Vincenzo Balzani (Ed.)

Electron Transfer in Chemistry

Volume 2

Vincenzo Balzani (Ed.)

# Electron Transfer in Chemistry

2  Organic Molecules

Organometallic and
Inorganic Molecules

**WILEY-VCH**

Weinheim · New York · Chichester
Brisbane · Singapore · Toronto

Prof. Vincenzo Balzani
Dipartimento de Chimica „G. Ciamician"
Università di Bologna
via Selmi 2
40126 Bologna
Italy

*100 2298754*

Library of Congress Card No.: applied for

A catalogue record for this book is available from the British Library.

Die Deutsche Bibliothek – CIP Cataloguing-in-Publication-Data
A catalogue record for this publication is available from Die Deutsche Bibliothek

ISBN 3-527-29912-2

© WILEY-VCH Verlag GmbH, D-69469 Weinheim (Federal Republic of Germany). 2001

Printed on acid-free paper.

Composition: Asco Typesetters, Hongkong.
Printing: betz-druck gmbh, 64291 Darmstadt.
Bookbinding: Wilhelm Osswald & Co., D-67433 Neustadt.
Printed in the Federal Republic of Germany.

# Contents

## Volume II

**Volume III**

# Volume IV

**Volume V**

Volume II




Part 1
Organic Molecules

Lennart Eberson (1933–2000)                    Eberhard Steckhan (1943–2000)

This section (Organic Molecules) is dedicated to the memory of Professor Lennart
Eberson and Professor Eberhard Steckhan, who both made fundamental contributions in this field, and who both died unexpectedly within few days in February
2000.

Lennart Eberson was one of the leading chemists in Sweden. He started to study
chemistry in 1950 and received his Ph.D. in 1959. His thesis dealt with the Kolbe
reaction and this was the start to his work in electrochemistry, a field where he
became an internationally recognized name. During a period in 1964–1965 he
worked with Saul Winstein's group at UCLA in Los Angeles. In 1979 he became
professor at Lund University where he created a research group working with
organic electrochemistry, silicon-organic chemistry and mechanistic studies of
radical reactions and electron transfer reactions. With skilful guidance and an endless flow of ideas, he brought 22 students to their Ph.D. degrees. In 1988 he became
a member of the Royal Swedish Academy of Sciences and on the occasion of his
death, Lennart Eberson was chairman of the Nobel Committee for Chemistry. For
his contributions he was recognized with several honors including the Norblad-
Ekstrand Medal and the Letterstedt Prize. He wrote several text-books including
"Electron Transfer Reactions in Organic Chemistry" and he was editor of the
organic chemistry section of Acta Chemica Scandinavica.

Eberhard Steckhan certainly belonged to the leading scientists in organic elec-
trochemistry. He started his career in Göttingen working as a graduate student in
Hans Schäfer's group at that time. In 1971 he finished his Ph.D. thesis and he
afterwards spent nearly two years as a postdoctoral fellow in the laboratory of
Ted Kuwana at the Ohio State University in Columbus, Ohio. After returning to
Germany, he followed his mentor of the Göttingen period to the University of
Münster, where he worked on his independent research projects and was appointed
Privatdozent in 1978. During that time he developed the fundamental criteria of
indirect electrolysis, which is now one of the well-established methods in organic
electrochemistry. Two years later he moved to the University of Bonn as Professor

of Organic Chemistry. Over the years he was active in various fields beside electrochemistry. For example, he was one of the pioneers who combined aspects of electrochemistry with biological applications. He also contributed fundamentally to the electron transfer induced Diels-Alder reactions and he recently made some spectacular applications to the synthesis of alkaloids.

Eberhard Steckhan was an ideal example of both academic mentor and friend. His warm personality and scientific integrity will be missed by the scientific community.

Bielefeld, April 2000                                          Jochen Mattay

# 1 Reactivity Patterns of Radical Ions—A Unifying Picture of Radical-anion and Radical-cation Transformations

*Michael Schmittel and Manas K. Ghorai*

## 1.1 Introduction

The chemistry of charged radicals, both radical anions and radical cations [1], has attracted much attention over the last two decades [2]. Despite a growing number of electron-transfer (ET)-promoted reactions that proceed with high selectivity and with unprecedented low activation barriers, however, most preparative chemists are still hesitant to use ET concepts when mapping out strategies for interesting synthetic target molecules. The reluctance to use radical-ion chemistry is firmly founded in the common notion that ET-initiated processes are mostly extremely complex transformations involving short-lived radical-ion intermediates the reactivity of which is hard to control and to predict. In our analysis covering several hundred transformations of odd-electron charged species [1], however, the majority of reactivity patterns proves to be extremely similar, irrespective of whether radical anions or radical cations are involved, making predictions on the *primary reaction channel* very reliable. Because a large body of kinetic data has recently become available on the reactivity of radical ions, it is now possible to map out the primary reaction channels not only conceptually but also on the basis of reliable data. We hope that the following unifying treatment, although it might not do justice to every possible individual case study in the literature, provides a helpful guide to the understanding of the fascinating chances of radical-ion transformations.

## 1.2 A Unifying Picture of Radical-anion and Radical-cation Chemistry

Radical ions are usually generated by one-electron reduction or oxidation processes starting from neutral compounds—an electron is either injected into the LUMO or removed from the HOMO. Even a simple MO picture of this process (Scheme 1)

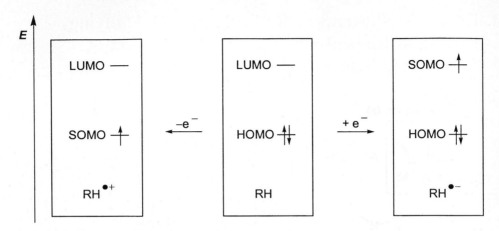

**Scheme 1.** One-electron-transfer reduction and oxidation of RH.

suggests that the two radical-ion intermediates should share, at least qualitatively, some common features that will determine their chemical fate:

1) Both species carry an unpaired electron and a charge: hence, they will behave as either nucleophiles (or electrophiles) and likewise as radicals [3].
2) By electron injection into an antibonding orbital or electron removal from a bonding orbital characteristic bonds are weakened: as a consequence a plethora of bond cleavage processes is typical for both radical anions and cations.

To comprehend our categorization and to conceive the similarity between radical-anion and radical-cation reactivity the concept of the electrophore is of paramount importance. In simple terms, the electrophore is the ET active part of the molecule which either accepts or ejects an electron. To illustrate this concept it is helpful to analyze the reaction behavior of complex molecules after one-electron-transfer activation.

The alkaloid **1** [4] (Scheme 2), for example, has four donor functionalities (**a**)–(**d**) that could potentially be involved in the one-electron oxidation. At what site do we expect the oxidation to occur? For rapid analysis we approximate the various sub-units (**a**)–(**d**) by small, structurally related molecules the potentials of which are known. Accordingly, triethylamine and methyl acetate can serve as models for (**c**) and (**d**). As donors (**a**) and (**b**) are linked by conjugation, we treat them together and, indeed, find with *N*-acetylindoline a good approximation.

From the oxidation potentials provided above we have to conclude that oxidation will originate from the donor site (**c**) with the lowest $E_{ox}$, which is indeed observed in the selective photoinduced electron-transfer (PET) cyanation of **1** using DAP$^{2+}$ (*N,N'*-dimethyl-2,7-diazapyreniumbistetrafluoroborate) as sensitizer [4].

As reversible redox potentials in solution are not available *in extenso* they can be replaced to a first approximation by ionization energies [5] or electron affinities [5, 6] either from experimental compilations or quantum-chemical calculations [7].

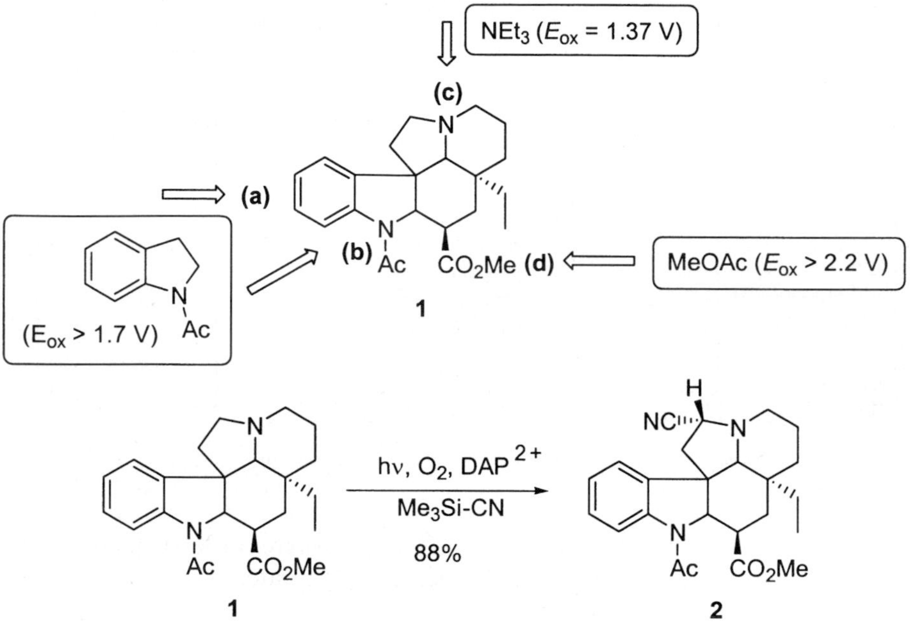

**Scheme 2.** Finding the relevant donor in alkaloid **1** containing several alternative electrophoric sites and the selective transformation of **1** [4] using PET.

Hence, a quick analysis might use LUMO energies from semi-empirical calculations for the analysis of the reduction of **3** (Scheme 3). Because the LUMO energy of acetone as model for site (**a**) in **3** is the lowest within the three electrophores, one must expect that an electron will be accepted at this site. This is indeed demonstrated in the PET reduction of **3** to a ketyl radical anionic intermediate using triethylamine as sensitizer [8].

While identification of the relevant electrophores in both examples above has enabled specification of the site at which the reaction will occur, it will be difficult for the non-expert reader to predict the primary reaction of the radical ion. Hence, we have written this review with the intention of providing *guidelines for the identification of primary reaction channels of radical ions and information about their kinetic reactivity*, which should enable the reader to set up defined ET-induced reactions even in polyfunctional molecules.

## 1.3 A Construction Set of Electrophores and its Relevance for Devising Selective Reactions *via* Radical Ions

The examples in Schemes 2 and 3 have illustrated the significance of identifying the electrophore with the lowest LUMO (or highest HOMO) energy for one-electron

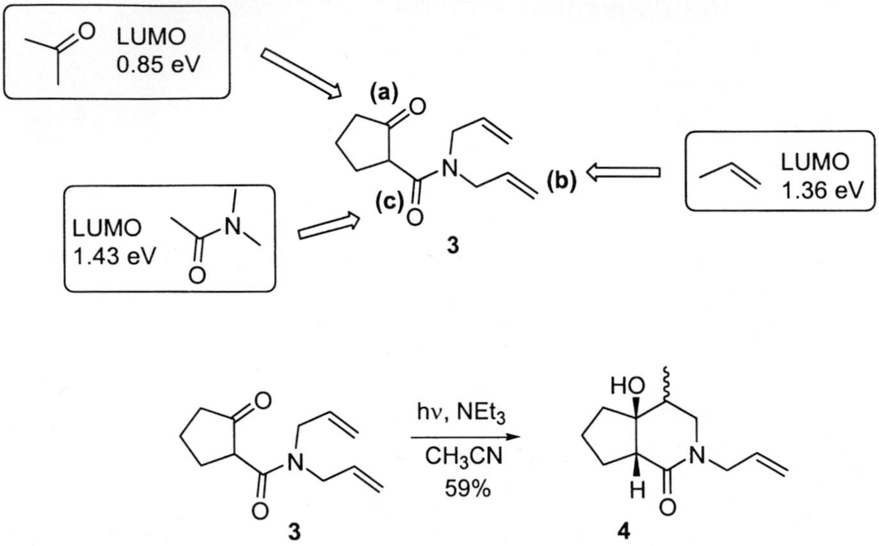

**Scheme 3.** LUMO energies calculated using AM1 [9a,c] and the selective transformation of **3** [4].

reduction (or oxidation) reactions. But that is not sufficient! In addition, it is necessary to categorize the relevant electrophore according to its main constituents, i.e. to check whether it has mainly $\sigma^*$ or $\pi^*$ (for reductions) or $\sigma$, $\pi$ or $n$ character (for oxidations). This categorization is helpful because it will enable us to make predictions on the follow-up reactions of the radical ions.

*If one wishes to induce a bond-cleavage reaction at the stage of the radical ion, one must choose either an acceptor whose LUMO contains a significant $\sigma^*$ component or a donor whose HOMO has high $\sigma$ participation.* In both, electron transfer will lead to a significant bond weakening, resulting in bond cleavage (Scheme 4).

*If we like to induce bond formation at the stage of the radical ion, it is advisable to select either an acceptor whose LUMO contains significant $\pi^*$ character or a donor whose HOMO has a predominant $\pi$ (or n) component.* Electron transfer in both will lead to either radical or ion-type bond-formation reactions (Scheme 5).

The above considerations have emphasized the need to establish the main character of the LUMO (of acceptors) and likewise of the HOMO (of donors). For that

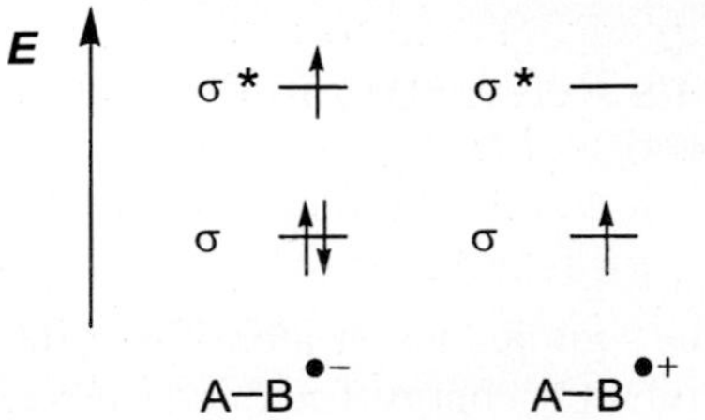

**Scheme 4.** Illustration of bond weakening in both $\sigma^*$ radical anions and $\sigma$ radical cations.

## Bond formation through

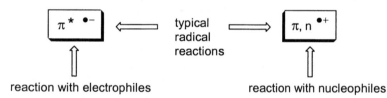

reaction with electrophiles          reaction with nucleophiles

**Scheme 5.** Illustration of the similar reactivity patterns of radical anions and radical cations in bond-formation reactions.

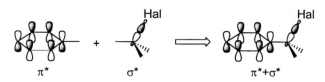

**Scheme 6.** Electrophores are often assembled from individual functional groups and bonds at the periphery.

purpose, the electrophore of interest should be analyzed with regard to the functional group(s) (halide, phenyl, carbonyl, amino etc.) and the peripheral bonds (such as C–C, C–H, C–Hal bonds) that have orbital overlap with the main functional group(s). It is indisputable that the acceptor (or donor) qualities of different groups and bonds will be different and that only those with the lowest individual LUMO (or highest individual HOMO) energies will be main contributors to the LUMO (or HOMO) of the electrophore (Scheme 6).

The main character of a composed electrophore will be determined by the character of the strongest contributor (acceptor or donor). Hence in the following, the acceptor properties of several common bonds and functional groups that are relevant for organic ET synthesis will be listed. Because of the lack of extensive electron affinity data, we have preferred to approximate the acceptor qualities by the LUMO energies as calculated by AM1 (Table 1) [9a, c, 10]. The calculations also enable categorization of acceptors as either $\sigma^*$ or $\pi^*$.

Likewise, we have established a qualitative order of the donor properties of several bonds and functional groups. The ordering was based on ionization potentials [5], because these are readily accessible for many molecules (Table 2).

It is obvious that a reduction of a pure $\sigma^*$ acceptor will lead to rather efficient bond weakening as much as oxidation of a $\sigma$ donor will entail bond cleavage (Scheme 4). The numbers [11, 12] in Table 3 underline the drastic effect of one-electron transfer on the bond strength of appropriate $\sigma^*$ acceptors and $\sigma$ donors. In all cases the bond-dissociation energies (BDE) of the radical ions are much lower than those of the neutral precursors (Table 3).

As mentioned above, however, most acceptors (or donors) will not be pure $\sigma^*$

**Table 1.** Acceptor qualities (LUMO energies) of a variety of $\sigma^*$ and $\pi^*$ acceptors as determined by AM1 semi-empirical calculations [9a, c, 10].

| Energy of LUMO | $\sigma^*$ Acceptor (energy of LUMO, eV) | $\pi^*$ Acceptor (energy of LUMO, eV) |
|---|---|---|
| +5 to +3 eV | C–H ethane (+4.1) | |
| | C–F fluoromethane (+3.8) | |
| | C–N dimethylamine (+3.5) | |
| | C–O oxirane (+3.4)[a] | |
| | C–O dimethyl ether (+3.2) | |
| +3 to +2 eV | C–C oxirane (+2.7) | C≡C ethyne (+2.1) |
| +2 to +1 eV | C–Cl chloromethane (+1.6) | C≡N acetonitrile (+1.7) |
| | C–F tetrafluoromethane (+1.2) | C=C ethylene (+1.4) |
| | | C=N acetonimine (+1.2) |
| | | C=O methyl acetate (+1.1) |
| +1 to 0 eV | C–S dimethylsulfide (+0.9) | C=O acetaldehyde (+0.8) |
| | C–Br bromomethane (+0.9) | Ph benzene (+0.6) |
| | C–I iodomethane (+0.5) | C=N azirine (+0.6) |
| | | Ar pyridine (+0.1) |
| 0 to −1 eV | C–Se dimethylselenide (−0.5)[b] | $NO_2$ nitromethane (−0.1) |
| | C–O C–S methyl mesylate (−0.7) | Ar pyrimidine (−0.2) |
| | | Ar phenanthroline (−0.7) |
| | | Ar anthracene (−0.8) |
| −1 to −3 eV | S–S dimethyl disulfide (−1.5) | Ar nitrobenzene (−1.1) |
| | C–Br tetrabromomethane (−2.1) | |
| | C–O, C–S methyl triflate (−2.4) | |

[a] Higher-lying unoccupied MO.
[b] Value calculated by use of PM3 [9b].

acceptors (or $\sigma$ donors). The relevant frontier orbital will be composed of two or more constituents, e.g. for benzyl halides and benzylsilanes (Scheme 7).

(σ*+π*) radical anion (benzyl halide)
•−
Hal β
Y
α
weakening of $C_α$-$Hal_β$ bond by overlap of
the σ*($C_α$-$Hal_β$) with the π* orbital (phenyl)
(σ+π) radical cation (benzylsilane)
•+
$SiMe_3$
β
α
weakening of $C_α$-$SiR_{3\,β}$ bond by overlap of
the σ ($C_α$-$SiR_{3\,β}$) with the π orbital (phenyl)

**Scheme 7.** The relevant frontier MO for benzyl halides and benzylsilane.

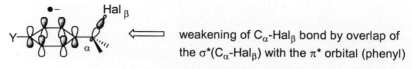

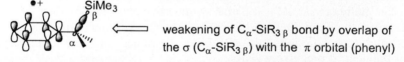

Let us examine the reduction of a benzyl halide, which is, according to our understanding a typical $(\sigma^* + \pi^*)$ acceptor. $(\sigma^* + \pi^*)$ radical anions are composed of $\sigma^*$ and $\pi^*$ components in the singly occupied orbital. If there is good overlap between them (stereoelectronic effect) then for a given C–X bond the weakening will be stronger the closer the energy of the $\sigma^*$ and the $\pi^*$ electrophore (cf. Scheme 8).

For the simplest case, i.e. compounds with the same kind of scissile bond (similar homolytic bond-dissociation energies and $\sigma^*$ energies!), the splitting $E_{(\sigma^* - \pi^*)}$ provides a good measure of the bond weakening. Along this line, substituted benzyl bromide radical anions have a lower $\Delta G_{BDE}$ when the energy of the $\pi^*$ LUMO is raised, because the $\sigma^*$ character in the SOMO is increased [13].

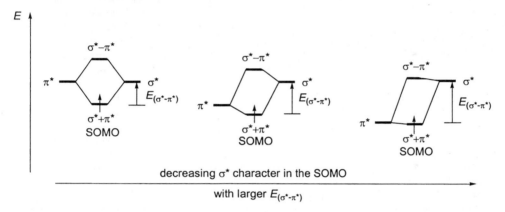

**Scheme 8.** In a radical anion with a constant $\sigma^*$(C–X) energy $E_{(\sigma^* - \pi^*)}$ becomes larger when the energy of the $\pi^*$ orbital is reduced. Hence, the $\sigma^*$ character of the SOMO decreases.

A particular situation is met when the observed bond cleavage in the radical anion is not expected because of the low $\sigma^*$ character of the SOMO, as observed in the reduction of aryl halides. For example, the analysis of 4-chloronitrobenzene based on AM1 [9a, c] calculations clearly indicates that the LUMO and the nLUMO do not share any $\sigma^*$(C–Cl) character. Actually, the lowest unoccupied orbital with a

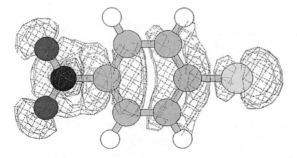

**Figure 1.** LUMO + 3 orbital of 4-chloronitrobenzene according to AM1 [9a, c] calculations.

**Table 2.** Donor qualities (HOMO energies) of a variety of $\sigma$, $\pi$, and $n$ donors as determined by experimental ionization energies from the NIST webbook [5].

| Energy of HOMO | $\sigma$ donor (energy of HOMO, eV) | $\pi$ or $n$ donor (energy of HOMO, eV) |
|---|---|---|
| $-7$ to $-8$ eV | Sn–Sn hexamethyldistannane ($-8.0$) | Ar anthracene ($-7.4$) |
| | | $n_N$ trimethylamine ($-7.9$) |
| | | Ar $p$-hydroquinone ($-7.9$) |
| $-8$ to $-9$ eV | Ge–Ge hexamethyldigermane ($-8.1$) | $n_P$ triethylphosphane ($-8.1$) |
| | Si–Si hexamethyldisilane ($-8.3$) | Ar pyrrole ($-8.2$) |
| | C–Sn tetramethylstannane ($-8.9$) | Ar anisole ($-8.2$) |
| | | $n_{Se}$ diethylselenide ($-8.3$) |
| | | $n_S$ diethylsulfide ($-8.4$) |
| | | Ar furan ($-8.9$) |
| $-9$ to $-10$ eV | C–Ge tetramethylgermane ($-9.3$) | Ar benzene ($-9.2$) |
| | C–C cyclopropane ($-9.9$) | $n_I$ iodomethane ($-9.5$) |
| | C–Si tetramethylsilane ($-9.8$) | $n_O$ diethyl ether ($-9.5$) |
| | | $n_{Se}$ H$_2$Se ($-9.9$) |
| $-10$ to $-11$ eV | C–H cyclopentane ($-10.3$) | $n_N$ ammonia ($-10.1$) |
| | | $n_S$ H$_2$S ($-10.5$) |
| | | $n_{Br}$ tribromomethane ($-10.5$) |
| | | $n_{Br}$ bromomethane ($-10.5$) |
| | | C=C ethene ($-10.5$) |
| $-11$ to $-12$ eV | C–H ethane ($-11.5$) | $n_{Cl}$ chloromethane ($-11.3$) |
| | | C≡C ethyne ($-11.4$) |
| | | $n_{Cl}$ tetrachloromethane ($-11.5$) |
| $-12$ to $-13$ eV | C–H fluoromethane ($-12.5$) | $n_O$ H$_2$O ($-12.6$) |
| | C–H methane ($-12.6$) | |

**Table 3.** Bond-dissociation energies (BDE) of $\sigma^*$ radical anions and $\sigma$ radical cations.

| Radical anion | BDE (kJ mol$^{-1}$) [11] | Radical cation | BDE (kJ mol$^{-1}$) [12] |
|---|---|---|---|
| Cl–CCl$_3$$^{\cdot-}$ | $-59$ (306)[a] | CH$_3$–H$^{\cdot+}$ | 172 (440)[a] |
| Cl–CHCl$_2$$^{\cdot-}$ | $-92$ (325)[a] | CH$_3$–CH$_3$$^{\cdot+}$ | 215 (377)[a] |
| Br–CBr$_3$$^{\cdot-}$ | $-13$ (235)[a] | (CH$_3$)$_3$C–CH$_3$$^{\cdot+}$ | 25 (364)[a] |
| Br–CHBr$_2$$^{\cdot-}$ | $-50$ (259)[a] | Cyclopropane$^{\cdot+}$ | 89 (251)[a] |

[a] Values in parentheses are the bond-dissociation energies of the neutral compounds.

sizeable $\sigma^*$(C–Cl) component is the LUMO+3 orbital (Figure 1). Nevertheless, C–Cl bond cleavage can be initiated by first forming an aromatic $\pi^*$ radical anion that undergoes activation by intramolecular ET to the LUMO+3 orbital (Scheme 9).

Conceptually seen, such an ET reaction is possible through the avoided crossing of both electronic states separated by the energy gap $\Delta E_{\sigma^*\pi^*}$. In a more pictorial description, the $\pi^*$ acceptor system serves as an intramolecular redox relay which helps to shuttle the electron to the $\sigma^*$(C–Cl) orbital which is not accessible by direct

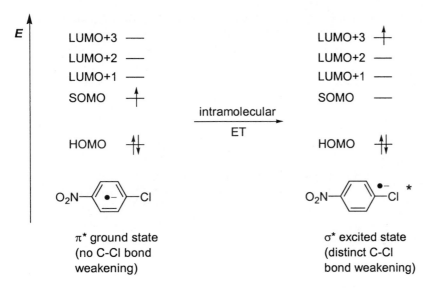

**Scheme 9.** 'Intramolecular ET' activates 4-chloronitrobenzene radical anion for bond cleavage.

reduction because of its higher energy. It is obvious that the $\pi^*$ radical anion will undergo bond cleavage only in the presence of a low-lying $\sigma^*(C–X)$; otherwise it will undergo typical $\pi^*$ reactions, such as transformations with either electrophiles or radical species.

What we have summed up for radical anions is likewise valid for radical cations, except that they might contain, in addition to the $\sigma$ and $\pi$ donor sites, also $n$ donor character. Hence, $\sigma$, $(\sigma + \pi)$, and $(\sigma + n)$ radical cations may undergo bond cleav-

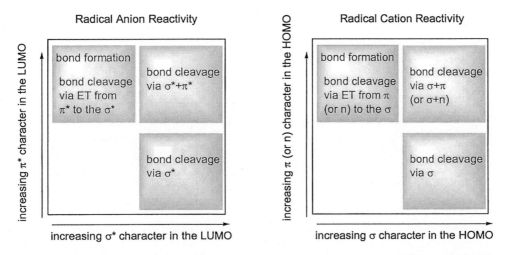

**Figure 2.** Radical anion and radical cation reactivity related to the character of the frontier MO of the redox-active molecule.

age because an electron is completely or partly removed from a bonding $\sigma$ type orbital. In contrast, pure $\pi$ and $n$ radical cations, will undergo bond-formation reactions. Hence, to sum up the following picture of radical-ion chemistry may be sketched (Figure 2).

The above categorization emphasizes the similarities of the reactivity patterns of radical anions and radical cations. On the basis of the bond and fragment electron affinities (Table 1) or ionization potentials (Table 2), we can now devise systems that should undergo, for example, facile bond cleavage. As a prerequisite the $\sigma^*$ (or $\sigma$) orbital representing the scissile bond has to be made a major contributor to the LUMO (or HOMO).

For example, C–Cl bonds can be easily cleaved by reduction because of their low-lying $\sigma^*$ orbitals, whereas C–N bonds are much more difficult to cleave. Hence, it is advisable to incorporate the $\sigma^*$(C–N) into a $(\sigma^* + \pi^*)$ system by overlap of $\sigma^*$(C–N) with a low-lying $\pi^*$ (e.g. from nitrobenzene). If overlap of the $\sigma^*$ with a $\pi^*$ is not possible, we still can set up a low-lying $\pi^*$ redox relay to shuffle the electron to the $\sigma^*$ by intramolecular ET.

The analogous situation can be arranged for bond cleavage in radical cations by using appropriately placed $\pi$ or $n$ donors to assist bond cleavage (Scheme 10).

To reveal the analogies in the primary reaction behavior of radical ions, we will use in this review the following structure:

• reductive and oxidative **bond-cleavage reaction**s,
• reductive and oxidative **bond-formation reaction**s, and
• **'pericyclic' processes** under reducing and oxidizing conditions.

The vast body of examples covering many decades can obviously not be presented in a comprehensive manner here. Thus, we have decided to focus on a few selected examples from the recent literature; this should give the interesting reader the chance to find analogous examples through cited references. We regret that this way of presenting the material will not do justice to the numerous pioneers of radical-ion chemistry, whose work is extensively quoted in older reviews [14] or monographs [2]. The synthetic examples will be accompanied by kinetic and thermodynamic data to provide a qualitative and quantitative picture of radical-ion reactivity. When devising new synthetic schemes using these odd-electron species the combination of synthetic, kinetic, and thermodynamic aspects hopefully will serve as guide to the achievement of highly selective transformations.

## 1.4  Reductive and Oxidative Bond-cleavage Reactions

### 1.4.1  General Principles of Bond Cleavage

As discussed above, bond cleavage in radical ions is a result of the weakening of a single bond after electron transfer.

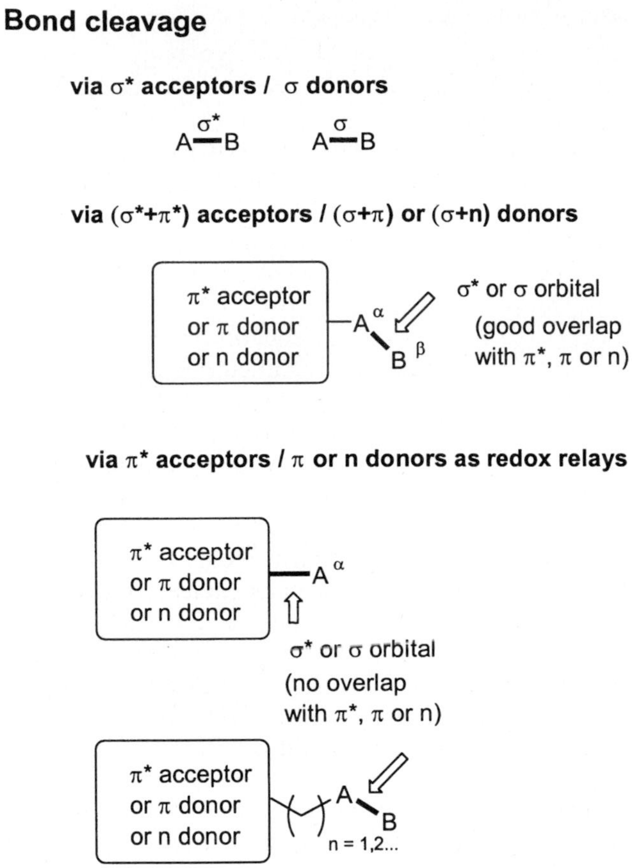

## Bond cleavage

### via σ* acceptors / σ donors

$$A \overset{\sigma^*}{\text{—}} B \qquad A \overset{\sigma}{\text{—}} B$$

### via (σ*+π*) acceptors / (σ+π) or (σ+n) donors

| π* acceptor or π donor or n donor |
|---|

$-A^{\alpha}$  $B^{\beta}$   σ* or σ orbital (good overlap with π*, π or n)

### via π* acceptors / π or n donors as redox relays

| π* acceptor or π donor or n donor |
|---|

$-A^{\alpha}$  ⇑

σ* or σ orbital (no overlap with π*, π or n)

| π* acceptor or π donor or n donor |
|---|

$A$  $B$   $n = 1,2...$

**Scheme 10.** Schematic view of bond cleavage in radical ions.

$$A\text{–}B \xrightarrow{+e^-} A^{\bullet} + B^-$$

There are three distinct cases to be recognized depending on the $\sigma^*$ (or $\sigma$) character of the LUMO (or HOMO).

1) If we deal with a pure $\sigma^*$ LUMO (or $\sigma$ HOMO), electron injection (or removal) will entail the strongest bond weakening effect. ET activation of such systems will always lead to bond dissociation.
2) In contrast, radical ions, whose LUMO has $(\pi^* + \sigma^*)$ character (or whose HOMO has $(\pi + \sigma)$ or $(n + \sigma)$ character), will experience reduced bond weakening after ET. Nevertheless, these systems still undergo efficient bond cleavage.
3) Molecules with pure $\pi^*$ LUMO ($\pi$ or $n$ HOMO) character might undergo—in competition to other reactions—efficient bond cleavage, if they contain $\sigma^*$ (or $\sigma$) orbitals which are energetically accessible by intramolecular ET.

**Table 4.** Acceptor type classification of aryl halides according to PM3 [9b, d] calculations.

| Molecule | LUMO (energy) | Energy of lowest $\sigma^*$ orbital |
|---|---|---|
| Fluorobenzene | $\pi^*$ (+0.03 eV) | +2.33 eV (LUMO + 2) |
| Chlorobenzene | $\pi^*$ (+0.06 eV) | +1.46 eV (LUMO + 2) |
| Bromobenzene | $\sigma^*$ (−0.05 eV) | |
| Iodobenzene | $\sigma^*$ (−0.43 eV) | |
| 1,4-Dibromobenzene | $\sigma^*$ (−0.32 eV) | |
| *p*-Cyanobromobenzene | $\pi^*$ (−0.86 eV) | −0.45 eV (C–Br) |
| | | +1.69 eV (C–CN) |
| *p*-Nitrobromobenzene | $\pi^*$ (−1.39 eV) | −0.72 eV (C–Br) |
| | | +1.10 eV (C–NO$_2$) |

To realize how the acceptor character can change from a $\sigma^*$ to $\pi^*$ system within a seemingly analogous series of compounds it is instructive to check the classification in Table 4.

While the cartoon-like depiction of the acceptor (or donor) in Scheme 10 emphasizes the analogies in radical-ion chemistry in a rather general manner, our compilation (Scheme 11) aims at providing more insight into some relevant functionalities involved in dissociation processes.

Irrespective of the electrophoric system involved in one-electron transfer-initiated bond dissociation one can easily derive the thermodynamic driving force for a such process by use of thermochemical cycle calculations [15]. Such estimates are particularly valuable as experimental numbers, because bond-dissociation data are scarce.

The cleavage of radical ions can proceed homolytically (or heterolytically) depending on whether the scission leaves (or does not leave) the charge mainly in the same region as in the radical ion. Change-over from homolytic to heterolytic bond cleavage can even occur in the same family of compounds, and the kinetic changes seem to be simply related to the thermodynamics of the bond cleavage and *not* to the mode of scission [16].

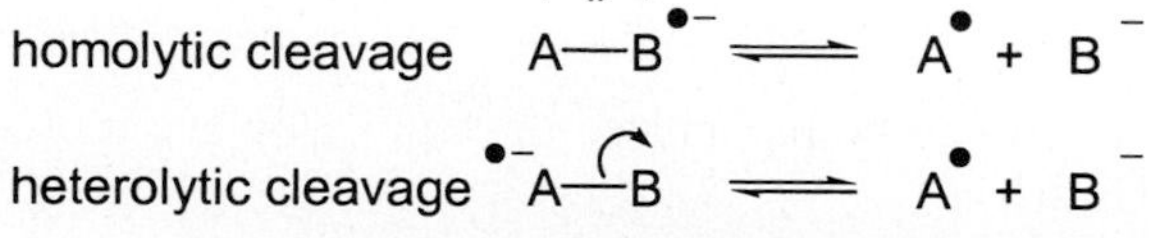

homolytic cleavage    A—B$^{\bullet-}$ ⇌ A$^{\bullet}$ + B$^{-}$

heterolytic cleavage    $^{\bullet-}$A—B ⇌ A$^{\bullet}$ + B$^{-}$

The kinetics of bond cleavage will be treated separately below according to the type of electrophore. Such data are important as cleavage rate constants can vary by more than ten orders of magnitude, depending on the nature of R and X in RX$^{\bullet-}$ [17].

The huge amount of bond-dissociation [18] examples makes this reactivity pattern one of the most prominent in radical-ion chemistry. We will therefore discuss this mode of reaction by providing thermodynamic and kinetic data along with some representative synthetic examples.

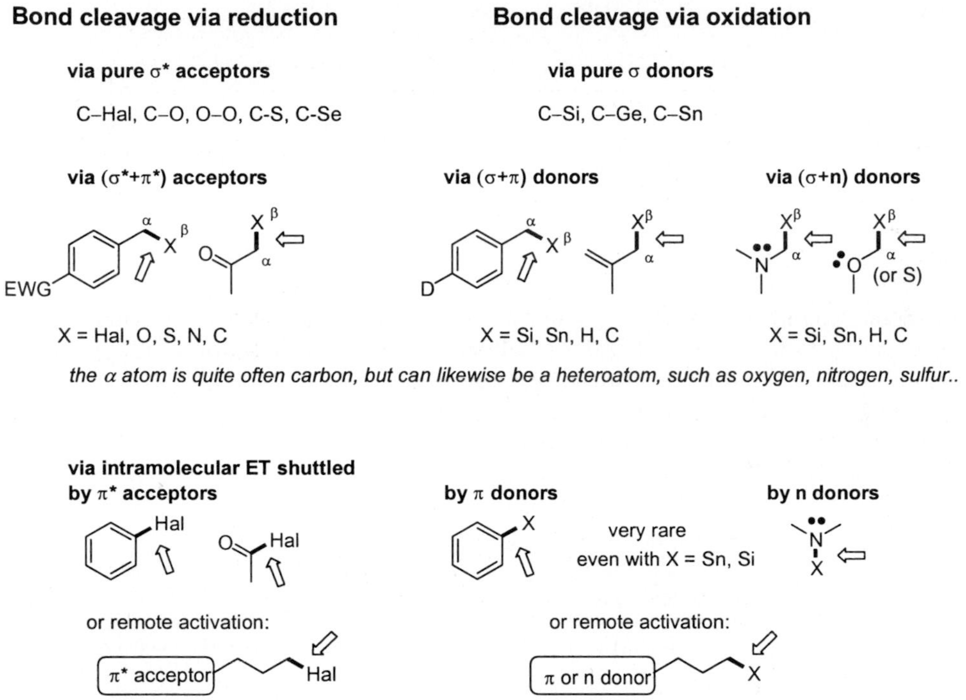

**Bond cleavage via reduction**

**via pure σ* acceptors**

C–Hal, C–O, O–O, C-S, C-Se

**via (σ*+π*) acceptors**

EWG

X = Hal, O, S, N, C

**Bond cleavage via oxidation**

**via pure σ donors**

C–Si, C–Ge, C–Sn

**via (σ+π) donors**

D

X = Si, Sn, H, C

**via (σ+n) donors**

(or S)

X = Si, Sn, H, C

*the α atom is quite often carbon, but can likewise be a heteroatom, such as oxygen, nitrogen, sulfur..*

**via intramolecular ET shuttled by π* acceptors**

or remote activation:

π* acceptor — Hal

**by π donors**

very rare
even with X = Sn, Si

or remote activation:

π or n donor — X

**by n donors**

**Scheme 11.** A compilation of scissile bonds often encountered in radical ion cleavage reactions.

## 1.4.2 Synthetic, Kinetic and Thermodynamic Aspects of Reductive Bond Cleavage

### Bond cleavage after reduction of σ* acceptors

*Synthetic aspects*

We encounter reductive bond cleavage of $\sigma^*$ acceptors when there are no readily reducible $\pi^*$ acceptors, such as aryl, carbonyl, and analogous functionalities present. Hence, this reaction pattern is commonly realized in aliphatic systems containing the electrophiles depicted in Scheme 12, such as in simple reductions R–Hal → R–H [19]. Equally of importance is the dehalogenation of geminal [20] and vicinal dihaloalkanes [21], which has also been described for $SO_2R$, SR, OAc, OMe, OH,

### Common σ* acceptors for reductive bond cleavage

C–Hal, C–O, C–S, C–Se, O–O, Si–Hal

**Scheme 12.** Frequently used $\sigma^*$ acceptors that undergo rapid bond cleavage after reduction.

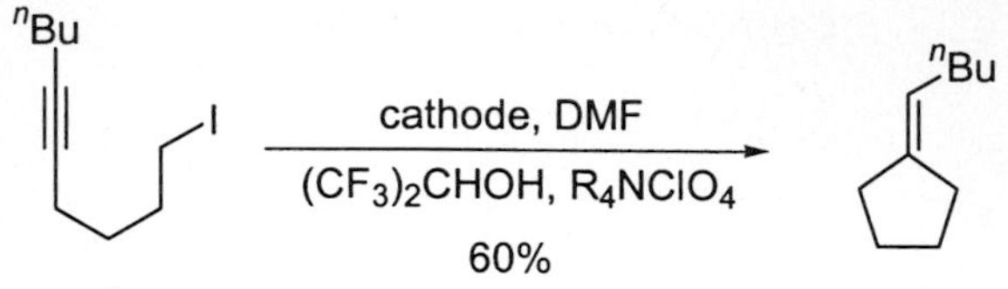

**Scheme 13.** 5-*exo-dig* cyclization initiated after reductive C–I bond cleavage [25].

and other leaving groups [21]. C–Hal bond cleavage is, moreover, frequently applied in $S_{RN}1$ reactions [2b] with aliphatic halides [22], in the generation of alkyl lithium compounds [23], and in cyclization reactions (Scheme 13) [24, 25].

C–O bond cleavage is most efficiently set up in strained systems, e.g. in oxiranes [26] and oxetanes [27], whereas O–O bond cleavage in cyclic peroxides [28] and C–S [29] and C–Se [30] bond rupture occur efficiently even in unstrained compounds. Si(Ge)–Hal bond cleavage [31] has become a valuable reaction, e.g. in the preparation of network polysilanes and polygermanes [32].

*Kinetic and thermodynamic aspects*

Pure $\sigma^*$ acceptors, e.g. aliphatic halides and peroxides, are mostly believed to undergo dissociative ET, i.e. concomitant with ET into the $\sigma^*$ orbital the bond elongates until it cleaves. Hence, over years it was believed that in such processes radical anions are generally not involved as distinct intermediates, although in very few examples parent anions had been observed, e.g. for $CF_3Cl$, $CF_3Br$, $CCl_4$ and $CBr_4$ [33]. Moreover, recent femtosecond studies revealed for the $Et_2S–I_2$ system after photoinduced charge transfer that reversible ET occurs in less than 150 fs (fastest trajectory) and is followed by the rupture of the I–I bond with the release of the first I atom in 510 fs [34]. Although semi-empirical calculations found both dissociative and non-dissociative ET with polyhalogen methanes [35], one must also take into consideration that the surrounding solvent will be important. Indeed, electron-transfer reduction of $CCl_4$ in the gas phase is predicted to yield an intermediate radical anion [36], whereas it is dissociative in $H_2O–CH_3CN$ (36:64 % $v/v$) [35]. As a consequence, the question of dissociative or non-dissociative ET has to be answered depending on the individual molecule and the reaction conditions.

For dissociative ET reactions thermodynamic bond dissociation data are not very meaningful, because they are intrinsically interwoven with the ET step. Nevertheless, some data are available, e.g. on halomethanes [37], and polyhalomethanes [11, 38].

Seminal work by Savéant and his group [39] has contributed to our understanding of the dynamics of dissociative ET, compared with the corresponding stepwise process [40]. Obviously, bond-dissociation energy, rather than bond dissociation free energy, represents the important contribution to the intrinsic barrier [41]. One must, moreover, take into account that dissociative ET can occur in the adiabatic (e.g. *tert*-butyl bromide) and in the non-adiabatic regime (di-*tert*-butyl peroxide)

[42]. Reorganization energies of such processes have been calculated and evaluated [43].

Irrespective of the mechanistic pathway, radical anions from $\sigma^*$ acceptors undergo very rapid bond cleavage, a process that might be related to the toxicity [44] and carcinogenic activity [45] of polyhalogen alkanes, and which is of huge importance in synthetic chemistry.

**Bond cleavage after reduction of ($\sigma^*$ + $\pi^*$) acceptors**

*Synthetic aspects*

Reductive bond cleavage in ($\sigma^*$ + $\pi^*$) radical anions can be accomplished under milder conditions than with $\sigma^*$ acceptors, because the $\pi^*$ component can be used to move the reduction potential towards less cathodic potentials. The most common examples involve C–Hal bond cleavage in (hetero)benzylic and (hetero)allylic positions. Even those bonds can, however, be cleaved whose $\sigma^*$ orbital is very high in energy, e.g. C–O, C–N, and C–C bonds (Scheme 14).

**Common ($\sigma^*$+$\pi^*$) acceptors for reductive bond cleavage.**

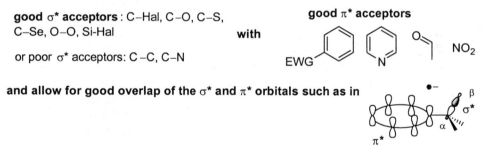

**Combine**

**good $\sigma^*$ acceptors**: C–Hal, C–O, C–S, C–Se, O–O, Si–Hal

**with**

**good $\pi^*$ acceptors**

EWG      N      O      NO$_2$

or poor $\sigma^*$ acceptors: C–C, C–N

**and allow for good overlap of the $\sigma^*$ and $\pi^*$ orbitals such as in**

$\pi^*$      $\alpha$      $\beta$      $\sigma^*$

**Scheme 14.** Frequently used ($\sigma^*$ + $\pi^*$) acceptors that undergo rapid reductive bond cleavage.

C–Hal bond cleavage is a main motif in simple reductions R–Hal → R–H [46] or for R–Hal → R–E by trapping the resulting anions with electrophiles E$^+$ [47]. Likewise, S$_{RN}$1 reactions [2b] involving benzyl halides [48], heteroarene methyl halides [49], and difluoromethyl quinones are frequently used [50]. Whereas bond cleavage of pure $\sigma^*$ (C–N, C–O, C–C) acceptors is rather difficult, C–N [51], C–O [52], C–S [53], and C–Se bond rupture [30] can be readily realized in the corresponding ($\sigma^*$ + $\pi^*$) systems. Even C–C bond cleavage is possible, for example in $\alpha$-cyclopropylketones [54], but is restricted to a few instances. Other types of bond cleavage include rupture of the disulfide bond [55] and Se–S [56], Se–Se [56], and O–Si bond scission [52].

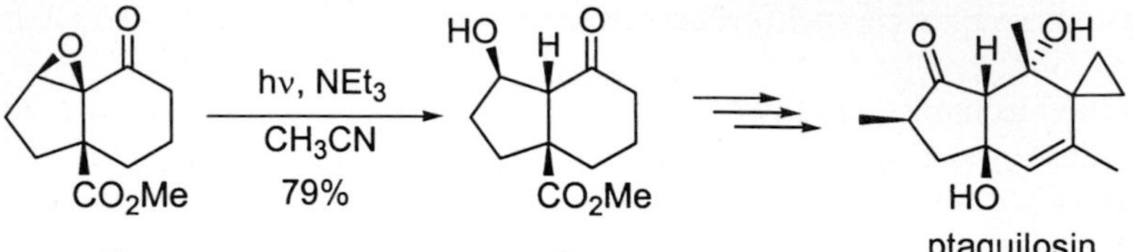

**Scheme 15.** Regioselective C–O bond cleavage as a key step in the synthesis of ptaquilosin [57].

## Kinetic and thermodynamic aspects

$(\sigma^* + \pi^*)$ acceptors are often involved in thermal $S_{RN}1$ reactions [2b], because dissociative ET, for example between 4-nitrocumylchloride ($E^\circ_{SCE} = -1.12$ V) and 2-nitropropanate ($E^\circ_{SCE} = 0.077$ V), overcomes the sluggishness of an outer-sphere ET followed by bond dissociation (kinetic amplification through the $S_{RN}1$ process) [58].

## C–Hal bond cleavage

When $\sigma^*, \pi^*$ overlap in the radical anion is comparable within a series of compounds then thermodynamic factors will govern the kinetics. Following the trend in $\Delta G_{het}(RX^{\cdot-})$, bond cleavage in benzyl chloride$^{\cdot-}$ is dissociative, whereas the corresponding *p*-nitrobenzyl chloride$^{\cdot-}$ undergoes bond scission at $4 \times 10^6$ s$^{-1}$ [59]. The situation becomes somewhat more complicated when we compare the C–X bond cleavage of different groups X, because then the different homolytic bond-dissociation energies and $\sigma^*$ energies also become important. For example, bond scission in benzyl halide$^{\cdot-}$ usually follows the order BzBr$^{\cdot-}$ > BzCl$^{\cdot-}$ > BzF$^{\cdot-}$ [60, 61]. Bond weakening is also responsible for the complete defluorination of trifluoromethylarenes—NC–C$_6$H$_4$–CH$_2$F$^{\cdot-}$ ($k = 7 \times 10^6$ s$^{-1}$) cleaves more rapidly than NC–C$_6$H$_4$–CHF$_2$$^{\cdot-}$ ($k = 4 \times 10^5$ s$^{-1}$) and NC–C$_6$H$_4$–CF$_3$$^{\cdot-}$ ($k = 3.8 \times 10^1$ s$^{-1}$) [62].

## C–O bond cleavage

After one-electron reduction of cyanoanisoles the C–O bond rupture is very slow [63]. Much more rapid and in line with the driving force $\Delta G_{het}(RX^{\cdot-})$ is bond cleavage in ArCOCH$_2$–OPh$^{\cdot-}$: Ar = *p*MeOPh at $k = 1.2 \times 10^7$ s$^{-1}$ compared with Ar = Ph at $k = 7.0 \times 10^5$ s$^{-1}$ and Ar = *p*CF$_3$Ph at $k = 1.0 \times 10^3$ s$^{-1}$ [64]. The role of the anionic leaving group Y has also been investigated in the bond scission of α-phenoxyacetophenones PhCOCH$_2$OPh–Y [65]. Here, the rate constant correlates well with $\sigma$ ($\rho = 3.1$) [66], because of the thermodynamic contribution to the activation barrier, as required by the Savéant model [17, 40, 67]. In this model the cleavage is viewed as an intramolecular dissociative ET, in which the electron is transferred from a $\pi^*$ to the $\sigma^*$ orbital of the bond to be cleaved.

## C–X bond cleavage

It has been demonstrated that in $(\sigma^* + \pi^*)$ acceptors even bonds (e.g. C–C) with very high-lying $\sigma^*$ orbitals, which are neither accessible in $\sigma^*$ acceptors by direct electron injection nor in $\pi^*$ acceptors by intramolecular ET, can be broken. To effect such difficult processes, the $(\sigma^* + \pi^*)$ acceptor must have a weak homolytic bond and a high reduction potential, and the anion produced should be very stable [68]. By following these rules, rate constants of C–C bond cleavage in 1,2-diarylethane·⁻ were increased from $8 \times 10^5$ to $>3 \times 10^{10}$ s⁻¹; this enabled the design of interesting self-destructive electron acceptors [68]. Analogous C–S bond scission is feasible, but much slower [69, 70].

**Bond cleavage after intramolecular ET via $\pi^*$ redox relays**

Many compounds undergo reductive C–X bond cleavage even though the LUMO does not have observable $\sigma^*$(C–X) character; examples are aromatic and vinylic halides [72] where the C–Hal bond is located orthogonal to the LUMO. It has been proposed [71] on the basis of semi-empirical calculations [72] and experimental evidence [73] that for aryl halides an intramolecular ET from the primarily formed $\pi^*$ to the $\sigma^*$(C–X) is responsible for the observed difference in cleavage rates $(k = 10^{10}$ s⁻¹ for phenyl halides compared with $k = 10^{-2}$ s⁻¹ for nitrophenylhalides). More precisely, intramolecular $\pi^* \rightarrow \sigma^*$ dissociative ET takes place when the energy of the $\sigma^*$ orbital is reduced (as a result of bond stretching) to match the energy of the $\pi^*$ LUMO to which the unpaired electron has initially been transferred [17].

$$\text{Ar–X} \xrightarrow{+e^-} \underset{\pi^*}{(\overset{\bullet\,-}{\text{Ar}})\text{–X}} \xrightarrow[\text{ET}]{\text{intramol.}} \left[\underset{\sigma^*}{\overset{\bullet\,-}{\text{Ar–X}}}\right]^{\ddagger} \longrightarrow \text{Ar}^{\bullet} + \text{X}^-$$

*Synthetic aspects*

As we have seen above, the rate of bond cleavage can be modulated very readily by the energy splitting between the $\pi^*$ LUMO and the unoccupied $\sigma^*$(C–X). Usually the $\pi^*$ LUMO belongs to an aromatic, heteroaromatic, vinylic, or heterovinylic (e.g. acyl groups) acceptor. Bonds that are frequently cleaved are depicted in Scheme 16.

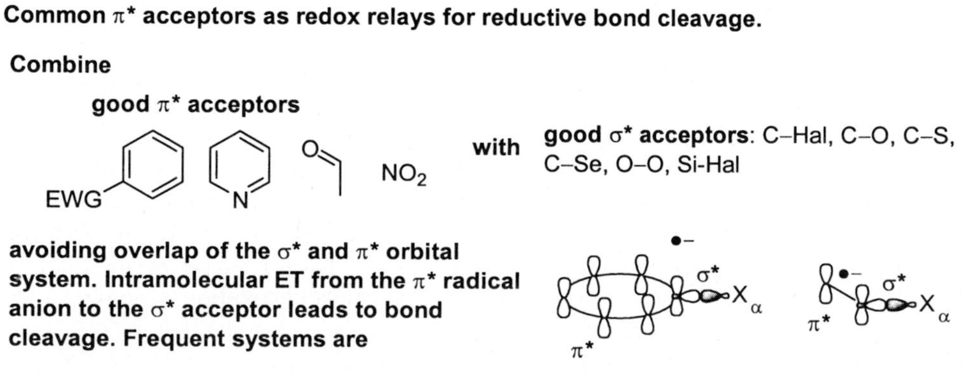

**Common $\pi^*$ acceptors as redox relays for reductive bond cleavage.**

**Combine**

**good $\pi^*$ acceptors**

**with**  **good $\sigma^*$ acceptors**: C–Hal, C–O, C–S, C–Se, O–O, Si–Hal

**avoiding overlap of the $\sigma^*$ and $\pi^*$ orbital system. Intramolecular ET from the $\pi^*$ radical anion to the $\sigma^*$ acceptor leads to bond cleavage. Frequent systems are**

**Scheme 16.** Frequently used $\pi^*$ acceptors that function as redox relays to shuttle electrons to the scissile C–X bonds.

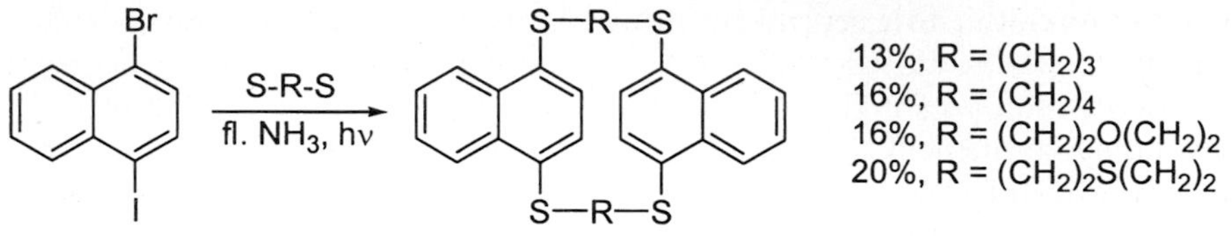

**Scheme 17.** Four reductive C–Hal bond-cleavage reactions are used in the quadruple $S_{RN}1$ reaction to assemble **10** [80].

Frequently, bond cleavage is used for R–Hal → R–H dehalogenations [74], and the formation of carbanions [75]. More seldom encountered are reactions of $\pi^*$ acceptors that have the $\sigma^*$ bond spatially well separated and that depend on long-range ET. Such a situation occurs in phenyl-substituted alkyl chlorides [76], and bridgehead halides [77] with various redox relay functions (e.g. a nitroaryl group) [78].

$\pi^*$ Haloacceptors and related systems play a prominent role in $S_{RN}1$ reactions [2b] of aryl halides [79, 80], hetaryl halides [79c, 81], and vinyl halides [82]. A spectacular example of the use of such acceptors is the macrocycle synthesis by a quadruple $S_{RN}1$ reaction depicted in Scheme 17 [80].

C–C bond cleavage is readily possible with nitrile radical anions as a shuttle [83]; C–N bond cleavage is rare, but has been effected in phenyl-substituted dimethylanilines [84], some benzenesulfonamides [52], and by using benzotriazole both as the $\pi^*$ acceptor and as the anionic leaving group [85]. C–O or C–S bond cleavage can readily be accomplished in reductions of the type R–O → R–H [86] and lithium carbanion generation [23]. C–S bond cleavage in aromatic polysulfones [87] and alkenyl phenyl sulfones [88] can, moreover, be regarded as originating from $\pi^*$ acceptors. Other bond-cleavage processes have been recognized—for example in the cathodic cleavage of S–P [89], N–S (in aryl azo sulfides) [90], and Si–O bonds [91].

*Kinetic and thermodynamic aspects*

C–Hal bond cleavage

Recent work [72b, d] has demonstrated that for a variety of halonitrobenzene radical anions there is a correlation between the experimental rate constants (spanning eight orders of magnitude) and $\Delta E_{\sigma^*\pi^*}$. If the scissile bond of interest does not change its $\sigma^*$ LUMO energy significantly within a series of analogous compounds, changes in the energy gap $\Delta E_{\sigma^*\pi^*}$ can be experimentally accessed through $E^{\circ}_{(RX, RX^{\cdot-})}$. As a result, linear correlations between $\log k$ and $E^{\circ}_{(RX, RX^{\cdot-})}$ were obtained for aryl chlorides and bromides [17]. Moreover, a correlation of $\log k$ (C–Br bond cleavage) with $\sigma$ was found in $YC_6H_4$–$Br^{\cdot-}$ ($\rho = -17.9$) [66].

With regard to different leaving groups and the substitution pattern at the aryl group several general statements are possible [71, 72a, 92]. The dissociation rate constant of Ar–$X^{\cdot-}$ varies considerably with the nature of the halogen (I > Br > Cl) and its position relative to a second substituent ($o > p > m$). Comparison of cleavage rates reveals the order of reactivity $CN > CH_3C(=O) > NO_2$. For a series of compounds with similar structures the intrinsic activation barrier changes with the thermodynamic component [93], in accordance with the Savéant model [17, 40]. Solvent effects on the cleavage rate constants of $\pi^*$ radical anions are much better understood, as a result of recent investigations [94]. Earlier results implied specific solvation of the transition state, but recent work has emphasized the dominant role of solvation of the leaving group. It was, moreover, shown that solvent effects, which can be described by the Pekar factor, determine the intrinsic rate constant [94].

By use of a series of 1-aroyl-$\omega$-haloalkanes [95a] and 1-aryl-$\omega$-haloalkanes [95b] intramolecular ET between the $\pi^*$ radical anion site and the distal $\sigma^*$ (C–X) acceptor was evaluated for longer distances. The correlation with the acceptor number of the solvent indicates that the negative charge is more localized in the transition state than in the initial radical anion [95b]. Notably, rate constants for the dissociation of Ar–Cl$^{\cdot-}$ show the same correlation with solvent polarity as aroyl-$\omega$-haloalkanes [95b]. Experimental and theoretical investigations on $\omega$-halo-1-alkene$^{\cdot-}$ [96] and phenyl-substituted 4-benzoyloxy-1-methylcyclohexylbromide$^{\cdot-}$ [97] shed further light on intramolecular dissociative ET processes.

## C–X bond cleavage

C–O bond cleavage rates have become accessible through kinetic investigations of diphenyl ether ($k = 4 \times 10^5$ s$^{-1}$) [98] and of trifluoromethoxybenzenes (4-CN: $k = 4 \times 10^3$ s$^{-1}$, 4-H: $k = 3 \times 10^7$ s$^{-1}$). Interestingly, when the leaving group properties of methoxide and chloride were compared [99], cleavage was rapid when attached to a phenyl ring, but much slower when attached to a 4-cyanophenyl group. The C–S thioether bond is more readily cleaved than the C–O bond [100].

## Cleavage of other bonds

Rate data for cleavage of other bonds are rare. According to electrochemical investigations, S–S bond cleavage in diphenyldisulfide$^{\cdot-}$ proceeds at $5 \times 10^8$ s$^{-1}$, whereas that of dialkyldisulfides is slightly slower [101].

### 1.4.3 Synthetic, Kinetic and Thermodynamic Aspects of Oxidative Bond Cleavage

**Bond cleavage after oxidation of $\sigma$ donors**

In comparison with the rich chemistry of $\sigma^*$ radical anions the synthetic utility of $\sigma$ radical cations is quite restricted, because pure $\sigma$ donors are mostly limited to Si–Si, C–Si, C–Ge, and C–Sn functionalities (Scheme 18). Only strained carbocyclic compounds [119] have low ionization energies (*IE*) that make them readily accessible $\sigma$(C–C) and $\sigma$(C–H) donors, but they are of limited synthetic use. In contrast, Me$_3$M–MMe$_3$ (M = Sn, Ge, Si; *IEs* ranging from 8.0 to 8.3 eV) [5] and Me$_4$M (M = Sn, Ge, Si; *IEs* ranging from 8.9 to 9.8 eV) [5] are, in general, better donors

**Common σ donors for oxidative bond cleavage**

C–Si, C-Ge, C–Sn, Si–Si, C–N (in azo compounds)

**Scheme 18.** Frequently encountered σ donors that undergo efficient oxidative bond cleavage.

than terminal alkenes (e.g. propene has a $IE = 9.73$ eV) [5], which makes them attractive for oxidative cleavage.

*Synthetic aspects*

Whereas C–H bond cleavage has been used only rarely, e.g. in adamantane [103], C–C bond rupture is more common, but only in strained carbocycles [104–106]. C–N bond cleavage is observed in azoalkane·+ decomposition [106d], although little synthetic use has been elaborated [107]. In contrast, C–Sn bond cleavage in alkylstannanes has been developed into a novel alkylation of α,β-unsaturated ketones [108] and a versatile alkylation of electron-poor olefins [109]. Oxidative Si–Si [110] and Ge–Ge bond cleavage [102] have also found use.

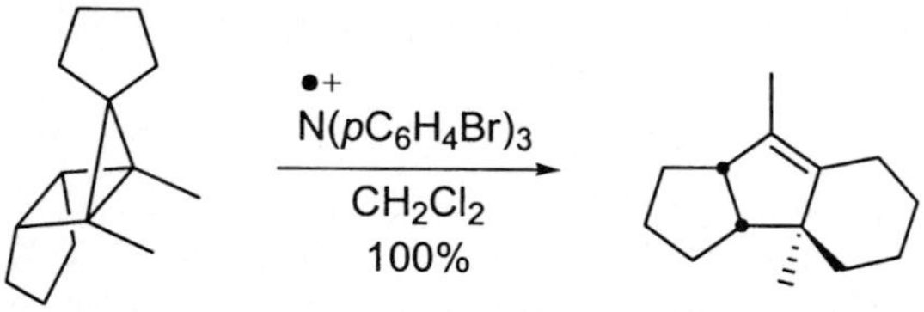

**Scheme 19.** Oxidative C–C bond cleavage in housane **11** leads to the formation of diquinane 12 [106c].

*Kinetic and thermodynamic aspects*

Whereas the rather strong donor qualities of the above electrophores can readily be quantified by the corresponding ionization energies [5, 111], very little is known about the thermochemical data, except as a result of calculation [112, 113]. Fragmentation of $Me_3MR^{·+}$ (M = Si, Sn; R = Me, *i*-Pr, allyl, benzyl, Ph) is endothermic for all R and occurs in the order *i*-Pr ≥ Me > allyl > benzyl > phenyl and M = Sn > Si [113]. C–Sn bonds can be cleaved efficiently through intramolecular assistance by a carbonyl group or heteroatoms [114], as is also known for Si–Si bond scission [115]. Sn–Sn [116], Sn–Ge [116], Sn–Si [116], and Ge–Ge [117] bond-cleavage reactions roughly follow the qualitative order: Sn–C ≫ Ge–Ge > Si–Si [117].

C–H and C–C bond cleavage in σ radical cations is feasible, but not a versatile option for synthetic use because of the very high oxidation potentials. Deprotonations in alkane radical cations have strongly negative Gibbs energies, with strong preference for tertiary over secondary or primary C–H bonds [118]. C–C bonds are selectively weakened in strained carbocycles [119].

**Bond cleavage after oxidation of ($\sigma + \pi$) or ($\sigma + n$) donors**

Analogous to the situation of ($\sigma^* + \pi^*$) acceptors, a donor can be composed of two interacting $\sigma$ and $\pi$ (or $n$) electrophore subunits [120]. For stereoelectronic reasons [121] bond weakening in ($\sigma + \pi$) or ($\sigma + n$) donors is most often encountered between the $\alpha$ and $\beta$ atoms (with respect to the $\pi$ or $n$ electrophore, Scheme 10).

*Synthetic aspects of ($\sigma + \pi$) donors*

C–H deprotonation is frequently encountered at the benzylic position of aromatic [122] and heteroaromatic [123] systems (e.g. in deprotections [124]), in bisallylic systems [125], but less often in allylic functionalization [126]. To compete with other pathways C–C bond scission often needs an extra driving force which is available in strained cyclic systems [127]. Unstrained systems require the assistance of radical- and cation-stabilizing groups to reduce the C–C bond-dissociation energy [128].

**Common (σ+π) donors for oxidative bond cleavage.**

**Combine**

**good σ donor**: C–Si, C–Ge, C–Sn, Si–Si,

or poor σ donors: C–C, C–H

**and allow for good overlap of the σ and π orbitals such as in**

**good π donors**

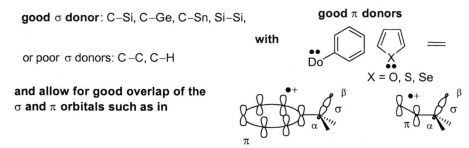

X = O, S, Se

**Scheme 20.** Frequently used ($\sigma + \pi$) donors that undergo oxidative bond cleavage.

The proper use of a strong electrofugal group $X^+$ ($X = SiR_3$, $SnR_3$, $SR$) facilitates highly chemoselective bond scission [129], as in many C–Si [130, 131] and C–Sn cleavage reactions [132]. When alternatives are lacking oxidation of sulfides often results in the oxidative cleavage of the C–S bond [133]. O–H deprotonation is a key step in the oxidative transformation of phenols [134] (Scheme 21) and enols [135]. Much less used are other scissions, e.g. Si–Si bond cleavage [136].

*Synthetic aspects of ($\sigma + n$) donors*

C–H deprotonation plays an important role in the $\alpha$-functionalization of alcohols [138], amines [139] + related *N*-functionalities [140], ethers [141], and sulfides [142]. *N,O*-acetals are versatile precursors to *N*-acyliminium intermediates [143].

Because cation stabilization makes an important contribution, N-, S-, and O-substituents are often used to support C–C bond cleavage [109, 144]. C–C bond scission is readily achieved in strained carbocyclic [145] or heterocyclic [146] ring systems, with many examples stemming from cyclopropyl sulfides [147].

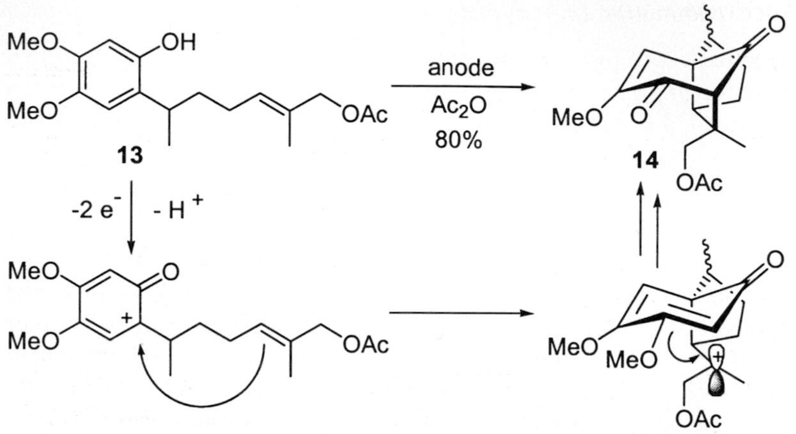
MeO
OH
MeO
OAc
13
anode
Ac2O
80%
-2 e-
- H+
14
OAc
MeO
O
OAc
MeO
MeO
OAc

**Scheme 21.** O–H Deprotonation is the key step in the oxidation of phenol **13** to the tricyclic product **14** [137].

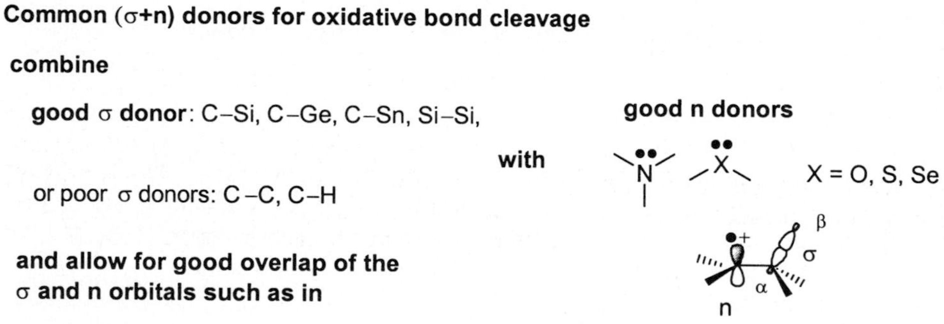
Common (σ+n) donors for oxidative bond cleavage
combine
good σ donor: C–Si, C–Ge, C–Sn, Si–Si,
or poor σ donors: C–C, C–H
with
good n donors
X = O, S, Se
and allow for good overlap of the
σ and n orbitals such as in
β
σ
α
n

**Scheme 22.** Frequently used $(\sigma + n)$ donors that undergo oxidative bond cleavage.

C–Sn [148] and C–Si [121, 149] bond cleavage has developed into a versatile synthetic method for adding functionality $\alpha$ to nitrogen, oxygen and sulfur centers. C–Si bond cleavage is known to be induced by nucleophiles such as pyridine [150] and methanol, and stereoselectivities have been interpreted as indicating that C–Si bonds can even be attacked by a double bond system [151].

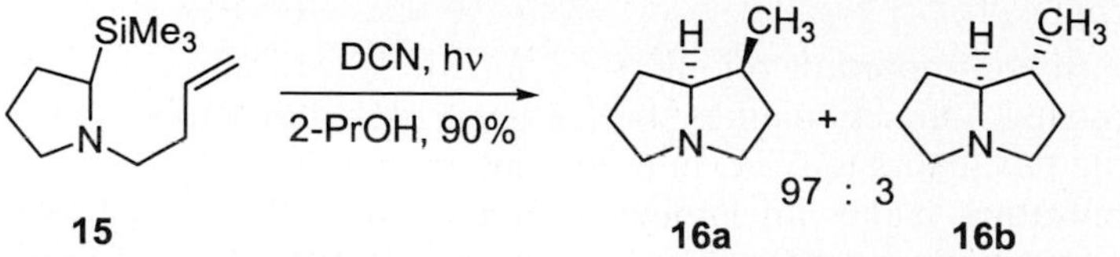
SiMe3
DCN, hν
2-PrOH, 90%
H
CH3
+
97 : 3
15
16a
16b

**Scheme 23.** C–Si bond cleavage assisted by the neighboring double bond [151b].

*Kinetic and thermodynamic aspects*

C–H bond cleavage

Removal of an electron generally brings about significant thermochemical C–H bond weakening both in $(\sigma + \pi)$ radical cations (i.e. deprotonation in the benzylic position [152]) and $(\sigma + n)$ radical cations (i.e. deprotonation at the $\alpha$-carbon of amines [153, 154], ethers [155], and thioethers [156]). The kinetic acidity of benzyl radical cations is rather well understood [152d, 157, 158], although occasionally an addition–elimination route was favored over direct proton transfer [159]. For alkyl arene radical cations [160] a linear relationship has been established between $\log k_{dep}$ and thermochemical acidity [157].

Likewise, the biologically important C–H deprotonation [161] of amine radical cations has been extensively studied [162], with the kinetic acidities paralleling the thermodynamic $pK_a$ values [163]. C–H deprotonation of thioethers has also been studied, but more rarely [164].

C–C bond cleavage

Bond dissociation data [165] indicate that C–C bond cleavage can only occur with appropriate substitution at the carbon centers (alkyl, aryl, hydroxy, siloxy, alkoxy) [166] to stabilize the resultant radical and cationic intermediates [152e]. Reasonable prediction of C–C bond cleavage of $(\sigma + \pi)$ radical cations can be made on the basis of linear correlation between kinetic and thermodynamic data for 1,2-biarylethanes [167]. Activation data correspond closely to the experimental $\Delta G^\circ$ values [60]. Importantly, C–C bond-cleavage reactions in strained-ring systems can be induced by nucleophiles leading to inversion of configuration at the carbon center [168]. $(\sigma + n)$ radical cations equally can undergo C–C bond cleavage at a sufficiently high rate [169] (even decarboxylation [170]).

C–X bond cleavage

Monomolecular C–X bond cleavage energies are accessible by thermochemical cycle calculations; such bond dissociation processes are often induced by nucleophiles, at least with X = $SiR_3$ [163, 171]. As a consequence, rates are much faster than expected on the basis of thermochemical data. Detailed data are available for aniline radical cation desilylation [170].

C–N bond cleavage in cyclic azoalkane$^{\bullet+}$ with expulsion of $N_2$ is apparently very rapid, because only follow-up products were detected [172]. In contrast, C–S bond cleavage in benzyl phenyl sulfide$^{\bullet+}$ is rather slow with $k \approx 10^3$ s$^{-1}$ [164].

O(N)–H bond cleavage

In such systems, C–H deprotonation—if feasible at all—cannot usually compete with O–H deprotonation. Whereas thermodynamic $pK_a$ values are available [173], kinetic data are rather scarce [174], despite the importance of such processes. Very recently, the rate constants for deprotonation of phenol radical cations by water were measured to be $(0.6-6) \times 10^8 \, \text{M}^{-1} \, \text{s}^{-1}$. Consistent with the $pK_a$ values, the 2-methoxyphenol radical cation is more reactive than the 4-methoxy derivative [175].

Cleavage of other bonds

Other bond cleavage (O–Si [176], O–Sn [177], O–C [178], O–P [179], O–Ti [176b, 180], O–Zr [181], S–C [182], Si–Si [183]) can be readily set up in radical cations for synthetic use and in several instances kinetic data are known. Rapid O–C bond cleavage might even occur in the antioxidant action of 7,8-diacetoxy-4-methylcoumarin [184].

**Bond cleavage after intramolecular ET via π or n redox relays**

Although the analogous reaction mode is frequently observed with radical anions, at present this seems to be largely a *terra incognita* in radical-cation chemistry. We would, therefore, like to propose the combination of readily scissile σ donors, e.g. C–Sn, C–Si, and the Group IV M–M bonds (M = Si, Ge, Sn), with good π donors in the same molecules. Because the π donor must be more electron-rich than the σ donor, electron-rich π donors such as ferrocene-, amino-, and alkoxy-substituted phenyl groups, and larger aromatic systems might be appropriate. Such a system is, for example, realized in the C–Sn bond cleavage of trimethylphenylstannane⁺ [113]. Other examples might include methylenecyclopropane [185], which is known to effect rapid cleavage of the C–C bond [186], and of the Ge–O bond in digermoxanes [187].

The elegant synthesis of the harringtonine alkaloid skeleton might involve the photoinduced oxidation of an electron-rich phenyl group that acts as a redox relay by accepting an electron from the σ(C–Si) donor [188].

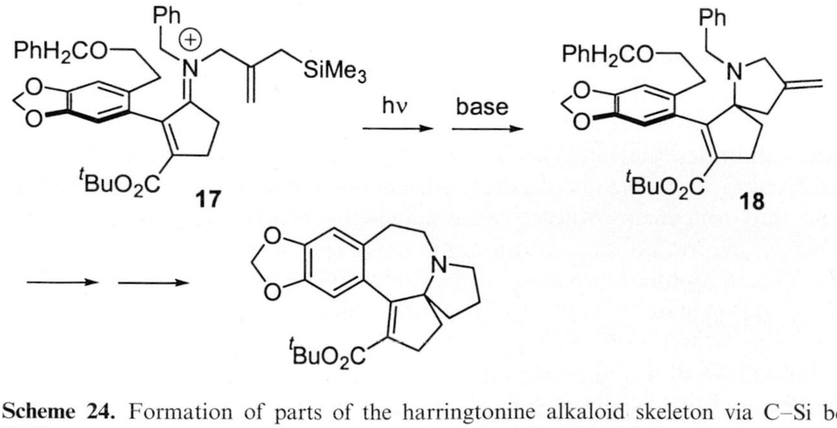

**Scheme 24.** Formation of parts of the harringtonine alkaloid skeleton via C–Si bond cleavage [188].

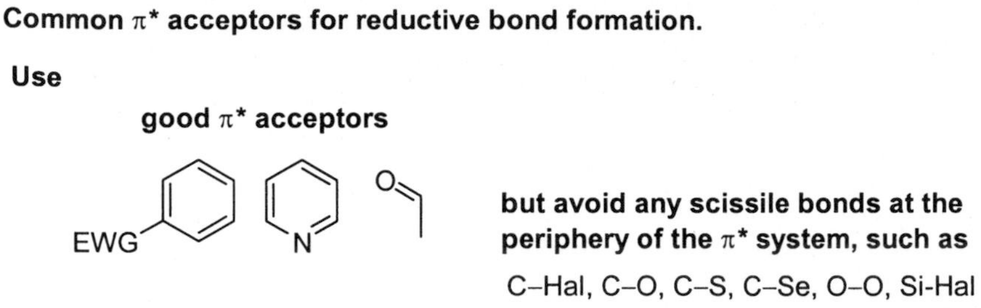

**Common π\* acceptors for reductive bond formation.**

**Use**

      **good π\* acceptors**

**but avoid any scissile bonds at the periphery of the π\* system, such as**

C–Hal, C–O, C–S, C–Se, O–O, Si–Hal

**Scheme 25.** Frequently used $\pi^*$ acceptors for reductive bond-formation reactions.

# 1.5 Reductive and Oxidative Bond-formation Reactions

## 1.5.1 General Principles of Bond Formation

Although the bond-cleavage reactions described above might also lead to bond formation *via* a follow-up intermediate, we would like to discuss in the ensuing section bond formation right at the stage of the radical ion. Again, we recognize very similar reaction motifs in both radical-anion and radical-cation chemistry (cf. Schemes 5 and 25).

## 1.5.2 Synthetic, Kinetic and Thermodynamic Aspects of Reductive Bond-formation

Bond formation at the radical anion stage (Scheme 25) usually occurs either by reaction with an electrophile (proton, cation, unsaturated electron-poor system) or with another odd-electron species (charged or neutral radical).

### Reactions of π * radical anions with closed-shell systems

*Synthetic aspects*

One of the most frequently encountered reactions is that with proton sources, as observed with arenes (Birch reduction) [189], aldehydes [190], alkynes [2d], fullerenes [191], ketones [192] (even enantioselective protonation of ketyl radical anions [193]), nitriles [194], nitro [195] and nitroso compounds [196], and olefins [197]. Protons are often replaced as electrophiles by trialkylsilyl chloride [198].

    C–C bond formation in inter- or intramolecular additions starting with olefin [199, 200], ketyl [201] (generated by PET [202], chemical or cathodic [24, 203] reduction) [204, 205] or imine radical anions [206] has become a versatile method. In general, the intramolecular addition is highly suitable for the construction of five membered rings, less so for six-, and not effective for seven-membered ring formation.

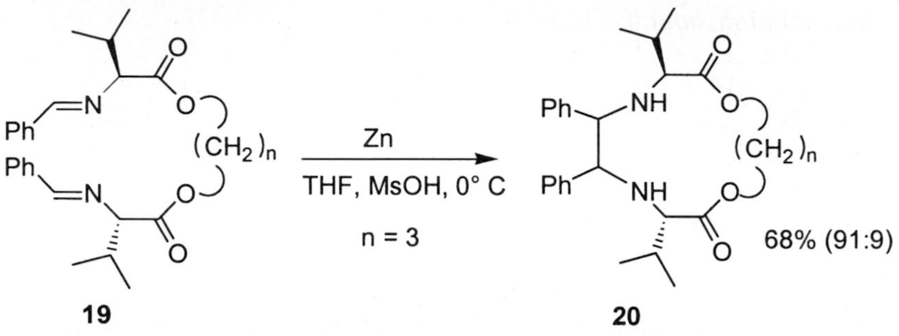

**Scheme 26.** Stereoselective bisimine cyclization via radical anion intermediates [206b].

*Kinetic and thermodynamic aspects*

Protonation

Protonation becomes a rapid reaction in protic solvents and in the presence of acids, as demonstrated for, e.g., *n*-butyl acrylate in aqueous solution [207], methyl acrylate in EtOH [208], cinnamates in the presence of phenol in DMF [209], and benzaldehyde in ethanolic buffer solution [210]. Rate constants for protonation of aromatic radical anions (anthracene [211], naphthalene, 2-methoxynaphthalene, 2,3-dimethoxynaphthalene) by a number of proton donors including phenols, acetic acid, and benzoic acids in aprotic DMF were found to vary from $5.0 \times 10^4$ $M^{-1}$ $s^{-1}$ (for anthracene, in the presence of *p*-chlorophenol) to $6.2 \times 10^7$ $M^{-1}$ $s^{-1}$ (for anthracene, in the presence of pentachlorophenol) [212]. For dimedone, PhOH, or $PhCO_2H$ the rate of protonation depends on the hydrogen-bond basicity of the solvent and increases in the order DMSO < DMF ≪ MeCN [213].

When the substrate was more difficult to reduce the rate of protonation of alkyl cinnamates in the presence of phenol in DMF [209] increased. For aryl cinnamates the competing rate-determining dimerization of the radical anions is predominant [214]. In MeOH proton transfer occurs as a monomolecular reaction with the hydrogen-bonded radical anion complex (first-order rate constant $k_{prot}$ varies from $2.9 \times 10^2$ $s^{-1}$ to $3.5 \times 10^3$ $s^{-1}$) [215].

Faster protonation rates were observed in the self-protonation of aromatic carboxylic radical anions ($k_{prot} > 10^6$ $M^{-1}$ $s^{-1}$) [216] and (*R*)-(−)-1,1′-biphenyl-2,2′-diyl hydrogen phosphate radical anion ($k_{prot} = 5.7 \times 10^9$ $M^{-1}$ $s^{-1}$) [217].

Reaction with other electrophiles

Kinetic studies of the reaction of $CO_2$ with radical anions generated from dialkyl fumarates and maleates showed that C–C bond formation was the rate-determining step. The pseudo first-order rate constants, $k_{CO2}$, for fumarate radical anions in $CO_2$-saturated DMF were found to vary between 0.35 and 1.5 $s^{-1}$ and to decrease in the same order as observed for dimerization [218]. Rate constants for maleates ($k_{CO2}$ varied from 32.0 $s^{-1}$ to 18.0 $s^{-1}$) were higher. Rather slow is the coupling of $CO_2$ with the 4-keto isophorone·− in MeCN ($k_{CO2} = 0.35$ $s^{-1}$) [219].

Aromatic radical anions and alkyl halides can react either by ET or direct $S_N2$ [95a, 220]. The $S_N2$ reaction depends on the steric requirements of the substrate and the magnitude of the driving force for the ET process [221]. In general, with increasing standard potential of the aromatic compound or decreasing ability of the substrate as electron acceptor, the $S_N2$ mechanism becomes more important.

**Reactions of $\pi^*$ radical anions with open-shell systems**

*Synthetic aspects*

The coupling of two radical anions is frequently realized in the reductive dimerization of olefins with different groups R, for example R = CN (e.g. acrylonitrile [234]), C(=O)OR [198b, 237], C(=O)X [222], Ph [223], $NO_2$ [224], and $SO_2Ph$ [225]. Diastereoselective and enantioselective dimerizations are reported for cinnamic esters [226] and oxazolidones [227], respectively.

On the other side, coupling of radical anions with radicals is a key step in the electrochemical reductive *t*-butylation of aromatic compounds [228], in the photo-NOCAS reaction [229], and for other combinations [87].

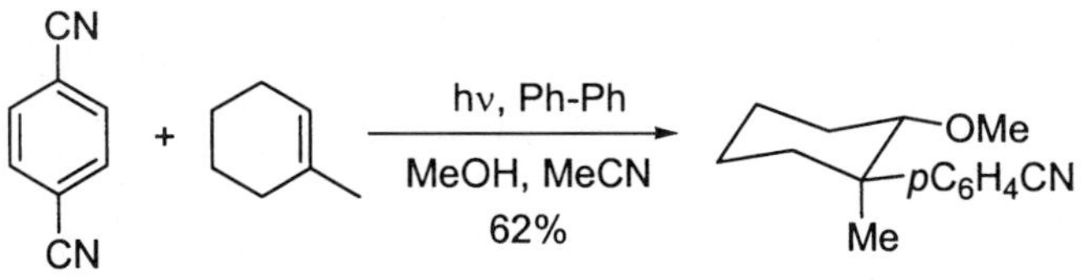

**Scheme 27.** Radical anion–radical coupling in the photo-NOCAS reaction [229].

Radical-anion and radical-cation intermediates, for example, react with each other after PET in donor–acceptor systems. After proton reorganization they undergo cyclization to provide a direct synthetic route to macrocycles and *N*-heterocycles with a variety of ring sizes [230]. Cycloadditions *via* radical ion pairs [231] and the C–C bond formation between $C_{60}$ and $N,O$-ketene acetals [232] also fit this category.

*Kinetic and thermodynamic aspects*

Intermolecular reactions of $\pi^*$ radical anions with electron-poor olefins often proceed quite sluggishly. As a consequence, radical anion electrohydrodimerization (EHD) and radical anion–radical reactions are frequently encountered during electrochemical reduction when the radical concentration is rather high.

Dimerization of activated olefins

In general, the cathodic reduction of electron-deficient olefins generates a $\pi^*$ radical anion which undergoes radical anion–radical anion dimerization to form the dianion dimer which undergoes protonation and/or cyclization. Several investigations

[233] have been performed on reductive dimerization; the best known example is the production of adiponitrile from acrylonitrile [234]. The initial step in the EHD is a rapidly established equilibrium between two radical anions and a dimer dianion; this has been extensively studied for cinnamates [235], fumarates [236], and maleates [237].

For dimerization of methyl and ethyl cinnamate the equilibrium constants were estimated to be 53.1 and 109.0 $M^{-1}$, respectively [238]. In solvents of low acidity (acetonitrile, DMF, alkaline ethanol) [235] the radical anion dimerization step was found to be rate-determining ($k \approx 5 \times 10^3$ $M^{-1}$ $s^{-1}$) depending on the structure of the alkyl group of the cinnamic ester moiety [239] (e.g., for electron-donating $R = Bu^t$, $k_{obs} = 4.1 \times 10^2$ $M^{-1}$ $s^{-1}$; for electron-withdrawing $R = 4\text{-}NC\text{-}C_6H_5$, $k_{obs} = 5.7 \times 10^4$ $M^{-1}$ $s^{-1}$). Rate studies indicate that complexation with water is crucial for the radical anion–radical anion coupling step of ethyl cinnamate ($k = 5.4 \times 10^2$ $M^{-1}$ $s^{-1}$ in DMF containing 0.28 $M$ $H_2O$; $k = 1.4 \times 10^2$ $M^{-1}$ $s^{-1}$ in dry DMF [236]). As a consequence, stereoselectivity in cinnamic ester coupling could be controlled by appropriate selection of the solvent. In slightly wet DMF dimerization is relatively slow and stereoselective—involving two radical anions with a hydrogen-bonded water molecule serving as a template—whereas in MeOH two neutral radicals are involved in the rapid and unselective dimerization.

The dimerization rate constants of dialkyl fumarates ($k = 25$–$120$ $M^{-1}$ $s^{-1}$) [236] decrease with increasing steric hindrance of the alkyl group and prove to be significantly smaller than those of fumaronitrile ($k = 7 \times 10^5$ $M^{-1}$ $s^{-1}$) [240]. The electrochemically generated dimethyl maleate$^{\bullet-}$ undergoes rapid *cis–trans* isomerization to the dimethyl fumarate$^{\bullet-}$ then rapid radical anion dimerization [241]. In comparison, the EHD rate constant for 4-methylcoumarin ($k = 1.8 \times 10^5$ $M^{-1}$ $s^{-1}$) proved similar to that of a relatively fast cinnamate (4-cyano; $k = 5.7 \times 10^4$ $M^{-1}$ $s^{-1}$) [242].

Dimerization of arenes and heteroarenes

The dimerization of anthracene$^{\bullet-}$ has been studied extensively [243, 244]. With strongly electron-withdrawing substituents at position 9 the radical anions undergo reversible dimerization in aprotic solvents such as DMF, MeCN, propylene carbonate, DMSO etc. followed by rate determining $\sigma$ bond formation to furnish the stable dimer dianion [245]. In DMF $k_{dim}$ were found to decrease in the order $NO_2$ ($k_{dim} = 1.6 \times 10^6$ $M^{-1}$ $s^{-1}$) > CHO ($k_{dim} = 3.1 \times 10^5$ $M^{-1}$ $s^{-1}$) > CN ($k_{dim} = 1.3 \times 10^5$ $M^{-1}$ $s^{-1}$).

Reductive dimerization of benzene [246], cyano biphenyl ether [247], pyridine [248], and acridine [249] derivatives has also been investigated. Radical anions of diesters of pyridine and benzene undergo rapid reversible dimerization ($k_{dim} = 10^3$–$10^4$ $M^{-1}$ $s^{-1}$) [250].

Dimerization of aldchydes, ketones and related systems

In general, reductive generation furnishes a ketyl radical anion which undergoes radical–radical coupling to form the dimer dianion [251]. This dimerization was found to be faster than the reaction of the radical anion with the parent molecule (e.g. for of *p*-cyanobenzaldehyde: 28.6 $M^{-1}$ $s^{-1}$ compared with 1.45 $M^{-1}$ $s^{-1}$) [252].

Kinetic studies have revealed that aliphatic ketyl radical anions are very short-lived compared with aromatic (half life of acetone$^{\bullet-}$ in aqueous 2-propanol is 72 μs, whereas that for acetophenone$^{\bullet-}$ is 1.5 ms) [253]. The reductive dimerization of simple aromatic aldehydes has been studied in aprotic solvents, with the second order rate constant being larger in acetonitrile than in DMF, because of ion-pair effects [254]. Electron-withdrawing substituents reduce the speed of dimerization (benzaldehyde$^{\bullet-}$: $k = 2.4 \times 10^3$ M$^{-1}$ s$^{-1}$, p-cyanobenzaldehyde$^{\bullet-}$: $k = 5$ M$^{-1}$ s$^{-1}$) [255], whereas protic solvents lead to protonation before dimerization [256].

In imine reductions rapid equilibration of dimeric dianions and the precursor radical anions leads to the thermodynamically more stable isomer [257]. In the reductive coupling of salicylideneanilines, however, rate-determining C–C bond formation is preceded by intramolecular H-bridging [258].

## Reaction with neutral radicals

Few rate data have been reported for the coupling of radical anions with alkyl radicals [259]. In general, coupling is rapid with rate constants approaching the diffusion limit (1-hexenyl radical + naphthalene$^{\bullet-}$: $k_2 = 2.4 \times 10^9$ M$^{-1}$ s$^{-1}$ in 1,2-dimethoxyethane) [260]. Although the rates of coupling of alkyl radicals with aromatic radical anions are found to vary slightly in the order primary > secondary > tertiary, the small activation energy is not sensitive to structural differences between the alkyl radical and in the aromatic substrate or to variations in SOMO–SOMO energies [261].

## 1.5.3 Synthetic, Kinetic, and Thermodynamic Aspects of Oxidative Bond-formation

### Reactions of π radical cations with closed-shell systems

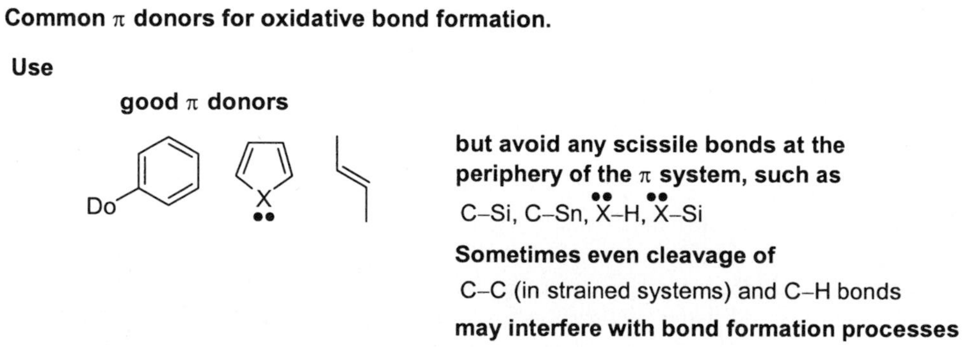

#### Common π donors for oxidative bond formation.

**Use**

**good π donors**

but avoid any scissile bonds at the periphery of the π system, such as
C–Si, C–Sn, X–H, X–Si

**Sometimes even cleavage of**
C–C (in strained systems) and C–H bonds
**may interfere with bond formation processes**

**Scheme 28.** Frequently used π donors for oxidative bond-formation reactions.

*Synthetic aspects*

Intermolecular reactions involving attack of n-nucleophiles (even F$^-$) [262] at π radical cations include reactions at alkenes [263], allenes [264], aromatic systems

[265], enol esters [266], enol ethers [267], fullerenes [268], benzofurans [269], dienes [270, 271], furans [272], pyrrolidines [273], and ketene imines [274]. These reactions usually proceed by *anti*-Markovnikov addition [275]. C–C bond formation is possible by reaction with cyanide [264]. Truly spectacular examples are derived from PET cascade cyclizations [276], as demonstrated in a short biomimetic route to the steroid system **25** (two out of 256 stereoisomers) [277].

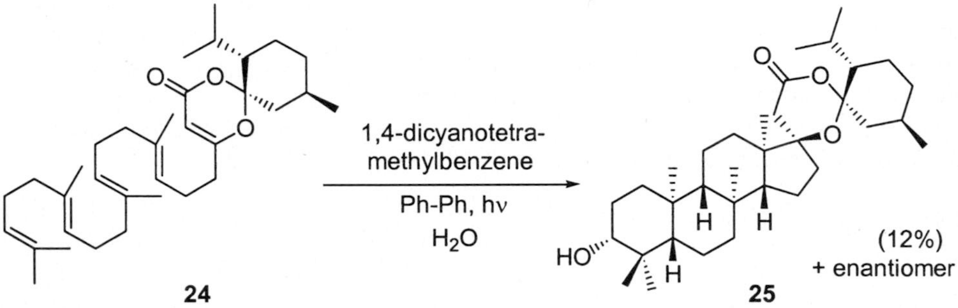

**Scheme 29.** Biomimetic cascade cyclization of **24** to steroid ring system **25** [277].

Likewise, *n*-nucleophiles react well in intramolecular cyclizations to trap olefin and arene radical cations [278].

Because both radical–nucleophile and radical–radical coupling (electropolymerization [316]) can occur under anodic conditions mechanistic differentiation is often difficult. Under chemical oxidation [279b] and PET conditions [279a], however, only radical–nucleophile coupling is encountered. Intermolecular [280] and intramolecular [281] C–C bond formation used for biomimetic atroposelective coupling [282] and for natural product synthesis [283] is nowadays a well-established synthetic tool.

*Kinetic and thermodynamic aspects*

Although addition of a nucleophile to a $\pi$ radical cation seems to be a straightforward process, several mechanistic scenarios [284] (e.g. the disproportionation, complexation, and half-regeneration pathways) that depend on the reactants need to be considered [285]. It has, moreover, been stressed that for protic nucleophiles such as alcohols and water, deprotonation of the primary adduct is important [286]. As a consequence, the rational design of bond-forming reactions requires deeper understanding of mechanistic matters.

Extensive kinetic data have been collected over the last decade especially for substituted arylanthracenes [287], dienes [288], styrenes [289], and psoralen [290]. Combinations of $\pi$ radical cations and *n*-type nucleophiles are very rapid processes that approach the diffusion controlled limit for anionic nucleophiles. Primary amines were found to add to 9-phenylanthracene radical cation with second-order rate constants in the range $8 \times 10^6$ to $3 \times 10^9$ $M^{-1}$ $s^{-1}$. The rates increase with increasing amine basicity and decreasing steric requirements [287]. If, however, the amines become too strong donors (e.g. tertiary amines and anilines) they might react with the radical cation by ET [291]. Less studied are reactions with protic nucleophiles that occur according to the deprotonation mechanism. In line with nucleophilicity data [292], bond formation with water and alcohols is much slower than with amines and pyridines [286, 289].

The reaction of $\pi$ radical cations with $\pi$ nucleophiles usually leads to C–C bond formation, a reaction that can be very fast (cf. pericyclic reactions also), as in the oxidative dimerization of triphenylamine ($k = 1$–$10 \times 10^7$ $M^{-1}$ $s^{-1}$) [293]. Hence, such a reaction mechanism can even operate in anodic oxidations (4-methoxybiphenyl [294], tetrahydrocarbazole [295], 4,4'-dimethoxystilbene [296] and 9-methoxyanthracene) [297], where the radical cation concentration is very high.

**Reactions of $\pi$ radical cations with open-shell systems**

*Synthetic aspects*

$\pi$ Radical cations react readily with stable radicals, e.g. oxygen [298], superoxide [299] or $NO_2$ [300], and short-lived radicals (as detected in the charge transfer nitration [301] of electron-rich aromatic systems with tetranitromethane) [302]. Moreover, PET in general leads to intra- [303] or intermolecular [304] radical ion pairs [305], but direct bond formation between organic radical cations and radical anions is rare. It might play a role in PET Paternò–Büchi reactions [306], and in related PET cycloadditions [307]. If the radical anion site of an intramolecular radical ion pair is designed to lose an anion, efficient protocols to radical cation–radical combinations can be installed.

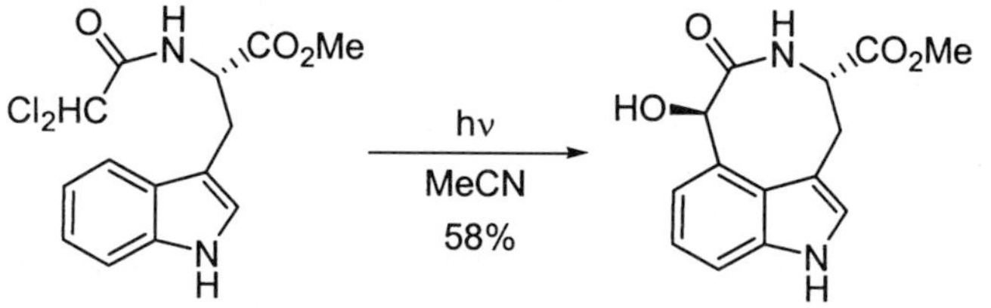

**Scheme 30.** Radical cation–radical coupling after intramolecular PET and loss of chloride at the radical anion site [305a].

Radical cation–radical cation coupling [308] is often observed during anodic oxidations when their concentration is sufficiently high; the capacity of $\pi$ radical cations to undergo atom abstraction reactions, e.g. of fullerene$^{\cdot+}$ [309], has little significance only.

*Kinetic and thermodynamic aspects*

Reactions with stable radical species, e.g. oxygen and $NO_2$, occur readily, although oxygenation rates of aryl olefin radical cations (ca $10^6$ $M^{-1}$ $s^{-1}$ [310]) are two or three orders of magnitude slower than those of neutral radicals with oxygen [311]. The polar effect in such reactions is remarkable. Whereas *trans*-stilbene$^{\cdot+}$ has little reactivity towards oxygen, derivatives with a *p*-MeO group react at $k = 1.2$–$4.5 \times 10^7$ $M^{-1}$ $s^{-1}$ [312]. Much more rapid is the reaction of stilbene$^{\cdot+}$ with superoxide ($k = 4.1 \times 10^{10}$ $M^{-1}$ $s^{-1}$) [310].

Reactions that include the dimerization of two radical cations are, for example, those of anethole [313], pyrroles [314], oligopyrroles [315], thiophenes [316], oligo-thiophenes [317], diphenylamine [318], triphenylamine [319], *N,N*-dimethylaniline [320], and diphenylpolyene [321].

**Reactions of *n* radical cations with closed-shell systems**

**Common n donors for oxidative bond formation.**

**Use**

    **good n donors**

**but avoid any scissile bonds at the periphery of the π system, such as**
C–Si, C–Sn, X–H, X–Si

**Sometimes even cleavage of**
C–C (in strained systems) and C–H bonds
**may interfere with bond formation processes**

**Scheme 31.** Commonly used *n* donors for oxidative bond formation.

*Synthetic aspects*

Whereas nitrogen-centered radical cations are prone to C–H, C–C bond-cleavage reactions, $R_3P^{\cdot+}$ and $R_2S^{\cdot+}$ suffer predominantly nucleophilic attack [322, 323]; few exceptions have been reported to date. This reaction plays an important part in the formation of oxygenated derivatives (P [324], S [325]), C–S [326], S–S, and Se–Se bond formation [327], and in removing sulfur-protecting groups [328–330].

Whereas intermolecular attack of a π nucleophile is rather rare [331] (e.g. disulfide$^{\cdot+}$ + arene [332], and diselenide$^{\cdot+}$ + dienes [333]), intramolecular cycliza-tions of amine radical cations have been useful in the preparation of five-membered rings [334].

Hydrogen abstraction is the key factor in the intramolecular radical functionali-zation step of the Hofmann–Löffler–Freytag reaction (Scheme 32) [335].

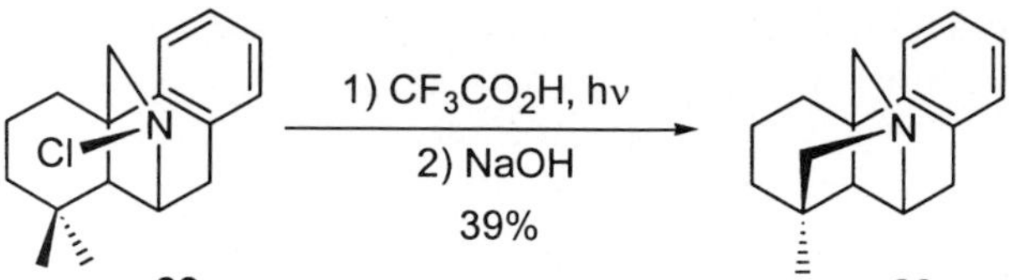

**Scheme 32.** Successful preparation of pentacycle **29** [336] via the Hofmann–Löffler–Freytag reaction.

*Kinetic and thermodynamic aspects*

Kinetic data are very scarce and available mostly for amine, phosphane [337], and sulfide [338] radical cation reactions with nucleophiles. In intermolecular reactions, phosphane radical cations react with sulfur nucleophiles [339], $k = 10^5$–$10^7$ $M^{-1}$ $s^{-1}$, following the same reactivity pattern as with $\pi$ radical cations [289]. Cyclization rate constants for the attack at double-bond systems are faster than those of the neutral radicals, as shown for dialkylamine radical cations [340] ($k = 6 \times 10^5$– $1 \times 10^{10}$ $M^{-1}$ $s^{-1}$).

**Reactions of *n* radical cations with open-shell systems**

*Synthetic aspects*

Oxygen reacts readily with aromatic sulfides [341], *N*-methionyl peptides [342], 1,3-dithianes [343], aziridines [344], and phosphorus [345] compounds after oxidation. On the other hand, aryl radicals are involved in the ET reaction of trivalent phosphorus compounds with aromatic diazonium salts [346].

*Kinetic aspects*

Although the homogeneous reaction of sulfide$^{\bullet+}$ with oxygen has been accomplished at high $O_2$ pressures only [347], the spontaneous reaction in a zeolite was recently monitored kinetically [348]. The very large rate constant for the reaction between dimethylsulfide$^{\bullet+}$ and superoxide, $k = 2.3 \times 10^{11}$ $M^{-1}$ $s^{-1}$, can be interpreted in terms of rapid ET rather than radical–radical coupling [349]. Rates are also known for the reaction of phosphane$^{\bullet+}$ with oxygen [339].

# 1.6 Pericyclic Reactions

## 1.6.1 General Principles of Pericyclic Reactions

*Per definitionem*, pericyclic reactions proceed in a concerted manner, i.e. bond cleavage and bond formation occur simultaneously. Recent high-level calculations on radical cation Diels–Alder reaction have, however, put the stepwise [350] and concerted [351] cycloadditions close in energy whereas experimental evidence (mostly [352] stereospecific! [353]) argues for concertedness. Stereospecific retro-Diels–Alder fragmentation has, moreover, been observed in the gas phase [354].

### 1.6.2 Synthetic, Kinetic and Thermodynamic Aspects

**Radical anion pericyclic processes**

*Synthetic aspects*

Although cycloadditions have frequently been observed in radical-cation chemistry, this reaction mode is apparently very rare in radical-anion chemistry because of the electron repulsion term. Few examples are known of Diels–Alder dimerizations [355], [2 + 2] cycloadditions [356], retro-[2 + 2] cycloadditions [357], and cyclo-trimerizations [358]. Equally, little is known about electrocyclic reactions, despite their interesting stereochemical course [359].

Although the synthetic utility of radical anion pericyclic processes is still to be explored, the recently disclosed intermediacy of radical anions in metathesis reactions with Grubb's catalyst [360] should ignite the search for further examples of this interesting class of reactions.

*Kinetic and thermodynamic aspects*

To the best of our knowledge no experimental kinetic and thermodynamic studies have been performed on radical anion pericyclic processes. Interesting structural information has, however, been obtained for some [2 + 2] cycloadduct·⁻, e.g. 4N/5e radical anions [361].

**Radical cation pericyclic processes**

*Synthetic aspects*

Since their original discovery [362] pericyclic radical cation reactions have been developed for various synthetic formats. The methodology nowadays is most advanced for intra- [363] and intermolecular Diels–Alder cycloadditions [364], and

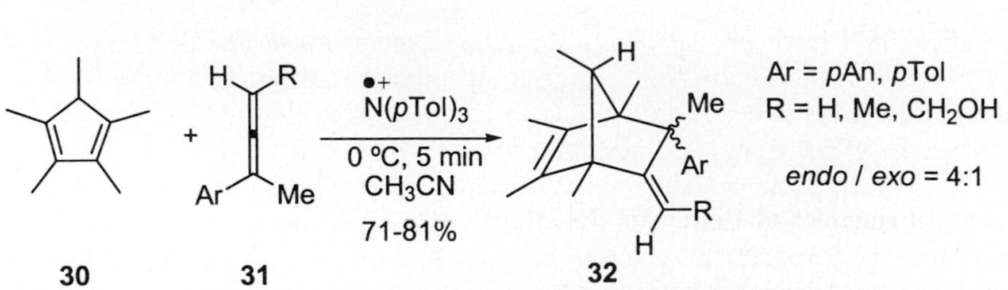

**Scheme 33.** Only two out of 32 possible isomers of **32** are formed in this highly peri-, chemo-, regio- and stereoselective radical cation Diels–Alder cycloaddition [365b, c].

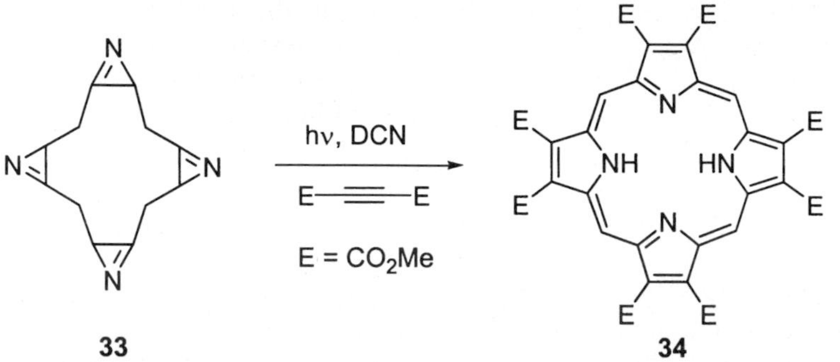

**33**                                                    **34**

**Scheme 34.** Preparation of porphyrin **34** initiated through electrocyclic ring opening of four azirine radical cations to 2-azaallenyl radical cations [373c].

wide variations in the dienophile [365] (cf. allene **31**) and the diene [366] have been realized. Likewise, several examples are known of other cycloadditions ([2 + 2] [367], with diazomethane [368] or osmium tetroxide [369], epoxidation [370], cyclopropanations [371]) and cycloreversions [372].

An exciting cycloaddition which is not pericyclic in nature has been developed on the basis of PET opening of azirines to 2-azaallenyl radical cations [373]; these add readily to substrates such as imines and acetylenes. This paves the way for the synthesis of imidazoles [373a, e], heterophanes [373b, e], and even porphyrins (cf. the preparation of **34**) [373c, e].

Many classical pericyclic reactions, for example the Cope [374, 375], Claisen [376], vinylcyclobutane [377], vinylcyclopropane [378], and vinylcyclobutanone [365a] rearrangements and the [1, 16] hydrogen shift in an A/D *seco* corrin radical cation [379], can be translated into the radical cation format equally easily. The stereochemical course of these rearrangements has been described as stereospecific in the vinylcyclobutane [377, 380] and Cope [374b] rearrangements but with retention, inversion, and partial loss of the stereochemical integrity in the rearrangement of vinylcyclopropane [381] and vinylcyclobutanone [365a] radical cations.

*Kinetic aspects*

Recent studies [382] have provided rate data for cycloaddition reactions. Accordingly, steric effects at the electrophile site, and the capacity of the added unsaturated component $RCH=CH_2$ to stabilize radicals and cations, play a vital role, the importance of which is reflected in a rate decrease for $R = Ph \approx OR >$ vinyl $>$ alkyl by a factor of 100–300. In principle, the observed trends follow those for addition of carbocations to alkenes [292]. A study of the [2 + 1] cycloaddition of 4-methoxystyrene also emphasizes the importance of the rapid one-electron reduction of the intermediate dimer radical cation [383]. A direct view of 4-center 3-electron cyclobutane [384] and bisdiazene-oxide [385] radical cations has been obtained with polycyclic, rigid systems.

## 1.7 Conclusion

In this brief review we have outlined a unifying concept to promote understanding of primary reaction pathways of radical ions in solution. From the analysis in this review, which is based on (i) identifying the electrophore in a complex molecule and (ii) identifying the character of the electrophore, it has become clear that both radical anions and radical cations undergo analogous reactions with regard to the electrophoric system. Our heuristic concept of relating the reactivity of a radical ion to the relevant electrophore is closely connected to the principle of localized charge as used in mass spectroscopy [386]. Hence, the interested reader will find many analogies in radical ion gas-phase chemistry, an area which unfortunately could not be covered in our present analysis [387].

It is hoped that our guide for predicting the reactivity of radical ions, whether generated by electrolysis, or by chemical or photochemical ET processes, will encourage scientists to devise novel radical-ion reactions for synthetic applications. Because our analysis has aimed at covering synthetically relevant radical-ion transformations, it should be noted that less frequently used reactions, such as *cis–trans* isomerizations, and ET oxidation or reduction of radical ions are not included. One should, moreover, bear in mind that the reactivity of radical ionic intermediates might be heavily influenced by counterion effects [388], a research area which still deserves major attention.

Finally, it can be stated that many reactivity patterns of free radical ions are equally found in oxidative and reductive transformations involving initial inner-sphere ET, such as in reactions with samarium iodide [389], low valent titanium [390] and titanocene complexes [391], manganese(III) [392], and CAN [393].

### Acknowledgments

M. K. G. is deeply indebted to the Alexander von Humboldt Stiftung, and M. S. would like to thank all his coworkers whose commitment has made possible the joint contributions in the field of electron transfer activation. Generous financial support over many years from the Deutsche Forschungsgemeinschaft (SFB 347, Schwerpunkt Normalverfahren, Graduiertenkolleg), the Volkswagen-Stiftung, and the Fonds der Chemischen Industrie is gratefully acknowledged.

### References

1. T. Linker, M. Schmittel, *Radikale und Radikalionen in der Organischen Synthese*, Wiley–VCH, Weinheim, **1998**.
2. (a) *Photoinduced Electron Transfer*, (Eds.: M. Chanon, M.-A. Fox), Parts A–D, Elsevier, Amsterdam, **1981**; (b) R. A. Rossi, R. H. de Rossi, *Aromatic Substitution by the $S_{RN}1$ Mechanism*, ACS 178, Washington DC, **1983**; (c) L. Eberson, *Electron Transfer Reactions in Organic Chemistry*, Springer, Berlin, **1987**; (d) *Organic Electrochemistry*, (Eds.: H. Lund, M. M. Baizer) Marcel Dekker, New York **1991**; (e) T. Shono, *Electroorganic Synthesis*, Acad. Press,

London, **1991**; (f) G. J. Kavarnos, *Fundamentals of Photoinduced Electron Transfer*, VCH, New York, **1993**; (g) K. Yoshida, *Electrooxidation in Organic Chemistry*, Krieger, Malabar, **1993**; (h) D. Kyriacou, *Modern Electroorganic Chemistry*, Springer, Berlin, **1994**; (i) J. Volke, F. Liška, *Electrochemistry in Organic Synthesis*, Springer, Berlin, **1994**.

3. V. D. Parker, *Acta Chem. Scand.* **1998**, *52*, 145–153.
4. J. Santamaria, M. T. Kaddachi, J. Rigaudy, *Tetrahedron Lett.* **1990**, *31*, 4735–4738.
5. NIST Chemistry WebBook by P. J. Linstrom; General Editor W. G. Mallard, NIST Standard Reference Database No 69 (November 1998 Release), [http://webbook.nist.gov/chemistry/]
6. P. Kebarle, S. Chowdhury, *Chem. Rev.* **1987**, *87*, 513–534.
7. J. A. Pople, *Angew. Chem.* **1999**, *111*, 2014–2023; *Angew. Chem. Int. Ed.* **1999**, *38*, 1894–1902.
8. J. Cossy, D. Belotti, J. P. Pete, *Tetrahedron Lett.* **1987**, *28*, 4545–4546.
9. (a) M. J. S. Dewar, E. G. Zoebisch, E. F. Healy, J. J. P. Stewart, *J. Am. Chem. Soc.* **1985**, *107*, 3902–3909; (b) J. J. P. Stewart, *J. Comput. Chem.* **1989**, *10*, 221–264; (c) A comparison of calculated and experimental electron affinity data [5] indicates that AM1 [9a] performs somewhat better than PM3. [9b] d) As ionization potentials and electron affinity data of iodo compounds are much better calculated by PM3 [9b] than by AM1 [9a] we used the PM3 semi-empirical method for the data in Table4.
10. A comparison of calculated and experimental ionization energies shows that AM1 [9a] and PM3 [9b] perform better than ab initio (STO-3G, 3–21 G, 6–31G*, 6–31+G**) and DFT (B3PW91, BLYP, B3LYP, B3P86) calculations, see P. Politzer, F. Abu-Awwad, *Theor. Chem. Acc.* **1998**, *99*, 83–87.
11. J.-P. Cheng, Z. Zheng, *Tetrahedron Lett.* **1996**, *37*, 1457–1460.
12. Determined by thermochemical cycle calculations using ionization energies from reference [5].
13. K. Daasbjerg, *J. Chem. Soc., Perkin Trans.* 2 **1994**, 1275–1277.
14. Electrochemistry I–VI, Ed. E. Steckhan, *Top. Curr. Chem.* **1987**, *142*; **1988**, *148* and *143*; **1990**, *152*; **1994**, *170*; **1997**, *185*; Electron Transfer I–II, Ed. J. Mattay, *Top. Curr. Chem.* **1994**, *169*; **1996**, *177*.
15. D. D. M. Wayner, V. D. Parker, *Acc. Chem. Res.* **1993**, *26*, 287–294.
16. Z.-R. Zheng, D. H. Evans, E. S. Chan-Shing, J. Lessard, *J. Am. Chem. Soc.* **1999**, *121*, 9429–9434.
17. J.-M. Savéant, *Tetrahedron* **1994**, *50*, 10117–10165.
18. T. W. Greene, P. G. M. Wuts, *Protective Groups in Organic Synthesis*, Wiley, New York, **1999**.
19. M. S. Mubarak, D. G. Peters, *J. Electrochem. Soc.* **1996**, *143*, 3833–3838; T. Fuchigami, M. Kasuga, A. Konno, *J. Electroanal. Chem.* **1996**, *411*, 115–119; A. A. Pud, G. S. Shapoval, O. E. Mikulina, L. L. Gervits, *Russ. J. Electrochem.* **1996**, *32*, 337–342; A. Kotsinaris, G. Kyriacou, C. Lambrou, *J. Appl. Electrochem.* **1998**, *28*, 613–616.
20. A. J. Fry, J. Touster, *Electrochim. Acta* **1997**, *42*, 2057–2063.
21. J. Simonet, *L'Actualite Chimique* **1998**, 4–42.
22. Some recent examples providing access to older references: M. Médebielle, M. A. Oturan, J. Pinson, J. M. Savéant, *J. Org. Chem.* **1996**, *61*, 1331–1340; M. A. Nazareno, R. A. Rossi, *J. Org. Chem.* **1996**, *61*, 1645–1649; M. C. Murguia, R. A. Rossi, *Tetrahedron Lett.* **1997**, *38*, 1355–1358; A. N. Santiago, A. E. Stahl, G. L. Rodriguez, R. A. Rossi, *J. Org. Chem.* **1997**, *62*, 4406–4411; A. E. Lukach, A. N. Santiago, R. A. Rossi, *J. Org. Chem.* **1997**, *62*, 4260–4265; Z. Y. Long, Q. Y. Chen, *J. Fluor. Chem.* **1998**, *91*, 95–98; E. Delli, S. Kouloumtzoglou, G. Kyriacou, C. Lambrou, *Chem. Comm.* **1998**, 1693–1694; A. E. Lukach, R. A. Rossi, *J. Org. Chem.* **1999**, *64*, 5826–5831.
23. M. Yus, *Chem. Soc. Rev.* **1996**, *25*, 155–161.
24. R. D. Little, M. K. Schwaebe, *Top. Curr. Chem.* **1997**, *185*, 1–48.
25. R. Shao, D. G. Peters, *J. Org. Chem.* **1987**, *52*, 652–657.
26. E. Bartmann, *Angew. Chem.* **1986**, *98*, 629–631; *Angew. Chem. Int. Ed. Engl.* **1986**, *25*, 653–655; T. Cohen, I.-H. Jeong, B. Mudryk, M. Bhupathy, M. A. Awad, *J. Org. Chem.* **1990**, *55*, 1528–1536.
27. B. Mudryk, T. Cohen, *J. Org. Chem.* **1989**, *54*, 5657–5659.
28. Y. Chen, J.-M. Zheng, S.-M. Zhu, H.-Y. Chen, *Electrochimica Acta* **1999**, *44*, 2345–2350.
29. W. E. Truce, D. P. Tate, D. N. Burdge, *J. Am. Chem. Soc.* **1960**, *82*, 2872–2876.

30. A. Krief, A. Nazih, *Tetrahedron Lett.* **1995**, *36*, 8115–8118.
31. Z. J. Jedliñski, A. Stolarzewicz, Z. Grobelny, M. Szwarc, *J. Phys. Chem.* **1994**, *88*, 6094–6095.
32. M. Okano, H. Fukai, M. Arakawa, H. Hamano, *Electrochem. Comm.* **1999**, *1*, 223–226.
33. A. Hasegawa, F. Williams, *Chem. Phys. Lett.* **1977**, *46*, 66–68; C. E. Clots, R. N. Compton, *J. Chem. Phys.* **1977**, *67*, 1779–1780; A. Kühn, E. Illenberger, *J. Phys. Chem.* **1989**, *93*, 7060–7061; A. Kühn, E. Illenberger, *J. Chem. Phys.* **1990**, *93*, 357–364; H. Muto, K. Nunome, *J. Chem. Phys.* **1991**, *94*, 4741–4748; R. A. Popple, X. D. Finch, K. A. Smith, F. B. Dunning, *J. Chem. Phys.* **1996**, *104*, 8485–8489.
34. D. P. Zhong, A. H. Zewail, *Proc. Natl. Acad. Sci.* **1999**, 96, 2602–2607.
35. V. A. Tikhomirov, E. D. German, *J. Electroanal. Chem.* **1998**, *450*, 13–20.
36. E. D. German, V. A. Tikhomirov, *J. Mol. Struct. (Theochem)* **1998**, *423*, 251–261.
37. (a) R. Benassi, F. Bernardi, A. Bottoni, M. A. Robb, F. Taddei, *Chem. Phys. Lett.* **1989**, *161*, 79–84; (b) T. Tada, R. Yoshimura, *J. Am. Chem. Soc.* **1992**, *114*, 1593–1595; (c) J. Bertran, I. Gallardo, M. Moreno, J.-M. Savéant, *J. Am. Chem. Soc.* **1992**, *114*, 9576–9583.
38. S. Roszak, W. S. Koski, J. J. Kaufman, K. Balasubramanian, *J. Chem. Phys.* **1997**, *106*, 7709–7713.
39. J.-M. Savéant, *J. Am. Chem. Soc.* **1987**, *109*, 6788–6795; J.-M. Savéant, in *Advances in Electron Transfer Chemistry*, ed. P. S. Mariano, JAI Press, Greenwich, **1994**, Vol. 4, p. 53–116.
40. J.-M. Savéant, *Acc. Chem. Res.* **1993**, *26*, 455–461.
41. C. P. Andrieux, J.-M. Savéant, C. Tardy, *J. Am. Chem. Soc.* **1998**, *120*, 4167–4175.
42. R. L. Donkers, F. Maran, D. D. M. Wayner, M. S. Workentin, *J. Am. Chem. Soc.* **1999**, *121*, 7239–7248.
43. E. D. German, A. M. Kuznetsov, V. A. Tikhomirov. *J. Electroanal. Chem.* **1997**, *420*, 235–241.
44. C. G. Fraga, B. E. Leibovitz, A. L. Tappel, *Free Rad. Biol. Med.* **1987**, *3*, 119–123.
45. Y. T. Woo, D. Y. Lai, J. C. Arcos, M. F. Argus in *Chemical Induction of Cancer*, Vol. IIIb, p. 71, Academic Press, New York, **1985**; J. J. Kaufman, W. S. Koski, S. Roszak, K. Balasubramanian, *Chem. Phys.* **1996**, *204*, 233–237.
46. M. F. Semmelhack, R. J. de Franco, J. Stock, *Tetrahedron Lett.* **1972**, 1371–1374; P. Martigny, H. Lund, *Acta. Chem. Scand.* **1979**, *B33*, 575–579; J. I. Lozano, F. Barba, *Electrochim. Acta* **1997**, *42*, 2173–2176.
47. I. Chiarotto, M. Feroci, C. Giomini, A. Inesi, *Bull. Soc. Chim. Fr.* **1996**, 133, 167–175; E. Oguntoye, S. Szunerits, J. H. P. Utley, P. B. Wyatt, Tetrahedron **1996**, 52, 7771–7778; G. Montero, G. Quintanilla, F. Barba, *Electrochim. Acta* **1997**, *42*, 2177–2180; K. Uneyama, G. Mizutani, *Chem. Comm.* **1999**, 613–614; B. Batanero, M. J. Perez, F. Barba, *J. Electroanal. Chem.* **1999**, *469*, 201–205.
48. M. P. Crozet, A. Gellis, C. Pasquier, P. Vanelle, J. P. Aune, *Tetrahedron Lett.* **1995**, *36*, 525–528; D. Witt, J. Rachon, *Heteroatom Chem.* **1996**, *7*, 359–364; S. A. Dandekar, S. N. Greenwood, T. D. Greenwood, S. Mabic, J. S. Merola, J. M. Tanko, J. F. Wolfe, *J. Org. Chem.* **1999**, 64, 1543–1553.
49. C. Roubaud, P. Vanelle, J. Maldonado, M. P. Crozet, *Tetrahedron* **1995**, *51*, 9643–9656; A. Gellis, P. Vanelle, M. Kaafarani, K. Benakli, M. P. Crozet, *Tetrahedron* **1997**, *53*, 5471–5484; P. Vanelle, P. Rathelot, J. Maldonado, M. P. Crozet, *Heterocyles* **1997**, *45*, 1519–1528; V. Beraud, P. Perfetti, C. Pfister, M. Kaafarani, P. Vanelle, M. P. Crozet, Tetrahedron **1998**, *54*, 4923–4934.
50. A. Giraud, L. Giraud, M. P. Crozet, P. Vanelle, *Synlett* **1997**, 1159–1160.
51. J. Almena, F. Foubelo, M. Yus, *Tetrahedron Lett.* **1993**, *34*, 1649–1652; J. Almena, F. Foubelo, M. Yus, *J. Org. Chem.* **1994**, *59*, 3210–3215; E. W. Oliver, D. H. Evans, J. V. Caspar, *J. Electroanal. Chem.* **1996**, 403, 153–158.
52. J. F. Gil, D. J. Ramón, M. Yus, *Tetrahedron* **1993**, *49*, 9535–9546; J. F. Gil, D. J. Ramón, M. Yus, *Tetrahedron* **1994**, *50*, 3437–3446; J. F. Gil, D. J. Ramón, M. Yus, *Tetrahedron* **1994**, *50*, 7307–7314; J. M. Saa, P. Ballester, P. M. Deya, M. Capo, X. Garcias, *J. Org. Chem.* **1996**, *61*, 1035–1046; U. Azzena, S. Demartis, G. Melloni, *J. Org. Chem.* **1996**, *61*, 4913–4919; J. F. Bunnett, J. Jenvey, *J. Org. Chem.* **1996**, *61*, 8069–8073; E. Alonso, D. J. Ramon, M. Yus, *Tetrahedron* **1997**, *53*, 14355–14368.

53. E. M. Kaiser, C. G. Edmonds, S. D. Grubb, J. W. Smith, D. Tramp, *J. Org. Chem.* **1971**, *36*, 330–335; T. Mandai, H. Irei, M. Kuwada, J. Otera, *Tetrahedron Lett.* **1984**, *25*, 2371–2374; S. D. Rychnovsky, A. J. Buckmelter, V. H. Dahanukar, D. J. Skalitzky, *J. Org. Chem.* **1999**, *64*, 6849–6860; K. Wakamatsu, J. Dairiki, T. Etoh, H. Yamamoto, S. Yamamoto, Y. Shigetomi, *Tetrahedron Lett.* **2000**, *41*, 365–369.
54. M. Fagnoni, P. Schmoldt, T. Kirschberg, J. Mattay, *Tetrahedron* **1998**, *54*, 6427–6444.
55. E. Shouji, N. Oyama, *J. Electroanal. Chem.* **1996**, *410*, 229–234; J. Ludvik, B. Nygard, *J. Electroanal. Chem.* **1997**, *423*, 1–11.
56. V. Jouikov, L. Grigorieva, *Electrochim. Acta* **1996**, *41*, 2489–2491.
57. J. Cossy, S. Ibhi, P. H. Kahn, L. Tacchini, *Tetrahedron Lett.* **1995**, *36*, 7877–7880.
58. C. Costentin, P. Hapiot, M. Médebielle, J.-M. Savéant, *J. Am. Chem. Soc.* **1999**, *121*, 4451–4460.
59. C. P. Andrieux, A. Le Gorande, J.-M. Savéant, *J. Am. Chem. Soc.* **1992**, *114*, 6892–6904.
60. P. Maslak, *Top. Curr. Chem.* **1993**, *168*, 1–46.
61. N. Kimura, S. Takamuku, *Bull. Chem. Soc. Jpn.* **1993**, *66*, 3613–3617.
62. C. P. Andrieux, C. Combellas, F. Kanoufi, J.-M. Savéant, A. Thiébault, *J. Am. Chem. Soc.* **1997**, *119*, 9527–9540.
63. M. D. Koppang, N. F. Woolsey, D. E. Bartak, *J. Am. Chem. Soc.* **1984**, *106*, 2799–2805.
64. M. L. Andersen, W. Long, D. D. M. Wayner, *J. Am. Chem. Soc.* **1997**, *119*, 6590–6595.
65. M. L. Andersen, N. Mathivanan, D. D. M. Wayner, *J. Am. Chem. Soc.* **1996**, *118*, 4871–4879.
66. J. S. Jaworski, *J. Chem. Res.* (S) **1997**, 412–413.
67. J.-M. Savéant, *J. Phys. Chem.* **1994**, *98*, 3716–3724.
68. P. Maslak, J. Kula, J. E. Chateauneuf, *J. Am. Chem. Soc.* **1991**, *113*, 2304–2306.
69. P. Maslak, J. Theroff, *J. Am. Chem. Soc.* **1996**, *118*, 7235–7236.
70. P. Maslak, J. N. Narvaez, T. M. Vallombroso, Jr. *J. Am. Chem. Soc.* **1995**, *117*, 12373–12379.
71. D. Behar, P. Neta, *J. Phys. Chem.* **1981**, *85*, 690–693; P. Neta, D. Behar, *J. Am. Chem. Soc.* **1981**, *103*, 103–106.
72. (a) J.-M. Savéant, *Adv. Phys. Org. Chem.* **1990**, *26*, 1; (b) A. B. Pierini, J. S. Duca, Jr. *J. Chem. Soc., Perkin Trans.* 2 **1995**, 1821–1828; (c) T. Underwood-Lemons, G. Sághi-Szabó, J. A. Tossell, J. H. Moore, *J. Chem. Phys.* **1996**, *105*, 7896–7903; (d) A. B. Pierini, J. S. Duca, Jr., D. M. A. Vera *J. Chem. Soc., Perkin Trans.* 2 **1999**, 1003–1009.
73. D. D. Clarke, C. A. Coulson, *J. Chem. Soc. A* **1969**, 169–172; M. C. R. Symons, *Pure Appl. Chem.* **1981**, *53*, 223–229; R. Dressler, M. Allan, E. Haselbach, *Chimia* **1985**, *39*, 385–389; M. C. R. Symons, *Acta Chem. Scand.* **1997**, *51*, 127–134.
74. M. S. Mubarak, D. G. Peters, *J. Electroanal. Chem.* **1997**, *425*, 13–17; C. Ji, D. G. Peters, *J. Electroanal. Chem.* **1998**, *455*, 147–152; A. I. Tsyganok, K. Otsuka, *Appl. Cat. B*, **1999**, *22*, 15–26; A. Profumo, E. Fasani, A. Albini, *Heterocycles* **1999**, *51*, 1499–1502; E. Fasani, F. F. Barberis Negra, M. Mella, S. Monti, A. Albini, *J. Org. Chem.* **1999**, *64*, 5388–5395.
75. D. J. Ramón, M. Yus, *Tetrahedron Lett.* **1993**, *34*, 7115–7118; A. Bachki, F. Foubelo, M. Yus, *Tetrahedron Lett.* **1994**, *35*, 7643–7646; F. F. Huerta, C. Gómez, M: Yus, *Tetrahedron* **1995**, *51*, 3375–3388.
76. J. S. Duca, M. H. Gallego, A. B. Pierini, R. A. Rossi, *J. Org. Chem.* **1999**, *64*, 2626–2629.
77. M. C. Harsanyi, P. A. Lay, R. K. Norris, P. K. Witting, *Aust. J. Chem.* **1996**, *49*, 581–597.
78. M. C. Harsanyi, P. A. Lay, R. K. Norris, P. K. Witting, *J. Org. Chem.* **1995**, *60*, 5487–5493.
79. (a) R. Beugelmans, M. Chbani, *Bull. Soc. Chim. Fr.* **1995**, *132*, 290–305; (b) C. G. Ferrayoli, S. M. Palacios, R. A. Alonso, *J. Chem. Soc., Perkin Trans. 1*, **1995**, 1635–1638; (c) J. W. Wong, K. J. Natalie, G. C. Nwokogu, J. S. Pisipati, P. T. Flaherty, T. D. Greenwood, J. F. Wolfe, *J. Org. Chem.* **1997**, *62*, 6152–6159; (d) M. Médebielle, J. Pinson, J.-M. Savéant, *Electrochim. Acta* **1997**, *42*, 2049–2055; (e) M. T. Baumgartner, M. H. Gallego, A. B. Pierini, *J. Org. Chem.* **1998**, *63*, 6394–6397.
80. R. Beugelmans, M. Chbani, M. Soufiaoui, *Tetrahedron* **1996**, *37*, 1603–1604.
81. M. Chbani, J. P. Bouillon, J. Chastanet, M. Soufiaoui, R. Beugelmans, *Bull. Soc. Chim. Fr.* **1995**, 132, 1053–1060; M. Médebielle, M. A. Oturan, J. Pinson, *New J. Chem.* **1995**, *19*, 349–352.
82. A. N. Santiago, G. Lassaga, Z. Rappoport, R. A. Rossi, *J. Org. Chem.* **1996**, *61*, 1125–1128.

83. S. D. Rychnovsky, S. S. Swenson, *J. Org. Chem.* **1997**, *62*, 1333–1340.
84. U. Azzena, F. Dessanti, G. Melloni, L. Pisano, *Tetrahedron Lett.* **1999**, *40*, 8291–8293.
85. Y. H. Kang, K. Kim, *Tetrahedron* **1999**, *55*, 4271–4286.
86. S. M. A. Jorge, N. R. Stradiotto, *J. Electroanal. Chem.* **1997**, 431, 237–241.
87. P. Cauliez, M. Benaskar, A. Ghanimi, J. Simonet, *New J. Chem.* **1998**, 253–261.
88. S. Prigent, P. Cauliez, J. Simonet, D. G. Peters, *Acta Chem. Scand.* **1999**, *53*, 892–900.
89. J. F. Pilard, J. Simonet, *Tetrahedron Lett.* **1997**, *38*, 3735–3738.
90. C. Dell'Erba, M. Novi, G. Petrillo, C. Tavani, *Gazz. Chim. Ital.* **1997**, *127*, 361–366.
91. S.-H. Chen, V. Farina, D. M. Vyas, T. W. Doyle, B. H. Long, C. Fairchild, *J. Org. Chem.* **1996**, *61*, 2065–2070.
92. J. G. Lawless, M. D. Hawley, *J. Electroanal. Chem.* **1969**, *21*, 365–375; J. C. Stellhammer, W. E. Wentworth, *J. Chem. Phys.* **1969**, *51*, 1802–1814; F. M'Halla, J. Pinson, J.-M. Savéant, *J. Electroanal. Chem.* **1978**, *89*, 347–361; G. J. Gores, C. E. Koeppe, D. E. Bartak, *J. Org. Chem.* **1979**, *44*, 380–385; F. M'Halla, J. Pinson, J.-M. Savéant, *J. Am. Chem. Soc.* **1980**, *102*, 4120–4127; V. D. Parker, *Acta Chem. Scand. Ser. B.* **1981**, *35*, 595–599 and 655–660.
93. J. S. Jaworski, P. Leszczynski, S. Filipek, *J. Electroanal. Chem.* **1997**, *440*, 163–167.
94. J. S. Jaworski, P. Leszczynski, J. Tykarski, *J. Chem. Res. (S)* **1995**, 510–511; J. S. Jaworski, *Tetrahedron Lett.* **1999**, *40*, 5771–5772; J. S. Jaworski, P. Leszczynski, *J. Electroanal. Chem.* **1999**, *464*, 259–262.
95. (a) N. Kimura, S. Takamuku, *J. Am. Chem. Soc.* **1994**, *116*, 4087–4088; (b) N. Kimura, S. Takamuku, *J. Am. Chem. Soc.* **1995**, *117*, 8023–8024.
96. T. Underwood-Lemons, G. Sághi-Szabó, J. A. Tossell, J. H. Moore, *J. Chem. Phys.* **1996**, *105*, 7896–7903.
97. S. Antonello, F. Maran, *J. Am. Chem. Soc.* **1998**, *120*, 5713–5722.
98. M. D. Koppang, N. F. Woolsey, D. E. Bartak, *J. Am. Chem. Soc.* **1986**, *108*, 6497–6502.
99. C. Combellas, F. Kanoufi, A. Thiébault, *J. Electroanal. Chem.* **1997**, *432*, 181–192.
100. A. Maercker, *Angew. Chem.* **1987**, *99*, 1002–1019; *Angew. Chem. Int. Ed.* **1987**, *26*, 972–989.
101. T. B. Christensen, K. Daasbjerg, *Acta Chem. Scand.* **1997**, *51*, 307–317.
102. K. Mochida, H. Watanabe, S. Murata, M. Fujitsuka, O. Ito, *J. Organomet. Chem.* **1998**, *568*, 121–125.
103. A. Bewick, J. M. Mellor, B. S. Pons, *J. Chem. Soc., Chem. Commun.* **1978**, 738; M. Mella, M. Freccero, A. Albini, *J. Chem. Soc., Chem. Commun.* **1995**, 41–42; M. Mella, M. Freccero, A. Albini, *Tetrahedron* **1996**, *52*, 5533–5548.
104. P. G. Gassman, B. A. Hay, *J. Am. Chem. Soc.* **1985**, *107*, 4075–4076; P. G. Gassman, B. A. Hay, *J. Am. Chem. Soc.* **1986**, *108*, 4227–4228; P. G. Gassman, S. J. Burns, *J. Org. Chem.* **1988**, *53*, 5576–5678.
105. C. J. Abelt, H. D. Roth, M. L. M. Schilling, *J. Am. Chem. Soc.* **1985**, *107*, 4148–4152; K. I. Booker-Milburn, *Synlett* **1992**, 809–810; M. Abe, A. Oku, *J. Org. Chem.* **1995**, *60*, 3065–3073; G. Maier, H. Rang, R. Emrich, S. Gries, H. Irngartinger, *Liebigs Ann.* **1995**, 161–167; B. Hong, M. A. Fox, G. Maier, C. Hermann, *Tetrahedron Lett.* **1996**, *37*, 583–586; M. Abe, M. Nojima, A. Oku, *Tetrahedron Lett.* **1996**, *37*, 1833–1836.
106. (a) W. Adam, H. Walter, G.-F. Chen, F. Williams, *J. Am. Chem. Soc.* **1992**, *114*, 3007–3014; (b) W. Adam, C. Sahin, J. Sendelbach, H. Walter, G.-F. Chen, F. Williams, *J. Am. Chem. Soc.* **1994**, *116*, 2576–2584; (c) W. Adam, T. Heidenfelder, C. Sahin, *Synthesis* **1995**, 1163–1170; (d) W. Adam, T. Heidenfelder, *Chem. Soc. Rev.* **1999**, *28*, 359–365.
107. T. Karatsu, Y. Ichino, A. Kitamura, W. H. Owens, P. S. Engel, *J. Chem. Res. (S)* **1995**, 440–441.
108. M. Fagnoni, M. Mella, A. Albini, *J. Phys. Org. Chem.* **1997**, *10*, 777–780.
109. M. Mella, M. Fagnoni, M. Freccero, E. Fasani, A. Albini, *Chem. Soc. Rev.* **1998**, *27*, 81–89.
110. Y. Nakadaira, N. Komatsu, H. Sakurai, *Chem. Lett.* **1985**, 1781–1782; H. Watanabe, M. Kato, E. Tabei, II. Kuwabara, N. Hirai, T. Sato, Y. Nagai, *J. Chem. Soc., Chem. Commun.* **1986**, 1662–1663; S. Kyushin, Y. Ehara, Y. Nakadaira, M. Ohashi, *J. Chem. Soc., Chem. Commun.* **1989**, 279–280.
111. J. M. White, *Aust. J. Chem.* **1995**, *48*, 1227–1251; H. Bock, B. Solouki, *Chem. Rev.* **1995**, *95*, 1161–1190.
112. C. Carra, F. Fiussello, G. Tonachini, *J. Org. Chem.* **1999**, *64*, 3867–3877.

113. G. L. Borosky, A. B. Pierini, *J. Mol. Struct. Theochem* **1999**, *466*, 165–175.
114. J. Yoshida, M. Izawa, *J. Am. Chem. Soc.* **1997**, *119*, 9361–9365.
115. Y. Nakadaira, S. Otani, S. Kyushin, M. Ohashi, H. Sakurai, Y. Funada, K. Sakamoto, A. Sekiguchi, *Chem. Lett.* **1991**, 601–602.
116. S. Fukuzumi, T. Kitano, K. Mochida, *J. Am. Chem. Soc.* **1990**, *112*, 3246–3247.
117. K. Mochida, C. Hodota, R. Hata, S. Fukuzumi, *Organomet.* **1993**, *12*, 586–588.
118. M. Mella, M. Freccero, A. Albini, *J. Chem. Soc., Chem. Commun.* **1995**, 41–42; M. Mella, M. Freccero, A. Albini, *Tetrahedron* **1996**, *52*, 5533–5548.
119. P. G. Gassman, R. Yamaguchi, G. F. Koser, *J. Org. Chem.* **1978**, *43*, 4392–4393.
120. Actually, there is an alternative option that is frequently encountered when a $\sigma$ donor is flanked on one side by a $\pi$ donor, on the other by an $n$ donor. In our analysis, we have included this option by assigning such systems either to the $(\sigma + \pi)$ or $(\sigma + n)$ donor class based on the better donor (either $\pi$ or $n$).
121. J. Yoshida, K. Nishiwaki, *J. Chem. Soc., Dalton Trans.* **1998**, 2589–2596.
122. A. J. Baggaley, R. Brettle, *J. Chem. Soc., Chem. Commun.* **1966**, 108; K. Ponsold, H. Kasch, *Tetrahedron Lett.* **1979**, 4463–4464; D. P. DeCosta, A. K. Bennett, J. A. Pincock, *J. Am. Chem. Soc.* **1999**, *121*, 3785–3786.
123. T. Thyrann, D. A. Lightner, *Tetrahedron Lett.* **1996**, *37*, 315–318.
124. W. Schmidt, E. Steckhan, *Angew. Chem.* **1978**, *90*, 717; *Angew. Chem., Int. Ed. Engl.* **1978**, *17*, 673–674; W. Schmidt, E. Steckhan, *Angew. Chem.* **1979**, *91*, 850–851; *Angew. Chem., Int. Ed. Engl.* **1979**, *18*, 801–802; W. Schmidt, E. Steckhan, *Angew. Chem.* **1979**, *91*, 851–852; *Angew. Chem., Int. Ed. Engl.* **1979**, *18*, 802–803.
125. Z. Yang, B. X. Wang, H. W. Hu, S. M. Zhu, *Syn. Commun.* **1998**, *28*, 3163–3171.
126. T. Shono, Y. Matsumura, Y. Nakagawa, *J. Am. Chem. Soc.* **1974**, *96*, 3532–3536; T. Chiba, M. Okimoto, H. Nagai, Y. Takata, *J. Org. Chem.* **1979**, *44*, 3519–3523; R. M. Borg, D. R. Arnold, T. S. Cameron, *Can. J. Chem.* **1984**, *62*, 1785–1802.
127. Y. Takahashi, H. Ohaku, S. Morishima, T. Suzuki, T. Miyashi, *Tetrahedron Lett.* **1995**, *36*, 5207–5210; C. Gaebert, J. Mattay, *Tetrahedron* **1997**, *53*, 14297–14316; W. Bergmark, S. Hector, G. Jones II, C. Oh, T. Kumagai, S. Hara, T. Segawa, N. Tanaka, T. Mukai, *Photochem. Photobiol.* **1997**, *109*, 119–124; T. Herbertz, H. D. Roth, *J. Am. Chem. Soc.* **1998**, *120*, 11904–11911; T. Herbertz, H. D. Roth, *J. Org. Chem.* **1999**, *64*, 3708–3713; R. S. Glass, *Top. Curr. Chem.* **1999**, *205*, 1–87.
128. A. L. Perrott, H. J. P. deLijser, D. R. Arnold, *Can. J. Chem.* **1997**, *75*, 384–397; M. Suzuki, T. Ikeno, T. Osoda, K. Narasaka, T. Suenobu, S. Fukuzumi, A. Ishida, *Bull. Chem. Soc. Jpn.* **1997**, *70*, 2269–2277; A. Anne, S. Fraoua, J. Moiroux, J.-M. Savéant, *J. Phys. Org. Chem.* **1998**, *11*, 774–780; J. L. Faria, R. A. McClelland, S. Steenken, *Chem. Eur. J.* **1998**, *4*, 1275–1280; W. Zhang, L. Yang, L.-M. Wu, Y. C. Liu, Z. L. Liu, *J. Chem. Soc., Perkin Trans. 2* **1998**, 1189–1193.
129. J. Yoshida, T. Murata, S. Isoe, *Tetrahedron Lett.* **1986**, *27*, 3373–3376; T. Koizumi, T. Fuchigami, T. Nonaka, *Bull. Chem. Soc. Jpn.* **1989**, *62*, 219–225.
130. P. S. Mariano, *Acc. Chem. Res.* **1983**, *16*, 130–137; A. J. Y. Lan, S. L. Quillen, R. O. Heuckeroth, P. S. Mariano, *J. Am. Chem. Soc.* **1984**, *106*, 6439–6440.
131. M. Freccero, E. Fasani, A. Albini, *J. Org. Chem.* **1993**, *58*, 1740–1745; K. Mizuno, G. Konishi, T. Nishiyama, H. Inoue, *Chem. Lett.* **1995**, 1077–1078; M. Kako, K. Hatakenaka, S. Kakuma, M. Ninomiya, Y. Nakadaira, M. Yasui, F. Iwasaki, M. Wakasa, H. Hayashi, *Tetrahedron Lett.* **1999**, *40*, 1133–1136.
132. S. Fukuzumi, K. Yasui, S. Itoh, *Chem. Lett.* **1997**, 161–162.
133. T. Yoshiyama, T. Fuchigami, *Chem. Lett.* **1992**, 1995–1998; T. Fuchigami, T. Fujita, *J. Org. Chem.* **1994**, *59*, 7190–7192; E. Baciocchi, C. Crescenzi, O. Lanzalunga, *Tetrahedron* **1997**, *53*, 4469–4478.
134. K. Chiba, M. Fukuda, S. Kim, Y. Kitano, M. Tada, *J. Org. Chem.* **1999**, *64*, 7654–7656; H. Takakura, S. Yamamura, *Tetrahedron Lett.* **1999**, *40*, 299–302.
135. M. Schmittel, A. Abufarag, O. Luche, M. Levis, *Angew. Chem.* **1990**, *102*, 1174–1176; *Angew. Chem. Int. Ed. Engl.* **1990**, *29*, 1144–1146; M. Schmittel, M. Levis, *Chem. Lett.* **1994**, 1935–1938.
136. M. Kako, H. Takada, Y. Nakadaira, *Tetrahedron Lett.* **1997**, *38*, 3525–3528; Y. Tajima, H. Ishikawa, Y. Shimanuki, T. Miyazawa, N. Mikami, M. Kira, *Chem. Lett.* **1998**, 415–416.

137. S. Yamamura, Y. Shizuri, H. Shigemori, Y. Okuno, M. Ohkubo, *Tetrahedron* **1991**, *47*, 635–644.
138. R. Suau, F. Nájera, R. Rico, *Tetrahedron* **1999**, *55*, 4019–4028.
139. G. Pandey, P. Y. Reddy, U. T. Bhalerao, *Tetrahedron Lett.* **1991**, *32*, 5147–5150; Furuta S, Fuchigami T, *Electrochim. Acta* **1998**, *43*, 3183–3191.
140. A. G. M. Barrett, D. Pilipauskas, *J. Org. Chem.* **1991**, *56*, 2787–2800; A. Papadopoulos, B. Lewall, E. Steckhan, K.-D. Ginzel, F. Knoch, M. Nieger, *Tetrahedron* **1991**, *47*, 563–572; K. D. Moeller, L. D. Rutledge, *J. Org. Chem.* **1992**, *57*, 6360–6363; V. N. Belevskii, D. A. Tyurin, N. D. Chuvylkin, *High Ener. Chem.* **1998**, *32*, 305–315; T.-S. Kam, T.-M. Lim, Y.-M. Choo, *Tetrahedron* **1999**, *55*, 1457–1468; J. Yoshida, S. Suga, S. Suzuki, N. Kinomura, A. Yamamoto, K. Fujiwara, *J. Am. Chem. Soc.* **1999**, *121*, 9546–9549.
141. V. N. Belevskii, S. I. Belopushkin, D. A. Tyurin, N. D. Chuvylkin, *High Ener. Chem.* **1999**, *33*, 77–86.
142. T. Fuchigami, K. Yamamoto, A. Konno, *Tetrahedron* **1991**, *47*, 625–634; T. Fuchigami, A. Konno, K. Nakagawa, M. Shimojo, *J. Org. Chem.* **1994**, *59*, 5937–5941.
143. F. Cornille, U. Slomczynska, M. L. Smythe, D. D. Beusen, K. D. Moeller, G. R. Marshall, *J. Am. Chem. Soc.* **1995**, *117*, 909–917; P. Brungs, K. Danielmeier, J. Jakobi, C. Nothhelfer, A. Stahl, A. Zietlow, E. Steckhan, *J. Chim. Phys.* **1996**, *93*, 575–590.
144. M. Kimura, N. Saitoh, H. Kawai, Y. Sawaki, *Novel Trends in Electroorganic Synthesis*, Ed. S. Torii, Springer, Tokyo **1998**, p. 73–76; W. A. McHale, A. G. Kutateladze, *J. Org. Chem.* **1998**, *63*, 9924–9931; M. Nakamura, M. Toganoh, H. Ohara, E. Nakamura, *Org. Lett.* **1999**, *1*, 7–9.
145. Y. Takemoto, T. Ohra, S. Furuse, H. Koike, C. Iwata, *J. Chem. Soc., Chem. Commun.* **1994**, 1529–1530.
146. A. P. Schaap, S. Prasad, S. D. Gagnon, *Tetrahedron Lett.* **1983**, 3047–3050; E. Hasegawa, S. Koshii, T. Horaguchi, T. Shimizu, *J. Org. Chem.* **1992**, *57*, 6342–6344; A. J. Highton, T. N. Majid, N. S. Simpkins, *Synlett* **1999**, 237–239.
147. Y. Takemoto, S. Furuse, H. Koike, T. Ohra, C. Iwata, H. Ohishi, *Tetrahedron Lett.* **1995**, *36*, 4085–4088.; M. Kimura, H. Kawai, Y. Sawaki, in *Novel Trends in Electroorganic Synthesis*, Ed. S. Torii, Springer, Tokyo **1998**, p. 77–78.
148. J. Yoshida, Y. Ishichi, S. Isoe, *J. Am. Chem. Soc.* **1992**, *114*, 7594–7595; K. Narasaka, Y. Kohno, *Bull. Chem. Soc. Jpn.* **1993**, *66*, 3456–3463; K. Narasaka, Y. Kohno, S. Shimada, *Chem. Lett.* **1993**, 125–128; S. Kyushin, S. Otani, Y. Nakadaira, M. Ohashi, *Chem. Lett.* **1995**, 29–30; T. Ikeno, M. Harada, N. Arai, K. Narasaka, *Chem. Lett.* **1997**, 169–170; K. Narasaka, *Pure Appl. Chem.* **1997**, *69*, 601–604; T. Mikami, M. Harada, K. Narasaka, *Chem. Lett.* **1999**, 425–426.
149. Y. S. Jung, W. H. Swartz, W. Xu, P. S. Mariano, N. J. Green, A. G. Schultz, *J. Org. Chem.* **1992**, *56*, 6037–6047; E. Meggers, E. Steckhan, S. Blechert, *Angew. Chem.* **1995**, *107*, 2317–2319; *Angew. Chem. Int. Ed. Engl.* **1995**, *34*, 2137–2139; E. Le Gall, J.-P. Hurvois, S. Sinbandhit, *Eur. J. Org. Chem.* **1999**, 2645–2653.
150. J. Yoshida, S. Suga, K. Fuke, M. Watanabe, *Chem. Lett.* **1999**, 251–252.
151. (a) G. Pandey, G. Kumaraswamy, U. T. Bhalerao, *Tetrahedron Lett.* **1989** *30*, 6059–6062; (b) G. Pandey, G. D. Reddy, *Tetrahedron Lett.* **1992**, *33*, 6533–6536; (c) G. Pandey, G. D. Reddy, G. Kumaraswamy, *Tetrahedron* **1994**, *50*, 8185–8194.
152. (a) F. G. Bordwell, J.-P. Cheng, M. J. Bausch *J. Am. Chem. Soc.* **1988**, *110*, 2872–2877; (b) F. G. Bordwell, J.-P. Cheng, *J. Am. Chem. Soc.* **1989**, *111*, 1792–1795; (c) F. G. Bordwell, A. V. Satish, *J. Am. Chem. Soc.* **1992**, *114*, 10173–10176; (d) M. Bietti, E. Baciocchi, S. Steenken, *J. Phys. Chem. A* **1998**, *102*, 7337–7342; (e) C. X. Zhao, Y. F. Gong, H. Y. He, X. K. Jiang, *J. Phys. Org. Chem.* **1999**, *12*, 688–694.
153. X. Zhang, F. G. Bordwell, *J. Org. Chem.* **1992**, *57*, 4163–4168.
154. M. Bietti, A. Cuppoletti, C. Dagostin, C. Florea, C. Galli, P. Gentili, H. Petride, C. R. Caia, *Eur. J. Org. Chem.* **1998**, 2425–2429.
155. E. Fasani, M. Mella, A. Albini, *J. Chem. Soc. Perkin Trans. 2* **1995**, 449–452.
156. J. Mönig, R. Goslich, K.-D. Asmus, *Ber. Bunsen Ges. Phys. Chem.* **1986**, *90*, 115–121.
157. E. Baciocchi, T. Del Giacco, F. Elisei, *J. Am. Chem. Soc.* **1993**, *115*, 12290–12295.
158. P. Hapiot, J. Moiroux, J.-M. Savéant, *J. Am. Chem. Soc.* **1990**, *112*, 1337–1343; A. Anne, P. Hapiot, J. Moiroux, P. Neta, J.-M. Savéant, *J. Phys. Chem.* **1991**, *95*, 2370–2377; G. N. R.

Tripathi, *Chem. Phys. Lett.* **1992**, *199*, 409–416; A. Anne, P. Hapiot, J. Moiroux, P. Neta, J.-M. Savéant, *J. Am. Chem. Soc.* **1992**, *112*, 4694–4701; V. D. Parker, Y. X. Zhao, Y. Lu, G. Zheng, *J. Am. Chem. Soc.* **1998**, *120*, 12720–12727.
159. V. D. Parker, E. T. Chao, G. Zheng, *J. Am. Chem. Soc.* **1997**, *119*, 11390–11394.
160. E. Baciocchi, M. Bietti, S. Steenken, *Chem. Eur. J.* **1999**, *5*, 1785–1793.
161. R. B. Silverman, *Adv. Electr. Trans. Chem.* **1992**, *2*, 177–213.
162. C. G. Shaefer, K. S. Peters, *J. Am. Chem. Soc.* **1980**, *102*, 7566–7567; A. Sinha, T. C. Bruice, *J. Am. Chem. Soc.* **1984**, *106*, 7291–7292; S. Das, C. von Sonntag, *Z. Naturforsch.* **1986**, *41b*, 505–513; F. D. Lewis, *Acc. Chem. Res.* **1986**, *19*, 401–405; J. P. Dinnocenzo, T. E. Banach, *J. Am. Chem. Soc.* **1989**, *111*, 8646–8653; W. Xu, X.-M. Zhang, P. S. Mariano, *J. Am. Chem. Soc.* **1991**, *113*, 8863–8878; W. Xu, P. S. Mariano, *J. Am. Chem. Soc.* **1991**, *113*, 1431–1432; M. Goez, I. Sartorius, *J. Am. Chem. Soc.* **1993**, *115*, 11123–11133; M. Goez, I. Sartorius, *Chem. Ber.* **1994**, *127*, 2273–2276; O. Brede, D. Beckert, C. Windolph, H. A. Göttinger, *J. Phys. Chem. A* **1998**, *102*, 1457–1464; K. S. Peters, A. Cashin, P. Timbers, *J. Am. Chem. Soc.* **2000**, *122*, 107–113.
163. X. Zhang, S.-R. Yeh, S. Hong, M. Freccero, A. Albini, D. E. Falvey, P. S. Mariano, *J. Am. Chem. Soc.* **1994**, *116*, 4211–4220.
164. M. Ioele, S. Steenken, E. Baciocchi, *J. Phys. Chem. A* **1997**, *101*, 2979–2987.
165. D. M. Camaioni, *J. Am. Chem. Soc.* **1990**, *112*, 9475–9483.
166. J. H. Penn, D.-L. Deng, K.-J. Chai, *Tetrahedron Lett.* **1988**, *29*, 3635–3638; D. R. Arnold, L. J. Lamont, *Can. J. Chem.* **1989**, *67*, 2119–2127; R. Popielarz, D. R. Arnold, *J. Am. Chem. Soc.* **1990**, *112*, 3068–3082; D. R. Arnold, L. J. Lamont, A. L. Perrott, *Can. J. Chem.* **1991**, *69*, 225–233; S. Perrier, S. Sankararaman, J. K. Kochi, *J. Chem. Soc. Perkin Trans. 2* **1993**, 825–837.
167. P. Maslak, T. M. Vallombroso, W. H. Chapman, Jr., J. N. Narvaez, *Angew. Chem.* **1994**, *106*, 110–113; *Angew. Chem. Int. Ed. Engl.* **1994**, *33*, 73–75.
168. J. P. Dinnocenzo, W. P. Todd, T. R. Simpson, I. R. Gould, *J. Am. Chem. Soc.* **1990**, *112*, 2462–2464; J. P. Dinnocenzo, D. R. Lieberman, T. R. Simpson, *J. Am. Chem. Soc.* **1993**, *115*, 366–367.
169. L. A. Lucia, R. D. Burton, K. S. Schanze, *J. Phys. Chem.* **1993**, *97*, 9078–9080; J. W. Leon, D. G. Whitten, *J. Am. Chem. Soc.* **1993**, *115*, 8038–8043.
170. Z. Su, P. S. Mariano, D. E. Falvey, U. C. Yoon, S. W. Oh, *J. Am. Chem. Soc.* **1998**, *120*, 10676–10686.
171. J. P. Dinnocenzo, S. Farid, J. L. Goodman, I. R. Gould, W. P. Todd, S. L. Mattes, *J. Am. Chem. Soc.* **1989**, *111*, 8973–8975; W. P. Todd, J. P. Dinnocenzo, S. Farid, J. L. Goodman, I. R. Gould, *Tetrahedron Lett.* **1993**, *34*, 2863–2866.
172. W. Adam, T. Kammel, M. Toubartz, S. Steenken, *J. Am. Chem. Soc.* **1997**, *119*, 10673–10676.
173. L. Qin, G. N. R. Tripathi, R. H. Schuler, *Z. Naturforsch.* **1985**, *40a*, 1026–1039; F. G. Bordwell, X.-M. Zhang, J.-P. Cheng, *J. Org. Chem.* **1993**, *58*, 6410–6416; J.-P. Cheng, Y. Zhao, *Tetrahedron* **1993**, *49*, 5267–5276; M. Jonsson, J. Lind, T. E. Eriksen, G. Merényi, *J. Am. Chem. Soc.* **1994**, *116*, 1423–1427.
174. O. Hammerich, V. D. Parker, A. Ronlán, *Acta Chem. Scand.* **1976**, *B30*, 89–90; V. D. Parker, Y. Chao, B. Reitstöen, *J. Am. Chem. Soc.* **1991**, *113*, 2336–2338; M. Schmittel, G. Gescheidt, M. Röck, *Angew. Chem.* **1994**, *106*, 2056–2058; *Angew. Chem. Int. Ed. Engl* **1994**, *33*, 1961–1963.
175. T. A. Gadosy, D. Shukla, L. J. Johnston, *J. Phys. Chem. A* **1999**, *103*, 8834–8839.
176. (a) M. Schmittel, M. Keller, A. Burghart, *J. Chem. Soc. Perkin Trans. 2*, **1995**, 2327–2333; (b) M. Schmittel, A. Burghart, H. Werner, M. Laubender, R. Söllner, *J. Org. Chem.*, **1999**, 64, 3077–3085.
177. Y. Kohno, K. Narasaka, *Bull. Chem. Soc. Jpn.* **1995**, *68*, 322–329.
178. M. Schmittel, J. Heinze, H. Trenkle, *J. Org. Chem.* **1995**, *60*, 2726–2733.
179. M Schmittel, J.-P. Steffen, A. Burghart, *Acta Chem. Scand.* **1999**, *53*, 781–791.
180. M. Schmittel, R. Söllner, *Angew. Chem.* **1996**, 108, 2248–2250, *Angew. Chem. Int. Ed. Engl.* **1996**, *35*, 2107–2109.
181. M. Schmittel, R. Söllner, *J. Chem. Soc., Perkin Trans. 2*, **1999**, 515–520.

182. E. Baciocchi, E. Fasella, O. Lanzalunga, M. Mattioli, *Angew. Chem.* **1993**, *105*, 1110–1112; *Angew. Chem. Int. Ed. Engl.* **1993**, *32*, 1071–1073.
183. K. Mizuno, T. Tamai, I. Hashida, Y. Otsuji, *J. Org. Chem.* **1995**, *60*, 2935–2937.
184. H. G. Raj, V. S. Parmar, S. C. Jain, K. I. Priyadarsini, J. P. Mittal, S. Goel, S. K. Das, S. K. Sharma, C. E. Olsen, J. Wengel, *Bioorg. Med. Chem.* **1999**, *7*, 2091–2094.
185. H. J. P. de Lijser, T. S. Cameron, D. R. Arnold, *Can. J. Chem.* **1997**, *75*, 1795–1809.
186. K. Komaguchi, M. Shiotani, A. Lund, *Chem. Phys. Lett.* **1997**, *265*, 217–223.
187. K. Mochida, K. Takekuma, H. Watanabe, S. Murata, *Chem. Lett.* **1998**, 623–624.
188. R. W. Kavash, P. S. Mariano, *Tetrahedron Lett.* **1989**, *30*, 4185–4188.
189. P. W. Rabideau, Z. Marcinow, *Org. React.* **1992**, *23*, 1–334.
190. T. Guena, D. Pletcher, *Acta Chem. Scand.* **1998**, *52*, 23–31.
191. G. E. Lawson, A. Kitaygorodskiy, Y.-P. Sun, *J. Org. Chem.* **1999**, *64*, 5913–5920.
192. D. C. de Azevedo, J. F. C. Boodts, J. C. M. Cavalcanti, A. E. G. Santana, A. F. dos Santos, E. S. Bento, J. Tonholo, M. O. F. Goulart, *J. Electroanal. Chem.* **1999**, *466*, 99–106.
193. G. Asensio, A. Cuenca, P. Gaviña, M. Medio-Simón, *Tetrahedron Lett.* **1999**, *40*, 3939–3940.
194. J. Volke, V. Skála, *J. Electroanal. Chem.* **1972**, *36*, 383–388.
195. F. Marken, S. Kumbhat, G. H. W. Sanders, R. G. Compton, *J. Electroanal. Chem.* **1996**, *414*, 95–105; P. A. Lay, R. K. Norris, P. K. Witting, *Aust. J. Chem.* **1997**, *50*, 999–1007.
196. F. Williot, M. Bernard, D. Lucas, Y. Mugnier, J. Lessard, *Can. J. Chem.* **1999**, *77*, 1648–1654.
197. A. Arranz, S. F. deBetoño, J. M. Moreda, A. Cid, J. F. Arranz, *Anal. Chim. Acta* **1997**, *351*, 97–103.
198. (a) T. Ohno, H. Nakahiro, K. Sanemitsu, T. Hirashima, I. Nishiguchi, *Tetrahedorn Lett.* **1992**, *33*, 5515–5516; (b) S. Kashimura, Y. Murai, M. Ishifune, H. Masuda, H. Murase, T. Shono, *Tetrahedron Lett.* **1995**, *36*, 4805–4808.
199. D. A. Tyssee, M. M. Baizer, *J. Org. Chem.* **1974**, *39*, 2819–2823; T. Ohno, Y. Ishino, Y. Tsumagari, I. Nishiguchi, *J. Org. Chem.* **1995**, *60*, 458–460; T. Ohno, H. Aramaki, H. Nakahiro, I. Nishiguchi, *Tetrahedron* **1996**, *52*, 1943–1952; G. Pandey, M. K. Ghorai, S. Hajra, *Tetrahedron Lett.* **1998**, *39*, 8341–8344; N. Kise, Y. Hirata, T. Hamaguchi, N. Ueda, *Tetrahedron Lett.* **1999**, *40*, 8125–8128.
200. H. E. Bode, G. Sowell, R. D. Little, *Tetrahedron Lett.* **1990**, *31*, 2525–2528; S. P. Chavan, K. S. Ethiraj, *Tetrahedron Lett.* **1995**, *36*, 2281–2284; G. Pandey, S. Hajra, M. K. Ghorai, K. Ravikumar, *J. Am. Chem. Soc.* **1997**, *119*, 8777–8787; G. Pandey, M. K. Ghorai, S. Hajra, *Tetrahedron Lett.* **1998**, *39*, 1831–1834; S. Kashimura, Y. Murai, M. Ishifune, H. Masuda, M. Shimomura, H. Murase, T. Shono, *Acta Chem. Scand.* **1999**, *53*, 949–951.
201. (a) G. Pattenden, G. M. Robertson, *Tetrahedron Lett.* **1983**, *24*, 4617–4620; (b) J. Cossy, S. BouzBouz, C. Mouza, *Synlett* **1998**, 621–622.
202. J. Cossy, J.-P. Pete, *Adv. Electron Transfer Chem.* **1996**, *5*, 141–195.
203. S. Hintz, A. Heidbreder, J. Mattay, *Top. Curr. Chem.* **1996**, *177*, 77–124.
204. D. P. Fox, R. D. Little, M. M. Baizer, *J. Org. Chem.* **1985**, *50*, 2202–2204; D. Beloti, J. Cossy, J. P. Pete, C. Portella, *J. Org. Chem.* **1986**, *51*, 4196–4200; R. D. Little, D. P. Fox, L. Van Hijfte, R. Dannecker, G. Sowell, R. L. Wolin, L. Moëns, M. M. Baizer, *J. Org. Chem.* **1988**, *53*, 2287–2294; N. Kise, T. Suzumoto, T. Shono, *J. Org. Chem.* **1994**, *59*, 1407–1413; G. H. Lee, E. B. Choi, E. Lee, C. S. Pak, *J. Org. Chem.* **1994**, *59*, 1428–1443; G. Pandey, S. Hajra, M. K. Ghorai, K. Ravikumar, *J. Org. Chem.* **1997**, *62*, 5966–5973; G. H. Lee, S. J. Ha, I. K. Yoon, C. S. Pak, *Tetrahedron Lett.* **1999**, *40*, 2581–2584.
205. R. Gorny, H. J. Schäfer, R. Fröhlich, *Angew. Chem.* **1995**, *107*, 2188–2191; *Angew. Chem. Int. Ed. Engl.* **1995**, *34*, 2007–2009.
206. (a) T. Shono, N. Kise, E. Shirakawa, H. Matsumoto, E. Okazaki, *J. Org. Chem.* **1991**, *56*, 3063–3067; (b) N. Kise, H. Oike, E. Okajaki, M. Yoshimoto, T. Shono, *J. Org. Chem.* **1995**, *60*, 3980–3992.
207. P. Kujawa, N. Mohid, C. K. Zaman, W. Manshol, P. Ulanski, J. M. Rosiak, *Radiat. Phys. Chem.* **1998**, *53*, 403–409.
208. G. Sun, J. Wu, X. Qin, X. Fang, W. Wang, *Radiat. Phys. Chem.* **1999**, *55*, 409–416.
209. V. D. Parker, *Acta Chem. Scand.* **1981**, *B35*, 295–301.
210. C. P. Andrieux, M. Grzeszczuk, J.-M. Savéant, *J. Am. Chem. Soc.* **1991**, *113*, 8811–8817.

211. M. F. Nielsen, O. Hammerich, *Acta. Chem. Scand.* **1992**, 46, 883–896.
212. M. F. Nielsen, O. Hammerich, *Acta. Chem. Scand.* **1987**, *B 41*, 668–678.
213. M. F. Nielsen, H. Eggert, O. Hammerich, *Acta. Chem. Scand.* **1991**, *45*, 292–301.
214. O. Hammerich, M. F. Nielsen, *Acta. Chem. Scand.* **1998**, *52*, 831–857.
215. I. Fussing, O. Hammerich, A. Hussain, M. F. Nielsen, J. H. P. Utley, *Acta. Chem. Scand.* **1998**, *52*, 328–337.
216. A. S. Mendkovich, O. Hammerich, T. Y. Rubinskaya, V. P. Gultyai, *Acta. Chem. Scand.* **1991**, *45*, 644–651.
217. L. C. T. Shoute, *J. Phys. Chem.* **1997**, *101*, 5335–5542.
218. L.-S. R. Yeh, A. J. Bard, *J. Electrochem. Soc.*, **1977**, *124*, 355–360.
219. A. R. de Andrade, J. F. C. Boodts, *J. Brazil. Chem. Soc.*, **1998**, *9*, 257–161.
220. Y. Huang, D. D. M. Wayner, *J. Am. Chem. Soc.* **1994**, *116*, 2157–2158.
221. H. S. Sørensen, K. Daasbjerg, *Acta Chem. Scand.* **1998**, *52*, 51–61.
222. F. Barba, J. L. de la Fuente, M. Galakhov, *Tetrahedron*, **1997**, *53*, 5831–5838; S. Kashimura, M. Ishifune, Y. Murai, H. Murase, M. Shimomura, T. Shono, *Tetrahedron Lett.* **1998**, *39*, 6199–6202; J. Gruber, F. F. Camilo, A. C. M. Arantes, *Helv. Chim. Acta* **1999**, *82*, 389–393.
223. M. Kimura, N. Moritani, Y. Swaki, *Electroorg. Synth.; M. M. Baizer Mem. Symp.*, Dekker, New York **1991**, p. 61.
224. P. Mikesell, M. Schwaebe, M. DiMare, R. D. Little, *Acta Chem. Scand.* **1999**, *53*, 792–799.
225. J. Delaunay, A. Orliac, J. Simonet, *J. Electrochem. Soc.* **1995**, *142*, 3613–3619.
226. J. H. P. Utley, M. Güllü, M. Motevalli, *J. Chem. Soc. Perkin Trans I* **1995**, 1961–1970.
227. N. Kise, M. Echigo, T. Shono, *Tetrahedron Lett.* **1994**, *35*, 1897–1900; N. Kise, S. Mashiba, N. Ueda, *J. Org. Chem.* **1998**, *63*, 7931–7938.
228. C. Degrand, H. Lund, *Acta Chem. Scand.* **1977**, *B31*, 593–598; C. Degrand, D. Jacquin, P.-L. Compagnon, *J. Chem. Res. (S)* **1978**, 246–247.
229. D. R. Arnold, K. A McManus, *Can. J. Chem.* **1998**, 1238–1248 and cited references.
230. M. Machida, H. Takechi, Y. Kanaoka, *Chem. Pharm. Bull.* **1982**, *30*, 1579–1587; W. Xu, P. S. Mariano, *J. Am Chem. Soc.* **1991**, *113*, 1431–1432; U. C. Yoon, P. S. Mariano, *Acc. Chem. Res.* **1992**, *25*, 233–240; A. G. Griesbeck, A. Henz, J. Hirt, *Synthesis* **1996**, 1261–1276; T. Hasegawa, Y. Yamazaki, *Tetrahedron* **1998**, *54*, 12223–12232.
231. D. R. Arnold, P. C. Wong, A. J. Maroulis, T. S. Cameron, *Pure Appl. Chem.* **1980**, *52*, 2609–2619; N. E. Polyakov, A. I. Kruppa, V. S. Bashurova, R. N. Musin, T. V. Leshina, L. D. Kispert, *J. Photochem. Photobiol. A* **1999**, *128*, 65–74.
232. Y. Rubin, P. S. Ganapathi, A. Franz, Y.-Z. An, W. Qian, R. Neier, *Chem. Eur. J.* **1999**, *5*, 3162–3184.
233. M. M. Baizer in *Organic Electrochemistry*, eds. H. Lund and M.M. Baizer, New York, Marcel Dekker, **1991**, chapter 22, 879–948.
234. M. M. Baizer *Chemtech* **1980**, *10*, 161–164; D. E. Danly *Chemtech* **1980**, *10*, 302–311.
235. E. Lamy, L. Nadjo, J.-M. Savéant *J. Electroanal. Chem.* **1973**, *42*, 189–221.
236. M. J. Hazelrigg Jr., A. J. Bard, *J. Electrochem. Soc.* **1975**, *122*, 211–220.
237. E. A. Casanova, M. C. Dutton, D. J. Kalota, J. H. Wagenknecht, *J. Electrochem. Soc.*, **1993**, *140*, 2565–2567.
238. B.M. Bezilla Jr., J. T. Maloy, *J. Electrochem. Soc.* **1979**, *126*, 579–583.
239. I. Fussing, M. Güllü, O. Hammerich, A. Hussain, M. F. Nielson, J. H. P. Utley, *J. Chem. Soc., Perkin Trans. 2* **1996**, 649–658.
240. V. J. Puglishi, A. J. Bard, *J. Electrochem. Soc.* **1972**, *119*, 829–833.
241. A. P. Doherty, K. Scott, *J. Electroanal Chem.* **1998**, *442*, 35–40.
242. M. F. Nielsen, B. Batanero, T. Löhl, H. J. Schäfer, E.-U. Würthwein, R. Fröhlich, *Chem. Eur. J.* **1997**, *3*, 2011–2024.
243. O. Hammerich, V. D. Parker, *Acta. Chem. Scand* **1981**, *B 35*, 341–347; O. Hammerich, V. D. Parker, *Acta. Chem. Scand.* **1983**, *B37*, 379–392.
244. C. Z. Smith, J. H. P. Utley, *J. Chem. Res. (S)* **1982**, 18–19; C. Amatore, J. Pinson, J.-M. Savéant, *J. Electroanal. Chem.* **1982**, *137*, 143–148; C. Amatore, D. Garreau, M. Hammi, J. Pinson, J.-M. Savéant, *J. Electroanal. Chem.* **1985**, *184*, 1–24; A. S. Mendkovich, L. V. Michalchenko, V. P. Gultyai, *J. Electroanal. Chem.* **1987**, *224*, 273–275.
245. R. Eliason, O. Hammerich, V. D. Parker, *Acta. Chem. Scand.* **1988**, *B46*, 7–10.

246. M. Sertel, A. Yildiz, R. Gambert, H. Baumgärtel, *Electrochim. Acta* **1986**, *31*, 1287–1292.
247. M. D. Koppang, N. F. Woolsey, D. E. Bartak, *J. Am. Chem. Soc.* **1985**, *107*, 4692–4700.
248. C. Degrand, D. Jacquin, P.-L. Compagnon, *J. Chem. Res. (S)* **1978**, 246–247; C. Degrand, D. Jacquin, P.-L. Compagnon, *J. Chem. Res. (M)* **1978**, 3272–3282.
249. R. M. Crooks, A. J. Bard, *J. Electroanal. Chem.* **1988**, *240*, 253–279.
250. R. D. Webster, *J. Chem. Soc., Perkin Trans. 2* **1999**, 263–269.
251. A less common route involves the further reduction of the ketyl radical anion to the corresponding dianion and subsequent nucleophilic addition to a carbonyl group.
252. L.-S. R. Yeh, *J. Electroanal. Chem.* **1977**, *84*, 159–168.
253. M. Grätzel, A. Henglein, K. M. Bansal, *Ber. Bunsenges. Phys. Chem.*, **1973**, *77*, 6–11.
254. W. R. Fawcett, T. M. Krygowski, *Can. J. Chem.* **1976**, *54*, 3283–3292; W. R. Fawcett, A. Lasia, *Can. J. Chem.* **1981**, *59*, 3256–3260.
255. N. R. Armstrong, N. E. Vanderborgh, R. K. Quinn, *J. Electrochem. Soc.* **1975**, *122*, 615–619.
256. C. P. Andrieux, M. Grzeszczuk, J.-M. Savéant, *J. Am. Chem. Soc.* **1991**, *113*, 8811–8817.
257. J. G. Smith, C. D. Veach, *Can. J. Chem.* **1966**, *44*, 2497–2502; J. H. Smith, I. Ho, *J. Org. Chem.* **1972**, *37*, 653–656.
258. A. A. Isse, A. M. Abdurahman, E. Vianello, *J. Electroanal. Chem.* **1997**, *431*, 249–255.
259. S. U. Pedersen, T. Lund, K. Daasbjerg, M. Pop, I. Fussing, H. Lund, *Acta Chem. Scand.* **1998**, *52*, 657–671.
260. J. F. Garst, F. E. Barton, *Tetrahedron Lett.* **1969**, 587–590.
261. C. Chatgilialoglu, K. U. Ingold, J. C. Scaiano, *J. Am. Chem. Soc.* **1981**, *103*, 7739–7742; S. U. Padersen, T. Lund, *Acta Chem. Scand.* **1991**, *45*, 397–402.
262. M. S. W. Chan, D. R. Arnold, *Can. J. Chem.* **1997**, *75*, 1810–1819.
263. M. Kojima, A. Ishida, S. Takamuku, Y. Wada, S. Yanagida, *Chem. Lett.* **1994**, 1897–1900; H. Weng, H. D. Roth, *J. Org. Chem.* **1995**, *60*, 4136–4145; H. J. P. de Lijser, D. R. Arnold, *J. Org. Chem.* **1997**, *62*, 8432–8438; H. J. P. de Lijser, D. R. Arnold, *J. Chem. Soc. Perkin Trans. 2* **1997**, 1369–1380; S. Swansburg, K. Janz, G. Jocys, A. Pincock, J. Pincock, *Can. J. Chem.* **1998**, *76*, 35–47; V. Nair, L. G. Nair, *Tetrahedron Lett.* **1998**, *39*, 4585–4586; V. Nair, S. B. Panicker, *Tetrahedron Lett.* **1999**, *40*, 563–564.
264. D. R. Arnold, K. A. McManus, M. S. W. Chan, *Can. J. Chem.* **1997**, *75*, 1055–1075.
265. C. P. Butts, L. Eberson, M. P. Hartshorn, W. T. Robinson, D. J. Timmerman-Vaughan, *Acta Chem. Scand.* **1997**, *51*, 73–87; L. Eberson, M. P. Hartshorn, J. O. Svensson, *Acta Chem. Scand.* **1997**, *51*, 279–288; F. Ciminale, A. Ciardo, S. Francioso, A. Nacci, *J. Org. Chem.* **1999**, *64*, 2459–2464.
266. T. Shono, S. Kashimura, *J. Org. Chem.* **1983**, *48*, 1939–1940.
267. L. Lopez, V. Calò, F. Stasi, *Synthesis* **1987**, 947–948.
268. G. Lem, D. I. Schuster, S. H. Courtney, Q. Lu, S. R. Wilson, *J. Am. Chem. Soc.* **1995**, *117*, 554–555; I. G. Safonov, S. H. Courtney, D. I. Schuster, *Res. Chem. Intermed.* **1997**, *23*, 541–548.
269. M. Cariou, J. Simonet, *Tetrahedron Lett.* **1991**, *32*, 4913–4916; C. P. Butts, L. Eberson, M. P. Hartshorn, W. T. Robinson, B. R. Wood, *Acta Chem. Scand.* **1996**, *50*, 587–595.
270. D. R. Arnold, M. S. W. Chan, K. A. McManus, *Can. J. Chem.* **1996**, *74*, 2143–2166.
271. H. Baltes, L. Stork, H. J. Schäfer, *Justus Liebigs Ann. Chem.* **1979**, 318–327.
272. As a keystep in the synthesis of a) maltol: T. Shono, Y. Matsumura, *Tetrahedron Lett.* **1976**, 1363–1364; (b) rethrolones: T. Shono, Y. Matsumura, H. Hamaguchi, K. Nakamura, *Chem. Lett.* **1976**, 1249–1252; (c) the flavouring agent cylotene: T. Shono, Y. Matsumura, H. Hamaguchi, *J. Chem. Soc., Chem. Commun.* **1977**, 712–713.
273. T. Shono, Y. Matsumura, K. Tsubata, Y. Sugihara, S. Yamane, T. Kanazawa, T. Aoki, *J. Am. Chem. Soc.* **1982**, *104*, 6697–6703.
274. J. Y. Becker, E. Shakkour, J. A. P. R. Sarma, *J. Chem. Soc., Chem. Commun.* **1990**, 1016–1017; J. Y. Becker, E. Shakkour, J. A. P. R. Sarma, *J. Org. Chem.* **1992**, *57*, 3716–3720.
275. R. A. Neunteufel, D. R. Arnold, *J. Am. Chem. Soc.* **1973**, *95*, 4080–4081; M. Kojima, A. Ishida, Y. Kuriyama, Y. Wada, H. Takeya, *Bull. Chem. Soc. Jpn.* **1999**, *72*, 1049–1055.
276. K. D. Warzecha, X. C. Xing, M. Demuth, *Pure Appl. Chem.* **1997**, *69*, 109–112; C. Heinemann, M. Demuth, *J. Am. Chem. Soc.* **1997**, *119*, 1129–1130; X. Xing, M. Demuth, *Synlett* **1999**, 987–990.

277. C. Heinemann, M. Demuth, *J. Am. Chem. Soc.* **1999**, *121*, 4894–4895.
278. G. Pandey, M. Sridar, U. T. Bhalerao, *Tetrahedron Lett.* **1990**, *31*, 5373–5376; A. K. Panfilov, G. V. Cherkaev, T. V. Magdesieva, N. M. Przhiyalgovskaya, *Zh. Org. Khim.* **1992**, *28*, 691–699; Y. Inoue, T. Okano, N. Yamasaki, A. Tai, *J. Chem. Soc., Chem. Commun.* **1993**, 718–720; A. Goosen, C. W. McCleland, F. C. Rinaldi, *J. Chem. Soc., Perkin Trans. 2* **1993**, 279–281; F. D. Lewis, G. D. Reddy, B. E. Cohen, *Tetrahedron Lett.* **1994**, *35*, 535–538.
279. (a) M. Fujita, A. Shindo, A. Ishida, T. Majima, S. Takamuku, S. Fukuzumi, *Bull. Chem. Soc. Jpn.* **1996**, *69*, 743–749; (b) V. Nair, J. Mathew, P. P. Kanakamma, S. B. Panicker, V. Sheeba, S. Zeena, G. K. Eigendorf, *Tetrahedron Lett.* **1997**, *38*, 2191–2194.
280. K. McMahon, D. R. Arnold, *Can. J. Chem.* **1993**, *71*, 450–468; D. R. Arnold, X. Du, H. J. P. de Lijser, *Can. J. Chem.* **1995**, *73*, 522–530; M. Tanaka, H. Nakashima, M. Fujiwara, H. Ando, Y. Souma, *J. Org. Chem.* **1996**, *61*, 788–792; T. Douadi, M. Cariou, J. Simonet, *Tetrahedron* **1996**, *52*, 4449–4456.
281. B. B. Snider, T. Kwon, *J. Org. Chem.* **1992**, *57*, 2399–2410; A. Heidbreder, J. Mattay, *Tetrahedron Lett.* **1992**, *33*, 1973–1976; G. Pandey, A. Krishna, K. Girja, M. Karthikeyan, *Tetrahedron Lett.* **1993**, *34*, 6631–6634; D. G. New, Z. Tesfai, K. D. Moeller, *J. Org. Chem.* **1996**, *61*, 1578–1598; H. E. Zimmerman, K. D. Hoffacker, *J. Org. Chem.* **1996**, *61*, 6526–6534; F. Ciminale, L. Lopez, G. M. Farinola, *Tetrahedron Lett.* **1999**, *40*, 7267–7270.
282. D. A. Evans, C. J. Dinsmore, D. A. Evrard, K. M. DeVries, *J. Am. Chem. Soc.* **1993**, *115*, 6426–6427.
283. R. B. Herbert, A. E. Kattah, A. J. Murtagh, P. W. Sheldrake, *Tetrahedron Lett.* **1995**, *36*, 5649–5650.
284. O. Hammerich, V. D. Parker, *Adv. Phys. Org. Chem.* **1984**, *20*, 55–189.
285. M. Schmittel, A. Burghart, *Angew. Chem.* **1997**, *109*, 2659–2699; *Angew. Chem. Int. Ed. Engl.* **1997**, *36*, 2550–2589.
286. M. Oyama, K. Nozaki, T. Nagaoka, S. Okazaki, *Bull. Chem. Soc. Jpn.* **1990**, *63*, 33–41; M. Oyama, K. Nozaki, S. Okazaki, *J. Electroanal. Chem.* **1991**, *304*, 61–73.
287. V. D. Parker, B. Reitstöen, M. Tilset, *J. Phys. Org. Chem.* **1989**, *2*, 580–584; F. Norrsell, K. L. Handoo, V. D. Parker, *J. Org. Chem.* **1993**, *58*, 4929–4932; M. S. Workentin, L. J. Johnston, D. D. M. Wayner, V. D. Parker, *J. Am. Chem. Soc.* **1994**, *116*, 8279–8287; H. J. Wang, G. Zheng, V. D. Parker, *Acta Chem. Scand.* **1995**, *49*, 311–312; M. S. Workentin, V. D. Parker, T. L. Morkin, D. D. M. Wayner, *J. Phys. Chem. A* **1998**, *102*, 6503–6512.
288. C. S. Q. Lew, J. R. Brisson, L. J. Johnston, *J. Org. Chem.* **1997**, *62*, 4047–4056.
289. L. J. Johnston, N. P. Schepp, *J. Am. Chem. Soc.* **1993**, *115*, 6564–6571; L. J. Johnston, N. P. Schepp, *Pure Appl. Chem.* **1995**, *67*, 71–78.
290. P. D. Wood, L. J. Johnston, *Photochem. Photobiol.* **1997**, *66*, 642–648.
291. M. S. Workentin, L. J. Johnston, D. D. M. Wayner, V. D. Parker, *J. Am. Chem. Soc.* **1994**, *116*, 8279–8287.
292. H. Mayr, M. Patz, *Angew. Chem.* **1994**, *106*, 990–1010; *Angew. Chem., Int. Ed. Engl.* **1994**, *33*, 938–958.
293. T. Sumiyoshi, *Chem. Lett.* **1995**, 645–646.
294. V. D. Parker, *Acta Chem. Scand.* **1983**, *B37*, 393–401.
295. C. L. Kulkarni, B. J. Scheer, J. F. Rusling, *J. Electroanal. Chem.* **1982**, *140*, 57–74.
296. B. Aalstad, A. Ronlán, V. D. Parker, *Acta Chem. Scand.* **1981**, *B35*, 247–257.
297. K. Nozaki, M. Oyama, H. Hatano, S. Okazaki, *J. Electroanal. Chem.* **1989**, *270*, 191–204.
298. L. Lopez, *Top. Curr. Chem.* **1990**, *156*, 117–166.
299. A. P. Schaap, K. A. Zaklika, B. Kaskar, L. W.-M. Fung, *J. Am. Chem. Soc.* **1980**, *102*, 389–391; L. T. Spada, C. S. Foote, *J. Am. Chem. Soc.* **1980**, *102*, 391–393; K. E. O'Shea, S. H. Jannach, I. Garcia, *J. Photochem. Photobiol. A* **1999**, *122*, 127–131.
300. E. Bosch, J. K. Kochi, *J. Am. Chem. Soc.* **1996**, *118*, 1319–1329.
301. M. Lehnig, *J. Chem. Soc., Perkin Trans. 2* **1996**, 1943–1948.
302. C. Amatore, J. K. Kochi, *Adv. Electron Transfer Chem.* **1991**, *1*, 55–148.
303. K. Mizuno, M. Ikeda, Y. Otsuji, *Tetrahedron Lett.* **1985**, *26*, 461–464.
304. R. J. Sundberg in *Organic Photochemistry*, Ed. A. Padwa, Marcel Dekker, New York, Vol. 6, p 121–176, **1983**.

305. M. Mascal, C. J. Moody, *J. Chem. Soc., Chem. Commun.* **1988**, 587–588; M. Mascal, C. J. Moody, *J. Chem. Soc., Chem. Commun.* **1988**, 589–590.
306. (a) J. Mattay, J. Gersdorf, K. Buchkremer, *Chem. Ber.* **1987**, *120*, 307–318; (b) J. Gersdorf, J. Mattay, H. Görner, *J. Am. Chem. Soc.* **1987**, *109*, 1203–1209; D. Sun, S. M. Hubig, J. K. Kochi, *J. Org. Chem.* **1999**, *64*, 2250–2258.
307. K. Maruyama, T. Otsuki, S. Tai, *J. Org. Chem.* **1985**, *50*, 52–60.
308. V. D. Parker, *Acta Chem. Scand.* **1998**, *52*, 154–159.
309. C. Siedschlag, H. Luftmann, C. Wolff, J. Mattay, *Tetrahedron* **1997**, *53*, 3587–3592; C. Siedschlag, H. Luftmann, C. Wolff, J. Mattay, *Tetrahedron* **1999**, *55*, 7805–7818.
310. M. Tsuchiya, T. W. Ebbesen, Y. Nishimura, H. Sakuragi, K. Tokumaru, *Chem. Lett.* **1987**, 2121–2124.
311. Although the bimolecular rate constant of radical cations with superoxide anion is higher, this process most often results in electron transfer rather than in oxygenation, see ref. [310].
312. S. Tojo, K. Morishima, A. Ishida, T. Majima, S. Takamuku, *J. Org. Chem.* **1995**, *60*, 4684–4685.
313. C. Demaille, A. J. Bard, *Acta Chem. Scand.* **1999**, *53*, 842–848.
314. C. P. Andrieux, P. Audebert, P. Hapiot, J.-M. Savéant, *J. Phys. Chem.* **1991**, *95*, 10158–10164.
315. P. Audebert, J. M. Catel, V. Duchenet, L. Guyard, P. Hapiot, G. Le Coustumer, *Syn. Met.* **1999**, *101*, 642–645; A. Merz, J. Kronberger, L. Dunsch, A. Neudeck, A. Petr, L. Parkanyi, *Angew. Chem.* **1999**, *111*, 1533–1538; *Angew. Chem. Int. Ed.* **1999**, *38*, 1442–1446.
316. A. F. Diaz in *Organic Electrochemistry*, H. Lund, M. M. Baizer, Eds., Chapter 33, Marcel Dekker, New York, **1991**; P. Hapiot, L. Gaillon, P. Audebert, J. J. E. Moreau, J. P. Lère-Porte, M. W. C. Man, *J. Electroanal. Chem.* **1997**, *435*, 85–94; J. J. Apperloo, R. A. J. Janssen, *Syn. Met.* **1999**, *101*, 373–374.
317. E. Levillain, J. Roncali, *J. Am. Chem. Soc.* **1999**, *121*, 8760–8765.
318. H. Yang, A. J. Bard, *J. Electroanal. Chem.* **1991**, *306*, 87–109.
319. S. C. Creason, J. Wheeler, R. F. Nelson, *J. Org. Chem.* **1972**, *37*, 4440–4446; R. F. Nelson, R. H. Philp Jr., *J. Phys. Chem.* **1979**, *83*, 713–716.
320. D. Larumbe, I. Gallardo, C. P. Andrieux, *J. Electroanal. Chem.* **1991**, *304*, 241–247.
321. A. Smie, J. Heinze, *Angew. Chem.* **1997**, *109*, 375–379; *Angew. Chem. Int. Ed. Engl.* **1997**, *36*, 363–367.
322. G. Pandey, D. Pooranchand, U. T. Bhalerao, *Tetrahedron* **1991**, *47*, 1745–1752.
323. H. J. Shine, *Phosporus Silicon* **1994**, *95–6*, 429–430.
324. S. Yasui, K. Shioji, A. Ohno, M. Yoshihara, *J. Org. Chem.* **1995**, *60*, 2099–2105.
325. T.-L. Ho, *Synthesis* **1973**, 347–354.
326. A. Houmam, D. Shukla, H.-B. Kraatz, D. D. M. Wayner, *J. Org. Chem.* **1999**, *64*, 3342–3345; W. K. Lee, B. Liu, C. W. Park, H. J. Shine, I. Y. Guzman-Jimenez, K. H. Whitmire, *J. Org. Chem.* **1999**, *64*, 9206–9210.
327. V. Jouikov, D. Fattahova, *Electrochim. Acta* **1998**, *43*, 1811–1819.
328. A. Lebouc, J. Simonet, J. Gelas, A. Debhi, *Synthesis* **1987**, 320–321; H. J. Cristau, B. Chabaud, C. Niangoran, *J. Org. Chem.* **1983**, *48*, 1527–1529.
329. M. Kimura, S. Matsubara, Y. Sawaki, H. Iwamura, *Tetrahedron Lett.* **1986**, *27*, 4177–4178.
330. T. Mandai, H. Yasunaga, M. Kawada, J. Otera, *Chem. Lett.* **1984**, 715–716; T. Mandai, H. Irei, M. Kawada, J. Otera, *Tetrahedron Lett.* **1984**, *25*, 2371–2374.
331. K. Narasaka, T. Okauchi, *Chem. Lett.* **1991**, 515–518.
332. H. Takeuchi, T. Hiyama, N. Kamai, H. Õya, *J. Chem. Soc., Perkin Trans. 2* **1997**, 2301–2305.
333. G. Pandey, R. Sochanchingwung, S. K. Tiwari, *Synlett* **1999**, 1257–1258.
334. M. Newcomb, T. M. Deeb, D. J. Marquardt, *Tetrahedron* **1990**, *46*, 2317–2328; S. Karady, E. G. Corley, N. L. Abrahamson, J. S. Amato, L. M. Weinstock, *Tetrahedron* **1991**, *47*, 757–766; M. Newcomb, K. A. Weber, *J. Org. Chem.* **1991**, *56*, 1309–1313; U. Jahn, S. Aussieker, *Org. Lett.* **1999**, *1*, 849–852.
335. M. Kimura, Y. Ban, *Synthesis*, **1976**, 201–202; M. W. Wolff, *Chem. Rev.* **1963**, *63*, 55–64.
336. Y. Shibanuma, T. Okamoto, *Chem. Pharm Bull.* **1985**, *33*, 3187–3194.
337. B. Merzougui, Y. Berchadsky, P. Tordo, G. Gronchi, *Electrochim. Acta* **1997**, *42*, 2445–2453; S. Yasui, K. Shioji, M. Tsujimoto, A. Ohno, *J. Chem. Soc., Perkin Trans. 2* **1999**, 855–862.

338. H. Yokoi, A. Hatta, K. Ishiguro, Y. Sawaki, *J. Am. Chem. Soc.* **1998**, *120*, 12728–12733.
339. S. Yasui, K. Shioji, M. Tsujimoto, A. Ohno, *Chem. Lett.* **1995**, 783–784.
340. J. H. Horner, F. N. Martinez, O. M. Musa, M. Newcomb, H. E. Shahin, *J. Am. Chem. Soc.* **1995**, *117*, 11124–11133.
341. P. H. Sackett, J. S. Mayausky, T. Smith, S. Kalus, R. L. McCreery, *J. Med. Chem.* **1981**, *24*, 1342–1347.
342. B. L. Miller, K. Kuczera, C. Schöneich, *J. Am. Chem. Soc.* **1998**, *120*, 3345–3356.
343. M. Kamata, M. Sato, E. Hasegawa, *Tetrahedron Lett.* **1992**, *35*, 5085–5088.
344. A. P. Schaap, S. Prasad, S. D. Gagnon, *Tetrahedron Lett.* **1983**, *24*, 3047–3050; A. P. Schaap, G. Prasad, S. Siddiqui, *Tetrahedron Lett.* **1984**, *25*, 3035–3038.
345. S. Yasui, K. Shioji, M. Tsujimoto, A. Ohno, *Tetrahedron Lett.* **1996**, *37*, 1625–1628.
346. S. Yasui, M. Fujii, C. Kawano, Y. Nishimura, A. Ohno, *Tetrahedron Lett.* **1991**, *32*, 5601–5604; S. Yasui, M. Fujii, C. Kawano, Y. Nishimura, K. Shioji, A. Ohno, *J. Chem. Soc., Perkin Trans. 2* **1994**, 177–183; S. Yasui, K. Shioji, A. Ohno, *Tetrahedron Lett.* **1994**, *35*, 2695–2698.
347. D. P. Riley, M. R. Smith, P. E. Correa, *J. Am. Chem. Soc.* **1988**, *110*, 177–180.
348. W. Zhou, E. L. Clennan, *Chem. Commun.* **1999**, 2261–2262.
349. B. L. Miller, T. D. Williams, C. Schöneich, *J. Am. Chem. Soc.* **1996**, *118*, 11014–11025.
350. U. Haberl, O. Wiest, E. Steckhan, *J. Am. Chem. Soc.* **1999**, *121*, 6730–6736.
351. M. Hofmann, H. F. Schaefer, III *J. Am. Chem. Soc.* **1999**, *121*, 6719–6729.
352. N. L. Bauld, J. Yang, *Tetrahedron Lett.* **1999**, *40*, 8519–8522.
353. N. L. Bauld, J. Yang, *Org. Lett.* **1999**, *1*, 773–774.
354. (a) G. Bouchoux, J. Y. Salpin., *Rapid Commun. Mass Spectrom.* **1994**, *8*, 325–328; (b) C. Denekamp, A. Weisz, A. Mandelbaum, *J. Mass Spectr.* **1996**, *31*, 1028–1032; (c) N. Morlender-Vais, A. Mandelbaum, *J. Mass Spectr.* **1998**, *33*, 229–241.
355. D. W. Borhani, F. D. Greene, *J. Org. Chem.* **1986**, *51*, 1563–1570.
356. J. Delaunay, G. Mabon, A. Orliac, J. Simonet, *Tetrahedron Lett.* **1990**, *31*, 667–668; R. G. Janssen, M. Motevalli, J. H. P. Utley, *Chem. Commun.* **1998**, 539–540.
357. A. Pezeshk, I. D. Podmore, P. F. Heelis, M. C. R. Symons, *J. Phys. Chem.* **1996**, *100*, 19714–19718; A. A. Voityuk, M.-E. MichelBeyerle, N. Rösch, *J. Am. Chem. Soc.* **1996**, *118*, 9750–9758; R. Epple, E.-U. Wallenborn, T. Carell, *J. Am. Chem. Soc.* **1997**, *119*, 7440–7451.
358. A. S. Kurbatova, Y. V. Kurbatov, *Russ. J. Org. Chem.* **1997**, *33*, 1050.
359. N. L. Bauld, J. Cessac, C.-S. Chang, F. R. Farr, R. Holloway, *J. Am. Chem. Soc.* **1976**, *98*, 4561–4567; M. A. Fox, J. R. Hurst, *J. Am. Chem. Soc.* **1984**, *106*, 7626–7627.
360. V. Amir-Ebrahimi, J. G. Hamilton, J. Nelson, J. J. Rooney, J. M. Thompson, A. J. Beaumont, A. D. Rooney, C. J. Harding, *Chem. Commun.* **1999**, 1621–1622.
361. K. Exner, D. Hunkler, G. Gescheidt, H. Prinzbach, *Angew. Chem.* **1998**, *110*, 2013–2016; *Angew. Chem. Int. Ed.* **1998**, *37*, 1910–1913.
362. F. A. Bell, R. A. Crellin, N. Fujii, A. Ledwith, *J. Chem. Soc., Chem. Commun.* **1969**, 251–252; A. Ledwith, *Acc. Chem. Res.* **1972**, *5*, 133–139.
363. D. B. Rusterholz, D. B. Gorman, P. G. Gassman, *Molecules* **1997**, *2*, 80–86.
364. N. L. Bauld, *Adv. Electron Transfer Chem.* **1992**, *2*, 1–66.
365. (a) M. Schmittel, H. von Seggern, *J. Am. Chem. Soc.* **1993**, *115*, 2165–2177; (b) M. Schmittel, C. Wöhrle, I. Bohn, *Chem. Eur. J.* **1996**, *2*, 1031–1040; (c) M. Schmittel, C. Wöhrle, *J. Org. Chem.* **1995**, *60*, 8223–8230; (d) U. Haberl, E. Steckhan, S. Blechert, O. Wiest, *Chem. Eur. J.* **1999**, *5*, 2859–2865; N. L. Bauld, J. Yang, D. Gao, *Perkin Trans. 2* **2000**, 207–210.
366. C. F. Gürtler, E. Steckhan, S. Blechert, *J. Org. Chem.* **1996**, *61*, 4136–4143; C. F. Gürtler, S. Blechert, E. Steckhan, *Chem. Eur. J.* **1997**, *3*, 447–452; T. Peglow, S. Blechert, E. Steckhan, *Chem. Eur. J.* **1998**, *4*, 107–112; J. Botzem, U. Haberl, E. Steckhan, S. Blechert, *Acta Chem. Scand.* **1998**, *52*, 175–193; T. Peglow, S. Blechert, E. Steckhan, *Chem. Commun.* **1999**, 433–434.
367. Y. Takahashi, M. Ando, T. Miyashi, *J. Chem. Soc., Chem. Commun.* **1995**, 521–522; L. Brancaleon, D. Brousmiche, V. J. Rao, L. J. Johnston, V. Ramamurthy, *J. Am. Chem. Soc.* **1998**, *120*, 4926–4933M; N. L. Bauld, D. Gao, *Perkin Trans. 2* **2000**, 191–192.
368. K. Ishiguro, M. Ikeda, Y. Sawaki, *J. Org. Chem.* **1992**, *57*, 3057–3066.
369. W. B. Motherwell, A. S. Williams, *Angew. Chem.* **1995**, *107*, 2207–2209; *Angew. Chem., Int. Ed. Engl.* **1995**, *34*, 2031–2033.

370. N. L. Bauld, G. A. Mirafzal, *J. Am. Chem. Soc.* **1991**, *113*, 3613–3614.
371. N. L. Bauld, G. W. Stufflebeme, K. T. Lorenz, *J. Phys. Org. Chem.* **1989**, *2*, 585–601.
372. R. Herges, F. Starck, T. Winkler, M. Schmittel, *Chem. Eur. J.* **1999**, *5*, 2965–2970; G. D. Reddy, O. Wiest, T. Hudlický, V. Schapiro, D. Gonzalez, *J. Org. Chem.* **1999**, *64*, 2860–2863.
373. (a) F. Müller, J. Mattay, *Angew. Chem.* **1991**, *103*, 1352–1353; *Angew. Chem., Int. Ed. Engl.* **1991**, *30*, 1336–1337; (b) F. Müller, J. Mattay, *Angew. Chem.* **1992**, *104*, 207–208; *Angew. Chem., Int. Ed. Engl.* **1992**, *31*, 209–210; (c) F. Müller, A. Karwe, J. Mattay, *J. Org. Chem.* **1992**, *57*, 6080–6082; (d) F. Müller, J. Mattay, S. Steenken, *J. Org. Chem.* **1993**, *58*, 4462–4464; (e) F. Müller, J. Mattay, *Chem. Ber.* **1993**, *126*, 543–549.
374. (a) K. Lorenz, N. L. Bauld, *Catalysis* **1985**, *95*, 613–616; (b) T. Miyashi, A. Konno, Y. Takahashi, *J. Am. Chem. Soc.* **1988**, *110*, 3676–3677; (c) H. Ikeda, T. Oikawa, T. Miyashi, *Tetrahedron Lett.* **1993**, *34*, 2323–2326.
375. H. Ikeda, T. Minegishi, H. Abe, A. Konno, J. L. Goodman, T. Miyashi, *J. Am. Chem. Soc.* **1998**, *120*, 87–95; H. Ikeda, T. Takasaki, Y. Takahashi, A. Konno, M. Matsumoto, Y. Hoshi, T. Aoki, T. Suzuki, J. L. Goodman, T. Miyashi, *J. Org. Chem.* **1999**, *64*, 1640–1649.
376. S. Dhanalekshmi, C. S. Venkatachalam, K. K. Balasubramanian, *J. Chem. Soc., Chem. Commun.* **1994**, 511–512.
377. D. W. Reynolds, B. Harirchian, H.-S. Chiou, B. K. Marsh, N. L. Bauld, *J. Phys. Org. Chem.* **1989**, *2*, 57–88.
378. J. P. Dinnocenzo, D. A. Conlon, *J. Am. Chem. Soc.* **1988**, *110*, 2324–2326.
379. B. Kräutler, A. Pfaltz, R. Nordmann, K. O. Hodgson, J. D. Dunitz, A. Eschenmoser, *Helv. Chim. Acta* **1976**, *59*, 924–937.
380. N. L. Bauld, *J. Comput. Chem.* **1990**, *11*, 896–898.
381. J. P. Dinnocenzo, D. A. Conlon, *Tetrahedron Lett.* **1995**, *36*, 7415–7418.
382. S. Tojo, S. Toki, S. Takamuku, *J. Org. Chem.* **1991**, *56*, 6240–6243; N. P. Schepp, L. J. Johnston, *J. Am. Chem. Soc.* **1994**, *116*, 6895–6903; N. P. Schepp, L. J. Johnston, *J. Am. Chem. Soc.* **1994**, *116*, 10330–10331; O. Brede, F. David, *Radiat. Phys. Chem.* **1996**, *47*, 53–58; N. P. Schepp, D. Shukla, H. Sarker, N. L. Bauld, L. J. Johnston, *J. Am. Chem. Soc.* **1997**, *119*, 10325–10334.
383. L. J. Johnston, N. P. Schepp, *Pure Appl. Chem.* **1995**, *67*, 71–78.
384. H. Prinzbach, G. Gescheidt, H. D. Martin, R. Herges, J. Heinze, G. K. S. Prakash, G. A. Olah, *Pure Appl. Chem.* **1995**, *67*, 673–682; G. Gescheidt, H. Prinzbach, A. G. Davies, R. Herges, *Acta Chem. Scand.* **1997**, *51*, 174–180; M. Etzkorn, F. Wahl, M. Keller, H. Prinzbach, F. Barbosa, V. Peron, G. Gescheidt, J. Heinze, R. Herges, *J. Org. Chem.* **1998**, *63*, 6080–6081.
385. K. Exner, H. Prinzbach, G. Gescheidt, B. Grossmann, J. Heinze, *J. Am. Chem. Soc.* **1999**, *121*, 1964–1965.
386. F. W. McLafferty, F. Turecek, *Interpretation of Mass Spectra*, 4th Ed., University Science Books, Mill Valley', 1993, Chapter 4.3.
387. M. Born, S. Ingemann, N. M. M. Nibbering, *Mass Spectr. Rev.* **1997**, *16*, 181–200.
388. F. Casado, L. Pisano, M. Farriol, I. Gallardo, J. Marquet, G. Melloni, *J. Org. Chem.* **2000**, *65*, 322–331.
389. G. A. Molander, *Org. React.* **1994**, *46*, 211–367; G. A. Molander, C. R. Harris, *Chem. Rev.* **1996**, *96*, 307–338; R. J. Enemærke, K. Daasbjerg, T. Skrydstrup, *Chem. Commun.* **1999**, 343–344.
390. J. E. McMurry, *Chem. Rev.* **1989**, *89*, 1513–1524.
391. A. Gansäuer, T. Lauterbach, H. Bluhm, M. Noltemeyer, *Angew. Chem.* **1999**, *111*, 3112–3114; *Angew. Chem. Int. Ed.* **1999**, *38*, 2909–2910 and cited references.
392. G. G. Melikyan, *Org. React.* **1996**, *49*, 427–675; B. B. Snider, *Chem. Rev.* **1996**, *96*, 339–363; T. Linker, *J. Prakt. Chem.* **1997**, *339*, 488–492.
393. V. Nair, J. Mathew, J. Prabhakaran, *Chem. Soc. Rev.* **1997**, 127–132.

# 2 Electron Transfer from Aliphatic and Alicyclic Compounds

*Heinz D. Roth*

## 2.1 Introduction

Electron-transfer processes have attracted considerable attention because of their central role in the chemistry of life and as a method for generating interesting intermediates, in the gas phase, in solution, or in the solid state. Photo-induced electron transfer is one of the very few fundamental reaction types known in excited-state organic chemistry. It occurs in the solid state, in solution, and in the gas phase, including the Earth's atmosphere and outer space. The products generated by electron transfer vary with the nature of the substrate and with the medium in which the reaction is conducted. They can be macromolecular 'charge-separated' entities (in the photosynthesis of the green plant), zwitterions, ion pairs, or 'free' radical ions, carrying a positive or negative charge in addition to an unpaired spin. These intermediates can be characterized by a range of spectroscopic techniques and their reactions can be studied conveniently. This chapter deals with electron transfer from saturated chain- or cyclic hydrocarbons, and the structures and reactions of their positively charged radical ions.

Alkanes (Section 2.2) are among the least reactive compounds; their C–H bonds are among the least acidic moieties (highest $pK_a$ values) known. Likewise, $n$-alkanes have very low electron affinities and very high ionization potentials; thus, they are poor electron acceptors and electron donors; it takes significant energies to activate them. For the small-ring compounds, cyclopropane and cyclobutane (Section 2.3), electron transfer occurs with relief of ring strain; these systems are more reactive. Strained bicyclic systems, particularly bicyclo[1.1.0]butane and bicyclo[2.1.0]pentane (Section 2.4), are more reactive still.

Electron transfer from methane is significant because it seems likely that the radical cation and secondary intermediates derived from it ($CH_3^{\cdot}$, $CH_3^+$) played a significant role in the chemical evolution preceding the origins of life. Methane is a probable constituent of early planetary atmospheres and its radical cation has potential significance as an interstellar species.

## 2.2 Electron-transfer Reactions of Aliphatic Compounds

Alkanes are among the least reactive classes of compound; their C–H bonds are among the least acidic moieties (highest $pK_a$ values) known. Likewise, $n$-alkanes have very low electron affinities and very high ionization potentials; thus, they are poor electron acceptors and electron donors. The molecular anions of $n$-alkanes are even less stable than are those of cycloalkanes (Section 2.3) and their temporary anion states are even shorter lived [1]. Non-dissociative electron attachment requires the dissipation of excess energy by radiation or collision; for simple alkanes, methane through butane, negative ion yields are ca $10^4$ times lower than positive ion yields. Typically, electron impact does not result in molecular ions; instead, small fragment ions such as $CH^-$, $CH_2^-$, $CH_3^-$, $C_2H^-$ predominate [2–4]. The adiabatic ionization potentials of $n$-alkanes decrease from 12.61 eV for methane and 11.52 eV for ethane to 10.25 eV for $n$-pentane and 10.13 eV for $n$-hexane and to 9.80 eV for $n$-octane and 9.65 eV for $n$-decane [5]. As a consequence, it takes significant energies to activate these compounds.

In the gas phase, $n$-alkanes are readily ionized upon electron impact (mass spectrometry, MS) or upon $He_\alpha$ impact (photoelectron spectroscopy, PES). The resulting spectra provide information about the molecular ions, for example, on their ease of fragmentation and their fragmentation patterns (MS) [6, 7], whereas the PE spectra reveal the ordering and energies of their molecular orbitals [8]. In solution the high oxidation potentials of $n$-alkanes pose significant difficulties for the generation of their radical cations and render the resulting species highly reactive, although the electron-donor properties improve somewhat with increasing branching (vide infra). Accordingly, these radical cations were long thought to be unstable and decompose by instantaneous deprotonation. Radical cations have become readily accessible only with the advent of halocarbon and noble gas matrix-isolation techniques. The development of these techniques was crucial for the detailed study of $n$-alkane radical cations.

### 2.2.1 High-energy Irradiation of Matrices

The pioneering work of W. H. Hamill and co-workers established that high-energy photons, such as X-rays or $\gamma$-rays, can eject electrons from halogen-containing matrix materials, viz., tetrachloromethane, chlorobutanes, and several Freon mixtures [9–11]. High-energy photons interact with these materials by generating 'holes' while ejecting electrons into the matrix (Eq. 1). In the presence of solute molecules (S) of lower oxidation potential than the host ($n$-alkane guests at low concentrations; ca 1 in $10^3$), the holes are unstable. A host molecule transfers an electron to the 'hole', regenerating the matrix material (Eq. 2). The ejected electrons are 'solvated' in the matrix or attach themselves to a halogen-containing molecule (Eq. 3), causing its fragmentation to a halide ion and a free radical (Eq. 4). These processes are without direct consequence for the radical cation; they are mentioned here because they can interfere with the observation of the species chosen for study, or because the halide ions can scavenge the radical cations to form free radicals (Eq. 5).

$$R-X + \gamma \rightarrow R-X^{\cdot +} + e^- \tag{1}$$

$$R-X^{\cdot +} + S \rightarrow R-X + S^{\cdot +} \tag{2}$$

$$R-X + e^- \rightarrow R-X^{\cdot -} \tag{3}$$

$$R-X^{\cdot -} \rightarrow R^{\cdot} + X^- \tag{4}$$

$$S^{\cdot +} + X^- \rightarrow S^{\cdot}-X \tag{5}$$

The overall procedure, based on a combination of cryogenic techniques, matrix isolation, ionizing radiation, and ESR detection, has become the method of choice for generating and studying alkane radical cations. One note of caution is in order, however. It is generally recognized that the barriers to the rearrangements of radical cations are considerably lower than are those on the parent energy surfaces. Accordingly, some radical cations are unstable in solid matrices and rearrange instantaneously under the conditions of radiolytic generation in frozen matrices. Applications of optical spectroscopy to the study of alkane radical cations are limited. Although this technique is of great importance for studying many organic radical ions, the spectra of alkane radical cations do not fall into an accessible spectral range.

An alternative method of radical cation generation involves photo- or electron-impact ionization of substrates at high dilution (ca 1 in $10^6$) during vapor deposition in neon matrices at (ca 5 K) [12]. The radical cations deposited in the matrix owe their stability to the high ionization potential of the matrix material, which is more difficult to oxidize than are $n$-alkanes. This technique has found significant applications and has provided invaluable insights in special cases (Section 2.3); it is not, however, generally accessible and offers only a very limited temperature range for study. Finally, some of the more stable radical cations of highly branched alkanes can be generated in zeolites.

The application of matrix isolation methods has facilitated the generation and study of many radical cations. Early examples of alkane radiolysis generating radical cations include $n$-octane, which produced a species '$n$-$C_8H_{18}+$' [13, 14], and the well characterized radical cation of hexamethylethane, reported independently by three different groups in 1980/81 [15–18]. Although the ESR spectrum of '$n$-$C_8H_{18}+$' provided little specific information about its structure and the latter might be considered a 'special case', the time was ripe for systematic investigation of electron transfer from alkanes. Thus, when Iwasaki and co-workers reported clear information on the structure of ethane radical cation, the study of alkane radical cations became a significant area of scientific investigation [19].

## 2.2.2 Electron Spin Resonance

Among the techniques available for the study of free radicals or radical ions in solution, electron spin resonance (ESR) stands out as a technique with sufficient resolution to provide detailed information about the identity of the intermediate in question. In an external magnetic field the unpaired electron of such a species can adopt either of two spin orientations, parallel or antiparallel to the field $H_0$. The two orientations are of slightly different energies. Transitions between the corresponding spin levels can be stimulated by applying radiation of a frequency satisfying the resonance condition:

$$hv = g\mu H_0 \qquad (6)$$

where $h$ is the Planck constant, $g$ is a parameter characteristic for the radical under scrutiny, $\mu$ is the Bohr magneton, and $H_0$ is the applied magnetic field strength [20, 21].

Experimentally, resonance is approached by sweeping the field in the range near 3400 G while holding the frequency constant, typically at ca 9.6 GHz. A key contribution to the identification of the radical (ion) is due to the interaction of the unpaired spin with nearby magnetic nuclei, the so called hyperfine interaction. This gives rise to a pattern of signals characteristic for the radical (ion). The spacing of the signals and their relative intensities identify the magnitude of the interaction between the electron and a group of equivalent nuclei as well as the nuclei so coupled; these results reveal the spin-density distribution in the paramagnetic intermediate under study [20, 21].

The hyperfine coupling constants of magnetic nuclei in organic radicals range from essentially 0 to 200 G (0 to 20 mT). The resulting spectra appear in absorption with intensities determined by the Boltzmann population of the states at thermal equilibrium. This feature limits the sensitivity of the ESR method and has restricted its application to relatively stable species with lifetimes greater than a few milliseconds. Another limitation has its origin in the tendency of free radicals to dimerize, thereby annihilating the unpaired spins. Nevertheless, this method has been the method of choice for the study of numerous families of radicals and radical ions; it has proved invaluable for probing their spin-density distributions and has provided detailed insights into many structures.

Appropriate modification of the ESR spectrometer and generation of free radicals by flash photolysis enables time-resolved (TR) ESR spectroscopy [22]. Spectra observed under these conditions are remarkable for their signal directions and intensities. They can be enhanced as much as one-hundredfold and appear as absorption, emission, or a combination of both. Effects of this type are a result of chemically induced dynamic electron polarization (CIDEP); these spectra indicate the intermediacy of radicals whose sublevel populations deviate substantially from equilibrium populations. Significantly, the splitting pattern characteristic of the spin-density distribution of the intermediate remains unaffected; thus, the CIDEP enhancement not only facilitates the detection of short-lived radicals at low concentrations, but also aids their identification. Time-resolved ESR techniques cannot be expected to be of much use for electron-transfer reactions from alkanes, because their oxidation potentials are prohibitively high. Even branched alkanes have oxidation potentials well above the excited-state reduction potential of typical photo-

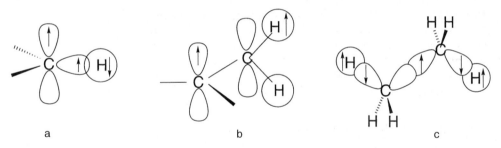

**Figure 1.** (a, left) Preferred configuration of electron spins in the $\sigma$ orbital connecting a $^1$H nucleus to an sp$^2$-hybridized C atom bearing unpaired $\pi$ spin density ($\pi,\sigma$-polarization); (b, center) "molecular $\pi$ orbital" consisting of two carbon p$_z$ orbitals and an H$_2$ "group orbital" due to hyperconjugation between an sp$^2$-hybridized C atom bearing unpaired spin and a CH$_2$-R group ($\pi,\sigma$-delocalization); (c, right) delocalization of an unpaired electron throughout the C–C $\sigma$-frame and extending directly into two chain-end in-plane C–H bonds ($\sigma$-delocalization).

sensitizers (see Section 2.3.3). They have, however, proved useful for some strained ring compounds (below).

The $^1$H hyperfine coupling constants are related to carbon spin densities by different mechanisms of interaction. For $\pi$ radicals, there are two principal mechanisms involving either induction ($\pi,\sigma$-polarization) or delocalization ($\pi,\sigma$-delocalization, hyperconjugation). Protons attached directly to carbon atoms bearing positive spin-density have negative hyperfine coupling constants because of the preferred exchange interaction between the unpaired $\pi$ spin-density and the carbon $\sigma$ electron (Figure 1a). Positive hyperfine coupling constants, on the other hand, are usually observed for protons which are one C–C bond removed from a carbon bearing positive spin-density. The positive sign is a result of a hyperconjugative interaction which delocalizes the $\pi$ spin-density on carbon into an H$_n$ 'group-orbital' (cf. Figure 1b). Finally, the radical cations formed upon electron transfer from $n$-alkanes are typically $\sigma$ radicals, in which the unpaired electron is delocalized throughout the carbon framework and extends directly into the two chain-end in-plane C–H bonds, resulting in large positive hyperfine coupling constants for these $^1$H nuclei (Figure 1c).

In the next section we discuss the radical cations generated upon electron transfer from simple $n$-alkanes to holes generated in various matrices upon radiolysis. The ESR spectra observed in these matrices reveal many interesting structural details. We will begin with the special case of the long-elusive methane radical cation, followed by the radical cations of $n$-alkanes and those of branched alkanes.

### 2.2.3 Electron-transfer Reactions of Methane

Electron transfer from methane is significant because it seems likely that CH$_4$$^{\bullet+}$ and secondary intermediates derived from it (CH$_3$$^{\bullet}$, CH$_3$$^+$) played a significant role in

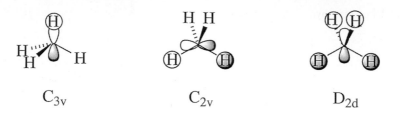

$C_{3v}$                    $C_{2v}$                    $D_{2d}$

**Figure 2.** Possible Jahn-Teller distorted geometries for methane radical cation.

the chemical evolution preceding the origins of life. Methane is a likely constituent of early planetary atmospheres and its radical cation has potential significance as an interstellar species [23]. In the laboratory, however, the ESR spectrum of this species has long been elusive, probably because methane is the alkane most difficult to oxidize and its radical cation is least stable. For example, radiolysis experiments in $SF_6$ only gave complexes of methyl radical with a fluorinated compound [24]. In 1984, Knight and co-workers succeeded in the ESR detection of $CH_4^{\cdot+}$ in a neon matrix at 4 K [25]. Neon discharge ionization of methane ($IP = 12.6$ eV) gave rise to a quintet ($a_H = 5.48$ mT) whereas methane-$D_2$ gave rise to a 1:2:1 triplet ($a_H = 12.17$ mT) of 1:2:3:2:1 quintets ($a_D = 0.222$ mT). The $a_D$ splitting corresponds to an $a_H$ splitting of 1.4 mT.

For a species with tetrahedral symmetry ($T_d$), Jahn–Teller distortion upon ionization can lead to radical cations of $C_{3v}$, $C_{2v}$, or $D_{2d}$ symmetry (Figure 2). The ESR spectrum observed for $CH_4^{\cdot+}$ suggests the presence of four equivalent protons, i.e., a species with $D_{2d}$ symmetry. On the other hand, the spectrum of $CH_2D_2^{\cdot+}$ clearly shows the presence of two non-equivalent pairs, i.e., a species with $C_{2v}$ symmetry. $C_{2v}$ symmetry was, therefore, assigned to both $CH_4^{\cdot+}$ and $CH_2D_2^{\cdot+}$; the perprotio species, $CH_4^{\cdot+}$, is assumed to undergo dynamic Jahn–Teller distortion (fluxional behavior), so that the spectrum reflects an averaged geometry. The 5.48 mT splitting of $CH_4^{\cdot+}$ is essentially equal to the average splitting of the $C_{2v}$ species, $[a_H + (a_D \times 6.5)]/2 = (12.17 - 1.46)/2 = 5.36$ [25].

Ab initio calculations on $CH_4^{\cdot+}$ at the UHF/6-31G* level indicated the presence of two elongated (116.4 pm) and two shortened (107.5 pm) C–H bonds with hyperfine coupling constants, $a_H = 13.7$ mT and $a_H = -1.7$ mT, respectively. For $CH_2D_2^{\cdot+}$, the deuterons preferentially occupy the shorter bonds [25]. The $C_{2v}$ symmetry assigned to $CH_4^{\cdot+}$ is the same as that of $BH_4^{\cdot}$ [26, 27], with which it is isoelectronic. The radical cation of $C_{2v}$ symmetry (two different types of C–H bond) was probed by Eriksson and collaborators by a wide range of different calculation methods; the results are summarized in Table 1 [28]. The nature of the (large amplitude) tunneling among Jahn–Teller distorted structures was elucidated in a group theoretical study [29] and by results obtained with various isotopomers [30].

It is noteworthy, that ab initio calculations even for small molecules give substantially different results depending on the basis set and the method used to take electron correlation into account. These difficulties are illustrated here with the results from three different calculations on two different methane radical cations. At

**Table 1.** Structures and isotropic hyperfine coupling constants (Gauss) of the methane radical cation.

| Species | | Method[a] | | | | | |
|---|---|---|---|---|---|---|---|
| | | LDA | BP | PWP | LDA//PF//D2P | CISP | Experiment |
| | C | 4.1 | 14.7 | 23.3 | – | – | – |
| $C_{2v}$ | $H_a$ | 118.3 | 128.7 | 133.6 | 121.1 | 137.0 | 121.7 |
| | $H_b$ | −16.4 | −20.1 | −18.6 | −16.3 | −17.0 | −14.6 |

[a] According to [28].

different levels of theory the relative energies of the different SOMOs change along with the structure parameters. Accordingly, the global minimum can depend on the method of calculation and more than one species might become local minima. For the methane radical cation calculations with the 6-31G* basis set at different levels of electron correlation gave considerably different results. At the UHF/6-31G* level, a $C_s$ species reported earlier [25] was the global minimum, whereas the $D_{2d}$ structure had one negative frequency. Using UB3LYP/6-31G*, the $D_{2d}$ structure was the global minimum, whereas the $C_s$ structure showed one negative frequency. Finally, the UMP2/6-31G* level gave both $D_{2d}$ and $C_s$ species as minima with similar energies, but different hyperfine coupling patterns (Table 1; Figure 3) [31]. The $^1H$ nuclei connected by long bonds have strong positive hyperfine coupling constants. Aside from the different $^1H$ hyperfine coupling patterns (which might be masked in the experiment by dynamic Jahn–Teller effect) the two structures have significantly different isotropic $^{13}C$ hyperfine couplings. The $D_{2d}$ structure has a very small $^{13}C$ hyperfine coupling (−0.36 to 0.42 mT), whereas the $C_s$ species has a more significant interaction (1.8–3.3 mT) [31]. Accordingly, the actual structure should be identified unambiguously by the isotropic $^{13}C$ hyperfine splitting. Unfortunately, the spectra of $^{13}CH_4^{·+}$ are anisotropic, and the isotropic coupling is not known unambiguously [30].

### 2.2.4 Electron-transfer Reactions of *n*-Alkanes

The ESR spectrum of ethane radical cation at 4 K in $SF_6$ contained a 1:2:1 triplet because of two strongly coupled protons (15.25 mT), corresponding to a spin-

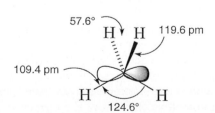

**Figure 3.** Possible Jahn-Teller distorted geometries for methane radical cation.

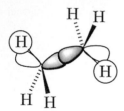

**Figure 4.** Schematic representation of the SOMO for ethane radical cation.

density of 0.3 in the $H_{1s}$ orbital. The hyperfine coupling of the remaining four protons is smaller than the linewidth ($\leq 1$ mT) [19]. The coupling documents an unambiguous example of $\sigma$-delocalization. Another important result is the direct detection of static Jahn–Teller distortion. These effects are usually based on indirect detection, i.e., via the temperature-dependence or isotope effects on hyperfine coupling [32–34]. In a few special cases, the distortion is believed to be introduced by the matrix [35, 36]. However, for most alkane radical cations (with the notable exception of $C_3H_8^{\cdot+}$, vide infra), the coordinate of distortion is independent of the matrix; the mode of distortion appears to be intrinsic to the radical cations. The matrix only serves to stabilize the distorted form [37].

These results suggest that the $C_2H_6$ molecule ($D_{3d}$ symmetry) is distorted to $D_{3d}$ symmetry in the electron-transfer process generating $C_2H_6^{\cdot+}$. The mode of deformation was associated with a Jahn–Teller active rocking vibration ($e_g$) of the methyl groups (Figure 4) [38]. The structure of the resulting radical cation resembles diborane, $B_2H_6$, rather than the diborane anion, $B_2H_6^-$, with which it is isoelectronic [39].

Upon raising the temperature to 77 K, the spectrum changes to a septet with a splitting of 5.04 mT, one third of the value at 4 K. This change was ascribed to a dynamic Jahn–Teller effect, viz., rapid equilibration of three equivalent distorted forms, each with two strongly coupled $^1H$ nuclei. An apparent activation energy, $E_a = 250$ cal mol$^{-1}$, was derived for the exchange; this value is one order of magnitude smaller than that for methyl rotation in the gas phase ($E_a = 2.8$ kcal mol$^{-1}$) [19].

Detailed calculations using ab initio or density functional theory methods revealed two different structures for $C_2H_6^{\cdot+}$, a 'diborane-type' structure and a 'long-bond' structure. In the diborane structure, the unpaired electron is delocalized between the C–C bond and two 'bridging' C–H bonds; whereas in the long-bond structure it resides principally in the C–C bond. The large positive hyperfine coupling constant observed at low temperatures [19] is reproduced well by the results of B3LYP calculations on the diborane structure [40].

The next higher homolog, propane, $C_3H_8$, poses an additional interesting problem. Electron transfer from $C_3H_8$ gave rise to three different radical cations, depending on the matrix [41]. Different structures obtained in different matrices are ascribed to the fact that the energies of several high-lying orbitals are very close to each other. Therefore, even small perturbations as a result of the matrix can alter the relative energies of these levels.

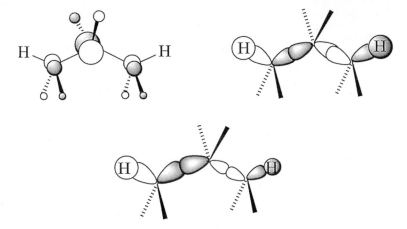

**Figure 5.** Schematic representation of three SOMOs corresponding to three different propane radical cations observed by ESR.

The species observed in F113 has two strong hyperfine interactions (10.5 mT) and four weaker ones (5.25 mT). This spectrum was assigned to a delocalized $C_3H_8^{\cdot+}$ species with a pseudo $\pi$ orbital, resulting from the antibonding combination of three pseudo $\pi$-CH$_2$ orbitals. The $^1$H nuclei of the center carbon are most strongly coupled whereas the two protons in the C–C–C plane have negligible hfcs. The magnitude of the strong hfcs (C-2) relative to the weaker ones (C-1,3) reflects the orbital coefficient at the corresponding carbons (Figure 5) [41].

Electron transfer from $C_3H_8$ in SF$_6$ generated a different ESR spectrum—a 1:2:1 triplet with two strongly coupled protons (9.8 mT). This spectrum is compatible with a SOMO consisting of two C–H orbitals, lying in the plane of two C–C $\sigma$-orbitals. The two in-plane $^1$H nuclei at C-1 and C-3 are strongly coupled [42]; the hyperfine interactions of the remaining protons were derived from line-shape analysis at higher temperatures. The out-of-plane $^1$H nuclei at C-1 and C-3 have a splitting constant, $a = -0.3$ mT, whereas those at C-2 are coupled less strongly. Compared with the parent, $C_3H_8$, the probable structure of the radical cation is one in which the rotational axes of the CH$_3$ groups are bent toward the central carbon and the two in-plane C–H bonds are lengthened ($^4b_1$ symmetry).

A third structure type of $C_3H_8^{\cdot+}$ was obtained in $C_3F_8$; the ESR spectrum of this species has only one strongly coupled ($a = 8.4$ mT) and one moderately coupled proton ($a = 1.8$ mT); the other splittings are considerably weaker ($a = 0.9, 0.6$ mT). Deuteration at C-2 identified the 0.6 mT splitting as due to the protons at C-2. The spectrum was assigned to a radical cation with one lengthened C–C bond. The two major splittings were assigned to the in-plane terminal protons at the lengthened bond and at the shorter one, respectively [43, 44].

Eriksson and colleagues used the $^4b_1$ species as a test case for a wide range of different calculation methods (Figure 5; Table 2) [28]. More recent detailed calculations revealed two different minima on the $C_3H_8^{\cdot+}$ potential surface, corresponding to the structure types with either two or one lengthened C–C bonds (Table 2).

**Table 2.** Structure and isotropic hyperfine coupling constants (Gauss) of propane radical cations.

| Species | | Method[a] | | | | | |
|---|---|---|---|---|---|---|---|
| | | LDA | BP | PWP | LDA//PF//D2P | CISP | Experiment |
| | C1,3 | 16.1 | 16.7 | 20.3 | – | – | – |
| | C2 | −3.3 | −2.9 | −0.11 | – | – | – |
| $C_S$ | $H1,3_{long}$ | 123.3 | 129.0 | 130.4 | 114.0 | 88.6 | 98.0 |
| | $H1,3_{short}$ | −2.7 | −4.2 | −4.1 | −3.9 | −5.8 | – |
| | H2 | −5.1 | −6.9 | −6.2 | −5.2 | −7.1 | – |

[a] According to [28].

Electron-transfer reactions of the longer *n*-alkanes generated the higher homologs of the σ-delocalized radical cation, $C_3H_8^{\bullet+}$ observed in F113. The resulting radical cations also have 1:2:1 ESR triplets with two strongly coupled protons [41, 45–47]. The splitting decreases sharply from 5.5 mT for $C_4H_{10}^{\bullet+}$ to 1.0 mT for $C_{10}H_{22}^{\bullet+}$. By analogy to the $C_3H_8^{\bullet+}$ structure in F113, the SOMOs are σ-orbitals spread over fully extended planar C–C systems; two C–H group-orbitals at the terminal carbons contain the strongly coupled ${}^1H$ nuclei (Figure 6). The steadily diminishing hfcs with increasing chain length can be attributed to the distribution of the unpaired electron spin over an ever-increasing number of carbons (Figure 7). The hyperfine coupling constants of the inner protons and the out-of-plane methyl protons are typically small and are resolved only in special cases [42, 48].

Actually, the ESR spectra are more complex than the above simple description based on the exclusive formation of fully extended radical cations. In most frozen matrices, depending on the chain length, radical cations in *gauche* conformations were also detected [41, 49–53]. In the *gauche* forms, the spin delocalization is ter-

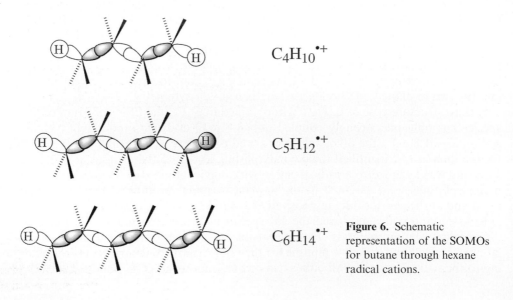

$C_4H_{10}^{\bullet+}$

$C_5H_{12}^{\bullet+}$

$C_6H_{14}^{\bullet+}$

**Figure 6.** Schematic representation of the SOMOs for butane through hexane radical cations.

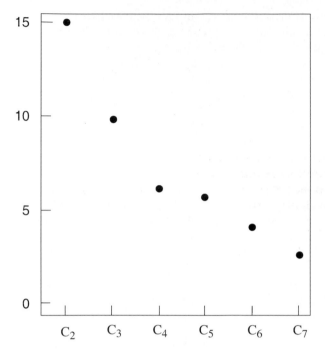

**Figure 7.** Hyperfine coupling constants (mT) of the in-plane chain-end $^1$H nuclei for extended $\sigma$ radical cations of ethane through heptane.

minated at the *gauche* carbon and the unpaired electron is confined mostly to the longer fragment. Therefore, the hfcs of the in-plane $^1$H nuclei become larger than for the fully extended conformers. The strongly coupled protons are one chain-end $^1$H nucleus and an additional in-plane $^1$H nucleus at the *gauche* carbon; the latter has a slightly larger hfc because the *gauche* carbon has slightly greater spin-density [41, 50]. The co-existence of two conformers is illustrated for *n*-pentane, which can exist in the *s-trans,trans,trans-* and the *s-trans,trans,gauche*-conformation (Figure 8). The observed ESR spectrum was dissected into two independent contributions, a

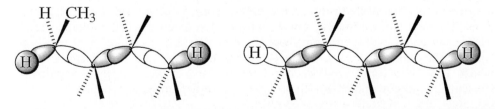

**Figure 8.** Schematic representation of the SOMOs for two conformers of pentane radical cation.

narrower one for the linear conformer and a spectrum with wider extension for the kinked one. The different spin and charge densities at the chain-end and the *gauche* carbon causes the $^1$H nuclei attached 'in-plane' to have different reactivities (see Section 2.2.5).

An interesting extension of these results was observed upon radiolysis inside pentasil zeolite (ZSM-5) [54]. The oxidation potentials of alkanes are too high to enable oxidation by the zeolite; indeed, incorporation of alkanes into Na- or HZSM-5 does not give rise to any absorption above 260 nm (cf. Section 2.2.6). Therefore, the neutral hydrocarbon molecules undergo electron transfer to holes generated in the zeolite upon radiolysis. At low alkane concentration (loading 0.5 % by weight), only the characteristic triplets of the extended radical cations, ($n = 3,5$), were observed (*n*-hexane, $a = 3.9$ mT; *n*-octane, $a = 2.9$ mT). Comparison of these spectra with those obtained in $SF_6$ and chlorofluorocarbons [19] revealed that only the fully extended conformers of the alkane radical cations are present in the zeolite enforced by the geometry of the zeolite channels [19].

Radiolysis at higher substrate loadings (3.0 % by weight), yielded different spectra, which were assigned to the corresponding primary alkyl radicals, ($n = 3,5$). Because the appearance of the new species was concentration-dependent, their formation was ascribed to an ion–molecule reaction, generating a protonated *n*-alkane, at the same time [54].

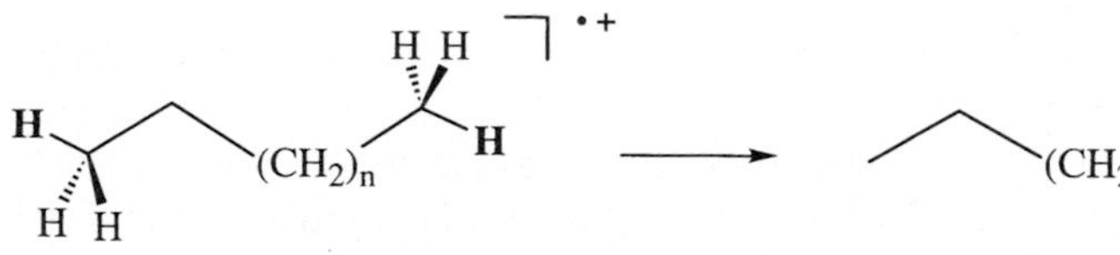

### 2.2.5 Deprotonation of *n*-Alkane Radical Cations

The reaction of the *n*-hexane and *n*-octane radical cations with their diamagnetic parents to generate alkyl radicals and protonated *n*-alkanes is one of the few reactions observed for *n*-alkane radical cations. Under the limited range of conditions in which they can be generated, only a limited range of reagents is available. The bimolecular H-atom transfer seems to be a general reaction for *n*-alkane radical cations. This reaction is not limited to zeolite media but occurs in halogen-containing matrices also. The mechanism of these conversions poses several interesting questions. The key step can involve transfer of a proton to the neutral *n*-alkane (Eq. 7) or abstraction of hydrogen atom from the *n*-alkane (Eq. 8). Additional questions concern the donor site, from which H· or H$^+$ originate, and the acceptor site to which H· or H$^+$ is transferred. The mechanism of this conversion was elucidated by Ceulemans and coworkers in elegant studies of *n*-heptane and *n*-octane in $CCl_3F$ [55, 56].

$$R_1H^{\bullet+} + R_2H \xrightarrow[\text{transfer}]{H^+} R_1^{\bullet} + R_2H_2^+ \tag{7}$$

$$R_1H^{\bullet+} + R_2H \xrightarrow[\text{abstraction}]{H^{\bullet}} R_1H_2^+ + R_2^{\bullet} \tag{8}$$

$$RH_2^+ + R_3^+Cl^- \longrightarrow R-Cl^+ + R_3^+ + H_2 \tag{9}$$

The prevailing conformers of alkane radical cations depend on specific inter-actions between the host matrix and the guest alkane. For example, $\gamma$-irradiation of *n*-heptane in $CCl_3F$ generated an extended all-*trans* conformer of $n$-$C_7H_{16}^{\bullet+}$, whereas *n*-octane gave rise to the *gauche* at-C2 conformer of $n$-$C_8H_{18}^{\bullet+}$. In experi-ments with $n$-$C_7H_{16}$ containing small amounts of $n$-$C_8D_{18}$, the spectrum of 1-heptyl radical, $n$-$C_7H_{15}^{\bullet}$, was superimposed on that of $n$-$C_7H_{16}^{\bullet+}$; this finding was ascribed to $H^+$ transfer from $n$-$C_7H_{16}^{\bullet+}$ to $n$-$C_8D_{18}$. The preferential formation of primary heptyl radical during the early phase of the reaction identified a chain-end carbon as the donor site. This conclusion is in agreement with the reactivity expected for a $\sigma$-delocalized radical cation; a proton of a lengthened and weakened C–H bond should be donated preferentially. This is the same proton which also has a strong hyperfine coupling constant. The formation of secondary heptyl radicals appearing at higher *n*-heptane concentrations was explained as a result of intermolecular radical site transfer.

Analogous experiments with *n*-octane revealed that the secondary octyl radical, 2-$C_8H_{17}^{\bullet}$, was present from the very onset of proton transfer, suggesting the 2-position as the donor site. $\gamma$-Irradiation of *n*-octane generates the *gauche* at-C2 conformer of $n$-$C_8H_{18}^{\bullet+}$, a species with large unpaired electron density on one in-plane chain-end C–H bond and on the in-plane C–H bond at C2 [55]. The forma-tion of secondary octyl radical clearly supports significant donor site selectivity for $H^+$ transfer from *n*-alkane radical cations to *n*-alkanes. The *gauche* at-C2 con-former of $C_8H_{18}^{\bullet+}$ reacted at the site of greater electron spin (and positive hole) density (C2) and generated the more stable free radical. Thus, this mode of reaction is favored by both kinetic and thermodynamic principles.

Deprotonation from the carbon of highest spin-density has been suggested as a general rule. This is strongly supported by the deprotonation of the propane radical cation in different matrices. The $C_3H_8^{\bullet+}$ species observed in F113, which has high spin-density on C-2 (Figure 5) and strongly coupled secondary protons (10.5 mT) gave rise to 2-propyl radical. In contrast, the $C_3H_8^{\bullet+}$ species formed in $SF_6$, which has high spin-density on C-1 and C-3 and in which the chain-end in-plane $^1H$ nuclei are strongly coupled, generated the 1-propyl radical [41]. Additional strong evi-dence for this principle was derived from the deprotonation of isobutane radical cation (Section 2.2.6). On the other hand, it has been noted that the spin-density–reactivity correlation might not hold universally [41, 46, 49, 51].

The $H^+$ acceptor site was identified by experiments on $n$-$C_7H_{16}$ containing dif-ferent amounts of $n$-$C_8D_{18}$ and a small percentage of 1-chlorohexane (e.g. Eq. 9). These experiments revealed the formation of small amounts of chlorooctanes, pre-sumably via neutralization of protonated octane molecules by chloride ions. Among the resulting chlorooctanes, the 2-isomer predominates, identifying C2 as the primary $H^+$ acceptor site [55].

### 2.2.6 Electron-transfer Reactions of Branched Alkanes

Branched alkanes are somewhat more easily oxidized than *n*-alkanes, although typically still quite unreactive. Their ionization potentials are barely lower than those of *n*-alkanes, ranging from 10.68 eV for 2-methylpropane to 9.89 eV for 2,2,4-trimethylpentane. As typical saturated hydrocarbons they are unreactive in the potential range that is experimentally accessible in non-aqueous solvents. Some branched alkanes with especially weak C–H bonds can, however, be oxidized under these conditions. For example, oxidation potentials (relative to the Ag/Ag$^+$ electrode) were assigned to 2,2-dimethylbutane (ca 3.3 V) or 2-methylpentane (ca 3.0 V) [57]. The radical cations of branched alkanes have interesting structural features. Electron transfer from the simplest branched alkane, 2-methylpropane, to electron holes generated in a halocarbon or $SF_6$ matrix gave rise to two different radical cations, of $C_s$ and $C_{3v}$ symmetry. Radiolysis in $SF_6$ at 4 K generated a radical cation, whose ESR spectrum contained a triplet ($a_H = 5.8$ mT). In this radical cation the unpaired spin is localized largely in one C–C bond ($C_s$ symmetry); two *trans-β*-protons, one on each adjacent methyl group, are strongly coupled [19, 41]. This species is deprotonated to form *iso*-butyl free radical by loss of $H^+$ from a methyl group. Radiolysis in Freon at 4 K generated the same $C_s$ radical cation. However, warming to 77 K caused irreversible conversion to a different species ($C_{3v}$ symmetry), in which the spin is localized largely in the tertiary C–H bond, causing the tertiary proton to be strongly coupled ($a_H = 25.0$ mT) (Figure 9) [41]. This species is deprotonated to form *t*-butyl free radical.

It was the oxidation of 2,2,3,3-tetramethylbutane (hexamethylethane), studied independently in three different laboratories, which might have signaled the onset of systematic investigation of electron-transfer reactions from alkanes and of the structures of the resulting radical cations. Oxidation of hexamethylethane generated a radical cation with a seven-line ESR spectrum with $a_H = 2.9$ mT. The unpaired spin of this radical cation is localized in the central C–C bond; six *trans-β*-protons, one on each adjacent methyl group, are strongly coupled by hyperconjugation [17]. The hyperfine couplings of the *gauche* β-hydrogens were also assigned ($a_H =$

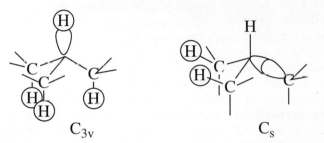

**Figure 9.** Schematic representation of σ- and π-SOMOs considered for isobutane radical cation.

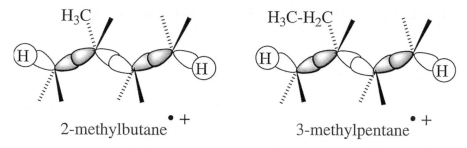

Figure 10. Schematic representation of SOMOs for *s-trans,trans*-2-methylbutane and *s-trans,trans*-3-methylpentane radical cations.

0.42 mT) [18]. Interestingly, the six methyl groups remain static even at 77 K [18] in considerable contrast to the facile rotation of the methyl groups of the ethane radical cation. The reason for the pronounced hindrance of methyl rotation must lie in the significant steric congestion of the tetramethylbutane framework.

Electron transfer from 2-methylbutane [58] and 3-methylpentane [58, 59] generated remarkably similar ESR spectra—both donors gave rise to four-line spectra with $a_H \approx 4.5$ mT. These patterns were rationalized in terms of radical cations, with the unpaired spin localized in the C3–C4 bond; the large hfcs arise from one proton in each of three methyl groups (Figure 10) [58, 59]. Radiolysis of specific deuterated derivatives and spectral simulations led Shiotani and coworkers to a refined assignment [58]. Details go beyond the scope of this review.

Many other branched alkanes undergo electron-transfer reactions in cryogenic matrices. Although the ESR spectra of the resulting radical cations reveal interesting structural features [60–62], detailed discussion would exceed the scope of this review.

### 2.2.7 Electron Transfer from Alkanes to Zeolites

Zeolites were introduced as catalysts for large-scale heterogeneous reactions in petrochemistry, because they have significant advantages for processes such as cracking, aromatic isomerization, or disproportionation [63–68]. One of the most intriguing properties of acid zeolites is their spontaneous generation of organic radical cations on adsorption of organic electron donors [69]. Molecules containing strained rings are especially readily converted to their radical cations on incorporation into zeolites [70], although radical cations have also been obtained from selected alkanes. Because radical cations sequestered in zeolite pores are protected from reagents that typically would cause their decay in solution, their lifetimes are increased so that they can be studied by conventional spectroscopic techniques. Also, the limiting geometry of the zeolite pores can manifest itself in two ways: (i) it might selectively incorporate substrates with suitable geometries (shape selectivity); and (ii) it might restrict the geometry (conformation) of a sequestered intermediate. Thus, *trans*-1,2-diphenylcyclopropane was incorporated readily into ZSM-5 whereas the *cis* isomer was not [70]. On the other hand, radiolysis of *n*-hexane and *n*-octane in pentasil zeolite generated solely the EPR spectra of the 'extended' radical cations (vide supra) [54].

Radical cations are also generated efficiently upon γ-radiolysis of appropriate 'guests' sequestered in zeolites. In this experiment, a 'hole' and a free electron are generated in the zeolite, and the guest transfers an electron to the 'hole'. Given the redox- and acid–base-active nature of zeolites it is not obvious whether the original substrate or a rearranged product is oxidized by interaction with the newly created 'hole'. In fact, many alkenes undergo significant reactions before the onset of radiolysis.

Chen and Fripiat succeeded in generating radical ions from saturated hydrocarbons, e.g., 3-methylpentane (**1**), upon sequestering them into H-mordenite and heating for several hours [71]. The resulting species were identified by their EPR spectra [72]. Adsorption of **1** and heating to 100 °C for 4 h produced a 'nine-line spectrum' ($a = 17.3$ G, $g = 2.003$), which was identified by Roduner and Crockett as that of the 2,3-dimethylbutene radical cation (**2·+**) [72]. Although an electron-transfer step must be involved at some stage of this conversion, it is not clear at what stage electron transfer to the zeolite occurs.

Significant applications of γ-radiolysis in the study of electron-transfer processes and radical cations in zeolites have emanated from the radiation group at Argonne National Laboratories [73]. For example, Barnabas et al. studied the fate of highly branched alkanes, 2,3-dimethylbutane, 2,2,3-trimethylbutane, and 2,2,3,3-tetramethylbutane, upon γ-radiolysis in pentasil zeolite (ZSM-5) [74, 75].

γ-Radiolysis of 2,3-dimethylbutane (**3**) in pentasil zeolite at 4 K forms the corresponding radical cation (**3·+**; a broad quintet, $a = 3.7$ mT); by analogy with typical radical cations of branched hydrocarbons, the quintet is caused by coupling of a single proton per methyl group. In addition, 2,3-dimethyl-2-butyl free radical (**4·**), recognized as a septet ($a = 2.34$ mT) of doublets ($a = 0.55$ mT), is formed to a lesser extent [75].

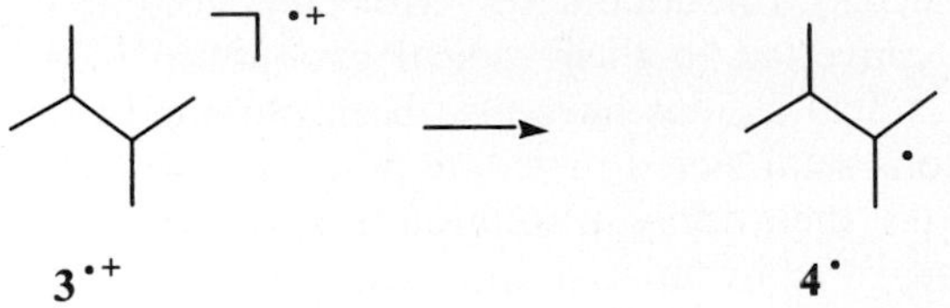

Radiolysis of ZSM-5 containing 2,2,3-trimethylbutane (**5**) at 77 K yielded a six-line spectrum characteristic of **5·+** ($a = 3.1$ mT, 5H) [76]. Upon raising the temperature, in the range 120–200 K, the well known 2,3-dimethyl-2-butene radical cation ($2^{\cdot+}$) was observed ($a = 1.7$ mT, nine lines observed, 12H). Its formation was ascribed to "thermal elimination of methane", a reaction similar to the above-mentioned elimination of $H_2$ [72].

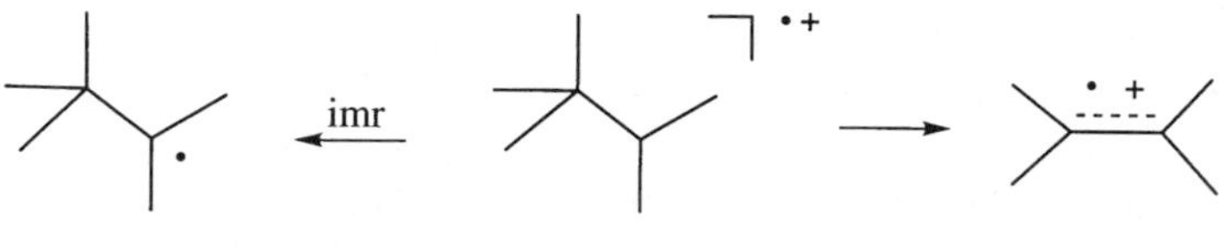

Above 200 K, an additional seven-line pattern ($a = 2.25$ mT, 6H) appeared, which was ascribed to 2,3,3-trimethyl-2-butyl radical, $6^{\cdot}$, apparently as a result of deprotonation of $5^{\cdot+}$. Because the abundance of this species increased with substrate loading, its formation was ascribed to an ion–molecule reaction (Section 2.2.5).

$\gamma$-Radiolysis of tetramethylbutane (7) in pentasil zeolite gave rise to a 'seven-line' spectrum ($a = 3.1$ mT, 6H; outer lines not distinct), identified as that of $7^{\cdot+}$ [72, 74]. The splitting is, once again, caused by a single (in-plane) proton per methyl group. An additional weak quartet ($a = 2.2$ mT) indicated formation of methyl radical, most probably formed by fragmentation of $7^{\cdot+}$. Upon annealing at 125 K, a new spectrum appeared ($a = 1.49$ mT, 2H; $a = 1.38$ mT, 2H; $a = 0.33$ mT, 3H); this was assigned to 2-methylpropenyl free radical, $7^{\cdot}$. This species was explained by rapid deprotonation of 2-methylallyl radical cation ($8^{\cdot+}$, not observed). This reaction, assigned by the authors to "scission of the central C–C bond" of $9^{\cdot+}$ [74], amounts to 'elimination' of 2-methylpropane ($C_4H_{10}$), similar to the loss of $CH_4$ [76] and $H_2$ [72] mentioned above.

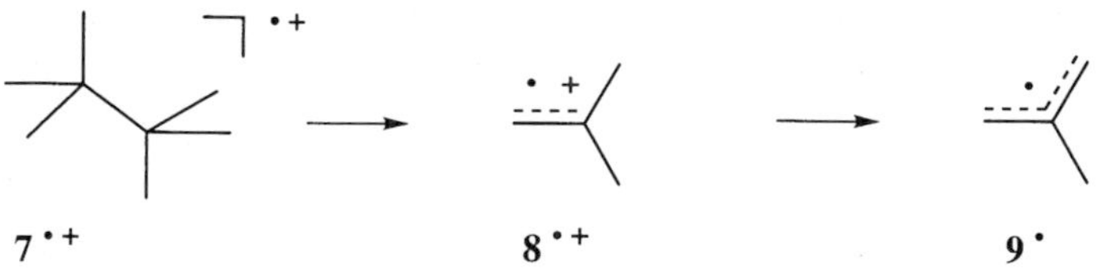

Among the various conversions discussed in this section, deprotonation is clearly a general reaction of alkane radical cations. The interesting elimination or fragmentation reactions, on the other hand, seem to be zeolite-specific reactions without precedent in halogen containing matrices.

## 2.3 Electron-transfer Reactions of Cycloalkanes

Similar to alkanes most cycloalkanes also are quite unreactive and have very high $pK_a$ values and low electron affinities, making them poor electron acceptors. Although the molecular anions of cycloalkanes are very unstable, temporary anion states with lifetimes shorter than picoseconds have been observed. The four smallest ring systems, cyclopropane through cyclohexane have temporary anion states at 5.29, 5.80, 6.14, and 4.11 eV [1]. The ionization potentials of cycloalkanes are somewhat lower than those of the low *n*-alkanes (the *IP* of cyclopropane through cyclooctane lie between 9.75 and 9.90 eV); they are still poor electron donors and are difficult to oxidize in solution or in solid matrices. The molecular ions are accessible in the gas phase (electron or $He_\alpha$ impact ionization), as documented by their mass and photoelectron spectra. Also, simple cycloalkanes and their derivatives readily undergo electron transfer to 'holes' generated upon radiolysis of halogen-containing matrices [9–11]. Some substituted cyclopropanes and cyclobutanes undergo photo-induced electron transfer to photo-excited electron acceptors in solution, generating radical cation–radical anion pairs (Section 2.3.3).

The basic cycloalkanes with three to eight carbons ($c$-$C_3H_6$–$c$-$C_8H_{16}$) undergo electron transfer upon radiolysis in halogen-containing matrices, generating the radical cations ($c$-$C_3H_6^{\cdot+}$–$c$-$C_8H_{16}^{\cdot+}$). The ESR spectra observed in these systems reveal interesting structural details. Some properties of these species are discussed below. Common features include delocalization of the unpaired electron throughout the molecular frame and a noticeable positive $g$-shift relative to alkane and aromatic radical cations. In addition, the overall width of the ESR spectra, i.e., the sum of all hyperfine coupling constants, increases with increasing ring size (Figure 11). In marked contrast, a steady decrease in spectral width with increasing chain length is observed for aliphatic radical cations (cf., Figure 7). For the typical $\sigma$-delocalized linear aliphatic radical cations, only C–H bonds in the molecular plane contribute to delocalizing the unpaired electron spin. Only the in-plane $^1H$ nuclei are strongly coupled; typically, two terminal protons satisfy this requirement. Because the spin-density in the terminal C–H bond decreases with increasing numbers of carbons delocalizing the electron spin, so does the magnitude of the $^1H$ hyperfine coupling.

Among the simple cycloalkanes, we first discuss electron transfer from the three- to eight-membered cycloalkane prototypes to electron holes generated by radiolysis in different matrices, giving rise to the simple cycloalkane radical cations. Because of the significant interest they have attracted, the electron-transfer reactions of cyclopropane and, to a lesser extent, cyclobutane derivatives will be treated separately. Finally, electron transfer from some bicyclic hydrocarbons and the resulting radical cations will be discussed in a separate section (Section 2.4).

### 2.3.1 Electron Transfer from $C_3$–$C_8$ Prototype Cycloalkanes

Several of the simple cycloalkane radical cations, generated by electron transfer, are Jahn–Teller active. Apart from the interesting small-ring radical cations, $c$-$C_3H_6^{\cdot+}$ and $c$-$C_4H_8^{\cdot+}$, one common-ring radical cation, $c$-$C_6H_{12}^{\cdot+}$, and one medium-ring species, $c$-$C_8H_{16}^{\cdot+}$, also show this interesting phenomenon. On the other hand, $c$-$C_5H_{10}^{\cdot+}$ and $c$-$C_7H_{14}^{\cdot+}$ seem to be Jahn–Teller inactive.

Cyclopropane, the simplest strained ring system, continues to be subject to intense and detailed scrutiny by a variety of techniques. Its photoelectron spectrum,

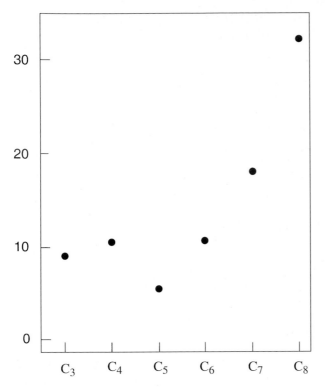

**Figure 11.** Sum of all hyperfine coupling constants (mT) for radical cations of $C_3$–$C_8$ cycloalkanes; the strained rings ($C_3$, $C_4$) clearly deviate from the trend of $C_5$–$C_8$.

investigated over thirty years ago, showed Gaussian shaped bands without fine structure. Molecular-orbital calculations suggest that the first two bands, a double peak centered near 11 eV, be assigned to a $^2E'$ state. The large splitting of these bands (0.8 eV), also observed for many derivatives, is ascribed to Jahn–Teller distortion of the ionic ground state [77]. The two states resulting from the distortion, $^2B_2$ and $^2A_1$, ($C_{2v}$ symmetry) correspond to two different molecular structures (Figure 12). In rigid matrices, cyclopropane readily undergoes electron transfer upon $\gamma$-radiolysis at 4.2 K; the ESR spectrum of the resulting radical cation has two strongly coupled $^1$H nuclei ($a_H = 2.04$ mT) and four less strongly coupled nuclei ($a_H = -1.17$ mT). These splittings were assigned to the $\beta$- and $\alpha$-protons, respectively, of a trimethylene ($^2A_1$)-type radical cation structure, an equilibrium structure with one lengthened C–C bond, because of static Jahn–Teller distortion of the cyclopropane ring ($D_{3h}$ symmetry) [78, 79]. At higher temperatures, an averaged spectrum, consisting of only a single line (FWHH = 1.5 mT), was observed. This spectrum was ascribed to an averaged structure resulting from dynamic Jahn–Teller distortion, interconverting three equivalent $^2A_1$ structures with three equivalent $^2B_2$

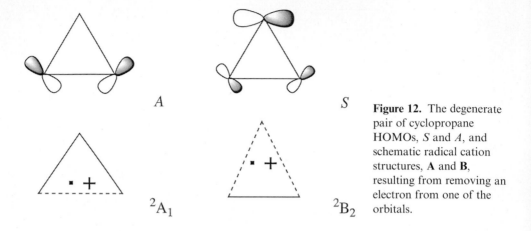

*A*    *S*

$^2A_1$    $^2B_2$

**Figure 12.** The degenerate pair of cyclopropane HOMOs, *S* and *A*, and schematic radical cation structures, **A** and **B**, resulting from removing an electron from one of the orbitals.

structures as transition states [37, 78–80]. The magnitude of $a_\beta$ (2.04 mT) and $a_\alpha$ (−1.17 mT) fortuitously averages to $a_{avg} \approx 0$ [80].

The structure of the cyclobutane radical cation also is of major interest. The parent cyclobutane system is known to have a puckered ring with $D_{2d}$ symmetry [81]. Electron transfer from one of the cyclobutane e orbitals is expected to lead to a Jahn–Teller unstable radical cation, which might distort to structures of $D_{2d}$ and $C_{2v}$ symmetry [81]. Ushida et al. studied the EPR spectra obtained upon X-irradiation of cyclobutane in frozen $CFCl_3$ solution [82]. At 4 K, a triplet of triplets appeared with hyperfine coupling constants, $a_{H1} = 4.9$ mT (2H), $a_{H2} = 1.4$ mT (2 H), and four weakly coupled protons ($a_{H1} = 0.5$ mT), suggesting an intermediate with $C_{2v}$ symmetry. These findings were interpreted in terms of a radical cation that had undergone static Jahn–Teller distortion to a rhombic structure of $C_{2v}$ symmetry (Figure 13, type D) [82]. Upon annealing the sample at 77 K, the spectrum changed irreversibly to a five-line pattern (measured at 4 K). This result was ascribed to an irreversibly reduced barrier to ring puckering and/or flattening. Finally, at temperatures above 77 K, a nearly isotropic nine-line spectrum with a hyperfine coupling constant, $a_H = 1.33$ mT was observed. This change was ascribed to a dynamic Jahn–Teller effect via pseudorotation. Additional structure types can be envisaged that are compatible with the low-temperature ESR pattern and cannot be rigorously eliminated [82].

The cyclobutane radical cation was calculated by several groups, who evaluated Jahn–Teller distorted structures potentially arising upon ionization of cyclobutane [79, 83–87]. Four distorted local minima were considered (Figure 13). A rectangular structure with two weakened C–C bonds (type A) and a rhomboidal structure with four weakened C C bonds (type B) result from first order Jahn–Teller distortion. Two further structures, a trapezoidal one (one weakened C–C bond; type C), and an irregular structure shaped like a kite (two lengthened C–C bonds; type D) can be envisaged as a consequence of second-order Jahn–Teller distortion [86]. A detailed recent study also considered puckered equivalents of the planar structure types, A–D

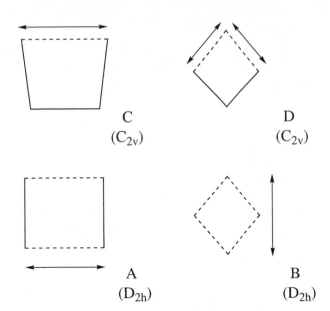

C
($C_{2v}$)

D
($C_{2v}$)

**Figure 13.** Possible structure types of cyclobutane radical cations.

A
($D_{2h}$)

B
($D_{2h}$)

[86]. At the QCISD-(T)/6-31G*//UMP2/6-31G* level of theory a rhombic structure, quite flexible to ring-puckering, emerged as the most stable.

The most recent study, using UHF, MP2, BLYP and B3LYP methodologies, reached somewhat different conclusions. Density functional theory calculations showed one imaginary frequency for the rhombic structure (four equivalent bonds, 157.3 pm), suggesting that it is a (very low-lying) transition structure between two parallelograms (two pairs of equivalent bonds, 149.5 and 169.5 pm, respectively) [87]. A calculation of the hyperfine coupling constants for the minimum yielded values, $a = 2.09$, 0.29 mT; the rather poor agreement with the experimental splittings, $a = 4.9$, 1.4 mT, is unsettling, particularly because hyperfine coupling constants calculated for many strained-ring systems are in significantly better agreement with experiment. Because hyperfine coupling patterns (and the related CIDNP patterns) are the only experimental data available for many strained ring systems, closer reproduction of these data by calculation seems desirable.

In contrast to the radical cations of strained-ring cycloalkanes, the cyclopentane radical cation, $c\text{-}C_5H_{10}{}^{\cdot+}$, formed by electron transfer to radiolysis-induced holes in halocarbon matrices, had a simpler spectrum. A triplet with $a_H = 2.5$ mT (2H) was attributed to a localized species with $C_s$ symmetry. The unpaired electron was assigned to a W-shaped $\sigma$-orbital, involving C5–C1–C2, and the two equatorial protons at C5 and C2 [80, 88, 89]. At temperatures above 77 K, all ring protons become equivalent, most probably as a result of processes such as ring inversion, or pseudo-rotation around the C5-axis [89].

Five-membered rings typically have non-planar conformers, either a puckered $C_2$ 'envelope' or a $C_s$ 'half chair' (Figure 14). The conformers are readily interconverted by flipping one CH$_2$ from above the approximate ring plane to below. Ring inversion converts the molecule into its mirror image, interchanging equatorial and

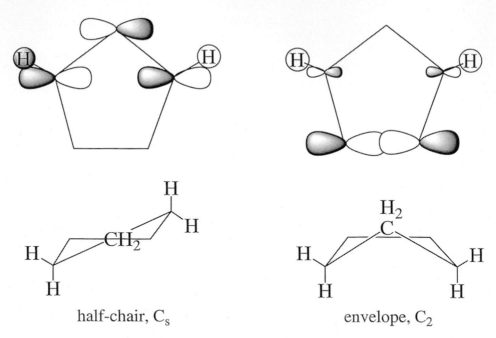

half-chair, $C_s$                          envelope, $C_2$

**Figure 14.** Conformers of cyclopentane and schematic representation of possible singly occupied molecular orbitals (SOMOs) of cyclopentane radical cation.

axial $^1H$ nuclei at a given carbon, whereas pseudo-rotation would move the unpaired electron spin to an adjacent set of carbon atoms. Spectra simulation led to estimates of 1.2 kcal mol$^{-1}$ for the barrier of ring inversion and 3.6 kcal mol$^{-1}$ for the barrier to pseudo-rotation [84]. The observed ESR pattern is consistent with either the $C_s$ or the $C_2$ structure. Ab initio calculations carried out at the time supported a $^2A''$ state with the $\sigma$-type W shaped SOMO [80, 88]; however, an element of ambiguity persisted, because the calculations did not reproduce the radical cation well. Alternatively, the unpaired electron may be localized mainly in one C–C bond (Figure 14) interacting mainly with two quasi-equatorial $\beta$ hydrogens. Recent ab initio calculations (6-31G* basis set, UHF, UB3LYP, or UMP2) favored the alternative structure. The C3–C4 bond was significantly lengthened (ca 210 pm) and the electron spin-density was almost entirely localized at C3,4 ($\rho_{3,4} = 0.46$, UB3LYP) [31]. At this level, only the quasi-equatorial $^1H$ nuclei had large hyperfine coupling constants ($a_{2,5} = 1.98$ mT), although still somewhat different from the experimental values.

Interestingly, the averaged hyperfine coupling of the eleven-line spectrum changes significantly in different matrices, ranging from 0.59 mT in $C_6F_{12}$ to 0.77 mT in CFCl$_3$. This result might suggest that the electronic ground state is influenced by the nature of the matrix. Other matrix-dependent ESR spectra will be discussed for the radical cations of the bicyclo[4.4.0]decane isomers (Section 2.4.2).

Electron transfer from $c$-C$_6$H$_{12}$ was studied by many groups as a convenient and simple target for pulse radiolysis [37, 90–93]. Radiolysis of $c$-C$_6$H$_{12}$ in Freon-113

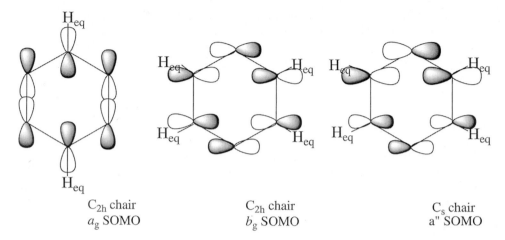

$C_{2h}$ chair
$a_g$ SOMO

$C_{2h}$ chair
$b_g$ SOMO

$C_s$ chair
a″ SOMO

**Figure 15.** Possible SOMOs of cyclohexane radical cations. Because the $a_g$ and $b_g$ SOMOs are incompatible with the observed hyperfine coupling pattern, further distortion of the $b_g$ SOMO to an a″ SOMO was suggested [94, 95].

produced a Jahn–Teller active radical cation, $c\text{-}C_6H_{12}^{\cdot+}$. The spectra observed at 4 or 77 K were broad and difficult to analyze. The ESR spectrum at 4 K contained three pairs of equivalent protons, $a_H = 8.5$ mT, $a_H = 3.4$ mT, $a_H = 1.4$ mT. Jahn–Teller distortion of the degenerate $E_g$ structure (a $D_{2d}$ chair) might lead to either a $^2A_g$ or a $^2B_g$ state ($C_{2h}$ symmetry; Figure 15) [92, 93]. The observed spectrum was assigned to a radical cation of $^2A_g$ structure, with two elongated C–C bonds and the splitting to the six equatorial protons of $c\text{-}C_6H_{12}^{\cdot+}$ [37, 92]. The simple species fitting this description would have four equivalent protons, however; an additional distortion, twisting two $^1H$ nuclei 'out of plane', must, therefore, be invoked to account for the spectrum [94, 95]. At temperatures $\geq 140$ K a well resolved seven-line spectrum appeared ($a_H = 4.3$ mT); this is readily explained as the average of three rapidly equilibrating $^2A_g$ structures with an estimated activation barrier between 170 and 240 cal mol$^{-1}$ [37].

Electron-transfer reactions of higher cycloalkanes were also studied. Electron transfer from $c\text{-}C_7H_{14}$ to unstable holes generated by radiolysis in Freon-113 gave rise to a stable radical cation, $c\text{-}C_7H_{14}^{\cdot+}$; its spectrum was interpreted in terms of a twisted chair form with $C_2$ symmetry [37]. Finally, radiolysis of $c\text{-}C_8H_{16}$ in a Freon-113 matrix generated a Jahn–Teller-active radical cation, $c\text{-}C_8H_{16}^{\cdot+}$, with three sets of non-equivalent protons [37]. A detailed discussion of these species exceeds the scope of this review.

### 2.3.2 Cyclopropane Radical Cations

Cyclopropane has a degenerate pair of in-plane e′ orbitals $(S, A)$. Accordingly, vertical ionization leads to a doubly degenerate $^2E'$ state. Jahn–Teller (JT) distortion of this state results in two non-degenerate electronic states, $^2A_1$ and $^2B_2$ ($C_{2v}$ symmetry) [96–104]. The $^2A_1$ component (orbital $S$ singly occupied) relaxes to an equilibrium structure with one lengthened C–C bond, which is the lowest energy species for many cyclopropane radical cations (Figure 12). This assignment is based unambiguously on ESR [78, 105, 106] and nuclear spin polarization (CIDNP) studies (cf. below) [107–110]. The two structure-types considered for the cyclopropane radical cation, $^2B_2$ ('type **B**') and $^2A_1$ ('type **A**') pose several interesting questions. Which structure is of lower energy? Can the $^2A_1$ species undergo ring-opening to trimethylene radical cation? How is the ring-opened species related to propene radical cation? How do substituents affect the relative stabilities of the $^2B_2$ and $^2A_1$ structures? These questions have been pursued in theoretical and experimental studies. We will review some molecular orbital calculations and discuss them in the light of experimental results.

**Molecular orbital calculations**

The radical cations of cyclopropane have long been the target of theoretical investigations) [96–104]. Here, we will discuss mainly a thorough ab initio study by Borden and co-workers [101], dealing with the cyclopropane radical cation, $10^{\cdot+}$, its potential ring opening to trimethylene radical cation, $11^{\cdot+}$, and the further rearrangement to propene cation radical $12^{\cdot+}$. In addition, we will briefly mention a study by Krogh-Jespersen and Roth [103], dealing specifically with the existence of the $^2B_2$ structure type and its potential stabilization by appropriate substituents, and two papers dealing with the ring-opened species [102, 104].

In view of the significant ring-strain of the cyclopropane system, it is hardly surprising, that the propene radical cation, $12^{\cdot+}$ was found to be the lowest energy isomer on the $C_3H_6^{\cdot+}$ potential surface. At the unrestricted Hartree–Fock (UHF/6-31G*//MP2/6-31G*) level ($C_s$ symmetry), $12^{\cdot+}$ lies ca 10 kcal mol$^{-1}$ below the $^2A_1$ radical cation with only one lengthened C–C bond '$10^{\cdot+}$'. A vibrational analysis of this structure showed only positive frequencies, thus identifying this species as a local minimum. In contrast, the ring-opened trimethylene radical cation, $11^{\cdot+}$, does not appear to be a minimum, because vibrational analysis in $C_{2v}$ symmetry showed one imaginary and one low frequency.

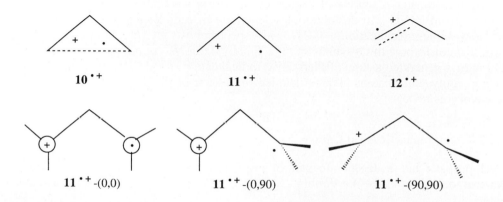

$10^{\cdot+}$          $11^{\cdot+}$          $12^{\cdot+}$

$11^{\cdot+}$-(0,0)          $11^{\cdot+}$-(0,90)          $11^{\cdot+}$-(90,90)

Additional geometries in which the terminal $CH_2$ groups were rotated relative to the plane of the three-carbon unit from orientation '0' (in plane) to orientation '90' (perpendicular) likewise failed to qualify as local minima. The calculations failed to indicate a chemically significant barrier for the conversion of $11^{\cdot+}$ to $12^{\cdot+}$. The essence of these findings was confirmed by UMP2/6-31G* calculations of Skancke [104], who identified the conversion of $10^{\cdot+}$ ($^2A_1$) to $12^{\cdot+}$ as a one-step reaction with a barrier of ca 30 kcal mol$^{-1}$, approximately one half that measured for the parent system [111, 112]. In summary, these calculations argue against the opening of $10^{\cdot+}$ to $11^{\cdot+}$ reported to occur in $CF_2Cl$–$CFCl_2$ matrices, even at cryogenic temperatures [105, 106]. On the other hand, it is possible that matrix forces affect the prevailing structure. We will discuss several examples of effects of this nature (vide infra).

**Photoinduced electron transfer of cyclopropane systems**

The ring-strain inherent in cyclopropane makes it a significantly better electron donor than non-strained cyclic hydrocarbons. Accordingly, a wide range of cyclopropane derivatives undergo electron transfer upon interacting with acceptor excited states. The resulting 'photoinduced electron transfer' (PET) is a mild and versatile method for the generation of radical cation–radical anion pairs in solution (Scheme 1) [113–115]. The PET method utilizes the fact that the oxidative power of an acceptor and the reductive power of a donor are substantially enhanced by photoexcitation. Thus, donor–acceptor pairs with negligible or weak interactions in the ground state, can readily undergo electron transfer, generating radical ion pairs, if either reactant is excited electronically. For the study of radical cations it is advantageous to excite the acceptor (Eq. 10). Depending on the nature of the acceptor and its lifetime, it will be quenched before or after intersystem crossing to the triplet state.

In Scheme 1, the reaction is formulated for triplet quenching (Eq. 11), which generates radical ion pairs of triplet spin multiplicity. Even so, the resulting radical ions have limited lifetimes, because the pairs readily undergo intersystem crossing

$$A \longrightarrow {}^1A^* \longrightarrow A^* \qquad (10)$$

$$^3A^* + D \longrightarrow \overline{{}^3 A^{\cdot-}\ D^{\cdot+}} \qquad (11)$$

$$\overline{{}^3 A^{\cdot-}\ D^{\cdot+}} \longrightarrow \overline{{}^1 A^{\cdot-}\ D^{\cdot+}} \qquad (12)$$

$$\overline{{}^1 A^{\cdot-}\ D^{\cdot+}} \longrightarrow A + D \qquad (13)$$

$$\overline{{}^3 A^{\cdot-}\ D^{\cdot+}} \longrightarrow {}^2A^{\cdot-} + {}^2D^{\cdot+} \qquad (14)$$

**Scheme 1.**

(Eq. 12), followed by recombination of the singlet pairs (Eq. 13); alternatively, separation by diffusion (Eq. 14) might generate 'free' radical ions.

The most common triplet-state electron acceptors are ketones and quinones, whereas aromatic hydrocarbons, often bearing one or more cyano groups, are the most frequently used singlet-state electron acceptors. For the generation of radical cations from a given donor it is important that the exothermicity of electron-transfer reactions can be adjusted to fall within an appropriate range, typically between 0.2 and 1.0 eV. The change in free energy ($\Delta G$) for an electron-transfer reaction is given by the Rehm Weller equation (Eq. 11) [116]

$$\Delta G = -E_T - E_{red} + E_{ox} - e^2/\varepsilon a \tag{15}$$

where $E_T$ is the excited state energy (0–0 transition), $E_{ox}$ is the one-electron oxidation potential of the donor, $E_{red}$ is the one-electron reduction potential of the acceptor, and $a$ is the distance (Å) between donor and acceptor. The term $e^2/\varepsilon a$ (Coulomb term) takes account of ion pairing. Alternatively, one can define the reduction potential of the acceptor excited state as

$$^*E_{red} = -E_T + E_{red} \tag{16}$$

The application of PET is limited to the oxidation of substrates with oxidation potentials well below the threshold value defined by $^*E_{red}$. According to Eq. 16 the change in free energy of the reaction can be tuned by variation of the solvent (polarity) and of the acceptor (reduction potential, excited-state energy). For a given class of acceptors the excited state energies typically vary over a narrow range, whereas the reduction potentials can be altered substantially by the introduction of appropriate substituents (Table 3). It is not, therefore, generally a problem to adjust the exothermicity to an appropriate range.

The parent cyclopropane system does not, in fact, readily undergo electron transfer in solution; apparently, the excited state reduction potentials of most sensitizers are too low (Table 3). However, introducing simple alkyl substituents increases the donor capacity of the cyclopropane system. This is aptly shown by the (gas-phase) ionization potential of 1,1-dimethylcyclopropane (9.0 eV) compared with that of cyclopropane (9.87 eV). PET from a series of methyl-substituted cyclopropanes to photoexcited chloranil was probed in solution. These experiments failed to provide evidence for electron transfer from *cis*- or *trans*-1,2-dimethylcyclopropane. On the other hand, 1,1,2-trimethyl- and 1,1,2,2-tetramethylcyclopropane were oxidized [108, 109].

Conjugation with one or two phenyl groups also converts the cyclopropane ring to an excellent electron donor. Although 1-phenyl- and 1,2-diphenylcyclopropane arguably belong into the category of aromatic compounds, their electron-transfer chemistry is included here, because their reactions are essentially those of cyclopropane compounds.

Two types of competing reaction pose potential drawbacks to the PET method. First, the principal types of electron acceptor are ketones and quinones, the triplet states of which are known to abstract hydrogen atoms with formation of neutral radicals. Second, many of the radical cations generated by PET are potential proton

**Table 3.** Excited state reduction potentials of selected electron acceptors.

| | $E_{A/A^-}$ [a] | $E_{(0,0)}$ [b] | $^3E^*$ [c] | $^*E_{A/A^-}$ [d] |
|---|---|---|---|---|
| *Singlet acceptors* | | | | |
| 2,4,6-Triphenylpyrylium tetrafluoroborate (TTF) | −0.29[e] | 2.8 | | 2.5 |
| 2,6,9,10-Tetracyanoanthracene (TCA) | −0.45 | 2.82 | | 2.35 |
| 1,2,4,5-Tetracyanobenzene (TCB) | −0.65 | 3.83 | | 3.2 |
| 9,10-Dicyanoanthracene (DCA) | −0.89 | 2.88 | | 2.0 |
| 1,4-Dicyanonaphthalene (DCN) | −1.28 | 3.45 | | 2.15 |
| 9-Cyanoanthracene (CA) | −1.39 | 2.96 | | 1.55 |
| *p*-Dicyanobenzene (p-DCB) | −1.60 | 4.29[f] | | 2.7 |
| 9-Cyanophenanthrene (CP) | −1.88[g] | 3.42 | | 1.55 |
| 1-Cyanonaphthalene (1-CN) | −1.98[h] | 3.75 | | 1.75 |
| 2-Cyanonaphthalene (2-CN) | −2.13[h] | 3.68 | | 1.55 |
| Naphthalene (N) | −2.50[i] | 3.97 | | 1.45 |
| Phenanthrene (P) | −2.45[i] | 3.58 | | 1.15 |
| Anthracene (A) | −1.96[i] | 3.28 | | 1.3 |
| *Triplet acceptors* | | | | |
| Chloranil (CA) | +0.02[i] | | 2.7[j] | 2.7 |
| Benzoquinone | −0.54[i] | | 2.95[j] | 2.4 |
| Naphthoquinone | −0.60[i] | | | |
| Anthraquinone | −0.94[k] | | 2.72[k] | 1.8 |
| Benzil | −1.50[k] | | 2.36[k] | 0.85 |
| Benzophenone | −2.16[k] | | 2.95[k] | 0.8 |

[a] Half wave reduction potential (V) vs. SCE; from Ref. [98], except as noted otherwise.
[b] Singlet energy (eV) from the 0,0 transition of the fluorescence spectrum; from Ref. [98], except as noted otherwise.
[c] Triplet energy (eV).
[d] Excited state reduction potential.
[e] 48. Saeva, F. D., Olin, G. R. *J. Am. Chem. Soc.* **1980**, *102*, 299.
[f] Arnold, D. R., Maroulis, A. J. *J. Am. Chem. Soc.* **1976**, *98*, 5931.
[g] Park, S.-M. Caldwell, R. A. *J. Electrochem. Soc.* **1977**, *124*, 1859.
[h] McCullough, J. L, Miller, R. C., Fung, D., Wu, W.-S. *J. Am. Chem. Soc.* **1975**, *97*, 5942.
[i] Mann, C. K., Barnes, K. K. *Electrochemical Reactions in Nonaqueous Systems*, Marcel Dekker, Inc., New York, **1970**.
[j] Kavarnos, G. L, Turro, N. *J. Chem. Rev.* **1986**, *86*, 40.
[k] Gersdorf, L, Mattay, L, Görner, H. *J. Am. Chem. Soc.* **1987**, *109*, 1203.

donors (Section 2.2.5) and the radical anions are comparably strong bases. Accordingly, proton transfer in the geminate radical ion pair might produce neutral radicals, and the potential involvement of two or more competing reactions might introduce mechanistic ambiguities. On the other hand, this feature has made it possible to study interesting electron-transfer–proton-transfer sequences.

**Electron transfer of cyclopropane systems—the CIDNP method**

As mentioned earlier, the cyclopropane radical cation, prepared by $\gamma$-radiolysis in rigid matrices, had an ESR spectrum compatible with the trimethylene structure ($^2A_1$) [78, 79]. Irradiation of several methyl-substituted derivatives at 77 K gave rise to a family of radical cations of the same structure type (type **A**) [105], which had been identified previously on the basis of chemically induced dynamic nuclear polarization (CIDNP) results [107, 110]. Because of the significance of this method for the study of electron-transfer reactions and for assigning radical cation structures, we briefly discuss the underlying basic principles.

Chemically induced dynamic nuclear polarization (CIDNP) is a nuclear magnetic resonance method based on the observation of transient signals, typically substantially enhanced, in either absorption of emission. These effects are induced as a result of magnetic interactions in radical or radical ion pairs on the nanosecond time scale. This method requires acquisition of an NMR spectrum during (or within a few seconds of) the generation of the radical ion pairs. The CIDNP technique is applied in solution, typically at room temperature, and lends itself to modest time resolution. The first CIDNP effects were reported in 1967, and their potential as a mechanistic tool for radical pair reactions was soon recognized [117, 118]. Nuclear spin polarization effects were discovered in reactions of neutral radicals and experiments in the author's laboratory established that similar effects could also be induced in radical ions [119–121].

The theory underlying this effect depends critically on two selection principles: the nuclear spin-dependence of intersystem crossing in a radical pair, and the electron spin-dependence of the rates of radical pair reactions. The combination of these selection principles causes a 'sorting' of nuclear spin states into different products, formed by geminate recombination (allowed for singlet pairs but spin-forbidden for triplet pairs) or by free-radical ('escape') products (whose formation is electron spin-independent). As a result, geminate reaction products are formed with characteristic non-equilibrium populations of nuclear spin levels, whereas 'escape' products show complementary non-equilibrium spin level populations.

The transitions between levels with non-equilibrium populations will be in the direction towards restoring the normal Boltzmann population; their signal intensities will depend on the extent of non-equilibrium population. The observed effects are optimal for radical pairs with lifetimes in the nanosecond range. On a shorter time-scale, hyperfine induced intersystem crossing is negligible whereas on a longer time-scale the polarization decays owing to spin–lattice relaxation in the radicals.

The quantitative theory of CIDNP enables one to compute the intensity ratios of CIDNP spectra on the basis of reaction and relaxation rates and characteristic parameters of the radical pair (initial spin multiplicity, $\mu$), the individual radicals (electron $g$ factors, hyperfine coupling constants, $a$), and the products (spin–spin coupling constants, $J$) [122–126]. Conversely, the patterns of signal directions and intensities observed for different nuclei of a reaction product can be interpreted in terms of the hyperfine coupling constants of the same nuclei in the radical cation intermediate. This feature has proved significant for the assignment of radical cation structures.

In a well designed experiment, the pattern of CIDNP signal directions and intensities observed for a diamagnetic product reveal the relative magnitude and the absolute sign of the hyperfine coupling constants of the corresponding nuclei in the paramagnetic intermediate. The hyperfine coupling constants, in turn, can be interpreted in terms of carbon spin densities and these reveal important structural features of the intermediates. These results often are quite unambiguous, because

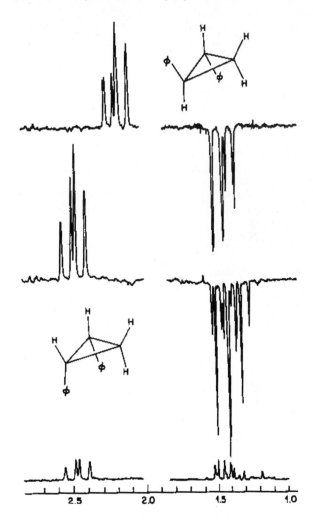

**Figure 16.** PMR spectra (90 MHz) observed during the irradiation of chloranil in acetonitrile-d$_3$ solutions containing *trans*- (top) or *cis*-1,2-diphenylcyclopropane (bottom), respectively [107].

NMR chemical shifts are usually well understood, and the identity of the coupled nuclei is clearly established. Combined with PET as a method of radical ion generation, the CIDNP technique has been the key to elucidating mechanistic details of important reactions and provided insight into many short-lived radical cations with unusual structures, many of which had previously eluded any other technique.

The nature of the cyclopropane radical cation was first characterized unambiguously by CIDNP effects of a 1,2-disubstituted derivative. The pattern of benzylic and geminal polarization observed during the reaction of chloranil with *cis*- and *trans*-1,2-diphenylcyclopropane (Figure 16) supported radical ions with spin-density on the benzylic carbons [107].

Strictly, the results do not differentiate a priori between a 'closed' and an 'open' radical cation. The 'closed' structure was assigned because the reaction did not cause geometric isomerization, suggesting that the stereochemistry at the key car-

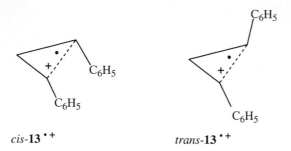

*cis*-**13**<sup></sup>     *trans*-**13**<sup></sup>

bons is preserved in the intermediate. These results establish local minima on the radical cation potential surface but do not eliminate the possibility of the existence of additional minima with different stereochemistry and allow no conclusions concerning the global minimum on the radical ion energy surface.

**'Standard' structure types of cyclopropane radical cations**

In this context we consider the potential effects of substituents on the structure of the cyclopropane radical cation. For molecules with a pair of degenerate HOMOs suitable substitution might be expected to lift the degeneracy and favor one structure over the other. Qualitative predictions of the favored structure can be based on a frontier (F) MO/perturbational (P) MO approach [110, 127]. The substrates are dissected into molecular fragments and the potential interactions of the component FMOs are considered. According to PMO theory [127] the strength of the fragment perturbation is approximately proportional to $S^2/\Delta E$, where $S$ is the overlap integral between the components and $\Delta E$ is the difference between the FMO orbital energies. For the $S^2$ term, three factors will be of primary importance: the FMO orbital symmetry (where present); the magnitude of the coefficients at the point(s) of union; and the orientation of the fragments relative to each other [110].

For cyclopropane, substituents at a single carbon might most effectively stabilize the anti-symmetrical HOMO, whereas substitution at two carbons is expected stabilize the symmetrical orbital. The radical ions of many derivatives also belong to the general structure type **A**. On the other hand, cyclopropane radical cations with the alternative, anti-symmetrical singly occupied (SO) MO should be of particular interest. The reversal of the ordering of structure types **A** and **B** can be envisaged via three different mechanisms, involving stabilization of structure type **B** by conjugation, homoconjugation, or hyperconjugation.

Because a structure of $^2A_1$ symmetry was established for the prototype [78], it is hardly surprising that the radical ions of many derivatives also belong to that general structure type. Radical cations of the same structure type as those derived from *cis*- and *trans*-1,2-diphenylcyclopropane were established for numerous cyclopropane derivatives, including 1,2-di-, 1,1,2-tri- and 1,1,2,2-tetramethylcyclopropane (Table 4) [104, 109]. Two of these systems provide a direct comparison between the results of CIDNP and ESR experiments. In both instances the ESR spectra observed by Williams and coworkers after $\gamma$-irradiation in frozen solutions [105, 106] contain splitting patterns supporting the presence of spin-density on two car-

**Table 4.** $^1$H Hyperfine coupling patterns constants for radical cations of selected cyclopropane systems.

| Radical cation | Calculation | CIDNP | ESR |
|---|---|---|---|
| 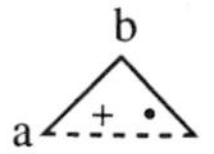 | | No result | a − 12.5<br>b + 21.0 |
| 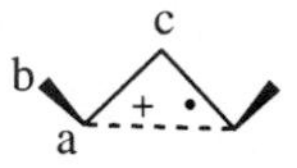 | | No result | a (−) 10.4<br>b (+) 20.5<br>c (+) 20.5 |
| 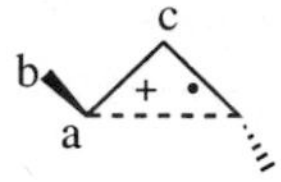 | | No result | a (−) 11.9<br>b (+) 21.8<br>c (+) 21.8 |
| 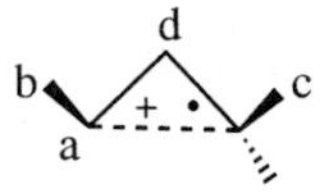 | | a −<br>b +<br>c +<br>d + | a (−) 9.8<br>b (+) 14.5<br>c (+) 20.6<br>d (+) 17.9 |
| 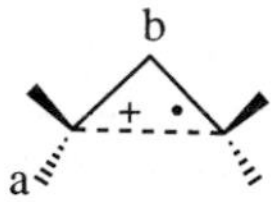 | | a +<br>b + | a (+) 15.0<br>b (+) 18.7 |
| 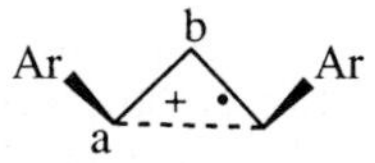 | | a −<br>b + | |
| 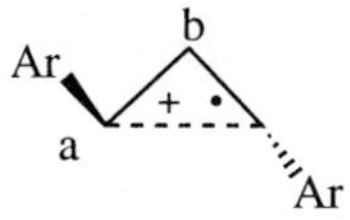 | | a −<br>b + | |
| 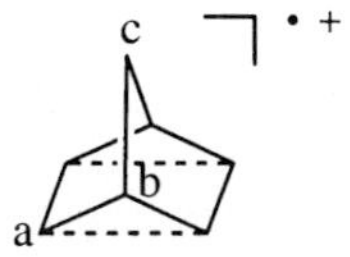 | a (−) 10.5<br>b (+) 7.9<br>c (−) 4.0 | a −<br>b +<br>c − | a (−) 5.1<br>b (+) 6.6<br>c (−) 2.0 |
| 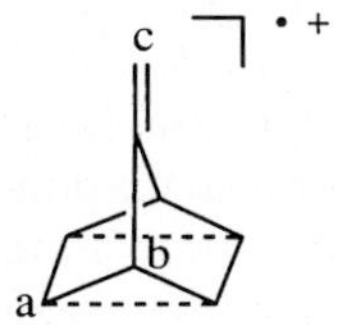 | a (−) 10.6<br>b (+) 9.9<br>c (+) 12.6 | a −<br>b +<br>c + | |

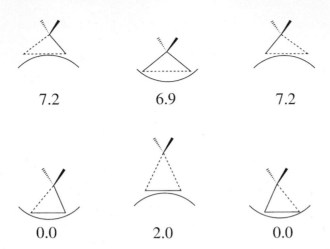

7.2                6.9                7.2

0.0                2.0                0.0

**Figure 17.** Minima and transition states on the potential energy surface of radical cation states of 1-methylcyclopropane (PMP4/6-311G*//UMP2/6-31G* + ΔZPE; relative energies in kcal/mol) [103].

bon centers and, thus, fully confirm the structure type (**A**) assigned on the basis of CIDNP results.

Stabilization by hyperconjugation was probed by ab initio calculations on the radical cations of methyl- and 1,1-dimethylcyclopropane [103]. Two sets of structures reflect the first-order Jahn–Teller distortion of the parent cation from the doubly degenerate $^2E'$ ($D_{3h}$ symmetry) ground state to non-degenerate states $^2A_1$ and $^2B_2$. States of type **A** (one long and two short ring C–C bonds) are always minima. For methyl and 1,1-dimethyl derivatives, type **B** structures are the preferred first-order Jahn–Teller type distorted structures. Although their energies lie below the type **A** structures, the type **B** structures are, however, transition states, undergoing second-order Jahn–Teller type distortions to unsymmetrical (scalene) structures with one very long C–C bond. These structures represent the absolute minima for 1-methyl- and 1,1-dimethylcyclopropane radical cations (Figure 17) [103]. The 'scalene' structures can be viewed as distorted type **B** structures or as unsymmetrical type **A** structures with substituents at one 'terminal' carbon.

The calculated hyperfine coupling constants (B3LYP/6-31G*//MP2/6-31G*) for the type **B** transition state and the distorted minimum clearly show that this species must be considered a type **A** structure. The hyperfine coupling pattern of the lowest-energy minimum ($a_2 = -1.43$ mT; $a_3 = 1.98$ mT) shows a trend similar to the experimental splittings of the *trans*-1,2-dimethylcyclopropane radical cation ($a_{1,2} = -1.19$ mT; $a_3 = 2.18$ mT), whereas the pattern calculated for the transition state ($a_{2,3} = 0.55$ mT) is incompatible with that model (Figure 18). The distorted structure type calculated for the methyl-substituted systems seems to prevail also under other conditions (see below).

Although the results indicated some stabilization for the type **B** structures, they clearly indicate that hyperconjugation is not sufficient to alter the 'natural' preference of cyclopropane radical cations for the type **A** structure. On the other hand, both conjugation and homoconjugation have been shown to reverse the stabilities

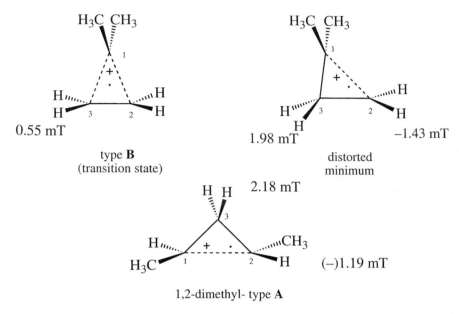

0.55 mT

type **B**
(transition state)

1.98 mT          −1.43 mT

distorted
minimum

2.18 mT

(−)1.19 mT

1,2-dimethyl- type **A**

**Figure 18.** Comparison between the hyperfine coupling constants calculated for the type **B** radical cation (top, left) and the distorted minimum (top, right), respectively, of 1,1-dimethylcyclopropane [31] and the experimental hyperfine coupling constants [105, 106] of 1,2-dimethylcyclopropane (bottom).

of the 'natural' structure types $^2B_2$ and $^2A_1$. Conjugation with a suitable $\pi$-system, either a vinyl [128] or a phenyl group [129, 130], is sufficient to stabilize structures of type **B**, whereas homoconjugation stabilized type **B** radical cations for substrates, such as norcaradiene and derivatives [108, 131] and spiro[cyclopropane-1,9'-fluorene] [110, 132]. These assignments are based on experimental results and born out by ab initio calculations.

The structure of the phenylcyclopropane radical cation, **14**·+, was based on CIDNP effects observed during the electron-transfer reaction from **14** to photo-excited chloranil. The results indicated that the unpaired electron spin was delocalized between the aromatic ring and the benzylic cyclopropane carbon [129]. $\pi,\sigma$-Polarization induces negative hyperfine coupling (hfc) constants in the aromatic *ortho* and *para* protons and in the benzylic cyclopropane protons. The secondary cyclopropane protons had significantly divergent hyperfine coupling constants, because of a pronounced stereoelectronic effect. The strongly polarized secondary protons confirm the presence of spin-density on the benzylic carbon; the emission supports positive hfcs, which arise typically via $\pi,\sigma$-delocalization (hyperconjugation) of spin-density on to the $^1$H nuclei. These results support a radical cation, **14**·+, R = H, in which spin and charge are delocalized between the phenyl ring and the benzylic cyclopropane carbon; two cyclopropane bonds are lengthened and weakened (type **B**). These conclusions are in full accord with the results of ab initio calculations (B3LYP/6-31G*), which delineated the principal structural features and the charge density distribution in the radical cation, **14**·+ [129, 130].

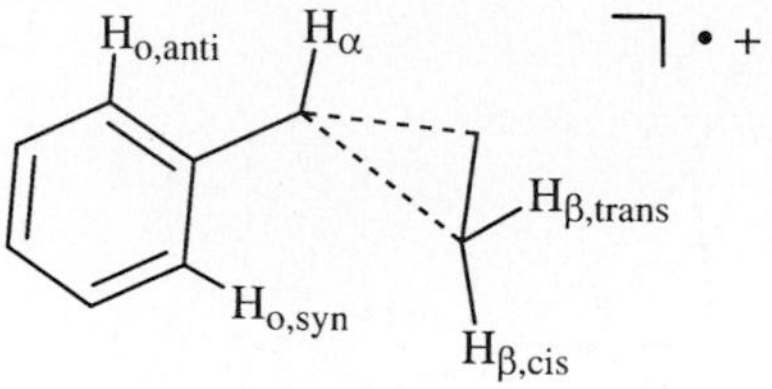

**14·⁺**

The vinylcyclopropane radical cation, **15·⁺**, is another radical cation of structure type **B**, which is stabilized by conjugation. Its proposed structure was based exclusively on ab initio calculations (B3LYP/6-31G*) because the electron-transfer photochemistry of this species failed to provide clear-cut CIDNP effects [128]. In this context it is worth noting that product studies cannot, in principle, establish the cyclopropane radical cation structure type. Irrespective of the structure, nucleophilic capture is expected to result in the cleavage of the strained ring.

Calculations using the 6-31G* basis set and employing UB3LYP//UB3LYP and UMP2//UB3LYP methodologies, respectively, indicated the existence of two conformers, s-anti-**1·⁺** and s-syn-**1·⁺**. Both have structures of type **B**, resembling a π-complex between vinylmethylene and ethene. The calculated bond lengths for the two conformers show similar trends; the allylic cyclopropane bonds are lengthened (+6 %) whereas the bond between the secondary carbons C2–C3 is shortened (−4 %). Also, the distinct difference between the (vinyl) double bond (134.0 pm) and the bond linking the cyclopropane and ethene functions (147.5 pm) is reduced; these bonds are of essentially equal length (139.8); in essence, the array, Cβ–Cα–C1, has been converted to an allyl moiety [128].

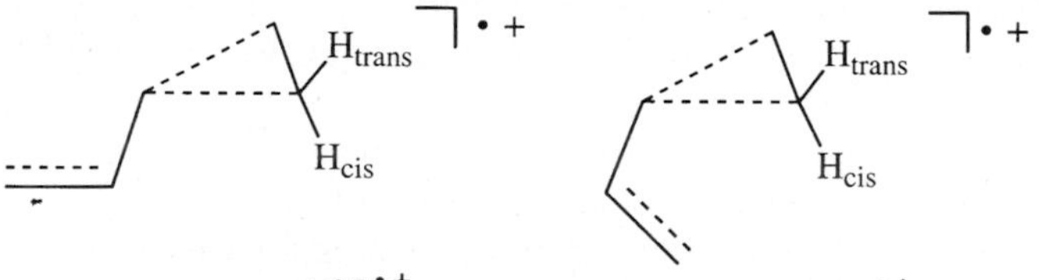

*anti-***15·⁺**          *syn-***15·⁺**

The calculated spin densities of the two conformers support the conclusions derived from the bond lengths—most of the unpaired electron density is located on the tertiary cyclopropane carbon (C1) and the terminal vinyl carbon (Cβ). The general type of spin-density distribution calculated for the two conformers of **15·⁺** has precedent in several vinylcyclopropane systems with 'locked' geometries (vide infra) [134, 135].

Perhaps the most interesting mechanism stabilizing radical cations of type **B** involves homoconjugation. The interaction with the butadiene frontier molecular orbital (FMO) can lift the degeneracy of the cyclopropane in-plane e' orbitals (S, A) and favor the type **B** structure. This principle was shown to stabilize substrates such as norcaradiene and derivatives [110, 131] and, to a lesser extent,

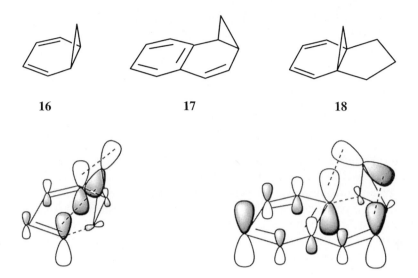

**16**                    **17**                    **18**

**Figure 19.** Structures of norcaradiene, **16**, and two derivatives, **17**, **18**, and schematic illustration of homoconjugation between the frontier molecular orbitals of butadiene (bottom left) and styrene (bottom right), respectively, with the anti-symmetrical cyclopropane HOMO.

spiro[cyclopropane-1,9'-fluorene] [110, 132]. The assignments are based on ab initio calculations on the parent system (**16**) and two derivatives, on CIDNP results observed for a benzo-annelated system (benzonorcaradiene, **17**), and on the electron-transfer photochemistry of a bridged tricyclic derivative, tricyclo[4.3.1.0$^{1,6}$]deca-2,4-diene (**18**) in the presence of methanol. This study identified **18** as a unique probe elucidating mechanistic features of the nucleophilic capture of radical cations (Figure 19).

The norcaradiene radical cation, **16$^{\cdot+}$**, has $C_s$ symmetry and a $^2A''$ electronic ground state. The bond lengths offer limited support for homoconjugation. Although the internal cyclopropane bond of **16$^{\cdot+}$** (C1–C6 = 153.8 pm) is slightly longer than the two lateral bonds (C1–C7 = C6–C7 = 153.3 pm), it is actually shortened upon ionization (−3.4 pm) while the lateral bonds are lengthened (+2.8 pm). The bonds between the pairs of olefinic carbons (C2–C3 = C4–C5 = 139.5 pm) are only marginally shorter than the intervening bond (C3–C4 = 140.5 pm). However, the delocalization of spin-density to C7 clearly supports the effect of homoconjugation. Although the (UMP2) spin-density at C7 ($\rho_7 = 0.246$) is lower than that at the terminal butadiene carbons, ($\rho_{2,5} = 0.359$), the delocalization of spin and charge on to C7 supports a structure of type **B**. The extent of delocalization depends on the level of perturbation theory; calculations at the UHF level show significantly less delocalization ($\rho_{2,5} = 0.61$, $\rho_7 = 0.231$) than do the UMP2 calculations. The hyperfine coupling pattern also reflects a type **B** structure. Two alkene protons (H2,5) and the geminal cyclopropane protons have significant negative coupling constants ($a_{2,5} = -0.091$ mT; $a_{7syn} = -0.057$ mT; $a_{7anti} = -0.063$ mT); sizeable positive coupling constants ($a_{1,6} = 1.36$ mT) are observed for the bridgehead carbons; the alkene protons near the nodal plane have negligible coupling constants ($a_{3,4} = -0.04$ G) [131].

Benzobicyclo[4.1.0]hepta-2,4-diene radical cation, **17·+** (no symmetry), was optimized at the UHF/6-31G* level of theory. The cyclopropane C–C bonds are subtly different; the lateral bond conjugated with the ethene function (C1–C7, 152.4 pm) is longer, whereas that conjugated with the benzene ring is shorter (C6–C7, 149.1 pm), than the internal bond (C1–C6, 151.3 pm). The bonds linking the strained ring to the styrene moiety are slightly longer (C1–C2, 145.4 pm; C5–C6, 148.5 pm) than those of the styrene function (C2–C3 138.5 pm; C4–C5 143.8 pm; C3–C4 139.9 pm) [131].

The assignment of an antisymmetric cyclopropane SOMO to the radical cation, **17·+**, was based on a comparison of CIDNP effects (Figure 20) with those for *cis*-1,2-diphenylcyclopropane. While the nuclei of the aromatic segments showed identical or very similar polarization, the cyclopropane protons show characteristic differences. This suggests significantly different spin-density distributions for the cyclopropane moieties of the two species and, thus, different structures [108]. The benzonorcaradiene radical cation should owe its structure to the symmetry of the fragment FMOs at the points of union. The styrene HOMO is antisymmetric at the positions of attachment, suggesting preferred interaction with the antisymmetric cyclopropane HOMO (Figure 19).

The calculated carbon spin densities of **17·+** document the extent of homo-conjugation; most of the spin is located on C2 ($\rho_2 = 0.355$), significantly less on C5 and C7 ($\rho_5 = 0.153$, $\rho_7 = 0.149$), whereas the tertiary cyclopropane carbons, C1 and C6 ($\rho_1 = -0.009$, $\rho_6 = -0.007$), have negative spin-density. The calculated hfcs are in qualitative agreement with the CIDNP effects observed during the electron transfer from benzonorcaradiene, **17**, to photo-excited chloranil [108]. The tertiary cyclopropane $^1$H nuclei, H1 and H6, have large positive hfcs ($a_1 = 0.93$ mT, $a_6 = 1.06$ mT), whereas the geminal cyclopropane nuclei (H7$_{s,a}$), have negative hfcs of moderate magnitude ($a_{7s} = -0.34$ mT, $a_{7a} = -0.28$ mT) [131].

Tricyclo[4.3.1.0$^{1,6}$]deca-2,4-diene radical cation, **18·+**, also has a $^2A''$ electronic ground state. When optimized at the UHF/6-31G* level of theory with imposed $C_s$ symmetry. The C–C bond lengths of the cyclopropane ring are slightly more divergent than are those of **16·+**; the internal bond (C1–C6 = 153.9 pm) is 3 pm longer than the lateral bonds (C1–C10 = C6–C10 = 150.8 pm). Both types of bonds are lengthened upon ionization, perhaps due to the release of strain. The bonds between the olefinic carbons are essentially equal in length (C2–C3 = C4–C5 = 139.3 pm; C3–C4 = 139.1 pm) [131].

The carbon spin-density on the bridge carbon is lower than that of **16·+** ($\rho_{10} = 0.203$), even considering the lower level of theory, whereas the terminal butadiene carbons have correspondingly higher spin densities ($\rho_{2,5} = 0.383$). The hyperfine coupling pattern of **18·+** shows minor changes relative to **16·+**: increased coupling constants for the olefinic protons H2,5 = $-0.96$ mT) and decreased coupling constants for the cyclopropane protons ($a_{10syn} = -0.54$ mT; $a_{10anti} = -0.48$ mT). These data support a radical cation related to structure type **B**. The extent of homo-conjugation, as judged by the extent of delocalization, is less than for **16·+**, possibly

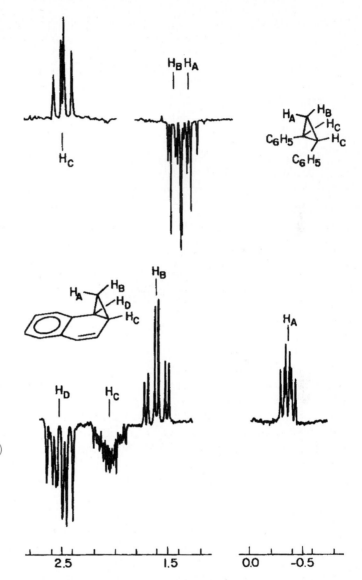

**Figure 20.** $^{1}$H CIDNP spectra (cyclopropane resonances only) observed during the photo-reaction of chloranil with *cis*-1,2-diphenylcyclopropane (top) and benzonorcaradiene (bottom). The opposite signal directions observed for analogous protons provide evidence that the two radical cations belong to two different structure types [107, 108].

because of strain introduced by the trimethylene bridge. This feature, though crucial for enforcing the norcaradiene structure, might limit the extent of homoconjugation because of steric factors. Still, the calculations predict a significant role of homo-conjugation in the radical cations, **16**$^{\cdot+}$ and **18**$^{\cdot+}$.

The radical cation of spirofluorene, **19**, is mentioned only in passing. The orientation of the fragments allows only the second highest MO of the biphenyl moiety to interact with a cyclopropane FMO (the *A* HOMO). Thus, the type B structure should be stabilized to a lesser extent. The extent of homoconjugation also is expected to be weak because of a more serious mismatch in orbital energies [110, 127].

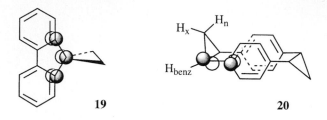

19                                      20

The interaction between a cyclopropane group and olefinic or aromatic systems conjugated to it depends on their relative orientation. Although the 3°,3°-cyclopropane bond of benzonorcaradiene (**17**) lies in a plane perpendicular to the aromatic π system, [1:2,9:10]bismethano[2.2]paracyclophane (**20**) features a parallel arrangement of these elements [133]. The photoreaction of this obviously strained compound with chloranil gives rise to CIDNP effects not unlike those observed for *cis*-diphenylcyclopropane (*cis*-**13**). The relative signal intensities of **20** are, however, noticeably distorted (Figure 21). The (secondary) endo proton (1.2 ppm) is more strongly enhanced (indicating a larger hyperfine coupling) than both the *exo* (2.0 ppm) and the benzylic protons.

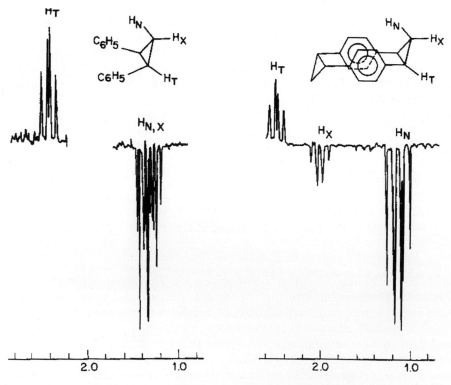

**Figure 21.** A comparison of the CIDNP effects observed for the cyclopropane protons of *cis*-diphenylcyclopropane (left) and [1:2.9:10]bismethanol[2.2]paracyclophane (right) [133].

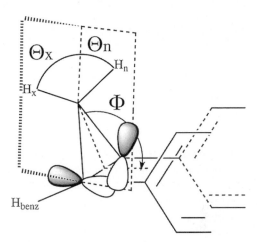

**Figure 22.** Partial view of [1:2.9:10] bismethano[2.2]paracyclophane showing the angles $\Phi$ and $\Theta$ discussed in the text.

This finding was interpreted in terms of different dihedral angles, $\Theta_x$ and $\Theta_n$, between the singly occupied orbital and the two geminal C–H bonds. Non-identical dihedral angles for the secondary cyclopropane protons can result if the angle, $\Phi$ (Figure 22), and the distance between the benzylic cyclopropane carbons is increased in the radical cation relative to the parent hydrocarbon. The endo proton, which in the diamagnetic molecule lies in the shielding cone of the aromatic moieties, has in the radical cation the smaller dihedral angle $\Theta$, with the benzylic '$\pi$' orbital (Figure 22) and, therefore, the greater hyperfine coupling [133].

The comparably weak enhancement of the benzylic protons and the relatively strong $^{13}$C polarization for the benzylic carbons are indicative of a pyramidal benzylic carbon. Although radicals containing pyramidal carbon are reasonably rare, existing examples are derived from strained ring systems [136–139]. For the radical cation **20**$^{\cdot+}$, the change in the angle, $\Phi$, the stretching of the doubly benzylic bond, and the adoption of a pyramidal structure apparently relieve some of the strain in the carbon skeleton, yet maintain reasonable overlap between the benzylic carbons and the benzene rings [133]. Possibly, the sum of these changes may amount to a 'ring-opened' cyclopropane radical cation (vide infra).

### Electron transfer and ring-opening—trimethylene radical cation

The potential 'ring-opening' of cyclopropane radical cations, 'breaking' the weakened bond of type **A** radical cations (**21**$^{\cdot+}$), has been a subject of both interest and controversy. The ESR spectra of cyclopropane radical cation and its methyl-substituted derivatives decayed at temperatures near 100 K. They were replaced by secondary spectra, in which the protons at one cyclopropane center do not interact with the electron spin. This coupling pattern was interpreted as evidence for a ring-opened trimethylene species (**22**$^{\cdot+}$) in which one terminal carbon has rotated into an orthogonal orientation [105, 106, 140].

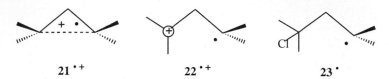

**21**·⁺                **22**·⁺                **23**·

Similar species were postulated to explain the geometric isomerization of 1-aryl-2-vinylcyclopropanes upon reaction with aminium radical cations [141]. As mentioned above, ab initio calculations fail to support the existence of this structure type for the parent system and simple derivatives. Another potential argument against the existence of species such as **22**·⁺ lies in its failure to undergo a hydrogen shift, generating the known and stable propene radical cation [142]. Hydrogen shifts have been observed in numerous electron-transfer-induced reactions [143–147], including in matrices at cryogenic temperatures, under the very conditions that give rise to the putative **22**·⁺ [143–146]. These considerations have led several authors to advance alternative explanations. They include structures, such as **23**· [99–101, 148], in which a chloride ion has captured the cationic center of either **21**·⁺ or **22**·⁺. Although interaction of **21**·⁺ or **22**·⁺ with one or more matrix molecules might form a chloronium-substituted free radical or a matrix cluster [149], ESR spectra failed to furnish direct evidence for coupling to the matrix.

Interestingly, the rearrangement was found to be limited to two matrices, $CFCl_2CF_2Cl$ and $CF_2ClCF_2Cl$, suggesting a major role of the matrix in the conversion. The putative nucleophilic ring opening by chloride ion, generating the β-chloroalkyl radical (**23**·), has precedent in many substitution of cyclopropane radical cations in solution (Section 2.4.3). In this instance the matrix might prevent the approach of the ion. Comparison of radical cation structures or reactions in matrices and in solution can, however, be misleading. Matrix forces are responsible for several reactions unprecedented in solution, such as the series of 'fragmentations' in zeolites (Section 2.2.7). Matrix forces have been invoked also to account for a 'reordering' of electronic states in matrices (Section 2.2.6) or zeolites (Section 2.4.2). In view of these considerations the discrepancy between the postulated structure (**22**·⁺) and the ab initio calculations is unsettling but not irreconcilable.

In this context, we mention two zeolite-induced conversions of cyclopropane derivatives. Incorporation of *trans*-1,2-diphenylcyclopropane (*trans*-**13**) and its 3,3-D₂-isotopomer into the channels of a redox-active pentasil zeolite (Na-ZSM-5) generated *exo,exo*-1,3-diphenylallyl radical (**24**·) and its 2-D₁-isotopomer. This conversion is a zeolite-specific reaction; it requires a series of reactions, including oxidation, ring opening, and deprotonation [70].

H
$C_6H_5$

$C_6H_5$        H

**13**

ZSM 5 →

H
|
C
$C_6H_5$—C ⋯⋯ C—$C_6H_5$
|                |
H              H

**24**·

Incorporation of arylcyclopropanes (**25**, R = H, $OCH_3$) into ZSM-5 also caused a matrix-specific conversion, generating *trans*-propenylbenzene radical cations

($26^{•+}$, R = H, OCH$_3$) [70]. The formation of $26^{•+}$ requires a series of reactions, including oxidation, ring opening, and a hydrogen shift. Interestingly, the 2,2-$d_2$-isotopomer of **25** (R = OCH$_3$) gave rise to three different isotopomers of $26^{•+}$ (R = OCH$_3$), containing two, one or no deuterium. This observation requires several competing mechanisms, including one proceeding with two D vs. H exchanges. This reaction also is zeolite-specific.

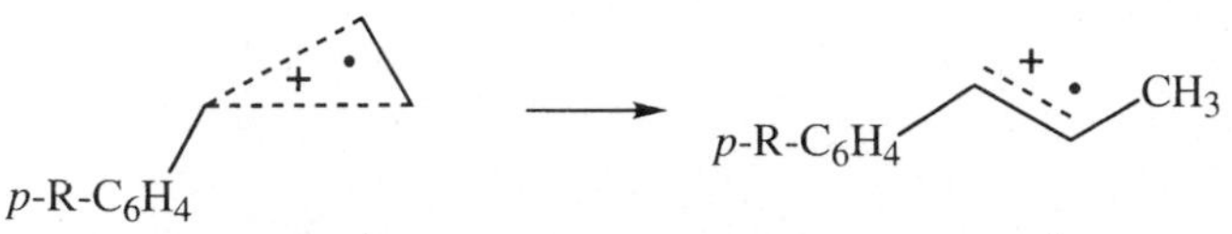

Ring-opened cyclopropane radical cations have been postulated also to account for the stereochemistry of the aminium radical cation-catalyzed rearrangement of 1-aryl-2-vinylcyclopropanes (**27**) [141]. These systems, of course, contain substituents that might veil the 'true' nature of the cyclopropane radical cation by delocalizing spin and charge.

The potential surface of the radical cations derived from **27** might contain several minima separated by low barriers; the ultimate reaction products might afford little information about any one minimum. Even ESR results might not differentiate unambiguously between a planar ring-opened radical cation ($29^{•+}$) and a rapidly equilibrating pair of orthogonal bifunctional intermediates (e.g., $28^{•+}$). Perhaps the most likely explanation involves a pair of isomeric cyclopropane radical cations with one very weak bond, viz., $27^{•+}$, which might enable either or both tertiary carbon atoms to rotate. The bifunctional structures shown below, either orthogonal (e.g. $28^{•+}$) or planar ($29^{•+}$) might signify minima or transition structures.

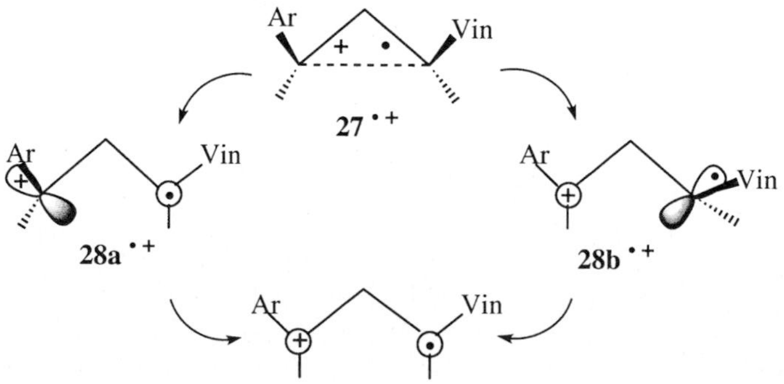

Additional insight into the 'ring-closed' or 'ring-opened' structure of cyclopropane radical cations can be derived unambiguously from the stereochemical course of their reactions. These reactions are discussed in detail in Section 2.4; here we briefly mention several illuminating examples of cyclopropane radical cation reactivity, chosen solely to illuminate the structure of cyclopropane radical cations.

Various substituted cyclopropanes undergo electron-transfer-induced nucleophilic addition of alcoholic solvents. For example, the electron-transfer reaction of phenylcyclopropane (**30**, R = H) with *p*-dicyanobenzene resulted in a ring-opened ether, **31** [150]. This reaction also produced an aromatic substitution product (**32**, R = H) formed by coupling with the sensitizer anion [150–152]. A related pair of products (**31**, **33**, R = H) was obtained upon photo-induced electron transfer between phenylcyclopropane and *N*-methylphthalimide [153].

More recently, Dinnocenzo and colleagues showed that a 2,3-dimethyl derivative (**30**, R = CH₃) and several 1-phenyl- and 1,1-diphenyl-2-alkyl-substituted cyclopropanes are captured with complete inversion of configuration [154, 155]. The observed stereochemistry requires an intermediate radical cation, **30**$^{\cdot+}$, with the unperturbed stereochemistry of the parent molecule. This result unambiguously rules out a ring-opened cyclopropane radical cation in solution.

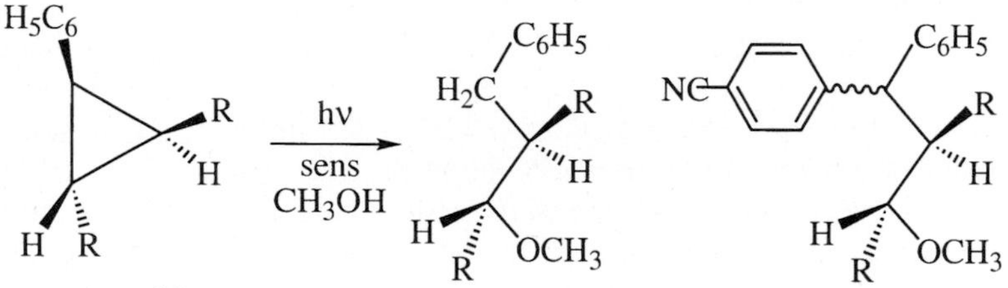

The nature of vinylcyclopropane radical cation was elucidated via the electron-transfer photochemistry of a simple vinylcyclopropane system, in which the two functions are locked in the *anti* configuration, viz. 4-methylene-1-isopropylbicyclo-[3.1.0]hexane (sabinene, **34**) [156]. The electron-transfer reaction of (1*R*, 5*R*)-(+)-**34** with 1,4-dicyanobenzene–phenanthrene in acetonitrile–methanol produced various optically active ring-opened products. The stereochemical relationship between **34** and the products requires that the radical cation, **34**$^{\cdot+}$, retain the three-dimensional integrity of **34**, i.e., that significant bonding is maintained in the three-membered ring. Any ring-opened radical cation, in which C1 would be planar, is clearly eliminated. Nucleophilic capture of **34**$^{\cdot+}$ by methanol generated free radical **35**$^{\cdot}$ in stereoselective fashion. In competition with this intermolecular reaction, an intramolecular hydride shift from C6 to C1 produced *β*-phellandrene radical cation, **36**$^{\cdot+}$, which is also captured by the nucleophile.

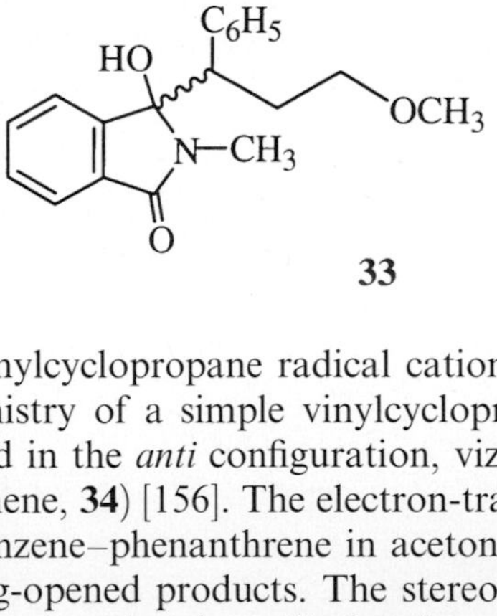

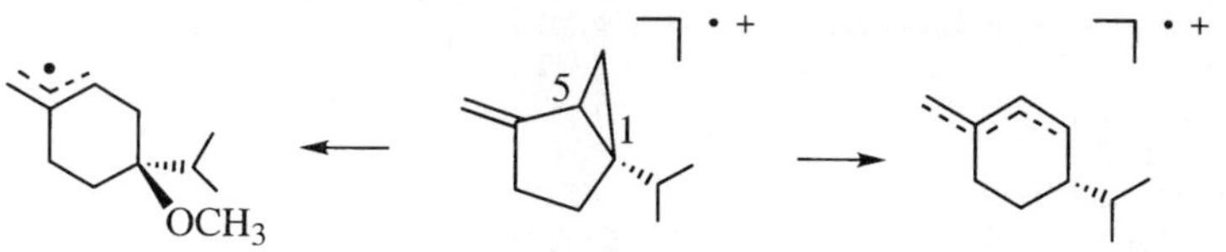

35•          34•+          36•+

### 2.3.3 Electron-transfer Reactions of Substituted Cyclobutane Systems

Having discussed the electron-transfer reactions of cyclopropane derivatives in some detail, we only touch briefly on the related reactions of the larger ring systems. Cyclobutane, cyclohexane, and cyclooctane have degenerate HOMOs (Section 2.3.1); thus, their radical cations are Jahn–Teller active. It is reasonable to expect that suitable substitution will lift the degeneracy and favor one radical cation structure over the other. Qualitative predictions of the preferred structure are readily derived.

Considering cyclobutane, one might expect that 1,2-disubstituted derivatives favor structure type C, in which the doubly substituted cyclobutane bond is weakened (and lengthened), or a related structure type, in which this bond is actually broken. Ab initio calculations at the QCISD-(T)/6-31G*//UMP2/6-31G* level of theory showed that the radical cation of *trans*-1,2-dimethylcyclobutane has a trapezoidal structure [86]. Although ESR results on simple derivatives are not available, this expectation is born out by several experimental observations.

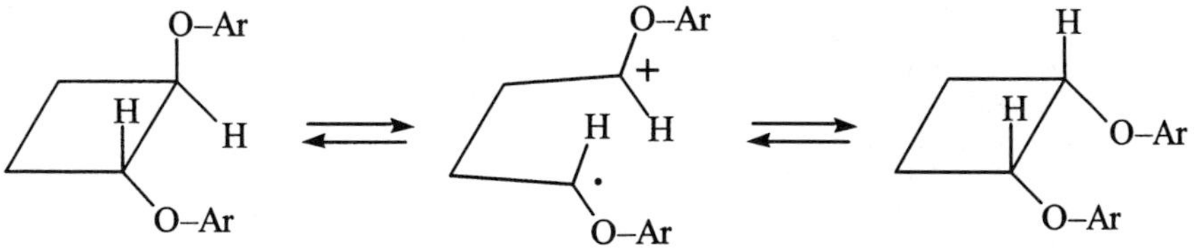

*trans*-37•+          38•+          *cis*-37•+

For example, the electron-transfer-induced geometric isomerization of 1,2-diaryloxycyclobutane, **37**, can be rationalized via the bifunctional radical cation, **38**•+, formed by ring opening of a type C radical ion [157]. Similarly, the fragmentation of the *anti*-head-to-head dimer of dimethylindene might involve consecutive cleavage of two cyclobutane bonds in a type C radical ion. Because of the low ionization potential of the aryl substituents (*IP* 9.25 eV) [8], the primary ionization is expected to occur from one of the aryl groups. CIDNP results observed during the dimer cleavage were rationalized in terms of an equilibrium between ring-closed (**39**•+) and ring-open (**40**•+) radical cations [158].

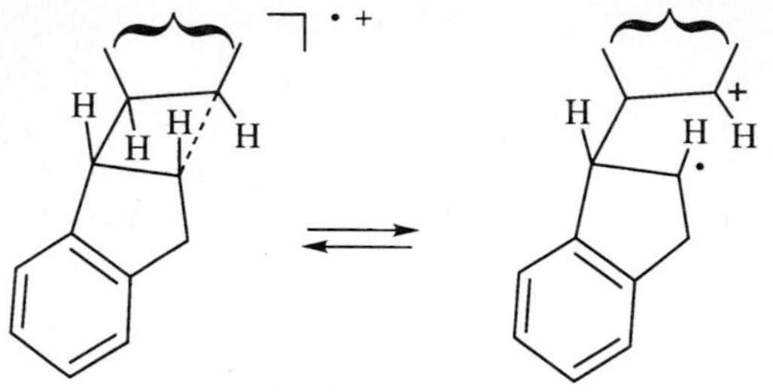

**39**·⁺                                    **40**·⁺

By analogy with substituent effects discussed for the cyclopropane system (Section 2.3.2), one might expect that 1,1-disubstitution might stabilize cyclobutane radical cations of the structure type resembling a kite (type D). As far as we are aware such a structure has not yet been established unambiguously, although the oxetane radical cation (**41**·⁺) clearly belongs to that structure type. The unpaired electron is localized on the oxygen and the adjacent $^1$H nuclei are strongly coupled ($a = 6.6$ mT, $a = 1.1$ mT) [159, 160].

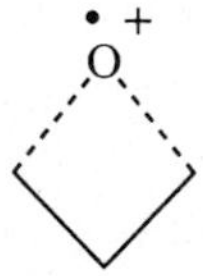

**41**·⁺

Cyclobutane radical cations can also be formed by addition of ethylene radical cations to a neutral ethylene. For example, vinyl amines [161] and other electron-rich alkenes [162] undergo [1 + 2] cycloadditions, which even might proceed as chain reactions [163]. Fluorescence-detected magnetic resonance effects observed during the pulse radiolysis of anthracene-$d_{10}$ in the presence of tetramethylethene generated an ESR spectrum compatible with eight equivalent methyl groups; the splitting, $a_d = 0.82$ mT, was approximately one half that of the monomer splitting, $a_m = 17.1$ mT [164–167]. The data were interpreted as evidence for a 'sandwich' dimer; higher cluster cations were also invoked [167]. Although the ESR spectrum is compatible with a radical cation with two weakened C–C bonds (type A; **42**·⁺), they do not rule out other structures.

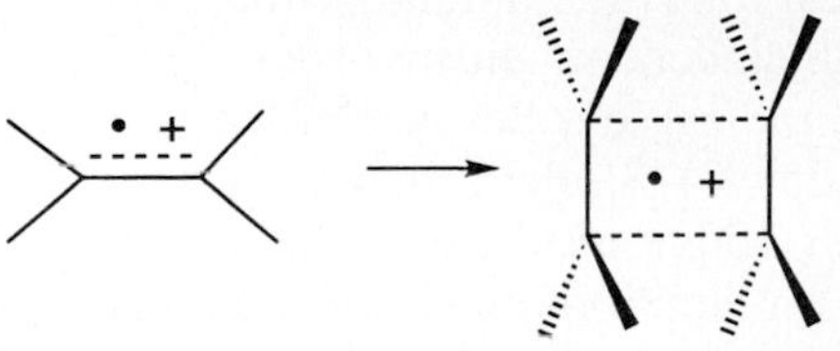

**42**·⁺

Ab initio calculations at the QCISD-(T)/6-31G*//UMP2/6-31G* level of theory failed to furnish evidence for the 'sandwich-type' $\pi$-complex; rather, they support the formation of an *'anti'*-$\pi$-complex between ethylene and its radical cation, in which the two components are connected by one long (190 pm) bond only. This complex is connected to two different transition states leading to the (rhombic) cyclobutane radical cation or, by 1,3-H-shift, to 1-butene radical cation [167].

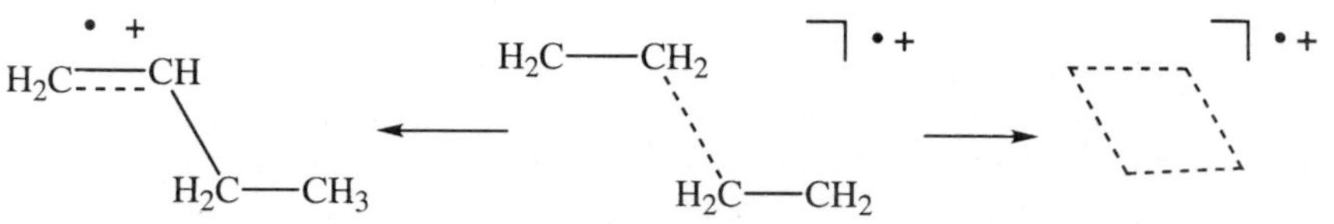

The density functional theory calculations mentioned earlier also modeled the ring-opening of cyclobutane radical cation [87]. This reaction proceeded via a distorted trapezoid transition state with two shorter (142.2, 147.1 pm) and two longer bonds (187.7, 208.4 pm), forming the *'anti'*-complex (one long bond; 227.6 pm), similar to that calculated earlier [168].

As for derivatives of unstrained ring systems, we mention—in passing—the electron-transfer oxidation of various mono-, di-, or trialkylcyclohexanes, which were studied in significant detail. Interestingly, the radical cations of 1-alkyl and 1,1-dialkyl derivatives have been assigned a SOMO resembling the $a_g$ SOMO of the cyclohexane radical cation [169–171]. A more detailed discussion would exceed the scope of this review.

## 2.3.4 Electron-transfer Reactions of Cycloalkanes in Zeolites

Similar to linear and branched alkanes, cycloalkanes also give rise to radical cations in zeolites, spontaneously or upon $\gamma$-radiolysis. This brief discussion of selected examples is intended only to give a flavor of the work being done. Thus, a 13-line radical cation spectrum ($a = 0.17$ mT, $g = 2.003$) obtained upon incorporation of 1-methylcyclohexane, **43**, into zeolites [71] was identified as 1,2-dimethylcyclopentene radical cation, **44**·+ (two sets of protons with hyperfine coupling constants in the ratio of ca 2:1; $a = 1.67$ mT, 2 CH$_3$; $a = 3.42$ mT, 4H) [72]. The formation of **44**·+ was rationalized by protonation of the 3° carbon of **43**, followed by loss of H$_2$. Loss of a proton from a rearranged carbocation may generate **44**, which is oxidized to **44**·+ by a Lewis site.

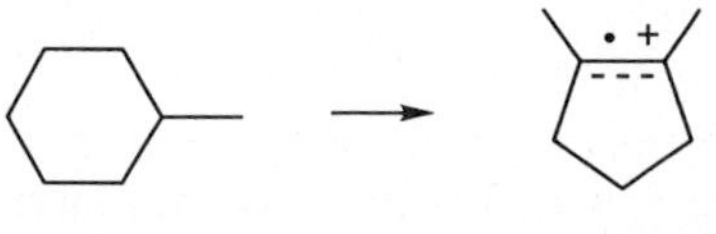

**43**                **44**·+

$\gamma$-Irradiation of cyclohexane on Na-ZSM-5 at 77 K generated the EPR spectrum of cyclohexyl free radical, **45·** [172]. The clear difference between two sets of $\beta$-protons ($a = 2.2$ mT, 1H$_\alpha$; $a = 4.0$ mT, 2H$_{ax}$; $a = 0.5$ mT, 2H$_{eq}$) indicates that at 77 K **45·** is conformationally rigid. A cyclohexane radical cation (Section 2.3.1) is a probable intermediate, but no evidence for this species was observed. The formation of **45·** was ascribed to 'spontaneous' proton transfer to the zeolite matrix, in contrast with previously discussed deprotonations (Section 2.2.5), which proceeded by ion–molecule reactions.

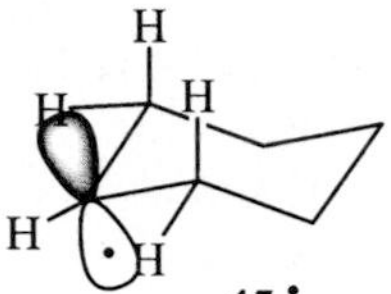

**45·**

$\gamma$-Radiolysis of 1,1,2,2-tetramethylcyclopropane, **21**, in zeolite Na–Y produced an EPR spectrum that was simulated with the known hyperfine coupling parameters of radical cation, **21·+**, with one lengthened cyclopropane bond (Section 2.3.4; $a = 1.87$ mT, 2H; $a = 1.49$ mT, 12H) [173]. In contrast with *trans*-1,2-diphenylcyclopropane (*trans*-**13**), which is spontaneously oxidized and undergoes ring opening and deprotonation (below), **45·+** remains unchanged. Apparently, its bulky shape isolates it from other molecules as well as precluding migration in the channels.

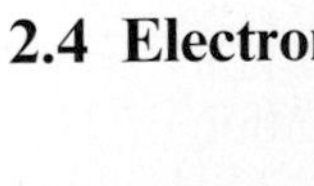

**21·+**

## 2.4  Electron-transfer Reactions of Bicyclic Systems

Many bicycloalkanes are as unreactive as alkanes or unstrained cycloalkanes. They have very high p$K_a$ values and their low electron affinities make them poor electron acceptors. The ionization potentials of unstrained bicycloalkanes are somewhat lower than those of unstrained cycloalkanes—bicyclo[2.2.2]octane (*IP* 9.47 eV), bicyclo[4.3.0]nonane (*IP* 9.46 eV), or bicyclo[4.4.0]decane (*IP* 9.32 eV) lie below the 9.75–9.90 eV range characteristic of cycloalkanes [5]. As a result, they are somewhat easier to oxidize in solution or in solid matrices. Strained bicycloalkanes have even lower ionization potentials (bicyclo[1.1.0]butane, *IP* 8.7 eV); the ring strain present in these systems provides an additional driving force for their oxidation, making them significantly better electron donors. As a result, they readily undergo photo-induced electron transfer to excited electron acceptors in solution. All bicycloalkanes are readily ionized upon electron or $He_\alpha$ impact in the gas phase. Because of the difference in reactivity, we will discuss strained bicycloalkanes separately from those with little or no strain.

### 2.4.1 Electron Transfer of Strained Bicyclic Systems

Bicyclic ring systems containing three or four-membered rings have special bonding forces, which are reflected in their unusual chemistry. Their electron-transfer reactions have been studied in the gas, liquid, and solid phases, and the structure and reactivity of their radical cations have been the target of intense scrutiny.

#### Bicyclobutane radical cations

The unique bonding in bicyclobutane, **46**, and its unusual chemistry resulted in early interest in its radical cation. The structure and reactivity of the parent radical cation and of various derivatives have been studied by ESR–ENDOR spectroscopy and CIDNP effects, by identification of representative reaction products, and by MNDO and ab initio calculations. The structure of the bicyclobutane radical cation was characterized by ESR–ENDOR spectroscopy [174–176]. The ESR spectrum contained a triplet (7.7 mT) of quintets (1.14 mT). These results were interpreted in terms of a radical cation, **46**$^{\cdot+}$, in which the bridgehead carbons bear spin-density and the transannular bond is lengthened. Given this spin-density distribution, the large hyperfine coupling was assigned to the axial protons ($H_{ax}$); the smaller quintet splitting suggests that the remaining two pairs of protons have identical hyperfine couplings. The bridgehead protons have negative coupling constants, because they are adjacent to centers of major spin-density, whereas the axial and equatorial protons ($H_{eq}$) must have positive hyperfine couplings, as a result of '$\pi,\sigma$-delocalization'. The large difference between axial and equatorial hyperfine coupling indicates a non-planar, puckered geometry for the radical cation. No evidence for interconversion is apparent up to 160 K; this finding requires an inversion barrier of at least 12 kJ mol$^{-1}$.

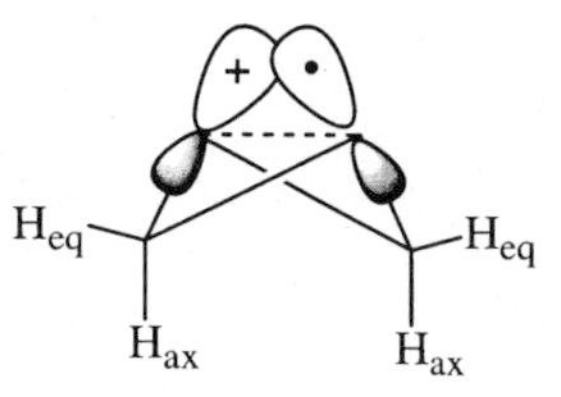

**46** $^{\cdot+}$

MNDO calculations [174, 175] identified the highest occupied MOs (Figure 23). The key feature of the HOMO is a bonding contribution to the transannular link-

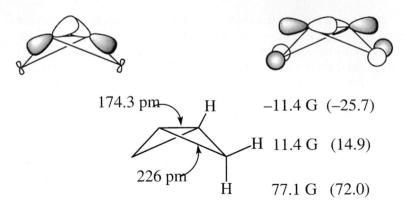

**Figure 23.** Schematic representation of two high lying MOs for bicyclobutane (top) and structure parameters of the bicyclobutane radical cation.

age. Removal of an electron from this HOMO yields radical cation, **46**[·+], with a lengthened transannular bond (178.6 pm, MNDO; 174.3 pm, B3LYP/6-31G*// MP2/6-31G*) and an increased flap angle (132°, MNDO; 133.6°, B3LYP/6-31G*// MP2/6-31G*; Figure 23). Coupling constants calculated at different levels of theory are summarized in Table 5.

The photoreactions of electron acceptors with several bridged bicyclobutane systems, e.g., tricyclo[4.1.0.0$^{2,7}$]heptane, **47**[·+], give rise to CIDNP effects supporting negative hfcs for H1 and H7 and positive hfcs for H2 and H6 and for H3,3' and H5,5'. These results would place electron spin-density at C1 and C7, in agreement with the involvement of the A$_1$ type HOMO (Figure 23) [77, 177]. The axial 'flag-pole' $^1$H nuclei of **46**[·+] are replaced in **47**[·+] by the trimethylene bridge. This structure type has been confirmed by CIDNP results for several derivatives [177, 178].

Although radical cation **47**[·+] is persistent in solution on the nanosecond time-scale, both **47**[·+] and dihydrobenzvalene radical cation, **48**[·+], the lower homolog

**Table 5.** Isotropic hyperfine coupling constants (Gauss) of the bicyclo[1.1.0]butane radical cation.

| Nucleus | Calculation method | | | Experiment[a] |
|---|---|---|---|---|
| | B3LYP | MP2 | MP2/B3LYP | |
| H$_{1,3}$ | −10.88 | −27.04 | −11.6 | −11.4 |
| H$_{2,4ax}$ | 79.83 | 76.35 | 79.28 | 77.1 |
| H$_{2,4eq}$ | 12.98 | 13.76 | 11.65 | 11.4 |

[a] ESR/ENDOR [174–176].

with a dimethylene bridge, undergo ring opening in cryogenic matrices, forming the radical cations of cycloheptadiene and cyclohexadiene, respectively (**49**$^{\cdot+}$; $n = 2, 3$). The identity of the product radical cations was established unambiguously by ENDOR, which gives particularly rich information about the cycloalkadiene radical cations [175, 176]. The formation of **49**$^{\cdot+}$ in cryogenic matrices stands in interesting contrast to the rearrangements of **47**$^{\cdot+}$ and of a benzolog of a dehydro derivative in solution (vide infra).

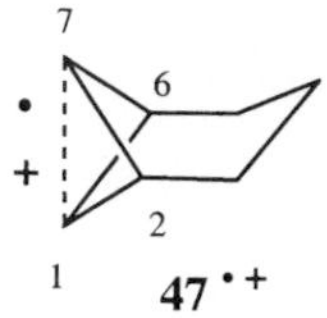

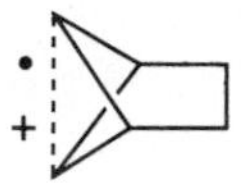

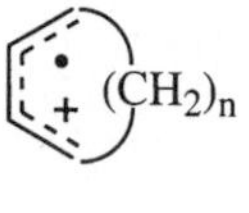

The electron-transfer-induced chemistry of bicyclobutane systems offer a rich variety of reactions. Irradiation of naphthalene in the presence of **46** resulted in rapid fluorescence quenching without rearrangement. In contrast, irradiation with either 1-cyanonaphthalene or 9,10-dicyanoanthracene in solutions containing derivatives of **46** resulted in product formation. The product distribution obtained under electron-transfer conditions is compatible with radical cations of structure type **46**$^{\cdot+}$, which is firmly established by ESR and CIDNP results. Nucleophilic capture of the 1,2,2-trimethyl derivative, **50**$^{\cdot+}$, led to cleavage of the transannular bond. The initial capture is followed by net addition, producing **51**, or dehydrogenation, yielding **52** [179].

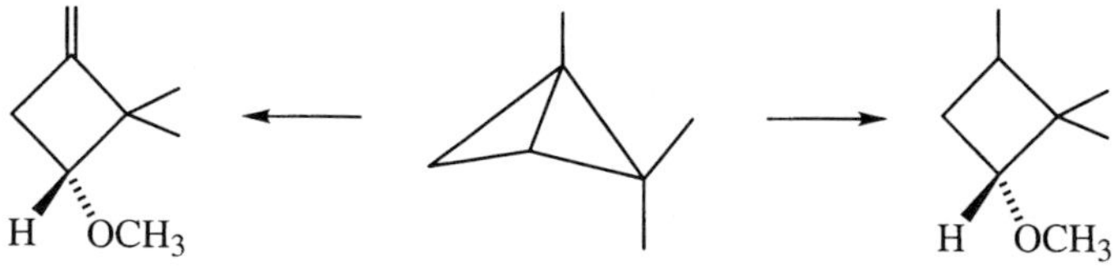

The most thoroughly studied bicyclobutane system is a bridged derivative, tricyclo[4.1.0.0$^{2,7}$]heptane (Moore's hydrocarbon, **47**), which has revealed many facets of radical cation reactivity. Nucleophilic solvents ($CH_3OH$, $H_2O$) or ionic nucleophiles ($CN^-$) capture **47**$^{\cdot+}$, leading to products formally derived by addition across the transannular bond (**53**, **54**, **55**). Significantly, each product is derived by backside attack, as indicated clearly for the monomethyl derivative (**54**, **55**, R = $CH_3$) [180]. In the absence of nucleophiles, a (dimeric) rearrangement product (**57**) was obtained (Scheme 2) [180, 181].

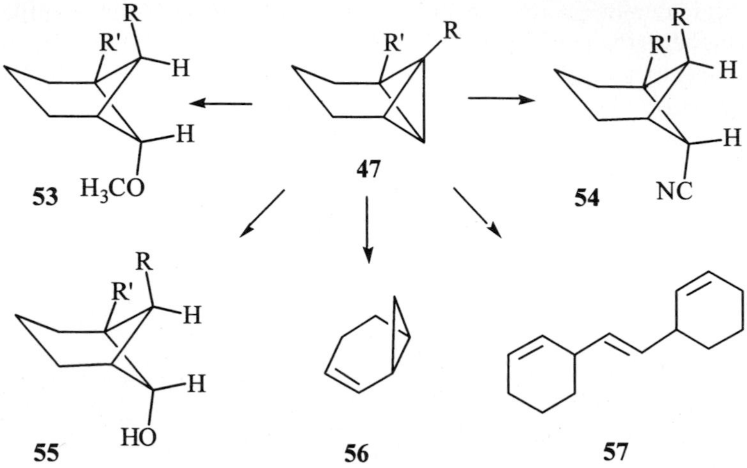

**Scheme 2.**

The bridged bicyclobutane, **47**, and the benzolog, **58**, of a dehydro derivative undergo an interesting rearrangement to norcarene systems; these conversions require a 1,3-hydrogen (hydride) migration or their intermolecular equivalent [177, 178]. The rearrangement products, **56** and **17** are well known and their NMR features are unmistakable. These results establish a striking difference between the rearrangements in cryogenic matrices [175, 176] and in solution [177, 178].

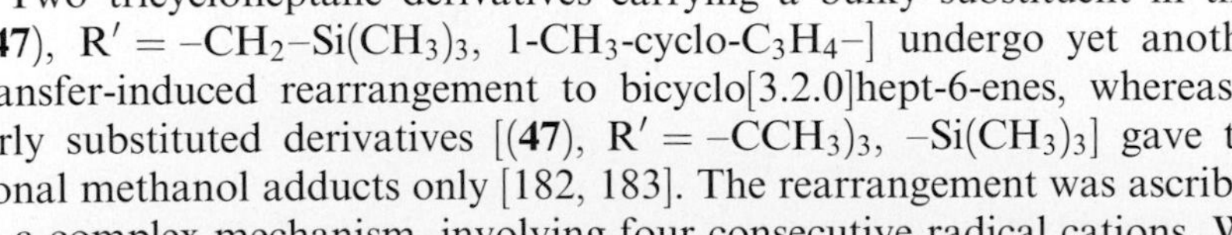

Two tricycloheptane derivatives carrying a bulky substituent in the 2-position [(**47**), $R' = -CH_2-Si(CH_3)_3$, 1-CH$_3$-cyclo-C$_3$H$_4-$] undergo yet another electron-transfer-induced rearrangement to bicyclo[3.2.0]hept-6-enes, whereas other similarly substituted derivatives [(**47**), $R' = -CCH_3)_3$, $-Si(CH_3)_3$] gave the unexceptional methanol adducts only [182, 183]. The rearrangement was ascribed originally to a complex mechanism, involving four consecutive radical cations. We have proposed that this reorganization might be interpreted, by analogy with the conversion of the neutral parent molecule [184], as a conrotatory ring-opening, generating **59**$^{\cdot+}$, followed by conrotatory ring-closure, yielding **60**$^{\cdot+}$ [185, 186]. Alternatively, the key intermediate **59**$^{\cdot+}$ might be trapped by electron-return, generating *cis,trans*-cycloheptadiene, a structure type which has been invoked as an intermediate in several bicyclobutane-to-cyclobutene rearrangements [187]. Although the revised explanation was accepted by the original authors [188], it remains to be established whether orbital symmetry plays a definite role in this radical cation reaction. Altogether, **47**

and derivatives undergo at least three types of rearrangement. Although the identity of the products is established beyond doubt, the mechanistic details remain to be established.

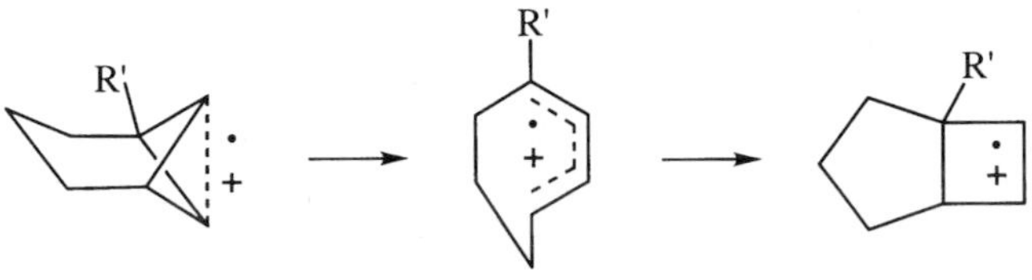

$47^{\cdot +}$          $59^{\cdot +}$          $60^{\cdot +}$

Benzvalene (**61**) is a tricyclic benzene isomer containing a bicyclobutane ring system bridged by an ethene moiety. The radical cation of this system is accessible by photo-induced electron transfer in solution or by radiolysis in cryogenic matrices. A CIDNP study indicated negative hfcs for the olefinic protons ($H_{ol}$), strong positive hfcs for the (non-allylic) bridgehead protons ($H_n$), and negligible hfcs for the allylic bridgehead protons ($H_a$) [189]. These results suggest that benzvalene radical cation has spin and charge essentially localized in the olefinic moiety, although with efficient spin delocalization on to the non-allylic bridgehead $^1H$ nuclei. The strong positive hfcs of the non-allylic bridgehead protons are evidence for a strong hyperconjugative interaction between these protons and the $\pi$-orbitals bearing the unpaired spin. These assignments were confirmed by the ESR spectrum of **61**$^{\cdot +}$, a triplet (+2.79 mT; $H_{ol}$) of triplets (−0.835 mT; $H_n$). A third triplet splitting, from the allylic $\beta$-protons (−0.158 mT; $H_n$), is revealed in the ENDOR spectrum [190]. The small negative value of the coupling constant reflects the position of the $\beta$-protons in the nodal plane of the $\pi$-system. This allows only spin polarization (inefficient for $\beta$-$^1H$-nuclei) as a coupling mechanism.

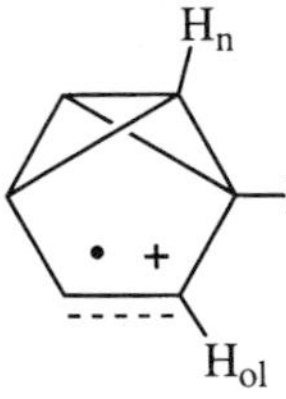

**61**$^{\cdot +}$

The above data are consistent with a radical cation corresponding to the HOMO established by photoelectron spectroscopy and simple theoretical models [191–193]. This assignment is also supported by theoretical calculations [31, 193]. Ab initio molecular orbital calculations at the MP2/6-31G* level support two low-lying radical cationic states of **61**$^{\cdot +}$, $^2B_1$ and $^2A_1$ in $C_{2v}$ symmetry [31]. The hyperfine coupling constants calculated for the $^2B_1$ state (B3LYP/6-31G*//MP2/6-31G*; Figure 24) are fully compatible with the observed CIDNP and ESR/ENDOR results. In contrast, those for the $^2A_1$ state (a single point calculation at the MP2/6-31G*//UHF/6-31G* level) show irreconcilable differences with the experimental findings. Cal-

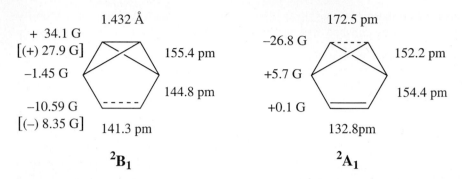

1.432 Å

+ 34.1 G
[(+) 27.9 G]        155.4 pm
−1.45 G
                    144.8 pm
−10.59 G
[(−) 8.35 G]    141.3 pm

$^2\boldsymbol{B_1}$

172.5 pm

−26.8 G        152.2 pm
+5.7 G
                    154.4 pm
+0.1 G
                    132.8pm

$^2\boldsymbol{A_1}$

**Figure 24.** Two structure types of benzvalene radical cation with spin and charge located mainly in the ethylene ($^2\boldsymbol{B_1}$; left) or the bicyclobutane function ($^2\boldsymbol{A_1}$; right). Hyperfine coupling constants calculated for the $^2\boldsymbol{B_1}$ structure (B3LYP/6-31G*//MP2/6-31G*) agree well with the experimental values (in brackets) and signs; in contrast, the $^2\boldsymbol{A_1}$ structure (MP2/6-31G*//UHF/6-31G*) shows irreconcilable differences.

culations at the UMP2/6-31G* level in general show excellent agreement for the positive hyperfine coupling constants but overestimate the negative ones significantly. Clearly, the $^2\boldsymbol{B_1}$ state is the ground state of $\boldsymbol{61}^{\cdot+}$; the bicyclobutane moiety is a 'pendant' group. Although spin-density is efficiently delocalized on to the $\gamma$-$^1$H nuclei, no spin-density is found in the $\gamma$-carbons, and the transannular C–C bond is slightly shortened relative to the parent molecule.

In this context, we mention the formation of a derivative of **61** by rearrangement of 3,3′-dimethylbicyclopropenyl (**62**). The electron-transfer chemistry of bicyclopropenyl was of interest, because its radical cation might be an intermediate in the rearrangement of prismane (vide infra), and also as a potential adduct between cyclopropenium cation and cyclopropenyl radical. For reasons of practicality the 3,3′-dimethyl-derivative, **62**, was chosen for study because it is not subject to prototropic rearrangements.

$\boldsymbol{62}^{\cdot+}$                $\boldsymbol{63}^{\cdot+}$

CIDNP results indicate that the radical cation, $\boldsymbol{62}^{\cdot+}$, is exceedingly short-lived and undergoes selective rearrangement to one dimethylbenzvalene radical cation ($\boldsymbol{63}^{\cdot+}$). This species was identified by its unmistakable chemical shift and polarization pattern. Because the CIDNP effects reflect the spin-density distribution of $\boldsymbol{63}^{\cdot+}$, any intermediate(s) preceding it cannot have lifetimes exceeding (fractions of) nanoseconds [194]. The interesting mechanistic challenge posed by this rearrangement exceeds the scope of this review.

**Bicyclopentane radical cation**

The highly strained bicyclo[2.1.0]pentane, **64**, undergoes electron transfer readily in solution and in cryogenic matrices. Electron transfer to radiolytically generated holes in $CF_3CCl_3$ generated an ESR spectrum with several strongly coupled nuclei, $a_1 = 4.49$ mT (1H), $a_2 = 3.35$ mT (2H), $a_3 = 1.17$ mT (2H). These data, especially the single strongly coupled proton, suggest a puckered conformation for the resulting radical cation, **64·+** [195, 196]. This assignment is unlikely to be compatible with planar radical centers at the bridgehead carbons. Ab initio calculations indicate that the transannular bond of **64·+**, similar to numerous cyclopropane systems, remains bonding and that the bridgehead carbons are still pyramidal [31]. Upon annealing the sample above 90 K, **64·+** rearranges to cyclopentene radical cation [195], a conversion previously believed to be spontaneous [196].

Additional support for the puckered structure of **64·+** comes from the interesting stereochemistry of the hydrogen or methyl migration in *syn*- and *anti*-5-methyl-**64**; the radical cation of *anti*-**65·+** rearranged to 1-methylcyclopentene radical cation, **66·+**, via a H-shift of the strongly coupled 'flagpole' *syn*-5-hydrogen [145]. In contrast, *syn*-**65·+** generated 3-methylcyclopentene, **67**, via a methyl shift, under identical conditions.

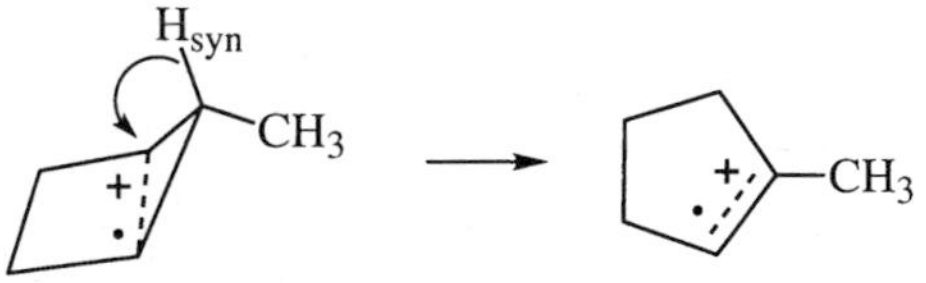

*anti*-**65·+**          **66·+**

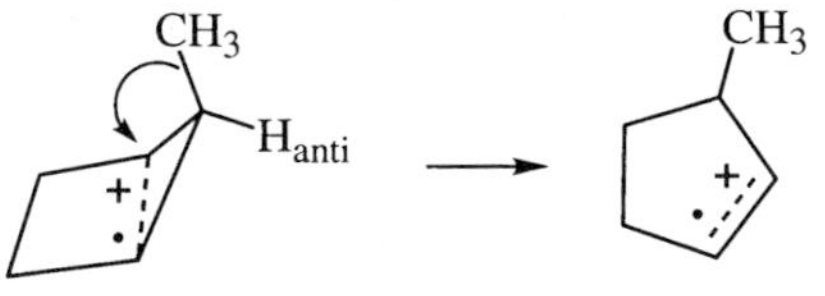

*syn*-**65·+**          **67·+**

Derivatives of **64** with a single bridgehead substituent rearranged with significant regiospecificity. Thus, the 1-methyl derivative, **68a**, yielded **67** exclusively, whereas the 1-phenyl derivative, **68b**, gave mainly 1-phenylcyclopentene, **69**. Radiolysis in cryogenic matrices failed to provide any evidence for the primary radical cations, **68·+**; only the ESR spectra of the cyclopentene radical cations were observed. In these systems, the rearrangement of the bicyclobutane radical cations may, indeed, be spontaneous [197].

These rearrangements have been probed by CASSCF calculations [198]. They show that the rearrangements have two components, bond breaking and atom or group migration. Although these are discussed as two separate steps, there is only

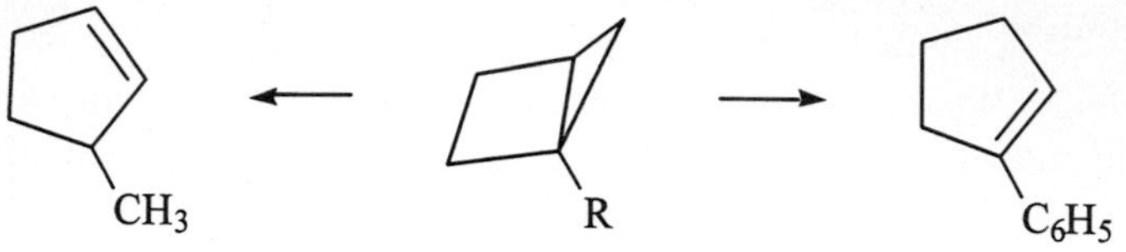

**68**

67

a) R = CH₃

b) R = C₆H₅

69

one minimum after the rate-determining transition state. In this respect, the potential surface of bicyclopentane radical cation is similar to that of cyclopropane radical cation, which also failed to show a minimum after the rate-determining transition state [148]. In view of this analogy, one might consider the rearrangements of *anti*-**65**·⁺ to **66**·⁺ and of *syn*-**65**·⁺ to **67**·⁺ as non-synchronous 1,3-sigmatropic shifts. All experimental results are fully compatible with this proposed assignment.

Radical cations of derivatives of **64**·⁺ are accessible also by irradiation of azobicycloalkanes, **70**. For a detailed description of the rich chemistry of **64** and its derivatives, and of a large family of azobicyclo[2.2.1]alkanes the reader is referred to a recent review [199].

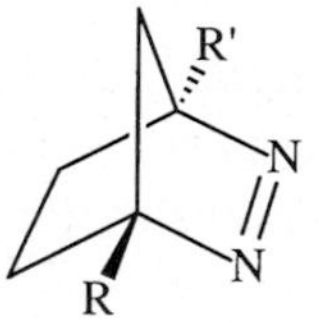

**70**

**Bicyclo[2.2.0]hexane**

Although less strained than bicyclobutane (**46**) and bicyclopentane (**64**), bicyclo[2.2.0]hexane, **71**, undergoes electron transfer not only in cryogenic matrices but also in solution. Electron transfer from this strained ring system to holes generated in halocarbon matrices, $CFCl_3$ or $CF_3CCl_3$, gave rise to an ESR spectrum ($g = 2.0026$; $a = 1.2$ mT), indicating the presence of six equivalent nuclei. This spectrum was identified as that of the ring-opened chair conformer of cyclohexane-1,4-diyl radical cation, **73**·⁺, a species in which the unpaired spin is delocalized between two 2p-orbitals at C1 and C4; two α-protons and four axial β-protons strongly interact with the unpaired spin [145]. No evidence was obtained for either the bicyclo[2.2.0]hexane radical cation, **71**·⁺, or the boat conformer of cyclohexane-1,4-diyl radical cation, **72**·⁺, which is the necessary intermediate, however fleeting, in the ring opening of **71**·⁺. At temperatures above 90 K, the spectrum of **73**·⁺ is replaced by that of cyclohexene radical cation, a conversion readily explained by a 1,3-H shift of an axial hydrogen.

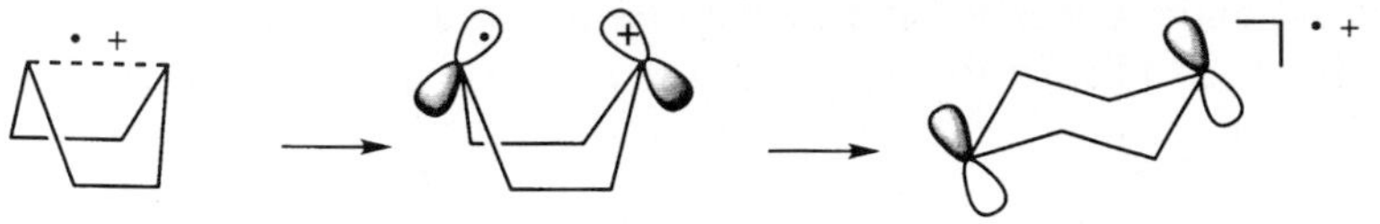

**71**·⁺        **72**·⁺        **73**·⁺

Some chemical evidence for the existence of a boat conformer was obtained in the electron-transfer reaction of hexamethylbicyclo[2.2.0]hexane, **74**. This reaction formed the *erythro*-(*E,E*)-diene, **76**, along with lower yields of hexamethylcyclohexene formed by hydrogen migration [200]. These results are compatible with the initial generation of the 1,4-diyl radical cation, **75**·⁺, which is partitioned between stereospecific ring opening, yielding **76**, and hydrogen migration, generating hexamethylcyclohexene. The cleavage of **75**·⁺ should be favored by the repulsive interaction between the four all-*cis* methyl groups. Of course, these results provide only indirect evidence for the boat conformer, **75**·⁺.

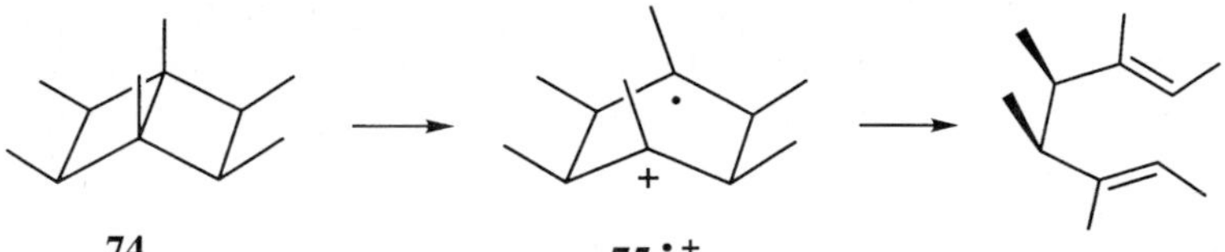

**74**        **75**·⁺        **76**

The 1,4-diyl species, **73**·⁺, is obtained also upon electron transfer from 1,5-cyclohexadiene, **77**, both in cryogenic matrices [143] and in solution [201]. Solution experiments provided chemical evidence for the cyclohexane-1,4-diyl structure type in an elegant study of the electron-transfer-initiated photochemistry of 2,5-diphenylhexa-1,5-diene, **78**, and derivatives in the presence of molecular oxygen [201–203]. The intermediate 1,4-cyclohexanediyls, **79**·⁺, were intercepted by $O_2$; the stereochemistry of the *endo*-peroxide products, **80**, showed that the initial cycloaddition occurred in the same stereospecific manner established for the thermal rearrangement of the neutral parent [204].

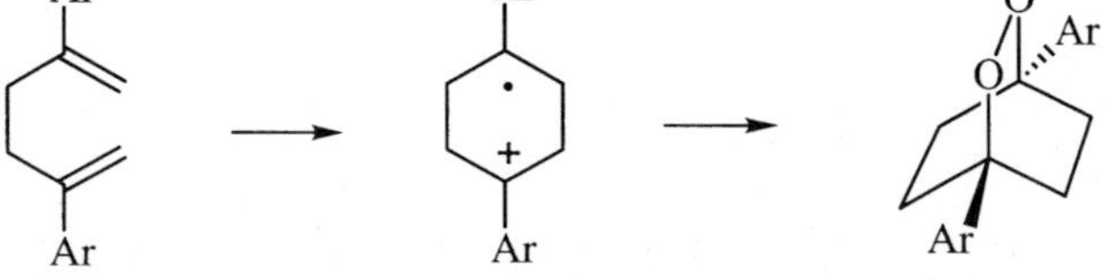

**78**        **79**·⁺        **80**

The formation of **79**·⁺ could be viewed as a stepwise Cope rearrangement, which is 'arrested' after the first (ring-closing) step. Radical cations of different hexadiene systems constitute an interesting family of intermediates, related to the potential mechanistic extremes of the Cope rearrangement. Possible pathways include an associative mechanism (addition precedes cleavage), a dissociative mechanism (cleavage precedes addition) and a concerted mechanism (addition and cleavage proceed in coordinated fashion). Radical cations corresponding to the three mechanistic extremes have been characterized, illustrating remarkable differences between the potential surfaces of radical cations and neutral precursors. On the precursor potential surface, states of intermediate geometry are saddle points (transition structures), whereas they are pronounced minima on the radical cation potential surface. In essence, the parent molecules undergo a concerted Cope rearrangement via a transition structure, whereas the radical cations undergo cycloaddition or cleavage reactions, which are 'arrested' at intermediate geometries [205–214].

It is an interesting question whether intermediates of type $79^{\cdot+}$ cleave to hexa-1,5-diene radical cations, i.e., whether they complete the radical cation Cope rearrangement. The hydrogen shift in cryogenic matrices [143] efficiently competes (and suppresses) any Cope rearrangement. On the other hand, the photoinduced electron-transfer reaction of **78** in polar solvents results in exchange of a deuterium label between the terminal olefinic and the allylic positions. These results are compatible with a cycloreversion [201]. Recent results suggest that the second, ring-opening step does not involve the radical cation, $79^{\cdot+}$, but occurs in a biradical, $79^{\cdot\cdot}$, generated by triplet recombination (Section 2.4.4) [202, 203].

**Bicyclo[*n*.1.0]alkane systems**

Although bicyclo[3.1.0]hexane and bicyclo[4.1.0]heptane are modestly efficient electron donors, the main value of these ring systems lies in the rigid framework they offer for the elucidation of the stereochemical course of radical cation reactions. We will mention several vinylcyclopropane systems and several tricyclic cyclopropane derivatives.

The tricyclanes, **81** and **82**, readily undergo electron transfer in solution to sensitizers in either the excited singlet [215] or triplet states [216]. The resulting radical cations undergo interesting ring-opening substitution reactions in the presence of nucleophiles. For $81^{\cdot+}$, the attack occurred exclusively at the tertiary, rather than quaternary, carbon [215]. The chiral isomer, **82**, has two 3°–4° bonds, either of which might be the site of spin and charge, possibly in an equilibrium. The attack of the nucleophile is less hindered at the carbon further removed from the (neopentyl type) dimethyl-substituted bridge (approach a). The isolated product, **83**, is optically active, and formed by backside attack on the less hindered carbon [216]. These results show that the nucleophilic substitution at the cyclopropane one-electron bond is subject to 'conventional' steric hindrance and does not proceed with 'inverse' steric effects [154].

An interesting electron-transfer reaction was observed for a bicyclo[4.1.0]heptane

**81**              **82** $^{\cdot+}$              **83**

system, **84**, bearing a (3-hydroxybutyl) substituent and a *p*-tolylthio moiety in the 1- and 6-positions, respectively. The compound readily undergoes electron transfer and the resulting radical cation, **84**·⁺, generates the spiro product **85** by regiospecific intramolecular nucleophilic capture [217]. This attack corresponds to a backside 'substitution' of an intramolecular leaving group.

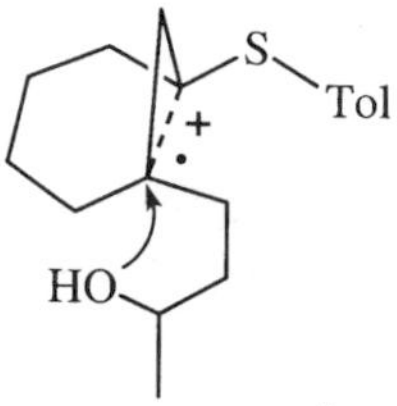

**84**·⁺ 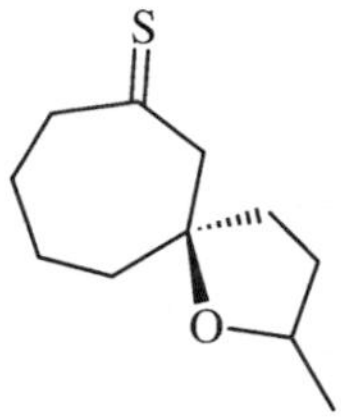           **85**

The CIDNP method was used to probe the hyperfine coupling patterns and structures of several rigid vinylcyclopropane radical cations, in which the two functionalities are locked in either the *syn-* (**86–88**) or the *anti-* (**35**) configuration. Spin polarization effects observed during electron-transfer reactions of these systems suggest significantly different structure types, in which either the internal cyclopropane bond or a lateral one is involved in delocalizing spin and charge [218]. In view of the fixed orientations of the vinyl group relative to the cyclopropane moiety and the different substitution patterns in these substrates, these findings elucidate the electronic and stereochemical requirements for conjugation between the two functionalities. They probe the significance of such factors as orbital overlap and charge stabilization in the radical cations and the release of ring strain in their formation or in their reactions. Three structure types were considered for the radical cations: the well-documented structure with one lengthened C–C bond (type **A**), the structure with two lengthened bonds (type **B**), and the ring-opened bifunctional structure.

The effects observed for bicyclo[3.1.0]hex-2-ene, **86**, were particularly clear-cut, because the ¹H spectrum is fully resolved [218]. The polarization pattern supported a species with spin-density on C3 and C6, indicating the delocalization of spin and charge into the lateral cyclopropane bond. The bicyclohexene system has limited mobility, enabling more significant orbital overlap of the lateral cyclopropane bond with the alkene p-orbitals (**86**·⁺). The participation of the lateral bicyclohexene bond is supported by ab initio calculations, carried to the MP2/6-31G* level of theory. The lateral cyclopropane bond is lengthened (C1–C6 = 1.748 Å), and carbons C3 and C6 carry prominent spin densities, whereas lower spin densities were found at C2 and C1 [220].

CIDNP effects observed for norcarene [147, 178] supported a radical cation, **87**·⁺, which spin and charge are delocalized between the olefinic group and the Walsh orbital of the (more highly substituted) internal cyclopropane bond. The bicycloheptene system, **87**·⁺, appears to be more flexible than **86**·⁺; either the internal or the lateral cyclopropane bond can align with the alkene p-orbitals. Delocalization of spin and charge into the more highly substituted bond is preferred.

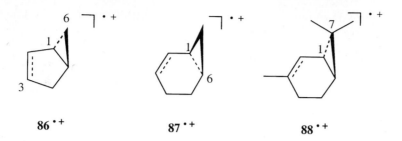

**86**·⁺                    **87**·⁺                    **88**·⁺

The spin polarization effects upon electron transfer from carene suggested a radical cation structure, **88**·⁺, in which spin and charge are delocalized between the olefinic group and the Walsh orbital of the adjacent lateral cyclopropane bond. The change in structure (**87**·⁺ compared with **88**·⁺), caused by the geminal methyl groups at C7, reflects stabilization of the radical cation as a result of hyperconjugative interaction with the methyl groups. For both **87**·⁺ and **88**·⁺, the more highly substituted cyclopropane bonds are involved in delocalizing spin and charge.

4-Methylene-1-isopropylbicyclo[3.1.0]hexane (sabinene, **34**) contains an alkene and a cyclopropane function locked in the *anti* orientation. CIDNP effects for **34** support a radical cation, in which the electron spin is delocalized between the olefinic π-system and the internal cyclopropane bond. The results are compatible with a vinylcyclopropane radical cation, **34**·⁺, with one weakened cyclopropane bond, which has retained the steric integrity of the parent molecule [219].

**34**·⁺

### Systems containing two adjacent cyclopropane rings

Some tri- and tetracyclic alkanes contain two cyclopropane moieties locked into a fixed orientation. These arrangements might give rise to interesting stereoelectronic effects in the interaction between the cyclopropane rings. For example, CIDNP effects during electron-transfer reactions of *syn-* and *anti*-tricyclo[5.1.0.0²,⁴]octane (*syn-*, *anti*-**89**) indicate significantly different structures for their radical cations.

*syn*-**89**                    *anti*-**89**

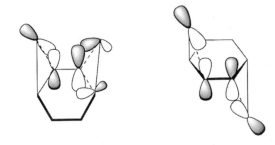

**Figure 25.** Schematic representation of cyclopropane MOs combining to form the SOMOs of *syn*- (left) and *anti*-tricyclo [5.1.0.0$^{2,4}$]octane (right) radical cations.

The polarization pattern for *anti*-**89** fits a structure, *anti*-**89**$^{\cdot+}$, in which spin and charge are localized in the lateral bonds (C1–C8, C2–C3; cf., Figure 25). The less substituted lateral bonds are involved because the orientation of the cyclopropane rings enables overlap between the Walsh orbitals of two lateral cyclopropane bonds [220]. Ab initio calculations (B3LYP/6-31G*//MP2/6-31G*) support this assignment. The carbons of the lateral cyclopropane bonds, C1,2 and C3,8, bear most of the spin-density ($\rho_{1,2} = 0.192$; $\rho_{3,8} = 0.322$) and the 3° cyclopropane protons (H4,7) show large positive hyperfine couplings ($a = 2.0$ mT). The geminal protons, H3,8s and H3,8a ($a_s = -0.72$ mT; $a_a = -0.05$ mT) have significantly different hyperfine coupling constants. The structure suggested for *anti*-**89**$^{\cdot+}$ has precedent in that of **86**$^{\cdot+}$.

The CIDNP spectrum of *syn*-**89** supported a quite different, although not obvious, structure. Ab initio calculations indicated a structure with dissimilar cyclopropane fragments and 12 distinct $^1$H nuclei. A lengthened internal bond (C1–C7) and spin densities $\rho_1 = 0.2$ and $\rho_7 = 0.31$ support a type **A** structure for one cyclopropane ring whereas lengthened C2–C3 and C2–C4 bonds and spin densities $\rho_2 = 0.24$, $\rho_3 = 0.15$, and $\rho_4 = 0.17$ suggest a type **B** structure for the second cyclopropane ring (Figure 25). The barrier between the structure shown and its mirror image is very low so rapid equilibrium between two equivalent structures is probable.

No discussion of cyclopropane radical cations would be complete without reference to the quadricyclane radical cation, **90**$^{\cdot+}$, and its valence isomer, the norbornadiene radical cation, **91**$^{\cdot+}$. These valence isomers have attracted considerable attention; one features two rigidly arranged adjacent cyclopropane rings, whereas the other contains two ethene $\pi$ systems uniquely suited to the probing of through-space interactions [221–225]. In addition, the potential energy surface of these radical cations poses the interesting question of whether there are two discrete minima or a single minimum, accessible upon oxidation of either parent. Although other methods had failed to provide evidence for more than one intermediate [225], the CIDNP technique furnished clear-cut evidence for the existence of two distinct species with characteristic spin-density distributions [226, 227]. This result is compatible with molecular orbital considerations, which suggest the antisymmetric combination of two ethene $\pi$ orbitals or cyclopropane Walsh orbitals as respective HOMOs of the two parent molecules (Figure 26). The radical ions have different state symmetries and their SOMOs have different orbital symmetries.

The structures of these ions (Table 6) rest on detailed CIDNP spectra delineating the hyperfine patterns of both ions [226, 227], ab initio calculations on both ions

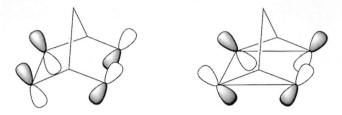

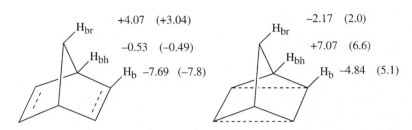

**Figure 26.** Schematic representation of SOMOs of bicyclo[2.2.1]heptadiene (**91**) and quadricyclane (**90**) and calculated hyperfine coupling constants (Gauss; B3LYP/6-31G*//MP2/6-31G*) [31]. Experimental values [230, 231] are given in parentheses.

[31, 228], and ESR and ENDOR data for norbornadiene [229, 230] and quadricyclane radical cations [231]. Early calculations at the UMP2/6-31G* level [228] showed excellent agreement for the positive hyperfine coupling constants but overestimated the negative ones significantly. More recent calculations, using the MP2 structure and calculating spin densities and hyperfine coupling constants with density functional methods (B3LYP/6-31G*//UMP2/6-31G*), produce plausible values for both positive and negative hyperfine interactions [31]. The CIDNP results indicate the absolute signs and relative magnitude of the hyperfine coupling constants; comparison with the calculated values (Table 6) shows satisfactory agreement [228].

**Table 6.** Hyperfine coupling constants (Gauss) of bicyclo[2.2.1]heptadiene and quadricyclane radical cations.

| Species | Quadricyclane | | | Norbornadiene | | | |
|---|---|---|---|---|---|---|---|
| | CIDNP[a] | Calc[b] | ESR[c] | CIDNP[a] | Calc[b] | ESR[d] | ESR[e] |
| $H_{ol}$ | −1 | −4.84 | −5.1 | −1 | −7.69 | −7.8 | 8 |
| $H_{bh}$ | +1 | +7.07 | +6.6 | −vs | −0.53 | −0.49 | |
| $H_b$ | −m | 2.17 | −2.0 | +s | +4.07 | +3.04 | 3.3 |

[a] According to [226, 228]; l = large, m = medium, s = small, vs = very small.
[b] B3LYP/6-31G*//MP2/6-31G* [31].
[c] According to [231].
[d] ESR/ENDOR [230].
[e] According to [229].

The calculations indicate that each radical cation is related uniquely to the geometry of one of the precursors.

For both species the unpaired spin-density resides on four equivalent carbons; $^1H$ nuclei attached to these centers ('$H_{ol}$') have sizeable hfcs, because of the familiar $\pi,\sigma$ spin polarization (cf. Figure 1a). The CIDNP effects of the bridgehead ($\beta$) protons ('$H_{bh}$') indicate major differences—a sizeable positive hfc of $90^{\cdot+}$ can be ascribed to hyperconjugation ($\pi,\sigma$ spin delocalization; cf. Figure 1b); in contrast only very weak and negative hyperfine coupling constants were observed for $91^{\cdot+}$. Because the $\beta$ protons of $91^{\cdot+}$ lie in the nodal plane of its SOMO, the hyperconjugative interaction is inefficient; the observed sign of the $\beta$ protons was, therefore ascribed to 'residual' $\pi,\sigma$ polarization [230]. This type of interaction is usually obscured by the typically much stronger hyperconjugative interaction.

The CIDNP effects for the protons of the methylene bridge ('$H_b$') again are quite different, suggesting a sizeable positive hyperfine coupling for $91^{\cdot+}$ and an even larger negative one for $90^{\cdot+}$. The positive sign for the $\gamma$ hyperfine coupling of $91^{\cdot+}$ can be ascribed reasonably to a 'long-range' $\pi,\sigma$ spin delocalization which is aided by an approximate W arrangement of the $\gamma$ C–H bond relative to the p-orbitals at the olefinic carbons. The bridge ($\gamma$) protons of $90^{\cdot+}$ lie in the nodal plane of the SOMO, rendering the $\pi,\sigma$ spin delocalization mechanism inefficient. The relatively large negative hyperfine coupling indicated by the CIDNP results might suggest a $\sigma,\sigma$ polarization mechanism operating between the bridgehead ($\beta$) and bridge ($\gamma$) protons. The dihedral angle ($H-C\beta-C\gamma-H \approx 60°$) is compatible with a sizeable interaction. This interesting assignment was confirmed independently by an ESR study [231].

Another system with two adjacent cyclopropane rings is prismane, **92**, the highly strained quadricyclic isomer of benzene first considered by Ladenburg [232]. Its radical cation has proved elusive. Even hexamethylprismane, **93**, is rapidly rearranged by chloranil in polar solvents at room temperature. CIDNP experiments provided tentative evidence for the fleeting existence of $93^{\cdot+}$ during the electron-transfer reaction of **93** with anthraquinone. The prismane polarization is very weak; apparently, only a small fraction of $93^{\cdot+}$ reverts back to **93**. The predominant reaction is ring opening to hexamethyl-Dewar-benzene, **94** [233]. In essence, **94** is related to **93** as **91** is related to **90**. Because of the high symmetry of **93**, the CIDNP results do not offer clues about the structure of the intermediate; the polarization observed for **94** implicates a radical cation in which spin and charge is located on four equivalent carbons.

**92**, R = H

**93**, R = $CH_3$

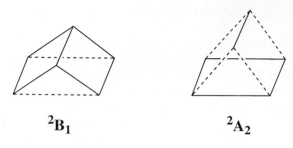

$^2\mathbf{B}_1$    $^2\mathbf{A}_2$

**Figure 27.** Schematic representation of two radical cation structures derived from prismane.

Ab initio calculations for $\mathbf{92}^{\cdot+}$ support the existence of $\mathbf{93}^{\cdot+}$ and help to delineate its structure. The highest occupied molecular orbital in prismane is $e''$, which has significant bonding character in the cyclopropane bonds and antibonding character in the transannular bonds connecting them. The next lower orbital ($e'$) has bonding character mainly in the three transannular bonds. Removal of an electron from the $e''$ orbital gives rise to a $^2E''$ state which undergoes Jahn–Teller distortion, yielding a $^2A_2$ and a $^2B_1$ state (in $C_{2v}$ symmetry).

Calculations (MP2/6-31 G) identified the $^2B_1$ state as lowest in energy. Two of its cyclopropane bonds are lengthened to 1.73 Å whereas the transannular bonds are shortened to 1.48 Å (Figure 27). These changes reflect the bonding pattern in the $e''$ orbital. The $^2A_2$ state lies close in energy to the $^2B_1$ state; four equivalent cyclopropane bonds are stretched to 1.63 Å and one transannular bond is shortened to 1.46 Å. This state has an imaginary $b_1$ frequency that converts it to the $^2B_1$ state through $C_s$ pathways. In essence, the $^2A_2$ state is a transition state in the pseudo-rotation interconverting two equivalent $^2B_1$ states [233].

The lowest state of prismane ($^2B_1$) radical cation lies 16 kcal mol$^{-1}$ above the $^2B_2$ state of the Dewar benzene radical cation (at the MP2/6-31G* level); the energy difference is considerably less than that between the ground states of the corresponding neutral systems (37 kcal mol$^{-1}$) [234]. The ground electronic states of prismane and Dewar benzene ions do not correlate; their interconversion is forbidden from both state-symmetry and orbital-symmetry considerations. Nevertheless, the CIDNP experiments indicate that the actual barrier is quite small [233].

The prismane system exemplifies several characteristics of strained-ring radical cations, including the low barrier to isomerization and the fact that its ion apparently has less strain energy than the parent molecule. Additional investigations of $\mathbf{92}^{\cdot+}$ or even $\mathbf{93}^{\cdot+}$ will be difficult, because of its kinetic instability and its high symmetry (one might expect a 13-line ESR spectrum showing an unexceptional coupling constant near 1.0 mT),

### 2.4.2 Electron Transfer of Unstrained Bicyclic Systems

In contrast with the highly strained bicyclic hydrocarbons containing three- and four-membered rings, the radical cations of bicyclic hydrocarbons containing five-membered or larger rings undergo electron transfer only reluctantly. As for simple

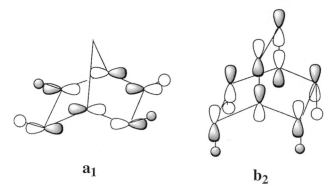

$\mathbf{a_1}$            $\mathbf{b_2}$

**Figure 28.** Schematic representation of SOMOs for bicyclo[2.2.1]heptane, **95**, and bicyclo[2.2.2] octane, **96**.

alkanes or cycloalkanes and their derivatives, electron transfer can be achieved in halogen-containing matrices upon radiolysis. Few systems are selected here; they were chosen to give a flavor of the research that has been performed. Among the species studied we mention bicyclo[2.2.1]heptane, **95**, and bicyclo[2.2.2]octane, **96**. Radiolysis of **95** generated the corresponding radical cation, **95**$^{\cdot+}$. The ESR spectrum contained a five-line pattern, $a_H = 6.5$ mT, indicating the presence of four equivalent strongly coupled protons. The spectrum was interpreted in terms of a structure in which the unpaired electron spin is delocalized over the four *exo*-C–H bonds (Figure 28) [229, 235].

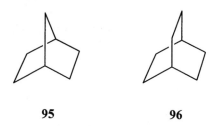

**95**             **96**

Radiolysis of **96** resulted in an interesting, temperature-dependent spectrum. At 4 K, the species is Jahn–Teller-active and exhibits a static distortion from $D_{3h}$ to $C_{2v}$ symmetry. In contrast to the bicyclo[2.2.1]heptane radical cation, the SOMO of **96**$^{\cdot+}$ involves four *endo*-C–H bonds, $a_H = 3.8$ mT. At 77 K in perfluorocyclohexane or Freon-113, the radical cation is dynamically averaged, with splitting from 12 equivalent protons [235].

We note that the simple symmetrical structures might be an over-simplified or idealized view of the structures of **95**$^{\cdot+}$, at least. More recent detailed investigations in other matrices and using line-shape analysis furnished evidence for less symmetrical species with several modes of distortion. For details, the reader is referred to a recent detailed description [236].

Two bicycloalkane isomers, *cis*- and *trans*-bicyclo[4.4.0]decane (decalin; *cis*-, *trans*-**97**) have been the focus of investigation by both radiolysis [237] and photolysis [238]. Radiolysis of *cis*-bicyclo[4.4.0]decane also gave rise to a five-line spectrum with $a_H = 5.0$ mT. Most of the spin-density is confined to the central C–C bond [237, 238]; the four protons in *trans* positions are strongly coupled [239]. Interestingly, significant matrix dependence was observed for the hyperfine pattern of the *trans* radical cation (*trans*-**97**$^{\cdot+}$) in halocarbon matrices [239]. Radiolysis of *cis*- and *trans*-**97** in several synthetic zeolites also gave rise to significantly different ESR spectra. Depending on the nature of the zeolite host and the temperature, the hyperfine splitting of the resulting spectra showed variations of a magnitude that supported the stabilization of two different structure types ('electronic states') for both *cis*- and *trans*-**97**$^{\cdot+}$ [240]. In some zeolites a single structure was observed whereas in others both structures were found.

*trans*-**97**                    *cis*-**97**

For *cis*-**97**$^{\cdot+}$, the authors discussed two radical cation structures, one corresponding to the $^2A_1$ state ($a = 4.95$ mT, 4H, in silicalite, 45 K), the other corresponding to the $^2A_2$ state ($a = 2.8$ mT, 4H, in silicalite, 95 K). Calculations predict the $^2A_2$ state to be higher in energy. For *trans*-**97**$^{\cdot+}$, they assigned radical cation structures corresponding to a (lower-energy) $^2A_g$ state, ($a = 5.05$ mT, in silicalite) and a higher-energy $^2B_g$ state ($a = 2.85$ mT, in Na–Y), respectively. The spectra contain five-line patterns, because of coupling with four equatorial protons, either those at the $\alpha$-carbons (adjacent to the transannular bond; $^2A_1$, $^2A_g$), or those at the $\beta$-carbons, one C–C bond further removed ($^2A_2$, $^2B_g$) [240]. Even more unusual is the interpretation of a nine-line spectrum ($a = 3.0$ mT) of *cis*-**97**$^{\cdot+}$ in offretite, which was interpreted in terms of a fast equilibrium between structures corresponding to the $^2A_1$ and $^2A_2$, states, a most interesting assignment.

### 2.4.3  Electron-transfer Chemistry of Bicyclic and Higher Ring Systems

Many radical cations derived from cyclopropane (or cyclobutane) systems undergo bond formation with nucleophiles, typically neutralizing the positive charge and generating addition products via free-radical intermediates [140, 147]. In one sense, these reactions are akin to the well known nucleophilic capture of carbocations, which is the second step of nucleophilic substitution via an $S_N1$ mechanism. The capture of cyclopropane radical cations has the special feature that an sp$^2$-hybridized carbon center serves as an (intramolecular) leaving group, which changes the reaction, in essence, to a second-order substitution. Whereas the $S_N1$ reaction involves two electrons and an empty p-orbital and the $S_N2$ reaction occurs with redistribution of four electrons, the related radical cation reaction involves three electrons.

The high regio- and stereoselectivity in the nucleophilic capture of the 'one-electron bonds' of **30**$^{\cdot+}$ and **34**$^{\cdot+}$ (above) was interpreted as significant because it indicated that the radical cations had retained the three-dimensional integrity of the parent molecules. Clearly, a significant degree of bonding is maintained in the three-membered ring. These results also established the second-order nature of nucleo-philic capture. Further, because the substitution reaction occurred even at cyclo-propane sites carrying bulky alkyl groups [154, 155], an early transition state was indicated for the nucleophilic substitution–ring opening on cyclopropane radical cations.

In this section, we consider the electron-transfer photochemistry of bi-, tri-, and quadricyclic cyclopropane systems to illuminate the general factors affecting the stereo- and regiochemistry of radical cation nucleophilic capture. Experimental results were interpreted in support of several governing factors. These include: (i) the spin and charge density distribution in the radical cation (the educt); (ii) the extent of conjugation in the educt and the free-radical formed (the product); (iii) the release of ring strain upon forming the product; (iv) steric factors; and (v) the selectivity (reactivity) of the nucleophile.

The spin and charge density distribution of the radical cation is an obvious (perhaps trivial) consideration, because it delineates the singly occupied molecular orbital (SOMO), which must be involved in the reaction. For example, the regio-selective capture of **30**$^{\cdot+}$ and **34**$^{\cdot+}$ occurs at positions of high spin and charge density. The reduced regioselectivity for benzobicyclo[3.1.0]hexene, **98** (comparable yields of **99** and **100**), suggests that both C5 and C6 have significant spin-density [241]. These results are compatible with a radical cation in which the Walsh orbitals of both the 3°–3° and the 3°–2° cyclopropane bonds overlap the benzene $\pi$ system [241].

**98**$^{\cdot+}$        **99**        **100**

The thermodynamic changes in the nucleophilic capture, i.e., the extent of con-jugation in radical cation educt and free-radical product and the release of ring strain in forming the product are also of obvious importance. For example, **34**$^{\cdot+}$ generates a conjugated radical cation, **36**$^{\cdot+}$, via a sigmatropic shift, and forms an allyl radical, **35**$^{\cdot}$, upon nucleophilic attack [156]. Both reactions form fully con-jugated 'products' with full relief of ring strain. Similarly, nucleophilic attack on 1-aryl-2-alkylcyclopropanes, **30**$^{\cdot+}$, forms benzyl radicals [154, 155]. The high re-gioselectivity in the nucleophilic capture of the 'one-electron bonds' of **30**$^{\cdot+}$ or **34**$^{\cdot+}$ reflects the unfavorable energetics for the formation of alternative products. Thus, attack at the benzylic position of **30**$^{\cdot+}$ or the *exo*-methylene position of **34**$^{\cdot+}$ (yielding **101**$^{\cdot}$) is energetically disfavored. Similarly, the nucleophilic attack on **98**$^{\cdot+}$ is regio-random, because it generates two benzyl radicals of comparable stability.

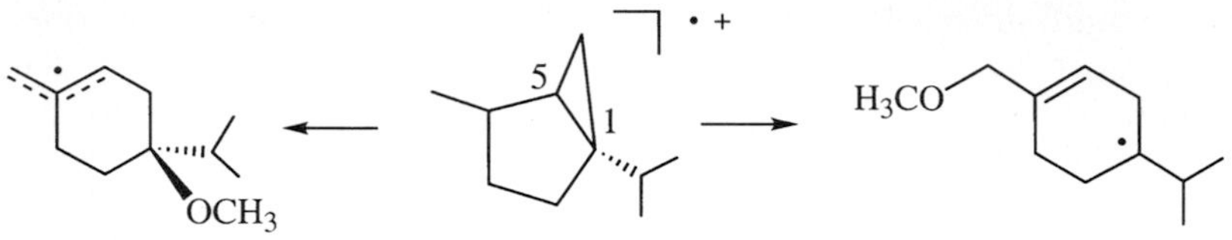

**35**·          **34**·+          **101**·

The nucleophilic addition of methanol to quadricyclane radical cation, **90**·+, produced two methanol adducts, **102**, having a 3-*exo*-methoxy group, and **103**, bearing a 7-*anti*-methoxy group. The stereochemistry of the methoxy groups in these structures identified the direction of nucleophilic attack upon **90**·+ as exclusively from the *exo* position [242]. It can be viewed as a backside attack of the nucleophile on the weakened cyclopropane bond with inversion of configuration. 7-Methylenequadricyclane also was attacked exclusively from the *exo* face [243].

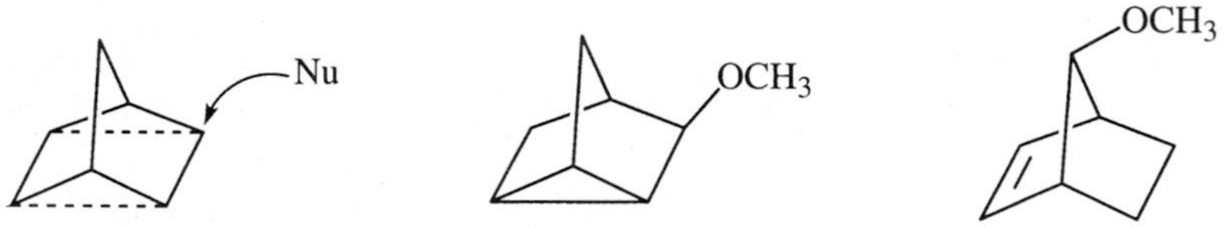

**90**·+          **102**          **103**

The regiospecific intramolecular capture of the internal cyclopropane bond of the bicyclo[4.1.0]heptane radical cation, **84**·+, by the pendant hydroxyalkyl group also amounts to backside 'substitution' of an intramolecular leaving group [217].

Steric factors have been mentioned; they are not expected to play a major role. In fact, several radical cations have been captured by attack on highly congested centers [154, 156]. The ring-opening substitution of tricyclane radical cation, **81**·+, occurred exclusively at the tertiary carbon [215] whereas that of **82**·+ occurred at the tertiary carbon further removed from the dimethyl-substituted bridge, i.e., at the less hindered of two tertiary carbons, not at the quaternary carbon [216]. These results clearly show that the nucleophilic substitution at the cyclopropane one-electron bond is subject to 'conventional' steric hindrance and is not subject to 'inverse' steric effects [154].

The selectivity (reactivity) of the nucleophile may play a role. Methanol seems to be a selective reagent, as it captures β-phellandrene radical cation, **36**·+, with high regiospecificity at the *exo*-methylene carbon, and sabinene radical cation, **35**·+, exclusively at the quaternary carbon [156].

Significant insights into the nature of the nucleophilic capture of radial cations is provided by the regiochemistry of the attack on the bridged norcaradiene radical cation, **18**·+. The products suggested regiospecific attack of methanol on **18**·+ with capture at C2 and C5 generating **104**· and **105**·. The attack occurs with limited stereoselectivity, because products derived from **104**· and **105**· were formed in comparable yields [131].

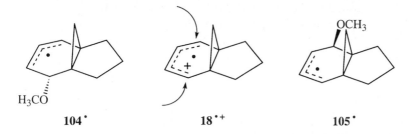

104 •          18 •+          105 •

Interestingly, no products were derived by nucleophilic attack on the cyclopropane ring of **18•+**, even though ab initio calculations suggest that spin and charge are delocalized on to the strained ring, notably on to C10 (Section 2.3.2) [131]. Contrary to ample precedent suggesting release of ring strain is important in radical cation reactions [154–156], particularly when leading to delocalized free-radicals or radical cations, **18•+** failed to generate products implying nucleophilic attack at the cyclopropane ring. While the failure of the nucleophile to attack at C1/C6 can be explained by the fact that the bridgehead carbons are nodal centers, attack at C10 remains feasible.

To eliminate unforeseen energetic issues, the relative energies of the four methoxy-substituted free radicals were confirmed by calculations on the free radicals, **106•**–**109•**, derived from the truncated parent system, **16•+**. The results (UHF/6-31G*) confirmed the expected ordering of energies (relative to the *anti*-methoxynorcarenyl radical, **106•**). The bicyclic radicals, **106•** and **107•**, retaining the strained ring lie significantly above the monocyclic ring-opened radicals, methoxycycloheptadienyl, **108•**, and methoxymethylcyclohexadienyl, **109•** (Scheme 3).

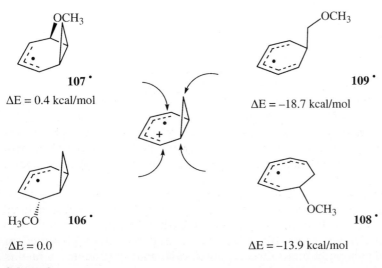

107 •
ΔE = 0.4 kcal/mol

109 •
ΔE = −18.7 kcal/mol

106 •
ΔE = 0.0

108 •
ΔE = −13.9 kcal/mol

**Scheme 3.**

Because the factors considered above cannot explain the failure of **18**·+ to undergo nucleophilic capture at the strained ring, we sought the key to the reactivity in the nature of its orbitals which might be involved in the reaction, i.e. its SOMO and LUMO. This approach has precedent in several theoretical treatments. Pross probed the capture of radical cations by nucleophiles with curve-crossing methodologies. The excited state involved in the curve crossing requires 'double excitation' [244]. Later, Shaik and Pross showed that the excitation energy could be small and the resulting barrier low [245]. This prediction was confirmed by results of Eberson and coworkers; nucleophiles attack the dibenzofuran radical cation at the site of the highest LUMO coefficient, which coincides with the site of highest spin-density in the dibenzofuran triplet state [246–248]. Similarly, Shaik and collaborators explained the well-documented stereochemical course of nucleophilic displacement of a $\sigma$ bond (with inversion of configuration) [156, 180, 181, 242, 243, 249–253] by involvement of the $\sigma^*$ orbital (the LUMO) of the weakened bond [254, 255].

The bridged norcaradiene, **18**·+, is well suited to probing the orbitals involved in the nucleophilic capture of radical cations. This becomes evident when considering the combined effects of free energy and molecular orbital contributions. We have mentioned the significant free energy differences between attack at the strained ring and addition to the diene moiety (cf., Scheme 3). Perhaps more importantly, the two MOs potentially involved in the reaction have a substantially different distribution of orbital coefficients. The SOMOs of **16**·+ and **18**·+ have large orbital coefficients at C2,5 and C7 (C10), which are reflected in the hyperfine pattern of these species (cf., Figure 29, bottom). In contrast, the principal orbital coefficients of the LUMOs are located at C2,5 and C3,4 and the orbital lobes at C7 (C10) offer no target for attack by the nucleophile (Figure 29, top). Because the orbital coefficients of SOMO and LUMO differ, the norcaradiene system will elucidate whether the regioselectivity of nucleophilic capture is governed by the SOMO, or by the nature and topology of the LUMO. The products derived from **18**·+ are clearly formed by attack at a center where both SOMO and LUMO have significant orbital coefficients. This assignment seems persuasive, because the reaction of **18**·+ fails to follow the direction of a significant advantage in driving force. For the norcaradiene system, molecular orbital arguments and free energy considerations predict different regioselectivities. Obviously, the molecular orbital requirements outweigh the significant free energy differences, because of the relief of ring strain and extended conjugation.

The product-determining involvement of the LUMO can also explain the regioselective capture of many other radical cations, including the nucleophilic attack on 1-aryl-2-alkylcyclopropanes, **30**·+. The high regioselectivity in the nucleophilic capture of the 'one-electron bonds' of **30**·+ was originally cited as evidence of inverse steric effects [154, 155]. However, the SOMO and LUMO of disubstituted cyclopropane radical cations (e.g. 1,2-dimethylcyclopropane; Figure 30) clearly suggest that the observed regioselectivity reflects electronic rather than steric factors. Capture at the unsubstituted cyclopropane carbon cannot be expected, because neither SOMO nor LUMO have orbital coefficients at that carbon [31].

A detailed discussion of the reactivity of organic radical ions is presented in the contribution of Schmittel and Ghorai (Volume II, Part 1, Chapter 1 in this handbook).

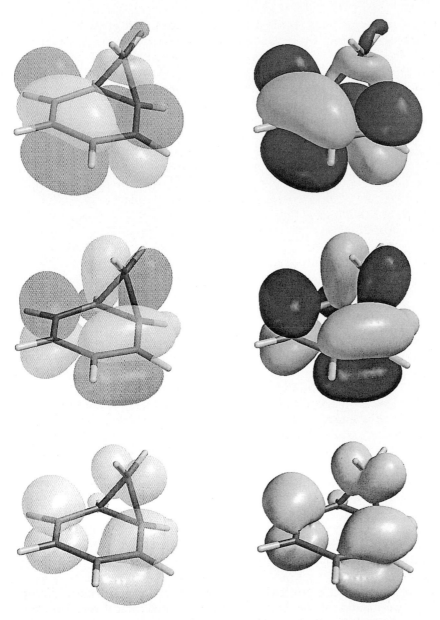

**Figure 29.** Pictorial representation (calculated by Spartan) of the spin density distribution of nor-caradiene radical cation, **18**$^{\cdot+}$, (bottom), and its SOMO (center) and LUMO (top).

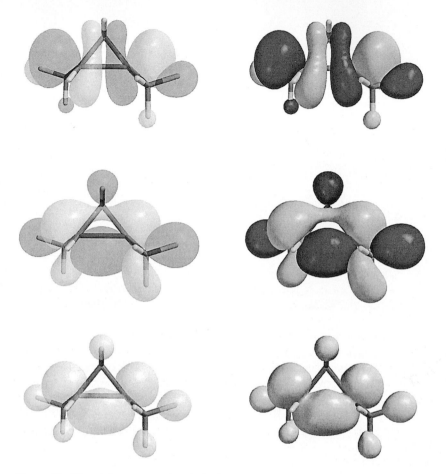

**Figure 30.** Pictorial representation (calculated by Spartan) of the spin density distribution of *trans*-1,2-dimethylcyclopropane radical cation (bottom), and its SOMO (center) and LUMO (top).

### 2.4.4 Triplet Recombination of Radical Ion Pairs

In the context of the potential Cope rearrangement of hexa-1,5-diene radical cations (Section 2.4.1), we mentioned the triplet recombination of radical ion pairs generating a biradical [202, 203]. Because of continuing interest in this type of reaction we briefly mention two additional examples involving radical cationic systems discussed in this review, viz., the isomeric 1,2-diphenylcyclopropane radical cations, *cis*- and *trans*-**13**$^{\cdot+}$, and norbornadiene radical cation, **91**$^{\cdot+}$.

The steric integrity of *cis*- and *trans*-**13**$^{\cdot+}$ is established clearly; it is a key argument for their ring-closed structure [223]. *cis*- and *trans*-**13**, however, undergo geometric isomerization when reacted with photo-excited singlet acceptors, such as 1,4-dicyanonaphthalene. The electron-transfer-induced conversion was ascribed to

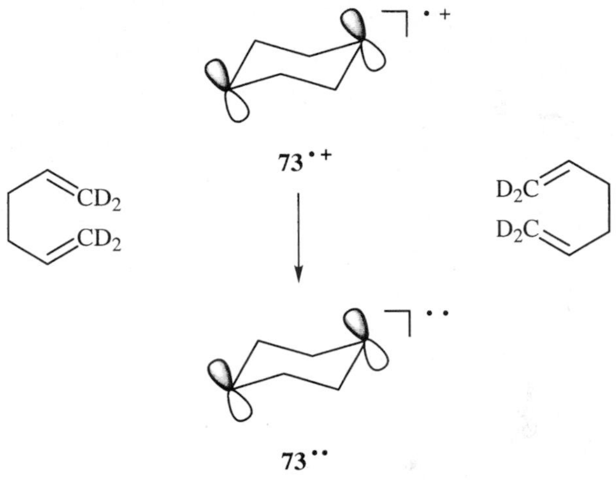

**Scheme 4.**

reverse electron transfer in pairs of triplet spin multiplicity, a process that populates a triplet state of **13** [107, 227]; accordingly, the overall reaction involves two consecutive intermediates, a radical cation and a triplet state.

The triplet recombination of cyclohexanediyl radical cations with a sensitizer radical anion (above) generates a triplet biradical with similar geometry. This conversion requires only that the biradical energy lies below the radical ion pair energy. The key structural changes amounting to a Cope rearrangement occur in two separate steps, the formation of **73**$^{\cdot+}$ and the conversion of **73**$^{\cdot\cdot}$ to the hexadienes. The detailed pathway of the formation of **73**$^{\cdot+}$ is not known. The conversion of **73**$^{\cdot\cdot}$ to the hexadiene isomers requires breaking one of two essentially equivalent doubly allylic C–C bonds (Scheme 4).

The rearrangements of **13** and **91** require specific energetic and structural features: (i) the energy of the radical ion pair must lie above an accessible triplet state and (ii) a mismatch must exist between the potential surfaces of ground and triplet state. Both requirements seem to be fulfilled. Because of the relatively high negative reduction potential of dicyanonaphthalene ($E_{A^-/A} \approx -1.6$ V relative to the SCE), the free energy of the radical ion pair generated by reaction with *cis*- or *trans*-**13** ($E_{D/D+} \approx 1.5$ V relative to the SCE; $\Delta G > 3.1$ eV) lies above an accessible triplet state of **13**. A ring-opened triplet state with two orthogonal p-orbitals, **13**$^{\cdot\cdot}$, was postulated [107, 227]; as its structure lies between those of *cis*- and *trans*-**13**, it can decay to either isomer. The geometry of minimum energy of the intermediate, **13**$^{\cdot\cdot}$, corresponds to (or lies near) a saddle point on the potential surface of the parent molecule, enabling decay to either isomer (Scheme 5). The biradical was observed recently and its free energy determined [256], confirming the mechanism previously assigned [107, 227]. Because a (partial) bond is broken in this step, the recombination was called 'dissociative' [256].

Another example of triplet recombination involves the formation of quadricyclane upon electron transfer from norbornadiene to 1-cyanonaphthalene (1-CNN) [226]. Once again, the radical ion pair energy lies above 3 eV, so an intermediate on the

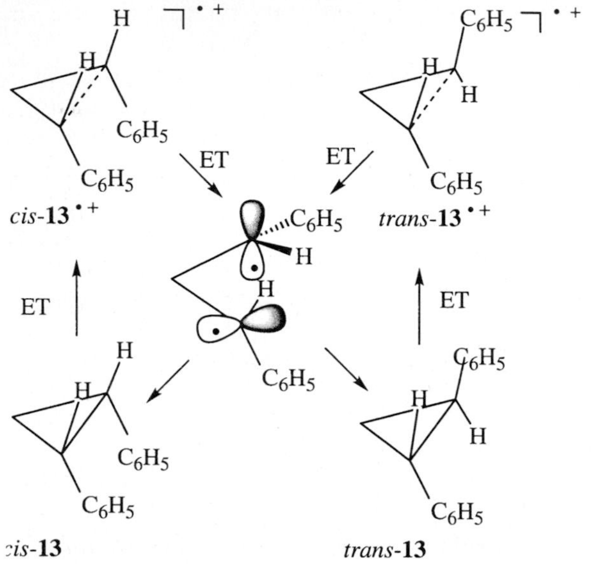

**Scheme 5.**

norbornadiene triplet potential surface is postulated. This species probably has a nortricyclanediyl structure, **109**$^{\cdot\cdot}$ [228], which enables decay to quadricyclane, **90**. By analogy with the nomenclature used above [256], the recombination of **91**$^{\cdot+}$ would be 'associative'.

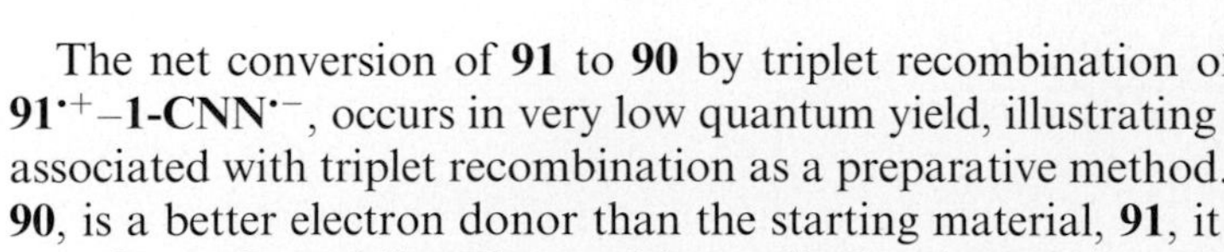

The net conversion of **91** to **90** by triplet recombination of the radical ion pair, **91**$^{\cdot+}$–**1-CNN**$^{\cdot-}$, occurs in very low quantum yield, illustrating the potential problem associated with triplet recombination as a preparative method. Because the product, **90**, is a better electron donor than the starting material, **91**, it will undergo electron transfer preferentially. The resulting radical cation, **90**$^{\cdot+}$, undergoes the well-known ring opening to **91**$^{\cdot+}$, causing the rapid loss of any product formed.

## 2.4.5  Concluding Remarks

The electron-transfer processes discussed in this chapter constitute an interesting facet of the overall field. Although the substrates discussed here do not play a role in the important chemistry of life, they do occur in the Earth's atmosphere and in outer space. Electron-transfer reactions of alkanes and cycloalkanes generate a plethora of interesting intermediates, and their study has revealed a wide range of interesting reactions. Some (unrecognized) electron-transfer reactions preceded the actual discovery of the electron [257], and some radical ions were observed long before their nature was recognized [258]. Given these early beginnings, the field of electron-transfer chemistry is well over one hundred years old. It is safe to predict significant and vigorous advances in this field for at least the first decade of the new century. In particular, time-resolved spectroscopic methods and their extension to as yet inaccessible spectral regions is expected to facilitate major new contributions.

# References

1. Jordan KD, Burrow PD, *Chem. Rev.* **1987**, *87*, 557–588.
2. Melton CE, in *Mass Spectrometry of Organic Ions*, McLafferty FW, Ed, Academic Press, **1963**, 163–205.
3. Melton CE, in *Principles of Mass Spectrometry and Negative Ions*, Marcel Dekker, New York, **1970**, 163–205.
4. Janousek BK, Brauman JI, in *Gas Phase Ion Chemistry*, Bowers MT, Ed, Academic Press, **1979**, Vol II, 53–86.
5. Lias SG, Ed., *J. Phys. Chem. Ref. Data* **1988**.
6. Howe I, Williams DH, Bowen RD, *Mass Spectrometry Principles and Applications*, Wiley, New York, **1981**.
7. Lehman TA, Bursey MM *Ion Cyclotron Resonance Spectrometry*, **1976**.
8. Turner DW, Baker AD, Baker C, Brundle CR, *Molecular Photoelectron Spectroscopy*, Wiley–Interscience, New York, **1970**.
9. Rao PS, Nash JR, Guarino JP, Ronayne MR, Hamill WH, *J. Am. Chem. Soc.* **1962**, *84*, 500–501.
10. Guarino JP, Ronayne MR, Hamill WH *J. Am. Chem. Soc.* **1962**, *84*, 4230–4235.
11. Louwrier PWF, Hamill WH, *J. Phys. Chem.* **1970**, *74*, 1418–1421.
12. Knight LB Jr, *Accounts Chem. Res.* **1986**, *19*, 313–321.
13. Nauwelaerts F, Ceulemans J, *Chem. Phys. Lett.* **1976**, *38*, 354–356.
14. Nauwelaerts F, Lemahieu M, Ceulemans J, *J. Chem. Phys.* **1977**, *66*, 140–142.
15. Smith IG, Symons MCR, *J. Chem. Res.* **1989**, 382.
16. Symons MCR, *Chem. Phys. Lett.* **1980**, *69*, 198–200.
17. Wang JT, Williams F, *J. Phys. Chem.* **1980**, *84*, 3156–3159.
18. Shida T, Kubodera H, Egawa Y, *Chem. Phys. Lett.* **1981**, *79*, 179–182.
19. Toriyama K, Nunome K, Iwasaki M, *J. Am. Chem. Soc.* **1981**, *103*, 3591–3592.
20. Carrington A, McLachlan AD, *Introduction to Magnetic Resonance*, Harper & Row, New York, **1967**.
21. Wertz JE, Bolton JR, *Electron Spin Resonance: Elementary Theory and Practical Applications*, McGraw–Hill, New York, **1972**.
22. McLauchlan KA, Stevens DG, *Accounts Chem Res.* **1988**, *21*, 54–59.
23. Miller SL, Urey HC, *Science* **1959**, *130*, 245.
24. Toriyama K, Nunome K, Iwasaki M, *J. Phys. Chem.* **1988**, *92*, 5097–5103.
25. Knight LB, Jr, Steadman J, Feller D, Davidson ER, *J. Am. Chem. Soc.* **1984**, *106*, 3700–3701.
26. Symons MCR, Chen T, Glidewell C, *J. Chem. Soc. Chem. Commun.* **1983**, 326–328.
27. Claxton TA, Chen T, Symons MCR, *Faraday Disc. Chem. Soc.* **1984**, *78*, 1.
28. Eriksson LE, Malkin VG, Malkina OL, Salahub DR, *Int. J. Quant. Chem*, **1994**, *52*, 879–901.
29. Matsushita M, Momose T, Shida T, Knight LB, Jr, *J. Chem. Phys.* **1995**, *103*, 3367–3376.
30. Knight LB, Jr, King GM, Petty JT, Matsushita M, Momose T, Shida T, *J. Chem. Phys.* **1995**, *103*, 3377–3385.

31. Herbertz T, Roth HD, **2000**, unpublished results.
32. Carter MK, Vincow G, *J. Chem. Phys.* **1967**, *47*, 292–302.
33. Kira M, Watanabe M, Sakurai H, *Chem. Letters* **1979**, 973–976.
34. Clark T, Chandrasekhar J, Jr, Schleyer PvR, *J. Chem. Soc., Chem. Commun.* **1980**, 26.
35. Liebling GR, McConnel HM, *J. Chem. Phys.* **1965**, *42*, 3931–3934.
36. Closs GL, Redwine OD, *J. Am. Chem. Soc.* **1986**, *108*, 506–507.
37. Iwasaki M, Toriyama K, Nunome K, *Faraday Disc. Chem. Soc.* **1984**, *78*, 19.
38. Herzberg G, *Electronic Spectra and Electronic Structure of Polyatomic Molecules*, Van Nostrand Reinhold, New York, **1966**, p. 50.
39. Claxton TA, Overill RE, Symons MCR, *Mol. Phys.* **1974**, *27*, 701–706.
40. Zuilhof H, Dinnocenzo JP, Reddy AC, Shaik SS, *J. Phys. Chem.* **1996**, *100*, 15774–15784.
41. Toriyama K, Nunome K, Iwasaki M, *J. Chem. Phys.* **1982**, *77*, 5891–5912.
42. Matsuura K, Nunome K, Toriyama K, Iwasaki M, *J. Phys. Chem.* **1989**, *93*, 149–154.
43. Toriyama K, *Chem. Phys. Lett.* **1991**, *177*, 39–44.
44. Toriyama K, Okazaki M, Nunome K, *J. Chem. Phys.* **1982**, *77*, 3955–3963.
45. Toriyama K, Nunome K, Iwasaki M, *J. Am. Chem. Soc.* **1987**, *109*, 3591–3592.
46. Toriyama K, Nunome K, Iwasaki M, *J. Phys. Chem.* **1981**, *85*, 2149–2152.
47. Wang JT, Williams F, *Chem. Phys. Lett.*, **1981**, *82*, 177–181.
48. Iwasaki M, Toriyama K, *J. Am. Chem. Soc.* **1986**, *108*, 6441–6443.
49. Tabata M, Lund A, *Rad. Phys. Chem.* **1984**, *23*, 545.
50. Lindgren M, Lund A, Dolivo G, *Chem. Phys.* **1985**, *99*, 103–110.
51. Lund A, Lindgren M, Dolivo G, Tabata M, *Rad. Phys. Chem.* **1985**, *26*, 491.
52. Dolivo G, Lund A, *J. Phys. Chem.* **1985**, *89*, 3977–3984.
53. Dolivo G, Lund A, *Z. Naturforsch.* **1985**, *40a*, 52–65.
54. Toriyama K, Nunome K, Iwasaki M, *J. Am. Chem. Soc.* **1987**, *109*, 4496–4500.
55. Demeyer A, Stienlet D, Ceulemans J, *J. Phys. Chem.* **1994**, *98*, 9530–9543.
56. Ceulemans J, *Spectrochim. Acta* A **1998**, *54*, 2359–2372.
57. Fleischmann M, Pletcher D, *Tetrahedron Lett.* **1968**, 6255–6258.
58. Toriyama K, Nunome K. Iwasaki, M. *Chem. Phys. Lett.* **1986**, *132*, 456–458.
59. Shiotani M, Yano A, Ohta N, Ichikawa T, *Chem. Phys. Lett.* **1988**, *147*, 38–42.
60. Cochran EL, Adrian FJ, Bowers, *J. Chem. Phys.* **1964**, *40*, 213–220.
61. Muto H, Iwasaki M, Takahashi Y, *J. Chem. Phys.* **1977**, *66*, 1943–1952.
62. Hulme R, Symons MCR, *J. Chem. Soc.* **1965**, 1120.
63. Wojciechowski BW, Corma A, *Catalytic Cracking, Catalysts Kinetics and Mechanisms*, Marcel Dekker: New York, **1984**.
64. Venuto PB, *Adv. Catal.* **1968**, *18*, 259–371.
65. Venuto PB, *Microporous Mater.* **1994**, *2*, 297–411.
66. Hölderich WF, Hesse M, Näumann F, *Angew. Chem., Int. Ed. Engl.* **1988**, *27*, 226–246.
67. Hölderich WF, *Stud. Surf. Sci. Catal.* **1989**, *49*, 69.
68. Corma A, García H, *Catal. Today* **1997**, *38*, 257–308.
69. Corma A, García H, *Top. Catal.* **1998**, *6*, 127–140.
70. Herbertz T, Lakkaraju PS, Blume F, Blume M, Roth HD, *Eur. J. Org. Chem.* **2000**, 467–472.
71. Chen FR, Fripiat JJ, *J. Phys. Chem.* **1993**, *97*, 5796–5797.
72. Roduner E, Crockett R, *J. Phys. Chem.* **1993**, *97*, 11853–11854.
73. Trifunac AD, Qin X-Z, *Appl. Magn. Resonance* **1990**, *1*, 29–40.
74. Barnabas MV, Werst DW, Trifunac AD, *Chem. Phys. Lett.* **1993**, *204*, 435–439.
75. Werst DW, Tartakovsky EE, Piocos EA, Trifunac AD, *J. Phys. Chem.* **1994**, *98*, 10249–10257.
76. Shiotani M, Lund A, *Radical Ionic Systems*, Lund A, Shiotani M, Eds; Kluwer Academic: Dordrecht, **1991**, p. 151–176.
77. Gleiter R, *Top. Curr. Chem.* **1980**, *86*, 197–285.
78. Iwasaki M, Toriyama K, Nunome K, *J. Chem. Soc. Chem. Comm.* **1984**, 202–203.
79. Shida T, Takemura Y, *Radiat. Phys. Chem.* **1983**, *21*, 157.
80. Ohta K, Nakatsuji H, Kubodera H, Shida T, *Chem. Phys.* **198**3, *76*, 271–281.
81. Almennigen A, Bastiansen O, Skancke PN, *Acta Chim. Scand.* **1961**, *15*, 711.

82. Ushida K, Shida T, Iwasaki M, Toriyama K, Numone K, *J. Am. Chem. Soc.* **1983**, *105*, 5496–5497.
83. Bauld NL, Bellville DJ, Pabon R, Chelsky R, Green G, *J. Am. Chem. Soc.* **1983**, *105*, 2378–2382.
84. Bouma WJ, Poppinger D, Radom L, *Isr. J. Chem.* **1983**, *23*, 21–36.
85. Dewar MS, Merz KM, Jr, *J. Mol. Struct. THEOCHEM* **1985**, *122*, 59–65.
86. Jungwirth P, Carsky P, Bally T, *J. Am. Chem. Soc.* **1993**, *115*, 5776–5782.
87. Wiest O, *J. Phys. Chem. A* **1999**, *103*, 7907–7911.
88. Sjöquist L, Lund A, Maruani J, *Chem. Phys.* **1988**, *125*, 293–298.
89. Huang MB, Lunell S, Lund A, *Chem. Phys. Letters* **1983**, *99*, 201–205.
90. Tabata M, Lund A, *Chem. Phys.* **1983**, *75*, 379–388.
91. Shida T, Takemura Y, *Rad. Phys. Chem.* **1983**, *21*, 157.
92. Toriyama K, Nunome K, Iwasaki M, *J. Chem. Soc., Chem. Commun.* **1984**, 143–145.
93. Lunell S, Huang MB, Claesson O, Lund A, *J. Chem. Phys.* **1985**, *82*, 5121–5126.
94. Shiotani M, Ohta N, Ichikawa T, *Chem. Phys. Letters* **1988**, *149*, 85–88.
95. Toriyama K, Nunome K, Iwasaki M, *J. Phys. Chem.* **1986**, *90*, 6836–6842.
96. Haselbach E, *Chem. Phys. Lett.* **1970**, *7*, 428–430.
97. Rowland CG, *Chem. Phys. Lett.* **1971**, *9*, 169–173.
98. Collins JR, Gallup GA, *J. Am. Chem. Soc.* **1982**, *104*, 1530–1533.
99. Wayner DDM, Boyd RJ, Arnold DR, *Can. J. Chem.* **1985**, *63*, 3283–3289.
100. Wayner DDM, Boyd RJ, Arnold DR, *Can. J. Chem.* **1983**, *61*, 2310–2315.
101. Du P, Hrovat DA, Borden WT, *J. Am. Chem. Soc.* **1988**, *110*, 3405–3412.
102. Lunell S, Yin L, Huang M-B, *Chem. Phys.* **1989**, *139*, 293–299.
103. Krogh-Jespersen K, Roth HD, *J. Am. Chem. Soc.* **1992**, *114*, 8388–8394.
104. Skancke A, *J. Phys. Chem.* **1995**, *99*, 13886–13889.
105. Qin XZ, Snow LD, Williams F, *J. Am. Chem. Soc.* **1984**, *106*, 7640–7641.
106. Qin XZ, Williams F, *Tetrahedron* **1986**, *42*, 6301–6314.
107. Roth HD, Schilling MLM, *J. Am. Chem. Soc.* **1980**, *102*, 7956–7958.
108. Roth HD, Schilling MLM, *Can. J. Chem.* **1983**, *61*, 1027–1035.
109. Roth HD, Schilling MLM, *J. Am. Chem. Soc.* **1983**, *105*, 6805–6808.
110. Haddon RC, Roth HD, *Croat. Chem. Acta* **1984**, *57*, 1165–1176.
111. Rabinovitch BS, Schlag EW, Wiberg K, *J. Chem. Phys.* **1958**, *28*, 504–505.
112. Schlag EW, Rabinovitch BS, *J. Am. Chem. Soc.* **1960**, *82*, 5996–6000.
113. Mattes SL, Farid S, *Org. Photochem.* **1983**, *6*, 233–326.
114. Mattay J, Vondenhof M, *Topics Curr. Chem.* **1991**, *159*, 219–255.
115. *Photoinduced Electron Transfer*, Fox MA, Chanon M, Eds, Elsevier, Amsterdam, **1988**.
116. Knibbe H, Rehm D, Weller A, *Ber. Bunsenges. Phys. Chem.* **1969**, *73*, 839–845.
117. Bargon J, Fischer H, Johnson U, *Z. Naturforsch.* **1967**, *A 22*, 1551–1556.
118. Ward HR, Lawler RG, *J. Am. Chem. Soc.* **1967**, *89*, 5518–5519.
119. Roth HD, Lamola AA, *J. Am. Chem. Soc.* **1972**, *94*, 1013–1014.
120. Roth HD, Lamola AA, *J. Am. Chem. Soc.* **1974**, *96*, 6270–6275.
121. Lamola AA, Roth HD, Schilling MLM, Tollin G, *Proc. Nat. Acad. Sci. USA* **1975**, *72*, 3265–3269.
122. Closs GL, *Adv. Magn. Reson.* **1974**, *7*, 157–229.
123. Kaptein R, *Adv. Free Radical Chem.* **1975**, *5*, 319–380.
124. Adrian FJ, *Rev. Chem. Intermed.* **1979**, *3*, 3–43.
125. Freed JH, Pedersen JB, *Adv. Magn Reson.* **1976**, *8*, 2–84,.
126. Roth HD, in *Encyclopedia of Nuclear Magnetic Resonance*, Grant DM, Harris RK, Eds, **1996**, vol. *2*, 1337–1350.
127. Dewar MJS, Dougherty RC, *The PMO Theory of Organic Chemistry*, Plenum Press, N.Y. **1975**.
128. Herbertz T, Roth HD, *J. Am. Chem. Soc.* **1998**, *120*, 11904–11911.
129. Herbertz T, Lakkaraju PS, Roth HD, Sluggett G, Turro NJ, *J. Phys. Chem. A* **1999**, *103*, 11350–11354.
130. Dinnocenzo JP, Zuilhof H, Lieberman DR, Simpson TR, McKechney MW, *J. Am. Chem. Soc.* **1997**, *119*, 994–1004.

131. Herbertz T, Blume F, Roth HD, *J. Am. Chem. Soc.* **1998**, *120*, 4591–4599.
132. Roth HD, Schilling MLM, Schilling FC, *J. Am. Chem. Soc.* **1985**, *107*, 4152–4158.
133. Roth HD, Schilling MLM, Hutton RS, Truesdale EA, *J. Am. Chem. Soc.* **1983**, *105*, 153–157.
134. Roth HD, Herbertz T, *J. Am. Chem. Soc.* **1993**, *115*, 9804–9805.
135. Roth HD, Weng H, Herbertz T, *Tetrahedron*, **1997**, *53*, 10051–10070.
136. Kawamura T, Tsumura M, Yokomichi Y, Yonegawa T, *J. Am. Chem. Soc.* **1977**, *99*, 8251–8256.
137. Fessenden RW, *J. Phys. Chem.* **1967**, *71*, 74.
138. Ohta K, Nakatsuji H, Hirao K, Yonezawa T, *J. Chem. Phys.* **1980**, *73*, 1770–1776.
139. Behrens G, Schulte-Frohlinde D, *Angew. Chem.* **1973**, *85*, 993.
140. Qin X-Z, Williams F, *Chem. Phys. Lett.* **1984**, *112*, 79–83.
141. Dinnocenzo JP, Schmittel M, *J. Am. Chem. Soc.* **1987**, *109*, 1561–1562.
142. Shiotani M, Nagata Y, Sohma J, *J. Phys. Chem.* **1984**, *88*, 4078–4082.
143. Guo QX, Qin XZ, Wang JT, Williams F, *J. Am. Chem. Soc.* **1988**, *110*, 1974–1976.
144. Williams F, Guo QX, Petillo PA, Nelsen SF, *J. Am. Chem. Soc.* **1988**, *110*, 7887–7888.
145. Williams F, Guo QX, Bebout DC, Carpenter BK, *J. Am. Chem. Soc.* **1989**, *111*, 4133–4134.
146. Adam W, Walter H, Chen G-F, Williams F, *J. Am. Chem. Soc.* **1992**, *114*, 3007–3014.
147. Roth HD, Schilling MLM, Gassman PG, Smith JL, *J. Am. Chem. Soc.* **1984**, *106*, 2711–2712.
148. Du P, Hrovat DA, Borden WT, *Chem. Phys. Lett.* **1986**, *123*, 337–340.
149. Symons MCR, *Chem. Phys. Lett.* **1978**, *117*, 381–382.
150. Rao VR, Hixson SS, *J. Am. Chem. Soc.* **1979**, *101*, 6458–6459.
151. Mizuno K, Ogawa J, Kagano H, Otsuji Y, *Chem. Lett.* **1981**, 437–438.
152. Mizuno K, Ogawa J, Otsuji Y, *Chem. Lett.* **1981**, 741–744.
153. Mazzocchi PH, Somich C, Edwards M, Morgan T, Ammon HL, *J. Am. Chem. Soc.* **1986**, *108*, 6828–6829.
154. Dinnocenzo JP, Todd WP, Simpson TR, Gould IR, *J. Am. Chem. Soc.* **1990**, *112*, 2462–2464.
155. Dinnocenzo JP, Lieberman DR, Simpson TR, *J. Am. Chem. Soc.* **1993**, *115*, 366–367.
156. Weng H, Sethuraman V, Roth HD, *J. Am. Chem. Soc.* **1994**, *116*, 7021–7025.
157. Evans TR, Wake RW, Jaenicke O, The Exciplex, Academic Press, New York, **1975**, 345–358.
158. Roth HD, Hutton RS, *J. Phys. Org. Chem.* **1990**, *3*, 119–125.
159. Symons, MCR, Wren BW, *Tetrahedron Lett.* **1983**, *24*, 2315–2318.
160. Snow LD, Wang JT, Williams F, *Chem. Phys. Lett.* **1983**, *100*, 193–197.
161. Ledwith A, *Accounts Chem. Res.* **1972**, *57*, 133–139.
162. Mattes SL, Farid S, *Accounts Chem. Res.* **1982**, *15*, 80–86.
163. Bauld NL, *Tetrahedron*, **1989**, *45*, 5307–5363.
164. Ichikawa T, Ohta N, Kajioka H, *J. Phys. Chem.* **1979**, *83*, 284–295.
165. Saik VO, Anisimov OA, Lozovoy VV, Molin YuN, *Z. Naturforsch.* **1985**, *40a*, 239–245.
166. Barnabas MW, Trifunac AD, *Chem. Phys. Lett.* **1992**, *193*, 298–304.
167. Desrosiers MF, Trifunac AD, *J. Phys. Chem* **1986**, *90*, 1560–1564.
168. Jungwirth P, Bally T, *J. Am. Chem. Soc.* **1993**, *115*, 5783–5789.
169. Lindgren M, Shiotani M, Ohta N, Sjoquist L, Lund A, *Chem. Phys. Lett.* **1989**, *161*, 127–130.
170. Shiotani M, Lindgren M, Ichikawa T, *J. Am. Chem. Soc.* **1990**, *112*, 967–973.
171. Lindgren M, Shiotani M, in *Radical Ionic Systems*, Lund A, Shiotani M, Eds; Kluwer Academic: Dordrecht, **1991**, 125–150.
172. Werst DW, Piocos EA, Tartakovsky EE, Trifunac AD, *Chem. Phys. Lett.* **1994**, *229*, 421–428.
173. Qin X-Z, Trifunac AD, *J. Phys. Chem.* **1990**, *94*, 4751–4754.
174. Gerson F, Qin X-Z, Ess C, Kloster-Jensen E, *J. Am. Chem. Soc.* **1989**, *111*, 6456–6457.
175. Arnold A, Burger U, Gerson F, Kloster-Jensen E, Schmidlin SP, *J. Am. Chem. Soc.* **1993**, *115*, 4271–4281.
176. Gerson F, *Accounts Chem. Res.* **1994**, *27*, 63–69.
177. Roth HD, Schilling MLM, Gassman PG, Smith JL, *J. Am. Chem. Soc.* **1984**, *106*, 2711–2712.
178. Abelt CJ, Roth HD, Schilling MLM, *J. Am. Chem. Soc.* **1985**, *10*, 4148–4152.
179. Gassman PG, Carroll GT, *Tetrahedron* **1986**, *42*, 6201–6206.
180. Gassman PG, Olson KD, Walter L, Yamaguchi R, *J. Am. Chem. Soc.* **1981**, *103*, 4977–4979.
181. Gassman PG, Olson KD *J. Am. Chem. Soc.* **1982**, *104*, 3740–3742.
182. Gassman PG, Hay BA, *J. Am. Chem. Soc.* **1985**, *107*, 4075–4078.

183. Gassman PG, Hay BA, *J. Am. Chem. Soc.* **1986**, *108*, 4227–4228.
184. Wiberg KB, Szeimies G, Tetrahedron Letters **1968**, 1235–1239.
185. Roth HD, *Accounts Chem. Res.* **1987**, *20*, 343–350.
186. Roth HD, *Topics Curr. Chem.* **1992**, *163*, 131–245.
187. Woodward RB, Hoffmann R, *The Conservation of Orbital Symmetry*, Verlag Chemie, Weinheim, **1971**.
188. Gassman PG (1988) in *Photoinduced Electron Transfer*, Fox MA, Chanon M, Eds, Elsevier, Amsterdam, **1988**, Vol. C, 70.
189. Abelt CJ, Roth HD, Schilling MLM, *J. Am. Chem. Soc.* **1985**, *107*, 4148–4152.
190. Gerson F, Arnold A, Burger U, *J. Am. Chem. Soc.* **1991**, *113*, 4359–4360.
191. Bischof P, Gleiter R, Mueller E, *Tetrahedron* **1976**, *32*, 2769–2773.
192. Harman PJ, Kent JE, Gan TH, Peel JB, Willett GD, *J. Am. Chem. Soc.* **1977**, *99*, 943–944.
193. Gleiter R, Gubernator K, Eckert-Maksic M, Spanget-Larsen J, Bianco B, Gandillion G, Burger U, *Helv. Chim. Acta* **1981**, *64*, 1312–1313.
194. Abelt CJ, Roth HD, *J. Am. Chem. Soc.* **1985**, *107*, 3840–3843.
195. Williams F, Guo QX, Kolb TM, Nelsen SF, *J. Chem. Soc., Chem. Commun.* **1989**, 1835–1836.
196. Ushida K, Shida T, Walton JC, *J. Am. Chem. Soc.* **1986**, *108*, 2805–2808.
197. Adam W, Sahin C, Sendelbach J, Walter H, Chen G-F, Williams F, *J. Am. Chem. Soc.* **1994**, *116*, 2576–3584.
198. Blancafort L, Adam W, González D, Olivucci M, Vreven T, Robb MA, *J. Am. Chem. Soc.* **1999**, *121*, 10583–10590.
199. Adam W, Heidenfelder T, *Chem. Soc. Rev.* **1999**, *28*, 359–365.
200. Tsuji T, Miura T, Sugiura K, Nishida S, *J. Am. Chem. Soc.* **1990**, *112*, 1998–1999.
201. Miyashi T, Konno A, Takahashi Y, *J. Am. Chem. Soc.* **1988**, *110*, 3676–3677.
202. Ikeda H, Minegishi T, Abe H, Konno A, Goodman JL, Miyashi T, *J. Am. Chem. Soc.* **1998**, *120*, 87–95.
203. Miyashi T, Ikeda H, Takahashi Y, *Accounts Chem. Res.* **1999**, *32*, 815–824.
204. Doering WvE, Roth WR, *Tetrahedron* **1962**, *18*, 67–74.
205. Roth HD, *Proc. IUPAC Symp. Photochem.* **1984**, *10*, 455–456.
206. Roth HD, Schilling MLM, *J. Am. Chem. Soc.* **1985**, *107*, 716–718.
207. Roth HD, Schilling MLM, Abelt CJ, *Tetrahedron* **1986**, *42*, 6157–6166.
208. Roth HD, Schilling MLM, Abelt CJ, *J. Am. Chem. Soc.* **1986**, *108*, 6098–6099.
209. Momose T, Shida T, Kobayashi T, *Tetrahedron* **1986**, *42*, 6337–6342.
210. Dai S, Wang JT, Williams F, *J. Am. Chem. Soc.* **1990**, *112*, 2835–2836.
211. Dai S, Wang JT, Williams F, *J. Am. Chem. Soc.* **1990**, *112*, 2837–2837.
212. Abelt CJ, *J. Am. Chem. Soc.* **1986**, *108*, 2013–2019.
213. Roth HD, *Z. Phys. Chem.* **1993**, *180*, 135–158.
214. Williams F, *J. Chem. Soc., Faraday Trans.*, **1994**, *90*, 1681–1687.
215. Arnold DR, Du X, *Can. J. Chem.* **1994**, *72*, 403–414.
216. Wlostowski M, Roth HD, unpublished results.
217. Takemoto Y, Ohra T, Koike H, Furuse S-I, Iwata C, Ohishi H, *J. Org. Chem.* **1994**, *59*, 4727–4729.
218. Roth HD, Herbertz T, *J. Am. Chem. Soc.* **1993**, *115*, 9804–9805.
219. Roth HD, Weng H, Zhou D, Herbertz T, *Pure Appl. Chem.* **1997**, *69*, 809–814.
220. Herbertz T, Roth HD, *J. Am. Chem. Soc.* **1997**, *119*, 9574–9575.
221. Hoffmann R, Heilbronner E, Gleiter R, *J. Am. Chem. Soc.* **1970**, *92*, 706–707.
222. Hoffmann R, *Accounts Chem. Res.* **1971**, *4*, 1–9.
223. Dewar MJS, Wasson JS, *J. Am. Chem. Soc.* **1970**, *92*, 3506–3508.
224. Heilbronner E, Schmelzer A, *Helv. Chim. Acta* **1975**, *58*, 936–967.
225. Haselbach E, Bally T, Lanyiova Z, Baertschi P *Helv. Chim. Acta* **1979**, *62*, 583–592.
226. Roth HD, Schilling MLM, Jones G, *J. Am. Chem. Soc.* **1981**, *103*, 1246–1248.
227. Roth HD, Schilling MLM *J. Am. Chem. Soc.* **1981**, *103*, 7210–7217.
228. Raghavachari K, Haddon RC, Roth HD, *J. Am. Chem. Soc.* **1983**, *105*, 3110–3114.
229. Toriyama K, Nunone K, Iwasaki M, *J. Chem. Soc., Chem. Commun.* **1983**, 1346–1347.
230. Gerson F, Qin X-Z, *Helv. Chim. Acta* **1989**, *72*, 383–390.

231. Ishiguro K, Khudyakov IV, McGarry PF, Turro NJ, Roth, HD, *J. Am. Chem. Soc.* **1994**, *116*, 6933–6934.
232. Ladenburg A, *Chem. Ber.* **1869**, *2*, 140.
233. Raghavachari K, Roth HD, *J. Am. Chem. Soc.* **1989**, *111*, 7132–7136.
234. Oth JMF, *Ang. Chem.* **1968**, *79*, 1102; *Ang. Chem. Int. Ed. Engl.* **1968**, *7*, 646.
235. Nunome K, Toriyama K, Iwasaki M, *Tetrahedron* **1986**, *42*, 6315–6323.
236. Toriyama K, Okazaki M, *Radiat. Phys. Chem.* **1989**, *33*, 505.
237. Lotais BC, Jonah CD, *Radiat. Phys. Chem.* **1989**, *33*, 505.
238. Sauer MCJ, Schmidt KH, *Scand. Chem. Acta.* **1996**, *51*, 167–173.
239. Melekhov VI, Anisimov OA, Sjoquist L, Lund A, *Chem. Phys. Lett.* **1990**, *174*, 95–102.
240. Barnabas MV, Trifunac AD, *Chem. Phys. Lett.* **1991**, *187*, 565–570.
241. Hixson SS, Xing Y, *Tetrahedron Lett.* **1991**, *32*, 173–174.
242. Weng H, Roth HD, *J. Org. Chem.* **1995**, *60*, 4136–4145.
243. Weng H, Du X-M, Roth HD, *J. Am. Chem. Soc.* **1995**, *117*, 135–140.
244. Pross A, *J. Am. Chem. Soc.* **1986**, *108*, 3537–3538.
245. Shaik SS, Pross A, *J. Am. Chem. Soc.* **1989**, *111*, 4306–4312.
246. Eberson L, Radner F, *Acta. Chem. Scand.* **1992**, *46*, 312–314.
247. Eberson L, Radner F, *Acta. Chem. Scand.* **1992**, *46*, 802–804.
248. Eberson L, Hartshorn MP, Radner F, Merchan M, Roos BO, *Acta. Chem. Scand.* **1993**, *47*, 176–183.
249. Dinnocenzo JP, Simpson TR, Zuilhof H, Todd WP, Heinrich T, *J. Am. Chem. Soc.* **1997**, *119*, 987–993.
250. Arnold DR, Humphreys RWR, *J. Am. Chem. Soc.* **1979**, *101*, 2743–2744.
251. Weng H, Sheik Q, Roth HD, *J. Am. Chem. Soc.* **1995**, *117*, 10655–10661.
252. Herbertz T, Roth HD, *J. Am. Chem. Soc.* **1996**, *118*, 10954–10962.
253. Gassman PG, Olson KD, *Tetrahedron Lett.* **1983**, *1*, 19–22.
254. Shaik SS, Dinnocenzo JP, *J. Org. Chem.* **1990**, *55*, 3434–3436.
255. Shaik SS, Reddy AC, Ioffe A, Dinnocenzo JP, Danovich D, Cho JK, *J. Am. Chem. Soc.* **1995**, *117*, 3205–3222.
256. Karki SB, Dinnocenzo JP, Farid S, Goodman JC, Gould I, Zona TA, *J. Am. Chem. Soc.* **1997**, *119*, 431–432.
257. Roth HD, *Topics Curr. Chem.* **1990**, *156*, 1–19.
258. Roth HD, *Tetrahedron* **1986**, *42*, 6097–6100.

# 3 The Electron-transfer Chemistry of Carbon–Carbon Multiple Bonds

*Nathan L. Bauld and Daxin Gao*

## 3.1 Introduction

The research of the Ledwith group at Liverpool on the cyclodimerization of *N*-vinylcarbazole (NVC) is the key to much of the modern electron transfer chemistry of carbon–carbon multiple bonds (Scheme 1) [1].

Not only was the overall reaction novel in its *connectivity*, i.e., the formation of cyclobutane rings (cyclobutanation), which is unusual outside of photochemistry, but even more importantly, the *cation radical chain mechanism* which was proposed and demonstrated for the reaction represented a fundamentally new reaction mechanism. Significantly for the present context, this mechanism involved not one but two electron transfer (ET) steps. The first, ionization of NVC to the corresponding cation radical ($NVC^{+\bullet}$) was effected by either a metal ion oxidant (e.g. $Fe^{3+}$) or by photosensitized electron transfer (PET) to excited state chloranil. This step is followed by cycloaddition of $NVC^{+\bullet}$ to neutral NVC to yield, in Ledwith's formulation, an acyclic 1,4-butanediyl cation radical. The latter then cyclizes to the cyclobutane cation radical, which is subsequently reduced to the neutral cyclobutane product by electron transfer from neutral NVC, thereby also regenerating $NVC^{+\bullet}$ and setting up the chain mechanism. Following the suggestion of Farid, this type of dimer is referred to more specifically as a *cyclobutadimer*. These latter three steps thus constitute a chain process. Although this mechanism *might* be valid in the case of NVC, one aspect of it appears to require revision for the cyclobutadimerizations of many other substrates. In particular, many of the cyclodimerizations appear to proceed in a concerted manner, directly forming a cyclobutane cation radical, and avoiding the formation of an intermediate acyclic cation radical (vide infra). This latter mechanism, in its simplest form, consists essentially of a *substrate ionization* step followed by a propagation cycle consisting of alternating *cation radical/neutral cycloaddition* and *electron transfer* steps. Incidentally, a *catalytic* (as opposed to a chain) *version* of this mechanism is also potentially available and can be realized when the cyclodimer cation radical is neutralized by the reduced form of

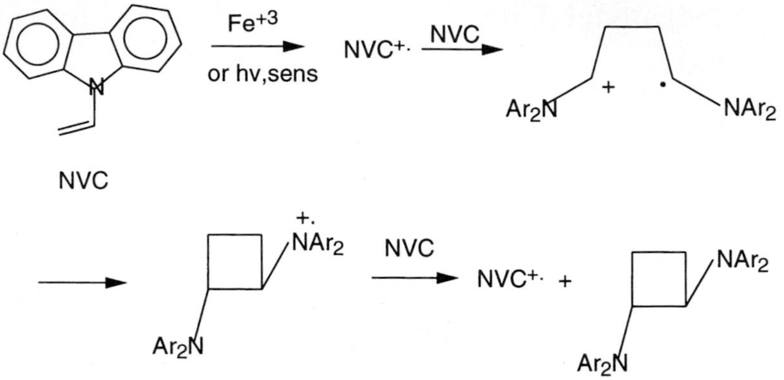

**Scheme 1.** The cation radical chain cyclodimerization of *N*-vinylcarbazole.

the species which initially oxidizes the substrate, instead of being neutralized by the substrate itself. It is evident that cation radicals (or radical cations) are key reactive intermediates in these ET reactions. However, analogous mechanisms exist, at least on paper, for the case in which the substrate accepts an electron to form an *anion radical* (radical anion). The ET chemistry of both organic cation and anion radicals will be considered in this chapter, beginning with the more abundant ET chemistry of cation radicals.

## 3.2 Electron Transfer Chemistry Involving C–C Multiple Bonds as Single-electron Donors

### 3.2.1 Discovery of Cation Radicals

The first stable cation radical salts (**1**) were obtained by Würster in 1879, simply by reacting the corresponding diamine with bromine in methanol/acetic acid solvent (Scheme 2) [2].

A stable triarylaminium salt (**2**) was isolated by Wieland in 1907 [3]. Neither **1** nor **2** was recognized as a cation radical salt, since this concept had not yet even begun to emerge. It was not until 1926 that Weitz formulated the corresponding perchlorate salt of **2** as a delocalized cation which is also a 'free ammonium radical' [4]. Michaelis then made Weitz's interpretation more certain via electrochemical oxidation studies, and he coined the term 'cationic free radical', which is the appropriate antecedent of the current term 'cation radical' [5]. Since that time, many other cation radical salts have been prepared, but the salts **3**$^{+\cdot}$ and **4** have an especially important role in the electron transfer chemistry of cation radicals [6, 7].

$\mathbf{1}^{+\bullet}$ $Br_3^-$

$\mathbf{2}^{+\bullet}$ $\mathbf{Br_3^-/ClO_4^-}$($A_1 = CH_3$, $A_2 = H$)

$\mathbf{3}^{+\bullet}$ $\mathbf{SbCl_6^-}$ ($A_1 = Br$; $A_2 = H$)

$\mathbf{4}^{+}$ $\mathbf{SbCl_6^-}$ ($A_1 = A_2 = Br$)

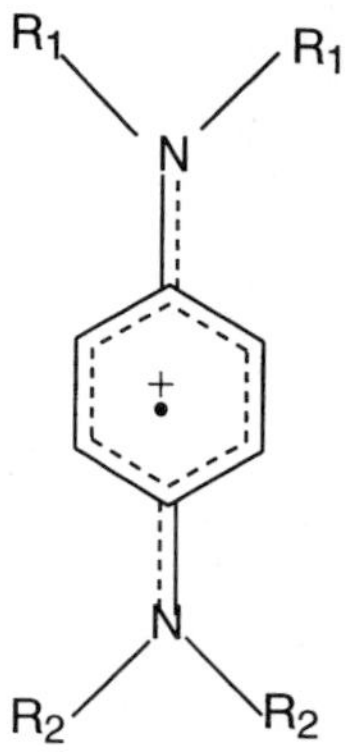

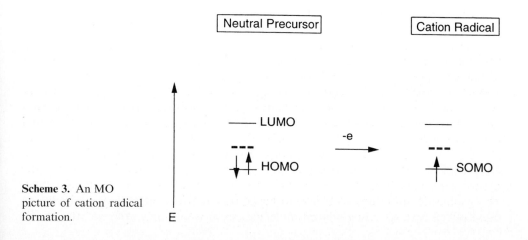

**Scheme 2.**

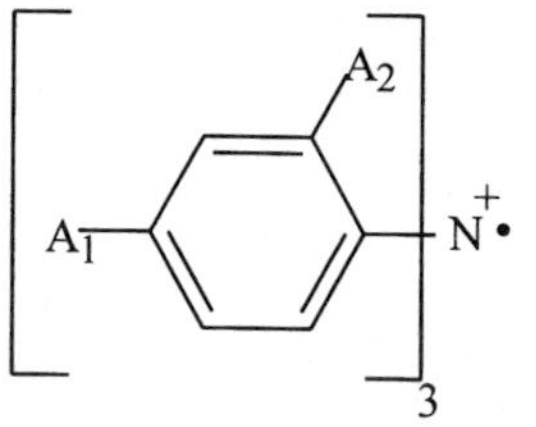

### 3.2.2 A Molecular-orbital Picture of Cation Radical Formation

Conceptually, cation radicals are related to the corresponding neutral molecules through removal (ionization) of one electron (Scheme 3).

In contrast to carbocation, radical, or carbanion formation, no bonds are broken in cation radical formation, although bonding is diminished by removal of an electron from a bonding MO. If the geometry of the resulting cation radical is not very different from that of the neutral precursor (by no means always the case!), the ground state cation radical can be considered to result from the removal of an electron from the highest energy MO (i.e., the HOMO) of the precursor, which MO then becomes the SOMO (singly occupied MO) of the cation radical. This latter MO then predominantly controls the spin distribution in the cation radical. If, in

**Scheme 3.** An MO picture of cation radical formation.

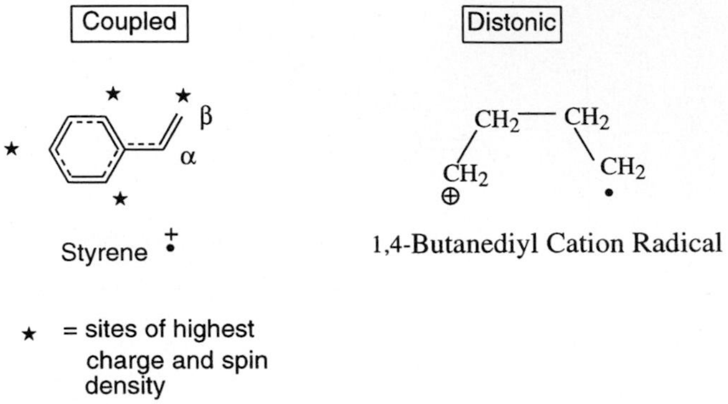

**Scheme 4.** Spin and Charge Coupling/Uncoupling in Cation Radicals.

addition, the original precursor molecule was nonpolar or at least not very polar, the same SOMO also primarily controls the (positive) charge density in the cation radical. For this reason, the spin and charge in many pi hydrocarbon cation radicals are said to be 'coupled'. Thus, to a first approximation, the spin and charge density at a given carbon atom are equal. We may say that spin and charge 'travel together'. We may even consider this spin/charge density to be 'hole density', which term implies both spin and positive charge. However, if the neutral precursor has substantial charge separation, the net charge density is a composite of the effects of the SOMO electron density and the density contributed by other occupied MOs. The spin and the charge densities are then said to be 'uncoupled'. A cation radical in which the spin and charge are essentially completely separated (i.e., uncoupled) is termed a 'distonic' cation radical. An example of a distonic cation radical which is particularly relevant to the reactions of cation radicals with pi bonds is the 1,4-butanediyl cation radical (Scheme 4).

In terms of reactivity, the importance of the spin density is that radicals, including cation radicals, tend to react at the position(s) of highest spin density, especially in coupling reactions with other radicals, but also in abstractions of atoms from neutral molecules. The importance of charge density is, of course, that in their reactions with nucleophiles (neutral or negatively charged), cation radicals tend to react at the position of highest charge density. For hydrocarbon cation radicals, in which spin and charge are coupled, both types of reaction tend to occur at the same position(s). This scenario also holds even for many non-hydrocarbon cation radicals. However, it should also be noted that when a neutral molecule is significantly dipolar, the charge distribution in the cation radical does not arise exclusively from the SOMO, so that spin and charge are partially uncoupled.

### 3.2.3 Cation Radical Structures

The geometric structures of cation radicals are often considered to be similar to those of the corresponding neutral molecules from which they are generated. Such

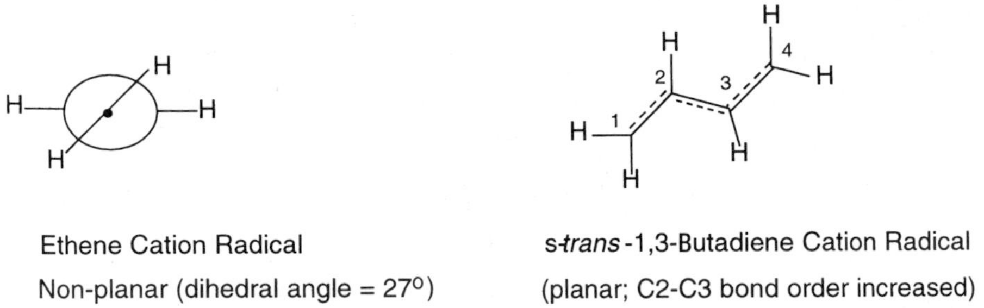

Ethene Cation Radical

Non-planar (dihedral angle = 27°)

*s-trans*-1,3-Butadiene Cation Radical

(planar; C2-C3 bond order increased)

**Scheme 5.** Ethene and butadiene cation radicals.

an approximation serves best when the charge and spin density (hole density) is extensively delocalized, so that no individual bond is sharply weakened. However, cation radical structures can and often do differ in very important ways from those of the corresponding neutrals. We will focus here upon just a few of the structural adjustments which are most pertinent to the ET chemistry of carbon–carbon multiple bonds.

**Pi cation radicals**

Pi cation radicals, that is cation radicals in which the SOMO is a pi type orbital, are obviously key intermediates in the ET chemistry of alkenes and alkynes. The ethene cation radical, which has the novel twisted structure shown in Scheme 5, provides an excellent example of a significant structural adjustment which can accompany the loss of an electron [8].

Since in a planar ethene cation radical of the pi type the SOMO would be completely localized in one bond (the pi bond), the latter is substantially weakened. Torsion around this weakened carbon–carbon bond relieves torsional strain (eclipsing of the C–H bonds) and allows additional delocalization of the SOMO to the C–H bonds via hyperconjugation. However, if the SOMO is delocalized over two pi bonds, as in the case of diene cation radicals (which are especially important intermediates in the ET chemistry of interest), the ground state cation radical structure remains planar [9]. Evidently neither of the pi bonds is weakened sufficiently to occasion torsion around the formal pi bonds. In the case of diene cation radicals, an important consequence of cation radical formation is the strengthening of the central (C2–C3) bond, so that torsion around this bond becomes much more difficult than in the parent diene. Although net bonding is diminished in the cation radical, the C2–C3 bond is actually strengthened when an electron is removed from the HOMO, which is antibonding between these two atoms. On the other hand, the C1–C2 and C3–C4 bonds are more strongly attenuated. The consequence is that all three C–C bonds in the butadiene cation radical have nearly equal stretching force constants [9]. *A direct consequence of this is that cation radicals corresponding to the s-cis and s-trans forms of the diene are not readily interconvertible.*

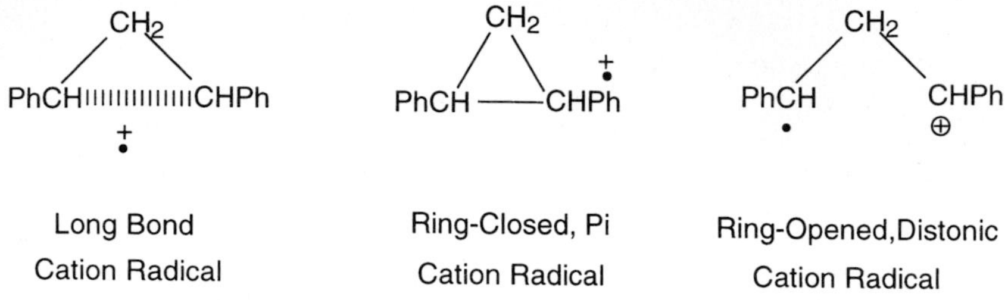

$$\underset{\substack{+\\ \bullet}}{PhCH}\diagdown\!\!\!\!\!\!\!\!\!\!\diagdown\!\!\!\!\!\!\!\!\!\! CH_2 \diagdown\!\!\!\!\!\!\!\! CHPh$$

CH₂ / PhCH⫼⫼⫼⫼⫼⫼CHPh

Long Bond
Cation Radical

CH₂ / PhCH——CHPh (+•)

Ring-Closed, Pi
Cation Radical

CH₂ / PhCH(•)   CHPh(⊕)

Ring-Opened, Distonic
Cation Radical

**Scheme 6.** The long, one-electron bond.

### Sigma cation radicals

The second type of structural adjustment which is pertinent to the present chemistry is the 'long bond'. An especially good example of this phenomenon is available in the case of the 1,2-diphenylcyclopropane cation radical [10]. Any of the three possible structures of this cation radical illustrated in Scheme 6 might be considered to be plausible, a priori.

The distonic cation radical appears plausible because in it a strained cyclopropane bond is broken and both the spin and charge are delocalized benzylically. The ring-closed structure in which the (pi) SOMO (i.e., the cation radical moiety) resides on one of the aryl rings is also plausible, since delocalization of the SOMO over an aromatic ring tends to stabilize it. However, the long bond structure, in which the SOMO is somewhat concentrated in a specific, much-lengthened cyclopropane bond, but with additional benzylic delocalization of the SOMO onto both phenyl rings, is preferred. This long, one electron bond is nevertheless sufficiently strong that the *cis* and *trans* isomers of this cation radical are not interconverting on the relevant time scale (the CIDNP time scale). Nevertheless, all three types of cation radical structure—ring-opened distonic, ring closed pi, and long bond—may play a role in the relevant ET chemistry and all deserve consideration. It should be noted that the long bond (or trapezium) structure is particularly favored by a symmetrical 1,2-disubstitution pattern, which is usually encountered in cation radical cyclobutanation reactions. This is especially the case if the substituents are cation stabilizing substituents, such as aryl or vinyl, but even 1,2-dimethylcyclobutane appears to favor the long bond structure [11]. The cyclobutane cation radical itself, however, has a novel rhomboid structure [11].

### 3.2.4 The Generation of Cation Radicals in Solution

#### The photosensitized electron transfer (PET) method

In their pioneering research on the cyclodimerization of *N*-vinylcarbazole, the Ledwith group used both metal ion oxidants (e.g., $Fe^{3+}$) and photosensitized electron transfer (PET) to generate substrate cation radicals. This latter method, previously

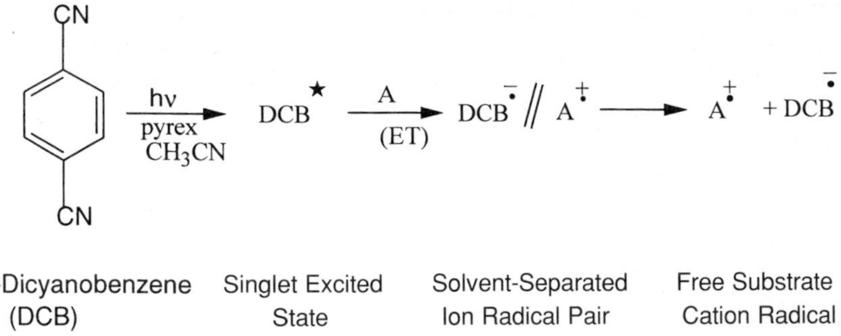

$$\text{DCB} \xrightarrow[\substack{\text{pyrex}\\ \text{CH}_3\text{CN}}]{hv} \text{DCB}^\ast \xrightarrow[\text{(ET)}]{A} \text{DCB}^{\overset{-}{\bullet}} /\!/ A^{\overset{+}{\bullet}} \longrightarrow A^{\overset{+}{\bullet}} + \text{DCB}^{\bullet}$$

| 1,4-Dicyanobenzene (DCB) | Singlet Excited State | Solvent-Separated Ion Radical Pair | Free Substrate Cation Radical |

**Scheme 7.** Photosensitized electron transfer ionization.

used by Ellinger in his own research on *N*-vinylcarbazole [12], has remained rather popular, and the basic processes involved in it are therefore briefly illustrated in Scheme 7.

Essentially, the photoexcited state of an electron deficient photosensitizer becomes a potent single electron acceptor (SEA), in part because of the electronic excitation energy available to drive the reaction, and in part because the electron withdrawing groups on the sensitizer strongly stabilize the resulting sensitizer anion radical. Although Ledwith's group used chloranil as a sensitizer, many other sensitizers are now in common use, including 1,4-dicyanobenzene (DCB),cyanonaphthalene (CN),dicyanonaphthalenes (DCN), and di- and tetracyanoanthracene (DCA and TCA, respectively). A key requirement in the PET method is that the ultraviolet light must be exclusively absorbed by the sensitizer, and not significantly by the substrate, so that direct photochemistry is circumvented. For many substrates it is sufficient merely to carry out the reaction in a Pyrex reaction vessel, but for substrates which have significant absorption at or above 290 nm appropriate filters (e.g., uranium filters) are required. Traditionally, these PET reactions are carried out in dry acetonitrile solvent, even though it is known that cation radicals sometimes react with such polar solvents. Further, hydrocarbon substrates may not be soluble in acetonitrile. To avert reactions of the ion radical intermediates with the solvent, and to provide solubility for most organic substrates, dichloromethane may sometimes be a preferable solvent. The efficiency of PET reactions in dichloromethane and other relatively less polar solvents can be substantially improved through the use of a cationic photosensitizer such as an *N*-methylacridinium salt [13]. When the excited state of such a cation accepts an electron from the substrate, a radical/cation radical pair is formed in contrast to the more usual anion radical/ cation radical pair. In the latter case but not in the former, diffusive separation of the cation radical is retarded by coulombic attractions. Slower diffusive separation, in turn, leads to more back electron transfer and to less efficient generation of free cation radicals. This would be especially true if the solvent is a relatively nonpolar one, such as dichloromethane. Another variation in the PET method consists of using a *co-sensitizer*, i.e., one which does not absorb the ultraviolet light but which, when present in excess, acts as the primary electron donor, forming a co-sensitizer cation radical [14]. The co-sensitizer, often biphenyl, is chosen so that the subsequent electron transfer from the ultimate substrate molecule is exergonic. A consequence of this method of generating substrate cation radicals is that presumably only free substrate cation radicals are formed, thus minimizing the chances of reaction with the anion radical of the sensitizer. Further, the ionization of a substrate or a mixture of substrates is considerably more selective in the much less exergonic process involving a co-sensitizer cation radical.

**The aminium salt method**

In 1981, Bauld and Bellville demonstrated that substrate cation radicals can be generated at a rate sufficient to produce efficient ET cycloaddition chemistry by using triarylaminium salts such as **3**$^{+\cdot}$ in catalytic amounts [15]. Since **3**$^{+\cdot}$ is commercially available and shelf-stable, this chemical method of oxidation is a particularly convenient method for generating cation radicals of substrates which have oxidation potentials up to about 1.6 V vs SCE and for studying their unique ET chemistry in solution. The oxidation potential of **3** is 1.05 V vs. SCE [7]. Other triarylaminium salts such as **4**$^{+\cdot}$ which have even higher oxidation potentials (the oxidation potential of **4** is 1.59 V [7]) are also available and have been employed in many successful ET reactions of less readily oxidizable substrates. A particular advantage of **4**$^{+\cdot}$ is its much greater solubility than **3**$^{+\cdot}$, in dichloromethane, so that reactions can be run at temperatures as low as $-78\,°C$ or even lower.

**Electrochemical and other methods**

Anodic electrochemical (EC) oxidation is also a very attractive procedure for inducing ET chemistry of pi cation radicals [16], and a variety of other ionization methods have also been employed, many of which involve the generation of cation radicals on surfaces. Examples include zeolites [17], clays [18], semiconductors [19], and cation radical polymer surfaces [20]. The generation of cation radicals in solution by photoionization and photosensitized electron transfer, in conjunction with time-resolved spectroscopic studies of the reactions of the cation radical represents a particularly powerful method for the quantitative study of cation radical chemistry, which will be extensively noted in this review [14, 21]. Finally, the generation of cation radicals in rigid matrices by gamma radiolysis has proved to be an excellent way of not only studying the structures of cation radicals, but even of revealing their intramolecular chemistry [22].

**General mechanisms of substrate ionization**

Using a familiar method of classifying the mechanisms of intermolecular electron transfers, the mechanism of substrate ionization may be of the *inner sphere* or the *outer sphere* type. *Outer sphere electron transfer* involves the transfer of an electron between two chemical species without the development of any significant covalent interaction between them, i.e., through space or through another molecule or molecules. In contrast, *inner sphere electron transfer* occurs with the assistance of at least some significant degree of covalent interaction. Typically, photosensitized electron transfer is highly exergonic and the ET is of the outer sphere variety. Evidence strongly suggest that, at least in acetonitrile, the most common solvent for PET reactions, the cation radical/anion radical pair is initially formed as a *solvent separated ion pair* [23]. In solvents which are much less polar, as for example dichloromethane, it is not certain whether solvent-separated or contact ion pairs are formed. A reliable method of generating contact ion pairs is available for instances in which the sensitizer and the substrate form a ground state donor/acceptor complex. Irradiation of this complex at the charge transfer frequency generates the contact cation radical/anion radical pair [23]. In the case of chemical ionization by an agent such as an aminium salt, which is typically an endergonic process, recent evidence indicates that many of these electron transfers occur by an inner sphere process which involves strong covalent interaction, the more precise details of which will be discussed further on [24].

**Electron transfer and hole transfer**

The term 'hole transfer' is frequently used to describe the specific type of electron transfer which occurs between a cation radical and a neutral molecule. The cation radical is considered to have a missing electron or 'hole', and transfers this to the neutral molecule, which then becomes a cation radical, while the original cation radical is neutralized. Hole transfer is of one of the simplest and most common types of reaction between cation radicals and neutral molecules.

### 3.2.5 Cation Radical Cyclobutanation

**Scope of the cation radical chain cyclobutanation reaction**

Subsequent to the work of the Ledwith group, cation radical cyclobutanation was extended to a variety of relatively readily ionizable substrates, including styrenes, indenes, and phenyl vinyl ether (Scheme 8) [25, 26].

Farid also reported the first example of an efficient cross addition, that occurring between dimethylindene and phenyl vinyl ether [27]. Essentially all of these reactions were carried out using the PET method.

**The mechanism of cation radical cyclobutanation**

The stereochemistry of cyclobutanation was first studied using the aminium salt (**3**$^{+\cdot}$) method. The cyclobutadimerizations of both *trans*- and *cis*-anethole were found (Scheme 9) to be stereospecific [28].

Subsequently the same reactions were studied using the PET method (acetonitrile solvent), and confirmed to be stereospecific [29]. The mechanistic significance of these results is that a mechanism analogous to the one proposed by Ledwith for the cyclobutadimerization of *N*-vinylcarbazole which involves a distonic cation radical,

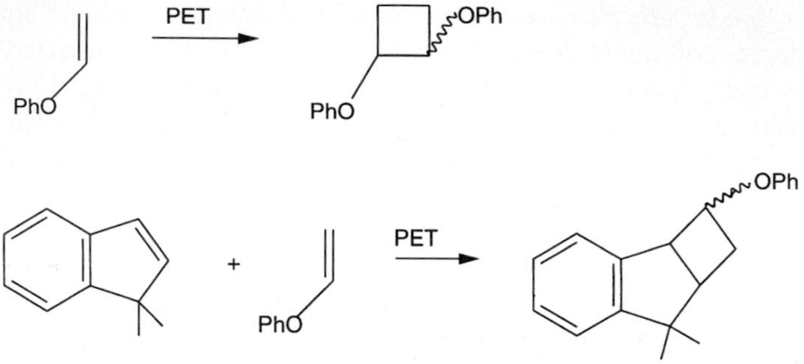

**Scheme 8.** Other early examples of cyclobutanation.

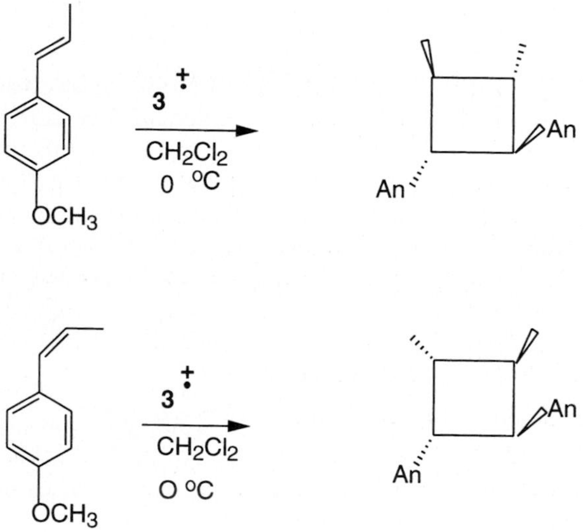

An = p-Anisyl = 4-Methoxyphenyl

**Scheme 9.** The cyclobutadimerization of *trans*- and *cis*-anethole.

is difficult to reconcile with stereospecificity. Since the cycloadditions were found to be completely stereospecific in both a polar (acetonitrile) and a relatively non-polar(dichloromethane) solvent, a more plausible mechanism involves the concerted addition of the *trans*-anethole cation radical to neutral *trans*-anethole (Scheme 10).

The resulting dimer cation radical could conceivably be of either the ring-closed, pi type or the long bond type. By analogy to the known structure of the 1,2-

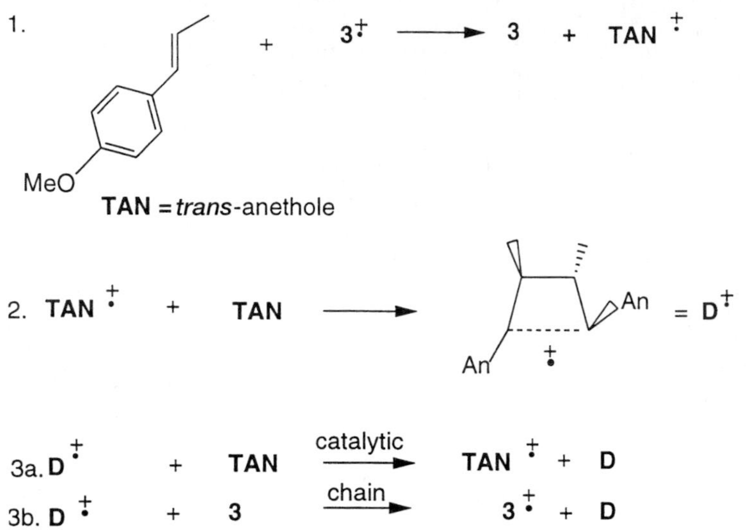

1.

$+$     $3\overset{+}{\cdot}$     $\longrightarrow$     3     +     TAN$\overset{+}{\cdot}$

MeO

**TAN** = *trans*-anethole

2.   **TAN**$\overset{+}{\cdot}$   +   **TAN**   $\longrightarrow$

An = **D**$^{+}$

An$\overset{+}{\cdot}$

3a. **D**$\overset{+}{\cdot}$     +     **TAN**   $\xrightarrow[\text{chain}]{\text{catalytic}}$   **TAN**$\overset{+}{\cdot}$ + **D**

3b. **D**$\overset{+}{\cdot}$   +   3   $\xrightarrow{}$   $3\overset{+}{\cdot}$ + **D**

**Scheme 10.** Mechanism of the aminium-salt induced cation radical cyclobutadimerization of *trans*-anethole.

diphenylcyclopropane cation radical and in view of the calculational results for the 1,2-dimethylcyclobutane cation radical, this dimer cation radical has been assumed to be of the long bond type.

**Selectivity characteristics of cation radical cyclobutanation**

In every case, including those of *trans*-anethole and phenyl vinyl ether, cyclo-butadimerization has proved to be completely head to head regiospecific. This is eminently plausible since in the case of either a styrene or an enol ether type cation radical, there is much more spin and charge density on the carbon beta to the aryl or alkoxy group than alpha to it. Reaction therefore preferentially occurs at the beta position. This tendency is further reinforced by the circumstance that when the covalent bond is established at the beta position of the cation radical, positive charge and/or spin will be generated at the alpha position, where it can be effec-tively stabilized by the aryl or heteroatom substituent. The cation radical also tends to add to the beta carbon of the neutral molecule for this latter reason. Although the thermodynamically more stable *anti* cyclobutane isomer is often favored over the *syn* isomer, *anti* diastereoselectivity is not necessarily high in all cases. An in-teresting case is that of *trans*-anethole dimerization. When the reaction is carried out at $0\,°C$, using the aminium salt catalyst, only the *trans,anti,trans* cyclobutadimer is obtained. However, if the reaction is carried at $-40\,°C$, an approximately 50:50 mixture of this isomer along with the *trans,syn,trans* isomer is obtained (Scheme 11) [28].

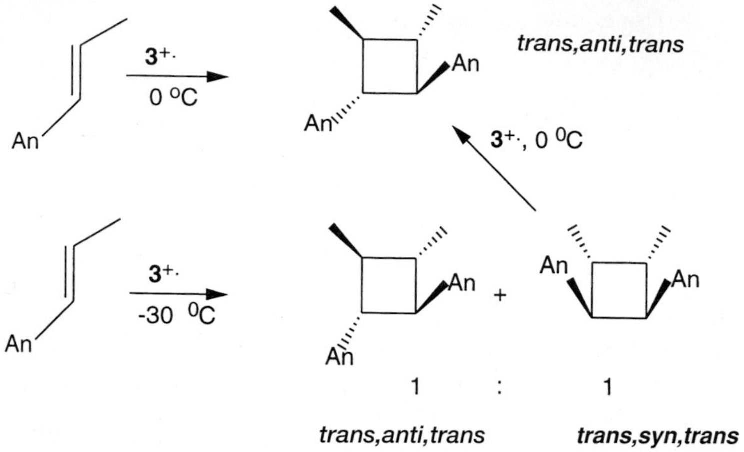

**Scheme 11.** *Syn/anti* diastereoselectivity in the dimerization of *trans*-anethole.

When the latter mixture is treated with the aminium salt at $0\,°C$, the mixture isomerizes completely to the more stable *trans,anti,trans* isomer. Evidently, the cyclodimerization is reversible at the higher temperature. It is also noteworthy that even the reversal is stereospecific, since none of the other possible diastereoisomers are formed. This result is nicely interpreted in terms of the ionization of the *trans,syn,trans* cyclobutadimer to a long bond cation radical, followed by concerted fragmentation to *trans*-anethole and the *trans*-anethole cation radical, avoiding the formation of any *cis*-anethole. The cyclodimerization of phenyl vinyl ether gives both *trans* (or *anti*) and *cis* (or *syn*) isomers initially, but subsequent reaction induces some isomerization of the less stable *cis* isomer to the more stable *trans* isomer. This is believed to occur through the intervention of a distonic 1,4-butanediyl type cation radical.

### Comparison of the selectivity of cation radical cyclobutadimerization with direct photocyclodimerization

The reaction of singlet excited *trans*-anethole with ground state *trans*-anethole is considered to occur *via* an excited state dimer (excimer), in a process which selectively gives rise to the *trans,syn,trans* cyclobutadimer, in direct contrast to the cation radical process [30]. This appears to be a rather general result for direct photodimerization *via* singlet excited states. In further contrast, triplet excited states of *cis* and *trans* isomers undergo extensive *cis/trans* isomerization (of the starting alkene), and moreover the cycloaddition is stepwise. Both of these processes are inconsistent with stereospecific addition. However, head to head regiospecificity is often the strong preference in both singlet and triplet photocyclodimerizations as well as in cation radical cyclodimerizations.

**Rates of cation radical cyclobutanation reactions**

The cation radicals of a substantial number of styrene derivatives have been generated both by photoionization and by photosensitized electron transfer and the absolute rates of their reactions with a variety of neutral styrenes, dienes, and electron rich alkenes measured by means of time resolved UV–visible spectroscopy. As an example, the rate of cycloaddition of the *p*-methoxystyrene cation radical to *p*-methoxystyrene is $1.4 \times 10^9$ $\mathrm{M}^{-1}$ $\mathrm{s}^{-1}$, only about a factor of ten less than the diffusion controlled rate in acetonitrile [31]. Incidentally, the rate of the reverse of this reaction is also rather high, at $8 \times 10^7$. The authors conclude, in agreement with the findings of both the Bauld and Lewis groups, that there is no evidence for a distonic cation radical intermediate, suggesting that the evidence is most consistent with the concerted formation of a long bond cyclobutane cation radical intermediate. Incidentally, the rate constant for the electron transfer reduction of this long bond cyclobutane cation radical is found to be $1.5 \times 10^{10}$ $\mathrm{M}^{-1}$ $\mathrm{s}^{-1}$, i.e., essentially the diffusion controlled rate. That terminal methyl groups attached to the double bond have a substantial steric effect on cation radical cycloadditions is evident from the rate constant for the cycloaddition of the *trans*-anethole cation radical to *trans*-anethole, which is only $8 \times 10^7$ $\mathrm{M}^{-1}$ $\mathrm{s}^{-1}$. This steric effect is also evident in the relative rates of addition of the vinylanisole and *trans*-anethole cation radicals to various alkenes, enol ethers, and styrenes [32]. The profound stabilizing effect of the *p*-methoxy function upon the cation radical moiety is also evident in comparing the relative rates of addition of the parent styrene cation radical and the vinylanisole cation radical to alkenes and enol ethers. Whereas the styrene cation radical reacts even with cyclohexene with a rate constant of $1.2 \times 10^9$, the reaction of the vinyl-anisole cation radical with this simple alkene is too slow to measure ($<5 \times 10^5$).

**Kinetics of the cyclobutadimerization of *trans*-anethole**

A kinetic study has been reported for the cyclodimerization of *trans*-anethole as initiated by $3^{+\cdot}$ in dichloromethane solution [33]. The kinetic rate law which applies over a range of concentrations and for four different temperatures ranging from 0 to 25 °C is second order in the concentration of *trans*-anethole and first order in the aminium salt concentration. These kinetics are consistent with a chain mechanism in which termination is a unimolecular reaction of the *trans*-anethole cation radical, quite possibly *via* de-protonation. The initiation step consists of the ionization of *trans*-anethole by the aminium salt. The second molecule of *trans*-anethole enters the rate law through the first propagation step, in which the *trans*-anethole cation radical reacts with *trans*-anethole. The activation parameters derived for the reaction are $\Delta G_a = 10.6$ kcal $\mathrm{mol}^{-1}$, $\Delta H_a = 2.1$ kcal $\mathrm{mol}^{-1}$, and $\Delta S_a = -29.85$ e.u.. The unusually low value of $\Delta H$ is reflected in the circumstance that the rate decreases by less than a factor of 2 from the highest to the lowest temperature studied. As expected for a reaction involving the generation of *trans*-anethole cation radicals, the reaction rate is slowed by the addition of large amounts of the neutral triarylamine **3**, revealing either reversal of the ionization step under these conditions or conversion of the more efficient chain mechanism, in part, to a catalytic one by

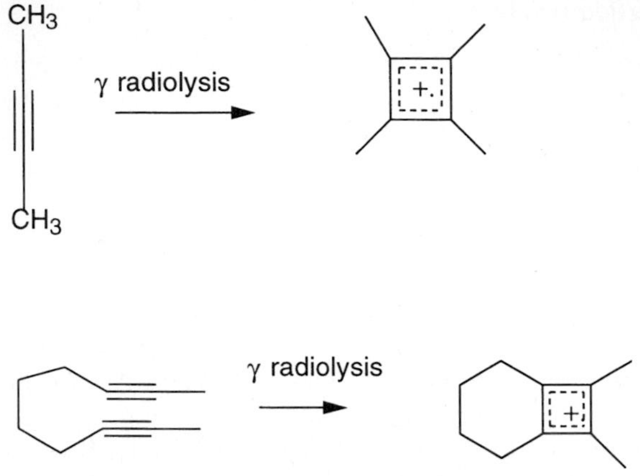

Scheme 12. Cation radical cyclo-additions of alkynes.

neutralizing the dimer cation radical with **3** instead of with *trans*-anethole. However, the effect is small, the addition of 100 mol % of **3** producing only a rate diminution of at most a factor of 4.

**Cation radical cyclobutanation reactions of alkynes, ketenes, and allenes**

The cation radical cyclodimerization of 2-butyne has been observed by ESR spectroscopy under matrix isolation conditions (Scheme 12) [34].

When the cation radical of this alkyne is generated by $\gamma$ radiolysis in a solid matrix at 77 K and then warmed to 150 K, the ESR spectrum of the 1,2,3,4-tetramethyl-1,3-butadiene cation radical is observed. An analogous intramolecular reaction was also observed even in a rigid matrix at 77 K. The feasibility of the cycloaddition step itself is therefore indicated, but little work has yet been done in respect of the aminium salt or PET induced cycloadditions of alkynes in solution at ambient or near-ambient temperatures. Whether a chain or catalytic alkyne cyclo-dimerization can be effected is yet unclear, as is the potential fate of the cyclo-butadiene products.

**Theoretical considerations in cyclobutanation**

A unique aspect of cation radical cycloaddition is the ability to form cyclobutane rings in a thermal reaction of amazing facility. The corresponding reaction on the neutral potential surface is extremely difficult for most unsaturated systems. Like the latter reaction, which is formally a [2 + 2] cycloaddition, the cation radical cy-clobutanation reaction, which is a [2 + 1] cycloaddition, is formally symmetry forbidden [35]. The kinetic driving force for cation radical cycloadditions, in general, as contrasted to neutral cycloadditions will be discussed in detail further on. Reaction path calculations suggest that cation radical cyclobutanations, in the simpler cases, have little or no activation energy and proceed via non-synchronous but, at

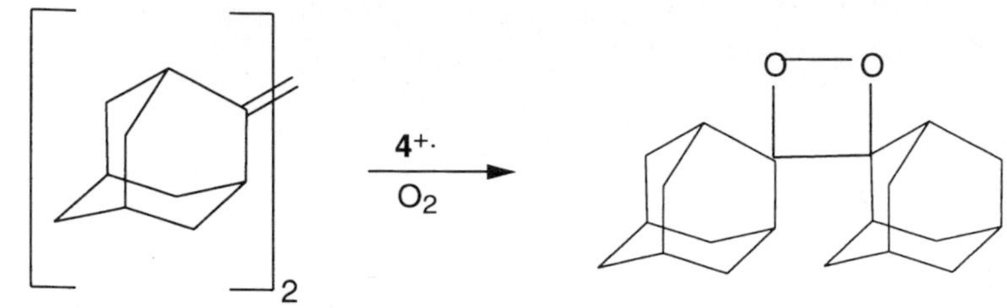

bis(adamantylidene)

**Scheme 13.** Oxetane formation by cation-radical cycloaddition to dioxygen.

least in some cases, concerted reaction paths. For example, the cycloaddition of the ethene cation radical both to ethene and to ethyne involves the formation of a loose, T-shaped ion/molecule complex intermediate which collapses smoothly to the cyclobutane or cyclobutene cation radical, respectively [11, 36]. Further results indicate that the ethyne cation radical also adds to neutral ethyne via an effectively concerted path [36]. In none of these cases do the calculations reveal the intervention of a distonic cation radical intermediate.

### 3.2.6 Formation of 1,2-Dioxetanes by Cation Radical Additions to Triplet Dioxygen

In a reaction quite analogous to cation radical cyclobutanation, cation radicals of some relatively ionizable, but sterically hindered alkenes have been found to add to ground state dioxygen in a very efficient cation radical chain process (Scheme 13) [37].

The reaction is somewhat limited in scope because cyclobutadimerization is a potentially competing reaction for substrates which are ionizable but not sterically hindered, and conjugated dienes react with oxygen under these same conditions to give 1,4-endoperoxides (vide infra). The reaction is best carried out using the more potent, hexabromo salt at $-78\,°C$, and has been found to be incompletely stereospecific. Consequently, a distonic cation radical intermediate has been postulated to play a role in the mechanism (Scheme 14).

### 3.2.7 Cation Radical Diels–Alder Cycloadditions

#### Historical

The $\gamma$ radiolysis of 1,3-cyclohexadiene was reported, in 1969, to yield two sets of dimers [38]. One set was the familiar pair of *syn* and *anti* cyclobutane dimers which had previously been obtained via triplet sensitized photodimerization. The other set

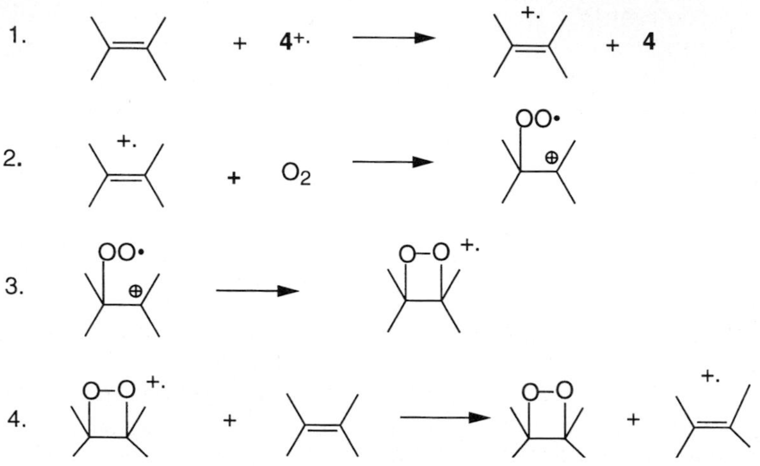

**Scheme 14.** Cation radical chain mechanism for oxetane formation.

of dimers was the *endo/exo* Diels–Alder dimer pair. The Diels–Alder dimers were shown to arise primarily through a cation radical chain mechanism, which was highly periselective for Diels–Alder cycloaddition as opposed to cyclobutanation. The cyclobutane dimers and a portion of the Diels–Alder dimers (primarily the *exo* dimer) were produced by triplet 1,3-cyclohexadiene chemistry.

**The aminium salt catalyzed cation radical Diels–Alder reaction**

In 1981, this cation radical Diels–Alder cyclodimerization of 1,3-cyclohexadiene was shown to be more cleanly (only 1 % of the cyclobutane dimers is produced), conveniently (in a synthetic organic context), and efficiently (70 % yield) carried out by chemical ionization of the diene, using **3**$^{+\cdot}$ (Scheme 15) [39].

The reaction was carried out in dichloromethane solvent at 0 °C for 5 min, using 5 mol % of the initiator. The *endo/exo* ratio (4.5:1) was similar to that found in the thermal Diels–Alder cyclodimerization of this same diene. This encouraging result and the convenience of the procedure led to the extensive study of the cation radical Diels–Alder reaction and its utility in synthetic organic chemistry [40]. It is of further note that the same cyclodimerization was subsequently carried out by the PET

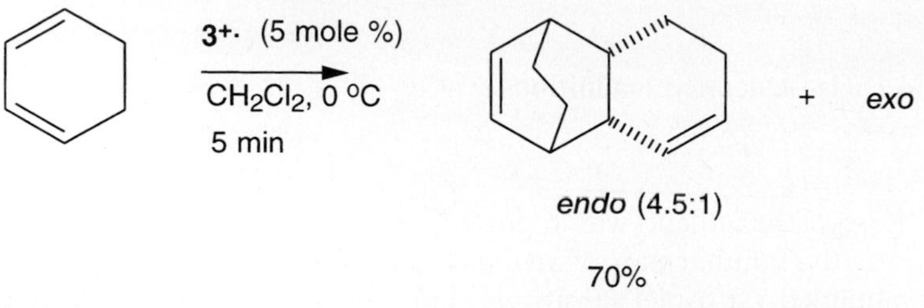

**Scheme 15.** The aminium salt initiated cation radical Diels–Alder reaction.

method, and the rate constant for the cycloaddition of the 1,3-cyclohexadiene cation radical to the neutral diene was estimated to be $2 \times 10^8$ s$^{-1}$, only a factor of
sixty less that the diffusion controlled rate [41]. Other relatively ionizable 1,3-
cyclohexadiene derivatives have more recently been found to undergo facile cation
radical Diels–Alder cyclodimerization. These include 1-methyl-1,3-cyclohexadiene
and 1-methoxy-1,3-cyclohexadiene. Readily ionizable acyclic dienes such as 2,4-
dimethyl-1,3-pentadiene and 1,1'-dicyclohexenyl also undergo such cyclodimerization, but dienes which ionize with more difficulty, such as 1,3-butadiene, isoprene,
2,3-dimethyl-1,3-butadiene, and 1,3-cyclopentadiene do not. Another very early
development in cation radical cycloaddition chemistry was the photosensitized
electron transfer cyclodimerization of 1,1-diphenylethene first reported in 1973 by
Neunteufel and Arnold [42]. These authors reported the efficient (70 %) formation
of a Diels–Alder type adduct using methyl 4-cyanobenzoate as the sensitizer. Subsequently, Mattes and Farid observed the formation of both the cyclobutadimer
and Diels–Alder type dimers, and proposed a common distonic cation radical intermediate (Scheme 16) [43].

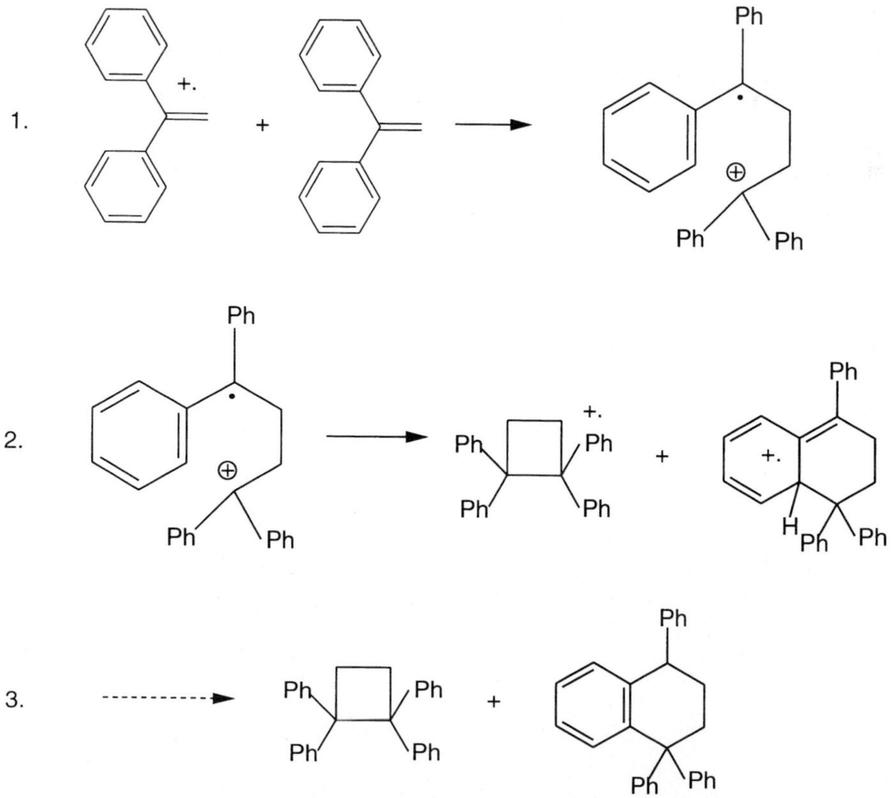

**Scheme 16.** The pet induced cyclodimerization of 1,1-diphenylethene.

Note that the DA dimers are formed in an unusual way, in effect, by using a dienic moiety which includes one of the benzene ring bonds. Additional important observations of Diels–Alder reactions induced under PET conditions were reported by Libman [44] and by Mizuno [45].

### Stereochemistry of the cation radical Diels–Alder reaction

The uncatalyzed Diels–Alder reaction is well known to be highly stereospecific, preferentially occurring via syn addition to both the diene and dienophilic components. Stereochemical studies of the cation radical Diels–Alder reaction have confirmed an analogous stereospecificity in two distinctly different systems. The initial study was carried out using the cycloaddition of the three geometric isomers of 2,4-hexadiene as dienophilic components and 1,3-cyclohexadiene as the diene component [39]. Each of the three isomers of the acyclic diene was found to add stereospecifically to cyclohexadiene. In a more recent study, the *cis* and *trans* isomers of 1,2-diaryloxyethenes were found to add stereospecifically to 1,3-cyclopentadiene (Scheme 17) and also to 2,3-dimethyl-1,3-butadiene [46].

### Regiochemistry of the cation radical Diels–Alder reaction

In most cases the cation radical Diels–Alder reaction has proved to be highly regiospecific with respect to the reaction of an unsymmetrical dienophile with an unsymmetrical diene. As examples, the cyclodimerizations of both 1-methyl-1,3-cyclohexadiene and 1-methoxy-1,3-cyclohexadiene are essentially regioexclusive (Scheme 18) [47–49].

It is noteworthy that the latter cyclodimerization cannot be carried out using the aminium salt method, because both the reactant and the Diels–Alder product react with the aminium salt in a destructive manner.

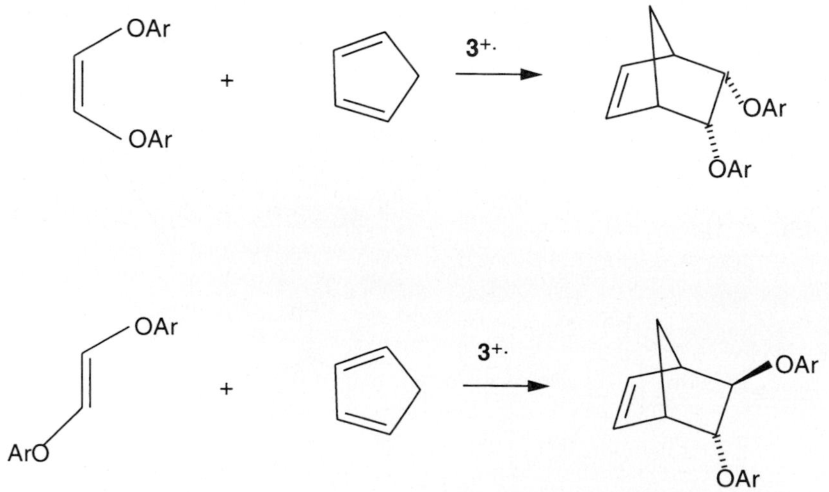

**Scheme 17.** The stereospecificity of the cation radical Diels–Alder reaction.

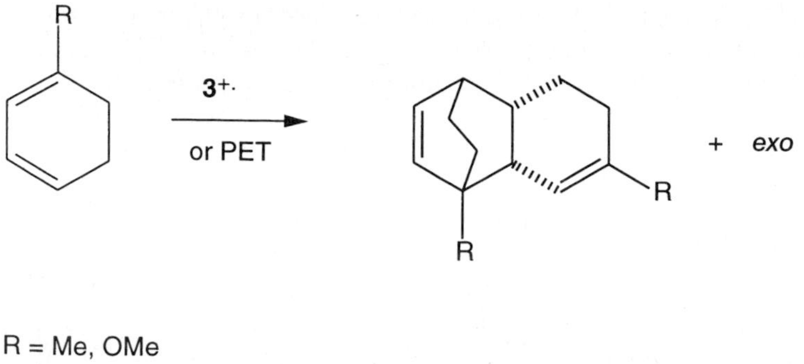

R = Me, OMe

**Scheme 18.** Regiospecificity of the cation radical Diels–Alder.

## *endo/exo* **Diastereoselectivity**

Besides being essentially 100 % stereospecific, the reaction of *trans,trans*-2,4-hexadiene with 1,3-cyclohexadiene is at least 98 % *endo* diastereoselective (Scheme 19) [39, 47, 48].

However, the cyclodimerization of 1,3-cyclohexadiene and also the addition of the *cis,cis* isomer of 2,4-hexadiene to 1,3-cyclohexadiene are only modestly stereoselective. The addition of *cis,trans*-2,4-hexadiene to 1,3-cyclohexadiene is highly stereoselective for the addition to the *trans*-propenyl group, but only modestly stereoselective for the addition to the *cis*-propenyl group. Further, the addition of a dienophile having a pendant, unsubstituted vinyl double bond to this diene is also highly *endo* stereoselective. The installation of a *cis* group at the terminus of the dienophilic moiety consistently appears to reduce the *endo* stereoselectivity to a more modest level. It has been proposed that the *cis* substituent attenuates the secondary interaction involving the *endo* double bond in the transition state for cycloaddition [47, 48]. The effect has been termed the '*cis*-propenyl effect'. The addition of the *trans*-anethole cation radical to both 1,3-cyclohexadiene and 1,3-cyclopentadiene is, however, only moderately diastereospecific (ca 3:1) [49].

## **Scope of the cation radical Diels–Alder reaction**

Although the (neutral) Diels–Alder reaction is one of the most useful in the repertoire of synthetic organic chemistry, it has, like any other reaction, some well

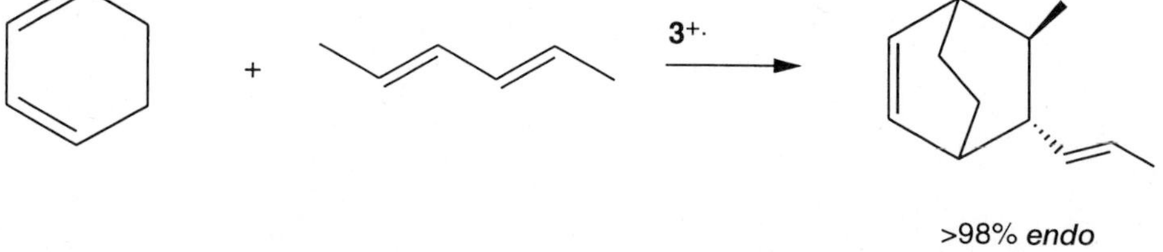

>98% *endo*

**Scheme 19.** *endo* Diastereoselectivity of the cation radical Diels–Alder.

defined limitations. In particular, it is rather sensitive to steric effects in both the dienophilic and dienic components, and it tends to work much more efficiently with electron deficient dienophiles than with electron rich ones. Although the cation radical Diels–Alder reaction has its own set of limitations, it is nicely complementary to the neutral Diels–Alder in being far less sensitive to steric effects, and in being most readily applicable to electron rich dienophiles. The range of substrate molecules which have oxidation potentials appropriate for use in cation radical cycloadditions induced by $3^{+\cdot}$ would appear to be approximately from about 1.2 V to about 1.6 V vs. SCE. Where the oxidation potential is substantially less than 1.2 V, the substrate ionization may be too rapid, resulting in cation radical/cation radical reactions (especially coupling). *Trans*-1,2-di-*p*-anisylethene ($E_{ox} = 1.20$ V vs SCE), which is readily ionized but fails to undergo either cyclodimerization or Diels–Alder addition, appears to be such a case. Stilbene, which has an ionization potential of 1.59 V, reacts cleanly but rather slowly, taking hours instead of the usual minutes to react when $3^{+\cdot}$ is used as the catalyst. Classes of ionizable substrates include many conjugated dienes (but not 1,3-butadiene and 1,3-cyclopentadiene, the oxidation potentials of which are much too high), electron rich styrene derivatives (especially those with beta alkyl substitution), aryl vinyl sulfides, aryl vinyl and aryl propenyl ethers, and 1,2-diaryloxyethenes. It appears to be rather generally the case that in cross additions (as opposed to dimerizations) the substrate which has the lower oxidation potential preferentially reacts as the cation radical component of the cation radical Diels–Alder reaction. Under PET and anodic oxidation conditions, enol alkyl ethers, *N*-vinylamides (enamides), and vinylindoles are also potentially viable substrates.

**Additions to sterically hindered dienophiles**

With respect to the matter of steric sensitivity, the reaction of 2,5-dimethyl-2,4-hexadiene with 1,3-cyclohexadiene is instructive [39]. Apparently, no Diels–Alder additions of this sterically hindered acyclic diene, either as the dienophilic or dienic component, have ever been reported. However, the cation radical Diels–Alder cycloaddition referred to above (Scheme 20) occurs smoothly, the readily ionizable acyclic diene serving as the dienophilic component.

Of further interest is the observation that the same reaction is not observed at all under PET conditions, which should reliably furnish the same diene cation radical [50]. Further, the normally efficient cation radical cyclodimerization of 1,3-cyclohexadiene is completely inhibited in the presence of the hindered diene, so that

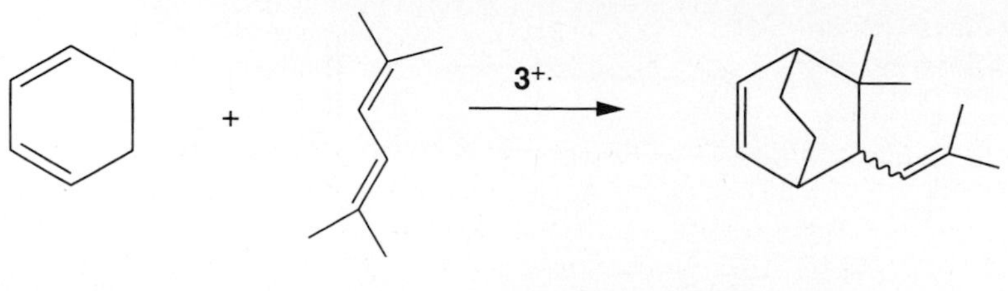

**Scheme 20.** Diels–Alder addition of a sterically hindered dienophile.

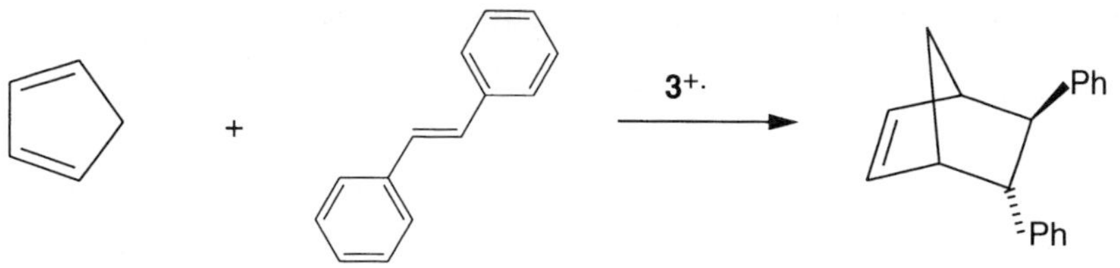

**Scheme 21.** Diels–Alder addition of stilbenes.

no cycloaddition at all occurs, even though the cation radicals of both dienes should be formed in the highly exergonic electron transfer to the excited state sensitizer. Evidently the rate of cycloaddition of the acyclic diene cation radical to the cyclic diene is indeed relatively slow as a result of steric repulsions, and the lifetime of this cation radical is therefore great enough that it is quenched by back electron transfer from the sensitizer anion radical more rapidly than it can add to 1,3-cyclohexadiene. The cation radical of the latter diene, assuming that it is indeed produced, evidently must undergo rapid, probably diffusion controlled, electron transfer from (hole transfer to) the acyclic diene, which has a much lower oxidation potential. Another reaction system which highlights the relatively lower level of sensitivity of the cation radical Diels–Alder reaction to steric effects is the stilbene/1,3-cyclopentadiene system (Scheme 21).

Although neutral Diels–Alder reactions of stilbene as a dienophile are unknown, the cation radical Diels–Alder reaction cited above occurs smoothly [51]. *Trans* stilbenes yields only *trans* adducts, while *cis*-stilbenes yield primarily, but not exclusively, *cis* adducts. The stereospecificity of the cycloaddition step in the latter reaction system was, however, not readily evaluated because of the occurrence of competing *cis* to *trans* isomerization of the starting material. It was established, however, that at low conversions, the reaction tends toward stereospecificity. Since *cis*-stilbene is much less readily ionizable than *trans*-stilbene, the Diels–Alder cycloadditions in the *cis*-stilbene system were studied using the $p,p'$-dimethyl derivative.

**Electron-rich alkenes as dienophiles in the cation radical Diels–Alder reaction**

In the present context, the term electron rich alkenes refers primarily to enol ethers, enol sulfides, and N-vinylamides or N-vinylamines. Such alkenes are typically much more readily ionizable than are simple alkenes. The conversion of these substrates to the corresponding (highly electron deficient) cation radicals represents a sharp Umpolung. The Diels–Alder additions of *trans*-anethole, phenyl vinyl ether, phenyl vinyl sulfide, 1,3-dioxole, and N-methylindole to 1,3-cyclohexadiene have been reported (Scheme 22) [49, 52].

It should also be born in mind that electron rich alkenes are also especially reactive neutral components of cation radical cycloaddition reactions, since they are also highly nucleophilic. Consequently, in appropriate instances, either role sense of the cation radical Diels–Alder reaction may be operative, i.e. either the diene or the electron rich alkene could be reacting as the cation radical.

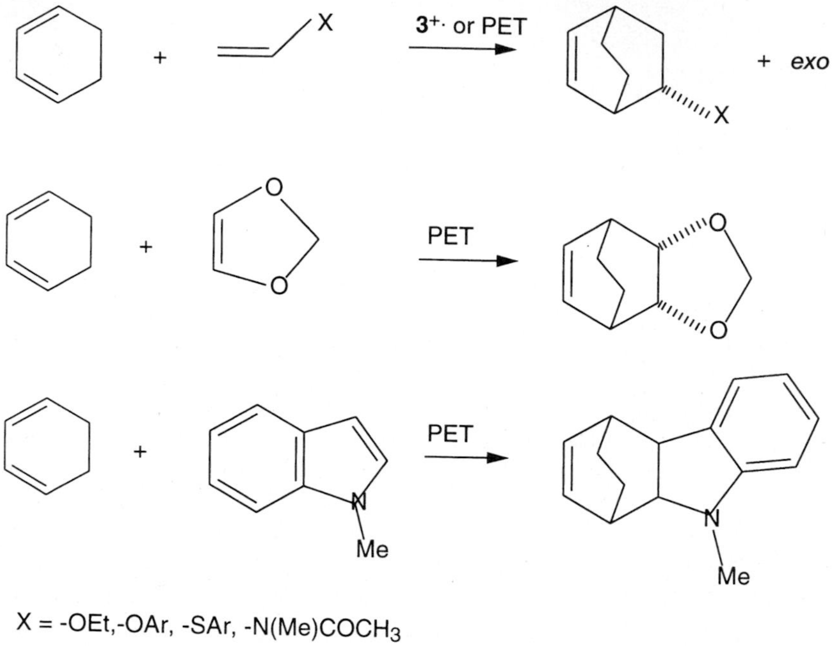

X = -OEt,-OAr, -SAr, -N(Me)COCH₃

**Scheme 22.** Diels–Alder additions to electron rich alkenes.

### Diels–Alder cycloadditions to ketenes and allenes

The facile Diels–Alder addition of certain ketenes and allenes to a relatively readily ionizable diene, 1,2,3,4,5-pentamethylcyclopentadiene, initiated by $3^{+\cdot}$ has also been observed (Scheme 23) [53, 54].

### Role-selectivity in the cation radical Diels–Alder reaction

In view of the demonstrated stereospecificity of at least some cation radical Diels–Alder reactions, it is at least possible that these reactions, like the neutral Diels–Alder, are true pericyclic reactions, i.e., they may occur via a concerted cyclo-addition. The results of a variety of calculations, however, make clear that the cycloadditions must at least be highly non-synchronous, so that the extent of the formation of the second bond, which completes the cyclic transition state, is no more than slight [55, 56]. If the cation radical Diels–Alder reaction is nevertheless interpreted as pericyclic and the concept of orbital correlation diagrams is applied to them, it emerges that the cycloaddition is symmetry allowed if the ionized (cation radical) component is the dienophile, but forbidden if it is the diene [39, 55]. The former mode of reaction has been referred to as the [4 + 1] mode, and the latter as the [3 + 2] mode. Interestingly, the great majority of cation radical Diels–Alder reactions thus far observed seem to represent the formally allowed [4 + 1] mode. An interesting case in point is the reaction of 1,1′-dicyclohexenyl with 2,3-dimethylbutadiene (Scheme 24) [57].

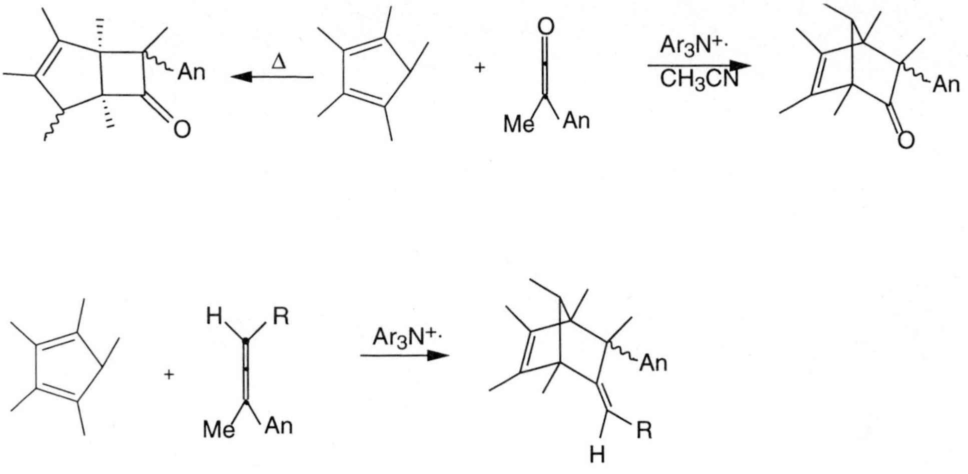

Ar = 4-methylphenyl
An = 4-methoxyphenyl

**Scheme 23.** Cation radical cycloadditions of ketenes and allenes.

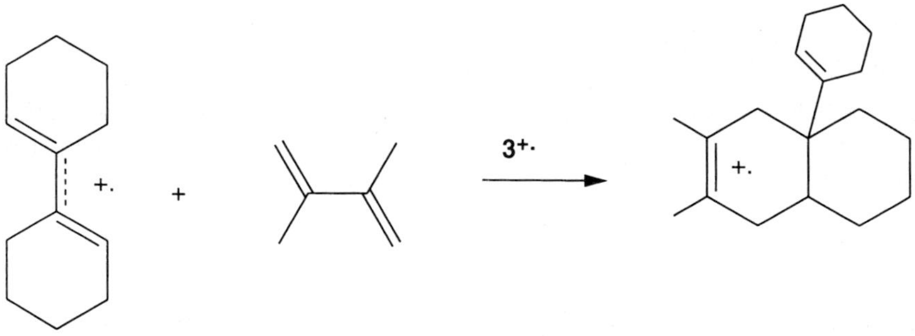

**Scheme 24.** Role preferences in the cation radical Diels–Alder.

Both of these dienes are conformationally flexible as regards the adoption of an *s-cis* conformation, but the former apparently has a somewhat higher *s-cis* content and might therefore be expected to preferentially adopt the dienic role. However, the product corresponds to the use of 2,3-dimethyl-1,3-butadiene in the dienic role. Since this latter diene is by far the more difficult to ionize of the two dienes, the observed reaction must be of the allowed [4 + 1] type. The possibility that this empirically observed sense of role selectivity has its origin in orbital symmetry allowedness/forbiddenness was tested in the manner shown in Scheme 25.

In each case the ionized component is readily identified by the large difference in oxidation potentials of the two reactants, but there is little, if any, preference for the

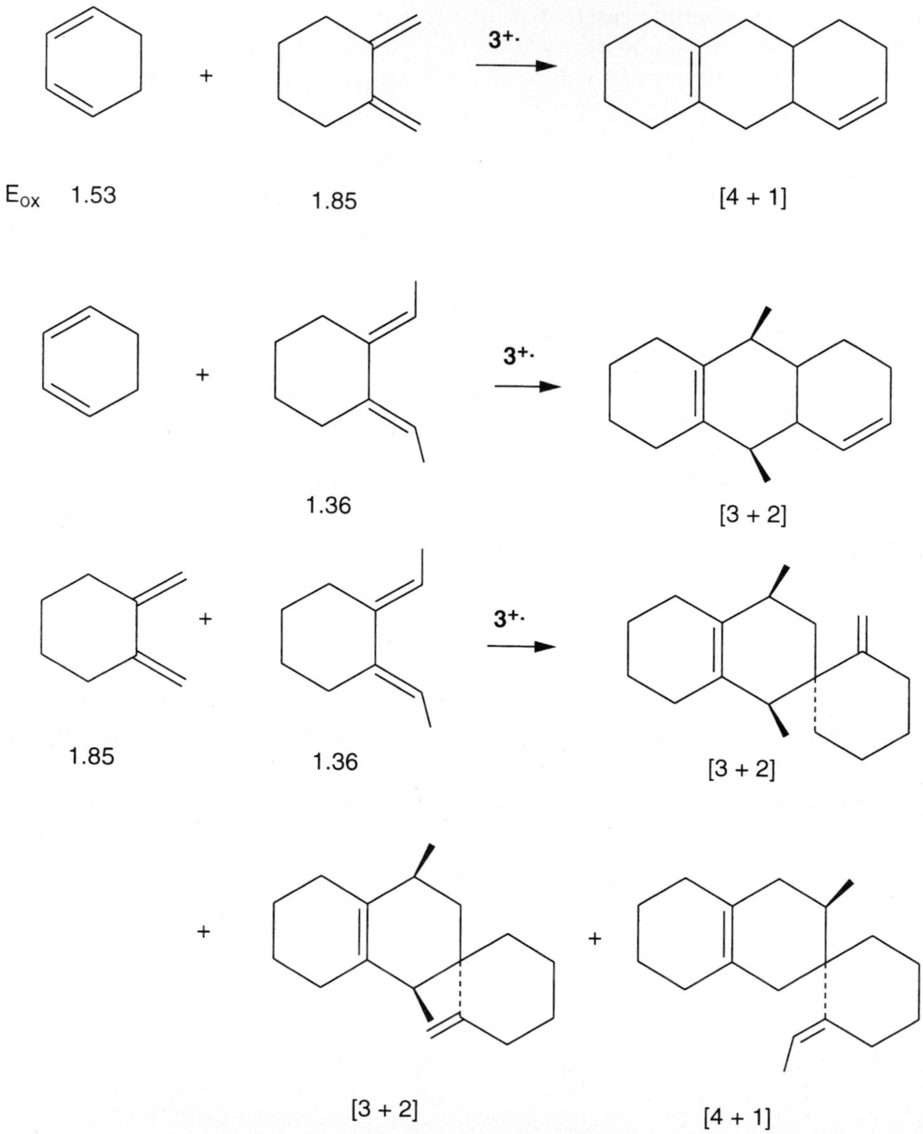

$E_{OX}$    1.53                1.85                                    [4 + 1]

1.36                                    [3 + 2]

1.85                1.36                                    [3 + 2]

[3 + 2]                                    [4 + 1]

**Scheme 25.** Test of orbital symmetry effects on role specificity.

ionized component to participate as the dienophilic as opposed to the dienic component. For instance, in the final example of Scheme 25, the ratio of [4 + 1]/[3 + 2] products is 1:1.2. This result is not especially surprising since, as mentioned previously, the highly non-synchronous transition state can have but little pericyclic character. It is of further interest that cation radical cyclobutanation, a formally [2 + 1] cycloaddition, is also formally symmetry forbidden, but is nevertheless ex-

tremely rapid in appropriate cases. A more appropriate explanation for the preference for [4 + 1] cycloaddition over [3 + 2] cycloaddition, based upon conformational effects, will be proposed further on.

**Periselectivity**

An intriguing competition arises in the context of cation radical cycloadditions (as
in the context of Diels–Alder cycloadditions) which involve at least one conjugated
diene component. Since both cyclobutanation and Diels–Alder addition are extremely facile reactions on the cation radical potential energy surface, it would
not be surprising to find a mixture of cyclobutane (CB) and Diels–Alder (DA) addition to the diene component in such cases. Even in the cyclodimerization of 1,3-
cyclohexadiene, *syn* and *anti* cyclobutadimers are observed as 1 % of the total
dimeric product. Incidentally, the DA dimers have been shown not to arise indirectly via the CB dimers in this case [58]. The cross addition of *trans*-anethole to
1,3-cyclohexadiene also proceeds directly and essentially exclusively to the Diels–
Alder adducts (*endo* > *exo*). Similarly, additions to 1,3-cyclopentadiene yield essentially only Diels–Alder adducts. However, additions to acyclic dienes, which
typically exist predominantly in the *s-trans* conformation which is inherently unsuitable for Diels–Alder cycloaddition, can yield either exclusively CB adducts, a
mixture of CB and DA adducts or essentially exclusively DA adducts (Scheme 26)
[59].

   Although clearly the *s-cis* conformational content of the diene component has a
major effect on the CB/DA periselectivity, a second effect related to role specificity
also appears to represent a significant factor (Scheme 27).

   If an *s-trans* diene component is ionized to the corresponding cation radical, it
cannot participate in a Diels–Alder cycloaddition directly, and the large bond order
between C2–C3 of the diene cation radical strongly prohibits equilibration with the
*s-cis* diene cation radical. When proceeding in this role sense, only CB adducts can
result. In contrast, if the dienophilic component is ionized, either DA or CB adducts
can result, since the *s-trans* neutral diene, at least in some cases, may be able to
equilibrate with the *s-cis* conformer at a rate faster than the cycloaddition rate. As a
further illustration, the reaction of 1,1-dicyclopentenyl with phenyl vinyl sulfide
yields an initial adduct distribution corresponding to a CB/DA ratio of 3.02. The
excess of CB products is plausible in view of the slightly lower oxidation potential
of this diene (1.34) than phenyl vinyl sulfide (1.42). If the oxidation potential of the
vinyl sulfide is increased by using a *p*-bromo substituent, the CB/DA ratio is increased to 8.74. On the other hand, if the oxidation potential is decreased by the use
of a *p*-ethyl substituent, the CB/DA is decreased to 0.85. This type of effect could
also explain, at least in part, the tendency toward the prevalence of [4 + 1] over
[3 + 2] cycloadditions. In each of the three cases described above, the rearrangement of the CB adducts to DA adducts via a cation radical mechanism (vide infra)
is relatively rapid, so that the initial product ratio at very low conversions must be
evaluated [60]. Finally, in connection with the CB vs DA competition represented
in Scheme 26, it should be noted that the oxidation potentials are in the order
*trans*-anethole (lowest) and vinyl ethers (highest), with phenylvinyl sulfide and 1,1-

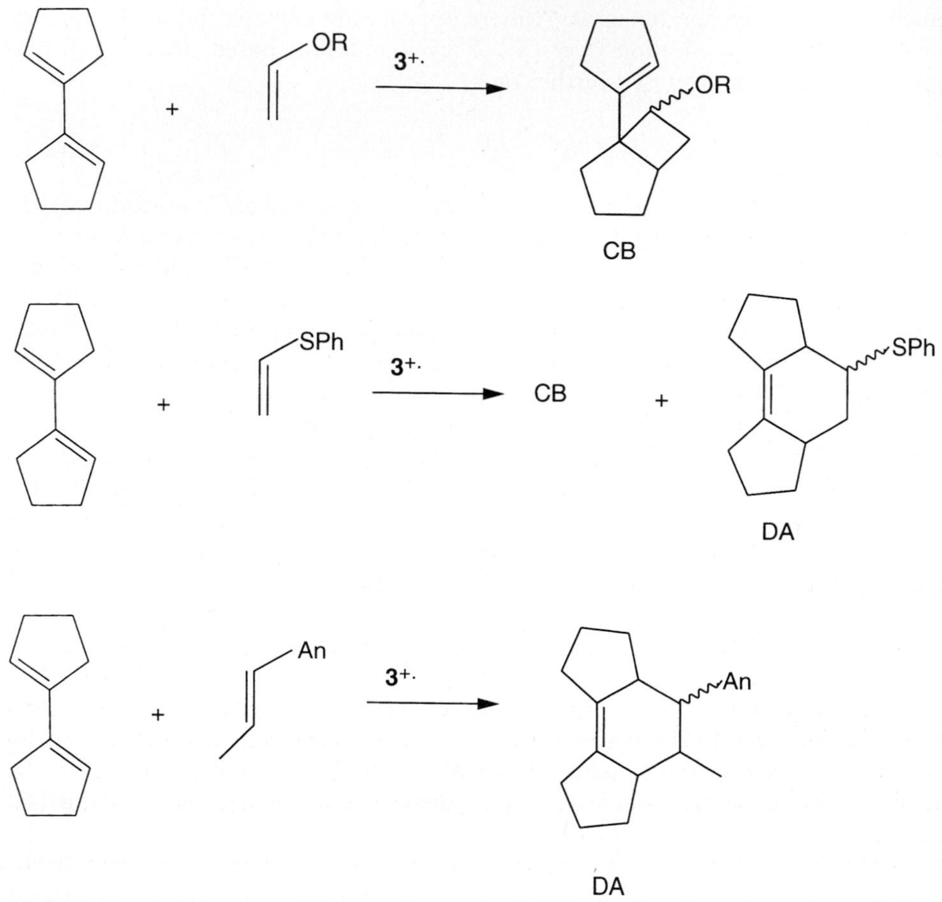

**Scheme 26.** Competition between Diels–Alder and cyclobutanation paths for acyclic diene components.

dicyclopentenyl being intermediate. This order of oxidation potentials is thus consistent with the expectation that if an acyclic diene is the more ionizable component, cyclobutanation (i.e. $[2 + 1]$ addition) is the predominant result, and a $[3 + 2]$ Diels–Alder addition is unlikely. On the other hand, if the dienophilic component is more oxidizable, the $[4 + 1]$ Diels–Alder product is likely to predominate.

**Periselectivity in cycloadditions of electron-rich alkenes**

The cation radical cycloaddition reactions of conjugated dienes which can adopt either the *s-cis* or *s-trans* conformation with a dienophile can proceed to yield either CB or DA adducts or an admixture of both, as noted above. In the case where the dienophile is an electron rich alkene, the tendency toward CB adduct formation appears to be even more pronounced than in the case of a dienophile of the diene or

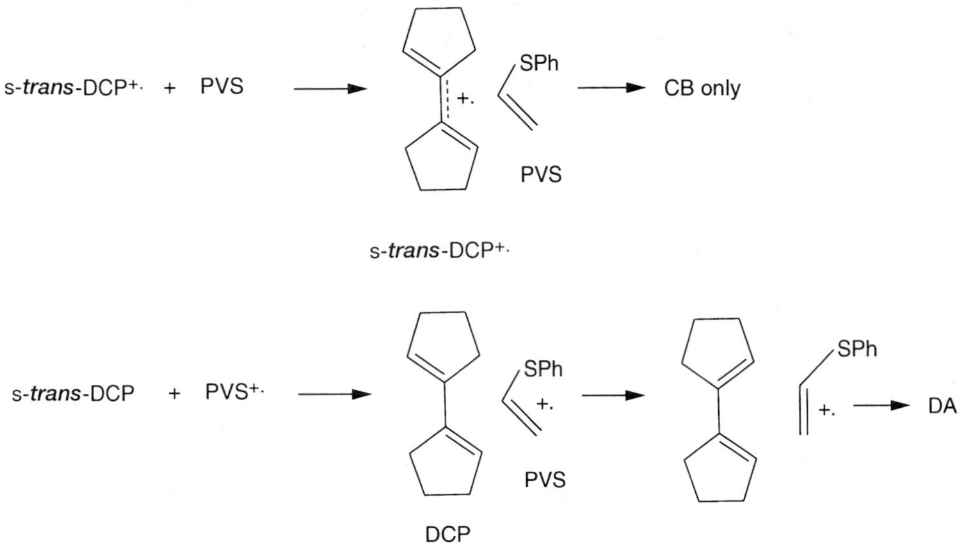

**Scheme 27.** Conformational effects on CB vs DA periselectivity.

styrene type [59a]. As will be noted further on, the tendency of these initially formed vinylcyclobutane adducts to quickly rearrange via a cation radical mechanism to the cyclohexene (i.e., Diels–Alder) adducts is also more pronounced. In the case of a very highly electron rich substrate such as N-methyl-N-vinylacetamide, even the reaction with the rigidly s-cis-1,3-cyclohexadiene is CB periselective (Scheme 28) [61].

Observations such as this tend to suggest that the TS for cation radical cyclo-butanation has the positive charge either more localized or more highly developed than in the Diels–Alder TS, so that strongly carbocation stabilizing functions tend to favor the cyclobutanation reaction over Diels–Alder addition. It is important to note that many electron rich alkenes are incompatible with the aminium salt cata-lyst, requiring that their cation radical cycloadditions be carried out using the PET method or anodic oxidation. Vinyl sulfides and enol phenyl ethers such as phenyl vinyl ether or phenyl propenyl ether are exceptions to this generalization, and can be used effectively in the context of aminium salt catalysis.

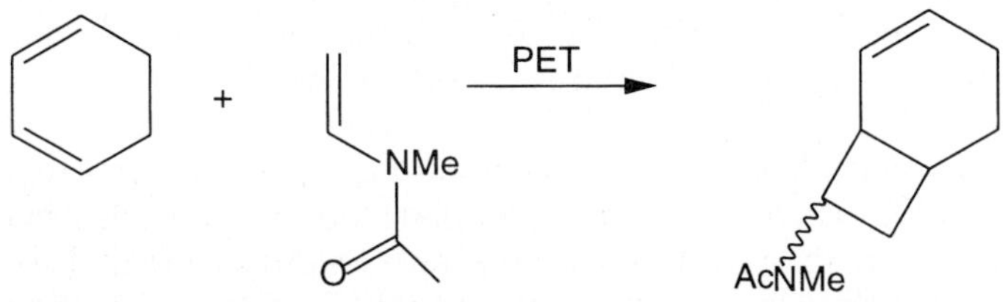

**Scheme 28.** Cyclobutane periselectivity in electron rich alkenes.

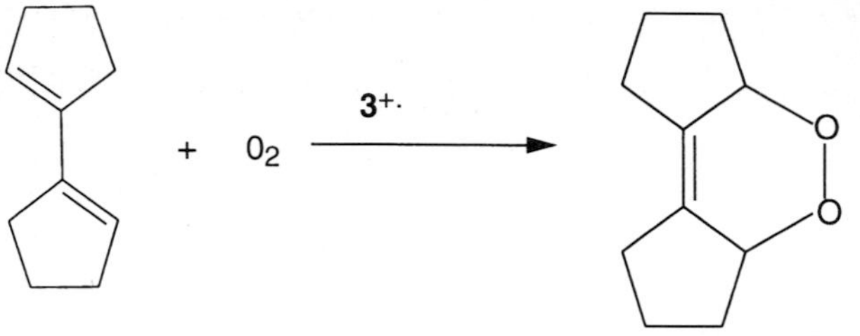

**Scheme 29.** Cation radical chain dioxygenation of conjugated dienes.

### 3.2.8 Cation Radical Diels–Alder Cycloadditions to Dioxygen

The cycloaddition of an *s-cis* conjugated diene to ground state (triplet) dioxygen to give the hetero Diels–Alder adduct, a 1,4-*endo* peroxide, in its ground electronic state is spin forbidden. In an early, ingenious attempt to break down the 'spin barrier' toward the addition of a singlet molecule to triplet oxygen, the triarylaminium salt was successfully employed as a catalyst (Scheme 29) [62].

Although a cation radical mechanism was not initially proposed, it was later established that these reactions proceed via a cation radical chain process involving the cation radical of the conjugated diene [63]. Since the *s-trans* cation radical can not directly undergo cyclization to the *endo* peroxide cation radical, the reaction either must proceed via the *s-cis* cation radical or perhaps, in the case of conformationally flexible dienes, indirectly via the vinyl dioxetane cation radical (the CB type product), which subsequently rearranges to the *endo* peroxide (the DA type product).

### 3.2.9 Mechanisms of Formation of Substrate Cation Radicals from the Aminium Salt

#### General considerations

The question of the nature of the process whereby substrate cation radicals are generated from the neutral substrate by electron transfer to an aminium salt (or other *chemical* agents) is of fundamental mechanistic interest. Moreover it is relevant to the *stereochemical outcome* of the reaction, as well as to the *selectivity* in ionizing one of two dissimilar reacting neutral molecules (as in the cation radical Diels–Alder reaction or in cross cyclobutanations) or when two or more functional groups are present in a single substrate molecule. Since substrate ionization via the aminium salt $3^{+\cdot}$ is virtually always endergonic, it is considered highly *unlikely* that solvent separated pairs would be formed directly, i.e. that electron transfer would occur through a solvent molecule. This conclusion follows because solvent reorganization and thus the intrinsic activation energy for the latter type of electron transfer is typically much greater than for electron transfers yielding intimate (i.e., contact) pairs. Thus, if electron transfer is of the outer sphere type, i.e., involves no

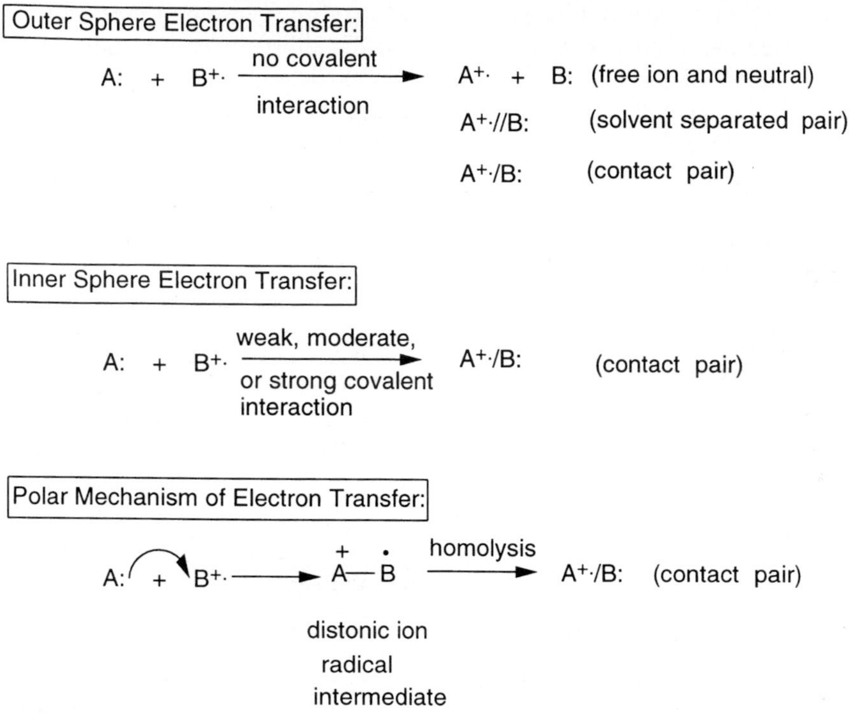

Outer Sphere Electron Transfer:

$$A{:} \;\; + \;\; B^{+\cdot} \xrightarrow[\text{interaction}]{\text{no covalent}} A^{+\cdot} \;\; + \;\; B{:} \quad \text{(free ion and neutral)}$$

$$A^{+\cdot}//B{:} \qquad \text{(solvent separated pair)}$$

$$A^{+\cdot}/B{:} \qquad \text{(contact pair)}$$

Inner Sphere Electron Transfer:

$$A{:} \;\; + \;\; B^{+\cdot} \xrightarrow[\substack{\text{or strong covalent} \\ \text{interaction}}]{\text{weak, moderate,}} A^{+\cdot}/B{:} \qquad \text{(contact pair)}$$

Polar Mechanism of Electron Transfer:

$$A{:} \; + \; B^{+\cdot} \longrightarrow \overset{+}{A}\!\!-\!\!\overset{\bullet}{B} \xrightarrow{\text{homolysis}} A^{+\cdot}/B{:} \quad \text{(contact pair)}$$

distonic ion
radical
intermediate

**Scheme 30.** Inner and outer sphere electron transfer mechanisms.

significant covalent interaction between the substrate and the aminium salt, it must presumably generate contact pairs. On the other hand, the range of possible inner sphere mechanisms is quite large, encompassing covalent interactions ranging from very weak to quite strong (Scheme 30).

The limit of the strong interaction case is presumably a stepwise electrophilic reaction establishing a full covalent bond in a distonic cation radical intermediate, followed by homolysis of this bond to yield the substrate cation radical and the neutral triarylamine (Scheme 31).

A plausible variation of this latter theme is the oxidation of the distonic cation

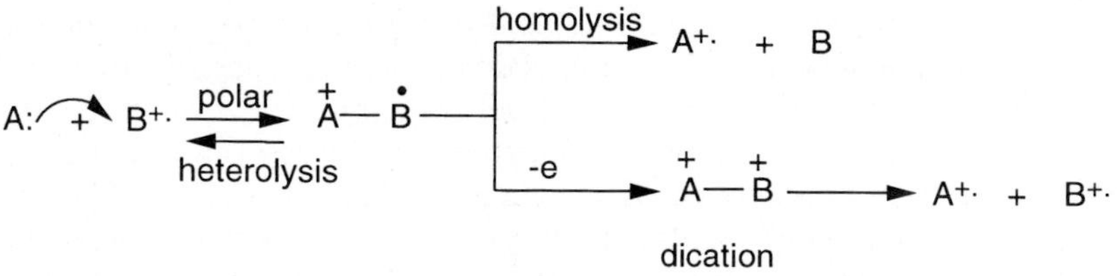

$$A{:} \; + \; B^{+\cdot} \underset{\text{heterolysis}}{\overset{\text{polar}}{\rightleftarrows}} \overset{+}{A}\!\!-\!\!\overset{\bullet}{B} \begin{cases} \xrightarrow{\text{homolysis}} A^{+\cdot} \; + \; B \\ \xrightarrow{-e} \overset{+}{A}\!\!-\!\!\overset{+}{B} \longrightarrow A^{+\cdot} \; + \; B^{+\cdot} \end{cases}$$

dication

**Scheme 31.** Variants of the polar or limiting inner sphere et mechanism.

radical produced by the electrophilic mechanism by another aminium ion to give a di-cation, followed by homolysis of the latter to yield a substrate cation radical and an aminium ion [64]. In the case of the cleavage of a distonic cation radical inter-mediate, it might be thought that this should preferentially cleave to the aminium ion and a neutral substrate molecule, since this process is typically exergonic, than cleavage to the substrate cation radical. However, it should be noted that the latter process involves a bond heterolysis, whereas the former one involves homolysis. Since the redistribution of charge is much smaller in a homolysis, it is conceivable that the endergonic homolytic route could be kinetically favored over the exergonic heterolytic route.

### Mechanistic criteria for the substrate ionization step

The mechanism of ionization of the two substrate molecules phenyl vinyl sulfide and phenyl *cis*-1-propenyl ether by the aminium salt $3^{+\cdot}$ has been studied in detail and a mechanistic criterion established for distinguishing outer sphere and inner sphere ionization in these and analogous substrate molecules [65]. Using this crite-rion, which is designated the $\sigma/\sigma^+$ criterion, it is also possible to distinguish between weak and strong inner-sphere ionization. The method takes advantage of the cir-cumstance that the ionization of *meta* and *para* substituted derivatives of these two types of substrate to the corresponding cation radicals correlates extremely well with the Hammett–Brown substituent parameters $\sigma^+$ whereas the addition of an electrophile to the double bond to yield an α-aryloxy or α-arylthiocarbocation, or to the heteroatom to yield an onium ion, correlates with the Hammett $\sigma$ substituent parameters (Scheme 32).

This dichotomy has been rigorously established by both theoretical methods (semi-empirical and ab initio calculations) and by experimental measurements for both cation radical and carbocation formation (Scheme 32) [66]. The kinetics of aminium salt-induced cation radical Diels–Alder reactions of these substrates were then studied, and an excellent correlation with the Hammett $\sigma$ values was observed for both the enol ethers and the enol sulfides. The reactions of the aryl propenyl ethers involved 2,3-dimethyl-1,3-butadiene as the dienic component [64], while the reactions of the aryl vinyl sulfides involved cyclopentadiene as the dienic component (Scheme 33) [65].

The distinctions between the correlations with $\sigma$ and $\sigma^+$ were both statistically significant at well above the 95 % confidence level. *These results suggest that the transition state for the ionization step, which was established as rate determining, closely resembles that for an electrophilic attack and has little if any cation radical (outer sphere) character.* This and other supporting evidence suggests that the transition states for these two reaction series closely resembles a *distonic* cation radical. Whether the site of the electrophilic attack is on the heteroatom or upon the β-carbon of the double bond is less certain, but stereochemical evidence (stereo-specific addition in the case of *cis*-β-deutero phenyl vinyl sulfide) is more consistent with attack at sulfur in the case of the aryl vinyl sulfides. In any case, these results are most consistent with a polar mechanism for ET, which involves a distonic cation radical intermediate (Scheme 34).

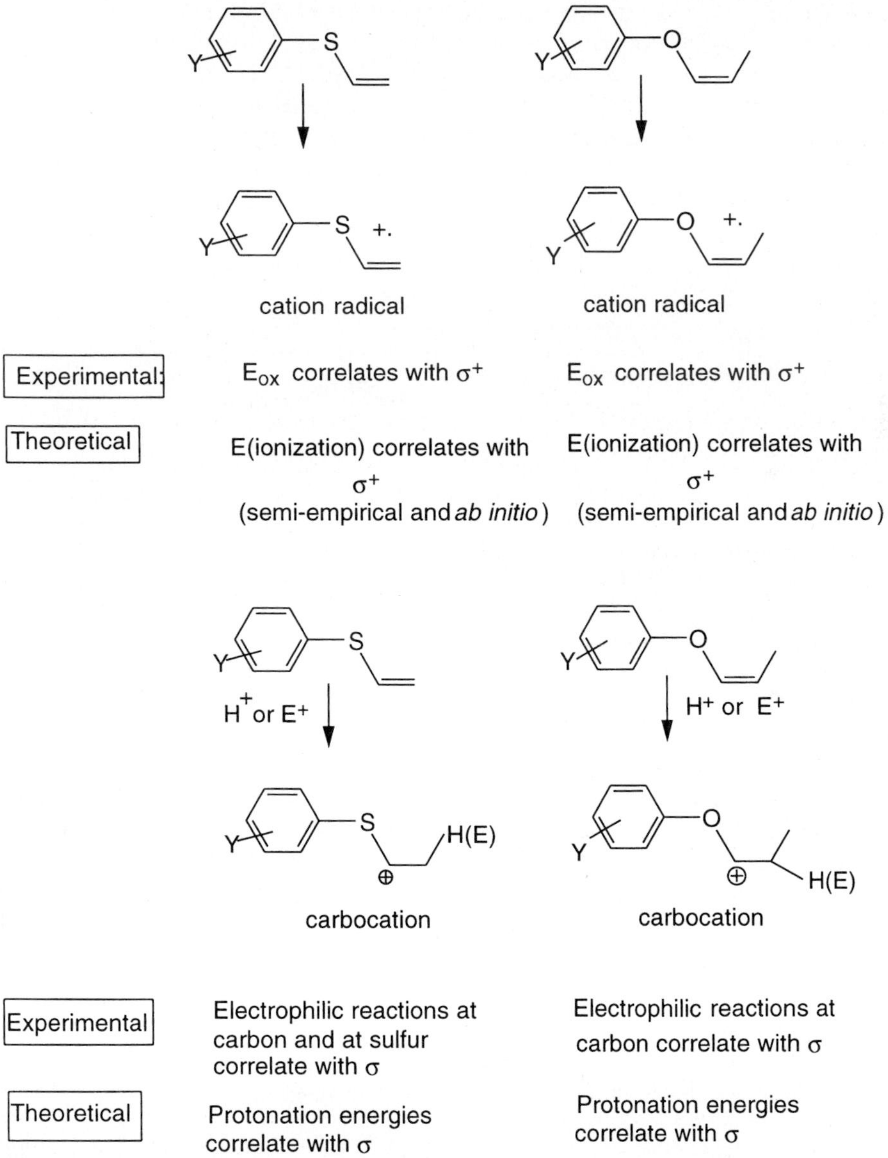
S
Y
O
Y
S +.
Y
O +.
Y
cation radical
cation radical
Experimental
Eox correlates with σ+
Eox correlates with σ+
Theoretical
E(ionization) correlates with σ+
(semi-empirical and ab initio)
E(ionization) correlates with σ+
(semi-empirical and ab initio)
S
Y
O
Y
H+ or E+
H+ or E+
S
H(E)
Y
⊕
O
Y
⊕
H(E)
carbocation
carbocation
Experimental
Electrophilic reactions at carbon and at sulfur correlate with σ
Electrophilic reactions at carbon correlate with σ
Theoretical
Protonation energies correlate with σ
Protonation energies correlate with σ

**Scheme 32.** The $\sigma/\sigma^+$ criterion for inner sphere vs. outer sphere ionization.

### 3.2.10 Neutralization of the Product Cation Radical

An electron transfer process is also involved in the second propagation step of a cation radical chain reaction, i.e., the step in which the product cation radical is neutralized by electron transfer from the neutral substrate molecule. For an efficient chain reaction, this step should, of course, be exothermic. If the assumption is made

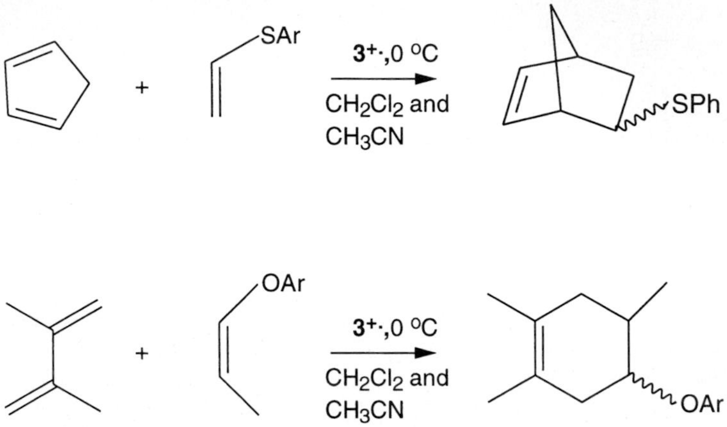
SAr
+
3+·,0 °C
CH2Cl2 and
CH3CN
SPh
OAr
+
3+·,0 °C
CH2Cl2 and
CH3CN
OAr

**Scheme 33.** Kinetic studies of the cation radical Diels–Alder reactions of aryl vinyl sulfides with cyclopentadiene and of aryl propenyl ethers with 2,3-dimethyl1,3-butadiene. In both solvents, both extended reaction series have rates which correlate excellently with the Hammett $\sigma$ parameters and poorly with the Hammett-Brown $\sigma^+$ values. The preference is statistically significant at or above the 95 % confidence level.

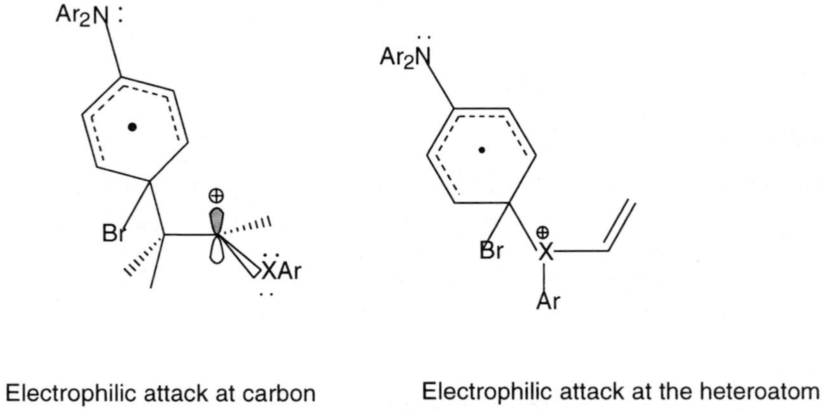
Ar2N
Br
XAr
Ar2N
Br
X
Ar
Electrophilic attack at carbon
Electrophilic attack at the heteroatom

**Scheme 34.** Possible structures of the distonic cation radical intermediate.

that the cation radical cycloaddition step is concerted, i.e., it does not proceed via an intermediate distonic cation radical, this requirement is relatively easily met, since the net conversion of a pi bond to a sigma bond usually will assure that the cation radical moiety (hole) is delocalized over a less extensive pi system and is therefore less stable in the cycloadduct than in the reactant. Interestingly, this is not necessarily the case if a distonic cation radical intermediate is involved. It has been plausibly argued by both D'Innocenzo [67] and by Nelsen [68] that the (vertical) neutralization of the carbocation site of a distonic cation radical, which would generate a diradical intermediate, is generally endothermic. Therefore, if a distonic

**Scheme 35.** Hypothetical neutralization of a distonic cation radical by the bond-coupled electron transfer mechanism.

cation radical intermediate is involved, it appears reasonable to assume that it must first cyclize to the ring-closed product cation radical before it can be neutralized by exothermic hole transfer. This was considered to be a plausible scenario in the cation radical chain addition to triplet oxygen to yield dioxetanes, which is a non-stereospecific but highly efficient chain process, and also in the vinylcyclopropane rearrangement [68]. More recently, the possibility of *bond-coupled electron transfer* has been considered for the neutralization of distonic cation radical intermediates [69]. Since concerted electron transfer and bond fragmentations have now been established, it appears quite reasonable to consider the possibility of bond-coupled electron (hole) transfer to neutralize the distonic cation radical intermediate, i.e., a step in which the neutral product is generated directly by simultaneous ET and covalent bond formation. This mechanism, which is at this point still hypothetical, is illustrated in Scheme 35 for a cation radical cyclobutanation reaction.

### 3.2.11 Mechanistic Diagnosis of Cation Radical Cycloadditions

**Qualitative diagnosis**

Evidence for the operation of cation radical mechanisms for cycloaddition has often been provided by means of a comparison of the results obtained for various methods of generating cation radicals. For example, in the Diels–Alder cycloaddition of phenyl vinyl sulfide to 1,3-cyclopentadiene (Scheme 36) the same adducts are formed whether the cation radicals are generated by chemical ionization (aminium salt), photochemical ionization (the PET method), or electrochemical ionization (anodic oxidation) [65].

No other adducts are formed, and the *endo/exo* diastereomeric ratio is essentially the same for all of these methods. Further, the existence of an acid catalyzed mechanism for cycloaddition can be explicitly excluded by using an excess of a hindered amine base (2,6-di-*tert*-butylpyridine, DTBP) in the aminium salt induced reaction and by examining the results of an authentic acid catalyzed reaction (using, for example, triflic acid). In the former case, the same *endo* and *exo* adducts are formed in virtually the same relative amounts, but in the latter case neither of these adducts is formed. It is worth noting that acid catalyzed reactions have indeed sometimes been observed under typical aminium salt conditions [70], but these have never been observed, nor would they be expected, under PET conditions. Finally, in the instance where cation radicals are generated by the aminium salt method, the intervention of substrate cation radicals can usually be verified by the addition of the reduced form of the catalyst, i.e., the neutral triarylamine, to the reaction mixture.

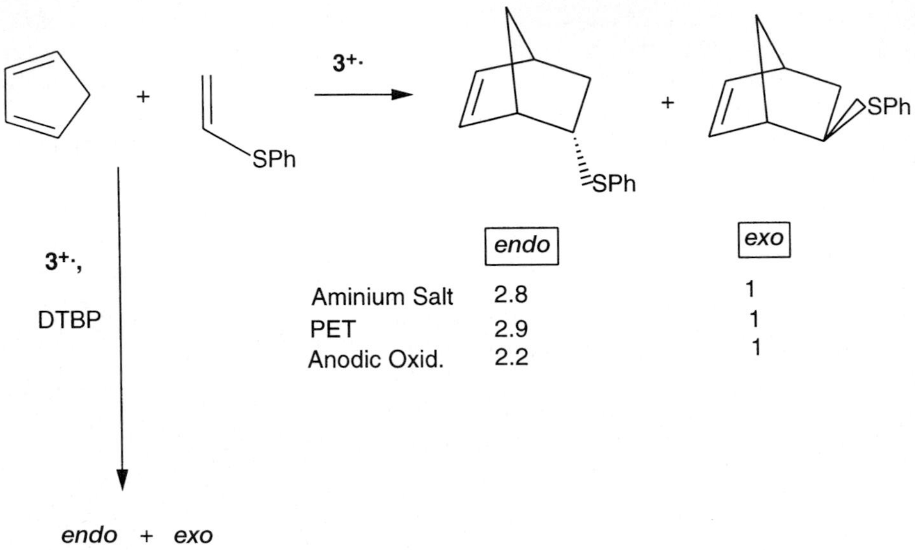

**Scheme 36.** Mechanistic diagnosis of cation radical cycloadditions.

The substrate ionization step, if it is not already reversible, often becomes reversible under these conditions, resulting in a modest to very pronounced rate diminution [40].

**Quantitative diagnosis**

A more quantitative criterion for the operation of a cation radical cycloaddition mechanism has been carried out for a Diels–Alder reaction system consisting of various substituted derivatives of *trans*-stilbene (as the ionizable dienophile) and 2,3-dimethyl-1,3-butadiene (Scheme 37) [71].

The criterion relies on the proposal that *for a symmetrical pi system (such as trans-stilbene), the development of a full unit of positive charge distributed symmet-*

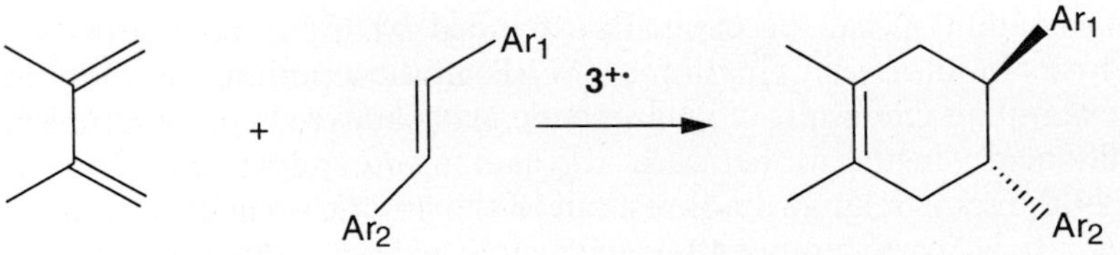

**Scheme 37.** The symmetry criterion for cation radical formation via outer sphere electron transfer. Substituent effects in mono- and disubstituted stilbenes are multiplicative; $\log k_{rel}$ correlates with $\sigma^+$ with $\rho = 4.16$; the oxidation potentials of these same stilbene derivatives correlate with the same substituent parameters with $\rho = -5.02$.

*rically over the system is a unique and positive indication of cation radical formation.*
The symmetrical distribution of charge in the TS was established by a competition
kinetic study which reveals that both mono- and disubstituted stilbenes are nicely
incorporated on the same Hammett–Brown plot, and that substituent effects are
essentially multiplicative for substitutions on one or both rings. The magnitude of
the positive charge development was assessed via the Hammett–Brown $\rho$ value and
a comparison with the $\rho$ value for the complete ionization of these same substrates,
in the same solvent, to the corresponding cation radicals. Thus the oxidation potentials for these substituted stilbenes correlate nicely with Hammett–Brown $\sigma^+$ values,
and the slope corresponds to a $\rho$ value of −5.02. In the same solvent (acetonitrile)
the $\rho$ value for the plot of the log of the relative rate constants for the Diels–Alder
cycloadditions vs $\sigma^+$ is −4.16, i.e., approximately 83 % of a unit of positive charge
is present in this transition state. It is of further interest that similar studies have
been carried out for the cation radical cyclopropanation of these same substrates,
and the $\rho$ value (−4.56) represents 91 % of the full unit of positive charge [72]. The
rate determining step of these reactions has been found to be the cycloaddition step,
so that even in the TS of cycloadditions, 83–91 % of the charge remains on the
stilbene moiety, consistently with a reaction having an early transition state. Incidentally, *cis*-stilbene has a much higher oxidation potential than *trans*-stilbene (1.70 V
vs 1.59 V vs SCE), so that its addition to dienes under these same conditions is too
slow to observe. However, 4,4′-dimethyl-*cis*-stilbene has been observed to undergo
the cation radical Diels–Alder reaction [73].

### 3.2.12 Absolute Reaction Rates and Cation Radical Probes

The rate constant for the Diels–Alder cyclization of substrate **5**, generated by photoionization and by photosensitized electron transfer, has been determined to be
$3 \times 10^8$ s$^{-1}$ (Scheme 38) [31].

This substrate, along with the corresponding substrate **6**, which has a cyclobutanation rate of $1.2 \times 10^9$, have been proposed as sensitive cation radical probes,
for detecting the presence of cation radical intermediates in various reactions [74].
An interesting example is the reaction of tetracyanoethylene (TCNE) with electron
rich alkenes, a reaction for which an electron transfer mechanism had been considered (Scheme 39) [75].

The reaction of **6** with TCNE yields only the conventional adduct corresponding
to the uncyclized probe and none of the product expected from the cation radical
cyclization. That the probe cyclization of the cation radical of **6** would have been
observed in the context of an ET mechanism, if it had been involved, was demonstrated by generating the contact ion radical pair of **6**$^{+•}$/TCNE$^{-•}$ via excitation of
the charge transfer complex of **6** and TCNE. The cyclobutane cyclization product
of the probe reaction was easily detected under these conditions. Consequently, an
ET mechanism for this reaction can be confidently excluded. In a similar manner,
the epoxidation of **5** and **6** by oxidized metalloporphyrins provides strong evidence
against a cation radical mechanism for these reactions [76].

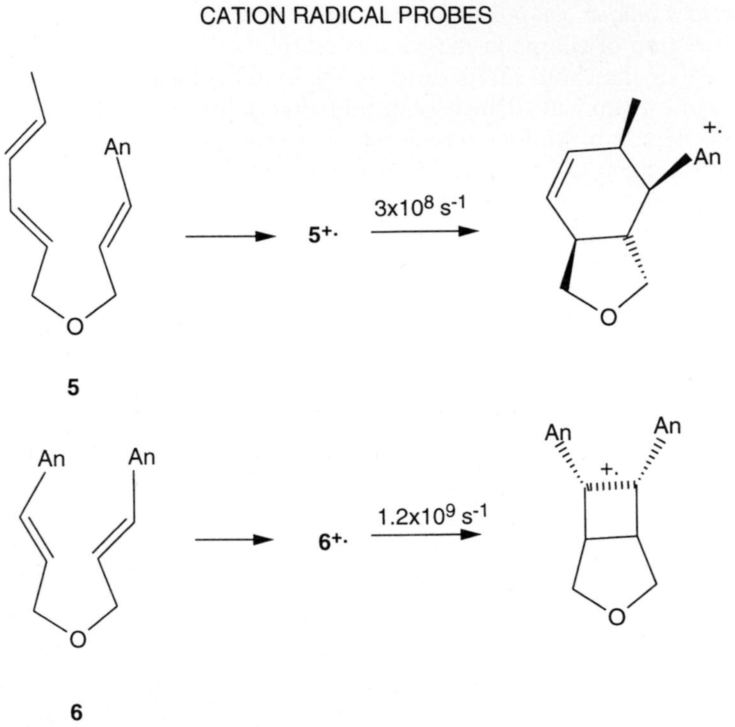

CATION RADICAL PROBES

**Scheme 38.** Cation radical probes.

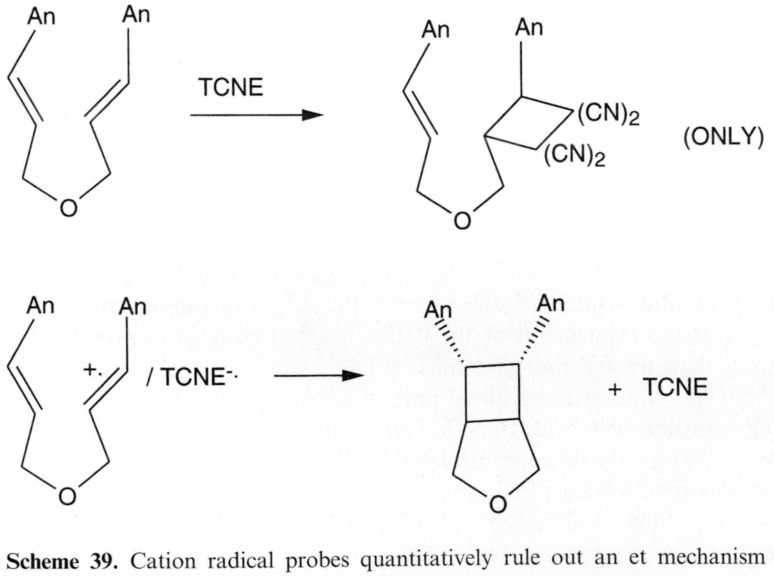

**Scheme 39.** Cation radical probes quantitatively rule out an et mechanism for the addition of tetracyano-ethylene to electron rich alkenes.

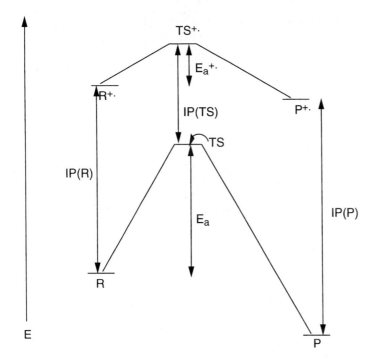

**Scheme 40.** The kinetic impetus for cation radical cycloadditions. R = reactant, P = product, R$^{+\cdot}$ = reactant cation radical, P$^{+\cdot}$ = product cation radical, IP = ionization potential, E$_a$ = activation energy.

### 3.2.13  General Theoretical Considerations in Cation Radical Cycloadditions

The facility of many cation radical cycloadditions is impressive, and it is clear that this is not primarily the result of a powerful thermodynamic driving force. In fact, the thermodynamic driving force for cation radical cycloadditions is often less than that for the corresponding neutral reactions [35, 77]. Thus, although cation radicals are indeed relatively high energy intermediates compared to the corresponding neutral compounds, both the reactants and the products of cation radical cyclo-additions exist upon the same high energy potential surface. It is therefore necessary to look for a transition state effect not related to product development as the common and primary basis for the rapidity of these reactions. One such effect is evident upon examination of the potential surface given in Scheme 40.

From this perspective, it emerges that *the kinetic driving force for the cation radical cycloaddition relative to the corresponding neutral one can be expressed as the difference in the ionizabilities of the reactant and the transition state.* As a specific example, we may consider a highly non-synchronous TS for cation radical Diels–Alder cycloaddition—which is supported by theoretical calculations (vide infra). This transition state is essentially a distonic cation radical, the radical site of which is easily ionized because the odd electron is in a non-bonding MO. In contrast, the ionization of the reactant involves the removal of an electron from an orbital which is strongly bonding. Further, transition states in general, because they are relatively less bonding and relatively higher in energy than reactant states, can be expected to have lower oxidation potentials. It also appears likely that the diminished electronic repulsions in a cation radical transition state are another intrinsic factor favoring these reactions not only over the corresponding neutral reactions, but also over the corresponding anion radical reactions.

Additional insights follow from an inspection of the curve shown in Scheme 40. First, for an effective chain process, *the oxidation potential of the product should be higher than that of the reactant*, so that the hole transfer process involved in the propagation cycle will be exergonic. This is normally the case in cycloadditions because the product typically has one less pi bond than the reactants, so that the former has a less highly delocalized cation radical moiety (hole) than the latter. Also, the use of the exergonicity associated with the neutralization of the product cation radical to ionize the reactant is relatively much more efficient than quenching the product by a molecule of the neutral triarylamine, followed by the ionization of the substrate by the aminium salt. This follows because in the latter scenario the difference in the oxidation potentials of the product cation radical and the neutral triarylamine represents wasted energy. Consequently, *the catalytic mechanism, which involves an endergonic ionization step, is intrinsically less energy efficient than the chain mechanism*, which involves an exergonic substrate ionization step. On the other hand, it also follows that *substrate ionization is more selective in the context of a catalytic mechanism than in the chain mechanism.*

### 3.2.14  Calculational Results

All calculations presented to date, including both semi-empirical and ab initio results, indicate that the cation radical DA reaction transition state is at least highly non-synchronous, somewhat resembling a distonic cation radical [35, 56]. The question of whether the reaction is even weakly concerted, i.e., whether the second C–C bond has begun to form at all in the transition state is of special interest in relation to reaction stereospecificity. Early results based upon MP2/6-31G* calculations for the DA reaction of the *s-cis*-1,3-butadiene cation radical with ethene suggested that the reaction might well be concerted [56], but more recent, and more sophisticated, calculations suggest that the reaction may be stepwise, involving a short-lived distonic cation radical intermediate which has essentially no covalent interaction between the distonic termini [78]. On the other hand, experimental results for three distinct cation radical DA reaction systems reveal highly stereo-specific reactions [79]. Of course, the reaction systems studied experimentally are not those for which the reaction path was calculated, nor are solvation effects included in the calculations. Further, it is formally possible that a distonic cation radical intermediate could cyclize to the DA adduct cation radical at a rate faster than bond rotation, and thus be stereospecific. It is also relevant to note that an intermediate distonic cation radical has been detected in the DA cyclodimerization of spiro[2,4]heptadiene, although it can not be definitely determined that this is the

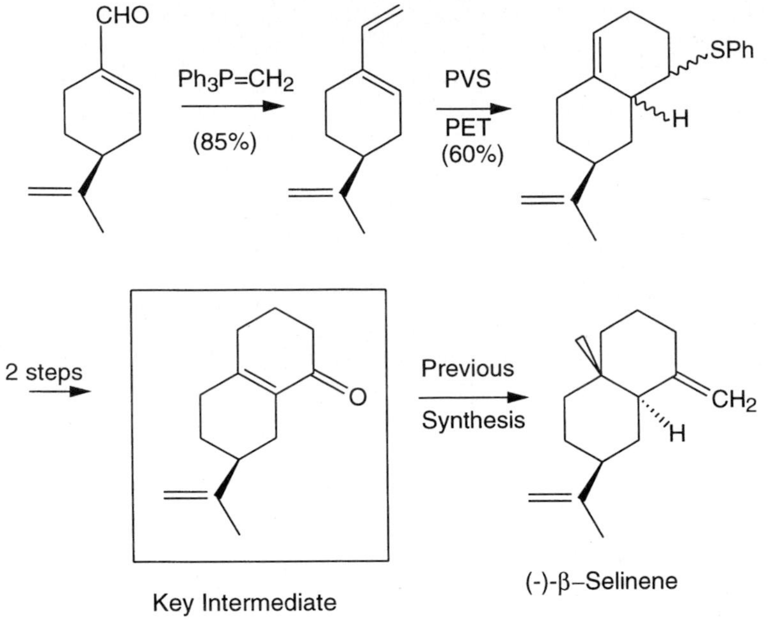

**Scheme 41.** The cation radical Diels–Alder reaction in a formal total synthesis of beta selinene.

sole or even the major pathway for this reaction [80]. Consequently, it is reasonable to assume a certain mechanistic diversity in the cation radical DA reaction with respect to its concerted vs. stepwise nature. Nevertheless, the observations of stereo-specificity in three structurally very different systems continues to support the concept of concerted cation radical Diels–Alder reactions.

### 3.2.15 Natural Product Synthesis and Synthetic Methodology using Cation Radical Cycloaddition Reactions

The usefulness of the cation radical Diels–Alder reaction for natural product synthesis has been illustrated in the synthesis of β-selinene (Scheme 41) [81], which utilizes phenyl vinyl sulfide as a Diels–Alder dienophile and the photosensitized electron transfer method as the preferable method for generating the corresponding cation radical.

It was demonstrated that the use of phenyl vinyl sulfoxide or sulfone in a thermal Diels–Alder reaction was highly unsatisfactory as an alternate route. The synthesis of the neolignins galbulin, isogalbulin [82], magnoshinin, and magnosalin [83] have all been accomplished by means of a cation radical cycloaddition (Scheme 42).

The syntheses of galbulin, isogalbulin, and magnoshinin involve cation radical Diels–Alder reactions, while that of magnosalin involves a cation radical cyclobutanation. The galbulin and isogalbulin syntheses involve aminium salt induced cycloadditions, whereas the magnoshinin and magnosalin syntheses were carried

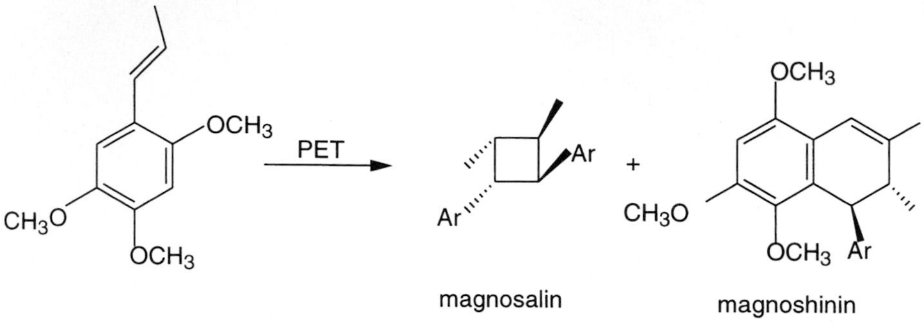

**Scheme 42.** Cation radical cycloadditions in the synthesis of magnosalin and magnoshinin.

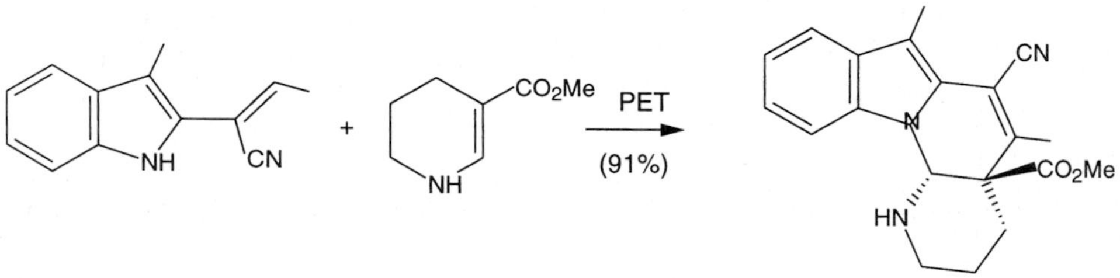

**Scheme 43.** Construction of the goniomitine skeleton using vinylindoles as diene components.

out using the PET method. Another especially impressive synthetic use of cation radical cycloadditions is the use of 2-vinylindoles as Diels–Alder dienes for the construction of alkaloid skeletons. As an example, the Goniomitine framework has been generated in this way, using the PET method (Scheme 43) [84].

## 3.2.16 Cation Radical Polymerization

The cation radical intermediate and the process of electron (hole) transfer have recently been shown to constitute the basis for a fundamentally new addition to the repertoire of polymerization methods [85]. Both cation radical chain cyclobutanation polymerization (Scheme 44) and Diels–Alder polymerization have been demonstrated under the typical aminium salt conditions.

Cycloaddition polymers having average molecular weights of up to 200 000 have been generated in this way using appropriately constituted difunctional monomers. The mechanistic advantage of polymerization over the familiar cycloaddition reactions of monofunctional substrates is considered to be that in the propagation cycle electron (hole) transfer is intramolecular in the difunctional context, but intermolecular in the monofunctional context. An intriguing aspect of these novel polymers is that they retain ionizable end groups, such that the polymerization process can be 'resurrected'. The ability to rather efficiently produce cyclobuta-

a "cyclobutapolymer"

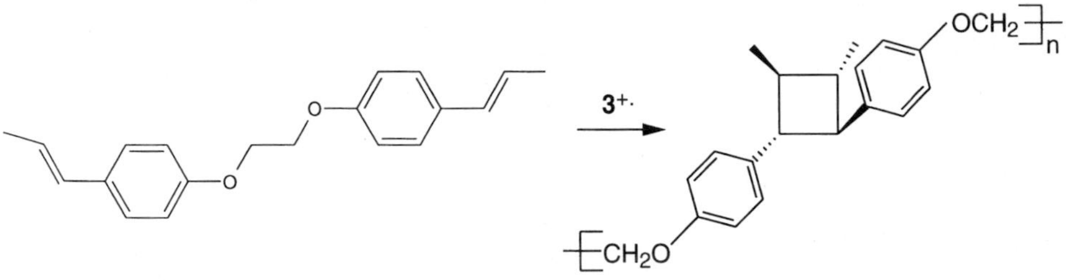

**Scheme 44.** Cation radical chain cycloaddition polymerization.

polymers under very mild thermal conditions is an especially novel aspect of this polymerization method.

### 3.2.17 The Cation Radical Vinylcyclobutane (VCB) Rearrangement

#### Scope and occurrence

A further complication in cation radical cycloadditions involving at least one con-jugated diene component is the often facile rearrangement of initially formed cyclo-butane (CB) adducts to cyclohexene (i.e., DA) adducts. However, this can also provide a synthetically useful route for *indirect, net Diels–Alder cycloaddition* to acyclic dienes [86]. Further, the cation radical vinylcyclobutane (VCB) rearrange-ment can usually be suppressed by carrying out the reaction using $3^{+\cdot}$ for a very short period of time at $0\,°C$, for a somewhat longer time at $-30\,°C$, or at $-78\,°C$ with the corresponding tris(2,4-dibromophenyl)aminium hexachloroantimonate salt ($4^{+\cdot}$) which is much more soluble than $3^{+\cdot}$. In this way, either the CB rich mixture or the DA adducts may be obtained. For example, the CB adduct strongly pre-dominates in the cycloaddition of *trans*-anethole with 1,3-butadiene at $-30\,°C$, but this is rapidly rearranged at $0\,°C$ in the presence of $3^{+\cdot}$ or by the PET method to the corresponding Diels–Alder adduct (Scheme 45).

Similarly, the cyclobutane adducts of phenyl vinyl ether and 1,1'-dicyclopentenyl are the predominant products under PET conditions, but these rearrange smoothly at $-45\,°C$ in the presence of $4^{+\cdot}$ to Diels–Alder adducts.

#### Mechanism of the cation radical VCB rearrangement

The cation radical VCB rearrangement has been found, at least in several cases, to occur intramolecularly, rather than by dissociation/recombination [86]. An espe-cially interesting case is the rearrangement of the CB dimers of 1,3-cyclohexadiene generated by triplet sensitized photochemistry. In the presence of the usual aminium salt catalyst ($3^{+\cdot}$), these CB dimers are quite stable, but if the more powerful hex-abromo aminium salt, $4^{+\cdot}$, is used, these individual dimers rearrange to the DA dimers via a stereospecific VCB rearrangement (Scheme 46).

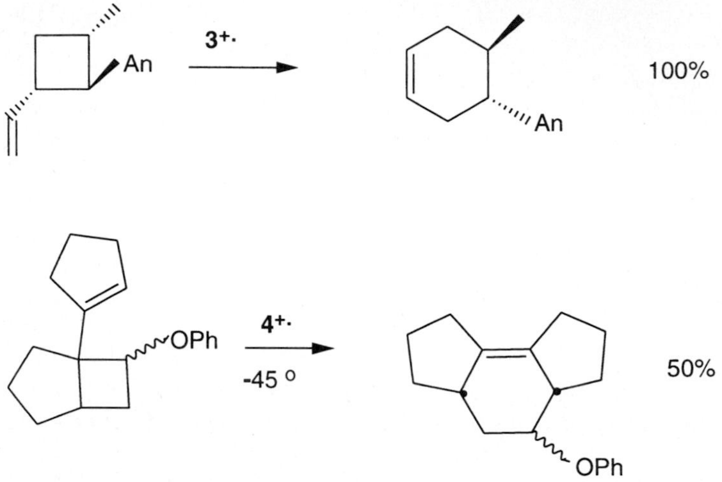

**Scheme 45.** The cation radical vinylcyclobutane (VCB) rearrangement.

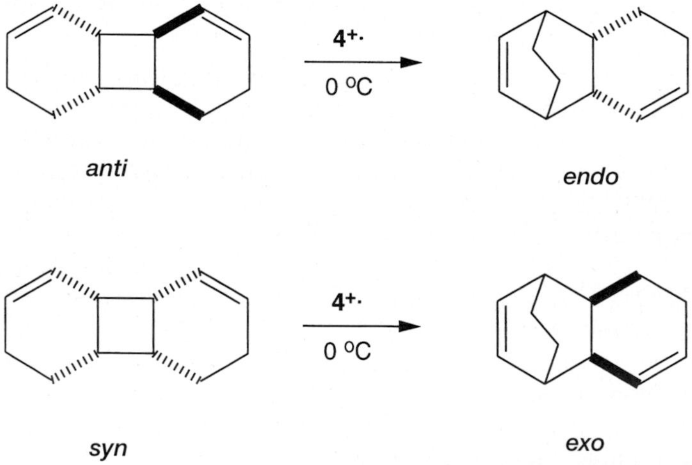

**Scheme 46.** Intramolecular vs. dissociation/recombination mechanisms for the VCB rearrangement.

Thus, the *anti* CB dimer rearranges to the *endo* DA dimer, and the *syn* CB dimer rearranges to the *exo* DA dimer by a 1,3-sigmatropic shift which occurs suprafacially with retention of configuration. The necessity for using the more powerful aminium salt is inherent in the difficulty of ionizing the CB dimers, which have only simple alkene moieties available for ionization. More generally, the feasibility of an efficient cation radical VCB rearrangement depends upon the presence of a readily ionizable functionality on the cyclobutane ring carbon vicinal to the carbon bearing the vinyl group. The anisyl, arylthio, and aryloxy groups are especially effective in this context. In the rearrangement of the CB adduct of *trans*-anethole and 1,3-butadiene it was further shown that the highly efficient rearrangement to the DA adduct did not involve dissociation/recombination. Thus, the addition of 800 mol % of 2,3-dimethyl-1,3-butadiene, which is three times as reactive toward the *trans*-anethole cation radical as is 1,3-butadiene, to the reaction mixture prior to the addition of the initiator, fails to generate even a trace of the adduct of *trans*-anethole and 2,3-dimethyl-1,3-butadiene.

### 3.2.18 The Cation Radical 'Phenylcyclobutane' Rearrangement

A reaction which is formally analogous to, and in a sense much more surprising than, the vinylcyclobutane rearrangement is a reaction in which one of the unsaturated bonds of an aryl ring fulfills the same role as an alkene double bond in the latter rearrangement. In view of the close analogy to the vinylcyclobutane rearrangement, it might be appropriate to designate this type of rearrangement as the phenylcyclobutane rearrangement. An example of this latter type of rearrangement which has been studied especially carefully occurs in the cyclobutadimerization of vinylanisole [87]. The initial cation radical cyclobutanation reaction occurs with a rate constant of $1.4 \times 10^9$ s$^{-1}$ to give the long bond cyclobutane cation radical. The latter is able to rearrange, via a phenylcyclobutane rearrangement, to a hexatriene type cation radical in which the aromaticity of one of the aryl rings has been disrupted (Scheme 47).

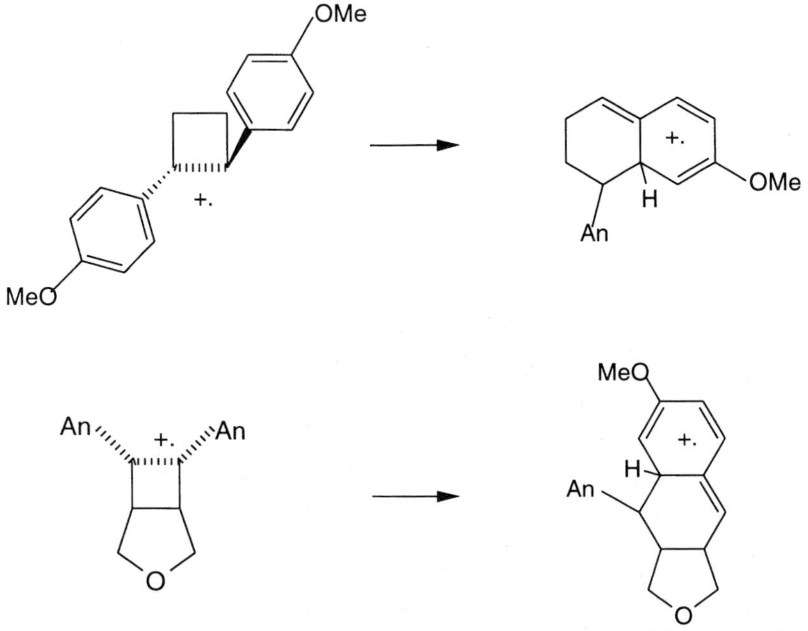

**Scheme 47.** Cation radical "phenylcyclobutane" rearrangements.

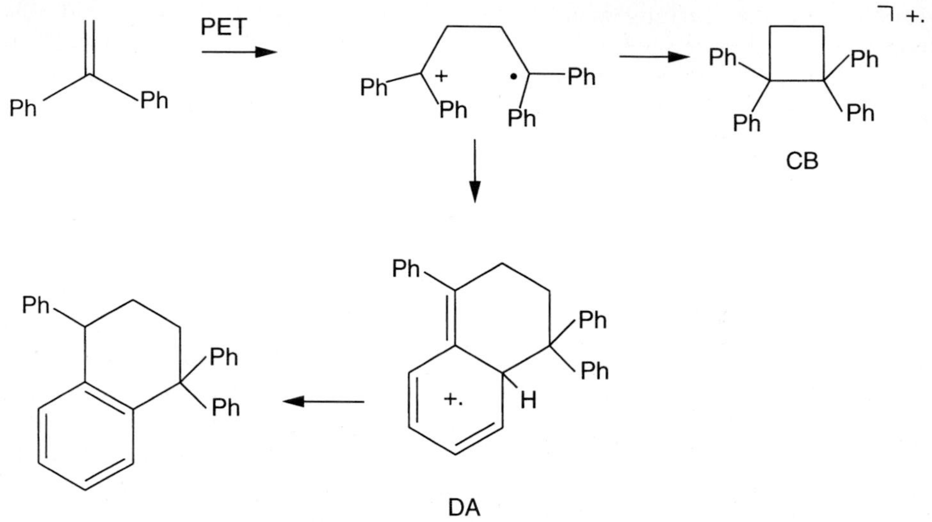

**Scheme 48.** The cyclodimerization of 1,1-diphenylethene.

A similar reaction has been observed in the case of an intramolecular cation radical cyclobutanation, and in this case the cyclized product having a re-aromatized aryl ring was isolated and characterized [31]. Although these reactions appear to be concerted rearrangements, not involving a distonic cation radical intermediate, analogous products could be formed from a distonic cation radical intermediate. The photosensitized cation radical cyclodimerization of 1,1-diphenylethene, originally reported by Arnold's group [42], and subsequently scrutinized by Farid [43], appears to involve an intermediate distonic cation radical which can close to give a cyclobutane or a hexatriene cation radical (Scheme 48).

### 3.2.19 The Cation Radical Vinylcyclopropane Rearrangement

Efficient and facile rearrangements analogous to the cation radical vinylcyclobutane rearrangement have also be identified in the vinylcyclopropane (VCP) series (Scheme 49).

In rather marked contrast to the vinylcyclobutane series, neither *cis*- nor *trans*-2-*p*-anisyl vinylcyclopropane undergoes ring expansion in the presence of aminium salts. However, several more highly alkylated derivatives undergo the reaction with impressive efficiency [88]. Evidently a stepwise pathway is involved in at least some cases, since non-stereospecificity has been demonstrated in one instance (Scheme 49) [89]. Interestingly, the intermediate distonic cation radical, in this latter case, does not re-cyclize to regenerate the original vinylcyclopropane cation radical, since the latter does not undergo *cis–trans* isomerization under the relevant conditions. In view of the cation radical cyclopropanation reaction to be discussed below, the

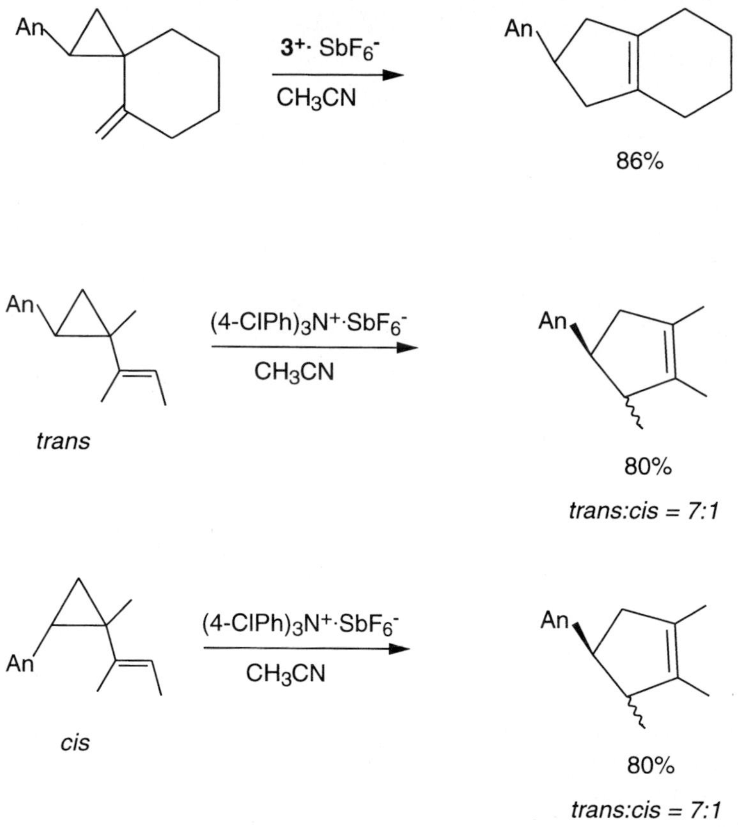

86%

*trans*

80%

*trans:cis* = 7:1

*cis*

80%

*trans:cis* = 7:1

**Scheme 49.** The cation radical vinylcyclopropane rearrangement.

cation radical vinylcyclopropane rearrangement is of further interest because it formally makes available cation radical routes to all ring sizes from three to six.

### 3.2.20 Cation Radical Chain Cyclopropanation

The chain addition of cation radicals of conjugated dienes and electron rich styrenes to ethyl diazoacetate has been found to be a relatively efficient method of generating cyclopropane derivatives (Scheme 50) [90].

An excess (5:1 mole ratio) of ethyl diazoacetate is used in these reactions to suppress cyclobutadimerization or Diels–Alder cyclodimerization. In difunctional molecules which have non-equivalent ionizable functionalities, cyclopropanation is highly selective for the more easily oxidized functionality. The latter selectivity is perhaps the most attractive aspect of the reaction. In contrast to transition metal (e.g. rhodium) catalyzed cyclopropanations, cation radical additions to electron deficient alkene moieties do not occur at all. The reaction is relatively sensitive to

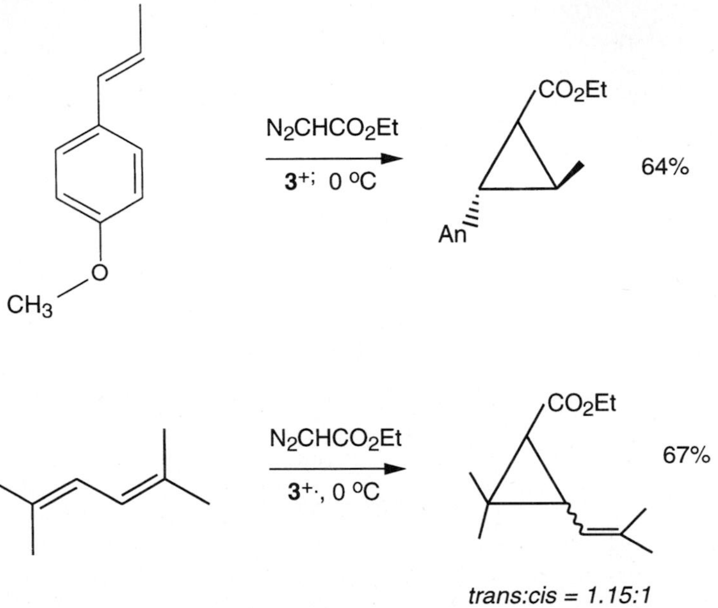

**Scheme 50.** Cation radical chain cyclopropanation.

steric effects, since α-terpinene is cyclopropanated at the less sterically hindered double bond with a very high degree of selectivity (Scheme 51).

Similarly, tetrasubstituted alkene functions can be cyclopropanated selectively in the presence of less highly substituted alkene moieties. These latter reactions require the more potent aminium salt **4⁺·** as the initiator, since simple alkene functions are

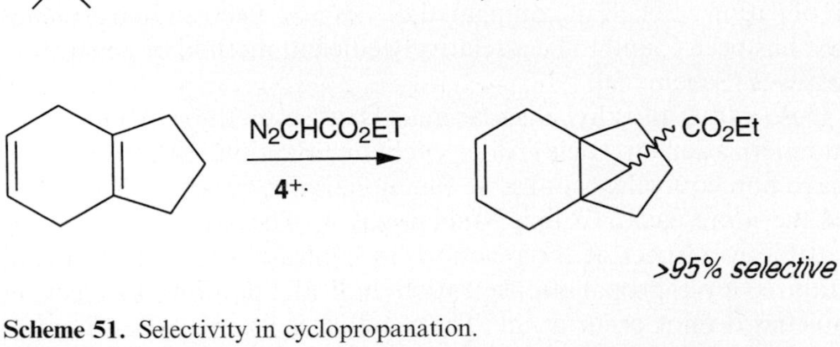

**Scheme 51.** Selectivity in cyclopropanation.

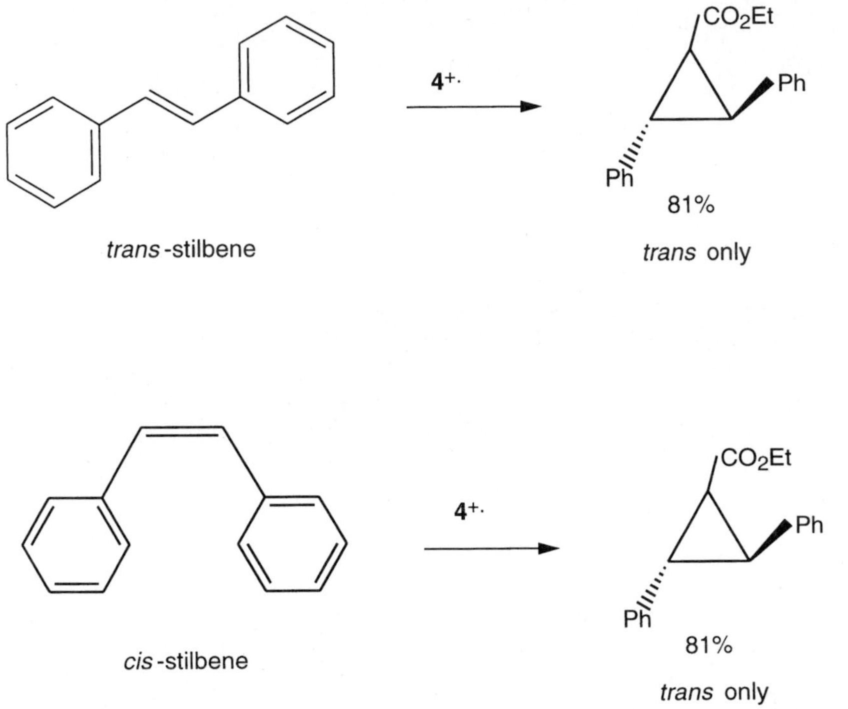

trans-stilbene → 4+· → CO₂Et, Ph, Ph 81% trans only

cis-stilbene → 4+· → CO₂Et, Ph, Ph 81% trans only

**Scheme 52.** Stereochemistry of cyclopropanation.

extremely difficult to ionize. The use of this more potent initiator also allows the reactions to be performed at −78 °C, thus enhancing the selectivity. Stilbenes and even tetrasubstituted alkenes can be cyclopropanated (Scheme 52).

The cyclopropanation of *cis-* and *trans*-stilbene is completely non-stereospecific, suggesting a stepwise process involving a distonic cation radical intermediate. In accord with this supposition, a long bond cyclopropane cation radical structure would seem to be rather unfavorable in this system, because of the presence of the electron withdrawing ester function at the site of the hypothetical long bond. Similarly, a ring closed structure for the cyclopropane cation radical would not appear to be especially appealing. This gives rise to the interesting question of how this distonic cation radical is neutralized, since electron transfer to the non-bonding orbital of a carbocation site would not appear to be energetically favorable. *In this case, it might be reasonable to postulate a bond-coupled ET process, in which ET is concerted with the formation of the cyclopropane bond of the neutral product.*

### 3.2.21 Cation Radical Cycloadditions Forming Five-membered Rings

When 2,3-diphenylazirine is ionized by the PET method, the resulting cation radical apparently undergoes rapid cleavage to a distonic cation radical (Scheme 53) [91].

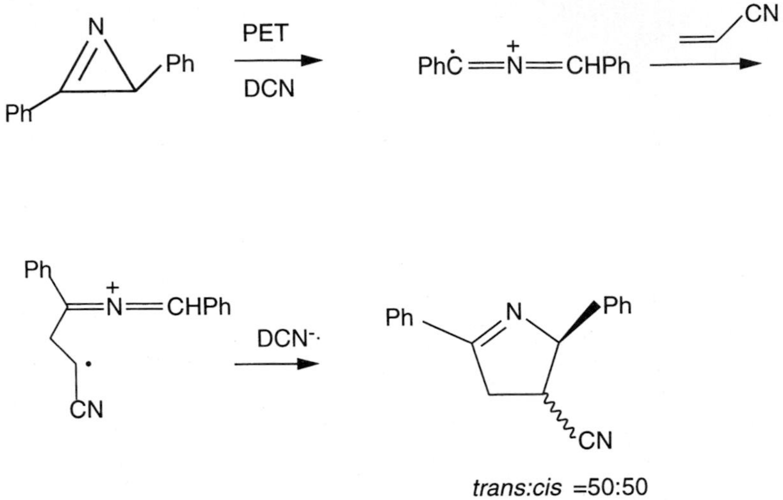

**Scheme 53.** [3 + 2] Cycloadditions with azirine cation radicals.

The cationic site is of the immonium type, and is relatively unreactive, but the radical site is reactive toward electron deficient alkenes such as acrylonitrile, yielding a 50:50 mixture of the diastereoisomeric pyrrolines, after being neutralized by back electron transfer from the sensitizer anion radical. The net result is an interesting example of a net 1,3-dipolar cycloaddition which, in the Huisgen method of classification of cycloadditions, is of the [3 + 2] type. The same general reaction had previously been carried out by Padwa, using direct photochemical excitation, a procedure which, in contrast, was highly diastereoselective (90:10, in favor of the *trans* isomer) [92].

### 3.2.22 The Cation Radical Cope Reaction

The cation radical version of the familiar Cope reaction has also received substantial attention. Semi-empirical calculations predict that the cation radical of 1,5-hexadiene should undergo cyclization to a 1,4-cyclohexanediyl cation radical with little or no activation, but that the subsequent ring opening cleavage in the opposite sense would necessarily be highly endothermic [55]. Consequently, the full Cope reaction can presumably not be realized in the simple case of 1,5-hexadiene. This prediction has been amply verified by generating this cation radical via gamma radiolysis in a rigid matrix and observing its closure to the expected 1,4-cyclohexanediyl cation radical, which is relatively stable under the conditions of its generation (Scheme 54) [93].

The difficulty in completing the full Cope reaction can, however, be circumvented by installing cation radical stabilizing groups at the 3 and 4 positions of a 1,5-hexadiene derivative [94] or by providing for the delocalization of the cation radical

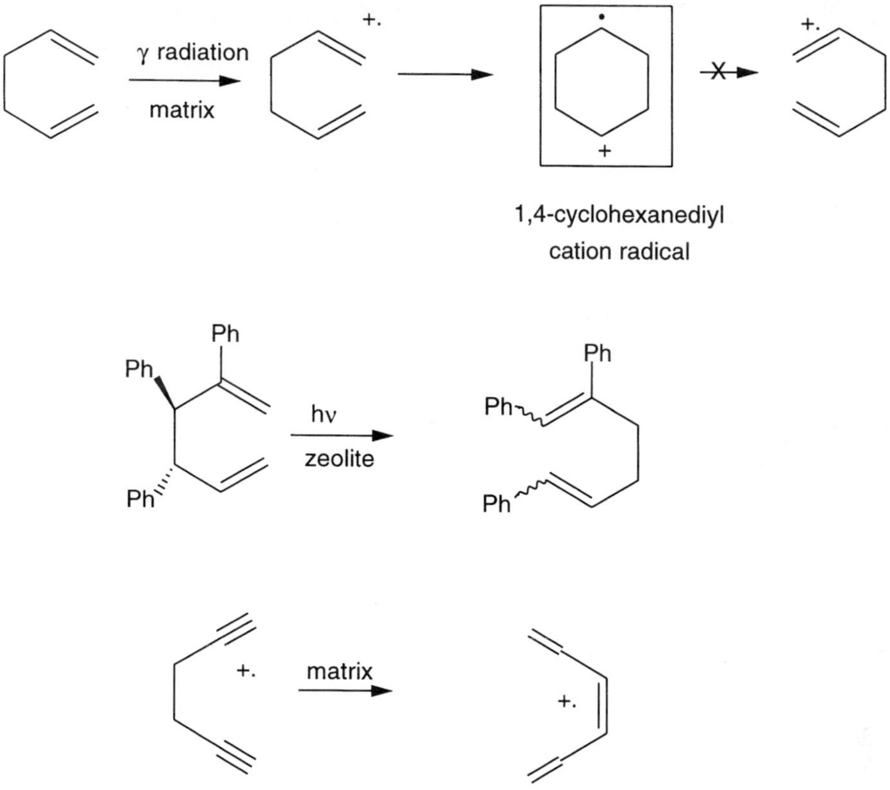

1,4-cyclohexanediyl
cation radical

**Scheme 54.** The cation radical cope reaction.

moiety of the product over a conjugated system consisting of the original C1, C2, C5, and C6 atoms (Scheme 54) [95]. Although the former case may well proceed by a stepwise, distonic cation radical, mechanism, the latter one does appear to be a bona fide example of a concerted [3,3] sigmatropic shift. An especially interesting way to bring about the final stage of the Cope reaction, that is the cleavage of an intermediate cyclohexanediyl cation radical to a 1,5-hexadiene derivative, has been discovered by Miyashi (Scheme 55) [96].

When the PET method is used to generated the 1,5-hexadiene-type cation radical, the latter cyclizes to the 1,4-cyclohexanediyl cation radical, which is then reduced by electron transfer from the sensitizer anion radical to the 1,4-cyclohexanediyl diradical. The cleavage of the latter to a neutral 1,5-hexadiene derivative is then thermodynamically favorable. The course of these reactions has been followed by the scrambling of deuterium from the 1- and 6-positions of the diene to the 3- and 4-positions, by trapping the intermediate cation radical with dioxygen, and by studying the corresponding reactions of the meso- and dl-3,4-dimethyl compounds. The latter reactions have been found to be stereospecific.

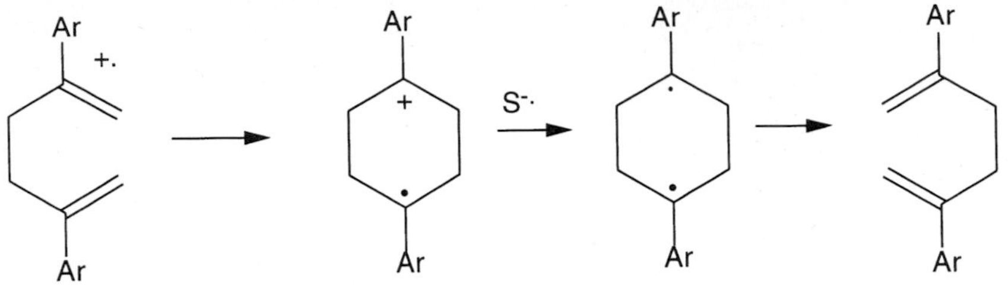

S⁻· = sensitizer anion radical in the PET method
Ar  = phenyl or 4-methoxyphenyl

**Scheme 55.** An indirect version of the cation radical cope reaction.

### 3.2.23  A [1,16] Sigmatropic Shift

One of the earliest, and certainly one of the more impressive examples of apparent cation radical sigmatropic shifts involves the corrin series [97]. Electrochemical oxidation of a nickel(II)-A/D-secocorrinate in acetonitrile containing a trace of water was found to provide an almost quantitative yield of a secocorrinoxide in which a [1,16] hydrogen shift, from the methylene group at C19 in the D ring to the methylidene carbon in the A ring, had occurred in the intermediate cation radical.

### 3.2.24  Electrocyclic Reactions of Cation Radicals

The retroelectrocyclic ring opening of cyclobutene cation radicals to the 1,3-butadiene type cation radicals has been studied both in the gas phase and in solution. A particularly elegant experiment is the demonstration of conrotatory stereo-specificity in the ring opening of the *cis-* and *trans*-1,2-diphenylbenzocyclobutene cation radicals, generated by charge transfer excitation of the electron donor–acceptor complexes of the corresponding neutral substrates with tetracyano-ethylene [98]. The formation of the tetracyanoethylene adducts of the *o*-xylylene type cation radical (Scheme 56) is stereospecific, the *cis* benzocyclobutene derivative yielding the *trans* adduct, and the *trans*-benzocyclobutene derivative yielding the *cis* adduct.

Interestingly, the retroelectrocyclic cleavage of the anion radicals of the same substrates had also been found to be symmetry allowed in the conrotatory mode, which was also experimentally observed (vide infra). The stereochemistry of both of these ring openings is the same as that found in the neutral molecules (conrotatory). The retroelectrocyclic cleavage of the parent cyclobutene cation radical, generated by gamma radiolysis of matrix isolated cyclobutene, has also been studied in elegant detail [99]. Very interestingly, the reaction does not appear to generate the *s-cis*-1,3-butadiene cation radical as would be naively expected, but appears to proceed directly to the *s-trans*-1,3-butadiene cation radical, which can be authentically gen-

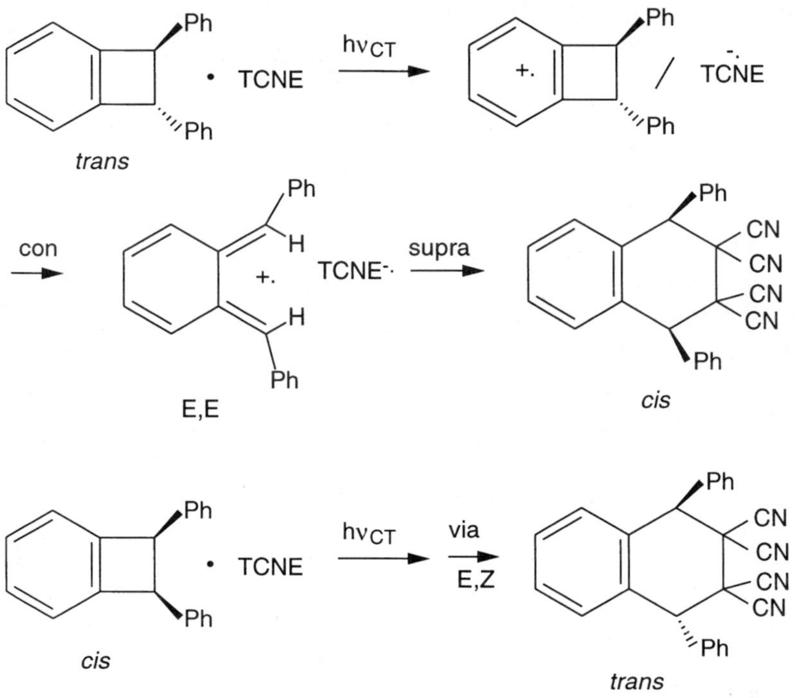

**Scheme 56.** Electrocyclic reactions of cation radicals.

erated directly from 1,3-butadiene. This result had previously been predicted by low level ab initio calculations which envisioned the formation of an intermediate distonic cyclopropylcarbinyl cation radical [100]. Subsequently, much more sophisticated calculations have revealed that a reaction path proceeding via the cyclopropylcarbinyl cation radical is indeed the preferred one (Scheme 57) [101].

However, the calculations indicate that this species is not an actual energy minimum but a flat portion of the potential surface. This reaction path, which has been termed a *non-electrocyclic reaction*, somewhat surprisingly is also predicted to be conrotatorily stereospecific.

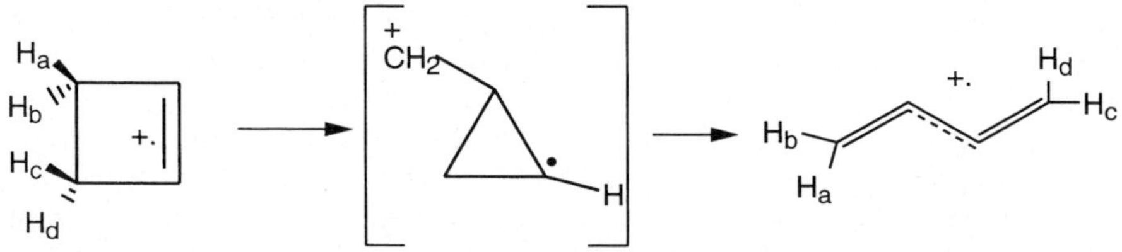

**Scheme 57.** The conrotatory, nonelectrocyclic path from the cyclobutene cation radical to 1,3-butadiene cation radical.

### 3.2.25 Cation Radical Cyclizations

#### Scope and general aspects of cation radical cyclizations

The distinction between the two terms 'cycloaddition' and 'cyclization' should be noted. In contrast to cycloadditions, in which a ring is formed by generating two new bonds either in an intermolecular or an intramolecular context, cyclization is an intramolecular reaction in which a ring is formed by generating only one new bond. The Diels–Alder reaction is a cycloaddition, while the cyclization of the 1,5-hexadiene cation radical to give the 1,4-cyclohexanediyl cation radical represents a cation radical cyclization. Moeller's group has been especially ingenious in developing the synthetic potential of cation radical cyclization reactions [102]. It should be born in mind that, unlike most of the previously discussed cycloadditions, these cyclization reactions are typically neither chain nor catalytic processes, but require the stoichiometric oxidation of an appropriate substrate to its cation radical. For this reason, among others, most of the efficient cyclizations already reported have involved the generation of substrate cation radicals by anodic oxidation. Normally the substrate is bifunctional, and contains one functionality which is especially easily oxidizable and one to which the cation radical moiety may readily add by an electrophilic or radical addition mode. The vinyl ether function plays an especially prominent role in much of the anodic cyclization chemistry reported thus far, especially as the oxidizable function, but also in some cases as the neutral component of the cyclization reaction (Scheme 58).

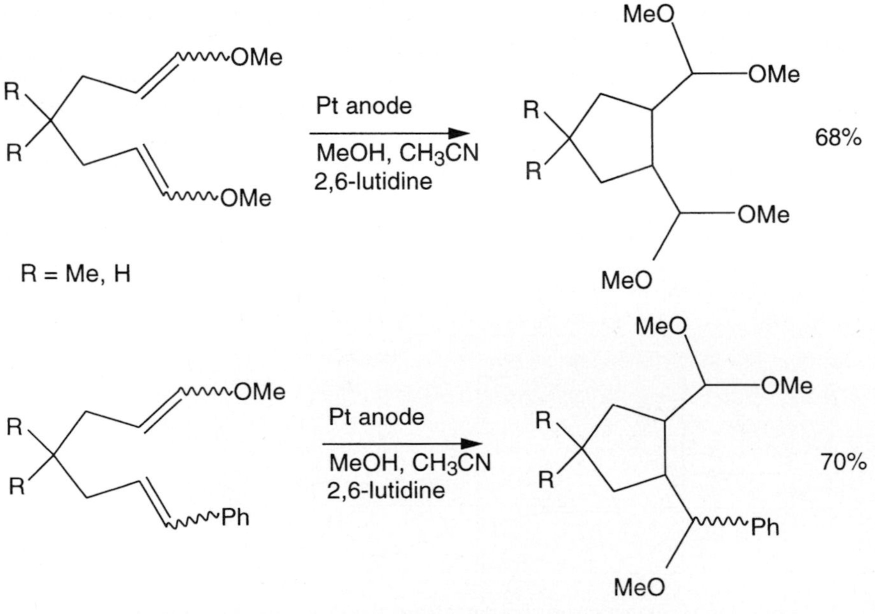

**Scheme 58.** Cation radical cyclization reactions.

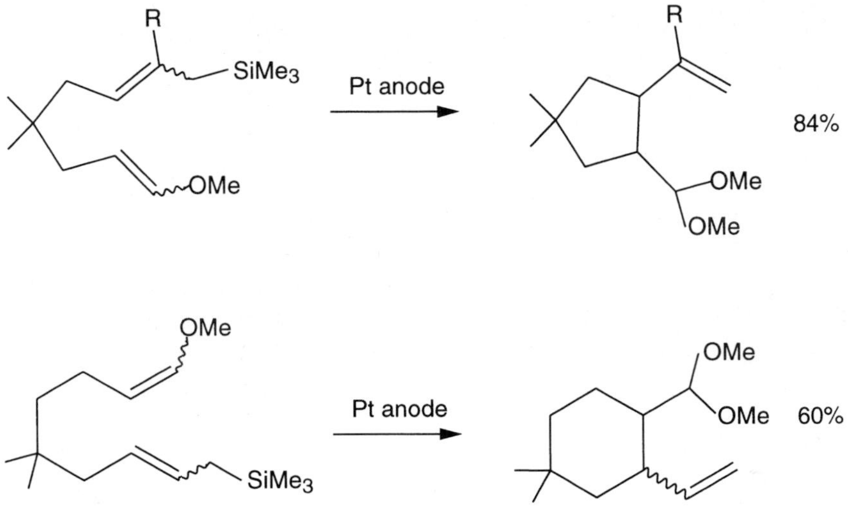

**Scheme 59.** Cation radical cyclizations using allylsilanes as neutral components.

As neutral components, simple styryl and even alkene functions have been employed in cyclizations generating 5–7 membered rings. Vinylsilane and allylsilane functionality have also been ingeniously employed, the silyl group providing an effective electrofugal leaving group (Scheme 59).

Most of these reactions take place in good yield and using relatively simple electrochemical apparatus. It is especially important in this context to note the initial example of such cation radical cyclizations, especially under anodic oxidation conditions, which was provided by Shono (Scheme 60) [103].

Although this reaction proceeded in a relatively low yield, it provides not only the prototype example, but an interesting exemplification of the use of a vinyl sulfide moiety as the ionizable group, and of an electron rich aromatic ring as the neutral moiety.

**Cation radical cyclizations in natural product synthesis**

The effective potential use of cation radical cyclization reactions for natural product synthesis has been proposed, and the syntheses of Crinipellin B and Scopadulcic

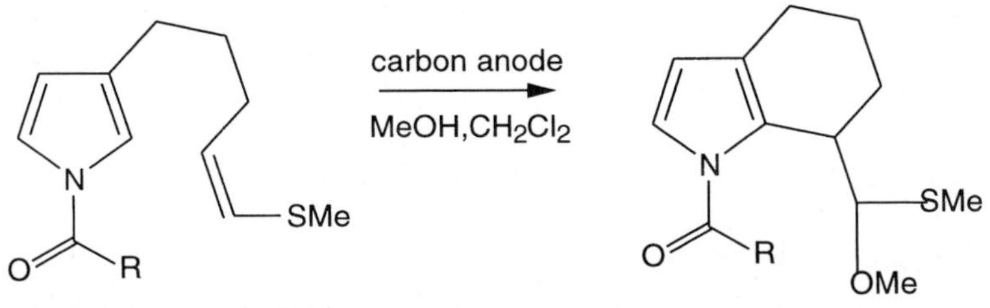

**Scheme 60.** The precedent for anodic cation radical cyclizations.

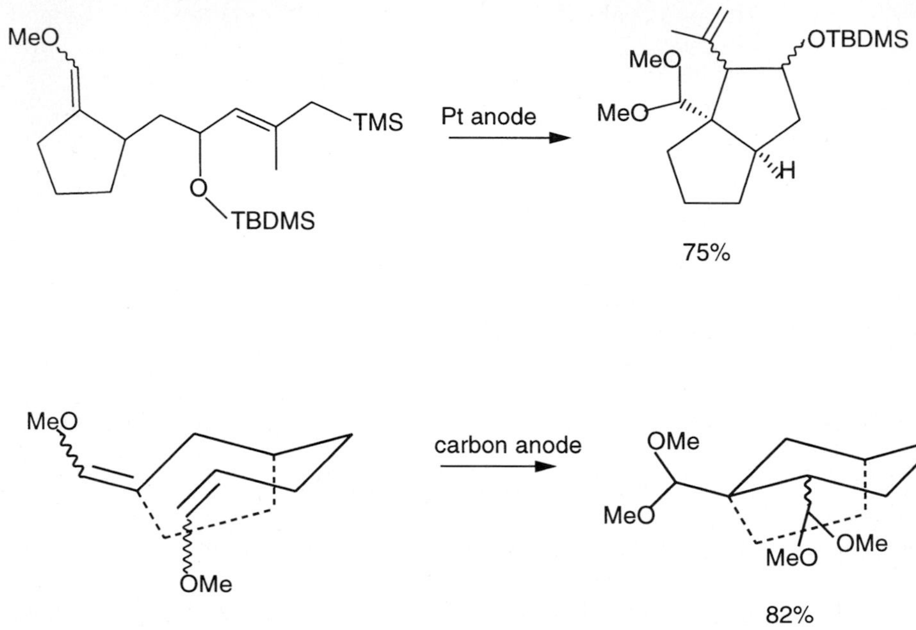

**Scheme 61.** Cation radical approaches to crinipellin B and scopadulcic acid B.

Acid B appear to be well underway, as illustrated by the published model studies illustrated in Scheme 61 [104].

### 3.2.26 Other Reactions of Cation Radicals

#### Coupling reactions

It is abundantly clear that *electron transfer, cycloaddition, cyclization*, and *re-arrangement* are characteristic and facile reactions of pi cation radicals. Equally as characteristic, even if not quite as useful, are the coupling reactions which frequently occur between cation radicals when they are generated in higher concentrations, e.g., at anodes or when an aminium salt initiator is used which too readily oxidizes the substrate. As radical species, cation radicals would be expected to participate in coupling reactions which afford di-cations. An especially interesting case of this type involves the coupling of vinylcarbazole cation radicals under certain conditions to give di-cations, which then propagate addition polymerization from both cationic sites [105]. Many other such couplings have been observed.

#### Cation radicals as Brønsted acids and electrophiles

As electrophilic species, cation radicals react readily with a variety of nucleophiles. The rates of such reactions, in fact, often approach diffusion control. For example, the 4-vinylanisole cation radical reacts with azide ion in trifluoroethanol solution at the diffusion controlled rate $(7 \times 10^9 \text{ M}^{-1} \text{ s}^{-1})$ [106]. Even the *trans*-anethole cation radical reacts at a rate which is only a factor of two less than the diffusion controlled rate. Finally, cation radicals which have allylic hydrogens are highly acidic, often being in the category of superacids, so that acid catalyzed chemistry is a potential concomitant of cation radical chemistry, especially under aminium salt conditions. Further, the di-cations which may be generated by cation radical/cation radical coupling may de-protonate to yield strong acids. This appears to be an especial caveat for aminium salt initiated chemistry, which can however, be controlled in many cases by the addition of a hindered amine base. Acid catalyzed, electrophilic chemistry appears not to be problematic in the case of the PET method of generating cation radicals, because the corresponding anion radicals of the sensitizer sweep out any protonic species. Further, it has not yet become evident that acid catalyzed processes are problematic under anodic oxidation conditions. A classic instance of the intrusion of acid-catalyzed electrophilic chemistry in the context of an aminium salt catalyzed reaction is available in the Diels–Alder cyclodimerization of 2,4-dimethyl-1,3-pentadiene (Scheme 62) [70].

The DA dimer obtained in the aminium salt reaction differs from that obtained by the PET method and from that obtained using the aminium salt/hindered base method, but is the same as that obtained by the Brønsted acid catalyzed reaction of this diene. The addition of insoluble bases like sodium carbonate is not sufficient to suppress the extremely facile, acid-catalyzed cyclodimerization of this particular diene, which yield a highly stabilized tetramethylallyl carbocation intermediate. Nevertheless, this results demonstrates the necessity for caution in assigning a cation radical mechanism to a cyclodimerization reaction observed under aminium salt conditions.

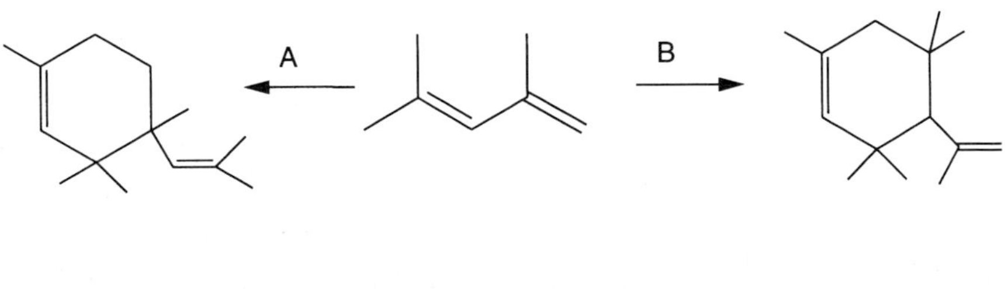

A = PET induced reaction of cation radicals or
    aminium salt induced reaction in the presence
    of a hindered base: cation radical mechanism

B = aminium salt induced reaction in the absence of
    soluble, hindered base or triflic acid catalyzed reaction:
    Bronsted acid catalyzed, carbocation mediated reaction.

**Scheme 62.** Cation radical vs. carbocationic Diels–Alder cycloaddition reactions.

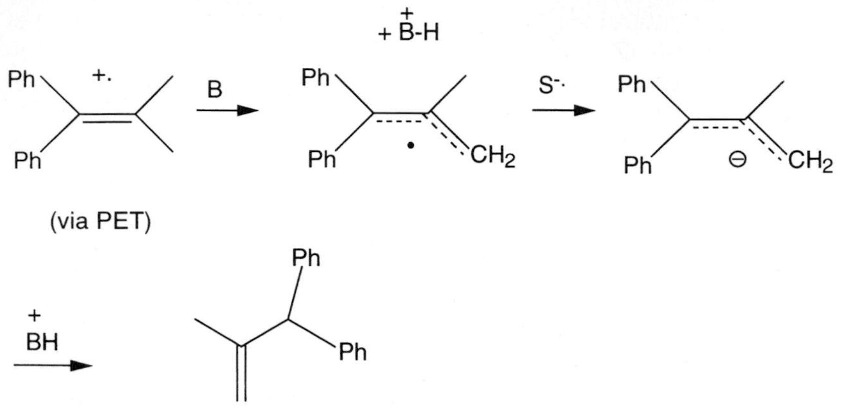

**Scheme 63.** Contrathermodynamic isomerization of alkenes.

## Contrathermodynamic alkene isomerization

Arnold's group has exploited the acidity of alkene cation radicals which have allylic hydrogens to isomerize an alkene function from a position in conjugation with an aryl ring to the position once removed from its original placement (Scheme 63) [107].

This reaction presumably relies upon the relative ease of ionization of the styrene type system to the corresponding cation radical using the PET method. This can be rapidly deprotonated by an appropriate base to the corresponding allylic radical, which is then reduced by the sensitizer anion radical to the corresponding allylic anion. This anion, in turn, can be protonated at either allylic terminus by the conjugate acid of the base which de-protonated the cation radical. When the non-conjugated alkene is generated, it is relatively much less ionizable by the excited state sensitizer, so that the photostationary state favors the less thermodynamically stable, non-conjugated alkene function.

## 3.3 Electron Transfer Chemistry Involving C–C Multiple Bonds as Single Electron Acceptors

### 3.3.1 The Three-electron Bond of the Ethene Anion Radical

The structure of the ethene anion radical remains unknown, but a simple HMO picture of this species (Scheme 64) reveals a three electron bond loosely analogous to those found in some radicals, such as the nitroxyl radicals.

Not only does the addition of an electron to the $\pi^*$ MO of ethene inherently greatly decrease the extent of bonding relative to the two electron bond present in ethene, but it also appears highly likely that because of the additional electronic

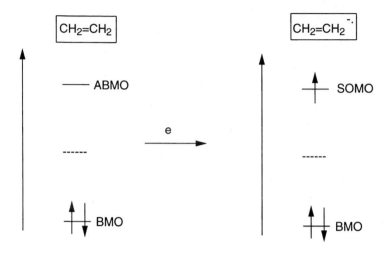

**Scheme 64.** The three electron bond of the ethene anion radical.

repulsions in a three electron pi system the bond energy is far less than that of even a one electron pi bond (as in the ethene cation radical). The severely weakened pi bond could thus be amenable to twisting in order to decrease the torsional strain present in a planar species, just as was the case with the corresponding cation radical. On the other hand, the cationic hyperconjugative interactions with the CH bonds which are present in the twisted but not in the planar form of the ethene cation radical are also likely to be less favorable in the anionic case. What does appear clear is that not only is ethene relatively difficult to reduce to the corresponding anion radical, in contrast to, e.g. benzene, but the resulting anion radical is also highly reactive and unstable.

### 3.3.2 The Shape of the SOMO of a Pi Anion Radical

The energy of the SOMO of an ethene derivative which is substituted with conjugating, electron withdrawing substituents is of course much lower than that of ethene itself. An extreme case of such an anion radical is that of tetracyanoethylene (TCNE), which has been isolated as the tetrabutylammonium salt. A reasonably direct, experimental examination of the SOMO distribution of this anion radical was possible through polarized single crystal neutron diffraction studies [108]. Interestingly, the pi SOMO was found not to be centered directly around the two alkene carbon nuclei, but rather to be bent back, away from the alkene C–C bond, as is theoretically expected for an MO which is anti-bonding between these two carbons (Scheme 65).

The spin densities found on the central alkene carbons, the cyano group carbons, and the terminal nitrogens are 0.33, −0.05, and 0.13, per atom, respectively. These experimental findings are in good agreement with the results of density functional

NODAL PLANE

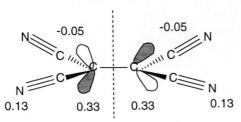

0.13        0.33        0.33        0.13

**Scheme 65.** Shape and spin densities in the SOMO of the TCNE anion radical.

theoretical (DFT) calculations. Thus, the alkene carbons have a total of 66 % of the spin, while the remainder of the positive spin is delocalized onto the terminal nitrogens of the cyano groups.

### 3.3.3 The Butadiene Anion Radical

The LUMOs of conjugated pi systems are typically much lower in energy than that of ethene, so that the reduction of these systems to the corresponding anion radicals is more facile. The electrochemical reduction of 1,3-butadiene in liquid ammonia solution at $-78\,^\circ$C, in fact, yields an anion radical which is stable enough to permit observation by ESR spectroscopy (Scheme 66) [109].

The hyperfine splitting arising from the four protons at the terminal carbon atoms is found to be $-0.76$ mT, while that of the two other protons is $-0.28$ mT, indicating that, as theory predicts, the SOMO of this diene anion radical is more heavily concentrated on C1 and C4, as opposed to C2 and C3. In contrast, the cation radical of 1,3-butadiene is apparently too short-lived in *fluid* solution to be observed by ESR spectroscopy.

### 3.3.4 Disproportionation of Anion Radicals to Dianions and Neutrals

The tendency of anion radicals to undergo an electron transfer reaction yielding a neutral molecule and a dianion (Scheme 67) is especially noteworthy.

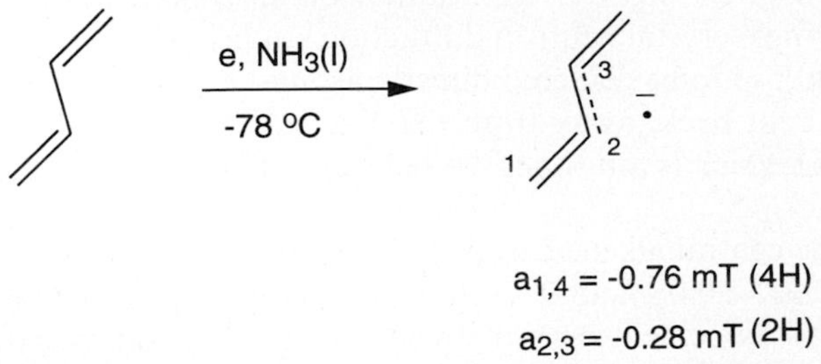

e, NH$_3$(l)

$-78\,^\circ$C

$a_{1,4} = -0.76$ mT (4H)

$a_{2,3} = -0.28$ mT (2H)

**Scheme 66.** Formation and hyperfine couplings in the 1,3-butadiene anion radical.

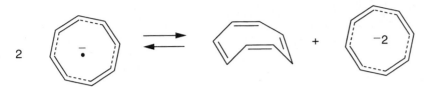

**Scheme 67.** Disproportionation of anion radicals.

Though this tendency to disproportionate is especially marked in relatively non-polar solvents, it can occur to a significant extent even in polar solvents. Apparently, the additional electronic repulsions incurred by adding a second electron to the SOMO of an anion radical are negated by the even stronger electrostatic attraction of the dianion to two unipositive metal ions. Besides the coordination of metal ions to the dianion, substantial structural changes incurred in converting the neutral species to the anion radical or in converting the anion radical to the dianion may affect the extent of disproportionation. In the cyclooctatetraene case, for example, the neutral molecule is tub-shaped, with the four alkene pi bonds essentially perpendicular to each other [110]. Formally, this molecule has a very high energy LUMO, comparable to that in ethene. However, the anion radical which is actually formed is planar, and the SOMO of this anion radical occupies a very low energy, non-bonding orbital. Consequently, the conversion from a tub shaped anion radical to a planar structure is highly favored by the dramatic lowering of the energy of the SOMO. On the other hand, the planar octagonal shape of the anion radical has a substantial amount of angle strain, owing to the wider than normal bond angles, along with some additional torsional strain. In contrast, the conversion of the anion radical to the dianion takes advantage of the lower SOMO energy without having to compensate for any additional strain, since the dianion is also planar. To put the argument more succinctly, on the anion radical side of the equilibrium, two species have ring strain, whereas on the dianion side, only one does. The tetraphenylethylene anion radical is another persistent anion radical for which disproportionation has been found to be extensive [111]. In this instance, a possible incremental contributing reason is the release of steric effects in the dianion, which could reasonably be expected to have a twisted structure in which the two trigonal planes of the alkene carbons are by no means coincident, and possibly might even be perpendicular.

### 3.3.5 Dianion Radicals, Trianion Radicals, and Multianion Radicals

Just as the one electron reduction of a neutral molecule produces an anion radical, the one electron reduction of an organic anion can yield a dianion radical, assuming that the LUMO of the anion is not too high in energy. Two anions familiar in organic chemistry are the dibenzoylmethide (an enolate of a $\beta$-diketone) [112] and fluorenide anions [113] (Scheme 68).

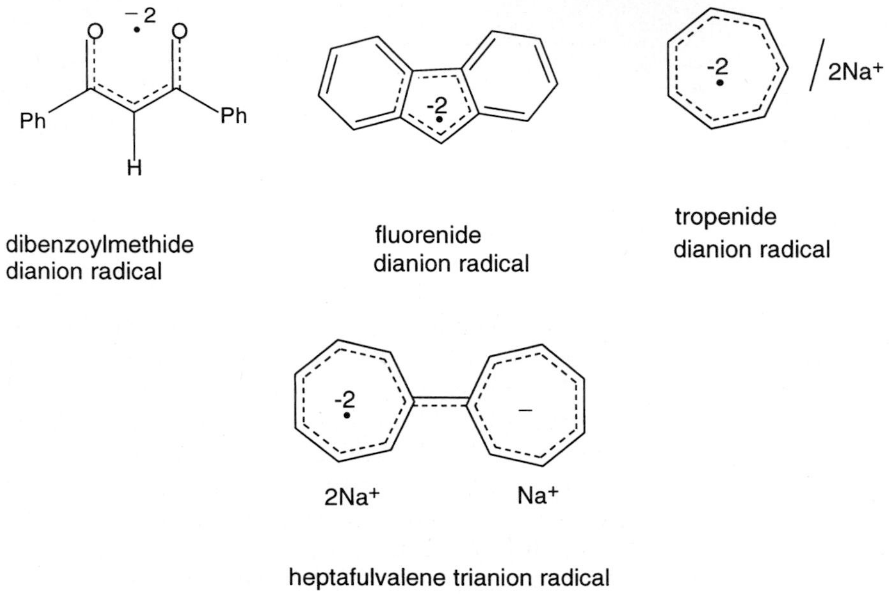

dibenzoylmethide
dianion radical

fluorenide
dianion radical

tropenide
dianion radical

heptafulvalene trianion radical

**Scheme 68.** Stable dianion radicals and trianion radicals.

These anions have both been converted to the corresponding persistent dianion radicals by alkali metal reduction. In the latter case, the agreement of the ESR hyperfine splitting constants with those predicted from the LUMO of the fluorenide anion were confirmed by deuterium labeling experiments. A simple hydrocarbon dianion radical was prepared by the reduction of the tropenide anion to the corresponding dianion radical [114]. In the case of the sodium salt of this dianion radical, the dianionic nature of the radical species was confirmed additionally by the observation of two equivalent sodium hyperfine splittings. A trianion radical was prepared by the alkali metal reduction of sesquifulvalene [115]. Unlike the corresponding anion radical, the trianion radical has the spin localized on only one of the seven-membered rings, presumably because of tight ion pairing of the other ring (which contains the dianion moiety) with two metal ions, while the ring which contains the anion radical moiety is ion paired to just one of the three metal ions.

### 3.3.6 Methods for Generating Anion Radicals

From the earliest times, anion radicals have been generated by the reaction of an appropriate neutral molecule with metals, especially, with sodium, potassium, or lithium in ethereal solvents. In sharp contrast to the corresponding cation radicals, these anion radicals have often proved to be quite persistent in the absence of moisture or air. This is particularly true for anion radicals of aromatic systems such

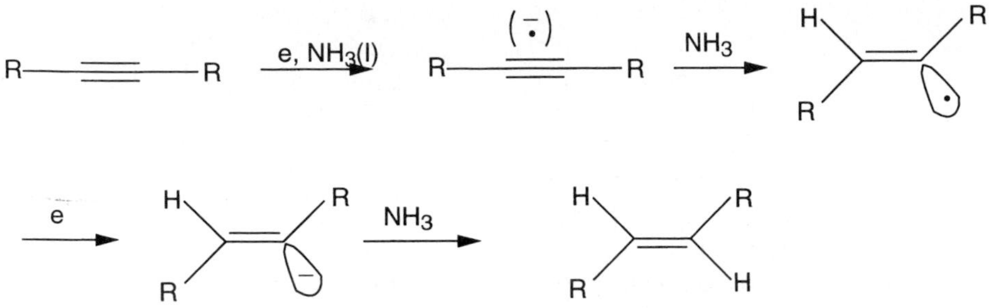

**Scheme 69.** Birch reduction of alkynes to trans-alkenes.

as naphthalene and biphenyl. The latter have often been prepared and used as single electron transfer (SET) agents for the homogeneous reduction of other substrates to their anion radicals. Further, there is a rich corpus of anion radical chemistry deriving from electroreduction of neutral organic substrates, which will be discussed further on. More recently, anion radicals have also been generated by the PET method, using excited states of electron rich sensitizers to reduce substrate molecules. Finally, anion radical/cation radical pairs can be generated in appropriate instances by SET between neutral, ground state molecules. Each of the above methods will be illustrated in the course of the following discussions of the ET chemistry of anion radicals.

### 3.3.7 The Birch Reduction of Non-Terminal Alkynes

Although alkali metal/liquid ammonia reductions (Birch reductions) of simple alkenes is difficult, presumably as a result of the very high energy of an ethene type LUMO, the corresponding reduction of non-terminal alkynes to *trans*-alkenes is an efficient and useful synthetic tool for accessing *trans*-alkenes [116]. The mechanism for this reaction (Scheme 69), involves the homogeneous reduction of the alkyne to the corresponding anion radical by the solvated electrons present in liquid ammonia solutions of alkali metals.

The anion radicals undergo protonation to give a vinyl radical, followed by homogeneous reduction of the latter to the corresponding vinyl anion, which is then protonated. The step in which the product stereochemistry is established is still not certain, but it is clear that both the vinyl radical and carbanion are bent species which should have a preference for *trans* R groups. More likely is the possibility that the vinyl radical is either initially formed with a preference for the *trans* bent structure in the protonation of the anion radical or that it rapidly equilibrates to this more favorable structure prior to reduction to the vinyl anion. The vinyl anion might be expected to be configurationally more stable than is the radical, but it is also possible that rapid equilibration of the anion yields the more stable *trans*-vinyl anion.

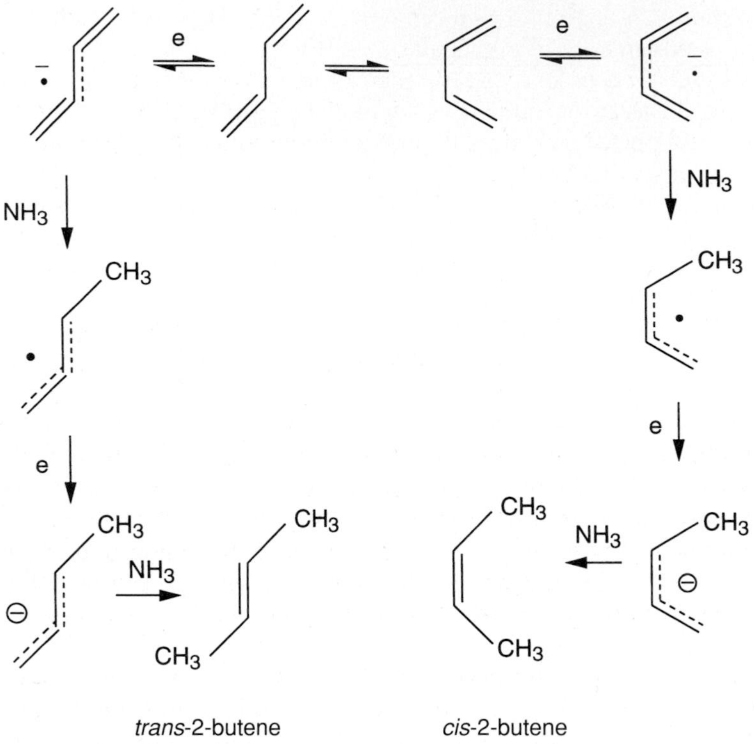

trans-2-butene          cis-2-butene

**Scheme 70.** Birch reduction of conjugated dienes.

### 3.3.8 Birch Reduction of Conjugated Dienes

The reaction of 1,3-butadiene with the solvated electrons of liquid ammonia results in homogeneous electron transfer to yield the diene anion radical, which is ultimately converted to a mixture of *cis*- and *trans*-2-butene [117]. The mechanism of this reduction (Scheme 70) involves protonation of the anion radical at C1 or C4 of the diene, where most of the negative charge resides, followed by homogeneous reduction of the resulting allylic radical by the solvated electrons and finally protonation of the allylic carbanion at the primary carbanion site.

An interesting aspect of this reaction is the formation of substantial amounts of *cis*-2-butene, which would appear to require the intermediacy of the *s-cis*-1,3-butadiene anion radical, even though butadiene exists almost exclusively in the *s-trans* conformation (98 %). At −33 °C, 13 % of the 2-butene mixture is the *cis* alkene, and at −78 °C 50 % of the mixture is *cis*-2-butene. In the case of 1,3-pentadiene, 68 % of the 2-pentene is the *cis* isomer. The most plausible explanation for these stereochemical results appears to be the reversible reduction of the diene to the diene anion radical at −78 °C by the pool of solvated electrons, which yields an equilibrium mixture of the *s-cis* and *s-trans*-anion radicals (ca. 50:50), which are then protonated by ammonia to yield the *cis-* or *trans* allylic radical. An alternative equilibration mechanism involving unimolecular *trans*-to-*cis* isomerization appears unlikely since the C2–C3 bond order is greatly increased in the anion radical. That the protonation of the diene anion radical by ammonia might be slow enough to permit equilibration is supported by the weak acidity of ammonia ($pK_a$ 38) and the soft basicity of the delocalized anion radical. It appears likely that once the allylic radical is formed, equilibration between the *cis* and *trans* allylic radicals or anions is highly unlikely, since this would involve the loss of allylic resonance. At $-33\,°C$, the protonation of the anion radical may be fast enough to prevent the complete equilibration of the *cis* and *trans* anion radicals.

### 3.3.9 Coupling of Anion Radicals

In the previously described reactions, the basicity of anion radicals and their reactions with proton donors was emphasized. In the absence of viable proton donors or even in their presence, if the anion radical is relatively stable, radical coupling may be the dominant reaction. Thus even in aqueous solution, the anion radicals of alkenes substituted with strongly electron withdrawing moieties may undergo coupling in preference to protonation. The synthesis of adiponitrile from acrylonitrile (Scheme 71) is an outstanding example [118].

This kind of reaction is effective with acrylate esters and even with 4-vinylpyridine. Somewhat surprisingly, it works well even with some terminally disubstituted acrylate esters. Intramolecular versions involving cyclization to five-membered rings are especially effective. Mechanistically, these reactions could involve the coupling of two anion radicals or the addition of an anion radical to a corresponding neutral, followed by reduction of the distonic anion radical to the dianion, which is then protonated twice, yielding the hydrodimer. Apparently, the distinction between these two plausible mechanistic types has not yet been decisively made in most cases (however, please see the discussion on intramolecular cyclizations immediately

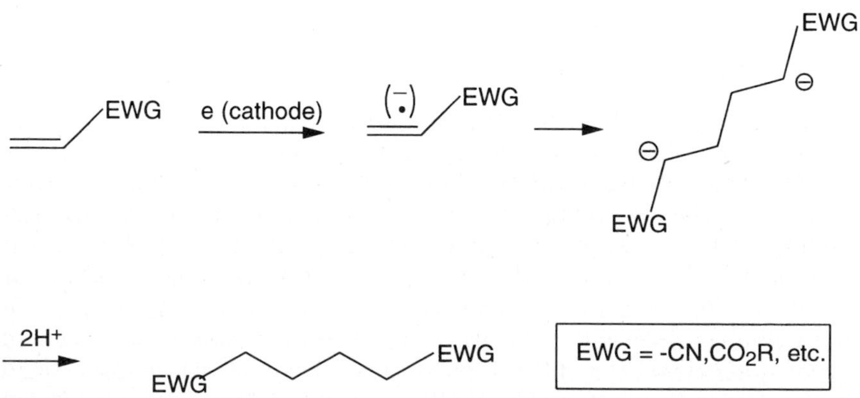

**Scheme 71.** The hydrodimerization of electron deficient alkenes.

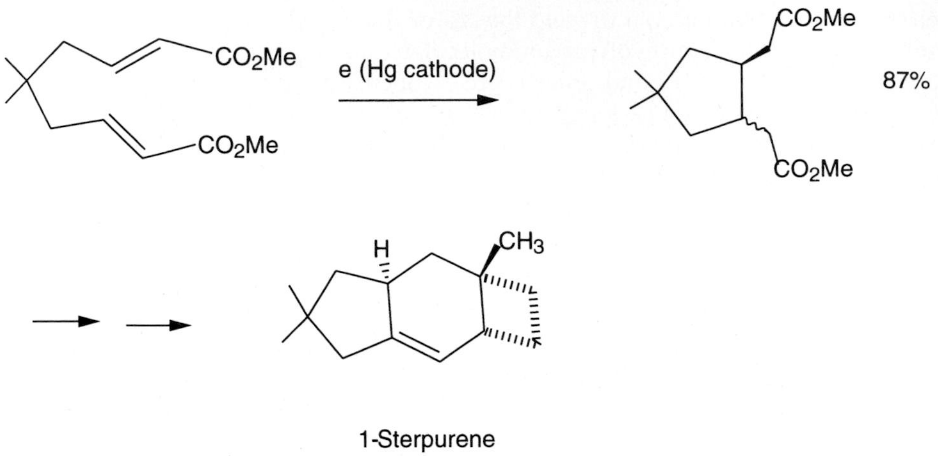

1-Sterpurene

**Scheme 72.** Total synthesis of sterpurene.

below). When substrates are reduced in the absence of any viable proton donor, as in the case of reactions carried out in an ethereal solvent, the dianion is the usual product. In the case of styrene, this dianion has been employed to propagate anion polymerization of styrene at both anionic centers of the dianion [119].

### 3.3.10 Intramolecular Cyclizations Involving Anion Radicals

An intramolecular version of the hydrodimerization reaction discussed in the preceding section has proved to be especially attractive and efficient for natural product synthesis. A total synthesis of the sesquiterpene sterpurene exploits just such a reaction, at a very early stage of the synthesis, for the closure of the five-membered ring (Scheme 72) [120].

The anion radicals were generated by cathodic reduction, providing yields of cyclized product as high as 87 %. In a total synthesis of quadrone, efficient reductive cyclizations were used at two different stages of the synthesis (Scheme 73) [121].

These cyclizations both involve the reductive intramolecular addition of an electron deficient alkene function to an aldehyde carbonyl function, and both are effected in ca 90 % yields. The mechanism of this latter type of electrochemically induced cyclizations of carbon–carbon double bonds to carbonyl double bonds have been studied rather extensively, with especial attention to the fundamental mechanistic question of whether the cyclization step involves an anion radical, radical, or anionic mechanism [122]. The latter two mechanisms would involve the protonation of the initially formed anion radical intermediate to form a radical, which could then cyclize or, alternatively, be further reduced to an anion, which could then cyclize. Extensive and elegant electrochemical and chemical studies have led to the formulation of these reactions as involving anionic cyclization (Scheme 74).

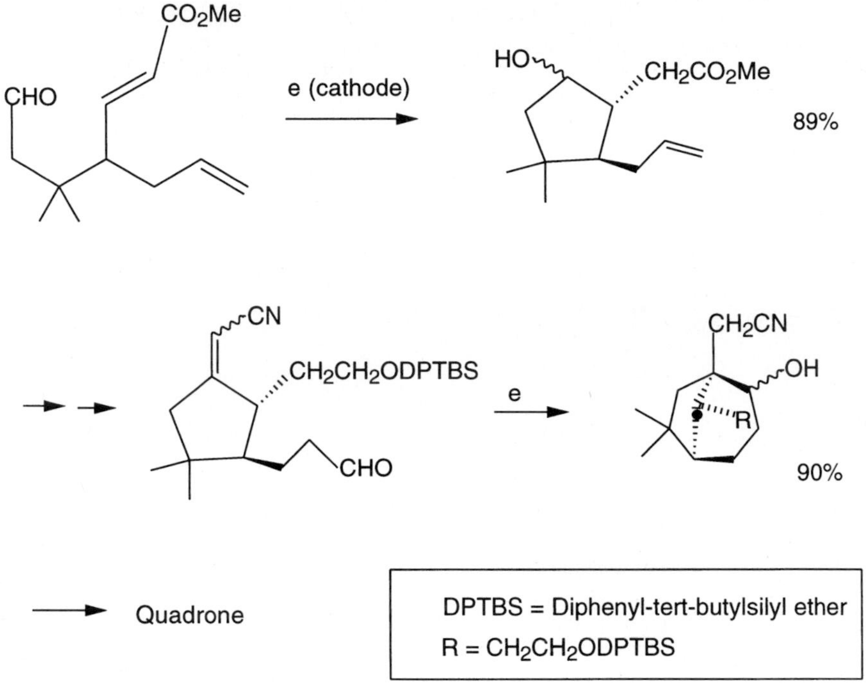

**Scheme 73.** Formal total synthesis of quadrone.

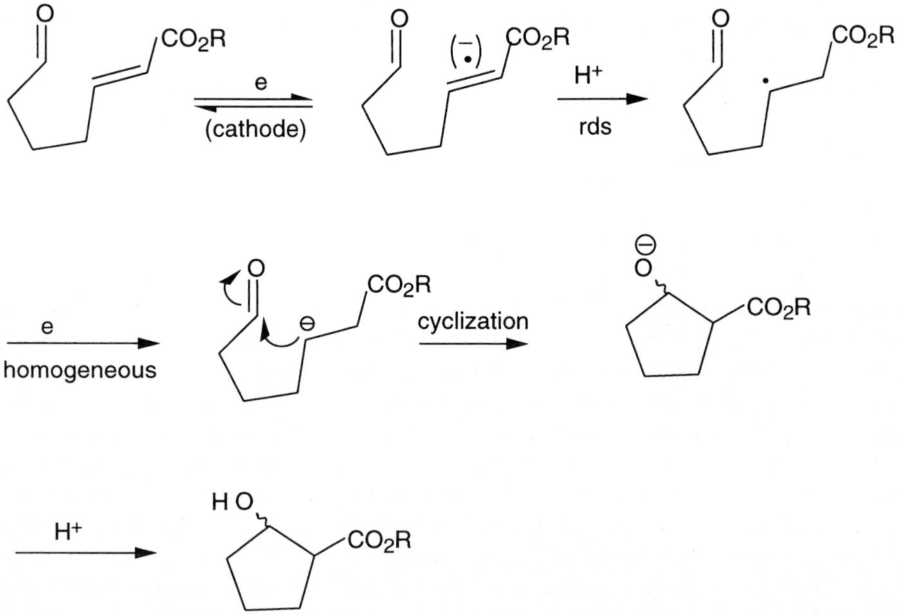

**Scheme 74.** The proposed anionic mechanism for cyclizations involving anion radical intermediates.

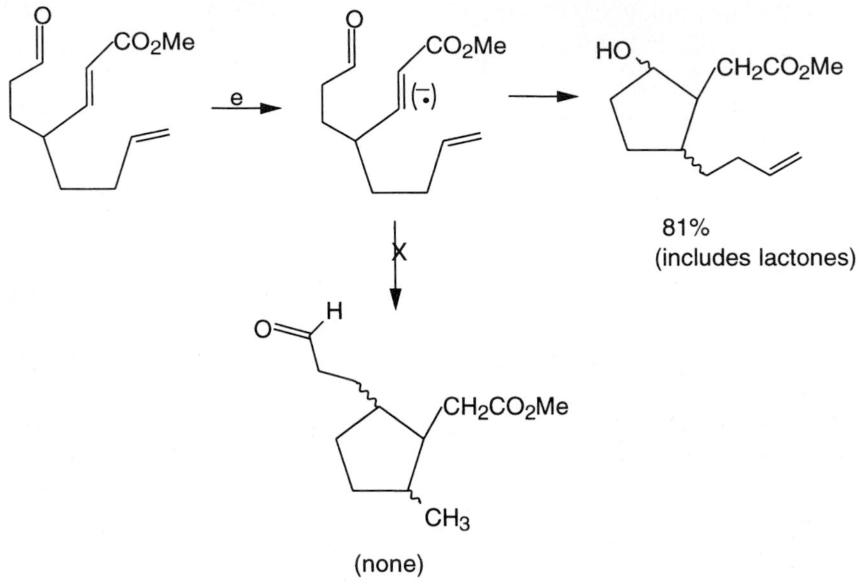

**Scheme 75.** A probe for distinguishing the radical and anionic mechanisms of cyclization.

Indeed, radical probe studies have very decisively excluded the radical cyclization mechanism in at least one typical case. This experiment relies upon the circumstance that the cyclization of a carbon-centered radical to an aldehyde carbon group is known to occur at approximately the same rate as the *exo*-trig cyclization of such a radical to a carbon–carbon double bond. In a probe molecule designed to provide an opportunity for a hypothetical radical intermediate to add to either or both of these functionalities, no addition to the vinyl double bond was observed (Scheme 75).

This indicates that the reaction must involve either the cyclization of an anionic intermediate, which obviously would add more rapidly to a carbonyl group than to an alkene double bond, or of the initially formed anion radical intermediate, which because of Umpolung might also prefer addition to the carbonyl group. The latter possibility was considered to be effectively ruled out because the electrochemical studies revealed that protonation of an intermediate was involved at or before the rate determining step. Thus, if cyclization had occurred at the anion radical stage, it would, according to this reasoning, be expected that this (cyclization) step would be rate determining, not the subsequent protonation of the cyclized intermediate, which would be an alkoxide anion. It might be considered, however, that if the anion radical cyclization were rapid and reversible and thermodynamically unfavorable (as appears likely), the subsequent protonation (especially by a weak acid or even a strong acid present at low concentration) could indeed be rate determining. The rapid, reversible, and selective cyclization of a delocalized and thus rather stabilized anion radical to a carbonyl group would not appear to be implausible. One further point should be made in relation to the proposed anionic cyclization mechanism.

The protonation of the anion radical intermediate proposed in Scheme 74 would appear to be highly unlikely to give the indicated radical as the primary product of protonation. Instead, protonation on oxygen would appear to be favored not only by the higher partial negative charge on oxygen, but also by the circumstance that such a protonation would yield a stabilized, allylic radical, in contrast to the simple primary radical proposed in Scheme 74. Moreover, the rearrangement of an initially formed allylic radical to the primary radical intermediate proposed would appear to be thermodynamically unfavorable, as well as kinetically and mechanistically unlikely. Of course, the allylic radical intermediate could be reduced to the corresponding allylic anion, which might then cyclize to give the (ester) enol form of the expected product. Consequently, it would appear reasonable, if the anionic cyclization mechanism is to be retained, to modify the structures of the proposed radical and anionic intermediates in this cyclization. On the other hand, as was noted above, it may be desirable to more rigorously examine the possibility of an anion radical cyclization mechanism.

### 3.3.11 Pericyclic Reactions of Anion Radicals

#### General aspects of anion radical pericyclic additions

Although the pericyclic chemistry of anion radicals has been much slower to emerge than that of cation radicals, the number of intriguing examples now available suggests that this could be an attractive area for future development in electron transfer chemistry. Reaction types which have been exemplified include cyclobutanation, retrocyclobutanation, Diels–Alder addition, electrocyclic reactions, and retroelectrocyclic reactions.

#### Anion radical cyclobutanation

An especially clean example of electron transfer chemistry involving carbon–carbon multiple bonds as electron acceptors, i.e., proceeding via anion radicals is the cyclobutanation of phenyl vinyl sulfone induced by reduction at a mercury pool cathode (Scheme 76) [122].

That the reaction is electrocatalytic is shown by the observation that only 0.2 F $mol^{-1}$ of electricity is consumed. The dimer is isolated in 66 % yield. The interesting question of whether the addition of the anion radical to the neutral sulfone is con-

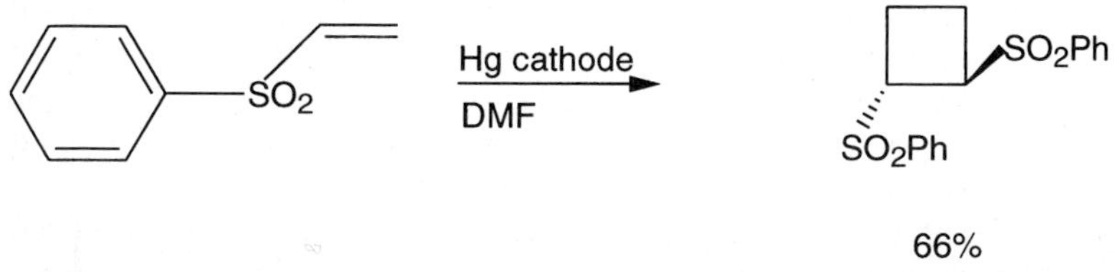

66%

Scheme 76. Anion radical cyclobutanation.

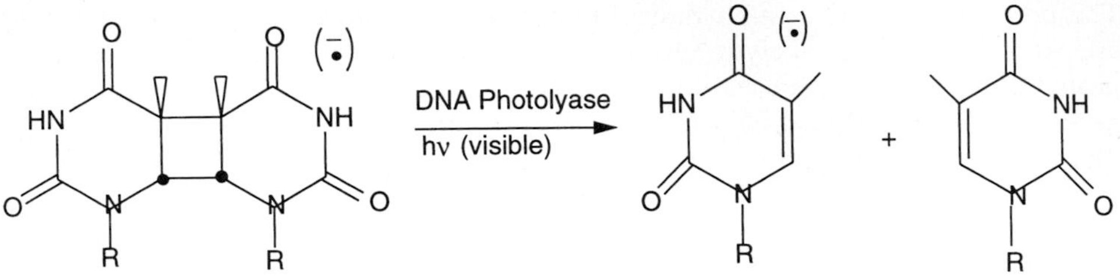

**Scheme 77.** Anion radical retrocyclobutanation in the repair of DNA.

certed or stepwise has not yet been addressed. Other examples of this type of reaction appear not to have been observed, but clearly this is an area where additional research efforts would be justified.

**Anion radical retrocyclobutanation**

The reversal of cyclobutanation, i.e., retrocyclobutanation has also been observed in the context of a reaction of substantial biological importance (Scheme 77) [123].

The predominant DNA photolesion engendered by long wave length UV light is pyrimidine cyclodimerization, a cyclobutanation reaction involving an excited state of the pyrimidine moiety. One of the processes available for the repair of such photolesions is an enzymatic retrocyclobutanation catalyzed by the enzyme DNA photolyase in the presence of visible light and one or more cofactors. Elegant work by Begley's group has demonstrated that this cleavage involves electron transfer to the pyrimidine dimer moiety to form the anion radical, which rapidly cleaves the dimer, giving one anion radical moiety and a neutral moiety.

**Anion radical electrocyclic reactions**

An unusually efficient example of an electron transfer catalyzed electrocyclic reaction proceeding via anion radicals has also been established [124]. The conversion of the bis(allene) shown in Scheme 78 to the corresponding cyclobutene derivative

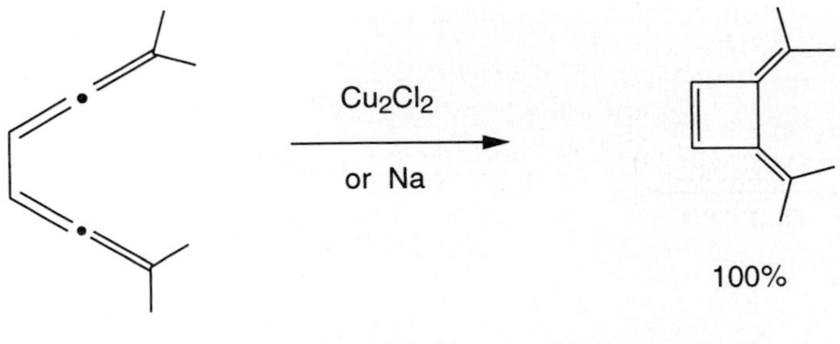

**Scheme 78.** An anion radical chain electrocyclic reaction.

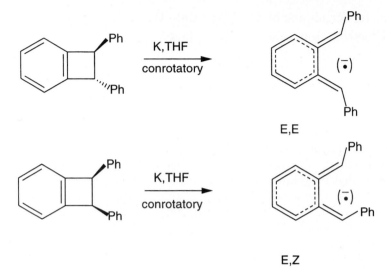

**Scheme 79.** A symmetry allowed anion radical retroelectrocylic reaction.

proceeds in quantitative yield in the presence of cuprous chloride, sodium metal, or the naphthalene anion radical, and is inhibited by molecular oxygen, indicating that it proceeds via a chain mechanism.

The retroelectrocyclic cleavage of the *cis*- and *trans*-diphenylbenzocyclobutene anion radicals, generated by alkali metal reduction, has been shown to occur in a conrotatorily stereospecific manner (Scheme 79) [125].

An electrocyclic closure to a six-membered ring system has also been reported (Scheme 80) [126].

Perhaps the most interesting reaction of all of the possible anion radical pericyclic reactions, a Diels–Alder cycloaddition, has not been definitively exemplified, but one potential example of such a reaction has been proposed [127].

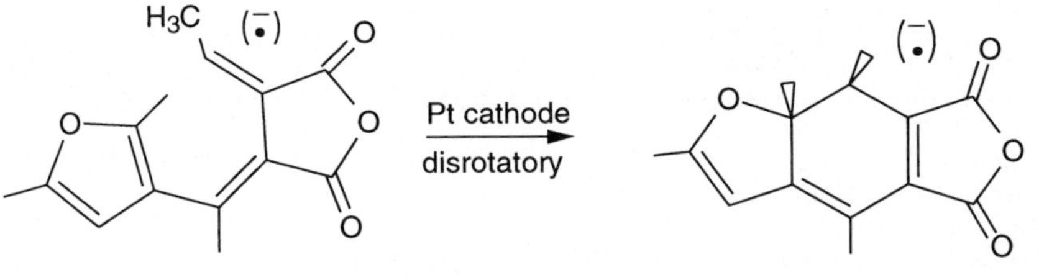

**Scheme 80.** A 1,3,5-hexatriene to 1,3-cyclohexadiene-type anion radical electrocyclic reaction.

## 3.4 Electron Transfer Reactions of C–C Multiple Bonds Which Involve Both Single Electron Donation and Acceptance

### 3.4.1 Reactions Involving both Anion Radicals and Cation Radicals

Reactions which involve thermal electron transfer (TET) between two neutral substrate molecules in their respective ground states are still relatively rare in the chemistry of alkenes and alkynes. An especially intriguing example has been provided, however, the reaction of nitroalkenes and other highly electron deficient alkenes with certain methylenecyclopropanone acetals [128]. Electron transfer between two such substrates appears to generate the anion radical/cation radical pair, the cation radical of which undergoes ring opening to form a dialkoxytrimethylenemethane cation radical. The latter then reacts with its anion radical partner *in a non-stereospecific manner* to yield largely *exo*-methylenecyclopentanone acetals. With less strongly electron deficient alkenes, e.g., those having a single ester substituent, a different reaction course is followed, in which the methylenecyclopropanone ketal initially cleaves to a zwitterion, which subsequently reacts in a *stereospecific manner* to yield only cyclopentanone ketal products.

### Acknowledgment

The authors wish to acknowledge the support of the Robert A. Welch Foundation (F-149) and the National Science Foundation (CHE-9610227).

### References

1. (a) F. A. Bell, A. Ledwith, D. C. Sherrington, *J. Chem. Soc.* **1969**, 2719; (b) R. A. Crellin, M. C. Lambert, A. Ledwith, *J. Chem. Soc., Chem. Commun.* **1970**, 682; A. Ledwith, *Accts. Chem. Res.* **1972**, *5*, 133.
2. C. Wurster, R. Sendtner, *Ber.* **1879**, *12*, 1803 and 2071.
3. H. Wieland, *Ber.* **1907**, *40*, 4260.
4. (a) E. Weitz, H. W. Schwechtin, *Ber.* **1926**, *59*, 2307; (b) *ibid.* **1927**, *60*, 545.
5. L. Michaelis, *Chem. Rev.* **1935**, *16*, 243.
6. R. I. Walter, *J. Am. Chem. Soc.* **1966**, *88*, 1923.
7. W. Schmidt, F. Steckhan, *Chem. Ber.* **1980**, *113*, 577.
8. (a) R. S. Mulliken, C. C. J. Roothan, *Chem. Rev.* **1947**, *41*, 219; (b) A. J. Meerer, L. Schoonveld, *J. Chem. Phys.* **1968**, *48*, 522; (c) H. Koppel, W. Domcke, L. S. Cederbaum, W. von Niessen, *J. Chem. Phys.* **1978**, *69*, 4252; (d) D. J. Bellville, N. L. Bauld, *J. Am. Chem. Soc.* **1982**, *104*, 294.
9. (a) W. Tang, X.-L. Zhang, T. Bally, *J. Phys. Chem.* **1993**, *97*, 4373; (b) T. Keszthelyi, R. Wilbrandt, J.-L. Roulin, T. Bally, *ibid.* **1996**, *100*, 16850.
10. H. D. Roth, M. L. Mannion, *J. Am. Chem. Soc.* **1981**, *103*, 7210.
11. P. Jungwirth, T. Bally, *J. Am. Chem. Soc.* **1993**, *115*, 5783–5789.
12. L. P. Ellinger, *Polymer* **1964**, *5*, 559.
13. W. P. Todd, J. P. Dinnocenzo, S. Farid, J. L. Goodman, I. R. Gould, *J. Am. Chem. Soc.* **1991**, *113*, 3601.

14. S. L. Mattes, S. Farid in *Organic Photochemistry*, Vol. 6 (Ed.: A. Padw (a), Marcel Dekker, New York, **1983**, p. 233.
15. D. J. Bellville, D. D. Wirth, N. L. Bauld, *J. Am. Chem. Soc.* **1981**, *103*, 718.
16. L. Cedheim, L. Eberson, *Acta Chem. Scand.* **1976**, *B30*, 527–532.
17. (a) D. Ghosh, N. L. Bauld, *J. Catalysis* **1985**, *95*, 300–304; (b) K. Lorenz, N. L. Bauld, J. Catalysis 1985, 95, 613–616.
18. P. Laszlo, J. Lucchetti, Tetrahedron Lett. 1984, 25, 1567.
19. M. A. Fox, D. D. Sackett, J. N. Younathan, *Tetrahedron* **1987**, *43*, 7.
20. N. L. Bauld, D. J. Bellville, S. A. Gardner, Y. Migron, G. Cogswell, *Tetrahedron Lett.* **1982**, *8*, 825–828.
21. N. P. Schepp, L. J. Johnston, *J. Am. Chem. Soc.* **1994**, *116*, 6895–6903.
22. T. Bally, *Chimia* **1994**, *48*, 378.
23. I. R. Gould, S. Farid, *Acct. Chem. Res.* **1996**, *29*, 522–528.
24. (a) N. L. Bauld, J. T. Aplin, W. Yueh, S. Endo, A. Loving, *J. Phys. Org. Chem.* **1998**, *11*, 15–24; (b) N. L. Bauld, J. T. Aplin, W. Yueh, A. Loinaz, *J. Am. Chem. Soc.* **1997**, *119*, 11381–11389.
25. (a) S. Kuwata, Y. Shigemitsu, Y. Odaira, *J. Org. Chem.* **1973**, *21*, 3803; S. Farid, S. E. Hartman, T. R. Evans in *The Exciplex* (Eds.: M. Gordon, T. R. Ware), Academic Press, New York, **1975**, p. 317.
26. T. R. Evans, R. W. Wake, O. Jaenicke, *ibid.*, p. 345.
27. S. Farid, S. E. Shealer, *J. Chem. Soc. Chem. Commun.* **1973**, 677.
28. N. L. Bauld, R. Pabon, *J. Am. Chem. Soc.* **1983**, *105*, 633–634.
29. F. D. Lewis, M. Kojima, *J. Am. Chem. Soc.* **1988**, *110*, 8664–8670.
30. H. Nozaki, I. Otani, R. Noyori, M. Kawanisi, *Tetrahedron* **1968**, *24*, 2183.
31. N. P. Schepp, D. Shukla, H. Sarker, N. L. Bauld, L. J. Johnston, *J. Am. Chem. Soc.* **1997**, *119*, 10325–10334.
32. (a) N. P. Schepp, L. J. Johnston, *J. Am. Chem. Soc.* **1996**, *118*, 2872–2881; (b) N. P. Schepp, L. J. Johnston, *ibid.* **1994**, *116*, 6895–6903.
33. K. T. Lorenz, N. L. Bauld, *J. Am. Chem. Soc.* **1987**, *109*, 1157–1160.
34. J. L. Courtneidge, A. G. Davies, S. M. Tollerfield, J. Rideout, M. C. R. Symons, *J. Chem. Soc. Chem. Commun.* **1985**, 1092–1093.
35. N. L. Bauld, D. J. Bellville, R. Pabon, R. Chelsky, G. Green, *J. Am. Chem. Soc.* **1983**, *105*, 2378–2382.
36. (a) G. N. Sastry, V. Hrouda, M. Ingr, P. Carsky, T. Bally, *J. Phys. Chem. A* **1998**, *102*, 9297; (b) V. Hrouda, M. Roeselova, *J. Phys. Chem. A* **1997**, *101*, 3925.
37. S. F. Nelsen, *Accts. Chem. Research* **1987**, *20*, 276.
38. (a) G. O. Schenck, S.-P. Mannsfeld, G. Schomburg, C. H. Krauch, *Z. Naturforsch.* **1964**, *19B*, 18; (b) R. Schutte, G. R. Freeman, *J. Am. Chem. Soc.* **1969**, *91*, 3715; (c) T. L. Penner, D. G. Whitten, G. S. Hammond, *J. Am. Chem. Soc.* **1970**, *92*, 2861.
39. D. J. Bellville, D. D. Wirth, N. L. Bauld, *ibid.* **1981**, *103*, 718–720.
40. (a) N. L. Bauld, D. J. Bellville, B. Harirchian, K. T. Lorenz, R. A. Pabon, Jr., D. W. Reynolds, D. D. Wirth, H.- S. Chiou, B. K. Marsh, *Accts. Chem. Research* **1987**, *20*, 371–378; (b) N. L. Bauld, *Tetrahedron* **1989**, *45*, 5307–5363; (c) N. L. Bauld in *Advances in Electron Transfer Chemistry*, Vol. 2 (Ed. P. S. Mariano), JAI Press, Greenwich, **1992**, pp. 1–65.
41. G. C. Calhoun, G. B. Schuster, *J. Am. Chem. Soc.* **1984**, *106*, 6870.
42. (a) R. A. Neunteufel, D. R. Arnold, *J. Am. Chem. Soc.* **1973**, *95*, 4080; (b) A. J. Maroulis, D. R. Arnold, *J. Chem. Soc. Chem. Commun* **1979**, 351.
43. S. L. Mattes, S. Farid, *J. Am. Chem. Soc.* **1983**, *105*, 1386
44. J. Libman, *J. Chem. Soc. Chem. Commun.* **1976**, 361.
45. K. Mizuno, R. Kaji, H. Okada, Y. Otsuji, *ibid.* **1978**, 594.
46. N. L. Bauld, J. Yang, unpublished work.
47. D. J. Bellville, N. L. Bauld, *J. Am. Chem. Soc.* **1982**, *104*, 2665–2667.
48. D. J. Bellville, N. L. Bauld, R. Pabon, S. A. Gardner, *ibid.* **1983**, *105*, 3584–3588.
49. R. A. Pabon, D. J. Bellville, N. L. Bauld, *J. Am. Chem. Soc.* **1983**, *105*, 5158–5159.
50. C. R. Jones, B. J. Allman, A. Mooring, B. Spahic, *ibid.* **1983**, *105*, 652.
51. (a) W. Yueh, N. L. Bauld, *J. Chem. Soc. Perkin Trans. 2*, **1995**, 871–873; (b) W. Yueh, N. L. Bauld, *ibid.*, **1996**, 1761–1766.

52. J. Mattay, J. Gersdorf, J. Mertes, *J. Chem. Soc. Chem. Commun.* **1985**, 1088.
53. M. Schmittel, C. Wöhrle, I. Bohn, *Chem. Eur. J.* **1996**, *2*, 1031–1040.
54. M. Schmittel, H. von Seggern, *Angew. Chemie Intl. Ed. Eng.* **1991**, *8*, 999–1001.
55. N. L. Bauld, D. J. Bellville, R. Pabon, R. Chelsky, G. Green, *J. Am. Chem. Soc.* **1983**, *105*, 2378–2382.
56. N. L. Bauld, *J. Am. Chem. Soc.* **1992**, *114*, 5800–5804.
57. Refernce 40c, p. 15.
58. Reference 40c, pp. 5–7.
59. (a) R. A. Pabon, D. J. Bellville, N. L. Bauld, *J. Am. Chem. Soc.* **1984**, *106*, 2730–2731; (b) D. W. Reynolds, N. L. Bauld, *Tetrahedron* **1984**, *42*, 6189–6194.
60. T. Kim, J. Pye, N. L. Bauld, *J. Am. Chem. Soc.* **1990**, *112*, 6285–6290.
61. N. L. Bauld, B. Harirchian, D. W. Reynolds, J. C. White, *ibid.* **1988**, *110*, 8111–8117.
62. D. H. R. Barton, G. Leclerc, P. D. Magnus, I. D. Menzies, *J. Chem. Soc. Chem. Commun.* **1972**, 447.
63. R. Tang, H. J. Yue, J. F. Wolf, F. Mares, *J. Am. Chem. Soc.* **1978**, *100*, 5248.
64. J. F. Evans, H. N. Blount, *J. Am. Chem. Soc.* **1978**, *100*, 4191–4196.
65. (a) N. L. Bauld, J. T. Aplin, W. Yueh, A. Loving, S. Endo, *J. Chem. Soc. Perkin Trans. 2* **1998**, 2733–2776; (b) N. L. Bauld, J. T. Aplin, W. Yueh, A. Loinaz, *J. Am. Chem. Soc.* **1997**, *119*, 11381–11389.
66. N. L. Bauld, J. T. Aplin, W. Yueh, S. Endo, A. Loving, *J. Phy. Org. Chem.* **1998**, *11*, 15–24.
67. J. P. Dinnocenzo, M. Schmittel, *J. Am. Chem. Soc.* **1987**, *109*, 1561–1562.
68. S. F. Nelsen, *Accounts Chem. Research* **1987**, 269–276.
69. S. B. Karki, J. P. Dinnocenzo, S. Farid, J. L. Goodman, I. R. Gould, T. A. Zona, *J. Am. Chem. Soc.* **1997**, *119*, 431–432.
70. (a) P. G. Gassman, D. A. Singleton, *ibid.*, **1984**, *106*, 7993; (b) D. W. Reynolds, K. T. Lorenz, H.-S. Chiou, D. J. Bellville, R. A. Pabon, Jr., N. L. Bauld, *J. Am. Chem. Soc.* **1987**, *109*, 4960.
71. W. Yueh, N. L. Bauld, *J. Chem. Soc. Perkin. Trans. 2* **1995**, 871–873.
72. W. Yueh, N. L. Bauldl, *J. Am. Chem. Soc.* **1995**, *117*, 5671–5676.
73. W. Yueh, N. L. Bauld, unpublished results.
74. G. A. Mirafzal, T. Kim, J. Liu, N. L. Bauld, *J. Am. Chem. Soc.* **1992**, *114*, 10968–10969.
75. T. Kim, G. A. Mirafzal, N. L. Bauld, *Tetrahedron Lett.* **1993**, *34*, 7201–7204.
76. G. A. Mirafzal, T. Kim, J. Liu, N. L. Bauld, *J. Am. Chem. Soc.* **1992**, *114*, 10968–10969.
77. Reference 40c, pp. 62–63.
78. U. Haberl, O. Wiest, E. Steckhan, *J. Am. Chem. Soc.*, **1999**, *121*, in press.
79. The three systems are: (1) 1,3-cyclohexadiene with all three geometric isomers of 2,4-hexadiene (2) 1,3-cyclopentadiene with *cis-* and *trans*-diaryloxyethenes and (3) 2,3-dimethyl-1,3-butadiene with *cis-* and *trans*-1,2-diaryloxyethenes. Further, stereospecific cyclobutanation has been observed in two distinct systems: (1) cyclobutadimerization of *cis-* and *trans*-anethole (in both dichloromethane and acetonitrile) and (2) 2,3-dimethyl-1,3-butadiene with *cis-* and *trans*-1,2-diaryloxyethanes.
80. H. D. Roth, M. L. Schilling, C. Abelt, *J. Am. Chem. Soc.* **1986**, *108*, 6098.
81. B. Harirchian, N. L. Bauld, *ibid.* **1989**, *111*, 1826–1828.
82. R. M. Wilson, J. G. Dietz, T. A. Shepherd, D. M. Ho, K. A. Schnapp, R. C. Elder, J. W. Watkins II, L. S. Geraci, C. F. Campana, *J. Am. Chem. Soc.* **1989**, *111*, 1749–1754.
83. S. Kadota, K. Tsubono, K. Makino, M. Takeshita, T. Kibuchi, *Tetrahedron Lett.* **1987**, *28*, 2857.
84. C. F. Gurtler, S. Blechert, E. Steckhan, *J. Org. Chem.* **1996**, *61*, 4136–4143.
85. N. L. Bauld, J. T. Aplin, W. Yueh, H. Sarker, D. J. Bellville, *Macromolecules* **1996**, *29*, 3661–3662.
86. D. W. Reynolds, B. Harirchian, H.-S. Chiou, B. K. Marsh, N. L. Bauld, *J. Phys. Org. Chem.* **1989**, *2*, 57–88.
87. N. P. Schepp, L. J. Johnston, *J. Am. Chem. Soc.* **1994**, *116*, 6895–6903.
88. J. P. Dinnocenzo, D. A. Conlon, *ibid.* **1988**, *110*, 2324–2326.
89. J. P. Dinnocenzo, D. A. Conlon, *Tetrahedron Lett.* **1995**, *41*, 7415–7418.
90. (a) G. Stufflebeme, K. T. Lorenz, N. L. Bauld, *J. Am. Chem. Soc.* **1986**, *108*, 4234–4235; (b) G. A. Mirafzal, A. M. Lozeva, J. A. Olson, *Tetrahedron Lett.* **1998**, *39*, 9323–9326.

91. F. Müller, J. Mattay, *Angew. Chem. Int. Ed. Eng.* **1991**, *103*, 1336–1337.
92. A. Padwa, J. Smolanoff, *J. Am. Chem. Soc.* **1971**, *93*, 548.
93. Q.-X. Guo, X.-Z. Quin, J. T. Wang, F. Williams, *J. Am. Chem. Soc.* **1988**, *110*, 1974–1976.
94. K. Lorenz, N. L. Bauld, *J. Catalysis* **1985**, *95*, 613–616.
95. S. Dai, R. S. Pappas, G.-F. Chen, Q.-X. Guo, J. T. Wang, *J. Am. Chem. Soc.* **1989**, *111*, 8759–8761.
96. T. Miyashi, A. Konno, Y. Takahashi, *J. Am. Chem. Soc.* **1988**, *110*, 3676.
97. B. Krautler, A. Pfaltz, R. Nordmann, K. O. Hodgson, J. D. Dunitz, A. Eschenmoser, *Helv. Chim. Acta* **1976**, *59*, 924–937.
98. Y Takahashi, J. K. Kochi, *Chem. Berichte* **1988**, *121*, 253–259.
99. F. Gerson, X. Z. Qin, T. Bally, J. N. Aebischer, *Helv. Chim. Acta* **1988**, *71*, 1069.
100. D. J. Bellville, R. Chelsky, N. L. Bauld, *J. Comput. Chem.* **1982**, *3*, 548.
101. (a) G. N. Sastry, T. Bally, V. Hrouda, P. Carsky, *J. Am. Chem. Soc.* **1998**, *120*, 9323–9334; (b) V. Barone, N. Rega, G. N. Sastra, T. Bally, *J. Phys. Chem. A* **1998**, *102*, in print.
102. (a) C. M. Hudson, M. R. Marzabadi, K. D. Moeller, D. G. New, *J. Am. Chem. Soc.* **1991**, *113*, 7372–7385; (b) K. D. Moeller, L. V. Tinao, *ibid.* **1992**, *114*, 1032–1041; (c) K. D. Moeller, *Top. Curr. Chem.* **1997**, *185*, 49–86.
103. T. Shono, I. Nishiguchi, S. Kashimura, M. Okawa, *Bull. Chem. Soc. Jpn* **1978**, *51*, 2181.
104. K. D. Moeller, D. Frey, L. Matson-Beal, S. H. K. Reddy, Y. Tong in *Novel Trends in Electroorganic Synthesis* (Ed.: S. Torii), Springer Verlag, Tokyo, **1998**, pp. 51–54.
105. H. Scott, G. A. Miller, M. M. Labes, *Tetrahedron Lett.* **1963**, *17*, 1073.
106. M. S. Workentin, N. P. Schepp, L. J. Johnston, D. D. M. Wayner, *J. Am. Chem. Soc.* **1994**, *116*, 1141.
107. D. R. Arnold, S. A. Mines, *Can. J. Chem.* **1987**, *65*, 2312–2314.
108. A. Zedulev, A. Grand, E. Ressouche, J. Schweizer, B. G. Morin, A. J. Epstein, D. A. Dixon, J. S. Miller, *J. Am. Chem. Soc.* **1994**, *116*, 7243.
109. D. H. Levy, R. J. Myers, *J. Chem. Phys.* **1964**, *41*, 1062.
110. (a) T. J. Katz, H. L. Strauss, *J. Chem. Phys.* **1960**, *32*, 1873; (b) H. L. Strauss, T. J. Katz, G. K. Fraenkel, *J. Am. Chem. Soc.* **1963**, *85*, 2360.
111. J. F. Garst in *Free Radicals* (Ed.: J. K. Kochi), Vol. 1, John Wiley & Sons, New York, **1973**, pp. 518–519.
112. N. L. Bauld, M. S. Brown, *J. Am. Chem. Soc.* **1967**, *89*, 5413–5417.
113. N. L. Bauld, J. H. Zoeller, Jr., *Tetrahedron Lett.* **1967**, *10*, 885–889.
114. (a) N. L. Bauld, M. S. Brown, *J. Am. Chem. Soc.* **1967**, *89*, 5417–5429; (b) F. Farr, Y. S. Rim, N. L. Bauld, *ibid.* **1971**, *93*, 6888–6890.
115. N. L. Bauld, C.-S. Chang, J. H. Eilert, *Tetrahedron Lett.* **1973**, *2*, 153–154.
116. H. O. House, *Modern Synthetic Reactions*, W. A. Benjamin, Menlo Park, **1972**, pp. 206–209.
117. N. L. Bauld, *J. Am. Chem. Soc.* **1962**, *84*, 4347.
118. (a) M. M. Baizer, *Tetrahedron Lett.* **1963**, 973; (b) M. M. Baizer, *J. Electrochem. Soc.* **1964**, *111*, 215.
119. M. Swarz in *Progress in Physical Organic Chemistry* Vol. 6 (Eds.: A. Streitweiser, Jr., R. W. Taft) Interscience, New York, **1968**, 323–438.
120. L. Moens, M. M. Baizer, R. D. Little, *J. Org. Chem.* **1986**, *51*, 4497.
121. (A) H. E. Bode, C. G. Sowell, R. D. Little, *Tetrahedron Lett.* **1979**, 2525; (b) R. D. Little, M. K. Schwaebe, *Top. Curr. Chem.* **1997**, *185*, 1–48.
122. J. Delaunay, G. Mabon, A. Orliac, J. Simonet, *Tetrahedron Lett.* **1990**, *31*, 667–668.
123. M. Witmer, E. Altmann, H. Young, A. Sancar, T. P. Begley, *J. Am. Chem. Soc.* **1989**, *111*. 9264.
124. D. J. Pasto, S.-H. Yang, *J. Org. Chem.* **1989**, *54*, 3544–3549.
125. (a) N. L. Bauld, C.-S. Chang, F. R. Farr, *J. Am. Chem. Soc.* **1972**, *94*, 7164; (b) N. L. Bauld, J. Cessac, C.-S. Chang, F. R. Farr, R. Holloway, *J. Am. Chem. Soc.* **1976**, *98*, 4561–4567.
126. M. A. Fox, J. R. Hurst, *J. Am. Chem. Soc.* **1984**, *106*, 7626–7627.
127. D. W. Borhani, F. D. Greene, *J. Org. Chem.* **1986**, *51*, 1563–1570.
128. S. Yamago, S. Ejiri, M. Nakamura, E. Nakamura, *J. Am. Chem. Soc.* **1993**, *115*, 5344.

# 4 Electron-transfer Reactions of Aromatic Compounds

*Georg Gescheidt and Md. Nadeem Khan*

## 4.1 Introduction

Aromatic molecules are inclined to undergo electron-transfer reactions. The additional charges introduced by the electron-transfer process are efficiently stabilized by delocalization. Nevertheless rearrangements of the molecular skeleton or follow-up reactions can occur. The course of the electron-transfer reactions and the properties of the species thus formed are the subject of this section.

In particular, we concentrate on the species formed after an one-electron transfer reaction. Starting from closed-shell diamagnetic precursors, paramagnetic stages, i.e. radical cations or radical anions are formed in this first step. To understand the properties and reactivity of these species, detailed knowledge in terms of charge and spin delocalization and electronic structure is an important prerequisite. This information is preferably derived from electronic and paramagnetic resonance (EPR and electron-nuclear multiple resonance-techniques such as ENDOR or Triple spectroscopy). It has recently been shown that the experimental results can be substantiated by the use of quantum-chemical calculations. In addition to Hartree–Fock-based ab initio procedures, calculations on the density-functional level of theory have been established as rather efficient tools. Therefore, before representing examples of electron-transfer-generated radicals a short survey of the computational methods is given.

Another substantial factor directing the kinetic and thermodynamic stability of charged radicals is ion pairing. This phenomenon, although well established for many years, is often not directly distinct in experiments. To establish the importance of ion pairing, or, in other words, supramolecular interactions, a separate introductory chapter is dedicated to these aspects.

The references are selected particularly from recent publications, without, however, ignoring some 'classical' contributions. In some sections topics are included which do not strictly represent aromatic molecules but involve interactions of $\pi$-type orbitals.

## 4.2 Computational Methods for Organic Radicals

The calculation of radical hyperfine properties has received enormous attention in recent years [1–4]. With rapid improvement in computer technology, combined with increasingly accurate computational schemes, theoretical predictions of radical hyperfine structures (hfs) are today serving an important role in the understanding of the properties of radicals and their reactions. Comparison of observed and computed hyperfine coupling constants (hfcc) leads to the assignment of plausible geometries and the identification of reaction products. Further analysis of the theoretical data also provides answers to questions about reaction barriers, transition states, charge and spin distributions, and a variety of other properties [1, 2, 5–7].

Of the many conventional ab initio approaches, multireference configuration interaction (MRC1), quadratic configuration interaction (QC1), and coupled cluster (CC) techniques, in conjunction with large basis sets, that consistently have proven able to generate hfcc of high accuracy [8–12]. Unfortunately, one problem with such approaches is that they are computationally quite expensive even for moderately sized systems, hence restricting studies to date to relatively small molecules. An alternative approach for calculating hfcc is represented by Density Functional Theory (DFT) [13], for which the computational cost and memory requirements are considerably less than those of conventional correlated ab initio procedures. As a consequence, the number of basis functions, and hence atoms, is not nearly as limiting a factor at the DFT level as it is for the earlier approaches. The use of DFT has increased tremendously as a serious and competitive alternative to more conventional ab initio approaches to elucidate molecular electronic structures. This increased interest has been further stimulated by the formulation of a 'Hartree–Fock type' formalism for DFT [13], by the development of accurate gradient correction schemes [14–19], and by the incorporation of DFT into widely used computational quantum chemistry programs [20, 21].

The performance of various exchange and correlation functions has been extensively tested [22–27] and number of books that review various applications of DFT has appeared in last few years [28–30]. The consensus at present appears to be that for the so-called hybrid or adiabatic connection method (ACM), functions containing a mixture of different exchange terms and gradient-corrected correlation constitutes the most accurate forms of DFT currently at hand. Method for converting DFT energies to enthalpies of formation is described and its performance, in conjunction with six DFT methods, are examined for 23 stable hydrocarbons [31]. The B3LYP atom equivalents of carbon and hydrogen are used without adjustment to calculate the enthalpies of formation of some free radicals and carbocations. The mean deviation between calculated and experimental results is found to be around 2 kcal mol$^{-1}$, which is of the same order as experimental uncertainties for these highly reactive species. In search of the optimal combination of basis set and exchange-correlation potential, the dependence of atomization energies ($D_0$) and reaction enthalpies ($\Delta H$) for a set of 44 molecules using gradient-corrected DFT has been investigated [32]. Of the six functions tested, those that include a portion of the exact (Hartree–Fock) exchange performed best and yielded $D_0$ values within 3–

5 kcal mol$^{-1}$ of the experimental value. The thermal motion of the •CCH radical embedded in matrix of solid argon is simulated at 4 and 40 K, using hybrid DFT–molecular dynamics (DFT–MD) approach [33]. The results reveal that •CCH when embedded in the argon matrix favors an oscillating, slightly bent geometric structure, whereas in vacuum the molecule is linear. The geometries and $^{17}$O hyperfine coupling constants in several alkyl peroxy radicals is determined by DFT [34]. Results comparable to experimental values are obtained for all the alkyl peroxy radicals but erratic and strongly fluctuating results are noticed for the fluoro peroxy radicals, apart from considerable spin contamination, multireference, and matrix effects are proposed for observed deviation. A total of 22 substituted benzene radicals are investigated, two different functional schemes, the B3LYP hybrid DFT–HF functional and the 'pure' gradient corrected DFT functional PWP86 are employed [35]. The mechanism of degenerate [5,5]-sigmatropic rearrangements of 5,5a,10,10a-tetrahydroheptalene and $(z,z)$-1,3,7,9-decatetraene are explored with restricted and unrestricted Becke 3LYP/6-31G* hybrid HF–DFT calculations [36]. Calculations predicted that the [5,5]-sigmatropic rearrangements occur via stepwise diradical mechanism with activation barriers of 29.1 and 34.3 kcal mol$^{-1}$, respectively. The [3,3]- (cope) rearrangement of 3,6-bismethylene-1,7-octadiene is also predicted to occur by a stepwise diradical mechanism. The barrier for this reaction is much less than those for the [5,5]- rearrangements due to greater stability of its constituent radical moieties and lack of unfavorable steric interactions in the transition state. The fully optimized potential energy curves for the unimolecular decomposition of the lowest singlet and triplet states of nitromethane through the C–NO$_2$ bond dissociation pathway is calculated using various DFT and high level ab initio electronic structure methods [37]. For the most simple $(H_2O)_2{}^+$ system, previous theoretical studies using Post Hartree–Fock methods [38], have shown that proton-transferred OH–H$_3$O$^+$ isomer is the ground state structure of the ionized water dimer, however, recent density functional calculation with exchange correlation gradient corrections predict the three-electron bond isomer to be the ground-state structure [39].

This anomaly is attributed to overestimation by the exchange functions of the self-interaction part of the exchange energy in the hemi-bond ion due to its delocalized electron hole [40]. It is cautioned that this behavior of the density functions for exchange, if unrecognized, may lead to wrong predictions for ground-state structures of systems with a three-electron bond. Corrected calculations showed that the proton-transferred OH–H$_3$O$^+$ isomer to be the ground state structure of the $(H_2O)_2{}^+$ dimer ion. The mechanism for the 1,3-dipolar cycloaddition of trifluoromethyl azomethine ylide with acetonitrile has been characterized using DFT methods with B3LYP functional and the 6-31G* and 6-31$^+$G** basis sets [41]. Density functional calculations at the B3LYP/6-31G* theory level provided data on transition-state energies for this 1,3-dipolar cycloaddition in full agreement with the stereo chemical outcome. More recently uncertainties in configurations of camphene hydrochloride and isobornyl chloride are laid to rest by a comparison of computed and experimental $^{13}$C and methyl proton chemical shifts [42]. There has been great deal of interest in short-strong or low-barrier hydrogen bonds [43–45], recently high level ab initio and DFT calculations have been employed to investigate these systems with reasonable accuracy [46]. The Hartree–Fock method failed to predict the relative energies of the non-ionized amino acid conformers correctly [47], however, recently DFT and MP2 methods are successfully employed for relative stability calculations of non-ionized Valine conformers [48]. Extensive density functional calculations are reported for the geometrical structures, thermochemistry, infrared, and hyperfine parameters of representative carbon-centered $\pi$-radicals [49]. In short, the merits of DFT, by far the most useful non-empirical alternative to conventional post Hartree–Fock methods for studying physicochemical properties of molecules, is now well recognized. Gradient-corrected functions have significantly increased the reliability of DFT methods, small modifications of the functional form or partial inclusion of the Hartree–Fock exchange provides even better results and contrary to methods based on an Hartree–Fock zero-order wave function, DFT approaches appear equally reliable for closed and open shell systems.

## 4.3 Ion Pairing

The fact that ions of opposite charges tend to pair is of great importance for the course of organic reactions [50]. Evidence for such an ion pairing is provided by hyperfine splittings from paramagnetic nuclei of alkali-metal cations associated with organic radical anions. EPR spectroscopy represents a method of choice for the study of ion-pairs. Since the first observation of a $^{23}$Na-hyperfine splitting in the EPR spectrum of naphthalene reduced to its radical anion with sodium in tetrahydrofuran (THF) or 2-methyltetrahydrofuran (MTHF) [51], a large number of papers describing similar findings appeared. The largest coupling constants of alkali-metal nuclei in counterions or organic radical anions were reported therein for cation attached in a chelate-like fashion to the lone electron pairs of the oxygen atoms in the radical anion. In this respect, alkali-metal cations associated with the radical anion of *o*-dimesitoylbenzene in 1,2-dimethoxyethene and THF served as a paradigm [52]. With 1,2-dimethoxyethane at room temperature the $^{7}$Li-, $^{23}$Na-, $^{39}$K-, $^{85}$Rb-, $^{87}$Rb-, and $^{133}$Cs-coupling constants amounted there to 0.375 [52], 0.695 [52], 0.491 [53], 1.66 [53], and 1.02 [52] mT respectively; such ion pairs are denoted as 'tight' or 'contact' ones. On the other hand, ion-pairs of alkali-metal cations associated with radical anions of hydrocarbons are considered as 'loose' or 'solvent-separated' ones, because the molecules of the ether solvent successfully compete with the radical anion for the positively charged counterion. Accordingly, in these ion-pairs, the hyperfine splittings from alkali-metal nuclei are smaller by 1–2 orders of magnitude than the values described above in the case of the *o*-dimesitoylbenzene radical anion [54]. Typically for the ion pairs of the naphthalene radical anion in DME at room temperature, the $^{7}$Li-, $^{23}$Na-, $^{39}$K-, $^{85}$Rb-, $^{87}$Rb-, and $^{133}$Cs-coupling constants amounted to <0.01, <0.0405, <0.01, 0.0095, 0.0316, and 0.1071 mT, respectively [55]. In addition to EPR, the ENDOR technique is successfully applied to study ion-pairing of some very persistent organic radical anions with alkali-metal cations [53, 56], the relative signs of the coupling constants of proton and metal nuclei could be determined by general triple resonance [57]. Absolute signs of these values are accessible by NMR spectroscopy also used for such studies [58], although the high concentration of the radical anion salts required by this method led to some deviations from the corresponding results obtained by EPR and ENDOR spectroscopy.

In ion pairs consisting of two radicals, the exchange interaction ($J$) is usually negative, however there have been reports that $J$ is positive in some radical ion pairs [59a,b]. A conflicting mechanism has also been proposed [59c]. Tero-Kubota et al. discussed indications of a positive $J$ for the radical ion pairs of benzophenone radical anions with the radical cation of DABCO (1,4-diazabicyclo[2,2,2]-octane) [59b]. They studied the effects of Lewis acid on the sign of $J$ of the radical ion pair as the charge of the radical pair seems to be an important factor in the inter-radical interactions. A recent FTEPR study revealed that the sign of the exchange interaction of the radical ion pair including the 4,4'-dimethoxybenzophenone radical anion and DABCO radical cation changes from positive to negative by addition of $BF_3$ [59d].

The strength of association with the counterion depends on several factors, such as solvating power of the solvent, the temperature, and the radius of the cation. The ion-pairing tightens and the coupling constant of alkali-metal nucleus increases with the decreasing solvating power parallel to raising the temperature [60], however, the dependence on the radius of the cation is less clear. For tight ion-pairs of radical anions containing heteroatoms, the association tends to weaken with increase in cation radius, in the order $Li^+$, $Na^+$, $K^+$, $Rb^+$, and $Cs^+$, because the contact of the counter ion with the lone electron pairs of the heteroatom diminishes in the same order. On the other hand, for loose ion pairs of radical anions without heteroatoms, an opposite sequence seems to prevail as the solvation of the counterion decreases with increasing radius of the cation. This behavior mirrors the hard–soft relationships between the radical ion, the counterion, and the solvent. The radical anion of buta-1,3-diene, 2,3-dimethylbuta-1,3-diene, and 1,1,4,4-tetramethylbuta-1,3-diene were generated from their neutral precursors by electrolytic reduction with THF replacing liquid ammonia as the solvent [61]. The buta-1,3-diene radical anion was also generated by electrolytic reduction of parent molecule at a Pt-wire cathode in liquid ammonia at 195 K [62]. A similar method is employed for 2-methylbuta-1,3-diene, 2,3-dimethylbuta-1,3-diene, and cyclohexa-1,3-diene radical anions [63]. However, in these experiments, a tetraalkylammonium cation served as the counterion, the nuclei of which did not give rise to additional hyperfine splittings in the EPR spectra. In general, reaction of buta-1,3-diene and its derivatives with an alkali-metal mirror in an ether as solvent failed to yield a persistent radical anion, because of rapid polymerization to rubber-like products [64]. EPR and ENDOR study of the radical anions of the 1,4-di-*tert*-butylbuta-1,3-diene, 2,3-di-*tert*-butylbuta-1,3-diene, its 1,1-dideutero derivative and of 1,4-di-*tert*-butylcyclohexa-1,3-diene is recently reported by Gerson et al. [65]. Characteristic of the hyperfine patterns of 1,4- and 2,3-di-*tert*-butylbuta-1,3-diene radical anions are very large coupling constants of the alkali-metal nuclei in the counter ion: $a(^{39}K) = 0.12$–$0.15$, $a(^{85}Rb) = 0.40$–$0.84$, $a(^{87}Rb) = 1.4$–$2.8$ and $a(^{133}Cs) = 0.70$–$2.6$ mT. Values of this magnitude, unusual for counterions of hydrocarbon radical anions, points to a tight or contact ion pairing. In addition to large hyperfine splittings, a striking feature of the tight ion pair of 2,3-di-*tert*-butylbuta-1,3-diene, is the coupling constant of two protons in the 1,4-positions.

The dynamic processes in tetrahydrofuran (THF) solutions of 2,5-di-*tert*-butyl-*p*-benzoquinone$^-$,Na$^+$ ion pairs obtained by reduction with a sodium mirror is investigated by two-dimensional (2D) exchange Fourier Transformation (FT) EPR [66]. Results indicate two types of ion pairs, in first the intramolecular Na$^+$ hopping is slow while in other it is fast and results in a selective smearing of some of the hyperfine lines. The origin of the latter is tentatively ascribed to complexation with OH$^-$ generated by water impurity. The work demonstrates the efficiency of the 2D exchange FT EPR method in elucidating mechanisms of dynamic processes and determining kinetic parameters, in particular when several such processes occur simultaneously. Radical anions and radical cations of tetraphenylethane and its cyclophane derivatives were generated by chemical methods and characterized by their EPR and simultaneously recorded optical spectra [67]. In radical anions no specific complexation of alkali-metal counterions, K$^+$ and Li$^+$, by the crown-ether moieties could be observed. The solution structures of the radical cations closely matched the geometry established by X-ray diffraction analysis of crystalline tetra anisylethene [68] where the ethenic bond is twisted by 30.5° and phenyl-ring torsion amounts to 33°.

The effect of Zeeman levels crossing in spin-correlated radical ion pairs (naphthalene)$^{\cdot+}$/(hexafluorobenzene)$^{\cdot-}$ was monitored as the influence of an external magnetic field on the solution fluorescence under X-irradiated (MARY spectrum) [69]. The results show the possibility of observing hyperfine structures of a counterion in short-lived radical ion pairs using level crossing technique. Increasing the technique sensitivity will certainly open new prospects for employing the level crossing phenomenon in studying hyperfine structures in short-lived radical ion-pairs.

# 4.4 Radical Cations

## 4.4.1 π Systems

One-electron oxidation of aromatic compounds (ArH) leads primarily to corresponding radical cation which exist either in monomeric (ArH$^{\cdot+}$) or dimeric form [(ArH)$_2{}^{\cdot+}$] the latter usually formulated as π-dimer [70]. However, radical cations are reactive species and can undergo further reaction yielding more persistent radical cations e.g. oxidation of *tert*-butylbenzene or of toluene or *o*-xylene yielded radical cation of 4,4'-di-*tert*-butyl biphenyl, 4,4'-bitoluene or 3,3',4,4'-tetramethyl biphenyls, products of further σ-coupling, proton loss and further one-electron oxidation [71]. This is a well-known pathway of biaryl dehydrodimerization, explored in anodic and metal-ion oxidation of ArH [72, 73]. Other compounds with high reactivity in σ-coupling are alkoxy and amino substituted ArH [73]. Thus a risk with characterization of radical cations is that "hardy survivors and not primary radical

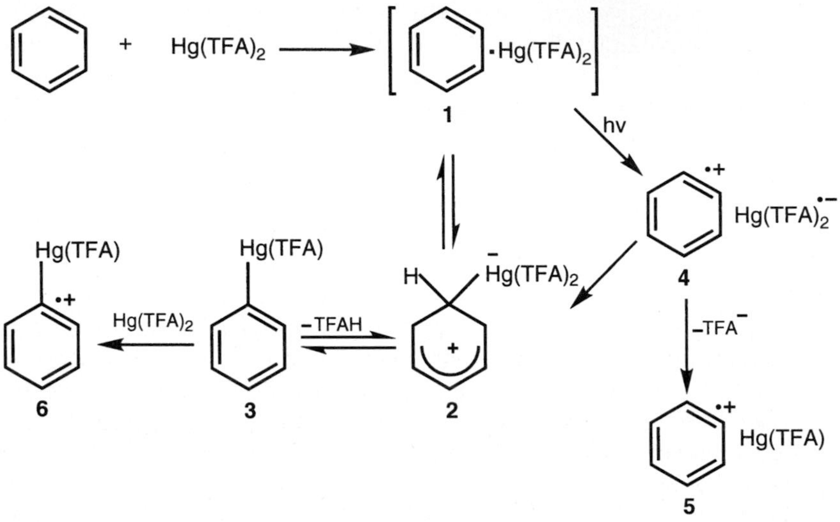

Scheme 1.

cations are detected" for example oxidation of 1-methoxynaphthalene by tetranitromethane [74]. Radical cations from oxidation of methylnaphthalenes are characterized by EPR spectroscopy [75] either as monomer ArH$^{\cdot+}$ or (ArH)$_2$$^{\cdot+}$ or radical cations (derived from perylene structure) resulting form the occurrence of the Scholl reaction [76]. Results from oxidation of 1-methyl-, 1,2-, 1,3-, 1,4-, 1,5-, 1,7-, 1,8-dimethylnaphthalene, and acenaphthene is reported [77]. Except in cases where the perimethyl group can interfere (1,4-, 1,5-dimethylnaphthalene) with $\sigma$-coupling at the 1-position, oxidation yielded radical cations of the corresponding 1,1'-binaphthalenes. A number of reactions can occur on oxidation of arene, represented by benzene in Scheme 1, on treatment with mercury(II) bis(trifluoroacetate) (TTFA) in trifluoroacetic acid (TFAH). The various possibilities are charge transfer complexation 1 [78], Wheeland complex 2, giving rise to arylmercury(II) trifluoroacetate 3 [79]. If charge transfer complex 1 is irradiated in its absorption band it may lead to radical cation 5 [80].

However, in certain cases under photolytic conditions, spectra of the corresponding arylmercury radical cations 6 developed, whereas no mercuration occurred in dark [81] signifying collapse of the ArH$^{\cdot+}$,Hg(TFA)$_2$$^{\cdot-}$ radical ion pair 4, provides an alternative path way to Wheeland complex 2 and hence to ArHg(TFA)$^{\cdot+}$ 6. Arene radical cations can also be generated from arene and thallium(III) tris-(trifluoroacetate) in trifluoroacetic acid [82], but with a different mechanism proposed by Eberson et al. [83]. Oxidation of anthracene showed 9-trifluoroacetoxy and 9,10-bis(trifluoroacetoxy)anthracene [84, 85], benzo[a]pyrene, 7-methylbenzo-[a]pyrene and 12-methylbenzo[a]pyrene yielded radical cations of 7- and/or 12-trifluoroacetates [86], triptycene (9,10-dihydro-9,10-[1,2]benzanthracene) showed

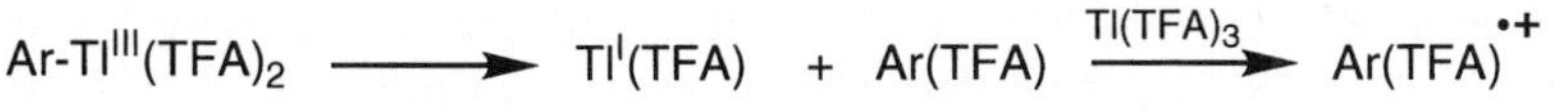

$$\text{Ar-Tl}^{\text{III}}\text{(TFA)}_2 \longrightarrow \text{Tl}^{\text{I}}\text{(TFA)} + \text{Ar(TFA)} \xrightarrow{\text{Tl(TFA)}_3} \text{Ar(TFA)}^{\bullet+}$$

**7**

**Scheme 2.**

tris(trifluoroacetoxy) triptycene radical cation [85], and it is suggested that these trifluoroacetates are formed via thalliated intermediate **7** (Scheme 2).

To date the mercurated arene radical cation is known for biphenylene [87], acenaphthene, pyracene, hexahydropyrene, triptycene, *p*-terphenyl, tetramethylnaphthopyran, anthracene, dibenzodioxin [89], and 4-*tert*-butylanisole [88]. In certain cases multiple mercuration is observed, for example in case of diphenylene [87] and dibenzodioxin [89]. Mercuration causes a decrease in *g*-value and always occurs at the site where the local coefficient of the Hückel HOMO of the hydrocarbon is greatest, and there is a constant ratio of about 20.6 between the hyperfine couplings by the $^{199}\text{Hg}$ ($I\frac{1}{2}$, abundance 16.84 %) which has been introduced, and by the proton which has been displaced [89]. EPR spectroscopic evidence is reported for **8**, **9**, **10**, **11**, **12**, **13**, **14**, **15** and **16** as new examples of recently recognized alternative mechanism of arene mercuration in which collapse of $\text{ArH}^{\bullet+}\text{Hg(TFA)}_2^{\bullet-}$ radical ion pair leads to arylmercury trifluoroacetate $\text{ArHg(TFA)}^{\bullet+}$ [90].

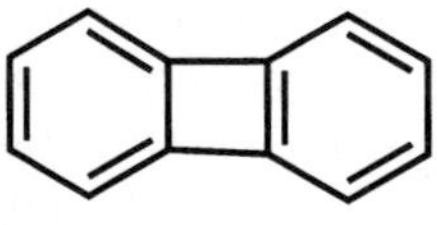

**8** 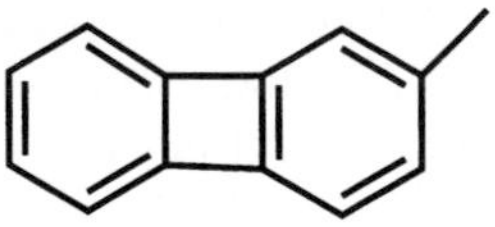           **9** 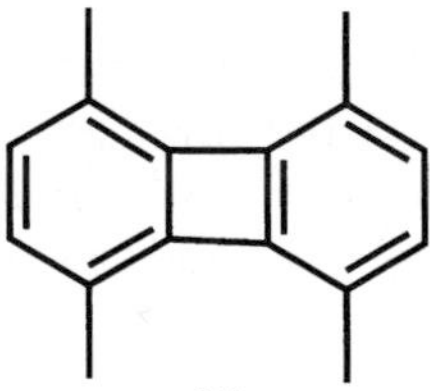           **10**

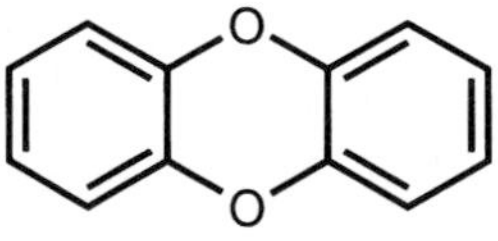

**11** 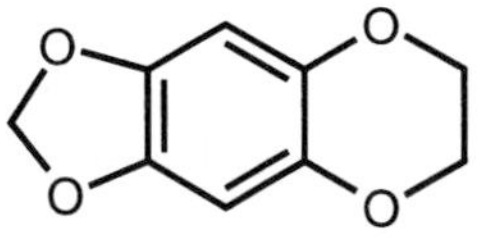           **12** 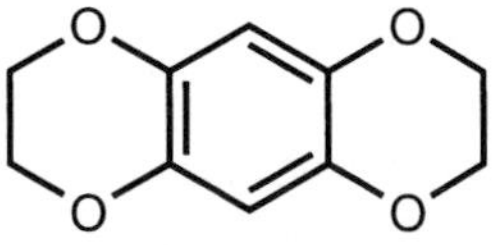           **13**

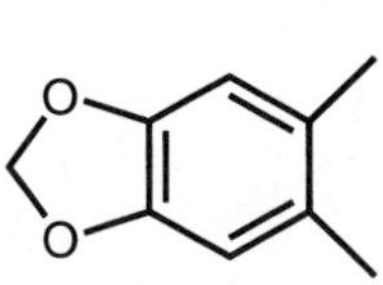

**14** 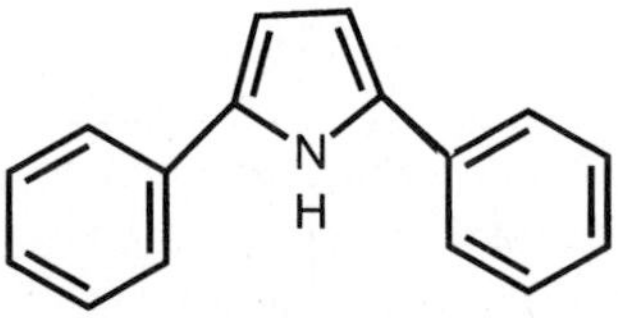           **15** 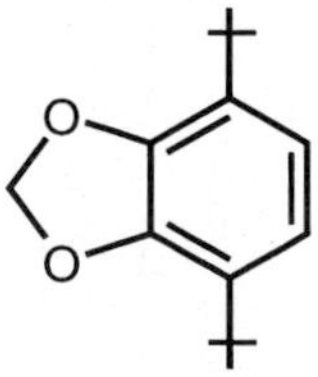           **16**

Spectra of **17**·⁺ and **18**·⁺ have been reported by Fleischhauer et al. using a cerium(IV) flow system [91], **19**·⁺ by Sullivan and co-workers using $AlCl_3$–$MeNO_2$ [92] and **20**·⁺ by Bubnov by using $Tl(TFA)_3$–TFAH reagent [93].

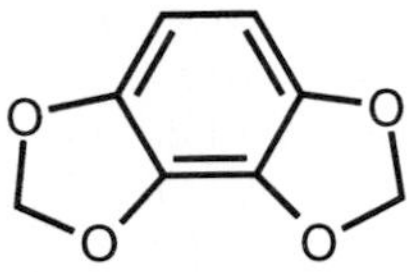

17

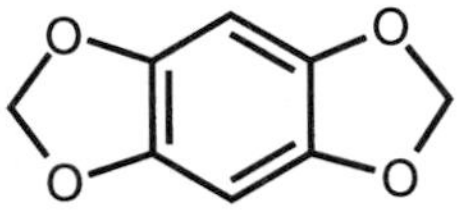

18

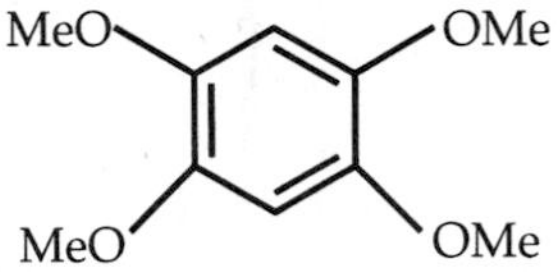

19

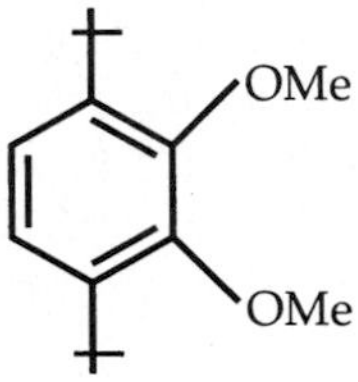

20

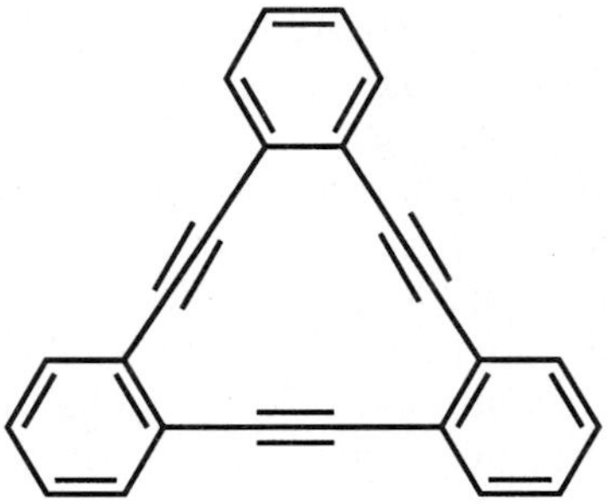

21

Dialkyl alkylene cations react with their parents to give tetraalkylcyclobutadiene cation [94] but diaryl alkynes on oxidation showed triarylazulene radical cations [95]; the radical anions of diaryl alkynes on the other hand do not rearrange. Reduction of **21** with potassium and small amount of benzo-18-crown-6 in THF gave septet of septet spectra with a(6H) 1.38, a(6H) 0.10 G, *g* 2.0026, while oxidation gave simple septet spectrum with a(6H) 1.29 G [90]. It is proposed that the small hyperfine coupling to aromatic protons in **12**·⁺, **13**·⁺, **14**·⁺, **18**·⁺, **19**·⁺, and relatively large coupling to aromatic protons in **16**·⁺ and **20**·⁺, reveal that the *ortho* oxygen substituents from the dioxine ring break the degeneracy of the orbitals of the benzene ring to confer $\psi_A$ character on the SOMO. Reaction between aromatic compounds, **22–25**, and several others with halogenating agents in 1,1,1,3,3,3-hexafluoropropan-

22

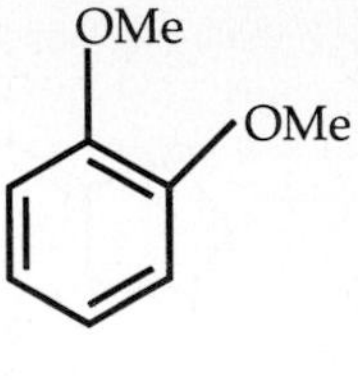

23

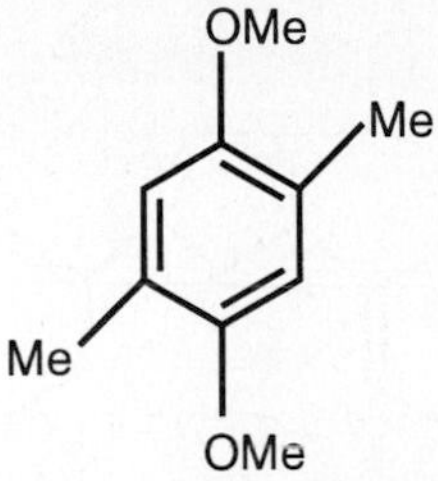

24

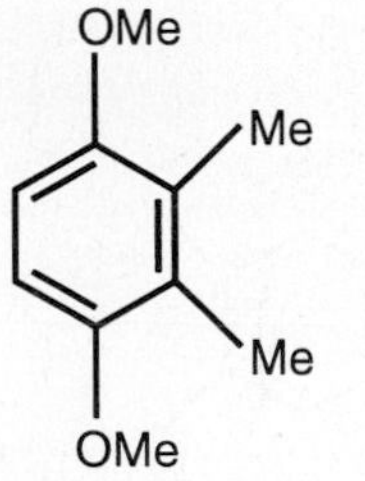

25

2-ol (HFP) was studied by Eberson et al. [96]. Results show that halogenating agents ICl, $Cl_2$, $Br_2$, $I_2$, *N*-bromosuccinimide and *N*-chlorosuccinimide gave persistent radical cations and their persistency in HFP is discussed.

Recently, HFP a non-nucleophilic and strongly hydrogen bonding solvent [97] is extensively employed in several cation studies, as it increases the stability of the cations in an unprecedented manner. To a large extent this effect is assumed to be due to the strong attenuation of reactivity experienced by many nucleophiles [98]. Dihydro[1,4]dioxines can readily be converted into their radical cations, the ring inversion occurs at a rate which can be measured by monitoring line shape effects on signals of axial and equatorial protons over a range of temperatures. Computer simulations provide rate constants and, hence the activation parameters for these reactions. This study is reported for **26–28** [99].

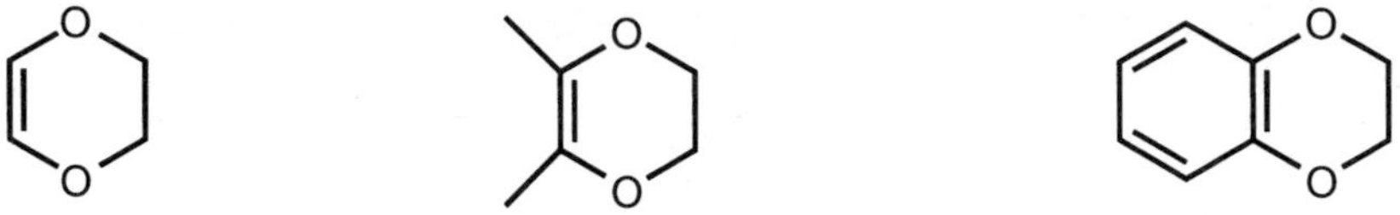

**26**          **27**          **28**

The same technique was employed to determine the kinetics of the inversion of the dioxine ring in the radical cations of **29–34** [100].

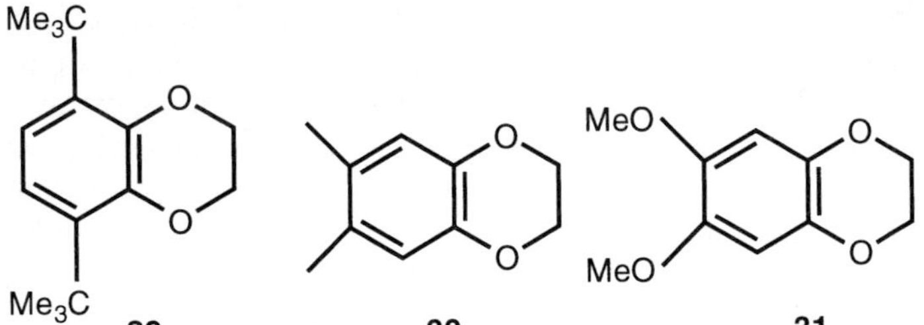

**29**          **30**          **31**

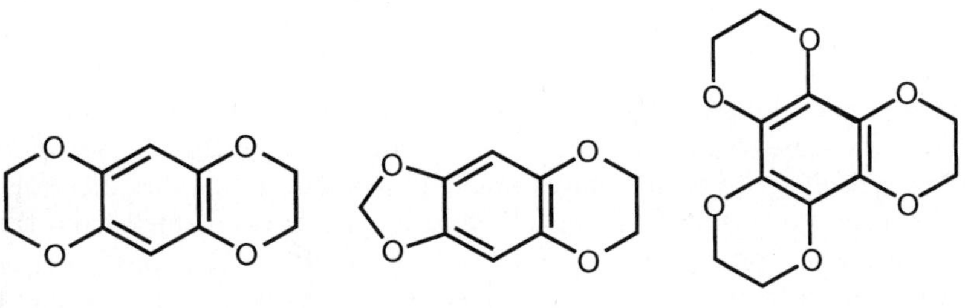

**32**          **33**          **34**

The factors in determining the barriers to ring-inversion in the spin paired and odd-electron molecules are proposed to be different bond lengths and angles associated with the oxygen atoms, related stretching and bending force constants, torsional and Van der Waals interactions and the nature of the wave function in the studied systems. Iwaizumi et al. considered two factors that might account for the barrier in the hexahydropyrene radical anion **35·⁻** being less than its radical cation. First, in the radical cation, the positive charge caused a contraction of the electron distribution, resulting in smaller overlap integrals and a decrease in the bending and stretching force constants. Second, hyperconjugative stabilization by the methylene groups, which was expected to be more important in the transition state than in the reactants, should be larger in the radical cations than in the anions [101].

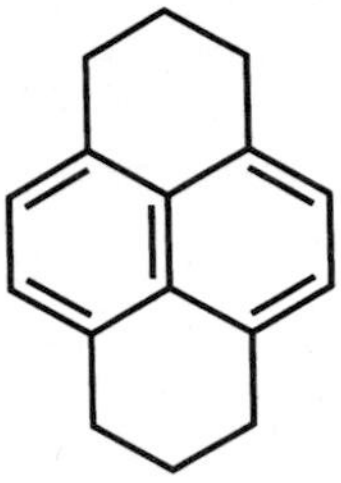

**35** 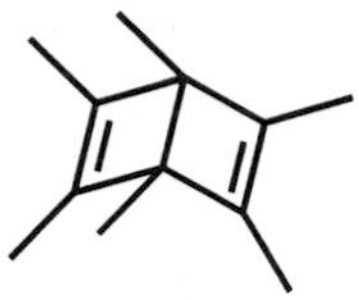         **36** 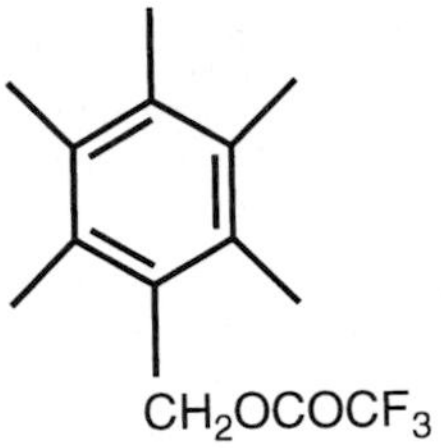         **37**

Several attempts have been made to characterize hexamethyl (Dewar benzene) radical cation **36·⁺** [102]. A 13-line EPR spectrum with hyperfine splitting constant of 0.98 mT was assigned [103] to the $^2B_2$ state of **36·⁺**, but as pointed out later [104], other interpretations are also possible. The reaction between Dewar benzene and TTFA in TFAH is investigated in detail by Eberson et al. [105]. It is proposed that the reaction involves a slow acid-catalyzed conversion of **36** into hexamethyl benzene (HMB) which is then oxidized by TTFA to pentamethylbenzyl trifluoroacetate **37**. Oxidation of benzene in Freon ($CCl_3F$) matrices at 4 K showed removal of orbital degeneracy and the unpaired electron occupied the $b_{2g}$ orbital with $D_{2h}$ symmetry, giving major spin densities on the C(1) and C(4) atoms [106]. Radical cations of a series of fluorinated benzenes, generated by $\gamma$-irradiation in halocarbon solid matrices were studied [107]. The ab initio results reveal that an unambiguous deformation in geometry is brought about by cationization and it is concluded that structure and symmetry of the SOMO of these radical cations are affected not only be the number of substitutions by fluorine but also by the position of the substitution.

An important step in making photochemically generated phenyl cations visible, is achieved by their addition to aromatics [108]. On the basis of their absorption spectra and their reactivity with typical bases/nucleophiles such as halides, alcohols, and ethers, the transients are identified as cyclohexadienyl cations formed from the photo produced 'invisible' phenyl cations by addition to the ring of added aromatics. With respect to both spectral and reactivity properties, the 'phenyl adducts' are very similar to the corresponding 'proton adducts' [109, 110]. Electrophilic reagents generally attack dibenzofuran **38** predominantly at 2-position [111], while nucleophiles such as acetate [112], cyanide [113] and trinitromethanide ion [114] attack at the 3-position, thus providing a tool for the testing of the Pross–Shaik CM model [115].

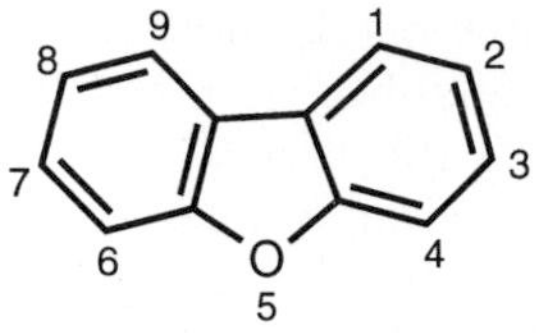

**38**

Radical cations of dibenzofuran and variety of symmetrically methylated dibenzofurans have been studied by cyclic voltammetry (CV) and EPR spectroscopy [116]. **38**$^{\cdot+}$ could not be generated in fluid solution due to its high reactivity however, monomeric compounds with 3- and 7-positions blocked, **39–42**, gave persistent cations.

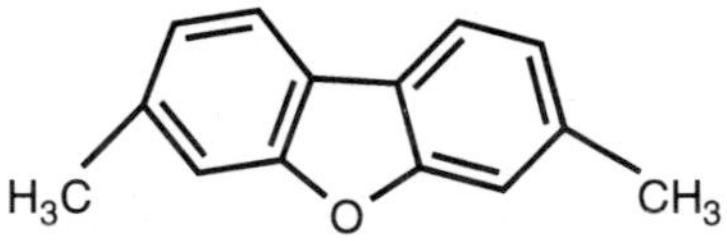

**39** 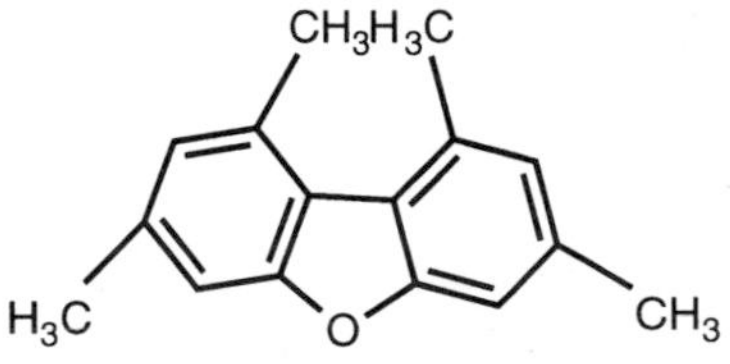 **40**

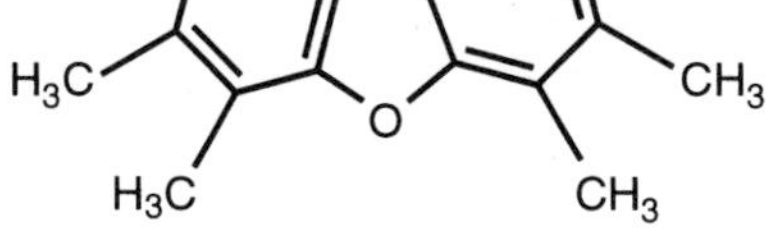

**41** 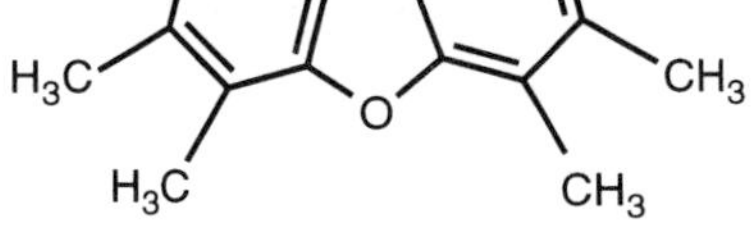 **42**

Both dehydro dimers **43**, **44** gave fairly persistent cations, whereas that of **45** could not be detected, presumably due to oligomerization.

Detailed investigation of the kinetic stability and electronic structure of **46a–d** is carried out by using EPR/ENDOR, CV and UV techniques [117].

The persistent benzofuran **46c**$^{\cdot+}$ and **46d**$^{\cdot+}$ showed reduced amount of spin population in the benzofuran $\pi$-systems due to delocalization into the aromatic substituents. This reduces the deprotonation reactions or nucleophilic attack [118]. Conjugative and homo conjugative interactions between strained ring moieties and olefinic fragments in organic radical cations have been the focus of much interest in recent years [119]. Changes in molecular geometry caused by one electron oxidation has been delineated for various substrates and the spin and charge density distributions in the resulting radical cations have been assessed [120]. Typically, the reactions of these species proceed with release of ring strain [121, 122], while in some

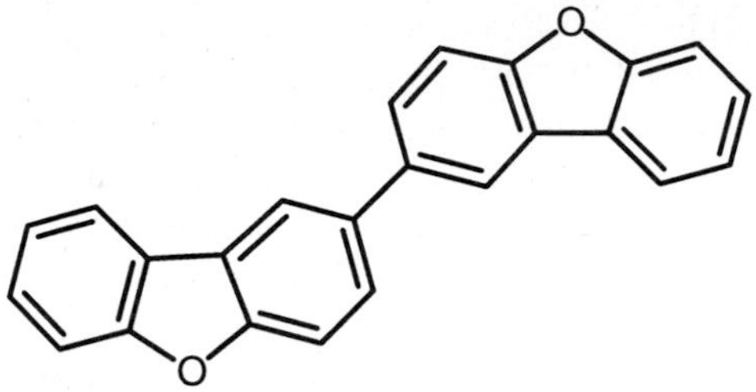

**43**

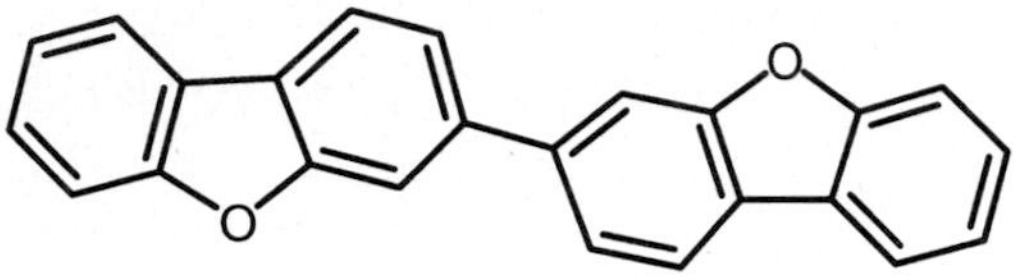

**44**

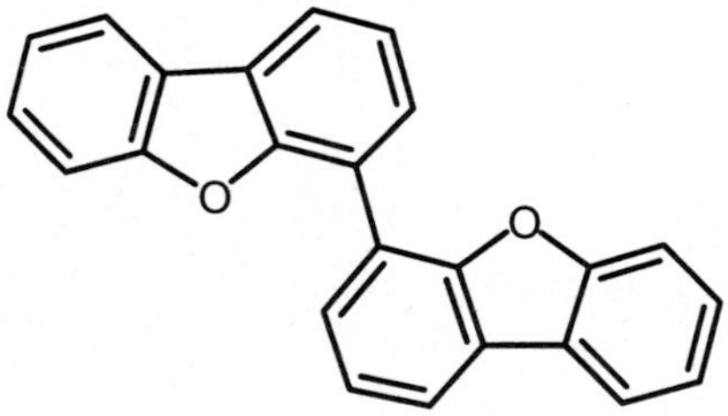

**45**

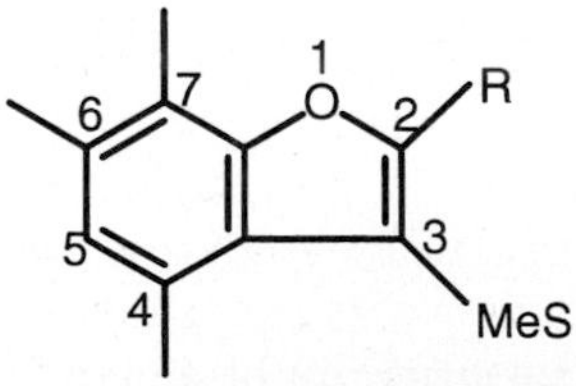

**46**

a  R = CH₃   b  R = Buᵗ

c  R = Ph    d  R = MeS = mesityl = 2,4,6-trimethylphenyl

systems ring opening is assisted by a nucleophile [123, 124]. Vinylcyclopropane radical cation, the simplest radical cation containing olefinic moiety and a cyclo-propane ring has not been characterized adequately, although several derivatives have been studied in recent years [125]. The ring opening and isomerization of cyclopropane is shown by ab initio calculations to be catalyzed by complexation with Be$^{•+}$ [126]. Only corner attack by the metal is found in the ring-opening of cyclopropane and a reactive metacyclic intermediate is suggested. Investigations showed a facile reaction pathway, for the catalysis of the ring-opening and isomerization of cyclopropane with group II metal radical cations. The mass spectrum of vinylcyclopropane showed evidence for ring opening to penta-1,3-diene radical cation [127]. Theoretical approaches include an early STO-3G calculation for a vinylcyclopropane radical cation with seriously restricted geometry [128], and more recent INDO, MP2/6-31G*, and B3LYP/6-31G* approaches [129]. Radical cations of bicyclo [3.1.0]hex-2-ene and bicyclo [4.1.0] hept-2-ene system, are characterized by CIDNP experiments [130].

Radical cations of three terpenes, sabinene **47**, α-thujene **48**, and β-thujene **49**, containing vinylcyclopropane functions held rigidly in either an *anti*- or *syn*-orientation, are elucidated by CIDNP results observed during there electron transfer reactions with photoexcited (triplet) chloranil [131].

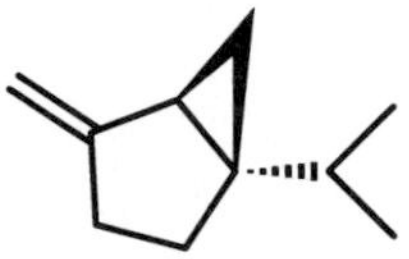

**47** 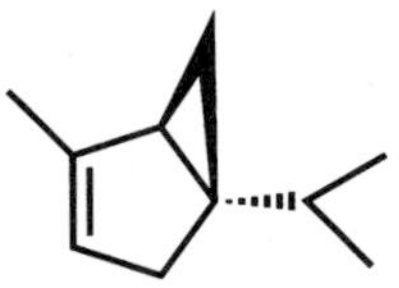  **48** 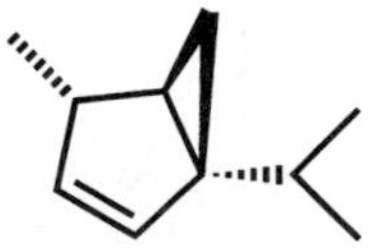  **49**

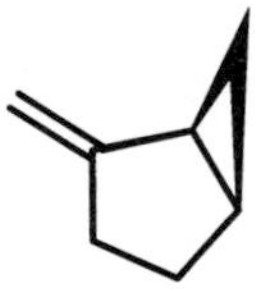

**50**   **51** 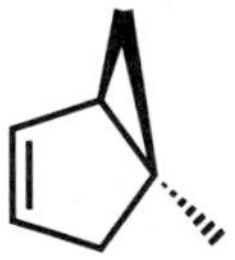  **52**

The resulting hyperfine couplings and spin densities are compared with the results of ab initio calculations for radical cations **50**, **51** and **52** [132]. In gas phase, the vinylcyclopropane radical cation, rearranges to the penta-1,3-diene radical cations [127]. Related rearrangements are documented for two rigidly linked vinyl-cyclopropane systems in solution: the electron-transfer-induced rearrangements of sabinene to β-phellandrene, and of α-thujene to α-phellandrene, are interpreted as novel examples of sigmatropic shifts in radical cations [133]. Products of **53–55**, generated by electron-transfer photochemistry of vinylcyclopropane are rationalized by attack of methanol on the radical cation, either at a secondary or the terminal vinyl carbon.

The potential surface of radical cations probed by ab initio calculations showed two minima, both belong to an unusual structure type with two lengthened cyclopropane bonds. Radical cations of **56–60** are studied by γ-irradiation in Freon matrices [134], the bicyclic cation **56**$^{•+}$, **57**$^{•+}$ and **58**$^{•+}$ were persistent and were characterized by EPR and ENDOR experiments. Tricyclic radical cations **59**$^{•+}$ and **60**$^{•+}$ readily underwent rearrangement to **61** and **62** respectively.

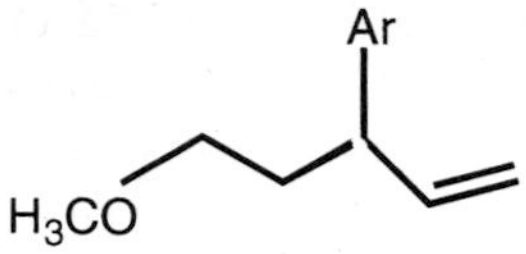

**53**

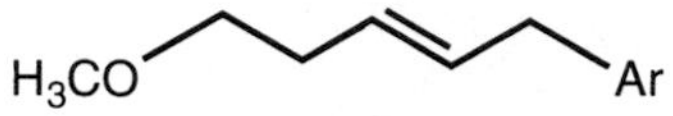

**54**

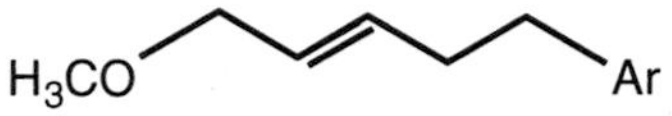

**55**

**56**
(R = H; R = D)

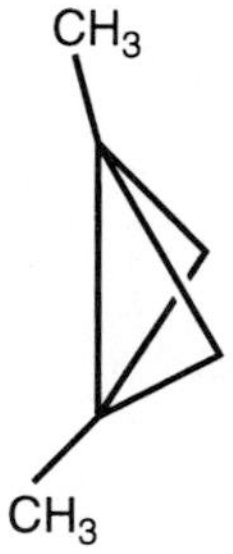

**57**

**58**
(R = H; R = D)

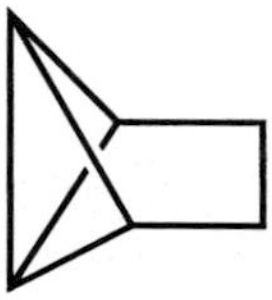

**59**

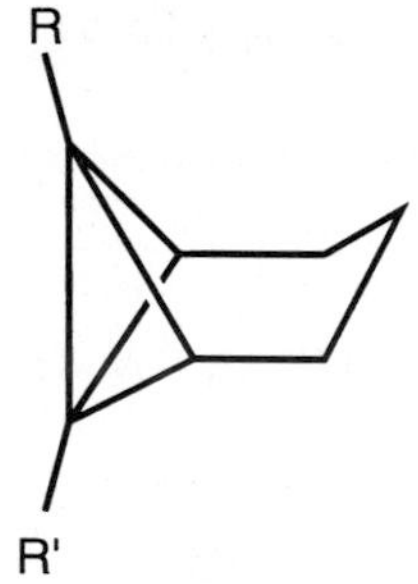

**60**
$\begin{pmatrix} R = R' = H \\ R = D, R' = H \\ R = R' = D \end{pmatrix}$

**61**

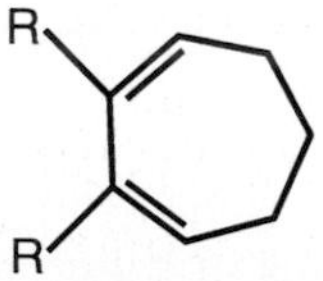

**62**
(R = H, R = D)

Photoinduced electron transfer (PET) is of current interest and numerous electron transfer sensitizers have been employed not only for mechanistic but also for synthetic purposes [135], effectiveness of PET being further improved by cosensitization technique [136]. PET and Chemical Electron Transfer (CET) have been conducted in solution and within zeolite cavities for the bicyclo [2.1.0] pentanes [137]. The results indicated the CET method to be comparatively better and providing mechanistic insights on the regioselectivity (effective charge localization) and

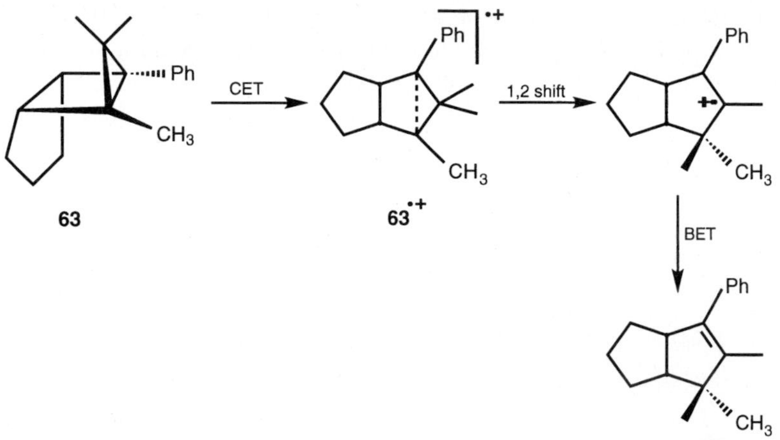

**Scheme 3.**

diastereoselectivity (conformational memory effect) in the rearrangement of the intermediary 1,3-diyl radical cations to the corresponding cyclopentenes. A study carried by Adam and coworkers [138] showed that 1,3-diyl radical cations generated from bicyclo [2.1.0] pentane derivatives, exhibit a high propensity to re-arrange by 1,2-shift to the corresponding 1,2 radical cations, which after electron back-transfer yield cyclopentenes. The tricyclo [3.3.0.0] octane (housane) [138, 139] **63** affords only the cyclopentene *exo*-**63** upon electron transfer oxidation, and thus, the 1,2 migration in the intermediary radical cation **63**$^{\cdot+}$ occurs exclusively to the methyl terminus (Scheme 3).

Further investigation by the Adam group [140] showed electronic substituent effects on the diyl sites profoundly influences the regioselectivities of the 1,2 shift. The data obtained illustrate the distinct electronic character of the cationic inter-mediates involved in the electron transfer oxidation vs acid catalysis of the housanes. The radical anions and cations of dipleiadiene **64** and its 12b,12c-homo derivatives **65** are characterized by EPR and ENDOR spectroscopy [141]. The $\pi$-spin distribu-tion over the perimeter is similar in the radical cations **64**$^{\cdot+}$, **65**$^{\cdot+}$, and an analogous statement holds for the corresponding radical anions.

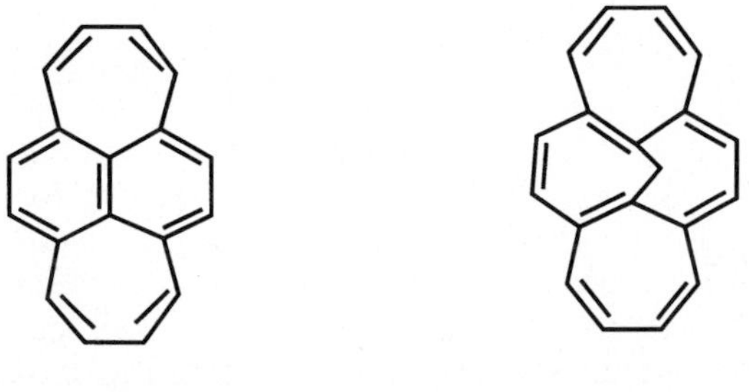

From the coupling constants of $CH_2$ protons, it is concluded that the cation $65^{\cdot+}$ exists in the methano-bridged from **66** of the neutral **64**, whereas the anion $65^{\cdot-}$ adopts the bisnorcaradiene form **67** of the dianion $65^{2-}$.

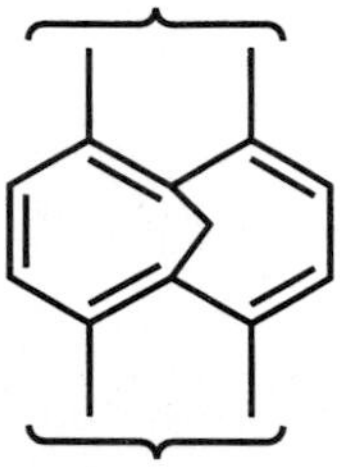

66 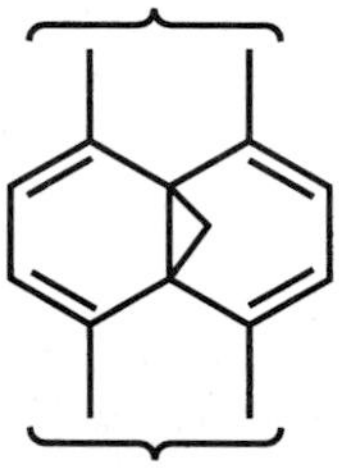                        67

Pioneering work of Dowd on trimethylene methane [142], 1,8-naphthoquinodimethane diradicals were among the first examples of non-Kekule $\pi,\pi$-diradicals that were found to be persistent in their triplet ground state at cryogenic temperatures and were thoroughly characterized by various spectroscopic methods [143]. Time resolved studies of diradicals have provided lifetimes and reactivities of numerous diradical intermediates in solution [144]. $\gamma$-irradiation of CAN, **68**, in haloalkane glasses at 77 K yielded radical cation $68^{\cdot+}$; its electronic absorption spectrum has been obtained and represents first report of absorption spectrum of an ionized diradical [145]. Irradiation at >640 nm converted it into an isomer identical with that formed by ionization of 1,4-dihydro-1,4-ethanonaphtho[1,8-*de*][1,2]diazepine **70**, and subsequent irradiation of the radical cation at $\lambda > 540$ nm. Results identified the photo isomer of $68^{\cdot+}$ to be $69^{\cdot+}$ (Scheme 4).

The molecular and electronic structures are discussed on the basis of B3LYP and CASPT2 quantum chemical calculations. Paracyclophanes are attractive model compounds for studying specific intramolecular interactions [146, 147]. Gerson et al. [148] showed the radical anion of [2,2] paracyclophane **71** is an unassociated specie, with the unpaired electron being equally distributed over both $\pi$-systems.

A range of paracyclophane radical cations has recently been studied by EPR and ENDOR spectroscopy [149]. The [5,5] and [7,7]paracyclophane showed localized radical cation at low temperatures, however at room temperature the higher molec-

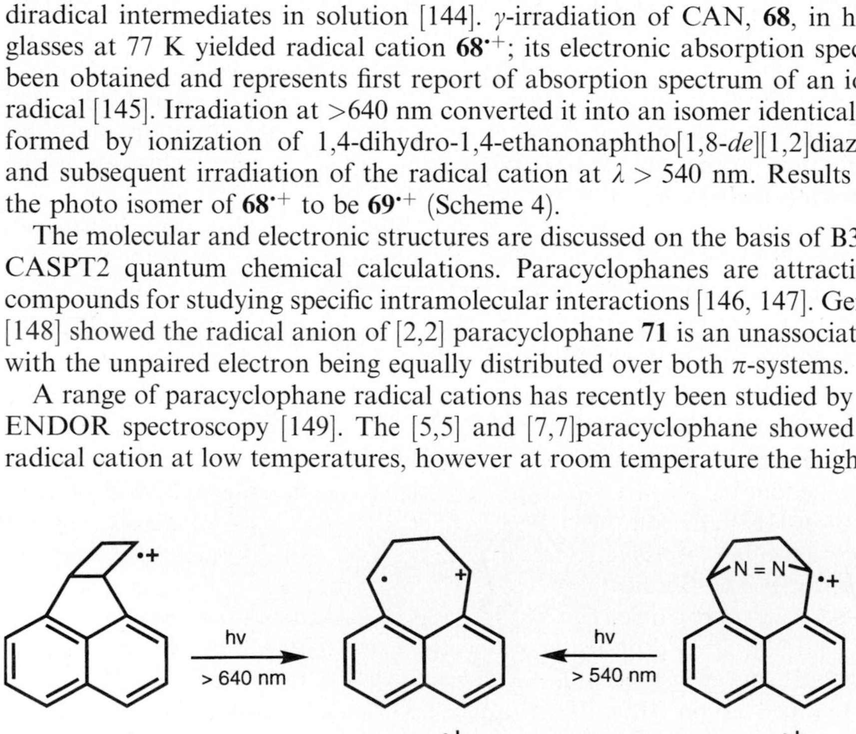

**Scheme 4.**

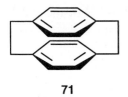

**71**

ular flexibility leads to a fast (EPR time scale) intramolecular electron transfer. In this series radical cations in which two terminal 1,4-dimethoxy benzene units are anellated to [2.2] paracyclophane, [2,2](1,4) naphthalenophane and anthracene bridges have been studied by the same group employing EPR and ENDOR spectroscopic techniques [150].

Single electron transfer (SET) oxidation seems to be the most important pathway for propellanic $\sigma_{c-c}$-bond interactions with $NO_2^+$, the same reactivity is expected for nitrosonium reagent ($NO^+$) which are effective in SET oxidations of organic sulfides [151], urazines [152], aromatic compounds [153] and cyclopropanes [154]. $NO^+$ containing reagents are also reactive towards saturated hydrocarbons [155]. Chemical ($NO^+BF_4^-$/EtOAc, $NO^+OAc^-$/$Ac_2O$, and $NO^+BF_4^-$/$CH_3CN$), photochemical (photoexcited 1,2,4,5-tetracyanobenzene), and electrochemical (Pt anode, $CH_3CN$, $NH_4BF_4$) oxidation of 3,6-dehydrohomoadamantane **72** is elucidated [156].

**72**

Results indicated that the activation of propellane $\sigma_{c-c}$ bonds with strong oxidizing electrophiles occurs by a sequence of SET steps. Ab initio {density functional theory (BLYP) and Møller–Plesset perturbation theory (MP2)} computations utilizing standard basis sets, 6-31G* and 6-311+G* (single-point energy evaluations), agreed with the experimental results implicating the involvement of the same radical cation intermediates in the activation process. Pagodane and (seco) dodecahedrane cage structures [157] allowed the study of unusual electronic phenomena [158] and of unusual molecular species (e.g. valence isomeric 4C/3e radical cations [159] and $\sigma$-bishomoaromatic 4C/2e dications [160]). The ionization and detection of [1.1.1.1] pagodane is successfully carried out by time-resolved fluorescence detected magnetic resonance [161], the first example in which the rate of conversion of two valence isomeric radical cations has been directly measured. The discovery of highly persistent 4C/3e radical cation [162], $\sigma$-bishomoaromatic 4C/2e dication [163] on oxidation of [1.1.1.1] pagodane triggered modification of the cage skeleton surrounding the central cyclobutane ring with respect to stability, persistence, and behavior of the respective radical cations and dications [164]. Synthesis, one and two electron oxidation of [2.2.2.2] pagodane is reported by Gescheidt et al. [165]. For the quantification of the factors determining the transannular electronic interactions, a continuous search for model structures with well-defined molecular dimensions is on the way [166].

Spontaneous generation of long-lived organic cation radicals on zeolite surfaces is known for over three decades [167], however, only during the last few years has this phenomenon been widely recognized and the technique has been developed into a routine laboratory tool to generate stable cation radicals [168]. Stable and ultrastable carbocations from 4-vinylanisole are generated within Zeolites, these reactive intermediates are found to be stable for several weeks within the confirmed environments of zeolites [169]. Results reveal that zeolites, when properly prepared, can generate and stabilize 'reactive' intermediates from neutral molecules. A new method for elucidating elementary reaction steps in zeolite catalysis is demonstrated for reactions of isobutene and other mono olefins on HZSM5 [170], one of the most active zeolite catalyst. Because of its strong Brønsted acidity it has found many critical applications in petroleum cracking, benzene alkylation, xylene isomerization and in aromatization of alkanes and alkenes [171]. The study demonstrates the use of EPR to detect and identify radiolytically generated radical cations and neutral radicals of product molecules of hydrocarbon catalysis on HZSM5. For isobutene on HZSM5, dimerization and isomerization even at cryogenic temperatures was shown [172]. Aromatic radical cations e.g. benzene$^{\bullet+}$, biphenyl$^{\bullet+}$, and naphthalene$^{\bullet+}$ being difficult to stabilize and studied in solution, are prepared in rigid matrices of certain halocarbons or on surfaces of zeolites or silica gel [173], however in certain cases EPR spectra from these disordered samples suffer from limited resolution. For biphenyl$^{\bullet+}$ the higher resolution of ENDOR greatly facilitated investigations of the ring proton hyperfine tensors in disordered solids [174].

The naphthalene radical cation generated by UV- or X-irradiation on HZSM5 zeolite and in frozen $CFCl_3$ is investigated by EPR, ENDOR and ESEEM spectroscopic methods [175]. Analysis of powder EPR line shape was made possible by the simultaneous use of ENDOR and ESEEM spectroscopies. The $g$ and ring-proton hyperfine (hf) tensors were determined and were found identical in the two media, within experimental error. Iwasaki et al. [176], stabilized benzene radical cations in a $CFCl_3$ matrix and obtained EPR evidence for the Jahn–Teller distortion at 4.2 K, with the unpaired electron density in the $^2b_{2g}$ orbital. EPR investigation of the benzene radical cation adsorbed on the HY molecular sieve at 203 K resulted in spectra characterized by axially symmetric $g$ and $A$ tensors with $g = 2.00238$, $A_{II} = -14.00$ MHz and $A_\perp = -11.34$ MHz, suggesting rapid rotation of the benzene cation about the molecular sixfold symmetry axis [177]. EPR, ENDOR and ESEEM studies of $C_6H_6^{\bullet+}$ and $C_6H_5D^{\bullet+}$ absorbed on HY molecular sieve and silica gel suggested that $C_6H_6^{\bullet+}$ undergoes pseudorotation at 3.5 K, whereas $C_6H_5D^{\bullet+}$ on silica gel showed slowing down of the molecular motion to a rigid or semi-rigid state [178]. Recent ENDOR and EPR investigations by Lund et al. [179] yielded evidence for a Jahn–Teller distortion of monomeric benzene radical cation $C_6H_6^{\bullet+}$ in a $CFCl_3$ matrix at 30 K, with unpaired electron density predominantly in the $^2b_{2g}$ orbital. Better resolution in case of ENDOR as compared to EPR enabled precise measurement of isotropic and dipolar hyperfine coupling constants in the

Jahn–Teller distorted state. ENDOR results of dimeric benzene cation formed after warming in $CFCl_3$ and present in $CF_3CCl_3$ at all temperatures indicate that the dimer had a sandwich structure in both matrices. An EPR study of the benzene radical cation in argon matrix generated by fast electron irradiation at 16 K showed favorable stabilization of $^2b_{1g}$ state rather than $^2b_{2g}$ state, in contrast to previous results found in Freon matrices [180].

Substitution of hydrogen by weakly perturbing alkyl groups like the methyl group in toluene and xylene anions, removes the twofold orbital degeneracy of the ground state of benzene radical cations/anions artificially [181]. The radical cations of toluene and xylene in silica gel and halocarbon matrices are reported [182], however hyperfine coupling constants determined in these investigations were different and detailed assignment of spectra is lacking. Komatsu et al. [183] studied toluene and partly deuterated toluene (toluene $\alpha$-d$_3$) on silica gel and Vycor glass and provided unambiguous evidence for the formation of toluene radical cation. EPR and ENDOR results on radical cations of methyl-substituted benzene, *p*-xylene, *o*-xylene, *m*-xylene and their deuterated analogs generated in $CFCl_3$ and $CF_3CCl_3$ matrices by X-ray irradiation at 77 K are investigated by Lund et al. [184]. Spectra in the radical cations are dominated by large axially symmetric hyperfine splitting due to methyl group protons. Experiments and theoretical calculations supported the stabilization of $^2b_{2g}$ type of state for the radical cations of toluene and *p*-xylene compared with the stabilization of $^2b_{1g}$ type of ground state for *o*-xylene and *m*-xylene due to lifting of the orbital degeneracy of the $e_{1g}$ orbital by lowering the symmetry obtained after substitution of hydrogen by methyl groups.

Eriksson [185] studied radiolytic reduction of acetylene adsorbed on HZSM5. Observance of hydrocarbon radical anion is a rare finding and contrasts with results for other unsaturated hydrocarbons adsorbed on HZSM5, for which radical cations are only observed. An EPR spectrum revealed radical anion to be in the *cis*-bent form. The energies, geometries, and hyperfine coupling constants are calculated for the acetylene, vinylidene, cyclobutadiene and butatriene radical anions and radical cations, using different levels of theory, including gradient-corrected density functional theory. Variable-temperature EPR study of various olefin radical cations, 1,4-cyclohexadiene, 1,3-cyclohexadiene, 3,3-dimethyl-1-butene, 2,3-dimethyl-1-butene and tetramethyl ethylene, generated radiolytically in non-acidic and acidic ZSM-5 zeolites showed isomerization reactions even at 4 K [186]. The observation of H-addition-type radicals indicates Brønsted acid-catalyzed rearrangements prior to irradiation on more acidic zeolites. Radical cations of *cis*- and *trans*-decalin are studied in Freon matrices and zeolites by EPR [187], in hydrocarbon solutions by ODEPR in CW [188] and time resolved variants [189]. ODEPR of *cis*-decalin radical cation in dilute cyclohexane and squalane [190] solution were observed at temperatures up to ambient, but signal of *trans*-decalin could be observed only below 40 K [189]. ODEPR and MARY (magnetic field effect on recombination luminescence yield) spectroscopy of spin-correlated radical ion pairs were used to study *cis*- and *trans*-decalin radical cations in nonpolar medium [191]. The analysis demonstrates that the peculiarities of the *trans*-decalin ODEPR and MARY spectra at high temperatures root in the quasi-degeneracy of electron levels and the related dynamic transitions and spin relaxation due to strong spin–orbital interactions.

### 4.4.2 Olefins

Radical cations of alkanes are extensively studied both in Freon matrices by EPR [192] and in hydrocarbon solutions by ODEPR technique in CW [193] and time-resolved FDMR [194] variants. An alkane with higher ionization potential is used as a matrix, and the substance of interest is added in small ($<10^{-2}$ M) concentrations as an admixture. ODEPR spectra of *cis*-decalin radical cation in cyclohexane [195] are observed for dilute solutions but efforts to observe signals of solvent radical cations (holes) in non-dilute solutions failed. This resulted in the assumption that some fast hole decay is responsible for the absence of any signal [196]. It has been demonstrated that the effect of level crossing in spin-correlated radical ion pairs [197] permits EPR spectra of radical cations without microwave pumping [198]. Recently Tadjikov et al. [199] succeeded in detecting squalane radical cation (hole) formed under the action of ionizing radiation at room temperature in a non-dilute solution by employing MARY–EPR technique. The spectrum showed one single line, homogeneously narrowed due to resonance charge transfer over the solute molecules with the rate constant exceeding the diffusion-controlled limit. The technique of radical ion pair level crossing spectroscopy revealed EPR spectrum parameters of solvent radical cations for a series of non-viscous alkanes, *n*-pentane, *n*-hexane, *n*-octane, *n*-decane, *n*-dodecane and *n*-hexadecane [200]. The rate constant for squalane hole scavenging by a diphenyl sulfide molecule is directly obtained using quantum-beats technique [201]. Results are in good agreement with that of pulse-radiolysis experiments [196], providing stronger background to the hypothesis about the hole nature of the highly mobile solvent cation in squalane. A number of EPR and theoretical studies are carried out for the cyclohexane and related radical cations [202], however, static and dynamic structures are not yet completely understood. Combined with theoretical calculations employing ab initio and DFT [203] methods, the radical cation is concluded to take an $^2A_g$ electronic ground state in $C_{2h}$ symmetry, in which two C–C bonds are elongated and other four compressed. An EPR study of radical cations of selectively deuterated cyclohexanes in a $cC_6F_{12}$ matrix, revealed that the geometrical structure of the radical cation is distorted into a $C_{2h}$ symmetry from the original $D_{3d}$ symmetry. The temperature-dependent EPR line shapes are analyzed in terms of dynamical averaging among the $C_{2h}$ structures with different zero-point vibrational energies due to deuterium mass effects under the assumption of Boltzmann distribution.

Photoheterolysis of benzylic chlorides [204] yielded results signifying that simple benzyl cations, such as cumyl and 1-phenylethyl cations, can exist in the solution as free ions; radicals arising from a competing photohomolysis are also observed frequently. Haloalkyl-carbocations are studied by heterolysis of the corresponding dihalides in super acid media [205]. $^{13}C$ NMR chemical shifts are interpreted as evidence for an interaction between the vacant orbital of cationic center of the haloalkyl carbocations with a lone electron pair of the halogen atom. 3-chloro-1-methylcyclopentyl cation **73**, thermally eliminates hydrogen chloride and yields 1-methyl-2-cyclopentyl cation **74**, a similar behavior reported for $\gamma$-chloroalkyl carbocations [206] (Scheme 5).

Flash photolysis experiments [207] of 1,3-dichloro-1,3-diphenyl propane **75** in

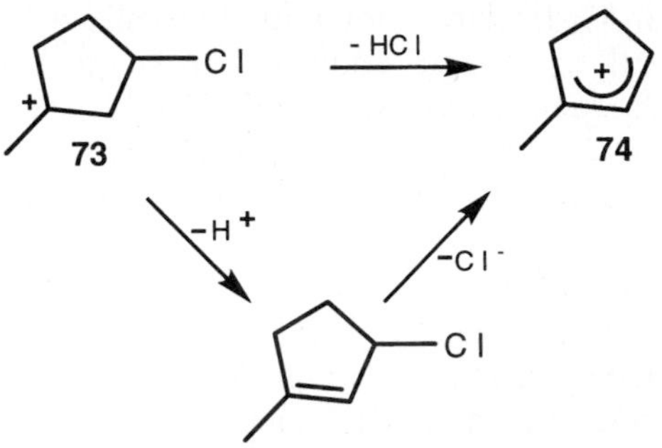

**Scheme 5.**

2,2,2-trifluoroethanol at 266 nm lead to γ-chloropropyl cation **76a**, kinetic results showed interconversion between the open chain cation and corresponding cyclic chloronium ion **76b**; loss of HCl yielded 1,3-diphenyl-2-propenyl cation **77** (Scheme 6).

Two-laser two-photon results revealed photoisomerization of the cation *E,E*-**77** to its stereoisomer *Z,E*-**77**, which undergoes thermal reversion with a lifetime of 3.5 μs at room temperature. Absolute rate constants for reaction of styrene, 4-methylstyrene, 4-methoxystyrene and β-methyl-4-methoxystyrene radical cation with a series of alkanes, dienes and enol ethers are measured by Laser flash photolysis [208]. The addition reactions are sensitive to steric and electronic effects on both the radical cation and the alkene or diene. Reactivity of radical cations follows the general trend of 4-H > 4-CH$_3$ > 4-CH$_3$O > 4-CH$_3$O-β-CH$_3$, while the effect of alkyl substitution on the relative reactivity of alkenes toward styrene radical cations may be summarized as 1,2-dialkyl < 2-alkyl < trialkyl ≤ 2,2-dialkyl < tetraalkyl.

There is a considerable current interest in synthetic applications of radical cation chemistry. Alkene radical cations have been studied extensively, notable examples include the *anti* Markovnikov addition of nucleophiles [209], and the photo-NO-CAS reaction [210]. The synthetic utility of radical cation mediated chemistry, and

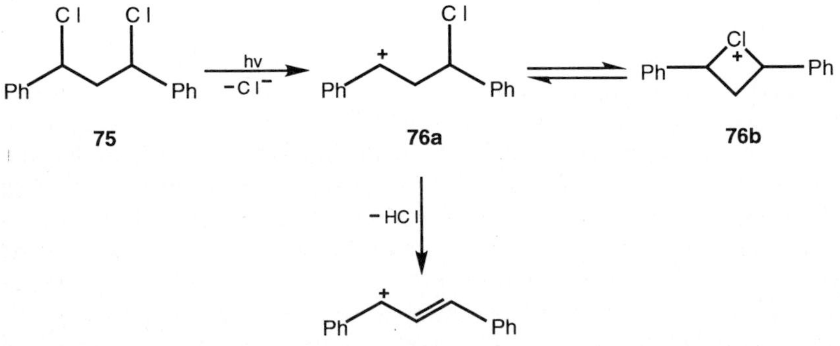

**Scheme 6.**

in particular cyclobutanation and Diels–Alder reactions, is well noted [211]. Radical cation cycloaddition can be carried out under relatively mild conditions and frequently results in product yields, stereoselectivity, and regioselectivity that equal or surpass their thermal counterparts. However, it should be noted that the rational design of synthetic strategies based on this chemistry requires detailed kinetic and mechanistic information for all the potential reaction pathways of the various intermediates. A number of conjugated diene radical cations have been generated photochemically by either photoionization or photosensitized electron transfer and have been spectroscopically and kinetically characterized using laser flash photolysis. Measured rate constants for addition to methanol demonstrate that the more highly alkyl-substituted radical cations are substantially less reactive toward nucleophilic addition reactions [212].

Fluorine substitution in the organic molecules has a significant effect on their geometry and electronic structure [213], in particular for the benzene derivatives, the structure and the symmetry of the radical cation are affected not only by the number of substitutions by fluorine but also by the position of the substitution [214]. Other examples studied are pyridine [215], butadiene [216], and ethylene [217]. Although the geometrical structure of neutral olefins are planar at the ground electronic state, non-polar structures in bent form or in twist form have been suggested for their radical cations and anions by EPR [218], photoelectron spectroscopy, and theoretical calculations [219]. EPR spectrum from $\gamma$-irradiated $CCl_2FCCl_2F$ solution containing $C_2H_4$, was attributed to ethylene radical cation with a torsional angle of 45° about the C–C bond [220a]. However, Fujisawa et al. [220b] suggested that the spectrum should be attributed to a propagating radical cation of $^+CH_2$–$(CH_2)_n$–$CH_2$. Lunell and group calculated the isotropic $^1H$ hf splitting by an accurate CI study and concluded that the splitting of $(-)0.3$ mT corresponds to the torsional angle of 28° [221]. Based on EPR results of the propene radical cation, $CH_2=CH–CH_3^+$ produced in a $CCl_3F$ matrix at 77 K, Toriyama et al. [222] analyzed that the rotation of the $CH_3$ group is hindered. However, study of partially deuterated propene $CH_3CH=CD_2$ revealed that the $CH_3$ group rotates freely even at 77 K [223]. INDO calculations suggested an energy minimum with a twist angle of 45°. On comparing the observed hf splitting of 0.70 mT for the CH proton and calculated value, it is concluded that the propene cation might interchange between two configurations corresponding to the energy minimum rapidly enough to give the apparent planarity on the EPR time scale because of its low barrier. Toriyama et al. reinvestigated propene cation, in addition to a study on a trimethylethylene cation, $(CH_3)_2C=CH(CH_3)^+$. Using a twisted model along the C=C bond and the bent structure in the group of $CHCH_3$ or $CH_2$ of the cations, they interpreted the observed hf splittings of 0.6 mT for the CH proton of the propene cation as well as a large non-equivalence of the two $CH_2$ proton coupling constants of 2.35 and 1.1 mT [224]. Radical cation of a series of fluorinated ethylenes and propenes, generated by irradiation with $\gamma$-rays in halocarbon solid matrices is recently being studied by Hasegawa et al. [225]. EPR spectra obtained for trifluoro olefin cations, $CF_2=CFX^+$ (X = H, $CH_3$ or $CF_3$) are successfully analyzed in terms of three $^{19}F$ nuclei with coaxial parallel components, thus suggesting a planar structure. The optimized geometry of the radical cations was calculated by ab initio MO method.

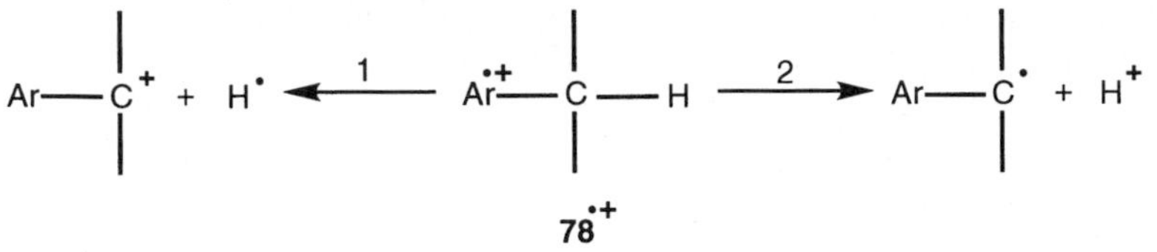

**Scheme 7.**

In contrast with the ethylene cation and the propene cation, having non-planar twisted structures, the fluorinated ethylene and propene cations are concluded to have planar structures. The cleavage of the C–H bond of alkyl aromatic radical cations usually occurs in a heterolytic fashion in solution, however, homolytic cleavage is equally possible [226] (Scheme 7).

Path 2 is generally observed in solution [227], and accordingly, radical cations $78^{\cdot+}$ represent a very interesting class of strong carbon acids, whose kinetic and thermodynamic acidity is the object of continuous investigation [228]. The possibility of the alternative path 1 involving a hydrogen atom transfer reaction is rarely observed [229]. Recently photoinduced electron transfer of bis(4-methoxyphenyl)-methane, sensitized by chloranil in $CH_3CN$ clearly indicated the occurrence of a hydrogen atom transfer from the radical cation to the chloranil radical anion $(CA^{\cdot-})$ [230] (Scheme 8).

The low basicity of $CA^{\cdot-}$ in $CH_3CN$ ($pK_a = 6.8$) and the substantial spin density on the oxygen atoms probably play a fundamental role in this respect, as well as, of course, the reduction potential of the formed carbocation. Trimethyl methane (TMM) is one of the most attractive organic molecules, investigated with respect to electronic structure because it is a rather small fundamental molecule with a high symmetry and a triplet ground state with a doubly degenerate HOMO [231]. The 4.2 K EPR showed a septet with isotropic $^1$H hf splittings of 0.93 mT (6H) in $CF_2ClCF_2Cl$ matrix. Ab initio MO calculations predicted that $TMM^{\cdot+}$ is a $^2A_2$ state in a distorted $C_{2v}$ structure, and the unpaired electron mainly resides in the $P_z$ orbitals of the two equivalent terminal carbons [232]. The spectra is explained by intramolecular dynamics between the three energetically equivalent $C_{2v}$ structures to average the structural distortion giving an apparent $D_{3h}$ structure even at low temperatures. Pulsed EPR technique is employed to measure electron spin–lattice $(T_1)$ and spin–spin relaxation times $(T_2)$ for the tetramethyl ethylene radical cation in ZSM5 zeolite [233]. Major conclusions are (a) methyl rotation is hindered by radical cation interactions with co-adsorbed argon, (b) the rotational potential of the methyl group is affected by radical cation–matrix interactions, (c) $TME^{\cdot+}$–TME interactions do not appear to occur to an appreciable extent at TME loadings less than 3 uc$^{-1}$ (molecules per unit cell), but a change in the TME loading causes a

$$CA^{\cdot-} + (4\text{-MeOPh})_2CH_2^{\cdot+} \longrightarrow CAH^- + (4\text{-MeOPh})_2CH^+$$

**Scheme 8.**

change in the radical cation–matrix interactions due to site heterogeneity and/or sorbate-induced lattice distortions.

### 4.4.3 Heteronuclear Radical Cations

Ground-state azoalkanes can serve as electron donors and thereby be converted to radical cations [234], a method frequently used to induce electron transfer in photosensitization [235]. Azoalkane radical cations have been studied by EPR directly under matrix isolation, e.g. 2,3-diazabicyclo[2.2.2]oct-2-ene [234a], and in solution, e.g. *trans-N,N'*-di(1-norbornyl)diazene [234b], however some azoalkanes experience rapid loss of nitrogen upon one-electron oxidation [234c–f]. Well known for $N_2$ extrusion upon direct and triplet-sensitized photolysis [236], 2,3-diazabicyclo[2.2.1]-hept-2-ene (DBH) and its derivatives denitrogenate effectively upon electron transfer to yield carbon-centered radical cations [234e, 234f, 237]. The azoalkanes **79a–c** showed nitrogen extrusion upon photosensitized electron transfer (PET) to yield 1,3-radical cation intermediate **80**$^{\bullet+}$, reverse electron transfer (RET) affords the unrearranged housanes **80**, but significant rearrangement to dicyclopentadiene derivative **81** was observed prior to RET [238] (Scheme 9).

The housanes **80a,b** are also oxidized by PET, but are more reluctant to rearrange, results illustrating the intriguing chemical fate of radical cations derived from dicyclopentadienes and their isomers. In continuation of their investigation, Adam et al. [239] studied oxidation of 2,3-diazabicyclo[2.2.1]hept-2-enes, **82a–d**,

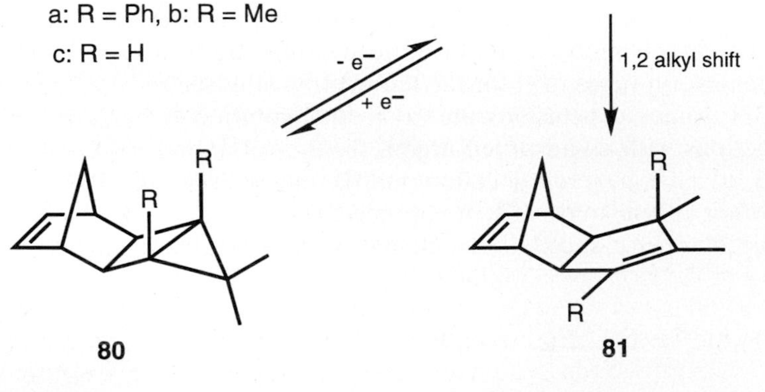

a: R = Ph, b: R = Me
c: R = H

**Scheme 9.**

and bicyclo[2.1.0]pentanes, **83a–d**, with catalytic amounts of tris(aryl)aminium hexachloroantimonates afforded the corresponding cyclopentenes, **84a–d**.

| 82 | 83 | 84 |

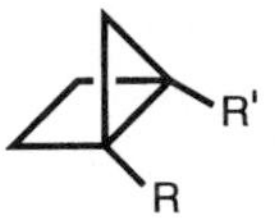

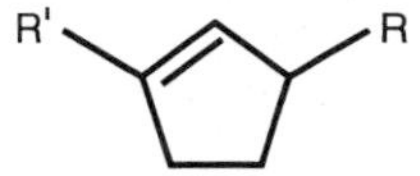

a (R = R' = H);  b (R = Me ; R' = H)

c (R = Ph ; R' = H);  d (R = R' = Me)

A reversal in the regioselectivity of the 1,2-migration was observed for the unsymmetrical derivative **80b**, **80c**, proposed to be due to delocalization of the positive charge into the aromatic ring for the 1,3-diyl radical cation, as corroborated by AM1 calculations. EPR spectra are reported for radical anion and for the first time radical cation derivatives of pyrazolines **85–88** [240].

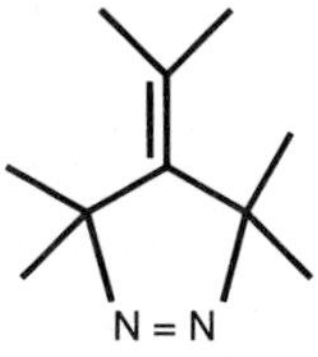

| 85 | 86 | 87 | 88 |

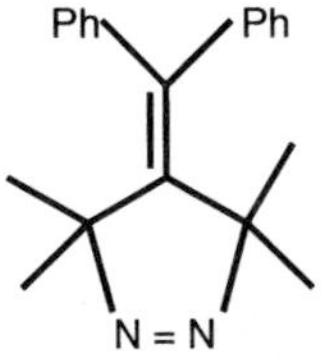

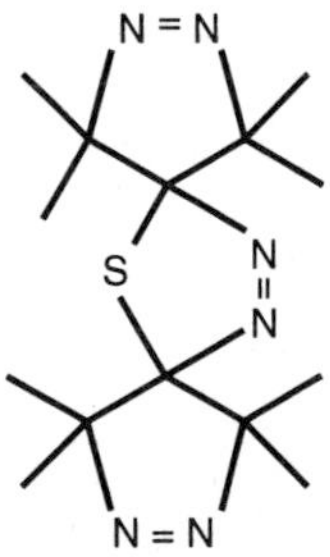

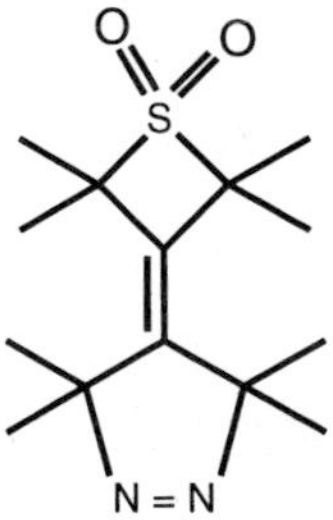

The radical anion spectra are consistent with the unpaired electron being localized principally in the N=N $\pi^*$-orbital, however, for the radical cation of 4-alkylidenepyrazolines, the near degeneracy of the N=N $n^-$ and C=C $\pi$-orbitals made the distribution of the unpaired electron difficult to predict. The radical cations of cyclopentane- and urazole-annelated azoalkanes and housanes were generated by pulse radiolysis and the intermediates characterized spectrally and kinetically by time-resolved optical monitoring. For the bridgehead-substituted diphenyl derivatives only the corresponding proximate 1,2-radical cation is detected, generated from the initially formed and too short-lived (<1 μs) distonic 1,3-radical cation by 1,2-methyl migration [241]. Triazenes increasingly find application as initiators of radical polymerization [242] and promoters of polymer ablation by ultraviolet irradiation [243]. The decomposition of symmetrical dialkyltriazenes in acidic aqueous solution proceeds with formation of alkylamines and alkyl alcohols, whereas, asymmetric 1,3-dialkyltriazenes preferably form different alkyldiazonium ions and alkyl amines due to two tautomeric forms in the starting triazenes [244]. In

cSiC5          cSiC5-2,2,6,6-d$_4$          1,1-Me$_2$-cSiC5          4,4-Me$_2$-cSiC5

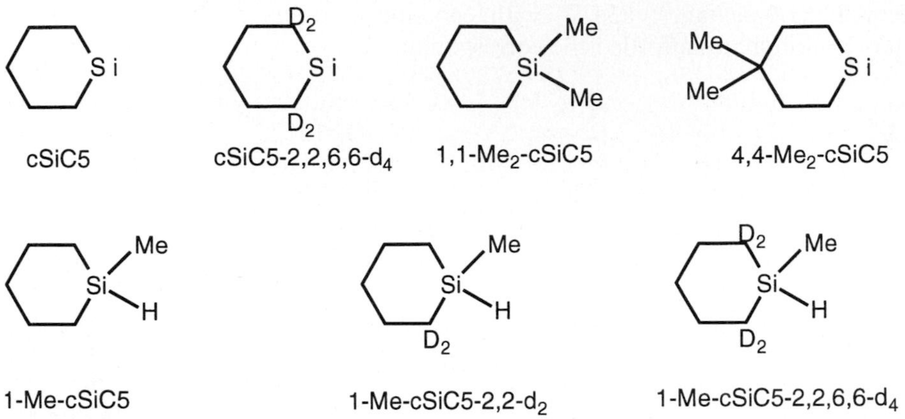

1-Me-cSiC5          1-Me-cSiC5-2,2-d$_2$          1-Me-cSiC5-2,2,6,6-d$_4$

**Scheme 10.**

electrochemical oxidation of 1,3-diphenyltriazenes, intermediate formation of the triazene cation radical is assumed, which after protonation with residual water rearranges to aniline and finally converts to *p*-aminoazobenzene [245]. Redox properties of asymmetric 1-phenyl-3-alkyltriazenes and their corresponding triazenidoplatinum complexes are characterized by CV and the structure of the radical intermediates formed in electrolytic reactions and in oxidation with peroxy compounds is investigated [246]. Hydrazines linked by a *p*-phenylene bridging unit are charge localized in the intervalence oxidation state, proper choosing of substituents allows the isolation of compounds in 0, 1$^+$, and 2$^+$ oxidation states which are stable enough for variety of physical studies [247]. The structure, dynamics and thermal reactions of the radical cation of silacyclohexanes (cSiC$_5$) containing one Si atom in the six-membered ring have been studied by EPR and ab initio MO calculations. The cations were generated in perfluoromethylcyclohexane matrix by ionization radiation at low temperatures. Results suggest that cSiC$_5$ radical cation takes an asymmetrically distorted $C_1$ structure with one of the Si–C bonds elongated in which the unpaired electron resides, supported by ab initio MO calculations (Scheme 10).

At temperatures above 140 K, the cSiC$_5$ radical cations were converted into the neutral radicals via geoselective deprotonations. In recent years, a number of compounds which show multiple bonding to silicon has been isolated and studied [248]. Monomeric silanones, **89**, have as yet been observed only in solid matrices [249], apparently the presence of two bulky groups is not sufficient to prevent oligomeri-

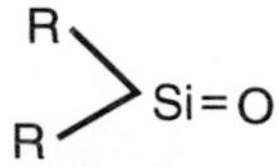

$$R_2Si=O$$

**89**

zation in fluid solutions, but formation of the silanones can be inferred from their reactions with traps such as dienes [250].

However, there is good evidence from EPR experiments for the existence of sterically stabilized silanone radical anions in solution [251]. EPR study of tris(trimethylsilyl)methylsilyl radicals $(Me_3Si)_3CSiH_{n-1}R_{3-n}$ (R = H, Et or F, $n = 1$, 2 or 3), prepared by hydrogen abstraction from the corresponding tris(trimethylsilyl)methylsilanes is reported [252]. Hydrogen abstraction from silanols $(Me_3Si)_3CSiH(R)OH$ (R = H, Me, Et, Bu, Ph, F) under neutral conditions yielded spectra of the corresponding hydroxysilyl radicals $(Me_3Si)_3CSi(R)OH^{\cdot}$, while reaction with potassium *tert*-butoxide gave silanone radical anions $(Me_3Si)_3CSi(R){=}O^{\cdot-}$. The electronic absorption spectra of polysilane radical ions have been studied by Ban et al. [253], Irie et al. [254] and by Ushida et al. [255]. They observed a UV band assigned to the HOMO → LUMO transition of neutral polysilanes (SOMO → LUMO for the cation and HOMO → SOMO for the anion). Ushida et al. assigned the near IR band as due to charge resonance band arising from the interaction between adjacent σ-conjugated polymer segments. Ichikawa et al. [256] studied the electronic structure of polysilane and oligosilane radical anions by electronic absorption and EPR spectroscopy. They concluded that the SOMO is a pseudo-π orbital composed of antibonding Si–Si and Si–C orbitals and the unpaired electron is not delocalized all over the polymer chain but is confined to a part of the chain. A similar study of oligosilane radical cations suggested that the unpaired electron of polymer radical cation is also not delocalized all over the Si–Si main chain [257]. Tetrathiafulvalene (TTF) and its derivative represents the most popular donor π-systems and have been used for the last two decades in conducting and supraconducting radical cation salts [258]. Radical cations of **90–92** are studied by CV, EPR and ENDOR spectroscopic techniques [259].

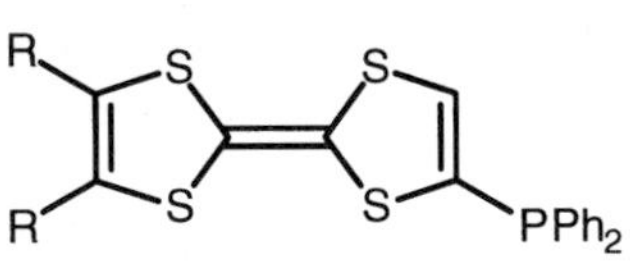

**90**

R = H, R = Me

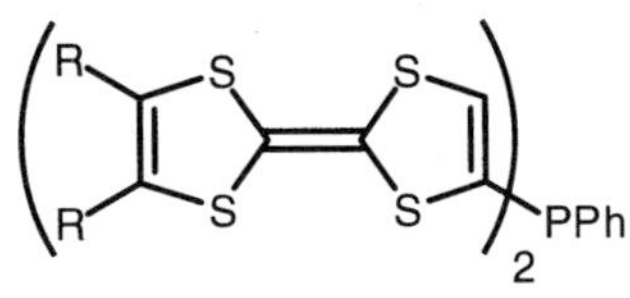

**91**

R = H, R = Me

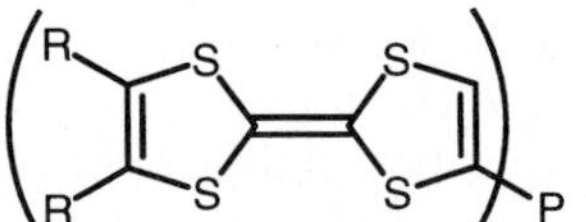

**92**

R = H, R = Me

During stepwise oxidation, each of the tetrathiafulvalene (TTF) or 4,5-dimethyltetrathiofulvalene (o-DMTTF) moieties in **90–92** donates two electrons so that exhaustive oxidation at potentials below IV (vs. SCE) leads to dications $\mathbf{90}^{2+}$, the tetracations $\mathbf{91}^{4+}$ and hexacations $\mathbf{92}^{6+}$. In the initially formed radical cations,

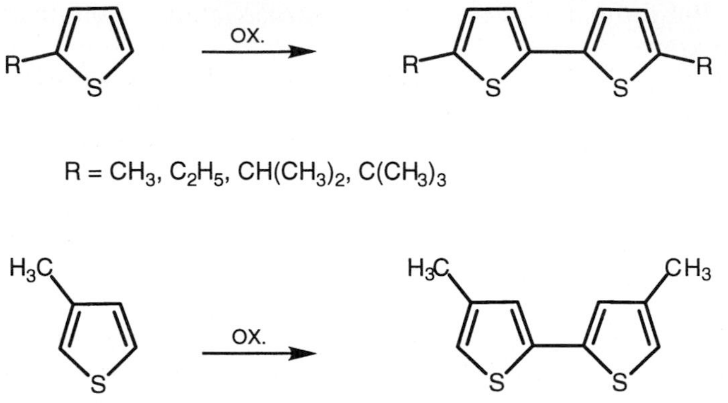

R = CH$_3$, C$_2$H$_5$, CH(CH$_3$)$_2$, C(CH$_3$)$_3$

**Scheme 11.**

the electron hole is delocalized over all the TTF or $o$-DMTTF moieties, two in **91**$^{\cdot+}$ and three in **92**$^{\cdot+}$. On the other hand, in the paramagnetic species produced by further oxidation, also giving rise to well defined EPR and ENDOR spectra, the unpaired electron appears localized in only one donor moiety on the hyperfine time-scale. The dications **91**$^{2+}$ and **92**$^{2+}$, as well as the tetracations **92**$^{4+}$ are proposed to have a triplet ground state, while the trications **92**$^{3+}$ are able to form quartet states. Oxidation of 2-alkylthiophenes by thallium(III) tris(trifluoroacetate) in HFP yielded the corresponding persistent 5,5′-dialkyl-2,2′-bithiophene radical cations, while 3-methylthiophene showed 3,3′-dimethyl-2,2′-bithiophene [260] (Scheme 11).

A combination of collisional activation, neutralization–reionization, and ion–molecule reaction experiments has been used to characterize various [C$_2$H$_4$OS]$^{\cdot+}$ radical cations derived from $s$-alkyl thiaformate precursors, and the results suggest the occurrence of distonic ions and ion-molecular complexes [261]. Experimental evidence for the existence of isomeric distonic HC$^+$(SH)OCH$_2^{\cdot}$ species is not found, but the calculation of the related [C$_2$H$_4$OS]$^{\cdot+}$ potential energy surface suggests that the latter ion is stable and likely to play a role in the fragmentation processes. Investigation by Asmus et al. showed that the radical cations of aliphatic sulfides are stabilized by forming the sulfur–sulfur two-center three-electron bonded dimers [262]. The nature of such 2C-3e S–S bonds has attracted considerable interest as possible intermediates in many enzymatic oxidations of organic sulfides [263]. However, radical cations of aromatic sulfides [264] are believed not to form any dimers because of the delocalization of positive charge/spin density over the aromatic rings. Recently, Sawaki et al. [265] studied aromatic sulfides by photochemical one-electron oxidation in acetonitrile. Irradiation of dicyanonaphthalene and thioanisole with nanosecond laser flash (308 nm), revealed two types of dimer radical cations at 470 ($\sigma$-type) and 800 ($\pi$-type) nm, at the expense of the monomer radical cation (520 nm). The density functional BLYP/6-31*G calculations on thioanisole predicted the existence of $\sigma$- and $\pi$-type dimer radical cations, in accordance with the experimental observation of approximately equal stability. In recent years, special interest has been given to cumulenes containing one or two atoms of

Group VB [266] (e.g. RP=C=PR), mainly due to considerable progress made in synthetic chemistry, especially in field of phosphorus-containing compounds [267], and to a better understanding of the redox properties of these compounds [268]. Crystal structure of diphosphaalenes stabilized by bulky substituents (Ar = 2,4,6-$(Bu^t)_3C_6H_2$), have shown that the P–C–P sequence is not linear and that these molecules adopt a structure close to $C_2$ symmetry with the substituents on the opposite sides of the PCP plane [269]. The structure of the first term of the series, HPCPH, has been the subject of several ab initio studies [270], all of which reproduce the characteristics of the X-ray measurements. Electrochemical oxidation of ArP=C=PAr and ArP=$^{13}$C=PAr is studied by EPR in tetrahydrofuran [271]. Taking HPCPH as a model compound, the structure of the radical cation assessed by extensive ab initio calculations revealed two rotamers with HPPH dihedral angles of 45° and 135°.

## 4.5 Radical Anions

### 4.5.1 π Systems

Radical anions of organic compounds have received much attention due to several reasons. The transition into a radical anion state activates organic molecules with respect to various transformations, which often are not typical for the starting systems and opens up new ways of their functionalization [272]. In particular, one electron reduction enhances nucleophilic and protophilic properties of organic compounds, facilitates their reactions with free radicals and makes their fragmentation and dimerization possible. There has been considerable interest in the design of organic materials that are suited in terms of electron transfer and electron storage [273]. An important aspect in the redox behavior of bis- and poly-electrophoric systems with vanishing conjugative interactions of π-subunits involves the possibility of intramolecular electron-hopping processes upon photoexcitation or upon thermal activation of a monocharged state [274]. Another factor controlling the electron transfer is the number of extra charges since successive reduction to di-, tri-, or even tetra-anion effects not only the prevailing ion-pair structure but also the spin multiplicity of the species and relative spatial arrangement of the subunits [275]. For charge storage purpose, a conjugative barrier with nearly independent redox active behavior of the subunits and density of localized charges is particularly favorable. Vanishing conjugative interaction of π-units and, thus, strongly inhibited spin–spin pairing, on the other hand also occur in organic high-spin systems, which play a central role in the search for organic ferromagnets [276]. While the 'unpairing' of the individual spins has usually been sought in π-conjugative compounds with the so-called 'non-Kekulé' structures [276, 277], approaches towards high-spin molecules include the almost orthogonal alignment of spin-carrying subunits [278]. In agreement with theoretical studies [278b, 278c, 279], ferromagnetic coupling should be possible in the triplet dianion of **93** and in the polyanions of poly(9,10-

anthrylenes) **94** as long as orthogonality (85–90°) of the subunits is retained upon charging [280].

EPR and ENDOR studies of the radical anions of **95**, **96** and **97** generated by chemical reduction are reported [281].

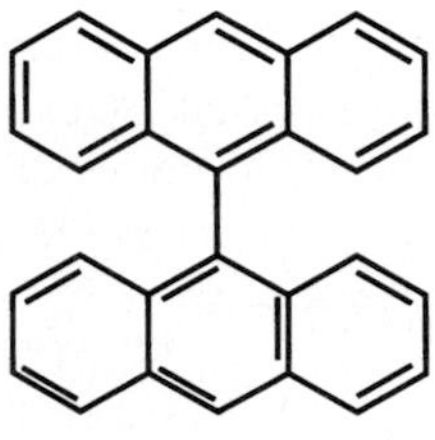

93

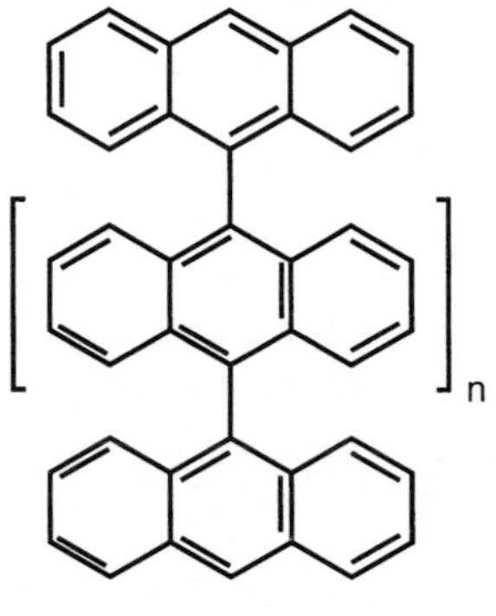

94

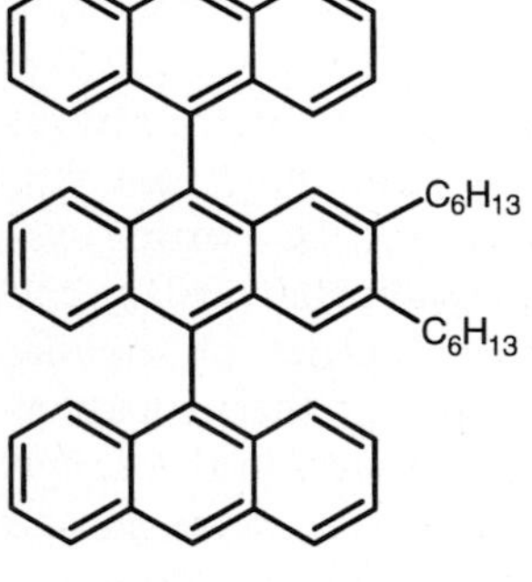

95

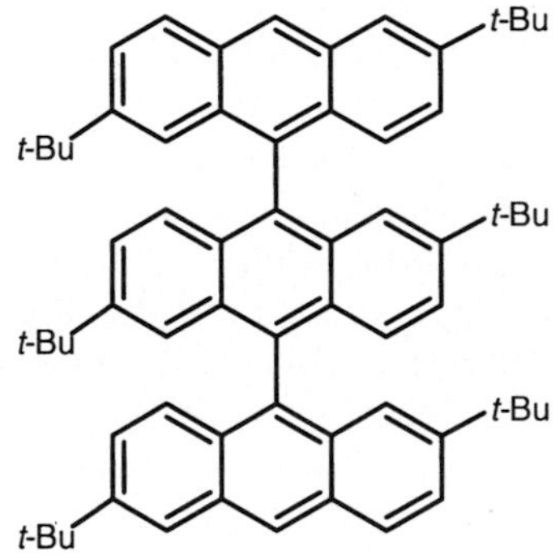

96

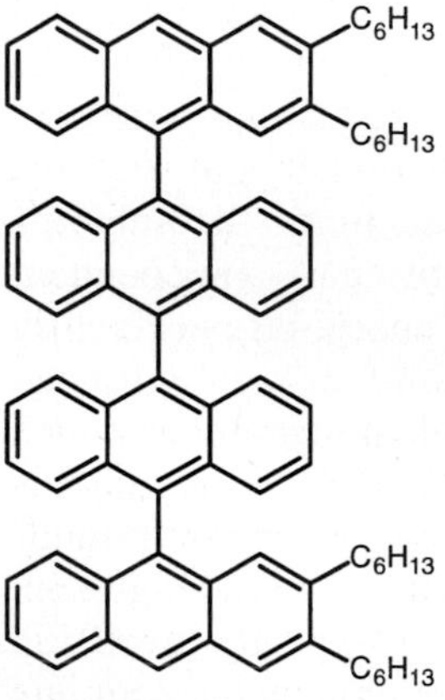

97

The intramolecular electron transfer in mono charged species of oligo(9,10-anthrylene)s between anthracene moieties sensitively depends on the special substitution pattern, on the radical concentration, and thus on the ion-pair conditions. The trianion of **95** and tetra anion of **96** are found in a quartet state and a quintet state. Temperature dependent EPR measurements of the dimer dianion revealed that the triplet state is thermally excited and the zero-field splitting looses its axial character (typically for orthogonal alignment) at low temperatures, pointing to increased orthorhombicity with decreasing temperature. The higher spin states of the oligomers are found to be thermally activated with 120 cal mol$^{-1}$ for the doublet–quartet ($S = 3/2$) transition and 180 cal mol for the quintet–singlet ($S = 2$) transitions. Electrochemically generated 9,9'-bianthryl (BA) and 10,10'-dimethoxy-9,9'-bianthryl (DA) radical anions have been investigated by means of UV–Vis and EPR spectroscopy. Unpaired electron of the radical anion is localized on one of the anthracene subunits with large counter ions, tetrabutyl ammonium. The rate constants for the intramolecular electron exchange reactions were measured by EPR line-broadening effects, $(4.2 \pm 0.3) \times 10^7$ s$^{-1}$ and $(2.2 \pm 0.3) \times 10^7$ s$^{-1}$ for BA$^{\bullet-}$ and DA$^{\bullet-}$ in DMF at 298 K, respectively [282].

UV–Vis, $^1$H and $^{13}$C NMR study of monometallic salts of 9,10-dihydroanthracene and its 9,10-disubstituted derivatives in THF, showed lithium 9-phenyl-9,10-dihydroanthracene-9-ide, lithium 9,10-dimethyl-9,10-dihydroanthracenide and lithium 9,10-diphenyl-9,10-dihydroanthracenide exist as a solvent separated ion pair (SSIP). Sodium, potassium, rubidium and cesium 9,10-dihydroanthracenides, 9-methyl-9,10-dihydroanthracene-10-ides and 9-cyano-9,10-dihydroanthracenides exist as contact ion pairs (CIP) in solution. A model, taking into account the geometry and charge distribution, for the transition of CIP of alkali metal salts of 9,10-dihydroanthracene and its derivatives into SSIP is proposed [283].

Reduction of cyclooctatetraenes by electrolysis or with alkali metals yields essentially planar radical anions [284], and dianions. X-ray crystallographic structure determination of the potassium salt of 1,3,5,7-tetramethylcyclooctatetraene dianion in diglyme revealed the eight membered ring in a planar conformation with an average C–C bond length of 1.407 Å [285], further substantiated by $^1$H NMR results [286]. In the electroreduction of dibenzo[a,e]cyclooctene [287], the first reduction step (corresponding to the conversion of the tub-shaped parent to planar radical anion) was found to be slower than the second step leading to the planar dianion. The $^1$H NMR spectra of the dianion in solution showed the protons at the cyclooctatetraene ring, shifted to lower filed than the corresponding proton peaks in the neutral compound [288], behavior characteristic of an induced ring current in the cyclooctatetraene ring, as occurs in aromatic molecules and suggests strong electron delocalization over the ring. EPR and thermodynamic studies for the reduction of benzocyclooctene and cycloocta[b]naphthalene [289], with potassium in hexamethylphosphoric triamide (HMPA), also point to the planarity of the corresponding radical anions and dianions. A set of structurally different radicals has been detected by EPR in the reduction of cycloocta[b]naphthalene **98** by potassium in 1,2-dimethoxyethane [290]. In the first step, an apparently symmetry-forbidden ring closure lead to benzo[b]biphenylene **99**$^{\bullet-}$, subsequent reduction lead to 2-phenyl-naphthalene radical anion **100**$^{\bullet-}$. Finally, the EPR spectrum of the radical anion of a compound tentatively proposed to be dipotassium salt of dibenzoheptafulvene was observed.

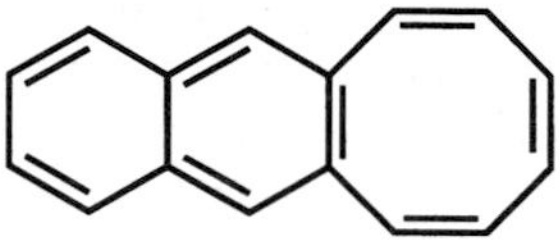

**98** 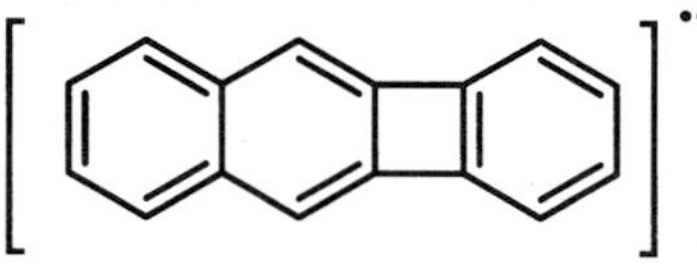                                    **99**$^{\cdot-}$

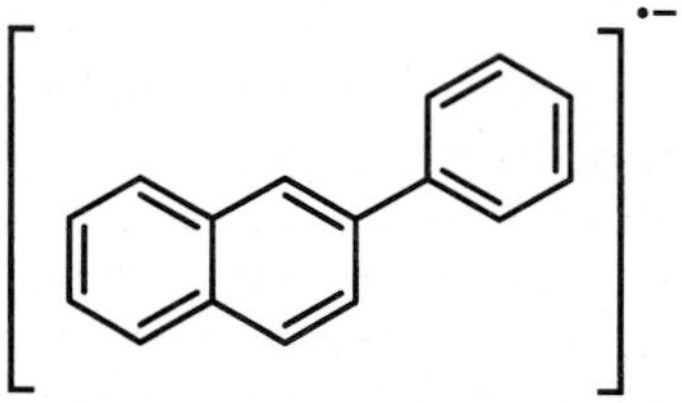

**100**$^{\cdot-}$

Several EPR investigations of the radical anion of double-layered [2.2]para-cyclophane for the study of electron transfer between the two aromatic rings and the effects of the counter ions on the spin distribution is reported [291]. In a thorough EPR study of conjugated quinone and imide radical anions, Miller et al. [292] generated electrochemically the alkyl substituted 1,4,8,11-pentacenetetrone radical anions **101**$^{\cdot-}$ in dimethylformamide and dichloromethane solutions containing 0.1 M tetrabutylammonium tetrafluoroborate. Results indicate that at room temperature the unpaired electron is equally displaced over both quinone units (electrophores) however, EPR spectra at lower temperature clearly indicated a localization of the unpaired electron on one quinone unit of the planar conjugated pentacene skeleton. As a result of this and other findings the question of whether these radical anions might be thought of as mixed-valence species was considered.

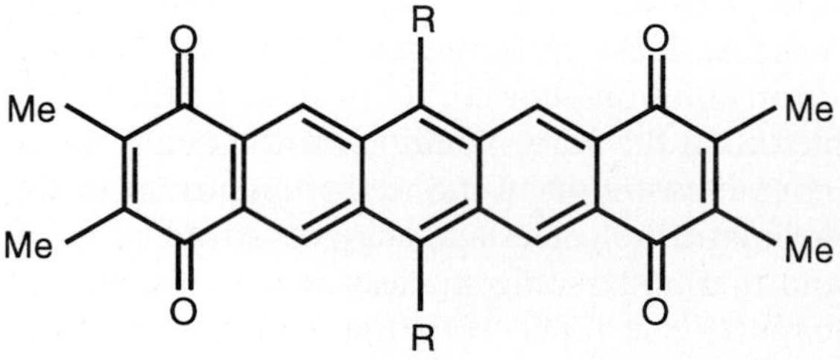

**101**

R = H, hexyl

The CV of the paracyclophanes containing one quinone unit showed reversible one-electron reduction step similar to the reference compound 2,5-dimethyl-1,4-

benzoquinone. The higher reduction potential of the paracyclophane compared to the parent quinone indicates that the linked 1,4-phenylene moiety acts as an additional electrophore [293]. Three types of tetrone radical anions in which two 1,4-benzoquinone units are connected by ethane, [2.2]paracyclophane, and anthracene bridges are studied by EPR and ENDOR spectroscopy. The displacement of the unpaired electron over the two $\pi$ moieties in the [2.2]cyclophane radical anions and a marked differences between the first and the second reduction potentials, are evidence for substantial intramolecular electronic interactions between the two electrophores. The existence of several types of ion pairs obtained by reduction of benzophenone and its derivatives by different methods has been evidenced by EPR [294]. In 4,4'-dinitrobenzophenone (DNBF) two types of anion radicals, characterized by different electron distributions have been observed [295]. A radical anion with delocalized odd electron is observed when the parent neutral molecule is reduced by the electrochemical method and by chemical methods with sodium methoxide in DMSO, on the other hand radical anion with localized electron ($p$-nitrophenyl fragment) is observed when the concentration of sodium methoxide is larger than $10^{-1}$ M in the same solvent and when the alkali metals in ethereal solution is employed. On the other hand, a mixture of electronically symmetric and asymmetric ion pairs is obtained by adding dimethylsulfoxide (40 %) to a solution of the ion-pair DNBF$^{\cdot+}$,K$^+$ in 1,2-dimethoxyethane and the superposition of the corresponding EPR spectra is observed. Radical anions of benzophenone, 4,4'-dinitrobenzophenone and 4-nitrobenzophenone generated by reduction with alkali metals (Li, Na, K) showed hindered rotation of the two rings on the EPR time-scale in the temperature range explored [296]. A four-site exchange model is employed to obtain the best fit of the experimental spectra for systems with symmetric spin distribution while a two-site exchange model is used for systems with spin-distribution localized only on one $p$-nitrophenyl fragment. Electron self-exchange rate constants of the nitrobenzene–nitrobenzene radical anion couple and their temperature dependence is measured by EPR-line broadening effects. The first order rate constants of these diffusionless reactions vary between $2.8 \times 10^8$ and $9.7 \times 10^8$ s$^{-1}$ within a temperature range of 296–353 K [297]. Electron self-exchange reactions are good examples for the application of Marcus theory, since for this type of reaction the driving force $\Delta G^\circ = 0$ and activation energy in the sense of Marcus reduces to $\lambda/4$ ($\lambda$ is the total reorganization energy). Different redox couples with radical cations [298] and radical anions [299] are known.

Radical anions of haloaromatic compounds are proposed to be intermediates in different type of reactions. Their fragmentation rates, determined electrochemically [300] or by pulse radiolysis [301] range from $10^{10}$ s$^{-1}$ for phenyl halides to $10^{-2}$ s$^{-1}$ for some halonitrobenzenes. The rate of the reaction for some aryl halide radical anions is too high to be measured electrochemically, the fragmentation of more stable radical anions such as those of 1-bromo- and 1-iodoanthraquinone [302], $p$- [303] and $m$-bromo- [304] and $p$- [303] and $m$-chloronitrobenzenes [304] occurs at considerably lower rates and the reaction is favored from their photoexcited state. Aryl halide radical anions may present $\sigma$–$\pi$ 'orbital isomerism' depending on the orbital symmetry of their singly occupied molecular orbital [305], a proposal derived from theoretical and experimental evidences [306]. The isomerism is possible when the excitation of the unpaired electron from the singly occupied to a low lying unoccupied MO of different symmetry, give rise to a species that differ in geometry from the initial state, both being different local minima on the ground state surface [307]. The $\pi-\sigma$ isomerization of these radical anions is interpreted in terms of an intramolecular electron transfer (intra-ET) from the $\pi$-system, corresponding to the generally more stable and initially formed radical anions, to the $\sigma^*$ C–halogen bond. From the two possible radical anions, it is proposed that only the $\sigma$ dissociates into an aromatic radical and the anion of the leaving group [306a]. A relatively limited number of theoretical studies are known for haloaromatic radical anions [308], recently, a theoretical inspection of the potential surface of the radical anions of halobenzenes, halobenzonitriles and o-, m-, and p-haloacetophenones is reported [309]. It is proposed that the difference in the energy between both radical anions could be correlated with their experimental fragmentation rates, considering the intra-ET from the $\pi$ to the $\sigma$ system as the limiting step of the cleavage reaction. An AM1 study of the radical anions of o-, m-, and p-halonitrobenzenes in relation to their $\sigma-\pi$ orbital isomerism and energy of their interconversion is carried out by Pierini et al. [310]. Based on the calculated energy of their interconversion, the intramolecular thermal electron transfer from the $\pi$-system to the $\sigma^*_{C-X}$ bond involved in the fragmentation of the intermediates into an aromatic radical and the anion of the leaving group occurs with considerable energy for the p-, m-, and o-chloronitrobenzenes (**102a–c**) and the p- and m-bromo (**103a,b**) derivatives. On the other hand, the intramolecular thermal electron transfer is favored for the p-, m-, o-iodo (**104a–c**) and *ortho* bromo (**103c**) derivatives.

| **102a** X = Cl | **102b** X = Cl | **102c** X = Cl |
| **103a** X = Br | **103b** X = Br | **103c** X = Br |
| **104a** X = I | **104b** X = I | **104c** X = I |

The loss of the halide anion from p-chloronitrobenzene and p-bromonitrobenzene is accomplished by dual photochemical and electrochemical activation [303], the mechanism being absorption of light by electrogenerated anion radicals [X-$C_6H_4$-$NO_2$]$^{\cdot-}$ (X = Br, Cl) followed by halide loss and, ultimately, the formation of nitrobenzene radical via photo ECE mechanism [311]. In contrast the dual activation of p-cyanonitrobenzene, p-dinitrobenzene, phenyl 4-nitrophenyl sulfone, methyl 4-nitrophenyl sulfone, phenyl 4-nitrophenyl sulfoxide or methyl 4-nitrophenyl sulfoxide failed to induce the leaving of any substituent [304]. It is presumed that in cases where no fragmentation occurs, the excitation energy of the radical anion is lost through collisional deactivation by the solvent. An interesting controversy has appeared concerning (in part) the mechanism of nucleophilic aromatic substitution reactions of aromatic compounds, as dinitrobenzenes, nitrobenzophenones, nitro-benzonitriles, polyfluoronitrobenzenes etc., able to stabilize radical anions [312]. Abe and Ikegami [313] observed radical anions in reaction of dinitrobenzenes with OH$^-$ in aqueous DMSO that led to the postulation of radical anions as inter-mediates. Sammes et al. [314] studied the displacement of nitro groups from *p*-dinitrobenzene and other nitro compounds by various phenoxide in DMSO and concluded the radical nature of these reactions. Shein et al. [315] demonstrated the formation of *p*-dinitrobenzene radical anion on reaction with a series of anion nucleophiles, among them phenoxide ion. Electrochemical studies by Gallardo et al. [316] indicate that the reactions of *p*-dinitrobenzene and *p*-nitrobenzonitrile with phenolate or phenol in DMF show radical features and cannot be attributed to the direct reaction of the nucleophile on the substrate radical anion. Reductive activation is feasible in reaction of *p*-dinitrobenzene with phenol, however, substrate radical anion formation is not responsible for it. In radicals derived from 2,4-dinitrophenol, the localization of the spin density is found to change from the *ortho* nitro group to the *para* nitro group on charging from the radical anion to radical dianion [317]. A significant change in electronic structure of radical anion may occur in transition from aprotic to protic media. In radical anions of 1,3-dinitrobenzenes [318] and 2,7-dinitronaphthalene [319] in protic solvents, the strong solvation of one nitro group through hydrogen bonding causes an asymmetric distribution of the spin densities on the nitrogen atoms, inducing slow intramolecular electron exchange between the two nitro groups. The rate constants determined from line broadening effects are in the fast region range ($K \approx 10^9$ s$^{-1}$) in aprotic solvents while in alcohols the exchange is slow ($K \approx 10^6$ s$^{-1}$) on the EPR time scale [318, 319]. The position of the OH group is a decisive factor in the variation of the rate constant with the deprotonation of the radical anion, variation being large when the OH group is in the *ortho* position to one nitro group. The EPR spectra of radicals derived from 4,6-dinitrobenzene-1,3-diol showed no line-broadening effects in the experimental range of temperatures [320]. In the radical-anion and radical-dianion, the electron is localized mainly in one nitro group, while in the radical trianion the spin density is evenly distributed over the two nitro groups. $\Delta G^*$ values range from 12.6 kJ mol$^{-1}$ for radical dianion of 2,6-dinitrophenol to 46.0 kJ mol$^{-1}$ for radical dianion of 3,5-dinitrophenol.

Stilbenes are an important class of compounds with a broad range of applications in basic and applied research [321]. The isoelectronic (*E*)-stilbene **105** and (*E*)-azobenzene **106** belong to the basic organic compounds of which radical anions were first investigated by EPR 30 years ago [322], and since then repeatedly studied by EPR and ENDOR techniques [323]. Radical anions of **107** and **108** are not persistent because they rapidly isomerize to **105** and **106** respectively. Nevertheless, **107**$^{•-}$ could be characterized by hyperfine data under specific conditions [324], whereas its azo counterpart **108**$^{•-}$ has hitherto escaped detection by EPR.

Sterically congested stilbenes such as 2,2,5,5-tetramethyl-3,4-diphenylhex-3-ene has received growing attention [325]. The preparation and characteristic properties of the radical anion and dianion is investigated by EPR and the spectra analyzed using DFT techniques [326]. In radical anion the steric effect (steric repulsion of *tert*-butyl groups) and electronic effects just about balance as rotation occurs about

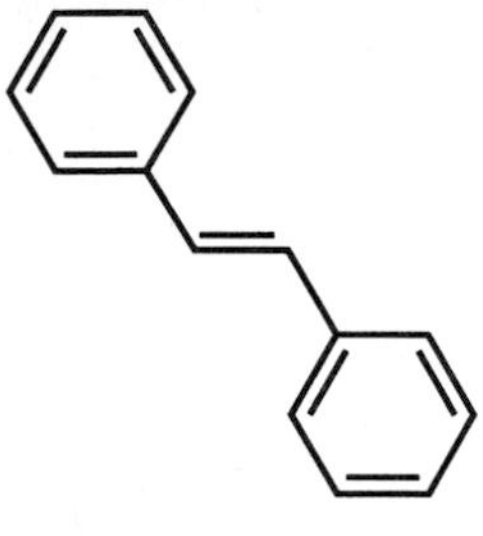

**105** 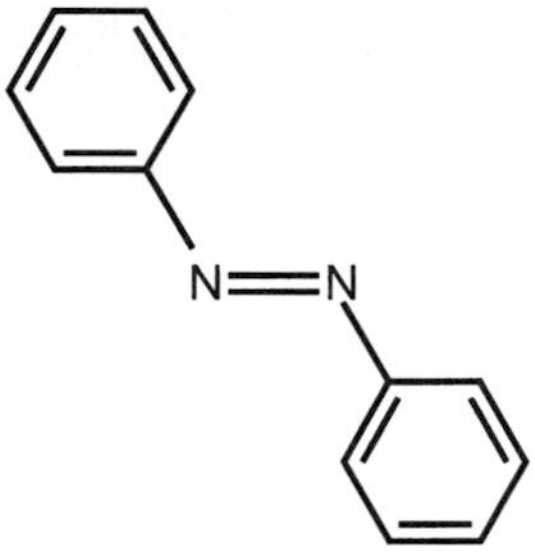         **106** 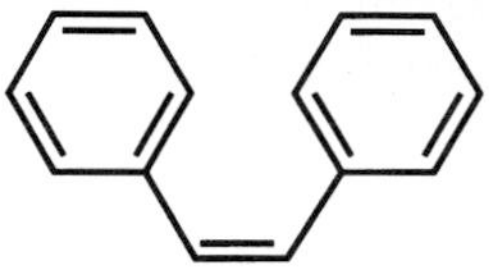         **107**

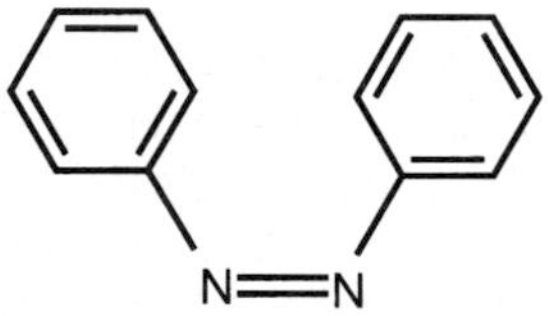

**108**

the central double bond, on the other hand, on addition of a second electron, no residual electronic stabilization exists and the most stable conformation is dictated by the steric effect to be the 90° twisted form. In 1,2-diphenylcycloalkene **109**, when the alkene is propene ($n = 1$) [327], butene ($n = 2$) [328], pentene ($n = 3$) [328b], or hexene ($n = 4$) [329], the (Z) configuration of $n = 3$ is fixed by incorporation of the central C=C bond into the cycloalkene ring.

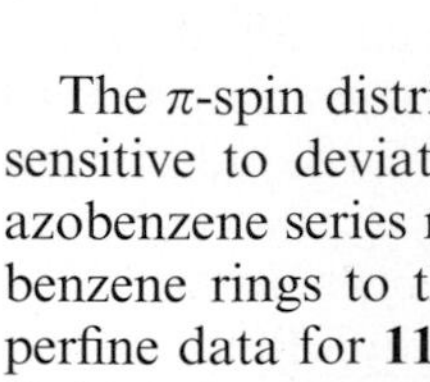

**109**

The $\pi$-spin distribution in the radical anions of stilbene series is only moderately sensitive to deviations of the $\pi$-system from planarity, the radical anions of the azobenzene series respond to steric strain by shifting the $\pi$-spin population from the benzene rings to the azo group. This is impressively demonstrated by similar hyperfine data for **110˙⁻** and **111˙⁻** which contrast with the strongly differing one for their azo counterparts **112˙⁻** and **113˙⁻**, as well as by the corresponding values for sterically highly hindered **114˙⁻** [330].

The redox properties of free radicals are essential parameters for predicting whether electron transfer to molecules may trigger radical or ionic chemistry [331]. If these redox properties can be converted into standard potentials, that may then

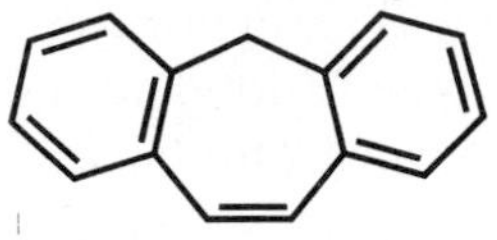

**110**

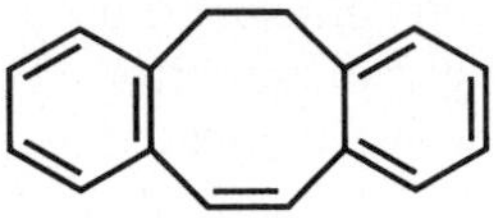

**111**

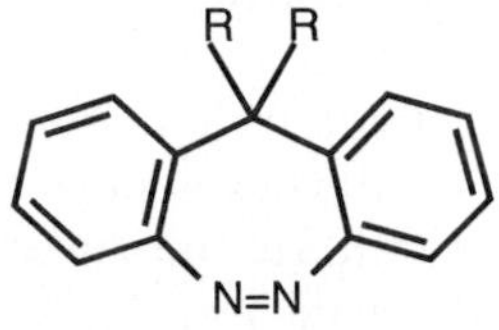

**112**

R = H, D

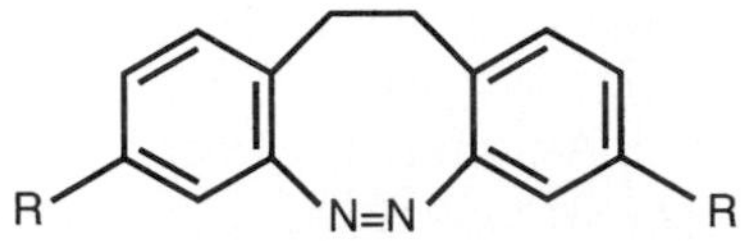

**113**

R = H, D

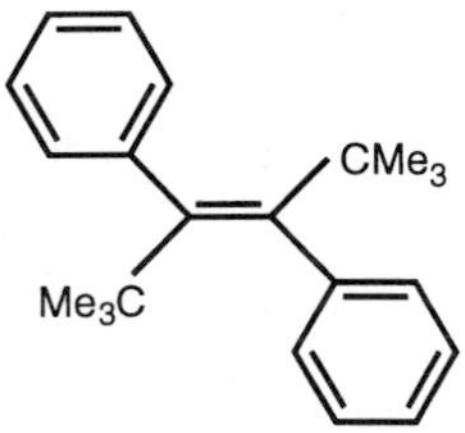

**114**

be used to estimate other thermodynamic parameters, such as p$K_a$ and bond-dissociation free energies [332]. Several direct or indirect electrochemical methods are used for investigating reduction characteristics of free radicals, the simplest ones involve the recording of the oxidation wave of the anion in steady-state or CV [333]. An indirect electrochemical method, based on redox catalysis [334] has also been applied to determine reduction potentials of transient radicals [335]. Two photo-electrochemical methods have been proposed for investigating the reduction characteristics of transient free radicals [336]. One may also produce the radicals by continuous irradiation and use fast electrochemical techniques, such as normal and reverse pulse voltammetry at an ultramicro electrode to obtain the radical standard potential, at least in favorable cases, such as the oxidation of the diphenylmethyl radical and the reduction of the diphenyl cyanomethyl radical [337]. However, one limitation of this method is the possible superposition of large photocurrents arising from the photo injection of electrons from the electrode into the solution. Theoretical expressions are known that relate half-wave potential and shape of the radical polarograms to the thermodynamics and kinetics of various reactions in which the radical and the anion resulting from its reduction may be engaged. The 9-anthrylmethyl, diphenylmethyl, benzyl, and 4-methylbenzyl radicals were generated by reduction of the corresponding chlorides by electron photoinjection in laser pulse experiments. The variation of the half-wave potential with time is used as a source of mechanism and reactivity information, and the shape of the polarograms was used as an additional diagnostic criterion. The results are compared with data obtained by other techniques and the reactivity parameters are discussed with the help of density functional quantum chemical calculations [338].

Organic high-spin molecules are basic materials for the design of organic ferromagnets [339]. Unfortunately, most of the known stable high-spin molecules are not suitable for obtaining organic based ferromagnetic materials due to antiferromagnetic intermolecular interactions which compensate the intramolecular spin alignment in the bulk material. A promising category of compounds for the development of macroscopic high-spin systems is the class of aromatic ketone radical anions of type **115**, since both intra- and intermolecular high-spin coupling seems possible.

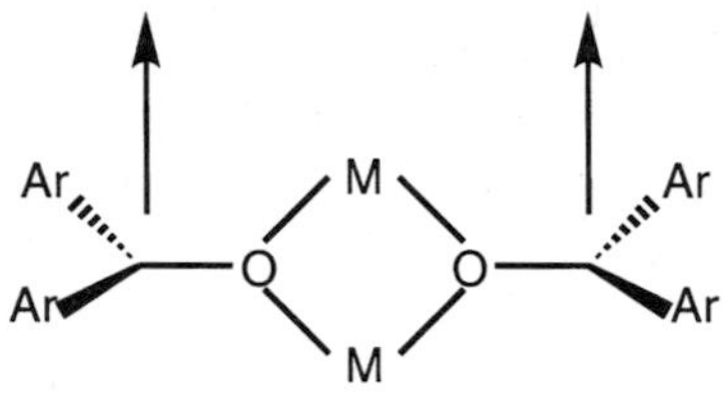

**115**

Hou et al. [340] confirmed earlier conclusions arrived at by Hirota and Weissman [341], that alkali metal aromatic ketone radical anions form strongly coupled biradicals in which the metal acts as a spin-carrying center. The combination of fluorenone and benzophenone entities within a molecule seems attractive in this regard, because of their well known tendency to form stable radical anions. The mono- and poly-anions of some mono-, di-, and tetraketones containing fluorenone and benzophenone moieties have been studied by NMR and CV [342]. Alkali metal anion radicals of fluorenone generated by electron transfer from an alkali metal naphthalene radical anion exhibit markedly lower molar paramagnetic solvent shifts than those generated by direct reduction with an alkali metal. NMR data combined with those obtained by CV, indicate that polyketones possessing fluorenone moieties connected through isophthaloyl 'spacers' are promising system for the preparation of high-spin organics and electrophores.

Electron transfer to endoperoxides and peroxides results in the cleavage of the weak oxygen–oxygen bond [343]. This process proceeds in peroxides and endoperoxides, generally by dissociative ET mechanism, in which ET and O–O bond fragmentation are concerted. However, in case of *tert*-butyl-*p*-cyanoperbenzoate there is evidence for transition to a stepwise mechanism with formation of the intermediate radical anion [343c]. Such mechanistic studies are important because ET processes of peroxides and endoperoxides play a key role in their activity in chemical and biological systems [344], for example, $Fe^{III}$-promoted electron transfer reduction of the O–O bond in the antimalarial endoperoxide artemisinin, its semi-synthetic derivatives has recently been shown to be the key step in its antimalarial activity [345]. A first report of the observed reactivity from the alkoxy radical in the ·O–R–R–O⁻ distonic radical anion formed in a heterogeneous dissociative ET, namely 1,2-phenyl migration (*o*-neophyl rearrangement), has appeared recently [346]. This *o*-neophyl-type rearrangement of a phenyl group to the alkoxy radical fragment occurs in the distonic radical anion formed by electrochemical single ET to the O–O bond in the endoperoxide 9,10-diphenyl-9,10-epidioxyanthracene. This rearrangement occurs at the expense of the reduction of the alkoxy radical portion of the distonic radical anion. Systems such as these can be developed as kinetic probes of the rate of second heterogeneous ET. Kinetics of both intra and intermolecular electron transfer reactions are of much interest and importance, as these are the fundamental chemical steps in numerous biological processes. In particular, the understanding of intramolecular ET has evolved to a high degree [347]. Most of the studies of intramolecular ET focus on photoinduced charge separation, involving different donor and acceptor moieties to achieve a suitable driving force $\Delta G°$ [348]. Nelsen et al. [349] studied the self-exchange within various hydrazine derivatives by optical and EPR methods. The kinetics of intramolecular electron exchange (IEE) in the radical anions of 1,3-dinitrobenzene (1,3-DNB) is studied by EPR spectroscopy. The rate constants, K, determined are ca $10^6$ s$^{-1}$ in alcohols and $10^9$ s$^{-1}$ in aprotic solvents and the reaction was found to be adiabatic and uniform in the sense of classic transition state theory [350]. It is usually believed that the absence of direct resonance between the donor and the acceptor is necessary for the occurrence of alternating line-broadening effects caused by IEE in the EPR spectra of free radicals. The examples known in the literature, where the two groups are either in non-conjugative *meta* positions, as in the radical anions of 1,3-dinitrobenzene and 2,7-dinitronaphthalene [351], or sterically forced out of planarity, as in the radical anion of dinitrodurene [352], seems to confirm this behavior. Line broadening effects due to intramolecular electron exchange are recently observed in the EPR spectra of 2,6-dinitrophenolate radical dianion, where the resonance structures can be drawn delocalizing the electron through the two nitro groups [353]. The rate constants of the intramolecular electron exchange reaction in the 1,4-dinitrobenzene radical anion (1,4-DNB) in linear alcohols are determined from the alternating line-broadening effects in EPR spectra [354]. Results demonstrate the role of solvent dynamics on the kinetics of intramolecular electron exchange in 1,4-DNB radical anion. It obeys a diffusive adiabatic regime, while the same reaction in the isomeric 1,3-DNB radical anion in alcohol follows a uniform adiabatic reaction behavior. This difference reflects a faster reaction in the 1,4-DNB radical, the ET occurs on a time scale where solvent-relaxation processes controls the reaction rate. On the other hand in 1,3-DNB radical, solvent relaxation being much faster than the ET step, does not effect the reaction dynamics, 1,4-DNB radical anion in aprotic solvents do not show evidence of alternating line broadening effects [355]. In aprotic medium, the fraction of transferred charge is substantially lower than in alcohols, due to a lower specific solvation. Since, in most aprotic solvents, the values of $1/\Gamma$ (where $\Gamma$ is the Longitudinal correlation time) are higher than in alcohols, the reaction is too fast to exert discernible alternating line-broadening effects on the EPR spectra.

Despite the simmering debate within the community of physical organic chemists over the scope and importance of electron transfer in nucleophilic aliphatic substitution [356], evidence continues to accumulate that electron transfer may play a role in the mechanism for this substitution [357], particularly when the electrophiles are sterically hindered alkyl iodides. Although $S_N2$ mechanisms are known to exhibit significant steric constraints, of which Walden inversion is the most profound, evidence exists that electron transfer processes may have steric components as well [357c]. For methyl iodide and 9-phenylfluorenyl anion, backside approach is sterically feasible; whereas with the same nucleophile and neopentyl iodide, it is not, and photo activation is required. Conversely, for the 9-mesitylfluorenyl anion, backside approach is possible but not to within an $S_N2$ bonding distance. Spectral evidence for single electron transfer in Nucleophilic aliphatic substitution is recently observed by Tolbert et al. [358]. The photochemical generation of closed-shell ions is a topic of current interest in organic photochemistry [359]. Carbanions are one of the most common reactive intermediates in organic synthesis [360], and hence the photogeneration of such ionic species contributes to progress in mechanistic organic chemistry. Only a few examples are known for the photochemical formation of carbanions [361], most of which are based on the photodissociation of carboxylic acids. A related reaction is the anodic oxidation of carboxylate ions ($RCO_2^-$), i.e. the Kolbe reaction [362], in which carbon centered radicals ($R^\bullet$) are formed via decarboxylation of carboxyl radicals ($RCO_2^\bullet$). These carbon radicals undergo dimerization or further oxidation, yielding R–R or $R^+$, respectively. Photoinduced electron transfer reaction of carboxylic acids also gives carbon-centered radicals via decarboxylation [363]. In this reactions, carbon radicals are formed together with sensitizer radical anion, and hence the formation of carbanion might be expected if an effective electron transfer between them takes place. However, the reported photoinduced electron transfer reactions yielded a mixture of radical–sensitizer adducts and other radical coupling products [364]. Direct photolysis of carboxylate anions yielded carbanions [365] but their spectroscopic observation has been achieved only for carbanions stabilized by electron-withdrawing substituents e.g. *p*-nitrobenzyl anion [366] or by resonance with a carbonyl group [367]. In order to develop a novel method for the photochemical generation of carbanions, photoinduced electron transfer reactions of carboxylate salts is recently studied by Sawaki et al. [368]. They succeeded in generating carbanion from carboxylate ions by photoinduced electron transfer. This reaction is potentially useful in the conversion of arylacetic acids ($RCO_2H$) to RH and also as a method to generate carbanions photochemically. The key process in the formation of carbanion is the electron transfer from sensitizer anion radical to radical intermediates proceeding in-cage and depending on the free energy changes. The influence of the solute–solvent interaction through hydrogen bonding was reported recently by Chan and Chan [369] and Tominaga et al. [370]. They reported that the hydrogen bonding between the OH or $NH_2$ group of a solute molecule and polar solvents makes the diffusion process very slow. Radicals are interesting systems for diffusion studies as well as for elucidating the mechanism [371] and dynamics [372] of chemical reactions. Unfortunately, until recently, only a few diffusion constants ($D$) of radicals have been reported [373], because of technical difficulties. However, Terazima et al. [374] succeeded in measuring the $D$ values of many intermediate radicals which appear during photochemical reactions by the transient grating (TG) method. Results indicate that transient neutral radicals created by photo induced hydrogen abstraction of quinones and *N*-heteroaromatic molecules diffuse much slower than their parent molecules. The same group also investigated the viscosity ($\eta$) dependence [374a], the solute radius ($r$) dependence [374c] and the temperature ($T$) dependence [374d] of the $D$ values of such radicals. The difference in $D$ between the radicals and the parent molecules became larger with increasing $\eta$, $1/r$, or $1/T$, tendency similar to those of ions [375]. Slow diffusion of radicals was observed not only in polar solvents but also in non-polar solvents and in aprotic solvents [374b], and hence hydrogen bonding between the OH or NH group of the radical and the solvents cannot be the origin of the slow diffusion. Therefore, contrary to ions or hydrogen bonding systems, the origin of the slow diffusion remains unclear. The mobilities of the photochemically produced intermediate radical cations and anions probed by the time of flight (TOF) technique is reported by Houser and Jarnagin [376], Freeman and coworkers [377], and Albrecht and coworkers [378]. Study revealed that $D$ of the ion radicals are smaller than those of neutral molecules of similar shapes and sizes. Freeman and coworkers attributed the origin of the slow diffusion to the electrostrictive drag by the charged species and dimerization for some compounds. Albrecht and coworkers found that $D$ of the charged radicals can be well reproduced by the Stokes–Einstein (SE) equation. The result is in good agreement with the SE relation and is similar to that found for neutral radicals. However, if one wants to extract the effect of the charge or the unpaired electron by comparison of $D$ of the charged radicals determined by this method with those of closed shell molecules, one should use $D$ of closed shell molecules measured by other methods under different conditions. Since $D$ is very sensitive to environmental and experimental conditions, it makes accurate comparisons very difficult. With the use of the TG method, $D$ values of stable molecules can be measured simultaneously with those of the transient species. For example, Terazima et al. determined $D$ of a cation radical and its parent molecule, $N,N,N',N'$-tetramethyl-$p$-phenylenediamine (TMPD), by the TG method under exactly the same conditions [379]. The results showed that the cation radicals diffuse only half as quickly as the parent molecule in ethanol, the contribution of the charge and the unpaired electron could not be separated from this measurement. The translational diffusion constants of the ketyl anion radicals, the neutral radicals, and the parent stable molecules is successfully measured under same conditions by the transient grating method [380]. Both the neutral and anion radicals diffuse slower than the parent molecules. The values are compared in detail in wide range of solvent viscosities, solute sizes and temperatures.

Reverse electron transfer (RET) process in photo induced ET is typically viewed as undesirable energy-wasting steps. Indeed, a considerable amount of work has been directed towards strategies to increase the efficiencies of photoinduced electron transfers by decreasing the rate of RET, or increasing the rates of useful competing processes such as ion-pair separation or follow-up ion radical interactions [381]. Recently an alternative approach enabling RET to give useful products, namely dissociative reverse electron transfer (DRET), is much under investigation. For efficient DRET, a triplet ion radical pair would be more useful than a singlet pair, several examples of reactions via triplet ion radical pairs are described in literature [382]. On the basis of CIDNP experiments, Wong and Arnold [383] proposed a mechanism involving ring opening of 1,2-diphenylcyclopropane (DPC) radical cation to a 1,3-cation radical, followed by RET to a triplet 1,3-biradical, with subsequent intersystem crossing and ring closure. On the basis of additional CIDNP experiments, Schilling and Roth showed that configurationally stable *cis* and *trans*-DPC cation radicals are involved in these reactions and that the isomerization involves RET in the triplet ion radical pair leading to formation of the triplet biradical [384]. However, biradical formation could occur by RET leading to a locally excited triplet DPC, which subsequently undergoes bond breaking to form the triplet biradical or alternatively, the biradical could be the direct product of RET. These two fundamentally different mechanisms cannot be distinguished by CIDNP. Recently Farid et al. provided direct identification of the intermediates involved in the electron transfer photosensitized isomerization of DPC and demonstrated that the isomerization occurs by DRET with several triplet sensitizers and bond fragmentation can occur with 100 % quantum efficiency [385].

In bimolecular ET between freely diffusing donors and acceptors in solution, the nuclear prearrangement of the reactants in the transition state with its critical donor/acceptor distance and orbital overlap-limits the intrinsic rate of the electron exchange [386]. As a result all theoretical calculations of ET rate constants invoke far-reaching assumptions on the relative orientation and electronic interaction of the donor and acceptor in the transition state [387]. Owning to intrinsic life time of transition states, their direct (spectroscopic) observation constitutes an experimental challenge [388]. However, attempts have been made to predict structures and degrees of donor/acceptor bonding in various ET transition states by different theoretical methods [389]. Electronic coupling that promotes electron transfer between redox partners is revealed experimentally by charge-transfer interactions extent in the donor/acceptor precursor or encounter complex prior to ET, and the degree of charge transfer as defined by Mulliken theory can be taken as a measure of the donor/acceptor bonding [390]. For example, electron transfer from arene donors to photoactivated quinones occurs via encounter complexes with substantial charge transfer bonding, postulated by the observation of near-IR absorption bands and relatively high formation constants [391]. Steric effects on the kinetics of ET from hindered and unhindered arene donors to quinone acceptors and their temperature, solvent, and driving-force dependence reveal a structure-induced (mechanistic) changeover [392]. Unhindered donors undergo inner-sphere electron transfer owning to the strong electronic coupling of donor and acceptor in a well defined encounter complex preceding the ET transition state. On the other hand, hindered donors show no (kinetic or spectroscopic) evidence for a discrete encounter complex in the preequilibrium step, and the kinetics follows outer-sphere ET behavior expected for weakly coupled donors and acceptors. Although the comparative study of hindered and unhindered electron donors establishes a clear cut (experimental) distinction between outer-sphere and inner-sphere electron transfers, we believe that there will generally be a broad borderline region between the two mechanisms.

## 4.5.2 Alkenes

The radical anions of buta-1,3-diene and its methyl substituted derivatives are studied electrolytically in liquid ammonia [393] or in tetrahydrofuran [394]. In general, reaction of these compounds with an alkali metal in ethereal solvents fails to yield a persistent radical anion, because of rapid polymerization to rubber-like products [395]. However, it has been shown recently [396] that these processes can be avoided by bulky substituents. Radical anions of various aliphatic $\alpha$-substituted nitro compounds have been proposed as reactive intermediates in radical–radical–anion chain-substitution reaction. The mechanism was termed $S_{RN}{}^1$ (substitution radical-nucleophilic, first order) by Bunnett [397] in his studies on aromatic substrates, while the same mechanism for aliphatic nitro compounds was elaborated by Russell [398] and Kornblum [399]. A growing number of anions have been shown to participate in $S_{RN}{}^1$ reactions with $\alpha$-substituted nitro compounds e.g. $R_2CNO_2{}^-$, $RSO_2{}^-$, $RS^-$, $N_3{}^-$, $SO_2{}^{2-}$ [400], $(RO)_2PO^-$ [401], and carbanions. Other mechanisms including these radical–anion intermediates have also been reported e.g. $S_{ET}{}^2$ (substitution, electron transfer, second order) [402], and reduction by dihydrobenzylnicotinamide [403], trialkyltin hydride [404], or methane thiolate [405]. All these mechanisms have an initial step of electron capture by the $\alpha$-substituted nitro compounds to yield an intermediate radical–anion, with sufficient life time to allow reaction with other species. From the viewpoint of EPR, electron capture by $RNO_2$ (R = alkyl, aryl) molecules to give $RNO_2{}^{\cdot-}$ radical anion is well established, both in solid state [406] and liquid phase [407] studies. A range of radical–anions, $[Me_2C(X)NO_2]^{\cdot-}$ with X = Br, Cl, SCN, $NO_2$, CN, $PO_3Et_2$, $CO_2Et$, COMe, $SO_2Me$, $SO_2Ar$, and Me have been identified by EPR and are found to be long lived at low temperatures [408]. Results indicate that one or more of at least three pathways are followed on the reaction of $Me_2C(X)NO_2$ with electrons; (a) electron-capture to yield a stable radical anion, $[Me_2C(X)NO_2]^{\cdot-}$, (b) dissociative electron capture to yield $Me_2C^{\cdot}NO_2$ and $X^-$ (for X = Br, Cl, SCN), (c) dissociative electron-capture to yield $Me_2C^{\cdot}X$ and $NO_2{}^-$ (for X = CN, $NO_2$, $PO_3Et_2$ and $CO_2Et$).

It has been proposed that the reactions which occur remote from and uninfluenced by the charge center (charge-remote reactions) can occur following collisional activation of even electron organic anions in the gas phase [409]. Charge remote loss of a radical from an $(M–H)^-$ ion is sometimes observed when the product formed is a stable radical anion [410], however, evidence in favor of the loss of (even electron) neutrals commonly occurring by charge-remote processes from even electron anions is not strong [411]. Acetylene, with its linear structure in the ground state is expected to have a *trans*-bent structure in its electronically excited state [412]. Theoretical calculations have predicted that the acetylene radical anion has a *trans*-bent structure, but the energy difference between the *trans* and *cis* is as small as 7.4 kcal mol$^{-1}$ [413]. Muto and his collaborators first reported the acetylene radical anion trapped in a 3-methylpentane matrix at 77 K. Consistent with the theoretical predictions, the *trans*-bent structure was concluded based on the EPR data [414]. On the other hand Manceron and Andrews reported an IR study on an Li–acetylene complex anion radical generated in an argon matrix and found that the acetylene moiety has a *cis*-bent structure [415]. Kasai observed the EPR spectrum of the same complex anion and confirmed the *cis*-bent structure [416]. An EPR and MO study by Itagaki et al. [417] to elucidate the electronic structure of methylacetylene radical anion generated in glassy 2-methyltetrahydrofuran matrix by ionization radiation at 77 K. The spectrum was dominated by a large and slightly anisotropic $^1H$ hyperfine splitting of ca 4.53 mT due to one ethynyl proton.

With the help of a selectively deuterated methylacetylene, the anisotropic hyperfine couplings are determined. Comparison of experimental values with the theoretical ones calculated by ab initio MO and INDO methods, a *trans*-bent structure is concluded. The formation of methylacetylene radical anion is also confirmed by an electron absorption spectroscopic study.

### 4.5.3 Heteronuclear Radical Anions

Polycyclic arenes, e.g. perylene, have been widely studied in the preparation of molecular conductors, some of the radical cations show semiconducting or metallic behavior [418]. Introduction of one or more sulfur atoms at the periphery of such systems, i.e. thia arene derivatives, generally imparts greater stability to the radical-cation salts, coupled with increase conductivity [419]. For compound **116**, X-ray structure studied have been reported on the pure donor and some radical-ion salts [420].

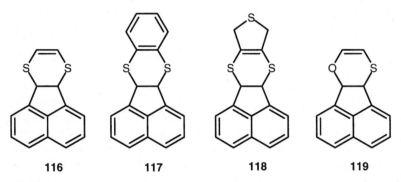

|      116      |      117      |      118      |      119      |

Radical cations and anions of compounds **116–119** are characterized by EPR and ENDOR techniques [421]. The *g*-factors of the radical cations comply with the values for oxidized S-donors, whereas those of the corresponding anions are typically of reduced π-systems without heteroatoms. The new derivative 8,9-bis(methylsulfanyl)-acenaphtho[1,2-*b*][1,4]dithiine has emerged as a promising electron donor for the formation of crystalline charge-transfer complexes and radical ion salts, X-ray crystal structures have been obtained for its 1:1 complexes with TCNQ and $Br_2TCNQ$. In his extensive study of the UV spectra of organic thiocarbonyl compounds, Janssen [422] included some sodium dithiocarbamates based on primary and secondary aliphatic amines. The spectra in ethanol showed one weak band near 350 nm and stronger bands at 288 and 253 nm. The bands were interpreted with aid of calculations of Hückel type as an $n \rightarrow \pi^*$ and two $\pi \rightarrow \pi^*$ transitions. Multiple $n \rightarrow \pi^*$ transitions have been observed in 1,2- and 1,3-dithiones, and up to four bands assignable to $n \rightarrow \pi^*$ transitions were observed in UV spectra of tetra-thiooxalates [423] and in CD spectra of chiral dithiooxamides [424]. In an attempt to locate $n_+ \rightarrow \pi^*$ transition in the dithiocarbamate anion and to obtain information about the polarization of the $\pi \rightarrow \pi^*$ transitions, Sandström et al. [425] studied UV and CD spectra of chiral dithiocarbamates. The UV spectrum is interpreted with the aid of CNDO/S calculations, and the CD spectrum with calculations by the Schellman matrix method. The CD spectrum showed a medium–strong positive band at 230 nm, however, the theoretical calculation predicted a negative sign for this band, and consequently no assignment for the 230 band was made. Except for 230 nm band, the calculations predict correct signs and qualitatively correct intensities for the CD bands between 200 and 400 nm in a narrow range of orientations of the 1-phenylethyl group with respect to the dithiocarbamate ion, and it is expected that the favored conformation falls in this conformational range. Low-coordinated trivalent phosphorus continues to attract attention in several fields of chemistry [426]. Reduction of systems containing a phenyl ring linked to the phosphaalkene carbon [427], leads to radical anion whose phosphorus spin density is <0.5, the unpaired electron being delocalized on both the phosphaethylenic bond and the phenyl ring. Owning to resonance stabilization of the cyclopentadienide ion, it is expected that incorporation of the phosphaalkenic carbon in a cyclopentadiene ring appreciably modified the spin distribution by increasing the contribution of the phosphinyl mesomeric structure **120**.

**120**

EPR spectra of phosphafulvene and dibenzophosphafulvene radical anions, formed by electrochemical reduction or by reaction on a potassium mirror are studied [428]. Ab initio calculations on model phosphaalkene and phosphafulvene radical anions show that, in accord with the experimental results, the electronic structures of these two species are quite different; whereas the unpaired electron is delocalized on the whole P=C(H)R moiety in the phosphaalkene anion, it is markedly localized on the phosphorus atom in the phosphafulvene anion. The isotropic coupling constants obtained for diphosphaallenic radical anion ArP=C=PAr$^{\cdot-}$ (Ar = 2,4,6-(Bu$^t$)$_3$C$_6$H$_2$) reflects the probability of unpaired electron mainly localized on two equivalent phosphorus atoms [429]. The interpretation is substantiated by considering the coupling constants calculated for the *trans*-like and *cis*-like isomers at the MP2 and MCSCF optimized geometries. In light of experimental and theoretical results it is proposed that the reduction of diphosphaallene gives a slightly asymmetric *trans*-like radical anion, which rapidly interconverts between two equivalent structures. Here, a comparison with the corresponding radical cation is worthwhile to be mentioned: Diphosphaallene derivatives [430] RP=C=PR are electrochemically oxidized to radical cations with the unpaired electron located in a *p*-orbital with a spin population of ca 20 % at each phosphorus (ca. 20 %). Ab initio calculation on HPCPH revealed that oxidation is accompanied by a modification of the molecular geometry. Whereas the HPPH dihedral angle, $\theta_{HPPH}$, is equal to 90° in the neutral molecule, the radical cation has two possible $\theta_{HPPH}$ values of ca 45° and ca 135°, corresponding to two isomers of similar energy.

Recently Alberti et al. [431] applied a multi-disciplinary approach involving CV, EPR, pulse radiolysis and MO calculations to investigate the radical ions resulting from the one-electron oxidation and one-electron reduction of bis(2,4,6-tri-*tert*-butylphenyl)-1,3-diphosphaallene. Results showed significant differences from those reported by Geoffroy et al. [429, 430] at both experimental (EPR) and theoretical (MO calculation) levels, the disagreement being particularly severe in the case of the reduction. A higher spin density on the phosphorus atoms is observed in the anion than in the cation, in agreement with the different nature of the SOMO in the two species predicted by UB3LYP calculations on the model compound diphenyl-1,3-diphosphaallene, which also predict for both ions the existence of *cis* and *trans* geometrical isomers, the latter being more stable. In both radical ions the unpaired electron is found mainly localized in the PCP moiety, namely in a $\pi$-allylic type MO in the cation and in a $\sigma$ MO in the anion. The paramagnetic species detected by EPR upon electrochemical reduction of bis(2,4,6-tri-*tert*-butylphenyl)-1,3-diphosphaallene in THF has been identified as the bis(2,4,6-tri-*tert*-butyl-phenyl)-1,3-diphosphaallyl radical, the identification is supported by DFT calculations [432]. One and two electron reduction of bisdiazenes has provided access to non-classical, cyclically delocalized 4N/5e radical anions and 4N/6e dianions of high persistence [433]. EPR investigation on the relatively facile one-electron reduction of cycloaliphatic monodiazenes is reported [434]. Neutral and anionic cyanocarbons and azacyanocarbons show interesting optical, electrical and/or magnetic properties [435]. Although a great deal of chemistry is known for cyanocarbanions but their exist very few reports on azacyanocarbanions [436]. Electrochemical, spectro-chemical and structural characterization of the tetraalkyl ammonium salts of the 1,1,2,5,6,6-hexacyano-3,4-diazahexa-1,5-dienedide anion $[(R_4N)_2(C_{10}N_8); R = Et, Bu]$ is reported [437]. 2,3-diazabicyclo[2.2.1]hept-2-ene and its derivatives have been the focus of numerous mechanistic studies concerned with the extrusion step (dinitrogen loss) and fragmentation, giving rise to radicals, diradicals and radical ions. The radical anions of both *E*- and cyclic *Z*-azoalkanes proved to be much more resistant to the extrusion of dinitrogen and so have been amply investigated by EPR and ENDOR spectroscopy [438]. Unlike azoarenes, such as azobenzene with a half-wave reduction potential of $-1.38$ V relative to the SCE, conversion of azoalkanes to their radical anions requires a highly negative voltage [439]. Radical anions of 2,3-diazabicyclo[2.2.1]hept-2-ene and its 18 substituted and tricyclic derivatives are characterized by EPR and ENDOR [440]. Structural modification reported lead to only minor changes in the geometry of the carbon framework of radical anion and do not substantially alter the $\pi$-spin distribution.

The voltammetric behavior of disubstituted alkyl pyridine and benzene esters **121** and several of their dithioic *S,S'*-diesters **122** analogs have recently been reported

**121**                                          **122**

Ar = pyridine or benzene, R = Me, Et or Pr

[441], as well as the identity and yields of the products obtained by bulk controlled potential electrolysis experiments in acetonitrile [442].

Bulk controlled-potential electrolysis experiments have shown that most of the (O) ester anion radicals decay via a simple bond cleavage mechanism to form carboxylate anions in very high yield, while the (S) ester radicals decay via a very complicated mechanism often involving aromatic substitution reactions [442]. Using CV, many of the compounds are shown to display chemically (and electrochemically) reversible behavior at slow scan rates, in the sense that the $i_p^{ox.}/i_p^{red.}$ ratios were close to one, indicating that the anion radicals formed were stable for at least several seconds and existence confirmed by EPR [442] and UV–Vis spectroscopy [443]. In contrast, other compounds, including the *meta*-substituted (O) and (S) diesters **123** and **124**, appeared to show chemically irreversible behavior at slow scan rates suggesting that the associated anion radicals of these compounds were much less stable and quickly decompose to other products.

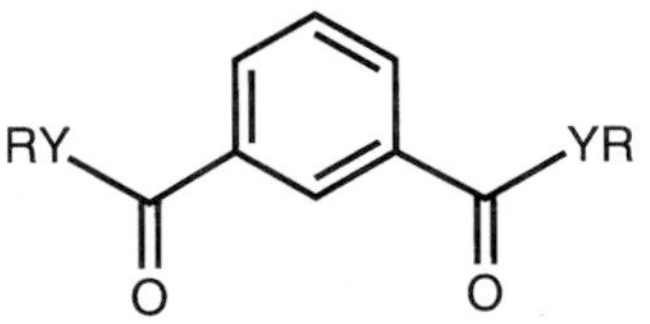

**123** 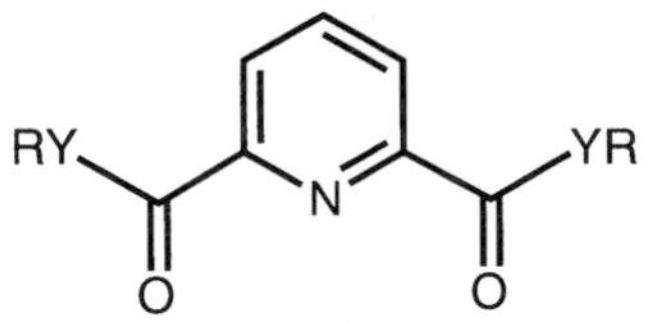                     **124**

Y = O, R = Me, dimethyl benzene-1,3-dicarboxylate

Y = S, R = Pr, S,S'-dipropyl benzene-1,3-dicarbothioate

Y = O, R = Me, dimethyl pyridine-2,6-dicarboxylate

Y = O, R = Pr, dipropyl pyridine-2,6-dicarboxylate

Y = S, R = Pr, S,S'-dipropyl pyridine-2,6-dicarbothioate

Cyclic voltammograms (CV) obtained at various concentrations and scan rates from several *meta*-disubstituted (O) and (S) diesters of pyridine and benzene [444]. The CV and EPR experiments revealed that a likely explanation for the surprising stability of the anion radicals of dialkylbenzene-1,3-dicarboxylates, dialkylpyridine-2,6-dicarboxylates, and their corresponding dithioic $S,S'$-diesters is due to a reversible dimerization mechanism. The cyclic voltammetric data were complicated and digital simulation for an $EC_{dim}$ mechanism required several homogeneous and heterogeneous rate constants in order to obtain a good theoretical match to the experimental data. Rate constants evaluated by simulation at scan rates between $0.1$–$50$ V s$^{-1}$ and substrate concentrations between 0.2 and 10 mM were estimated to be ca $10^3$–$10^4$ L mol$^{-1}$ s$^{-1}$ for the dimerization reaction and ca $10^{-1}$–$10^0$ s$^{-1}$ for the monomerization reaction. Only approximate rate constants could be derived from the experimental curves because a large number of variables needed to be included in the simulations, meaning that no unique combination of variables would give a reasonable data fit.

## 4.6 So what?—Conclusions and Outlook

So what? This is an often-asked question after listening to a lecture or reading a publication about a topic, which is not too familiar to ones own field of work. Our answer contains the following aspects:

The aim of this contribution is to represent the widespread occurrence of $\pi$-type radical ions in a variety of electron-transfer processes; it should help to illustrate the state of the art in terms of the identification and characterizations of aromatic compounds in a variety of redox stages predominately by spectroscopic techniques. Aromatic (radical) polyions were upon the first species inspected by such techniques. Their structures were generally interpreted by the simplistic Hückel model and, in many cases, an astonishing agreement between the experimental results and models were found.

Nowadays, sophisticated experimental techniques and quantum-mechanical calculations allow generating and describing even rather short-lived molecules and their reactivity. Spectroscopic methods allow the detection of radicals on the ns time scale (magnetic resonance) and beyond (optical spectroscopy). Thus, it is becoming more and more practicable to establish electron-transfer generated species as intermediates in a variety of reactions. This is quite rewarding since many chemical, biochemical, and technical transformations involve electron-transfer steps: Several examples exist in which $\pi$ systems undergo electron transfer and form key intermediates, e.g., in coupling reactions or isomerizations. Extensive studies exist in which quinone radical anions are established as substantial participants in photosynthesis, as taking part in the NADH pathways, or being involved in radiation damage of tissue. Moreover reduced or oxidized $\pi$ systems are shown to be able to work as (semi)conductors or serve as building blocks of molecular magnets. Such systems are very helpful to gain information about electron delocalization and (para)magnetic coupling. These highly charged systems, yet, have to become considerably more stable to be useful for technical applications; even the fullerenes which are able to accept up to six electrons ($C_{60}$, $C_{70}$), do not form salts which are stable enough to be used as materials [445].

For essentially all $\pi$ systems, electron transfer is connected with particular changes in the electronic absorption spectra. Thus, the usefulness of aromatic (poly)-anions and cations should particularly come to pass in cases when electron-transfer generated stages occur as short-lived mediators or as optoelectronic devices and sensors.

The references given below should serve as a guide to the fields mentioned above when connected with an appropriate online data base. Moreover, some of the above aspects are covered in the other chapters of this series.

### References

1. A. Lund, M. Lindgren, S. Lunell, J. Maruani, *Molecules in Physics, Chemistry and Biology* (Ed.: J. Maruani), Kluwer Academic Publishers, Dordrecht, 1989, Vol. 3.
2. D. Feller, E. R. Davidson, *Theoretical Models of Chemical Bonding* (Ed.: Z. B. Maksic), Springer, Berlin, 1991, Part 3.

3. V. G. Malkin, O. L. Malkina, D. R. Salahub, L. A. Eriksson, *Theoretical and Computational Chemistry* (Eds.: P. Politzer, J. M. Seminario), Elsevier, Amsterdam, 1995, Vol. 2.
4. D. M. Chipman, *Quantum Mechanical Electronic Structure Calculations With Chemical Accuracy* (Ed.: S. R. Langhoff), Kluwer Academic Publishers, Dordrecht, 1995.
5. A. Lund, M. Shiotani, *Radical Ionic Systems. Properties In Condensed Phases*, Kluwer, Dordrecht, 1991.
6. L. B. Knight, Jr., *Acc. Chem. Res.* **1986**, *19*, 313.
7. V. Barone, *Recent Advances In Density Functional Methods* (Ed.: D. P. Chong), World Scientific, Singapore 1995, Part 1.
8. (a) B. Engels, *Chem. Phys. Lett.* **1991**, *179*, 398; (b) K. Funken, B. Engels, S. D. Peyerimhoff, F. Grein, *Chem. Phys. Lett.* **1990**, *172*, 180; (c) J. Kong, R. J. Boyd, L. A. Eriksson, *J. Chem. Phys.* **1995**, *102*, 3674; (d) B. Engels, S. D. Peyerimhoff, *J. Phys. B.* **1988**, *21*, 3459.
9. (a) D. Feller, *J. Chem. Phys.* **1990**, *93*, 579; (b) D. Feller, E. Glendening, E. A. McCullough Jr., R. J. Miller, *J. Chem. Phys.* **1993**, *99*, 2829.
10. V. Barone, C. Adamo, A. Grand, R. Subra, *Chem. Phys. Lett.* **1995**, *242*, 351.
11. (a) I. Carmichael, *J. Phys. Chem.* **1991**, *95*, 6198; **1991**, *95*, 108; *J. Chem. Phys.* **1990**, *93*, 863; (b) D. M. Chipman, I. Carmichael, D. Feller, *J. Phys. Chem.* **1991**, *95*, 4702; (c) I. Carmichael, ibid. **1994**, *98*, 5044; **1995**, *99*, 6832.
12. S. A. Perera, J. D. Watts, R. J. Bartlett, *J. Chem. Phys.* **1994**, *100*, 1425.
13. (a) P. Hohenberg, W. Kohn, *Phys. Rev. B* **1964**, *136*, 864; (b) W. Kohn, L. Sham, *J. Phys. Rev. A* **1965**, *140*, 1133.
14. A. D. Becke, *Phys. Rev. A* **1988**, *38*, 3098; *J. Chem. Phys.* **1992**, *96*, 2155.
15. J. P. Perdew, Y. Wang, *Phys. Rev. B* **1986**, *33*, 8800.
16. J. P. Perdew, J. A. Chevary, S. H. Vosko, K. A. Jackson, M. R. Pederson, J. D. Singh, C. Fiolhas, *Phys. Rev. B* **1992**, *46*, 6671; (b) ibid **1992**, *48*, 4978; (c) J. P. Perdew, *Physica B* **1991**, *172*, 1.
17. A. D. Becke, *J. Chem. Phys.* **1993**, *98*, 1372; **1993**, *98*, 5684.
18. C. Lee, W. Yang, R. G. Parr, *Phys. Rev. B* **1988**, *37*, 785.
19. G. J. Laming, V. Termath, N. C. Handy, *J. Chem. Phys.* **1993**, *99*, 8765.
20. M. J. Frisch, G. W. Trucks, H. B. Schlegel, P. M. W. Gill, B. G. Johnson, M. A. Robb, J. R. Cheeseman, T. A. Keith, A. G. Peterson, J. A. Montgomery, K. Raghavachari, M. A. Al-Laham, V. G. Zakrevwski, J. V. Ortiz, J. B. Foresman, J. Cioslowski, B.B. Stefanov, A. Nanayakkara, M. Challacombe, C. Y. Peng, P. Y. Ayala, W. Chen, M. W. Wong, J. L. Andres, E. S. Replogle, R. Gomperts, R. L. Martin, D. J. Fox, J. S. Binkley, D. J. Defrees, J. Baker, J. P. Stewart, Head-M. Gordon, Gonzalez, J. A. Pople, *GAUSSIAN 94, Revision A.1*, Gaussian Inc., Pittsburgh, PA, 1995.
21. R. D. Amos, I. L. Alberts, J. S. Andrews, S. M. Colwell, N. C. Handy, D. Jayatilaka, P. J. Knowles, R. Kobayashi, G. J. Laming, A. M. Lee, P. E. Maslen, C. W. Murray, P. Palmieri, J. E. Rice, E. D. Simandiras, A. J. Stone, M. D. Su, D. J. Tozer, *The Cambridge Analytic Derivatives Package Issue 6.0*, University of Cambridge, Cambridge, U.K 1995.
22. (a) J. Andzelm, E. Wimmer, *J. Chem. Phys.* **1992**, *96*, 1280; (b) B. G. Johnson, P. M. W. Gill, J. A. Pople, *J. Chem. Phys.* **1993**, *98*, 5612.
23. N. Oliphant, R. J. Barlett, *J. Chem. Phys.* **1994**, *100*, 6550.
24. (a) C. W. Bauschlicher, Jr., *Chem. Phys. Lett.* **1995**, *246*, 40; (b) C. W. Bauschlicher, Jr., H. Patridge, *ibid* **1995**, *240*, 533.
25. R. Neumann, N. C. Handy, *Chem. Phys. Lett.* **1995**, *246*, 381.
26. (a) L. A. Eriksson, L. G. M. Pettersson, P. E. M. Siegbahn, U. Wahlgren, *J. Chem. Phys.* **1995**, *102*, 872; (b) P. A. Stewart, P. W. M. Gill, *J. Chem. Soc. Faraday Trans.* **1995**, *91*, 4337; (c) J. A. Altmann, N. C. Handy, V. E. Ingamells, *Int. J. Quantum Chem.* **1996**, *57*, 533; (d) E. I. Proynov, E. Ruiz, A. Vela, D. R. Salahub, *Int. J. Quantum Chem. Symp.* **1995**, *29*, 61.
27. V. Barone, C. Adamo, F. Mele, *Chem. Phys. Lett.* **1996**, *249*, 290.
28. *Density Functional Methods in Chemistry* (Eds.: J. Labanowski, J. Andzelm), Springer, New York, 1991.
29. *Theoretical and Computational Chemistry, Modern Density Functional Theory—A Tool for Chemistry* (Eds.: J. M. Seminario, P. Politzer), Elsevier, Amsterdam 1995, Vol. 2.

30. *Density Functional Methods: Applications in Chemistry and Materials Science* (Ed.: M. Springborg), Wiley, New York 1996.
31. S. J. Mole, X. Zhou, R. Liu, *J. Phys. Chem.* **1996**, *100*, 14665–14671.
32. J. M. Martell, J. D. Goddard, L. A. Eriksson, *J. Phys. Chem. A* **1997**, *101*, 1927–1934.
33. L. A. Eriksson, A. Laaksonen, *J. Chem. Phys.* **1996**, *105*, 8195–8202.
34. S. D. Wetmore, R. J. Boyd, L. A. Eriksson, *J. Chem. Phys.* **1997**, *106*, 7738–7748.
35. L. A. Eriksson, *Molecular Physics* **1997**, *91*, 827–833.
36. B. R. Beno, J. Fennen, K. N. Houk, H. J. Lindner, K. Hafner, *J. Am. Chem. Soc.* **1998**, *120*, 10490–10493.
37. (a) A. M. Mebel, A. Luna, M. C. Lin, K. Morokuma, *J. Chem. Phys.* **1996**, *105*, 6439; (b) E. Sicilia, M. Toscano, T. Mineva, N. Russo, *Int. J. Quantum Chem.* **1997**, *61*, 571; (c) J. J. Queralt, V. S. Safont, V. Moliner, J. Andres, *Theor. Chim. Acta* **1996**, *94*, 247; (d) A. Irigoras, J. M. Ugalde, X. Lopez, C. Sarasola, *Can. J. Chem.* **1996**, *74*, 1824; (e) M. R. Manaa, L. E. Fried, *J. Phys. Chem. A* **1998**, *102*, 9884–9889.
38. (a) P. M. Gill, L. Radom, *J. Am. Chem. Soc.* **1988**, *110*, 4931; (b) M. Sodupe, A. Oliva, J. Bertran, *J. Am. Chem. Soc.* **1994**, *116*, 8249.
39. R. N. Barnett, U. Landman, *J. Phys. Chem.* **1995**, *99*, 17305; *J. Phys. Chem. A* **1997**, *101*, 164.
40. M. Sodupe, J. Bertran, L. Rodríguez-Santiago, E. J. Baerends, *J. Phys. Chem. A* **1999**, *103*, 166–170.
41. L. R. Domingo, *J. Org. Chem.* **1999**, *64*, 3922–3929.
42. W. B. Smith, *J. Org. Chem.* **1999**, *64*, 60–64.
43. (a) M. Garcia-Viloca, A. Gonzalez-Lafont, J. M. Lluch, *J. Am. Chem. Soc.* **1997**, *119*, 1081; (b) M. E. Tuckerman, D. Marx, M. L. Klein, M. Parrinello, *Science* **1997**, *275*, 817; (c) S. Shan, D. Herschlag, *J. Am. Chem. Soc.* **1996**, *118*, 5515; (d) B. Schwartz, D. G. Drueckhammer, *J. Am. Chem. Soc.* **1995**, *117*, 11902; (e) J. A. Gerlt, M. M. Kreevoy, W. W. Cleland, P. A. Frey, *Chem. Biol.* **1997**, *4*, 259.
44. (a) J. A. Gerlt, P. G. Gassman, *J. Am. Chem. Soc.* **1993**, *115*, 11552; (b) J. P. Guthrie, *Chem. Biol.* **1996**, *3*, 163; (c) Y. Kato, L. M. Toledo, J. Rebek Jr., *J. Am. Chem. Soc.* **1996**, *118*, 8575; (d) G. A. Jeffrey, *An Introduction to Hydrogen Bonding*, Oxford University Press, New York 1997.
45. (a) Y. Pan, M. A. McAllister, *J. Am. Chem. Soc.* **1997**, *119*, 7561; (b) M. A. McAllister, *Can. J. Chem.* **1997**, *75*, 1195; (c) Y. Pan, M. A. McAllister, *J. Org. Chem.* **1997**, *62*, 8176; (d) C. J. Smallwood, M. A. McAllister, *J. Am. Chem. Soc.* **1997**, *119*, 11277.
46. Y. Pan, M. A. McAllister, *J. Am. Chem. Soc.* **1998**, *120*, 166–169.
47. A. Császár, *J. Mol. Struct.* **1995**, *346*, 141; *J. Am. Chem. Soc.* **1992**, *114*, 9568.
48. S. G. Stepanian, I. D. Reva, E. D. Radchenko, L. Adamowicz, *J. Phys. Chem. A* **1999**, *103*, 4404–4412.
49. C. Adamo, V. Barone, A. Fortunelli, *J. Chem. Phys.* **1995**, *102*, 384–393.
50. *Ions and Ion Pairs in Organic Reactions* (Ed.: M. Szwarc), Wiley–Interscience, New York, 1972, Vol. 1; 1974, Vol. 2.
51. N. M. Atherton, S. I. Weissman, *J. Am. Chem. Soc.* **1961**, *83*, 1330–1334.
52. B. J. Herold, A. F. Neiva Correia, J. dos Santos Veiga, *J. Am. Chem. Soc.* **1965**, *87*, 2661–2665.
53. H. Van Willigen, M. Plato, R. Biehl, K. P. Dinse, K. Möbius, *Mol. Phys.* **1973**, *26*, 793–809.
54. (a) F. Gerson, J. Jachimowicz, M. Nakagawa, M. Iyoda, *Helv. Chim. Acta* **1974**, *57*, 2141–2148; (b) W. Huber, *ibid* **1985**, *68*, 1140–1148.
55. C. L. Dodson, A. H. Reddoch, *J. Chem. Phys.* **1968**, *48*, 3226–3234.
56. H. Kurreck, B. Kirste, W. Lubitz, *Electron Nuclear Double Resonance Spectroscopy of Radicals in Solution*, VCH, New York, 1988, Ch. 4.7, 99–100.
57. H. Kurreck, B. Kirste, W. Lubitz, *Electron Nuclear Double Resonance Spectroscopy of Radicals in Solution*, VCH, New York, 1988, Ch. 2.2, 20–24.
58. E. De Boer, J. L. Sommerdijk, In *Ions and Ion Pairs in Organic Reactions* (Ed.: M. Szwarc), Wiley–Interscience, New York, 1972, Vol. 1, Ch. 7, 289–309.
59. (a) S. N. Batchelor, H. Heikkila, C. W. M. Kay, K. A. McLauchlan and I. A. Shkrob, *Chem. Phys.* **1992**, *162*, 29; (b) S. Sekiguchi, K. Akiyama, S. Tero-Kubota, *Chem. Phys. Lett.* **1996**, *263*, 161; (c) N. J. Avdievich, A. S. Jeevarajan, M. D. E. Forbes, *J. Phys. Chem.* **1996**, *100*,

5334; (d) S. Sekiguchi, K. Akiyama, S. Tero-Kubota, *J. Chem. Soc., Perkin Trans. 2*, **1997**, 1619–1620.

60. (a) J. H. Sharp, M. C. R. Symons, *Ions and Ion Pairs in Organic Reactions* (Ed.: M. Szwarc), Wiley–Interscience, New York, 1972, Vol. 1, Ch. 5, 177–262; (b) F. Gerson, *High-Resolution ESR Spectroscopy*, Wiley and Verlag Chemie, New York and Weinheim, 1970, appendix 2.2, 137–143.

61. W. M. Tolles, D. W. Moore, *J. Chem. Phys.* **1967**, *46*, 2102–2106.

62. D. H. Levy, R. J. Myers, *J. Chem. Phys.* **1964**, *41*, 1062–1065.

63. D. H. Levy, R. J. Myers, *J. Chem. Phys.* **1966**, *44*, 4177–4180.

64. N. Sommer, *Kautsch. Gummi Kunstst.* **1975**, *28*, 131–135.

65. F. Gerson, H. Hopf, P. Merstetter, C. Mlynek, D. Fischer, *J. Am. Chem. Soc.* **1998**, *120*, 4815–4824.

66. S. Kababya, Z. Luz, D. Goldfarb, *J. Am. Chem. Soc.* **1994**, *116*, 5805–5813.

67. F. Barbosa, V. Péron, G. Gescheidt, A. Fürstner, *J. Org. Chem.* **1998**, *63*, 8806–8814.

68. R. Rathore, S. V. Lindeman, A. S. Kumar, J. K. Kochi, *J. Am. Chem. Soc.* **1998**, *120*, 6931.

69. B. M. Tadjikov, D. V. Stass, Yu. N. Molin, *Chem. Phys. Lett.* **1996**, *260*, 529–532.

70. (a) O. Hammerich, V. D. Parker, *Adv. Phys. Org. Chem.* **1994**, *20*, 1; (b) A. J. Bard, A. Ledwith, H. J. Shine, *Adv. Phys. Org. Chem.* **1976**, *12*, 155.

71. (a) J. K. Kochi, R. T. Tang, T. Bernath, *J. Am. Chem. Soc.* **1973**, *95*, 7114; (b) I. H. Elson, J. K. Kochi, *J. Am. Chem. Soc.* **1973**, *95*, 5060.

72. (a) K. Nyberg, *Acta Chem. Scand.* **1970**, *24*, 1609; **1971**, *25*, 2499; **1971**, *25*, 2983; **1971**, *25*, 3770; **1971**, *25*, 534; (b) L. Eberson, K. Nyberg, H. Sternerup, *Acta Chem. Scand.* **1973**, *27*, 1679; *Acc. Chem. Res.* **1973**, *6*, 106; (c) A. McKillop, A. G. Turrell, D. W. Young, E. C. Taylor, *J. Am. Chem. Soc.* **1980**, *102*, 6504; (d) L. Eberson, F. Radner, *Acta Chem. Scand.* **1992**, *46*, 630; (e) R. Sebastiano, J. D. Krop, J. K. Kochi, *J. Chem. Soc., Chem. Commun.* **1991**, 1481.

73. (a) A. Ronlan, K. Bechgaard, V. D. Parker, *Acta Chem. Scand.* **1973**, *27*, 2375; (b) R. N. Adams, *Acc. Chem. Res.* **1966**, *2*, 175.

74. (a) J. K. Kochi, *Acc. Chem. Res.* **1992**, *25*, 39; (b) L. Eberson, M. P. Hartshorn, F. Radner, J. O. Svensson, *J. Chem. Soc., Perkin Trans. 2* **1994**, 1719.

75. (a) A. Terahara, H. Ohya-Nishiguchi, N. Hirota, A. Oku, *J. Phys. Chem.* **1986**, *90*, 1564; (b) M. G. Bakker, R. F. Claridge, C. M. Kirk, *J. Chem. Soc., Perkin Trans. 2*, **1986**, 1735.

76. A. T. Balaban, C. D. Nenitzescu, *Friedel–Crafts and Related Reactions* (Ed.: G. A. Olah), Wiley, New York, 1963–1965, Vol. II, Ch. 23.

77. L. Eberson, M. P. Hartshorn, O. Persson, *J. Chem. Soc., Perkin Trans. 2* **1995**, 409–416.

78. W. Lau, J. C. Huffman, J. K. Kochi, *J. Am. Chem. Soc.* **1982**, *104*, 5515.

79. R. Taylor, *Electrophilic Aromatic Substitution*, Wiley, Chichester, 1990.

80. (a) W. Lau, J. K. Kochi, *J. Org. Chem.* **1986**, *51*, 1801; (b) A. G. Davies, *Chem. Soc. Rev.* **1993**, 299.

81. (a) J. L. Courtneidge, A. G. Davies, P. S. Gregory, D. C. McGuchan, S. N. Yazdi, *J. Chem. Soc., Chem. Commun.* **1987**, 1192; (b) A. G. Davies, C. J. Shields, J. C. Evans, C. C. Rowlands, *Can. J. Chem.* **1989**, *67*, 1748.

82. I. H. Elson, J. K. Kochi, *J. Am. Chem. Soc.* **1973**, *95*, 5060.

83. L. Eberson, M. P. Hartshorn, O. Persson, J. O. Stevensson, *J. Chem. Soc., Perkin Trans. 2* **1995**, 1253.

84. (a) P. D. Sullivan, E. M. Menger, A. H. Reddoch, D. H. Paskovich, *J. Phys. Chem.* **1978**, *82*, 1158; (b) J. Eloranta, S. Kasa, *Acta Chem. Scand., Ser. A* **1985**, *399*, 63.

85. A. G. Davies, D. C. McGuchan, *Organometallics*, **1991**, *10*, 329.

86. X. H. Chen, P. D. Sullivan, *J. Magn. Reson.* **1989**, *83*, 484.

87. J. L. Courtneidge, A. G. Davies, D. C. McGuchan, S. N. Yazdi, *J. Organomet. Chem.* **1988**, *341*, 63.

88. L. Eberson, M. P. Hartshorn, O. Persson, *J. Chem. Soc., Perkin Trans. 2* **1995**, 1735.

89. D. V. Avila, A. G. Davies, *J. Chem. Soc., Perkin Trans. 2* **1991**, 1111.

90. A. G. Davies, K. M. Ng, *J. Chem. Soc., Perkin Trans. 2*, **1998**, 2599–2607.

91. J. Fleischhauer, S. Ma, W. Schleker, K. Gersonde, H. Twilfer, F. Dallacker, *Z. Naturforsch, Teil A*, **1982**, *37*, 680.

92. P. D. Sullivan, N. A. Brette, *J. Phys. Chem.* **1975**, *79*, 474.
93. N. A. Malysheva, A. I. Prokof'ev, N. N. Bubnov, S. P. Solodovnikov, T. I. Prokof'ev, V. B. Vol'eva, V. V. Ershov, M. I. Kabachnik, *Izvest. Akad. Nauk SSSR, Ser. Khim.* **1988**, 1040.
94. (a) J. L. Courtneidge, A. G. Davies, E. Lisztyk, J. Lusztyk, *J. Chem. Soc., Perkin Trans. 2* **1984**, 155; (b) J. L. Courtneidge, A. G. Davies, S. M. Tollerfield, J. Rideout, M. C. R. Symons, *J. Chem. Soc., Chem. Commun.* **1985**, 1092.
95. C. J. Cooksey, J. L. Courtneidge, A. G. Davies, J. C. Evans, P. S. Gregory, C. C. Rowlands, *J. Chem. Soc., Perkin Trans. 2* **1988**, 807.
96. L. Eberson, M. P. Hartshorn, F. Radner, O. Persson, *J. Chem. Soc., Perkin Trans. 2* **1998**, 59–70.
97. (a) L. Eberson, M. P. Hartshorn, O. Persson, *J. Chem. Soc., Chem. Commun.* **1995**, 1131; *J. Chem. Soc., Perkin Trans. 2* **1995**, 1735; *Angew. Chem. Int. Ed. Engl.* **1995**, *34*, 2268; *Res. Chem. Intermed.* **1996**, *22*, 799; (b) L. Eberson, M. P. Hartshorn, O. Persson, F. Radner, *J. Chem. Soc., Chem. Commun.* **1996**, 2105.
98. (a) L. Eberson, M. P. Hartshorn, O. Persson, *J. Chem. Soc., Perkin Trans. 2* **1996**, 141: (b) L. Eberson, M. P. Hartshorn, O. Persson, F. Radner, *J. Chem. Soc., Chem. Commun.* **1996**, 215.
99. A. G. Davies, C. J. Shields, J. C. Evans, C. C. Rowlands, *Can. J. Chem.* **1989**, *67*, 1748.
100. D. V. Avila, A. G. Davies, R. Lapouyade, K. M. Ng, *J. Chem. Soc., Perkin Trans. 2* **1998**, 2609–2615.
101. M. Iwaizumi, T. Isobe, *Mol. Phys.* **1975**, *29*, 549.
102. (a) H. Hogeveen, H. C. Volger, *Recl. Trav. Chim. Pays-Bas* **1968**, *87*, 385; (b) L. A. Paquette, G. R. Krow, J. M. Bollinger, G. A. Olah, *J. Am. Chem. Soc.* **1968**, *90*, 7147; (c) T. R. Evans, R. W. Wake, M. M. Sifain, *Tetrahedron Lett.* **1973**, 701; (d) N. Peacock, G. B. Schuster, *J. Am. Chem. Soc.* **1983**, *105*, 3632.
103. (a) C. J. Rhodes, *J. Am. Chem. Soc.*, **1988**, *110*, 4446; (b) X. Z. Qin, D. W. Werst, A. D. Trifunac, *J. Am. Chem. Soc.* **1990**, *112*, 2026; (c) F. Williams, Q. X. Guo, S. F. Nelsen, *J. Am. Chem. Soc.* **1990**, *112*, 2028.
104. (a) H. D. Roth, *Top. Curr. Chem.* **1992**, *163*, 131; (b) H. D. Roth, P. Lakkaraju, J. Zhang, *J. Chem. Soc., Chem. Commun.* **1994**, 1969.
105. L. Eberson, M. P. Hartshorn, O. Persson, J. O. Svensson, *J. Chem. Soc., Perkin Trans. 2* **1995**, 1253–1262.
106. (a) M. Iwasaki, K. Toriyama, K. Nunome, *J. Chem. Soc., Chem. Commun.* **1983**, 320–322; (b) K. Raghavachari, R. C. Haddon, T. A. Miller, V. E. Bondybey, *J. Chem. Phys.* **1983**, *79*, 1387.
107. A. Hasegawa, Y. Itagaki, M. Shiotani, *J. Chem. Soc., Perkin Trans. 2* **1997**, 1625–1631.
108. S. Steenken, M. Ashokkumar, P. Maruthamuthu, R. A. McClelland, *J. Am. Chem. Soc.* **1998**, *120*, 11925–11931.
109. (a) H. J. Bakoss, R. J. Ranson, R. M. G. Roberts, A. R. Sadri, *Tetrahedron* **1982**, *38*, 623; (b) G. P. Smith, A. S. Dworkin, R. M. Pagni, S. P. Zingg, *J. Am. Chem. Soc.* **1989**, *111*, 525; (c) For Review see: V. A. Koptyuk, *Arenium Ions—Structure and Reactivity, In Topics in Current Chemistry* (Ed.: F. L. Boschke), Springer, Berlin, **1984**, *122*, 96.
110. S. Steenken, R. A. McClelland, *J. Am. Chem. Soc.* **1990**, *112*, 9648.
111. M. V. Sargent, F. M. Dean, *Comprehensive Heterocyclic Chemistry* (Eds.: A. J. Boulton, A. McKillop), Pergamon Press, Oxford, **1984**, *3*, 599.
112. L. Eberson, F. Radner, *Acta Chem. Scand.* **1992**, *46*, 802.
113. L. Eberson, F. Radner, *Acta Chem. Scand.* **1992**, *46*, 312.
114. L. Eberson, M. P. Hartshorn, F. Radner, M. Merchàn, B. O. Roos, *Acta Chem. Scand.* **1993**, *47*, 176.
115. (a) S. S. Shaik, A. Pross, *J. Am. Chem. Soc.* **1989**, *111*, 4306; (b) S. S. Shaik, E. J. Canadell, *ibid* **1990**, *112*, 1452; (c) S. S. Shaik, *J. Org. Chem.* **1990**, *55*, 3434.
116. L. Eberson, M. P. Hartshorn, O. Persson, F. Radner, C. J. Rhodes, *J. Chem. Soc., Perkin Trans. 2* **1996**, 1289–1295.
117. M. Schmittel, G. Gescheidt, L. Eberson, H. Trenkle, *J. Chem. Soc., Perkin Trans. 2* **1997**, 2145–2150.
118. L. Eberson, R. González-Luque, M. Merchàn, F. Radner, B. O. Roos, S. Shaik, *J. Chem. Soc., Perkin Trans. 2* **1997**, 463.

119. (a) R. A. Forrester, K. Ishizu, G. Kothe, S. F. Nelsen, H. Ohya-Nishiguchi, K. Watanabe, W. Wilker, *Organic Cation Radicals and Polyradicals. In Landolt Börnstein, Numerical Data and Functional Relationships in Science and Technology*, Springer, Heidelberg, 1980, IX, Part d2; (b) K. Yoshida, *Electrooxidation in Organic Chemistry: The Role of Cation Radicals as Synthetic Intermediates*, Wiley, New York, 1984; (c) *Radical Ionic Systems* (Eds.: A. Lund, M. Shiotani), Kluwer Academics, Dordrecht, 1991; (d) H. D. Roth, *Top. Curr. Chem.* **1992**, *163*, 133.

120. (a) A. Ledwith, *Acc. Chem. Res.* **1972**, *5*, 133; (b) T. Shida, E. Haselbach, T. Bally, *Acc. Chem. Res.* **1984**, *17*, 180–186; (c) S. F. Nelsen, *Acc. Chem. Res.* **1987**, *20*, 269–276; (d) H. D. Roth, *Acc. Chem. Res.* **1987**, *20*, 343–350; (e) G. A. Mirafzal, J. Liu, N. Bauld, *J. Am. Chem. Soc.* **1993**, *115*, 6072; (f) F. Gerson, *Acc. Chem. Res.* **1994**, *27*, 63.

121. (a) H. D. Roth, M. L. M. Schilling, G. Jones II, *J. Am. Chem. Soc.* **1981**, *103*, 1246–1248; (b) H. D. Roth, M. L. M. Schilling, *J. Am. Chem., Soc.* **1981**, *103*, 7210–7217; (c) Y. Takahashi, T. Mukai, T. Miyashi, *J. Am. Chem. Soc.* **1983**, *105*, 6511–6513.

122. (a) P. G. Gassman, B. A. Hay, *J. Am. Chem. Soc.* **1985**, *107*, 4075; (b) A. Arnold, U. Burger, F. Gerson, E. Kloster-Jensen, S. P. Schmidlin, *J. Am. Chem. Soc.* **1993**, *115*, 4271–4281; (c) J. P. Dinnocenzo, M. Schmittel, *J. Am. Chem. Soc.* **1987**, *109*, 1561–1562; (d) J. P. Dinnocenzo, D. A. Conlon, *J. Am. Chem. Soc.* **1988**, *110*, 2324–2326.

123. (a) J. P. Dinnocenzo, W. P. Todd, T. R. Simpson, I. R. Gould, *J. Am. Chem. Soc.* **1990**, *112*, 2462–2464; (b) S. S. Hixson, Y. Xing, *Tetrahedron Lett.* **1991**, *32*, 173–174; (c) J. P. Dinnocenzo, D. R. Lieberman, T. R. Simpson, *J. Am. Chem. Soc.* **1993**, *115*, 366–367.

124. P. G. Gassman, K. D. Olson, *J. Am. Chem. Soc.* **1982**, *104*, 3740.

125. (a) H. D. Roth, T. Herbertz, *J. Am. Chem. Soc.* **1993**, *115*, 9804–9805; (b) D. R. Arnold, X. Du, H. J. P. de Lijser, *Can. J. Chem.* **1995**, *73*, 522; (c) H. D. Roth, H. Weng, T. Herbertz, *Tetrahedron* **1997**, *53*, 10051–10070 and references therein.

126. A. Alex, T. Clark, *J. Am. Chem. Soc.* **1992**, *114*, 10897–10902.

127. C. Dass, D. A. Peake, M. L. Gross, *Org. Mass. Spectrom.* **1986**, *21*, 741–746.

128. L. T. Scott, I. Erden, W. R. Brunsvold, T. H. Schultzu, K. N. Houk, M. N. Paddon-Row, *J. Am. Chem. Soc.* **1982**, *104*, 3659.

129. I. Yu. Shchapin, V. I. Fel'dman, V. N. Belevskii, N. A. Donskaya, N. D. Chuvylkin, *Russ. Chem. Bull.* **1995**, *44*, 203–227.

130. H. D. Roth, T. Herbertz, *J. Am. Chem. Soc.* **1993**, *115*, 9804–9805.

131. H. D. Roth, H. Weng, T. Herbertz, *Tetrahedron* **1997**, *53*, 10051–10070.

132. W. J. Hehre, L. Radom, J. A. Pople, P. V. R. Schleyer, *Ab Initio Molecular Orbital Theory*, Wiley Interscience, New York, 1986.

133. H. Weng, Q. Sheik, H. D. Roth, *J. Am. Chem. Soc.* **1995**, *117*, 10655–10661.

134. A. Arnold, U. Burger, F. Gerson, E. Kloster-Jensen, S. P. Schmidlin, *J. Am. Chem. Soc.* **1993**, *115*, 4271–4281.

135. (a) G. J. Kavarnos, N. J. Turro, *Chem. Rev.* **1986**, *86*, 401–449; (b) T. Oguchi, T. Arai, H. Sakuragi, K. Tokumaru, *Bull. Chem. Soc. Jpn.* **1987**, *60*, 2395–2399; (c) *Photoinduced Electron Transfer* (Eds.: M. A. Fox, M. Chanon), Elsevier, Amsterdam, 1988; (d) G. J. Kavarnos, *Top. Curr. Chem.* **1990**, *156*, 21–58.

136. (a) T. Majiama, C. Pac, A. Makasone, H. Sakurai, *J. Am. Chem. Soc.* **1981**, *103*, 4499–4508; (b) M. Julliard, In *Photoinduced Electron Transfer* (Eds.: M. A. Fox, M. Chanon), Elsevier, Amsterdam, 1988, Part B, 216–313; (c) I. R. Gould, D. Ege, J. E. Mooser, S. Farid, *J. Am. Chem. Soc.* **1990**, *112*, 4290–4301.

137. W. Adam, A. Corma, M. A. Miranda, M. Sabater-Picot, C. Sahin, *J. Am. Chem. Soc.* **1996**, *118*, 2380–2386.

138. (a) W. Adam, C. Sahin, J. Sendelbach, H. Walter, G. F. Chen, F. Williams, *J. Am. Chem. Soc.* **1994**, *116*, 2576; (b) W. Adam, T. Heidenfelder, C. Sahin, *Synthesis* **1995**, 1163; (c) W. Adam, A. Corma, M. A. Miranda, M. J. Sabater-Picot, C. Sahin, *J. Am. Chem. Soc.* **1996**, *118*, 2380.

139. W. Schmidt, E. Steckhan, *Chem. Ber.* **1980**, *113*, 577.

140. W. Adam, V. Handmann, F. Kita, T. Heidenfelder, *J. Am. Chem. Soc.* **1998**, *120*, 831–832.

141. R. Bachmann, F. Gerson, P. Merstetter, E. Vogel, *Helvetica Chem. Acta* **1996**, *79*, 1627–1634.

142. (a) P. Dowd, *J. Am. Chem. Soc.* **1966**, *88*, 2587; (b) P. Dowd, M. Chow, *Tetrahedron* **1982**, *38*, 799.
143. (a) M. S. Platz In *Diradicals* (Ed.: W. T. Borden), Wiley, New York, **1982**; (b) J. Wirz, *Pure Appl. Chem.* **1984**, *56*, 1289.
144. M. C. Biewer, M. S. Platz, M. Roth, J. Wirz, *J. Am. Chem. Soc.* **1991**, *113*, 8069.
145. Z. Zhu, T. Bally, J. Wirz, M. Fülscher, *J. Chem. Soc., Perkin Trans. 2* **1998**, 1083–1091.
146. (a) W. Rebafka, H. A. Staab, *Angew. Chem.* **1973**, *85*, 831–832; *Chem. Ber.* **1977**, *110*, 3333–3350; (b) H. A. Staab, C. P. Herz, C. Krieger, M. Rentea, *Chem. Ber.* **1983**, *116*, 3813–3830; (c) H. A. Staab, A. Döhling, C. Kreiger, *Liebigs Ann. Chem.* **1981**, 1052–1064.
147. (a) H. A. Staab, B. Starker, C. Krieger, *Chem. Ber.* **1983**, *116*, 3831–3845; (b) F. Gerson, *Top. Curr. Chem.* **1983**, *115*, 57–105.
148. F. Gerson, W. B. Martin, *J. Am. Chem. Soc.* **1969**, *91*, 1883–1891.
149. A. R. Wartini, J. Valenzuela, H. A. Staab, F. A. Neugebauer, *Eur. J. Org. Chem.* **1998**, 139–148.
150. A. R. Wartini, H. A. Staab, F. A. Neugebauer, *Eur. J. Org. Chem.* **1998**, 1161–1170.
151. (a) L. Eberson, F. Radner, *Acc. Chem. Res.* **1988**, *20*, 53–60; (b) B. Boduszek, H. J. Shine, *J. Org. Chem.* **1988**, *53*, 5142–5147.
152. S. F. Nelsen, Y. Kim, *J. Org. Chem.* **1991**, *56*, 1045–1049.
153. (a) E. K. Kim, J. K. Kochi, *J. Am. Chem. Soc.* **1991**, *113*, 4962–4967; (b) M. Lehning, K. Schürmann, *Eur. J. Org. Chem.* **1998**, 913–917.
154. K. Mizuno, N. Ichinose, T. Tamai, Y. Otsuji, *J. Org. Chem.* **1992**, *57*, 4669–4673.
155. G. A. Olah, P. Ramaiah, C. B. Rao, G. Sandford, R. Golam, N. J. Trivedi, J. A. Olah, *J. Am. Chem. Soc.* **1993**, *115*, 7246–7249.
156. A. A. Fokin, P. A. Gunchenko, S. A. Peleshanko, P. V. R. Schleyer, P. R. Schreiner, *Eur. J. Org. Chem.* **1999**, 855–860.
157. H. Prinzbach, K. Weber, *Angew. Chem.* **1994**, *106*, 2329; *Angew. Chem. Int. Ed. Engl.* **1994**, *33*, 2239.
158. (a) H. D. Martin, B. Mayer, K. Weber, H. Prinzbach, *Liebgis Ann.* **1995**, 2019; (b) K. Weber, G. Lutz, L. Knothe, J. Mortensen, J. Heinze, H. Prinzbach, *J. Chem. Soc., Perkin Trans. 1* **1995**, 1991; (c) H. Prinzbach, G. Gescheidt, H. D. Martin, R. Herges, J. Heinze, G. K. S. Prakash, G. A. Olah, *Pure Appl. Chem.* **1995**, *67*, 673.
159. A. D. Trifunac, D. W. Werst, R. Herges, H. Neumann, H. Prinzbach, M. Etzkorn, *J. Am. Chem. Soc.* **1996**, *118*, 9444.
160. (a) R. Herges, P. von R. Schleyer, M. Schindler, W. D. Fessner, *J. Am. Chem. Soc.* **1991**, *113*, 3649; (b) G. K. S. Prakash in *Stable Carbocation Chemistry* (Eds.: G. K. S. Prakash, P. von R. Schleyer), Wiley, New York, 1996.
161. A. D. Trifunac, D. W. Werst, *J. Am. Chem. Soc.* **1996**, *118*, 9444–9445.
162. H. Prinzbach, B. A. R. C. Murty, W. D. Fessner, J. Mortensen, J. Heinze, G. Gescheidt, F. Gerson, *Angew. Chem. Int. Ed. Engl.* **1987**, *26*, 457.
163. G. K. S. Prakash, V. V. Krishnamurthy, R. Herges, R. Bau, H. Yuan, G. A. Olah, W. D. Fessner, H. Prinzbach, *J. Am. Chem. Soc.* **1988**, *110*, 7764.
164. A. D. Trifunac, D. W. Werst, R. Herges, H. Neumann, H. Prinzbach, M. Etzkorn, *J. Am. Chem. Soc.* **1996**, *118*, 9444.
165. M. Etzkorn, F. Wahl, M. Keller, H. Prinzbach, F. Barbosa, V. Peron, G. Gescheidt, J. Heinze, R. Herges, *J. Org. Chem.* **1998**, *63*, 6080–6081.
166. (a) M. Bertau, F. Wahl, A. Weiler, K. Scheumann, J. Wörth, M. Keller, H. Prinzbach, *Tetrahedron* **1997**, *53*, 10029–10040; (b) K. Scheumann, E. Sackers, M. Bertau, J. Leonhardt, D. Hunkler, H. Fritz, J. Wörth, H. Prinzbach, *J. Chem. Soc., Perkin Trans. 2* **1998**, 1195–1210.
167. D. N. Stamires, J. Turkevich, *J. Am. Chem. Soc.* **1964**, *86*, 749.
168. (a) P. S. Lakkaraju, D. Zhou, H. D. Roth, *Chem. Commun.* **1996**, 2605; (b) E. A. Piocos, D. W. Werst, A. D. Trifunac, *J. Phys. Chem.* **1996**, *100*, 8408; (c) V. Ramamurthy, J. V. Caspar, D. R. Corbin, *J. Am. Chem. Soc.* **1991**, *113*, 594; **1991**, *113*, 600.
169. V. J. Rao, N. Prevost, V. Ramamurthy, M. Kojima, L. J. Johnston, *Chem. Commun.* **1997**, 2209–2210.
170. E. A. Piocos, P. Han, D. W. Werst, *J. Phys. Chem.* **1996**, *100*, 7191–7199.

171. (a) G. T. Kerr, *Sci. Am.* **1989**, 100; (b) A. Corma, *Chem. Rev.* **1995**, *95*, 559; (c) F. R. Ribeiro, F. Alvarez, C. Henriques, F. Lemos, J. M. Lopes, M. F. Ribeiro, *J. Mol. Catal. A* **1995**, *96*, 245.
172. J. P. Lange, A. Gutsze, H. G. Karge, *J. Catal.* **1988**, *114*, 136.
173. (a) T. Komatsu, A. Lund, *J. Phys. Chem.* **1972**, *76*, 1727; (b) A. M. Volodin, V. A. Boloshov, T. A. Konovalova, *Radicals on Surfaces* (Eds.: A. Lund, C. Rhodes), Kluwer, Dordrecht, **1995**, 201–226.
174. R. Erickson, A. Lund, M. Lindgren, *Chem. Phys.* **1995**, *193*, 89.
175. R. Erickson, N. P. Benetis, A. Lund, M. Lindgren, *J. Phys. Chem.* A **1997**, *101*, 2390–2396.
176. M. Iwasaki, K. Toriyama, K. Nunoma, *J. Chem. Soc., Chem. Commun.* **1983**, 320.
177. T. Komatsu, A. Lund, *J. Phys. Chem.* **1972**, *76*, 1727.
178. R. Erickson, M. Lindgren, A. Lund, L. Sjöqvist, *Colloids Surf. A* **1993**, *72*, 207.
179. R. M. Kadam, R. Erickson, K. Komaguchi, M. Shiotani, A. Lund, *Chem. Phys.* Lett. **1998**, *290*, 371–378.
180. V. I. Feldman, F. F. Sukhov, A. Yu. Orlov, *Chem. Phys. Lett.* **1999**, *300*, 713–718.
181. (a) T. R. Tuttle, S. I. Weissman Jr., *J. Am. Chem. Soc.* **1958**, *80*, 5342; (b) C. D. Stevenson, E. P. Wagner, R. C. Reiter, *J. Phys. Chem.* **1993**, *97*, 10587.
182. (a) M. C. R. Symons, L. Harris, *J. Chem. Res. (S)* **1982**, 268; (b) M. Tabata, A. Lund, *Chem. Phys.* **1983**, *78*, 379.
183. T. Komatsu, A. Lund, *J. Phys. Chem.* **1972**, *76*, 1721.
184. R. M. Kadam, Y. Itagaki, R. Erickson, A. Lund, *J. Phys. Chem. A* **1999**, *103*, 1480–1486.
185. E. A. Piocos, D. W. Werst, A. D. Trifunac, L. A. Eriksson, *J. Phys. Chem.* **1996**, *100*, 8408–8417.
186. D. W. Werst, E. E. Tartakovsky, E. A. Piocos, A. D. Trifunac, *J. Phys. Chem.* **1994**, *98*, 10249–10257.
187. (a) T. Shida, Y. Takemura, *Radiat. Phys. Chem.* **1983**, *21*, 157; (b) M. Iwasaki, K. Toriyama, K. Nunome, *Faraday Discuss. Chem. Soc.* **1984**, *78*, 19; (c) M. V. Barnabas, A. D. Trifunac, *Chem. Phys. Lett.* **1991**, *187*, 565.
188. (a) V. I. Melekhov, O. A. Anisimov, A. V. Veselov, Yu. N. Molin, *Chem. Phys. Lett.* **1986**, *127*, 97; (b) B. M. Tadjikov, V. I. Melekhov, O. A. Anisimov, Yu. N. Molin, *Radiat. Phys. Chem.* **1989**, *34*, 353.
189. (a) D. W. Werst, L. T. Percy, A. D. Trifunac, *Chem. Phys. Lett.* **1988**, *153*, 45; (b) D. W. Werst, M. G. Bakker, A. D. Trifunac, *J. Am. Chem. Soc.* **1990**, *112*, 40.
190. D. W. Werst, A. D. Trifunac, *J. Phys. Chem.* **1988**, *92*, 1093.
191. B. M. Tadjikov, D. V. Stass, Y. N. Molin, *J. Phys. Chem. A* **1997**, *101*, 377–383.
192. (a) T. Shida, Y. Takemura, *Radiat. Phys. Chem.* **1983**, *21*, 157; (b) M. Iwasaki, K. Toriyama, K. Nunome, *Faraday. Discuss. Chem. Soc.* **1984**, *78*, 19; (c) V. I. Melekhov, O. A. Anisimov, L. Sjöqvist, A. Lund, *Chem. Phys. Lett.* **1990**, *174*, 95.
193. (a) V. I. Melekhov, O. A. Anisimov, A. V. Veselov, Yu. N. Molin, *Chem. Phys. Lett.* **1986**, *127*, 97; (b) B. M. Tadjikov, V. I. Melekhov, O. A. Anisimov, Yu. N. Molin, *Radiat. Phys. Chem.* **1989**, *34*, 353.
194. (a) A. D. Trifunac, D. W. Werst, L. T. Percy, *Radiat. Phys. Chem.* **1989**, *34*, 547; (b) D. W. Werst, M. G. Bakker, A. D. Trifunac, *J. Am. Chem. Soc.* **1990**, *112*, 40.
195. D. W. Werst, A. D. Trifunac, *J. Phys. Chem.* **1988**, *92*, 1093.
196. I. A. Shkrob, M. C. Sauer, A. D. Trifunac, *J. Phys. Chem.* **1996**, *100*, 5993.
197. D. V. Tass, B. M. Tadjikov, Yu. N. Molin, *Chem. Phys. Lett.* **1995**, *235*, 511.
198. B. M. Tadjikov, D. V. Tass, Yu. N. Molin, *Chem. Phys. Lett.* **1996**, *260*, 529.
199. B. M. Tadjikov, D. V. Tass, O. M. Usov, Yu. N. Molin, *Chem. Phys. Lett.* **1997**, *273*, 25–30.
200. F. B. Sviridenko, D. V. Tass, Yu. N. Molin, *Chem. Phys. Lett.* **1998**, *297*, 343–349.
201. O. M. Usov, D. V. Tass, B. M. Tadjikov, Yu. N. Molin, *J. Phys. Chem. A* **1997**, *101*, 7711–7717.
202. (a) M. Lindgren, M. Shiotani In *Radical Ionic Systems* (Eds.: A. Lund, M. Shiotani), Kluwer, Dordrecht, **1991**, 115; (b) M. Lindgren, M. Matsumoto, M. Shiotani, *J. Chem. Soc., Perkin Trans. 2* **1992**, 1397; (c) P. V. Schastnev, L. N. Shchegoleva, *Molecular Distortions in Ionic and Excited States*, CRC Press, Boca Raton, FL. 1995.
203. T. Fängström, S. Lunell, B. Engels, L. Eriksson, M. Shiotani, K. Komaguchi, *J. Chem. Phys.* **1997**, *107*, 297.

204. (a) T. L. Amyes, J. P. Richard, *J. Am. Chem. Soc.* **1990**, *112*, 9507; (b) J. P. Richard, T. L. Amyes, L. Bei, V. Stubblefield, *J. Am. Chem. Soc.* **1990**, *112*, 9513.
205. P. M. Henrichs, P. E. Peterson, *J. Am. Chem. Soc.* **1973**, *95*, 7449; *J. Org. Chem.* **1976**, *41*, 362.
206. D. Farcasiu, *J. Chem. Soc., Chem. Commun.* **1977**, 394.
207. M. A. Miranda, J. Péret-Prieto, E. Font-Sanchis, K. Kónya, J. C. Scaiano, *J. Phys. Chem. A* **1998**, *102*, 5724–5727.
208. N. P. Schepp, L. J. Johnston, *J. Am. Chem. Soc.* **1996**, *118*, 2872–2881.
209. (a) K. Mizuno, I. Nakanishi, N. Ichinose, Y. Otsuji, *Chem. lett.* **1989**, 1095–1098; (b) D. R. Arnold, X. Du, K. M. Hensleit, *Can. J. Chem.* **1991**, *69*, 839–852; (c) D. R. Arnold, X. Du, J. Chen, *Can. J. Chem.* **1995**, *73*, 307–318.
210. K. A. McManus, D. R. Arnold, *Can. J. Chem.* **1994**, *72*, 2291–2304; ibid **1995**, *73*, 2158–2169.
211. (a) F. D. Lewis In *Photoinduced Electron Transfer* (Eds.: M. A. Fox, M. Chanon), Elsevier, Amsterdam, 1988, Part C, Ch. 4.1; (b) N. L. Bauld, *Tetrahedron* **1989**, *45*, 5307–5363; (c) N. L. Bauld, *Adv. Electron Transfer Chem.* **1992**, Vol. 2, 1–66; (d) D. A. Connor, D. R. Arnold, P. K. Bakshi, T. S. Cameron, *Can. J. Chem.* **1995**, *73*, 762–771.
212. C. S. Q. Lew, J. R. Brisson, L. J. Johnston, *J. Org. Chem.* **1997**, *62*, 4047–4056.
213. (a) C. R. Brundle, M. B. Robin, N. A. Kuebler, H. Basch, *J. Am. Chem. Soc.* **1972**, *94*, 1451; (b) C. R. Brundle, M. B. Robin, N. A. Kuebler, *J. Am. Chem. Soc.* **1972**, *94*, 1466.
214. A. Hasegawa, Y. Itagaki, M. Shiotani, *J. Chem. Soc., Perkin Trans. 2* **1997**, 1625.
215. M. Shiotani, H. Kawazoe, J. Sohma, *J. Phys. Chem.* **1984**, *88*, 2220.
216. M. Shiotani, H. Kawazoe, J. Sohma, *Chem. Phys. Lett.* **1984**, *111*, 254.
217. K. Ohta, M. Shiotani, J. Sohma, *Chem. Phys. Lett.* **1987**, *140*, 148.
218. A. Hasegawa, M. C. R. Symons, *J. Chem. Soc., Faraday Trans. 1* **1983**, *79*, 1565.
219. (a) S. Marry, C. Thomson, *Chem. Phys. Lett.* **1981**, *82*, 373; (b) M. Kira, H. Nakagawa, H. Sakurai, *J. Am. Chem. Soc.* **1983**, *105*, 6983.
220. (a) M. Shiotani, Y. Nagata, J. Sohma, *J. Am. Chem. Soc.* **1984**, *106*, 4640; (b) J. Fujisawa, S. Sato, K. Shimokoshi, *Chem. Phys. Lett.* **1986**, *124*, 391.
221. S. Lunell, M. B. Huang, *Chem. Phys. Lett.* **1990**, *168*, 63.
222. K. Toriyama, K. Nunome, M. Iwasaki, *J. Chem. Phys.* **1982**, *77*, 5981.
223. M. Shiotani, Y. Nagata, J. Sohma, *J. Phys. Chem.* **1984**, *88*, 4078.
224. K. Toriyama, K. Nunome, M. Iwasaki, *Chem. Phys. Lett.* **1984**, *107*, 86.
225. Y. Itagaki, M. Shiotani, A. Hasegawa, H. Kawazoe, *Bull. Chem. Soc. Jpn.* **1998**, *71*, 2547–2554.
226. E. Baciocchi, *Acta Chem. Scand.* **1990**, *44*, 645 and references therein.
227. X. M. Zhang, F. G. Bordwell, *J. Am. Chem. Soc.* **1994**, *116*, 904–908 and references therein.
228. (a) A. Anne, P. Hapiot, J. Moiroux, P. Neta, J. M. Savéant, *J. Am. Chem. Soc.* **1992**, *114*, 4694–4701; (b) E. Baciocchi, T. Del Giacco, F. Elisei, *J. Am. Chem. Soc.* **1993**, *115*, 12290.
229. A. Anne, J. Moiroux, J. M. Savéant, *J. Am. Chem. Soc.* **1993**, *115*, 10224–10230.
230. E. Baciocchi, T. D. Giacco, F. Elisei, O. Lanzalunga, *J. Am. Chem. Soc.* **1998**, *120*, 11800–11801.
231. (a) J. H. Davis, W. A. Goddard III, *J. Am. Chem. Soc.* **1977**, *99*, 4242; (b) H. J. P. Lijser, D. R. Arnold, *J. Phys. Chem.* **1996**, *100*, 3996.
232. K. Komaguchi, M. Shiotani, A. Lund, *Chem. Phys. Lett.* **1997**, *265*, 217–223.
233. S. D. Chemerisov, D. W. Werst, A. D. Trifunac, *Chem. Phys. Lett.* **1998**, *291*, 262–268.
234. (a) F. Gerson, X. Z. Qin, *Helv. Chim. Acta* **1988**, *71*, 1498; (b) G. Gescheidt, A. Lamprecht, C. Rüchardt, M. Schmittel, *Helv. Chim. Acta* **1992**, *75*, 351; (c) P. S. Engel, D. M. Robertson, J. N. Scholz, H. J. Shine, *J. Org. Chem.* **1992**, *57*, 6178; (d) J. L. Goodman, T. A. Zona, *Tetrahedron Lett.* **1992**, 6093–6096; (e) W. Adam, G. F. Chen, H. Walter, F. Williams, *J. Am. Chem. Soc.* **1992**, *114*, 3007; (f) W. Adam, U. Denninger, R. Finzel, F. Kita, H. Platsch, H. Walter, G. Zang, *J. Am. Chem. Soc.* **1992**, *114*, 5027.
235. (a) N. J. Turro, G. J. Kavarnos, *J. Chem. Rev.* **1986**, *86*, 401; (b) G. J. Kavarnos, *Top. Curr. Chem.* **1990**, *156*, 20.
236. W. Adam, S. Grabowski, R. M. Wilson, *Acc. Chem. Res.* **1990**, *23*, 165.
237. (a) W. Adam, M. Dörr, *J. Am. Chem. Soc.* **1987**, *109*, 1570; (b) W. Adam, M. A. Miranda, *J. Org. Chem.* **1987**, *52*, 5498.

238. W. Adam, J. Sendelbach, *J. Org. Chem.* **1993**, *58*, 5310–5315.
239. W. Adam, C. Sahin, *Tetrahedron Lett.* **1994**, *35*, 9027–9030.
240. R. J. Bushby, K. M. Ng, *J. Chem. Soc., Perkin Trans. 2* **1996**, 1053–1056.
241. W. Adam, T. Kammel, M. Toubartz, S. Steenken, *J. Am. Chem. Soc.* **1997**, *119*, 10673–10676.
242. F. A. Benson, *The High Nitrogen Compounds*, Wiley, New York, **1984**.
243. M. Bolle, K. Luther, *J. Troe. Appl. Surf. Sc.* **1990**, *46*, 279.
244. R. H. Smith Jr., B. D. Wladkowski, J. A. Herling, T. D. Pfaltzgraff, B. Pruski, J. Klose, Ch. J. Michejda, *J. Org. Chem.* **1992**, *57*, 654.
245. (a) J. Huguet, M. Libert, C. Caullet, *Bull. Soc. Chim. Fr.* **1972**, *12*, 4860; (b) L. Dunsch, B. Gollas, A. Neudeck, A. Petr, B. Speiser, H. Stahl, *Chem. Ber.* **1994**, *127*, 2423.
246. P. Rapta, L. Omelka, A. Stasko, J. Dauth, B. Deubzer, J. Weis, *J. Chem. Soc., Perkin Trans. 2* **1995**, 255–261.
247. K. Komaguchi, M. Shiotani, *J. Phys. Chem. A* **1997**, *101*, 6983–6990.
248. (a) Y. Apeloig In *The Chemistry of Organosilicon Compounds* (Eds.: S. Patai, Z. Rappoport), Wiley, Chichester, 1989, 57–225; (b) J. Chojnowski, W. Stanczyk, *Adv. Organomet. Chem.* **1990**, *30*, 243; (c) N. C. Norman, *Polyhedron* **1993**, *12*, 2431.
249. V. N. Khabashesku, Z. A. Kerzina, A. K. Maltsev, O. M. Nefedov, *J. Organomet. Chem.* **1989**, *364*, 301.
250. (a) G. Manuel, G. Bertrand, F. El Anba, Organometallics **1983**, *2*, 391; (b) C. A. Arrington, R. West, J. Michl, *J. Am. Chem. Soc.* **1983**, *105*, 6176; (c) V. N. Khabashesku, Z. A. Kerzina, E. G. Baskir, A. K. Maltsev, O. F. Nefedov, J. Organomet. Chem. **1988**, *347*, 277.
251. A. G. Davies, A. G. Neville, *J. Organomet. Chem.* **1992**, *436*, 255.
252. A. G. Davies, C. Eaborn, P. D. Lickiss, A. G. Neville, *J. Chem. Soc., Perkin Trans. 2* **1995**, 163–169.
253. H. Ban, A. Tanaka, N. Hayashi, S. Tagawa, Y. Tabata, *Radiat. Phys. Chem.* **1989**, *34*, 587.
254. (a) S. Irie, K. Oka, R. Nakao, M. Irie, *J. Organomet. Chem.* **1990**, *388*, 253; (b) S. Irie, M. Irie, *Macromolecules* **1992**, *25*, 1766.
255. K. Ushida, A. Kira, S. Tagawa, Y. Yoshida, H. Shibata, *Polym. Mater. Sci. Eng.* **1992**, *66*, 299.
256. J. Kumagai, H. Yoshida, H. Koizumi, T. Ichikawa, *J. Phys. Chem.* **1994**, *98*, 13117.
257. J. Kumagai, H. Yoshida, T. Ichikawa, *J. Phys. Chem.* **1995**, *99*, 7965–7969.
258. (a) M. Adam, K. Müllen, *Adv. Mater.* **1994**, *6*, 439; (b) M. R. Bryce, *J. Mater. Chem.* **1995**, *5*, 1481.
259. F. Gerson, A. Lamprecht, M. Fourmigué, *J. Chem. Soc., Perkin Trans. 2* **1996**, 1409–1414.
260. F. Barbosa, L. Eberson, G. Gescheidt, S. Gronowitz, A. Hörnfeldt, L. Juliá, O. Persson, *Acta Chem. Scand.* **1998**, *52*, 1275–1284.
261. D. Lahem, R. Flammang, H. T. Le, T. L. Nguyen, M. T. Nguyen, *J. Chem. Soc., Perkin Trans. 2* **1999**, 821–826.
262. (a) *Sulfur-Centered Reactive Intermediates in Chemistry and Biology* (Eds.: C. Chatgilialoglu, K. D. Asmus), Plenum Press, New York 1990; (b) S. A. Chaudhri, H. Mohan, E. Anklam, K. D. Asmus, *J. Chem. Soc., Perkin Trans. 2* **1996**, 383–390.
263. (a) C. von Sonntag, *The Chemical Basis of Radiation Biology*, Taylor and Francis, London, **1987**; (b) P. Wardman, In *Sulfur-Centered Reactive Intermediates in Chemistry and Biology* (Eds.: C. Chatgilialoglu, K. D. Asmus), NATO Ser. A, 197, Plenum Press, New York **1990**, 415; (c) C. Schöneich, K. D. Asmus, M. Bonifacic, *J. Chem. Soc., Faraday Trans.* **1995**, *91*, 1923–1930.
264. (a) L. Engman, J. Lind, G. Merényi, *J. Phys. Chem.* **1994**, *98*, 3174–3182; (b) M. Ioele, S. Steenken, E. Baciocchi, *J. Phys. Chem. A* **1997**, *101*, 2979–2897.
265. H. Yokoi, A. Hatta, K. Ishiguro, Y. Sawaki, *J. Am. Chem. Soc.* **1998**, *120*, 12728–12733.
266. (a) N. J. Fitzpatrick, D. F. Brougham, P. J. Goarke, M. T. Nguyen, *Chem. Ber.* **1994**, *127*, 969; (b) K. Toyota, M. Shibata, M. Yoshifuji, *Bull. Chem. Soc. Jpn.* **1995**, *68*, 2633.
267. *Multiple Bonds and Low Coordination in Phosphorus Chemistry* (Eds.: M. Regitz, O. Scherer), Thieme, New York, **1990**.
268. (a) A. Jouaiti, M. Geoffroy, G. Terron, G. Bernardinelli, *J. Am. Chem. Soc.* **1995**, *117*, 2251; (b) W. W. Schoeller, W. Haug, J. Strutwolf, T. Busch, *J. Chem. Soc., Faraday Trans.* **1996**, *92*, 1751.

269. H. H. Karsh, H. U. Reisacher, G. Müller, *Angew. Chem. Int. Ed. Engl.* **1994**, *23*, 618.
270. M. T. Nguyen, A. F. Hegarty, *J. Chem. Soc., Perkin Trans. 2* **1985**, 2005.
271. M. Chentit, H. Sidorenkova, A. Jouaiti, G. Terron, M. Geoffroy, Y. Ellinger, *J. Chem. Soc., Perkin Trans. 2* **1997**, 921–925.
272. I. I. Bilkis, V. D. Shteingarts, *Izv. SO Akad. Nauk. SSSR, Ser. Khim. Nauk*, **1987**, *5*, 105.
273. (a) A. Bohnen, J. Räder, K. Müllen, *Synth. Met.* **1992**, *47*, 37; (b) M. Baumgarten, U. Müller, K. Müllen, *AIP Conf. Proc. Ser.*, St. Thomas **1992**, 68.
274. (a) A. Farazdel, M. Dupuis, E. Clementi, A. Aviram, *J. Am. Chem. Soc.* **1990**, *112*, 4206; (b) N. Liang, J. R. Miller, G. L. Closs, *J. Am. Chem. Soc.* **1990**, *112*, 5353.
275. (a) J. Fielder, W. Huber, K. Müllen, *Angew. Chem., Int. Ed. Engl.* **1986**, *25*, 443; (b) G. L. Closs, P. Piotrowiak, J. M. MacInnes, G. R. Flemming, *J. Am. Chem. Soc.* **1988**, *110*, 2652.
276. (a) J. S. Miller, A. J. Epstein, *Angew. Chem., Int. Ed. Engl.* **1994**, *33*, 385; (b) M. Baumgarten, K. Müllen, *Top. Curr. Chem.* **1994**, *169*, 1; (c) A. Rajca, *Chem. Rev.* **1994**, *94*, 871.
277. *Proceedings of the Symposium on Chemistry and Physics of Molecular Based Magnetic Materials* (Eds.: H. Iwamura, J. S. Miller), Mol. Cryst. Liq. Cryst. **1993**, Vol. 232, 1–360; Vol. 233, 1–366.
278. (a) D. A. Dougherty In *Research Frontiers in Magnetochemistry* (Ed.: C. O. Connor), World Scientific Publishing, River Edge, NJ, 1992; (b) M. Baumgarten, U. Müller, A. Bohnen, K. Müllen, *Angew. Chem., Int. Ed. Engl.* **1992**, *31*, 448; (c) M. Baumgarten, U. Müller, *Synth. Met.* **1993**, *57*, 4755.
279. (a) M. H. Whangbo, *Acc. Chem. Res.* **1983**, *16*, 95; (b) L. Salem, C. Rowland, *Angew. Chem., Int. Ed. Engl.* **1972**, *11*, 91.
280. (a) M. Dietrich, J. Mortensen, J. Heinze, *Angew. Chem., Int. Ed. Engl.* **1985**, *24*, 508; (b) K. Meehrholz, J. Heinze, *J. Am. Chem. Soc.* **1990**, *112*, 5142.
281. U. Müller, M. Baumgarten, *J. Am. Chem. Soc.* **1995**, *117*, 5840–5850.
282. G. Grampp, A. Kapturkiewicz, J. Salbeck, *Chem. Phys.* **1994**, *187*, 391–397.
283. R. E. Hoffman, M. Nir, I. O. Shapiro, M. Rabinovitz, *J. Chem. Soc., Perkin Trans. 2*, **1996**, 1225–1233.
284. J. H. Hammons, C. T. Kresge, L. A. Paquette, *J. Am. Chem. Soc.* **1976**, *98*, 8172.
285. S. Z. Goldberg, K. N. Raymond, C. A. Harmon, D. H. Templeton, *J. Am. Chem. Soc.* **1974**, *96*, 1348.
286. T. J. Katz, *J. Am. Chem. Soc.* **1960**, *82*, 3784.
287. H. Kojima, A. J. Bard, H. N. C. Wong, F. Sondheimer, *J. Am. Chem. Soc.* **1976**, *98*, 5560.
288. T. J. Katz, M. Yoshida, L. C. Siew, *J. Am. Chem. Soc.* **1965**, *87*, 4516.
289. G. R. Stevenson, M. Colón, I. Ocasio, J. G. Concepcion, A. McB. Block, *J. Phys. Chem.* **1975**, *79*, 1685.
290. M. Celina. R. L. R. Lazana, M. Luisa T. M. B. Franco, B. J. Herold, *J. Chem. Soc., Perkin Trans. 2* **1991**, 1791–1795.
291. M. Iwaizumi, S. Kita, T. Isobe, M. Kohna, T. Yamamoto, H. Horita, T. Otsubo, S. Misumi, *Bull. Chem. Soc. Jpn.* **1977**, *50*, 2074–2083.
292. S. F. Rak, L. L. Miller, *J. Am. Chem. Soc.* **1992**, *114*, 1388–1394.
293. A. R. Wartini, J. Valenzuela, H. A. Staab, F. A. Neugebauer, *Eur. J. Org. Chem.* **1998**, 221–227.
294. (a) K. Maruyama, T. Otsuki, *Bull. Chem. Soc. Jpn.* **1968**, *41*, 444; (b) T. Takeshita, N. Hirota, *J. Chem. Phys.* **1969**, *51*, 2146.
295. M. Barzaghi, P. L. Beltrame, A. Gamba, M. Simonetta, *J. Am. Chem. Soc.* **1978**, *100*, 251.
296. M. Barzaghi, S. S. Mlertus, C. Oliva, E. Ortoleva, M. Simonetta, *J. Phys. Chem.* **1983**, *87*, 881–888.
297. G. Grampp, Y. A. Khan, H. Larsen, *J. Chem. Soc., Perkin Trans. 2* **1997**, 2555–2557.
298. (a) G. Grampp, W. Jaenicke, *Ber. Bunsenges. Phys. Chem.* **1991**, *95*, 904; (b) G. Grampp, W. Jaenicke, *J. Chem. Soc., Faraday Trans. 2* **1985**, *81*, 1035.
299. (a) M. J. Weaver, *Chem. Rev.* **1992**, *92*, 463; (b) H. Larsen, St. U. Pedersen, J. A. Pedersen, *J. Electroanal. Chem.* **1992**, *331*, 971; (c) C. D. Stevenson, C. V. Rice, *J. Am. Chem. Soc.* **1995**, *117*, 10551.
300. J. M. Savéant, *Adv. Phys. Org. Chem.* **1990**, *26*, 1.
301. (a) P. Neta, D. Behar, *J. Am. Chem. Soc.* **1981**, *103*, 103; (b) M. MeotNer, P. Neta, R. K. Norris, K. Wilson, *J. Phys. Chem.* **1986**, *90*, 168.

302. D. Bethell, R. G. Compton, R. G. Welling, *J. Chem. Soc., Perkin Trans. 2* **1992**, 147.
303. (a) R. G. Compton, R. A. W. Dryfe, A. C. Fisher, *J. Electroanal. Chem.* **1993**, *361*, 275; *J. Chem. Soc., Perkin Trans. 2* **1994**, 1581; (b) R. G. Compton, R. A. W. Dryfe, *J. Electroanal. Chem.* **1994**, *375*, 247.
304. R. G. Compton, R. A. W. Dryfe, J. C. Eklund, S. D. Page, J. Hirst, L. B. Nei, G. W. J. Fleet, K. Y. Hsia, D. Bethell, L. Martingale, *J. Chem. Soc., Perkin Trans. 2* **1995**, 1673.
305. M. J. S. Dewar, S. Kirschner, H. W. Kollmar, *J. Am. Chem. Soc.* **1974**, *96*, 5242.
306. (a) R. Dressler, M. Allan, E. Haselbach, *Chimia* **1985**, *39*, 385; (b) M. C. R. Symons, *Acta Chem. Scand.* **1997**, *51*, 127; *Pure Appl. Chem.* **1981**, *53*, 223.
307. M. J. S. Dewar, A. H. Pakiari, A. B. Pierini, *J. Am. Chem. Soc.* **1982**, *104*, 3242.
308. (a) H. Villar, E. A. Castro, R. A. Rossi, *Can. J. Chem.* **1982**, *60*, 2525; *Z. Naturforsch. Teil A* **1984**, *39*, 49; (b) R. Benassi, C. Bertarini, F. Taddei, *Chem. Phys. Lett.* **1996**, *257*, 633.
309. A. B. Pierini, J. S. Duca, Jr., *J. Chem. Soc., Perkin Trans. 2* **1995**, 1821.
310. A. B. Pierini, J. S. Duca, Jr., D. M. A. Vera, *J. Chem. Soc., Perkin Trans. 2* **1999**, 1003–1009.
311. R. G. Compton, R. A. W. Dryfe, J. Hirst, *J. Phys. Chem.* **1994**, *98*, 10497.
312. (a) J. F. Bunnett, E. Mitchell, C. Galli, *Tetrahedron* **1985**, *41*, 4119; (b) C. Galli, *ibid* **1988**, *44*, 5205; (c) D. B. Denney, D. Z. Denney, *ibid* **1991**, *47*, 6577; (d) D. B. Denney, D. Z. Denney, A. J. Perez, *ibid* **1993**, *49*, 4463; (e) X. M. Zhang, D. L. Yang, X. Q. Jia, Y. C. Liu, *J. Org. Chem.* **1993**, *58*, 7350.
313. T. Abe, Y. Ikegami, *Bull. Chem. Soc. Jpn.* **1976**, *49*, 3227; **1978**, *51*, 196.
314. P. G. Sammes, D. Thetford, M. Voyle, *J. Chem. Soc., Perkin Trans. 1* **1988**, 3229.
315. (a) S. M. Shein, L. V. Bryukhovetskaya, F. V. Pishchugin, V. F. Starichenco, V. N. Panfilov, V. V. Voevodskii, *J. Struct. Chem.* **1970**, *11*, 228; (b) I. I. Bilkis, S. M. Shein, *Tetrahedron* **1975**, *31*, 969.
316. M. Mir, M. Espín, J. Marquet, I. Gallardo, C. Tomasi, *Tetrahedron Lett.* **1994**, *35*, 9055–9058.
317. I. I. Bilkis, V. D. Shteingarts, *Zh. Org. Khim.* **1982**, *18*, 359.
318. G. Grampp, M. C. B. L. Shohoji, B. Herold, S. Steenken, *Ber. Bunsenges. Phys. Chem.* **1990**, *94*, 1507.
319. J. P. Telo, M. C. B. L. Shohoji, B. J. Herald, G. Grampp, *J. Chem. Soc., Faraday Trans.* **1992**, *88*, 47.
320. J. P. Telo, M. C. B. L. Shohoji, *J. Chem. Soc., Perkin Trans. 2* **1998**, 711–714.
321. (a) H. Meier, *Angew. Chem., Int. Ed. Engl.* **1992**, *31*, 1399–1420; (b) H. Goerner, H. J. Kuhn In *Ad. Photochem.* (Eds.: D. C. Neckers, D. H. Volman, G. V. Von Buenau), Wiley, New York, **1995**, Vol. 19, 1–118 and references cited therein.
322. (a) R. Chang, C. S. Johnson, *J. Chem. Phys.* **1965**, *43*, 3183; (b) N. M. Atherton, F. Gerson, N. J. Ockwell, *J. Chem. Soc. A* **1966**, 109.
323. (a) J. Higuchi, K. Ishizu, F. Nemoto, K. Tajima, H. Suzuki, K. Ogawa, *J. Am. Chem. Soc.* **1984**, *106*, 5403; (b) U. Buser, C. H. Ess, F. Gerson, *Magn. Reson. Chem.* **1991**, *29*, 721.
324. F. Gerson, H. Ohya-Nishiguchi, M. Szwarc, G. Levin, *Chem. Phys. Lett.* **1977**, *52*, 587.
325. (a) J. E. Gano, G. Subramaniam, R. Birnbaum, *J. Org. Chem.* **1990**, *55*, 4760–4763; (b) K. K. Laali, J. E. Gano, D. Lenoir, C. W. I. Gundlach, *J. Chem. Soc., Perkin Trans. 2* **1994**, 2169–2173 and references therein.
326. J. E. Gano, E. J. Jacob, P. Sekher, G. Subramaniam, L. A. Eriksson, D. Lenoir, *J. Org. Chem.* **1996**, *61*, 6739–6743.
327. S. Konishi, A. H. Reddoch, *J. Magn. Reson.* **1978**, *29*, 113.
328. F. Gerson, W. B. Martin, Jr., C. Wydler, *Helv. Chim. Acta* **1979**, *62*, 2517.
329. F. Gerson, W. B. Martin, Jr., *Helv. Chim. Acta* **1987**, *70*, 1558.
330. F. Gerson, A. Lamprecht, M. Scholz, H. Troxler, D. Lenoir, *Helv. Chim. Acta* **1996**, *79*, 307–318.
331. C. P. Andrieux, I. Gallardo, J. M. Savéant, *J. Am. Chem. Soc.* **1989**, *111*, 1620.
332. (a) D. M. Wayner, V. D. Parker, *Acc. Chem. Res.* **1993**, *26*, 287; (b) F. G. Bordwell, X. M. Zhang, *ibid* **1993**, *26*, 510.
333. (a) F. G. Bordwell, M. J. Bausch, *J. Am. Chem. Soc.* **1986**, *108*, 1985; (b) C. P. Andrieux, P. Hapiot, J. Pinson, J. M. Savéant, *ibid* **1993**, *115*, 7783.
334. (a) C. P. Andrieux, J. M. Savéant In *Electrochemical Reactions in Investigation of Rates and Mechanisms of Reactions, Techniques of Chemistry* (Ed.: C. F. Bernasconi), Wiley, New York,

**1986**, *VI/4E*, Part 2, 305–390; (b) C. P. Andrieux, P. Hapiot, J. M. Savéant, *Chem. Rev.* **1990**, *90*, 723.

335. H. Lund, K. Daasbjerg, T. Lund, D. Occhialini, S. U. Pedersen, *Acta Chem. Scand.* **1997**, *51*, 135.

336. (a) T. Nagaoka, D. Griller, D. D. M. Wayner, *J. Phys. Chem.* **1991**, *95*, 6264; (b) D. D. M. Wayner, A. Houman, *Acta Chem. Scand.* **1998**, *52*, 377.

337. P. Hapiot, V. V. Konovalov, J. M. Savéant, *J. Am. Chem. Soc.* **1995**, *117*, 1428 and references therein.

338. J. Gonzalez, P. Hapiot, V. Konovalov, J. M. Savéant, *J. Am. Chem. Soc.* **1998**, *120*, 10171–10179.

339. (a) J. S. Miller, A. J. Epstein, *Angew. Chemie.* **1994**, *106*, 399; *Angew. Chemie Int. Ed. Engl.* **1994**, *33*, 385; (b) M. Baumgarten, K. Müllen, *Top. Curr. Chem.* **1994**, *169*, 1.

340. Z. Hou, A. Fujita, H. Yamazaki, Y. Wakatsuki, *J. Am. Chem. Soc.* **1996**, *118*, 2503.

341. (a) N. Hirota, S. I. Weissman, *Mol. Phys.* **1962**, *5*, 537; (b) N. Hirota, *J. Am. Chem. Soc.* **1967**, *89*, 32.

342. A. Behrendt, C. G. Screttas, D. Bethell, O. Schiemann, B. R. Steele, *J. Chem. Soc., Perkin Trans. 2* **1998**, 2039–2045.

343. (a) M. S. Workentin, R. L. Donkers, *J. Am. Chem. Soc.* **1998**, *120*, 2664; *J. Phys. Chem. B* **1998**, *102*, 4061; (b) S. Antonello, M. Musumeci, D. D. M. Wayner, F. Maran, *J. Am. Chem. Soc.* **1997**, *119*, 9541; (c) S. Antonello, F. Maran, *J. Am. Chem. Soc.* **1997**, *119*, 12595.

344. (a) *Organic Peroxides* (Ed.: W. Ando), Wiley, Chichester, England, 1992; (b) *Active Oxygen in Chemistry*, *Search Series* (Eds.: C. S. Foote, J. S. Valentine, A. Greenberg, J. F. Lieban), Blackie, New York, 1995, Vol. 2 ; (c) *Active Oxygen in Biochemistry*, *Search Series* (Eds.: C. S Foote, J. S. Valentine, A. Greenberg, J. F. Lieban), Blackie, New York, 1995, Vol. 3.

345. (a) W. M. Wu, Y. Wu, Y. L. Wu, Z. J. Yao, C. M. Xhou, Y. Li, F. Shan, *J. Am. Chem. Soc.* **1998**, *120*, 3316; (b) J. N. Cumming, D. Wang, S. B. Park, T. A. Shapiro, G. H. Posner, *J. Med. Chem.* **1998**, *41*, 952 and references therein; (c) A. Robert, B. Meunier, *Chem. Soc. Rev.* **1998**, *27*, 273.

346. R. L. Donkers, J. Tse, M. S. Workentin, *Chem. Commun.* **1999**, 135–136.

347. M. J. Weaver, *Chem. Rev.* **1992**, *92*, 463.

348. P. F. Barbara, F. J. Meyer, M. A. Ratner, *J. Phys. Chem.* **1996**, *100*, 13148.

349. (a) S. F. Nelsen, R. F. Ismagilov, D. R. Powell, *J. Am. Chem. Soc.* **1997**, *119*, 10213; (b) S. F. Nelsen, R. F. Ismagilov, D. A. Trieber II, *Science* **1997**, *278*, 846.

350. G. Grampp, M. C. B. L. Shohoji, B. Herold, S. Steenken, *Ber. Bunsenges. Phys. Chem.* **1990**, *94*, 1507.

351. J. P. Telo, M. C. B. L. Shohoji, B. J. Herold, G. Grampp, *J. Chem. Soc., Faraday Trans.* **1992**, *88*, 47.

352. S. Mahmood, B. J. Tabner, V. A. Tabner, *J. Chem. Soc., Faraday Trans.* **1990**, *86*, 3253.

353. J. P. Telo, M. C. B. L. Shohoji, *J. Chem. Soc., Perkin Trans. 2* **1998**, 711.

354. J. P. Telo, G. Grampp, M. C. B. L. Shohoji, *Phys. Chem. Chem. Phys.* **1999**, *1*, 99–104.

355. *Landolt-Börnstein New Series, Magnetic Properties of Free Radicals*, Springer, Berlin, **1980**, group II, *9d1*, 634–639.

356. (a) M. Newcomb, D. Curran, *Acc. Chem. Res.* **1988**, *21*, 206; (b) M. Newcomb, T. R. Varick, S. Y. Choi, *J. Org. Chem.* **1992**, *57*, 373.

357. (a) M. Ahbala, P. Hapiot, A. Houmam, M. Jouini, J. Pinson, J. M. Savéant, *J. Am. Chem. Soc.* **1995**, *117*, 11488; (b) K. Daasbierg, T. B. Christensen, *Acta Chem. Scand.* **1995**, *49*, 128; (c) E. C. Ashby, *Acc. Chem. Res.* **1988**, *21*, 414.

358. L. M. Tolbert, J. Bedlek, M. Terapane, J. Kowalik, *J. Am. Chem. Soc.* **1997**, *119*, 2291–2292.

359. (a) P. K. Das, *Chem. Rev.* **1993**, *93*, 119–144; (b) R. A. McClelland, *Tetrahedron* **1996**, *52*, 6823–6858 and references therein.

360. E. Buncel, T. Durst, *Comprehensive Carbanion Chemistry*, Elsevier, Amsterdam, 1980.

361. D. Budac, P. Wan, *Can. J. Chem.* **1996**, *74*, 1447–1464.

362. (a) S. Torii, *Electro-Organic Syntheses*, Kodansha, Tokyo, **1985**, Ch. 2, 51–74; (b) T. Ohno, T. Fukumoto, T. Hirashima, I. Nishiguchi, *Chem. Lett.* **1991**, 1085–1088.

363. P. R. Bowers, K. A. McLauchlan, R. C. Sealy, *J. Chem. Soc., Perkin Trans. 2* **1976**, 915–921.

364. (a) J. Libman, *J. Am. Chem. Soc.* **1975**, *99*, 4139–4141; (b) M. Novak, A. Miller, T. C. Bruice, *ibid* **1980**, *102*, 1465–1467; (c) Y. Kurauchi, H. Nobuhara, K. Ohga, *Bull. Chem. Soc. Jpn.* **1986**, *59*, 897–905; (d) K. Tsujimoto, N. Nakao, M. Ohashi, *J. Chem. Soc., Chem. Commun.* **1992**, 336–337; (e) M. Mella, M. Freccero, T. Soldi, E. Fasani, A. Albini, *J. Org. Chem.* **1996**, *61*, 1413–1422.
365. (a) T. O. Meiggs, S. I. Miller, *J. Am. Chem. Soc.* **1972**, *94*, 1989–1996; (b) J. D. Coyle, *Chem. Rev.* **1978**, *78*, 97–123.
366. J. D. Margerum, *J. Am. Chem. Soc.* **1965**, *87*, 3772–3773.
367. J. L. Martínez, J. C. Scaiano, *J. Am. Chem. Soc.* **1997**, *119*, 11066–11070.
368. H. Yokoi, T. Nakano, W. Fujita, K. Ishiguro, Y. Sawaki, *J. Am. Chem. Soc.* **1998**, *120*, 12453–12458.
369. M. L. Chan, T. C. Chan, *J. Phys. Chem.* **1995**, *99*, 5765.
370. T. Tominaga, S. Tenma, H. Watanabe, *J. Chem. Soc., Faraday Trans.* **1996**, *92*, 1863.
371. (a) M. Lehni, H. Schuh, H. Fischer, *Int. J. Chem. Kinet.* **1979**, *11*, 705; (b) M. Sitarski, *ibid* **1981**, *13*, 125; (c) M. Lehni, H. Fischer, *ibid* **1983**, *15*, 733.
372. (a) H. J. Werner, Z. Schulten, K. Schulten, *J. Chem. Phys.* **1977**, *67*, 646; (b) R. Kaptein, *J. Am. Chem. Soc.* **1972**, *94*, 6251.
373. (a) R. D. Burkhart, R. J. Wong, *J. Am. Chem. Soc.* **1973**, *95*, 7203; (b) P. P. Levin, I. V. Khudyakov, V. A. Kuzumin, *J. Phys. Chem.* **1989**, *93*, 208.
374. (a) M. Terazima, K. Okamoto, N. Hirota, *Laser Chem.* **1994**, *13*, 169; (b) M. Terazima, K. Okamoto, N. Hirota, *J. Phys. Chem.* **1993**, *97*, 13387; (c) *J. Chem. Phys.* **1995**, *102*, 2506; (d) K. Okamoto, M. Terazima, N. Hirota, *J. Chem. Phys.* **1995**, *103*, 10445.
375. (a) F. D. Evans, C. Chan, B. C. Lamartine, *J. Am. Chem. Soc.* **1977**, *28*, 6492; (b) D. F. Evans, T. Tominaga, C. Chan, *J. Solution Chem.* **1979**, *8*, 461.
376. N. Houser, R. C. Jarnagin, *J. Chem. Phys.* **1970**, *52*, 1069.
377. (a) J. P. Dodelet, G. R. Freeman, *Can. J. Chem.* **1975**, *53*, 1263; (b) S. S. Sam, G. R. Freeman, *J. Chem. Phys.* **1979**, *70*, 1538; **1980**, *72*, 1989.
378. (a) S. K. Lim, M. E. Burba, A. C. Albrecht, *Chem. Phys. Lett.* **1993**, *216*, 405; (b) *J. Phys. Chem.* **1994**, *98*, 9665.
379. M. Terazima, T. Okazaki, N. Hirota, *J. Photochem. Photobiol.* **1995**, *92*, 7.
380. K. Okamoto, N. Hirota, M. Terazima, *J. Chem. Soc., Faraday Trans.* **1998**, *94*, 185–194.
381. I. R. Gould, S. Farid, *Acc. Chem. Res.* **1996**, *29*, 522.
382. (a) H. D. Roth, M. L. M. Schilling, *J. Am. Chem. Soc.* **1980**, *102*, 4303; (b) H. D. Roth, M. L. M. Schilling, G. Jones, *J. Am. Chem. Soc.* **1981**, *103*, 1246.
383. P. C. Wong, D. R. Arnold, *Tetrahedron Lett.* **1979**, 2101.
384. M. L. M. Schilling, H. D. Roth, *J. Am. Chem. Soc.* **1980**, *102*, 7956; **1981**, *103*, 7210.
385. S. B. Karki, J. P. Dinnocenzo, S. Farid, J. L. Goodman, I. R. Gould, T. A. Zona, *J. Am. Chem. Soc.* **1997**, *119*, 431–432.
386. (a) H. Eyring, M. Z. Polanyi, *Phys. Chem.* **1931**, *B12*, 279; (b) K. Wynne, C. Galli, R. M. Hochstrasser, *J. Chem. Phys.* **1994**, *100*, 4797; (c) T. Hannappel, B. Burfeindt, W. Storck, F. Willig, *J. Phys. Chem.* **1997**, *B101*, 6799.
387. (a) L. Eberson, *New J. Chem.* **1992**, *16*, 151; (b) H. Zipse, *Angew. Chem. Int. Ed. Engl.* **1997**, *36*, 1697; (c) H. Tributsch, L. Pohlmann, *Science* **1998**, *279*, 1891.
388. D. Zhong, A. H. Zewail, *J. Phys. Chem.* **1998**, *A102*, 4031.
389. (a) G. N. Sastry, S. Shaik, *J. Am. Chem. Soc.* **1998**, *120*, 2131; (b) J. T. Su, A. H. Zewail, *J. Phys. Chem.* **1998**, *A102*, 4082.
390. (a) R. S. Mulliken, *J. Am. Chem. Soc.* **1950**, *72*, 600; (b) R. S. Mulliken, W. M. Person, *Molecular Complexes*, Wiley, New York, 1969; (c) M. Tamres, M. Brandon, *J. Am. Chem. Soc.* **1960**, *82*, 2134.
391. R. Rathore, S. M. Hubig, J. K. Kochi, *J. Am. Chem. Soc.* **1997**, *119*, 11468.
392. S. M. Hubig, R. Rathore, J. K. Kochi, *J. Am. Chem. Soc.* **1999**, *121*, 617–626.
393. D. H. Levy, R. J. Myers, *J. Chem. Phys.* **1964**, *41*, 1062; **1966**, *44*, 4177.
394. W. M. Tolles, D. W. Moore, *J. Chem. Phys.* **1967**, *46*, 2102.
395. N. Sommer, *Kautschuk, Gummi Kunstst.* **1975**, *28*, 131.
396. F. Gerson, H. Hopf, P. Merstetter, C. Mlynek, D. Fischer, *J. Am. Chem. Soc.* **1998**, *120*, 4815.

397. J. K. Kim, J. F. Bunnett, *J. Am. Chem. Soc.* **1970**, *92*, 7463.
398. (a) G. A. Russell, W. C. Danen, *J. Am. Chem. Soc.* **1966**, *88*, 5663; **1968**, *90*, 347; (b) G. A. Russell, *Chem. Soc. Special Publication* **1970**, No. 24, 271; *Pure Appl. Chem.* **1971**, *4*, 67.
399. (a) N. Kornblum, R. E. Michael, P. C. Kerber, *J. Am. Chem. Soc.* **1966**, *88*, 5660; (b) N. Kornblum, *Angew. Chem., Int. Ed. Engl.* **1975**, *14*, 734.
400. G. A. Russell, A. R. Metcalfe, *J. Am. Chem. Soc.* **1979**, *101*, 2359.
401. G. A. Russell, J. Hershberger, *J. Chem. Soc., Chem. Commun.* **1980**, 216.
402. G. A. Russell, M. Jawdosiuk, M. Makosza, *J. Am. Chem. Soc.* **1979**, *101*, 2355.
403. N. Ono, R. Tamura, A. Kagi, *J. Am. Chem. Soc.* **1980**, *102*, 2851.
404. D. D. Tanner, E. V. Blackburn, G. E. Diaz, *J. Am. Chem. Soc.* **1981**, *103*, 1557.
405. N. Kornblum, S. C. Carlson, R. G. Smith, *J. Am. Chem. Soc.* **1979**, *101*, 647.
406. (a) C. Chacaty, *Compt. Rend. Ser. C* **1966**, *262*, 680; (b) M. C. R. Symons, J. H. Sharp, *Nature* (London) **1969**, *224*, 5226.
407. (a) G. A. Russell, R. K. Norris, E. J. Panek, *J. Am. Chem. Soc.* **1971**, *93*, 5839; (b) H. Sayo, M. Masui, *Tetrahedron* **1968**, *24*, 5075.
408. W. R. Bowman, M. C. R. Symons, *J. Chem. Soc., Perkin Trans. 2* **1983**, 25.
409. (a) J. Adams, M. L. Gross, *J. Am. Chem. Soc.* **1989**, *111*, 435; (b) M. L. Gross, *Int. J. Mass Spectrom Ion Process* **1992**, *118/119*, 137; (c) J. Adams, *Mass Spectrom. Rev.* **1990**, *9*, 141.
410. P. C. H. Eichinger, J. H. Bowie, *Int. J. Mass Spectrom. Ion Process* **1991**, *110*, 123.
411. (a) V. H. Wysocki, M. H. Bier, R. G. Cooks, *Org. Mass. Spectrom.* **1988**, *23*, 627; (b) V. H. Wysocki, M. M. Ross, S. R. Horning, R. G. Cooks, *Rapid Commun. Mass. Spectrom.* **1988**, *2*, 214; (c) V. H. Wysocki, M. M. Ross, *Int. J. Mass Spectrom. Ion Process* **1991**, *104*, 179.
412. (a) C. K. Ingold, G. W. King, *J. Chem. Soc.* **1953**, 2702; (b) K. K. Innes, *J. Chem. Phys.* **1954**, *22*, 863.
413. S. Sakai, K. Morokuma, *J. Phys. Chem.* **1987**, *91*, 3661.
414. (a) K. Matsuura, H. Muto, *J. Chem. Phys.* **1991**, *94*, 4078; (b) *J. Phys. Chem.* **1993**, *97*, 8842.
415. (a) L. Manceron, L. Andrews, *J. Am. Chem. Soc.* **1985**, *107*, 563; (b) *J. Phys. Chem.* **1985**, *89*, 4094.
416. (a) P. H. Kasai, *J. Phys. Chem.* **1982**, *86*, 4092; (b) *J. Am. Chem. Soc.* **1992**, *114*, 3299.
417. Y. Itagaki, M. Shiotani, H. Tachikawa, *Acta Chem. Scand.* **1997**, *51*, 220–223.
418. (a) C. Krohnke, V. Enkelmann, G. Wegner, *Angew. Chem., Int. Ed. Engl.* **1980**, *19*, 912; (b) M. Almeida, R. T. Henriques In *Handbook of Organic Conductive Molecules and Polymers* (Ed.: H. S. Nalwa), John Wiley and Sons, Chichester, 1997, Vol. 1, Ch. 2, 87.
419. (a) K. Bechgaard In *Structure and Properties of Molecular Crystals* (Ed.: M. Pierrot), Elsevier, Amsterdam, 1990, 235; (b) C. Heywang, S. Roth, *Angew. Chem., Int. Ed. Engl.* **1991**, *30*, 176; (c) J. Morgado, I. C. Santos, L. F. Veiros, R. T. Henriques, M. T. Duarte, M. Almeida, L. Alcácer, *J. Mater. Chem.* **1997**, *7*, 2387.
420. H. Tani, Y. Kamada, N. Azuma, N. Ono, *Tetrahedron Lett.* **1994**, *35*, 7051; *Mol. Cryst. Liq. Cryst.* **1996**, *278*, 131.
421. M. R. Bryce, A. K. Lay, A. Chesney, A. S. Batsanov, J. A. K. Howard, U. Buser, F. Gerson, P. Merstetter, *J. Chem. Soc., Perkin Trans. 2* **1999**, 755–763.
422. (a) M. J. Janssen, *The Electronic Structure of Organic Thione Compounds*, Thesis, Utrecht University, 1959; (b) *Recl. Trav. Chim. Pays-Bas* **1960**, *79*, 454, 464, 1067.
423. L. Fälth, U. Håkansson, J. Sandström, *J. Mol. Struct. (THEOCHEM)* **1989**, *185*, 239.
424. A. Mannschreck, A. Talvitie, W. Fischer, G. Snatzke, *Monatsh. Chem.* **1983**, *114*, 101.
425. K. Rang, J. Sandström, *J. Chem. Soc., Perkin Trans. 2* **1999**, 827–832.
426. (a) D. Bourissou, G. Bertrand, *C. R. Acad. Sci. Paris* **1996**, *322*, 489; (b) M. Yoshifuji, Y. Ichikawa, K. Toyota, E. Kasashima, Y. Okamoto, *Chem. Lett.* **1997**, 87.
427. M. Geoffroy, A. Jouaiti, G. Terron, M. Cattani-Lorente, Y. Ellinger, *J. Phys. Chem.* **1992**, *96*, 8241.
428. A. Al Badri, M. Chentit, M. Geoffroy, A. Jouaiti, *J. Chem. Soc., Faraday Trans.* **1997**, *93*, 3631–3635.
429. H. Sidorenkova, M. Chentit, A. Jouaiti, G. Terron, M. Geoffroy, Y. Ellinger, *J. Chem. Soc., Perkin Trans. 2* **1998**, 71–74.

430. M. Chentit, H. Sidorenkova, A. Jouaiti, G. Terron, M. Geoffroy, Y. Ellinger, *J. Chem. Soc., Perkin Trans. 2* **1997**, 921.
431. A. Alberti, M. Benaglia, M. D' Angelantonio, S. S. Emmi, M. Guerra, A. Hudson, D. Macciantelli, F. Paolucci, S. Roffia, *J. Chem. Soc., Perkin Trans. 2*, **1999**, 309–323.
432. A. Alberti, M. Benaglia, M. Guerra, A. Hudson, D. Macciantelli, *J. Chem. Soc., Perkin Trans. 2*, **1999**, 1567–1568.
433. K. Exner, D. Hunkler, G. Gescheidt, H. Prinzbach, *Angew. Chem. Int. Ed.* **1998**, *37*, 1910–1912.
434. (a) C. H. Ess, F. Gerson, W. Adam, *Helv. Chim. Acta* **1992**, *75*, 335; (b) S. F. Gerson, A. Lamprecht, M. Scholz, H. Troxler, *ibid* **1996**, *79*, 307; (c) G. Gescheidt, A. Lamprecht, C. Rüchardt, M. Schmittel, *ibid* **1992**, *75*, 351.
435. (a) A. J. Fatiadi, *The Chemistry of Triple Bonded Groups*, Wiley, New York, **1983**; *Synthesis* **1986**, 249; **1987**, 749, 959; (b) W. Kaim, M. Moscherosch, *Coord. Chem. Rev.* **1994**, *129*, 157; (c) W. E. Buschmann, A. M. Arif, J. S. Miller, *J. Chem. Soc., Chem. Commun.* **1995**, 2343; (d) H. Yamochi, S. Horiuchi, G. Saito, M. Kusonoki, K. Sakaguchi, T. Kikuchi, S. Sato, *Synth. Met.* **1993**, *56*, 2096.
436. (a) J. E. Lind Jr., R. M. Fuoss, *J. Am. Chem. Soc.* **1961**, *83*, 1828; (b) A. L. Farragher, F. M. Page, *Trans. Faraday Soc.* **1967**, *63*, 10; (c) M. L. Bruce, R. C. Wallis, B. W. Skelton, A. H. White, *J. Chem. Soc., Dalton Trans.* **1981**, 1205.
437. M. Decoster, F. Conan, M. Kubicki, Y. Le Mest, P. Richard, J. S. Pala, L. Toupet, *J. Chem. Soc., Perkin Trans. 2*, **1997**, 265–271.
438. (a) G. Gescheidt, A. Lamprecht, C. Rüchardt, M. Schmittel, *Helv. Chim. Acta*, **1991**, *74*, 2094; (b) R. J. Bushby, K. M. Ng, *J. Chem. Soc., Perkin Trans. 2* **1996**, 1053, 1525.
439. U. Buser, C. H. Ess, F. Gerson, *Magn. Reson. Chem.* **1991**, *29*, 721.
440. F. Gerson, C. Sahin, *J. Chem. Soc., Perkin Trans. 2* **1997**, 1127–1132.
441. R. D. Webster, A. M. Bond, R. G. Compton, *J. Phys. Chem.* **1996**, *100*, 10288.
442. (a) R. D. Webster, A. M. Bond, T. Schmidt, *J. Chem. Soc., Perkin Trans. 2* **1995**, 1365; (b) R. D. Webster, A. M. Bond, *J. Org. Chem.* **1997**, *62*, 1779; *J. Chem. Soc., Perkin Trans. 2* **1997**, 1079.
443. R. D. Webster, A. M. Bond, D. C. Coomber, *J. Electroanal. Chem.* **1998**, *422*, 217.
444. R. D. Webster, *J. Chem. Soc., Perkin Trans. 2* **1999**, 263–269.
445. see, e.g., D. M. Guldi, *Chem. Commun.* **2000**, 321–327; L. Echegoyen, L. E. Echegoyen, *Acc. Chem. Res*, **1998**, 593–601.

# 5 Electron-transfer Chemistry of Fullerenes

*Shunichi Fukuzumi and Dirk M. Guldi*

## 5.1 Introduction

The spherical shape of buckminsterfullerenes ($C_{60}$) makes these carbon allotropes ideal probes for the investigation of electron-transfer reactions, especially in relation to Marcus's electron-transfer theory [1]. In the reduction of $C_{60}$, the first electron is added to a triply degenerate $t_{1u}$ unoccupied molecular orbital resulting in a maximum delocalization [2]. In fact, the remarkable cyclic voltammogram and differential pulse voltammogram of $C_{60}$ reveal six reversible reduction steps, all equally separated from each other by ca 450 mV (Figure 1) [3–5]. When, however, $C_{60}$ is derivatized by, for example, introducing suitable addends to the fullerene core, the multi-electron reduction usually becomes thermodynamically more difficult than for the non-derivatized $C_{60}$ [2]. Nonetheless the multi-electron reduction of $C_{60}$ derivatives (i.e. up to six electrons) has still been observed [6–9]. Multi-electron reduction is also possible for higher fullerenes including $C_{70}$, $C_{76}$, $C_{78}$, and $C_{84}$, but the reduction becomes thermodynamically easier as the HOMO–LUMO gap gradually decreases for the higher fullerenes [10–12]. Thus, fullerenes have quite different electron acceptor abilities—the higher fullerenes are markedly better electron acceptors than, for example, $C_{60}$. Encapsulating metals inside the skeleton of fullerenes to form endohedral fullerene metal complexes further attenuates the oxidizing behavior of fullerenes [13–15].

In comparison with the facile reduction of $C_{60}$, the oxidation of $C_{60}$ is rendered more difficult with a reported one-electron oxidation potential of 1.26 V relative to ferrocene/ferricenium [16]. Nonetheless the electrochemical oxidation of $C_{60}$ has shown that multi-electron oxidation occurs at ambient potentials [17]. As for $C_{60}{}^{\bullet-}$, the charge of $C_{60}{}^{\bullet+}$ is highly delocalized, because the five HOMO orbitals of $C_{60}$ ($h_u$) are degenerate [2]. Both reductive and oxidative electron transfer to or from $C_{60}$ are expected to be quite efficient because minimal changes of structure and solvation are associated with the respective electron-transfer events.

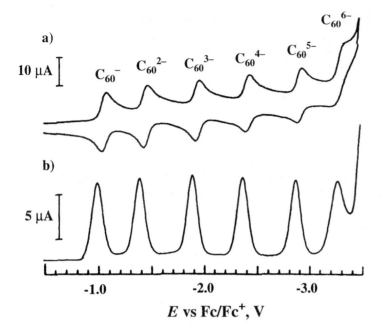

**Figure 1.** Cyclic voltammogram and differential pulse voltammogram of $C_{60}$ [3].

Use of photoexcited fullerenes (i.e., the singlet or triplet excited state) widens the scope of electron-transfer reactions. This assumption is because excitation of fullerenes enhances both the electron-acceptor and -donor behavior of the photoexcited fullerenes. For example, the triplet excited state of $C_{60}$, which is formed by efficient intersystem crossing (i.e. with a quantum yield close to unity) [18, 19] has a reduction potential of $E°_{red} = 1.14$ V relative to the SCE [18, 19]. This potential is clearly more positive than the reduction potential of the ground state $(-0.43$ V) [20]. Thus, the triplet excited state of $C_{60}$ can be reduced with a variety of organic compounds yielding the $C_{60}$ radical anion and the oxidized donor [18].

The rich redox properties of fullerenes and their derivatives, in the ground and excited states, have rendered them important three-dimensional components for the design of novel artificial photosynthesis systems. This chapter focuses on both the fundamental electron-transfer properties of fullerenes and the chemical processes associated with their electron-transfer reactions. Recent advances in time-resolved spectroscopic techniques, including laser flash photolysis and pulse radiolysis, provide a means of monitoring fast electron-transfer reactions and detecting unstable reactive intermediates of fullerene radical anions and cations as typically produced during the electron-transfer reactions. In the first part, the fundamental electron-transfer properties of fullerenes are highlighted. This section is then followed by a detailed description of the mechanistic viability of electron-transfer reactions by summarizing the numerous kinds of research currently ongoing in the electron-transfer chemistry of fullerenes.

## 5.2 Fundamental Electron-transfer Properties of Fullerenes

Photoinduced electron transfer from different electron donors to the triplet excited states of $C_{60}$ and $C_{70}$ occurs efficiently and is typically associated with a small reorganization energy [18, 19, 21–27]. Consequently, the occurrence of fast electron-transfer events involving the fullerene excited states has been well established as giving rise to small intrinsic barriers. In contrast with the fast electron-transfer reactions of the triplet excited state of $C_{60}$, an extremely slow electron-transfer reaction has been reported for the reaction of $C_{60}$ in its ground state with 1,8-diazabicyclo[5.4.0]undec-7-ene (DBU) to produce $C_{60}^{\bullet-}$ in benzonitrile. The latter can be followed even by conventional Vis–NIR spectroscopy [28]. In this instance, however, it is not clear whether the generation of $C_{60}^{\bullet-}$ is directly related to electron transfer from DBU to $C_{60}$, or if $C_{60}^{\bullet-}$ evolves from the product of a secondary reaction.

The efficiency of electron-transfer reduction of $C_{60}$ can be expressed by the self-exchange rates between $C_{60}$ and the radical anion ($C_{60}^{\bullet-}$), which is the most fundamental property of electron-transfer reactions in solution. In fact, an electrochemical study on $C_{60}$ has indicated that the electron transfer of $C_{60}$ is fast, as one would expect for a large spherical reactant. This conclusion is based on the electroreduction kinetics of $C_{60}$ in a benzonitrile solution of tetrabutylammonium perchlorate at ultramicroelectrodes by applying the ac admittance technique [29]. The reported standard rate constant for the electroreduction of $C_{60}$ ($0.3$ cm s$^{-1}$) is comparable with that known for the ferricenium ion ($0.2$ cm s$^{-1}$) [22], whereas the self-exchange rate constant of ferrocene in acetonitrile is reported as $5.3 \times 10^6$ M$^{-1}$ s$^{-1}$, far smaller than the diffusion limit [30, 31].

In general, linewidth variations of the ESR signal of radical anions in the presence of the corresponding neutral compounds at different concentrations are typically used to determine fast exchange rate constants, which are close to the diffusion limit [32, 33]. Unfortunately, this method cannot be applied to $C_{60}^{\bullet-}$, because of the absence of hyperfine structure in its ESR spectrum [34–37]. In contrast, spin polarization and hyperconjugative effects in mono-alkyl adduct radicals of the type $RC_{60}^{\bullet}$ give rise to resolvable hyperfine interactions between R and $RC_{60}^{\bullet}$ [38–42]. The linewidth of the ESR spectrum of $t$-BuC$_{60}^{\bullet}$ increases with increasing concentration of $t$-BuC$_{60}^{-}$ added to a benzonitrile solution of $RC_{60}^{\bullet}$, due to the electron-exchange reaction between $t$-BuC$_{60}^{-}$ and $t$-BuC$_{60}^{\bullet}$ (Eq. 1) [43].

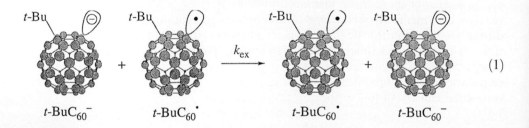

$$t\text{-BuC}_{60}^{-} \quad + \quad t\text{-BuC}_{60}^{\bullet} \quad \xrightarrow{k_{\text{ex}}} \quad t\text{-BuC}_{60}^{\bullet} \quad + \quad t\text{-BuC}_{60}^{-} \qquad (1)$$

The rate constant ($k_{ex}$) of the electron-exchange reaction between $t\text{-BuC}_{60}^-$ and $t\text{-BuC}_{60}^{\bullet}$ was determined as $1.9 \times 10^8$ $\text{M}^{-1}$ $\text{s}^{-1}$ at 298 K by analyzing the linewidth variations of the ESR spectra [43]. Although $C_{60}$ is perturbed by the $t$-Bu addend, the much larger $k_{ex}$ value, compared with the value for the ferrocene/ferricenium exchange ($5.3 \times 10^6$ $\text{M}^{-1}$ $\text{s}^{-1}$), corroborates the high efficiency of electron transfer of $C_{60}$.

The electron transfer reactivity of $C_{60}$ has been compared with that of $p$-benzoquinone which has a slightly more negative one-electron reduction potential ($E^{\circ}_{red}$ relative to the SCE = $-0.50$ V) [44] than $C_{60}$ ($E^{\circ}_{red}$ $-0.43$ V). The rate constants of electron transfer from $C_{60}^{\bullet-}$ and $C_{60}^{2-}$ to electron acceptors such as allyl halides and manganese(III) dodecaphenylporphyrin [45] correlate well with those from semiquinone radical anions and their derivatives. Linear correlations are obtained between logarithms of rate constants and the oxidation potentials of $C_{60}^{\bullet-}$, $C_{60}^{2-}$ and semiquinone radical anions for the different electron acceptors [43]. Such correlations clearly indicate that the reorganization energies for the electron-transfer reactions of $C_{60}^{\bullet-}$ and $C_{60}^{2-}$ are essentially the same as those of semiquinone radical anions. The self-exchange rate constants ($k_{ex}$) between $p$-benzoquinone (Q) and the semiquinone radical anion ($Q^{\bullet-}$) in benzonitrile have been determined independently at different temperatures by analyzing the linewidth variations of the ESR spectra [43]. Arrhenius plots of $\log k_{ex}$ against $1/T$ gave the activation enthalpy ($\Delta H^{\neq}$) and activation entropy ($\Delta S^{\neq}$) which are 3.0 kcal $\text{mol}^{-1}$ and $-3.9$ cal $\text{K}^{-1}$ $\text{mol}^{-1}$, respectively [43]. The small $\Delta S^{\neq}$ value is consistent with adiabatic outer-sphere electron transfer [1]. The reorganization energy ($\lambda$) of the $Q^{\bullet-}/Q$ system in benzonitrile at 298 K is obtained as 16.9 kcal $\text{mol}^{-1}$; this can be regarded as the $\lambda$ value for electron exchange between $C_{60}^{\bullet-}$ and $C_{60}$ (Eq. 2) and between $C_{60}^{2-}$ and $C_{60}^{\bullet-}$.

$$C_{60}^{\bullet-} + C_{60} \xrightarrow{k_{ex}} C_{60} + C_{60}^{\bullet-} \qquad (2)$$

This $\lambda$ value is larger than the value reported for the $Q^{\bullet-}/Q$ system in DMF (13.1 kcal $\text{mol}^{-1}$) but smaller than the value in DMF–$H_2O$ (9:1; 17.7 kcal $\text{mol}^{-1}$) [46]. Because the larger $\lambda$ value in DMF–$H_2O$ is ascribed to the larger solvation of $Q^{\bullet-}$ in the presence of $H_2O$ [46], the solvation of $Q^{\bullet-}$ in benzonitrile might be slightly smaller than that in DMF–$H_2O$ (9:1).

The reorganization energy ($\lambda$) consists of (i) the reorganization of the inner coordination spheres ($\lambda_i$) associated with the structural change upon the electron-transfer reduction of $C_{60}$ in the gas phase and (ii) the solvent reorganization of the outer coordination spheres ($\lambda_0$) [1]. The $\lambda_i$ value for the one-electron reduction of $C_{60}$ has been calculated theoretically as 0.001 kcal $\text{mol}^{-1}$ [43]. Thus, the small $\lambda_i$ value as compared with the observed $\lambda$ values, which include the solvent reorganization energy, indicate that the solvent reorganization plays a major role in determining the intrinsic barrier to the electron-transfer reduction of $C_{60}$.

When the reorganization energy ($\lambda$) of electron transfer is small, the Marcus inverted region ($\lambda < -\Delta G^{\circ}_{et}$) where the rate constant of electron transfer decreases with increasing driving force ($-\Delta G^{\circ}_{et}$) is reached at smaller $-\Delta G^{\circ}_{et}$ values [1, 47]. To date, experimental verification of the existence of such an inverted region has only been achieved by investigating highly exergonic ($-\Delta G^\circ_{\mathrm{et}} \gg 0$) electron-transfer reactions [48–61]. The maximum value of the rate constant for electron transfer is obtained when $-\Delta G^\circ_{\mathrm{et}}$ equals $\lambda$. Thus, the $\lambda$ value of electron transfer can be evaluated from the dependence of the rate constant ($k_{\mathrm{et}}$) for electron transfer on the driving force ($-\Delta G^\circ_{\mathrm{et}}$) provided that the $\lambda$ value is constant for a series of electron-transfer reactions. To obtain the accurate reorganization energy it is, however, necessary to study a system with a small $\lambda$ value and the driving-force range sufficiently wide to probe the behavior far in the inverted region ($\lambda < -\Delta G^\circ_{\mathrm{et}}$). For this purpose the electron-transfer reactions of fullerenes seem to be an ideal system, because the $\lambda$ value is small.

In fact, a Marcus inverted type dependence of the rate constants on $-\Delta G^\circ_{\mathrm{et}}$ has been reported for electron-transfer oxidation of $C_{76}$ and $C_{78}$ with a series of arene radical cations produced upon radiolysis of the $C_{76}/C_{78}$-arene systems [62]. A series of arene radical cations was generated as electron acceptors, and probed in intermolecular electron-transfer reactions with different fullerenes with ionization potentials between 7.59 and 7.1 eV (Eq. 3) [62, 63].

$$(\text{arene})^{\bullet+} + \text{fullerene} \rightarrow \text{arene} + (\text{fullerene})^{\bullet+} \tag{3}$$

The kinetic studies make use of the unequally sized reaction partners (e.g. a large electron donor and a small electron acceptor couple) and benefit from the low viscosity of dichloromethane (DCM), both of which elevate the diffusion-controlled limit. To study the electron transfer, deoxygenated DCM solutions of, for example, *m*-terphenyl at high concentrations (0.02 M) were irradiated in the presence of different concentrations of fullerene (ca $10^{-5}$ M) [62]. This resulted in accelerated decay of (arene)$^{\bullet+}$ UV–Vis absorption, with rates linearly depending on fullerene concentration [62]. Formation of the electron-transfer product, fullerene$^{\bullet+}$, was confirmed spectroscopically by measurement of the NIR fingerprint ($\lambda_{\mathrm{max}} = 1080$ nm) [62, 65].

The rate constants for the investigated arenes vary between $7.8 \times 10^8$ and $4.5 \times 10^{10}$ $M^{-1}$ $s^{-1}$ [62]. The driving forces ($-\Delta G^\circ_{\mathrm{et}}$) calculated on the basis of the difference in the respective arene and fullerene ionization potentials ($-\Delta G^\circ_{\mathrm{et}} = \Delta IP = IP_{\mathrm{arene}} - IP_{\mathrm{fullerene}}$) have a pronounced parabolic dependency on the measured rate constants for the electron-transfer reactions, as shown in Figure 2.

It is important to note that the different solvation of the arene radical cation and the fullerene radical cation is largely canceled out. It is interesting to note that these results indicate a decrease of the rate constant with larger free energy towards the highly exothermic region. From the maximum value of the rate constant of electron-transfer oxidation of $C_{76}$ and $C_{78}$ is obtained the $\lambda$ value (13.8 kcal mol$^{-1}$) [62]. These relatively low values and minor vibrational differences between the fullerene oxidized and ground states are both consequences of the rigid structure and extensive delocalization within the resonance structure of the fullerene $\pi$-system. The $\lambda$ value for the electron-transfer oxidation of fullerenes in dichloromethane (13.8 kcal mol$^{-1}$) is smaller than that for the electron-transfer reduction of $C_{60}$ (16.9 kcal mol$^{-1}$) [43]. This difference might result from the smaller solvent reorganization energy in dichloromethane compared with that in benzonitrile, because

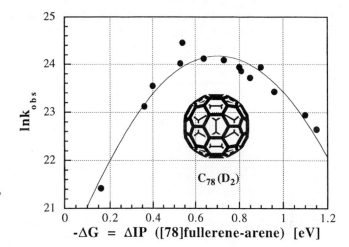

**Figure 2.** Plot of ln $k_{obs}$ for electron transfer from [78]fullerene ($D_2$) to (arene)$^{\cdot+}$, in dichloromethane at room temperature, as a function of the free energy change.

the reorganization of inner coordination spheres ($\lambda_i$) associated with the structural change upon the electron-transfer reduction or oxidation of fullerenes should be close to zero (vide supra).

## 5.3 Thermal Electron Transfer

### 5.3.1 Radiolytically Generated Radical Species

**Non-aqueous media**

Suitable conditions for the study of radiation-induced reduction experiments are usually found in alcohols ($\alpha$-hydroxyl radicals) or aqueous solutions (hydrated electrons). Fullerenes are, however, poorly soluble in polar alcohols and almost insoluble in aqueous solutions. As a consequence of dissolving fullerenes in polar solvents, the spontaneous and irreversible formation of variable-sized clusters and aggregates is observed. Thus to perform radiation-induced reduction studies toluene was employed as a fullerene-dissolving medium [65, 66]. Methodical fullerene reduction was then achieve by addition of adequate co-solvents, namely acetone and 2-propanol. Acetone was chosen as an efficient electron scavenger to avoid generation of excited states, stemming from a reaction between solvated electrons and toluene. After the fast protonation of the initially formed $(CH_3)_2{}^{\cdot}CO^-$ species the $(CH_3)_2{}^{\cdot}COH$ radical is produced; this has exclusively reducing properties and is identical with the main product of the radiolysis of the second co-solvent (2-propanol). Upon pulse radiolysis of an oxygen-free solution of $C_{60}$ in 8:1:1 ($v/v$) toluene–acetone–2-propanol a characteristic NIR spectrum was obtained [65, 66]. In particular, the detection of a maximum around 1080 nm is in excellent agreement

**Table 1.** Redox potentials, rate constants for the radiolytic reduction and $\pi$-radical anion absorption maxima of fullerenes [11, 65, 66, 68–70].

| Compound | $E_{1/2}$ fullerene/fullerene$^{\bullet-}$ (V, relative to Fc/Fc$^+$ [11]) | $k$ ((CH$_3$)$_2$$^{\bullet}$COH) (M$^{-1}$ s$^{-1}$) | $\lambda_{max}$ (nm) |
|---|---|---|---|
| [60]fullerene | −1.06 | $8.5 \times 10^8$ | 1080 |
| [70]fullerene | −1.02 | $8.0 \times 10^8$ | 880 |
| [76]fullerene | −0.83 | $1.3 \times 10^9$ | 905 |
| [78]fullerene | −0.77 | $1.6 \times 10^9$ | 975 |
| [78]fullerene | −0.67 | $1.0 \times 10^9$ | 960 |

with an independently performed CNDO/S calculation of the electronic structure of the fullerene $\pi$-radical anion (i.e., C$_{60}$$^{\bullet-}$) [67]. This suggests successful fullerene reduction as a result of a nearly diffusion-controlled reaction between the fullerene core and (CH$_3$)$_2$$^{\bullet}$COH radicals with an intermolecular rate constant of $8.5 \times 10^8$ M$^{-1}$ s$^{-1}$ (Eq. 4) [65, 66]:

$$(CH_3)_2{}^{\bullet}COH + C_{60} \rightarrow (CH_3)_2CO + C_{60}{}^{\bullet-} + H^+ \qquad (4)$$

A qualitative similar picture was established for the reduction of higher fullerenes (e.g., C$_{70}$, C$_{76}$, C$_{78}$, and C$_{84}$) [68–70]. The associated symmetries of these higher fullerenes were expected to have an impact on the reduction potentials and, equally important, on the relative position of the corresponding $\pi$-radical anion absorption bands (Table 1).

The faster rate constants of $1.3 \times 10^9$ and $1.6 \times 10^9$ M$^{-1}$ s$^{-1}$, respectively, for the radiation-induced reduction of C$_{76}$ and C$_{78}$ with (CH$_3$)$_2$$^{\bullet}$COH radicals [70] compared with that for C$_{60}$ ($8.5 \times 10^8$ M$^{-1}$ s$^{-1}$) [65, 66] support the view that higher fullerenes are better electron-acceptor moieties than C$_{60}$. Also, the position of the absorption maxima was effected by the absolute fullerene size [68–70].

Fullerene reduction was also studied by study of electron transfer from a series of reduced metalloporphyrin $\pi$-radical anions at high porphyrin concentrations (ca $5 \times 10^{-4}$ M) [63]. This enables the fine-tuning of the reducing strength of the reducing radical species, with reduction potentials typically varying between −1.35 V relative to the SCE (ZnP/ZnP$^{\bullet-}$ couple) and −0.8 V relative to the SCE (SnP/SnP$^{\bullet-}$ couple). In this study the same solvent mixture as described above for the direct reduction of fullerenes (i.e. 8:1:1 ($v/v$) toluene–acetone–2-propanol) was used [65, 66, 68–70]. Again, the solely reducing species is the (CH$_3$)$_2$$^{\bullet}$COH radical, which reduces a number of metalloporphyrins at the macrocyclic ligand to yield the corresponding metalloporphyrin $\pi$-radical anions (MP$^{\bullet-}$). Whereas in the absence of an electron acceptor (i.e. fullerene) the MP$^{\bullet-}$ decay slowly via a sequence of disproportionation and protonation reactions, addition of various concentrations of C$_{60}$ ($0.2$–$5 \times 10^{-5}$ M) resulted in an accelerated decay of the absorption between 600 and 800 nm. The linear dependence of the observed rates on the C$_{60}$ concentration support the view that the MP$^{\bullet-}$ indeed react with C$_{60}$ via electron transfer [63]. To confirm the possible fullerene reduction, the formation of the NIR absorption around 1080 nm was monitored. Most importantly, the grow-in rate of the $C_{60}^{\bullet-}$ absorption at different wavelengths in the 980–1100 nm range was nearly identical with the decay rate of the $MP^{\bullet-}$ absorption between 600 and 800 nm. For example, for $ZnP^{\bullet-}$ an intermolecular rate constant of $(2.5 \pm 1.0) \times 10^9 \, M^{-1} \, s^{-1}$ was derived from the $ZnP^{\bullet-}$ decay and $(1.4 \pm 1.0) \times 10^9 \, M^{-1} \, s^{-1}$ from the $C_{60}^{\bullet-}$ formation [63]. These two values are in reasonable agreement and confirm unmistakably the following electron transfer from the one-electron reduced metalloporphyrin $\pi$-radical anion to the fullerene (Eq. 5):

$$ZnP^{\bullet-} + C_{60} \rightarrow ZnP + C_{60}^{\bullet-} \tag{5}$$

Surprisingly, the rate constants for electron transfer between all the investigated $MP^{\bullet-}$ and $C_{60}$ were found to be in the range $(1–3) \times 10^9 \, M^{-1} \, s^{-1}$, despite the large variation in the one-electron reduction potentials of the examined metalloporphyrins [63]. The lack of dependence of the rate constant on the free energy change of the reaction has been ascribed to weak electronic interactions between the metalloporphyrins and $C_{60}$ in the ground state. It should be noted, however, that no ground state charge-transfer interactions were detectable.

With the objective of oxidizing the fullerene core, radiolysis of any chlorinated hydrocarbon solvent provides the means of forming strongly oxidizing radical species [71]. For example, the radiation-induced ionization of dichloroethane (DCE) yields the short-lived and highly reactive solvent radical cation. In general, the electron affinity of $[DCE]^{\bullet+}$ is sufficient to initiate one-electron oxidation of the fullerene moiety (Eq. 6) [72–76].

$$[DCE]^{\bullet+} + C_{60} \rightarrow DCE + C_{60}^{\bullet+} \tag{6}$$

Pulse radiolytic experiments with $C_{60}$ in nitrogen-saturated or aerated DCE solutions yielded a doublet with maxima at 960 and 980 nm [64, 65]. This fingerprint is identical with that computed in CNDO/S calculations [66]. Because of the short-lived nature of the $[DCE]^{\bullet+}$ radical, the rate constants for the $C_{60}$ oxidation could only be estimated; values were ca $2 \times 10^{10} \, M^{-1} \, s^{-1}$ [64, 65].

**Aqueous media**

Viable means of overcoming the insolubility of pristine fullerenes in polar solvents involve (i) incorporation of pristine fullerenes into the hydrophobic cavity of water-soluble host structures (e.g. cyclodextrin [77, 78], surfactants [79–82], and vesicles [81, 83, 84], or (ii) functionalization of the fullerene core with hydrophilic addends (e.g. $C_{60}C(COO^-)_2$, $C_{60}[C(OCH_2CH_2)_3CH_3]_2$, $C_{60}(C_4H_{10}N^+)$, etc.) [2]

*Cyclodextrin*

Cyclodextrins are water-soluble cyclic oligosaccharides with hydrophilic cavities of different sizes. For incorporation of $C_{60}$, thereby facilitating its dissolution in aqueous media, $\gamma$-CD is the most promising candidate. Molecular modeling (i.e., balancing of cavity radius and fullerene size) suggests that a full inclusion in the form of a $C_{60}/\gamma$-CD 1:1 complex is almost impossible to achieve [77, 78]. Despite this apparent size mismatch, incorporation of a single $C_{60}$ molecule between the cavities of two $\gamma$-CD molecules has been unequivocally demonstrated by use of different spectroscopic techniques [77, 78]. On the other hand, incorporation of $C_{60}$ into the smaller analogs of the cyclodextrin family ($\alpha$- and $\beta$-CD) failed; attempts to host the slightly larger $C_{70}$ were similarly unsuccessful. Reaction of the 1:2 $C_{60}/\gamma$-CD complex with radiolytically generated hydrated electrons ($e^-_{aq}$) led to strong NIR changes, with the $C_{60}^{\bullet-}$ fingerprint absorption at 1080 nm [64, 85, 86] (Eq. 7) resembling those described for reduction experiments in 8:1:1 ($v/v$) toluene–2-propanol–acetone [65, 66].

$$C_{60}/\gamma\text{-CD} + e^-_{aq} \rightarrow C_{60}^{\bullet-}/\gamma\text{-CD} \tag{7}$$

To obtain further evidence of the fullerene reduction, different $\alpha$-hydroxyl radicals (i.e. $^\bullet CH_2OH$, $CH_3{}^\bullet CHOH$ and $(CH_3)_2{}^\bullet COH$) were probed, as were hydrated electrons [65, 66, 85, 86]. In this context the different reducing strength of the generated radicals should be considered. In fact, the rate constants associated with the $C_{60}/\gamma$-CD reduction track the strength of these $\alpha$-hydroxyl radicals. In addition, the reduction experiments disclose a noticeable slow-down of the $C_{60}/\gamma$-CD reduction ($2.7 \times 10^8$ $M^{-1}$ $s^{-1}$) [85, 86], initiated by the bulky $(CH_3)_2{}^\bullet COH$ radicals, relative to the intermolecular reduction in homogeneous systems ($8.5 \times 10^8$ $M^{-1}$ $s^{-1}$) [65, 66]. This clearly points to the heterogeneity of the $C_{60}/\gamma$-CD 1:2 complex in which the fullerene core is separated from the surrounding aqueous phase. It is notable that the shielding is not complete and, as a matter of fact, efficient fullerene reduction occurs via hydrated electrons and significantly larger $\alpha$-hydroxyl radicals.

### Surfactants

In contrast with cyclodextrins, amphiphilic micellar assemblies are a class of host molecules with flexible cavities. In a simple view, these molecules bear a hydrophilic head group and a hydrophobic tail and aggregate as 3-dimensional surface-active assemblies. To dissolve fullerenes in aqueous solutions it is important to note that surfactants create hydrophobic micro-domains capable of accommodating water-insoluble entities. A very intriguing aspect of micellar assemblies is their structural flexibility, which enables them to host not only $C_{60}$ but also the larger $C_{70}$, $C_{76}$ and $C_{78}$. A variety of surfactants, e.g. Triton X-100, Tween 20, BRIJ 35, and cetyltrimethylammonium chloride, were successfully shown to host hydrophobic fullerene moieties [80–82, 87]. This assumption is based on (i) the well-resolved absorption bands, which are in excellent resemblance with those reported for organic solutions (e.g., cyclohexane and dichloromethane) and (ii) the lack of any significant light scattering [80–82, 87]. These are attributes that point to the monomeric dissolution of the fullerene moieties within freshly prepared surfactant ensembles. The pulse radiolytically induced reduction of the fullerene was investigated in aqueous solution containing the respective fullerene–surfactant assembly [87]. The differential absorption spectrum recorded after reaction with hydrated electrons (i.e. a NIR fingerprint maximum at 1075 nm) suggests formation of the $C_{60}^{\bullet-}$ radical anion generated with an intermolecular rate constant of $2.6 \times 10^{10}$ M$^{-1}$ s$^{-1}$ [87]. Complementary reduction experiments with $(CH_3)_2{}^\bullet COH$ radicals led to spectral changes superimposable on those described for the hydrated electrons (Eq. 8). In the absence of oxygen $C_{60}{}^{\bullet-}$ is stable on the experimental time scale.

$$C_{60}/\text{surfactant} + e^-{}_{aq} \rightarrow C_{60}{}^{\bullet-}/\text{surfactant} \qquad (8)$$

*Vesicular systems*

Incorporation of $C_{60}$ into different vesicular media was confirmed by monitoring the characteristic absorption bands at 220, 260, and 340 nm [81, 83, 88]. Surprisingly, all lipid bilayer membrane solutions in, for example, toluene or DCM were yellow–brownish in color rather than the magenta color of $C_{60}$. The UV bands are generally red-shifted with increasing fullerene concentration and the spectral shifts are accompanied by light scattering and significantly decreasing extinction coefficients [81, 88]. All these observations point to the same phenomenon—fullerene aggregation within the interior of the lipid bilayer membranes. Measurements of the fullerene triplet lifetime in positively charged DODAB, negatively charged DHP and zwitterionic lecithin systems revealed significantly reduced triplet lifetimes ($<0.2$ μs), relative to that of the fullerene monomer (ca 100 μs) [88]. Because these triplet lifetimes are strongly effected by triplet–triplet or triplet–singlet ground-state annihilation, the shorter lifetimes provide further proof of the proposed aggregation concept. Detection of a NIR band for $C_{60}$, incorporated into the three types of vesicle discussed, confirms the successful reduction, for example, via hydrated electrons $(e^-{}_{aq})$ (Eq. 9) [81, 88].

$$C_{60}/\text{vesicle} + e^-{}_{aq} \rightarrow C_{60}{}^{\bullet-}/\text{vesicle} \qquad (9)$$

Interestingly, the characteristic NIR band of $C_{60}{}^{\bullet-}$ within these lipid materials has a blue shift with a maximum centered at ca 1020 nm [88]; for organic solutions of $C_{60}$ the maximum was recorded at 1080 nm [65, 66]. In addition, the 1020 nm band is substantially broadened, which again suggests fullerene aggregation.

In summary, the incorporation of $C_{60}$ into artificial bilayer membranes, despite being successful in principle, gives rise to a number of unexpected complications. In consideration of the strong aggregation forces among fullerene cores, it is imperative to separate the individual fullerene moieties. Only the adequate hydrophilic–hydrophobic balance of the host matrix is an appropriate means to hinder the spontaneous cluster formation [88].

*Monofunctionalized fullerene derivatives*

Experiments on the radiation-induced reduction of the functionalized fullerene carboxylate, $C_{60}C(COO^-)_2$, in aqueous media were unsuccessful, despite the use of a concentration range that should guarantee rapid fullerene reduction (e.g., as found for $C_{60}$–γ-CD, $C_{60}$–surfactant, and $C_{60}$–vesicle) [89, 90]. In particular, the absorption at ca 700 nm, attributable to the initially generated hydrated electron, remained virtually unaffected on addition of $C_{60}C(COO^-)_2$ and the characteristic NIR absorption of the fullerene radical anion was completely lacking.

Micellar aggregation of the $C_{60}C(COO^-)_2$ derivative in water, a consequence of its amphiphilic structure (i.e., hydrophobic fullerene core and hydrophilic carboxylates), is responsible for the lack of reduction [89, 90]. These clusters are considered to contain an inward oriented hydrophobic fullerene moiety and a hydrophilic layer of carboxylate head groups, which prevent the negatively charged electrons from approaching the fullerene core.

Evidence of clustering in aqueous and alcoholic systems is also provided by the ground state spectra [89, 90]. In particular, the visible region is dominated by strong dynamic light scattering and significant broadening of the absorption bands. An elegant way of avoiding aggregation of $C_{60}C(COO^-)_2$ implies the incorporation of the functionalized fullerene into a water-soluble host molecule (e.g., $\gamma$-cyclodextrin, surfactants, or vesicle [77–83]). The ground-state spectrum of this guest–host complex contains the same narrow bands as, for example, those of monomeric $C_{60}C(COOEt)_2$ or $C_{60}-\gamma$-CD and clearly differs from the presumed $\{C_{60}C(COO^-)_2\}_n$ cluster. The resulting non-aggregated fullerene complex, $C_{60}C(COO^-)_2-\gamma$-CD, gives rise to a rapid reduction via hydrated electrons and other reducing $\alpha$-hydroxyl radicals (i.e., $^\bullet CH_2OH$, $CH_3{}^\bullet CHOH$ and $(CH_3)_2{}^\bullet COH$) [89, 90].

Varying the fullerene concentration always resulted in accelerated decay of the hydrated electron absorption at ca 700 nm. In addition, the characteristic fullerene $\pi$-radical anion absorption, which maximizes for $C_{60}C(COO^-)_2$ at 1040 nm, grows in simultaneously. The rate constant for the fullerene reduction discloses an interesting observation: the value $9.8 \times 10^9$ $M^{-1}$ $s^{-1}$ [89, 90] is slightly lower than that for the $C_{60}-\gamma$-CD complex ($1.8 \times 10^{10}$ $M^{-1}$ $s^{-1}$) [85, 86], paralleling the trend noted for the electrochemically determined reduction potentials of $C_{60}$ ($E_{1/2} = -0.55$ V relative to the SCE; in 4:1 ($v/v$) toluene–acetonitrile) [91] and various mono-functionalized fullerene derivatives ($E_{1/2} = -0.64$ V relative to the SCE; in 4:1 ($v/v$) toluene–acetonitrile) [91]. It is also in line with the observation that mono-functionalization increases the LUMO energy and, in turn, shifts the reduction potentials to more negative values [92].

A fullerene derivative, $C_{60}$ $[C(OCH_2CH_2)_3CH_3]_2$, bearing a non-ionic triethylene glycol chain rather than carboxylates, was investigated to discover potential contributions from repelling effects between the hydrated electrons and the negatively charged fullerene core [93]. It turned out, however, that clustering has the strongest effect on the fullerene reactivity. This conclusion was drawn from strong light scattering noticed in the ground state absorption spectrum; very short triplet excited state lifetimes substantiate fullerene clustering. Only the surfactant-embedded or $\gamma$-CD-incorporated complex revealed the expected ease of radiolytic reduction [93].

In the next step a positively charged pyrrolidinium salt ($C_{60}(C_4H_{10}N^+)$) was probed, not only to overcome the Coulombic charge repulsion, but, more importantly, to attract a reaction with the hydrated electrons (Eq. 10) [90, 93, 94].

$$\{C_{60}(C_4H_{10}N^+)\}_n + e^-{}_{aq} \rightarrow \{(C_{60}{}^{\bullet-})(C_4H_{10}N^+)\}_n \tag{10}$$

**Table 2.** Redox potentials and rate constants for the radiolytic reduction of some water-soluble fullerene derivatives [95].

| Compound | $E_{1/2}$ fullerene/fullerene$^{\cdot-}$ (V, relative to SCE) | $k_{\text{(hydrated electrons)}}$ ($M^{-1}$ $s^{-1}$)[a] |
|---|---|---|
| $C_{60}$ | $-0.35$[b] | $1.8 \times 10^{10}$ |
| $C_{60}C(COO^-)_2$ | $-0.46$[c,d] | $9.8 \times 10^9$ |
| $C_{60}[C(OCH_2CH_2)_3CH_3]_2$ | $-0.46$[e] | Not measured |
| $C_{60}(C_4H_{10}N^+)$ | $-0.29$[b] | $3.5 \times 10^{10}$ |

[a] Measured for the $\gamma$-CD-incorporated complexes [85, 86, 89, 94].
[b] Measured in THF solutions [95].
[c] Recalculated from values, determined in toluene–acetonitrile (4:1 $v/v$) [91].
[d] Measured for the analogous $C_{60}C(COOEt)_2$ [91].
[e] Unpublished result.

Despite the also unequivocally occurring clustering in aqueous media, attachment of a pyrrolidinium group facilitated the reduction of these clusters by hydrated electrons, compared with the negatively charged system $\{C_{60}C(COO^-)_2\}_n$. Comparison of the NIR fingerprints of the reduced cluster ($\{(C_{60}^{\cdot-})(C_4H_{10}N^+)\}_n$) and the reduced $\gamma$-CD-incorporated monomer ($(C_{60}^{\cdot-})(C_4H_{10}N^+)$–$\gamma$-CD) revealed several significant differences [94]. Most importantly, the spectral features of the fullerene cluster are much broader. This is further accompanied by lower yields of the reduced cluster compared with the reduced monomer.

In conclusion, reduction of monomeric fullerenes in aqueous media (e.g. $\gamma$-CD-incorporated or surfactant-embedded [77–83]) by hydrated electrons and uncharged $(CH_3)_2\cdot COH$ radicals reveals significant rate enhancement for the pyrrolidinium salt $C_{60}(C_4H_{10}N^+)$ [94] compared with $C_{60}$ [85, 86] and negatively charged carboxylates $C_{60}C(COO^-)_2$ [89, 90]. This suggests, in line with independently determined quenching rates, an anodic shift of the reduction potential of $C_{60}(C_4H_{10}N^+)$ relative to $C_{60}C(COO^-)_2$, $C_{60}$ [C(OCH_2CH_2)_3CH_3]_2, and $C_{60}$ (Table 2).

*Bisfunctionalized fullerene derivatives*

Because monofunctionalization of fullerenes shows that a single hydrophilic addend is insufficient to prevent the strong hydrophobic interactions among the compounds [89, 90, 93, 94], multiple functionalized derivatives were examined as water-soluble probes. In particular, introduction of a second hydrophilic ligand (e.g., pyrrolidinium salts or carboxylates) to the fullerene core enhances the surface coverage of the hydrophobic fullerene surface. In turn, it was expected that fullerene aggregation might be suppressed. It should be stated that these water-soluble derivatives are important alternatives to the $\gamma$-CD-incorporated and surfactant-embedded fullerenes.

A series of water-soluble bis-carboxylates and bis-pyrrolidinium salts (e.g. $C_{60}$ $[C(COO^-)_2]_2$ and $C_{60}(C_4H_{10}N^+)_2$) were studied in aqueous solutions and compared with $[(C_{60})C(COO^-)_2]_n$ and $[(C_{60}{}^{\bullet -})(C_4H_{10}N^+)]_n$ clusters [96–98]. In the ground state, the fact that none of these derivatives deviates significantly from the Lambert–Beer law is been used as a first argument in support of the absence of fullerene clusters [96]. Furthermore, triplet lifetimes of ca 40 µs [96–98] comparred with 0.4 µs and less (i.e. for fullerene clusters) [89, 90, 94] is extra evidence for the truly monomeric occurrence of these bis-functionalized fullerenes in aqueous solutions.

Radiation-induced reduction of bis-carboxylates or bis-pyrrolidinium salts in $N_2$-purged aqueous solution showed that the expected formation of the diagnostic NIR transition band occurs synchronously with the decay of the hydrated electron absorption at ca 700 nm [96]. The bis-carboxylates ($C_{60}[C(COO^-)_2]_2$) were reduced by hydrated electrons and $(CH_3)_2{}^{\bullet}COH$ radicals with rate constants of $(0.75 - 3.4) \times 10^9$ and $(0.9 - 2.2) \times 10^8$ $M^{-1}$ $s^{-1}$, respectively [96]. These values are, nevertheless, significantly lower than those for the reduction of $C_{60}$–surfactant, $C_{60}C(COO^-)_2$–surfactant, and the respective $\gamma$-CD-incorporated complexes. Such an effect reflects the perturbation of the fullerene $\pi$-system caused by placing two functional negatively charged appendices on to the fullerene core. Because of the electron-withdrawing nature of the pyrrolidinium groups, the electron-acceptor properties of the bis-pyrrolidinium salts ($C_{60}(C_4H_{10}N^+)_2$) [97] are markedly improved relative to the bis-carboxylates ($C_{60}[C(COO^-)_2]_2$) [96] and also relative to pristine $C_{60}$ [85, 86]. For example, the rate constants for the fullerene reduction with hydrated electrons ($(0.88 - 2.2) \times 10^{10}$ $M^{-1}$ $s^{-1}$) and $(CH_3)_2{}^{\bullet}COH$ radicals ($(4.7 - 7.1) \times 10^8$ $M^{-1}$ $s^{-1}$) [97] are clearly faster than those noted for any other functionalized fullerene derivative. In conclusion, the extended and highly delocalized $\pi$-system experiences gradual destruction. Specifically, perturbation is observed which depends strongly on (i) the extent of functionalization, (ii) the relative distance of the individual addends from each other, and (iii) the electronic structure of the substituent.

### 5.3.2 Electron-transfer Reactions of Fullerene Anions

#### Derivatization of fullerene anions

Fullerenes such as $C_{60}$ are easy to reduce [1], but difficult to oxidize electrochemically [2], and thus are generally regarded as electrophiles, or electron acceptors, rather than nucleophiles or electron donors. The derivatization of fullerenes has therefore been achieved by polar reactions with a variety of nucleophiles [99–104], e.g. electron-rich olefins [105–114], carbenes [115–122], carbanions [123–129], alkoxides [130, 131], and organometallic reagents [132–138]. When, on the other hand, electrons are chemically or electrochemically added to $C_{60}$, the resulting anions are expected to behave as strong nucleophiles or electron donors.

The two-electron reduction of $C_{60}$ to produce $C_{60}{}^{2-}$ can be achieved electrochemically in the presence of methyl iodide to yield the dimethyl adduct, $Me_2C_{60}$ [139]. In this context, catalytic currents are observed for the electrochemical reduction of $C_{60}$ or $C_{70}$ to produce $C_{60}{}^{2-}$ or $C_{70}{}^{2-}$, respectively, in the presence of vicinal dibromides [140] or $\alpha,\omega$-dihaloalkanes [141, 142]. Electrosynthesis of methanofullerenes has been achieved by the reaction of $C_{60}{}^{2-}$ with *ipso*-brominated and iodinated reagents [143]. The reaction of $C_{60}{}^{2-}$ with halogenated compounds was performed with no potential applied after the completion of the electrochemical conversion of $C_{60}$ to $C_{60}{}^{2-}$ [143].

The dianion, $C_{60}{}^{2-}$, is also been produced by chemical reduction of $C_{60}$ with thiolate anions [144] or with the trimethyl-*p*-benzoquinone dianion [145]. For example, the reaction of $C_{60}{}^{2-}$ with two equivalents of benzyl bromide yields the dibenzyl adduct, $(PhCH_2)_2C_{60}$ (Eq 11).

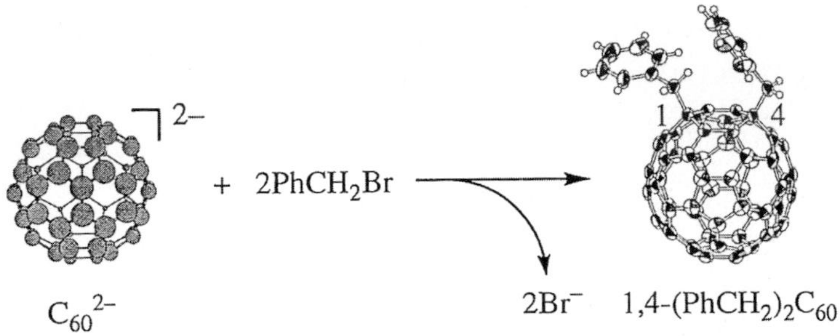

$$C_{60}{}^{2-} \quad + \quad 2PhCH_2Br \quad \longrightarrow \quad 2Br^- \quad 1,4\text{-}(PhCH_2)_2C_{60} \tag{11}$$

The major isomer isolated, and characterized by single-crystal X-ray diffraction, was found to be the 1,4 isomer rather than the 1,2 isomer (Eq. 11) [145]. In a 1,4 adduct of $C_{60}$ the double bond in a pentagon increases the energy of the $C_{60}$ cage by about 8 kcal mol$^{-1}$ compared with that of the 1,2 adduct [146, 147]. However, this disfavored energy increase is compensated by a decrease in steric hindrance when two bulky groups are placed in a 1,4-position. Thus, 1,4 adducts are usually the dominant products when bulky groups are added to $C_{60}$ [145, 148]. It has been shown that the 1,2 and 1,4 adducts of $C_{60}$ can be identified from their different absorption spectra [148]—the 1,2 adducts have a weak spike at ca 432 nm [149] and the 1,4 adducts have a broad absorption band at ca 445 nm [145, 148, 150].

A small amount of 1,2 adduct for the reaction of $C_{60}{}^{2-}$ with benzyl bromide was isolated and also characterized spectroscopically [151]. The mono- and dianions of the 1,2- and 1,4-$(PhCH_2)_2C_{60}$ derivatives were then generated by controlled-potential electrolysis of the organofullerenes in benzonitrile. The visible and near-IR spectrum of $[1,2\text{-}(PhCH_2)_2C_{60}]^{\bullet-}$ has a strong absorption band at 1030 nm and a weaker band at 906 nm, whereas that of $[1,4\text{-}(PhCH_2)_2C_{60}]^{\bullet-}$ has two well-defined moderate absorption bands at 1498 and 989 nm and much weaker bands at 874 and 1237 nm [151]. Doubly reduced 1,2-$(PhCH_2)_2C_{60}$ has one strong absorption band at 905 nm and weaker bands at 801 and 1028 nm, whereas doubly reduced 1,4-$(PhCH_2)_2C_{60}$ has two strong absorption bands at 1294 and 903 nm and two much weaker bands at 806 and 1087 nm [151]. These spectroscopic data can be used to differentiate between the 1,2 and 1,4 isomers of $C_{60}$.

**Electron transfer or the $S_N2$ pathway**

The conversion of $C_{60}{}^{2-}$ to $R_2C_{60}$ passes through a $RC_{60}{}^-$ intermediate [145]. This alkylation could proceed via two probable mechanistic pathways. One is an electron

**First Addition**

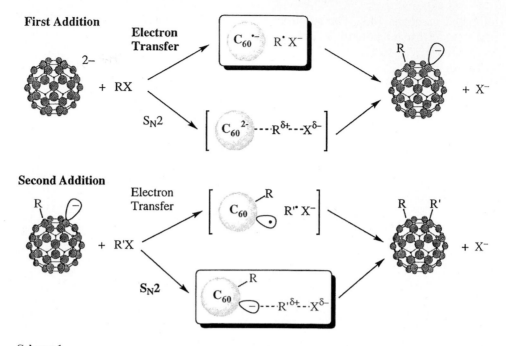

**Scheme 1.**

transfer mechanism where $C_{60}^{2-}$ acts as an electron donor and the other is an $S_N2$ mechanism where $C_{60}^{2-}$ acts as a nucleophile (Scheme 1).

Such an electron-transfer rather than nucleophilic process has been one of the most central propositions in reaction mechanism [152–160]. The second step in the formation of the final bis-adduct may also proceed by an electron transfer or alternatively by an $S_N2$ reaction (Scheme 1). An electron transfer from $C_{60}^{2-}$ to RX would give a radical pair ($C_{60}^{\bullet-}$ R$^{\bullet}$ X$^-$) where the R–X bond is cleaved upon dissociative electron transfer [160]. A facile radical coupling in the radical pair may subsequently afford the same $RC_{60}^-$ product as expected also from the alternative $S_N2$ pathway shown in the first R group addition step of Scheme 1. The addition of a second alkyl group to $RC_{60}^-$ may also proceed via electron transfer from $RC_{60}^-$ to a different alkyl halide (R′X) to give a radical pair ($RC_{60}^{\bullet}$R′$^{\bullet}$X$^-$), followed by a fast radical coupling to yield the final product, $RR'C_{60}$. As shown in Scheme 1, the electron transfer and $S_N2$ pathways in each step would give the same products. However, the operating mechanism has been distinguished by evaluating the reactivity of RX in each step [161].

The reactivity of RX toward $C_{60}^{2-}$ is found to be insensitive to the steric hindrance of the alkyl group of RX. This is however in contrast to the normal reactivity in $S_N2$ reactions [162]. For example, the largest rate constant is seen for a sterically hindered $CCl_4$, while the also sterically hindered *t*-BuI gives rise to a larger $k_{obs}$ value ($4.7 \times 10^{-2}$ M$^{-1}$ s$^{-1}$) than the much less sterically hindered MeI ($3.5 \times 10^{-2}$ M$^{-1}$ s$^{-1}$) [161]. Such an insensitivity of the rate on the steric hindrance

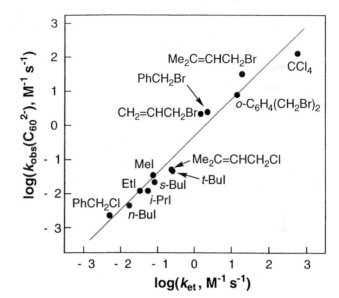

**Figure 3.** Correlation between $\log k_{obs}$ for the reaction of $C_{60}^{2-}$ with alkyl halides and $\log k_{et}$ for the electron-transfer reaction of $Me_4Q^{\cdot-}$ with the same alkyl halides in degenerated PhCN at 298 K [161].

of RX is a known characteristic of electron-transfer reductions of RX [163–165]. The electron transfer pathway in the reaction of $C_{60}^{2-}$ with RX was confirmed by comparing the observed rate constants ($k_{obs}$) for formation of $RC_{60}^-$ and the rate constants ($k_{et}$) of electron-transfer reactions from tetramethylsemiquinone radical anion ($Me_4Q^{\cdot-}$) to the same series of RX [161]. The radical anion $Me_4Q^{\cdot-}$ was chosen to model the electron-transfer reactions of $C_{60}^{2-}$ with RX, since $Me_4Q^{\cdot-}$ has nearly the same oxidation potential ($E_{1/2} = -0.84$ V relative to the SCE in MeCN) [166] as that of $C_{60}^{2-}$ ($E_{1/2} = -0.87$ V relative to the SCE in PhCN) [20]. The plot between the experimentally determined $\log k_{obs}(C_{60}^{2-})$ and $\log k_{et}$ values in Figure 3 shows a good linear correlation with a slope of unity. This is a clear demonstration that a reaction of $C_{60}^{2-}$ with RX proceeds via the rate-determining electron transfer from $C_{60}^{2-}$ to RX as the authentic electron-transfer reaction of $Me_4Q^{\cdot-}$ with the same series of RX [161].

The reactivity order of RX for the second addition step to yield a dialkyl adduct of $C_{60}$ is quite different from that for the first addition step [161]. For example, $CCl_4$ is the most reactive RX group in the first addition step, but it is much less reactive in the second addition step. For example, the $k_{obs}$ value of $3.0 \times 10^{-5}$ $M^{-1}$ $s^{-1}$ is three orders of magnitude smaller than the most reactive $Me_2C=CHCH_2Br$ ($2.7 \times 10^{-2}$ $M^{-1}$ $s^{-1}$) [161]. The sterically hindered $t$-BuI and $s$-BuI reveal both good reactivities in the first R group addition, but show no reactivity at all in the second R group addition. Such sensitivity to the steric hindrance of the RX alkyl groups is characteristic of $S_N2$ reactions involving RX and nucleophiles. A typical example for this reaction type is the conversion of cobalt(I) tetraphenylporphyrin anion, $Co^ITPP^-$, to the $\sigma$-bonded complex, $RCoTPP$ [162]. A good linear correlation with a slope of unity is observed between $\log k_{obs}$ of $t$-$BuC_{60}^-$ and $\log k_{S_N2}$ of $CoTPP^-$

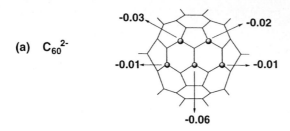

**(a)** $C_{60}^{2-}$

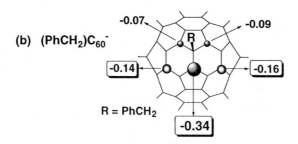

**(b)** $(PhCH_2)C_{60}^-$

R = PhCH₂

Figure 4. Calculated charge distribution on (a) $C_{60}^{2-}$ and (b) $(PhCH_2)C_{60}^-$ [145].

[162]. Such a correlation clearly indicates that the alkylation of $t$-BuC$_{60}^-$ with RX proceeds via an $S_N2$ pathway rather than via an electron transfer pathway.

It is interesting to discuss the reason why the reactions of $C_{60}^{2-}$ and $t$-BuC$_{60}^-$ with RX proceed via two different mechanisms, namely, an electron transfer or an $S_N2$, respectively. The high delocalization of negative charge on the $C_{60}$ skeleton should be noted with the most negative charge (i.e. $-0.06$) at a $C_{60}^{2-}$ carbon atom (see for illustration Figure 4) [145].

$C_{60}^{2-}$ with such a small negative charge at each carbon cannot act, however, as an effective nucleophile in the reaction with RX. In such a case, an electron transfer pathway is the only choice for the reaction of $C_{60}^{2-}$ to proceed. A nucleophile, which is also an electron donor, is generally forced to undergo an electron transfer pathway when the steric hindrance at the reaction center prevents the nucleophile from interacting strongly enough to undergo an $S_N2$ reaction [152–160, 167]. Thus, the $C_{60}^{2-}$/RX system provides a unique example for an electron transfer pathway of a nucleophile which has a highly delocalized negative charge. In contrast to $C_{60}^{2-}$, the negative charge on $(PhCH_2)C_{60}^-$ becomes significantly localized at the C(2) position ($-0.34$) as well as at the C(4) and C(11) positions ($-0.15$) [145]. Such a localization of negative charge is also observed for $t$-BuC$_{60}^-$ [168]. Addition at the C(2) position may be difficult because of steric repulsion between the proximate benzyl or $t$-Bu groups and the nucleophilic addition of RX may then occur at the C(4) or C(11) positions, where the negative charge is the second largest [156].

In conclusion, the drastic difference in charge distribution between $C_{60}^{2-}$ and $RC_{60}^-$ causes significant differences in their nucleophilic reactivities so that the reactions of $C_{60}^{2-}$ and $RC_{60}^-$ with R'X proceed via different pathways, namely, an electron transfer in the first step and an $S_N2$ pathway in the second step. A summary of the mechanism is shown in Scheme 2 for the formation of $t$-Bu$(PhCH_2)C_{60}$ as a typical example [161].

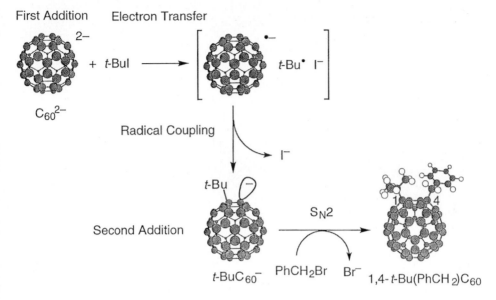

**Scheme 2.**

The initial electron transfer from $C_{60}^{2-}$ to $t$-BuI gives the $C_{60}^{\cdot-}/t$-Bu$^{\cdot}$ radical pair following an instantaneous cleavage of the C–I bond upon the dissociative electron-transfer reduction of $t$-BuI. A fast radical coupling between $C_{60}^{\cdot-}$ and $t$-Bu$^{\cdot}$ occurs in the radical pair to produce the monoadduct anion, $t$-BuC$_{60}^{-}$. The further addition of a second sterically hindered group does not occur via an $S_N2$ reaction. $t$-BuC$_{60}^{-}$ then undergoes an $S_N2$ reaction with less sterically hindered PhCH$_2$Br to yield the final product with two different alkyl groups such as $t$-Bu(PhCH$_2$)C$_{60}$ [161]. Such a sequential process for the derivatization of $C_{60}$ is made possible by differences in alkyl halide reactivity in the first electron transfer and the second $S_N2$ step of the reaction.

Only a few examples of organofullerenes with multiple addends involving a 1,4-addition pattern have been reported [133, 169, 170]. The electron transfer and $S_N2$ sequence in Scheme 2 can be repeated by starting from the dianion of 1,4-(PhCH$_2$)$_2$C$_{60}$ to produce the corresponding tetrakisadduct, (PhCH$_2$)$_4$C$_{60}$ (Scheme 3) [148, 171].

The two isomers were isolated by HPLC from the products and the X-ray crystal structure of each isomer is shown in Scheme 3, while a schematic representation of each compound is illustrated in Figure 5 [171].

Each isomer has four benzyl groups on the $C_{60}$ cage, with a 1,4;1,4-addition pattern for one compound (Figure 5a) and a 1,4;1,2-addition pattern for the other (Figure 5b). The positions of the two benzyl groups in the starting compound are labeled as 'A' and 'B' while the two added benzyl groups in the products are labeled as 'C' and 'D' (see Figure 5). The two isomers have three benzyl groups at identical positions. The regiochemistry of the two isomers can be rationalized by the initial

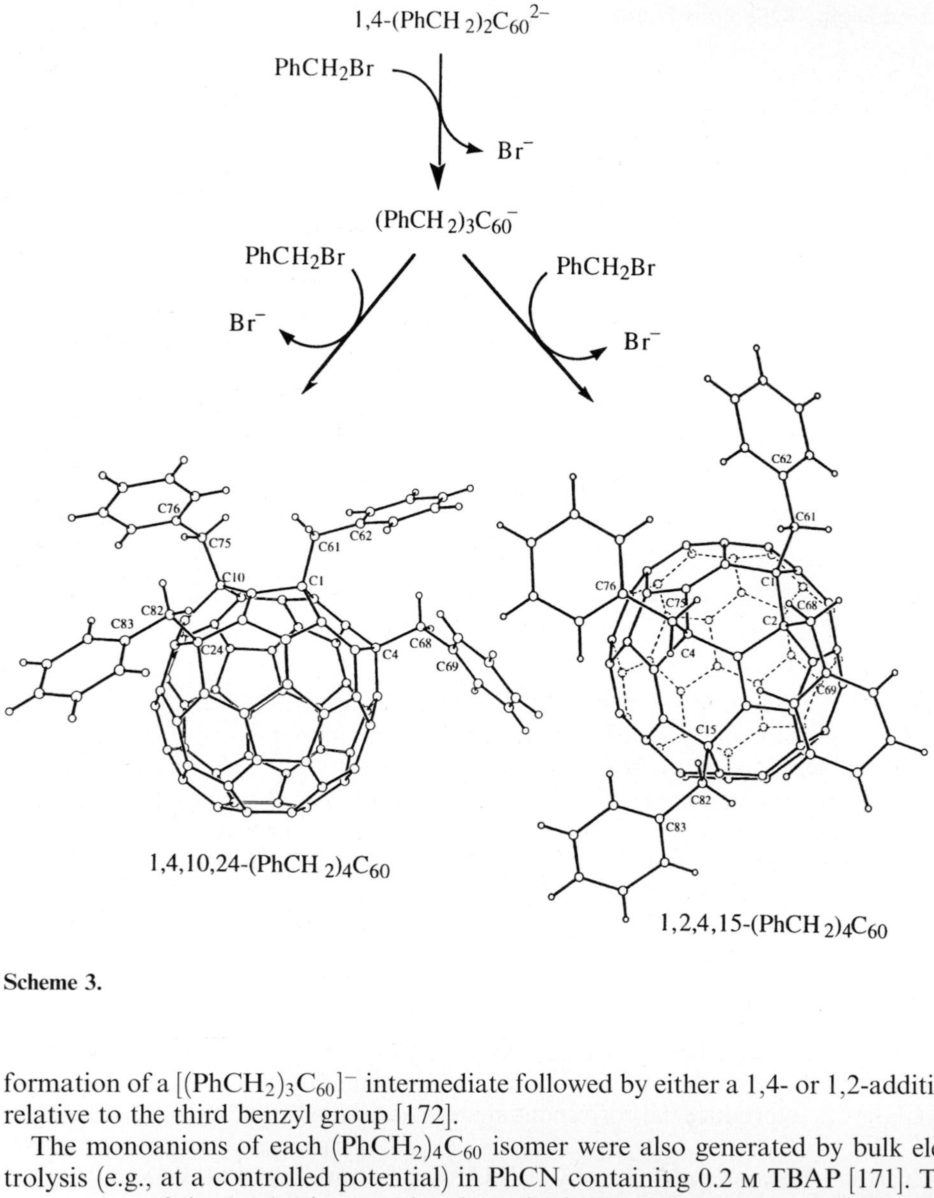

**Scheme 3.**

formation of a $[(PhCH_2)_3C_{60}]^-$ intermediate followed by either a 1,4- or 1,2-addition relative to the third benzyl group [172].

The monoanions of each $(PhCH_2)_4C_{60}$ isomer were also generated by bulk electrolysis (e.g., at a controlled potential) in PhCN containing 0.2 M TBAP [171]. The monoanion of the 1,4;1,4 isomer absorbs at both 801 and 1374 nm and this result contrasts with the monoanion of the 1,4;1,2 isomer which exhibits only a single absorption band at 1084 nm [171].

The $I_h$ symmetry of the parent $C_{60}$ is lowered by the introduction of two benzyl groups to $C_{60}$ to give $1,4-(C_6H_5CH_2)_2C_{60}$, which leads to the larger $g$ value (2.0004) of $1,4-(C_6H_5CH_2)_2C_{60}^{\cdot-}$ than that of $C_{60}^{\cdot-}$ ($g = 1.9984$) and a much smaller line-width ($\Delta H_{msl} = 2.5$ G at 213 K) which is temperature independent [172]. An even smaller $\Delta H_{msl}$ value (0.17 G) and a larger $g$ value (2.0011) are observed for the

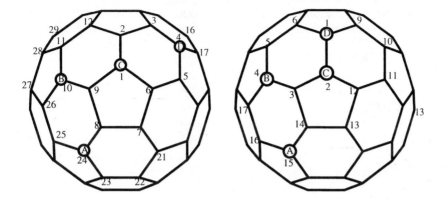

(a) 1,4,10,24-(PhCH$_2$)$_4$C$_{60}$ (1,4;1,4-isomer)  (b) 1,2,4,15-(PhCH$_2$)$_4$C$_{60}$ (1,4;1,2-isomer)

**Figure 5.** Schematic representations of (a) the 1,4;1,4 isomer and (b) the 1,4;1,2 isomer of (PhCH$_2$)$_4$C$_{60}$. The positions of substitution on the C$_{60}$ molecule are indicated by the circles A, B, C, and D. Selected carbon atoms are labeled in accordance with IUPAC numbering [171].

tetrabenzyl C$_{60}$ adduct radical anion, 1,4,10,24-(C$_6$H$_5$CH$_2$)$_4$C$_{60}$$^{\cdot-}$ [172]. This is ascribed to a large splitting of the degenerate $t_{1u}$ orbitals as a result of introducing four benzyl groups to C$_{60}$ [172]. In this case a hyperfine structure due to two non-equivalent protons of only one benzyl group ($a$H$_1$ = 0.31 G, $a$H$_2$ = 0.11 G) is observed. This is consistent with the predicted localized spin density at the C2 position next to the C1 carbon to which a benzyl group is attached [172].

Dibenzyl adducts of C$_{70}$ were also electrosynthesized from C$_{70}$$^{2-}$ in the presence of PhCH$_2$Br and purified by HPLC [173]. Mass spectral results for the largest HPLC fraction confirm formation of the [(PhCH$_2$)$_2$C$_{70}$] compound while $^1$H NMR spectroscopy suggests that a mixture of three isomers is present in this fraction, all of which are 1,4-addition products [173].

**Base-induced electron transfer**

C$_{60}$ is known to undergo reduction to C$_{60}$$^{\cdot-}$ by addition of a MeO$^-$ solution. This reduction is, nevertheless, accompanied by the formation of the adduct anions, C$_{60}$(OMe)$_n^-$ (n = 1, 3, 5, 7) [174]. Electron transfer from MeO$^-$ to C$_{60}$ is thought to result in the formation of C$_{60}$$^{\cdot-}$ and the corresponding adduct anions [174]. $p$-Benzoquinone is also known to be reduced to semiquinone radical anion in a reaction with OH$^-$ in acetonitrile [175, 176]. The hydroxide ion is a much stronger electron donor in aprotic solvents such as acetonitrile than in water, since the solvation energy for OH$^-$ is less in aprotic solvents than in water [175]. However, no oxidized products of OH$^-$ were found in the reaction of $p$-benzoquinone with OH$^-$ [176]. The only oxidized product detected evolved from $p$-benzoquinone itself, namely, the rhodizonate dianion that is the ten-electron oxidized product of $p$-benzoquinone [176]. Thus, the one-electron reduction of ten equivalents of $p$-benzoquinone is accompanied by the ten electron oxidation of one equivalent of $p$-benzoquinone. In this case

$OH^-$ is not the electron donor, but instead *p*-benzoquinone itself acts as the electron donor in the presence of $OH^-$. The addition of $OH^-$ to *p*-benzoquinone initiates an electron transfer from the $OH^-$ adduct of *p*-benzoquinone to *p*-benzoquinone leading to the noble disproportionation. This yields ten equivalents of semiquinone radical anion and rhodizonate dianion [176].

In the reaction of $C_{60}$ with $MeO^-$, the ultimate electron source to reduce $C_{60}$ should be $C_{60}$ itself, since $C_{60}$ is oxidized to $C_{60}(OMe)_n^-$ accompanied by the reduction of $C_{60}$ to $C_{60}^{\cdot -}$ [174]. The detailed spectroscopic and kinetic studies have revealed that a methoxy adduct anion of $C_{60}$ is the true electron donor and that $MeO^-$ is acting as a very strong base or nucleophile rather than an electron donor in PhCN [177]. The reaction sequence is shown in Scheme 4 [177]. $(MeO)C_{60}^-$ becomes a much stronger reductant than the parent $C_{60}$ once the methoxide ion adduct is formed. Then, an electron transfer from $(MeO)C_{60}^-$ to $C_{60}$ occurs to give a methoxy adduct radical of $C_{60}$ $((MeO)C_{60}^{\cdot})$ and $C_{60}^{\cdot -}$. Although the electron transfer from $(MeO)C_{60}^-$ $(E^p_{ox} = 0.14$ V) to $C_{60}$ $(E^{\circ}_{red} = -0.43$ V) [20] is thermodynamically unfavorable as indicated by the difference in the redox potentials the follow-up reaction involving an addition of the second $MeO^-$ to $(MeO)C_{60}^{\cdot}$, is a highly exothermic process to give $(MeO)_2C_{60}^{\cdot -}$. Since $(MeO)_2C_{60}^{\cdot -}$ is a much stronger electron donor than the parent $C_{60}^{\cdot -}$, an electron transfer from $(MeO)_2C_{60}^{\cdot -}$ to $C_{60}$ occurs readily to yield the dimethoxy adduct of $C_{60}$ $[(MeO)_2C_{60}]$ and $C_{60}^{\cdot -}$ [177] (Scheme 4).

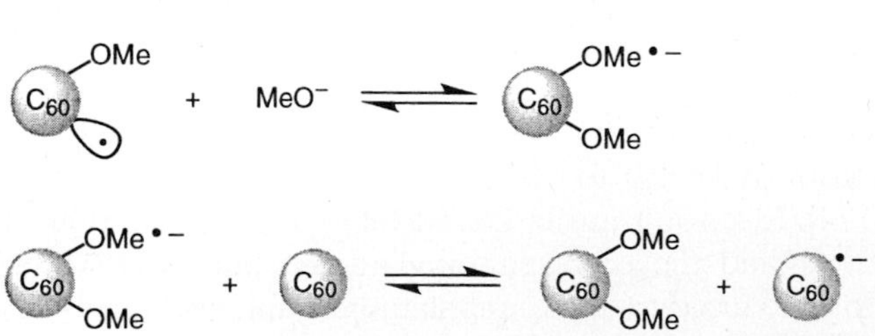

Scheme 4.

## 5.4 Intermolecular Photoinduced Electron Transfer

Since the absorption features of fullerenes extend over the UV and visible spectrum, light of different frequency can be used to excite fullerenes [19, 178]. Depending on the light frequency used, vibrational levels of the first singlet excited state or higher singlet excited states are populated. After a series of ultra-fast, cascade-like relaxation processes, the population of the lowest vibrational level of the first singlet excited state is completed before a quantitative intersystem crossing ($\tau \approx 1.3$ ns) to the triplet excited state takes place [179]. The latter is then the starting point for a slow regeneration ($\tau \approx 100$ μs) of the singlet ground state [19]. Promoting an electron from a low-energy bonding orbital to a high-energy anti-bonding orbital leads to two profound effects. It increases the electron affinity ($C_{60}/C_{60}{}^{\bullet-} = -0.43$ V relative to the SCE; in benzonitrile) [20] and, at the same time, reduces the oxidation potential of the fullerene ($C_{60}/C_{60}{}^{\bullet+} = 1.76$ V relative to the SCE; in benzonitrile) [16, 92] by precisely the amount of the excited state energy. Thus, it makes the fullerene excited states stronger oxidants ($E_{1/2}$ ($^1C_{60}{}^*/C_{60}{}^{\bullet-}$) $= 1.56$ V relative to the SCE; $E_{1/2}$ ($^3C_{60}{}^*/C_{60}{}^{\bullet-}$) $= 1.13$ V relative to the SCE) and stronger reductants ($E_{1/2}$ ($^1C_{60}{}^*/C_{60}{}^{\bullet+}$) $= -0.23$ V relative to the SCE; $E_{1/2}$ ($^3C_{60}{}^*/C_{60}{}^{\bullet+}$) $= 0.19$ V relative to the SCE). In principle, both excited states can be probed in inter- and intramolecular charge transfer reactions with adequate electron acceptor and electron donor moieties. An intermolecular reaction with the initially formed singlet excited state stands, however, in direct competition with the rapidly occurring intersystem crossing and is, in essence, overshadowed by the latter. Thus, intermolecular electron-transfer reactions, as outlined in the following section, usually occur with the much longer-lived triplet excited state. An intermolecular electron-transfer reaction, yielding a charge separated radical pair, involves a sequence of events. In particular, the initial photoexcitation process is succeeded by a diffusion- or activation-controlled intermolecular electron transfer, which, in turn, creates a primary contact radical pair. In the current context, it is the one-electron reduced fullerene π-radical anion (or the one-electron oxidized fullerene π-radical cation) and the oxidized form of the applied electron donor (or the reduced form of the applied electron acceptor). In the final step the diffusional break-up of the contact pair to give the products completes the sequence. The dissociation depends on the ability of the solvent to retard the exergonic back electron transfer to the ground state or excited state reactant pair and stabilize the radical pair, for example, via strong dipole–dipole interactions.

### 5.4.1 Formation of Fullerene Radical Anions

The high reduction potential of the fullerene triplet excited state permits an efficient reductive quenching with a large number of tertiary amines to produce $C_{60}{}^{\bullet-}$ (Eq. 12) [18, 23, 180–191].

$$^3C_{60}{}^* + \text{amine} \rightarrow C_{60} + (\text{amine})^{\bullet+} \tag{12}$$

Selection of a polar solvent, such as different alcohols or benzonitrile, is crucial in view of accomplishing an efficient charge separation, which is, under these circumstances, typically confirmed through formation of $C_{60}{}^{\bullet-}$ at $\lambda_{max} = 1080$ nm [65]. In contrast, the low dielectric constant of toluene prevents the successful charge separation. Ferrocene (Fc) [192, 193], tetrathiafulvalene (TTF) [194, 195], bis-(ethylenedithio)tetrathiafulvalene (BEDT-TTF) [194, 208], aromatic thiols [196], tetra-selenafulvalenes [197], tetraethoxyethene [198], phenothiazines [199], 3,3′,5,5′-tetramethylbenzidine (NTMB) [26, 200–202], 1,4-diazabicyclo[2.2.2]octane (DABCO) [190, 203], stable nitroxide radicals [204], tetraphenylborate [205], triphenylbutylborate [205], retinol [205–207], $\beta$-carotene [209, 210], polysilane [201, 211], polygermane [201, 212], oligo-thiophene [213–216], tetrathiophene [213–216], zinc tetraphenyl porphyrin (ZnP) [217, 218] and zinc phthalocyanine (ZnPc) [27], to name a few representative cases, are all excellent electron donors. These compounds have therefore been probed as sacrificial electron donors in intermolecular quenching reactions with triplet excited fullerenes. Again, polar solutions play a central role in creating conditions which support an efficient charge separation and give rise to diffusion-controlled charge recombination ($\approx 10^{10}$ $\mathrm{M}^{-1}$ $\mathrm{s}^{-1}$).

Beside the photoinduced electron transfer in homogeneous solutions, fullerenes have also been shown to play a key function in mediating electron transfer through an artificial lipid bilayer membrane [84, 219, 220]. For example, the fullerene moiety was embedded in form of aggregates within the membrane interior of a three-component cell [220]. A strong electron donor (e.g., ascorbate) was placed on one side of the lipid bilayer, while the three-component cell was completed with an adequate electron acceptor on the other side of the membrane (e.g., an aqueous solution containing anthraquinone 2-sulfonate). In an oxygen-free environment, illumination generated a measurable photocurrent between the two aqueous phases [220]. To confirm a possible electron hopping mechanism or, alternatively, a diffusion of an one-electron reduced fullerene moiety, the same studies were carried out in blank experiments without a bilayer incorporated fullerene moiety and without removal of oxygen. While the earlier set-up is meant to interrupt the redox gradient across the cell, oxygen quenches the excited and reduced states. Indeed, no photocurrent was observed in the two blanks [220]. A similar across-membrane electron transport was reported focusing on an intermolecular electron transport between a porphyrin, attached to the bilayer-water interface, and fullerenes, embedded in the lipid bilayer membrane.

ZnO or TiO$_2$ semiconductor clusters are excellent probes to perform electron transfer experiments since their conduction bands ($E_{CB} = -0.5$ V relative to the NHE) favor only one-electron reduction of, for example, C$_{60}$, avoiding, however, multiple reduction steps [221–223]. Pioneering experiments focusing on the reduction of C$_{60}$ were conducted via exclusive excitation of ZnO colloids.

As an alternative to an intermolecular quenching reaction involving the fullerene triplet excited state, strong interactions between the fullerene and a donor, in form of a ground state complex or excited state complex ('exciplex'), are also an appropriate means to accelerate transfer dynamics evolving from the singlet excited state [23, 180–183, 188]. Formation of fullerene charge transfer complexes has been reported for a large variety of substrates, such as *N,N*-dimethylaniline (DMA) [180, 188], *N,N*-diethylaniline (DEA) [180, 183, 188], diphenylamines (DPA) [23], triphenylamine (TPA) [23], triethylamine (TEA) [23, 181, 182], and a number of substituted naphthalenes [224]. Generally, the resulting complexes are weak and high amine concentrations are required to study the characteristics and properties of fullerene containing charge-transfer complexes. In aliphatic hydrocarbon solvents such as hexane or methylcyclohexane, the quenching of the monomer excited states occurs via the transient formation of 'exciplexes'.

A long-lived transient of $C_{60}^{\bullet-}$ is formed in photoinduced electron transfer from ZnO or $TiO_2$ semiconductor colloids to $C_{60}$ [222]. In homogeneous systems described above, however, the lifetime of the generated $C_{60}^{\bullet-}$ is generally extremely short due to fast back electron transfer to the reactant pair, resulting in no net formation of $C_{60}^{\bullet-}$. This is not the case for a photoinduced electron-transfer reaction from an NADH analog, 1-benzyl-1,4-dihydronicotinamide (BNAH), and the dimer analog [(BNA)$_2$] to the triplet excited state of $C_{60}$ ($^3C_{60}^*$). A stable $C_{60}^{\bullet-}$ was found in benzonitrile solution with a surprisingly high quantum yield exceeding unity in the latter case; $\Phi_\infty = 1.3$ (Eq. 13) [225].

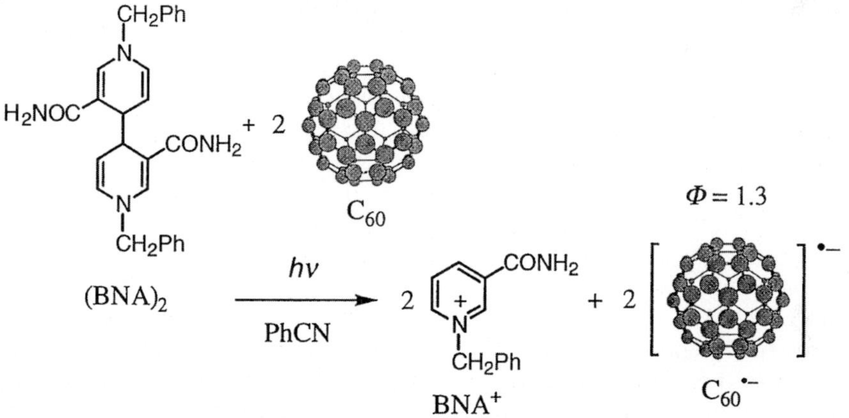

$$\Phi = 1.3 \tag{13}$$

A quantum yield exceeding unity is consistent with the stoichiometry in Eq. 13, where (BNA)$_2$ can reduce two equivalents of $C_{60}$. The reaction mechanism for the formation of a stable $C_{60}^{\bullet-}$ is shown in Scheme 5 [225].

The triplet excited state, namely, $^3C_{60}^*$ generated by the efficient intersystem crossing upon photoexcitation of $C_{60}$, is quenched by electron transfer from (BNA)$_2$ to $^3C_{60}^*$ to give the radical ion pair [(BNA)$_2^{\bullet+}$ and $C_{60}^{\bullet-}$] in competition with the decay to the ground state. This step is followed by a fast cleavage of the C–C bond of the dimer ($k_c$) to produce *N*-benzylnicotinamide radical (BNA$^\bullet$) and BNA$^+$ [226]. The subsequent second electron transfer from BNA$^\bullet$ to $C_{60}$ should be by far faster than the first, as BNA$^\bullet$ is a strong reductant ($E^\circ_{ox} = -1.08$ V relative to the SCE) [44]. Thus, once photoinduced electron transfer from (BNA)$_2$ to $^3C_{60}^*$ occurs, two $C_{60}^{\bullet-}$ molecules are produced. The generation of stable $C_{60}^{\bullet-}$ in the photoreduction of $C_{60}$ by BNAH also proceeds via photoinduced electron transfer from BNAH to $^3C_{60}^*$ [225]. In this case, the deprotonation of BNAH$^{\bullet+}$ gives BNA$^\bullet$, which can further reduce $C_{60}$ to $C_{60}^{\bullet-}$ [225].

The selective one-electron reduction of $C_{70}$ to $C_{70}^{\bullet-}$ is also attained through photoinduced electron transfer from BNAH or (BNA)$_2$ to the triplet excited state of $C_{70}$ [227]. The limiting quantum yields for formation of $C_{70}^{\bullet-}$ in the photoreduction

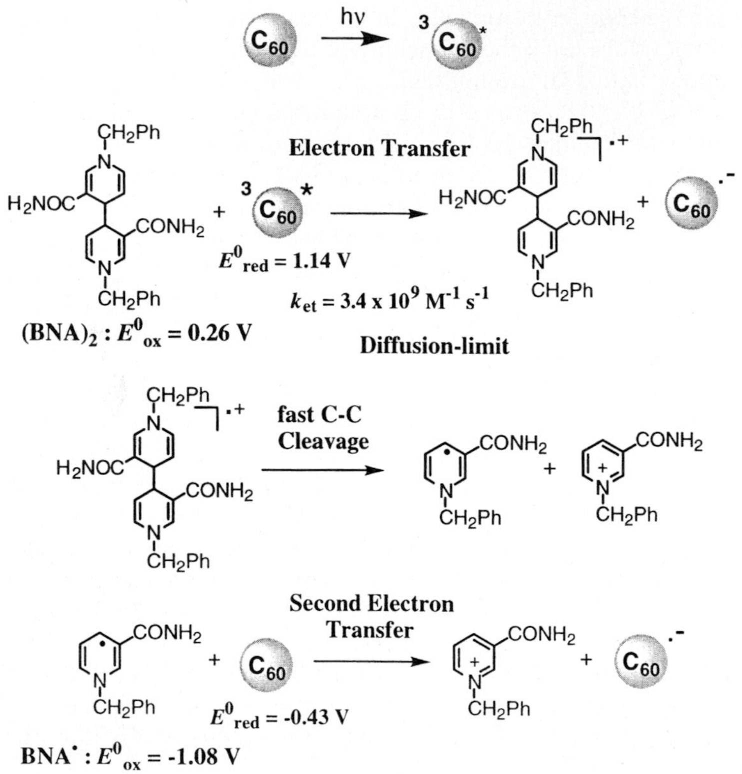

$$C_{60} \xrightarrow{h\nu} {}^3C_{60}{}^*$$

**Electron Transfer**

$$\left[ \text{(BNA)}_2 \right]^{\bullet+} + {}^3C_{60}{}^* \longrightarrow \left[ \text{(BNA)}_2 \right]^{\bullet+} + C_{60}{}^{\bullet-}$$

$E^0{}_{red} = 1.14$ V

$k_{et} = 3.4 \times 10^9$ M$^{-1}$ s$^{-1}$

(BNA)$_2$ : $E^0{}_{ox} = 0.26$ V

**Diffusion-limit**

**fast C-C Cleavage**

**Second Electron Transfer**

$E^0{}_{red} = -0.43$ V

BNA$^{\bullet}$ : $E^0{}_{ox} = -1.08$ V

**Scheme 5.**

of C$_{70}$ by BNAH and (BNA)$_2$ exceed unity; $\Phi_\infty = 2.0$ and 1.9. Both values are larger than the corresponding values for formation of C$_{60}{}^{\bullet-}$ ($\Phi_\infty = 1.3$ and 0.80, respectively) [227]. The formation of stable C$_{70}{}^{\bullet-}$ is readily detected by the typical NIR spectrum ($\lambda_{max} = 1374$ nm) [228, 229]. The enhanced reactivity of C$_{70}$ as compared to C$_{60}$ is ascribed to a more localized unpaired electron and negative charge in C$_{70}{}^{\bullet-}$ due to loss of symmetry. This facilitates the follow-up reaction in competition with the back electron transfer to the ground state reactant pair.

### 5.4.2 Selective Two-electron Reduction

The reduction of C$_{60}$ and C$_{70}$ has normally been achieved by the use of strong reductants such as BH$_3$, which yield not only C$_{60}$H$_2$ and C$_{70}$H$_2$ but also polyhydride mixtures [104, 230–237]. The use of the triplet excited state of C$_{60}$ has made it possible to attain the selective two-electron reduction of C$_{60}$ to 1,2-C$_{60}$H$_2$ by 10-methyl-9,10-dihydroacridine (AcrH$_2$) which is a mild hydride donor (Eq. 14) [225, 238].

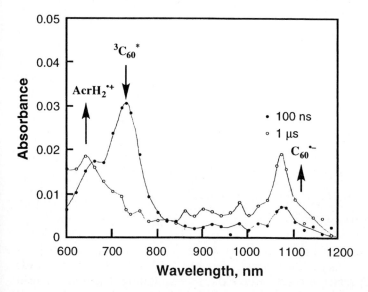

Irradiation of a deaerated PhCN solution of $C_{60}$, $AcrH_2$ and $CF_3COOH$ with a Xe lamp equipped a UV-cut filter ($\lambda < 540$ nm) resulted in exclusive formation of 1,2-$C_{60}H_2$ in approximately 70 % yield [238]. In the dark, however, no reaction occurred even at high temperatures (e.g., 373 K) [238]. The transient absorption spectra of the radical ion pair ($AcrH_2^{\bullet+}$ $C_{60}^{\bullet-}$) in the visible and near-IR region are observed by laser flash photolysis of a deaerated PhCN solution of $C_{60}$ in the presence of $AcrH_2$ and $CF_3COOH$ with 532 nm laser light as shown in Figure 6 [225].

The absorption band at 640 nm in the visible region agrees with that reported for $AcrH_2^{\bullet+}$ observed as a transient spectrum in the electron-transfer oxidation of $AcrH_2$ by $Fe(phen)_3^{3+}$ (phen = 1,10-phenanthroline) [239]. The decay of the ab-

**Figure 6.** Transient absorption spectra observed in the photoreduction of $C_{60}$ ($1.0 \times 10^{-4}$ M) by $AcrH_2$ ($1.0 \times 10^{-3}$ M), 100 ns (●) and 1 μs (○) after laser excitation in deaerated PhCN at 295 K [225].

sorbance at 740 nm due to $^3C_{60}^*$ coincides with the rise of the absorbance at 640 nm due to $AcrH_2^{\bullet+}$ as well as the absorbance at 1080 nm due to $C_{60}^{\bullet-}$ [64, 225]. Thus, the photochemical two-electron reduction of $C_{60}$ by $AcrH_2$ proceeds via electron transfer from $AcrH_2$ to $^3C_{60}^*$, which gives the radical ion pair $(AcrH_2^{\bullet+}C_{60}^{\bullet-})$ in competition with the decay to the ground state. The $pK_a$ of singly reduced $C_{60}$ ($C_{60}H^{\bullet}$) encapsulated in $\gamma$-cyclodextrin ($\gamma$-CD) and dissolved in water/propan-2-ol has been determined as 4.5 on the basis of a specific IR absorption band for $C_{60}^{\bullet-}$-$\gamma$-CD [85]. On the other hand, the $pK_a$ of $AcrH_2^{\bullet+}$ in water has previously been determined as 2.0 [44, 239]. The proton transfer from $AcrH_2^{\bullet+}$ to $C_{60}^{\bullet-}$ is therefore exergonic, and it should occur efficiently in the radical ion pair in competition with the back electron transfer to the reactant pair to give $C_{60}H^{\bullet}$ which is converted to 1,2-$C_{60}H_2$ by the fast electron transfer from $AcrH^{\bullet}$ in the presence of $CF_3COOH$. Thus, the selective two-electron reduction of $C_{60}$ to $C_{60}H_2$ occurs via sequential electron–proton–electron–proton transfer.

### 5.4.3 C–C Bond Formation via Photoinduced Electron Transfer

The carbon–carbon bond formation via photoinduced electron transfer has recently attracted considerable attention from both synthetic and mechanistic viewpoints [240–243]. In order to achieve efficient C–C bond formation via photoinduced electron transfer, the choice of an appropriate electron donor is essential. Most importantly, the donor should be sufficiently strong to attain efficient photoinduced electron transfer. Furthermore, the bond cleavage in the donor radical cation produced in the photoinduced electron transfer should occur rapidly in competition with the fast back electron transfer. Organosilanes that have been frequently used as key reagents for many synthetically important transformations [244–247] have been reported to act as good electron donors in photoinduced electron-transfer reactions [248, 249]. The one-electron oxidation potentials of ketene silyl acetals (e.g., $E°_{ox}$ relative to the SCE = 0.90 V for $Me_2C=C(OMe)OSiMe_3$) [248] are sufficiently low to render the efficient photoinduced electron transfer to $^3C_{60}^*$ [22], which, after the addition of ketene silyl acetals, yields the fullerene with an ester functionality (Eq. 15) [250, 251].

$$ (15) $$

$$C_{60}\text{-1,2-RH}$$
$$R = CMe_2COOMe$$

The initial monoadduct reacts further to give the corresponding bisadduct at prolonged irradiation times particularly at higher concentration of the ketene silyl acetal (10 equiv.) [22]. When the unsubstituted ketene silyl acetal ($H_2C=C(OEt)OSiMe_3$) is employed, only the monoadduct, ethyl 1,2-dihydro[60]fullerene-1-acetate, is obtained in quantitative yields based on recovered $C_{60}$ [22]. Under these conditions, no polyadducts were obtained even after prolonged reaction times [22]. Such a difference in the reactivity between $Me_2C=C(OMe)OSiMe_3$ and $H_2C=C(OEt)OSiMe_3$ can be well accounted for by the decrease in the electron donor ability of the latter [248]. The laser flash photolysis study showed that the disappearance of the band due to $^3C_{60}*$ at 740 nm is accompanied by the appearance of a new absorption band at 1080 nm due to $C_{60}{}^{\bullet-}$ and that the decay rate of $^3C_{60}*$ coincides completely with the rate of formation of $C_{60}{}^{\bullet-}$ [252]. The reaction mechanism of the photo-addition of ketene silyl acetals to $C_{60}$ can be summarized as shown in Scheme 6 which is essentially the same as that of the photoaddition of ketene silyl acetals to 10-methylacridone [22, 249, 253]. The photoinduced electron transfer from the ketene silyl acetal to $^3C_{60}*$ gives the radical ion pair in competition with the decay to the ground state. The spin density of the ketene silyl acetal radical cation thus produced is nearly localized on the terminal carbon atom [248]. Thus, the ketene silyl acetal radical cation is coupled efficiently with the $C_{60}$ radical anion in competition with the back electron transfer to the reactant pair. The resulting zwitterionic intermediate converts eventually to the product after an efficient protonation step is completed (Scheme 6) [22, 252].

The photoinduced electron-transfer reactions are usually performed in polar solvents such as acetonitrile and benzonitrile, in which the solvation energy of free ions is greatly increased compared to non-polar solvents such as benzene. Such a difference in the solvation energy of ketene silyl acetal radical cation in benzene and

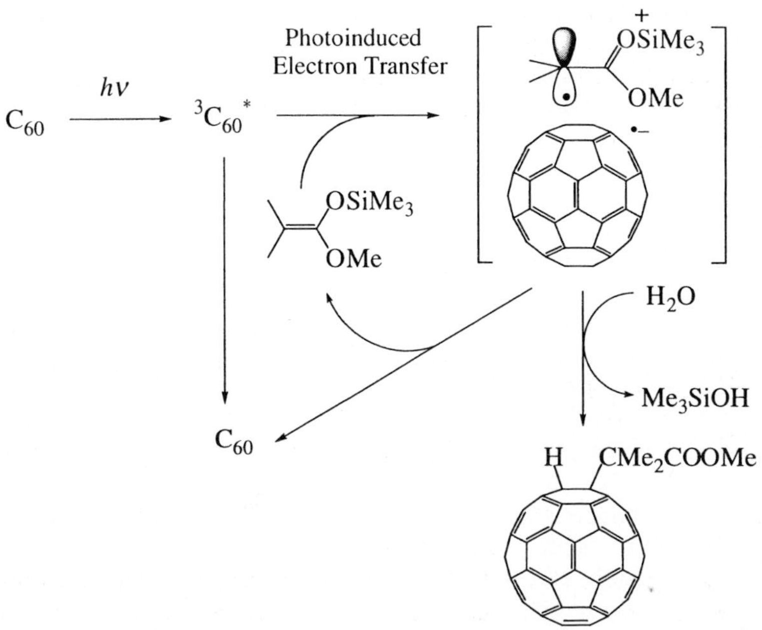

**Scheme 6.**

benzonitrile has been estimated as 0.86 eV [248, 249]. On the other hand, the difference in the solvation energy of $C_{60}^{\cdot-}$ in benzene and benzonitrile may be negligible because of the highly delocalized negative charge. The overall difference in the free energy change of electron transfer to produce free ions in benzene and benzonitrile is thereby estimated as 0.86 eV [22]. The difference of 0.86 eV in the $\Delta G^{\circ}_{et}$ values between benzene and benzonitrile may well be canceled by the large Coulombic interaction in benzene as compared with that in benzonitrile, since the difference in the values in benzene and benzonitrile with the mean separation of 6.7 Å is equal to 0.86 eV [22]. Thus, the photoinduced electron transfer from ketene silyl acetals to $^{3}C_{60}*$ can also occur in benzene leading to the same product as obtained in benzonitrile, although the rate becomes slower [22].

There is an interesting case in which a C–C bond formation via photoinduced electron transfer yields a product which gives rise to a different regioselectivity than that predicted for a thermal polar reaction [254]. The selective $\gamma$-addition of prenyltributyltin with 10-methylacridinium ion proceeds via a polar mechanism in which the C–C bond formation should occur prior to the cleavage of the Sn–C bond. In contrast, the selective $\alpha$-addition instead of $\gamma$-addition occurs through a photoinduced electron transfer from prenyltributyltin to the singlet excited sate of 10-methylacridinium ion [254]. The photoinduced electron transfer from prenyltributyltin to $^{3}C_{60}*$ also leads to the selective formation of $C_{60}$-1,2-(H)CH$_2$CH=CMe$_2$ in which the allylic group is introduced selectively at the $\alpha$ position and no $\gamma$ adduct has been formed [255]. In an electron-transfer reaction, the Sn–C bond may be significantly lengthened in the prenyltributyltin radical cation generated upon the electron transfer [256]. In such a case the C–C bond formation may occur after the elongation or the cleavage of the Sn–C bond. This leads to the more favorable $\alpha$-addition than the $\gamma$-addition because of the steric hindrance of two methyl groups of prenyltributyltin radical cation as shown in Scheme 7.

The free energy change of photoinduced electron transfer from electron donors to $^{3}C_{60}*$ ($\Delta G^{\circ}_{et}$) is given by Eq. 16:

$$\Delta G^{\circ}_{et} = F(E^{\circ}_{ox} - E^{\circ}_{red}) \tag{16}$$

where $E^{\circ}_{ox}$ and $E^{\circ}_{red}$ are the one-electron oxidation potentials of the electron donor and the one-electron reduction potential of $^{3}C_{60}*$, respectively. The dependence of the activation free energy of photoinduced electron transfer $\Delta G^{\neq}$ on the free energy change of electron transfer ($\Delta G^{\circ}_{et}$) has well been established as given by the Rehm–Weller free energy relation, Eq. 17:

$$\Delta G^{\neq} = (\Delta G^{\circ}_{et}/2) + [(\Delta G^{\circ}_{et}/2)^2 + (\Delta G^{\neq}_0)^2]^{1/2} \tag{17}$$

where $\Delta G^{\neq}_0$ is the intrinsic barrier that represents the activation free energy when the driving force of electron transfer is zero, i.e., $\Delta G^{\neq} = \Delta G^{\neq}_0$ at $\Delta G^{\circ}_{et} = 0$ [257]. The $\Delta G^{\neq}$ values are related to the rate constant of electron transfer ($k_{et}$) as given by Eq. 18:

$$\Delta G^{\neq} = 2.3RT \log[Z(k_{et}^{-1} - k_{diff}^{-1})] \tag{18}$$

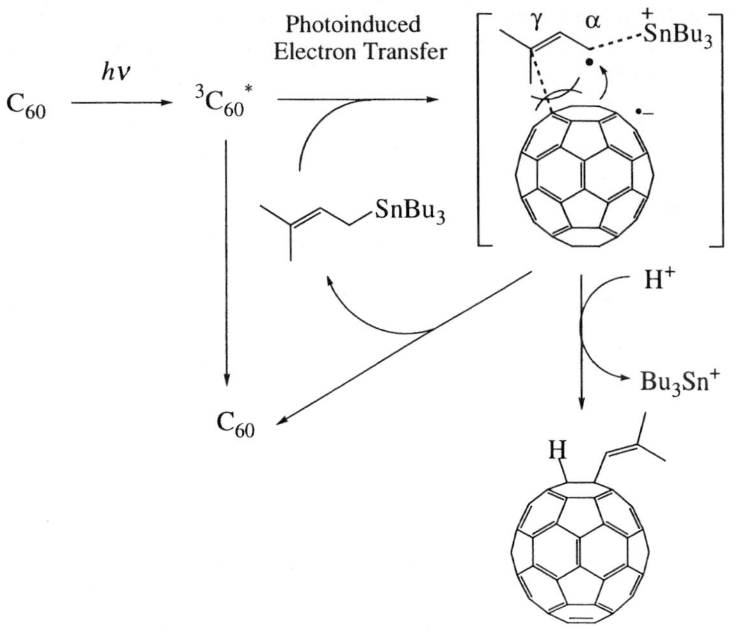

**Scheme 7.**

where Z is the collision frequency (i.e., $1 \times 10^{11}$ $M^{-1}$ $s^{-1}$) and $k_{diff}$ is the diffusion rate constant (i.e., $5.6 \times 10^9$ $M^{-1}$ $s^{-1}$) in benzonitrile. The $k_{et}$ values can be calculated from the $\Delta G°_{et}$ and $\Delta G^{\neq}_0$ values by using Eqs. 16 and 17. The $k_{et}$ values of photoinduced electron transfer from 14 different organometallic electron donors are listed in Table 3, together with the free energy change of electron transfer ($\Delta G°_{et}$), observed rate constants $k_{obs}$ derived from the dependence of quantum yields on the donor concentration, the quenching rate constants $k_q$ of $^3C_{60}*$, and the limiting quantum yields ($\Phi_\infty$).

The $k_{et}$ value increases with an increase in the driving force ($-\Delta G°_{et}$) and they agree with the $k_q$ and $k_{obs}$ values (Table 3). Such agreement confirms the validity of the photoinduced electron transfer mechanisms shown in Schemes 6 and 7.

As mentioned earlier, organometallic reagents which are strong nucleophiles have usually been used for the alkylation of fullerenes [132–138]. On the other hand, 4-alkylated NADH analogs can be used as a mild alkylating reagent in the photochemical reactions with $C_{60}$ and $C_{70}$ [225, 227]. The one-electron oxidation of 4-*tert*-butylated BNAH (*t*-BuBNAH) results in the selective C(4)–C bond cleavage of *t*-BuBNAH$^{\cdot+}$ [258, 259]. Although there are two possible modes of the carbon–carbon bond cleavage in such reactions to generate (i) *t*-Bu$^{\cdot}$ and BNA$^+$ or (ii) *t*-Bu$^+$ and BNA$^{\cdot}$, the formation of *t*-Bu$^{\cdot}$ in the one-electron oxidation of *t*-BuBNAH has been confirmed by applying a rapid mixing flow ESR technique [259]. Thus, the photoinduced electron transfer from *t*-BuBNAH to $^3C_{60}*$ is followed by the facile C(4)–C bond cleavage of *t*-BuBNAH$^{\cdot+}$. The resulting *t*-Bu$^{\cdot}$ is coupled immediately

**Table 3.** Free energy change, $\Delta G^{\circ}{}_{et}$, and rate constants, $k_{et}$, of photoinduced electron transfer from group 14 organometallic electron donors to $^{3}C_{60}{}^{*}$, observed rate constants, $k_{obs}$, triplet quenching rate constants, $k_{q}$, and limiting quantum yields, $\Phi_{\infty}$, in the photoaddition of the donors to $C_{60}$ in benzonitrile at 298 K [212].

| Electron donor | $\Delta G^{\circ}{}_{et}$ (kcal mol$^{-1}$) | $k_{obs}$ ($M^{-1}$ s$^{-1}$) | $k_{q}$ ($M^{-1}$ s$^{-1}$) | $k_{et}$ ($M^{-1}$ s$^{-1}$) | $\Phi_{\infty}$ |
|---|---|---|---|---|---|
| 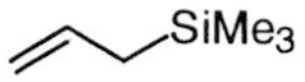 SiMe$_3$ | 8.3 | No reaction | No reaction | $2.3 \times 10^{3}$ | No reaction |
| 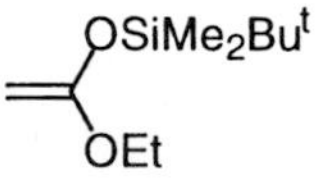 OSiMe$_2$Bu$^t$ / OEt | 3.2 | $2.5 \times 10^{6}$ | – | $1.3 \times 10^{6}$ | 0.14 |
|  SnBu$_3$ | −1.8 | $5.0 \times 10^{7}$ | $7.8 \times 10^{7}$ | $4.5 \times 10^{7}$ | 0.13 |
| 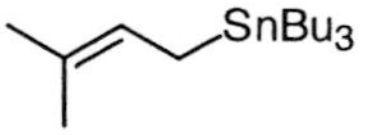 SnBu$_3$ | −5.8 | $9.8 \times 10^{8}$ | – | $1.7 \times 10^{9}$ | 0.13 |
| 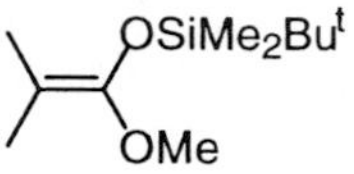 OSiMe$_2$Bu$^t$ / OMe | – | $2.1 \times 10^{8}$ | $3.7 \times 10^{8}$ | – | 0.19 |
| 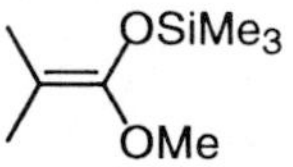 OSiMe$_3$ / OMe | −5.5 | $2.2 \times 10^{8}$ | $4.0 \times 10^{8}$ | $4.7 \times 10^{8}$ | 0.26 |

with $C_{60}{}^{\bullet-}$ to yield the final product ($t$-BuC$_{60}{}^{-}$) in competition with the back electron transfer as shown in Scheme 8 [225].

The formation of $C_{60}{}^{\bullet-}$ produced as a reactive intermediate in the photochemical reaction of $C_{60}$ with $t$-BuBNAH is also detected by the laser flash photolysis of a deaerated PhCN solution of $C_{60}$ in the presence of $t$-BuBNAH [225]. This type of photochemical reaction provides a unique and new way to prepare $t$-BuC$_{60}{}^{-}$ by using $t$-BuBNAH, which is a mild alkylating reagent under neutral conditions. The subsequent trap of the photoproduct by $CF_3COOH$ and $PhCH_2Br$ gave 1,2-$t$-BuC$_{60}$H and 1,4-$t$-Bu(PhCH$_2$)C$_{60}$, respectively as reported for the reactions of $t$-BuC$_{60}{}^{-}$ with electrophiles [260–262].

Strained Si–Si $\sigma$ bonds are also known to act as good electron donors [263–265]. Photoinduced electron transfer from strained disiliranes such as 1,1,2,2-tetramesityl-1,2-disilirane to $^{3}C_{60}{}^{*}$ occurs in benzonitrile to give a transient absorption spectrum at 1080 nm due to $C_{60}{}^{\bullet-}$ obtained by 532 nm laser flash photolysis of $C_{60}$ in the presence of disilirane [266]. The photochemical reaction of $C_{60}$ with the disilirane in benzonitrile afforded the 1:1 and 1:2 adducts of disilirane and benzonitrile in good yields. The structures of the corresponding adducts were determined by X-ray crystallographic analysis [266]. These adducts were also obtained in good yields in the presence of catalytic amounts of $C_{60}$. Thus, $C_{60}$ acts as a photoinduced electron-transfer catalyst as shown in Scheme 9.

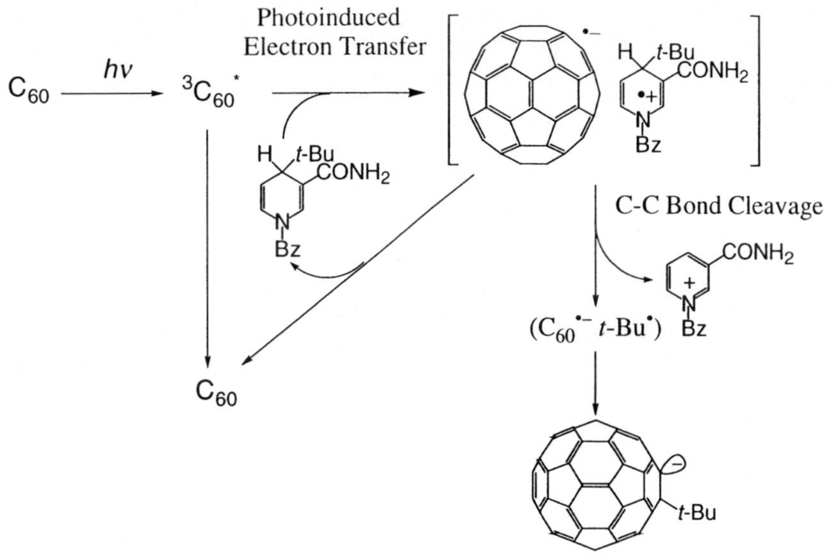

**Scheme 8.**

Photoinduced electron transfer from the silirane to $^3C_{60}$* gives the radical ion pair of $C_{60}^{\cdot-}$ and the disilirane radical cation. The Si–Si bond of the disilirane radical cation may be cleaved and trapped by benzonitrile. This sequence is followed by back electron transfer from $C_{60}^{\cdot-}$ to the adduct cation to yield the corre-

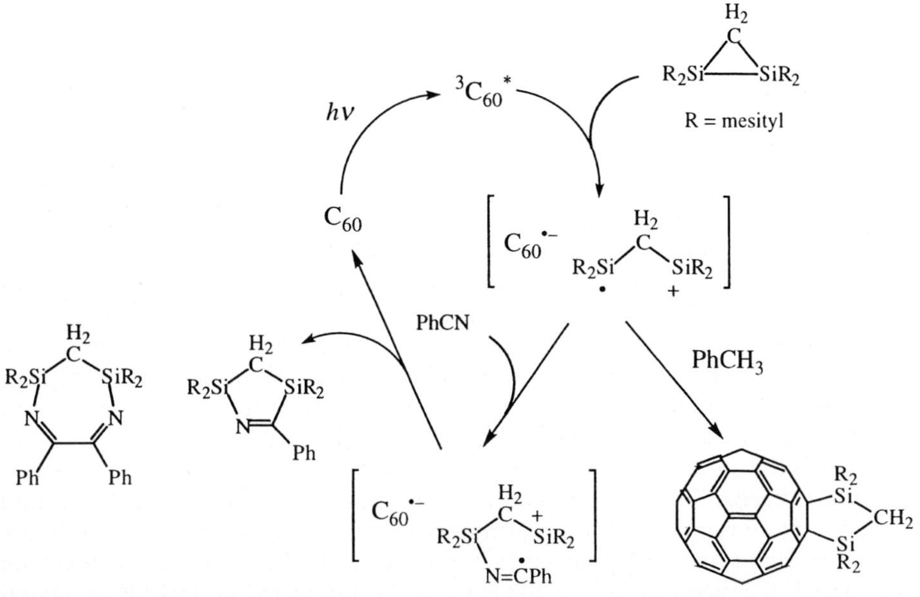

**Scheme 9.**

sponding adducts when $C_{60}$ is regenerated (Scheme 9). Irradiation of a toluene solution of the disilirane and $C_{60}$ with a high-pressure mercury lamp ($\lambda > 300$ nm) resulted, however, in the formation of 1,1,3,3-tetramesityl-1,3-disilolane [1,2-$(Mes_4Si_2CH_2)_2C_{60}$] (Scheme 9) [267]. In a nonpolar solvent, the transient absorption band due to $C_{60}^{\bullet-}$ was not observed although appreciable acceleration of the decay of $^3C_{60}^*$ at 740 nm was observed [266]. This suggests that the decay of the radical ion pair due to the Si–C bond formation is too fast to be detected or alternatively $^3C_{60}^*$ forms an exciplex with the disilirane.

When endohedral metallofullerenes which are fullerenes with metal(s) inside the hollow spherical cage are employed, the addition of disilirane to La@$C_{82}$ and Gd@$C_{82}$ (Eq. 19) occurs in both thermal and photochemical ways [268, 269].

$$
\begin{array}{c}
\left(\!\!\left( M \right)\!\!\right) \;+\; \underset{R_2Si\!-\!\!-\!SiR_2}{\overset{\overset{\displaystyle H_2}{C}}{\triangle}} \quad \xrightarrow[\Delta]{h\nu} \quad \left(\!\!\left( M \right)\!\!\right)\!\!\begin{array}{l} \overset{R_2}{Si}\\ \quad\!\! CH_2 \\ \underset{R_2}{Si} \end{array}
\end{array}
\qquad (19)
$$

M@$C_{82}$: M = La, Gd

The thermal addition of digermirane to La@$C_{82}$ has been reported to occur even at 20 °C [270]. This is in sharp contrast with the fact that empty fullerenes react only photochemically. Such an enhanced reactivity of endohedral metallofullerenes may be ascribed to a significantly larger electron affinity (3.22 eV) of La@$C_{82}$ as compared to $C_{60}$ (2.57 eV) and $C_{70}$ (2.69 eV) [271].

Irradiation of a benzene solution of 1,1,2,2-tetraphenyl-1,2-di-*tert*-butyl-1,2-disilane and $C_{60}$ with a low-pressure mercury lamp in a quartz tube resulted in formation of the bissilylated adduct (Eq. 20) [272].

$$
C_{60} + (t\text{-BuPh}_2\text{Si})_2 \xrightarrow[\text{benzene}]{h\nu} (t\text{-BuPh}_2\text{Si})_2 C_{60} \qquad (20)
$$

In this case the decay of $^3C_{60}^*$ was not accelerated upon the addition of the disilane. The electron transfer from the disilane to $^3C_{60}^*$ may not occur because of the higher oxidation potential of disilanes and the large reorganization energy associated with electron-transfer reaction of disilanes [273] as compared with the strained disilirane. Thus, the bissilylated adduct is suggested to be formed via the intermediacy of $t$-BuPh$_2$Si$^{\bullet}$ radical generated by the photochemical Si–Si cleavage of the disilane. The UV photolysis of the disilane is known to produce the $t$-BuPh$_2$Si$^{\bullet}$ radical almost exclusively [274]. In fact, none of the bissilylated adducts were produced upon irradiation at $\lambda > 300$ nm where the cleavage of the disilane does not take place [272].

Fullerene amino acid derivatives have merited special attention in relation with the biological application [275–280]. The photochemical reactions of $C_{60}$ with amino acid esters have been utilized to obtain fullerene amino acid derivatives [281, 282]. For example, the photochemical addition of iminodiacetic methyl ester to $C_{60}$ provides a pyrrolidine ring-fused fullerene carboxylate, $C_{60}$(MeOOCCHNHCHCOOMe) [281]. Although the mechanistic detail has yet to be elucidated, a possible sequence

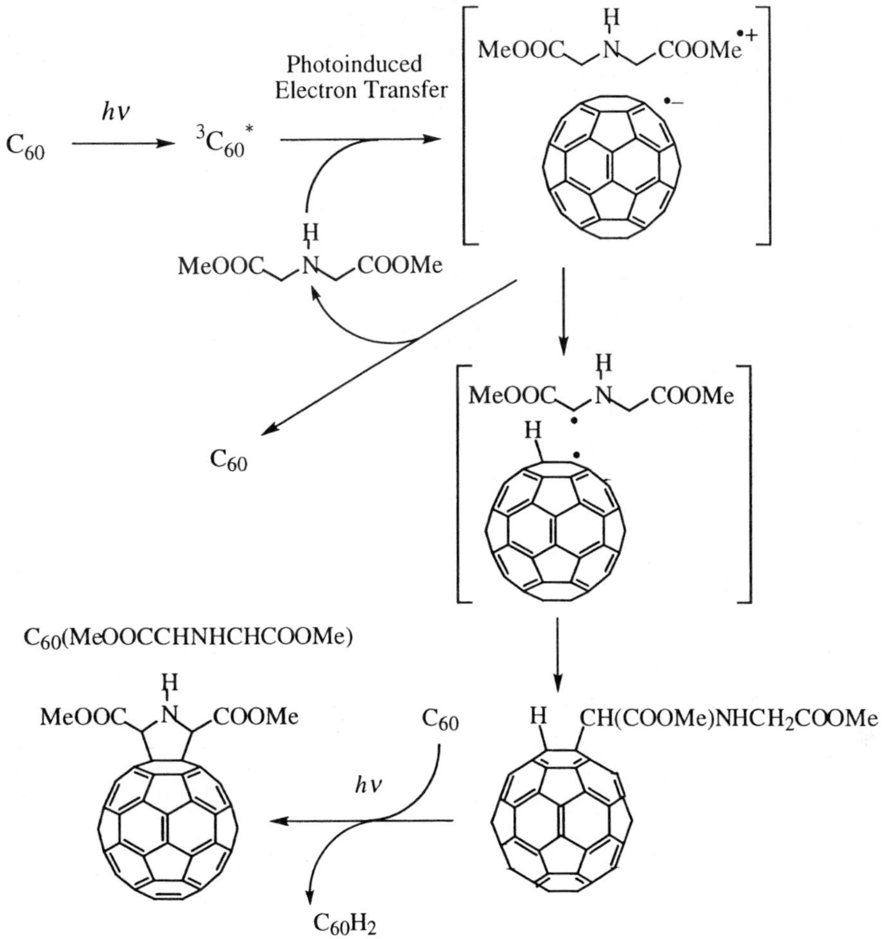

**Scheme 10.**

of reactions via photoinduced electron transfer from iminodiacetic methyl ester to $^3C_{60}^*$ is suggested, as shown in Scheme 10. The key step is the formation of the $\alpha$-carbon-centered radical produced by photoinduced electron transfer followed by the deprotonation [281, 282]. The radical coupling between the $\alpha$-carbon-centered radical and $C_{60}H^{\bullet}$ gives a 1,2 adduct which may be oxidized by $C_{60}$ photochemically to yield the final product, $C_{60}(MeOOCCHNHCHCOOMe)$ in Scheme 10.

Diels–Alder reactions of $C_{60}$ are generally believed to proceed via a thermally allowed concerted (suprafacial) process or a photochemical concerted (antarafacial) process [283–286]. However, an alternative stepwise (open-shell) mechanism for the Diels–Alder reaction has recently merited increasing attention [287–294]. Along this line several reports describe an electron transfer with the formation of radical ion pairs as primary step of the Diels–Alder reactions, followed by a stepwise bond formation [295–301]. The photochemical Diels–Alder reaction of $C_{60}$ with an-

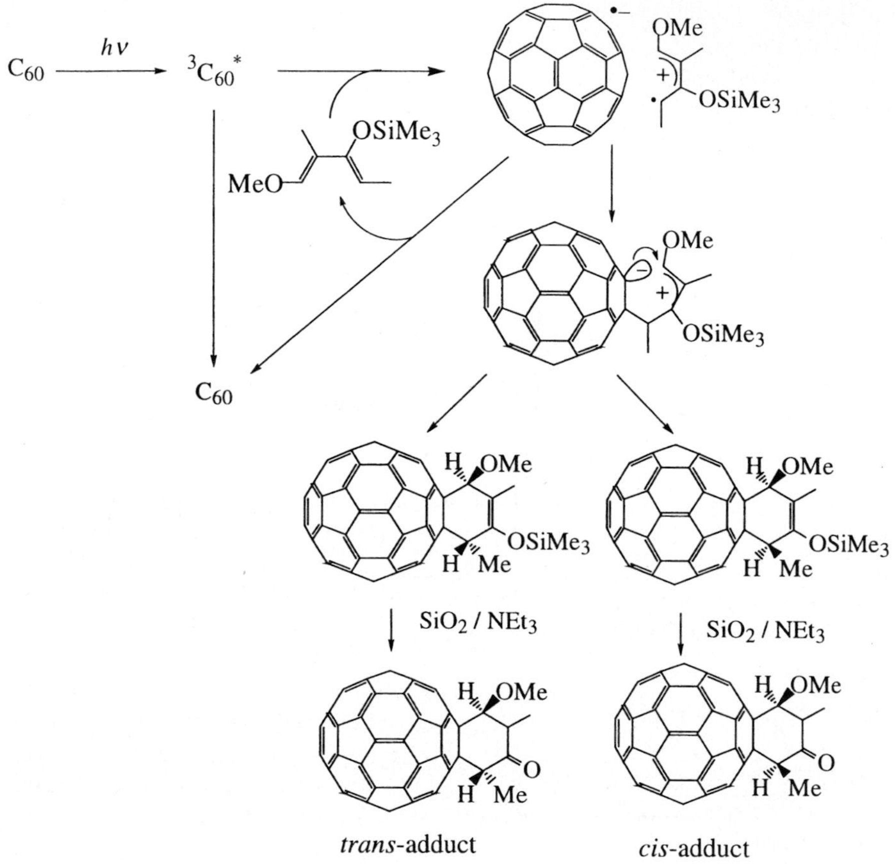

**Scheme 11.**

thracenes proceeds in the solid state via a photoinduced electron transfer within a charge-transfer complex comprising $C_{60}$ and anthracene [302]. Also the thermal Diels–Alder reaction of $C_{60}$ with condensed aromatics has been reported. Interestingly, in the solid state, using a high-speed vibration milling technique, the reaction is clearly accelerated [303].

The photoinduced electron transfer pathway has been unequivocally demonstrated in a photochemical Diels–Alder reactions between $C_{60}$ and Danishefsky's dienes (Scheme 11) [304, 305].

The transient spectra of $C_{60}^{\bullet-}$ at $\lambda_{max} = 1080$ nm formed in photoinduced electron transfer from Danishefsky's diene to $^3C_{60}^*$ have been detected, accompanied by the decay of $^3C_{60}^*$ in a laser flash photolysis of the reaction system [305]. The observed rate constant agrees well with the predicted rate constant on the basis of an electron transfer mechanism shown in Scheme 11 [305]. In the photochemical Diels–Alder reaction of $C_{60}$, a stereochemically defined $(1E,3Z)$-1,4-disubstituted Danishefsky's diene is used as a stereochemical probe. In the photochemical Diels–

Alder reaction the major Diels–Alder product is the *trans* adduct rather than the *cis* adduct (Scheme 11). Such a stereochemistry indicates that the Diels–Alder reactions proceed by a stepwise mechanism rather than a concerted mechanism. The photo-induced electron transfer from Danishefsky's diene to $^3C_{60}{}^*$ gives the triplet radical ion pair. The triplet radical ion pair is then converted to the singlet radical ion pair to give a zwitterionic intermediate (or a diradical intermediate) in competition with the back electron transfer to the reactant pair. The bond formation occurs stepwise with no symmetry restriction for the bond formation. Thus, both *trans* and *cis* adducts were obtained as the final products [305].

## 5.5 Intramolecular Photoinduced Electron Transfer

The most important classes of monofunctionalized fullerene derivatives give rise to a cancellation of the fivefold degeneration of their HOMO and threefold degeneration of their LUMO levels [92, 306]. In view of electron transfer processes, it should be noted that functionalization leads to wider HOMO–LUMO gaps and, consequently, to more negative reduction potentials [306]. Typically, saturation of a double bond causes a negative shift of $\approx 100$ mV relative to pristine $C_{60}$. Similarly the excited states energies are shifted to lower energies. Taken the reduction potentials and the excited state energies in concert, significant consequences for the reduction potentials of the excited states are expected. For example, for methano- and pyrrolidinofullerenes values of $+1.16$ and $+0.86$ V relative to the SCE are reported for the $^1C_{60}$-R$^*$/$C_{60}$-R$^{\cdot-}$ and $^3C_{60}$-R$^*$/$C_{60}$-R$^{\cdot-}$ couples, respectively [307].

### 5.5.1 Fullerene Electron-donor Systems

More negative reduction potentials [306] and lower excited state energies of the functionalized fullerene derivatives [307] effect the activation energy for an intramolecular electron transfer. Despite this general disadvantage, methano- and pyrrolidinofullerenes have been employed as an electron acceptor unit. In line with the concept of using the fullerene core as the primary photosensitizer, a series of $C_{60}$-donor dyads have been reported [306, 308–312]. The general reaction scheme in these fullerene systems can be summarized as follows: Excitation of the fullerene core in the respective dyad yields the fullerene singlet excited state, from which an intramolecular electron transfer process evolves.

**Ferrocene**

A systematic fluorescence and flash photolytic investigation of a series of covalently linked $C_{60}$-based ferrocene dyads (**1**, **2**) is reported as a function of the nature of the bridge (i.e., flexible or rigid) and the separation (i.e., relative distance between the donor and acceptor site) [192, 193]. The fate of the fullerene singlet excited state was found to be governed by rapid intramolecular electron transfer, an assumption that was supported by monitoring the characteristic fullerene $\pi$-radical anion band in the NIR [192, 193]. The nature of the hydrocarbon bridge, weak electronic ground state interactions, steady-state fluorescence, picosecond-resolved photolysis and thermo-dynamic relation suggest two different quenching mechanisms, namely, a 'through-bond' electron transfer for rigidly-spaced dyads (**2**) and formation of a transient intramolecular 'exciplex' for flexibly-spaced dyads (**1**). With the view to improving the lifetime of the charge-separated radical pair, the flexibly-spaced dyads, in which the fullerene is connected to a ferrocene donor by extended hydrocarbon chains containing either ten $\sigma$-bonds or six $\sigma$-bonds deserve special mentioning [192, 193]. In polar solvents, the charge-separated radical pair is then regarded to diffuse semi-freely, almost like two different entities in solution. The general flexibility in these dyads promotes lifetimes of up to 2.6 $\mu$s and 3.6 $\mu$s in polar benzonitrile solutions for the $C_{60}$-donor dyad with 10 $\sigma$-bonds (**1**) and 6 $\sigma$-bonds (not shown), respectively [192, 193]. Non-polar solvents lack the beneficial role of stabilizing the radical pair. Therefore, charge recombination to yield the singlet ground state or an excited state product is usually very fast.

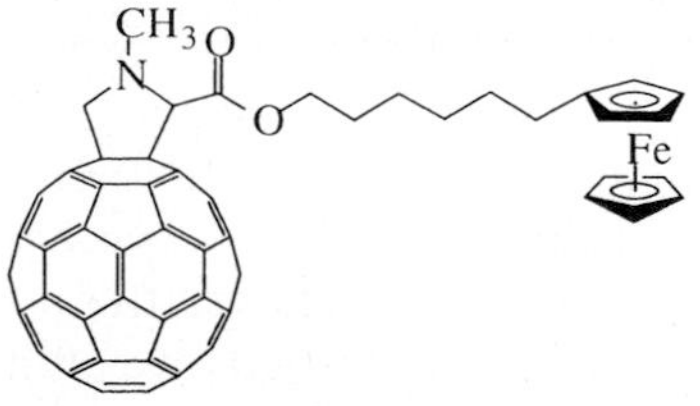

**1** 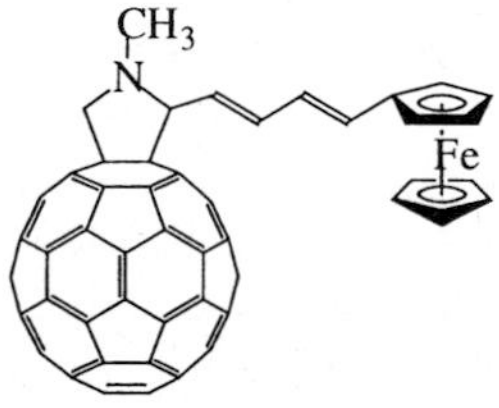                              **2**

### Tetrathiafulvalene

The organic donor tetrathiafulvalene (TTF) is an appealing compound as it is a $\pi$-electron donor in the field of organic molecular materials [313]. More importantly, TTF is a non-aromatic molecule in its singlet ground state configuration. In contrast to most of the donor units used in the preparation of fullerene containing dyads, it converts into the thermodynamically very stable 6 $\pi$-electron heteroaromatic 1,3-dithiolium cation upon oxidation [313]. $C_{60}$-based dyads (**3–5**), in which the full-erene is covalently linked to the electron donor TTF, have been synthesized by 1,3-dipolar cycloadditions of 'in situ' generated TTF-containing azomethine ylides to $C_{60}$ [6, 306, 314–317].

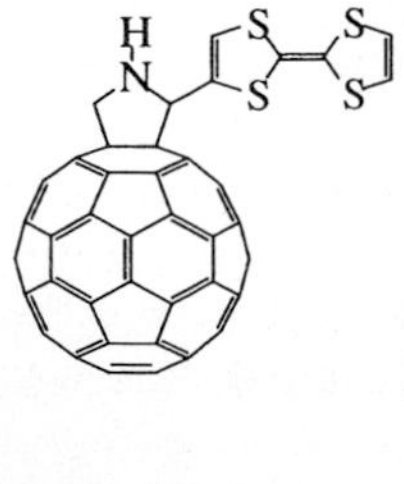

**3** 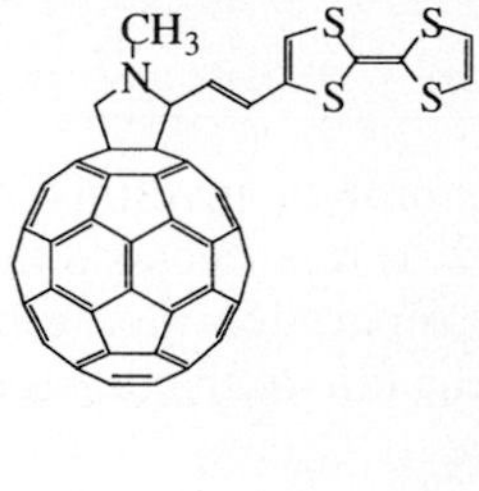                    **4** 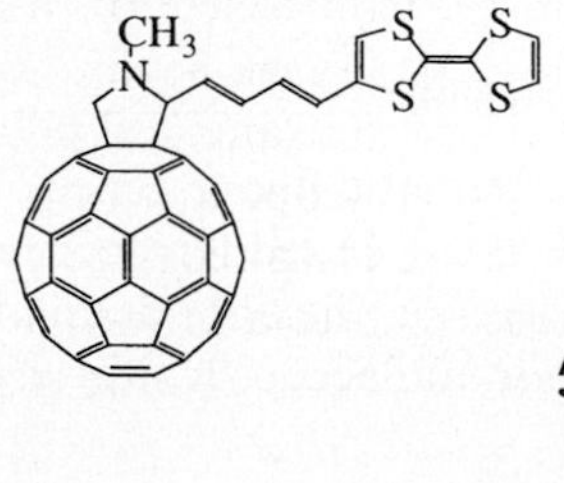                    **5**

Steady-state and time-resolved photolysis reveal that the fullerene singlet excited states in $C_{60}$-TTF dyads are subject to rapid intramolecular electron-transfer events yielding a charge-separated radical pair, namely, $C_{60}^{\bullet-}$-TTF$^{\bullet+}$ [306, 314, 315]. The gain of aromaticity, as a result of the formation of the TTF radical cation, is a promising concept to increase the stabilization of the charge-separated state. In fact, the lifetime of the charge-separated state in $C_{60}$-TTF dyads was increased by a factor of $\approx 4$ relative to comparable systems that do not contain TTF (e.g., ferrocene, aniline, etc.) [306, 314]. A radical pair lifetime of ca 2 ns was observed for the closely spaced $C_{60}$-TTF dyad (**3**), while much shorter lifetimes of 0.526 ns, 0.045 ns and 0.294 ns are reported for similarly spaced Carotene–$C_{60}$, ZnP–$C_{60}$ (**22**) and $H_2P$–$C_{60}$ dyads (**23**) (see Section 5.5.3), respectively.

In contrast to $C_{60}$-ferrocene and $C_{60}$-aniline dyads, back electron transfer proceeds mainly via formation of the fullerene triplet excited state [306, 314, 315]. The energy level of the latter (ca. 1.50 eV) is still sufficient to activate a second, intramolecular electron transfer, as noticed in an intermolecular quenching between the triplet excited state of $C_{60}$ and TTF. In a different family of $C_{60}$-TTF dyads (**6** and **7**) ESR values and $^{32}S$ hyperfine coupling constants give rise to a spin density distribution located on the TTF and fullerene moieties, for the cationic and anionic species, respectively [318]. Complementary performed flash photolysis substantiates the rapid quenching of the fullerene excited states yielding long-lived ($\tau \approx 75$ μs) charge-separated states with an open-shell character [318].

**6**          **7**

**Aniline**

One of the first examples regarding the design of $C_{60}$-based donor–acceptor dyads encompasses an electron donating *N,N*-dimethylaniline (DMA) group [186, 319–322]. In polar solvents, a significant quenching of the fullerene fluorescence was interpreted in terms of an intramolecular electron transfer process along a saturated heterocyclic bridge. In non-polar solvents, this deactivation route is, however, not observed, a consequence that stems from an overall less exergonic reaction [319].

Replacing the weak electron donor DMA with a *N,N,N',N'*-tetramethyl-*p*-phenylenediamine (TMPD) donor increases the energy gap between the fullerene singlet excited state and the charge-separated state [323]. As a consequence of a more exergonic electron transfer, the $C_{60}$-TMPD dyad (**8**) gives rise to fast intramolecular charge separation, irrespective of the solvent polarity (i.e., non-polar methylcyclohexane and polar benzonitrile) [323].

With the scope to retard the energy-wasting back electron transfer the separation between the two redox active moieties (i.e., fullerene and aniline) was systematically increased up to a 11-$\sigma$-bond containing 'norbornylogous' bridge (**9**) [323–325]. The reactivity of this 11-$\sigma$-bonds system followed the first class of shorter DMA hybrids

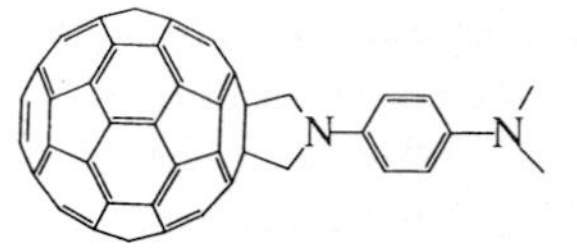

**8**

[319]. Successful electron transfer occurs, however, with much slower dynamics, and furthermore only in polar solutions. On the other hand, the 11-$\sigma$-bond 'norbornylogous' bridge has a remarkable effect on the back electron transfer kinetics ($\tau \approx 250$ ns) [323]. Strong electron coupling between the hydrocarbon bridge and the fullerene HOMO facilitates the fast charge separation, while the relevant electronic coupling with the fullerene LUMO, which determines the back electron transfer dynamics, remains small [323].

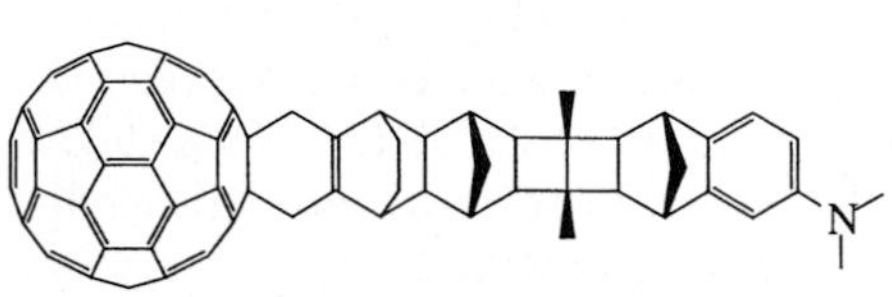

**9**

A topographically controlled electron transfer process was the subject in another generation of *ortho*- and *para*-substituted fullerene-aniline dyads (**10** and **11**) [326]. Minimum energy conformations yielded a folded configuration for the *ortho*-substituted dyad **11** (e.g., minimum separation between the fullerene and aniline) while the *para*-substituted analog **10** adapts a stretched one (e.g., maximum separation between the fullerene and aniline) [326]. The implemented decrease in the spatial separation between the electron donating aniline and electron accepting fullerene was beneficial in light of an efficient electron transfer. Dynamics of the singlet excited state deactivation and quantum yield of charge separation supports the computational view: In particular, the close proximity between the aniline group and the fullerene $\pi$-system in the *ortho*-substituted dyad (**11**) accelerates the electron transfer process and enhances its efficiency relative to the *para*-substituted dyad (**10**) [326].

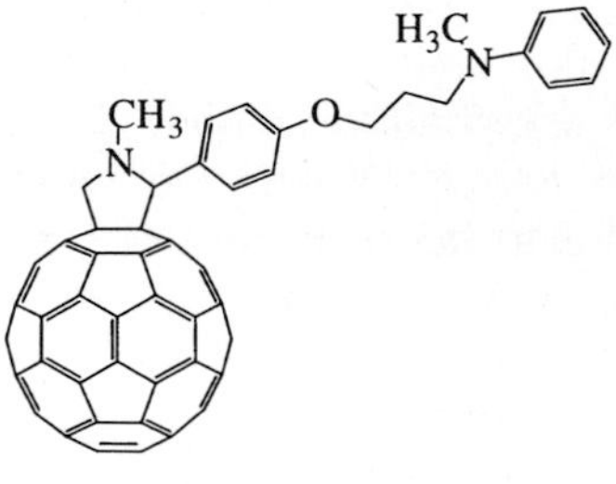

**10** 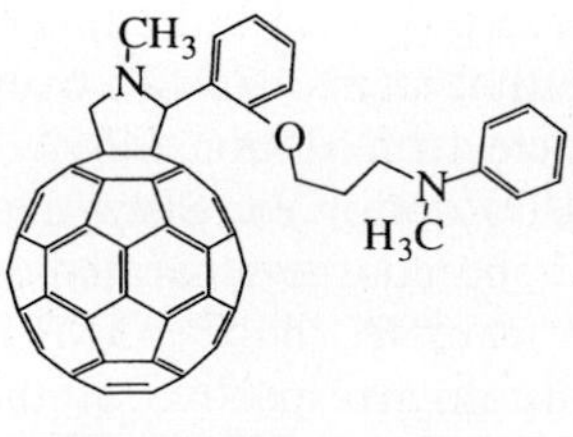                    **11**

A different objective was pursued in a recent work that reported the cluster formation of a $C_{60}$-based aniline dyad (**10**) [327] The feasibility to enhance the electron delocalization from a single: fullerene molecule to a fullerene cluster was probed under the aspect of improving charge-separation. For example, the *para*-substituted fullerene–aniline dyad **10** (i.e., the one that lacks any electron transfer activity) [327] forms stable clusters in polar solvents. The latter give rise to a strong quenching of the fullerene fluorescence. Identification of the fullerene π-radical anion in the NIR provides unambiguous evidence for the charge-transfer interactions between the photoexcited fullerene and aniline moieties [327]. Far more important is, however, the lifetime of the charge-separated state, being on the order of several hundred nanoseconds. This documents that formation of fullerene clusters is a viable alternative for the stabilization of charge-separated radical-pairs.

## 5.5.2 Fullerene Electron-acceptor Systems

The well-documented feature of fullerenes to form stable multi-anions is in sharp contrast to the poor ability to generate a stable cationic species [92]. Nevertheless, along the last several years preparation of fullerene derivatives in which $C_{60}$ is co-valently attached to an electron acceptor unit have been reported [315, 328].

Linking the fullerene core to a tetracyanoanthraquinodimethane (TCAQ) ($E_{1/2} = -0.38$ V relative to the SCE) (**12**) or a 2,3-dibromonaphthoquinone acceptor ($E_{1/2} = -0.21$ V relative to the SCE) [315] (**13**) afforded two rare examples of dyads in which an organic addend is covalently attached to the fullerene core that exhibits better electron-acceptor properties than the parent fullerene core [329].

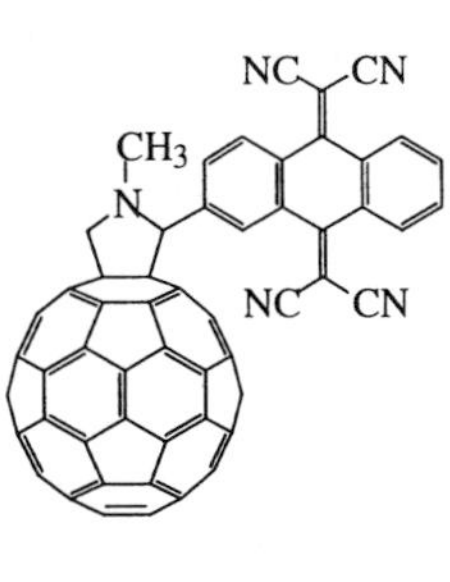

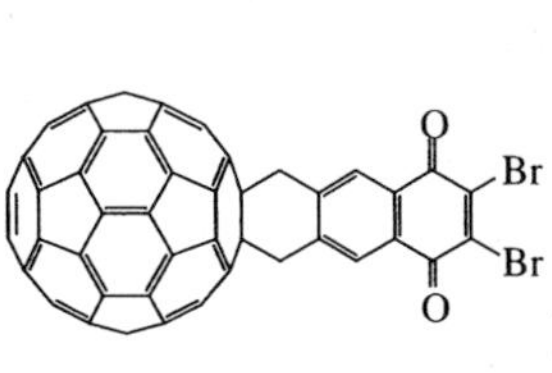

In the light of a possible electron transfer from the photoexcited fullerene to the adjacent electron acceptor, the fullerene emission was probed in a variety of solvents. Only a moderate fluorescence quenching was noted in non-polar toluene and polar benzonitrile. Considering the energetics ($E_{1/2}$ ($^1C_{60}$-R*/$C_{60}$-R$^{\cdot+}$) = −0.15 relative to the SCE) of the charge-separated radical pair small thermodynamic driving forces for an associated electron transfer were determined [315]. The main reason therefore lies in the unfavorable oxidation potential of the electron donor (i.e., fullerene). Further confirmation for the inadequate mixing of redox potentials and the inefficient quenching resulting therefrom, stems from complementary pico- and nanosecond resolved photolysis. Similar to the reference compound, illumination of the $C_{60}$-acceptor dyads (**12** and **13**) leads to the instantaneous formation of the fullerene singlet excited state absorption [315]. These transients exhibit short-ened singlet lifetimes, which fall in line with the above trend of the fluorescence intensities [315]. Despite the shorter singlet excited state lifetimes, the differential absorption spectrum, recorded after the completion of the reaction, shows spectral characteristics of the fullerene triplet excited state, rather than those of a transient product that evolves from a possible intramolecular electron transfer.

### 5.5.3 Fullerene Chromophore Systems

The moderate visible absorption of $C_{60}$-donor dyads (see above) prompts to $C_{60}$-chromophore systems as suitable models for the mimicry of the primary events in natural photosynthesis [330–332]. To enhance the absorption of the fullerene containing dyads in the visible region strong chromophoric units, such as (i) ruthenium(II) polypyridyl complexes or (ii) metalloporphyrins have been implemented as antenna systems. In these $C_{60}$-chromophore systems the role of the fullerene changed dramatically, namely, it only accepts an electron or energy from the photoexcited chromophore.

### Ruthenium complexes

Ruthenium(II) polypyridyl complexes are employed as antenna molecules [311, 333–337] for a variety of reasons: [338, 339]: Ruthenium(II) polypyridyl complexes (i) are photostable under illumination, (ii) have strong metal-to-ligand-charge-transfer absorption (MLCT) in the visible, (iii) have high lying MLCT excited states ($E^{\circ}_0 \approx 2.0$ eV), and (iv) are powerful electron donors in the MLCT excited state ($E_{1/2}$ (*[Ru(bpy)$_3$]$^{2+}$/[Ru(bpy)$_3$]$^{3+}$) = $-0.63$ V relative to the SCE; in dichloromethane). Thus, the latter should be able to generate the one-electron reduced fullerene $\pi$-radical anion ($E_{1/2} = -0.41$ V relative to the SCE; in dichloromethane) [338, 339].

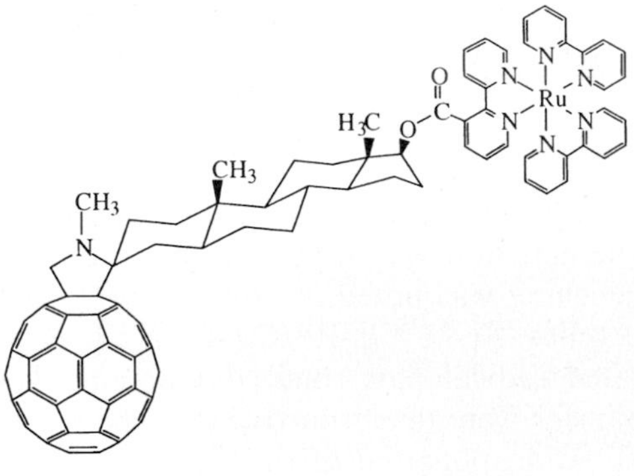

**14**

Since a general principle predicts that a larger separation evokes a slow-down of the back electron transfer, an androstane bridge was selected to separate the fullerene core from a [Ru(bpy)$_3$]$^{2+}$ complex (bpy = 2,2'-bipyridine) with an edge-to-edge distance of 11.5 Å (dyad **14**) [336]. The ruthenium MLCT excited state was found to convert, upon excitation, into the [Ru(bpy)$_3$]$^{3+}$–$C_{60}^{\bullet-}$ state with intramolecular rates of $0.69 \times 10^9$ s$^{-1}$ (DCM–toluene, 1:1, $v/v$) and $5.1 \times 10^9$ s$^{-1}$ (CH$_3$CN) [336]. The photochemically formed radical pair displays the superimposed features of the oxidized ruthenium complex ([Ru(bpy)$_3$]$^{3+}$) and of the reduced fullerene moiety (C$_{60}$·$^-$). Lifetimes of 304 and 169 ns in deoxygenated DCM and CH$_3$CN [336], respectively, are good indicators for the usefulness of this C$_{60}$-chromophore system.

Molecular recognition has been utilized as a motif to control the dynamics of intra- and intermolecular electron-transfer events and the efficiency of charge separation in a [Ru(bpy)$_3$]$^{2+}$–C$_{60}$ donor acceptor assembly (**15**) [337]. Specifically, hydrogen bonding in polypeptide systems, with a $3_{10}$ α-helix supporting α-butyric acid is selected to ensure tunable separations of the donor (i.e. [Ru(bpy)$_3$]$^{2+}$) and acceptor (i.e., fullerene) moieties. This factor is expected to have profound effects on intramolecular electron transfer dynamics between the photoexcited [Ru(bpy)$_3$]$^{2+}$ and the electron accepting fullerene. Structures, such as the $3_{10}$ α-helix in non-protic solvents, are prone to configurational changes upon addition of protic solvents (i.e., a denaturation of the helical structure is observed). In non-protic solvents the two redox active moieties are locked in well-defined positions, with an edge-to-edge distance of ca 12 Å, and electron transfer occurs from the ruthenium MLCT excited state to the electron accepting fullerene [337]. In contrast, protic solvents interfere with the intramolecular hydrogen bonding of the peptide backbone and, in turn, the relative distance between the [Ru(bpy)$_3$]$^{2+}$ and fullerene is markedly increased. Despite the general flexibility of the peptide backbone, the encountered scenario prevents either a 'through-bond' or 'exciplex' mechanism to support the electron transfer. This deactivation is reversible and repetitive addition or removal of the protic component interrupts or activates the electron transfer mechanism, respectively.

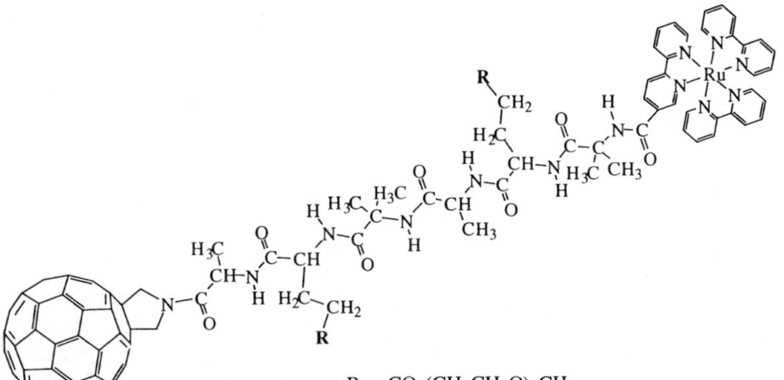

R = CO$_2$(CH$_2$CH$_2$O)$_3$CH$_3$

**15**

The design of a dinuclear ruthenium chromophore, in which the mononuclear [Ru(bpy)$_3$]$^{2+}$ antenna molecule was replaced by a dinuclear [(bpy)$_2$Ru(BL)Ru(bpy)$_2$]$^{4+}$ complex (BL = 2,3-bis-(2-pyridyl)quinoxaline) led to a red-shift of the MLCT transition from 460 nm to 625 nm [311]. The luminescence quantum yield of the dinuclear dyad **16** gives rise to a very inefficient intramolecular

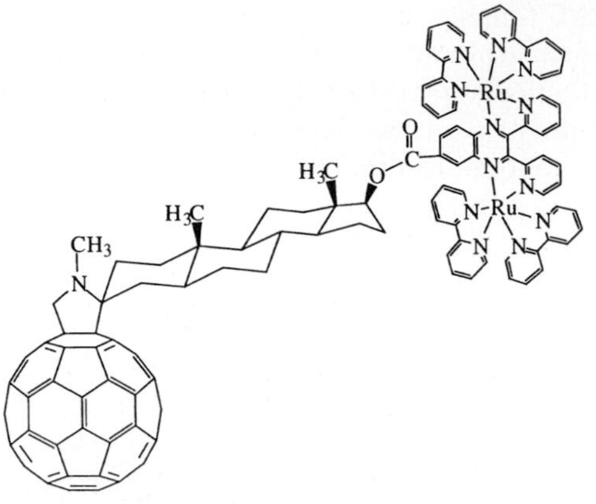

**16**

quenching reaction. In line with the luminescence quenching, picosecond and nanosecond resolved experiments revealed that the photoexcited MLCT $[(bpy)_2Ru(BL)Ru(bpy)_2]^{4+}$ state transforms slowly $(4.5 \times 10^7 \ s^{-1})$ into the fullerene triplet excited state [311]. The difference in reactivity relative to the $[Ru(bpy)_3]^{2+}$ dyad stems from the markedly different MLCT excited state energies of the $[(bpy)_2Ru(BL)Ru(bpy)_2]^{4+}$ antenna molecule.

The systematic extension of the dyad concept aimed at the design of a molecular triad by attaching an additional electron donor to the $C_{60}$-chromophore dyad [340]. Selection of the second electron donor should guarantee an electron relay along a vertical energy gradient before charge recombination takes over to dominate the fate of the first couple. As a secondary electron donor unit phenothiazine (PTZ) was chosen and covalently linked to the ruthenium complex (PTZ-$[Ru(bpy)_3]^{2+}$–$C_{60}$) **17**. In $CH_3CN$ solutions, first the formation of the PTZ-$[Ru(bpy)_3]^{3+}$–$C_{60}^{\bullet-}$ pair

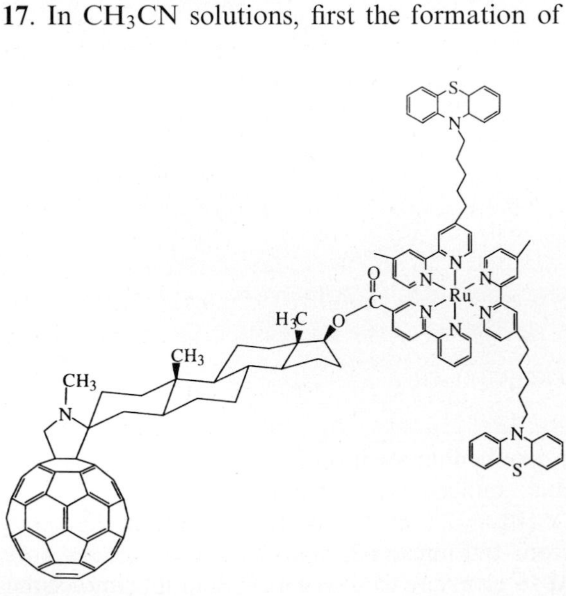

**17**

develops with intramolecular kinetics $(2.8 \times 10^9 \text{ s}^{-1})$ that are virtually identical to those of the precursor dyad. After its completion a follow-up process leads to the grow-in of the charge-separated state $PTZ^{\bullet+}\text{-}[Ru(bpy)_3]^{2+}\text{-}C_{60}^{\bullet-}$. The lifetimes observed (1.29 μs in DCM and 304 ns in $CH_3CN$) point to a significant improvement over the dyad system **14** and corroborate the value of the synthetic strategy [340].

**Metalloporphyrins**

By far the largest number of $C_{60}$-based donor–acceptor systems studied to date utilize porphyrins as antennas for efficient light capture in the visible region of the spectrum [306, 308–310, 324, 329, 341]. In the following section systems are presented for which time-resolved and steady-state absorption and emission measurements are available, which document the intramolecular electron transfer processes under a variety of experimental conditions. But we also like to acknowledge the purely synthetic work, for which no experimental data is obtainable [342–351].

Up front, a general comment seemed necessary: Photophysical studies indicate that, when structurally possible, dyads adopt conformations in which the porphyrin and fullerene moieties are in close proximity, thus facilitating through-space interactions of photoexcited porphyrins with the fullerenes. A similar conclusion stems from a recent work that demonstrates impressively weak ligand interactions between $C_{60}$ and a hard metal such as iron(III) in an iron(III)tetraphenylporphyrin π-radical cation $(FeP^{\bullet+})$ [352].

*Dyads*

The first direct evidence for photoinduced electron transfer in donor-linked fullerenes comes from photophysical studies regarding a $ZnP$–$C_{60}$ dyad (**18**), in which two chromophores are separated by a phenyl spacer including an amido group, using picosecond time-resolved transient absorption spectroscopy [353]. As the singlet excited state absorption of the porphyrin decayed after excitation of the porphyrin, simultaneous rise and decay of the transient bands due to $ZnP^{\bullet+}$ and $C_{60}^{\bullet-}$ were observed clearly. In summary, photoinduced electron transfer evolves from the excited singlet state of the porphyrin to $C_{60}$ in dyad **18**. To evaluate the three-dimensional effect of $C_{60}$ in photoinduced electron transfer, a porphyrin–quinone dyad **19** (ZnP–Q) was prepared which gives rise to driving forces and electronic coupling interactions that are quite similar to the $ZnP$–$C_{60}$ dyad [354]. In THF photoinduced charge separation and charge recombination in the $ZnP$–$C_{60}$ dyad (**18**) is accelerated and retarded, respectively, compared to that of the ZnP–Q (**19**). The acceleration and deceleration effect of $C_{60}$ in intramolecular electron transfer has been attributed to the smaller reorganization energy of $C_{60}$ relative to those of typical two-dimensional acceptors such as quinones and imides [308, 309, 353–356, 358]. A combination of small reorganization energy of fullerenes with multistep electron transfer strategy is a fascinating proposal for the construction of artificial photosynthesis.

One of the first reported $C_{60}$-based donor–acceptor systems is also the $H_2P$–$C_{60}$ and the analog $ZnP$–$C_{60}$ dyad (**20**), in which $H_2P$ denotes the metal-free base tetraphenyl porphyrin and ZnP is the corresponding zinc tetraphenylporphyrin

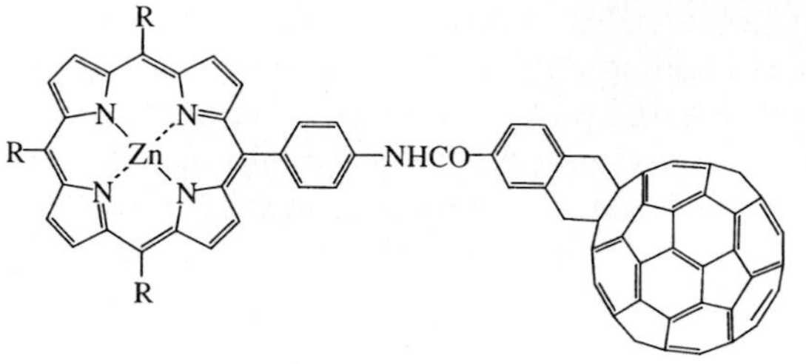

**18**

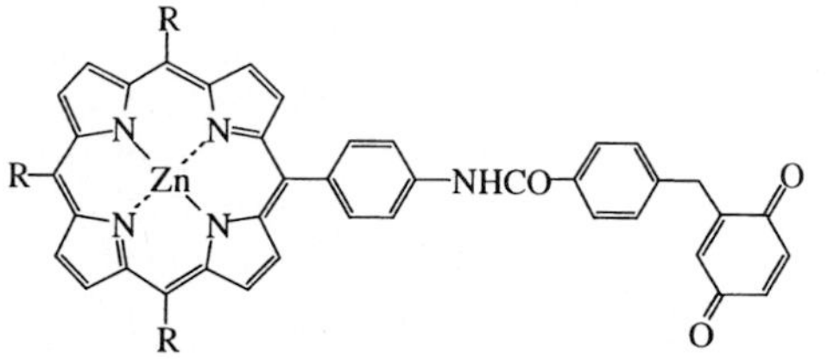

**19**

complex [360]. In polar benzonitrile the singlet excited porphyrin undergoes photo-induced electron transfer to give $H_2P^{\bullet+}-C_{60}^{\bullet-}$ and $ZnP^{\bullet+}-C_{60}^{\bullet-}$, with intramolec-ular rates of $\approx 2 \times 10^{11}$ s$^{-1}$ [360]. Unfavorable thermodynamics render an electron transfer in non-polar toluene difficult. While the lower oxidation potential of ZnP ($\approx 200$ mV) still facilitates an electron transfer, the markedly shifted oxidation po-tential of $H_2P$ gives rise to singlet–singlet energy transfer in $H_2P-C_{60}$ from the porphyrin to the fullerene.

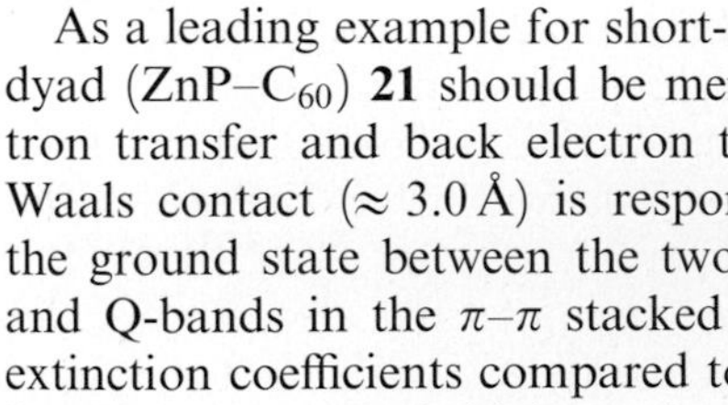

**20**

As a leading example for short-spaced dyads, a $\pi$–$\pi$ stacked porphyrin–fullerene dyad (ZnP–$C_{60}$) **21** should be mentioned, which was probed in light of their elec-tron transfer and back electron transfer dynamics [361, 362]. The close van der Waals contact ($\approx 3.0$ Å) is responsible for pronounced electronic interactions in the ground state between the two $\pi$-chromophores. For example, the ZnP Soret- and Q-bands in the $\pi$–$\pi$ stacked dyad **21** show a bathochromic shift and lower extinction coefficients compared to free ZnP [361]. In the $\pi$–$\pi$ stacked dyad **21** the linkage of the two bridging units occurs in the *trans*-2 position at the fullerene. A charge-separated radical pair evolves from a rapid intramolecular electron transfer ($\approx 35$ ps) between the photoexcited metalloporphyrin and the fullerene core in a variety of solvents (i.e., ranging from toluene to benzonitrile). Remarkably, the lifetimes in tetrahydrofuran ($\tau = 385$ ps) and DCM ($\tau = 122$ ps) are markedly in-creased relative to the more polar solvents dichloromethane ($\tau = 61$ ps) and ben-zonitrile ($\tau = 38$ ps) [362]. This dependency prompts to an important conclusion:

The rates of the back electron transfer processes, from the $ZnP^{\bullet+}-C_{60}^{\bullet-}$ state to the ground state, are clearly in the 'Marcus-inverted' region. This is one of the important examples that provide experimental support to electron transfer kinetics clearly occurring in the inverted region. Using the same synthetic methodology, a similar family of $\pi-\pi$ stacked porphyrin–fullerene dyads (i.e., $ZnP-C_{60}$ and $H_2P-C_{60}$), in which the porphyrin chromophore is attached by a malonate bridge to the *trans*-1 position of the fullerene sphere, has been reported [363].

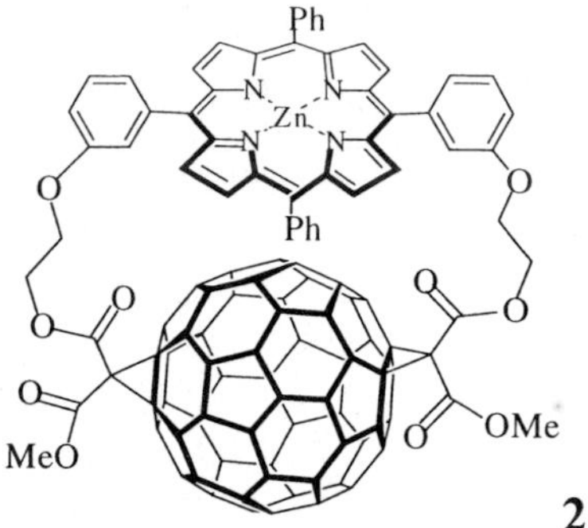

**21**

Increasing the distance between the donor and acceptor, by means of using spacer units of increasing size, is one approach to slow down the charge recombination [364, 365]. In this context, a fulleropyrrolidine should be mentioned which is been attached to one of the pyrrole units of $H_2P$ (**23**) or alternatively $ZnP$ (**22**) [365]. Based on the close proximity between the donor and acceptor efficient through-space interactions are postulated. In benzonitrile intramolecular electron-transfer events prevail yielding the $H_2P^{\bullet+}-C_{60}^{\bullet-}$ and $ZnP^{\bullet+}-C_{60}^{\bullet-}$ pairs with a quantum efficiency near unity and lifetimes of 290 ps and 50 ps, respectively [364]. The spatial separation impacts, at the same time, the thermodynamic driving force $(-\Delta G^{\circ}_{et})$ for an intramolecular electron transfer. In particular, increasing the donor acceptor separation lowers the driving force. Thus, an electron transfer that is exothermic in a closely spaced dyad becomes less exothermic in a widely spaced dyad and, in turn, is deactivated. For example, in toluene a rapid singlet–singlet energy transfer governs the deactivation of the photoexcited chromophore complex [364].

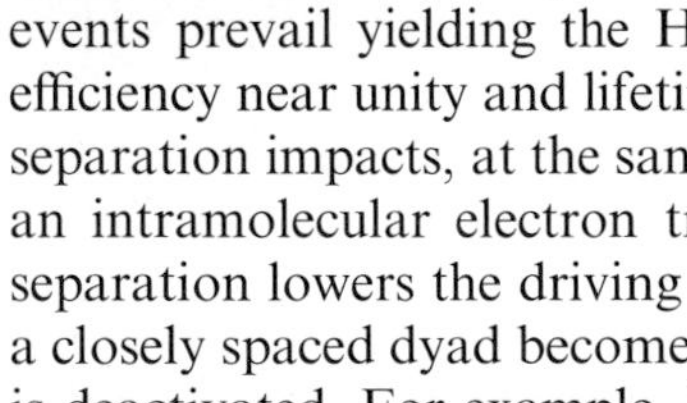

**22** 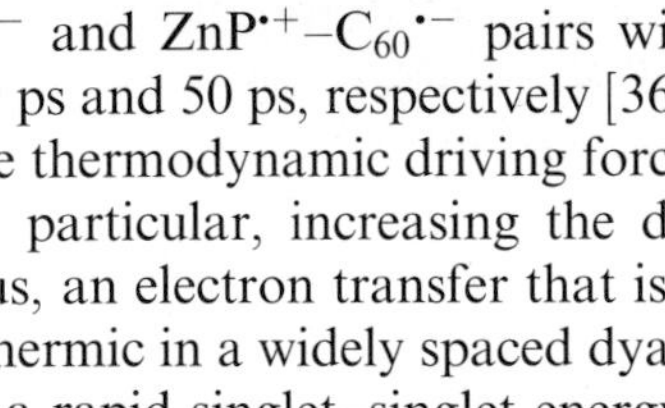   **23**

Further advances with respect to the separation principle were made via some work that reports on a long-range photoinduced electron transfer in a $ZnP-C_{60}$ dyad **24**, in which the two redox active moieties are separated by a saturated norbornylogous bridge nine sigma bonds in length [324, 366]. The lifetime of the radical ion pair in the 'norbornylogous' bridged dyad is impressive (420 ns) [366] and clearly discloses a significant improvement relative to the short-spaced analogs ($\pi$–$\pi$ stacked dyad: 38 ps [362]; pyrrole-linked dyad: 50 ps [364]).

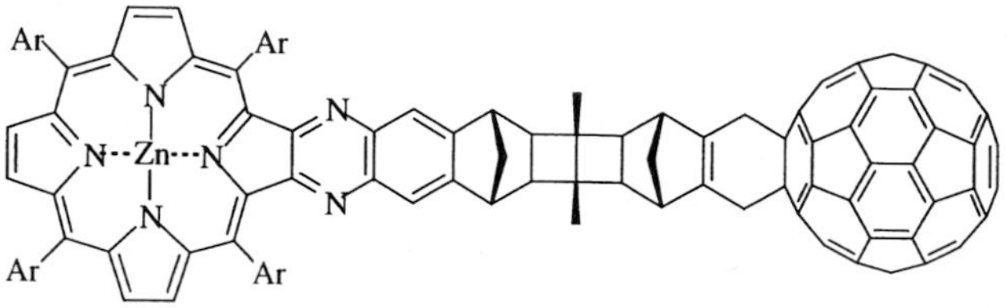

**24**

A phenyl group has also been employed to link the electron acceptor ($C_{60}$) to a ZnP, which, in addition, enabled a systematic variation of the linkage (e.g., in *ortho* (**26**), *meta* (**25**) or *para* position (**27**) of the phenyl group) [353–359]. Regardless of the specific position, the initial photoexcitation event is followed by a rapidly occurring charge separation to yield the $ZnP^{\bullet+}$–$C_{60}^{\bullet-}$ with rates between $2.2 \times 10^{10}$ s$^{-1}$ and $0.45 \times 10^{10}$ s$^{-1}$ [355]. The final step in the sequence is the charge recombination process, which results in the complete restoration of the ground state. For example, in THF the radical ion pair evolves from the singlet excited state of both moieties (i.e., fullerene and porphyrin). On the other hand, in nonpolar solvents such as benzene the charge-separated state, generated by photoinduced electron transfer from the $^1$ZnP* to the $C_{60}$, recombines to yield the $^1C_{60}$* or energetically equilibrates with the $^1C_{60}$*. The photodynamic behavior in nonpolar solvents is in sharp contrast with that in the other porphyrin–$C_{60}$ dyads where direct singlet–singlet energy transfer takes place from the porphyrin to the $C_{60}$ [360, 364, 366, 376]. In light of the intramolecular dynamics (i.e., forward and back electron transfer), it is important to note that both processes are markedly slowed-down by the *meta*-linkage (**25**) [355].

Work from a different laboratory focuses on the synthesis and photophysical properties of $C_{60}$-based porphyrin dyads with a variety of linkers [367–371]. These include new types of flexible polyether linkers (**30**) [367, 368], rigid steroid linkers [370] as well as novel doubly linked porphyrins [369, 371], which provide the means for a parachute-shaped dyad (e.g., ZnP–$C_{60}$ (**29**) and $H_2$P–$C_{60}$ (**28**)). In both para-

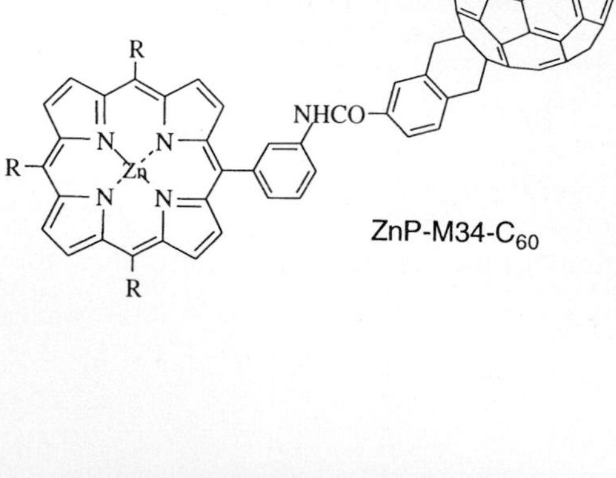

ZnP-M34-C$_{60}$

**25**

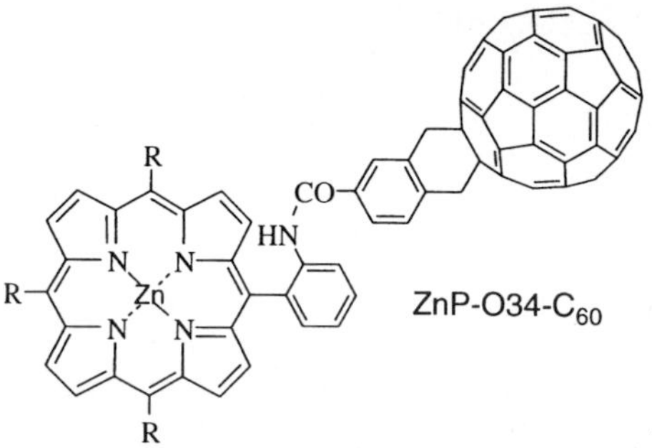

ZnP-O34-C$_{60}$

**26**

ZnP-P23-C$_{60}$

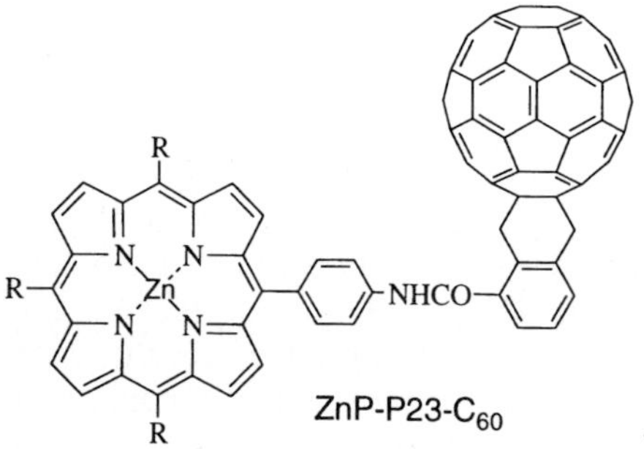

**27**

chute compounds, the porphyrin fluorescence is almost completely quenched with fluorescence lifetimes of 13 ps ($^{1}$ZnP*–C$_{60}$) and 23 ps ($^{1}$H$_{2}$P*–C$_{60}$) [369, 371]. It should be noted that in the ZnP–C$_{60}$ (**29**) and H$_{2}$P–C$_{60}$ dyads (**28**) the porphyrin is enforced in close proximity to the fullerene core. With the help of transient absorption studies the dynamics of the interactions in these systems were elucidated, especially to differentiate between electron and energy transfer as the primary process. The unique orientation of the porphyrin and fullerene moieties in the parachute-shaped dyad (**28** and **29**) seems to facilitate electron transfer in both non-polar

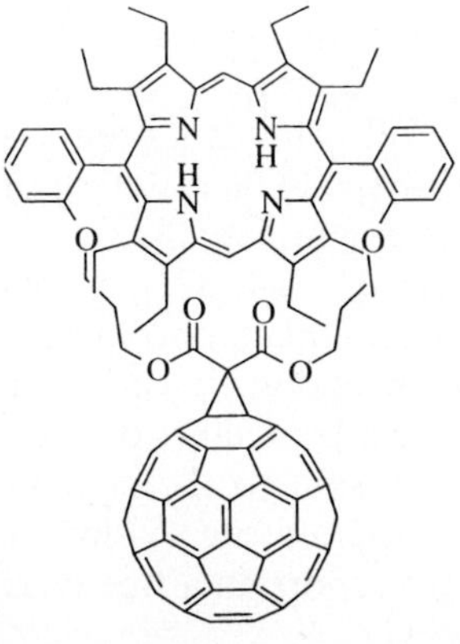

**28**

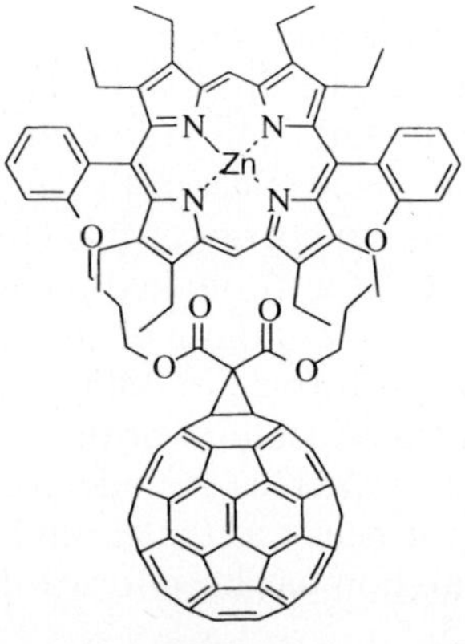

**29**

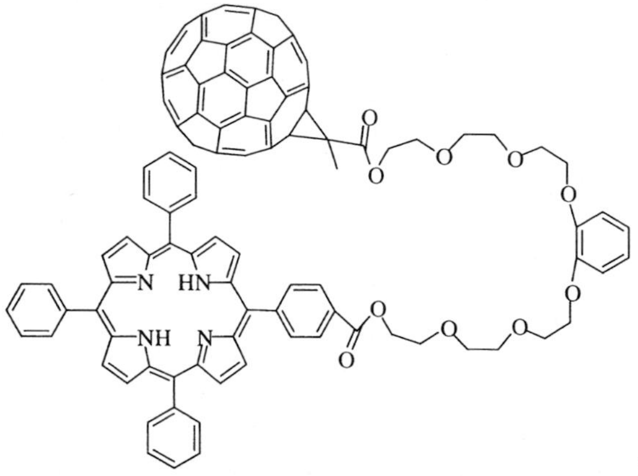

**30**

and polar media to yield the $ZnP^{\bullet+}-C_{60}^{\bullet-}$ and $H_2P^{\bullet+}-C_{60}^{\bullet-}$ pair, respectively. It is curious that, for example, the lifetime of the $ZnP^{\bullet+}-C_{60}^{\bullet-}$ state is longer in toluene ($\tau = 1015$ ps) than in the more polar solvents THF ($\tau = 99$ ps) or benzonitrile ($\tau = 69$ ps) [371]. It has been concluded that these solvent effect results from changes in the energetics of the ion pair state as well as the reorganization energy in these three solvents, being well in the Marcus inverted region. The same quenching effect is also seen in a series of flexible-linked dyads (**30**), indicating that non-covalent interaction between the porphyrin and fullerene moieties bring them into close proximity [367, 368].

A fundamental challenge addresses the question to what extent the choice of a higher fullerene impacts the electron transfer dynamics in fullerene-based donor–acceptor systems [372]. To shed light onto this important aspect and to evaluate the intrinsic properties of $C_{70}$ in electron-transfer reactions a $ZnP-C_{70}$ dyad was designed, resembling the basic features of the analogous $ZnP-C_{60}$ system [372]. In this context, the reduction potentials of fullerenes should be considered: In general higher fullerenes become better electron acceptors as the fullerene size increases. Despite this general trend, the difference between $C_{60}$ and $C_{70}$ is reported to be marginally small. Still, the intramolecular electron transfer dynamics in $ZnP-C_{70}$ ($3.5 \times 10^{10}$ s$^{-1}$) following the initial excitation of the ZnP chromophore, were found to be nearly twice as fast as in $ZnP-C_{60}$ ($1.9 \times 10^{10}$ s$^{-1}$) [372]. This has been ascribed to the inherent size and shape effect of the rugby ball-shaped $C_{70}$ resulting in a somewhat smaller reorganization energy of this larger fullerene moiety [372].

The photophysical properties of a $C_{60}$-linked phytochlorin (**31**) [344, 373], a porphyrin analog, are quite different from those of conventional $C_{60}$-linked porphyrins. The phytochlorin–$C_{60}$ exciplex is formed in both toluene and benzonitrile via either the singlet excited states of the phytochlorin or the $C_{60}$. The exciplex relaxes directly to the ground state in toluene, whereas it undergoes a conversion to the charge-separated state, followed by the decay to the ground state in benzonitrile. A similar proposal for the exciplex formation has been forwarded by the report on a free-base porphyrin $C_{60}$ dyad [374, 375].

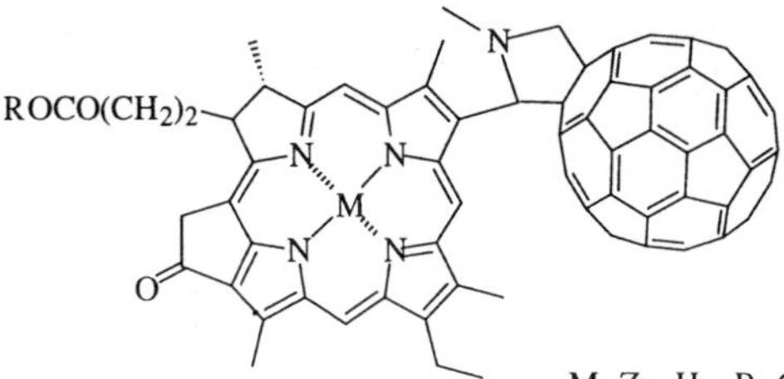

ROCO(CH$_2$)$_2$

M=Zn, H$_2$; R=CH$_3$, H  **31**

*Triads*

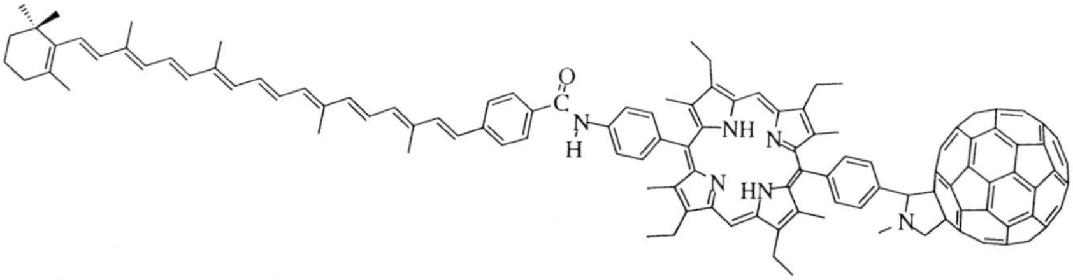

Using the above-summarized C$_{60}$-based porphyrin dyads as a work base, a carotenoid polyene (C) [331, 332] has been covalently linked to a H$_2$P–C$_{60}$ dyad, constituting the first molecular triad **32** published (e.g., C-H$_2$P–C$_{60}$) [376, 377]. Upon photoexcitation, a sequence of intramolecular electron-transfer events were found for C-P–C$_{60}$ triad **32**, tracking the energy gradient C-H$_2$P*–C$_{60}$ → C–(H$_2$P$^{•+}$)–(C$_{60}$$^{•-}$) → (C$^{•+}$)–H$_2$P–(C$_{60}$$^{•-}$). The final (C$^{•+}$)–H$_2$P–(C$_{60}$$^{•-}$) pair, which was produced with an overall quantum yield of 0.14 decays via charge recombination ($\tau = 170$ ns) to afford the carotenoid triplet state, rather than the singlet ground state [376]. Even in a glassy matrix at 77 K the radical pair is generated with a still remarkable quantum yield of 0.1. Spectral support for this recombination route was presented in terms of time-resolved EPR spectroscopy (i.e., correlating the spin-polarized radical pair and carotenoid triplet state) [377]. The sequence of (i) generating a long-lived charge separated state and (ii) forming a triplet state via charge recombination is a phenomenon heretofore observed almost exclusively in photosynthetic reaction centers.

**32**

A viable alternative to the C–H$_2$P–C$_{60}$ triad **32** [376, 377] is the design of a sequential electron transfer relay system in which the porphyrin–fullerene pair (ZnP–C$_{60}$) is bridged via an electroactive pyromellitimide functionality (In) (**33**) [358, 378]. The photochemical results clearly show a photoinduced two-step electron transfer sequence. Starting with the initial excited porphyrin state the bridging pyromellitimide accepts an electron from the photoexcited ZnP chromophore affording a charge-separated radical pair. This transient state subsequently transfers the accepted charge from the pyromellitimide to the end terminus, namely, the fullerene moiety. The resulting spatially separated (ZnP$^{•+}$)–In–(C$_{60}$$^{•-}$) radical, formed with an overall quantum yield of 0.46, undergoes a remarkably slow back electron transfer with a rate of $7.7 \times 10^8$ s$^{-1}$ [378].

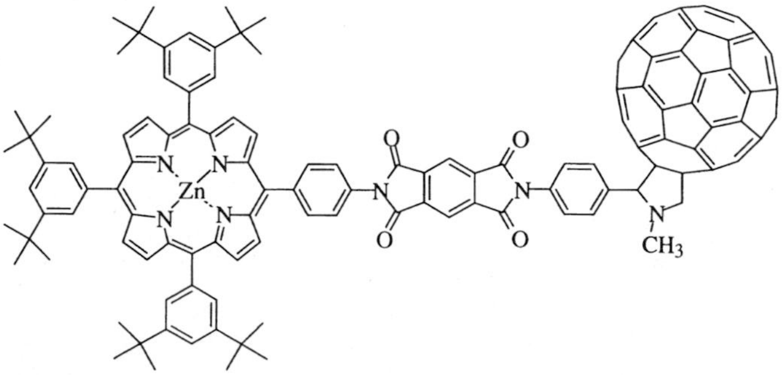

**33**

Stabilization of the charge-separated state is the focus of a more recent study regarding a Fc–ZnP–C$_{60}$ triad **34**, whose conceptional idea is similar to that described for the C–P–C$_{60}$ triad **32** [376, 377] (i.e., a controlled redox gradient Fc–*ZnP–C$_{60}$ → Fc–(ZnP$^{\bullet+}$)–(C$_{60}^{\bullet-}$) → (Fc$^+$)–ZnP–(C$_{60}^{\bullet-}$) [358, 379–381]. The improvement is, however, based on increasing the spatial separation between the individual building blocks (e.g., Fc–ZnP and ZnP–C$_{60}$ couples, respectively). Via utilizing of a nearly rigid bridge, comprising two phenyl groups linked by an amide functionality, a center-to-center distance of $\approx 18$ Å is achieved between each of the two couples [358, 380]. Upon excitation, efficient multi-step electron transfer governs the deactivation of the photoexcited ZnP chromophore, producing a long-lived charge-separated state (Fc$^{\bullet+}$)–ZnP–(C$_{60}^{\bullet-}$) with a center-to-center distance of 34.2 Å. In DMF a remarkable lifetime of 15.6 μs is observed [358, 380].

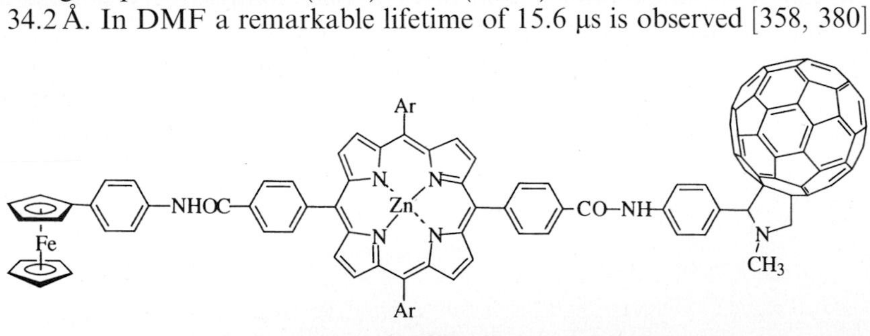

**34**

A novel molecular triad **35**, representing an artificial reaction center, was synthesized via linking a fullerene moiety to an array of two porphyrins (i.e., ZnP and H$_2$P) [382, 383]. In this ZnP–H$_2$P–C$_{60}$ triad **35**, the ZnP performs as an antenna molecule, mediating its singlet excited state energy to the energetically lower lying H$_2$P. This energy transfer ($k = 1.49 \times 10^{10}$ s$^{-1}$) is followed by a sequential electron transfer relay evolving from the generated singlet excited state of H$_2$P to yield ZnP–(H$_2$P$^{\bullet+}$)–(C$_{60}^{\bullet-}$) and subsequently (ZnP$^{\bullet+}$)–H$_2$P–(C$_{60}^{\bullet-}$) with rate constants of $6.99 \times 10^9$ s$^{-1}$ and $2.15 \times 10^9$ s$^{-1}$, respectively [383]. The final charge-separated state, formed in high yield (0.4), gives rise to a remarkable lifetime of 21 μs in deoxygenated benzonitrile and decays directly to the singlet ground state [383]. In contrast, in non-polar toluene solutions the deactivation of the porphyrin chromophores (ZnP and H$_2$P) takes place via singlet–singlet energy transfer leading to the fullerene singlet excited state. This stems from the unfavorable free energy changes for an intramolecular electron-transfer event from the singlet excited state of H$_2$P to the adjacent fullerene acceptor.

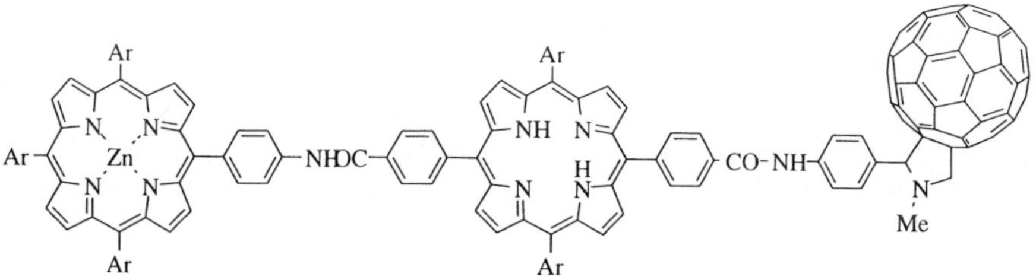

**35**

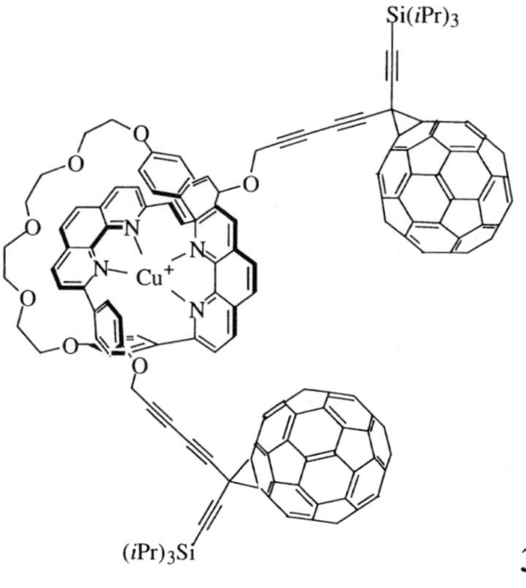

**36**

A novel bis-fullerene rotaxane **36** structure was successfully synthesized via oxidative coupling between two methanofullerenes and a Cu(I)–bisphenanthroline complex [384]. This supramolecular assembly (**36**) provides well-defined inter-component separations. Interestingly, perturbation of the absorption spectrum of rotaxane, relative to the sum of the individual moieties, indicates strong electronic communication between the two components in the ground state [384]. More importantly, the excited states of the two fullerenes and also the Cu(I)–bisphenanthroline complex are subject to a rapid deactivation. In particular, a sequence of energy and electron-transfer events lead to a low-lying charge-separated state. However, no clear evidence for the charge-separated state is been found and its existence is only been inferred. The lack of spectral identification has been rationalized in terms of a fast back electron transfer prevailing over a slower forward electron transfer [384].

*Hexad*

The idea of an artificial reaction center (e.g., the sequence of energy and electron transfer) was realized in a recent report of a $C_{60}$-based hexad, $(ZnP)_3$–ZnP–$H_2P$–$C_{60}$, employing an array of star-shaped aligned porphyrins (i.e., four zinc porphyrins linked to a free base porphyrin) as a multichromophoric antenna system [385]. Excitation of the peripheral zinc porphyrins lead to rapid singlet–singlet energy transfer to the central zinc porphyrin ($\approx 50$ ps) [385]. This excited state energy is passed on to the free porphyrin, yielding the singlet excited state, which, in turn, decays via an intramolecular electron-transfer event to the adjacent fullerene moiety (3 ps). The charge-separated state formed (i.e., $(ZnP)_3$–ZnP–$(H_2P^{\bullet+})$–$(C_{60}^{\bullet-})$) has a lifetime of 1330 ps and is generated with a quantum yield of 0.70 based on light absorbed by the ZnP antenna [385].

*Supramolecular complexes*

The following summarizes recent studies, whose aim is to retard the fast back electron transfer, commonly observed in supermolecular systems [386–389]. In this context, a reversible coordination of the acceptor moiety (ligand or substrate) to the donor (coordination center or receptor) rather than a covalent linkage was pursued [386–389]. This was expected to enable the diffusional splitting of the charge-separated radical pair after the initial electron transfer takes place.

Photoprocesses associated with the complexation of a pyridine-functionalized fullerene derivative to a ruthenium tetraphenyl porphyrin (Ru(CO)P) and to a ZnP, yielding the ZnP–$C_{60}$ (**37**) and RuP–$C_{60}$ dyads, respectively, have been studied by time resolved optical and transient EPR spectroscopies [386–389].

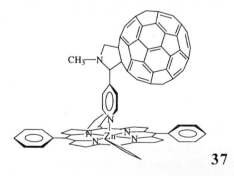

**37**

It is found that, upon irradiation in toluene, a highly efficient triplet–triplet energy transfer governs the deactivation of the RuP, while electron transfer from the porphyrin to the fullerene (RuP$^{\bullet+}$–$C_{60}^{\bullet-}$) prevails in polar solvents [389]. Complexation of ZnP by the fullerene derivative is reversible and, following excitation of the ZnP, gives rise to very efficient charge separation. In fluid polar solvents like THF and benzonitrile radical ion pairs (ZnP$^{\bullet+}$–$C_{60}^{\bullet-}$) are generated both by intramolecular electron transfer in the complex and by intermolecular electron transfer in the uncomplexed form [386]. In the latter case radical pairs live about 10 μs in THF and about 50 μs in benzonitrile at room temperature [386]. Thus, complexation of the donor–acceptor couple appears a viable alternative to super-molecular polyads (e.g., triads, tetrads, etc.) involving covalent links between the components for increasing the lifetime of the charge-separated state. The coordination concept, namely, complexation of a fullerene–pyridine ligand by a macrocyclic $\pi$-system that bears a potential utilization as a chromophore system, is very general and can be employed successfully to metallophthalocyanines and structural porphyrin isomers as has been demonstrated recently [310, 388].

## 5.6 Oxidation of Fullerenes

### 5.6.1 Gas-phase Oxidation

The oxidation of fullerenes is much more difficult than their reduction because of the highly positive oxidation potentials (1.26 V relative to $Fc/Fc^+$ for $C_{60}$) [16]. Since the five HOMO orbitals of $C_{60}$ ($h_u$) are degenerate and fully occupied, $C_{60}$ may be oxidized up to $C_{60}^{10+}$. The formation of $C_{60}^{7+}$ in which 7 electrons were removed from the ocean of 240 valence electrons that surround its hollow cage of $C_{60}$ has been detected in the mass spectrum [390]. This has been achieved in an ion source that delivers high electron currents at high electron energies. Singly, doubly and triply charged $C_{60}^{n+}$ (n = 1, 2, 3) cations can be readily formed in the conventional electron-impact ion source that delivers ions into a selected-ion flow tube (SIFT) apparatus [391]. $C_{60}^{\bullet+}$ is generally quite unreactive towards nucleophiles such as alcohols under SIFT conditions in the gas phase [392, 393]. Electron transfer is rare because of the low electron recombination energy of $C_{60}^{\bullet+}$ (7.6 eV) [394]. Addition of nucleophiles such as ammonia, amines and iron pentacarbonyl to $C_{60}^{\bullet+}$ has been observed rather than electron transfer [395, 396]. Diels–Alder additions of 1,3-cyclopentadiene, 1,3-cyclohexadiene, anthracene and corannulene to $C_{60}^{\bullet+}$ have also been reported to form the $C_{60}$ adduct radical cation [397]. $C_{60}^{2+}$ is much more reactive than $C_{60}^{\bullet+}$ in both electron transfer and addition reactions [388]. Electron transfer becomes a dominant process only when $C_{60}^{\bullet+}$ reacts with molecules with sufficiently low ionization energies and it must be exothermic by at least 1.8 eV due to the Coulombic repulsion between the charged products in order to compete with the addition pathway effectively [398, 399]. Electron transfer with $C_{60}^{3+}$ becomes even more competitive than with $C_{60}^{2+}$. However, the ionization-energy threshold for electron transfer to $C_{60}^{3+}$ lies 4.4 eV below the thermodynamic threshold of the electron recombination energy of $C_{60}^{3+}$ (15.6 eV) because of the larger Coulombic repulsion between the charged products as compared with the case of $C_{60}^{2+}$ [400].

### 5.6.2 Oxidation of Fullerenes in Solution

Electron acceptors such as tetracyanoquinodimethane (TCNQ) [401–404], tetra-cyanoethylene (TCNE) [405, 407] and chloroaniline (ClA) [24, 406] have been employed in photolysis experiments regarding the oxidative quenching of triplet excited fullerenes. In general, very fast quenching reactions are reported, but these processes are not accompanied by the separation of the free radical pairs. A triplet 'exciplex' formation has been proposed as a possible mechanism, which, in turn, would prevent the escaping from the cage. Only, in the case of TCNE the 'exciplex' has been considered as an ion–radical pair [405, 406]. Alternatively, semiconductor supports (i.e., $TiO_2$) have been used to initiate photochemical processes on their surfaces, including oxidation of surface-adsorbed substrates [408, 409]. In this context, the reader is reminded of that band-gap excitation of anatase $TiO_2$ ($E_g = 3.2$ eV) requires highly energetic UV light. On the other hand, illumination of the adsorbed substrate with visible light avoids direct excitation of the semiconductor support. Irreversible chemical changes are reported following the fullerene excitation on a $TiO_2$ support [408, 409]. The photoejection of electrons from the photoexcited $C_{60}$ into the $TiO_2$ semiconductor is considered to be the primary step of the photooxidation process. The oxidation potential of the fullerene singlet excited state renders an electron injection from the singlet excited state into $TiO_2$ particles thermodynamically impossible. Rationalization of the experimental observation was forwarded in form of a hypothesis that suggests a biphotonic process (Eq. 21). In support of this hypothesis is the linear dependence of the transient oxidation product formed on the $TiO_2$ surface on the square of the laser dose.

$$(C_{60}/TiO_2)-2h\nu \rightarrow C_{60}{}^{\bullet+} + TiO_{2(e)} \tag{21}$$

The one-electron oxidation potential of $C_{60}$ is highly positive ($E°_{ox} = 1.26$ V relative to ferrocene/ferricenium, which is equivalent to 1.63 V relative to the SCE) [16]. The singlet excited states of 9,10-dicyanoanthracene ($^1DCA*$: $E°_{red} = 1.91$ V relative to the SCE) [410] and 10-methylacridinium ion ($^1AcrH^+*$: $E°_{red} = 2.32$ V relative to the SCE) [411, 412] have both higher one-electron reduction potentials than that of $C_{60}$. Thus, both of them can be used as effective electron acceptors to oxidize $C_{60}$ by photoinduced electron transfer [413]. The strong oxidizing ability of $^1AcrH^+*$ in the photoinduced electron-transfer reactions with a variety of electron donors has been well established [409, 410, 414–417]. Although $C_{60}{}^{\bullet+}$ is unreactive toward alcohols in the gas phase (vide supra) [390, 391], $C_{60}{}^{\bullet+}$ produced by photoinduced electron transfer from $C_{60}$ to $^1DCA*$ or $^1AcrH^+*$ can react with alcohols and hydrogen donors to afford the alkylated fullerenes [418–420]. Such a different reactivity of $C_{60}{}^{\bullet+}$ (i.e., gas phase or solution) can be ascribed to the strong solvation of the product cation. The addition of biphenyl (BP) as a cosensitizer to the DCA–$C_{60}$ system in the presence of a hydrogen donor resulted in a significant increase in the reaction efficiency via the electron transfer catalysis as shown in Scheme 12 [417, 418].

The photochemical reaction is started by a photoinduced electron transfer from BP (used in excess to $C_{60}$) to $^1DCA*$ producing $DCA^{\bullet-}$ and $BP^{\bullet+}$. Since electron transfer from $C_{60}$ to $BP^{\bullet+}$ is exothermic, $C_{60}{}^{\bullet+}$ is formed as a free ion by electron transfer from $C_{60}$ to $BP^{\bullet+}$ to regenerate BP. Without BP the facile back electron transfer from $DCA^{\bullet-}$ to $C_{60}{}^{\bullet+}$ in the radical ion pair produced direct electron transfer between $C_{60}$ and $^1DCA*$ would proceed leading to no product formation. In contrast

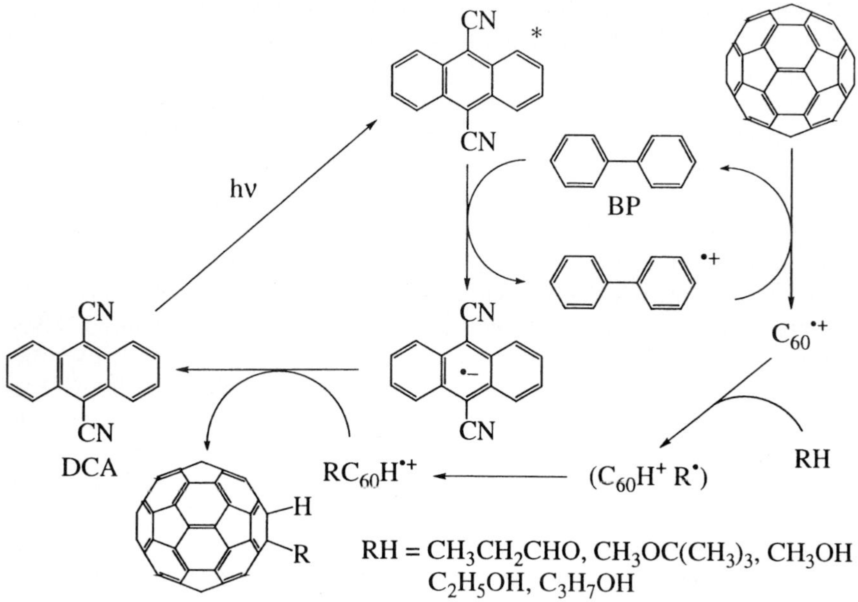

**Scheme 12**

to nucleophilic attacks of alcohols to the radical cation as expected from a common reaction of radical cations with nucleophiles [421, 422], the C–C bond formation with the alcohol takes place rather than the C–O bond formation (Scheme 12). Such a bizarre reaction of $C_{60}^{\bullet+}$ may be ascribed to the low electrophilicity of $C_{60}^{\bullet+}$ due to the highly delocalized positive charge. This reactivity is reminiscent of the low nucleophilicity of $C_{60}^{2-}$ due to the highly delocalized negative charge in the reactions with alkyl halides in Scheme 2. As is the case of $C_{60}^{2-}$, the only reaction pathway left for $C_{60}^{\bullet+}$ is an electron transfer from RH to $C_{60}^{\bullet+}$ to give $RH^{\bullet+}$ and $C_{60}$. This may be followed by a rapid deprotonation of $RH^{\bullet+}$ to produce $R^{\bullet}$, since $RH^{\bullet+}$ becomes an extremely strong acid as compared to RH [423]. The addition of $R^{\bullet}$ to $C_{60}$ followed by back electron transfer from $DCA^{\bullet-}$ leads to the final product, namely, 1-substituted 1,2-dihydro [60]fullerene (Scheme 12).

## 5.7 Summary

This review has described the fundamental electron-transfer properties of fullerenes together with a variety of electron-transfer reactions of fullerenes, which often lead to a new type of derivatization of fullerenes. The small reorganization energy of fullerenes, especially in electron-transfer reactions, and the long triplet excited sate lifetimes has rendered fullerenes as useful components in the design of novel electron transfer systems in both the ground and excited states. Most studies have so far dealt with $C_{60}$ and a variety of $C_{60}$ derivatives, the electron-transfer chemistry of higher fullerenes and endohedral metallofullerenes has remained to be explored further. There are many common mechanistic aspects in electron-transfer chemistry of fullerenes with that of other organic compounds described in other chapters. However, the highly delocalized negative and positive charges of fullerene anions and cations produced in the electron-transfer reduction and oxidation result in peculiar reactivities. Therefore fullerene anions and cations react in favor of an electron transfer pathway rather than a polar two-electron pathway. There have been cases where the regio- and stereo-selectivities in the reactions of fullerenes via electron transfer are different from those expected from the concerted reactions. Thus, electron-transfer chemistry of fullerenes provides an unique opportunity to synthesize new fullerene derivatives which would otherwise be impossible to obtain. It is hoped that this overview of electron-transfer chemistry of fullerenes facilitates to expand this area from both fundamental and applicational points of view.

The systematic investigation of fullerene chemistry has led to the development of useful molecular donor acceptor dyads. The selected examples are a clear demonstration of the continuing interest in fullerenes as novel, multifunctional electron storage moieties in well-ordered multicomponent composites. In this context, the unique delocalization, provided by the fullerene core, in combination with the small reorganization energy are key factors that prevent a fast back electron transfer process in fullerene-containing systems. Thus, the contributed systems are promising candidates for the design of artificial photosynthetic systems with efficient and long-lived charge separation but fewer electron transfer steps and less energy loss.

## Acknowledgment

The authors are deeply indebted to the work of all collaborators and co-workers whose names are listed in the references of this chapter (in particular, S.F.: Prof. K. M. Kadish, Prof. K. Mikami and Prof. O. Ito; DMG: Prof. K.-D. Asmus, Prof. M. Maggini, Prof. N. Martin and Prof. M. Prato). S.F. acknowledges continuous support of his study on electron-transfer chemistry by a Grant-in-Aid from the Ministry of Education, Science, Culture and Sports, Japan. D.M.G. acknowledges the support by the Office of Basic Energy Sciences of the Department of Energy. This is document NDRL-4267 from the Notre Dame Radiation Laboratory). We would like to thank Drs. H. Imahori and C. Luo for their helpful discussion.

## References

1. R. A. Marcus, *Annu. Rev. Phys. Chem.* **1964**, *15*, 155; R. A. Marcus, *Angew. Chem. Int. Ed. Engl.* **1993**, *32*, 1111.
2. A. Hirsch, *The Chemistry of the Fullerenes*, Thieme, New York, **1994**.
3. Q. Xie, E. Perez-Cordero, L. Echegoyen, *J. Am. Chem. Soc.* **1992**, *114*, 3978
4. Y. Ohsawa, T. Saji, *J. Chem. Soc. Chem. Commun.* **1992**, 781.
5. F. Zhou, C. Jehoulet, A. J. Bard, *J. Am. Chem. Soc.* **1992**, *114*, 11004.

6. M. Prato, M. Maggini, C. Giacometti, G. Scorrano, C. Sandona, G. Farnia, *Tetrahedron* **1996**, *52*, 5221.
7. M. Keshavarz-K., B. Knight, R. C. Haddon, F. Wudl, *Tetrahedron* **1996**, *52*, 5149.
8. S. Yamago, M. Yanagawa, H. Mukai, E. Nakamura, *Tetrahedron* **1996**, *52*, 5091.
9. H. Nagashima, M. Saito, Y. Kato, H. Goto, E. Osawa, M. Haga, K. Itoh, *Tetrahedron*, **1996**, *52*, 5053.
10. Q. Li, F. Wudl, C. Thilgen, R. L. Whetten, F. Diederich, *J. Am. Chem. Soc.* **1992**, *114*, 3994.
11. Y. Yang, F. Arias, L. Echegoyen, L. P. F. Chibante, S. Flanagan, A. Robertson, L. J. Wilson, *J. Am. Chem. Soc.* **1995**, *117*, 7801.
12. M. D. Diener, J. M. Alford, *Nature* **1998**, *393*, 668.
13. T. Suzuki, Y. Maruyama, T. Kato, K. Kikuchi, Y. Achiba, *J. Am. Chem. Soc.* **1993**, *115*, 11006.
14. K. Kikuchi, Y. Nakao, S. Suzuki, Y. Achiba, T. Suzuki, Y. Maruyama, *J. Am. Chem. Soc.* **1994**, *116*, 9367.
15. T. Suzuki, K. Kikuchi, F. Oguri, Y. Nakao, S. Suzuki, Y. Achiba, K. Yamamoto, H. Funasaka, T. Takahashi, *Tetrahedron* **1996**, *52*, 4973.
16. Q. Xie, F. Arias, L. Echegoyen, *J. Am. Chem. Soc.* **1993**, *115*, 9818.
17. D. Dubois, K. M. Kadish, S. Flanagan, L. J. Wilson, *J. Am. Chem. Soc.* **1991**, *113*, 7773.
18. J. W. Arbogast, C. S. Foote, M. Kao, *J. Am. Chem. Soc.* **1992**, *114*, 2277.
19. C. S. Foote, *Top. Curr. Chem.* **1994**, *169*, 347.
20. D. Dubois, G. Moninot, W. Kutner, M. T. Jones, K. M. Kadish, *J. Phys. Chem.* **1992**, *96*, 7137.
21. T. Osaki, Y. Tai, M. Tazawa, S. Tanemura, K. Inukai, K. Ishiguro, Y. Sawaki, Y. Saito, H. Shinohara, H. Nagashima, H. *Chem. Lett.* **1993**, 789.
22. K. Mikami, S. Matsumoto, A. Ishida, S. Takamuku, T. Suenobu, S. Fukuzumi, *J. Am. Chem. Soc.* **1995**, *117*, 11134.
23. H. N. Ghosh, H. Pal, A. V. Sapre, J. P. Mittal, *J. Am. Chem. Soc.* **1993**, *115*, 11722.
24. C. A. Steren, H. van Willigen, L. Biczók, N. Gupta, H. Linschitz, *J. Phys. Chem.* **1996**, *100*, 8920.
25. L. Biczók, N. Gupta, H. Linschitz, *J. Am. Chem. Soc.* **1997**, *119*, 12601.
26. O. Ito, Y. Sasaki, Y. Yoshikawa, A. Watanabe, *J. Phys. Chem.* **1995**, *99*, 9838.
27. T. Nojiri, M. M. Alam, H. Konami, A. Watanabe, O. Ito, *J. Phys. Chem. A* **1997**, *101*, 7943.
28. A. Skiebe, A. Hirsch, H. Klos, B. Gotwschy, *Chem. Phys. Lett.* **1994**, *220*, 138.
29. W. R. Fawcett, M. Opallo, M. Fedurco, J. W. Lee, *J. Am. Chem. Soc.* **1993**, *115*, 196.
30. E. S. Yang, M.-S. Chan, A. C. Wahl, *J. Phys. Chem.* **1980**, *84*, 3094.
31. S. Fukuzumi, S. Mochizuki, T. Tanaka, *Inorg. Chem.* **1989**, *28*, 2459.
32. K. S. Cheng, N. Hirota, In *Investigation of Rates and Mechanisms of Reactions*, (Ed: G. G. Hammes), Wiley–Interscience, New York, **1974**, Vol. VI, p 565.
33. R. Chang, *J. Chem. Educ.* **1970**, *47*, 563.
34. C. A. Reed, R. D. Bolskar, *Chem. Rev.* **1999**, *99*, 1075.
35. P.-M. Allemand, G. Srdanov, A. Koch, K. Khemani, F. Wudl, Y. Rubin, F. Diederich, M. M. Alvarez, S. J. Anz, R. L. Whetten, *J. Am. Chem. Soc.* **1991**, *113*, 2780.
36. D. Dubois, K. M. Kadish, S. Flanagan, R. E. Haufler, L. P. F. Chibante, L. J. Wilson, *J. Am. Chem. Soc.* **1991**, *113*, 4364.
37. J. Stinchcombe, A. Pénicaud, P. Bhyrappa, P. D. W. Boyd, C. A. Reed, *J. Am. Chem. Soc.* **1993**, *115*, 5212.
38. J. R. Morton, F. Negri, K. F. Preston, *Acc. Chem. Res.* **1998**, *31*, 63.
39. J. R. Morton, K. F. Preston, P. J. Kruisic, S. A. Hill, E. Wasserman, *J. Am. Chem. Soc.* **1992**, *114*, 5454.
40. J. R. Morton, K. F. Preston, P. J. Kruisic, S. A. Hill, E. Wasserman, *J. Phys. Chem.* **1992**, *96*, 3576.
41. P. J. Krusic, E. Wasserman, P. N. Keizer, J. R. Morton, K. F. Preston, *Science* **1991**, *254*, 1183.
42. P. J. Krusic, E. Wasserman, B. A. Parkinson, B. Malone, E. R. Holler Jr., P. N. Keizer, J. R. Morton, K. F. Preston, *J. Am. Chem. Soc.* **1991**, *113*, 6274.
43. S. Fukuzumi, I. Nakanishi, T. Suenobu, K. M. Kadish, *J. Am. Chem. Soc.* **1999**, *121*, 3468.

44. S. Fukuzumi, K. Koumitsu, K. Hironaka, T. Tanaka, *J. Am. Chem. Soc.* **1987**, *109*, 305.
45. S. Fukuzumi, I. Nakanishi, J.-M. Barbe, R. Guilard, E. Van Caemelbecke, G. Ning, K. M. Kadish, *Angew. Chem. Int. Ed. Engl.* **1999**, *38*, 964.
46. T. Layloff, T. Miller, R. N. Adams, H. Fäh, A. Horsfield, W. Proctor, *Nature* **1965**, *205*, 382.
47. R. A. Marcus, N. Sutin, N. *Biochim. Biophys. Acta* **1985**, *811*, 265.
48. J. R. Miller, L. T. Calcaterra, G. L. Closs, *J. Am. Chem. Soc.* **1984**, *106*, 3047.
49. G. L. Closs, J. R. Miller, *Science* **1988**, *240*, 440.
50. N. Liang, J. R Miller, G. L. Closs, *J. Am. Chem. Soc.* **1990**, *112*, 5353.
51. N. Mataga, T. Asahi, Y. Kanda, T. Okada, T. Kakitani, *Chem. Phys.* **1988**, *127*, 249.
52. I. R. Gould, S. Farid, *Acc. Chem. Res.* **1996**, *29*, 522.
53. J. R. Winkler, H. B. Gray, H. B. *Chem. Rev.* **1992**, *92*, 369.
54. I. R. Gould, J. E. Moser, B. Armitage, S. Farid, *J. Am. Chem. Soc.* **1989**, *111*, 1917.
55. L. S. Fox, M. Kozik, J. R. Winkler, H. B. Gray, *Science* **1990**, *247*, 1069.
56. P. Chen, R. Duesing, G. Tapolsky, T. J. Meyer, *J. Am. Chem. Soc.* **1989**, *111*, 8305.
57. E. H. Yonemoto, R. L. Riley, Y. I. Kim, S. J. Atherton, R. H. Schmehl, T. E. Mallouk, *J. Am. Chem. Soc.* **1992**, *114*, 8081.
58. M. R. Wasielewski, M. P. Niemczyk, W. A. Svec, E. B. Pewitt, *J. Am. Chem. Soc.* **1985**, *107*, 1080.
59. D. B. MacQueen, K. S. Schanze, *J. Am. Chem. Soc.* **1991**, *113*, 7470.
60. T. M. McCleskey, J. R. Winkler, H. B. Gray, *J. Am. Chem. Soc.* **1992**, *114*, 6935.
61. C. Turro, J. M. Zaleski, Y. M. Karabatsos, D. G. Nocera, *J. Am. Chem. Soc.* **1996**, *118*, 6060.
62. D. M. Guldi, K.-D. Asmus, *J. Am. Chem. Soc.* **1997**, *119*, 5744.
63. D. M. Guldi, P. Neta, K.-D. Asmus, *J. Phys. Chem.* **1994**, *98*, 4617.
64. D. R. Lawson, D. L. Feldheim, C. A. Foss, P. K. Dorhout, C. M. Elliot, C. R. Martin, B. Parkinson, *J. Electrochem. Soc.* **1992**, *139*, L68.
65. D. M. Guldi, H. Hungerbühler, E. Janata, K. D. Asmus, *J. Phys. Chem.* **1993**, *97*, 11258.
66. D. M. Guldi, H. Hungerbühler, E. Janata, K. D. Asmus, *J. Chem. Soc. Chem. Commun* **1993**, *6*, 84.
67. T. Kato, T. Kodama, T. Shida, T. Nakagawa, Y. Matsui, S. Suzuki, H. Shiromaru, K. Yamauchi, Y. Achiba, *Chem. Phys. Lett.* **1991**, *180*, 446.
68. P. V. Kamat, G. Sauvé, D. M. Guldi, K.-D. Asmus, *Res. Chem. Intermed.* **1997**, *23*, 575.
69. D. M. Guldi, H. Hungerbühler, M. Wilhelm, K.-D. Asmus, *J. Chem. Soc. Faraday Trans.* **1994**, *90*, 1391.
70. D. M. Guldi, D. Liu, P. V. Kamat, *J. Phys. Chem. A* **1997**, *101*, 6195.
71. N. E. Shank, L. M. Dorfman, *J. Chem. Phys.* **1970**, *52*, 4441.
72. N. M. Dimitrijevic, *Chem. Phys. Lett.* **1992**, *194*, 457.
73. Z. R. Lian, S. D. Yao, W. Z. Lin, W. F. Wang, N. Y. Lin, *Radiat. Phys. Chem.* **1997**, *50*, 245.
74. H. Hou, C. Luo, Z. Liu, D. Mao, Q. Qin, Z. R. Lian, S. Yao, W. Wang, J. Zhang, N. Lin, *Chem. Phys. Lett.* **1993**, *203*, 555.
75. S. D. Yao, Z. R. Lian, W. F. Wang, J. S. Zhang, N. Y. Lin, H. Hou, Z. Zhang, Q. Qin, *Chem. Phys. Lett.* **1995**, *239*, 112.
76. S. D. Yao, W. Z. Lin, Z. R. Lian, W. F. Wang, N. Y. Lin, *Radiat. Phys. Chem.* **1997**, *50*, 249.
77. M. Sundahl, T. Andersson, K. Nilsson, O. Wennerstrom, G. Westman, *Synth. Met.* **1993**, *55*, 3252.
78. T. Andersson, K. Nilsson, M. Sundahl, G. Westman, O. Wennerström, *J. Chem. Soc. Chem. Commun.* **1992**, 604.
79. A. Beeby, J. Eastoe, E. R. Crooks, *Chem. Comun.* **1996**, 901.
80. J. Eastoe, E. R. Crooks, A. Beeby, R. K. Heenan, *Chem. Phys. Lett* **1995**, *245*, 571.
81. H. Hungerbühler, D. M. Guldi, K.-D. Asmus, *J. Am. Chem. Soc.* **1993**, *115*, 3386.
82. Y. N. Yamakoshi, T. Yagami, K. Fukuhara, S. Sueyoshi, N. Miyata, *J. Chem. Soc. Chem. Commun.* **1994**, 517.
83. R. V. Bensasson, E. Bienvenue, M. Dellinger, S. Leach, P. Seta, *J. Phys. Chem.* **1994**, *98*, 3492.
84. S. Niu, D. Mauzerall, *J. Am. Chem. Soc.* **1996**, *118*, 5791.
85. V. Ohlendorf, A. Willnow, H. Hungerbühler, D. M. Guldi, K.-D. Asmus, *J. Chem. Soc. Chem. Commun.* **1995**, 759.

86. K. I. Priyadarsini, H. Mohan, J. P. Mittal, D. M. Guldi, H. Hungerbühler, K.-D. Asmus, *J. Phys. Chem.* **1994**, *98*, 9565.
87. D. M. Guldi, *J. Phys. Chem. B* **1997**, *101*, 9600.
88. D. M. Guldi, H. Hungerbühler, *Res. Chem. Intermed.* **1999**, *25*, 615.
89. D. M. Guldi, H. Hungerbühler, K.-D. Asmus, *J. Phys. Chem.* **1995**, *99*, 13487.
90. D. M. Guldi, *Res. Chem. Intermed.* **1997**, *23*, 653.
91. D. M. Guldi, H. Hungerbühler, K.-D. Asmus, *J. Phys. Chem.* **1995**, *99*, 9380.
92. L. Echegoyen, L. E. Echegoyen, *Acc. Chem. Res.* **1998**, *31*, 593.
93. D. M. Guldi, *J. Phys. Chem. A* **1997**, *101*, 3895.
94. D. M. Guldi, H. Hungerbühler, K.-D. Asmus, *J. Phys. Chem. A* **1997**, *101*, 1783.
95. T. Da Ros, M. Prato, M. Carano, P. Ceroni, F. Paolucci, S. Roffia, *J. Am. Chem. Soc.* **1998**, *120*, 11645.
96. D. M. Guldi, H. Hungerbühler, K.-D. Asmus, *J. Phys. Chem. B* **1999**, *103*, 1444.
97. D. M. Guldi, *J. Phys. Chem. A* **2000**, *104*, 1484.
98. D. M. Guldi, M. Prato, *to be published*.
99. F. Wudl, *Acc. Chem. Res.* **1992**, *25*, 157.
100. (a) F. Diederich, C. Thilgen, *Science* **1996**, *271*, 317; (b) A. Hirsch, *Angew. Chem. Int. Ed. Engl.* **1993**, *32*, 1138.
101. R. Taylor, D. R. M. Walton, *Nature* **1993**, *363*, 685.
102. P. J. Fagan, J. C. Calabrese, B. Malone, *Science* **1991**, *252*, 1160.
103. P. J. Fagan, J. C. Calabrese, B. Malone, *Acc. Chem. Res.* **1992**, *25*, 134.
104. C. C. Henderson, P. A. Cahill, *Science* **1993**, *259*, 1885.
105. R. Schwenninger, T. Müller, B. Kräutler, *J. Am. Chem. Soc.* **1997**, *119*, 9317.
106. J. Llacay, M. Mas, E. Molins, J. Veciana, D. Powell, C. Rovira, *Chem. Commun.* **1997**, 659.
107. M. Ohno, T. Azuma, S. Kojima, Y. Shirakawa, S. Eguchi, *Tetrahedron* **1996**, *52*, 4983.
108. B. Kräutler, T. Müller, J. Maynollo, K. Gruber, C. Kratky, P. Ochsenbein, D. Schwarzenbach, H.-B. Bürgi, *Angew. Chem. Int. Ed. Engl.* **1996**, *35*, 1204.
109. L. Isaacs, R. F. Haldimann, F. Diederich, *Angew. Chem. Int. Ed. Engl.* **1994**, *33*, 2339.
110. P. Belik, A. Gügel, J. Spickermann, K. Müllen, *Angew. Chem. Int. Ed. Engl.* **1993**, *32*, 78.
111. S. Yamago, A. Takeichi, E. Nakamura, *J. Am. Chem. Soc.* **1994**, *116*, 1123.
112. M. Prato, H. Suzuki, H. Foroudian, Q. Li, K. Khemani, F. Wudl, J. Leonetti, R. D. Little, T. White, B. Rickborn, S. Yamago, E. Nakamura, *J. Am. Chem. Soc.* **1993**, *115*, 1594.
113. P. Belik, A. Gügel, J. Spickermann, K. Müllen, *Angew. Chem. Int. Ed. Engl.* **1993**, *32*, 78.
114. A. Gügel, A. Kraus, J. Spickermann, P. Belik, K. Müllen, *Angew. Chem. Int. Ed. Engl.* **1994**, *33*, 559.
115. T. Suzuki, Q. Li, K. C. Khemani, F. Wudl, Ö. Almarsson, *Science* **1991**, *254*, 1186.
116. T. Suzuki, Q. Li, K. C. Khemani, F. Wudl, *J. Am. Chem. Soc.* **1992**, *114*, 7300.
117. S. Shi, K. C. Khemani, Q. Li, F. Wudl, *J. Am. Chem. Soc.* **1992**, *114*, 10656.
118. R. Sijbesma, G. Srdanov, F. Wudl, J. A. Castoro, C. Wilkins, S. H. Friedman, D. L. DeCamp, G. L. Kenyon, *J. Am. Chem. Soc.* **1993**, *115*, 6510.
119. M. Prato, Q. Li, F. Wudl, V. Lucchini, *J. Am. Chem. Soc.* **1993**, *115*, 1148.
120. J. E. Chateauneuf, *J. Am. Chem. Soc.* **1995**, *117*, 2677.
121. M. Eiermann, F. Wudl, M. Prato, M. Maggini, *J. Am. Chem. Soc.* **1994**, *116*, 8364.
122. A. F. Kiely, R. C. Haddon, M. S. Meier, J. P. Selegue, C. P. Brock, B. O. Patrick, C.-W. Wang, Y. Chen, *J. Am. Chem. Soc.* **1999**, *121*, 7971.
123. P. J. Fagan, P. J. Krusic, D. H. Evans, S. A. Lerke, E. Johnston, *J. Am. Chem. Soc.* **1992**, *114*, 9697.
124. A. Hirsch, A. Soi, H. R. Karfunkel, *Angew. Chem. Int. Ed. Engl.* **1992**, *31*, 766.
125. A. Hirsch, T. Grösser, A. Skiebe, A. Soi, *Chem. Ber.* **1993**, *126*, 1061.
126. K. Komatsu, Y. Murata, N. Takimoto, S. Mori, N. Sugita, T. S. M. Wan, *J. Org. Chem.* **1994**, *59*, 6101.
127. Y. Murata, K. Motoyama, K. Komatsu, T. S. M. Wan, *Tetrahedron*, **1996**, *52*, 5077.
128. H. Nagashima, H. Terasaki, E. Kimura, K. Nakajima, K. Itoh, *J. Org. Chem.* **1994**, *59*, 1246.
129. H. Nagashima, M. Saito, Y. Kato, H. Goto, E. Osawa, M. Haga, K. Itoh, *Tetrahedron* **1996**, *52*, 5053.
130. S. R. Wilson, Y. Wu, *J. Am. Chem. Soc.* **1993**, *115*, 10334.

131. G.-W. Wang, L.-H. Shu, S.-H. Wu, H.-M. Wu, X.-F. Lao, *J. Chem. Soc. Chem. Commun.* **1995**, 1071.
132. J. M. Hawkins, *Acc. Chem. Res.* **1992**, *25*, 150.
133. M. Sawamura, H. Iikura, E. Nakamura, *J. Am. Chem. Soc.* **1996**, *118*, 12850.
134. G.-W. Wang, Y. Murata, K. Komatsu, T. S. M. Wan, *Chem. Commun.* **1996**, 2059.
135. M. Maggini, A. Karlsson, G. Scorrano, G. Sandona, G. Farnia, M. Prato, *J. Chem. Soc. Chem. Commun.* **1994**, 589.
136. S. Ballenweg, R. Gleiter, W. Krätschmer, *J. Chem. Soc. Chem. Commun.* **1994**, 2269.
137. A. L. Balch, V. J. Catalano, J. W. Lee, M. M. Olmstead, *J. Am. Chem. Soc.* **1992**, *114*, 5455.
138. A. L. Balch, V. J. Catalano, L. W. Lee, M. M. Olmstead, S. R. Parkin, *J. Am. Chem. Soc.* **1991**, *113*, 8953.
139. C. Caron, R. Subramanian, F. D'Souza, J. Kim, W. Kutner, M. T. Jones, K. M. Kadish, *J. Am. Chem. Soc.* **1993**, *115*, 8505.
140. Y. Huang, D. D. M. Wayner, *J. Am. Chem. Soc.* **1993**, *115*, 367.
141. F. D'Souza, J.-p. Choi, W. Kutner, *J. Phys. Chem. B* **1997**, *102*, 4247.
142. F. D'Souza, J.-p. Choi, Y.-Y. Hsieh, K. Shriver, W. Kutner, *J. Phys. Chem. B* **1998**, *102*, 212.
143. P. L. Boulas, Y. Zuo, L. Echegoyen, *Chem. Commun.* **1996**, 1547.
144. R. Subramanian, P. Boulas, M. N. Vijayashree, F. D'Souza, M. T. Jones, K. M. Kadish, *J. Chem. Soc. Chem. Comun.* **1994**, 1847.
145. R. Subramanian, K. M. Kadish, M. N. Vijayashree, X. Gao, M. T. Jones, M. D. Miller, K. L. Krause, T. Suenobu, S. Fukuzumi, *J. Phys. Chem.* **1996**, *100*, 16327.
146. N. Matsuzawa, D. Dixon, T. Fukunaga, *J. Phys. Chem.* **1992**, *96*, 7594.
147. P. A. Cahill, C. M. Rohlfing, *Tetrahedron* **1996**, *52*, 5247.
148. K. M. Kadish, X. Gao, E. Van Caemelbecke, T. Hirasaka, T. Suenobu, S. Fukuzumi, *J. Phys. Chem.* **1998**, *102*, 3898.
149. A. B. Smith, III, M. Strongin, L. Brard, G. T. Furst, J. Romanow, K. G. Owens, R. J. Goldschmidt, R. C. King, *J. Am. Chem. Soc.* **1995**, *117*, 5492.
150. S. Miki, M. Kitao, K. Fukunishi, *Tetrahedron Lett.* **1996**, *37*, 2042.
151. K. M. Kadish, X. Gao, E. Van Caemelbecke, T. Suenobu, S. Fukuzumi, *J. Phys. Chem. A* **2000**, *104*, 3878.
152. L. Eberson, *Electron-transfer reactions in Organic Chemistry*, Springer, Tokyo, **1987**.
153. N. Kornblum, *Angew. Chem. Int. Ed. Engl.* **1975**, *14*, 734.
154. J. F. Bunnett, *Acc. Chem. Res.* **1992**, *25*, 2.
155. M. Chanon, M. L. Tobe, *Angew. Chem. Int. Ed. Engl.* **1982**, *21*, 1.
156. S. Fukuzumi, in *Advances in Electron-transfer chemistry*, Vol. 2, (Ed.: P. S. Mariano), JAI Press, Greenwich, CT, **1992**, p. 65.
157. M. Patz, S. Fukuzumi, *J. Phys. Org. Chem.* **1997**, *10*, 129.
158. J. K. Kochi, *Angew. Chem. Int. Ed. Engl.* **1988**, *27*, 1227.
159. E. C. Ashby, *Acc. Chem. Res.* **1988**, *21*, 414.
160. M. Chanon, M. Rajzmann, F. Chanon, *Tetrahedron* **1990**, *46*, 6193.
161. S. Fukuzumi, T. Suenobu, T. Hirasaka, R. Arakawa, K. M. Kadish, *J. Am. Chem. Soc.* **1998**, *120*, 9220.
162. S. Fukuzumi, J. Maruta, *Inorg. Chim. Acta* **1994**, *226*, 145.
163. M. Ishikawa, S. Fukuzumi, *J. Am. Chem. Soc.* **1990**, *112*, 8864.
164. J.-M. Savéant, *J. Am. Chem. Soc.* **1990**, *109*, 6788.
165. C. P. Andrieux, A. L. Gorande, J.-M. Savéant, *J. Am. Chem. Soc.* **1990**, *114*, 6892.
166. S. Fukuzumi, T. Yorisue, *Bull. Chem. Soc. Jpn.* **1992**, *65*, 715.
167. S. Fukuzumi, T. Kitano, M. Ishikawa, Y. Matsuda, *Chem. Phys.* **1993**, *176*, 337.
168. Y. Murata, K. Komatsu, T. S. M. Wan, *Tetrahedron Lett.* **1996**, *37*, 7061.
169. Y. Murata, M. Shiro, K. Komatsu, *J. Am. Chem. Soc.* **1997**, *119*, 8117.
170. P. R. Birkett, A. G. Avent, A. D. Darwish, H. W. Kroto, R. Taylor, D. R. M. Walton, *J. Chem. Soc. Perkin Trans. 2* **1997**, 457.
171. K. M. Kadish, X. Gao, E. Van Caemelbecke, T. Suenobu, S. Fukuzumi, *J. Am. Chem. Soc.* **2000**, *122*, 563.
172. S. Fukuzumi, T. Suenobu, X. Gao, K. M. Kadish, *J. Phys. Chem. A* **2000**, *104*, 2908.

173. K. M. Kadish, X. Gao, O. Gorelik, E. Van Caemelbecke, T. Suenobu, S. Fukuzumi, *J. Phys. Chem. A* **2000**, *104*, 2902.
174. S. R. Wilson, Y. Wu, *J. Am. Chem. Soc.* **1993**, *115*, 10334.
175. D. T. Sawyer, J. L. Roberts, Jr. *Acc. Chem. Res.* **1988**, *21*, 469.
176. S. Fukuzumi, T. Yorisue, *J. Am. Chem. Soc.* **1991**, *113*, 7764.
177. S. Fukuzumi, I. Nakanishi, J. Maruta, T. Yorisue, T. Suenobu, S. Itoh, R. Arakawa, K. M. Kadish, *J. Am. Chem. Soc.* **1998**, *120*, 6673.
178. Y.-P. Sun, in *Molecular and Supramolecular Photochemistry*, Vol. 1, (Eds.: V. Ramamurthy, K. S. Schanze), Marcel Dekker, New York, **1997**, p. 325–389.
179. T. W. Ebbesen, K. Tanigaki, S. Kuroshima, *Chem. Phys. Lett.* **1991**, *181*, 501.
180. R. Seshadri, C. N. R. Rao, H. Pal, T. Mukherjee, J. P. Mittal, *Chem. Phys. Lett.* **1993**, *205*, 395.
181. B. Ma, G. E. Lawson, C. E. Bunker, A. Kitaygorodskiy, Y.-P. Sun, *Chem. Phys. Lett.* **1995**, *247*, 51.
182. Y.-P. Sun, B. Ma, G. E. Lawson, *Chem. Phys. Lett.* **1995**, *233*, 57.
183. Y. Wang, *J. Phys. Chem.* **1992**, *96*, 764.
184. E. Schaffner, H. Fischer, *J. Phys. Chem* **1993**, *97*, 13149.
185. M. Fujitsuka, C. Luo, O. Ito, *J. Phys. Chem. B* **1999**, *103*, 445.
186. C. Luo, M. Fujitsuka, C.-H. Huang, O. Ito, *J. Phys. Chem. A* **1998**, *102*, 8716.
187. C. Luo, M. Fujitsuka, C.-H. Huang, O. Ito, *Phys. Chem. Chem. Phys.* **1999**, *1*, 2923.
188. M. C. Rath, H. Pal, T. Mukherjee, *J. Phys. Chem. A* **1999**, *103*, 4993.
189. S. Komamine, M. Fujitsuka, O. Ito, *Phys. Chem. Chem. Phys.* **1999**, *1*, 4745.
190. D. M. Guldi, R. E. Huie, P. Neta, H. Hungerbühler, K.-D. Asmus, *Chem. Phys. Lett.* **1994**, *223*, 511.
191. V. A. Nadtochenko, F. F. Brazgun, *Russ. Chem. Bull. (Transl. of Izv. Akad. Nauk, Ser. Khim)* **1997**, *46*, 1074.
192. D. M. Guldi, M. Maggini, G. Scorrano, M. Prato, *Res. Chem. Intermed.* **1997**, *23*, 561.
193. D. M. Guldi, M. Maggini, G. Scorrano, M. Prato, *J. Am. Chem. Soc.* **1997**, *119*, 974.
194. M. M. Alam, A. Watanabe, O. Ito, *J. Photochem. Photobiol. A: Chem.* **1997**, *104*, 59.
195. A. Graja, M. A. Tanatar, Y. L. Li, D. B. Zhu, *Polish J. Chem.* **1998**, *72*, 869.
196. M. Alam, M. Sato, A. Watanabe, T. Akasaka, O. Ito, *J. Phys. Chem. A* **1998**, *102*, 7447.
197. A. Alam, O. Ito, N. Sakurai, H. Moriyama, *Fullerene Science Technology* **1998**, *6*, 1007.
198. O. Ito, Y. Sasaki, A. Watanabe, R. Hoffmann, C. Siedschlag, J. Mattay, *J. Chem. Soc. Perkin Trans. 2* **1997**, 1007.
199. H. N. Ghosh, D. K. Palit, A. V. Sapre, J. P. Mittal, *Chem. Phys. Lett.* **1997**, *265*, 365.
200. Y. Sasaki, Y. Yoshikawa, A. Watanabe, O. Ito, *J. Chem. Soc. Faraday Trans.* **1995**, *91*, 2287.
201. O. Ito, *Res. Chem. Intermed.* **1997**, *23*, 389.
202. S. Michaeli, V. Meiklyar, B. Endeward, K. Möbius, H. Levanon, *Res. Chem. Intermed.* **1997**, *23*, 505.
203. M. Fujitsuka, O. Ito, Y. Maeda, M. Kako, T. Wakahara, T. Akasaka, *Phys. Chem. Chem. Phys.* **1999**, *1*, 3527.
204. A. Samanta, P. V. Kamat, *Chem. Phys. Lett.* **1992**, *199*, 635.
205. T. Konishi, Y. Sasaki, M. Fujitsuka, Y. Toba, H. Moriyama, O. Ito, *J. Chem. Soc. Perkin Trans. 2* **1999**, 551.
206. M. Fujitsuka, C. Luo, O. Ito, Y. Murata, K. Komatsu, *J. Phys. Chem. A* **1999**, *103*, 7155.
207. Y. Sasaki, T. Konishi, M. Yamazaki, M. Fujitsuka, O. Ito, *Phys. Chem. Chem. Phys.* **1999**, *1*, 4555.
208. A. Alam, O. Ito, N. Sakurai, H. Moriyama, *Res. Chem. Intermed.* **1999**, *25*, 323.
209. Y. Sasaki, M. Fujitsuka, A. Watanabe, O. Ito, *J. Chem. Soc. Faraday Trans.* **1997**, *93*, 4275.
210. C. Luo, M. Fujitsuka, A. Watanabe, O. Ito, L. Gan, Y. Huang, C.-H. Huang, *J. Chem. Soc. Faraday Trans.* **1998**, *94*, 527.
211. A. Watanabe, O. Ito, *J. Phys. Chem.* **1994**, *98*, 7736.
212. A. Watanabe, O. Ito, K. Mochida, *Organometallics* **1995**, *14*, 4281.
213. M. Bennati, A. Grupp, M. Mehring, K. P. Dinse, J. Fink, *Chem. Phys. Lett.* **1992**, *200*, 440.
214. M. Bennati, A. Grupp, P. Bäuerle, M. Mehring, *Chem. Phys.* **1994**, *185*, 221.
215. M. Bennati, A. Grupp, P. Bäuerle, M. Mehring, *Mol. Cryst. Liq. Cryst.* **1994**, *256*, 751.

216. M. Bennati, A. Grupp, M. Mehring, *J. Chem. Phys.* **1995**, *102*, 9457.
217. J.-I. Fujisawa, Y. Ohba, S. Yamauchi, *Chem. Phys. Lett.* **1998**, *294*, 248.
218. J.-I. Fujisawa, Y. Ohba, S. Yamauchi, *Chem. Phys. Lett.* **1998**, *282*, 181.
219. K. C. Hwang, D. Mauzerall, *J. Am. Chem. Soc.* **1992**, *114*, 9705.
220. K. C. Hwang, D. C. Mauzerall, *Nature* **1993**, *361*, 138.
221. P. V. Kamat, I. Bedja, S. Hotchandani, *J. Phys. Chem.* **1994**, *98*, 9137.
222. P. V. Kamat, *J. Am. Chem. Soc.* **1991**, *113*, 9705.
223. A. Stasko, V. Brezova, S. Biskupic, K.-P. Dinse, P. Schweitzer, M. Baumgarten, *J. Phys. Chem.* **1995**, *99*, 8782.
224. R. D. Scurlock, P. R. Ogilby, *J. Photochem. Photobiol. A: Chem.* **1995**, *91*, 21.
225. S. Fukuzumi, T. Suenobu, M. Patz, T. Hirasaka, S. Itoh, M. Fujitsuka, O. Ito, *J. Am. Chem. Soc.* **1998**, *120*, 8060.
226. M. Patz, Y. Kuwahara, T. Suenobu, S. Fukuzumi, *Chem. Lett.*, **1997**, 567.
227. S. Fukuzumi, T. Suenobu, T. Hirasaka, N. Sakurada, R. Arakawa, M. Fujitsuka, O. Ito, *J. Phys. Chem. A* **1999**, *103*, 5935.
228. D. R. Lawson, D. L. Feldheim, C. A. Foss, P. K. Dorhout, C. M. Elliot, C. R. Martin, B. Parkinson, B. *J. Phys. Chem.* **1992**, *96*, 7175.
229. A. Watanabe, O. Ito, M. Watanabe, H. Saito, M. Koishi, *J. Phys. Chem.* **1996**, *100*, 10518.
230. L. Becker, T. P. Evans, J. L. Bada, *J. Org. Chem.* **1993**, *58*, 7630.
231. S. Ballenweg, R. Gleiter, W. Krätschmer, *Tetrahedron Lett.* **1993**, *34*, 3737.
232. T. F. Guarr, M. S. Meier, V. K. Vance, M. Clayton, *J. Am. Chem. Soc.* **1993**, *115*, 9862.
233. D. Mandrus, M. Kele, R. L. Hettich, G. Guiochon, B. C. Sales, L. A. Boatner, *J. Phys. Chem. B* **1997**, *101*, 123.
234. M. S. Meier, B. R. Weedon, H. P. Spielmann, *J. Am. Chem. Soc.* **1996**, *118*, 11682.
235. R. G. Bergosh, M. S. Meier, J. A. L. Cooke, H. P. Spielmann, B. R. Weedon, *J. Org. Chem.* **1997**, *62*, 7667.
236. M. S. Meier, P. S. Corbin, V. K. Vance, M. Clayton. M. Mollman, *Tetrahedron Lett.* **1994**, *35*, 5789.
237. H. P. Spielmann, G.-W. Wang, M. S. Meier, B. R. Weedon, *J. Org. Chem.* **1998**, *63*, 9865.
238. S. Fukuzumi, T. Suenobu, S. Kawamura, A. Ishida, K. Mikami, *Chem. Commun.* **1997**, 291.
239. S. Fukuzumi, Y. Tokuda, T. Kitano, T. Okamoto, J. Otera, *J. Am. Chem. Soc.* **1993**, *115*, 8960.
240. M. A. Fox, N. Chanon, (Eds.), *Photoinduced Electron Transfer*, Elsevier, Amsterdam, **1988**.
241. S. Fukuzumi, S. Itoh, in *Advances in Photochemistry*, Vol. 25, (Eds.: D. Volman, G. S. Hammond, D. C. Neckers), Wiley, New York, **1998**, pp 107–172.
242. F. Müller, J. Mattay, *Chem. Rev.* **1993**, *93*, 99.
243. G. J. Kavarnos, N. J. Turro, *Chem. Rev.* **1986**, 86, 401.
244. T. Mukaiyama, *Org. Reac.* **1982**, *28*, 203.
245. A. Hosomi, *Acc. Chem. Res.* **1988**, *21*, 200.
246. I. Fleming, *Comprehensive Organic Synthesis*, Vol. 2, Pergamon, London, **1991**.
247. H. Sakurai, *Synlett* **1989**, 1.
248. S. Fukuzumi, M. Fjita, J. Otera, Y. Fujita, *J. Am. Chem. Soc.* **1992**, *114*, 10271.
249. S. Fukuzumi, M. Fujita, J. Otera, *J. Org. Chem.* **1993**, *58*, 5405.
250. H. Tokuyama, H. Isobe, E. Nakamura, *J. Chem. Soc. Chem. Commun.* **1994**, 2753.
251. K. Mikami, S. Matsumoto, *Synlett* **1995**, 229.
252. S. Fukuzumi, T. Suenobu, M. Fujitsuka, O. Ito, T. Tonoi, S. Matsumto, K. Mikami, *J. Organomet. Chem.* **1999**, *574*, 32.
253. S. Fukuzumi, *Res. Chem. Intermediat.* **1997**, *23*, 519.
254. S. Fukuzumi, M. Fujita, J. Otera, *J. Chem. Soc. Chem. Commun.* **1993**, 1536.
255. K. Mikami, S. Matsumoto, T. Tonoi, T. Suenobu, A. Ishida, S. Fukuzumi, *Synlett* **1997**, 85.
256. E. Butcher, C.J. Rhodes, M. Standing, R.S. Davidson, R. Bowser, *J. Chem. Soc. Perkin Trans. 2* **1992**, 1469.
257. A. Rehm, A. Weller, *Isr. J. Chem.* **1970**, *8*, 259.
258. A. Anne, J. Moiroux, J.-M. Savéant, *J. Am. Chem. Soc.* **1993**, *115*, 10224.
259. N. Takada, S. Itoh, S. Fukuzumi, *Chem. Lett.* **1996**, 1103.
260. A. Hirsch, A. Soi, H. R. Karfunkel, *Angew. Chem. Int. Ed. Engl.* **1992**, *31*, 766.

261. T. Kitagawa, T. Tanaka, Y. Takata, K. Takeuchi, *J. Org. Chem.* **1995**, *60*, 1490.
262. T. Tanaka, T. Kitagawa, K. Komatsu, K. Takeuchi, *J. Am. Chem. Soc.* **1997**, *119*, 9313.
263. H. Sakurai, M. Kira, T. Uchida, *J. Am. Chem. Soc.* **1973**, *95*, 6826.
264. Y. F. Traven, R. West, *J. Am. Chem. Soc.* **1973**, *95*, 6824.
265. W. Ando, M. Kako, T. Akasaka, S. Nagase, *Organometallics* **1993**, *12*, 1514.
266. Y. Maeda, T. Wakahara, T. Akasaka, M. Fujitsuka, O. Ito, M. Kato, Y. Nakadaira, K. Kobayashi, S. Nagase, *Proceedings of 45th Symposium on Organometallic Chemistry*, Tolyo, **1998**, pp 42–43.
267. T. Akasaka, W. Ando, K. Kobayashi, S. Nagase, *J. Am. Chem. Soc.* **1993**, *115*, 10366.
268. T. Akasaka, T. Kato, K. Kobayashi, S. Nagase, K. Yamamoto, H. Funasaka, T. Takahashi, *Nature* **1995**, *374*, 600.
269. T. Akasaka, S. Nagase, K. Kobayashi, *J. Syn. Org. Chem. Jpn.* **1996**, *54*, 580.
270. T. Akasaka, K. Kato, S. Nagase, K. Kobayahsi, K. Yamamoto, H. Funasaka, T. Takahashi, *Tetrahedron* **1996**, *52*, 5015.
271. S. Nagase, K. Kobayahsi, *J. Chem. Soc. Chem. Commun.* **1994**, 1837.
272. T. Akasaka, T. Suzuki, Y. Maeda, M. Ara, T. Wakahara, K. Kobayashi, S. Nagase, M. Kako, Y. Nakadaira, M. Fujitsuka, O. Ito, *J. Org. Chem.* **1999**, *64*, 566.
273. S. Fukuzumi, T. Kitano, K. Mochida, *Chem. Lett.* **1990**, 1741.
274. F. W. Sluggett, W. J. Leigh, *Organometallics* **1992**, *11*, 3731.
275. Y. Z. An, L. J. Anderson, Y. Rubin, *J. Org. Chem.* **1993**, *58*, 4799.
276. M. Prato, A. Bianco, M. Magini, G. Scorrano, C. Toniolo, F. Wudl, *J. Org. Chem.* **1993**, *58*, 5578.
277. A. Skiebe, A. Hirsch, *J. Chem. Soc. Chem. Commun.* **1994**, 335.
278. L. Isaacs, A. Wehrsig, F. Diederich, *Helv. Chim. Acta* **1993**, *76*, 1231.
279. L. Isaacs, F. Diederich, *Helv. Chim. Acta* **1993**, *76*, 2454.
280. D. J. Zhou, L. B. Gan, L.B. Xu, C. P. Luo, C. H. Huang, *Fullerene Sci. Technol.* **1995**, *3*, 127.
281. L. Gan, D. Zhou, C. Luo, H. Tan, C. Huang, M. Lü, J. Pan, Y. Wu, *J. Org. Chem.* **1996**, *61*, 1954.
282. L. Gan, J. Jiang, W. Zhang, Y. Su, Y. Shi, C. Huang, J. Pan, M. Lü, Y. Wu, *J. Org. Chem.* **1998**, *63*, 4240.
283. M.-J. Arce, A. L. Viado, Y.-Z. An, S. I. Khan, Y. Rubin, *J. Am. Chem. Soc.* **1996**, *118*, 3775.
284. M. J. Arce, A. L. Viado, S. I. Khan, Y. Rubin, *Organometallics*, **1996**, *15*, 4340.
285. S. R. Wilson, Q. Y. Lu, J. R. Cao, Y. H. Wu, C. J. Welch, D. I. Schuster, *Tetrahedron* **1996**, *52*, 5131.
286. M. Ohno, N. Koide, H. Sato, S. Eguchi, *Tetrahedron* **1997**, *53*, 9075.
287. K. N. Houk, J. González, Y. Li, *Acc. Chem. Res.* **1995**, *28*, 81.
288. K. N. Houk, J. D. Evanseck, *Angew. Chem. Int. Ed. Engl.* **1992**, *31*, 682.
289. M. J. S. Dewar, C. Jie, *Acc. Chem. Res.* **1992**, *25*, 537.
290. B. A. Horn, J. L. Herek, A. H. Zewail, *J. Am. Chem. Soc.* **1996**, *118*, 8755.
291. R. Sustmann, S. Tappanchai, H. Bandmann, *J. Am. Chem. Soc.* **1996**, *118*, 12555.
292. B. R. Beno, S. Wilsey, K. N. Houk, *J. Am. Chem. Soc.* **1999**, *121*, 4816.
293. J. S. Chen, K. N. Houk, C. S. Foote, *J. Am. Chem. Soc.* **1998**, *120*, 12303.
294. E. Goldstein, B. Beno, K. N. Houk, *J. Am. Chem. Soc.* **1996**, *118*, 6036.
295. S. Fukuzumi, J. K. Kochi, *Tetrahedron* **1982**, *38*, 1035.
296. R. Sustmann, K. Lücking, G. Kopp, M. Rese, *Angew. Chem. Int. Ed. Engl.* **1989**, *28*, 1713.
297. R. Sustmann, H.-G. Korth, U. Nüchter, I. Siangouri-Feulner, W. Sicking, *Chem. Ber.* **1991**, *124*, 2811.
298. S. Fukuzumi, T. Okamoto, *J. Am. Chem. Soc.* **1993**, *115*, 11600.
299. S. Yamago, S. Ejiri, M. Nakamura, E. Nakamura, *J. Am. Chem. Soc.* **1993**, *115*, 5344.
300. D. L. Boger, C. E. Brotherton, in *Advances in Cycloaddition*, Vo. 2, (Ed.: D. P. Curran), JAI Press, Greenwich, CT, **1990**, p. 147.
301. M. Schmittel, C. Wöhrle, I. Bohn, *Chem. Eur. J.* **1996**, *2*, 1031.
302. K. Mikami, S. Matsumoto, T. Tonoi, Y. Okubo, T. Suenobu, S. Fukuzumi, *Tetrahedron Lett.* **1998**, *39*, 3733.
303. Y. Murata, N. Kato, K. Fujiwara, K. Komatsu, *J. Org. Chem.* **1999**, *64*, 3483.
304. K. Mikami, S. Matsumoto, Y. Okubo, T. Suenobu, S. Fukuzumi, *Synlett* **1999**, 1130.

305. K. Mikami, S. Matsumoto, Y. Okubo, M. Fujitsuka, O. Ito, T. Suenobu, S. Fukuzumi, *J. Am. Chem. Soc.* **2000**, *122*, 2236.
306. N. Martin, L. Sánchez, B. Illescas, I. Pérez, *Chem. Rev.* **1998**, *98*, 2527.
307. D. M. Guldi, K.-D. Asmus, *J. Phys. Chem. A* **1997**, *101*, 1472.
308. H. Imahori, Y. Sakata, *Adv. Mater.* **1997**, *9*, 537.
309. H. Imahori, Y. Sakata, *Eur. J. Org. Chem.* **1999**, 2445.
310. D. M. Guldi, *Chem. Commun.* **2000**, 321.
311. D. M. Guldi, M. Maggini, N. Martin, M. Prato, *Carbon* **2000**, *38*, 1615.
312. F. Diederich, M. Gomez-Lopez, *Chem. Soc. Rev.* **1999**, *28*, 263.
313. M. R. Bryce, W. Devonport, L. M. Goldenberg, C. Wang, *Chem. Commun.* **1998**, 945.
314. N. Martin, L. Sanchez, M. A. Herranz, D. M. Guldi, *J. Phys. Chem. A* **2000**, *104*, 4648.
315. N. Martin, L. Sanchez, B. Illescas, S. Gonzalez, M. A. Herranz, D. M. Guldi, *Carbon* **2000**, *38*, 1577.
316. K. B. Simonsen, V. V. Konovalov, T. A. Konovalov, T. Kawai, M. P. Cava, L. D. Kispert, R. M. Metzger, J. Becher, *J. Chem. Soc. Perkin Trans. 2* **1999**, 657.
317. N. Martin, L. Sanchez, C. Seoane, R. Andreu, J. Garin, J. Orduna, *Tetrahedron Lett.* **1996**, *37*, 5979.
318. J. Llacay, J. Veciana, J. Vidal-Gancedo, J. L. Bourdelande, R. Gonzalez-Moreno, C. Rovira, *J. Org. Chem.* **1998**, *63*, 5201.
319. R. M. Williams, J. M. Zwier, J. W. Verhoeven, *J. Am. Chem. Soc.* **1995**, *117*, 4093.
320. K. George Thomas, V. Biju, M. V. George, D. M. Guldi, P. V. Kamat, *J. Phys. Chem. A* **1998**, *102*, 5341.
321. S.-G. Liu, L. Shu, J. Rivera, H. Liu, J.-M. Raimundo, J. Roncali, A. Gorgues, L. Echegoyen, *J. Org. Chem.* **1999**, *64*, 4884.
322. S. I. Khan, A. M. Oliver, M. N. Paddon-Row, Y. Rubin, *J. Am. Chem. Soc.* **1993**, *115*, 4919.
323. R. M. Williams, M. Koeberg, J. M. Lawson, Y.-Z. An, Y. Rubin, M. N. Paddon-Row, J. W. Verhoeven, *J. Org. Chem.* **1996**, *61*, 5055.
324. M. N. Paddon-Row, *Fullerene Science Technology* **1999**, *7*, 1151.
325. M. J. Shephard, M. N. Paddon-Row, *Aust. J. Chem.* **1996**, *49*, 395.
326. K. George Thomas, V. Biju, D. M. Guldi, P. V. Kamat, M. V. George, *J. Phys. Chem. A* **1999**, *103*, 10755.
327. K. George Thomas, V. Biju, M. V. George, D. M. Guldi, P. V. Kamat, *J. Phys. Chem. B* **1999**, *103*, 8864.
328. M. Iyoda, F. Sultana, S. Sasaki, M. Yoshida, *J. Chem. Soc. Chem. Commun.* **1994**, 1929.
329. M. Diekers, A. Hirsch, S. Pyo, J. Rivera, L. Echegoyen, *Eur. J. Org. Chem.* **1998**, 1111.
330. D. Gust, T. A. Moore, A. L. Moore, *Pure Appl. Chem.* **1998**, *70*, 2189.
331. A. L. Moore, T. A. Moore, D. Gust, J. J. Silber, L. Sereno, F. Fungo, L. Otero, G. Steinberg-Yfrach, P. A. Liddell, S. -C. Hung, H. Imahori, S. Cardoso, D. Tatman, A. N. Macpherson, *Pure Appl. Chem.* **1997**, *69*, 2111.
332. H. Imahori, S. Cardoso, D. Tatman, S. Lin, L. Noss, G. R. Seely, L. Sereno, J. C. de Silber, T. A. Moore, A. L. Moore, D. Gust, *Photochem. Photobiol.* **1995**, *62*, 1009.
333. D. Armspach, E. C. Constable, F. Diederich, C. E. Housecraft, J.-F. Nierengarten, *J. Chem. Soc. Chem. Commun.* **1996**, 2009.
334. D. Armspach, E. C. Constable, F. Diederich, C. E. Housecroft, J.-F. Nierengarten, *Chem. Eur. J.* **1998**, *4*, 723.
335. M. Maggini, A. Dono, G. Scorrano, M. Prato, *J. Chem. Soc. Chem. Commun.* **1995**, 845.
336. M. Maggini, D. M. Guldi, S. Mondini, G. Scorrano, F. Paolucci, P. Ceroni, S. Roffia, *Chem. Eur. J.* **1998**, *4*, 1992.
337. A. Polese, S. Mondini, A. Bianco, C. Toniolo, G. Scorrano, D. M. Guldi, M. Maggini, *J. Am. Chem. Soc.* **1999**, *121*, 3456.
338. A. Juris, V. Balzani, F. Barigelletti, S. Campagna, P. Belser, A. Zelewsky, *Coord. Chem. Rev.* **1988**, *84*, 85.
339. V. Balzani, A. Juris, M. Venturi, S. Campagna, S. Serroni, *Chem. Rev.* **1996**, *96*, 759.
340. M. Maggini, D. M. Guldi, S. Mondini, G. Scorrano, F. Paolucci, P. Ceroni, S. Roffia, unpublished results.
341. D. Gust, T. A. Moore, A. L. Moore, *Res. Chem. Intermed.* **1997**, *23*, 621.

342. H. Imahori, Y. Sakata, *Chem. Lett.* **1996**, 199.
343. J.-F. Nierengarten, C. Schall, J.-F. Nicoud, *Angew. Chem. Int. Ed.* **1998**, *37*, 1934.
344. J. Helaja, A. Y. Tauber, Y. Abel, N. V. Tkachenko, H. Lemmetyinen, I. Kilpeläinen, P. H. Hynninen, *J. Chem. Soc. Perkin Trans. 1* **1999**, 2403.
345. H. Maruyama, M. Fujiwara, K. Tanaka, *Chem. Lett.* **1998**, 805.
346. K. Tashiro, T. Aida, J.-Y. Zheng, K. Kinbara, K. Saigo, S. Sakamoto, K. Yamaguchi, *J. Am. Chem. Soc.* **1999**, *121*, 9477.
347. E. Dietel, A. Hirsch, J. Zhou, A. Rieker, *J. Chem. Soc. Perkin Trans. 2* **1998**, 1357.
348. J.-F. Nierengarten, L. Oswald, J.-F. Nicaud, *Chem. Commun.* **1998**, 1545.
349. M. Wedel, F.-P. Montforts, *Tetrahedron Lett.* **1999**, *40*, 7071.
350. X. Camps, E. Dietel, A. Hirsch, S. Pyo, L. Echegoyen, S. Hackbarth, B. Röder, *Chem. Eur. J.* **1999**, *5*, 2362.
351. S. Higashida, H. Imahori, T. Kaneda, Y. Sakata, *Chem. Lett.* **1998**, 605.
352. D. R. Evans, N. L. P. Fackler, X. Zuowei, C. E. F. Rickard, P. D. W. Boyd, C. A. Reed, *J. Am. Chem. Soc.* **1999**, *121*, 8466.
353. H. Imahori, K. Hagiwara, T. Akiyama, S. Taniguchi, T. Okada, Y. Sakata, *Chem. Lett.* **1995**, 265.
354. H. Imahori, K. Hagiwara, T. Akiyama, M. Aoki, S. Taniguchi, T. Okada, M. Shirakawa, Y. Sakata, *Chem. Phys. Lett.* **1996**, *263*, 545.
355. H. Imahori, K. Hagiwara, T. Akiyama, M. Aoki, S. Taniguchi, T. Okada, M. Shirakawa, Y. Sakata, *J. Am. Chem. Soc.* **1996**, *118*, 11771.
356. Y. Sakata, H. Imahori, H. Tsue, S. Higashida, T. Akiyama, E. Yoshizawa, M. Aoki, K. Yamada, K. Hagiwara, S. Taniguchi, T. Okada, *Pure Appl. Chem.* **1997**, *69*, 1951.
357. K. Yamada, H. Imahori, Y. Nishimura, I. Yamazaki, Y. Sakata, *Chem. Lett.* **1999**, 895.
358. H. Imahori, K. Tamaki, H. Yamada, K. Yamada, Y. Sakata, Y. 'Nishimura, I. Yamazaki, M. Fujitsuka, O. Ito, *Carbon*, **2000**, 1599.
359. N. V. Tkachenko, C. Guenther, H. Imahori, K. Tamaki, Y. Sakata, H. Lemmetyinen, S. Fukuzumi, *Chem. Phys. Lett.* **2000**, *326*, 344.
360. P. A. Liddell, J. P. Sumida, A. N. MacPherson, L. Noss, G. R. Seely, K. N. Clark, A. L. Moore, T. A. Moore, D. Gust, *Photochem. Photobiol.* **1994**, *60*, 537.
361. E. Dietel, A. Hirsch, E. Eichborn, A. Rieker, S. Hackbarth, B. Röder, *Chem. Commun.* **1998**, 1981.
362. D. M. Guldi, C. Luo, M. Prato, E. Dietel, A. Hirsch, *Chem. Commun.* **2000**, 373.
363. J.-P. Bourgeois, F. Diederich, L. Echegoyen, J.-F. Nierengarten, *Helv. Chim. Acta* **1998**, *81*, 1835.
364. D. Kuciauskas, S. Lin, G. R. Seely, A. L. Moore, T. A. Moore, D. Gust, T. Drovetskaya, C. A. Reed, P. D. W. Boyd, *J. Phys. Chem.* **1996**, *100*, 15926.
365. T. Drovetskaya, C. A. Reed, P. Boyd, *Tetrahedron Lett.* **1995**, *36*, 7971.
366. T. D. M. Bell, T. A. Smith, K. P. Ghiggino, M. G. Ranasinghe, M. J. Shephard, M. N. Paddon-Row, *Chem. Phys. Lett.* **1997**, *268*, 223.
367. I. G. Safonov, P. S. Baran, D. I. Schuster, *Tetrahedron Lett.* **1997**, *38*, 8133.
368. P. S. Baran, R. R. Monaco, A. U. Khan, D. I. Schuster, S. R. Wilson, *J. Am. Chem. Soc.* **1997**, *119*, 8363.
369. P. Cheng, S. Wilson, D. I. Schuster, *Chem. Commun.* **1999**, 89.
370. R. II Fong, D. I. Schuster, S. R. Wilson, *Organic Letters* **1999**, *1*, 729.
371. D. I. Schuster, P. Cheng, S. R. Wilson, V. Prokhorenko, M. Katterle, A. R. Holzwarth, S. E. Braslavsky, G. Klihm, R. M. Williams, C. Luo, *J. Am. Chem. Soc.* **1999**, *121*, 11599.
372. K. Tamaki, H. Imahori, Y. Nishimura, I. Yamazaki, A. Shimomura, T. Okada, Y. Sakata, *Chem. Lett.* **1999**, 227.
373. N. V. Tkachenko, L. Rantala, A. Y. Tauber, J. Helaja, P. H. Hynninen, H. Lemmetyinen, *J. Am. Chem. Soc.* **1999**, *121*, 9378.
374. H. Imahori, S. Ozawa, K. Ushida, M. Takahashi, T. Azuma, A. Ajavakom, T. Akiyama, M. Hasegawa, S. Taniguchi, T. Okada, Y. Sakata, *Bull. Chem. Soc. Jpn.* **1999**, *72*, 485.
375. T. Akiyama, H. Imahori, A. Ajavakom, Y. Sakata, *Chem. Lett.* **1996**, 907.
376. P. A. Liddell, D. Kuciauskas, J. P. Sumida, B. Nash, D. Nguyen, A. L. Moore, T. A. Moore, D. Gust, *J. Am. Chem. Soc.* **1997**, *119*, 1400.

377. D. Carbonera, M. Di Valentin, C. Corvaja, G. Agostini, G. Giacometti, P. A. Liddell, D. Kuciauskas, A. L. Moore, T. A. Moore, D. J. Gust, *J. Am. Chem. Soc.* **1998**, *120*, 4398.
378. H. Imahori, K. Yamada, M. Hasegawa, S. Taniguchi, T. Okada, Y. Sakata, *Angew. Chem. Int. Ed. Engl.* **1997**, *36*, 2626.
379. H. Imahori, H. Yamada, S. Ozawa, K. Ushida, Y. Sakata, *Chem. Commun.* **1999**, 1165.
380. M. Fujitsuka, O. Ito, H. Imahori, K. Yamada, H. Yamada, Y. Sakata, *Chem. Lett.* **1999**, 721.
381. H. Imahori, H. Yamada, Y. Nishimura, I. Yamazaki, Y. Sakata, *J. Phys. Chem. B*, **2000**, *104*, 2099.
382. K. Tamaki, H. Imahori, Y. Nishimura, I. Yamazaki, Y. Sakata, *Chem. Commun.* **1999**, 625.
383. C. Luo, D. M. Guldi, H. Imahori, K. Tamaki, Y. Sakata, *J. Am. Chem. Soc.* **2000**, *122*, 6535.
384. N. Armaroli, F. Diederich, C. O. Dietrich-Buchecker, L. Flamigni, G. Marconi, J.-F. Nierengarten, J.-P. Sauvage, *Chem. Eur. J.* **1998**, *4*, 406.
385. D. Kuciauskas, P. A. Liddell, S. Lin, T. E. Johnson, S. J. Weghorn, J. S. Lindsey, A. L. Moore, T. A. Moore, D. Gust, *J. Am. Chem. Soc.* **1999**, *121*, 8604.
386. T. Da Ros, M. Prato, D. Guldi, E. Alessio, M. Ruzzi, L. Pasimeni, *Chem. Commun.* **1999**, 635.
387. N. Armaroli, F. Diederich, L. Echegoyen, T. Habicher, L. Flamigni, G. Marconi, J.-F. Nierengarten, *New J. Chem.* **1999**, *23*, 77.
388. F. D'Souza, G. R. Deviprasad, M. S. Rahman, J. P. Choi, *Inorg. Chem.* **1999**, *1999*, 2157.
389. T. Da Ros, M. Prato, D.M. Guldi, M. Ruzzi, L. Pasimeni, *Chem. Eur. J.* **2000**, *6*, in press.
390. P. Scheier, B. Düner, R. Wörgötter, S. Matt, D. Muigg, G. Senn, T. D. Märk, *Int. Rev. Phys. Chem.* **1996**, *15*, 93.
391. D. K. Böhme, *Can. J. Chem.* **1999**, *77*, 1453.
392. G. Javahery, S. Petrie, J. Wang, H. Wincel, D. K. Böhme, *J. Am. Chem. Soc.* **1993**, *115*, 9701.
393. G. Javahery, S. Petrie, H. Wincel, J. Wang, D. K. Böhme, *J. Am. Chem. Soc.* **1993**, *115*, 6295.
394. D. L. Lichtenburger, M. E. Rempe, S. B. Gogosha, *Chem. Phys. Lett.* **1992**, *198*, 454.
395. G. Javahery, S. Petrie, H. Wincel, J. Wang, D. K. Böhme, *J. Am. Chem. Soc.* **1993**, *115*, 5716.
396. V. Baranov, D. K. Böhme, *Int. J. Mass Spect. Ion Processes* **1997**, *165/166*, 249.
397. H. Becker, G. Javahery, S. Petrie, D. K. Böhme, *J. Phys. Chem.* **1994**, *98*, 5591.
398. S. Petrie, G. Javahery, J. Wang, D. K. Böhme, *J. Am. Chem. Soc.* **1992**, *114*, 6268.
399. S. Petrie, J. Javahery, J. Wang, D. K. Böhme, *J. Phys. Chem.* **1992**, *96*, 6162.
400. D. K. Böhme, *Int. Rev. Phys. Chem.* **1994**, *13*, 163.
401. V. A. Nadtochenko, N. N. Denisov, I. V. Rubtsov, A. S. Lobach, A. P. Moravskii, *Chem. Phys. Lett.* **1993**, *208*, 431.
402. V. A. Nadtochenko, I. V. Vasil'ev, N. N. Denisov, I. V. Rubtsov, A. S. Lobach, A. P. Moravskii, A. F. Shestakov, *J. Photochem. Photobiol., A* **1993**, *70*, 153.
403. V. A. Nadtochenko, N. N. Denisov, I. V. Rubtsov, A. S. Lobach, A. P. Moravsky, *Russ. Chem. Bull.* **1993**, *42*, 1171.
404. V. A. Nadtochenko, N. N. Denisov, A. S. Lobach, A. P. Moravskii, *Zh. Fiz. Khim.* **1994**, *68*, 228.
405. C. A. Steren, H. Van Willigen, *Proc. Indian Acad. Sci., Chem. Sci.* **1994**, *106*, 1671.
406. C. A. Steren, P. R. Levstein, H. van Willigen, H. Linschitz, L. Biczok, *Chem. Phys. Lett.* **1993**, *204*, 23.
407. S. Michaeli, V. Meiklyar, M. Schulz, K. Möbius, H. Levanon, *J. Phys. Chem.* **1994**, *98*, 7444.
408. P. V. Kamat, M. Gevaert, K. Vinodgopal, *J. Phys. Chem. B* **1997**, *101*, 4422.
409. M. Gevaert, P. V. Kamat, *J. Chem. Soc. Chem. Commun.* **1992**, 1470.
410. J. Eriksen, C. S. Foote, *J. Phys. Chem.* **1978**, *82*, 2659.
411. S. Fukuzumi, T. Tanaka in *Photoinduced Electron Transfer, Part C* (Eds.: M. A. Fox, M. Chanon), Elsevier, Amsterdam, **1988**, pp. 578.
412. S. Fukuzumi, S. Kuroda, T. Tanaka, *J. Chem. Soc. Chem. Commun.* **1986**, 1533.
413. S. Nonell, J. W. Arbogast, C. S. Foote, *J. Phys. Chem.* **1992**, *96*, 4169.
414. S. Fukuzumi in *Advances in Electron-transfer chemistry*, *Vol. 2* (Ed.: P. S. Mariano), JAI Press, Greenwich, **1992**, pp. 67–175.
415. M. Fujita, S. Fukuzumi, *J. Chem. Soc. Perkin Trans. 2* **1993**, 1915.
416. M. Fujita, A. Ishida, S. Takamuku, S. Fukuzumi, *J. Am. Chem. Soc.* **1996**, *118*, 8566.

417. S. Fukuzumi, T. Kitano, T. Tanaka, *Chem. Lett.* **1989**, 1231.
418. G. Lem, D. I. Schuster, H. Courtney, Q. Lu, S. R. Wilson, *J. Am. Chem. Soc.* **1995**, *117*, 554.
419. C. Siedschlag, H. Luftmann, C. Wolff, J. Mattay, *Tetrahedron* **1997**, *53*, 3587.
420. C. Siedschlag, H. Luftmann, C. Wolff, J. Mattay, *Tetrahedron* **1999**, *55*, 7805.
421. A. J. Maroulis, D. R. Arnold, *Synthesis* **1979**, 819.
422. D. R. Arnold, X. Du, J. Chen, *Can. J. Chem.* **1995**, *73*, 307.
423. F. G. Bordwell, J.-P. Cheng, M. J. Bausch, J. E. Bases, *J. Phys. Org. Chem.* **1988**, 1, 209.

# 6 Electron-transfer Reactions of Heteroaromatic Compounds

*Angelo Albini and Maurizio Fagnoni*

## 6.1 Introduction

Most practitioners of heterocyclic chemistry will concur with the statement by Newkome and Paudler that "the chemistry of heterocyclic compounds encompasses most, if not all, of the general reactions of organic chemistry. The presence of heteroatoms merely endows the heterocyclic systems with some *additional*, theoretical predictable properties" [1]. Thus we have here the whole body of organic chemistry, plus something peculiar to this (very large) class of compounds.

Indeed this is true also for the class of reactions treated in this book, viz. electron-transfer reactions, which also with these compounds already have an important role and are expected to have a much larger one in the near future. The number of chemical reactions more or less unambiguously characterized as electron-transfer processes in heteroaromatic chemistry in the current literature is so large that an exhaustive presentation is far beyond the space allotted for this contribution. This is, therefore, limited to discussion of the most characteristic chemical paths—or at least those which seem to answer this requirement in the opinion of these authors—with a choice of appropriate examples.

Because it is felt that particular attention should be given to the electron-transfer step, after a brief discussion of the structure of the radical ions of heteroaromatic molecules, the reactions are discussed in two sections, the first concerning the reactions in which the heterocycle is the donor, the latter reactions in which it is the acceptor. When both donor and acceptor are heterocycles, the reaction is presented in the most appropriate section.

The discussion is limited to the reactions of heteroaromatic molecules, non-aromatic heterocycles are discussed elsewhere in this book in accordance with the appropriate heteroatom-containing functionality. The most important electron-transfer phenomena involving heteroaromatics are to be found on the one hand in biochemistry, where some heterocycles have a key role in metabolism, and on the other in material science, the applications of conductive polymers derived from heterocycles. Neither of these topics is discussed in this chapter, which is rather concerned with typical 'organic' reactions. Furthermore, many electrochemical reaction involving a two (or $2n$)-electron oxidation or reduction of heterocycles are skipped or barely mentioned (while being exhaustively discussed in electrochemistry textbooks and in some reviews quoted in this text) and the attention is concentrated on processes where the key chemical step occurs after single-electron transfer and involves odd-electron species.

## 6.2 Structure of the Radical Ions of Heteroaromatic Compounds

The generation and structure of heterocyclic radical ions has been discussed previously [2] and needs not be addressed at length in this context, because the main concern here is with reactivity. A few examples of the main structural differences are, however, reported in this section to facilitate subsequent discussion of chemical reactions.

### 6.2.1 Radical Cations

Ionization potentials have been measured for all reasonably simple heterocycles by photoelectron spectroscopy, and the bands have been assigned by comparison with calculated molecular orbital energies. Photoelectron angular distribution studies gave further support for the assignment. A more detailed description of the structure can be obtained by EPR spectroscopy of the relaxed radical cation in matrices or fluid solutions. Many heterocycles are good donors, and radical cations are often formed by mere dissolution in a Brønsted acid (e.g. standing of alkylthiophenes in hexafluoropropanol–3 % methanesulfonic acid leads to the development of the EPR spectra of the corresponding radical cations) [3]. The process occurring is illustrated in Scheme 1a. Quantitative conversion can be obtained in the presence of an oxidant, e.g. by dissolving the heterocycle in methanesulfonic acid in the presence of chloranil, according to Scheme 1b [4]. Convenient methods for the characterization of radical cations are $^{60}$Co $\gamma$ irradiation (in matrix) or, in solution, UV irradiation in acids (trifluoroacetic acid, sulfuric acid) or protic media (e.g. hexafluoropropan-2-ol, HFP), if appropriate in the presence of Hg(II) or Tl(III) salts) [5] or thermal oxidation by thallium(III) or mercury trifluoroacetates or nitrosonium tetrafluoroborate [6, 7] or by strong organic oxidants such as 2,3-dichloro-4,5-dicyanobenzoquinone (DDQ) in trifluoroacetic acid or in HFP [8, 9].

As expected from simple MO considerations, the radical cations of five-membered heterocycles, e.g. the blue species formed from furan, pyrrole, thiophene, and their alkyl derivatives, are $\pi$ ions. The semi-occupied orbital is the $\pi$ orbital with the heteroatom in the nodal plane ($a_1$), see Scheme 2, structure **1**. In radical cations of $\alpha,\omega$-bis-(1-pyrrolyl) alkanes the charge remains localized on a single ring, rather than being delocalized over both units [5, 10, 11].

a)

$$Het + H^+ \longrightarrow HetH^+$$

$$HetH^+ + Het \longrightarrow Het^{+\bullet} + HetH^\bullet$$

$$HetH^\bullet + HetH^+ \longrightarrow Het + HetH_2^{+\bullet}$$

$$HetH_2^{+\bullet} + Het \longrightarrow Het^{+\bullet} + HetH_2$$

$$2\,Het + HetH^+ + H^+ \longrightarrow 2\,Het^{+\bullet} + HetH_2$$

b)

$$2\,Het + 2\,H^+ \longrightarrow 2\,HetH^+$$

$$2\,HetH^+ + 2\,Het \longrightarrow 2\,Het^{+\bullet} + 2\,HetH^\bullet$$

$$2\,HetH^\bullet + 2\,Chl \longrightarrow 2\,Het + 2\,ChlH^\bullet$$

$$2\,ChlH^\bullet \longrightarrow ChlH_2 + Chl$$

$$2\,Het + 2\,H^+ + Chl \longrightarrow 2\,Het^{+\bullet} + ChlH_2 \qquad \textbf{Scheme 1.}$$

Interestingly, it has been observed that the ionization potentials (determined by the electron-impact technique) of several substituted five-membered heterocycles correlate with the $\sigma_p^+$ constants. The slopes obtained give a value of $\rho = -20.2$ for furan, $-18.2$ for pyrrole, $-16.5$ for thiophene, and a value close to the last for selenophene. Thus, the sensitivity of ionization to substituent effects is larger than for benzene ($\rho = -14.7$) and larger than that measured for the most sensitive electrophilic substitution (bromination in acetic acid, which gives $\rho = -10.0$ for thiophene). As pointed out by Linda et al. [12], the sensitivity to substituent effects

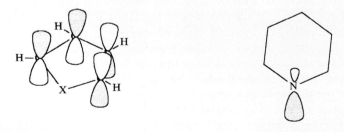

**1**                                **2**        **Scheme 2.**

varies in the reverse order to that of the ground state aromaticity of the hetero-cycles.

With azines the situation is varied. In the radical cations of pyridine and diazines the semi-occupied orbital is largely confined to the $n_N$ orbital(s) (see Scheme 2, structure **2**), while the radical cation is of the $\pi$ type with monoazanaphthalenes, -phenanthrenes and -anthracenes. The situation might change with substitution. As an example, alkylpyridine radical cations are of the n type, like the parent compound, whereas for the 2,5-dimethyl, 2-chloro, and 2-bromo derivatives the structure is of the $\pi$ type [13]. Likewise, with benzo[c]cinnoline the parent compound and its alkyl derivatives give an n radical cation, but with some dimethoxy derivatives a $\pi$ structure is found [14] and a switch from n to $\pi$ structure occurs also in passing from 1,2,4,5-tetrazine to its 3,6-diamino derivatives [15].

Illustrative examples of the ionization potential of heteroaromatics and of the structure of the corresponding radical cations are reported in Table 1 [5, 16–24].

## 6.2.2 Radical Anions

The radical anions have been likewise characterized by methods ranging from the various electrochemical techniques supplemented by EPR/ENDOR measurements to electron transmission spectroscopy. Radical anions can be conveniently generated in a matrix by $^{60}$Co $\gamma$ irradiation, e.g. in CD$_3$OD, or in the liquid phase by potassium metal reduction in solvents such as hexamethylphosphoric triamide or dimethoxyethane. These ions have the $\pi^*$ structure expected from calculations [14, 15, 25, 26].

**Table 1.** Ionization potential and structure of the radical cations of some representative hetero-aromatic compounds.

| Compound | Ionization potential, eV[a] | Structure | Other methods of generation |
|---|---|---|---|
| Furan | 8.77 [16] | $\pi$ | $\gamma$ irradiation [17]; UV irradiation in CF$_3$CO$_2$H [5] |
| Pyrrole | 8.90 [18] | $\pi$ | $\gamma$ irradiation [17]; irradiation in CF$_3$CO$_2$H/Hg(CF$_3$CO$_2$) [5] |
| Thiophene | 8.91 [19] | $\pi$ | $\gamma$ irradiation [17]; irradiation in H$_2$SO$_4$ [5] |
| Oxazole | 9.83 [20] | $\pi$ | |
| Isoxazole | 10.15 [21] | $\pi$ | |
| Indole | 8.20 [22] | $\pi$ | |
| Pyridine | 9.23 [19] | $n$ | $\gamma$ irradiation [13] |
| Pyridazine | 9.31 [23] | $n$ | $\gamma$ irradiation [24] |
| Pyrimidine | 9.73 [23] | $n$ | $\gamma$ irradiation [24] |
| Pyrazine | 9.63 [23] | $n$ | $\gamma$ irradiation [24] |
| Quinoline | 8.62 [25] | $\pi$ | $\gamma$ irradiation [24] |
| Isoquinoline | 8.50 [25] | $\pi$ | $\gamma$ irradiation [24] |

[a] Measured by photoelectron spectroscopy.

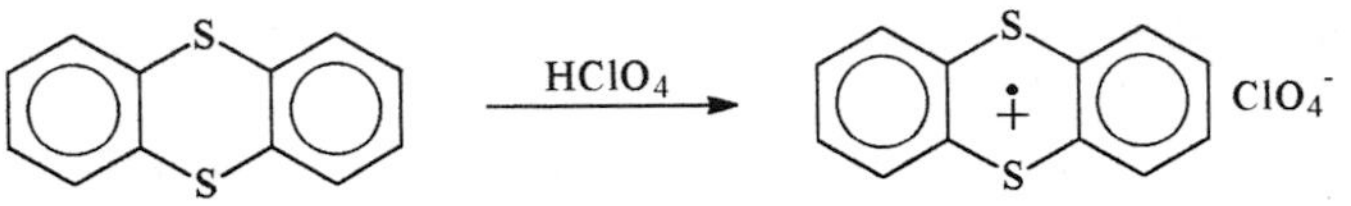

**3**                                    Scheme 3.

Comparison of the $\pi^*$ electron affinities of pyridine, phospha-, arsa-, and stilbabenzenes showed that these quantities, and the ionization potentials, are correlated both with heteroatom electronegativities and with the lengths of the C–X bonds, in keeping with the expectation that lengthening of the bond destabilizes the $b_1$ $(\pi)$ orbital and stabilizes the $b_1$ $(\pi^*)$ orbital [27].

### 6.2.3 Stable Radical Ions

Finally, one should mention that several heteroaromatic radical cations have been prepared as stable, crystalline salts with weakly nucleophilic anions. Some of these count among the earliest known examples—for example the dark red phenothiazine radical cation tribromide obtained, and correctly identified, at the beginning of the 20th century [28]. The electrochemical or chemical preparation of the radical cations of related heterocycles such as phenoxathiin, thianthrene [7], and selenium analogs [29] have afforded many well-characterized examples (*caution*: preparation of these salts might be not free from hazard: the explosion of a dry sample of thianthrenium radical cation perchlorate (**3**, Scheme 3), apparently caused by friction, has been reported) [30]. Other chalcogen heterocycles also give stable derivatives, such as the purple or blue hexafluorophosphates of the radical cations of some benzotrithiols, **4** (see Scheme 4) [31], benzotriselenols [32], and phenanthrotrithiol [33]. Both with these compounds and with six-membered heterocycles (**5**), the one-electron redox process can be performed either electrochemically, with close to 100 % efficiency, or equally well by chemical means, oxidizing with nitrosonium hexafluorophosphate and reducing by means of samariun iodide.

Examples of stable salts of heteroaromatic radical anions isolated as crystalline compounds are the lithium, sodium, and potassium salts of 2,4,6-triphenylpyridine (**6**, Scheme 5) [34].

## 6.3 Chemical Reactions via Electron Transfer—The Heteroaromatic Compound is the Donor

### 6.3.1 General Scheme

Radical cations are produced under a variety of conditions, from gas-phase to rigid matrix or to solution, and obviously their chemical behavior varies accordingly. The main preparative interest is centered on the solution phase, although the reac-

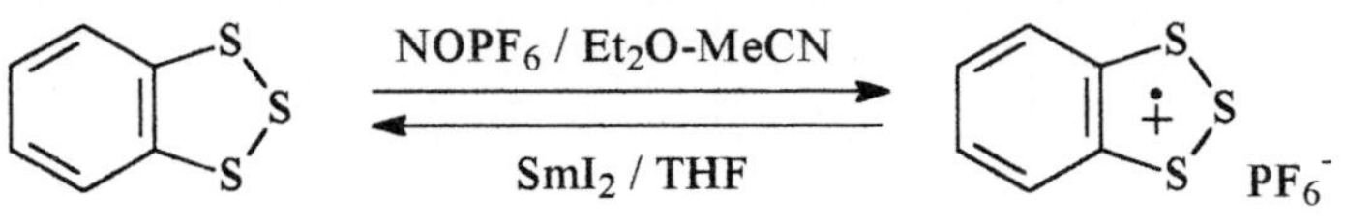

**4**

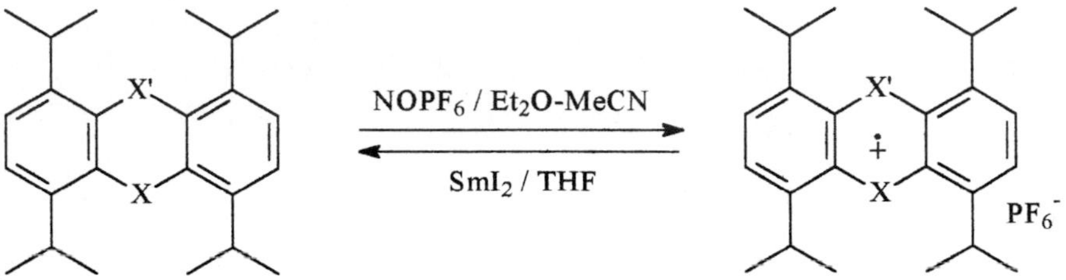

**5**

**5a** X = X' = S, **5b** X = X' = Se

**Scheme 4.**

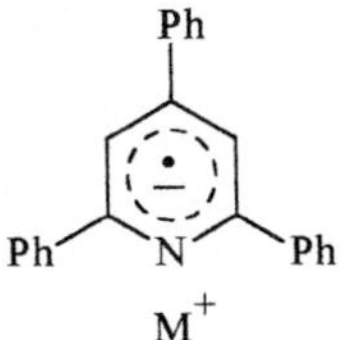

M = Li, Na, K

**6**

**Scheme 5.**

tivity of the radical cations also depends on the mode of generation and on the medium characteristics. One-electron oxidation occurs either at the anode, or chemically by reaction with a variety of species, including $Tl^{III}$, $Hg^{II}$, $PhI^{III}$, $NOBF_4$, persistent organic radical cations such as amminium cations or hydroquinone radical cations, or by the action of acids or light. The general pattern of reactivity of radical cations is discussed elsewhere in this book, and certainly includes useful processes. Quite appealing are electron transfer catalytic processes, although, particularly with (hetero)aromatic substrates, it is not trivial to distinguish reactions involving radical cations as the key species from processes where a diamagnetic proton adduct is involved, i.e. 'traditional' acid catalysis. As recently pointed out by Kochi, apart from not being readily identified, paramagnetic radical cations and diamagnetic cations are at any rate expected to have a similar chemistry, and the two classes of reaction might not be completely clearly differentiated. It also true that non-acidic media in which radical cations have longer lifetimes are now avail-

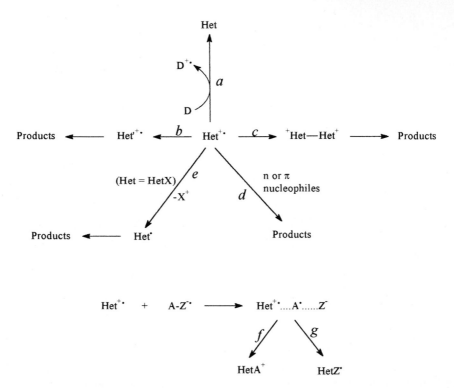

**Scheme 6.**

able, and this enables deeper investigation of such species and their involvement in chemical reactions [35, 36].

The main mechanistic paths for heteroaromatic radical cations are shown in Scheme 6. Besides electron transfer regenerating the starting material and re-arrangement (paths *a* and *b*), these involve radical reactions, such as coupling (path *c*), cationic reactions, viz. reaction with a nucleophile (path *d*), and cleavage of an electrofugal group giving a heteroaryl radical (path *e*). An important group of reactions involves fragmentation of the radical anion concomitantly formed in the initial electron-transfer (ET) step, followed by recombination of the heteroaryl radical cation either with the radical or with the anion fragment (paths *f* and *g*, see Scheme 6, bottom).

These mechanistic alternatives are recognized in the reactions discussed below, which are, however, grouped according to overall chemical transformation occurring rather than to the specific mechanism involved.

### 6.3.2 Electron and Proton Transfer

An important aspect of rationalization of the reaction between the radical cations of aromatics and nucleophiles (see below) is the possibility of an electron-transfer

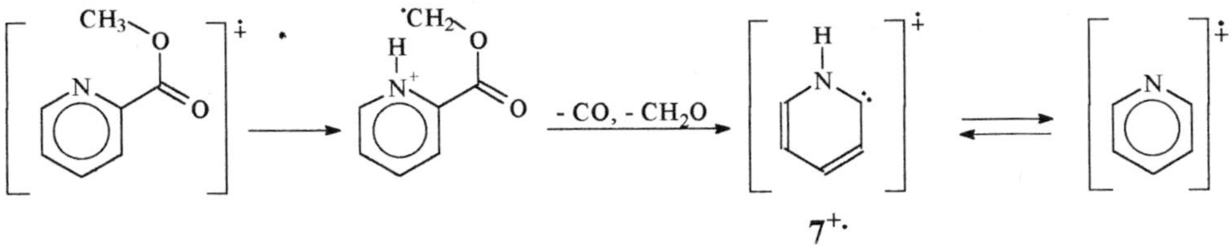

**Scheme 7.**

step. As an example the role of an ET step in the interaction between dibenzodioxin and thianthrene radical cation with aromatic donors such as anisole and anthracene has been evaluated through kinetic studies [37].

A related question is proton transfer, where two aspects are relevant, viz. proton transfer from a heterocyclic radical cation after an ET step (e.g. after photoinduced ET from indole to anthracenes or pyrene) [38] and the possible role of ET in proton transfer from aromatic radical cations and heterocyclic nucleophiles, in particular to pyridine derivatives [37b, 39].

### 6.3.3 Ring Opening and Rearrangement

The form of the potential hypersurface for a molecular species changes drastically upon ionization. As an example, whereas azacyclohexatriene-2-ylidene **7** is largely destabilized (by ca 50 kcal mol$^{-1}$) relative to isomeric pyridine, it has been calculated that the difference is reduced to a few kcal mol$^{-1}$ for the corresponding radical cation **7$^{+\cdot}$** (see Scheme 7) [40]. Isomerization of the ions is prohibited by an energy barrier, evaluated at 40–60 kcal mol$^{-1}$, which is lower, however, than the dissociation threshold, so interconversion does occur under mass spectrometric conditions. The effect of aromaticity even at the radical cation stage can, on the other hand, be evaluated because the furan radical cation is the main fragment formed (along with a minor amount of vinylketene) from the decarbonylation of the 2- and 4-pyrone radical cation (**8$^{+\cdot}$**, see Scheme 8) [41].

Fragmentation of 2,3-diazabicyclo[2.2.1]hept-2-enes via photoinduced electron transfer offers a convenient route to cyclopentane-1,3-diyl radical cations, in turn undergoing interesting and selective rearrangements [42].

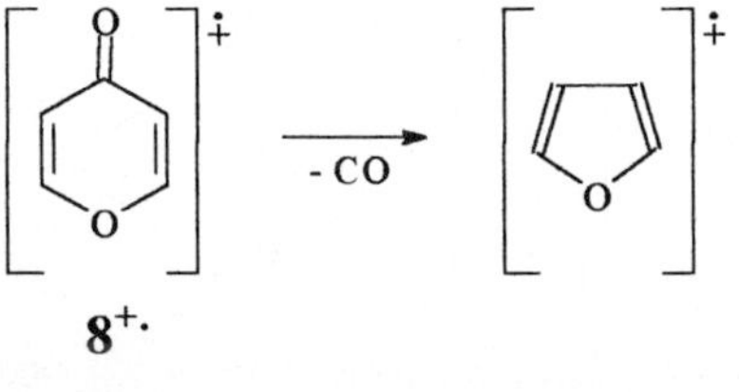

**8$^{+\cdot}$**

**Scheme 8.**

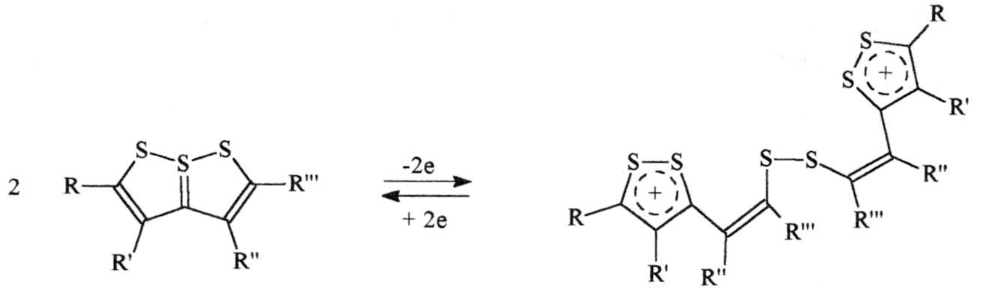

**9**

**Scheme 9.**

Ring-opening then dimerization to give bis (dithiolyium) di-cations linked by a disulfide bond occurs upon anodic oxidation of some $1,6,6a\lambda^4$-trithiapentalenes (**9**) and the starting material can be regenerated upon reduction (see Scheme 9) [43].

A single electron transfer- (SET) induced ring enlargement has been found in the conversion of substituted 2-furylmethanols (**10**) into 6-hydroxytetrahydropyran-3-ones (**12**) by the action of the binary reagent $PhI(OAc)_2$–$Mg(ClO_4)_2$; it has been suggested that the reaction occurs via a furan radical cation (**11**) as illustrated in Scheme 10 [44].

### 6.3.4 Coupling, Dimerization, and Polymerization

As mentioned in Section 6.3.1, the radical cations of many heteroaromatics have been formed anodically, thermally, or photochemically. With electron-rich heterocycles, one-electron oxidation is often followed by dimerization, either via coupling of two radical cations or via radical cation–substrate reaction, which in turn is most often followed by a second electron-transfer step leading to the same non-radical di-cation formed by the previous mechanism (see Scheme 11). The course of the reac-

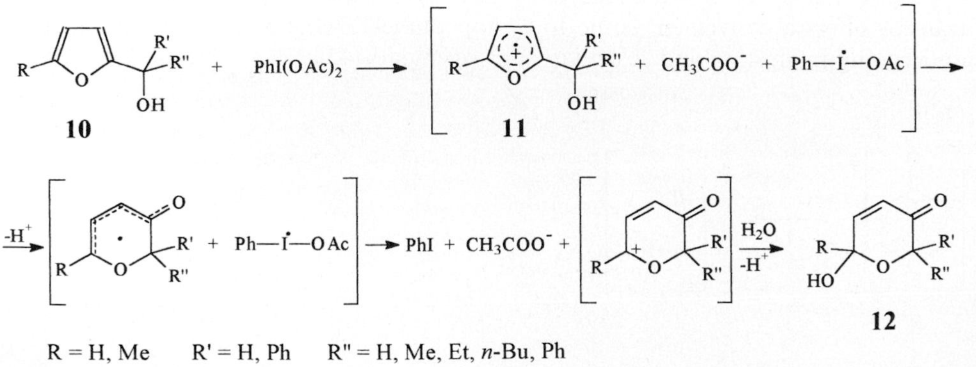

R = H, Me    R' = H, Ph    R" = H, Me, Et, n-Bu, Ph

**Scheme 10.**

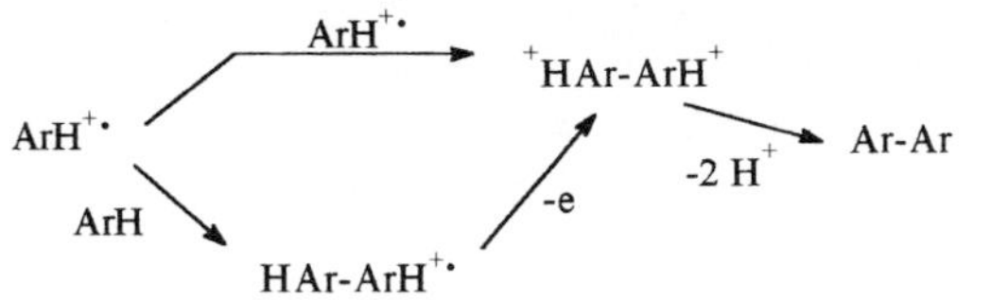

**Scheme 11.**

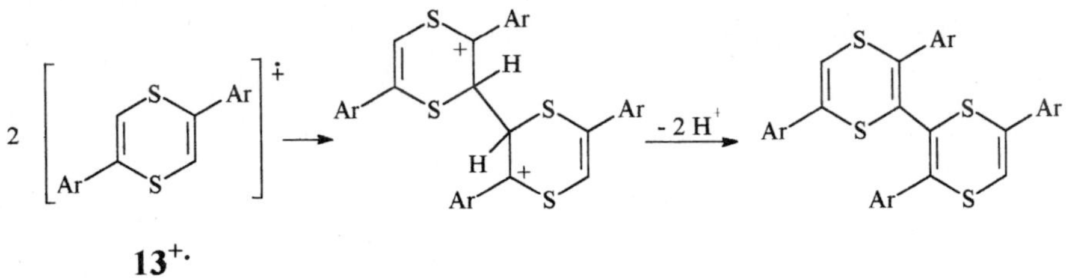

**13$^{+\cdot}$**

**Scheme 12.**

tion then depends on the paths enabling proton transfer. The mechanism of anodic dimerization has long been a subject of extensive investigation [45].

A typical example in heteroaromatic chemistry is the preparative electrolysis of some 2,5-diaryl-1,4-dithiins **13**, which give two well resolved quasi-reversible one-electron transfer waves to the radical cation and the di-cation, respectively, in cyclic voltammetry. The electrolysis gives low yields of the 2,2′ dimers (see Scheme 12) via the radical cation coupling mechanism [46].

The 2-(4′-aminophenyl)thiophene **14** is converted in 71 % yield to dimer **15** by preparative electrolysis. The process is effected by oxidation at 0.9 V (1.5 F mol$^{-1}$), involving formation of the radical cation of the substrate (**14$^{+\cdot}$**) dimerization, de-protonation, and further oxidation of dimer **15** (occurring at a potential lower than the applied potential), followed by controlled cathodic reduction at 0.0 V of the resulting di-cation (0.5 F mol$^{-1}$, see Scheme 13) [47].

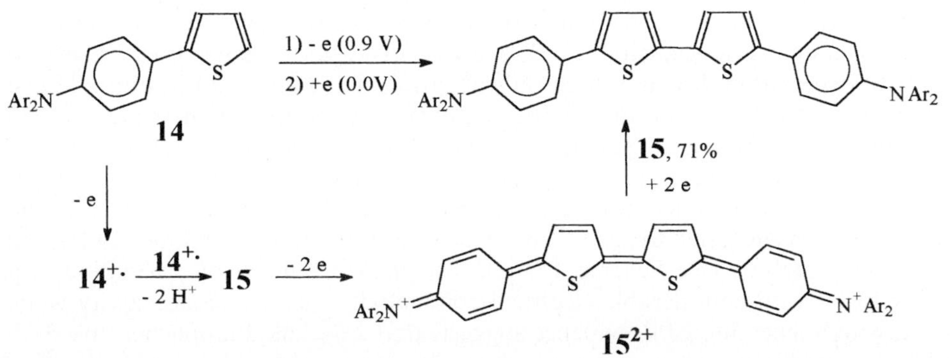

**Scheme 13.**

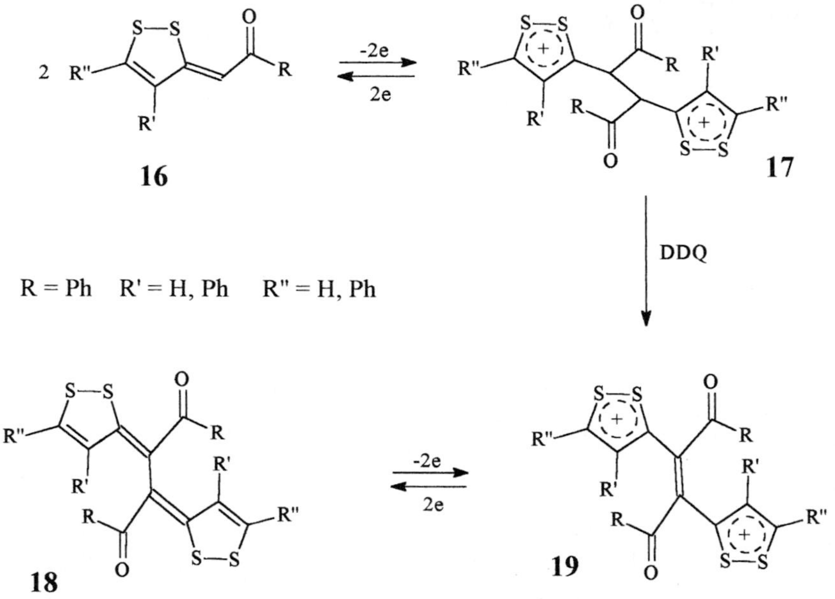

R = Ph    R' = H, Ph    R" = H, Ph

**Scheme 14.**

Coupling can also occur via a substituent, as for 1,2-dithiol-3-thiones, which are anodically oxidized to bis(dithiolyium)disulfides [48]. The related α-(1',2'-dithiol-3'-ylidene)acetophenones **16** are analogously oxidized to a di-cation (**17**) that deprotonates to give the uncharged dimer **18**. When, however, the electrolysis was conducted in the presence of an oxidant such as DDQ, or by further oxidizing dimer **18**, a new di-cation (**19**) was formed (Scheme 14) [49].

Most chemical or photochemical oxidations were initially conducted under acidic conditions, but use of other non-nucleophilic protic solvents such as HFP enables the oxidation to be performed under neutral conditions. The result of coupling reactions depends on conditions. As an example, oxidation of 2-alkylthiophenes (**20**) by thallium tris(trifluoroacetate) leads to the corresponding 5,5'-dialkyl-2,2'-bithiophene radical cations (**21**$^{+\bullet}$), which are quite persistent species (hours) [3]. Unsubstituted thiophene and 3-alkyl derivatives (**22**) react in the same way, although the dimeric radical cations (**23**$^{+\bullet}$), are short-lived (see Scheme 15). This is in line with expectation, because these radical cations have a free α position and fast follow-up reactions can occur, leading to oligomers or to polymers. In fact, in reactions starting from authentic 4,4'-dialkyl-2,2'-thiophene EPR spectra attributed to a tetrathiophene were observed. Under acidic conditions, however, different radical cations are obtained. Structure **25**$^{+\bullet}$ has been attributed to these species and these have been suggested to arise from oxidation of the thiophene dimer **24** formed under acid catalysis (see Scheme 15) [3]. The final product from oxidation is a dark powder with considerable electrical conductivity and semiconductivity when doped with iodine [50]. EPR evidence suggests that 2,5-dialkylthiophenes give 3,3'-bithiophenes (mixture of *cis* and *trans* isomers).

With the radical cations of benzoannelated heterocycles, various paths are

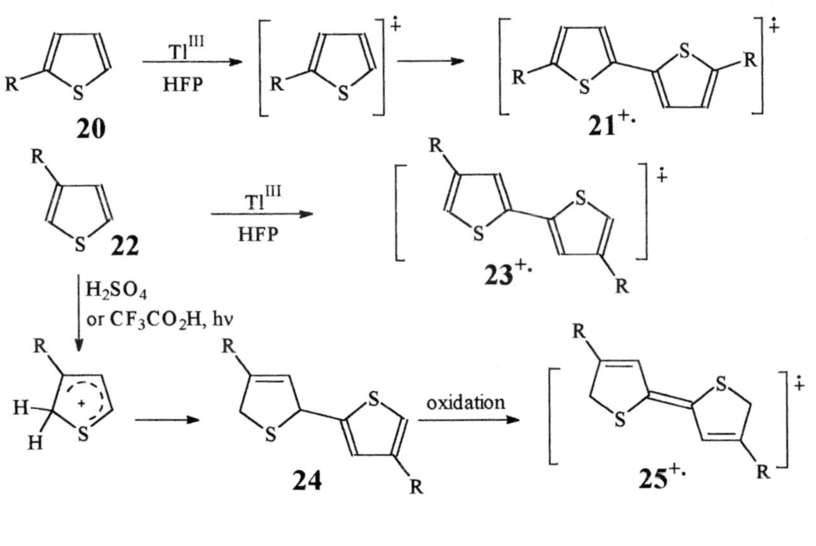

R = Me, Et, *i*-Pr, *t*-Bu

**Scheme 15.**

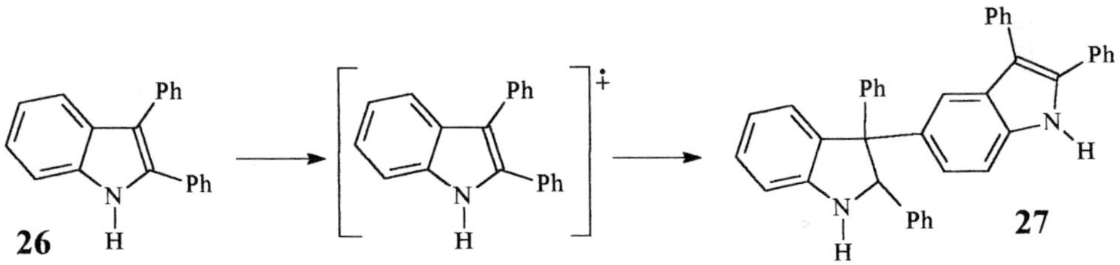

**Scheme 16.**

followed and C–C coupling is occasionally observed. As an example, it has been reported that a 5-[3′-(2′,3′-dihydroindolyl)]indole (**27**) is formed from 2,3-diphenylindole (**26**) [51]. 3,3′-Dimers (**29**) are efficiently formed as the corresponding radical cation tetrafluoroborates by treatment of carbazole and *N*-ethylcarbazole (**28**) with nitrosonium tetrafluoroborate (97–98 %; Scheme 17) [52]; electrochemically, the *N,N′* dimer seems to be formed initially, but the 3,3′ isomer is finally obtained [53]. Likewise, it has been suggested that dibenzothiophene gives the 2,2′ dimer [54], but the heterocycle often functions as a heteroatom-centered nucleophile. As an example, phenothiazine, phenoxazine, and phenoselenine all give the 3,10′ dimers (**30**) as the main products on oxidation by $I_2$ in dimethyl sulfoxide. The intermediate radical cation has been characterized; it reacts with the substrate which behaves as an *N* nucleophile (Scheme 18) [55]. Similar reactions have been performed on the same substrates under different oxidizing conditions [56]. With 1,2-dihydro-2,2-dialkyl-3-oxo-3*H*-indoles (indoxyl), a class of derivatives the radical

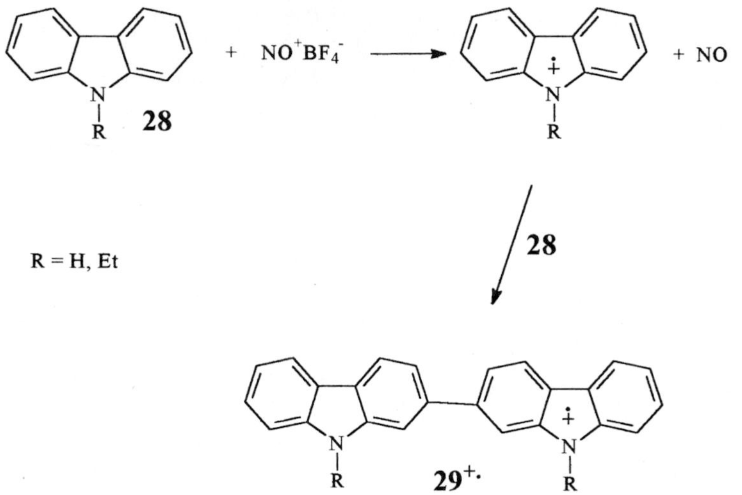

Scheme 17.

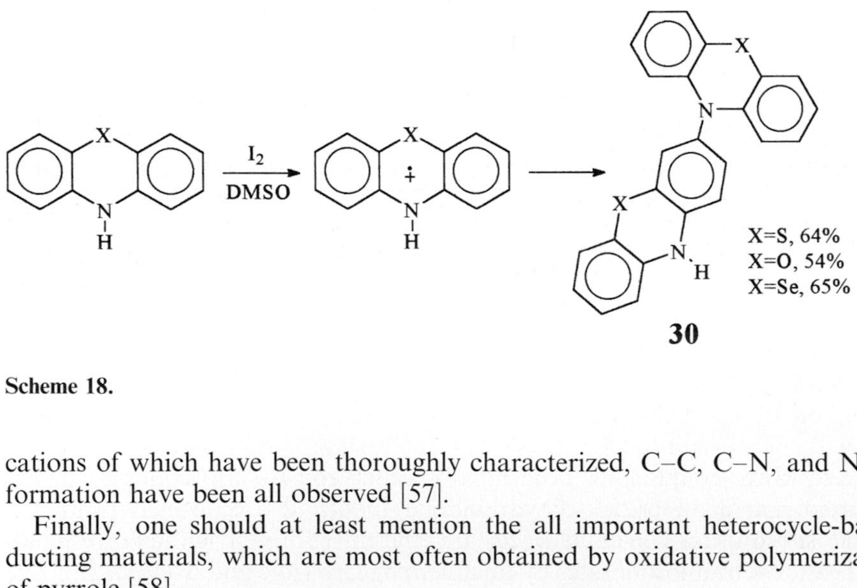

X=S, 64%
X=O, 54%
X=Se, 65%

**30**

Scheme 18.

cations of which have been thoroughly characterized, C–C, C–N, and N–N bond formation have been all observed [57].

Finally, one should at least mention the all important heterocycle-based conducting materials, which are most often obtained by oxidative polymerization, e.g. of pyrrole [58].

### 6.3.5  Functionalization of the Ring

#### Halogenation

The possibility that some electrophilic aromatic substitutions occur via a SET mechanism has long been debated in the recent literature, as reported in more detail elsewhere in this book.

$$HetH + ICl \longrightarrow HetH^{+\bullet} + I^\bullet + Cl^-$$

**31**

$$HetH^{+\bullet} + Cl^- \longrightarrow HetHCl^\bullet \longrightarrow HetCl$$

$$HetH^{+\bullet} + I^\bullet (or\ I_2) \longrightarrow HetHI^+ \longrightarrow HetI$$

$$HetH + Br_2 \longrightarrow (HetH^{+\bullet}\ Br^-\ Br^\bullet) \xrightarrow{in\ cage} HetHBr^\bullet$$

$$\Big\downarrow diffusion$$

**Scheme 19.**

$$HetH^{+\bullet} + Br^- + 1/2\ Br_2$$

The halogenation of electron-rich heteroaromatics, as an example, can proceed either via the normal addition–elimination mechanism via a cation or via a SET mechanism, because the halogen, or a mixed halogen, e.g. ICl, can act as an oxidant, and the radical cation then add the halide anion. Indeed, in HFP heteroaromatics with $E°$ (Het$^{\bullet+}$/Het) < 1.5 V relative to Ag/AgCl, e.g. thianthrene ($E° = 1.32$ V), are oxidized by ICl, and the radical cation can be expediently characterized by EPR. The same holds for chlorine, bromine, and iodine with heterocycles with $E°$ (Het$^{\bullet+}$/Het) < 1.0, 1.4, and 1.1 V, respectively, as observed for $N$-methylbenzothiazine ($E° = 0.73$ V). Under these conditions, electron transfer is followed by fragmentation of the reduced acceptor to give a triad (e.g. **31**). The final course of the substitution reaction depends on the structure, involving, e.g., ionic or radical coupling within the triad (Scheme 19), and on conditions, which affect the competition between coupling within the triad or diffusion out of cage and subsequent coupling of the free solvated species (Scheme 19, bottom) [59].

Bromination of 10-phenylphenothiazine (**32a**; Scheme 20) and 10-phenylphenoxazine with bromine in acetic acid gives the corresponding 3-bromo and 3,7-dibromo derivatives via the radical cation, whereas the use of a milder reagent such as pyridinium bromide perbromide gives predominantly the 4'-bromo derivative via electrophilic substitution [60]. The preformed phenothiazine radical cation indeed adds bromide (and the nitrite and thiocyanate anions) to give the corresponding 3-substituted derivatives [56]. With 10-alkylphenothiazines (**32b**) the brominated derivatives are further oxidized to the radical cations and these undergo dealkylation. In the limiting case, 10-benzylphenothiazine (**32c**) is debenzylated, rather than brominated, apparently because of the combined effect of retardation of ring-bromination by the bulky substituent and easy detachment of the good electrofugal group (Scheme 20) [61]. The kinetic role of the radical cations of thianthrene, phenoxathiin, and phenothiazine derivatives has been investigated in detail and reviewed [7]. It should also been considered that the heterocycle itself can act as a nucleophile, leading to dimers, as mentioned in Section 6.3.4 The view has been

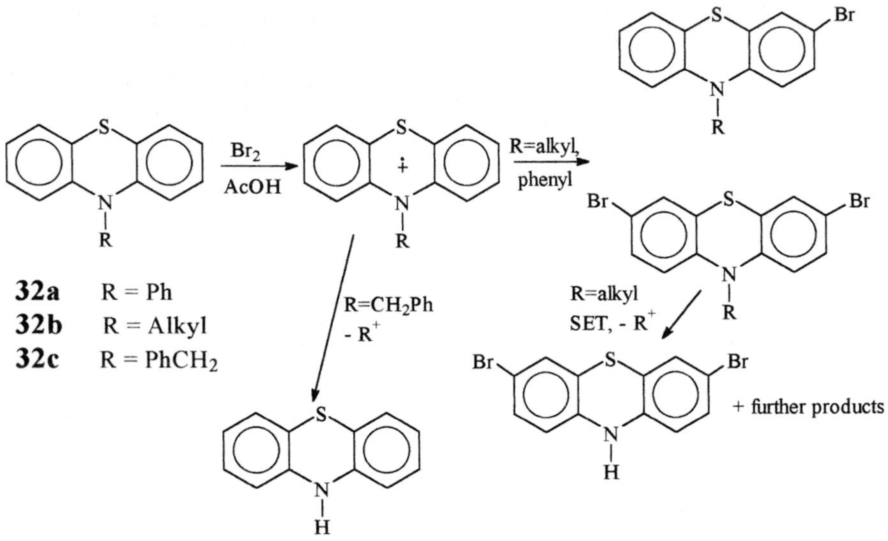

**Scheme 20.**

expressed that, particularly with *N*-heterocycles, the localization of the charge on the nitrogen atom makes reaction with nucleophiles slower than for the radical cations arising from the carbocyclic analogs, thus making coupling more important; this is certainly true with indoxyl (for example) [62].

Iodination of thiophene occurs quite cleanly (92 % of the 2-substituted, 2 % of the 2,5-disubstituted derivatives) in the presence of nitrosonium tetrafluoroborate [63]. The oxidative fluorination of heterocycles is of some interest, because of the limited accessibility of such compounds, despite the fact that mixtures are usually obtained. Anodic oxidation in a mixture of triethylamine and hydrogen fluoride, for example, leads to unstable dihydrodifluoro derivatives from furan and benzofuran. If, however, 1,10-phenanthrolines are used the dihydrodifluoro adducts can be conveniently dehydrofluorinated by bases; repeating the oxidation–rearomatization sequence eventually furnishes 5,6-dihydro-5,5,6,6,-tetrafluoro-1,10-phenanthroline in 34 % yield. High-valence metal fluorides can also be used. As an example, fluorination of pyridine by cesium cobalt(III) fluoride leads to polyfluoropyridines (**33**) and a perfluoropiperidine (**35**) arising via 2,5-dihydro-2,2,5,5-tetrafluoropyridine (**34**; Scheme 21). Similar results are obtained with benzofuran [64].

## Nitration

The intervention of a SET path in (hetero)aromatic nitration has been the subject of an extended debate [65]. The evidence is usually based either on the detection of intermediates or on the regioselectivity of the attack in relation to the spin and charge density on the radical cation. Limiting the attention to heteroaromatic substrates, it should be noted that the radical cations were detected spectroscopically [66] or even isolated as a salt [67] during nitration of, e.g., phenothiazine, phenox-

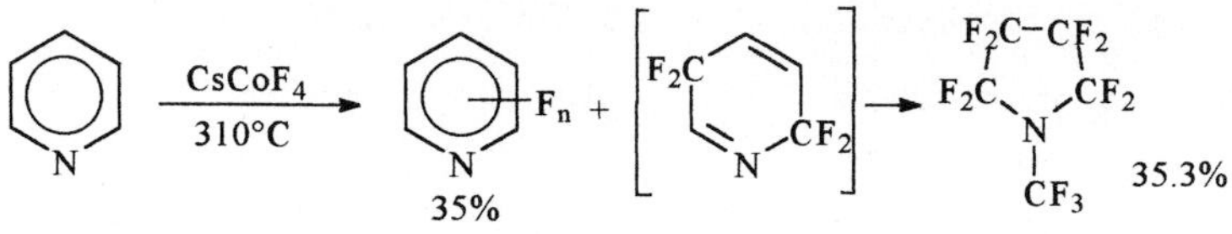

35%         35.3%

**33**     **34**     **35**

(of which 13.6% F$_5$)

**Scheme 21.**

azine, or dibenzo-1,4-dioxin with HNO$_3$. With the last substrate, and use of 0.3–0.5 mol of the acid, the blue radical cation is the final product in 90 % yield whereas the use of 2 mol leads to isomeric dinitrodibenzodioxins [68]. With *N*-methylphenoxazine the use of a milder reagent such as tetranitromethane enables monitoring of the conversion of the initially formed radical cation into the nitro derivative [69].

When thermal reaction with tetranitromethane is too slow, photochemical excitation in the CT band induces the SET step and dihydroaromatics arising from the addition of the nitro and the trinitromethyl groups are formed (although the final products can result from further transformations). The first step here is attack of the trinitromethanide ion on the aromatic radical cation and the regioselectivity is in keeping with theoretical predictions, e.g. with predominant attack of the trinitromethide ion on positions 1 and 3 in dibenzofuran [70]. The regiochemistry of the reactions of benzofuran (Scheme 22) [71] and dibenzothiophene [72] has been similarly rationalized. Another method for SET nitration is the reaction of preformed, or in-situ formed, radical cations with NO$_2$. In this reaction the regiochemistry is

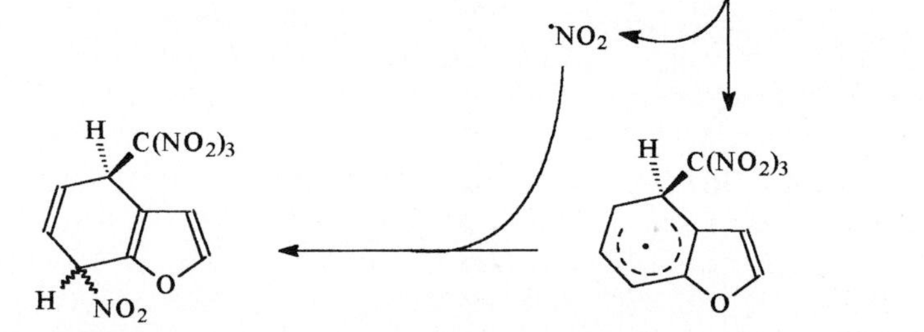

**Scheme 22.**

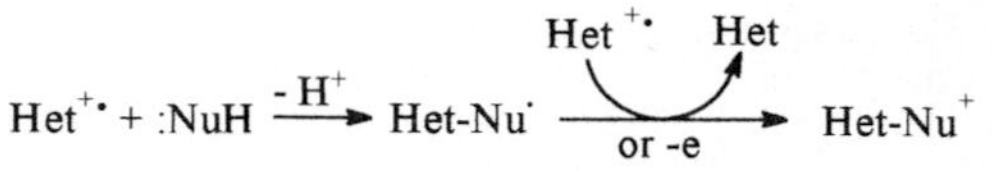

$$\text{Het}^{+\cdot} + :\text{NuH} \xrightarrow{-\text{H}^+} \text{Het-Nu}^\cdot \xrightarrow[\text{or } -e]{\overset{\text{Het}^{+\cdot} \quad \text{Het}}{\curvearrowright}} \text{Het-Nu}^+ \qquad \textbf{Scheme 23.}$$

controlled by the spin density distribution in the radical cation, because this is a coupling between two radicals, as indeed has been shown for some compounds (e.g. benzofuran [71] and phenothiazine, where the main product is, however, the *S*-oxide [73]).

**Reaction with heteroatom- or carbon-centered nucleophiles**

A variety of functional groups has been obtained by reaction of radical cations with neutral nucleophiles. The initial reaction produces a radical more easily oxidized than the substrate from which the radical cation arises. A further oxidation then intervenes leading to the final cationic product (Scheme 23). When the reaction is performed electrochemically, a further ET occurs at the anode, whereas in homogenous solution the oxidant is usually a second equivalent of cation radical that is reduced to the starting substrate. The stoichiometry of the reaction thus generally requires 2 mol of the radical cation, although the mechanism might actually be different from the oversimplified delineation of the scheme and involve further intermediates, e.g. di-cations.

Many reactions involving formation of a sulfur heteroatom or a sulfur–carbon bond have been developed with the radical cations of thianthrene and analogs. As an example, anodic oxidation of phenoxathiin [74] or of thianthrene in wet acetonitrile or reaction with water of the preformed thianthrene radical cation [30, 75, 76] lead to formation of the corresponding *S*-oxides. The same is true of phenoxathiin derivatives, as has been shown for the drug chlorpromazine [77]. When treated with ammonia and amines, the thianthrene, phenoxathiin, and phenoxazine radical cation perchlorates give the corresponding sulfinimine perchlorates (**36**; Scheme 24). With ammonia, these can be deprotonated to give the neutral sulfinimines (**37**); if ammonia is slowly added to the initial radical cation salt the dimeric iminum salt can be prepared [52, 78]. Thianthrene radical cation salts have been similarly reacted with a variety of nucleophiles. Carbazole reacts at the nitrogen atom to give

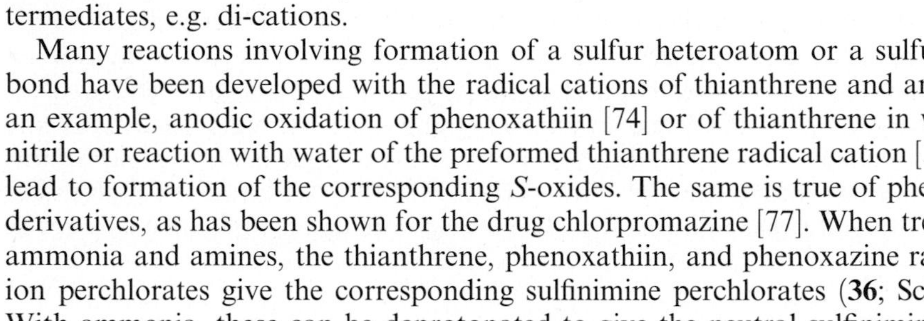

X = S, O, NR    **36**    **37**

R = Me, Ph

**Scheme 24.**

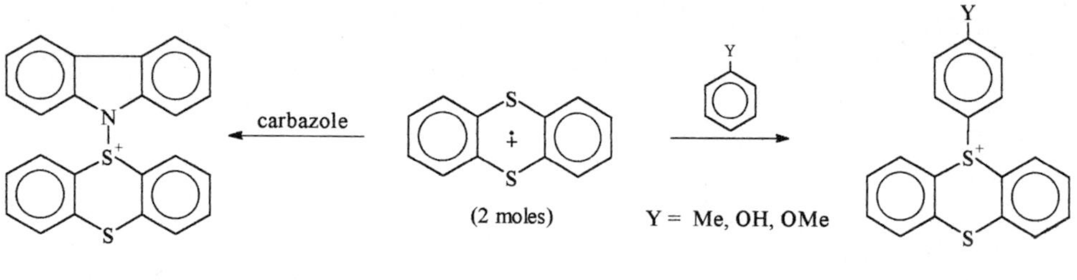

**38**

Scheme 25.

product **38** (Scheme 25) [79], but with phenol [80], toluene, and anisole [81] the reaction is at the *para* carbon atom and yields products **39**. A second mole of the thianthrene radical cation always participates into the reaction as an oxidant (compare Scheme 23).

Further interesting examples of C–S bond formation involve the reaction of previously prepared (or in situ anodically generated) thianthrene or phenothiazine radical cations with alkenes or alkynes, to give 1,2-bis(hetaryl) alkanes (or the respective alkenes) [82]. With cyclooctene a 1:1 adduct is obtained instead. Another valuable application is the smooth reaction with ketones (Scheme 26). The thianthrenium salts (**40**) now obtained are readily deprotonated to the corresponding ylides (**41**) [83]. The latter compounds are directly obtained when β-dicarbonyls are used.

A ring carbon can also be involved, however, as in the reaction of the thianthrene and phenothiazine radical cations in neat pyridine or with pyridine in an anhydrous solvent. In this reaction the 1-pyridinium group is inserted on to the benzo ring (**43**), apparently via nucleophilic attack on di-cations **42**, in turn resulting from oxidation of the initially formed radical cation adducts (Scheme 27). In the presence of moisture the sulfoxides are again formed [84].

Other reactions such as the anodic cyanation or alkoxylation of electron-rich heterocycles such as pyrroles and indoles [85] or of electron-donating substituted azines [86] are important, but are not discussed in detail here, because this subject is well covered in electrochemistry textbooks.

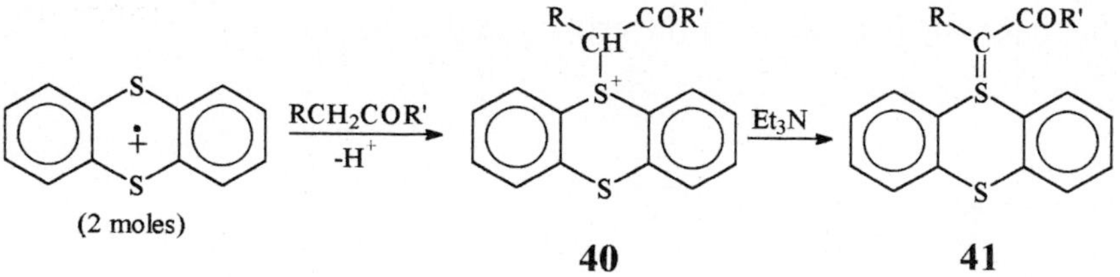

Scheme 26.

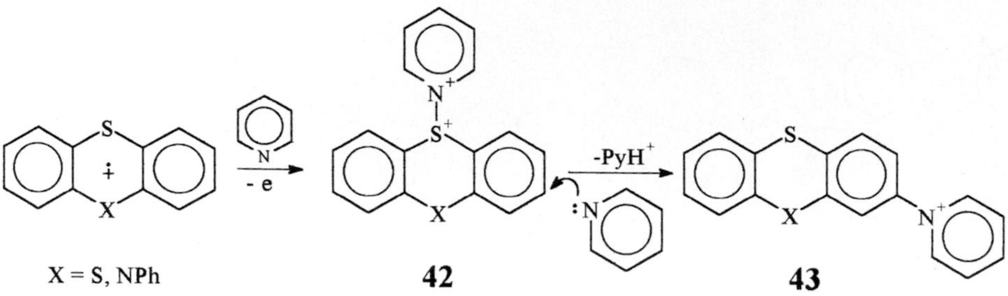

X = S, NPh                    **42**                          **43**

**Scheme 27.**

## Alkylation and arylation via ET

In the gas phase, alkylation of five-membered heterocycles by alkyl cations usually occurs via the usual addition–elimination mechanism of aromatic electrophilic substitution. The phenyl cation behaves differently, however; although its substrate discrimination is limited, in accord with its exceedingly high reactivity, it has marked selectivity for the α position, which does not conform with the 'hard' character of this cation. It has, therefore, been suggested that an electron-transfer mechanism is followed; this is thermodynamically allowed for the phenylium, and likewise for the methyl cation, but not for other alkyl cations (Scheme 28). This SET mechanism applies also for acyl cations [87].

Solution-phase studies are more important preparatively. Two main mechanisms seem to operate in solution. The first is attack of the radical cation of a heteroaromatic donor on a π nucleophile, as happens in the arylation reactions reported above. Other examples include photochemical reactions in which the heterocycle participates as a donor—for example the formation of 2- and 3-(1,2-diphenylethyl)pyrroles (yield 44 and 10 %, respectively) from the irradiation of (E)-stilbene in the presence of pyrrole, a reaction which evidence implies is initiated by SET from pyrrole [88]. 2-(2′,2′-Diphenylethyl)furans are cleanly formed on irradiation of the corresponding furans in the presence of 1,1-diphenylethylene and an electron-accepting sensitizer [89]. Likewise, irradiation of naphthalene and benzothiophene in the presence of pyrrole results in electron transfer from the latter and leads eventually to pyrrolyldihydronaphthalene or benzothiophene, **44**, respectively (Scheme 29) [90].

A partially related reaction is the photochemical synthesis of bis(pyrrolyl)- and bis(indolyl)methanes by irradiation of (hetero)aryl aldehydes in the presence of the

Y = NH, NMe, O, S

**Scheme 28.**

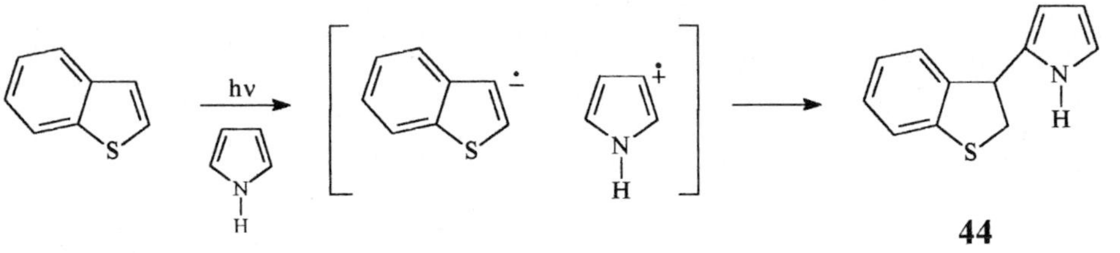

**44**

**Scheme 29.**

respective heterocycles; it is suggested that this involves SET to the excited aldehyde then coupling of the radical ions [91].

The second mechanism involves radical or ionic coupling when at least one of the radical ions formed by thermal or photoinduced electron transfer fragments with sufficient efficacy (compare paths *f* and *g* in Scheme 6). In this case, coupling of the radical or charged species can lead to the formation of a new bond. A typical example is the addition of tetranitromethane fragments across a (hetero)aromatic molecule; for convenience this was mentioned above along with the nitration reaction. Other processes resulting in alkylation or arylation of the heterocyclic ring are reported below. A synthetically useful reaction of this class is the perfluoroalkylation of heterocycles by use of bis(perfluoroalkanoyl)peroxides. The radical anion of the peroxide formed in the initial SET step decomposes to give a perfluoroalkyl radical, perfluoroalkanoate, and carbon dioxide. The radical couples with the donor radical cation leading to the alkylated compound (Scheme 30). In this way 2-perfluoropropylfuran and thiophene are obtained smoothly (40 °C, degassed Freon solution) in quantitative yield [92]. The reaction is equally effective with pyrrole at a lower temperature (−30 °C), with some substituted pyrroles, and with indole (with increased selectivity for the 2 position at −80 °C). The reaction fails with pyridine, but is successful with 2-pyridone or when the nitrogen lone pair is shielded, as in 2,6-di-*tert*-butylpyridine [93].

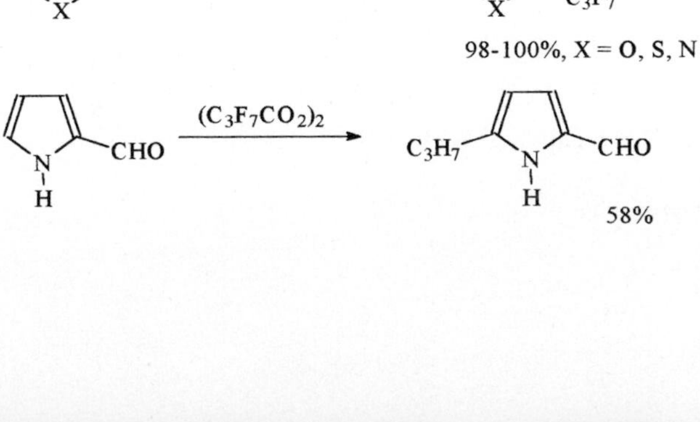

**Scheme 30.**

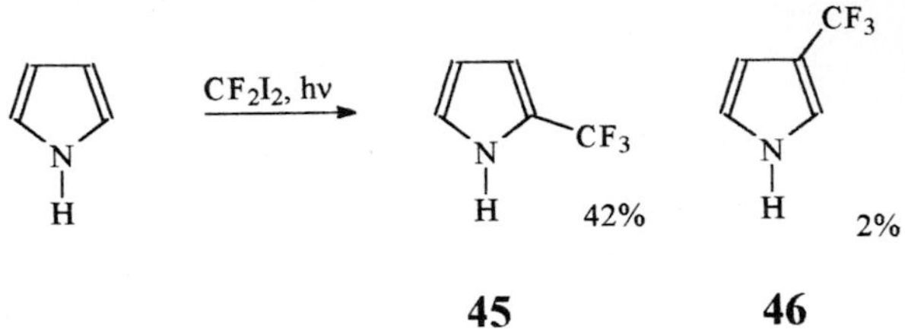

**45**        **46**

**Scheme 31.**

Trifluoromethylation of pyrrole (and indole and imidazole) occurs on irradiation of a mixture of the compound with difluorodiiodomethane, again via a SET mechanism involving fragmentation of the radical anion (Scheme 31). The presence of the $CF_3$ group in the final products, **45** and **46**, is a result either of secondary decomposition of the initially formed—but not isolated—difluoroiodomethyl derivatives, or of formation of the $CF_3$ anion or radical in situ [94]. Perfluoroalkylation of pyrroles can also be achieved by an $S_{RN}1$ mechanism—by reaction with perfluoroalkyl iodides in the presence of magnesium or zinc [95]. Indole, on the other hand, gives a mixture of the seven possible alkylated derivatives when irradiated in the presence of ethyl chloroacetate [96].

Examples of related photochemical (hetero)arylations are the pentafluoralkylation of pyrroles and indoles by irradiation in the presence of pentafluoroiodobenzene or of pentafluorophenyl perfluoroalkanesulfonates [97], and presumably also the acetone-sensitized photocoupling of 5-bromo-1,3-dimethyluracil (**47**) with indoles. The latter reactions apparently follow an ET path, rather than the homolytic debromination which occurs on direct irradiation (Scheme 32) [98]. A different arylation leads to 2,5-dihydro-2-(4′cyanophenyl)-5-methoxyfuran on irradiation with 1,4-dicyanobenzene in methanol [99].

An important reaction is that with haloketones. 2,5-Dibromo-1,4-benzoquinone

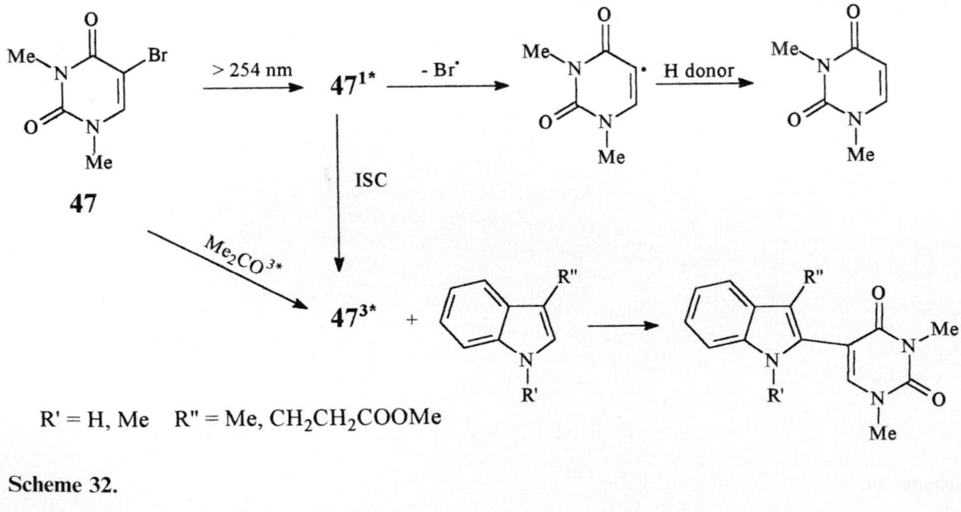

R′ = H, Me   R″ = Me, $CH_2CH_2COOMe$

**Scheme 32.**

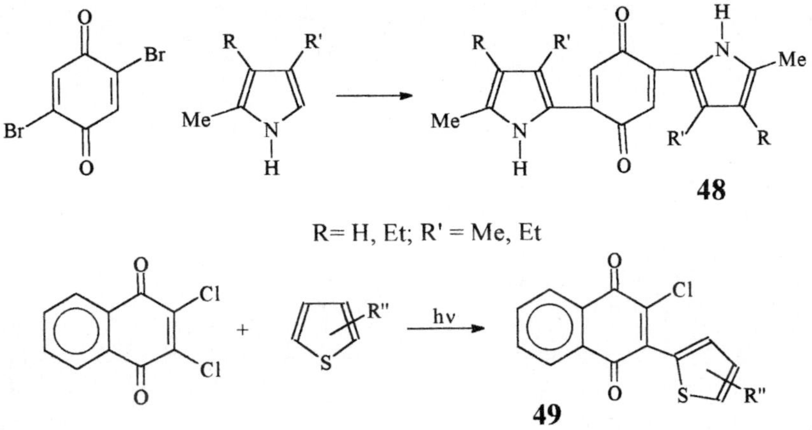

**48**

R= H, Et; R' = Me, Et

**49**

**Scheme 33.**

gives dyes of structure **48** (Scheme 33) smoothly on addition to pyrrole solutions; the reaction can be envisaged as being initiated by a SET step [100] (it should be noticed that 3,4,5,6-tetrachloro-1,2-benzoquinone gives enolic adducts rather than eliminating hydrogen chloride, a reaction which is envisaged as Michael addition [101]).

Analogous reactions occur smoothly upon photochemical excitation with halo-quinones; these have much greater scope. As an example, furans, thiophenes, indoles, and pyrroles are readily functionalized by irradiation in the presence of haloquinones (see, e.g., product **49** in Scheme 33) [102]. Maruyama's group has also used the SET photoinduced reaction between halonaphthoquinones (or halo-methoxynaphthoquinones) and 1,1-diarylethylenes for the synthesis of a variety of heterocycles, for example compounds **50** and **51** in Scheme 34 [103].

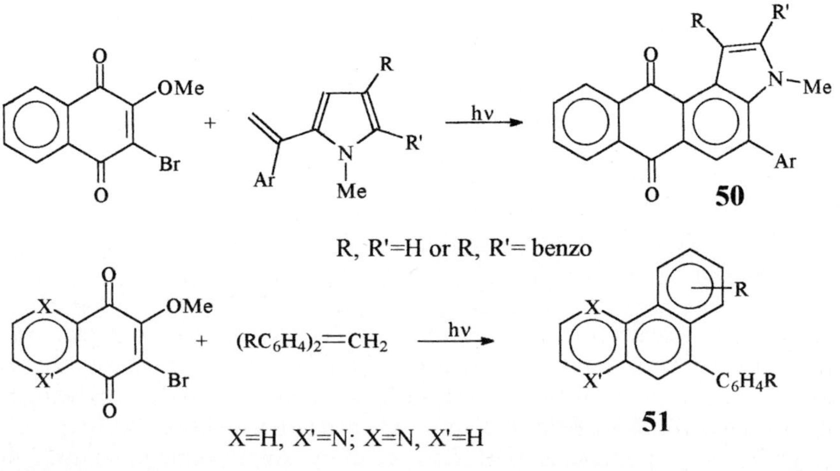

**50**

R, R'=H or R, R'= benzo

**51**

X=H, X'=N; X=N, X'=H

**Scheme 34.**

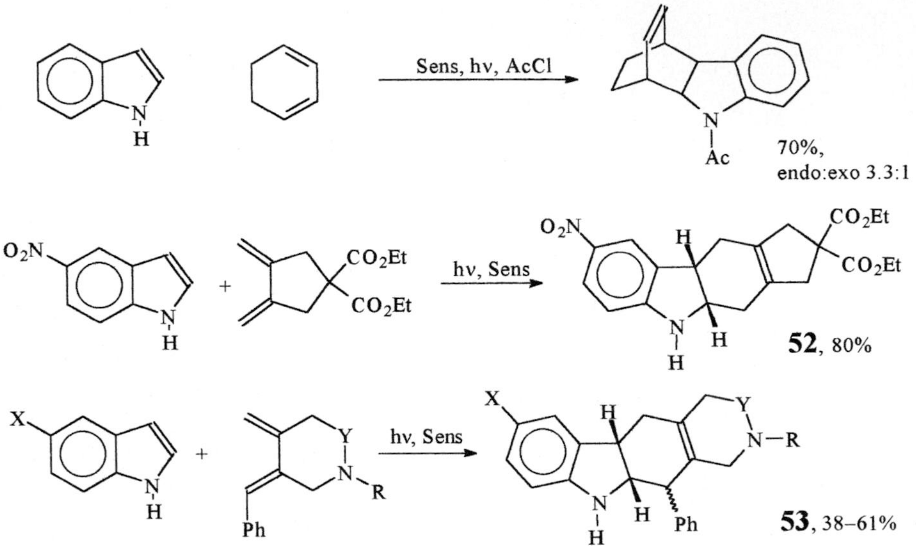

70%,
endo:exo 3.3:1

**52**, 80%

**53**, 38–61%

X=H, CO$_2$Me, CN, NO$_2$; Y=O, NR; R=Ts, Ac, CO$_2$Et

**Scheme 35.**

### 6.3.6 Cycloadditions

The radical cation Diels–Alder reaction has been the subject of many mechanistic and theoretical investigations and has been shown to have much synthetic potential. With regard to heteroaromatics, the reaction has been exploited by Steckhan in the cycloaddition of indoles and 1,3-dienes. This reaction occurs smoothly upon photosensitization by triarylpyrrilium tetrafluoroborates. The reaction is satisfactory rationalized as involving addition of the indole radical cation to electron-rich dienes (Scheme 35), and the regioselectivity is in accord with theoretical predictions [104]. The reaction with exocyclic dienes has been developed for the synthesis of carbazole derivatives such as **52** and **53** [105].

### 6.3.7 Oxygenation

The possibility that an electron-transfer path is involved in photo-sensitized oxygenation has been considered on several occasions. This is relevant in several fields of application, from the biomimetic oxygenation of indole and flavin derivatives [106] to pollutant control. With reference to latter, it has been suggested that SET occurs in heterogeneous photosensitized oxidation by solid semiconductors, in which the adsorbed substrate donates an electron to the photogenerated hole and

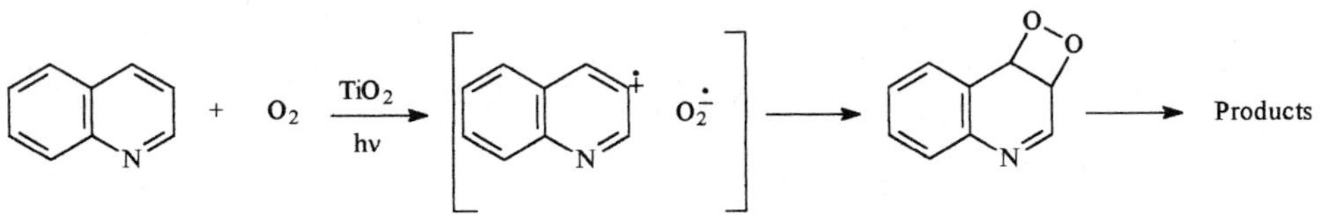

**Scheme 36.**

oxygen is reduced to the superoxide anion. This mechanism has been probed, e.g. by use of quinoline as a model (Scheme 36) [107].

## 6.4 Chemical Reactions via Electron Transfer—The Heteroaromatic is the Acceptor

### 6.4.1 General Scheme

The general pattern of reactions is quite similar (with inverted charge) to that occurring after electron transfer in the reverse direction (see Scheme 37). Thus, apart from electron transfer and rearrangement or ring cleavage (path *a*), the reactions are those expected from radical (coupling, path *b*), from an anion (addition of an

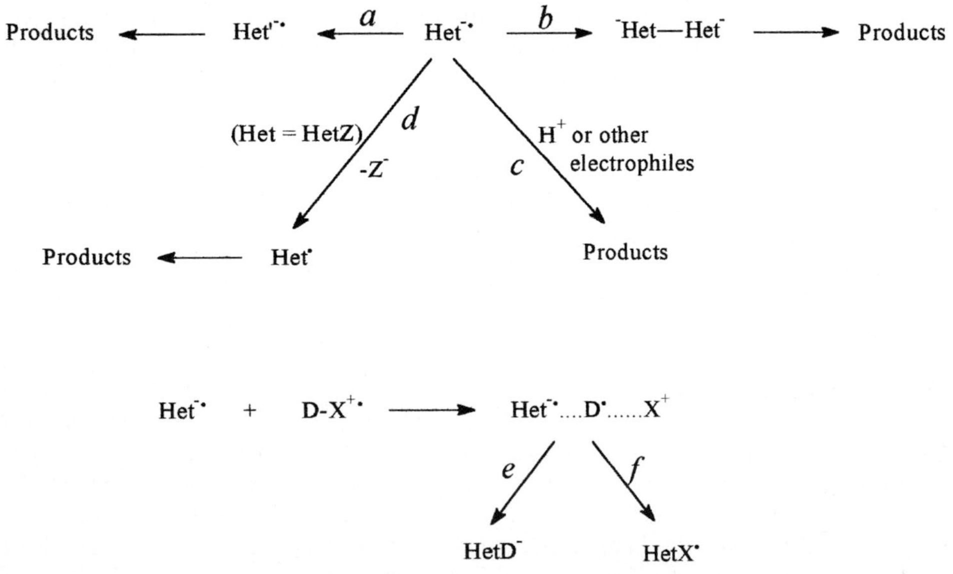

**Scheme 37.**

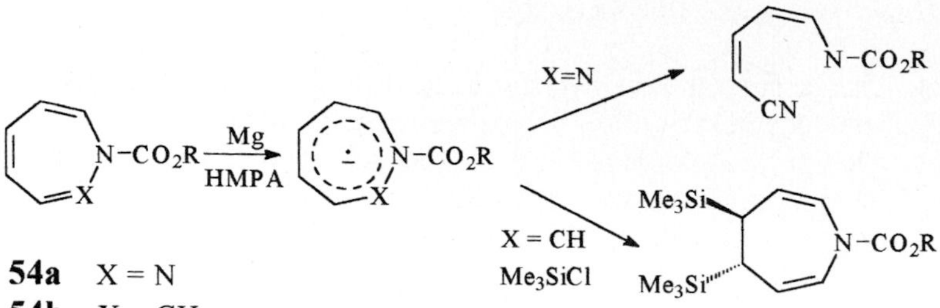

**54a**  X = N
**54b**  X = CH

**Scheme 38.**

electrophile, here quite often the proton, path *c*), or cleavage of a nucleofugal group (path *d*). Here again, fragmentation of the concomitantly formed radical cation can lead to addition of the heteroaromatic radical anion either with a radical (path *e*) or with a cation (path *f*).

### 6.4.2 Ring Opening and Rearrangement

The ET-induced desulfurization of thiophene and related sulfur heteroaromatics by the action of metal naphthalenides [108], metal hydride–metal complexes [109], or transition metals [110] is a process with high practical significance, in view of its implication in the desulfurization of coal.

Many 1,2-diazines undergo N–N-bond cleavage upon electroreduction, eventually resulting either in reductive ring opening or in ring contraction, e.g. of the type pyridazine → pyrrole [111] or phthalazine → isoindole [112].

1*H*-1,2-Diazepincarboxylates (**54a**) undergo ring cleavage upon reduction by magnesium in hexamethyl phosphoramide, whereas the corresponding azepines (**54b**) preserve the ring and are reduced under these conditions, as evidenced by the formation of a disilyl derivative (Scheme 38) [113]. The 2,3,5-triphenyltetrazolium cation (**55**) undergoes one-electron reduction by electrolysis in a strictly apolar medium, but in the presence of water the resulting radical (**56**) undergoes protonation and ring-opening to yield a formazane in an overall 2-electron process (Scheme 39) [114].

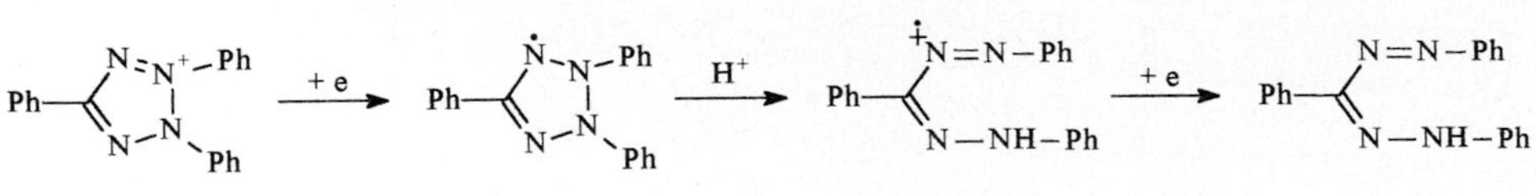

**55**                    **56**

**Scheme 39.**

### 6.4.3 Coupling, Dimerization and Polymerization

With some heterocycles, separated one-electron waves are observed upon cathodic reduction, an important phenomenon i.a. with sensitizing dyes such as porphyrins [115] and cyanins. Occasionally dimers may be formed by preparative electro-reduction competitively with the formation of partially or totally ring-hydrogenated heterocycles. This happens. e.g.. with acridine (giving biacridane) [116], pyrimidine [117], phthalazine [118], quinazoline [119], 4-cinnolinone [116], and 4-quinazolone [120]. Cathodic coupling of 2- and 4-vinylpyridine gives the corresponding 1,4-dipyridylbutanes in good yield [121]. The dehalogenative coupling of pentafluoro-pyridine is discussed below.

The quaternary salts of *N*-heterocycles are readily reduced and often give reversible one-electron reduction waves, as, for example, with pyridium salts [122]. Dimers have been obtained by this pathway e.g. from acridizinium and benzothiazolium salts [116]. Such processes are particularly important both with biological substrates (e.g. NAD is reported to give essentially 4,4'-dimers on reduction) [123] and in the monoelectronic reduction of di-cations such as dipyridinium and a variety of diaza polycyclic aromatic di-cations [124].

Alternatively, the radical anions of pyridine and other azines are obtained on reduction by metals or other reducing agents, e.g. lithium diethylamide. Under suitable conditions the salts of the radical anions can be obtained as crystalline materials (Section 6.2.3). Alternatively, dimerization follows; as an example, treatment of pyridine and other azines with 1 equiv. sodium in HMPA gives the well characterized radical anion, whereas in tetrahydrofuran the dimeric dianion is formed [125]. A later study with pyridine showed that treatment with sodium leads to a tetrahydro-4,4-bipyridine dianion (**57**), which is rearomatized to yield **58** in the presence of excess sodium (Scheme 40) [126]. Treatment with LiNEt$_2$ gives 2,2'-bipyridine, however, possibly because of stronger coordination with the lithium cation [127]. 2,2'-Biquinoline and 1,1'-biisoquinoline are similarly obtained [128].

The procedure has been extended to heterocoupling. As an example, heating quinoline and indole in the presence of a metal gives biquinolines and indolylqui-nolyines (occasionally accompanied by the corresponding 1,2-dihydroderivatives; Scheme 41). The result depends on the metal chosen, e.g. with copper only 2,2'-biquinoline is formed whereas with lithium a 2-(3-indolyl)quinoline is by far the main product [129]. A SET step might be involved in other hetero-coupling reactions via metalated heterocycles [130].

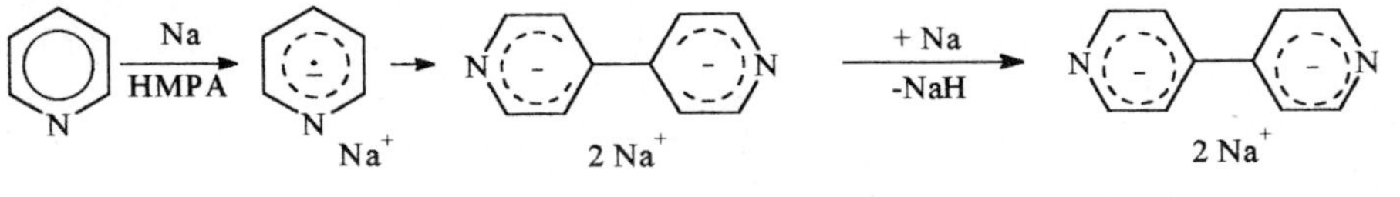

**57**                    **58**

**Scheme 40.**

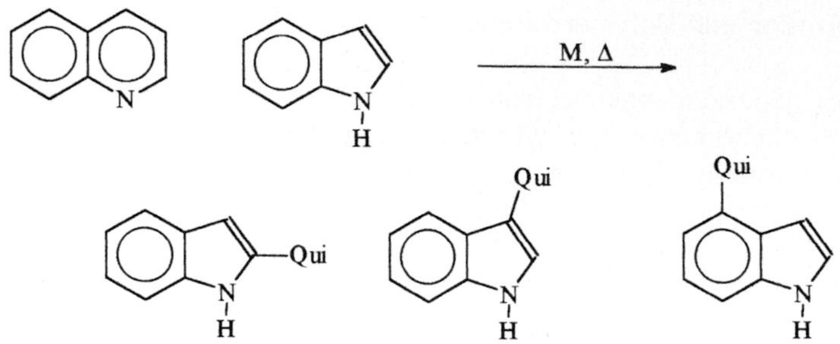

| | | |
|---|---|---|
| M = Li | 13% | 90 | - |
| M = Na | 3.5% | 26.5 (+ 12.5 1,2-dihydro) | 9 |

(+ biquinolynes)

**Scheme 41.**

A peculiar example of heterocoupling has been found in the electroreduction of a mixture of pyrrole (**60**) and 2,4,7-trinitrofluoren-9-one (**59**), which leads to the formation of two Meisenhemer anions (the major isomer (structure **61**) from attack at position 3, the minor one from attack at position 1) arguably via coupling of the two radical anions (see Scheme 42) [131]. Electro-oxidation of such anions gives a black polypyrrole film which differs from that obtained by the normal electro-oxidative polymerization of pyrrole (compare Section 6.3.4).

### 6.4.4  Direct Ring Functionalization

ET-initiated alkylation of azines in the presence of alkyl halides has been found to occur under electrochemical conditions according to Scheme 43. Either alkylated dihydroheterocycles or the rearomatized products are obtained. As an example, electroreduction of quinoline in the presence of 1-bromoadamantane gives alkylated quinolines (10 % in position 2, 5 % in position 7) and 2-methyl- and 2-methoxyquinoline give the 7-adamantyl derivative in 20 and 23 % yield, respectively. 1,10-Phenanthroline gives bis(adamantyl) derivatives, whereas isoquinoline and phenanthridine give the 6-adamantyl-5,6-dihydro and the 9-adamantyl-9,10-dihydro derivatives, respectively [132]. Reaction with *tert*-butyl chloride also gives alkylated dihydroheterocycles [133]. Arylation has also been performed, e.g. in the synthesis of pyrazolophenanthridines (**62**, Scheme 44) [134].

Pyridinium and related cations are well known as excellent radical traps, yielding, e.g., conveniently alkylated derivatives by addition of radicals generated by oxidants from the appropriate precursors [135]. Arylation of such derivatives can be efficiently performed at the cathode, as shown in Scheme 45 with an intermolecular example affording a precursor of cularine (**63**) [136] and with an intramolecular example giving aporphine (**64**) [137].

Under irradiation the cation itself can function as acceptor and generate the radicals. Addition of the radicals, particularly bulky *tert*-butyl radicals, is reversible and the yield of the end-products depends on competition between subsequent re-

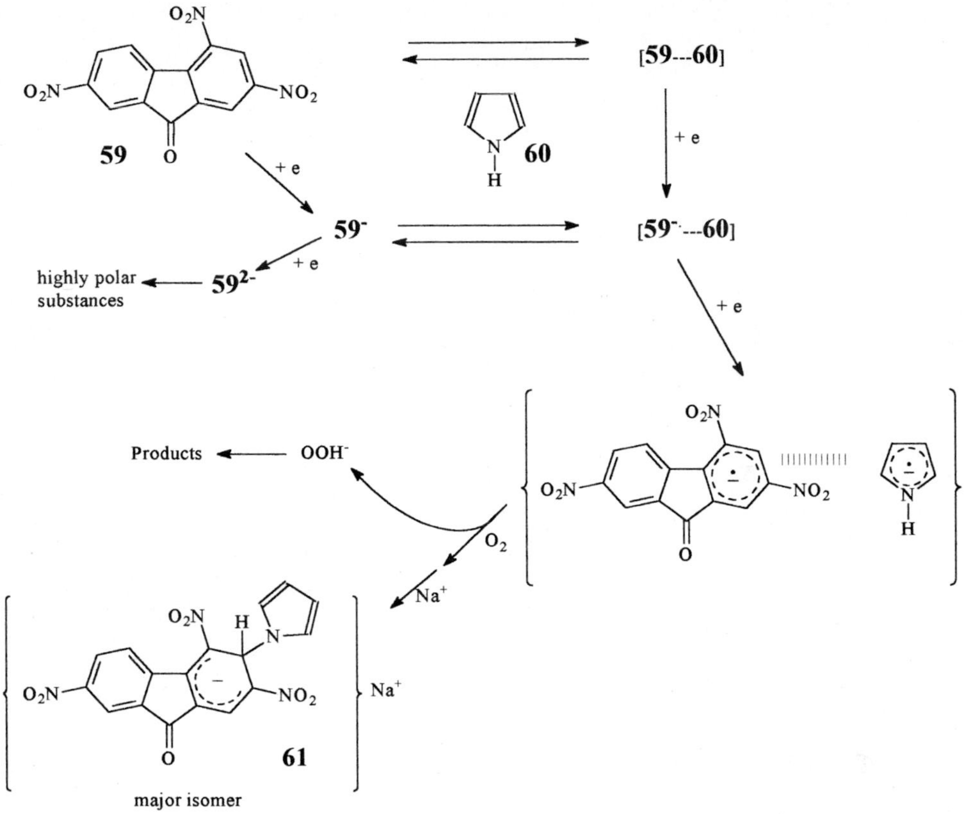

**Scheme 42.**

**Scheme 43.**

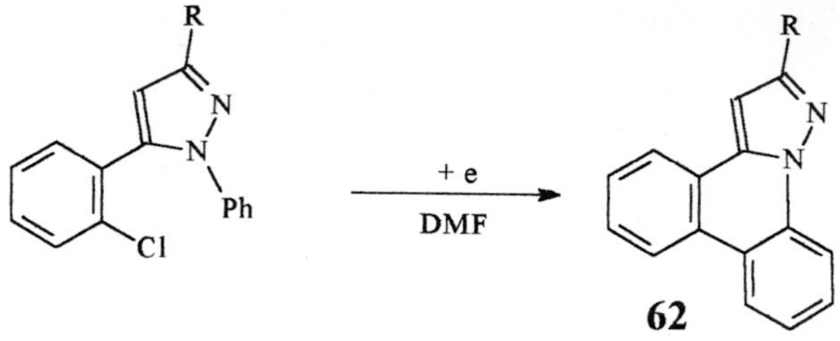

**62**

**Scheme 44.**

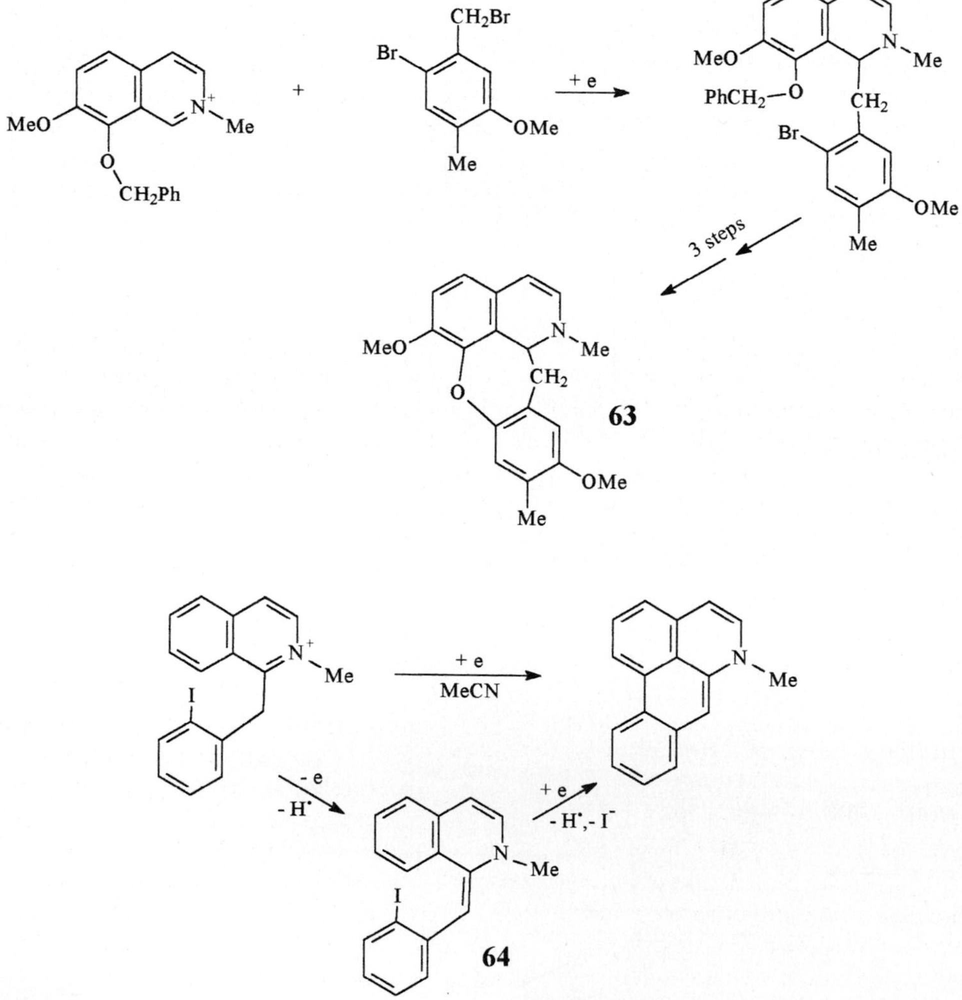

3 steps

**63**

**64**

**Scheme 45.**

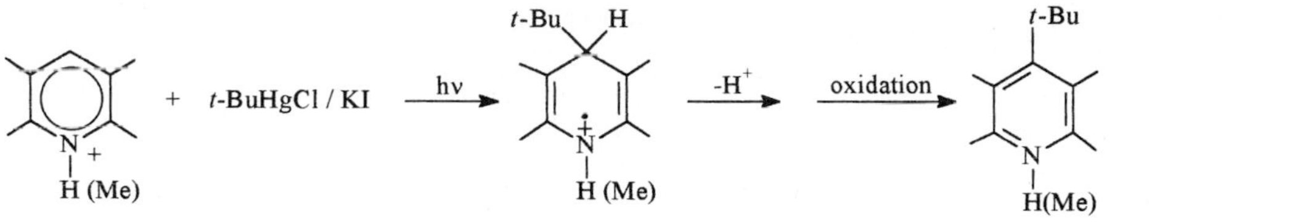

**Scheme 46.**

actions, viz. proton loss and oxidation of the resulting radical to give the rearomatized heterocycle or reduction to give a dihydroheterocycle. Under these conditions good yields of alkylated pyridines, quinolines, isoquinolines, or acridines have been obtained from the heterocycles by irradiation in the presence of alkyl (usually *tert*-butyl) mercury halides and acids (Scheme 46) [138]. Interestingly, 4-vinylpyridine gives 2-(2-alkylethyl) derivatives under these conditions. Pyrilyum salts are also conveniently alkylated to the corresponding 2- or 4-alkylpyrans by irradiation in the presence of Group-4 organometallic compounds [139].

It has been suggested that the photochemical hydroxyalkylation of pyridine and other azines in acidic alcohols likewise proceeds through an ET path [140].

## 6.4.5 Cleavage of a Group

SET promoted dehalogenation is occasionally of interest. Treatment of 2-chloro-, 2-bromo- or 2-iodopyridine with sodium in liquid ammonia gives directly the EPR-detectable radical anion of the parent molecule, dehalogenation being rapid on the time-scale of the experiment [141]. The same is true for a variety of dichloro- and dibromopyridines and pyrimidines, but with 2-fluoropyridine the radical anion is more persistent and only the spectrum of the starting material is registered. For the cathodic reduction of pentachloropyridine, see Section 6.4.6.

In this connection it is noteworthy that smooth defluorination of some fluoroquinolones (used as antibacterial drugs) is obtained both at the cathode and by photolysis in the presence of sodium sulfite in aqueous solution, as it is shown for enoxacin (**65**) in Scheme 47 [142]. The 6,8-difluoro derivative lomefloxacin is selectively defluorinated at position 8 by cathodic reduction. Electron capture of electrons by 5-halouracils in neutral organic and aqueous glasses leads to species characterized as $\sigma^*$ radical anions (more advanced in the path towards cleavage) with the iodo and bromo derivatives, whereas $\pi^*$ radical anions are formed from the chloro and fluoro derivatives [143]. The radical anion of 2-chlorothiophene (obtained by irradiation in the crystal state) is of the $\sigma^*$ type [144].

The cleavage of a side-chain can be useful. As an example, alkyl 1-pyrrolecarboxylates and the corresponding pyrazoles (**66a, b**; Scheme 48) lose the alkoxycarbonyl group upon reduction by magnesium [145] and the corresponding trityl derivatives (**66c**) are conveniently dealkylated upon Birch reduction [146]. The latter group has an important protecting–directing role on the chemistry of pyrrole derivatives; thus the availability of simple deprotections is preparatively useful.

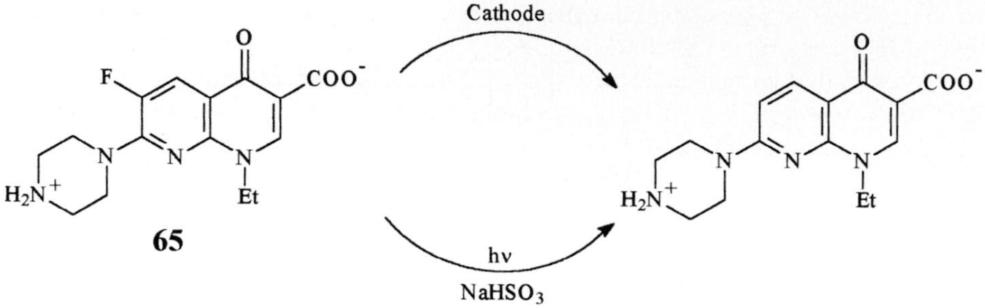

**Scheme 47.**

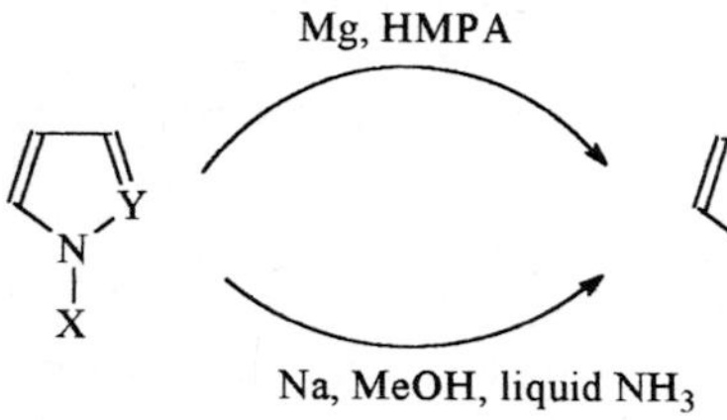

**66a**  Y = CH, X = COOMe
**66b**  Y = N, X = COOMe
**66c**  Y = CH,  X = CPh$_3$

**Scheme 48.**

An interesting fragmentation process involving an ET step is the reaction of the antibiotic toxoflavine (**67**) and some analogs with nucleophiles. This involves transfer of an *N*-methyl group to the nucleophile and formation in equilibrium of the toxoflavine radical anion, which has been characterized by EPR (see Scheme 49) [147].

A particular example of reductive group elimination is elimination of a *N*-oxide function. Because of the rich chemistry of azine *N*-oxide it is often expedient to carry over useful functionalizations at the *N*-oxide level and then to reduce such a

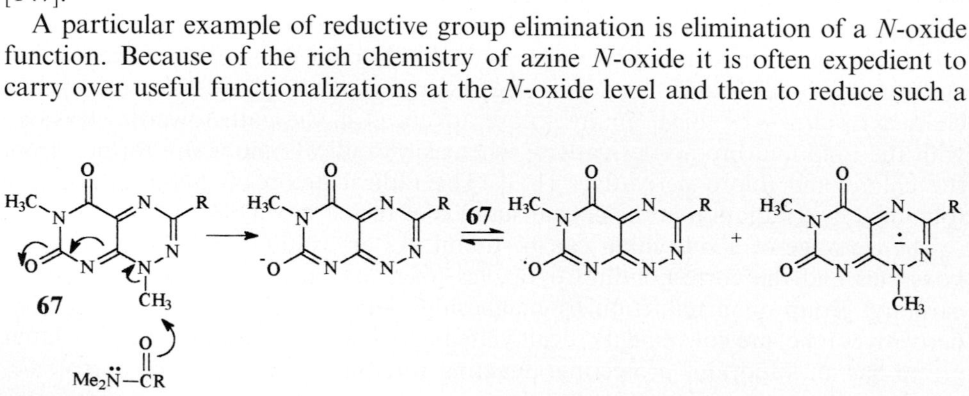

**Scheme 49.**

function selectively. One expedient method for *N*-deoxygenation is photoinduced ET to a donor such as an aliphatic amine, an alkene, or triethyl phosphite, a quite general procedure that usually leaves unchanged ring substituents, including the nitro group and suppresses the ring rearrangement otherwise observed on irradiation of the *N*-oxides [148].

### 6.4.6 Substitution of a Group

Most reactions involve halogen substitution. Many preparatively useful reactions are initiated by reductive dehalogenation of heterocycles. Cathodic reduction leads to different results with the pentafluoro- (**68a**) and pentachloropyridine (**68b**). In the first, radical anion coupling at the position of highest spin density ensues and finally gives octafluoro-4,4′-bipyridine (**69**), although in the presence of a proton source (hydroquinone) dehalogenation to **70** results (Scheme 50). With the latter substrate, fragmentation of the weaker C–Cl bond is the fastest process and thus 2,3,5,6-tetrachloropyridine (**71**) is the main product, although trapping by carbon dioxide is successful [149].

The $S_{RN}1$ reaction has been applied to heterocycles. Among five-membered ring compounds, halothiophenes have been the most studied; they have been shown to be susceptible to both electron-stimulated and photostimulated reactions, and have been converted to the corresponding acetonitriles [150], acetones [151], and phenyl-sulfides [152] in low to medium yields. In the study with the benzenethiolate anion it has been shown that the yield is low because of fragmentation of the adduct radical anion; it can be increased by adding an electron acceptor, e.g. benzonitrile which prevents decomposition. Further applications include the thermally activated $S_{RN}1$ reaction between 3-iodobenzothiophene and enolates [153] and the photo-stimulated reaction of 3-halo-2-aminobenzothiophenes [154].

The reaction has been more extensively used for azines, however. As an example,

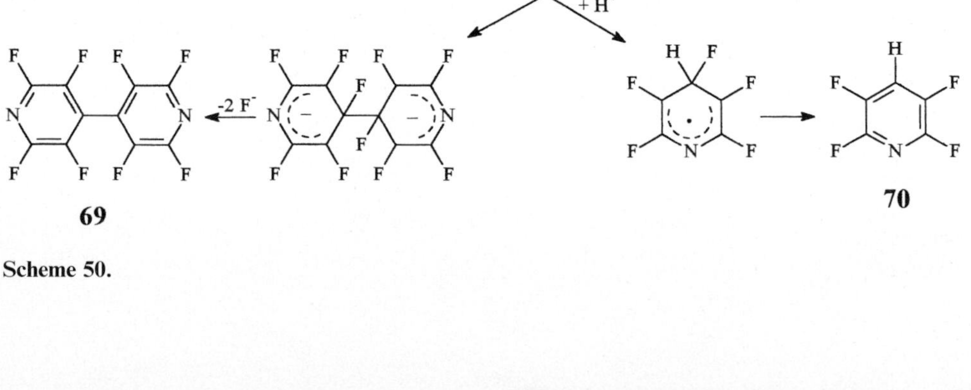

**68a** X = F
**68b** X = Cl

**71**

**69**

**70**

**Scheme 50.**

medium to good yields of substituted pyridylmethyl ketones are obtained by photostimulated reaction of 2-chloro-, 2-bromo-, and (although at a much lower rate) 2-fluoropyridine with the corresponding enolates [155]. Photostimulation is also effective in promoting the reaction of 2-chloroquinoline with enolates [156]. Heteroarylketones have been similarly prepared from chloropyrimidines [157] and pyrazines [158] and phenylsulfides and selenides have been prepared similarly from haloquinolines [159] and haloisoquinolines [160] via an $S_{RN}1$ reaction.

Non-chain functionalization of five-membered heterocycles to halogen derivatives via photoinduced SET has been developed by an Italian group [161]. In a typical reaction, irradiation of acylhalofurans [162], thiophenes [163], and pyrroles [164] in the presence of aromatic substrates gives rise to the corresponding arylated heterocycles (**72**; Scheme 51). The reaction is though to involve SET and fragmentation of the heteroaryl radical anion. Heteroarylation is also effective through this path and offers an expedient entry to bithiophenecarboxyaldehydes (**73**), in turn useful synthons for naturally occurring bithiophenes [165] and for other coupled heterocycles such as compound **74** [166] (Scheme 51).

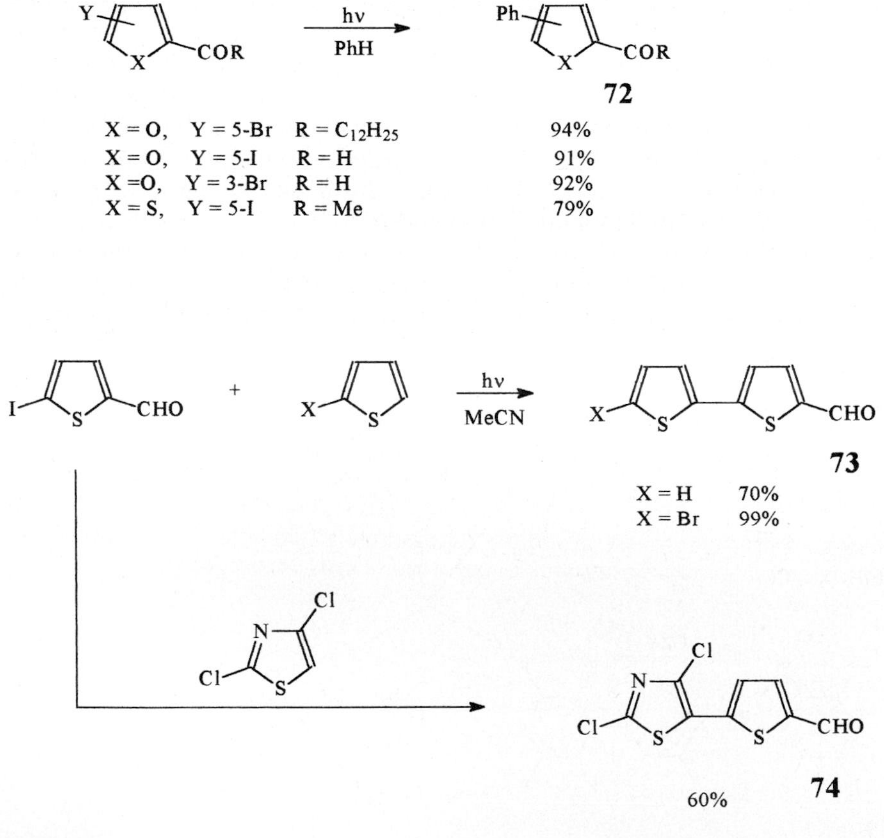
Y
COR
X
hν
PhH
Ph
COR
X
72
X = O, Y = 5-Br R = C12H25 94%
X = O, Y = 5-I R = H 91%
X = O, Y = 3-Br R = H 92%
X = S, Y = 5-I R = Me 79%
I
S
CHO
+
X
S
hν
MeCN
X
S
S
CHO
73
X = H 70%
X = Br 99%
Cl
N
Cl
S
Cl
N
Cl
S
S
CHO
60%
74

Scheme 51.

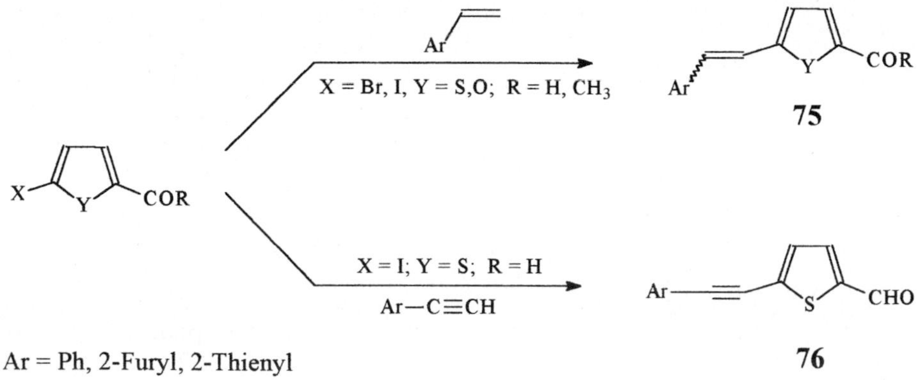

X = Br, I, Y = S,O; R = H, CH₃

$X = Br, I, \; Y = S,O; \; R = H, CH_3$

**75**

$X = I; \; Y = S; \; R = H$

Ar—C≡CH

**76**

Ar = Ph, 2-Furyl, 2-Thienyl

**Scheme 52.**

Irradiation of the same acylhaloheterocycles in the presence of (hetero)-arylalkenes and alkynes (**75**) leads to the corresponding bis(hetero)arylalkenes and alkynes (**76**) (Scheme 52) [167].

An ET mechanism is involved in the photochemical reaction of 2- and 4-cyanopiridine and other cyanoazines with alcohols, alkenes, or amines; substitution products are usually obtained. As an example, allylpyridines (and quinolines) are obtained from the acetone-sensitized reaction of the corresponding cyanohetero-cycles in the presence of alkenes (Scheme 53). The reactions of various dicyano-pyridines have also been studied in detail and the distribution of the products obtained is well rationalized with an ET mechanism [168].

Other good acceptors are functionalized by a similar mechanism when irradiated in the presence of donors; as an example, aminonitrophenazines are formed when nitrophenazines are irradiated in the presence of aliphatic amines [169].

### 6.4.7 Cycloaddition

Single electron donors such as sodium metal, sodium iodide, or sodium naph-thalenide (the last in an amount as low as 5 ‰ molar) convert triazolinediones to deaza dimers **77** via a reaction that has been formulated as a radical anion chain

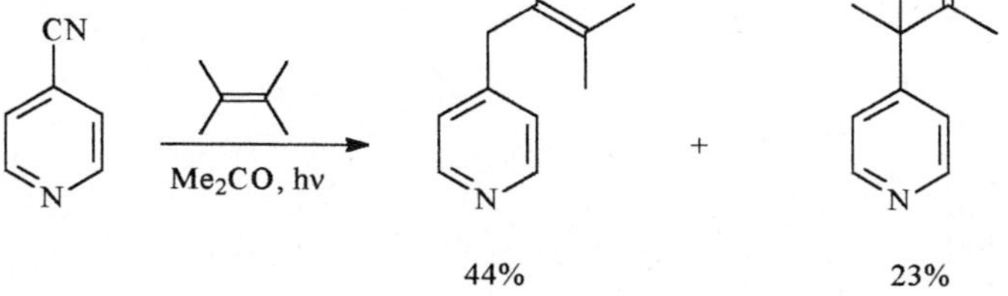

44%                23%

**Scheme 53.**

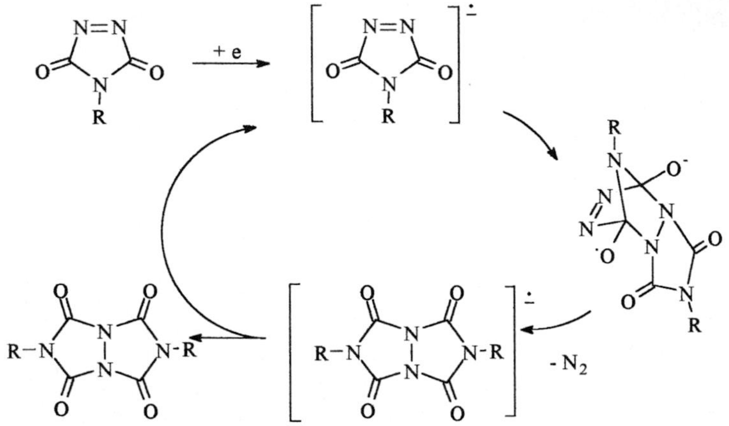

**77**

**Scheme 54.**

process based on a radical anion Diels–Alder reaction (Scheme 54). A similar re-action occurs with azadicarboxylates [170].

### 6.4.8 Ring Reduction

Selective cathodic reduction of pyridines is a process of great industrial significance [171]. As an example, dimethyl pyridinedicarboxylates undergo a highly selective electroreduction in methanol by use of a divided cell. The product obtained depends on the position of the substituents. Thus, the 2,3- and 2,5-dicarboxylates give the 1,2-dihydropyridines whereas the 2,6-, 3,4- and 2,4-dicarboxylates the 1,4-dihydropyridines [172]. Many other ring reductions of azines (often with dimeriza-tion as a side-reaction, see Section 6.4.3) are discussed in electrochemistry texts and reviews [116]. Attempts have been made to rationalize the herbicidal properties of dipyridinium salts in terms of their cathodic behavior [173].

Another process worth mentioning is the cathodic reduction of the bis-(quinoxalino)cyclobutane **78** (Scheme 55) [174].

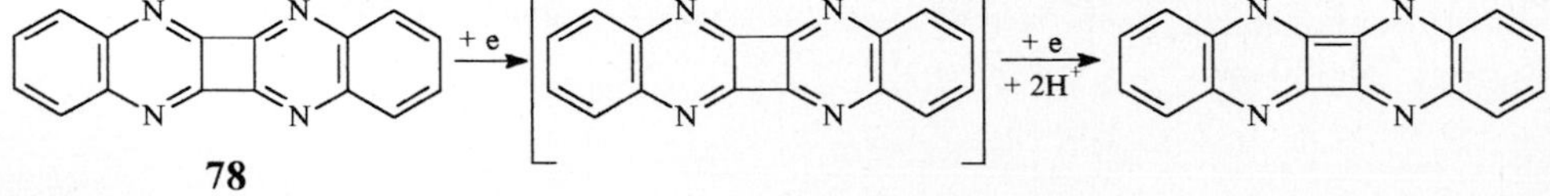

**78**

**Scheme 55.**

## 6.5 Conclusions and Outlook

We believe it is apparent from the discussion above that heteroaromatic compounds are in no way minor players in the field of electron transfer. These compounds are more variable in their redox properties than their homocyclic counterparts and the presence of heteroatoms has a substantial effect on the electron distribution of the resulting odd-electron species, and this is an important source of selectivity in the ensuing chemical reactions. Perhaps in this field more than in any other field of organic chemistry it is apparent that the study of reactions involving electron transfer has been confronted in different laboratories and times, with different techniques and intellectual approaches, hardly realizing the benefit of an interdisciplinary approach. It is hoped that this contribution centered on the reactions of organic molecules, coupled with other, more physically oriented, contributions in this book, might help foster new interest in this topic. The impression is that we are under-exploiting the possibilities of electron transfer in the chemistry of heterocyclic compounds and this should be one of the main topics of development in the near future, not least because of its relevance to biological systems and its technical application.

### Acknowledgment

We are grateful to Drs L. Cermenati and M. G. Uggetti for their help in the literature search. The work from our laboratory quoted here was supported by CNR, Rome, and MURST, Rome).

### References

1. G. R. Newkome, W. W. Paudler, *Contemporary Heterocyclic Chemistry*, Wiley, New York, **1982**, p. 1.
2. (a) P. Hanson, *Adv. Heterocycl. Chem.* **1979**, *25*, 205–301; (b) P. Hanson, *ibid.*. **1980**, *27*, 31–149; (c) A. S. Morkovnik, O. Y. Okhlobystin, *Khim. Geterotsikl. Soedin.* **1980**, 1011–1029.
3. F. Barbosa, L. Eberson, G. Gescheidt, S. Gronowitz, A. B. Hörnfeldt, L. Julia, O. Persson, *Acta Chem. Scand.* **1998**, *52*, 1275–1284.
4. R. Rathore, J. K. Kochi, *Acta Chem. Scand.* **1998**, *52*, 114–130.
5. A. G. Davies, L. Julia, S. N. Yazdi, *J. Chem. Soc., Perkin Trans. 2* **1989**, 239–244.
6. A. S. Markovnick, *Zh. Obshsh. Khim.* **1982**, *52*, 1877–1883.
7. (a) A. J. Bard, A. Ledwith, H. J. Shine, *Adv. Phys. Org. Chem.* **1976**, *13*, 155–278; (b) H. J. Shine, *ACS Symp. Ser.* **1978**, *69*, 359–375.
8. W. Lau, J. K. Kochi, *J. Am. Chem. Soc.* **1984**, *106*, 7100–7112.
9. L. Eberson, M. P. Hartshorn, O. Persson, *J. Chem. Soc., Perkin Trans. 2* **1995**, 1735–1744.
10. A. G. Davies, *Chem. Soc. Rev.* **1993**, *22*, 299–304.
11. L. Eberson, M. P. Hartshorn, O. Persson, F. Radner, *Chem. Commun.* **1996**, 2105–2112.
12. P. Linda, G. Marino, S. Pignataro, *J. Chem. Soc. (B)* **1971**, 1585–1587.
13. D. N. R. Rao, G. W. Eastland, M. C. R. Symons, *J. Chem. Soc., Faraday Trans. 1* **1984**, *80*, 2803–2815.
14. H. Fischer, F. A. Neugebauer, H. Chandra, M. C. R. Symons, *J. Chem. Soc., Perkin Trans. 2* **1989**, 727–730.

15. H. Fischer, T. Müller, I. Umminger, F. A. Neugebauer, H. Chandra, M. C. R. Symons, *J. Chem. Soc. Trans. 2* **1988**, 413–421.
16. C. Aussems, S. Jaspers, G. Leroy, F. Van Remoortere, *Bull. Soc. Chim. Belges* **1969**, *78*, 407–420.
17. D. N. Ramakrishna Rao, M. C. R. Symons, *J. Chem. Soc., Perkin Trans. 2* **1983**, 135–137.
18. D. W. Turner, *Adv. Phys. Org. Chem.* **1966**, *4*, 31–71.
19. K. Watanabe, *J. Chem. Phys.* **1957**, *26*, 542–547.
20. M. H. Palmer, R. H. Findlay, R. G. Edgell, *J. Mol. Struct.* **1977**, *40*, 191–210.
21. T. Kobayashi, T. Kubota, K. Ezumi, C. Utsunomiya, *Bull. Chem. Soc. Jpn.* **1982**, *55*, 3915–3919.
22. C. Aussems, S. Jaspers, G. Leroy, F. Van Remoortere, *Bull. Soc. Chim. Belges* **1969**, *78*, 479–485.
23. J. R. Gleiter, E. Heilbronner, V. Hornung, *Helv. Chim. Acta* **1972**, *55*, 255–274.
24. F. Brogli, E. Heilbronner, T. Kobayashi, *Helv. Chim. Acta* **1972**, 274–288. T. Kato, T. Shida, *J. Am. Chem. Soc.* **1979**, *101*, 6879–6876.
25. J. Chaudhuri, S. Kuhe, J. Jagur-Grodzinki, M. Szwarc, *J. Am. Chem. Soc.* **1968**, *90*, 6421–6425.
26. A. Modler-Spreitzer, A. Mannschreck, M. Scholz, G. Gescheidt, H. Spreitzer, J. Daub, *J. Chem. Res. (S)* **1995**, 180–181, *(M)*, 1229–1251.
27. P. D. Burrow, A. J. Ashe III, D. J. Bellville, K. D. Jordan, *J. Am. Chem. Soc.* **1982**, *104*, 425–429.
28. (a) R. Pummerer, S. Gaßner, *Ber. Dtsch. Chem. Ges.* **1913**, *46*, 2310–2327; (b) F. Kehrmann, L. Diserens. *Ber. Dtsch. Chem. Ges.* **1915**, *48*, 318–328.
29. S. Ogawa, M. Sugawara, Y. Kawai, S. Niizuma, T. Kimura, R. Sato, *Tetrahedron Lett.* **1999**, *40*, 9101–9106.
30. Y. Murata, H. J. Shine *J. Org. Chem.* **1969**, *34*, 3368–3372.
31. S. Ogawa, S. Saito, T. Kikuchi, Y. Kawai, S. Niizuma, R. Sato, *Chem. Lett.* **1995**, 321–322.
32. (a) S. Ogawa, S. Nobuta, R. Nakayama, Y. Kawai, S. Niizuma, R. Sato, *Chem. Lett.* **1996**, 757–758. (b). S. Ogawa, T. Ohmiya, T. Kikuchi, Y. Kawai, S. Niizuma, R. Sato, *Heterocycles* **1996**, *43*, 1843–1846.
33. S. Ogawa, T. Kikuchi, S. Niizuma, R. Sato, *J. Chem. Soc., Chem. Commun.* **1994**, 1593–1594.
34. (a) K. Lühdler, *Z. Chem.* **1967**, *7*, 198–199; (b) K. Lühdler, H. Langanke, *ibid.* **1970**, *10*, 74–75.
35. R. Rathore, J.–K. Kochi, *Acta Chem, Scand.* **1998**, *52*, 114–130.
36. L. Eberson, M. P. Hartshorn, O. Persson, F. Radner, *Chem. Commun.* **1996**, 2105–2112.
37. (a) U. Svanholm, V. D. Parker, *J. Chem. Soc., Perkin Trans. 2* **1976**, 1567–1574; (b) V. D. Parker, *Acc. Chem. Res.* **1984**, *17*, 243.
38. (a) A. I. Novaira, C. D. Borsarelli, J. J. Cosa, C. M. Previtali, *J. Photochem. Photobiol. A*, **1998**, *115*, 43–47; (b) C. D. Borsarelli, H. A. Montejano, J. J. Cosa, C. M. Previtali, *ibid.* **1995**, *91*, 13–19; (c) H. A. Montejano, J. J. Cosa, H. A. Garrera, C. M. Previtali, *ibid.* **1995**, *86*, 115–120.
39. (a) V. D. Parker, Y. T. Chao, G. Zheng, *J. Am. Chem. Soc.* **1997**, *119*, 11390. (b) B. Reitstöen, V. D. Parker, *J. Am. Chem. Soc.* **1991**, *113*, 6954–6958; (c)V. D. Parker, Y. Zhao, Y. Lu, G. Zheng, *ibid.* **1998**, *120*, 12720–12727.
40. D. Lavorato, J. K. Terlouw, T. K. Dargel, W. Koch, G. A. McGibbon, H. Schwartz, *J. Am. Chem. Soc.* **1996**, *118*, 11898–11904.
41. J. H. Holmes, J. K. Terlouw, *J. Am. Chem. Soc.* **1979**, *101*, 4973–4975.
42. (a) W. Adam, C. Sahin, J. Sendelbach, H. Walter, G. F. Chen, F. Williams, *J. Am. Chem. Soc.* **1994**, *116*, 2576–2584; (b) W. Adam, U. Denninger, R. Finzel, F. Kita, H. Platsch, H. Walter, G. Zang, *J. Am. Chem. Soc.* **1992**, *114*, 5027–5035; (c) W. Adam, T. Heidenfelder, *Chem. Soc. Rev.* **1999**, *28*, 359–365.
43. C. T. Pedersen, O. Hammerich, V. D. Parker, *J. Electroanal. Chem.* **1972**, *38*, 479–481.
44. A. De Mico, R. Margarita, G. Piancatelli, *Tetrahedron Lett.* **1995**, *36*, 3553–3556.
45. (a) C. P. Andrieux, L. Nadjo, J. M. Savéant, *J. Electroanalytical. Chem.* **1970**, *26*, 147–186; (b) L. Nadjo, J. M. Savéant, *J. Electroanal. Chem.* **1973**, *44*, 327–366.
46. M. L. Andersen, M. F. Nielsen, O. Hammerich, *Acta Chem. Scand.* **1997**, *51*, 94–107.
47. I. Tabakovic, Y. Kunugi, A. Canavesi, L. L. Miller, *Acta Chem. Scand.* **1998**, *52*, 131–136.

48. C. T. Pedersen, V. D. Parker, *Tetrahedron Lett.* **1972**, 771–772.
49. C. T. Pedersen, V. D. Parker, O. Hammerich, *Acta Chem. Scand.* **1976**, *B 30*, 478–84.
50. J. Tormo, F. J. Moreno, J. Ruiz, L. Fajari, L. Julia, *J. Org. Chem.* **1997**, *62*, 878–884.
51. G. T. Cheek, R. F. Nelson, *J. Org. Chem.* **1978**, *43*, 1230–1232.
52. B. K. Bandlish, H. J. Shine, *J. Org. Chem.* **1977**, *42*, 561–563.
53. J. F. Ambrose, R. F. Nelson, *J. Electrochem. Soc.* **1968**, *115*, 1159–1164.
54. L. Eberson, M. P. Hartshorn, O. Persson, F. Radner, *Acta Chem Scand.* **1997**, *51*, 492–500.
55. Y. Tsujino, *Tetrahedron Lett.* **1969**, 763–766.
56. (a) H. J. Shine, S. M. Wu, *J. Org. Chem.* **1979**, *44*, 3310–3316; B) H. Musso, *Chem. Ber.* **1959**, *92*, 2862–2873; (c) H. Musso, *ibid.* **1959**, *92*, 2873–2871; (d) Y. Tsujino, *Tetrahedron Lett.* **1968**, 4111–4114.
57. C. Berti, L. Greci, R. Andruzzi, A. Trazza, *J. Org. Chem.* **1985**, *50*, 368–373.
58. (a) A. F. Diaz, K. K. Kanazawa, G. P. Gardini, *J. Chem. Soc., Chem. Commun.* **1979**, 635–636; (b) K. K. Kanazawa, A. F. Diaz, R. H. Geiss, W. D. Gill, J. F. Kwak, J. A. Logan, J. F. Rabolt, G. B. Street, *ibid.* **1979**, 854–855.
59. L. Eberson, M. P. Hartshorn, F. Radner, O. Persson, *J. Chem. Soc., Perkin Trans. 2* **1998**, 59–70.
60. M. V. Jovanovic, E. R. Biehl, *J. Org. Chem.* **1984**, *49*, 1905–1908.
61. H. Chiou, P. C. Reeves, E. R. Biehl, *J. Heterocycl. Chem.* **1976**, *13*, 77–82.
62. P. Carloni, E. Damiani, L. Greci, P. Stipa, *Acta Chem. Scand.* **1998**, *52*, 137–140.
63. F. Radner, *J. Org. Chem.* **1988**, *53*, 3548–3553.
64. J. Burden, I. W. Parsons *Tetrahedron* **1980**, *36*, 1423–1433.
65. (a) J. H. Ridd, *Chem. Soc. Rev.* **1991**, *20*, 149–165; (b) A. S. Morkovnik, Uspekhi Khimii **1988**, *57*, 254–280; (c) L. Eberson, F. Radner, *Acta Chem. Scand.* **1984**, *B 38*, 861–870; (d) L. G. Gorb, I. A. Sbronini, L. Korsukov, G. M. Zhidemirov, J. P. Litvinov, *Isv. Ak. Nauk SSSR, Ser. Khim.*, **1984**, 1079–1085.
66. (a) A. S. Morkovnick, N. M. Dobaeva, V. B. Panov, O. Y. Okhlobystin, *Dokl. Akad. Nauk SSSR* **1980**, *251*, 125–8; (b) V. G. Koshechko, A. N. Inozemtsev, V. D. Pokhodenko, *Zhur. Org. Khim.* **1981**, *17*, 2608–2612.
67. A. S. Morkovnick, N. M. Dobaeva, O. Y. Okhlobystin, *Khim. Geterotsikl. Soedin.* **1981**, 1214–1216.
68. A. S. Morkovnick, E. Y. Belinskii, N. M. Dobaeva, O. Y. Okhlobystin, *Zhur. Org. Khim.* **1982**, *18*, 378–386.
69. A. S. Morkovnick, N. M. Dobaeva, O. Y. Okhlobystin, *Khim. Geteotsikl. Soed.* **1983**, 122–123.
70. (a) C. P. Butts, L. Eberson, M. P. Hartshorn, W. T. Robinson, B. R. Wood, *Acta Chem. Scand.* **1996**, *50*, 587–595; (b) C. P. Butts, L. Eberson, M. P. Hartshorn, F. Radner, B. R. Wood, *ibid.* **1997**, *51*, 476–482.
71. C. P. Butts, L. Eberson, R. Gonzalez-Luque, C. M. Hartshorn, M. P. Hartshorn, M. Merchan, W. T. Robinson, B. O. Roos, C. Vallance, B. R. Wood, *Acta Chem. Scand.* **1997**, *51*, 984–999.
72. C. P. Butts, L. Eberson, M. P. Hartshorn, F. Radner, W. T. Robinson, B. R. Wood, *Acta Chem. Scand.* **1997**, *51*, 839–848.
73. A. S. Morkovnik, E. Y. Belinski, O. Y. Okhlobystin, *Zhur. Org. Khim.* **1979**, *15*, 1565.
74. C. Barry, G. Cauquis, M. Maurey, *Bull. Soc. Chim. Fr.* **1966**, 2510–2516.
75. H. J. Shine, Y. Murata, *J. Am. Chem. Soc.* **1969**, *91*, 1872–1874.
76. V. D. Parker, L. Eberson, *J. Am. Chem. Soc.* **1970**, *92*, 7488–7489.
77. H. J. Cheng, P. H. Sackett, R. L. McCreery, *J. Am. Chem. Soc.* **1978**, *100*, 962–967.
78. S. R. Mani, H. J. Shine, *J. Org. Chem.* **1975**, *40*, 2756–2758.
79. K. Kim, H. J. Shine, *J. Org. Chem.* **1974**, *39*, 2537–2539.
80. U. Svanholm, V. D. Parker, *J. Am. Chem. Soc.* **1976**, *98*, 997–1001.
81. J. J. Silber, H. J. Shine, *J. Org. Chem.* **1971**, *36*, 2923–2926.
82. (a) K. Iwai, H. J. Shine, *J. Org. Chem.*, **1981**, *46*, 271–276; (b) H. J. Shine, B. K. Bandlish, S. R. Mani, A. G. Padilla, *ibid.*, **1979**, *44*, 915–917; (c) W. K. Lee, B. Liu, C. W. Park, H. J. Shine, J. Y. Guzman-Jimenez, K. H. Whitmire, *ibid.*, **1999**, *64*, 9206–9210, (d) A. Hoomam, D. Shukla, H. B. Kraatz, D. D. M. Wayner, *ibid.* **1999**, *64*, 3342–3345.

83. K. Kim, H. J. Shine, *Tetrahedron Lett.* **1974**, 4413–4416.
84. (a) H. J. Shine, J. J. Silver, R. J. Bussey, T. Okuyama, *J. Org. Chem.* **1972**, *37*, 2691–2697; (b) J. F. Evans, H. N. Blount, *ibid.* **1977**, *42*, 976–982; (c) J. F. Evans, J. R. Lenhard, H. N. Blount, *ibid.* **1977**, *42*, 983–988.
85. K. Yoshida, *J. Am. Chem. Soc.* **1979**, *101*, 2116–2121.
86. N. L. Weinberg, E. A. Brown, *J. Org. Chem.* **1966**, *31*, 4054–4058.
87. (a) A. Filippi, G. Occhiucci, M. Speranza, *Can. J. Chem.* **1991**, *69*, 732–739; (b) A. Filippi, G. Occhiucci, C. Sparapani, M. Speranza, *Can. J. Chem.* **1991**, *69*, 740–748.
88. (a) T. Kubota, H. Sakurai, *Chem. Lett.* **1972**, 923–926; (b) T. Kubota, H. Sakurai, *ibid..* **1972**, 1249–1250.
89. K. Mizuno, M. Ishii, Y. Otsuji, *J. Am. Chem. Soc.* **1981**, *103*, 5570–5572.
90. (a) J. J. McCullough, W. S. Wu, C. W. Huang, *J. Chem. Soc., Perkin Trans. 2* **1972**, 370–375; (b) P. Grandclaudon, A. Lablache-Combier, C. Parkanyi, *Tetrahedron* **1973**, *29*, 651–658.
91. (a) M. D'Auria, E. De Luca, V. Esposito, G. Mauriello, R. Racioppi, *Tetrahedron* **1997**, *53*, 1157–1166; (b) M. D'Auria, *Tetrahedron* **1991**, *47*, 9225–9230.
92. H. Sawada, M. Yoshida, H. Hagii, K. Aoshima, M. Kobayashi, *Bull. Chem. Soc. Jpn.* **1986**, *59*, 215–219.
93. M. Yoshida, T. Yoshida, M. Kobayashi, N. Kamigata, *J. Chem. Soc., Perkin Trans. 1* **1989**, 909–914.
94. Q. Y. Chen, Z. T. Li, *J. Chem. Soc., Perkin Trans 1* **1993**, 645–648.
95. (a) Q. Y. Chen., Z. M. Qiu, *J. Fluorine Chem.* **1988**, *39*, 289–292; (b) Q. Y. Chen, Z. M. Qiu, *Hyaxue Xuebao*, **1988**, *46*, 258–263 through *Chem. Abstr.* **1989**, *110*, 75218a.
96. S. Naruto, O. Yonemitsu, *Tetrahedron Lett.* **1971**, 2297–2300.
97. (a) Q. Y. Chen, Z. T. Li, *J. Chem. Soc., Perkin Trans 1* **1993**, 1705–1710; (b) Q.Y. Chen, Z. T. Li, *J. Org. Chem.* **1993**, *58*, 2599–2604.
98. S. Ito, I. Saito, T. Matsuura, *J. Am. Chem. Soc.* **1980**, *102*, 7535–7541.
99. T. Majima, C. Pac, A. Nakasone, H. Sakurai, *J. Am. Chem. Soc.* **1981**, *103*, 4499–4508.
100. H. Fischer, A. Treibs, E. Zaucker, *Chem. Ber.* **1959**, *92*, 2026–2029.
101. K. Saito, Y. Horie, *Heterocycl.* **1986**, *24*, 579–582.
102. (a) K. Maruyama, T. Otsuki, *Chem. Lett.* **1977**, 851–852; (b) K. Maruyama, T. Otsuki, *Bull. Chem. Soc. Jpn.* **1977**, *50*, 3429–3430; (c) K. Maruyama, T. Otsuki, H. Tamiaki, *ibid.* **1985**, *58*, 3049–3050.
103. (a) K. Maruyama, K. Mitsui, T. Otsuki, *Chem. Lett.* **1978**, 323–324; (b) K. Maruyama, T. Otsuki, S. Tai, *J. Org. Chem.* **1985**, *50*, 52–60.
104. (a) A. Gieseler, E. Steckhan, O. Wiest, F. Knoch, *J. Org. Chem.* **1991**, *56*, 1405–1411; (b) O. Wiest, E. Steckhan, F. Grein, *ibid.* **1992**, *57*, 4034–4037; (c) A. Gieseler, E. Steckhan, O. Wiest, *Synlett.* **1990**, 275–277.
105. O. Wiest, E. Steckhan, *Tetrahedron Lett.* **1993**, 6391–6394; T. Peglow, S. Blechert, E. Steckhan, *Chem. Commun.* **1999**, 433–434.
106. Y. Yoshioka, S. Yamanaka, S. Yamada, T. Kawakami, M. Nishino, K. Yamaguchi, A. Nishinag, *Bull. Chem. Soc. Jpn.* **1996**, *69*, 2701–2722.
107. L. Cermenati, P. Pichat, C. Guillard, A. Albini, *J. Chem. Phys. B* **1997**, *101*, 2650–2658.
108. (a) T. Ignasiak, A. V. Kemp-Jones, O. P. Stausz, *J. Org. Chem.* **1977**, *42*, 312–320; (b) K. Chatterjee, L. M. Stock, M. L. Gorbaty, G. N. George, S. R. Kelemen, *Energy Fuels* **1991**, *5*, 771–773.
109. J. J. Eish, L. E. Hallenbeck, K. I. Han, *J. Am. Chem. Soc.* **1986**, *108*, 7763–7767.
110. (a) A. N. Statrsev, *Kinetic Catal.* **1995**, *36*, 471–478; (b) S. W. Oliver, T. D. Smith, J. R. Pilbrow, K. C. Pratts, V. Christov, *J. Catal.* **1988**, *111*, 88–93.
111. H. Lund, P. Lunde, *Acta Chem. Scand.* **1967**, *21*, 1067–1080.
112. H. Lund, E. T. Jensen, *Acta Chem. Scand.* **1970**, *24*, 1867–1877.
113. K. Saito, H. Kojima, T. Okudaira, K. Takahashi, *Bull. Chem. Soc. Jpn.* **1983**, *56*, 175–179.
114. I. Tabakovic, M. Trkovnik, Z. Grujic, *J. Chem. Soc., Perkin Trans. 2,* **1979**, 166–171.
115. A. Stanienda, G. Biebl, *Z. Phys. Chem.* **1967**, *52*, 254–275.
116. (a) H. Lund, *Adv. Heterocycl. Chem.* **1970**, *12*, 213–316; (b) H. Lund, I Tabakovic, *ibid.* **1984**, *36*, 237–316.
117. D. L. Smith, P. J. Elving, *J. Am. Chem. Soc.* **1962**, *84*, 2741–2747.

118. H. Lund, *Österr. Chemiker Z.* **1967**, *68*, 43–53.
119. H. Lund, *Acta Chem. Scand.* **1964**, *18*, 1984–1995.
120. P. Pflegel, G. Wagner, *Z. Chem.* **1969**, *9*, 151–152.
121. T. Nonaka, T. Kato, T. Fuchigani, T. Sekine, *Electrochim. Acta* **1981**, *26*, 887–92.
122. S. I. Zhdanov, L. S. Mirkin, *Coll. Czech. Chem. Commun.* **1961**, *26*, 370–379.
123. H. Jaegfeldt, *Bioelectrochem. Bioenerg.* **1981**, *8*, 355–370.
124. S. Hünig, J. Gross, *Tetrahedron Lett.* **1968**, 2599–2604.
125. J. Chaudhuri, S. Kume, J. Jagur-Grodzinski, M. Szwarc, *J. Am. Chem. Soc.* **1968**, *90*, 6421–6425.
126. K. Lühdler, H. Füllbier, *Z. Chem.* **1988**, *28*, 402–404.
127. G. R. Newkome, D. C. Hager, *J. Org. Chem.* **1982**, *47*, 599–601.
128. A. J. Clarke, S. McNamara, O. Meth-Cohn, *Tetrahedron Lett.* **1974**, 2373–2376.
129. A. K. Sheinkman, V. A. Ivanov, N. A. Klyuev, G. A. Mal'tseva, *Zhur. Org. Khim.* **1973**, *9*, 2550–2560.
130. T. Kauffmann, *Angew. Chem. Int. Ed. Engl.* **1979**, *18*, 1–19.
131. K. Oshino, N. Ozawa, H. Kokado, H. Seki, T. Tokunaga, T. Ishikawa, *J. Org. Chem.* **1999**, *64*, 4572–4573.
132. U. Hess, D. Huhn, H. Lund, *Acta Chem. Scand.* **1980**, *B34*, 413–417.
133. C. Degrand, H. Lund, *Acta Chem. Scand.* **1977**, *B31*, 593–598.
134. J. Grimshaw, J. Trocha-Grimshaw, *Tetrahedron Lett.* **1974**, 993–996; *ibid.* **1975**, 2601–2604.
135. (a) F. Minisci, R. Galli, M. Cecere, V. Malatesta, T. Caronna, *Tetrahedron Lett.* **1968**, 5609–5612; (b) F. Minisci, A. Citterio, C. Giordano, *Acc. Chem. Res.* **1983**, *16*, 27–32.
136. T. Shono, T. Miyamoto, M. Mizukami, H. Hamaguchi, *Tetrahedron Lett.* **1981**, 2385–2388.
137. R. Gottlieb, J. L. Neumeyer, *J. Am. Chem. Soc.* **1976**, *98*, 7108–7109.
138. (a) G. A. Russel, R. Rajaratnam, L. Wang, B. Z. Shi, B. H. Kim, C. F. Yao, *J. Am. Chem. Soc.* **1993**, *115*, 10596–10604; (b) G. A. Russel, D. Guo, W. Baik, S. J. Herron, *Heterocycl.* **1989**, *28*, 143–146.
139. (a) S. Kyushin, Y. Nakadaira, M. Ohashi, *Chem. Lett.* **1990**, 2191–2194; (b) E. Baciocchi, G. Doddi, M. Ioele, G. Ercolani, *Tetrahedron* **1993**, *49*, 3793–3800.
140. (a) A. Castellano, J. P. Catteau, A. Lablache-Combier, *Tetrahedron* **1975**, *31*, 2255–2261; (b) A. Albini, G. F. Bettinetti, M. De Bernardi, S. Pietra, *Gazz. Chim. Ital.* **1970**, *100*, 700–702.
141. A. R. Buick, T. J. Kemp, G. T. Neal, T. J. Stone, *J. Chem. Soc. (A)* **1969**, 666–669.
142. (a) A. Profumo, E. Fasani, A. Albini, *Heterocycl.* **1999**, *51*, 1499–1502; (b) E. Fasani, F. F. Barberis Negra, M. Mella, S. Monti, A. Albini, *J. Org. Chem.* **1999**, *64*, 5388–5395.
143. H. Riederer, J. Hüttermann, M. C. R. Symons, *J. Chem. Soc., Chem. Commun.* **1978**, 313–314.
144. S. Nagai, T. Gillbro, *J. Phys. Chem.* **1977**, *81*, 1793–1794.
145. K. Saito, Y. Horie, T. Murase, K. Takahashi, *Heterocycl.* **1989**, *29*, 1545–1550.
146. D. J. Chadwick, S. T. Hodgson, *J. Chem. Soc., Perkin Trans. 1*, **1983**, 93–102.
147. F. Yoneda, T. Nagamatsu, *J. Am. Chem. Soc.* **1973**, *95*, 5735–5737.
148. (a) E. Fasani, A. M. Amer, A. Albini, *Heterocycl.* **1994**, *37*, 985–992; (b) S. Pietra, G. F. Bettinetti, A. Albini, E. Fasani, R. Oberti, *J. Chem. Soc., Perkin Trans. 2* **1978**, 185–189.
149. R. D. Chambers, W. K. R. Musgrave, C. R. Sargent, F. G. Drakesmith, *Tetrahedron* **1981**, *37*, 591–595.
150. Y. L. Goldfarb, A. P. Yakubov, L. I. Belen'kii, *Khim. Geterotsikl. Soedin.* **1979**, 1044–1046.
151. J. F. Bunnet, B. F. Gloor, *Heterocycl.* **1976**, *5*, 377–399.
152. M. Novi, G. Garbarino, G. Petrillo, C. Dell'Erba, *J. Org. Chem.* **1987**, *52*, 5382–5386.
153. M. Prats, C. Galvez, L. Beltran, *Heterocycl.* **1992**, *34*, 1039–1046.
154. L. Beltran, C. Galvez, M. Prats, J. Salgado, *J. Heterocycl. Chem.* **1992**, *29*, 905–909.
155. A. P. Komin, J. F. Wolfe, *J. Org. Chem.* **1977**, *42*, 2481–2486.
156. (a) J. V. Hay, T. Hulicky, J. F. Wolfe, *J. Am. Chem. Soc.* **1975**, *97*, 374–377; (b) J. V. Hay, J. F. Wolfe, *ibid.* **1975**, *97*, 3702–3706.
157. E. A. Oostreen, H. C. van der Plas, *Rec. Trav. Chim. Pays-Bas* **1979**, *98*, 441–444.
158. D. R. Carver, A. P. Komin, J. S. Hubbard, J. F. Wolfe, *J. Org. Chem.* **1981**, *46*, 294–299.
159. (a) A. B. Pierini, R. A. Rossi, *J. Org. Chem.*, **1979**, *44*, 4667–4673; (b) R. A. Rossi, S. M. Palacios, *ibid.* **1981**, *46*, 5300–5304.
160. J. A. Zoltewicz, T. M. Oestreich, *J. Am. Chem. Soc.* **1973**, *95*, 6863–6864.

161.  M. D'Auria, *Gazz. Chim. It.* **1989**, *119*, 419–433.
162.  (a) R. Antonioletti, M. D'Auria, A. De Mico, G. Piancatelli, A. Scettri, *Tetrahedron* **1985**, *41*, 3441–3446; (b) R. Antonioletti, M. D'Auria, A. De Mico, G. Piancatelli, A. Scettri, *J. Chem. Soc., Perkin Trans. 1* **1985**, 1285–1288; (c) M. D'Auria, R. Antonioletti, A. De Mico, G. Piancatelli, *Heterocycl.* **1986**, *24*, 1575–1578.
163.  R. Antonioletti, M. D'Auria, F. D'Onofrio, G. Piancatelli, A. Scettri, *J. Chem. Soc., Perkin Trans. 1* **1986**, 1755–1758.
164.  M. D'Auria, E. De Luca, V. Esposito, G. Mauriello, R. Racioppi, *Tetrahedron* **1997**, *53*, 1157–1166.
165.  (a) M. D'Auria, A. De Mico, F. D'Onofrio, G. Piancatelli, *J. Org. Chem.* **1987**, *52*, 5243–5247; (b) M. D'Auria, A. De Mico, F. D'Onofrio, G. Piancatelli, *J. Chem. Soc., Perkin Trans. 1* **1987**, 1777–1780.
166.  M. D'Auria, A. De Mico, F. D'Onofrio, G. Piancatelli, *Gazz. Chim. It.* **1989**, *119*, 381–384.
167.  M. D'Auria, G. Piancatelli, T. Ferri, *J. Org. Chem.* **1990**, *55*, 4019–4025.
168.  (a) R. Bernardi, T. Caronna, D. Coggiola, F. Ganazzoli, *J. Org. Chem.* **1986**, *51*, 1045–1050; (b) R. Bernardi, T. Caronna, S. Morrocchi, P. Traldi, B. M. Vittimberga, *J. Chem. Soc., Perkin Trans. 1* **1981**, 1607–1609; (c) T. Caronna, B. M. Vittimberga, M. E. Kernn, W. G. McGimpsey, *J. Photochem. Photobiol. A* **1995**, *90*, 137–140; (d) R. Bernardi, T. Caronna, S. Morrocchi, M. Ursini, B. M. Vittimberga, *J. Chem. Soc., Perkin Trans. 1* **1990**, 97–100; (e) R. Bernardi, T. Caronna, G. Poggi, B. M. Vittimberga, *J. Heterocycl. Chem.* **1994**, *31*, 903–908.
169.  A. Albini, G. F. Bettinetti, E. Fasani, G. Minoli, *J. Chem. Soc., Perkin Trans. 1*, **1978**, 299–303.
170.  D. W. Borhani, F. D. Greene, *J. Org. Chem.* **1986**, *51*, 1563–1570.
171.  J. E. Toomey, *Adv. Heterocycl. Chem.* **1984**, *37*, 167–215.
172.  Y. Kita, H. Maekawa, Y. Yamasaki, I. Nishiguchi, *Tetrahedron Lett.* **1999**, *40*, 8587–8590.
173.  J. Volke, *Collect. Czech. Chem. Commun.* **1968**, *33*, 3044–3048.
174.  K. Hesse, S. Hünig, H. J. Bestmann, G. Schmid, E. Wilhelm, G. Seitz, R. Matusch, K. Mann, *Chem. Ber.* **1982**, *115*, 795–797.

# 7 Electron-transfer Reactions of Amines

*Suresh Das and V. Suresh*

## 7.1 Introduction

Amines and their derivatives are more widely distributed in nature than any other functional group family. This combined with the ease of oxidation of amines because of the lone pair of electrons on the nitrogen atom makes electron-transfer reactions of amines important in several electrochemical [1–3], photochemical [4–14], and biochemical [15–18] redox processes. One-electron oxidation of amines leads to the formation of radical intermediates that can be used for the synthesis of amino acids, alkaloids, and several other nitrogen-containing compounds of biological and pharmaceutical importance [19–23]. Electron-transfer reactions of amines are also important in several technological applications such as imaging [24], photopolymerization [25–27], and fading of textile dyes [28]. More recently the electron-donating capacity of the amino functionality has been extensively used for designing new materials such as fluoroionophores [29, 30], organic conductors [31], electroluminiscent materials [32–34], photovoltaics [35, 36], and materials with non-linear optical activity [37–39].

In this chapter, we focus on the mechanistic and synthetic aspects of electron-transfer reactions of amines. The role of electron transfer of amines in enzyme-catalyzed oxidations of amines is also briefly discussed.

## 7.2 Mechanistic Studies

The mechanism of electron-transfer reactions of amines has been studied by different methods, including the thermochemical, electrochemical, photochemical, and radiation chemical techniques.

## 7.2.1 Thermal Oxidation

Electron-transfer reactions of amines can be initiated by a variety of chemical oxidants. One-electron oxidation has been observed, for example, in the reactions of amines with metal salts such as ceric ammonium nitrate (CAN) [23, 40–42], manganese oxalate [23], alkaline ferricyanide [43–46], phenanthroline complexes of iron [46], octacyanomolybdate [45]. The mechanism of electron-transfer-catalyzed reactions of amines by chlorine dioxide [45, 47] and permanganate [48] have been intensively investigated in aqueous solution. In non-aqueous solvents, *N*-bromosuccinimide in carbon tetrachloride [49] and *N*-chlorobenzotriazole in benzene [50] are reported to react with amines via single-electron transfer (SET). Metal ion-catalyzed oxidation of vinyl amines by molecular oxygen [51, 52] and of aromatic amines by nitrogen dioxide have been reported [53]. In the presence of transition metal ions, hydrogen peroxide and peracids are known to liberate hydroxyl radicals [54–58] which can react with amines either via electron transfer or hydrogen-atom transfer. It has been proposed that oxidation of amines catalyzed by enzymes such as amine oxidases [16] and cytochrome P-450 [18] occurs via single-electron-transfer processes. Reagents such as hydrogen peroxide, peroxy acid [22], and ozone [22] bring about two-electron oxidation of amines, usually resulting in the formation of amine oxides or hydroxylamines.

Mechanistic studies of the chemical oxidation of aliphatic amines have been reviewed extensively by Chow et al. [22]. Many studies of the mechanism of oxidation of amines have been performed with chlorine dioxide or ferricyanide as oxidants, because they have absorption bands with maxima at 357 and 420 nm, respectively. Changes in the absorbance at these wavelengths for the respective oxidants can be conveniently used to follow the kinetics of the reactions. On the basis of these studies, the electron-transfer mechanism shown in Scheme 1 has been proposed for amine oxidation.

The amine radical cation formed in the initial one-electron-transfer process deprotonates at the $\alpha$-carbon and the amino alkyl radical formed is oxidized to the iminium salt which hydrolyzes to the dealkylated amine and a carbonyl compound. With the use of benzoyl peroxide as oxidant the aminium radical is also believed to

$$\text{Ox}^{n+} + \text{R}_2\overset{\cdot\cdot}{\text{N}}-\overset{\overset{\text{H}}{|}}{\text{C}}\text{R}'_2 \longrightarrow \text{Ox}^{(n-1)+} + \text{R}_2\overset{\oplus\bullet}{\text{N}}-\overset{\overset{\text{H}}{|}}{\text{C}}\text{R}'_2 \qquad (1)$$

$$\text{R}_2\overset{\oplus\bullet}{\text{N}}-\overset{\overset{\text{H}}{|}}{\text{C}}\text{R}'_2 \longrightarrow \text{R}_2\overset{\cdot\cdot}{\text{N}}-\overset{\bullet}{\text{C}}\text{R}'_2 + \text{H}^+ \qquad (2)$$

$$\text{R}_2\overset{\cdot\cdot}{\text{N}}-\overset{\bullet}{\text{C}}\text{R}'_2 + \text{Ox}^{n+} \longrightarrow \text{Ox}^{(n-1)+} + \text{R}_2\overset{\oplus}{\text{N}}=\text{CR}'_2 \qquad (3)$$

$$\text{R}_2\overset{\oplus}{\text{N}}=\text{CR}'_2 \xrightarrow{\text{H}_2\text{O}} \text{R}_2\text{NH} + \text{R}'_2\text{C}=\text{O} \qquad (4)$$

**Scheme 1.**

be formed. In many of these reactions hydrogen-atom abstraction to yield the aminoalkyl radical directly is also feasible. Chlorine dioxide, for example, reacts with dibenzyl amine by 35 % hydrogen abstraction and 65 % electron transfer [59]. Permanganate, on the other hand, reacts with triethylamine solely by electron transfer [48], whereas it reacts with benzyl amine predominantly via hydrogen-atom abstraction [60].

In the oxidation of benzylamines, apart from formation of the dealkylated amine and benzaldehyde [61–64], formation of imines [65, 66], benzonitrile [64], diazines [62] anilines [67], and *N*-benzylidene benzyl amines [64, 66, 67] has also been observed.

## 7.2.2 Electrochemical Oxidation

Electrochemical techniques are a convenient means of studying one-electron oxidations of amines. The reaction pattern of the anodic oxidation of amines depends greatly on the reaction conditions, including the nature of the electrode and the nucleophilicity of the solvent [1–3]. A major drawback of electrode oxidations is that unwanted secondary electron-transfer reactions can occur at the electrode surface. Also in electrochemical processes the effective reaction volume is limited at the electrode surface, thereby creating a high local concentration of reactive intermediates which can lead to dimerization and disproportionation reactions. These factors have to some extent, limited the synthetic utility of the anodic oxidation of amines. Because of this the anodic oxidation of amines has been intensively studied, although mainly from a mechanistic standpoint.

### Aliphatic amines

Electrolytic oxidation of amines results in the formation of an amine radical cation by one-electron transfer to the electrode. Scheme 2 shows the various reaction pathways observed on electrochemical oxidation of aliphatic amines [1–3, 68–71]. The radical cation **2** can undergo C–N bond cleavage yielding a relatively stable carbonium ion **3**, which can undergo solvolysis, add to a nucleophile, or polymerize. As observed in the thermal oxidation reaction deprotonation at the α-carbon to yield the α-aminoalkyl radical, followed by its rapid oxidation to the iminium ion **7**, forms a major reaction pathway. The iminium ion can undergo hydrolysis or addition to a nucleophile to yield substituted product (**11**). Under strongly basic conditions, nitrogen deprotonation can occur in primary and secondary amines. The resulting nitrogen-centered radicals can couple to form dimers or can undergo further oxidation to yield nitrenium ions ($>N:^+$). In the presence of silver [70] or nickel [71] electrodes, the nitrogen-centered radicals can be further oxidized to yield the nitrile **16**.

### Aromatic amines

Anodic oxidation of aromatic amines has been well studied [1–3, 68, 72, 73]. These reactions are rather complex and are substantially affected by the reaction condition.

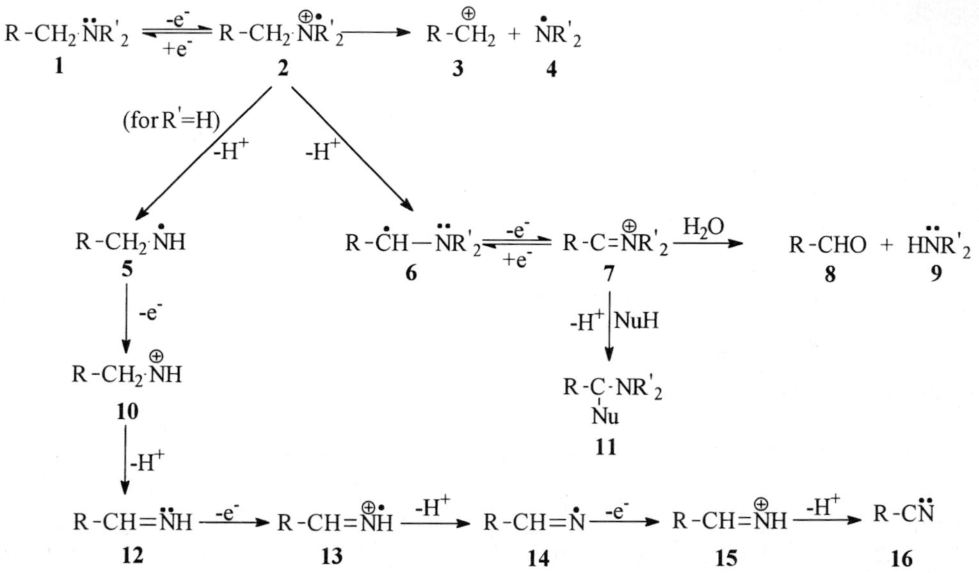

**Scheme 2.**

Scheme 3 shows the different reaction pathways observed in the anodic oxidation of aniline and its derivatives [74]. The cation radical formed initially can undergo dimerization or deprotonation. Dimerization is favored under strongly acidic conditions and high current densities. Subsequent deprotonation of the dimer yields the benzidine **18**. Benzidine, being more easily oxidizable than the parent amine, is converted to the diimine species. In basic solutions at low current densities and high concentrations of the starting material, deprotonation is favored, yielding a resonance-stabilized radical. This undergoes C–N coupling with a parent molecule,

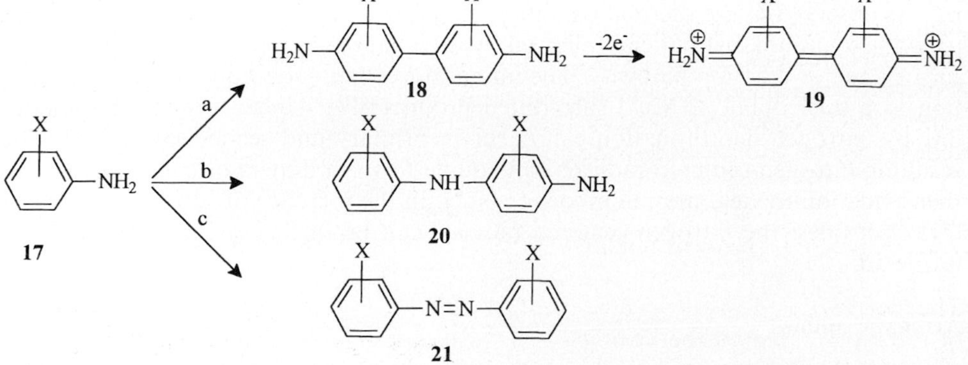

**Scheme 3.**

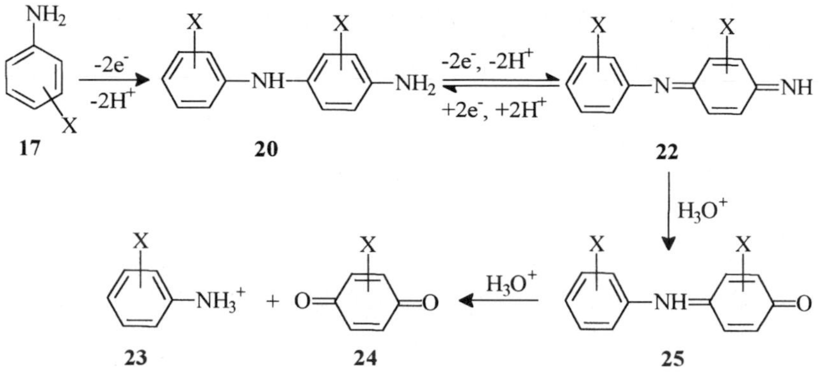

**Scheme 4.**

leading to the formation of 4-aminodiphenylamine **20**. Further oxidation of **20** yields *p*-benzoquinone and the parent amine as shown in Scheme 4.

In basic solution head-to-head coupling of the deprotonated radicals result in the formation of azo compounds (**21**) from *N*-unsubstituted anilines. This occurs via the formation of hydrazo derivatives. *p*-Substituted anilines also preferentially undergo head-to-head coupling to yield azo derivatives.

The influence of reaction conditions on the electrochemical oxidation of anilines and *N*-alkylanilines is shown in Table 1 [74]. It has been proposed that electrochemical formation of the conducting polyaniline polymers occurs via the formation of *p*-aminodiphenylamine [31, 75–77].

Diphenylamines on electrooxidation also undergo aryl–aryl, *N*–aryl or N–N coupling as shown in Scheme 5 [1–3].

Tail-to-tail coupling of radicals obtained in the anodic oxidation of triphenylamines results in the formation of tetraphenylbenzidines. Oxidation of triarylamines to the di-cation results in the formation of the carbazoles, as observed for *N*-alkyl-*p,p'*-disubstituted diphenylamines [1–3, 78]. The cation radicals of triarylamines with substituents in the *para* position of the aryl groups, which can protect them against nucleophilic attack, are very stable and can be used as organic redox catalysts for indirect electrochemical oxidation reactions. Depending on the substitution pattern on the phenyl group the oxidation potentials of the triarylamines can be tuned over a wide range ($E_{ox} = 0.7$–$2.0$ V) and many of these have been used as redox catalysts in numerous indirect electrochemical reactions [1–3, 79–83].

### 7.2.3 Radiation Chemical Studies of Amines

The mechanistic details of amine oxidation has also been extensively studied by use of radiation chemical methods [84–90]. In the radiolysis of dilute solutions, interaction of the ionizing radiation occurs predominantly with the solvent molecules resulting in the formation of reactive intermediates derived from the solvent [91].

**Table 1.** Influence of electrolysis conditions on product distribution in the anodic oxidation of aniline and *N*-alkylanilines.[a,b]

| Compound | Electrolysis conditions | Products (%)[c] |
|---|---|---|
| Aniline | MeCN–TEAP, low current density | 4-Aminodiphenylamine (40) |
| Aniline | 0.05 M $H_2SO_4$, high or low current density | *p*-Benzoquinone (90–100) |
| Aniline | 6 M $H_2SO_4$, high current density | *p*-Benzoquinone (80–90)<br>Benzidine (10–20) |
| *N*-Ethylaniline | MeCN–TEAP, low current density, low parent concentration | *N,N′*-Diethylbenzidine (60–70) |
| *N*-Ethylaniline | MeCN–TEAP, high current density | *N,N′*-Diethylbenzidine (70–80) |
| *N*-Ethylaniline | MeCN–TEAP, low current density, high parent concentration | *N*-Ethyl-4-ethylaminodiphenylamine (70) |
| *N*-Ethylaniline | 6 M $H_2SO_4$, high current density | *p*-Benzoquinone (40)<br>*N,N′*-Diethylbenzidine (60) |
| *N-tert*-Butylaniline | MeCN–TEAP, high current density | *N,N′*-di-*tert*-Butylbenzidine (100) |

[a] Controlled-potential electrolyses.
[b] According to [1].
[c] Based on consumed parent compound.

Thus, the radiolysis of dilute aqueous solutions results in the formation of solvated electrons ($e_{aq}^-$) hydroxyl radicals, and hydrogen atoms (Eq. 5). In neutral and alkaline solutions, H atom formation is negligible.

$$H_2O \rightsquigarrow \ ^\bullet OH, H^\bullet, e^-_{aq}, H^+, H_2O_2, H_2 \qquad (5)$$

Oxidizing or reducing conditions can be created by selective scavenging of the intermediates. In nitrous oxide saturated solutions, for example, reaction of the solvated electron with $N_2O$ leads to further generation of hydroxyl radicals. Alternatively, reducing conditions can be achieved by using tertiary butanol radicals to scavenge the hydroxyl radicals. Oxidizing species other than hydroxyl radicals that have normally been generated in aqueous solutions to study amine oxidation reactions include the azide radical ($N_3^\bullet$) [88] and halogen radicals [89]; these are normally generated by reaction of hydroxyl radicals with azide and halogen ions. Sulfate radicals ($SO_4^{-\bullet}$) generated by reaction of electrons with persulfate anions, have also been used to study the oxidation of amines [92]. An oxidizing environment can also be generated in non-aqueous solvents. Radiolysis of dichloromethane solutions, for example, results in the formation of $CH_2ClO_2^\bullet$ and $CHCl_2O_2^\bullet$ radicals, both of which are oxidizing in nature [89].

The formation and subsequent reaction of amine radical cations by radiolysis can be studied by pulse radiolysis in combination with optical [84, 87–90], conductivity

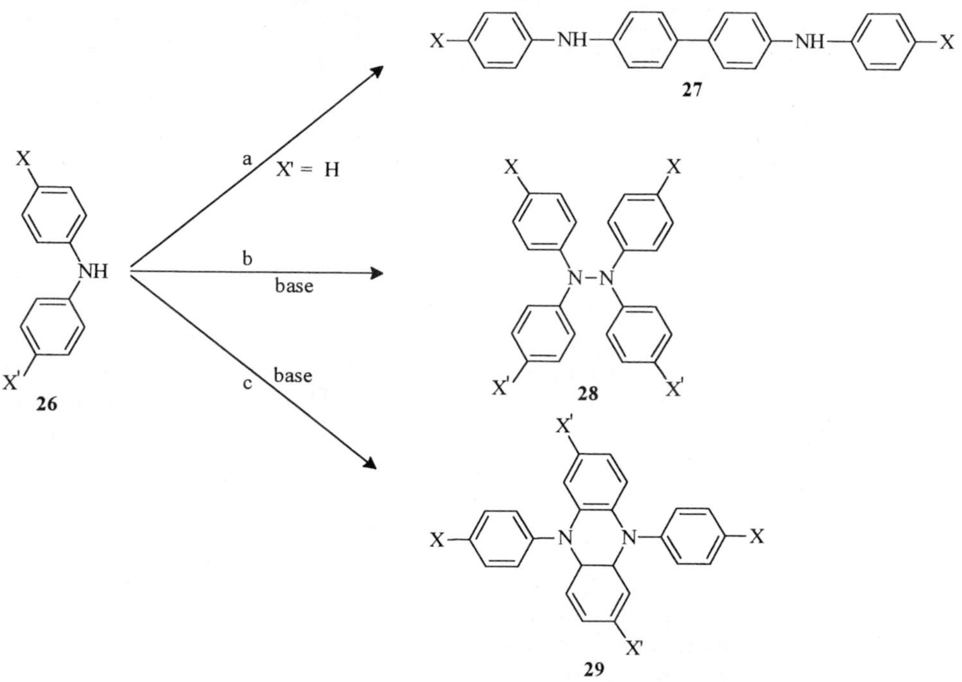

**Scheme 5.**

[85], and ESR [93–95] detection techniques. A major advantage of these methods over flash photolysis is that the transient signals from the amine radicals are not complicated by the presence of radical anions, which are normally generated along with amine radical cations in photoinduced electron-transfer processes. Aminoalkyl radical intermediates formed in the process can be monitored optically, because of strong absorption in the 220–450 nm region.

Apart from being good electron donors, the lability of the hydrogen atoms attached to the $\alpha$-carbon of amines also make the amines very good hydrogen-atom donors. The very high rate constant ($k = 1.2 \times 10^{10}$ $M^{-1}$ $s^{-1}$) for the reaction of hydroxyl radicals with amines is, however, strongly indicative of a predominantly electron-transfer mechanism [84]. Rate constants for normal hydrogen-atom-abstraction reactions of OH$^{\cdot}$ with good hydrogen-atom donors such as formate ($k = 3 \times 10^{9}$ $M^{-1}$ $s^{-1}$) and isopropanol ($k = 3 \times 10^{9}$ $M^{-1}$ $s^{-1}$) [96] are generally much lower. The radical cation formed in the electron-transfer reaction is an oxidizing species whereas the $\alpha$-aminoalkyl radical formed via hydrogen abstraction is reducing in nature. Formation of substantial amounts of oxidizing amine radical cation even at high pH ($>$10.0) in the reaction of hydroxyl radicals with the amine was indicated by the reaction of the radical intermediates with $N,N'$-tetramethyl-$p$-phenylenediamine (TMPD), which results in the formation of TMPD$^{+\cdot}$. This confirms that the reaction of hydroxyl radicals with amine occurs predominantly via electron transfer [84].

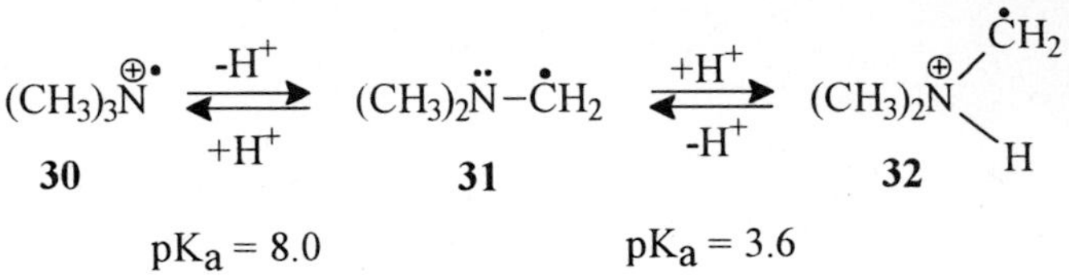

$$(CH_3)_3\overset{\oplus\bullet}{N} \quad \underset{+H^+}{\overset{-H^+}{\rightleftharpoons}} \quad (CH_3)_2\overset{\cdot\cdot}{N}-\overset{\bullet}{C}H_2 \quad \underset{-H^+}{\overset{+H^+}{\rightleftharpoons}} \quad (CH_3)_2\overset{\oplus}{N}\overset{\overset{\overset{\bullet}{C}H_2}{\diagup}}{\underset{\diagdown}{H}}$$

$$\textbf{30} \qquad\qquad\qquad \textbf{31} \qquad\qquad\qquad \textbf{32}$$

$$pK_a = 8.0 \qquad\qquad\qquad pK_a = 3.6$$

**Scheme 6.**

The amine radical cation of triethylamine formed in these reactions can exist in an acid–base equilibrium, as shown in Scheme 6.

Pulse radiolysis studies using optical detection suggested the main species in equilibrium to be the $\alpha$-aminoalkyl radical (**31**) and the *N*-protonated $\alpha$-aminoalkyl radical (**32**) whereas results from ESR studies were indicative of protonation at the $\alpha$-carbon site to form the *N*-centered radical cation (**30**). A subsequent study showed that these results could be attributed to kinetic and thermodynamic factors [84]. Thus *N*-protonation resulting in the formation of **32** is kinetically favored and is hence observed in the short time-scales involved in pulse-radiolysis systems. In the longer times involved in ESR measurements the thermodynamically more stable *N*-centered radical cation will be observed. The $pK_a$ of **30** and **32** were estimated as 8.0 and 3.6 by use of pulse radiolysis [84]. Using the equation for similar keto $\leftrightarrow$ enol tautomerism of barbituric acid [97], the ratio of **30/32** was estimated to be $10^{-3.6}/10^{-8} = 10^{4.4}$, indicating the N-centered radical (**30**) to be the predominant protonated species.

The rate constants for deprotonation of the trimethylamine radical cation by the parent amine and by hydroxyl radicals were estimated as $1 \times 10^9$ and $1 \times 10^{10}$ $M^{-1} s^{-1}$, respectively [84]. The rate constants for the different steps involved in the reaction of oxygen with $\alpha$-aminoalkyl radical **31** (Scheme 7) were determined by pulse radiolysis with conductivity detection [85].

The $\alpha$-aminoalkyl radical (**31**) reacts rapidly with oxygen ($k = 3.5 \times 10^9$ $M^{-1} s^{-1}$) to form $O_2^{-\bullet}$ possibly via a short-lived ($\tau = 10^{-6}$ s) peroxy radical **33**. Elimination

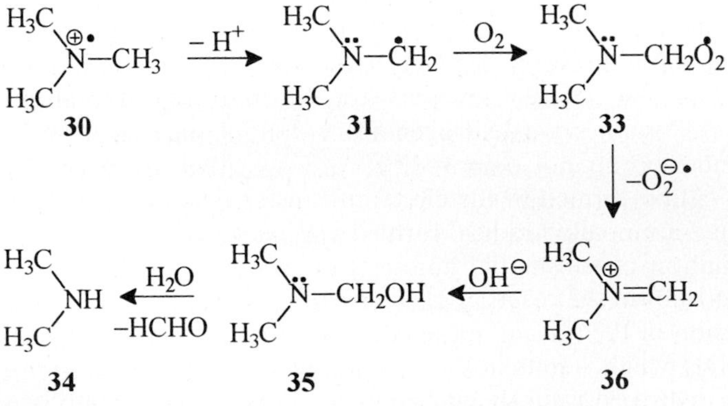

**Scheme 7.**

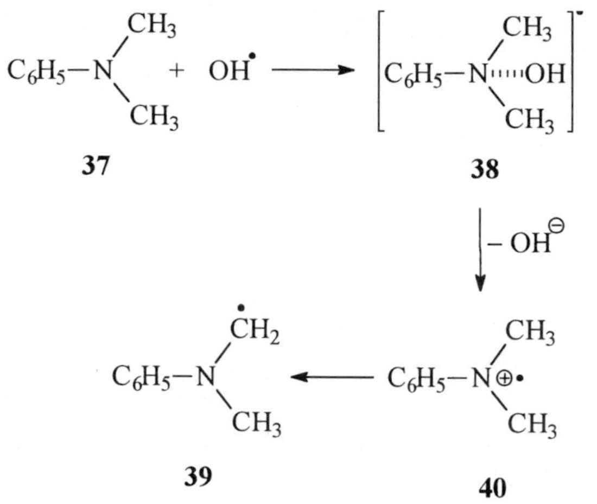

$$C_6H_5-N\overset{CH_3}{\underset{CH_3}{}} + \overset{\bullet}{OH} \longrightarrow \left[ C_6H_5-N\overset{CH_3}{\underset{CH_3}{\cdots\cdots OH}} \right]^{\cdot}$$

**37**                                **38**

$$\Big\downarrow - OH^{\ominus}$$

$$C_6H_5-N\overset{\overset{\bullet}{CH_2}}{\underset{CH_3}{}} \longleftarrow C_6H_5-N\overset{CH_3}{\underset{CH_3}{\oplus\bullet}}$$

**39**                                **40**

**Scheme 8.**

of peroxide ion from **33** will lead to **36** (Scheme 7). The dimethyl iminium ion (**36**) adds OH⁻ to form (hydroxymethyl)dimethylamine (**35**) ($k = 2.8 \times 10^8$ M$^{-1}$s$^{-1}$). This is hydrolyzed much more slowly ($k_{obs} = 4.0$ s$^{-1}$) to yield formaldehyde and the dealkylated amine (**34**).

One-electron oxidation of aniline by hydroxyl radicals yields the anilinium radical [C$_6$H$_5$NH$_2$]$^{+\bullet}$, which loses a proton over several microseconds to give the neutral radical C$_6$H$_5$NH$^\bullet$ [98, 99]. The deprotonation was followed at 400–440 nm where the radical cation absorbs more strongly than the neutral radical. Because the p$K_a$ of parent aromatic amines are generally lower than those of the corresponding radical cations, an external base is required to effect deprotonation of the anilinium radical.

On the basis of studies using nanosecond time resolution, Holcman and Sehested suggested that the reaction of hydroxyl radicals with dimethylaniline occurs via the intermediate formation of a hydroxyl radical–amine adduct [100], as described in Scheme 8.

The transient absorption spectrum of the OH adduct of dimethylaniline has a broad band centered at ca 380 nm. Decay of this species results in a growth of absorption in the 450 nm region where the radical cation has a strong absorption. Assignment of the radical cation spectrum was supported by comparison with the spectra obtained in the electron-transfer reaction of sulfate radicals with DMA, and with DMA$^{+\bullet}$ spectra generated by flash photolysis and electrochemical methods [101, 102]. The spectral characteristics of the observed transient species are summarized in Table 2.

The mechanism of hydroxyl-radical-mediated oxidation of aromatic amines has recently been reinvestigated by Tripathi and Sun, who used time-resolved Raman spectroscopy as a diagnostic tool. Their studies suggest that the initial transient formed in the reaction of hydroxyl radicals with the neutral *p*-phenylene diamine is the amine radical cation and not the OH-adduct of the amine as originally believed.

**Table 2.** Spectral characteristic of the radicals formed in the reaction of H and OH radicals with DMA.[a]

| Radical | $\lambda_{max}$ (nm) | $\varepsilon$ ($M^{-1}$ cm$^{-1}$) |
|---|---|---|
| $[C_6H_5N(CH_3)_2]H^{\cdot}$ | 380 | |
| $[C_6H_5N(CH_3)_2]OH^{\cdot}$ | 380 | $6500 \pm 1000$ |
| | <270 | |
| $C_6H_5N(CH_3)CH_2^{\cdot}$ | 330 | $10500 \pm 500$ |
| | 450 | $1400 \pm 100$ |
| $[C_6H_5N(CH_3)_2]^{+\cdot}$ | 270 | |
| | 465 | $4500 \pm 400$ |

[a] According to [100].

An adduct-mediated electron-transfer pathway was, however, observed for the protonated amine [103].

The electronic absorption maxima of the few other aminium ions that have been investigated are given in Table 3 [22, 56].

The acidities of aminium ions derived from primary and secondary amines have also been measured; a few representative examples are listed in Table 4, with the acidities of the parent compounds.

The p$K_a$ of the aminium ions is generally lower than those of the corresponding amine. The greater acidity of the aminium ion can be attributed to the greater amount of 's' character in the lone-pair orbital of the amine radical, $(CH_3)_2N^{\cdot}$, which is sp$^2$-hybridized compared, with the more basic lone pair orbital in $(CH_3)_2NH$, which is sp$^3$-hybridized. In contrast, the p$K_a$ of the anilinium radical is much higher than that of the protonated aniline. The basicity of the anilinium radical is comparable with that of $(CH_3)_2NH^{+\cdot}$, reflecting the sp$^2$-hybrid character of the N–H bonds in the two species. The low basicity of the parent aniline results from

**Table 3.** Electronic absorption spectra of radical cations.[a]

| Radical cation precursor | Solvent | $\lambda_{max}$ (nm) |
|---|---|---|
| $(CH_2)_5NH$ (piperidine) | Acidic water | 295 |
| DABCO | Aqueous acetonitrile | 465 ($\varepsilon = 2000$) |
| $HONH_2$ | Acidic water | 220 |
| Aniline | Water or EPA | 425 ($\varepsilon = 2000$) |
| N-Methylaniline | Acidic water | 455 |
| N,N'-Dimethylaniline | Water or Acetonitrile | 460 |
| Diphenylamine | Acetonitrile | 660 |
| Triphenylamine (TPA) | Acetonitrile or EPA | 655 ($\varepsilon = 10^4$) |
| *tris-p*-Methoxy TPA | Acetonitrile | 714 |
| 2-Aminonaphthalene | Water | 525 |

[a] According to [12].

**Table 4.** Acidities of radical cations and protonated parent compounds.[a]

| Radical cation precursor | p$K_a$ value | |
| --- | --- | --- |
| | $-N^{+\bullet}$ | $-N-H^+$ |
| Ammonia | 6.7 | 9.2 |
| Diethylamine | 7.0 | 10.7 |
| Hydroxylamine | 4.2 | 6.1 |
| Methoxylamine | 2.9 | 4.6 |
| Aniline | 7.0 | 4.6 |

[a] According to [12].

delocalization of the lone-pair electrons and the loss in resonance stabilization upon protonation.

ESR spectra of a wide variety of aminium radicals have been studied [22, 93–95, 104–114]. The radical cations were generated for this purpose by UV photolysis or γ-radiolysis in strongly acidic solutions for unstable radical cations [93–95, 104–106]. More stable radical cations such as those of triisopropylamine and 9-*tert*-butylazabicyclo(3.3.1)nonane were prepared by oxidation in dichloromethane by SbF$_5$ or ($p$-BrC$_6$H$_4$)$_3$N$^{+\bullet}$/SbCl$_6^-$ [107–109]. γ-Radiation in Freon matrices at 77 K has also been used to study the ESR spectra of amine radical cations [109–114]. With the exception of a few oligocyclic amines, for which flattening is impaired by the rigid molecular framework, aminoalkyl radicals are generally planar. Various parameters related to the structure and reactivity of aminoalkyl radical cations have been extensively reviewed [22].

### 7.2.4 Photoinduced Electron Transfer

Photoinduced electron transfer (PET) processes can be used to bringing about one-electron oxidation and reduction of organic molecules [115–118]. In these processes, photoexcitation of an electron acceptor or donor leads to enhancement of their electron-accepting and -donating properties, respectively. When the excited state molecule comes into contact with a ground-state electron donor or acceptor within the excited state lifetime, electron transfer can occur. The feasibility of producing radical ions via these processes can be predicted by use of the Rehm–Weller equation [119], which is given in Eq. 6 [120]:

$$\Delta G^\circ = -23.06[E^\circ{}_{A/A}{}^{\ominus\bullet} - E_{D/D}{}^{\oplus\bullet}] - \frac{e_o^2}{\varepsilon_A} - \Delta E_{0,0} \tag{6}$$

where $\Delta G^\circ$ is the change in free energy associated with the electron transfer. The figure 23.06 enables $\Delta G^\circ$ to be expressed in kcal, $E^\circ{}_{A/A}{}^{-\bullet}$ is the reduction potential in volts of the acceptor A in the ground state, $E^\circ{}_{D/D}{}^{+\bullet}$ is the oxidation potential of the donor; $\varepsilon_A$ is the dielectric constant of the solvent and the term $e_o^2/\varepsilon_A$ represents

**Table 5.** Redox potentials and free energy changes for electron transfer between the ground state of selected group of amines and the excited states of anthraquinone and dicyanoanthracene.[a]

| Donor (D) | Acceptor (A) | $\Delta E_{0,0}(A)$ | $E_0(D/D^{+\bullet})$[b] | $E_0(A/A^{-\bullet})$[b] | $\Delta G_{el}$ |
|---|---|---|---|---|---|
| N,N-Diethylaniline | Anthraquinone | 3.1 | 0.76 | −0.94 | 36.10 |
| N,N-Dimethylaniline | Anthraquinone | 3.1 | 0.81 | −0.94 | 37.26 |
| Diethylamine | Anthraquinone | 3.1 | 0.78 | −0.94 | 36.56 |
| Triethylamine | Anthraquinone | 3.1 | 0.76 | −0.94 | 36.10 |
| Triphenylamine | Anthraquinone | 3.1 | 0.98 | −0.94 | 41.18 |
| N,N-Diethylaniline | Dicyanoanthracene | 2.86 | 0.76 | −0.89 | 35.19 |
| N,N-Dimethylaniline | Dicyanoanthracene | 2.86 | 0.81 | −0.89 | 36.34 |
| Diethylamine | Dicyanoanthracene | 2.86 | 0.78 | −0.89 | 35.65 |
| Triethylamine | Dicyanoanthracene | 2.86 | 0.76 | −0.89 | 35.19 |
| Triphenylamine | Dicyanoanthracene | 2.86 | 0.98 | −0.89 | 40.26 |

[a] According to [13].
[b] Redox potentials are reported relative to SCE in polar solvents.

the free enthalpy associated with the Coulombic interaction between D and A. $\Delta E_{0,0}$ is the electronic excitation energy of the partner whose redox properties have been enhanced by excitation.

The Coulombic term in Eq. 6 becomes negligible in polar solvents owing to 'shielding' of the electrostatic interaction between the radical ions generated. Table 5 lists representative redox potentials and free energies for photo-mediated electron transfer between a selected group of amines and photosensitizers such as anthraquinone (AQ) and dicyanoanthracene (DCA).

The problem of secondary electron transfer normally observed in thermal and electrochemical oxidation can be partially circumvented by using photo- and radiation chemical methods, because in such processes the oxidizing species is generated as a transient and the steady-state concentration of radical intermediates is usually very low ($\leq 10^{-6}$ M).

## Deprotonation at the α-carbon of aminium radicals produced in SET reactions

As discussed earlier, deprotonation of α–carbon forms a major reaction pathway for the disappearance of the amine radical cation. Studies of photoinduced electron-transfer reactions of tertiary amines by Lewis [7, 11] and by Mariano [5, 10] have contributed significantly towards our understanding of the factors that control this process. Lewis and coworkers used product-distribution ratios of stilbene–amine adducts to elucidate the stereoelectronic effects involved in the deprotonation process [5, 10, 121, 122]. In non-polar solvents, the singlet excited state of trans-stilbene forms non-reactive but fluorescent exciplexes with simple trialkylamines. Increasing solvent polarity brings about a decrease in the fluorescence intensity and an increase in adduct formation. For non-symmetrically substituted tertiary amines two types of stilbene–amine adduct can be formed, as is shown in Scheme 9, depending on whether the aminoalkyl radical adding to the stilbene radical is formed by de-

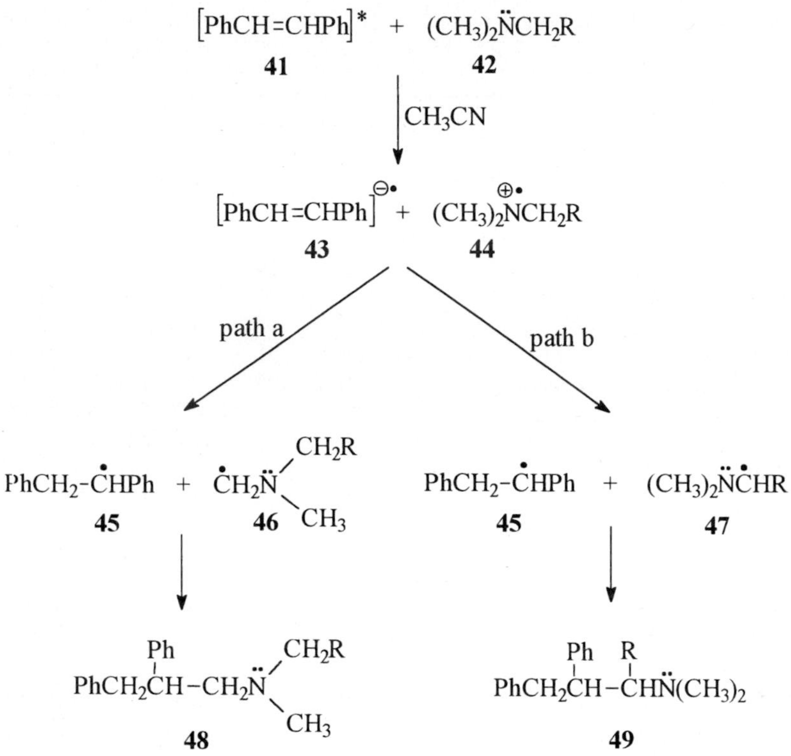

$$[PhCH=CHPh]^* \quad + \quad (CH_3)_2\ddot{N}CH_2R$$

**41**                                    **42**

$$\downarrow CH_3CN$$

$$[PhCH=CHPh]^{\ominus\bullet} \quad + \quad (CH_3)_2\overset{\oplus\bullet}{N}CH_2R$$

**43**                          **44**

path a                              path b

$$PhCH_2\text{-}\overset{\bullet}{C}HPh \quad + \quad \overset{\bullet}{C}H_2\ddot{N}\overset{CH_2R}{\underset{CH_3}{\diagdown}}$$

**45**                **46**

$$PhCH_2\text{-}\overset{\bullet}{C}HPh \quad + \quad (CH_3)_2\ddot{N}\overset{\bullet}{C}HR$$

**45**                        **47**

$$\downarrow$$                                    $$\downarrow$$

$$\overset{Ph}{\underset{}{PhCH_2CH}}\text{-}CH_2\ddot{N}\overset{CH_2R}{\underset{CH_3}{\diagdown}}$$

**48**

$$PhCH_2\overset{Ph}{\underset{}{CH}}\text{-}\overset{R}{\underset{}{CH}}\ddot{N}(CH_3)_2$$

**49**

**Scheme 9.**

protonation of the less substituted (path a) or more substituted (path b) $\alpha$-CH bond. The yields and ratios of the two types of adduct formed from a variety of unsymmetrical amines are summarized in Table 6 [122].

These studies indicate that the selectivity of $\alpha$-aminoalkyl radical formation is Me > Et $\gg$ *i*-Pr, which is the opposite of that expected on the basis of radical stability. For example, in diisopropylmethylamine, methyl adducts are formed exclusively. Similar selectivity has been observed for oxidation of unsymmetrical amines by ferricyanide [43, 123] and in anodic processes [124]. This selectivity has been attributed to requirement of overlap between the half-vacant nitrogen *p* orbital and the $\alpha$-CH orbital of the $\alpha$-carbon. From the Newman projections below it can be seen that the conformation necessary for methyl deprotonation (a) is lower in energy than that for the isopropyl deprotonation (b) [5, 122].

Selective formation of the more substituted $\alpha$-aminoalkyl radical was, however, observed for amines of the type $Me_2NCH_2R$, where R is a radical stabilizing group with minimal steric requirements ($R=CH_2=CH-$, $Ph-$, $HC\equiv C-$, and $C_2H_5CO_2$) [125]. It has been proposed that the base strength can also affect the selectivity of deprotonation, as observed in the formation of the less substituted alkene as the base strength is increased in E1 elimination reactions [122, 126, 127].

**Table 6.** *trans*-Stilbene–amine adduct yields and ratios.[a]

| Amine | % of **48** | % of **49** | **48/49** corrected |
|---|---|---|---|
| Diisopropylmethyl | >95 | <5 | >20 |
| Isopropyldimethyl | >95 | <5 | >20 |
| Ethyldiisopropyl | 92 | 8 | 12 |
| Diethylmethyl | 63 | 37 | 2.3 |
| *n*-Butyldimethyl | 86 | 14 | 2.0 |
| Ethyldimethyl | 84 | 16 | 1.8 |
| Dimethylbenzyl | 77 | 23 | 1.1 |
| Dimethylallyl | 87 | 13 | 2.2 |
| Dimethylglycine ethyl ester | 49 | 51 | 0.48 |
| 1-(Dimethylamine)-2-butyne dimethyl propargyl | <5 | >95 | <0.01 |

[a] According to [122].

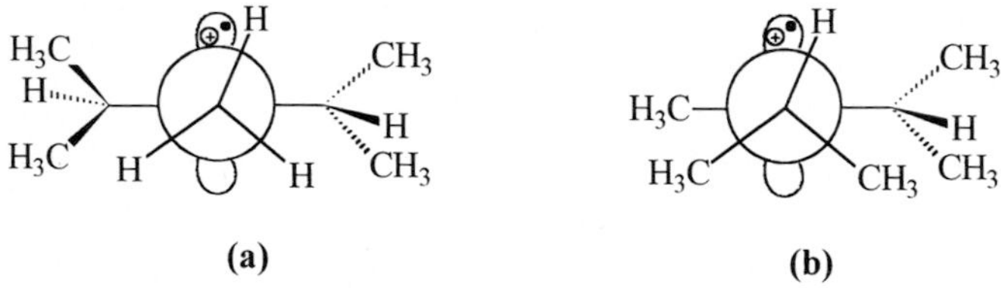

(a)                          (b)

**Chart 1**

Mariano and coworkers have investigated the effect of substituents on the $\alpha$-CH kinetic acidity of several tertiary aromatic amine radical cations generated by electron transfer from the parent amines to the excited state of dicyanobenzene [128]. Laser excitation of 60:40 methanol–acetonitrile solutions of dimethylaniline (DMA) and dicyanobenzene (DCB) result in the formation of the DMA cation radical with an absorption maximum at ca 460 nm and the DCB radical anion with an absorption maximum at ca 340 nm, which decay by back-electron-transfer at diffusion-controlled rates ($k = 1.1 \times 10^{10}$ $\text{M}^{-1}$ $\text{s}^{-1}$). Bases such as tetrabutyl-ammonium acetate (*n*-Bu$_4$NOAc) and tetrabutylammonium trifluoroacetate (*n*-Bu$_4$NOCCF$_3$) were observed to deprotonate the DMA radical cation with rate constants of $3.1 \times 10^5$ and $8 \times 10^4$ $\text{M}^{-1}$ $\text{s}^{-1}$ respectively.

The rate constant for *n*-Bu$_4$NOAc-catalyzed deprotonation increases with *para*-substitution of the amine in the order *p*-OMeC$_6$H$_4$NMe$_2$ < *p*-MeC$_6$H$_4$NMe$_2$ < *p*-CF$_3$C$_6$H$_4$NMe$_2$, which is consistent with the effects of *para* substituents, because electron-donating groups stabilize the radical cation leading to an increase in its p$K_a$. The effects of $\alpha$-substituents on the rates of *n*-Bu$_4$NOAc-induced $\alpha$-CH deprotonation of tertiary amines was studied for the amines Ph$_2$N(CHR$^1$R$^2$) (Table 7).

The results of these studies were compared with those obtained earlier for the stilbene amine photoaddition [7, 11] and amine–enone photoelectron-transfer-catalyzed cyclizations (Table 7). The effects of $\alpha$-substituents on the $\alpha$-CH deprotonation rates of aminium radicals measured in the stilbene–amine photoaddition

**Table 7.** Relative rates of deprotonation of the α-substituted tertiary amine cation radical from product distribution and laser spectroscopic studies.[a]

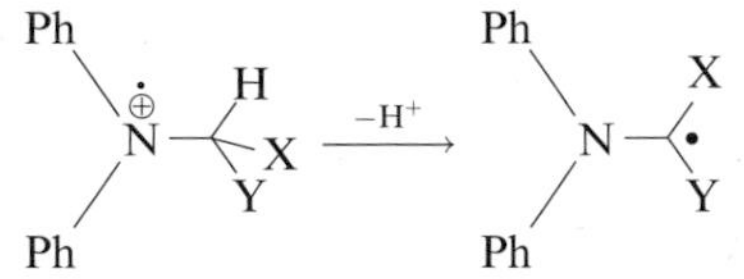

| Entry | X | Y | From stilbene–amine SET system | From enone–amine SET photocyclizations | From rate constant for acetate promoted deprotonation of diphenylaminium radical (laser spectroscopy) |
|---|---|---|---|---|---|
| a | H | H | 1.1 | 1.1 | 0.3 |
| b | H | $CH_3$ | 0.5 | 0.2 | 0.1 |
| c | $CH_3$ | $CH_3$ | <0.05 | – | 0.05 |
| d | H | Ph | 1.0 | 1.0 | 1.0 |
| e | H | $CH=CH_2$ | 0.5 | 1.9 | 0.8 |
| f | H | $C≡CH$ | 111 | 3.9 | 22 |

[a] According to [128].

reactions and by laser spectroscopic methods show the same trend, although in the stilbene–amine photoaddition the steric effect on α-CH deprotonation is much more pronounced. It has been suggested that the difference might arise from the different nature of the deprotonation in the two systems [129]. Proton transfer in the stilbene–amine system occurs within the contact ion-pair [7, 121, 122], the structure of which might be governed by the need for charge stabilization. This effect could make kinetic factors more important in such systems than those determined by the acetate-promoted deprotonation of the free radical cation. The differences could also be attributed to the weak nature of the base used in the laser spectroscopic studies. The second-order rate constants for the deprotonation of the aminium radical catalyzed by acetate are, in fact, much lower than diffusion-controlled processes. Thus in this instance electronic factors, i.e. the capacity of the substituents to stabilize the resulting α-amino radical could play a greater role in deciding the nature of the α-CH deprotonation.

For enone–amine photocyclization reactions the α-CH kinetic acidity seems to be mainly dependent on electronic factors and this has been attributed to the intramolecular nature of proton transfer in the enone–amine zwitterionic radicals (diagram below), which would disfavor overlap of the half vacant 'p' orbital on nitrogen with the α-CH bond [128, 130, 131].

**Photoinduced bond-cleavage reactions**

More recently, Mariano and coworkers have investigated the effects of substituents and medium on desilylation, decarboxylation, and retro-aldol cleavage reactions

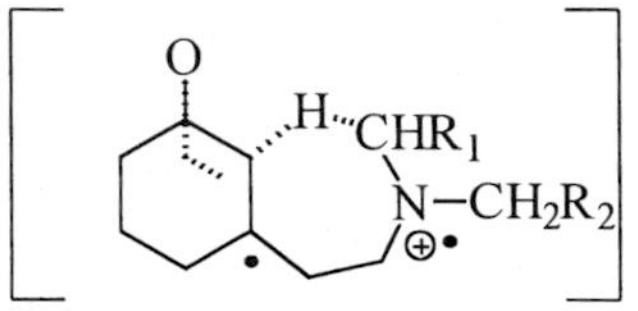

Enone-Amine

**Chart 2**

(Scheme 10) of anilinium radicals by laser-flash photolysis [129]. 308-nm laser excitation of the anilinocarboxylate in the presence of dicyanobenzene leads to electron transfer from the excited state of anilinocarboxylate to the ground state of DCB (Eq. 7) as evidenced by the transient absorption maxima at 340 nm (DCB$^{-\cdot}$) and 450–480 nm (anilinium radical). The decay of the anilinium radical occurs at a much faster rate than that of DCB$^{-\cdot}$ indicating that back-electron-transfer is not a major reaction pathway unlike in the SET reaction of dimethyl aniline. The fast decay of the anilinium radical has been attributed to unimolecular $\alpha$-decarboxylation, as shown in Eq. 8 (Scheme 10).

The unimolecular rates of reactions (Eqs. 8 and 9) were found to depend on solvent polarity and substituents on the aniline ring, nitrogen, and $\alpha$-carbon. The second-order rate constants for methanol-promoted desilylation (Eq. 10) of the aminium radical generated via SET reactions of $\alpha$-anilinosilanes increased tenfold on changing the substituent on nitrogen from an alkyl group to acyl group.

The competition between fragmentation of the C–C bond and deprotonation (Scheme 11) of the aminium radical has been extensively probed by Whitten and others [132–139]. Cleavage of the C–C bond can occur to produce either carbocations (Eq. 11) or radicals (Eq. 12).

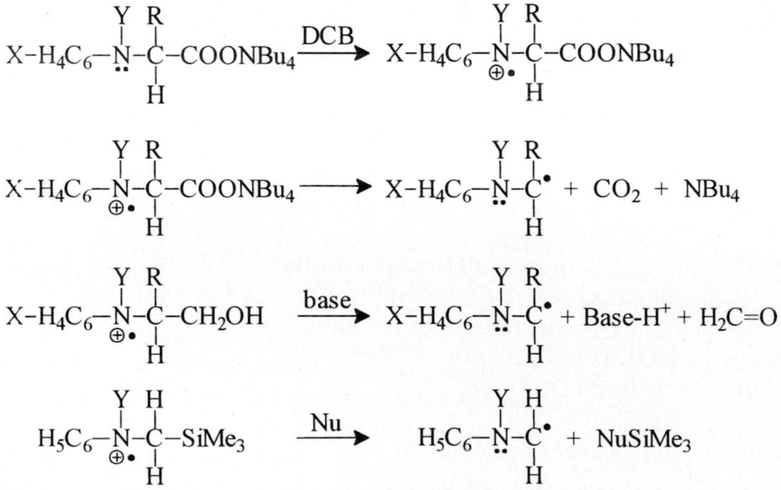

$$X\text{-}H_4C_6\text{-}\overset{Y}{\underset{H}{N}}\text{-}\overset{R}{\underset{}{C}}\text{-COONBu}_4 \xrightarrow{\text{DCB}} X\text{-}H_4C_6\text{-}\overset{Y}{\underset{H}{\overset{\oplus\cdot}{N}}}\text{-}\overset{R}{\underset{}{C}}\text{-COONBu}_4 \qquad (7)$$

$$X\text{-}H_4C_6\text{-}\overset{Y}{\underset{H}{\overset{\oplus\cdot}{N}}}\text{-}\overset{R}{\underset{}{C}}\text{-COONBu}_4 \longrightarrow X\text{-}H_4C_6\text{-}\overset{Y}{\underset{H}{N}}\text{-}\overset{R}{\underset{}{\overset{\cdot}{C}}} + CO_2 + NBu_4 \qquad (8)$$

$$X\text{-}H_4C_6\text{-}\overset{Y}{\underset{H}{\overset{\oplus\cdot}{N}}}\text{-}\overset{R}{\underset{}{C}}\text{-CH}_2OH \xrightarrow{\text{base}} X\text{-}H_4C_6\text{-}\overset{Y}{\underset{H}{N}}\text{-}\overset{R}{\underset{}{\overset{\cdot}{C}}} + \text{Base-H}^+ + H_2C=O \qquad (9)$$

$$H_5C_6\text{-}\overset{Y}{\underset{H}{\overset{\oplus\cdot}{N}}}\text{-}\overset{H}{\underset{}{C}}\text{-SiMe}_3 \xrightarrow{\text{Nu}} H_5C_6\text{-}\overset{Y}{\underset{H}{N}}\text{-}\overset{H}{\underset{}{\overset{\cdot}{C}}} + \text{NuSiMe}_3 \qquad (10)$$

**Scheme 10.**

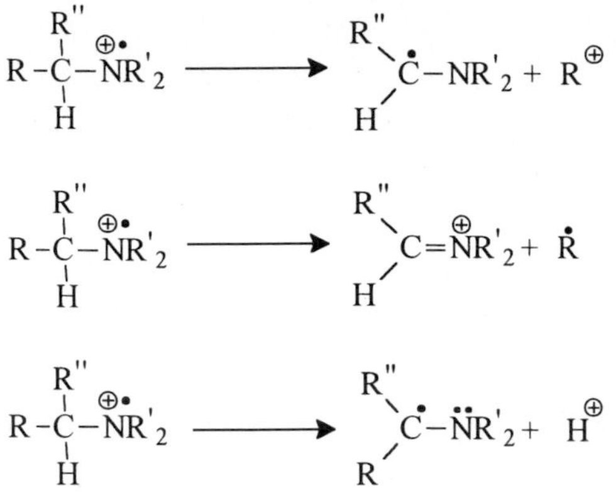

$$R-\overset{\overset{R''}{|}}{\underset{\underset{H}{|}}{C}}-\overset{\oplus\bullet}{N}R'_2 \longrightarrow \overset{R''}{\underset{H}{\diagdown}}\overset{\bullet}{C}-NR'_2 + R^{\oplus} \tag{11}$$

$$R-\overset{\overset{R''}{|}}{\underset{\underset{H}{|}}{C}}-\overset{\oplus\bullet}{N}R'_2 \longrightarrow \overset{R''}{\underset{H}{\diagdown}}C=\overset{\oplus}{N}R'_2 + \overset{\bullet}{R} \tag{12}$$

$$R-\overset{\overset{R''}{|}}{\underset{\underset{H}{|}}{C}}-\overset{\oplus\bullet}{N}R'_2 \longrightarrow \overset{R''}{\underset{R}{\diagdown}}\overset{\bullet}{C}-\overset{\bullet\bullet}{N}R'_2 + H^{\oplus} \tag{13}$$

**Scheme 11.**

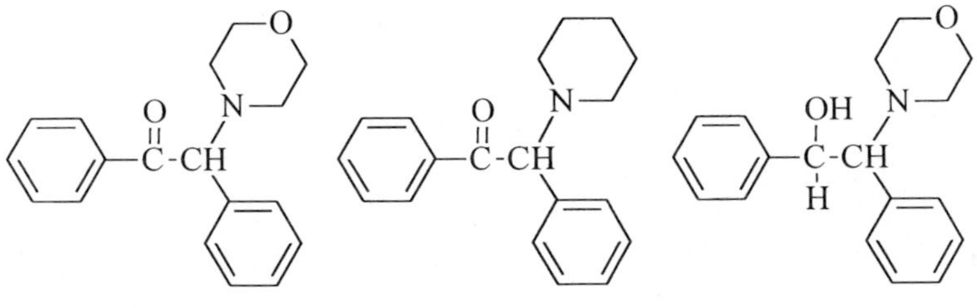

**50**          **51**          **52**

**Chart 3**

Photoinduced oxidation of the aminoketones **50** and **51** results predominantly in the formation of benzoyl radicals indicating Eq. 12 in Scheme 11 to be the predominant reaction pathway for the corresponding aminium radicals.

A general mechanism for fragmentation of aminium radicals derived from $\beta$-aminoalcohols in the presence of an electron acceptor (A*) has been proposed (Scheme 12) on the basis of studies of a variety of substituted 2-morpholino-1,2-diphenylethanol derivatives.

The fragmentation has been proposed to occur via proton loss from the alcohol. Studies of diastereomeric isomers of 2-morpholino-1,2-diphenyl ethanol **52** indicated a striking preference for the reaction of the *erythro* over *threo* isomers. This has been attributed to two factors. The first is the different efficiency of the quenching step (Eq. 15, Scheme 12)—it has been suggested that efficient intramolecular hydrogen-bonding of the *threo* isomer reduces the electron-donating capacity of the amine. A second and more important factor explaining the different yields is the competition between fragmentation (Eqs 17 and 18) and back-electron-transfer (Eq. 16), which yields the starting compounds. It has been proposed that the requirement of a co-planar, *anti*-parallel relationship between the alcohol, C–C bond, and amine radical cation is more easily obtained in the *erythro* isomer than for the *threo* compound.

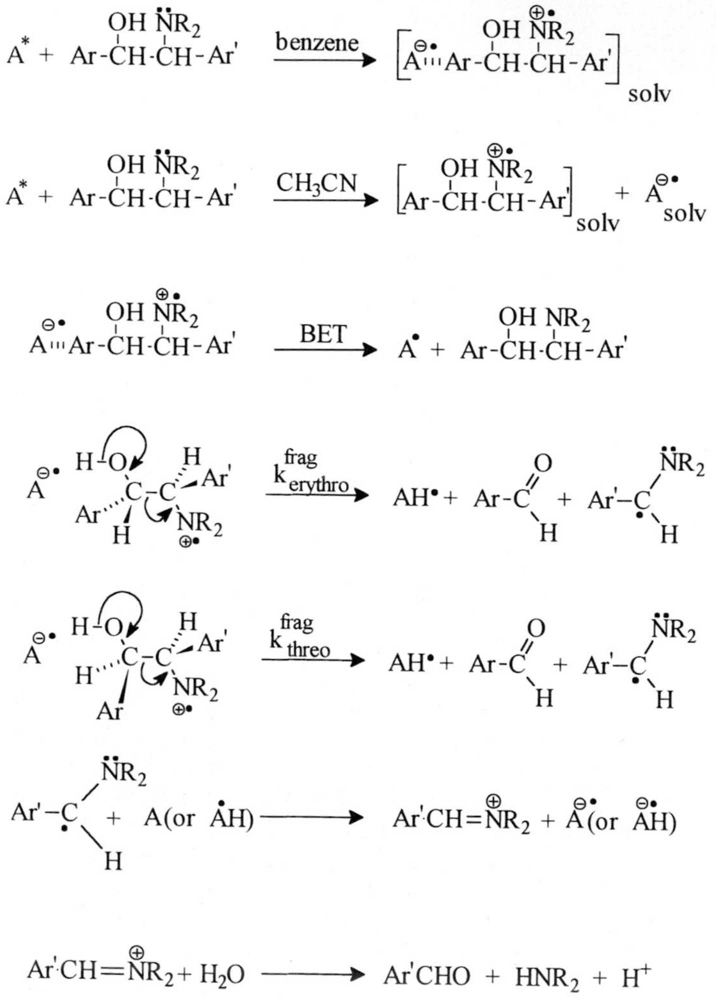

$$A^* + \text{Ar-}\overset{\text{OH}}{\underset{}{\text{CH}}}\cdot\overset{\ddot{\text{N}}\text{R}_2}{\underset{}{\text{CH}}}\text{-Ar}' \xrightarrow{\text{benzene}} \left[ A^{\ominus\bullet}{}^{\text{"}}\text{Ar-}\overset{\text{OH}}{\underset{}{\text{CH}}}\cdot\overset{\overset{\oplus\bullet}{\text{N}}\text{R}_2}{\underset{}{\text{CH}}}\text{-Ar}' \right]_{\text{solv}} \tag{14}$$

$$A^* + \text{Ar-}\overset{\text{OH}}{\underset{}{\text{CH}}}\cdot\overset{\ddot{\text{N}}\text{R}_2}{\underset{}{\text{CH}}}\text{-Ar}' \xrightarrow{\text{CH}_3\text{CN}} \left[ \text{Ar-}\overset{\text{OH}}{\underset{}{\text{CH}}}\cdot\overset{\overset{\oplus\bullet}{\text{N}}\text{R}_2}{\underset{}{\text{CH}}}\text{-Ar}' \right]_{\text{solv}} + A^{\ominus\bullet}_{\text{solv}} \tag{15}$$

$$A^{\ominus\bullet}{}^{\text{"}}\text{Ar-}\overset{\text{OH}}{\underset{}{\text{CH}}}\cdot\overset{\overset{\oplus\bullet}{\text{N}}\text{R}_2}{\underset{}{\text{CH}}}\text{-Ar}' \xrightarrow{\text{BET}} A^{\bullet} + \text{Ar-}\overset{\text{OH}}{\underset{}{\text{CH}}}\cdot\overset{\text{NR}_2}{\underset{}{\text{CH}}}\text{-Ar}' \tag{16}$$

$$A^{\ominus\bullet} \quad \text{H-O} \quad \overset{\text{frag}}{\xrightarrow{k_{\text{erythro}}}} \quad \text{AH}^{\bullet} + \text{Ar-}\overset{\text{O}}{\underset{\text{H}}{\overset{\|}{\text{C}}}} + \text{Ar}'\text{-}\overset{\ddot{\text{N}}\text{R}_2}{\underset{\text{H}}{\text{C}}}^{\bullet} \tag{17}$$

$$A^{\ominus\bullet} \quad \text{H-O} \quad \overset{\text{frag}}{\xrightarrow{k_{\text{threo}}}} \quad \text{AH}^{\bullet} + \text{Ar-}\overset{\text{O}}{\underset{\text{H}}{\overset{\|}{\text{C}}}} + \text{Ar}'\text{-}\overset{\ddot{\text{N}}\text{R}_2}{\underset{\text{H}}{\text{C}}}^{\bullet} \tag{18}$$

$$\text{Ar}'\text{-}\overset{\ddot{\text{N}}\text{R}_2}{\underset{\text{H}}{\text{C}}}^{\bullet} + A(\text{or } \dot{A}\text{H}) \longrightarrow \text{Ar}'\text{CH}=\overset{\oplus}{\text{N}}\text{R}_2 + \overset{\ominus\bullet}{A}(\text{or } \overset{\ominus\bullet}{A}\text{H}) \tag{19}$$

$$\text{Ar}'\text{CH}=\overset{\oplus}{\text{N}}\text{R}_2 + \text{H}_2\text{O} \longrightarrow \text{Ar}'\text{CHO} + \text{HNR}_2 + \text{H}^+ \tag{20}$$

**Scheme 12.**

## Study of excited-state and radical-ion intermediates in SET reactions using time-resolved techniques

Nanosecond and picosecond laser-flash photolysis techniques have been used by different groups to elucidate the various intermediate stages involved in the photo-induced reactions of amines. The overall mechanism involving the electron-transfer process in a fluid medium is illustrated in Scheme 13. The dynamics of the process involve the formation of an encounter complex between the excited-state molecule and the ground-state molecule [117, 140, 141]. The encounter complex can be described as an intermolecular ensemble of excited- and ground-state molecules, separated by a small distance (ca 7 Å) and surrounded by solvent molecules. During

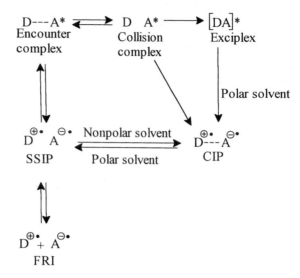

**Scheme 13.**

the lifetime of the encounter complex, the reactants undergo mutual collisions inside the solvent cage and as a result of these collisions, a stage is reached where the reactants are in contact to form what is called the collision complex. If the interaction between the reactants is strong enough (ca 5–20 kcal mol$^{-1}$), the collision complex can rapidly change to a new intermediate called an exciplex, which has partial charge transfer character and a large dipole moment. Electron transfer can occur at any of these stages [117]. Electron transfer from the collision complex or from the exciplex leads to the charge-transfer species called a contact ion pair (CIP). The contact ion pair can undergo slight separation in the solvent cage to generate a solvent-separated ion pair (SSIP). Alternatively, electron transfer from the encounter complex can directly lead to the SSIP. The solvent-separated ion pairs can then diffuse apart from the solvent cage and become separated to form the free solvated radical ions (FRI), which are analogous to free radicals and can undergo chemical reactions to yield products.

All these processes, namely, the formation of the encounter complex, collision complex, contact ion pairs, solvent-separated ion pairs, and free-radical ion pairs, are reversible. For generation of free radicals in good yields, forward electron-transfer processes have to compete efficiently with the energy wasting back-electron-transfer processes [142].

Picosecond and nanosecond transient spectroscopic studies to elucidate the nature of various intermediates formed in the photoinduced electron-transfer reaction of amines have been most extensive for the ketone–amine systems. These studies have been described in detail in a recent review by Yoon et al. [10]. Some aspects of these studies are briefly described here. In the earlier studies of Cohen and coworkers, the photochemical reactions of benzophenone with aliphatic amines were probed by fluorescence quenching, determination of product quantum yields, and nanosecond laser-flash photolysis [143–147]. They proposed that the reactions of amines with the triplet states of benzophenone proceed via the formation of an intermediate charge-transfer complex ($^3$CTC*), although no direct evidence for the formation of such complexes was available from their studies. Laser-flash photolysis of benzophenone in the presence of a variety of aliphatic amines indicates the formation of ketyl radicals ($\lambda_{max}$ 555 nm) with a quantum yield of unity. Ketyl radicals could be generated via hydrogen-atom abstraction from amines rather than via electron transfer followed by proton transfer between the radical ion pair. However the extremely fast reaction between amine and the triplet excited state of benzophenone ($k = 5 \times 10^{10}$ s$^{-1}$) supports the view that electron-transfer reactions are involved. Direct proof of the electron-transfer mechanism was obtained from their studies on the flash photolysis of benzophenone–DABCO systems [146]. In these systems the formation of the ketyl radical anion as a transient species ($\lambda_{max}$ 660 nm) could be observed, because of the increased stability of DABCO$^{+\bullet}$ and because deprotonation of the $\alpha$-CH is not facile for this constrained amine [147].

Photoreduction of benzophenone by primary and secondary amines leads to the formation of benzpinacol and imines [145]. Quantum yields greater than unity for reduction of benzophenone indicated that the $\alpha$-aminoalkyl radical could further reduce the ground state of benzophenone. Bhattacharyya and Das confirmed this in a laser-flash photolysis study of the benzophenone–triethylamine system, which showed that ketyl radical anion formation occurs by a fast and a slow process wherein the slow process corresponds to the reaction of $\alpha$-aminoalkyl radical in the ground state of benzophenone [148]. Direct evidence for similar secondary reduction of benzil [149] and naphthalimides [150] by the $\alpha$-aminoalkyl radical have also been reported. The secondary dark reaction of $\alpha$-aminoalkyl radicals in photoinduced electron-transfer reactions with a variety of quinones, dyes, and metal complexes has been studied by Whitten and coworkers [151].

Formation of the ketyl radical anion in photoinduced electron-transfer reactions of benzophenone with butylamine and triethylamine was observed by Shizuka and coworkers in neat amines at low temperatures [14, 152]. At room temperatures they observed the formation of the ketyl radical ($\lambda_{max}$ 555 nm), whereas on reducing the temperature to 120 K the transient absorption spectrum showed a decrease in intensity at 555 nm and a subsequent increase at 660 nm. This has been attributed to the slowing down of proton transfer between the radical ion-pairs at low temperatures, making it possible to measure ketyl radical anion absorption at 660 nm [152]. At very low temperatures however (ca 120 K) only the triplet state of benzophenone was observed ($\lambda_{max}$ 525 nm), indicating thereby that under these conditions the triplet state of benzophenone does not undergo electron-transfer reactions with amines.

Direct evidence for the formation of radical ion intermediates in the benzophenone–amine system was also obtained by Peters et al. [153–156]. Picosecond laser-flash photolysis studies have indicated the formation of ketyl radical anions concomitant with the decay of the benzophenone triplet. For 1.0 M dimethylaniline and diethylaniline the rate of electron transfer to the benzophenone triplet was $3.6 \times 10^{10}$ and $4.2 \times 10^{10}$ M$^{-1}$ s$^{-1}$, respectively. On the basis of their studies Peters et al. proposed a mechanism in which a solvent-separated ion pair

(SSIP) is formed first; this subsequently collapses into a contact ion pair (CIP). This mechanism was based on the observation of a blue shift of the absorption maximum of the ketyl radical anion from 715 to 690 nm which occurs with a half-life of $200 \pm 50$ ps. The SSIP, being more solvated than the CIP, is expected to have its absorption spectrum red-shifted compared with that of CIP. Subsequent studies by Devadoss and Fessenden on the benzophenone–DABCO system indicated, however, that the spectrum of initial transient has an absorption maximum at 700 nm, which shifts to the red (720 nm) in the picosecond time domain [157, 158]. These results seem to suggest that the initial species to be formed is the CIP which eventually separates to yield the SSIP and proton transfer occurs in the SSIP.

They also observed that the species absorbing at 720 nm which decays on the microsecond time-scale by first-order kinetics is a long-lived SSIP. Evolution of the SSIP to free-radical ions was ruled out, because free-radical ions should decay by second-order kinetics. Later observations by Mataga [159, 160] and by Haselbach [161], on the basis of photoconductivity measurements and transient absorption spectroscopy of the benzophenone–DABCO system, showed that the species absorbing at 710 nm decays not by first-order kinetics but by second-order kinetics; this is consistent with the formation of free-radical ions.

The mechanism of the photoreduction of benzophenone by dimethylaniline (DMA) has subsequently been probed by Mataga and coworkers by use of femtosecond and picosecond absorption spectroscopy [160]. For the first time they considered the role in the photoreduction process of the first excited singlet state ($^1$BP*) and the triplet excited state ($^3$BP*) of benzophenone. Mataga's investigation revealed that three radical ion pairs are formed in the reduction process. In the ground state, benzophenone forms a ground state charge-transfer complex with dimethylaniline. Thus 355 nm irradiation leads to excitation of both benzophenone and the charge-transfer complex. The charge-transfer complex on excitation directly forms the contact ion pair, with an absorption maximum at 740 nm; this decays by back-electron-transfer in 85 ps to reform the ground-state reactants. During the decay process, there is no change in the absorption maximum, and proton transfer does not occur. Excitation of the uncomplexed benzophenone produces $^1$BP*, which undergoes intersystem crossing to $^3$BP* with a rise time of $9 \pm 2$ ps. Both $^1$BP* and $^3$BP* are quenched by electron transfer to form the singlet radical ion pair and triplet radical ion pair with absorption maximum at 710 nm. These studies suggest that the electron-transfer process produces the SSIP, and that it is within the $^1$BP and $^3$BP SSIP that proton transfer occurs. There is no evidence based on spectral shifts for the conversion of CIP to SSIP or SSIP to CIP. Recent studies by Peters and coworkers support Mataga's results [162, 163].

Mataga and coworkers have also confirmed that ion-pair separation was highly dependent on the energy gap of charge separation ($-\Delta G_{cs}$) [164]. With increasing $-\Delta G_{cs}$, proton transfer rates within the radical ion-pair decreased rapidly, whereas the rates for dissociation increased slightly. These results indicated that the charge-separation process occurred at larger encounter distances for larger $-\Delta G_{cs}$ leading to a decrease in proton transfer within the radical ion pair and increase in ion-pair dissociation.

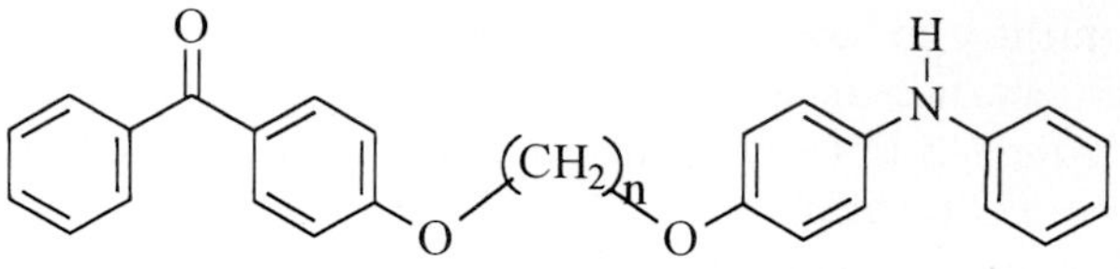

**53**

**Chart 4**

The effects of chain-length, solvent, and temperature on the intramolecular photoreduction of benzophenone linked to diphenylamine (**53**) has also been explored in detail [165, 166].

For these systems, direct hydrogen abstraction by benzophenone triplets was observed in benzene whereas in a polar solvent electron transfer and hydrogen-atom abstraction were observed. Electron transfer followed by an intramolecular proton transfer was observed in these systems although such proton-transfer reactions are not observed in unlinked systems of primary and secondary amines. The observed differences between the linked and unlinked systems have been attributed to the dependence of electron transfer, proton transfer, and hydrogen transfer on mutual distance and orientation. In the unlinked systems, rotational and translational motion of two reacting molecules are usually much faster than those in linked systems.

In the benzophenone-photosensitized reactions of $N,N'$-dialkyl-1-naphthylamines (DANA) in acetonitrile, Kiyota et al. observed that the initial process was a triplet energy transfer, both in the absence and presence of water or methanol [167]. The triplet aminonaphthalene so formed reacts with ground-state benzophenone to yield a triplet exciplex $^3$(DANA–BP)*. In the absence of water or methanol, the exciplex decays back to DANA and benzophenone. In the presence of water or methanol however, charge separation to yield DANA$^{+\cdot}$ and BP$^{-\cdot}$ was observed. It has been proposed that the driving force for intra-exciplex electron transfer is the enlarged reduction potential of benzophenone in the exciplex, because of hydrogen-bonding by water or methanol to the carbonyl group.

Hamanoue et al. have investigated electron-transfer reactions of several amines with the triplet state of anthraquinones [168, 169]. The photochemistry of anthraquinone is expected to vary depending on the nature of the lowest excited state. It has been observed that the lowest $n$–$\pi^*$ state abstracts a hydrogen atom, whereas the lowest $\pi$–$\pi^*$ state reacts via electron transfer. By introducing substituents on the anthraquinone chromophore, the nature of the lowest triplet state can be changed from $n$–$\pi^*$ to $\pi$–$\pi^*$. Picosecond laser-flash photolysis studies of the mixtures of anthraquinone and 1-chloroanthraquinone with amines reveal that the anthraquinone triplet forms a triplet exciplex with triethylamine. In polar solvents the exciplex changes to a contact ion pair which disappears by proton exchange between the radical ions. In acetonitrile, two transients from the SSIP and the triplet exciplex were observed. Among these transients, the SSIP is assumed to be produced by electron transfer from triethylamine to a higher triplet ($T_2$) of anthraquinone upon laser excitation, whereas the exciplex is produced by the reaction of the lowest triplet ($T_1$) of anthraquinone with triethylamine. It was proposed that deprotonation of the aminium cation, produced after the electron-transfer reaction with anthraquinone, occurred in the contact ion pair. Recent CIDNP studies of the deprotonation of the aminium cation, generated from the anthraquinone sensitized electron-transfer reaction reveal that the relative group reactivity or substituent effect plays a key role in determining the time of proton elimination [170]. These studies show that deprotonation of the methyl, ethyl and isopropyl substituents occurs exclusively in the solvent-separated ion pair whereas the deprotonation of the allyl substituents occurs within the contact ion pair. This difference in behavior can be explained on the basis of an increase in the rate of in-cage proton transfer relative to cage life with increasing $\Delta G$ of the reaction.

The photoreduction of 9,10-anthraquinone-1,5-disulfonate by 2,2,6,6-tetramethyl piperidine in aqueous media has been studied in the nanosecond and microsecond time domains by use of time-resolved optical and ESR measurements [171]. Electron transfer from the amine to the excited state of the anthraquinone derivative occurs with a rate constant of $5.7 \times 10^8$ $M^{-1}$ $s^{-1}$. The aminyl radicals formed via deprotonation of the aminium radicals are long-lived (ca 0.5 ms), because the steric hindrance of these radicals slows down recombination reactions. The aminyl radicals formed in these systems have been characterized by ESR.

Another area where the mechanism of intramolecular photoinduced charge separation involving amines has attracted much attention is the phenomenon of dual luminescence observed in dimethylaminobenzonitrile (DMABN) and related molecules. Dual luminescence of DMABN, first observed by Lippert [172, 173] has been extensively studied [174–183]. Grabowski proposed that the anomalous long-wavelength emission of DMABN and related molecules could be assigned to a rotational isomer namely a twisted intramolecular charge-transfer state (TICT), in which a full charge is transferred from the dimethylamino moiety to benzonitrile. The process is depicted in Scheme 14 where A and D represent the acceptor and donor moieties, respectively. It is proposed that in this state the amine and benzonitrile orbitals are completely decoupled in a 90° twisted conformation. Although many other mechanisms have been proposed to explain the phenomenon of dual luminescence, the TICT model has attracted most attention. Since then the idea of TICT has been extended to other classes of molecule with anomalous dual fluorescence [174, 179].

Recently Zachariasse and coworkers have questioned the role of TICT in DMABN and related molecules [180–183]. They observed an absence of a linear correlation between the energy of the anomalous fluorescence band and the redox potential of the donor and acceptor groups and also observed dual emission in

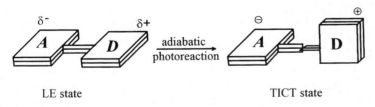

LE state                                    TICT state

**Scheme 14.**

derivatives where 90° twist of the amino group is not possible. Zachariasse et al. propose a new model, involving a planar intramolecular charge-transfer state (PICT). They proposed that the dual fluorescence in these classes of molecules arise as a result of two close-lying excited states [$S_1$ and $S_2$ (CT)], which have an energy gap small enough to enable vibronic coupling of the two states. It has been suggested that *N*-inversion of the amino group acts as a promoting mode between the two levels.

### 7.2.5 Electron-transfer Reactions of Amines in Biochemical Systems

Electron-transfer reactions of amines are of significant importance in biochemical systems. Enzymes known to catalyze the oxidative dealkylation of amines include monoamine oxidase [16, 17], cytochrome-P450 [18, 184–186], horseradish peroxidase [187], hemoproteins [188, 189], and chloroperoxidase [187, 188]. N-dealkylation of amines by peroxidases are generally accepted to occur via one-electron transfer, whereas the role of electron transfer in reactions catalyzed by enzymes such as monoamine oxidase [16, 17] and cytochrome P-450 [18, 184, 185] is currently a topic of debate.

### Amine oxidations catalyzed by monoamine oxidase

Monoamine oxidase, which exists in two distinct forms, referred to as MAO A and MAO B, is one of the enzymes responsible for the degradation of biologically important amines. Compounds that block the catalytic action of MAO A, which is selective for the degradation of norepinephrine and serotinin, have antidepressant effects whereas compounds that inhibit MAO B, which degrades dopamine in the brain, are useful for treating Parkinson's disease [190, 191]. Both MAO A and MAO B contain flavin co-enzyme attached at the 8-α-position to an enzyme-active cysteine residue (**54**). A one-electron transfer mechanism (Scheme 15) for the oxidations catalyzed by MAO was first proposed by Silverman [192] and Krantz [193, 194].

One-electron transfer from the substrate amino group to flavin (Fl) results in the formation of the aminium radical and the flavin radical anion (Fl$^{-\cdot}$) (Scheme 15). Deprotonation of the aminium radical to yield an α-aminoalkyl radical followed by a second electron transfer to the flavin radical anion will result in the formation of the reduced flavin and iminium ion. Alternatively the iminium ion can be formed by path d in Scheme 15; this involves formation of a covalent adduct which can

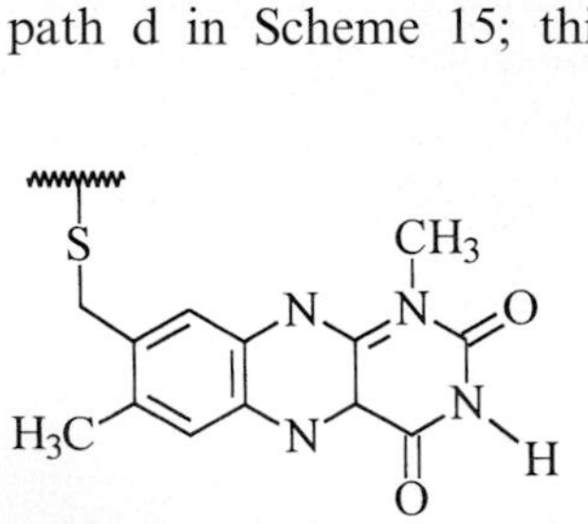

**54**

**Chart 5**

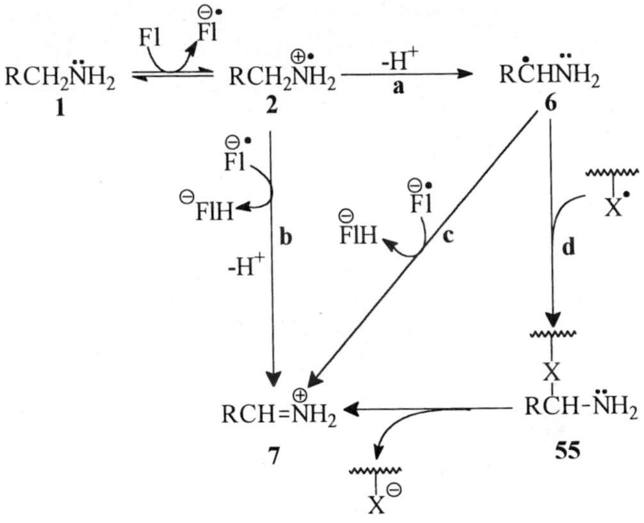

**Scheme 15.**

subsequently decompose to give the iminium ion. The 'X' in Scheme 15 can be the flavin radical anion or an amino acid radical formed by hydrogen-atom abstraction from the amino acid by the flavin radical anion. Instead of deprotonation the aminium radical could also undergo hydrogen abstraction by the flavin radical anion (path b) which would directly result in the formation of the iminium ion.

The transfer of a first electron from amine to the flavin is a thermodynamically uphill process, being endothermic by about 1.75 V for a primary amine. It has been proposed that intrinsic binding, which could distort the bonds of the amine and flavin, results in a lowering of the redox potentials of the amines and the flavin, making electron transfer feasible [16, 17].

Extensive studies by Silverman and coworkers on the mechanism of inactivation of the enzyme have provided strong support for the role of electron transfer in the catalytic mechanism of MAO [16, 17]. Laser-flash photolysis of cyclopropylamine **56** showed that its aminyl radical undergoes a very fast ring opening ($>5 \times 10^8$ M$^{-1}$ s$^{-1}$) to yield the primary radical **58** (Scheme 16) [195].

Using similar systems Silverman and coworkers have explored the oxidation of a variety of cyclopropyl amine substrates by MAO [16, 17, 196, 197]. Involvement of electron transfer in such processes would lead to the formation of the ring-opened alkyl radical and subsequent attachment of the resultant alkyl radical to the enzyme would result in its inactivation. Scheme 17 shows the proposed mechanism for enzyme deactivation by 1-phenylcyclopropylamine.

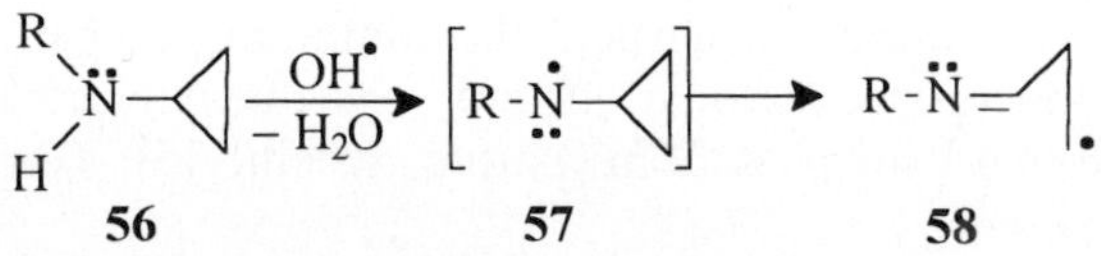

**Scheme 16.**    **56**    **57**    **58**

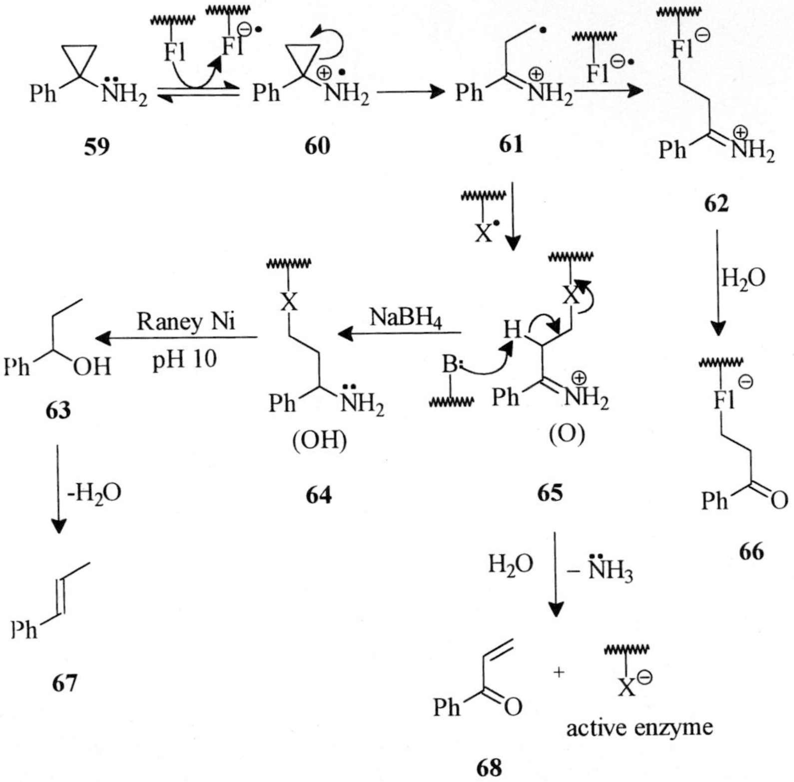

**Scheme 17.**

Electron transfer from the amine to flavin would result in the aminium radical which is expected to rearrange rapidly to radical **61**. Inactivation of the enzyme would then occur via coupling of the radical with the flavin radical anion resulting in the formation of **66**. Coupling of the aminium radical with an amino acid radical would result in the formation of **65**. By use of radioactive labeling techniques Silverman et al. have confirmed the formation of **65** and **66**; this confirms the role of electron transfer in the oxidation process. Similar studies have been performed using 1-phenylcyclobutylamine (Scheme 18) [198].

Identification of **74** on incubation of MAO with **69** confirmed the initial formation of the 1-phenylcyclobutylaminium radical. That a variety of cyclopropylamines and cyclobutylamines inactivate MAO and result in ring-opened adducts is consistent with the intermediacy of the aminyl radicals. These results strongly suggest that the MAO-catalyzed oxidation of amines involves one-electron transfer as the first step.

As discussed earlier, the aminium radical formed in this process can either undergo deprotonation and subsequent electron transfer (path a, Scheme 15) or lose a hydrogen atom (path c, Scheme 15). To differentiate between these processes

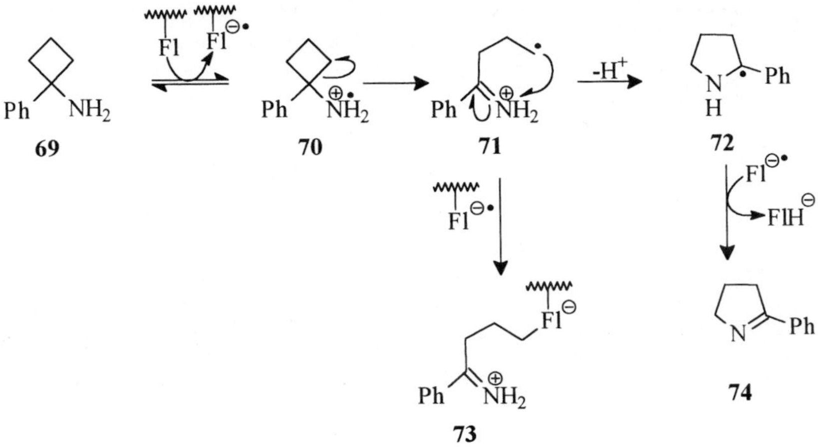

**Scheme 18.**

Zelechnak and Silverman [199] incubated MAO with *trans*-1-aminomethyl-2-phenylcyclopropane (**75**) (Scheme 19). Newcomb et al. had previously shown that the *trans*-2-phenyl(cyclopropenyl)carbinyl radical undergoes very fast ring cleavage ($k = 3 \times 10^{11}$ s$^{-1}$) [200]. By comparison the $\alpha$-aminoalkyl radical (**78**) if generated in these systems would be expected to form 2-hydroxy-5-phenyltetrahydrofuran (**76**). However **76** could not be detected, suggesting that the reaction occurred by path b (Scheme 19) or that the second electron transfer (path c) was much faster than the ring-opening reaction. It has been proposed that the cleavage reaction of

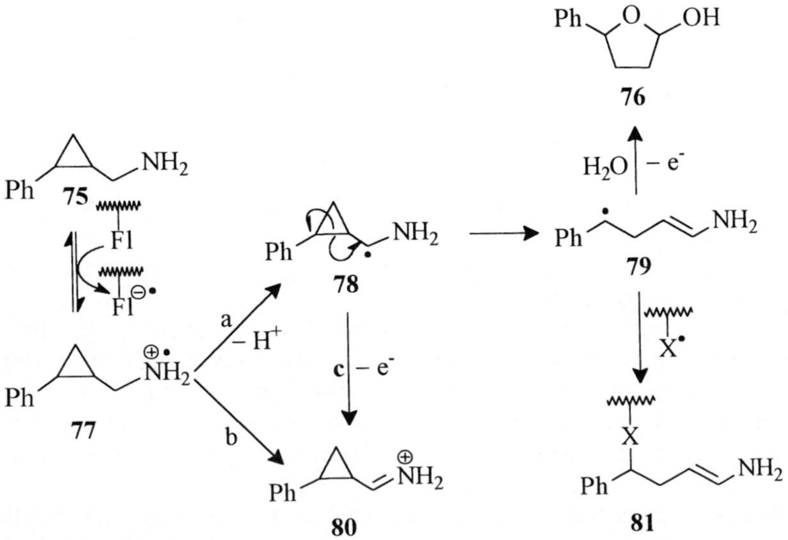

**Scheme 19.**

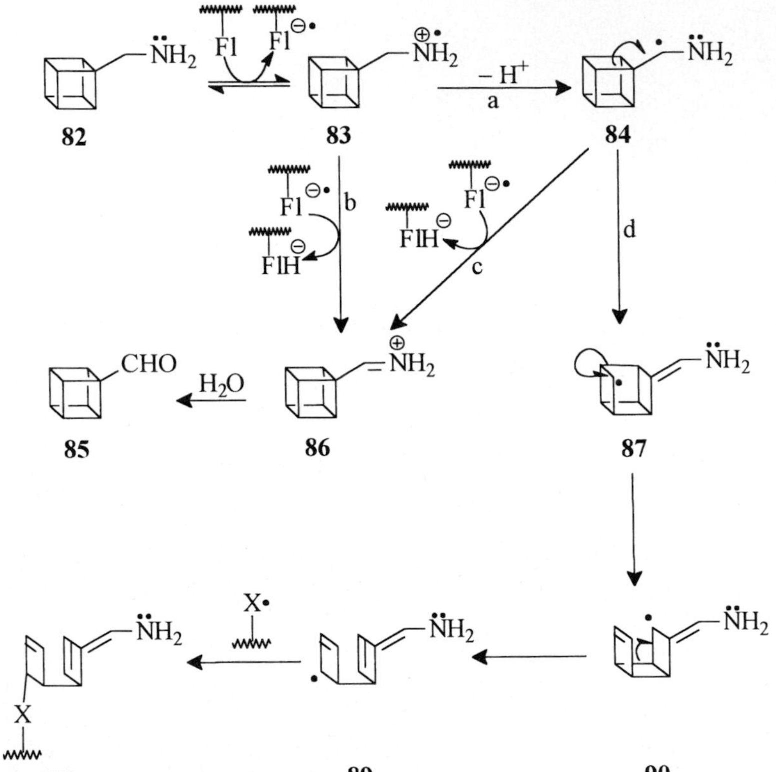

**Scheme 20.**

**78** (Scheme 19) could be slowed because of stabilization by the amine group and during enzyme binding the free rotation of the molecule is frozen preventing overlap of the $\alpha$-carbon radical orbital and the cyclopropyl radicals.

The occurrence of deprotonation/second electron transfer in such systems was confirmed by use of aminomethylcubane **82** (Scheme 20) [201]. Treatment of MAO with aminomethylcubane led to time-dependent inactivation of the enzyme and the formation of products arising from ring cleavage. If the hydrogen abstraction by flavinium radical anion (path b) is the predominant pathway the major product formed would be iminium ion **86**. Ring cleavage products cannot arise from the iminium ion **86**, because cubyl carbonyl cations are well known to undergo exclusive rearrangement to homocubanes. Decomposition of the cubane nucleus indicates that deprotonation of the aminium radical (path d) is the major reaction pathway in MAO catalyzed oxidation of amines.

Destruction of the cubane structure could also however occur via carbanion intermediates. To rule out this possibility Silverman et al. studied the MAO-catalyzed reactions of cinnamylamine-2,3-oxide [202].

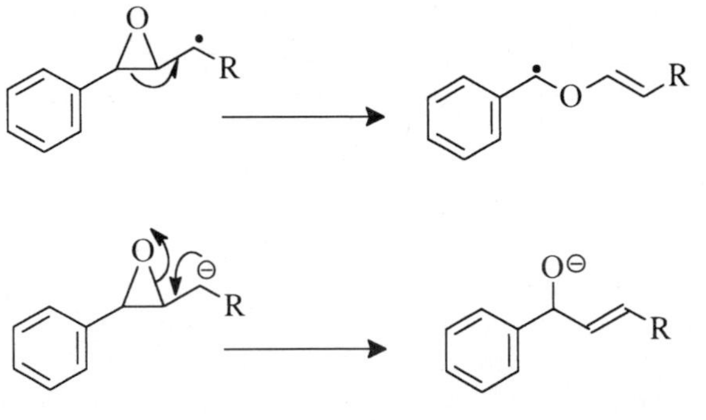

(21)

(22)

**Scheme 21.**

1-Substituted-2,3-epoxy-3-phenylpropyl radicals have been reported to cleave the C–C bond exclusively (Eq. 21, Scheme 21) whereas, the corresponding carbanion undergoes exclusive C–O bond cleavage (Eq. 22, Scheme 21) [203, 204]. Formation of benzaldehyde and glycolaldehyde and the absence of cinnamaldehyde, a product expected via C–O bond cleavage supports the mechanism shown in Scheme 22 for the MAO catalyzed oxidation of cinnamylamine-2,3-oxide (**91**).

Silverman's studies on mechanism based MAO inactivation have provided overwhelming support for the role of electron transfer in the MAO catalyzed dealkylation of amines. It must be mentioned however that spectroscopic attempts for detecting the radical ion intermediates have hitherto been unsuccessful. Yasanobu and coworkers could not find EPR spectral evidence for radical intermediates in MAO-catalyzed oxidation of benzylamine [205]. Miller et al. failed to observe the flavin semiquinone or an amine–flavin adduct in rapid-scan-stopped flow spectroscopy [206]. The only time-dependent absorption change observed in this study was the bleaching of the oxidized flavin. Furthermore, no influence of a magnetic field up to 6500 G was observed on the rate of MAO B reduction. The reaction rates of systems with kinetically significant radical pair intermediates are known to be altered

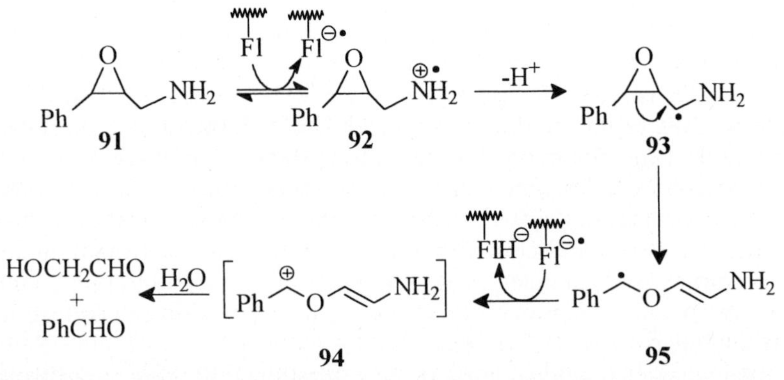

**Scheme 22.**

by exogenous magnetic fields ($B = 10$–$3000$ G) [207, 208]. A possible explanation proposed by Silverman for the lack of observation of the flavin semiquinone is that the first electron transfer is reversible and that the back-electron-transfer is much faster than the forward reaction leading to very short lifetimes for the flavin semi-quinone. In the ESR measurements this would lead to very low steady state concentrations of the free radicals and hence would not be observable.

Search for new mechanism based investigations for deducing the mechanism of the enzyme catalyzed activity continues to be active area of research. Mariano and coworkers have used activated flavins such as 5-ethylflavinium perchlorate, whose ground state reduction potentials are high enough to promote oxidative dealkylation of amines, as enzyme models [209]. Studies on the inactivation of the model enzymes by cyclopropylamines and α-silylamines suggest a polar mechanistic model. Silverman attributes this result to the drastically altered nature of the flavin used in these studies, which could favor a nucleophilic mechanism [16].

In a more recent study, Mariano and coworkers have probed the use of the reported retro-aldol type fragmentation of aminium radicals generated from *tert-β*-allylic and -propargylic *β*-aminoalcohols as MAO inactivators [210]. Photoexcitation of a model flavin, [3-methyllumiflavin (3MLF)], generates the triplet excited state of 3MLF which undergoes SET reactions with *β*-aminoalcohols. Analysis of the photoproducts, namely 4a adducts of 3MLF are indicative of the formation of aminium radicals which undergo facile retro-aldol-like fragmentation. These studies as well as the studies on MAO A inactivation by the *β*-hydroxylamines suggest that the enzymatic sequence is most likely initiated by one-electron transfer within the enzyme–inactivator complex which is then followed by a retro-aldol fragmentation of the intermediate tertiary aminium radical.

### Amine oxidations catalyzed by cytochrome P-450

Cytochrome P-450 and hemoproteins are known to be involved in the oxidation of a broad variety of drugs, pesticides, carcinogens, steroids and fat soluble vitamins [18, 184, 185]. Amine oxidation by P-450 has been proposed to proceed either via electron/proton transfer mechanism or via hydrogen-atom abstraction mechanism (Scheme 23).

Studies using isotopic labeling as well as mechanism based on inactivation such as 4-alkyldihydropyridines and cycloalkylamines have supported the view that the first step involves an electron-transfer process (path a, Scheme 23) [18, 184–186, 211]. Deprotonation of the resultant aminium radical would yield the α-aminoalkyl radical. The formation of the dealkylated amine and carbonyl derivative has been proposed to occur via a second electron transfer to the enzyme and a nonenzymatic hydrolysis of the imine formed. In the P-450 catalyzed reaction, however this process is proposed to occur via a radical recombination process to yield a carbinolamine (**99**) which then decomposes to the dealkylated amine and the corresponding carbonyl derivative. Evidence for this was obtained by the incorporation of label from $^{18}O_2$, into the carbonyl derivative [212–214].

In a recent study, Dinnocenzo and coworkers have questioned the role of electron

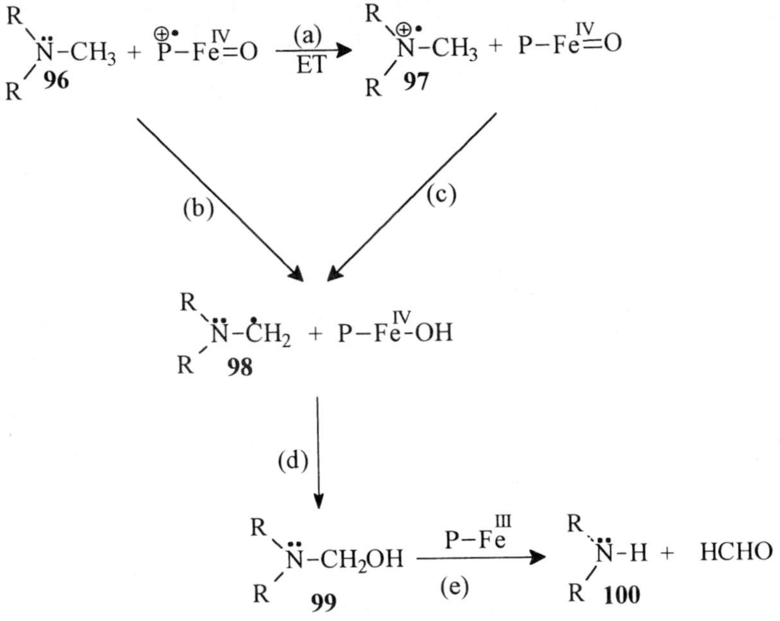

$$R_2\overset{..}{N}-CH_3 + \overset{\oplus\bullet}{P}-\overset{IV}{Fe}=O \xrightarrow[\text{ET}]{(a)} R_2\overset{\oplus\bullet}{N}-CH_3 + P-\overset{IV}{Fe}=O$$

**96**  →  **97**

(b)        (c)

$$R_2\overset{..}{N}-\overset{\bullet}{C}H_2 + P-\overset{IV}{Fe}-OH$$

**98**

(d) ↓

$$R_2\overset{..}{N}-CH_2OH \xrightarrow[(e)]{P-\overset{III}{Fe}} R_2\overset{..}{N}-H + HCHO$$

**99**                        **100**

**Scheme 23.**

transfer in these processes [215, 216]. They have studied the isotope effect profiles for *para*-substituted *N*-methyl-*N*-(trideuteromethyl)aniline cation radicals by pyridine and for the hydrogen abstraction from their parent amines by *tert*-butoxyl radicals. The isotope effect profiles for the P-450 catalyzed oxidation of these amines (increased isotope effects with increasing electronegativity of the *para*-substituent) were found to be indistinguishable from the hydrogen abstraction profile and distinctly different from the deprotonation profile.

In a subsequent study Guengerich et al. did not observe a continuous trend for an increase in isotopic effects with increase in electron withdrawing nature of the *para* substituent [217]. The larger value observed in their study for the 4-nitro derivative has been attributed to either a reversible electron transfer step proceeding to deprotonation or to a hydrogen-atom abstraction mechanism for this molecule. The higher oxidation potential of 4-nitro-*N,N*-dimethylaniline could be a contributing factor.

Recent studies by Baciocchi et al. on kinetic deuterium isotope effect profiles and substituent effects in the oxidizing N-demethylation of *N,N*-dimethylanilines catalyzed by tetrakis (pentafluorophenyl)porphyrin also supports the electron-transfer mechanism and exclude the hydrogen-atom transfer mechanism in such processes [218].

Evidence in favor of the electron-transfer mechanism has also come from studies on substrates such as *N*-ethyl-*N*-methylaniline which show a demethylation/

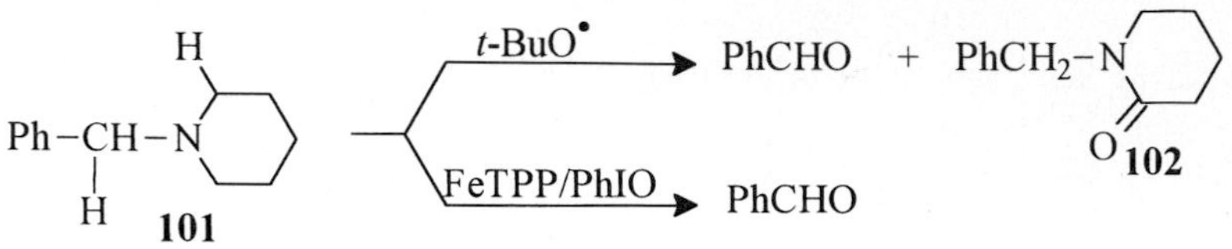

**Scheme 24.**

deethylation ratio of 16 to 20 [219, 220]. These results can be attributed to steric preference for deprotonation of the methyl proton [7, 221] of the aminium radicals. In *tert*-butoxyl radical systems the rate of hydrogen-atom abstraction was twice as that for the $N$-methyl group [222].

Recently Bietti et al. have investigated the use of $N$-benzylpiperidine for differentiating between the electron transfer and hydrogen-atom transfer (HAT) mechanisms [223]. Reaction of the substrate with $t$-BuO$^•$ produced benzaldehyde and 1-benzyl-2-piperidone indicating hydrogen-atom abstraction both from the benzylic and piperidine ring $\alpha$-methylene groups (Scheme 24). From the product ratio a relative rate of $k_{benzyl}/k_{alkyl}$ of 1.4 was obtained taking into account the number of equivalent positions. Under biomimetic conditions using iodosobenzene and tetraphenylporphyrin iron (III) chloride as the catalyst only benzaldehyde was obtained (Scheme 24). These differences can be attributed to the preferential deprotonation of the benzylic proton from aminium radicals, for stereoelectronic reasons, thereby confirming the electron-transfer mechanism for the reaction.

Baciocchi and Lapi have investigated the oxidation of $N$-methyl-9-$t$-butylacridane **103** by iodosylbenzene catalyzed by tetrakis(pentafluorophenyl)porphyrin iron (III). (FeTPFPP) [224]. It has been reported earlier by Anne and coworkers that the radical cation of **103** generated electrochemically undergoes a very fast C–C bond cleavage to form the $t$-butyl radical and $N$-methylacridinium cation [225, 226]. In the presence of a very strong base such as 2,4,6-trimethylpyridine (p$K_a$ ~ 15.6) deprotonation of the $N$-methyl group occurs leading to the formation of acridine. The biomimetic oxidation of **103** with iodosylbenzene in the presence of FeTPFPP in $CH_2Cl_2$ at room temperature using a substrate/oxidant/catalyst ratio of 50:10:1 led exclusively to the formation of 9-$t$-butylacridane and formaldehyde [224]. On decreasing the substrate oxidant ratio, formation of acridine was also observed. The mechanism shown in Scheme 25 was proposed to explain the results observed.

The low kinetic isotope effect using $N$-*tri*-deuteromethyl-9-$t$-butylacridane indicated that a hydrogen-atom abstraction was not operative in this case.

Dealkylation of the $N$-methyl group indicates that the initially formed radical cation **104** (Scheme 25) undergoes preferential deprotonation of the $N$-methyl group instead of C–C bond cleavage. In view of the earlier report on electrochemical oxidation of **104**, this would suggest that P–Fe(IV)=O is a relatively strong base (p$K$ ≥ 16). This observation is interesting in view of the earlier studies where significant basicity of P–Fe(IV)=O has been implied [227].

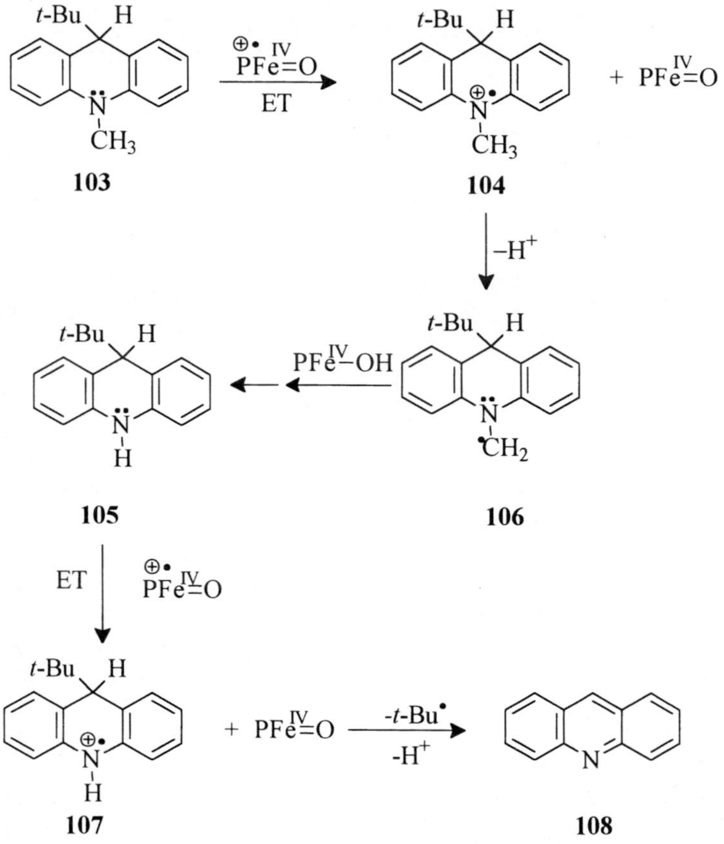

**Scheme 25.**

## 7.3 Synthetic Applications

The α-aminoalkyl radicals as well as iminium ions generated as intermediates in electron-transfer reactions of amines can be used for bringing about synthetically useful transformations of amines. The synthetic applications of amine oxidation reactions brought about by thermal, electrochemical and photochemical methods as discussed below.

### 7.3.1 Thermal Methods

#### Chemical oxidation

Because of the relative instability of amine radicals, their synthetic applications in thermal one-electron-catalyzed reactions are rare. In contrast free radicals generated in the oxidation of amides are more common [19–23, 228]. Aminium radicals gen-

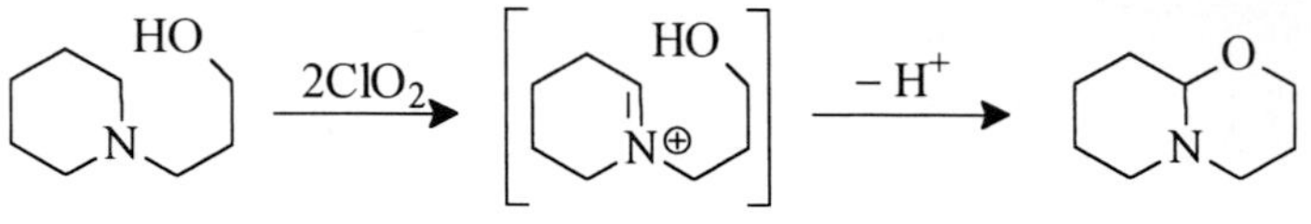

**Scheme 26.**

erated via metal catalyzed degradation of chloro- and hydroxylamines can undergo a variety of synthetically useful reactions such as inter- and intramolecular addition to olefins and in aromatic amination reactions [22, 55].

Chlorine dioxide catalyzed cyclization of tertiary aminoalcohols (Scheme 26) to oxazilidines and tetrahydro-1,3-oxazines in basic aqueous media have been reported [229].

Similar reactions, were earlier reported by Audeh and Lindsay Smith in ferricyanide catalyzed reactions [44], although in much lower yields. The $ClO_2$ catalyzed cyclization showed a greater regiochemical tendency for ring closure to occur at the less substituted α-carbon of the piperidine moiety compared to $Hg(OAc)_2$ catalyzed reactions of the same compounds.

α-Cyanoamines are powerful synthons for the preparation of a variety of substituted piperidines. Reactions of $ClO_2$ with tertiary amines in the presence of 5–7 mol equivalent of aqueous sodium cyanide afforded 53–88 % of the α-cyano-substituted tertiary amine. This strategy has been utilized for the synthesis of the alkaloid (±)-elaeocarpidine (**115**) in moderate yields (Scheme 27) [229].

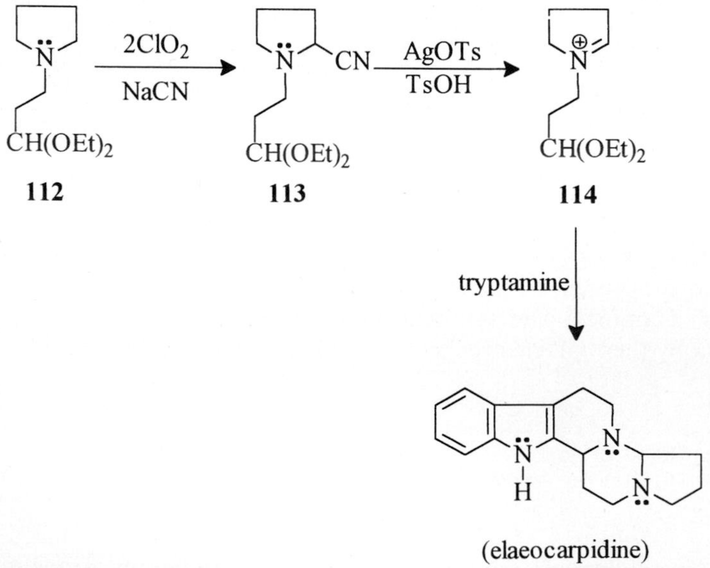

(elaeocarpidine)
**115**

**Scheme 27.**

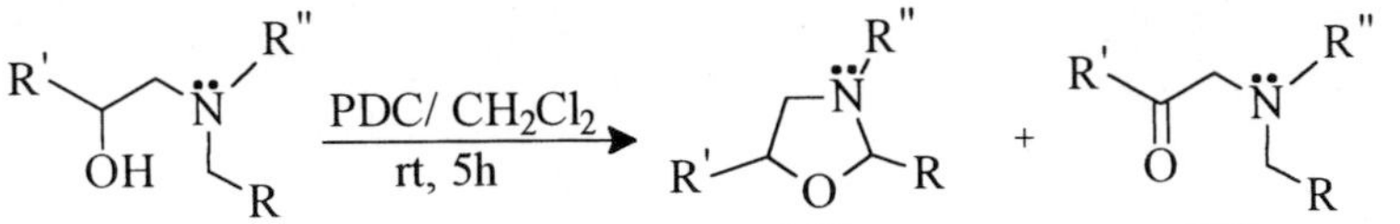

116 117 118

Where R = H
R' = Ph–$CH_2$ ; alkyl
R" = aryl

Scheme 28.

Oxidation of *N*-aryl-*N*-methyl-substituted $\beta$-aminoalcohols using pyridinium dichromate (PDC) has recently been reported to give moderate to excellent yields of oxazolidines (Scheme 28) [230].

Hoornaert and coworkers have used the regioselective oxidation of piperidine-3 derivatives by mercuric acetate for the preparation of $\alpha$-cyanoamines [231]. Mercuric acetate oxidation of cyclic amines to enamines is believed to occur with a two electron reduction of mercuric ion to metallic mercury which is subsequently oxidized to mercuric ion by mercurous ions [232, 233]. 1-Benzyl-3,3-(ethylenedioxy)piperidine (**119**) and 3-$CO_2$Et and 3-$CH_2$OH-substituted piperidines (**121**) were regioselectively oxidized at the $\alpha$-position and trapping of the resulting 6-iminium ions with cyanide yielded the corresponding 5-substituted-2-piperidinecarbonitriles (**120**, **122**) (Scheme 29).

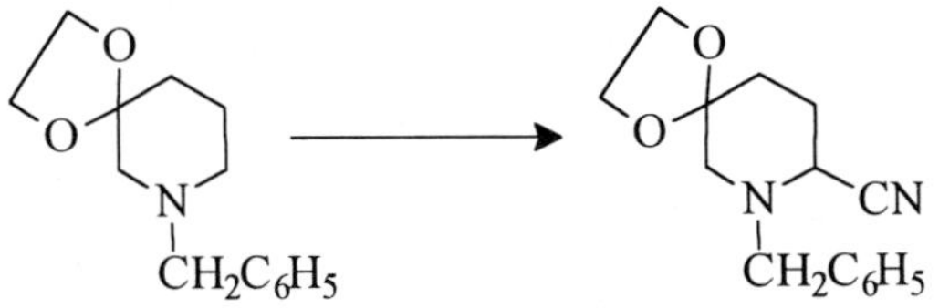

119 120

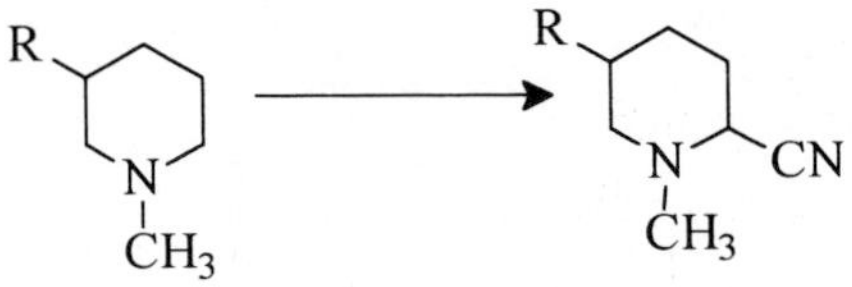

121 122

Where R = $CO_2$Et ; $CH_2$OH

Scheme 29.

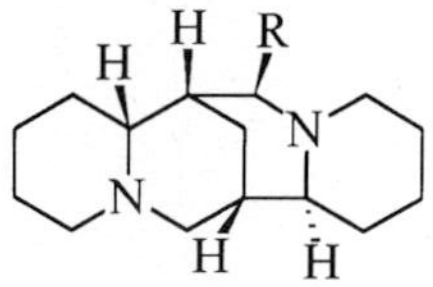

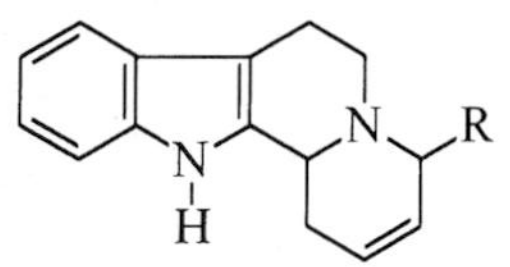

**127**  R = H indoloquinolizidine
**128**  R = CN 17-cyanoindolo-
quinolizindine

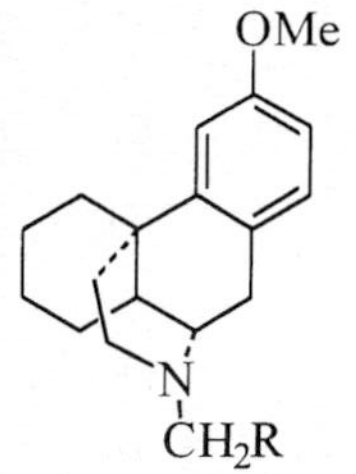

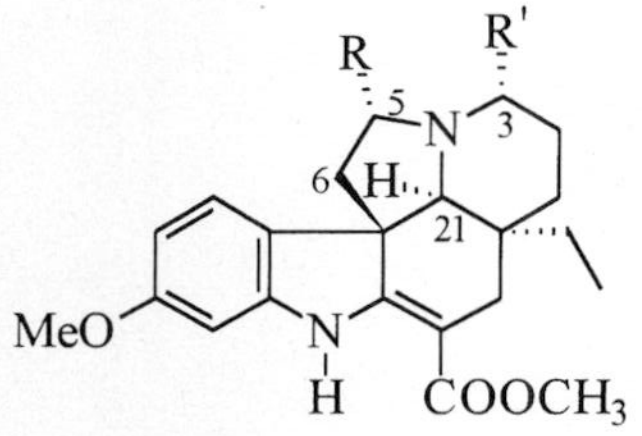

**125**  R = H dextromethorphan
**126**  R = CN N-cyanomethyl-
dextromethorphan

**123**  R = H sparteine
**124**  R = CN 17b cyanosparteine

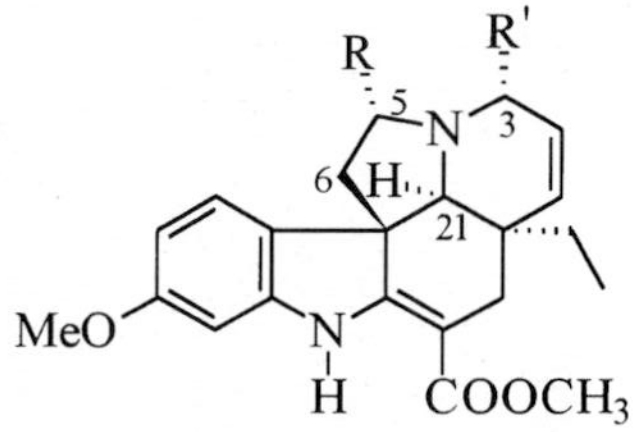

**132**  R = R' = H tabersonine
**133**  R = CN; R' = H 5A-cyanotabersonine
**134**  R = H;  R' = CN 3a=cyanotabersonine

**129**  R = R' = H vincadifformine
**130**  R = CN; R' = H 5A-cyanovincadifformine
**131**  R = H;  R' = CN 3a=cyanovincadifformine

**Chart 6**

The observed regioselectivity was attributed to a transition state in which the mercuric ion is coordinated to both the N atom and the axially oriented O atom. Using this method Hoornaert and coworkers have synthesized several substituted piperidines and their bicyclic piperazine analogs [231, 234, 235].

Singlet oxygen has been reported to react with amines via an electron-transfer mechanism [236]. Electron-transfer-catalyzed reactions of alkaloids in the presence of cyanide ions, by singlet oxygen generated thermally from 1,4-naphthaleneendoperoxide led to the formation of the corresponding α-cyano-aminated products in good yields [237].

Ferric ion-induced coupling of catharanthine (135) and vindoline (140) in aqueous acidic media to produce 3,4'-anhydrovinblastine has been proposed to occur via the formation of a cation radical (136) of the tertiary amine of catharanthine (Scheme 30). Rearrangement and subsequent fragmentation between C16 and C21 leads to ring opening. A second oxidation followed by nucleophilic attack of the diiminium (137) by vindolene (140) results in the formation of iminium (139), which on borohydride reduction yields 3,4'-anhydrovinblastine (77 %) [238].

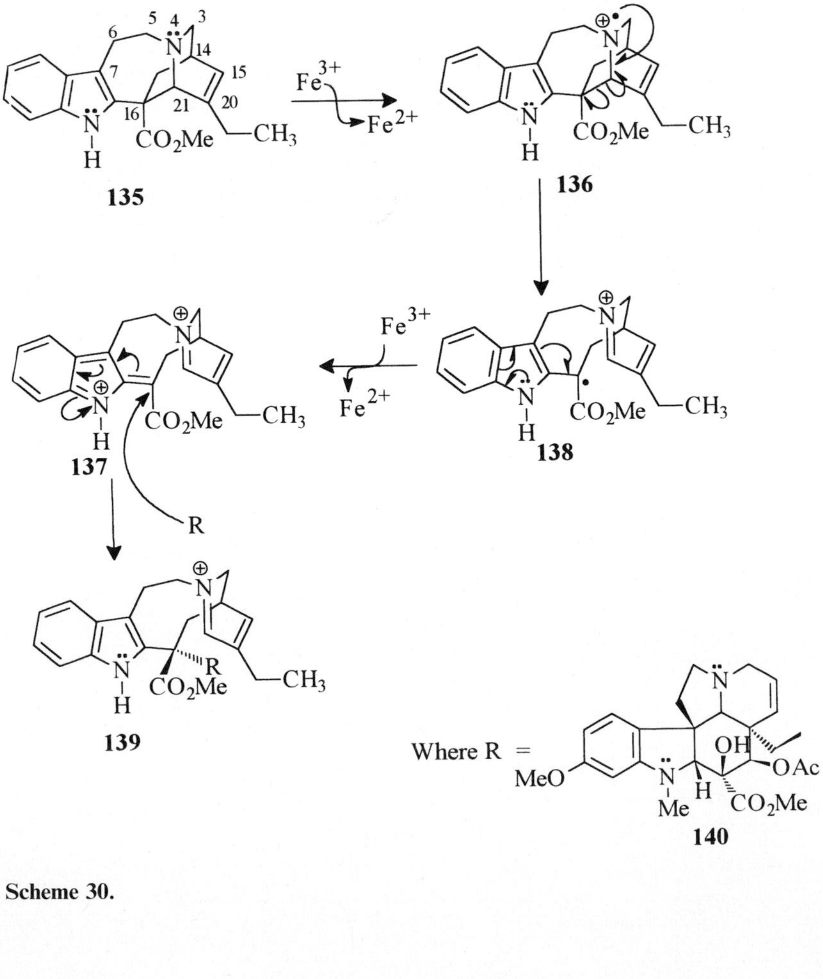

**Scheme 30.**

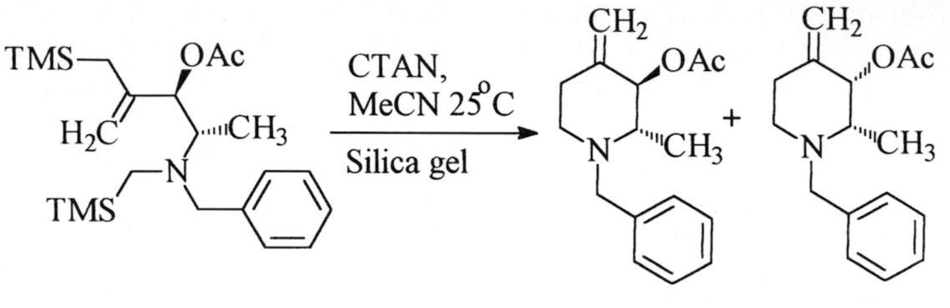

**141**          **142** (41%)   **143** (34%)

**Scheme 31.**

Fragmentation of the C16–C21 bond of catharanthine following formation of the radical cation has also been proposed in the oxidation reaction with dichloro-dicyanoquinone [239].

Ceric ammonium nitrate (CAN) and ceric tetrabutylammonium nitrate (CTAN) [40–42, 240] oxidation of α-silylamine and α-silylamide have been utilized for the synthesis of hydropyridines (Scheme 31).

Unlike the amine analogs α-silylamido (*E*)-vinylsilanes undergo cyclization to produce tetra hydropyridines with retention of absolute and relative stereo-chemistry. This method has been utilized for the synthesis of the aza-sugars, (−)-1-deoxymannojirimycin and (+)-1-deoxyallonojirimycin.

Metal oxide catalyzed reactions of anilines and benzylamines have been utilized to synthesize azobenzenes and *N*-benzylidenebenzyl amines in good yields (Scheme 32) [67, 241, 242].

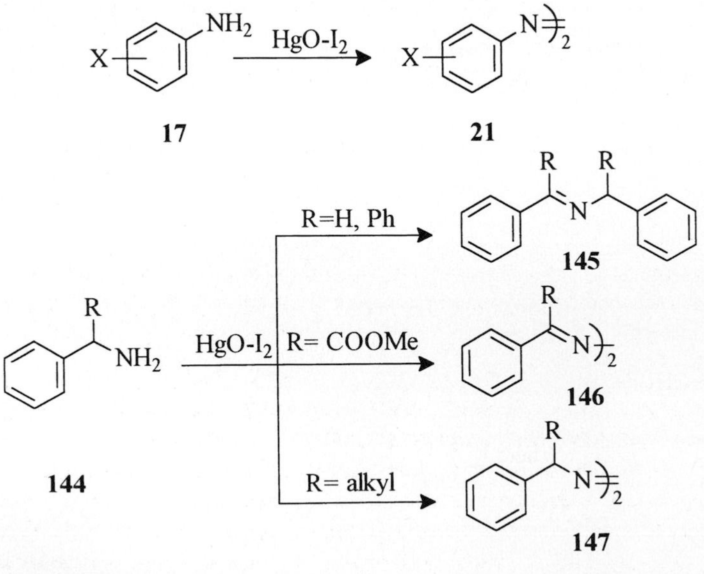

**17**                **21**

**144**

**145**

**146**

**147**

**Scheme 32.**

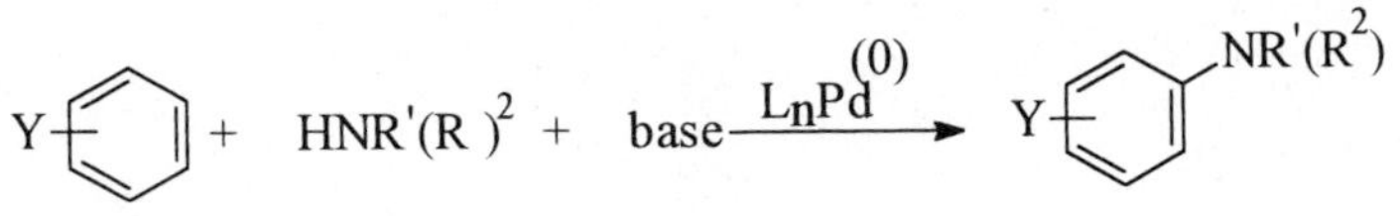

$$Y \text{—} \langle \bigcirc \rangle + \text{HNR}'(\text{R})^2 + \text{base} \xrightarrow{\underset{\text{L}_n\text{Pd}}{(0)}} Y \text{—} \langle \bigcirc \rangle \text{—NR}'(\text{R}^2)$$

**148**          **149**                          **150**

**Scheme 33.**

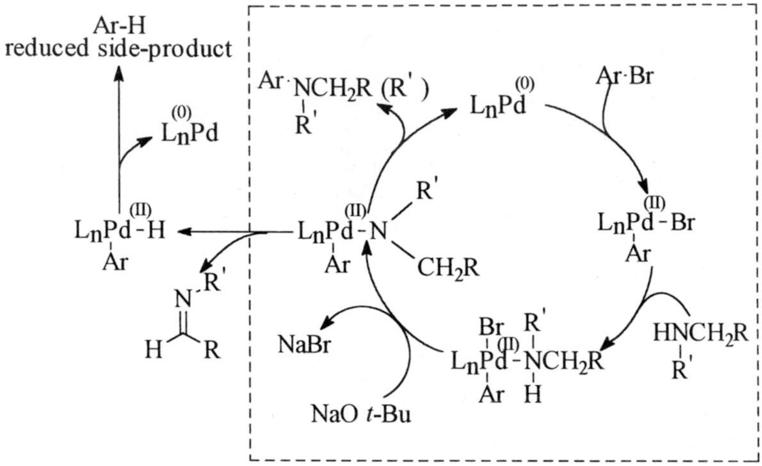

Ar-H
reduced side-product

*[reaction cycle diagram]*

Ar·NCH₂R (R')    L$_n$Pd$^{(0)}$    Ar·Br

L$_n$Pd$^{(0)}$

L$_n$Pd$^{(II)}$-H / Ar

L$_n$Pd$^{(II)}$-N / Ar / CH₂R

L$_n$Pd$^{(II)}$-Br / Ar

NaBr

Br R' / L$_n$Pd$^{(II)}$-NCH₂R / Ar H    HNCH₂R / R'

NaO *t*-Bu

**Scheme 34.**

Palladium catalyzed aminations of aryl halides and triflates have become a very efficient and useful method for the synthesis of aromatic amines (Scheme 33) [243–248].

The currently accepted mechanism for these transformations is shown in Scheme 34.

Oxidative addition of the aryl bromide to $L_n\text{Pd}^0$ gives the $\text{Pd}^{II}$ complex. Metathesis of amine from bromide gives the aryl amido intermediate. This can reductively eliminate to produce the aniline and regenerate $\text{Pd}^0$. Current research efforts in this area have concentrated on improving the nature of the palladium catalyst.

**Enzyme-catalyzed reactions**

Enzyme catalysis also provides a convenient method for bringing about synthetically useful oxidizing transformations of amines [249].

Scheme 35 shows the multistep synthesis of optically pure tritium-labeled histamine, which involves an amine oxidase-catalyzed reaction as the first step [250].

Diamine oxidase (DAMOX) from pea seedlings catalyze the oxidative deamination of diamines. DAMOX-catalyzed reactions of diamines in the synthesis of phenacyl azaheterocycles is shown in Scheme 36 [251–255].

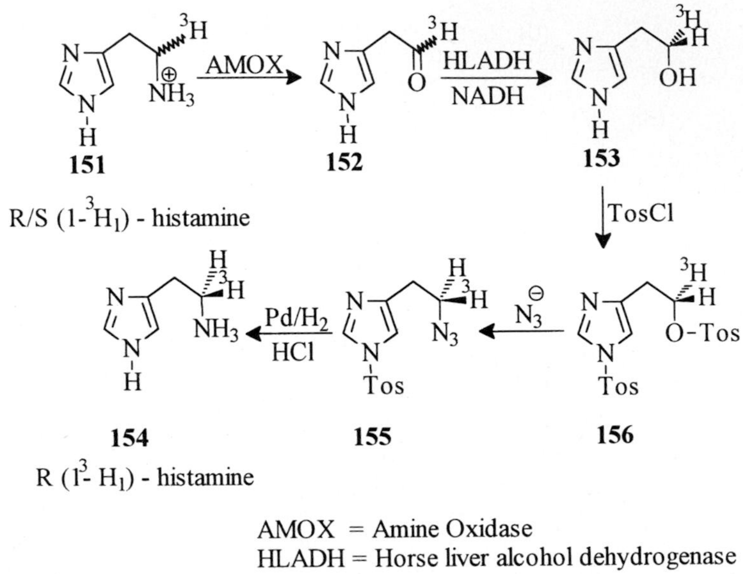

**151**    **152**    **153**

R/S $(1\text{-}^3H_1)$ - histamine

**154**    **155**    **156**

R $(1\text{-}^3H_1)$ - histamine

AMOX  = Amine Oxidase
HLADH = Horse liver alcohol dehydrogenase

**Scheme 35.**

Enantioselective oxidation of the D-amino acid by D-amino acid oxidase has been utilized to transform racemic mixtures of methionine and thionine to the corresponding L-amino acid [256, 257]. The separation of L-amino acid from a racemic β-hydroxyamino acid mixture using D-amino acid oxidase (D-ASOX), was used as a key step in the synthesis of bleomycine (Scheme 37) [258].

The resolution of racemic mixtures by amino acid oxidases also forms an important step in the synthesis of isotope labeled sugars and amino acids [259].

### 7.3.2  Electrochemical Methods

Dehydrogenation of primary amines to yield nitriles can be carried out with high efficiency at nickel hydroxide electrodes [1–3, 260]. Short chain and reactive amines are easily oxidized at low temperatures, whereas higher temperatures are required for long-chain amines (Table 8) [260].

The electrochemical dealkylation of aliphatic amines is a useful way of mimicking enzymatic dealkylation. This has been effectively used for the synthesis of N-dealkylated metabolites of drugs with much better efficiencies than the enzyme catalyzed reactions (Scheme 38) [2].

Nucleophilic substitution of the iminium ions formed in the electrochemical process provides a convenient route for the synthesis of α-substituted amino derivatives. For example the anodic oxidation of N,N-dimethylbenzylamine in methanol containing tetrabutyl ammonium fluoroborate or potassium hydroxide gives rise to methoxylated products (Scheme 39) [261].

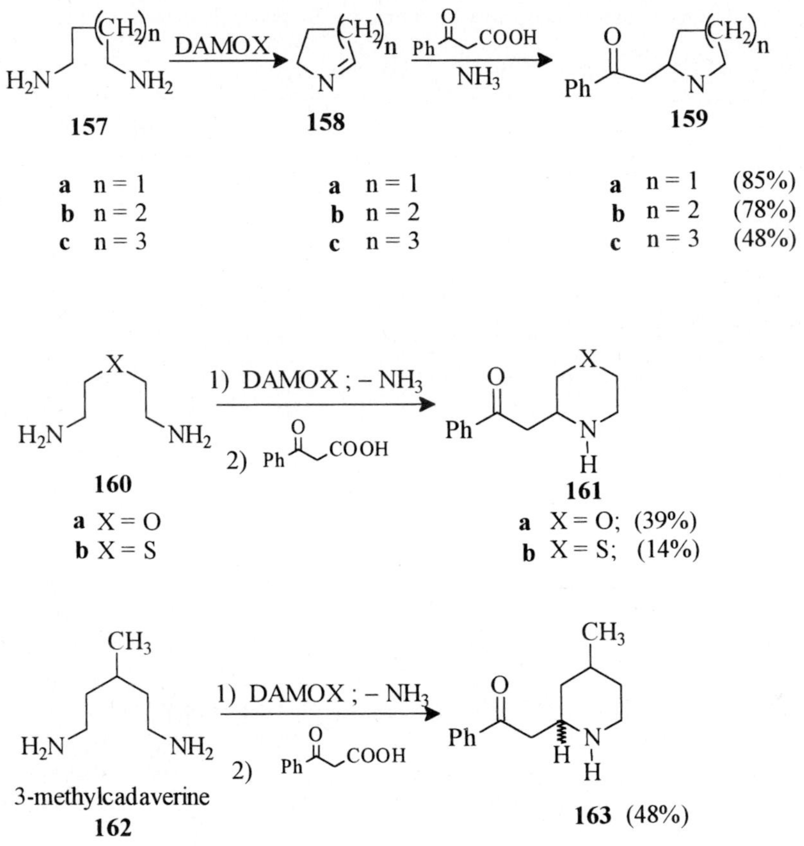
DAMOX
NH3
157
158
159
a n = 1
b n = 2
c n = 3
a n = 1
b n = 2
c n = 3
a n = 1 (85%)
b n = 2 (78%)
c n = 3 (48%)
1) DAMOX ; – NH3
2)
160
a X = O
b X = S
161
a X = O; (39%)
b X = S; (14%)
1) DAMOX ; – NH3
2)
3-methylcadaverine
162
163 (48%)

**Scheme 36.**

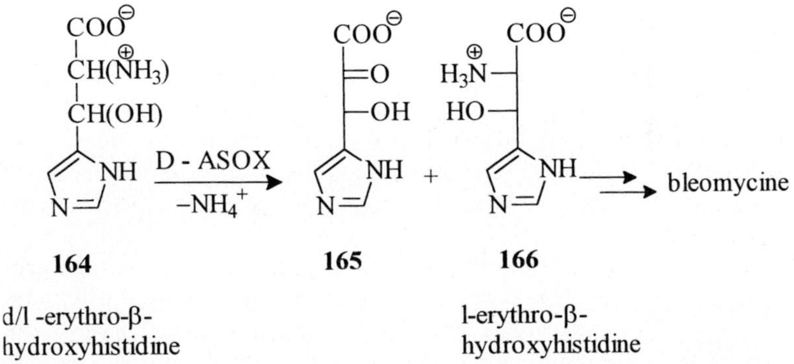
D - ASOX
–NH4+
bleomycine
164
165
166
d/l -erythro-β-
hydroxyhistidine
l-erythro-β-
hydroxyhistidine

**Scheme 37.**

**Table 8.** Oxidation of primary α-unbranched amines to nitriles at the nickel hydroxide electrode.[a]

$$RCH_2NH_2 \xrightarrow{\text{NiOOH},-e^-} RC\equiv N$$

| Amine | Electrolyte[b] | T (°C) | Yield (%) of nitrile or dinitrile |
|---|---|---|---|
| Ethylamine | – | – | 68 |
| 1-Propylamine | – | – | 84 |
| 1-Butylamine | A | 30 | 85 |
| 1-Hexylamine | B | 5 | 72 |
| 1-Octylamine | C | 40 | 95 |
| 1-Decylamine | C | 60 | 91 |
| Benzylamine | C | 40 | 90 |
| Furfurylamine | A | 5 | 86 |
| 1,6-Diaminohexane | C | 40 | 93 |
| 1,10-Diaminodecane | C | 40 | 90 |
| 1,12-Diaminododocane | C | 40 | 88 |
| 6-Aminohexanoic acid | D | 40 | 97 |

[a] According to [260].
[b] A = 0.1 M aqueous potassium hydroxide; B = 0.1 M potassium hydroxide in acetonitrile–water (1:1, $v/v$); C = 0.1 M potassium hydroxide in $t$-butanol–water (1:1, $v/v$); D = 0.3 M potassium hydroxide in $t$-butanol–water (1:1, $v/v$).

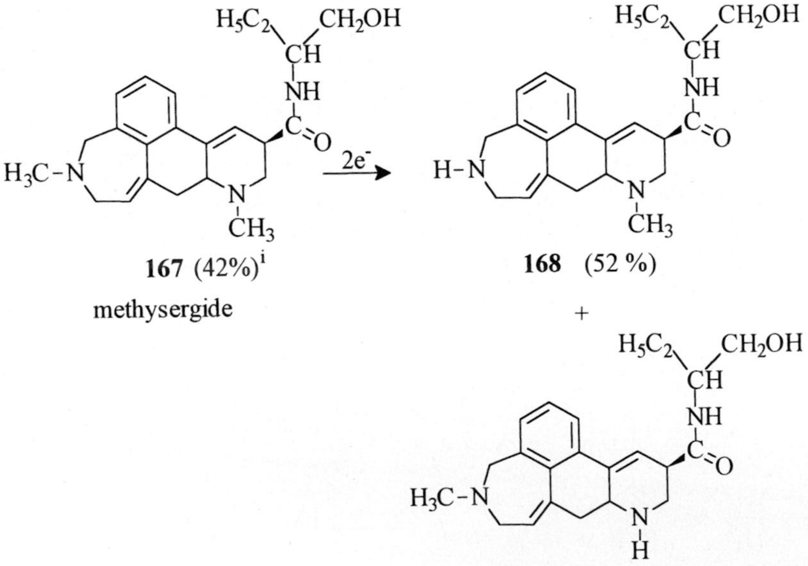

**167** (42 %)[i]
methysergide

**168** (52 %)

+

**169** (37 %)

(i) Percentage of starting
material consumed

**Scheme 38.**

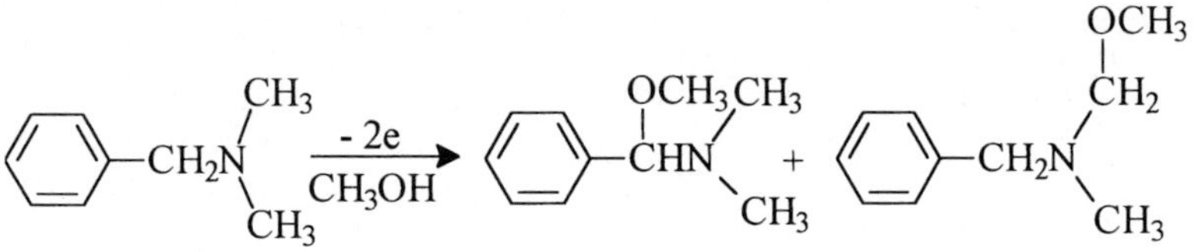


**170** **171** (20%) **172** (80%)

**Scheme 39.**

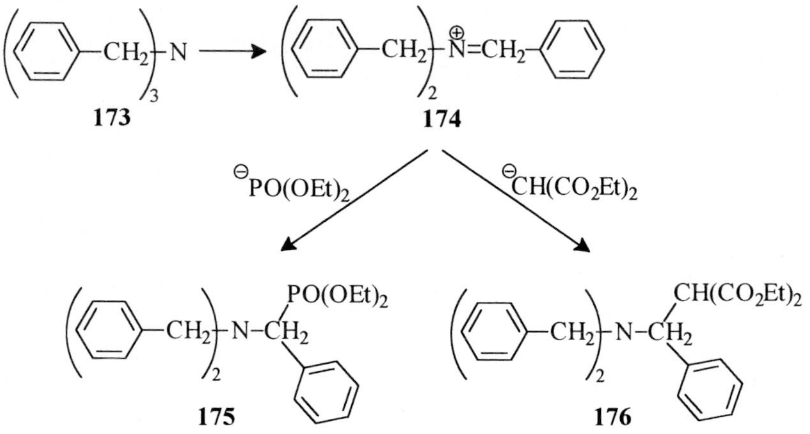


**173** **174** **175** **176**

**Scheme 40.**

The iminium cation can also react with nucleophiles such as diethyl malonate or diethyl phosphonate as shown in Scheme 40 for tribenzylamine [262].

Anodic oxidation of *N*,*N*-dimethyl, or *N*-methyl-*N*-alkylanilines in basic methanol leads predominantly to the methoxylation of the *N*-methyl group (Scheme 41) [263–265].

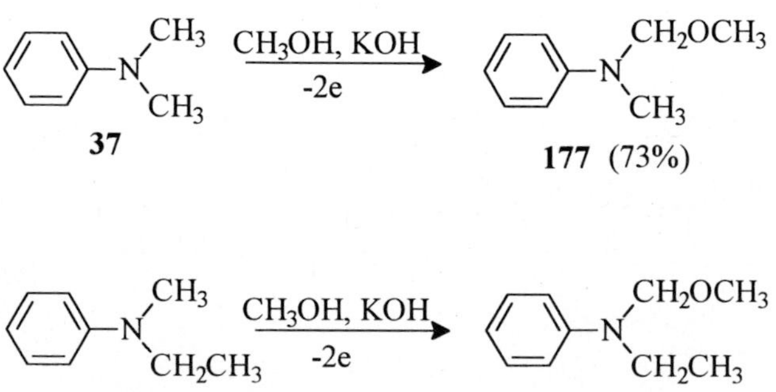


**37** **177** (73%) **178** **179** (85%)

**Scheme 41.**

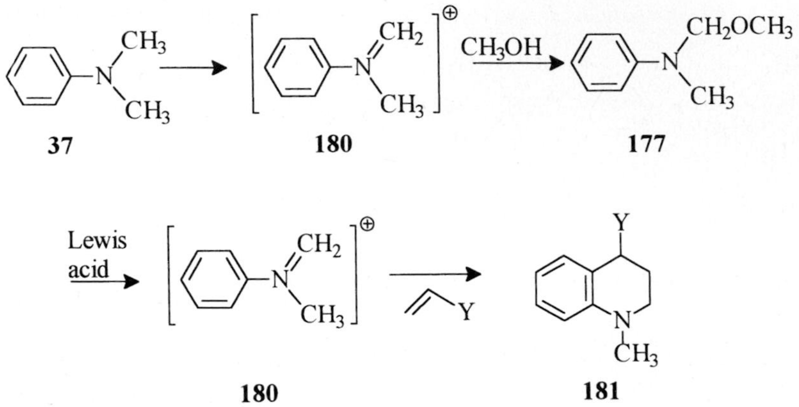

**Scheme 42.**

α-Methoxylated amines serve as useful synthons as they can be converted to iminium cations by Lewis acid, which can be trapped in situ by nucleophiles such as electron rich olefins as shown in Schemes 42 and 43, for the synthesis of tetra-hydroquinolines and julolidine derivatives (Table 9) [265].

The nucleophilic addition of cyanide ions to the electrochemically generated iminium cation is a synthetically very useful reaction and has been studied for a variety of symmetric and unsymmetric ions (Table 10) [266].

In unsymmetrical amines this process generally leads to α-cyanation of the less substituted carbon. It was earlier proposed that the regioselectivity of cyanation was decided by the conformation of the amine adsorbed on the anode. However

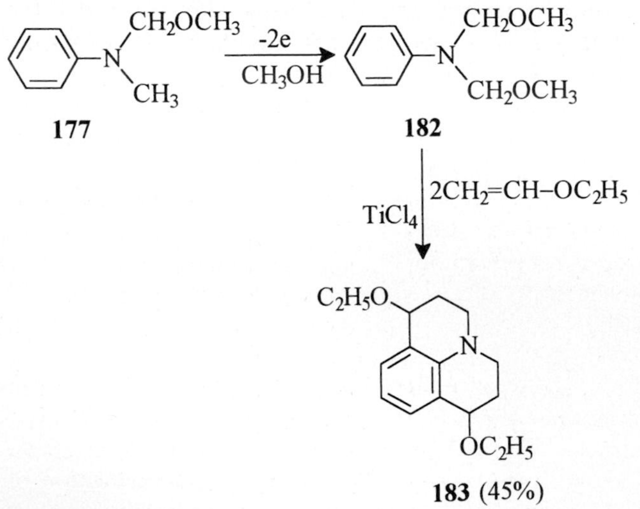

**Scheme 43.**

**Table 9.** Synthesis of tetrahydroquinolines from $N$-methoxymethyl-$N$-methylaniline using $TiCl_4$ as a catalyst.[a]

| Electron-rich olefin | Product | | | | Yield (%) |
|---|---|---|---|---|---|
| | | X | X' | X" | |
| $C_6H_5CH=CH_2$ | | $C_6H_5$ | H | H | 84 |
| $C_6H_5C(CH_3)=CH_2$ | | $C_6H_5$ | $CH_3$ | H | 89 |
| $C_6H_{13}CH=CH_2$ | | $C_6H_{13}$ | H | H | 58 |
| $C_2H_5OCH=CH_2$ | | $OC_2H_5$ | H | H | 64 |
| $C_2H_5OCH=CHOSi(CH_3)$ | | $OSi(CH_3)_3$ | H | $C_2H_5$ | 61 |
| $(CH_3)_3SiO$ $\diagdown$ $C=CH_2$ $C_6H_5$ | | $OSi(CH_3)_3$ | $C_6H_5$ | H | 31 |
| $C_4H_9$ $\diagdown$ $C=CH_2$ $(CH_3)_3SiO$ | | $OSi(CH_3)_3$ $OCH_3$ $OH$ | $C_4H_9$ $C_4H_9$ $C_4H_9$ | H H H | 29 11 21 |
| $C_6H_5CH=CH-N\bigcirc O$ 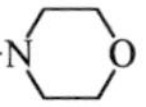 | $N\bigcirc O$ | | H | $C_2H_5$ | 81 |
| $CH_2=CHOAc$ | | $OCH_3$ $OH$ | H H | H H | 11 69 |

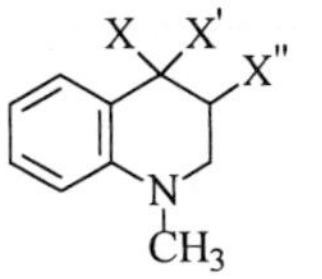

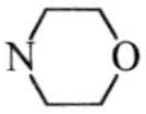

[a] According to [265].

efficiency of deprotonation depending upon the ability of $\alpha$-CH orbitals to overlap with the half vacant 'p' orbitals of the aminium radical is better able to explain the observed results [7, 9, 121, 122].

Recently Le Gall et al. have been able to bring about *regio-* and *diastereo-* selective synthesis of $\alpha$-cyanoamines by anodic oxidation of cyclic $\alpha$-silylamines (Scheme 44) [267].

Anodic cyanation of aromatic amines can lead to cyanation on the phenyl or methyl groups as shown in Scheme 45 [268, 269].

Anodic oxidation of amides and carbamates on the other hand are far more useful than those of amines. Scheme 46 shows the synthesis of a pharaoh ant trail pheromone in which the anodic oxidation of 2-pyrrolidone is an important step [270].

The use of electrochemical oxidations of amides and carbamates as a key step in the synthesis of alkaloidal compounds has been recently reviewed by Shono [271].

The electrochemical oxidations of several fluorinated amines have been investigated with a view to synthesizing fluoroorganic compounds [272]. Substitution of the hydrogen on the carbon of amines by chloromethyl and fluoromethyl groups slightly increases their oxidation potential [273]. Anodic oxidation of fluoromethyl

**Table 10.** Anodic cyanation of tertiary amines.[a,b]

| Amine | Anode potential (V vs SCE) | Product | Yield (%)[c] |
|---|---|---|---|
| $(CH_3)_3N$ | 1.2–1.4 | $(CH_3)_2NCH_2CN$ | 53 |
| $(C_2H_5)_3N$ | 1.1–1.2 | $(C_2H_5)_2NCH(CH_3)CN$ | 53 |
| $(C_2H_5)_2NCH_3$ | 1.0–1.2 | $(C_2H_5)_2N(CH_2)CN +$ | |
| | | $\begin{array}{c} H_5C_2 \\ \diagdown \\ \diagup \\ H_3C \end{array} N{-}CHCN \\ \qquad\quad | \\ \qquad\quad CH_3$ | 37 |
| $(i\text{-}C_3H_7)_2NCH_3$ | 0.9–1.1 | $(i\text{-}C_3H_7)_2NCH_2CN$ | 55 |
| 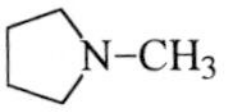 [pyrrolidine]N–CH₃ | 1.0–1.3 | 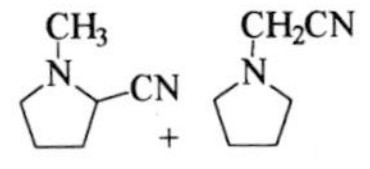 $CH_3$ / CN  +  $CH_2CN$ | 59 |
| 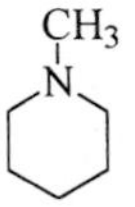 [piperidine] CH₃ | 1.1–1.4 | 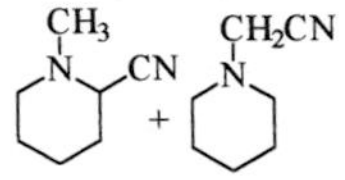 $CH_3$ / CN  +  $CH_2CN$ | 66 |
| 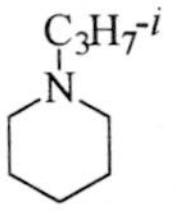 [piperidine] $C_3H_7{-}i$ | 1.0–1.2 | 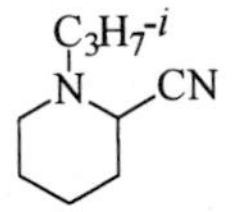 $C_3H_7{-}i$ / CN | 62 |

[a] According to [266].
[b] Analyte: amine (0.10 mol) + NaCN (0.15 mmol) in $CH_3OH{-}H_2O$ (1:1, 75 mL).
[c] GLC.

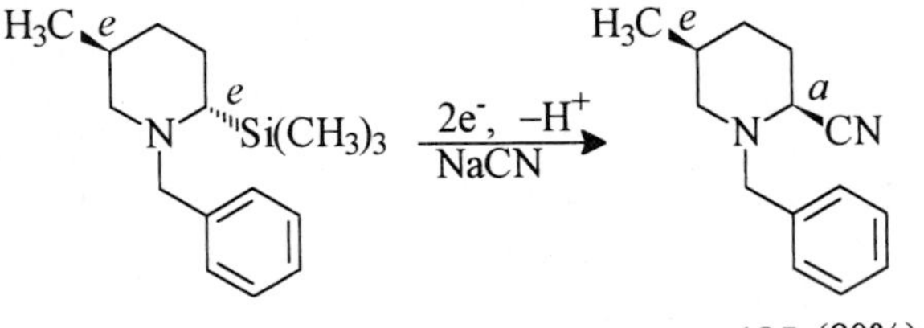

**Scheme 44.**

amines, such as $ArN(R)CH_2X$ ($X = CF_3$, $CHF_2$, $CH_2F$) in alkaline methanol leads to preferential addition of a methoxy group at a position $\alpha$ to the fluoromethyl group, as shown in Scheme 47 for an *N*-methylaniline derivative [265, 273, 274].

Whereas in *N*-ethyl and *N*-methylaniline methoxylation occurs only on the methyl groups in the fluoromethyl substituted compounds, the regioselectivity is drastically changed. The $\alpha$-methoxylated fluoroalkylamine have been used as precursors for the synthesis of fluorinated amino acids and dihydroquinolines [272].

In contrast to the anodic methoxylation of *N*-(2,2,2-trifluoroethyl)amines, anodic cyanation does not occur at the position $\alpha$ to the trifluoroalkyl group (Scheme 48) [275].

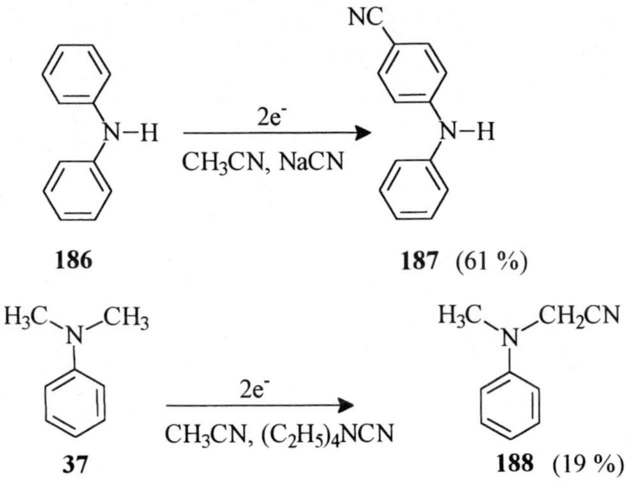

**186**  **187** (61 %)

**37**  **188** (19 %)

**Scheme 45.**

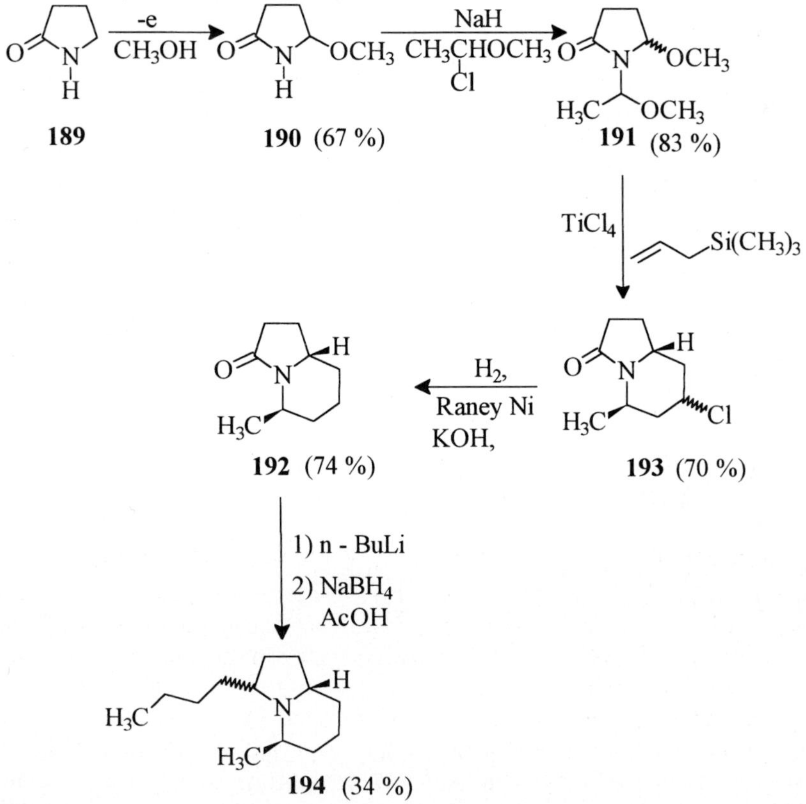

**189**  **190** (67 %)  **191** (83 %)

**192** (74 %)  **193** (70 %)

1) n - BuLi
2) NaBH₄
   AcOH

**194** (34 %)

**Scheme 46.**

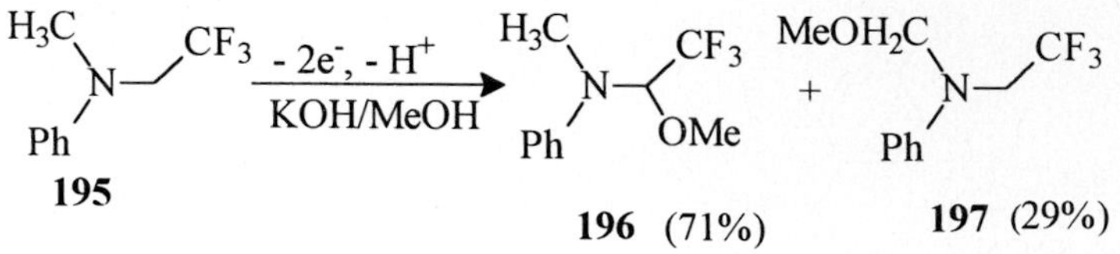

**Scheme 47.**

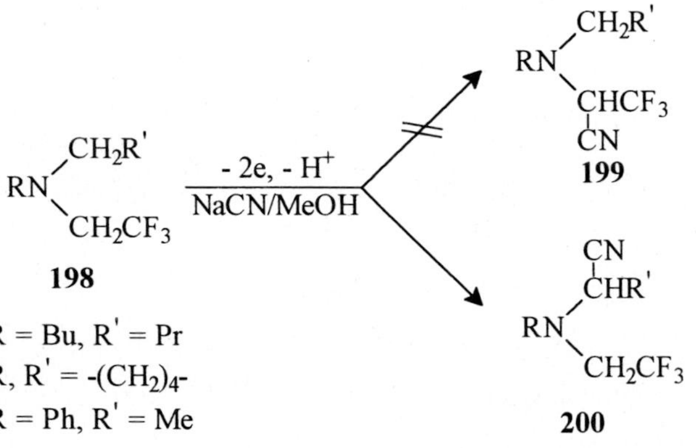

**a** R = Bu, R' = Pr
**b** R, R' = -(CH$_2$)$_4$-
**c** R = Ph, R' = Me

**a** R = Bu, R' = Pr   (54%)
**b** R, R' = -(CH$_2$)$_4$-   (40%)
**c** R = Ph, R' = Me   (9%)

**Scheme 48.**

Trifluoromethylated sulfenimines which are useful starting materials for the preparation of *N*-fluoroalkylamines, aminoketones and aminoalkanoates, have been prepared by the anodic oxidation of 2,2,2-trifluoroethylamine and diaryl sulfides in MeCN/Et$_4$NClO$_4$ with MgBr$_2$ as the redox mediator (Scheme 49) [276].

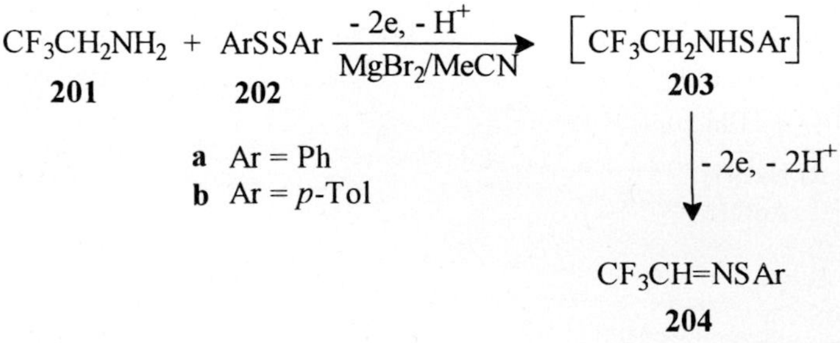

**a** Ar = Ph
**b** Ar = *p*-Tol

CF$_3$CH=NSAr
**204**

**a** Ar = Ph   (72%)
**b** Ar = *p*-Tol   (58%)

**Scheme 49.**

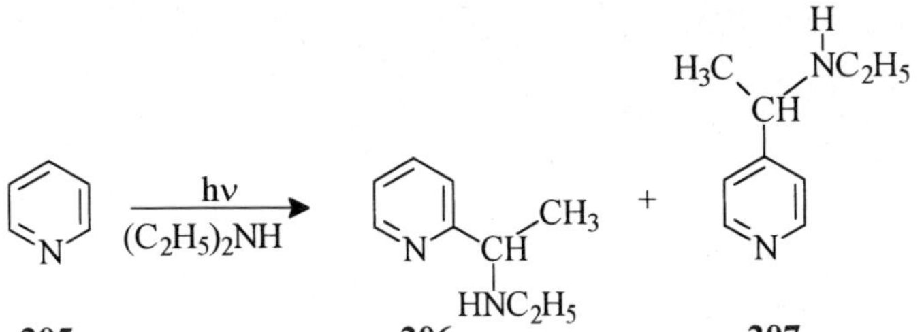

Scheme 50.            205              206              207

## 7.3.3 Photoinduced Electron-transfer Reactions

Both direct and sensitized photoelectron-transfer reactions of amines have been utilized for bringing about synthetically useful transformations of amines.

**Direct irradiation of amines**

The synthetic and mechanistic aspects of photoaddition reactions of amines with excited states of arenes and aryl olefins have been extensively investigated [10–12, 277].

The reactions of excited states of arenes and arenealkenes with primary and secondary amines result in products arising out of aminyl radical addition, whereas $\alpha$-aminoalkyl radical addition occurs with tertiary amines [11, 278, 279].

N-Heteroarenes such as pyridine however photochemically add amines to yield $\alpha$-aminoalkyl substituted products even with secondary amines (Scheme 50) [280].

Intramolecular photoaddition reactions of aminoarenes which are of interest from the synthetic point of view, was first reported by Bryce-Smith et al., for (N,N-dimethylaminoalkyl)benzenes (Scheme 51) [281].

Aoyama and coworkers have synthesized several pyrrolidine and piperidine derivatives via intramolecular photoaddition reactions of tertiary (aminoalkyl) styrene (Scheme 52) [282, 283].

Similar intramolecular photoaddition reactions of styrylamides have also been reported by Aoyama and coworkers [282–286].

Lewis and coworkers have also observed efficient intramolecular photoaddition for several (N,N-dimethylaminoalkyl)styrenes, leading to the formation of nitrogen heterocycles or aminocycloalkanes (Scheme 53) [287].

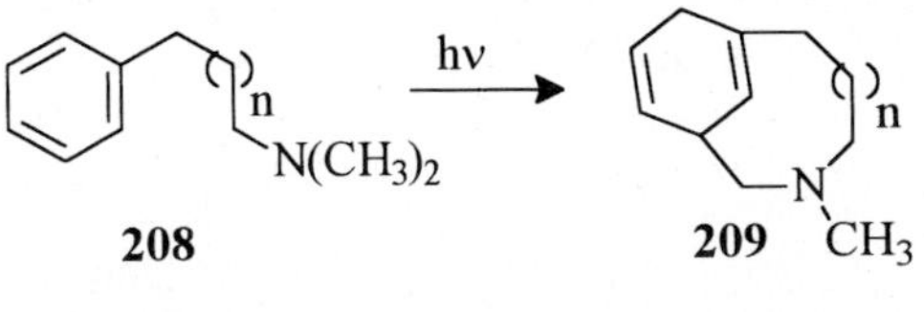

208                     209  CH₃

a  n = 1               a  n = 1
b  n = 2               b  n = 2

Scheme 51.

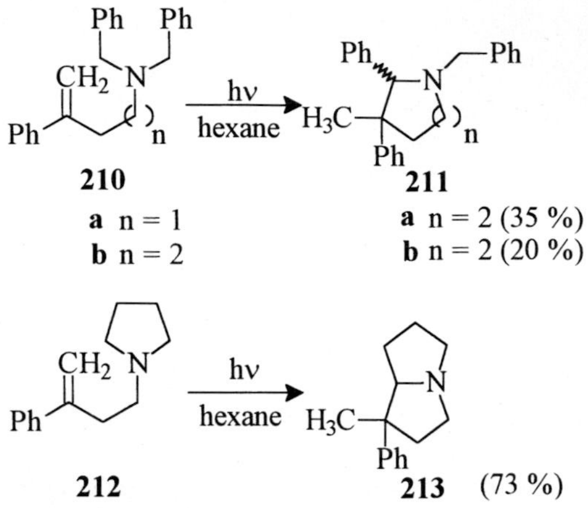

**Scheme 52.**

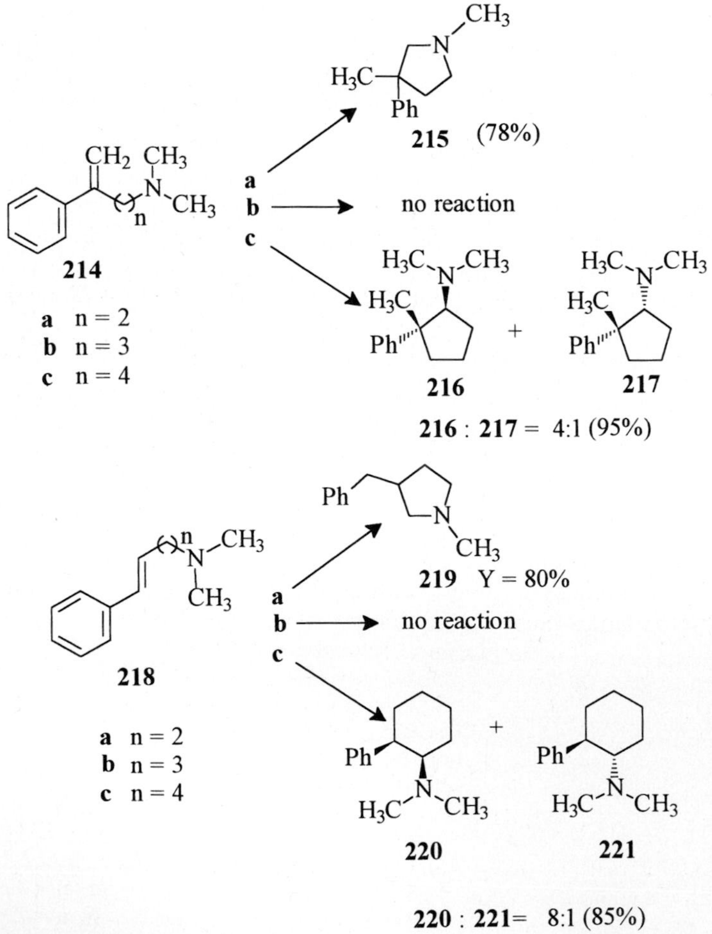

216 : 217 = 4:1 (95%)

220 : 221= 8:1 (85%)

**Scheme 53.**

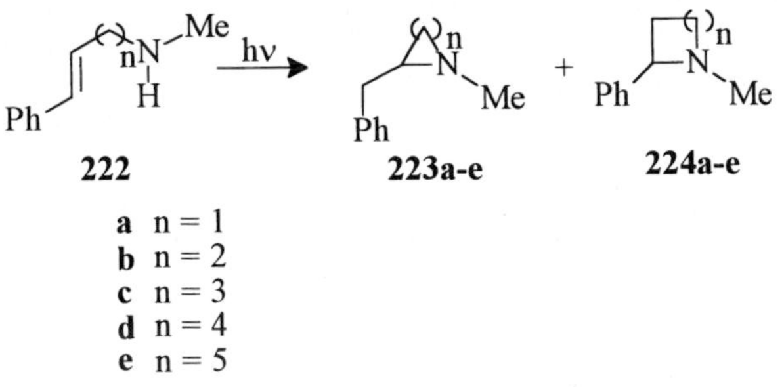

**Scheme 54.**

Intramolecular photoaddition reactions of secondary amines linked to styrenes can result in the formation of C–N bonds at the $\alpha$- or $\beta$-carbon (Scheme 54) [288–290]. The product yields and the ratio of the two products are shown in Table 11.

Intramolecular photoaddition of secondary amines linked to various other arenes and stilbenes also provide an efficient method for the synthesis of *N*-heterocycles (Scheme 55) [291–293].

Recently, Lewis et al. have reported the synthesis of larger rings via the photo-induced intramolecular addition of (aminoalkyl)styrylamides and *N*-(aminoalkyl)-2-stilbene carboxamides (Scheme 56) [294, 295].

Intramolecular photoaddition reactions of amines for the construction of *N*-heterocycles has also been successfully explored by Sugimoto and coworkers (Scheme 57) [296–302].

Mariano and coworkers have shown that photoreactions of silylated tertiary amines such as **242,** with a cyclohexenone derivative **241** give two types of adduct, the trimethylsilyl (TMS) containing adduct **243** and the non-TMS adduct **244** (Scheme 58) [303–305].

The distribution of **243** and **244** varied, depending on the nature of the solvent [306]. The TMS adduct was found to predominate in less polar aprotic solvents

**Table 11.** Product yields for *N*-methyl $\omega$-styrylaminoalkenes.[a]

| Styrylamine | $\Phi_{add}$[b] | Yield (%)[c] | 224/223[d] |
|---|---|---|---|
| **222a** | – | 15 | >10 |
| **222b** | 0.024 | 63 | 14 |
| **t-222c** | 0.011 | 57 | 2.4 |
| **c-222c** | 0.019 | 62 | 2.0 |
| **222d** | 0.050 | 82 | 0.15 |
| **222e** | – | 30 | >10 |

[a] According to [291].
[b] Quantum yield for total adduct formation using monochromatic 288 nm irradiation of 0.005 M styrylamine in nitrogen-purged acetonitrile solution.
[c] Yield of the major adduct (**223** or **224**) determined by GC analysis at conversion less than 20 %.
[d] Product ratio determined by GC analysis.

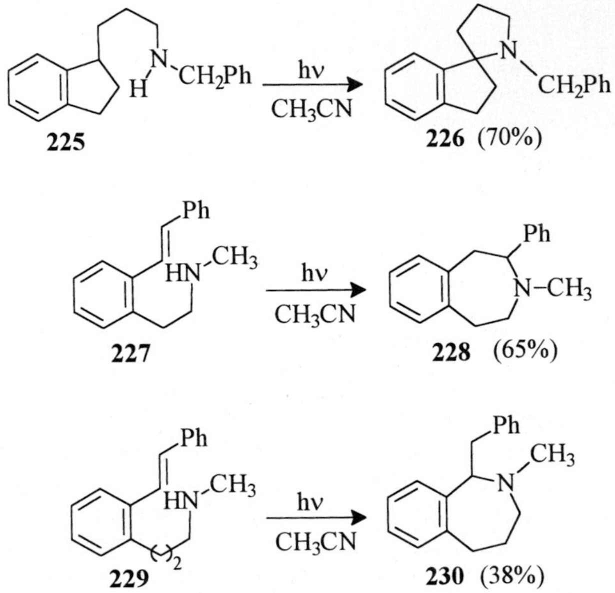

225 → 226 (70%)

227 → 228 (65%)

229 → 230 (38%)

**Scheme 55.**

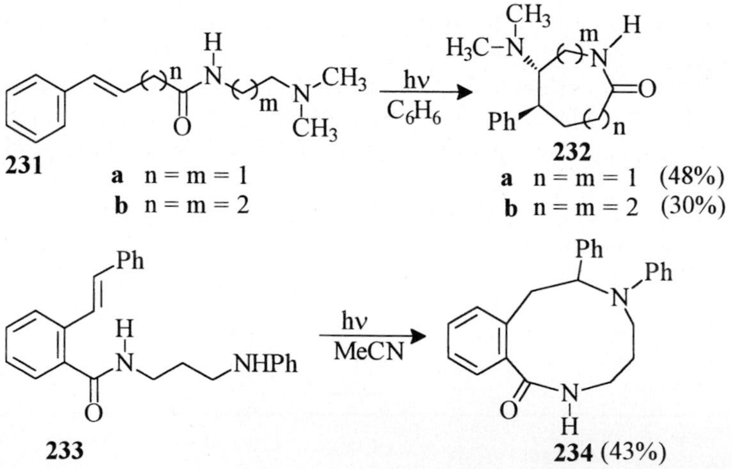

231

a n = m = 1
b n = m = 2

232

a n = m = 1 (48%)
b n = m = 2 (30%)

233

234 (43%)

**Scheme 56.**

such as acetonitrile, whereas, the non-TMS adduct was formed in higher yields in polar protic solvents such as methanol.

A higher chemoselectivity was obtained in the intramolecular photoreaction of α-silylamines (Scheme 59). Irradiation of **246** and **249** in methanol leads to the cyclized non-TMS products **247** and **250**, respectively, whereas, TMS containing

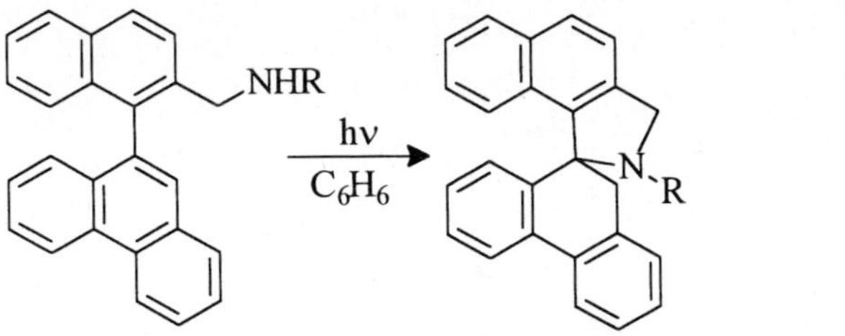

**235**
**a** R = Me,
**b** R = *t*-Bu,
**c** R = Ph,

**236**
**a** R = Me, (30%)
**b** R = *t*-Bu,(84%)
**c** R = Ph, (93%)

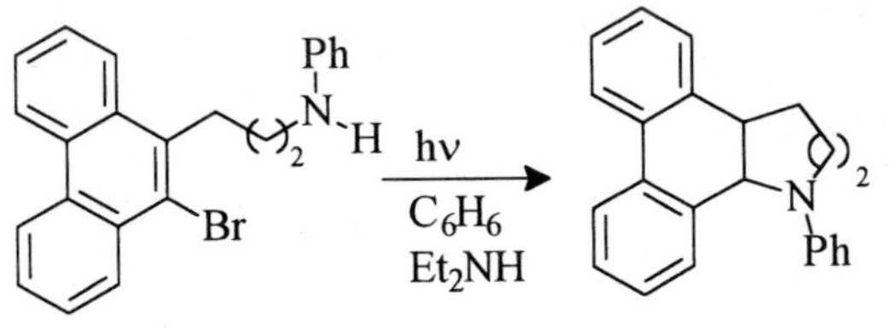

**237**

**238** (39%)

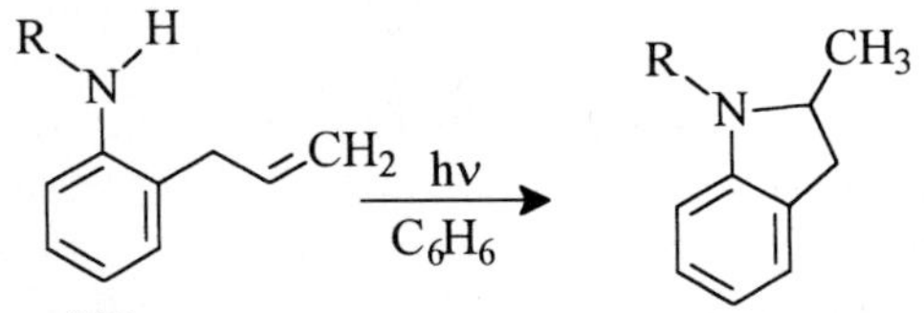

**239**
**a** R = H,
**b** R = Me,
**c** R = Ph,

**240**
**a** R = H, (40%)
**b** R = Me,(75%)
**c** R = Ph, (80%)

**Scheme 57.**

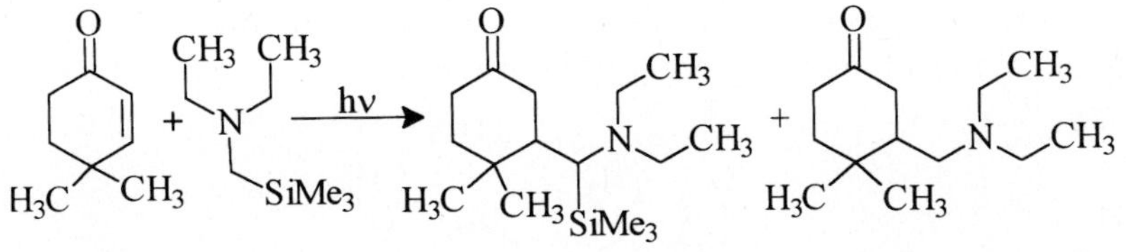

**241** **242** **243** (70% (i) ; 30% (ii)) **244** (9% (i) ; 60% (ii))

(i) in MeCN
(ii) in MeOH

**Scheme 58.**

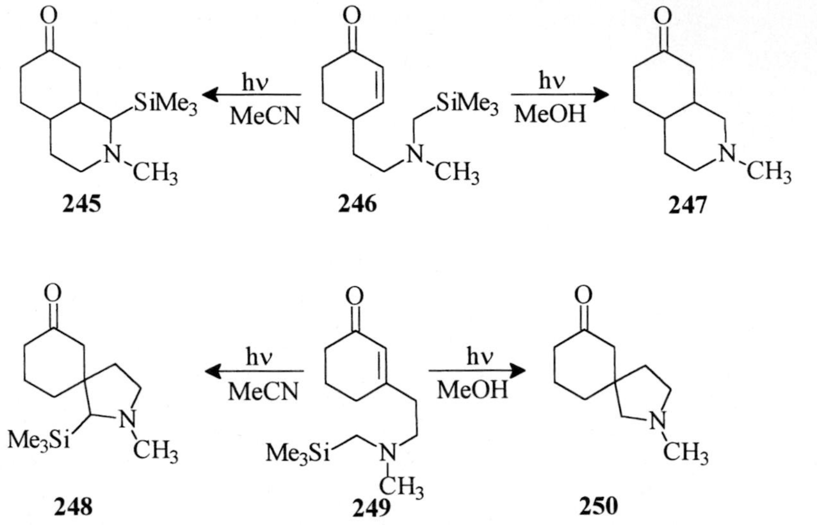

**Scheme 59.**

products **245** and **248** were formed when irradiation was performed in acetonitrile
[131].

Mattay and coworkers reported that direct irradiation ($\lambda > 300$ nm) of 2-
cyclohexene-1-one (**251**) in the presence of *N,N*-dimethylallylamine (**252**) leads to
2,4 *β*-dimethyl-*cis*-decahydro-5-isoquinolineone (**253**) in yields up to 20 % along
with trace amounts of **254** (Scheme 60), depending on the polarity of the medium
(Table 12) [307]. In polar solvents like acetonitrile containing 10 % water the PET
promoted addition–cyclization process competes efficiently with the [2 + 2] cycloaddition pathway leading to dimers (Table 12). Irradiation in the presence of
PET sensitizer such as dicyanonaphthalene (DCN) led to only trace amounts of
products.

### Sensitized photoreactions of amines

In photosensitized reactions, electron transfer occurs between the excited state of
the sensitizer and amine resulting in the formation of the radical ion-pair. The sen-

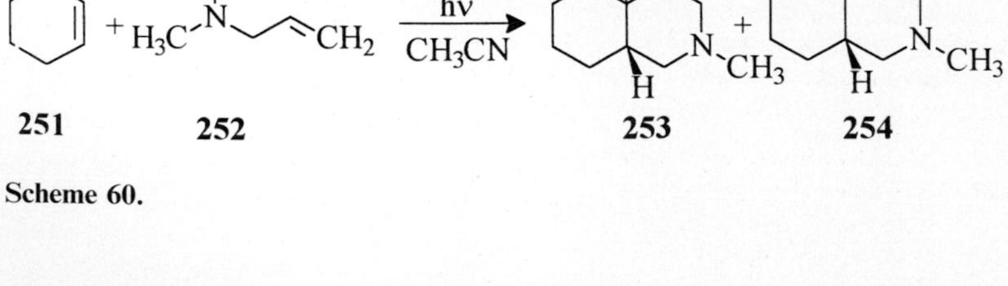

**Scheme 60.**

**Table 12.** Solvent effect on the product yields in various photoaddition of *N*,*N*-dimethylallylamine (**252**) to 2-cyclohexen-1-one (**251**).[a]

| Solvent | 253 | 254 | [2 + 2] dimer | Cyclohexanone |
|---|---|---|---|---|
| Cyclohexene | 2.6 % | traces | 3.5 % | |
| THF | 8.2 % | 3.9 % | 2.9 % | – |
| MeOH, 20 °C | 7.6 % | traces | 5.5 % | – |
| MeOH, 25 °C | 5.8 % | 3.3 % | 2.4 % | – |
| CH₃CN | 3.7 % | 10.4 % | traces | – |
| CH₃CN, 0.1 eq. LiClO4 | 7.4 % | traces | traces | 7.5 % |
| CH₃CN, DCN | traces | 3.8 % | 2.2 % | – |
| CH₃CN–10 % H₂O | 20 % | traces | traces | – |

[a] According to [307].

sitizer is used in low concentrations as it is usually regenerated during the course of the reaction and hence acts in a catalytic manner. The commonly used sensitizers for the generation of aminium ions are cyanoaromatics such as dicyanobenzene (DCB), dicyanonaphthalene (DCN) and dicyanoanthracene (DCA). Ketonic sensitizers such as anthraquinone and benzophenone have also been used.

Mariano and coworkers have investigated the DCA sensitized photocyclization of silylamino–enone derivatives (Scheme 61) [130, 131, 308]. DCA-sensitized photoreactions of **255** in acetonitrile–methanol mixture gave the substituted piperidines **256** in high yields along with the pyrrolidine ester **257** in the case of **255b**. DCA sensitized reaction of **246** resulted in the formation of a diastereomeric mixture (6:1) of hydrosoquinolines [212, 213] and <2 % of the hydrosoindoline, **258**.

Khim and Mariano have used the same strategy for the synthesis of several functionalized piperidines (Scheme 62) [309].

Sensitized photocyclization was used as a key step in the synthesis of the *E*-ring functionalized alkaloid, yohimbane **267** (Scheme 63) [310].

Pandey and coworkers have reported several DCN-sensitized photocyclization of α-silylamines connected to unactivated olefins [311–314]. Thus, the silyl amines **268** (Scheme 64) were reported to proceed in a *regio*- and *stereo*-selective manner yielding the cyclized products **269** and **270** [311].

Similarly the biologically active natural products (±)-epilupine (**273**) and (±)-isoretronecanol (**276**) were formed in the DCN-sensitized photoreactions of the silyl amines **271** and **274**, respectively (Scheme 64) [312].

Mariano and coworkers have shown that DCN sensitized photoreaction of **268b** in non-deoxygenated solutions leads to the efficient formation (75 %) of the corresponding pyrrolidone and not the azabicyclic product [315]. Earlier studies by Padwa et al. have shown that α-aminoalkyl radicals, unlike their α-amido analogs, do not add efficiently to unactivated olefins (Scheme 65) [316]. The α-silyl amines **277** connected to electron deficient olefins on the other hand, were found to undergo efficient DCN photosensitized cyclization in deaerated solutions to yield **278** and

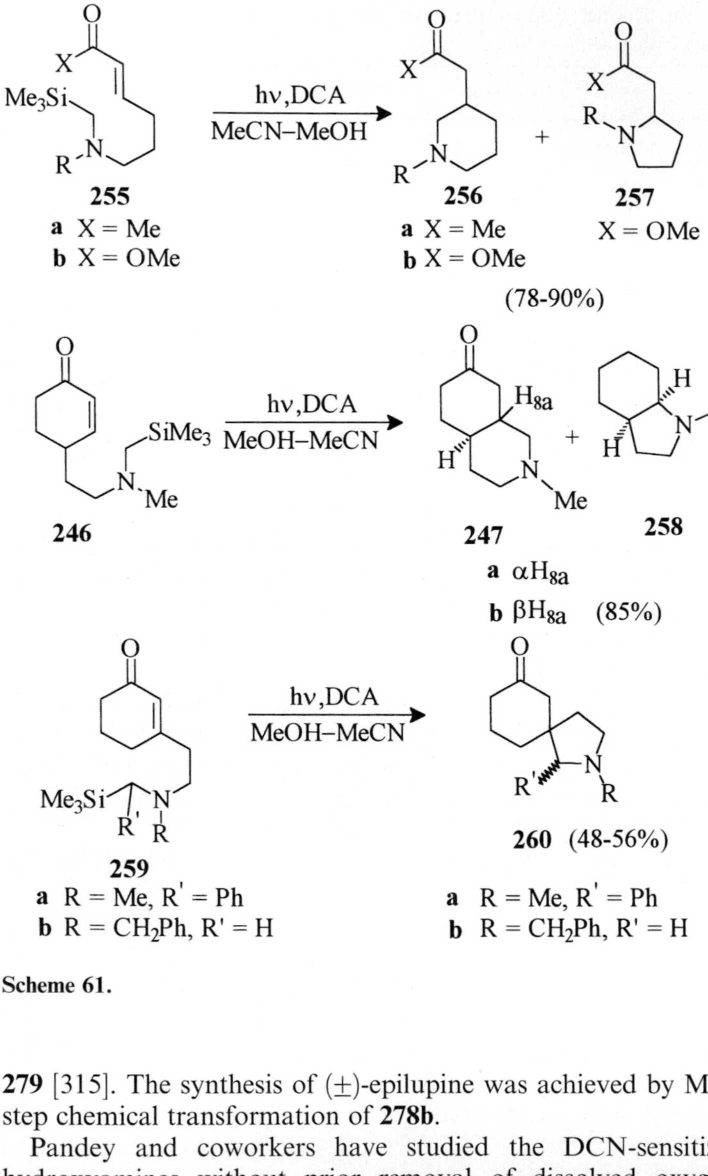

**Scheme 61.**

**279** [315]. The synthesis of (±)-epilupine was achieved by Mariano et al. by a two step chemical transformation of **278b**.

Pandey and coworkers have studied the DCN-sensitized oxidation of *N*-hydroxyamines without prior removal of dissolved oxygen, to nitrone intermediates. Trapping the nitrones by dimethylfumarate (**282**) gave the corresponding cycloadducts in good yields (Scheme 66) [317].

Synthesis of oxazabicycloalkanes and related products was achieved by a one pot electron–proton–electron (EPE) transfer mediated reactions of the amine moiety [317]. Here the iminium cation is formed from the second electron oxidation of the α-aminoalkyl radical, generated via the α-deprotonation of the planar aminium radical owing to their low ionization potential. The iminium cation thus formed can

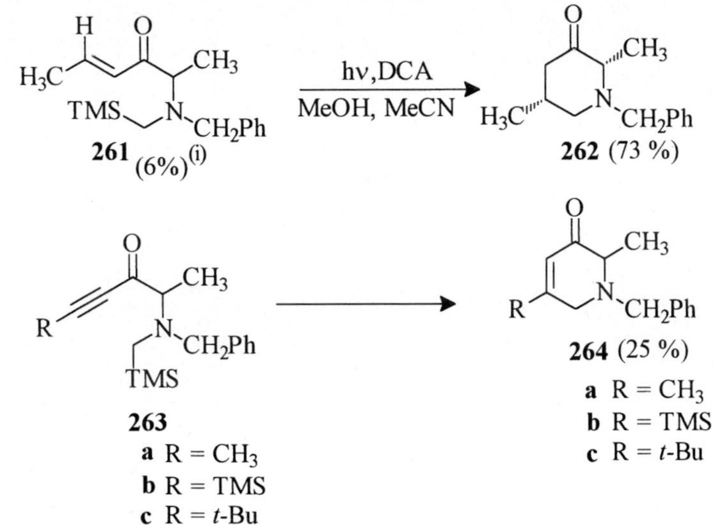

**261** (6%)$^{(i)}$    →    hv,DCA / MeOH, MeCN    →    **262** (73 %)

**263**

a R = CH$_3$
b R = TMS
c R = t-Bu

**264** (25 %)

a R = CH$_3$
b R = TMS
c R = t-Bu

(i) Percentage of starting material consumed

**Scheme 62.**

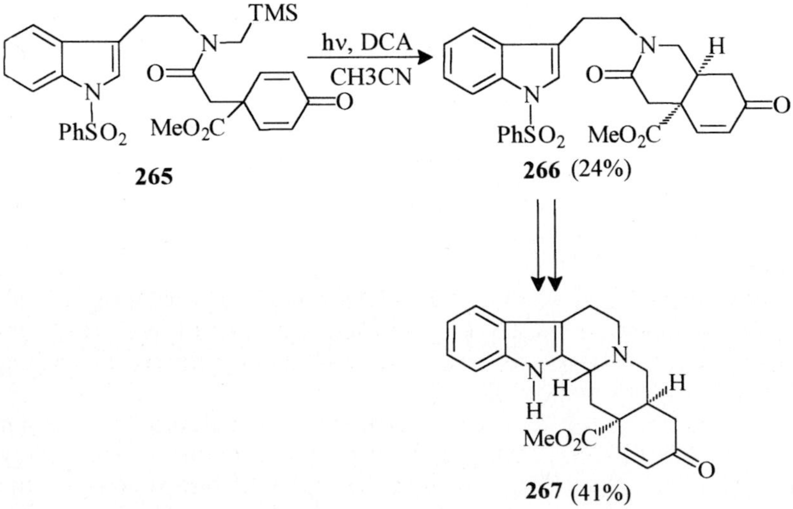

**265**    →    hv, DCA / CH3CN    →    **266** (24%)

**267** (41%)

**Scheme 63.**

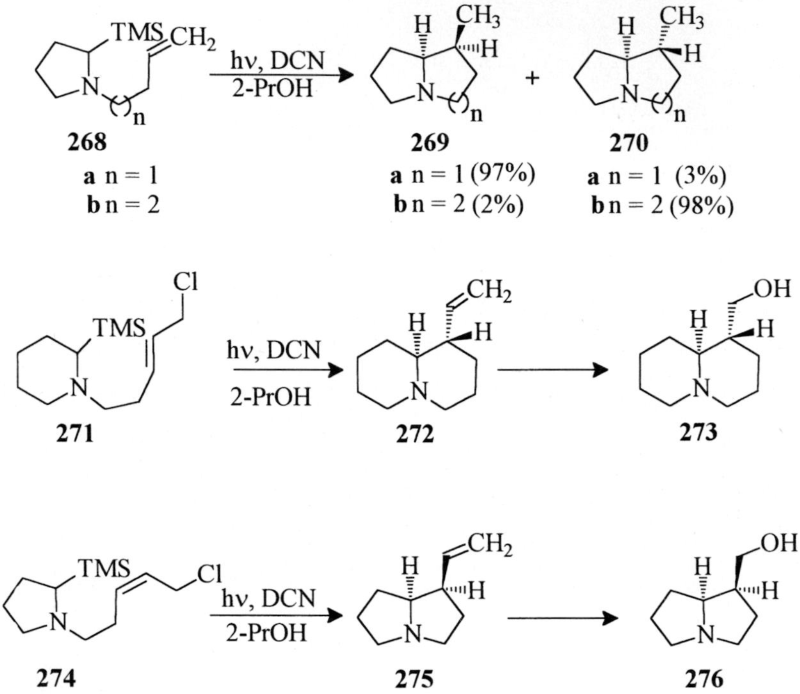
TMS
$CH_2$
hν, DCN
2-PrOH
H
$CH_3$
H
n
+
268
a n = 1
b n = 2
269
a n = 1 (97%)
b n = 2 (2%)
270
a n = 1 (3%)
b n = 2 (98%)
Cl
TMS
hν, DCN
2-PrOH
$CH_2$
OH
271
272
273
TMS
Cl
hν, DCN
2-PrOH
$CH_2$
OH
274
275
276

Scheme 64.

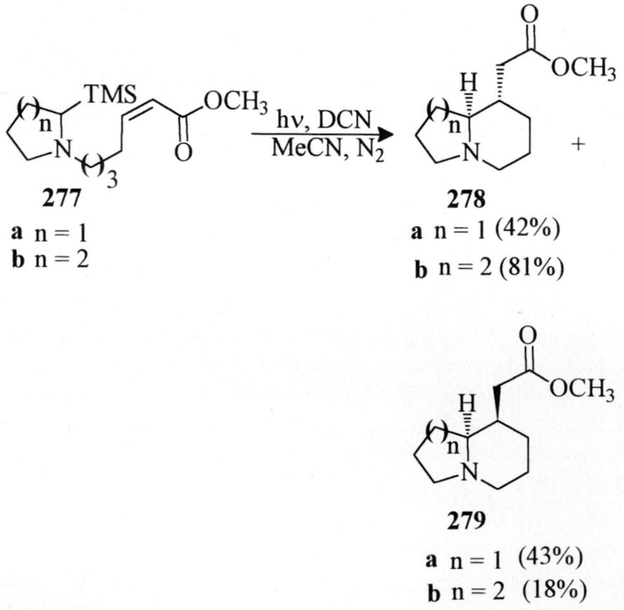
TMS
$OCH_3$
n
3
hν, DCN
MeCN, $N_2$
O
$OCH_3$
H
+
277
a n = 1
b n = 2
278
a n = 1 (42%)
b n = 2 (81%)
279
a n = 1 (43%)
b n = 2 (18%)

Scheme 65.

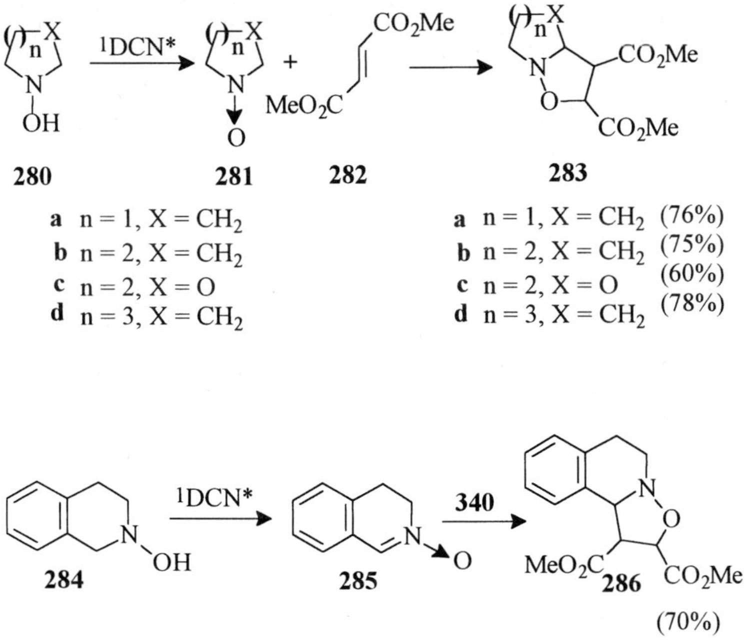

a  n = 1, X = CH$_2$
b  n = 2, X = CH$_2$
c  n = 2, X = O
d  n = 3, X = CH$_2$

a  n = 1, X = CH$_2$ (76%)
b  n = 2, X = CH$_2$ (75%)
c  n = 2, X = O     (60%)
d  n = 3, X = CH$_2$ (78%)

(70%)

**Scheme 66.**

undergo nucleophilic intramolecular cyclization to form the corresponding bicyclic products (Scheme 67) [318].

Lewis and coworkers have studied the sensitized photocyclization of stilbene-linked primary amines **300** which gave **302**, in good yields (Scheme 68) [293].

Similarly, the photosensitized electron transfer mediated reactions of several 1- and 9-(aminoalkyl)phenanthrenes using DCB have been investigated by Lewis and coworkers (Scheme 69) [319]. These reactions provide an efficient method of construction of skeletal structures of aporphine, phenanthropiperidine, phenanthroazepine and related alkaloids. The mechanism of these reactions involves an initial electron transfer from the ground state amine to the singlet arene followed by N–H proton transfer and biradical cyclization.

Photoamination of a variety of aromatic compounds with ammonia and alkyl-amines under photosensitized electron transfer mediated reaction conditions has been extensively investigated by Yasuda and coworkers (Scheme 70) [320–324]. The photoamination reactions of stilbene derivatives with ammonia have been utilized for the synthesis of a variety of isoquinoline derivatives [324]. The photoamination is initiated by photochemical electron transfer from the arenes to the electron acceptor followed by nucleophilic attack of ammonia or primary amines on the aromatic radical cations (Scheme 70).

Anthraquinone photosensitized intermolecular carbon–carbon bond forming reactions of several primary, secondary and tertiary amines with electron deficient olefins have been investigated. The addition of α-aminoalkyl radical, generated

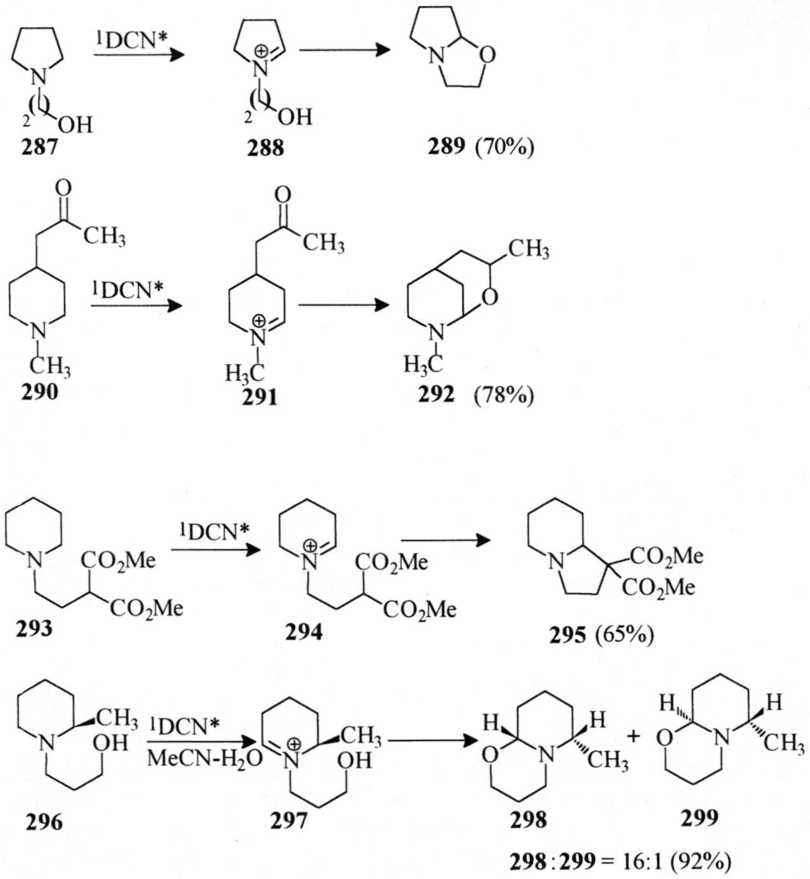

**287**    **288**    **289** (70%)

**290**    **291**    **292** (78%)

**293**    **294**    **295** (65%)

**296**    **297**    **298**    **299**

**298** : **299** = 16:1 (92%)

**Scheme 67.**

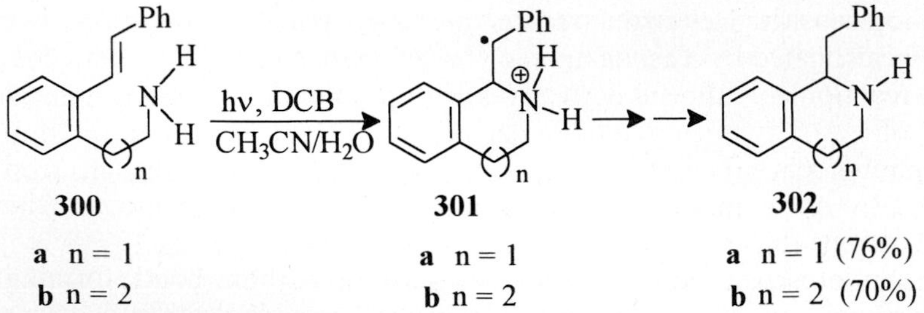

**300**    **301**    **302**

**a** n = 1    **a** n = 1    **a** n = 1 (76%)
**b** n = 2    **b** n = 2    **b** n = 2 (70%)

**Scheme 68.**

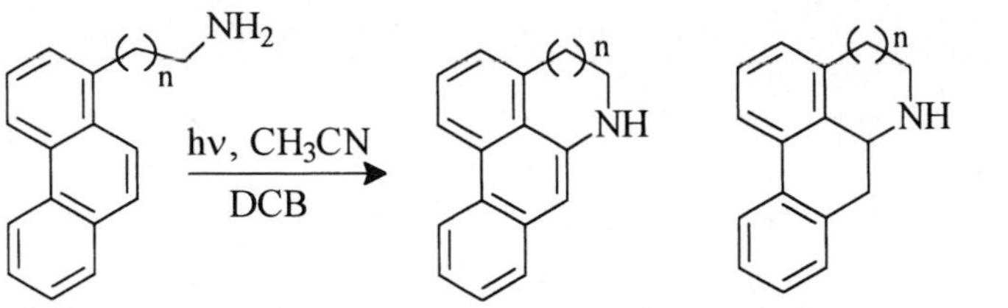

**303**

**a** n = 1

**b** n = 2

**304**

**a** n = 1 (15%)

**b** n = 2 (12%)

**305**

**a** n = 1 (70%)

**b** n = 2 (68%)

**Scheme 69.**

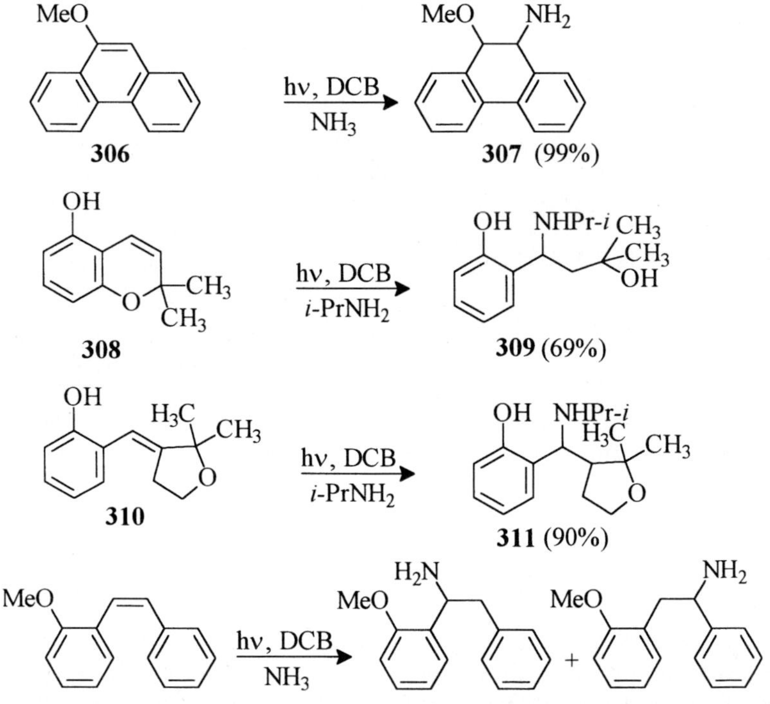

**306**

**307** (99%)

**308**

**309** (69%)

**310**

**311** (90%)

**312**

**313** (42%)

**314** (17%)

**Scheme 70.**

from triethylamine to electron deficient olefins, is followed by an intramolecular 1,5-hydrogen abstraction of the adduct radical. These reactions lead to the formation of products arising out of multiple addition of the olefinic substrate [325].

In the anthraquinone photosensitized reactions of primary and secondary amines with methyl methacrylate, the adducts formed undergo thermal cyclization to the corresponding bicyclolactams and spirolactams (Scheme 71) [325].

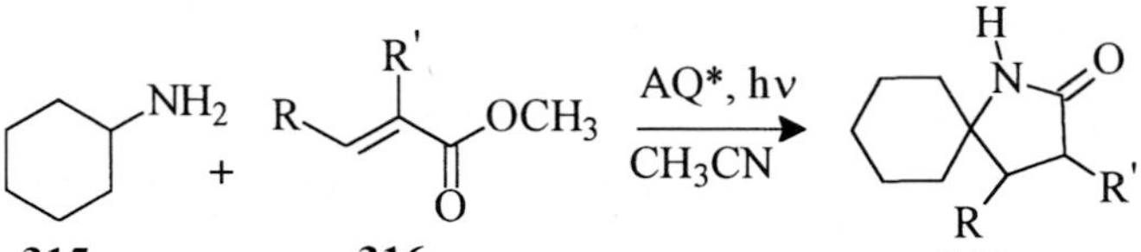

**315**    **316**    **317**

a R, R' = H        (35%) (i)        a R = R' = H        (80%)
b R = H, R' = CH₃ (30%) (i)        b R = H, R' = CH₃ (100%)
c R = CH₃, R' = H (20%) (i)        c R = CH₃, R' = H (50%)

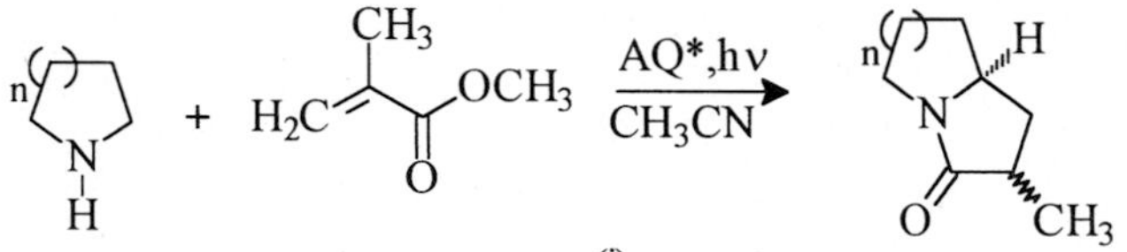

**318**        **316b** (15%) (i)        **319**
a  n = 1                                a  n = 1 (65%)
b  n = 2                                b  n = 2 (70%)

(i) Percentage of starting material consumed

**Scheme 71.**

Most, α,β-unsaturated esters other than methyl methacrylate undergo facile Michael type addition with primary and secondary amines to yield N-addition products. By carrying out the anthraquinone photosensitized addition of several secondary amines with α,β-unsaturated esters at low temperatures (~0 °C), the thermal reactions could be controlled without adversely affecting the photochemical free radical reactions. Using this procedure, indolizidone, pyrrolizidone and a mixture of heliotridone and pseudoheliotridone was synthesized (Scheme 72) [326].

Sensitizers such as benzophenone, anthrone and xanthene were also observed to catalyze these reactions [327]. Anthraquinone sensitized photoreactions of N-allylamines with α,β-unsaturated esters also yield lactams as the main product. Minor amounts of tandem radical addition products were also formed [328].

Recently Betrand et al. have reported the sensitized intermolecular addition of tertiary amines derived from pyrrolidines to (5R)-5-menthyloxy-2[5H]-furanone (**328**) using ketonic sensitizers (Scheme 73, Table 13) [329].

The addition occurs with complete facial selectivity on the furanone ring. The method has been generalized to the addition of tertiary amines and N-protected secondary amines to electron-deficient olefins (Table 13).

The electron donating ability of amines, has been utilized in several photochemical reactions for the generation of reactive radical ions which can subsequently undergo synthetically important reactions.

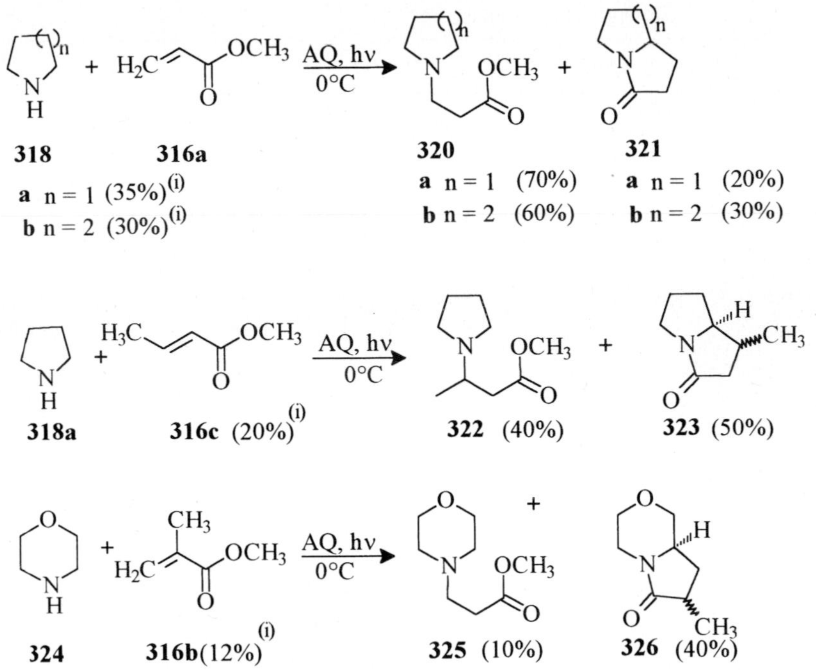

**318**      **316a**                        **320**                  **321**

**a** n = 1 (35%)[(i)]                **a** n = 1   (70%)     **a** n = 1   (20%)

**b** n = 2 (30%)[(i)]                **b** n = 2   (60%)     **b** n = 2   (30%)

**318a**      **316c** (20%)[(i)]         **322** (40%)       **323** (50%)

**324**      **316b**(12%)[(i)]         **325** (10%)      **326** (40%)

(i) Percentage of starting material consumed

**Scheme 72.**

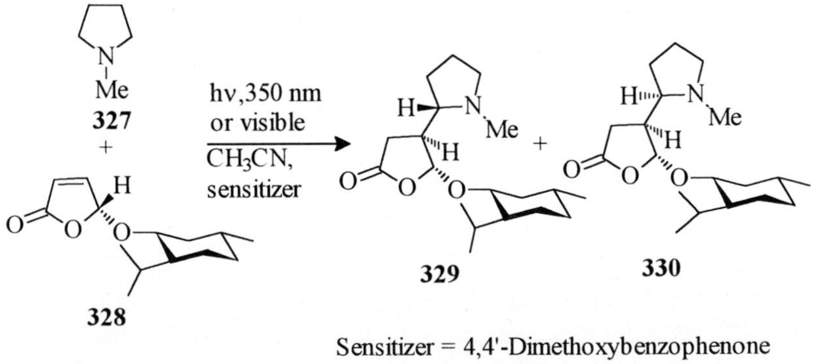

Sensitizer = 4,4'-Dimethoxybenzophenone

**Scheme 73.**

Cossy and coworkers have reported the stereoselective synthesis of several bicyclic pentanols in good yields, by the intramolecular addition of the ketyl radicals of $\delta,\varepsilon$-unsaturated ketones, generated via photoinduced electron transfer, using triethyl-amine (TEA) or hexamethylphosphonic triamide (HMPA), as donors in acetonitrile

**Table 13.** Addition of **327** to electron-deficient alkenes.[a]

| Alkene | Irradiation Time (min) | Product | Yield (%) |
|---|---|---|---|
| **331** A = CN<br>**332** A = CO$_2$Me | 60 | **333** A = CN<br>**334** A = CO$_2$Me | 80[b]<br><br>87[b] |
| **335** = *cis*<br>**336** = *trans* | 10 | **337** | 69<br>73 |
| **338** | 7 | **339** | 76 |
| **340** | 5 | **341** | 86 |
| **342** | 120 | **343** | 71[d] |
| **344** | 180 | **345** | 90 |

[a] According to [329].
[b] The reaction was performed in pure *N*-methylpyrrolidine.
[c] Men = 1*R*-(+) = menthyl.
[d] Conversion = 55 %.

(Scheme 74) [330]. The use of photoreductive cyclization for the total synthesis of actinidine and isooxyskytanthine has been reported [331].

Bischof and Mattay have shown that radical anion cyclization leading to spirocyclic products compete effectively with intramolecular [2 + 2]-cycloaddition on photoexcitation of olefinic enones in the presence of triethylamine [332, 333]. The [2 + 2] cycloadducts could be converted to the corresponding spiro compounds under PET conditions (Scheme 75) [307].

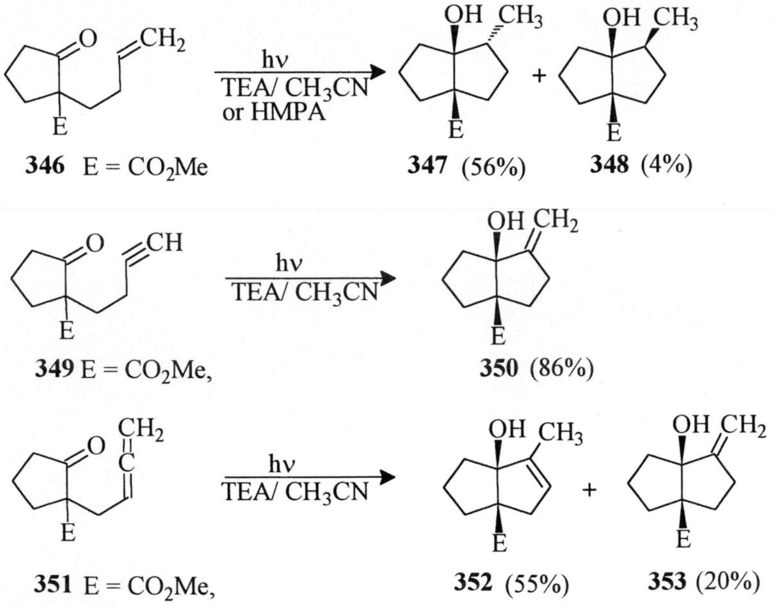

**346** E = CO₂Me

**347** (56%)    **348** (4%)

**349** E = CO₂Me,

**350** (86%)

**351** E = CO₂Me,

**352** (55%)    **353** (20%)

**Scheme 74.**

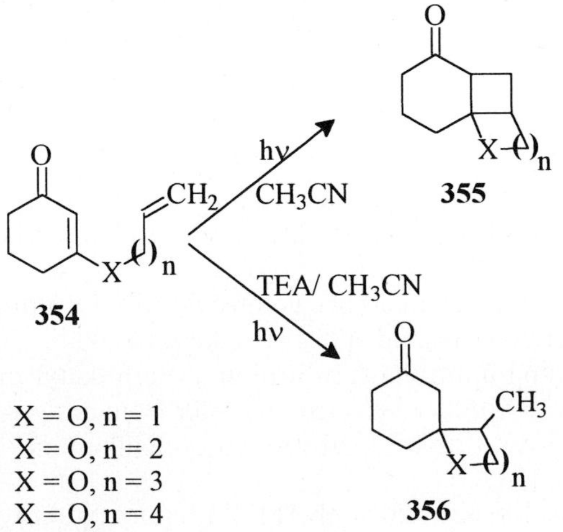

**354**

**355**

**356**

a   X = O, n = 1
b   X = O, n = 2
c   X = O, n = 3
d   X = O, n = 4
e   X = CH₂, n = 1
f   X = CH₂, n = 2
g   X = CH₂, n = 3
h   X = CH₂, n = 4

**Scheme 75.**

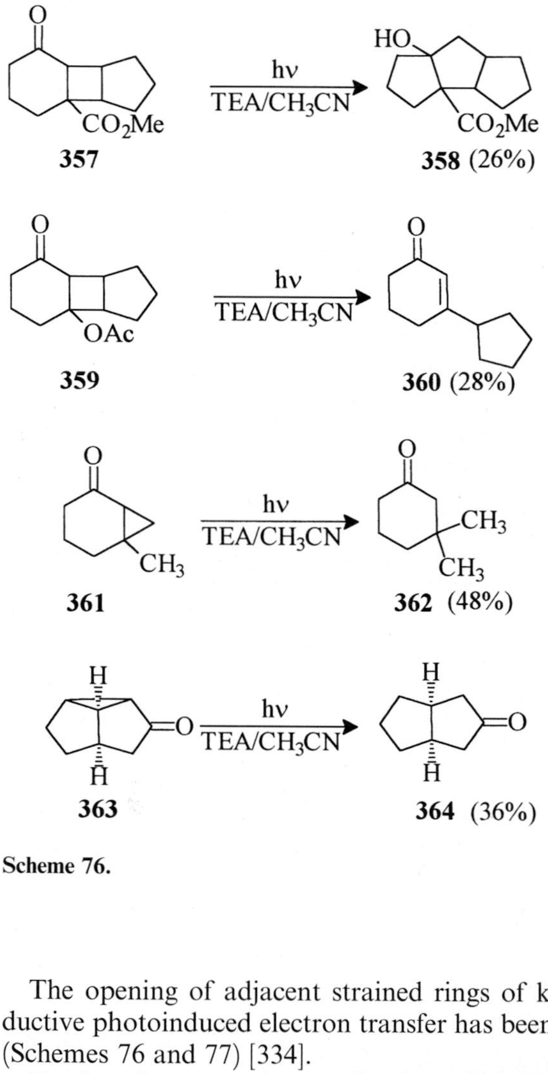

**357**  →  **358** (26%)

**359**  →  **360** (28%)

**361**  →  **362** (48%)

**363**  →  **364** (36%)

**Scheme 76.**

The opening of adjacent strained rings of ketyl radical anions generated in reductive photoinduced electron transfer has been explored by Mattay and coworkers (Schemes 76 and 77) [334].

Tandem fragmentation–cyclization of bicyclic ketones connected to unsaturated side chains have been used to construct bi-, tri-, and spirocyclic ketones [334].

Cossy et al. have shown that depending upon the substitution pattern, ketyl radical anions obtained from photochemically induced electron transfer from amines to cyclopropylketones lead either to the formation of 3-substituted cycloalkanones or to ring expanded products (Scheme 78) [335].

Intramolecular trapping of the radical produced by the PET fragmentation was utilized by Cossy et al. for the construction of bicyclic ring systems (Scheme 79) [335].

Bond cleavage reactions of cyclopropyl and cyclobutyl rings by adjacent ketyl radical anions generated via photoinduced electron transfer from amines have also been reported by Pandey and coworkers (Scheme 80) [336, 337].

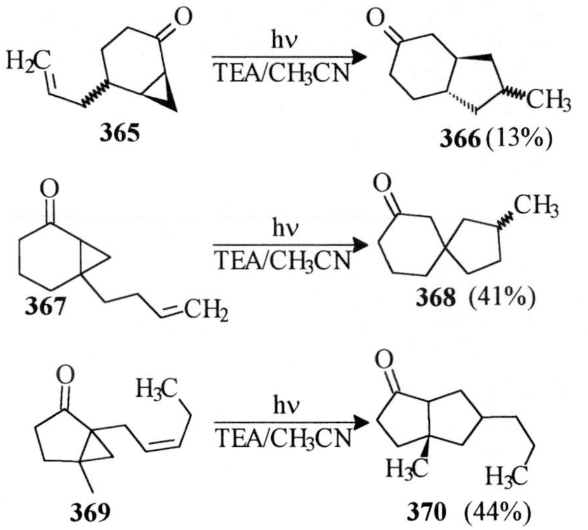

**Scheme 77.**

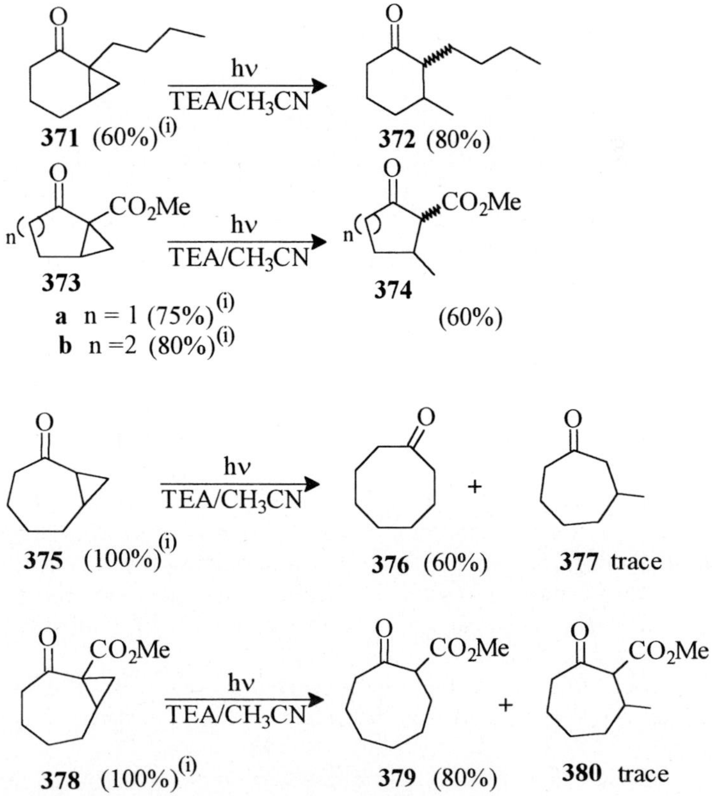

**Scheme 78.**  (i) Percentage of starting material consumed

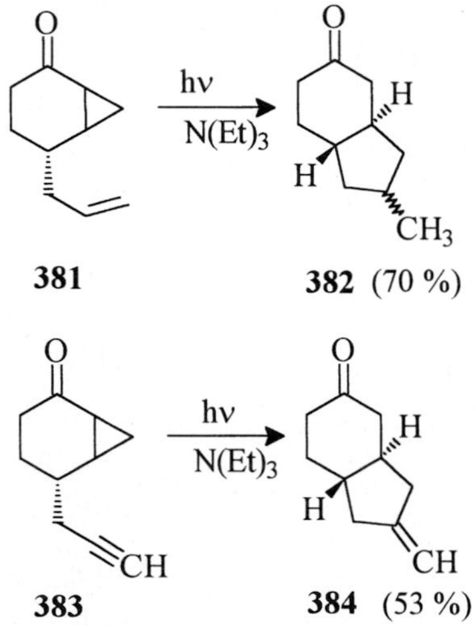

**381**    **382** (70 %)

**383**    **384** (53 %)

**Scheme 79.**

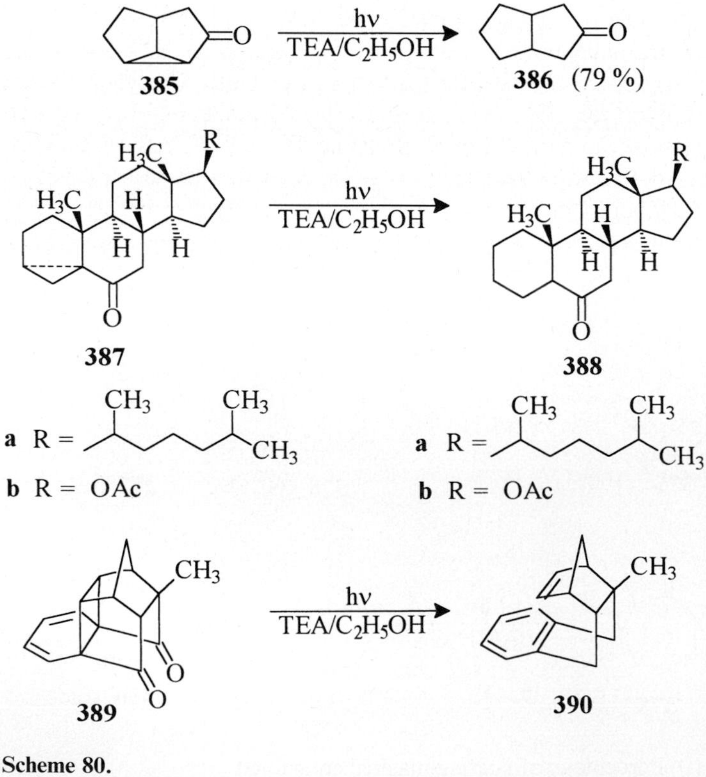

**385**    **386** (79 %)

**387**    **388**

**a** R =    **a** R =

**b** R = OAc    **b** R = OAc

**389**    **390**

**Scheme 80.**

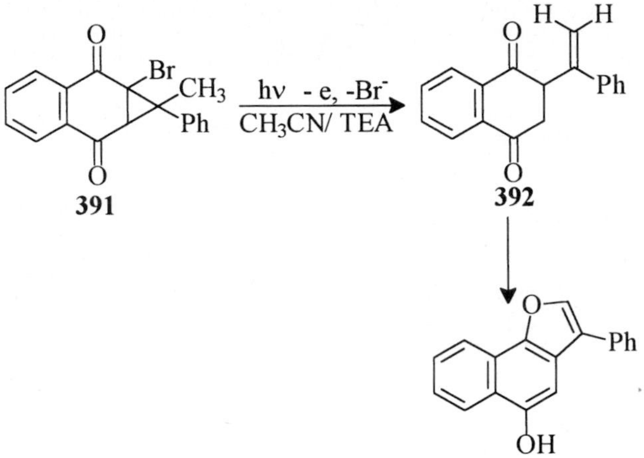


**391**

**392**

**393** (40%)

**Scheme 81.**

Reductive ring opening of the cyclopropyl ring in the homonaphthaquinone (**391**), gave the naphthafuran derivative **393** via the photocyclization of the intermediate 2-(α-phenylvinyl-1,4-naphthaquinone (**392**) (Scheme 81) [338].

Photoinduced electron transfer from tertiary amines to aromatic acid chlorides leads to the formation of acyl radicals. Acyl radicals generated from *O*-allyl salicyloyl chloride (**394**) undergoes intramolecular cyclization to form 3-methyl-4-chromanone (**395**) along with formation of the amide and *O*-allylsalicylic acid (Scheme 82) [339].

Wu et al. have reported that direct excitation of amines in the presence of 3, *β*-hydroxy-5-α-androstan-17-one, under conditions where light is exclusively absorbed by the amine leads to the reduction of the 17-keto group with high stereoselectivity (Scheme 83) [340].

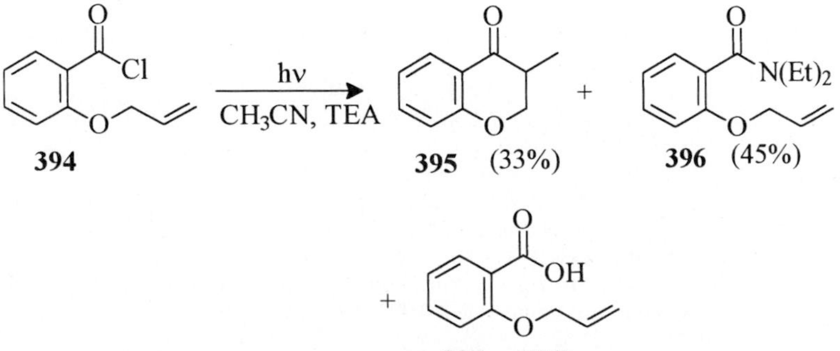


**394**

**395** (33%)

**396** (45%)

**397** (7%)

**Scheme 82.**

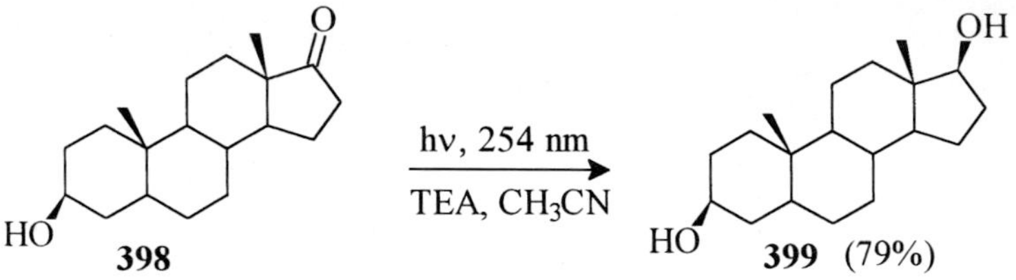

**Scheme 83.**

## Acknowledgment

Financial support from the Council of Scientific and Industrial Research, Government of India and the Volkswagen Foundation, Germany is gratefully acknowledged. This is the contribution No. RRLT-PRU-122 from the Regional Research Laboratory (CSIR), Trivandrum.

## References

1. E. Steckhan, in *Organic Electrochemistry* (Ed.: H. Lund, M. M. Baizer) **1990**, 581–613.
2. T. Shono in *Electroorganic Chemistry as a New Tool in Organic Synthesis*, Springer, Berlin, **1984**, 54–62.
3. Albert J. Fry in *Synthetic Organic Chemistry II nd Ed.*, **1989**, 274–287.
4. J. Mattay, *Synthesis* **1989**, 233–252.
5. P. S. Mariano, J. L. Stavinoha, in *Synthetic Organic Photochemistry* (Ed.: W. M. Horspool) Plenum Press, New York **1983**, 145–147.
6. U. C. Yoon, P. S. Mariano, *Acc. Chem. Res.* **1992**, *25*, 233–240.
7. F. D. Lewis, *Acc. Chem. Res.* **1986**, *19*, 401–405.
8. G. Pandey, *Synlett.* **1994**, 546–552.
9. A. Albini, M. Mella, M. Freccero, *Tetrahedron* **1994**, *50*, 575–607.
10. U. C. Yoon, P. S. Mariano, R. S. Givens, R. W. Atwater, in *Advances in Electron Transfer Chemistry* (Ed.: P. S. Mariano) *Vol. 4*, JAI Press, Greenwich, CT, **1994**, 117–205.
11. F. D. Lewis, in *Advances in Electron Transfer Chemistry* (Ed.: P. S. Mariano) *Vol. 5*, JAI Press, Greenwich, CT, **1996**, 1–40.
12. N. J. Pienta, in *Photoinduced Electron Transfer Part C* (Ed.: M. A. Fox, M. Chanon) Elsevier, Amsterdam, **1988**, 421–486.
13. J. S. D. Kumar, S. Das, *Res. Chem. Intermed.* **1997**, *23*, 755–800.
14. M. Hoshino, H. Shizuka, in *Photoinduced Electron Transfer Part C* (Ed.: M. A. Fox, M. Chanon) Elsevier, Amsterdam **1988**, 313–371.
15. J. Stubbe, W. A. van der Donk, *Chem. Rev.* **1998**, *98*, 705–762.
16. R. B. Silverman, *Acc. Chem. Res.* **1995**, *28*, 335–342.
17. R. B. Silverman, in *Advances in Electron Transfer Chemistry* (Ed.: P. S. Mariano) *Vol. 2*, JAI Press, Greenwich, Connecticut **1992**, 177–213.
18. F. P. Guengerich, T. L. MacDonald, in *Advances in Electron Transfer Chemistry* (Ed.: P. S. Mariano) *Vol. 3*, JAI Press, Greenwich, Connecticut **1993**, 191–241.
19. C. J. Easton, *Chem. Rev.* **1997**, *97*, 53–82.
20. P. Renaud, L. Giruad, *Synthesis* **1996**, 913–926.
21. B. Giese in *Radicals in Organic Synthesis: Formation of Carbon–Carbon Bonds*, Pergamon Press, New York, **1986**.

22. Y. L. Chow, W. C. Danen, S. F. Nelson, D. H. Rosenblatt, *Chem. Rev.* **1978**, *78*, 243–274.
23. P. I. Dalko, *Tetrahedron* **1995**, *51*, 7579–7653.
24. R. D. Theys, G. Sosnovsky, *Chem. Rev.* **1997**, *97*, 83–132.
25. L.-A. Linden, *Proc. Indian Acad. Sci. (Chem. Sci.)* **1993**, *105*, 405–419.
26. N.S. Allen, G. Pullen, M. Edge, I. Weddell, R. Swart, F. Catalina, *J. Photochem. Photobiol. A: Chem.* **1997**, *109*, 71–75.
27. G.K. Pullen, N.S. Allen, M. Edge, I. Weddell, R. Swart, F. Catalina, S. Navaratnam, *European Polymer Journal* **1996**, *32*, 943–955.
28. K. Venkataraman, *The Chemistry of Synthetic Dyes Vol. I–VIII* **1953–1978**,.
29. A. P. de Silva, H. Q. N. Gunaratne, T. Gunnlaugsson, A. J. M. Huxley, C. P. McCoy, J. T. Rademacher, T. R. Rice, *Chem. Rev.* **1997**, *97*, 1515–1566.
30. A. W. Czarnik, in *Fluorescent Chemosensors for Ion and Molecule Recognition* American Chemical Society, Washington D. C. **1992**.
31. P. Novàk, K. Müller, K. S. V. Santhanam, O. Haas, *Chem. Rev.* **1997**, *97*, 207–281.
32. T. Tsutsui, C. Adachi, S. Saito, *Electroluminescence in Organic Thin Films*, in *Photochemical Progress in Organized Molecular Systems* (Ed.: K. Honda) Elsevier Science Publishers, B. V., Amsterdam **1991**, 437–450.
33. J. Kido, M. Kimura, K. Nagai, *Science* **1996**, *275*, 1267–1332.
34. K. Sakanoue, M. Motoda, M. Sugimoto, S. Sakaki, *J. Phys. Chem. A* **1999**, *103*, 5551–5556.
35. D. L. Morel, E. L. Stogryn, A. K. Ghosh, T. Feng, P. E. Purwin, R. F. Shaw, C. Fishman, G. R. Bird, A. P. Piechowski, *J. Phys. Chem.* **1984**, *88*, 923–933.
36. A. P. Piechowski, G. R. Bird, D. L. Morel, E. L. Stogryn, *J. Phys. Chem.* **1984**, *88*, 934–950.
37. S. R. Marder, J. E. Sohn, G. D. Stucky in *Materials for Nonlinear Optics: Chemical Perspectives, ACS Symp. Ser.*, **1991**.
38. D. S. Chemla, J. Zyss in *Non-linear Optical Properties of Organic Molecules and Crystals, Vol. 1 and 2*, Academic Press, New York, **1987**.
39. D. J. Williams in *Nonlinear Optical Properties of Organic and Polymeric Materials, ACS Symp. Ser., Vol. 233*, Washington D. C., **1983**, 251pp.
40. X-D. Wu, S-K. Khim, X. Zhang, E. M. Cederstrom, P. S. Mariano, *J. Org. Chem.* **1998**, *63*, 841–859.
41. H. J. Kim, U. C. Yoon, Y-S Jung, N. S. Park, E. M. Cederstrom, P. S. Mariano, *J. Org. Chem.* **1998**, *63*, 860–863.
42. X. Zhang, Y-S Jung, P. S. Mariano, M. A. Fox, P. S. Martin, J. Merkert, *Tetrahedron Lett.* **1993**, *34*, 5239–5242.
43. C. A. Audeh, J. R. Lindsay Smith, *J. Chem. Soc. B* **1970**, 1280–1285.
44. C. A. Audeh, J. R. Lindsay Smith, *J. Chem. Soc. B* **1971**, 1745–1747.
45. L. A. Hull, G. T. Davis, D. H. Rosenblatt, *J. Am. Chem. Soc.* **1969**, *91*, 6247–6250.
46. K. S. Shukla, P. C. Mathur, O. P. Bansal, *J. Inorg. Nucl. Chem.* **1973**, *35*, 1301–1307.
47. G. T. Davis, M. M. Demek, D. H. Rosenblatt, *J. Am. Chem. Soc.* **1972**, *94*, 3321–3325.
48. D. H. Rosenblatt, G. T. Davis, L. A. Hull, G. D. Forberg, *J. Org. Chem.* **1968**, *33*, 1649–1650.
49. H. J. Dauben Jr, L.L. McCoy, *J. Am. Chem. Soc.* **1959**, *81*, 5404–5409.
50. J. R. Lindsay Smith, J. S. Sadd, *J. Chem. Soc. Perkin Trans. 2* **1976**, 741–747.
51. S. K. Malhotra, J. J. Hostynek, A. F. Lundin, *J. Am. Chem. Soc.* **1968**, *90*, 6565–6566.
52. H. Weingarten, J. S. Wager, *J. Org. Chem.* **1970**, *35*, 1750–1753.
53. L. Horner, C. Betzel, *Justus Leibigs Ann. Chem.* **1953**, *579*, 175–192.
54. J. K. Kochi, in *Free Radicals* (Ed.: J. K. Kochi) *Vol. 2*, John Wiley and Sons, New York **1973**, 665–710.
55. J. K. Kochi, in *Free Radicals* (Ed.: J. K. Kochi) *Vol. 1*, Interscience, New York **1973**, 591–683.
56. Y. L. Chow in *Reactive Intermediates*, Plenum Press, **1980**, 417–441.
57. Y. L. Chow, *Acc. Chem. Res.* **1973**, *6*, 354–360.
58. D. H. Rosenblatt, E. P. Burrows, in *Chemistry of Amino, Nitroso, Nitro Compounds and Their Derivatives* (Ed.: S. Patai) *Vol. 2*, Wiley, New York **1982**, 1085–1149.
59. L. A. Hull, G. T. Davis, D. H. Rosenblatt, H. K. R. Williams, R. C. Weglein, *J. Am. Chem. Soc.* **1967**, *89*, 1163–1170.

60. M. Wei, R. Stewart, *J. Am. Chem. Soc.* **1966**, *88*, 1974–1979.
61. H. Firouzabadi, A. R. Sardarian, M. Naderi, B. Vessal, *Tetrahedron* **1984**, *40*, 5001–5004.
62. H. Firouzabadi, B. Vessal, M. Naderi, *Tetrahedron Lett.* **1982**, *23*, 1847–1850.
63. H. Firouzabadi, Z. Mostafavipoor, *Bull. Chem. Soc. Jpn.* **1983**, *56*, 914–917.
64. L. Castedo, R. Riguera, M. J. Rodriguez, *Tetrahedron* **1982**, *38*, 1569–1570.
65. D. S. Kashdan, J. A. Schwartz, H. Rapoport, *J. Org. Chem.* **1982**, *47*, 2638–2643.
66. R. Neumann, M. Levin, *J. Org. Chem.* **1991**, *56*, 5707–5710.
67. K. Orito, T. Hatakakeyama, M. Takeo, S. Uchiito, M. Tokuda, H. Suginome, *Tetrahedron* **1998**, *54*, 8403–8410.
68. C. K. Mann, K. A. Barnes in *Electrochemical Reactions in Nonaqueous Systems*, Dekker, New York, **1970**, 259–296.
69. K. K. Barnes, C. K. Mann, *J. Org. Chem.* **1967**, *32*, 1474–1479.
70. N. A. Hampson, J. B. Lee, J. R. Morley, K. I. MacDonald, B. F. Scanlon, *Tetrahedron* **1970**, *26*, 1109–1114.
71. P. M. Robertson, F. Scwager, N. Ibl, *Electrochim. Acta.* **1973**, *18*, 923–924.
72. R. F. Nelson, *Tech. Electroorg. Synth. Part 2*, in *Tech. Chem.* (Ed.: N. L. Weinberg) *Vol. V*, Wiley–Interscience, New York **1974**, 269–395.
73. N. L. Weinberg, H. R. Weinberg, *Chem. Rev.* **1968**, *68*, 449–523.
74. R. L. Hand, R. F. Nelson, *J. Am. Chem. Soc.* **1974**, *96*, 850–860.
75. D. M. Mohilner, R. N. Adams, W. J. Argersinger Jr., *J. Am. Chem. Soc.* **1962**, *84*, 3618–3622.
76. M. Breitenbach, K. H. Heckner, *J. Electroanal. Chem. Interfacial Electrochem.* **1971**, *29*, 309–323.
77. J. Heinze, *Top. Curr. Chem.* **1990**, *152*, 1–47.
78. R. Reynolds, L.L. Line, R. F. Nelson, *J. Am. Chem. Soc.* **1974**, *96*, 1087–1092.
79. E. Steckhan, *Angew. Chem.* **1986**, *25*, 681–699.
80. E. Steckhan, *Top. Curr. Chem.* **1987**, *142*, 1–69.
81. W. Schmidt, E. Steckhan, *Angew. Chem. Int. Ed. Engl.* **1978**, *17*, 673–674.
82. N. Schultz, S. Töteberg-Kaulen, S. Dapperheld, J. Heyer, M. Platen, K. Schumacher, E. Steckhan, in *Recent Advances in Electro Organic Synthesis, Proceedings of the 1st International Symposium on Electroorganic Synthesis* (Ed.: S. Torii) Elsevier, Amsterdam **1987**, 127–135.
83. K.-H. Grosse Brinkhaus, E. Steckhan, D. Degner, *Tetrahedron* **1986**, *42*, 553–560.
84. S. Das, C. von Sonntag, *Z. Naturforsch.* **1986**, *41b*, 505–513.
85. S. Das, M. N. Schuchmann, H.-P. Schuchmann, C. von Sonntag, *Chem. Ber.* **1987**, *120*, 319–323.
86. S. Das, O. J. Mieden, X.-M. Pan, M. Pepas, M. N. Schuchmann, H.-P. Schuchmann, C. von Sonntag, H. Zegota, in *Oxygen Radicals in Biology and Medicine* (Ed.: M. G. Simic, K. A. Taylor, J. F. Ward, C. von Sonntag) Plenum Press, New York **1988**, 55–58.
87. M. Simic, P. Neta, E. Hayon, *Int. J. Radiat. Phys. Chem.* **1971**, *3*, 309–320.
88. Z. B. Alfassi, R. H. Schuler, *J. Phys. Chem.* **1985**, *89*, 3359–3363.
89. Z. B. Alfassi, S. Mosseri, P. Neta, *J. Phys. Chem.* **1989**, *93*, 1380–1385.
90. S. Solar, W. Solar, N. Getoff, *Radiat. Phys. Chem.* **1986**, *28*, 229–234.
91. J. W. T. Spinks, R. J. Woods in *An Introduction to Radiation Chemistry*, John Wiley and Sons, New York, **1990**, 592pp.
92. S. Das, P. V. Kamat, S. Padmaja, V. Au, S. A. Madison, *J. Chem. Soc. Perkin Trans. 2* **1999**, 1219–1223.
93. R. O. C. Norman, N. H. Anderson, *J. Chem. Soc. B* **1971**, 993–1003.
94. R. W. Fessenden, P. Neta, *J. Phys. Chem.* **1972**, *76*, 2857–2859.
95. P. Wardman, D. R. Smith, *Can. J. Chem.* **1971**, *49*, 1880–1887.
96. Farhataziz, A. B. Ross, *U. S. Department of Commerce Washington NSRDS-NBS* **1977**, *59*, 113pp.
97. M. Eigen, G. Ilgenfritz, W. Kruse, *Chem. Ber.* **1965**, *98*, 1623–1638.
98. L. Qin, G. N. R. Tripathi, R. H. Schuler, *Z. Naturforsch.* **1985**, *40a*, 1026–1039.
99. P. S. Rao, E. Hayon, *J. Phys. Chem.* **1975**, *79*, 1063–1066.
100. J. Holcman, K. Sehested, *J. Phys. Chem.* **1977**, *81*, 1963–1966.
101. J. Wheeler, R. Nelsen, *J. Phys. Chem.* **1973**, *77*, 2490–2492.

102. E. Zador, J. M. Warman, A. Hummel, *J. Chem. Soc. Faraday Transactions 1* **1976**, *72*, 1368–1376.
103. G. N. R. Tripathi, Q. Sun, *J. Phys. Chem. A* **1999**, *103*, 9055–9060.
104. W. C. Danen, R. C. Rickard, *J. Am. Chem. Soc.* **1972**, *94*, 3254–3256.
105. V. Malatesta, K. U. Ingold, *J. Am. Chem. Soc.* **1973**, *95*, 6400–6404.
106. W. C. Danen, R. C. Rickard, *J. Am. Chem. Soc.* **1975**, *97*, 2303–2304.
107. H. Bock, I. Goebel, Z-Havlas, S. Liedle, H. Oberhammer, *Angew. Chem. Int. Ed. Engl.* **1991**, *30*, 187–190.
108. S. F. Nelsen, C. R. Kessel, *J. Chem. Soc. Chem. Commun.* **1977**, *30*, 490–491.
109. A. de Meijere, V. Chaplinski, F. Gerson, P. Merstetter, E. Haselbach, *J. Org. Chem.* **1999**, *64*, 6951–6959.
110. T. Shida, E. Haselbach, T. Bally, *Acc. Chem. Res.* **1984**, *17*, 180–186.
111. M.C.R. Symons, *Chem. Soc. Rev.* **1984**, *13*, 393–439.
112. X.-Z. Qin, F. Williams, *J. Phys. Chem.* **1986**, *90*, 2292–2296.
113. A. de. Meijere, V. Chaplinski, H. Winsel, M. A. Kusnetsov, P. Rademacher, R. Boese, T. Haumann, M. Traetteberg, P. V. Schleyer, T. Zyweitz, H. Jiao, P. Merstetter, F. Gerson, *Angew. Chem. Int. Ed. Engl.* **1999**, *38*, 2430–2433.
114. M. B. Khusidman, V. P. Vyatkin, N. V. Grogor'eva, S. L. Dobychin, *Zh. Prikl. Khim. (Leningrad)* **1983**, *56*, 222–224 [Chem. Abstr. **1983**, *98*, 160106r].
115. *Photoinduced Electron Transfer Part A–D*, (Ed.: M. A. Fox, M. Chanon) Elsevier, New York **1988**.
116. M. Julliard, M. Chanon, *Chem. Rev.* **1983**, *83*, 425–506.
117. G. J. Kavarnos, N. J. Turro, *Chem. Rev.* **1986**, *86*, 401–449.
118. J. Mattay, M. Vondenhof, *Top. Curr. Chem.* **1991**, 219–255.
119. D. Rehm, A. Weller, *Isr. J. Chem.* **1970**, *8*, 259–271.
120. M. Chanon, M. D. Hawley, M. A. Fox, in *Photoinduced Electron Transfer Part A* (Ed.: M. A. Fox, M. Chanon) **1988**, 1–59.
121. F. D. Lewis, T.-I Ho, J. T. Simpson, *J. Am. Chem. Soc.* **1982**, *104*, 1924–1929.
122. F. D. Lewis, T.-I Ho, J. T. Simpson, *J. Org. Chem.* **1981**, *46*, 1077–1082.
123. J. R. Lindsay Smith, L. A. V. Mead, *J. Chem. Soc. Perkin Trans. 2* **1973**, 206–210.
124. P. J. Smith, C.K. Mann, *J. Org. Chem.* **1969**, *34*, 1821–1826.
125. F. D. Lewis, J. T Simpson, *J. Am. Chem. Soc.* **1980**, *102*, 7593–7595.
126. H. C. Brown, I. Moritani, *J. Am. Chem. Soc.* **1955**, *77*, 3607–3610.
127. D. J. Cram, M. R. V. Sahyun, *J. Am. Chem. Soc.* **1963**, *85*, 1257–1263.
128. X. Zhang, S.-R. Yeh, S. Hong, M. Freccero, A. Albini, D. E. Falvey, P. S. Mariano, *J. Am. Chem. Soc.* **1994**, *116*, 4211–4220.
129. Z. Su, P. S. Mariano, D. E. Falvey, U. C. Yoon, S. W. Oh, *J. Am. Chem. Soc.* **1998**, *120*, 10676–10686.
130. W. Xu, P. S. Mariano, *J. Am. Chem. Soc.* **1991**, *113*, 1431–1432.
131. W. Xu, X. M. Zhang, P. S. Mariano, *J. Am. Chem. Soc.* **1991**, *113*, 8863–8878.
132. E. R. Gaillard, D. G. Whitten, *Acc. Chem. Res.* **1996**, 292–297.
133. X. Ci, D. G. Whitten, *J. Phys. Chem.* **1991**, *95*, 1988–1993.
134. C. M. Haugen, W. R. Bergmark, D. G. Whitten, *J. Am. Chem. Soc.* **1992**, *114*, 10293–10297.
135. X. Ci, M. A. Kellet, D. G. Whitten, *J. Am. Chem. Soc.* **1993**, *113*, 3893–3904.
136. X. Ci, D. G. Whitten, *J. Am. Chem. Soc.* **1989**, *111*, 3459–3461.
137. X. Ci, D. G. Whitten, *J. Am. Chem. Soc.* **1987**, *109*, 7215–7217.
138. L. A. Lucia, R. D. Burton, K. S. Schanze, *J. Phys. Chem.* **1993**, *97*, 9078–9080.
139. Y. Wang, B. T. Hauser, M. M. Rooney, R. D. Burton, K. S. Schanze, *J. Am. Chem. Soc.* **1993**, *115*, 5675–5683.
140. N. Sutin, *Acc. Chem. Res.* **1982**, *15*, 275–282.
141. *Electron-transfer Reactions*, (Ed.: R. D. Cannon) Butterworths, Sevenoaks, London **1981**, 368pp.
142. M. D. Mauzerall, in *Photoinduced Electron Transfer Part A*, (Ed.: M. A. Fox, M. Chanon) Elsevier, Amsterdam **1988**, 228–244.
143. S. G. Cohen, J. I. Cohen, *J. Phys. Chem.* **1968**, *72*, 3782–3793.
144. S. Inbar, S. G. Cohen, *J. Am. Chem. Soc.* **1978**, *100*, 4490–4495.

145. S. G. Cohen, H. M. Chao, *J. Am. Chem. Soc.* **1968**, *90*, 165–173.
146. S. Inbar, H. Linschitz, S. G. Cohen, *J. Am. Chem. Soc.* **1980**, *102*, 1419–1421.
147. S. Inbar, H. Linschitz, S. G. Cohen, *J. Am. Chem. Soc.* **1981**, *103*, 1048–1054.
148. K. Bhattacharyya, P. K. Das, *J. Phys. Chem.* **1986**, *90*, 3987–3993.
149. J. C. Scaiano, *J. Phys. Chem.* **1981**, *85*, 2851–2855.
150. A. Demeter, L. Biczók, T. Bérces, V. Wintgens, P. Valat, J. Kossanyi, *J. Phys. Chem.* **1993**, *97*, 3217–3224.
151. X. Ci, D. G. Whitten, in *Photoinduced Electron Transfer Part C* (Ed.: M. A. Fox, M. Chanon) Elsevier, Amsterdam **1988**, 553–577.
152. M. Hoshino, H. Shizuka, *J. Phys. Chem.* **1987**, *91*, 714–718.
153. J. D. Simon, K. S. Peters, *J. Am. Chem. Soc.* **1981**, *103*, 6403–6406.
154. J. D. Simon, K. S. Peters, *Acc. Chem. Res.* **1984**, *17*, 277–283.
155. J. D. Simon, K. S. Peters, *J. Am. Chem. Soc.* **1982**, *104*, 6542–6547.
156. K. S. Peters, E. Pang, J. Rudzki, *J. Am. Chem. Soc.* **1982**, *104*, 5535–5577.
157. C. Devadoss, R. W. Fessenden, *J. Phys. Chem.* **1990**, *94*, 4540–4549.
158. C. Devadoss, R. W. Fessenden, *J. Phys. Chem.* **1991**, *95*, 7253–7260.
159. H. Miyasaka, K. Morita, K. Kamada, N. Mataga, *Chem. Phys. Lett.* **1991**, *178*, 504–510.
160. H. Miyasaka, K. Morita, K. Kamada, N. Mataga, *Bull. Chem. Soc. Jpn.* **1990**, *63*, 3385–3397.
161. E. Haselbach, P. Jacques, D. Pilloud, P. Suppan, E. Vauthey, *J. Phys. Chem.* **1991**, *95*, 7115–7117.
162. K. S. Peters, J. Lee, *J. Phys. Chem.* **1993**, *97*, 3761–3764.
163. K. S. Peters, J. Lee, *J. Phys. Chem.* **1992**, *96*, 8941–8945.
164. H. Miyasaka, K. Morita, K. Kamada, T. Nagata, M. Kiri, N. Mataga, *Bull. Chem. Soc. Jpn.* **1991**, *64*, 3229–3244.
165. H. Miyasaka, M. Kiri, K. Morita, N. Mataga, Y. Tanimoto, *Chem. Phys. Lett.* **1992**, *199*, 21–28.
166. H. Miyasaka, M. Kiri, K. Morita, N. Mataga, Y. Tanimoto, *Bull. Chem. Soc. Jpn.* **1995**, *68*, 1569–1821.
167. T. Kiyota, M. Yamaji, H. Shizuka, *J. Phys. Chem.* **1996**, *100*, 672–679.
168. K. Hamanoue, K. Yokoyama, Y. Kajiwara, M. Kimoto, T. Nakayama, H. Teranishi, *Chem. Phys. Lett.* **1985**, *113*, 207–212.
169. M. Hamanoue, K. Sawada, K. Yokoyama, T. Nakayama, S. Hirase, H. Teranishi, *J. Photochem.* **1986**, *33*, 99–111.
170. M. Goez, I. Frisch, *J. Photochem. Photobiol. A: Chem.* **1994**, *84*, 1–12.
171. D. Beckert, R. W. Fessenden, *J. Phys. Chem.* **1996**, *100*, 1622–1629.
172. E. Lippert, W. Luder, H. Boos, in *Advances in Molecular Spectroscopy* (Ed.: A. Mangini) Pergamon Press, Oxford **1962**, 443–457.
173. E. Lippert, W. Lüder, F. Moll, W. Naegele, H. Boos, H. Prigge, I. S. Blankenstein, *Angew. Chem.* **1961**, *73*, 695–706.
174. Z. R. Grabowski, J. Dobkowski, *Pure. Appl. Chem.* **1983**, *55*, 245–252.
175. K. Rotkiewicz, K. H. Grellmann, Z. R. Grabowski, *Chem. Phys. Lett.* **1973**, *19*, 315–318.
176. W. Rettig, *Angew. Chem.* **1986**, *98*, 969–976.
177. E. Lippert, W. Rettig, V. B. Koutecky, F. Heisel, J. A. Miehe, *Adv. Chem. Phys.* **1987**, *68*, 1–173.
178. W. Rettig, W. Baumann, J. F. Rabek, in *Photochem. Photophys.* (Ed.: J. F. Rabek) *Vol. 6*, CRC Press, Boca Raton **1992**, 79–134.
179. K. Bhattacharyya, M. Chowdhury, *Chem. Rev.* **1993**, *93*, 507–535.
180. K. A. Zachariasse, Th. von der Haar, A. Hebecker, U. Leinhos, W. Kuhnle, *Pure. Appl. Chem.* **1993**, *65*, 1745–1750.
181. Th. von der Haar, A. Hebecker, Y. V. Il'ichev, Y. B. Jiang, W. Kühnle, K. A. Zachariasse, *Recl. Trav. Chim. Pays-Bas* **1995**, *114*, 430–442.
182. K. A. Zachariasse, M. Grobys, Th. von der Haar, A. Hebecker, Y. V. Il'ichev, Y.-B Jiang, O. Morawski, W. Kuhnle, *J. Photochem. Photobiol. A: Chem.* **1996**, *102*, 59–70.
183. K. A. Zachariasse, M. Grobys, Th. von der Haar, A. Hebecker, Y. V. Il'ichev, O. Morawski, I. Ruckert, W. Kühnle, *J. Photochem. Photobiol. A: Chem.* **1997**, *105*, 373–383.
184. F. P. Guengerich, T. L. MacDonald, *Acc. Chem. Res.* **1984**, *17*, 9–16.

185. F. P. Guengerich, *Crit. Rev. Biochem. Mol. Biol.* **1990**, *25*, 97–153.
186. *Cytochrome P-450: Structure, Mechanism and Biochemistry*, Plenum Press, New York **1995**.
187. G. L. Kedderis, P. F. Hollenberg, *J. Biol. Chem.* **1983**, *258*, 8129–8138.
188. G. L. Kedderis, D. R. Koop, P. F. Hollenberg, *J. Biol. Chem.* **1980**, *255*, 10174–10182.
189. B. W. Griffin, P. L. Ting, *Biochemistry* **1978**, *17*, 2206–2211.
190. V. W. Tetrud, J. W. Langston, A. J. Ruttenber, *Science* **1989**, *245*, 519–532.
191. K. F. Tipton, P. Dostert, M. S. Ben, in *Monoamine Oxidase and Disease* Academic Press, New York **1984**.
192. R. B. Silverman, S. J. Hoffman, W. B. Catus III, *J. Am. Chem. Soc.* **1980**, *102*, 7126–7128.
193. J. T. Simpson, A. Krantz, F. D. Lewis, B. Kokel, *J. Am. Chem. Soc.* **1982**, *104*, 7155–7161.
194. A. Krantz, B. Kokel, Y. P. Sachdeva, J. I. Salach, K. Detmer, A. Claesson, C. Sahlberg, in *Monoamine Oxidase: Struct. Funct. Altered Funct. (Proc. Symp.)* (Ed.: T. P. Singer, R. W. von Korff, D. L. Murphy) Academic Press, New York **1979**, 51–70.
195. Y. Maeda, K. U. Ingold, *J. Am. Chem. Soc.* **1980**, *102*, 328–331.
196. R. B. Silverman, S. J. Hoffman, *Biochem. Biophys. Res. Commun.* **1981**, *101*, 1396–1401.
197. R. B. Silverman, J. M. Cesarone, X. Lu, *J. Am. Chem. Soc.* **1993**, *115*, 4955–4961.
198. R. B. Silverman, P. A. Zeiske, *Biochemistry* **1986**, *25*, 341–346.
199. Y. Zelechonok, R. B. Silverman, *J. Org. Chem.* **1992**, *57*, 5787–5790.
200. M. Newcomb, C. C. Johnson, M. B. Manek, T. R. Varick, *J. Am. Chem. Soc.* **1992**, *114*, 10915–10921.
201. R. B. Silverman, J. P. Zhou, P. E. Eaton, *J. Am. Chem. Soc.* **1993**, *115*, 8841–8842.
202. R. B. Silverman, X. Lu, J. J. P. Zhou, A. Swihart, *J. Am. Chem. Soc.* **1994**, *116*, 11590–11591.
203. J. M. Dickinson, J. A. Murphy, C. W. Patterson, N. F. Wooster, *J. Chem. Soc. Perkin Trans. 1* **1990**, 1179–1184.
204. A. Johns, J. A. Murphy, C. W. Patterson, N. F. Wooster, *J. Chem. Soc. Chem. Commun.* **1987**, 1238–1240.
205. A. Tan, M. D. Glantz, L. H. Piette, K. T. Yasunobu, *Biochem. Biophys. Res. Commun.* **1983**, *117*, 517–523.
206. J. R. Miller, D. Edmondson, C. B. Grissom, *J. Am. Chem. Soc.* **1995**, *117*, 7830–7831.
207. C. B. Grissom, *Chem. Rev.* **1995**, *95*, 3–24.
208. J. C. Scaiano, F. L. Cozens, J. McLean, *Photochem. Photobiol.* **1994**, *59*, 585–589.
209. J-M. Kim, S. E. Hoegy, P. S. Mariano, *J. Am. Chem. Soc.* **1995**, *117*, 100–105.
210. K. A. Van Houten, J-M. Kim, M. A. Bogdan, D. C. Ferri, P. S. Mariano, *J. Am. Chem. Soc.* **1998**, *120*, 5864–5872.
211. J. R. Lindsay Smith, D. N. Martimer, *J. Chem. Soc. Perkin Trans. 2* **1986**, 1743–1749.
212. H. Kurebayashi, *Arch. Biochem. Biophys.* **1989**, *270*, 320–329.
213. C. J. Parli, N. Wang, R. E. McMahon, *Biochem. Biophys. Res. Commun.* **1971**, *43*, 1204–1209.
214. R. E. McMahon, H. W. Culp, J. C. Craig, N. Ekwuribe, *J. Med. Chem.* **1979**, *22*, 1100–1103.
215. J. P. Dinnocenzo, S. B. Karki, J. P. Jones, *J. Am. Chem. Soc.* **1993**, *115*, 7111–7116.
216. S. B. Karki, J. P. Dinnocenzo, J. P. Jones, K. R. Korzekwa, *J. Am. Chem. Soc.* **1995**, *117*, 3657–3664.
217. F. P. Guengerich, C-H. Yun, T. L. MacDonald, *J. Biol. Chem.* **1996**, *271*, 27321–27329.
218. E. Baciocchi, O. Lanzalunga, A. Lapi, L. Manduchi, *J. Am. Chem. Soc.* **1998**, *120*, 5783–5787.
219. T. Shono, T. Toda, N. Oshino, *J. Am. Chem. Soc.* **1982**, *104*, 2639–2641.
220. Y. Seto, F. P. Guengerich, *J. Biol. Chem.* **1993**, *268*, 9986–9997.
221. V. D. Parker, M. Tilset, *J. Am. Chem. Soc.* **1991**, *113*, 8778–8781.
222. D. Griller, J. A. Howard, P. R. Mariott, J. C. Scaiano, *J. Am. Chem. Soc.* **1981**, *103*, 619–623.
223. M. Bietti, A. Cuppoletti, C. Dagastin, C. Florea, C. Galli, P. Gentili, H. Petride, C. R. Caia, *Eur. J. Org. Chem.* **1998**, 2425–2529.
224. E. Baciocchi, A. Lapi, *Tetrahedron Lett.* **1999**, *40*, 5425–5428.
225. A. Anne, S. Fraoua, J. Moiroux, J.M. Saveant, *J. Phy. Org. Chem.* **1998**, *11*, 774–780.
226. A. Anne, S. Fraoua, J. Moiroux, J. M. Saveant, *J. Am. Chem. Soc.* **1996**, *118*, 3938–3945.
227. O. Okazaki, F. P. Guengerich, *J. Biol. Chem.* **1993**, *268*, 1546–1552.
228. F. E. McDonald, A. K. Chatterjee, *Tetrahedron Lett.* **1997**, *38*, 7687–7690.

229. C-K. Chen, A. G. Hortmann, M. R. Marzabadi, *J. Am. Chem. Soc.* **1988**, *110*, 4829–4831.
230. J. T. Yli-Kauhaluoma, C. W. Harwig, P. Wentworth Jr., K. D. Janda, *Tetrahedron Lett.* **1998**, *39*, 2269–2272.
231. F. Compernolle, M. A. Saleh, S. Van den Branden, S. Toppet, G. Hoornaert, *J. Org. Chem.* **1991**, *56*, 2386–2390.
232. N. J. Leonard, D. F. Morrow, *J. Am. Chem. Soc.* **1958**, *80*, 371–375.
233. N. J. Leonard, R. R. Sauers, *J. Am. Chem. Soc.* **1957**, *79*, 6210–6214.
234. F. Compernolle, M. A. Saleh, S. Toppet, G. Hoornaert, *J. Org. Chem.* **1991**, *56*, 5192–5196.
235. S. Van den Branden, F. Compernolle, G. Hoornaert, *Tetrahedron* **1992**, *48*, 9753–9766.
236. A. P. Darmanyan, W. S. Jenks, P. Jardon, *J. Phys. Chem.* **1998**, *102*, 7420–7426.
237. C. Ferroud, P. Rool, J. Santamaria, *Tetrahedron Lett.* **1998**, *39*, 9423–9426.
238. J. Vukovic, A. E. Goodbody, J. P. Kutney, M. Misawa, *Tetrahedron* **1988**, *44*, 325–331.
239. R. J. Sundberg, P. J. Hunt, P. Desos, K. G. Gadamasetti, *J. Org. Chem.* **1991**, *56*, 1689–1692.
240. S-K. Khim, X. Wu, P. S. Mariano, *Tetrahedron Lett.* **1996**, *37*, 571–574.
241. M. Hirano, S. Yakabe, H. Chikamori, J. H. Clark, T. Morimoto, *J. Chem. Res. (S)* **1998**, 770–771.
242. M. Hudlicky, *Oxidations in Organic Chemistry*, in *ACS Monograph Vol. 186*, **1990**, 234–242.
243. J. P. Wolfe, S. Wagaw, J.-F. Marcoux, S. L. Buchwald, *Acc. Chem. Res.* **1998**, *31*, 805–818.
244. D. W. Old, J. P. Wolfe, S. L. Buchwald, *J. Am. Chem. Soc.* **1998**, *120*, 9722–9723.
245. J. P. Wolfe, S. Wagaw, S. L. Buchwald, *J. Am. Chem. Soc.* **1996**, *118*, 7215–7216.
246. V. V. Grushin, H. Alper, *Chem. Rev.* **1994**, *94*, 1047–1062.
247. B. C. Hamann, J.F. Hartwig, *J. Am. Chem. Soc.* **1998**, *120*, 7369–7370.
248. I. P. Beletskaya, A. G. Bessmertnykh, R. Guilard, *Tetrahedron Lett.* **1999**, *40*, 6393–6397.
249. *Enzyme Catalysis in Organic Synthesis, A Comprehensive Handbook*, (Ed. K. Drauz, H. Waldman) *Vol. II*, VCH, **1995**, 774–780.
250. A. R. Battersby, *in Ciba Foundation Symposium* **1985**, 22–30.
251. J. E. Cragg, R. B. Herbert, M. M. Kgaphola, *Tetrahedron Lett.* **1990**, *31*, 6907–6910.
252. W. D. Fessner, G. Sinerius, *Angew. Chem. Int. Ed. Engl.* **1994**, *33*, 209–212.
253. G. Hilt, E. Steckhan, *J. Chem. Soc. Chem. Commun.* **1993**, 1706–1707.
254. A. M. Equi, A. M. Brown, A. Cooper, S. K. Ner, A. B. Watson, D. J. Robins, *Tetrahedron* **1991**, *47*, 507–518.
255. E. Santaniello, A. Manzocchi, P. A. Bondi, C. Secchi, T. Simanic, *J. Chem. Soc. Chem. Commun.* **1984**, 803–804.
256. R. Parkin, H. O. Hultin, *Biotech. Bioeng.* **1979**, 939–953.
257. N. Nakajima, D. Conrad, H. Sumi, K. Suzuki, N. Esaki, C. Wandrey, K. Soda, *Ferment. and Bioeng.* **1990**, *70*, 322–325.
258. S. M. Hecht, K. M. Rupprecht, P. M. Jacobs, *J. Am. Chem. Soc.* **1979**, *101*, 3982–3983.
259. Y. Asada, K. Tanizawa, S. Sawada, T. Suzuki, H. Misono, K. Soda, *Biochemistry* **1981**, *20*, 6881–6886.
260. H. J. Schäfer, *Top. Curr. Chem.* **1987**, *142*, 101–129.
261. J. E. Barry, M. Finkelstein, E. A. Mayeda, S. D. Ross, *J. Org. Chem.* **1974**, *39*, 2695–2699.
262. G. Bidan, M. Genies, *Tetrahedron* **1981**, *37*, 2297–2301.
263. N. L. Weinberg, E. A. Brown, *J. Org. Chem.* **1966**, *31*, 4054–4058.
264. N. L. Weinberg, T. B. Reddy, *J. Am. Chem. Soc.* **1968**, *90*, 91–93.
265. T. Shono, Y. Matsumura, K. Inoue, H. Ohimzu, S. Kashimura, *J. Am. Chem. Soc.* **1982**, *104*, 5753–5757.
266. T. Chiba, Y. Takata, *J. Org. Chem.* **1977**, *42*, 2973–2977.
267. E. Le Gall, J-P. Hurvois, S. Sinbandhit, *Eur. J. Org. Chem.* **1999**, 2645–2633.
268. S. Andreades, E. W. Zahnow, *J. Am. Chem. Soc.* **1969**, *91*, 4181–4190.
269. K. Yoshida, T. Fueno, *J. Org. Chem.* **1976**, *41*, 731–732.
270. T. Shono, Y. Matsumura, K. Uchida, H. Kobayashi, *J. Org. Chem.* **1985**, *50*, 3243–3245.
271. T. Shono, *Top. Curr. Chem.* **1988**, *148*, 131–151.
272. T. Fuchigami, in *Advances in Electron Transfer Chemistry* (Ed.: P. S. Mariano) *Vol. 6*, JAI Press Inc., Stanford, Connecticut **1999**, 41–130.
273. M. Kimura, K. Koie, S. Matsubara, Y. Sawaki, H. Iwamura, *J. Chem. Soc. Chem. Commun.* **1987**, 122–123.

274. S. Furuta, T. Fuchigami, *Electrochim. Acta.* **1998**, *43*, 3153–3157.
275. A. Konno, T. Fuchigami, Y. Fujita, T. Nonaka, *J. Org. Chem.* **1990**, *55*, 1952–1954.
276. T. Fuchigami, S. Ichikawa, A. Konno, *Chem. Lett.* **1992**, 2405–2408.
277. N. J. Bunce, in *Handbook of Organic Photochemistry and Photobiology* (Ed.: W. M. Horspool, P.-S. Song) CRC Press, Boca Raton Fl **1995**, 266–279.
278. F. D. Lewis, T.-I. Ho, *J. Am. Chem. Soc.* **1977**, *99*, 7991–7996.
279. R. C. Cookson, S. M. de B. Costa, J. Hudec, *J. Chem. Soc. D* **1969**, 753–754.
280. A. Gilbert, S. Krestnosich, *J. Chem. Soc. Perkin Trans. 1* **1980**, 2531–2534.
281. D. Bryce-Smith, A. Gilbert, G. Klunklin, *J. Chem. Soc. Chem. Commun.* **1973**, 330–331.
282. H. Aoyama, Y. Arata, Y. Omote, *J. Chem. Soc. Chem. Commun.* **1985**, 1381–1382.
283. H. Aoyama, J. Sugiyama, M. Yoshida, H. Hatori, A. Hosomi, *J. Org. Chem.* **1992**, *57*, 3037–3041.
284. H. Aoyama, Y. Inoue, Y.Omote, *J. Org. Chem.* **1991**, *46*, 1965–1967.
285. H. Aoyama, Y. Arata, Y. Omote, *J. Chem. Soc. Perkin Trans. 1* **1986**, 1165–1169.
286. H. Aoyama, Y. Arata, Y. Omote, *J. Org. Chem.* **1987**, *52*, 4639–4640.
287. W. Hub, S. Schneider, F. Doerr, J. D. Oxman, F. D. Lewis, *J. Am. Chem. Soc.* **1984**, *106*, 701–708.
288. F. D. Lewis, G. D. Reddy, D. Bassani, S. Schneider, M. Gahr, *J. Photochem. Photobiol A: Chem.* **1992**, *65*, 205–220.
289. F. D. Lewis, D. M. Bassani, *J. Photochem. Photobiol. A: Chem.* **1992**, *66*, 43–52.
290. F. D. Lewis, G. D. Reddy, S. Schneider, M. Gahr, *J. Am. Chem. Soc.* **1991**, *113*, 3498–3506.
291. F. D. Lewis, D. M. Bassani, G. D. Reddy, *J. Org. Chem.* **1993**, *58*, 6390–6393.
292. F. D. Lewis, D. M. Bassani, E. L. Burch, B. E. Cohen, J. A. Engleman, G. D. Reddy, S. Schneider, W. Jaeger, P. Gedeck, M. Gatr, *J. Am. Chem. Soc.* **1995**, *117*, 660–669.
293. F. D. Lewis, G. D. Reddy, *Tetrahedron Lett.* **1992**, *33*, 4249–4252.
294. F. D. Lewis, J. M. Wagner-Brennan, J. M. Denari, *J. Photochem. Photobiol. A: Chem.* **1998**, *112*, 139–143.
295. F. D. Lewis, S. G. Kultgen, *J. Photochem. Photobiol. A: Chem.* **1998**, *112*, 159–164.
296. A. Sugimoto, R. Sumida, N. Tamai, H. Inoue, Y. Oytsuji, *Bull. Chem. Soc. Jpn.* **1981**, *54*, 3500–3504.
297. A. Sugimoto, S. Yoneda, *J. Chem. Soc. Chem. Commun.* **1982**, 376–377.
298. A. Sugimoto, K. Sumi, K. Urakawa, M. Ikemura, S. SaKamoto, S. Yoneda, Y. Otsuji, *Bull. Chem. Soc. Jpn.* **1988**, *56*, 3118.
299. A. Sugimoto, R. Hiraoka, H. Inoue, T. Adachi, *J. Chem. Soc. Perkin Trans. 1* **1992**, 1559–1560.
300. A. Sugimoto, N. Fukada, T. Adachi, H. Inoue, *J. Chem. Soc. Perkin Trans. 1* **1995**, 1597–1602.
301. A. Sugimoto, R. Hiraoka, N. Fukuda, H. Kosaka, H. Inoue, *J. Chem. Soc. Perkin Trans. 1* **1992**, 2871–2875.
302. A. Sugimoto, J. Yamano, K. Suyuma, S. Yonada, *J. Chem. Soc. Perkin Trans. 1* **1989**, 483–487.
303. U. C. Yoon, J. U. Kim, E. Hasegawa, P. S Mariano, *J. Am. Chem. Soc.* **1987**, *109*, 4421–4423.
304. E. Hasegawa, W. Xu, P. S. Mariano, U. C. Yoon, J. U. Kim, *J. Am. Chem. Soc.* **1988**, *110*, 8099–8111.
305. X-M. Zhang, P. S. Mariano, *J. Org. Chem.* **1991**, *56*, 1655–1660.
306. U. C. Yoon, Y. C. Kim, J. Choi, D. U. Kim, P. S. Mariano, I. S. Cho, Y. T. Jeon, *J. Org. Chem.* **1992**, *57*, 1422–1428.
307. J. Mattay, A. Banning, E. W. Bischof, A. Heidbreder, J. Runsink, *Chem. Ber.* **1992**, *125*, 2119–2127.
308. Y. T. Jeon, C. P. Lee, P. S. Mariano, *J. Am. Chem. Soc.* **1991**, *113*, 8847–8863.
309. S. K. Khim, P. S. Mariano, *Tetrahedron Lett.* **1994**, *35*, 999–1002.
310. Y. S. Jung, P. S. Mariano, *Tetrahedron Lett.* **1993**, *34*, 4611–4614.
311. G. Pandey, G. R. Reddy, *Tetrahedron Lett.* **1992**, *33*, 6533–6536.
312. G. Pandey, G. D. Reddy, D. Chakrabarti, *J. Chem. Soc. Perkin Trans. 1* **1996**, 219–224.
313. G. Pandey, D. Chakrabarti, *Tetrahedron Lett.* **1996**, *37*, 2285–2288.

314. G. Pandey, D. Chakrabarti, *Tetrahedron Lett.* **1998**, *39*, 8371.
315. S. E. Hoegy, P. S. Mariano, *Tetrahedron Lett.* **1994**, *35*, 8319–8322.
316. A. Padwa, H. Nimmesgern, G.S.K. Wong, *J. Org. Chem.* **1985**, *50*, 5620–5627.
317. G. Pandey, G. Kumaraswamy, *Tetrahedron Lett.* **1988**, *29*, 4153–4156.
318. G. Pandey, G. D. Reddy, U. T. Bhalerao, *Tetrahedron Lett.* **1991**, *32*, 5147–5150.
319. F. D. Lewis, G. D. Reddy, B. E. Cohen, *Tetrahedron Lett.* **1994**, *35*, 535–538.
320. M. Yasuda, C. Pac, H. Sakurai, *J. Org. Chem.* **1981**, *46*, 788–793.
321. M. Yasuda, T. Yamashita, K. Shima, *J. Org. Chem.* **1987**, *52*, 753–759.
322. M. Yasuda, Y. Watanabe, K. Tanabe, K. Shima, *J. Photochem. Photobiol. A: Chem.* **1994**, *79*, 61–65.
323. M. Yasuda, T. Isami, J.-I. Kubo, M. Mizutani, T. Yamashita, *J. Org. Chem.* **1992**, *57*, 1351–1354.
324. M. Yasuda, T. Sone, K. Tanabe, K. Shima, *Chemistry Lett.* **1994**, 453–456.
325. S. Das, J.S.D. Kumar, K. George Thomas, K. Shivaramayya, M. V. George, *J. Org. Chem.* **1994**, *59*, 624–628.
326. S. Das, J.S.D. Kumar, K. Shivaramayya, M. V. George, *J. Chem. Soc. Perkin Trans. 1* **1995**, 1797–1799.
327. S. Das, J.S.D. Kumar, K. Shivaramayya, M. V. George, *Tetrahedron* **1996**, *52*, 3425–3434.
328. S. Das, J.S.D. Kumar, K. Shivaramayya, M. V. George, *J. Photochem. Photobiol. A: Chem.* **1996**, *97*, 139–150.
329. S. Bertrand, C. Glapski, N. Hoffmann, J. P. Pete, *Tetrahedron Lett.* **1999**, *40*, 3169–3172.
330. D. Belotti, J. Cossy, J. P. Pete, C. Portella, *J. Org. Chem.* **1986**, *51*, 4196–4200.
331. J. Cossy, D. Belotti, C. Leblanc, *J. Org. Chem.* **1993**, *58*, 2351–2354.
332. E. W. Bischof, J. Mattay, *Tetrahedron Lett.* **1990**, *31*, 7137–7140.
333. E. W. Bischof, J. Mattay, *J. Photochem. Photobiol. A: Chem.* **1992**, *63*, 249–251.
334. T. Kirschberg, J. Mattay, *J. Org. Chem.* **1996**, *61*, 8885–8891.
335. J. Cossy, N. Furet, S. BouzBouz, *Tetrahedron* **1995**, *51*, 11751–11764.
336. B. Pandey, A. T. Rao, P. V. Dalvi, P. Kumar, *Tetrahedron* **1994**, *50*, 3835–3842.
337. B. Pandey, A. T. Rao, P. V. Dalvi, P. Kumar, *Tetrahedron* **1994**, *50*, 3843–3848.
338. H. Moriwaki, T. Oshima, T. Nagai, *J. Chem. Soc. Perkin Trans. 1* **1995**, 2517–2523.
339. S. Das, C. S. Rajesh, T. L. Thanulingam, D. Ramaiah, M. V. George, *J. Chem. Soc. Perkin Trans. 2* **1994**, 1545–1547.
340. Z.-Z. Wu, G. L. Hug, H. Morrison, *J. Am. Chem. Soc.* **1992**, *114*, 1812–1816.

# 8 Electron-Transfer Reactions of Carbonyl Compounds

*Axel G. Griesbeck and Stefan Schieffer*

## 8.1 Generation of Carbonyl Radical Ions by Electron Transfer

The carbonyl group and other organic functional groups which can be described as analogous (imines, iminium salts, thiocarbonyls) often behave as chromophores and at the same time as electrophores. Because of the functional group, most of these compounds have moderate to strong absorption in the UV to visible region of the electromagnetic spectrum and can be electronically excited to give singlet and (subsequently) triplet states with a broad range of excitation energies and lifetimes. These compounds usually also have relatively low reduction potentials and can be reduced either by chemical methods (metals, solvated electron, organic electron donors), photochemical methods (direct or mediated photoinduced electron transfer), or electrochemically by cathodic reduction. These two distinct properties make carbonyl and carbonyl analogous compounds reducible even in their ground states and even more efficiently reducible in their first electronically excited singlet and triplet states. Many processes proceed with full electron transfer, although partial electron transfer might also be responsible for modified reactivity, e.g. in Lewis acid–base interactions, charge-transfer complex formation in ground state chemistry, or contact ion pair formation in photoinduced electron-transfer processes.

Single electron oxidation of the non-activated carbonyl group, e.g. in aliphatic or aromatic aldehydes, ketones and carboxylic acid derivatives, is, on the other hand, much less feasible and only a handful of methods and synthetic applications are known. Useful methods for synthetic applications are chemical modifications to lower the oxidation potentials by peripheral donor substitution and α-silylation, or redox umpolung via oxidation of the corresponding carbonyl enols or enol ethers.

### 8.1.1 Electrochemical Generation of Carbonyl Radical Anions

The electrophoric carbonyl group present in all compounds makes these substrates electroreducible under a broad variety of conditions [1]. Aliphatic ketones have very

**Table 1.** Reduction potentials of some aliphatic ketones in EtOH–H$_2$O.[a]

| Ketone | Reduction potential (V, relative to SCE) |
|---|---|
| Acetone | −2.57 |
| Cyclopentanone | −2.46 |
| Cyclohexanone | −2.40 |

[a] According to [2].

high reduction potentials and can hardly be selectively reduced in aprotic solvents (Table 1).

Under protic conditions, however, the potentials are somewhat lower and selective reduction can be achieved. The control of the pH is critical for the chemo-selectivity of the reduction: at low pH pinacols are often formed because of hydroxycarbinyl radical coupling, a process which is also favored by high substrate concentrations and high hydrogen overvoltage cathodes. Alkaline solutions, on the other hand, favor the formation of alcohols as the primary reduction products. Further reduction to give hydrocarbons can be achieved in acidic media [3]. The intermediacy of pinacols, which are further reduced at more negative potentials, has been demonstrated for several aromatic ketones. Aliphatic ketones are preferentially reduced to the hydrocarbons via the corresponding alcohols (Scheme 1). Protonated carbonyl substrates have much lower reduction potentials, e.g. acetone has a reduction potentials of −1.2 V in 0.5 M sulfuric acid whereas the unprotonated substrate has a reduction potential of −2.57 V (relative to the SCE).

The various stages of reduction and protonation during the course of the electrochemical reduction have been carefully investigated and are described in several

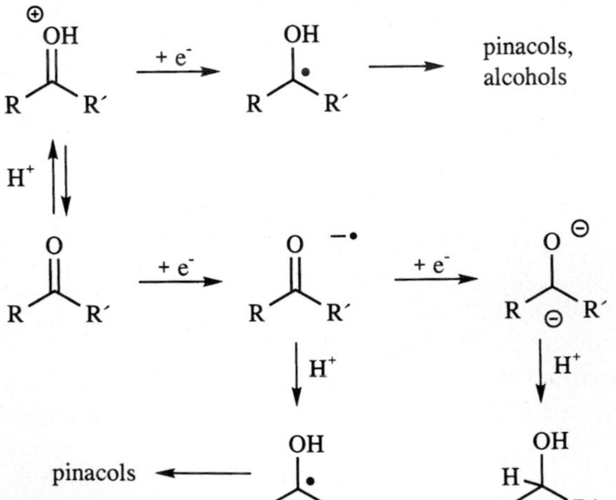

Scheme 1.

**Table 2.** Reduction potentials of some saturated and unsaturated aliphatic aldehydes and ketones.[a]

| Compound | Reduction potential (V, relative to SCE) |
| --- | --- |
| $CH_3CH_2CHO$ | −1.8 |
| $PhCH_2CH_2CHO$ | −1.8 |
| $CH_2CH_2COCH_3$ | −2.25 |
| Cyclohexanone | −2.45 |
| Acrolein | −1.5 |
| Cinnamyl aldehyde | −0.8 |
| Vinyl methyl ketone | −1.42 |
| Cyclohex-2-enone | −1.55 |

[a] According to [7].

reviews. The reduction to give alcohols from ketones can be used for preparative purposes with the use of trialkylammonium halides as electrolytes in mixtures of organic solvents and water [4, 5]. Cyclohexanone and alkylated derivatives are reduced at a potential of ca −2.9 V (relative to the SCE) to give the corresponding alcohols and pinacols in good yields [6]. Aliphatic aldehydes have much lower reduction potentials than ketones and these can be further reduced by conjugation to double bonds (Table 2).

The effective electroreductive pinacolization has been exemplified in the terpene series, i.e. aldehydes such as retinal and ketones such as α- or β-ionone give the corresponding pinacols in good yields when electrolyzed in a DMF–sodium perchlorate–Hg system or with a Pt cathode in the presence of tin [8, 9].

The intermediacy of carbonyl radical ions in electrochemical reductions can be demonstrated by trapping reactions, e.g. by intramolecular cyclization of unsaturated ketones. These ring formations can occur at the electrode surface (graphite cathode) or at a mercury cathode in the presence of catalytic amounts of dimethylpyrrolidinium salts. The cyclization of 6-heptene-2-one to give *cis*-1,2-dimethylcyclopentan-1-ol was the first example in this series. These reactions have been thoroughly investigated and stereochemical features, in particular, have been elucidated by the groups of Shono, Pattenden, and Kariv-Miller [10–15]. Ketones with arene substituents at position C4 can be electrochemically cyclized to give hexahydronaphthalenes in good yields [12]. In the absence of the catalyst, the corresponding alcohols were formed with good chemoselectivity.

The structure of the tetraalkylammonium electrolyte salt is crucial to the result of the reaction, i.e. the control of two-electron-transfer steps rather than one-electron-transfer steps followed by subsequent non-reductive reactions. The alkenone

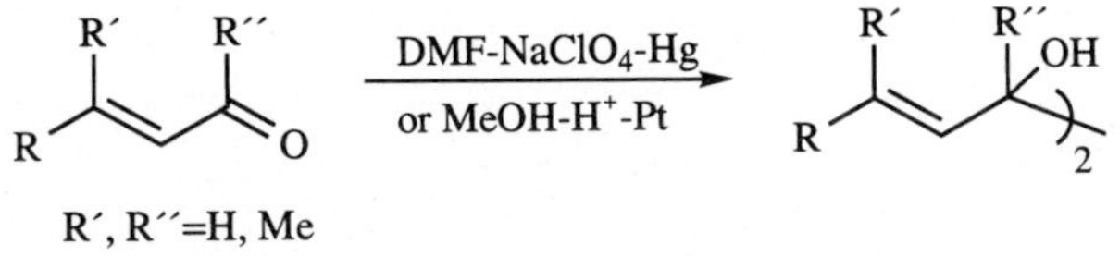

R′, R″=H, Me

**Scheme 2.**

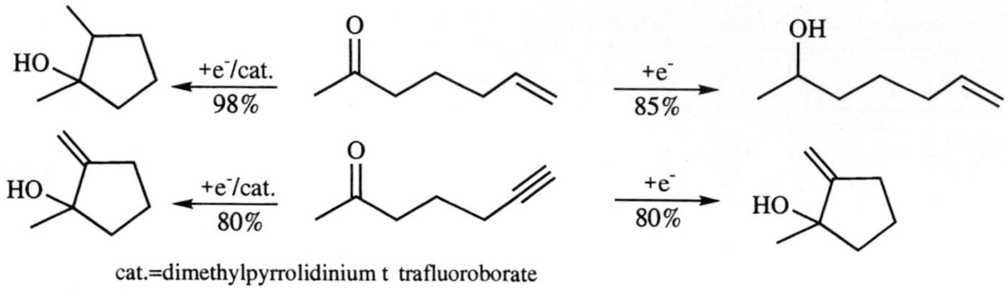

cat.=dimethylpyrrolidinium t trafluoroborate

**Scheme 3.**

(Scheme 3) was preferentially reduced to the alcohol in the absence of the pyrrolidinium catalyst whereas in the presence of this salt, 5-*exo-trig* cyclization prevailed.
The corresponding alkyneone, on the other hand, gave the 5-*exo-dig* cyclization
product in 80 % yield even in the absence of the catalyst [6]. To achieve multiple
cyclization reactions, *bis*-unsaturated ketones were cathodically reduced to give the
bicyclo[3.3.0]octane skeleton. The electrolyses were performed in one-compartment
cells in DMF with several allyl pentenyl ketones as substrates. The first reductive
step is followed by two subsequent 5-*exo-trig* cyclizations to give stereoselectively
the corresponding *cis*-fused bicyclooctanols in good yields (Scheme 4) [16].

Intramolecular trapping of the electroreduced carbonyl group is also possible
with cyano substituents; when $\gamma$- and $\delta$-cyano ketones are reduced in i-PrOH with a
Sn cathode the corresponding $\alpha$-hydroxy ketones are formed in high yields. This
reaction proceeds via two subsequent electron-transfer steps generating the ketyl
dianions which act as highly reactive nucleophiles and attack the cyano group. The
imine intermediate can be hydrolyzed to give the $\alpha$-hydroxy ketone or dehydrated
and subsequently electroreduced to give the corresponding ketone. Thus, the overall
reaction can be described as cyclization coupled with 1,2-carbonyl transposition
(Scheme 5) [17].

Likewise, intermolecular reactions are possible and lead to coupling products
which correspond retrosynthetically to the addition of an acyl anion synthon to a
ketone. The presence of a proton-donor cosolvent is crucial, otherwise $\beta$-hydroxy
nitriles are formed preferentially. The nitrile addition reaction proceeds with good
stereoselectivity, e.g. preferentially one diastereoisomer is formed from the electroreductive addition of acetonitrile (which can advantageously be used as solvent) to
dihydrocarvone.

The inherent problem of electrochemical reduction (and oxidation) of carbonyl
and carbonyl-analogous compounds is that the electron transfer constitutes an

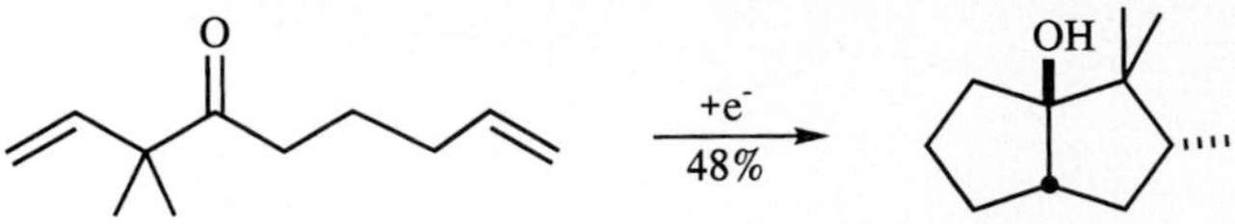

**Scheme 4.**

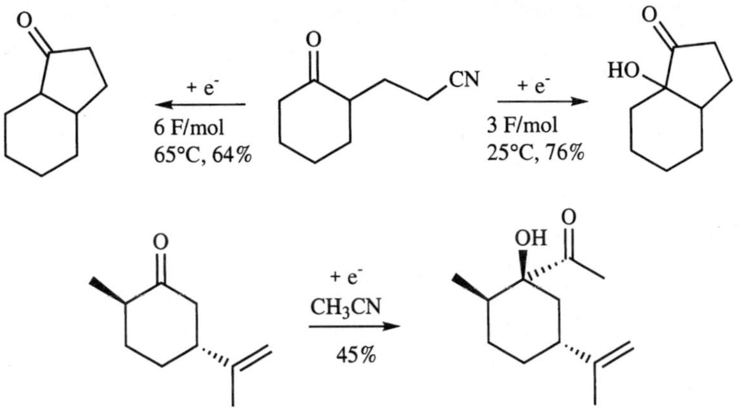

**Scheme 5.**

heterogeneous step and large overpotentials often have to be applied to effect the reaction [18]. A solution to this problem can sometimes be the use of an indirect electrosynthesis, i.e. two coupled heterogeneous–homogeneous electron-transfer steps [19]. In the first electron-transfer step, a catalyst is cathodically reduced (or anodically oxidized) and then diffuses into the reaction medium. In the second, homogeneous, step the catalyst is regenerated and transfers an electron (or a hole) to the substrate. The catalyst thus acts as electron-transfer *mediator* and can also be used in photoinduced electron-transfer reactions (vide infra). A series of aromatic mono- and polycyclic hydrocarbons has been intensively investigated as mediators for indirect reductive electrosynthesis; their reduction potentials are given in Table 3.

**Table 3.** Reduction potentials of some aromatic compounds.

| Compound | Reduction potential (V, relative to SCE) |
|---|---|
| Perylene | −1.67 |
| Phthalonitrile | −1.69 |
| Anthracene | −1.96 |
| Pyrene | −2.09 |
| Methyl benzoate | −2.17 |
| Benzonitrile | −2.24 |
| Chrysene | −2.24 |
| 3-Toluonitrile | −2.27 |
| 4-Toluonitrile | −2.34 |
| Phenanthrene | −2.45 |
| Naphthalene | −2.50 |
| Biphenyl | −2.70 |

[a] According to [20–22].

## 8.1.2 Chemical Generation of Carbonyl Radical Anions

The classical method of chemical one-electron reduction of carbonyl substrates is treatment with low-valent metals, e.g. alkali metals, in inert solvents. Less reactive radical anions can be characterized by UV absorption and emission spectroscopy, e.g. the benzophenone radical anion. This species has been generated also by low-temperature $\gamma$-ray irradiation and absorbs visible light in the region of 630 nm (in ethanol) to 800 nm (in 2-methyltetrahydrofuran) [23]. The benzophenone radical anions generated by reduction with alkali metals in tetrahydrofuran have similar absorption peaks at 654 nm (for the lithium compound) up to 714 nm (for the potassium compound) [24]. In hydrogen-donating solvents, the ketyl radical anions ($\lambda_{max}$ ca 620–660 nm) are rapidly protonated to give the corresponding ketyl radicals ($\lambda_{max}$ ca 540 nm). The role of active metals in reductive processes involving carbonyl and carbonyl analogous compounds has been extensively investigated in recent decades and is summarized in excellent reviews [25–27].

The reductive coupling of aldehydes and ketones to give 1,2-diols (pinacol coupling) is an important reactivity pathway for ketyl radical anions and an important method for C–C bond formation. A multitude of reagents has recently been developed for stoichiometric reductive dimerization, e.g. diverse alkali and earth alkali metals, low-valent metal complexes with Ce, Ti, V, Zr, Sn, Nb or Sm. Some of these reagents are prepared before use, others in situ [28, 29]. Stereoselective pinacol coupling is often achieved when chelating interactions direct the C–C formation steps. The homocoupling of aromatic carbonyl groups can be directed to high D,L-selectivities when performed with samarium metal in the presence of diethyl aluminum iodide [30]. The latter reagent generates the samarium (II) species active in electron transfer and probably also serves as chelating element in the C–C coupling step (Scheme 6).

To avoid secondary reductive steps, low concentrations of the active low-valent metal are advantageous, i.e. the use of catalytic one-electron reduction cycles [31]. Low-valent vanadium species are efficient catalysts for pinacolization. After the primary electron transfer, the oxidized vanadium catalyst can be reduced by Zn(0) or by aluminum [32]. Trialkylsilanes have to be added to decomplex the oxidized vanadium catalyst from the pinacol [33]. The same catalytic system can also be used for the synthesis of 1,2-diamines via coupling of aldimines [34]. Alternatively, a titanium (II)–samarium system gives the 1,2-diamines with moderate D,L-selectivity [35].

Ketyl radical anions can be easily trapped by electron-deficient alkenes, e.g. acrylates, to give enolate radicals. Further reduction of these species results in enolates which can be used as nucleophiles in alkylation reactions. A multi-component

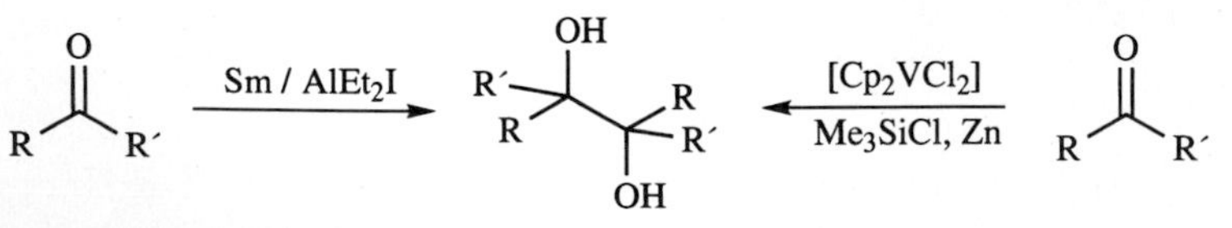

**Scheme 6.**

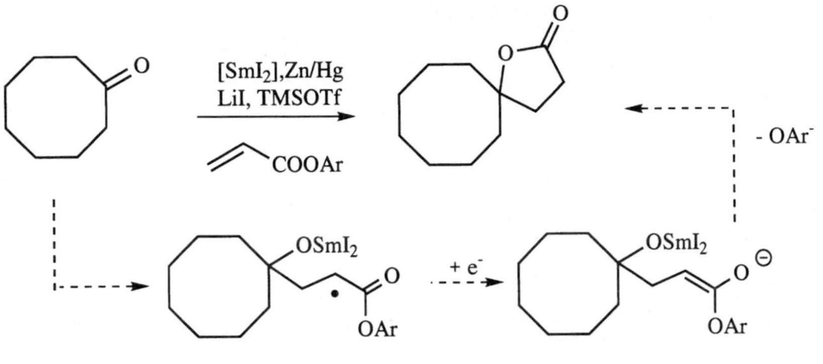

**Scheme 7.**

catalytic reductive system consisting of samarium (II), Zn/Hg (for reduction of the consumed samarium reagent), LiI, and TMSOTf has been developed by Corey and Zheng for the spirolactonization of cycloalkanones (Scheme 7) [36].

*Raney nickel* can be used for the chemoselective reduction of α,β-unsaturated ketones, esters, acids, nitriles, and nitroalkenes to give the corresponding saturated carbonyl compounds and carbonyl analogs in excellent yields. From trapping experiments it became evident that electron transfer from nickel to give the enone radical anion initiates the reaction which then proceeds via proton transfer and second electron–proton transfer cycle (Scheme 8) [37].

The reductive coupling of carbonyl compounds with formation of C–C double bonds was developed in the early seventies and is now known as McMurry reaction [38, 39]. The active metal in these reactions is *titanium* in a low-valent oxidation state. The reactive Ti species is usually generated from Ti(IV) or Ti(III) substrates by reduction with Zn, a Zn–Cu couple, or lithium aluminum hydride. A broad variety of dicarbonyl compounds can be cyclized by means of this reaction, unfunctionalized cycloalkenes can be synthesized from diketones, enolethers from ketone–ester substrates, enamines from ketone–amide substrates [40–42]. Cycloalkanones can be synthesized from external keto esters ($X = OR'$) by subsequent hydrolysis of the primary formed enol ethers (Scheme 9).

Several variations of this exceedingly important reaction have been reported in recent years, e.g. the use of *titanium–graphite* from Ti(III) and $C_8K$, especially

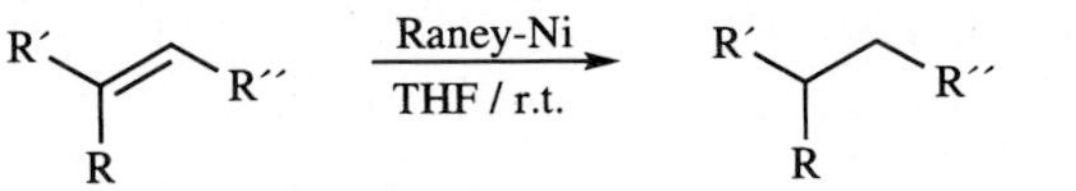

R=H, R´=Ph, R´´=COOEt 93%; R,R´=H, R´´=COOMe 90%;
R,R´=H, R´=CN 93%, R,R´=Me, R´´=COOH 93%.

**Scheme 8.**

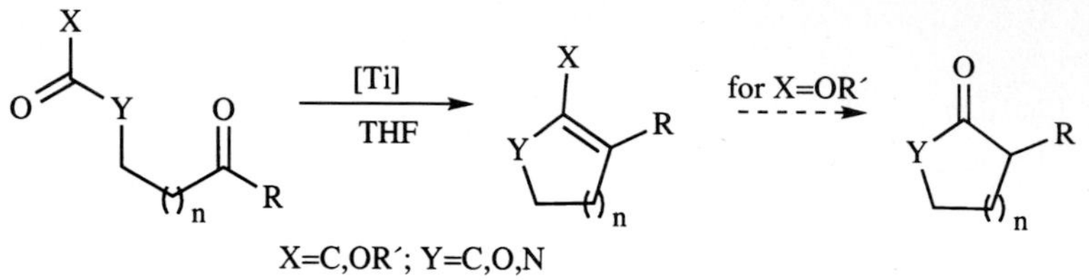

X=C,OR′; Y=C,O,N

**Scheme 9.**

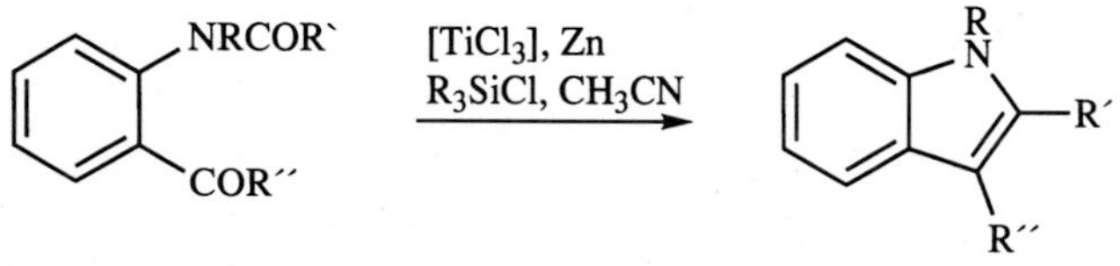

**Scheme 10.**

for the synthesis of unsaturated heterocycles [43–46]. Catalytic versions of the McMurry reaction have been developed for the synthesis of carbo- and heterocyclic ring systems. As in the samarium- and vanadium-catalyzed pinacolizations, Zn is used as the electron source for reloading the low-valent titanium catalyst, e.g. titanium trichloride, in the presence of trialkyl silylchlorides (Scheme 10) [47].

By use of weaker reductants, the C–C coupling can be stopped at the stage of the pinacols [48]. The best metals for achieving chemical pinacolization from ketones are magnesium in the form of its amalgam and mixtures of *Zn–Hg* with titanium tetrachloride (Scheme 11). In the latter reaction the Ti(II) species presumably initiates electron transfer [49].

Another important coupling reaction uses esters as the electron-accepting species and leads to α-hydroxy ketones (acyloin coupling). *Sodium, potassium* (less frequently) or *sodium–potassium alloys* are commonly used as electron donors in nonpolar solvents such as toluene or xylene. The first detectable reaction intermediate after the primary reductive step is the enediolate which can be trapped with trialkylsilyl chloride. This method is widely used to synthesize highly nucleophilic alkenes and/or protected acyloins (Scheme 12) [50, 51].

In the acyloin protocol, the alkoxides serve as leaving groups after the C–C coupling step. If appropriate leaving groups (OR, SR, Hal) are localized at the α-

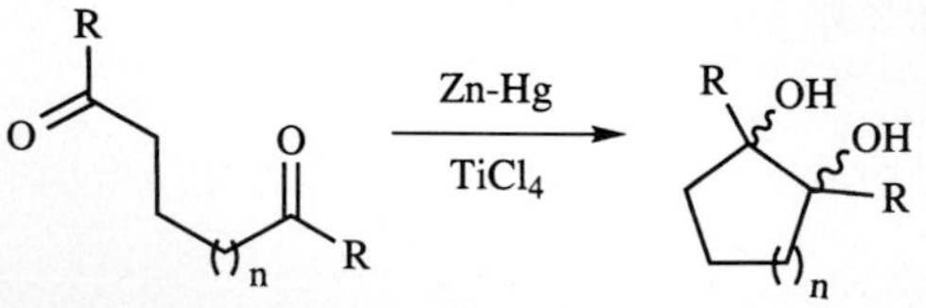

**Scheme 11.**

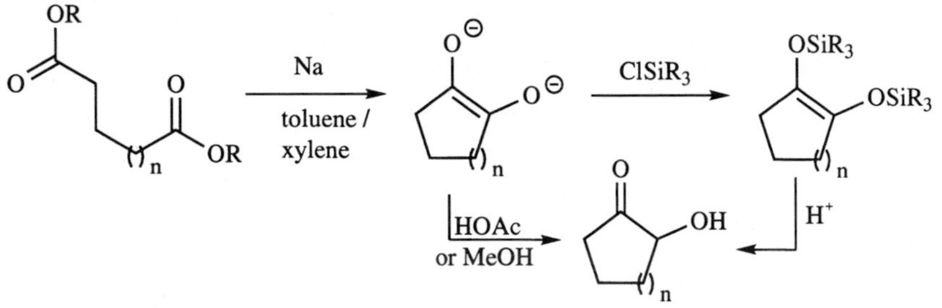

**Scheme 12.**

position with respect to the carbonyl group, they eliminate rapidly after two-electron transfer reduction and the corresponding enolates are formed [52]. Acyloins can also be used as substrates for this route; they are usually transformed into ketones with reductive elimination of the $\alpha$-hydroxy group. As reductant, metallic *zinc* is usually used as the ideal two-electron donor. The vinylogous substrates ($\gamma$-activated $\alpha,\beta$-unsaturated carbonyl compounds) can also be reductively cleaved with possible migration of the double bond (Scheme 13) [53]. Lactones acylated at the $\omega$-position (e.g. $\gamma$-acyl $\gamma$-butyrolactones) are cleaved into the corresponding dicarboxylic acids [54].

*Solvated electrons* can be generated by dissolving alkali metals in liquid ammonia or similar solvents [55]. They are widely used for the reduction of organic compounds, e.g. in the Birch reduction of aromatic substrates. Alternatively, by using larger current densities than in direct cathodic reductions, solvated electrons can also be generated under electrochemical conditions. This methodology is useful for avoiding side-reactions derived from the use of alkali metals and electrode reactions. Even benzene can be hydrogenated by solvated electrons, a process which is not possible by direct cathodic reduction. Chiral and prochiral carbonyl substrates can be reduced to alcohols by solvated electrons with high diastereofacial selectivity. This behavior was reported for cyclic alkanones which were reduced in ethanol predominantly to give the *trans* products [56]. The selectivities are higher than those obtained from direct electroreduction of the ketones and thus the process via solvated electrons is advantageous (Scheme 14) [57].

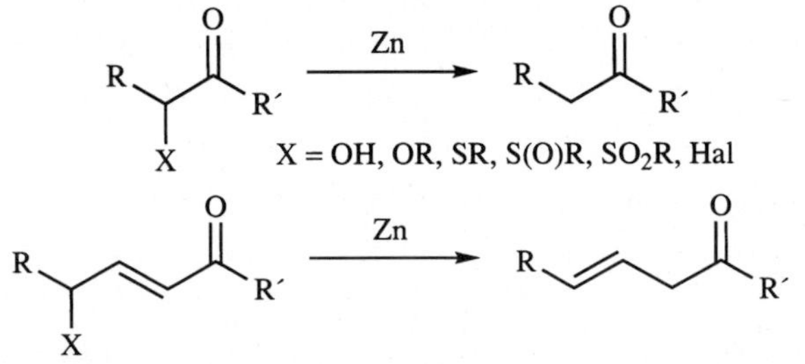

**Scheme 13.**

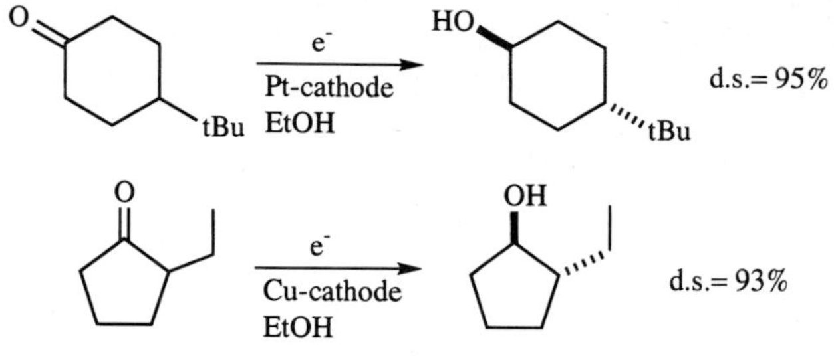

**Scheme 14.**

An intramolecular coupling reaction of macrodithionolactones has been developed by analogy with the C=O–C=O coupling reactions after the acyloin and McMurry protocols. *Sodium naphthalide* (vide infra), used as chemical reductant, subsequently generates two thiocarbonyl radical anions. After C–C coupling the resulting dithiolate is alkylated with methyl iodide. The methylthio groups are easily replaced by hydrogen via elimination–hydrogenation or direct substitution (Scheme 15) [58].

One of the synthetically most powerful and versatile homogeneous reductants is *samarium diiodide* (SmI$_2$); it was first described by Kagan et al. [59] and in recent decades has acquired the status of miracle reagent. Carbonyl compounds are ideal candidates for reductive activation by SmI$_2$, because of its enormous reducing power. Reactions initiated by SmI$_2$ can be described as Sm(III) radical anion chemistry, i.e. the samarium (III) can alter the reactivity by complexation with other heteroatoms and eventually is destroyed by hydrolysis of the Sm(III) alkoxide. This means, on the other hand, that stoichiometric amounts of the SmI$_2$ reagent must be used in nearly all the applications yet developed, because of deactivation of the reagent by the product, e.g. two-electron reduction of $\beta$-hydroxy ketones with three equivalents of SmI$_2$ gives the corresponding 1,3-diols with high *anti*-selectivities and excellent yields [60]. In recent years, however, several regeneration systems for SmI$_2$ have been developed: Mg for reductive pinacolizations [61], Zn–Hg for spirolactonization [36], and Sm/Mischmetall (an alloy of light lanthanides) for Barbier reactions, pinacolization, and acyloin coupling reactions [62].

The ring-opening reaction of three-membered rings (cyclopropanes and oxiranes, respectively) adjacent to the carbonyl group is a straightforward process for the

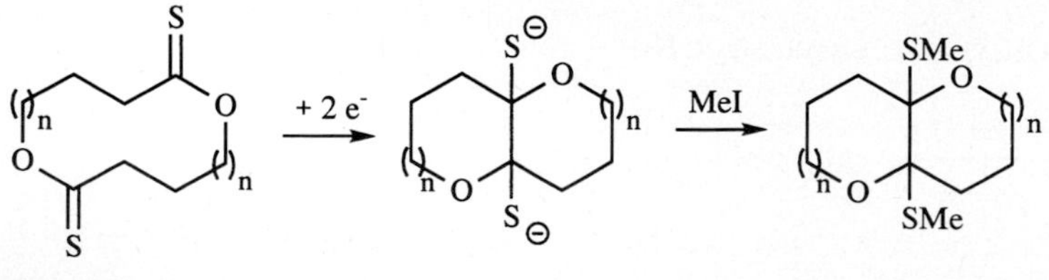

**Scheme 15.**

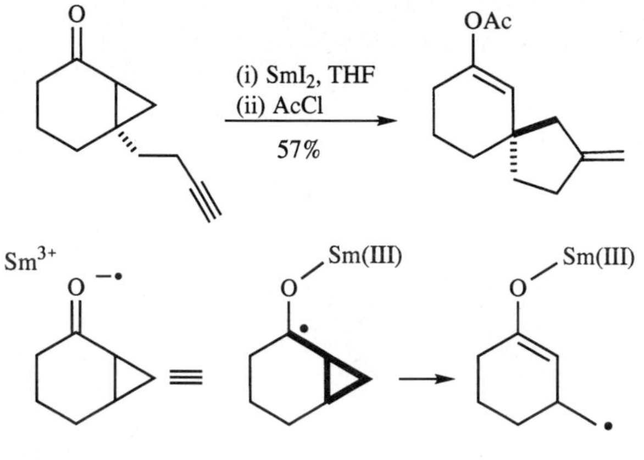

**Scheme 16.**

generation of alkyl radicals which can undergo further radical cyclization reactions [63–65]. Five-membered rings are easily formed via a 5-*exo-trig* or 5-*exo-dig* process. The resulting alkyl or vinyl radicals are subsequently reduced by SmI$_2$ to give the corresponding carbanions which are eventually protonated. The enolate anion, on the other hand, formed after ring-opening can be trapped by electrophiles to give enol ethers or enol acetates (Scheme 16).

By analogy, the ring-opening of epoxy ketones gives rise to 1,3-alkoxy radical anions which are stabilized by Sm(III) [66–68]. These reactions are performed in the presence of protic solvents and thus, after enolate protonation, the reformed carbonyl group is reduced again to give the reactive species which initiates radical reactions. As typical of free radical cyclizations, 5-*exo-trig* or 6-*exo-trig* reactions dominate the course of the reaction and further reductive steps lead to the formation of five-membered or six-membered monocyclic or annulated ring systems (Scheme 17).

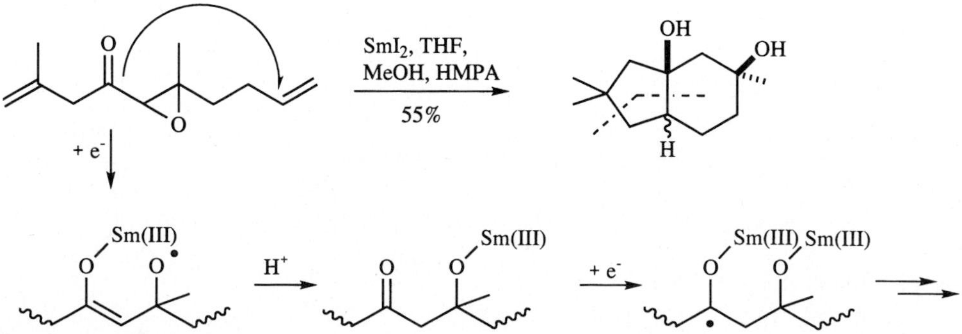

**Scheme 17.**

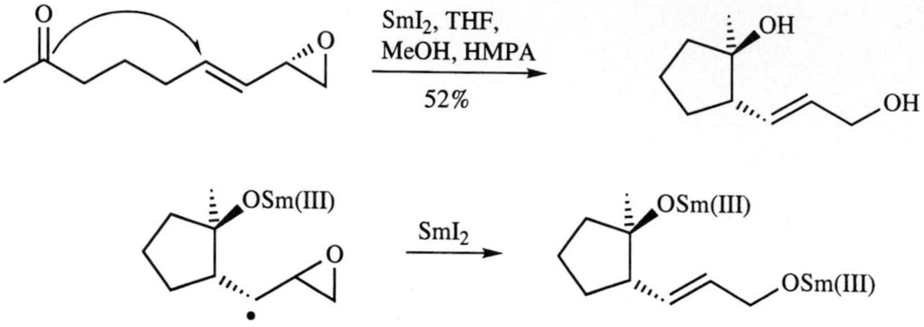

**Scheme 18.**

The termination step in these sequences turns out to be the second reduction step of the ultimate carbon radical. The carbanions formed thereby can be protonated or, in the presence of an appropriate leaving group, undergo $\beta$-elimination to give alkenes. Acetate, halide, and alkoxide have been reported to be appropriate leaving groups [69, 70]. Oxirane ring opening can also terminate the cyclization sequence to give hydroxyalkyl-substituted ring systems (Scheme 18) [71].

By analogy with cyclopropanes and oxiranes, 2-acylaziridines, aziridine-2-carboxylic esters and amides, are also suitable substrates for one-electron reductive ring cleavage. This is an efficient and highly regioselective method for the synthesis of $\beta$-amino carbonyl compounds. Vinylogous substrates are, furthermore, transformed into $\delta$-amino $\beta,\gamma$-unsaturated carbonyl derivatives; azetidines can also be used to achieve $\gamma$-amino functionalization (Scheme 19) [72].

The secondary reduction of the terminal radical by $SmI_2$ generates samarium alkyl species which are suitable for classical organometallic reactions, e.g. protonation, acylation, reactions with carbon dioxide, disulfides, diselenides, or the Eschenmoser salt. A broad variety of products is available (hydroxy-substituted alkanes, esters, carboxylic acids, thioethers, selenoethers, tertiary amines) by use of the double-redox four-step (reduction–radical reaction–reduction–anion reaction) route (Scheme 20) [73].

The chain linking the carbonyl group and the alkenyl part can also be functionalized with heteroatoms. Thus, ether-linked $\delta,\varepsilon$- or $\varepsilon,\zeta$-unsaturated carbonyl substrates are converted into tetrahydrofurans and tetrahydropyrans, respectively, via 5-*exo-trig* or 6-*exo-trig* cyclizations [74]. The radicalophilic part of the molecule can

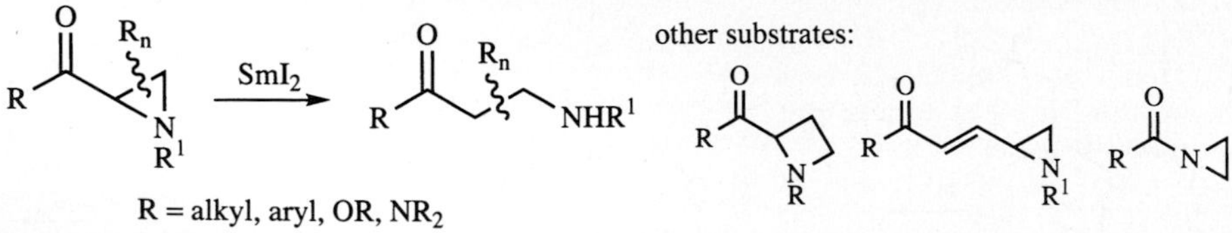

$R$ = alkyl, aryl, OR, $NR_2$
$R^1$ = Boc, Tr, Ts, (CO)alkyl

**Scheme 19.**

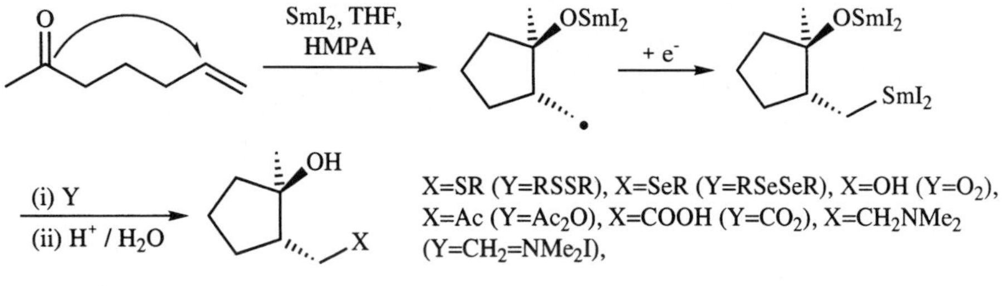

**Scheme 20.**

also be in conjugation with an arene group, i.e. vinyl- or alkynyl-substituted benzene rings are attacked by the ketyl radical anions [75]. The regioselectivity of this reaction, however, depends on the substituent at the alkene–alkyne termini— unsubstituted alkenes and alkynes underwent 8-*endo-trig* and 8-*endo-dig* cyclizations, respectively, whereas alkynes with substituents at the termini gave Birch-type products derived from radical attack at the *ortho'* position of the arene skeleton, with very high stereoselectivity (Scheme 21) [76].

Carbonyl substrates with appropriate leaving groups can be reduced with $SmI_2$ to give acyl radicals. In the presence of radical-trapping reagents these reactive intermediates can be transformed into stable products, e.g. via cyclization to give mono- or bicyclic ring systems [77]. In the absence of trapping reagents further reduction of the acyl radicals leads to acyl anions which are capable of nucleophilic addition reactions [78]. Fine tuning of the reaction conditions enables the sequential reaction of acyl anion and ketyl radical anion. Thus, bicyclization via anionic and subsequent radical reactions can be induced and proceed highly stereoselectively (Scheme 22) [79].

The mechanism of the intramolecular samarium-initiated Barbier reaction is still a matter of debate [80–82]. One of several mechanistic possibilities is primary reductive generation of the ketyl radical anion which can subsequently initiate a second

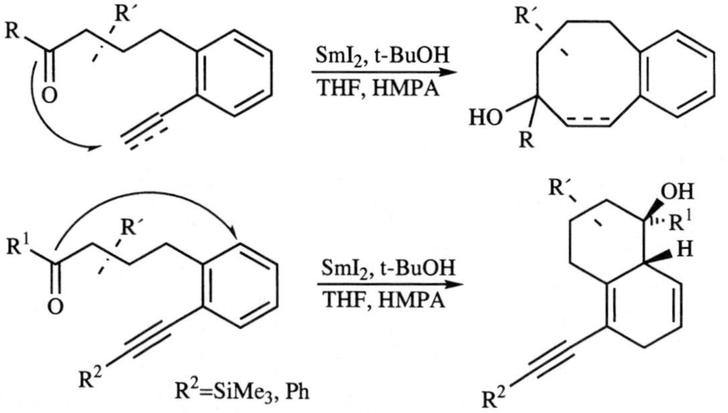

**Scheme 21.**

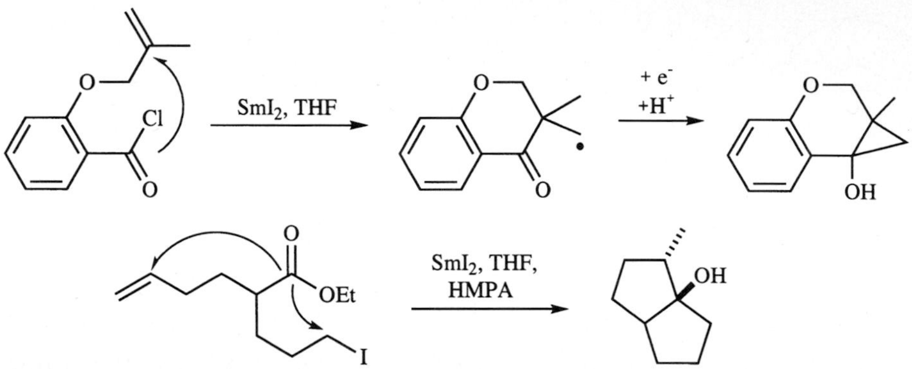

**Scheme 22.**

electron transfer to give the primary alkyl radical, or direct attack with substitution of the iodide leaving group (Scheme 23).

The classical (Grignard-like) mechanism for Barbier reactions involves the primary formation of an Sm-alkyl species via halogen abstraction and subsequent reduction of the alkyl radical formed after the first electron transfer. Be that as it may, the Barbier reaction can be used to construct complex polycyclic target molecules, e.g. the synthesis of tetraquinanes from diquinane precursors by two independent intramolecular cyclization steps (Scheme 24) [83].

Intramolecular coupling of carbonyl groups with vinylbromides after an addition–elimination sequence involves two electron-transfer steps. In the first step the ketone is reduced to the ketyl radical anion. After 5-*exo-trig* radical addition of the vinylbromide group, a primary radical is generated which is subsequently reduced to give an alkyl samarium species which eliminates bromide. This reaction

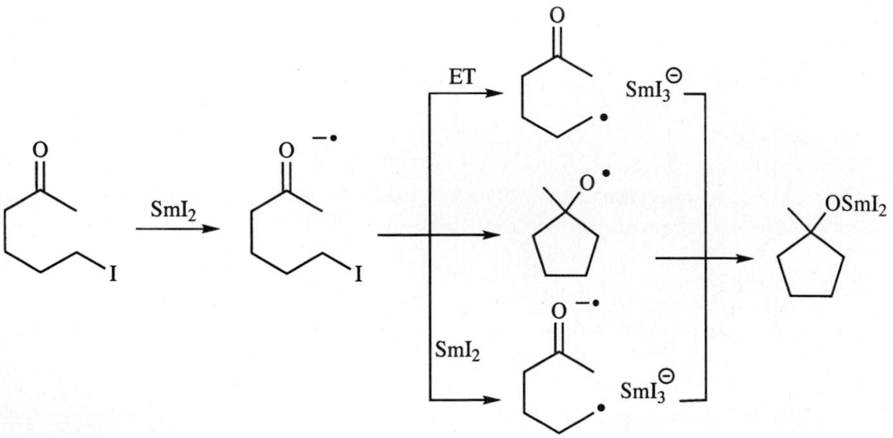

**Scheme 23.**

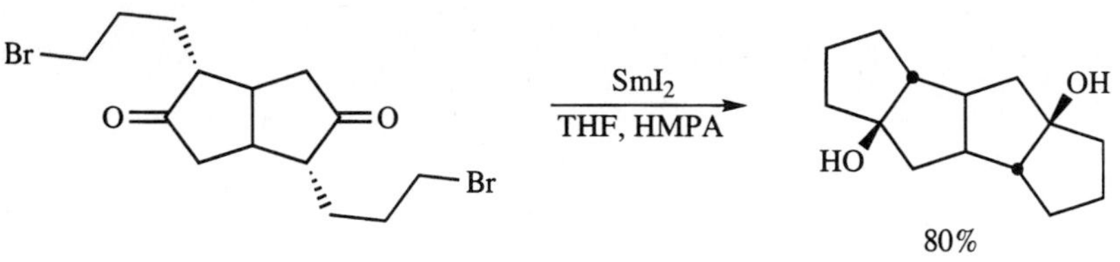

80%

**Scheme 24.**

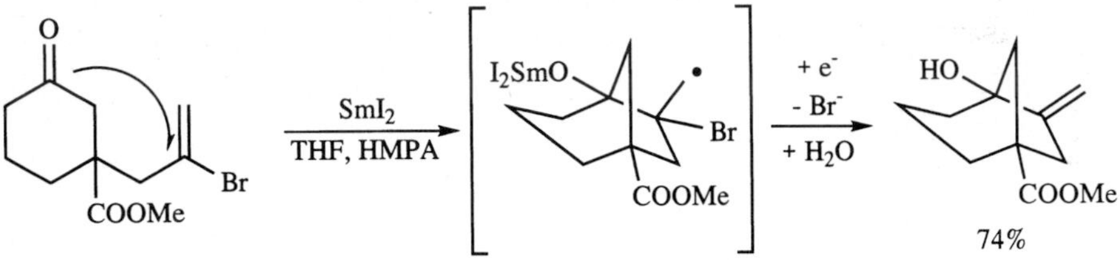

74%

**Scheme 25.**

is equivalent to the intramolecular addition of the ketyl radical anion to an alkyne group (Scheme 25) [84].

Allylic substituents (X = SR, $SO_2R$) at the radical trapping alkene can be used for radical coupling–elimination sequences leading to vinyl-substituted cyclo-alkanes [85]. Even non-activated aliphatic aldehydes can be reduced with $SmI_2$ and 5-*exo-trig* and 6-*exo-trig* radical cyclizations, respectively, (Scheme 26) and tandem cyclization processes can also be initiated by this route, leading to polyquinane structures [86].

An intramolecular reductive coupling of ketones with nitriles has been reported for acyclic and monocyclic substrates; $\beta$-amino nitriles were isolated from acyclic malononitrile adducts [87]. The $SmI_2$-initiated reductive cyclization of cyclic $\alpha$-cyanoalkyl-substituted ketones leads to acyloin products in high yields. In this instance further irradiation of the reaction mixtures was performed to afford complete conversion (Scheme 27) [88]. More applications have been collected in several excellent reviews [89–92].

Another versatile chemical reductant which has not been widely applied in electron-transfer chemistry is sodium naphthalide. This arene radical anion is easy to generate and capable of reducing carbonyl compounds to the corresponding ketyl

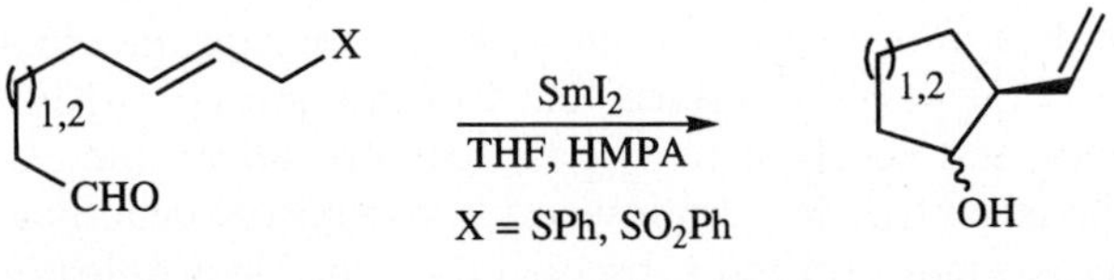

**Scheme 26.**

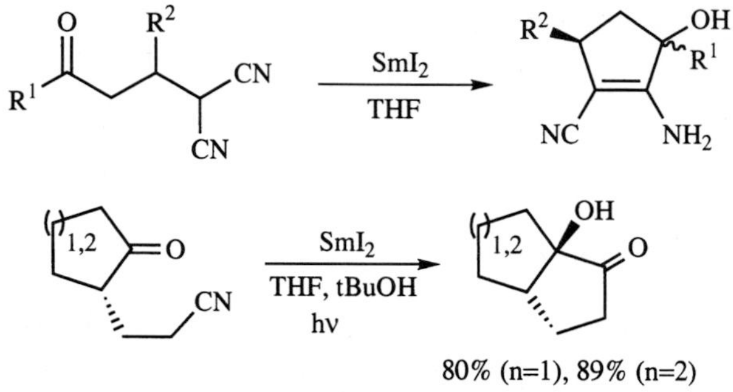

**Scheme 27.**

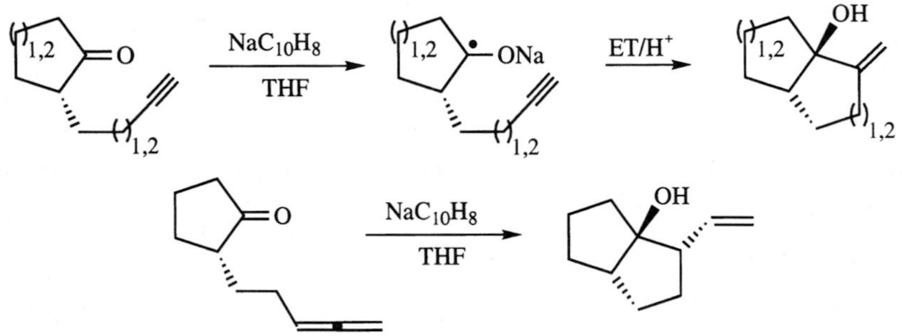

**Scheme 28.**

radical anions under mild conditions. In this reaction the naphthalene serves as electron-transfer mediator. Applications have been reported for 5-*exo-dig* cyclizations in polyquinane and steroid syntheses (Scheme 28) [93, 94].

Allenyl-substituted cycloalkanones have also been studied; the one-electron reduction of these resulted in 5-*exo-trig* or 5-*endo-trig* cyclizations (depending on the relative position of the allene group) [95]. The reductive cyclization of ketones in the presence of acrylate side-chains have been investigated; the reductant was magnesium metal in methanol [96].

### 8.1.3 Photochemical Generation of Carbonyl Radical Anions

One of the most important means of inducing electron-transfer processes is photoexcitation of a substrate or a catalyst molecule in the presence of appropriate electron donor or acceptor molecules (PET = photoinduced electron transfer) [97–108].

In the simplest example, a donor and acceptor pair is activated by electronic excitation of either the donor or the acceptor. In addition to photophysical deactivation or energy transfer, two processes can proceed subsequently—the electronically excited donor donates an electron from its SOMO′ into the acceptor LUMO or the electronically excited acceptor takes up an electron from the donor HOMO into the acceptor SOMO. The free energy change for these photoinduced electron-transfer processes can be calculated from the Weller equation (Eq. 1) [109, 110]:

$$\Delta G_{ET} = \Delta E_{1/2}(\text{Donor}) - \Delta E_{1/2}(\text{Acceptor}) - \Delta E_{00} + \Delta E_{coul}. \qquad (1)$$

The rates of electron transfer correlate with the free energy change as a parabolic relationship and approach the maximum values (i.e. diffusion-controlled rates for bimolecular PET reactions) for exergonic processes with negative free energy changes larger than the reorganization energy (following the Marcus equation) and decreasing rate constants in the Marcus inverted region [111]. This quadratic free energy relationship has been found for numerous photoinduced electron-transfer processes and serves nowadays as a useful tool for predicting reactivity and selectivity in PET chemistry. The secondary processes after the electron-transfer step can be categorized into three principle types:

*Type I*: Acceptor radical anion and donor radical cation undergo subsequent reactions leading to uncharged closed-shell molecules which cannot be photoexcited again.

*Type II*: Either the acceptor radical anion or the donor radical cation is retransformed into its original state during subsequent reaction steps. If this molecule corresponds to the light-absorbing species, it serves as *sensitizer* for PET processes and can be used in catalytic amounts.

*Type III*: A non-absorbing molecule transports the electron or the hole after the first electron-transfer step to the target species and thus enable separation of the originally formed acceptor radical anion–donor radical cation pair. This *mediator* is often used in catalytic amounts but can also be consumed during the PET reaction. In many of these cases, mediators serve as the terminal proton–hydrogen donors.

Simple carbonyl compounds such as *aliphatic aldehydes or ketones* have singlet excitation energies of ca 3.2 to 3.5 eV and reduction potentials of ca $-2.0$ to $-2.5$ V (see Tables 1 and 2). Thus, the first singlets of these species have excited state reduction potentials of ca 0.7 to 1.5 V. Conjugation of the carbonyl group to arenes, C–C multiple bonds, and additional C=O or C=X groups reduce the corresponding excitation energies and the ground-state reduction potentials. Tuning of chromophoric and electrophoric groups enables the design of electron-accepting substrates with reduction potentials as large as 2 V. Thus, PET with these electron acceptors can lead to oxidation of organic substrates with oxidations potentials of $+2$ V or even higher (Table 4).

A broad variety of organic functionality has been described as donor groups— heteroatoms (N, O, S, Se), C–C double bonds (donor-substituted or conjugated), alkynes, and arenes. The oxidation potentials of the electron donor can be altered by modification of possible leaving groups for mesolytic bond cleavage, e.g. the use of silyl or stannyl groups instead of protons as electrofugs. In carbonyl photochemistry, amines have been thoroughly investigated as efficient electron donors. The photoreduction of triplet benzophenone by triethylamine leads to a caged radical ion pair which rapidly undergoes mesolytic bond cleavage of the amine radical cation to give an α-amino radical and the benzhydrol radical (Scheme 29) [113].

**Table 4.** Reduction potentials and triplet energies of aromatic carbonyl compounds.[a]

| Compound | Reduction potential, $E_{red}$ (V, relative to SCE) | Triplet energies, $E(T_1)$ (eV) |
|---|---|---|
| Benzophenone | −2.16 | 2.95 |
| Benzil | −1.50 | 2.36 |
| Anthraquinone | −0.94 | 2.72 |

[a] According to [112].

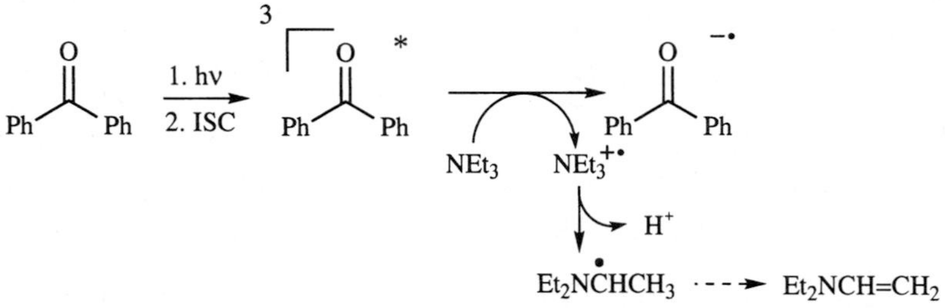

**Scheme 29.**

The α-amino radical can undergo cage-combination with the hydroxyalkyl radicals but can also transfer a hydrogen atom which leads to neutral photoreduction products. This procedure enables the reductive photocyclization of simple carbonyl compounds in the presence of radicalophilic groups. This has been described for a series of γ,δ- and δ,ε-unsaturated ketones. These substrates were irradiated in the presence of electron-donating amines such as triethylamine or in neat HMPT. This methodology has been used for the synthesis, in moderate to high yields, of natural products and interesting unnatural biologically active molecules such as diquinanes [114], triquinanes [115], and heteroindanes such as actinidine [116] and iridoids [117] (Scheme 30).

Cyclopropyl-substituted ketones are suitable substrates for generating distonic radical anions from ketyl radical anions. A series of cycloalkanone substrates with unsaturated side-chains, to trap the primary radical formed after cyclopropylcarbinyl ring opening, has been investigated (Scheme 31) [118, 119]. For the first electron-transfer step triethylamine is used as electron donor. The reaction sequence is terminated by proton or hydrogen transfer from the solvent or the α-amino radical formed after deprotonation of the amine radical cation.

The radical anions of carbonyl groups can also be generated via PET from activated alkenes, e.g. allylic silanes or stannanes. Triplet excited aromatic ketones, α-dicarbonyls and Michael systems are suitable substrates for oxidizing allylic Group 14 organometallic compounds with subsequent formation of homoallylic alcohols or β-allylated ketones (Scheme 32) [120–122].

The nitrile group serves as an universal electron-accepting substituent, activating olefinic and aromatic substrates for electron-transfer steps. Cyano-substituted

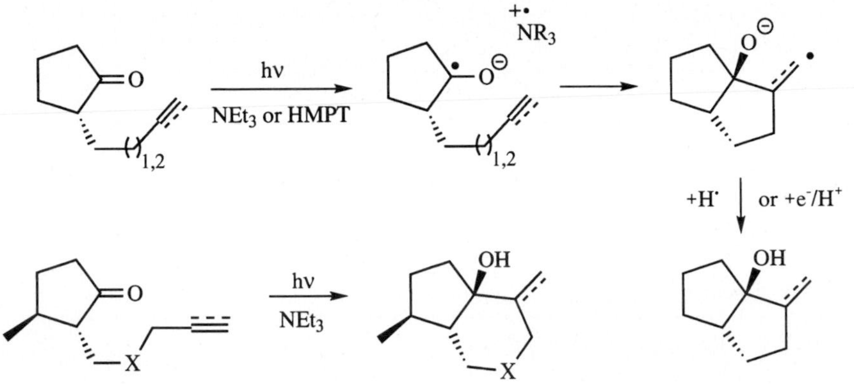

**Scheme 30.**

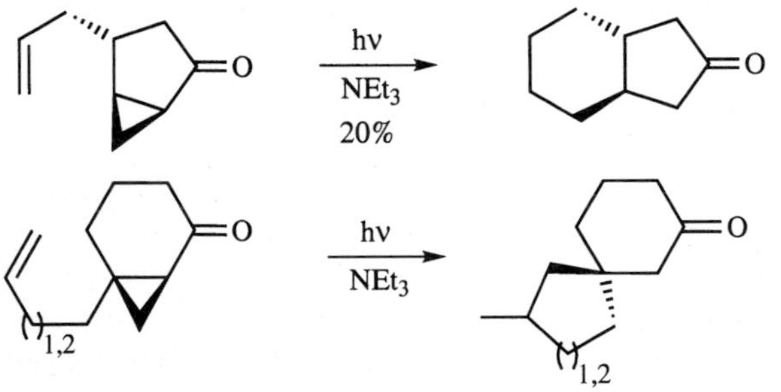

**Scheme 31.**

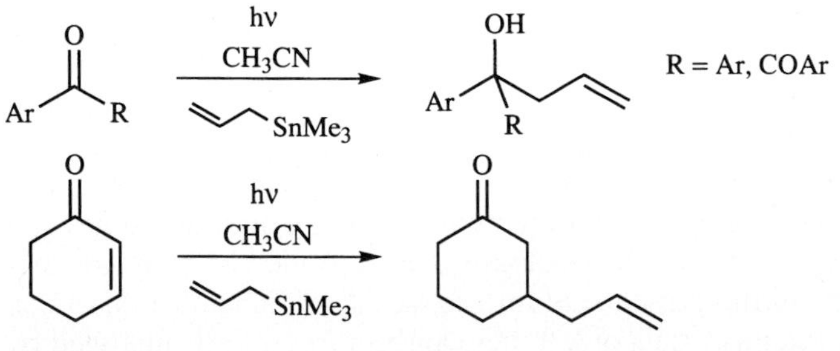

**Scheme 32.**

alkenes are easily alkylated in the presence of Group 14 organometallic reagents such as benzylic or allylic silanes and stannanes (Scheme 33) [123–125]. The radical anion formed after PET from an electron-donating substrate to a cyano-substituted aromatic hydrocarbon can be highly stabilized by spin- and ion-dilution or by steric effects of arene substituents. This enables the generation of long-lived radical anions

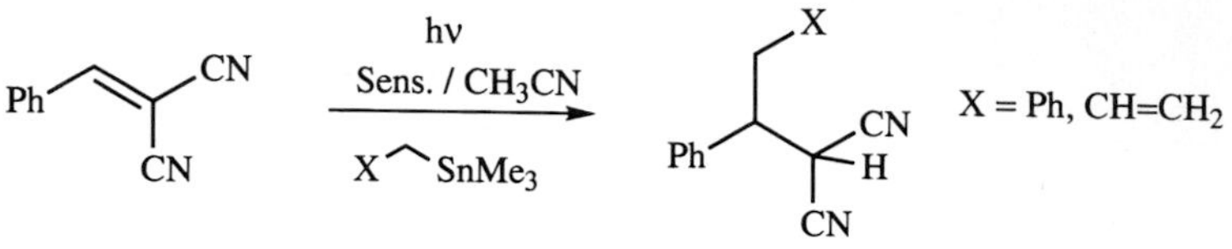

**Scheme 33.**

**Table 5.** Reduction potentials and singlet energies of cyanoaromatic compounds.[a]

| Compound[b] | Reduction potential, $E_{red}$ (V, relative to SCE) | Singlet energies, $E(S_1)$ (eV) |
| --- | --- | --- |
| 2,6,9,10-TCA | −0.45 | 2.82 |
| 1,2,4,5-TCB | −0.65 | 3.83 |
| 9,10-DCA | −0.89 | 2.88 |
| 1,4-DCN | −1.28 | 3.45 |
| 9-CA | −1.39 | 2.96 |
| 1,4-DCB | −1.60 | 4.29 |

[a] According to [126].
[b] TCA = tetracyanoanthracene, TCB = tetracyanobenzene, DCA = dicyanoanthracene, DCN = dicyanonaphthalene, CA = cyanoanthracene, DCB = dicyanobenzene.

which are chemically reluctant, and thus serve as electron source in subsequent reaction steps. These compounds are often used as electron-accepting sensitizers in PET processes (Table 5). The sensitizers most often used are 9,10-dicyano-anthracene (DCA) and 1,4-dicyanonaphthalene (DCN).

Radical anions of carbonyl groups and imines also seem to be produced in the presence of *titanium(IV) chloride* in methanol as solvent. Consecutive oxidation and deprotonation of methanol leads to hydroxymethyl radicals which combine with the carbonyl radical anions to give 1,2-diols and 1,2-aminoalcohols, respectively. The synthesis of the pheromone frontalin has been achieved in a one-pot reaction by hydroxy-methylation of a diketone [127–129]. Likewise triplet sensitizers [130] can be used for direct excitation of the substrate in methanol [131]. Chiral aldimines can be conveniently hydroxymethylated with moderate diastereoselectivity by irradiation of methanolic solutions in the presence of an excess $TiCl_4$ (Scheme 34) [132].

The high negative reduction potentials of ketones can be substantially changed by transforming them into *iminium salts*. These carbonyl derivatives absorb light in similar wavelength regions but are much more easily reduced to neutral α-amino radicals. A multitude of electron-donating reagents has been investigated [133–136]. Arenes are oxidized by electronically excited iminium salts to give arene radical cations. When benzylic CH groups are present deprotonation to give benzylic radicals is rapid and subsequent radical combination occurs. Pyrrolidinium salts are attractive building blocks for syntheses of heterocyclic target molecules (Scheme 35) [137]. The fluorescence quenching of singlet excited iminium salts by allylsilanes or

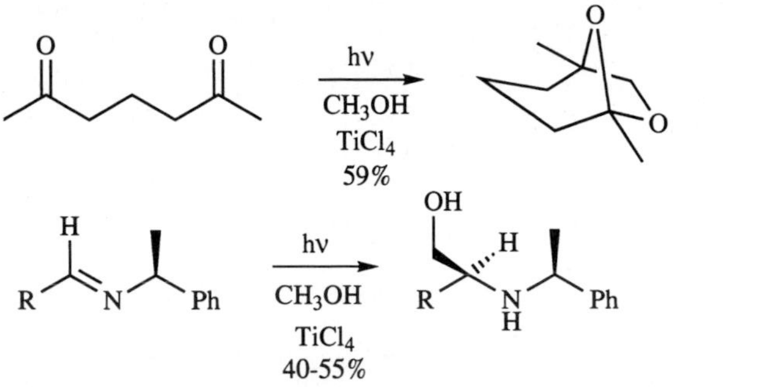

**Scheme 34.**

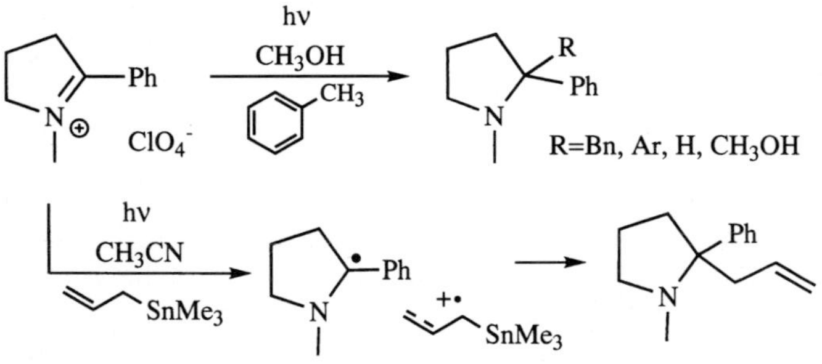

**Scheme 35.**

allylstannanes indicates rapid electron transfer to give the alkene radical cations. In the presence of nucleophilic solvents, allylic radicals are produced via mesolytic cleavage of the radical cations and combine with the α-amino radicals [138, 139].

These two concepts can be combined by using benzyl trialkylsilanes as electron donors in intramolecular PET-reactions. Excitation of the iminium chromophore and electron transfer leads to arene radical cations which rapidly cleave the C–Si bond to give a 1,6-biradical which combines to give the indolizidine skeleton (Scheme 36) [140].

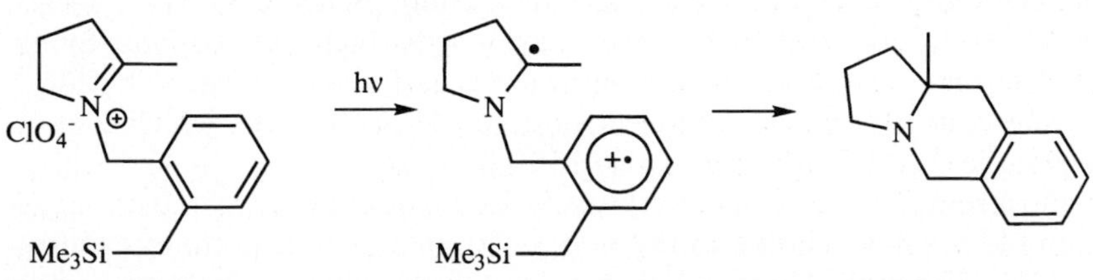

**Scheme 36.**

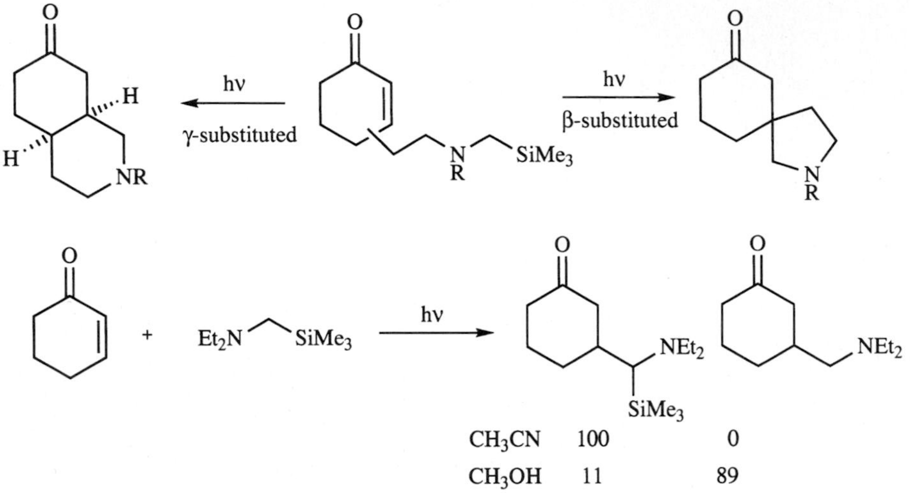

**Scheme 37.**

Conjugated carbonyl groups can also serve as electron acceptors in intramolecular PET reactions. Tertiary amines are versatile donor groups and can be additionally activated by α-silylation. Cyclohexene-2-ones can be coupled at the 3- or 4-position with alkylamino groups. After direct excitation of this donor–acceptor pair, electron transfer generates the enone radical anion–amine radical cation pair. Starting with the β-branched cyclohexenone, after desilylation a 1,5-biradical anion is formed which combines to give the spiro heterocyclic product. Likewise, the 1,6-biradical from the γ-branched cyclohexenone gives a perhydroisoquinoline [141, 142]. An electron transfer sensitizer such as DCN can also be used to initiate the N-oxidation. An important feature is the regiochemistry of the mesolytic bond cleavage step after the oxidation of α-trialkylsilyl amines—in methanol desilylation dominates whereas in acetonitrile only deprotonation is observed (Scheme 37).

In the presence of alkenyl side chains, electronically excited cycloalkene-3-ones readily undergo intramolecular [2 + 2] cycloadditions [143]. If, however, the radical anion of the enone is generated via electron transfer, 5- or 6-*exo-trig* cyclizations can be induced; these lead to primary carbon radicals which can be trapped by hydrogen donors or reduced to the corresponding carbanions which are subsequently protonated to give spirocyclic products. Alternatively, the photocycloadducts can be reductively cleaved by cyclobutylcarbinyl ring-opening (Scheme 38) [144, 145].

An elegant electron relay system has been developed which uses triphenylphosphine as the electron source and dicyanoanthracene as PET-sensitizer. The DCA radical anion subsequently transfers an electron to a Michael system which eventually undergoes radical cyclization chemistry [146].

Aliphatic and aromatic *imides* have relatively low reduction potentials (e.g. *N*-methylphthalimide: −1.4 V relative to the SCE [147]) and absorb in the near ultraviolet region (260–320 nm). These properties have been widely used in synthetic transformations of imides and in catalytic processes in which imides serve as redox

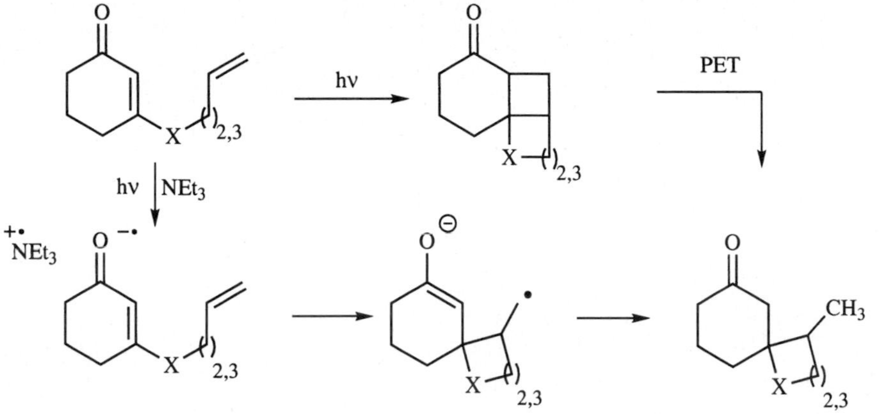

**Scheme 38.**

sensitizers or mediators. A photodecarboxylation protocol has been developed for *N-acyloxyphthalimides* [148, 149]. The light-absorbing species in this process is 1,6-bis(dimethylamino)pyrene which is oxidized by the phthalimide to give the phthalimide radical anion. Mesolytic cleavage results in the carboxylate radical and the phthalimide anion. Alternatively, *phthalimides* can be electronically excited directly in the presence of external alkylcarboxylates and give PET with formation of the phthalimide radical anion and the carboxylate radical [150, 151]. In both reactions decarboxylation results in the formation of free radicals which can be trapped by hydrogen-donating reagents, alkenes, solvent molecules, or phthalimide radical anions (Scheme 39).

Cycloadditions with electronically excited carbonyl groups (or homologs) usually follow biradical mechanisms. Alternatively, PET-pathways can be observed either for energetically favorable donor–acceptor situations or in highly polar solvents. For

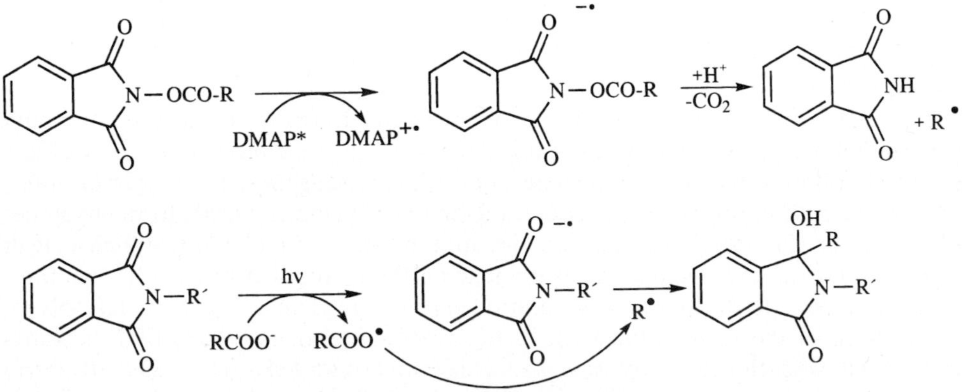

**Scheme 39.**

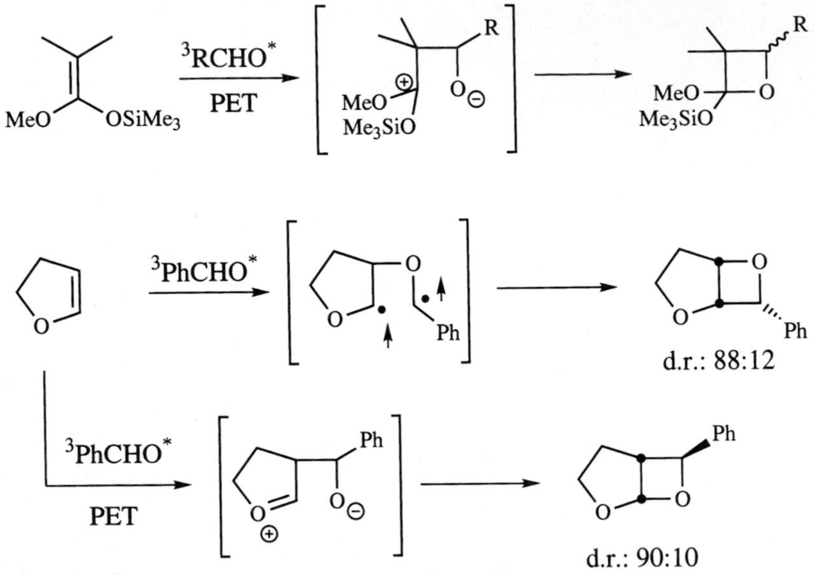

**Scheme 40.**

*Paternò–Büchi reactions* [152] this competition has been investigated for electron-rich alkene substrates for several combinations of carbonyl compounds and electron-donors, e.g. α-diketones and ketene acetals [153], aromatic aldehydes and silyl ketene acetals, and enol ethers. In polar solvents, the assumption of a 1,4-zwitterion as decisive intermediate is reasonable. This situation then resembles the sequence observed for ET-induced thermal [2 + 2]-cycloaddition reactions [154]. Both regio- and diastereoselectivity are influenced by this mechanistic scenario. The regioselectivity is now a consequence of maximum charge stabilization and no longer a consequence of the primary interaction between excited carbonyl compound and alkene. Whereas 3-alkoxyoxetanes are preferentially formed from triplet excited aldehydes and enolethers, 2-alkoxyoxetanes result from the reaction of triplet excited ketones or aldehydes and highly electron-rich ketene silylacetals (Scheme 40) [155].

In the latter reaction photoinduced electron transfer (PET) to give the carbonyl radical anion and the ketene acetal radical cation is energetically feasible [156]. PET might be followed by ISC and formation of a highly stabilized 1,4-zwitterion intermediate (aldol intermediate). C–O bond formation eventually leads to the oxetanes with correct (in respect of the experimental results) regiochemistry. Substituent tuning is not the only possibility of influencing the regioselectivity of the Paternò–Büchi reaction. If the photocycloaddition is performed in a highly polar solvent which reduces the Coulombic term in the Rehm–Weller equation, PET becomes compatible with radical pathways. This effect was observed with 2,3-dihydrofuran as electron-rich substrate which gave selectively the 3-alkoxyoxetane when added to triplet excited aliphatic aldehydes in non-polar solvents. In acetonitrile, however,

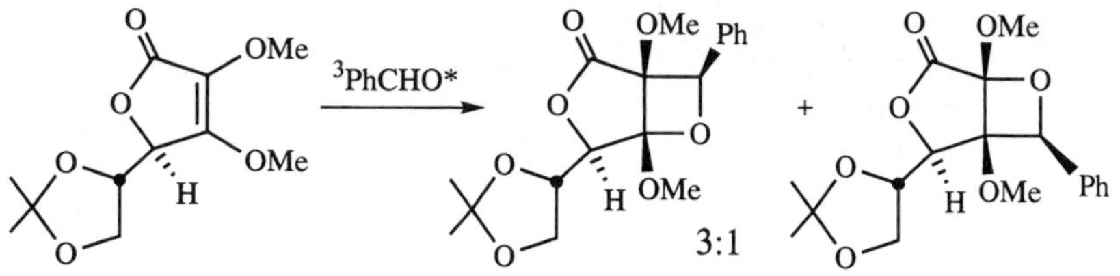

**Scheme 41.**

the corresponding 2-alkoxyoxetane was also detected. The relative amount of this product correlated with solvent polarity parameters thus indicating PET as the mechanism responsible [157].

Abe and coworkers have also observed this stereochemical effect in the Paternò–Büchi reaction of aromatic aldehydes with cyclic ketene acetals [158]. The addition of benzaldehyde to the 5-silyloxy-substituted 2,3-dihydrofuran resulted in the *exo*-phenyl product, albeit in low yields. Higher yields and selectivity were obtained with naphthaldehydes and acceptor-substituted benzaldehydes as carbonyl addends. Interesting substrates in this context were also the 2,3-dialkylated ascorbic acid acetonides investigated by Kulkarni and coworkers [159]. Photocycloaddition of benzaldehyde with these substrates resulted in the formation of two regioisomeric products (Scheme 41).

Both oxetanes were formed with exclusively the *exo*-phenyl configuration. The regio- and diastereoselectivity observed are in accord with the assumption of a PET process involving the oxidation of the ascorbic acid derivatives and the formation of the carbonyl radical anions. In these special instances 1,4-biradical and 1,4-zwitterion stabilization result in similar product regiochemistry. The relative configuration of the products favors the assumption of a PET-process.

*Macrocyclization reactions* are an important class of reactions involving photo-induced electron-transfer steps. Donor–acceptor pairs linked by a flexible hydrocarbon chain are used as starting materials. Mesolytic cleavage of a CH bond proximate to the radical cation (i.e. oxidized heteroatom, alkene, or arene group) usually precedes the primary electron transfer [160]. The resulting (1,$n$)-biradicals combine to give medium- to large-sized ring systems. A very useful carbonyl chromophore which has been intensively investigated in the last two decades is the phthalimido group [161, 162]. The reduction potential (vide supra) is remarkably low in comparison with aromatic carbonyl compounds [163]. In the presence of electron-donor groups (thioether, amines, alkenes, arenes) exergonic electron transfer can occur after electronic excitation. Applications of these PET-macrocyclizations are described with reference to the electron-donating groups.

The course of the intramolecular photoreaction of carbonyl compounds with electron-rich *alkenyl- or aryl-substituents* in the side-chain is dictated essentially by the thermodynamics of the electron-transfer step. This relationship has been intensively studied for phthalimides. When $\Delta G^\circ_{ET}$ is positive, $[\pi^2 + \sigma^2]$ cycloaddition reactions were observed with alkenyl substituents and classical Norrish II chemistry for aryl-substituted substrates. When $\Delta G^\circ_{ET}$ was negative, electron transfer prod-

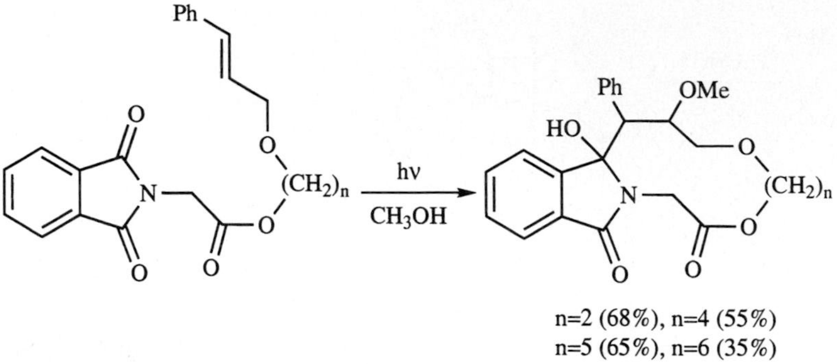

n=2 (68%), n=4 (55%)
n=5 (65%), n=6 (35%)

**Scheme 42.**

ucts predominate. Macrocyclization reactions of phthalimides with remote styryl-substituents were studied by Kubo and coworkers (Scheme 42) [164–166].

When the photolyses were performed in methanol as solvent the styrene radical cations were trapped by methanol in an *anti*-Markovnikov fashion and the resulting biradicals combined to give macrocyclic lactones in moderate to good yields. Spiro-annulated products were formed by irradiation of the corresponding indenyl-substituted starting materials in methanol [167]. The yields for the heterocyclic products were excellent even for the 13-membered representative (86 %).

Intramolecular PET cyclizations of *aminoalkyl-substituted* phthalimides were investigated by the groups of Kanaoka and Coyle [168–171]. Because of the relatively low oxidation potentials of tertiary amines, these electron-transfer reactions are highly exergonic [172, 173]. The regioselectivity of the CH-activation step was remarkably high for the *N*-methyl, *N*-phenyl substituted substrates (Scheme 43).

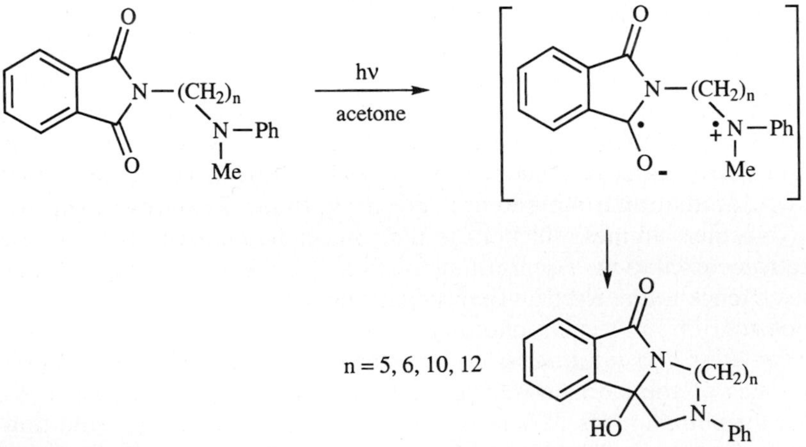

n = 5, 6, 10, 12

**Scheme 43.**

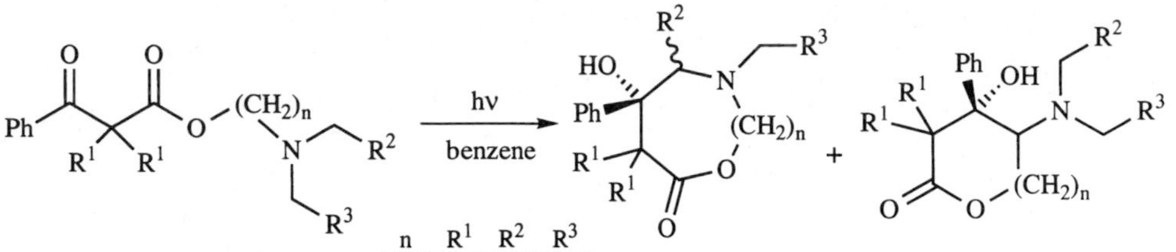

| n | R$^1$ | R$^2$ | R$^3$ | | |
|---|---|---|---|---|---|
| 2 | H | H | H | 33 | - |
| 2 | H | Ph | Ph | 52 | - |
| 2 | Me | Ph | Ph | 63 | - |
| 3 | H | H | H | 34 | 7 |
| 4 | H | Ph | Ph | 25 | 6 |

**Scheme 44.**

This effect has been rationalized by the higher kinetic acidity of *N*-methyl compared with *N*-methylene groups in amine radical cations. The chemical yields for the hydroxyisoindolinones were, however, always relatively poor (ring sizes between 5 and 16). The yields of these cyclization reactions could be improved by use of aromatic carbonyl groups as electron acceptors instead of phthalimides. Hasegawa and coworkers used aromatic *β*-oxoesters as substrates. Thus, medium-sized azalactones and lactones were available via photolysis of *N,N*-benzylaminoalkyl benzoylacetates in moderate yields [174]. Charge-transfer complexes were postulated as precursors and quenching experiments showed that the $\pi\pi^*$ triplet states are the reactive species. The regioselectivity of these cyclization reactions could be improved by use of twofold *N*-benzyl-substituted substrates which were exclusively activated by CH mesolysis of one of the benzylic C–H bonds (Scheme 44) [175].

PET-activation using the widely investigated cyclohexenone chromophore as the electron accepting group has been developed with remote amino substituents as electron-donor groups [176]. The regioselectivity of the mesolytic CH-cleavage was again controlled by use of the *N,N*-dimethylamino group. In contrast to the reactions described above, conjugate addition was observed.

*N*-alkyl substituted phthalimides with *alkylthio groups* in the side chain undergo facile photocyclization to give a variety of azathiacyclols. This reaction is, in contrast to the analogous transformations described above, remarkably efficient and tolerates many different functional groups in the hydrocarbon spacer. The scope and limitation of this important method were intensively investigated by Kanaoka and coworkers [177–182]. Limitations concerning the maximum ring size of the macrocycles have not yet been exactly determined. The chemical yields of unbranched azathiacyclols decreased significantly with increasing spacer length in the substrates (Scheme 45).

This process enabled the synthesis of cyclopeptide [183] and macrolide [184] model compounds with ring sizes up to 37. The acyclic starting materials were conveniently cyclized, with yields between 25 and 50 %. One major problem connected with this thioether method is the incorporation of the electron-donating group into the macrocyclic ring systems. A practical way of overcoming this dis-

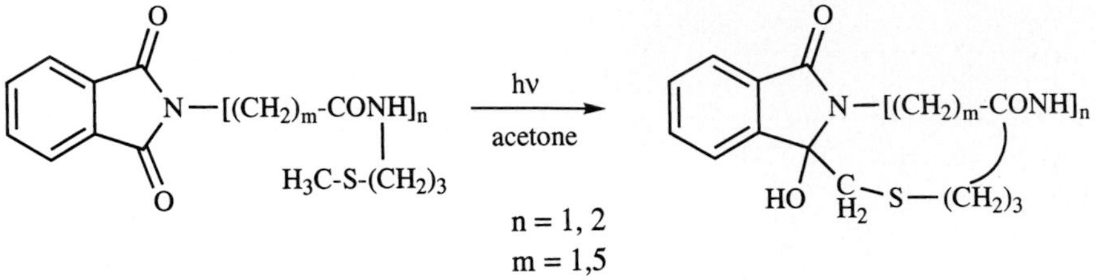

**Scheme 45.**

advantage was developed by *Kanaoka* et al. using the 1,3-dithiolanyl group as electron-donor substituent; this could be removed to give the sulfur-free compounds by treatment with Raney Ni [185]. PET-induced cyclization reactions were also investigated with enantiomerically pure phthalimides from the amino acid pool [186].

The molecular systems for the remote photocyclizations, as described in the previous chapters, consist of an electron-accepting and an electron-donating group connected by a flexible hydrocarbon chain. Both donor and acceptor group are inserted into the newly formed macrocyclic ring system and must be eliminated in a second reaction step. This detour makes these reactions less attractive for C–C coupling steps, e.g. for macrolide synthesis. An improved route has been developed using $\omega$-aminocarboxylic acids as substrates [187]. The photodecarboxylation of *N*-phthaloyl $\alpha$-aminocarboxylic acids is an efficient reaction when conducted in acetone ($\lambda = 300$ nm), in acetonitrile, and even when solid state irradiation is used [188]. This reaction also proceeds with high regioselectivity in the presence of another carboxyl group in the $\beta$- or $\gamma$-position, i.e. aspartic and glutamic acid derivatives. Electron-transfer activation from remote positions became possible when the corresponding potassium carboxylates were used (Scheme 46).

That the electron-transfer step occurs predominantly in an *intramolecular* fashion is probably because of complexation effects in the ground state of the substrates. The synthesis of cyclopeptide model compounds was intensively investigated by use of this photo-decarboxylation route. The yields of these macrocyclizations did not depend on spacer length but on the position of the amide group relative to

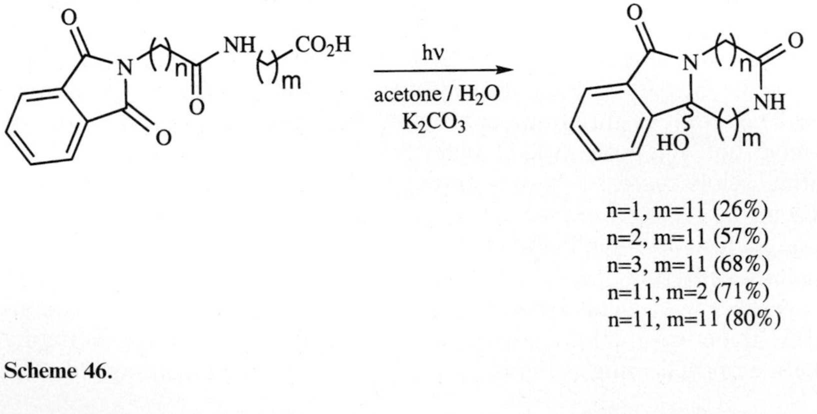

n=1, m=11 (26%)
n=2, m=11 (57%)
n=3, m=11 (68%)
n=11, m=2 (71%)
n=11, m=11 (80%)

**Scheme 46.**

**Table 6.** Oxidation potentials of some aliphatic and aromatic ketones.[a]

| Ketone | Oxidation potential (V, relative to SCE) |
|---|---|
| Acetone | 3.06 |
| Cyclohexanone | 2.71 |
| Propiophenone | 2.86 |
| Acetophenone | 2.82 |
| *p*-Methoxyacetophenone | 2.09 |

[a] According to [190, 191].

the phthalimide chromophore. The maximum yield (80 %) was achieved for a 26-membered cyclopeptide from the acyclic dipeptide [189].

## 8.1.4 Generation of Carbonyl Radical Cations

In general, the electrochemical oxidation potentials of carbonyl compounds are very high (Table 6) and oxidative activation of aliphatic or aromatic carbonyl compounds is problematic whether by anodic oxidation or by photochemical methods. Oxidation at very high positive potentials is circumvented by at least three chemical modifications of the carbonyl group which enable the subsequent oxidation by chemical, electrochemical, or photoinduced electron-transfer processes.

The first modification is peripheral substitution to reduce the oxidation potential of the carbonyl group—a method which often switches the reactivity because of hole localization at positions remote with respect to the C=O group (e.g. acetophenone or 4-methoxyacetophenone). Actually, the electrophore is extended by additional conjugation which reduces the oxidation potential and the HOMO–LUMO energy gap (i.e. the effect of auxochromic groups on the chromophore skeleton).

The second modification is *Umpolung* of the carbonyl group via conversion into the corresponding enol ethers or enols and subsequent oxidation to give the radical cations of enol ether and enol, respectively [192]. The oxidation potentials of these substrates are approximately 1 V (relative to the SCE) and thus oxidation is feasible even with moderately active oxidants or via anodic oxidation. Subsequent reactions of the enol radical cations and radical cations of enol ethers can result in α-substitution products (e.g. by running the reaction in the presence of nucleophiles) and re-formation of the carbonyl group (Scheme 47). Thus, the overall process corresponds to α-activation of carbonyl substrates via intermediate tautomeric enols (and sometimes also enolates) [193, 194].

The corresponding silyl enol ethers are likewise readily available carbonyl *umpolung* substrates which can be oxidized by a variety of chemical oxidants and also by cathodic oxidation. If not trapped by nucleophiles, the radical cations can dimerize and subsequently hydrolyze to give 1,4-dicarbonyl (homo)coupling products [195].

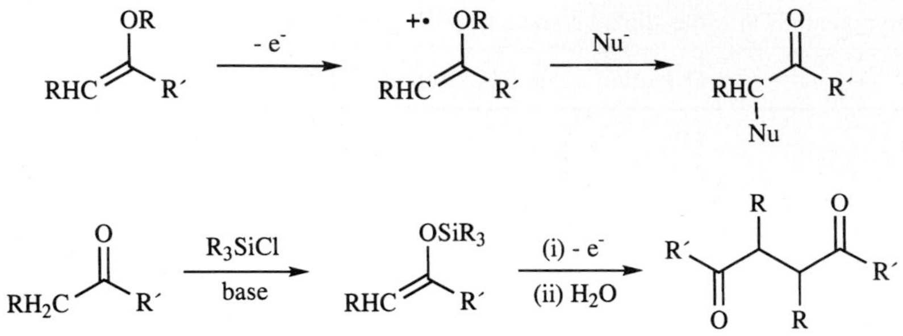

**Scheme 47.**

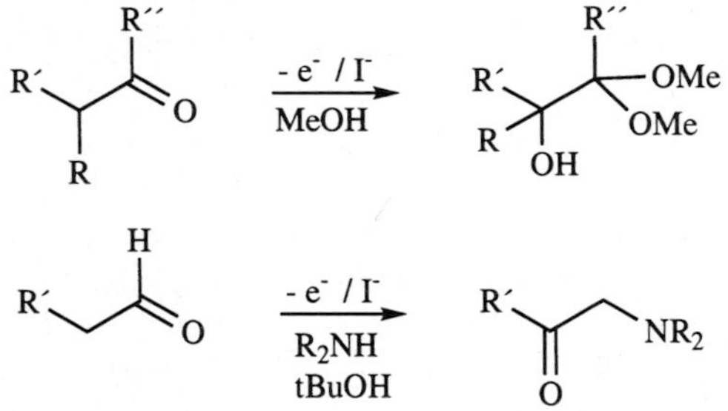

**Scheme 48.**

Intermediate enols can also be oxidized by indirect electrosynthesis (vide supra); halide anions are used as oxidation mediators. A multistep reaction converts ketones into α-hydroxylated acetals when oxidized electrochemically in the presence of iodine as the redox catalyst (Scheme 48) [196].

An analogous process has been developed for unbranched aldehydes which can be transformed into α-amino ketones when oxidized in the presence of an secondary amine and iodine, as the mediator, in aqueous *tert*-butanol. The actual reactive species is probably the enamine which is attacked by iodine cations and subsequently by water. Carbonyl transposition reaction releases iodine anions which can be anodically reoxidized [197].

A third possibility of chemical modification is conversion into an acylsilane which reduces the oxidation potential of the corresponding ketone by approximately 1 V. A peak potential of 1.45 V (relative to Ag/AgCl) for the oxidation of undecanoyltrimethylsilane has been reported. Preparative electrochemical oxidations of acylsilanes proceed in methanol to give the corresponding methyl esters. A two-step oxidation process must be assumed because of the reaction stoichiometry —oxidation of the acylsilane results in the carbonyl radical cation which is mesolytically cleaved to give the silyl cation and the acyl radical, which is subsequently oxidized to give the acyl cation as ultimate electrophile which reacts with the solvent. A variety of other nucleophiles have been used and a series of carboxylic acid derivatives are available via this pathway (Scheme 49) [198].

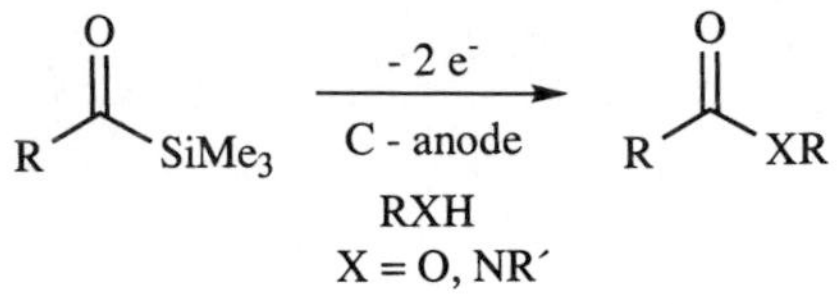

$$R \overset{O}{\underset{}{\|}} SiMe_3 \quad \xrightarrow[\substack{RXH \\ X = O, NR'}]{\substack{-2\,e^- \\ C\text{ - anode}}} \quad R \overset{O}{\underset{}{\|}} XR$$

**Scheme 49.**

## 8.2 Reactivity Pattern of Carbonyl Radical Ions

This section summarizes the reactivity pattern of carbonyl radical ion reactions which are described in numerous applications in Section 8.1. After the first one-electron-transfer step, the carbonyl radical ions can, in principle, undergo three modes of reaction—second electron transfer, bond cleavage, and bond formation.

Carbonyl radical anions are potential *electron donors* and are capable of reducing other carbonyl groups, electron acceptors with lower reduction potentials, or the primary electron source which was oxidized during the first electron-transfer step. The latter reaction is an important side-reaction in photoinduced electron-transfer processes, reconstituting the original electron donor–acceptor pair. Also carbon radicals can be reduced by carbonyl radical anions to give carbanions which can be protonated in the termination step of complex reaction cycles [199]. Further reduction of carbonyl radical anions can also proceed under strong reducing conditions to give 1,2-dianions (Scheme 50).

*Bond cleavage reactions* are important unimolecular processes for carbonyl radical anions which lead to radical–anion separation and stabilization. Two important cleavage modes are known: α- and β-cleavage. In the presence of good anionic leaving groups, mesolytic cleavage leads to acyl radicals which can be further reduced to acyl anions. The latter can also be directly formed via heterolytic cleavage from carbonyl dianions. In the presence of good radical leaving groups X in α-position with respect to the C=O group, rapid mesolytic cleavage leads to enolate anions and X radicals which can undergo further radical reactions. This reaction is synthetically highly useful in its intramolecular version, e.g. in cyclopropyl- or cyclobutylcarbinyl ring opening reactions. In the presence of good anionic leaving

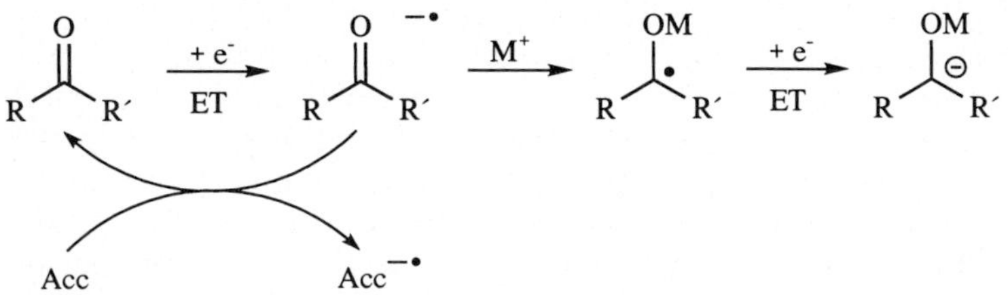

**Scheme 50.**

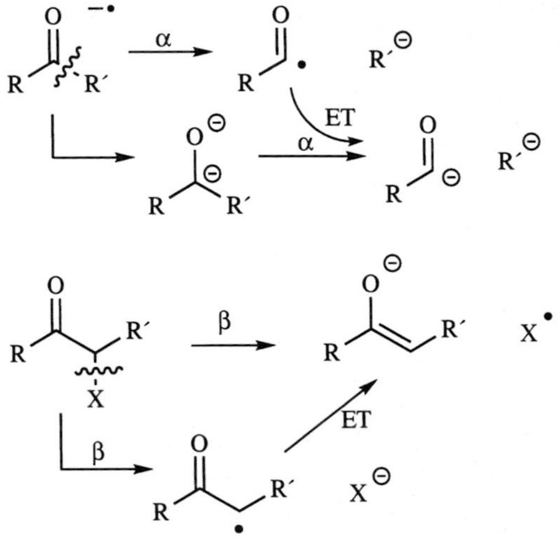

**Scheme 51.**

groups X in the $\beta$ position, the mesolysis leads to enolate radicals which can be further reduced to enolate anions or undergo radical addition reactions (Scheme 51). Bond-cleavage reactions of radical anions and of the corresponding radical cations are often coupled, especially in photoinduced electron-transfer reactions where radical ion pairs are produced in the first electron-transfer step [200, 201].

From a synthetic point of view, *bond forming steps* are the most important reactions of radical ions [202]. Several principle possibilities have been described in Section 8.1 and are summarized in Scheme 52. Many carbo- and heterocyclic ring systems can be constructed by (inter- and intramolecular) radical addition to alkenes, alkynes, or arenes. Coupling of carbonyl radical anions leads to pinacols either intra- or inter-molecular which can be further modified to give 1,2-diols, acyloins or alkenes. Radical combination reactions with alkyl radicals afford the opportunity to synthesize macrocyclic rings. These radical ion–radical pairs can be generated most efficiently by inter- or intramolecular photoinduced electron transfer.

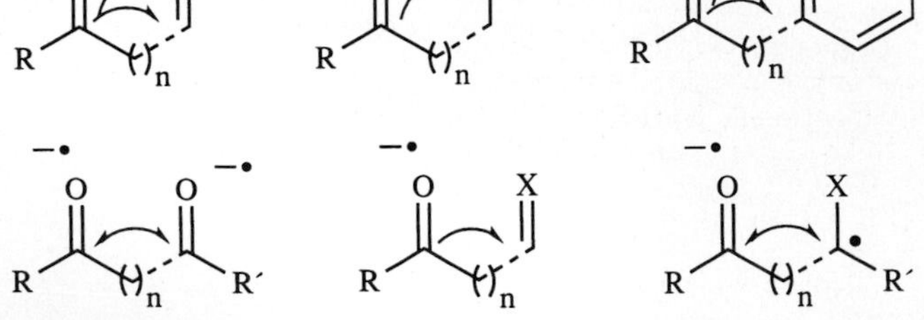

**Scheme 52.**

# References

1. H. Lund, M. M. Baizer in *Organic Electrochemistry*, 3. ed. (Eds.: H. Lund, M. M. Baizer), Marcel Dekker, New York, **1991**, 433.
2. P. Kabasakalian, J. McGlotten, *Anal. Chem.* **1959**, *31*, 1091.
3. M.-A. Michel, J. Simonet, G. Mousset, H. Lund, *Electrochim. Acta* **1975**, *20*, 143.
4. J. P. Coleman, R. J. Holman, J. H. P. Utley, *J. Chem. Soc., Perkin Trans. II* **1976**, 879.
5. R. J. Holman, J. H. P. Utley, *J. Chem. Soc., Perkin Trans. II* **1976**, 884.
6. E. Kariv-Miller, T. J. Mahachi, *J. Org. Chem.* **1986**, *51*, 1041.
7. J. W. Hayes, I. Ruzic, D. E. Smith, G. L. Booman, J. R. Delmastro, *J. Electroanal. Chem.* **1974**, *51*, 269.
8. D. W. Sopher, J. H. P. Utley, *J. Chem. Soc., Perkin Trans. II* **1984**, 1361.
9. K. Uneyama, T. Tokunaga, S. Torii, *Bull. Chem. Soc. Jpn.* **1987**, *60*, 3427.
10. T. Shono, M. Mitani, *J. Am. Chem. Soc.* **1971**, *93*, 5284.
11. T. Shono, I. Nishiguchi, H. Ohmizu, M. Mitani, *J. Am. Chem. Soc.* **1978**, *100*, 545.
12. T. Shono, N. Kise, T. Suzumoto, T. Morimoto, *J. Am. Chem. Soc.* **1986**, *108*, 4676.
13. G. Pattenden, G. M. Robertson, *Tetrahedron Lett.* **1983**, *24*, 4617.
14. G. Pattenden, G. M. Robertson, *Tetrahedron* **1985**, *41*, 4001.
15. J. E. Swartz, E. Kariv-Miller, S. J. Harrold, *J. Am. Chem. Soc.* **1989**, *111*, 1211.
16. E. Kariv-Miller, H. Maeda, F. Lombardo, *J. Org. Chem.* **1989**, *54*, 4022.
17. T. Shono, N. Kise, T. Fujimoto, N. Tominaga, H. Morita, *J. Org. Chem.* **1992**, *57*, 7175.
18. E. Kariv-Miller, R. I. Pacut, G. K. Lehmann, *Top. Curr. Chem.* **1988**, *148*, 97.
19. E. Steckhan, *Top. Curr. Chem.* **1987**, *142*, 2.
20. C. K. Mann, K. K. Barnes, in *Electrochemical Reactions in Nonaqueous Systems*, Marcel Dekker: New York, **1970**.
21. C. P. Andrieux, C. Blocman, J. M. Dumas-Bouchiat, J. M. Savéant, *J. Am. Chem. Soc.* **1979**, *101*, 3431.
22. C. P. Andrieux, C. Blocman, J. M. Dumas-Bouchiat, F. M'Halla, J. M. Savéant, *J. Am. Chem. Soc.* **1980**, *102*, 3806.
23. M. Hoshino, S. Arai, M. Imamura, A. Namiki, *Chem. Phys. Lett.* **1974**, *26*, 582.
24. B. J. McClelland, *Trans. Faraday Soc.* **1961**, *57*, 1458.
25. A. Fürstner, *Angew. Chem.* **1993**, *105*, 171.
26. P. W. Rabideau, Z. Marcinow, *Org. React.* **1992**, *42*, 1.
27. A. Fürstner (ed.), *Active Metals—Preparation, Characterization, Applications*, VCH: Weinheim, New York, **1996**.
28. T. Wirth, *Angew. Chem. Int. Ed. Engl.* **1996**, *35*, 61.
29. A. Gansäuer, *Synlett*, **1998**, 801.
30. Y. Nishiyama, E. Shinomiya, S. Kimura, K. Itoh, N. Sonoda, *Tetrahedron Lett.* **1998**, *39*, 3705.
31. T. Hirao, *Synlett* **1999**, 175.
32. T. Hirao, B. Hatano, Y. Imamoto, A. Ogawa, *J. Org. Chem.* **1999**, *64*, 7665.
33. T. Hirao, B. Hatano, M. Asahara, Y. Muguruma, A. Ogawa, *Tetrahedron Lett.* **1998**, *39*, 5247.
34. B. Hatano, A. Ogawa, T. Hirao, *J. Org. Chem.* **1998**, *63*, 9421.
35. P. Liao, Y. Huang, Y. Zhang, *Synth. Commun.* **1997**, *27*, 1483.
36. E. J. Corey, G. Z. Zheng, *Tetrahedron Lett.* **1997**, *38*, 2045.
37. A. F. Barrero, E. J. Alvarez-Manzaneda, R. Chahboun, R. Meneses, *Synlett* **1999**, 1663.
38. J. E. McMurry, M. P. Fleming, *J. Am. Chem. Soc.* **1974**, *96*, 4708.
39. T. Mukaiyama, T. Sato, J. Hanna, *Chem. Lett.* **1973**, 1041.
40. J. E. McMurry, *Chem. Rev.* **1989**, *89*, 1513.
41. J. E. McMurry, *Acc. Chem. Res.* **1983**, *16*, 405.
42. D. Lenoir, *Synthesis* **1989**, 883.
43. A. Fürstner, H. Weidmann, *Synthesis* **1987**, 1071.
44. G. P. Boldrini, D. Savoia, E. Tagliavini, C. Trombini, A. Umani-Ronchi, *J. Organomet. Chem.* **1985**, *280*, 307.

45. P. Burger, H. H. Brintzinger, *J. Organomet. Chem.* **1991**, *407*, 207.
46. D. L. J. Clive, C. Zhang, K. S. K. Murthy, W. D. Hayward, S. Daigneault, *J. Org. Chem.* **1991**, *56*, 6447.
47. A. Fürstner, A. Hupperts, *J. Am. Chem. Soc.* **1995**, *117*, 4468.
48. E. J. Corey, R. L. Danheiser, S. Chandrasekaran, *J. Org. Chem.* **1976**, *41*, 260.
49. J. E. McMurry, N. O. Siemers, *Tetrahedron Lett.* **1993**, *34*, 7891.
50. J. J. Blomfield, D. C. Owsley, J. M. Nelke, *Org. React.* **1976**, *23*, 259.
51. K. Rühlmann, *Synthesis* **1971**, 236.
52. A. C. Cope, J. W. Barthel, R. D. Smith, *Org. Synth.* **1963**, *IV*, 218.
53. L. E. Overman, C. Fukaya, *J. Am. Chem. Soc.* **1980**, *102*, 1454.
54. P. A. Grieco, E. Williams, H. Tanaka, S. Gilman, *J. Org. Chem.* **1980**, *45*, 3537.
55. N. M. Alpatova, L. I. Krishtalik, Y. V. Pleskov, *Top. Curr. Chem.* **1987**, *138*, 149.
56. J. P. Coleman, R. J. Kobylecky, J. H. P. Utley, *Chem. Commun.* **1971**, 104.
57. S. E. Zabusova, A. P. Tomilov, L. F. Filimonova, N. M. Alpatova, *Elektrokhimiya* **1980**, *16*, 970.
58. K. C. Nicolaou, C.-K. Hwang, M. E. Duggan, K. B. Reddy, B. E. Marron, D. G. McGarry, *J. Am. Chem. Soc.* **1986**, *108*, 6800.
59. P. Girard, J. L. Namy, H. B. Kagan, *J. Am. Chem. Soc.* **1980**, *102*, 2693.
60. G. E. Keck, C. A. Wagner, T. Sell, T. T. Wager, *J. Org. Chem.* **1999**, *64*, 2172.
61. R. Nomura, T. Matsuno, T. Endo, *J. Am. Chem. Soc.* **1996**, *118*, 11666.
62. F. Hélion, J.-L. Namy, *J. Org. Chem.* **1999**, *64*, 2944.
63. R. A. Batey, W. B. Motherwell, *Tetrahedron Lett.* **1991**, *32*, 6649.
64. R. A. Batey, J. D. Harling, W. B. Motherwell, *Tetrahedron* **1996**, *52*, 11421.
65. G. A. Molander, C. Alonso-Alija, *Tetrahedron* **1997**, *53*, 8067.
66. G. A. Molander, C. P. Losada, *J. Org. Chem.* **1997**, *62*, 2935.
67. G. A. Molander, J. A. McKie, *J. Org. Chem.* **1995**, *60*, 872.
68. J. L. Chiara, S. Martinez, M. Bernabé, *J. Org. Chem.* **1996**, *61*, 6488.
69. G. A. Molander, J. A. McKie, *J. Org. Chem.* **1994**, *59*, 3186.
70. G. A. Molander, S. R. Shakya, *J. Org. Chem.* **1996**, *61*, 5885.
71. D. P. Curran, B. Yoo, *Tetrahedron Lett.* **1992**, *33*, 6931.
72. G. A. Molander, P. J. Stengel, *Tetrahedron* **1997**, *53*, 8887.
73. G. A. Molander, J. A. McKie, *J. Org. Chem.* **1992**, *57*, 3132.
74. G. A. Molander, G. Hahn, *J. Org. Chem.* **1986**, *51*, 2596.
75. F. A. Khan, R. Czerwonka, R. Zimmer, H.-U. Reißig, *Synlett* **1997**, 995.
76. C. U. Dinesh, H.-U. Reißig, *Angew. Chem. Int. Ed. Engl.* **1999**, *38*, 789.
77. M. Sasaki, J. Collin, H. B. Kagan, *Tetrahedron Lett.* **1988**, *29*, 6105.
78. J. Collin, F. Dallemer, J.-L. Namy, H. B. Kagan, *Tetrahedron Lett.* **1989**, *30*, 7407.
79. G. A. Molander, J. A. McKie, *J. Org. Chem.* **1993**, *58*, 7216.
80. P. Girard, J. L. Namy, H. B. Kagan, *J. Am. Chem. Soc.* **1980**, *102*, 2693.
81. D. P. Curran, T. L. Fevig, C. P. Jasperse, M. Totleben, *Synlett* **1992**, 943.
82. D. P. Curran, X. Gu, W. Zhang, P. Dowd, *Tetrahedron* **1997**, *53*, 9023.
83. G. Lannoye, K. Sambasivarao, S. Wehrli, J. Cook, *J. Org. Chem.* **1988**, *53*, 2327.
84. D. P. Curran, B. Yoo, *Tetrahedron Lett.* **1992**, *33*, 6931.
85. T. Kan, S. Nara, S. Ito, F. Matsuda, H. Shirahama, *J. Org. Chem.* **1994**, *59*, 5111.
86. T. L. Fevig, R. L. Elliott, D. P. Curran, *J. Am. Chem. Soc.* **1988**, *110*, 5064.
87. L. Zhou, Y. Zhang, D. Shi, *Tetrahedron Lett.* **1998**, *39*, 8491.
88. G. A. Molander, C. N. Wolfe, *J. Org. Chem.* **1998**, *63*, 9031.
89. G. A. Molander, C. R. Harris, *Tetrahedron* **1998**, *54*, 3321.
90. M. Sasaki, J. Collin, H. B. Kagan, *New J. Chem.* **1992**, *16*, 89.
91. G. A. Molander, C. R. Harris, *Chem. Rev.* **1992**, *92*, 29.
92. G. A. Molander, *Acc. Chem. Res.* **1998**, *31*, 603.
93. G. Pattenden, S. J. Teague, *Tetrahedron Lett.* **1982**, *23*, 4571.
94. G. Pattenden, G. M. Robertson, *Tetrahedron* **1985**, *41*, 4001.
95. G. Pattenden, G. M. Robertson, *Tetrahedron Lett.* **1983**, *24*, 4617.
96. G. H. Lee, E. B. Choi, E. Lee, C. S. Pak, *J. Org. Chem.* **1994**, *59*, 1428.
97. K. Mizuno, Y. Otsuji, *Top. Curr. Chem.* **1994**, *169*, 301.

98. G. Pandey, *Top. Curr. Chem.* **1993**, *168*, 175.
99. M. Chanon, M. D. Hawley, M. A. Fox, in *Photoinduced Electron Transfer*, (Eds.: M. A. Fox, M. Chanon), Part A, Elsevier: Amsterdam, **1988**, 1.
100. M. Chanon, L. Eberson, in *Photoinduced Electron Transfer*, (Eds.: M. A. Fox, M. Chanon), Part A, Elsevier: Amsterdam, **1988**, 409.
101. G. J. Kavarnos, *Fundamentals of Photoinduced Electron Transfer*, VCH: Weinheim, New York, **1993**.
102. H. D. Roth, *Top. Curr. Chem.* **1990**, *156*, 1.
103. J. Mattay, *Top. Curr. Chem.* **1991**, *159*, 219.
104. J. Mattay, *Angew. Chem. Int. Ed. Engl.* **1987**, *26*, 825.
105. S. L. Mattes, S. Farid, *Org. Photochem.* **1983**, *6*, 233.
106. M. Chanon, M. Rajzmann, F. Chanon, *Tetrahedron* **1990**, *46*, 6193.
107. J. K. Kochi, *Angew. Chem. Int. Ed. Engl.* **1988**, *27*, 1227.
108. G. J. Kavarnos, N. J. Turro, *Chem. Rev.* **1986**, *86*, 401.
109. D. Rehm, A. Weller, *Ber. Bunsenges. Phys. Chem.* **1969**, *73*, 834
110. D. Rehm, A. Weller, *Z. Phys. Chem.* **1970**, *69*, 183.
111. I. R. Gould, J. E. Moser, B. Armitage, S. Farid, J. L. Goodman, M. S. Herman, *J. Am. Chem. Soc.* **1989**, *111*, 1917.
112. J. Gersdorf, J. Mattay, H. Görner, *J. Am. Chem. Soc.* **1987**, *109*, 1203.
113. J. B. Guttenplan, S. G. Cohen, *J. Am. Chem. Soc.* **1972**, *94*, 4040.
114. D. Belotti, J. Cossy, J. P. Pete, C. Portella, *J. Org. Chem.* **1986**, *51*, 4196.
115. J. Cossy, D. Belotti, J. P. Pete, *Tetrahedron Lett.* **1987**, *28*, 4545.
116. J. Cossy, D. Belotti, *Tetrahedron Lett.* **1988**, *29*, 6113.
117. J. Cossy, *Tetrahedron Lett.* **1989**, *30*, 4113.
118. T. Kirschberg, J. Mattay, *Tetrahedron Lett.* **1994**, *35*, 7217.
119. M. Fagnoni, P. Schmoldt, T. Kirschberg, J. Mattay, *Tetrahedron* **1998**, *54*, 6427.
120. K. Maruyama, H. Imahori, A. Osuka, A. Takuwa, H. Tagawa, *Chem. Lett.* **1986**, 1719.
121. A. Takuwa, Y. Nishigaichi, K. Yamashita, H. Iwamoto, *Chem. Lett.* **1990**, 639.
122. A. Takuwa, Y. Nishigaichi, H. Iwamoto, *Chem. Lett.* **1991**, 1013.
123. K. Mizuno, M. Ikeda, Y. Otsuji, *Chem. Lett.* **1988**, 1507.
124. K. Mizuno, K. Nakanishi, A. Tachibana, Y. Otsuji, *J. Chem. Soc., Chem. Commun.* **1991**, 344.
125. K. Mizuno, Y. Otsuji, *Top. Curr. Chem.* **1994**, *169*, 301.
126. S. L. Mattes, S. Farid, in *Organic Photochemistry* (Ed.: A. Padwa), Marcel Dekker: New York, **1983**, *6*, 233.
127. T. Sato, G. Izumi, T. Imamura, *Tetrahedron Lett.* **1975**, 2191.
128. T. Sato, S. Yamaguchi, H. Kaneko, *Tetrahedron Lett.* **1979**, *21*, 1863.
129. T. Sato, H. Kaneko, S. Yamaguchi, *J. Org. Chem.* **1980**, *45*, 3778.
130. B. Fraser-Reid, R. C. Anderson, D. R. Hicks, D. L. Walker, *Can. J. Chem.* **1977**, *55*, 3986.
131. N. Reineke, N. A. Zaidi, M. Mitra, D. O'Hagan, A. S. Batsanov, J. A. K. Howard, D. Y. Naumov, *J. Chem. Soc., Perkin Trans 1* **1996**, 147.
132. A. G. Griesbeck, S. Buhr, J. Lex, *Tetrahedron Lett.* **1998**, *39*, 2535.
133. P. S. Mariano, *Acc. Chem. Res.* **1983**, *16*, 130.
134. P. S. Mariano, J. L. Stavinoha, in *Synthetic Organic Photochemistry*, (Ed.: W. M. Horspool), Plenum Press: New York, **1983**, 145.
135. U. C. Yoon, P. S. Mariano, *Acc. Chem. Res.* **1992**, *25*, 233.
136. P. S. Mariano, in *Photoinduced Electron Transfer*, (Eds.: M. A. Fox, M. Chanon), Part C, Elsevier: Amsterdam, **1988**, 372.
137. R. M. Borg, R. O. Heuckeroth, A. J. Y. Lan, S. L. Quillen, P. S. Mariano, *J. Am. Chem. Soc.* **1987**, *109*, 2728.
138. K. Ohga, P. S. Mariano, *J. Am. Chem. Soc.* **1982**, *104*, 617.
139. R. M. Borg, P. S. Mariano, *Tetrahedron Lett.* **1986**, *27*, 2821.
140. A. J. Y. Lan, S. L. Quillen, R. O. Heuckeroth, P. S. Mariano, *J. Am. Chem. Soc.* **1984**, *106*, 6439.
141. W. Xu, Y. T. Jeon, E. Hasegawa, U. C. Yoon, P. S. Mariano, *J. Am. Chem. Soc.* **1989**, *111*, 406.

142. W. Xu, X. M. Zhang, P. S. Mariano, *J. Am. Chem. Soc.* **1991**, *113*, 8863.
143. M. T. Crimmins, *Chem. Rev.* **1988**, *88*, 1453.
144. E. W. Bischof, J. Mattay, *Tetrahedron Lett.* **1990**, *31*, 7137.
145. J. Mattay, A. Banning, E. W. Bischof, A. Heidbreder, J. Runsink, *Chem. Ber.* **1992**, *125*, 2119.
146. G. Pandey, S. Hajra, *Angew. Chem. Int. Ed. Engl.* **1994**, *33*, 1169.
147. D. W. Leedy, D. L. Muck, *J. Am. Chem. Soc.* **1971**, *93*, 4264.
148. K. Okada, K. Okamoto, M. Oda, *J. Am. Chem. Soc.* **1988**, *110*, 8736.
149. K. Okada, K. Okamoto, N. Morita, M. Oda, K. Okubo, *J. Am. Chem. Soc.* **1991**, *113*, 9401.
150. A. G. Griesbeck, M. Oelgemöller, *Synlett* **1999**, 492.
151. A. G. Griesbeck, W. Kramer, M. Oelgemöller, *Synlett* **1999**, 1169.
152. J. Mattay, R. Conrads, R. Hoffmann, in *Methoden der Organischen Chemie* (Houben–Weyl) (Eds.: G. Helmchen, R. W. Hoffmann, J. Mulzer, E. Schaumann), Thieme Verlag: Stuttgart, **1995**, E21c, 3133.
153. J. Mattay, J. Gersdorf, K. Buchkremer, *Chem. Ber.* **1987**, *120*, 307.
154. T. Kim, H. Sarker, N. L. Bauld, *J. Chem. Soc. Perkin Trans. 2* **1995**, 577.
155. M. Abe, Y. Shirodai, M. Nojima, *J. Chem. Soc., Perkin Trans. 1* **1998**, 3253.
156. D. L. Sun, S. M. Hubig, J. K. Kochi, *J. Org. Chem.* **1999**, *64*, 2250.
157. A. G. Griesbeck, S. Buhr, M. Fiege, H. Schmickler, J. Lex, *J. Org. Chem.* **1998**, *63*, 3847.
158. M. Abe, I. Masayuki, M. Nojima, *J. Chem. Soc., Perkin Trans. 1* **1998**, 3261.
159. R. S. Thopate, M. G. Kulkarni, V. G. Puranik, *Angew. Chem. Int. Ed. Engl.* **1998**, *37*, 1110.
160. P. Maslak, J. N. Navaraez, *Angew. Chem. Int. Ed. Engl.* **1990**, *29*, 302.
161. Y. Kanaoka, *Acc. Chem. Res.* **1978**, *11*, 407.
162. A. G. Griesbeck, H. Mauder in *CRC Handbook of Organic Photochemistry and Photobiology* (Eds.: W. M. Horspool, P.-S. Song), CRC Press: New York, **1995**, 513.
163. H. Hoshino, H. Shizuka, in *Photoinduced Electron Transfer*, (Eds.: M. A. Fox, M. Chanon), Part C, Elsevier: Amsterdam, **1988**, 313.
164. K. Maruyama, Y. Kubo, *J. Am. Chem. Soc.* **1978**, *100*, 7772.
165. K. Maruyama, Y. Kubo, *Chem. Lett.* **1978**, 851.
166. K. Maruyama, Y. Kubo, M. Machida, K. Oda, Y. Kanaoka, K. Furuyama, *J. Org. Chem.* **1978**, *43*, 2303.
167. M. Machida, K. Oda, Y. Kanaoka, *Tetrahedron* **1985**, *41*, 4995.
168. M. Machida, H. Takechi, Y. Kanaoka, *Heterocycles* **1980**, *14*, 1255.
169. M. Machida, H. Takechi, Y. Kanaoka, *Heterocycles* **1977**, *7*, 273.
170. J. D. Coyle, G. L. Newport, *Tetrahedron Lett.* **1977**, 899.
171. J. D. Coyle, G. L. Newport, *J. Chem. Soc. Perkin Trans. 1* **1980**, 93.
172. M. Machida, H. Takechi, Y. Kanaoka, *Chem. Pharm. Bull.* **1982**, *30*, 1579.
173. M. Machida, H. Takechi, Y. Kanaoka, *Heterocycles* **1980**, *14*, 1255.
174. T. Hasegawa, K. Miyata, T. Ogawa, N. Yoshihara, M. Yoshioka, *J. Chem. Soc. Chem. Commun.* **1985**, 363.
175. T. Hasegawa, T. Ogawa, K. Miyata, A. Karakizawa, M. Komiyama, Y. Nishizawa, M. Yoshioka, *J. Chem. Soc. Perkin Trans. 1* **1990**, 901.
176. G. A. Kraus, L. Chen, *Tetrahedron Lett.* **1991**, *32*, 7151.
177. Y. Kanaoka, H. Migita, Y. Sato, H. Nakai, *Tetrahedron Lett.* **1973**, 51.
178. Y. Hatanaka, Y. Sato, H. Nakai, M. Wada, T. Mizoguchi, Y. Kanaoka, *Liebigs Ann. Chem.* **1992**, 1113.
179. Y. Sato, H. Nakai, T. Mizoguchi, Y. Kanaoka, *Tetrahedron Lett.* **1976**, 1889.
180. H. Takechi, S. Tateuchi, M. Machida, Y. Nishibata, K. Aoe, Y. Sato, Y. Kanaoka, *Chem. Pharm. Bull.* **1986**, *34*, 3142.
181. Y. Sato, H. Nakai, M. Wada, T. Mizuguchi, Y. Hatanaka, H. Migita, Y. Kanaoka, *Liebigs Ann. Chem.* **1985**, 1099.
182. Y. Sato, H. Nakai, M. Wada, T. Mizuguchi, Y. Hatanaka, Y. Kanaoka, *Chem. Pharm. Bull.* **1992**, *40*, 3174.
183. Sato, Y.; Nakai, H.; Mizoguchi, T.; Hatanaka, Y.; Kanaoka, Y. *J. Am. Chem. Soc.* **1976**, *98*, 2349.

184. M. Wada, H. Nakai, K. Aoe, K. Kotera, Y. Sato, Y. Hatanaka, Y. Kanaoka, *Tetrahedron* **1983**, *39*, 1273.
185. M. Wada, H. Nakai, Y. Sato, Y. Hatanaka, Y. Kanaoka, *Tetrahedron* **1983**, *39*, 2691.
186. A. G. Griesbeck, H. Mauder, I. Müller, E.-M. Peters, K. Peters, H. G. von Schnering, *Tetrahedron Lett.* **1993**, *34*, 453.
187. A. G. Griesbeck, A. Henz, K. Peters, E.-M. Peters, H. G. von Schnering, H. G. *Angew. Chem. Int. Ed. Engl.* **1995**, *34*, 474.
188. A. G. Griesbeck, A. Henz, *Synlett* **1994**, 931.
189. A. G. Griesbeck, A. Henz, W. Kramer, J. Lex, F. Nerowski, M. Oelgemöller, K. Peters, E.-M. Peters, *Helvetica Chim. Acta*, **1997**, *80*, 912–933.
190. F. G. Bordwell, J. A. Harrelson, Jr., *Can. J. Chem.* **1990**, *68*, 1714.
191. J. Y. Becker, L. L. Miller, T. M. Siegel, *J. Am. Chem. Soc.* **1975**, *97*, 849.
192. M. Schmittel, *Top. Curr. Chem.* **1994**, *169*, 183.
193. M. Schulz, R. Kluge, L. Sivilai, B. Kamm, *Tetrahedron* **1990**, *46*, 2371.
194. M. Schmittel, A. Abufarag, O. Luche, M. Levis, *Angew. Chem. Int. Ed. Engl.* **1990**, *29*, 1144.
195. H. J. Schäfer, *Angew. Chem. Int. Ed. Engl.* **1981**, *20*, 911.
196. T. Shono, Y. Matsumura, K. Inoue, F. Iwasaki, *J. Chem. Soc., Perkin Trans. I* **1986**, 73.
197. T. Shono, Y. Matsumura, J. Hayashi, M. Usui, S. Yamane, K. Inoue, *Acta Chem. Scand. B37* **1983**, 491.
198. J. Yoshida, M. Itoh, S. Matsunaga, S. Isoe, *J. Org. Chem.* **1992**, *57*, 4877.
199. J.-P. Soumillion, *Top. Curr. Chem.* **1993**, *168*, 93.
200. N. J. Pienta, in *Photoinduced Electron Transfer*, (Eds.: M. A. Fox, M. Chanon), Part C, Elsevier: Amsterdam, **1988**, 421.
201. F. D. Saeva, *Top. Curr. Chem.* **1990**, *156*, 59.
202. S. Hintz, A. Heidbreder, J. Mattay, *Top. Curr. Chem.* **1996**, *177*, 77.

# 9 Electron Transfer in Radicals

*Massimo Bietti and Steen Steenken*

## 9.1 Introduction

Radical chemistry has long been an exotic (and allegedly 'dirty') area of chemistry. However, because of (i) contributions to the understanding of radical *reaction mechanisms*, mainly from *radiation*-chemical studies in aqueous solution [1–7], (ii) growing awareness of the occurrence and importance of *radical reactions in biological systems* such as enzymes [8, 9] or photosystem II [10–12] or *biological processes* such as diseases, cancer, or aging [13, 14], and, finally, (iii) the recognition of the *synthetic potential* of radicals [15–20] radical chemistry has succeeded in gaining legitimacy as a proper area of chemistry.

A particularly simple approach with respect to synthetic applications of radical reactions is 'electron transfer activation' which consists in producing radical *ions* as the reactive species. The activation can proceed by electron *addition* to the molecule to be activated, when a radical *anion* is produced, or by electron *removal* which leads to a radical *cation*. If there is more than one electron or hole trap on a molecule, such as in A–spacer–A', the electron or hole can transfer to the other (secondary) trap.

$$A^{\cdot-/+}\text{–spacer–}A' \rightarrow A\text{–spacer–}A'^{\cdot-/+} \tag{1}$$

Intramolecular electron- or hole-transfer processes of this type have been intensively studied [21–23], because of their obvious relevance to photosynthesis [24–26] or to ET processes in other biological systems such as proteins [27–32] or DNA [33–35]. These reactions as well as *inter*molecular electron transfer (ET) processes, which consist just in *exchange* of charge between the reaction partners, will *not* be discussed here, although radicals do have a strong tendency to engage in one-electron transfer processes (since this is the simplest way for these species to lose their radical nature, i.e. to become closed-shell molecules) [36]. What *will* be discussed (in Sections 9.2 and 9.3) are reactions where a radical X$^{\cdot}$ oxidizes or reduces a substrate S, Eq. 2, in which charged species are *produced de novo*:

$$X^{\cdot} + S \rightarrow X^- + S^{\cdot+} \tag{2a}$$

$$X^{\cdot} + S \rightarrow X^+ + S^{\cdot-} \tag{2b}$$

In reactions of this type, the *direction* of the electron flow between the radical $X^{\cdot}$ and the substrate molecule S depends, of course, on the oxidizing or reducing properties of $X^{\cdot}$ and the ability of S either to donate or to accept an electron. Irrespective of electron flow *direction*, transfer between $X^{\cdot}$ and S can proceed by the 'outer-sphere' mechanism [37], Eq. 2, or by 'inner-sphere' or 'bonded' paths (Eqs 3 and 4) [38, 39]. The stoichiometry in Eqs 3 and 4 is the same as in Eq. 2. In the bonded mechanism, however, a heterolytic fragmentation of the bond joining X and S is the step in which the electron transfer actually occurs:

$$X^{\cdot} + S \rightarrow X_{\beta}\text{--}S_{\alpha}^{\cdot} \qquad \text{bond formation} \tag{3}$$

S is *oxidized*

$$X_{\beta}\text{--}S_{\alpha}^{\cdot} \rightarrow X^- + S^{\cdot+} \qquad \text{bond heterolysis } (\beta\text{-elimination}) \tag{4a}$$

S is *reduced*

$$X_{\beta}\text{--}S_{\alpha}^{\cdot} \rightarrow X^+ + S^{\cdot-} \qquad \text{bond heterolysis } (\beta\text{-elimination}) \tag{4b}$$

For the heterolysis reactions, Eq. 4, to occur it is, of course, irrelevant *how* $X\text{--}S^{\cdot}$ was *formed*; e.g., rather than by addition (Eq. 3), it can be produced by H-abstraction from $X\text{--}S\text{--}H$:

$$X\text{--}S\text{--}H \xrightarrow{\ -H^{\cdot}\ } X_{\beta}\text{--}S_{\alpha}^{\cdot} \tag{5}$$

The reaction depicted by Eq. 5 can be called 'activation by H-abstraction', which is often analogous to 'electron-transfer activation'. Examples of the latter (in which $X\text{--}S^{\cdot+}$ or $X\text{--}S^{\cdot-}$ are produced) will also be presented (Section 9.5) but only in heterolysis is ET taking place between X and S [40, 41].

## 9.2 Oxidation of S by $X^{\cdot}$ (ET by Addition–Elimination (The 'ad–el' Mechanism)

Condition: $(E(X^{\cdot}/X^-) > E(S^{\cdot+}/S))$

### 9.2.1 S = alkene

The reaction starts with addition to the double bond (rate constant $k_{ad}$ ($\text{M}^{-1}\,\text{s}^{-1}$)):

$$X^{\cdot} + {>}C{=}C{<} \rightarrow X\text{--}\overset{|}{\underset{|}{C}}\text{--}C^{\cdot}{<} \tag{6}$$

If the so-formed $\beta$-X–alk–$(\alpha)$yl radical undergoes rapid C–X heterolysis with the rate constant $k_{het}$ $(s^{-1})$ (Eq. 7):

$$X-\overset{|}{\underset{|}{C}}-\overset{|}{C}< \;\rightarrow\; X^- + {}^+\overset{|}{C}-\overset{|}{C}< \tag{7}$$

the reaction will *seem* as if it is electron transfer to all detection methods whose time resolution $\tau$ is larger than $1/k_{het}$. In Eqs 6 and 7 the stoichiometry is such that the electrophilic radical X$^{\bullet}$ is the electron *acceptor* (which yields X$^-$) and the alkene is the electron *donor* (which is thereby converted to the alkene radical cation). On the basis of this concept it can be predicted that the rate of X$^{\bullet}$ addition, $k_{ad}$, and the rate of X$^-$ elimination, $k_{het}$, should increase with increasing reduction potential of X$^{\bullet}$, $E(X^{\bullet}/X^-)$, and, of course, with increasing ease of oxidation of the alkene, which should depend strongly on the kind and number of substituents on the alkene function.

As shown in the next section, these ideas are borne out by the reactions of alkenes with the strong oxidizing radical $SO_4^{\bullet-}$ $(E(SO_4^{\bullet-}/SO_4^{2-}) = 2.4\text{–}3.1$ V$)$ [42–44].

## X = $SO_4^{\bullet-}$

The oxidation of alkenes by $SO_4^{\bullet-}$ has been studied in aqueous acetonitrile solution, using time-resolved (a) optical and (b) conductance methods [45]. $SO_4^{\bullet-}$ was produced in the presence of the alkenes by (i) laser photolysis (20 ns pulse, $\lambda_{excitation} = 248$ nm)

$$S_2O_8^{2-} \xrightarrow[\text{248 nm}]{h\nu} 2SO_4^{\bullet-} \tag{8}$$

or (ii) pulse radiolysis

$$S_2O_8^{2-} + e_{aq}^- \rightarrow SO_4^{\bullet-} + SO_4^{2-} \tag{9}$$

In (a), the optical experiments, the reaction of $SO_4^{\bullet-}$ with the alkenes was studied by monitoring the rate of decay of $SO_4^{\bullet-}$ as a function of the concentration of alkene. It was found that the bimolecular rate constant for reaction of $SO_4^{\bullet-}$ (obtained from the plot of $k_{obs(erved)}$ as a function of [alkene]) increases strongly with increasing water content of the solvent, as shown in Figure 1, where allyl alcohol is used as an example. This indicates that the transition state for the reaction (which consists in addition to the C=C double bond, Eq. 10; see later and Refs [46] and [47]) is highly ionic [48].

$$SO_4^{\bullet-} + >C=C< \;\rightarrow\; {}^-O_3SO\overset{|}{\underset{|}{C}}-C^{\bullet}< \tag{10}$$

It was observed that, after $SO_4^{\bullet-}$ had completely disappeared, due to reaction with the alkene (using concentrated or saturated solutions of the alkenes), with electron-rich alkenes, there was a first-order increase of conductance of the aqueous

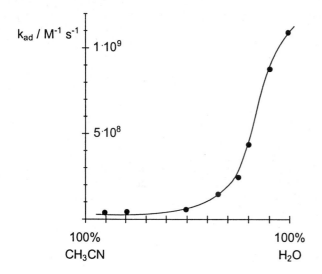

**Figure 1.** Rate constant $k_{ad}$ ($M^{-1}$ $s^{-1}$) for the reaction $SO_4^{·-} + HOCH_2CH=CH_2 \rightarrow$ $HOCH_2C^·HCH_2OSO_3^-$ as a function of solvent ($CH_3CN-H_2O$) composition.

solution at pH 4–6. The rate of this conductance increase (which could be reversibly changed into a conductance *decrease* in basic solution, indicating that the conductance change is due to the production of $H^+$) was independent of [alkene] and pH but dependent on temperature. The reaction responsible is identified [49] in terms of $\beta$-fragmentation (heterolysis) of the $SO_4^-$ adduct to give a (short-lived) [50] alkene radical cation, followed by its rapid reaction with water.

$$>C^·-\underset{|}{\overset{|}{C}}-OSO_3^- \xrightarrow[-SO_4^{2-}]{k_{het}} [>C^·-C^+<] \xrightarrow{H_2O} >C^·-C(OH)< + H^+ \qquad (11)$$

An example is the cyclohexene derivative shown below. Figure 2 depicts the dependence of the rate of heterolysis $k_{het}$ on solvent composition for this system. The fact that $k_{het}$ increases strongly with increasing water content is indicative of the strongly ionic nature ($S_N1$-type) of the transition state of the C–O heterolysis.

Table 1 lists the rate constants for reaction 11 in aqueous solution measured via the first-order conductance increase in $H_2O$ at pH 4–6 and $20 \pm 2\,°C$.

As is apparent from Table 1, the rate of heterolysis is drastically *in*creased by introducing methyl groups at C=C. If, in contrast, the electron density is somewhat *reduced*, e.g. by the inductive effect of an OH group at $C_\gamma$, $C_\delta$, or even $C_\varepsilon$, $k_{het}$ decreases [51].

These results indicate that the reaction of $SO_4^{·-}$ with alkenes changes from pure addition to what *seems* to be electron transfer, followed by, or concerted with, addition of water to the radical cation, as the electron density of the olefinic bond is increased. One consequence is that at the one extreme (characterized by *low* electron density), the reaction $SO_4^{·-}$ + alkene stops at the *adduct* stage. At the other extreme (high electron availability) the reaction probably proceeds by outer-sphere ET. In the in-between situation, addition is followed by elimination (rates strongly depen-

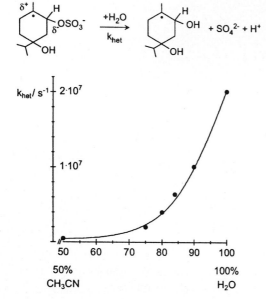

**Figure 2.** Effect of solvent (CH$_3$CN–H$_2$O) composition on the rate constant for heterolysis, $k_{het}$, of the $\alpha$-methyl-$\beta$-sulfatocyclohexyl-type radical shown above.

dent on solvent), and this constitutes an inner-sphere ET mechanism. The solvent has an analogous effect: in(de)creasing the polarity of the solvent (dis)favors the 'electron transfer' path.

Concerning (semi)quantitative conclusions from the data in Table 1, the following statements can be made:

1) A methyl group at C$_\alpha$ increases $k_{het}$ by ca 4000 (a consequence is that extrapolation to 1,2-dimethylcyclohexenes leads to the prediction that for these compounds $k_{het} \approx 2 \times 10^{11}$ s$^{-1}$) [52];
2) CH$_2$OH is more electron-donating than H;
3) OH (at C$_\gamma$ or further removed from the reaction center (C$_\alpha$/C$_\beta$)) *reduces* $k_{het}$ (–I effect); and
4) the cyclopentyl system is less 'electron rich' than the cyclohexyl.

## X = Cl$^\cdot$

A further example of Eqs 6 and 7 has been studied using the time-resolved conductance technique [53]. It involves the addition to alkenes of Cl$^\cdot$, e.g. Scheme 1, in which the intermediate $\beta$-chloroalkyl radical undergoes C–Cl heterolysis with $k_{het} = 3.1$–$3.5 \times 10^4$ s$^{-1}$ to give the short-lived ($\tau < 20$ ns) alkene-type radical cation >C$^+$–C$^\cdot$H$_2$ which is trapped by the nucleophile water yielding >C$^\cdot$–CH$_2$OH/ HO–C$|$–CH$_2^\cdot$ and H$^+$. It is noteworthy that in this system, Cl$^-$ is a much better leaving group than SO$_4{}^{2-}$ (see Table 1, entry 2), as reflected by the ratio of the two heterolysis rate constants, $k_{het}$(Cl$^-$)/$k_{het}$(SO$_4{}^{2-}$) $\geq 300$ [53]. Further examples of C–Cl heterolysis reactions will be presented in Section 9.2.2.

**Table 1.** The dependence of $k_{het}{}^a$ on alkene structure for the reaction depicted by Eq. 11.

| $>\!\overset{\mid}{\underset{\mid}{\text{C}^\bullet\!-\!\text{C}}}\!-\!OSO_3{}^-$ | $k_{het}$ (s$^{-1}$) at 20 °C |
|---|---|
| $HOCH_2C^\bullet HCH_2OSO_3{}^-$ | $<10^2$ |
| $Me_2C^\bullet CH_2OSO_3{}^-$ | $\leq 10^3$ |
| $Me_2C^\bullet CH(OSO_3{}^-)CH_2OH$ | $4 \times 10^4$ |
| (cyclopentane ring, H, H, $OSO_3{}^-$) | $9.3 \times 10^3$ |
| (cyclopentane ring, $CH_2OH$, H, $OSO_3{}^-$) | $2.9 \times 10^5$ |
| (cyclohexane ring, H, H, $OSO_3{}^-$) | $3.0 \times 10^{4\,b}$ |
| (cyclohexane ring, H, H, $OSO_3{}^-$, HO) | $1.4 \times 10^3$ |
| (cyclohexane ring, H, H, $OSO_3{}^-$, OH) | $8.4 \times 10^3$ |
| (cyclohexane ring, Me, H, $OSO_3{}^-$) | $\geq 5 \times 10^7$ |
| (cyclohexane ring, Me, H, $OSO_3{}^-$, OH) | $4 \times 10^7$ |

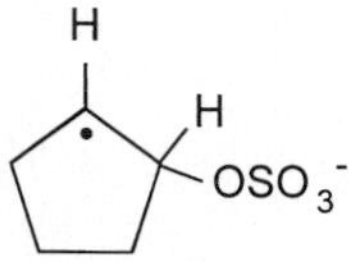

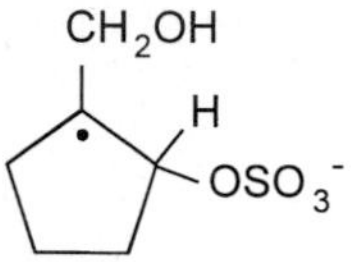

For the cyclohexane entry ($3.0 \times 10^{4\,b}$): $\Delta H^{\#} = 17$ kJ mol$^{-1\,c}$, $\Delta S^{\#} = -103$ J mol$^{-1}$ K$^{-1\,c}$

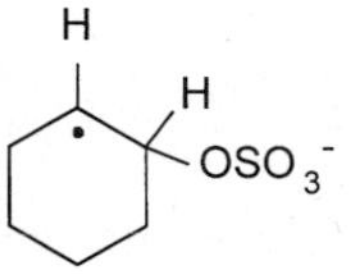

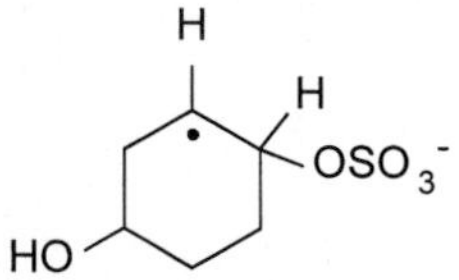

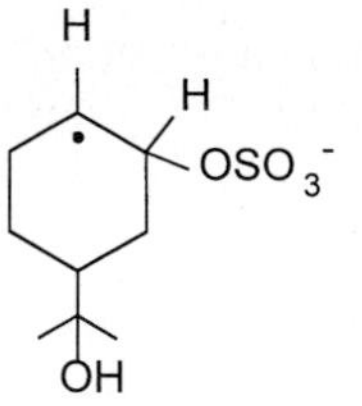

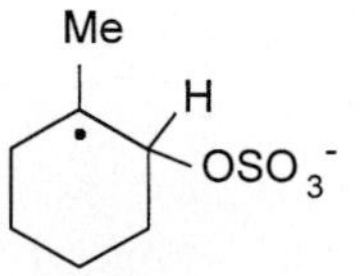

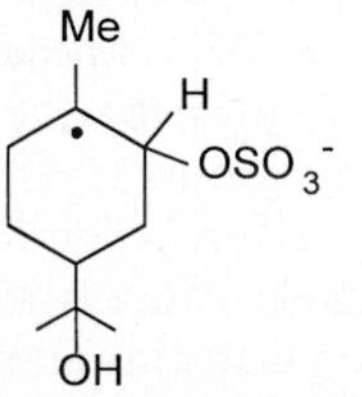

**Table 1** (*continued*)

| $>C^{\cdot}-\underset{\vert}{\overset{\vert}{C}}-OSO_3^-$ | $k_{het}$ (s$^{-1}$) at 20 °C |
| --- | --- |

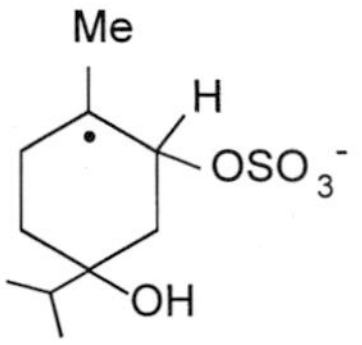

$2 \times 10^7$

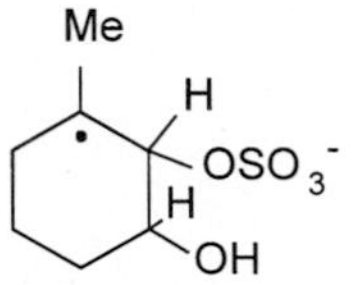

$8.4 \times 10^4$

[a] Measured using the time-resolved AC or DC (time resolution ≤ 10 ns) conductance technique. In all cases the concentration of the alkene was varied to make sure that the *formation* of the $SO_4^-$ adduct was *not* rate-limiting.

[b] The $\beta$-sulfatoalkyl radical was produced independently from the cyclohexylsulfate by abstraction of H by OH$^{\cdot}$ (see Eq. 5).

[c] The activation parameters were determined by measuring the rate constants in the temperature range 5–50 °C (in 5° steps). Good Arrhenius behavior was observed.

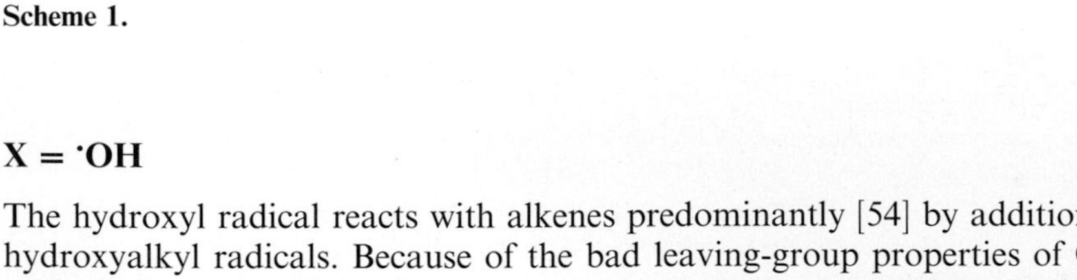

**Scheme 1.**

## X = $^{\cdot}$OH

The hydroxyl radical reacts with alkenes predominantly [54] by addition to give $\beta$-hydroxyalkyl radicals. Because of the bad leaving-group properties of OH$^-$, these radicals do *not* usually undergo spontaneous $\beta$-elimination of OH$^-$. If, however, the leaving group OH is changed by protonation into the better leaving group $H_2O$, radical cations can be produced, indicating that electron transfer has proceeded. Examples of this type of reaction will be given in Section 9.2.2, because the $\beta$-hydroxyalkyl radicals involved are more readily accessible by H-abstraction reactions from aliphatic systems.

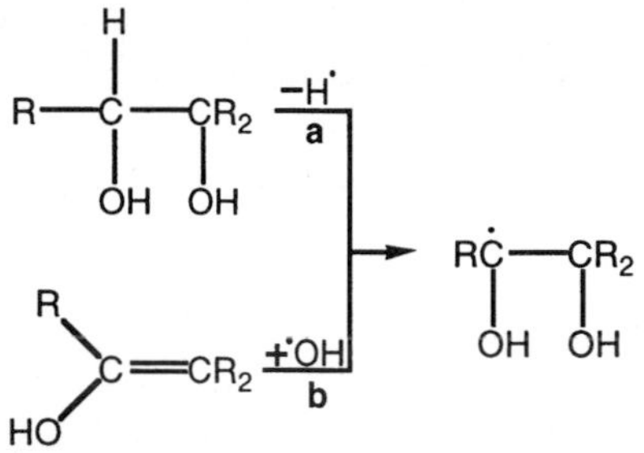

**Scheme 2.**

## 9.2.2 S = X–C–C–H (activation by H-abstraction)

### X = OH

The reactions of polyhydric alcohols with the hydroxyl radical in aqueous solution have been extensively studied (e.g. in radiolytic and biomimetic systems), mainly because of their suitability as models for more complicated carbohydrate substrates [55] or enzymatic systems involving glycol-type radicals [56, 57]. Because there are no double bonds to which •OH could add, only H-abstraction reactions are possible. Because the C–H bond energy is significantly lower than the O–H bond energy, it is the carbons from which H are abstracted and not the alcohol function. In this type of reaction, $\alpha,\beta$-dihydroxyalkyl radicals are formed. The same radicals could, in principle, be produced by *addition* of •OH to enols, see Scheme 2, lower part. This shows the complementarity of H-abstraction and OH-addition and thereby the relevance of the former to one-electron oxidation of olefinic bonds (Scheme 2).

A particularly important finding, first revealed by ESR studies on ethane-1,2-diol (R = H) at low pH [58, 59], and confirmed by product analysis [55, 60] and pulse-radiolysis studies [61], is that the initial $\alpha,\beta$-dihydroxyalkyl radical from this substrate undergoes an acid-catalyzed dehydration reaction to give the carbonyl-conjugated radical •CH₂CHO. It has been suggested [61, 62] on the basis of a wide range of experiments that this process occurs via a radical-*cation* (Scheme 3).

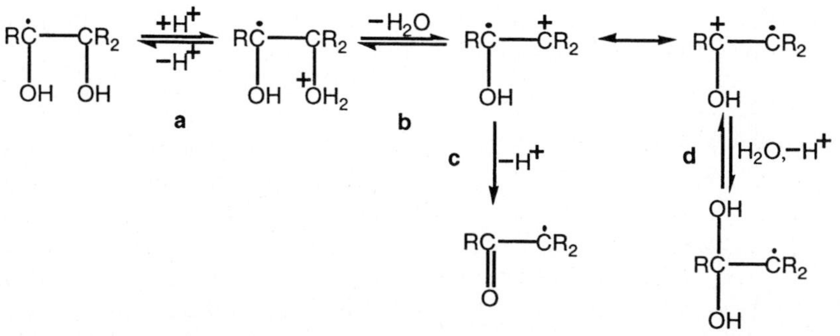

**Scheme 3.** H⁺-catalyzed dehydration.

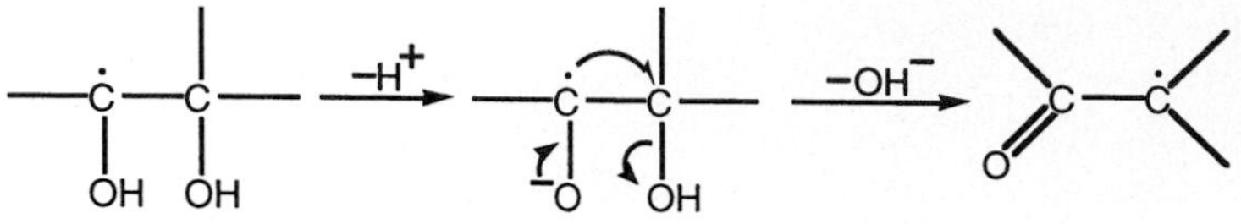

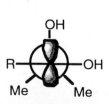

**Scheme 4.** OH$^-$-catalyzed dehydration.

A related base-catalyzed reaction has been shown to give the same carbonyl-con-jugated radical, via a radical–*anion* intermediate [61, 63] (Scheme 4), and rate con-stants for the loss of OH$^-$ (Scheme 4) have been determined [63].

The presence of alkyl substituents can markedly increase the rate of the acid-catalyzed reaction [64] (cf. second-order rate constants for dehydration of the radicals from ethane-1,2-diol [61] and butane-2,3-diol [65] of $9 \times 10^6$ and $1.3 \times 10^8$ M$^{-1}$ s$^{-1}$, respectively). Alkyl substituents also increase the rate of loss of OH$^-$ in their ionized counterparts (Scheme 4) [63].

To rationalize the effect of structure and substituents on the acid- or base-catalyzed loss of water from $\alpha,\beta$-dihydroxy-substituted radicals, the complementary techniques of pulse-radiolysis and ESR spectroscopy have been employed in a kinetic study of the dehydration of a variety of $\alpha,\beta$-dihydroxyalkyl radicals [·CR$^1$(OH)CR$^2$R$^3$OH] into the corresponding carbonyl-conjugated radicals [·CR$^2$R$^3$C(O)R$^1$]. The overall rates of proton-catalyzed dehydration, as revealed by steady-state (ESR) and time-resolved (pulse-radiolysis) experiments, indicate the importance of the electronic effects of substituents (contrast values of $1.2 \times 10^9$ and $9.8 \times 10^8$ M$^{-1}$ s$^{-1}$ for the radicals from cyclohexane-1,2-diol and butane-2,3-diol, respectively, with that for the radicals from erythritol of $4.2 \times 10^6$ M$^{-1}$ s$^{-1}$). Time-resolved experiments enable information to be obtained about the generation of the protonated species [·CR$^1$(OH)CR$^2$R$^3$OH$_2^+$] and the loss of water from this intermediate.

For ·CR(OH)CMe$_2$OH (R = H and Me) a rapid *un*catalyzed dehydration reac-tion occurs, $(k = 1–2 \times 10^4$ s$^{-1})$, which is believed to be assisted by the steric effect of the *gem*-dimethyl group and the polarity of the solvent. For the latter sub-strate, the reaction is characterized by a strongly negative activation entropy $(-93$ J mol$^{-1}$ K$^{-1})$. These results have been interpreted in terms of Schemes 5 and 6 [66], the latter providing an explanation of the negative activation entropy of the spontaneous dehydration in terms of 'freezing' of water molecules in the transition state, because of solvation of the (incipient) H$^+$ and OH$^-$ ions:

**Scheme 5.**

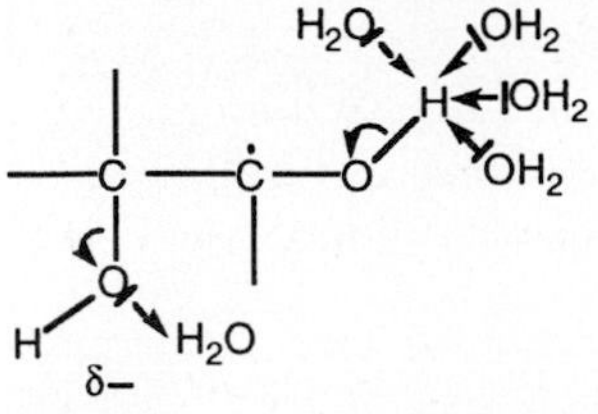

**Scheme 6.**

Mechanistically, the acid-catalyzed dehydration, Scheme 3, involves conversion of a *bad* nucleofugal leaving group, i.e. $OH^-$, into the much better one, $H_2O$. This difference is (quantitatively) reflected in the corresponding $pK_a$ values, i.e. $pK_a(H_3O^+) = -1.7$ compared with $pK_a(H_2O) = 15.7$. Qualitatively, protonation of the $\beta$-OH group increases the *pull* on the $C_\beta$–O bond, such that the $H_2O$ molecule is 'pulled out'. The opposite of this happens when the $\alpha$-OH group is *de*protonated by a Brønsted base such as $OH^-$, as in the *base*-catalyzed dehydration [63], Scheme 4. Here, the increased electron density at $C_\alpha$ caused by deprotonation at $C_\alpha$–OH finally ends up *pushing* the OH group out from $C_\beta$. The rate constants for elimination of $OH^-$ from the ionized 1,2-dihydroxyalkyl radicals $R_1C\cdot(O^-)CH(OH)R_2$ are between $3 \times 10^6$ s$^{-1}$ (for $R_1 = R_2 = H$ and $R_1$, $R_2 = H$ or hydroxyalkyl) and $\geq 8 \times 10^6$ s$^{-1}$ (for $R_1 = R_2 = Me$). Clearly, the *spontaneous* dehydration, Scheme 6, should contain elements of both push and pull.

It may not be surprising that $\alpha,\beta$-dihydroxyalkyl radicals are able to achieve one-electron reduction of oxidants such as benzoquinone or methyl viologen, with the rate constants strongly dependent on the nature of the substituents, whereby reducing power of the radical and its rate of dehydration are correlated with each other [66]. This indicates that the push-component might be more important than the pull-component.

## X = Oalk

A $\beta$-elimination reaction also occurs with linear and cyclic glycol mono-alkyl *ether* radicals. In this case the nucleofugal leaving group is the alkoxy function, RO, as shown in Scheme 7 for base catalysis and a cyclic system in which alkoxide elimination leads to ring-opening [63]:

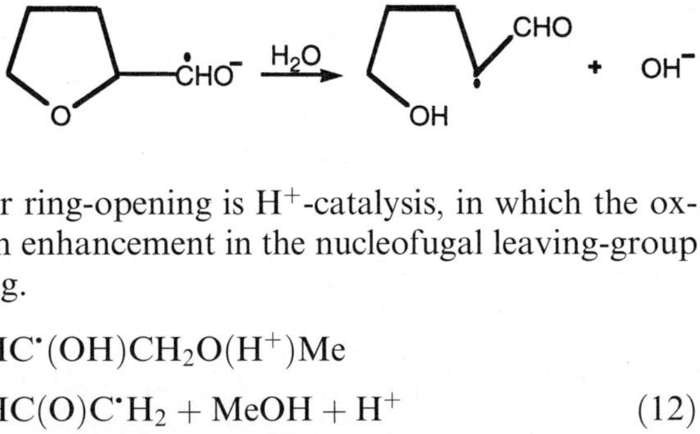

**Scheme 7.**

The reciprocal [67] mechanism for ring-opening is $H^+$-catalysis, in which the oxygen at $C_\beta$ is protonated, resulting in enhancement in the nucleofugal leaving-group properties of the alkoxy function, e.g.

$$HC\cdot(OH)CH_2OMe \overset{H^+}{\rightleftharpoons} HC\cdot(OH)CH_2O(H^+)Me$$

$$\longrightarrow HC(O)C\cdot H_2 + MeOH + H^+ \qquad (12)$$

As the leaving group properties of X are enhanced (equivalent to reducing the $pK_a$ values of the conjugate acids H–X) compared with X = OH or OR (Schemes 1–7), the rates of $\beta$-elimination increase considerably such that support of heterolysis by deprotonation of $C_\alpha$–OH or by protonation of X to achieve complete reaction tends to become unnecessary (see later).

## X = NH$_3$

Examples with $NH_3$ as the leaving group are $\alpha$-hydroxyalkyl-type radicals from $\beta$-amino alcohols [68] or from serine (R = H) or threonine (R = Me):

$$RC^{\bullet}(OH)CH(CO_2{}^-)\overset{+}{N}H_3 \rightarrow RC(O)C^{\bullet}HCO_2{}^- + NH_3 \tag{13}$$

$$R = H, \text{ Me}$$

When $R = $ Me (threonine), the breakage of the C–N bond proceeds from the *neutral* α-hydroxyalkyl, as in Eq. 13; however, with the less electron-rich serine radical ($R = $ H), elimination of $NH_3$ requires prior deprotonation from OH, i.e., the ketyl radical, $>C^{\bullet}–O^-$, is needed to *push* the $NH_3$ group out [69].

## X = OAc

An example of further increase in the acidity of H–X are β-acetato radicals [70]:

$$HOC^{\bullet}HCH_2OCOMe \rightarrow O{=}CHC^{\bullet}H_2 + MeCO_2{}^- + H^+ \tag{14}$$

where the rate of C–O heterolysis in aqueous solution at room temperature has been determined as $5.5 \times 10^5$ s$^{-1}$ [71–74].

## X = Hal

An acid of roughly similar strength as acetic acid is HF. In fact, elimination of F$^-$ from α-hydroxy-β-fluoroalkyl radicals has been observed [69], e.g.:

$$HOC^{\bullet}HCF_3 \rightarrow O{=}CHC^{\bullet}F_2 + H^+ + F^- \tag{15}$$

Moving from fluorine to chlorine, the acidity of H–X is drastically increased $(pK_a(HF) = 3.16, pK_a(HCl) = -7)$. As a reflection of this, the rate constant for $C_\beta$–Cl bond heterolysis in $HOC^{\bullet}HCH_2Cl$ in aqueous solution is very fast, $k_{het} \geq 10^6$ s$^{-1}$ [71].

As a matter of fact, Cl$^-$ is such an excellent nucleofugal leaving group (from $C_\beta$) that activation at the adjacent carbon (i.e., at $C_\alpha$) by an electron-donating substituent (such as OH) is not necessary for the reaction to occur rapidly and efficiently. For instance, the β-chloroalkyl radical formed by H-abstraction from 1-chloro-2-methylpropane, $Me_2C^{\bullet}CH_2Cl$ (this is the same radical as that formed in Scheme 1, Section 9.2.1), undergoes C–Cl heterolysis in aqueous solution with $k_{het} = 3.4 \times 10^4$ s$^{-1}$ [75].

$$Me_2C^{\bullet}CH_2Cl \xrightarrow[-Cl^-]{k_{het}} [Me_2C^{\bullet}CH_2{}^+] \xrightarrow{H_2O} Me_2C^{\bullet}CH_2OH + H^+ \tag{16}$$

This rate is ca $5 \times 10^6$ times larger [75] than that for the C–Cl heterolysis from the analogous ('classical') *non*-radical system, i.e. $Me_3CCl$, the cornerstone of $S_N1$ mechanistic chemistry. This demonstrates the dramatic acceleration of heterolysis *as a result of radical formation* at the site adjacent to the leaving group [76]. With this species, as with *t*-BuCl, the 'hydrolysis' is of the $S_N1$-type which means that the reaction proceeds via a cationic intermediate, a *radical* cation, which, however, was not *directly* seen, because its lifetime is too short.

Of the many interesting examples of these heterolysis reactions [75], that involving a *cyclic* system should be specifically mentioned, with $k_{het} \geq 10^6$ s$^{-1}$, as derived

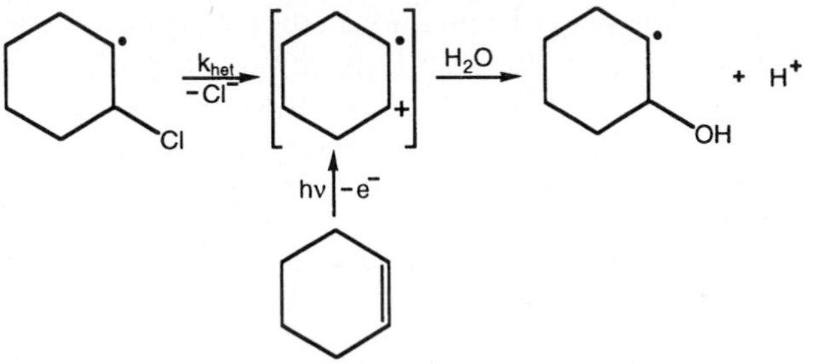

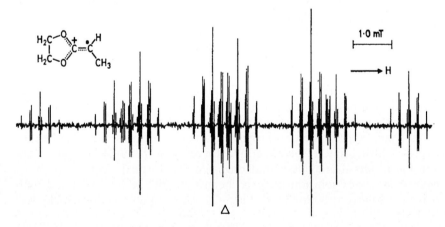

**Scheme 8.**

from time-resolved conductance measurements [77]. This is an example of the relevance of the olefin-type radical cation intermediate in one-electron oxidation of alkenes (Scheme 8).

It should be possible to make visible the radical cation-type intermediate postulated as in Eq. 16 by stabilizing the positive charge on the carbon skeleton. This was successfully done [78, 79] by introducing *two* alkoxy substituents at that position, e.g.:

$$(MeO)_2CHCH_2Cl \xrightarrow[-H_2O]{•OH} (MeO)_2C•CH_2Cl$$

$$\xrightarrow{k_{het}} (MeO)_2C^+C•H_2 + Cl^- \qquad (17)$$

with $k_{het} \geq 5 \times 10^7 \text{ s}^{-1}$ [80].

The ESR spectrum of a cyclic and methylated derivative of the enol ether-type radical cation formed in Eq. 17 is shown in Figure 3.

**Figure 3.** ESR spectrum of the 2-ethylidene-1,3-dioxolan radical cation; center indicated by a triangle. The radical cation has large $CH_3$ splittings; note, e.g., the well resolved second-order components. The size of the $CH_3$ coupling, 2.511 mT, indicates that the unpaired spin resides mainly on the carbon atom next to the $CH_3$ group. Experimental conditions: photolytic flow system, aqueous solutions of pH 5 at 3 °C containing 0.3 M acetone, 0.02 M $K_2S_2O_8$, and 0.03 M 2-(1-bromoethyl)-1,3-dioxolan. From Ref. [79].

Many of the radical cations studied [79] are very long-lived (ms range), because they decay only bimolecularly (by disproportionation or dimerization). Occasionally, however, a first-order reaction with water was observed, with $k$ between $10^2$ and $10^4$ $s^{-1}$ [79]; this involves (nucleophilic) *addition* of water to the radical cation, e.g.:

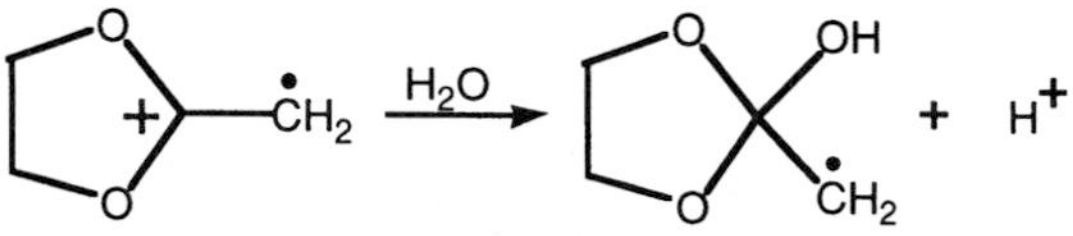

$$(18)$$

Nucleophiles stronger than water, e.g., $OH^-$ or $HPO_4^{2-}$, react with the radical cations, with $k = 0.2–6 \times 10^9$ or $2 \times 10^6$ to $3 \times 10^8$ $M^{-1}$ $s^{-1}$, producing alkoxy-carbonylalkyl radicals or $(RO)_2P(O)OC-C^\bullet H_2$-type radicals, respectively [81, 82]. These radical cations seem to be exceptional in so far as they show a relatively high reactivity with $O_2$ ($k > 10^8$ $M^{-1}$ $s^{-1}$) [79].

Heterolytic $\beta$-fragmentation reactions of the type in Eq. 17 have also been observed for the leaving groups $F^-$ and $Br^-$ [82]. For the latter, elimination is fast even when there is a methylene group in-between the radical site ($\equiv C_\alpha$) and the leaving group, i.e., if the reaction is a $\gamma$-elimination, e.g.:

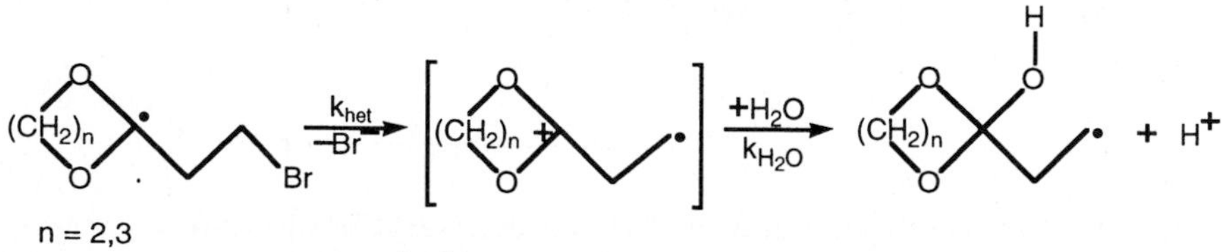

$$(19)$$

$$n = 2,3$$
$$k_{het} = 1.2 \times 10^5 \text{ s}^{-1}, \qquad k_{H2O} \approx 10^8 \text{ s}^{-1}$$

which gives rise to a *distonic* radical cation with a characteristically *short* lifetime compared with the *delocalized* type radical cation produced in, e.g., Eq. 17.

### $\alpha$-Mono*alkoxy-$\beta$-chloroalkyl radicals*

The reactions of these radicals in aqueous solution are particularly interesting, because of their model character with respect to deoxyribose-derived radicals in DNA [83], which lead to strand breaks of this macromolecule. These model reactions have been studied in detail [84], by use of a large number of substrates, with the help of in-situ photolysis ESR, time-resolved conductance, and product-analysis techniques. From the results it is evident that the primarily formed $\alpha$-alkoxy-$\beta$-chloroalkyl radicals in aqueous solution undergo heterolysis of the $\beta$-C–Cl bond with rates $k_{het} > 10^8$ $s^{-1}$ to give rise, finally, to the $\beta$-OH-substituted analogs which were identified by ESR.

$$CH_3O-C^\bullet HCH_2-Cl$$

$$(20)$$

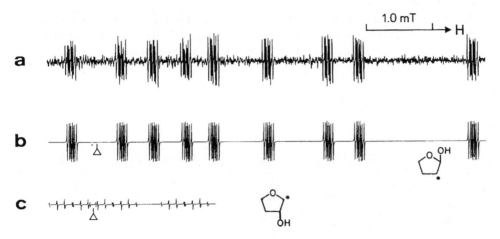

**Figure 4.** (a) High-field part of an ESR spectrum obtained after H-abstraction from 3-chlorotetrahydrofuran, 0.1 M, in aqueous solution, at pH 4, upon UV irradiation in a flow system at 276 K, 3 mM $K_2S_2O_8$, and 0.3 M acetone as sensitizer; (b) and (c) are simulated spectra calculated with a line width of 0.005 mT, centers marked by triangles [84].

Analogous results are shown in Figure 4 for the 2-yl radical derived from 3-chlorotetrahydrofuran.

From substituent and solvent effects on reactions such as Eq. 20 it was concluded [84] that these reactions are of the $S_N1$ type, i.e. that alkoxyalkene (enol ether) type *radical cations* are intermediates. The lifetimes of these radical cations were estimated [84] to be of the order of nanoseconds, *much* shorter than those [78, 79, 81] of the corresponding 1,1-*di*-alkoxyalkene radical cations. This shows the importance of the additional (second) alkoxy group in stabilizing the positive charge on the carbon skeleton. On the basis of these mechanistic model studies, very detailed suggestions could be made [84] regarding the deoxyribose-derived radical reactions that lead to chain breaks in DNA (see below).

## X = $OSO_n^-$ (n = 2, 3)

The results described in this chapter are complementary to those given above. On the basis of the acidity of sulfuric acid, the sulfonate group $ROSO_3^-$ is a leaving group at least as efficient as $Cl^-$ or $Br^-$. This is reflected in the rate constants for $C_\beta$–$OSO_3R$ heterolysis of alkyl radicals [75], e.g.:

$$MeC^{\bullet}HCH_2-Br \xrightarrow[-Br^-]{k_{het} = 7 \times 10^3 \text{ s}^{-1}}$$

$$MeC^{\bullet}HCH_2-OSO_3n-Pr \xrightarrow[k_{het} > 10^6 \text{ s}^{-1}]{-n-PrOSO_3^-} \longrightarrow [MeC^{\bullet}HCH_2^+ \leftrightarrow MeCH^+CH_2^{\bullet}]$$

$$\xrightarrow{H_2O} MeC^{\bullet}HCH_2OH + H^+ \quad (21)$$

On the basis of 193 nm photoionization experiments [45], the lifetime of the propene radical cation formed in the heterolysis step is <20 ns. As expected on the basis of the lower acidity of sulfonic acid compared with sulfuric, sulfonates are weaker leaving groups than sulfates, e.g.:

$$MeC^{\cdot}HCH_2-OSO_2Me \xrightarrow[-MeOSO_2^-]{k_{het}} [MeC^{\cdot}HCH_2^+]$$

$$\xrightarrow{H_2O} MeC^{\cdot}HCH_2OH + H^+ \tag{22}$$

$k_{het} = 2 \times 10^5$ s$^{-1}$ [75], compared with $k_{het} > 10^6$ s$^{-1}$ for the reaction depicted by Eq. 21.

The one-electron oxidation of cyclohexenes by $SO_4^{\cdot-}$ in aqueous solution has been studied from kinetic and stereochemical standpoints [49]. It was found that the alkene oxidation proceeds by an addition–elimination mechanism with $SO_4^{\cdot-}$ adding to the C=C double bond in the first step followed by $C_\beta$–$OSO_3^-$ heterolysis to give a solvent-separated alkene radical cation–sulfate ion pair ($S_N1$ mechanism) (Scheme 9).

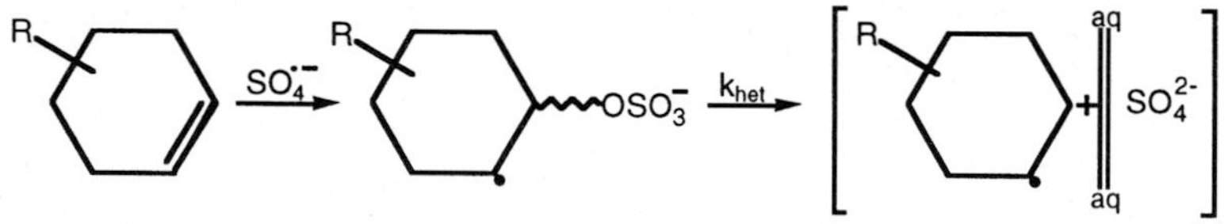

**Scheme 9.**

Attack of water on the radical cation occurs before the sulfate group has completely departed, the sulfate group hindering the approach of water from one side of the cyclohexene skeleton. The lifetime of the solvent-separated radical cation was estimated to be in the 10–100 ps range [49].

The rate constant for heterolysis of the '$SO_4^-$ adduct' depends on the nature of R (Scheme 9) and it increases with additional alkyl (methyl) groups on the double bond from $k_{het} = 3 \times 10^4$ s$^{-1}$ for R = H and *no* methyl groups to >$5 \times 10^6$ s$^{-1}$ for R = $Me_2COH$ and 1 or 2 methyl groups, in agreement with the $S_N1$-character of the reaction.

## X = OP(O)(OR)$_2$ (phosphate as leaving group) [85]

The reactions of alkyl radials carrying phosphate as substituent in the $\beta$-position are particularly interesting because they are close models for those deoxyribose radicals that cause strand breaks in DNA [83, 86]. Historically speaking, the compounds studied first were $\beta$-phosphato *alcohols*, where H-abstraction (by $^{\cdot}OH$) from C atoms leads to $\alpha$-*hydroxy*-$\beta$-phosphatoalkyl radicals, e.g. for glycerol-2-phosphate, to $HOC^{\cdot}HCH(OPO_3H^-)CH_2OH$ which was found to undergo a rapid elimination of (inorganic) phosphate ($k_{het} > 10^6$ s$^{-1}$) to yield a $\beta$-oxoalkyl radical, Eq. 23 [87, 88]:

$$HOC^{\cdot}HCH(OPO_3H^-)CH_2OH \xrightarrow[-H_2PO_4^-]{k_{het}} HC(O)C^{\cdot}HCH_2OH \tag{23}$$

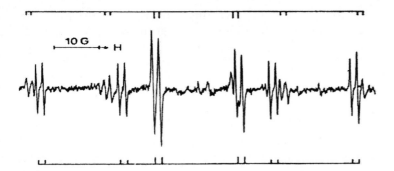

**Figure 5.** Spectra obtained on UV irradiation (at 2 °C) of an aqueous solution of $H_2O_2$ (3 %) in the presence of 0.04 M glycerol-2-phosphate, pH 1.6. Lines not assigned by the stick spectra are from impurity radicals. The spectrum assigned by the upper lines is from the *cis* form of HOCH₂C·HCHO; that assigned by the lower lines is from the *trans* form [88].

which exists as *cis* and *trans* forms, owing to partial double bond character around the $C_\alpha$–$C_\beta$ bond, as shown in Figure 5. It is likely that in this fragmentation reaction a radical cation is formed which should, however, be extremely short-lived because of very rapid deprotonation from the α-OH group.

To obtain an even better model compound for the corresponding deoxyribose-centered radical in DNA, *ether* phosphates were used instead of the *alcohol* phosphates described in reaction 23. For example, on reaction of ·OH with 2-methoxyethylphosphate at pH 4.5 the 2-yl radical is the most prominent species (see Scheme 10), as observed by ESR [89]. By measuring the coupling constants as a function of pH, the $pK_a$ of the radical was determined to be 6.5. On reducing the pH, the 2-yl radical disappeared giving rise to two product radicals, as shown below (Scheme 10).

This mechanism is analogous with that described for the 2-methoxyethyl chloride 2-yl radical (Eq. 20). By performing product-analysis studies under conditions of radical scavenging with $O_2$ it was possible to arrive at the rate constants for heterolysis of the C–OPO₃ bond as a function of the protonation state of the phosphate leaving group, from –OPO₃H₂ to –OPO₃H⁻ to –OPO₃²⁻. The rate constants

$$MeOC·HCH_2OPO_3{}^{2-} \overset{H^+}{\underset{pK=6.5}{\rightleftarrows}} MeOC·HCH_2OPO_3H^- \overset{H^+}{\underset{pK\approx1.5}{\rightleftarrows}}$$

$$MeOC·HCH_2OPO_3H_2 \xrightarrow[-H_2PO_4{}^-]{k_{het}} [MeOC·HCH_2{}^+]$$

$$\begin{array}{c} \longrightarrow MeOC·HCH_2OH \\ \overset{H_2O}{\underset{-H^+}{\phantom{x}}} \Big| \\ \longrightarrow MeOCH(OH)C·H_2 \end{array}$$

**Scheme 10.**

obtained (at $0\,^\circ$C) are $3 \times 10^6$, $10^3$, and $0.1–1$ s$^{-1}$, respectively [89]. $H_2PO_4^-$ is thus seen to be a *much* better (by a factor of ca $10^7$) leaving group than $PO_4^{3-}$, which reflects the difference between the p$K_a$ values of the conjugate acids, i.e., p$K_a$ ($H_3PO_4$) $= 2.1$, p$K_a$ ($HPO_4^{2-}$) $= 12.7$. The effect of alkylation of the phosphate group is similar to that of protonation, thus, for the relevant *model* reaction (that mimicking the C4$'$ radical of DNA (Scheme 11)) [83], i.e., the reaction which constitutes the breakage of the DNA chain, the rate $k_{het}$ is $10^3–10^4$ s$^{-1}$ [90]. Of course, it would be interesting to know the *rate* of this type of reaction *in DNA*. In this polymer, the deoxyribose and the phosphate groups protrude into the aqueous phase and, correspondingly, they are strongly hydrogen-bonded to the adjacent water molecules, but they are *not immersed* in the aqueous phase. Thus, their environment is not identical with that of bulk water. The expected effect (by analogy with the $\alpha$-methoxy-$\beta$-chloroalkyl radicals such as in Eq. 20, see also Ref. [73]) of a decrease of polarity of the environment is a *de*crease of the rate constant of heterolysis. This is, however, probably compensated by the effect of *alkylation* on $C_\alpha$ and $C_\beta$: Alkylation leads to a drastic ($\geq 2$ orders of magnitude) *in*crease of the heterolysis rate [89]. On the basis of the experiments of Giese et al [91], this type of reaction also occurs from the C4$'$ radical of a nucleotide containing the thymine moiety dissolved in acetonitrile, in which the rate constant $k_{het}$ can be estimated to be ca $1.8 \times 10^6$ s$^{-1}$ [92].

Going one step further in the direction of DNA one arrives at *oligo*nucleotides [86, 93] where, again, the C4$'$-type radical is found to undergo rapid $C_\beta$–OPO$_3^-$R heterolysis ($k_{het} \geq 10^2–10^3$ s$^{-1}$) [93] to yield the enol ether-type (deoxy)ribose-derived radical cation already suggested [83, 94] for the DNA *polymer*. Concerning the further fate of this radical cation, two major reaction channels have been suggested: (i) reaction as an *electrophile*, whereby the solvent water acts as the nucleophile, which attacks at either position of the 'olefinic' bond, see Scheme 11, and (ii) electron transfer from an 'adjacent' [95] guanine moiety in the DNA chain.

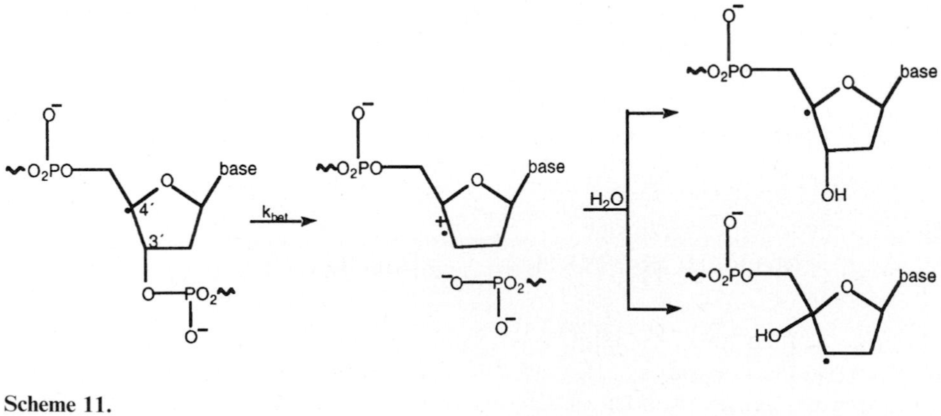

**Scheme 11.**

From scavenging experiments, the *rate* of electron transfer from a guanine moiety (G) in oligonucleotides to the enol ether-type radical cation (see below) can be estimated to be $\geq 10^6$ s$^{-1}$ [96]. This is apparently fast enough to compete with the electrophilic reaction of the radical cation with water. It is interesting that a *third* characteristic [62] reaction of enol ether-type radical cations, i.e. deprotonation to yield an allyl-type radical, Eq. 24, has (so far) *not* been observed [86, 94].

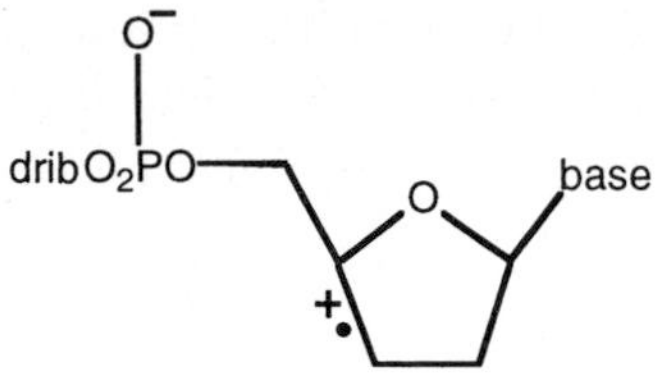

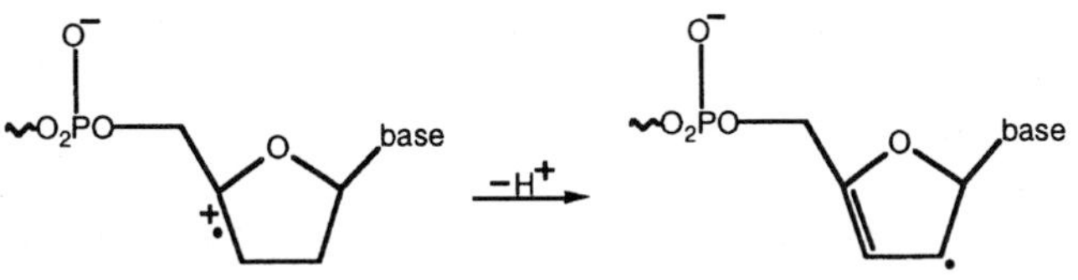

$$\xrightarrow{-H^+} \tag{24}$$

### 9.2.3  S = Aromatic (Ar)

## X = OSO$_3{}^-$

The strongly oxidizing radical $SO_4{}^{\bullet-}$ is often considered to be a 'natural' one-electron (outer-sphere) oxidant, Eq. 25:

$$SO_4{}^{\bullet-} + Ar \rightarrow SO_4{}^{2-} + Ar^{\bullet+} \tag{25}$$

mainly, because the final (non-radical) products of its reaction with aromatics are derived from the *radical cations* of the aromatics [97–99]. This mechanistic assumption is in fact supported by time-resolved studies in which the expected radical cations (or their characteristic neutral (radical-type) reaction products) could be identified [46, 100–102]. In principle, however, and by analogy to the olefinic systems discussed above (Section 9.2.1), there is the possibility that the reaction proceeds by an *inner*-sphere path, i.e.:

$$SO_4{}^{\bullet-} + Ar \rightarrow [^-O_3SO{-}Ar^\bullet] \xrightarrow{k_{het}} SO_4{}^{2-} + Ar^{\bullet+} \tag{26}$$

If this is the mechanism, to explain the results presently available [46, 100, 101] it is necessary to assume that $k_{het}$ (Eq. 26) $\geq 10^7$ s$^{-1}$. Rates of this order of magnitude are difficult to measure (under conditions of bimolecular formation of the hypothetical adduct $^-O_3SO{-}Ar^\bullet$), and it would therefore be desirable to *reduce* $k_{het}$, e.g. by decreasing the electron density of Ar. As long as experiments of this type have not been reported the question of the outer- or inner-sphere nature of oxidation of Ar by $SO_4{}^{\bullet-}$ remains unanswered.

In addition to reactions of $SO_4^{\cdot-}$ with *homo*cyclic Ar, those with *hetero*cyclic systems have also been studied [103–105]. For these systems also, however, solid evidence for adduct formation has *not* been obtained [106]. In sharp contrast to this are the results with the strong oxidant $\cdot OH$ (below).

## X = OH

The OH radical is (one of) the most reactive species around. The reaction of $\cdot OH$ usually leads to the one-electron *oxidation* of the substrate, so a common impression is that $\cdot OH$ reacts by electron transfer. This is understandable because thermodynamically the OH radical is a very powerful one-electron oxidant. Its reduction potential at pH 0 is 2.7 V relative to the NHE [107, 108], and at pH 7 it is still 2.3 V. That this number indicates strong oxidizing power is evident on comparing it with those of some well known oxidants such as $IrCl_6^{2-}$ (0.87 V) or $Tl^{2+}$ (2.2 V). Despite this, $\cdot OH$ does usually *not* react by electron transfer but by addition, not only with organic substrates (containing double bonds) [109], but also with anions [110] and even metal ions [111].

This tendency to add rather than to oxidize is probably a result of stabilization of the transition state for addition by contributions from bond *making*, whereas electron transfer requires pronounced bond and solvent reorganization with a correspondingly large free energy change to reach the transition state.

Addition of $\cdot OH$ to a substrate S leads to the '$\cdot OH$-adduct' $HO–S^\cdot$ (Scheme 12).

For $HO–S^\cdot$ to yield electron transfer products, heterolysis of the bond joining HO and Y must occur. Because $OH^-$ is a very bad leaving group (as evidenced by the high $pK_a$ (15.7) of its conjugate acid, $H_2O$), however, the rate of the spontaneous heterolysis, $k_{het}$ (Scheme 12, upper part), is very low (very often $\ll 10^2$ $s^{-1}$). As a consequence, the final (non-radical) products from $\cdot OH$ reactions with S are typically derived from dimerization or disproportionation of $HOS^\cdot$.

One-electron oxidation of S by $\cdot OH$ is, however, possible if the leaving group capacities of the adduct components, $HO–$ and $S^\cdot$ are changed. As for aliphatic systems (Scheme 3), protonation of $HO–$ to give $H_2O^+–$ converts the bad leaving group $OH^-$ into the excellent $H_2O$ ($pK_a(H_3O^+) = -1.7$). If it is assumed that the Brønsted catalysis law is applicable and that the Brønsted coefficient, $\alpha$, is equal to 0.5, a rate enhancement of $10^{8.7}$ induced by protonation of the leaving group $–OH$ is calculated from the difference in the $pK_a$ values of $H_2O$ and $H_3O^+$. A somewhat similar number ($10^7$) is obtained by considering the difference in the reduction potentials of $\cdot OH$ at pH 7 and at pH 0 [4].

Protonation is a very effective means of improving the *nucleo*fugacity of the leaving group OH. It is not restricted to *radical* systems, as shown below (Scheme 13).

The alternative method, which can be even more efficient, is to improve the *electro*fugacity of $S^\cdot$. A means of achieving this is to put electron-donating sub-

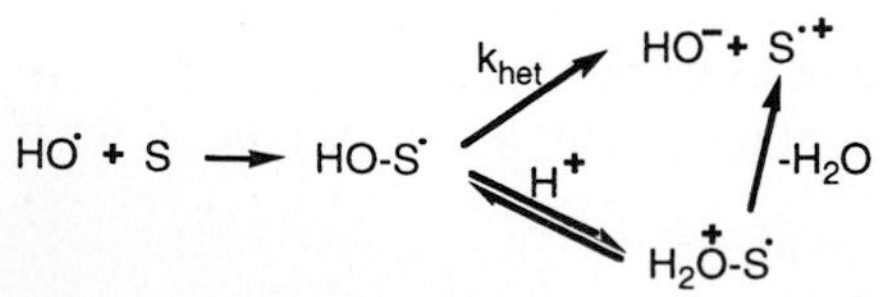

Scheme 12.

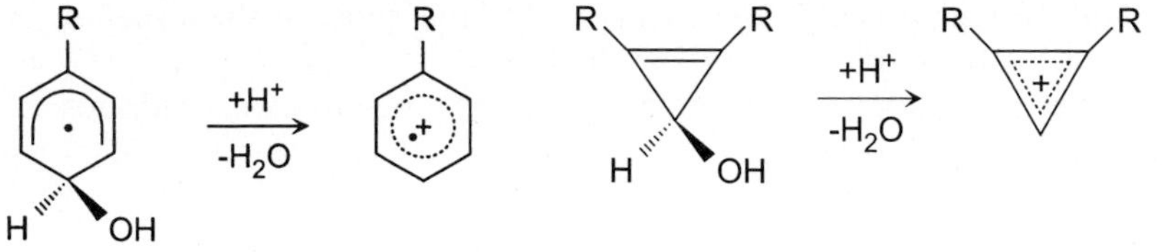

**Scheme 13.**

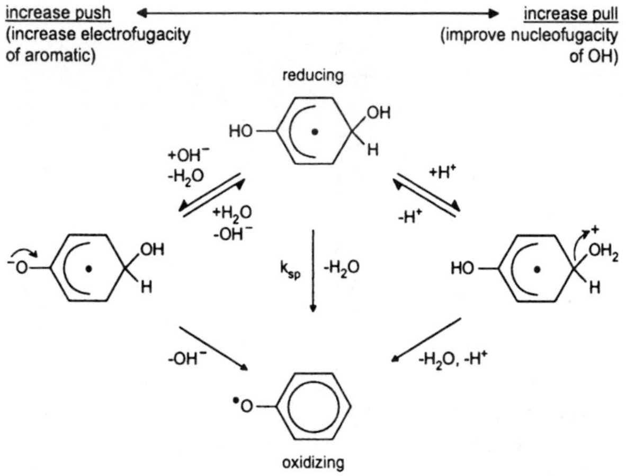

**Scheme 14.**

stituents into S·. An elegant method (of practical importance) is to increase the electron density on S· by ionization of a substituent which is a Brønsted acid. An example of this, which serves also to summarize the mechanisms of H+- and OH−-supported dehydration, is shown in Scheme 14. This scheme is analogous to Schemes 3 and 4 which represent the situation in the 'aliphatic–olefinic world' [112].

Another example relates to OH−-aided one-electron oxidation of cytosine. With cytosine, the OH reaction proceeds by addition to C5, a process that has a selectivity of 90 % [113]. The 5-hydroxy-6-yl radical is an excellent *reductant,* and the same is true for the ionized 6-yl radical formed by deprotonation from N1. This radical anion now contains sufficient electron density to eliminate the OH group at C5 as OH−. (This is analogous to Scheme 4, as an example from the aliphatic world). The result is the cytosine-1-yl radical which is *oxidizing,* probably because of appreciable spin density on the hetero atoms N1 and O2 [113].

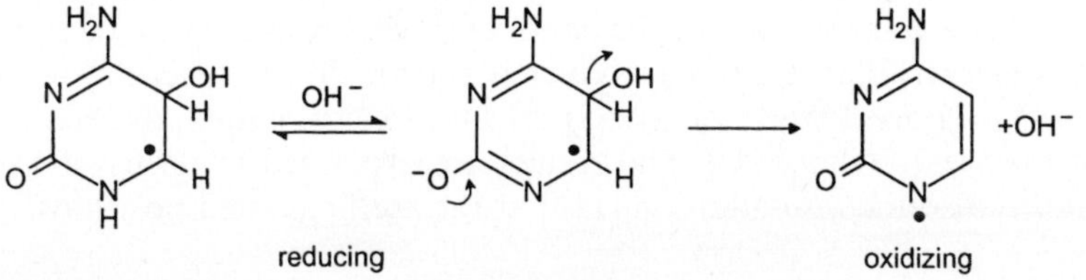

(27)

Schemes 14 and 15 and Eq. (27) are examples for the general phenomenon of 'redox-inversion' [4] by dehydration of OH adducts.

Another example from heterocyclic chemistry is the one-electron oxidation of guanine-derivatives by the OH-radical.

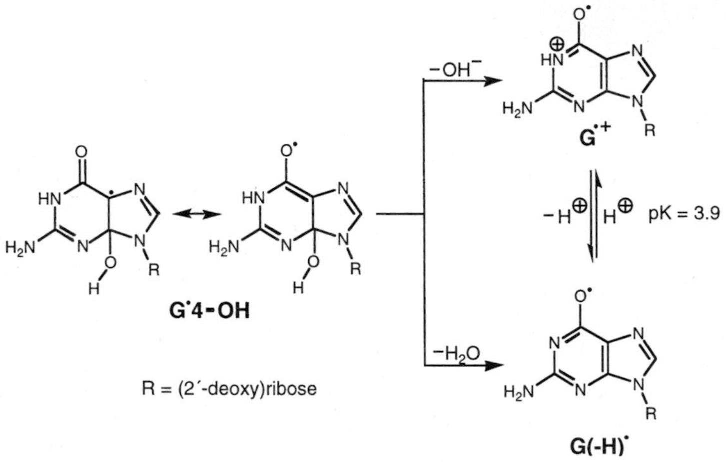

**Scheme 15.**

The reaction between ˙OH and phenol (see Scheme 14) can be analyzed in terms of its thermochemistry. On the basis of the potential at pH 7, $E_7(˙OH) = 2.3$ V relative to the NHE and $E_7(PhO˙) = 0.97$ V [114], the formation of PhO˙ and $H_2O$ via an electron-transfer mechanism is exothermic by 1.33 V = 31 kcal mol$^{-1}$. Despite this, the reaction proceeds by addition, as outlined in Scheme 14. Thus, again, the propensity of ˙OH to add rather than to oxidize can be understood in terms of the transition state for addition being stabilized by contributions from bond *making*, in contrast to electron transfer which requires pronounced bond and solvent reorganization which results in a large (entropy-caused) free energy change.

It might be expected that not only the rate of addition of ˙OH [115] to the substrate S but also, particularly, the rate of heterolysis of HO–S˙ will increase with increasing electron density of S. So, as one goes to the extreme of *very electron-rich* aromatics, the ad–el mechanism might become indistinguishable from an outersphere ET mechanism. Examples where one is close to this situation are the *tetra*-alkoxybenzenes **a** and **b** (Scheme 16).

On the basis of pulse radiolysis experiments in aqueous solution [45], the rate constant for reaction of ˙OH with **a** has the very high value $1.5 \times 10^{10}$ M$^{-1}$ s$^{-1}$. If performed in *basic* solution (pH ≈ 10), an *increase* of conductance is observed which results from the formation of OH$^-$ (this ion is neutralized in *acidic* solution by H$^+$ which leads to a *decrease* of the conductance). From the *rate* of conductance increase it is evident that if an OH adduct is formed in the reaction, it eliminates OH$^-$ with $k_{het} > 1.5 \times 10^6$ s$^{-1}$ (to give rise to the radical cation, as seen by optical detection), the yield being 47 %. The results are explained in terms of Scheme 17 which contains also the *ipso*-addition [116] at the methoxylated positions.

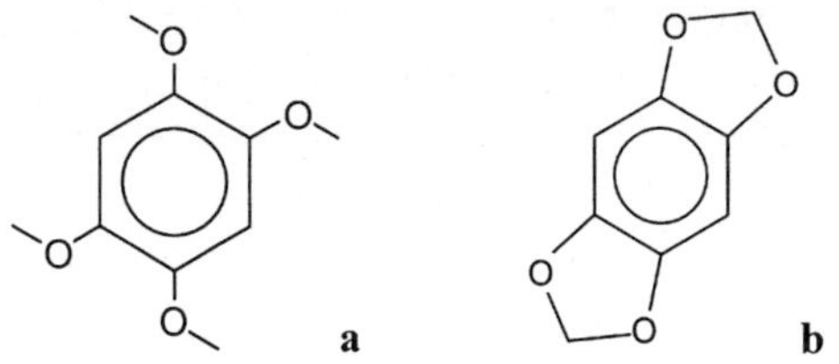

**Scheme 16.**

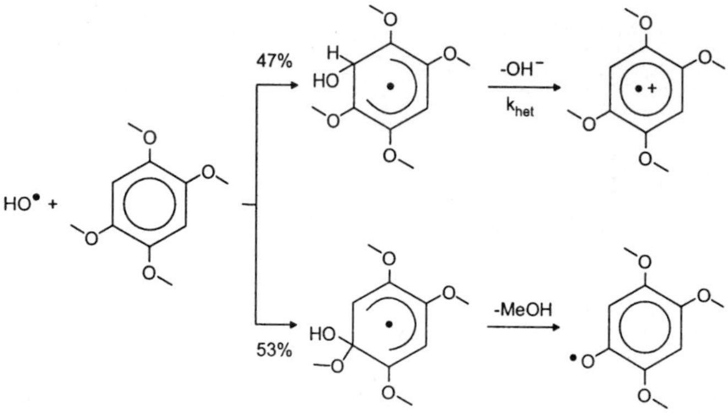

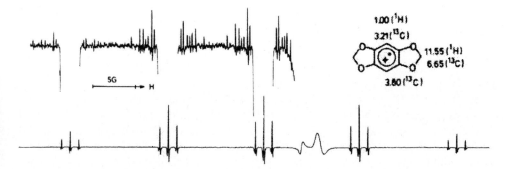

**Scheme 17.**

The ESR spectrum of the radical cation of the tetraalkoxy-type benzene **b**, as produced by reaction with ·OH at pH 2, is shown in Figure 6.

An analogous, possibly even more extreme example has recently been studied by time-resolved Raman spectroscopy [117]; it was concluded that addition of ·OH [118] and elimination of OH⁻ from all the ring positions are equally fast. This in effect means that the reaction is indistinguishable from an outer-sphere ET (Scheme 18) [119].

**Figure 6.** ESR spectrum recorded on reaction of ·OH with 0.2 mM compound **b** in an $N_2O$-saturated aqueous solution at pH 2 and ca 5 °C. The experimental spectrum from the four equivalent methylene hydrogens and the two equivalent aromatic hydrogens is shown in the lower part; the upper part contains the lines from $^{13}C$ in natural abundance. S. Steenken, unpublished material.

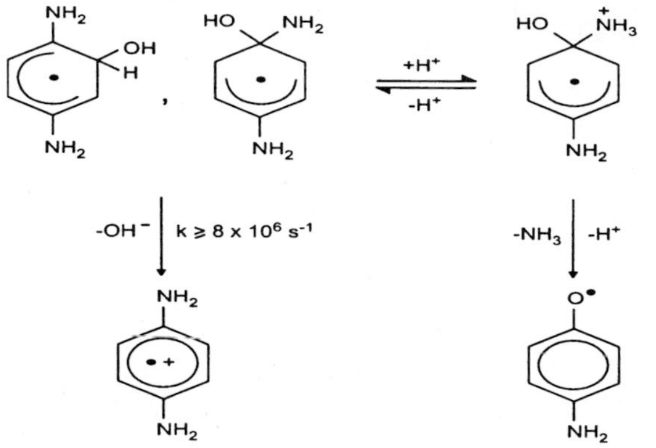

**Scheme 18.**

Other electron-rich systems are 1,3-dioxols and dioxenes. With the former, ˙OH adds to the 4,5-C–C bond to give α-alkoxy-β-hydroxyalkyl radicals, Scheme 19. With the fully methylated system (left side of Scheme 19), at pH 2 the OH adduct is quantitatively converted (by $H^+$-induced dehydration) to the radical cation, as judged by ESR, whereas with only hydrogens as substituents (right side), at pH 2 dehydration is too slow to lead to a visible decrease of the (stationary) concentration of the OH adduct (Scheme 19) [45].

Even a small increase in the electron density of the system, by introducing a methylene group, i.e., by going from the 1,3-dioxol to the 1,4-dioxene system, leads, however, to complete $H^+$-induced conversion of the OH adduct into the radical cation (Figure 7 and Scheme 20) [45].

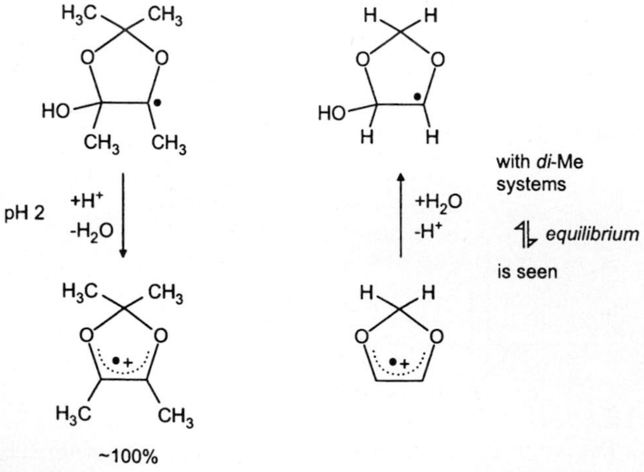

**Scheme 19.**

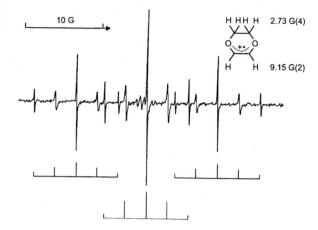

**Figure 7.** ESR spectrum observed on photolysis of an aqueous 0.1 M $H_2O_2$ solution containing ca 1 mM 1,4-dioxene at pH 2. In the quintet, there are line-broadening effects caused by partial non-equivalence of the four methylene protons.

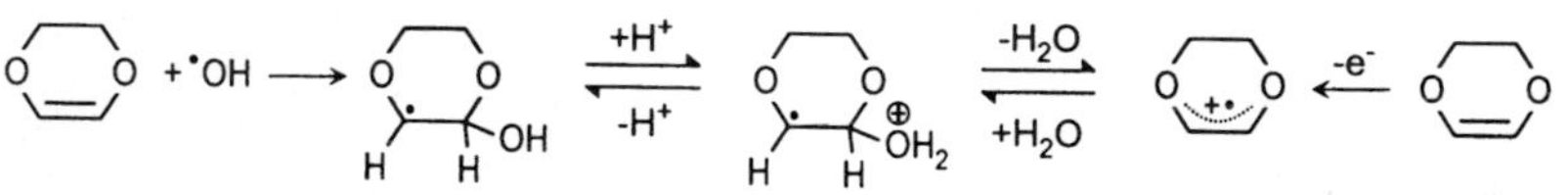

**Scheme 20.**

## 9.3 Intramolecular Electron Transfer with Bond Formation

There are so far only a few examples belonging to this category. By use of pulse radiolysis with optical and with ESR detection it was found [45] that the reaction of ˙OH with 4-nitrobenzaldehyde in aqueous solution leads quantitatively to the radical *anion* of 4-nitrobenzoate, which is explained in terms of H-abstraction from the CHO function followed by rapid addition of water ($k \geq 1.2 \times 10^6$ s$^{-1}$) (Scheme 21).

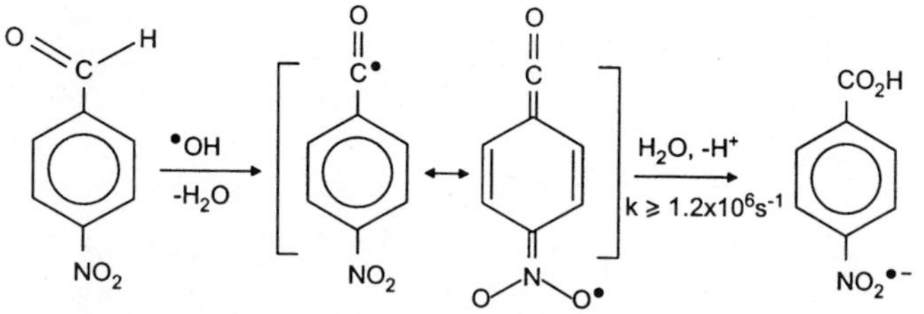

**Scheme 21.**

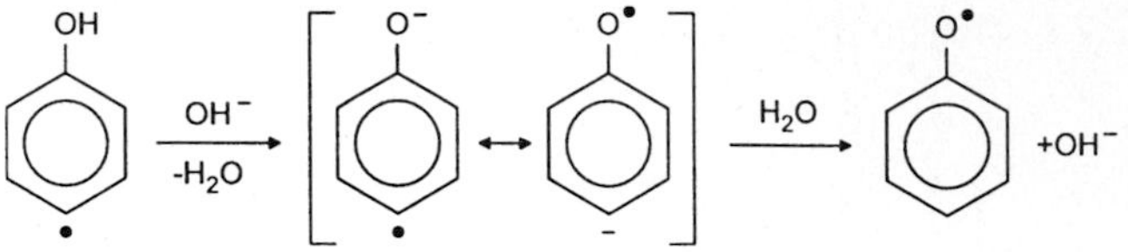

**Scheme 22.**

Possibly, the reaction is driven by the clectrophilicity of the ketenic and nitroxide-type mesomer. In effect, it leads to the *oxidation* of the aldehyde and the *reduction* of the nitro function.

Another example of solvent-assisted intramolecular ET has been reported by Schuler [120]. It involves the transformation of a ($\sigma$-type) phenyl radical into a much more stable ($\pi$-type) phenoxyl radical (Scheme 22).

Protonation by water of the ring carbon occurs only if the phenol function is *de*-protonated, apparently because only then is the electron density at the ring sufficient for protonation by the weak acid $H_2O$.

In Scheme 22 another reaction is described in which a C–H bond is *formed* by proton transfer from the solvent ($H_2O$) to a carbon atom of the substrate. For this reaction to occur, a *high* electron density is required at the site of protonation. This reaction is the opposite to proton-transfer reactions that electron *deficient* species (e.g. radical cations) undergo, namely proton transfer *from a carbon* **to** the solvent water, as discussed in Section 9.5.1. There are some interesting examples, of biological relevance, of proton transfer **to** carbon from heterocyclic systems (Schemes 23 and 24):

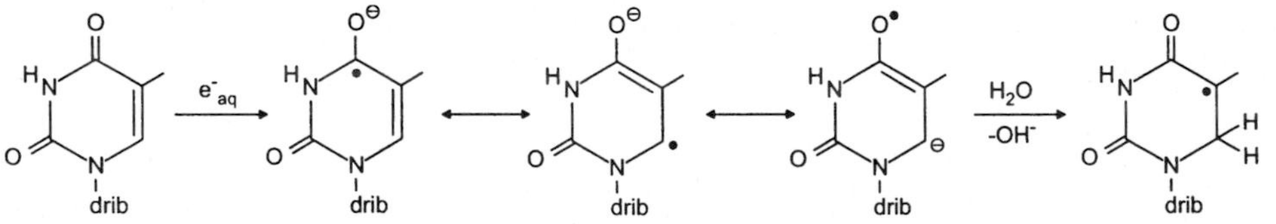

**Scheme 23.** Electron addition to thymidine followed by protonation on carbon [121, 122].

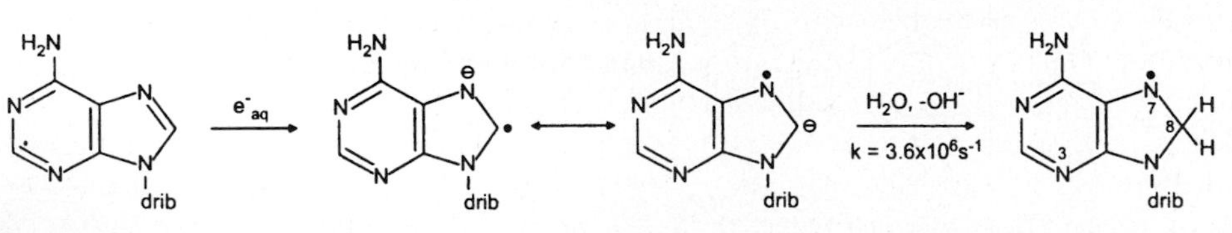

**Scheme 24.** Electron addition to 2′-deoxyadenosine [123, 124].

## 9.4 Reduction of S by X˙, X = E–C˙< (E = Hetero Atom)

The examples so far described always involve a carbon-centered radical which is substituted at $C_\alpha$ by a hetero atom such as O, S, or N. S is always a functional group that contains a double bond, usually an oxygen, i.e. the group =O. X˙ then adds to S according to:

$$X^\cdot + S \xrightarrow{k_{ad}} X\text{–}S^\cdot \xrightarrow{k_{het}} X^+ + S^{\cdot-} \tag{4b}$$

As in the opposite (or complementary) case (Eq. 4a), the mechanism thus involves the formation of X–S˙ by covalent bond formation between X˙ and S, with the second-order rate constant $k_{ad}$, followed by heterolysis of X–S˙ with the first-order rate constant $k_{het}$, to give X$^+$ and S$^{\cdot-}$. As described above, the kinetic condition for the detection of X–S˙ is $k_{het} < k_f$ [S] (under the (usual) condition that [X˙] ≪ [S]). If $k_{het} > k_{ad}$ [S], the reaction will appear to be 'non-bonded' (i.e. outer-sphere ET). Since $k_{ad}$ is bimolecular (i.e. $10^9$ M$^{-1}$ s$^{-1}$) and almost always [S] < 1 M, $k_{het}$ values ≥ $10^9$ s$^{-1}$ are not usually experimentally accessible.

The conditions to be fulfilled for the heterolysis of X–S˙ to be rapid can easily be defined: the reduction potential of X$^+$, E(X$^+$/X˙) should be negative whereas that of S, E(S/S˙$^-$) should be positive. A prediction is that above some (large) value for the *difference* in redox potentials of X$^+$ and S the species X–S˙ will be difficult to detect, because of the rapid rate of its heterolysis. Under these conditions the reaction will *seem* to proceed by outer-sphere electron transfer. Again, as in Eq. 4a, although X–S˙ is a radical and the overall reaction results in the transfer of a *single* electron, in the actual electron-transfer step an electron *pair* is transferred (heterolysis or electron pair transfer), rather than a single electron. Examples—in many cases involving molecules relevant for biological systems—of reduction of S by carbon-centered radicals substituted at $C_\alpha$ by a hetero atom, in most cases O or N, have already been discussed in detail [53]. The reader is referred to this literature.

## 9.5 Intramolecular Electron Transfer Accompanied by Bond Cleavage in Radical Ions

Although a previous chapter in this volume provides a broader perspective on the reactivity of radical cations, in this section we will examine intramolecular electron-transfer reactions coupled with or followed by cleavage of a bond in odd electron species such as radical cations, radical zwitterions and radical anions. In particular, this paragraph will be divided in oxidative and reductive bond-cleavage processes. Because this field is however too large to be covered extensively here, the discussion will be limited to selected examples—for oxidative cleavages, side-chain fragmentation reactions of alkylaromatic radical cations and decarboxylation reactions of radical zwitterions derived from benzoic and arylalkanoic acids, and for reductive

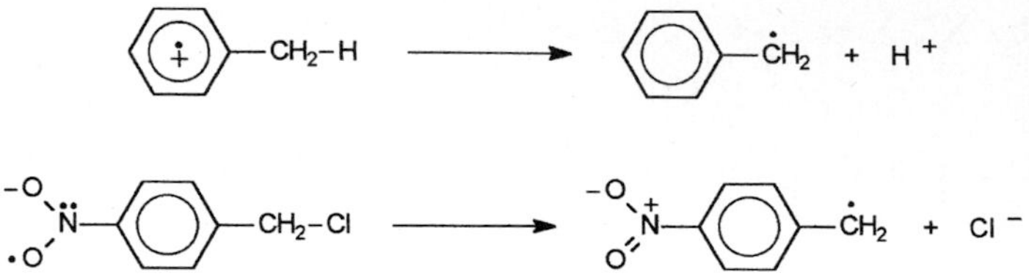

Scheme 25.

cleavages, fragmentation reactions of radical anions, such as those derived from aryl and benzyl halides and 1,2-diphenylethanes.

The occurrence of intramolecular electron transfer in radical ion bond cleavage reactions is probably not straightforward and in this respect, two examples may help the reader—C–H deprotonation in the toluene radical cation and C–Cl bond cleavage in the 4-nitrobenzyl chloride radical anion (Scheme 25).

In the toluene radical cation the electron hole is delocalized over the $\pi$-system and the deprotonation reaction is coupled with intramolecular electron transfer from the scissible bond to the aromatic ring, leading to the benzyl radical. In the 4-nitrobenzyl chloride radical anion, the unpaired electron resides in a $\pi^*$ orbital that does not belong to the leaving group and C–Cl bond cleavage occurs simultaneously with an intramolecular electron transfer from this orbital to the $\sigma^*$ orbital of the scissible bond.

### 9.5.1 Oxidative Bond-cleavage Processes

#### Alkylaromatic radical cations

The one-electron oxidation of side-chain-substituted aromatics to give alkylaromatic radical cations often leads to oxidative transformation of the side-chain even though the primary oxidation step involves removal of an electron from the aromatic part of the molecule. Side-chain reactions of alkylaromatic radical cations mostly involve the cleavage of bonds between the $\alpha$ and $\beta$ atoms, because the considerable overlap achievable between these bonds and the $\pi$-orbitals of the aromatic ring results in their significant weakening as compared with the neutral substrate. Many side-chain reactions are thus possible depending on the nature of the bond being cleaved—C–H bond cleavage is the most common fragmentation pattern of these reactive intermediates but fragmentation of C–C and C–X bonds (X = S, Si, Sn) can also occur. This discussion will be limited to the cleavage of side-chain C–H and C–C bonds in aromatic radical cations [125].

Usually the initially generated radical cations have the electron hole delocalized over the $\pi$-system and the fragmentation reaction is accompanied by an intramolecular electron transfer from the scissible bond to the aromatic ring, ultimately resulting in two formal ways of electron apportionment between the fragments, which depends on their relative thermodynamic stability. In the *heterolytic* mode the charge is transferred through the scissible bond while in the *homolytic* mode the spin is transferred across the bond being cleaved (Scheme 26) [126]. Such cleavage modes have been named *mesolytic* to account for both mechanistic possibilities [127].

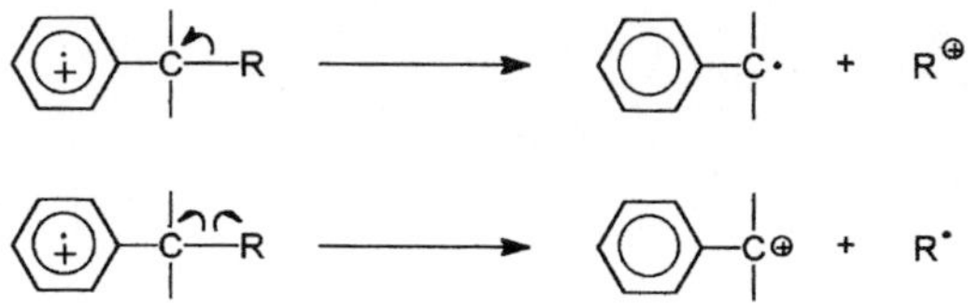

**Scheme 26.**

The free energy for radical cation bond cleavage, $\Delta G_f$, can be estimated by means of simple thermodynamic cycles [126, 128], where, regardless of the cleavage mode, $\Delta G_f$ depends on the free energy of homolysis of the radical cation precursor ($\Delta G_h$) and the difference between the reduction potentials ($\Delta E$) of the radical cation and the ionic fragment formed—$\Delta G_f = \Delta G_h - \Delta E$.

The rate of side-chain fragmentation of an alkylaromatic radical cation can be influenced by the relative orientation of the scissible bond and the aromatic $\pi$ system (stereoelectronic effect). The orientation most suited for cleavage is that where the dihedral angle between the plane of the $\pi$ system and the plane defined by the scissible bond and the atom of the $\pi$ system to which this bond is connected is 90°. Scheme 27 shows the conformation most suited for C–H bond cleavage.

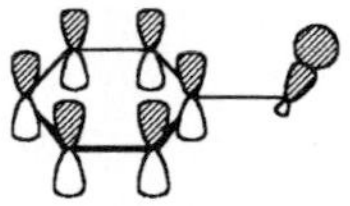

**Scheme 27.**

In such conformation the scissible bond is collinear with the $\pi$ orbitals of the aromatic system bearing the unpaired electron, and the best orbital overlap for intramolecular ET, required for bond cleavage, can be achieved. Interestingly, when competition between C–H and C–C bond cleavage is possible, stereoelectronic effects generally play a significant role, because the former process is depressed owing to its steric requirements, which are much smaller than those of a C–C bond. In such circumstances, therefore, the conformation with the latter bond collinear with the $\pi$ system is the most favored (see later).

*C–H bond cleavage*

The most simple and common side-chain reaction of alkylaromatic radical cations is the cleavage of a $C_\alpha$–H (from now on simply indicated as C–H) bond. These reactions have features of great theoretical interest, because C–H bond cleavage must be accompanied by extensive electronic reorganization, because one of the electrons of the $\sigma$ C–H bond has to be transferred to the aromatic $\pi$ system. In solution such a cleavage generally occurs via the heterolytic mode, owing to the relative stability of the benzyl radical and the large negative free energy of solvation of the proton (in water $\Delta H(H^+_{gas} \rightarrow H^+_{soln}) = 260$ kcal mol$^{-1}$ [129]). Accordingly, alkylaromatic radical cations behave as carbon acids with a carbon radical as the conjugate base (Scheme 28; where B is a Brønsted base, including the solvent).

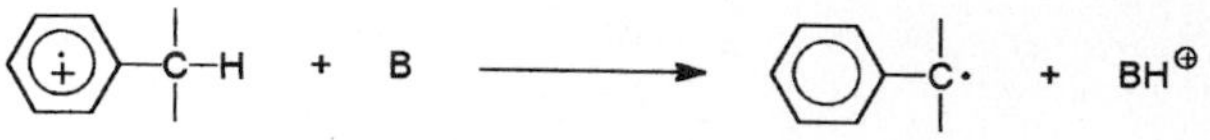

Scheme 28.

$pK_a$ values for alkylaromatic radical cations can be calculated by the use of appropriate thermodynamic cycles and by this approach a $pK_a$ value between $-11$ and $-13$ has been estimated in acetonitrile for the deprotonation of the toluene radical cation [130]. On the basis of extrapolation to this radical cation, the acidities of methylbenzene radical cations have been estimated and show the expected decrease in acidity on increasing radical cation stability; this is reflected in the values of $E°_{Ar}$ which progressively decrease with increasing number of methyl substituents. Accordingly, the difference of 440 mV in $E°_{Ar}$ between *p*-xylene and hexamethyl-benzene [131], corresponds to $pK_a$ values for the two radical cations of $-8$ and $+1$ respectively [132]. The same trend is observed for the kinetic acidity, the rate constants for deprotonation of methylbenzene radical cations in acidic aqueous solution decreasing by three orders of magnitude as the number of methyl groups increases from one to five (Table 2) [133, 134].

When considering the rate constants for deprotonation reported in Table 2 one should, however, take into account that in methylbenzene radical cations with $\leq 3$ methyl groups, nucleophilic attack of water on the aromatic ring can compete efficiently with side-chain deprotonation. An elegant explanation which accounts for this competition has been provided for the toluene radical cation on the basis of the three-electron three-orbital three-configuration approach [135, 136]. Three electrons are involved in the deprotonation reaction—the unpaired electron delocalized over

**Table 2.** Rate constants for the deprotonation of methylbenzene radical cations in acidic aqueous solution.

| Radical cation | $k \ (\mathrm{s}^{-1})^a$ | $k \ (\mathrm{s}^{-1})^b$ |
|---|---|---|
| Toluene | $1.0 \times 10^7$ | $\geq 5 \times 10^{7c}$ |
| *o*-Xylene | $2.0 \times 10^6$ | $1.3 \times 10^7$ |
| *m*-Xylene | $2.0 \times 10^6$ | $3.9 \times 10^7$ |
| *p*-Xylene | $1.4 \times 10^6$ | $2.1 \times 10^{6d}$ |
| Mesitylene | $1.5 \times 10^6$ | $\geq 5 \times 10^7$ |
| Hemimellitene | $1.5 \times 10^6$ | $\geq 5 \times 10^7$ |
| Isodurene | $1.0 \times 10^5$ | – |
| Durene | $2.7 \times 10^4$ | $4.3 \times 10^4$ |
| Pentamethylbenzene | $1.6 \times 10^4$ | – |

[a] Radical cations generated by pulse radiolysis (PR) (Section 9.2.3), pH $\leq 3$. Conditions under which exclusive side-chain deprotonation is reported to occur [133].
[b] Radical cations generated by laser-flash photolysis (LFP), pH $= 4.5$. Conditions under which water addition to the radical cation competes with side-chain deprotonation [134].
[c] Addition $> 90\%$.
[d] Deprotonation $> 60\%$.

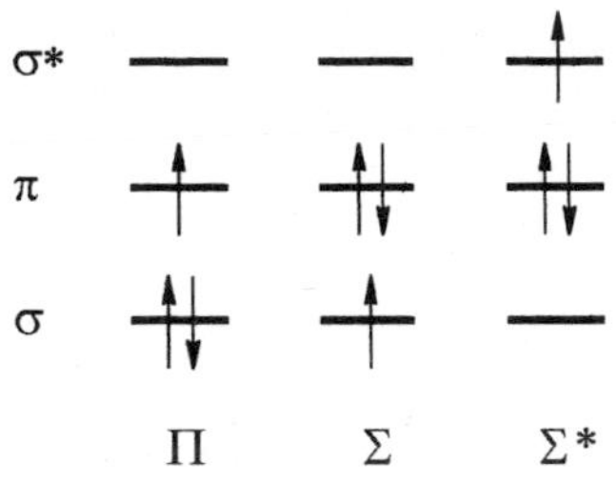

$$\sigma^* \quad \underline{\phantom{x}} \quad \underline{\phantom{x}} \quad \underline{\uparrow}$$
$$\pi \quad \underline{\uparrow} \quad \underline{\uparrow\downarrow} \quad \underline{\uparrow\downarrow}$$
$$\sigma \quad \underline{\uparrow\downarrow} \quad \underline{\uparrow} \quad \underline{\phantom{x}}$$

**Scheme 29.**        $\Pi \qquad \Sigma \qquad \Sigma^*$

the aromatic ring and the two $\sigma$ electrons of the scissible C–H bond, while the orbitals of interest are the SOMO of the phenyl ring ($\pi$) and the bonding and antibonding orbitals of the C–H bond ($\sigma$ and $\sigma^*$).

Three configurations are important for description of the cleavage (Scheme 29): $\Pi$, representing the reactant radical cation, and $\Sigma$ and $\Sigma^*$ which make little contribution to the ground state of the radical cation but become important as the C–H bond elongates, accounting for the significant weakening of this bond compared with neutral toluene.

According to this picture the toluene radical cation should react predominantly by nucleophilic attack at the aromatic ring, side-chain deprotonation being a minor reactive pathway, in full agreement with the experimental results (Table 2). Interestingly, substitution of electron-donating groups at the *para* position will stabilize the $\Pi$ state more than $\Sigma$ and $\Sigma^*$, leading to an overall decrease in reactivity which is, however, associated with an increase in the relative importance of the deprotonation pathway, as clearly shown by comparing the experimental results for toluene and *p*-xylene radical cations in Table 2.

The kinetic acidity of alkylaromatic radical cations has received great attention. Savéant and his associates studied the C–H deprotonation of NADH analog radical cations and showed that the intrinsic *kinetic* acidities of the radical cations do not correlate with the *thermodynamic* acidities while a good correlation is instead observed between the intrinsic barrier and the homolytic C–H bond dissociation energy [137–139]. Thus, in such radical cations the dynamics of proton transfer are governed by the homolytic cleavage of the C–H bond, suggesting that C–H deprotonation of an aromatic radical cation can be better described as a concerted H-atom electron transfer rather than a direct proton transfer. Intrinsic barriers as high as 0.5–0.6 eV have been determined; these are comparable with those generally observed for acid–base reactions involving carbon acids [140, 141]. The behavior of NADH analog radical cations can be explained in terms of the mesomeric structures shown in Scheme 30 for the transition state of proton transfer from the radical cation to a pyridine base, where the positive charge is delocalized from the departing proton and is mostly borne by the nitrogen atom.

Whereas within the precursor complex the presence of the base does not influence the covalent state of the radical cation when the C–H bond elongates, the ionic state is highly stabilized by the presence of the base as the C–H distance increases [137]. This model implies that formation of the N–H bond does not interfere to a large extent in the dynamics of the reaction, the deprotonation of the radical cation

covalent                    ionic                **Scheme 30.**

involving three concerted steps—homolytic cleavage of the C–H bond, electron transfer, and heterolytic formation of the N–H bond, and proceeding through an early transition state in which the N–H distance is larger than the C–H distance [139].

Interestingly, such a model leads to similar results when applied to the deprotonation reactions of methylarene radical cations by a series of pyridine bases [132] and to the deprotonation of the $(4\text{-MeOC}_6\text{H}_4)_2\text{NCH}_3$ radical cation by a series of quinuclidine bases [142].

An early transition state has also been suggested for the side-chain deprotonation of alkylaromatic radical cations in the seminal studies of Kochi [132, 143, 144] and Baciocchi [145]. Kochi determined the rates of proton transfer from various methylarene radical cations to substituted pyridine bases in acetonitrile generating the radical cations by oxidation with *tris*(phenanthroline)iron complexes [132], or charge-transfer excitation of the electron donor–acceptor complex of the methylarenes with tetranitromethane [146]. The former method leads to an indirect evaluation of the rate constants for deprotonation of the radical cations whereas the latter enables direct measurement by observation of the radical cation by means of time-resolved techniques. The deprotonation rate constants follow a general Brønsted relationship leading to $\beta$ values between 0.30 and 0.15 [143, 144]. If this free energy relationship for proton transfer is considered in the context of the Marcus equation, the magnitude of these Brønsted coefficients corresponds to an overall free energy change lying in the exoergonic region of the driving force. The same conclusion is obtained from deuterium kinetic isotope effects which decrease by increasing driving force (base strength). These observations are in line with an early transition state in which proton transfer has not proceeded beyond the symmetrical situation. Such a qualitative description is also in agreement with the relatively low sensitivity of proton transfer rates to the steric effects of the pyridine bases. The Marcus approach leads to an intrinsic barrier of 0.6 eV for deprotonation of methylarene radical cations by substituted pyridine bases [147].

Similar results have been obtained by Baciocchi for the deprotonation of $\alpha$-substituted 4-methoxytoluenes by 2,6-lutidine and $\text{NO}_3^-$ in acetonitrile [145]. In this study, the same values of the Brønsted coefficient ($\alpha = 0.24$), and of the deuterium kinetic isotope effect ($k_\text{H}/k_\text{D} = 2.0$ for 4-methoxytoluene radical cation) have been obtained with the two bases; these results point again towards a highly asymmetric transition state with a very small amount of C–H bond cleavage. Moreover, values of 0.53 and 0.66 eV have been calculated for the intrinsic barrier of the reactions of the radical cations with $\text{NO}_3^-$ and 2,6-lutidine, respectively, again comparable with those observed for acid–base reactions involving carbon acids [140, 141].

**Table 3.** Relative reactivities for the deprotonation of $4\text{-MeOC}_6H_4CH_2X$ and $4\text{-MeC}_6H_4CH_2X$ radical cations.

| X | $4\text{-MeOC}_6H_4CH_2X^{\cdot+a}$ | | $4\text{-MeC}_6H_4CH_2X^{\cdot+b}$ $k(CH_2X)/k(CH_3)$ |
|---|---|---|---|
| | $(NO_3^-)$ | (2,6-Lutidine) | |
| CN | 183 | – | 9.3 |
| OMe | 20.5 | 13.3 | >100 |
| OH | 15 | 19 | >100 |
| OAc | 3.4 | 11.7 | 1.8 |
| Me | 2.4 | 2 | 14.1 |
| H | 1 | 1 | 1 |
| D | 0.5 | 0.5 | – |
| tBu | – | – | 0.01 |

[a] Intermolecular, from LFP in MeCN, $T = 22\,°C$ [145].
[b] Intramolecular, from anodic oxidations in AcOH–MeCN (3:1), $T = 25\,°C$ [148].

Useful information about the structure of the transition state for C–H deprotonation can be obtained by comparing the relative reactivities of two series of alkylaromatic radical cations $4\text{-MeOC}_6H_4CH_2X^{\cdot+}$ [145] and $4\text{-MeC}_6H_4CH_2X^{\cdot+}$ [148] (Table 3). In both instances all $\alpha$-substituents (with the exception of X = tBu, see later) increase the deprotonation rate compared with X = H, although there are some discrepancies between the two series of radical cations—$\alpha$-substituents of the +R type (OH, OMe, Me) have a much larger kinetic effect in the deprotonation of $4\text{-MeC}_6H_4CH_2X^{\cdot+}$ than in that of $4\text{-MeOC}_6H_4CH_2X^{\cdot+}$, whereas the opposite behavior is observed with a –I, –R substituent such as CN.

It is possible to interpret these results in terms of the model proposed by Savéant for the deprotonation of NADH analog radical cations, where the dynamics of the deprotonation process in the radical cation are controlled by the homolytic C–H bond dissociation energy [137, 139].

The transition state for the deprotonation of an alkylaromatic radical cation might be described in terms of the mesomeric structures in Scheme 31. In structure **I** the positive charge is delocalized on the aromatic ring whereas structures **II** and **III** represent the homolytic and heterolytic C–H bond cleavage modes, respectively.

If the homolytic bond-dissociation energy (BDE) is the main factor governing the dynamics of proton transfer, the transition state is better represented by structures **I** and **II**. In an alkylaromatic radical cation the build-up of positive charge at the $\alpha$-carbon requires extensive charge delocalization from the aromatic ring and this is likely to occur with greater efficiency in 4-Me rather than 4-MeO substituted radical

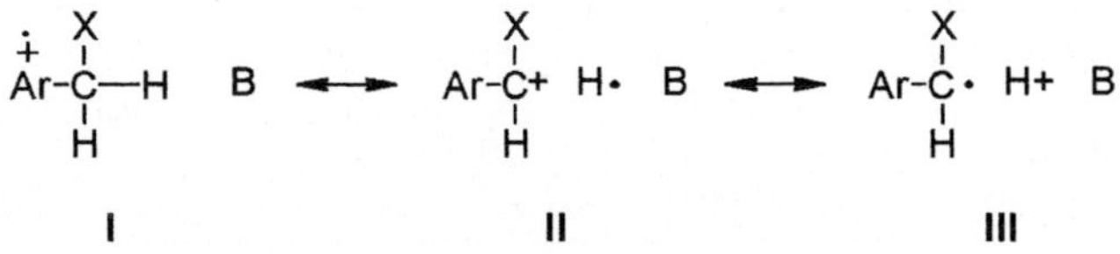

**Scheme 31.**  **I**  **II**  **III**

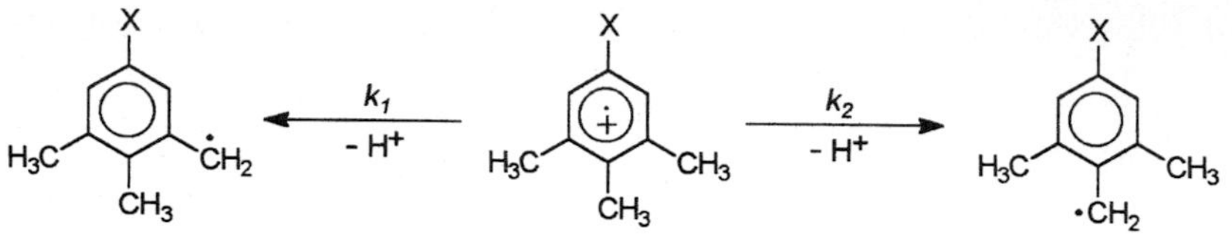

**Scheme 32.**

cations, because the latter group is very effective in stabilizing a positive charge. In $4\text{-MeOC}_6\text{H}_4\text{CH}_2\text{X}^{\bullet+}$, the favorable kinetic effect of $+R$ $\alpha$-substituents in reducing the homolytic BDE might be compensated by the presence of the 4-MeO group which opposes charge transfer from the ring to the $\alpha$-carbon, thus explaining the higher relative reactivities observed with $4\text{-MeC}_6\text{H}_4\text{CH}_2\text{X}^{\bullet+}$. On the other hand, whereas the $\alpha$-CN group is again expected to reduce the homolytic BDE, resulting in an increase of the deprotonation rate [145], this favorable effect is likely to be felt less in the reaction of $4\text{-MeC}_6\text{H}_4\text{CH}_2\text{CN}^{\bullet+}$, because of the larger fraction of positive charge residing on the $\alpha$-carbon in the transition state.

The intramolecular selectivity in the deprotonation of alkylaromatic radical cations has been investigated in a series of 5-X-1,2,3-trimethylbenzene radical cations; the results showed that spin and/or charge density at the scissible C–H bond can strongly influence the kinetic acidity, which is, accordingly, very sensitive to the nature and position (*meta* or *para*) of ring substituents (Scheme 32) [149].

Very high selectivity $(k_2/k_1)$ is observed in the presence of $+R$ substituents (X = Br, Me, OAc, *t*-Bu, OMe), deprotonation at the 2-methyl (*para* to the X group, the site bearing the largest charge/spin density) being strongly favored over that at the methyl groups in the 1 and 3 positions (Table 4).

As remarked previously, the deprotonation rate of an alkylaromatic radical cation can be influenced by the stereoelectronic effect (Scheme 27). In this respect, Tolbert provided convincing evidence for the operation of stereoelectronic effects in the deprotonation of 9-alkylanthracene and 9,10-dialkylanthracene radical cations

**Table 4.** Intramolecular selectivities $(k_2/k_1)$ in the deprotonation of 5-X-1,2,3-trimethylbenzene radical cations generated by anodic or CAN-promoted oxidation in acetic acid.

| X | $k_2/k_1$ | |
|---|---|---|
| | CAN | Anodic |
| COOMe | 2.0 | 1.5 |
| H | 3.4 | 3.0 |
| *t*-Bu | 24 | 16 |
| Br | 42 | 31 |
| Me | 55 | 37 |
| OAc | 110 | 59 |
| OMe | >200 | – |

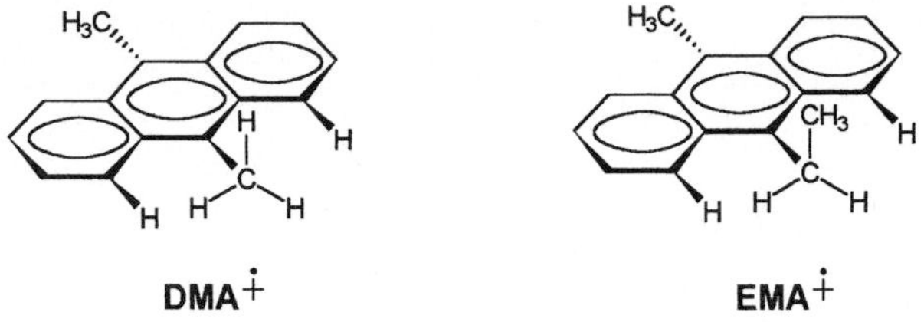

**Scheme 33.**                          DMA$\overset{+}{\cdot}$                          EMA$\overset{+}{\cdot}$

[150, 151]. Whereas the 9,10-dimethylanthracene (DMA) radical cation undergoes side-chain deprotonation leading to the oxidized product 9-hydroxymethyl-10-methylanthracene [152], the 9-ethyl-10-methylanthracene (EMA) radical cation forms 9-ethyl-10-hydroxymethylanthracene as the exclusive deprotonation product and no deprotonation products are observed with the 9,10-diethylanthracene (DEA) radical cation [151], despite the relatively high thermodynamic acidity of these species ($pK_a = -6$ for DMA$^{\cdot+}$ [150]). These observations can be explained in terms of stereoelectronic control of the deprotonation reaction—whereas in DMA$^{\cdot+}$ the C–H bond can easily assume the conformation required for deprotonation, where the scissible C–H bond is collinear with the $\pi$ orbitals of the aromatic system, in the presence of an ethyl substituent as in EMA$^{\cdot+}$ the analogous conformation is prevented by the unfavorable interactions with the *peri* hydrogens, which force the CH$_2$–CH$_3$ bond in a conformation perpendicular to the anthracene plane (Scheme 33, in which the unpaired electron is not shown).

Interestingly, when the C–H bond is forced into a conformation in which it is almost collinear with the $\pi$-system as in 1,9-ethanoanthracene (EA), a ratio for deprotonation of 28:1 is observed between the ethano group in EA$^{\cdot+}$ and the methyl group in 9-methylanthracene radical cation [150].

Additional information on the role of stereoelectronic effects comes from the anodic oxidation of $\alpha$-substituted *p*-xylenes [148], and from the oxidation of alkyl- and 1,4-dialkylbenzenes with a variety of one-electron oxidants [153]. The *p*-neopentyltoluene radical cation, generated by anodic oxidation, undergoes deprotonation almost exclusively at the methyl group, a behavior which has been attributed to the greater stability of the conformation having the *t*Bu group perpendicular to the plane of the aromatic ring (Scheme 34, structure **I**) compared with that suitable for C–H bond cleavage from the CH$_2$*t*Bu group (structure **II**), which is destabilized by the interaction of the *t*Bu group with the *ortho* hydrogens of the ring [148].

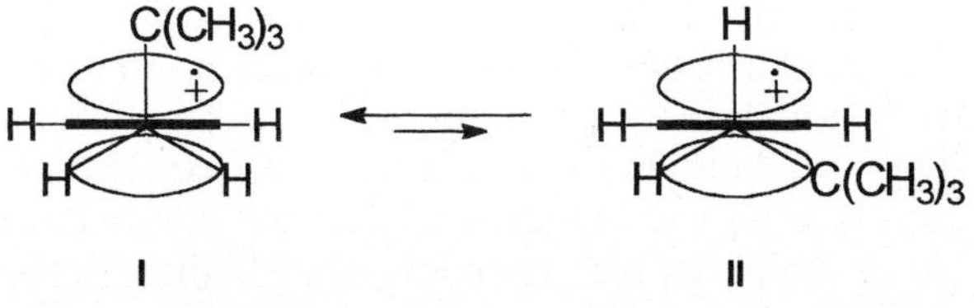

**Scheme 34.**                    **I**                            **II**

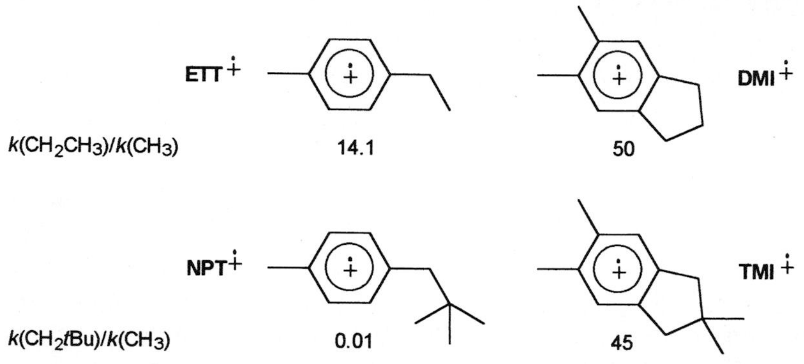

ETT$\overset{\centerdot}{\underset{}{+}}$     $k(CH_2CH_3)/k(CH_3)$     14.1

DMI$\overset{\centerdot}{\underset{}{+}}$     50

NPT$\overset{\centerdot}{\underset{}{+}}$     $k(CH_2tBu)/k(CH_3)$     0.01

TMI$\overset{\centerdot}{\underset{}{+}}$     45

**Scheme 35.**

The importance of stereoelectronic effects has been assessed quantitatively by comparing the $CH_3/CH_2CH_3$ and $CH_3/CH_2tBu$ reactivity ratios (statistically corrected) in the radical cations of *p*-ethyltoluene (ETT), *p*-neopentyltoluene (NPT), 5,6-dimethylindane (DMI), and 2,2,5,6-tetramethylindane (TMI) (Scheme 35) [148]. In the two indane derivatives the methylene groups in positions 1 and 3 bear a substituent which resembles an ethyl and a *t*-butyl group, respectively. In the *p*-xylene series, on going from ETT$^{\cdot+}$ to NPT$^{\cdot+}$ a dramatic 1400-fold decrease in reactivity is observed for deprotonation from the methylene position, behavior which is readily explained on the basis of a stereoelectronic effect (see above and Scheme 34).

This effect disappears as we move to the indane system, where comparable relative reactivities (50 and 45, respectively) are observed for DMI$^{\cdot+}$ and TMI$^{\cdot+}$, in which stereoelectronic effects cannot operate, because the scissible C–H bonds in the 1- and 3-positions are kept almost collinear with the aromatic system, because of the rigidity imposed by the cyclopentane ring. It thus seems, by comparing NPT$^{\cdot+}$ with TMI$^{\cdot+}$, that stereoelectronic control of deprotonation determines a decrease in the deprotonation rate from the $CH_2tBu$ as large as 4500-fold. Interestingly, comparison of ETT$^{\cdot+}$ and DMI$^{\cdot+}$ shows that a stereoelectronic effect is operating in the former radical cation also. An important conclusion that comes from the study of the indane system is that steric hindrance to the approach of the base is of limited importance, a result which is in line with the early transition state proposed for deprotonation of alkylaromatic radical cations. Because stereoelectronic effects always operate in the direction of depressing the deprotonation rate, the observation that for all X substituents $k(CH_2X)/k(CH_3)$ is always greater than unity (Table 3) clearly indicates a major role for electronic effects.

In another study, tertiary:secondary:primary relative reactivity data (TSP selectivity) for the deprotonation reactions of alkylbenzene radical cations [153] showed that with both intra- and intermolecular TSP selectivity the order $S > T > P$ is usually observed, suggesting that the combination of steric and stereoelectronic effects makes an *i*Pr group less reactive than Et, but still more reactive than Me.

$$\overset{\cdot+}{ArCH_2\text{-}H} \quad + \quad B \quad \rightleftharpoons \quad (ArCH_2\text{-}H)\overset{\cdot}{-}B^+$$

**Scheme 36.**

$$(ArCH_2\text{-}H)\overset{\cdot}{-}B^+ \quad \longrightarrow \quad Ar\overset{\cdot}{C}H_2 \quad + \quad BH^+$$

The mechanism discussed above for the deprotonation of alkylaromatic radical cations, involving a bimolecular reaction between the radical cation and the base (B), leading to a carbon centered neutral radical and the conjugated acid of the base (BH$^+$) as described in Scheme 28, has been recently questioned by Parker who provided evidence for an alternative mechanism in proton-transfer reactions between methylanthracene radical cations and pyridine bases [154]; this involved reversible covalent adduct formation between the radical cation and the base followed by elimination of BH$^+$ (Scheme 36).

Resolution of the rate constants for the individual steps of this mechanistic scheme under non-steady-state conditions was also achieved [155].

Product studies of the reaction of the 9-methylanthracene radical cation with 2,6-lutidine in acetonitrile show **X** to be the exclusive reaction product; it results from proton transfer from the methyl substituent (Scheme 37).

An Arrhenius activation energy, $E_a$, of $-1.3$ kcal mol$^{-1}$ has been measured for this process. Substitution of the hydrogen at the 10-position with deuterium leads to a characteristic inverse deuterium kinetic isotope effect—$k_H/k_D = 0.83$. Both values are typical of radical cation–nucleophile combination reactions [156], and the latter is inconsistent with direct deprotonation from the 9-methyl group. Replacement of the 9-methyl with CD$_3$ results, moreover, in a primary deuterium kinetic isotope effect, $k_H/k_D = 5.9$. A mechanism which can account for all these observations involves nucleophilic attack of 2,6-lutidine at the 10-position of 9-methylanthracene radical cation followed by a unimolecular elimination of the 2,6-lutidinium ion from the adduct leading to the 9-anthracenylmethyl radical from which the observed reaction product can be formed after oxidation. Observation of an inverse deuterium kinetic isotope effect implicates concerted bond breaking and formation, with an early transition state for the elimination reaction; this is also supported by the large primary deuterium kinetic isotope effect observed on deuterium substitution at the 9-methyl group. The transition state for the elimination reaction is thus suggested to involve a boat-like conformation of the central ring of the anthracene moiety bearing both the 9-methyl and 10-lutidinium groups in axial positions (Scheme 38, showing in the transition state only the central anthracene ring).

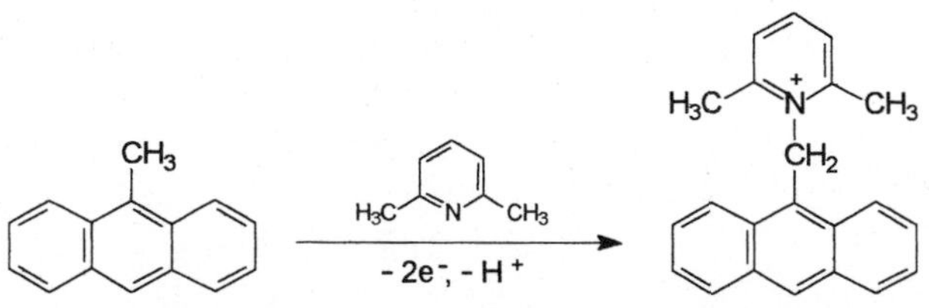

**Scheme 37.**

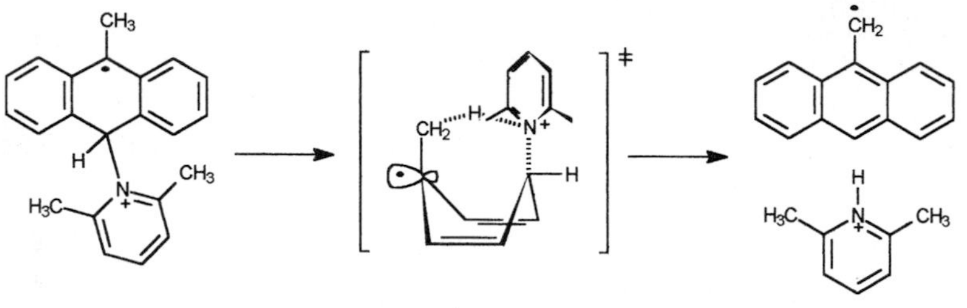

**Scheme 38.**

According to this picture, the elimination reaction occurs from a distonic radical cation in which the positive charge is localized on the nitrogen atom and is maintained at this site throughout the reaction. Thus, the formal deprotonation step can be described as an hydrogen-atom transfer from the 9-methyl group to an incipient 2,6-lutidine radical cation.

Comparison of the reactivities of both 9-methylanthracene and 9,10-dimethylanthracene radical cations with those of pyridine and 2,6-lutidine shows that in the absence of severe steric effects radical cation–nucleophile combination is kinetically favored over direct proton transfer, even though the latter process is highly exergonic. Because the addition step cannot usually be readily detected, however, experimental evidence does not usually enable distinction between addition–elimination and direct proton transfer mechanisms.

Another mechanistic ambiguity which has attracted considerable interest is the possible intermediacy of radical cations in the oxidative functionalization of alkylaromatic substrates with different organic and inorganic oxidants, a process involving C–H bond cleavage which can occur in a single step by direct hydrogen-atom transfer (HAT) from the neutral substrate to the oxidant (Scheme 39, path **a**) or by deprotonation of a first-formed aromatic radical cation (paths **b** and **c**).

Whereas in organic solvents the oxidation of toluene with permanganate ($MnO_4^-$) and chromyl chloride ($CrO_2Cl_2$) has been established to occur by HAT [157], Kochi showed that in the reaction of methylarenes with photoactivated quinones such as chloranil intermediate aromatic radical cations are formed which then undergo side-chain deprotonation [158].

Primary deuterium kinetic isotope effects have always been observed, ruling out their use as a mechanistic tool to distinguish between the two possibilities. The different behavior observed with these systems can be explained on the basis of the

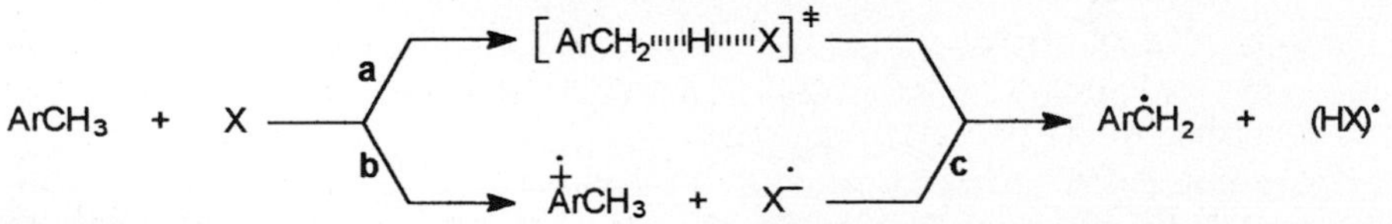

**Scheme 39.**

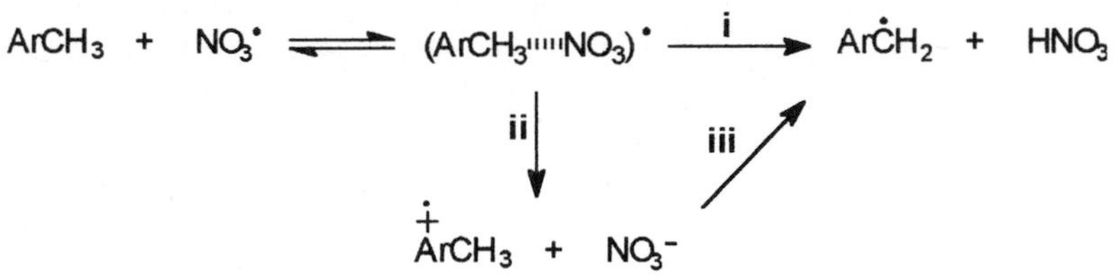

$$ArCH_3 + NO_3^{\bullet} \rightleftharpoons (ArCH_3^{\cdots\cdots}NO_3)^{\bullet} \xrightarrow{\;\;i\;\;} Ar\overset{\bullet}{C}H_2 + HNO_3$$

Scheme 40.

much higher oxidizing power of photoactivated chloranil (for which $E^*_{red} = 2.38$ V relative to the NHE in MeCN [159];) compared with inorganic oxidants (i.e. for $MnO_4^-$ $E° = 0.56$ V in $H_2O$ [160]), and also by considering the hydrogen-abstracting strength of both $MnO_4^-$ and $CrO_2Cl_2$ [157].

A different mechanistic picture has been proposed for the oxidation of alkylbenzenes by $NO_3^{\bullet}$ generated by LFP in MeCN; this involves reversible formation of a weak adduct between the aromatic compound and $NO_3^{\bullet}$ (Scheme 40) [161].

With monoalkylbenzenes and *m*-xylene no evidence has been obtained for formation of an intermediate radical cation, the complex decomposing to a benzyl radical (path **i**) either by a direct HAT or via a concerted electron–proton transfer mechanism. The latter seems more likely on the basis of the observation of the order of reactivity ethylbenzene > isopropylbenzene > toluene (statistically corrected) for reaction **i**, in line with the stereoelectronic effects observed for the deprotonation of alkylbenzene radical cations [153]. With more readily oxidizable substrates such as *p*-xylene and methylbenzenes with $\geq 3$ methyl groups, an analogous complex is formed; this, however, is characterized by a lifetime of $\leq 20$ ns and decomposes into the intermediate radical cation (experimentally observed) from which the benzyl radical is formed by deprotonation (paths **ii** and **iii**).

Along this line, much effort has been devoted to understanding the mechanism of hydroxylation of organic substrates promoted by cytochrome P-450 [162]. Whereas for unactivated alkanes experimental evidence is strongly in favor of a HAT mechanisms [163], for alkylaromatic compounds an electron-transfer mechanism can, in principle, occur, because of the similar oxidation potentials of these compounds [131, 164] and the active oxidant of cytochrome P-450, for which a value of 1.7–2.0 V relative to the SCE has been estimated [165]. Given the experimental difficulties in the direct detection of reactive intermediates in enzymatic reactions, valuable mechanistic information can be obtained, considering that with suitably chosen substrates, reactions involving radical ions can have intramolecular or intermolecular selectivity or lead to product patterns which differ from those obtained in HAT processes. Thus, by comparing the products and selectivity of the enzymatic reactions of these compounds with those obtained using well known electron transfer oxidants it is possible to distinguish between the two mechanistic pathways described in Scheme 39 [166]. When this approach was applied to a study of the cytochrome P-450-promoted oxidation of alkylaromatic compounds such as 4-substituted 1,2-dimethylbenzenes [167], 1,2-dialkylbenzenes [153], and α-alkyl-4-methoxybenzyl alcohols [168], no evidence was found for the occurrence of an

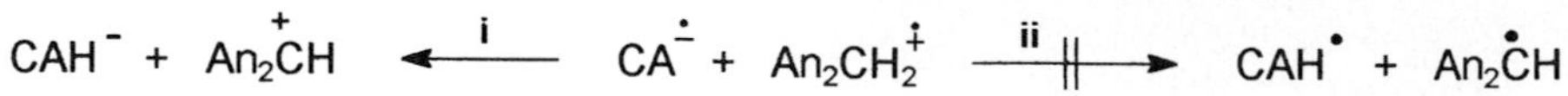

$$CAH^- + An_2\overset{+}{C}H \quad \overset{i}{\longleftarrow} \quad CA^{\cdot -} + An_2CH_2^{\cdot +} \quad \overset{ii}{\longrightarrow}\!\!\!+\!\!\!\rightarrow \quad CAH^{\cdot} + An_2\overset{\cdot}{C}H$$

**Scheme 41.**

electron-transfer mechanism; the results obtained suggested that these reactions proceed through a HAT mechanism.

Whereas in solution alkylaromatic radical cations undergo predominantly heterolytic C–H bond cleavage, evidence for homolytic fragmentation has recently been provided by a LFP study of the photoinduced electron-transfer reaction of *bis*(4-methoxyphenyl)methane ($An_2CH_2$) sensitized by chloranil (CA) in MeCN [169]. 355-nm laser excitation results in the formation of the excited triplet chloranil, $^3CA^*$; this reacts with $An_2CH_2$ by electron transfer leading to $An_2CH_2^{\cdot +}$ and $CA^{\cdot -}$. Quite surprisingly, from this radical ion pair $An_2CH^+$ and $CAH^-$ are formed indicating the occurrence of a HAT reaction (Scheme 41, path **i**).

$\Delta G^\circ$ for this reaction has been determined, on the basis of the appropriate thermodynamic cycle, to be $\leq -29$ kcal mol$^{-1}$ a value which is significantly more negative than that ($\Delta G^\circ = -16$ kcal mol$^{-1}$) calculated for the alternative deprotonation reaction (heterolytic C–H bond cleavage) which leads to $CAH^{\cdot}$ and $An_2CH^{\cdot}$ (path **ii**), thus suggesting that homolytic C–H bond cleavage reactions of alkylaromatic radical cations might be more frequent than hitherto thought.

## Comparison of oxygen and carbon acidity

A major achievement in the study of the acidity of aromatic radical cations is the recent discovery of a pH-dependent mechanistic dichotomy for 1-arylalkanol radical cations in water. Whereas in acidic solution (pH $\leq$ 5) these species undergo C–H deprotonation, in basic media deprotonation of the alcoholic O–H group is observed. Clear evidence in this respect has been provided by a study of the reactivity of the radical cations of 4-methoxybenzyl alcohol (MBA) and its methyl ether (MBAME) generated by pulse radiolysis in aqueous solution [170, 171]. At pH $\leq$ 5, both radical cations decay at the same rate by an $H_2O$-induced C–H deprotonation process, well supported by primary deuterium kinetic isotope effects ($k_H/k_D$ between 4.5 and 5.0) measured with the corresponding $\alpha$-deuterated substrates (Table 5). Interestingly, whereas the effect of an $\alpha$-OH group on the rate of deprotonation of alkylaromatic radical cations is clearly shown by comparing the reactivity of MBA$^{\cdot +}$ with that of 4-methoxytoluene (MT) radical cation—a 40-fold difference— [172], there is no significant difference between the $\alpha$-OH- and $\alpha$-OMe-substituted radical cations.

In alkaline solution MBA$^{\cdot +}$ reacts instead with $^-$OH at a diffusion-controlled rate and is at least 50 and 210 times more reactive than MBAME$^{\cdot +}$ and MT$^{\cdot +}$, respectively, differences which cannot be explained on the basis of the electronic effects of the $\alpha$-OH and $\alpha$-OMe groups. Moreover, no deuterium kinetic isotope effect is observed for MBA$^{\cdot +}$ under these conditions, in line with the diffusion control of the reaction, whereas a $k_H/k_D$ value of 1.8 has been measured for MBAME$^{\cdot +}$. These observations point towards a key role of the $\alpha$-OH group in the

**Table 5.** Rate constants for the uncatalyzed ($k_{dec}$) and $^-$OH-catalyzed ($k_{-OH}$) decay of ArCH$_2$X$^{\cdot+}$ (Ar = 4-MeOC$_6$H$_4$, 3,4-(MeO)$_2$C$_6$H$_3$; X = H, OH, OMe) generated by pulse radiolysis of the parent substrate in aqueous solution, measured at $T = 25\,°C$.

| Substrate | Radical cation | $k_{dec}$ (s$^{-1}$) | $k_{-OH}$ (M$^{-1}$ s$^{-1}$) |
|---|---|---|---|
| 4-MeOC$_6$H$_4$CH$_2$OH[a] | MBA$^{\cdot+}$ | $1.5 \times 10^4$ | $1.2 \times 10^{10}$ |
| 4-MeOC$_6$H$_4$CD$_2$OH[a] | MBA-$d_2$$^{\cdot+}$ | $3.3 \times 10^3$ | $1.1 \times 10^{10}$ |
| 4-MeOC$_6$H$_4$CH$_2$OMe[a] | MBAME$^{\cdot+}$ | $1.5 \times 10^4$ | $2.5 \times 10^8$ |
| 4-MeOC$_6$H$_4$CD$_2$OMe[a] | MBAME-$d_2$$^{\cdot+}$ | $3.0 \times 10^3$ | $1.4 \times 10^8$ |
| 4-MeOC$_6$H$_4$CH$_3$[b] | MT$^{\cdot+}$ | $4.0 \times 10^2$ | $5.5 \times 10^7$ |
| 3,4-(MeO)$_2$C$_6$H$_3$CH$_2$OH[c] | VA$^{\cdot+}$ | 17 | $1.3 \times 10^9$ |
| 3,4-(MeO)$_2$C$_6$H$_3$CH$_2$OMe[c] | VAME$^{\cdot+}$ | 20 | $2.0 \times 10^7$ |
| 3,4-(MeO)$_2$C$_6$H$_3$CH$_3$[c] | DMT$^{\cdot+}$ | 1.1 | $2.1 \times 10^6$ |

[a] According to [170].
[b] According to [172].
[c] According to [102].

deprotonation of 1-arylalkanol radical cations in alkaline aqueous solution, as described in Scheme 42 (An = 4-MeOC$_6$H$_4$).

At pH $\leq 5$, where H$_2$O is the proton-abstracting base, MBA$^{\cdot+}$ undergoes C–H deprotonation (path **a**) whereas when the base is $^-$OH deprotonation from the alcoholic O–H group occurs, the radical cation behaving as an oxygen acid. Because under the latter conditions the rate constant for deprotonation is diffusion controlled, formation of the encounter complex (path **b**) is the rate-determining step [173]. Proton transfer occurs within the complex leading either directly (intramolecular electron transfer coupled with deprotonation, path **f**) or via an intermediate radical zwitterion (paths **c** and **d**), to a benzyloxyl radical which eventually undergoes a formal 1,2-H atom shift (path **e**) leading to the same α-hydroxy-4-methoxybenzyl radical observed at pH $\leq 5$. The mechanism is the same [170, 171] when R = Me, but for R = Et, $i$Pr, C–C bond cleavage (see later) starts to compete (paths **g** and **h**).

This shift from carbon to oxygen acidity is particularly interesting, because only the C–H but not the O–H bond can overlap efficiently with the π system of the

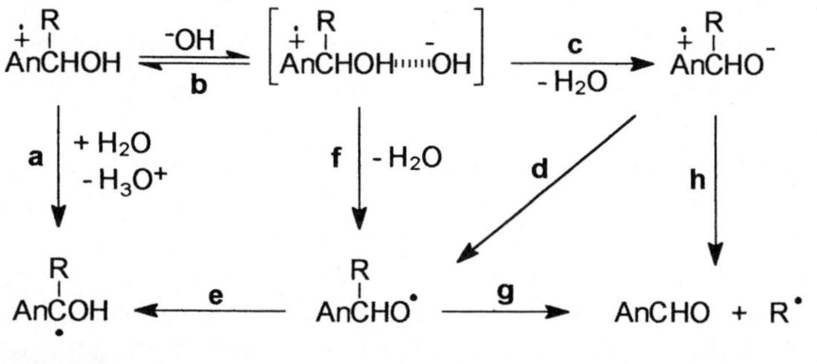

**Scheme 42.**

aromatic radical cation and, moreover, C–H deprotonation is largely favored thermodynamically over O–H deprotonation.

It is, however, well known that carbon acids have much larger intrinsic barriers for proton transfer than oxygen acids [140, 141, 173], thus when a base such as $H_2O$ is involved, the much larger driving force predominates and C–H deprotonation is observed. When, on the other hand, the base is stronger ($^-OH$) both deprotonations are thermodynamically feasible and O–H deprotonation, characterized by a smaller intrinsic barrier, is favored over C–H deprotonation. An explanation in terms of the hard and soft acids and bases concept is also possible—O–H is a much *harder* acid center than C–H, thus in the presence of the charged *hard* base $^-OH$ an effective electrostatic interaction might be particularly favorable. With the uncharged and weaker base $H_2O$, no favorable electrostatic interaction can occur and reaction takes place at the *softer* acid center. Evidence in this respect has been provided by MO calculations showing that in MBA$^{•+}$ the hydrogen bound to the oxygen atom bears a large fraction of the positive charge whereas the benzylic hydrogens are characterized by relatively large spin densities [174].

Analogous behavior is also observed with the corresponding 3,4-dimethoxy derivatives [102]; with these derivatives, however, because of the increased stability of the radical cations, in acidic solution a decrease in reactivity of three orders of magnitude for the deprotonation reaction is observed on going from MBA$^{•+}$ to the 3,4-dimethoxybenzyl alcohol radical cation (veratryl alcohol, VA). In the presence of $^-OH$ the deprotonation site shifts from carbon to oxygen, as previously described for the 4-methoxy derivatives, and a dramatic increase in reactivity is observed, VA$^{•+}$ reacting with $^-OH$ with $k = 1.3 \times 10^9$ M$^{-1}$ s$^{-1}$ (Table 5), a value which is lower than the diffusion limit and which, compared with that measured for MBA$^{•+}$, rather than the decreased O–H acidity of VA$^{•+}$, presumably reflects the increased electron density on the aromatic system which disfavors the intramolecular (side-chain to nucleus) electron transfer eventually leading to the benzyloxyl radical.

*Comparison of C–H and C–C bond cleavage*

When an α-carbon is bonded to both an hydrogen atom and a saturated carbon, $C_\alpha$–$C_\beta$ (from now on indicated as C–C) bond cleavage can start to compete with C–H deprotonation (Scheme 43).

Thermodynamically cleavage of a C–H bond in an alkylaromatic radical cation is strongly favored over C–C fragmentation due to the much higher solvation free energy of the proton as compared to a carbocation. However, the former process is characterized by significantly higher intrinsic barriers (0.5–0.6 eV for C–H

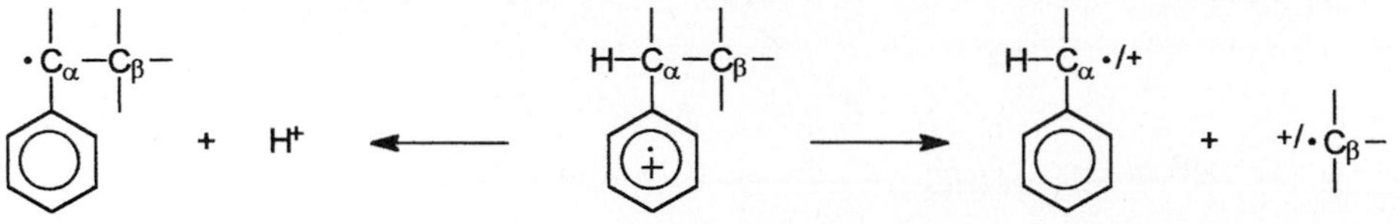

**Scheme 43.**

(see above), compared with 0.1–0.2 eV for C–C bond cleavage [175, 176]) and this difference may play a major role in the competition between the two cleavages. As an example, whereas in solution $PhCH_2–CH_2Ph^{•+}$ undergoes exclusive deprotonation [177–179], C–C bond cleavage is the only fragmentation pathway for $Ph_2CH–CHPh_2^{•+}$, in spite of the fact that for this radical cation the C–C BDE (9.3 kcal mol$^{-1}$ [180]) is still much higher than that for cleavage of the C–H bond ($-17$ kcal mol$^{-1}$ [128]). This behavior may be rationalized in terms of the differences in intrinsic barriers discussed above, and also by considering that the rate of C–H bond cleavage can be strongly depressed by stereoelectronic effects, which always disfavor this process when in competition with the cleavage of other bonds. Evidence in favor of stereoelectronic control of fragmentation in C–C bond cleavage reactions of alkylaromatic radical cations has been provided by Arnold for a variety of systems. Accordingly, while 2,2-diphenylethyl methyl ether radical cation undergoes C–C bond cleavage [181], no products deriving from this pathway but only from a reversible C–H deprotonation reaction are observed with the radical cations of the cyclic analogs 3-phenyl-2,3-dihydrobenzofuran, 5-methyl-3-phenyl-2,3-dihydrobenzofuran, *cis*- and *trans*-2-methoxy-1-phenylindane, all incorporating the $\beta,\beta$-diphenylethyl methyl ether moiety, where collinearity of the scissible C–C bond with the $\pi$-system (Scheme 27) is prevented by the presence of the five-membered ring (Scheme 44) [182].

Additional evidence comes from the study of the radical cations of *trans*- and *cis*-methyl-2-phenycyclopentyl ether and *trans*- and *cis*-1-methyl-2-phenylcyclopentane, where only the first radical cation undergoes C–C bond cleavage since in this case in addition to the overlap between the scissible bond and the $\pi$-system, the conformation of the methoxy group allows interaction of a nonbonding electron pair on the oxygen with the $\sigma^*$ orbital of this bond [183, 184].

In contrast with these findings are however the similar rates of C–C bond cleavage calculated for the radical cations of $PhCH_2CH_2OH$ and 2-indanol [185, 186], indicating that further studies are needed in order to assess the importance of stereoelectronic effects on C–C bond fragmentation reactions of alkylaromatic radical cations.

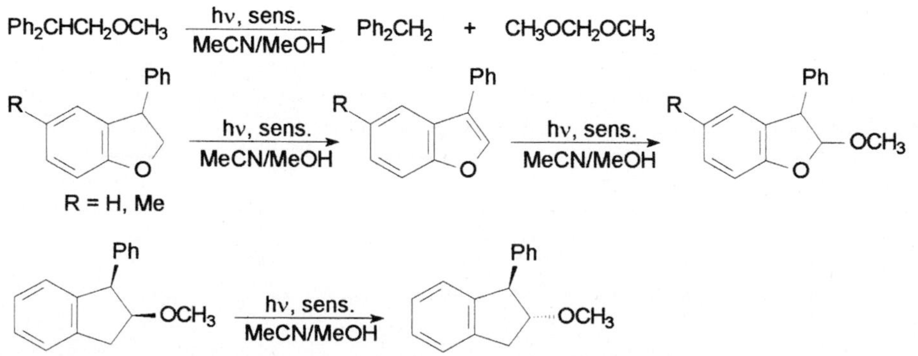

**Scheme 44.**

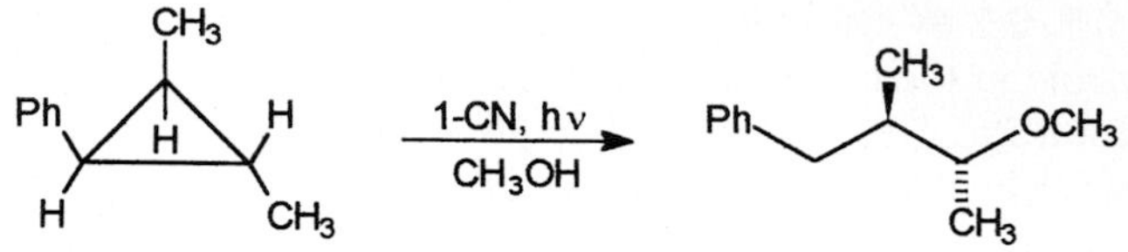

**Scheme 45.**

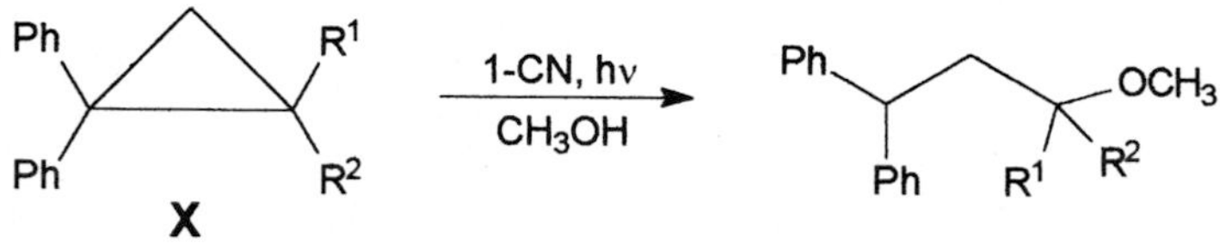

**Scheme 46.**

Interestingly, C–C bond cleavage can successfully compete with C–H deprotonation in systems where fragmentation leads to a very stable fragment, as in the one electron oxidation of phenylacetic acids followed by rapid decarboxylation (see for example ref [187].).

C–C bond cleavage reactions are in most cases unimolecular processes, however, the possibility that C–C bond fragmentation occurs through a nucleophilically assisted pathway has been clearly shown in the three-electron $S_N2$ reactions between arylcyclopropane radical cations and nucleophiles which proceed stereospecifically with inversion of configuration at the carbon atom undergoing nucleophilic substitution (Scheme 45, 1-CN = 1-cyanonaphthalene) [188, 189].

Quite surprisingly, substitution occurs at the more hindered carbon atom (Scheme 46), even when this atom bears a *tert*-butyl or two methyl substituents.

A quantitative evaluation of the steric effects has been obtained by measuring the rate constants for reaction of MeOH with $X^{\cdot+}$ (Table 6), which decrease on going from $R^1 = H$ to $R^1 = t$Bu, showing however effects that are much smaller than those generally observed in four electron $S_N2$ reactions [190, 191]. Moreover, an increase in rate constants is observed with increasing alkyl substitution at the reaction center.

**Table 6.** Rate constants for reaction of MeOH with $X^{\cdot+}$ in $CH_2Cl_2$ at $T = 22\,°C$.

| $R^1$ | $R^2$ | $k$ ($M^{-1}\,s^{-1}$) | $k_{rel}$ |
|-------|-------|------------------------|-----------|
| H     | H     | $1.7 \times 10^7$      | 1         |
| Me    | H     | $1.5 \times 10^8$      | 8.8       |
| Et    | H     | $8.3 \times 10^7$      | 4.9       |
| *i*Pr | H     | $3.0 \times 10^7$      | 1.8       |
| *t*Bu | H     | $4.8 \times 10^6$      | 0.3       |
| Me    | Me    | $3.2 \times 10^8$      | 18.8      |

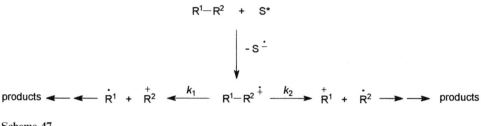

**Scheme 47.**

From these results it appears that the substitution regiochemistry is not determined by steric effects but rather by the ability of the alkyl group at $C_\beta$ to stabilize a positive charge in the substitution transition state, indicating that three-electron $S_N2$ reactions of arylcyclopropane radical cations are dominated by electronic rather than steric effects (see later).

Evidence for nucleophile-assisted C–C bond cleavage has also been found in the nitrate-induced fragmentation reaction of 2-X-2-(4-methoxybenzyl)-1,3-dioxolane (X = H, Me) radical cations [192].

C–C bond cleavage can be heterolytic or homolytic depending on whether the charge or the spin is transferred across the scissible bond (Scheme 26). In this respect, Arnold studied the regioselectivity of the C–C bond-cleavage process in radical cations of diphenylethane derivatives (tri- and tetraarylalkanes) generated by photoinduced electron transfer (PET), showing that C–C bond cleavage may occur if the radical cation BDE is $\leq 15$ kcal mol$^{-1}$ (with higher BDE values fragmentation cannot efficiently compete with back electron transfer within the radical ion pair) and that the product distribution is dependent upon the relative oxidation potentials of the two fragments formed, the carbocation deriving from the radical fragment which has the lower oxidation potential [193]. Since different products arise from the alternative fragmentation pathways of the radical cation, and given that the product distribution is determined within the radical ion pair (Scheme 47), if the product ratio from the cleavage reaction can be measured, the relative oxidation potentials of the two radical fragments can be determined.

It has been suggested that the extent of charge delocalization in the transition state may be greater for the heterolytic than for the homolytic fragmentation mode, since in the former the departing fragment is predisposed to accommodate a net charge [126]. If charge delocalization decreases the transition state energy, the heterolytic cleavage may have an intrinsic kinetic advantage over the homolytic one. However, Savéant proposed that for endoergonic cleavages, C–C bond fragmentation is reversible with the rate kinetically controlled by the diffusion of the two fragments from the solvent cage rather than by activation parameters, a behavior associated with small intrinsic barriers ($\leq 0.2$ eV), indicating that the reorganization energies for these processes are small [175]. Therefore, the reactivity of the fragments may be important with respect to the C–C bond cleavage rate.

The importance of through bond delocalization in C–C bond cleavage reactions of alkylaromatic radical cations has been established in both theoretical and experimental studies. In a MO-theoretical investigation, Kikuchi showed that a drastic change in the electronic structure of phenylethane radical cation occurs during (homolytic) C–C bond cleavage, discussing the results in terms of the three electron three orbital three configuration approach, as described in Scheme 29 [136]. In the radical cation both charge and spin are highly localized on the aromatic ring; as the C–C bond elongates a change in electronic structure occurs (attributed to the avoided crossing between the $\Pi$ and $\Sigma$ states), the unpaired electron density rapidly decreases on the phenyl ring, increases at $C_\beta$ (due to the contribution of the $\Sigma^*$ state) and transiently increases at $C_\alpha$. Transient accumulation of positive charge at $C_\beta$ is also observed, irrespective of the homolytic cleavage, finally leading to a benzyl cation and a methyl radical. Substitution of electron releasing groups at $C_\alpha$ and/or $C_\beta$ stabilizes both $\Sigma$ and $\Sigma^*$ states favoring C–C bond cleavage, while substitution of electron releasing groups at the *para* position will mainly stabilize the $\Pi$ state decreasing the importance of this reaction.

In another study, Camaioni showed that bibenzylic radical cations are $\pi$-type structures with the charge mainly localized in one of the aromatic rings and that the barrier for C–C bond cleavage is mainly associated with establishing orbital overlap between the $\pi$- and $\sigma$-systems by stretching the ethylenic bond, thus enabling through bond delocalization [194]. When sufficient overlap is achieved, the charge developing on $C_\alpha$ and $C_\beta$ can be stabilized by delocalization on the aromatic rings as well as on the side-chain substituents. The presence of electron releasing groups such as OH and Me on the phenyl ring does not stabilize significantly the long bond structure, while substitution on the ethylenic bond leads to a change in the potential energy surface for C–C bond cleavage. Accordingly, bibenzyl, 4-methylbibenzyl and 4-hydroxybibenzyl radical cations have similar BDEs, and long-bond structures with charge and spin delocalized through the scissible C–C bond are not predicted to be stable minima. On the other hand, the presence of $\alpha$-methyl or $\alpha$-hydroxy groups leads to a significant decrease in radical cation BDE, and accordingly C–C bond cleavage in bicumyl radical cation is predicted to be exothermic with an early transition state and only partial delocalization into the $\beta$-ring, with fragmentation occurring when the C–C bond is elongated by 10 % of its equilibrium length. Within bicumyl radical cations, where C–C bond cleavage is the only possible fragmentation pathway, an increase in ring electron density is predicted to result in later transition states characterized by a larger extent of through bond delocalization [194].

Taken together, these studies suggest that the transition state for C–C bond cleavage in alkylaromatic radical cations can be conveniently described by the following mesomeric structures (Scheme 48), where the contribution of structure **I** is

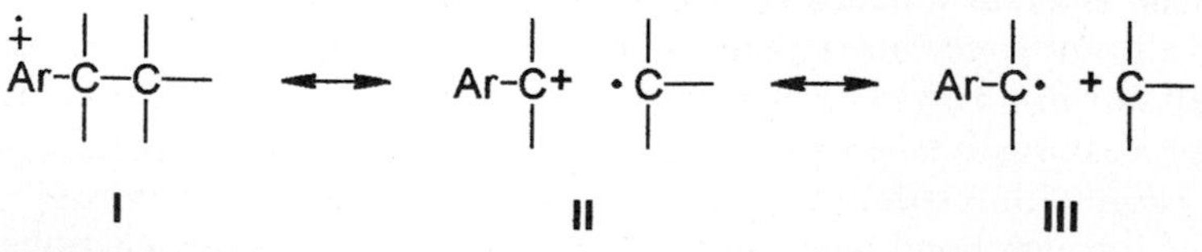

Scheme 48.

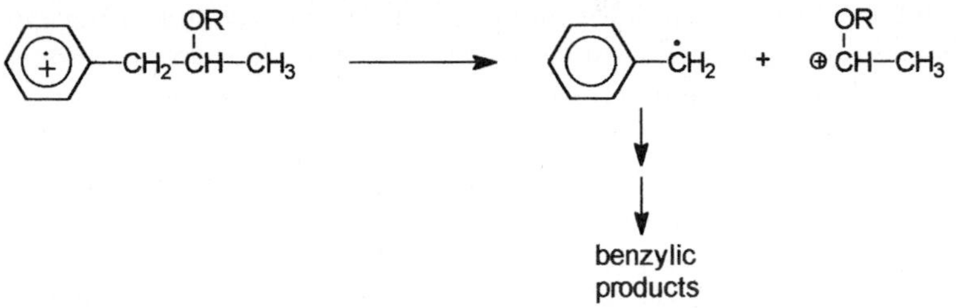

$$\text{[ring}^{+}\text{]}-CH_2-\overset{\displaystyle OR}{CH}-CH_3 \longrightarrow \text{[ring]}-\overset{\bullet}{C}H_2 \;+\; \overset{\oplus}{C}H-\overset{\displaystyle OR}{}CH_3$$

benzylic
products

**Scheme 49.**

predominant for exothermic scissions with early transition states, while structures **II** and **III**, accounting for through bond delocalization, give an increasing contribution as the transition states become later.

These *theoretical* studies fully support the *experimental* results obtained for side-chain fragmentation reactions of aromatic radical cations. Accordingly, product studies have shown that while bibenzyl, 4-methyl-, 4-methoxy- and 4,4′-dimethoxybibenzyl radical cations undergo exclusive C–H bond cleavage [177–179, 195], C–C bond cleavage is favored by $\alpha$ substitution of electron releasing groups [179], as well as by $\beta$ substitution [196].

Additional information on the importance of through bond delocalization and on the competition between C–H and C–C bond cleavage comes from studies on arylalkane radical cations bearing OH or OR substituents at $C_\beta$, generated under a variety of conditions. While fragmentation of $PhCH_2CH(OR)CH_3^{\bullet+}$ (R = H, Me) [98, 180, 196, 197] and $PhCH_2CH(OEt)_2^{\bullet+}$ [198] lead to predominant or exclusive formation of benzylic products deriving from heterolytic C–C bond cleavage (Scheme 49), a different reactivity pattern is observed with the corresponding ring methoxylated radical cations. $4\text{-MeOC}_6H_4CH_2CH(OMe)CH_3^{\bullet+}$ undergoes exclusive C–H deprotonation, while the product distribution changes with conditions for $4\text{-MeOC}_6H_4CH_2CH(OH)CH_3^{\bullet+}$ [196]. At 50 °C, in the absence of added base only products of C–H bond cleavage have been observed, the importance of the C–C bond cleavage pathway increasing in the presence of base. At 110 °C in the absence of base C–H fragmentation is still predominating but C–C bond cleavage competes significantly.

On the basis of these results several conclusions can be drawn:

The relative importance of C–C bond cleavage significantly decreases by adding a 4-methoxy substituent to the ring, which indicates the existence of different electronic requirements for C–C and C–H bond cleavage. Probably, the extent of positive charge which has to accumulate on the scissible bond in the transition state (through bond delocalization), is larger for C–C than for C–H bond cleavage, as also suggested by the early transition state of the latter process. Thus, in the radical cation, the presence of the 4-methoxy group, which stabilizes the positive charge on the aromatic ring, opposing the intramolecular side-chain to nucleus electron transfer, may exert an unfavorable effect more on C–C than on C–H bond cleavage.

Replacement of the $\beta$-OMe group by OH has no influence on the reactivity of the ring-unsubstituted radical cations, while it leads to a significant increase in the importance of the C–C bond cleavage pathway in the corresponding ring methoxylated ones, this reaction however being favored by the presence of a base (see later).

C–C bond cleavage can be thermally activated, as also shown by Arnold in the fragmentation of 1,1,2,2-tetraphenylethane radical cation [181]. In this respect, Maslak measured the activation parameters for a series of bicumene radical cations showing that activation enthalpies ($\Delta H^{\neq}$) for mesolysis are between 8 and 16 kcal mol$^{-1}$, i.e. 30–35 kcal mol$^{-1}$ lower than the values for homolysis of the corresponding neutral substrate, while activation entropies ($\Delta S^{\neq}$) are very small or negative (between 2 and $-27$ eu) in spite of the production of fragments in the cleavage process, suggesting the involvement of a highly ordered solvation of the transition state [199, 200]. A similar trend has been observed for C–C bond fragmentation in a series of 4-methoxybenzylic radical cations ($4\text{-MeOC}_6\text{H}_4\text{CH}_2\text{R}^{\cdot+}$), where, moreover, the study of the temperature effect on the competition between C–H and C–C bond cleavage has lead to the conclusion that $\Delta S^{\neq}$ contributes more to the C–C than to the C–H fragmentation pathway while the opposite holds for $\Delta H^{\neq}$ [192].

In the same study it has also been shown that when the nucleophilicity of the system (neat MeCN) is increased by addition of $NO_3^-$, certain radical cations which in the absence of $NO_3^-$ undergo C–C fragmentation with $\Delta G^{\neq} > 11$ kcal mol$^{-1}$ (R = $CMe_2OMe$, $CPh_2OMe$, 1,3-dioxolan-2-yl), change their reactivity pattern to C–H deprotonation, suggesting that this process requires a more efficient base than the solvent. Attack of $NO_3^-$ is expected to depend on the charge density ($Q$) at the reaction center (benzylic hydrogens or $\beta$-carbon) and accordingly calculations show that while on the benzylic hydrogens $Q$ is almost constant for all radical cations (between 0.12 and 0.16), different values are obtained for the $\beta$-carbon depending on the nature of R: $Q \leq -0.02$ for R = $CMe_2OMe$ and $CPh_2OMe$, $Q = 0.16$ for R = 1,3-dioxolan-2-yl, thus providing a rationale for the observed reactivity.

The presence of electron releasing groups on both $C_\alpha$ and $C_\beta$ as in pinacols, pinacol ethers and bicumyls strongly favors C–C bond cleavage in the corresponding radical cation. Kochi studied the mesolytic C–C bond cleavage in pinacol radical cations, $(\text{ArC(OH)R})_2^{\cdot+}$, generated by PET [201, 202], or charge transfer excitation of the EDA complexes between the pinacol donor ($D_2$) and an acceptor (A = quinones [201, 202], or A = methyl viologen di-cation ($MV^{2+}$) [203]) (Scheme 50, where $D_2 = (\text{ArC(OH)R})_2$).

Direct kinetic measurements by time-resolved fs spectroscopy, following charge-transfer activation with $MV^{2+}$ as the acceptor, lead to $k_C$ values between $3 \times 10^9$

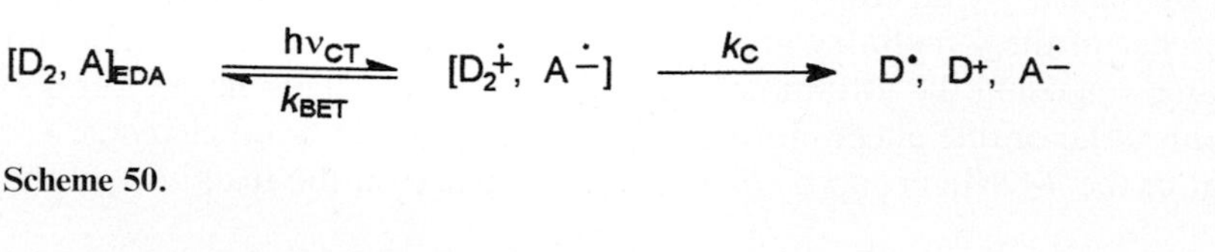

$$A \xrightarrow{h\nu_A} A^* \underset{k_{BET}}{\overset{D_2}{\rightleftharpoons}} [D_2^{\dot{+}}, A^{\dot{-}}] \xrightarrow{k_C} D^{\cdot}, D^+, A^{\dot{-}}$$

$$[D_2, A]_{EDA} \underset{k_{BET}}{\overset{h\nu_{CT}}{\rightleftharpoons}} [D_2^{\dot{+}}, A^{\dot{-}}] \xrightarrow{k_C} D^{\cdot}, D^+, A^{\dot{-}}$$

**Scheme 50.**

**Table 7.** Unimolecular rate constants for C–C bond cleavage of $(ArC(OH)R)_2^{\bullet+}$ generated by charge-transfer activation of EDA complexes in MeCN at $T = 25\,°C$.

| Ar | R | $k_C$ (s$^{-1}$) |
|---|---|---|
| $C_6H_5$ | $C_6H_5$ | $1.4 \times 10^{10}$ |
| $3\text{-}C_6H_5\text{-}C_6H_4$ | Me | $3 \times 10^9$ |
| $4\text{-}C_6H_5\text{-}C_6H_4$ | Me | $5 \times 10^9$ |
| $4\text{-}MeOC_6H_4$ | Me | $5.3 \times 10^{10}$ |
| $4\text{-}MeOC_6H_4$ | $C_6H_5$ | $4.9 \times 10^{10}$ |
| $4\text{-}MeOC_6H_4$ | $4\text{-}MeOC_6H_4$ | $8.0 \times 10^{10}$ |

and $8.0 \times 10^{10}$ s$^{-1}$ (Table 7), indicating ultrafast fragmentation rate constants for all radical cations with activation energies $< 5$ kcal mol$^{-1}$ [203].

The rate of fragmentation is enhanced by electron-releasing ring substituents as shown by the analysis of two series of structurally related pinacol radical cations: $(X\text{-}C_6H_4C(OH)Me)_2^{\bullet+}$ with X = 3-Ph, 4-Ph, OMe (Table 7), from which a Hammett correlation is observed between $\log k_C$ and $\sigma$ with $\rho = -3.8$ [203]; and $(X\text{-}C_6H_4C(OSiMe_3)Me)_2^{\bullet+}$ with X = 2,4-(MeO)$_2$, 4-MeO, H, 4-Br showing the following relative reactivities, 5000, 70, 1.0, 0.2, respectively [201]. A behavior which has been attributed to the increased stability of the fragmentation products as the ring substituents are changed, pointing towards a late transition state for the C–C bond cleavage reaction which can be described by structures **II** and **III** in Scheme 48.

Similar behavior has been observed by Maslak in the C–C bond cleavage of 4-methoxy- [204–206] and 4-dimethylaminobicumene [199, 200] radical cations (4MBC$^{\bullet+}$ and 4DMABC$^{\bullet+}$, respectively, Scheme 51).

Accordingly, $\rho^+$ values of $-2.2$ and $-0.8$ have been obtained for mesolytic C–C bond cleavages of 4MBC$^{\bullet+}$ [205], and 4DMABC$^{\bullet+}$ [200], respectively, indicating buildup of a partial positive charge on C$_\beta$ in the transition state.

Moreover, the unimolecular rate constants for mesolytic cleavage ($k_m$) for both 4MBC$^{\bullet+}$ [204], and 4DMABC$^{\bullet+}$ [199, 200] significantly increase by substitution of electron releasing groups at the 4-position of the $\beta$-phenyl ring as well as by replacing the side-chain Me groups with Et, while electron withdrawing substituents (X = CF$_3$, CN) slow down the fragmentation reaction (Table 8). These observations point again towards a polarized transition state in which a substantial amount of positive charge has been transferred across the scissible bond (Scheme 48).

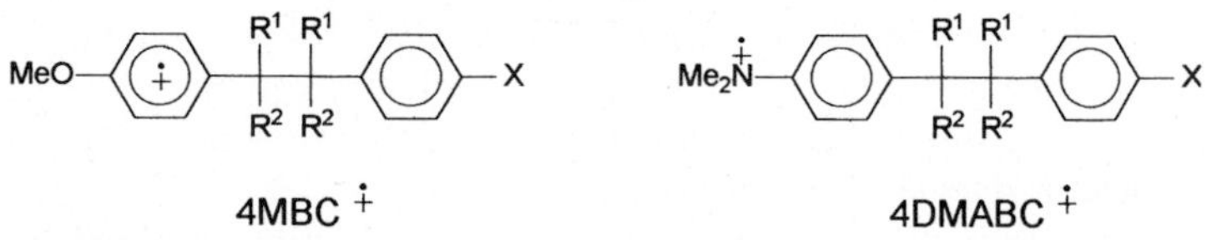

**Scheme 51.**

**Table 8.** Unimolecular rate constants for mesolytic cleavage of 4MBC$^{\bullet+}$ and 4DMABC$^{\bullet+}$ in CH$_2$Cl$_2$–MeOH$^a$ at room temperature.

| R$^1$ | R$^2$ | X | $k_m$ (s$^{-1}$) |
|---|---|---|---|
| *4MBC$^{\bullet+b}$* | | | |
| Me | Me | H | $2.4 \times 10^7$ |
| Me | Me | MeO | $6.0 \times 10^8$ |
| Me | Me | Me | $1.2 \times 10^8$ |
| Me | Me | CF$_3$ | $1.6 \times 10^6$ |
| Me | Me | CN | $2.6 \times 10^6$ |
| Me | Et | MeO | $1.2 \times 10^9$ |
| Et | Et | MeO | $4.4 \times 10^9$ |
| *4DMABC$^{\bullet+c}$* | | | |
| Me | Me | H | 0.7 |
| Me | Et | H | 20 |
| Et | Et | H | $1.6 \times 10^5$ |

$^a$ MeOH 0.5–5 %.
$^b$ According to [204].
$^c$ According to [199].

On the basis of the comparison between the reactivities of the different series of pinacol and bicumyl radical cations discussed above (Tables 7 and 8), some conclusions can be drawn:

Increased electron density on the aromatic ring bearing the unpaired electron allows a smaller extent of charge transfer to the scissible C–C bond (compare the $\rho^+$ values for 4DMABC$^{\bullet+}$ and 4MBC$^{\bullet+}$), by opposing intramolecular electron transfer, reflected in a dramatic decrease in $k_m$ (Table 8).

Increased electron density on the $\beta$-aromatic ring (not bearing the unpaired electron) favors through bond delocalization ($\rho^+ < 0$) enhancing the fragmentation rate.

Increased electron density on both aromatic rings enhances the fragmentation rate, suggesting that in such cases stabilization of the transition state by through bond delocalization is more important than stabilization of the reactant radical cation.

Replacement of side-chain Me groups by Et or OH leads to an increase in the rate of C–C bond cleavage, as clearly shown by comparing the reactivities of three structurally related radical cations: (4-MeOC$_6$H$_4$CMe$_2$)$_2$$^{\bullet+}$, (4-MeOC$_6$H$_4$CEt$_2$)$_2$$^{\bullet+}$, undergoing fragmentation with $k_m = 6.0 \times 10^8$ and $4.4 \times 10^9$ s$^{-1}$, respectively (Table 8) and (4-MeOC$_6$H$_4$C(OH)Me)$_2$$^{\bullet+}$, for which $k_m = 5.3 \times 10^{10}$ s$^{-1}$ (Table 7), in line with ability of these groups in stabilizing a positive charge on the adjacent carbon atom. Interestingly, this rate enhancement is significantly more pronounced when the $\beta$-ring lacks electron releasing substituents.

However, in partial contrast with these conclusions are the results obtained by Whitten for the fragmentation of substituted pinacol radical cations (ArC(OH)R)$_2$$^{\bullet+}$,

**Table 9.** Unimolecular rate constants for C–C bond cleavage of $(ArC(OH)R)_2{}^{\cdot+}$ generated by PET in MeCN at room temperature.

| Ar | R | $E_p$ (V relative to the NHE) | $k_{BF}$ (s$^{-1}$) |
|---|---|---|---|
| 4-MeOC$_6$H$_4$ | Me | 1.49 | $>1 \times 10^7$ |
| 2,4-(MeO)$_2$C$_6$H$_3$ | Me | 1.27 | $7.5 \times 10^6$ |
| 2,5-(MeO)$_2$C$_6$H$_3$ | Me | 1.24 | $4.5 \times 10^6$ |
| 2,4,5-(MeO)$_3$C$_6$H$_2$ | Me | 0.99 | $2.5 \times 10^6$ |
| 4-Me$_2$NC$_6$H$_4$ | H | 0.95 | $<5 \times 10^3$ |
| 4-Me$_2$NC$_6$H$_4$ | Me | 0.81 | $1.8 \times 10^5$ |
| 4-Me$_2$NC$_6$H$_4$ | Ph | 0.95 | $4 \times 10^4$ |

where the rate constants for C–C bond cleavage ($k_{BF}$) follow the stability of the corresponding radical cation (Table 9) [207–209].

An explanation accounting for the different behavior observed is not easy, but a more conclusive characterization of the radical cations by time-resolved spectroscopic techniques is certainly needed [203, 208]. It is moreover important to consider that if the reduced sensitizer is sufficiently basic, deprotonation from the side-chain OH group could also occur, followed by C–C fragmentation from an intermediate oxyl radical (see later) [201, 210].

The unexpected decrease in reactivity observed on going from (4-Me$_2$NC$_6$H$_4$C(OH)Me)$_2{}^{\cdot+}$ to (4-Me$_2$NC$_6$H$_4$C(OH)Ph)$_2{}^{\cdot+}$, has been instead attributed to conformational requirements in the transition state for C–C bond cleavage, based on the negative $\Delta S^{\neq}$ values ($-6.9$ and $-10.6$ eu, respectively) determined for this process, which suggest an ordered transition state [209]. Accordingly, calculations have shown that for (4-Me$_2$NC$_6$H$_4$C(OH)Me)$_2$ and (4-Me$_2$NC$_6$H$_4$C(OH)Ph)$_2$ the most stable conformations are those depicted in Scheme 52 (which should reasonably hold also for the corresponding radical cations), where the dimethylaminophenyl groups are in the correct orientation for C–C bond cleavage (face to face, where the $\pi$ orbitals are collinear with the scissible $\sigma$ bond) only in the former substrate.

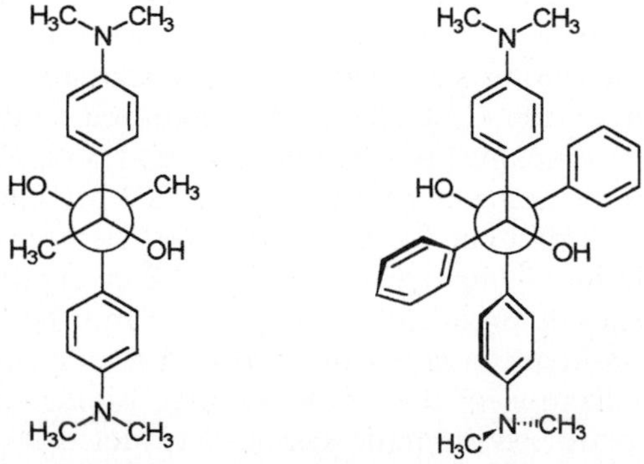

**Scheme 52.**

PCP $\overset{\cdot}{+}$          cMPCP $\overset{\cdot}{+}$          **Scheme 53.**

An analogous trend has been observed by Kochi for the corresponding ring methoxylated pinacols [203], where however the effects are much smaller (Table 7). Replacing the side-chain phenyl groups with 4-MeOC$_6$H$_4$ results however in a significant increase in reactivity, suggesting that in this case electronic effects overcome the steric ones.

It is also interesting to examine the nucleophilically assisted C–C bond cleavage in substituted arylcyclopropane radical cations previously described, in terms of the concepts discussed above. The experimental data obtained have shown that three-electron S$_N$2 reactions of arylcyclopropane radical cations are dominated by electronic rather than steric effects, the substitution regiochemistry being determined by the ability of the alkyl group at C$_\beta$ to stabilize a positive charge in the transition state [188, 189, 191]. Accordingly, quantum chemical calculations have shown that phenylcyclopropane radical cation (PCP$^{\cdot+}$) adopts a delocalized bisected conformation with two equally lengthened C–C bonds (Scheme 53) characterized by a significant interaction between the phenyl and cyclopropyl moieties [190, 211]. Addition of methyl groups at C$_\beta$ leads to a significant increase in the C$_\alpha$–C$_\beta$ bond length and to a change in the relative orientation of the two rings, with the plane of the phenyl ring almost completely aligned with the C$_\alpha$–C$_\gamma$ bond (Scheme 53, showing *cis*-1-methyl-2-phenylcyclopropane (*c*MPCP) radical cation), allowing efficient overlap between the $\pi$-orbitals and the long C$_\alpha$–C$_\beta$ bond.

Alkyl substitution also leads to a significant redistribution of the positive charge which decreases on the phenyl ring simultaneously increasing at C$_\beta$. Semi-empirical calculations of the transition state structures for backside nucleophilic attack of methanol on PCP$^{\cdot+}$ predict a significant increase in the C$_\alpha$–C$_\beta$ bond length and of the charge at C$_\beta$ at the expense of the charge on the phenyl ring, on going from the reactants to the transition state. In agreement with this shift in charge is the value of $\rho^+ = 2.2$ obtained from the correlation between $\sigma^+$ and the rate constants for nucleophilic substitution on *para*-substituted PCP$^{\cdot+}$, reflecting the decreased ease of intramolecular electron transfer as the ring electron density increases [190]. Moreover, the C$_\beta$–O distance is predicted to decrease by increasing alkyl substitution, accompanied by charge increase at both C$_\beta$ and O, a behavior which suggests that transition states for these processes become later with increasing alkyl substitution.

A product-like transition state has been proposed by Tanko for the nucleophile-induced ring opening of 1- and 2-cyclopropylnaphthalene radical cations, with the spin density delocalized over C$_\alpha$ and the aromatic system, and a large fraction of positive charge localized at both C$_\beta$ and O (Scheme 54) [212], suggesting that stabilization of the positive charge by the naphthyl group is of limited importance. Thus, the aromatic system is expected to stabilize the reactant radical cation sig-

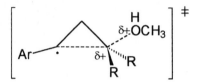

**Scheme 54.**

nificantly more than the transition state for nucleophilic substitution. In agreement with this hypothesis is the observation that no direct cyclopropyl ring opening is observed in the reaction of 9-cyclopropylanthracene radical cation with MeOH [213, 214].

Oxygen acidity and C–C bond cleavage

The favorable effect of side-chain OH groups on the C–C bond cleavage reactions of alkylaromatic radical cations has been shown in a number of studies (see above) [179, 186, 196, 203, 207]. Along this line, in acidic aqueous solution both $4\text{-MeOC}_6\text{H}_4\text{CH(OH)CH}_2\text{C}_6\text{H}_5{}^{\bullet+}$ and $4\text{-MeOC}_6\text{H}_4\text{CH}_2\text{CH(OH)C}_6\text{H}_5{}^{\bullet+}$ undergo C–C bond cleavage as the major reaction pathway with similar rate constants ($k = 3.2 \times 10^4$ and $2.0 \times 10^4$ s$^{-1}$, respectively) suggesting that at least for these systems the position of the OH group ($\alpha$ or $\beta$) does not play a significant role [196]. In both fragmentation reactions a small solvent kinetic isotope effect ($k(\text{H}_2\text{O})/k(\text{D}_2\text{O})$ between 1.2 and 1.4) and negative activation entropies have been observed, suggesting that hydrogen bonding plays a major role in the $\alpha$- and $\beta$-OH assisted C–C bond cleavage. Moreover, the OH group is much more effective than OMe in promoting C–C bond cleavage, as clearly shown by the observation that in acidic aqueous solution $4\text{-MeOC}_6\text{H}_4\text{CH(OH)}t\text{Bu}^{\bullet+}$ undergoes exclusive C–C bond cleavage with a rate ($k = 1.5 \times 10^5$ s$^{-1}$) which is about four orders of magnitude higher than that of the corresponding methyl ether, in spite of the fact that both radical cations are characterized by very similar values of C–C BDE. As a matter of comparison, under identical conditions no difference in the rate of C–H deprotonation was observed between the radical cations of 4-methoxybenzyl alcohol and its methyl ether (Table 5) [170].

On the basis of these observations, the transition state for C–C bond cleavage in 1-arylalkanol radical cations can be conveniently described by mesomeric structures **I–IV** (Scheme 55, Ar = $4\text{-MeOC}_6\text{H}_4$), where structure **III**, with the positively charged oxygen and the fully formed C=O double bond, strongly contributes to the transition state structure, profiting from stabilization by hydrogen bonding with the solvent water.

In contrast, in the transition state for C–H deprotonation (Scheme 56, also showing the proton-accepting base H$_2$O), the most important structure is now **VIII**, representing heterolytic C–H bond cleavage, whereas structure **VII** (corresponding to **III** in Scheme 55) should be of minor importance, since it involves the very unfavorable homolytic C–H bond cleavage.

A similar transition state stabilization by hydrogen bonding can be envisaged also for 2-arylalkanol radical cations, as described by structure **IX** (Scheme 57).

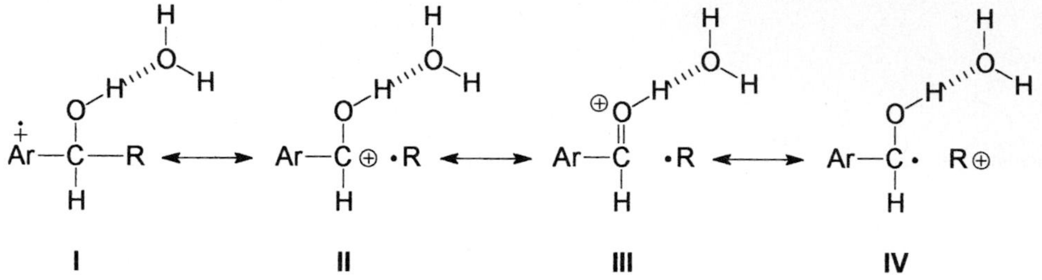

**Scheme 55.**

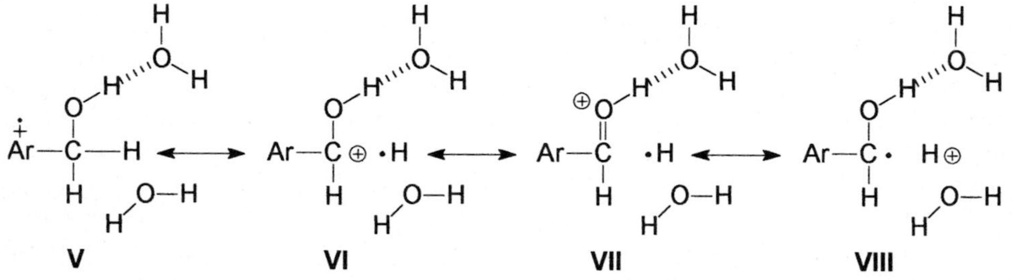

**I**    **II**    **III**    **IV**

**Scheme 56.**

**V**    **VI**    **VII**    **VIII**

**IX**

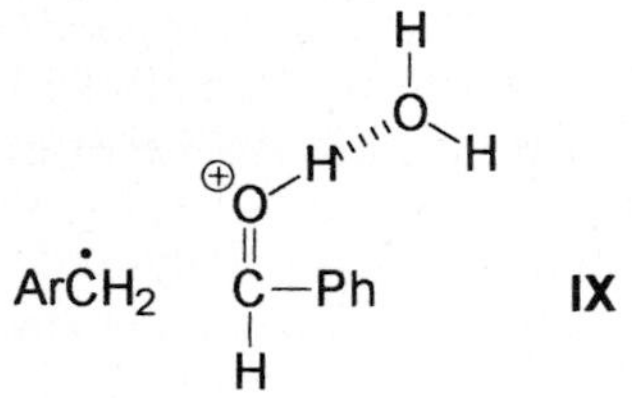

**Scheme 57.**

In both 1- and 2-(4-methoxyphenyl)-alkanol radical cations, C–C bond cleavage is strongly favored by the presence of $^-$OH. Accordingly, pulse radiolysis studies have shown that 4-MeOC$_6$H$_4$CH(OH)R$^{\cdot+}$ (R = Et, $i$Pr) undergo exclusive or predominant C–H bond cleavage at pH $\leq$ 5, while in the presence of $^-$OH C–C bond cleavage becomes the major reaction pathway [170]. Under these conditions, base-induced C–C bond cleavage occurs with a diffusion controlled rate ($k_{-OH} \approx 1 \times 10^{10}$ M$^{-1}$ s$^{-1}$) an observation which suggests that this reaction proceeds through an analogous mechanism as that proposed for the deprotonation of 4-methoxybenzyl alcohol radical cations (Scheme 42). Namely, 4-MeOC$_6$H$_4$CH(OH)R$^{\cdot+}$ behave as oxygen acids undergoing O–H deprotonation leading to an intermediate oxyl radical (4-MeOC$_6$H$_4$CH(R)O$^{\cdot}$), which can follow two reactive pathways: C–C bond cleavage and/or 1,2-H atom shift, depending on the nature of R (Scheme 42, paths **e** and **g**). When R = H or Me, the 1,2 shift is the exclusive pathway, whereas when

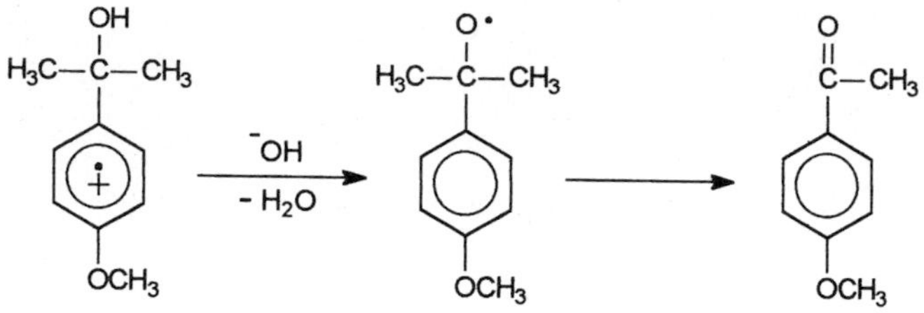

**Scheme 58.**

R = $t$Bu, only $\beta$-cleavage occurs. Intermediate situations apparently hold for R = Et and $i$Pr. Formation of an intermediate oxyl radical followed by $\beta$-fragmentation has been proposed in the reactions of photogenerated pinacol radical cations [201, 210].

Direct evidence for the formation of the oxyl radical was indeed obtained in a pulse radiolysis study of the ⁻OH-induced decay of 4-methoxycumylalcohol radical cation where, accordingly the 4-methoxycumyloxyl radical was observed by time-resolved UV–Vis spectroscopy on its way to form 4-methoxyacetophenone (Scheme 58) [215].

It is however important to point out that in this case the oxyl radical is relatively long-lived, since fragmentation leads to 4-methoxyacetophenone and Me·, the least stable alkyl radical. In the presence of alkyl groups which can produce, after $\beta$-fragmentation, radical fragments more stable than Me·, one cannot exclude that such process proceeds through a mechanism involving concerted C–C bond cleavage and intramolecular electron transfer in the radical zwitterion (Scheme 42, path **f**).

When the OH group is in the $\beta$ position, the mechanistic dichotomy between carbon and oxygen acidity is again observed. Accordingly, while at pH $\leq$ 5, 4-MeOC$_6$H$_4$CH$_2$CH$_2$OH·$^+$ undergoes exclusive $\alpha$-C–H deprotonation with $k = 5.2 \times 10^2$ s$^{-1}$, a rate constant very similar to that observed for 4-methoxytoluene radical cation (Table 5), in the presence of ⁻OH only C–C bond cleavage is observed with $k_{-OH} = 8.3 \times 10^9$ M$^{-1}$ s$^{-1}$, a value very close to the diffusion limit [172]. Thus, in the presence of ⁻OH, a mechanism resembling the one described in Scheme 42 for 1-arylalkanol radical cations can be proposed also for 2-arylalkanol radical cations (Scheme 59: An = 4-MeOC$_6$H$_4$).

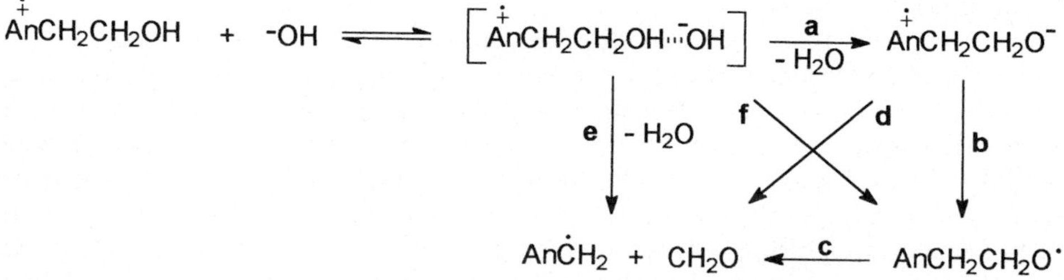

**Scheme 59.**

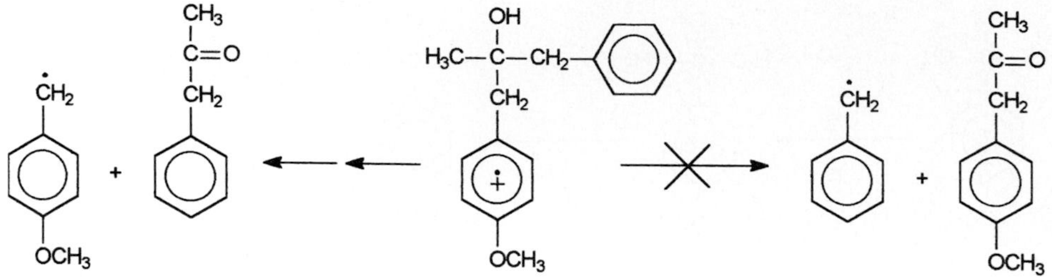

**Scheme 60.**

Proton transfer in the encounter complex may lead, either directly (path **f** ) or via a radical zwitterion (paths **a** and **b**), to an alkoxyl radical which then undergoes C–C bond $\beta$-cleavage (path **c**). As an alternative, two concerted processes are also possible; involving O–H deprotonation and C–C bond cleavage (E2 mechanism, path **e**) or C–C bond cleavage and intramolecular electron transfer in the radical zwitterion (E1cB mechanism, paths **a** and **d** ).

Useful information in this respect has been obtained through the product study of the $^-$OH-induced decay of 4-MeOC$_6$H$_4$CH$_2$(Me)C(OH)CH$_2$Ph$^{\cdot+}$ [172]. If alkoxyl radicals are involved in the fragmentation process, products deriving from both the benzyl and 4-methoxybenzyl radical should be observed (Scheme 60). Since only products deriving from the latter radical have been observed, the intermediacy of the alkoxyl radical in the $^-$OH-induced fragmentation of $\beta$-OH substituted alkyl-aromatic radical cations can be reasonably excluded, thus the reaction proceeds through one of the concerted processes (paths **d** or **e**) described in Scheme 59.

These two mechanisms (E1cB and E2) have been proposed by Schanze and Whitten, respectively to rationalize the effect of the $\beta$-OH group on the side-chain C–C bond cleavage in the radical cations of 2-(arylamino)-1,2-diphenylethanols [216], 1,2-aminoalcohols [217], and 2-(4-$N,N$-dimethylaminophenyl)-1-phenyl-ethanol [218].

In particular, Schanze proposed that the mechanism of base-catalyzed fragmentation in 2-(4-X-phenylamino)-1,2-diphenylethanol radical cations (Scheme 61) changes with the oxidation potential of the substrate, going from an E1cB-like mechanism (as in Scheme 59, paths **a** and **d** ) with the more stable radical cations (X = OMe and Me), to an E2-like mechanism (as in Scheme 59, path **e**) when X = H, Cl, CN.

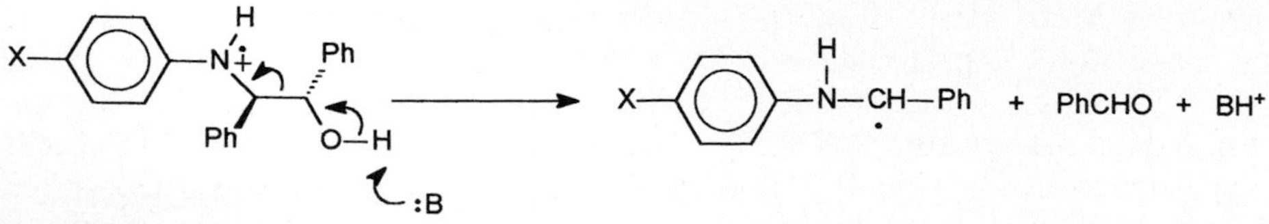

**Scheme 61.**

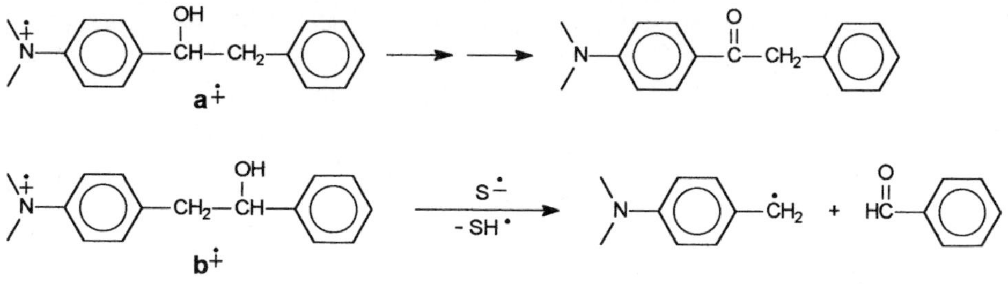

**Scheme 62.**

Moreover, a Brønsted correlation with $\alpha = 0.62$ has been obtained for 2-phenyl-amino-1,2-diphenylethanol radical cation [219]. Comparison of this value with that obtained for the C–C bond cleavage of 4-MeOC$_6$H$_4$CH(OH)$t$Bu$^{\cdot+}$ catalyzed by a series of bases in aqueous solution ($\beta = 0.4$) [170], suggests that localization of the positive charge at nitrogen results in a significantly later transition state for fragmentation.

It is also interesting to compare the results for 4-MeOC$_6$H$_4$CH(OH)CH$_2$C$_6$H$_5$$^{\cdot+}$ and 4-MeOC$_6$H$_4$CH$_2$CH(OH)C$_6$H$_5$$^{\cdot+}$, undergoing C–C bond cleavage as the major or exclusive reaction pathway in both acidic and basic aqueous solution [170, 172], with those obtained by Whitten for the fragmentation of structurally related aminoalcohol radical cations: 4-Me$_2$NC$_6$H$_4$CH(OH)CH$_2$C$_6$H$_5$$^{\cdot+}$ (**a**$^{\cdot+}$) and 4-Me$_2$NC$_6$H$_4$CH$_2$CH(OH)C$_6$H$_5$$^{\cdot+}$ (**b**$^{\cdot+}$), in MeCN or benzene [218]. While **a**$^{\cdot+}$ undergoes C–H bond cleavage leading to the corresponding aminoketone, **b**$^{\cdot+}$ gives clean C–C fragmentation (Scheme 62, S = sensitizer).

This peculiar behavior can be explained in terms of the already discussed through bond delocalization. Positive charge development at C$_\beta$ seems to be a prerequisite for C–C bond cleavage to occur. In aminoalcohol radical cations, the positive charge is strongly localized on the dimethylamino moiety thus opposing intra-molecular electron transfer from the side-chain scissible bond. In **a**$^{\cdot+}$, the presence of a phenyl group at C$_\beta$ is not sufficient to stabilize an incipient positive charge and accordingly no homolytic C–C fragmentation is observed. On the other hand, the presence of an additional $\beta$-OH group as in **b**$^{\cdot+}$ promotes through bond delocalization, making heterolytic (O–H assisted) C–C bond cleavage an efficient process. In line with this picture is the lower reactivity observed for N-aromatic substituted amino alcohol radical cations, as compared to N-alkyl substituted ones [217].

Finally, oxygen acidity extends also to 3-arylalkanol radical cations as clearly shown by 4-MeOC$_6$H$_4$(CH$_2$)$_3$OH$^{\cdot+}$ which undergoes $\alpha$-C–H deprotonation at pH $\leq 5$ and O–H deprotonation in the presence of $^-$OH, with a rate constant, $k_{-OH} = 1.7 \times 10^9$ M$^{-1}$ s$^{-1}$, which is 25 times higher than that measured for the corresponding methyl ether [172]. In basic solution, no products of C–C bond cleavage are observed, but only 3-(4-methoxyphenyl) propanal, deriving from 1,2-H shift in the intermediate oxyl radical, since $\beta$-cleavage would lead to a relatively unstable primary alkyl radical (Scheme 63, An = 4-MeOC$_6$H$_4$).

AnCH₂ĊH₂ + H₂C=O ◄──╳── AnCH₂CH₂CH₂Ȯ ────► AnCH₂CH₂ĊHOH ──►──► AnCH₂CH₂CHO

**Scheme 63.**

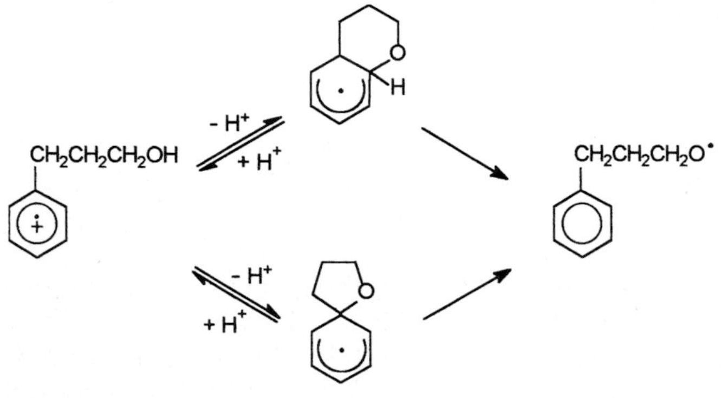

**Scheme 64.**

The value of $k_{-OH}$ is significantly below the diffusion limit and thus of the values measured for 1- and 2-arylalkanol radical cations, indicating that formation of the encounter complex is no longer the rate determining step. This behavior reflects the decreased oxygen acidity as well as the decreased ease of intramolecular electron transfer as the distance between the OH group and the aromatic ring bearing the positive charge is increased. An alternative mechanism has been however proposed by Gilbert accounting for the observed reactivity of 3-phenylalkanol radical cations in acidic solution, where intramolecular electron transfer proceeds through an inner-sphere mechanism by ring-closure via intramolecular nucleophilic attack of the γ-OH followed by ring opening, leading to the oxyl radical (Scheme 64) [186, 220].

Clearly, with this radical cation oxygen acidity is observed at much lower pH (between 1 and 3) suggesting that in the presence of a 4-MeO group, intramolecular electron transfer can efficiently occur only if the O–H group is deprotonated.

If the distance between the aromatic ring and the side-chain OH function is further increased as with 4-MeOC₆H₄(CH₂)₄OH, oxygen acidity in the corresponding radical cation is no longer observed [172].

### Aromatic radical zwitterions

As previously discussed, removal of an electron from the π-system of alkylaromatic substrates leads to a dramatic increase in the acid strength of the $C_\alpha$–H bonds, as clearly shown by the $pK_a$ between −11 and −13 estimated for the deprotonation of toluene radical cation in acetonitrile [130]. The presence of a positive charge on the aromatic ring can also influence, however to a much smaller extent, the acidity of groups which are further spaced from the aromatic ring such as OH and $CO_2H$ as

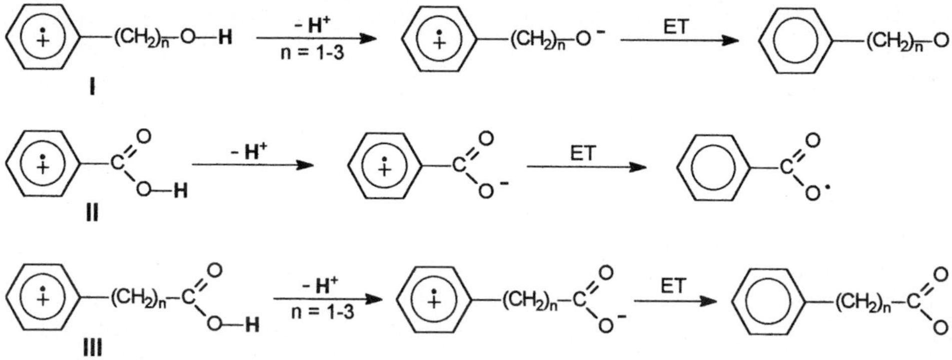

I

II

III

**Scheme 65.**

in arylalkanols, benzoic and arylalkanoic acids. Where possible, side-chain $C_\alpha$–H deprotonation in the corresponding radical cations is strongly favored thermodynamically over O–H deprotonation. However, carbon acids exhibit much larger intrinsic barriers for proton transfer than oxygen acids [140, 141, 173] and thus, in the presence of a sufficiently strong base which overcomes the thermodynamic gap, O–H deprotonation can take over (see above). No C–H deprotonation is usually observed with arylethanoic acid radical cations (in Scheme 65: structure **III** with $n = 1$), since in this case C–C bond cleavage leads to a very stable fragment ($CO_2$) and a relatively stable benzyl radical (see later).

O–H deprotonation of arylalkanol (**I**), benzoic acid (**II**) and arylalkanoic acid (**III**) radical cations eventually leads to radical zwitterions where the electron hole is localized on the aromatic system and the negative charge resides on the oxygen atom originally bonded to the acidic proton (Scheme 65). From the radical zwitterions an oxygen centered radical can be formed by intramolecular electron transfer from the side-chain $-O^-$ to the aromatic $\pi$-system, whose chemistry will be discussed below.

Since in these systems intramolecular electron transfer involves orbitals which cannot directly overlap, a point which has attracted great interest is whether or not the radical zwitterion and the oxygen centered radical are different representations of a single species. In addition, the mechanistic picture can be further complicated since intramolecular electron transfer in the radical zwitterion can be coupled with or followed by other processes (i.e. $C_\alpha$–$C_\beta$ bond cleavage or 1,2-carbon to oxygen H atom shift with **I**, and decarboxylation with **II** and **III**). Finally, it is also interesting to understand if in systems such as **II** and **III**, decarboxylation requires or not initial O–H deprotonation.

The mechanistic implications of carbon and oxygen acidity in arylalkanol radical cations have been discussed extensively in the previous section and will not be further considered here [170, 172, 185, 186, 215].

In this respect however, one point should be briefly considered. It has been suggested that some internal charge transfer may occur in arylcarbinyloxyl radicals, accounting for the very pronounced red-shift ($\approx 100$ nm) observed for the visible

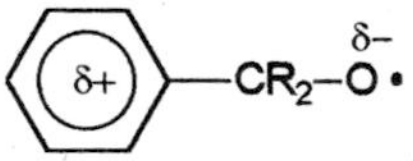

**Scheme 66.**

absorption band of these radicals by replacement of H with a 4-MeO group [221]. Moreover, calculations indicate that the unpaired electron is largely localized in the oxygen 2p orbital which is perpendicular to the plane of the aromatic ring and that in the ground state the oxygen atom of the benzyloxyl radical bears a significant amount of negative charge, which is increased by the $\pi \to \pi^*$ transition responsible for the visible absorption band [222]. These results suggest that arylcarbinyloxyl radicals display a certain zwitterionic character (Scheme 66), which should increase in the presence of electron releasing groups on the aromatic ring.

*Radical zwitterions derived from benzoic acids*

The reaction of benzoate anion with $SO_4^{\cdot-}$ has been established to proceed by electron transfer from the aromatic ring leading to the corresponding radical zwitterion [223], followed by decarboxylation of an intermediate benzoyloxyl radical to yield the phenyl radical (Scheme 67). Both benzoyloxyl and phenyl radicals have been identified by ESR spectroscopy through their adducts with $CH_2NO_2^-$ [224, 225], and in particular, in the case of polycarboxylated benzenes, the (carboxylated) phenyl radicals have been *directly* seen by ESR [224].

For the 4-methyl derivatives, the different nature of the radical zwitterion and the neutral benzoyloxyl radical, generated by $SO_4^{\cdot-}$ oxidation of 4-methylbenzoate and reaction of Ti(III) with 4-methylperoxybenzoic acid, respectively, has been clearly shown, since the former undergoes deprotonation and the latter hydrogen atom abstraction (by 4-methylbenzoyloxyl and/or 4-methylphenyl radicals) from the 4-methyl group. Moreover, in the presence of a 4-MeO substituent, decarboxylation is observed only in the Ti(III)–4-methoxyperoxybenzoic acid system, while in the $SO_4^{\cdot-}$–4-methoxybenzoate one this reaction becomes too slow and reduction of the radical zwitterion can take over, suggesting that this group significantly stabilizes the radical zwitterion by opposing intramolecular electron transfer leading to the 4-methoxybenzoyloxyl radical [223].

This hypothesis is supported by the results of a pulse radiolysis and ESR study of methoxylated benzoic acids in water, where again no decarboxylation was observed from the corresponding radical zwitterions [226]. The zwitterionic nature of these radicals, with no contribution from an aroyloxyl radical structure was shown on the

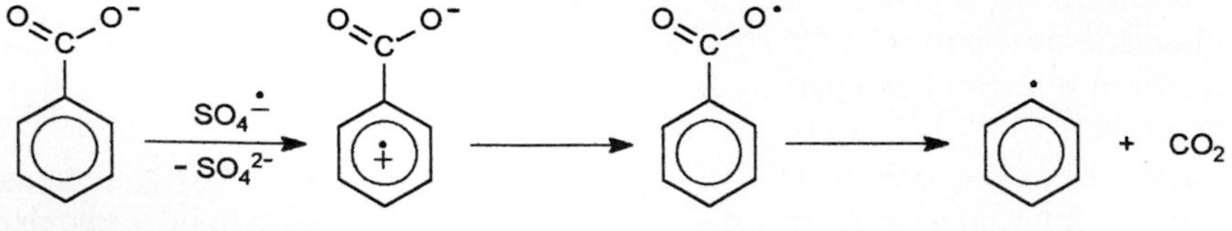

**Scheme 67.**

**Table 10.** Rate constants for the decarboxylation of $ArCO_2^•$ in $CCl_4$ at $T = 24\,°C$.

| Ar | $k\ (\mathrm{s}^{-1})$ |
|---|---|
| $C_6H_5$ | $2.0 \times 10^6$ |
| $4\text{-}ClC_6H_4$ | $1.4 \times 10^6$ |
| $4\text{-}MeOC_6H_4$ | $3.4 \times 10^5$ |

basis of the following observations. The coupling constants for the one-electron oxidized methoxylated benzoic acids are very similar to those obtained from the corresponding methoxybenzene radical cations and are constant between pH 2 and 10, pointing towards an analogous charge and spin distribution in the two series of radical cations, and indicating moreover that the ionization state is the same over this pH range and thus that the carboxyl group of methoxylated benzoic acid radical cations is ionized at pH $\geq 2$.

Additional information on the nature of the radical zwitterion and the oxyl radical comes from a kinetic study of the decarboxylation of aroyloxyl radicals ($ArCO_2^•$) generated by LFP of the parent diaroyl peroxide in $CCl_4$ [227], showing that the presence of a 4-MeO substituent leads to a significant decrease in the rate of decarboxylation (Table 10).

This behavior has been rationalized in terms of the relative arrangement of the aryl and $CO_2$ moieties, pointing towards a situation where these groups are coplanar (Scheme 68, structures **I** and **II**, where the unpaired electron is localized on the oxygen atom) rather than perpendicular (structures **III** and **IV**, where the arrangement of the two moieties allows through space electron transfer). Electron releasing substituents contribute to structure **II** increasing the double bond character of the $Ar–CO_2^•$ bond thus decreasing the decarboxylation rate. However, theoretical studies suggest that there is no strong preference for either the planar or perpendicular structure [227], and thus that intermediate structures may play a role in the behavior of aroyloxyl radicals.

It is to be expected that the solvent should exert a strong influence on the decarboxylation rate via stabilization of (zwitter)ionic structures (**II** and **IV**). As mentioned above [226], in water no decarboxylation was observed from the radical zwitterion of 4-methoxybenzoic acid (i.e. $k \leq 10^2\ \mathrm{s}^{-1}$).

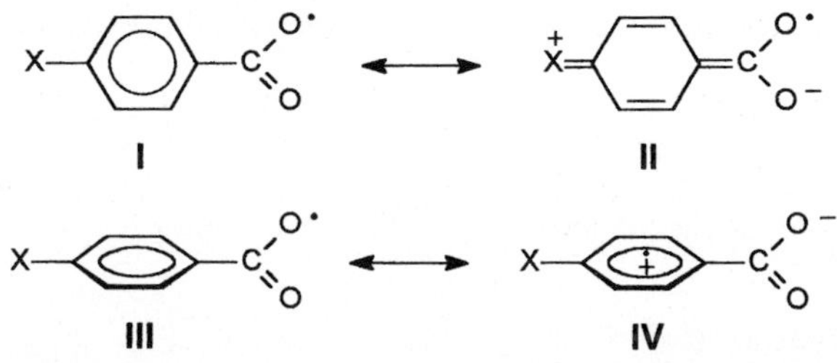

**Scheme 68.**

$$X-\underset{}{\boxed{\bigodot}}-CH_2-CO_2H/CO_2^- \longrightarrow X-\underset{}{\boxed{\bigodot}}-\dot{C}H_2 \; + \; CO_2 \;\; (+ H^+)$$

**Scheme 69.**

*Radical zwitterions derived from arylalkanoic acids*

The intermediacy of aromatic radical cations in the oxidative decarboxylation of arylethanoic acids has been supported by a number of studies [228–231]. Direct evidence for the involvement of radical cations in these processes has been provided by a LFP study where the radical cations of the parent acids have been produced in aqueous solution by photoionization or by reaction with $SO_4^{\cdot-}$ [187]. 4-Methylphenylethanoic acid radical cation undergoes decarboxylation leading to the 4-methylbenzyl radical with $k = 1.5 \times 10^7$ s$^{-1}$ at pH $= 0.3$ (Scheme 69, X = Me).

Interestingly, the rate constant increases by increasing pH ($k \geq 5 \times 10^7$ s$^{-1}$ at pH $\geq 2.5$), indicating that decarboxylation is faster when the carboxyl group is ionized. A similar reactivity has been observed with 4-methoxyphenylethanoic acid, where the corresponding radical zwitterion undergoes decarboxylation with $k \geq 5 \times 10^7$ s$^{-1}$ (Scheme 69, X = MeO). Unfortunately, instrumental limits did not allow a direct comparison between the reactivities of the methyl and methoxy substituted radical zwitterions.

When the reactions of 1- and 2-naphthylethanoic acids with $SO_4^{\cdot-}$ have been studied analogously, a characteristic naphthalene-type radical cation spectrum with absorption bands centered at 310, 380, 630 and 680 nm has been observed only for the radical cation derived from the latter acid, while the former gives a spectrum with a main absorption at 330 nm attributed to the 1-naphthylmethyl radical formed by decarboxylation ($k = 5 \times 10^5$ s$^{-1}$ at pH $= 3.5$) of a first formed radical zwitterion [187]. This behavior has been explained in terms of the much larger positive charge density at position 1 rather than 2 in the naphthalene radical cation (0.1809 and 0.0691, respectively from HMO calculations). Thus, the relatively larger accumulation of positive charge on the ring carbon atom bonded to the $CH_2CO_2H/CO_2^-$ group in 4-X-phenylethanoic acid radical cations as compared with 1-naphthylethanoic acid radical cation, reasonably accounts for the much faster rates of decarboxylation measured for the former.

More recently, Kochi studied the decarboxylation of acyloxyl radicals generated by PET in charge transfer ion pairs between benzilate anions ($Ar_2C(OH)CO_2^-$) and a cationic acceptor (methylviologen di-cation ($MV^{2+}$) or *N*-methyl-4-cyanopyridinium ($NCP^+$)) in aqueous solution (Scheme 70) [232, 233].

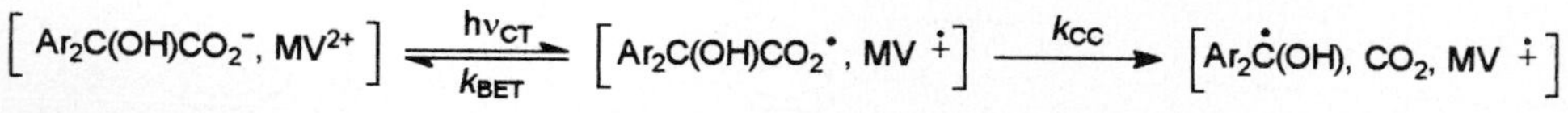

$$\left[ Ar_2C(OH)CO_2^-, MV^{2+} \right] \underset{k_{BET}}{\overset{h\nu_{CT}}{\rightleftharpoons}} \left[ Ar_2C(OH)CO_2^{\cdot}, MV^{\cdot+} \right] \overset{k_{CC}}{\longrightarrow} \left[ Ar_2\dot{C}(OH), CO_2, MV^{\cdot+} \right]$$

**Scheme 70.**

**Table 11.** Rate constants for decarboxylation of $MV^{2+}$–benzilate and $MV^{2+}$–arylacetate ion pairs in aqueous solution.

| Donor | $k_{CC}$ (s$^{-1}$) |
|---|---|
| *Benzilate* | |
| Benzilate | $8 \times 10^{11}$ |
| 4,4'-Dimethylbenzilate | $5 \times 10^{11}$ |
| 4-Methoxybenzilate | $1 \times 10^{11}$ |
| 4,4'-Dimethoxybenzilate | $2 \times 10^{11}$ |
| 2,2',5,5'-Tetramethoxybenzilate | $4 \times 10^{10}$ |
| *Arylacetate* | |
| Diphenylacetate | $6.1 \times 10^{9}$ |
| Phenylacetate | $1.8 \times 10^{9}$ |
| 4-Chlorophenylacetate | $1.6 \times 10^{9}$ |
| 4-Methoxyphenylacetate | $1.5 \times 10^{9}$ |
| 4-Biphenylacetate | $1.5 \times 10^{9}$ |
| 1-Naphthylacetate | $<2 \times 10^{8}$ |

The rate constants for decarboxylation of $Ar_2C(OH)CO_2^{\bullet}$ ($k_{CC}$) are collected in Table 11 together with those obtained analogously for a series of arylacetoxyl radicals ($ArCH_2CO_2^{\bullet}$).

Benziloxyl radicals undergo very fast decarboxylation ($k_{CC} > 10^{10}$ s$^{-1}$) indicating activation barriers for C–C bond cleavage of only 1–2 kcal mol$^{-1}$. Interestingly, the presence of electron donating substituents on the aromatic ring leads to a decrease in $k_{CC}$.

To explain this observation, it has been suggested that the benziloxyl radical can be thought of as a resonance hybrid between the two structures depicted in Scheme 71: a radical zwitterion (**I**) where the electron hole is localized on the aromatic ring, and the acyloxyl radical (**II**) having the unpaired electron on the carboxyl group [232]. In the presence of electron releasing substituents (**X**) on the aromatic ring, the relative importance of structure **I** increases and such structure is expected to be less prone than **II** to undergo decarboxylation.

However, the possibility that **I** and **II** are different species, with the benziloxyl radical deriving from intramolecular electron transfer in a first formed radical

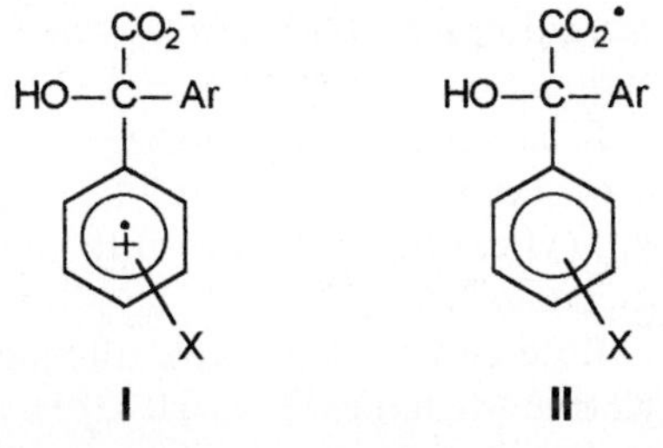

**Scheme 71.**                **I**                **II**

zwitterion cannot be ruled out, the observed differences in reactivity reflecting in this case the decreased ease of intramolecular electron transfer induced by the presence of electron releasing substituents on the aromatic ring. An additional possibility is that in the least stabilized systems, intramolecular electron transfer in the zwitterion (**I**) is coupled with decarboxylation.

With arylacetoxyl radicals, relatively slower decarboxylation rate constants have been measured (Table 11), pointing in this case towards higher activation barriers. Two main factors have been suggested to account for such behavior:

1) the stability of the radical formed by decarboxylation (ketyl, $Ar_2C(OH)^{\bullet}$, relative to benzyl, $ArCH_2^{\bullet}$), because the presence of two aromatic rings rather than one allows a more extensive delocalization of the incipient radical resulting in a greater stabilization of the transition state for decarboxylation (also compare in Table 11 phenylacetate with diphenylacetate). In this respect, the presence of an $\alpha$-OH group also appears to play a major role as shown by the 130-fold increase in $k_{CC}$ on going from diphenylacetate to benzilate.
2) Positive charge stabilization by the aromatic system, as shown in Scheme 71, structure **I**, which disfavors decarboxylation, reflected in the relatively low $k_{CC}$ values observed for 2,2′,5,5′-tetramethoxybenzilate and 1-naphthylacetate [232].

The former explanation contrasts however, especially in the benzilate system, with the very early transition state of the decarboxylation process and thus positive charge stabilization by the aromatic system (factor **ii**) rather than the stability of the radical formed by decarboxylation (factor **i**) may play a greater role. On the other hand the very similar values of $k_{CC}$ measured for the phenylacetoxyl radicals may reflect a relatively later transition state and a self-compensating stabilization by ring substituents of both reactant and product along the series.

When the distance between the carboxyl group and the aromatic ring is increased by the presence of additional methylene groups, as in 3-phenylpropanoic ($Ph(CH_2)_2CO_2H$) and 4-phenylbutanoic ($Ph(CH_2)_3CO_2H$) acids, decarboxylation from the radical zwitterion is still observed but $C_\alpha$–H deprotonation can now compete efficiently. Accordingly, when $Ph(CH_2)_2CO_2H^{\bullet+}$ is generated by reaction with $SO_4^{\bullet-}$ in aqueous solution, the ESR signals of $PhCH_2CH_2^{\bullet}$ are observed between pH 2 and 9, which are replaced below pH = 2 by those of $PhCH^{\bullet}CH_2CO_2H$ [220, 228]. Similar results are obtained for $Ph(CH_2)_3CO_2H^{\bullet+}$ where, however, decarboxylation is observed at pH $\geq$ 3. Interestingly, no decarboxylation is observed with $Ph(CH_2)_4CO_2H^{\bullet+}$ but only $C_\alpha$–H deprotonation, a result which confirms that electron transfer to $SO_4^{\bullet-}$ takes place from the aromatic ring leading to an aromatic radical cation (zwitterion) which can then undergo deprotonation and for shorter chains (in Scheme 72, $n = 0$–2) decarboxylation.

If decarboxylation occurs following intramolecular electron transfer, via an acyloxy radical, this is more likely to occur from the radical zwitterion rather than the radical cation, in line with the change in mechanism observed as the pH is reduced (Scheme 72).

Rate constants for overall side-chain to nucleus electron transfer and decarboxylation (Scheme 72, $n = 0$–2) have been calculated as $k \geq 10^9$ s$^{-1}$ [228].

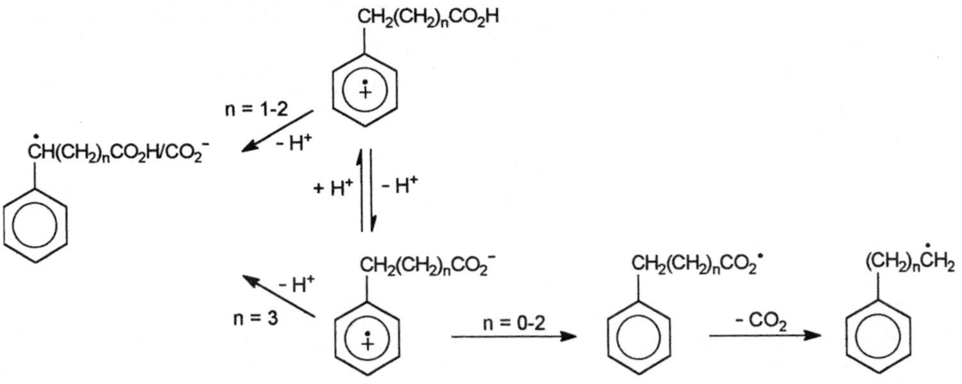

**Scheme 72.**

A somewhat different behavior has been instead observed for the radical zwitterions of indol-3-ylacetic and 3-indol-3-ylpropionic acids generated by pulse radiolysis in aqueous solution (through $Br_2^{\cdot-}$ oxidation), the former undergoing decarboxylation in acidic solution with $k = 1.6 \times 10^4$ s$^{-1}$, while no decarboxylation has been observed with the latter (Scheme 73) [234].

An explanation can be found in terms of the much lower reduction potential of the indole-type radical cations as compared to the phenylalkanoic acid ones, which results in a greater stabilization of the positive charge on the aromatic system thus opposing intramolecular electron transfer. In other words, a later transition state is expected in the decarboxylation of indole-type radical zwitterions as compared to the phenylalkanoic acid ones, with an increased importance of the stability of the carbon centered radical.

Interestingly, when the reactivity of phenylalkanoic acid radical cations was studied in non-aqueous solvents such as acetonitrile and acetic acid, decarboxylation is still the major path for $Ph(CH_2)_2CO_2H^{\cdot+}$ while exclusive C–H deprotonation is observed with $Ph(CH_2)_3CO_2H^{\cdot+}$ [230, 235], a behavior which could

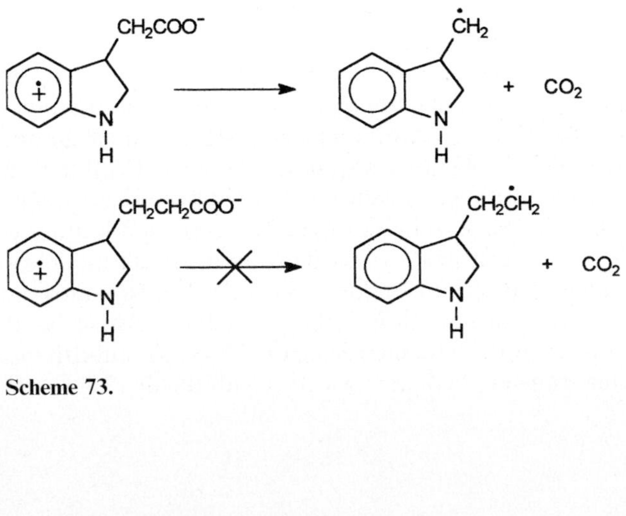

**Scheme 73.**

indicate a change in the acid–base properties of the radical cations under these conditions.

## 9.5.2 Reductive Bond-Cleavage Processes

Electron transfer to a neutral organic molecule is often accompanied by the cleavage of a bond, a process which can be coupled to or follow the electron transfer step (Scheme 74).

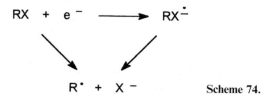

Scheme 74.

In the former case, the initial electron transfer process is dissociative, directly leading to the radical and anionic fragmentation products, a typical example being the reduction of unstrained aliphatic halides where the electron is added to the $\sigma^*$ orbital of the carbon–halogen bond [236, 237]. In the latter case, an intermediate radical anion is formed which then undergoes fragmentation coupled with intramolecular electron transfer (i.e. dissociative intramolecular electron transfer). With molecules containing $\pi^*$ orbitals, able to host transitorily the unpaired electron, the question which arises is whether reductive cleavage occurs via a concerted or stepwise mechanism. Distinction between these two possibilities is not easy when the lifetime of the intermediate radical anion is too short to allow detection, i.e. the kinetics of fragmentation are controlled by the initial electron transfer of the stepwise pathway or by the dissociative electron transfer of the concerted pathway. On the other hand, when the intermediate is sufficiently long lived, recognition of a stepwise mechanism is relatively straightforward. It is however important to point out that a change in the driving force of the reaction can induce a transition between the two mechanisms.

The problem of distinguishing between stepwise and concerted mechanisms has been addressed by Savéant who provided a thorough description of the dynamics of dissociative electron transfer by developing a model based on a Morse potential for the cleaving bond in the reactant, and on the assumption that the repulsive interaction of the two fragments formed upon electron transfer is the same as the repulsive part of the reactant potential [238, 239]. An expression where activation and reaction free energies ($\Delta G^{\neq}$ and $\Delta G_0$) follow a quadratic Marcus type relationship $[\Delta G^{\neq} = \Delta G^{\neq}_0(1 + \Delta G_0/4\Delta G^{\neq}_0)^2]$ is thus obtained, where however, in addition to the solvent reorganization free energy term ($\lambda_0$), as in the Marcus–Hush model of outer-sphere electron transfer, the intrinsic barrier $\Delta G^{\neq}_0$ also contains a contribution of bond breaking from the bond dissociation enthalpy of the scissible bond (BDE) $[\Delta G^{\neq}_0 = (BDE + \lambda_0)/4]$. The quadratic character of the activation-driving force relationship implies that the transfer coefficient $\alpha$ varies with the driving force:

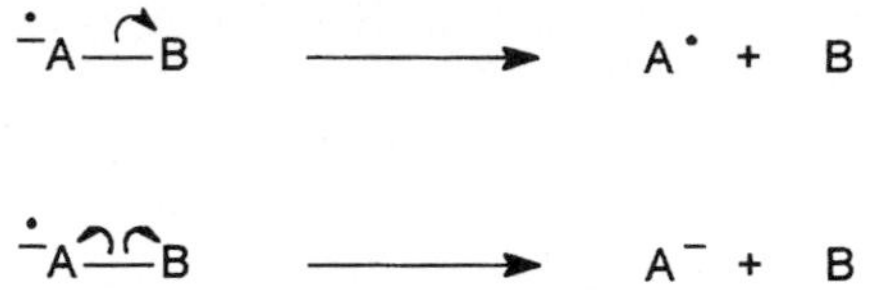

$$\overset{\cdot}{\phantom{|}}\overline{A}\text{——}B \quad\longrightarrow\quad A^{\bullet} \;+\; B^{-}$$

**Scheme 75.**

$$\overset{\cdot}{\phantom{|}}\overline{A}\text{——}B \quad\longrightarrow\quad A^{-} \;+\; B^{\bullet}$$

$\alpha = \partial\Delta G^{\neq}/\partial\Delta G_0$, and thus, the value of $\alpha$ may allow discrimination between the concerted and stepwise mechanisms. In the former case, a value of $\alpha$ significantly lower than 0.5 is expected. On the other hand, in the stepwise mechanism, given that the electron transfer step is fast compared to the following fragmentation step, $\alpha$ values close to or larger than 0.5 are expected [238, 240, 241].

Along these lines, new insights into the dynamics of dissociative electron transfer have been recently provided by a number of studies [240, 242–246].

In this section we will examine selected examples of reductive fragmentation processes, namely the reductive C–X bond cleavage accompanying reduction of aryl and benzyl halides and C–C bond cleavage in the radical anions of diphenylethane derivatives. In this respect, in analogy with the side-chain fragmentation reactions of alkylaromatic radical cations previously described (Scheme 26), when an intermediate radical anion is formed, fragmentation leading to radical and anionic fragments can occur in two formal ways: a heterolytic and a homolytic mode (Scheme 75) [126].

The following discussion will be focused on the effect of structural variations on the kinetics of fragmentation, the cleavage mode, and the stepwise or concerted nature of the process. Similar concepts have also been applied to the reductive cleavage reactions of other series of substrates, i.e. $\alpha$-substituted acetophenones [246–250], perbenzoates [240, 251], peroxides [242, 252–256], sulfides [244, 257, 258], sulfonium salts [259–263], nitrocumenes [264, 265], arylmethyl aryl ethers and thioethers [266–270].

Even though fragmentation of radical anions represents a key step in $S_{RN}1$ reactions (Scheme 76) and in aliphatic nucleophilic substitution reactions ($S_N2$) proceeding via single electron transfer (Scheme 77), such processes and their mechanistic implications will not be discussed in this section (several reviews are available [271–277]).

*Reductive dehalogenation of aryl and benzyl halides*

One-electron reduction of aryl (ArX) and benzyl halides (ArCH$_2$X) bearing electron withdrawing ring substituents eventually produces the corresponding radical

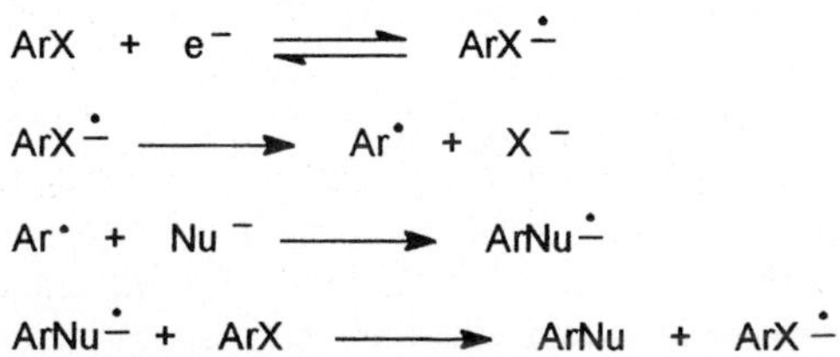

$$ArX \;+\; e^{-} \;\rightleftharpoons\; ArX^{\overset{\cdot}{-}}$$

$$ArX^{\overset{\cdot}{-}} \;\longrightarrow\; Ar^{\bullet} \;+\; X^{-}$$

$$Ar^{\bullet} \;+\; Nu^{-} \;\longrightarrow\; ArNu^{\overset{\cdot}{-}}$$

**Scheme 76.**

$$ArNu^{\overset{\cdot}{-}} \;+\; ArX \;\longrightarrow\; ArNu \;+\; ArX^{\overset{\cdot}{-}}$$

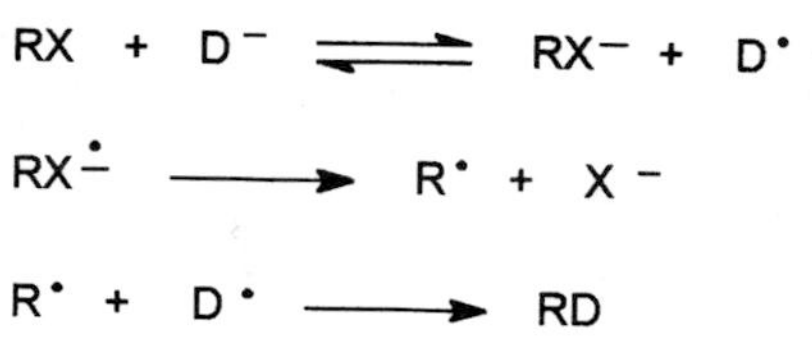

RX + D$^-$ ⇌ RX$^-$ + D$^{\bullet}$

RX$^{\bullet}{}^-$ ⟶ R$^{\bullet}$ + X$^-$

R$^{\bullet}$ + D$^{\bullet}$ ⟶ RD          **Scheme 77.**

anion which can undergo an intramolecular electron transfer process coupled to fragmentation leading to the halide anion and the aryl or benzyl radical (Scheme 74). The intramolecular electron transfer can be conveniently described in terms of a $\sigma$–$\pi$ radical anion isomerism with the initially formed $\pi$ radical anion ($\pi$ RA) undergoing intramolecular electron transfer from the $\pi$ system to the $\sigma^*$ orbital of the C–X bond leading to a $\sigma$ RA (a three-electron bond radical anion) which dissociates in a carbon centered radical and the anion of the leaving group (X$^-$) (Scheme 78) [278].

In this respect, theoretical studies have shown that in halobenzenes, halobenzonitriles, haloacetophenones and halonitrobenzenes, by considering the intramolecular electron transfer as the rate limiting step of the cleavage reaction, the energy difference between the two isomeric radical anions ($\Delta E_{\sigma\pi}$) correlates with the experimental fragmentation rates [279, 280].

Neta and Behar measured the rate constants for intramolecular electron transfer for an extended series of aryl (ArX) and benzyl halide (ArCH$_2$X) radical anions bearing an electron withdrawing ring substituent (NO$_2$, CN, COCH$_3$), generated by pulse radiolysis in aqueous solution (Table 12), conditions under which spectroscopic evidence for formation of the intermediate radical anions was obtained [281–284].

By comparing the data reported in Table 12 several conclusions can be drawn:

The rates of dehalogenation strongly depend on the nature of the halogen, increasing with decreasing C–X bond strength (I > Br > Cl).

The relative position of the two ring substituents (the electron withdrawing Z group and the X or CH$_2$X group) has a strong influence on the rate of fragmentation. The *ortho* isomer reacts always faster than the *para* isomer, a behavior which suggests the occurrence, at least in part, of a direct (through space) electron transfer from the Z group to the scissible bond in the former radical anions. In this respect, it has been suggested that in the radical anions of *ortho*-nitroarylhalides steric effects,

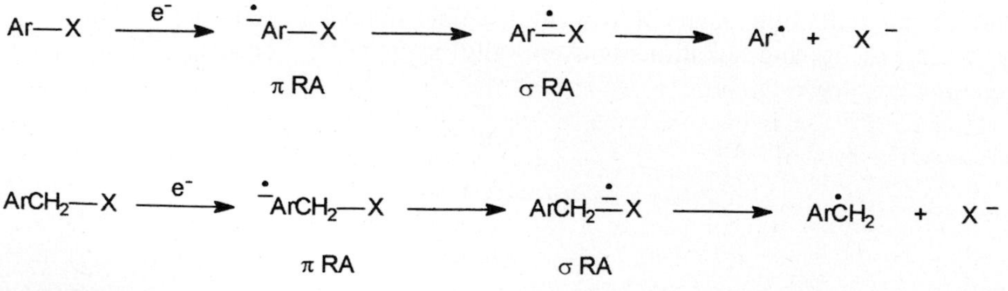

Ar—X $\xrightarrow{e^-}$ $^{\bullet-}$Ar—X ⟶ Ar$\overset{\bullet}{\doteq}$X ⟶ Ar$^{\bullet}$ + X$^-$

    $\pi$ RA                $\sigma$ RA

ArCH$_2$—X $\xrightarrow{e^-}$ $^{\bullet-}$ArCH$_2$—X ⟶ ArCH$_2\overset{\bullet}{\doteq}$X ⟶ ArĊH$_2$ + X$^-$

      $\pi$ RA                    $\sigma$ RA

**Scheme 78.**

**Table 12.** Rate constants for fragmentation of $ArX^{•-}$ and $ArCH_2X^{•-}$ in aqueous solution at room temperature.

| Ar | pH | $k$ (s$^{-1}$) (X = Cl) | $k$ (s$^{-1}$) (X = Br) | $k$ (s$^{-1}$) (X = I) |
|---|---|---|---|---|
| *ArX$^{•-}$* | | | | |
| 4-CNC$_6$H$_4$ | 12 | $5 \times 10^6$ | $>3 \times 10^7$ | – |
| 3-CNC$_6$H$_4$ | 12 | $4.2 \times 10^4$ | $8 \times 10^6$ | – |
| 2-CNC$_6$H$_4$ | 12 | $9 \times 10^6$ | – | – |
| 4-CH$_3$COC$_6$H$_4$ | 12 | $\sim 10^2$ | $5 \times 10^3$ | $1.4 \times 10^5$ |
| | 7 | – | – | $4 \times 10^3$ |
| 3-CH$_3$COC$_6$H$_4$ | 12 | – | $\sim 10^2$ | – |
| 2-CH$_3$COC$_6$H$_4$ | 12 | $1.5 \times 10^3$ | $5 \times 10^5$ | – |
| | 7 | – | $9 \times 10^3$ | – |
| 4-NO$_2$C$_6$H$_4$ | 7 | $<1$ | – | $<1$ |
| 3-NO$_2$C$_6$H$_4$ | 7 | – | $<1$ | – |
| 2-NO$_2$C$_6$H$_4$ | 7 | – | – | $<1$ |
| *ArCH$_2$X$^{•-}$* | | | | |
| 4-CNC$_6$H$_4$ | 12 | $>3 \times 10^7$ | $>6 \times 10^7$ | – |
| 3-CNC$_6$H$_4$ | 12 | – | $1.3 \times 10^7$ | – |
| 3-CH$_3$COC$_6$H$_4$ | 12 | $1.5 \times 10^4$ | – | – |
| | 7 | $4 \times 10^3$ | $1.7 \times 10^5$ | $5.7 \times 10^5$ |
| 4-NO$_2$C$_6$H$_4$ | 2 | – | $2 \times 10^4$ | – |
| | 0.6 | – | $3 \times 10^3$ | – |
| 3-NO$_2$C$_6$H$_4$ | 7 | $<5$ | 60 | $3 \times 10^3$ |
| 2-NO$_2$C$_6$H$_4$ | 7 | $1.0 \times 10^4$ | $4.0 \times 10^5$ | – |

which progressively increase with the size of the halogen atom, prevent conjugation of the nitro group with the aromatic ring thus reducing the stabilization of the $\pi^*$ orbital, favoring dehalogenation (see later) [285, 286]. On the other hand, fragmentation from the *meta* isomer is always significantly slower than from the *para*, because of the very low spin density at the *meta* position which disfavors intramolecular electron transfer to the halide.

Aryl halide radical anions ($^{•-}$ArX) undergo fragmentation much more slowly than the homologous benzyl halide ones ($^{•-}$ArCH$_2$X) in spite of the increased availability of the unpaired electron to the halogen in the former ones, due to reduced distance of the halogen from the ring. Such behavior reflects the importance of the C–X bond strength (which is significantly lower for ArCH$_2$X as compared to ArX [287, 288]) and thus the stability of the carbon centered radical formed by fragmentation. Interestingly, recent thermodynamic calculations have shown that while the C–X BDE is lowered by a factor of ca 4 on going from C$_6$H$_5$-Br to the corresponding radical anion, a 14-fold decrease is instead observed in the case of C$_6$H$_5$CH$_2$-Cl [289]. An additional factor accounting for this difference in reactivity is represented by the orientation of the C–X bond, since only in the benzyl halide radical anions the most favorable conformation for dissociative intramolecular electron transfer, where the C–X bond is perpendicular to the plane of the aromatic ring, allowing the most efficient orbital overlap, can be achieved [290]. In this re-

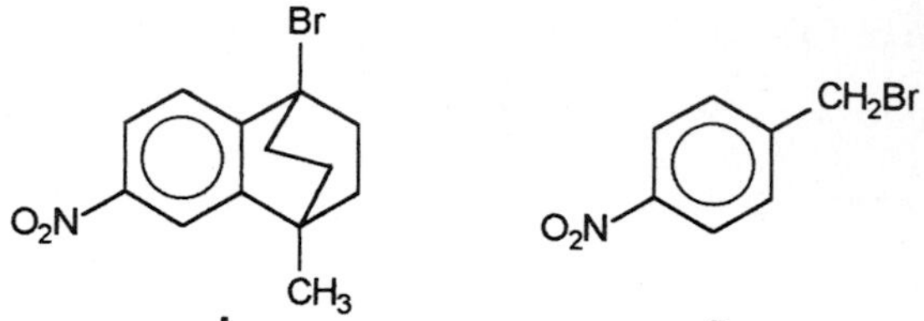

**Scheme 79.**

spect, it has been suggested that the transition state for cleavage of $\cdot^-$ArX must include a large intramolecular $\pi^* \rightarrow \sigma^*$ electron transfer component [291].

Clear evidence for orbital symmetry restrictions in dissociative intramolecular electron-transfer reactions has been provided by Savéant [292], by comparing the cleavage kinetics of radical anions generated from substrates **I** and **II** in DMF (Scheme 79).

In **I**$\cdot^-$, the rigid bicyclooctane structure precludes overlap between the $\pi^*$ orbital, bearing the unpaired electron, and the C–Br $\sigma^*$ orbital where the electron should be transferred concertedly with bond cleavage and moreover, fragmentation leads in this case to a bridgehead unconjugated sp$^3$ radical which is significantly less stable that the conjugated 4-nitrobenzyl radical formed by cleavage of **II**$\cdot^-$. Accordingly, **I**$\cdot^-$ undergoes cleavage with $k = 9.3 \times 10^{-3}$ s$^{-1}$, a value which is dramatically lower than that estimated for **II**$\cdot^-$ ($k \geq 7.9 \times 10^8$ s$^{-1}$) [292, 293].

No evidence for C–X bond cleavage has been found for the halonitrobenzene radical anions [283], while relatively fast dehalogenation occurs when Z = CN and COCH$_3$ [281, 282], due to the much higher electron affinity of the NO$_2$ group as compared to CN and COCH$_3$, which diminishes to a larger extent the overall ring charge density. Accordingly, the following order of reactivity is observed for dehalogenation of $\cdot^-$ArX: NO$_2$ < COCH$_3$ < CN (Table 12).

Protonation of the radical anion (Scheme 80) results in a significant decrease in the fragmentation rate, as clearly shown for 4-BrC$_6$H$_4$CH$_2$NO$_2$$\cdot^-$ (for which a p$K_a$ = 2.8 has been determined [284]), 2-Br- and 4-IC$_6$H$_4$COCH$_3$$\cdot^-$ [281].

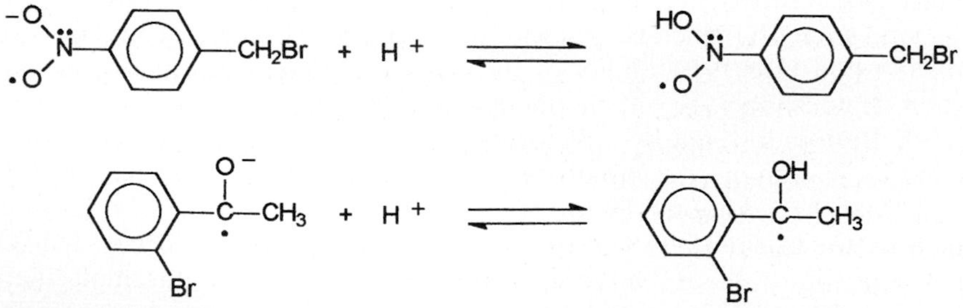

**Scheme 80.**

**Table 13.** Rate constants for fragmentation of $4\text{-RC}_6\text{H}_4\text{NO}_2^{\bullet-}$ in aqueous solution at room temperature.

| R | $k\ (\text{s}^{-1})$ |
|---|---|
| $CH_2Cl$ | $4 \times 10^3$ |
| $CH(CH_3)Cl$ | $9.7 \times 10^4$ |
| $CH(C(CH_3)_3)Cl$ | $4.0 \times 10^2$ |
| $CH_2Br$ | $1.7 \times 10^5$ |
| $CH(CH_3)Br$ | $3.5 \times 10^6$ |
| $CH(C(CH_3)_3)Br$ | $6.2 \times 10^4$ |
| $CH_2CH_2Br$ | $<1$ |
| $COCH_2Br$ | $4.2 \times 10^4$ |
| $CH{=}CHBr$ | $\leq 4$ |
| $CH{=}CHCH_2Br$ | $1 \times 10^5$ |

### Structural effects

To gain additional information on these fragmentation reactions, the reactivity of a wide variety of structurally modified nitrobenzyl halide radical anions has been investigated (Table 13) [283, 290, 294].

Addition of an $\alpha$-methyl substituent to $4\text{-XCH}_2\text{C}_6\text{H}_4\text{NO}_2^{\bullet-}$ (X = Cl, Br) reduces the C–X bond dissociation energy by stabilizing the resulting benzyl radical thus increasing the dehalogenation rate by a factor of $\approx 20\text{–}25$ [294].

On the other hand, $\alpha$-substitution with a *tert*-butyl group decreases the rate by a factor of $\approx 3\text{–}10$, indicating that this group exerts a steric effect on the orientation of the C–X bond by preventing the system from assuming a conformation in which the C–X bond is perpendicular to the plane of the aromatic ring [290]. Interestingly, the presence of an $\alpha$-*tert*-butyl group appears also to decrease the rate of intramolecular through space electron transfer as shown by the $\approx 50$–fold decrease in reactivity observed on going from $2\text{-ClCH}_2\text{C}_6\text{H}_4\text{NO}_2^{\bullet-}$ to $2\text{-ClCH}(\text{C}(\text{CH}_3)_3)\text{C}_6\text{H}_4\text{NO}_2^{\bullet-}$.

The presence of an additional methylene group as in $4\text{-BrCH}_2\text{CH}_2\text{C}_6\text{H}_4\text{NO}_2^{\bullet-}$ disfavors intramolecular electron transfer by isolating the halogen from the unpaired electron, and accordingly for this radical anion fragmentation is estimated to occur with $k < 1\ \text{s}^{-1}$ [283]. When the additional methylene group is replaced by a carbonyl group ($4\text{-BrCH}_2\text{COC}_6\text{H}_4\text{NO}_2^{\bullet-}$), C–Br bond cleavage is much more efficient occurring with $k = 4.2 \times 10^4\ \text{s}^{-1}$ [283], due to weakening of the scissible bond and increased ease of intramolecular electron transfer (as compared to $4\text{-BrCH}_2\text{CH}_2\text{C}_6\text{H}_4\text{NO}_2$) provided by the presence of the carbonyl spacer.

It is also interesting to compare the reactivity of radical anions bearing a vinyl and an allyl spacer with that of $4\text{-BrCH}_2\text{C}_6\text{H}_4\text{NO}_2^{\bullet-}$ (Table 13). While with $4\text{-BrCH}{=}\text{CHC}_6\text{H}_4\text{NO}_2^{\bullet-}$ a dramatic decrease in reactivity is observed, reflecting the relatively high C-Br bond strength and thus the reduced stability of a vinyl radical as compared with benzylic; with $4\text{-BrCH}_2\text{CH}{=}\text{CHC}_6\text{H}_4\text{NO}_2^{\bullet-}$ no significant decrease is observed indicating the effectiveness of the allyl group in channeling the electron to bromine [294].

In a later study, the temperature dependence on the unimolecular dehalogenation reaction of nitrobenzyl halide and haloacetophenone radical anions has been investigated [278], showing that the variation of the unimolecular rate constant with substrate structure depends in a complex way on the combination of $E_a$ and $\log A$, suggesting that comparisons between rate constants for fragmentation measured at a single temperature may lead to fortuitous conclusions. Interestingly, $\alpha$-substitution with methyl increases $A$ while ethyl and *tert*-butyl decrease it, suggesting that the presence of bulky substituents prevents the system from assuming a conformation suitable for intramolecular electron transfer [290].

### Comparison of concerted and stepwise cleavage

The problem of distinguishing between concerted and stepwise mechanisms in the reductive cleavage of both aryl and benzyl halides has been addressed by Savéant. While aliphatic halides undergo dehalogenation through the concerted pathway, with aryl halides intermediate radical anions are generally formed [274, 295]. This different behavior finds an explanation in the higher C–X bond dissociation energy of aryl rather than alkyl halides [287]. Moreover, in aryl halides, the incoming electron can be accommodated transitorily in the relatively low energy $\pi^*$ orbital of the aromatic moiety, while with simple aliphatic halides the C–X $\sigma^*$ orbital energy is so high that electron transfer concerted with bond cleavage is energetically more favorable. However, in a recent electrochemical study it has been found that the reductive cleavage of aromatic iodides in acetonitrile or DMF may follow a concerted mechanism at low driving forces [241]. For both iodobenzene and 3-methyliodobenzene an increase in the transfer coefficient $\alpha$ is observed on increasing the potential scan rate in cyclic voltammetry (CV) experiments, a behavior which indicates a transition from a concerted mechanism to a stepwise one as the driving force is increased [296]. An analogous behavior is observed at higher temperature, a condition which can play in favor of the concerted pathway [260]. Interestingly, no evidence for a change of mechanism has been observed with bromobenzene which undergoes a stepwise reductive cleavage even at the lower end of the scan rate domain. Such behavior has been explained in terms of the intrinsic barrier for cleavage since the concerted pathway is favored for iodobenzene as compared to bromobenzene by ca 0.17 eV [241]. From this example it follows that what governs the nature of the reductive cleavage is not the existence or non-existence of the radical anion but rather the energetic advantage offered by one mechanism over the other.

A borderline case is instead represented by benzyl halides, where a $\pi^*$ orbital is available for the incoming electron as in aryl halides, but the C–X bond is now significantly weaker. In this respect, Savéant studied the kinetics of the electrochemical reduction of a series of benzyl halides in acetonitrile and DMF, showing that while nitrobenzyl halides undergo a stepwise reductive cleavage via an intermediate radical anion, with weaker electron withdrawing ring substituents such as CN and $CO_2CH_3$, as well as with benzyl chloride and bromide or 9-anthracenylmethyl chloride, the reaction occurs via a concerted pathway [243, 293]. Apparently with nitrobenzyl halides, the $\pi^*$ orbital is too low in energy to allow a transition to a concerted mechanism at low CV scan rates (low driving forces), while the opposite situation holds for the cyanobenzyl halides where the $\pi^*$ energy is too high to allow a transition to a stepwise mechanism at high scan rates (high driving forces). In contrast with these conclusions is however the suggestion that, in the reaction of benzyl chloride with photoexcited $SmI_2$ in THF, reductive dehalogenation leading to a benzyl radical and $Cl^-$ occurs via a stepwise mechanism with formation of an intermediate radical anion [297].

### Effects of the medium

It is also interesting to compare the kinetic data for reductive dehalogenation in DMF obtained by Savéant ($k = 80$ and $4.0 \times 10^6$ s$^{-1}$ for 3-nitro- and 4-nitrobenzyl chloride, respectively and $k \gg 4.0 \times 10^6$ s$^{-1}$ for 4-nitrobenzyl bromide) [293] and Danen ($k = 8 \times 10^4$ s$^{-1}$ for 2-nitroiodobenzene measured chronoamperometrically) [286] with those obtained by Neta in aqueous solution for the same radical anions (Table 12: $k < 5$ s$^{-1}$, $k = 4 \times 10^3$ s$^{-1}$, $k = 1.7 \times 10^5$ s$^{-1}$ and $k < 1$ s$^{-1}$, respectively) [284], showing that the radical anions are significantly longer lived in water than in DMF. In nitrobenzyl and nitroaryl halide radical anions, the negative charge can be localized on the oxygen atoms of the nitro group, thus allowing a strong interaction with water. Two main consequences result from this stabilization: a lowering in the energy of the $\pi^*$ orbital, corresponding to a decrease in driving force of the cleavage reaction, and a large solvent reorganization energy, since solvation of the negatively charged nitro group has to be rearranged into solvation of the leaving halide anion. Accordingly, very large pre-exponential factors ($\log A$ between 12.8 and 17.1) have been measured for the dehalogenation of a series of nitrobenzyl halide radical anions in water, pointing towards a positive activation entropy [278, 293].

Moreover, 3-cyanobenzyl bromide radical anion has been *directly* observed by UV–Vis spectroscopy following pulse radiolytic generation in water, and a fragmentation rate constant $k = 1.3 \times 10^7$ s$^{-1}$ has been measured (Table 12) [282]; whereas, during electrochemical reduction in DMF, electron transfer and bond cleavage appear to be concerted. This different behavior can be explained on the basis of the solvation effect discussed above and also by considering that the driving force of the reaction (i.e. the energy of the incoming electron) is much larger when the radical anion is generated radiolytically [261, 293].

Solvation effects thus appear to play an important role in governing the dichotomy between concerted and stepwise mechanisms. In this respect, a detailed study of solvation and ion-pairing effects on the dynamics of intramolecular dissociative electron transfer in the cleavage of radical anions, generated by CV in MeCN or DMF, has been carried out [298]. Two families of compounds have been studied: the first one (group A) includes 4-chlorobenzophenone, 3-nitrobenzyl chloride and bromide, where the negative charge in the corresponding radical anions is localized on the oxygen atoms; the second one (group B) instead includes 2-chloro-, 9-fluoro- and 9-chloroanthracene, where the negative charge is delocalized over the entire system. In group A, the addition of water or of cations such as $Li^+$ or $Mg^{2+}$ strongly stabilizes the radical anion (corresponding to an increase in $E^\circ{}_{RX/RX \cdot -}$), thereby significantly decreasing its fragmentation rate constant. The effect of

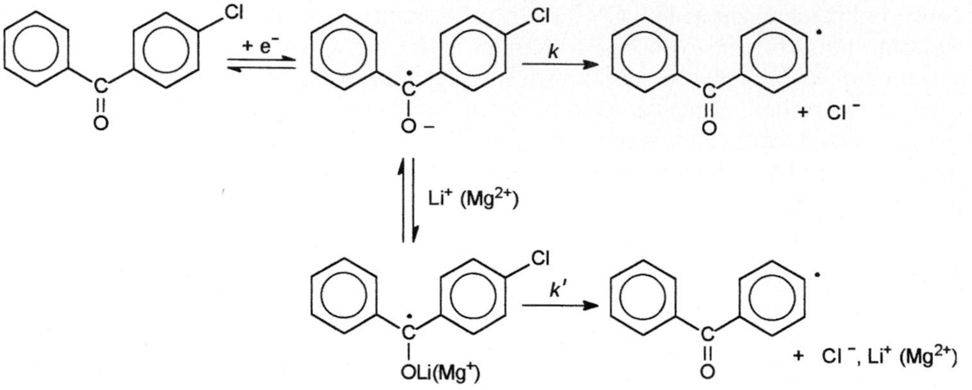

**Scheme 81.**

the addition of ion pairing cations is shown in Scheme 81 in the case of 4-chlorobenzophenone. Formation of the $RX^{\cdot-}Li^+$ adduct leads to a decrease in the $\pi^*$ orbital energy and thus in cleavage driving force, and for 4-chlorobenzophenone radical anion a ratio $k/k' = 20$ can be estimated. Even higher $k/k'$ ratios can be instead expected for cleavage of nitrobenzyl halide radical anions where, due to the weaker C–X bond as compared to aryl halides, smaller intrinsic barriers for cleavage are predicted.

In group B a slight increase in the fragmentation rate constant but no significant change in $E^\circ_{RX/RX\cdot-}$ is observed as the water concentration is raised, reflecting a specific solvation and thus a greater stabilization of the halide ion (corresponding to an increase in $E^\circ_{X\cdot/X-}$) which results in a decrease in the standard free energy of the reaction. On the other hand, addition of $Li^+$ has in this case no noticeable effect on the cleavage rate.

It is important to point out that stabilization of the halide ion by the presence of water occurs also in group A radical anions where, however, it is overcompensated by the lower energy of the $\pi^*$ orbital (reflected in the less negative $E^\circ_{RX/RX\cdot-}$) which disfavors intramolecular electron transfer to the $\sigma^*$ orbital of the scissible bond.

These observations clearly indicate that (in these cases) solvation and ion pairing effects on the cleavage reactivity of radical anions strongly depend upon the localization of the negative charge.

In another study, the reductive C–F bond cleavage of fluoromethylarenes has been investigated in liquid ammonia and DMF by CV and/or redox catalysis [299]. Within a series of 4-cyanotoluenes where the $\alpha$-carbon bears one, two or three fluorine atoms, the rate of radical anion cleavage increases on going from the trifluoro to the monofluoro derivative, a behavior which reflects the decrease in both C–F BDE and standard potential, $E^\circ_{RX/RX\cdot-}$, as the number of fluorine atoms is diminished (Table 14).

Along this line, within a series of trifluoromethylarenes, the cleavage rate constants are mainly governed by $E^\circ_{RX/RX\cdot-}$ which varies substantially as the aryl moiety is changed.

**Table 14.** Standard potentials $(E^\circ{}_{RX/RX\cdot-})$ and cleavage rate constants $(k)$ for 4-cyano-fluorotoluenes (4-CNC$_6$H$_4$X) and trifluoromethylarenes (4-XC$_6$H$_4$CF$_3$). Standard free energies $(\Delta G^\circ)$ and intrinsic barrier free energies $(\Delta G^{\neq}{}_0)$ for 4-CNC$_6$H$_4$X.

| X | DMF $(T = 20\,°C)$ | | | | NH$_3$ $(T = -38\,°C)$ | | | |
|---|---|---|---|---|---|---|---|---|
| | $E^{\circ\,a}$ | $k\ (s^{-1})$ | $\Delta G^\circ$ (eV) | $\Delta G^{\neq}{}_0$ (eV) | $E^{\circ\,b}$ | $k\ (s^{-1})$ | $\Delta G^\circ$ (eV) | $\Delta G^{\neq}{}_0$ (eV) |
| *4-CNC$_6$H$_4$X* | | | | | | | | |
| CH$_2$F | −2.020 | $7 \times 10^6$ | −0.93 | 0.85 | −1.540 | $2.2 \times 10^7$ | −1.35 | 0.91 |
| CHF$_2$ | −1.890 | $4 \times 10^5$ | −0.68 | 0.75 | −1.450 | $1 \times 10^6$ | −1.14 | 0.88 |
| CF$_3$ | −1.785 | $3.8 \times 10^1$ | −0.41 | 0.805 | −1.345 | $3.5 \times 10^2$ | −0.88 | 0.92 |
| *4-XC$_6$H$_4$CF$_3$* | | | | | | | | |
| CN | −1.785 | $3.8 \times 10^1$ | | | −1.345 | $3.5 \times 10^2$ | | |
| CO$_2$CH$_3$ | | | | | −1.330 | 7.5 | | |
| CO$_2t$Bu | | | | | −1.405 | $2.5 \times 10^1$ | | |
| CO$_2{}^-$ | | | | | −1.835 | $1.1 \times 10^4$ | | |
| H | | | | | −2.075 | $2.8 \times 10^8$ | | |

[a] V relative to SCE.
[b] V relative to 0.01 M Ag$^+$/Ag.

The cleavage rate constants for the 4-cyanofluorotoluenes measured in NH$_3$ are higher than those measured in DMF, in spite of the much lower temperature of the former system, reflecting the better stabilization of F$^-$ by this solvent as compared to DMF, as also indicated by the activation free energies $(\Delta G^\circ)$. Interestingly, the intrinsic barrier free energies $(\Delta G^{\neq}{}_0)$ are similar in the two solvents, due to a compensation between two contributions: nuclear reorganization, which is more important in DMF and solvent reorganization, which is instead more costly in NH$_3$.

### Mesolytic cleavage in radical anions of 1,2-diphenylethane derivatives

Diphenylethane derivatives bearing electron withdrawing ring substituents are suitable compounds for the study of mesolytic C–C bond fragmentation reactions in the corresponding radical anions. This problem has been addressed by Maslak using a series of sterically crowded 4-nitrodiphenylethane derivatives, **1** [300], where the incoming electron is added to the π-system of the nitrophenyl moiety leading to a radical anion (**1·−**), which can then undergo homolytic C–C bond cleavage to give an α,α-dialkyl-4-nitrobenzyl anion and an α,α-dialkylbenzyl radical (Scheme 82, X = H).

X-ray studies on strained 1,2-diphenyltetraalkyldiphenylethanes have shown that due to steric crowding, these molecules have a significantly elongated central C–C bond, and in the solid state they adopt a conformation where the torsional angles between this bond and the aromatic ring planes are very close to 90° [301], allowing the best overlap between the σ* orbitals of this bond and both aromatic π-systems. ESR studies suggest that an analogous conformation is also observed for the corresponding radical anions [291], thus allowing delocalization of the electron density

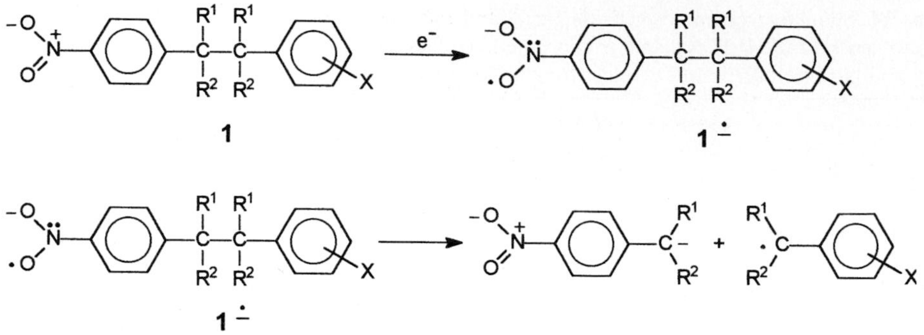

**Scheme 82.**

(initially localized over the nitroaryl moiety on one side of the scissible bond) over the entire molecule as the reaction progresses.

The unimolecular rate constants for C–C bond cleavage of $\mathbf{1}^{\cdot-}$ measured by ESR in DMSO are collected in Table 15, together with the corresponding activation free energies for mesolytic cleavage, $\Delta G^{\neq}_{m}$ [300].

The decay rates of the radical anions have been found to depend on the steric crowding. No fragmentation products but only that deriving from reduction of the parent compound has been detected from the relatively unstrained $\mathbf{1a}^{\cdot-}$ which decays with a measurable rate only at $T \geq 80\,°C$, exhibiting however second-order kinetics, a behavior attributed to disproportionation of $\mathbf{1a}^{\cdot-}$. The importance of C–C bond cleavage increases with increasing steric crowding and accordingly, while $\mathbf{1b}^{\cdot-}$ undergoes cleavage in competition with reduction, exclusive unimolecular C–C fragmentation has been observed for all the other radical anions in the series.

All cleavages are endoergonic and the activation free energies of mesolysis are on average 12.5 kcal mol$^{-1}$ lower than those for homolysis indicating significant C–C

**Table 15.** Unimolecular rate constants ($k_m$) and activation free energies ($\Delta G^{\neq}_{m}$) for mesolytic cleavage of 4-nitrodiphenylethane derivative radical anions $\mathbf{1}^{\cdot-}$ (X = H), in DMSO at $T = 27\,°C$.

| Radical anion | $R^1$ | $R^2$ | $k_m$ (s$^{-1}$) | $\Delta G^{\neq}_{m}$ (kcal mol$^{-1}$) |
|---|---|---|---|---|
| $\mathbf{1a}^{\cdot-}$ | Me | Me | b | $\geq 30.0$ |
| $\mathbf{1b}^{\cdot-}$ [a] | Me | Et | $4.4 \times 10^{-8}$ | 28.0 |
| $\mathbf{1c}^{\cdot-}$ [a] | Me | Pr | $1.0 \times 10^{-7}$ | 26.5 |
| $\mathbf{1d}^{\cdot-}$ [a] | Me | iBu | $2.0 \times 10^{-5}$ | 23.5 |
| $\mathbf{1e}^{\cdot-}$ | Et | Et | $6.3 \times 10^{-4}$ | 21.4 |
| $\mathbf{1f}^{\cdot-}$ | Pr | Pr | $2.0 \times 10^{-3}$ | 20.6 |
| $\mathbf{1g}^{\cdot-}$ | Bu | Bu | $3.1 \times 10^{-3}$ | 20.6 |

[a] *Erythro* diastereoisomer.
[b] Second-order kinetics.

**Table 16.** Unimolecular rate constants $(k_m)$ for mesolytic cleavage of $2^{\cdot-}$ (Scheme 82, $R^1 =$ $R^2 = Et$), in DMSO at $T = 20\,^{\circ}C$.

| Radical anion | X | $k_m$ $(s^{-1})$ |
|---|---|---|
| $2a^{\cdot-}$ | 4-CN | $2.3 \times 10^{-2}$ |
| $2b^{\cdot-}$ | 4-CF$_3$ | $9.3 \times 10^{-3}$ |
| $2c^{\cdot-}$ | 4-CH$_3$CO | $9.3 \times 10^{-3}$ |
| $2d^{\cdot-}$ | 4-NMe$_3{}^+$ | $4.0 \times 10^{-3}$ |
| $2e^{\cdot-}$ | 3-CF$_3$ | $2.3 \times 10^{-3}$ |
| $2f^{\cdot-}$ | 3-CN | $1.8 \times 10^{-3}$ |
| $2g^{\cdot-}$ | 4-Cl | $1.3 \times 10^{-3}$ |
| $2h^{\cdot-}$ | H | $6.5 \times 10^{-4}$ |
| $2i^{\cdot-}$ | 4-Me$_2$N | $5.4 \times 10^{-4}$ |
| $2j^{\cdot-}$ | 4-F | $4.5 \times 10^{-4}$ |
| $2k^{\cdot-}$ | 4-OMe | $3.8 \times 10^{-4}$ |

bond activation upon one-electron reduction. Such large activation has been shown to have thermodynamic origins, i.e. the transition state for mesolysis is lowered in energy as compared to that for homolysis by the difference in their free energies $(\Delta G^{\neq}{}_h - \Delta G^{\neq}{}_m \approx \Delta G_h - \Delta G_m)$.

The progress of charge delocalization on going from the radical anion to the cleavage transition state can be evaluated from the effect of ring substituents X on the rate of unimolecular C–C bond cleavage. The pertinent rate constants $k_m$ for a series of 4-nitrodiphenylethane derivative radical anions $2^{\cdot-}$ (Scheme 82, $R^1 = R^2 = Et$), measured by ESR in DMSO, are reported in Table 16 [291, 302].

The rate of C–C bond cleavage is increased by electron withdrawing X substituents while electron releasing ones determine a slight retarding effect. The very similar values of activation and reaction free energies $(\Delta G^{\neq}{}_m$ and $\Delta G_m$, respectively) observed for cleavage of $2^{\cdot-}$, point towards a late product-like transition state, suggesting that the relative cleavage rates reflect the radical stabilizing effect of the X substituent.

Quite surprisingly, however, no correlation has been observed between the relative cleavage rates and $\sigma^{\cdot}$ while a good correlation with $\sigma^-$ exists for the radical anions bearing strongly electron withdrawing substituents, a trend which is consistent with the development of a significant amount of negative charge on the $\beta$-carbon, in contrast with the predictions based on the thermodynamic stability of the fragments.

These results indicate that the fragmentation is kinetically controlled by the ability of the system to delocalize the negative charge across the scissible bond, suggesting that the transition state for C–C bond cleavage in $2^{\cdot-}$ can be described in terms of two electronic configurations: one, corresponding to relatively early transition states, where the $\beta$-carbon supports a significant amount of negative charge, becomes important with strongly electron withdrawing X substituents $(2a$–$e^{\cdot-})$; and

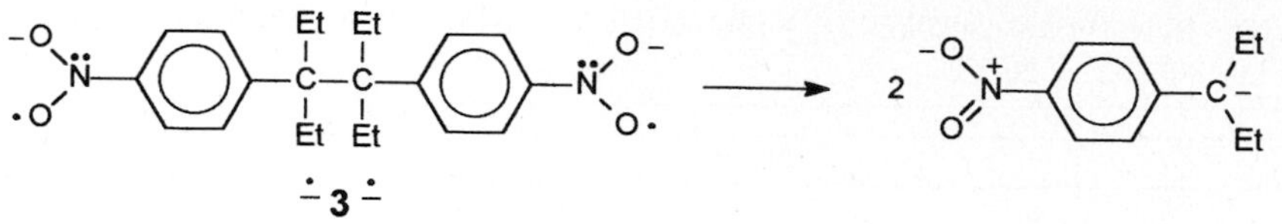

**Scheme 83.**

a second one represented by a $\sigma^*$ radical anion (three electron C–C bond) where the additional electron is shared by both benzylic carbons, which becomes increasingly important as the ability of the radical anions to delocalize the negative charge across the scissible bond is decreased, corresponding to progressively later transition states (**2f–k$^{\cdot-}$**).

Along this line, the importance of through bond delocalization has been clearly shown for the dinitro derivative (4-$O_2NC_6H_4CEt_2$)$_2$ (**3**) [291, 303], whose reduction leads to a dianion ($^{-\cdot}$**3$^{\cdot-}$**) having each of the two unpaired electrons delocalized over one nitroaryl moiety (Scheme 83). $^{-\cdot}$**3$^{\cdot-}$** undergoes a bimolecular fragmentation reaction (rationalized in terms of disproportionation of **3$^{\cdot-}$** followed by cleavage of $^{-\cdot}$**3$^{\cdot-}$**) which is at least 3 orders of magnitude faster than that (unimolecular) from **2a$^{\cdot-}$**.

In contrast with **2a$^{\cdot-}$** for which all of the thermodynamic advantage is expressed in the lowering of transition state energy (i.e. $\Delta G_m \approx \Delta G^{\neq}_m$), the driving force for cleavage of $^{-\cdot}$**3$^{\cdot-}$** is more negative by ca 14 kcal mol$^{-1}$ exhibiting however a significantly larger intrinsic barrier, suggested to derive from the inability of the dianion to delocalize the negative charge across the scissible bond due to charge repulsion. As a result, for $^{-\cdot}$**3$^{\cdot-}$** $\Delta G_m \ll \Delta G^{\neq}_m$, which shows that the possibility of through bond delocalization constitutes a kinetic advantage in the cleavage reactions of diphenylethane derivative radical anions.

Additional information comes from the study of the heterolytic C–C bond cleavage reactions of 4-cyanodiphenylethane derivative radical anions (**4$^{\cdot-}$**) generated by PET in MeCN (Scheme 84) [304]. The rate constants for fragmentation are collected in Table 17.

A dramatic increase in reactivity is observed on going from **1$^{\cdot-}$** and **2$^{\cdot-}$** (Tables 15 and 16, respectively) to **4$^{\cdot-}$**, indicating that the presence of an electron withdrawing substituent on the $\beta$-carbon strongly stabilizes the incipient negative charge which develops on this atom in the transition state for heterolytic fragmentation.

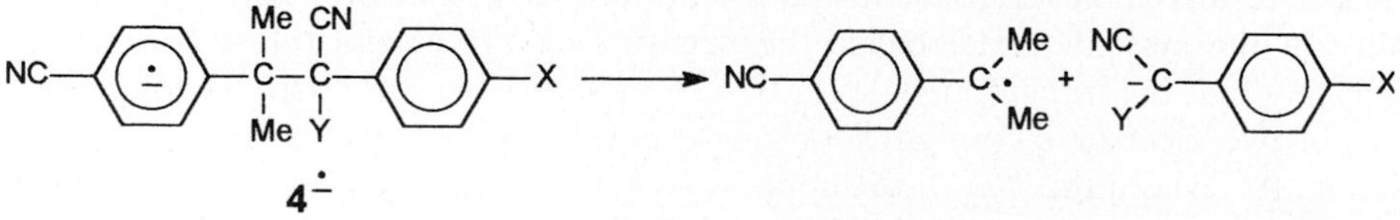

**Scheme 84.**

**Table 17.** Unimolecular rate constants $(k_m)$ for mesolytic cleavage of $\mathbf{4^{\cdot-}}$, in MeCN at $T = 27\,^{\circ}\mathrm{C}$.

| Radical anion | X | Y | $k_m$ $(s^{-1})$ |
|---|---|---|---|
| $\mathbf{4a^{\cdot-}}$ | OMe | Me | $8 \times 10^5$ |
| $\mathbf{4b^{\cdot-}}$ | H | Me | $9 \times 10^6$ |
| $\mathbf{4c^{\cdot-}}$ | $CF_3$ | Me | $4 \times 10^{8\,a}$ |
| $\mathbf{4d^{\cdot-}}$ | OMe | CN | $4 \times 10^{9\,a}$ |
| $\mathbf{4e^{\cdot-}}$ | H | CN | $>3 \times 10^{10\,a}$ |

[a] Estimated value.

# References

1. P. Neta, *Adv. Phys. Org. Chem.* **1976**, *12*, 223–297.
2. A. Henglein, *Electroanal. Chem.* **1976**, *9*, 163–244.
3. A. J. Swallow, *Progr. Reaction Kin.* **1978**, *9*, 195–366.
4. S. Steenken, *J. Chem. Soc., Faraday Trans. 1* **1987**, *83*, 113–124.
5. Farhataziz, M. A. J. Rodgers, VCH Publishers, Weinheim 1987.
6. S. Steenken, *Chem. Rev.* **1989**, *89*, 503–520.
7. C. von Sonntag, H. P. Schuchmann, *Angew. Chem. Int. Ed. Engl.* **1991**, *30*, 1229–1253.
8. J. A. Stubbe, W. A. van der Donk, *Chem. Rev.* **1998**, *98*, 705.
9. H. Eklund, M. Fontecave, *Structure With Folding & Design* **1999**, *7*, R257–R262.
10. B. A. Andersson, S. Styring, in C. P. Lee (Ed.): *Current Topics in Bioenergetics, Vol. 16*, Academic Press, San Diego, CA 1991, p. 1–81.
11. C. W. Hoganson, M. Sahlin, B.-M. Sjöberg, G. T. Babcock, *J. Am. Chem. Soc.* **1996**, *118*, 4672–4679.
12. S. Kim, B. A. Barry, *Biochemistry* **1998**, *37*, 13882–13892.
13. H. Sies, *Oxidative Stress*, Academic Press, Orlando, FL 1985.
14. K. C. Cundy, R. Kohen, B. N. Ames, in M. G. Simic, K. A. Taylor, J. F. Ward, C. von Sonntag (Eds.): *Oxygen Radicals in Biology and Medicine, Vol. 49*, Plenum Press, New York, London 1988, p. 479–482.
15. B. Giese, *Radicals in Organic Synthesis: Formation of Carbon–Carbon Bonds*, Pergamon Press, Oxford 1986.
16. M. Regitz, B. Giese, in Houben–Weyl (Ed.): *Methoden der Organischen Chemie, Vol. E19a*, Georg Thieme Verlag, Stuttgart 1989.
17. W. B. Motherwell, D. Crich, *Free Radical Chain Reactions in Organic Synthesis*, Academic Press, London 1992.
18. M. J. Perkins, *Radical Chemistry*, Ellis Horwood, New York 1994.
19. J. Fossey, D. Lefort, J. Sorba, *Free Radicals in Organic Chemistry*, Wiley, Chichester 1995.
20. T. Linker, M. Schmittel, *Radikale und Radikalionen in der Organischen Synthese*, Wiley–VCH, Weinheim 1998.
21. B. Paulson, K. Pramod, P. Eaton, G. Closs, J. R. Miller, *J. Phys. Chem.* **1993**, *97*, 13042–13045.
22. J. W. Verhoeven, In J. Jortner, M. Bixton (Eds.): *Electron Transfer—From Isolated Molecules To Biomolecules, Pt 1, Vol. 106*, John Wiley & Sons, New York Et Al. 1999, P. 603–644.
23. J. M. Warman, M. P. De Haas, J. W. Verhoeven, M. N. Paddon-Row, In J. Jortner, M. Bixton (Eds.): *Electron Transfer—From Isolated Molecules To Biomolecules, Pt 1, Vol. 106*, John Wiley & Sons, New York Et Al. 1999, P. 571–601.
24. A. J. Bard, M. A. Fox, *Acc. Chem. Res.* **1995**, *28*, 141–145.
25. T. J. Meyer, *Acc. Chem. Res.* **1989**, *22*, 163–170.
26. M. Grätzel, *Acc. Chem. Res.* **1981**, *14*, 376.
27. G. McLendon, *Acc. Chem. Res.* **1988**, *21*, 160–167.

28. S. S. Isied, M. Y. Ogawa, J. F. Wishart, *Chem. Rev.* **1992**, *92*, 381–394.
29. R. J. P. Williams, *J. Biol. Inorg. Chem.* **1997**, *2*, 373–377.
30. S. S. Skourtis, D. N. Beratan, *J. Biol. Inorg. Chem.* **1997**, *2*, 378–386.
31. C. C. Moser, C. C. Page, X. Chen, P. L. Dutton, *J. Biol. Inorg. Chem.* **1997**, 393–398.
32. J. R. Winkler, H. B. Gray, *J. Biol. Inorg. Chem.* **1997**, *2*, 399–404.
33. E. D. A. Stemp, J. K. Barton, *Metal Ions in Biol. Systems* **1996**, *33*, 325–365.
34. J. Jortner, M. Bixon, T. Langenbacher, M. E. Michel–Beyerle, *Proc. Natl. Acad. Sci. USA* **1998**, *95*, 12759–12765.
35. B. Giese, S. Wessely, M. Spormann, U. Lindemann, E. Meggers, M. E. Michel-Beyerle, *Angew. Chem. Int. Ed. Engl.* **1999**, *38*, 996–998.
36. Also (electron transfer) disproportionation is a way for radicals to lose their radical nature. With respect to disproportionation, the reader is referred to Alfassi, Z.B. (ed), General Aspects of the Chemistry of Radicals, Wiley, New York 1999.
37. R. A. Marcus, *Ann. Rev. Phys. Chem.* **1964**, *15*, 155–196.
38. J. S. Littler, In R. O. C. Norman (Ed.): *Essays On Free-Radical Chemistry, Special Publication 24*, The Chemical Society, London 1970.
39. L. Eberson, *Electron-transfer reactions in Organic Chemistry, Vol. 25*, Springer, Berlin 1987.
40. *Homolytic β-eliminations* are also known, see, e.g., Kim, S.; Cheong, J. H. *Chem. Commun.* **1998**, 1143–1144 and references therein.
41. The reverse of eq 3 is an example of a *homolytic* reaction.
42. L. Eberson, *Adv. Phys. Org. Chem.* **1982**, *18*, 79.
43. D. M. Stanbury, *Adv. Inorg. Chem.* **1989**, *33*, 69–138.
44. R. E. Huie, C. L. Clifton, P. Neta, *Radiat. Phys. Chem.* **1991**, *38*, 477–481.
45. S. Steenken, unpublished material.
46. O. P. Chawla, R. W. Fessenden, *J. Phys. Chem.* **1975**, *79*, 2693–2700.
47. M. J. Davies, B. C. Gilbert, *J. Chem. Soc. Perkin Trans. 2* **1984**, 1809–1815.
48. The rate constants also increase with alkyl substitution at the C=C double bond (see Table 1), again evidence for the polar nature of the transition state of the interaction of $SO_4{}^{\bullet-}$ with C=C.
49. G. Koltzenburg, E. Bastian, S. Steenken, *Angew. Chem. Int. Ed. Engl.* **1988**, *27*, 1066–1067.
50. The lifetime of the radical cation is <20 ns, as deduced from 193 nm photoionization experiments of the cyclohexene in aqueous solution.
51. If there are *two* methyl groups at C=C, the reaction with $SO_4{}^{\bullet-}$ leads 'directly', i.e. with a delay of <10 ns, to the β-hydroxyalkyl radical and $H^+$.
52. With this number for $k_{het}$, the distinction between addition–elimination and outer-sphere electron transfer becomes a borderline case.
53. S. Steenken, in J. Mattay (Ed.): *Topics in Current Chemistry 177, Electron Transfer II*, Springer, Berlin **1996**, p. 125–145.
54. H-abstraction, particularly from allylic positions, is also observed, but is typically a minor (<10 %) process.
55. C. von Sonntag, *Adv. Carbohydr. Chem. Biochem.* **1980**, *37*, 7–77.
56. J. Stubbe, W. A. Van der Donk, *Chemistry & Biology* **1995**, *2*, 793–801.
57. R. Lenz, B. Giese, *J. Am. Chem. Soc.* **1997**, *119*, 2784–2794.
58. A. L. Buley, R. O. C. Norman, R. J. Pritchett, *J. Chem. Soc. (B)* **1966**, 849–852.
59. R. Livingston, H. Zeldes, *J. Am. Chem. Soc.* **1966**, *88*, 4333–4336.
60. C. von Sonntag, E. Thoms, *Z. Naturf. B* **1970**, *25b*, 1405.
61. K. M. Bansal, M. Grätzel, A. Henglein, E. Janata, *J. Phys. Chem.* **1973**, *77*, 16–19.
62. B. C. Gilbert, R. O. C. Norman, P. S. Williams, *J. Chem. Soc. Perkin Trans. 2* **1980**, 647–656.
63. S. Steenken, *J. Phys. Chem.* **1979**, *83*, 595–599.
64. B. C. Gilbert, J. P. Larkin, R. O. C. Norman, *J. Chem. Soc. Perkin Trans. 2* **1972**, 794–802.
65. C. Walling, R. A. Johnson, *J. Am. Chem. Soc.* **1975**, *97*, 2405.
66. S. Steenken, M. J. Davies, B. C. Gilbert, *J. Chem. Soc., Perkin Trans. 2* **1986**, 1003–1010.
67. B. C. Gilbert, R. O. C. Norman, R. C. Sealy, *J. Chem. Soc. Perkin Trans. 2* **1974**, 824–830.
68. T. Foster, P. R. West, *Can. J. Chem.* **1973**, *51*, 4009–4017.
69. G. Behrens, G. Koltzenburg, *Z. Naturforsch.* **1985**, *40c*, 785–797.

70. A computational study on the mechanism of rearrangements of $\beta$-acyloxyalkyl radicals has been performed: Zipse, H. *J. Am. Chem. Soc.* **1997**, *119*, 1087.
71. T. Matsushige, G. Koltzenburg, D. Schulte-Frohlinde, *Ber. Bunsenges. Phys. Chem.* **1975**, *79*, 657–661.
72. G. Koltzenburg, T. Matsushige, D. Schulte-Frohlinde, *Z. Naturforsch.* **1976**, *31b*, 960–964.
73. G. Behrens, D. Schulte-Frohlinde, *Ber. Bunsenges. Phys. Chem.* **1976**, *80*, 429–436.
74. A further example for heterolysis of the acetate function are radicals from glycerol esters, see ref 75.
75. G. Koltzenburg, G. Behrens, D. Schulte-Frohlinde, *J. Am. Chem. Soc.* **1982**, *104*, 7311–7312.
76. This is certainly related to the fact that in the (*non*-radical) reaction involving *t*-BuCl a *localized* carbocation ($Me_3C^+$) is formed, whereas in the radical system (Eq. 16) the electrofuge is the *de*localized radical cation, $Me_2C^+CH_2^{\bullet}$.
77. The radical cation formed in the heterolysis can also be produced by 193 nm photoionization of cyclohexene in aqueous solution, see Scheme 8: S. Steenken, unpublished material.
78. G. Behrens, E. Bothe, J. Eibenberger, G. Koltzenburg, D. Schulte-Frohlinde, *Angew. Chem.* **1978**, *90*, 639.
79. G. Behrens, E. Bothe, G. Koltzenburg, D. Schulte-Frohlinde, *J. Chem. Soc. Perkin Trans. 2* **1980**, *3*, 883–889.
80. $(MeO)_2C^+CH_2^{\bullet}$ is able to oxidize N,N,N′,N′-tetramethyl-*p*-phenylenediamine (TMPD) to $TMPD^{\bullet+}$ with the rate constant $2.9 \times 10^9$ $\text{M}^{-1}$ $\text{s}^{-1}$: S. Steenken, unpublished.
81. G. Behrens, E. Bothe, G. Koltzenburg, D. Schulte-Frohlinde, *J. Chem. Soc. Perkin Trans. 2* **1981**, 143–154.
82. G. Koltzenburg, G. Behrens, D. Schulte-Frohlinde, *Angew. Chem. Int. Ed. Engl.* **1983**, *22*, 500–1.
83. M. Dizdaroglu, C. von Sonntag, D. Schulte-Frohlinde, *J. Am. Chem. Soc.* **1975**, *97*, 2277–2278.
84. G. Behrens, G. Koltzenburg, D. Schulte-Frohlinde, *Z. Naturforsch.* **1982**, *37c*, 1205–1227.
85. A computational study of the mechanism of fragmentation of $\beta$-phospatoalkyl radicals has been published: Zipse, H. *J. Am. Chem. Soc.* **1997**, *119*, 2889. See also Newcomb, M.; Horner, J.H.; Whitted, P.O.; Crich, D.; Huang, X.; Yao, Q.; Zipse, H. *J. Am. Chem. Soc.* **1999**, *121*, 10685.
86. B. Giese, X. Beyrich-Graf, P. Erdmann, M. Petretta, U. Schwitter, *Chemistry and Biology* **1995**, *2*, 367–375.
87. A. Samuni, P. Neta, *J. Phys. Chem.* **1973**, *77*, 2425–2429.
88. S. Steenken, G. Behrens, D. Schulte-Frohlinde, *Int. J. Radiat. Biol.* **1974**, *25*, 205–210.
89. G. Behrens, G. Koltzenburg, A. Ritter, D. Schulte-Frohlinde, *Int. J. Radiat. Biol.* **1978**, *33*, 163–171.
90. In oligonucleotide C4′-type radicals, the rate of heterolytic phosphate elimination has been estimated as $10^2$–$10^3$ $\text{s}^{-1}$: B. Giese, A. Dussy, E. Meggers, M. Petretta, U. Schwitter, *J. Am. Chem. Soc.* **1997**, *119*, 11130. On the other hand, with the diethyl *ester* of phosphoric acid, the rate constant for $\beta$-elimination from a 4-methyl-2′-deoxyribos-4′-yl radical is $\geq 3 \times 10^9$ $\text{s}^{-1}$: A. Gugger, R. Batra, P. Rzadek, G. Rist, B. Giese, *J. Am. Chem. Soc.* **1997**, *119*, 8740.
91. B. Giese, P. Erdmann, L. Giraud, T. Göbel, A. M. Petretta, T. Schäfer, M. Von Raumer, *Tetrahedron. Lett.* **1994**, *35*, 2683–2686.
92. The number is taken from Figure 2 of ref. 91.
93. A. Gugger, R. Batra, P. Rzadek, G. Rist, B. Giese, *J. Am. Chem. Soc.* **1997**, *119*, 8740–8741.
94. C. von Sonntag, U. Hagen, A. Schön-Bopp, D. Schulte-Frohlinde, *Adv. Radiat. Biol.* **1981**, *6*, 109–142.
95. The base may be on the same ('intra-strand') or the complementary ('inter-strand' electron transfer) strand or it may be up to 5 nucleotide units below or above the radical cation [35].
96. The ET rates are distance-dependent, for details see E. Meggers, A. Dussy, T. Schäfer, B. Giese, *Chem. Eur. J.* **2000**, *6*, 485.
97. C. Walling, D. M. Camaioni, *J. Am. Chem. Soc.* **1975**, *97*, 1603.
98. C. Walling, G. M. El-Taliawi, C. Zhao, *J. Org. Chem.* **1983**, *48*, 4914–4917.
99. C. Walling, C. X. Zhao, G. M. El-Taliawi, *J. Org. Chem.* **1983**, *48*, 4910–4914.

100. P. O'Neill, S. Steenken, D. Schulte-Frohlinde, *J. Phys. Chem.* **1975**, *79*, 2773–2779.
101. P. Neta, V. Madhavan, H. Zemel, R. W. Fessenden, *J. Am. Chem. Soc.* **1977**, *99*, 163–164.
102. M. Bietti, E. Baciocchi, S. Steenken, *J. Phys. Chem. A* **1998**, *102*, 7337–7342.
103. K. M. Bansal, R. W. Fessenden, *Radiat. Res.* **1978**, *75*, 497–507.
104. H. M. Novais, S. Steenken, *J. Phys. Chem.* **1987**, *91*, 426–433.
105. D. J. Deeble, M. N. Schuchmann, S. Steenken, C. von Sonntag, *J. Phys. Chem.* **1990**, *94*, 8186–8192.
106. The assignment of the reaction products of $SO_4^{\cdot-}$ with N1-alkylated thymines in aqueous solution in terms of $SO_4^-$ adducts (Lomoth, R.; Naumov, S.; Brede, O. *J. Phys. Chem. A* **1999**, *103*, 6571) is in disagreement with the conductance results reported in ref 105.
107. H. A. Schwarz, R. W. Dodson, *J. Phys. Chem.* **1984**, *88*, 3643–3647.
108. U. K. Kläning, K. Sehested, J. Holcman, *J. Phys. Chem.* **1985**, *89*, 760–763.
109. The tendency of ˙OH to add to double bonds has recently been confirmed by MO calculations: M. Peräkylä, T.A. Pakkanen, *J. Chem. Soc. Perkin Trans. 2* **1995**, 1405.
110. P. Fournier de Violet, *Rev. Chem. Intermediates* **1981**, *4*, 121–169.
111. K.-D. Asmus, M. Bonifacic, P. Toffel, P. O'Neill, D. Schulte-Frohlinde, S. Steenken, *J. Chem. Soc. Faraday Trans. 1* **1978**, *74*, 1820–1826.
112. A further example for the ad–el mechanism is the reaction of ˙OH with the strong reductant ascorbic acid. This reaction, which leads to the oxidation of ascorbic acid to yield the ascorbate radical, proceeds by addtion to the olefinic bond of the reductone function followed by elimination of $OH^-$: Abe, A.; Okada, S.; Nakao, R.; Horii, T.; Inoue, H.; Taniguchi, S.; Yamabe, S. *J. Chem. Soc. Perkin Trans. 2* **1992**, 2221.
113. D. K. Hazra, S. Steenken, *J. Am.Chem. Soc.* **1983**, *105*, 4380–4386.
114. J. Lind, X. Shen, T. E. Eriksen, G. Merenyi, *J. Am. Chem. Soc.* **1990**, *112*, 479–482.
115. M. Anbar, D. Meyerstein, P. Neta, *J. Phys. Chem.* **1966**, *70*, 2660–2662.
116. P. O'Neill, D. Schulte-Frohlinde, S. Steenken, *J. Chem. Soc. Faraday Disc.* **1977**, *63*, 141–148.
117. G. N. R. Tripathi, *J. Am. Chem. Soc.* **1998**, *120*, 4161–4166.
118. The formation of the 4-aminophenoxyl radical (see Scheme 18) *proves* that addition has occurred. The protonated amino function is expected to be a powerful leaving group, cf eq 13. If, alternatively, $OH^-$ was the leaving group, the radical cation would be formed.
119. In order to explain the results, the term 'adduct-mediated electron transfer' (AMET) was introduced. This term implies that it is the adduct which *mediates* electron transfer. However, the adduct does not *mediate* ET but its heterolysis *constitutes* the ET. Furthermore, the allegation of a distinction between ET and 'AMET' creates a mechanistic dichotomy which is unnecessary. It is possible to explain all the results on the basis of a single (unified), i.e., the ad–el mechanism.
120. R. H. Schuler, P. Neta, H. Zemel, R. W. Fessenden, *J. Am. Chem. Soc.* **1976**, *98*, 3825–3831.
121. D. J. Deeble, S. Das, C. von Sonntag, *J. Phys. Chem.* **1985**, *89*, 5784–5788.
122. H. M. Novais, S. Steenken, *J. Am. Chem. Soc.* **1986**, *108*, 1–6.
123. L. P. Candeias, S. Steenken, *J. Phys. Chem.* **1992**, *96*, 937–944.
124. L. P. Candeias, P. Wolf, P. O'Neill, S. Steenken, *J. Phys. Chem.* **1992**, *96*, 10302–7.
125. For a discussion on side-chain fragmentation reactions involving other C–X β-bonds, a number of reviews are available (E. Baciocchi, M. Bietti, O. Lanzalunga, *Acc. Chem. Res.* **200**, *33*, 243–251. K. Mizuno, T. Tamai, A. Sugimoto, H. Maeda, *Advances in Electron Transfer Chemistry* **1999**, *6*, 131–165. R. S. Glass, *Topics in Current Chemistry* **1999**, *205*, 1–87. M. Schmittel, A. Burghart, *Angew. Chem. Int. Ed. Engl.* **1997**, *36*, 2550–2589. E. Baciocchi, *Acta Chem. Scand.* **1990**, *44*, 645–652.
126. P. Maslak, *Topics in Current Chemistry* **1993**, *168*, 1–46.
127. P. Maslak, J. N. Narvaez, *Angew. Chem. Int. Ed. Engl.* **1990**, *29*, 283–285.
128. D. D. M. Wayner, V. D. Parker, *Acc. Chem. Res.* **1993**, *26*, 287–294.
129. H. L. Friedman, C. V. Krishnan, in E. Franks (Ed.): *Water, A Comprehensive Treatise*, Plenum Press, New York **1973**.
130. A. M. D. Nicholas, D. R. Arnold, *Can. J. Chem.* **1982**, *60*, 2165–2179.
131. J. O. Howell, J. M. Goncalves, C. Amatore, L. Klasinc, R. M. Wightman, J. K. Kochi, *J. Am. Chem. Soc.* **1984**, *106*, 3968–3976.
132. C. J. Schlesener, C. Amatore, J. K. Kochi, *J. Am. Chem. Soc.* **1984**, *106*, 7472–7482.

133. K. Sehested, J. Holcman, *J. Phys. Chem.* **1978**, *82*, 651–653.
134. C. Russo-Caia, S. Steenken, to be published.
135. M. Schmittel, A. Burghart, *Angew. Chem. Int. Ed. Engl.* **1997**, *36*, 2551–2589.
136. O. Takahashi, O. Kikuchi, *Tetrahedron Lett.* **1991**, *32*, 4933–4936.
137. A. Anne, S. Fraoua, V. Grass, J. Moiroux, J.-M. Savéant, *J. Am. Chem. Soc.* **1998**, *120*, 2951–2958.
138. A. Anne, S. Fraoua, P. Hapiot, J. Moiroux, J.-M. Savéant, *J. Am. Chem. Soc.* **1995**, *117*, 7412–7421.
139. A. Anne, P. Hapiot, J. Moiroux, P. Neta, J.-M. Savéant, *J. Am. Chem. Soc.* **1992**, *114*, 4694–4701.
140. R. A. Marcus, *J. Phys. Chem.* **1968**, *72*, 891–899.
141. A. O. Cohen, R. A. Marcus, *J. Phys. Chem.* **1968**, *72*, 4249–4256.
142. J. P. Dinnocenzo, T. E. Banach, *J. Am. Chem. Soc.* **1989**, *111*, 8646–8653.
143. C. Amatore, J. K. Kochi, *Advances in Electron Transfer Chemistry* **1991**, *1*, 55–148.
144. C. J. Schlesener, C. Amatore, J. K. Kochi, *J. Phys. Chem.* **1986**, *90*, 3747–3756.
145. E. Baciocchi, T. Del Giacco, F. Elisei, *J. Am. Chem. Soc* **1993**, *115*, 12290–12295.
146. J. M. Masnovi, S. Sankararaman, J. K. Kochi, *J. Am. Chem. Soc.* **1989**, *111*, 2263–2276.
147. This value was determined by Savéant (A. Anne, S. Fraoua, V. Grass, J. Moiroux, J.-M. Savéant, *J. Am. Chem. Soc.* **1998**, *120*, 2951–2958) using Kochi's data, while Kochi obtained 0.3 eV.
148. E. Baciocchi, M. Mattioli, R. Romano, R. Ruzziconi, *J. Org. Chem.* **1991**, *56*, 7154–7160.
149. E. Baciocchi, A. Dalla Cort, L. Eberson, L. Mandolini, C. Rol, *J. Org. Chem.* **1986**, *51*, 4544–4548.
150. L. M. Tolbert, Z. Z. Li, S. R. Sirimanne, D. G. Vanderveer, *J. Org. Chem.* **1997**, *62*, 3927–3930.
151. L. M. Tolbert, R. K. Khanna, A. E. Popp, L. Gelbaum, L. A. Bottomley, *J. Am. Chem. Soc.* **1990**, *112*, 2373–2378.
152. L. M. Tolbert, R. K. Khanna, *J. Am. Chem. Soc.* **1987**, *109*.
153. E. Baciocchi, F. D'Acunzo, C. Galli, O. Lanzalunga, *J. Chem. Soc., Perkin Trans. 2* **1996**, 133–140.
154. V. D. Parker, E. T. Chao, G. Zheng, *J. Am. Chem. Soc.* **1997**, *119*, 11390–11394.
155. V. D. Parker, Y. X. Zhao, Y. Lu, G. Zheng, *J. Am. Chem. Soc.* **1998**, *120*, 12720–12727.
156. B. Reitstoen, V. D. Parker, *J. Am. Chem. Soc.* **1991**, *113*, 6954–6958.
157. J. M. Mayer, *Acc. Chem. Res.* **1998**, *31*, 441–450.
158. T. M. Bockman, S. M. Hubig, J. K. Kochi, *J. Am. Chem. Soc.* **1998**, *120*, 2826–2830.
159. S. M. Hubig, R. Rathore, J. K. Kochi, *J. Am. Chem. Soc.* **1999**, *121*, 617–626.
160. K. A. Gardner, L. L. Kuehnert, J. M. Mayer, *Inorganic Chemistry* **1997**, *36*, 2069–2078.
161. T. Delgiacco, E. Baciocchi, S. Steenken, *J. Phys. Chem.* **1993**, *97*, 5451–56.
162. P. R. Ortiz De Montellano, *Cytochrome P-450: Structure, Mechanism And Biochemistry*, Plenum Press, New York **1995**.
163. T. G. Traylor, K. W. Hill, W. P. Fann, S. Tsuchiya, B. E. Dunlap, *J. Am. Chem. Soc.* **1992**, *114*, 1308–1312.
164. L. Eberson, *J. Am. Chem. Soc.* **1983**, *105*, 3192–3199.
165. T. L. Macdonald, W. G. Gutheim, R. B. Martin, F. P. Guengerich, *Biochemistry* **1989**, *28*, 2071–2077.
166. E. Baciocchi, *Xenobiotica* **1995**, *25*, 653–666.
167. R. Amodeo, E. Baciocchi, M. Crescenzi, O. Lanzalunga, *Tetrahedron Lett.* **1990**, *31*, 3477–3480.
168. E. Baciocchi, S. Belvedere, M. Bietti, O. Lanzalunga, *Eur. J. Org. Chem.* **1998**, 299–302.
169. E. Baciocchi, T. Del Giacco, F. Elisei, O. Lanzalunga, *J. Am. Chem. Soc.* **1998**, *120*, 11800–11801.
170. E. Baciocchi, M. Bietti, S. Steenken, *Chem. Eur. J.* **1999**, *5*, 1785–1793.
171. E. Baciocchi, M. Bietti, S. Steenken, *J. Am. Chem. Soc.* **1997**, *119*, 4078–4079.
172. E. Baciocchi, M. Bietti, L. Manduchi, S. Steenken, *J. Am. Chem. Soc.* **1999**, *121*, 6624–6629.
173. M. Eigen, *Angew. Chem., Int. Ed. Engl.* **1964**, *3*, 1–19.
174. V. Bachler, E. Baciocchi, M. Bietti, S. Steenken, to be published.

175. A. Anne, S. Fraoua, J. Moiroux, J.-M. Savéant, *J. Am. Chem. Soc.* **1996**, *118*, 3938–3945.
176. P. Maslak, T. M. Vallombroso, W. H. Chapman, Jr., J. N. Narvaez, *Angew. Chem. Int. Ed. Engl.* **1994**, *33*, 73–75.
177. L. Bardi, E. Fasani, A. Albini, *J. Chem. Soc. Perkin Trans. 1* **1994**, 545–549.
178. E. Baciocchi, D. Bartoli, C. Rol, R. Ruzziconi, G. V. Sebastiani, *J. Org. Chem.* **1986**, *51*, 3587–3593.
179. D. M. Camaioni, J. A. Franz, *J. Org. Chem.* **1984**, *49*, 1607–1613.
180. D. R. Arnold, L. J. Lamont, *Can. J. Chem.* **1989**, *67*, 2119–2127.
181. A. Okamoto, M. S. Snow, D. R. Arnold, *Tetrahedron* **1986**, *42*, 6175–87.
182. D. R. Arnold, B. J. Fahie, L. J. Lamont, J. Wierzchowski, K. M. Young, *Can. J. Chem.* **1987**, *65*, 2734–2743.
183. A. L. Perrott, D. R. Arnold, *Can. J. Chem.* **1992**, *70*, 272–279.
184. D. R. Arnold, L. J. Lamont, A. L. Perrott, *Can. J. Chem.* **1991**, *69*, 225–233.
185. M. J. Davies, B. C. Gilbert, *Advances In Detailed Reaction Mechanisms* **1991**, *1*, 35–81.
186. B. C. Gilbert, C. J. Warren, *Res. Chem. Intermediat.* **1989**, *11*, 1–17.
187. S. Steenken, C. J. Warren, B. C. Gilbert, *J. Chem. Soc. Perkin Trans. 2* **1990**, 335–342.
188. J. P. Dinnocenzo, T. R. Simpson, H. Zuilhof, W. P. Todd, T. Heinrich, *J. Am. Chem. Soc.* **1997**, *119*, 987–993.
189. J. P. Dinnocenzo, W. P. Todd, T. R. Simpson, I. R. Gould, *J. Am. Chem. Soc.* **1990**, *112*, 2462–2464.
190. J. P. Dinnocenzo, H. Zuilhof, D. R. Lieberman, T. R. Simpson, M. W. Mckechney, *J. Am. Chem. Soc.* **1997**, *119*, 994–1004.
191. J. P. Dinnocenzo, D. R. Lieberman, T. R. Simpson, *J. Am. Chem. Soc.* **1993**, *115*, 366–367.
192. M. Freccero, A. Pratt, A. Albini, C. Long, *J. Am. Chem. Soc.* **1998**, *120*, 284–297.
193. R. Popielarz, D. R. Arnold, *J. Am. Chem. Soc.* **1990**, *112*, 3068–3082.
194. D. M. Camaioni, *J. Am. Chem. Soc.* **1990**, *112*, 9475–9483.
195. However, when bibenzylic radical cations are generated by photoinduced electron transfer, no deprotonation is observed since this process is relatively inefficient compared to back electron transfer (R. Popielarz, D. R. Arnold, *J. Am. Chem. Soc.* **1990**, *112*, 3068–3082. H. F. Davis, P. K. Das, L. W. Reichel, G. W. Griffin, *J. Am. Chem. Soc.* **1984**, *106*, 6968–6973
196. E. Baciocchi, M. Bietti, L. Putignani, S. Steenken, *J. Am. Chem. Soc.* **1996**, *118*, 5952–5960.
197. E. Baciocchi, C. Rol, G. V. Sebastiani, L. Taglieri, *J. Org. Chem.* **1994**, *59*, 5272–5276.
198. S. Steenken, R. A. McClelland, *J. Am. Chem. Soc.* **1989**, *111*, 4967–4973.
199. P. Maslak, W. H. J. Chapman, T. M. J. Vallombroso, B. A. Watson, *J. Am. Chem. Soc.* **1995**, *117*, 12380–12389.
200. P. Maslak, S. L. Asel, *J. Am. Chem. Soc.* **1988**, *110*, 8260–8261.
201. S. Perrier, S. Sankararaman, J. K. Kochi, *J. Chem Soc. Perkin Trans. 2* **1993**, 825–837.
202. S. Sankararaman, S. Perrier, J. K. Kochi, *J. Am. Chem. Soc.* **1989**, *111*, 6448–6449.
203. T. M. Bockman, S. M. Hubig, J. K. Kochi, *J. Am. Chem. Soc.* **1998**, *120*, 6542–6547.
204. P. Maslak, W. H. Chapman, *J. Org. Chem.* **1996**, *61*, 2647–2656.
205. P. Maslak, W. H. Chapman, *J. Org. Chem.* **1990**, *55*, 6334–6347.
206. P. Maslak, W. H. Chapman, *Tetrahedron* **1990**, *46*, 2715–2724.
207. E. R. Gaillard, D. G. Whitten, *Acc. Chem. Res.* **1996**, *29*, 292–297.
208. L. H. Chen, M. S. Farahat, E. R. Gaillard, S. Farid, D. G. Whitten, *J. Photochem. Photobiol. A: Chem.* **1996**, *95*, 21–25.
209. H. Gan, U. Leinhos, I. R. Gould, D. G. Whitten, *J. Phys. Chem.* **1995**, *99*, 3566–3573.
210. A. Albini, M. Mella, *Tetrahedron* **1986**, *42*, 6219–6224.
211. J. P. Dinnocenzo, M. Merchan, B. O. Roos, S. Shaik, H. Zuilhof, *J. Phys. Chem. A* **1998**, *102*, 8979–8987.
212. Y. H. Wang, J. M. Tanko, *J. Am. Chem. Soc.* **1997**, *119*, 8201–8208.
213. Y. H. Wang, J. M. Tanko, *J. Chem. Soc. Perkin Trans. 2* **1998**, 2705–2711.
214. Y. H. Wang, K. H. Mclean, J. M. Tanko, *J. Org. Chem.* **1998**, *63*, 628–635.
215. E. Baciocchi, M. Bietti, O. Lanzalunga, S. Steenken, *J. Am. Chem. Soc.* **1998**, *120*, 11516–11517.

216. R. D. Burton, M. D. Bartberger, Y. Zhang, J. R. Eyler, K. S. Schanze, *J. Am. Chem. Soc.* **1996**, *118*, 5655–5664.
217. X. H. Ci, M. A. Kellett, D. G. Whitten, *J. Am. Chem. Soc.* **1991**, *113*, 3893–3904.
218. X. Ci, D. G. Whitten, *J. Am. Chem. Soc.* **1989**, *111*, 3459–3461.
219. L. A. Lucia, R. D. Burton, K. S. Schanze, *J. Phys. Chem.* **1993**, *97*, 9078–9080.
220. M. J. Davies, B. C. Gilbert, C. W. Mccleland, C. B. Thomas, J. Young, *J. Chem. Soc. Chem. Commun.* **1984**, 966–967.
221. D. V. Avila, J. Lusztyk, K. U. Ingold, *J. Am. Chem. Soc.* **1992**, *114*, 6576–6577.
222. D. V. Avila, K. U. Ingold, A. A. Di Nardo, F. Zerbetto, M. Z. Zgierski, J. Lusztyk, *J. Am. Chem. Soc.* **1995**, *117*, 2711–2718.
223. B. Ashworth, B. C. Gilbert, R. G. G. Holmes, R. O. C. Norman, *J. Chem. Soc., Perkin Trans. 2* **1978**, 951–956.
224. H. Zemel, R. W. Fessenden, *J. Phys. Chem.* **1975**, *79*, 1419–1427.
225. B. C. Gilbert, J. P. Larkin, R. O. C. Norman, *J. Chem. Soc. Perkin Trans. 2* **1972**, 1272–1279.
226. S. Steenken, P. O'neill, D. Schulte-Frohlinde, *J. Phys. Chem.* **1977**, *81*, 26–30.
227. J. Chateauneuf, J. Lusztyk, K. U. Ingold, *J. Am. Chem. Soc.* **1988**, *110*, 2877–2885.
228. B. C. Gilbert, C. J. Scarratt, C. B. Thomas, J. Young, *J. Chem. Soc. Perkin Trans. 2* **1987**, *2*, 371–80.
229. L. Jönsson, *Acta Chem. Scand.* **1983**, *B37*, 761–768.
230. R. M. Dessau, E. I. Heiba, *J. Org. Chem.* **1975**, *40*, 3647–3649.
231. W. S. Trahanowsky, J. Cramer, D. W. Brixius, *J. Am. Chem. Soc.* **1974**, *96*, 1077–1081.
232. T. M. Bockman, S. M. Hubig, J. K. Kochi, *J. Org. Chem.* **1997**, *62*, 2210–2221.
233. T. M. Bockman, S. M. Hubig, J. K. Kochi, *J. Am. Chem. Soc.* **1996**, *118*, 4502–4503.
234. L. K. Mehta, M. Porssa, J. Parrick, L. P. Candeias, P. Wardman, *J. Chem. Soc., Perkin Trans. 2* **1997**, 1487–1491.
235. C. Walling, G. M. El-Taliawi, K. Amarnath, *J. Am. Chem. Soc.* **1984**, *106*, 7573–7578.
236. J.-M. Savéant, *J. Am. Chem. Soc.* **1992**, *114*, 10595–10602.
237. J.-M. Savéant, *J. Am. Chem. Soc.* **1987**, *109*, 6788–6795.
238. J.-M. Savéant, In P. S. Mariano (Ed.): *Advances In Electron Transfer Chemistry, Vol. 4*, Jai Press Inc., Greenwich, Connecticut and London, England 1994, P. 53–116.
239. J.-M. Savéant, *Acc. Chem. Res.* **1993**, *26*, 455–461.
240. S. Antonello, F. Maran, *J. Am. Chem. Soc.* **1999**, *121*, 9668–9676.
241. L. Pause, M. Robert, J. M. Saveant, *J. Am. Chem. Soc.* **1999**, *121*, 7158–7159.
242. R. L. Donkers, F. Maran, D. D. M. Wayner, M. S. Workentin, *J. Am. Chem. Soc.* **1999**, *121*, 7239–7248.
243. C. Costentin, P. Hapiot, M. Medebielle, J.-M. Savéant, *J. Am. Chem. Soc.* **1999**, *121*, 4451–4460.
244. K. Daasbjerg, H. Jensen, R. Benassi, F. Taddei, S. Antonello, A. Gennaro, F. Maran, *J. Am. Chem. Soc.* **1999**, *121*, 1750–1751.
245. C. P. Andrieux, J.-M. Savéant, C. Tardy, *J. Am. Chem. Soc.* **1998**, *120*, 4167–4175.
246. C. P. Andrieux, J.-M. Savéant, A. Tallec, R. Tardivel, C. Tardy, *J. Am. Chem. Soc.* **1997**, *119*, 2420–2429.
247. M. L. Andersen, W. Long, D. D. M. Wayner, *J. Am. Chem. Soc.* **1997**, *119*, 6590–6595.
248. M. L. Andersen, N. Mathivanan, D. D. M. Wayner, *J. Am. Chem. Soc.* **1996**, *118*, 4871–4879.
249. C. P. Andrieux, J.-M. Savéant, A. Tallec, R. Tardivel, C. Tardy, *J. Am. Chem. Soc.* **1996**, *118*, 9788–9789.
250. D. D. Tanner, J. J. Chen, L. Chen, C. Luelo, *J. Am. Chem. Soc.* **1991**, *113*, 8074–8081.
251. S. Antonello, F. Maran, *J. Am. Chem. Soc.* **1997**, *119*, 12595–12600.
252. R. L. Donkers, J. Tse, M. S. Workentin, *Chem. Commun.* **1999**, 135–136.
253. M. S. Workentin, R. L. Donkers, *J. Am. Chem. Soc.* **1998**, *120*, 2664–2665.
254. R. L. Donkers, M. S. Workentin, *J. Phys. Chem. B* **1998**, *102*, 4061–4063.
255. S. Antonello, M. Musumeci, D. D. M. Wayner, F. Maran, *J. Am. Chem. Soc.* **1997**, *119*, 9541–9549.
256. M. S. Workentin, F. Maran, D. D. M. Wayner, *J. Am. Chem. Soc.* **1995**, *117*, 2120–2121.

257. S. Jakobsen, H. Jensen, S. U. Pedersen, K. Daasbjerg, *J. Phys. Chem. A* **1999**, *103*, 4141–4143.
258. M. G. Severin, G. Farnia, E. Vianello, M. C. Arevalo, *J. Electroanal. Chem.* **1988**, *251*, 369–382.
259. X. Z. Wang, F. D. Saeva, J. A. Kampmeier, *J. Am. Chem. Soc.* **1999**, *121*, 4364–4368.
260. C. P. Andrieux, J.-M. Savéant, C. Tardy, *J. Am. Chem. Soc.* **1997**, *119*, 11546–11547.
261. C. P. Andrieux, M. Robert, F. D. Saeva, J.-M. Savéant, *J. Am. Chem. Soc.* **1994**, *116*, 7864–7871.
262. F. D. Saeva, In P. S. Mariano (Ed.): *Advances In Electron Transfer Chemistry, Vol. 4*, Jai Press Inc., Greenwich, Connecticut and London, England 1994, P. 1–25.
263. F. D. Saeva, *Top. Curr. Chem.* **1990**, *156*, 59–92.
264. Z. R. Zheng, D. H. Evans, E. S. Chan-Shing, J. Lessard, *J. Am. Chem. Soc.* **1999**, *121*, 9429–9434.
265. F. F. Wu, T. F. Guarr, R. D. Guthrie, *J. Phys. Org. Chem.* **1992**, *5*, 7–18.
266. P. Maslak, J. Theroff, *J. Am. Chem. Soc.* **1996**, *118*, 7235–7236.
267. R. D. Guthrie, M. Patwardhan, J. E. Chateauneuf, *J. Phys. Org. Chem.* **1994**, *7*, 147–152.
268. R. D. Guthrie, B. C. Shi, *J. Am. Chem. Soc.* **1990**, *112*, 3136–3139.
269. P. Maslak, R. D. Guthrie, *J. Am. Chem. Soc.* **1986**, *108*, 2637–2640.
270. P. Maslak, R. D. Guthrie, *J. Am. Chem. Soc.* **1986**, *108*, 2628–2636.
271. H. Lund, K. Daasbjerg, T. Lund, S. U. Pedersen, *Acc. Chem. Res.* **1995**, *28*, 313–319.
272. J.-M. Savéant, *Tetrahedron* **1994**, *50*, 10117–10165.
273. R. A. Rossi, A. B. Pierini, S. M. Palacios, *Advances In Free Radical Chemistry* **1990**, *1*, 193–252.
274. J.-M. Savéant, *Adv. Phys. Org. Chem.* **1990**, *26*, 1–130.
275. R. A. Rossi, R. H. De Rossi, *Aromatic Substitution By The $S_{rn}1$ Mechanism, Vol. 178*, American Chemical Society, Washington D.C. 1983.
276. R. A. Rossi, *Acc. Chem. Res.* **1982**, *15*, 164–170.
277. J. F. Bunnett, *Acc. Chem. Res.* **1978**, *11*, 413–420.
278. M. Meot-Ner, P. Neta, R. K. Norris, K. Wilson, *J. Phys. Chem.* **1986**, *90*, 168.
279. A. B. Pierini, J. S. Duca, D. M. A. Vera, *J. Chem. Soc., Perkin Trans. 2* **1999**, 1003–1009.
280. A. B. Pierini, J. S. Duca, *J. Chem. Soc., Perkin Trans. 2* **1995**, 1821–1828.
281. D. Behar, P. Neta, *J. Am. Chem. Soc.* **1981**, *103*, 2280–2283.
282. P. Neta, D. Behar, *J. Am. Chem. Soc.* **1981**, *103*, 103–106.
283. D. Behar, P. Neta, *J. Phys. Chem.* **1981**, *85*, 690–693.
284. P. Neta, D. Behar, *J. Am. Chem. Soc.* **1980**, *102*, 4798–4802.
285. C. Galli, *Tetrahedron* **1988**, *44*, 5205–5208.
286. W. C. Danen, T. T. Kensler, J. G. Lawless, M. F. Marcus, M. D. Hawley, *J. Phys. Chem.* **1969**, *73*, 4389–4391.
287. *Handbook of Chemistry and Physics*, CRC Press, Boca Raton, Fl 1998–1999.
288. D. A. Pratt, J. S. Wright, K. U. Ingold, *J. Am. Chem. Soc.* **1999**, *121*, 4877–4882.
289. C. Galli, T. Pau, *Tetrahedron* **1998**, *54*, 2893–2904.
290. R. K. Norris, S. D. Barker, P. Neta, *J. Am. Chem. Soc.* **1984**, *106*, 3140–3144.
291. P. Maslak, J. N. Narvaez, J. Kula, D. S. Malinski, *J. Org. Chem.* **1990**, *55*, 4550–4559.
292. W. Adcock, C. P. Andrieux, C. I. Clark, A. Neudeck, J.-M. Savéant, C. Tardy, *J. Am. Chem. Soc.* **1995**, *117*, 8285–8286.
293. C. P. Andrieux, A. Le Gorande, J.-M. Savéant, *J. Am. Chem. Soc.* **1992**, *114*, 6892–6904.
294. J. P. Bays, S. T. Blumer, S. Baral-Tosh, D. Behar, P. Neta, *J. Am. Chem. Soc.* **1983**, *105*, 320–324.
295. J.-M. Savéant, *J. Phys. Chem.* **1994**, *98*, 3716–3724.
296. The driving force for R–X$^{\cdot-}$ bond cleavage may be expressed as a function of the bond dissociation energy of the neutral compound, $BDE_{RX}$, the standard potentials for radical anion formation and oxidation of the leaving anion X$^-$, $E^0_{RX/RX^{\cdot-}}$ and $E^0_{X^\cdot/X^-}$, respectively, and an entropic term, $\Delta S = S_{R^\cdot} + S_X - S_{RX}$: $\Delta G^0 = BDE_{RX} + E^0_{RX/RX^{\cdot-}} - E^0_{X^\cdot/X^-} - T\Delta S$.
297. W. G. Skene, J. C. Scaiano, F. L. Cozens, *J. Org. Chem.* **1996**, *61*, 7918–7921.
298. C. P. Andrieux, M. Robert, J.-M. Savéant, *J. Am. Chem. Soc.* **1995**, *117*, 9340–9346.

299. C. P. Andrieux, C. Combellas, F. Kanoufi, J.-M. Savéant, A. Thiebault, *J. Am. Chem. Soc.* **1997**, *119*, 9527–9540.
300. P. Maslak, J. N. Narvaez, T. M. Vallombroso, *J. Am. Chem. Soc.* **1995**, *117*, 12373–12379.
301. P. Maslak, J. N. Narvaez, M. Parvez, *J. Org. Chem.* **1991**, *56*, 602–607.
302. P. Maslak, J. N. Narvaez, *J. Chem. Soc., Chem. Commun.* **1989**, 138–139.
303. P. Maslak, J. Kula, J. N. Narvaez, *J. Org. Chem.* **1990**, *55*, 2277–2279.
304. P. Maslak, J. Kula, J. E. Chateauneuf, *J. Am. Chem. Soc.* **1991**, *113*, 2304–2306.

Volume II

Part 2

Organometallic and Inorganic Molecules

# 1 Reflections on the Two-state Electron-transfer Model

*Bruce S. Brunschwig and Norman Sutin*

## 1.1 Introduction

There is general agreement that the two most important factors determining electron transfer rates in solution are the degree of electronic interaction between the donor and acceptor sites, and the changes in the nuclear configurations of the donor, acceptor, and surrounding medium that occur upon the gain or loss of an electron [1–5]. The electronic interaction of the sites will be very weak, and the electron transfer slow, when the sites are far apart or their interaction is symmetry or spin forbidden. Since electron motion is much faster than nuclear motion, energy conservation requires that, prior to the actual electron transfer, the nuclear configurations of the reactants and the surrounding medium adjust from their equilibrium values to a configuration (generally) intermediate between that of the reactants and products. In the case of electron transfer between two metal complexes in a polar solvent, the nuclear configuration changes involve adjustments in the metal–ligand and intra-ligand bond lengths and angles, and changes in the orientations of the surrounding solvent molecules. In common with ordinary chemical reactions, an electron transfer reaction can then be described in terms of the motion of the system on an energy surface from the reactant equilibrium configuration (initial state) to the product equilibrium configuration (final state) via the activated complex (transition state) configuration.

This chapter will focus on the predictions of the traditional two-state electron transfer model. Only the ground and lowest excited state of the system are considered and contributions from higher electronic states are ignored. Thermal and optical electron transfers in both weakly and strongly interacting systems are discussed. The treatment is not intended to be exhaustive but instead will focus on certain features of the model that may be less familiar but which nevertheless have important implications.

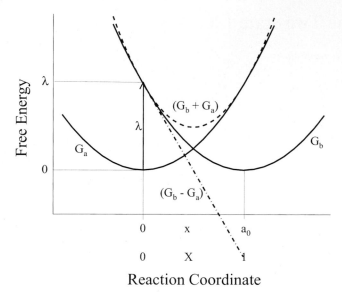

**Figure 1.** Plot of the diabatic free-energies of the reactants (left-hand curve, $G_a$) and products (right-hand curve, $G_b$) against the reaction coordinate for an electron transfer reaction with $\Delta G° = 0$. The sum (dashes) and difference (dot–dash) of the reactant and product free-energies are also plotted.

## 1.2  Zero-order Energy Surfaces

Provided a hypothetical change in the charge on the reactants produces a proportional change in the dielectric polarization of the surrounding medium, the distortions of the reactants and products from their equilibrium configurations can be described in terms of displacements on harmonic free-energy curves with identical force constants [1–5]. This is illustrated in Figures 1 and 2 where the free energy of the close-contact reactants plus surrounding medium (Curve $G_a$) and the free energy of the close-contact products plus surrounding medium (Curve $G_b$) are plotted against the reaction coordinate for a self-exchange reaction.

The free-energy curves depict the zero-order or diabatic states of the system. Figure 1 shows the diabatic free-energy curves for a self-exchange reaction (Eq. 1a, $\Delta G° = 0$) and Figure 2 the curves for an electron transfer reaction accompanied by a net chemical change (Eq. 1b, $\Delta G° < 0$ for an exergonic reaction).

$$\text{Fe}(\text{H}_2\text{O})_6{}^{2+} + \text{Fe}(\text{H}_2\text{O})_6{}^{3+} = \text{Fe}(\text{H}_2\text{O})_6{}^{3+} + \text{Fe}(\text{H}_2\text{O})_6{}^{2+} \tag{1a}$$

$$\text{Fe}(\text{H}_2\text{O})_6{}^{2+} + \text{Ru}(\text{bpy})_3{}^{3+} = \text{Fe}(\text{H}_2\text{O})_6{}^{3+} + \text{Ru}(\text{bpy})_3{}^{2+} \tag{1b}$$

The curves have identical force constants $f$ and their minima are separated by $a_0$. The vertical difference between the free energies of the reactants and products of a self-exchange reaction at the reactants' (or products') minimum (equilibrium configuration) is the reorganization parameter $\lambda = fa_0{}^2/2$. Denoting the displacement along the reaction coordinate by $x$, a dimensionless reaction coordinate $X$ may be defined as $x/a_0$: $X$ varies from 0 to 1 as the reaction proceeds and, with $X$ as the

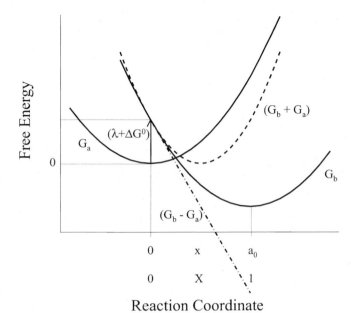

**Figure 2.** Plot of the diabatic free-energies of the reactants $(G_a)$ and products $(G_b)$ against the reaction coordinate for an electron transfer reaction with $\Delta G° < 0$. The sum (dashes) and difference (dot–dash) of the reactant and product free-energies are also plotted.

coordinate, the force constants of the parabolas are equal to $2\lambda$.

$$G_a = \frac{f}{2}x^2 = \lambda X^2 \tag{2a}$$

$$G_b = \frac{f}{2}(x - a_o)^2 + \Delta G° = \lambda(X - 1)^2 + \Delta G° \tag{2b}$$

The sum and difference of the zero-order free energies are given by:

$$G_b + G_a = 2\lambda(X - 1/2)^2 + \lambda/2 + \Delta G° \tag{3a}$$

$$G_b - G_a = \lambda(1 - 2X) + \Delta G° \tag{3b}$$

and the free-energy curves defined by these equations are included in Figures 1 and 2. The sum of the free energies of the reactants and products is a parabola [6] with force constant $4\lambda$ centered at $X = 1/2$ with its minimum vertically displaced relative to the reactant minimum by $\lambda/2 + \Delta G°$. Similarly, the dependence of the average diabatic energy $(G_b + G_a)/2$ on $X$ also is harmonic but with force constant $2\lambda$, identical to that of the separated reactants and product curves. The parabola defined by the average energies *is still centered at* $X = 1/2$ but with its minimum vertically displaced relative to the reactant minimum by $\lambda/4 + \Delta G°/2$. Since the difference between the diabatic free-energies of the reactants and products $(G_b - G_a)$ is linearly related to $X$, this difference affords a measure of the progress of the reaction [7, 8] and, as a consequence, it provides an alternate definition of the reaction coordinate. For both self-exchange reactions and reactions accompanied by a net chemical change, the slope of a plot of $(G_b - G_a)$ against $X$ is $-2\lambda$.

The free energy of activation for the electron transfer is the difference between the free energies of the transition-state configuration and the equilibrium configuration of the reactants.

$$\Delta G^* = G_X{}^* - G_{a, eq} \tag{4}$$

The equilibrium configuration of the reactants in the zero-interaction limit is located at $X = 0$ with $G_{a, eq} = 0$. At the transition state, $G_a{}^* = G_b{}^*$ so that $X^*$ and the free energy of activation in the zero-interaction limit are given by Eqs 5a and 5b, respectively.

$$X^* = \frac{1}{2}\left(1 + \frac{\Delta G^\circ}{\lambda}\right) \tag{5a}$$

$$\Delta G^* = \lambda(X^*)^2 = \frac{\lambda}{4}\left(1 + \frac{\Delta G^\circ}{\lambda}\right)^2 \tag{5b}$$

Evidently $\Delta G^* = \lambda/4$ for $\Delta G^\circ = 0$. Three free-energy regimes can be distinguished depending on the relative magnitudes of $\lambda$ and $\Delta G^\circ$. When $-\Delta G^\circ < \lambda$ the reaction is in the normal regime where $\Delta G^*$ decreases, and the rate constant increases, with increasing driving force. The reaction becomes barrierless ($\Delta G^* = 0$) when $-\Delta G^\circ = \lambda$ and $\Delta G^*$ is then insensitive to changes in $\Delta G^\circ$. If the driving force is increased even further then $-\Delta G^\circ > \lambda$ and $\Delta G^*$ increases, and the rate constant decreases, with increasing driving force. This is the counter-intuitive inverted regime.

By using the Gibbs–Helmholtz equations, it follows from Eq. 5b that the activation enthalpy and entropy are given by:

$$\Delta H^* = \frac{\Delta H_\lambda}{4}\left[1 - \left(\frac{\Delta G^\circ}{\lambda}\right)^2\right] + \frac{\Delta H^\circ}{2}\left(1 + \frac{\Delta G^\circ}{\lambda}\right) \tag{6a}$$

$$\Delta S^* = \frac{\Delta S_\lambda}{4}\left[1 - \left(\frac{\Delta G^\circ}{\lambda}\right)^2\right] + \frac{\Delta S^\circ}{2}\left(1 + \frac{\Delta G^\circ}{\lambda}\right) \tag{6b}$$

where $\Delta H_\lambda = \partial(\lambda/T)/\partial(1/T)$ and $\Delta S_\lambda = \partial(\lambda)/\partial(T)$ [9]. Eq. 5b for the free energy of activation can be rewritten as:

$$\Delta G^* = \frac{\lambda}{4}\left[1 - \left(\frac{\Delta G^\circ}{\lambda}\right)^2\right] + \frac{\Delta G^\circ}{2}\left(1 + \frac{\Delta G^\circ}{\lambda}\right) \tag{6c}$$

which is the same form as the activation enthalpy and entropy expressions.

The above equations give the activation parameters derived for the classical model. Departures are expected, and observed, for non-parabolic surfaces that are

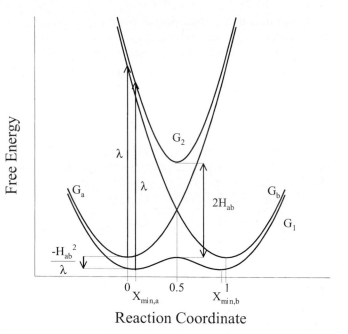

**Figure 3.** Plot of the diabatic $(G_a, G_b)$ and adiabatic $(G_1, G_2)$ free-energies of the reactants and products against the reaction coordinate for an electron transfer reaction with $\Delta G° = 0$. $H_{ab}$ is the electronic coupling element between the diabatic states of the reactants and products and $\lambda$ is the reorganization energy for the reaction.

very weakly coupled and/or when the experimental activation parameters contain contributions from other sources [10].

## 1.3 Semiclassical Treatment

Electronic interaction of the reactants gives rise to the first-order energy surfaces shown as $G_1$ and $G_2$ in Figure 3. The splitting at the intersection of the zero-order energy surfaces in Figure 3 is equal to $2H_{ab}$, where $H_{ab}$ is the electronic matrix element. We will treat $H_{ab}$ as a positive quantity.

### 1.3.1 First-order Energy Surfaces

If $\psi_a$ and $\psi_b$ denote the wave functions of the zero-order initial (reactant) and final (product) states, their interaction gives rise to two linear combinations, the first-order or adiabatic states

$$\psi_1 = c_a \psi_a + c_b \psi_b \tag{7a}$$

$$\psi_2 = c_a \psi_b - c_b \psi_a \tag{7b}$$

where $\psi_1$ is the wave function for the lower (ground) and $\psi_2$ is the wave function for the upper (excited) adiabatic state (energies $G_1$ and $G_2$, respectively) when the overlap integral $S_{ab}$ is neglected (or is zero by construction [11]), and the mixing coefficients are normalized, i.e., $c_a^2 + c_b^2 = 1$. The energies of the adiabatic states are obtained by solving the two-state secular determinant

$$\begin{vmatrix} G_a - G & H_{ab} \\ H_{ab} & G_b - G \end{vmatrix} = 0 \tag{8}$$

where, as before, $G_a = H_{aa} = \langle \psi_a | H | \psi_a \rangle$ and $G_b = H_{bb} = \langle \psi_b | H | \psi_b \rangle$ are the energies of the diabatic states. $H$ is the total Hamiltonian operator of the system including the interaction terms. The roots of the determinant are

$$G_1 = \tfrac{1}{2}\{(G_b + G_a) - [(G_b - G_a)^2 + 4H_{ab}^2]^{1/2}\} \tag{9a}$$

$$G_2 = \tfrac{1}{2}\{(G_b + G_a) + [(G_b - G_a)^2 + 4H_{ab}^2]^{1/2}\} \tag{9b}$$

The difference between the adiabatic energies is given by Eq. 10 while their sum is given by Eq. 11.

$$(G_2 - G_1) = [(G_b - G_a)^2 + 4H_{ab}^2]^{1/2} \tag{10a}$$

$$= \lambda \left[ \left(1 - 2X + \frac{\Delta G^\circ}{\lambda}\right)^2 + \frac{4H_{ab}^2}{\lambda^2} \right]^{1/2} \tag{10b}$$

$$(G_2 + G_1) = (G_b + G_a) \tag{11a}$$

$$= \frac{\lambda}{2}[(2X - 1)^2 + 1] + \Delta G^\circ \tag{11b}$$

*Evidently the average adiabatic energy $(G_2 + G_1)/2$, like the average diabatic energy, is described by a parabola with force constant $2\lambda$ centered at $X = 1/2$ with its minimum vertically displaced by $\lambda/4 + \Delta G^\circ/2$ relative to the diabatic minimum.*

The product of the adiabatic energies is given by Eq. 12a, the product of the mixing coefficients is given by Eq. 12b, and $(1 - 2c_b^2)$ is given by Eq. 12c [11].

$$G_1 G_2 = G_a G_b - H_{ab}^2 \tag{12a}$$

$$c_a c_b = H_{ab}/(G_2 - G_1) \tag{12b}$$

$$(1 - 2c_b^2) = \left(\frac{G_b - G_a}{G_2 - G_1}\right) \tag{12c}$$

The dependence of $c_b^2$ on the reaction coordinate is given by:

$$c_b^2 = \frac{1}{2}\left[ 1 - \frac{(1 - 2X)}{\{[(1 - 2X) + \Delta G^\circ/\lambda]^2 + 4H_{ab}^2/\lambda^2\}^{1/2}} \right] \tag{13}$$

The squares of $c_a$ and $c_b$ are the fraction of the charge of the transferring electron that is on the donor and acceptor, respectively, at any given nuclear configuration.

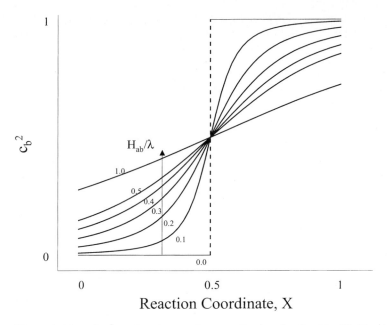

**Figure 4.** Plot of $c_b{}^2$ against the reaction coordinate using Eq. 13 with $H_{ab}/\lambda$ varying from 0 to 0.5.

Thus $c_b{}^2$ also provides a measure of the progress of the electron transfer. *However, unlike X, which is a nuclear configuration coordinate, $c_b{}^2$ is an electronic configuration coordinate.* Figure 4 shows plots of $c_b{}^2$ vs $X$ for various values of $H_{ab}/\lambda$ and $\Delta G° = 0$. As is evident from Eq. 13 and Figure 4, the two coordinates are not linearly related except at very large $H_{ab}$. In the very weak interaction limit (diabatic curves, $H_{ab} = 0$) no electron density is transferred until $X = 0.5$ when the electron 'suddenly' jumps from the donor to the acceptor. In this case $c_b{}^2$ is not a continuous function of $X$: instead $c_b{}^2 = 0$ for all $X < 1/2$ and $c_b{}^2 = 1$ for $X > 1/2$. As $H_{ab}$ increases, charge density is transferred more gradually (with more delocalization present in the initial reactant configuration) and $c_b{}^2$ approaches linearity in $X$ when $H_{ab} \geq \lambda$. Figure 4 shows that for typical symmetrical Class II systems most of the charge density is transferred between $X = 0.4$–$0.6$.

**Symmetrical systems**

As shown in Figure 3, the splitting at the intersection of the diabatic energy curves lowers the barrier by $H_{ab}$. Further, as $H_{ab}$ increases, the reactant and product minima of the adiabatic curves move closer together. The positions of the minima (reactant's and product's equilibrium configurations) are given by

$$X_{\text{min, a}} = \frac{1}{2}\left[1 - \left(1 - \frac{4H_{ab}^2}{\lambda^2}\right)^{1/2}\right] \tag{14a}$$

$$X_{min,b} = \frac{1}{2}\left[1 + \left(1 - \frac{4H_{ab}^2}{\lambda^2}\right)^{1/2}\right]$$ (14b)

and their energies are lowered by $H_{ab}^2/\lambda$ relative to the diabatic minima [4]. In view of these changes the free energy of activation for a self-exchange reaction with appreciable coupling of the reactants is given by

$$\Delta G^* = \lambda/4 - H_{ab} + H_{ab}^2/\lambda$$ (15a)

$$= \frac{\lambda}{4}\left(1 - \frac{2H_{ab}}{\lambda}\right)^2$$ (15b)

The second and third terms on the right of Eq. 15a are due to the lowering of the barrier and the stabilization of the reactants, respectively [4].

Three classes of symmetrical systems may be distinguished depending on the magnitude of the electronic coupling of the donor and acceptor sites [12–14]. In Class I systems the coupling is very weak (Figure 1) and the properties of Class I systems are essentially those of the separate reactants (i.e., the adiabatic energy curves are very close to the diabatic curves). Activated electron transfer either does not occur at all or it occurs only very slowly (because of its high non-adiabaticity) with $\Delta G^* = \lambda/4$ and optical electron transfer can not occur. Class II systems ($0 < H_{ab} < \lambda/2$, Figure 3) possess new optical and electronic properties in addition to those of the separate reactants. They remain valence trapped or charge localized: the electron transfers range from non-adiabatic ($H_{ab} < 10$ cm$^{-1}$) to strongly adiabatic ($H_{ab} > 200$ cm$^{-1}$) with $\Delta G^*$ given by Eq. 15. Eqs 14 and 15 hold as long as the self-exchange reaction is described by a double-well potential, i.e., as long as the system remains valence trapped. In Class III systems the interaction of the donor and acceptor sites has become so large that two separate minima are no longer discernible and the lower energy surface features a single well at $X = 1/2$ (Figure 5, bottom curve). This is the delocalized case which occurs when $H_{ab} \geq \lambda/2$. The latter condition follows readily from the zero barrier limit ($\Delta G^* = 0$) of Eq. 15.

From Eq. 3b the vertical difference between the *diabatic* energies at the equilibrium configuration (adiabatic minimum) of the reactants for $\Delta G^\circ = 0$ is given by

$$(G_b - G_a)_{eq} = \lambda(1 - 2X_{min})$$ (16a)

$$= \lambda\left(1 - \frac{4H_{ab}^2}{\lambda^2}\right)^{1/2}$$ (16b)

It therefore follows from Eq. 10a that the vertical difference between the adiabatic energies at the reactants' equilibrium configuration for $\Delta G^\circ = 0$ is given by:

$$(G_2 - G_1)_{eq} = \lambda$$ (17)

This result is independent of $H_{ab}$ for $H_{ab} \leq \lambda/2$. In other words, the vertical difference between the free energies of the reactants and products of a symmetrical reaction remains equal to $\lambda$ at the equilibrium configuration of the reactants (or

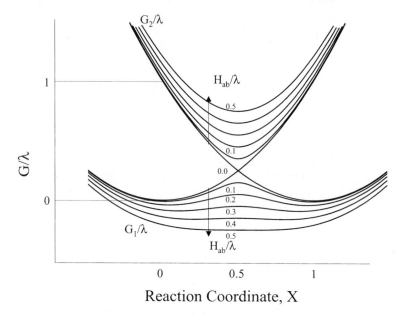

**Figure 5.** Plot of the adiabatic free-energy surfaces against the reaction coordinate for an electron transfer reaction with $\Delta G° = 0$ and $H_{ab}/\lambda$ varying from 0 to 0.5.

products) *regardless of the magnitude of the electronic coupling* as long as the system remains valence trapped [11]. Although the repulsion of the reactant and product curves increases with increasing $H_{ab}$, this is compensated for by the reactant and product minima moving closer together [15]. The net effect is that the adiabatic energy difference at $X_{min}$ remains equal to $\lambda$.

It follows from Eqs 12c and 17 that, for $H_{ab} \leq \lambda/2$, $(c_b^2)_{eq}$ is given by:

$$(c_b^2)_{eq} = \frac{1}{2}\left[1 - \left(1 - \frac{4H_{ab}^2}{\lambda^2}\right)^{1/2}\right] \tag{18}$$

Comparison with Eq. 14a shows that, for a symmetrical system with $H_{ab} > 0$, $(c_b^2)_{eq} = X_{min,a}$. For $H_{ab}/\lambda = 0.3$ this corresponds to $X_{min,a} = 0.10$. Moreover, at the transition state for a symmetrical system $c_b^2 = X^* = 1/2$. The equilibrium and transition-state configurations are the only configurations at which $X$ and $c_b^2$ for a symmetrical system are equal.

Values of $(G_2 - G_1)/\lambda$ calculated from Eq. 10b are plotted against $X$ for various $H_{ab}/\lambda$ values in Figure 6. The adiabatic energy difference flattens with increasing $H_{ab}$ and becomes essentially independent of $X$ when $H_{ab} \geq 2\lambda$. Under these conditions the system is deeply into the Class III regime. Further, it is evident from Eq. 9 that, except for extreme values of $X$, the force constants of the vertically aligned *adiabatic* surfaces in very strongly coupled symmetrical systems ($H_{ab} \geq 2\lambda$) are equal to that of the original *diabatic* parabolas (c.f. discussion of Eq. 11).

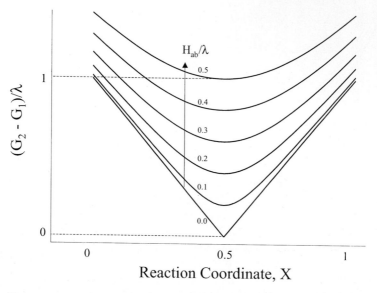

**Figure 6.** Plot of the differences between the adiabatic free-energy curves shown in Figure 5 vs the reaction coordinate for an electron transfer reaction with $\Delta G^{\circ} = 0$ and $H_{ab}/\lambda$ varying from 0 to 0.5.

### Unsymmetrical systems

As for symmetrical systems, the properties of an unsymmetrical Class I system are essentially those of the separate reactants. Although Class II systems are valence trapped, sufficiently endergonic reactions can exhibit a single minimum close to the non-interacting reactant minimum. This minimum shifts to $X^* = 0.5$ only when $H_{ab}$ becomes very large. Provided that $H_{ab} < (\lambda + \Delta G^{\circ})/2$ and $|\Delta G^{\circ}| < \lambda$, the positions of the reactant and product minima are given by Eqs 19a and 19b, while the location of the transition state is given by Eq. 19c.

$$X_{\min,a} = \frac{H_{ab}^2/\lambda^2}{(1 + \Delta G^{\circ}/\lambda)^2} \tag{19a}$$

$$X_{\min,b} = 1 - \frac{H_{ab}^2/\lambda^2}{(1 - \Delta G^{\circ}/\lambda)^2} \tag{19b}$$

$$X^* = \frac{(1 + \Delta G^{\circ}/\lambda - 2H_{ab}/\lambda)}{2(1 - 2H_{ab}/\lambda)} \tag{19c}$$

The free energy of activation is given by

$$\Delta G^* = \frac{\lambda}{4} + \frac{\Delta G^{\circ}}{2} + \frac{(\Delta G^{\circ})^2}{4(\lambda - 2H_{ab})} - H_{ab} + \frac{H_{ab}^2}{(\lambda + \Delta G^{\circ})} \tag{20}$$

In the above equations $-\Delta G°$ is the driving force in the *non-interacting* $(H_{ab} = 0)$ system [15].

## 1.3.2 Rate Constant Expressions

The first-order rate constant for intramolecular electron transfer or for electron transfer within the precursor complex formed from the reactants in a bimolecular reaction is given by

$$k_{el} = \kappa_{el} \nu_n \exp(-\Delta G^*/RT) \tag{21}$$

where $\kappa_{el}$ is the electronic transmission coefficient, $\nu_n$ is the nuclear vibration frequency that takes the system through the intersection region and $\Delta G^*$ is the free energy of activation for the electron transfer [4].

The electronic transmission coefficient is the probability that electron transfer will occur once the system has reached the intersection region (transition state). Provided that the electronic interaction of the reactants is sufficiently strong $\kappa_{el} \approx 1$ and the electron transfer will occur with near unit probability in the intersection region: the electron transfer reaction is *adiabatic* with the system remaining on the lower energy surface on passing through the intersection region. Under these conditions $k_{el}$ is given by

$$k_{el} = \nu_n \exp(-\Delta G^*/RT) \tag{22}$$

On the other hand, for a *non-adiabatic* reaction, $\kappa_{el} \ll 1$, $\kappa_{el}\nu_n = \nu_{el}$ and the rate constant is given by Eq. 23 where $\nu_{el}$ is the electron hopping frequency in the activated complex. The Landau–Zener treatment yields Eq. 24 for $\nu_{el}$ [16, 17].

$$k_{el} = \nu_{el} \exp(-\Delta G^*/RT) \tag{23}$$

$$\nu_{el} = \frac{2H_{ab}^2}{h} \left(\frac{\pi^3}{\lambda RT}\right)^{1/2} \tag{24}$$

In effect, the adiabatic and non-adiabatic limits of the transition state formalism correspond to $\nu_{el} \gg \nu_n$ and $\nu_{el} \ll \nu_n$, respectively.

The frequency of electron hopping in the activated complex may be estimated from $2H_{ab}/h$, the oscillating frequency of the two degenerate diabatic states [16]. Evidently $\nu_{el} \approx 10^{13}$ s$^{-1}$ for interaction energies of only a few hundred cals. A similar estimate is obtained from the Landau–Zener treatment of the intersection region [16]. Since the system typically spends about $10^{-13}$ s in the intersection region (i.e., $\nu_n \approx 10^{13}$ s$^{-1}$), the electron transfer will generally be adiabatic for interaction energies larger than about 100–300 cal (30–100 cm$^{-1}$).

### 1.3.3 Reorganization Parameters

The reorganization parameter is usually broken down into inner-shell (vibrational) and outer-shell (solvational) components.

$$\lambda = \lambda_{\text{in}} + \lambda_{\text{out}} \tag{25}$$

The inner-shell reorganization energy is generally treated within an harmonic approximation [18]. The outer-shell reorganization energy depends upon the properties of the solvent. When a continuum model for the solvent is used $\lambda_{\text{out}}$ is a function of the dielectric properties of the medium, the distance separating the donor and acceptor sites, and the shape of the reactants.

### Inner-shell reorganization energy

To illustrate the approach used to calculate the inner-shell contribution to the reorganization barrier we consider the symmetrical stretching vibrations of the two reactants in the $Fe(H_2O)_6{}^{2+}$–$Fe(H_2O)_6{}^{3+}$ self-exchange reaction (Eq. 1a). The inner-shell reorganization term is the sum of the reorganization parameters of the individual reactants, i.e.:

$$\lambda_{\text{in}} = \lambda_2(d_2{}^\circ \rightarrow d_3{}^\circ) + \lambda_3(d_3{}^\circ \rightarrow d_2{}^\circ) \tag{26a}$$

The first term on the right of Eq. 26a is the energy required to change the Fe–O distance in $Fe(H_2O)_6{}^{2+}$ from its equilibrium value $d_2{}^\circ$ to the equilibrium value $d_3{}^\circ$ in $Fe(H_2O)_6{}^{3+}$ and the second term is the energy required to change the Fe–O distance in $Fe(H_2O)_6{}^{3+}$ from $d_3{}^\circ$ to $d_2{}^\circ$. Denoting $(d_2{}^\circ - d_3{}^\circ)$ by $\Delta d^\circ$, the vertical reorganization energy is given by Eqs 26b and 26c where $f_2$ and $f_3$ are the respective breathing force constants.

$$\lambda_{\text{in}} = \frac{6f_2(\Delta d^\circ)^2}{2} + \frac{6f_3(\Delta d^\circ)^2}{2} \tag{26b}$$

$$= 3(f_2 + f_3)(\Delta d^\circ)^2 \tag{26c}$$

Evidently the contributions of the $Fe(H_2O)_6{}^{2+}$ and $Fe(H_2O)_6{}^{3+}$ breathing modes to $\lambda_{\text{in}}$ are *directly* proportional to their respective force constants.

In the activation process, the energy required to reach the transition state configuration is given by:

$$\Delta G_{\text{in}}^* = 3f_2(d_2^\circ - d_2^*)^2 + 3f_3(d_3^\circ - d_3^*)^2 \tag{27a}$$

Energy conservation requires that the Fe–O distances in the $Fe(H_2O)_6{}^{2+}$ and $Fe(H_2O)_6{}^{3+}$ adjust to a common value $d^*$ before electron transfer.

$$d_2{}^* = d_3{}^* = d^* \tag{27b}$$

Minimizing the resulting reorganization energy expression yields Eq. 27c and substitution into Eq. 27a gives Eq. 27d.

$$d^* = \frac{f_2 d_2^\circ + f_3 d_3^\circ}{f_2 + f_3} \tag{27c}$$

$$\Delta G_{in}^* = \frac{3 f_2 f_3 (\Delta d^\circ)^2}{f_2 + f_3} \tag{27d}$$

$$(d_2^\circ - d^*) = \frac{f_3 \Delta d^\circ}{f_2 + f_3} \tag{27e}$$

$$(d^* - d_3^\circ) = \frac{f_2 \Delta d^\circ}{f_2 + f_3} \tag{27f}$$

The ratio of the amounts that the $Fe(H_2O)_6^{2+}$ and $Fe(H_2O)_6^{3+}$ ions reorganize is equal to $f_3/f_2$ i.e., *inversely* proportional to their force constants. Since $f_3$ is larger than $f_2$, the $Fe(H_2O)_6^{2+}$ ion reorganizes more than the $Fe(H_2O)_6^{3+}$ ion. Note also that $\Delta G_{in}^* < \lambda_{in}/4$ because the free-energy surfaces are not harmonic along the reaction coordinate [15].

Considerable simplification results from using a common, reduced value $f_{in}$ for the force constant of the $Fe(H_2O)_6^{2+}$ and $Fe(H_2O)_6^{3+}$ symmetrical stretching vibrations.

$$f_{in} = \frac{2 f_2 f_3}{(f_2 + f_3)} \tag{28}$$

Under these conditions:

$$d^* = \frac{(d_2^\circ + d_3^\circ)}{2} \tag{29a}$$

$$\Delta G_{in}^* = \frac{3 f_{in} (\Delta d^\circ)^2}{2} \tag{29b}$$

$$\bar{\lambda}_{in} = 2\bar{\lambda}_2 = 2\bar{\lambda}_3 = 6 f_{in} (\Delta d^\circ)^2 \tag{29c}$$

where $\bar{\lambda}_2 = \bar{\lambda}_3 = 3 f_{in} (\Delta d^\circ)^2$ with the two reactants now reorganizing to the same extent and $\Delta G_{in}^* = \bar{\lambda}_{in}/4$. Moreover, $\bar{\lambda}_{in} = 4\lambda_2\lambda_3/(\lambda_2 + \lambda_3)$.

The relationship between the vertical reorganization parameter and the activation energy and the effect of using different criteria for the inner-shell reorganization have recently been considered in some detail [15]. The reorganization energy and the contributions of the individual reactants turn out to be quite sensitive to the model used.

### Solvent reorganization energy

Because of Coulomb interaction terms the solvent reorganization energy is not as readily broken down into contributions from the separate reactants. We illustrate the approach used to calculate the solvent reorganization energy by using the zero-electronic-interaction, two-sphere model developed by Marcus [1, 19, 20].

The familiar Born expression for the free energy of equilibrium solvation of a charged sphere is

$$\Delta G_{eq} = -\frac{(qe)^2}{2a}\left[1 - \frac{1}{D_s}\right] \tag{30}$$

where $qe$ is the charge on the ion, $a$ is its radius and $D_s$ is the static dielectric constant of the medium. The equilibrium solvation energy can be resolved into two contributions

$$\Delta G_{eq} = -\frac{(qe)^2}{2a}\left[1 - \frac{1}{D_{op}}\right] - \frac{(qe)^2}{2a}\left[\frac{1}{D_{op}} - \frac{1}{D_s}\right] \tag{31}$$

where the first contribution is the equilibrium solvation due to the electronic polarization of the medium and the second is the contribution from its orientational–vibrational polarization. $D_{op}$ is the optical dielectric constant of the medium. Note that the orientational–vibrational polarization term contains the Pekar factor $(1/D_{op} - 1/D_s)$. The electronic polarization is assumed to be rapid and capable of keeping up with the transferring electron. The orientational–vibrational polarization is much slower and lags behind. Energy conservation requires that the orientation–vibrational polarization adjust to a nonequilibrium value prior to the electron transfer.

Marcus devised a two-step path for calculating the reversible work required to establish a nonequilibrium orientational–vibrational polarization of the medium. In the first step the orientational–vibrational and electronic polarization of the medium is changed from being in equilibrium with the initial charges $q_2°$ and $q_3°$ to being in equilibrium with the (hypothetical) charges $q_2{}^*$ and $q_3{}^*$. In the second step the orientational–vibrational polarization remains appropriate to $q_2{}^*$ and $q_3{}^*$ but the electronic polarization is changed back to being in equilibrium with $q_2°$ and $q_3°$. The energy required to reorganize the solvent to the nonequilibrium configuration appropriate to charges $q_2{}^*$ and $q_3{}^*$ is then the sum of the work done in these two paths.

$$\Delta G^*_{out} = e^2\left(\frac{[q_2^* - q_2°]^2}{2a} + \frac{[q_3^* - q_3°]^2}{2a} + \frac{(q_2^* - q_2°)(q_3^* - q_3°)}{r}\right)\left(\frac{1}{D_{op}} - \frac{1}{D_s}\right) \tag{32}$$

The reactants are treated as rigid spheres and their radii are not allowed to change during the reorganization process: the radii in Eq. 32 are the average radii defined by $1/a = (1/a_2 + 1/a_3)/2$ and $r$ is the distance between the centers of the spheres. *Note that the numerators in Eq. 32 contain the square of the difference of the charges (or, in the case of the electrostatic interaction term, the product of charge differences) and are not simply differences between the squares of charges, as might have been expected on the basis of the Born equation.*

Analogous to the case of the inner-shell reorganization, energy conservation requires that the transition-state charges for the solvent reorganization be equal.

$$q_2^* = q_3^* = q^* \tag{33a}$$

Minimizing the resulting solvent reorganization expression yields:

$$q^* = (q_2^\circ + q_3^\circ)/2 \tag{33b}$$

and substitution into Eq. 32 yields Eq. 34a for the free energy of activation:

$$\Delta G_{\text{out}}^* = \frac{e^2}{4} \left( \frac{1}{2a_2} + \frac{1}{2a_3} - \frac{1}{r} \right) \left( \frac{1}{D_{\text{op}}} - \frac{1}{D_s} \right) \tag{34a}$$

where it has been assumed that the zero-interaction donor and acceptor sites differ by a single electron, i.e., $(q_3^\circ - q_2^\circ) = 1$. Similarly, substitution into Eq. 32 of $q_2^* = q_3^\circ$ and $q_3^* = q_2^\circ$ yields Eq. 34b for the solvent reorganization energy in a vertical one-electron transition with $\lambda_{\text{out}} = 4\Delta G_{\text{out}}^*$.

$$\lambda_{\text{out}} = e^2 \left( \frac{1}{2a_2} + \frac{1}{2a_3} - \frac{1}{r} \right) \left( \frac{1}{D_{\text{op}}} - \frac{1}{D_s} \right) \tag{34b}$$

If there is appreciable delocalization in the initial (equilibrium) state, then less than a unit of charge will be transferred from the donor to the acceptor. In terms of the mixing coefficients, the zero-interaction charge difference $(q_2^\circ - q_3^\circ)$ needs to be scaled by $(c_a^2 - c_b^2)_{\text{eq}} = (1 - 2c_b^2)_{\text{eq}}$ to obtain the 'real' charge transferred. We thus obtain

$$\Delta q = (c_a^2 - c_b^2)_{\text{eq}} = (1 - 2c_b^2)_{\text{eq}} \tag{35}$$

At the minimum of the adiabatic curve, i.e., at the equilibrium configuration of the reactants, $(c_b^2)_{\text{eq}}$ is given by Eq. 18, so that:

$$[(1 - 2c_b^2)_{\text{eq}}]^2 = \left( 1 - \frac{4H_{\text{ab}}^2}{\lambda^2} \right) \tag{36}$$

Electron delocalization in the initial state thus scales the solvent activation barrier by $(1 - 4H_{\text{ab}}^2/\lambda^2)$.

The vertical reorganization parameter $\lambda$ is a property of the diabatic states, $(H_{\text{ab}} = 0)$ and $\lambda_{\text{out}}$ continues to be given by Eq. 34b *regardless of the degree of initial state delocalization*. When appreciable delocalization is present, we add a prime to indicate that $\lambda$ has been modified to allow for the reduction in the charge transferred [15, 21]. In other words, $\lambda'$ denotes a reorganization energy that has been scaled by $(1 - 4H_{\text{ab}}^2/\lambda^2)$. (The parameter $\lambda'$ used here and in [15] corresponds to $\lambda_{\text{mod}}$ introduced earlier [21].)

$$\lambda'_{\text{out}} = \lambda_{\text{out}}\left(1 - \frac{4H_{\text{ab}}^2}{\lambda^2}\right) \tag{37a}$$

Similar considerations apply to the inner-shell reorganization. When initial-state delocalization is present $(d_2^\circ - d_3^\circ)$ is scaled by $(1 - 2c_b^2)_{\text{eq}}$ giving

$$\lambda'_{\text{in}} = \lambda_{\text{in}}\left(1 - \frac{4H_{\text{ab}}^2}{\lambda^2}\right) \tag{37b}$$

Consequently:

$$\lambda' = \lambda'_{\text{in}} + \lambda'_{\text{out}} = \lambda\left(1 - \frac{4H_{\text{ab}}^2}{\lambda^2}\right) \tag{37c}$$

In a sense the primed (scaled) quantities are the 'actual' vertical reorganization energies since their values are determined by the actual charge transferred. While it would be convenient if $\lambda'$ was the separation between the diabatic energy curves at the reactant's equilibrium configuration $(X_{\text{min}})$, it is not. The diabatic curves $(H_{\text{ab}} = 0)$ correspond to a charge transfer of one electron with this charge abruptly transferring at the transition state: delocalization is *not* incorporated into the diabatic surfaces. This topic is discussed further under optical charge transfer in Section 1.3.4.

## Time-scales for solvent electronic polarization and electron transfer

The above treatment is based upon the traditional Born–Oppenheimer approximation which states that, when nuclei move, the electrons can almost instantaneously adjust to their new positions. Another relevant time frame is the time required to establish the electronic polarization of the medium. To characterize this time frame, Kim and Hynes consider the ratio of $v_{\text{el}}$, the electron hopping frequency, to $v_{\text{ep}}$, the frequency characteristic of the solvent electronic polarization. The Born–Oppenheimer-based treatment is valid provided that this ratio is much less than unity, i.e., the time scale for the adjustment of the electronic polarization is much shorter than that for the transferring electron [22–26].

The electron hopping frequency may be estimated from time-dependent perturbation theory. If $H_{\text{ab}}$ is treated as a constant perturbation, the system will start to oscillate between the two diabatic states once the perturbation is turned on. In a bimolecular reaction, for example, the perturbation is turned on upon formation of the precursor complex, while in a covalently attached (bridged) binuclear system it can be turned on upon reduction (oxidation) of one end of the fully oxidized (reduced) system by an external reagent or by photoexcitation. If the system is in the diabatic reactant state at $t = 0$, then the probability of it being in the product state at some later time $t$ is given by the Rabi formula [27].

$$P_2 = \left[\frac{4H_{\text{ab}}^2}{(G_2 - G_1)^2}\right]\sin^2\left[\frac{(G_2 - G_1)}{h}\pi t\right] \tag{38}$$

Consider first the case where the system is initially at the nuclear configuration of the adiabatic minimum, i.e. $(G_2 - G_1) = \lambda$. The system will start to oscillate between the two diabatic states with a frequency equal to $\lambda/h$ which corresponds to $\sim 5 \times 10^{14}$ s$^{-1}$ for $\lambda = 40$ kcal mol$^{-1}$. The maximum value of the probability of finding the system in the product state is $4H_{ab}^2/\lambda^2$ or $2 \times 10^{-3}$ for $H_{ab}/\lambda = 2 \times 10^{-2}$. There is thus only a very small probability that weak coupling will drive the system into the product state at a nuclear configuration near the initial state minimum. Since the frequency with which the system oscillates under the influence of the perturbation is $\lambda/h$, the maximum frequency of attaining the product state (*i.e.*, the maximum probability per unit time) is $4H_{ab}^2/h\lambda$. In other words, $\nu_{el}$ at the adiabatic minimum is estimated to be $\approx 10^{12}$ s$^{-1}$ for a moderately coupled Class II system ($H_{ab} \approx 100$ cm$^{-1}$, $\lambda \approx 10$ kcal mol$^{-1}$). Since $\nu_{ep} \approx 10^{15}$ s$^{-1}$ or higher for most colorless solvents [25], the ratio $\nu_{el}/\nu_{ep}$ is much less than unity for a weakly or moderately coupled Class II system near the adiabatic minimum.

We turn next to the frequency of electron hopping in the transition state. As is evident from Eq. 38 with $(G_2 - G_1) = 2H_{ab}$, the frequency of electron hopping in the transition state is equal to $2H_{ab}/h$ (see also the discussion following Eq. 24. Thus the transition-state hopping frequency $\nu_{el}$ is $\leq 10^{13}$ s$^{-1}$ for a weakly or moderately coupled Class II system and $\nu_{el}/\nu_{ep} \ll 1$ at the transition state. Thus the condition for the validity of the Born–Oppenheimer approximation will be satisfied by most weakly and moderately coupled Class II systems. For symmetrical Class II systems, the free energy of activation will then be given by the traditional Eq. 15 [22–26] except that a correction for the so-called exchange field may be needed under certain circumstances. The exchange field arises from the overlap charge distribution $(e\psi_a\psi_b)$ and serves to lower $\lambda_{out}$ (more correctly, to stabilize the transition state) and to reduce the effective $H_{ab}$ [22, 25]. However, there is no exchange field when the diabatic wave functions are appropriately chosen, i.e., when they are based on the exchange dipole moment ($\mu_{ab}$, see below) being zero [28] and no exchange-field correction to the traditional expression for the free energy of activation is then required.

In the Born–Oppenheimer limit, the electrons of the surrounding medium equilibrate to the instantaneous position of the transferring electron while the orientations of the medium dipoles, which occur much more slowly, adjust to the smeared-out charge distribution of the transferring electron. When the time scale for electronic polarization is slower than, or comparable to, the time scale of the transferring electron, it becomes necessary to use a self-consistent treatment in which both the electronic polarization and the orientational polarization respond to the smeared-out charge distribution of the transferring electron [25]. Including the interaction of this charge distribution with the electronic polarization gives rise to a nonlinear Schrödinger equation in which the Hamiltonian depends on the wave function for the donor–acceptor pair. Such a treatment becomes increasingly important as the electronic interaction increases and introduces terms into the free energy of activation that have the net effect of *increasing* the activation energy beyond that given by the Born–Oppenheimer limit [25]. In the limit that the time scale for the solvent electronic polarization becomes very long the electronic polarization can no longer keep up with the transferring electron and the electronic polarization will contribute to the activation barrier in much the same way as the orientational–vibrational polarization.


### 1.3.4 Optical Charge Transfer

In addition to thermal activation, electron transfer between the donor and acceptor sites can also be effected by the absorption of light. As a consequence, $\lambda$ and $H_{ab}$ can be obtained from spectroscopic properties.

**Transition energies**

The energy of the light-induced charge-transfer transition in a symmetrical double-well system is given by Eq. 39 [29, 30].

$$h\nu_{max} = \lambda \tag{39}$$

Because $\lambda$ for a symmetrical localized system is *independent* of $H_{ab}$, Eq. 39 holds throughout the double-well regime [11]. Further insight into Eq. 39 can be obtained by noting that $h\nu_{max}$ is also given by:

$$h\nu_{max} = \lambda' + 4H_{ab}^2/\lambda \tag{40}$$

The first term on the RHS is the scaled reorganization energy and the second term is a further quantum-mechanical contribution to the transition energy. Although the scaled reorganization energy associated with the charge transfer is reduced by the delocalization, and reaches zero when $2H_{ab} = \lambda$, the decrease in the reorganization energy is compensated for by the repulsion of the curves. The net effect is that $h\nu_{max}$ remains constant and equal to $\lambda$ for a symmetrical system in the double-well regime. Thus, even when appreciable delocalization is present, $h\nu_{max}$ will still exhibit the full solvent dependence predicted for the very weakly interacting system.

The energy of the optical transition in a symmetrical Class III system is given by

$$h\nu_{max} = 2H_{ab} \tag{41}$$

so that $H_{ab}$ for symmetrical Class III complexes can be obtained directly from the energy of the optical transition [29]. Note that the optical transition in a Class III system no longer involves charge transfer: the transition occurs between delocalized molecular orbitals of the complex and is not accompanied by a net dipole-moment change.

The energy of the charge transfer transition in an unsymmetrical double-well system is given by:

$$h\nu_{max, a} = (\lambda + \Delta G^\circ)\left[1 + \frac{2H_{ab}^2\Delta G^\circ}{(\lambda + \Delta G^\circ)^3}\right] \tag{42}$$

assuming $H_{ab} < (\lambda + \Delta G^\circ)/2$ [15]. When the $H_{ab}{}^2$ contribution may be neglected, the energy of the charge transfer transition in an unsymmetrical double-well system is given by the familiar Eq. 43:

$$hv_{max,a} = \lambda + \Delta G^\circ \tag{43}$$

Finally, although $hv_{max}$ for a symmetrical double-well system is independent of the degree of electronic interaction, the free energy of activation does depend on $H_{ab}$. Thus when $\Delta G^\circ$ may be neglected, the ratio $hv_{max}/\Delta G^*$ for a double-well system is given by:

$$\frac{hv_{max,a}}{\Delta G^*} = \frac{4}{(1 - 2H_{ab}/\lambda)^2} \tag{44a}$$

while, when the electronic interaction may be neglected, the ratio is given by

$$\frac{hv_{max,a}}{\Delta G^*} = \frac{4}{1 + \Delta G^\circ/\lambda} \tag{44b}$$

Evidently $hv_{max}/\Delta G^*$ is $\leq 4$ for a weakly coupled, endergonic charge-transfer reaction and $\geq 4$ for a weakly coupled, exergonic charge-transfer reaction or for charge transfer in a moderately coupled symmetrical double-well system. The value of $hv_{max}/\Delta G^*$ can thus provide information about the degree of electronic interaction. However, in practice the latter is more readily obtained from the intensity of the charge transfer transition.

**Intensities and dipole-moment changes**

Using the Mulliken formalism, Hush [29] showed that the electronic coupling element is related to the intensity of the charge transfer transition by:

$$H_{ab} = 2.06 \times 10^{-2} \frac{(v_{max}\varepsilon_{max}\Delta v_{1/2})^{1/2}}{r_{ab}} \tag{45}$$

where $v_{max}$ and $\Delta v_{1/2}$ are the band maximum and width in wave numbers, $r_{ab}$ is the distance separating the donor and acceptor charge centroids in Ångströms, and the band is Gaussian shaped [11]. *Equation 45 is exact within a two-state model and is applicable to both symmetrical and unsymmetrical Class II and Class III systems [11].*

The Mulliken–Hush expression is a particular form of the more general equation

$$H_{ab} = \left|\frac{v_{max}\mu_{12}}{\mu_b - \mu_a}\right| \tag{46}$$

where $\mu_{12}$ is the transition dipole and $(\mu_b - \mu_a)$ is the difference between the dipole moments of the initial and final diabatic (localized) states [11, 31]. In the generalized Mulliken–Hush treatment formulated by Cave and Newton [32, 33], the diabatic states are obtained by applying the transformation that diagonalizes the adiabatic dipole-moment matrix. Because $\mu_{ab}$, the transition moment connecting the diabatic states, is zero, the value of $(\mu_b - \mu_a)$ is maximized. With this definition of the diabatic states, the diabatic dipole-moment difference is related to the measured dipole-moment change $(\mu_2 - \mu_1)$ by Eq. 47. The diabatic dipole-moment difference can thus be obtained from measurable quantities [32].

$$\mu_b - \mu_a = [(\mu_2 - \mu_1)^2 + 4\mu_{12}^2]^{1/2} \tag{47}$$

Equation 45 follows from Eq. 46 by noting that $r_{ab} \equiv |(\mu_b - \mu_a)/e|$ and that the transition dipole is given by Eq. 48:

$$\mu_{12} = \sqrt{\frac{f_{os}}{1.08 \times 10^{-5} \nu_{max}}} \tag{48a}$$

$$f_{os} = 4.61 \times 10^{-9}(\varepsilon_{max}\Delta\nu_{1/2}) \tag{48b}$$

where $f_{os}$ is the oscillator strength of the transition [11, 31]. Eq. 41 is obtained by noting that $(\mu_2 - \mu_1)$ is zero for a delocalized system and therefore, from Eq. 47, $(\mu_b - \mu_a) = 2\mu_{12}$. Finally, because the adiabatic and diabatic dipole-moment changes are related by:

$$(\mu_2 - \mu_1) = (\mu_b - \mu_a)(1 - 2c_b^2) \tag{49}$$

it follows from Eq. 12c that:

$$\frac{\mu_2 - \mu_1}{\mu_b - \mu_a} = \frac{G_b - G_a}{G_2 - G_1} \tag{50}$$

There is thus an inverse relationship between the ratio of the adiabatic and diabatic dipole-moment changes and the ratio of the corresponding free-energy differences within the two-state model.

## Optical band shapes

In the high-temperature or classical limit the molar absorptivity at a given transition energy is obtained by weighting the transition probability [34, 35], $p$, by the number of systems, $n(v)$, having a nuclear configuration appropriate to the particular transition energy.

$$\varepsilon_{cl}(v) = pn(v) \tag{51}$$

The transition probability is assumed to be energy independent with $p = \varepsilon_{max}/n_0$ where $n_0$ is the population of the ground-state energy minimum. Assuming a

Boltzmann distribution over the energies of the ground-state configurations gives:

$$\varepsilon_{cl}(v) = \varepsilon_{max} \exp[-(G_1 - G_{1,eq})/RT] \tag{52a}$$

where $G_1$ is the energy of the system on the lower (ground-state) free-energy surface.

We first consider systems in the zero-interaction limit i.e., $H_{ab} = 0$, $G_a = G_1$ and $G_b = G_2$. Substituting Eq. 2a into Eq. 52a yields Eq. 52b for the molar absorptivity for an optical transition at the nuclear configuration defined by $X$.

$$\varepsilon_{cl}(v) = \varepsilon_{max} \exp[-\lambda X^2/RT] \tag{52b}$$

From Eq. 3b the ground- excited-state energy difference $(G_2 - G_1)$ is related to $X^2$ by

$$X^2 = [\lambda + \Delta G^\circ - (G_2 - G_1)]^2/4\lambda^2 = [\lambda + \Delta G^\circ - hv]^2/4\lambda^2 \tag{53}$$

Substituting into Eq. 52b and recalling that $\lambda + \Delta G^\circ = hv_0$ gives

$$\varepsilon_{cl}(v) = \varepsilon_{max} \exp\left[\frac{-(hv_0 - hv)^2}{4\lambda RT}\right] \tag{54}$$

Eq. 54 predicts a Gaussian shape for the band obtained by plotting $\varepsilon(v)$ against $hv$. The band maximum occurs at $hv_0 = hv_{max} = \lambda + \Delta G^\circ$. The half-width of the band is given by:

$$\Delta v_{1/2} = [16\ln(2)\lambda RT]^{1/2} \tag{55a}$$

so that the expression for the molar absorptivity in terms of the band width is:

$$\varepsilon_{cl}(v) = \varepsilon_{max} \exp[-4\ln(2)(hv_0 - hv)^2/(\Delta v_{1/2})^2] \tag{55b}$$

From a semiclassical viewpoint the assumption underlying Eq. 51 is not correct. It can be shown [36] that:

$$\varepsilon_{sc}(v)/v = \frac{hn(v)B}{cn_t} \tag{56a}$$

where $B$, $c$, and $n_t$ are the Einstein coefficient of stimulated absorption, the speed of light and the total concentration of absorbers, respectively. The Einstein coefficient is given by [37]:

$$B = \frac{2\pi|\mu_{12}|^2}{3\hbar^2} \tag{56b}$$

where $\mu_{12}$ is the transition dipole moment. The above two equations yield:

$$\varepsilon_{sc}(v)/v = \frac{4\pi^2 |\mu_{12}|^2 n(v)}{3hcn_t} = p_{12}n(v) \tag{56c}$$

Again assuming a Boltzmann distribution for $n(v)$ over the ground-state energies and that $p_{12}$ is energy independent (i.e., that the transition dipole moment $\mu_{12} = \langle \psi_2 | \mu | \psi_1 \rangle$ does not depend on the vibrational levels) yields Eq. 57 for the optical band shape in the zero-interaction limit

$$\varepsilon_{sc}(v) = \left(\frac{\varepsilon}{v}\right)_{max} v \exp[-(\lambda + \Delta G^\circ - hv)^2 / 4\lambda RT] \tag{57a}$$

$$= \left(\frac{\varepsilon}{v}\right)_{max} v \exp[-(hv_0 - hv)^2 / 4\lambda RT] \tag{57b}$$

where $hv_0$ now corresponds to the maximum in the plot of $\varepsilon(v)/hv$ against $hv$ and is again the energy of the vertical transition from the ground-state minimum.

To the extent that $\Delta v_{1/2} \ll v_0$ the classical and semiclassical approaches yield the same results since then:

$$\frac{\varepsilon_{sc}(v)}{v} \approx \frac{\varepsilon_{sc}(v)}{v_0}$$

However, in practice the band width is about 20 % of the energy of the maximum and so the two approaches yield different results. In the classical formalism a plot of $\varepsilon_{cl}$ against $hv$ is Gaussian shaped while in the semiclassical formalism (Eq. 56), a plot of $\varepsilon_{sc}/hv$ against $hv$ is Gaussian with a maximum at $\lambda + \Delta G^\circ$; however, a plot of $\varepsilon_{sc}$ against $hv$ is not Gaussian and its maximum occurs at higher energy.

The situation is more complicated when the electronic coupling is appreciable. As discussed earlier, the energy surfaces are then no longer parabolic and the minimum of the initial state no longer occurs at $X = 0$. Moreover, the low energy side of the absorption band is cut off at $hv = 2H_{ab}$, the minimum difference between the energies of the ground and excited states. The relevant energy expressions for a symmetrical system are:

$$\frac{G_1}{\lambda} = \left(\frac{2X^2 - 2X + 1}{2}\right) - \frac{\left[(1 - 2X)^2 + \left(\frac{2H_{ab}}{\lambda}\right)^2\right]^{1/2}}{2} \tag{58a}$$

$$\frac{G_{1,eq}}{\lambda} = -\frac{H_{ab}^2}{\lambda^2} \tag{58b}$$

and

$$\frac{G_2 - G_1}{\lambda} = \left[(1 - 2X)^2 + \left(\frac{2H_{ab}}{\lambda}\right)^2\right]^{1/2} \tag{58c}$$

Assuming a Boltzmann distribution over the energies on the adiabatic ground-state surface yields the following expression for the band shape

$$\varepsilon_{cl}(v) = \varepsilon_{max} \exp\left[-\frac{(hv_0 - [(hv)^2 - 4H_{ab}^2]^{1/2})^2}{4\lambda RT}\right]$$

(58d)

$$\varepsilon_{sc}(v) = \left(\frac{\varepsilon}{v}\right)_{max} v \exp\left[-\frac{(hv_0 - [(hv)^2 - 4H_{ab}^2]^{1/2})^2}{4\lambda RT}\right]$$

where $v_0$ is defined above.

Plots of $\varepsilon_{cl}$ and $\varepsilon_{sc}$ vs $hv$ are presented in Figure 7 for $\Delta G° = 0$ and in Figure 8 for $\Delta G°/\lambda = 2$. One sees that the maximum of $\varepsilon_{sc}$ is always at higher energy than the maximum of $\varepsilon_{cl}$. For the zero driving force case, as $H_{ab}$ increases the spectrum becomes truncated at low energy. For $H_{ab}/\lambda > 1/2$ the truncated spectrum starts to shift to higher energy and becomes narrower; finally, at $H_{ab}/\lambda \gg 1$ only a sharp line remains at $hv = 2H_{ab}$. When $\Delta G° > 0$ the spectra do not show the low energy cut-off. However, when $H_{ab}/\lambda > 1/2$ the absorption maximum shifts to higher energy and narrows. Ultimately, for $H_{ab}/\lambda \gg \Delta G°/\lambda$, the spectrum again becomes a sharp line at $hv = 2H_{ab}$.


## 1.4 Quantum Mechanical Treatment


Although the semiclassical expressions work well at high temperatures, they break down at low temperatures and/or at high reaction exergonicities. Nuclear tunneling contributions to the rate can become very important under such conditions. Although corrections for nuclear tunneling can be introduced into the semiclassical treatment, tunneling enters naturally into a quantum mechanical treatment.

The quantum mechanical treatment of non-adiabatic electron transfers are normally considered in terms of the formalism developed for multiphonon radiationless transitions. This formalism starts from Fermi's golden rule for the probability of a transition from a vibronic state $A_v$ of the reactant (electronic state A with vibrational level v) to a set of vibronic levels $\{B_w\}$ of the product

$$W_{A_v} = \frac{4\pi^2 H_{ab}^2}{h} \rho_w$$

(59a)

$$\rho_w = \sum_w |\langle \chi_{A_v} | \chi_{B_w} \rangle|^2 \delta(\varepsilon_{A_v} - \varepsilon_{B_w})$$

(59b)

where $\rho_w$ is the density of final states, $\varepsilon_{Av}$ and $\varepsilon_{Bw}$ are the unperturbed energies of the vibronic levels and $\delta$ is the delta function that ensures energy conservation. To

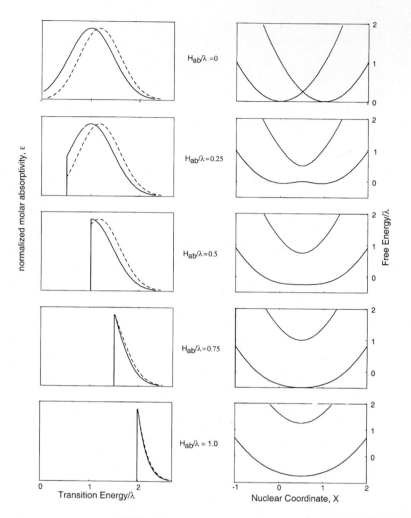

**Figure 7.** On the left side of the figure are plots of the normalized molar absorptivities, $\varepsilon_{cl}$ (solid line) and $\varepsilon_{sc}$ (dashed line), vs transition energy/$\lambda$ for $\Delta G° = 0$ and $H_{ab}/\lambda$ values of 0, 0.25, 0.5, 0.75 and 1, respectively. On the right side are plots of the free energy/$\lambda$ vs the nuclear coordinate $x$ for the same values of $\Delta G°$ and $H_{ab}/\lambda$.

obtain the thermally averaged probability per unit time, $k$, of passing from a set of vibrational levels $\{A_v\}$ of the reactant to a set of vibrational levels $\{B_w\}$ of the product we assume a Boltzmann distribution over the vibrational levels of the reactants and sum over these levels

$$k = \frac{4\pi^2 H_{ab}^2}{h}(FC) \tag{60a}$$

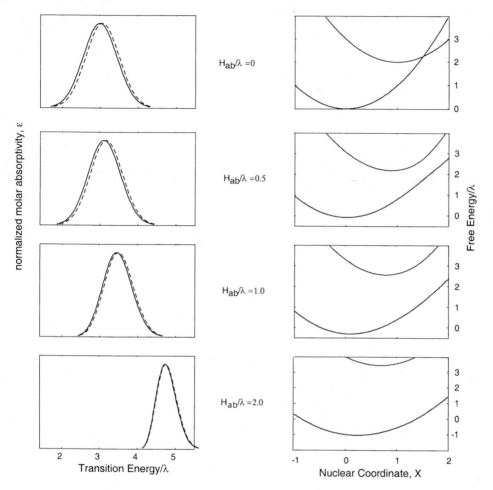

**Figure 8.** On the left side of the figure are shown plots of the normalized molar absorptivities, $\varepsilon_{cl}$ (solid line) and $\varepsilon_{sc}$ (dashed line), vs energy for $\Delta G°/\lambda = 2$ and $H_{ab}/\lambda$ values of 0, 0.5, 1.0 and 2, respectively. On the right side are plots of the free energy/$\lambda$ vs the nuclear coordinate $x$ for the same values of $\Delta G°/\lambda$ and $H_{ab}/\lambda$.

$$FC = \frac{1}{Q}\sum_{v}\sum_{w}\exp\left(\frac{-\varepsilon_{A_v}}{RT}\right)S^2_{A_v, B_w}\delta(\varepsilon_{A_v} - \varepsilon_{B_w}) \tag{60b}$$

$$Q = \sum_{v}\exp\left(\frac{-\varepsilon_{A_v}}{RT}\right) \tag{60c}$$

where $FC$ is the thermally averaged Franck–Condon factor and $S^2_{A_v, B_w}$ is the square of the overlap integral of the vibration wavefunctions of the reactants and products. If the reactant and product energy surfaces are approximated as harmonic functions, the $FC$ factors can be explicitly calculated [38, 39].

In the following discussion we first consider the form of $H_{ab}$ and then expressions for the thermally averaged Franck–Condon factor, *FC*.

### 1.4.1 The Electronic Coupling Element

In the present context the rate constant defined by Eq. 60 refers to a radiationless transition within a supramolecular complex that includes the precursor complex (donor–acceptor pair or excited molecule) and the surrounding solvent. Two limiting classes of radiationless transitions are of interest. The first involves the redistribution of charge within the supramolecular complex so that the transition is accompanied by a dipole-moment change. This is the type of CT (charge transfer) process that we have been considering. In the second class, there is no net charge transfer from one part of the supramolecular complex to another and thus there is no significant dipole-moment change in the transition. This type of radiationless transition is observed in the deactivation of excited states in aromatic molecules $(\pi^* \to \pi)$ and in spin conversion systems. For example, in $d^6$ Fe(II) octahedral complexes, either a low- or high-spin isomer is more stable depending on the whether the ligand-field splitting or the interelectronic repulsion energies are larger. Interconversions between such spin states are generally not accompanied by appreciable charge transfer and so are not considered as CT transitions.

In a radiationless CT transition, the transferring charge initially strongly interacts with the donor's nuclear charge (and electrons) and only weakly interacts with that of the acceptor; at the end of the transition, the transferred charge strongly interacts with the acceptor's nuclear charge (and electrons) and only weakly with that of the donor [40]. Thus the perturbation that promotes the radiationless transition is the coulombic interaction between the transferring charge and the acceptor nuclei (and electrons), $V_{e,ac}$. The diabatic potential energies and wavefunctions constitute the zero-order states for the perturbation calculation. Unfortunately, the two diabatic wavefunctions are not solutions of the same zero-order Hamiltonian, $H^\circ$, but rather solve Hamiltonians that ignore the interaction of the transferring charge with either the acceptor or the donor and are thus not orthogonal [40, 41]. Despite this complication [11] the perturbation is just the $V_{e,ac}$ interaction (or the coulombic interaction between the oxidized donor and the transferred charge, $V_{e,do}$) and $H_{ab}$ is given by

$$H_{ab} = \langle \psi_b | V_{e,ac} | \psi_a \rangle = \langle \psi_a | V_{e,dc} | \psi_b \rangle \tag{61a}$$

We note that, for CT transitions, the expression for $H_{ab}$ is not needed if the Mulliken–Hush approach is used to calculate $H_{ab}$ from experimental quantities as discussed in Section 1.3.4. Also, the generalized Mulliken–Hush treatment [32, 33] allows the calculation of $H_{ab}$ from the adiabatic wavefunctions and the complete Hamiltonian; the extension of Eq. 56 to include more than two states is then used to obtain $H_{ab}$.

In the non-CT radiationless transition the change in electronic charge interacts with the nuclei in a similar manner both before and after the transition. Two types of processes can be identified: internal conversion processes in which the transition is between spin states of the same multiplicity and intersystem crossing process in which the transition is between states of different spin multiplicity. For non-CT internal conversion processes the full BO (Born–Oppenheimer) adiabatic wavefunctions for the supramolecular complex are used as the zero-order basis [42–44]. The perturbations that cause the transition are the vibronic coupling between the nuclear and electron motions. These are just the terms that are neglected in the BO approximation [45]. The terms are expanded (normally to first order) in the normal vibrational coordinates of the nuclei as is customarily done for optical vibronic transitions. Thus one obtains Eq. 61b for cases when only one normal mode couples the two states

$$H_{ic} = \frac{\langle\psi_2|(\partial U/\partial Q)_0|\psi_1\rangle}{(E_2 - E_1)} \tag{61b}$$

In this equation the $\psi_i$ and $E_i$ are the adiabatic BO wavefunctions and energies (at the a particular nuclear configuration), respectively, $U$ is the electronic potential energy and $Q$ is the normal coordinate for the vibration that couples the electronic states $\psi_2$ and $\psi_1$.

By contrast, the pure spin BO states need to be mixed for intersystem crossing or spin-conversion processes. The BO wavefunctions are mixed by the spin–orbit coupling operator, $H_{so}$

$$\psi'_j = \psi_j + \sum_{k \neq j} \frac{\langle\psi_k|H_{so}|\psi_j\rangle}{E_j - E_k}\psi_k \tag{61c}$$

where the $\psi'_j$ are the mixed wavefunctions, the $\psi_j$ are the pure spin wavefunctions and the energies and spin–orbit matrix element are evaluated at the equilibrium configuration of the initial (or final) state. The spin–orbit coupling also promotes the radiationless transition between the $\psi'_j$ states with the intersystem crossing Hamiltonian is given by [46]:

$$\begin{aligned} H_{sc} &= \langle\psi'_1|H_{so}|\psi'_2\rangle \\ &\approx \langle\psi_1|H_{so}|\psi_2\rangle \\ &+ \sum_{j \neq 1,2} \langle\psi'_1|H_{so}|\psi'_j\rangle\langle\psi'_j|H_{so}|\psi'_2\rangle\left(\frac{1}{E_1 - E_j} + \frac{1}{E_2 - E_j}\right) \end{aligned} \tag{61d}$$

We note that for spin conversion processes that involve the flip of a single electron ($\Delta S = 1$) the first term in Eq. 61d is dominant whereas for processes in which the spins of several electrons change ($\Delta S > 1$) the first term is zero.

For most charge transfer reactions the electronic coupling element is simply given by the coulomb matrix element defined by Eq. 61a.

### 1.4.2 The Thermally Averaged Franck–Condon Factor

Three broad classes of vibrational modes may contribute to the thermally averaged Franck–Condon factor: the high-frequency (fast) modes ($hv > 1000$ cm$^{-1}$) which are mainly intraligand vibrations, intermediate modes ($1000$ cm$^{-1} > hv > 100$ cm$^{-1}$) that typically include the metal–ligand stretching vibrations and higher frequency solvent orientational–vibrational modes, and the low-frequency (slow) modes ($hv < 100$ cm$^{-1}$) which are primary solvent modes but can include low-frequency intramolecular modes. At ordinary temperatures $hv_v \gg kT \approx hv_c \gg hv_s$ and the low-frequency modes can be treated using classical (continuum) expressions.

### Two-mode systems

We first consider the case with one high-frequency mode and one low-frequency mode. When the high-frequency mode ($v_v$, with reorganization energy of $\lambda_v$) is in the low temperature limit and the low-frequency mode ($v_s, \lambda_s$) is treated classically, the rate constant for electron transfer is given by:

$$k_{el} = \frac{H_{ab}^2}{h} \left( \frac{4\pi^3}{\lambda_s RT} \right)^{1/2} \sum_{j=0}^{\infty} F_j \exp\left[ -\frac{(jhv_v + \Delta G^\circ + \lambda_s)^2}{4\lambda_s RT} \right] \tag{62}$$

where $F_j = e^{-S}S^j/j!$, $S = \lambda_v/hv_v$ is the Huang–Rhys factor and $F_j H_{ab}^2 < hv_s\sqrt{\lambda_s RT}/\pi^3$ [41, 47, 48]. Because the solvent (or other low-frequency) mode behaves classically while the high-frequency mode can tunnel it is most efficient for the solvent modes to use enough of the driving force to reduce the solvent barrier significantly with the remaining driving force absorbed by the high-frequency modes. Moreover, since $hv_v \gg kT$ all of the reaction occurs from the lowest vibrational level of the initial state, i.e., only $A_0 \rightarrow \{B_j\}$ vibronic transitions are considered. The exponential term in Eq. 62 is a Gaussian that describes the rate constant reduction deriving from the solvent reorganization. The Gaussian is peaked at $(jhv + \Delta G^\circ + \lambda_s) \approx 0$ and has a width of $2\sqrt{4\lambda_s RT}$. The transition with $j^* \approx -(G^\circ + \lambda_s)/hv$ will normally dominate the sum. The rate constant will be maximized when the solvent reorganization is barrierless. This occurs when the effective driving force for the solvent reorganization, $-(\Delta G^\circ + jhv_v)$ is approximately equal to $\lambda_s$, i.e., when $j^* \approx -(\Delta G^\circ + \lambda_s)/hv_v$. The effective energy gap for the high-frequency mode is $-(\Delta G^\circ + \lambda_s)$. The energy change of the reactant/product and the solvent for the single largest term in the sum of Eq. 62 is plotted against driving force is illustrated in Figure 9. The solvent accepts an amount of energy that is close to the $\lambda_s$ for the system while the high-frequency mode will accept no energy for very low driving forces and the majority of the energy change when $|\Delta G^\circ| \gg \lambda_s + \lambda_v$.

A convenient closed-form expression for the rate can be derived using Eq. 62

$$k_{el} = \frac{4\pi^2 H_{ab}^2 F_{j^*}}{h^2 v_v} \exp\left[ -\frac{(j^*hv_v + \Delta G^\circ + \lambda_s)^2}{4\lambda_s RT} \right] \tag{63a}$$

where

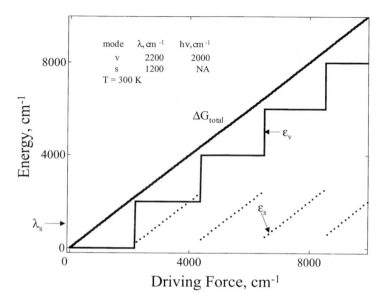

**Figure 9.** Plot of the energy in a particular mode for an electron transfer reaction with two active modes. $\Delta\varepsilon_v$ and $\Delta\varepsilon_s$ are the differences between the energies of the products and reactants in the high- and low-frequency modes, respectively; $\lambda_i$, $(cm^{-1})$ and $h\nu_i$ $(cm^{-1})$ are (2000, 2000) and (1200, −) for the high- and low-frequency modes and the temperature is 300 K. The calculations were done using Eq. 62. The straight line, the stepped solid line and the dotted lines are the total energy difference $(\Delta G^\circ)$ and the differences between the energies of the products and reactants in the high- and low-frequency modes, respectively.

$$j^* \approx -\frac{(\Delta G^\circ + \lambda_s)}{h\nu_v} - \frac{2\lambda_s RT(\gamma + 1)}{(h\nu_v)^2} \tag{63b}$$

$$\gamma = \ln\left[-\frac{(\Delta G^\circ + \lambda_s)}{\lambda_v}\right] - 1 \tag{63c}$$

The rate constants in the inverted region calculated from Eq. 59 are almost independent of temperature and decrease much less rapidly with driving force than predicted by classical models [49].

**Three-mode systems**

Next we consider a reaction that contains an active mode in each of the regions outlined above. The expression for the three-mode case is:

$$k_{el} = \frac{H_{ab}^2}{h}\left(\frac{4\pi^3}{\lambda_s RT}\right)^{1/2} \exp\left[-S_c \coth\left(\frac{h\nu_c}{2kT}\right)\right] \sum_{j_v=0}^{\infty} \sum_{j_c=-\infty}^{\infty} F_{j_v} \exp\left(\frac{j_c h\nu_c}{2kT}\right)$$

$$\times I_{j_c}\left(S_c \, \mathrm{csch}\left(\frac{h\nu_c}{2kT}\right)\right) \exp\left[-\frac{(\Delta G^\circ + \lambda_s + j_v h\nu_v + j_c h\nu_c)^2}{4\lambda_s RT}\right] \tag{64}$$

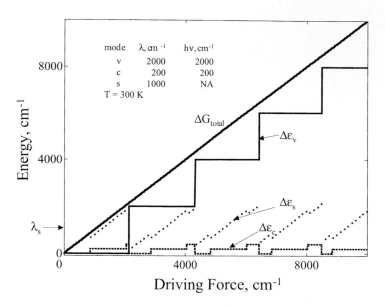

**Figure 10.** Plot of the energy in a particular mode for an electron transfer reaction with three active modes. $\Delta\varepsilon_v$, $\Delta\varepsilon_c$ and $\Delta\varepsilon_s$ are the differences between the energies of the products and reactants in the high-, intermediate- and low-frequency modes, respectively; $\lambda_i$, (cm$^{-1}$) and $h\nu_i$ (cm$^{-1}$) are (2000, 2000); (200, 200); and (1000, $-$) for the high-, intermediate- and low-frequency modes and the temperature is 300 K. The calculations were done using Eq. 64. The straight line, the stepped solid line, the dashed line and the dotted lines are the total energy difference ($\Delta G^\circ$) and the differences between the energies of the products and reactants in the high-, intermediate- and low-frequency modes, respectively.

where $j_v$ and $j_c$ are the changes in the vibrational quantum numbers for the high and intermediate frequency modes, respectively [49]. Again the last exponential term in Eq 64 is Gaussian peaked at $(\Delta G^\circ + \lambda_s + j_v h\nu_v + j_c h\nu_c) = 0$ with a width of $2\sqrt{4\lambda_s RT}$. In this case energy sharing can take place between the high-, intermediate- and low-frequency modes; the possibility of energy borrowing is increased but again the low-frequency mode is required to pass over its barrier while the other two modes can tunnel. Figures 10 and 11 show the energy distribution for the dominant contribution to the double sum. The low-frequency mode receives $\sim\lambda_s$ of energy to minimize its barrier; the intermediate mode receives $\sim\lambda_c$, and the bulk of the energy for large driving forces is deposited in the high-frequency mode. Only very seldom is the low or intermediate-frequency energy of the product less than that of the reactant. This is shown in Figure 11 where $\Delta\varepsilon_s$ is negative.

These expressions show that normally most of the excess energy is acquired by the high-frequency mode and that the intermediate-frequency mode receives an amount of energy that is less than one high-frequency vibrational quantum. Only when $\lambda_c \gg \lambda_s$ does the intermediate-frequency mode receive significantly more than a single high-frequency quantum. The effect of an intermediate mode on the rate constant for the reaction is relatively modest in the normal region but becomes important in the inverted region where the initial state needs to dispose of significantly more energy (Figure 12). In this region systems that have both high- and interme-

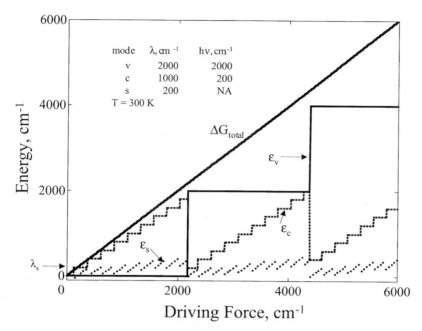

**Figure 11.** Plot of the energy in a particular mode for an electron transfer reaction with three active modes. $\Delta\varepsilon_v$, $\Delta\varepsilon_c$ and $\Delta\varepsilon_s$ are the difference between the energies of the products and reactants in the high-, intermediate- and low-frequency modes $\lambda_i$, (cm$^{-1}$) and $hv_i$ (cm$^{-1}$) are (2000, 2000); (1000, 200); and (200, −) for the high-, intermediate- and low-frequency modes, respectively, and the temperature is 300 K. The calculations were done using Eq. 64. The straight line, the stepped solid line, the dashed line and the dotted lines are the total energy difference ($\Delta G^\circ$) and the differences between the energies of the products and reactants in the high-, intermediate- and low-frequency modes, respectively.

diate-frequency modes exhibit significant rate enhancements due to tunneling and the decrease of the rate constant with increasing driving force is attenuated. Also, due to the intermediate-frequency mode the sinusoidal quantum beat effect observed for the dependence of the rate constant on driving force in the inverted region is significantly attenuated.

The three-mode expression is most useful when discussing the rates of non-radiative deactivation of excited states in the inverted region. In this region, where $-\Delta G^\circ \gg \lambda_v + \lambda_c + \lambda_s$, a much simpler expression can be used since the product is created with a high vibrational quantum number in the high-frequency mode. This expression is Eq. 65 provided that $5\lambda_c$ and $10\lambda_s$ are each $<|\Delta G^\circ|$.

$$k_{el} = \frac{4H_{ab}^2}{h}\left[-\frac{\pi^3}{2hv_v(\Delta G^\circ + \lambda_c + \lambda_s)}\right]^{1/2}$$

$$\times \exp\left(-\frac{\lambda_v - (\gamma_0+1)(\lambda_c+\lambda_s) - \gamma_0\Delta G^\circ - \frac{(\gamma_0+1)^2}{hv_v}\left(\lambda_s RT + \frac{\lambda_c hv_c}{2}\coth\left(\frac{hv_c}{2kT}\right)\right)}{hv_v}\right)$$

$$(65a)$$

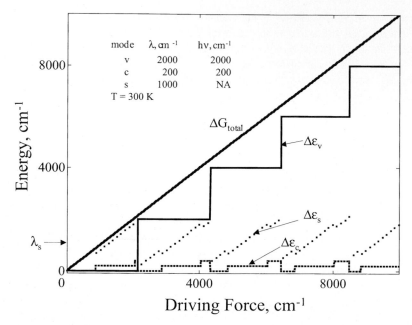

**Figure 12.** Plot of the logarithm of the Franck–Condon factors for the electron transfer reaction calculated using the classical expression, Eq. 23; two-mode expression, Eq. 62; three-mode expression, Eq. 64; and the approximate three-mode expression, Eq. 65 vs driving force. The parameters used $(\lambda_s, \lambda_c, h\nu_c, \lambda_h, h\nu_h$ in cm$^{-1}$) for the calculations are classical: (3200); two-mode, (1200, 2000, 600); three-mode (600, 600, 200, 2000, 600) and the temperature is 80 K. The solid line (inverted parabola), dotted line, oscillating solid line and the dashed line are for the classical expression, Eq. 23, the two-mode expression, Eq. 62, the full three-mode expression, Eq. 64, and the approximate three-mode expression, Eq. 65, respectively.

where

$$\gamma_0 = \ln\left(\frac{-\Delta G^\circ}{\lambda_v}\right) - 1 \tag{65b}$$

The above three-mode expression approximates the more exact expression Eq. 54 very well, but does not show the quantum beat effect.

## 1.5 Conclusions

The expressions derived from the traditional two-state model are useful in rationalizing a variety of electron transfer processes. Both thermal and optical charge transfer can be treated and, although not discussed here, electrochemical processes as well. The two-state model neglects contributions from higher electronic states in calculating the energies of the zero-order ground states of the reactants and products. Contributions from higher electronic states are, however, frequently needed in calculating electronic coupling elements. Mixing with such states leads to modification of the ground-state energies when the excited states are sufficiently low lying. Such perturbations are absent in the zero-interaction limit.

Some key features of the two-state model are summarized here:

(1) Although the reaction coordinate for charge transfer is not uniquely defined, the vertical difference between the zero-order reactant and product free energies is related to the degree of nuclear reorganization and consequently this difference provides a useful measure of the progress of the reaction (Section 1.2).

(2) The degree of charge transfer is *not* linearly related to the reaction coordinate defined above (Section 1.3.1).

(3) The splitting at the intersection of the adiabatic curves for a self-exchange reaction, $2H_{ab}$, enters into the expression for the free energy of activation for the exchange reaction analogous to the manner in which the driving force, $-\Delta G°$, enters into the expression for the free energy of activation for a marginally adiabatic net reaction (Section 1.3.1).

(4) The vertical difference between the free energies of the reactants and products of a self exchange reaction remains equal to $\lambda$ at the equilibrium configuration of the reactants (or products) regardless of the magnitude of the electronic coupling as long as the system remains valence trapped (Sections 1.3.1 and 1.3.4).

(5) The frequency of electron hopping in the transition state is equal to $2H_{ab}/h$ (Section 1.3.3).

(6) The electron transfer distance is defined by the difference between the dipole moments of the localized (diabatic) reactant and product states (Section 1.3.4).

(7) The electronic coupling element for a charge transfer reaction $H_{ab}$ is equal to $\langle\psi_a|V_{e,ac}|\psi_b\rangle$ where $V_{e,ac}$ is the coulomb interaction between the transferring electron (initially on the donor) and the acceptor (Section 1.4.1).

(8) At low temperatures and/or at high reaction exergonicities nuclear tunneling contributions to the rate and other quantum effects become important. Two- and three-mode expressions are presented that allow for tunneling of the higher frequency modes (Section 1.4.2).

Overall, the two-state model is remarkably successful in interpreting electron transfer and related properties and forms the cornerstone for interpreting a variety of complex physical, photosynthetic, catalytic and biological processes.

**Acknowledgments**

We wish to acknowledge helpful discussions with Rudolph A. Marcus. This research was carried out at Brookhaven National Laboratory under contract DE-AC02-98CH10886 with the U.S. Department of Energy and was supported by its Division of Chemical Sciences, Office of Basic Energy Sciences.

## References

1. R. A. Marcus, *J. Chem. Phys.* **1957**, *26*, 867–871.
2. R. A. Marcus, *Disc. Faraday. Soc.* **1960**, *29*, 21–31.
3. R. A. Marcus, *Rev. Modern Phys.* **1993**, *65*, 599–610.
4. N. Sutin, *Prog. Inorg. Chem.* **1983**, *30*, 441–498.
5. R. A. Marcus, N. Sutin, *Biochim. Biophys. Acta* **1985**, *811*, 265–322.
6. The sum of any two parabolas gives a parabola that has a force constant equal to the sum of the force constants of the two original parabolas and with its minimum located above the two original minima and horizontally between them.
7. J.-K. Hwang, A. Warshel, *J. Am. Chem. Soc.* **1987**, *109*, 715–720.
8. G. King, A. Warshel, *J. Chem. Phys.* **1990**, *93*, 8682–8692.
9. R. A. Marcus, N. Sutin, *Inorg. Chem* **1975**, *14*, 213–216.
10. L. W. Ungar, M. D. Newton, G. A. Voth, *J. Phys. Chem. B.* **1999**, *103*, 7367–7382.
11. C. Creutz, M. D. Newton, N. Sutin, *J. Photochem. Photobiol. A: Chem.* **1994**, *82*, 47–59.
12. M. B. Robin, P. Day, *Adv. Inorg. Chem. Radiochem.* **1967**, *10*, 247–422.
13. C. Creutz, *Prog. Inorg. Chem.* **1983**, *30*, 1–73.
14. R. J. Crutchley, *Adv. Inorg. Chem.* **1994**, *41*, 273–325.
15. B. S. Brunschwig, N. Sutin, *Coord. Chem. Rev.* **1999**, *187*, 233–254.
16. N. Sutin, in *Bioinorganic Chemistry, Vol. 2*, (Ed.: G. L. Gunther), Elsevier, New York **1973**, pp. 611–653.
17. B. S. Brunschwig, J. Logan, M. D. Newton, N. Sutin, *J. Am. Chem. Soc.* **1980**, *102*, 5798–5809.
18. R. A. Marcus, *Ann. Rev. Phys. Chem.* **1964**, *15*, 155–196.
19. R. A. Marcus, *J. Chem. Phys.* **1956**, *24*, 966–978.
20. R. A. Marcus, *J. Chem. Phys.* **1965**, *43*, 679–701.
21. B. S. Brunschwig, C. Creutz, N. Sutin, *Coord. Chem. Rev.* **1998**, *177*, 61–79.
22. H. J. Kim, J. T. Hynes, *J. Chem. Phys.* **1990**, *93*, 5194–5210.
23. J. N. Gehlen, D. Chandler, H. J. Kim, J. T. Hynes, *J. Phys. Chem.* **1992**, *96*, 1748–1753.
24. R. A. Marcus, *J. Phys. Chem.* **1992**, *96*, 1753–1757.
25. H. J. Kim, J. T. Hynes, *J. Chem. Phys.* **1992**, *96*, 5088–5110.
26. J. N. Gehlen, D. Chandler, *J. Chem. Phys.* **1992**, *97*, 4958–4963.
27. P. W. Atkins, R. S. Friedman, *Molecular Quantum Mechanics*, Oxford University Press, New York **1997**, p. 187.
28. H. J. Kim, R. Bianco, B. J. Gertner, J. T. Hynes, *J. Phys. Chem.* **1993**, *97*, 1723–1728.
29. N. S. Hush, *Prog. Inorg. Chem.* **1967**, *8*, 391–444.
30. N. S. Hush, *Electrochim. Acta* **1968**, *13*, 1005–1023.
31. Y.-G. K. Shin, B. S. Brunschwig, C. Creutz, N. Sutin, *J. Phys. Chem.* **1996**, *100*, 8157–8169.
32. R. J. Cave, M. D. Newton, *Chem. Phys. Lett.* **1996**, *249*, 15–19.
33. R. J. Cave, M. D. Newton, *J. Chem. Phys.* **1997**, *106*, 9213–9226.
34. C. Lambert, G. Nöll, *J. Am. Chem. Soc.* **1999**, *121*, 8434–8442.
35. S. F. Nelsen, R. F. Ismagilov, D. A. T. II, *Science* **1997**, *278*, 846–849.
36. P. W. Atkins, R. S. Friedman, *Molecular Quantum Mechanics*, Oxford University Press, New York **1997**, p. 509.
37. P. W. Atkins, R. S. Friedman, *Molecular Quantum Mechanics*, Oxford University Press, New York **1997**, p. 196.
38. The overlap integrals $S_{Av, Bw}$ can be calculated if the reactant and produce surfaces are assumed to be harmonic with the same force constants.
39. T. Terasaka, T. Matsushita, *Chem. Phys. Lett.* **1981**, *80*, 306–310.
40. N. R. Kestner, J. Logan, J. Jortner, *J. Phys. Chem.* **1974**, *78*, 2148.
41. J. Ulstrup, *Charge Transfer Processes in Condensed Media*, Springer, New York **1979**, pp. 71–79.
42. M. Bixon, J. Jortner, *J. Chem. Phys.* **1968**, *48*, 715–726.
43. K. F. Freed, J. Jortner, *J. Chem. Phys.* **1970**, *52*, 6272–6291.
44. R. Englman, J. Jortner, *Mol. Physics* **1970**, *18*, 145–164.

45. P. W. Atkins, R. S. Friedman, *Molecular Quantum Mechanics*, Oxford University Press, Oxford **1997**, 241.
46. E. Buhks, G. Navon, M. Bixon, J. Jortner, *J. Am. Chem. Soc.* **1980**, *102*, 2918–2923.
47. J. Ulstrup, J. Jortner, *J. Chem. Phys.* **1975**, *63*, 4358–4368.
48. J. Jortner, *J. Chem. Phys.* **1976**, *64*, 4860–4867.
49. B. S. Brunschwig, N. Sutin, *Comments Inorg. Chem.* **1987**, *6*, 209–235.

# 2 Charge-transfer Interactions and Electron-transfer-activated Reactions of Organometallic Complexes

*Stephan M. Hubig and Jay K. Kochi*

## 2.1 Introduction

Organometallic complexes display a rich electron-transfer chemistry as a consequence of their unique redox properties [1–3]. First, owing to their manifold oxidation states, the metal centers can undergo a variety of redox transformations involving single to multiple electron transfers that depend on the redox counterpart and the reaction conditions. Second, the coordination of the metal with various organic ligands (of different donor or acceptor properties) allows the selective stabilization of certain oxidation levels and a fine-tuning of their redox potentials. Thus, depending on their oxidation states and the ligand environment, organometallic complexes may act as electron donors or acceptors, which is discussed in detail in Section 2.2. Accordingly, they are ideal redox reagents to promote electron-transfer activated reactions of organic substrates which are (a) directly bound to the metal as ligands or (b) encounter the organometallic complex in a bimolecular (diffusional) reaction mode [4–6].

Bimolecular electron-transfer (ET) reactions in solution require the freely diffusing donor and acceptor molecules to achieve a nuclear pre-arrangement in the precursor or encounter complex [7–9] which enables electron exchange from the HOMO of the donor to the LUMO of the acceptor moiety. The distance and relative orientation of the ET partners in this precursor complex control the degree of orbital overlap or electronic coupling [10] of the redox partners in the ET transition state, which limits the rate of the intrinsically ultrafast electron jump [11–13]. As such, electron donor–acceptor (EDA) complexes represent precursor complexes with a rather strong electronic (charge-transfer) coupling [14–16] and thus play a critical role in the predisposition of redox reactants toward electron transfer. A variety of such EDA complexes involving organometallic donors and acceptors are introduced in Section 2.3 and their crystal structures and spectroscopic properties are discussed.

Depending on the energetic conditions, electron transfer may occur spontaneously in the EDA complex as a *thermal* reaction (see reaction path **A** in Scheme 1), or

**(A) Thermal ET:**

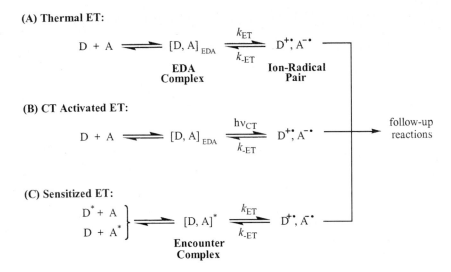

**(B) CT Activated ET:**

**(C) Sensitized ET:**

**Scheme 1.**

it may require additional (light) energy in a *photoactivated* reaction (see reaction **B** in Scheme 1) [5, 17]. In the latter case, irradiation of the charge-transfer (CT) absorption band of the (thermally unreactive) EDA complex effects a spontaneous electron transfer from the donor HOMO to the acceptor LUMO [18], which represents the initial step in a *CT-activated* photoreaction.

An alternative possibility to initiate electron transfer between two unreactive substrates is the photoactivation of one of the redox partners. In such a *photosensitized* reaction (path **C** in Scheme 1), photoexcitation of one of the redox partners enhances its oxidation or reduction potential substantially. As a consequence, the reactant in the excited state is now able to undergo spontaneous electron transfer upon the formation of an encounter complex (or exciplex) [19, 20] with a suitable redox partner. Owing to the enhanced redox potentials, the encounter complex exhibits even stronger charge-transfer character as compared to the (ground-state) EDA complex [20] and facilitates electron transfers that do not occur in the ground state owing to unfavorable (endergonic) energetics. Most importantly, all three types of electron transfer described above, viz. thermal, sensitized, and CT activated pathways, ultimately lead to ion–radical pairs as the critical reaction intermediates [21] that subsequently undergo various follow-up reactions to the final products. The various methods to generate and detect organometallic ion radicals will be described in Section 2.4.

Ion–radical pairs may undergo back electron transfer [22–24] ($k_{-ET}$ in Scheme 1) in competition with the desired follow-up reactions, which limits the yields of the final reaction products. In fact, the more endergonic the initial electron-transfer step ($k_{ET}$ in Scheme 1) is, the faster is the corresponding (exergonic) back electron transfer, which shifts the ET equilibrium towards the precursor (EDA) complex. However, even in such cases of highly endergonic electron transfer, a net chemical reaction to the final products may still be observed if the rate of the follow-up reaction is competitive with that of the back electron transfer in Scheme 1 [25]. This is demonstrated with many examples of follow-up reactions of organometallic ion radicals in Section 2.4.

Kinetic studies and the resulting correlation between the rate constants and the free energy change of the reaction are useful tools to elucidate mechanistic aspects of the electron-transfer step. Traditionally, electron-transfer reactions are classified as outer-sphere or inner-sphere processes [26–30]. Marcus theory [7, 8] describes the free-energy correlation of outer-sphere ET rate constants. Accordingly, free-energy correlations may be used to distinguish between outer-sphere and inner-sphere reactions [31–33]. This concept is demonstrated for various electron-transfer reactions of organometallic complexes in Section 2.5. Most importantly, steric effects introduced by bulky ligands [32, 33] are shown to induce a changeover between the two electron-transfer mechanisms as observed in organic electron transfers involving sterically encumbered electron donors [31].

This review illustrates the above delineated characteristics of electron-transfer activated reactions by analyzing some representative thermal and photoinduced organometallic reactions. Kinetic studies of thermal reactions, time-resolved spectroscopic studies of photoinduced reactions, and free-energy correlations are presented to underscore the unifying role of ion-radical intermediates [29] in—at first glance—unrelated reactions such as additions, insertions, eliminations, redox reactions, etc. (Photoinduced electron-transfer reactions of metal porphyrin and polypyridine complexes are not included here since they are reviewed separately in Chapters 2.2.16 and 2.2.17, respectively.)

## 2.2 Organometallic Complexes as Electron Donors or Acceptors

Because metals have low electronegativities and most frequently exist in low oxidation states when coordinated with organic ligands, organometallic complexes are most commonly employed as donors in electron-transfer reactions (see Section 2.2.1.). However, if the metal center is oxidized to its cationic state or (formally) promoted to a high oxidation level by coordination with electronegative ligands, the organometallic complex is transformed to an electron acceptor (see Section 2.2.2). A quantitative classification of organometallic complexes according to their electron donor or acceptor strength is generally based on measurements of their (vertical) ionization potentials ($IP$) and electron affinities ($EA$) in the gas phase or oxidation ($E^\circ_{ox}$) and reduction ($E^\circ_{red}$) potentials in solution. Thus, $IP$ and $E^\circ_{ox}$ values represent measures for the energy required to detach an electron from a donating substrate (Eq. 1):

$$D \xrightarrow[(E^\circ_{ox})]{IP} D^{+\bullet} + e^- \tag{1}$$

and electron affinities and reduction potentials are measures for the energetics of electron attachment to an acceptor (Eq. 2):

$$A + e^- \xrightarrow[(E^\circ_{red})]{EA} A^{-\bullet} \tag{2}$$

## 2.2.1 Organometallic Donors

### Alkylmetals

The coordination of alkyl groups R around a metal center M generates a simple organometallic complex $MR_n$ the electron-donor properties of which are readily accounted for by the electron-richness of the metal center due to the positive inductive effect of R. The (vertical) ionization potentials and (irreversible) oxidation potentials (vide infra) listed in Table 1 for a variety of polyalkyl complexes of Group IV and Group IIb metals establish the donor strength of these organometallic substrates to be comparable with those of organic aromatic donors such as naphthalene [34] or alkyl-substituted benzenes [36]. The data in Table 1 clearly show that with decreasing electronegativity of the (Group IV) metals (e.g., $C \gg Ge > Sn > Pb$) and with increasing electron donicity of the alkyl ligand (e.g., $CH_3 < C_2H_5 < i\text{-}C_3H_7 < t\text{-}C_4H_9$), the donor properties of the coordination complex improve substantially as illustrated by the ionization potentials which vary by more than 1 eV. In addition, stepwise substitution of methyl ligands by ethyl groups (see entries 1–5 in Table 1) results in incremental decrease of the ionization potential. The latter two observations are in excellent agreement with the MO description

**Table 1.** Vertical ionization potentials (*IP*) and oxidation (peak) potentials ($E_{ox}$) of alkylmetals ($R_nM$).

| $R_nM$ | $IP$ (eV) | Ref. | $E_{ox}$ (V relative to the SCE) | Ref. |
|---|---|---|---|---|
| PbMe$_4$ | 8.90 | 38 | 0.98 | 39 |
| PbMe$_3$Et | 8.65 | 38 | 0.84 | 39 |
| PbMe$_2$Et$_2$ | 8.45 | 38 | 0.73 | 39 |
| PbMeEt$_3$ | 8.26 | 38 | 0.63 | 39 |
| PbEt$_4$ | 8.13 | 38 | 0.55 | 39 |
| SnMe$_4$ | 9.70 | 40 | 1.42 | 39 |
| SnEt$_4$ | 8.93 | 40 | 1.65 | 41 |
| Sn($n$-)Pr$_4$ | 8.82 | 42 | | |
| Sn($n$-)Bu$_4$ | 8.83 | 40 | 0.90 | 39 |
| Sn($neo$-)Pent$_4$ | 8.67 | 42 | | |
| GeEt$_4$ | 9.41 | 42 | | |
| Neopentane[a] | 10.90 | 43 | | |
| ZnMe$_2$ | 9.4 | 44 | | |
| ZnEt$_2$ | 8.6 | 44 | | |
| CdMe$_2$ | 8.8 | 44 | | |
| CdEt$_2$ | 8.2 | 44 | | |
| Cd($n$-)Pr$_2$ | 8.2 | 44 | | |
| HgMe$_2$ | 9.33 | 42 | 1.46 | 42 |
| | 9.30 | 44 | | |
| HgEt$_2$ | 8.45 | 42 | 0.97 | 42 |
| | 8.90 | 44 | | |
| Hg($n$-)Bu$_2$ | 8.35 | 42 | | |

[a] For comparison.

of alkylmetal complexes which derives the HOMO from the $\sigma_{M-C}$ bonding orbitals [37].

Electrochemical oxidations of alkylmetal donors are irreversible in all cases studied owing to rapid homolytic cleavage of the metal–carbon bond at the cation-radical stage (Eq. 3):

$$R_nM \xrightarrow{-e^-} R_nM^{+\cdot} \xrightarrow{fast} R_{n-1}M^+ + R^\cdot \tag{3}$$

The facile metal–carbon bond cleavage upon one-electron oxidation is not surprising considering the $\sigma_{M-C}$ character of the HOMO (vide supra) [37], and the irreversible fragmentation of the cation radical allows even highly endergonic oxidations of alkylmetals to occur at reasonable rates and with good yields. In other words, even electron-transfer equilibria shifted far toward the side of the starting materials can be effectively 'drained' towards the electron-transfer products as demonstrated by the efficient insertion of tetracyanoethylene into the metal–carbon bond of various tetraalkyltin compounds (see Section 2.5).

**Metallocenes and (arene) metal sandwich complexes**

Metallocenes are useful electron donors as judged by their low (vertical) ionization potentials in the gas phase and oxidation potentials in solution (see Table 2). In fact, the electron-rich (19 $e^-$) cobaltocene with an oxidation potential of $E^\circ_{ox} = -0.9$ V relative to the SCE [45] is commonly employed as a very powerful reducing agent in solution. Unlike the alkylmetals (vide supra), the HOMOs of metallocenes reside at the metal center [46] which accounts for two effects: (i) Removal of an electron from the HOMO requires minimal reorganization energy which explains the facile oxidative conversion from metallocene to metallocenium. (ii) The metal–carbon bonding orbitals are little affected by the redox process, and thus the resulting metallocenium ions are very stable and can be isolated as salts.

Replacing cyclopentadienyl ligands in the metallocenes by arene ligands results in

**Table 2.** Vertical ionization potentials (*IP*) and oxidation potentials ($E^\circ_{ox}$) of metallocenes and (arene) metal sandwich complexes.

| Sandwich complex | *IP* (eV) | Ref. | $E^\circ_{ox}$ (V relative to the SCE) | Ref. |
|---|---|---|---|---|
| $Cp_2Fe$ | 6.86 | 47 | 0.41 | 48 |
| $Cp_2Co$ | 5.55 | 45 | -0.90 | 45 |
| $Cp_2Ru$ | 7.45 | 45 | | |
| $(\eta^6\text{-}C_6H_6)_2Cr$ | | | -0.80 | 49, 50 |
| $CpFe(HMB)^a$ | 4.68 | 45 | -1.41 | 45 |
| $(\eta^6\text{-}C_6H_6)Cr(CO)_3$ | | | 0.84 | 51 |
| | | | 0.71 | 52 |
| $(\eta^6\text{-}TOL)Cr(CO)_3{}^b$ | | | 0.82 | 51 |
| | | | 0.68 | 52 |

[a] HMB = hexamethylbenzene.
[b] TOL = toluene.

**Table 3.** Vertical ionization potentials ($IP$) and oxidation (peak) potentials ($E_{ox}$) of metal carbonyls.

| Metal carbonyl | $IP$ (eV) | Ref. | $E_{ox}$ (V relative to the SCE) | Ref. |
|---|---|---|---|---|
| $Cr(CO)_6$ | 8.40 | 55 | 1.53 | 53 |
| | | | 1.09[a] | 53 |
| $Mo(CO)_6$ | 8.50 | 55 | 1.53 | 53 |
| $Co(CO)_6$ | 8.56 | 55 | 1.53 | 53 |
| $Fe(CO)_5$ | 8.60 | 56 | 1.51 | 53 |
| $Ni(CO)_4$ | 8.93 | 56 | 1.26 | 53 |
| $Mn_2(CO)_{10}$ | 8.02 | 55 | 1.55 | 53 |
| $Re_2(CO)_{10}$ | 8.07 | 55 | 1.55 | 53 |

[a] Reversible potential in trifluoroacetic acid.

metal sandwich complexes with enhanced reducing properties (see entries 1 and 5 in Table 2). On the other hand, removal of one arene ligand and replacement by carbonyl ligands results in half-sandwich complexes with substantially higher oxidation potentials and thus diminished reducing capabilities.

**Metal carbonyls**

The ionization and oxidation (peak) potentials [53] of metal carbonyls listed in Table 3 establish their mild donicity despite their (formally) neutral oxidation state. However, partial replacement of carbonyls by stronger donor ligands (such as phosphine, sulfide, etc.) effects an incremental increase in their reducing properties [54]. Such a fine-tuning of the oxidation potentials of structurally similar organometallic donors is ideal for studies of the correlation between rate constants and the electron-transfer driving force (see Section 2.5).

**Metal hydrides**

Metal hydrides represent another type of electron donating organometallic complexes which exhibit ionization and oxidation potentials similar to alkyl or aryl coordinated metals (compare entries 2 and 3 or 2 and 4 in Table 4).

**Enhanced donicity in anionic organometallic complexes (metallates)**

The electron donicity of alkylmetal complexes is strongly enhanced when a negative charge is introduced. Thus, polyalkylmetallates such as borates or aurates exhibit oxidation (peak) potentials much lower than neutral polyalkylmetal complexes (compare Tables 1 and 5). An extreme case represents dimethylaurate(I) with an oxidation (peak) potential of 0.1 V (relative to the SCE) [64] which is more than 1 V lower than that of the isoelectronic dimethylmercury(II) complex ($E_{ox} = 1.46$ V relative to the SCE) [42]. Similarly, the oxidation potential of tetramethylborate ($E_{ox} = 0.58$ V relative to the SCE) [65] is substantially lower than that of the isoelectronic neopentane which exceeds 3 V [66].

**Table 4.** Vertical ionization potentials (*IP*) and oxidation potentials ($E°_{ox}$) of metal hydrides.

| Metal hydride | *IP* (eV) | Ref. | $E_{ox}$ (V relative to the SCE) | Ref. |
|---|---|---|---|---|
| $Cp_2WH_2$ | 6.4 | 57 | 0.01 | 58 |
| $Cp_2MoH_2$ | 6.4 | 57 | 0.13 | 58 |
| $Cp_2MoH(C_6H_6)$ | | | −0.06 | 59 |
| $Cp_2Mo(CH_3)_2$ | 6.1 | 57 | | |
| $Et_3SnH$ | 9.06 | 60 | | |
| $n\text{-}Bu_3SnH$ | 8.72 | 61 | | |
| $n\text{-}Bu_3GeH$ | 9.62 | 61 | | |
| $Re(CO)_5H$ | 8.89 | 62 | | |
| $Co(CO)_4H$ | 8.90 | 63 | | |

**Table 5.** Oxidation (peak) potentials ($E_{ox}$) of alkylmetallates.

| Alkyl metallate | $E_{ox}$ (*V* relative to the SCE) | Ref. |
|---|---|---|
| $BMe_4^-$ | 0.58 | 65 |
| $BMePh_3^-$ | 0.96 | 65 |
| $BPh_4^-$ | 0.99 | 65 |
| | 0.76[a] | 67 |
| $BTol_4^-$ | 0.55[a] | 67 |
| $AuMe_2^-$ | 0.10 | 64 |
| $AuMe_4^-$ | 0.42 | 64 |

[a] $E°_{ox}$ values calculated from electron-transfer rate constants [67].

Similarly, the mild reducing power of metal carbonyls can be effectively enhanced by conversion of the neutral complex to the corresponding carbonylmetallate. For example, hexacarbonylvanadate and pentacarbonylmanganate exhibit oxidation potentials about 1 V lower than those of the isoelectronic hexacarbonylchromium and pentacarbonyliron complexes, respectively (see Table 6). Similar trends apply to neutral and negatively charged half-sandwich complexes such as the isoelectronic $CpCr(CO)_3^-$ and $CpMn(CO)_3$ in Table 6.

The increase in the donor strength of negatively charged organometallic complexes as compared to their isoelectronic neutral counterparts is readily explained by also considering the different Coulombic work terms ($\omega$) involved in the one-electron oxidation processes. The removal of an electron from an anion generates a neutral oxidation product which does not attract the leaving electron ($\omega = 0$), whereas electron detachment from a neutral donor requires additional energy to overcome the strong Coulombic (attractive) forces ($\omega = e^2/r$) in the resulting electron–cation pair. The opposite effect on the donicity is observed when positive charges are introduced. Thus, neutral organometallic donors lose their reducing power when oxidized to the cationic state, and therefore cationic organometallic complexes are rarely used as electron donors in electron-transfer reactions [73].

**Table 6.** Comparison of isoelectronic neutral and anionic metal carbonyl complexes.

| Neutral complex | $E_{ox}$ (V relative to the SCE) | Ref. | Isoelectronic metallate | $E_{ox}$ (V relative to the SCE) | Ref. |
|---|---|---|---|---|---|
| $Cr(CO)_6$ | 1.09 | 53 | $V(CO)_6^-$ | 0.23 | 68 |
| $Fe(CO)_5$ | 1.5 | 53 | $Mn(CO)_5^-$ | 0.08 | 69 |
| $CpMn(CO)_3$ | 0.79 | 53 | $CpCr(CO)_3^-$ | −0.18 | 70 |
| $CpCo(CO)_2$ | −0.06 | 71 | $CpFe(CO)_2^-$ | −1.18 | 72 |

### 2.2.2 Organometallic Acceptors

Both the low electronegativity of the metallic core and the generally low (formal) oxidation state of the metal center make most organometallic complexes unsuitable as electron *acceptors*. In fact, *neutral* organometallic acceptors are very rare, however the few complexes which all contain metals in high (formal) oxidation states are very powerful oxidants. For example, $[(CF_3)_4C_2S_2]_2Ni(IV)$ and $Cp^*VCl_3$ have reduction potentials of $E^\circ_{red} = 1.0$ V [74] and 0.6 V [75], respectively, relative to the SCE. Thallium(III) trifluoroacetate and mercury(II) trifluoroacetate readily oxidize arenes (see Section 2.4.2), and we wish to mention several other high-valent metal complexes, e.g. $OsO_4$, $SnCl_4$, $TiCl_4$, which do not classify as organometallic acceptors, but act as acceptors in various electron-transfer reactions with organic donors and thus form organometallic donor–acceptor complexes (see Section 2.3) and covalently bound intermediates (see Section 2.4).

Most organometallic acceptors are positively charged complexes which show highly increased reduction potentials as compared to the isoelectronic neutral complexes. Several examples of such cationic organometallic acceptors (obtained by 'umpolung') and their isoelectronic neutral counterparts which act as donors are summarized in Table 7.

## 2.3 Electron Donor–Acceptor Interactions

The formation of charge-transfer complexes between organic electron donors and acceptors has been studied in great detail [85–87]. Similarly, the organometallic donors and acceptors introduced in Section 2.2 form electron donor–acceptor (EDA) complexes among themselves or when mixed together with an appropriate (organic or inorganic) donor or acceptor [88–90]. This association of electron donor and acceptor is critical for facilitating electron transfer between the two redox partners since it affects both the thermodynamics as well as the kinetics of the electron transfer. In fact, the formation of the EDA complex in Eq. 4, i.e.:

$$D + A \overset{K_{EDA}}{\rightleftharpoons} [D, A]_{EDA} \tag{4}$$

**Table 7.** Redox potentials of cationic electron acceptors and their isoelectronic neutral counterparts.

| Cationic complex[a] | $E^\circ_{red}$ (V relative to the SCE) | Ref. | Isoelectronic neutral complex | $E^\circ$ (V relative to the SCE) | Ref. |
|---|---|---|---|---|---|
| $Cp_2Co^+$ | −0.90 | 79 | $Cp_2Fe$ | $E^\circ_{ox} = 0.41$ | 48 |
| | | | | $E^\circ_{red} = -2.8$ | 80 |
| $(\eta^6\text{-MES})_2Fe^{2+}$ | −0.08 | 81 | $(\eta^6\text{-}C_6H_6)_2Cr$ | $E^\circ_{ox} = -0.8$ | 49 |
| $(\eta^6\text{-DUR})_2Fe^{2+}$ | −0.17 | 81 | | | |
| $(\eta^6\text{-HMB})_2Fe^{2+}$ | −0.27 | 81 | | | |
| $CpFe(C_6H_6)^+$ | −1.4 | 82 | $CpMn(C_6H_6)$ | $E^\circ_{ox} = 0.79$ | 53 |
| | | | | $E^\circ_{red} < -2.4$ | 53 |
| $(CpCF_3)_2Fe^+$ | +1.05 | 76 | | | |
| $Cp_2Fe^+$ | +0.41 | 48 | | | |
| $Cp^*_2Fe^+$ | −0.18 | 6 | | | |
| $(phen)_3Fe^{3+}$ | +0.98 | 77 | | | |
| $(phen)_3Ru^{3+}$ | +1.28 | 78 | | | |
| $CpCo(C_6H_6)^{2+}$ | +0.4 | 83 | | | |
| $\eta^5\text{-cyclohexadienyl}$ $Fe(CO)_3^+$ | −0.31 | 84 | | | |

[a] MES = mesitylene, DUR = durene, HMB = hexamethylbenzene.

effects a predisposition of the donor–acceptor couple toward electron transfer which can be quantified by the degree of charge transfer [91] or the orbital overlap [10] in the complex (see Section 2.3.2). In the following sections, we will first discuss the formation of organometallic EDA complexes and its energetics, and then present spectroscopic and structural evidence for charge-transfer interactions involving organometallic donors or acceptors.

## 2.3.1 Formation of Organometallic EDA Complexes

Organometallic EDA complexes can be roughly divided into two types based on the nature of the forces that stabilize the donor–acceptor association. In *molecular (outer-sphere or second-sphere)* [88–90] complexes with neutral organometallic donors and/or acceptors, the complex formation is caused by weak Coulombic interactions such as dipole–dipole [94], induced dipole [95], or dispersion forces [96] between readily polarizable substrates. In addition, hydrogen bonding may also play a stabilizing role [97]. Since the charge-transfer interactions occur between the ligated metal center and a donor (or acceptor) that is *not* a ligand in the first (or inner) coordination sphere of the metal, such weak complexes are commonly referred to as outer-sphere or second-sphere complexes [88–90]. The weak stabilization of these complexes is reflected in the rather low formation constants ($K_{EDA}$ in Eq. 4) which typically cover a range between 0.1 and 10 $M^{-1}$. Such low formation constants translate into free formation enthalpies of −2 kcal $mol^{-1}$ < $\Delta G^\circ_f$ < +2 kcal $mol^{-1}$. Extreme cases of such weak interactions are complexes with formation constants of $K_{EDA}$ approaching 0.01 M$^{-1}$. If diffusion-limited rates are assumed for the association of donor and acceptor, the low equilibrium constants of about 0.01 indicate lifetimes of about 0.5 ps for the EDA complex, which is in the time domain of molecular collisions. Thus, such weak complexes are also referred to as collision complexes [98] or contact charge-transfer complexes [99, 100].

Various examples for molecular complexes including organometallic donors or acceptors are listed in Tables 8 and 9, respectively, together with their formation constants (which are usually determined by spectrophotometric methods [101]). EDA complexes between organometallic donors and tetracyanoethylene (TCNE) may exhibit formation constants $K_{EDA} \gg 10$ (see entry 8 in Table 8) [110]. In these cases, the strong acceptor TCNE effects a high degree of charge transfer in the ground state, and thus EDA complexes containing TCNE are stabilized not only by the above mentioned dipole and dispersion forces, but also by charge resonance, i.e. $[D, A] \leftrightarrow [D^+, A^-]$ [102]. Such charge resonance in the ground state is frequently evidenced by (i) unusual bond lengths in the X-ray structure of the (crystalline) EDA complex as compared to the structures of the single (uncomplexed) donor or

**Table 8.** Molecular EDA complexes with organometallic donors.

| Organometallic donor | Acceptor[a] | $K_{EDA}$ (M$^{-1}$) | Ref. |
|---|---|---|---|
| Co(acac)$_2$(py)$_2$ | CHCl$_3$ | 3.7 | 88 |
| Co(acac)$_2$(py)$_2$ | CH$_2$Cl$_2$ | 3.0 | 106 |
| Fe(acac)$_3$ | CH$_3$CN | 0.6 | 107 |
| Cp$_2$Fe | CCl$_4$ | 1.5 | 108 |
| Cp$_2$Fe | CH$_2$I$_2$ | 0.1 | 109 |
| Cp$_2$MoH$_2$ | MA | 5.6 | 58 |
| (C$_6$H$_6$)Cr(CO)$_3$ | TNB | 0.42 | 110 |
| (C$_6$H$_6$)Cr(CO)$_3$ | TCNE | 1070 | 110 |
| Et$_4$Sn | I$_2$ | 3.6 | 111 |
| Et$_4$Pb | I$_2$ | 3.1 | 111 |
| $n$-Pr$_2$Hg | I$_2$ | 1.5 | 111 |

[a] MA = maleic anhydride; TNB = trinitrobenzene; TCNE = tetracyanoethylene.

**Table 9.** Molecular EDA complexes with organometallic acceptors.

| Organometallic acceptor | Donor | $K_{EDA}$ (M$^{-1}$) | Ref. |
|---|---|---|---|
| Ni[S$_2$C$_2$(CF$_3$)$_4$]$_2$ | Perylene | [b] | 112 |
| Hg(O$_2$CCF$_3$)$_2$ | Mesitylene | 5.1 | 113 |
| OsO$_4$[a] | Durene | 0.47 | 114 |
| SnCl$_4$[a] | Mesitylene | 0.4 | 115 |
| SnCl$_4$[a] | 1-Methoxynaphthalene | 1.3 | 115 |
| TiCl$_4$[a] | Naphthalene | 0.34 | 116 |

[a] High-valent metals that form organometallic EDA complexes.
[b] Unknown.

**Table 10.** Dissociation constants of charge-transfer ion pairs in different solvents.

| Donor anion | Acceptor cation | Solvent | $K_{diss}$ (m) | Ref. |
|---|---|---|---|---|
| $Co(CO)_4^-$ | $Cp_2Co^+$ | MeCN | 0.011 | 118 |
| | | $CH_2Cl_2$ | 0.00015 | 118 |
| $Co(CO)_4^-$ | $Q^{+a}$ | MeCN | 0.081 | 118 |
| | | THF | 0.00012 | 118 |
| | | $CH_2Cl_2$ | 0.000015 | 118 |
| $Co(CO)_4^-$ | $PPN^{+b}$ | THF | 0.000094 | 119 |
| $V(CO)_6^-$ | $PPN^+$ | THF | 0.00012 | 119 |
| $BPh_4^-$ | $PPN^+$ | THF | 0.000091 | 119 |
| $C(CN)_3^-$ | $(HMB)_2Fe^{2+c}$ | $CH_3CN$ | $0.01^d$ | 120 |

[a] *N*-methyl-1-quinolinium.
[b] *Bis*(triphenylphosphoranylidene)ammonium.
[c] HMB = hexamethylbenzene.
[d] Equilibrium constant ($K_{EDA,2}$) for *association* of the second $C(CN)_3^-$ ion to form the 2:1 complex, i.e. $[(HMB)_2Fe^{2+}, 2C(CN)_3^-]$. See [120].

acceptor components (see Section 2.3.3), (ii) unusual electrical and magnetic properties, and (iii) substantial changes in the UV–Vis and IR spectra (see Section 2.3.2).

The second type of EDA complexes involving organometallic donors or acceptors includes pairs of opposite charged ions in close contact, which are referred to as *contact ion pairs* (CIP) [103, 104] or *charge-transfer ion pairs* (CTIP) [105]. Since negative charges improve the donicity and positive charges enhance the acceptor strength, CT ion pairs usually consist of an anionic donor and a cationic acceptor, which are held together by strong Coulombic forces. Since Coulombic attraction between ions in solution strongly depends on the solvent polarity [104], the formation constants $K_{EDA}$ (see Eq. 4) for charge-transfer ion pairs may vary over orders of magnitude in solvents as different as water ($K_{EDA} \approx 1–100$ $M^{-1}$) and dichloromethane ($K_{EDA} > 10^5$ $M^{-1}$). The selected examples in Table 10 illustrate this dramatic solvent dependence for the equilibrium constant ($K_{diss}$) of the reverse process, viz., dissociation of the contact (charge-transfer) ion pairs into free (solvated) ions [117] (Eq. 5):

$$[D^-, A^+] \underset{}{\overset{K_{diss}}{\rightleftharpoons}} D^- + A^+ \qquad (5)$$
$$\text{(CIP or CTIP)} \qquad \text{(free ions)}$$

### 2.3.2 UV–Vis Spectroscopic Evidence for Charge-transfer Interactions in Organometallic EDA Complexes

**The Mulliken correlation**

The most characteristic spectroscopic features of electron donor–acceptor complexes are the new absorption bands in the electronic (UV–Vis) spectra, which are not found in the spectra of the single (donor or acceptor) components. Such absorption bands are universally observed with organic [85–87], inorganic [121], and organometallic substrates [122], for both molecular complexes [85] and contact ion pairs [103, 104], and even in the case of short-lived collision complexes with $K_{EDA} \ll 1$ (vide supra) [98]. Mulliken theory [14–16] provides the theoretical basis for the assignment of these new electronic transitions via the description of the wave function $\Psi_{EDA}$ of the EDA complex as the sum of a no-bond $(D, A)$ and a dative $(D^+, A^-)$ wave function (in first approximation), (Eq. 6):

$$\Psi_{EDA} = \mathbf{a}\,\psi_0(D, A) + \mathbf{b}\,\psi_1(D^+, A^-) + \cdots \tag{6}$$

For weakly interacting electron donors and acceptors, the wave function $\Psi_{EDA}$ for the *ground state* basically describes a no-bond situation between donor and acceptor $(D, A)$ in which the charge-transfer state $(D^+, A^-)$ does not play a significant role, i.e. $\mathbf{a} \gg \mathbf{b}$. In contrast, the ratio of the mixing coefficients $\mathbf{a}$ and $\mathbf{b}$ for no-bond and dative wave function is inverted in the *excited state*, i.e. the charge-transfer state $(D^+, A^-)$ is now predominant. The new absorption bands in the UV–Vis spectra of the EDA complexes are assigned to the electronic transition from the ground state (no-bond) to the excited (charge-transfer) state, and they are thus referred to as charge-transfer (CT) transitions. In other words, the population of the excited state of the EDA complex by photoexcitation of its CT absorption band is tantamount to a photoinduced electron transfer from a donor (HOMO) orbital to an acceptor (LUMO) orbital which yields the ion-pair state, (Eq. 7):

$$[D, A]_{EDA} \xrightarrow{h\nu_{CT}} [D^{+\bullet}, A^{-\bullet}] \tag{7}$$

According to Mulliken theory [14–16], the energy gap $(E_{CT})$ of the charge-transfer transition from the ground state to the excited ion-pair state determines the wavelength position of the CT absorption band $(\lambda_{CT})$, i.e. $E_{CT} = h\nu_{CT} = hc/\lambda_{CT}$. This energy gap directly depends on the ionization potential $(IP)$ of the donor and the electron affinity $(EA)$ of the acceptor, (Eq. 8):

$$E_{CT} = h\nu_{CT} = IP - EA + \omega + \text{const.} \tag{8}$$

where $\omega$ represents the Coulombic work term for the association of $D^{+\bullet}$ and $A^{-\bullet}$ to form an ion pair, and the constant term includes orbital-coupling energies which are negligible in weak EDA complexes. For a series of EDA complexes of various donors with the same acceptor, the Mulliken correlation in Eq. 8 predicts a unit slope for plots of $E_{CT}$ against the ionization potentials of the donors. This is demonstrated in Figure 1 for the EDA complexes of *bis*(durene)iron(II) acceptor with ferrocene and a series of organic donors [124]. Similarly, the charge-transfer absorption bands of EDA complexes of Ni(IV)dithiolene [112] with various (organic and organometallic) donors follow the Mulliken correlation in Eq. 8. The linearity of the plot of $E_{CT}$ against the ionization potentials of the donors has also been confirmed for a variety of other organometallic EDA complexes including ferrocene–alkyl halide [109], alkylmetal–iodine [111], alkylmetal–TCNE [123], metal hydride–TCNE [123], and (arene)metal tricarbonyl–nitroarene complexes [110].

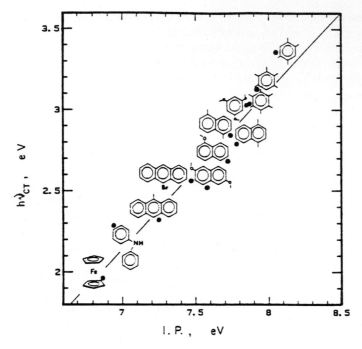

**Figure 1.** Mulliken correlation of the charge-transfer transition energy ($E_{CT} = h\nu_{CT}$) with the ionization potential (*IP*) of ferrocene and arene donors in *bis*(durene)iron(II) complexes. The straight line is arbitrarily drawn with a slope of unity [124].

Since ionization potentials of anionic donors and electron affinities of cationic acceptors are not readily available, Mulliken correlations for charge-transfer ion pairs are generally presented in a modified form using electrochemical oxidation or reduction potentials, respectively. A typical example of such a modified Mulliken plot with unit slope is shown in Figure 2 for the CT ion pairs of TpMo(CO)$_3^-$ [Tp = hydrido-*tris*-(3,5-dimethylpyrazolyl)borate] as the donor and various pyridinium acceptors [127]. Similar (modified) Mulliken correlations with unit slopes have been found for numerous other ion pairs with pyridinium acceptors and Mn(CO)$_5^-$ [126], Co(CO)$_4^-$ [118], or V(CO)$_6^-$ [118] as donors. It is important that the Coulombic work term ($\omega$) in Eq. 8 is explicitly included in all Mulliken evaluations of ion pairs with different structures since $\omega$ reflects the electrostatic energy of the (ground-state) ion pair which strongly depends on the inter-ionic distance [125].

**Solvatochromism of charge-transfer absorption bands**

Molecular EDA complexes as well as charge-transfer ion pairs show (negative) solvatochromism [128], i.e. the charge-transfer absorption maxima ($\lambda_{CT}$) undergo hypsochromic shifts with increasing solvent polarity. The solvatochromic effect is readily explained on the basis of the Marcus correlation for charge-transfer energies in solution [129], (Eq. 9):

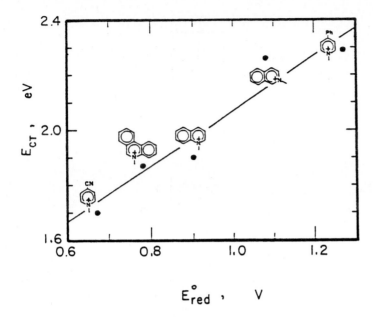

**Figure 2.** Variation of the charge-transfer transition energy ($E_{CT}$) with the reduction potential ($E°_{red}$) of pyridinium acceptors in CT ion pairs with TpMo(CO)$_3^-$ as the donor [127].

$$E_{CT} = \lambda_p + \lambda_i + \lambda_0 + \Delta G_0 \qquad (9)$$

with $\lambda_p$ being the energy of first-shell solvation of the ion-pair state, $\lambda_i$ the vibrational reorganization energy of the ion pair, $\lambda_o$ the solvent reorganization energy excluding first-shell solvent molecules, and $\Delta G°$ the free energy gap for the transition from the ground state to the ion-pair state in Eq. 7. Since $\lambda_p$ and $\lambda_o$ increase with increasing polarity of the solvent [129], $E_{CT}$ also increases with the solvent polarity which causes moderate, but sizable blue-shifts of the CT absorption bands of molecular EDA complexes [130].

In CT ion pairs the solvent-dependent color changes are much more pronounced as compared to molecular EDA complexes and hypsochromic shifts of more than 100 nm are frequently observed on going from a nonpolar solvent such as diethyl ether to the polar solvent acetonitrile. For example, the color of *N*-methylquinolinium tetracarbonylcobaltate [118] changes from green to purple to orange in diethyl ether, tetrahydrofuran, and acetonitrile, respectively. Such a strong (negative) solvatochromism is comparable with that of betaine dyes which are used as indicators to quantify solvent polarity [128]. In fact, the zwitterionic ground state of the betaine dye—like the contact ion pair—is more stabilized in polar solvents, and the biradical excited state of the dye—like the radical pair generated by CT excitation of the contact ion pair, (Eq. 10):

$$[D^-, A^+] \xrightarrow{h\nu_{CT}} [D^{\bullet}, A^{\bullet}] \qquad (10)$$

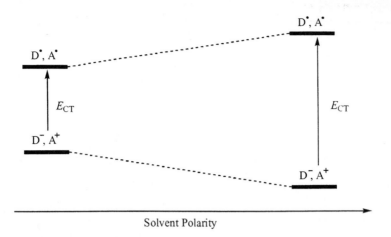

Solvent Polarity

**Scheme 2.**

is more destabilized in polar environment. Thus the energy gap between charged ground state and biradical excited state increases with increasing solvent polarity as illustrated in Scheme 2 [131].

Since contact ion pairs can be separated by the deliberate addition of an inert salt (such as tetra-*n*-butylammonium perchlorate or hexafluorophosphate) [132], (Eq. 11):

$$[D^-, A^+] + TBA^+PF_6^- \rightleftharpoons A^+PF_6^- + D^-TBA^+ \qquad (11)$$

the charge-transfer absorption bands of contact ion pairs are very sensitive to the presence of added salt. In other words, CT ion pairs are readily identified by the bleaching of their color due to added 'inert' salt [118].

### 2.3.3 Structural Changes due to Charge Transfer and/or Ion Pairing in Organometallic EDA Complexes

A number of the above described molecular complexes and CT ion-pair salts have been isolated in crystalline form for X-ray crystallographic studies. In general, the observation of significant changes in the structures of such charge-transfer crystals (as compared to the crystal structures of the separate components) will depend on the degree of charge transfer [91] between donor and acceptor in the ground state.

#### Weak molecular complexes

Weak (molecular) complexes in solution do not show any IR or NMR spectroscopic evidence for charge-transfer, i.e. the spectra of the complexes represent merely the superposition of the spectra of the separate (uncomplexed) donor and acceptor components [133]. Accordingly, X-ray crystal structures show the donors and acceptors with unperturbed bond lengths and angles [134, 135]. However, a

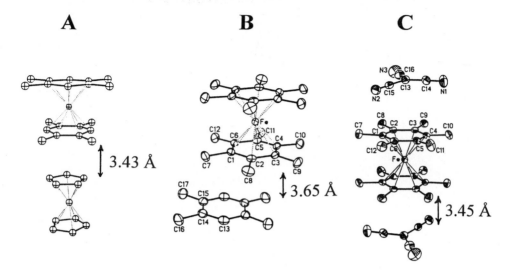

**Figure 3.** ORTEP diagrams of donor–acceptor pairs in charge-transfer crystals of (**A**) ferrocene and *bis*(durene)iron(II) [124], (**B**) durene and *bis*(hexamethylbenzene) iron(II) [124], and (**C**) tricyano-methide and *bis*(hexamethylbenzene)iron(II) [120].

characteristic crystal packing is generally observed which allows close (van der Waals) contact between the donor and acceptor molecules. For example, the distance between the Cp ring of ferrocene and the durene ring of *bis*-(durene)iron(II) in the 1:1 charge-transfer crystal (see Figure 3A) amounts to 3.43 Å [124], which is comparable to the interplanar distance of 3.65 Å between the hexamethylbenzene ring of *bis*-(hexamethylbenzene)iron(II) and durene in the CT crystal structure in Figure 3B [124]. Both distances lie within the range of donor–acceptor distances found in crystalline organic $(\pi, \pi)$ EDA complexes [134, 135]. Similarly, tricyano-methide donor and the hexamethylbenzene ring of *bis*-(hexamethylbenzene)iron(II) in the [(HMB)$_2$Fe$^{2+}$, 2C(CN)$_3^-$] CT crystal are separated by about 3.45 Å (see Figure 3C) [120]. The intermolecular distance of about 3.5 Å is very critical for charge-transfer interactions to occur between the $\pi$-orbitals of cofacially stacked donors and acceptors. A recent study shows that a mere increase by 1 Å in the closest interplanar distance between $\pi$-donors and acceptors (due to steric encumbrance) results in the charge-transfer interactions to be completely suppressed [136].

In contrast, the charge-transfer interactions in the charge-transfer salt of cationic ferrocenyl donor (CpFeCpCH$_2$NMe$_3^+$) with polyoxomolybdate (Mo$_6$O$_{19}^{2-}$) [137], as revealed by diffuse-reflectance UV–Vis spectroscopy, are not readily 'localized' in the crystal structure. Thus, the closest donor–acceptor contacts are located between three hydrogen atoms of the cyclopentadienyl ligands and (mostly bridged) oxygen atoms of the molybdate ion ($d = 2.66$–$2.76$ Å), whereas the closest distance between the iron redox center and the molybdate ion is about 4.2 Å (see Figure 4) [137]. Although the oxygen–hydrogen distances correspond to tight (van der Waals) contacts, the question remains open as to which orbitals are involved in the charge-transfer interactions.

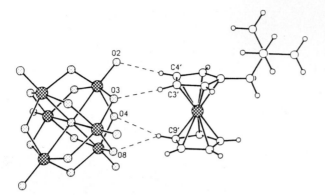

**Figure 4.** ORTEP diagram of the CT salt of ferrocenyl donor $(FeCpCH_2N^+Me_3)$ with poly-oxomolybdate $(Mo_6O_{19}^{2-})$ showing the closest donor–acceptor contacts [137].

## Charge-transfer ion pairs

In CT ion pairs, a wide variety of donor–acceptor interactions is observed—from weak interactions with little structural evidence for charge transfer, as observed in methylviologen dithiolene metallate salts [138, 139], to strong interactions in pyridinium carbonyl metallate salts which are readily recognized by the significant changes in the IR spectra and X-ray structures. In fact, carbonyl metallates themselves show varying degrees of charge-transfer depending on the acceptor strength of the cationic moiety in the ion-pair salt. For example, the infrared spectrum of the ion-pair salt of the weak tetraphenylphosphonium acceptor with tetracarbonylcobaltate does not show any distortion of the tetrahedral symmetry of the cobaltate anion [140]. However, contact ion pairs derived from strong acceptors such as cobaltocenium or pyridinium with tetracarbonylcobaltate exhibit a strong distortion of the tetrahedral structure of the cobaltate anion as revealed by IR spectroscopy in solution and X-ray crystallography in the crystalline state [118]. Generally, the smaller and stronger the acceptor is, the greater are the structural changes induced by the formation of intimate ion pairs. For example, the X-ray structures in Figure 5A and 5B show the close approach of the carbonyl ligands perpendicular to the aromatic plane of the pyridinium acceptor, which allows optimal orbital overlap. In fact, the cationic acceptor clearly penetrates into the carbonyl ligand shell and causes a significant distortion of the symmetry of the metallate complex. A similar penetration of the pyridinium acceptor into the ligand shell of the metallate is observed in the crystal structure of cyanopyridinium tetraphenylborate (see Figure 5C) [65]. Such *inner-sphere* charge-transfer ion pairs are clearly to be distinguished from the *outer-sphere* complexes [88–90] frequently observed between weakly coupling uncharged donors and acceptors (vide supra). Similarly, the CT ion-pair salt of *N*-methyl-4-cyanopyridinium with $TpMo(CO)_3^-$ shows significant distortion of the octahedral symmetry of the molybdate ion due to contact ion pairing (see Figure 5D) [127]. It is worth mentioning that the effects of contact ion pairing strongly depend on the solvent environment as illustrated with thallium(I) tetracarbonylco-

**A**     **B**

**C**     **D**

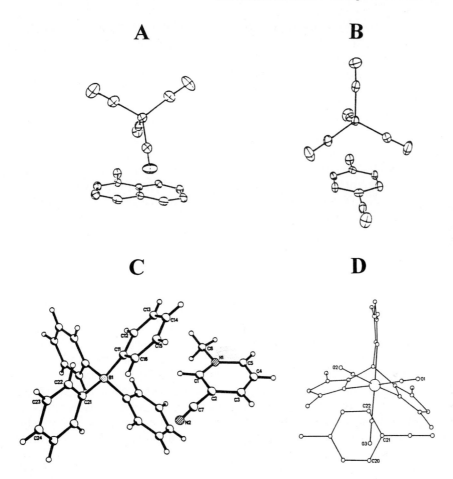

**Figure 5.** Intimate donor–acceptor orientation in the *inner-sphere* CT salts of (**A**) quinolinium with tetracarbonylcobaltate [118], (**B**) *N*-methyl-4-cyanopyridinium with tetracarbonylcobaltate [118], (**C**) *N*-methyl-3-cyanopyridinium with tetraphenylborate [65], and (**D**) *N*-methyl-4-cyanopyridinium with $TpMo(CO)_3^-$ [127].

baltate [141], for which a solvatotropic collapse of the CT ion pair (in dichloromethane) to a covalently-linked thallium–cobalt compound (in benzene) is observed.

**Strong charge-transfer interactions**

In coordinately saturated donor–acceptor complexes, the formation of new bonds as described with the thallium(I) tetracarbonylcobaltate (vide supra) is not readily accomplished. However, strong charge-transfer interactions between donor and acceptor moiety may lead to partial or complete electron transfer, which can be revealed by (UV–Vis or IR) spectroscopic and X-ray crystallographic techniques. For example, CT crystals of nitrosonium ($NO^+$) with aromatic donors show various degrees of charge transfer as detected by gradual shifts in the N–O stretch (IR) frequencies of the nitrosyl moiety approaching those of the NO radical, as well as by several X-ray crystallographic studies which clearly demonstrate the distortions of the aromatic ring of the donor that approach bond lengths and angles of the corresponding arene cation radical [142–145]. Tetracyanoethylene (TCNE) with a reduction potential of $E°_{red} = +0.24$ V relative to the SCE [146] represents a similarly strong and compact electron acceptor capable of fully oxidizing various electron donors to the cation-radical state. As a result, ion-pair salts of the type [D$^{+•}$, TCNE$^{-•}$] have been isolated in crystalline form, and the crystal structures clearly confirm complete electron transfer. For example, the X-ray structure of CT crystals prepared from Cp*$_2$Fe and tetracyanoethylene reveals one-dimensional stacks of Cp*$_2$Fe$^{+•}$ and TCNE$^{-•}$ units in cofacial orientation [147]. Interestingly, the corresponding CT crystals of TCNE with the weaker donor ferrocene (Cp$_2$Fe) crystallize as molecular complexes in indefinite stacks of uncharged ferrocene and TCNE [148, 149], owing to the endergonic ($\Delta G_{ET} = +0.17$ eV) energetics of the electron transfer from ferrocene ($E°_{ox} = 0.41$ V relative to the SCE) [48] to TCNE ($E°_{red} = 0.24$ V relative to the SCE) [146, 150]. The ion-pair character of the crystalline Cp*$_2$Fe–TCNE complex is quite similar to that in the CT crystals of Cp*$_2$Fe with 7,7,8,8-tetracyano-*p*-quinodimethane (TCNQ) as acceptor [151]. Both ion-pair salts show ferromagnetic properties (as revealed by Mössbauer spectroscopy), which is due to strong orbital overlap and spin coupling in the intimate ion-pairs [147, 151]. It is noteworthy that strong orbital interactions similar to those observed between oxidized organometallic donors and TCNE$^{-•}$ anion radical (generated in a *thermal* electron-transfer reaction) are also found in *photogenerated* ion–radical pairs observed upon CT excitation of EDA complexes of TCNE with olefinic donors [152]. The intimate ('inner-sphere') character of the latter ion–radical pairs is manifested in the ultrafast rates ($k \cong 5 \times 10^{11}$ s$^{-1}$) of back electron transfer (to restore the original molecular EDA complex) and their lack of driving-force dependence [152].

## 2.4 Electron-transfer Intermediates in Organometallic Reactions

In the previous section, we have shown that—owing to their unique donor or acceptor properties—organometallic compounds readily form electron donor–acceptor complexes in the presence of a suitable counterpart. Depending on the nature of the complex (viz., molecular complex or ion-pair salt) and the extent of charge transfer, the structures of these donor–acceptor assemblies reveal various degrees of predisposition to electron transfer between the two redox centers. Whether or not electron transfer occurs spontaneously in the EDA complex will depend on the free energy and the activation energy of the redox process. Favorable energetics will result in rapid thermal electron transfer which generally initiates a series of subsequent reactions. In this case, EDA complex formation and the generation of (usually short-lived) electron-transfer intermediates often evades its observation, and the electron-transfer mechanism in Scheme 1(A) can merely be inferred to on the basis of the donor–acceptor properties of the reactants and the

nature of the final products. Time-resolved spectroscopy [153, 154] applying ultrashort laser pulses makes it possible to generate and observe electron-transfer intermediates on an ultrafast (fs, ps) time scale in accord with Mulliken's charge-transfer
formulation [14–16] in Eqs 6 and 7. However, to observe EDA complex formation
and photoactivated electron transfers, the thermal electron transfer must be brought
to a standstill or at least, its rate needs to be substantially diminished, which can be
achieved by lowering of the temperature. The electron transfer in the thermally
stable EDA complex can then be *photoactivated* by deliberate irradiation of the
charge-transfer absorption band (as described in Scheme 1(B)). The steady-state
charge-transfer irradiation will lead to photoproducts which are unambiguously the
result of an initial photoinduced electron transfer from the donor to the acceptor
moiety in the EDA complex as monitored by time-resolved (laser) spectroscopy.

Depending on the starting materials, different electron-transfer intermediates
may be observed. Thus, CT excitation of a molecular complex leads to the generation of an ion–radical pair, (Eq. 12):

$$[D, A] \xrightarrow{\text{hv}_{CT}} [D^{+\bullet}, A^{-\bullet}] \tag{12}$$

whereas CT excitation of an ion-pair salt generates a geminate radical pair, (Eq. 13):

$$[D^-, A^+] \xrightarrow{\text{hv}_{CT}} [D^\bullet, A^\bullet] \tag{13}$$

There are also several mixed (charged–neutral) organometallic EDA complexes
known, the CT excitation of which results in the formation of ion-radical–radical
pairs, (Eq. 14):

$$[D, A^+] \xrightarrow{\text{hv}_{CT}} [D^{+\bullet}, A^\bullet] \tag{14}$$

Electron transfer induced by charge-transfer irradiation (Eqs 12–14) might or
might not lead to permanent chemical transformations that yield isolable photoproducts. The fate of the initial electron-transfer intermediate (viz., ion–radical pair
or radical pair) is mainly determined by the competition of two pathways, i.e. back
electron transfer or the follow-up reaction (see Scheme 1). If the former pathway
predominates and the follow-up reaction cannot efficiently compete, the ion-radical
or radical pair returns back to the original EDA complex and no net reaction is
observed. On the other hand, if the rate of the follow-up reaction is in the same
range as that of the back electron transfer, new intermediates and ultimately photoproducts will be formed that do not convert back to the starting materials, and
thus an electron-transfer activated reaction is obtained.

### 2.4.1 Charge-Transfer Excitation of Organometallic EDA Complexes Followed by Efficient Back Electron Transfer

There are many examples of organometallic EDA complexes that are thermally
stable and do not show permanent chemical transformations even after prolonged

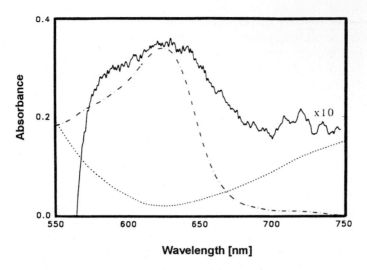

**Figure 6.** Transient (diffuse-reflectance) absorption spectrum (—) obtained at 25 ps following the 532-nm laser excitation of charge-transfer crystals of $[CpFeCpCH_2N^+Me_3]_2$ $Mo_6O_{19}$ dispersed in neutral alumina (10 %). The spectrum is deconvoluted as the (1:1) sum of reduced hexamolybdate acceptor ($Mo_6O_{19}{}^{3-}$, ----) and oxidized ferrocenyl donor $CpFe^+CpCH_2N^+Me_3$, $- \cdot - \cdot -$) [137].

charge-transfer irradiation. In these cases, CT excitation generates an ion-radical or radical pair which solely decays by back electron transfer to restore the original EDA complex, (Eq. 15):

$$[D,A]_{EDA} \underset{k_{-ET}}{\overset{h\nu_{CT}}{\rightleftharpoons}} [D^{+\cdot}, A^{-\cdot}] \tag{15}$$

Typical examples are the CT ion pairs of bipyridinium di-cations (viologens, $V^{2+}$) with metal (M = Zn, Cd, Hg) dithiolene (DT) anions [155], which form reduced viologen ($V^{+\cdot}$) and oxidized metal dithiolene complex upon CT excitation in dimethylsulfoxide, (Eq. 16):

$$[V^{2+}, M(DT)_2^{2-}]_{EDA} \underset{k_{-ET}}{\overset{h\nu_{CT}}{\rightleftharpoons}} [V^{+\cdot}, M(DT)_2^{-\cdot}] \tag{16}$$

The ion–radical pair subsequently decays as a result of back electron transfer ($k_{-ET}$, Eq. 16), which restores the original CT ion pair. In polar solvents such as dimethyl sulfoxide, dissociation of the ion–radical pair into free, solvated ion radicals can somewhat compete with the rapid back electron transfer ($k_{-ET} \approx 10^{10}$ s$^{-1}$). Thus, ion radicals are observed in small yields ($\Phi_{ion} \approx 0.05$) on the μs time scale upon 10-ns laser excitation [155]. They ultimately decay by (diffusional) back electron transfer.

Complete back electron transfer upon charge-transfer excitation has also been observed in crystalline EDA complexes. For example, laser excitation of the charge-transfer crystals of ferrocenyl donor and polyoxomolybdate acceptor results in short-lived (ps) transients the diffuse-reflectance absorption spectrum of which is shown in Figure 6 [137]. This transient spectrum can be deconvoluted as the sum of the absorption spectra of the oxidized ferrocenyl donor (ferrocenium) and the re-

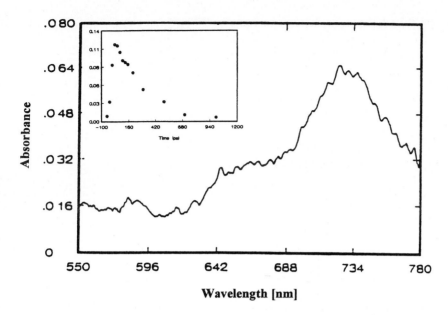

**Figure 7.** Transient absorption spectrum of 9-bromoanthracene cation radical obtained at 25 ps following the 532-nm CT excitation of the bromoanthracene–TiCl$_4$ complex in cyclohexane. The inset shows the complete decay of the cation radical within 1 ns due to back electron transfer [116].

duced hexamolybdate acceptor, which confirms the photoinduced electron transfer predicted by Mulliken theory, (Eq. 17):

$$\text{(17)}$$

Most organometallic EDA complexes of arenes with titanium tetrachloride [116] in solution also follow the general reaction scheme in Eq. 15 in that no net chemical reaction is observed upon charge-transfer irradiation due to rapid back electron transfer ($k_{-ET} \approx 10^{10}$ s$^{-1}$). For example, the transient absorption spectrum of bromoanthracene (BrAnt) cation radical generated by 532-nm laser excitation of the [BrAnt, TiCl$_4$] complex in cyclohexane (see Figure 7) decays completely to the spectral baseline within about 1 ns (see inset) due to back electron transfer [116], (Eq. 18):

$$[\text{BrAnt}, \text{TiCl}_4]_{\text{EDA}} \underset{k_{-ET}}{\overset{h\nu_{CT}}{\rightleftharpoons}} [\text{BrAnt}^{+\bullet}, \text{TiCl}_4^{-\bullet}] \tag{18}$$

However, if 9,10-anthracene dimer (dianthracene) or bicumene is used as the aromatic donor, the simple reaction scheme in Eq. 15 does not apply any longer, but a follow-up reaction is observed that efficiently competes with back electron transfer. Various examples for this case are described in the following section.

### 2.4.2 Charge-Transfer Activated Reactions of Organometallic EDA Complexes

**Arene fragmentation via TiCl₄ complexes**

A net chemical reaction upon charge-transfer irradiation of [arene, TiCl$_4$] complexes is obtained if rapid fragmentation of the arene cation radical competes with the back electron transfer in Eq. 18 [116]. Thus, dianthracene (Ant$_2$) cleaves into two anthracene moieties (Eqs 19–21) and bicumene cation radical fragments into cumyl cation and cumyl radical (Eqs. 22, 23), which ultimately form cumene and 1-phenyl-1,3,3-trimethylindane, respectively, as final products [116].

$$[\text{Ant}_2, \text{TiCl}_4] \xrightarrow{h\nu_{\text{CT}}} [\text{Ant}_2^{+\cdot}, \text{TiCl}_4^{-\cdot}] \tag{19}$$

$$[\text{Ant}_2^{+\cdot}, \text{TiCl}_4^{-\cdot}] \xrightarrow{\text{fast}} [\text{Ant}^{+\cdot}, \text{Ant}, \text{TiCl}_4^{-\cdot}] \tag{20}$$

$$[\text{Ant}^{+\cdot}, \text{Ant}, \text{TiCl}_4^{-\cdot}] \xrightarrow{k_{-\text{ET}}} [\text{Ant}, \text{TiCl}_4] + \text{Ant} \tag{21}$$

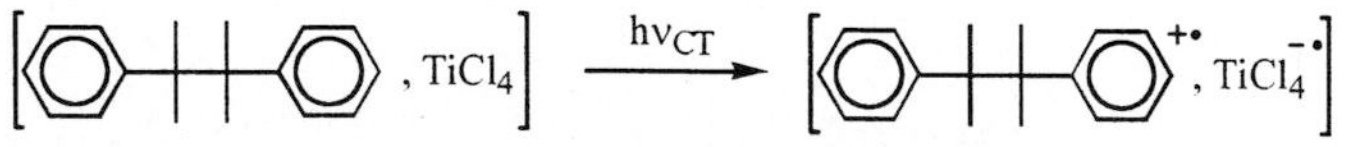

$$\tag{22}$$

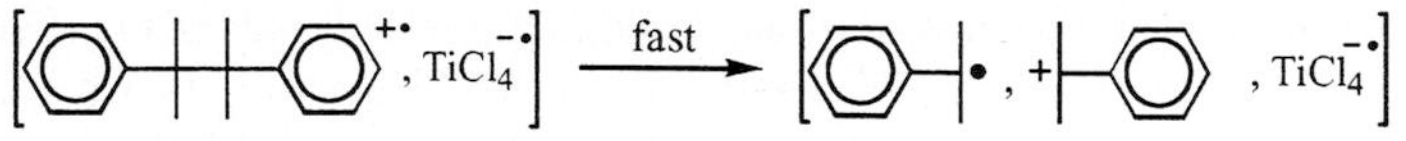

$$\tag{23}$$

**Mercuration and thallation of benzenes**

Methyl-substituted benzenes react with mercury(II) and thallium(III) electrophiles in thermal reactions to yield mercuration and thallation products, respectively [113], (Eqs 24, 25):

$$\text{ArH} + \text{Hg}^{\text{II}}(\text{O}_2\text{CCF}_3)_2 \rightarrow \text{ArH–Hg}(\text{O}_2\text{CCF}_3) + \text{CF}_3\text{CO}_2\text{H} \tag{24}$$

$$\text{ArH} + \text{Tl}^{\text{III}}(\text{O}_2\text{CCF}_3)_3 \rightarrow \text{ArH–Tl}(\text{O}_2\text{CCF}_3)_2 + \text{CF}_3\text{CO}_2\text{H} \tag{25}$$

During these reactions, transient charge-transfer absorption is observed; this is ascribed to the EDA complexes of the arene donors with the metal trifluoroacetate [113], (Eq. 26):

$$\text{ArH} + \text{Hg}^{\text{II}}(\text{O}_2\text{CCF}_3)_2 \rightleftharpoons [\text{ArH}, \text{Hg}^{\text{II}}(\text{O}_2\text{CCF}_3)_2]_{\text{EDA}} \tag{26}$$

Accordingly, a reaction mechanism has been proposed [113] which includes one-electron transfer from the arene donor to the high-valent metal within the EDA complex as the critical first step toward mercuration or thallation, (Eq. 27):

$$[\text{ArH}, \text{Hg}^{\text{II}}(\text{O}_2\text{CCF}_3)_2]_{\text{EDA}} \xrightleftharpoons{k_{\text{ET}}} [\text{ArH}^{+\cdot}, \text{Hg}^{\text{I}}(\text{O}_2\text{CCF}_3)_2^{-\cdot}] \tag{27}$$

In fact, the postulated arene cation-radical intermediate (ArH$^{+\cdot}$) has been observed recently [156] by picosecond time-resolved spectroscopy following the 355-nm laser excitation of the EDA complex of pentamethylbenzene with

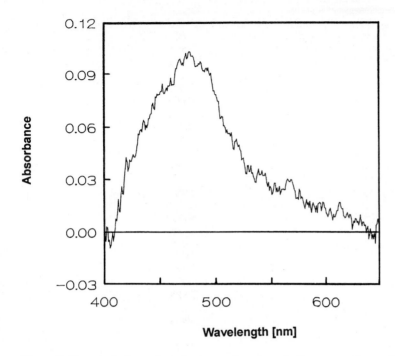

**Figure 8.** Transient absorption spectrum of pentamethylbenzene cation radical obtained at 25 ps following the 355-nm charge-transfer excitation of the pentamethylbenzene–Hg(CF$_3$CO$_2$)$_2$ complex in dichloromethane–CF$_3$COOH (2 %) mixture at −50 °C [156].

Hg$^{II}$(O$_2$CCF$_3$)$_2$ in dichloromethane (containing 2 % ($v/v$) trifluoroacetic acid) at low (−50 °C) temperature at which the thermal reaction is sufficiently suppressed (see Figure 8).

**Osmylation with osmium(VIII) tetroxide**

Osmium(VIII) tetroxide is an efficient reagent for *cis*-hydroxylations of alkenes [157, 158] via the osmium(VI) cycloadduct [159], which is stabilized by bases such as pyridine (pyr) [157, 160], (Eq. 28):

$$\text{OsO}_4 \;+\; \text{alkene} \;\xrightarrow{\;(Py)\;}\; \text{cycloadduct} \;\longrightarrow\; \text{diol} \tag{28}$$

In contrast, aromatic compounds are in general not oxidized by the high-valent osmium oxide despite their lower oxidation potentials as compared to those of olefinic substrates. Thus, the donor strength of the hydrocarbon is apparently not a critical factor for efficient oxidation or oxygen transfer. In fact, olefinic and aromatic donors (Ar) behave very similarly in the formation of organometallic EDA complexes [114], (Eq. 29):

$$[\text{olefin}, OsO_4]_{EDA} \xrightleftharpoons{+\text{olefin}} OsO_4 \xrightleftharpoons{+Ar} [Ar, OsO_4]_{EDA} \tag{29}$$

Molecular complexes of $OsO_4$ with various substituted benzenes, naphthalenes, and anthracenes have been identified by their charge-transfer absorption, which follows the Mulliken correlation in Eq. 8 [114, 161]. The arene–$OsO_4$ complexes are quite stable when kept in the dark and only very slowly form osmium(VI) cycloadducts by thermal osmylation (Eq. 30).

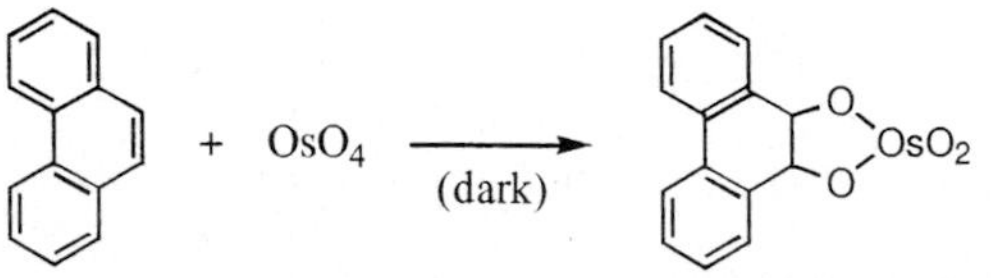

$$\tag{30}$$

On the other hand, charge-transfer irradiation of the same molecular complexes in *n*-hexane solution rapidly yields dark brown solids which—after dissolving in pyridine—have been identified by X-ray crystallography (see Figure 9), IR and NMR spectroscopy as the pyridine-ligated analogs of the osmium(VI) cycloadduct in Eq. 30 [114, 161], (Eq. 31):

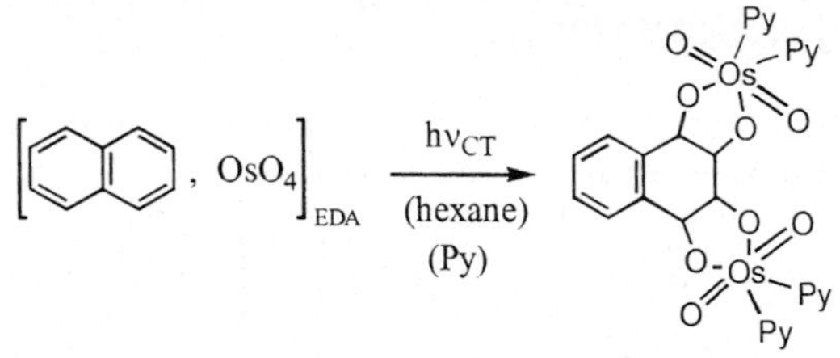

$$\tag{31}$$

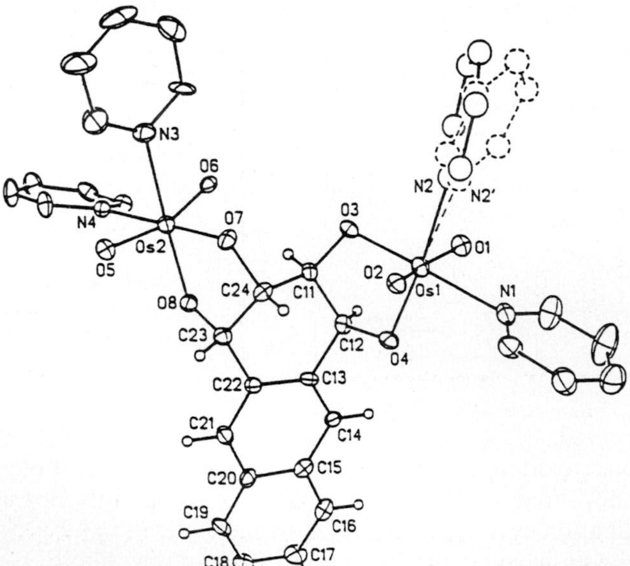

**Figure 9.** ORTEP diagram of the 2:1 adduct of pyridine-ligated $OsO_4$ with anthracene [161].

Interestingly, substituted anthracenes show different oxidation products depending on the solvent (Eq. 32):

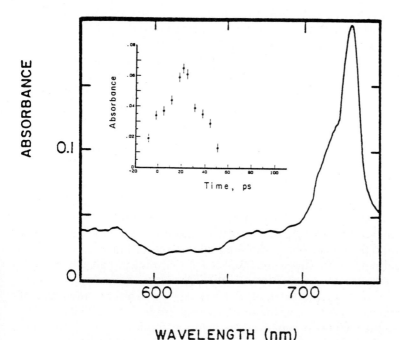

$$(32)$$

To identify the reactive intermediates in charge-transfer activated osmylations (Eqs 30–32), the EDA complexes of $OsO_4$ with various arene donors are photo-excited with the 25-ps laser pulse of a mode-locked Nd:YAG laser [161]. For example, the second harmonic laser output at 532 nm solely excites the charge-transfer absorption of the anthracene–$OsO_4$ complex and local excitation of (complexed or uncomplexed) anthracene or osmium tetroxide is avoided. The resulting transient spectrum at 20 ps after the laser pulse (see Figure 10) shows a strong, narrow ab-

**Figure 10.** Transient absorption spectrum of anthracene cation radical obtained at 35 ps following the 532-nm CT excitation of the anthracene–$OsO_4$ complex in dichloromethane. The inset shows the formation and decay of the cation radical within about 30 ps [161].

sorption band centered at 740 nm which has been assigned to the cation radical of anthracene by comparison with the spectrum of the electrochemically generated authentic species. Similar laser photolysis experiments with other arenes (Ar) result in the observation of the corresponding cation radicals ($Ar^{+\bullet}$) on the picosecond time scale. As such, charge-transfer excitation of the [Ar, $OsO_4$] complexes effects a spontaneous ($\tau < 20$ ps) electron transfer from the arene donor to the $OsO_4$ acceptor, which results in the formation of an ion–radical pair in accord with Mulliken theory [14–16] (Eq. 33):

$$[Ar, OsO_4]_{EDA} \xrightarrow{h\nu_{CT}} [Ar^{+\bullet}, OsO_4^{-\bullet}] \tag{33}$$

(Note that the spectral absorption of the reduced acceptor $OsO_4^{-\bullet}$ is obscured by the arene absorption.) This ion–radical pair is formed unambiguously as the first reactive intermediate upon charge-transfer excitation, and its collapse ultimately leads to the formation of the osmium(VI) cycloadduct observed upon steady-state CT irradiation (vide supra), as depicted for the simple benzene donor in Eq. 34:

$$\left[ \bigcirc, OsO_4 \right]_{EDA} \underset{k_{-ET}}{\overset{h\nu_{CT}}{\rightleftharpoons}} \left[ \bigcirc^{+\bullet}, OsO_4^{-\bullet} \right] \xrightarrow{k_C} \text{(cycloadduct)} \tag{34}$$

The coupling of the ion pair to form the cycloadduct is a very fast process as judged by the ultrashort lifetimes found for various arene ion radicals ($\tau \approx 20$ ps, see inset in Figure 10). However, the low quantum efficiencies ($\Phi_C < 0.01$) [114, 161] of the cycloadditions indicate that the predominant decay pathway of the ion radical pair in Eq. 34 is not the coupling step ($k_C$), but back electron transfer ($k_{-ET}$) to restore the original EDA complex. Quantitative evaluation of kinetics and quantum efficiencies yields the relatively fast cycloaddition rate constants of $k_C \approx 10^9$ s$^{-1}$ characteristic for highly exergonic bond formation.

The electron-transfer mechanism (Eq. 34) also explains the various regioselectivities observed for different arenes as the direct result of the symmetry of the arene HOMOs involved [161]. Moreover, the solvent effect on the oxidation products (Eq. 32) is now explicable on the basis of MO considerations. Thus, the ion–radical pair is very short-lived in hexane and collapses at the 9,10-positions where the anthracene HOMO is centered. The 9,10-cycloadduct is subsequently further oxidized to the anthraquinone product. In the more polar dichloromethane, the ion–radical pair is better stabilized and its longer lifetime allows relaxation of the original HOMO ion–radical pair to the subjacent (HOMO-1) ion–radical pair which leads to cycloaddition on the terminal ring (Eq. 32) [161].

Most importantly, both in the thermal and the charge-transfer osmylation of anthracene, identical cycloadducts on the terminal ring are observed, which underscores the close relationship between the two reaction modes. Thus, a unifying electron-transfer mechanism is proposed for both thermal and photoactivated osmylation, which reveals the ion–radical pair [$Ar^{+\bullet}$, $OsO_4^{-\bullet}$] as the common (primary) reactive intermediate [161].

**Deligation of bis(arene)iron(II) acceptors**

Organometallic acceptors of the type $Ar_2Fe^{2+}$ (Ar = mesitylene, durene, hexamethylbenzene, etc.) form EDA complexes with various organic and organometallic donors including molecular complexes with benzenes, naphthalenes, anthracenes, and ferrocene [124] and charge-transfer salts with tricyanomethide, tetraphenylborate, and tetracarbonylcobaltate (see Figure 3) [120]. As observed with $OsO_4$ complexes (vide supra), thermal and charge-transfer activation of the *bis*(arene)iron(II) complexes with donors (D) effect the same chemical transformations. Thus, a deligation process is observed in which the aromatic ligands of the acceptor are replaced by solvent (acetonitrile) molecules (Eq. 35).

$$[D, Ar_2Fe^{2+}]_{EDA} \xrightarrow[\text{(MeCN)}]{h\nu_{CT} \text{ or } \Delta} D + 2Ar + Fe(NCMe)_6^{2+} \tag{35}$$

Whether the deligation process occurs spontaneously in a thermal reaction or requires photoactivation depends on the donors employed. For example, most EDA complexes of *bis*(arene)iron(II) acceptors with aromatic donors such as anthracene or durene or with ferrocene are thermally stable and can be isolated in crystalline form (see Figure 3) [124]. Photoactivation of these complexes in acetonitrile by the selective irradiation of their charge-transfer absorption bands uniformly results in the deligation of the acceptor moiety (Eq. 36).

$$[Cp_2Fe, Ar_2Fe^{2+}]_{EDA} \xrightarrow[\text{(MeCN)}]{h\nu_{CT}} Cp_2Fe + 2Ar + Fe(NCMe)_6^{2+} \tag{36}$$

The reactive intermediates leading to the (charge-transfer) photodecomposition of the *bis*(arene)iron(II) acceptor are revealed by picosecond time-resolved spectroscopy. For example, photoexcitation of the CT absorption band of the ferrocene–$(HMB)_2Fe^{2+}$ complex (HMB = hexamethylbenzene) with the second harmonic output (at 532 nm) of a mode-locked Nd:YAG laser (25-ps pulse width) generates a transient spectrum with an absorption maximum at 580 nm (see Figure 11A). Careful deconvolution of this absorption spectrum reveals the superposition of the absorption bands of ferrocenium ($\lambda_{max} = 620$ nm, $\varepsilon = 360$ $M^{-1}$ $cm^{-1}$ [162]) and $(HMB)_2Fe^+$ ($\lambda_{max} = 580$ nm, $\varepsilon = 604$ $M^{-1}$ $cm^{-1}$ [163]).

The two intermediates are clearly the result of photoinduced electron transfer from the ferrocene donor to the *bis*(arene)iron(II) acceptor, which constitutes the first reaction step toward deligation (Eq. 37).

$$[Cp_2Fe, (HMB)_2Fe^{2+}]_{EDA} \xrightarrow{h\nu_{CT}} [Cp_2Fe^+, (HMB)_2Fe^+] \tag{37}$$

Similar electron-transfer intermediates are observed with other (aromatic) donors that promote deligation of $Ar_2Fe^{2+}$ acceptors. For example, charge-transfer laser excitation of the EDA complex of $(HMB)_2Fe^{2+}$ with 9-methylanthracene (MeANT) generates the 9-methylanthracene cation radical with its characteristic absorption centered at 700 nm (see Figure 11B; Eq. 38).

$$[MeANT, (HMB)_2Fe^{2+}]_{EDA} \xrightarrow{h\nu_{CT}} [MeANT^{+\bullet}, (HMB)_2Fe^+] \tag{38}$$

**A**

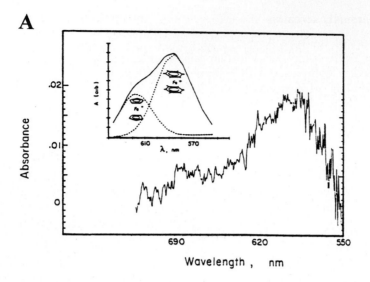

**B**

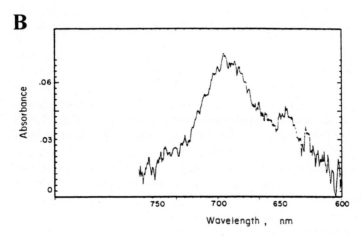

**Figure 11.** Transient absorption spectra obtained at 40 ps following the 532-nm CT excitation of the EDA complexes of *bis*(hexamethylbenzene) iron(II) with (**A**) ferrocene and (**B**) 9-methylanthracene. The spectrum in (**A**) is deconvoluted as the sum of reduced acceptor $(HMB)_2Fe^+$ and oxidized donor $Cp_2Fe^+$ (see inset) [124].

The monocationic intermediate $Ar_2Fe^+$ subsequently looses its two aromatic ligands by replacing them with solvent (acetonitrile) molecules ($k_1$), and then reacts ($k_2$) with another *bis*(arene)iron(II) acceptor in an electron-transfer chain (ETC) process which results in a catalytic deligation of $Ar_2Fe^{2+}$ [164, 165] (Eqs 39, 40).

$$Ar_2Fe^+ \xrightarrow[\text{(MeCN)}]{k_1} 2Ar + Fe(NCMe)_6^+ \qquad (39)$$

$$Fe(NCMe)_6^+ + Ar_2Fe^{2+} \xrightarrow{k_2} Fe(NCMe)_6^{2+} + Ar_2Fe^+ \qquad (40)$$

Whereas most EDA complexes with $Ar_2Fe^{2+}$ acceptors are stable in the dark, there are some examples where spontaneous thermal reactions are observed. For example, the color of the EDA complex of ferrocene with the strongest acceptor (mesitylene)$_2$iron(II) with $E°_{red} = -0.06$ V relative to the SCE [124] bleaches within 15 min upon mixing of the components, whereas the corresponding complexes with the weaker acceptors (durene)$_2Fe^{2+}$ ($E°_{red} = -0.16$ V relative to the SCE) and (HMB)$_2Fe^{2+}$ ($E°_{red} = -0.26$ V relative to the SCE) are thermally stable [124]. Moreover, replacing ferrocene ($E°_{ox} = 0.41$ V relative to the SCE [48]) by even stronger donors such as tetracarbonylcobaltate ($E°_{ox} = 0.32$ V relative to the SCE [118, 166]) or pentacarbonylmanganate ($E°_{ox} = 0.05$ V relative to the SCE [167]) also results in rapid thermal deligation of the (HMB)$_2Fe^{2+}$ acceptor. In fact, the complete kinetics of the catalytic electron-transfer activated deligation of (HMB)$_2Fe^{2+}$ (Eqs 39, 40) in the presence of the cobaltate and the manganate donor has been analyzed recently by digital simulation of the concentration–time profiles of the monocationic (HMB)$_2Fe^+$ intermediate for varying initial concentrations of the reactants [120].

### Ion-pair annihilation in carbonylmetallate salts

The anionic carbonylmetallate donors tetracarbonylcobaltate $[Co(CO)_4^-]$, hexacarbonylvanadate $[V(CO)_6^-]$, and pentacarbonylmanganate $[Mn(CO)_5^-]$ form a variety of (charge-transfer) ion-pair salts with cationic (organic or organometallic) acceptors the structures and spectroscopic properties of which have been described in Section 2.3. Most of these charge-transfer salts are thermally stable in dichloromethane solution, and even prolonged irradiation of their CT absorption bands does not result in any chemical transformations [118]. For example, the CT salt $Cp_2Co^+Co(CO)_4^-$ does not show any chemical change after 8 h irradiation at wavelengths $\lambda > 520$ nm where the single cationic and anionic components do not absorb. However, in the presence of triphenylphosphine, CT irradiation effects spontaneous evolution of carbon monoxide accompanied by the disappearance of the cobaltate anion as monitored by IR spectroscopy [118]. Thus, the characteristic carbonyl absorption at $\nu_{CO} = 1887$ cm$^{-1}$ of the cobaltate decays, and concomitantly the growth of a new IR absorption band is observed at $\nu_{CO} = 1958$ cm$^{-1}$ which is assigned to the dimeric $Co_2(CO)_6(PPh_3)_2$ [118]. The complete stoichiometry of the reaction is described in Eq. 41.

$$Cp_2Co^+Co(CO)_4^- + PPh_3 \xrightarrow[\text{(CH}_2\text{Cl}_2)]{h\nu_{CT}} Cp_2Co + 1/2\,Co_2(CO)_6(PPh_3)_2 + CO$$

$$(41)$$

Similar photoinduced dimerizations and ligand substitutions in the presence of additives such as triphenylphosphine are observed with ion-pairs salts of $Mn(CO)_5^-$ and $V(CO)_6^-$ with cobaltocenium or other cationic acceptors such as $Ph_2Cr^+$, pyridinium, quinolinium, etc [118]. Most importantly, all photochemical transformations of the various carbonyl metallate salts are initiated by actinic light that solely excites the charge-transfer absorption bands of the contact ion pairs whereas local excitation of the separate ions is deliberately excluded.

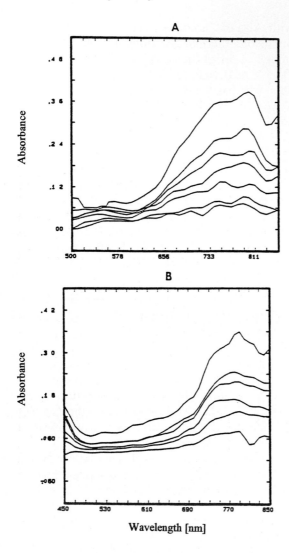

**Figure 12.** Transient absorption spectra of (**A**) $Mn(CO)_5^{\cdot}$ and (**B**) $Co(CO)_4^{\cdot}$ radicals recorded on the microsecond time scale upon 532-nm CT excitation of the cobaltocenium salts of $Mn(CO)_5^-$ and $Co(CO)_4^-$, respectively [118].

The reactive intermediates in the charge-transfer photoreactions of carbonyl metallate salts are examined by time-resolved spectroscopy applying a 10-ns pulse of a Q-switched Nd:YAG laser at 532 nm [118]. Thus, the charge-transfer excitation of various manganate and cobaltate salts results in the formation of intense transient absorption centered at 800 nm and 780 nm, respectively, which are readily assigned to the 17-electron radicals $Mn(CO)_5^{\cdot}$ and $Co(CO)_4^{\cdot}$, respectively (see Figure 12) [168–170].

Moreover, when pyridinium or quinolinium is used as the cationic partner of the carbonyl metallates, additional absorption bands (at shorter wavelengths) are observed which are readily assigned to the pyridinyl [171] and quinolinyl [172] radicals, respectively (see Figure 13).

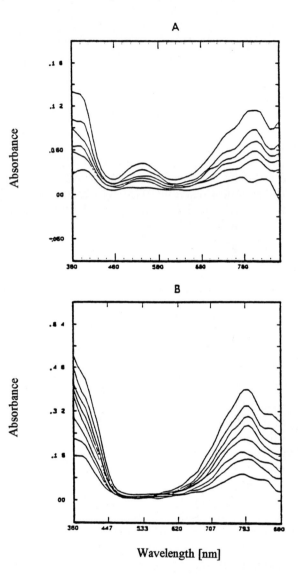

**Figure 13.** Transient absorption spectra recorded on the microsecond time scale upon 532-nm CT excitation of (**A**) quinolinium tetracarbonylcobaltate and (**B**) $N$-methyl-4-cyanopyridinium tetracarbonylcobaltate in acetone showing $Co(CO)_4{}^{\bullet}$ radical (at 790 nm) and quinolinyl (at 540 nm) or cyanopyridinyl (at 370 nm) radicals, respectively [118].

All laser experiments point uniformly to an initial photoinduced electron transfer from the carbonylmetallate donor $[M(CO)_n{}^-]$ to the cationic acceptor $A^+$ which results in the formation of a radical pair (Eq. 42).

$$A^+ M(CO)_n{}^- \xrightarrow{\text{h}\nu_{\text{CT}}} A^{\bullet}, M(CO)_n^{\bullet} \tag{42}$$

In the absence of triphenylphosphine, the photogenerated radicals undergo back electron transfer $(k_{-\text{ET}})$ either within the solvent cage of the geminate radical pair or upon diffusional encounter $(k_{\text{ass}})$ of the cage-escaped free radicals (Eq. 43).

$$A^+ M(CO)_n^- \xrightleftharpoons[k_{-ET}]{h\nu_{CT}} A^{\bullet}, M(CO)_n^{\bullet} \xrightleftharpoons[k_{ass}]{k_{diss}} A^{\bullet} + M(CO)_n^{\bullet} \qquad (43)$$

The latter process results in an overall second-order annihilation of the radicals as observed in the complete decays of the transient absorption to the spectral baseline on the microsecond time scale (see Figures 12 and 13). Since dimerization of the 17-electron radicals is orders of magnitude slower than the highly exothermic back electron transfer, no net photochemical transformations are observed even after prolonged charge-transfer irradiation.

However, even small amounts of additives (L) such as triphenylphosphine induce permanent photochemical changes due to redox processes, ligand exchange, and dimerization reactions [118]. The critical reaction step common to all these photo-reactions is the incorporation of the additive L as a ligand in the metal complex. This process must occur in a reactive stage of the metal complex since carbonyl metallates (in the ground state) do not undergo thermal ligand substitutions [173–175]. The time-resolved spectroscopic studies reveal the 17-electron radical $M(CO)_n^{\bullet}$ as the reactive intermediate that rapidly undergoes ligand substitution ($k_{sub}$) and other subsequent reactions such as homolytic dimerization ($k_{dim}$) (Eqs 44, 45).

$$Mn(CO)_5^{\bullet} + PPh_3 \xrightarrow{k_{sub}} Mn(CO)_4(PPh_3)^{\bullet} + CO \qquad (44)$$

$$2\ Mn(CO)_4(PPh_3)^{\bullet} \xrightarrow{k_{dim}} Mn_2(CO)_8(PPh_3)_2 \qquad (45)$$

Thus, a careful analysis of the rate constants for back electron transfer, ligand substitution, and dimerization leads to the conclusion that ligand exchange in the 17-electron radical ($k_{sub}$ in Eq. 44) lowers the rate of back electron transfer from the acceptor radical ($A^{\bullet}$) ($k_{-ET}$ in Eq. 43) to such an extent that dimerizations (and other possible follow-up reactions [118]) now become competitive and effect permanent photochemical transformations. The decrease of the back electron transfer rates is due to the attenuated reduction potentials of the phosphine-substituted radicals [176].

The isostructural pentacarbonyl metallates $Mn(CO)_5^-$ and $Re(CO)_5^-$ form a series of thermally or photochemically unstable charge-transfer salts with *N*-methylpyridinium cations. For example, *N*-methylacridinium reacts with penta-carbonylrhenate immediately upon mixing in acetonitrile to form the (*N*-methyl-9-acridanyl)pentacarbonylrhenium(I) adduct in 90 % yield [126] (Eq. 46).

$$(46)$$

X-ray crystallographic examination of the adduct reveals a folded acridanyl structure and a distorted square-pyramidal orientation of the carbonyl ligands around the rhenium(I) center (see Figure 14).

This adduct is very stable at room temperature and only decomposes upon heating at 80 °C in benzene to form dirheniumdecacarbonyl and biacridanyl [126] (Eq. 47).

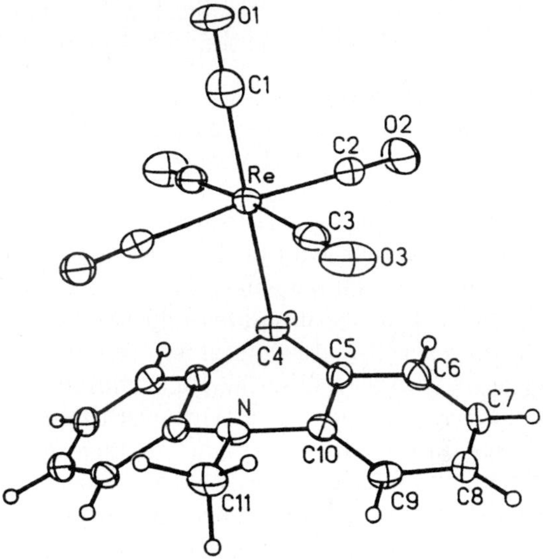

**Figure 14.** ORTEP diagram of (*N*-methyl-9-acridanyl)pentacarbonylrhenium(I) adduct [126].

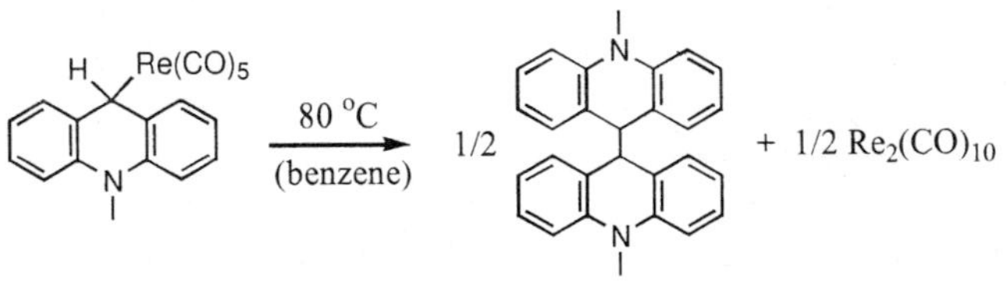

$$\text{(47)}$$

In contrast, the 9-phenylacridinium acceptor undergoes a thermal electron-transfer reaction upon mixing with the pentacarbonylrhenate(I) donor in acetonitrile to form dirheniumdecacarbonyl and the stable 9-phenylacridanyl radical (Eq. 48).

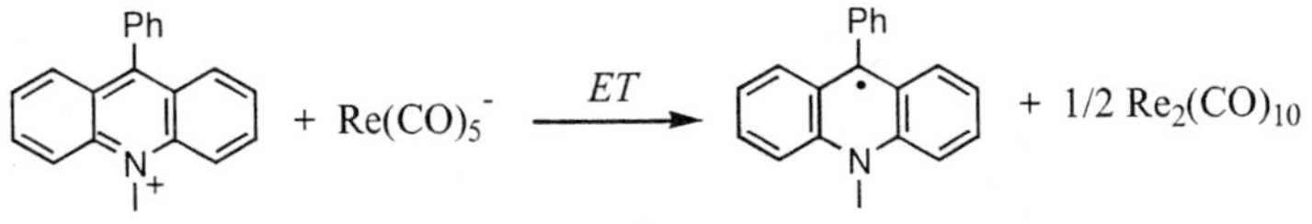

$$\text{(48)}$$

Other pyridinium acceptors yield either products of nucleophilic coupling (Eq. 46) or of electron transfer (Eq. 48). The partitioning between the two pathways is different for the two metal centers, and also strongly depends on the stereochemistry and electronic structure of the pyridinium substrate. Rhenate donors prefer adduct formation (Eq. 46), and pyridinium acceptors that form stable, delocalized pyranyl radicals favor the electron-transfer pathway (Eq. 48) [126]. The rate constants for both reactions correlate with the electron-transfer driving force in the same way, and electron transfer and nucleophilic addition cannot be distinguished on the basis of their linear free-energy relationships [177], which indicates that the transition states of both pathways are closely related. Accordingly, the question arises as to whether (i) the electron-transfer products are formed as secondary products via a metastable adduct that undergoes homolytic cleavage ($k_{CL}$ in Eq. 49) or (ii) adduct formation is the result of radical–radical coupling ($k_{RC}$) as the secondary step following an initial electron transfer ($k_{ET}$ in Eq. 50), i.e.

$$[Py^+, M(CO)_5^-] \xrightarrow{k_C} Py\text{---}M(CO)_5 \xrightarrow{k_{CL}} [Py^\bullet, M(CO)_5^\bullet] \qquad (49)$$

$$\text{or:} \quad [Py^+, M(CO)_5^-] \xrightarrow{ET} [Py^\bullet, M(CO)_5^\bullet] \xrightarrow{k_{RC}} Py\text{---}M(CO)_5 \qquad (50)$$

First, the inner-sphere formulation [26–30] for the net electron transfer in Eq. 49 seems unlikely in light of the relative rates of electron transfer and adduct decomposition. For example, biacridanyl and $Re_2(CO)_{10}$ are formed immediately as a 5 % by-product to the formation of the acridanylpentacarbonylrhenium(I) adduct in Eq. 46; however, the adduct itself is very stable at room temperature (vide supra). Thus, electron transfer and nucleophilic addition do not seem to be sequential, but rather concurrent reaction steps. However, the linear free-energy correlations [177] with slopes substantially less than the value of unity predicted for purely outer-sphere electron transfer do indicate strong inner-sphere character for the electron-transfer step [126]. On the other hand, the inverted reaction sequence in Eq. 50 is experimentally confirmed by laser flash photolysis experiments with the isoquinolinium–pentacarbonylmanganate ion pair. Charge-transfer excitation of the ion pair with a 10-ns laser pulse at 532 nm generates initially the $Mn(CO)_5^\bullet$ radical with its characteristic absorption at 780 nm (see Figure 15) [168–170]. The absorption band of

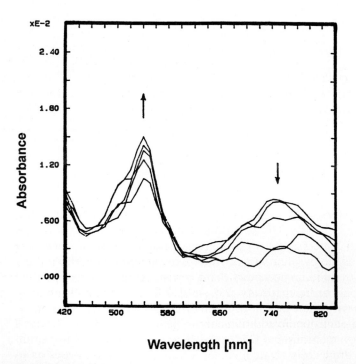

**Figure 15.** Transient absorption spectra recorded on the microsecond time scale upon 532-nm CT excitation of the isoquinolinium pentacarbonylmanganate ion pair showing the decay of $Mn(CO)_5^\bullet$ radical (at 750 nm) and the formation of the (1:1) $\sigma$-adduct (at 540 nm) [126].

the 17-electron radical decays on the microsecond time scale and a new absorption band appears (centered at 540 nm) which is readily assigned to the radical–radical coupling product based on the comparison with the spectrum of the acridanyl-analog adduct. However, the quinolyl adduct is very unstable and ultimately cleaves again to form the electron-transfer products [126].

In general, nucleophilic addition and electron transfer require the same structural changes in the pentacarbonyl metallate anion from its original trigonal-bipyramidal geometry to the square-pyramidal shape of both the 17-electron radical and the adduct (see Figure 14). A partial disposition toward this structural change is already apparent in the precursor charge-transfer ion pair which shows significant inner-sphere character as revealed by the distortion of the carbonyl metallate symmetry [178] similar to those described for cobaltate ion pairs (see Section 2.3.3). The distorted carbonyl metallate is thus activated towards both electron transfer and nucleophilic addition since both pathways exhibit structurally and energetically similar transition states. As a consequence, the final outcome of the reaction mostly depends on the energetics of the bond formation leading to the adduct.

A similar case of concurrence of one-electron transfer and nucleophilic addition is observed in the thermal ion-pair annihilation of $CpMo(CO)_3^-$ anion with (dienyl)$Fe(CO)_3^+$ cations [84, 179]. Thus, spontaneous electron transfer ($k_{ET}$) occurs upon mixing of ($\eta^5$-cyclohexadienyl)$Fe(CO)_3^+$ with $CpMo(CO)_3^-$ in acetonitrile to afford the transient 19-electron iron radical and the 17-electron molybdenum radical which both rapidly dimerize (Eq. 51).

$$\text{[diagram] } Fe(CO)_3^+ \ + \ CpMo(CO)_3^- \ \xrightarrow{k_{ET}} \ \left[ \text{[diagram] } Fe(CO)_3^{\bullet} \ , \ CpMo(CO)_3^{\bullet} \right] \qquad (51a)$$

$$\left[ \text{[diagram] } Fe(CO)_3^{\bullet} \ , \ CpMo(CO)_3^{\bullet} \right] \ \xrightarrow{fast} \ 1/2 \ (OC)_3Fe\text{[diagram]}Fe(CO)_3 \qquad (51b)$$
$$+ \ 1/2 \ [CpMo(CO)_3]_2$$

In contrast, the open-chain analog ($\eta^5$-hexadienyl)$Fe(CO)_3^+$ mainly (75 %) forms the nucleophilic adduct with the molybdate anion (Eq. 52).

$$\text{[diagram] } Fe(CO)_3^+ \ + \ CpMo(CO)_3^- \ \xrightarrow{k_N} \ \text{[diagram] } (OC)_3Fe \text{[diagram]} MoCp(CO)_3 \qquad (52)$$

Interestingly, the new bond in the adduct is formed at the terminal carbon of the dienyl ligand as established by X-ray crystallography (see Figure 16).

At room temperature, this adduct readily decomposes by homolytic cleavage to the 19-electron–17-electron radical pair which subsequently forms homodimers of the iron and the molybdenum complexes similar to those in Eq. 51. An overall electron transfer via the nucleophilic adduct as intermediate is observed as the favored pathway of iron complexes with acyclic dienyl ligands. However, the formation of the electron-transfer products, viz., homodimers of iron and molybdenum complexes, are also obtained directly (in 25 % yield) during the reaction. The nucleophilic adduct is not a prerequisite for the electron transfer from the molybdate

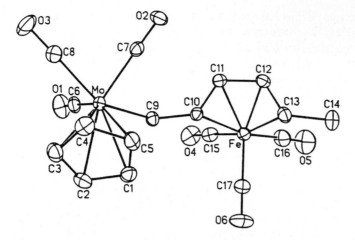

**Figure 16.** ORTEP diagram of the $\sigma$-adduct CpMo(CO)$_3$Fe(CO)$_3$($\eta^4$-hexadiene) [84].

donor to the cationic iron acceptor. In other words, direct *outer-sphere* electron transfer (Eq. 51) and *inner-sphere* electron transfer via the nucleophilic adduct (Eq. 52) are concurrent rather than sequential processes, and the partitioning between the two pathways is controlled by stereochemistry and the reorganization energies involved in the transformation from the charge-transfer ion pair to the radical pair [179].

In contrast to the inner-sphere charge-transfer salts of pyridinium acceptors with carbonyl metallates (vide supra), contact ion pairs of tetraphenylphosphonium or triphenylsulfonium acceptors with Co(CO)$_4^-$, Mn(CO)$_5^-$, and HFe(CO)$_4^-$ donors do not exhibit any distortions in the ($T_d$, $D_{3h}$, or $C_{3v}$) symmetry of the carbonyl metallate as confirmed by IR spectroscopy in solution and in the solid state [140]. Moreover, the strongly blue-shifted (as compared to the pyridinium analogs) CT absorption bands of these typical *outer-sphere* charge-transfer salts point to very weak donor–acceptor interactions owing to the rather low reduction potentials ($E_{red}^P \approx -2$ V relative to the SCE [140]) of the onium acceptors. Nonetheless, a rich photochemistry is observed upon charge-transfer activation of these salts, which is the direct result of the spontaneous fragmentation ($k_F$) of the labile sulfuranyl and phosphoranyl radicals [180] upon photoinduced electron transfer from the carbonylmetallate (Eq. 53).

$$[Ph_3S^+, Mn(CO)_5^-] \underset{k_{-ET}}{\overset{h\nu_{CT}}{\rightleftharpoons}} [Ph_3S^{\bullet}, Mn(CO)_5^{\bullet-}] \overset{k_F}{\longrightarrow} [Ph_2S, Ph^{\bullet}, Mn(CO)_5^{\bullet-}]$$

$$(53)$$

As a result, the energy-wasting back electron transfer (to regenerate the original charge-transfer salt) is partially suppressed, and the reactive phenyl radical couples with the 17-electron carbonylmetal radical within the solvent cage to form the phenyl-substituted metal complex (Eq. 54).

$$[Ph_2S, Ph^{\bullet}, Mn(CO)_5^{\bullet-}] \rightarrow Ph_2S + Ph\!-\!Mn(CO)_5 \qquad\qquad (54)$$

Similar homolytic combinations also apply to the radical pair photogenerated from the phosphonium salt. However, triphenylphosphine as a highly effective $\sigma$-donor replaces a CO ligand in the metal complex prior to phenyl coupling (Eqs 55, 56).

$$[Ph_4P^+, Mn(CO)_5{}^-] \underset{k_{-ET}}{\overset{h\nu_{CT}}{\rightleftharpoons}} [Ph_4P^{\bullet}, Mn(CO)_5^{\bullet}] \xrightarrow{k_F} [Ph_3P, Ph^{\bullet}, Mn(CO)_5^{\bullet}]$$

(55)

$$[Ph_3P, Ph^{\bullet}, Mn(CO)_5^{\bullet}] \rightarrow [Ph^{\bullet}, Mn(CO)_4(PPh_3)^{\bullet}, CO]$$

$$\rightarrow PhMn(CO)_4(PPh_3) + CO$$

(56)

The photochemical activation of the phosphonium salt (Eq. 55) has its *thermal* counterpart in the facile (dark) conversion of the iodonium salt $Ph_2I^+ Mn(CO)_5{}^-$. Owing to the greatly enhanced reduction potential of diphenyliodonium ($E_{red}{}^p \approx 0$ V relative to the SCE [181]) as compared to tetraphenylphosphonium ($E_{red}{}^p \approx -2.3$ V relative to the SCE [140]), electron transfer in the iodonium–metallate ion pair is now energetically feasible ($\Delta G_{ET} \approx 0$ eV). As a result, complete conversion of the charge-transfer salt to the electron-transfer products $Mn_2(CO)_{10}$ and $PhMn(CO)_5$ is observed within minutes upon mixing of the two components [140].

The formation of 17-electron radicals as the result of *thermal* electron transfer from a carbonyl metallate to a cationic pyrylium acceptor ($P^+$) is also observed with the charge-transfer salts of $[TpM(CO)_3{}^-]$ (Tp = hydrido-*tris*-(3,5-dimethylpyrazolyl)-borate; M = Mo, W, Cr) [127]. Owing to the stability of the $TpM(CO)_3{}^{\bullet}$ radical [182], electron-transfer equilibria ($K_{ET}$) are established by UV–Vis and IR spectroscopy which generally favor the intimate ion-pair state (see Structure **D** in Figure 5) in accord with the electron-transfer energetics [127] (Eq. 57).

$$[P^+, TpM(CO)_3^-]_{CT} \underset{\xleftarrow{\hspace{1cm}}}{\overset{k_{ET}}{\longrightarrow}} P^{\bullet} + TpMn(CO)_3^{\bullet}$$

(57)

However, with triarylpyrylium acceptors the stable 17-electron radicals are generated in high yields due to a shift of the endergonic equilibrium by the homolytic dimerization of the pyranyl radicals ($2 P^{\bullet} \rightarrow P_2$) [183, 184].

### Hydrometallation and hydrogenation of olefins with metal hydrides

Metal catalysis of olefin and arene hydrogenation is critically dependent on the reactive hydridometal intermediates. However, little is known about the mechanism of hydrogen transfer and hydrometallation and the reactive intermediates involved. The observation of transient charge-transfer absorption during hydrogenation and hydrometallation of olefins with tungsten or molybdenum hydrides [185], opens up the question of a possible electron-transfer mechanism in which the overall hydrogen-atom transfer is the result of a two-step electron-transfer–proton-transfer (ET–PT) process (Eq. 58).

$$DH + A \xrightarrow{ET} [DH^{+\bullet}, A^{-\bullet}] \xrightarrow{PT} D^{\bullet}, AH^{\bullet} \longrightarrow \text{etc.}$$

(58)

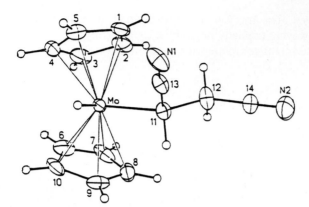

**Figure 17.** ORTEP diagram of the σ-adduct formed upon charge-transfer irradiation of the Cp$_2$MoH$_2$– fumaronitrile complex in toluene [58].

In this context, the transition metal hydrides Cp$_2$MoH$_2$, Cp$_2$WH$_2$, and Cp$_2$ReH$_2$ are of particular interest since they are excellent electron donors (see Table 4, Section 2.2.1) [57–59], and they form EDA complexes with various activated olefins such as fumaronitrile or maleic anhydride [58]. Charge-transfer activation of these EDA complexes in toluene by the deliberate irradiation of their charge-transfer absorption bands leads to either hydrometallation or hydrogenation [58]. For example, irradiation of the Cp$_2$MoH$_2$–fumaronitrile complex with visible light ($\lambda > 550$ nm) where neither the metal complex nor the olefin absorbs, yields the σ-adduct (see Figure 17) in excellent (95 %) yield (Eq. 59).

$$\text{(59)}$$

Under the same conditions, the analogous tungsten hydride reacts with fumaronitrile to form succinonitrile (95 %) and a mixture of fumaronitrile and maleonitrile complexed with tungstenocene (Eq. 60).

$$\text{(60)}$$

The two reaction pathways are not unique to the corresponding molybdenum or tungsten centers. For example, both metal hydrides effect hydrogenation of maleic anhydride [58]. Thus, the hydrogenation path may involve prior hydrometallation followed by reductive elimination of the hydridoalkyltungsten adduct (Eqs 61, 62).

$$\text{(61)}$$

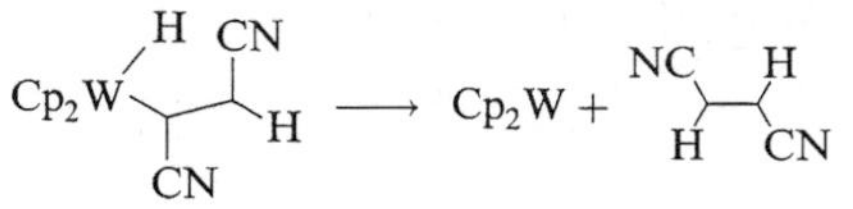

$$Cp_2W \underset{CN}{\overset{H\ CN}{\diagdown}}H \longrightarrow Cp_2W + \underset{H\ CN}{\overset{NC\ H}{\diagdown}} \tag{62}$$

Alternatively, hydrogenation is simply explained by the sequential electron-transfer–proton-transfer reaction followed by disproportionation of the resulting radical pair (Eqs 63, 64).

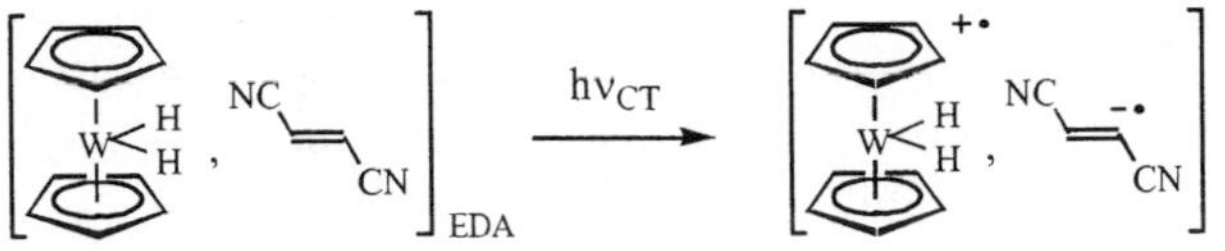

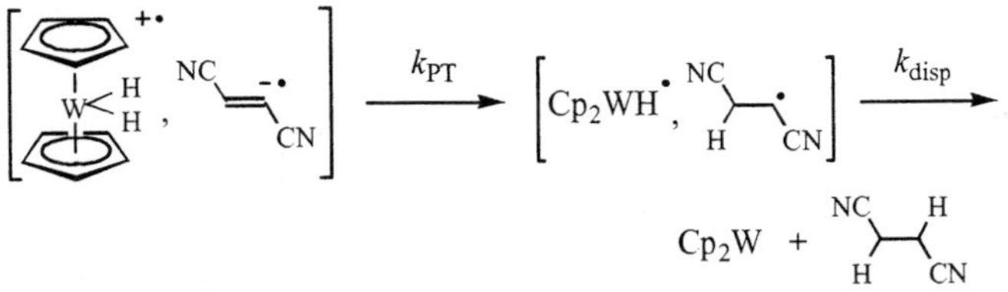

$$\left[ \begin{array}{c} W \overset{H}{\underset{H}{\diagup}} \ , \ NC \diagdown_{CN} \end{array} \right]_{EDA} \xrightarrow{h\nu_{CT}} \left[ \begin{array}{c} W \overset{H}{\underset{H}{\diagup}} \ , \ NC \diagdown_{CN} \end{array} \right] \tag{63}$$

$$\left[ \begin{array}{c} W \overset{H}{\underset{H}{\diagup}} \ , \ NC \diagdown_{CN} \end{array} \right] \xrightarrow{k_{PT}} \left[ Cp_2WH \ , \ \overset{NC}{\underset{H}{\diagup}} \overset{\cdot}{\underset{CN}{\diagdown}} \right] \xrightarrow{k_{disp}} \tag{64}$$

$$Cp_2W + \underset{H\ CN}{\overset{NC\ H}{\diagdown}}$$

Deprotonation of $Cp_2WH_2^{+\cdot}$ and $Cp_2MoH_2^{+\cdot}$ clearly occurs upon electrochemical generation of the cation radicals, and subsequent dimerization of the resulting hydridometallocene radicals has been observed [59]. Finally, the *charge-transfer* hydrogenation and hydrometallation of olefins closely parallels the *thermal* pathways in that the same type of $\sigma$-adducts are found [185].

**Alkylation with organoborates**

Tetraalkylborates are mild and selective alkylation reagents [186, 187], and they are commonly considered as sources of nucleophilic alkyl groups $(R^-)$ just as borohydrides are depicted as hydride $(H^-)$ sources. However, since organoborates represent excellent electron donors (see Table 5, Section 2.2.1), the question arises as to what role electron donor–acceptor interactions play in the nucleophilic alkyl transfer. Phenyl- and alkyl-substituted borate ions form highly colored charge-transfer salts with a variety of cationic pyridinium acceptors [65], which represent ideal substrates to probe the methyl-transfer mechanisms. Most pyridinium borate salts are quite stable in crystalline form (see for example Figure 5C), but decompose rapidly when dissolved in tetrahydrofuran to yield methylated hydropyridines (Eq. 65).

$$\underset{N^+}{\bigcirc\!\!\!\bigcirc} \ \ BMe_4^- \longrightarrow \underset{N}{\bigcirc\!\!\!\bigcirc} \ + \ BMe_3 \tag{65}$$

In some cases (e.g., 3-cyano-*N*-methylpyridinium tetramethylborate), this formal methide transfer is so fast that the ion-pair precursor cannot be isolated as a crystalline salt. Other charge-transfer salts such as 4-phenyl-*N*-methylpyridinium tetramethylborate do not react at all at room temperature. In these cases, methylation can still be achieved by deliberate irradiation of the charge-transfer absorption band of the ion-pair salt in solution (Eq. 66).

$$[Py^+, BMe_4^-] \xrightarrow{h\nu_{CT}} Py\text{—}Me + BMe_3 \tag{66}$$

Charge-transfer activated alkyl transfer can also be carried out with the thermally unstable salts at low ($-78\,°C$) temperature at which the thermal decomposition is suppressed. In these cases, the similarity in the regioselectivity of photoinduced and thermal alkyl transfer points to the critical role of the donor–acceptor orientation in the ion-pair precursor which ultimately determines the alkylation site. In mixed alkylborates containing methyl, *n*-butyl, and *sec*-butyl groups, the alkyl transfer rates are found to increase substantially from methyl to *sec*-butyl transfer, the latter being 67 times faster than the former [65].

Charge-transfer activation of the charge-transfer salts effects a spontaneous electron transfer [18] from the borate donor to the pyridinium acceptor which results in the formation of a radical pair (Eq. 67).

$$[Py^+, BMe_4^-] \xrightarrow{h\nu_{CT}} [Py^\bullet, BMe_4^\bullet] \tag{67}$$

The unstable $BR_4^\bullet$ radical subsequently either transfers ($k_{MT}$) an alkyl radical to the pyridinyl radical or undergoes back electron transfer ($k_{-ET}$) to restore the original charge-transfer salt (Eq. 68).

$$[Py^+, BR_4^-] \xleftarrow{k_{-ET}} [Py^\bullet, BR_4^\bullet] \xrightarrow{k_{MT}} Py\text{—}R + BR_3 \tag{68}$$

It is the competition between these two pathways that determines the thermal and photostability of the charge-transfer salt. For example, no photochemistry is observed even after prolonged CT irradiation of the charge-transfer salt of *tetrakis*[3,5-*bis*(trifluoromethyl)phenyl]borate with 4,4′-bipyridinium cation [188], which leads to the conclusion that the transfer of a *bis*(trifluoromethyl)-substituted phenyl group is too slow to efficiently compete with the (generally ultrafast [22–24]) back electron transfer. However, the photoinduced alkylations described above demonstrate that alkyl-transfer rates can be comparable or even faster than back electron transfer rates, which can only be explained by considering the alkyl transfer as an in-cage process. (The fact that no pyridine dimers are found, which readily form by coupling of free pyridinyl radicals, also supports the conclusion that no cage escape has taken place.) Accordingly, a complete mechanism for charge-transfer activated alkyl transfer can be formulated in three steps including photoinduced electron transfer ($h\nu_{CT}$) from the borate donor to the pyridinium acceptor followed by fragmentation ($k_F$) of the boranyl radical followed by coupling ($k_C$) of the alkyl and the pyridinyl radicals (Eqs 69–71).

$$[Py^+, BR_4^-] \xrightarrow{h\nu_{CT}} [Py^\bullet, BR_4^\bullet] \tag{69}$$

$$[Py^\bullet, BR_4^\bullet] \xrightarrow{k_F} [Py^\bullet, R^\bullet, BR_3] \tag{70}$$

$$[Py^\bullet, R^\bullet, BR_3] \xrightarrow{k_C} Py\text{—}R + BR_3 \tag{71}$$

The thermal reactions of the pyridinium borate salts are likely to follow the same electron-transfer path. Experimental evidence for this conclusion is the fact that the *sec*-butyl transfer is substantially faster than methyl transfer although a nucleophilic substitution mechanism would predict the less hindered group to be transferred preferentially. The fast rates of *sec*-butyl transfer can be readily explained on the basis of the electron-transfer mechanism (Eqs. 69–71) by considering the different boron–carbon bond strength [189, 190] for the various alkylborates. The boron–carbon bond cleavage (Eq. 70) is apparently the critical step, and its relative rate [191] as compared to that of the back electron transfer determines the overall rate for thermal alkyl transfers in pyridinium tetraalkylborate salts.

Alkyl transfer is also observed as the result of ion-pair annihilation of organoborate anions by carbonylmanganese(I) cations [192]. Thus, the manganese cations $[Mn(CO)_5(L)]^+$ (L = acetonitrile, pyridine, triphenylphosphine, methyldiphenylphosphine) readily react with tetramethylborate, methyltriphenylborate, and tetraphenylborate in tetrahydrofuran to yield a mixture of alkylmanganese carbonyls and/or dimanganese carbonyls and hydridomanganese carbonyls. Since in more polar solvents no such thermal reactions are observed, ion-pairing is apparently critical for all these chemical transformations. The fact that tetramethylborate (with the lowest oxidation potential) and the manganese cations $[Mn(CO)_5(acetonitrile)]^+$ and $[Mn(CO)_5(pyridine)]^+$ (with the highest reduction potentials) are the most reactive substrates, points to a redox-driven reaction scheme.

However, a general electron-transfer mechanism as formulated for the pyridinium borate salts (vide supra) is not applicable in this case owing to the difference in the redox products with varying manganese ligands (L). For example, the acetonitrile-substituted manganese cation affords methylmanganese pentacarbonyl as the sole product upon reaction with tetramethylborate or methyltriphenylborate (Eq. 72).

$$[Mn(CO)_5(MeCN)]^+ + [B(Me)R_3]^- \rightarrow MeMn(CO)_5 + BR_3 + MeCN$$
(72)

Similarly, tetraphenylborate transfers a phenyl group to form the phenyl-substituted manganese carbonyl as the only product. In contrast, pyridine- or phosphine-ligated manganese cations form dimanganese carbonyls and hydrido-manganese products besides varying amounts of alkylated manganese carbonyls depending on the manganese ligand (L). Interestingly, methyltriphenylborate does not form any alkylated or phenylated manganese carbonyls at all when reacted with pyridine- or phosphine-ligated manganese cations, and tetraphenylborate yields phenyl-substituted manganese carbonyl, dimanganese carbonyl, and hydrido manganese carbonyl in about equal amounts [192].

At first glance, all manganese products observed can readily be derived from the 19-electron carbonyl manganese radical $[Mn(CO)_5(L)]^{\bullet}$ which is generated by electron transfer from the borate donor (Eq. 73).

$$[Mn(CO)_5(L)]^+ + BR_4^- \rightarrow [Mn(CO)_5(L)^{\bullet}, BR_4^{\bullet}]$$
(73)

The 19-electron species either loses a CO or L ligand to form the corresponding 17-electron radical or abstracts hydrogen from the solvent and then extrudes a CO ligand to yield the hydridotetracarbonyl product (Eq. 74).

$$[Mn(CO)_5(L)]^{\bullet}$$

$$-CO \qquad \qquad +H^{\bullet} \qquad \qquad -L$$

$$[Mn(CO)_4(L)]^{\bullet} \quad H(CO)Mn(CO)_4(L) \quad [Mn(CO)_5]^{\bullet}$$

$$-CO$$

$$HMn(CO)_4(L) \tag{74}$$

Alternatively, both $[Mn(CO)_5]^{\bullet}$ or $[Mn(CO)_5(L)]^{\bullet}$ radicals may couple with an alkyl or phenyl radical ($R^{\bullet}$) which are present due to the fragmentation of the unstable boranyl radical (Eqs 75–77).

$$BR_4^{\bullet} \rightarrow R^{\bullet} + BR_3 \tag{75}$$

$$[Mn(CO)_5]^{\bullet} + R^{\bullet} \rightarrow R-Mn(CO)_5 \tag{76}$$

$$[Mn(CO)_5(L)]^{\bullet} + R^{\bullet} \rightarrow R-Mn(CO)_5(L) \rightarrow R-Mn(CO)_5 + L \tag{77}$$

Thus, all products may be derived from the primary radical pair generated by the initial electron transfer (Eq. 73). However, the acetonitrile-ligated carbonylmanganese cation yields $CH_3Mn(CO)_5$ and $C_6H_5Mn(CO)_5$ as the only carbonyl manganese products, and no other typical products from a 17-electron radical such as the dimer are observed. Moreover, butyltrimethyl borate transfers preferentially a methyl group to the manganese center and not the more stable butyl radical which would be the expected fragmentation product of the butyltrimethylboranyl radical. Both observations cannot be reconciled with the above delineated electron-transfer mechanism, but are satisfactorily explained on the basis of a direct nucleophilic attack of the borate anion on the manganese cation. Thus partitioning between the two pathways, i.e. electron transfer or nucleophilic addition, is controlled by the manganese ligand (L) which affects the electrophilicity of the carbonyl ligands as well as the electron affinity of the manganese center as follows [192]: acetonitrile as a poor $\sigma$-donor and a good $\pi$-acceptor withdraws electron density from the carbonyl ligands which enhances the nucleophilic attack of the borate anion on the carbonyl carbon atom. In contrast, pyridine and phosphine ligands are good $\sigma$-donors and poor $\pi$-acceptors and thus partially transfer electron density to the carbonyl ligands which decreases their electrophilicity to the point that electron transfer is a viable alternative to nucleophilic addition. Thus, manganese pentacarbonyl cations, which follow both electron-transfer and nucleophilic reaction pathways for ion-pair annihilation, are ideal substrates to demonstrate the dichotomy of the two related reactivities.

**Metal–carbon bond cleavage of dimethylaurate(I) and tetramethylaurate(III)**

The mechanistic dichotomy between nucleophilic addition and electron transfer may also apply to alkylmetallates derived from transition metals. For example, an electron-transfer mechanism has been discussed for alkylations with alkylcuprates [193, 194]. However, general mechanistic conclusions for transition alkylmetallates are hampered by the fact that thermal instability and air-sensitivity greatly limit their electron-transfer reactivity. Since alkylaurates act as alkylating reagents [195], but rapidly decompose under oxidative conditions [196], a detailed mechanistic study of the oxidative decomposition of dimethylaurate(I) and tetramethylaurate(III) has been undertaken recently [64].

Dimethylaurate(I) which can be isolated as the tetrabutylammonium salt, reacts with one-electron oxidants such as ferrocenium, decamethylferrocenium, or 2,4,6-trichlorobenzenediazonium in tetrahydrofuran to yield metallic gold and ethane [64] (Eq. 78).

$$Au^I(CH_3)_2^- + Cp_2Fe^+ \rightarrow C_2H_6 + Au^\circ + Cp_2Fe \qquad (78)$$

In the presence of 9,10-phenanthrenequinone (PQ), a chelating ligand with good $\sigma$-donor and $\pi$-acceptor properties, oxidation with ferrocenium at low temperature ($-78\,^\circ$C) leads to the formation of a green solid which decomposes at room temperature to form ethane, metallic gold, and two equivalents of phenanthrenequinone [64] (Eqs 79, 80).

$$Au^I(CH_3)_2^- + Cp_2Fe^+ + 2PQ \xrightarrow{(-78\,^\circ C)} (CH_3)_2Au(PQ)_2 + Cp_3Fe \qquad (79)$$

$$(CH_3)_2Au(PQ)_2 \xrightarrow{(25\,^\circ C)} C_2H_6 + Au^\circ + 2PQ \qquad (80)$$

The ESR spectrum of the trapped dimethylgold intermediate indicates that the unpaired electron is primarily localized on the quinone ligand, which suggests the polar formula $(CH_3)_2Au^+(PQ^{-\bullet})(PQ)$ for the green adduct [64]. The trapping experiment points to the following mechanism of oxidative decomposition of dimethylaurate(I) (Eqs 81–83).

$$Au^I(CH_3)_2^- + A^+ \rightarrow Au^{II}(CH_3)_2 + A \qquad (81)$$

$$Au^{II}(CH_3)_2 \rightarrow C_2H_6 + Au^\circ \qquad (82)$$

$$Au^{II}(CH_3)_2 + 2PQ \rightarrow (CH_3)_2Au^{II}(PQ)_2 \qquad (83)$$

The analogous oxidation of tetramethylaurate(III) with the same one-electron oxidants ($A^+$) yields methane, ethane, and metallic gold in the following stoichiometry (Eq. 84).

$$Au^{III}(CH_3)_4^- + A^+ \rightarrow CH_4 + 3/2C_2H_6 + Au^\circ + A \qquad (84)$$

In the presence of triphenylphosphine, a stable trimethylgold adduct $(CH_3)_3Au(PPh_3)$ is formed. The stoichiometry in Eq. 84, which is applicable to various one-electron oxidants, and the phosphine-ligated Au(III) intermediate strongly suggest the following electron-transfer mechanism with a tetramethylgold(IV) species as the initial (putative) reactive intermediate, which undergoes rapid homolytic cleavage to form trimethylgold(III) (Eqs 85–88).

$$Au^{III}(CH_3)_4{}^- + A^+ \rightarrow Au^{IV}(CH_3)_4 + A \qquad (85)$$

$$Au^{IV}(CH_3)_4 \rightarrow CH_4 + Au^{III}(CH_3)_3 \qquad (86)$$

$$Au^{III}(CH_3)_3 \rightarrow 3/2\,C_2H_6 + Au^\circ \qquad (87)$$

$$Au^{III}(CH_3)_3 + PPh_3 \rightarrow (CH_3)_3Au^{III}(PPh_3) \qquad (88)$$

Thus, dimethylaurate(I) and tetramethylaurate(III) exhibit an interesting difference in the fragmentation pattern upon one-electron oxidation which clearly requires further (theoretical) exploration.

## 2.5 Inner-sphere and Outer-sphere Mechanisms for Electron Transfer

The distinction between inner-sphere and outer-sphere electron transfer (ET) is originally based on redox reactions of inorganic coordination complexes [26]. *Inner-sphere* electron transfer is accompanied by ligand exchange and the formation of bridged intermediates, whereas no bond formation or breakage is observed in *outer-sphere* electron transfers [2, 26–30]. However, because this chemically based differentiation has often proved ambiguous [2, 197], we refer to a more general definition of inner-sphere and outer-sphere electron transfer [31] which considers the degree of electronic coupling of the donor and acceptor orbitals in the ET transition state as sole criterion. Owing to the technical difficulties in directly observing and characterizing of transition states [198, 199], we focus on the electron donor–acceptor precursor complex prior to electron transfer (which is assumed to be structurally and electronically similar to the ET transition state [7–9]) in which the electronic coupling (or charge-transfer interaction) is experimentally confirmed by X-ray structural and (UV–Vis or IR) spectroscopic evidence.

According to Mulliken theory [14–16], the degree of charge-transfer [91] in the precursor complex is critically dependent on the intermolecular donor–acceptor distance and is a direct measure of the bonding or electronic coupling of the electron-transfer partners, which is quantitatively described by the electronic coupling matrix element $V$ (or $H_{AB}$) [10]. Thus, *outer-sphere* electron transfer implies weak coupling ($V < 1$ kcal mol$^{-1}$ or 350 cm$^{-1}$ [200]) of the donor and acceptor orbitals in the ET transition state, whereas *inner-sphere* mechanisms should be considered whenever strong charge-transfer complexes (with $V > 1$ kcal mol$^{-1}$ or 350 cm$^{-1}$) are observed as reaction intermediates [29, 201, 202].

Organometallic donors and acceptors are particularly suitable to elucidate outer-sphere and inner-sphere ET processes as follows: First, they exist as neutral molecules as well as cationic or anionic species, and thus form both molecular EDA complexes and ion-pair salts (see Section 2.3). Molecular complexes with organometallic donors and acceptors exhibit a wide range of donor–acceptor interaction from very weak (contact) charge-transfer complexes such as the ferrocene–diiodomethane complex to the strong ionic character of TCNE complexes (see Section 2.3). In general, weak charge-transfer interaction is observed between ligated metal centers and donors (or acceptors) that are *not* ligands in the first (inner) coordination sphere of the metals. Such complexes are referred to as *outer-sphere* complexes [88–90]. In contrast, strong charge-transfer interaction is accompanied by partial penetration of the donor (or acceptor) into the inner coordination sphere of the metal. Such effects are most frequently observed in ion-pair salts. For example, the distortions in the symmetry of tetracarbonylcobaltate, tetraphenylborate, or $TpMo(CO)_3^-$ by ion pairing with pyridinium or metallocenium (see Section 2.3.3) are clear signs of a penetration of the cationic acceptor into the inner coordination sphere of the organometallic donor. The resulting inner-sphere character of such charge-transfer salts is highly solvent dependent and can in its extreme develop into a covalent bond as observed with thallium(I) tetracarbonylcobaltate in benzene [141].

Inner-sphere and outer-sphere electron transfers are not only distinguishable by the structures and charge-transfer interactions of the corresponding donor–acceptor precursor complexes, but also by the free-energy correlation of the ET rate constants. The free energy change ($\Delta G_{ET}$) accompanying outer-sphere electron transfers is simply defined by the potential gap between electron donor and acceptor and a coulombic work term ($\omega$) [203] (Eq. 89).

$$\Delta G_{ET} = E^\circ_{ox} - E^\circ_{red} + \omega \tag{89}$$

According to Marcus theory [7, 8], the activation free energy ($\Delta G^{\neq}$) of outer-sphere electron transfers depends on the free energy ($\Delta G_{ET}$) and the reorganization energy ($\lambda$) as follows (Eq. 90).

$$\Delta G^{\#} = \lambda/4[1 + \Delta G_{ET}/\lambda]^2 \tag{90}$$

As a consequence, the rate constants for *outer-sphere* electron transfers depend on the free energy of the reaction as a characteristic (quadratic) function, and *inner-sphere* ET reactions are readily revealed by substantial deviations from the Marcus behavior described in Eq. 90.

The charge-transfer concept for organic donors and acceptors derived from Mulliken theory [14–16] and the free-energy correlation of electron transfers in inorganic redox systems based on Marcus theory [7–9] are historically and methodologically separate theoretical approaches. Thus, Mulliken theory explains electronic (charge-transfer) transitions in EDA complexes, but does not consider their chemical consequences. On the other hand, Marcus theory predicts the correlation of kinetics and thermodynamics in electron-transfer reactions in the limit of weak coupling of donor and acceptor orbitals, but does not take into account energetic and electronic predisposition to electron transfer due to charge-transfer interactions in the precursor complex. However, both theoretical approaches find a common ground in the inner-sphere–outer-sphere distinction which applies to charge-transfer complexes as well as electron-transfer processes. The close relationship between charge transfer and electron transfer becomes most obvious in the study of sterically encumbered donors and acceptors. Thus, the increase in the donor–acceptor distance caused by steric hindrance effects a diminution of the donor–acceptor orbital overlap to such an extent that a changeover from an inner-sphere to an outer-sphere mechanism of electron transfer is obtained [29, 31–33].

Organometallic donors and acceptors are particularly suitable to demonstrate this mechanistic changeover since (i) their steric encumbrance can be readily modulated by the introduction of bulky ligands around the metal centers, and (ii) their oxidation and reduction potentials can be easily tuned over a wide range by varying the ligands (see Section 2.2), which is important for free-energy correlation studies. As a typical example, electron-transfer reactions of alkylmetals are discussed in Section 2.5.1. However, the steric control of electron-transfer mechanisms is not limited to organometallic redox systems, and examples of purely organic electron-transfer reactions are given in Section 2.5.2 for comparison.

### 2.5.1 Electron-transfer Reactions of Alkylmetals

Tetraalkyltin compounds ($R_4Sn$) react with tetracyanoethylene (TCNE) to form an insertion product [33] (Eq. 91).

$$\text{Me}_4\text{Sn} + \begin{array}{c} \text{NC} \quad \text{CN} \\ \diagdown \diagup \\ \diagup \diagdown \\ \text{NC} \quad \text{CN} \end{array} \xrightarrow{k_T} \text{Me}_3\text{Sn} \begin{array}{c} \text{NC} \quad \text{CN} \\ \diagup \diagdown \\ \end{array} \text{Me} \tag{91}$$

This reaction is rather slow in the absence of light ($k_T \approx 10^{-4}$ M$^{-1}$ s$^{-1}$ [33]), but can be photoactivated by deliberate irradiation of the charge-transfer absorption band of the EDA complex which forms upon mixing of the reactants in solution. The thermal as well as the photochemical pathway for TCNE insertion have been shown to occur via an initial electron transfer from the alkylmetal donor to the TCNE acceptor followed by rapid fragmentation of the resulting alkylmetal cation radical (see Section 2.2.1) (Eqs 92–94).

$$R_4\text{Sn} + \text{TCNE} \xrightleftharpoons{K_{EDA}} [R_4\text{Sn, TCNE}]_{EDA} \tag{92}$$

$$[R_4\text{Sn, TCNE}]_{EDA} \xrightarrow[\text{or } k_{ET}]{h\nu_{CT}} [R_4\text{Sn}^{+\cdot}, \text{TCNE}^{-\cdot}] \xrightarrow{fast} [R^\cdot, R_3\text{Sn}^+, \text{TCNE}^{-\cdot}] \tag{93}$$

$$[R^\cdot, R_3\text{Sn}^+, \text{TCNE}^{-\cdot}] \xrightarrow{fast} R_3\text{Sn} \begin{array}{c} \text{NC} \quad \text{CN} \\ \diagup \diagdown \\ \text{NC} \quad \text{CN} \end{array} R \tag{94}$$

Both, the fragmentation of $R_4\text{Sn}^{+\cdot}$ (Eq. 93) and the in-cage collapse of the triad (Eq. 94) are fast processes which allows the back electron transfer within the ion-radical pair to be omitted in Eq. 93 and the electron transfer ($k_{ET}$ (Eq. 93)) to be considered as the rate-determining step [33]. As a result, the rate constants for TCNE insertion can be directly compared with rate constants of other electron-transfer activated cleavages of alkyl tin donors, and in all cases the rate constants reflect the rate-determining initial electron transfer. Thus, a series of tetraalkyltin compounds of different donor strengths and steric encumbrance are reacted with TCNE, hexachloroiridate(IV) and *tris*(phenanthroline)-iron(III), $(FeL_3^{3+})$ as oxidants, and the second-order rate constants are determined by UV–Vis spectroscopic observation of the disappearance of TCNE and $IrCl_6^{2-}$ and the appearance of *tris*(phenanthroline)iron(II), respectively [32]. Moreover, the temperature dependence of the rate constants is obtained to calculate activation free energies $(\Delta G^{\neq})$ for the various electron-transfer reactions. The logarithms of the rate constants are plotted against the ionization potentials of the $SnR_4$ donors in Figure 18. Whereas the rate constants with the iron(III) acceptor $(FeL_3^{3+})$ are linearly correlated with the ionization potentials, the TCNE and $IrCl_6^{2-}$ data do not follow a linear relationship. In the case of the iron(III) acceptor, the linear correlation of $\log k$ with the ionization potentials also includes sterically encumbered alkyltin compounds such as di-*tert*-butyldimethyltin or tetra-*neo*-pentyltin (data points 17 and 18 in Figure 18). In contrast, the strongest deviations from the linear correlation are observed for electron transfer from the most sterically encumbered tetraalkyltin donors to TCNE and iridate(IV) acceptors. A similar trend applies to the activation free energies $(\Delta G^{\neq})$. Thus, the $\Delta G^{\neq}$ values for $FeL_3^{3+}$ reactions increase linearly with increasing ionization potentials of the tetraalkyltin donors, whereas the free activation enthalpies of the TCNE and iridate(IV) acceptors do not vary significantly with the ionization potentials.

A quasi-linear correlation of $\log k$ or $\Delta G^{\neq}$ with the ionization potentials of the electron donors as observed in the $FeL_3^{3+}$ reactions is predicted by Marcus theory for *outer-sphere* electron transfers. Accordingly, the free-energy dependence of $\Delta G^{\neq}$ can be satisfactorily simulated with the Marcus equation (Eq. 90), taking a (constant) value of $\lambda = 41$ kcal $mol^{-1}$ as reorganization energy for all tetraalkyltin compounds (see Figure 19A) [32].

In contrast, similar plots of $\Delta G^{\neq}$ against $\Delta G_{ET}$ for the oxidants TCNE and hexachloroiridate(IV) deviate substantially from the simulated curve for outer-sphere electron transfer (see Figures 19B, C). Moreover, the most pronounced deviations are observed with the least hindered tetraalkyltin donors. The fact that steric effects are only observed in the latter cases, but not with the $FeL_3^{3+}$ acceptor, leads to the conclusion that the inner coordination spheres of the tetraalkyltin donors are perturbed by TCNE and hexachloroiridate in the ET transition state. In other words, the electron transfers to TCNE and iridate(IV) exhibit strong inner-sphere character and thus occur from *quasi*-five-coordinate precursor complexes reminiscent of a variety of trigonal-bipyramidal structures known for tin(IV) derivatives, i.e.

In contrast, the outer-sphere iron(III) oxidant with its bulky phenanthroline ligands does not form such intimate precursor complexes owing to steric hindrance.

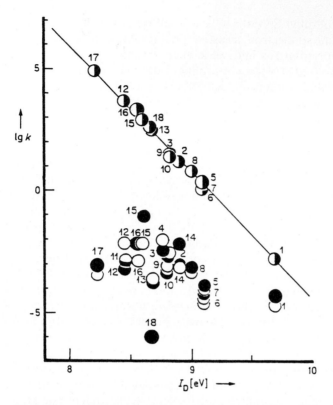

**Figure 18.** Contrasting behavior of $IrCl_6^{2-}$ ($\bullet$) and TCNE ($\bigcirc$) relative to *tris*(phenanthroline) iron(III) ($\circleddash$) acceptors in the correlation of the rates of oxidation with the ionization potentials of various alkylmetal donors: $1 = Me_4Sn$, $2 = Et_4Sn$, $3 = nPr_4Sn$, $4 = nBu_4Sn$, $5 = EtSnMe_3$, $6 = nPrSnMe_3$, $7 = nBuSnMe_3$, $8 = Et_2SnMe_2$, $9 = nPr_2SnMe_2$, $10 = nBu_2SnMe_2$, $11 = iPr_4Sn$, $12 = sBu_4Sn$, $13 = iBu_4Sn$, $14 = iPrSnMe_3$, $15 = tBuSnMe_3$, $16 = iPr_2SnMe_2$, $17 = tBu_2SnMe_2$, $18 = (tBuCH_2)_4Sn$ [32].

The inner-sphere complexes of tetraalkyltin donors with TCNE are spectroscopically manifested in the charge-transfer absorption bands observed for the EDA complexes such as in Eq. 92 [32, 33]. Quantitative evaluation of the charge-transfer energies according to Mulliken theory (Eq. 8) [14–16] leads to the determination of the coulombic work terms ($\omega$) and their dependence on steric encumbrance [39]. If the electron-transfer rate constants are corrected for the coulombic work terms of sterically more or less hindered donors [39], the corresponding free-energy correlations are linear for all three acceptors, however with different Brønsted slopes which point to different degrees of charge-transfer or inner-sphere character of the donor–acceptor pairs [32, 39]. In other words, the inner-sphere–outer-sphere distinction based on electronic coupling of donor and acceptor orbitals allows for a continuum of intermediate cases to exist between the two idealized extremes, which depends on the donor–acceptor distance and thus on the degree of steric hindrance [32, 204].

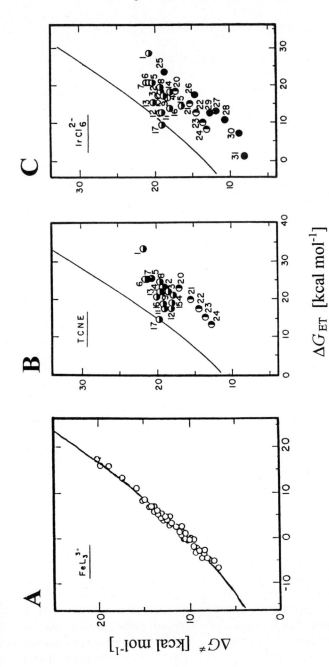

**Figure 19.** Correlation of the activation energy ($\Delta G^{\neq}$) with the free energy ($\Delta G_{ET}$) for electron transfer from alkylmetals (as identified in Figure 18) to (**A**) *tris*(phenanthroline) iron(III), (**B**) TCNE, and (**C**) $IrCl_6^{2-}$ acceptors. The solid lines represent the correlations for outer-sphere electron transfer according to Marcus theory [32].

### 2.5.2 Steric Control of Inner-sphere or Outer-sphere Electron Transfers

The general distinction of inner-sphere and outer-sphere electron transfers on the basis of electronic coupling of the donor and acceptor orbitals [31] goes beyond its original definition that is largely based on inorganic coordination complexes [26], and thus provides a universal terminology for electron-transfer mechanisms in all branches of chemistry including inorganic, organic, and organometallic chemistry. Accordingly, the examples of steric effects on the mechanism of organometallic electron transfers are to be complemented by analogous phenomena in organic chemistry [29]. Again, the comparison of electron transfers from sterically hindered and unhindered redox partners is the method of choice to reveal the relationship between electron transfer and charge transfer and to characterize inner-sphere and outer-sphere electron transfers in organic chemistry.

Thus, electron transfers from a series of unhindered, partially hindered, and heavily hindered aromatic electron donors (with matched oxidation potentials) to photoactivated quinone acceptors are kinetically examined by laser flash photolysis, and the free-energy correlations of the ET rate constants are scrutinized [31]. The second-order rate constants of electron transfers from hindered donors such as hexaethylbenzene or tri-*tert*-butylbenzene strongly depend on the temperature, the solvent polarity and salt effects, and they follow the free-energy correlation predicted by Marcus theory (see Figure 20A). Moreover, no spectroscopic or kinetic evidence for the formation of encounter complexes (exciplexes) with the photoactivated quinones prior to electron transfer is observed.

In contrast, electron transfers from unhindered (or partially hindered) donors such as hexamethylbenzene, mesitylene, di-*tert*-butyltoluene, etc. to photoactivated quinones exhibit temperature-independent rate constants that are up to 100 times faster than predicted by Marcus theory, poorly correlated with the accompanying free-energy changes (see Figure 20A), and only weakly affected by solvent polarity and salt effects. Most importantly, there is unambiguous (NIR) spectroscopic and kinetic evidence for the pre-equilibrium formation ($K_{EC}$) of long-lived encounter complexes (exciplexes) between arene donor (ArH) and photoexcited quinone acceptor (Q*) prior to electron transfer ($k_{ET}$) [20] (Eq. 95).

$$\text{ArH} + \text{Q}^* \underset{}{\overset{K_{EC}}{\rightleftharpoons}} \underset{\substack{\textbf{Encounter}\\\textbf{Complex}}}{[\text{ArH}, \text{Q}^*]} \xrightarrow{k_{ET}} [\text{ArH}^{+\bullet}, \text{Q}^{-\bullet}] \qquad (95)$$

The encounter complexes exhibit high degrees of charge-transfer [20, 91], and on the basis of absorption and emission data electronic coupling matrix elements for similar complexes (exciplexes) have been determined [205] which are comparable to those of mixed-valence metal complexes commonly used as prototypical models for the bridged-activated complex in inner-sphere electron transfers [2, 26, 197]. Accordingly, we ascribe the unusually high rate constants, their temperature-independence, and their *non*-Marcus behavior to an *inner-sphere* electron transfer process [31].

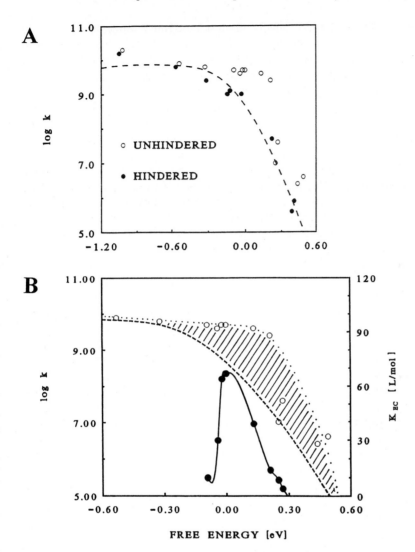

**Figure 20.** (**A**) Free-energy dependence of the second-order rate constants ($\log k$) of the electron transfer from hindered (●) and unhindered (○) arene donors to photo-activated quinones. The dashed line represents the best fit of the data points of the hindered donors to the Marcus correlation [31]. (**B**) Superposition of (**A**) and the free-energy dependence of the formation constants ($K_{EC}$) for encounter complexes of unhindered arenes with photoactivated quinones showing the coincidence of maximum encounter-complex formation (●) and maximum deviation of the electron-transfer rate constants of the unhindered donors (○) from Marcus behavior [31].

The difference in the ET mechanisms of hindered and unhindered donor–acceptor pairs is most evident in the comparison of the free-energy correlations in Figure 20A. Thus, the rate constants of the unhindered donors are faster than those of the hindered analogs, and they are faster than predicted by Marcus theory. In fact, the greatest deviations from Marcus behavior is observed for donor–acceptor couples that form the strongest encounter complexes (exciplexes) as gauged by their formation constants ($K_{EC}$ in Eq. 95). Figure 20B illustrates this finding by the superposition of the Marcus-type plots and the free-energy dependence of the formation constants $K_{EC}$. It is evident that the strongest complex formation is obtained in the isergonic and slightly endergonic free-energy region in which also the greatest deviations of the rate constants from Marcus behavior are observed. We note that the tight encounter complexes with strong charge-transfer character must experience a significant predisposition to electron transfer [20] which allows even endergonic electron transfers to occur at diffusion-limited rates (see Figure 20). In contrast, sterically encumbered donor–acceptor couples undergo 'loose' diffusive encounters for electron transfer which do not exhibit any detectable charge-transfer character. Thus, the latter redox partners are good models for the weak coupling limit in *outer-sphere* electron transfers [31].

In analogy to the findings for organometallic electron transfers (vide supra), we establish steric encumbrance as the critical effect in controlling the changeover from the outer-sphere to the inner-sphere pathway in electron transfers of organic donors. Both pathways ultimately lead to same electron-transfer products (viz. ion–radical pairs (Eq. 95)) without any overall changes in the intermolecular chemical bonds. Thus, the mechanistic difference between the two pathways solely lies in the degree of bonding in the precursor complex immediately preceding the ET transition state. Although this comparative study of hindered and unhindered organic donors suggests a clear-cut (experimental) distinction between inner-sphere and outer-sphere electron transfers, we believe there will generally be a broad borderline region between the two mechanisms. Thus, the idealized inner-sphere–outer-sphere descriptions should be taken as the two extremes in a continuum of electron transfer behaviors that are tuned by the magnitude of electronic coupling in the donor–acceptor pairs [28, 29, 32, 204].

## 2.6 Concluding Remarks

Owing to their remarkable redox properties, organometallic compounds readily form donor–acceptor assemblies with various organic, inorganic, and other organometallic substrates. X-ray crystallographic and UV–Vis or IR spectroscopic studies show that these donor–acceptor complexes cover a wide spectrum of varying degrees of charge transfer—from weakly-bound (outer-sphere) molecular complexes to strongly-bound (inner-sphere) ion pairs. This review identifies the effects of such donor–acceptor association on the chemical reactivity. Thus, high degrees of charge transfer imply an effective predisposition toward electron transfer which is recognized as the common (initial) step in various different organometallic reactions such as adduct formations, insertions, eliminations, or redox transformations. Depending on the energetic conditions, the critical electron transfer occurs either spontaneously in a thermal reaction or requires photoactivation, however both activation methods generally lead to identical products. Time-resolved (laser) spectroscopy is a powerful tool to detect the relevant (short-lived) reaction intermediates and to reveal the crucial role of ion-radical species in various organometallic reactions.

Most importantly, the organometallic donor–acceptor complexes and their electron-transfer activated reactions discussed in this review are ideal subjects to link together two independent theoretical approaches, viz. the charge-transfer concept derived from Mulliken theory [14–16] and the free-energy correlation of electron-transfer rates based on Marcus theory [7–9]. A unifying point of view of the inner-sphere–outer-sphere distinction applies to charge-transfer complexes as well as electron-transfer processes in organometallic chemistry.

## Acknowledgment

We thank the National Science Foundation and the Robert A. Welch Foundation for financial support.

## References

1. J. K. Kochi, *Organometallic Mechanisms and Catalysis*, Academic Press, New York, **1978**.
2. D. Astruc, *Electron Transfer and Radical Processes in Transition-Metal Chemistry*, VCH, New York, **1995**.
3. T. M. Bockman, J. K. Kochi in *Photosensitization and Photocatalysis Using Inorganic and Organometallic Compounds* (Eds.: K. Kalyanasundaram, M. Grätzel), Kluwer Academic, Amsterdam, **1993**, p. 407.
4. K. H. Dötz, R. W. Hoffmann (Eds.), *Organic Synthesis via Organometallics*, Vieweg, Braunschweig, **1991**.
5. J. K. Kochi, *Adv. Phys. Org. Chem.* **1994**, *29*, 185.
6. N. G. Connelly, W. E. Geiger, *Chem. Rev.* **1996**, *96*, 877.
7. R. A. Marcus, *J. Chem. Phys.* **1956**, *24*, 966.
8. R. A. Marcus, *Angew. Chem. Int. Ed. Engl.* **1993**, *32*, 1111, and references therein.
9. N. Sutin, *Acc. Chem. Res.* **1968**, *1*, 225.
10. Orbital overlap is commonly described by the electronic coupling matrix element V (or $H_{AB}$), which is assumed (within the limit of weak coupling) to exhibit an exponential falloff with the donor–acceptor distance. See: J. F. Endicott, K. Kumar, T. Ramasami, F. P. Rotzinger, *Prog. Inorg. Chem.* **1983**, *30*, 141, and references therein.
11. K. Wynne, C. Galli, R. M. Hochstrasser, *J. Chem. Phys.* **1994**, *100*, 4797.
12. T. Asahi, N. Mataga, *J. Phys. Chem.* **1989**, *93*, 6575.
13. T. Hannappel, B. Burfeindt, W. Storck, F. Willig, *J. Phys. Chem.* **1997**, *B101*, 6799.
14. R. S. Mulliken, *J. Am. Chem. Soc.* **1950**, *72*, 600.
15. R. S. Mulliken, *J. Am. Chem. Soc.* **1952**, *74*, 811.
16. R. S. Mulliken, W. M. Person, *Molecular Complexes*, Wiley, New York, **1969**.
17. J. K. Kochi in [4], p. 95.
18. Charge-transfer excitation of an electron donor–acceptor complex effects the transfer of an electron from the donor to the acceptor in less than 500 fs [11, 12].

19. G. J. Kavarnos, N. J. Turro, *Chem. Rev.* **1986**, *86*, 401.
20. R. Rathore, S. M. Hubig, J. K. Kochi, *J. Am. Chem. Soc.* **1997**, *119*, 11468.
21. Note that the ion–radical pairs may have different degrees of solvation depending on the electron-transfer pathway in Scheme 1. Thus, they may exist as contact ion pairs, solvent-separated ion pairs, or fully solvated ion radicals.
22. S. Ojima, H. Miyasaka, N. Mataga, *J. Phys. Chem.* **1990**, *94*, 7534.
23. K. Kikuchi, Y, Takahashi, K. Koike, K. Wakamatsu, H. Ikeda, T. Miyashi, *Z. Phys. Chem. N. F.* **1990**, *167*, 27.
24. I. R. Gould, S. Farid, *Acc. Chem. Res.* **1996**, *29*, 522.
25. D. C. Mauzerall in *Photoinduced Electron Transfer, Part A* (Eds.: M. A. Fox, M. Chanon), Elsevier, New York, **1988**, p. 228.
26. H. Taube, *Electron-Transfer Reactions of Complex Ions in Solution*, Academic, New York, **1970**.
27. R. D. Cannon, *Electron-Transfer Reactions*, Butterworth, London, **1980**.
28. L. Eberson, *Adv. Phys. Org. Chem.* **1982**, *18*, 79.
29. J. K. Kochi, *Angew. Chem. Int. Ed. Engl.* **1988**, *27*, 1227.
30. H. Zipse, *Angew. Chem. Int. Ed. Engl.* **1997**, *36*, 1697.
31. S. M. Hubig, R. Rathore, J. K. Kochi, *J. Am. Chem. Soc.* **1999**, *121*, 617.
32. S. Fukuzumi, C. L. Wong, J. K. Kochi, *J. Am. Chem. Soc.* **1980**, *102*, 2928.
33. S. Fukuzumi, K. Mochida, J. K. Kochi, *J. Am. Chem. Soc.* **1979**, *101*, 5961.
34. The ionization potential of naphthalene is $IP = 8.12$ [35].
35. R. C. Weast (Ed.), *CRC Handbook of Chemistry and Physics*, CRC Press, Boca Raton, FL, **1989**, p. E-87, and references therein.
36. Ionization potentials of polyalkylbenzenes lie between 7.8 and 8.8 [35].
37. See [1], p. 451.
38. H. C. Gardner, J. K. Kochi, *J. Am. Chem. Soc.* **1975**, *97*, 1855.
39. S. Fukuzumi, J. K. Kochi, *Bull. Chem. Soc. Jpn.* **1983**, *56*, 969.
40. V. F. Traven, R. West, *J. Gen. Chem. USSR* **1974**, *44*, 1803.
41. R. J. Klingler, J. K. Kochi, *J. Am. Chem. Soc.* **1981**, *103*, 5839.
42. See [32].
43. G. Bieri, F. Burger, E. Heilbronner, J. P. Maier, *Helv. Chim. Acta* **1977**, *60*, 2213.
44. D. K. Creber, G. M. Bancroft, *Inorg. Chem.* **1980**, *19*, 643.
45. S. Evans, J. C. Green, P. J. Joachim, A. F. Orchard, D. W. Turner, J. P. Maier, *J. Chem. Soc., Faraday Trans. 2* **1972**, *68*, 905.
46. T. A. Albright, J. K. Burdett, M.-H. Whangbo, *Orbital Interactions in Chemistry*, Wiley, New York, **1985**, p. 393.
47. J. W. Rabalais, L. O. Werme, T. Bergmark, L. Karlsson, M. Hussain, K. Siegbahn, *J. Chem. Phys.* **1972**, *57*, 1185.
48. A. M. Stolzenberg, M. T. Stershic, *J. Am. Chem. Soc.* **1988**, *110*, 6391.
49. N. Ito, T. Saji, K. Suga, S. Aoyagui, *J. Organomet. Chem.* **1982**, *229*, 43.
50. L. P. Yureva, L. N. Peregudova, L. N. Nekrasov, A. P. Korotkov, N. N. Zaitseva, N. V. Zakarin, A. Yu. Vasilkov, *J. Organomet. Chem.* **1981**, *219*, 43.
51. J. O. Howell, J. M. Goncalvez, C. Amatore, L. Klasinc, R. M. Wightman, J. K. Kochi, *J. Am. Chem. Soc.* **1984**, *106*, 3968.
52. M. K. Lloyd, J. A. McCleverty, J. A. Connor, E. M. Jones, *J. Chem. Soc., Dalton Trans.* **1973**, 1768.
53. Metal carbonyls are irreversibly oxidized in solution. See: C. J. Pickett, D. Pletcher, *J. Chem. Soc., Dalton Trans.* **1976**, 636.
54. S. Evans, A. F. Orchard, D. W. Turner, *Int. J. Mass Spectrom. Ion Phys.* **1971**, *7*, 261. See also: C. Amatore, D. J. Kuchynka, J. K. Kochi, *J. Electroanal. Chem.* **1988**, *241*, 181.
55. B. R. Higginson, D. R. Lloyd, P. Burroughs, D. M. Gibson, A. F. Orchard, *J. Chem. Soc., Faraday Trans. 2* **1973**, *69*, 1659.
56. D. R. Lloyd, E.W. Schlag, *Inorg. Chem.* **1969**, *8*, 2544.
57. J. C. Green, S. E. Jackson, B. Higginson, *J. Chem. Soc., Dalton Trans.* **1975**, 403.
58. J. J. Ko, T. M. Bockman, J. K. Kochi, *Organometallics* **1990**, *9*, 1833.
59. R. J. Klingler, C J. Huffman, J. K. Kochi, *J. Am. Chem. Soc.* **1980**, *102*, 208.

60. T. P. Fehlner, J. K. Kochi, unpublished results.
61. B. R. Higginson, D. R. Lloyd, S. Evans, A. F. Orchard, *J. Chem. Soc., Faraday Trans. 2* **1975**, *71*, 1913.
62. S. Cradock, E. A. V. Ebsworth, A. Robertson, *J. Chem. Soc., Dalton Trans.* **1973**, 23.
63. R. J. Klingler, K. Mochida, J. K. Kochi, *J. Am. Chem. Soc.* **1979**, *101*, 6626.
64. D. Zhu, S. V. Lindeman, J. K. Kochi, *Organometallics* **1999**, *18*, 2241.
65. D. Zhu, J. K. Kochi, *Organometallics* **1999**, *18*, 161.
66. L. L. Miller, G. D. Nordblom, E. A. Mayeda, *J. Org. Chem.* **1972**, *37*, 916.
67. S. Murphy, G. B. Schuster, *J. Phys. Chem.* **1995**, *99*, 511.
68. A. M. Bond, R. Colton, *Inorg. Chem.* **1976**, *15*, 2036.
69. D. J. Kuchynka, J. K. Kochi, *Inorg. Chem.* **1989**, *28*, 855.
70. T. Madach, H. Vahrenkamp, *Z. Naturforsch.* **1979**, *34b*, 573.
71. C. S. Ilenda, N. E. Schore, R. G. Bergman, *J. Am. Chem. Soc.* **1976**, *98*, 255.
72. D. Miholova, A. A. Vlcek, *Inorg. Chim. Acta* **1980**, *41*, 119.
73. For example, the cationic $[Co_2(fulvalene)_2]^+$ complex is readily oxidized to the dication by carbon tetrachloride. See: S. F. Clark, R. J. Watts, D. L. Dubois, J. S. Connolly, J. C. Smart, *Coord. Chem. Rev.* **1985**, *64*, 273.
74. J. A. McCleverty, *Prog. Inorg. Chem.* **1968**, *10*, 49.
75. D. B. Morse, D. N. Hendrickson, T. B. Rauchfuss, S. R. Wilson, *Organometallics* **1988**, *7*, 496.
76. P. G. Gassman, C. H. Winter, *J. Am. Chem. Soc.* **1986**, *108*, 4228.
77. C. L. Wong, J. K. Kochi, *J. Am. Chem. Soc.* **1979**, *101*, 5593.
78. K. L. Rollick, J. K. Kochi, *J. Am. Chem. Soc.* **1982**, *104*, 1319.
79. W. E. Geiger, Jr., D. E. Smith, *J. Electroanal. Chem. Interfac. Electrochem.* **1974**, *50*, 31.
80. Y. Mugnier, C. Moise, J. Tirouflet, E. Laviron, *J. Organomet. Chem.* **1980**, *186*, C49.
81. Z. J. Karpinski, J. K. Kochi, *J. Organomet. Chem.* **1992**, *437*, 211.
82. D. Astruc, *Top. Curr. Chem.* **1991**, *160*, 49.
83. U. Koelle, B. Fuss, M. V. Rajasekharan, B. L. Ramakrisna, J. H. Ammeter, M. C. Böhm, *J. Am. Chem. Soc.* **1984**, *106*, 4152.
84. R. E. Lehmann, T. M. Bockman, J. K. Kochi, *J. Am. Chem. Soc.* **1990**, *112*, 458.
85. L. J. Andrews, R. M. Keefer, *Molecular Complexes in Organic Chemistry*, Holden–Day, San Francisco, **1964**.
86. R. Foster, *Organic Charge-Transfer Complexes*, Academic, New York, **1969**.
87. G. Briegleb, *Elektronen-Donator–Acceptor Komplexe*, Springer, Heidelberg, **1961**.
88. V. M. Nekipelov, K. I. Zamaraev, *Coord. Chem. Rev.* **1985**, *61*, 185.
89. M. T. Beck, *Coord. Chem. Rev.* **1968**, *3*, 91.
90. V. Balzani, N. Sabbatini, F. Scandola, *Chem. Rev.* **1986**, *86*, 319.
91. On the basis of Mulliken theory [14–16], the degree of charge transfer is defined as the ratio $(\mathbf{b}/\mathbf{a})^2$ of the mixing coefficients $\mathbf{a}$ and $\mathbf{b}$ of the no-bond and the dative wave functions, respectively. For the experimental determination of $(\mathbf{b}/\mathbf{a})^2$, see [87, 92, 93].
92. J. A. A. Ketelaar, *J. Phys. Radium* **1954**, *15*, 197.
93. M. Tamres, M. Brandon, *J. Am. Chem. Soc.* **1960**, *82*, 2134.
94. P. Debye, *Z. Physik* **1920**, *21*, 178.
95. W. J. Duffin, *Electricity and Magnetism*, McGraw Hill, New York, **1973**, p. 79.
96. F. London, *Trans. Faraday Soc.* **1937**, *33*, 8.
97. See, for example: J. P. Fackler, T. S. Davis, I. D. Chawla, *Inorg. Chem.* **1965**, *4*, 130.
98. N. Mataga, T. Kubota, *Molecular Interactions and Electronic Spectra*, Dekker, New York, **1970**, p. 219.
99. M. Tamres, R. L. Strong in *Molecular Association Vol. 2* (Ed.: R. Foster), Academic, London, **1979**, p. 331.
100. D. F. Evans, *J. Chem. Soc.* **1953**, 345.
101. H. A. Benesi, J. H. Hildebrand, *J. Am. Chem. Soc.* **1949**, *71*, 2703.
102. The term 'charge resonance' is adopted from studies with mixed-valence metal complexes and cation-radical–neutral $\pi$-dimers $(D^+, D \leftrightarrow D, D^+)$ [150].
103. C. W. Davis, *Ion Association*, Butterworth, London, **1962**.
104. M. Szwarc (Ed.), *Ions and Ion Pairs in Organic Reactions, Vols. 1 and 2*, Wiley, New York, **1972**, **1974**.

105. A. Vogler, H. Kunkely, *Top. Curr. Chem.* **1990**, *158*, 1.
106. K. I. Zamaraev, A. N. Kitaigorodskii, *Coord. Chem.* **1980**, *6*, 563.
107. O. V. Nesterov, V. M. Nekipelov, Yu. N. Chirkov, A. N. Kitaigorodskii, S. G. Entelis, *Kinet. Katal.* **1980**, *21*, 1238.
108. J. C. D. Brand, W. Snedden, *Trans. Faraday Soc.* **1957**, *53*, 894.
109. F. K. Velichko, L. V. Balabanova, T. T. Vasileva, O. P. Bondarenko, G. A. Shvekhgeimer, *Isv. Akad. Nauk. SSSR, Ser. Khim.* **1988**, 711.
110. H. Kobayashi, M. Kobayashi, Y. Kaizu, *Bull. Chem. Soc. Jpn.* **1973**, *46*, 3109.
111. S. Fukuzumi, J. K. Kochi, *J. Am. Chem. Soc.* **1980**, *102*, 2141.
112. R. D. Schmitt, R. M. Wing, A. H. Maki, *J. Am. Chem. Soc.* **1969**, *91*, 4394.
113. W. Lau, J. K. Kochi, *J. Am. Chem. Soc.* **1986**, *108*, 6720.
114. J. M. Wallis, J. K. Kochi, *J. Org. Chem.* **1988**, *53*, 1679.
115. K. Brüggermann, J. K. Kochi, *J. Org. Chem.* **1992**, *57*, 2956.
116. K. Brüggermann, R. S. Czernuszewicz, J. K. Kochi, *J. Phys. Chem.* **1992**, *96*, 4405.
117. Note that the Drago–Rose determination (R. S. Drago, N. J. Rose, *J. Am. Chem. Soc.* **1959**, *81*, 6138) of $K_{diss}$ does not distinguish as to whether the products of the dissociation are free ions or solvent-separated ion pairs. See: H. C. Wang, G. Levin, M. Szwarc, *J. Am. Chem. Soc.* **1978**, *100*, 6137.
118. T. M. Bockman, J. K. Kochi, *J. Am. Chem. Soc.* **1989**, *111*, 4669.
119. M. Darensbourg, H. Barros, C. Bormann, *J. Am. Chem. Soc.* **1977**, *99*, 1647.
120. J. K. Kochi, C. H. Wei, *J. Organomet. Chem.* **1993**, *451*, 111.
121. See [105].
122. A comprehensive list of charge-transfer absorption bands of various organometallic EDA complexes is given in [3].
123. H. C. Gardner, J. K. Kochi, *J. Am. Chem. Soc.* **1976**, *98*, 2460.
124. R. E. Lehmann, J. K. Kochi, *J. Am. Chem. Soc.* **1991**, *113*, 501.
125. R. Billing, D. Rehorek, H. Hennig, *Top. Curr. Chem.* **1990**, *58*, 166.
126. T. M. Bockman, J. K. Kochi, *J. Phys. Org. Chem.* **1997**, *10*, 542.
127. T. M. Bockman, J. K. Kochi, *New J. Chem.* **1992**, *16*, 39.
128. C. Reichardt, *Solvents and Solvent Effects in Organic Chemistry*, VCH, 2nd ed., Weinheim, **1988**.
129. R. A. Marcus, *J. Phys. Chem.* **1989**, *93*, 3078.
130. S. M. Hubig, J. K. Kochi, *J. Phys. Chem.* **1995**, *99*, 17578.
131. C. Reichardt, *Chem. Soc. Rev.* **1992**, 147.
132. A. Loupy, B. Tchoubar, *Effets de Sel en Chimie Organique et Organometallique*, Dunod, Paris, **1988**.
133. J. Yarwood, R. Arndt in *Molecular Association Vol. 2* (Ed.: R. F. Foster), Academic, New York, **1979**, p. 267.
134. Z. G. Soos, D. J. Klein in *Molecular Association Vol. 1* (Ed.: R. F. Foster), Academic, New York, **1975**, p. 2.
135. C. K. Prout, B. Kamenar in *Molecular Complexes Vol. 1* (Ed.: R. F. Foster), Elek, London, **1969**, p. 49.
136. R. Rathore, S. V. Lindeman, J. K. Kochi, *J. Am. Chem. Soc.* **1997**, *119*, 9393.
137. P. L. Veya, J. K. Kochi, *J. Organomet. Chem* **1995**, *488*, C4.
138. H. Kisch, A. Fernández, Y. Wakatsuki, H. Yamazaki, *Z. Naturforsch.* **1985**, *40b*, 292.
139. M. Lemke, F. Knoch, H. Kisch, J. Salbeck, *Chem. Ber.* **1995**, *128*, 131.
140. C.-H. Wei, T. M. Bockman, J. K. Kochi, *J. Organomet. Chem.* **1992**, *428*, 85.
141. C. Schramm, J. I. Zink, *J. Am. Chem. Soc.* **1979**, *101*, 4554.
142. E. K. Kim, J. K. Kochi, *J. Am. Chem. Soc.* **1991**, *113*, 4962.
143. S. Brownstein, E. Gabe, F. Lee, A. Piotrowski, *Can. J. Chem.* **1986**, *64*, 1661.
144. R. Rathore, S. V. Lindeman, J. K. Kochi, *Angew. Chem. Int. Ed. Engl.* **1998**, *37*, 1585.
145. G. I. Borodkin, I. R. Elanov, M. M. Shakirov, V. G. Shubin, *J. Phys. Org. Chem.* **1993**, *6*, 153.
146. M. E. Peover, *Trans. Faraday Soc.* **1966**, *62*, 3535.
147. J. S. Miller, J. C. Calabrese, A. J. Epstein, J. H. Zhang, W. M. Reiff, *J. Chem. Soc., Chem. Commun.* **1986**, 1026.

148. E. Adman, M. Rosenblum, S. Sullivan, T. N. Margulis, *J. Am. Chem. Soc.* **1967**, *89*, 4540.

149. M. Rosenblum, R. W. Fish, C. Bennett, *J. Am. Chem. Soc.* **1964**, *86*, 5166.

150. Note that the CT absorption band of the complex in solution and in the solid state at $\lambda_{CT} \approx 1100$ nm [149] is in the same wavelength range as that of charge-resonance transitions in $\pi$-dimers of aromatic cation radicals with their neutral counterparts, which points to an electron-transfer driving force close to zero. See: B. Badger, B. Brocklehurst, *Trans. Faraday Soc.* **1969**, *65*, 2576, 2582, 2588.

151. J. S. Miller, J. H. Zhang, W. M. Reiff, D. A. Dixon, L. D. Preston, A. H. Reis, Jr., E. Gebert, M. Extine, J. Troup, A. J. Epstein, M. D. Ward, *J. Phys. Chem.* **1987**, *91*, 4344.

152. S. M. Hubig, T. M. Bockman, J. K. Kochi, *J. Am. Chem. Soc.* **1996**, *118*, 3842.

153. W. Kaiser, *Ultrashort Laser Pulses and Applications*, Springer, New York, **1988**.

154. J.-L. Martin, A. Migus, G. A. Mourou, A. H. Zewail, *Ultrafast Phenomena VIII*, Springer, New York, **1993**.

155. H. Kisch, W. Dümler, C. Chiorboli, F. Scandola, J. Salbeck, J. Daub, *J. Phys. Chem.* **1992**, *96*, 10323.

156. S. M. Hubig, unpublished results.

157. R. Criegee, *Liebigs Ann. Chem.* **1936**, *522*, 75.

158. R. Criegee, B. Marchand, H. Wannowius, *Liebigs Ann. Chem.* **1942**, *550*, 99.

159. R. J. Collin, J. Jones, W. P. Griffith, *J. Chem. Soc., Dalton Trans.* **1974**, 1094.

160. R. L. Clark, E. J. Behrmann, *Inorg. Chem.* **1975**, *14*, 1425.

161. J. M. Wallis, J. K. Kochi, *J. Am. Chem. Soc.* **1988**, *110*, 8207.

162. G. Wilkinson, M. Rosenblum, M. C. Whiting, R. B. Woodward, *J. Am. Chem. Soc.* **1952**, *74*, 2125.

163. E. O. Fischer, F. Röhrscheid, *Z. Naturforsch.* **1962**, *17b*, 483.

164. Z. J. Karpinski, J. K. Kochi, *J. Organomet. Chem.* **1992**, *437*, 211.

165. Z. J. Karpinski, J. K. Kochi, *Inorg. Chem.* **1992**, *31*, 2767.

166. Y. Mugnier, P. Reeb, C. Moise, E. Laviron, *J. Organomet. Chem.* **1983**, *254*, 111.

167. D. J. Kuchynka, J. K. Kochi, *Inorg. Chem.* **1989**, *28*, 855.

168. W. L. Waltz, O. Hackelberg, L. M. Dorfman, A. Wojcicki, *J. Am. Chem. Soc.* **1978**, *100*, 7259.

169. S. P. Church, M. Poliakoff, J. A. Timney, J. J. Turner, *J. Am. Chem. Soc.* **1981**, *103*, 7515.

170. H. W. Walker, R. S. Herrick, R. J. Olsen, T. L. Brown, *Inorg. Chem.* **1984**, *23*, 3748.

171. M. Itoh, S. Nagakura, *Bull. Chem. Soc. Jpn.* **1966**, *39*, 369.

172. R. F. Cozzens, T. A. Gover, *J. Phys. Chem.* **1970**, *74*, 3003.

173. J. A. S. Howell, P. M. Burkinshaw, *Chem. Rev.* **1983**, *83*, 557.

174. A. Davison, J. E. Ellis, *J. Organomet. Chem.* **1971**, *31*, 239.

175. F. Ungváry, A. Wojcicki, *J. Am. Chem. Soc.* **1987**, *109*, 6848.

176. D. J. Kuchynka, C. Amatore, J. K. Kochi, *J. Organomet. Chem.* **1987**, *328*, 133.

177. F. G. Bordwell, M. J. Bausch, *J. Am. Chem. Soc.* **1986**, *108*, 1979.

178. N. M. Boag, H. D. Kaesz in *Comprehensive Organometallic* Chemistry *Vol. 4* (Eds.: G. Wilkinson, F. G. A. Stone, E. W. Abel), Pergamon, Oxford, **1982**, p. 1161.

179. R. E. Lehmann, J. K. Kochi, *Organometallics* **1991**, *10*, 190.

180. F. D. Saeva, *Top. Curr. Chem* **1990**, *156*, 59.

181. A. N. Nesmeyanov, Yu. A. Chapovsky, I. V. Polovyanyuk, L. G. Makarova, *J. Organomet. Chem.* **1967**, *7*, 329.

182. M. D. Curtis, K. B. Shiu, W. M. Butler, J. C. Huffman, *J. Am. Chem. Soc.* **1986**, *108*, 3335.

183. A. T. Balaban, C. Bratu, C. N. Rentea, *Tetrahedron* **1964**, *20*, 265.

184. V. Wintgens, J. Pouliquen, J. Kossanyi, M. Heintz, *Nouv. J. Chim.* **1986**, *10*, 345.

185. A. Nakamura, S. Otsuka, *J. Am. Chem. Soc.* **1973**, *95*, 7262 and **1972**, *94*, 1886.

186. E. Negishi, *J. Organomet. Chem.* **1976**, *108*, 281.

187. R. Hunter, J. P. Michael, G. D. Tomlinson, *Tetrahedron* **1994**, *50*, 871.

188. T. Nagamura, *Pure & Appl. Chem.* **1996**, *68*, 1449.

189. S. Chatterjee, P. D. Davis, P. Gottschalk, M. E. Kurz, B. Sauerwein, X. Yang, G. B. Schuster, *J. Am. Chem. Soc.* **1990**, *112*, 6329.

190. G. B. Schuster, *Pure & Appl. Chem.* **1990**, *62*, 1565.

191. Cleavage rates for the boron–carbon bond in boranyl radicals have been estimated to exceed $k = 10^{11}$ s$^{-1}$ [189, 190].

192. D. Zhu, J. K. Kochi, *J. Organomet. Chem.* **1999**, *580*, 295.
193. H. O. House, *Acc. Chem. Res.* **1976**, *9*, 59.
194. G. Hallnemo, C. Ullenius, *Tetrahedron* **1983**, *39*, 1621.
195. D. Zhu, unpublished results.
196. S. Komiya, T. A. Albright, R. Hoffmann, J. K. Kochi, *J. Am. Chem. Soc.* **1977**, *99*, 8440.
197. A. Haim, *Prog. Inorg. Chem.* **1983**, *30*, 273.
198. J. C. Polanyi, A. H. Zewail, *Acc. Chem. Res.* **1995**, *28*, 119.
199. D. Zhong, A. H. Zewail, *J. Phys. Chem.* **1998**, *A102*, 4031.
200. W. L. Reynolds, R. W. Lumry, *Mechanisms of Electron Transfer*, Ronald Press, New York, **1966**, p. 12.
201. M. Juillard, M. Chanon, *Chem. Rev.* **1983**, *83*, 425.
202. L. Eberson, *Electron Transfer Reactions in Organic Chemistry*, Springer, New York, **1987**, p. 18.
203. The free energy ($\Delta G_{ET}$) for the transfer of an electron from a single donor to an acceptor molecule is given in eV, and thus the Faraday constant is omitted in Eq. 89.
204. D. R. Rosseinsky, *Chem. Rev.* **1972**, *72*, 215.
205. I. R. Gould, R. H. Young, L. J. Mueller, A. C. Albrecht, S. Farid, *J. Am. Chem. Soc.* **1994**, *116*, 8166.

# 3 The Thermodynamics of Organometallic Systems Involving Electron-transfer Paths

*Mats Tilset*

## 3.1 Introduction

Kochi's book from 1978 [1a] helped to establish electron-transfer and radical reactions as a crucial part of mainstream organometallic chemistry. The importance of such reactions is evident from Astruc's book [1b], still the most comprehensive and authoritative book in the area, and from several reviews and review collections [2] on aspects of organometallic electron-transfer reactivity. This chapter will be fully devoted to the use of electrochemical techniques to obtain bond-energy data for organometallic complexes, a topic that has not been previously reviewed. Aspects of the energetics of redox-induced structural changes and isomerizations, a thoroughly pursued topic, has been reviewed [2o] and will not be included here.

Bonding energetics is of fundamental importance to modern chemistry, as will be apparent from any contemporary general, organic, or inorganic chemistry text book. Knowledge of bond energies between metal and ligands and between atoms within ligands in organotransition-metal complexes is crucial to the understanding of stoichiometric and catalytic reactions. Insight into the often complex reaction mechanisms based on a quantitative understanding of strengths of bonds being formed and broken in the reaction steps involved might ultimately help in the design of new and improved processes [3]. Numerous methods, all with their particular scopes and limitations, are available for experimental estimation of the energetics of homolytic and heterolytic bond-cleavage reactions.

Electrochemical techniques, in particular cyclic voltammetry, have proven valuable for obtaining the electrode-potential data needed to obtain bond-energy data from thermochemical cycles. In a cyclic voltammetry experiment, a species generated at an electrode surface during the forward scan is detected again at the same electrode on the reverse scan. For a Nernstian, or nearly Nernstian, chemically reversible or partially reversible cyclic voltammogram, the midpoint between the anodic and cathodic peak potentials is a good estimate of the standard redox potential, $E°$. The time window of the experiment is adjustable through the voltage scan rate $v$ (V s$^{-1}$). In principle, $v$ can be varied by many orders of magnitude and, therefore, $E^\circ$ can be determined for very stable and for rather short-lived species. Further details about this experimental technique will not be discussed here, because it is well covered in electrochemistry textbooks [4].

## 3.2 Introduction to Thermochemical Cycles in the Study of Bond-cleavage Reactions of Molecules and Ions

A covalently bonded species R–X can be envisaged as undergoing cleavage of the R–X bond by the three modes depicted in Eqs 1–3. Here, Eq. 1 represents the homolytic cleavage reaction in which the X group departs as X$^{\bullet}$, along with the corresponding homolytic bond-dissociation energy (BDE) which is normally enthalpy-based. On the other hand, in Eqs 2 and 3, X departs as X$^+$ or X$^-$, respectively. The energetics of these three reactions are interrelated by the electron affinities ($EA$) and ionization potentials ($IP$) of the R$^{\bullet}$ and X$^{\bullet}$ radicals, and this relationship has formed the basis of extensive studies of species in the gas phase by mass spectrometric techniques [5]. In solution, the electron affinities and ionization potentials can be replaced by the corresponding standard electrode potentials ($E^\circ$) for the pertinent redox processes.

| Reaction | Energy change | |
|---|---|---|
| R–X $\rightarrow$ R$^{\bullet}$ + X$^{\bullet}$ | BDE(R–X) | (1) |
| R–X $\rightarrow$ R$^-$ + X$^+$ | BDE(R–X) $-\ EA$(R$^{\bullet}$) $+\ IP$(X$^{\bullet}$) | (2) |
| R–X $\rightarrow$ R$^+$ + X$^-$ | BDE(R–X) $+\ IP$(R$^{\bullet}$) $-\ EA$(X$^{\bullet}$) | (3) |

The concept 'bond-dissociation energy' is used to describe the energy change that accompanies the breakage of a specific bond in a molecule while the remaining molecular fragments are allowed to relax to their equilibrium geometries. Thus, the bond-dissociation energy includes two important terms. The first is the energy needed to cleave the bond and remove the two fragments from each other while retaining their respective geometries exactly as they were in the molecule. The second is the 'reorganization energy', a term that is necessarily exothermic, which is the energy that is gained when both fragments are allowed to attain their equilibrium geometries by relaxation of bond distances and angles [6]. This term can amount to a significant fraction of the total bond-dissociation energy. It is important to keep this in mind when bond-dissociation energies and trends are discussed in terms of the molecular structures and properties of the molecules that undergo cleavage. The contributions from reorganization energies might be particularly important for sterically encumbered systems, where the spatial demands of the ligands perturb the structure away from an otherwise ideal bonding geometry. As an example, consider the homolytic cleavage of the Fe–H bond in the organometallic

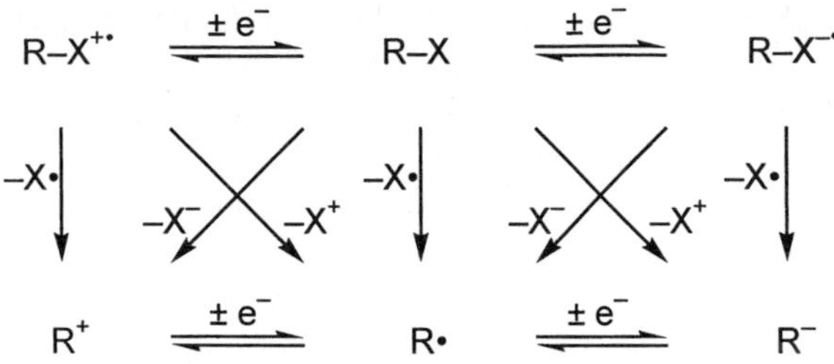

$$R-X^{+\cdot} \quad \xrightleftharpoons{\pm e^-} \quad R-X \quad \xrightleftharpoons{\pm e^-} \quad R-X^{\cdot-}$$

**Scheme 1.** Relationship between homolytic and heterolytic bond cleavage and electron-transfer reactions.

$$R^+ \quad \xrightleftharpoons{\pm e^-} \quad R\cdot \quad \xrightleftharpoons{\pm e^-} \quad R^-$$

complex Cp*Fe(dppe)H. The deuteride analog Cp*Fe(dppe)D was characterized by X-ray crystallography [7a] and has a three-legged piano-stool structure. Homolysis of the Fe–H bond, Eq. 4, generates a hydrogen atom and the Cp*Fe(dppe)$^\cdot$ radical with the same bond angles and distances as in the original molecule. This fragment then relaxes to the ground-state structure, which is a two-legged piano stool structure with the Fe and P atoms and the Cp* centroid lying in one plane, as verified by X-ray crystallography [7b].

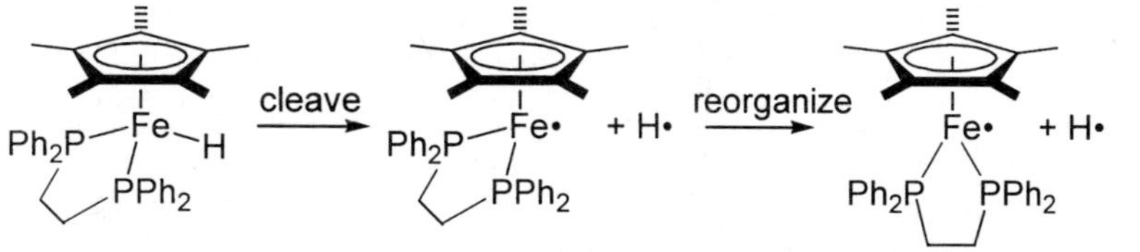

$$(4)$$

It is easy to envisage homolytic and heterolytic processes analogous to those in Eqs 1–3 that involve the substrate R–X in other oxidation states, i.e. its cation, dication, anion, etc. Wayner and Parker [8] have summarized the relationships between homolytic and heterolytic bond energies and redox processes. A mnemonic that shows available R–X cleavage patterns for a substrate R–X in three different oxidation states is shown in Scheme 1. Horizontal reactions involve the transfer of an electron, vertical reactions involve transfer of an X$^\cdot$ radical, and diagonal reactions involve the transfer of X$^+$ or X$^-$. The energy changes for any given reaction can be written as the sum of the changes for other reactions that add up to the same overall reaction. The use of thermochemical cycles based on this scheme will usually lead to *relative* rather than *absolute* bond energies. Absolute values can, however, be obtained when one or more reliable 'anchoring points' are available (see Section 3.3).

In a landmark paper, Breslow and coworkers described the determination of p$K_a$ values of weak hydrocarbon acids by use of thermochemical cycles involving electrochemical reduction data for triarylmethyl, cycloheptatrienyl, and triphenyl- and trialkylcyclopropenyl cations and radicals [9a]. Later, they derived p$K_a$ data from standard oxidation potentials and bond-dissociation energies [9b, c]. The methodology was further developed by Nicholas and Arnold [10a] for the determination of cation radical acidities, and later modified and extensively used by Bordwell and coworkers [10b, c] so that homolytic bond-dissociation energies and cation radical acidities could be determined in solution from a knowledge of pertinent electrode potential and $pK_a$ data. Since then the application of thermochemical cycles that incorporate electrode-potential data as a means of determining *absolute or relative bond energies* that are difficult or impossible to obtain by direct methods has gained widespread popularity. Electrochemical methods have been employed to determine bond-dissociation energy, acidity, and related data for organic molecules, radicals, ions, and ion radicals, and the methods and results have been thoroughly reviewed [8, 11].

During the last decade, extensive use has also been made of such thermochemical cycles to probe the bonding energetics in organometallic complexes. The rest of this chapter will describe efforts made by us and others towards this goal. We will first describe a thermochemical cycle that can be successfully used to determine *absolute* bond-dissociation energies for M–H and C–H bonds of coordinated ligands. Thereafter, the effects of single-electron-transfer processes on the energetics of homolytic and heterolytic bond-cleavage reactions will be described.

## 3.3  Absolute Bond-dissociation Energies for M–H, Coordinated Ligand C–H, and M–M Bonds

Many transition-metal hydrides exhibit significant equilibrium acidities [12]. This applies in particular to those that have electron-withdrawing ancillary ligands and/or a positive charge. This section presents a thermochemical cycle that uses these acidity ($pK_a$) data, when available, as an anchoring point to derive absolute BDE data. In principle the method is applicable to the investigation of BDEs of any M–X bond. In practice, applications have been limited to M–H because data on the prerequisite equilibrium dissociation of M–X to M$^-$ and X$^+$ in solution have been limited to metal hydrides.

### 3.3.1  Thermochemical Cycle for Determination of Absolute M–H Bond-dissociation Energies

By adapting the method of Bordwell for determination of C–H BDEs [10b], we arrived at a very useful method for estimating M–H bond-dissociation energies in solution (Scheme 2) [13]. The combined reactions represented by Eqs 5–9 give that depicted by Eq. 10 and constitute a thermochemical cycle which gives M–H BDE$_G$ expressed as Eq. 11. The subscript signifies that the derived BDE$_G$ quantity is *free-energy-based*, rather than enthalpy-based. Here, $pK_a$(M–H) is the Brønsted acidity of the metal hydride in the chosen solvent (solv), and $E^\circ_{ox}$(M$^-$) is the standard electrode potential for the oxidation of the conjugate base M$^-$ of the metal hydride. The electrode reaction depicted by Eq. 7 refers to the reduction of H$^+$ in the solvent to which all other reactions are referred.

| reaction | | free energy change $\Delta G°$ |
|---|---|---|
| M–H(solv) $\longrightarrow$ | M$^-$(solv) + H$^+$(solv) | $2.301RTpK_a$(M–H, solv)    (5) |
| M$^-$(solv) $\longrightarrow$ | M•(solv) + e$^-$ | $-FE°_{ox}$(M$^-$, solv)    (6) |
| H$^+$(solv) $\longrightarrow$ | 0.5H$_2$(g) | $FE°$(H$^+$/H$_2$, solv)    (7) |
| 0.5H$_2$(g) $\longrightarrow$ | H•(g) | $\Delta G°_f$(H•, gas)    (8) |
| H•(g) $\longrightarrow$ | H•(solv) | $\Delta G°_{solv}$(H•, solv)    (9) |
| M–H(solv) $\longrightarrow$ | M•(solv) + H•(solv) | $BDE_G$    (10) |

$$BDE_G = 2.301RTpK_a(\text{M–H, solv}) + F[E°_{ox}(\text{M}^-, \text{solv}) - E°(\text{H}^+/\text{H}_2, \text{solv})] +$$
$$\Delta G°_f(\text{H•, gas}) + \Delta G°_{solv}(\text{H•, solv}) \quad (11)$$

$$BDE_G = 2.301RTpK_a(\text{M–H, solv}) + FE°_{ox}(\text{M}^-, \text{solv}) + C_G \quad (12)$$

$$BDE = 2.301RTpK_a(\text{M–H, solv}) + FE°_{ox}(\text{M}^-, \text{solv}) + C \quad (13)$$

**Scheme 2.** Thermochemical cycle for determination of metal–hydride bond-dissociation energies.

It is convenient to gather all the constant terms in Eq. 11 to a constant $C_G$ to yield the simplified Eq. 12. The value of $C_G$ depends on the solvent chosen and on the reference electrode against which $E°_{ox}$(M$^-$, solv) and $E°$(H$^+$/H$_2$, solv) are referred. Compilations of metal-hydride p$K_a$ data in acetonitrile, mostly determined by Norton and coworkers, are available [14]. Acetonitrile is also an excellent solvent for many organometallic complexes, unreactive towards many (but far from all!) hydrides, and is also a superb solvent for non-aqueous electrochemistry. The high dielectric constant of this solvent ensures that ion-pairing effects do not strongly influence acid–base or electrochemical measurements. For these reasons, it was our preferred solvent. The free energy of solvation for the hydrogen atom, $\Delta G°_{solv}$(H•), was assumed to be equal to that of the hydrogen molecule, which varies very little in several organic solvents [15]. This term was taken to be $+21.4$ kJ mol$^{-1}$ in acetonitrile as calculated from reported $\Delta H°$ and $\Delta S°$ values [16]. The value for $\Delta G°_f$(H•)$_g$ was taken as 203.3 kJ mol$^{-1}$ [17]. In practice, the H$^+$/H$_2$ electrode, whether aqueous (NHE) or non-aqueous, is rarely used as an experimental reference. We therefore adhere to the recommended [18] use of the Fc/Fc$^+$(solv) system where Fc refers to ferrocene. When Fc is used as the reference, a quantity $FE°$(Fc/ Fc$^+$, solv) must be added to the constant terms in Eq. 11. Two presumably reliable values for $E°$(Fc/Fc$^+$, MeCN) relative to $E°$(H$^+$/H$_2$, MeCN) are $+0.034$ and $+0.068$ V [19]. Using the average value, 4.9 kJ mol$^{-1}$ must be added to $C_G$. This leads to $C_G = 229.7$ kJ mol$^{-1}$ for *free-energy-based data* when acetonitrile is the solvent for electrochemical and p$K_a$ measurements, and the electrochemical reference is Fc.

To convert the free energy data to enthalpy-based values that can be directly compared with quantities determined by calorimetric methods, the entropy contribution to the M–H cleavage reaction must be included. This contribution amounts to the entropy of the hydrogen atom in solution if one assumes that the entropies of solvation for M–H and M$^{\bullet}$ are equal. (This approximation should be valid if the two species have nearly identical shapes and charge distributions, and has been found to hold well for organic radicals R$^{\bullet}$ compared with hydrocarbons R–H [20]). The entropy of the hydrogen atom in solution, $\Delta S°(\text{H}^{\bullet}, \text{solv})$, is the sum of the entropy of formation of the hydrogen atom in the gas phase (114.6 J K$^{-1}$ mol$^{-1}$ [17]) and the entropy of solvation of the hydrogen atom, taken to equal that of the H$_2$ molecule ($-49.8$ J K$^{-1}$ mol$^{-1}$ [16]). Thus, $\Delta S°(\text{H}^{\bullet}, \text{solv}) = 64.9$ J K$^{-1}$ mol$^{-1}$, and $T\Delta S°(\text{H}^{\bullet}) = 19.3$ kJ mol$^{-1}$ at 298.15 K. This quantity must be added to the free-energy-based BDE$_G$, and thence to the constant $C_G$, to obtain enthalpy-based BDE data. When acetonitrile is the solvent for electrochemical and p$K_a$ measurements and the electrochemical reference is Fc, this gives $C = 249$ kJ mol$^{-1}$ for *enthalpy-based data*, Eq. 14. Similar treatment can be used for other solvents. For example [13b], $C = 308$ kJ mol$^{-1}$ in DMSO, Eq. 15.

$$\text{BDE} = 2.301RT\text{p}K_a(\text{M–H}, \text{MeCN})$$
$$+ FE°_{\text{ox}}(\text{M}^-, \text{MeCN}) + 249 \text{ kJ mol}^{-1} \tag{14}$$

$$\text{BDE} = 2.301RT\text{p}K_a(\text{M–H}, \text{DMSO})$$
$$+ FE°_{\text{ox}}(\text{M}^-, \text{DMSO}) + 308 \text{ kJ mol}^{-1} \tag{15}$$

Many assumptions go into the calculation of $C$ in Eq. 13 as described above, and the validity of these, and estimates of their uncertainties, have been discussed in detail [13b]. It is, however, desirable to have an independent check of the validity of the resulting data. Fortunately, M–H bond strengths have been directly determined by calorimetric investigation of the heat of hydrogenation of CpCr(CO)$_3$$^{\bullet}$ and related persistent radicals [21]. These systems are also amenable to the electrochemical determination of the Cr–H BDEs. Combining the calorimetrically determined BDEs and measured p$K_a$(M–H) and $E°_{\text{ox}}$(M$^-$) data for M–H = CpCr(CO)$_3$H and CpCr(CO)$_2$(PPh$_3$)H led, by insertion into Eq. 13, to calculated $C$ values of 248 and 250 kJ mol$^{-1}$, respectively. There is perfect agreement between the average of these values and the independently determined $C$ value of 249 kJ mol$^{-1}$ described above. This suggests that very reliable M–H BDE data can be obtained from this thermochemical cycle when p$K_a$ data in acetonitrile are available and accurate $E°_{\text{ox}}$(M$^-$) data can be obtained from reversible voltammograms recorded in acetonitrile. If other solvents are used, great care must be taken to eliminate errors caused by conversion of p$K_a$ scales for different solvents and reference-electrode systems. It is always desirable to have independent verification of some data when making approximations or assumptions such as these.

### 3.3.2 M–H Bond-dissociation Energies Determined in Acetonitrile

We have applied the thermochemical cycle that results in Eq. 13 to a wide range of organotransition-metal hydrides that have thermodynamic acidities suitable for de-

**Table 1.** Metal-hydride acidities, oxidation potentials, and bond-dissociation energies in acetonitrile.

| Entry | Compound MH | $pK_a(MH)^a$ | $E_{ox}(M^-)^b$ (V relative to Fc) | $BDE(MH)^c$ (kJ mol$^{-1}$) |
|---|---|---|---|---|
| 1 | CpCr(CO)$_3$H | 13.3 | −0.69 | 257 |
| 2 | CpCr(CO)$_2$[P(OMe)$_3$]H | 21.1$^e$ | −1.11 | 262$^f$ |
| 3 | CpCr(CO)$_2$(PPh$_3$)H | 21.8$^g$ | −1.29 | 250 |
| 4 | CpCr(CO)$_2$(PEt$_3$)H | 25.8$^e$ | −1.51 | 251$^f$ |
| 5 | Cp*Cr(CO)$_3$H | 16.1$^e$ | −0.83 | 261$^f$ |
| 6 | TpCr(CO)$_3$H | <8$^g$ | −0.82 | <215 |
| 7 | Tp'Cr(CO)$_3$H | –$^g$ | −0.86 | – |
| 8 | CpMo(CO)$_3$H | 13.9 | −0.50$^d$ | 290 |
| 9 | Cp*Mo(CO)$_3$H | 17.1 | −0.71$^d$ | 287 |
| 10 | TpMo(CO)$_3$H | 10.7$^h$ | −0.52 | 260 |
| 11 | Tp'Mo(CO)$_3$H | 9.7$^h$ | −0.58 | 248 |
| 12 | CpW(CO)$_3$H | 16.1 | −0.49$^d$ | 303 |
| 13 | CpW(CO)$_2$(PMe$_3$)H | 26.6 | −1.23$^d$ | 291 |
| 14 | CpW(CO)$_2$(PMe$_3$)H$_2^+$ | >9 | 0.16 | >316 |
| 15 | TpW(CO)$_3$H | 14.4$^h$ | −0.58 | 275 |
| 16 | Tp'W(CO)$_3$H | 12.9$^h$ | −0.65 | 260 |
| 17 | Mn(CO)$_5$H | 14.1 | −0.56$^d$ | 285 |
| 18 | Mn(CO)$_4$(PPh$_3$)H | 20.4 | −0.87$^d$ | 286 |
| 19 | Re(CO)$_5$H | 21.1 | −0.69$^d$ | 313 |
| 20 | CpFe(CO)$_2$H | 19.4 | −1.35$^d$ | 239 |
| 21 | Fe(CO)$_4$H$_2$ | 11.4 | −0.40$^d$ | 283 |
| 22 | CpRu(CO)$_2$H | 20.2 | −1.06$^d$ | 272 |
| 23 | Cp*$_2$OsH$^+$ | 9.9$^i$ | −0.06 | 298 |
| 24 | Co(CO)$_4$H | 8.3 | −0.27$^d$ | 278 |
| 25 | Co(CO)$_3$(PPh$_3$)H | 15.4 | −0.72$^d$ | 272 |
| 26 | Co(CO)$_3$[P(OPh)$_3$]H | 11.3 | −0.49$^d$ | 273 |

[a] According to [14] unless otherwise noted.
[b] $E°$ data estimated from reversible cyclic voltammograms unless otherwise noted.
[c] Calculated from Eq. 14 unless otherwise noted.
[d] Oxidation peak potentials from irreversible voltammograms [13a]. Kinetic potential shifts were applied before use of Eq. 14, as discussed elsewhere [13].
[e] $pK_a$ data were calculated [22a] from calorimetrically determined BDE data [21] by use of Eq. 14.
[f] According to [21].
[g] According to [13b].
[h] According to [22b].
[i] According to [22c].

termination by proton-transfer equilibrium measurements in acetonitrile [13, 22]. Table 1 lists the results currently available. Note that *reversible* voltammograms were obtained for entries 1–7, 10–11, 14–16, and 23 only. For the other species, irreversible peak potentials were measured for the anion oxidations, and the $E°$ values were estimated by correcting for the kinetic potential shift that is caused by the rapid follow-up reactions of the electrode-generated metalloradicals. These reactions are, under these conditions, second-order dimerization processes, often at rates approaching diffusion-control. Although kinetic potential shift corrections can easily be applied [23], the occurrence of irreversible oxidation waves does introduce an added uncertainty into the measurements. If electron transfer itself is not rate-limiting, i.e. the electron-transfer step is Nernstian, the maximum kinetic potential shift is $E_p - E° = 128$ mV for a diffusion-controlled ($10^{10}$ $M^{-1}$ $s^{-1}$) second-order dimerization at a substrate concentration of $10^{-3}$ M and a voltage scan rate $v$ of 1 V $s^{-1}$ which is quite typical of cyclic voltammetry experiments. The kinetic shift for an anodic wave will be to a potential less positive than $E°$. The 128 mV shift therefore results in the BDE being underestimated by 12 kJ $mol^{-1}$ unless a correction is applied. These and other complicating factors, such as non-Nernstian behavior arising from irreversible cyclic voltammograms, have been thoroughly discussed in this context [13b].

After the first systematic study of M–H BDEs [13a], we noted trends that seemed particularly interesting: (i) BDEs always *increase* when going down a group in the periodic table, i.e. in the order 1st row metal <2nd row <3rd row when ligands are equal; (ii) BDEs are relatively insensitive to a simple ligand substitutions in which a two-electron donor CO ligand is replaced by a tertiary phosphine; and (iii) BDEs are essentially unaffected when a Cp ligand is replaced by a Cp* ligand. The first trend is in accord with a substantial amount of accumulated data [3]. The ligand effects will be further discussed in Section 3.3.4.

A compilation of absolute BDE data is valuable in that the data might provide 'anchoring points' for relative BDEs available from other measurements. For example, Nolan and coworkers used the BDE for CpRu(CO)$_2$H (entry 22 in Table 1) with the assumption that Ru–H BDEs in CpRu(CO)$_2$H and Cp*Ru(CO)$_2$H are nearly equal, to estimate absolute Ru–X (X = Cl, Br, I) BDEs in Cp*Ru(CO)$_2$X complexes from calorimetry data that alone would have provided relative BDEs only [24].

### 3.3.3 M–H Bond-dissociation Energies Determined in Solvents other than Acetonitrile

Despite the favorable properties of acetonitrile as a solvent, its use for equilibrium acidity measurements has its definite limitations. The p$K_a$ range that is tolerable is limited at the high end by onset of solvent deprotonation, and at the low end by substrate autodissociation, as has been implicated for HCo(CO)$_4$ [14a] and TpCr(CO)$_3$H [22b]. These limitations can be overcome by the choice of a less polar solvent, e.g. 1,2-dichloroethane (DCE), dichloromethane, or THF. To make reliable, quantitative comparisons of thermodynamic data obtained in different solvents, it is necessary to link the acidity scales and electrode potential references in the different solvents. This has all too often proven to be a far from trivial task. Although, in principle, 1:1 relationship between the acidity scales in different solvents never exists, p$K_a$ differences between closely related compounds are often almost constant when compared in different solvents. This is because their solvation properties are similar, because of similarities in size and charge distribution. In less polar solvents, further complications result from extensive ion-pairing effects that will perturb acidity and electrochemical measurements. Further errors, unfortunately abundant in the literature, sometimes arise when electrode potentials measured against a certain reference in a certain solvent are converted to potentials referred against another reference, perhaps in another solvent. IUPAC has recommended [18] the use of Fc as an internal reference standard—it is commonly used not because its oxidation potential is solvent-independent (it is not!) but rather out of convenience—$Cp_2Fe$ is inexpensive, readily available, stable in two oxidation states, and soluble in most solvents. Ruiz and Astruc [25] have recently convincingly argued that decamethylmetallocenes and $Cp*Fe(C_6Me_6)$ would be better choices for standards, because their interactions with different solvents will be much more similar, simply because these permethylated transition-metal sandwiches are much bigger and the charge in the corresponding cations is better shielded against solvent interactions.

**Heat of protonation and electrochemical measurements in 1,2-dichloroethane**

Angelici [26] has reviewed a systematic study of relative acidities of *cationic* metal hydrides in DCE by measurement of the heats of protonation ($\Delta H_{MH+}$) of neutral metal precursors with triflic acid, Eq. 16. More recently, Wang and Angelici have used these data in a thermochemical cycle derived from Eq. 13 to obtain BDE data for these hydrides [27]. The advantage of the reaction calorimetry technique is that relative acidities can be obtained for compounds that are far too weakly acidic to be studied by proton-transfer equilibrium measurements in acetonitrile. Calorimetry does not, however, provide thermodynamically significant $pK_a$ data directly. We have determined the acidity of *one* of the compounds investigated calorimetrically by Angelici, $Cp*_2OsH^+$, in acetonitrile [22c]. Differences between $pK_a$ values in acetonitrile for this compound and others for which calorimetry data are available can be approximated by use of the relationship $2.301RT\Delta pK_a = -\Delta\Delta H_{MH+} + T\Delta\Delta S_{MH+}$. With the assumption that $\Delta\Delta S_{MH+} = 0$, this gives $\Delta pK_a = -\Delta\Delta H_{MH+}/2.301RT$, or $-\Delta\Delta H_{HM+} = 5.70\ pK_a$ kJ mol$^{-1}$ at 25 °C. This relationship predicts a slope of 5.70 for a plot of heat of protonation against $pK_a$ data; a slope of 6.36 (1.52 on a kcal mol$^{-1}$ basis) was found for such a correlation of data for a series of nitrogen bases with known aqueous $pK_a$ values.

This might be an indication that $\Delta\Delta S_{MH+} \neq 0$ in DCE, perhaps as a consequence of systematic covariance of $\Delta H_{MH+}$ and $\Delta S_{MH+}$ because of ion pairing in DCE. It is not known whether the same applies to metal-hydride data. Nevertheless, a systematic study of $\Delta H_{MH+}$ and $E°_{ox}(M)$ was undertaken for a wide range of complexes. On the assumption that the ideal linear relationship with a slope of 5.70 exists between $\Delta H_{MH+}$ (DCE) and $pK_a$ (acetonitrile) data, the complex $Cp*_2OsH^+$ can be used as an 'anchoring point' to relate DCE calorimetry data to acetonitrile $pK_a$ data. Eq. 17 was empirically derived using this 'anchor' and gives the relationship between BDE, $\Delta H_{MH+}$ (in DCE), and $E°_{ox}(M)$ (V relative to the SCE in DCE) [27]. Electrochemical measurements were obtained by cyclic voltammetry, second harmonic AC voltammetry, and/or Osteryoung square-wave voltammetry. For more than half of the compounds irreversible voltammograms were obtained

and because little is quantitatively known about the nature of the follow-up reactions, no kinetic corrections have been applied to the measured potentials.

$$L_nM \xrightarrow[\text{DCE}]{\text{CF}_3\text{SO}_3\text{H}} L_nM\text{–H}^+ \qquad \Delta H_{\text{rxn}} = \Delta H_{\text{MH+}} \qquad (16)$$

$$\text{BDE}(\text{MH}^+) = -\Delta H_{\text{MH+}} + FE^\circ_{\text{ox}}(M) + 139 \text{ kJ mol}^{-1} \qquad (17)$$

The resulting bond-energy data [27] are summarized in Table 2. As in Table 1,
there is an unambiguous trend that BDEs increase down a group in the periodic
table. Ligand effects are less clear-cut and will be discussed in Section 3.3.4.

**Aqueous $pK_a$ and electrochemical measurements in THF, dichloromethane, and
acetone**

Morris and coworkers have described the protonation reactions of a wide range of
Group 8 metal hydrides, Eq. 18, as part of their investigation of cationic metal dihydride and dihydrogen complexes [28]. The relative acidities of a series of cationic
dihydride or dihydrogen complexes were determined by proton-transfer equilibrium
measurements in dichloromethane, THF, and acetone [29]. The absolute $pK_a$ of
$CpRu(PPh_3)_2H_2^+$ was estimated by measuring the proton-transfer equilibrium
constant between $CpRu(PPh_3)_2H$ and $HPCy_3^+$ in dichloromethane and THF. The
results were converted into a pseudo-aqueous $pK_a$ scale by assuming constant
acidity differences between $CpRu(PPh_3)_2H_2^+$ and $HPCy_3^+$ in the solvents used
[29a], with the aqueous $pK_a = 9.7$ of $HPCy_3^+$ as the anchoring point. The constant-
acidity-difference assumption was then applied to the whole series of complexes.

$$L_nM\text{–H} \xrightarrow{\text{H}^+} L_nM(\text{H})_2^+ \text{ and/or } L_nM(\text{H}_2)^+ \qquad (18)$$

A $C$ value of 247 kJ mol$^{-1}$ for the use in Eq. 13, referring to the aqueous $pK_a$ scale
and electrode potentials relative to the Fc in the organic solvents, was originally
used, apparently [29d] on the assumption that Ru–H BDEs of 272 kJ mol$^{-1}$ were
reasonable estimates [29b, c]. We later presented evidence that the Ru–H BDE of
$CpRu(PPh_3)_2H_2^+$ might be substantially higher [30], and Morris and coworkers
revised the value of $C$ under these conditions to be 276 kJ mol$^{-1}$ [29d]. This $C$ value
is based on assumed constant acidity differences between $HPCy_3^+$ and the structurally very different cationic metal dihydrides in water and THF or dichloromethane. Therefore, in our view it is highly desirable to have a direct link to BDE
data independently determined under different conditions, preferably the acetoni-
trile-based data in Table 1. Unfortunately, thus far no such anchoring point exists.

The oxidation potentials for the conjugate bases, the neutral monohydrides, were
obtained by cyclic voltammetry in THF or dichloromethane. Irreversible waves
were often reported, also for $CpRu(PPh_3)_2H$, $CpRu(dppe)H$, and $CpRu(dppp)H$
for which other workers had reported reversible voltammetry [27, 30, 31]. It is not
clear to what extent, if any, these differences will affect the calculated BDEs, and for
the sake of internal consistency of data we will mostly use Morris's data. The data

**Table 2.** Heats of protonation, oxidation potentials, and bond-dissociation energies of metal hydrides in DCE [26].

| Entry | Compound M | $-\Delta H_{MH+}$ (kJ mol$^{-1}$) | $E^{\circ}{}_{ox}(M)^a$ (V relative to the SCE) | BDE(MH$^+$)$^b$ (kJ mol$^{-1}$) |
|---|---|---|---|---|
| 1 | Cr(CO)$_2$(dppm)$_2$ | 107 | −0.12 | 234 |
| 2 | Mo(CO)$_2$(arphos)$_2$ | 100 | 0.28 | 266 |
| 3 | Mo(CO)$_2$(dppe)$_2$ | 115 | 0.24 | 277 |
| 4 | Mo(CO)$_2$(dppm)$_2$ | 124 | 0.18 | 281 |
| 5 | W(CO)$_3$(PMePh$_2$)$_3$ | 63 | 0.48$^c$ | 249 |
| 6 | W(CO)$_3$(PEtPh$_2$)$_3$ | 71 | 0.45$^c$ | 254 |
| 7 | W(CO)$_3$(PEt$_2$Ph)$_3$ | 77 | 0.41$^c$ | 255 |
| 8 | W(CO)$_3$(PMe$_3$)$_3$ | 82 | 0.40$^c$ | 259 |
| 9 | W(CO)$_3$(PEt$_3$)$_3$ | 105 | 0.28$^c$ | 271 |
| 10 | W(CO)$_2$(dppm)$_2$ | 132 | 0.14 | 285 |
| 11 | W(CO)$_3$(tripod) | 44 | 0.72 | 253 |
| 12 | W(CO)$_3$(triphos) | 70 | 0.63 | 270 |
| 13 | Cp*Re(CO)$_2$(PMe$_2$Ph) | 77 | 0.84$^c$ | 297 |
| 14 | Cp*Re(CO)$_2$(PMe$_3$) | 84 | 0.80$^c$ | 300 |
| 15 | Fe(CO)$_3$(PPh$_3$)$_2$ | 59 | 0.55$^c$ | 251 |
| 16 | Fe(CO)$_3$(PMePh$_2$)$_2$ | 74 | 0.49$^c$ | 260 |
| 17 | Fe(CO)$_3$(PMe$_2$Ph)$_2$ | 89 | 0.45$^c$ | 272 |
| 18 | Fe(CO)$_3$(PMe$_3$)$_2$ | 97 | 0.41$^c$ | 277 |
| 19 | Fe(CO)$_3$(dppp) | 88 | 0.31$^c$ | 257 |
| 20 | Fe(CO)$_3$(dppm) | 100 | 0.40$^c$ | 278 |
| 21 | Cp*$_2$Ru | 79 | 0.68$^c$ | 285 |
| 22 | CpRu(PMe$_3$)$_2$I | 86 | 0.56 | 279 |
| 23 | CpRu(dppm)H | 121 | 0.37 | 296 |
| 24 | CpRu(dppe)H | 121 | 0.31 | 290 |
| 25 | CpRu(PPh$_3$)$_2$H | 124 | 0.23$^c$ | 286 |
| 26 | Cp*$_2$Os | 111 | 0.51$^c$ | 300 |
| 27 | CpOs(PPh$_3$)$_2$Br | 68 | 0.59$^c$ | 264 |
| 28 | CpOs(PPh$_3$)$_2$Cl | 82 | 0.58$^c$ | 278 |
| 29 | CpOs(PPh$_2$Me)$_2$Br | 85 | 0.51$^c$ | 273 |
| 30 | CpOs(PMe$_3$)$_2$Br | 123 | 0.34$^c$ | 295 |
| 31 | CpOs(PPh$_3$)$_2$H | 156 | 0.13$^c$ | 308 |
| 32 | CpIr(CO)[P($p$-C$_6$H$_4$CF$_3$)$_3$] | 117 | 0.60 | 314 |
| 33 | CpIr(CO)[P($p$-C$_6$H$_4$F)$_3$] | 125 | 0.53 | 315 |
| 34 | CpIr(CO)(PPh$_3$) | 126 | 0.50 | 313 |
| 35 | CpIr(CO)(PPh$_2$Me) | 132 | 0.45 | 315 |
| 36 | CpIr(CO)(PMe$_2$Ph) | 136 | 0.41 | 315 |
| 37 | CpIr(CO)(PMe$_3$) | 139 | 0.37 | 314 |
| 38 | CpIr(CO)(PEt$_3$) | 138 | 0.35 | 311 |
| 39 | CpIr(CO)(PCy$_3$) | 137 | 0.35 | 310 |
| 40 | CpIr(CS)(PPh$_3$) | 111 | 0.51 | 300 |
| 41 | CpIr(COD) | 95 | 0.69 | 301 |
| 42 | (C$_5$H$_4$Me)Ir(COD) | 101 | 0.61 | 299 |
| 43 | (1,2,3-C$_5$H$_2$Me$_3$)Ir(COD) | 110 | 0.54 | 302 |
| 44 | (C$_5$Me$_4$H)Ir(COD) | 115 | 0.47 | 300 |

**Table 2** (*continued*)

| Entry | Compound M | $-\Delta H_{MH+}$ (kJ mol$^{-1}$) | $E^{\circ}{}_{ox}(M)^a$ (V relative to the SCE) | BDE(MH$^+$)$^b$ (kJ mol$^{-1}$) |
|---|---|---|---|---|
| 45 | Cp*Ir(COD) | 119 | 0.45 | 302 |
| 46 | Cp*Ir(CO)[P($p$-C$_6$H$_4$CF$_3$)$_3$] | 141 | 0.30 | 310 |
| 47 | Cp*Ir(CO)[P($p$-C$_6$H$_4$Cl)$_3$] | 154 | 0.20 | 313 |
| 48 | Cp*Ir(CO)(PPh$_3$) | 155 | 0.09 | 303 |
| 49 | Cp*Ir(CO)(PPh$_2$Me) | 155 | 0.08 | 303 |
| 50 | Cp*Ir(CO)(PMe$_3$) | 159 | 0.07 | 305 |
| 51 | Cp*Ir(CO)$_2$ | 90 | 0.72 | 298 |

$^a$ Irreversible voltammograms unless otherwise noted.
$^b$ Calculated from Eq. 17.
$^c$ Reversible voltammogram.

were originally used in conjunction with a rearranged Eq. 13 to discuss correlations between oxidation potentials and acidities of metal hydrides on the assumption that M–H BDEs were constant for a series of CpRu(PR$_3$)$_2$H$_2$$^+$ complexes and the Cp* analogs [29a, c]. We have now discarded this assumption and instead used the published $pK_a$(MH$_2$$^+$) and $E_{ox}$(MH) data to derive the corresponding BDE values for a number of derivatives of CpRu(PR$_3$)$_2$H$_2$$^+$. Morris' revised value [29d] for $C$ of 276 kJ mol$^{-1}$ in Eq. 14 is used. The data are shown in Table 3.

Morris also investigated dihydrogen complexes of the type (diphosphine)$_2$M(X)(H$_2$)$^+$, obtained by protonation of *trans*-(diphosphine)MHX, Eq. 19 [29d–f]. Acidity and electrochemical measurements were performed under conditions identical to those above, and BDE determinations were performed analogously. These data are also included in Table 3.

$$ \tag{19} $$

Finally, electrode potential and acidity data were used by Ng et al. to study ligand effects on CpM(PR$_3$)$_2$H$_2$$^+$ and TpM(PPh$_3$)$_2$($\eta^2$-H$_2$)$^+$ (M = Ru, Os) complexes under similar conditions [31]; these data are included in Table 3. The Ru–H BDE determined by Ng for CpRu(PPh$_3$)$_2$H$_2$$^+$ (Table 3, entry 19; 289 kJ mol$^{-1}$) is in near perfect agreement with that of Angelici (Table 2, entry 25; 286 kJ mol$^{-1}$) but somewhat different from the value derived from Morris' data (Table 3, entry 1; 303 kJ mol$^{-1}$). The source of the discrepancy might partly be the assumptions and conversions made to align the $pK_a$ and electrochemical reference scales of the different solvents used.

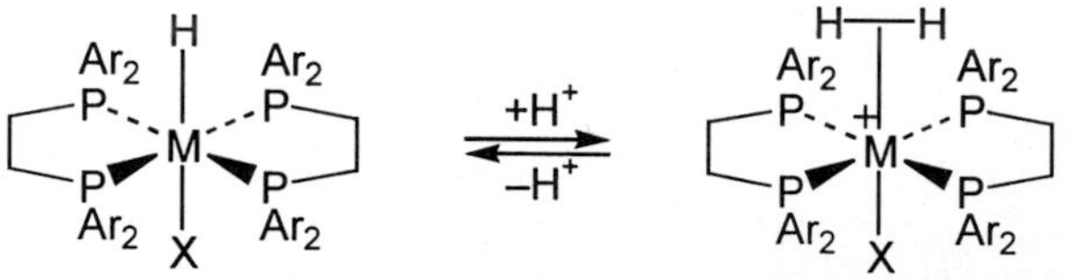

**Table 3.** Pseudo-aqueous $pK_a$ data and metal-hydride oxidation potential data in THF or dichloromethane.[a]

| Entry | Compound $MH_2^+$ | $pK_a(MH_2^+)$ | $E_{ox}(MH)$ (V relative to Fc) | $BDE(MH_2^+)$ (kJ mol$^{-1}$) |
|---|---|---|---|---|
| 1 | $CpRu(PPh_3)_2H_2^+$ | 8.0 | −0.20 | 303 |
| 2 | $CpRu(dppe)H_2^+$ | 7.5 | −0.09 | 310[b] |
| 3 | $CpRu(dtfpe)H_2^+$ | 4.6 | 0.17 | 319[b] |
| 4 | $CpRu(dppp)H_2^+$ | 8.6 | −0.22 | 304[b] |
| 5 | $CpRu(dape)H_2^+$ | 9.0 | −0.22 | 306[b] |
| 6 | $Cp^*Ru(PMe_3)_2H_2^+$ | 16.3 | −0.62 | 310[b] |
| 7 | $Cp^*Ru(PMe_2Ph)_2H_2^+$ | 14.3 | −0.57 | 303[b] |
| 8 | $Cp^*Ru(PMePh_2)_2H_2^+$ | 12.2 | −0.49 | 299[b] |
| 9 | $Cp^*Ru(PPh_3)_2H_2^+$ | 11.1 | −0.46 | 295[b] |
| 10 | $Cp^*Ru(dppm)H_2^+$ | 8.8 | −0.25 | 302 |
| 11 | $Cp^*Ru(dppm)(\eta^2\text{-}H_2)^+$ | 9.2 | −0.25 | 306 |
| 12 | $Cp^*Ru(dppp)H_2^+$ | 10.4 | −0.44 | 293[b] |
| 13 | *trans*-$CpOs(dppm)H_2^+$ | 10.0 | −0.16 | 318[c] |
| 14 | *cis*-$CpOs(dppm)H_2^+$ | 10.9 | −0.16 | 323[c] |
| 15 | *trans*-$CpOs(dppe)H_2^+$ | 11.8 | −0.17 | 327[c] |
| 16 | *cis*-$CpOs(dppe)H_2^+$ | 9.9 | −0.17 | 316[c] |
| 17 | *trans*-$CpOs(dppp)H_2^+$ | 13.4 | −0.35 | 319[c] |
| 18 | *trans*-$CpOs(PPh_3)_2H_2^+$ | 13.4 | −0.38 | 316[c] |
| 19 | $CpRu(PPh_3)_2H_2^+$ | 8.3 | −0.36 | 289[c] |
| 20 | $TpRu(PPh_3)_2(\eta^2\text{-}H_2)^+$ | 7.6 | −0.12 | 308[c] |
| 21 | $TpOs(PPh_3)_2(\eta^2\text{-}H_2)^+$ | 8.9 | −0.22 | 306[c] |
| 22 | *trans*-$HRu(dtfpe)_2(\eta^2\text{-}H_2)^+$ | 10.0 | 0.40 | 372 |
| 23 | *trans*-$HRu(dppe)_2(\eta^2\text{-}H_2)^+$ | 15.0 | −0.20 | 343 |
| 24 | *trans*-$ClRu(dppe)_2(\eta^2\text{-}H_2)^+$ | 6.0 | −0.12 | 299 |
| 25 | *trans*-$HRu(depe)_2(\eta^2\text{-}H_2)^+$ | 17.5 | −0.40 | 338 |
| 26 | *trans*-$HRu(dppf)_2H_2^+$ | 4.4 | −0.63 | 240 |
| 27 | *trans*-$HOs(dtfpe)_2(\eta^2\text{-}H_2)^+$ | 9.2 | 0.10 | 339 |
| 28 | *trans*-$HOs(dppe)_2(\eta^2\text{-}H_2)^+$ | 13.6 | −0.20 | 335 |
| 29 | *trans*-$(MeCN)Os(dppe)_2(\eta^2\text{-}H_2)^{2+}$ | −2.0 | 0.58 | 321 |
| 30 | *trans*-$ClOs(dppe)_2(\eta^2\text{-}H_2)^+$ | 7.4 | −0.13 | 306 |
| 31 | *trans*-$BrOs(dppe)_2(\eta^2\text{-}H_2)^+$ | 5.4 | −0.11 | 296 |
| 32 | *trans*-$HOs(depe)_2(\eta^2\text{-}H_2)^+$ | 17.3 | −0.60 | 317 |

[a] Data from [29] unless otherwise noted.
[b] Calculated from $E_{ox}$ and $pK_a$ data [29a, c] as mentioned in the text.
[c] Data from [31].

$$L_nM^{\pm}\!\!-\!\!\overset{\displaystyle H}{\underset{\displaystyle H}{|}} \longrightarrow L_nM^{\pm\cdot}\!\!-\!\!H + H^{\cdot} \qquad (20)$$

The variations in M–H BDEs for *dihydrogen* species are particularly intriguing. When a hydrogen atom is removed from a dihydrogen ($\eta^2$-$H_2$) ligand, Eq. 20, it is not only the M–H bond that is broken. The H–H bond to the other hydrogen atom in the $\eta^2$-H$_2$ ligand is also cleaved. Simultaneously, the M–H bond strength to the hydrogen atom that remains on the metal will change (this is part of the reorganization process, see Section 3.2).

The H–H BDE in H$_2$(g) is 435 kJ mol$^{-1}$ [17]. Inspection of the data for $\eta^2$-H$_2$ species in Table 3 reveals that the H–H bond is always considerably weakened as a result of the complexation. This is in agreement with spectroscopic and structural data for $\eta^2$-H$_2$ complexes [28, 32], i.e. $^1$H NMR $J_{HD}$ coupling constants, $r_{H-H}$ distances inferred from $T_1$ measurements, IR $v_{H-H}$ stretching frequencies, etc. The data in Table 3 show that the BDE values for $\eta^2$-H$_2$ complexes are *strikingly lower* for most of the Os complexes than for the Ru analogs, contrasting the trends for 'normal' M–H bonds. This result might be readily understood by taking into account that these $\eta^2$-H$_2$ BDE data reflect the combined effect of M–H and H–H bonding. It is well established that H–H cleavage is more pronounced for 3rd row metals than for 2nd row metals in $\eta^2$-H$_2$ complexes, where the ancillary ligands are identical. This is because the 3rd row metals are generally more capable of donating electron density to the $\sigma^*$ orbital of the H$_2$ ligand, ultimately leading to H–H cleavage and the formal oxidative addition of H$_2$ to the metal [28, 32]. It might, therefore, be argued that there is more H–H bonding left in Ru($\eta^2$-H$_2$) complexes than in analogous Os($\eta^2$-H$_2$) complexes. The compromise between a stronger M–H component and a weaker H–H component for Os can cause the apparent reversal of expected BDEs seen in Table 3.

### 3.3.4 Discussion of Ligand Effects on M–H BDE Data

The data in Tables 1–3 probably constitute the most extensive overview available for solution M–H BDE data that have been obtained by a single method. The effect of ancillary ligands on the M–H BDEs in series of related complexes can be studied in some detail. It was pointed out in Section 3.3.1 that the BDEs are remarkably insensitive to certain changes in the ligands. For example, within each of the following groups from Table 1, BDE differences are smaller than 8 kJ mol$^{-1}$: CpCr(CO)$_2$(L)H (entries 1–4); CpW(CO)$_2$(L)H (entries 12 and 13); Mn(CO)$_4$(L)H (entries 17 and 18); Co(CO)$_3$(L)H (entries 24–26); CpM(CO)$_3$H compared with Cp*M(CO)$_3$H (M = Cr, entries 1 and 5; Mo, entries 8 and 9). Similar effects are found in Table 2: less than 5 kJ mol$^{-1}$ differences within the series CpIr(CO)(L)H$^+$ (entries 32–39); Cp'Ir(COD)H$^+$ (entries 41–45); Cp*Ir(CO)(L)H$^+$ (entries 46–51, where the dicarbonyl differs somewhat more from the substituted compounds).

The essential lack of ligand effects in the above series can be qualitatively appreciated by considering that from Eq. 13 the BDE depends on contributions from the hydride p$K_a$ and the oxidation potential of the corresponding anion. The introduction of a better donor ligand will make $E^{\circ}_{ox}(M^-)$ less positive and contribute to a lower BDE, but will normally also lead to a p$K_a$ increase contributing to a higher BDE. Judging from the BDE data for the series mentioned above, these two opposing factors tend to cancel each other almost exactly.

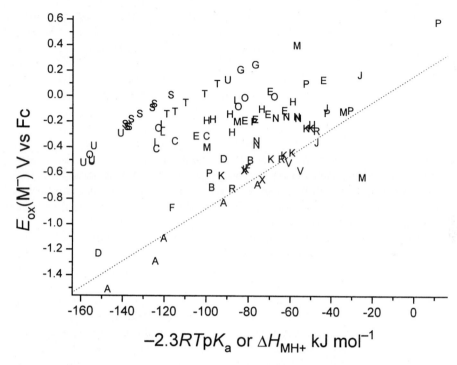

**Figure 1.** Plot of metal hydride acidity against oxidation potentials of their conjugate base. Legend (where L = CO or tertiary phosphine, P = tertiary phosphine, X = H or halide or MeCN$^+$, Cp″ = substituted or unsubstituted Cp, Tp″ = unsubstituted or substituted Tp): A = Cp″Cr(CO)$_2$(L)H (Table 1, 1–5); B = Cp″Mo(CO)$_3$H (Table 1, 8 and 9); C = Mo(CO)$_2$(P$_2$)$_2$H$^+$ (Table 2, 2–4); D = CpW(CO)$_2$(L)H (Table 1, 12 and 13); E = W(CO)$_3$(P)$_3$H$^+$ (Table 2, 5–9, 11, 12); F = Mn(CO)$_4$(P)H (Table 1, 17 and 18); G = Cp*Re(CO)$_2$(P)H$^+$ (Table 2, 13 and 14); H = Fe(CO)$_3$(P)$_2$H$^+$ (Table 2, 15–20); J = CpRu(P)$_2$H$_2^+$ (Table 3, 2–5, 19); K = Cp*Ru(P)$_2$H$_2^+$ (Table 3, 6–12); L = CpRu(P)$_2$(L)H$^+$ (Table 2, 23–25); M = Ru(P$_2$)$_2$(X)(H$_2$)$^+$ (Table 3, 22–26); N = CpOs(P)$_2$H$_2^+$ (Table 3, 13–18); O = CpOs(P)$_2$(X)H$^+$ (Table 2, 27–31); P = Os(P$_2$)$_2$(X)(H$_2$)$^+$ (Table 3, 27–32); R = Co(CO)$_3$(P)H (Table 1, 24–26); S = CpIr(CO)(P)H$^+$ (Table 2, 32–39); T = Cp″Ir(COD)H$^+$ (Table 2, 42–45); U = Cp*Ir(CO)(P)H$^+$ (Table 2, 46–51); V = Tp″Mo(CO)$_3$H (Table 1, 10 and 11); X = Tp″W(CO)$_3$H (Table 1, 13 and 14).

In an attempt to shed additional light on the observed ligand effects, Figure 1 summarizes many of the data in Tables 1–3. It shows a plot of $E_{ox}(M^-)$ against 2.3$RT$p$K_a$(M–H) (or, for Table 2, $-\Delta H_{MH+}$) for compounds that can be more or less reasonably grouped together according to metal and ligand similarities. The energy scales are different for data from the three different tables—for Tables 1 and 3 because p$K_a$ values refer to a different solvent, and for Table 2 because the data are enthalpy-based. Therefore, absolute energy values cannot be compared. This not important here, because the following discussion will focus on the *slope* of lines connecting data points for related complexes evaluated under the same conditions. (The slopes should be comparable and this was the basis, through the constant-entropy-assumption, for determining the BDEs in Table 2). The electrode-potential data from Table 2 were converted from the SCE to the Fc scale by addition of 0.59 V [27]. The dotted line in Figure 1 has the slope $1/F = 0.01036$ that would result if the ligand-induced changes in $E°_{ox}(M^-)$ and $pK_a$ contributions in Eq. 13 to the BDE data exactly cancelled each other, i.e. the BDEs within a series of related compounds remain completely insensitive to ligand variations. Data series with slopes smaller than that of the dotted line therefore show BDE variations that originate from large variations of M–H acidity rather than of $M^-$ oxidation potentials. To a first approximation this effect might be expected for sterically demanding complexes. Protonation or deprotonation of a metal, generating or removing a coordination site, will be more sensitive to steric effects than the removal or addition of an electron where no change in coordination number is involved. Stereoelectronic effects might, of course, perturb this rather simplistic picture.

In Figure 1, it is evident that the data for the series with BDEs that were insensitive to modest changes in the ligand environment are aligned in parallel with the dotted line. All these complexes are sterically relatively unencumbered species with no more than one phosphine ligand.

The other series presented in Figure 1 represent complexes that are sterically more crowded. With an exception for the Tp-substituted complexes (V and X labels) all contain at least two tertiary phosphine ligands, or at least one diphosphine. The data points for several of these series have a smaller slope than the dotted line, and for many there is much more scatter in the data than for the sterically undemanding complexes. It is possible that this pronounced difference between ligand sensitivity for mono-phosphine derivatives and the more highly substituted congeners might be traced to steric and/or ring-strain (for chelating diphosphines) effects. The BDE variations that result might then be traced to significant differences between the contributions of reorganization energies to the BDEs.

The complexes $LM(CO)_3H$ (L = Cp, Tp, Tp′; M = Cr, Mo, W) in Figure 2 constitute an interesting series of related complexes which can be compared [22b,

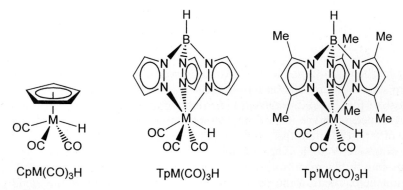

**Figure 2.** Cp, Tp, and Tp′ metal carbonyl hydrides of the Group 6 metals.

33]. The Tp and Cp ligand types are both monoanionic, six-electron donors that occupy three coordination sites [34]. Anion oxidation potential data (Table 1) and IR $\nu_{CO}$ data show that ligand donor power decreases in the order Tp$'$ > Tp > Cp. Metal-hydride acidity data determined by proton-transfer equilibrium measurements in acetonitrile suggest the opposite trend, however, because p$K_a$ decreases in the order Cp > Tp > Tp$'$. Thus, the better donor ligand—which should provide the least stable anion—nevertheless provides the stronger acid. As a result, the contributions from p$K_a$ and $E_{ox}(M^-)$ to Eq. 13 do not compensate. In fact, the slopes connecting the data points in Figure 1 (X–X—nearby D for W; V–V—nearby B for Mo) are negative, and significant C–H bond weakening is seen in the series, Cp > Tp > Tp$'$. Although the origin of this bond weakening can in principle be electronic (the Tp system is considered to be an 'octahedral enforcer' ligand [35]) or steric in nature, we consider the latter to be the predominant effect because it offers the best explanation for the distinct difference between Tp and Tp$'$. The added steric demand of Tp$'$ causes destabilization of the hepta-coordinate hydrides Tp$'$M(CO)$_3$H relative to the hexa-coordinate radical Tp$'$M(CO)$_3^\bullet$ and the anion Tp$'$M(CO)$_3^-$ that is not as strongly felt in the Tp series, and even less in the Cp series.

Other ligand effects may be interpreted by use of Figure 1 on the basis of ring-strain effects, *cis/trans* geometry effects, halide effects, and so on, but will not be further discussed here.

### 3.3.5 C–H Bond-dissociation Energies of Coordinated Ligands

One important reason that organotransition-metal chemistry has reached the importance it has today is that the binding of a ligand at a metal center alters the properties of the ligand. The coordination of a ligand can alter bond strengths within a ligand significantly, and this will have a pronounced effect on the reactivity of the ligand. The thermochemical cycle that led to Eq. 13 was a modification of one that was originally used to investigate C–H BDEs in hydrocarbons. The method is, of course, also well suited to investigation of the effects of metal coordination on the C–H BDEs of coordinated compared with free ligands. Although there are very few systematic, quantitative studies of this type in the literature, a few interesting examples that demonstrate its usefulness will be discussed.

#### $a$-C–H bonds in an $\eta^2$-acyl complex

Bruno and coworkers [36] probed the heterolytic (p$K_a$) and homolytic (BDE) bond energy of the $\alpha$-C–H bond in the $\eta^2$-acyl complex Cp$'_2$NbCl($\eta^2$-COCHEtPh)$^+$. Its acetonitrile p$K_a$ of 10.4 demonstrates that it is much more acidic than regular ketones. The deprotonation, Eq. 21, yields the $\eta^2$(C,O)-ketene complex Cp$'_2$NbCl($\eta^2$-COCEtPh). The irreversible oxidation of the ketene complex occurred at 0.24 V relative to the Fc in acetonitrile, and application of Eq. 14 led to an estimated C–H BDE in Cp$'_2$NbCl($\eta^2$-COCHEtPh)$^+$ of 343 kJ mol$^{-1}$, with some uncertainty because of the unknown kinetic potential shift. This value is in the range of C–H BDEs for comparably substituted ketones [37]. Thus, although the metal fragment has significantly changed the ligand acidity, it has hardly influenced the BDE of the C–H bond, and the stability of the radical at the α carbon is comparable with that of radicals derived from phenyl alkyl ketones. One possible interpretation is that delocalization of unpaired spin density on to Nb, the left structure in Eq. 21, is insignificant in the cation radical.

$$\left\{ \begin{array}{c} \text{Cp'}_2\overset{+\bullet}{\text{Nb}} \\ \text{Cl} \\ \text{O} \\ \text{C} \\ \text{Et}\quad\text{Ph} \end{array} \right\} \quad \begin{array}{c} \text{Cl} \\ \text{Cp'}_2\overset{+}{\text{Nb}} \\ \text{O} \\ \text{C} \\ \text{Et}\quad\text{Ph} \end{array} \quad \xleftarrow{-\text{H}\bullet} \quad \begin{array}{c} \text{Cl} \\ \text{Cp'}_2\overset{+}{\text{Nb}} \\ \text{O} \\ \text{C} \\ \text{Et}\quad\text{Ph} \\ \text{H} \end{array} \quad \xrightarrow{-\text{H}^+} \quad \begin{array}{c} \text{Cl} \\ \text{Cp'}_2\text{Nb} \\ \text{O} \\ \text{C} \\ \text{Et}\quad\text{Ph} \end{array} \tag{21}$$

### α-C–H bonds in an alkoxide complex

The methoxide complex $\text{Cp'}_2\text{Nb(Cl)(OCH}_3)^+$ has a $pK_a$ of 10.1 in acetonitrile. It is ca 50 $pK_a$ units more acidic than the model compound $\text{PhOCH}_3$, demonstrating a dramatic metal effect on the heterolytic C–H cleavage [36]. The methoxide complex is obtained by protonation of the formaldehyde complex, Eq. 22. The formaldehyde complex was reversibly oxidized at $E^\circ_{ox} = -0.07$ V relative to the Fc. The C–H BDE of the methoxy group is obtained from Eq. 14 as 301 kJ mol$^{-1}$. Because the C–H bond in methanol has a BDE of 393 kJ mol$^{-1}$ [38], the cationic Nb center has weakened the C–H bond by a substantial 92 kJ mol$^{-1}$. The cation radical that results from H$^\bullet$ cleavage gains additional stability through the formation of a Nb–C bond that was not present in the methoxide; resonance stabilization by delocalization of spin density on to Nb, the left structure in Eq. 22, was considered less important. The new Nb–C bond can be viewed as an extreme form of contribution to the *reorganization energy* discussed in Section 3.2.

$$\left\{ \begin{array}{c} \text{Cp'}_2\overset{+\bullet}{\text{Nb}} \\ \text{Cl} \\ \text{O} \\ \text{C} \\ \text{H}_2 \end{array} \right\} \quad \begin{array}{c} \text{Cl} \\ \text{Cp'}_2\text{Nb}\overset{+\bullet}{-}\text{O} \\ \text{C} \\ \text{H}_2 \end{array} \quad \xleftarrow{-\text{H}\bullet} \quad \begin{array}{c} \text{Cl} \\ \text{Cp'}_2\overset{+}{\text{Nb}} \\ \text{O} \\ \text{CH}_3 \end{array} \quad \xrightarrow{-\text{H}^+} \quad \begin{array}{c} \text{Cl} \\ \text{Cp'}_2\text{Nb}{-}\text{O} \\ \text{C} \\ \text{H}_2 \end{array} \tag{22}$$

### Benzylic C–H bonds in coordinated arenes

Zhang and Bordwell [39] looked at the effects of $\eta^6$ π-coordination on the acidities and C–H BDEs of fluorene ligands at the $\text{Cr(CO)}_3$, $\text{Mn(CO)}_3^+$, and $\text{CpFe}^+$ metal fragments in DMSO, Eq. 23. The metals increased the fluorene C–H acidity by 5.6, 16.9, and 8.0 $pK_a$ units, respectively, but application of Eq. 15 had no discernible effect on C–H BDE, which remained relatively constant at 335 kJ mol$^{-1}$ and are unchanged from that of the free ligand.

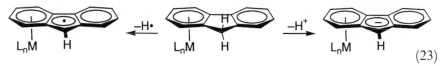

(23)

Astruc and co-workers [40] obtained similar results in an extensive study on how benzylic C–H acidities and BDEs are affected by arene complexation to the $CpFe^{n+}$ and $Cp*Fe^{n+}$ fragments in DMSO, Eq. 24. The effects on the coordinately saturated, 18-electron complex $CpFe(C_6Me_6)^+$ will be briefly mentioned here; the full investigation will be more thoroughly discussed in Section 3.8., and all data are summarized in Scheme 8. Coordination of hexamethylbenzene at the $CpFe^+$ fragment led to an estimated 14-unit increase in acidity compared with free hexamethylbenzene. The C–H BDE was calculated as 362 kJ mol$^{-1}$ from Eq. 15. This represents essentially no change when compared with the C–H BDE for free hexamethylbenzene.

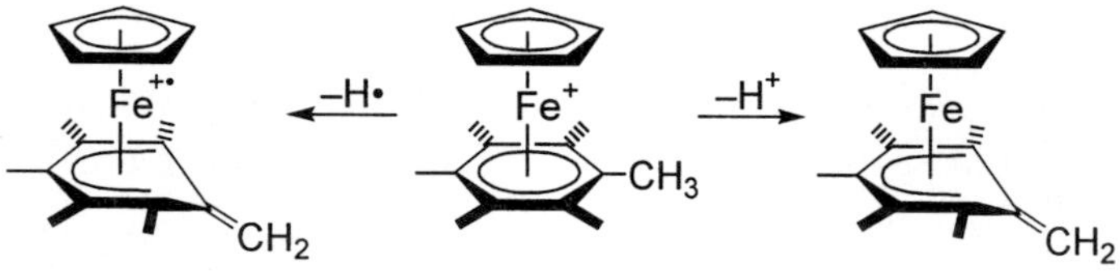

(24)

### 3.3.6 Metal–Metal BDEs from Redox Equilibrium and Electrode Potential Measurements

Dinuclear complexes such as $Mn_2(CO)_{10}$, $Cp_2Mo_2(CO)_6$, and $Cp_2Fe_2(CO)_4$ serve as precursors for many organometallic complexes via reaction sequences involving formal oxidative addition across the M–M bonds, Eq. 25. Access to M–M BDE data can provide a convenient anchoring point for the determination of other M–X data. Pugh and Meyer [41] have used a redox equilibration technique that enables the determination of M–M bond strengths.

$$L_nM–ML_n + X–X \rightarrow 2\, L_nM–X \qquad (25)$$

The thermochemical cycle in Scheme 3 was used for this purpose for the Mn, Mo, and Fe dimers. The formal potentials $E^{\circ\prime}$ ($M_2/2M^-$) were derived from equilibrium constant measurements of the redox equilibria, Eqs 26 and 27, by use of suitable reducing agents. When these formal potentials were combined with the anion oxidation potentials, the M–M BDE$_G$ could be calculated from Eq. 28 as 117 kJ mol$^{-1}$ for $Mn_2(CO)_{10}$, 92 kJ mol$^{-1}$ for $Cp_2Mo_2(CO)_6$, and 105 kJ mol$^{-1}$ for $Cp_2Fe_2(CO)_4$. For the Mn dimer, estimates for the entropy change were available and eventually led to an estimated Mn–Mn BDE of ca 159 kJ mol$^{-1}$ in enthalpy terms.

$$L_nM–ML_n + Red \underset{}{\overset{K_{eq}}{\rightleftharpoons}} 2\, L_nM^- + Red^+ \qquad (26)$$

$$E^{\circ\prime}(M_2/2M^-) = E^{\circ}(Red/Red^+) + (RT/2F) \ln K_{eq} \qquad (27)$$

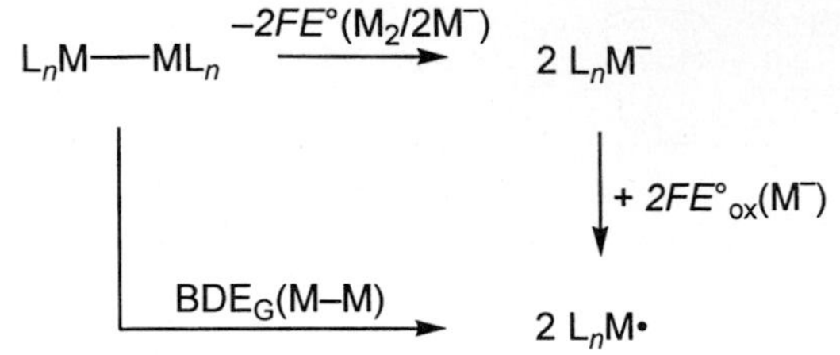

$$L_nM\text{——}ML_n \xrightarrow{-2FE°(M_2/2M^-)} 2\ L_nM^-$$

$$\downarrow + 2FE°_{ox}(M^-)$$

$$\text{BDE}_G(M\text{–}M) \longrightarrow 2\ L_nM\bullet$$

$$\text{BDE}_G(M\text{–}M) = 2FE°_{ox}(M^-) - 2FE°'(M_2/2M^-) \tag{28}$$

**Scheme 3.** Thermochemical cycle for the determination of M–M bond energies.

## 3.4 Metal–Hydride Cation Radical Acidities

The thermochemical cycle in Scheme 4 can be used to estimate the effect of one-electron oxidation on metal-hydride acidities. The method is analogous to one that has been extensively used to investigate organic cation radicals [10c]. Eq. 29 shows that measurements of $E°_{ox}(MH)$ and $E°_{ox}(M^-)$ provide *relative* $pK_a$ data for metal hydrides and their cation radicals. Absolute values for $pK_a(MH^{\bullet+})$ are obtained if the acidities of the neutral hydrides are known. The oxidation potentials of 18-electron hydrides can be readily obtained by cyclic voltammetry. In our experience, the waves that are obtained are frequently chemically irreversible, even at rather high scan rates. Consequently, the oxidation peak potentials will be kinetically shifted and represent *minimum* values for the true $E°_{ox}(MH)$ data, the estimates for $pK_a(MH^{\bullet+})$ represent *maximum* values, and calculated $\Delta pK_a$ are *minimum* values.

We have applied Eq. 29 to a series of Cp and Tp Group 6 metal carbonyl hydrides in acetonitrile, mostly on the basis of irreversible metal-hydride oxidation

51

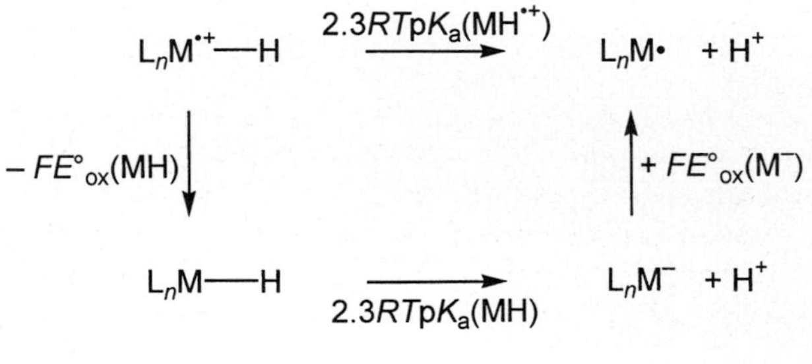

$$L_nM^{\bullet+}\text{——}H \xrightarrow{2.3RTpK_a(MH^{\bullet+})} L_nM\bullet\ +H^+$$

$$-FE°_{ox}(MH)\downarrow \qquad \uparrow +FE°_{ox}(M^-)$$

$$L_nM\text{——}H \xrightarrow{2.3RTpK_a(MH)} L_nM^-\ +H^+$$

$$\Delta pK_a = pK_a(MH^{\bullet+}) - pK_a(MH) = (F/2.3RT)\ [E°_{ox}(M^-) - E°_{ox}(MH)] \tag{29}$$

**Scheme 4.** Thermochemical cycle for the determination of metal-hydride cation radical acidities.

**Table 4.** $pK_a$ and BDE data for selected metal hydrides and their cation radicals in acetonitrile.

| Compound MH | $pK_a$(MH) | $pK_a$(MH$^{\cdot+}$) | $\Delta pK_a$ | BDE(MH) | BDE(MH$^{\cdot+}$) | $\Delta$BDE |
|---|---|---|---|---|---|---|
| CpCr(CO)$_3$H[a] | 13.3 | −9.5 | 22.8 | 257 | 233 | 24 |
| CpCr(CO)$_2$[P(OMe)$_3$]H[b] | 21.1 | −2.4 | 23.5 | 262 | 216 | 46 |
| CpCr(CO)$_2$(PEt$_3$)H[b] | 25.8 | 0.5 | 25.3 | 251 | 213 | 38 |
| CpCr(CO)$_2$(PPh$_3$)H[b] | 21.8 | −2.1 | 23.9 | 250 | 208 | 42 |
| Cp*Cr(CO)$_3$H[b] | 16.1 | −7.2 | 23.3 | 261 | 227 | 34 |
| TpCr(CO)$_3$H[c] | <8 | – | – | <215 | – | – |
| Tp'Cr(CO)$_3$H[c] | <8 | – | – | <212 | – | – |
| CpMo(CO)$_3$H[a] | 13.9 | −6.0 | 19.9 | 290 | – | – |
| TpMo(CO)$_3$H[c] | 10.7 | −8.2 | 18.9 | 260 | 233 | 27 |
| Tp'Mo(CO)$_3$H[c] | 9.7 | −9.5 | 19.2 | 248 | 223 | 25 |
| CpW(CO)$_3$H[a] | 16.1 | −3.0 | 19.1 | 303 | – | – |
| CpW(CO)$_2$(PMe$_3$)H[a] | 26.6 | 5.1 | 21.5 | 291 | – | – |
| TpW(CO)$_3$H[c] | 14.4 | −5.4 | 19.8 | 275 | 241 | 34 |
| Tp'W(CO)$_3$H[c] | 12.9 | −6.7 | 19.6 | 260 | 231 | 29 |
| *trans*-Cr(CO)$_2$(dppm)$_2$H$^+$[d] | – | – | 30.0 | – | – | – |
| *trans*-Cr(CO)$_2$(dppe)$_2$H$^+$[d] | – | – | 29.5 | – | – | – |
| *trans*-Mo(CO)$_2$(dppm)$_2$H$^+$[d] | – | – | 27.9 | – | – | – |
| *trans*-Mo(CO)$_2$(dppe)$_2$H$^+$[d] | – | – | 26.2 | – | – | – |
| *trans*-W(CO)$_2$(dppm)$_2$H$^+$[d] | – | – | 28.8 | – | – | – |
| *trans*-W(CO)$_2$(dppe)$_2$H$^+$[d] | – | – | 25.4 | – | – | – |
| Cp*Fe(dppe)H[e] | – | – | – | – | – | 51 |

[a] According to [42].
[b] According to [22a].
[c] According to [22b].
[d] According to [43].
[e] According to [48].

waves [22a, b, 42]. Bond and coworkers [43] more recently determined the effects of oxidations of Group 6 *trans*-M(CO)$_2$(diphosphine)$_2$H$^+$ (M = Cr, Mo, W) cations to di-cations, on the basis of *reversible* electrode processes in acetonitrile. The pertinent acidity data derived from Eq. 29 are summarized in Table 4 and, when absolute values are available, graphically represented in Figure 3. The data establish that the 17-electron cation radicals are more acidic than their 18-electron neutral parents by a relatively constant difference of 19–23 $pK_a$ units, translating to ca 120 kJ mol$^{-1}$ in bond weakening with respect to deprotonation. For the 17-electron di-cations, the bond activation is somewhat greater, 25–30 $pK_a$ units. Interestingly, many of these cation radicals can be termed *organometallic superacids*. (Acids are generally less acidic in acetonitrile than in water, primarily as a consequence of the significantly poorer solvation of the proton in acetonitrile, which alone corresponds to 7.8 $pK_a$ units [19]).

As might be expected, the 17-electron metal-hydride cation radicals typically undergo proton-transfer reactions [30, 42–44]. The 17-electron radical that results from the deprotonation is usually oxidized, presumably via solvent coordination, to

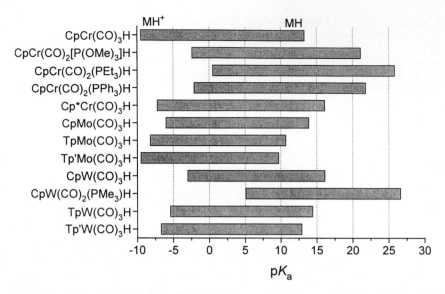

**Figure 3.** Acetonitrile $pK_a$ data for selected neutral 18-electron hydrides (right end of bars) and their 17-electron cation radicals (left end).

generate a more readily oxidized [45] 19-electron radical [1b, 2c–e] to give cationic 18-electron solvento complexes. The solvent/electrolyte system is often sufficiently basic. For the less acidic species, the unoxidized metal hydride can function as the base, Eq. 30. Alternatively, external bases such as amines can be added, as in Eq. 31. Note that the change of base changes the stoichiometry from 1 faraday mol$^{-1}$ in Eq. 30 to 2 faraday mol$^{-1}$ in Eq. 31.

$$\text{(30)}$$

$$\text{(31)}$$

## 3.5 Metal–Hydride Bond-dissociation Energies in Cation Radicals

Scheme 5 depicts a thermochemical cycle which can be used to determine metal-hydride cation radical BDEs. Adding the energy terms in the cycle gives Gibbs energy $\text{BDE}_G$ data. If the assumption is made that the solvation properties of MH$^{\bullet+}$

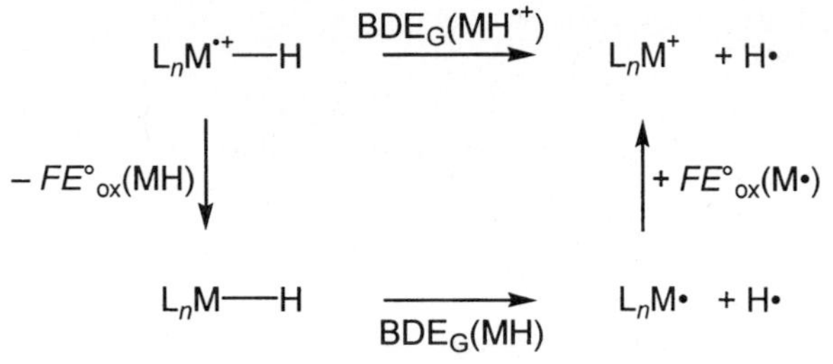

$$L_nM^{\bullet+}\!\!-\!\!H \xrightarrow{\;BDE_G(MH^{\bullet+})\;} L_nM^+ \; + \; H\bullet$$

$$-FE^\circ_{ox}(MH) \downarrow \qquad\qquad\qquad \uparrow +FE^\circ_{ox}(M\bullet)$$

$$L_nM\!\!-\!\!H \xrightarrow[\;BDE_G(MH)\;]{} L_nM\bullet \; + \; H\bullet$$

$$\Delta BDE = BDE(MH^{\bullet+}) - BDE(MH) = F[E^\circ_{ox}(M\bullet) - E^\circ_{ox}(MH)] \tag{32}$$

**Scheme 5.** Thermochemical cycle for the determination of metal-hydride cation radical bond-dissociation energies.

and $M^+$ are essentially identical, as are those of MH and $M^{\bullet}$ (this apparently works well for organic molecules [20]) then the solvation entropy terms will cancel and Eq. 32 applies to the desired enthalpy-based calculations. In the absence of BDE data for the parent metal hydride, Eq. 32 still provides useful insight into *relative* BDEs for the cation radical and the neutral parent.

Unfortunately, it is far from trivial to obtain oxidation potentials for commonly encountered 17-electron metalloradicals $M^{\bullet}$, because many such radicals dimerize at rates approaching diffusion-control, rendering it nearly impossible to observe such species by cyclic voltammetry. The use of ultramicroelectrodes was shown [41] to give a reversible signal for the oxidation of $Mn(CO)_5^-$ at scan rates of ca 5000 V s$^{-1}$, but the further oxidation of this radical to the 16-electron cation was not reported. There are, however, certain frequently encountered systems for which such radicals are stable at least on the time-scale of normal voltammetric measurements. Figure 4 shows an example, the oxidation of $CpCr(CO)_3^-$ in acetonitrile.

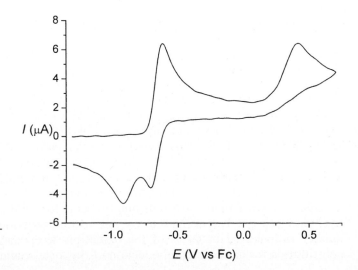

**Figure 4.** Cyclic voltammogram of the oxidation of $CpCr(CO)_3^-Et_4N^+$ in acetonitrile/0.1 M $Bu_4N^+PF_6^-$ at $\nu = 1.0$ V s$^{-1}$.

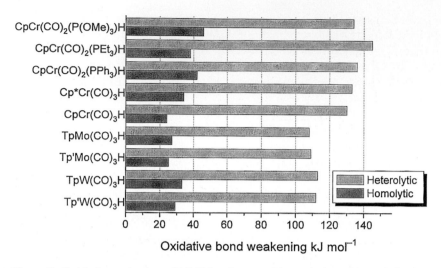

**Figure 5.** Oxidative activation of M–H bonds toward heterolytic and homolytic cleavage.

The anion is reversibly oxidized to the radical CpCr(CO)$_3$˙ at $-0.68$ V relative to the Fc. Furthermore, an irreversible oxidation occurs at $+0.42$ V. The irreversibility is caused by the extremely fast reaction of the unsaturated, 16-electron cation CpCr(CO)$_3{}^+$ with the solvent to give CpCr(CO)$_3$(NCMe)$^+$. The latter is reduced at $-0.92$ V during the reverse scan. Analogous behavior has been observed for several other complexes.

Table 4 includes p$K_a$ and BDE data for several metal hydrides and their cation radicals. In all instances where $\Delta$BDE data are given, except for Cp*Fe(dppe)H, oxidation of MH and M˙ was irreversible and therefore subject to unknown kinetic potential shifts. From these data it seems a consistent trend that the one-electron oxidation leads to a bond weakening with respect to homolytic M–H cleavage. Homolytic activation is much less pronounced than heterolytic, as depicted in Figure 5. This might not come as a surprise, because proton removal from a cationic species should be more facile than from a closely related neutral species. Heterolytic bond activation of organic molecules is usually even more pronounced, as seen in references cited in Refs [8] and [11]. Heterolytic bond weakening can be expressed as $-\Delta\Delta G^\circ{}_{het} = 2.3RT\Delta pK_a = F[E_{ox}(MH) - E_{ox}(M^-)]$ and homolytic as $-\Delta\Delta G^\circ{}_{hom} = -\Delta BDE = F[E_{ox}(MH) - E_{ox}(M˙)]$. Subtraction gives the difference between heterolytic and homolytic bond weakening, $-(\Delta\Delta G^\circ{}_{het} - \Delta\Delta G^\circ{}_{hom}) = F[E_{ox}(M˙) - E_{ox}(M^-)]$. Thus, heterolytic bond weakening will exceed homolytic bond weakening if $E_{ox}(M˙) > E_{ox}(M^-)$. This will generally be true, because the electron-rich 18-electron anion M$^-$ should be more readily oxidized than the electron-deficient 17-electron radical M˙. This does not, however, explain why bond weakening (rather than a strengthening!) occurs for the homolytic process. Oxidatively-induced bond weakening for a pure $\sigma$ donor ligand like the hydride can be naively rationalized in terms of the oxidation-state formalism. M–H homolysis constitutes a formal 1-electron reduction of the metal center and should be favored by a higher oxidation state. This is, however, probably too much of a simplification, as suggested by a recent contribution from Poli and coworkers [46].

The homolytic bond activation energies given in Table 4 for the Group 6 metal complexes were based on irreversible electrode processes. It is therefore conceivable that kinetic potential shifts might have reversed the order of the $M^{0/+}$ and $MH^{0/+}$ redox couples (note that a 30 kJ mol$^{-1}$ difference corresponds to as much as 0.3 V). This would have reversed what should have been a bond strengthening to an apparent bond weakening. Poli has argued [46], on MO and electrostatic grounds, that the oxidation processes should lead to bond strengthening, rather than bond weakening, for a $M–H^{\delta-}$ system with a very electron-rich metal fragment when the bonding energetics are dominated by the covalent component. Such an effect was inferred from an observed increase in the IR $\nu_{M-H}$ stretching frequencies when $Cp^*M(dppe)H_3$ (M = Mo, W) were oxidized to their relatively stable cation radicals. The inferred bond strengthening was further supported by results from DFT calculations. It has not yet been possible to apply thermochemical cycles to these systems. We recently investigated the effects of the oxidation of $Cp^*Fe(dppe)H$, for which the $M^{0/+}$ and $MH^{0/+}$ redox couples are both reversible [47]. The thermochemical cycle is indicative of homolytic bond *weakening* of 51 kJ mol$^{-1}$ in this instance. The bond weakening has been supported by DFT calculations (J.-Y. Saillard and K. Costuas, personal communication). These results are not necessarily at odds with each other because they arise from quite different molecules; rather, they tell us that our picture of bonding even in relatively simple M–H bonds is not yet complete.

## 3.6 Metal–Halide Bond Energies

The thermochemical cycle, Scheme 5, for determination of relative BDEs in metal hydrides in different oxidation states can be extended in a straightforward manner to any $\sigma$-bonded ligand X in any two adjacent oxidation states as shown in Scheme 6 and Eq. 32. We have recently [47] investigated the energetics of metal-halide bonding in $Cp^*Fe(dppe)X$ compounds (X = F, Cl, Br, I) in addition to the hydride that was discussed in Section 3.5. The sterically crowded and electron-rich $Cp^*Fe(dppe)X$ moiety supports metal complexes in a great number of oxidation states, with electron counts ranging from 15 to 19 [48]. Cyclic voltammetry in THF/ 0.2 M $Bu_4N^+PF_6^-$ shows that the couples $Cp^*Fe(dppe)^{0/+}$ and $Cp^*Fe(dppe)^{+/2+}$ are chemically reversible on the measurement time scale, as are $Cp^*Fe(dppe)X^{0/+}$ and $Cp^*Fe(dppe)X^{+/2+}$ for X = F, Cl, Br, and I. The electrode-potential data and the derived $\Delta$BDE data are listed in Table 5.

In the electrode-potential data for the halides, it is noteworthy that *the first oxidation is most facile for the most electronegative halide* and becomes increasingly more difficult in the series F < Cl < Br < I, with a particularly large jump (greater than 0.2 V) from F to Cl. The fluoride is also easiest to oxidize to a di-cation. The

**Table 5.** Electrode potential and ΔBDE data for Cp*Fe(dppe) derivatives.

| Compound MX | $E^{\circ}(MX/MX^+)^a$ | $E^{\circ}(MX^+/MX^{2+})^a$ | ΔBDE(MX/MX$^+$) (kJ mol$^{-1}$) | ΔBDE(MX$^+$/MX$^{2+}$) (kJ mol$^{-1}$) |
|---|---|---|---|---|
| Cp*Fe(dppe)˙ | −1.27 | −0.29 | – | – |
| Cp*Fe(dppe)H | −0.75 | 0.75$^b$ | 51 | 100 |
| Cp*Fe(dppe)F | −0.82 | 0.69 | 43 | 94 |
| Cp*Fe(dppe)Cl | −0.62 | 0.82 | 63 | 107 |
| Cp*Fe(dppe)Br | −0.58 | 0.81 | 67 | 106 |
| Cp*Fe(dppe)I | −0.54 | 0.78 | 71 | 103 |

$^a$ V relative to Fc in THF/0.2 M Bu$_4$N$^+$PF$_6^-$. Reversible voltammograms unless otherwise noted.
$^b$ Peak potential for irreversible oxidation.

trend is the opposite of that predicted on the basis of halide electronegativities alone. The electrode-potential data translate to an oxidatively induced bond weakening for all X. The weakening is smallest for X = F and increases in the order F < Cl < Br < I. A further weakening occurs as a consequence of the second oxidation. The origins of these effects are not yet well understood. Recent studies have shown that the nature of the bonding between organotransition-metal centers and electronegative σ-bonded ligands such as halide, alkoxide, and amido groups is more complex than usually thought [49]. Such ligands form not only 'normal' covalent M–X bonds, but can also act as π donors towards the metal. A ligand $p_\pi$ to metal $d_\pi$ electron-pair donation might destabilize electronically saturated complexes via repulsive filled–filled interactions, whereas electronically unsaturated species might be stabilized by partial π bond formation. IR $\nu_{CO}$ spectroscopy data, electrode potentials, theoretical calculations, and other evidence cited in [47] suggest that among the halides the fluoride ligand is the most efficient electron-pair donor towards the metal. It must, however, be emphasized that the full explanation of this behavior might be more complex. An alternative interpretation based on Drago's

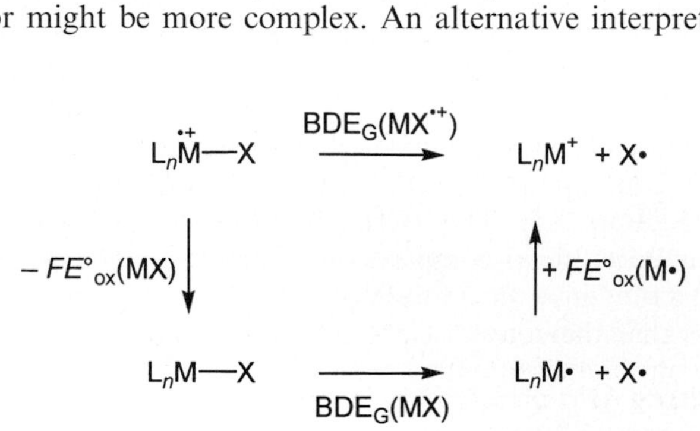

$$
\begin{array}{ccc}
L_n\overset{\bullet+}{\ddot{M}}{-}X & \xrightarrow{\;BDE_G(MX^{\bullet+})\;} & L_nM^+ + X\bullet \\[2mm]
{\scriptstyle -FE^{\circ}_{ox}(MX)}\Big\downarrow & & \Big\uparrow{\scriptstyle +FE^{\circ}_{ox}(M\bullet)} \\[2mm]
L_nM{-}X & \xrightarrow[\;BDE_G(MX)\;]{} & L_nM\bullet + X\bullet
\end{array}
$$

$$BDE(MX^{\bullet+}) - BDE(MX) = F\,[E^{\circ}_{ox}(M\bullet) - E^{\circ}_{ox}(MX)] \tag{33}$$

**Scheme 6.** Thermochemical cycle for determination of oxidatively induced M–X BDE changes in metal halides.

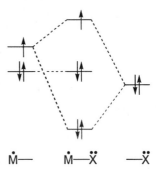

**Figure 6.** Simplified MO diagram to illustrate $p_\pi$–$d_\pi$ interaction in an M–X bond.

ECT model [50a–c] has also been proposed [50d–e] and theoretical studies have shown that covalent $\pi$-effects and $\sigma$-effects, and electrostatic effects and the ionicity of the M–X bond, might have a significant effect on trends in $v_{CO}$ and other observed parameters. The physical manifestations resemble those expected on the basis of $\pi$ donation, but the explanation may be much more complex; it might more appropriate to discuss the halide effects in terms of an *apparent* $\pi$ donor strength.

The presence of (apparent) halide-to-metal $\pi$ bonding might be reflected in BDE changes when the repulsive $p_\pi$–$d_\pi$ interactions in saturated complexes change to attractive interactions through the generation of coordinative unsaturation by dissociation of a ligand L or by other means. 17-electron cation radicals might exhibit behavior reflecting *partial* unsaturation, and 16-electron di-cations *full* unsaturation. The bond to the better $\pi$ donor should then be most strengthened, alternatively least weakened, when coordinative unsaturation is generated. This appears to be the case for the Cp*Fe(dppe)X series, with F being the best 'apparent $\pi$ donor'. The presence of apparent $\pi$ donation even in the 17-electron species is reminiscent of the 17/19-electron equilibria in organometallic radical chemistry: the halide, by virtue of a suitable $p_\pi$ orbital, acts as an 'intramolecular two-electron donor' that interacts with the metal SOMO in much the same sense as an incoming two-electron donor interacts with a 17-electron radical (Figure 6). Preliminary results from DFT calculations (J.-Y. Saillard and K. Costuas, personal communication) indicate that reorganization energies are relatively constant in the halogen series, so the jump in the $\Delta$BDE values from F to the other halogens appears to be associated with variations in the Fe–X bonds. Here, X-to-Fe $\pi$ donation is one of several contributing factors.

## 3.7 Bonding of 2-Electron Donor Ligands in 18- and 19-Electron Complexes

The association and dissociation of ligands at metal centers is a key reaction type in organometallic chemistry, including organometallic radical chemistry. Typically,

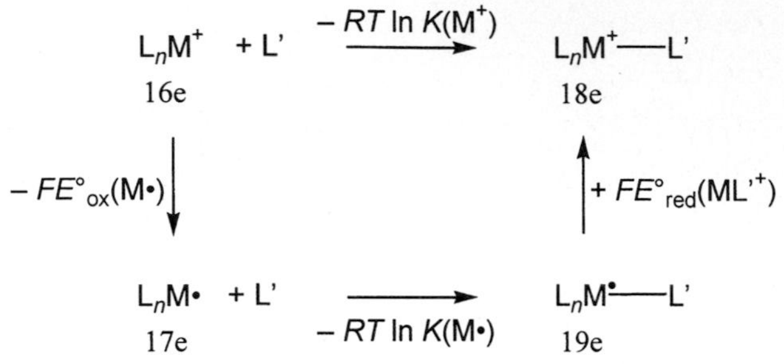

$$L_nM^+ \quad + L' \xrightarrow{\;-RT\ln K(M^+)\;} L_nM^+\!\!-\!\!L'$$

16e                                    18e

$-FE^\circ_{ox}(M\bullet) \downarrow \qquad\qquad\qquad\qquad \uparrow + FE^\circ_{red}(ML'^+)$

$$L_nM\bullet \;\; + L' \xrightarrow[\;-RT\ln K(M\bullet)\;]{} L_nM^{\bullet}\!\!-\!\!L'$$

17e                                    19e

$$\ln[K(M^+)/K(M\bullet)] \;=\; F/RT\,[E^\circ_{ox}(M\bullet) - E^\circ_{red}(ML'^+)] \tag{34}$$

**Scheme 7.** Thermochemical cycle for the determination of relative metal–ligand bond-dissociation energies in 18- and 19-electron complexes.

17-electron species undergo ligand substitution reactions many orders of magnitude faster than closely related 18-electron species. These rapid reactions are thought to occur by associative mechanisms that involve 19-electron intermediates [2c, e, h]. The 19th electron normally is considered to reside in a largely antibonding orbital, resulting in such species being rather labile with respect to ligand dissociation. Very little information is available concerning the relative M–L bond strengths in 18- and 19-electron species that only differ in the number of valence electrons and, consequently, charge. Scheme 7 shows a thermochemical cycle which may be useful in order to obtain such data. Eq. 34 requires knowledge about the reduction potential of the 18-electron species, taken to be a cationic complex $L_nM\!-\!L'^+$, and the oxidation potential for the 17-electron neutral species $L_nM^\bullet$. The use of this method is unfortunately severely limited by the short lifetimes of most 17-electron radicals that are of interest in this context. Additionally, both redox processes will normally be irreversible due to rapid follow-up reactions (ligand loss from 19-electron $L_nM\!-\!L'^\bullet$, ligand or solvent association at 16-electron $L_nM^+$) but at least some semi-quantitative information may be obtained.

As seen in Figure 4 and discussed in Section 3.5, $CpCr(CO)_3^\bullet$ undergoes oxidation at 0.42 V whereas $CpCr(CO)_3(NCMe)^+$ is reduced at $-0.92$ V relative to the Fc. This translates to $K(M^+)/K(M^\bullet) = 5 \times 10^{22}$ or $\Delta BDE = 130$ kJ mol$^{-1}$. For acetonitrile coordination at $CpCr(CO)_2[P(OMe)_3]^{0/+}$, the corresponding values are $-0.21$ and $-1.48$ V relative to the Fc, resulting in $\Delta BDE = 123$ kJ mol$^{-1}$. The *one-electron reductively induced* M–L bond activation for this heterolytic M–L cleavage in which the departing ligand retains both electrons is of the same magnitude as the *one-electron oxidatively induced* M–H bond activation for the heterolytic M–H cleavage (deprotonation) in which the metal retains both electrons (Section 3.4).

## 3.8 Changes in Ligand C–H Bond Strengths Induced by Electron Transfer

In Sections 3.5–3.7, one-electron changes in the oxidation state of a complex were shown to have profound effects on the energetics of metal–ligand bonding. Similarly, dramatic changes can occur for bonds within a coordinated ligand when a metal complex is subjected to electron-transfer processes. With the exception of a very recent contribution by Astruc and coworkers, which will be discussed first, very few studies have been performed in order to establish and to quantify such effects. Useful electrode-potential data that may be combined to yield such information are probably abundant, but scattered, in the literature. The discussion here will serve to show through selected examples how quantitative bond-energy data can be obtained, and is by no means intended to include all the pertinent data and reactions in the literature.

The $H^+$ and $H^·$ transfer reactions at the coordinated ligands that will be described in the following are frequently, but not always, accompanied by a metal–ligand hapticity change (for $\pi$-bonded ligands), a metal–ligand bond-order change (for $\sigma$-bonded ligands), or a C–C bond-order change (within a coordinated ligand). The thermodynamics of the reactions are often qualitatively understood when this is taken into account. The formation of new bonds, or more multiple character into existing ones, may significantly contribute to the observed changes in $pK_a$ or BDE data when the oxidation state of a complex is changed. The influence of the hapticity and bond-order changes are extreme manifestations of the importance of reorganization energies that were discussed in Section 3.2.

### 3.8.1 Benzylic C–H Bonds of $\pi$-Coordinated Arenes

Astruc and coworkers [40a] recently reported a comprehensive investigation of the bonding energetics of benzylic C–H bonds in 16–20-electron complexes $Fe(C_5R_5)(\eta^6\text{-arene})^n$ ($n = -1$ to $+3$; R = H, Me; arene = $C_6Me_6$, $C_6H_5CH_2Ph$, $C_6H_5CHPh_2$) where the entire analysis was based on variations on the thermochemical cycles in Schemes 2, 4, and 5. Acidity measurements were performed in DMSO. Electrode potential measurements were also conducted in DMSO, with a few exceptions. When electrode potentials were not accessible in DMSO, great care was taken to reliably convert data from other solvents. The most thorough investigation was done for compounds derived from the 18-electron species $CpFe^{II}(C_6Me_6)^+$ and $Cp^*Fe^{II}(C_6Me_6)^+$. The data that were obtained for these complexes are best summarized as in Scheme 8, where electrode potentials are referenced against Fc in DMSO. The relationship to Scheme 1 is obvious. Equilibrium $pK_a$ measurements on the 18-electron cation $CpFe^{II}(C_6Me_6)^+$ (top center) serves as the anchoring point for the other $pK_a$ and BDE data. The $pK_a$ of the $C_6Me_6$ ligand is ca 14 $pK_a$ units lower than for free $C_6Me_6$. Factors that may contribute to this bond activation are distribution of charge onto the ligand, and stabilization of the

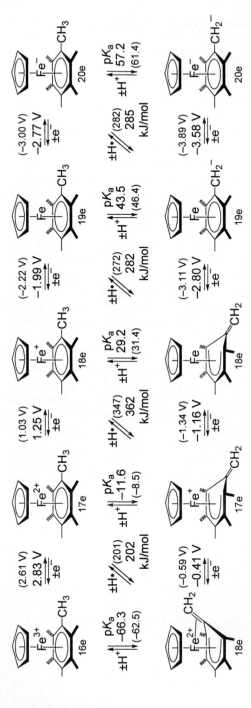

**Scheme 8.** Energetics of redox, proton transfer, and hydrogen atom transfer reactions of $CpFe(C_6Me_6)^n$ complexes in five different oxidation states. Numbers in parentheses are for the analogous $Cp^*Fe(C_6Me_6)^n$ complexes.

deprotonated ligand through the formation of a cyclohexadienyl ligand with an exocyclic C–C double bond. Reduction of $CpFe^{II}(C_6Me_6)^+$ gives the 19-electron 'electron reservoir' complex [1b, 2m, 51] $CpFe^{II}(C_6Me_6)^{\bullet}$ which at a $pK_a$ of 43.5 is ca 14 $pK_a$ units less acidic than the cation. In other words, the coordinated $C_6Me_6$ is approximately as poorly acidic as the free ligand—there is no charge stabilization upon complexation nor cyclohexadienyl formation in this case. For the $Cp^*Fe^{II}(C_6Me_6)^{\bullet}$ analog, the $pK_a$ was even somewhat *higher* than for the free ligand. The reduction of $CpFe^{I}(C_6Me_6)^{\bullet}$ is experimentally observable but the reduction of the conjugate base, 19-electron $CpFe^{I}(C_6Me_5CH_2^-)$, is not. Its reduction potential was approximated, based on the assumption that the added electron goes into an Fe-centered antibonding orbital of roughly known energy. The result is a further 14-unit $pK_a$ increase to 57.2 which was attributed to electrostatic effects. On the other side, oxidation of 18-electron $CpFe^{II}(C_6Me_6)^+$ to the 17-electron $Fe^{III}$ di-cation leads to an acidity enhancement of about 40 units to a $pK_a$ of $-11.6$. Interestingly, the $SbCl_6^-$ salt of the analogous $Cp^*$ complex, $Cp^*Fe^{III}(C_6Me_6)^{2+}$, with an estimated $pK_a$ in DMSO of $-8.5$, can be isolated (!) and is the most strongly oxidizing thermally stable organometallic complex known. Estimates for the experimentally inaccessible electrode potentials for the further oxidation of $CpFe^{III}(C_6Me_6)^{2+}$ to the $Fe^{IV}$ tri-cation, and for the oxidation of $CpFe^{III}(C_6Me_5CH_2)^+$ to the $Fe^{IV}$ di-cation, were obtained through certain not unreasonable assumptions, and further decreased the $pK_a$ of the $C_6Me_6$ ligand to $-66.3$.

The benzylic C–H BDEs of the coordinated arenes were also highly dependent on the oxidation state of the complex. From Eq. 15, the benzylic C–H BDE of $CpFe^{II}(C_6Me_6)^+$ was determined as 362 kJ mol$^{-1}$, essentially unchanged from that in the free ligand. The BDE dropped to 292 kJ mol$^{-1}$ for the neutral 19-electron $CpFe^{I}(C_6Me_6)^{\bullet}$, readily understood considering that an electron in the 19-electron complex resides in an antibonding orbital and that the H$^{\bullet}$ abstraction product is a stable, closed-shell, 18-electron species. No further BDE change occurred for the 20-electron $Fe^0$ arene complex. However, the BDE of 202 kJ mol$^{-1}$ in the 17-electron di-cation $CpFe^{III}(C_6Me_6)^{2+}$ makes the C–H bond 160 kJ mol$^{-1}$ weaker than that of its 18-electron parent. The H$^{\bullet}$ transfer generates a 16-electron cyclohexadienyl complex $CpFe^{IV}(C_6Me_5=CH_2)^{2+}$, and it is thought that coordination of the exocyclic double bond to give an 18-electron product contributes to the lowering of the BDE in this case.

### 3.8.2 α- and β-C–H Bonds in Metal Alkyls

The trityl cation $(Ph_3C^+)$ is a commonly used reagent to effect the abstraction of a hydride from coordinated ligands. For example, the preparation of metal alkylidene and alkene complexes by α- and β-hydride abstraction from metal alkyls using trityl salts is a well-established synthetic method, Eqs 35 and 36.

α abstraction

$$L_nM\overset{+}{=}CHCH_2R \qquad (35)$$

$$L_nM\!-\!CH_2CH_2R \quad \overset{+\,Ph_3C^+}{\underset{-\,Ph_3CH}{}}$$

$$L_n\overset{+}{M}\!-\!\overset{CH_2}{\underset{CHR}{\|}} \qquad (36)$$

β abstraction

It has been proposed that hydride abstraction reactions with trityl reagents may in many cases be two-step reactions that occur via a one-electron oxidation followed by a hydrogen atom transfer from the incipient organometallic cation radical and trityl radical, as shown for a simple α abstraction from a methyl group in Eq. 37 [52]. It is difficult to distinguish between the two-step reaction and a direct, one-step hydride abstraction mechanism. If the second step in the two-step mechanism is rate limiting, then the two mechanisms may in principle pass through identical transition states for the rate-limiting step and will be kinetically indistinguishable. Trapping and/or observation of the trityl radical may be a good indication for an electron-transfer pathway, but it should still be demonstrated that the electron transfer reaction is not just a non-productive side equilibrium.

$$\begin{array}{ccc} L_nM\!-\!CH_3 & L_n\overset{\bullet+}{M}\!-\!CH_3 & L_n\overset{+}{M}\!=\!CH_2 \\ + Ph_3C^+ & + Ph_3C^\bullet & + Ph_3CH \end{array} \qquad (37)$$

Bodner et al. [52a] presented strong evidence that α and β hydride abstractions from rhenium alkyls CpRe(NO)(PPh$_3$)R follow the two-step mechanism. Neutral Re alkyls that did not undergo reactions with Ph$_3$C$^\bullet$ were found to quickly react with Ph$_3$C$^+$. The thermochemical cycle in Scheme 9 may be employed to estimate how a one-electron oxidation affects the C–H BDE of a metal alkyl to produce an alkylidene complex. Eq. 38 quantifies the BDE change; the assumption is again made that entropy contributions cancel so that the relationship provides enthalpy-based data. For the benzyl complex CpRe(NO)(PPh$_3$)CH$_2$Ph, $E°_{ox}(MCH_2Ph) = 0.17$ V relative to Ag/Ag$^+$, and for CpRe(NO)(PPh$_3$)(=CHPh)$^+$, $E°_{red}(MCHPh^+) = -1.30$ V. Both electrode processes were reversible [52a], and translates to an oxidatively induced bond weakening $\Delta BDE = 142$ kJ mol$^{-1}$. For CpRe(NO)(PPh$_3$)CH$_3$, $E°_{ox}(MCH_3) = 0.04$ V (reversible) relative to Ag/Ag$^+$ and $E_{red}(MCH_2^+) = -1.49$ V relative to Ag/Ag$^+$ (irreversible) [53]. Neglecting kinetic potential shifts on the latter this results in a bond weakening for the methyl C–H bond of 150 kJ mol$^{-1}$ as a result of the oxidation. These data establish beyond doubt that CpRe(NO)(PPh$_3$)CH$_2$R$^{\bullet+}$ are much more prone to transfer H$^\bullet$ to the trityl radical than are their neutral parents CpRe(NO)(PPh$_3$)CH$_2$R. The stabilization gained by formation of a Re=C double bond following H$^\bullet$ abstraction from the cation radical must significantly contribute to the bond weakening effect. The Re–CH$_2^\bullet$ radical would give a 19-

56

$$L_nM{-}CH_2R \xrightarrow{\text{BDE(MCH}_2\text{R)}} L_nM{-}CHR\bullet + H\bullet$$

$+ FE°_{ox}(MCH_2R) \Big\downarrow \qquad\qquad\qquad \Big\uparrow - FE°_{red}(MCHR^+)$

$$L_n\overset{\bullet+}{M}{-}CH_2R \xrightarrow{\text{BDE(MCH}_2\text{R}^{\bullet+}\text{)}} L_n\overset{+}{M}{=}CHR + H\bullet$$

$$\Delta BDE = BDE(MCH_2R) - BDE(MCH_2R^{\bullet+}) = F[E°_{ox}(MCH_2R) - E°_{red}(MCHR^+)] \quad (38)$$

**Scheme 9.** Thermochemical cycle for determination of $\alpha$-C–H BDE changes in coordinated alkyl groups caused by a one-electron oxidation.

electron metal center if a double bond were present, and therefore lacks this stabilization.

Similar considerations may be applied to $\beta$-hydride transfer processes. For example, $E°_{ox}(ReCH_2CHMe_2) = +0.06$ V relative to Ag/Ag$^+$ (reversible) [52a] and $E_{red}(Re(CH_2{=}CMe_2)^+) = -1.34$ V (irreversible) which translates to a 135 kJ mol$^{-1}$ bond weakening of the $\beta$ C–H bond upon oxidation. In this case, H$^\bullet$ loss from the Re-alkyl cation radical generates an alkene complex in which a new Re–C bond has been formed and this contributes substantially to the thermodynamic weakening of the C–H bond.

### 3.8.3 Bridging Alkylidyne Ligands in Dinuclear Complexes

Substantial effects of one-electron redox processes on BDEs of C–H bonds in coordinated ligands can also be seen in dinuclear complexes and probably in higher clusters. For example, the reduction of the $\mu_2$-ethylidyne complexes $Cp_2M_2(CO)_3(\mu\text{-}CCH_3)^+$ (M = Fe, Ru) led to disproportionation of the ethylidyne groups to yield the vinylidene and ethylidene complexes $Cp_2M_2(CO)_3(\mu\text{-}C{=}CH_2)$ and $Cp_2M_2(CO)_3(\mu\text{-}CHCH_3)$ [54]. The reactions were proposed to take place via hydrogen atom transfer between two $Cp_2M_2(CO)_3(\mu\text{-}CCH_3)^\bullet$ radicals in a fashion reminiscent of disproportionation of caged organic radicals. In this case, the thermochemical cycle in Scheme 10 can be used to demonstrate that the $\beta$-C–H bond in the ethylidyne cations becomes substantially weakened following a one-electron reduction. Only irreversible electrode-potential data are available for the pertinent processes, but the semi-quantitative result is obtained that the reduction leads to an ethylidyne C–H BDE decrease of ca 140 kJ mol$^{-1}$ for Fe and ca 175 kJ mol$^{-1}$ for Ru. In this case, the bond weakening is facilitated by the formation of a strong C=C bond and two closed-shell metal centers when C–H scission takes place from the radical; such is not the case when the reaction occurs from the cation.

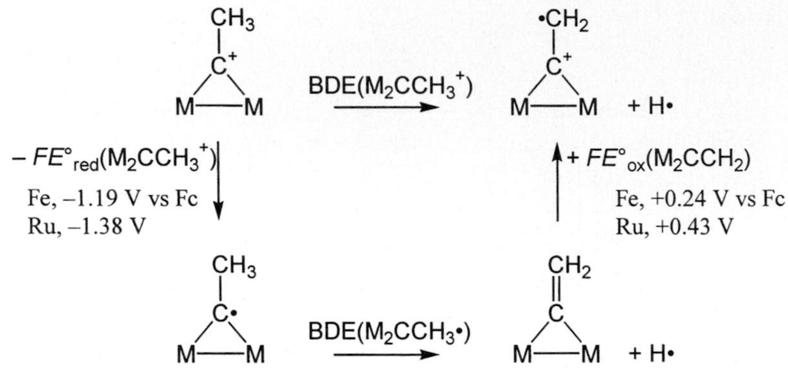

$$\Delta BDE = BDE(M_2CCH_3^+) - BDE(M_2CCH_3\bullet) =$$
$$F\,[E^{\circ}_{ox}(M_2CCH_2) - E^{\circ}_{red}(M_2CCH_3^+)]$$

(39)

**Scheme 10.** Thermochemical cycle for determination of reductively induced C–H bond activation in dinuclear ethylidyne complexes.

## 3.9 Concluding Remarks

Thermochemical cycles that incorporate electrode potentials provide a means for obtaining reasonable estimates of bond energies and other thermodynamic quantities that are difficult or impossible to obtain directly. Absolute bond energies can be determined when suitable anchoring points are available. Otherwise, important and interesting information can be obtained from cycles that provide relative bond-energy data. The selected examples have focused mostly on M–H and C–H bonding, but the methodology can be applied to any X–Y bond where the suitable redox partners are available for electrochemical studies.

The effects of one-electron redox processes on various bond strengths is often substantial and can often be qualitatively understood by appropriate consideration of pertinent Lewis structures for each side of the redox couples. Significant $pK_a$ or BDE changes can be ascribed to changes in metal–ligand hapticities, metal–ligand bond orders, or C–C bond orders.

It is not uncommon that one-electron redox processes cause bond energy changes in the range 100–150 kJ mol$^{-1}$. Looking at these numbers as relative activation energies for chemical reactions, one obtains relative rates of ca $10^{17}$:1–$10^{26}$:1 for the pertinent bond-cleavage reactions at ambient temperature! From this, it becomes clear that electron transfer pathways can open new reaction pathways, and that in many cases initially uphill electron transfer equilibria can be productive due to extremely fast follow-up reactions. This is particularly important in the context of understanding and exploiting efficient electron transfer catalyzed chain reactions [1b, 2i].

# Abbreviations

| | |
|---|---|
| arphos | $Ph_2PCH_2CH_2AsPh_2$ |
| BDE | Bond-dissociation energy (in enthalpy terms) |
| $BDE_G$ | Bond-dissociation energy (in free energy terms) |
| COD | 1,5-Cyclooctadiene |
| Cp | $\eta^5$-Cyclopentadienyl, $C_5H_5$ |
| Cp* | $\eta^5$-Pentamethylcyclopentadienyl, $C_5Me_5$ |
| Cp$'$ | Any substituted Cp |
| $Cp^{TMS}$ | $\eta^5$-Trimethylsilylcyclopentadienyl, $C_5H_4SiMe_3$ |
| Cy | Cyclohexyl |
| DCE | 1,2-Dichloroethane |
| dape | $(p\text{-}MeOC_6H_4)_2PCH_2CH_2P(p\text{-}MeOC_6H_4)_2$ |
| depe | $Et_2PCH_2CH_2PEt_2$ |
| dppe | $Ph_2PCH_2CH_2PPh_2$ |
| dppf | $(C_5H_4PPh_2)_2Fe$, 1,1$'$-bis(diphenylphosphino)ferrocene |
| dppm | $Ph_2PCH_2PPh_2$ |
| dppp | $Ph_2PCH_2CH_2CH_2PPh_2$ |
| dtfpe | $(p\text{-}CF_3C_6H_4)_2PCH_2CH_2P(p\text{-}CF_3C_6H_4)_2$ |
| EA | Electron affinity |
| Fc | Ferrocene, $Cp_2Fe$ |
| IP | Ionization potential |
| Tp | Hydridotris(pyrazolyl)borate |
| Tp$'$ | Hydridotris(3,5-dimethylpyrazolyl)borate |
| tripod | $MeC(CH_2PPh_2)_3$ |
| triphos | $PhP(CH_2CH_2PPh_2)_2$ |

# References

1. (a) J. K. Kochi, *Organometallic Mechanisms and Catalysis*, Academic Press, New York, **1978**. (b) D. Astruc, *Electron Transfer and Radical Processes in Transition-Metal Chemistry*, VCH, Weinheim, **1995**.
2. (a) N. G. Connelly, W. E. Geiger, *Chem. Rev.* **1996**, *96*, 877–910. (b) D. Astruc, *Acc. Chem. Res.* **1991**, *24*, 36–42. (c) D. R. Tyler, *Acc. Chem. Res.* **1991**, *24*, 325–331. (d) *Organometallic Radical Processes* (Ed: W. C. Trogler), Elsevier, Amsterdam, **1990**. (e) D. R. Tyler, F. Mao, *Coord. Chem. Rev.* **1990**, *97*, 119–140. (f) N. G. Connelly, *Chem. Soc. Rev.* **1989**, *18*, 153–185. (g) *Paramagnetic Organometallic Species in Activation/-Selectivity, Catalysis* (Eds.: M. Chanon, M. Julliard, J. C. Poite), Kluwer Academic, Dordrecht, **1989**. (h) D. R. Tyler, *Progr. Inorg. Chem.* **1988**, *36*, 125–194. (i) D. Astruc, *Angew. Chem., Int. Ed. Engl.* **1988**, *27*, 643–660. (j) D. Astruc, *Chem. Rev.* **1988**, *88*, 1189–1216. (k) M. C. Baird, *Chem. Rev.* **1988**, *88*, 1217–1227. (l) J. C. Kotz in *Topics in Electrochemistry* (Eds.: A. J. Fry, W. E. Britton), Plenum, New York, **1986**, Chapter 3. (m) D. Astruc, *Acc. Chem. Res.* **1986**, 19, 377–383. (n) W. E. Geiger, *Progr. Inorg. Chem.* **1985**, *33*, 275–351. (o) N. G. Connelly, W. E. Geiger, *Adv. Organomet. Chem.* **1984**, *23*, 1–93.
3. (a) J. A. M. Simões, J. L. Beauchamp, *Chem. Rev.* **1990**, *90*, 629–688. (b) *Energetics of Organometallic Species* (Ed.: J. A. M. Simões), Kluwer Academic, Dordrecht, **1992**. (c) *Bonding Energetics in Organometallic Compounds*, ACS Symposium Series No. 428 (Ed.: T. J. Marks), American Chemical Society, Washington, DC, **1990**. (d) J. Halpern, *Inorg. Chim. Acta* **1985**, *100*, 41–48. (e) J. A. Connor, *Top. Curr. Chem.* **1977**, *71*, 71–110.

4. (a) D. T. Sawyer, A. Sobkowiak, J. L. Roberts, Jr., *Electrochemistry for Chemists*, 2nd ed., Wiley, New York, **1995**. (b) A. J. Bard, L. R. Faulkner, *Electrochemical Methods: Fundamentals and Applications*, 2nd ed., Wiley, New York, **1998**. (c) D. Pletcher, *A First Course in Electrode Processes*, The Electrochemical Consultancy, Romsey, UK, **1991**. (d) D. K. Gosser, Jr., *Cyclic Voltammetry. Simulation and Analysis of Reaction Mechanisms*, Verlag Chemie, Weinheim, **1993**.

5. (a) S. G. Lias, J. E. Bartmess, J. F. Liebman, J. L. Holmes, R. D. Levin, W. G. Mallard, *J. Phys. Chem. Ref. Data* **1988**, *17*, Suppl. 1. (b) *NIST Chemistry WebBook, NIST Standard Reference Database Number 69* (Eds.: W. G. Mallard, P. J. Linstrom), National Institute of Standards and Technology, Gaithersburg, MD 20899, **1998** (http://webbook.nist.gov/chemistry/).

6. (a) A. S. Carson, in ref. [3b], p. 131–158. (b) J. A. M. Simões, in ref. [3b], p. 197–232.

7. (a) P. Hamon, L. Toupet, J.-R. Hamon, C. Lapinte, *Organometallics* **1992**, *11*, 1429–1431. (b) P. Hamon, L. Toupet, J.-R. Hamon, C. Lapinte, *Organometallics* **1996**, *15*, 10–12.

8. D. D. M. Wayner, V. D. Parker, *Acc. Chem. Res.* **1993**, *26*, 287–294.

9. (a) R. Breslow, W. Chu, *J. Am. Chem. Soc.* **1973**, *95*, 411–418. (b) R. Breslow, J. Grant, *J. Am. Chem. Soc.* **1977**, *99*, 7745–7746. (c) B. Jaun, J. Schwarz, R. Breslow, *J. Am. Chem. Soc.* **1980**, *102*, 5741–5748.

10. (a) A. M. de Nicholas, D. R. Arnold, *Can. J. Chem.* **1982**, *60*, 2165–2179. (b) F. G. Bordwell, J.-P. Cheng, J. A. Harrelson, *J. Am. Chem. Soc.* **1988**, *110*, 1229–1231. (c) F. G. Bordwell, J.-P. Cheng, *J. Am. Chem. Soc.* **1989**, *111*, 1792–1795.

11. (a) E. M. Arnett, R. A. Flowers, R. T. Ludwig, A. Meckhof, S. Walek, *Pure Appl. Chem.* **1995**, *67*, 729–734. (b) F. G. Bordwell, A. V. Satish, S. Zhang, X.-M. Zhang, *Pure Appl. Chem.* **1995**, *67*, 735–740. (c) E. M. Arnett, R. A. Flowers, *Chem. Soc. Rev.* **1993**, *22*, 9–15. (d) F. G. Bordwell, X.-M. Zhang, *Acc. Chem. Res.* **1993**, *26*, 510–517.

12. S. S. Kristjánsdóttir, J. R. Norton in *Transition Metal Hydrides* (Ed.: A. Dedieu), VCH, Weinheim, **1992**, p. 309–359.

13. (a) M. Tilset, V. D. Parker, *J. Am. Chem. Soc.* **1989**, *111*, 6711–6717; **1990**, *112*, 2843 (corrigendum). (b) V. D. Parker, K. L. Handoo, F. Roness, M. Tilset, *J. Am. Chem. Soc.* **1991**, *113*, 7493–7498.

14. (a) R. F. Jordan, J. R. Norton, *J. Am. Chem. Soc.* **1982**, *104*, 1255–1263. (b) E. J. Moore, J. M. Sullivan, J. R. Norton, *J. Am. Chem. Soc.* **1986**, *108*, 2257–2263. (c) S. S. Kristjánsdóttir, A. E. Moody, R. T. Weberg, J. R. Norton, *Organometallics* **1988**, *7*, 1983–1987.

15. *Solubility Data Series: Hydrogen and Deuterium, Vol. 5/6* (Ed.: C. L. Young), Pergamon, Oxford, **1981**.

16. E. Brunner, *J. Chem. Eng. Data* **1985**, *30*, 269–273.

17. *CRC Handbook of Chemistry and Physics* (Ed.: R. C. Weast), CRC Press, Boca Raton, 1987.

18. G. Gritzner, J. Kuta, *Pure Appl. Chem.* **1984**, *56*, 461–466.

19. I. M. Kolthoff, M. K. Chantooni Jr., *J. Phys. Chem.* **1972**, *76*, 2024–2034.

20. D. D. M. Wayner, D. J. McPhee, D. Griller, *J. Am. Chem. Soc.* **1988**, *110*, 132–137.

21. G. Kiss, K. Zhang, S. L. Mukerjee, C. D. Hoff, *J. Am. Chem. Soc.* **1990**, *112*, 5657–5658.

22. (a) M. Tilset, *J. Am. Chem. Soc.* **1992**, *114*, 2740–2741. (b) V. Skagestad, M. Tilset, *J. Am. Chem. Soc.* **1993**, *115*, 5077–5083. (c) A. Pedersen, V. Skagestad, M. Tilset, *Acta Chem. Scand.* **1995**, *49*, 632–635.

23. Ref. [4b], Chapter 11.

24. L. Luo, C. Li, M. E. Cucullu, S. P. Nolan, *Organometallics* **1995**, *14*, 1333–1338.

25. J. Ruiz, D. Astruc, *C. R. Acad. Sci., Ser. IIc: Chim.* **1998**, *1*, 21–28.

26. R. J. Angelici, *Acc. Chem. Res.* **1995**, *28*, 51–60.

27. D. Wang, R. J. Angelici, *J. Am. Chem. Soc.* **1996**, *118*, 935–942.

28. P. G. Jessop, R. H. Morris, *Coord. Chem. Rev.* **1992**, *121*, 155–284.

29. (a) G. Jia, R. H. Morris, *J. Am. Chem. Soc.* **1991**, *113*, 875–883. (b) R. Morris, *Inorg. Chem.* **1992**, *31*, 1471–1478. (c) G. Jia, A. J. Lough, R. H. Morris, *Organometallics* **1992**, *11*, 161–171. (d) E. P. Cappellani, S. D. Drouin, G. Jia, P. A. Maltby, R. H. Morris, C. T. Schweitzer, *J. Am. Chem. Soc.* **1994**, *116*, 3375–3388. (e) B. Chin, A. J. Lough, R. H. Morris, C. T. Schweitzer, C. D'Agostino, *Inorg. Chem.* **1994**, *33*, 6278–6288. (f) M. Schlaf, A. J. Lough, P. A. Maltby, R. H. Morris, *Organometallics* **1996**, *15*, 2270–2278.

30. K.-T. Smith, C. Rømming, M. Tilset, *J. Am. Chem. Soc.* **1993**, *115*, 8681–8689.

31. W. S. Ng, G. Jia, M. Y. Hung, C. P. Lau, K. Y. Wong, L. Wen, *Organometallics* **1998**, *17*, 4556–4561.
32. (a) D. M. Heinekey, W. J. Oldham, Jr., *Chem. Rev.* **1993**, *93*, 913–926. (b) Crabtree, R. H. *Angew. Chem., Int. Ed. Engl.* **1993**, *32*, 789–805.
33. J. D. Protasiewicz, K. H. Theopold, *J. Am. Chem. Soc.* **1993**, *115*, 5559–5569.
34. S. Trofimenko, *Scorpionates*, Imperial College Press, London, 1999.
35. (a) M. D. Curtis, K.-B. Shiu, *Inorg. Chem.* **1985**, *24*, 1213–1218. (b) M. D. Curtis, K.-B. Shiu, W. M. Butler, J. C. Huffman, *J. Am. Chem. Soc.* **1986**, *108*, 3335–3343. (c) Y. Alvarado, O. Boutry, E. Gutiérrez, A. Monge, M. C. Nicasio, M. L. Poveda, P. J. Pérez, C. Ruiz, C. Bianchini, E. Carmona, *Chem. Eur. J.* **1997**, *3*, 860–873.
36. M. E. Kerr, X.-M. Zhang, J. W. Bruno, *Organometallics* **1997**, *16*, 3249–3251.
37. F. G. Bordwell, X.-M. Zhang, R. Filler, *J. Org. Chem.* **1993**, *58*, 6067–6071.
38. D. F. McMillen, D. M. Golden, *Annu. Rev. Phys. Chem.* **1982**, *33*, 493–532.
39. S. Zhang, F. G. Bordwell, *Organometallics* **1994**, *13*, 2920–2921.
40. (a) H. A. Trujillo, C. M. Casado, J. Ruiz, D. Astruc, *J. Am. Chem. Soc.* **1999**, *121*, 5674–5686. (b) H. A. Trujillo, C. M. Casado, D. Astruc, *J. Chem. Soc., Chem. Commun.* **1995**, 7–8.
41. J. R. Pugh, T. J. Meyer, *J. Am. Chem. Soc.* **1992**, *114*, 3784–3792.
42. O. B. Ryan, M. Tilset, V. D. Parker, *J. Am. Chem. Soc.* **1990**, *112*, 2618–2626.
43. F. Marken, A. M. Bond, R. Colton, *Inorg. Chem.* **1995**, *34*, 1705–1710.
44. (a) O. B. Ryan, M. Tilset, V. D. Parker, *Organometallics* **1991**, *10*, 298–304. (b) O. B. Ryan, M. Tilset, *J. Am. Chem. Soc.* **1991**, *113*, 9554–9561. (c) K.-T. Smith, M. Tilset, *J. Organomet. Chem.* **1992**, *431*, 55–64. (d) M. Tilset, A. Zlota, K. G. Caulton, *Inorg. Chem.* **1993**, *32*, 3816–3821. (e) K.-T. Smith, M. Tilset, R. Kuhlman, K. G. Caulton, *J. Am. Chem. Soc.* **1995**, *117*, 9473–9480. (f) K.-T. Smith, M. Tilset, S. S. Kristjánsdóttir, J. R. Norton, *Inorg. Chem.* **1995**, *34*, 6497–6504. (g) J. C. Fettinger, H.-B. Kraatz, R. Poli, E. A. Quadrelli, R. C. Torralba, *Organometallics* **1998**, *17*, 5767–5775.
45. (a) M. Tilset, *Inorg. Chem.* **1994**, *33*, 3121–3126. (b) C. G. Zoski, D. A. Sweigart, N. J. Stone, P. H. Rieger, E. Mocellin, T. F. Mann, D. R. Mann, D. K. Gosser, M. M. Doeff, A. M. Bond, *J. Am. Chem. Soc.* **1998**, *110*, 2109–2116. (c) Y. Zhang, D. K. Gosser, P. H. Rieger, D. A. Sweigart, *J. Am. Chem. Soc.* **1991**, *113*, 4062–4068.
46. B. Pleune, D. Morales, R. Meunier-Prest, P. Richard, E. Collange, J. C. Fettinger, R. Poli, *J. Am. Chem. Soc.* **1999**, *121*, 2209–2225.
47. M. Tilset, J.-R. Hamon, P. Hamon, *Chem. Commun.* **1998**, 765–766.
48. (a) C. Roger, P. Hamon, L. Toupet, H. Rabaâ, J.-Y. Saillard, J.-R. Hamon, C. Lapinte, *Organometallics* **1991**, *10*, 1045–1054. (b) P. Hamon, L. Toupet, J.-R. Hamon, C. Lapinte, *Organometallics* **1992**, *11*, 1429–1431. (c) P. Hamon, J.-R. Hamon, C. Lapinte, *J. Chem. Soc., Chem. Commun.* **1992**, 1602–1603. (d) P. Hamon, L. Toupet, J.-R. Hamon, C. Lapinte, *Organometallics* **1996**, *15*, 10–12.
49. (a) K. G. Caulton, *New J. Chem.* **1994**, *18*, 25–41. (b) N. M. Doherty, N. W. Hoffman, *Chem. Rev.* **1991**, *91*, 553–573.
50. (a) R. S. Drago, N. M. Wong, D. C. Ferris, *J. Am. Chem. Soc.* **1992**, *114*, 91–98. (b) R. S. Drago, *Inorg. Chem.* **1990**, *29*, 1379–1382. (c) R. S. Drago, *Inorg. Chem.* **1995**, *34*, 3543–3548. (d) P. L. Holland, R. A. Andersen, R. G. Bergman, J. Huang, S. P. Nolan, *J. Am. Chem. Soc.* **1997**, *119*, 12800–12814. (e) P. L. Holland, R. A. Andersen, R. G. Bergman, *Comments Inorg. Chem.* **1999**, *21*, 115–129.
51. D. Astruc, J.-R. Hamon, G. Althoff, E. Román, P. Batail, P. Michaud, J.-P. Mariot, F. Varret, D. Cozak, *J. Am. Chem. Soc.* **1979**, *101*, 5445–5447.
52. (a) G. S. Bodner, J. A. Gladysz, M. F. Nielsen, V. D. Parker, *J. Am. Chem. Soc.* **1987**, *109*, 1757–1764. (b) D. Mandon, D. Astruc, *Organometallics* **1989**, *8*, 2372–2377. (c) V. Guerchais, C. Lapinte, *J. Chem. Soc., Chem. Commun.* **1986**, 663–664. (d) D. Mandon, L. Toupet, D. Astruc, *J. Am. Chem. Soc.* **1986**, *108*, 1320–1322. (e) J. C. Hayes, N. J. Cooper, *J. Am. Chem. Soc.* **1982**, *104*, 5570–5572.
53. M. Tilset, G. S. Bodner, D. R. Senn, J. A. Gladysz, V. D. Parker, *J. Am. Chem. Soc.* **1987**, *109*, 7551–7553.
54. T. Aase, M. Tilset, V. D. Parker, *Organometallics* **1989**, *8*, 1558–1563.

# 4 Electron-transfer Reactions of Electron-reservoir Complexes and other Monoelectronic Redox Reagents in Transition-metal Chemistry

*Didier Astruc*

## 4.1 Introduction

Redox reactions are a characteristic of transition-metal complexes because transition metals can readily vary their number of valence electrons. Indeed, the large majority of redox reagents is based on transition metals [1, 2]. This feature is also encountered in biochemistry with ferredoxines and cytochromes [3–5], in catalysis which involves the well-known oxidative addition and reductive elimination reactions [6, 7], in photochemistry where the remarkably long life-times of the excited states of transition-metal complexes enable exergonic electron-transfer reactions from these excited states [8, 9], in various parts of supramolecular chemistry [10], especially molecular electronics [11], where devices almost systematically involve transition metal components, and in solid-state chemistry and physics (non-linear optics, conductors, and superconductors) [12]. This chapter concerns monoelectronic redox reagents with emphasis on organometallic electron-reservoir complexes which are so useful in this context both as oxidants and reductants. Although most redox reagents are inorganic or organometallic, a minority of organic reagents are sometimes used and will be included in the review. The subject of redox chemistry has already been extensively treated in the literature from various standpoints. We intend to give here only a subjective and personal view of this immense area. In organic chemistry, asymmetric catalysis of oxidation [13, 14] and reduction [15, 16] reactions is a broad, popular field which we are not treating here. Electrochemistry [17] and photochemistry [8, 9] are also sub-disciplines of chemistry systematically involving redox reactions which are not dealt with in this chapter, except for thermodynamic redox potentials and redox catalysis. Likewise, organometallic oxidative additions and reductive elimination [6, 7] will not be discussed except when induced or catalyzed by monoelectronic reagents. We will not include biochemical redox reactions [3–5] either despite their importance.

## 4.2 Standard Redox Potentials and Complementary References

We will mostly deal with non-aqueous solutions. The IUPAC has recommended the reporting of standard redox potential $E^{o'}$ relative to the ferrocene/ferrocenium $(FeCp_2^{0/+})$ redox couple rather than the aqueous saturated calomel electrode [18] which is only modestly stable in non-aqueous solutions [19, 20]. In early studies, redox potentials in non-aqueous solutions were reported relative to the SCE, but recent studies systematically report redox potentials relative to $FeCp_2^{0/+}$. Conversion of values reported relative to the SCE to values reported relative to $FeCp_2^{0/+}$ have been made [21]. Unfortunately, despite the IUPAC recommendation, the absolute redox potential value of the couple $FeCp_2^{0/+}$ is somewhat dependent on the nature of the solvent and the supporting electrolyte, because the iron center which is the subject of redox change is not too well protected by the ligand in ferrocenium [22–24]. Indeed, the nucleophiles can interact from the sides with the cationic iron center. The energy of this interaction shifts the redox potential and depends on the nature of the nucleophiles. Therefore, we have selected the permethylated transition-metal sandwich redox systems decamethylferrocene/decamethylferrocenium $[FeCp*_2]^{0/+}$, decamethylcobaltocene/decamethylcobaltocenium $[CoCp*_2]^{0/+}$ and $[FeCp(\eta^6\text{-}C_6Me_6)]^{0/+}$ as alternative and complementary reference systems. In these electron-reservoir complexes the transition-metal center is really shielded from external nucleophiles (solvent, counter-anion of the electrolyte) so that their nature does not interfere with the redox potential value.

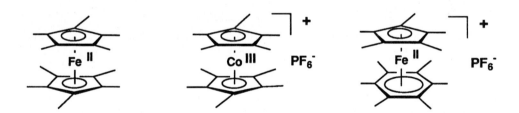

The redox potential values of these redox couples are reported in Table 1 relative to the SCE in different solvents. The independence of the redox couples of the permethylated redox couples was be verified by comparing the values of the redox potentials recorded relative to $[FeCp*_2]^{0/+}$, taken as the reference, which are reported in Table 2 for a variety of solvents. Only the values of the $[FeCp_2]^{0/+}$ couple varies, not those of the permethylated redox couples. This result is consistent with the fact that the redox potentials of the permethylated redox couples are independent of the nature of the solvent and supporting electrolyte. That the $[FeCp_2]^{0/+}$ reference is not quite sufficiently accurate does not mean that the many results that have been reported in the recent literature relative to $[FeCp_2]^{0/+}$ have to be re-measured relative to the more accurate $[FeCp*_2]^{0/+}$ reference. Indeed, it is possible to take advantage of all the results obtained relative to $[FeCp_2]^{0/+}$ using the conversion of

**Table 1.** $E_{1/2}$ values determined for the transition-metal sandwich complexes (V relative to the SCE).

| Complex | $E_{1/2}$ in the solvents: | | | | | |
|---|---|---|---|---|---|---|
| | DMF | CH$_3$CN | THF | CH$_2$Cl$_2$ | DMSO | DME |
| FeCp$_2{}^{0/+}$ | 0.470 | 0.382 | 0.547 | 0.475 | 0.435 | 0.580 |
| FeCp*$_2{}^{0/+}$ | −0.125 | −0.125 | 0.102 | −0.070 | −0.030 | 0.140 |
| CoCp*$_2{}^{0/+}$ | −1.402 | −1.525 | −1.295 | −1.497 | −1.425 | −1.260 |
| FeCp*(C$_6$Me$_6$)$^{0/+}$ | −1.762 | −1.865 | −1.645 | – | −1.775 | −1.605 |

**Table 2.** $E_{1/2}$ values determined for the transition-metal sandwich complexes (V relative to FeCp*$_2$).

| Complex | $E_{1/2}$ in the solvents: | | | | | |
|---|---|---|---|---|---|---|
| | DMF | CH$_3$CN | THF | CH$_2$Cl$_2$ | DMSO | DME |
| FeCp*$_2{}^{0/+}$ | 0 | 0 | 0 | 0 | 0 | 0 |
| FeCp$_2{}^{0/+}$ | 0.480 | 0.510 | 0.440 | 0.545 | 0.470 | 0.440 |
| CoCp*$_2{}^{0/+}$ | 1.390 | 1.390 | 1.400 | 1.390 | 1.395 | 1.400 |
| FeCp*(C$_6$Me$_6$)$^{0/+}$ | 1.750 | 1.745 | 1.750 | – | 1.750 | 1.750 |

$E_{1/2}$ values were determined using the cationic form: $E° \approx E_{1/2} = (E_{pa} + E_{pc}/2)$. $n$-Bu$_4$N$^+$PF$_6{}^-$ (0.1 M); cationic complex: $10^{-3}$ M; $v = 400$ mV s$^{-1}$, 20 °C; working and counter electrodes: Pt; reference electrode: SCE; $\Delta E_p = E_{pa} - E_{pc} = 40$–45 mV (DMF), 45–50 mV (MeCN), 50–55 mV (THF), 50–60 mV (CH$_2$Cl$_2$), 50–60 mV (DMSO), 60–70 mV (DME). The shifts versus the theoretical value of 58 mV at 20 °C are due to variations in ohmic compensation. $E_{1/2}$ values were recorded with an accuracy of ±0.005 V except the $E_{1/2}$ value of [FeCp*(C$_6$Me$_6$)]$^{0/+}$ in CH$_2$Cl$_2$, which is not accurate because of the interference with the reduction of the solvent. In CH$_2$Cl$_2$, $n$-Bu$_4$N$^+$ salts with various counter-anions and concentrations (vide infra) were used, but the difference between the $E_{1/2}$ values of FeCp*$_2{}^{+/0}$ and CoCp*$_2{}^{+/0}$ remained constant (1.390 V). The difference between the $E_{1/2}$ values of FeCp$_2{}^{+/0}$ and FeCp*$_2{}^{+/0}$ varied, however, between 0.460 V and 0.545 V. The $E_{1/2}$ values relative to the SCE obtained for FeCp$_2{}^{+/0}$ and FeCp*$_2{}^{+/0}$, respectively, are (V): $n$-Bu$_4$N$^+$PF$_6{}^-$ (0.1 M: 0.475 and −0.070; 0.4 M: 0.425 and −0.100); $n$-Bu$_4$N$^+$BF$_4{}^-$ (0.1 M: 0.585 and 0.045; 0.4 M: 0.555 and 0.035); $n$-Bu$_4$N$^+$Br$^-$ (0.1 M: 0.490 and −0.010; 0.4 M: 0.480 and 0.000); $n$-Bu$_4$N$^+$HSO$_4{}^-$ (0.1 M: 0.545 and 0.065; 0.4 M: 0.515 and 0.055); $n$-Bu$_4$N$^+$Cl$^-$ (0.1 M: 0.590 and 0.070; 0.4 M: 0.570 and 0.070).

values relative to [FeCp$_2$]$^{0/+}$ to values relative to [FeCp*$_2$]$^{0/+}$ of Table 2 in various solvents. Likewise, it is possible to continue using ferrocene as the internal reference, and yet to have a solvent- and electrolyte-independent redox potential by conversion with Table 2. The standard redox potentials of the main reductants and oxidants $E°'$ relative to [FeCp$_2$]$^{0/+}$ [FeCp$_2$],$^{0/+}$, SCE and the normal hydrogen electrode (NHE) are given in Tables 3 and 4, respectively [23].

**Table 3.** Formal potentials (V) of the main reductants.

| Reductant | Solvent | $E^{o\prime}$ relative to: | | | |
|---|---|---|---|---|---|
| | | FeCp*$_2$ | SCE | NHE | FeCp$_2$ |
| [C$_{10}$H$_8$]$^{\cdot-}$ | THF | −2.66 | −2.76 | −2.52 | −3.10 |
| – | DMF | −2.56 | −2.68 | −2.44 | −3.05 |
| Na | THF, DME | −2.60 | −2.72 | −2.58 | −3.04 |
| Li(Hg) | H$_2$O | −2.13 | −2.16 | −1.92 | −2.60 |
| [Anthracene]$^{\cdot-}$ | THF | −2.08* | −1.98 | | |
| FeCp*(C$_6$Me$_6$) | CH$_2$Cl$_2$ | −1.75* | −1.85 | −1.61 | −2.30 |
| – | DMF | −1.75* | −1.76 | −1.62 | −2.24 |
| [Perylene]$^{\cdot-}$ | THF | −1.75* | | | |
| [Benzophenone]$^{\cdot-}$ | THF | −1.75 | −1.86 | −1.62 | −2.30 |
| – | MeCN | 1.68* | −1.69 | −1.45 | −2.17 |
| FeCp(C$_6$Me$_6$) | DMF | −1.54 | −1.55 | −1.31 | −2.02 |
| – | THF | −1.53* | | | |
| – | DMSO | −1.52* | −1.56 | −1.32 | −1.99 |
| C$_{60}$$^{3-}$ | MeCN–PhMe | −1.41* | −1.36 | 1.12 | −1.87 |
| CoCp*$_2$ | DMF | −1.39* | −1.40 | −1.16 | −1.87 |
| FeCp(C$_6$H$_6$) | THF | −1.34* | | | |
| C$_{60}$$^{2-}$ | MeCN–PhMe | −0.915* | −0.91 | −0.67 | −1.37 |
| CoCp$_2$ | DME | −0.74* | −0.86 | −0.62 | −1.33 |
| C$_{60}$$^-$ | MeCN–PhMe | −0.515* | −0.51 | −0.27 | −0.98 |
| FeCp*$_2$ | DMF | 0 | −0.012 | +0.23 | −0.49 |
| – | CH$_2$Cl$_2$ | 0 | −0.105 | +0.13 | −0.55 |
| Hydrazine | DMSO | +0.06 | −0.03 | +0.21 | −0.41 |

## 4.3 Reductants

### 4.3.1 Alkali- and other Metals; Amalgams and Alloys

**Alkali metals**

Alkali metals (Li, Na, K) have long been used as strong reductants, *but their use is dangerous* and should be limited to occasions when other reductants are not efficient. *These metals should be used only by very experimented chemists.* They are the strongest reductants, with standard potentials of ca 3 V relative to the SCE in aqueous systems:

$$M \rightarrow M^+ + e^-$$

Their reducing power increases and electropositivity increases from lithium to rubidium. For instance, lithium reacts slowly with water, sodium reacts vigorously, potassium ignites, and cesium and rubidium react explosively. It is difficult to de-

**Table 4.** Formal potentials (V) of the main oxidants.

| Oxidant | Solvent | $E^{o\prime}$ relative to | | | |
|---|---|---|---|---|---|
| | | FeCp*$_2$ | SCE | NHE | FeCp$_2$ |
| $[N(C_6H_2Br_3\text{-}2,4,6)_3]^+$ | MeCN | 1.87 | 1.74 | 1.98 | 1.36 |
| $[N(C_6H_3Br_2\text{-}2,4)_3]^+$ | MeCN | 1.65 | 1.52 | 1.76 | 1.14 |
| $[NO]^+$ | CH$_2$Cl$_2$ | 1.55 | 1.44 | 1.68 | 1.00 |
| $[RuL_3]_{3+}$, L = bpy or phen | MeCN | 1.40* | 1.25 | 1.49 | 0.87 |
| $[\text{Thianthrene}]^{\cdot+}$ | MeCN | 1.32* | 1.24 | 1.48 | 0.86 |
| $[N(C_6H_4Br\text{-}4)_3]^{\cdot+}$ | MeCN | 1.21* | 1.05 | 1.29 | 0.70 |
| $[Fe(bipy)_3]^{3+}$ | MeCN | 1.17 | 1.04 | 1.28 | 0.66 |
| $Ag^+$ | CH$_2$Cl$_2$ | 1.16 | 1.05 | 1.29 | 0.65 |
| $FeCp(\eta^5\text{-}C_5H_4COMe)^+$ | MeCN | 0.75* | 0.71 | 0.95 | 0.27 |
| $Cl_2$ | MeCN | 0.69 | 0.56 | 0.80 | 0.18 |
| DDQ | MeCN | 0.635* | 0.51 | 0.75 | 0.13 |
| $Br_2$ | MeCN | 0.58 | 0.45 | 0.69 | 0.07 |
| FeCp$_2$ | MeCN | 0.51* | 0.38 | 0.62 | 0. |
| – | CH$_2$Cl$_2$ | 0.55* | 0.42 | 0.66 | 0. |
| $[N_2C_6H_4F\text{-}4]^+$ | MeCN | 0.44 | 0.31 | 0.55 | −0.07 |
| $Fe[C_5(CH_2Ph)_5]_2$ | CH$_2$Cl$_2$ | 0.49* | 0.50 | 0.74 | −0.06 |
| – | THF, MeCN | 0.41* | | | |
| – | DMF | 0.34* | 0.215 | 0.455 | −0.14 |
| $I_2$ | MeCN | 0.37 | 0.24 | 0.48 | −0.14 |
| TCNE | MeCN | 0.345* | 0.11 | 0.35 | −0.27 |
| TCNQ | MeCN | 0.32* | 0.08 | 0.32 | −0.30 |
| FeCp*$_2$ | MeCN | 0 | −0.13 | 0.11 | −0.51 |
| – | CH$_2$Cl$_2$ | 0 | −0.105 | 0.13 | −0.55 |

Footnote to Tables 3 and 4. Direct measurements relative to [FeCp*$_2$]. are denoted *. The mono, di- and trianion of C$_{60}$ are accessible by reduction of C$_{60}$ using [FeCp(C$_6$Me$_6$)] [25]. The other values relative to [FeCp$_2$*] are: C$_{60}^{3-/4-}$: −1.895 V, C$_{60}^{4-/5-}$: −2.39 V and C$_{60}^{5-/6-}$: −2.39 V. The two later values are deduced from the values relative to [FeCp$_2$]. reported in Ref. [25]. Other standard redox potentials $E^{o\prime}$ (V) of oxidants and reductants relative to various references extracted from Ref. [21] and corrected from the [FeCp$_2$] to the [FeCp*$_2$] reference using the $\Delta E^{o\prime}$ values of Table 2.

termine their redox potentials precisely, however, because they vary considerably with the nature of the solvent. Differences of up to 1 V are encountered because of variations in the solvation of the cation, a notion already discussed in Section 4.2. For instance, with NH$_3$, the redox potential of Na/Na$^+$ is ca 1 V less negative (approximately −2 V relative to the SCE) than in DMF (ca −3 V). Indeed, Na reacts with NH$_3$ to give solvated Na$^+$ and the stable blue solution of solvated electrons which have frequently been used as reducing agents in organic chemistry (for instance to reduce aromatic compounds to cyclohexadienes). The influence of ion pairs on redox potentials has been reviewed [26].

$$Na + (n + m)NH_3 \xrightarrow{NH_3} [Na(NH_3)_n]^+ + [e(NH_3)_m]^-$$

**Table 5.** Standard reduction potentials $E^{\circ\prime}$, relative to $FeCp_2$, of metal ions in water.

| Half reaction (acidic solution) | Standard reduction potential, $E^{\circ\prime}$ (V) |
|---|---|
| $Li^+(aq) + e^- \leftrightarrow Li(s)$ | −3.045 |
| $K^+(aq) + e^- \leftrightarrow K(s)$ | −2.925 |
| $Na^+(aq) + e^- \leftrightarrow Na(s)$ | −2.714 |
| $Al^{3+}(aq) + 3e^- \leftrightarrow Al(s)$ | −1.66 |
| $Zn^{2+}(aq) + 2e^- \leftrightarrow Zn(s)$ | −0.763 |
| $Fe^{2+}(aq) + 2e^- \leftrightarrow Fe(s)$ | −0.44 |
| $Sn^{2+}(aq) + 2e^- \leftrightarrow Sn(s)$ | −0.14 |

The redox potentials of the metals in water are given in Table 5 to provide an indication of the order of the reducing power of the main metals used as reductants [27].

Another major drawback is that metals must be used in excess of the necessary stoichiometric amount because it is impossible to know how much of the metal sample is contaminated with metal oxide and metal hydroxide covering the surface after contact with air. *Then excess metal must be destroyed very slowly by addition of absolute ethanol with the most extreme care, otherwise fire subsequent to reaction will occur almost systematically upon contact with air or water.* Metals freshly cut into pieces are particularly reactive. Even more reactive are the metallic sands obtained by heating pieces of the metal above its melting point in a suitable solvent, e.g. dry toluene, and the metallic mirrors formed by subliming a piece of metal (Na or more often K) from the bottom of a Schlenk flask on to its walls under vacuum. THF and DME are the solvents most often used for reductions with these metals, for example:

$$[Ce(\eta^8\text{-}C_8H_8)] \xrightarrow{\text{K, DME}} K[Ce(\eta^8\text{-}C_8H_8)] \xrightarrow{\text{K, DME}} K_2[Ce(\eta^8\text{-}C_8H_8)] \qquad [28]$$

$$[VCp_2] \xrightarrow{\text{K, THF}} [K(THF)(OEt_2)][VCp_2]$$
$$E^{\circ\prime} = -3.3 \text{ V vs. } FeCp_2 \quad \text{in THF} \qquad [29a]$$

$$[V(\eta^6\text{-arene})_2] \xrightarrow{\text{K, THF}} K[V(\eta^6\text{-arene})_2]$$
$$E^{\circ\prime} = -3.2 \text{ V vs. } FeCp_2 \quad \text{in THF} \qquad [29b]$$

$$[Co(\eta^2\text{-}C_2H_4)(PMe_3)_3] \xrightarrow{\text{K, pentane, cyclopentene}} K[Co(\eta^2\text{-}C_2H_4)(PMe_3)_3] \quad [30]$$

$$[FeCp^*(CO)_2]_2 \xrightarrow{\text{K, THF}} 2\,K[FeCp^*(CO)_2] \qquad [31]$$

$$[MoCp^*Cl_4] \xrightarrow{\text{Na, THF, nL}} [MoCp^*L_nCl] \qquad n = 2 \text{ or } 3, L = PMe_3 \qquad [32]$$

$$[TaCp(CO)_3] \xrightarrow{\text{Na, NH}_3} Na_2[TaCp(CO)_3] \quad \text{also Nb} \qquad [33]$$

$$[BMe_2Ph]^- Na^+ \xleftarrow{\text{Na, THF}} [BMe_2Ph] \xrightarrow{\text{K, THF}} K_2[BMe_2Ph] \qquad [34]$$

**Alkali metals in the presence of a macrocycle or a cryptand**

The presence of a polyether macrocycle or cryptand adapted to the size of the alkali cation [10] facilitates the reduction of molecules which are otherwise difficult to reduce by the alkali metal; it stabilizes and solubilizes the reduced molecules and induces the formation of crystals. $Li^+$ salts are stabilized by the presence of 12-crown-4 or 2,2,2-cryptand; $Na^+$ salts are stabilized by 12-crown-5 or 2,2,2-cryptand; and $K^+$ salts are stabilized by 18-crown-6 [10, 35], for example:

$$[Ga_2L_4] \xrightarrow{Li, 12\text{-crown-4}, Et_2O} [Li(12\text{-crown-4})_2][Ga_2L_4] \qquad [36]$$

$$[CoL_2] \xrightarrow{K, 18\text{-crown-6}, THF} [K(18\text{-crown-6})(THF)_2][CoL_2]$$

$$L = \eta^4\text{-}1,4\text{-}t\text{-}Bu_2\text{butadiene} \qquad [37]$$

**Less reducing metals—Zn powder**

Zn powder is a useful reducing agent with a less negative redox potential than the alkali metals (Table 5), and it is not dangerous.

$$[TcCl_4\{P(p\text{-tol})_3\}_2] \xrightarrow{Zn, P(p\text{-tol})_3, MeCN, reflux} [TcCl_3(MeCN)\{P(p\text{-tol})_3\}_2]$$
$$[38]$$

$$[Fe\{\eta^5\text{-}C_5(p\text{-tol})_5\}(CO)_2Br] \xrightarrow{Zn, THF} 1/2 \; trans\text{-}[Fe\{\eta^5\text{-}C_5(p\text{-tol})_5\}(CO)_2]_2$$
$$[39]$$

Transmetallation reactions are also known:

$$[CuL]^{2+} + Zn \rightarrow [ZnL]^{2+} + Cu \qquad L = \text{macrocycle} \qquad [40]$$

**Amalgams and alloys**

Amalgams and alloys, especially those of Na and K are very popular and have been extensively used for reductions. *Again, the potentials users of Na–K alloys (which are commercially available) should be warned that these reagents are extreme fire hazards when exposed to moist air, and should therefore be used and with extreme caution and carefully destroyed after the experiments*, for example:

$$[HfCl_4] \xrightarrow{Na/K, PEt_3, toluene} 1/2[Hf_2Cl_6(PEt_3)_4] \qquad [41]$$

Us of Na–Hg amalgams involves the handling of mercury, the vapor of which is toxic. Na–Hg amalgams are, on the other hand, milder reducing agents than Na metal. The redox potential of Na–Hg is probably ca 1 V less negative than that of sodium (despite a smaller difference in Table 6, see discussion below). It can even be stored in air without too much damage, its reaction with water being only slow and that with moist air extremely slow. This difference is because of its heat of formation from Na and Hg, a very exothermic reaction (preparation must take this into account and *be conducted under a well-ventilated hood*). If the amount of sodium is less than 1.25 %, the Na–Hg amalgam is in a liquid or semi-liquid (0.9 % is a good

**Table 6.** Standard redox potentials, $E^{\circ\prime}$, relative to FeCp$_2$ of alkali metals and amalgams in different solvents.

| Reductant | Solvent | $E^{\circ\prime}$ (V) |
|---|---|---|
| Na | THF, diglyme | −3.04 |
| Li | NH$_3$ | −2.64 |
| Li–Hg | H$_2$O | −2.60 |
| K | NH$_3$ | −2.38 |
| Na–Hg | Non-aqueous | −2.36 |
| Na | NH$_3$ | −2.25 |

amount for a convenient liquid state). The liquid state of the reductant is another significant advantage over pure sodium metal, because much better contact is obtained between the reductant with the substrate in solution. Na–Hg is most often used in THF or DME but diethyl ether, benzene, and toluene are also used. The redox potentials of the alkali metals and their main amalgams are compared in Table 6. The redox potentials of the alkali metals were measured by cathodic reduction of their cations and it is possible that adsorption of alkali-metal atoms on the metal cathode is also exothermic and results in some reduction of the absolute value of the redox potentials determined in this way. For instance, in Table 6, all the redox potentials of the metals are less negative than that of the naphthalene–naphthalene radical anion in the same solvent, even though the naphthalene radical anion is obtained by alkali-metal reduction of naphthalene in the same solvents! Ion pairing is stronger between alkali metals and nucleophiles (solvent, anion of the electrolyte) present in the electrochemical cell, which also contributes substantially to furnishing a less negative redox potential for the alkali metals. Thus the values of Table 6 should be regarded with caution in the light of the discussion above.

The di- and polynuclear metal-carbonyl complexes are readily reduced by Na–Hg to monometallic metal-carbonyl anions, their redox potentials usually being between −0.7 V and −1.7 V relative to FeCp$_2$ [42], i.e. well below the redox potential of Na/Hg (−2.36 V relative to FeCp$_2$ in non-aqueous solvents [43]). Redox potentials of a number of these complexes are listed in Table 7.

Binuclear complexes also containing Cp ligands are reduced at more negative potentials which are still accessible for Na–Hg, as are nitrosyl complexes, for example:

$$[Co_2(CO)_8] \xrightarrow{\text{Na/Hg, THF}} 2\,Na[Co(CO)_4] \qquad [45]$$

$$[M_2Cp_2(CO)_6] \xrightarrow{\text{Na/Hg, THF}} 2\,Na[MCp(CO)_3] \quad M = Cr, Mo, W \qquad [45a]$$

$$[Co_2Cp_2(NO)_2] \xrightarrow{\text{Na/Hg, THF}} 2\,Na[CoCp(NO)] \qquad [46]$$

$$[Fe_2Cp_2(CO)_4] \xrightarrow{\text{Na/Hg, THF}} 2\,Na[FeCp(CO)_2] \qquad [47]$$

$$[MCp_2Cl_2] \xrightarrow{\text{Na/Hg, THF}} 1/2\,[M_2Cp_4(\mu_2 - Cl)_2] \quad M = Zr, Hf \qquad [48]$$

**Table 7.** Two-electron reduction potential for selected bi- and trinuclear metal-carbonyl complexes (V relative to the SCE; obtained from reactivity studies).

| Compound | Product | Potential (V)[a] |
|---|---|---|
| $[CO_2(CO)_8]$ | $2\,[Co(CO)_4]^-$ | $-0.15$ |
| $[Cr_2Cp_2(CO)_6]$ | $2\,[CrCp(CO)_3]^-$ | $-0.70$ |
| $[Mo_2Cp_2(CO)_6]$ | $2\,[MoCp(CO)_3]^-$ | $-0.79$ |
| $[Fe_3(CO)_{12}]$ | $[Fe_3(CO)_{11}]^{2-}$ | $-0.80$ |
| $[W_2Cp_2(CO)_6]$ | $2\,[WCp(CO)_3]^-$ | $-0.80$ |
| $[Ru_3(CO)_{12}]$ | $[Ru_3(CO)_{11}]^{2-}$ | $-0.90$ |
| $[Mn_2(CO)_{10}]$ | $2\,[Mn(CO)_5]^-$ | $-0.97$ |
| $[Os_3(CO)_{12}]$ | $[Os_3(CO)_{11}]^{2-}$ | $-1.10$ |
| $[Re_2(CO)_{10}]$ | $2\,[Re(CO)_5]^-$ | $-1.20$ |
| $[FeCp_2(CO)_4]$ | $[FeCp(CO)_2]^-$ | $-1.70$ |

[a] The $2e^-$ oxidation of the metal-carbonyl anion to the dimer ($2M^- \rightarrow M_2 + 2e^-$) is evaluated as a combination of the oxidation potential of the anion and the metal–metal bond strength [44].

These are only simple, classical examples, but many more studies are known. The monoelectronic reduction of inorganic or organometallic complexes leads to radicals which are usually unstable [10]. For instance, $MCp(CO)_n$ radicals dimerize by formation of a metal–metal bond subsequent to the loss of a CO ligand as exemplified above [44, 45, 47]. The monoelectronic reduction of cationic complexes bearing a hydrocarbon ligand different from Cp often produces radicals whose SOMO has a strong ligand character. In these circumstances, ligand–ligand coupling occurs, leading to ligand-bridged dimers [49–52], for example (Scheme 1) [49, 50].

Coupling of Cp ligands also sometimes occurs with second- or third-row transition-metal complexes. For instance, reduction of the rhodocenium cation [53, 54] and even that of its Cp* [55] derivatives leads to rapid ligand–ligand coupling (equation and Scheme 2):

$$[Rh(\eta^5\text{-}C_5R_5)_2][PF_6] \xrightarrow{\text{Na/Hg, THF}} [Rh_2\{\mu_2,\eta^8\text{-}(C_5R_5)_2\}] \quad R = H \quad [53, 54]$$

$$\text{or Me} \quad [55]$$

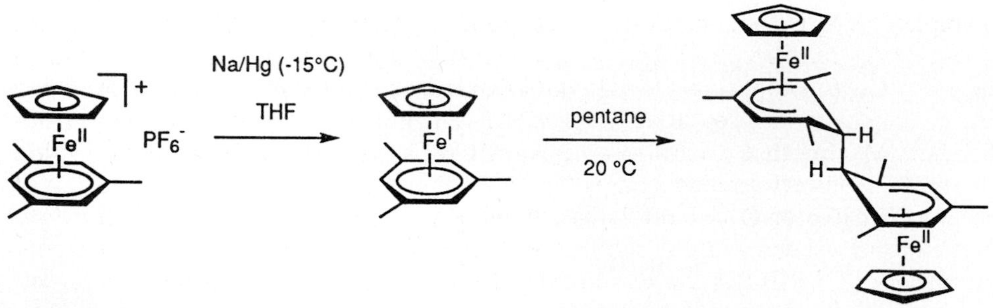

**Scheme 1.**

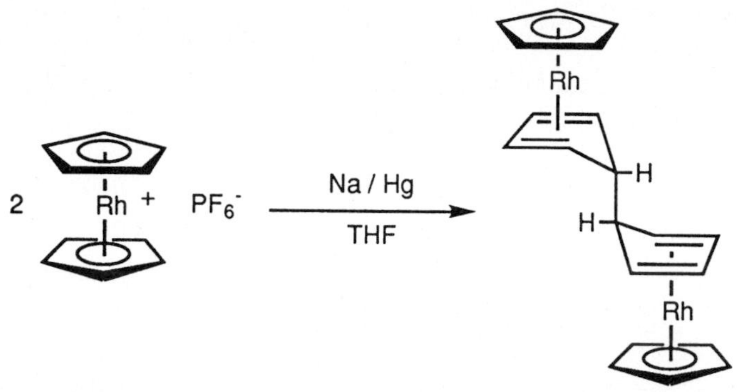

**Scheme 2.**

The other amalgams Mg–Hg [56] and Zn–Hg [57] and the commercially available alloy Na–Pb [21] have also occasionally been used.

### 4.3.2 Aromatic Radical Anions and Anions

**Naphthalene radical anion**

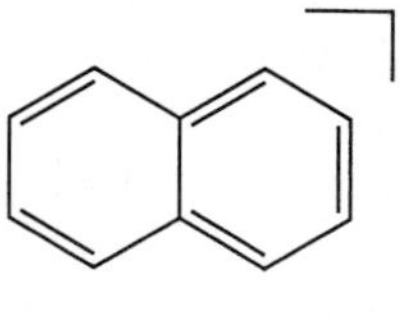

NpH $\cdot^-$

With a redox potential of $-3.10$ V relative to FeCp$_2$ in THF, the radical anion of naphthalene (NpH$^{\cdot-}$) is, in theory, the most powerful reducing agent used (see the discussion above, however):

$$M + NpH \xrightarrow{\text{THF or DME}} M^+, NpH^{\cdot-} \quad M = Li, Na, K$$

The problem is that, because it is extremely electron rich, this species is also extremely air sensitive, and the real quantity which is being used is not known. In addition, it reacts to give neutral naphthalene which is difficult to separate from the desired product. Finally, another problem (encountered, for instance, in cluster chemistry [58]) is that it can over-reduce the substrate and lead to decomposition whereas less powerful reducing agents are not marred by this problem. Addition of pieces of Li, Na, or K to naphthalene in THF or DME under extremely dry reaction conditions gives a green solution of the naphthalene radical anion, and concentrations of 0.1–0.5 M are employed [59]. The formation of the dianion in this reaction has also been proposed, but its formation is uncertain. It is, in any case, presumably present at low levels only, if at all.

$$M^+, ArH^{\cdot -} \rightarrow M^{2+}, Ar^{2-}$$

An advantage of naphthalene radical anion is that it is soluble in THF or DME, which facilitates the reduction, for example:

$$[U(\eta^8\text{-}C_8H_8)_2] \xrightarrow{\text{LiNp, THF}} [Li, THF][U(\eta^8\text{-}C_8H_8)_2] \tag{60}$$

$$[M(CO)_5(NMe_3)] \xrightarrow{\text{LiNp, THF}} Li_2[M(CO)_5] \quad M = Mo \text{ or } W \tag{61}$$

$$[Mn(\eta^8\text{-}C_6H_6)(CO)_3] \xrightarrow{\text{NaNp, THF}} Na_2[Mn(\eta^4\text{-}C_6H_6)(CO)_3] \tag{62}$$

Finally, these reactions are sometimes marred by coordination of naphthalene on to the transition metal, as, for example, for the last reaction [62].

### Cyclooctatetraene dianion

Aromatic salts $M_2COT$ resulting from the two-electron reduction of cyclo-octatetraene (COT) by alkali metals (Li, Na, K) have been used several times as reductants, although COT remains in solution which makes the separation difficult, for example:

$$[Zr\{\eta^5\text{-}C_5H_3(SiMe_3)_2\}_2Cl_2] \xrightarrow{\text{Li}_2[COT], CO(1 \text{ atm}), Et_2O} [Zr\{\eta^5\text{-}C_5H_3(SiMe_3)_2\}_2] \tag{63}$$

$$[MCp^*_2Cl_2] \xrightarrow{\text{M}'_2[COT], THF} [MCp^*_2] \quad M = Sn \text{ or } Ge; M' = Li \text{ or } K \tag{64}$$

### Benzophenone radical anion

The blue benzophenone radical anion (the first recognized radical anion [65]) is well-known as a color test for the absence of water and oxygen from ether solvents over Na. In the presence of water, it forms the colorless pinacol by dimerization of the ketyl radical $Ph_2C(O)H^\cdot$ (i.e. the protonated form of the radical anion) [66]. In the presence of dioxygen it is rapidly oxidized back to benzophenone, and thus removes traces of $O_2$ from $N_2$ [67]. It has been very little used as a stoichiometric reductant, probably because it is not easy to separate and use in precise amounts, the benzophenone obtained after reduction would not be easy to separate from products, and its redox potential ($E^{\circ\prime} = -2.2$ V relative to FeCp$_2$) is very similar to that of Na in NH$_3$ and Na–Hg which are often used instead, for example:

$$[W(CO)_5(NMe_3)] \xrightarrow{\text{Na[Ph}_2\text{CO]}} 1/2\ Na_2[W_2(CO)_{10}] \quad M = Mo\ or\ W \quad [61]$$

For this latter reaction, it should be noted that the use of potassium naphthalide leads to further reduction to K$_2$[W(CO)$_5$]; thus, the milder reductant Na[Ph$_2$CO] selectively leads to a less reduced complex [61]. M[Ph$_2$CO] salts have instead been used in catalytic amounts as mediators (see Section 4.4). They are prepared by reduction of benzophenone in THF, DME, or ammonia by pieces of alkali metals (Li, Na, K) [68]:

$$Ph_2CO \xrightarrow{\text{M, THF}} M[Ph_2CO] \quad M = Li, Na, K$$

Their reducing power is in the order K[Ph$_2$CO] > Na[Ph$_2$CO] > Li[Ph$_2$CO] because the strength of the ion pairing is in the opposite order [69]. This sequence is the opposite of that for polyaromatic anions such as naphthalide in ethers. These salts of Ph$_2$CO$^{\bullet-}$ can be reduced by excess metal to the purple salts of the dianion [70, 71], because the standard redox potential of the redox system Ph$_2$CO$^{\bullet-}$/Ph$_2$CO$^{2-}$ is less negative ($E^{\circ\prime} = -2.9$ V relative to FeCp$_2$ in THF [72, 73]) than those of the alkali metals in this solvent. The purple dianion is extremely sensitive to impurities, however, and is therefore not often seen:

$$M[Ph_2CO] \xrightarrow{\text{M, THF}} M_2[Ph_2CO] \quad M = Li, Na, K$$

**Other polycyclic aromatic radical anions**

Given the degeneracy of the LUMO in polycyclic hydrocarbons, addition of an electron into these orbitals is more or less facile and leads to thermally stable radical anions; redox potentials for a number of polycyclic hydrocarbon-radical anions are listed in Table 8.

**Table 8.** Reduction potentials of polycyclic hydrocarbon-radical anions relative to FeCp$_2$ in DMF[a].

| Compound | $E^{\circ\prime}$ (V relative to the SCE) | Compound | $E^{\circ\prime}$(V) |
|---|---|---|---|
| Biphenyl | −3.16 | Dibenz[a, h]anthracene | −2.47 |
| Naphthalene | −3.07 | Coronene | −2.46 |
| Phenanthrene | −2.99 | Perylene | −2.04 |
| Chrysene | −2.78 | Tetracene | −1.88 |
| Pyrene | −2.56 | | |

[a]Adapted from Refs [74] (biphenyl/biphenyl$^{\bullet-}$ in DMF) and [75] (others relative to biphenyl/biphenyl$^{\bullet-}$). Corrections into $E^{\circ\prime}$ values relative to FeCp$_2$ in DMF is made using Table 1.

Sodium anthracenide and potassium acenaphthylenide have been used. They are much milder reducing agents than naphthalides, for example:

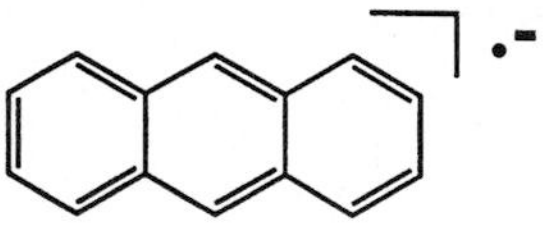

**anthracenide
radical anion** 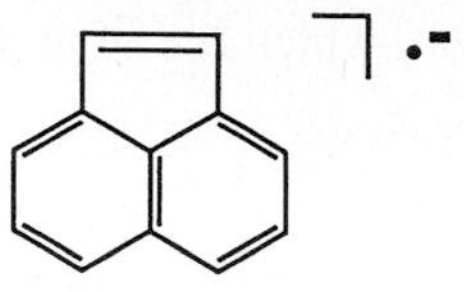                         **acenaphthylenide
radical anion**

$$[Fe^{III}Cl(TPP)] \xrightarrow{2\ Na[anthracene],\,THF} [Na(THF)_3][Fe^{I}(TPP)] \qquad [76]$$

$$[Fe^{III}Cl(TPP)] \xrightarrow{3\ Na[anthracene],\,THF} [Na(THF)_3]_2[Fe^{0}(TPP)] \qquad [76]$$

$$[Ni_3Cp_3(\mu^3\text{-}CO)_2] \xrightarrow{K[acenaphtylene]} K[Ni_3Cp_3(\mu^3\text{-}CO)_2]$$

$$[58;\ E^{\circ\prime}_{cluster} = -1.4\ V]$$

If the latter reaction is performed with the very strong reductant potassium naphthalide instead of potassium acenaphthylenide which is milder, it leads to decomposition as a result of over-reduction [58].

The radical anion of 2,2′-bipyridine is also a strong reductant which has been used as a mediator (see Section 4.3).

## Alkali-metal carbides

Volpin has noted the similarity between potassium salts of polycyclic aromatics and potassium graphite $KC_8$, an intercalation compound [75, 77, 78]. Table 7 shows that the reduction potentials of the radical anions become less negative as the number of fused rings increases. Graphite, which has an infinite number of fused rings, is thus easily reduced. The redox potential of $KC_8$ should thus be close to those of anthracene, pyrene, and coronene, i.e. ca $-2.5$ V relative to $FeCp_2$ in DMF, which also close to that of Na–Hg, for instance [75]. This compound is best described by the ionic form $K^+C_8^-$, because of the transfer of an electron from potassium to the conduction band of graphite. $KC_8$ is prepared from graphite powder and K at 350 °C [79], which is certainly a limitation for the community of molecular chemists. It has been used as a reductant on a few occasions [79, 80]. The insoluble by-product, carbon, is eliminated by filtration, for example:

$$[M(CO)_6] \xrightarrow{KC_8,\,THF} 1/2\ K_2[M_2(CO)_{10}] \quad M = Cr, Mo, W \qquad [80]$$

Sodium carbide, $Na_2C_2$, has also been used as a mild reductant, especially to reduce mono-and polynuclear metal carbonyl complexes [81]. It is relatively stable toward moist air, for example:

$$[Co_4(CO)_{12}] \xrightarrow{Na_2C_2,\,THF} 1/2\ Na_4[Co_6(CO)_{14}] + \cdots \qquad [81]$$

### 4.3.3 Hydrides, Grignard Reagents, Metal-alkyls and Metal-aryls, and Other Reductants

#### Hydrides

Commercial B and Al hydrides, e.g. NaH, LiAlH$_4$, NaBH$_4$, NH$_4$BH$_4$, and K[BH(CHMeEt)$_3$], are classic sources of hydride in organic, inorganic, and organometallic chemistry. The first example of their use as a monoelectronic reductant in transition-metal chemistry was disclosed with a variety of 18-electron cationic complexes [Fe$^{II}$Cp(arene)][PF$_6$] [82]. Although these complexes are reduced with overall hydride transfer onto the benzene ligand, it can be shown that this reduction comprises a first step, a single-electron transfer, then a second step, an H-atom transfer (Scheme 3). The 19-electron Fe$^I$ intermediates resulting from the initial electron transfer were isolated or generated and characterized as intermediates by use of ESR, UV–Vis, and Mössbauer spectroscopy and by the observation of their characteristic forest-green color.

Very unstable Fe$^I$ complexes such as [Fe$^I$Cp($\eta^6$-C$_6$H$_5$F)] and [Fe$^I$Cp($\eta^5$-C$_4$Me$_4$S)] could not be synthesized by Na–Hg reduction of their cationic precursor at $-10\,°$C in THF, because decomposition was rapid. They could be rapidly generated by use of LiAlH$_4$ in THF at $-95\,°$C and $-50\,°$C respectively. The parent complex [Fe$^I$Cp($\eta^6$-C$_6$H$_6$)] was generated at $-60\,°$C under these conditions. Even the complexes [Fe$^I$Cp($\eta^6$-C$_6$R$_6$)], R = Me or Et are generated rapidly by use of LiAlH$_4$). The electron-transfer reactions of LiAlH$_4$ are highly solvent-dependent. They were observed in THF and DME, but not in Et$_2$O. Although the standard redox potential has been estimated as $E^{o\prime} = -0.7$ V [83], such a value does not give an idea of the reducing power of the hydride. Even if it is correct, the effective reduction potential of LiAlH$_4$ is more negative, because the electron-transfer reactions of LiAlH$_4$ are shifted towards the electron-transfer products by rapid decomposition of the radical AlH$_4$• to $\frac{1}{2}$Al$_2$H$_6$ and H$_2$. Given the high rate and low temperatures above, one might conclude, however, that the effective reduction potential of LiAlH$_4$ is more negative than $-1.2$ V to $-1.5$ V relative to FeCp$_2$, the standard

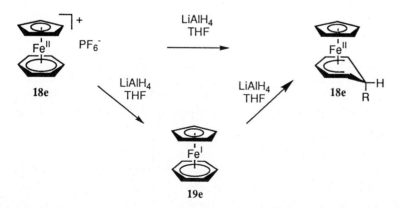

**Scheme 3.**

redox potentials of $[Fe^{I/II}Cp(\eta^6\text{-arene})]^{0/+}$ redox systems [84]. The hydrides $LiAlH_4$ and $NaBH_4$ also react with the complex $[Fe^{II}(\eta^5\text{-}C_6Me_5H)(\eta^6\text{-}C_6Me_6)][PF_6]$ as monoelectronic reductants in THF at $-35\,°C$ and $50\,°C$ respectively, this difference in reaction temperature illustrating that $NaBH_4$ is a much weaker reductant than $LiAlH_4$. The relative tendencies of different hydrides to transfer an electron parallels those to transfer $H^-$, as both depend on the relative energy levels of the HOMOs of the hydrides. This different reactivity is not reflected by the differences between reported standard redox potentials, because that of $NaBH_4$, $E^{o\prime} = -0.6$ V in diglyme [85], is almost the same as that of $LiAlH_4$. Whether electron transfer or hydride transfer occurs from a hydride depends, of course, on the comparative electron acceptor and $H^-$ acceptor properties of the substrate and on the solvent, concentrations, and reaction temperature. In the example above, the complexes $[FeCp(\eta^6\text{-arene})][PF_6]$ and $[Fe^{II}(\eta^5\text{-}C_6Me_5H)(\eta^6\text{-}C_6Me_6)][PF_6]$ can accept either an electron (to give 19-electron complexes) or a hydride (to give mostly 18-electron cyclohexadienyl complexes), which makes the competition uncertain. Indeed, before the study, the organometallic community believed direct hydride transfer was occurring [86]. One might think in terms of a continuum of situations between the direct hydride transfer mechanism seemingly occurring with $LiAlH_4$ and the complexes $[FeCp(\eta^6\text{-arene})][PF_6]$ in ether and the clear stepwise electron transfer plus H-atom transfer mechanism occurring between the same substrates in THF and DME (suffices therefore to continuously vary the proportion of these two solvents). Experimental probes can be difficult to establish in such borderline situations, especially because the final products are the same with both mechanisms. There are instances, however, for which a clear-cut electron-transfer mechanism can be established even for potential hydride acceptors. For instance, two-hydride reduction of the complex $[Fe^{II}(\eta^6\text{-}C_6Me_6)_2][PF_6]_2$ gives $[Fe^0(\eta^4\text{-}C_6Me_6H_2)(\eta^6\text{-}C_6Me_6)]$.

It was possible to show that the mechanism proceeds by the transfer sequence: $H^- + e^- + H^{\cdot}$. The monoelectronic reduction of the di-cation by use of Na–Hg in THF gives the 19-electron complex $[Fe^I(\eta^6\text{-}C_6Me_6)_2][PF_6]$ which reacts with either $LiAlH_4$ or $NaBH_4$ to give the 20-electron complex $[Fe^0(\eta^6\text{-}C_6Me_6)_2]$ [82, 87], but the reaction does not continue to give the di-hydrogenated complex obtained by hydride reduction starting from the di-cation. Thus, the direct hydride-transfer pathway and the electron-transfer pathways are clearly distinct when one starts from the di-cation. This shows that the electron-transfer pathway is no more like for a di-cation than for a mono-cation. The situation is, in fact, quite the opposite (Scheme 4) [82].

Similarly, the $LiAlH_4$ reduction of the piano-stool complex $[FeCp^*(\eta^2\text{-dppe})(CO)][PF_6]$ in THF was also reported to proceed by single-electron transfer, leading to the isostructural 19-electron species $[FeCp^*(\eta^2\text{-dppe})(CO)]$ which rapidly decoordinates to the 17-electron radical $[FeCp^*(\eta^1\text{-dppe})(CO)]$ [88]. Finally, H-atom transfer onto the $Fe^I$ center of this 17-electron species occurs to give the hydride $[FeCp^*(\eta^1\text{-dppe})(CO)(H)]$ (Scheme 5) [88, 89].

The $LiAlH_4$ reduction of the homologous complex $[MoCp^*(CO)_3(L)][PF_6]$ (L = phosphine) proceeded similarly to give a neutral radical resulting from electron transfer; this was followed by the formation of $[MoCp^*(CO)(L)_2(H)]$ or $[MoCp^*(CO)_2(L)(H)]$ [90]. $LiAlH_4$ also reduces polyaromatics and diarylketones

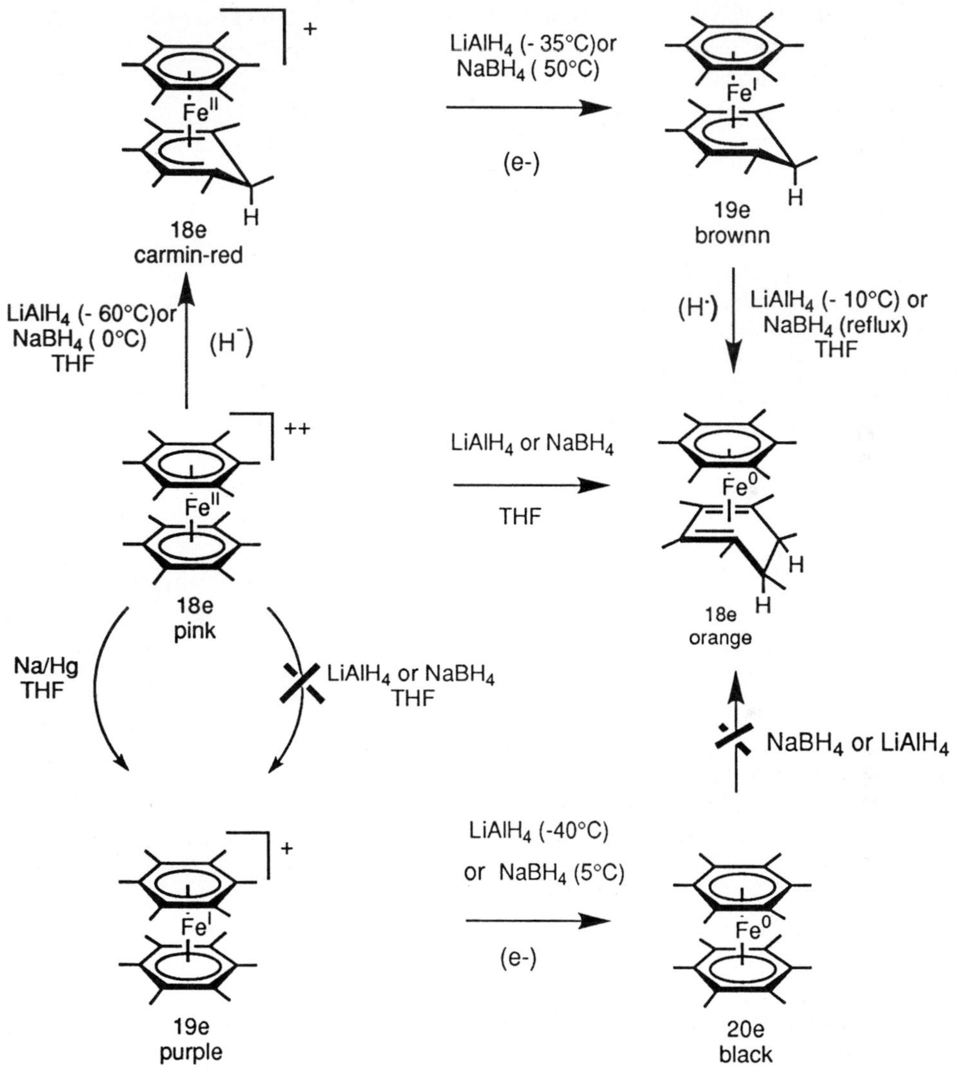

**Scheme 4.**

by a single-electron-transfer pathway [91, 92]. Other single-electron reduction of cationic transition-metal organometallic complexes to their neutral counterparts by $NaBH_4$ in THF–pentane [93] or by $[n\text{-}Bu_4][BH_4]$ in $CH_2Cl_2$ [94] are known. Electron-rich transition-metal hydrides such as, for instance, $[VCp(CO)_3(H)]^-$ can also behave as single-electron reductants towards electron-poor derivatives such as metal carbonyls [95]. Such hydrides are too sophisticated to be regarded as classic redox reagents, however. One might also expect that single-electron-transfer from transition-metal hydride complexes toward single-electron oxidants is a general trend. Such single-electron transfer steps generate 17-electron species which can

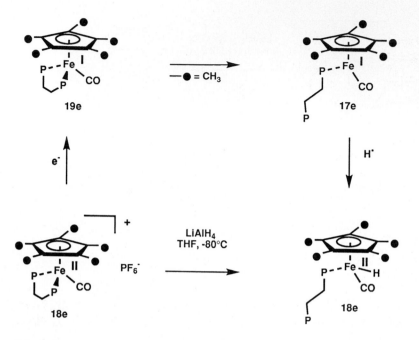

**Scheme 5.**

subsequently be subjected to fast ligand exchange and/or dimerization and/or reduction reactions [2, 96].

### Grignard reagents and alkali-metal alkyls and aryls

Grignard reagents have long been known to be reductants of metal halides for the synthesis of metal carbonyls in the presence of carbon monoxide [97]:

$$2CrCl_3 + 6PhMgBr + 12CO \rightarrow 2Cr(CO)_6 + 3Ph\text{–}Ph + 3MgCl_2 + 3MgBr_2$$
[97]

Wilkinson has also reported clear-cut examples of the use of RMgX as a single-electron reductant, although the oxidation products were not identified [98]:

$$[Ru_2(\mu\text{-}O_2CMe)_4Cl] + Me_3SiCH_2MgCl \xrightarrow{\text{THF}} [Ru_2(\mu\text{-}O_2CMe)_4(THF)_2]$$
[98]

There are many examples of the use of Grignard reagents or alkyllithium or aryllithium and other alkali-metal alkyls or aryls to alkylate, and also reduce, metal halides:

$$[(CoCp^*)_2(\mu^2\text{-}Cl)_2] + 2\,PhLi \rightarrow [(CoCp^*)_3(\mu^3, \eta^{12}\text{-}C_{12}H_{10})]$$
[99]

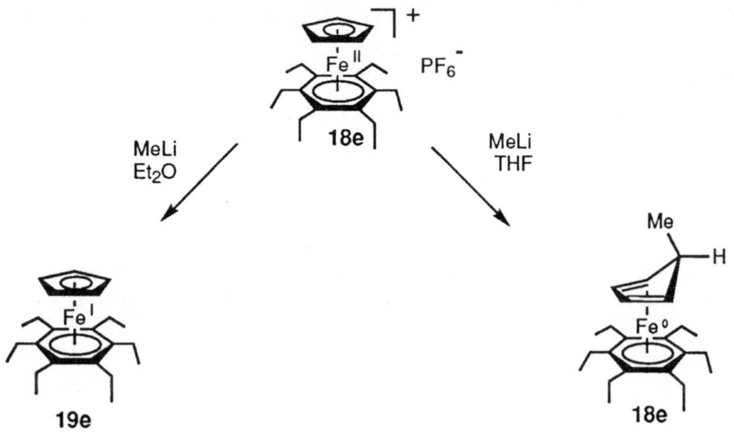

MeLi
Et₂O

18e

MeLi
THF

Me
H

Fe¹

19e

Fe⁰

18e

**Scheme 6.**

The solvent can have a dramatic effect on the electron transfer or alkylation properties of alkali-metal alkyls. Whereas MeLi reacts with $[Fe^{II}Cp(\eta^6\text{-}C_6Et_6)][PF_6]$ in THF to afford methylation of the Cp ring with regioselectivity opposite to that expected from charge control, because of the steric effect, reaction in ether gives the single-electron transfer product only, i.e. the green 19-electron complex. This solvent effect is the opposite of that observed for reaction with $NaBH_4$ and $LiAlH_4$, and noted above (Scheme 6) [100].

## 4.3.4 Miscellaneous Reductants

There are many reductants (CO, amines, phosphines, alcohols, sulfides, carboxylic acids, carboxylates, hydroquinones, electron-rich organometallics (low-oxidation state transition-metal complexes including those involved in oxidative addition), and organometallic anions) which are not single-electron reductants except towards single-electron oxidants, or are not commonly used as single-electron redox reagents (mostly inner-sphere reductants) and will, therefore, not be treated in detail here. We will only summarize these and give a few examples:

– When they are heated in the presence of CO under pressure, transition metal halides react to give metal carbonyls. For instance, $RuCl_3$ gives $[Ru_3(CO)_{12}]$ under a 65-atmosphere pressure of CO [101].
– Ascorbic acid (vitamin C) is a well-known reductant in biochemistry, but it can also reduce transition-metal complexes (for instance $Ru^{IV}$ to $Ru^{II}$) [102].
– Ethanol reduces, for example, $[PtCl_6]^{2-}$ to Zeise's salt $K[PtCl_3(\eta^2\text{-}C_2H_4)]$ [103].
– Reduction of $Au^{III}$ to $Au^I$ by sulfides has been known for a century [104] and is still useful [105].
– Sodium dithionite $Na_2S_2O_3$ reduces, inter alia, $Fe^{III}$ [106] and $Tc^{VII}$ [107].

- Derivatives of $Sn^{II}$, $Ti^{III}$, $Cr^{II}$, $V^{II}$, $Cu^{I}$, $Sm^{II}$ are well-known as reductants [108, 109]. In particular, Kagan's reagent $SmI_2$ reduces acyl chlorides to diketones, aldehydes, and ketones to pinacols in aprotic media and to alcohols in protic media [110]. It has also been applied to many other organic reactions [110]. $Cr^{II}$ salts have been extensively used to reduce alkyl- and aryl halides [109]. These reduced metal complexes have been known as reductants of transition metal ions since Taube's pioneering work [108, 109].
- Some reduced organometallic derivatives; e.g. $Na[FeCp(CO)_2]$ ($E^{\circ\prime} = -1.74$ V relative to $FeCp_2$ [111]) and $[Cp^*Cr(\mu\text{-}OPR_2)_3CoCp]$ ($E^{\circ\prime} = -2.0$ V relative to $FeCp_2$ [112]), have been used in single occasions [112, 113] to reduce other organometallic complexes.
- Amines are weak reductants. For instance, triethylamine ($E_{pa} = +0.47$ V relative to $FeCp_2$ in MeCN [114]) has been used as sacrificial reductant in photocatalyzed water photosplitting [115]. Hydrazine $NH_2\text{-}NH_2$ is more common in aqueous systems than in non-aqueous ($E_{pa} = -0.4$ V relative to $FeCp_2$ in DMSO [116]). It has occasionally been used to reduce cationic organometallic complexes to their neutral counterparts, the oxidized product ($N_2$) being readily removed [117]. Tetrakis(dimethylamino)ethylene, $C_2(NMe_2)_4$, is a convenient, commercially available two-electron reducing agent, soluble in organic solvents, which reduces metal carbonyls such as $[Co_2(CO)_8]$ to $[C_2(NMe_2)_4][Co(CO)_4]_2$ [118].
- Phosphines are deoxygenating reagents which form phosphoranes, the driving force being the strong P=O bond in the latter. Thus, they deoxygenate various metal–oxo complexes and oxygen-containing ligands [119].

### 4.3.5 Metallocenes and bis-Arene Metal Sandwiches

**Cobaltocene**

Cobaltocene is the most frequently used single-electron reductant although its standard oxidation potential is not very negative ($E^{\circ\prime} = -1.31$ V relative to $FeCp_2$ in DME). Thus, it can only reduce organometallic compounds which are very easy to reduce. It is readily prepared under an inert atmosphere, but another drawback is that it is air-sensitive in both solution and the solid state and must be sublimed just before use [120]. Its properties as a reductant have been reviewed [21, 121, 122]. Noteworthy applications of cobaltocene are single-electron reductions of neutral complexes to their monoanions giving ion pairs with cobaltocenium as the counter-cation:

$$[CoCp_2] + A \rightarrow [CoCp_2][A]$$

For instance, the $d^1$–$d^1$ dinuclear rhenium complex $[\{Re(O)Me_2\}(\mu_2\text{-}O)_2]$ is reduced in this way to the mixed-valence $Re^V$–$Re^{VI}$ monoanion by cobaltocene [123]; several other examples are known [21]. Cobaltocene is not always an 'innocent' reductant. For instance, it undergoes ring coupling with the 19-electron complex $[NiCp(\eta^4\text{-}C_4Ph_4)]$ when used in excess to reduce the 18-electron cationic precursor $[NiCp(\eta^4\text{-}C_4Ph_4)][PF_6]$ [124]. It also reacts with alkyl halides to give ring-coupling products and must therefore be used very rapidly in halogenated solvents such as chloroform and carbon tetrachloride [125]. Occasionally, the follow-up reaction of the reduced organometallic derivative can drive the endergonic electron-transfer reaction of cobaltocene [126].

### Decamethylcobaltocene

Decamethylcobaltocene [CoCp*$_2$], [126], has a standard redox potential ca 0.6 V more negative than cobaltocene ($-1.9$ V relative to FeCp$_2$ in MeCN) and it has, therefore, been used when cobaltocene was not an efficient single-electron-transfer reductant [127–129]. The precipitation of the salt [CoCp*$_2$][A] can be an additional driving force when the reduction of a neutral derivative A is slightly endergonic. For instance [W($\eta^5$-C$_5$HPh$_4$)(CO)$_2$(NO)] ($E^{o\prime} = -2.2$ V relative to FeCp$_2$) could be reduced by [CoCp*$_2$] to such a salt, despite the endergonicity of 0.3 V [129].

### Ferrocene

Ferrocene can be used as a weak reductant, but it is so weak that it has very rarely been used, for example:

$$[Pt^{IV}Cl_4(NH_3)_2] + 2[Fe^{II}Cp_2] \rightarrow [Pt^{II}Cl_2(NH_3)_2] + 2[Fe^{III}Cp_2]Cl \qquad [130]$$

It is useful for titraton of strong oxidants, for example:

$$[Fe^{III}Cp^*(C_6Me_6)][SbCl_6]_2 + Fe^{II}Cp_2$$
$$\xrightarrow{\text{THF}} [Fe^{II}Cp^*(C_6Me_6)][SbCl_6] + [Fe^{III}Cp_2][SbCl_6] \qquad [131]$$

The reduction potential of ferrocene can be increased by introduction of methyl groups or electron-releasing substituents on the rings [132]. In particular, decamethylferrocene [FeCp*$_2$], ($E^{o\prime} = -0.59$ V relative to FeCp$_2$ in CH$_2$Cl$_2$), is a single-electron reductant which is used as a reductant much more frequently than is ferrocene. For instance, it can reduce TCNE to form [FeCp*$_2$][TCNE], a ferromagnet [133]. The use of decamethylferrocene as a mild single-electron reductant is an especially practical means of reducing ferrocene derivatives. The many ferrocenium derivatives which have a redox potential close to that of ferrocene cannot be reduced efficiently by ferrocene because the redox reaction would be equilibrated rather than complete as hoped for. For instance, ferrocenium dendrimers containing theoretical numbers of ferrocenium units up to 243 were generated from ferrocenyl dendrimers by oxidation with [NO][PF$_6$], and could be reduced back to ferrocenyl dendrimers by use of decamethylferrocene in THF. [FeCp*$_2$][PF$_6$], which was generated in such reactions, precipitated and was easily removed by filtration of the THF solution in which the ferrocenyl dendrimers are soluble [134, 135].

### Bis(arene) chromium complexes

The 18-electron sandwich complexes [Cr($\eta^6$-arene)$_2$] have a range of standard redox potentials from $E^{o\prime} = -1.15$ V relative to FeCp$_2$ in CH$_2$Cl$_2$ (arene = C$_6$H$_6$) to $E^{o\prime} = -1.45$ V (arene = C$_6$Me$_6$) [136] and even $-1.5$ V (arene = C$_6$H$_5$NMe$_2$) [137], but have been very little used as reductants, for example:

$$2[Cr^0(C_6H_6)_2] + [Co_2(CO)_8] \rightarrow 2[Cr^I(C_6H_6)_2][Co(CO)_4] \qquad [138]$$

**The $d^7$ 19-electron organometallic electron-reservoir complexes**
**$[Fe^I(\eta^5\text{-}C_5R_5)(\eta^6\text{-}C_6R'_6)]$**

The $d^7$ 19-electron sandwich complexes $[Fe^I(\eta^5\text{-}C_5R_5)(\eta^6\text{-}C_6R'_6)]$ [139–143] (R = H or Me, R' = Me) are, on the basis of the values of their ionization potentials measured by He(I) photoelectron spectroscopy [143] the most electron-rich neutral molecules known. Accordingly, their redox potentials are very negative. For instance the standard oxidation potential of $[Fe^ICp(\eta^6\text{-}C_6Me_6)]$ ($E^{o\prime} = -2.02$ in DMF) is 0.7 V more negative than that of cobaltocene, and that of $[Fe^ICp^*(\eta^6\text{-}C_6Me_6)]$ ($E^{o\prime} = -2.30$ V in DMF) is almost 1 V more negative than that of cobaltocene and as negative as that of the benzophenone radical anion. Other advantages of these $Fe^I$ complexes as reductants are that they are crystalline and can be weighed accurately; they are also soluble in pentane, THF, and other organic solvents. They are thermally stable up to >100 °C and can be stored under an inert atmosphere; $[Fe^ICp(\eta^6\text{-}C_6Me_6)]$ sublimes at 70 °C ($2 \times 10^{-4}$ mmHg). A Mössbauer spectrum of this complex has even recently been recorded after storage for twenty years in a sealed tube.

They are Jahn–Teller active $d^7$ complexes subject to rhombic distortion, as has been shown by the ESR and Mössbauer spectroscopy. For instance, the values of the $g$ tensor recorded in the ESR spectra are $g = 2.063, 2.000, 1.864$ at 77 K (frozen DME) [144] and the temperature-dependent Mössbauer parameters in the solid state are I.S = 0.74 mm s$^{-1}$ and Q.S. = 1.6 mm s$^{-1}$ at 4 K, 0.5 mm s$^{-1}$ at 293 K for $[Fe^ICp(\eta^6\text{-}C_6Me_6)]$, indicative of a Boltzmann thermal population of the upper Kramers doublet [139, 145]. This indicates that the antibonding e$^*_1$ orbital of the classic metallocene MO diagram is singly occupied, and has a high metal character (70 %) as confirmed by recent accurate DFT MO calculations [146]. These physical properties show that the redox center which undergoes the redox change is located on Fe at the center of the molecular framework. This ligand shell protects it from the classic breakdown processes encountered in molecular chemistry upon redox change such as radical-type reactions of the ligands known for the second- and third-row transition-metal sandwich complexes for which single-electron reduction has been attempted [147, 148]. At the same time, the SOMO is only slightly antibonding, so the destabilization provoked by its single occupancy is low enough for thermal stability of the hexamethylbenzene $Fe^I$ complexes. These complexes were, therefore, denoted *electron-reservoir complexes* [139–141], which implies several properties [2]: strong single-electron stoichiometric reductants, redox catalysts, initiators of electron-transfer chain reactions, redox sensors and references for the determination of standard redox potentials by electrochemistry. Recently, these concepts were applied to the oxidation side with the isolation of the thermally stable $d^5$ 17-electron complex $[Fe^{III}Cp^*(\eta^6\text{-}C_6Me_6)][SbX_6]_2$ (X = Cl or F), and the finding that $[Fe^{III}Cp(\eta^6\text{-}C_6Me_6)]^{2+}$ is also a redox catalyst [131].

The $Fe^I$ complexes are synthesized by Na–Hg reduction in THF or DME of their $d^6$ 18-electron precursors $[Fe^{II}Cp(\eta^6\text{-arene})][PF_6]$ and $[Fe^{II}Cp^*(\eta^6\text{-arene})][PF_6]$ at

**Table 9.** First ionization energy and $E^{o'}$ of stable 19-electron complexes.

| Complex | Ionization energy (eV) | $(E^{o'}$ relative to FeCp$_2)^a$ | Ref. |
|---|---|---|---|
| [Cp*Fe($\eta^6$-C$_6$Me$_6$)] | 4.21 | −2.24 V | 142a[b] |
| K metal | 4.34 | −2.38 V | 142b |
| [CpFe($\eta^6$-C$_6$Et$_6$)] | 4.34 | −2.14 V | 142a[b] |
| [CpFe($\eta^6$-C$_6$Me$_6$)] | 4.68 | −2.02 V | 142a[b] |
| [CpFe($\eta^6$-1,3,5-$t$-Bu$_3$C$_6$H$_3$)] | 4.72 | −1.87 V | 142a[b] |
| [Cp*$_2$Co] | 4.8 | −1.84 V | 142c[b] |
| [Cp$_2$Co] | 5.3 | −1.20 V | 144[c] |
| [CpCo($\eta^6$-C$_5$H$_5$BMe)] | 6.5 | −0.80 V | 144[c] |
| [CpCo($\eta^6$-C$_5$H$_5$BPh)] | 6.6 | −0.78 V | 144[c] |
| [Co($\eta^6$-C$_5$H$_5$BMe)$_2$] | 7.1 | −0.36 V | 144[c] |
| Co(C$_5$H$_5$BPh)$_2$ | 7.2 | | 144[c] |

[a] $E^{o'}$ values are in DMF except K in NH$_3$. Stable, neutral 19-electron complexes having an even more negative redox potential than [Cp*Fe(C$_6$Me$_6$)] are [Cp*Fe$^I$(C$_6$Et$_5$H)] ($E^{o'} = -2.27$ V) and [Fe$^I{}_2(\mu_2, \eta^{10}$-fulvalene)(C$_6$Me$_6$)$_2$] ($E^{o'} = -2.30$ V).
[b] Determined by He(I) photoelectron spectroscopy; see Green et al. [143a].
[c] Determined by mass spectrometry; see Herberich et al. [144]; $E^{o'}$ values are estimated in DMF from this reference.

ambient temperature for a few hours [139–142] (see Section 4.9, Experimental). The precursor salts [Fe$^{II}$Cp($\eta^6$-arene)][PF$_6$] are easily synthesized by use of the ligand-exchange reaction of ferrocene with the appropriate arene in the presence of AlCl$_3$, Al, H$_2$O (4:1:1) at 80 °C, in the arene as solvent (or neat in the melt) or in an inert solvent such as heptane, cyclohexane, or decalin [100, 141, 149, 150] (Section 4.9). The precursor salts [Fe$^{II}$Cp*($\eta^6$-arene)][PF$_6$] are synthesized likewise from [FeCp*(CO)$_2$Br], AlCl$_3$ and the arene under analogous conditions [141, 148] (Section 4.9). The presence of the arene ligand has opened access to a large variety of complexes with different substituents which can influence the redox potential and the ionization potential. Table 9 gathers the standard redox potentials and ionization potentials of stable 19-electron Fe$^I$ and Co$^{II}$ complexes.

Many more [Fe$^I$Cp(arene)] complexes are known with numbers of methyl- or alkyl groups on the arene ring between one and six; this enables progressive variation of the redox potential. These Fe$^I$ complexes are only stable at room temperature when the arene ring is bulky, i.e. C$_6$Me$_6$, C$_6$Et$_6$, C$_6$Et$_5$H, or C$_6$H$_3$($t$-Bu)$_3$. The parent complex [Fe$^I$Cp($\eta^6$-C$_6$H$_6$)] [154] and the complexes [Fe$^I$Cp($\eta^6$-C$_6$H$_{6-n}$ Me$_n$)] [140] ($0 < n < 6$) are stable at −10 °C and can be synthesized at −10 °C by Na–Hg reduction of the PF$_6{}^-$ precursor salts in DME or THF. They are best be used as single-electron reductants if the solution prepared in this way is used directly at this temperature without isolating the complex in the solid state (the solutions are easily titrated by the oxidant since the color of the iron complex color changes from forest-green (Fe$^I$) to colorless (Fe$^{II}$, yellow precipitate) [155]. These complexes are soluble in pentane and organic solvents and more or less progressively dimerize in these solvents by coupling of the benzene ligand to bicyclohexadienyl between 0 °C and 25 °C. This reaction is logically faster in the solid state. The complex [Fe$^I$Cp*($\eta^6$-C$_6$H$_6$)] dimerizes much faster than the Cp analogs when the solvent is removed even at low temperature, and cannot even be obtained in the solid state. It is believed that this trend is due to an equilibrium between the 19-electron and the 18-electron form of the species. Alternatively, if only one potential well accounts for the structure of these complexes, the weight of the 18-electron form among the mesomer forms could be more pronounced with the Cp* complex. The spin density on the benzene ligand is lower in the Cp* complex than in the Cp complex, however, which would favor the tautomerism view, at least in this case [100]. It is possible to introduce a carboxylate group on the Cp or on the $C_6Me_6$ ligand of the [FeCp(arene)] structure in order to solubilize these complexes in water under both the $Fe^{II}$ and $Fe^I$ forms. Indeed the $Fe^I$ complex with a Cp-$CO_2^-$ ligand has been isolated and characterized as a genuine 19-electron complex. It is purple and very reactive, although thermally stable. It is very useful as a strong single-electron reductant in water, especially if kinetic measurements are desired, since the aqueous solution containing the reductant is then homogeneous.

The most frequently used electron-reservoir complex for stoichiometric single-electron transfer reactions is $[Fe^ICp(\eta^6\text{-}C_6Me_6)]$, because of its stability and ease of preparation, and since it has one of the most negative redox potentials in the series. It can reduce most inorganic and organometallic cations [2]. For instance, it is very useful to synthesize neutral 19-electron complexes (C in the equation below) such as other $[Fe^ICp(\eta^6\text{-arene})]$ complexes and $[Fe^I(\eta^5\text{-}C_6Me_6H)(\eta^6\text{-}C_6Me_6)]$ from the 18-electron cationic precursors $C[PF_6]$:

$$[Fe^ICp(\eta^6\text{-}C_6Me_6)](19e) + C[PF_6](18e)$$

$$\xrightarrow{\text{THF, RT}} [Fe^{II}Cp(\eta^6\text{-}C_6Me_6)][PF_6](18e) + C\ (19e)$$

It can reduce a variety of neutral compounds A to their monoanion $A^-$ in pentane, diethyl ether or THF; this has been shown with some polycyclic hetero-aromatics such as phenazine [156].

$$[Fe^ICp(\eta^6\text{-}C_6Me_6)](19e) + A \rightarrow [Fe^ICp(\eta^6\text{-}C_6Me_6)][A]\ (18e)$$

Compounds whose radical anions give follow-up reactions can be reduced, and the reaction medium (aqueous or non-aqueous) eventually governs the nature of the final products. A number of organic substrates such as aldehydes, ketones and alkynes have been reduced in this way. It is also possible to reduce substrates which have a potential much more negative than its standard redox potential. For instance, nitrate, nitrite and carbon dioxide are reduced [2]. In the latter case [157], the overpotential reaches 0.6 V and the reaction is still rapid because the rate of the follow-up reaction, i.e. the dimerization of $CO_2^{\bullet-}$, is extremely high ($k = 10^9$ $M^{-1}$ $s^{-1}$) [158].

$$[Fe^ICp(\eta^6\text{-}C_6Me_6)] + CO_2 \leftrightarrow [Fe^ICp(\eta^6\text{-}C_6Me_6)][CO_2^{\bullet-}]$$

$$\rightarrow \tfrac{1}{2}[Fe^ICp(\eta^6\text{-}C_6Me_6)][C_2O_4]$$

Scheme 7 shows representative examples of stoichiometric reductions using $[Fe^ICp(\eta^6\text{-}C_6Me_6)]$.

$O_2$

——————————→ $O_2^{\cdot-}$ → etc.    (Refs.159-162)

$C_{60}$

——————————→ $C_{60}^{\cdot-}$    (Ref. 163)

$C_{60}^{\cdot-}$

——————————→ $C_{60}^{2-}$    (Ref.163)

$C_{60}^{2-}$

——————————→ $C_{60}^{3-}$    (Ref.163)

phenazine

——————————→ [phenazine]$^{\cdot-}$    (Ref.156)

$CO_2$

——————————→ $\frac{1}{2}$ CO + $\frac{1}{2}$ $CO_3^{2-}$    (Ref.157)

$[Fe^ICp(C_6Me_6)]^-$ ⋯ PhCO*R*

——————————→ PhCHOH*R*   (*R* = H, CH$_3$) (Ref.164)

PhC≡CH

——————————→ PhCH = CH$_2$    (Ref.164)

$[Cp^*Fe^{IV}(dtc)_2]^+$

——————————→ $[Cp^*Fe^{III}(dtc)_2]$    (Ref.165)

$[Fe^{II}(C_6Me_6)(\eta^5\text{-}C_6Me_6H)]^+$

——————————→ $[Fe^I(C_6Me_6)(\eta^5\text{-}C_6Me_6H)]^{\cdot}$   (Ref.166)

$[Fe^{III}Cp^*(PMe_3)_2CH_3]^{\cdot+}$

——————————→ $[Fe^{II}Cp^*(PMe_3)_2CH_3]$    (Ref.167,168)

$[ML(C_4Ph_4)]^+$

——————————→ $[ML(C_4Ph_4)]^{\cdot}$    (Ref. 124)

*M* = NiCp or Co(C$_6$H$_3$Me$_3$)

**Scheme 7.**

It is interesting to compare the relative reduction power of cobaltocene, deca-methylcobaltocene and $[Fe^ICp(\eta^6\text{-}C_6Me_6)]$ towards $C_{60}$. The first of these studies was carried out with $[Fe^ICp(\eta^6\text{-}C_6Me_6)]$. It was shown that the mono-, di- and tri-anions of $C_{60}$ were obtained as salts of $[Fe^{II}Cp(\eta^6\text{-}C_6Me_6)]^+$ depending on the stoichiometry of the reaction. The titration of $C_{60}$ by $[Fe^ICp(\eta^6\text{-}C_6Me_6)]$ is conveniently carried out since the solution remains deep forest-green when all the $C_{60}$ is transformed into its trianion. The salt of the trianion $[Fe^{II}Cp(\eta^6\text{-}C_6Me_6)]_3$ $[C_{60}]$ is of special interest given the superconducting properties of the alkali-metal salts of this trianion [169]. In the present case, however, the lattice arrangement is governed by the organometallic cation which is larger than the anion, and the salt $[Fe^{II}Cp(\eta^6\text{-}C_6Me_6)]_3$ $[C_{60}]$ is paramagnetic [163]. In subsequent studies, it was shown that cobaltocene could only reduce $C_{60}$ to its monoanion [170] whereas decamethylcobaltocene could lead to the dianion, but not the trianion (Scheme 8) [171].

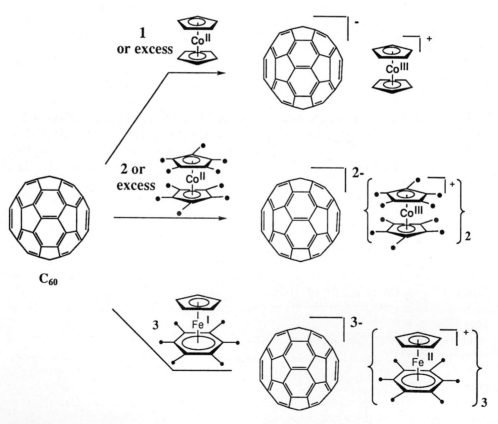

**Scheme 8.**

The reaction of $[Fe^ICp(\eta^6\text{-}C_6Me_6)]$ with $O_2$ also proceed by a very favorable electron transfer to generate superoxide radical anion $O_2^{\bullet-}$ (the electron-transfer process is exergonic by about 0.85 V, i.e. 20 kcal $mol^{-1}$ or 83.6 kJ $mol^{-1}$), but it is very peculiar [159–162], and the follow-up reaction considerably depends on the presence of an inorganic salt, the salt effect being maximum with $NaPF_6$ [172] In pentane or THF in the absence of a salt, the reaction is fast even at $-80\,°C$ and yields the deep-red 18-electron complex $[Fe^{II}Cp(\eta^5\text{-}C_6Me_5CH_2)]$ resulting from an overall removal of a H atom from the 19-electron complex; $\frac{1}{2}$ equiv. $H_2O$ is produced at 20 °C whereas $\frac{1}{2}$ $H_2O_2$ is formed at $-80\,°C$ (Scheme 9):

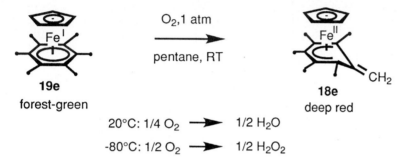

**19e**
forest-green

**18e**
deep red

20°C: 1/4 $O_2$ → 1/2 $H_2O$

-80°C: 1/2 $O_2$ → 1/2 $H_2O_2$

**Scheme 9.**

Superoxide radical anion can be characterized by ESR as an intermediate at low temperature. It is also possible to obtain the reaction product by reaction of $KO_2$ with $[Fe^{II}Cp(\eta^6\text{-}C_6Me_6)][PF_6]$ in DMSO in the presence of 18-crown-6. The overall H-atom abstraction in the above reaction of the 19e complex is thus decomposed into an electron transfer followed by benzylic deprotonation by $O_2^{\bullet-}$ in the ion pair $[Fe^{II}Cp(\eta^6\text{-}C_6Me_6)^+,O_2^{\bullet-}]$. This great reactivity of $O_2^{\bullet-}$ in the ion pair is due to the small size of this anion and to the fact that it is not strongly solvated in pentane or even in THF. It illustrates the dramatic reactivity of superoxide in a hydrophobic environment. When the reaction is carried out in THF, it is possible to do it in the presence of an inorganic salt such as $NaPF_6$ which completely changes its course: none of the deep-red complex $[Fe^{II}Cp(\eta^5\text{-}C_6Me_5CH_2)]$ is obtained (the resulting solution is colorless!) and a yellow precipitate of $[Fe^{II}Cp(\eta^6\text{-}C_6Me_6)][PF_6]$ together with $Na_2O_2$ (resulting from the disproportionation of $O_2^{\bullet-}$) are obtained (Scheme 10) [172].

The efficiency of the simple $Na^+$ salts to inhibit the reactivity of $O_2^{\bullet-}$ is very spectacular, and it matches the reactivity of superoxide dismutase enzymes in biological systems [173]. It is thus a unique property of the electron-reservoir complexes which lets us investigate the electron-transfer to $O_2$ in various media. The follow-up reactions of superoxide radical anion are reminiscent of its damage for cells in the aging processes which is well-known but little understood [173].

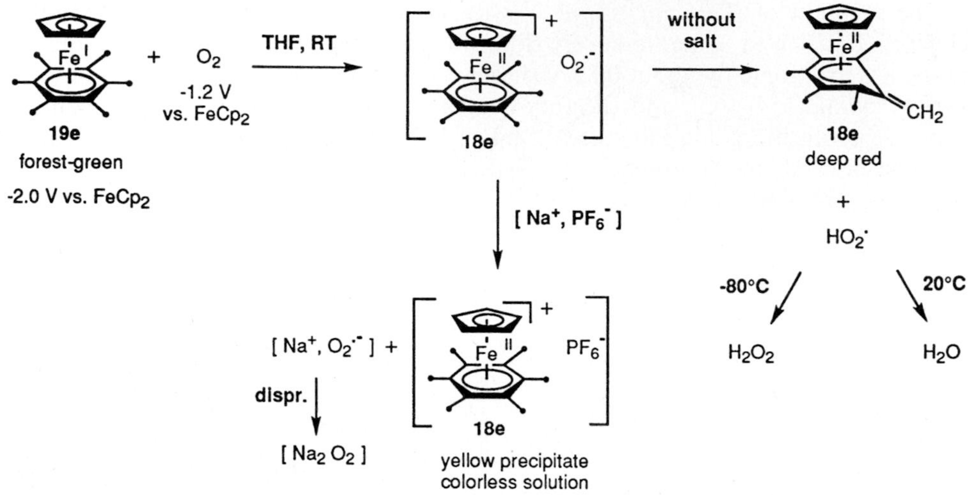

**Scheme 10.**

In the absence of benzylic proton, the reaction of $Fe^I$ complexes with $O_2$ leads to peroxo-bridged dimers resulting from nucleophilic attack of superoxide radical anion with the 18-electron cationic sandwich in the ion pair. The mechanism is rather similar to the preceding one except for the reaction following the primary electron transfer. Superoxide was also characterized at low temperature, and a stoichiometric amount of $NaPF_6$ inhibits its reaction likewise, leading to $[Fe^{II}Cp(\eta^6\text{-}C_6Me_6)][PF_6]$ and $Na_2O_2$. This thoroughly studied example illustrates the reactivity of the two partners of an ion cage in solvents such as pentane or THF in the absence of ions or solvating species, and indicates that a stoichiometric amount of $NaPF_6$ must be used in THF if this reactivity is not desired [162]. Thus, the reaction of the proto-type electron-reservoir complex $[Fe^ICp(\eta^6\text{-}C_6Me_6)]$ with neutral substrates were later carried out in the presence of $NaPF_6$ in stoichiometric amount when the aim was to reduce the substrate without being marred by the cage reactions. This can be easily done by using the THF solution used for the generation of the $Fe^I$ complex from Na–Hg and the precursor $PF_6$ salt. Indeed, these 'quantitative' salt effects best provided by $NaPF_6$ are very useful in organometallic reactions [174, 175] On the other hand, besides the interest in biomimetism, the useful product of the reaction of $O_2$ performed in the absence of a salt is not the reduced oxygen species, but the activated complex $[Fe^{II}Cp(\eta^5\text{-}C_6Me_5CH_2)]$. Indeed, this complex reacts smoothly with virtually all the electrophiles leading to the formation of carbon–element bonds. This has been shown with organic and inorganic halides (scheme below), $CO_2$ and $CS_2$, metal carbonyls and organometallic complexes, which led for instance to the formation of bonds between the benzylic carbon and C, Si, P, Fe, Mn and Mo (Scheme 11) [161].

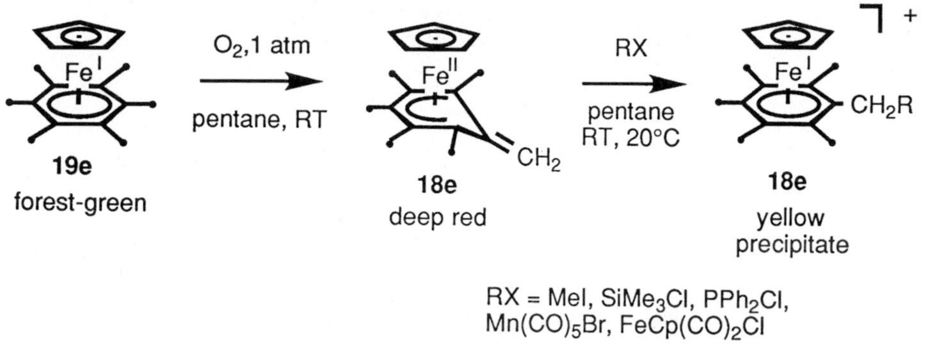

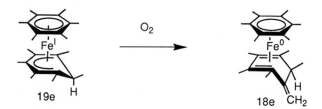

**Scheme 11.**

The reaction of $O_2$ with neutral 19-electron complexes bearing methyl groups on the rings is general and has been also found in other cases such as decamethyl-cobaltocene or others. Example (Scheme 12):

**Scheme 12.**

It is also possible to remove two H atoms using the 20-electron complex $[Fe^0(\eta^6\text{-}C_6Me_6)_2]$. The latter has a standard redox potential $E^{\circ\prime}(Fe^{0/I}) = -1.85$ V relative to $FeCp_2$ in DMF, i.e. also much more negative than that of $O_2/O_2^{\cdot-}$ ($-1.15$ V relative to $FeCp_2$ in DMF). The intermediate $[Fe^I(\eta^5\text{-}C_6Me_5CH_2)(\eta^6\text{-}C_6Me_6)]$ resulting from removal of the first benzylic hydrogen has a structure very close that of $[Fe^I(\eta^5\text{-}C_6Me_6H)(\eta^6\text{-}C_6Me_6)]$ (above) whose standard redox potential value is also very negative ($E^{\circ\prime}(Fe^{I/II}) = -1.92$ V relative to $FeCp_2$ in DMF) [166]. The same sequence of electron transfer + deprotonation in the cage is thus occurring twice at this time, yielding the *ortho*-xylylene complex $[Fe^0\{\eta^4\text{-}C_6Me_4(CH_2)_2\}\text{-}(\eta^6\text{-}C_6Me_6)]$, stable at $-40\,^\circ C$ and reactive towards electrophiles such as benzoyl chloride (Scheme 13) [166, 176].

The above examples shows that multiple H atoms may be removed from benzylic positions as long as the complexes resulting from overall H-atom abstraction have a standard redox potential more negative than that of $O_2/O_2^{\cdot-}$. During our attempt to extend this reaction to the functional complex $[Fe^I(\eta^5\text{-}C_5H_4CH_2NHn\text{-}Pr)\text{-}(\eta^6\text{-}C_6Me_6)]$ in order to carboxylate the hexamethylbenzene ligand, we found that the stoichiometry in dioxygen at $25\,^\circ C$ was not 0.25 equiv. as expected, but 0.75

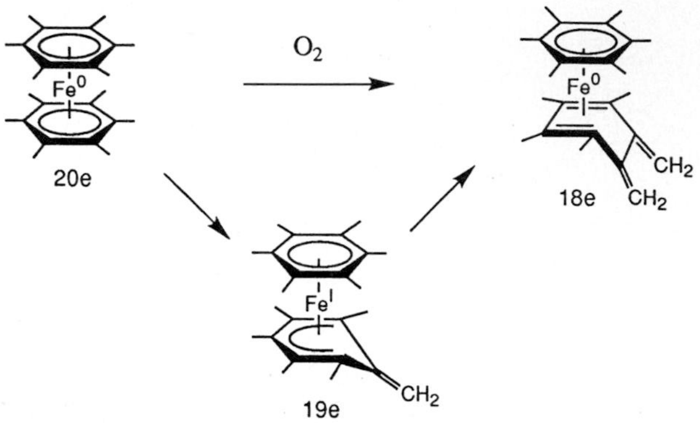

**Scheme 13.**

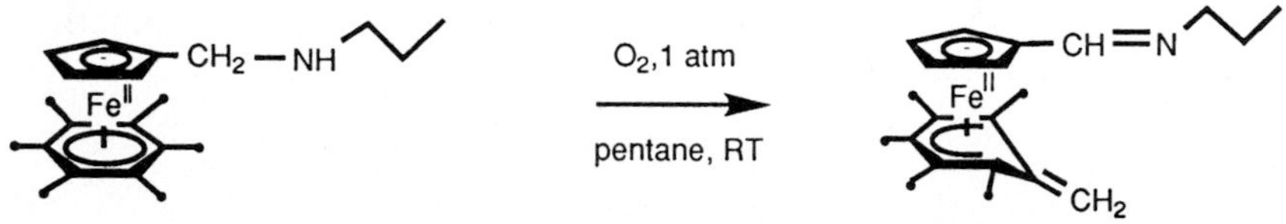

**Scheme 14.**

equiv., i.e. three times more. The resulting red product $[Fe^{II}(\eta^5\text{-}C_5H_4CH=Nn\text{-}Pr)\text{-}(\eta^5\text{-}C_6Me_5CH_2)]$ which was obtained after a short contact with air under ambient conditions in pentane resulted in fact from the removal of three H atoms from this $Fe^I$ complex (Scheme 14) [177].

Intriguingly, the reaction proceeds with oscillatory color changes from green to red. When it is carried out by stepwise addition of $\frac{1}{4}$ equiv. dioxygen, the color indeed changes from green ($Fe^I$) to red ($Fe^{II}$), but the red complexes are not stable and turn green again except after the third addition of dioxygen. The final red complex is stable. Its formation can be taken into account according to H-atom transfer from the amine to the methylenecyclohexadienyl ligand to form again the hexamethylbenzene ligand which in turn reacts again with dioxygen (Scheme 15).

The complex resulting from triple C–H/N–H activation can be heterobifunctionalized by successive addition of $CO_2$ (carboxylation of the hexamethylbenzene ligand) and water (acidic hydrolysis of the imine to the aldehyde on the Cp substituent). This heterobifunctional complex is very useful for the branching to nanoscopic substrates and yet keep the solubility in aqueous media by means of the carboxylate group. Use of this functionalized electron-reservoir system in redox catalysis is promising. Thus, from the starting 18-electron complex, the overall heterobifunctionalization can be carried out in one pot as shown in Scheme 16 [177, 178].

The reactions of organic halides RX with the 19-electron complex $[Fe^ICp\text{-}(\eta^6\text{-}C_6Me_6)]$ itself lead to coupling of the R group onto the Cp ring, producing

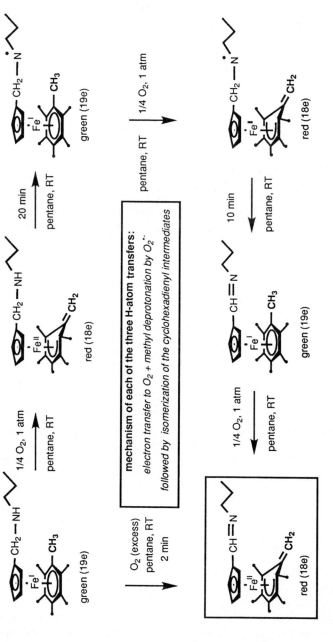

Scheme 15.

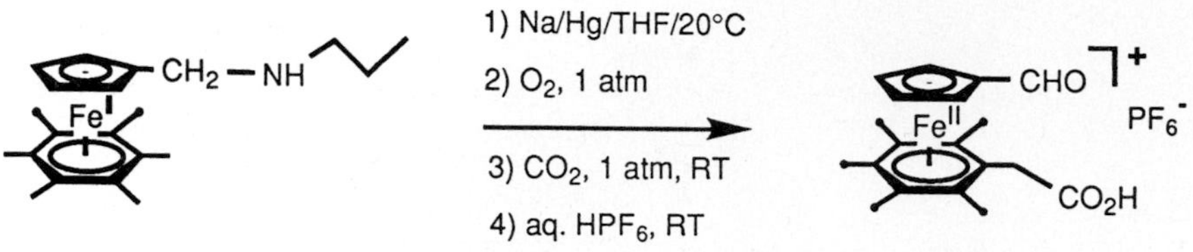

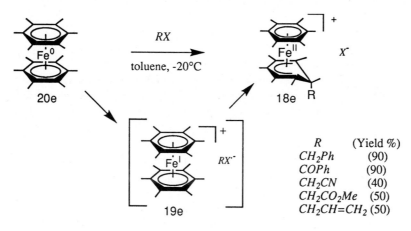

**Scheme 16.**

an equimolecular amount of $[Fe^ICp(\eta^6-C_6Me_6)]X$ and $[Fe^0(\eta^4-exo-RC_5H_5)-(\eta^6-C_6Me_6)]$, the latter adduct resulting from reaction between the starting $Fe^I$ complex of the radical $R^\bullet$ formed by reduction of $RX$ by another $Fe^I$. This reaction resembles that known with cobaltocene, and is not very productive for a synthetic purpose. On the other hand, the 20-electron complex $[Fe^0(\eta^6-C_6Me_6)_2]$ becomes a radical cation after having transferred one electron to $RX$, and the radical generated radical $R^\bullet$ can then couple to the 19-electron radical cation intermediate $[Fe^0(\eta^6-C_6Me_6)_2]^{+\bullet}$ to produce cationic 18-electron complexes with a functional cyclohexadienyl ligand. These reactions have been carried out with various functional halides (Scheme 17):

**Scheme 17.**

| R | (Yield %) |
|---|---|
| $CH_2Ph$ | (90) |
| $COPh$ | (90) |
| $CH_2CN$ | (40) |
| $CH_2CO_2Me$ | (50) |
| $CH_2CH=CH_2$ | (50) |

Such ligands can be deprotonated on a methyl substituent in $\alpha$ position with respect to the ring carbon bearing the functionalized group to give $Fe^0$ complexes and be subsequently removed by oxidation.

Starting from the parent 19-electron complex $[Fe^ICp(\eta^6-C_6H_6)]$, it is possible to generate very strong single-electron reducing agents by reaction with phosphines such as $PMe_3$. Indeed, this reaction leads to the exchange of the 6-electron ligand $C_6H_6$ with three 2-electron $PMe_3$ ligands. This provides the transient 19-electron species $[Fe^ICp(PMe_3)_3]$, whose standard redox potential is close to $-2.4$ V relative to

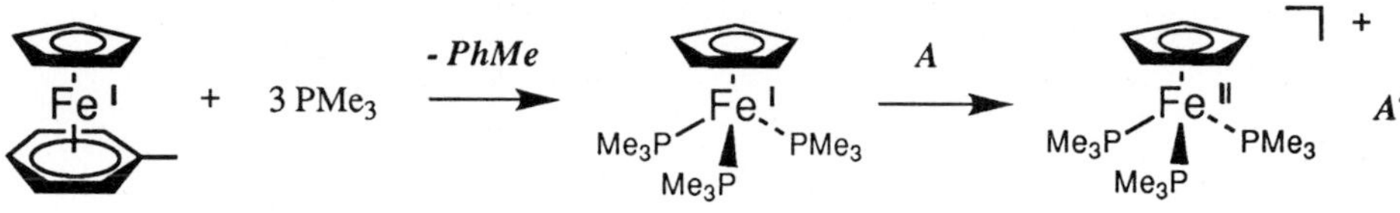


*A* = *p-dicyanobenzene* ($E^{0\prime}$ = - 2.04 vs. $FeCp_2$ in THF)
= *p-methylbenzoate* ($E^{0\prime}$ = - 2.16 vs. $FeCp_2$ in THF)

**Scheme 18.**

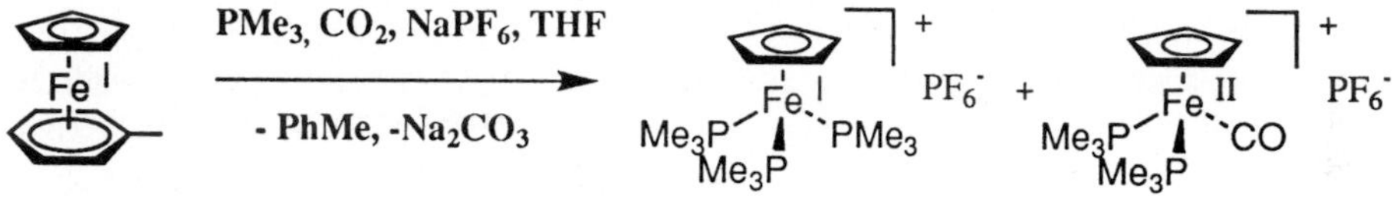


**Scheme 19.**

$FeCp_2$ in THF. Aromatic substrates such as *p*-dicyanobenzene and *p*-methylbenzoate whose standard redox potentials are more negative than −2 V relative to $FeCp_2$ can be reduced (Scheme 18).

Even $CO_2$ can be rapidly reduced to CO and carbonate, and CO incorporates in the coordination sphere of the metal at the 19-electron level whose very favorable ligand substitution properties are well known (Scheme 19).

The reducing power of the 19-electron species $[Fe^{I}Cp(PMe_3)_3]$, is so great that, in the absence of substrate, the 19-electron complex $[Fe^{I}Cp(\eta^6\text{-}C_6H_6)]$, itself is reduced to the unstable 20-electron anionic species which undergoes fast decomplexation under the influence of the ion pair $[Na^+, PF_6^-]$. This process corresponds to the disproportionation of the $Fe^I$ complex to $Fe^0$ and $Fe^{II}$ (Scheme 20).

In the absence of $PMe_3$ $[Na^+, PF_6^-]$, catalyzes this same process, except that it is slower. In this case, the salt effect is catalytic simply because all the species remain in solution [179, 180]. Thus, the presence of $[Na^+, PF_6^-]$ is important to favor electron-transfer processes from the 19-electron complexes, whether they are stable ($[Fe^{I}Cp(\eta^6\text{-}C_6Me_6)]$) or not ($[Fe^{I}Cp(\eta^6\text{-}C_6H_6)]$ or $[Fe^{I}Cp(PMe_3)_3]$). Salt effects are important in chain reactions as well [180–185] (see Section 4.5).

## 4.4 Oxidants

In the preceding section, we have shown the electron-reservoir properties of the redox series $[Fe(\eta^5\text{-}C_5R_5)(\eta^6\text{-}C_6Me_6)]^{2+/+/0}$ on the reduction side with the neutral 19-electron $Fe^I$ form. The 17-electron dicationic $Fe^{III}$ form can be isolated as the $SbX_6$ salts (R = Me, X = F or Cl)) and used as very strong oxidants, with a stan-

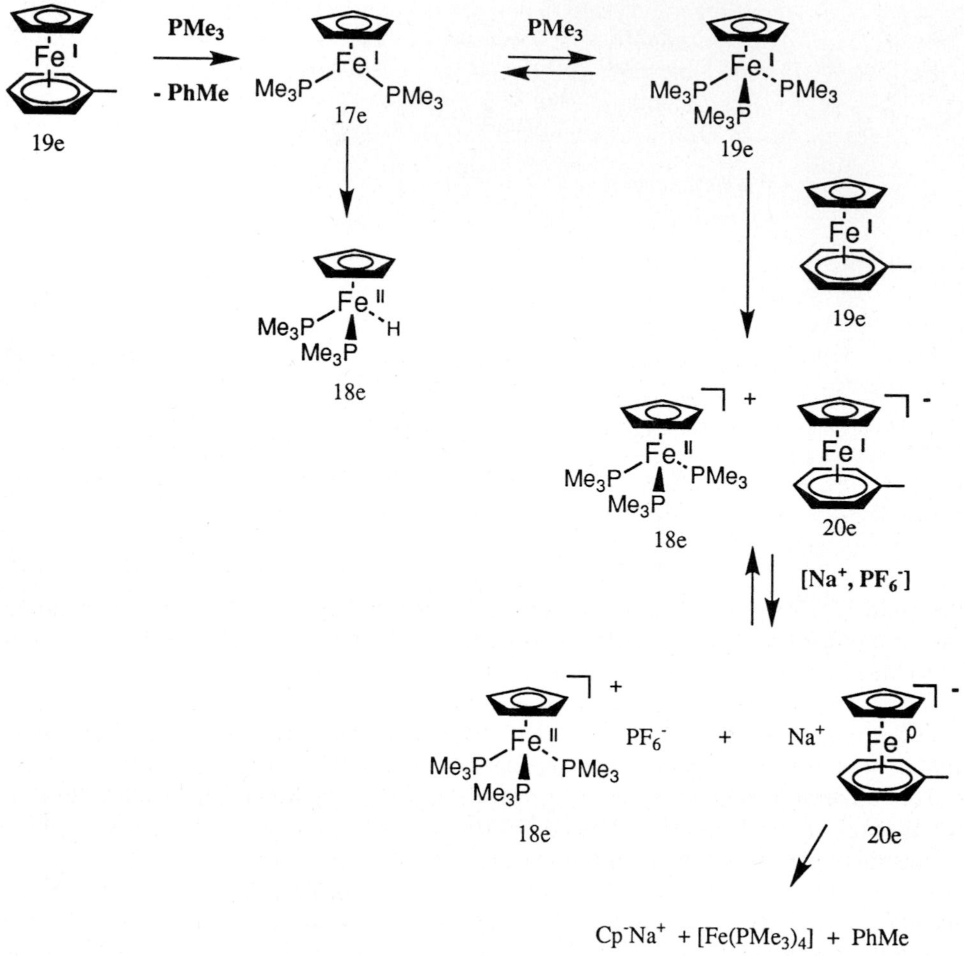

**Scheme 20.**

dard redox potential $Fe^{II/III}$ about 1 V more positive than that ferrocenium to which it is isoelectronic. Monoelectronic oxidants have been reviewed [21]. We will only summarize them in this section, and give the most noteworthy recent findings.

### 4.4.1 $Fe^{III}$ Sandwich Complexes: Ferrocenium Salts $[FeCp_2][X]$ and $[FeCp^*(\eta^6\text{-}C_6Me_6)][SbX_6]_2$

$[FeCp_2][BF_4]$ and $[FeCp_2][PF_6]$ are easily prepared by oxidation of ferrocene using a variety of oxidants (sulfuric acid [185], *p*-benzoquinone [21], $NO^+$, iron trichloride, etc. [186–191].) and commonly used in organometallic chemistry as very useful, almost innocent single-electron oxidants. They are thermally stable and can be handled in air in the solid state. They are more or less soluble in all the organic solvents, but must be used under an inert atmosphere in solution. They can also be used in the solid state (suspension) if the substrate is soluble in the solvent used. The reaction product, ferrocene, is soluble in non-polar solvents, which provides an easy means of separation. On the other hand, if excess ferrocenium salt is used, it is difficult to remove. These salts are non-innocent when the cation obtained upon ferrocenium oxidation is extremely reactive towards nucleophiles (for instance silicenium cation, early transition metal complexes). Fluorination of very reactive cations occurs indeed with the counter-anions $BF_4$ or $PF_6$, and other counter-anions such as $BPh_4^-$, $B(C_6F_5)_4^-$, $B\{3,5-C_6H_3(CF_3)_2\}_4^-$, or even carborane or perfluorocarborane anions must then be used [192, 193].

Substituents on the ferrocene rings lead to variations of the redox potentials [194]. Monoacetylferrocenium $[FeCp(\eta^5-C_5H_4COMe)]$, $[BF_4]$ $(E^{\circ\prime} = 0.27$ V in $CH_2Cl_2)$ is of particular interest because it is a stronger oxidant than parent ferrocenium salts. It can be prepared from $[FeCp(\eta^5-C_5H_4COMe)]$ and $AgBF_4$ (or other oxidants) in $CH_2Cl_2$ followed by filtration of colloidal Ag and precipitation by addition of ether [21, 195]. It should be used as freshly prepared because it is moisture sensitive and slowly deteriorates. 1,1'-diacetylferrocenium $[FeCp(\eta^5-C_5H_4(COMe)_2]$, $[BF_4]$ $(E^{\circ\prime} = 0.49$ V relative to $FeCp_2$ in $CH_2Cl_2)$ is an even stronger oxidant which is prepared likewise and is more air-sensitive [195]. These salts are also presumably easily synthesized from $[NO][BF_4]$ in $CH_2Cl_2$ followed by precipitation using ether as other ferrocenium salts, and have already been used when ferrocenium salts were not strong enough oxidants [196].

$[FeCp^*(\eta^6-C_6Me_6)][SbX_6]_2$ is a very strong oxidant $(E^{\circ\prime} = 1.03$ V relative to $FeCp_2$ in $CH_2Cl_2)$. Its standard redox potential is much more positive than that of ferrocenium, although it is isoelectronic to it, simply because of the additional positive charge. It is obtained by single- electron oxidation of $[FeCp^*(\eta^6-C_6Me_6)][PF_6]$ using $SbCl_5$ in $CH_2Cl_2$ or $SO_2$ or $SbF_5$ in $SO_2$ and can be recrystallized from $SO_2$. It can oxidize a wide range of compounds such as $[Ru(bpy)_3]^{2+}$ $[Cr(arene)(CO)_3]$, and transition-metal clusters which cannot be oxidized by ferrocenium salts. It is a rare innocent, very strong oxidant (Scheme 21). It is soluble in acetone and insoluble in $CH_2Cl_2$ but, given its strong reducing power, it is best to use it in the solid state provided the substrate to oxidize is soluble in the solvent used [131].

### 4.4.2 Silver(I) Salts

Silver(I)salts are commercially available with a variety of counter-anions which made them popular oxidants in both organic [197] and inorganic chemistry [198]. They are typical non-innocent oxidants in transition-metal chemistry, hygroscopic and photosensitive, and their standard redox potential very much depends on the solvent [199]. Thus Ag(I) salts are strong oxidants in 'non-coordinating' solvents such as $CH_2Cl_2$ $(E^{\circ\prime} = 0.65$ V relative to $FeCp_2)$, but their standard redox potential is all the less positive as the solvent is a better ligand. In MeCN $(E^{\circ\prime} = 0.04$ V relative to $FeCp_2)$, its oxidizing power is apparently not superior to that of ferrocenium salt [199]. $AgPF_6$ can oxidize, in $CH_2Cl_2$, substrates such as $[N(C_6H_4Br-4)_3]$

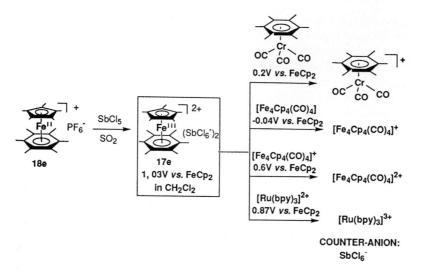

**Scheme 21.**

($E^{o\prime} = 0.70$ V relative to $FeCp_2$ in $CH_2Cl_2$), whose standard redox potential is slightly more positive than its own redox potential, because the redox equilibrium is shifted towards the products by the irreversible agglomeration and precipitation of silver particles. The silver ion Ag(I) has been used for almost a century as oxidant. For instance, in 1923, $AgNO_3$ was reported as an oxidant of metallic copper in acetonitrile to give the complex $[Cu(MeCN)_6][NO_3]$ [200]. Although they are soluble in organic solvents, silver (I) salts are best used in the solid state (fresh samples) because of their sensitivity. Filtration of colloidal silver using Celite or Kieselguhr is advised after the reaction. Toluene is a practical solvent in order to remove excess reactant as well as silver metal. In $CH_2Cl_2$ and $CHCl_3$, Ag(I) catalyzes the hydrolysis of the $[PF_6]^-$ counter-anion to $[PO_2F_2]^-$ in the presence of traces of water [201–203]. Side reactions of Ag(I) salts upon oxidation (non-innocent behavior) often involve the binding of reactive silver atoms with organometallic radicals generated by oxidation [21].

### 4.4.3 Copper(II) Salts

Copper(II) salts are familiar oxidants in organic chemistry [204, 205], but they have been little used by transition-metal chemists. They are commercially available with various counter-anions as anhydrous or hydrates ($CuCl_2$ is also simply obtained by heating $CuCl_2 \cdot 2H_2O$ in thionyl chloride) [206]. They have been used in $CH_2Cl_2$ and MeCN to cleave metal-carbon bonds in 18-electron iron (II) and ruthenium (II) complexes. The first step of this reaction is a single-electron oxidation [207].

$$[FeCp(CO)_2R] + 2CuCl_2 \rightarrow [FeCp(CO)_2Cl] + RX + 2CuCl$$

CuCl$_2$ is well-known as the oxidant of Pd(0) in the catalytic Wacker process (aerobic oxidation of ethylene to acetaldehyde) [208].

$$2CuCl_2 + Pd(0) \rightarrow 2CuCl + PdCl_2$$

### 4.4.4 Iron(III) Chloride

Iron(III) trichloride is commercially available both as the anhydrous or hydrate form which can also be refluxed in thionyl chloride to generate the anhydrous form (purified by sublimation) most often used [209]. Anhydrous iron trichloride is about as strong as ferrocenium salts, but it is not necessarily innocent. It is sometimes used to recover functionalized metal-free hydrocarbon derivatives which have been transformed by temporary complexation and activation on a transition-metal center [210].

$$[Fe\{\eta^4\text{-cis-1,2-}C_6H_6(CN)(CH_2Ph)\}(\eta^6\text{-}C_6H_6)]$$

$$\xrightarrow{FeCl_3,\ Et_2O} cis\text{-1,2-}C_6H_6(CN)(CH_2Ph + \cdots$$

### 4.4.5 Cerium(IV) Ammonium Nitrate (CAN)

Cerium (IV) ammonium nitrate (CAN) [NH$_4$]$_2$[Ce(NO$_3$)$_6$], is a water-soluble oxidant, classical in organic chemistry [197], which has sometimes been used in organometallic chemistry to disengage ligands from metal centers [211]. It can be used (i) in solution in an aqueous solvent such as ethanol, acetone and acetonitrile [212], (ii) with a phase-transfer catalyst such as [NBu$_4$][HSO$_4$] [213] or (iii) in the solid state (suspension) in, for instance, THF [214], for example (Scheme 22):

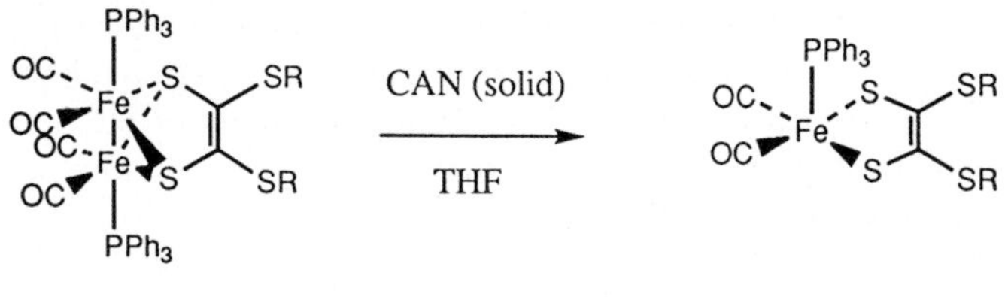

Scheme 22.

### 4.4.6 Miscellaneous High-oxidation-state Inorganic Complexes

M$^{III}$(L–L)$_3$]$^{3+}$ (M = Cr, Fe, Ru, Os, Co, Ni; L–L = bipy, phen, etc.) have been widely used in inorganic chemistry as outer-sphere single-electron oxidants [215], unfortunately as the potentially very hazardous and explosive perchlorate salts which should be replaced by $PF_6^-$ or $AsF_6^-$ salts. The tetracationic salt $[Ni^{IV}(bipy)_3][AsF_6]_4$, with a standard redox potential $E^{\circ\prime} = 2.06$ V, is a very strong oxidant which has been isolated [216]. $[Fe^{III}(bipy)_3]^{3+}$ has been used as an oxidant to determine the number of hydride ligands in transition-metal hydride complexes [217].

The anionic complex $M_2[IrCl_6]$ (M = Na and K) are commercially available and have been used as outer-sphere single-electron oxidants in mechanistic studies of the cleavage of metal-carbon bonds [218]. The isostructural oxidant $[PtCl_6]^{2-}$ is also known, especially its ability to oxidize saturated hydrocarbon by electrophilic C–H activation (inner-sphere two-electron oxidant) [219].

The neutral dithiolene complexes $[Ni(S_2C_2(CF_3)_2]$ and $[Mo(S_2C_2(CF_3)_2]$ are known as strong oxidants whose anions are stable, but their synthesis from the metal carbonyls [220] is tedious and dangerous in the case of the Ni complex because of the very high toxicity of $[Ni(CO)_4]$. Therefore, they have been little used [221] for this reason and also because some other oxidants with analogous oxidizing properties are commercial or easily accessible.

The metal hexahalides $MoF_6$, $WF_6$, $UF_6$ and $WCl_6$ are strong oxidants which are commercially available and soluble in various organic solvents, but the hexafluorides are readily hydrolyzed to the extremely corrosive and dangerous HF, and have therefore been little used. The standard redox potential of $WCl_6$ is around 1.1 V relative to $FeCp_2$ [222, 223] and it oxidizes $[N(C_6H_4Br-4)_3]$ to $[N(C_6H_4Br-4)_3][WCl_6]$ [224].

### 4.4.7 Nitrosonium Salts, [NO]X

The nitrosonium salts, [NO]X, are strong oxidants whose oxidizing power is solvent dependent ($E^{\circ\prime} = 0.56$ V in DMF, 0.87 V in MeCN and 1.00 V relative to $FeCp_2$ in $CH_2Cl_2$). They have been widely used in organic [225] and inorganic chemistry [12, 226, 227] and reviewed. They are commercially available (X = $BF_4$, $PF_6$, $SbF_6$) and can be easily prepared and purified [226–231], for instance by recrystallization from MeCN at $-20\,^\circ$C [229, 231]. They are very moisture sensitive, decomposing in air to higher oxides. They are stored in plastic containers under an inert atmosphere because the corrosive oxidation products attack glass. It has been reported that $[NO][BF_4]$ and $[NO][PF_6]$ can be rapidly weighted in air, however, and added as solids to substrates in a dry and deoxygenated solvent [21]. They are normally weighed in a dry-box for a more rigorous stoichiometry. In addition, a careful selection of solvents is necessary. Dichloromethane, alkanes, toluene, acetonitrile, nitromethane, ethyl acetate and carbon tetrachloride do not react, when they are carefully purified, with [NO]X salts in an inert atmosphere. Acetone, pyridine, ethers, alcohols and water, even when they are carefully purified, react with [NOX] salts whether the atmosphere is inert or not. Since the preparation of $Na[SbCl_6]$ from sodium and $[NO][SbCl_6]$ in liquid $SO_2$ in 1950 [232], the [NO]X salts have been used in the oxidation of metals for the preparation of homoleptic solvento complexes $[M(solvent)_n][X]_p$ (M = Cu, Ag, Au, Eu, Ni, Pd, etc.; solvent = MeCN, etc.; $n = 4$ or 6; X = $BF_4$ or $ClO_4$; $p = 1$–3) [233]. The hazardous, potentially explosive perchlorate salts must be avoided and obvious alternatives ($BF_4$ or $PF_6$) easily used instead. The [NO]X salts have been used in organometallic chemistry, but they are non-innocent oxidants, the most frequent reaction subsequent to oxidation being coordination of NO which leads to the formation of nitrosyl complexes. As in organic chemistry [225, 234], it is likely that such reactions proceed by the inner-sphere pathways [21]. The [NO]X salts are very useful for the oxidation, by an outer-sphere pathway, of ferrocene derivatives to the ferrocenium analogs, however, because nitrosyl complexes cannot coordinate in robust 18-electron complexes or in ferroceniums [134, 135]. NO derivatives have attracted a lot of interest recently since the discovery of the role of NO as a neurotransmitter.

### 4.4.8 Phosphorus, Arsenic, and Antimony Pentahalides and Aluminum Trichloride

Phosphorus pentahalides $PX_5$ (X = Cl and Br) are mild oxidants which were used already more than a century ago to oxidize platinum black to bridged dimers $[Pt_2(\mu_2\text{-}X)_2X_2(PX_3)_2]$ [235a]. Oxidative chlorination of $Pt^{II}$ to $Pt^{IV}$ complexes by $PCl_5$ is classic, and compares favorably with use of $Cl_2$ which is common to carry out this reaction [235b].

The arsenic and antimony pentahalides $EX_5$ (E: group 15 element As or Sb; X = F or Cl) are strong, irreversible oxidants; the gas $AsF_5$ has little been used, but $SbCl_5$ and $SbF_5$ are commercially available, very air-sensitive liquids which are used in dry and deoxygenated dichloromethane and liquid sulfur dioxide respectively. $SbCl_5$ is easier to handle than $SbF_5$ which gives the dangerous HF by reaction with moist air. Moreover, $SbCl_5$ is conveniently used in dichloromethane whereas $SbF_5$ is best used in liquid $SO_2$. On the other hand, the side products (halogenation) are more frequently encountered with $SbCl_5$ than with $SbF_5$. The redox process follows:

$$3EX_5 + 2e^- \rightarrow 2[EX_6]^- + EX_3$$

$[EX_6]^-$ can itself be an oxidant, for instance with ferrocene, iodides and aromatic amines [236]:

$$[SbCl_6]^- + 2e^- \rightarrow [SbCl_4]^- + 2Cl^-$$

Although the standard redox potentials of $SbX_5$ (X = F or Cl) are unknown because of the irreversibility of their redox reactions, it was possible to estimate rather accurately their oxidizing power by the investigation of their redox reactions with a series of 18-electron complexes $[FeCp^*(\eta^6\text{-arene})][PF_6]$ of known standard redox potentials. In these complexes, the arene ligands are polymethylaromatics in which the addition of each methyl group decreases the standard redox potential of the complex by 60 mV. For arene = hexamethylbenzene, pentamethylbenzene and tetramethyl-1,2,4,5-benzene (durene), the color change from yellow to purple observed upon reaction of $SbX_5$ (X = F or Cl) with the complex $[FeCp^*(\eta^6\text{-arene})][PF_6]$ indicates that electron transfer occurs, and it is also possible to isolate the salts $[FeCp^*(\eta^6\text{-arene})][SbX_6]_2$. For mesitylene and less methylated aromatics, no color change was observed, and the electron-transfer does not proceed. This is true for SbCl$_5$ in CH$_2$Cl$_2$ at 20 °C and SbF$_5$ in SO$_2$ at $-40°$ to $-10$ °C. Although it is known that SbF$_5$ is a better oxidant than SbF$_5$, they have the same oxidizing power in the conditions used here, because the solvent, temperature and rate of the follow-up reactions also play roles. Thus, these pentahalides can oxidize substrates which have standard redox potentials up to 1.045 V relative to FeCp$_2$, the standard redox potential of [FeCp*($\eta^6$-durene)][PF$_6$] in liquid SO$_2$.

AlCl$_3$ is a strong oxidant whose oxidation potential has been estimated to be about 1.1 V, but the mechanism by which oxidation reactions proceed is mysterious. It has essentially been used (even without sublimation, the method of purification) for spectroscopic (ESR) studies to generate radical cations [237, 238].

### 4.4.9 Dioxygenyl cation, O$_2$$^{\cdot+}$, dioxygen O$_2$, and superoxide anion O$_2$$^{\cdot-}$

The most oxidizing species ever isolated is Bartlett's dioxygenyl salt [O$_2$][PtF$_6$] made from O$_2$ and PtF$_6$ and reported in 1962. Since dioxygen has a higher ionization energy (12.1 eV) than xenon (11.6 eV), this synthesis of [O$_2$][PtF$_6$] led Bartlett to the idea that it would be possible to oxidize xenon using PtF$_6$. It is in this way that Bartlett isolated the first rare-gas compound [Xe][PtF$_6$] [239]. From this very high ionization potential, it could be calculated that [O$_2$][PtF$_6$] has a standard redox potential of 4.9 V relative to FeCp$_2$, although such high redox potentials cannot be reached in any solvent using electrochemistry. On may also imagine that considerable solvation energy of the dioxygenyl cation is released in any solvent, so the actual potential, still extremely high, is considerably less positive whenever a solvent is involved. The salts [O$_2$][EF$_6$] (E = As or Sb) are indeed air-sensitive, especially when the temperature is increased towards room temperature. Therefore, it has been used, especially in low-temperature Freon mixtures to oxidize several substrates which are difficult to oxidize [240–244] including organic compounds to their radical cations. Thus, for instance [O$_2$], [AsF$_6$] reacts with NMe(C$_6$H$_4$OMe-4)$_2$ in CHF$_2$Cl at $-130$ °C to give [NMe(C$_6$H$_4$OMe-4)$_2$][AsF$_6$] and O$_2$ which is released together with the solvent, leaving the salt [241].

Dioxygen O$_2$ is the most known and most common oxidant in combustion processes and in industry (Wacker process). It is not used in single-electron oxidation reactions, however, because the standard redox potential O$_2$/O$_2$$^{\cdot-}$ is relatively negative ($E^{\circ\prime} = -1.2$ V relative to FeCp$_2$ in DMF), making O$_2$ a mediocre single-electron acceptor, and superoxide radical anion is very reactive (although thermodynamically very stable). Thus, O$_2$ in a non-innocent oxidant, giving coordination of O$_2$ or superoxide onto transition-metal centers (formation of dioxygen-, super-oxo- and peroxo complexes MO$_2$, oxo complexes M=O and peroxide-bridged complexes MOM) and giving well-known radical reactions (R$^{\cdot}$ + O$_2$ → ROO$^{\cdot}$). In fact, it is a major concern for organometallic chemists to avoid contact of the reaction media with O$_2$ from air. Dioxygen can be used as a clean single-electron oxidant in reactions which are carried out in THF in the presence of one equiv. NaPF$_6$ in order to avoid the cage reactions of the intermediate ion pairs. This is a fairly general technique when the oxygen species cannot attack the metal center which is coordinatively saturated.

$$[Fe^ICp(arene)] + 1/2\,O_2 + [Na][PF_6] \xrightarrow{THF} [Fe^{II}Cp(arene)][PF_6] + Na_2O$$

The interesting cage reactions of superoxide anion subsequent to the reaction of $O_2$ in the absence of a salt have been reviewed in the preceding section.

Superoxide radical anion, $O_2{}^{\bullet-}$, a weak reductant, is also an extremely weak oxidant as its standard redox potential of reduction to peroxide, $O_2{}^{2-}$, is about $-2.4$ V relative to $FeCp_2$ [245]. First made by Gay Lussac almost two centuries ago by reaction of K with $O_2$, $KO_2$ is insoluble except in DMSO in the presence of 18-crown-6. Alternatively, soluble $O_2{}^{\bullet-}$ salts are generated cathodically in pyridine, DMF or DMSO in the presence of a solubilizing electrolyte of the type $[R_4N]Cl$ (R = Me or *n*-Bu), which provides the superoxide salt $[R_4N][O_2]$ [245].

## 4.4.10 Halogens: $Cl_2$, $Br_2$, and $I_2$

The halogens $Cl_2$, $Br_2$ and $I_2$ are commercially available (cheap) common mild (but non-innocent) oxidants which are soluble in organic solvents including non-polar ones. The order of their oxidizing powers follows that of their standard oxidation potentials $E^{\circ\prime}$ relative to $FeCp_2$ in MeCN: $Cl_2$ (0.18 V) > $Br_2$ (0.07 V) > $I_2$ ($-0.14$ V):

$$X_2 + 2e^- \leftrightarrow 2X^-$$

Oxidation of neutral organometallic complexes $ML_n$ using $I_2$ often leads to the salts $[ML_n]I$ (which can be transformed into $[ML_n][PF_6]$ by addition of $[NH_4][PF_6]$ in water), but the triiodides $I_3{}^-$ are also formed if excess $I_2$ is present in the reaction medium.

$$[Mo(\eta^5\text{-}C_7H_9)(\eta^7\text{-}C_7H_7)] + 1/2\,I_2 \xrightarrow{toluene} [Mo(\eta^5\text{-}C_7H_9)(\eta^7\text{-}C_7H_7)]I$$

The use of $Cl_2$ or $Br_2$ and $AlCl_3$ together in a mixture of dichloromethane and carbon tetrachloride provides a very strong oxidant which can oxidize complexes whose oxidation potential is as high as 1.4 V relative to $FeCp_2$ such as $[Pt(C_6Cl_5)_4]^-$ [245]. Halogens cleave metal–carbon bonds of metal–alkyl complexes with halogenation of both the alkyl and the metal fragments (the first step is an outer-sphere electron transfer) [246].

$$[FeCp(CO)_2R] + Cl_2 \rightarrow [FeCp(CO)_2Cl] + RCl$$

## 4.4.11 Arenediazonium, $[N_2aryl]^{\bullet+}$

The arenediazonium salts $[N_2aryl]X$ (X = $BF_4$ or $PF_6$) have been very much used in organic synthesis [247]. They are known with a variety of *para* substituents R, which allows variations of the standard redox potential $E^{o\prime}$ relative to $FeCp_2$: R = H (−0.10 V), Me (−0.15 V), Cl (0.01 V), $NO_2$ (0.05 V) and $NMe_2$ (−0.50 V) [248, 249]. Except the latter, these values are almost the same as that of ferrocenium. The redox reaction of aryldiazonium cations are irreversible, however, which shift the redox equilibrium towards the products:

$$[N_2aryl]^+ + A \leftrightarrow {}^\bullet N_2aryl + A^+$$

$${}^\bullet N_2aryl \rightarrow N_2 + aryl^\bullet$$

Therefore, they are better oxidants than ferrocenium salts. Indeed, they have been currently used in $CH_2Cl_2$ and THF to oxidize 18-electron complexes to their 17-electron counterparts [21], although the advantage of the irreversible oxidation has not been quantified. They are easily prepared [249, 250] and $[N_2C_6H_4NEt_2$-4][$BF_4$] is commercially available. They are stable, can be handled in air, and are soluble in polar solvents ($Me_2CO$, MeCN) but not in $CH_2Cl_2$, and decompose in THF and $CHCl_3$ [21]. They often behave as non-innocent oxidants in reactions of transition-metal complexes (H-atom abstraction, radical coupling, coordination and insertion in metal–ligand bonds are known) [21].

### 4.4.12 Acids

The reaction of concentrated sulfuric acid with neat ferrocene is a method of preparation of ferrocenium salts:

$$2FeCp_2 + H_2SO_4 \rightarrow [FeCp_2]_2[SO_4]$$

Cobaltocene [$Cr(arene)_2$], and [$FeCp^*_2$]$_2$ are also oxidized to the monocations by acids. It was shown in some cases that the first step consists in the protonation of the electron-rich metal center, and the likely follow-up reaction is oxidation of the neutral sandwich complex by the protonated complex [251].

$$Cp_2M + H^+ \rightarrow Cp_2MH^+$$

$$Cp_2MH^+ + Cp_2M \rightarrow Cp_2M^+ + Cp_2MH$$

$$Cp_2MH \rightarrow Cp_2M + \tfrac{1}{2}H_2$$

A very interesting example is the oxidation of biferrocenylene Fc–Fc by protic acids, because dihydrogen readily evolves at the two-metal centers and the biferrocenylene di-cation is reduced to the neutral compound at low cathodic potential providing a catalyst for the reduction of protons [252, 253]:

$$Fc\text{–}Fc + 2H^+ \rightarrow [HFc\text{–}FcH]^{2+} \rightarrow [Fc\text{–}Fc]^{2+} + H_2$$

However, the reaction of a protic acid with transition-metal complexes bearing non-bonding valence electrons (Lewis bases) usually leads to transition-metal hydrides complexes resulting in a formal oxidation of the metal center by two units of oxidation states. Various strengths of acids are required since the range of basicity of transition-metal hydrides is extremely large. With $d^0$ complexes, protonation of a metal–ligand bond can be achieved since the metal cannot be oxidized. Thus, $d^0$ transition-metal-hydride complexes as well as electron-poor transition-metal hydrides are protonated at the M–H bond to give dihydrogen complexes [254, 255]. This type of reaction is in competition with the electron-transfer reaction, but is predominant most of the time.

The most currently chosen acids are $HBF_4$ and $HPF_6$ which can be used either in water or in ether, this second possibility having provided other cases of single-electron oxidation of electron-rich neutral organometallic complexes to their mon-ocations as $BF_4$ or $PF_6$ salt [244]. Triflic acid, $HO_3SCF_3$ [256], and $[NH_4][PF_6]$ [257] can also be used in THF.

### 4.4.13 Triarylaminium Cations, $[N(aryl)_3]^{•+}$

The triarylaminium salts $[N(aryl)_3][X]$ in which the *para* position is substituted (in order to inhibit nucleophilic reactions at that position) are among the most useful oxidants because they are strong, almost innocent oxidants known for a range of standard redox potentials from 0.16 V to −1.76 V relative to $FeCp_2$ depending of the nature and number of substituents (Table 10), and they are easily accessible by oxidation of the triarylamine in $CH_2Cl_2$ using [NO] salts, silver salts in the presence

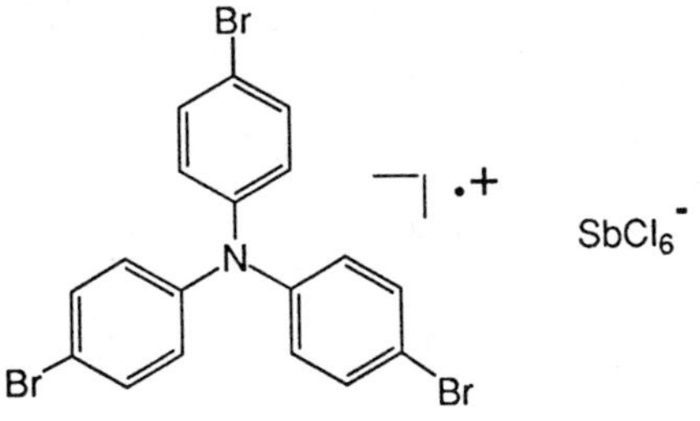

**MAGIC BLUE**

**Table 10.** Formal potentials (V, relative to Fc) of triarylaminium cations $[N(aryl)_3]^{+a}$.

| Cation | $E^{o\prime}$ (V) | Cation | $E^{o\prime}$ (V) |
|---|---|---|---|
| $C_6H_4OMe$-4 | 0.16 | $C_6H_4NO_2$-4 | 1.20 |
| $C_6H_4Me$-4 | 0.40 | $C_6H_4Br_2$-2,4 | 1.14 |
| $C_6H_4(COMe)$-4 | 0.90 | $C_6H_4Br_3$-2,4,6 | 1.36 |
| $C_6H_4Br$-4 | 0.70 | $C_6Cl_5$ | 1.72[b] |
| $C_6H_4CN$-4 | 1.08 | | |

[a] Standard redox potentials from Ref. [258] unless otherwise stated.
[b] From Ref. [21].

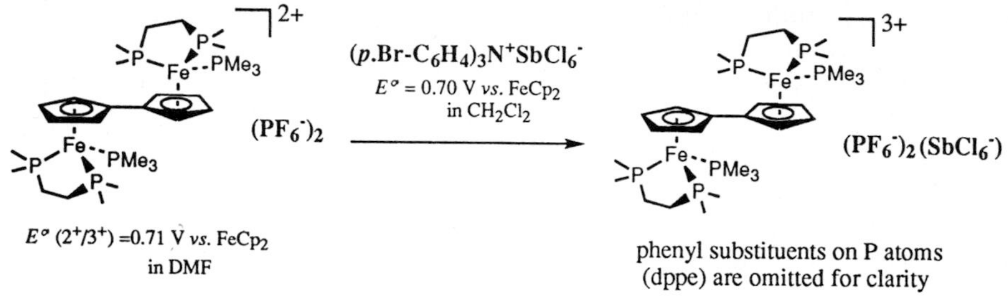
2+
PMe3
(p.Br-C6H4)3N+SbCl6-
E° = 0.70 V vs. FeCp2
in CH2Cl2
(PF6-)2
E° (2+/3+) =0.71 V vs. FeCp2
in DMF
3+
(PF6-)2 (SbCl6-)
phenyl substituents on P atoms
(dppe) are omitted for clarity

**Scheme 23.**

of iodine or $SbCl_5$ [258–260]. In addition, the most used one $[N(C_6H_4Br\text{-}4)_3]$, $[SbCl_6]$, named Magic Blue because of its intense royal blue color, is commercially available ($E^{\circ\prime} = 0.70$ V relative to $FeCp_2$ in MeCN).

The most frequent counter-anions are $BF_4^-$ (for the modest oxidants of this series), $PF_6^-$ (almost no side reactions), $SbCl_6^-$ (may give side nucleophilic reactions with the strongest oxidants of this series) and, for the strongest oxidants of this series, $SbF_6^-$. The $BF_4^-$ salts are the most fragile of the group and should be used as freshly prepared. $CH_2Cl_2$ is a good solvent for their use (the $SbCl_6$ salts are only sparingly soluble) and MeOH must be avoided with, for instance, Magic Blue, because of *para*-methoxylation.

An example of the use of Magic Blue for the oxidation of a binuclear 36-electron complex to the 35-electron complex as shown in Scheme 23 (this complex is a delocalized mixed-valence compound on the Mössbauer time scale) [261].

Since the oxidized species is sensitive to nucleophiles, even weakly nucleophilic counter-anions of strong oxidant can react with very sensitive species. This problem has been addressed for instance in the case of the oxidation of the fullerene $C_{76}$ ($E^{\circ\prime} = 0.81$ V relative to $FeCp_2$ for $C_{76}/C_{76}^+$) which is easier to oxidize than $C_{60}$ ($E^{\circ\prime} = 1.26$ V relative to $FeCp_2$ for $C_{60}/C_{60}^+$) and $C_{70}$ ($E^{\circ\prime} = 1.20$ V relative to $FeCp_2$ for $C_{70}/C_{70}^+$). The salt $[Ar_3N^{\bullet+}][CB_{11}H_6Br_6^-]$ (with Ar = 2,4-dibromophenyl; $E^{\circ\prime} = 1.16$ V relative to $FeCp_2$), a green solid, has been successfully used to synthesize the stable salt of the radical cation of $C_{76}$ because the counter-anion was carefully designed as one of the weakest nucleophilic anions known [262]:

$$Ar_3N + Ag(CB_{11}H_6Br_6) + 1/2\ Br_2 \xrightarrow{CH_2Cl_2} [Ar_3N^{\bullet+}][CB_{11}H_6Br_6^-] + AgBr$$

$$C_{76} + [Ar_3N^{\bullet+}][CB_{11}H_6Br_6^-] \xrightarrow{o\text{-dichlorobenzene}} [C_{76}{}^{\bullet+}][CB_{11}H_6Br_6^-] + Ar_3N$$

The strongest oxidants such as $[N(C_6H_4Br_3\text{-}2,4,6)_3]^{\bullet+}$ ($E^{\circ\prime} = 1.36$ V relative to $FeCp_2$ in MeCN]) and $[N(C_6Cl_5)_3]^{\bullet+}$ ($E^{\circ\prime} = 1.84$ V relative to $FeCp_2$ in MeCN [21]) can be generated using the appropriate amine and oxidant without the need to isolate them. These series of salts have been extensively used in organic chemistry and reviewed, and their use in transition-metal chemistry is spreading [21].

## 4.4.14 Thianthrene Radical Cation

Thianthrene can be oxidized by SbCl$_5$ or [NO][BF$_4$] in CH$_2$Cl$_2$ to give the red–brown salts [thianthrene][SbCl$_6$] and [thianthrene][BF$_4$] (purple in solution) which are stable in dry air but react with water. They are also stable for two weeks in MeCN in the absence of excess [NO]X salt [263, 264]. The $E^{o\prime}$ value of 0.86 V relative to FeCp$_2$ in MeCN [265] makes them attractive relatively strong oxidants which have been little used.

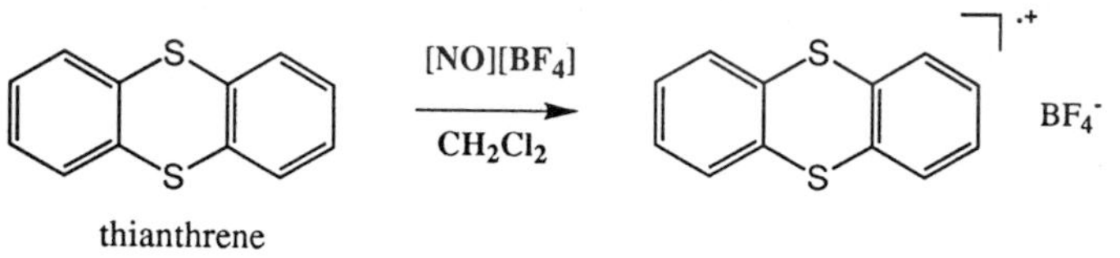

thianthrene

$E^{\alpha} = 0.86$ V *vs.* FeCp$_2$ in MeCN

## 4.4.15 Carbocations: Trityl, [C(aryl)$_3$]$^+$, Salts and Salts of other Carbocations

These carbocations can undergo three reactions: (i) single-electron oxidation, (ii) hydride abstraction and (iii) electrophilic addition. Thus, these compounds behave as mild single-electron oxidants towards reductants which do not react by other pathways. Actually, what is most interesting in these carbocations is the competition between electron-transfer and the other reactions. For instance, a hydride transfer can occur either directly or via the electron-transfer pathway using a trityl salt. From a synthetic standpoint, a hydride transfer which cannot be achieved directly for steric reasons may be attempted by means of the electron-transfer pathway.

The triphenylcarbenium ion (trityl cation) Ph$_3$C$^+$ ($E^{o\prime} = -0.11$ V relative to FeCp$_2$ in MeCN [266]) is commercially available as the yellow–orange BF$_4$$^-$, PF$_6$$^-$, AsF$_6$$^-$, SbCl$_6$$^-$ and CF$_3$SO$_3$$^-$ salts which are also easily synthesized, as other substituted trityl salts, for instance from Ph$_3$COH (or Ar$_3$COH) and the acid in acetic or propionic anhydride [267]. They are moisture sensitive in the solid state and sometimes light sensitive, and can be purified by recrystallization from MeCN at low temperature or from CH$_2$Cl$_2$–hexane or MeCN–ether mixtures. They can be used in CH$_2$Cl$_2$, MeCN and liquid SO$_2$ (they are insoluble in alkane and arene solvents and slowly react with ether [267]). An excellent alternative to the Ph$_3$C$^+$ salts (and weaker oxidant) is [Ph$_2$C(C$_6$H$_4$OMe-4)][BF$_4$] ($E^{o\prime} = -0.32$ V relative to FeCp$_2$ in MeCN) which is much more stable even in the presence of moist air, can be weighed and handled in air, and decompose in aqueous acetonitrile 100 times more slowly than [Ph$_3$C][BF$_4$] [268].

In addition to its classic use as hydride abstractor in organic [267] and organometallic chemistry [21], [Ph$_3$C][BF$_4$] has also been used as a single-electron oxidant [269] in a number of organometallic examples despite the side reactions. Indeed, the redox reaction of Ph$_3$C$^+$ salts produces the Ph$_3$C$^{\cdot}$ radical which dimerizes, but also can couple to organometallic substrates or products or abstract a H atom from some solvents and substrates. The electron-transfer properties of $Ph_3C^+$ could be used in a strategy to remove a hydride for a synthetic purpose as follows. The trityl salt $[Ph_3C][BF_4]$ abstracts a hydride from the cyclohexadiene complex $[Fe(\eta^4\text{-}C_6H_8)(CO)_3]$, giving the cyclohexadienyl complex $[Fe(\eta^5\text{-}C_6H_7)(CO)_3][BF_4]$, which opens the route to the syntheses of *exo*-functional cyclohexadiene complexes by reaction with zinc or cadmium reagents. $[Ph_3C][BF_4]$ does not react with these functional complexes for steric reasons, however, so that reiteration of this process leading to heterobifunctional complexes in not possible [270]. With the more electron-rich isolobal analog $[Fe(\eta^5\text{-}C_6H_7)(\eta^6\text{-}C_6H_6)][PF_6]$, directly accessible by reaction of $NaBH_4$ with $[Fe(\eta^6\text{-}C_6H_6)_2][PF_6]_2$, reaction of a variety of carbanionic nucleophile RM as sodium or potassium reagents cleanly gives $[Fe(\eta^4\text{-}exo\text{-}RC_6H_7)\text{-}(\eta^6\text{-}C_6H_6)]$. Reactions of these electron-rich organometallic compounds with $[Ph_3C][BF_4]$ at $-40\,°C$ proceeds according to an exergonic electron transfer followed by an H-atom transfer between the two radicals $[Fe(\eta^4\text{-}exo\text{-}RC_6H_7)(\eta^6\text{-}C_6H_6)]^{•+}$ and $Ph_3C^•$. Although the direct hydride transfer was not possible for steric reasons, an overall hydride transfer of the *endo* hydride was achieved in this two-step way producing the functional complex $[Fe(\eta^5\text{-}exo\text{-}RC_6H_6)(\eta^6\text{-}C_6H_6)]$. Further reactions with functional carbanions produced *cis*-heterobifunctional complexes $[Fe(\eta^4\text{-}exo, exo\text{-}RR'C_6H_6)(\eta^6\text{-}C_6H_6)]$ which could be decomposed using the oxidant $FeCl_3$ (stronger than $[Ph_3C][BF_4]$) to the free *cis*-heterobifunctional cyclohexadienes (Scheme 24).

Other noteworthy examples of hydride transfer reactions of $[Ph_3C][BF_4]$ proceeding by an electron-transfer path in organo-transition metal chemistry are the formation of $[WCp_2(H)(\eta^2\text{-}C_2H_4)]$ from $[WCp_2Me_2]$ [271] and of $[ReCp(PPh_3)\text{-}$

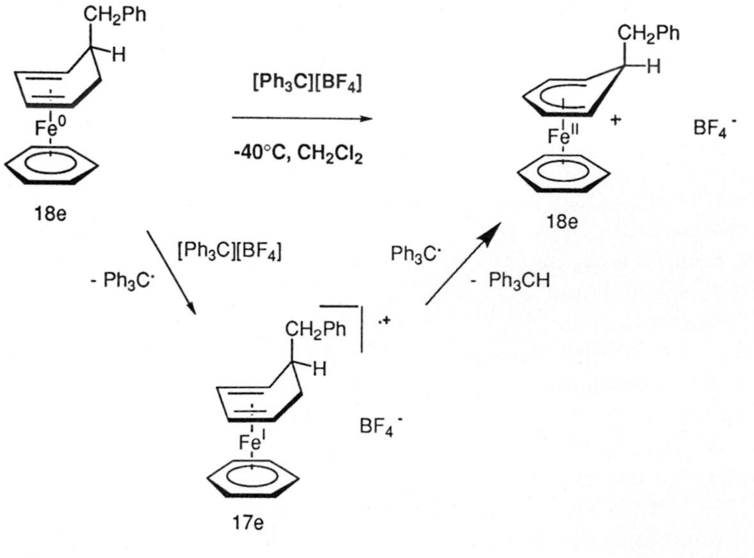

**Scheme 24.**

$(NO)(=CHCH_2R)(PPh_3)(NO)]$ from $[ReCp(PPh_3)(NO)(CH_2CH_2R)(PPh_3)(NO)]$ [272]. The early examples of the use of triphenylcarbonium salts in organometallic chemistry have been reviewed [273].

Other commercially available salts of carbocations such as $MeOSO_2F$ [274] and $Et_3OBF_4$ [275, 276] have been used in one-electron oxidation reactions:

$$[WCp(NO)_2\{P(OMe)_3\}] + MeOSO_2F \xrightarrow{CH_2Cl_2}$$
$$[WCp(NO)_2\{P(OMe)_3\}][OSO_2F] + C_2H_6$$

$$trans\text{-}[MoCl_2(dppe)_2] + Et_3OBF_4 \xrightarrow{CH_2Cl_2} trans\text{-}[MoCl_2(dppe)_2][BF_4]$$
$$+ Et_2O + n\text{-}C_4H_{10}$$

$$[ReCl(CO)_3(PPh_3)] + Et_3OBF_4 \xrightarrow{CH_2Cl_2} [ReCl(CO)_3(PPh_3)][BF_4]$$
$$+ Et_2O + n\text{-}C_4H_{10}$$

The tropylium cation, $C_7H_7^+$, is a very mild oxidant ($E^{\circ\prime} = -0.65$ V relative to $FeCp_2$ in MeCN [266]) as well as a mild hydride abstractor and a mild electrophile. The colorless $BF_4$ and $PF_6$ salts are commercially available and readily prepared, stable in air, non hygroscopic, soluble in $Me_2CO$, MeCN and liquid $SO_2$ and insoluble in $CH_2Cl_2$. There are several known examples of the use of these salts in $CH_2Cl_2$ or THF (suspension) for the one-electron oxidation of electron-rich (especially anionic) organometallic compounds, sometimes with coordination of the cycloheptatriene ligand.

### 4.4.16 TCNE, TCNQ, and other Cyanocarbons

TCNE ($E^{\circ\prime} = -0.27$ and $-1.27$ V relative to $FeCp_2$ in MeCN) and TCNQ ($E^{\circ\prime} = -0.30$ V and $-0.88$ V relative to $FeCp_2$ in MeCN) are commercially available mild one-electron acceptors which have attracted, as such, a considerable interest for the construction of low-dimensional organic and organometallic charge-transfer solids with magnetic [277] and conducting properties [278]. For instance, the salt

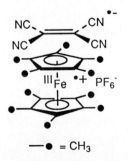

**molecular ferromagnet**

[TTF][TCNQ] was the first organic conductor [279] and the salt [FeCp*$_2$][TCNE] was one of the very first molecular ferromagnets [277, 280].

Thus, the interest of TCNE and TCNQ is more concentrated on their use as precursors of such materials than on their use as redox tools to obtain oxidized transition-metal compounds. Nevertheless, both aspects are somewhat connected, and transition-metal derivatives of TCNE and TCNQ have also been reviewed [281]. The most interesting aspect probably is the ability of TCNE to both oxidize and bind metal centers, but an important related problem is that **TCNE is toxic**.

TCNQ is reduced to the monoanion or dianion by the electron-reservoir complex [Fe$^I$Cp($\eta^6$-C$_6$Me$_6$)] depending on the stoichiometry and the respective redox potentials can let predict the same results with TCNE.

$$[Fe^ICp(\eta^6\text{-}C_6Me_6)] + TCNQ \xrightarrow{\text{THF}} [Fe^{II}Cp(\eta^6\text{-}C_6Me_6)][TCNQ]$$

$$[Fe^ICp(\eta^6\text{-}C_6Me_6)] + [Fe^{II}Cp(\eta^6\text{-}C_6Me_6)][TCNQ]$$
$$\xrightarrow{\text{THF}} [Fe^{II}Cp(\eta^6\text{-}C_6Me_6)]_2[TCNQ]$$

Another noteworthy example is the oxidation of the clusters [Co$_3$(CO)$_9$E] (E = S or PPh) by TCNQ to the salts [Co$_3$(CO)$_9$E][TCNQ] in various solvents [282]. Other cyanocarbons behave as single-electron oxidants: *n*-hexacyanobutadiene, *n*-C$_4$(CN)$_6$, the radical anion tris(dicyanomethylene) cyclopropane, [C$_6$(CN)$_6$]$^{\cdot-}$ (better formulated as [C$_3${C(CN)$_2$}$_3$]$^{\cdot-}$) [283].

### 4.4.17 Miscellaneous Organic Oxidants

Commercially available 1,2 and 1,4 quinones are easily handled. The formal potential of the 1,4-benzoquinone/hydroquinone couple at pH 0 is ca 0.30 V relative to FeCp$_2$. A variety of ferrocene derivatives including 1,1′-diacetylferrocene have been oxidized by 1,4-benzoquinone in the presence of HBF$_4$, and the ferrocenium salts have been prepared using this reaction [284].

Diphenyliodonium salts [IPh$_2$]X ($E^{o\prime} = -0.9$ V relative to FeCp$_2$ [285]) are commercially available (X = Cl, Br, I) and the BF$_4$, PF$_6$, AsF$_6$ and SbF$_6$ can be made [286]. As aryldiazonium salts, they decompose, after electron transfer, to the phenyl radical which dimerizes to biphenyl and abstracts a H atom from the solvent.

Iodosobenzene PhIO is a yellow amorphous powder which explodes above 200 °C, and has been used to oxidize [FeCp(dppe)(C=CHMe)]$^+$ to [{FeCp(dppe)}$_2$-{$\mu_2$-C=C(Me)C(Me)=C}]$^{2+}$ [287].

4-pyridinecarboxaldehyde, NC$_5$H$_4$CHO-4, is a very weak oxidant ($E^{o\prime} = $ ca $-1.0$ V relative to FeCp$_2$ [288]) which is a commercially available liquid. It must be distilled before use and stored under inert atmosphere at $-10$ °C. It may be useful to oxidize electron-rich organometallic complexes such as [Cr($\eta^6$-arene)$_2$] which are good monoelectronic reductants, for instance for the purpose of ESR studies [289].

# 4.5 Electron-transfer-chain (ETC) Catalysis

## 4.5.1 Historical Background, Main Types of Reaction, and Principle

Electron-transfer-chain catalysis is the catalysis of reactions by electrons or electron holes. Most organic and transition-metal reactions catalyzed in this way do not involve an overall redox change. The ETC catalyzed reactions belong to the general group of chain reactions [181, 182] which started in the 1920s with the chain reaction between $H_2$ and $Cl_2$ initiated by sodium in the gas phase [183]. This latter reaction, as many others discovered since, is of the type of atom-transfer-chain (ATC) catalysis, which can be considered as the inner-sphere version of ETC catalysis [184]. In transition-metal chemistry, the first ETC- and ATC-catalyzed reactions were disclosed by Taube in 1952 who reported chlorine exchange in $[AuCl_4]^-$ initiated by the reductant $[Fe^{II}(CN)_6]^{4-}$ [290]. Electrochemistry plays an important role in this field because electrodes are obvious sources of electrons and electron holes (although not the most practical ones for synthetic purpose). It is essential for the kinetic and thermodynamic investigations of the ETC systems, and this technique serves as an analytical tool for understanding the mechanism, thus improve the processes. This analytical approach of ETC mechanisms in molecular chemistry was first carried out by Feldberg who established the theoretical basis for the so called 'ECE' mechanism [291]. Once these features are determined, the ETC reactions are best carried out using redox reagents and especially electron-reservoir complexes as catalysts (initiators) given their very negative- and very positive redox potentials. ETC catalysis has spread over inorganic [292], organic [293] and organometallic chemistry [290, 294], although each sub-community seems to have developed very independently [292–294]. The first ETC catalyzed organic reactions were found by Russell [295], Kornblum [296] and Bunnett [297], and include the popular ETC catalyzed nucleophilic substitution in aromatic derivatives ($S_{NAr}1$) rationalized from an electrochemical standpoint by Savéant [298, 299]. The first ETC catalyzed reactions in organometallic chemistry are Feldberg's careful studies of the ligand exchange in the 15-electron species $[Cr^{III}(CN)_6]^{3-}$ reported in 1972 [291] and Rieke's studies of the anodically catalyzed *trans* to *cis* isomerization of bis-carbene $Mo^0$ and $W^0$ complexes reported in 1976. Rieke's studies were also thorough kinetic and thermodynamic analyzes and involved electrochemistry as well as preparative aspects [300, 301]. In the early 80's, there was an explosion of the field [294]. For instance, the group of Bruce reported an impressive number of ligand-exchange reactions of carbonyls in clusters [302] and Kochi published several in-depth studies and thoroughly analyzed cases of ligand substitution in mono- and polynuclear organometallic complexes [303]. Not only ligand substitution and isomerization of inorganic–organometallic compounds can be catalyzed by ETC chain but also migratory insertion (for instance migratory insertion of CO into a Fe–C bond in [FeCp(CO)L(Me)], L = CO [304] or $PPh_3$ [305]) and extrusion (for instance extrusion of CO from a Mn–formyl bond in $[Re_2(CO)_9(CHO)]^-$) [306].

The principle of ETC catalysis consists in carrying out the reaction at a kinetically labile redox state of the reactant when the ground state is not enough reactive.

This can be achieved by addition or removal of one electron. However, this type of procedure can also be stoichiometric. Then, a stoichiometric amount of oxidant or reductant is required to bring the substrate into the reactive state. For instance, the oxidative decomplexation of organometallic compound for the purpose of organic synthesis often falls into this category. The reaction only becomes catalytic in redox reagent if the product reacts with the substrate in a redox (cross electron transfer) reaction, thus entering a chain process. For instance, in a ligand substitution reaction (the most frequent ETC catalyzed reaction), a complex A reacts with a ligand C to give the new complex B and the ligand D according to a chain process, the overall course of which is represented by:

$$A + C \rightarrow B + D$$

There are two ways to try to catalyze the reaction: oxidation or reduction. Most of the time, only one way is efficient catalytically for kinetic reasons. The propagation cycle contains two reactions: the cross redox (cross electron-transfer) reaction and the non-redox ('chemical') reaction. Since the non-redox reaction proceeds in the most reactive state of the substrate, it is usually difficult to obtain information on such species and the ergonicity of this step is thus usually obscure. On the other hand, it is very easy to obtain information on the redox step. On may just have a look at the relative standard redox potentials of the substrate and product, or even only to watch these two compounds and consider if the entering ligand is more or less electron-releasing than the leaving ligand for the metallic redox center. Since only one ligand is usually exchanged, the molecular orbital diagram is approximately the same for both complexes except that the set of orbitals of the more electron-rich compound is at a slightly higher energy level than the same set of orbitals of the other, less electron-rich complex. It is thus easy to forecast the ergonicity (exergonic or endergonic) of the cross redox step as a function of the fact that oxidation or reduction is used as indicated on Scheme 25:

**B LESS ELECTRON-RICH THAN A**
$$(E^{o'}_{B/B^+} > E^{o'}_{A/A^+})$$

**B MORE ELECTRON-RICH THAN A**
$$(E^{o'}_{B/B^-} < E^{o'}_{A/A^-})$$

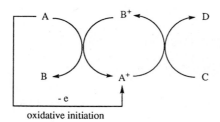

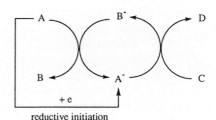

**Scheme 25.**

The scheme shows that only one way provides an exergonic redox step: if the entering ligand is more electron-rich than the leaving ligand, reduction provides an exergonic redox step whereas oxidation provides an endergonic redox step. On the other hand, if the entering ligand is less electron-releasing than the leaving ligand, it is the opposite which is true. According to Marcus theory, an exergonic electron transfer is fast and an endergonic electron transfer is slow[3]. The radical intermediates are the subject of side reactions which favorably compete with an endergonic redox reaction. Thus, the redox propagation step should be exergonic in order to proceed, i.e. the wrong type of initiation should be most of the time avoided. All the very efficient (high coulombic efficiency), ETC catalyzed reactions proceed according to an exergonic redox propagation step. There are exceptions, however, because a slightly endergonic reaction can still be driven by an extremely fast follow-up 'chemical' propagation step. Usually, the redox initiator must be chosen in order to involve an exergonic redox step, and its strength should be such that the initiation step between the redox reagent and the substrate complex be exergonic, isoergonic or not more than very slightly endergonic (but this latter solution requires that the follow-up propagation reaction be fast).

The redox initiators are easily chosen in the collection of oxidants and reductants described in the previous sections. In organic chemistry, very strong oxidants or reductants are required, because it is usually difficult to oxidize or reduce organic compounds, i.e. the organic compounds are oxidized or reduced at extreme potentials. Thus the triarylaminium salts on the oxidation side [258, 259] and the alkali metals (usually Na) on the reduction side are currently used. For instance, the nucleophilic aromatic substitution of halides by other nucleophiles is catalyzed by Na in $NH_3$ [297]. On the other hand, the redox potentials of inorganic and organometallic compounds are easily accessible, so that most oxidative initiations can be carried out using ferrocenium salts $[Fe^{III}Cp_2][X]$, $X = BF_4$ or $PF_6$, on the oxidation side and the electron-reservoir complex $[Fe^ICp(\eta^6\text{-}C_6Me_6)]$ on the reduction side. Not only these initiators cover most of the redox scale, but substrates with oxidation potential more positive than that of ferrocenium salts or reduction potentials more negative than that of $[Fe^ICp(\eta^6\text{-}C_6Me_6)]$ can also react when their redox reaction is followed by a fast 'chemical' reaction, which is often the case.

### 4.5.2 The Electron-reservoir Complexes $[Fe^ICp(\eta^6\text{-arene})]$ as ETC Catalysts

**Substitution of the arene ligand by three two-electron donors in the complexes $[Fe^{II}Cp(\eta^6\text{-arene})][PF_6]$**

This type of reaction was first reported to be induced cathodically, the coulombic efficiency being variable as a function of the coordinating solvent or ligand [307, 308]. In all the cases, these ligands bind the metal via an heteroatom (O, N) which is not the subject of back-bonding. Thus, the overall electron-releasing character of three such ligand is larger than that of the arene ligand which receives back-bonding from the transition metal. Consequently, reduction is the appropriate mode of initiation in order to provide an exergonic redox reaction. J. Ruiz used the 19-electron complex $[Fe^ICp(\eta^6\text{-arene})]$ as an initiator for the ETC catalyzed ligand substitution reaction of the arene ligand with various phosphanes and diphosphanes in the iso-structural complex $[Fe^{II}Cp(\eta^6\text{-arene})][PF_6]$, and the reaction is over in a few second at ambient temperature in THF (suspension) or MeCN (solution). Acetonitrile is also one of the three entering ligands L when the large PPh$_3$, dppe or dppm is used, because only two coordination sites can be filled with such large ligands [180, 310] In this case, the initiator and reduced form of the substrate have an identical struc-ture, and this is a unique example of such a situation. The arene could be any of the methylated benzene derivatives $C_6H_6Me_{6-n}$ ($n = 0\text{--}6$) and the P donor could be PMe$_3$, P(OMe)$_3$, dppm or dppe. With P(OPh)$_3$, the donicity of this ligand is not good enough to provide an exergonic electron-transfer step. This catalysis is re-markably efficient since an amount of 1 % of Fe$^I$ catalyst is sufficient to carry out the reaction quantitatively. The reaction of P donors with these 19-electron com-plexes was shown to be associative [180]. Indeed, these 19-electron Fe$^I$ complexes are either in rapid equilibrium with the 17-electron form in which one double bond of the arene ligand is decoordinated or this 17-electron $[Fe^ICp(\eta^4\text{-arene})]$ form simply is a mesomer of the 19-electron one (equation and Scheme 26) [146].

$$[Fe^{II}Cp(\eta^6\text{-arene})][PF_6] + 3\,L \xrightarrow{\text{cat } [Fe^ICp(\eta^6\text{-arene})]} [Fe^{II}CpL_3][PF_6] + \text{arene}$$

**Disproportionation of bimetallic fulvalene carbonyl complexes to bimetallic zwitterions**

A rare redox reaction carried out by ETC catalysis is the disproportionation of bimetallic metal carbonyl complexes such as $[MoCp(CO)_3]_2$ containing a metal–metal bond, in the presence of a two-electron donor such as PMe$_3$. In this typical example, the reaction leads to the ion pair $[Mo^{II}Cp(CO)_2(PMe_3)_2][Mo^0Cp(CO)_2]$. The photolysis of the dimer gives the 17-electron iron radical $[Mo^ICp(CO)_3]$ re-

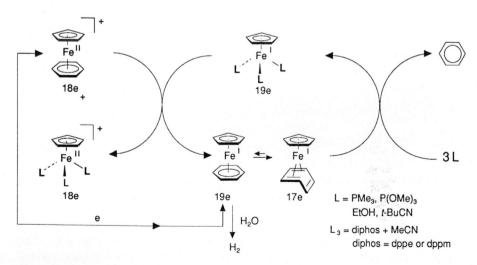

**Scheme 26.**

sulting from the cleavage of the Mo–Mo bond. In the presence of $PMe_3$, rapid uptake of this ligand by the 17-electron iron radical leads to the 19-electron radical $[Mo^ICp(CO)_2(PMe_3)]$. These 17- and 19-electron radicals are in fast equilibrium resulting from ligand exchange until a side follow-up reaction shift the overall equilibrium. At the level of the 19-electron radical $[Mo^ICp(CO)_2(PMe_3)_2]$, reduction of the dimer by this electron-rich species is exergonic, thus fast, which generates the reaction product and the 17-electron radical $[Mo^ICp(CO)_3]$, and a new propagation cycle can start [311, 312]:

$$\text{initiation:} \quad [MoCp(CO)_3]_2 \xrightarrow{h\nu} 2\,[MoCp(CO)_3]$$
$$\underset{18e}{} \qquad\qquad \underset{17e}{}$$

$$\text{propagation:} \quad [MoCp(CO)_3] + PMe_3 \longrightarrow [Mo^ICp(CO)_3(PMe_3)]$$
$$\underset{17e}{} \qquad\qquad\qquad \underset{19e}{}$$

$$[Mo^ICp(CO)_3(PMe_3)] \xrightarrow{-CO} [Mo^ICp(CO)_2(PMe_3)]$$
$$\underset{19e}{} \qquad\qquad\qquad \underset{17e}{}$$

$$\xrightarrow{PMe_3} [Mo^ICp(CO)_2(PMe_3)_2]$$
$$\underset{19e}{}$$

$$[Mo^ICp(CO)_2(PMe_3)_2] + [MoCp(CO)_3]_2$$
$$\underset{19e}{} \qquad\qquad\qquad \underset{18e}{}$$

$$\rightarrow [Mo^{II}Cp(CO)_2(PMe_3)_2]\,[Fe^0Cp(CO)_2] + [MoCp(CO)_3]$$
$$\underset{\text{18e cation}}{} \qquad\quad \underset{\text{18e anion}}{} \qquad \underset{17e}{}$$

Analogous chain reactions proceed with the other dimers $[Mn(CO)_5]_2$ and $[FeCp(CO)_2]_2$ [311, 312].

When the dimers are held together by a fulvalene bridging ligand instead of two Cp ligands, the cationic and anionic parts of the product are not separated from each other, i.e. a zwitterion forms. The MoMo and WW fulvalene hexacarbonyl dimers are reduced at $-1.22$ V and $-1.34$ V relative to $FeCp_2$ in THF respectively [313], and these potentials are less negative than the oxidation potentials of the 19-electron complexes $[Fe^ICp(\eta^6\text{-arene})]$. Thus the complexes $[Fe^ICp(\eta^6\text{-}C_6R_6)]$, R = H or Me are both efficient catalysts (0.1–0.2 equiv.) for the ETC disproportionation of the MoMo and WW fulvalene hexacarbonyl complexes in the presence of $PR_3$ (R = Me or OMe) to the corresponding zwitterions $[\mu_2, \eta^{10}\text{-fulvalene}]$-$\{[M(CO)_2(PR_3)_2\,[M(CO)_3]\}$. These reactions were also carried out thermally in the absence of catalyst, which gave lower reaction yields and required heating for days instead of a few minutes at ambient temperature in the presence of the catalyst. The mechanism is similar to the chain mechanism above, and the electron transfer between the two metal centers is now intramolecular and extremely fast, being facilitated through the delocalized bridging fulvalene ligand (Scheme 27) [314].

Similar reactions catalyzed by the 19-electron complexes $[Fe^ICp(\eta^6\text{-}C_6R_6)]$ were carried out with heterobinuclear FeW, RuMo and RuW fulvalene pentacarbonyl complexes. Since the two metals are different, it was of interest to look at the re-

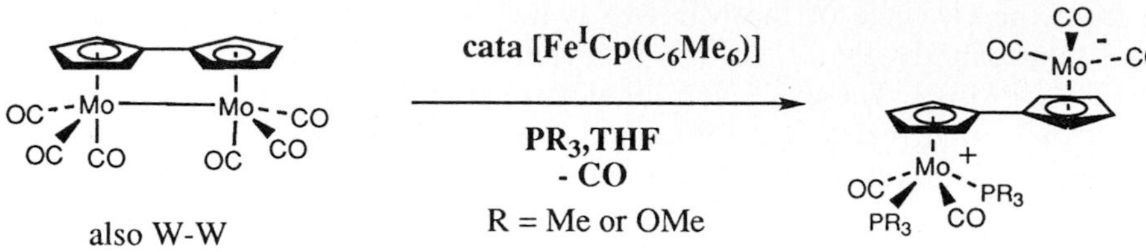

**Scheme 27.**

gioselectivity of the reaction. In the case of the FeW complex, the cathodic reduction of the Fe and W metal centers are separated, and the W center ($E_{p1} = -1.45$ V relative to FeCp$_2$) is more easily reduced than the Fe center ($E_{p2} = -2.0$ V relative to FeCp$_2$) [315, 316]. Both [Fe$^I$Cp($\eta^6$-C$_6$H$_6$)] and [Fe$^I$Cp($\eta^6$-C$_6$Me$_6$)] have oxidation potentials more negative than the first reduction wave, that of the W center, and indeed are good catalysts for the formation of the zwitterions in the presence of either PMe$_3$ or P(OMe)$_3$. Since the reduction occurs first on the W center, cleavage of the metal–metal bond in the substrate which has accepted an electron leave an anionic W center and a neutral radical Fe center. The rapid ligand addition and substitution occurs on this latter center, finally leading to [$\mu_2, \eta^{10}$-fulvalene]-\{[Fe(CO)(PR$_3$)$_2$ [W(CO)$_3$]\}. Likewise, the tetracarbonyl mono-trimethylphosphite non-zwitterionic complex [$\mu_2, \eta^{10}$-fulvalene][$\mu_2$CO]$_2$\{[Fe(PR$_3$)$_2$ [W(CO)$_2$]\} resulting from photolysis of the substrate in the presence of P(OMe)$_3$ leads to the same zwitterions upon ETC reactions catalyzed by [Fe$^I$Cp($\eta^6$-C$_6$Me$_6$)] as those yielded from the pentacarbonyl complex. This means that substitution of P(OMe)$_3$ is preferred to that of CO in this complex (Scheme 28) [317].

The RuMo and RuW complexes are reduced in a two-electron waves without differentiation of the two metals. This make the result of the regioselectivity of the

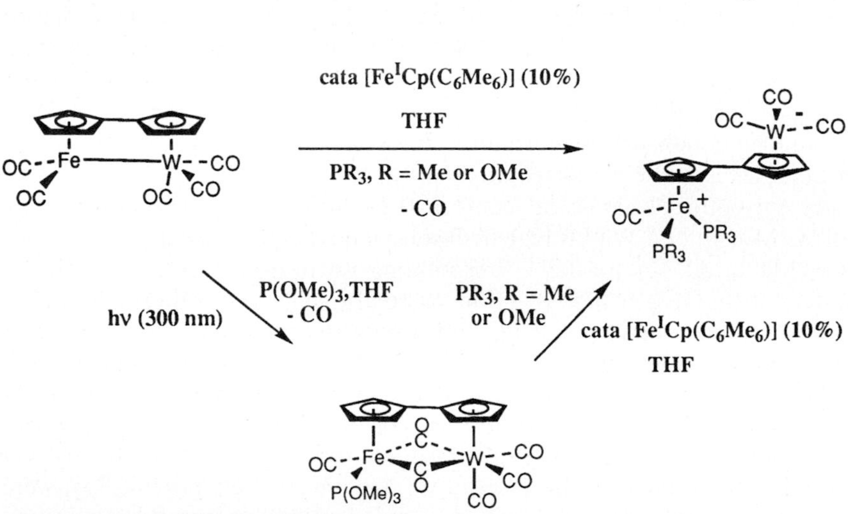

**Scheme 28.**

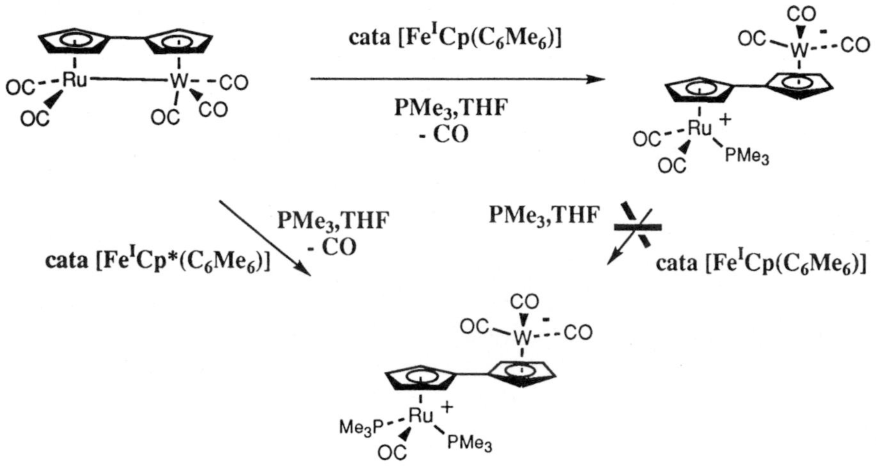

cata [Fe$^I$Cp(C$_6$Me$_6$)]

PMe$_3$,THF
- CO

OC--Ru$^+$
OC   PMe$_3$

CO
OC—W---CO

cata [Fe$^I$Cp*(C$_6$Me$_6$)]

PMe$_3$,THF
- CO

PMe$_3$,THF

cata [Fe$^I$Cp(C$_6$Me$_6$)]

CO
OC—W---CO

Me$_3$P--Ru$^+$
OC   PMe$_3$

**Scheme 29.**

ETC catalyzed reaction of more interest. It is again the Ru center which undergoes the ligand substitution and become positive, which means that reduction again occurs first on Mo or W.

In the case of the RuW complex, the double ligand substitution occurs stepwise, indicating that the first ligand substitution is much easier than the second one. At this time, the driving force of the initiation redox reaction of the Fe$^I$ complex is decisive concerning the number of COs which are substituted by PMe$_3$. Thus, in THF, catalysis by [Fe$^I$Cp($\eta^6$-C$_6$Me$_6$)] of ligand substitution in [$\mu_2,\eta^{10}$-fulvalene]-{[Ru(CO)$_2$][W(CO)$_3$]} in the presence of PMe$_3$ leads to the monosubstituted zwitterion [$\mu_2,\eta^{10}$-fulvalene]{[Ru$^{II}$(CO)$_2$(PMe$_3$)][W$^0$(CO)$_3$]}. On the other hand, the same reaction catalyzed by [Fe$^I$($\eta^5$-C$_5$Me$_5$)($\eta^6$-C$_6$Me$_6$)] under the same conditions gives the disubstituted zwitterion [$\mu_2,\eta^{10}$-fulvalene]{[Ru$^{II}$(CO)(PMe$_3$)$_2$][W$^0$(CO)$_3$]}. The 300 mV of additional driving force provided by the use of the Fe$^I$ complex containing the permethylated Cp* ligand appears to provoke the second ligand substitution. The monosubstituted complex cannot be reduced by [Fe$^I$Cp($\eta^6$-C$_6$Me$_6$)], but can be so by [Fe$^I$($\eta^5$-C$_5$Me$_5$)($\eta^6$-C$_6$Me$_6$)] (Scheme 29).

The relative values of the oxidation potentials of the catalysts and of the reduction potentials of the starting and monosubstituted complexes are in accord with the observed selectivity. This also means that the second ligand substitution is slow compared to the cross redox reaction of the propagation step, and that the zwitterionic monosubstituted complex is thus an intermediate in the ETC catalyzed synthesis of the disubstituted zwitterion. The precipitation of the monosubstituted zwitterion also precludes the second substitution, since the latter occurs in MeCN even with [Fe$^I$Cp($\eta^6$-C$_6$Me$_6$)]. Ion-pairing may play a substantial role in the balance between the various reaction pathways, and especially here in the competition between cross redox reaction, ligand substitution, side radical reactions and precipitation. In the present case, the counter-cation of the anionic species of the propagation cycle is the large 18-electron sandwich [Fe$^{II}$Cp($\eta^6$-C$_6$Me$_6$)]$^+$ (Scheme 30).

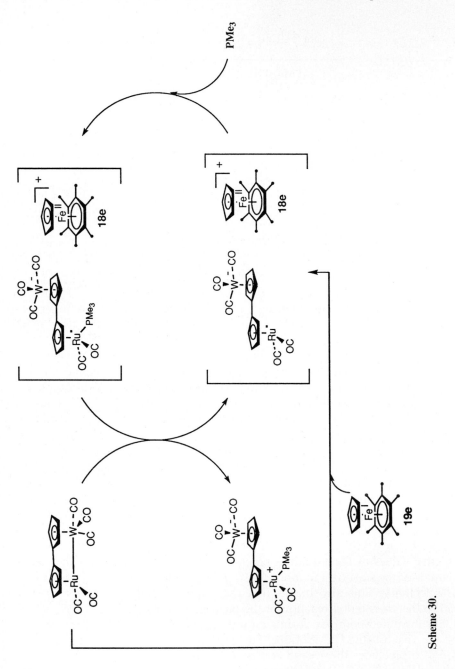

Scheme 30.

One of the conclusions of this study is that the first ligand substitution proceeds with a much higher coulombic efficiency than the second one, simply because of the increasing bulk at the metal center [317]. This trend was known from previous studies of carbonyl substitution by phosphanes in transition-metal carbonyl clusters catalyzed by benzophenone radical anion [318, 319]. In such a case, selective ETC catalysts may be designed as a function of the required exergonicity of the desired initiation step [317].

**Substitution of carbonyls by phosphines in transition-metal clusters and ETC-catalyzed synthesis of dendrimers with cluster termini**

Many phosphane-substituted transition-metal clusters have been synthesized from late transition-metal carbonyl clusters and the appropriate phosphane using reductive ETC catalysis with reductive initiation [318–333]. Indeed such an initiation provides an exergonic cross electron-transfer propagation step. Most syntheses were carried out using a cathodic initiation or sodium benzophenone radical anion. The method was successful because it turned out that the first substitution of a carbonyl by a phosphane proceeds with high yield and coulombic efficiency in homoleptic metal carbonyl clusters and some others.

The family of 19-electron complexes $[Fe^ICp(\eta^6\text{-arene})]$ has also been investigated as catalysts for the substitution of one, two and three carbonyls in metal-carbonyl clusters as shown below.

$$[M_3(CO)_m] + nL \xrightarrow{\text{cata } [FeCp(C_6Me_6)]} [M_3(CO)_{m-n}L_n] + nCO$$

$$M = Fe: n = 1; M = Ru \text{ or } M_3 = CH_3CCo_3: n = 1, 2 \text{ or } 3$$

The standard mechanism is shown in Scheme 31 and the catalysts and products appear in Scheme 32.

The inconvenient of the sodium benzophenone catalyst is that it is not easy to determine the quantity of catalyst used since it is never isolated but simply generated from sodium and benzophenone in THF and the THF solution is used as it is. On the other hand, the quantities of $Fe^I$ complexes are exactly known since they

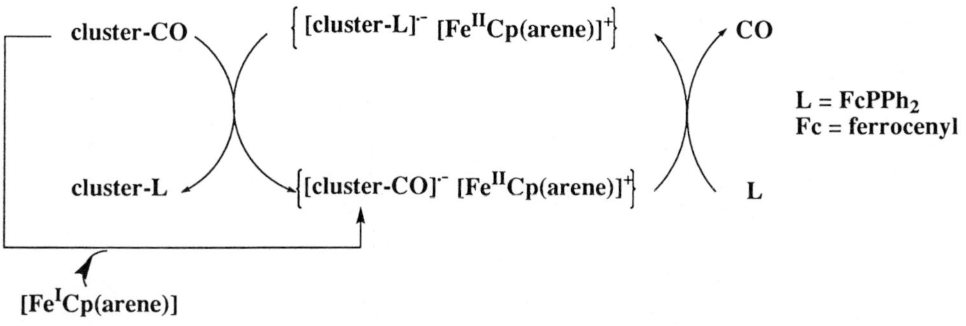

**Scheme 31.**

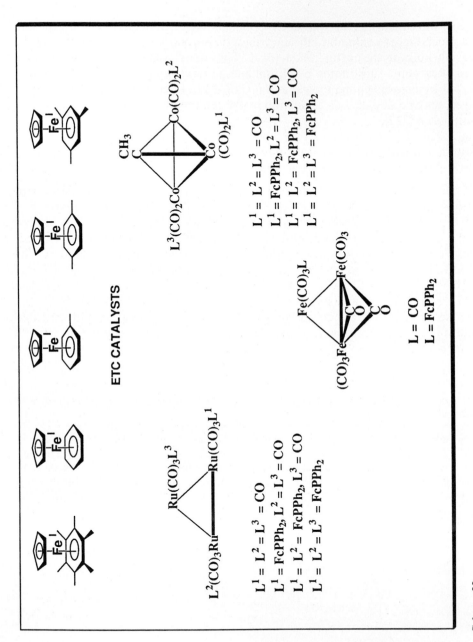

Scheme 32.

can usually be weighted. In addition, they can be used in a variety of solvents and not only in THF. The goal was also partly synthetic and partly mechanistic, and the link between these two aspects is that mechanistic information leads to synthetic improvements. It was desirable to know if the $Fe^I$ catalysts are as efficient as sodium benzophenone radical anion. First, the catalysis using $[Fe^ICp(\eta^6\text{-}C_6Me_6)]$ was carried out. Then, the influence of the driving force of the initiator on the selectivity and efficiency of the catalytic process were also examined. Indeed, the change of driving force is easily reached with these 19-electron complexes by variation of the number of methyl groups on the ligands as we know from the preceding section. In particular, the progressive variation of the number of methyl group from 0 to 6 on the arene ligand is easily achieved. Each methyl group on the arene ligand makes the redox potential more negative by about 0.03 V to 0.04 V so that the standard redox potential of this series of initiators spreads from $-1.75$ V for $[Fe^ICp(\eta^6\text{-}C_6H_6)]$ to $-2.0$ V relative to $FeCp_2$ for $[Fe^ICp(\eta^6\text{-}C_6Me_6)]$ in THF. The results of the catalyzed ETC reactions are summarized below. Table 11 shows the data using only $[Fe^ICp(\eta^6\text{-}C_6Me_6)]$, and the Table 12 compares the different $[Fe^ICp(\eta^6\text{-arene})]$ catalysts.

The quantity of catalyst used was either 1 % or 10 %, and no reaction takes place under ambient condition in the absence of catalyst. When only one equiv. of phos-

**Table 11.** Amount (%) of substituted cluster in the mixture mono- + di- + trisubstituted products obtained using $[Fe^ICpC_6Me_6]$ as catalyst at 20 °C (30 min) in THF on a one-millimolar scale (determined by $^1H$ NMR).

| Starting cluster | Amount of L (no. of equiv.) | Amount of catalyst (%) | Number, $n$, of ligand(s) in the final product | | | Stability of products in solution |
|---|---|---|---|---|---|---|
| | | | $n = 1$ | $n = 2$ | $n = 3$ | |
| $Fe_3(CO)_{12}$ | 1 | 1 | 100 | 0 | 0 | Mediocre |
| (Reaction at $-40$ °C) | 1 | 10 | 100 | 0 | 0 | Mediocre |
| | 2 | 10 | – | – | – | Decomposition |
| | 2 | 10 | – | – | – | Decomposition |
| $Ru_3(CO)_{12}$ | 1 | 1 | 100 | 0 | 0 | Stable |
| | 1 | 10 | 80 | 20 | – | Stable |
| | 2 | 10 | – | 90 | 10 | Stable |
| | 3 | 1 | 10 | 60 | 30 | Stable |
| | 3 | 10 | – | 20 | 80 | Stable |
| $Ru_3(CO)_{10}L_2$ | 1 | 10 | – | – | 100 | |
| $(\mu\text{-}CCH_3)Co_3(CO)_9$ | 1 | 1 | 100 | – | – | |
| | 1 | 10 | 85 | 15 | – | Stable |
| | 2 | 10 | 15 | 80 | trace | Mediocre |
| | 3 | 1 | 70 | 25 | 0 | Mediocre |
| | 3 | 10 | 15 | 30 | 55 | Mediocre |

**Table 12.** Amount (%) of mono-, di- and trisubstituted ruthenium carbonyl clusters obtained by reaction of $[Ru_3(CO)_{12}]$ with 3 equiv. FDPP and use of 10 % iron(I) electrocatalyst. The standard redox potentials of $[Fe^ICp(C_6H_{6-n}Me_n)]$ in THF + $n$-Bu$_4$NPF$_6$, 0.1 M, are $E^{o\prime} = -(1.745 + 0.032n)$ V relative to $[FeCp_2]$ [84].

| | Monosubstitution | Disubstituted | Trisubstituted |
|---|---|---|---|
| $[FeCp(C_6Me_6)]$ | 0 | 0 | 100 |
| $[FeCp(C_6H_6)]$ | 10 | 70 | 20 |
| $[FeCp(toluene)]$ | 5 | 65 | 30 |
| $[FeCp(p\text{-xylene})]$ | 0 | 60 | 40 |
| $[FeCp(mesitylene)]$ | 0 | 55 | 45 |

phine per cluster is used, the syntheses of monosubstituted clusters proceeds in high yield even with only 1 % catalyst except for the Fe$_3$ cluster which is slightly unstable in solution and partly decomposed on a chromatographic column. With two and three equiv. phosphine, the Ru$_3$ and Co$_3$ clusters could be obtained, but the reaction with 10 % catalyst gives much better yields of the desired products. This shows that side reactions are more competitive as substitution increases as expected because the bulk slows down the ligand exchange propagation step. The influence of the driving force of the catalyst is evident from the second table. The first substitution proceeds quantitatively even with the catalyst of weakest driving force $[Fe^ICp(\eta^6\text{-}C_6H_6)]$. As the number of substitutions increases, however, the influence of the driving force of the catalyst becomes more dramatic. The largest influence is obtained for the third substitution as evidenced by the data in the table. The yield of trisubstituted cluster regularly increases, when 10 % catalyst is used, from 20 % with $[Fe^ICp(\eta^6\text{-}C_6H_6)]$ to 100 %, obtained only with $[Fe^ICp(\eta^6\text{-}C_6Me_6)]$. The conclusions of such a study are multiple. It confirms the crucial influence of the driving force of the initiator on the yield and efficiency in ETC catalyzed reactions. It indicates that a maximum efficiency requires a high driving force in the initiation when slow propagation steps are involved, which is the case in multiple substitution. It also indicates that, if a given number of substitutions is desired, the use of a catalyst of intermediate driving force is optimal in order to avoid the formation of more substituted compounds. Thus chemioselectivity is improved in this way. Another difference with sodium benzophenone radical anion is that the counter cation of the anionic cluster intermediates in the propagation ETC cycle is not Na$^+$, but the large cation $[Fe^{II}Cp(\eta^6\text{-arene})]^+$. This makes the anion less strongly bound to the cation in the large organometallic ion pair than in the tight ion pair containing the sodium cation. Finally, the 19-electron complexes $[Fe^ICp(\eta^6\text{-arene})]$ and especially $[Fe^ICp(\eta^6\text{-}C_6Me_6)]$ are excellent catalysts for ETC reactions, and the variation of driving force by variation of the number of methyl groups offers an additional flexibility as a function of the requirements [334]. For instance, this study was very useful to determine the optimal conditions for more delicate substrates as follows.

## Introduction of the cluster fragment $Ru_3(CO)_{11}$ at the termini of polyphosphine dendrimers

The clean introduction of clusters on to the termini of polyphosphine dendrimers is a real challenge because of the current interest of dendritic clusters in catalysis and the mixtures usually obtained in thermal reactions. The diphosphine $CH_3(CH_2)_2N(CH_2PPh_2)_2$ (abbreviated P-P below) was used as a simple, model ligand. The reaction between P-P and $[Ru_3(CO)_{12}]$ [11] (molar ratio 1:1.05) in the presence of 0.1 equiv. $[Fe^ICp(\eta^6-C_6Me_6)]$. in THF at $20\,^{\circ}C$ led to the complete disappearance of $[Ru_3(CO)_{12}]$ in a few minutes and the appearance of a mixture of chelate $[P-P.Ru_3(CO)_{10}]$, monodentate $[P-P.Ru_3(CO)_{11}]$ and bis-cluster $[P-P.\{Ru_3(CO)_{11}\}_2]$. This type of reaction was reported by Bruce with simple diphosphines [318, 319]. On the other hand, the reaction of P-P with $[Ru_3(CO)_{12}]$ in excess (1:4) and only 0.01 equiv. $[Fe^ICp(\eta^6-C_6Me_6)]$. in THF at $20\,^{\circ}C$ led, in 20 min, to the formation of the air-stable, light-sensitive bis-cluster $[P-P.\{Ru_3(CO)_{11}\}_2]$ as the only reaction product (Scheme 33):

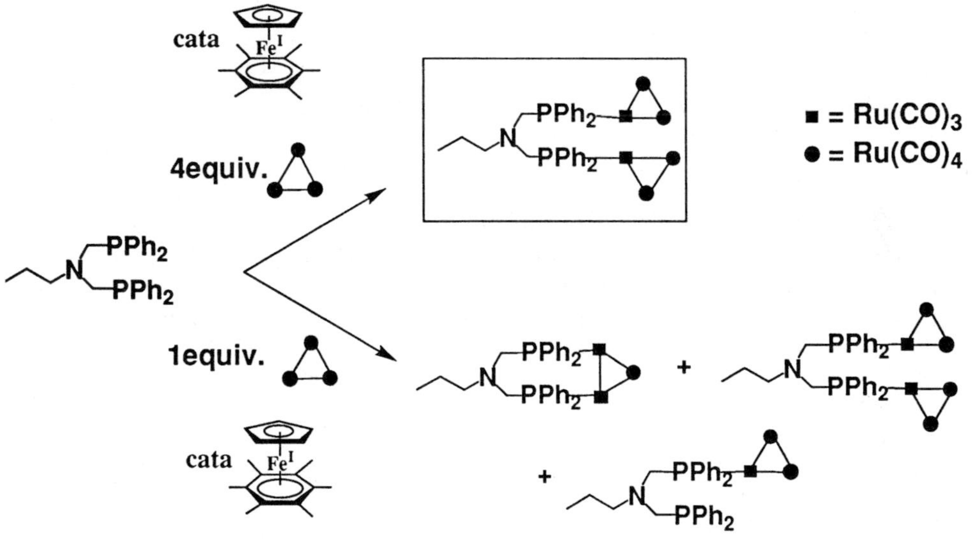

**Scheme 33.**

Given the simplicity of the above characterization of the reaction product by $^{31}P$ NMR [13] and the excellent selectivity of this model reaction when excess $[Ru_3(CO)_{12}]$ was used, the same reaction between the phosphine dendrimers and $[Ru_3(CO)_{12}]$ could be more confidently envisaged. This reaction, catalyzed by 1 % equiv. $[Fe^ICp(\eta^6-C_6Me_6)]$ was carried out in THF at $20\,^{\circ}C$. The dendrimer-cluster assembly was obtained in 50 % yield. This shows the selectivity and completion of the coordination of each of the 32 phosphino ligands of P-P to a $Ru_3(CO)_{11}$ cluster fragment (Scheme 34).

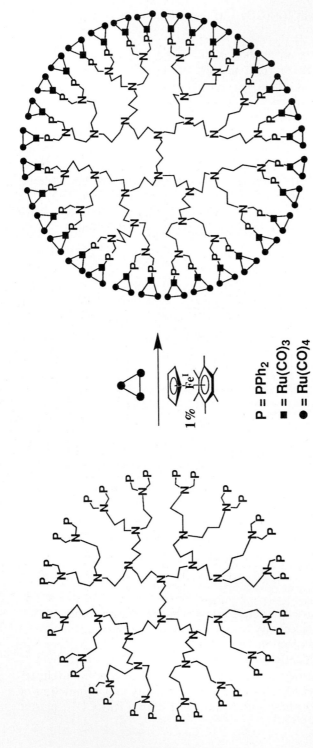

P = PPh₂

■ = Ru(CO)₃

● = Ru(CO)₄

Scheme 34.

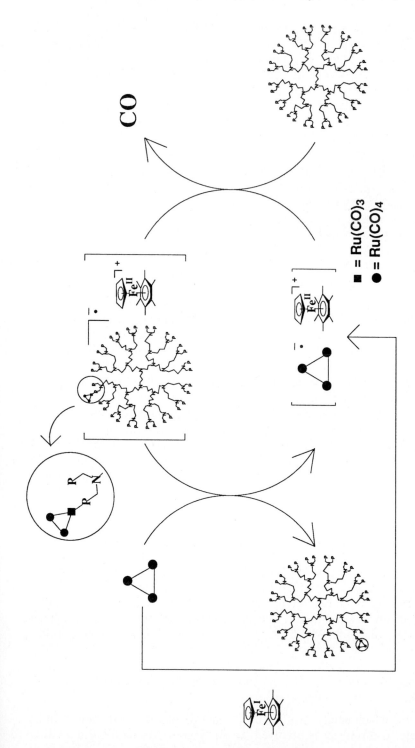

**dendr.phos. $\longrightarrow$ dendr.phos. (Ru$_3$) $\longrightarrow$ dendr.phos.(Ru$_3$)$_2$ $\longrightarrow$ $\cdots$ $\longrightarrow$ dendr.phos.(Ru$_3$)$_{32}$**

Scheme 35.

ETC mechanism proceeds for the introduction of the 32 cluster fragments in the dendrimer for ligation of the first $Ru_3(CO)_{11}$ fragment to the dendritic phosphine. Then, this first complex [dendriphosphine.$Ru_3(CO)_{11}$] would undergo the same ETC cycle as [$Ru_3(CO)_{12}$] initially does to generate the bis-cluster complex [dendriphosphine.$\{Ru_3(CO)_{11}\}_2$], and so on (Scheme 35).

Finally, the 64-branch phosphine DAB-*dendr*-G4-[N(CH_2PPh_2)_2]_{32} analogously reacts with [$Ru_3(CO)_{12}$] and 1 % [$Fe^ICp(\eta^6\text{-}C_6Me_6)$] (20 °C, THF, 20 min) to give the dark-red 192-Ru dendrimer. Characterization of the purity of these dendrimer-cluster assemblies is conveniently monitored by $^{31}P$ NMR. This application should find extension to other metal-carbonyl clusters and other families of phosphine dendrimers.

### 4.5.3 The Ferrocenium Salts as ETC Catalysts

#### The classical case of [MnCp(CO)_2(MeCN)]

ETC-catalyzed reactions which required an oxidative initiation are often catalyzed by a ferrocenium salt in inorganic and organometallic chemistry despite the modest value of the reduction potential of ferrocenium cation. This is especially the case when the substrate is a neutral complex. Most ligand exchange reactions are those involving the substitution of CO. Thus, since CO is one of the less electron-donating ligand, substitution by other ligands systematically requires a cathodic initiation if an exergonic cross redox step is required. On the other hand, substitution of inorganic ligands such as nitriles, pyridine or ethers is favorably induced by an oxidizing initiation if it does require ETC catalysis at all to proceed. Indeed, these inorganic ligands are also very often easy to replace thermally. The classic case is that of [MnCp(CO)_2(MeCN)] in which substitution by various phosphane ligands is favorably induced anodically or using ferrocenium salt as the ETC catalyst. The largest coulombic efficiencies (up to than 1000) ever found in ETC catalysis were indeed disclosed with this type of reaction (Scheme 36).

$$[MnCp(CO)_2(MeCN)] + PR_3 \xrightarrow{\text{cata } [FeCp_2]^+} [MnCp(CO)_2(PR_3)] + MeCN$$

Since they are so clean, these reactions were the subject of in-depth mechanistic studies by electrochemistry [336–338], including by derivative cyclic voltammetry (DCV) [337, 338]. Acetonitrile and other solvento complexes are accessible by photolysis of transition-metal carbonyl complexes in the desired solvent. Such ETC catalyzed reactions are also of interest because they are practical and easily generalized, for instance to the series $M(CO)_3L_3$ (M = Cr, Mo, W; L = MeCN, pyridine) [339].

#### Initiation by ferrocenium salts of an ETC-catalyzed chelation involving an endergonic cross propagation step

There are cases for which catalysis of an ETC reaction can proceed even if an endergonic cross electron transfer is involved. In such cases, the required conditions

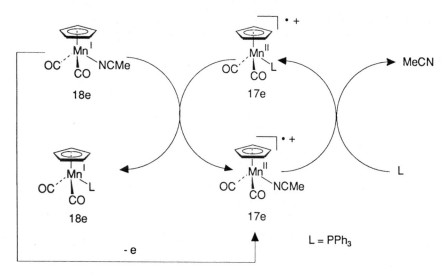

**Scheme 36.**

are (i) that the endergonicity be weak and (ii) that the chemical propagation step be irreversible and fast in order to shift the unfavorable redox propagation step. Even if these conditions are fulfilled, the resulting coulombic efficiencies are weak (i.e. there are only small turnover numbers). The catalysis by ferrocenium salts of the chelation of the monodentate dithiocarbamate complexes $[Fe(\eta^5-C_5R_5)(CO)_2(\eta^1-S_2CNMe_2]$ (R = H or Me) to $[Fe(\eta^5-C_5R_5)(CO)(\eta^2-S_2CNMe_2]$ disclosed in our group by Catheline [340] and Madonik [341] represents a typical example. The chelation of the $Fe^{III}$ intermediate generated by single-electron oxidation of the substrate is so fast that this intermediate could not be detected using fast electrochemical techniques (cyclic voltammetry with sweep rates up to 5000 V s$^{-1}$) [341]. The yield could not overtake a certain limit even if the quantity of ferrocenium used was optimized (0.2 equiv. was the optimum as shown in Figure 1) because of the side reaction (CO exchange with a solvent molecule and precipitation of the generated 17-electron complex) [341]. The ferrocenium counter-anion influences the reaction because the ion pairing plays a crucial role in the competition between the cross redox propagation step and the side reactions of the 17-electron chelated cation intermediate (Scheme 37). Thus, the larger SbCl$_6^-$ salt gave a better yield of chelated product $[FeCp(CO)(\eta^2-S_2CNMe_2]$ (63 %) than the PF$_6^-$ (52 %) and BF$_4^-$ (47 %) salts.

If the reaction is initiated by reduction, the very strong reductant naphthylsodium is required, and the cross redox step is now exergonic (thus all the starting material in consumed). The yields are not higher than with ferrocenium, however, because, reduction of the $Fe^{II}$ substrate generates an inorganic anionic Fe$^I$ intermediate. Such a low oxidation state is not favorable in inorganic chemistry. Thus, this Fe$^I$ species rapidly decomposes to the dithiocarbamate anion S$_2$CNMe$_2^-$ and the radical $[FeCp(CO)_2]$ which dimerizes to $[FeCp(CO)_2]_2$ (Scheme 38). This efficient side reaction competes with the chemical propagation step, chelation. The driving force

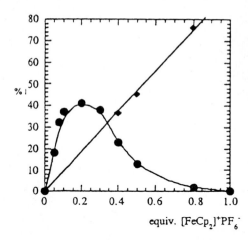

**Figure 1.** Yields of the ETC catalysis product [FeCp(CO)($\eta^2$-dtc)] and stoichiometrically oxidized complex [FeCp(CO)($\eta^2$-dtc)][PF$_6^-$] (circles) as a function of the amount (%) of [FeCp$_2$][PF$_6^-$] added to the starting complex [FeCp(CO)$_2$($\eta^1$-dtc)] (see Scheme 37).

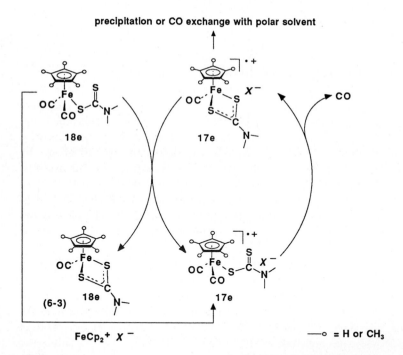

**Scheme 37.**

of the initiator and the nature of the counter-cation also influence this competition, and thus the reaction yield, in a way similar to that detailed in Section 4.5.2. This case is probably a unique one for which both oxidative and reductive initiation could be used [342].

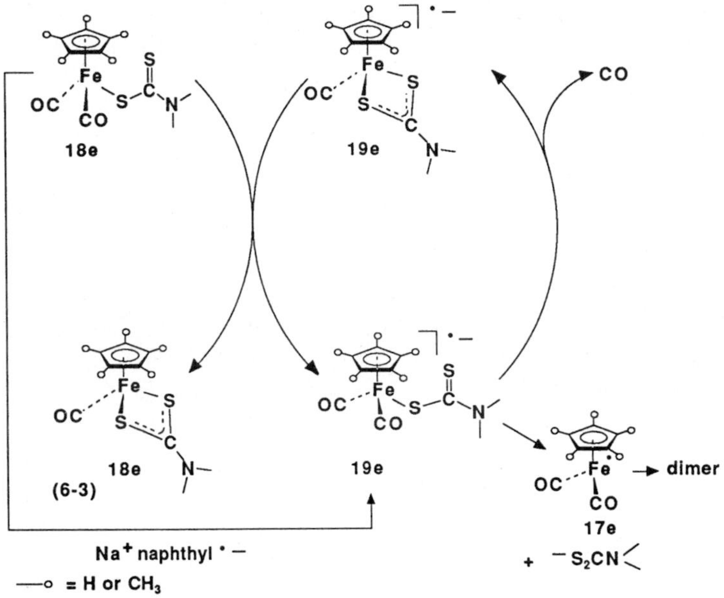

**Scheme 38.**

### Coupling oxidative ETC catalysis using ferrocenium cation with organometallic catalysis: polymerization of terminal alkynes

A major drawback in homogeneous catalysis is sometimes the need to heat 18-electron transition-metal complexes, the catalyst precursors, so that they generate catalytically active species with vacant coordination sites. Since 18-electron complexes are most of the time inert to substitution reactions, they usually need to be heated to quite high temperature in order to generate efficient catalytically active species. Our concept consisted in generating catalytically active species in which the substrates are ligated to the transition-metal centers using the fast ETC catalyzed ligand substitution process under ambient conditions [185, 343]. Since 17-and 19-electron species react in ligand substitution reactions with rates in the range of $10^6$ to $10^9$ times higher than their 18-electron isostructural precursors [344, 345], the kinetic gain thereof could be used to perform homogeneous transition-metal catalysis under ambient conditions in cases where extensive heating was required. In short, the idea was to couple ETC catalyzed ligand substitution with homogeneous transition-metal catalysis. We choose the polymerization of terminal alkynes which was known to proceed at 100 °C according to the Katz mechanism with the 18-electron catalyst [W(CO)$_6$] [346]. It was also known that [W(CO)$_3$(MeCN)$_3$] could undergo oxidatively induced ETC catalyzed ligand substitution reaction with phosphanes [339]. We needed to bring two alkyne molecules on to the W center in order to achieve the catalytic alkyne polymerization according to the Katz mechanism [346]. We therefore envisaged that the substitution of two MeCN ligands by alkyne ligands would also be possible by oxidative ETC catalysis, since alkynes are not as good donors as MeCN. What was needed was to use ferrocenium cation in catalytic amounts relative to the W catalyst itself, i.e. the amount of required ferrocenium salt was extremely minute. The use of ferrocenium in stoichiometric amount relative to the W catalyst would quantitatively generate catalytically inactive $W^I$. Also the reaction should be done in a solvent which would not be a good ligand to compete with alkynes. This catalytic engineering was successful and led to the rapid polymerization (10 min) of terminal alkynes under ambient conditions in toluene, THF, $CH_2Cl_2$ and neat but not in MeCN solvent. The optimum amount of $[FeCp_2][PF_6]$ was 20 % relative to the W catalyst in THF whereas this yield dropped continuously as this amount was increased, and reached only a few % when a stoichiometric amount of $[FeCp_2][PF_6]$ was used. The results were rather similar for different terminal alkynes, but the yield was only 6 % with the internal alkyne 2-hexyne. The yields, molecular weights and polydispersities with 10 % $[FeCp_2][PF_6]$ and without it are gathered in the Table 13.

The above data are consistent with a mechanism involving double ETC catalyzed substitution of two MeCN ligands by two terminal alkynes followed by isomerization of one alkyne ligand to a vinylidene ligand and Chauvin-type [347] square metathesis-like formation of a tungstacyclobutene, then continuation of the alkyne polymerization. It is assumed that the third MeCN ligand is only very slowly exchanged if at all, and probably not before polymerization (ancillary role, but no bulk interference). Indeed, the complex $[W(CO)_3(PPh_3)(MeCN)_2]$ is about ten times less active than $[W(CO)_3(MeCN)_3]$, presumably because of the large bulk of the $PPh_3$ ligand compared to MeCN [185] (Scheme 39). Obviously, this type of process combining ETC catalysis and homogeneous transition-metal catalysis has a future for other comparable applications.

Besides ETC catalysis, atom-transfer chain (ATC) catalysis has been reviewed [2]. The initiation involves radicals and the redox propagation step exchanges an

**Table 13.** Solvent effect on the polymerization of $PhC_2H$ by $W(CO)_3(NCMe)_3$ and 0.1 equiv. $[FeCp_2]^+PF_6^-$ per W complex.

| Solvent | THF | $CH_2Cl_2$ | $CH_3CN$ | Toluene | Bulk |
|---|---|---|---|---|---|
| Polymer yield (%)[a] | 45 (15)[b] | 42 (45)[b] | 14 (12.6) | 36 (13) | 13 (7) |
| $Mw^c$ | 32000 (21350) | 30200 (22250) | 2070 (1530) | 7000 (30600) | 21400 (16800) |
| $Mn^c$ | 13200 (10650) | 11600 (6250) | 800 (600) | 3400 (7000) | 6400 (5600) |
| Polydispersity index[c] | 2.42 (2.00) | 2.61 (3.56) | 2.57 (2.63) | 2.06 (4.37) | 3.34 (2.99) |

[a] Polymerized at RT, in the dark, under argon for 10 min; 50 equiv. alkyne per W complex was used.

[b] As indication: numbers in brackets are the yields and characterizations obtained without addition of the cocatalyst $[FeCp_2]^+PF_6^-$ and after one week reaction time.

[c] Determined by gel-permeation chromatography. The polydispersity index, close to 2, is indicative either of slow initiation or of random disruption of chain growth.

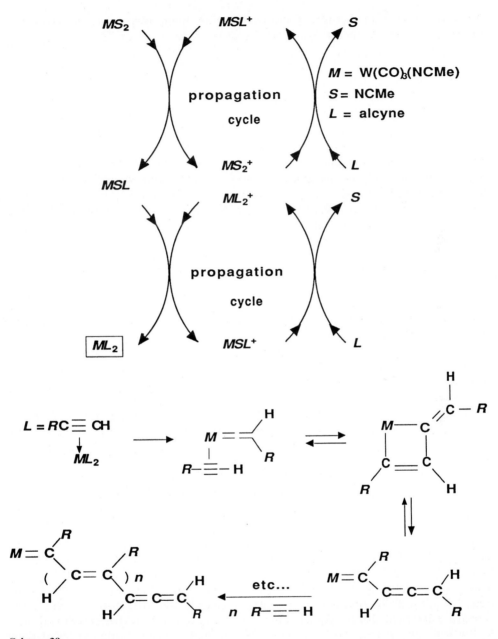

Scheme 39.

atom (H, halogen) or a group of atoms. There are many ways to generate radicals and single-electron reduction could be one of them [2].

## 4.6 Redox Catalysis

Redox catalysis is the catalysis of redox reactions and constitutes a broad area of chemistry embracing biochemistry (cytochromes, iron–sulfur proteins, copper proteins, flavodoxins and quinones), photochemical processes (energy conversion), electrochemistry (modified electrodes, organic synthesis) and 'chemical' processes (Wacker-type reactions). It has been reviewed altogether relatively recently [2]. We will essentially review here the redox catalysis by electron reservoir complexes and give a few examples of the use of ferrocenium derivatives.

### 4.6.1 Principle of Redox Catalysis of Electrochemical Reactions

The electrochemical redox reaction of a substrate resulting from the heterogeneous electron transfer from the electrode to this substrate (cathodic reduction) or the opposite (anodic oxidation) is said to be electrochemically reversible if it occurs at the Nernstian redox potential without surtension (overpotential). This is the case if the heterogeneous electron transfer is fast, i.e. there must not be a significant structural change in the substrate upon electron transfer. Any structural change slows down the electron transfer. When the rate of heterogeneous electron transfer is within the time scale of the electrochemical experiment, the electrochemical process is fast (reversible). In the opposite case, it appears to be slow (electrochemically irreversible). Structural transformations are accompanied by a slow electron transfer (slow E), except if this transformation occurs after electron transfer (EC mechanism).

Redox mediators or catalysts should facilitate the heterogeneous electron transfer from the electrode to the substrate (or the opposite) and partly cancel the required surtension observed in the absence of mediator or redox catalyst. Mediators or redox catalysts must undergo a redox change without significant structural change at a milder potential than the substrate. In this way, the heterogeneous electron transfer between the electrode and the mediator is Nernstian (fast i.e. electrochemically reversible). Electron transfer proceed more readily between the mediator or redox catalyst in solution than between the electrode and the substrate because the substrate and mediator or redox catalyst can move easily in the tridimensional bulk of the solution so that they find the right, optimal respective orientation of their orbitals for electron transfer. Such selective orientation of both partners of the electron transfer is much less easy with the electrode surface which is only two-dimensional. This is the reason for the kinetic advantage provided by the mediator or redox catalyst for a given driving force and is illustrated in Scheme 40.

The nomenclature [348] is such that a *mediator* exchanges electrons with substrates by an outer-sphere electron-transfer mechanism (for instance ferrocene/

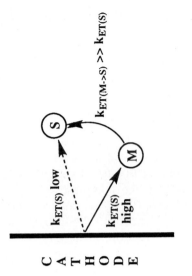

$k_{ET(M \to S)} >> k_{ET(S)}$

$k_{ET(S)}$ low

$k_{ET(S)}$ high

C
A
T
H
O
D
E

S = substrate
M = mediator

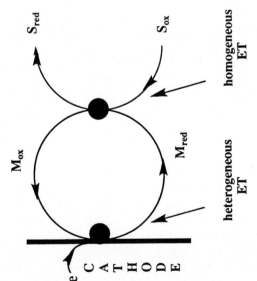

$S_{red}$

$S_{ox}$

$M_{ox}$

$M_{red}$

homogeneous
ET

heterogeneous
ET

e

C
A
T
H
O
D
E

Scheme 40.

ferrocenium [349]) whereas a *redox catalyst* does so by an inner-sphere electron-transfer mechanism (for instance $Ni^0$ catalysts for the cathodic transformation of aryl halides [350, 351]). Redox catalysts provide a larger kinetic gain than redox mediators because inner-sphere electron transfers are much faster than outer-sphere ones.

### 4.6.2  Catalysis of Cathodic Reduction of Nitrates and Nitrites by the Electron-reservoir Complexes [$Fe^I(\eta^5$-$C_5H_4R)(\eta^6$-arene)]

The electroreduction of the complexes [$Fe^{II}Cp(\eta^6$-arene)][$PF_6$] was first studied in basic aqueous medium and ethanol. It was found that the complexes [$Fe^I(\eta^5$-$C_5H_4R)(\eta^6$-$C_6Me_6$)] (R = H or $CO_2^-$) are both initiators for ETC catalyzed self-decomposition and redox catalysts for the reduction of water to dihydrogen on mercury cathode [352]. The reduction of water is a side reaction of the ETC catalytic process and limits its coulombic efficiency (Scheme 41). On such a cathode, the surtension of the reduction of water is very high, which makes this electrode especially suitable for the study of metal ions in such a medium. During this study, it was found that the most stable $Fe^I$ complex [$Fe^I(\eta^5$-$C_5H_5)(\eta^6$-$C_6Me_6$)], also catalyzes the cathodic reduction of nitrates to ammonia in the same basic aqueous medium (pH 13) [353, 354].

This result was all the more interesting as nitrate is not electroactive in these conditions in the absence of catalyst and the catalyst is stable after a large turnover number. The catalytic reduction of nitrate has been known for a long time [355], but this was the first example of organometallic catalyst for this reaction. Electro-

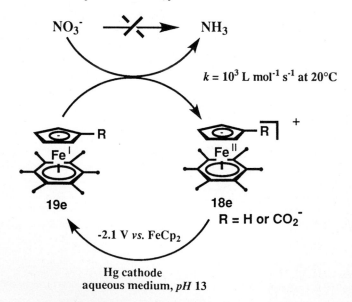

Scheme 41.

syntheses catalyzed by $[Fe^I(\eta^5\text{-}C_5H_4CO_2^-)(\eta^6\text{-}C_6Me_6)]$ under these conditions showed that $NH_3$ was produced in 63 % chemical yield and 57 % electrical yield with $R = CO_2^-$. Nitrite, hydroxylamine, hydrazine and dinitrogen (minute amounts) were intermediates towards the formation of ammonia. The cathodic reduction stopped at the level of hydroxylamine when the experiment was carried out at pH 7 instead of 13. Kinetic studies in homogeneous basic aqueous solution using a polarographic method were carried out with $[Fe^{II}(\eta^5\text{-}C_5H_4R)(\eta^6\text{-}arene)]$, $R = CO_2^-$, or when only the cationic $Fe^{II}$ form was soluble in the medium ($R = H$; arene = benzene, $m$-xylene, hexamethylbenzene). These studies led to the conclusion that the rate of redox catalysis was independent of the nature of the catalyst within this series [353]. This study has recently been reconsidered, however, with the series of complexes $[Fe^{II}(\eta^5\text{-}C_5H_4CO_2^-)(\eta^6\text{-arene})][PF_6]$, arene = $C_6Me_{6-n}$, $n = 0$ to 6 [356]. The polarographic, cyclic voltammetry and chronoamperometry techniques were used to investigate the kinetics of the redox catalysis. The three techniques provided similar results. Thus, the rate constant of the redox catalysis can be calculated from the enhancement of the intensity of the cyclovoltammogram wave observed upon addition of the nitrate or nitrite salt into the electrochemical cell (Figure 2). A Marcus-type linear relationship was found between the logarithms of the rate constants and the standard redox potentials of the catalysts, indicating that the electron transfer in solution between the 19-electron $Fe^I$ complex and nitrate or nitrite ion is rate limiting.

To investigate whether the mechanism proceeds by inner-sphere or outer-sphere electron transfer, other catalysts of the type $[Fe^{II}(\eta^5\text{-}C_5H_4CO_2^-)(\eta^6\text{-arene})][PF_6]$ were synthesized with bulky arenes such as 1,3,5-tris-*tert*-butylbenzene and $C_6(CH_2CH_2p\text{-}C_6H_4OH)_6$. The rate constants were found to be one to two orders of

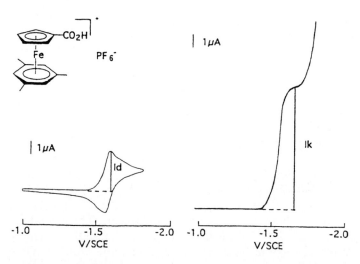

**Figure 2.** Redox catalysis of the reduction of nitrate on Hg cathode: cyclovoltammograms of $[Fe^{II}(\eta^5\text{-}C_5H_4CO_2^-)(\eta^6\text{-}1,3,5\text{-}Me_3\text{-}C_6H_3)]$ 1.27 M in aq. LiOH 0.1 M: a) in the absence of nitrate; b) in the presence of nitrate $1.18 \times 10^{-2}$ M; scan rate: 0.1 V s$^{-1}$.

magnitude lower than what would be expected for the same driving force if the steric effect was not interfering, by comparison with the above series of catalysts with polymethylbenzene ligands. This showed a significant inner-sphere component to the electron-transfer process, although the kinetic drop would have been even more dramatic with a fully inner-sphere electron-transfer mechanism. It is likely that nitrate and nitrite ions coordinate to the 17-electron form of the $Fe^I$ catalysts and that electron transfer proceeds in such intermediates rather than by outer sphere. However, the bond between an oxygen atom of nitrate or nitrite to such a low oxidation-state center must be very loose and long because $\pi$-back-bonding is impossible with these ligands [356].

### 4.6.3 Mediation by Ferrocenium Cation: Derivatized Electrodes

We briefly mention here the use of the ferrocene/ferrocenium redox couple to mediate electron transfer on the oxidation (anodic) side, especially in derivatized electrode. This broad area has been reviewed [349]. For instance, polymers and dendrimers containing ferrocene units have been used to derivatize electrodes and mediate electron transfer between a substrate and the anode. Recently, ferrocene dendrimers up to a theoretical number of 243 ferrocene units were synthesized, reversibly oxidized, and shown to make stable derivatized electrodes. Thus, these polyferrocene dendrimers behave as molecular batteries (Scheme 42). These modified electrodes are characterized by the identical potential for the anodic and cathodic peak in cyclic voltammetry and by a linear relationship between the sweep rate and the intensity [134, 135]. Electrodes modified with ferrocene dendrimers were shown to be efficient mediators [357–359]. For the sake of convenience, the redox process of a smaller ferrocene dendrimer is represented below.

Electrical communication between redox centers of enzymes and electrodes is mediated by ferrocene links in amperometric redox biosensors. Direct communication with large enzymes is not possible because of the thick insulating protein layers around the redox centers of the proteins. This is the case, for instance, with the glycoprotein shell of glucose oxidase. Ferrocene units with low molecular weights can diffuse freely in and out the protein and mediate electron transfer. For instance, attachment of ferrocene units to amino groups of enzymes such as lysine, tyrosine and histidine led to increase of anodic current with glucose concentration when gold, platinum and carbon electrodes modified with such ferrocenylated enzymes [360–365].

$$glucose + O_2 \xrightarrow{\text{glucose oxidase}} gluconolactone + H_2O_2$$

An average of one mediator was bound per 12–75 kDa of enzyme. The mediators used have redox potentials 0.07 V to 0.55 V positive of the FAD/FADH enzyme's redox potential ($-0.05$ V relative to the NHE at pH 7), but those which have a redox potential more negative than glucose oxidase enzyme do not mediate electron transfer. Electrons can be relayed both by tunneling and by motion of the mediator in and out of the protein chains. For distances $>8$ Å, tunneling rates decrease ex-

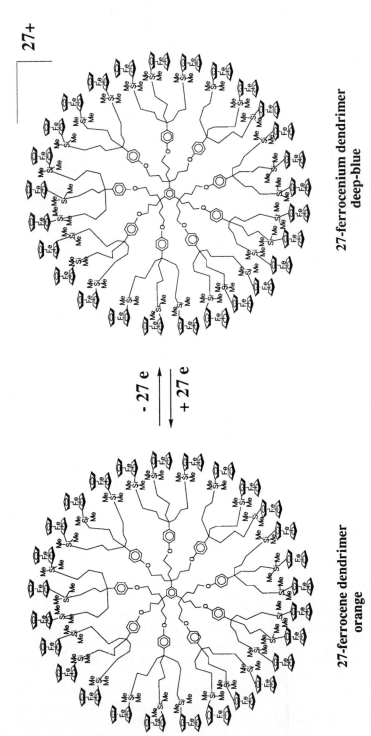

27+

27-ferrocenium dendrimer
deep-blue

– 27 e
+ 27 e

27-ferrocene dendrimer
orange

Scheme 42.

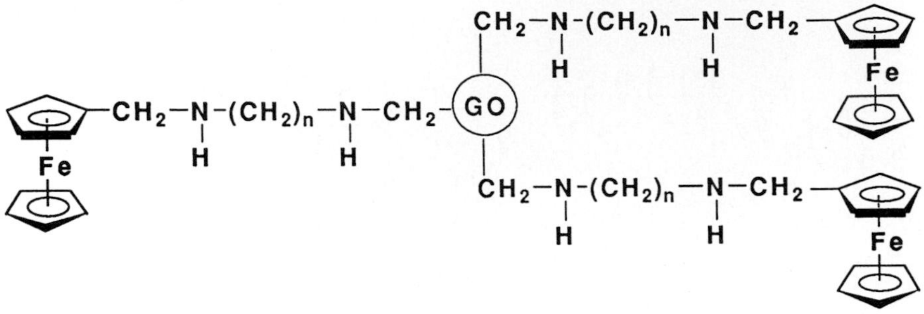

$$CH_2-N\text{-}(CH_2)_n-\overset{\displaystyle |}{\underset{\displaystyle H}{N}}-CH_2-\boxed{GO}$$

**Scheme 43.**

ponentially with the electron transfer distance. Communication between glucose oxidase and a vitreous carbon electrode was effective when the chain between the enzyme and the ferrocene unit was long (>10 bonds), but not when the chain was short. A long flexible chain allows the ferrocene mediator unit to penetrate the enzyme to a sufficient depth to reduce the electron-transfer distance (Scheme 43) [364, 365]. Several variations are known [360–365].

## 4.7 Sensors

The binding of a redox system such as ferrocene to endoreceptors has been used to recognize particular guests. The first example concerned with the recognition of alkali-metal cations by a crown ether incorporating a ferrocenophane moiety [366] as shown on Figure 3.

The recognition of anions using amidometallocenes bound to endoreceptors such as crown ethers and calixarenes has been efficiently developed by Beer [368–371]. Our group has synthesized and examined dendrimers whose branches are terminated by amidoferrocene [372, 373] and amidocobalticinium [374, 375] units; they were shown to be very excellent exoreceptors for hydrogen sulfate and dihydrogen phosphate [373–375]. The variation of redox potential of the redox metallocene centers were much larger than that of monoamidometallocenes or tripodal amidometallocenes. The equivalence point determined by cyclic voltammetry was found to be one equiv. per dendritic branch in all the cases. Moreover, large dendritic effects were found, i.e. the variation of the redox potential obtained after addition of one equiv. anion per dendritic branch was all the larger as the dendritic generation increased (3 → 9 → 18 branches; Scheme 44).

These amidometallocenes do not significantly recognize halides, however, the redox potential shifts being very small [372]. Chloride and bromide were specifically recognized by 24-cationic dendrimers bearing iron sandwich units at the termini of the branches [376, 377]. The sandwich units was $[FeCp^*(\eta^6\text{-}N\text{-alkylaniline})]^+$.

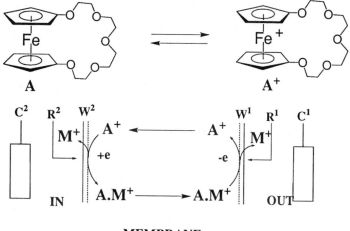

**MEMBRANE**

**Figure 3.** Amperometric sensory device: scheme of electrochemical ion transport with a redox-active crown ether. $W^1$, $W^2$: minigrid platinum working electrodes. $C^1$, $C^2$:platinum plate counter electrode. $R^1$, $R^2$:SCE reference.

Although these dendrimers have a different topology compared to the former, the corresponding 24-amidoferrocene dendrimers of the new type also recognize hydrogen sulfate and dihydrogen phosphate, but not the halides (Scheme 45).

Bromide and chloride were sensed by the variation of the chemical shift of the NH proton in $^1$H NMR. The equivalence point for bromide was also one equiv. bromide per branch, but the recognition of chloride was sharper and different; the equivalence point for this anion was one equiv. chloride per dendritic tripod, i.e. eight equiv. chloride per dendrimer [376]. This complementarity between the two families of dendrimers shows that dendritic variation can be searched to recognize a given anion. Nevertheless, the recognition of oxoanions by amido- or amino-metallocene dendrimers is based on the synergy between three effects: (i) the electrostatic effect with the ferrocenium form, (ii) the H bonding (simple or double) and (iii) especially the topology effect. In the case of the 24-branch polycationic dendrimers, only one H bonding per branch is occurring.

## 4.8 Conclusion

Redox reagents are useful for stoichiometric electron-transfer reactions, catalytic initiation of electron-transfer-chain (ETC) catalyzed reactions, as redox mediators (by outer-sphere electron transfer) or redox catalysts (by inner-sphere electron transfer) of redox reactions, as part of sensory devices when then are attached to

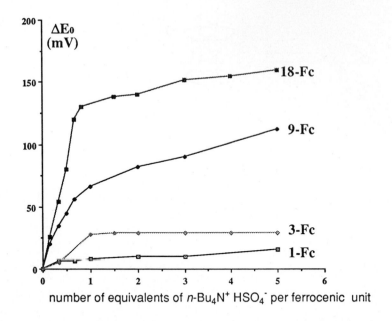

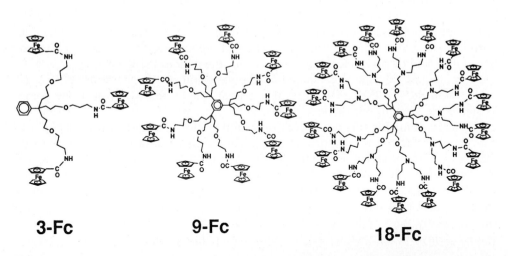

**3-Fc**          **9-Fc**                    **18-Fc**

**Scheme 44.**

*endo-* or *exo*-receptors and as references for the determination of redox potentials by cyclic voltammetry.

Electron-reservoir complexes were so named because they have the ability to transfer electrons without breakdown at very negative or very positive redox potentials, can be synthesized in large quantities, are crystalline and thermally robust and can be used in various solvents. The 19-electron complexes $[Fe^I(\eta^5\text{-}C_5R_5)(\eta^6\text{-}C_6Me_6)]$ (R = H or Me) are the most electron-rich neutral molecules known and the recently

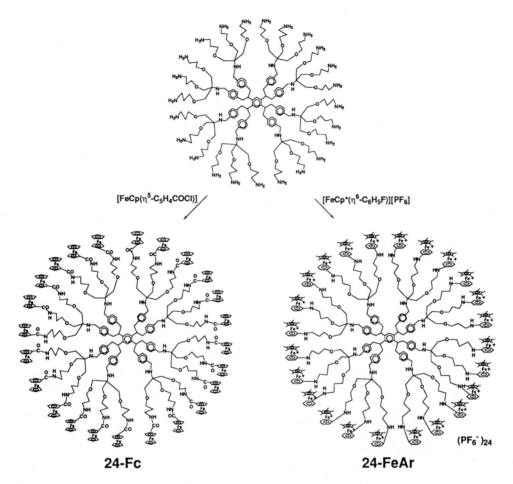

$[FeCp(\eta^5-C_5H_4COCl)]$ $[FeCp*(\eta^6-C_6H_5F)][PF_6]$

**24-Fc** **24-FeAr**

$(PF_6^-)_{24}$

**Scheme 45.**

disclosed 17-electron complex $[Fe^{III}(\eta^5-C_5Me_5)(\eta^6-C_6Me_6)][SbX_6]_2$, X = Cl or F (isoelectronic to ferrocenium but with a redox potential $Fe^{II/III}$ one volt more positive), is the strongest known stable organometallic oxidant. Thus, they are ideal redox reagents in many respects, as largely exemplified in this review. Their driving force can be changed at will by variation of the number of methyl groups on the ligands, which leads to noteworthy selectivities in ETC chain reactions, for instance.

These electron-reservoir iron-sandwich complexes have been compared in this chapter on the reduction side with other redox reagents: alkali metals and their amalgams and alloys, radical anions of aromatic such as those of naphthalene (extremely strongly reducing) and benzophenone (strongly reducing), hydrides (electron-transfer rather than hydride transfer), cobaltocene and decamethyl-cobaltocene (good and strong reductants). On the oxidation side, ferrocenium salts (mild oxidants) have been most used and discussed together with the other one-electron oxidants: $NO^+$, $Ag^I$, $Cu^{II}$, $FeCl_3$, $Ce^{IV}$ (CAN), halogens, aryl diazonium salts, acids, the very useful triarylaminium radical cations, carbocations (trityl and derivatives), TCNE, TCNQ and quinones. Most of these redox reagents (except the metallocenes) are non innocent by contrast with the electron-reservoir complexes in which the permethylation of the ring(s) protects the metal center from external influences to a great extent.

The well-known metal polypyridine [378, 379] and metal-porphyrin [380–382] and phthalocyanine complexes also have extended redox series which make them good candidates as electron-reservoirs. They are thoroughly reviewed and discussed in other chapters of this volume.

## 4.9 Experimental: Syntheses of the Electron-reservoir Complexes $[Fe^{II}(\eta^5\text{-}C_5R_5)(\eta^6\text{-}C_6Me_6)][PF_6]$, $[Fe^I(\eta^5\text{-}C_5R_5)(\eta^6\text{-}C_6Me_6)]$, and $[Fe^{III}(\eta^5\text{-}C_5R_5)(\eta^6\text{-}C_6Me_6)][SbCl_6]_2$ (R = H or Me)

### 4.9.1 $[Fe^{II}Cp(\eta^6\text{-}C_6Me_6)][PF_6]$

The reaction is conducted in a 500-mL three-necked, round-bottomed flask equipped with a nitrogen inlet and a reflux condenser topped with a gas outlet. Ferrocene (18.6 g, 100 mmol), aluminum powder (2.7 g, 100 mmol), and $C_6Me_6$ (16.2 g, 100 mmol) are added, together, to degassed decalin (200 mL) under nitrogen or argon. After stirring for 5 min, $AlCl_3$ (40 g, 300 mmol) is added to the solution. Finally, after another 5 min, $H_2O$ (1.8 mL, 100 mmol) is introduced by syringe to the reaction medium. The mixture is heated to 190 °C with a oil bath for 12 h. The reaction mixture is hydrolyzed at 50 °C (to prevent solidification) very slowly under nitrogen or argon with degassed ice–water (100 mL). After filtration, the two layers are separated and the aqueous layer is washed with diethyl ether (3 × 50 mL). $Al(OH)_3$ is precipitated by the addition of concentrated aqueous $NH_4OH$ until pH 9. Filtration into aqueous $HPF_6$ solution (100 mmol) gives a yellow precipitate of $[Fe^{II}Cp(\eta^6\text{-}C_6Me_6)][PF_6]$. The latter is removed by filtration, dissolved in acetone (500 mL), and dried for 1 h over $MgSO_4$. Addition of ethanol (400 mL) to the yellow filtered solution then slow evaporation of the acetone by rotary evaporation gives, after standing overnight at −25 °C, 27.8 g (65 %) of yellow microcrystals. The other complexes $[FeCp(\eta^6\text{-arene})][PF_6]$ are prepared similarly from the arene (neat) by reaction at 100 °C or under reflux overnight, then the above work-up $[Fe^{III}(\eta^5\text{-}C_5Me_5)(\eta^6\text{-}C_6Me_6)]$. $[PF_6]$ is analogously prepared from $[Fe^{III}(\eta^5\text{-}C_5Me_5)(CO)_2Br]$ by reaction with $C_6Me_6$ and $AlCl_3$ overnight without water at 100 °C, then similar work-up [50].

### 4.9.2 $[Fe^ICp(\eta^6\text{-}C_6Me_6)]$

A sample of the yellow crystalline salt $[Fe^{II}Cp(\eta^6\text{-}C_6Me_6)][PF_6]$:$PF_6^{-1}$ (8.56 g, 20 mmol) in dimethoxyethane (30 mL) is stirred with Na–Hg (230 g, 0.9 %, 100 mmol) for 1 h at ambient temperature under argon. The DME is then removed in vacuo, and the residue is extracted with pentane ($3 \times 50$ mL) and filtered. The deep green solution is concentrated to 40 mL and cooled to $-80\,°C$. Filtration of this solution provides 4.53 g, 80 % yield, of $[Fe^{I}Cp(\eta^{6}\text{-}C_{6}Me_{6})]$ as green microcrystals. Recrystallization can also be performed overnight in the freezer of the dry-lab ($-40\,°C$). Concentration and treatment of subsequent crops gives an overall 92 % yield of crystals.

### 4.9.3 $[Fe^{I}(\eta^{5}\text{-}C_{5}Me_{5})(\eta^{6}\text{-}C_{6}Me_{6})]$

A sample of $[Fe^{I}(\eta^{5}\text{-}C_{5}Me_{5})(\eta^{6}\text{-}C_{6}Me_{6})][PF_{6}]$ (0.498 g, 1 mmol) in THF (10 mL) was stirred on sodium amalgam (0.8 %, 10 g). The reduction was achieved in 1 h. THF was then removed in vacuo, and the residue was extracted with pentane ($2 \times 15$ mL). Pentane was then removed in vacuo to leave a dark-green product. Recrystallization from pentane overnight in the freezer of the dry-lab ($-40\,°C$) gave dark-green crystals of $[Fe^{I}(\eta^{5}\text{-}C_{5}Me_{5})(\eta^{6}\text{-}C_{6}Me_{6})]$ (0.307 g, 87 % yield) [50].

### 4.9.4 $[Fe^{III}(\eta^{5}\text{-}C_{5}Me_{5})(\eta^{6}\text{-}C_{6}Me_{6})][SbCl_{6}]_{2}$

$SbCl_{5}$ (1 M solution in $CH_{2}Cl_{2}$, 5 mmol) was added with magnetic stirring to $[Fe^{III}(\eta^{5}\text{-}C_{5}Me_{5})(\eta^{6}\text{-}C_{6}Me_{6})][PF_{6}]$ (0.498 g, 1 mmol) in $CH_{2}Cl_{2}$ (10 mL) under argon. Formation of a purple precipitate from the initially yellow solution was instantaneous. The purple solid was removed by filtration, washed with $CH_{2}Cl_{2}$ ($4 \times 15$ mL) until the latter was colorless, then recrystallized from $SO_{2}$ at $-40\,°C$ (0.624 g, 71 % yield). Analysis for $[Fe^{III}(\eta^{5}\text{-}C_{5}Me_{5})(\eta^{6}\text{-}C_{6}Me_{6})][SbCl_{6}]_{2}$: calc. for $C_{22}H_{33}FeSb_{2}Cl_{12}$: C 25.85, H 3.25. Found C 25.84, H 3.47.

Alternatively, $[Fe^{III}(\eta^{5}\text{-}C_{5}Me_{5})(\eta^{6}\text{-}C_{6}Me_{6})][SbCl_{6}]_{2}$ can also be synthesized from $[Fe^{I}(\eta^{5}\text{-}C_{5}Me_{5})(\eta^{6}\text{-}C_{6}Me_{6})]$. The complex $[Fe^{I}(\eta^{5}\text{-}C_{5}Me_{5})(\eta^{6}\text{-}C_{6}Me_{6})]$ (0.300 g, neat) was cooled to $-80\,°C$ and $SbCl_{5}$ (1 M solution in $CH_{2}Cl_{2}$, 5 mmol) at $-80\,°C$ were added with magnetic stirring. The reaction mixture was then slowly warmed to room temperature and a brown solution and a purple precipitate appeared. The solution was then transferred by cannula, and the solid residue was washed with $CH_{2}Cl_{2}$ ($4 \times 15$ mL) and recrystallized from $SO_{2}$ at $-40\,°C$ to yield the complex $[Fe^{III}(\eta^{5}\text{-}C_{5}Me_{5})(\eta^{6}\text{-}C_{6}Me_{6})][SbCl_{6}]_{2}$ as a microcrystalline purple powder (0.694 g, 78 % yield). UV–Vis for 1 $[SbF_{6}]_{2}$: $\lambda_{max}$ (MeCN) = 487 nm ($\varepsilon = 228$ L $M^{-1}$ $cm^{-1}$); $\lambda_{max}$ (MeCN) = 525 nm ($\varepsilon = 213$ L $M^{-1}$ $cm^{-1}$).

## References

1. H. Taube, *Electron Transfer Reactions of Complex Ions in Solution*, Academic Press, New York, **1970**.
2. D. Astruc, *Electron Transfer and Radical Processes in Transition Metal Chemistry*, VCH, New York, **1995**.

3. R. A. Marcus, N. Sutin, *Biophys. Biochem. Acta* **1985**, *811*, 265.
4. I. Bertini, H. B. Gray, S. J. Lippard, J. S. Valentine, *Bioinorganic Chemistry*, University Science Books, Mill Valley, **1994**.
5. L. Stryer, Biochemistry, 2nd Ed., Freeman, New York, **1981**.
6. J. K. Kochi, *Organometallic Mechanisms and Catalysis*, Academic Press, New York, **1981**.
7. R. Crabtree, *The Organometallic Chemistry of the Transition Metals*, 2nd Ed., Wiley, New York, **1994**.
8. V. Balzani, F. Scandola, *Supramolecular Photochemistry*, Ellis Hordwood, New York, **1991**.
9. *Homogeneous and Heterogeneous Photocatalysis* (Ed.: V. Balzani, A. Juris, F. Scandola, E. Pelizzetti, N. Serpone), Reidel, Dordrecht, **1986**.
10. J.-M. Lehn, *Supramolecular Chemistry: Concepts and Perspectives*, VCH, Weinheim, **1995**.
11. *Molecular Electronics* (Guest Ed.: J.-P. Launay), Gauthier–Villars, Paris, *New J. Chem.* **1991**, *15*, 97. See also Ref. [2] (Chapter 4), Ref. [8] (Chapter 12) and Ref. [10] (Chapter 8).
12. *Electron Transfer in Biology and in the Solid State* (Eds: M. K. Johnson et al.), *Adv. Chem. Ser.* No. 226, American Chemical Society, Washington, DC, **1990**.
13. R. A. Sheldon, J. K. Kochi, *Metal-Catalyzed Oxidations of Organic Compounds*, Academic Press, New York, **1981**.
14. M. Hudlicky, *Oxidations in Organic Chemistry*, A.C.S Monograph No. 186, American Chemical Society, Washington, DC, **1990**.
15. M. Hudlicky, *Reductions in Organic Chemistry*, Ellis Horwood, Chichester, **1984**.
16. J. Seyden-Penne, *Reductions by the Alumino- and Borohydrides in Organic Synthesis*, 2nd Ed.; Wiley, VCH, New York, **1997**.
17. A. J. Bard, L. R. Faulkner, *Electrochemical Methods*, Wiley, New York, **1980**.
18. G. J. Hills in *Reference Electrodes* (Eds: D. Ives, J. G. Janz), Academic Press, New York, **1961**, Chapter 10.
19. J. F. Coetzee, G. R. Padmanablan, *J. Phys. Chem.* **1962**, *66*, 7708.
20. V. D. Parker, *Adv. Phys. Org. Chem.* **1983**, *19*, 131.
21. N. Connelly, W. E. Geiger, *Chem. Rev.* **1996**, *96*, 877.
22. J. T. Hupp, *Inorg. Chem.* **1990**, *29*, 5010.
23. J. Ruiz, D. Astruc, *C. R. Acad. Sci. Paris, Sér. IIc*, **1998**, *1*, 27.
24. K. N. Brown, D. S. Fleming, P. T. Gulyas, P. A. Lay, A. F. Masters, *XVIIth Intern. Conf. Organomet. Chem.* Brisbane, Australia, **1996**, Abstr. OA12.
25. Q. Xie, E. Pérez-Cordero, L. Echegoyen, *J. Am. Chem. Soc.* **1992**, *114*, 3978.
26. A. Loupy, B. Tchoubar, D. Astruc, *Chem. Rev.* **1992**, *92*, 1141.
27. J. A. Davies, C. M. Hockensmith, V. Yu Kukushkin, Yu N. Kukushkin, Synthetic Coordination Chemistry, World Scientific, London, **1996**, Chapters 8 and 9.
28. K. Jonas, V. Z. Wiskamp, *Naturforsch.* **1983**, *38B*, 1113.
29. (a) J. D. L. Holloway, W. E. Geiger, *J. Am. Chem. Soc.* **1979**, *101*, 2038; (b) G. Fochi, X. Runjuan, A. Colligiani, *J. Chem. Soc. Dalton Trans.* **1990**, 2551.
30. H.-F. Klein, J. Gross, J. M. Bassett, U. Schubert, *Z. Naturforsch.* **1980**, 35B, 614.
31. D. Catheline, D. Astruc, *Organometallics* **1984**, *3*, 1094.
32. F. Abugideiri, D. W. Keogh, R. Poli, *J. Chem. Soc., Chem. Commun.* **1994**, 2317.
33. K. M. Phahl, J. Ellis, *Organometallics* **1984**, *3*, 230.
34. K. Okada, T. Kawata, M. Oda, *J. Chem. Soc. Chem. Commun.* **1995**, 233.
35. J. S. Bradshaw, K. E. Krakoviak, R. M. Izatt, *Aza-crown Macrocycles*, Wiley, New York, **1993**, Chapter 1.
36. M. M. Olmstead, P. P. Power, *J. Am. Chem. Soc.* **1986**, *108*, 4235.
37. F. G. N. Cloke, P. B. Hitchcock, A. McCamley, *J. Chem. Soc. Chem. Commun.* **1993**, 248.
38. C. M. Archer, J. R. Dilworth, R. M. Thompson, M. McPartlin, D. C. Povey, J. D. Kelly, *Dalton Trans.* **1993**, 461.
39. L. D. Field, A. F. Masters, M. Gibson, D. R. Latimer, T. W. Hambley, I. E. Buys, *Inorg. Chem.* **1993**, *32*, 211.
40. B. K. Coltrain, S. C. Jackels, *Inorg. Chem.* **1981**, *20*, 2032.
41. M. E. Riehl, S. R. Wilson, G. S. Girolami, *Inorg. Chem.* **1993**, *32*, 218.
42. M. S. Corraine, J. Atwood, *Organometallics 10*, 2647.
43. J. Balej, *Electrochemica Acta* **1976**, *21*, 953.

44. M. S. Corraine, J. D. Atwood, *Organometallics* **1991**, *10*, 2985.
45. (a) M. Pfeffer, J. Fisher, A. Mitschler, *Organometallics* **1984**, *3*, 1531; (b) V. G. Syrkin, *Metal Carbonyls* (in Russian), Khimia, Moscow, **1984**.
46. W. P. Weiner, F. J. Hollander, R. G. Bergman, *J. Am. Chem. Soc.* **1984**, *106*, 7462.
47. J. Li, A. D. Hunter, R. McDonald, B. D. Santarsiero, S. G. Bott, J. L. Atwood, *Organometallics* **1992**, *11*, 3050.
48. T. Cuenca, P. Royo, *J. Organomet. Chem.* **1985**, *293*, 61.
49. C. Moinet, E. Román, D. Astruc, *J. Organomet. Chem.* **1977**, *128*, C45.
50. J.-R. Hamon, D. Astruc, P. Michaud, *J. Am. Chem. Soc.* **1981**, *103*, 758.
51. N. G. Connelly in *Paramagnetic Species in Activation, Selectivity, Catalysis* (Eds: M. Chanon, M. Juliard, J. C. Poite), Kluwer, Dordrecht, **1989**, p. 71.
52. N. G. Connelly, *Chem. Soc. Rev.* **1989**, *18*, 153.
53. E. O. Fischer, H. Waversik, *J. Organomet. Chem.* **1966**, *5*, 559.
54. H. J. Keller, H. Waversik, *J. Organomet. Chem.* **1967**, *8*, 185.
55. O.V. Gusev, L. I. Denisovich, M. G. Peterleitner, A. Z. Rubezhov, N. A. Ustynyuk, P. M. Maitlis, *J. Organomet. Chem.* **1993**, *452*, 219.
56. D. J. Cole-Hamilton, G. Wilkinson, *J. Chem. Soc. Dalton Trans.* **1979**, 1283.
57. M. Green, *Polyhedron* **1986**, *5*, 427.
58. J. J. Maj, A. D. Rae, L. F. Dahl, *J. Am. Chem. Soc.* **1982**, *104*, 3054.
59. *Radical Ions* (Ed.: E. T. Kaiser, L. Kevan), Interscience, New York, **1968**.
60. F. Billiau, G. Folcher, H. Marguet-Ellis, P. Rigny, E. Saito, *J. Am. Chem. Soc.* **1981**, *103*, 5603.
61. J. M. Maher, R. P. Beatty, N. J. Cooper, *Organometallics* **1982**, *1*, 215 and **1985**, *4*, 1354.
62. R. L. Thompson, S. J. Geib, N. J. Cooper, *J. Am. Chem. Soc.* **1991**, *113*, 8961.
63. A. Antinolo, M. F. Lappert, D. J. W. J. Winterborn, *J. Organomet. Chem.* **1985**, *272*, C37.
64. P. Jutzi, B. J. Heilscher, *Organomet. Chem.* **1985**, *291*, C25.
65. W. Schlenk, T. Weickel, *Chem. Ber.* **1911**, *44*, 1182.
66. *Electrochemistry of the Elements* (Eds: A. J. Bard, H. Lund), Marcel Dekker, New York, **1978**, *Vol XII*, p. 209.
67. L. F. Fieser, *Experiments in Organic Chemistry*, 3rd Ed., Heath & Co, Boston, p. 209.
68. M. I. Bruce, B. K. Nicholson, M. L. Williams, *Inorg. Synth.* **1989**, *26*, 273.
69. N. Hirota in *Radical Ions* (Eds: E.T. Kaiser, L. Kevan), Wiley, New York, **1968**, p. 65.
70. C. B. Wooster, *J. Am. Chem. Soc.* **1928**, *50*, 1388.
71. D. H. Geske, A. H. Maki, *J. Am. Chem. Soc.* **1960**, *82*, 2671.
72. B. S. Jensen, V. D. Parker, *J. Chem. Soc., Chem. Commun.* **1974**, 367.
73. J. Grimshaw, R. Hamilton, *Electroanal. Chem.* **1980**, *106*, 339.
74. M. Grezeszczuck, D. E. Smith, *J. Electroanal. Chem.* **1983**, *157*, 205.
75. N. Novikov, Yu., M. E. Volpin, *Upsekhi Khim.* **1971**, *40*, 1568.
76. T. Mashiko, C. A. Reed, K. J. Haller, W. R. Scheidt, *Inorg. Chem.* **1984**, *23*, 3192.
77. W. Rüdorff, *Adv. Inorg. Chem. Radiochem.* **1959**, *1*, 223.
78. A. Fürstner, *Advan. Organomet. Chem.* **1988**, *28*, 58.
79. K. A. Jensen, B. Nygaard, G. Elisson, P. H. Nielsen, *Acta Chem. Scand.* **1965**, *19*, 768.
80. C. Ungurenasu, M. Palie, *J. Chem. Soc., Chem. Commun.* **1975**, 388.
81. M. C. Manning, W. C. Trogler, *Inorg. Chim. Acta* **1981**, *50*, 247.
82. P. Michaud, D. Astruc, J. H. Ammeter, *J. Am. Chem. Soc.* **1982**, *104*, 3755.
83. L. Eberson, *Acta Chem. Scand., Ser. B* **1984**, *38*, 439.
84. D. Astruc, *Chem. Rev.* **1988**, *88*, 1189.
85. P. Massur, Yu., I. S. Antonov, A. P. Tomilov, L. N. Ovsyannikov, *Soviet Electrochemistry* **1969**, *5*, 452.
86. S. G. Davies, M. L. H. Green, D. M. P. Mingos, *Tetrahedron* **1978**, *34*, 3047.
87. P. Michaud, J.-P. Mariot, F. Varret, D. Astruc, *J. Chem. Soc., Chem. Commun.* **1982**, 1383.
88. C. Lapinte, D. Catheline, D. Astruc, *Organometallics* **1984**, *3*, 817.
89. P. Michaud, C. Lapinte, D. Astruc, *Annals New York. Acad. Science* **1983**, *415*, 97.
90. M.-J. Tudoret, M.-L. Robo, C. Lapinte, *Organometallics* **1992**, *11*, 1419.
91. E. C. Ashby, A. B. Goel, R. N. de Priest, *J. Am. Chem. Soc.* **1980**, *102*, 7779.
92. E. C. Asby, A. B. Goel, R. N. de Priest, N. S. Prasad, *J. Am. Chem. Soc.* **1981**, *103*, 973.

93. N. G. Connelly, J. D. Payne, W. E. Geiger, *J. Chem. Soc., Dalton Trans.* **1983**, 295.
94. D. C. Boyd, N. G. Connelly, G. G. Herbosa, M. G. Hill, K. R. Mann, C. Mealli, A. G. Orpen, K. Richardson, P. H. Rieger, *Inorg. Chem.* **1994**, *33*, 960.
95. W. D. Jones, J. M. Huggins, R. G. Bergman, *J. Am. Chem. Soc.* **1981**, *103*, 4415.
96. D. R. Tyler in *Prog. Inorg. Chem.* (Ed.: S. J. Lippard), Wiley, New York, **1988**, *35*, 125.
97. K. A. Kocheshkov, A. N. Nesmeyanov, M. M. Nad', *Dokl. Akad. Nauk. SSSR* **1940**, *26*, 53.
98. A. J. Linsay, R. P. Tooze, M. Motevalli, M. B. Hurthouse, G. Wilkinson, *J. Chem. Soc., Chem. Commun.* 799 and 1383.
99. H. Lehmkuhl, H. Nehl, R. Benn, R. Mynott, *Angew. Chem.* **1986**, *98*, 628.
100. D. Astruc, *Tetrahedron* **1983**, *39*, 4027.
101. M. I. Bruce, C. M. Jensen, N. L. Jones, *Inorg. Syn.* **1989**, *26*, 259.
102. C.-M. Che, M. Jamal, C.-K. Poon, W.-C. Chung, *Inorg. Chem.* **1985**, *24*, 2868.
103. *Synthesis of Complex Compounds of the Platinum Metals* (Ed.: I. I. Cheryaev), (in Russian), Nauka, Moscow, **1964**, p. 24.
104. F. Herrmann, *Ber.* **1905**, *38*, 2813.
105. R. Uson, A. Laguna, M. Laguna, *Inorg. Syn.* **1989**, *26*, 85.
106. R. S. Wade, C. E. Castro, *Inorg. Chem.* **1985**, *24*, 2862.
107. D. Brenner, A. Davison, J. Lister-James, A. G. Jones, *Inorg. Chem.* **1984**, *23*, 3793.
108. H. Taube, *Chem. Rev.* **1951**, *50*, 69.
109. Ref. [2], Chapter 5, p. 393.
110. H. Kagan, J.-L. Namy, *Tetrahedron* **1980**, *42*, 6573.
111. J. R. Pugh, T. J. Meyer, *J. Am. Chem. Soc.* **1992**, *114*, 3784.
112. M. Green, N. K. Jetha, R. J. Mercer, N. C. Norman, A. G. Orpen, *J. Chem. Soc., Dalton Trans.* **1988**, 1843.
112. U. Kölle, *Coord. Chem. Rev.* **1994**, *135*, 623.
114. S. F. Nelsen, P. J. Hintz, *J. Am. Chem. Soc.* **1972**, *94*, 7114.
115. J. Kiwi, K. Kalyanasundaram, M. Grätzel, *Struct. Bond.* **1981**, *49*, 37.
116. F. A. Cotton, G. Wilkinson, *Advanced Inorganic Chemistry*, 5th Ed., Wiley, New York, **1988**, p. 316.
117. G. A. Carriedo, M. C. Crespo, V. Riera, M. G. Sanchez, M. L. Valin, D. Moreiras, X. Solans, *J. Organomet. Chem.* **1986**, *302*, 47.
118. R. B. King, *Inorg. Chem.* **1965**, *4*, 1518.
119. G. W. Parshall, *Inorg. Syn.* **1977**, *17*, 110.
120. R. B. King, *Organometallic Syntheses* Academic Press, New York, **1965**, *Vol. 1*, p. 70.
121. J. E. Sheat, *J. Organomet. Chem. Library* **1977**, *7*, 461.
122. R. D. Kemmit, D. R. Russell in *Comprehensive Organometallic Chemistry* (Eds: G. Wilkinson, F. G. A. Stone, E. W. Abel), Pergamon, Oxford, **1982**, *Vol. 5*, Chapter 34, p. 244.
123. R. W. Albach, U. Kürstardt, J. Behm, B. Ebert, M.-H. Delville, D. Astruc, *J. Organomet. Chem.* **1993**, *450*, 165.
124. G. E. Herberich, W. Klein, U. Kölle, D. Spiliotis, *Chem. Ber.* **1992**, *125*, 1589.
125. G. E. Herberich, E. Bauer, J. Schwarzer, *J. Organomet. Chem.* **1969**, *17*, 445.
126. S. Nlate, V. Guerchais, C. Lapinte, *J. Organomet. Chem.* **1992**, *434*, 89.
127. F. Calderazzo, G. Pampaloni, U. Englert, J. Strahle, *J. Organomet. Chem.* **1990**, *383*, 45.
128. W. J. Bowyer, W. E. Geiger, *J. Am. Chem. Soc.* **1985**, *107*, 5657.
129. P. Legzdins, R. Reina, M. J. Shaw, R. J. Batchelor, F. W. B. Einstein, *Organometallics* **1993**, *12*, 1029.
130. A. Peloso, *J. Chem. Soc., Dalton Trans.* **1984**, 249.
131. J. Ruiz, F. Ogliaro, J.-Y. Saillard, J.-F. Halet, F. Varret, D. Astruc, *J. Am. Chem. Soc.* **1998**, *120*, 11693.
132. K. N. Brown, P. T. Gulyas, P. A. Lay, N. S. McAlpine, A. F. Masters, L. Phillips, *J. Chem. Soc., Dalton Trans.* **1993**, 835.
133. J. S. Miller, A. J. Epstein, W. M. Reiff, *Chem. Rev.* **1988**, *88*, 201.
134. S. Nlate, J. Ruiz, J.-C. Blais, D. Astruc, *Chem. Commun.* 2000, 417.
135. S. Nlate, J. Ruiz, V. Sartor, R. Navarro, J.-C. Blais, D. Astruc, *Chem. Eur. J.* **2000**, *6*, 2544.
136. R. Davis, L. A. P. Kane-Maguire in *Comprehensive Organometallic Chemistry* (Eds: G. Wilkinson, F. G. A. Stone, E. W. Abel), Pergamon, Oxford, **1982**, *Vol. 3*, Chapter 26.2, p. 975.

137. H. Brunner, H. Koch, *Chem. Ber.* **1982**, *115*, 65.
138. T. Bockman, J. K. Kochi, *J. Am. Chem. Soc.* **1989**, *111*, 4669.
139. D. Astruc, J.-R. Hamon, G. Althoff, E. Román, P. Batail, P. Michaud, J.-P. Mariot, F. Varret, D. Cozak, *J. Am. Chem. Soc.* **1979**, *101*, 5445.
140. (a) D. Astruc, *Acc. Chem. Res.* **1986**, *19*, 377; (b) D. Astruc, *Comments Inorg. Chem.* **1987**, *6*, 61.
141. D. Astruc, J.-R. Hamon, M. Lacoste, M.-H. Desbois, E. Román, *Organomet. Synthesis* (Ed.: R.B. King), *Vol. IV*, **1988**, p. 172–187.
142. (a) J. C. Green, M. R. Kelly, M. P. Payne, E. A. Seddon, D. Astruc, J.-R. Hamon, P. Michaud, *Organometallics* **1983**, *2*, 211; (b) D. Briggs Handbook of X-ray and Photoelectron Spectroscopy, Heyden, London, **1977**; (c) C. Cauletti, J. C. Green, M. R. Kelly, P. Powell, J. Van Tilborg, J. Robbin, J. Smart *Electron Spectrosc. Relat. Phenom.* **1980**, *19*, 327.
143. (a) G. E. Herberich, G. Greiss, H. F. Heil, J. Müller *J. Chem. Soc. Chem. Commun.* **1971**, 1328; (b) G. Herberich In *Comprehensive Organometallic Chemistry* (Eds.: G. Wilkinson, F. G. A. Stone, W. E. Abel), Pergamon, London, **1982**, Vol 5, p. 381.
144. M. V. Rajasekharan, S. Giezynski, J. H. Ammeter, J.-R. Hamon, P. Michaud, D. Astruc, *J. Am. Chem. Soc.*, **1982**, *104*, 2400.
145. J.-P. Mariot, P. Michaud, S. Lauer, D. Astruc, A. X. Trautwein, F. Varret, *J. Physique* **1983**, *44*, 1377.
146. F. Ogliaro, J.-Y. Saillard, J.-F. Halet, D. Astruc, *New J. Chem.* **2000**, *24*, 257.
147. E. O. Fisher, H. Waversik, *J. Organomet. Chem.* **1967**, *8*, 185.
148. E. O. Fischer, M. F. Schmidt, *Chem. Ber.* **1966**, *99*, 2206.
149. A. N. Nesmeyanov, A. N. Vol'kenau, I. N. Bolesova, *Tetrahedron Let.* **1963**, 1725.
150. I. U. Khand, P. L. Pauson, W. E. Watts, *J. Chem. Soc.* **1968**, C 2257.
151. D. Briggs, *Handbook of X-Ray and Photoelectron Spectroscopy*, Heyden, London, **1977**.
152. C. Cauletti, J. C. Green, M. R. Kelly, P. Powell, J. Van Tilborg, J. Robbins, J. Smart, *J. Electron Spectros. Relat. Phenom.* **1980**, *39*, 1.
153. J. E. Herberich, G. Greiss, H. F. Heil, J. Müller, *J. Chem. Soc. Chem. Commun.* **1971**, 1328.
154. A. N. Nesmeyanov, N. A. Vol'kenau, L. S. Shilovtseva, V. A. Petrakova, *J. Organomet. Chem.* **1973**, *61*, 329.
155. E. Alonso, J. Ruiz, D. Astruc, *J. Clust. Sci.* **1998**, *9*, 271.
156. M.-H. Desbois, P. Michaud, D. Astruc, *J. Chem. Soc., Chem. Commun.* **1985**, 450.
157. J. Ruiz, V. Guerchais, D. Astruc, *J. Chem. Soc., Chem. Commun.* **1989**, 812.
158. J.-C. Gressin, D. Michelet, L. Nadjo, J. M. Savéant, *Nouv. J. Chim.* **1979**, 545.
159. D. Astruc, E. Román, J.-R. Hamon, P. Batail, *J. Am. Chem. Soc.* **1979**, *101*, 2240.
160. J.-R. Hamon, D. Astruc, E. Román, P. Batail, J. J. Meyerle, *J. Am. Chem. Soc.* **1981**, *103*, 2431.
161. D. Astruc, J.-R. Hamon, E. Román, P. Michaud, *J. Am. Chem. Soc.* **1981**, *103*, 7502.
162. J.-R. Hamon, D. Astruc, *Organometallics* **1988**, *7*, 1036.
163. C. Bossard, S. Rigaut, D. Astruc, M.-H. Delville, G. Félix, A. Février-Bouvier, J. Amiell, S. Flandrois, P. Delhaès, *J. Chem. Soc., Chem. Commun.* **1993**, 333.
164. D. Astruc in *Mechanisms and Processes in Molecular Chemistry* (Ed.: D. Astruc), Gauthier–Villars, Paris, *New J. Chem.* **1992**, *16*, 305.
165. M.-H. Desbois, D. Astruc, *Angew. Chem. Int. Engl.* **1989**, *101*, 459.
166. D. Astruc, D. Mandon, A. M. Madonik, P. Michaud, N. Ardoin, F. Varret, *Organometallics* **1990**, *9*, 2155.
167. J. Morrow, D. Catheline, M.-H. Desbois, J. M. Manriquez, J. Ruiz, D. Astruc, *Organometallics* **1987**, *6*, 2605.
168. J. R. Morrow, D. Astruc, *Bull. Soc. Chim. Fr.* **1992**, *129*, 319.
169. A. F. Hebrard, M. J. Roseinsky, R. C. Hadden, D. W. Murphy, S. H. Glarum, T. T. M. Palbtra, A. P. Ramirez, A. R. Kortan, *Nature* **1991**, *350*, 600.
170. J. Stinchcombe, A. Pénicaud, P. Bhyrappa, P. D. W. Boyd, C. A. Reed, *J. Am. Chem. Soc.* **1993**, *115*, 5212.
171. P. D. W. Boyd, P. Bhyrappa, P. Paul, J. Stinchcombe, R. D. Bolskar, Y. Sun, C. A. Reed, *J. Am. Chem. Soc.* **1995**, *117*, 2907.
172. J.-R. Hamon, D. Astruc, *J. Am. Chem. Soc.* **1983**, *105*, 5951.

173. I. Fridovitch in *Free Radicals in Biology* (Ed.: W. A. Pryor), Academic Press, New York, **1976**, pp. 239–277.
174. D. Astruc, M.-H. Delville, M. Lacoste, J. Ruiz, F. Moulines, J.-R. Hamon in *Recent Advances in the Chemistry of the Main-Group Elements* (Ed.: N. Hosmane), *Phosphorus, Sulfur, Silicon and the Related Elements*, **1994**, *87*, 11.
175. A. Loupy, B. Tchoubar, *Salt Effects in Organic and Organometallic Chemistry*, VCH, Weinheim, **1992**.
176. A. M. Madonik, D. Astruc, *J. Am. Chem. Soc.* **1984**, *106*, 2437.
177. S. Rigaut, M.-H. Delville, D. Astruc, *J. Am. Chem. Soc.* **1997**, *119*, 11132.
178. D. Astruc, S. Rigaut, E. Alonso, M.-H. Delville, J. Ruiz, *Hung. Chim. Acta.* **1998**, *135*, 751.
179. J. Ruiz, M. Lacoste, D. Astruc, *J. Chem. Soc., Chem. Commun.* **1989**, 813.
180. J. Ruiz, M. Lacoste, D. Astruc, *J. Am. Chem. Soc.* **1990**, *112*, 5471.
181. F. R. Mayo, *Acc. Chem. Res.* **1968**, 1.
182. A. G. Davies, *J. Organomet. Chem.* **1980**, *200*, 87.
183. *Foundations of Chemical Kinetics*, E. N. Mir, Moscow, **1979** (English translation, p. 247).
184. Ref. [2], Chapter 6: Chain reactions.
185. M.-H. Desbois, D. Astruc, *New J. Chem.* **1989**, *13*, 595.
186. M. J. Carney, J. S. Lesniak, M. D. Likar, J. R. Pladziewicz, *J. Am. Chem. Soc.* **1984**, *106*, 2565.
187. J. C. Smart, B. L. Pinsky, *J. Am. Chem. Soc.* **1980**, *102*, 1009.
188. D. N. Hendrickson, Y. S. Sohn, H. B. Gray, *Inorg. Chem.* **1971**, *10*, 1559.
189. H. J. Schumann, *Organomet. Chem.* **1986**, *304*, 341.
190. R. F. Jordan, R. E. Lapointe, C. S. Bajgur, S. F. Echols, R. Willett, *J. Am. Chem. Soc.* **1987**, *109*, 4111.
191. a) R. P. Aggarwal, N. G. Connelly, M. C. Crespo, B. J. Dunne, P. M. Hopkins, A. G. Orpen, *J. Chem. Soc., Dalton Trans.* **1992**, 655; b) J. M. Forward, D. M. P. Mingos, A. V. Powell, *J. Organomet. Chem.* **1994**, *465*, 251.
192. C. A. Reed, *Acc. Chem. Res.* **1998**, *31*, 133.
193. B. T. King, Z. Janusek, B. Grüner, M. Trammel, B. C. Noll, J. Michl, *J. Am. Chem. Soc.* **1996**, *118*, 3313; S. H. Strauss, *Chem. Rev.* **1993**, *93*, 927.
194. W. E. Geiger in *Organometallic Radical Processes*, *J. Organomet. Chem.* Library 22 (Ed.: W. C. Trogler), Elsevier, Amsterdam, **1990**, pp. 142–172.
195. C. Guillon, P. Vierling, *J. Organomet. Chem.* **1994**, *464*, C42.
196. A. Pedersen, M. Tilset, K. Folting, K. G. Caulton, *Organometallics* **1995**, *14*, 875.
197. M. Hudlicky *Oxidation in Organic Chemistry*; ACS Monograph 186, American Chemical Society: Washington, DC, **1990**.
198. M. Bjorgvinsson, T. Heinze; H. W. Roesky, F. Pauer, D. Stalke, G. M. Sheldrick, *Angew. Chem., Int. Ed. Engl.* **1991**, *30*, 1677.
199. a) L. Song, W. C. Trogler, *Angew. Chem., Int. Ed. Engl.* **1992**, *31*, 770; b) F. A. Cotton, L. Falvello, T. Ren, K. Vidyasagar, *Inorg. Chim. Acta* **1992**, *194*, 163.
200. H. H. Morgan, *J. Chem. Soc.* **1923**, 2901.
201. R. Fernandez-Galan, B. R. Manzano, A. Otero, M. Lanfranchi, M. A. Pellinghelli, *Inorg. Chem.* **1994**, *33*, 2309.
202. C. White, S. J. Thompson, P. M. Maitlis, *J. Organomet. Chem.* **1977**, *134*, 319.
203. N. G. Connelly, T. Einig, G. Garcia Herbosa, P. M. Hopkins, C. Mealli, A. G. Orpen, G. M. Rosair, F. Viguri, *J. Chem. Soc., Dalton Trans.* **1994**, 2025.
204. C. L. Jenkis, *Chem. Soc.* **1972**, *94*, 843, 856.
205. A. Heumann, K.-J. Jens, M. Reglier, *Prog. Inorg. Chem.* **1994**, *42*, 505.
206. A. R. Pray, *Inorg. Synth.* **1957**, *5*, 153.
207. M. F. Joseph, J. A. Page, M. C. Baird, *Organometallics* **1984**, *3*, 1749.
208. Ref. [13], Chapter 7.
209. S. P. Schmidt, F. Basolo, W. C. Trogler, *Inorg. Chim. Acta* **1987**, *131*, 181.
210. D. Mandon, L. Toupet, D. Astruc, *J. Am. Chem. Soc.* **1986**, *108*, 1320.
211. A. J. Pearson in *Comprehensive Organometallic Chemistry* (Eds: G. Wilkinson, F. G. A. Stone, E. W. Abel), Pergamon, Oxford, **1982**, *Vol. 8*, Chapter 58, p. 939.
212. M. W. Droege, W. D. Harman, H. Taube, *Inorg. Chem.* **1987**, *26*, 1309.

213. W. Adam, M. A. Miranda, F. Mojarrad, H. Sheikh, *Chem. Ber.* **1994**, *127*, 875.
214. D. Touchard, J.-L. Fillaut, D. U. Khasnis, P. H. Dixneuf, C. Meali, D. Massi, L. Toupet, *Organometallics* **1988**, 7, 67.
215. K. L. Rollick, J. K. Kochi, *J. Am. Chem. Soc.* **1982**, *104*, 1319.
216. J. B. Chlistunoff, A. J. Bard, *Inorg. Chem.* **1992**, *31*, 4582.
217. T. H. Lemmen, E. G. Lundquist, L. F. Rhodes, B. R. Sutherland, D. E. Werterberg, K. G. Caulton *Inorg. Chem.* **1986**, *25*, 3915.
218. Ref. [6], Chapter 16.
219. A. E. Shilov *Metal Complexes in Biomimetic Chemical Reactions*, CRC Press, New York, **1997**.
220. A. Davison, R. H. Holm, *Inorg. Synth.* **1967**, *10*, 18; A. Davison, R. H. Holm, *Inorg. Synth.* **1967**, *10*, 22.
221. P. Baird, J. A. Bandy, M. L. H. Green, A. Hamnett, E. Marseglia, D. S. Obertelli, K. Prout, J. Qin, *J. Chem. Soc., Dalton Trans.* **1991**, 2377.
222. G. Olah, *Acc. Chem. Res.* **1980**, *13*, 330.
223. M. Gilet, A. Mortreux, J.-C. Folest, F. Petit, *J. Am. Chem. Soc.* **1983**, *105*, 3876.
224. L. Eberson, L. Jonsson, O. Sanneskog, *Acta Chem. Scand., Ser. B.* **1985**, *39*, 113.
225. L. Eberson, F. Radner, *Acc. Chem. Res.* **1987**, *20*, 53.
226. M. T. Mocella, M. S. Okamoto, E. K. Barefield, *Synth. React. Inorg. Met.-Org. Chem.* **1974**, *4*, 69.
227. K. G. Caulton, *Coord. Chem. Rev.* **1975**, *14*, 317.
228. N. G. Connelly, P. T. Draggett, M. Green, T. A. Kuc, *J. Chem. Soc., Dalton Trans.* **1977**, 70.
229. C. C. Addison, J. Lewis, *Quater. Rev.* **1955**, *9*, 115.
230. E. K. Kim, J. K. Kochi, *J. Am. Chem. Soc.* **1991**, *113*, 4962.
231. W. K. Musker, T. L. Wolford, P. B. Roush, *J. Am. Chem. Soc.* **1978**, *100*, 6416.
232. F. Seel, *Z. Anorg. Chem.* **1950**, *261*, 75.
233. R. R. Thomas, A. Sen, *Inorg. Synth.* **1990**, *28*, 63.
234. J. K. Kochi, *Acc. Chem. Res.* **1992**, *25*, 39.
235. a) A. E. Arbuzov, V. M. Zoroastrova, *Izv. Akad. Nauk SSSR, Ser. Khim.* **1952**, 818; b) V. Yu. Kukushin, N. P. Kiseleva *Koord. Khim.* **1988**, *14*, 334.
236. G. W. Cowell, A. Ledwith, A. C. White, H. J. Woods, *J. Chem. Soc. B* **1970**, 227.
237. W. Kaim, *Acc. Chem. Res.* **1985**, *18*, 160.
238. H. Bock, W. Kaim, *Acc. Chem. Res.* **1982**, *15*, 9.
239. N. Barlett, D. H. Lohmann, *J. Chem. Soc.* **1962**, 5253.
240. a) T. J. Richardson, N. Bartlett, *J. Chem. Soc. Chem. Commun.* **1974**, 427; b) K. Zücher, T. J. Richardson, O. Glemser, N. Bartlett, *Angew. Chem. Int. Ed. Engl.* **1980**, *19*, 944.
241. a) J. P. Dinnocenzo, T. E. Banach, *J. Am. Chem. Soc.* **1986**, *108*, 6063; b) J. P. Dinnocenzo, T. E. Banach *J. Am. Chem. Soc.* **1988**, *110*, 911.
242. J. P. Dinnocenzo, T. E. Banach, *J. Am. Chem. Soc.* **1989**, *111*, 8646.
243. D. T. Sawyer, J. S. Valentine, *Acc. Chem. Res.* **1981**, *14*, 393.
244. M. L. H. Green, D. K. P. Ng, R. C. Tovey, A. N. Chernega, *J. Chem. Soc., Dalton Trans.* **1993**, 3203.
245. F. G. N. Cloke, P. J. Fyne, V. C. Gibson, M. L. H. Green, M. J. Ledoux, R. N. Perutz, A. Dix, A. Gourdon, K. Prout, *J. Organomet. Chem.* **1984**, *277*, 61.
246. D. A. Slack, M. C. Baird, *J. Am. Chem. Soc.* **1976**, *98*, 5539.
247. *The Chemistry of Diazonium and Diazo Groups* (Ed.: S. J. Patai), John Wiley & Sons, New York, **1978**.
248. R. M. Elofson, F. F. Gadallah, *J. Org. Chem.* **1969**, *34*, 854.
249. M. P. Doyle, J. K. Guy, K. C. Brown, S. N. Mahapatro, C. M. van Zyl, J. R. Pladziewicz, *J. Am. Chem. Soc.* **1987**, *109*, 1536.
250. A. Roe, *Org. React.* **1949**, *5*, 193.
251. U. Kölle in *Mechanism and Processes in Molecular Chemistry* (Guest Ed.: D. Astruc), Gauthier–Villars, Paris, *New. J. Chem.* **1992**, *16*, 157.
252. T. E. Bitterwolf, A. C. Ling, *J. Organomet. Chem.* **1973**, *57*, C17.
253. U. T. Mueller-Westerhoff, *Angew. Chem. Int. Ed. Engl.* **1986**, *25*, 702.

254. R. H. Crabtree *The Organometallic Chemistry of Transition Metals*, 2nd Ed., Wiley, New York, **1994**, Chapters 3, 6 and 15.
255. J. Norton, *Acc. Chem. Res.* **1979**, *12*, 139.
256. J. R. Bleeke, W.-J. Peng, *Organometallics* **1986**, *5*, 635.
257. D. O'Hare, V. J. Murphy, N. J. Kaltsoyannis, *J. Chem. Soc. Dalton Trans.* **1993**, 383.
258. E. Steckhan, *Top. Curr. Chem.* **1987**, *142*, 1.
259. E. Steckhan, *Angew. Chem., Int. Ed. Engl.* **1986**, *25*, 683.
260. L. Eberson, B. Larsson, *Acta Chem. Scand., Ser. B* **1986**, *40*, 210.
261. (a) M. Lacoste, D. Astruc, M.-T. Garland, F. Varret, *Organometallics*, **1988**, *7*, 2253; (b) M. Lacoste, M.-H. Delville, N. Ardoin, D. Astruc, *Organometallics* **1997**, *16*, 2343.
262. (a) R. D. Bolskar, R. S. Mathur, C. A. Reed, *J. Am. Chem. Soc.* **1996**, *118*, 13093; (b) C. Reed, R. D. Bolskar, *Chem. Rev.* **2000**, *100*, 1075.
263. B. Boduszek, H. J. Shine, *J. Org. Chem.* **1988**, *53*, 5142.
264. S. Lochynski, H. J. Shine, M. Siroka, T. K. Venkatachalam, *J. Org. Chem.* **1990**, *55*, 2702.
265. O. Hammerich, V. D. Parker, *Electrochim. Acta* **1973**, *18*, 537.
266. H. Volz, W. Lotsch, *Tetrahedron Lett.* **1969**, 2275.
267. H. J. Dauben Jr, L. R. Honnen, K. M. Harmon, *J. Org. Chem.* **1960**, *25*, 1442.
268. O. B. Ryan, M. Tilset, V. D. Parker, *J. Am. Chem. Soc.* **1990**, *112*, 2618.
269. L. R. Eberson, *Adv. Phys. Org. Chem.* **1982**, *18*, 79.
270. a) A. J. Birch, K. B. Chamberlain, M. A. Hass, D. J. Thompson, *J. Chem. Soc., Perkin Trans.* **1973**, *1*, 1882; b) A. J. Pearson, *Acc. Chem. Res.* **1980**, *13*, 463.
271. J. C. Hayes, G. D. N. Pearson, N. J. Cooper, *J. Am. Chem. Soc.* **1981**, *103*, 4648.
272. G. S. Bodner, J. A. Gladysz, M. F. Nielsen, V. D. Parker, *J. Am. Chem. Soc.* **1987**, *109*, 1757.
273. E. W. Abel, S. P. Tyfield, *Adv. Organomet. Chem.* **1970**, 126.
274. Y. S. Yu, R. A. Jacobson, R. J. Angelici, *Inorg. Chem.* **1982**, *21*, 3106.
275. C. Eaborn, N. Farrell, J. L. Murphy, A. Pidcock, *J. Chem. Soc., Dalton Trans.* **1976**, 58.
276. N. G. Connelly, P. M. Hopkins, A. G. Orpen, J. Slater, *J. Chem. Soc., Dalton Trans.* **1992**, 3303.
277. O. Kahn, *Molecular Magnetism*, VCH, New York, **1993**.
278. J. B. Torrance, *Acc. Chem. Res.* **1979**, *12*, 79.
279. L. R. Melby, R. J. Harder, W. R. Hertler, R. E. Mahler, R. E. Benson, W. E. Mochel, *J. Am. Chem. Soc.* **1962**, *84*, 3374.
280. J. S. Miller, A. J. Epstein, W. M. Reiff, *Chem. Rev.* **1988**, *88*, 201.
281. W. Kaim, M. Moscherosh, *Coord. Chem. Rev.* **1994**, *129*, 157.
282. U. Honrath, H. Vahrenkamp, *Z. Naturforsch.* **1984**, *39b*, 555.
283. J. S. Miller, D. T. Glatzhofer, D. M. O'Hare, W. M. Reiff, A. Chakraborty, A. J. Epstein, *Inorg. Chem.* **1989**, 28, 2930.
284. C. Guillon, P. Vierling, *J. Organomet. Chem.* **1994**, *464*, C42.
285. M. D. Haley in *Encyclopedia of Electrochemistry of the Elements* (Ed.: A. J. Bard), Marcel Dekker, New York, **1980**, *Vol. IV*, p. 283.
286. J. V. Crivello, J. H. W. Lam, *J. Org. Chem.* **1978**, *43*, 3055.
287. R. S. Iyer, J. P. Selegue, *J. Am. Chem. Soc.* **1987**, *109*, 910.
288. D. H. Evans in *Encyclopedia of Electrochemistry of the Elements* (Ed.: A. J. Bard), Marcel Dekker, New York, **1978**, *Vol. XII*, p. 563.
289. C. Elschenbroich, G. Heikenfeld, M. Wunsch, W. Massa, G. Baum, *Angew. Chem. Int. Ed. Engl.* **1988**, *27*, 414.
290. R. L. Rich, H. Taube, *J. Am. Chem. Soc.* **1954**, *76*, 2608.
291. S. W. Feldberg, L. Jeftic, *J. Phys. Chem.* **1972**, *76*, 2439.
292. M. Chanon, *Acc. Chem. Res.* **1987**, *20*, 214.
293. L. Eberson *"Electron Transfer in Organic Chemistry"* Springer Verlag, Berlin, **1987**, Chapter X.
294. E. Román, D. Astruc, A. Darchen, *J. Chem. Soc., Chem. Commun.* **1976**, 512–513.
295. G. A. Russell, *Spec. Publ. Chem. Soc.* **1970**, *24*, 271.
296. N. Kornblum, *Angew. Chem.* **1975**, *87*, 797; *Angew. Chem. Int. Ed. Engl.* **1975**, *14*, 734.
297. J. F. Bunnett, *Acc. Chem. Res.* **1978**, *11*, 413; *Ibid* **1992**, *25*, 2.
298. J.-M. Savéant, *Acc. Chem. Res.* **1980**, *13*, 323

299. J.-M. Savéant in *Mechanisms and Processes in Molecular Chemistry* (Guest Ed.: D. Astruc), Gauthier–Villars, Paris, *New J. Chem.* **1992**, *16*, 131.
300. R. D. Rieke, H. Kojima, K. Öfele, *J. Am. Chem. Soc.* **1976**, *98*, 6735.
301. R. D. Rieke, H. Kojima, K. Öfele, *Angew. Chem. Int. Ed. Engl.* **1980**, *19*, 538.
302. M. I. Bruce, *Coord. Chem. Rev.* **1987**, *76*, 1.
303. J. K. Kochi, *J. Organomet. Chem.* **1986**, *300*, 139.
304. R. H. Magnuson, R. Meirowitz, S. J. Zulu, W. P. Giering, *Organometallics* **1983**, *2*, 460.
305. D. Miholová, A. A. Vlček, *J. Organomet. Chem.* **1982**, *240*, 413.
306. B. A. Narayanan, C. Amatore, C. P. Casey, J. K. Kochi, *J. Am. Chem. Soc.* **1983**, *105*, 6351.
307. C. Moinet, E. Román, D. Astruc, *J. Electroanal. Chem. Interfacial Electrochem.* **1981**, *241*, 121.
308. A. Darchen, *J. Chem. Soc. Chem., Commun.* **1983**, 768; *J. Organomet.* **1986**, *302*, 389.
309. P. Boudeville, A. Darchen, *Inorg. Chem.* **1991**, *30*, 1663.
310. D. Astruc, M.-H. Delville, J. Ruiz in *Molecular Electrochemistry of Inorganic, Bioinorganic and Organometallic Compounds* (Eds: A. J. L. Pombeiro, J. A. Mc Cleverty), NATO ASI Series, Kluwer, Dordrecht, **1993**, *385*, 277.
311. D. R. Tyler, *Coord. Chem. Rev.* **1985**, *63*, 217.
312. D. R. Tyler, *Comments Inorg. Chem.* **1986**, *5*, 215.
313. R. Moulton, T. W. Weidman, K. P. C. Vollhardt, A. J. Bard, *Inorg. Chem.* **1986**, *25*, 1846.
314. D. S. Brown, M.-H. Delville, R. Boese, K. P. C. Vollhardt, D. Astruc, *Angew. Chem.* **1994**, *106*, 715; *Angew. Chem. Int. Engl. Ed.* **1994**, *33*, 661.
315. M.-H. Delville, D. S. Brown, K. P. C. Vollhardt, D. Astruc, *J. Chem. Soc., Chem. Commun.*, **1991**, 1355.
316. D. S. Brown, M.-H. Delville, K. P. C. Vollhardt, D. Astruc, *New. J. Chem.* **1992**, *16*, 899.
317. D. S. Brown, M.-H. Delville, K. P. C. Vollhardt, D. Astruc, *Organometallics* **1996**, *15*, 2360.
318. M. I. Bruce, D. C. Kehoe, J. G. Matisons, B. K. Nicholson, P. H. Rieger, M. L. J. Williams, *J. Chem. Soc. Chem. Commun.* **1982**, 442.
319. M. I. Bruce, J. G. Mattisons, B. K. Nicholson, *J. Organomet. Chem.* **1983**, *247*, 321.
320. M. G. Richmond, J. K. Kochi, *Inorg. Chem.* **1986**, *25*, 656 and 1334; *Organometallics* **1987**, *6*, 254.
321. H. H. Ohst, J. K. Kochi, *J. Chem. Soc. Chem. Commun. 186*, 121; *Inorg. Chem.* **1986**, *25*, 2066.
322. M. Arewgoda, B. H. Robinson, J. Simpson, *J. Am. Chem. Soc.* **1983**, *105*, 1893.
323. A. J. Downard, B. H. Robinson, J. Simpson, *Organometallics* **1986**, *5*, 1122, 1132 and 1140.
324. S. Aime, M. Botta, R. Gobetto, D. Osella, *Organometallics* **1985**, *4*, 1475; *Inorg. Chim. Acta* **1986**, *115*, 129.
325. E. Cabrera, J. C. Daran, Y. Jeannin, *J. Chem. Soc. Chem. Commun.* **1988**, 607.
326. F. Richter, H. Varenkamp, *Angew. Chem. Int. Ed. Engl.* **1978**, *17*, 864.
327. M. O. Albers, N. J. Coville, E. Singleton, *J. Organomet. Chem.* **1987**, *323*, 37 and *326*, 229.
328. N. G. Connelly, S. J. Raven, G. A. Carriedo, V. Riera, *J. Chem. Soc. Chem. Commun.* **1986**, 992.
329. T. Venäläinen, T. Pakkanen, *J. Organomet. Chem.* **1984**, *266*, 269.
330. J. Pursiainen, T. A. Pakkanen, J. Jääskeläinen, *J. Organomet. Chem.* **1985**, *290*, 85.
331. J. Rimmelin, P. Lemoine, M. Gross, A. A. Bahsoun, J. A. Osborn, *Nouv. J. Chim.* **1985**, *9*, 181.
332. P. Lahuerta, J. Latorre, M. Sanau, H. Kisch, *J. Organomet. Chem.* **1985**, *286*, C27.
333. E. K. Lahdi, C. Mahé, H. Patin, A. Darchen, *J. Organomet. Chem.* **1983**, *246*, C61 and *259*, 189.
334. E. K. Lahdi, C. Mahé, H. Patin, A. Darchen, *Nouv. J. Chim.* **1982**, *6*, 539; *Organometallics* **1984**, *3*, 1128.
335. E. Alonso, D. Astruc, *J. Am. Chem. Soc.* **2000**, *122*, 3222.
336. J. W. Hershberger, R. J. Klingler, J. K. Kochi, *J. Am. Chem. Soc.* **1983**, *105*, 61.
337. M. Tilset in *Energetics of Organometallic Species* (Ed.: S. Martinho), Kluwer, Dordrecht, **1992**, p. 109.

338. M. Tilset in *Molecular Electrochemistryof Inorganic, Bioinorganic and Organometallic Compounds*, NATO ASI Series C385 (Eds.: A. J. L. Pombeiro, J. A. Mc Cleverty), Kluwer, Dordrecht, **1993**, p. 269.
339. J. W. Hershberger, R. J. Klingler, J. K. Kochi, *J. Am. Chem. Soc.* **1982**, *104*, 3034.
340. D. Catheline, D. Astruc, Proceedings of the XXIIth Intern. Conf. Coord. Chem. Budapest, *Coord. Chem.* **1982**, *23*, F41.
341. J.-N. Verpeaux, M.-H. Desbois, A. Madonik, C. Amatore, D. Astruc, *Organometallics* **1990**, *9*, 630.
342. M.-H. Desbois, D. Astruc, *J. Chem. Soc., Chem. Commun.* **1990**, 943.
343. M.-H. Desbois, D. Astruc, *J. Chem. Soc., Chem. Commun.*, **1988**, 472.
344. W. C. Trogler in *Organometallic Radical Processes, J. Organomet. Chem. Library 22* (Ed.: W. C. Trogler), Elsevier, Amsterdam, **1990**, p. 306.
345. H. Taube, *J. Chem. Ed.* **1986**, *45*, 452.
346. T. J. Katz, T. H. Ho, N. Y. Shih, V. Ying, Y. I. W. Stuart, *J. Am. Chem. Soc.* **1984**, *106*, 2659.
347. J. L. Hérisson, Y. Chauvin, *Makromol. Chem.* **1970**, *141*, 161.
348. J.-M. Savéant, *Acc. Chem. Res.* **1980**, *13*, 323.
349. R. W. Murray in *Electroanalytical Chemistry* (Ed.: A. J. Bard), Marcel Dekker, New York, **1984**, *Vol. 13*, pp. 191–357.
350. J.-F. Fauvarque, M. A. Petit, F. Pfluger, A. Jutand, C. Chevrot, M. Troupel, *Makromol. Chem. Commun.* **1983**, *4*, 455.
351. Y. Rollin, G. Meyer, M. Troupel, J.-F. Fauvarque, J. Périchon, *J. Chem. Soc., Chem. Commun.* **1983**, 793.
352. C. Moinet, E. Román, D. Astruc, *J. Am. Chem. Soc.* **1981**, *103*, 2431.
353. A. Buet, A. Darchen, C. Moinet, *J. Chem. Soc., Chem. Commun.* **1979**, 447.
354. E. Román, R. Dabard, C. Moinet, D. Astruc, *Tetrahedron Lett.* **1979**, *16*, 1433.
355. M. Tokuaka, *Collect. Czech. Chem. Commun.* **1932**, *4*, 444; **1934**, *6*, 339.
356. S. Rigaud, Ph. D. thesis, University Bordeaux I, **1997**.
357. J. Losada, I. Cuadrado, M. Morán, C. M. Casado, B. Alonso, M. Barranco, *Anal. Chim. Acta*, **1996**, *251*, 5.
358. I. Cuadrado, M. Morán, J. Losada, C. M. Casado, C. Pascual, B. Alonso, F. Lobete In *Advances in Dendritic Macromolecules* (Ed.: G. R. Newkome), JAI Press, Vol 3, Greenwich, Connecticut, **1996**, p. 151.
359. C. M. Casado, I. Cuadrado, M. Morán, B. Alonso, B. Garcia, B. Gonzales, J. Losada, *Coord. Chem. Rev.* **1999**, *185–186*, 53; I. Cuadrado, M. Morán, C. M. Casado, B. Alonso, J. Losada, *Coord. Chem. Rev.*, **1999**, *189*, 123.
360. A. Heller, *Acc. Chem. Res.* **1990**, *23*, 128.
361. Y. Degani, A. Heller, *J. Am. Chem. Soc.* **1988**, *110*, 2615.
362. I. Taniguchi, S. Miyanoto, S. Tomimura, F. M. Hawkridge, *J. Electroanal. Chem.* **1988**, *240*, 33.
363. A. L. Crumbliss, H. A. O. Hill, D. Page, *J. Electroanal. Chem.* **1986**, *206*, 327.
364. Y. Degani, A. Heller, *J. Am. Chem. Soc.* **1989**, *111*, 2357, "*Proceedings of the Third International Symposium on Redox Mechanism and Interfacial Properties of Molecules of Biological Importance*" (Eds: G. Dryhurst, K. Niki, H. I. Honolulu), Plenum, New York, **1988**, p. 151.
365. C. Bourdillon, M. Majda, *J. Am. Chem. Soc.* **1990**, *112*, 1795.
366. T. E. Edmonds in *Chemical Sensors* (Ed.: T. E. Edmonds), Blackie, Glasgow, **1988**, Chapter 8, p. 193.
367. T. Saji, I. Kinoshita, *J. Chem. Soc., Chem. Commun.* **1986**, 716.
368. P. D. Beer, *Chem. Soc. Rev.* **1989**, *18*, 409.
369. P. D. Beer, *Advan. Inorg. Chem.* **1992**, *39*, 79.
370. P. D. Beer, *J. Chem. Soc., Chem. Commun.* **1996**, 689.
371. P. D. Beer, *Acc. Chem. Res.* **1998**, *31*, 71.
372. C. Valério, J.-L. Fillaut, J. Ruiz, J. Guittard, J.-C. Blais, D. Astruc, *J. Am. Chem. Soc.*, **1997**, *119*, 2588; D. Astruc, *Acc. Chem. Res.* **2000**, *33*, 287.
373. J. Guittard, J.-C. Blais, C. Valério, E. Alonso, J. Ruiz, J.-L. Fillaut, D. Astruc, *Pure Appl. Chem* **1998**, *70*, n° 4, 809; Top. Curr. Chem. **2000**, *210*, 229.
374. E. Alonso, C. Valério, J. Ruiz, D. Astruc, *New J. Chem.* **1997**, *21*, 1139–1141.

375. C. Valério, J. Ruiz, J.-L. Fillaut, D. Astruc, *C. R. Acad. Sci.* Paris, **1999**, *2, Série II c*, 79.
376. C. Valério, E. Alonso, J. Ruiz, J.-C. Blais, D. Astruc, *Angew. Chem. Int. Ed. Engl.*, **1999**, *38,* 1747.
377. C. Valério, F. Moulines, J. Ruiz, J.-C. Blais, D. Astruc, *J. Org. Chem.* **2000**, *65*, 1996.
378. A. Vlček Jr., see this Section 2.2, Chapter 2.2.5.
379. W. Kaim, see this Section 2.2, Chapter 2.2.9.
380. S. Fukuzumi, see this Section 2.2, Chapter 2.2.8.
381. S. Fukusumi, the Phorphyrin Handbook, Vol 8 (Eds.: K. M. Kadish, K. Smith, R. Guillard), Academic Press, San Diego, CA, 2000, pp. 115–152.
382. R. Guillard, K. M. Kadish, Chem. Rev. **1993**, **1988**, *88*, 1121.

# 5 Electron-transfer Processes in Mononuclear Polypyridine Metal Complexes*

*Antonín Vlček, Jr*

## 5.1 Introduction

Transition metal complexes of 2,2′-bipyridine and other polypyridine ligands (N,N) show an exceptionally rich electron-transfer chemistry which makes them a clearly distinct class of coordination compounds. Detailed investigations of their electrochemistry, redox chemistry, and photochemistry have much strengthened our understanding of electron-transfer processes in general and enabled many interesting applications. Such important concepts as localization of a redox change [1–5], redox orbital [1, 3, 6], redox series [4, 7, 8], relations between electrochemistry and electronic spectroscopy [6, 9], or analogy between non-radiative decay of charge-transfer excited states and intramolecular electron transfer [10–14] were either discovered or deeply elaborated by use of metal–polypyridine complexes. In general, metal–polypyridines are known to exist in unusual oxidation states, to undergo sequential electron transfer, and/or to couple electron-transfer with bond activation. Consequently, they can act as redox catalysts [24, 30, 31], electron-transfer mediators [32], redox sensors [33–35], or electrochromic materials [36]. Moreover, their electron transfer reactivity is often retained upon electronic excitation, opening the field of photochemical excited-state electron transfer [37–39], with potential applications in light energy conversion [36, 40–43], photocatalysis [42, 44], fluorescence sensors [45], or in molecular electronic and photonic devices [38, 46].

The great structural variability of polypyridine complexes enable us to tune and control their redox properties over a very broad range. For example, variations in polypyridine ligand structure can shift the reduction or oxidation potential within a range almost 2 V wide. Further control can be achieved by changing the nature of the metal and, in mixed-ligand complexes, of ancillary ligands. Polypyridine com-

---

*This chapter is dedicated to the memory of my father, Professor Antonín A. Vlček (1927–1999), who has contributed much to the electrochemistry of metal–polypyridine complexes and to our understanding of underlying principles [1, 4–8, 15–29, 209].

plexes can be prepared stereochemically pure, with a predetermined chirality [47–50]. Moreover, it is relatively easy to connect metal–polypyridine units to form larger electro- and photo-active supermolecular assemblies, including dendrimers [38, 51–53]. The structural versatility and role of polypyridine complexes in processes and devices based on electron transfer are unique, paralleled perhaps only by metal porphyrins.

Transition metal polypyridine complexes have attracted much research interest for several decades. The first low-valent 2,2′-bipyridine (bpy) complexes, $[Cr(bpy)_3]^+$ and $[Cr(bpy)_3]^0$, were prepared by Herzog et al. as long ago as the fifties [54]. On the basis of magnetochemical data they were formulated as $Cr^I$ and $Cr^0$ species, respectively. Soon after, facile electrochemical reduction of polypyridine complexes to their low-valent forms was observed by A. A. Vlcek. This was first demonstrated [22, 23] by reduction of $[Co^{II}(bpy)_3]^{2+}$ to $[Co^I(bpy)_3]^+$. Electrochemical reduction of $Ni^{II}$, $Mn^{II}$, and $Cr^{III}$ complexes [25, 27] was described next. $[Cr(bpy)_3]^{3+}$ is especially important [2, 25, 55] because this complex shows a six-step sequential reversible reduction, with a single electron being transferred in each step without a change in complex composition. A 7-membered redox series $[Cr(bpy)_3]^z$; $z = 3+, 2+, 1+, 0, 1-, 2-, 3-$; was thus established and all its members isolated. The existence of such a redox series is a typical feature of metal–polypyridine chemistry. At about the same time, a redox-catalytic reactivity of polypyridine complexes was demonstrated by reduction of aromatic nitro compounds catalyzed by cobalt–bipyridine complexes [24]. These initial studies sparked a series of investigations of polypyridine (initially mostly bpy) complexes of other transition metals. Electrochemistry (polarography and, later, cyclic voltammetry) became the most important technique used to detect polypyridine complexes in unusual oxidation states and to assess their chemical stability. In parallel, many low-valent polypyridine complexes of almost all the first-row transition metals were prepared and isolated by chemical means, and their UV–Vis and IR absorption spectra determined, along with their magnetochemical properties [2]. The problem of redox change localization became soon apparent, because these studies revealed that it is often the polypyridine ligand which is reduced, instead of the metal atom. For example, $[Cr(bpy)_3]^0$, originally believed to contain $Cr^0$, was re-formulated as a $Cr^{III}$ complex with three radical-anionic $bpy^{•−}$ ligands, possibly with some electronic delocalization [2]. In the 80s, Bard et al. turned the attention to oxidation of bipyridine complexes at very positive potentials, finding that bpy and its analogs can also stabilize metal atoms in unusually high oxidation states [56–58].

A real explosion of research interest in metal polypyridines followed Adamson's and Gafney's discovery [59] that, upon irradiation with visible light, the long-lived excited state of $[Ru(bpy)_3]^{2+}$ is oxidized in a bimolecular reaction with $[Co(NH_3)_5Br]^{2+}$ to $[Ru(bpy)_3]^{3+}$. The great potential of bimolecular excited-state electron transfer for photochemical conversion and storage of light (solar) energy was immediately recognized [37]. The number of photo- and redox-active metal–polypyridine complexes was further extended by Wrighton's investigations [60–62] of the photochemistry of carbonyl complexes $[Re(L)(CO)_3(N,N)]$. Photochemical research focused mostly on polypyridine complexes of heavy $d^6$ metals, especially $Ru^{II}$, $Os^{II}$, $Re^I$ and, to a lesser extent, $Rh^{III}$ and $Ir^{III}$. Many new polypyridine ligands and their complexes were prepared. Their photochemical, photophysical, and electrochemical properties were investigated in much depth, especially by Balzani, Meyer, Elliott, DeArmond, Lever, Sutin, Creutz, Rillema, Scandola, and other research groups. The wealth of data available provided further insight into the molecular mechanism of metal–polypyridine electron-transfer reactions and revealed general structure–reactivity relationships and patterns. Sequential reduction by, at least, six one-electron steps have been observed [63–65] for tris-bipyridine complexes of $Os^{II}$ and $Ru^{II}$, in addition to their metal-centered oxidation. On the basis of spectroscopic studies of reduction products it was soon realized that the electrons gained in the reduction of metal tris-bipyridine complexes are localized on individual ligands [66–69], instead of being delocalized over the whole complex molecule. The chemistry of redox series was deeply elaborated through the work of Aoyagui, DeArmond, Elliott, Roffia, A. A. Vlcek, and others. These observation led to detailed examination of the redox orbital concept [1, 6] by De Armond [3, 65, 69]. Redox potential patterns in metal–polypyridine redox series were analyzed and explained by A. A. Vlcek [4, 5, 7, 8]. In the early 80s much attention was attracted by an observation [70–72] of a close similarity between spectra of reduced forms and excited states of $[Ru(bpy)_3]^{2+}$ and its analogs. This finding surprisingly implied that the excited electron is localized at a single bpy ligand, similarly to the localization of the extra electron in the respective reduced complexes.

In the 70s and 80s much was learned about the mechanism of excited-state electron transfer, control of excited-state lifetime, and on separation of products of photochemical redox reactions to prevent their recombination [32, 37, 39, 73, 74]. Although several schemes for light-energy conversion based on $[Ru(bpy)_3]^{2+}$ have been proposed, none was practically applicable [40, 41, 75, 76]. Only the in late 80s was it found that $Ru^{II}$ polypyridine complexes efficiently sensitize $TiO_2$ electrodes, leading to their use in a solar cell developed by Grätzel et al. [36, 43, 77–82]. The search for an optimum sensitizer of photoelectrochemical light energy conversion still continues and $Ru^{II}$ polypyridines seem to be the most promising candidates [36].

Another important research area, linking metal–polypyridine and supramolecular chemistry, was opened in mid-eighties, initially by Balzani and Scandola. Metal–polypyridine units were recognized as excellent building blocks of oligonuclear complexes and supramolecular assemblies, because they can be linked together either through bridging ancillary ligands (e.g. $CN^-$) or polypyridine ligands themselves [38, 51, 52, 83], often with controlled stereochemistry [48–50]. For example, a dendrimer composed of twenty-two $Ru^{II}(N,N)_3{}^{2+}$ units has been made [53, 84]. In other types of supermolecular complexes, metal–polypyridine units are linked with other functionalities, for example receptor groups (crown ethers, calixarenes, etc.) or different electron- or energy-acceptors or -donors (porphyrins, $C_{60}$, etc.) [85–89]. Metal–polypyridine fragments are also used as redox- and/or photo-active components of catenanes and rotaxanes [90–92]. The supramolecular chemistry, electrochemistry, and photochemistry of metal–polypyridines is a very promising research area which is undergoing rapid expansion. Questions of electronic communication between individual active centers, and possible emergence of collective redox, photophysical or photochemical behavior, are very intriguing, as are possible applications in sensors and molecular electronic and photonic devices [38, 46].

Currently, metal–polypyridine units are the molecular building blocks of choice whenever a compound with special electro- and/or photo-activity is to be designed. The research emphasis has somewhat shifted from fundamental studies of electron transfer reactivity and excited state properties of individual complexes to the design of new functional molecules and supermolecules with predetermined properties.

The amount of research performed and literature published on electron-transfer reactions of metal–polypyridine complexes is enormous. Several excellent reviews [42, 74, 93–97] and books [38, 62, 98, 99] deal with polypyridine complexes, their redox chemistry, photochemistry, and applications. Hereinafter, the most prominent aspects of electron transfer reactivity of mononuclear metal–polypyridine complexes will be surveyed without attempting to cover exhaustively the vast original literature. Instead, the main purpose of this chapter is to single out the structural, thermodynamic, and kinetic factors which enable and control the special and diverse electron-transfer behavior of metal–polypyridine complexes in their electronic ground and excited states. Although supramolecular electron-transfer chemistry of metal–polypyridines is not discussed here in detail, because it is covered in Volume 3 of this monograph, links connecting the redox behavior of mononuclear polypyridine complexes and their supramolecular counterparts will be briefly outlined.

## 5.2 Electron-transfer Properties of Ground-state Polypyridine Complexes: A Survey

### 5.2.1 The 2,2′-Bipyridine and other Polypyridine Ligands

Formulas of selected polypyridine ligands and the abbreviations used are shown in the Appendix. 2,2′-Bipyridine (bpy) is the generic member of the polypyridine family and the most common polypyridine-type ligand. Many bpy ligands are derived from bpy merely by attaching different functional groups to its pyridine rings. These groups range from strong electron acceptors ($-CF_3$) or donors ($-NEt_2$) to bulky aliphatic substituents ($-Bu^t$) or polyether chains. The choice of substituent enables us to control many of the properties of bpy complexes—their redox potentials, excited state energy and lifetime, chemical stability of reduction and oxidation products, and their solubility. Closing the central aromatic ring in 2,2′-bipyridine by connecting the 3,3′ positions with a $-CH=CH-$ group leads to the 1,10-phenanthroline ligand, phen. Again, the properties of phenanthroline complexes can be controlled by use of functional groups. Linking three pyridine rings leads to 2,2′:6′,2″-terpyridine (tpy) which gives rise to another family of polypyridine complexes. Tpy complexes are especially important as building blocks of supermolecules [87].

Still more polypyridine-type ligands, better called polyazines, are derived from bpy and phen by replacing some of the CH groups of their aromatic rings with N atoms. This strongly affects the redox properties of both the ligands and their complexes. A great variety of polyazine- and polypyridine-type bridging ligands [100] is, moreover, available for construction of polynuclear complexes.

α-Diimine ligands, e.g. 2-pyridine-*N*-carbaldehydes (PyCa) or 1,4-diazabutadienes (dab), should also be considered here, because the redox behavior of their complexes often resembles that of metal–polypyridines (polypyridines are, in fact, sometimes regarded as aromatic α-diimines).

Most of the interesting redox properties of metal–polypyridine complexes originate from the electron-transfer activity of polypyridine ligands themselves. The free bpy ligand and its analogs are sequentially reduced in two one-electron electrochemically reversible steps, producing the corresponding radical-anion and dianion, respectively:

$$\text{bpy} \xleftrightarrow{e^-} \text{bpy}^{\cdot-} \xleftrightarrow{e^-} \text{bpy}^{2-} \tag{1}$$

The respective redox potentials of the first and second reduction step in THF [18, 21] are −2.68 and −3.31 V. The corresponding values in DMF [101] are 2.57 and −3.13 V. (***All redox potentials in this chapter are reported relative to the ferrocene/ferricinium couple.***) Both reduced forms of bpy are intrinsically stable in strictly aprotic and anaerobic media. For bpy and some other polypyridines, the radical anions and/or dianions have been characterized by UV–Vis absorption, resonance Raman, and/or EPR spectroscopy [18, 21, 70, 71, 102, 103]. Bipyridine derivatives with strongly electron-accepting groups –C(O)OEt or –Ph in 4,4′ positions have a third reduction step at very negative potentials [18, 21]. Free phen [104] and tpy [101] are reduced by one electron at potentials similar to that of the first bpy reduction. The second reduction steps of phen and tpy are chemically irreversible.

Electrochemical oxidation of free bpy was observed as an irreversible process at ca +1.8 V in liquid $SO_2$ [105] and at ca +2 V in acetonitrile [106].

The strong dependence of the redox potentials of both reduction steps on substituents attached to aromatic rings of polypyridine ligands provides a very convenient means of tuning the redox properties, because the reduction potentials of polypyridine complexes reflect those of the free ligands. Generally, the first reduction of free bpy-type ligands occurs in the range from ca −2.8 to −2.20 V. This is significantly more negative than the reduction of most transition metals in common oxidation states. On the other hand, some polyazines and polypyridine derivatives such as abpy are reduced at much more positive potentials (e.g. −1.37 V for abpy [107]).

Polypyridines form stable complexes with most transition metals. Complexes with metals in oxidation states which correspond to the $d^6$ configuration seem to be most abundant. Polypyridines are bidentate ligands. Tpy is generally tridentate although it can coordinate as a bidentate ligand through two pyridine rings only. M(N,N) chelate rings are rather stable, although a polypyridine ligand can dissociate from highly reduced complexes, especially of first-row transition metals.

Polypyridines are good σ-donors. Importantly, they also have low-lying, unoccupied $\pi^*$ orbitals which can accept electrons on reduction. The extent of d → $\pi^*$

back-donation is small for most metal–polypyridines, except for some complexes of first-row metals and some special ligands, e.g. abpy. Hence, most reduced poly-pyridine complexes are not complexes of metals in unusually low oxidation states, stabilized by $\pi$ back-donation. Instead, they behave as complexes of metals in common oxidation states (usually with a $d^6$ configuration) with polypyridine radi-cal-anions or dianions as ligands.

Important steric effects can be induced by attaching bulky substituents to posi-tions adjacent to the nitrogen donor atoms, that is 6,6'-bpy and 2,9-phen. Resulting ligands strongly favor tetrahedral over octahedral or square-planar coordination. This effect was employed in the construction of entwined catenate ligands (catphen)$_2$ in which two phen ligands are connected by interlocked polyether rings attached to their 2,9 positions (see Appendix). Catphen-type ligands form com-plexes with most first-row transition metals [108]. They impose a pseudotetrahedral coordination geometry on the metal atoms.

An octahedral cage is obtained by connecting three bpy ligands together through their 5,5'-positions by amide links [109, 110].

### 5.2.2 Polypyridine Complexes

**Titanium**

Two successive, reversible one-electron reductions in the range from $-2.0$ to $-3.1$ V are observed for the [Ti(bpy)$_3$] complex [111]; these are attributed to the formation of [Ti(bpy)$_3$]$^z$, $z = 1-, 2-$. These species seem to undergo partial bpy dissociation in solution. Oxidation-state assignment to Ti and the bpy ligands is not completely clear. Both reductions are assumed to be Ti-localized [111]. Indeed, the $z = 1-, 2-$ complexes have spectroscopic features [112, 113] of a bpy$^{\bullet-}$ ligand whereas [Ti(bpy)$_3$]$^{2-}$ probably contains a bpy$^{2-}$ ligand also [112].

**Vanadium**

[V(bpy)$_3$]$^z$ complexes form [111, 114] a six-membered redox series $z = 2+, 1+, 0,$ $1-, 2-, 3-$. The corresponding one-electron redox steps are reversible. They occur in the range from $-1.48$ to $-3.2$ V. Correlation between spectroscopic and electro-chemical data and dependence of redox potentials on methylation of bpy ligands indicate that the [V(bpy)$_3$]$^{2+}$/[V(bpy)$_3$]$^+$ and [V(bpy)$_3$]$^+$/[V(bpy)$_3$] redox changes are metal-localized, whereas the steps $0/-1$, $-1/-2$, and $-2/-3$ seem to be ligand-localized. This suggests [111] the formulations [V$^{II}$(bpy)$_3$]$^{2+}$, [V$^I$(bpy)$_3$]$^+$, [V$^0$(bpy)$_3$], [V$^0$(bpy)$_2$(bpy$^{\bullet-}$)]$^-$, [V$^0$(bpy)(bpy$^{\bullet-}$)$_2$]$^{2-}$, and [V$^0$(bpy$^{\bullet-}$)$_3$]$^{3-}$. A more recent EPR study [115] has indicated that [V(bpy)$_3$] has a highly delocalized struc-ture, the unpaired electron occupying a delocalized orbital of ca 30 % metal 3d character. UV–Vis spectra of [V$^0$(bpy)$_3$] and [V$^0$(bpy)$_2$(bpy$^{\bullet-}$)]$^-$ have features arising from V $\rightarrow$ bpy metal-to-ligand charge transfer (MLCT) and bpy$^{\bullet-}$ intra-ligand transitions, respectively [114]. Nevertheless, it seems that the bonding in the first 3 or 4 members of the [V(bpy)$_3$]$^z$ redox series is rather delocalized.

**Chromium, molybdenum, and tungsten**

Chromium forms [2, 25, 55, 111, 113, 116] a seven-membered redox series $[Cr(bpy)_3]^z$; $z = 3+, 2+, 1+, 0, 1-, 2-, 3-$. All its members have been synthesized in the solid state. Corresponding one-electron redox steps are reversible, occurring in the range from $-0.70$ to $-3.0$ V. Some of the more reduced species ($z < 1+$) are partially labile in solution. A combination of electrochemical [25, 111] and spectroscopic (UV–Vis [2, 111, 117] and IR [113]) data indicates that the $3+/2+$ step is metal-localized $Cr^{III}/Cr^{II}$, although some electronic delocalization can occur at the $2+$ level. The $z = 1+$ and 0 members look rather delocalized. $[Cr(bpy)_3]^0$ could be viewed as $[Cr^{III}(bpy^{\cdot -})_3]^0$ with a strong antiferromagnetic coupling between unpaired electrons of the Cr atoms and three $bpy^{\cdot -}$ ligands. Nevertheless, the use of integer metal and ligand oxidation states in the $z = 1+$ and 0 complexes seems inappropriate, owing to electron delocalization. Further reduction steps to $z = 1-, 2-, 3-$ are bpy-localized. Five members of an analogous $[Mo(bpy)_3]^z$ redox series are known [111]: $z = 1+, 0, 1-, 2-, 3-$. The $1+/0$, $0/1-$, and $1-/2-$ reduction steps are reversible.

Complexes of the type $[M(CO)_4(N,N)]$; M = Cr, Mo, W; have been prepared for many polypyridine ligands and $\alpha$-diimines PyCa or dab. These compounds are very important spectroscopically and photochemically, because of their low-lying MLCT electronic transitions [96, 118]. Polypyridine complexes $[M(CO)_4(N,N)]$ are reduced in two successive one-electron steps which are localized on the polypyridine ligand [20, 21, 119–122]. (Four reduction steps have been found [21] for N,N = 4,4'-Ph$_2$-bpy.) A chemically partly reversible oxidation occurs at positive potentials [17].

Investigations of $[M(CO)_4(N,N)]$ reduction were important for our understanding of the effects of metal coordination on the redox behavior of polypyridine ligands. Its simple molecular composition makes $[M(CO)_4(N,N)]$ ideal for such studies, because the single M(N,N) chelate ring is the only redox-active unit present. The most typical features of $[M(CO)_4(N,N)]$ electrochemistry are: (i) The M(N,N) chelate ring has the same reduction pattern (number of steps, number of electrons exchanged) as the free N,N ligand itself. Compared with the free ligands, reductions of the corresponding complexes are shifted positively by ca 0.5–0.8 V. Hence even the most negative ligand-based reductions become observable upon coordination (e.g. the 3rd and 4th reduction of 4,4'-Ph$_2$-bpy). The positive shift is even larger for a polyazine ligand bridging between two metal atoms [120] in complexes of the type $[\{M(CO)_4\}_2(\mu\text{-}N,N)]$. (ii) Reduction potentials of the complexes correlate with reduction potentials of the free N,N ligand. (iii) Despite the first reduction being essentially N,N-localized, its potential is weakly dependent [20] on M: W < Mo < Cr. (iv) EPR [119–122] and IR [121, 122] spectra, and DFT calculations [122] confirm the predominant N,N-localization of the first reduction. Importantly, they show that reduction to $[M(CO)_4(N,N^{\cdot -})]$ affects also the electron density distribution within the M(CO)$_4$ fragment by increasing the total electron density on all four CO ligands (on COs *trans* to N,N more than on *cis*) and by a spin-delocalization to axial (*cis*) CO ligands. (v) Reduction labilizes an axial M–CO bond toward substitution [20, 21]. The rate of axial CO dissociation strongly increases on going from

$[M(CO)_4(N,N^{\bullet-})]$ to $[M(CO)_4(N,N^{2-})]$. (vi) Potentials of both reduction steps of complexes $[Mo(PBu_3)_n(CO)_{4-n}(bpy)]$ or $[Mo(PBu_3)_n(CO)_{4-n}(tBu\text{-}dab)]$ shift slightly negatively when CO is replaced by a stronger electron donor $PBu_3$: $n = 0 > 1 > 2$ [123] (vii) Oxidation labilizes a M–CO bond [17].

## Manganese and rhenium: coordination compounds

$[Mn^{II}(bpy)_3]^{2+}$ is reduced in three one-electron steps to $[Mn^{II}(bpy)_3]^z$; $z = 1+, 0, 1-$; which are accompanied by a partial loss of a bpy ligand [27, 116, 124]. Compared with other first-row metal bpy complexes the reductions occur rather negatively: $E_{1/2}(2+/1+) = -1.75$ V. Localization of the electrons added is uncertain. Magnetochemical data [125] suggest that the $z = 0$ and $1-$ species should be viewed as $[Mn^{II}(bpy)(bpy^{\bullet-})_2]$ and $[Mn^{II}(bpy^{\bullet-})_3]^-$, essentially in agreement with IR spectra [113]. An irreversible oxidation of $[Mn^{II}(bpy)_3]^{2+}$ and $[Mn^{II}(phen)_3]^{2+}$ occurs at $+0.97$ V, producing, in the presence of traces of water, $[(N,N)_2Mn(\mu\text{-}O)_2\text{-}Mn(N,N)_2]^{4+}$. These, formally $Mn^{IV}Mn^{IV}$ complexes, are strong oxidants, able to oxidize $OH^-$, $Cl^-$, and catechol [126]. $[(phen)_2Mn(\mu\text{-}O)_2Mn(phen)_2]^{4+}$ oxidizes toluene by a unique hydride abstraction mechanism [127]. Manganese–polypyridine–oxo complexes have also been employed as models of the water-oxidation catalyst of plant photosynthesis [128].

The complex $[Re^{II}(bpy)_3]^{2+}$ and its tpy and phen analogs undergo a Re-localized reduction to $[Re^I(bpy)_3]^+$, followed by two, presumably bpy-localized, one-electron reductions. Oxidation of $[Re^{II}(bpy)_3]^{2+}$ is irreversible, because of the formation of 7-coordinated Re(III) complexes [129].

## Rhenium and manganese: organometallic compounds

$[Re^I(E)(CO)_3(N,N)]^{n+}$ complexes form a special class of organometallic compounds with distinct electrochemical, photochemical, and photophysical properties [97]. Their electrochemical behavior epitomizes many general aspects of the redox chemistry of metal–diimine units. Neutral ($n = 0$) complexes exist with anionic axial ligands E = halides (X), $CN^-$, $CF_3SO_3^-$, alkyl or aryl group, metal fragment like $Mn(CO)_5$, $Re(CO)_5$, $Ph_3Sn$, etc. Cationic ($n = 1$) complexes are known for E = neutral Lewis base, e.g. pyridine derivatives, phosphines, phosphites, nitriles, isonitriles, CO, imidazole, histidine, etc. The nature of the N,N ligand can also be varied broadly: bpy- or phen-type polypyridines, polyazines, and dab and PyCa-type ligands.

In principle, these complexes undergo a one-electron oxidation and two successive one-electron reductions. The potential and reversibility of the oxidation step depend very much on the axial ligand E. Oxidation of halide and cyanide complexes is followed by rapid loss of the oxidized axial ligand [130, 131]:

$$[Re^I(E)(CO)_3(bpy)] \xrightarrow{e^-} [Re(E)(CO)_3(bpy)]^{\bullet+}$$
$$\xrightarrow{S} [Re^I(S)(CO)_3(bpy)]^+ + E^\bullet \qquad (2)$$

$E_{ox} = +1.00$ V and $+1.17$ V for E = Cl and CN, respectively.

Oxidation of a metal–metal bonded complex [Re$^I$(Ph$_3$Sn)(CO)$_3$(bpy)] is also chemically irreversible. It occurs at less positive potential, +0.39 V [61]. On the other hand, cationic complexes [Re$^I$(CH$_3$CN)(CO)$_3$(bpy)]$^+$ [130] and [Re$^I$(imidazole)-(CO)$_3$(bpy)]$^+$ [132] are oxidized reversibly in CH$_3$CN at +1.40 and 1.46 V, respectively, producing the corresponding Re$^{II}$ complexes. Oxidation is, however, irreversible in water. No oxidation was observed for [Re(CO)$_4$(bpy)]$^+$ and its bpym and phen analogs up to +1.6 V in acetonitrile [133].

Reduction of [Re$^I$(E)(CO)$_3$(N,N)]$^{n+}$ occurs in two successive one-electron steps which are predominantly localized on the N,N ligand, with a partial, N,N-dependent, delocalization of the extra electron density over the Re$^I$(E)(CO)$_3$$^{n+}$ fragment [134]. The respective products, [Re$^I$(E)(CO)$_3$(N,N$^{\bullet-}$)]$^{(n-1)+}$ and [Re(E)(CO)$_3$(N,N$^{2-}$)]$^{(n-2)+}$, are amenable to dissociation of the axial ligand E, whose rate and extent depends on the N,N and E ligands, and on the temperature and the excess of free ligand E added:

$$[Re(X)(CO)_3(bpy)] \xleftarrow{e^-} [Re(X)(CO)_3(bpy)]^{\bullet-}$$
$$\xrightarrow{S} [Re(S)(CO)_3(bpy)]^{\bullet} + X^- \tag{3a}$$

X = Cl, Br

$$[Re(E)(CO)_3(bpy)]^+ \xleftarrow{e^-} [Re(E)(CO)_3(bpy)]^{\bullet}$$
$$\xrightarrow{S} [Re(S)(CO)_3(bpy)]^{\bullet} + E \tag{3b}$$

E = phosphines, phosphites, nitriles, THF, CO

$$[Re(X)(CO)_3(bpy)]^{\bullet-} \xleftarrow{e^-} [Re(X)(CO)_3(bpy)]^{2-}$$
$$\longrightarrow [Re(CO)_3(bpy)]^- + X^- \tag{4a}$$

$$[Re(S)(CO)_3(bpy)]^{\bullet} \xleftarrow{e^-} [Re(S)(CO)_3(bpy)]^-$$
$$\longrightarrow [Re(CO)_3(bpy)]^- + S \tag{4b}$$

For example, [Re$^I$(CN)(CO)$_3$(bpy)] in strictly aprotic solvents at low temperature is reduced in two chemically reversible one-electron steps at −1.77 and −2.42 V, which correspond to formation of [Re$^I$(CN)(CO)$_3$(bpy)]$^z$; z = 1− and 2−, respectively [130]. The first reduction of [Re(CO)$_4$(bpy)]$^+$ or [Re(P(OEt)$_3$))(CO)$_3$(bpy)]$^+$ is chemically reversible even at room temperature [133, 135]. Halide complexes [Re$^I$(X)(CO)$_3$(bpy)] (X = Cl, Br) have a complicated reduction pattern because of further reactions of the reduced products. Several reaction intermediates and products have been characterized spectroelectrochemically [131, 135–137] and rate constants of the main steps have been determined [130] by low-temperature cyclic voltammetry.

The rate and extent of dissociation of the axial ligand E upon uptake of the first electron (Eq. 3) depend on the nature of the axial ligand [134, 136], decreasing in the order Br$^-$ > Cl$^-$ ≫ THF > PPh$_3$, $n$PrCN > P(OMe)$_3$, CO. The structure of the polypyridine N,N ligand also has a profound effect. For series of complexes

$[Re(Br)(CO)_3(N,N)]^-$, $X = Br$, $Cl$, the reactivity orders are $t$Bu-dab $> $ bpz $\approx$ bpdz $>$ bpym $\gg$ bpm, abpy and bpy $>$ iPr-PyCa $>$ 2,3-dpp $>$ abpy [134, 136]. The reactivity decreases in the same order as the $\pi^*$-orbital coefficient at the N donor atoms [134].

For example, reduced bromo complexes are much more labile than their chloro counterparts [134]. In contrast, the solvento species $[Re(S)(CO)_3(bpy)]^{\bullet}$ are relatively stable in coordinating solvents S for S $= n$PrCN, but not for S $=$ THF, in which they dimerize to $[Re(CO)_3(bpy)]_2$. Reduction of $[Re(PPh_3)(CO)_3(bpy)]^+$ generates $[Re^I(PPh_3)(CO)_3(bpy)]^{\bullet}$, which is stable in $n$PrCN even at room temperature. As an exception, $[Re^I(P(OEt)_3)(CO)_3(bpy)]^{\bullet}$ slowly reacts [135] with excess P(OEt)$_3$ to give $[Re^I(P(OEt)_3)_2(CO)_2(bpy)]^{\bullet}$. The stabilizing effect of the abpy ligand is manifested by $[Re(Br)(CO)_3(abpy)]^{\bullet-}$ which undergoes only a very slow Br$^-$ substitution by a solvent molecule. The complexes $[Re(L)(CO)_3(abpy)]^{\bullet}$; and $[Re(L)(CO)_3(bpym)]^{\bullet}$ L $=$ THF, $n$PrCN, or PPh$_3$ are stable in solution.

The rate of axial ligand dissociation increases dramatically with the number of electrons added, being much faster (Eq. 4) for $[Re(X)(CO)_3(bpy)]^{2-}$ than $[Re(X)(CO)_3(bpy)]^-$; $X = Cl$ or $Br$. Five-coordinated species $[Re(CO)_3(bpy)]^-$ are formed. The same reaction occurs for other polypyridine ligands. Only $[Re(Br)(CO)_3(abpy^{\bullet-})]^{2-}$ reacts via slow substitution of Br$^-$ by a solvent (THF) molecule instead of dissociation, and full axial ligand labilization requires addition of one more (third) electron [136]. The complexes $[Re(S)(CO)_3(bpym)]^-$; S $= n$PrCN, THF; are also stable in solvent S even at room temperature [137]. For other polypyridines, the two-electron reduced solvento species $[Re(S)(CO)_3(N,N)]^-$; S $=$ CH$_3$CN, $n$PrCN; were observed [137] only at low temperature, in the solvent S, in equilibrium with $[Re(CO)_3(N,N)]^-$.

The pentacoordinated complexes $[Re(CO)_3(N,N)]^-$ have highly delocalized electronic structures and assignment of integer oxidation states to Re and N,N seems inappropriate. The complex $[Re(CO)_3(bpy)]^-$ undergoes two further reductions [130] at more negative potentials producing $[Re(CO)_3(bpy)]^{2-}$ and $[Re(CO)_3(bpy)]^{3-}$. Reoxidation of $[Re(CO)_3(bpy)]^-$ is rather complicated [130]; it reacts in with the parent complex $[Re(Cl)(CO)_3(bpy)]$ to give $[Re(S)(CO)_3(bpy)]^{\bullet}$ and $[Re(Cl)(CO)_3(bpy)]^{\bullet-}$. In a parallel reaction, $[Re(CO)_3(bpy)]^-$ dimerizes in the presence of Cl$^-$ to give $[\{Re(CO)_3(bpy)\}_2(\mu\text{-}Cl)]^{3-}$, which is electrochemically reoxidized to $[Re(S)(CO)_3(bpy)]^{\bullet}$.

Both $[Re(S)(CO)_3(bpy)]^{\bullet}$ (for weakly coordinating S) and $[Re(CO)_3(bpy)]^-$ react with CO$_2$. Hence, $[Re(Cl)(CO)_3(bpy)]$ and related complexes are catalysts of electrochemical CO$_2$ reduction to CO and/or formate [131, 135, 138–141].

The redox behavior of complexes of the type $[Re^I(X)(CO)_3(R\text{-}dab)]$; $X =$ halide; is similar to that of their polypyridine counterparts—one oxidation and two successive reductions which ultimately produce pentacoordinated $[Re^I(CO)_3(R\text{-}dab)]^-$ species [134, 136, 142]. There are, however, some remarkable differences between the redox chemistry of Re dab and polypyridine complexes: (i) Oxidation [134] to $[Re^{II}(X)(CO)_3(R\text{-}dab)]^+$ is reversible and less positive (+0.58 V for $X = Cl$, R $= t$Bu), indicating the greater stability of Re$^{II}$-dab complexes and smaller involvement of E ligand orbitals in the HOMO of the Re$^I$ complex. (ii) The rate of dissociation of the axial ligand after the first reduction to $[Re^I(Br)(CO)_3(R\text{-}dab^{\bullet-})]^-$

depends very much on the *N*-substituent R; dissociation of an axial $Br^-$ ligand is much faster for *i*Pr-dab than for aryl-dab ligands (aryl = *p*-tolyl or *p*-anisyl), which are strong $\pi$-acceptors [142].

Interestingly, the first reduction of Re-alkyl complexes [142] $[Re(R)(CO)_3-(iPr\text{-}dab)]$; R = Me, Et, Bz; is reversible, occurring some 0.34 V more negatively than reduction of $[Re(Br)(CO)_3(R\text{-}dab)]$. This might be caused by a mixing between the Re-C $\sigma$ and $\pi^*(dab)$ orbitals. The second reduction of $[Re(R)(CO)_3(iPr\text{-}dab)]$ is irreversible. The same behavior was found for $[Re(R)(CO)_3(bpy)]$.

Manganese complexes $[Mn^I(E)(CO)_3(iPr\text{-}dab)]^{n+}$; E = Br ($n = 0$) or a donor solvent molecule ($n = 1$); are reduced [143] to a five-coordinated complex $[Mn(CO)_3(iPr\text{-}dab)]^-$ in a two-electron process by an ECE mechanism: The first reduction to $[Mn(E)(CO)_3(iPr\text{-}dab)]^{\cdot(n-1)+}$ is followed by a very fast dissociation of the axial ligand E. The $[Mn(CO)_3(iPr\text{-}dab)]^{\cdot}$ product is reduced to $[Mn(CO)_3-(iPr\text{-}dab)]^-$ at a more positive potential than that of the first reduction. Hence, an apparently two-electron reduction wave results. Reduction products are engaged in complicated coupling and reoxidation reactions. Only the complex $[Mn^I(P(OEt)_3)-(CO)_3(iPr\text{-}dab)]^+$ behaves differently [143], because its reduction produces the stable radical $[Mn^I(P(OEt)_3)(CO)_3(iPr\text{-}dab^{\cdot-})]$.

### Iron

$[Fe(N,N)_3]^z$ complexes (N,N = bpy, phen, bpym) form [101, 144–147] a six-membered redox series $z = 3+, 2+, 1+, 0, 1-, 2-$. The 3+/2+ couple is metal-localized—$Fe^{III}/Fe^{II}$. For N,N = bpy, it occurs at +0.64 V. (Three more one-electron oxidations were found [57] at positive potentials in liquid $SO_2$.) Reduction steps are predominantly N,N-localized. They occur at rather negative potentials and are accompanied by partial ligand dissociation. For N,N = bpy, the 2+/1+ and 1+/0 couples were observed at $-1.73$ and $-1.91$ V, respectively. The presence of radical-anionic $N,N^{\cdot-}$ ligand(s) in the reduced Fe complexes ($z = 1+, 0, 1-$) was confirmed by UV–Vis spectroelectrochemistry [144], EPR [69], and resonance Raman [146] spectroscopy. Two more reduction steps were observed [148] for $[Fe(4,7\text{-}Ph_2\text{-}phen)_3]^{2+}$, which is reduced in six one-electron steps to $z = 4-$. On the other hand, only three reductions up to $z = 1-$ were found [144] for $[Fe(tpy)_2]^{2+}$.

The catphen ligand forms tetracoordinated, encaged complex $[Fe^{II}(catphen)_2]^{2+}$ which cannot be oxidized to $Fe^{III}$. In contrast, reversible reductions to $[Fe(catphen)_2]^+$ and $[Fe(catphen)_2]^0$ occur relatively positively, at $-1.17$ and $-1.71$ V, respectively [108]. The locations of these redox changes are unclear, although the positive potentials indicate Fe-reduction in the first step at least. The entwined geometry of the catphen ligand prevents low-valent Fe complexes from dissociation. The topological effect of catphen becomes even more obvious in comparison with sterically hindered, but not entwined, danphen ligand, which does not form complexes with Fe(II) [108].

### Ruthenium and osmium

In aprotic solvents the complexes $[Ru(bpy)_3]^{2+}$ and $[Os(bpy)_3]^{2+}$ and their analogs form extensive redox series [4, 63–67, 101, 149, 150] comprising at least eight members $[M(N,N)_3]^z$: $z = 3+, 2+, 1+, 0, 1-, 2-, 3-, 4-$. The oxidation to $[M^{III}(N,N)_3]^{3+}$ is metal-localized, occurring at = +0.85 and +0.42 V for Ru and Os, respectively (N,N = bpy) [117]. Further oxidations of $[M^{III}(bpy)_3]^{3+}$; M = Ru, Os; were found [56, 57] in liquid $SO_2$ solutions at very positive potentials, extending the redox series by three more members: $z = 6+, 5+, 4+$. They were described as complexes of $Ru^{III}$ and $Os^{IV}$ containing bpy and bpy$^{•+}$ ligands.

All six reductions of $[M(N,N)_3]^{2+}$ are invariably ligand-localized. They occur in two triplets, that is in two groups of three closely-spaced steps, the potential separation between the triplets (0.65 V for bpy) being much larger than that within the triplets, ca 0.2 V [4]. Characterization of reduction products by UV–Vis absorption [63, 66, 67, 150], resonance Raman [68], and EPR spectroscopy [69], and detailed analysis of the reduction pattern [4, 65] (Section 5.3.2) have shown that all reduction steps can be attributed to sequential reductions of individual ligands, which behave as (nearly) electronically isolated [3, 4, 65]. This implies the existence of complexes containing simultaneously N,N ligands in different oxidation states, for example $[Ru^{II}(bpy)_2(bpy^{•-})]^+$ or $[Ru^{II}(bpy^{•-})_2(bpy^{2-})]^{2-}$, etc. Reduction potentials of $[M(N,N)_3]^{2+}$ complexes decrease linearly with decreasing reduction potential of free N,N ligands with a slope of 0.76–1.03, depending on the potential range examined [67, 74, 151]. Besides providing evidence of polypyridine-localization of the reduction steps, this dependence also enables us to tune the reduction potential of $[Ru(N,N)_3]^{2+}$ complexes over a very broad range, from ca −0.45 to −3.0 V for Ru (Section 5.3.3). Most Ru complexes show the first reduction between −1.0 and −2.0 V, however. The first reduction potential is, moreover, shifted positively by some 0.7–0.9 V relative to the free ligand values. This enables even the most negative ligand-localized steps to be observed for ligands like 4,4′-Ph$_2$-bpy [149] or 4,4′-(EtC(O)O)$_2$-bpy [65]. Thus, $[Ru(4,4'-(EtC(O)O)_2-bpy)_3]^{2+}$ shows [65] ten reduction steps, all the way down to $[Ru(4,4'-(EtC(O)O)_2-bpy)_3]^{8-}$! Surprisingly, even the oxidation potential of the $Ru^{II}/Ru^{III}$ couple is linearly dependent on the reduction potential of the ligand, albeit with a smaller slope than that found for the reduction-potential dependence [67]. Together, these two dependencies imply that the oxidation and reduction potentials in a series of $[Ru(N,N)_3]^{2+}$ complexes are linearly dependent on each other, and this has, indeed, been found [29] for many N,N ligands (Sections 5.3.2 and 5.3.3). Overall, the oxidation potential of $[Ru(N,N)_3]^{2+}$ can be tuned by the choice of the N,N ligand from ca +0.15 to +1.8 V, although most Ru complexes have oxidation between +0.7 and +1.4 V. Os complexes are oxidized ca 0.4 V more negatively than their Ru counterparts.

Reduction patterns of heteroleptic $Ru^{II}$ polypyridine complexes $[Ru(A)_n(B)_{3-n}]^{2+}$ and $[Ru(A)(B)(C)]^{2+}$, where A, B and C, are different polypyridine ligands, consist of separate reductions of individual ligands [5, 16, 149, 152, 153]. The localization problem, i.e. the order in which the ligands are reduced, is rather complicated. It depends on the energy of the redox orbitals of individual ligands (i.e. the reduction potentials of the free ligands), on electron–electron repulsion, on interligand interaction energy, and on solvation energy [5, 16]. For example, the strongly electron-accepting 5dcebpy = 5,5′-(EtC(O)O)$_2$-bpy ligand in $[Ru(bpy)_2(5dcebpy)]^{2+}$ is reduced first. Hence, the first two reductions of this complex produce $[Ru(bpy)_2(5dcebpy^{•-})]^+$ and $[Ru(bpy)_2(5dcebpy^{2-})]$, respectively, followed by a more negative set of bpy-localized reductions [16]. A general theory of reduction patterns in heteroleptic complexes $[Ru(A)_n(B)_{3-n}]^{2+}$, which accounts well for experimental data, has been developed [5]. It is also applicable to dinuclear (and, possibly, polynuclear) complexes in which both the terminal and the bridging polypyridine ligands are reducible. The 8-step reduction of $[\{Ru(bpym)_2\}_2(\mu\text{-}bpym)]^{4+}$ is an example studied in much detail [19].

Substituted complexes of the type $[M^{II}(N,N)_2XY]^n$; $M = Ru$, Os, X,Y = halides, $CN^-$, $C_2O_4{}^{2-}$, py, en, $NH_3$, etc.; show, in general, two doublets of N,N-localized reductions [154–156]. The reduction behavior is complicated by loss of an ancillary ligand X upon the second and, more rapidly, the 3rd bpy-localized reduction. The reduction potentials are only slightly X-dependent. On the other hand, the $M^{II}/M^{III}$ oxidation potential is highly dependent on X. Extensive tables of oxidation and the first reduction potentials of these complexes are available [15, 28, 74, 157]. Their values can be predicted by use of electrochemical ligand parameters [15, 28, 157].

Labilization of an ancillary ligand X on reduction indicates the possibility of employing Ru and Os polypyridine complexes as redox catalysts (Section 5.3.5). Indeed, electrocatalysis of $CO_2$ reduction to CO or formate by $[Ru(bpy)_2(CO)_2]^{2+}$, $[Ru(bpy)_2(CO)Cl]^+$, or $[Os(bpy)_2(CO)H]^+$ has been reported [158, 159]. The electrocatalytically most active Ru-bpy species is, however, a film of a –Ru–Ru– bonded polymer $[-\{Ru(bpy)(CO)_2\}_n-]$. It is formed on an electrode surface by reduction of various mono- or bis-bpy $Ru^{II}$ carbonyl or carbonyl–chloro complexes [160–163]. Dissociation of a CO ligand from the polymer upon further bpy-localized reduction is the crucial step which enables $CO_2$ coordination and reduction.

Organometallic complexes $[Ru(E)(E')(CO)_2(N,N)]$ form another interesting class of redox-active complexes [164]. They undergo two one-electron reductions, usually accompanied by loss of an axial ligand E. Five-coordinated species $[Ru(E)(CO)_2(N,N)]^-$ are the ultimate products of the second reduction. The overall mechanism involves coupling reactions between intermediates and electron-transfer catalyzed reduction of the parent complex. This reactivity resembles reduction of $[Re(Cl)(CO)_3(N,N)]$, discussed above. $[Ru(SnPh_3)_2(CO)_2(iPr\text{-}dab)]$ is rather exceptional, because it undergoes reversible reduction to a stable $[Ru(SnPh_3)_2(CO)_2(iPr\text{-}dab^{\bullet-})]^-$ complex [165]. Spectra and DFT calculations of this radical-anion imply significant delocalization of the spin density to axial $SnPh_3$ ligands.

Osmium and ruthenium in high oxidation states (III, IV, or V) form polypyridine (bpy, phen or tpy) hydroxo and oxo complexes, which have a rich redox chemistry [166, 167]. Redox changes are metal-localized and accompanied by reactions of M=O or M–OH bonds. These complexes are active as electrocatalysts of the oxidation of water to $O_2$, or of $Cl^-$ to $Cl_2$, or they can transfer oxygen atoms and oxidize organic substrates like $(CH_3)_2CHOH$ or $C_6H_5\text{-}CH(CH_3)_2$.

## Cobalt

Bipyridine forms a very stable $[Co(bpy)_3]^{3+}$ complex with $Co^{III}$. Its sequential electrochemical reduction produces [22, 23, 147, 168] $[Co(bpy)_3]^{2+}$, $[Co(bpy)_3]^+$, and $[Co(bpy)_3]^-$:

$$[Co(bpy)_3]^{3+} \xleftrightarrow{e^-} [Co(bpy)_3]^{2+} \xleftrightarrow{e^-} [Co(bpy)_3]^+ \xleftrightarrow{2e^-} [Co(bpy)_3]^- \quad (5)$$

The corresponding redox potentials are: $-0.15$, $-1.38$, and $-2.02$ V in acetonitrile [147, 168]. Identical electrochemical behavior was found for cobalt complexes of phen, dmb, and 4,4'-Ph$_2$-bpy [147].

The complexes [Co(bpy)$_3$]$^+$ and [Co(bpy)$_3$]$^-$ undergo partial dissociation of a bpy ligand, so fully reversible reduction is observed only in the presence on an excess of free bpy. The two-electron reduction to the anion was observed only in CH$_3$CN, whereas the monocation is stable even in water. Interestingly, the neutral [Co(bpy)$_3$]$^0$ complex was prepared [169] chemically in THF, although it was not detected electrochemically. [Co(bpy)$_3$]$^+$ is a high-spin d$^8$ Co$^I$ complex, stabilized by a Co $\rightarrow$ bpy $\pi$-back-donation [170]. Besides electrochemical and chemical reduction, e.g. by BH$_4^-$, [Co(bpy)$_3$]$^+$ can be produced by pulse radiolysis [171] or flash photolysis [172, 173]. Release of a bpy ligand from reduced complexes makes cobalt–bipyridine complexes active redox catalysts. Substrates are activated by coordination to low-valent Co$^I$ and undergo chemical transformation with concomitant regeneration of [Co(bpy)$_3$]$^{2+}$. Besides the first reported [24] reduction of aromatic nitro compounds by BH$_4^-$ catalyzed by [Co(bpy)$_3$]$^{2+}$, it was found that cobalt–bipyridine complexes catalyze reduction of CO$_2$ to CO or of H$^+$ to H$_2$ [30, 31, 44]. These reactions involve coordination of CO$_2$ or oxidative addition of H$^+$ to the Co$^I$ center, respectively [174]. [Co(bpy)$_3$]$^+$ also mediates reduction of PhC≡CH to PhCH=CH$_2$ [175].

The mechanism and localization of reduction of the Co$^I$ complex are subtly dependent on the character of diimine ligand [147]. Whereas a two-electron reduction to [Co(N,N)$_3$]$^-$ has been observed [147, 168] for N,N = bpy, phen, 4,4'-Me$_2$-bpy and 4,4'-Ph$_2$-bpy, for the [Co(tpy)$_2$]$^+$ complex [176] a one-electron reduction to a formally Co$^0$ complex, [Co(tpy)$_2$], is followed by another reduction to [Co(tpy)$_2$]$^-$. The 1+/0 and 0/1− couples are separated by 0.36 V.

Sterically hindered danphen and entwined catphen ligands form [108] tetracoordinated, pseudotetrahedral complexes [Co$^{II}$(danphen)$_2$]$^{2+}$ and [Co$^{II}$(catphen)$_2$]$^{2+}$ which cannot be oxidized to Co$^{III}$. The reduction mechanism is different from that of [Co(bpy)$_3$]$^{2+}$ or [Co(phen)$_3$]$^{2+}$ (Eq. 5), because it consists [108] of two successive one-electron steps producing corresponding Co$^I$ and Co$^0$ complexes at $-0.99$ and $-1.67$ V for [Co$^{II}$(danphen)$_2$]$^{2+}$ and at $-0.99$ and $-1.71$ V for [Co$^{II}$(catphen)$_2$]$^{2+}$.

## Rhodium

The stable [Rh(bpy)$_3$]$^{3+}$ complex undergoes reduction to (formally) Rh$^{II}$, Rh$^I$ and Rh$^0$ species [177–180]. This redox chemistry is complicated by substitutional lability of the reduced species and their high reactivity, e.g. with H$^+$ [181]. In general, [Rh(bpy)$_3$]$^{3+}$ is reduced in a series of one-electron steps between $-1.2$ and $-2.1$ V. The complexes [Rh(bpy)$_3$]$^{2+}$, [Rh(bpy)$_2$]$^+$, [Rh(bpy)$_2$], and [Rh(bpy)$_2$]$^-$ are produced. Reduction of [Rh(bpy)$_2$Cl$_2$]$^+$ leads to the same species via Cl$^-$ dissociation. The mechanism of reduction of Rh(III) phenanthroline complexes is similar, although the ligand-dissociation steps are slower. Reduction to [Rh(bpy)$_3$]$^{2+}$ seems to be predominantly bpy-localized, although the possibility of relatively strong Rh-bpy $\pi$-delocalization cannot be excluded [180]. The possibility of a redox equilibrium between [Rh$^{III}$(bpy)$_2$(bpy$^{•-}$)]$^{2+}$ and [Rh$^{II}$(bpy)$_3$]$^{2+}$ has been proposed [180] to account for the solution reactivity of $[Rh(bpy)_3]^{2+}$. The complexes $[Rh(bpy)_2]$ and $[Rh(phen)_2]$ have been characterized by EPR as distorted-planar, electronically highly delocalized systems [179]. The substitutional lability of reduced Rh–polypyridine complexes [180] suggests that they can be active redox catalysts; although this role has been explored only very little, electrocatalytic reduction of $CO_2$ has been reported [30, 182].

An organometallic complex $[(C_5Me_5)Rh(bpy)(Cl)]^+$ is an efficient catalyst of electrochemical $CO_2$ and $H^+$ reduction [183–185]. Its redox behavior is similar to that of the analogous Ir complex (see below) and of $[(C_6Me_6)M(bpy)(Cl)]^+$; M = Ru, Os [184].

**Iridium**

Similarly to polypyridine complexes of other $d^6$ metals, Ir(III)–bipyridine [106, 186] and phenanthroline [186, 187] complexes form extensive ligand-based redox series. The $[Ir(bpy)_3]^z$ series comprises eight members related by one-electron steps [106]. $[Ir(bpy)_3]^{3+}$ is reduced in six successive one-electron steps which occur in two groups of closely spaced three reductions spanning the range from $-1.22$ to $-2.76$ V. It also has an one-electron irreversible oxidation at ca $+1.7$ V, presumably to $[Ir(bpy)_3]^{4+}$. The substituted complex $[Ir(bpy)_2(Cl)_2]^+$ has an irreversible oxidation and four one-electron reductions, presumably ligand-based. The reduction mechanism is complicated by dissociation of $Cl^-$ ligand(s) on electron uptake [186].

The organometallic complex $[(C_5Me_5)Ir(bpy)(Cl)]^+$ undergoes an irreversible two-electron reduction with a concomitant loss of the $Cl^-$ ligand [184, 185]. On the basis of spectroscopic data the resulting $[(C_5Me_5)Ir(bpy)]$ was formulated [185] as an $Ir^{II}(bpy^{·-})$ species. $[(C_5Me_5)Ir(bpy)]$ can be protonated to yield photoactive $[(C_5Me_5)Ir(bpy)(H)]^+$. $[(C_5Me_5)Ir(bpy)]$ is also further reduced at $-3.29$ V to $[(C_5Me_5)Ir^I(bpy^{2-})]^-$. The redox chemistry of $[(C_5Me_5)Ir(bpy)(Cl)]^+$ is relevant to electrocatalysis of $CO_2$ reduction [183] and photocatalysis of the water-gas-shift reaction [44, 188–190]. Further applications of this complex and of its Rh analog in electrocatalysis can be expected.

**Nickel**

Two redox series of Ni–bpy complexes are known [191]: $[Ni(bpy)_3]^z$ and $[Ni(bpy)_2]^z$. The complexes $[Ni(bpy)_3]^{2+}$ and $[Ni(bpy)_2]$ are the respective entries to the Ni–bpy chemistry, the latter complex having been prepared by a metal-vapor synthesis [191]. The electrochemical behavior (including redox potentials) is almost identical for both series. In $CH_3CN$ both complexes form a four-membered redox series $z = 3+, 2+, 0, 1-$. $Ni^{II}/Ni^{III}$ oxidation is rather positive, $+1.3$ V Further oxidations of $[Ni(bpy)_3]^{3+}$ to $[Ni^{IV}(bpy)_3]^{4+}$, $[Ni^{IV}(bpy^{·+})(bpy)_2]^{5+}$, and, finally, to the extremely unstable $[Ni^{IV}(bpy^{·+})_2(bpy)]^{6+}$ were observed [58] in liquid $SO_2$. The $[Ni^{IV}(bpy)_3]^{4+}$ complex seems to be in a temperature-dependent equilibrium with its electronic isomer $[Ni^{III}(bpy^{·+})(bpy)_2]^{4+}$.

Reduction of $Ni^{II}$ complexes $[Ni(bpy)_3]^{2+}$ and $[Ni(bpy)_2]^{2+}$ occurs in a two-electron step [27, 191], at $-1.64$ V producing $[Ni(bpy)_3]$ and $[Ni(bpy)_2]$, respectively. These complexes are further reduced at more negative potentials to the corresponding anions. $[Ni(bpy)_3]$, $[Ni(bpy)_3]^-$, and $[Ni(bpy)_2]^-$ are prone to dissociation of a bpy ligand. $[Ni(bpy)_2]^-$ is reduced by one electron to Ni metal and 2bpy$^{\cdot -}$. In parallel, $[Ni(bpy)(CH_3CN)]^-$ is formed, which is further reduced to $[Ni(bpy)(CH_3CN)]^{2-}$. No direct information is available on the localization of these redox changes. Presumably, all anionic Ni complexes contain reduced bpy ligands [191]. Reduced Ni–bpy complexes undergo oxidative addition of organic halides RX. This reaction is the basis of coupling of R groups electrocatalyzed [192] by $[Ni(X)_2(bpy)]$.

The reduction mechanism changes for complexes containing sterically hindered danphen or catphen ligands. The 4-coordinated species $[Ni(danphen)_2]^{2+}$ and $[Ni(catphen)_2]^{2+}$ are reduced by one electron to the corresponding Ni$^I$ complexes [108]. The distorted tetrahedral coordination required by these ligands stabilizes Ni$^I$ complexes both thermodynamically (reduction occurs at ca $-0.57$ V) and kinetically towards ligand loss and, for $[Ni(catphen)_2]^{2+}$, even towards reoxidation by $O_2$ [108]. $[Ni(catphen)_2]^{2+}$, on the other hand, cannot be oxidized to a Ni$^{III}$ complex.

**Platinum**

The $[Pt(bpy)_2]^{2+}$ complex [193] is reduced in a sequence of three reversible one-electron steps in the range between $-1.14$ and $-2.07$ V, demonstrating the existence of a redox series $[Pt(bpy)_2]^z$; $z = 2+, 1+, 0, 1-$. The first reduction product $[Pt(bpy)_2]^+$ is very reactive. Depending on reaction conditions, it dimerizes to give $[Pt_2(bpy)_3]^{2+}$ with a highly unusual structural unit: a Pt–Pt bond supported by a bridging bpy ligand [193]. Alternatively, electrically-conductive crystals of stacked Pt(I) complex $\{[Pt(bpy)_2]_n\}_n$ can be grown electrolytically [194]. Complexes of the type $[Pt^{II}(bpy)(L)_2]^n$; $n = 2$, L = pyridine, $NH_3$, $PMe_3$, and $n = 0$, L = aryl, Me, $Cl^-$, $CN^-$, etc.; generally undergo two successive one-electron reductions [195–200]. The first step is localized on the bpy ligand, producing $[Pt^{II}(bpy^{\cdot -})(L)_2]^{(n-1)}$. The redox potential of this step and the extent of Pt–bpy electronic delocalization depend on the ancillary ligand L, because of strong interaction between the ligands mediated by the Pt atom. The second reduction is Pt-localized. It leads to $[Pt^I(bpy^{\cdot -})(L)_2]^{(z-1)}$, except for L = mesityl, for which the formula $[Pt^{II}(bpy^{2-})(mes)_2]^{2-}$ has been suggested [197]. Oxidation of $[Pt^{II}(bpy)(L)_2]$ is generally irreversible, again with the exception of $[Pt^{II}(bpy)(mes)_2]$ for which the steric protection of the Pt atom enables reversible oxidation to Pt$^{III}$ [197]. The influence of steric strain on redox behavior has been investigated for the series $[Pt^{II}(R-phen)(X)_2]$; X = halide, R-phen = phenanthroline derivatives [201]. Reversible one-electron reduction was found only for complexes with undistorted planar geometry.

In another electrochemically and photochemically important group of Pt–polypyridine complexes [202, 203], an electron-accepting polypyridine ligand is combined with an electron-donating dithiolate dianion or dithiocarbamate anion: $[Pt^{II}(S,S)(N,N)]^n$; $n = 0$ or $1+$, respectively. These complexes are oxidized in an irreversible or partly reversible one-electron step and reduced in two successive, reversible one-electron steps between ca $-1.5$ and $-2.4$ V. Presumably, both reductions are N,N-localized.

**Copper**

The $Cu^I$ complexes $[Cu(bpy)_2]^+$ and $[Cu(phen)_2]^+$ undergo electrochemically quasi-reversible or irreversible (phen) oxidation ($-0.21$ V, $-0.35$ V, respectively) to the corresponding $Cu^{II}$ species. The reduction is accompanied by decomposition to Cu metal. This all changes when sterically hindered ligands 6,6'-Me$_2$-bpy, 2,9-Me$_2$-phen, 2,9-Ph$_2$-phen, danphen, or catphen are used [108, 204]. These ligands favor the tetrahedral geometry of $Cu^I$ but inhibit the octahedral or planar geometry of $Cu^{II}$. Hence, $Cu^I$ complexes are strongly stabilized and the oxidation potential shifts positively to the range $+0.29$ to $+0.34$ V. Although reduction of $[Cu^I(6,6'-Me_2-bpy)_2]^+$ is still irreversible, the more rigid, encaged geometry of $[Cu^I(N,N)_2]^+$; N,N = 2,9-Me$_2$-phen, 2,9-Ph$_2$-phen, danphen, or catphen; prevents the reduced complex from dissociation and $[Cu^I(N,N)(N,N^{\bullet-})]$ species are formed by reversible reduction which occurs between $-2.03$ and $-2.12$ V. For danphen and catphen even the second reduction to $[Cu^I(N,N^{\bullet-})_2]^-$ was observed ca 0.21 V more negatively [108, 204].

## 5.3 Electron-transfer Properties of Ground-state Polypyridine Complexes: Phenomena

The systematic survey given above enables us to single out general features of the redox chemistry of transition metal polypyridine complexes:

- multiple reductions and oxidations in sequences of reversible one-electron redox steps;
- the existence of redox series, that is of polypyridine complexes of the same composition but with different numbers of electrons (i.e. overall charge);
- for most polypyridine complexes, the redox steps are distinctly localized at the metal atom or a polypyridine ligand;
- for some complexes, reduction labilizes an ancillary ligand or a polypyridine ligand.

### 5.3.1 Localization of Redox Steps

In principle, redox changes of polypyridine complexes can be localized either on the metal atom, on a polypyridine ligand or delocalized over the whole complex molecule. Reduction steps of complexes containing more than one polypyridine ligand are localized on individual ligands instead of being delocalized over the ligand cluster. Hence, in heteroleptic complexes, it is necessary to determine the order in which the ligands are reduced. In this section metal and ligand localization will be discussed. Successive ligand-localized reductions will be treated in Section 5.3.2. The localization of a redox change is usually decided by use of the following experimental criteria:

1) Spectroelectrochemical (EPR, UV–Vis, res. Raman, and to a lesser extent IR) characterization of redox products often reveals features characteristic of reduced polypyridine ligands for ligand-localized reductions or of oxidized metal atoms for metal-centered oxidations. Stretching CO frequencies, obtained by IR spectroelectrochemistry, are an especially useful marker of a metal oxidation state in carbonyl–polypyridine complexes.

2) Potentials of polypyridine-localized redox couples depend linearly on the reduction potential of free polypyridines with a slope close to unity, when measured for homologous series of structurally related complexes [26, 67, 74, 101, 151, 205, 206]. This argument should be used with caution because even the potentials of metal-centered oxidation depends on the free polypyridine reduction potential, albeit with a significantly lower slope [29, 67] (Sections 5.3.2 and 5.3.3). Nevertheless, correlations of redox potentials in a series of $[M(N,N)_3]^z$ complexes, N,N = bpy, 4,4′-Me$_2$-bpy, and 5,5′-Me$_2$-bpy, with those of free ligands were often sufficient to identify N,N-localized reductions [101, 111, 117].

3) The energy of a spectroscopic metal-to-ligand charge-transfer (MLCT) electronic transition (if present) depends linearly on the difference between the oxidation and reduction potentials, if the oxidation and reduction are predominantly metal- and ligand-localized, respectively [6, 9, 74, 111, 117, 149, 152, 153, 205, 207–211]. This criterion can be used in various forms. The proper procedure uses the oxidation and reduction potential difference in series of complexes in which the polypyridine and/or ancillary ligands are varied. However, the oxidation potential alone can be correlated with spectroscopic energies in series of $[M(W)(X)(Y)(Z)(N,N)]$ or $[M(X)(Y)(N,N)_2]$ complexes, in which only the ancillary ligands are varied. If the reduction step is difficult to observe, or its localization is unclear, differences between the complex oxidation potential and the free ligand reduction potential can be correlated with the spectroscopic transition energy to decide on the localization of the oxidation step.

4) Experimentally observed redox patterns can be compared with those calculated by quantum-chemical techniques for a particular localization [5, 212]. This approach is especially useful for assignment of ligand-localized reductions in heteroleptic complexes.

5) The lack of correlations (2) or (3) and the appearance of spectroscopic features that do not resemble those of reduced polypyridine or an oxidized metal atom indicate a delocalized redox change.

Polypyridine complexes of second and third-row transition metals of a $d^6$ electron configuration are chemically stable and readily accessible, enabling the most convenient entry into metal–polypyridine redox chemistry and photochemistry. The corresponding metal atoms ($Mo^0$, $W^0$, $Re^I$, $Ru^{II}$, $Os^{II}$, $Rh^{III}$, $Ir^{III}$, and $Cr^0$ in tetracarbonyls) are much harder to reduce than polypyridine ligands and the $\pi$ delocalization within the corresponding M(N,N) chelate rings is, in most complexes, weak. At the same time, the first oxidation of second- and third-row transition $d^6$ metal atoms always occurs well before oxidation of polypyridine ligands. Consequently, the formal oxidation states of the metal atom and polypyridine ligands are well defined in the parent complexes and in their reduced forms and the first oxidation product. Reductions are predominantly ligand-localized and reduced complexes can best be described as containing a $d^6$ metal atom and reduced polypyridine ligand(s). One-electron oxidized complexes are typical polypyridine complexes of $d^5$ metals (e.g. $Ru^{III}$, $Re^{II}$, etc.). Only the products of most positive oxidations in $SO_2$ possibly contain coordinated polypyridine cations.

Delocalized redox steps are rare among the polypyridine complexes of second- and third-row transition metals. Two-electron reduction of the organometallic species $[(C_5Me_5)M(bpy)(Cl)]^+$; $M = Rh$, $Ir$; and $[(C_6Me_6)M(bpy)(Cl)]^+$; $M = Ru$, $Os$; complexes [184, 185, 213, 214] is rather delocalized. Specifically, reduction of $[(C_5Me_5)Ir(Ar\text{-}dab)(Cl)]^+$ leads to $[(C_5Me_5)Ir(Ar\text{-}dab)]$ which contains a delocalized, aromatic $Ir(dab)$ chelate ring [214]. The second reduction (Eq. 4) of $[Re(E)(CO)_3(N,N)]^{\cdot n}$ to pentacoordinated species $[Re(CO)_3(N,N)]^-$ is also largely delocalized. Few $\alpha$-diimine and polypyridine-type ligands, especially abpy, bpm or Ar-dab favor metal–ligand delocalization, which arises from a relatively strong $\pi$-back-bonding. This is manifested, e.g., in the electrochemistry of $[Ru(abpy)_3]^{2+}$ [107], see Section 5.3.3.

The localization problem (metal or ligand-centered) is more complicated in poly- pyridine complexes of first-row transition metals which are relatively easy to reduce and/or oxidize. Hence, potentials of metal-centered redox couples can occur in the range of polypyridine-localized reductions. Occupied d-orbitals of low-valent metal atoms can lie energetically close to the polypyridine $\pi^*$ orbital, rising the possibility of delocalized redox changes. Because there is no simple rule enabling the predic- tion of the localization of redox steps in polypyridine complexes of first-row tran- sition metals, their electron-transfer reactivity must be discussed individually (Sec- tion 5.2.2), with the aid of a general knowledge of the redox chemistry of the metal atoms and polypyridine ligands involved. Clear-cut metal-centered redox couples are known for polypyridine complexes of Cr, Fe, Co, and Cu: $Cr^{II}/Cr^{III}$, $Fe^{II}/Fe^{III}$, $Co^{II}/Co^{III}$, $Co^{I}/Co^{II}$, $Cu^{I}/Cu^{II}$. Well-defined polypyridine-localized reductions are known for $[Fe(N,N)_3]^{2+}$. Reduction of sterically confined $[Cu(catphen)_2]^+$ and related Cu(I) complexes is also clearly catphen-localized, although reduction of sterically unhindered $[Cu(N,N)_2]^+$ species leads to decomposition and formation of metallic Cu. The most negative reductions of the first-row metal polypyridine complexes seem to be predominantly ligand-localized, producing formally $N,N^{2-}$ complexes. This is, for example, observed for the last three reductions in the $[Cr(bpy)_3]^z$ series, i.e. $z = 0/1-$, $1-/2-$, and $2-/3-$. The localization of redox couples occurring at intermediate potentials, which correspond to electron exchange between the species in the middle of the respective redox series, are hard to assign for first-row transition metals. Very often, these changes in the electron number are delocalized, affecting the metal atom and polypyridine ligands. The use of integer formal oxidation states is then inappropriate. This is, for example, observed for the $[Cr(bpy)_3]^{2+}/[Cr(bpy)_3]^+/[Cr(bpy)_3]^+$ couples, and for the reduction of Ti and V and, presumably, also Mn complexes. Delocalized redox changes are manifested by complicated UV–Vis and EPR spectra of the reduced forms which were, for Ti, V, Cr, and Fe, analyzed with the help of MO calculations [215]. Interesting experi- mental criteria for delocalized structures are also provided by IR spectra of the re- duced bpy complexes measured in the range of bpy intraligand vibrations [113].

## 5.3.2 Redox Patterns

Typical polypyridine complexes of second- or third-row $d^6$ transition metals are characterized by metal-localized oxidation and a series of predominantly polypyridine-localized one-electron reductions. Their redox patterns are schematically shown in Figure 1.

Coordination to a metal atom shifts N,N-localized reductions positively by some 0.4–1 V, relative to the free ligand values. The magnitude of this shifts increases with the positive charge on the metal atom. For example, the first reduction of $[M^{III}(bpy)_3]^{3+}$; M = Rh, Ir; occurs almost 0.5 V more positively than for their Ru and Os analogs $[M^{II}(bpy)_3]^{2+}$ and the first reduction of $[Re(CO)_4(bpy)]^+$ is by 0.23 V more positive than that of isoelectronic $[W(CO)_4(bpy)]$. This positive potential shift is a result of stabilization of the reduced ligand by coordination [4]. Higher positive shifts can occur for polyazine ligands bridging between two metal atoms. The actual magnitude of this effect is a result of bonding and solvation effects [4, 212].

The number of reduction steps depends on the number and type of polypyridine ligands present. In principle, a complex with $m$ polypyridine ligands, each of which carries $k$ redox orbitals, will show a maximum of $2mk$ one-electron reductions, grouped in $2k$ separate sets each with $m$ members [4, 212] (or, more simply, the maximum number of polypyridine-localized reduction steps equals the number or reduction steps of a free ligand multiplied by the number of ligands present). The bpy ligand and most polypyridines have only one low-lying, unoccupied $\pi^*$ orbital, which acts as a redox orbital. Hence, usually $k = 1$ and the number of reduction steps is twice the number of polypyridine ligands, $2m$, see Figure 1. Polypyridines with electron-accepting substituents (e.g. 4,4'-Ph$_2$-bpy, or 4,4'-(COOEt)$_2$-bpy) have two low-lying $\pi^*$ orbitals and, hence, their complexes have more than $2m$ reduction steps, although the full number of $2mk$ steps is usually not observed because the most negative lie outside the experimentally accessible potential range.

Figure 1 shows the origin of individual polypyridine-localized reduction steps of mono-, bis-, and tris-polypyridine complexes. It follows that a mono-polypyridine complex $[M(W)(X)(Y)(Z)(N,N)]$ will have the same number of reduction steps as the free N,N ligand itself. The potentials are shifted positively relative to those of the free ligand, as was discussed above. The first reduction leads to $[M(W)(X)(Y)(Z)(N,N^{\bullet-})]^-$. The actual values of the corresponding reduction potential are a function of the N,N ligand and the metal fragment M(W)(X)(Y)(Z). They can be calculated empirically, by use of electrochemical parameters, for the M(N,N) fragment, ancillary ligands W, X, Y, Z, and the parameter $S_L$ which describes the coupling between the ancillary ligands and the reducible N,N ligand [9, 15]. This coupling has been amply demonstrated experimentally, e.g. by the dependence of the first reduction potential of $[Mo(CO)_{4-n}(PBu_3)_n(bpy)]$ on the number of PBu$_3$ ligands ($n = 0, 2, 4$) [123] or by EPR spectra of $[Re(E)(CO)_3(N,N)]^{\bullet n}$ complexes; E = Cl ($n = 1-$) or phosphine ($n = 0$) [134]. The effects of bpy-reduction on the Cr(CO)$_4$ fragment in $[Cr(CO)_4(bpy)]$, analyzed by IR and EPR spectroelectrochemistry [121] and DFT calculations [122], also point to important interactions between bpy and CO ligands.

The second reduction of a mono-polypyridine complex corresponds to formation

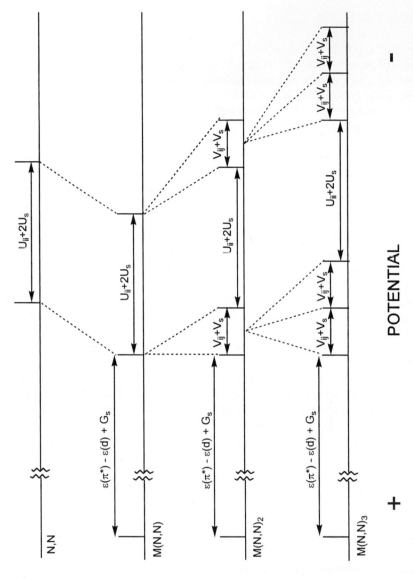

**Figure 1.** Redox patterns of a free polypyridine ligand and typical mono-, bis- and tris-polypyridine complexes [4, 8, 9, 212]. $\varepsilon$ is the orbital energy, $U_{ii}$ and $V_{ij}$ are pair interaction energies between electrons placed on the same and different polypyridine ligands, respectively. $U_s$, $V_s$, and $G_s$ are solvation energy terms. In the first approximation, interaction and solvation terms are assumed to be independent of the number of electrons, that is to be constant along the redox series. In reality, they show small changes. Interaction energies are positive and solvation terms are negative. Hence, solvation diminishes the redox potential differences. Dashed lines show the origin of reduction doublets and triplets in bis- and tris-polypyridine complexes, respectively.

of a dianion $[M(W)(X)(Y)(Z)(N,N^{2-})]^{2-}$. The difference between the first and second reduction potential is equal to the energy of a pair interaction between the two electrons placed on *the same* N,N ligand (i.e. between the two electrons in the $\pi^*$ redox orbital), $U_{ii}$, plus a solvation term [4, 8, 212]. Experimental values of the potential difference (0.6–0.7 V for bpy complexes) are very close to those the free ligands and almost independent on the metal fragment.

The two polypyridine ligands in bis-polypyridine complexes are reduced sequentially, one by one, at different potentials, instead of a single 2-electron step. Hence, the number of reduction steps doubles on going from mono- to bis-polypyridine complexes (Figure 1). Bis-polypyridine complexes are reduced in four one-electron reduction steps, grouped into two doublets [155]. The first two closely-spaced reductions (the first doublet) correspond to $M(N,N)(N,N)/M(N,N)(N,N^{\bullet-})/$ $M(N,N^{\bullet-})(N,N^{\bullet-})$ couples while the reductions comprising the second doublet are due to $M(N,N^{\bullet-})(N,N^{\bullet-})/M(N,N^{\bullet-})(N,N^{2-})/M(N,N^{2-})(N,N^{2-})$ couples. The fact that the two N,N ligands are reduced separately is the consequence of an interligand electronic interaction, due to which the reduction of the second N,N ligand is more difficult if the first one was already reduced. The potential difference between the sequential reductions of the individual N,N ligands (i.e. within the doublet, ca 0.2 V) is then equal to the energy of a pair interaction between two electrons localized on *different* polypyridine ligands $V_{ij}$, plus a solvation term [4, 8, 212]. The potential difference between the doublets (0.6–0.7 V) corresponds to the energy of pair interaction between two electrons on *the same* N,N ligand, $U_{ii}$, plus a solvation term, as was discussed above for mono-polypyridine complexes. This description of the reduction pattern of metal bis-polypyridines is valid both for 6-coordinate species $[M(X)(Y)(N,N)_2]$ and for pseudotetrahedral complexes $[Cu(catphen)_2]$. The magnitude of the difference between the reduction potentials in $[Cu(catphen)_2]$, that is of the interaction energies between the extra electrons on the two phen moieties, is comparable with that in 6-coordinated bis-phen complexes, regardless of the different spatial orientation of the phen units. As for the mono-polypyridine complexes, potentials of ligand-localized reductions in 6-coordinate $[M(X)(Y)(N,N)_2]$ are slightly dependent on the nature of ancillary ligands X, Y and the metal atom M [216]. The magnitude of this effect can be estimated using Lever electrochemical parameters [9, 15].

The reduction pattern of tris-polypyridine complexes (Figure 1) can be easily deduced by considering interactions between the three N,N ligands [4, 5, 8, 149, 212]. They lead to sequential reduction of the individual N,N ligands which occurs in two triplets. The first triplet corresponds to reductions up to $[M(N,N^{\bullet-})_3]^z$ while the next three reduction steps, grouped in the second triplet, lead all the way down to $[M(N,N^{2-})_3]^{z-3}$. The potential difference between successive reductions within the triplets corresponds to the energy of interaction between electrons on different ligands $V_{ij}$, plus a solvation term. Corresponding values are typically between 0.13 and 0.25 V. Larger values (e.g. 0.36 found [107, 217] for $[Ru(abpy)_3]^{2+}$) indicate that the reductions can no longer be regarded as fully localized, most probably due to a metal-ligand $\pi$ delocalization, which can mediate an interligand interaction. Even in well-behaved systems, the splitting between the most negative reductions within the second triplet is sometimes larger (ca. 0.3 V) than within the first triplet

[4, 5]. Apparently, this is caused by a change of the interligand interaction energy with the total number of electrons. The potential difference between the triplets is equal to the usual energy of interaction between the two electrons placed at *the same* N,N ligand, $U_{ii}$, plus a solvation term [4, 212]. Corresponding values are very similar for free polypyridines and their tris-complexes.

Heteroleptic complexes which contain chemically different polypyridine ligands give rise to complicated reduction patterns [5, 16, 149, 153]. The ligand with the most stable (lowest lying) $\pi^*$ orbital gets reduced first. The following reduction sequence depends on the relative magnitude of the differences between the redox orbital energies (i.e. reduction potentials of the corresponding free ligands) and various pair interaction energies involved [5]. Hence, the ligands need not to be reduced in the order of the increasing energy of their redox orbital, that is in the order the decreasing reduction potentials of their free forms. It can even happen that several different electron configurations of the complex molecule, which differ in electron localization on different ligands, have very similar energies. Such complexes may exist as equilibrium mixtures of 'electronic isomers' [5].

The same approach to the interpretation and prediction of reduction patterns is valid also for polynuclear polypyridine complexes in which metal atoms are bridged by reducible polypyridine or polyazine ligand(s). Polyazines are generally reduced at more positive potentials than polypyridines and, moreover, reduction of a bridging ligand can be subjected to a larger positive shift than that of the terminal ligands. Therefore, bridging ligands are usually reduced first, before the terminal polypyridines. This is, for example, the case of the dinuclear complex $[(bpy)_2Ru(\mu\text{-bpym})\text{-}Ru(bpy)_2]^{4+}$ in which the bpym ligand is reduced to $\mu\text{-bpym}^{\cdot-}$ and $\mu\text{-bpym}^{2-}$ before the first reduction of the bpy ligands [19]. Because of a very small interaction energy between the electrons on the two bpy ligands coordinated to different metal atoms, they are reduced at very similar or even identical potentials [19]. This type of redox pattern (viz. bridging ligands reduced first, terminal ligands bound to different metal atoms next, at similar potentials) was found [53, 84, 218] also for large supermolecular polypyridine complexes, including dendrimers (Section 5.3.6.).

An interpretation [4, 5, 8, 149, 212] of reduction sequences of $[M(N,N)_3]^z$ complexes requires to use redox orbitals localized on individual N,N ligands, instead of the $a_2+e$ set of delocalized molecular orbitals of the $(N,N)_3$ ligand cluster, pertinent to the parent, unreduced complexes of a $D_3$ symmetry. The $[M(N,N)_3]^z$ complexes are then treated as a set of weakly interacting redox centers, that is individual N,N ligands. Qualitatively, this situation can be described using the redox orbital concept [1, 4, 6, 8] and assuming [3, 65, 69, 219] that the redox orbital is 'spatially isolated' at a single ligand. A quantum chemical model of sequential ligand-localized reductions, which includes solvation, has been developed for $[M(N,N)_3]^z$ as well as for heteroleptic complexes $[M(A)_{3-m}(B)_m]^z$, where A and B represent different reducible polypyridine ligands [5, 212]. It uses a unitary transformation of delocalized molecular orbitals into a set of localized redox orbitals and a specially developed form of the Hubbard Hamiltonian. This model allows us to express individual redox potentials and their differences in terms of redox orbital energies, inter- and intra-ligand electron pair interaction energies and solvation terms. A very good agreement between redox patterns predicted by this model and those observed experimentally has been found. It follows that the localization of redox changes on individual N,N ligands, as opposed to a delocalization over a $(N,N)_3$ ligand cluster, is driven by energetically favorable solvation and structural reorganization of the reduced ligand that accompany localization of an extra electron at a single ligand [212].

Most polypyridine complexes of second- and third-row transition metals also display a predominantly metal-localized oxidation at positive potentials which are chemically either reversible or partly reversible. Further one-electron oxidations often occur at more positive potentials in liquid $SO_2$ [57]. The first oxidation potential depends on the metal atom (for example, Ru > Os), the ancillary ligands in $[M(W)(X)(Y)(Z)(N,N)]$ or $[M(X)(Y)(N,N)_2]$ and, also, on the structure of the polypyridine ligand. Empirically, oxidation potentials can be calculated using additive Lever electrochemical parameters which quantify the influence of the metal atom and individual ligands on metal-centered redox couples [9, 157, 220].

The dependence of the first oxidation potential on the nature of the polypyridine ligand is intriguing. For example, a linear correlation between the potential of the $Ru^{II}/Ru^{III}$ redox couple and the reduction potential of the free N,N ligand has been established for a series of $[Ru(N,N)_3]^{2+}$ complexes [67]. The slope of this correlation is smaller than that of the correlation between the reduction potentials of the complex and the ligand. However, the very existence of these two correlations implies an existence of a linear correlation between the first reduction and oxidation potentials in a series of $[Ru(N,N)_3]^{2+}$ complexes. Indeed, such a correlation has been found [29] for a large (52) set of these compounds with a slope of 1.16. It follows that the increase of the polypyridine $\pi^*$ LUMO energy, which is responsible for the negative shift of the reduction potential, is accompanied by changes in other structural factors which stabilize the $Ru^{III}$ oxidation state relative to $Ru^{II}$ [29]. These include: (i) increase of $\sigma$-basicity [205, 210, 217], (ii) increase in the energy of occupied $\pi$ polypyridine orbitals resulting in a $\pi$-donation to $Ru^{III}$ [29], and, (iii) decrease of $\pi$-back donation from $Ru^{II}$.

Linear correlations between the first reduction and oxidation potentials were also found in series of $[M^{II}(bpy)_2(X)(Y)]^n$; M = Ru, Os; where the ancillary ligands are varied [152, 216], and for heteroleptic tris-polypyridine complexes [210]. The existence of such correlation elegantly follows from the theory of electrochemical parameters [9]. Since the ancillary ligand nature affects the $M^{II}/M^{III}$ potential much more than the $bpy^{\cdot-}/bpy$ potential, the slope of this correlation is only about 0.25, i.e. much smaller than that found for the $[Ru(N,N)_3]^{2+}$ series, 1.16 [29].

The difference between the first reduction and oxidation potentials is an interesting parameter [6, 9, 207, 208]. Approximately, it corresponds to the difference between the energies of the metal-localized HOMO and polypyridine-localized $\pi^*$ LUMO orbitals, plus a solvation term, Figure 1. This is the basis for the above-mentioned correlation with the spectroscopic MLCT transition energy (Section 5.4.3, Eqs 16 and 17) [6, 9, 207, 208]. The existence of a linear correlation between the oxidation and reduction potentials of $[Ru(N,N)_3]^{2+}$ complexes with a near-unity slope implies, however, that their difference and, hence, the MLCT energy are only little dependent on the nature of the polypyridine ligand.

Redox patterns of first-row transition metal polypyridine complexes are very much dependent on the metal atom, see Section 5.2.2. Redox behavior of $[Fe(N,N)_3]^{2+}$ complexes is similar to that of its Ru and Os counterparts. Tris-polypyridine complexes of Ti, V, Cr, Mn, and Fe show sequences of one-electron reductions and oxidations. However, there is no regular grouping of the redox steps into triplets because of delocalization of the redox changes whose extent, moreover, changes along redox series. Hence, delocalized redox orbitals, whose very nature changes with the number of electron added, have to be considered. $Co^I$ and $Ni^I$ polypyridine complexes are exceptional since they are reduced in two-electron steps.

### 5.3.3  Control of Redox Potentials by the Polypyridine Ligand Structure

Variations in the structure of a polypyridine ligand provide a means of control of the number of reduction steps of polypyridine complexes, of their reduction and oxidation potentials, and of the extent of $\pi$-delocalization between the metal atom and a polypyridine ligand. These aspects were discussed above in Sections 5.3.1 and 5.3.2. Here, the polypyridine influence on redox potentials will be discussed in more detail.

Potentials of ligand-localized reductions of polypyridine complexes depend on the ligand structure in the same way as free-ligand reduction potentials [26, 67, 74, 101, 151, 205]. Similar, but slightly less prominent correlation also exists between the potential of the first metal-localized oxidation and the free polypyridine reduction potential, see Section 5.3.2 [29, 67]. The first reduction steps of the generic bpy, phen and tpy ligands occur at similar potentials, between $-2.4$ and $-2.6$ V for the free ligands. The first reduction of the $[Ru(bpy)_3]^{2+}$, $[Ru(phen)_3]^{2+}$, and $[Ru(tpy)_2]^{2+}$ complexes occur at $-1.76$, $-1.74$, and $-1.74$ V, respectively. The reduction potentials can be controlled either by introducing electron-accepting or donating substituents at pyridine rings of the generic polypyridine ligands or by introducing more N atoms, that is by going from polypyridines to polyazines. (Note that taphen and dppz complexes deviate from the correlations between ligand and complex reduction potentials [74].)

The first approach, substituting the H atoms on polypyridine ligands by electron-accepting or donating groups, affects reduction potentials without changing significantly the extent of a metal/polypyridine $\pi$-delocalization. A great variety of substituents were used to derivatize the bpy ligand, ranging from electron acceptors like $CF_3$, $C(O)OEt$, $C(O)Et$, Ph, Cl, Br, to electron donors Me, OMe, $NH_2$, or $NR_2$. In most cases, the bpy ligand is substituted symmetrically, either at $4,4'$ or $5,5'$ positions. The effect of substituents at $5,5'$ positions is by $0.06$–$0.25$ V larger than at the positions $4,4'$ [111, 221]. Redox potentials of bpy complexes can be tuned by substituents over a range of almost 1 V broad [74]. For example, $[Ru(5,5'-(CF_3)_2-bpy)_3]^{2+}$ is reduced at $-1.16$ V while the reduction of $[Ru(4,4',5,5'-(Me)_4-bpy)_3]^{2+}$ occurs at $-1.96$ V. The same complexes are oxidized at $+1.30$ and $+0.67$ V, respectively [74, 222]. Oxidation of the complex $[Ru(4,4'-(NMe_2)_2-bpy)_3]^{2+}$ is even more negative, $-0.24$ V [223]. Similar substituent-dependence occurs also for redox potentials of phen and tpy complexes [74, 224]. In general, the dependence of redox potentials on ligand substituents is well predictable. Both oxidation and reduction potentials increase as the substituents become more electron accepting and decrease (become more negative) for electron-donating substituents. Linear correlations between redox potentials of polypyridine complexes and substituent Hammett parameters have been found [200, 206, 223–226].

Annealing two benzene rings to the bpy framework leads either to biquinoline or isomeric *iso*-biquinoline. Their respective complexes differ significantly in their redox behavior [74, 153, 227, 228]: The first reduction of $[Ru(biq)_3]^{2+}$ is by 0.77 V more positive than that of $[Ru(i\text{-}biq)_3]^{2+}$, in accordance with the trend in the reduction potentials of the free ligands. The difference between the oxidation and reduction potential for $[Ru(biq)_3]^{2+}$ is unusually small, 2.20 V, relative to $[Ru(i\text{-}biq)_3]^{2+}$, 2.63 V, and most other Ru-polypyridines.

Annealing one or two pyridine rings to the bpy framework forms the pynapy and dinapy ligands, respectively. They give rise to complexes that are relatively easy to oxidize and reduce [229]. The oxidation and the first reduction of $[Ru(pynapy)_3]^{2+}$ occur at +0.69 and −1.38 V, respectively. The corresponding values for $[Ru(dinapy)_3]^{2+}$ are +0.58 and −1.22 V. The differences between the oxidation and reduction potentials of these complexes are unusually low: 2.07 and 1.8 V, respectively. At the same time, differences between the first and second reduction potentials are rather high: about 0.24 V for both complexes. This behavior can be, as in the case of $[Ru(biq)_3]^{2+}$ complexes, the result of a steric strain.

Introducing more N atoms into the framework of a bpy or phen ligand, that is on going from polypyridines to polyazines, changes not only redox potentials but also the localization of the $\pi^*$ LUMO on N donor atoms and the $\sigma$-basicity [119, 217]. Generally, polyazine complexes are oxidized and reduced at more positive potentials than their polypyridine counterparts. This is demonstrated by the values for several representative complexes collected in Table 1. These data also demonstrate the parallel trend of the oxidation and reduction potential as a function of the ligand (see Section 5.3.2) which keeps the difference between them nearly constant, the only exceptions being bpm and taphen. The abpy ligand is clearly different: the

**Table 1.** The first oxidation ($E_{1/2}{}^{ox}$) and reduction ($E_{1/2}{}^{red}$) potentials of selected $[Ru(N,N)_3]^{2+}$ complexes. $\Delta E_{1/2}{}^{red1,2}$ denotes the difference between the first and second reduction potential. All values are relative to the Fc/Fc$^+$ couple.

| Complex | $E_{1/2}{}^{ox}$ | $E_{1/2}{}^{red}$ | $\Delta E_{1/2}{}^{red1,2}$ | $E_{1/2}{}^{ox} - E_{1/2}{}^{red}$ | Ref. |
|---|---|---|---|---|---|
| bpy | +0.85 | −1.76 | 0.19 | 2.61 | 74 |
| bpdz | +1.19 | −1.39 | 0.25 | 2.58 | 217 |
| bpym | +1.30 | −1.30 | 0.17 | 2.60 | 217 |
| bpz | +1.59 | −1.07 | 0.19 | 2.66 | 217 |
| bpm | +1.36 | −0.97 | 0.15 | 2.33 | 217 |
| phen | +0.88 | −1.74 | 0.11 | 2.62 | 205 |
| tap | +1.55 | −1.14 | 0.13 | 2.69 | 211 |
| taphen | +1.21 | −1.09 | – | 2.30 | 288 |
| abpy | – | −0.41 | 0.36 | – | 107, 217 |

reduction potential of $[Ru(abpy)_3]^{2+}$ is very positive and the splitting between the first and second potential is rather large. These effects signal a much larger Ru-abpy $\pi$-delocalization as compared with other complexes. An exceptionally strong electron-acceptor ability of the abpy ligand was noted [136] also for $[Re(L)(CO)_3(abpy)]^+$; L = THF or $PPh_3$.

A comparison of electrochemistry of some of the free polypyridine ligands and their $Ru^{II}$, $Re^{I}$ and $Mo^0$ complexes shows [97, 123, 134, 136, 137, 142, 230, 231] that, all other factors being constant, the potential of the first N,N-localized reduction increases in the order:

$$\text{phen, bpy} < \text{R-PyCa} < \text{2,3-dpp} < \text{R-dab} \approx \text{bpym} < \text{bpdz} < \text{bpz}$$

$$\approx \text{dpq} < \text{tap} < \text{bpm} < \text{Ar-dab} \ll \text{abpy} \qquad (6)$$

$$(\text{R, Ar} = \text{alkyl, aryl})$$

This implies that the $\pi^*$ LUMO energy decreases in the same order, that is from bpy to Ar-dab and abpy. The Re $\rightarrow$ diimine $\pi$ back bonding strengthens in a similar order:

$$\text{phen, bpy} < \text{R-PyCa} < \text{2,3-dpp} < \text{bpym}$$

$$\leq \text{dpq} \leq \text{R-dab} < \text{Ar-dab} < \text{abpy} \qquad (7)$$

R-dab ligands are somewhat stronger $\pi$-acceptors than aromatic diimines of comparable $\pi^*$ orbital energy like dpp, bpym or dpq. This is caused by a larger contribution from the $N-2p_z$ orbitals to the $\pi^*$ orbital of the dab ligands, which, in turn, results in a larger $d\pi - \pi^*$ overlap. The very strong $\pi$ acceptor ability of Ar-dab should be noted.

### 5.3.4 Electron-transfer Kinetics

Given the importance and great variety of electron-transfer reactions of polypyridine complexes, systematic kinetic studies are surprisingly scarce and only few kinetic data are available. Some representative rate constant values for homogeneous self-exchange redox reactions were reported:

bpy/$K^+$bpy$^{\bullet-}$: $3.2 \times 10^6$ M s$^{-1}$ in 1,2-dimethoxyethane [232]
$[Co^{III}(bpy)_3]^{3+}/[Co^{II}(bpy)_3]^{2+}$: 18 M s$^{-1}$ in water [170]
$[Co^{II}(bpy)_3]^{2+}/[Co^{I}(bpy)_3]^{+}$: $10^9$ M s$^{-1}$ in water [170]
$[Cr^{II}(bpy)_3]^{+}/[Cr^{I}(bpy)_3]$: $1.5 \times 10^9$ M s$^{-1}$ in DMF [233]
$[Rh^{III}(4,4'-Me_2bpy)_3]^{3+}/[Rh^{II}(4,4'-Me_2bpy)_3]^{2+}$: $\geq 10^9$ M s$^{-1}$ in water or $CH_3CN$ [180]
$[Fe^{III}(bpy)_3]^{3+}/[Fe^{II}(bpy)_3]^{2+}$: $>10^8$ M s$^{-1}$ in water [234]

The apparent slowness of the bpy/$K^+$bpy$^{\cdot-}$ rate could be caused by a strong ion-pairing with the $K^+$ ion upon reduction.

The spectacular $5 \times 10^7$-fold acceleration on going from the $Co^{III}/Co^{II}$ to $Co^{II}/Co^I$ couple can easily be accounted for by different inner-sphere reorganization energies, according to the Marcus theory. Electron uptake by a $t_{2g}^6$-$Co^{III}$ complex is accompanied by a large structural change, since two electrons are placed in a $\sigma$-antibonding $e_g$ orbital of the $t_{2g}^5e_g^2$-$Co^{II}$ configuration. Accordingly, the average Co–N distance elongates [170] from 1.93 to 2.13 Å. On the other hand the extra electron in $[Co^I(bpy)_3]^+$ is placed in a $t_{2g}$ orbital and its structural effects are largely compensated for by $\pi$-back bonding to bpy ligands. Consequently, the Co–N bond distance [170] is nearly the same (2.11 Å) as in the $Co^{II}$ complex, reorganization energy is small and electron transfer fast. The solution kinetics of electron-transfer reaction involving $[Co^I(bpy)_3]^+$ are complicated by concurrent bpy-dissociation equilibria [171, 172]. They give rise to various cross-reactions between $[Co^I(bpy)_m]^+$ and $[Co^{II}(bpy)_3]^{2+}$ complexes, which are all very fast $>10^8$ M s$^{-1}$.

Notably, the self-exchange electron transfer $[Co^{III}(bpy)_3]^{3+}/[Co^{II}(bpy)_3]^{2+}$ is some 4–8 orders of magnitude faster than that between $Co^{III}$ and $Co^{II}$ amine complexes of a comparable size [235]. This demonstrates that bpy ligands are better able to compensate for redox-induced structural changes and to distribute the charge-change over a large molecular surface. Moreover, aromatic bpy ligands provide much better possibility for electronic coupling between the reactants. This can make the $[Co^{III}(bpy)_3]^{3+}/[Co^{II}(bpy)_3]^{2+}$ electron exchange partly adiabatic.

The much faster electron self-exchange rate observed for Rh than for Co reflects different localization of the corresponding redox couples, at the bpy ligand and the Co metal, respectively.

The heterogeneous rate constant [236] of electrochemical reduction of $[Co^{III}(bpy)_3]^{3+}$ to $[Co^{II}(bpy)_3]^{2+}$ is relatively slow, about 0.1 cm s$^{-1}$ in $CH_3CN$ or $CH_2Cl_2$. Detailed studies of solvent and pressure effects [236, 237] have revealed that the rate of heterogeneous electron transfer is controlled by solvent dynamics. This implies that the electron transfer is adiabatic.

Electrochemical (heterogeneous) rate constants $k^\circ$ have been reported for a large set of $[M(bpy)_3]^z/[M(bpy)_3]^{z-1}$ redox couples [238], hereinafter denoted as $M^z/M^{z-1}$. According to the values obtained, the complexes fall into two categories:

1) Relatively fast reacting ($0.6 < k^\circ < 1.3$ cm s$^{-1}$) complexes in which the redox change involves $t_{2g}$ orbitals and/or systems with a delocalized $\pi$ bonding due to a strong $t_{2g}$-$\pi^*$(bpy) interaction: Ti/Ti$^-$, (V$^{2+}$/V$^+$), V$^+$/V, V/V$^-$, Cr$^{2+}$/Cr$^+$, Cr$^+$/Cr, Mo$^+$/Mo, Fe$^{3+}$/Fe$^{2+}$, Os$^{3+}$/Os$^{2+}$. The V$^{2+}$/V$^+$ couple, although falling into this category, is rather slow, 0.35 cm s$^{-1}$. Interestingly, the rate of electrochemical reduction of $[Fe^{III}(bpy)_{3-n}(CN)_n]^z$ complexes gradually decreases as the bpy ligands are substituted by CN$^-$. The corresponding $k^\circ$ values (obtained in water) range [239] from 0.88 cm s$^{-1}$ for $[Fe^{III}(bpy)_3]^{3+}$ to 0.15 cm s$^{-1}$ for $[Fe^{III}(bpy)(CN)_4]^-$. This could be caused by diminishing the size of the molecule and, hence, increasing outer-sphere reorganization energy.

2) Relatively slow reacting $(0.1 < k° < 0.3$ cm s$^{-1})$ complexes in which the redox change is localized on a single bpy ligand. These rate constants are comparable to that of the reduction of free bpy, i.e. of the bpy/bpy$^{\cdot-}$ redox couple, 0.13–0.21 cm s$^{-1}$, depending on the electrode material. This behavior is pertinent to following couples [238]: Cr$^0$/Cr$^-$, Mo$^0$/Mo$^-$, Fe$^{2+}$/Fe$^+$, Ru$^{2+}$/Ru$^+$, Ru$^+$/Ru$^0$, Ru$^0$/Ru$^-$, Os$^{2+}$/Os$^+$, Os$^+$/Os$^0$, Os$^0$/Os$^-$, and to the [Ir(bpy)$_3$]$^{z+}$ series [106]: Ir$^{3+}$/Ir$^{2+}$, Ir$^{2+}$/Ir$^+$, Ir$^+$/Ir$^0$, Ir$^0$/Ir$^-$, Ir$^-$/Ir$^{2-}$. Bpy-localized reductions of bis-bpy complexes [Ir(bpy)$_2$Cl$_2$]$^+$, and [Fe$^{II}$(bpy)$_2$(CN)$_2$] also belong to this class [106, 240]. It is interesting to note that the electron transfer rate constant is virtually independent of the oxidation state, that is of the number of electrons already present. The Ir series even shows [106] that the reduction steps in which a bpy ligand is reduced to bpy$^{\cdot-}$ (i.e. Ir$^{3+}$/Ir$^{2+}$, Ir$^{2+}$/Ir$^+$, Ir$^+$/Ir$^0$) have similar rate constants as the steps which correspond to bpy$^{\cdot-}$/bpy$^{2-}$ ligand-localized couples (i.e. Ir$^0$/Ir$^-$, Ir$^-$/Ir$^{2-}$).

The relatively slow rate of electrochemical bpy-localized reduction steps is rather puzzling. Possibly, it can be explained by large reorganization energy which arises from the fact that the redox change is localized on a single bpy ligand, instead of being delocalized over the whole complex ions. The internal reorganization energy would then consist of contributions from structural changes within the M(bpy) moiety affected. More importantly, localization of a redox change on a single ligand would lead to a significant dipole moment change, and, therefore, a large outer sphere reorganization energy.

Electrochemical rate constants are, generally, much lower than would correspond to the very fast homogeneous self-exchange rate constants observed for the Fe$^{3+}$/Fe$^{2+}$ or Cr$^+$/Cr$^0$ couples. (For example, the $1.5 \times 10^9$ M s$^{-1}$ rate constant measured [233] for Cr$^+$/Cr$^0$ implies a heterogeneous rate constant of about $10^3$ cm s$^{-1}$, in a sharp contrast to the experimental value of 1 cm s$^{-1}$.) Also, the slow electrochemical rates of bpy-localized redox couples do not agree with the ultrafast electron injection rates observed at semiconductor electrodes, see Section 5.4.6. To address these problems, it would be worthwhile to reexamine some of the electrochemical kinetic data using ultramicroelectrodes and to compare electrochemical and homogeneous electron transfer constants for a larger number of complexes. Moreover, the important questions of adiabaticity and possible kinetic solvent control were never addressed for bpy-localized redox couples.

A different type of electron transfer occurs in semi-reduced polypyridine complexes which contain bpy ligands in different oxidation states. Electron hopping between N,N and N,N$^{\cdot-}$ ligands can take place in such species. This was indeed manifested [69, 241] by EPR line-broadening for several complexes [Fe(N,N)$_3$]$^+$ and [Ru(N,N)$_3$]$^z$; $z = 1+, 0$; N,N = bpy, phen, and their Me$_2$ derivatives. Electron hopping occurs over an energy barrier [69] of 740–960 cm$^{-1}$ for [M(N,N$^{\cdot-}$)(N,N)$_2$]$^+$; M = Fe, Ru; and 300–440 cm$^{-1}$ for [Ru(N,N$^{\cdot-}$)$_2$(N,N)]$^0$. Electron hopping between (EtOOC)$_2$-4,4'-bpy and (EtOOC)$_2$-4,4'-bpy$^{\cdot-}$ ligands in [Ru((EtOOC)$_2$-4,4'-bpy)$_2$(X)$_2$]$^{\cdot-}$ is manifested even electrochemically, by the redox potential temperature dependence [242]. In accordance with its formulation as [Fe$^{II}$(bpy$^{\cdot-}$)$_3$]$^-$, no evidence for electron hopping was found for this complex [241].

### 5.3.5 Ligand Labilization

Polypyridine complexes are usually remarkably stable toward ligand dissociation upon reduction or oxidation. This is because polypyridines are very good ligands in their neutral, radical-anionic, as well as dianionic forms. However, in some cases (Section 5.2.2), a polypyridine ligand or, more often, an ancillary ligand gets labilized by reduction, with important chemical consequences and applications in catalysis.

Tris-polypyridine complexes of first row transition metals undergo polypyridine dissociation upon reduction. Hence, their highly reduced forms has to be studied in the presence of an excess of free ligand, which complicates electrochemical investigations. For example, $[Co^{III}(bpy)_3]^{3+}$, which has the $d^6$ electron configuration, is completely stable, while a bpy-dissociation from $[Co^{II}(bpy)_3]^{2+}$ takes place with a rate constant of 3 $s^{-1}$. The $Co^I$ complex $[Co^I(bpy)_3]^+$ reacts faster [171] (10–300 $s^{-1}$), giving rise to highly reactive coordinatively unsaturated $Co^I$ species which activate organic substrates as well as $CO_2$ [30]. Similar reductively induced bpy dissociation was found [180] for $[Rh(bpy)_3]^{3+}$ whose primary reduction product, $[Rh(bpy)_3]^{2+}$ loses a bpy ligand with a rate constant of 1 $s^{-1}$. Further reduction to $[Rh(bpy)_3]^+$ accelerates the bpy loss to $5 \times 10^4$ $s^{-1}$. Moreover, $[Rh(bpy)_3]^{2+}$ undergoes several redox-disproportionation equilibria whereby $[Rh(bpy)_3]^{3+}$ and $[Rh(bpy)_2]^+$ are produced. The reason for a bpy loss from $[Rh(bpy)_3]^{2+}$ is not quite clear since, unlike its $Co^{II}$ analog, it was formulated as a $d^6$ $Rh^{III}$ complex $[Rh^{III}(bpy)_2(bpy^{\bullet-})]^{2+}$. A close energetic proximity of the singly occupied $\pi^*(bpy)$ orbital and the empty $e_g(Rh)$ orbital could be the reason [180]. Except for Rh, even the most reduced complexes of second- and third-row transition metals are mostly stable, at least on the time scale of cyclic voltammetry at low-temperatures. A partial loss of a bpy ligand upon the fourth reduction has been noted [154] for $[Os(bpy)_3]^{2+}$.

Ancillary ligands in substituted polypyridine complexes are often labilized upon reduction. This is, for example, the case of $[M(CO)_4(bpy)]$; M = Cr, Mo, W; which undergo electron transfer catalyzed substitution of an axial CO ligand [20]. Dissociation of a $Cl^-$ ligand was observed [154, 155, 186, 216] on reduction of $[M(bpy)_2(L)(Cl)]^{n+}$; M = Ru, Os; or on the second reduction of $[Ir(bpy)_2(Cl)_2]^{2+}$. For $[Ru(4,4'-(EtOOC)_2-bpy)_2(X)_2]$, the following order of the rate of ligand dissociation upon the first reduction was found [242]: X = $CN^- < NCS^- < Cl^- < I^-$. Similarly, $[Re(E)(CO)_3(N,N)]^{n+}$ complexes undergo reductively-induced loss of the axial ligand E = halide ($n = 0$) and, more slowly, of neutral ligands THF, $PPh_3$, or RCN [130, 134–137, 243]. Dissociation of an electron accepting ligand E = $P(OR)_3$ requires an addition of a second electron [135]. The organometallic complexes $[(C_5Me_5)M(bpy)(Cl)]^+$; M = Rh, Ir; and $[(C_6Me_6)M(bpy)(Cl)]^+$; M = Ru, Os; undergo a one-step, two-electron reduction, concerted with $Cl^-$ dissociation [184, 185, 213].

Generally, the rate and extent of the reductively-induced ancillary ligand dissociation increase with the number of electrons added and decrease with increasing delocalization within the M(N,N) chelate ring. Hence, reduced bpy or phen complexes are much more reactive than those of bpym, Ar-dab or, especially, abpy.

Labilization of the ancillary ligands in mono-reduced complexes appears to be linked with the extent of delocalization of unpaired electron density from the diimine $\pi^*$-orbital to the labilized ligand [121, 122, 134].

Reductively induced dissociation of an ancillary ligand from polypyridine complexes has a great importance for redox catalysis:

$$M(X)(N, N) \xrightarrow{\ e^-\ } M(X)(N, N^{\boldsymbol{\cdot}}) \xrightarrow{\ e^-\ } M(N, N^{2-}) + X^- \tag{8}$$

$$M(N, N^{2-}) + A \longrightarrow M(A)(N, N^{2-}) \xrightarrow{\ +X^-\ } M(X)(N, N) + A_{red} \tag{9}$$

From this simplified scheme, it follows that the reductively-induced ligand labilization creates a vacant site to which a substrate A can coordinate. At the same time, the N,N ligand acts as an electron reservoir, accommodating electrons which are used to transform the coordinated substrate to $A_{red}$. A similar scheme can be drawn for one-electron substrate reduction following a ligand loss upon the first reduction. Indeed, both one- and two-electron catalytic reductions based on polypyridine complexes have been observed [135, 141].

An ancillary ligand in substituted polypyridine complexes can also be labilized by one-electron oxidation. The cases in which the oxidized radical of the ancillary ligand (e.g. $CN^{\boldsymbol{\cdot}}$, $Cl^{\boldsymbol{\cdot}}$) dissociates are most interesting and need further mechanistic investigation. This is, for example, the case of [130, 131] [Re$^I$(X)(CO)$_3$(bpy)]; X = Cl, CN; Eq. 2, or [244] [Ru(4,4'-(EtOC(O))$_2$-bpy)$_2$(X)$_2$]; X = CN, NCS$^-$, Cl$^-$, I$^-$.

### 5.3.6 Redox Properties of Supermolecules Containing Metal–Polypyridine Units

Depending on the particular structure, the following supramolecular effects can be distinguished: (i) encapsulation of a metal–polypyridine unit, (ii) interaction between linked metal–polypyridine units, (iii) interactions between metal–polypyridine units and chemically different redox centers, and, (iv) host–guest interactions where a metal–polypyridine unit is the part of the host.

Encapsulation. The electrochemistry of a caged Ru$^{II}$ complex in which all three bpy ligands are covalently linked together hardly differs from structurally similar tris-bpy complexes [109, 110]. A chemical stabilization of reduced first-row metal complexes by caged ligands can be expected. However, such species were not investigated. A spectacular encapsulation effect is shown by entwined catphen ligands which impose the coordination number four and a pseudotetrahedral geometry on first-row metal atoms, determining thus the redox behavior [108, 204] (Section 5.2.2). A strong enhancement of the stability of the reduced catenate complexes both toward decomposition (Cu) and reoxidation by $O_2$ (Ni) relative to their bis-danphen analogs should be noted.

A redox-induced translocation of an Fe atom occurs in complexes with tripodal, hemicaged ligands, each branch of which contains a bpy and hydroxamate [245] or a salicylate [246] binding sites. The Fe atom shuttles between the O-coordinating and bpy sites upon its oxidation to Fe$^{3+}$ and reduction to Fe$^{2+}$, respectively.

Another type of redox-induced movement was found for $Cu^{2+/+}$-catenates which contain both phen and tpy units in the catenate rings [247].

Dendrimers with a $[Ru(bpy)_3]^{2+}$ core show the typical electrochemical behavior of the $[Ru(bpy)_3]^{2+}$ unit, at least as the first reduction and oxidation are concerned [248]. Encapsulation by the dendrimer, which branches from the 4,4' positions of the bpy ligand(s) shifts the first oxidation and reduction potentials positively by ca 0.1 V, while the corresponding redox couples are no longer fully electrochemically reversible.

Metal atoms in polynuclear polypyridine complexes [51–53, 84, 87, 88] are usually bridged by another polypyridine or a polyazine ligand or by an ancillary ligand like $CN^-$. The use of a 2,3-dpp bridge leads to large dendrimers formed of interconnected $Ru^{II}(dpp)_3$ moieties and terminal $Ru^{II}(dpp)_2(bpy)_2$ units. Dendrimers prepared so far contain up to 22 Ru centers [84]. Electronic interaction between metal atoms mediated by the bridge causes neighboring metal atoms to be oxidized at different potentials even if they are chemically identical. In large dendrimeric species, the peripheral metal atoms are oxidized first in a single multi-electron step, followed by an oxidation of the inner metal atoms at more positive potentials. Metal atoms of each layer are thus oxidized together and the number of electrons exchanged corresponds to the number of metal atoms present in the layer [249]. Polypyridine ligands are reduced sequentially as in mononuclear complexes, but the overall redox pattern is complicated by weak (few tens of mV) interactions between the dpp ligands at the neighboring centers, and by large positive shifts of reduction steps localized at the bridging ligands (Section 5.3.2). Generally, bridging polypyridine or polyazine ligands are reduced first: a group of closely spaced reductions converts all bridging ligands to anions, one by one. A stepwise reduction of all the bridging ligands to dianions comes next at more negative potentials. A group of very closely spaced reductions, in which one electron is added to each ligand of each terminal group, follows. In the next manifold of closely-spaced reductions, the second ligand of each terminal group is reduced. At the end of these reduction sequences, all bridging ligands are present as dianions and all terminal ligands as anions. Then, two sets of reductions of the terminal ligands to dianions occur. This type of redox pattern is quite general [53, 218, 250]. For example, 26 ligand-localized reduction steps have recently been distinguished and assigned for a hexanuclear Ru-polypyridine dendrimer [218]. The general reduction pattern described above may differ if the bridging ligands are difficult to reduce. Electrochemistry emerges as an important technique to characterize supramolecular species and to map electronic interactions between their components.

In other systems [51, 52, 87, 88], the tpy ligands of neighboring metal-containing units are connected through their 4' positions, either directly or by a spacer (e.g. 1,4-phenylene, or $-C\equiv C-$). Most of homometallic dinuclear complexes of this type show a single two-electron, metal-centered oxidation, indicating that the metal–metal interaction is negligible. The ligands are reduced sequentially. Localization of reduced steps was not studied in detail, but a perusal of trends in reported redox potentials [52] indicates that the interconnected ligands are reduced first.

Another type of supramolecular systems contain a metal–polypyridine unit ($Ru^{II}$ or $Os^{II}$ bis- or tris-polypyridines, $Re^{I}(CO)_3(N,N)^+$, or $Cu^{I}(catphen)_2^+$) linked covalently with chemically different redox- and/or photo-active groups: alkylviologens ($MV^{2+}$ and derivatives) [91, 251], $C_{60}$ [252], porphyrins [86, 253], phenothiazine (PTZ) [254, 255] etc. (See Appendix for $MV^{2+}$ and PTZ.) These groups can be bound either as ancillary ligands, or covalently attached to the pyridine rings of N,N ligands, or inserted into catenate rings. Rotaxanes with a $Cu^I(-9,7-phen)_2{}^+$ central unit and $Ru(bpy)_3{}^{2+}$, $C_{60}$ or porphyrin stoppers are also known [92]. Supermolecules of this type are designed and synthesized especially for photochemical applications. Their electrochemical behavior is, in most cases, a simple sum of that of their components. No electrochemically important synergetic effects or interactions between the components were observed. (Positive shifts of metal-centered oxidations observed in some dinuclear Cu-catenanes are due to a steric strain at the metal centers [256].) UV–Vis absorption spectra also show only sums of the features due to the components.

It is also possible to build metal–polypyridines into supramolecular hosts and receptors. For example, one of the bpy ligands may be a part of a crown ether ring [34], or it may be capped with a calixarene [33, 257] attached at the 4,4′ positions. Electrochemical behavior of the metal–polypyridine unit is then affected by an interaction with a 'guest', e.g. an alkali metal cation.

### 5.3.7 Applications of Ground-state Electron Transfer

Most chemical applications of metal–polypyridine complexes are based on the high reactivity of their unusual oxidation states. Their use as stoichiometric reagents [258] seems to be largely limited to outer-sphere oxidations with tris-polypyridine complexes of $Fe^{III}$, $Ru^{III}$, and of the less oxidizing $Co^{III}$ or $Cr^{III}$. Redox potentials of Fe and Ru complexes can be finely tuned, Section 5.3.3. By contrast, reduced polypyridine complexes are usually not employed as stoichiometric reductants, probably because of an abundance of other reducing agents which are more easily handled.

One of the most prominent feature of polypyridine complexes is that they can be easily oxidized and reduced both photochemically and electrochemically. This opens an enormous potential for applications in photocatalysis (Section 5.4.9) and electrocatalysis.

Most electrocatalytic processes studied so far are reductive, taking an advantage of a reductive ligand labilization, Section 5.3.5. They are essentially based on Eqs 8 and 9, but the real mechanisms are more complex. Electrocatalysis of $CO_2$ reduction is the most studied process [131, 135, 138, 139, 141, 158–163, 182, 183, 259] with $[Re^I(Cl)(CO)_3(N,N)]$, $[Ru^{II}(N,N)_3]^{2+}$, $[Ru^{II}(bpy)_2(CO)(Cl)]^+$, $[Ru^{II}(bpy)_2(CO)_2]^{2+}$, $Os^{II}$ complexes, or $[(C_5Me_5)M(bpy)(Cl)]^+$; M = Rh, Ir; serving as catalysts. The catalyst is either homogeneous, in a solution phase, or deposited on the electrode surface as a polymer film [160–162, 259]. Depending on the catalyst and conditions, $CO_2$ reduction electrocatalyzed by $[Re(Cl)(CO)_3(bpy)]$ can follow either one- or two-electron mechanism [135, 141]. Electrogenerated $[Re(CO)_3(bpy)]^{•-}$ and $[Re(CO)_3(bpy)]^{2-}$, are the respective active species.

Other examples of electrocatalysis include reduction of $H^+$ to $H_2$ catalyzed [185] by [(C$_5$Me$_5$)Ir(bpy)(Cl)]$^+$, coupling of organic groups (with a C–C bond formation) catalyzed [192] by [Ni(X)$_2$(bpy)], or coupling of alkynes and CO$_2$ catalyzed by [Ni(bpy)$_3$]$^{2+}$ in the presence of Mg$^{2+}$ ions [260]. [Co(bpy)$_3$]$^{2+}$ is another useful reduction catalyst [24, 30, 31]. The active form, [Co(bpy)$_3$]$^+$, is usually generated chemically or photochemically [173].

An electrocatalytic oxidation of guanine in oligonucleotides and DNA using the [Ru(bpy)$_3$]$^{2+/3+}$ redox couple has been observed and its mechanism investigated [261]. Metal–polypyridine complexes with oxo ligands act as electrochemical catalysts of water, Cl$^-$, or hydrocarbon oxidations [128, 166, 167, 262].

It is surprising that the great potential of metal–polypyridine complexes as reduction catalysts and electrocatalysts has not been investigated in more detail and extended to other types of substrates.

Supramolecules containing metal–polypyridine units, especially the Ru(dpp)-based dendrimers, could be used as electron reservoirs or components of molecular-electronic devices. Supramolecules in which an electroactive M(N,N)$_n$ group is attached to a receptor capable of molecular recognition (crown ethers, calixarenes, cryptands etc.) can work as electrochemical sensors. Electrochemical recognition of cations as well as anions has been reported [33–35, 257, 263].

Redox series of metal–polypyridines still await their practical exploration. The existence of multistep, reversible, sequential reduction processes, each step occurring at a defined potential and being localized at a specific molecular site, is very promising for possible applications in molecular electronics. This would require to organize the active complexes in films, polymers or supermolecules. Up to now, only the electrochromic behavior of some [Ru(N,N)$_3$]$^{2+}$ complexes has been explored with potential applications in electrochromic glasses, displays and redox sensors [206, 262, 264].

## 5.4 Electron-transfer Properties of Electronically Excited Polypyridine Complexes

This area was opened [59] in 1972 by the observation of a bimolecular reaction between the long-lived MLCT excited state of [Ru(bpy)$_3$]$^{2+}$ and [Co(NH$_3$)$_5$Br]$^{2+}$:

$$[Ru(bpy)_3]^{2+} \xrightarrow{h\nu} {}^*[Ru(bpy)_3]^{2+} \tag{10}$$

$$\begin{aligned} &{}^*[Ru(bpy)_3]^{2+} + [Co(NH_3)_5Br]^{2+} \\ &\rightarrow [Ru(bpy)_3]^{3+} + Co^{2+} + 5NH_3 + Br^- \end{aligned} \tag{11}$$

Since then, excited state electron transfer of metal polypyridines has developed into a major research area which is covered by many excellent reviews and books [32, 37, 38, 42, 43, 62, 74, 83, 95, 98, 99, 265–270]. Herein, some general aspects and more recent developments will be discussed.

The prominent electron transfer photochemistry of polypyridine complexes stems from a unique combination of their ground state redox reactivity and excited state properties:

1) Fast one-electron oxidation and/or reduction of ground-state complexes at suitable potentials. (Out of all the multiple redox steps, it is only the first oxidation and the first reduction which are photochemically relevant.)
2) Chemical stability of oxidized and reduced complexes.
3) Presence of low-lying excited states which can be efficiently populated by optical excitation.
4) Excited state lifetimes are sufficiently long-lived to allow for an efficient bimolecular electron transfer.
5) Photochemical stability.

These conditions are met by excited states of many polypyridine complexes of $d^6$ metals and, in part, of $d^3$ Cr$^{III}$. Electron transfer reactivity is then retained upon electronic excitation, giving rise to important photochemistry and possible applications.

### 5.4.1 Basic Thermodynamic and Kinetic Aspects of Excited-state Electron Transfer

Optical excitation occurs as a vertical transition, without a change in molecular geometry or in the arrangement of the surrounding medium. It prepares a so-called Franck–Condon excited state. Since the molecular geometry and orientation of solvent molecules are not adapted to the changed electron density distribution, Franck–Condon excited states are not equilibrated. Hence, ultrafast relaxation to a thermally equilibrated excited state follows [118, 271]. For most polypyridine complexes, relaxation of a Franck–Condon state is a convoluted process that includes redistribution of electron density, change of a spin state (i.e. intersystem crossing), vibrational relaxation, and solvent relaxation. It is usually completed within a few hundreds of femtoseconds, although in some cases a full relaxation can take as long as few picoseconds [272, 273].

Relaxation of a Franck–Condon state produces an energetically low-lying, thermally equilibrated excited state [274–276]. In most cases, this is the lowest excited state. A molecule in a thermally equilibrated excited state is a distinct chemical species in its own right, which has defined equilibrium geometry, thermodynamic properties, and lifetime. It can be viewed as an energy-rich isomer of the respective ground state. Indeed, most of photochemical electron-transfer reactions of polypyridine complexes involve equilibrated states, whose lifetimes are long enough so that electron transfer can occur competitively with unproductive decay to the ground state. Direct electron transfer from Franck–Condon state is very rare, albeit possibly very important in photoelectrochemical light–energy conversion (Section 5.4.6).

Redox behavior of electronically excited molecules is derived from that of the

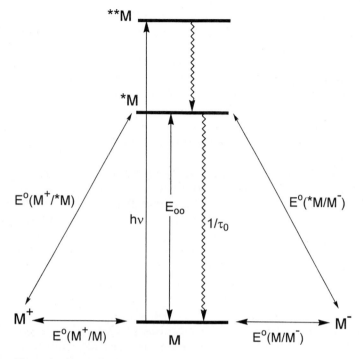

**Figure 2.** Generalized excited state diagram. M represents a ground-state molecule, **M and *M are the Franck–Condon and thermally equilibrated excited states, respectively. $h\nu$ is the excitation energy, $E_{00}$ the excited state energy, and $\tau_0$ is the inherent excited state lifetime. Relevant ground- and excited state redox couples are shown.

corresponding ground state species (Figures 2 and 5). Electronic excitation activates electron transfer reactivity in two ways: thermodynamically by shifting the redox potentials and kinetically by changing the electron distribution. Excited state redox potentials [37] are determined by the corresponding ground state potentials and the excited state energy, $E_{00}$:

$$E^\circ(M^+/*M) = E^\circ(M^+/M) - E_{00} \tag{12}$$

$$E^\circ(*M/M^-) = E^\circ(M/M^-) + E_{00} \tag{13}$$

where M and *M represent the ground and excited state, respectively, see Figure 2. (The potentials $E^\circ(M^+/M)$ and $E^\circ(M/M^-)$ correspond to $E_{1/2}^{ox}$ and $E_{1/2}^{red}$, as used in Table 1.) Since redox potentials are thermodynamic quantities, they can be defined only for thermally equilibrated excited states. The excited state energy, $E_{00}$, is the energy difference between the minima of the ground and excited state potential energy surfaces. The entropy change between the ground and excited state is neglected. Eqs 12 and 13 show that, on excitation, the oxidation potential shifts negatively and the reduction potential positively. This means that excited state of a molecule which, in its ground state, can be both reversibly oxidized and reduced (as many polypyridine complexes are) is simultaneously a very strong reductant and oxidant. This can be demonstrated using the best-known polypyridine photosensitizer, $[Ru(bpy)_3]^{2+}$: In its ground state, it is neither an oxidant nor a reductant since its reduction and oxidation occurs at rather negative and positive potentials, $-1.76$ and $+0.85$ V, respectively. However, its MLCT excited state lies 2.12 eV above the ground state. This shifts the corresponding reduction and oxidation potentials to $+0.36$ and $-1.27$ V, making the MLCT-excited $[Ru(bpy)_3]^{2+}$ a very strong oxidant *and* reductant.

The respective products of oxidation and reduction of an excited molecule, that is the species $M^+$ and $M^-$ in their electronic ground states are themselves strong oxidants and reductants, respectively. They can return to the initial ground state molecule M by further electron-transfer reactions, as shown in Figure 3. Hence, photochemical electron transfer can operate as a catalytic cycle. (In the $[Ru(bpy)_3]^{2+}$ example, $M^+$ and $M^-$ represent $[Ru(bpy)_3]^{3+}$ and $[Ru(bpy)_3]^{+}$, respectively.)

### 5.4.2 Excited States of Polypyridine Complexes

Three basic types of excited states need to be considered here: (i) metal to ligand charge transfer excited states (MLCT) in which an electron is excited from a $d\pi$

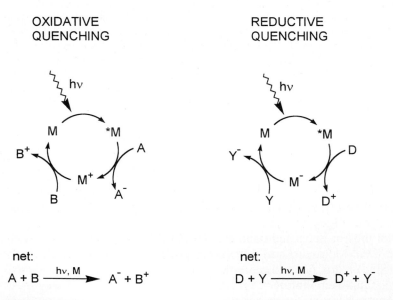

**Figure 3.** Photocatalytic cycles based on an oxidative (left) and reductive (right) quenching of an excited polypyridine complex M. (The photosensitizer M in these cycles has been called 'light absorption sensitizer' (LAS), since it enables a photochemical reactions between chemical species which do not absorb light [74, 266].)

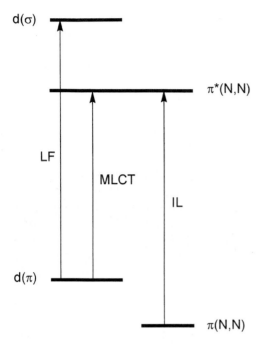

**Figure 4.** Typical electronic transitions of metal polypyridine complexes. LF = ligand field, MLCT = metal to ligand charge transfer, IL = intraligand, d($\pi$) and d($\sigma$) are metal d orbitals of a $t_{2g}$ and $e_g$ origin, respectively. $\pi$(N,N) and $\pi^*$(N,N) are the respective highest occupied and lowest unoccupied polypyridine $\pi$ orbitals. LF states become the lowest-lying in complexes of first-row metals with a small d-orbital splitting. The IL state can lie below MLCT states for highly delocalized polypyridine ligands with a small $\pi^*$–$\pi$ gap, or if d($\pi$) orbitals lie low in energy due to high positive charge on the metal atom (Rh$^{III}$, Ir$^{III}$) or electron-accepting ancillary ligands (CO, RNC…).

(essentially $t_{2g}$) metal orbital to a $\pi^*$ orbital localized predominantly on the polypyridine ligand, (ii) ligand-field, (LF) states which correspond to electron excitation within the predominantly metal-localized d-orbital manifold, and, (iii) intraligand excited states (IL) which originate in electron excitation from an occupied $\pi$ polypyridine orbital to a corresponding $\pi^*$ orbital. These three types of excitation are shown schematically in Figure 4.

The character of the lowest excited state depends on the nature of the metal atom, the polypyridine ligand and the ancillary ligands. MLCT excited states are the lowest excited states of those complexes which contain easily-oxidized low-valent metal atoms: Cr$^0$, Mo$^0$, W$^0$, Mn$^I$, Re$^I$, Fe$^{II}$, Ru$^{II}$, Os$^{II}$, Pt$^{II}$, and Cu$^I$. These are the polypyridine complexes whose ground-state reduction and oxidation is predominantly localized on the metal atom and the polypyridine ligand, respectively (Section 5.3.1). MLCT electronic transitions are strongly allowed and usually occur in the visible spectral region. Hence, irradiation into MLCT absorption bands of polypyridine complexes provides a very efficient way of collecting light energy. Many complexes which possess a lowest lying MLCT state also meet other practical conditions for a good photochemical sensitizer [277], namely those of convenient redox potentials of both oxidation and reduction and chemical stability of the excited, oxidized, and reduced forms.

Since the ligand-field splitting of the d-orbital manifold of low-valent $d^6$ metal atoms is relatively large, LF excited states of $d^6$-metals occur at higher energy than MLCT states, with the only exception of $Fe^{II}$. Hence, LF states are not involved in electron transfer reactivity but they can provide a non-radiative deactivation pathway for the reactive MLCT state, shortening its lifetime. LF states do not exist for $d^{10}$ $Cu^I$. The only polypyridine complexes with a redox-active LF state are $[Cr(N,N)_3]^{3+}$, whose $^2T/^2E$ LF states are strong oxidants [278, 279].

Intraligand excited states become the lowest in complexes of metal atoms which are difficult to oxidize, that is $Rh^{III}$ and $Ir^{III}$. Moreover, IL excited states are relatively low-lying for those polypyridine ligands which have highly delocalized, extended $\pi$ system, for example dppz or some phen derivatives. Hence, complexes of these ligands often have close-lying MLCT and IL states. This situation has interesting photophysical consequences [280–282]. The much stronger medium-dependence of the MLCT than IL excited state energy provides a way of controlling their relative energetic order. Generally, rigid or highly polar media shift MLCT states to higher energies. The order between MLCT and IL states also depends on the nature of the ancillary ligands, whereby the MLCT states are stabilized in energy by strong electron donors and destabilized by electron acceptors. This effect is, for example, demonstrated by $[Re(Cl)(CO)_3(bpy)]$ and $[Re(CO)_4(bpy)]^+$, whose lowest-lying excited states are of a MLCT and IL origin, respectively [97, 133, 283]. Polypyridine complexes with lowest IL excited states are mostly poor excited state reductants. Moreover, their oxidized forms are often chemically unstable, precluding their use in photocatalytic cycles shown in the Figure 3, left. On the other hand, these species can be strong excited state oxidants. Photochemistry of complexes with lowest-lying IL excited states usually requires UV irradiation.

### 5.4.3 Control of the Electron-transfer Properties of MLCT Excited States of Polypyridine Complexes

It follows from the discussion in the Section 5.4.1 that excited state redox properties can be controlled through ground-state redox potentials (Sections 5.3.2. and 5.3.3.) and excited state energy $E_{00}$. The excited state energy is dependent on the molecular structure, but in a way that is interrelated with the structural dependence of the redox potentials. In Section 5.3.1 it was mentioned that the MLCT excitation energy, $hv$, (Figure 2) depends linearly on the difference between oxidation and reduction potentials. This dependence stems from the intramolecular charge separation that occurs in MLCT excited states:

$$[M(B)_n(N, N)] \xrightarrow{hv, \text{MLCT}} [M^+(B)_n(N, N^{\cdot -})] \tag{14}$$

where B represents ancillary ligands. For complexes containing two or three polypyridine ligands, the excited electron is localized [70–72, 219, 284] on a single N,N ligand:

$$[M(N, N)_3] \xrightarrow{hv, \text{MLCT}} [M^+(N, N)_2(N, N^{\bullet-})] \tag{15}$$

in the same way as in reduced species. Indeed, the MLCT-excited and reduced polypyridine complexes show the same spectroscopic patterns (UV–Vis, res. Raman, IR), since they contain the same chromophore, viz. $N,N^{\bullet-}$. The localized formulation (Eq. 15) implies that electron hopping between N,N ligands ($N,N^{\bullet-} \leftrightarrow N,N$) should occur in MLCT-excited tris-polypyridine complexes. Indeed, such interligand electron transfer was observed for $[Os(bpy)_3]^{2+}$. It occurs in the picosecond time domain. The actual rate and mechanism depend on the solvent relaxation time [285, 286].

Energies of optical MLCT transitions ($hv$) and energy of a thermally equilibrated MLCT state ($E_{00}$) depend linearly on the difference between the reduction and oxidation potential [6, 9, 28, 149, 207, 208]:

$$hv = E^\circ(M^+/M) - E^\circ(M/M^-) + \chi_i + \chi_0 + C \tag{16}$$

$$E_{00} = E^\circ(M^+/M) - E^\circ(M/M^-) - D \tag{17}$$

Reorganization energies $\chi_i$ and $\chi_0$ correspond to the inner-sphere (vibrational) and outer-sphere (solvent) reorganization that follows excitation into the Franck–Condon MLCT excited state (Figure 2). The parameters $C$ and $D$ include further electronic and solvation terms [9, 28]. Eqs 16 and 17 have been tested in numerous studies using absorption or emission energies (for example [9, 149, 152, 153, 205, 210, 287–290]) and, indeed, the predicted linear relationships between the transition or excited state energies and ($E^\circ(M^+/M) - E^\circ(M/M^-)$) were found. The experimentally determined slopes are usually slightly less than 1, due to a partial delocalization of redox changes and of MLCT excitation between the metal atom and the polypyridine ligand [9]. (*Note that $E_{00}$ is defined here as an energy difference between fully equilibrated excited and ground state. Hence, it is different from the emission energy as used in ref. [9].)

Combining Eqs 12, 13, and 17 yields new expressions for excited state redox potentials [28]:

$$E^\circ(M^+/^*M) = E^\circ(M/M^-) + D \tag{18}$$

$$E^\circ(^*M/M^-) = E^\circ(M^+/M) - D \tag{19}$$

Now, the excited state oxidation potential is related to the potential of the ground-state ligand-localized redox couple and the excited state reduction potential is related to the potential of the ground-state metal-localized redox couple, see Figure 5. These relations are very logical since oxidation of an MLCT-excited polypyridine complex actually amounts to oxidation of the reduced polypyridine ligand $N,N^{\bullet-}$. Similarly, reduction of an MLCT-excited polypyridine complex corresponds to re-

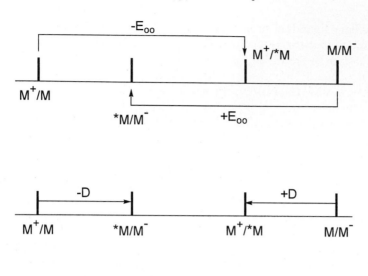

**Figure 5.** Relationship between ground- and excited state redox potentials. The upper diagram is valid generally, regardless of the excited state character. The lower diagram is applicable only to predominantly localized MLCT excited states of those complexes whose first oxidation and reduction are predominantly metal- and ligand-localized, respectively. See Figure 2 and Eqs 12, 13, 18, and 19.

duction of the oxidized central metal atom $M^+$. (Note the formulation of a MLCT excited state as $[M^+(B)_n(N,N^{\cdot-})]$, Eqs 14 and 15) The parameter $D$ is characteristic of a given homologous series of structurally related complexes. For example, an average value of 0.48 V (with a standard deviation of 0.09 V) was determined [28] for a group of 70 tris- and bis-polypyridine $Ru^{II}$ complexes. $Os^{II}$ complexes gave [28] a value of 0.54 V, with a standard deviation of 0.1 V. Importantly, Eqs 18 and 19 imply that excited state redox potentials are uniquely determined by ground state redox potentials, Figure 5. Hence, with the knowledge of $D$ for the particular set of complexes, the excited state redox potential values can be estimated using electrochemical parameters [9, 15, 28, 157, 220]. Obviously, complexes with hard-to-reduce polypyridine ligands will be strong excited state reductants and complexes of hard-to-oxidize metals will be strong excited state oxidants.

It is often important to control the MLCT excitation energy, $h\nu$. Generally, it is advantageous if a photocatalytic reaction can be driven by light of as low energy as thermodynamically possible. Photosensitizers for solar energy conversion should ideally absorb throughout the whole visible spectral region. Hence, photoactive polypyridine complexes with rather low-lying MLCT states are needed. Eqs 16, 18, and 19 provide a clue how to tune MLCT excitation energy without losing valuable redox properties. It has been shown in Sections 5.3.2. and 5.3.3. that the difference between the ground-state oxidation and reduction potentials changes only little in extended series of tris-polypyridine complexes, since the reduction and oxidation potentials are linearly dependent on each other with a slope close to one (1.16 for $Ru^{II}$) [29]. Hence, structural variations of a polypyridine ligand in tris-polypyridine complexes of the same metal atom have only limited effect on MLCT excitation energy, Eq. 16. In contrast, the redox-potential difference is strongly dependent on the nature of the ancillary ligands in mono- and bis-polypyridine complexes since the dependence between the reduction and oxidation potentials has a much smaller slope (ca 0.25 for $[Ru^{II}(bpy)_2XY]$). Hence, varying the nature of ancillary ligands is a much more effective way to shift the MLCT excitation energy to the desired spectral range than varying the polypyridine ligand. Ancillary ligand variations are especially advantageous for designing complexes that should act as excited state reductants, since the corresponding excited state redox potential $E°(M^+/*M)$ is determined by the ground-state reduction potential $E°(M/M^-)$, which, in turn, is only little dependent on the nature of the ancillary ligands. Therefore, changing the nature of ancillary ligands allows us to tune the MLCT excitation energy while keeping the excited state oxidation potential nearly constant, essentially determined by the polypyridine ligand chosen. Indeed, the most promising sensitizers for photoelectrochemical light energy conversions, $[Ru(4,4'-(COOH)_2-bpy)_2(SCN)_2]$ and $[Ru(4,4',4''-(COOH)_3-tpy)(NCS)_3]^-$ are mixed-ligand complexes [36]. Another way to control excitation energy is to change the metal atom. For example, MLCT transitions of $Os^{II}$ complexes occur at lower energies than for their $Ru^{II}$ counterparts, since the redox potential difference is smaller.

Excited state lifetime $\tau_0$ is another important parameter to be controlled, especially if the photoactive complex is intended for bimolecular photochemical electron transfer. MLCT excited states of most polypyridine complexes decay both radiatively and non-radiatively, with the respective rate constants $k_r$ and $k_{nr}$. The inherent excited state lifetime is defined as $\tau_0 = 1/(k_r + k_{nr})$. The non-radiative decay pathway in most cases prevails: $k_{nr} \gg k_r$. Hence: $\tau_0 \cong 1/k_{nr}$. Non-radiative decay of MLCT excited states can be treated as intramolecular electron transfer in the Marcus inverted region:

$$[M^+(B)_n(N, N^{\cdot-})] \xrightarrow{k_{nr}} [M(B)_n(N, N)] \tag{20}$$

using the energy-gap law [10–12, 14, 291–295]. Generally, MLCT excited state lifetime is expected to increase with increasing excited state energy and decreasing structural distortion between the ground and excited state. Eqs 16 and 17 show that MLCT excited state energy $E_{00}$ can be tuned by structural variations in the same way as the excitation energy, vide supra. For a given excited state energy, MLCT lifetime will increase with increasing structural rigidity and, namely, with increasing delocalization of the excited electron over the polypyridine ligand, which spreads the structural effects of excitation over a large number of bonds. The use of polypyridine ligands with extended delocalized $\pi$ system can even compensate for a small energy gap, making it possible to design polypyridine complexes which absorb light throughout much of the visible spectral region, while still possesssing relatively long-lived MLCT excited states [293, 294].

The presence of a short-lived LF excited state that lies close above the MLCT state can seriously diminish MLCT excited state lifetimes of $Ru^{II}$ complexes. Thermally activated population of such LF states from an MLCT state provides an efficient deactivation pathway and can lead to photochemical decomposition [152, 296, 297]. The energy gap between the MLCT and deactivating LF excited states can be increased by a judicious choice of the polypyridine and ancillary ligands, thereby increasing the MLCT lifetime and improving photostability [74, 152, 297, 298]. LF states do not intervene with a non-radiative decay of $Os^{II}$-polypyridine MLCT excited states since they are too high in energy. On the other hand, LF states lie below MLCT states in $Fe^{II}$ complexes, providing an ultrafast deactivation pathway (e.g. $\tau_0 = 0.84$ ns for $[Fe(bpy)_3]^{2+}$) [270].

Increasing the complex rigidity can prolong the MLCT lifetime and improve the complex photostability. This is the case of rigid caged [109, 110] or hemi-caged [299] derivatives of $[Ru^{II}(bpy)_3]^{2+}$, in which all three bpy ligands are covalently linked together. The structural rigidity and electron delocalization have a profound effect on MLCT lifetimes of $[Cu(N,N)_2]^+$ complexes [204, 268, 300–302]. Recently, it was found that MLCT excited state lifetime can be extended significantly by coupling with a long-lived IL state. For example, excited state lifetime of $[Ru^{II}(bpy)_3]^{2+}$ derivatives in which a pyrene unit is covalently attached to one bpy ligand is 42 μs long [282].

MLCT excited state lifetime decreases when polypyridine ligands are replaced by PyCa or dab-type ligands. This has been amply manifested by the photophysical behavior of $[Re(E)(CO)_3(N,N)]^{n+}$ complexes [97] and by the series [231] $[Ru(bpy)_{3-n}(Me-PyCa)_n]^{2+}$; $n = 0, 1, 2, 3$. The MLCT lifetime decreases from 860 ns to 58 ns on going from the $[Ru(bpy)_3]^{2+}$ to $[Ru(bpy)_2(Me-PyCa)]^{2+}$, since the localization of the excited electron changes from bpy to PyCa. Replacement of further bpy ligands have almost no influence. Values of 62 and 67 ns were measured for $[Ru(bpy)(Me-PyCa)_2]^{2+}$ and $[Ru(Me-PyCa)_3]^{2+}$, respectively.

Finally, it should be noted that an enormous research effort have been spent on the development of photo-redox active polypyridine complexes and optimization of their excited state redox properties. The vast literature on this topic offers discussions and theories of the relations between molecular structure, medium and excited state redox properties and lifetimes, as well as a broad choice of polypyridine complexes suitable as photosensitizers, especially those based on $Ru^{II}$ [9, 14, 28, 74, 95, 98, 151–153, 205, 217, 222, 224, 226, 287, 289, 290, 295, 297].

### 5.4.4 Properties of Some Common Photo-redox-active Metal–Polypyridine Chromophores

**Chromium(III)**

The lowest excited states of $[Cr(N,N)_3]^{3+}$; N,N = bpy, phen and their derivatives; have a LF $^2T_1/^2E$ character [278, 279]. Excited state lifetimes are very long (63 μs for bpy, 270 µs for phen). $[Cr(N,N)_3]^{3+}$ complexes are strong excited state oxidants [265, 278]: $E(*Cr^{3+}/Cr^{2+}) \cong +1.3$ V. However, they did not find much use in photochemistry, perhaps due to poor absorption in the visible spectral region.

**Rhenium(I)**

The character and behavior of the lowest excited state of $[Re^I(E)(CO)_3(N,N)]^{n+}$ complexes is strongly dependent on the E and N,N ligands [97]. Complexes which have lowest $^3$MLCT excited state are very useful in excited state electron-transfer reactions [61, 62, 303]. This situation occurs for E = Cl ($n = 0$), pyridine derivatives, nitriles or phosphines ($n = 1$), N,N = bpy or phen derivatives. Such species are photochemically stable and their excited state lifetimes range from tens to hundreds of nanoseconds, being controlled by the energy gap law [12]. Excited state redox properties can be broadly tuned by variations of the axial (E) and polypyridine ligands. Thermodynamically, $[Re^I(E)(CO)_3(N,N)]^{n+}$ complexes are both strong excited state oxidants and reductants. Their use in bimolecular excited state electron transfer is somewhat complicated by limited stability of their redox products, especially of oxidized $Re^{II}$ species. Hence, they are mostly employed as excited state oxidants. The $Re^I(CO)_3(N,N)^+$ unit can be linked with other metal-containing groups either through the axial ligand E (e.g. $\mu$-CN$^-$) or through a bridging polyazine ligand ($\mu$-N,N). The $Re^I(CO)_3(N,N)^+$ moiety in polynuclear complexes often acts as an energy donor. It is also used to construct various chromophore-quencher complexes, in which the reducing or oxidizing unit is attached to the axial position. These species undergo intramolecular photochemical electron-transfer reactions, Section 5.4.7. Intraligand excited states become the lowest for those $Re^I$ complexes in which the N,N ligand is highly delocalized (dppz, phenyl-substituted phen, etc.) and/or the axial ligand is a strong electron acceptor (CO, RNC). Such species are very powerful excited state oxidants.

**Iron(II)**

Polypyridine complexes of Fe(II) are strongly colored due to MLCT transitions. However, their MLCT excited states are very short-lived because of an efficient deactivation through lower-lying LF states. MLCT excited state lifetimes of $[Fe(N,N)_3]^{2+}$ range from 2.54 ns for tpy to 0.8 ns for bpy and phen [270]. Therefore, $Fe^{II}$ complexes were deemed unsuitable for electron-transfer reactions. Recently, there is a renewed interest in excited state properties of $Fe^{II}$-polypyridines caused by the observation [304] of an ultrafast electron transfer from $[Fe(4,4'-(COOH)_2-bpy)_2(CN)_2]$ ($\tau_0 = 330$ ps) to $TiO_2$, see Section 5.4.6.

**Ruthenium(II)**

Bis- and tris-polypyridine complexes of $Ru^{II}$ are the best known and most useful photo-redox active coordination compounds. The possibility to vary the structure and number of polypyridine ligands and of the ancillary ligands in mono- and bis-polypyridine complexes gives rise to a virtually endless number of photoactive species and provides many opportunities for controlling their properties. Photophysical, photochemical and redox properties of Ru-polypyridines were collected in a very comprehensive review [74]. The lowest excited state of Ru$^{II}$ complexes has a $^3$MLCT character. In most case, it is sufficiently long-lived for bimolecular excited state reactions (0.5–1.1 µs for [Ru(bpy)$_3$]$^{2+}$ and [Ru(phen)$_3$]$^{2+}$). MLCT lifetimes are strongly dependent on complex composition, the medium, and temperature. Their values are determined by the energy gap law [11, 74, 298] and by a thermally activated population of close-lying, upper LF states [74, 295]. A tuning of the LF—MLCT—ground state energy gaps is the key to control photophysical properties of Ru$^{II}$ polypyridines [152, 297, 298]. Complexes in which the MLCT–LF gap is small have short MLCT lifetimes and are prone to slow photochemical substitution of a polypyridine ligand. As is discussed in the Section 5.4.3, the MLCT excitation and excited state energies are best controlled by variations of ancillary ligands in bis-polypyridine Ru$^{II}$ complexes. Thermodynamically, Ru$^{II}$-polypyridines are both strong excited state oxidants and reductants. Their excited state redox potentials can be broadly tuned by structural variations in the same way as ground state redox potentials. Variations in the polypyridine ligand affect both the excited state reduction and oxidation potential while ancillary ligands have a much bigger effect on the excited state reduction potential E(*Ru$^{2+}$/Ru$^+$). Ru$^{II}$(N,N)$_2$$^{2+}$ and Ru$^{II}$(N,N)$_3$$^{2+}$ units are used as building blocks of photo-redox active polynuclear complexes [51, 83] and supermolecular assemblies, in which the Ru-polypyridine chromophore can act as an electron donor or acceptor and energy donor or acceptor. [Ru(tpy)$_2$]$^{2+}$ is only a very weak light emitter and its MLCT excited state lifetime is extremely short (250 ps). Fortunately, photophysical and photo-redox properties of [Ru(tpy)$_2$]$^{2+}$ can be much improved by placing electron-accepting or donating substituents at the 4′ position of the tpy ligand [87, 224]. Groups which extend de-localization of the excited electron (Me$_2$SO$_2$, Ph, –C≡C–, etc.) are especially useful to prolong the MLCT lifetime. Depending on the substituents, [Ru(4′-X-tpy)$_2$]$^{2+}$ and [Ru(4′-X-tpy)(4′-Y-tpy)]$^{2+}$ are strong excited state reductants, oxidants, or energy donors [87, 224]. For geometric and synthetic reasons, [Ru(4′-X-tpy)(4′-Y-tpy)]$^{2+}$ is an excellent building block for photo-redox active supramolecular systems [87]. Here, the (-tpy)Ru$^{II}$(tpy-) chromophores are linked with other molecular units through their 4′ positions, either directly or by –C≡C– or 1,4-phenylene bridges [87, 305]. Besides connecting the components, the substitution at the 4′ position also improves the photophysical properties of the Ru(tpy)$_2$ units [87, 305].

## Osmium(II)

MLCT states are the lowest excited states for Os$^{II}$ polypyridine complexes. Compared with the corresponding Ru$^{II}$ species, the MLCT states of Os$^{II}$-polypyridines lie at lower energies (e.g. 1.66 eV for [Os(bpy)$_3$]$^{2+}$). Thermodynamically, they are comparably strong excited state reductants as Ru$^{II}$ complexes, but weaker oxidants. LF excited states are high in energy and do not provide a decay pathway for MLCT states, whose lifetime is thus governed by the energy gap law [10]. Due to a smaller energy gap and larger spin–orbit coupling, MLCT lifetimes of Os$^{II}$ complexes are shorter than those of their Ru$^{II}$ analogs of the same molecular composition. As members of polynuclear assemblies, Os$^{II}$-polypyridine units usually play the role of an energy acceptor. Because of the absence of close-lying LF states, $[Os(tpy)_2]^{2+}$ has a long MLCT excited state lifetime, 269 ns [87]. This chromophore finds much use as an energy acceptor in supramolecular systems.

### Rhodium(III) and iridium(III)

The lowest-lying excited states of tris-polypyridine $Rh^{III}$ and $Ir^{III}$ complexes have intraligand $\pi\pi^*$ character [37, 269]. These states are populated by UV irradiation. They are long lived ($\mu$s range) and very strong oxidants (e.g. +1.42 V for $*[Ir(bpy)_3]^{3+}/[Ir(bpy)_3]^{2+}$), but relatively weak reductants. In supramolecular systems, $Ir^{III}$ and $Rh^{III}$ usually function as excited-state energy donors or ground-state electron acceptors.

### Platinum(II)

The $[Pt^{II}(S,S)(N,N)]^n$; complexes which combine an electron-accepting polypyridine ligand with an electron-donating dithiolate dianion ($n = 0$) or dithiocarbamate anion ($n = 1+$) are important chromophores and thermodynamically strong excited state oxidants and reductants, whose photochemistry still awaits deeper investigation [202, 203]. The much better chemical stability of their reduced than oxidized forms would favor their use as photochemical oxidants. The interesting photochemistry and emission of these species arise from low-lying ligand to ligand charge transfer excited states, in which an electron is excited from an occupied $\pi$ orbital of the S,S ligand to a $\pi^*$ polypyridine orbital.

### Copper(I)

The properties of low-lying MLCT excited states of $[Cu^I(N,N)_2]^+$ complexes are extremely dependent on the molecular geometry and the medium [300, 302, 306]. Unless protected by bulky ligands, the $Cu^{II}$ atom in an MLCT exited state $[Cu^{II}(N,N)(N,N)^{\cdot-}]^+$ coordinates a donor solvent molecule. Resulting 5-coordinate exciplexes decay rapidly to the ground state [307]. Hence, long-lived MLCT excited states (tens to hundreds of nanoseconds), applicable in photochemical electron transfer, are found mostly for complexes of phen substituted at 2,9 positions with bulky substituents, especially aromatic ones, which force the excited molecule to maintain a pseudotetrahedral geometry and prevent solvent coordination. This is the case of $N,N = 2,9\text{-}Ph_2\text{-phen}$, danphen, or, especially, catphen. Notably, these are the complexes which, in their respective ground states, show reversible one-electron reduction and rather positive oxidation. Resulting large redox-potential difference implies large MLCT excited state energy and, hence, longer lifetimes. $[Cu(R_2\text{-phen})_2]^+$ and related complexes are strong excited state reductants, able to reduce substrates in a homogeneous solution as well as inject electron into a semiconductor electrode [301, 306]. Their role as excited state oxidants, although thermodynamically possible, is hampered by slow electron transfer kinetics, caused by a high reorganization energy of the metal-localized $Cu^{II}/Cu^I$ couple. The $Cu(catpen)_2^+$ unit is used as the active part of various photo-redox active supramolecular assemblies.

### 5.4.5 Bimolecular Electron-transfer Reactions

The photocatalytic cycles shown in Figure 3 are based on oxidative (left) and reductive (right) quenching of electronically excited polypyridine complexes. Such cycles can operate in a homogeneous solution, provided that the excited state electron transfer is much faster than concurrent, unproductive decay to the ground state: $k_{et} \gg 1/\tau_0$, where $\tau_0$ is the inherent excited state lifetime at given experimental conditions, but in the absence of the excited state chemical reaction: $1/\tau_0 = k_r + k_{nr}$, vide supra. Obviously, an electron-transfer reaction should be as fast as possible in order to utilize even short-lived excited states. However, diffusion restricts the maximum rate of bimolecular reactions: as the reaction rate increases, the kinetics become diffusion controlled. In practice, this means that only complexes with relatively long-lived excited states ($\tau_0 > 10$ ns) are useful in bimolecular excited state electron transfer. An occurrence of electron transfer shortens the actual excited state lifetime from $\tau_0$ to $\tau$. $\tau_0/\tau = 1 + k_q \tau_0$ [Q], where Q is the species (quencher) reacting with the excited molecule and $k_q$ is the reaction rate constant (not necessarily equal to $k_{et}$) [37].

Generally, electron-transfer reactions of MLCT excited states of polypyridine complexes occur with small reorganization energies on the side of the excited complex molecules. These reactions can be treated as non-adiabatic, using the Marcus theory. Reaction rates of bimolecular excited state electron transfer, as studied for many polypyridine complexes of $d^6$ metal atoms, are very fast, approaching the diffusion limit. When measured in series involving structurally related polypyridine complexes or quenchers, the logarithm of the rate constant ($\ln k_{et}$) usually increases with increasing driving force ($-\Delta G^\circ$) until it levels off between $10^9$ and $10^{10}$ M s$^{-1}$. The decrease of the reaction rate at very high driving force values, predicted by the Marcus theory, was not seen in most of the reaction series examined. Therefore, kinetics of bimolecular electron transfer quenching of excited states of polypyridine complexes are often described by the Rehm–Weller formalism which replaces the Marcus-type quadratic dependence of the activation free energy on the driving force by empirical relationships. The Rehm–Weller kinetics and underlying mechanisms of bimolecular electron transfer quenching of excited polypyridine complexes were discussed in detail in several excellent articles [37, 38, 42, 265, 308].

The lack of observations of the predicted fall of electron transfer rate at very high driving force values is due to several factors: (i) examination of only limited driving force ranges, (ii) diffusion control of fast reaction rates, and, (iii) formation of redox products in electronically excited states that diminishes the actual driving force, are the most important reasons. However, despite of all these effects, the Marcus theory is perfectly valid for bimolecular electron-transfer reactions of excited polypyridine complexes. This has been demonstrated [309] by the observation of the full bell-shaped Marcus curve (that is the dependence of $\ln k_{et}$ on $-\Delta G^\circ$) for both oxidative and reductive quenching of a series of excited [Ru(N,N)$_3$]$^{2+}$ complexes by Fe$^{III}$cytochrome $c$ and Fe$^{II}$cytochrome $c$, respectively. In this case, the maximum of the Marcus curve was shifted below the diffusion limit, because of a weak electronic coupling and, at the same time, to relatively low driving force values because of small reorganization energy. Moreover, very strong excited state oxidant [Ru(4,4'-(CF$_3$)$_2$-bpy)$_3$]$^{2+}$ and reductant [Ru(4,7-(MeO)$_2$-phen)$_3$]$^{2+}$ were employed.

Oxidative and reducing quenching of excited polypyridine complexes produce strong ground state oxidants and reductants, respectively (Figure 2). For example, $[Ru(bpy)_3]^{3+}$ $(E_{1/2}(3+/2+) = +0.85$ V) and $[Ru(bpy)_3]^+$ $(E_{1/2}(2+/+) = -1.76$ V) are formed by oxidative and reductive quenching, respectively, of *$[Ru(bpy)_3]^{2+}$. Very often, these primary products of excited state electron transfer are the proper oxidants and reductants for the substrate, that is B or Y in Figure 3.

Inspection of the cycles shown in Figure 3 points to possible shortcut reactions which compete with the desired photocatalytic process. Besides the excited state decay, these are the back-reactions $M^+/A^-$ or $M^-/D^+$ and product recombinations $A^-/B^+$ or $D^+/Y^-$. All these reactions regenerate the starting species without accomplishing any useful chemical transformation. They diminish the yield of the redox products and, hence, the efficiency of light energy utilization. One possible solution is to use sacrificial electron acceptors (A) or donors (D), which decompose on reduction or oxidation, respectively, preventing thus the back reaction. $[Co(NH_3)_5Br]^{2+}$ is an example of a sacrificial electron acceptor, widely used in oxidative quenching of excited polypyridine complexes, Eq. 11. Organic amines, whose cations undergo fragmentation, are typical sacrificial electron donors used in reductive quenching. Another strategy is to use reversible donors and acceptors (e.g. the very convenient methylviologen $MV^{2+}$), but to run photocatalytic reactions in organized or microheterogeneous systems (membranes, micelles, vesicles, etc.) which allow for a separation of redox products.

Since its first discovery [59] in 1972, bimolecular electron transfer reactivity of metal polypyridine complexes has been much developed and used in many photocatalytic processes, including, for example, water gas shift reaction, $CO_2$ reduction or transformation of organic substrates. However, the ultimate, but still elusive, goal is to use polypyridine complexes as molecular sensitizers of solar energy conversion, especially of photochemical water splitting. For example, $[Ru(bpy)_3]^{2+}$ in its $^3$MLCT excited state has enough energy and the right redox potentials to reduce *and* oxidize water to $H_2$ and $O_2$, respectively [40, 41, 75, 76]. However, although many possible schemes based on a bimolecular excited state electron transfer have been developed, none of them shows the necessary efficiency and long-term stability, mostly due to short-cut reactions and chemical instability of the species involved. Nevertheless, much knowledge on tuning excited state properties and redox potentials of polypyridine complexes, and on handling electron-transfer reactions in homogeneous and microheterogeneous systems has been obtained in the course of this research. Bimolecular electron-transfer reactions of polypyridine complexes are amply covered by the original literature and many reviews [32, 37, 39, 42, 62, 74, 95, 98, 99, 265–268, 270, 301, 310].

## 5.4.6 Ultra-fast Electron Injection to Semiconductor Electrodes

At present, photoelectrochemistry on semiconductor $(TiO_2)$ electrodes seems to be the most promising way forward towards light energy conversion by chemical means [36, 76–78]. Figure 6 shows schematically a photoelectrochemical cell. In principle, an electron can be excited directly from semiconductor's valence band into its conduction band, followed by the passage of current to a counter electrode.

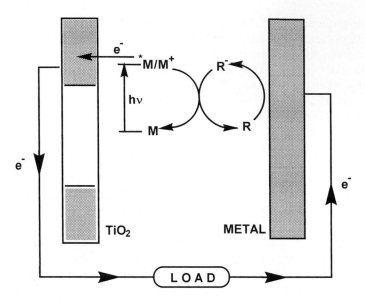

**Figure 6.** Schematic diagram of a photoelectrochemical cell based on a sensitized semiconductor electrode. M = metal–polypyridine sensitizer, R = reversible electron-relay system, e.g. $I_2/I_3^-$.

The circuit is closed with a suitable redox mediator R (e.g. $I_3^-/I_2$ couple) that transfers the electron back to the semiconductor. In this way, light energy is converted to electrical energy. However, $TiO_2$ does not absorb light in the visible spectral region, being thus unsuitable for applications in solar energy conversion. This drawback can be overcome (Figure 6) by using a sensitizer M, that is a photoactive compound which is excited by visible light and then transfers an electron directly from its excited state to the conduction band of the semiconductor electrode. Several polypyridine transition metal complexes proved to be good and efficient sensitizers. The best ones known [36] at the time of writing are [Ru(4,4'-$(COOH)_2$-bpy)$_2(NCS)_2$] and [Ru(4,4',4''-$(COOH)_3$-tpy)$(NCS)_3$]$^-$. The -COOH groups guarantee a strong adsorption to the surface of a $TiO_2$ electrode, while the NCS$^-$ ligands shift light absorption to long wavelengths and provide good photochemical stability.

Importantly, it was found [80–82, 311] that interfacial electron transfer from MLCT-excited Ru$^{II}$ polypyridine complexes to $TiO_2$ is an ultrafast process, completed in 25–150 fs! This groundbreaking discovery implies that the search for new sensitizers need not to be limited to complexes with long-lived excited states. Indeed, [Fe(4,4'-$(COOH)_2$-bpy)$_2(CN)_2$], whose MLCT excited state lifetime is only ca 330 ps, was found [304] to act as a sensitizer in a $TiO_2$-based solar cell. In fact, even the classical Grätzel cell [36, 77, 78] would not operate as well as it does, were the interfacial electron transfer not ultrafast, since the [Ru(4,4'-$(COOH)_2$-bpy)$_2$-$(NCS)_2$] sensitizer has an inherent excited state lifetime of only 50 ns.

In particular, it was found that electron transfer from an MLCT excited state of [Ru(4,4'-$(COOH)_2$-bpy)$(NCS)_2$] into the conduction band of colloidal or thin-film

TiO$_2$ takes place with a time constant of about 50 fs in a colloidal solution [80–82], and less than 25 fs in high vacuum [311]. Some studies have indicated biphasic kinetics, with a minor component of about 1 ps. Similar ultrafast electron injection was found to occur also from an inherently short-lived MLCT excited state of [Fe(bpy-4,4'-(COOH)$_2$)$_2$(CN)$_2$] [304], and [Fe(CN)$_6$]$^{4-}$ [312], or from various organic dyes [313–317]. The rate is independent of temperature over a broad range (22–300 K, in vacuum [314]). Common to all these systems is a strong adsorption of the sensitizer molecule to the TiO$_2$ surface via various anchoring groups (–COOH, –P(O)(OH)$_2$, –CN, –OH, etc.) which also provides an electronic coupling by mixing between the bpy $\pi^*$ orbital, into which the electron is excited, and the ultimate TiO$_2$ acceptor orbitals [82].

Typical electron injection times are faster than, or comparable with, relaxation of the optically prepared Franck–Condon $^1$MLCT excited states. Hence, the electron injection can actually occur directly from the Franck–Condon state. This is the case of [Fe(4,4'-(COOH)$_2$-bpy)$_2$(CN)$_2$] which reacts from its optically prepared $^1$MLCT state [304], before its deactivation through the lower-lying LF states can occur. The relaxation time of the $^1$MLCT state of the actual [Ru(4,4'-(COOH)$_2$-bpy)$_2$(NCS)$_2$] sensitizer was determined as <75 fs [82], only a little slower than electron injection itself. Hence, it is possible that electron transfer occurs from both the optically prepared $^1$MLCT and relaxed $^3$MLCT states.

The high degree of electronic coupling and ultrafast rates of electron injection imply that conventional electron transfer theories can hardly be used. It was suggested [311] that the electron injection rate is controlled "by the electron tunneling barrier, essentially determined by the anchor and spacer group, by the Franck–Condon overlap of the respective vibrational states of reactant and product, and by the escape time for the initially prepared wave packet describing the hot electron". The high density of acceptor states in TiO$_2$ can also contribute to ultrafast injection rates [82]. More detailed quantitative theory is definitely needed.

Much attention has been devoted to the development of optimal photosensitizers of semiconductor electrodes [36, 43]. Ruthenium(II) polypyridine complexes are especially well suited for this purpose. They are strong light absorbers in the visible spectral region and bpy or tpy ligands can be easily derivatized with anchoring groups. Moreover, localization of the excited electron on the ligand which is attached to the semiconductor surface facilitates the electron injection.

### 5.4.7 Excited-state Electron-transfer in Supermolecules Containing Metal–polypyridine Chromophores

Several kinds of supramolecular effects on the redox behavior of metal–polypyridine units were mentioned in Section 5.3.6. Besides influencing photophysical properties, incorporation of metal–bipyridine chromophores into supramolecular structures enables new electron-transfer reactions. Since these processes are dealt with in detail in other chapters, only basic principles and links between the behavior of isolated and supramolecular metal–polypyridine units will be mentioned here.

Encapsulation of a $Ru^{II}$ atom into a caged [109, 110] or hemicaged [299] tris-bipyridine ligand extends the MLCT excited state lifetime and improves photo-stability relative to $[Ru(bpy)_3]^{2+}$, while retaining the fast (diffusion controlled) bimolecular excited state electron transfer reactivity. In contrast, the $[Ru(bpy)_3]^{2+}$ in the core of a dendrimer [248] has about the same inherent lifetime as the free complex but the rate of electron transfer quenching rapidly decreases with increasing the number and size of dendrimer branches.

Intramolecular excited state electron transfer in bi- and polynuclear metal poly-pyridine complex is a much studied process, especially in relation with the development of molecular wires [38, 51, 52, 83, 87, 88, 318]. The rates are dependent on the nature of the bridging groups linking the metal–polypyridine units. These process will be discussed in detail elsewhere in this book.

Multicomponent assemblies in which a photoredox-active metal–polypyridine unit is combined with electron acceptors and/or donors show very rich photo-induced electron transfer reactivity [38]. Such species are often called molecular dyads (triads, tetrads ...). Electron transfer usually occurs from an excited metal–polypyridine unit M to an attached acceptor:

$$M-A \xrightarrow{h\nu} {}^*M-A \xrightarrow{k_{et}} M^+-A^- \tag{21}$$

The back reaction $M^+-A^- \rightarrow M-A$, which regenerates the initial state, often occurs in the inverted Marcus region, which makes it much slower than the forward electron transfer. In this situation, the charge-separated state can be utilized in follow-up reactions (energy conversion, catalysis) or as a bit of chemical information. A long-lived, long-distance charge separation can be produced in molecular triads in which an electron donor and acceptor are attached simultaneously to the photo-active center:

$$D-M-A \xrightarrow{h\nu} D-{}^*M-A \xrightarrow{k_{et,1}} D-M^+-A^- \xrightarrow{k_{et,2}} D^+-M-A^- \tag{22}$$

The electron injection from the MLCT-excited central metal–polypyridine unit M to an acceptor A seems to be kinetically preferred over an alternative $D-{}^*M-A \rightarrow D^+-M^--A$ step, even if the energetics are similar. This is because the excited electron resides at the complex periphery, that is at the polypyridine ligand to which the acceptor group is attached.

The maximum intramolecular electron transfer rates in dyads and triads are not diffusion-limited and very fast electron transfer rates can be expected. The rate constant $k_{et}$ depends on the nature of the link between the active components, which not only hold them together but also provides an electronic coupling. In reality, most of the reactions studied occur on the time scale of a few tens of pico-seconds and longer. The ultrafast rates known from metal–polypyridine/TiO$_2$ interfaces have not been attained in molecular systems. This can be caused by weaker electronic coupling and lower density of acceptor states. In some supramolecules, intramolecular electron transfer occurs concurrently with energy transfer. The overall photodynamics of molecular dyads and triads can thus be quite complicated, involving several excited and charge-separated states. Structural variations in the bridging group can discriminate between electron- and energy-transfer [85, 88].

An occurrence of intramolecular electron transfer shortens the MLCT excited state lifetime: $1/\tau = 1/\tau_{et} + 1/\tau_0$ where $\tau_{et} = 1/k_{et}$ This is demonstrated by the the $^3$MLCT excited state of $[Ru(bpy)_3]^{2+}$ [251], the ca 1 μs lifetime of which is shortened to a few tens of picoseconds when a methylviologen di-cation ($MV^{2+}$) is attached to the 4-position of the bpy ligand by a $-(CH_2)_n-$ chain; $n = 1-8$. This is caused by a bpy$^{\bullet-} \to MV^{2+}$ intramolecular electron transfer which occurs from the $^3$MLCT excited state. The rate constant of $5.9 \times 10^{10}$ s$^{-1}$ ($\tau_{et} = 17$ ps) has been determined for $n = 1$, decreasing with increasing $n$.

There is a great number of photo-redox active supramolecular systems which utilize metal–polypyridine moieties. A considerable knowledge about electron transfer mechanism was obtained by studying so-called chromophore-quencher complexes of the type $[Re(Q)(CO)_3(bpy)]^{n+}$ in which Q is an electron acceptor, e.g. *N*-Me-4,4'-bipyridinium ($MQ^+$, $n = 2$) [319–322], or an electron donor, for example phenothiazine-4-pyridine (py-PTZ, $n = 1$) [254, 255, 323–325], see Appendix for formulas. Thus, Re $\to$ bpy excitation of $[Re^I(MQ^+)(CO)_3(bpy)]^{2+}$ prepares a $[Re^{II}(MQ^+)(CO)_3(bpy^{\bullet-})]^{2+}$ MLCT state which undergoes a bpy$^{\bullet-} \to MQ^+$ electron transfer in 8–15 ps, depending on the solvent relaxation time [322]. Excitation of $[Re(py-PTZ)(CO)_3(bpy)]$ produces initially an MLCT state $[Re^{II}(py-PTZ)(CO)_3(bpy^{\bullet-})]^+$ which is rapidly ($4 \times 10^9$ s$^{-1}$) converted to $[Re(py-PTZ^{\bullet+})(CO)_3(bpy^{\bullet-})]^+$ by a PTZ $\to$ Re$^{II}$ electron transfer. Following inter-ligand electron transfer bpy$^{\bullet-} \to$ PTZ$^{\bullet+}$ regenerates the ground state. It is relatively slow ($4 \times 10^7$ s$^{-1}$) because of a highly inverted character. Detailed investigation of its kinetics helped to understand the dynamical influence of structural and medium factors and to highlight similarities between inverted electron transfer and non-radiative decay of MLCT excited states. $[Re(py-D)(CO)_3(bpy)]^+$ complexes in which D is an electron donor that undergoes fragmentation on oxidation (e.g. α-aminoalcohols or 1,2-diamines) show an irreversible photochemistry, whereby the D $\to$ Re$^{II}$ intramolecular electron transfer, that follows Re $\to$ bpy MLCT excitation, causes bond breaking within the oxidized D group [326–329].

An interesting case of a charge separation across an oligopeptide chain has been found [330] for the triad $[PTZ-Ru^{2+}-MV^{2+}]$ shown in Figure 7. Excitation of the Ru(bpy) chromophore triggers a series of intramolecular electron-transfer reactions that eventually produce a charge separated state $[PTZ^{\bullet+}-Ru^{2+}-MV^{\bullet+}]$, which stores 1.17 eV and lives for 108 ns. Another interesting example of a charge separated state is provided by $[Ru(bpy)_2(bpy-C_{60})]^{2+}$ where a rigid steroidal spacer links bpy and $C_{60}$ [252]. A charge separated state $Ru^{3+}-C_{60}^{\bullet-}$ is formed after MLCT excitation of the Ru(bpy) center. In $CH_3CN$, the charge-separated state undergoes recombination to produce first a $Ru^{2+}-^3C_{60}$ excited state which then decays to the ground state. In contrast, a dynamic equilibrium between a $Ru^{3+}-C_{60}^{\bullet-}$ charge separated state and a *Ru–$C_{60}$ excited state exists in $CH_2Cl_2$. This is a very rare example of a reversible excited state transfer.

Intriguing excited state electron transfer dynamics are displayed by Cu$^I$-complexed rotaxanes which combine several photo-redox active units [92, 331, 332]. For example, a rotaxane $[Cu(catphen)(phen-9,7-(C_{60})_2)]^+$ based on a Cu$^I$(phen)$_2$

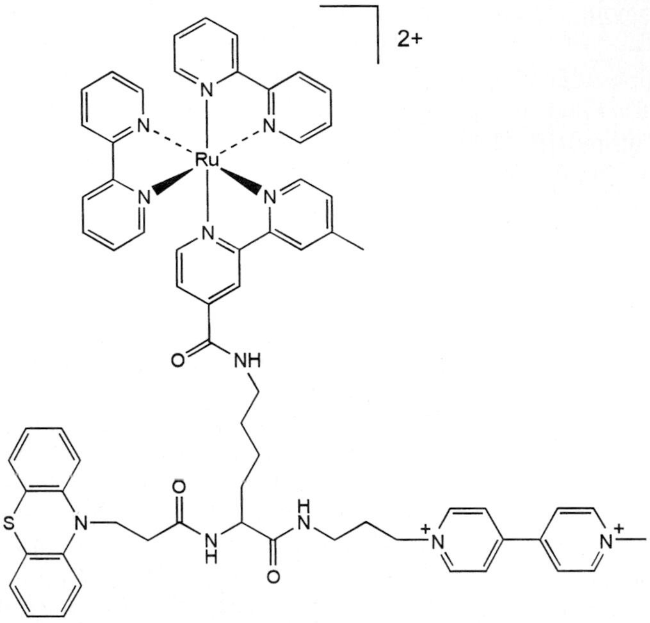

**Figure 7.** Structure of the triad [PTZ–Ru²⁺–MV²⁺]. A charge separated state [PTZ˙⁺–Ru²⁺–MV˙⁺] is formed upon excitation of the Ru(bpy)₃²⁺ unit [330].

central unit and two $C_{60}$ stoppers, shows quenching of both the $^1C_{60}$ and $^3$MLCT Cu(phen)₂ excited states: the $^1C_{60}$ state rapidly decays by an energy transfer ($k_{ent} = 1.6 \times 10^9$ s⁻¹) to the [Cu(phen)₂]⁺ unit while the $^3$MLCT excited state of Cu(phen)₂ undergoes an electron transfer ($k_{et} = 5.8 \times 10^9$ s⁻¹) to $C_{60}$. In another example, fast electron transfer between *[Ru(bpy)₃]²⁺ and attached electron acceptor (-4,4′-bipyridinium or -2,7-diazapyrenium) is responsible for a light-driven threading and dethreading of a pseudorotaxane [332].

Dyads composed of a linked porphyrin and Ru(bpy)₃²⁺ show a very interesting photochemistry [85, 86, 89]. In assemblies with Zn-porphyrins (ZnP), the [Ru(bpy)₃]²⁺ unit plays the role of a ground state electron acceptor:

$$ZnP–Ru \xrightarrow{h\nu} {}^*ZnP–Ru \xrightarrow{k_{et}} ZnP^+–Ru^- \qquad (23)$$

For a ZnP-[Ru(bpy)₃]²⁺ dyad where the linkage is provided by an amide group, intramolecular electron transfer is fast enough ($k_{et} = 6 \times 10^{11}$ s⁻¹; $\tau_{et} = 1.7$ ps) to occur from the upper, $S_2$ excited state of ZnP ($\tau_0 = 1.6$ ps) prior to its electronic relaxation to the long-lived $S_1$ state [253]. This is a very rare example of a capture of a high-lying excited state by a chemical reaction. The relaxed $S_1$-excited ZnP unit also reduces attached [Ru(bpy)₃]²⁺, but with a 100-times smaller rate [253]. In other systems, Zn and Au porphyrins were used as terminal groups of complexed rotaxanes based on a central [Cu(catpen)₂]⁺ unit [92, 333]. Excitation of these species triggers several intramolecular electron transfers between the porphyrin units, as well as those which change the Cu oxidation state. These reactions occur in the picosecond time domain. Intramolecular energy transfer is a common competing process in porphyrin–(metal–polypyridine assemblies) [86].

Finally, it is important to note (Section 5.3.6) that electrochemistry and UV–Vis absorption spectra of molecular dyads or triads based on metal polypyridines show that electronic interactions between the components of the systems discussed above are too small to influence ground-state behavior. Nevertheless, they are sufficient to allow for very fast intramolecular electron transfer when electronically excited. In fact electronic coupling of 0.002–0.005 eV would be quite enough, but hardly detectable electrochemically. Detailed studies of electrochemistry and spectroscopy of these supramolecular systems and their components are, nevertheless, essential for the understanding of the energetics of photoinduced intramolecular electron and energy transfer reactions.

### 5.4.8 Chemiluminescence

If an electron-transfer reaction occurs with a driving force that is higher than the excited state energy of its product, that species can be formed in an excited state, instead of the ground state. Formation of an excited state occurs with a smaller driving force $(-\Delta G° - E_{00})$ than of the corresponding ground state $(-\Delta G°)$. For large $-\Delta G°$ values, the ground state formation is slow because of the Marcus inverted effect. Reaction to the excited state then becomes kinetically competitive with the ground state formation. It is accompanied by light emission from the excited state formed. The chemiluminescence quantum yield increases with increasing ratio between the rate constants of the excited and ground state formation. Strong chemiluminescence can be expected for complexes which have high emission quantum yields and whose oxidized and/or reduced forms are chemically stable. This often is the case of polypyridine complexes of $d^6$-metals ($Re^I$, $Ru^{II}$, $Os^{II}$) and of $Cr^{III}$ whose excited state energies are in the range ca 1.5–2.3 eV, accessible by homogeneous or electrochemical redox reactions [32, 266, 277, 334–339].

For example, an intense orange chemiluminescence from the $^3$MLCT state of $[Ru(bpy)_3]^{2+}$ has been observed during $[Ru(bpy)_3]^{3+}$ reduction [334] by $C_2O_4^{2-}$ or $[Ru(bpy)_3]^+$ oxidation [340] by $S_2O_8^{2-}$. These reactions follow complex multi-step mechanisms. The actual steps responsible for chemiluminescence are:

$$[Ru(bpy)_3]^{3+} + CO_2^{·-} \rightarrow {}^*[Ru(bpy)_3]^{2+} + CO_2 \tag{24}$$

$$[Ru(bpy)_3]^+ + SO_4^{·-} \rightarrow {}^*[Ru(bpy)_3]^{2+} + SO_4^{2-} \tag{25}$$

The $[Ru(bpy)_3]^{3+}$ and $[Ru(bpy)_3]^+$ complexes can be generated by another chemical reaction from $[Ru(bpy)_3]^{2+}$, for example with $PbO_2$ or Mg, respectively. The whole chemiluminescent reaction can then be run cyclic, as a catalytic process, shown Figure 8. Alternatively, the oxidized or reduced Ru complex is generated electrochemically at an electrode inserted to the solution, giving rise to electrochemiluminescence.

MLCT excited states of metal–polypyridine complexes are also produced by recombination of their oxidized and reduced forms. For example, the reaction

$$[Ru(bpy)_3]^{3+} + [Ru(bpy)_3]^+ \rightarrow 2[Ru(bpy)_3]^{2+} \tag{26}$$

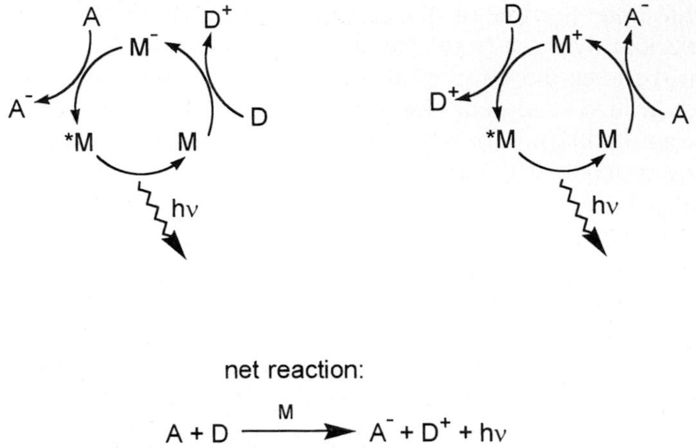

net reaction:

$$A + D \xrightarrow{M} A^- + D^+ + h\nu$$

**Figure 8.** Chemiluminescence in redox-catalytic cycles. The metal–polypyridine complex M acts as light-emission sensitizer, LES [74, 266]. D and A can be either a chemical reductant and oxidant, respectively, or an electrode polarized at appropriate potential.

occurs with a driving force $-\Delta G° = 2.61$ eV, well above the 2.12 eV energy of the emissive $^3$MLCT state. Hence, a parallel reaction takes place:

$$[\text{Ru(bpy)}_3]^{3+} + [\text{Ru(bpy)}_3]^{+} \rightarrow {}^*[\text{Ru(bpy)}_3]^{2+} + [\text{Ru(bpy)}_3]^{2+} \qquad (27)$$

with a nearly 100 % efficiency [341, 342]. Similar chemiluminescent reactions were observed for $Os^{II}$ [336] and $Re^{I}$ [60, 335] polypyridine complexes. Technically, they are performed at an electrode whose potential is rapidly switched between the values at which oxidation and reduction of the parent complex occurs.

Chemiluminescence quantum yields were studied as a function of the driving force for series of reactions between $[\text{Ru(N,N)}_3]^{3+}$ and $[\text{Co(N,N)}_3]^{+}$. The yield of $^*[\text{Ru(N,N)}_3]^{2+}$ emission decreases from 0.31 to 0.07 with decreasing driving force [339].

### 5.4.9 Applications of Photochemical Electron Transfer

The strong oxidizing and reducing power of MLCT excited states of many metal ($Ru^{II}$, $Os^{II}$, $Re^{I}$, $Cu^{I}$) polypyridine complexes is employed in various photocatalytic processes, shown schematically in Figure 3. The excited polypyridine complex molecule can react directly with the substrate. More often, the excited state is first quenched by a fast reaction with a sacrificial electron acceptor or donor to form a powerful ground-state oxidant or reductant $M^+$ and $M^-$, respectively (Figure 3). These species then react with the substrate in a subsequent step. In this case, the photochemical step serves to invest an extra energy and to generate the reactive species in a sufficient concentration. General aspects of homogeneous photocatalytic reactions and the interference with detrimental short-cut reactions were discussed in the Section 5.4.5.

Many photocatalytic cycles were developed in relation to light energy conversion [40, 41], especially aiming at water photodecomposition to $H_2$ and $O_2$. However, the low efficiency and poor long-term stability make most of homogeneous and microheterogeneous photocatalytic cycles unsuitable for practical energetic or synthetic applications. Sensitization of semiconductors (Section 5.4.6) is a much more promising way toward photochemical light energy conversion [36, 43, 76–78]. Photoelectrochemical cells operating on the principle shown in Figure 6 which, using the sensitizer $[Ru(4,4'-(COOH)_2-bpy)_2(NCS)_2]$, achieve an efficiency of nearly 10 % and a reasonable long-term stability [36]. The ultrafast rate of electron injection from excited sensitizers to $TiO_2$ effectively prevents any photochemical decomposition and minimizes losses caused by excited state decay. Out of many different sensitizers examined, Ru polypyridine complexes show the biggest promise [36, 43]. The quest for more efficient and stable sensitizers still continues. The discovery of ultrafast rates of electron injection has directed researcher's attention also towards cheaper and more environmentally friendly Fe complexes [304].

Photochemical generation of strong reductants have been employed in photocatalytic reduction of $CO_2$ [30, 31, 44, 139, 343–346] to CO or formate, reduction of $H^+$ [30] to $H_2$, or in photocatalyzed water-gas-shift reaction [44]. The mechanisms are very similar to those of the corresponding electrocatalyzed reactions (Section 5.3.7), only the active reductants are generated photochemically, instead of electrochemically. For example, $[Re(Cl)(CO)_3(bpy)]$ is reduced photochemically by the sacrificial electron donor triethanolamine to $[Re(Cl)(CO)_3(bpy)]^{\cdot-}$ which loses $Cl^-$ to give $[Re(CO)_3(bpy)]^{\cdot}$. The latter species coordinates and reduces $CO_2$, with an eventual photocatalyst regeneration. On the other hand, $[Ru(bpy)_3]^+$, formed by photoreduction of $[Ru(bpy)_3]^{2+}$, is substitutionally inert. Hence, $CO_2$ reduction photosensitized by $[Ru(bpy)_3]^+$, involves two catalytic cycles, Figure 9. In the first cycle, the active catalyst of $CO_2$ reduction is photocatalytically reduced using $[Ru(bpy)_3]^{2+}$ and $CO_2$ is then reduced in the second cycle, which involves ground-state reactions only.

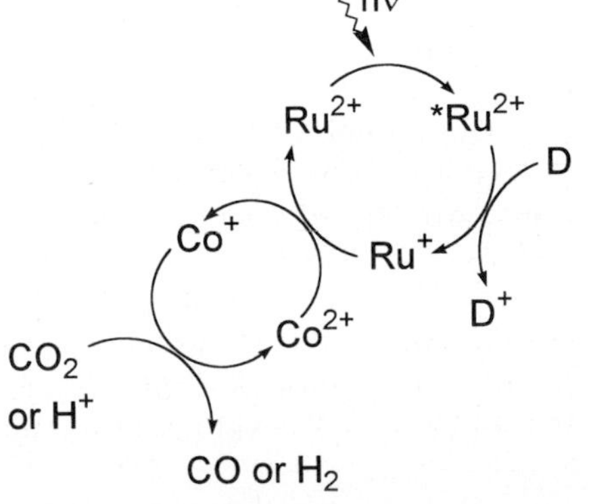

**Figure 9.** Photocatalytic reduction of $CO_2$ or $H^+$ using $[Ru(bpy)_3]^{2+}$ photosensitizer and $[Co(bpy)_3]^{2+}$ catalyst. The same scheme applies to photocatalytic reduction of $CO_2$ with different catalysts, e.g. $[Ru(bpy)_2XY]^n$ (X, Y = CO, $Cl^-$), or $[Ni(cyclam)]^{2+}$, instead of Co.

Electron-transfer reactions triggered by optical excitation of metal–polypyridines have very important applications in studying proteins and DNA. They are the basis of the flash-quench technique [347–351] used to investigate kinetics of electron-transfer reactions of metalloproteins, for example cytochromes, myoglobins or azurines: The complex $[Ru^{II}(bpy)_2(im)(His-)]^{2+}$; im = imidazole is attached to a specific site at a protein surface though a histidine group (His-) and excited by a laser flash. The MLCT excited state created is then rapidly quenched by an electron-transfer reaction with a Fe or Cu center of the protein. The rate of this electron transfer and/or of the back reaction is then measured. Alternatively, the excited $*[Ru^{II}(bpy)_2(im)(His-)]^{2+}$ unit at the protein surface is quenched by an oxidizing or reducing quencher presents in the solution. Again, various forward- and back electron-transfer reactions between the oxidized or reduced surface complex and the metalloprotein occur and their kinetics are measured. Since the Ru complex can be attached at different positions of the protein, it is possible to follow the electron-transfer kinetics as a function of the redox pathway through the protein. Derivatives of $[Ru^{II}(bpy)_2(im)(His-)]^{2+}$ containing substituted bpy or phen ligands can be used to vary redox potentials and, hence, the electron transfer energetics [349]. Besides that, the strong excited-state oxidant $Re^{I}(CO)_3(phen)$ can be attached to surface histidine groups of metalloproteins, e.g. azurines [132]. Another strategy uses $[Ru(bpy)_3]^{2+}$ attached covalently to an amino acid chain of cytochrome c through an -alanine-4-bpy link [352]. Electron injection from the MLCT excited state of $[Ru(bpy)(bpy-alanin-]^{2+}$ to cyt-$Fe^{3+}$ and the back reaction were found to proceed about ten times faster than those observed for histidine-bound $[Ru^{III}(bpy)_2(im)(His-)]^{2+}$. This is because the electron transfer now occurs directly from/to the Ru(bpy) chelate ring, without the intervening imidazole group. The flash-quench and direct injection techniques are also used to induce protein folding [353].

In a very interesting extension of the flash-quench technique, electron transfer probes have been developed [354] (Figure 10) to oxidize or reduce an active site of cytochrome P450, which is deeply buried within the protein. Reaction between the probe and P450 forms a complex in which the alkyl chain penetrates through the protein and the end-group of the probe interacts with the active site. Laser excitation of this $Ru^{II}$-P450 complex in the presence of the quencher $[Co^{III}(NH_3)_5Cl]^{2+}$ oxidizes the $Ru(bpy)_3^{2+}$ unit at the protein surface and triggers a rapid $(6 \times 10^3 \text{ s}^{-1})$ electron transfer P450 $\rightarrow Ru^{III}$, which occurs along the alkyl chain. The structure and the chemistry of the oxidized cytochrome P450 active site is then studied spectroscopically. Similarly, chemistry of the reduced active site of cytochrome P450 can be explored if a reducing quencher (*para*-methoxy-$N,N$-dimethylaniline) is used. Electron injection $Ru^+ \rightarrow$ P450 then occurs with a rate constant of $2 \times 10^4 \text{ s}^{-1}$.

Polypyridine complexes containing a phen ligand or its derivatives (4,7-$Ph_2$-phen), dppz, tap, hat or other extended polypyridines or polyazines intercalate into DNA strands at specific sites. If the DNA-bound polypyridine complex is a sufficiently strong excited-state oxidant, a photoinduced guanine oxidation takes place with a concomitant DNA strand cleavage. This reaction has indeed been observed for $[Ru(tap)_3]^{2+}$ or $[Ru(hat)_3]^{2+}$ and some of their derivatives [355]. Photoexcited $[Rh(phi)_2(bpy)]^{3+}$ oxidizes DNA by an H-atom abstraction from a sugar moiety,

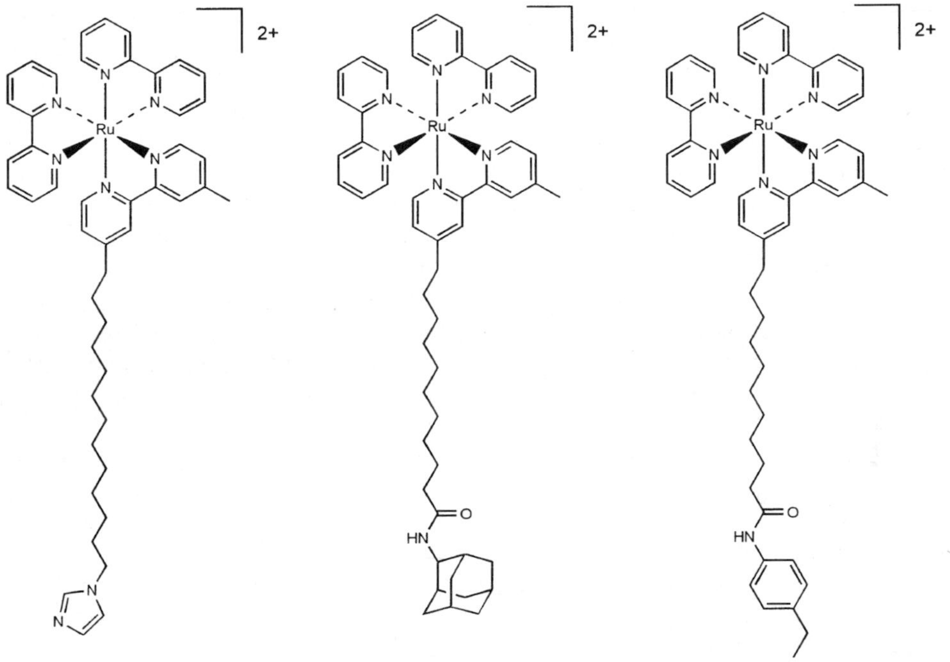

**Figure 10.** Structures of electron-transfer probes used to examine the chemistry of enzyme active sites which are deeply buried within the protein, in particular the cytochrome P450 [354].

again causing a strand cleavage [356]. However, in the latter case, the reaction occurs from the dpi → Rh ligand to metal charge transfer excited state, without a direct involvement of the bpy ligand. Interactions between DNA and metal–polypyridines can find application as medical diagnostic tools or light-activated anti-cancer drugs. For example, a photoinduced thymine repair in the DNA helix catalyzed by $[Rh(phi)_2(4,4'-Me_2-bpy)]^{3+}$ has been observed [357].

Electron injection from MLCT-excited Ru-polypyridine complexes are used to investigate electron transfer along DNA strands, that is to decide whether DNA can behave as a molecular wire [358–360]. In these studies, derivatives of $[Ru(phen)_2(dppz)]^{2+}$ act as excited-state electron donors and $[Rh^{III}(phi)_2(bpy)]^{3+}$ as a ground-state electron acceptor. Both complexes are anchored at different DNA sites and the rate of *Ru → Rh photoinduced electron transfer is measured. In another study [361], a $[Ru^{II}(bpy)_2(im)(NH_2-)]^{2+}$ unit attached to a terminal ribose of a DNA duplex acted as an excited-state oxidant toward a $[Ru^{II}(NH_3)_4(py)(NH_2-)]^{2+}$ unit attached at the other end.

A tremendous research effort is being devoted to intramolecular electron transfer in all kinds of supermolecules containing metal–polypyridine components, Section 5.4.7. At this moment, their applications in luminescence sensors is developed most.

It is based on quenching of the emission from a $Ru^{II}$ or $Re^{I}$ polypyridine part of the supermolecule by electron transfer with a redox-active species recognized and bound at the receptor site. Excited state electron transfer can be utilized even in pH-sensors, provided that pH affects the redox potential of the receptor group, e.g. a phenolate. Supermolecules in which a Ru-polypyridine chromophore is co-valently linked with a redox-catalyst can utilize photoinduced excited state electron transfer to run energetically up-hill reactions. This is demonstrated, for example, by a $[Ru(phen)_2(phen-CH_2-cyclam-Ni^{II})]^{4+}$ complex [362]. Upon excitation, the Ru(phen) center injects an electron into the Ni(cyclam) moiety, at which catalytic reduction of $CO_2$ proceeds. The $Ru^{II}$ oxidation state is regenerated by reduction with ascorbate present in the solution. Hence, the overall reaction amounts to reduction of $CO_2$ to CO by ascorbate, photocatalyzed by $[Ru(phen)_2(phen-CH_2-cyclam-Ni^{II})]^{4+}$. The CO yield is much larger than if isolated $[Ru(phen)_3]^{2+}$ photosensitizer and $[Ni(cyclam)]^{2+}$ catalysts are used in a mixture (Figure 9). The oxidizing power of MLCT-excited $Ru(bpy)_3^{2+}$ unit has been explored in an assembly where a mono- or a tri-nuclear Mn phenolate unit is linked to one of the bpy ligands [363]. This system is studied as a model of the photosystem II, since the Mn part is a potential water oxidation catalyst.

Polynuclear complexes, molecular dyads, triads, and other supermolecules composed of redox- and photo-active metal polypyridine units have a great promise as components of future molecular electronic or photonic devices as optical switches, relays, memories, etc. [38, 46].

## 5.5  Conclusions

Transition metal polypyridine complexes are highly redox-active, both in their electronic ground- and excited states. Their electron transfer reactivity and properties can be fine-tuned by variations in the molecular structure and composition. They are excellent candidates for applications in redox-catalysis and photocatalysis, conversion of light energy into chemical or electrical energy, as sensors, active components of functional supramolecular assemblies, and molecular electronic and photonic devices.

The rich electron-transfer behavior of transition metal polypyridine complexes originates in a unique combination of several factors:

– Complexes of neutral polypyridine ligands with metals in 'usual' oxidation states (e.g. $Cr^{III}$, $Mn^{II}$, $Fe^{II}$, $Co^{III}$, $Co^{II}$, $Ni^{II}$, $Cu^{I}$, $Ru^{II}$, $Rh^{III}$, $Re^{I}$, $Os^{II}$, $Ir^{III}$, $Pt^{II}$) are stable and rather unreactive. This makes them convenient entries into the rich redox chemistry and photochemistry.
– Free polypyridine ligands are reduced in (at least) two successive one-electron steps.
– Redox activity of free polypyridines is retained upon coordination.

- Polypyridines are strongly coordinating ligands in all their oxidation states (i.e. as neutral species, radical anions or dianions.)
- Electron delocalization between the metal atom and a polypyridine ligand and between polypyridine ligands within the coordination sphere is weak. (Complexes of first-row transition metals in intermediate oxidation states and complexes of a few unusual polypyridines (abpy) are the only exceptions.)
- Oxidation of polypyridine complexes of $d^6$-metal atoms is predominantly metal-localized while the reductions are polypyridine-localized. This defined localization, together with a small inter-ligand electronic interaction, gives rise to extensive redox series, that is series of polypyridine complexes of the same chemical composition, but different number and localization of electrons.
- Labilization of an ancillary ligand occurs upon reduction of some substituted polypyridine complexes, allowing for a substrate coordination and catalytic reduction.
- The predominant metal- and polypyridine-localization of ground-state oxidation and reduction, respectively, is paralleled by the localized character of metal $\rightarrow$ polypyridine MLCT electronic excitation, whereby the excited electron resides on a single polypyridine ligand and the positive hole on the metal atom (more precisely, on the metal-ancillary ligands moiety).
- Excited state redox properties reflect those of the ground state. Excited states of polypyridine complexes are thermodynamically strong oxidants and reductants.
- The localized character of the first oxidation, reduction, and MLCT excitation allows the formulation of linear relationships between (i) the complex and free ligand reduction potentials, (ii) ground- and excited-state redox potentials, and, (iii) the difference between the ground-state first oxidation and first reduction potentials and MLCT excited-state energy. These relations are of a great importance in the design of polypyridine complexes for specific purposes.
- Electron-transfer reactions of polypyridine complexes occur with a small reorganization energy. They are very fast, depending on the reaction energetics and electronic coupling.
- In supramolecular systems, electronic interactions between metal–polypyridine and other redox-active or units are too small to perturb ground-state electrochemical and spectroscopic properties but are sufficient to enable very fast intramolecular electron-transfer reactions upon excitation.
- When connected to other components of a supramolecular assembly or to a semiconductor electrode through a polypyridine ligand, a metal–polypyridine unit is kinetically especially suited to inject an electron from its MLCT excited state, that is to act as an excited state reductant. Ultrafast rates can be reached.

**Appendix**

Formulas of selected polypyridine ligands and the abbreviations used.

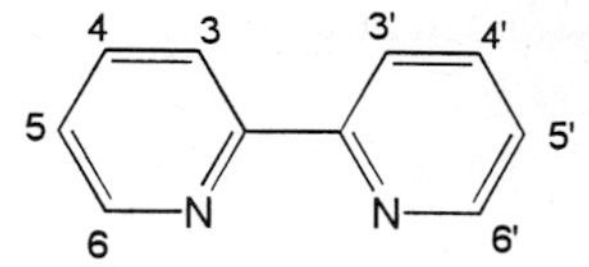

bpy

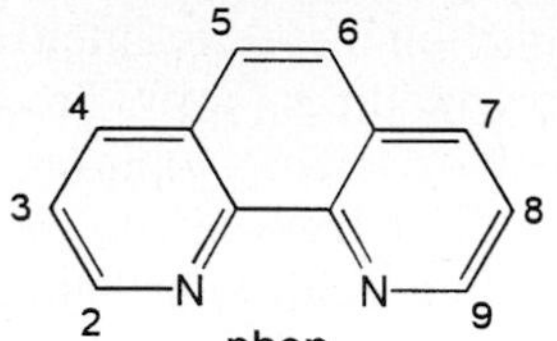

phen

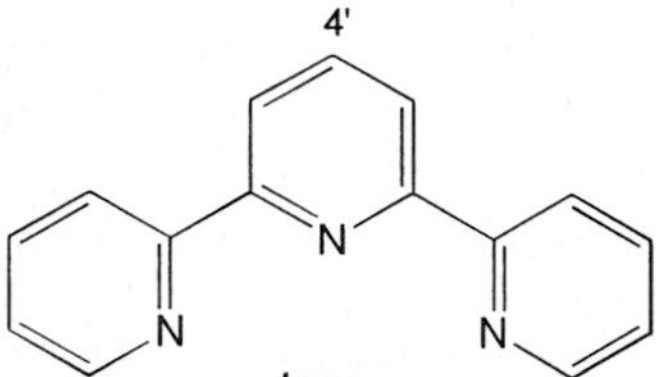

trpy

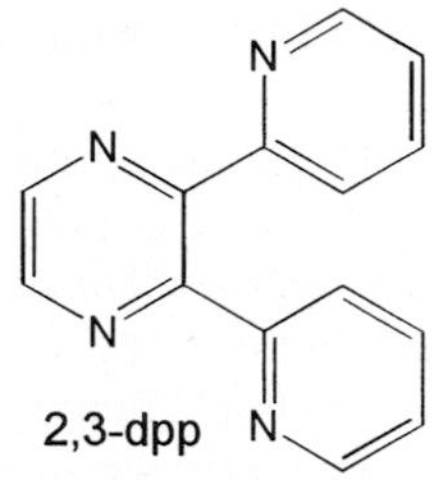

2,3-dpp

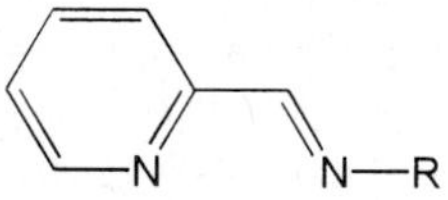

R-PyCa

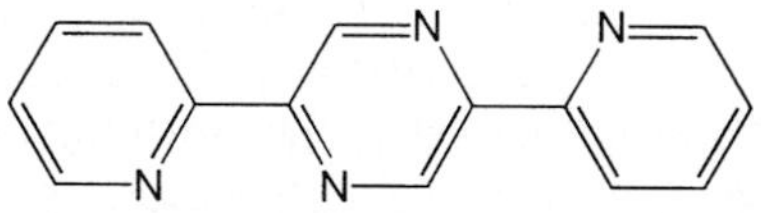

2,5-dpp

R-DAB

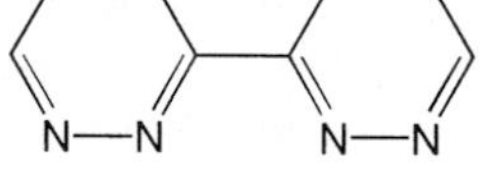

bpdz

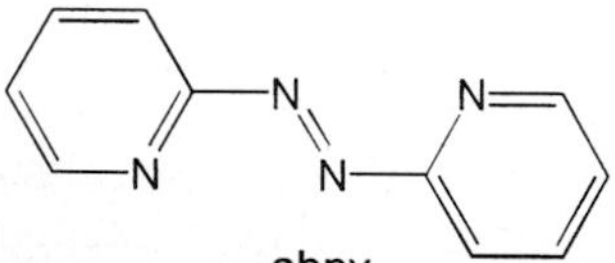

abpy

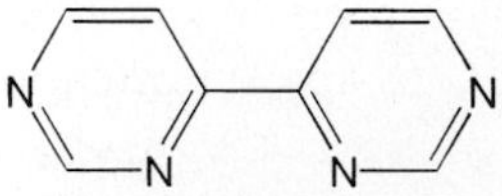

bpm

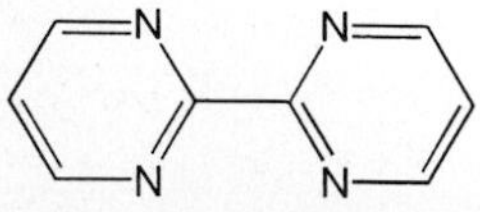

bpym

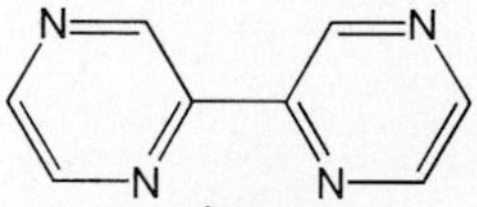

bpz

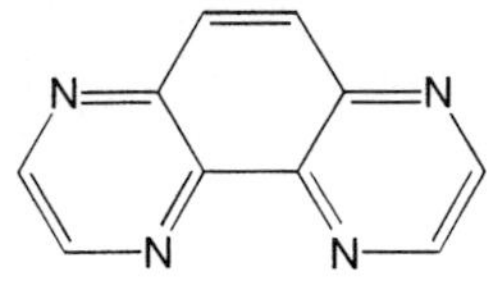

tap

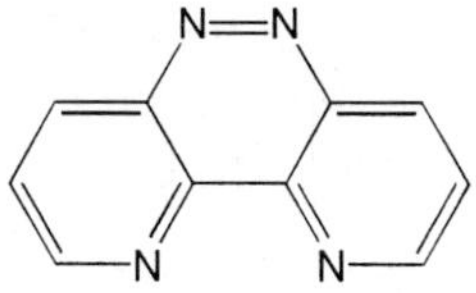

taphen

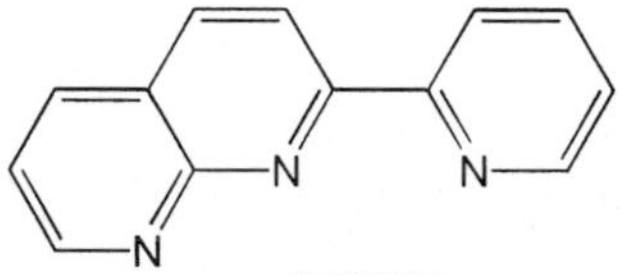

pynapy

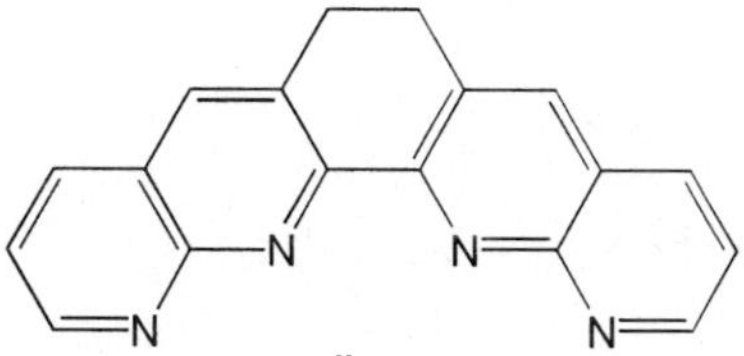

dinapy

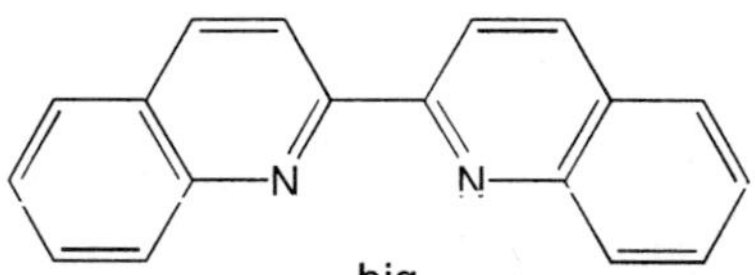

biq

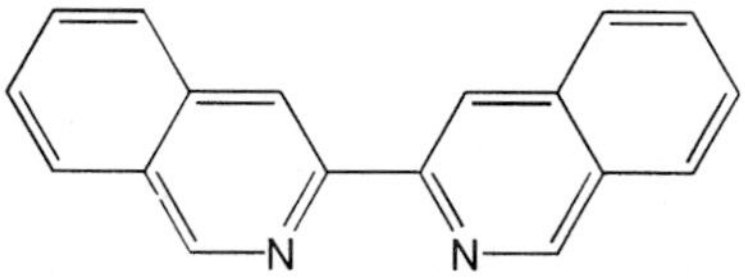

*i*-biq

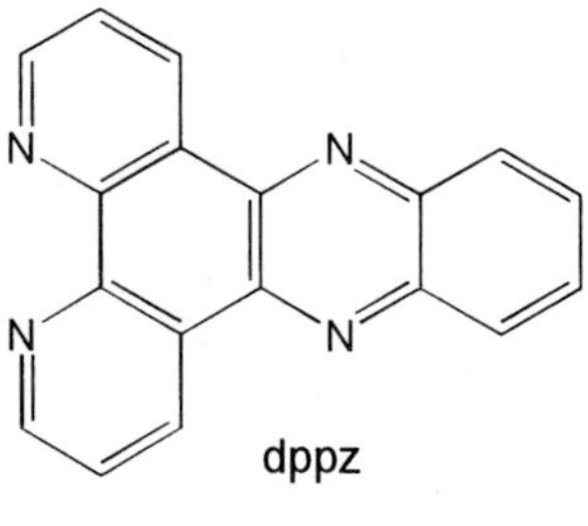

dppz

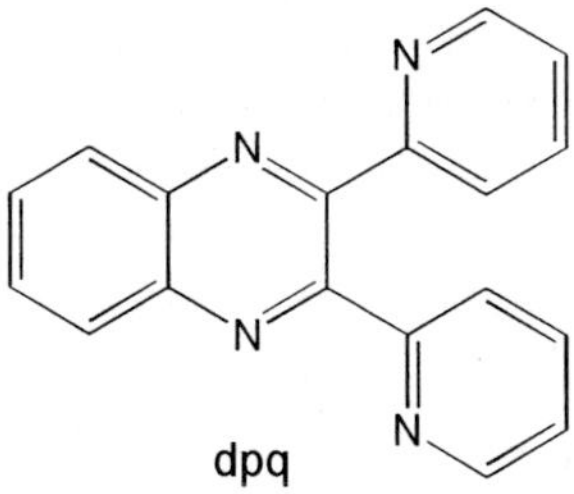

dpq

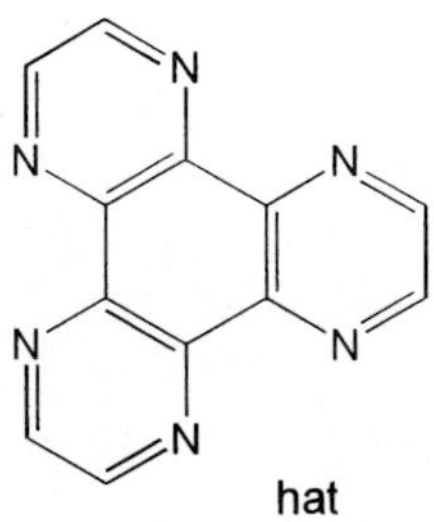

hat

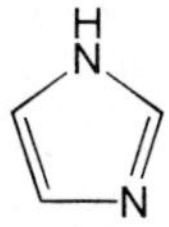

im

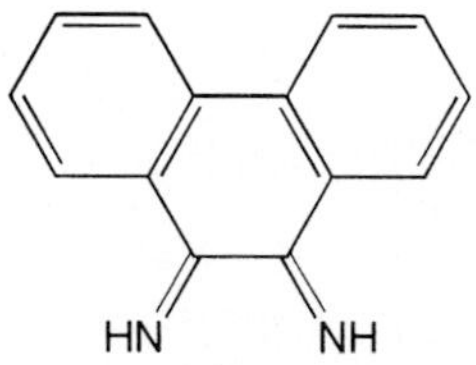

dpi

danphen

catphen
(The actual number of O–O units may differ.)

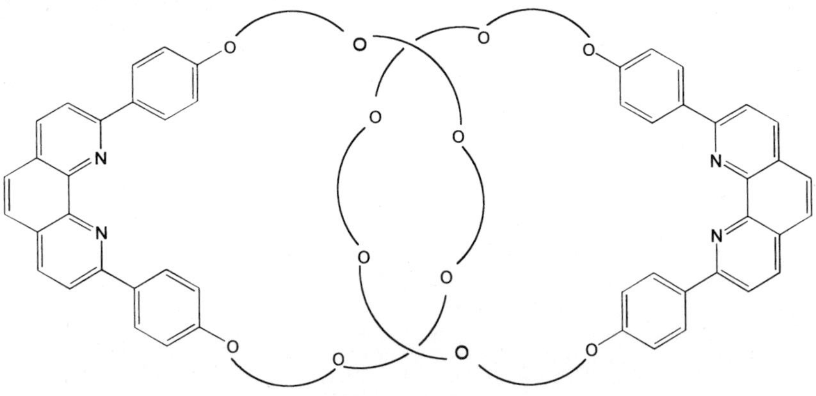

Two entwinned catphen ligands

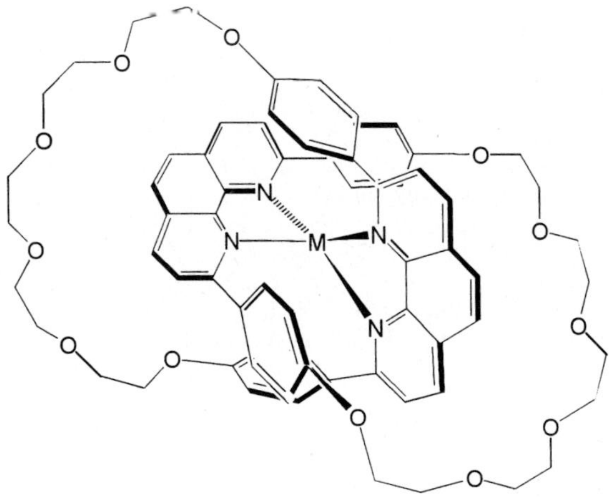

Pseudotetrahedral [M(catphen)₂] complex

PTZ                    py-PTZ

MV$^{2+}$                   MQ$^{+}$

# References

1. A. A. Vlček, *Rev. Chim. Minérale* **1968**, *5*, 299–316.
2. E. König, S. Herzog, *J. Inorg. Nucl. Chem.* **1970**, *32*, 585–599.
3. M. K. DeArmond, K. W. Hanck, D. W. Wertz, *Coord. Chem. Revs.* **1985**, *64*, 65–81.
4. A. A. Vlček, *Coord. Chem. Rev.* **1982**, *43*, 39–62.
5. S. Záliš, M. Krejčík, V. Drchal, A. A. Vlček, *Inorg. Chem.* **1995**, *34*, 6008–6014.
6. A. A. Vlček, *Electrochim. Acta* **1968**, *13*, 1063–1078.
7. A. A. Vlček, *IUPAC—Coord. Chem.* **1981**, *21*, 99–112.
8. A. A. Vlček, *Rev. Chim. Minérale* **1983**, *20*, 612–627.
9. A. B. P. Lever, E. S. Dodsworth, in E. I. Solomon, A. B. P. Lever (Eds.): *Inorganic Electronic Structure and Spectroscopy, Volume II: Applications and Case Studies,* J. Wiley&Sons, Inc. 1999, p. 227–289.
10. E. M. Kober, J. V. Caspar, R. S. Lumpkin, T. J. Meyer, *J. Phys. Chem.* **1986**, *90*, 3722–3734.
11. J. V. Caspar, T. J. Meyer, *Inorg. Chem.* **1983**, *22*, 2444–2453.
12. L. A. Worl, R. Duesing, P. Chen, L. Della Ciana, T. J. Meyer, *J. Chem. Soc. Dalton Trans.* **1991**, 849–858.
13. B. S. Brunschwig, S. Ehrenson, N. Sutin, *J. Phys. Chem.* **1986**, *90*, 3657–3668.
14. P. Chen, T. J. Meyer, *Chem. Rev.* **1998**, *98*, 1439–1477.
15. E. S. Dodsworth, A. A. Vlček, A. B. P. Lever, *Inorg. Chem.* **1994**, *33*, 1045–1049.
16. B. Gaš, J. Klíma, S. Záliš, A. A. Vlček, *J. Electroanal. Chem.* **1987**, *222*, 161–171.
17. J. Hanzlík, L. Pospíšil, A. A. Vlček, M. Krejčík, *J. Electroanal. Chem.* **1992**, *331*, 831–844.
18. M. Krejčík, A. A. Vlček, *J. Electroanal. Chem.* **1991**, *313*, 243–257.
19. M. Krejčík, A. A. Vlček, *Inorg. Chem.* **1992**, *31*, 2390–2395.
20. D. Miholová, A. A. Vlček, *J. Organomet. Chem.* **1985**, *279*, 317–326.
21. D. Miholová, B. Gaš, S. Záliš, J. Klíma, A. A. Vlček, *J. Organomet. Chem.* **1987**, *330*, 75–84.
22. A. A. Vlček, *Nature* **1957**, *180*, 753.
23. A. A. Vlček, *Z. Phys. Chem.* **1958**, 143–151.
24. A. A. Vlček, A. Rusina, *Proc. Chem. Soc.* **1961**, 161.
25. A. A. Vlček, *Nature* **1961**, *189*, 393–394.
26. A. A. Vlček, *Progr. Inorg. Chem.* **1963**, *5*, 211–384.

27. A. A. Vlček, *Pure and Appl. Chem.* **1965**, *10*, 61–70.
28. A. A. Vlček, E. S. Dodsworth, W. J. Pietro, A. B. P. Lever, *Inorg. Chem.* **1995**, *34*, 1906–1913.
29. A. A. Vlček, *Coord. Chem. Rev.* **2000**, *200–202*, 979–990.
30. N. Sutin, C. Creutz, E. Fujita, *Comments Inorg. Chem.* **1997**, *19*, 67–92.
31. F. R. Keene, C. Creutz, N. Sutin, *Coord. Chem. Rev.* **1985**, *64*, 247–260.
32. V. Balzani, F. Bolletta, *Comments Inorg. Chem.* **1983**, *2*, 211–226.
33. P. D. Beer, S. W. Dent, T. J. Wear, *J. Chem. Soc., Dalton Trans.* **1996**, 2341–2346.
34. V. W.-W. Yam, V. W.-M. Lee, F. Ke, K.-W. M. Siu, *Inorg. Chem.* **1997**, *36*, 2124–2129.
35. P. D. Beer, P. A. Gale, G. Z. Chen, *J. Chem. Soc., Dalton Trans.* **1999**, 1897–1909.
36. K. Kalyanasundaram, M. Grätzel, *Coord. Chem. Rev.* **1998**, *177*, 347–414.
37. V. Balzani, F. Bolletta, M. T. Gandolfi, M. Maestri, *Top. Curr. Chem.* **1978**, *75*, 1–64.
38. V. Balzani, F. Scandola, *Supramolecular Photochemistry*, Ellis Horwood, Chichester 1991.
39. V. Balzani, M. Maestri, in K. Kalyanasundaram, M. Grätzel (Eds.): *Photosensitization and Photocatalysis Using Inorganic and Organometallic Compounds*, Kluwer Academic Publishers, Dordrecht 1993, p. 15–50.
40. E. Amouyal, *Solar Energy Materials and Solar Cells* **1995**, *38*, 249–276.
41. A. J. Bard, M. A. Fox, *Acc. Chem. Res.* **1995**, *28*, 141–145.
42. K. Kalyanasundaram, in K. Kalyanasundaram, M. Grätzel (Eds.): *Photosensitization and Photocatalysis Using Inorganic and Organometallic Compounds*, Kluwer Academic Publishers, Dordrecht 1993, p. 113–160.
43. M. Grätzel, K. Kalyanasundaram, in K. Kalyanasundaram, M. Grätzel (Eds.): *Photosensitization and Photocatalysis Using Inorganic and Organometallic Compounds*, Kluwer Academic Publishers, Dordrecht 1993, p. 247–272.
44. R. Ziessel, in K. Kalyanasundaram, M. Grätzel (Eds.): *Photosensitization and Photocatalysis Using Inorganic and Organometallic Compounds*, Kluwer Academic Publishers, Dordrecht 1993, p. 217–245.
45. J. N. Demas, B. A. DeGraff, P. B. Coleman, *Anal. Chem.* **1999**, *71*, 793A–800A.
46. V. Balzani, F. Scandola, in D. N. Reinhoudt (Ed.): *Comprehensive Supramolecular Chemistry*, Vol. 10, Elsevier Science Ltd., Oxford 1996, p. 687.
47. A. von Zelewsky, *Coord. Chem. Rev.* **1999**, *190–192*, 811–825.
48. A. von Zelewsky, P. Belser, P. Hayoz, R. Dux, X. Hua, A. Suckling, H. Stoeckli-Evans, *Coord. Chem. Rev.* **1994**, *132*, 75–85.
49. P. Belser, S. Bernhard, E. Jandrasics, A. von Zelewsky, L. De Cola, V. Balzani, *Coord. Chem. Rev.* **1997**, *159*, 1–8.
50. U. Knof, A. von Zelewsky, *Angew. Chem. Int. Ed.* **1999**, *38*, 302–322.
51. V. Balzani, A. Juris, M. Venturi, S. Campagna, S. Serroni, *Chem. Rev.* **1996**, *96*, 759–833.
52. L. De Cola, P. Belser, *Coord. Chem. Rev.* **1998**, *177*, 301–346.
53. V. Balzani, S. Campagna, G. Denti, A. Juris, S. Serroni, M. Venturi, *Acc. Chem. Res.* **1998**, *31*, 26–34.
54. F. Hein, S. Herzog, *Z. anorg. allg. Chem.* **1952**, *267*, 337.
55. Y. Sato, N. Tanaka, *Bull. Chem. Soc. Jpn.* **1969**, *42*, 1021–1024.
56. J. G. Gaudiello, P. R. Sharp, A. J. Bard, *J. Am. Chem. Soc.* **1982**, *104*, 6373–6377.
57. E. Garcia, J. Kwak, A. J. Bard, *Inorg. Chem.* **1988**, *27*, 4377–4382.
58. J. B. Chlistunoff, A. J. Bard, *Inorg. Chem.* **1992**, *31*, 4582–4587.
59. H. D. Gafney, A. W. Adamson, *J. Am. Chem. Soc.* **1972**, *94*, 8238–8239.
60. J. C. Luong, L. Nadjo, M. S. Wrighton, *J. Am. Chem. Soc.* **1978**, *100*, 5790.
61. J. C. Luong, R. A. Faltynek, M. S. Wrighton, *J. Am. Chem. Soc.* **1980**, *102*, 7892.
62. G. L. Geoffroy, M. S. Wrighton, *Organometallic Photochemistry*, Academic Press, New York 1979.
63. C. M. Elliott, *J.C.S. Chem. Comm.* **1980**, 261–262.
64. S. Roffia, M. A. Raggi, M. Ciano, *J. Electroanal. Chem.* **1980**, *108*, 69–76.
65. Y. Ohsawa, M. K. DeArmond, K. W. Hanck, D. E. Morris, D. G. Whitten, P. E. Neveux, Jr., *J. Am. Chem. Soc.* **1983**, *105*, 6522–6524.
66. G. A. Heath, L. J. Yellowlees, P. S. Braterman, *J.C.S. Chem. Comm.* **1981**, 287–289.
67. C. M. Elliott, E. J. Hershenhart, *J. Am. Chem. Soc.* **1982**, *104*, 7519–7526.

68. S. M. Angel, M. K. DeArmond, R. J. Donohoe, K. W. Hanck, D. W. Wertz, *J. Am. Chem. Soc.* **1984**, *106*, 3688–3689.
69. D. E. Morris, K. W. Hanck, M. K. DeArmond, *J. Am. Chem. Soc.* **1983**, *105*, 3032–3038.
70. P. G. Bradley, N. Kress, B. A. Hornberger, R. F. Dallinger, W. H. Woodruff, *J. Am. Chem. Soc.* **1981**, *103*, 7441–7446.
71. M. Forster, R. E. Hester, *Chem. Phys. Lett.* **1981**, *81*, 42–47.
72. P. S. Braterman, A. Harriman, G. A. Heath, L. J. Yellowlees, *J. Chem. Soc. Dalton Trans.* **1983**, 1801–1803.
73. J. K. Nagle, D. M. Roundhill, *Chemtracts—Inorg. Chem.* **1992**, *4*, 141–155.
74. A. Juris, V. Balzani, F. Barigelletti, S. Campagna, P. Belser, A. von Zelewsky, *Coord. Chem. Rev.* **1988**, *84*, 85–277.
75. E. Borgarello, J. Kiwi, E. Pelizzetti, M. Visca, M. Grätzel, *Nature* **1981**, *289*, 158–160.
76. M. Grätzel, *Acc. Chem. Res.* **1981**, *14*, 376–384.
77. M. Grätzel, *Coord. Chem. Rev.* **1991**, 111, 167–174.
78. B. O'Regan, M. Grätzel, *Nature* **1991**, *353*, 737.
79. M. K. Nazeeruddin, A. Kay, I. Rodicio, R. Humphry-Baker, E. Müller, P. Liska, N. Vlachopoulos, M. Grätzel, *J. Am. Chem. Soc.* **1993**, *115*, 6382–6390.
80. Y. Tachibana, J. E. Moser, M. Grätzel, D. R. Klug, J. R. Durrant, *J. Phys. Chem.* **1996**, *100*, 20056–20062.
81. R. J. Ellingson, J. B. Asbury, S. Ferrere, H. N. Ghosh, J. R. Sprague, T. Lian, A. J. Nozik, *J. Phys. Chem. B.* **1998**, *102*, 6455–6458.
82. J. B. Asbury, R. J. Ellingson, H. N. Ghosh, S. Ferrere, A. J. Nozik, T. Lian, *J. Phys. Chem. B* **1999**, *103*, 3110–3119.
83. F. Scandola, C. A. Bignozzi, M. T. Indelli, in K. Kalyanasundaram, M. Grätzel (Eds.): *Photosensitization and Photocatalysis Using Inorganic and Organometallic Compounds*, Kluwer Academic Publishers, Dordrecht 1993, p. 161–216.
84. S. Campagna, G. Denti, S. Serroni, A. Juris, M. Venturi, V. Ricevuto, V. Balzani, *Chem. Eur. J.* **1995**, 1, 211.
85. R. Ziessel, M. Hissler, A. El-ghayoury, A. Harriman, *Coord. Chem. Rev.* **1998**, *178–180*, 1251–1298.
86. L. Flamigni, F. Barigelletti, N. Armaroli, J.-P. Collin, I. M. Dixon, J.-P. Sauvage, J. A. Gareth Williams, *Coord. Chem. Rev.* **1999**, *190–192*, 671–682.
87. J.-P. Sauvage, J.-P. Collin, J.-C. Chambron, S. Guillerez, C. Coudret, V. Balzani, F. Barigelletti, L. De Cola, L. Flamigni, *Chem. Rev.* **1994**, *94*, 993–1019.
88. A. Harriman, R. Ziessel, *Chem. Commun.* **1996**, 1707–1716.
89. J.-P. Collin, P. Gaviña, V. Heitz, J.-P. Sauvage, *Eur. J. Inorg. Chem.* **1998**, 1–14.
90. J.-M. Kern, J.-P. Sauvage, J.-L. Weidmann, N. Armaroli, L. Flamigni, P. Ceroni, V. Balzani, *Inorg. Chem.* **1997**, *36*, 5329–5338.
91. P. R. Ashton, V. Balzani, A. Credi, O. Kocian, D. Pasini, L. Prodi, N. Spencer, J. F. Stoddart, M. S. Tolley, M. Venturi, A. J. P. White, D. J. Williams, *Chem. Eur. J.* **1998**, *4*, 590–607.
92. M.-J. Blanco, M. C. Jiménez, J.-C. Chambron, V. Heitz, M. Linke, J.-P. Sauvage, *Chem. Soc. Rev.* **1999**, *28*, 293–305.
93. W. R. McWhinnie, J. D. Miller, *Adv. Inorg. Chem. Radiochem.* **1969**, *12*, 135–215.
94. E. C. Constable, *Adv. Inorg. Chem.* **1989**, *34*, 1–63.
95. K. Kalyanasundaram, *Coord. Chem. Rev.* **1982**, *46*, 159–244.
96. D. J. Stufkens, *Coord. Chem. Rev.* **1990**, *104*, 39–112.
97. D. J. Stufkens, A. Vlček, Jr., *Coord. Chem. Rev.* **1998**, *177*, 127–179.
98. K. Kalyanasundaram, *Photochemistry of Polypyridine and Porphyrin Complexes*, Academic Press, London 1992.
99. D. M. Roundhill, *Photochemistry and Photophysics of Metal Complexes*, Plenum Press, New York 1994.
100. P. J. Steel, *Coord. Chem. Rev.* **1990**, *106*, 227–265.
101. T. Saji, S. Aoyagui, *J. Electroanal. Chem.* **1975**, *58*, 401–410.
102. P. S. Braterman, J.-I. Song, *J. Org. Chem.* **1991**, *56*, 4678–4682.
103. B. C. Noble, R. D. Peacock, *Spectrochimica Acta* **1990**, *46A*, 407–412.

104. A. Klein, W. Kaim, E. Waldhör, H.-D. Hausen, *J. Chem. Soc. Perkin Trans. 2* **1995**, 2121–2126.
105. J. B. Chlistunoff, A. J. Bard, *Inorg. Chem.* **1993**, *32*, 3521–3527.
106. J. L. Kahl, K. Hanck, K. DeArmond, *J. Phys. Chem.* **1978**, *82*, 540–545.
107. M. Krejčík, S. Záliš, J. Klíma, D. Sýkora, W. Matheis, A. Klein, W. Kaim, *Inorg. Chem.* **1993**, *32*, 3362–3368.
108. C. Dietrich-Buchecker, J.-P. Sauvage, J.-M. Kern, *J. Am. Chem. Soc.* **1989**, 111, 7791–7800.
109. L. De Cola, F. Barigelletti, V. Balzani, P. Belser, A. von Zelewsky, F. Vögtle, F. Ebmeyer, S. Grammenudi, *J. Am. Chem. Soc.* **1988**, *110*, 7210–7212.
110. F. Barigelletti, L. De Cola, V. Balzani, P. Belser, A. von Zelewsky, F. Vögtle, F. Ebmeyer, S. Grammenudi, *J. Am. Chem. Soc.* **1989**, 111, 4662–4668.
111. T. Saji, S. Aoyagui, *J. Electroanal. Chem.* **1975**, *63*, 405–419.
112. E. König, S. Herzog, *J. Inorg. Nucl. Chem.* **1970**, *32*, 613–617.
113. Y. Saito, J. Takemoto, B. Hutchinson, K. Nakamoto, *Inorg. Chem.* **1972**, 11, 2003–2011.
114. E. König, S. Herzog, *J. Inorg. Nucl. Chem.* **1970**, *32*, 601–611.
115. A. L. Rieger, J. L. Scott, P. H. Rieger, *Inorg. Chem.* **1994**, *33*, 621–622.
116. M. C. Hughes, J. M. Rao, D. J. Macero, *Inorg. Chim. Acta* **1979**, *35*, L321–324.
117. T. Saji, S. Aoyagui, *J. Electroanal. Chem.* **1975**, *60*, 1–10.
118. A. Vlček, Jr., *Coord. Chem. Rev.* **1998**, *177*, 219–256.
119. S. Ernst, W. Kaim, *J. Am. Chem. Soc.* **1986**, *108*, 3578–3586.
120. W. Kaim, S. Kohlmann, *Inorg. Chem.* **1987**, *26*, 68–77.
121. A. Vlček, Jr., F. Baumann, W. Kaim, F.-W. Grevels, F. Hartl, *J. Chem. Soc., Dalton Trans.* **1998**, 215–220.
122. S. Záliš, C. Daniel, A. Vlček, Jr., *J. Chem. Soc., Dalton Trans.* **1999**, 3081–3086.
123. H. tom Dieck, K.-D. Franz, F. Hohmann, *Chem. Ber.* **1975**, *108*, 163–173.
124. Y. Sato, N. Tanaka, *Bull. Chem. Soc. Jpn.* **1968**, *41*, 2064–2066.
125. M. Inoue, K. Hara, T. Horiba, M. Kubo, *Bull. Chem. Soc. Jpn.* **1974**, *47*, 2137–2140.
126. M. M. Morrison, D. T. Sawyer, *Inorg. Chem.* **1978**, *17*, 333–337.
127. M. A. Lockwood, K. Wang, J. M. Mayer, *J. Am. Chem. Soc.* **1999**, *121*, 11894–11895.
128. G. W. Brudvig, H. H. Thorp, R. H. Crabtree, *Acc. Chem. Res.* **1991**, *24*, 311–316.
129. L. E. Helberg, S. D. Orth, M. Sabat, W. D. Harman, *Inorg. Chem.* **1996**, *35*, 5584–5594.
130. F. Paolucci, M. Marcaccio, C. Paradisi, S. Roffia, C. A. Bignozzi, C. Amatore, *J. Phys. Chem. B* **1998**, *102*, 4759–4769.
131. P. Christensen, A. Hamnett, A. V. G. Muir, J. A. Timney, *J. Chem. Soc. Dalton Trans.* **1992**, 1455–1463.
132. W. B. Connick, A. J. Di Bilio, M. G. Hill, J. R. Winkler, H. B. Gray, *Inorg. Chim. Acta* **1995**, *240*, 169–173.
133. R. J. Shaver, D. P. Rillema, *Inorg. Chem.* **1992**, *31*, 4101–4107.
134. A. Klein, C. Vogler, W. Kaim, *Organometallics* **1996**, *15*, 236–244.
135. F. P. A. Johnson, M. W. George, F. Hartl, J. J. Turner, *Organometallics* **1996**, *15*, 3374–3387.
136. G. J. Stor, F. Hartl, J. W. M. van Outersterp, D. J. Stufkens, *Organometallics* **1995**, *14*, 1115–1131.
137. J. W. M. van Outersterp, F. Hartl, D. J. Stufkens, *Organometallics* **1995**, *14*, 3303–3310.
138. J. Hawecker, J.-M. Lehn, R. Ziessel, *J. Chem. Soc., Chem. Commun.* **1984**, 328–330.
139. J. Hawecker, J.-M. Lehn, R. Ziessel, *Helv. Chim. Acta* **1986**, *69*, 1990–2009.
140. K. Koike, H. Hori, M. Ishizuka, J. R. Westwell, K. Takeuchi, T. Ibusuki, K. Enjouji, H. Konno, K. Sakamoto, O. Ishitani, *Organometallics* **1997**, *16*, 5724–5729.
141. B. P. Sullivan, C. M. Bolinger, D. Conrad, W. J. Vining, T. J. Meyer, *J. Chem. Soc., Chem. Commun.* **1985**, 1414–1416.
142. B. D. Rossenaar, F. Hartl, D. J. Stufkens, *Inorg. Chem.* **1996**, *35*, 6194–6203.
143. B. D. Rossenaar, F. Hartl, D. J. Stufkens, C. Amatore, E. Maisonhaute, J.-N. Verpeaux, *Organometallics* **1997**, *16*, 4675–4685.
144. P. S. Braterman, J.-I. Song, R. D. Peacock, *Inorg. Chem.* **1992**, *31*, 555–559.
145. N. Tanaka, T. Ogata, S. Niizuma, *Bull. Chem. Soc. Jpn.* **1973**, *46*, 3299–3301.
146. S. M. Angel, M. K. DeArmond, R. J. Donohoe, D. W. Wertz, *J. Phys. Chem.* **1985**, *89*, 282–285.

147. J. M. Rao, M. C. Hughes, D. J. Macero, *Inorg. Chim. Acta* **1979**, *35*, L369–L373.
148. T. Saji, S. Aoyagui, *J. Electroanal. Chem.* **1980**, *110*, 329–334.
149. Y. Ohsawa, K. W. Hanck, M. K. DeArmond, *J. Electroanal. Chem.* **1984**, *175*, 229–240.
150. C. D. Tait, D. B. MacQueen, R. J. Donohoe, M. K. De Armond, K. W. Hanck, D. W. Wertz, *J. Phys. Chem.* **1986**, *90*, 1766–1771.
151. F. Barigelletti, A. Juris, V. Balzani, P. Belser, A. von Zelewsky, *Inorg. Chem.* **1987**, *26*, 4115–4119.
152. H. B. Ross, M. Boldaji, D. P. Rillema, C. B. Blanton, R. P. White, *Inorg. Chem.* **1989**, *28*, 1013–1021.
153. A. Juris, S. Campagna, V. Balzani, G. Gremaud, A. von Zelewsky, *Inorg. Chem.* **1988**, *27*, 3652–3655.
154. S. Roffia, M. Ciano, *J. Electroanal. Chem.* **1979**, *100*, 809–817.
155. S. L. Tan, M. K. De Armond, K. W. Hanck, *J. Electroanal. Chem.* **1984**, *181*, 187–197.
156. T. Matsumura-Inoue, T. Tominaga-Morimoto, *J. Electroanal. Chem.* **1978**, *93*, 127–139.
157. A. B. P. Lever, *Inorg. Chem.* **1990**, *29*, 1271–1285.
158. H. Ishida, K. Tanaka, T. Tanaka, *Organometallics* **1987**, *6*, 181–186.
159. M. R. M. Bruce, E. Megehee, B. P. Sullivan, H. H. Thorp, T. R. O'Toole, A. Downard, J. R. Pugh, T. J. Meyer, *Inorg. Chem.* **1992**, *31*, 4864–4873.
160. M. N. Collomb-Dunand-Sauthier, A. Deronzier, R. Ziessel, *J. Chem. Soc., Chem. Comm.* **1994**, 189–191.
161. S. Chardon-Noblat, M. N. Collomb-Dunand-Sauthier, A. Deronzier, R. Ziessel, D. Zsoldos, *Inorg. Chem.* **1994**, *33*, 4410–4412.
162. M. N. Collomb-Dunand-Sauthier, A. Deronzier, R. Ziessel, *Inorg. Chem.* **1994**, *33*, 2961–2967.
163. S. Chardon-Noblat, A. Deronzier, R. Ziessel, D. Zsoldos, *Inorg. Chem.* **1997**, *36*, 5384–5389.
164. M. P. Aarnts, F. Hartl, K. Peelen, D. J. Stufkens, C. Amatore, J.-N. Verpeaux, *Organometallics* **1997**, *16*, 4686–4695.
165. M. P. Aarnts, M. P. Wilms, K. Peelen, J. Fraanje, K. Goubitz, F. Hartl, D. J. Stufkens, E. J. Baerends, A. Vlček, Jr., *Inorg. Chem.* **1996**, *35*, 5468–5477.
166. T. J. Meyer, *J. Electrochem. Soc.* **1984**, *131*, 221C–228C.
167. J. A. Gilbert, D. S. Eggleston, W. R. Murphy, Jr., D. A. Geselowitz, S. W. Gersten, D. J. Hodgson, T. J. Meyer, *J. Am. Chem. Soc.* **1985**, *107*, 3855–3864.
168. N. Tanaka, Y. Sato, *Bull. Chem. Soc. Jpn.* **1968**, *41*, 2059–2064.
169. S. Herzog, R. Klausch, J. Lantos, *Z. Chem.* **1964**, *4*, 150.
170. D. J. Szalda, C. Creutz, D. Mahajan, N. Sutin, *Inorg. Chem.* **1983**, *22*, 2372–2379.
171. H. A. Schwarz, C. Creutz, N. Sutin, *Inorg. Chem.* **1985**, *24*, 433–439.
172. P. Biagini, T. Funaioli, A. Juris, G. Fachinetti, *J. Organomet. Chem.* **1990**, *390*, C61–C63.
173. C. Creutz, N. Sutin, *Coord. Chem. Rev.* **1985**, *64*, 321–341.
174. C. Creutz, H. A. Schwarz, N. Sutin, *J. Am. Chem. Soc.* **1984**, *106*, 3036–3037.
175. D. A. Reitsma, F. R. Keene, *Organometallics* **1994**, *13*, 1351–1354.
176. J. M. Rao, M. C. Hughes, D. J. Macero, *Inorg. Chim. Acta* **1976**, *16*, 231–236.
177. G. Kew, K. DeArmond, K. Hanck, *J. Phys. Chem.* **1974**, *78*, 727–734.
178. G. Kew, K. Hanck, K. DeArmond, *J. Phys. Chem.* **1975**, *79*, 1828–1835.
179. H. Câldãraru, M. K. De Armond, K. W. Hanck, V. E. Sahini, *J. Am. Chem. Soc.* **1976**, *98*, 4455–4457.
180. C. Creutz, A. D. Keller, H. A. Schwarz, N. Sutin, A. P. Zipp, in D. B. Rorabacher, J. F. Endicott (Eds.): *Mechanistic Aspects of Inorganic Reactions. ACS Symposium Series 198*, ACS, Washington, D.C. 1982, p. 385–402.
181. Q. G. Mulazzani, S. Emmi, M. Z. Hoffman, M. Venturi, *J. Am. Chem. Soc.* **1981**, *103*, 3362–3370.
182. C. M. Bolinger, B. P. Sullivan, D. Conrad, J. A. Gilbert, N. Story, T. J. Meyer, *J. Chem. Soc., Chem. Commun.* **1985**, 796–797.
183. C. Caix, S. Chardon-Noblat, A. Deronzier, *J. Electroanal. Chem.* **1997**, *434*, 163–170.
184. W. Kaim, R. Reinhardt, M. Sieger, *Inorg. Chem.* **1994**, *33*, 4453–4459.
185. M. Ladwig, W. Kaim, *J. Organomet. Chem.* **1992**, *439*, 79–90.
186. S. Roffia, M. Ciano, *J. Electroanal. Chem.* **1978**, *87*, 267–274.

187. J. L. Kahl, K. Hanck, K. DeArmond, *J. Phys. Chem.* **1979**, *83*, 2606–2611.
188. R. Ziessel, *J. Chem. Soc., Chem. Commun.* **1988**, 16–17.
189. R. Ziessel, *Angew. Chem. Int. Ed. Engl.* **1991**, *30*, 844–847.
190. R. Ziessel, *J. Am. Chem. Soc.* **1993**, *115*, 118–127.
191. B. J. Henne, D. E. Bartak, *Inorg. Chem.* **1984**, *23*, 369–373.
192. Y. Rollin, M. Troupel, D. G. Tuck, J. Perichon, *J. Organomet. Chem.* **1986**, *303*, 131–137.
193. A. R. Brown, Z. Guo, F. W. J. Mosselmans, S. Parsons, M. Schröder, L. J. Yellowlees, *J. Am. Chem. Soc.* **1998**, *120*, 8805–8811.
194. R. Palmans, D. B. MacQueen, C. G. Pierpont, A. J. Frank, *J. Am. Chem. Soc.* **1996**, *118*, 12647–12653.
195. P. S. Braterman, J.-I. Song, C. Vogler, W. Kaim, *Inorg. Chem.* **1992**, *31*, 222–224.
196. P. S. Braterman, J.-I. Song, F. M. Wimmer, S. Wimmer, W. Kaim, A. Klein, R. D. Peacock, *Inorg. Chem.* **1992**, *31*, 5084–5088.
197. A. Klein, W. Kaim, *Organometallics* **1995**, *14*, 1176–1186.
198. L. Yang, F. L. Wimmer, S. Wimmer, J. Zhao, P. S. Braterman, *J. Organomet. Chem.* **1996**, *525*, 1–8.
199. D. Collison, F. E. Mabbs, E. J. L. McInnes, K. J. Taylor, A. J. Welch, L. J. Yellowlees, *J. Chem. Soc., Dalton. Trans.* **1996**, 329–334.
200. E. J. L. McInnes, R. D. Farley, C. C. Rowlands, A. J. Welch, L. Rovatti, L. J. Yellowlees, *J. Chem. Soc., Dalton Trans.* **1999**, 4203–4208.
201. F. P. Fanizzi, G. Natile, M. Lanfranchi, A. Tiripicchio, F. Laschi, P. Zanello, *Inorg. Chem.* **1996**, *35*, 3173–3182.
202. W. Paw, S. D. Cummings, M. A. Mansour, W. B. Connick, D. K. Geiger, R. Eisenberg, *Coord. Chem. Rev.* **1998**, *171*, 125–150.
203. J. A. Zuleta, M. S. Burberry, R. Eisenberg, *Coord. Chem. Rev.* **1990**, *97*, 47–64.
204. P. Federlin, J.-M. Kern, A. Rastegar, C. Dietrich-Buchecker, P. A. Marnot, J.-P. Sauvage, *New. J. Chem.* **1990**, *14*, 9–12.
205. Y. Kawanishi, N. Kitamura, S. Tazuke, *Inorg. Chem.* **1989**, *28*, 2968–2975.
206. F. Pichot, J. H. Beck, C. M. Elliott, *J. Phys. Chem. A* **1999**, *103*, 6263–6267.
207. E. S. Dodsworth, A. B. P. Lever, *Chem. Phys. Lett.* **1985**, *119*, 61–66.
208. E. S. Dodsworth, A. B. P. Lever, *Chem. Phys. Lett.* **1986**, *124*, 152–158.
209. S. I. Gorelsky, E. S. Dodsworth, A. B. P. Lever, A. A. Vlček, *Coord. Chem. Rev.* **1998**, *174*, 469–494.
210. D. P. Rillema, G. Allen, T. J. Meyer, D. Conrad, *Inorg. Chem.* **1983**, *22*, 1617–1622.
211. A. Masschelein, L. Jacquet, A. Kirsch-De Mesmaeker, J. Nasielski, *Inorg. Chem.* **1990**, *29*, 855–860.
212. S. Záliš, V. Drchal, *Chem. Phys.* **1987**, *118*, 313–323.
213. W. Kaim, R. Reinhardt, E. Waldhör, J. Fiedler, *J. Organomet. Chem.* **1996**, *524*, 195–202.
214. S. Greulich, W. Kaim, A. F. Stange, H. Stoll, J. Fiedler, S. Záliš, *Inorg. Chem.* **1996**, *35*, 3998–4002.
215. I. Hanazaki, S. Nagakura, *Bull. Chem. Soc. Jpn.* **1971**, *44*, 2312–2321.
216. B. P. Sullivan, D. Conrad, T. J. Meyer, *Inorg. Chem.* **1985**, *24*, 3640–3645.
217. S. D. Ernst, W. Kaim, *Inorg. Chem.* **1989**, *28*, 1520–1528.
218. M. Marcaccio, F. Paolucci, C. Paradisi, S. Roffia, C. Fontanesi, L. J. Yellowlees, S. Serroni, S. Campagna, G. Denti, V. Balzani, *J. Am. Chem. Soc.* **1999**, *121*, 10081–10091.
219. M. K. De Armond, M. L. Myrick, *Acc. Chem. Res.* **1989**, *22*, 364–370.
220. A. B. P. Lever, *Inorg. Chem.* **1991**, *30*, 1980–1985.
221. Y. Ohsawa, M.-H. Whangbo, K. W. Hanck, M. K. DeArmond, *Inorg. Chem.* **1984**, *23*, 3426–3428.
222. M. Furue, K. Maruyama, T. Oguni, M. Naiki, M. Kamachi, *Inorg. Chem.* **1992**, *31*, 3792–3795.
223. M. K. Nazeeruddin, S. M. Zakeeruddin, K. Kalyanasundaram, *J. Phys. Chem.* **1993**, *97*, 9607–9612.
224. M. Maestri, N. Armaroli, V. Balzani, E. C. Constable, A. M. W. Cargill Thompson, *Inorg. Chem.* **1995**, *34*, 2759–2767.
225. J. K. Hino, L. Della Ciana, W. J. Dressick, B. P. Sullivan, *Inorg. Chem.* **1992**, *31*, 1072–1080.

226. V. Skarda, M. J. Cook, A. P. Lewis, G. S. G. McAuliffe, A. J. Thomson, D. J. Robbins, *J. Chem. Soc. Perkin Trans. II* **1984**, 1309–1311.
227. P. Bugnon, R. E. Hester, *Chem. Phys. Lett.* **1983**, *102*, 537–543.
228. P. Belser, A. von Zelewsky, A. Juris, F. Barigelletti, A. Tucci, V. Balzani, *Chem. Phys. Lett.* **1982**, *89*, 101–104.
229. E. Binamira-Soriaga, S. D. Sprouse, R. J. Watts, W. C. Kaska, *Inorg. Chim. Acta* **1984**, *84*, 135–139.
230. J. A. Baiano, R. J. Kessler, R. S. Lumpkin, M. J. Munley, W. R. Murphy, Jr., *J. Phys. Chem.* **1995**, *99*, 17680–17690.
231. M. Maruyama, Y. Kaizu, *Inorg. Chim. Acta* **1996**, *247*, 155–159.
232. W. L. Reynolds, *J. Phys. Chem.* **1963**, *67*, 2866–2868.
233. T. Saji, S. Aoyagui, *Bull. Chem. Soc. Jpn.* **1973**, *46*, 2101–2105.
234. R. Stasiw, R. G. Wilkins, *Inorg. Chem.* **1969**, *8*, 156–157.
235. P. Hendry, A. Ludi, *Adv. Inorg. Chem.* **1990**, *35*, 117–198.
236. R. Pyati, R. W. Murray, *J. Am. Chem. Soc.* **1996**, *118*, 1743–1749.
237. Y. Fu, A. S. Cole, T. W. Swaddle, *J. Am. Chem. Soc.* **1999**, *121*, 10410–10415.
238. T. Saji, S. Aoyagui, *J. Electroanal. Chem.* **1975**, *63*, 31–37.
239. T. Saji, T. Yamada, S. Aoyagui, *J. Electroanal. Chem.* **1975**, *61*, 147–153.
240. T. Saji, T. Yamada, S. Aoyagui, *Bull. Chem. Soc. Jpn.* **1975**, *48*, 1641–1642.
241. A. G. Motten, K. Hanck, M. K. DeArmond, *Chem. Phys. Lett.* **1981**, *79*, 541–546.
242. G. Wolfbauer, A. M. Bond, D. R. MacFarlane, *J. Chem. Soc., Dalton Trans.* **1999**, 4363–4372.
243. W. Kaim, S. Kohlmann, *Inorg. Chem.* **1990**, *29*, 2909–2914.
244. G. Wolfbauer, A. M. Bond, D. R. MacFarlane, *Inorg. Chem.* **1999**, *38*, 3836–3846.
245. L. Zelikovich, J. Libman, A. Shanzer, *Nature* **1995**, *374*, 790.
246. T. R. Ward, A. Lutz, S. P. Parel, J. Ensling, P. Gütlich, P. Buglyó, C. Orvig, *Inorg. Chem.* **1999**, *38*, 5007–5017.
247. D. J. Cárdenas, A. Livoreil, J.-P. Sauvage, *J. Am. Chem. Soc.* **1996**, *118*, 11980–11981.
248. F. Vögtle, M. Plevoets, M. Nieger, G. C. Azzellini, A. Credi, L. De Cola, V. De Marchis, M. Venturi, V. Balzani, *J. Am. Chem. Soc.* **1999**, *121*, 6290–6298.
249. P. Ceroni, F. Paolucci, C. Paradisi, A. Juris, S. Roffia, S. Serroni, S. Campagna, A. J. Bard, *J. Am. Chem. Soc.* **1998**, *120*, 5480–5487.
250. M. Venturi, A. Credi, V. Balzani, *Coord. Chem. Rev.* **1999**, *185–186*, 233–256.
251. E. H. Yonemoto, G. B. Saupe, R. H. Schmehl, S. M. Hubig, R. L. Riley, B. L. Iverson, T. E. Mallouk, *J. Am. Chem. Soc.* **1994**, *116*, 4786–4795.
252. M. Maggini, D. M. Guldi, S. Mondini, G. Scorrano, F. Paolucci, P. Ceroni, S. Roffia, *Chem. Eur. J.* **1998**, *4*, 1992–2000.
253. D. LeGourriérec, M. Andersson, J. Davidsson, E. Mukhtar, L. Sun, L. Hammarström, *J. Phys. Chem. A.* **1999**, *103*, 557–559.
254. P. Chen, T. D. Westmoreland, E. Danielson, K. S. Schanze, D. Anthon, P. E. Neveux, Jr., T. J. Meyer, *Inorg. Chem.* **1987**, *26*, 1116–1126.
255. P. Chen, R. Duesing, D. K. Graff, T. J. Meyer, *J. Phys. Chem.* **1991**, *95*, 5850–5858.
256. C. O. Dietrich-Buchecker, J.-F. Nierengarten, J.-P. Sauvage, N. Armaroli, V. Balzani, L. De Cola, *J. Am. Chem. Soc.* **1993**, *115*, 11237–11244.
257. P. D. Beer, Z. Chen, A. J. Goulden, A. Grieve, D. Hesek, F. Szemes, T. Wear, *J. Chem. Soc., Chem. Commun.* **1994**, 1269.
258. N. G. Connelly, W. E. Geiger, *Chem. Rev.* **1996**, *96*, 877–910.
259. T. R. O'Toole, L. D. Margerum, T. D. Westmoreland, W. J. Vining, R. W. Murray, *J. Chem. Soc., Chem. Commun.* **1985**, 1416–1417.
260. S. Dérien, E. Duñach, J. Périchon, *J. Am. Chem. Soc.* **1991**, *113*, 8447–8454.
261. M. F. Sistare, R. C. Holmberg, H. H. Thorp, *J. Phys. Chem. B* **1999**, *103*, 10718–10728.
262. S. Gould, R. M. Leasure, T. J. Meyer, *Chem. Brit.* **1995**, *31*, 891–893.
263. P. D. Beer, P. A. Gale, G. Z. Chen, *Coord. Chem. Rev.* **1999**, *185–186*, 3–36.
264. C. M. Elliott, J. G. Redepenning, *J. Electroanal. Chem.* **1986**, *197*, 219–232.
265. V. Balzani, F. Bolletta, F. Scandola, R. Ballardini, *Pure & Appl. Chem.* **1979**, *51*, 299–311.
266. V. Balzani, F. Barigelletti, L. De Cola, *Top. Curr. Chem.* **1990**, *158*, 31–71.

267. N. Serpone, in M. A. Fox, M. Chanon (Eds.): *Photoinduced Electron Transfer. Part D. Photoinduced Electron-transfer reactions: Inorganic Substrates and Applications*, Elsevier, Amsterdam 1988, p. 47.
268. M. Ruthkosky, F. N. Castellano, G. J. Meyer, *Inorg. Chem.* **1996**, *35*, 6406–6412.
269. V. Balzani, F. Scandola, in J. S. Connolly (Ed.): *Photochemical Conversion and Storage of Solar Energy*, Academic Press, New York 1981, p. 97–129.
270. C. Creutz, M. Chou, T. L. Netzel, M. Okumura, N. Sutin, *J. Am. Chem. Soc.* **1980**, *102*, 1309–1319.
271. A. Vlček, *Coord. Chem. Rev.* **2000**, *200–202*, 933–977.
272. N. H. Damrauer, G. Cerullo, A. Yeh, T. R. Boussie, C. V. Shank, J. K. McCusker, *Science* **1997**, *275*, 54–57.
273. N. H. Damrauer, J. K. McCusker, *J. Phys. Chem. A* **1999**, *103*, 8440–8446.
274. A. W. Adamson, *Pure & Appl. Chem.* **1979**, *51*, 313–329.
275. P. D. Fleischauer, A. W. Adamson, G. Sartori, *Progr. Inorg. Chem.* **1972**, *17*, 1.
276. A. W. Adamson, in K. Kalyanasundaram, M. Grätzel (Eds.): *Photosensitization and Photocatalysis Using Inorganic and Organometallic Compounds*, Kluwer Academic Publishers, Dordrecht 1993, p. 1–14.
277. V. Balzani, A. Juris, F. Scandola, in E. Pelizzetti, N. Serpone (Eds.): *Homogeneous and Heterogeneous Photocatalysis*, D. Reidel Publishing Company 1986, p. 1–27.
278. N. Serpone, M. A. Jamieson, M. S. Henry, M. Z. Hoffman, F. Bolletta, M. Maestri, *J. Am. Chem. Soc.* **1979**, *101*, 2907–2916.
279. N. Serpone, M. A. Jamieson, R. Sriram, M. Z. Hoffman, *Inorg. Chem.* **1981**, *20*, 3983–3988.
280. A. I. Baba, J. R. Shaw, J. A. Simon, R. P. Thummel, R. H. Schmehl, *Coord. Chem. Rev.* **1998**, *171*, 43–59.
281. W. E. Ford, M. A. J. Rodgers, *J. Phys. Chem.* **1992**, *96*, 2917–2920.
282. A. Harriman, M. Hissler, A. Khatyr, R. Ziessel, *Chem. Commun.* **1999**, 735–736.
283. G. F. Strouse, H. U. Gudel, *Inorg. Chim. Acta* **1995**, *240*, 453–464.
284. H. Riesen, E. Krausz, *Comments Inorg. Chem.* **1995**, *18*, 27–63.
285. J. L. Pogge, D. F. Kelley, *Chem. Phys. Letters* **1995**, *238*, 16–24.
286. J. P. Cushing, C. Butoi, D. F. Kelley, *J. Phys. Chem. A.* **1997**, *101*, 7222–7230.
287. D. P. Rillema, D. G. Taghdiri, D. S. Jones, C. D. Keller, L. A. Worl, T. J. Meyer, H. A. Levy, *Inorg. Chem.* **1987**, *26*, 578–585.
288. A. Juris, P. Belser, F. Barigelletti, A. von Zelewsky, V. Balzani, *Inorg. Chem.* **1986**, *25*, 256–259.
289. S. R. Johnson, T. D. Westmoreland, J. V. Caspar, K. R. Barqawi, T. J. Meyer, *Inorg. Chem.* **1988**, *27*, 3195–3200.
290. E. M. Kober, J. L. Marshall, W. J. Dressick, B. P. Sullivan, J. V. Caspar, T. J. Meyer, *Inorg. Chem.* **1985**, *24*, 2755–2763.
291. J. V. Caspar, T. J. Meyer, *J. Phys. Chem.* **1983**, *87*, 952–957.
292. J. V. Caspar, E. M. Kober, B. P. Sullivan, T. J. Meyer, *J. Am. Chem. Soc.* **1982**, *104*, 630–632.
293. G. F. Strouse, J. R. Schoonover, R. Duesing, S. Boyde, W. E. Jones, T. J. Meyer, *Inorg. Chem.* **1995**, *34*, 473–487.
294. J. A. Treadway, B. Loeb, R. Lopez, P. A. Anderson, F. R. Keene, T. J. Meyer, *Inorg. Chem.* **1996**, *35*, 2242–2246.
295. T. J. Meyer, *Pure & Appl. Chem.* **1986**, *58*, 1193–1206.
296. B. Durham, J. V. Caspar, J. K. Nagle, T. J. Meyer, *J. Am. Chem. Soc.* **1982**, *104*, 4803–4810.
297. D. P. Rillema, C. B. Blanton, R. J. Shaver, D. C. Jackman, M. Boldaji, S. Bundy, L. A. Worl, T. J. Meyer, *Inorg. Chem.* **1992**, *31*, 1600–1606.
298. G. H. Allen, R. P. White, D. P. Rillema, T. J. Meyer, *J. Am. Chem. Soc.* **1984**, *106*, 2613–2620.
299. R. F. Beeston, W. S. Aldridge, J. A. Treadway, M. C. Fitzgerald, B. A. DeGraff, S. E. Stitzel, *Inorg. Chem.* **1998**, *37*, 4368–4379.
300. M. T. Miller, P. K. Gantzel, T. B. Karpishin, *Inorg. Chem.* **1999**, *38*, 3414–3422.
301. M. Ruthkosky, C. A. Kelly, F. N. Castellano, G. J. Meyer, *Coord. Chem. Rev.* **1998**, *171*, 309–322.

302. M. T. Miller, T. B. Karpishin, *Inorg. Chem.* **1999**, *38*, 5246–5249.
303. K. Kalyanasundaram, *J. Chem. Soc. Faraday Trans. 2* **1986**, *82*, 2401–2415.
304. S. Ferrere, B. A. Gregg, *J. Am. Chem. Soc.* **1998**, *120*, 843–844.
305. L. Hammarström, F. Barigelletti, L. Flamigni, M. T. Indelli, N. Armaroli, G. Calogero, M. Guardigli, A. Sour, J.-P. Collin, J.-P. Sauvage, *J. Phys. Chem. A* **1997**, *101*, 9061–9069.
306. C. Kutal, *Coord. Chem. Revs.* **1990**, *99*, 213–252.
307. D. R. McMillin, J. R. Kirchhoff, K. V. Goodwin, *Coord. Chem. Rev.* **1985**, *64*, 83–92.
308. F. Scandola, V. Balzani, G. B. Schuster, *J. Am. Chem. Soc.* **1981**, *103*, 2519–2523.
309. C. Turró, J. M. Zaleski, Y. M. Karabatsos, D. G. Nocera, *J. Am. Chem. Soc.* **1996**, *118*, 6060–6067.
310. D. G. Whitten, *Acc. Chem. Res.* **1980**, *13*, 83–90.
311. T. Hannappel, B. Burfeindt, W. Storck, F. Willig, *J. Phys. Chem. B* **1997**, *101*, 6799–6802.
312. H. N. Ghosh, J. B. Asbury, Y. Weng, T. Lian, *J. Phys. Chem. B* **1998**, *102*, 10208–10215.
313. J. M. Rehm, G. L. McLendon, Y. Nagasawa, K. Yoshihara, J. Moser, M. Grätzel, *J. Phys. Chem.* **1996**, *100*, 9577–9578.
314. B. Burfeindt, T. Hannappel, W. Storck, F. Willig, *J. Phys. Chem.* **1996**, *100*, 16463–16465.
315. N. J. Cherepy, G. P. Smestad, M. Grätzel, J. Z. Zhang, *J. Phys. Chem. B* **1997**, *101*, 9342–9351.
316. M. Hilgendorff, V. Sundström, *J. Phys. Chem. B* **1998**, *102*, 10505–10514.
317. H. N. Ghosh, J. B. Asbury, T. Lian, *J. Phys. Chem. B* **1998**, *102*, 6482–6486.
318. P. Belser, S. Bernhard, C. Blum, A. Beyeler, L. De Cola, V. Balzani, *Coord. Chem. Rev.* **1999**, *190–192*, 155–169.
319. P. Chen, M. Curry, T. J. Meyer, *Inorg. Chem.* **1989**, *28*, 2271–2280.
320. P. Chen, E. Danielson, T. J. Meyer, *J. Phys. Chem.* **1988**, *92*, 3708–3711.
321. J. R. Schoonover, P. Chen, W. D. Bates, R. B. Dyer, T. J. Meyer, *Inorg. Chem.* **1994**, *33*, 793–797.
322. D. J. Liard, A. Vlček, Jr., *Inorg. Chem.* **2000**, *39*, 485–490.
323. N. E. Katz, S. L. Mecklenburg, D. K. Graff, P. Chen, T. J. Meyer, *J. Phys. Chem.* **1994**, *98*, 8959–8961.
324. P. Chen, S. L. Mecklenburg, T. J. Meyer, *J. Phys. Chem.* **1993**, *97*, 13126–13131.
325. J. R. Schoonover, G. F. Strouse, P. Chen, W. D. Bates, T. J. Meyer, *Inorg. Chem.* **1993**, *32*, 2618–2619.
326. Y. Wang, B. T. Hauser, M. M. Rooney, R. D. Burton, K. S. Schanze, *J. Am. Chem. Soc.* **1993**, *115*, 5675–5683.
327. L. A. Lucia, Y. Wang, K. Nafisi, T. L. Netzel, K. S. Schanze, *J. Phys. Chem.* **1995**, *99*, 11801–11804.
328. Y. Wang, L. A. Lucia, K. S. Schanze, *J. Phys. Chem.* **1995**, *99*, 1961–1968.
329. Y. Wang, K. S. Schanze, *J. Phys. Chem.* **1996**, *100*, 5408–5419.
330. S. L. Mecklenburg, B. M. Peek, J. R. Schoonover, D. G. McCafferty, C. G. Wall, B. W. Erickson, T. J. Meyer, *J. Am. Chem. Soc.* **1993**, *115*, 5479–5495.
331. N. Armaroli, F. Diederich, C. O. Dietrich-Buchecker, L. Flamigni, J.-F. Nierengarten, J.-P. Sauvage, *Chem. Eur. J.* **1998**, *4*, 406–416.
332. P. R. Ashton, R. Ballardini, V. Balzani, E. C. Constable, A. Credi, O. Kocian, S. J. Langford, J. A. Preece, L. Prodi, E. R. Schofield, N. Spencer, J. F. Stoddart, S. Wenger, *Chem. Eur. J.* **1998**, *4*, 2413–2422.
333. J.-C. Chambron, A. Harriman, V. Heitz, J.-P. Sauvage, *J. Am. Chem. Soc.* **1993**, *115*, 6109–6114.
334. I. Rubinstein, A. J. Bard, *J. Am. Chem. Soc.* **1981**, *103*, 512–516.
335. M. M. Richter, J. D. Debad, D. R. Striplin, G. A. Crosby, A. J. Bard, *Anal. Chem.* **1996**, *68*, 4370–4376.
336. H. D. Abruña, *J. Electroanal. Chem.* **1984**, *175*, 321–326.
337. A. Vogler, L. El-Sayed, R. G. Jones, J. Namnath, A. W. Adamson, *Inorg. Chim. Acta* **1981**, *53*, L35–L37.
338. V. Balzani, F. Bolletta, *J. Photochem.* **1981**, *17*, 479–485.
339. D. K. Liu, B. S. Brunschwig, C. Creutz, N. Sutin, *J. Am. Chem. Soc.* **1986**, *108*, 1749–1755.
340. F. Bolletta, M. Ciano, V. Balzani, N. Serpone, *Inorg. Chim. Acta* **1982**, *62*, 207–213.

341. N. E. Tokel-Takvoryan, R. E. Hemingway, A. J. Bard, *J. Am. Chem. Soc.* **1973**, *95*, 6582–6589.
342. W. L. Wallace, A. J. Bard, *J. Phys. Chem.* **1979**, *83*, 1350–1357.
343. J.-M. Lehn, R. Ziessel, *J. Organomet. Chem.* **1990**, *382*, 157–173.
344. C. Kutal, A. J. Corbin, G. Ferraudi, *Organometallics* **1987**, *6*, 553–557.
345. C. Kutal, M. A. Weber, G. Ferraudi, D. Geiger, *Organometallics* **1985**, *4*, 2161–2166.
346. J. Hawecker, J.-M. Lehn, R. Ziessel, *J. Chem. Soc., Chem. Commun.* **1983**, 536–538.
347. J. R. Winkler, H. B. Gray, *Chem. Rev.* **1992**, *92*, 369–379.
348. D. S. Wuttke, M. J. Bjerrum, J. R. Winkler, H. B. Gray, *Science* **1992**, *256*, 1007–1009.
349. G. A. Mines, M. J. Bjerrum, M. G. Hill, D. R. Casimiro, I.-J. Chang, J. R. Winkler, H. B. Gray, *J. Am. Chem. Soc.* **1996**, *118*, 1961–1965.
350. L. K. Skov, T. Pascher, J. R. Winkler, H. B. Gray, *J. Am. Chem. Soc.* **1998**, *120*, 1102–1103.
351. D. R. Casimiro, J. H. Richards, J. R. Winkler, H. B. Gray, *J. Phys. Chem.* **1993**, *97*, 13073–13077.
352. D. S. Wuttke, H. B. Gray, S. L. Fisher, B. Imperiali, *J. Am. Chem. Soc.* **1993**, *115*, 8455–8456.
353. J. R. Telford, P. Wittung-Stafshede, H. B. Gray, J. R. Winkler, *Acc. Chem. Res.* **1998**, *31*, 755–763.
354. J. J. Wilker, I. J. Dmochowski, J. H. Dawson, J. R. Winkler, H. B. Gray, *Angew. Chem. Int. Ed.* **1999**, *38*, 90–92.
355. I. Ortmans, C. Moucheron, A. Kirsch-De Mesmaeker, *Coord. Chem. Rev.* **1998**, *168*, 233–271.
356. J. K. Barton, in I. Bertini, H. B. Gray, S. J. Lippard, J. S. Valentine (Eds.): *Bioinorganic Chemistry*, University Science Books, Mill Valley, CA 1994, p. 455–503.
357. P. J. Dandliker, R. E. Holmlin, J. K. Barton, *Science* **1997**, *275*, 1465–1467.
358. C. J. Murphy, M. R. Arkin, Y. Jenkins, N. D. Ghatlia, S. H. Bossmann, N. J. Turro, J. K. Barton, *Science* **1993**, *262*, 1025–1029.
359. M. R. Arkin, E. D. A. Stemp, R. E. Holmlin, J. K. Barton, A. Hörmann, E. J. C. Olson, P. F. Babara, *Science* **1996**, *273*, 475–480.
360. E. D. A. Stemp, R. E. Holmlin, J. K. Barton, *Inorg. Chim. Acta* **2000**, *297*, 88–97.
361. T. J. Meade, J. F. Kayyem, *Angew. Chem. Int. Ed. Engl.* **1995**, *34*, 352–354.
362. E. Kimura, X. Bu, M. Shionoya, S. Wada, S. Maruyama, *Inorg. Chem.* **1992**, *31*, 4542–4546.
363. D. Burdinski, K. Wieghardt, S. Steenken, *J. Am. Chem. Soc.* **1999**, *121*, 10781–10787.

# 6 Electron Transfer in Catalytic Dinitrogen Reduction

*Alexander E. Shilov*

## 6.1 Introduction

Dinitrogen, $N_2$, is chemically very inert, yet its hydrogenation to ammonia is exothermic, i.e. thermodynamically favored at low temperatures. In practice, however, the hydrogenation requires high temperatures and pressures because of kinetic obstacles to consecutive cleavage of the $N\equiv N$ triple bond.

Hydrogenation is closely related to $N_2$ reduction, because the same result is achieved when $N_2$ receives six electrons from an electron donor and six protons from the medium:

$$N_2 + 6e^- + 6H^+ \rightarrow 2NH_3$$

The reduction of dinitrogen to ammonia constitutes the first observed step in biological nitrogen fixation. Because nature might have chosen and developed an optimum means of reducing $N_2$ by use of comparatively mild electron donors in protic surroundings at ambient temperature and pressure, it has long seemed very attractive to try to understand the mechanism of biological $N_2$ fixation and to use the knowledge of this mechanism for construction of synthetic systems capable of performing a similar process in a chemistry laboratory.

Progress in both understanding biological dinitrogen reduction and in the search for purely chemical systems which reduce $N_2$ at low temperatures was initially very slow, but in recent decades there have been very important developments in investigations of enzymatic $N_2$ reduction and at the same time in low-temperature dinitrogen chemistry including $N_2$ reduction in solution. We now know both stoichiometric and catalytic systems reducing $N_2$ to hydrazine and ammonia in solution, and we understand how to make $N_2$ unusually chemically active under mild conditions. Polynuclear metal complexes have been found to be catalysts for chemical reduction of $N_2$ in protic media when bound to the surface of such donors as titanium(III) hydroxides, metals amalgams, or mercury cathodes in electrochemical reduction. Recently the FeMo cofactor of nitrogenase, which was isolated for the first time in 1977 and which is also a polynuclear complex of composition $Fe_7MoS_9$ (homocitrate), was found to be an active catalyst of acetylene reduction and of reduction of other nitrogenase substrates; it also coordinates dinitrogen in non-protein surroundings. For the first time, we can now compare catalytic functions of FeMoco in natural and non-protein surrounding on the one hand and also on the other hand FeMoco and synthetic complexes active to dinitrogen and other nitrogenase substrates when bound to the surface of the same artificial electron donor.

Peculiarities of the $N_2$ molecule make it necessary to use special means of electron transfer to and inside the active center containing the substrate. The mechanism of the catalysis in protic surroundings, at least for dinitrogen reduction, presumably necessarily includes coupled one-electron transfer from an external electron donor and multi-electron transfer to the substrate coordinated in the polynuclear complex. The coupled electron transfer helps to activate and reduce the 'difficult' substrate dinitrogen at ambient temperatures.

In this article, the reasons for this mechanism in catalytic reduction of $N_2$ will be considered, and results for electron transfer in biological $N_2$ reduction and model synthetic systems will be compared. Hopefully, this consideration will lead to understanding how inert dinitrogen can be turned into a very active substrate readily reacting in solution in the presence of comparatively mild reducing agents.

## 6.2 Peculiarities of the $N_2$ Molecule

The inertness of dinitrogen is very well known and reflected in its properties (Table 1). Dinitrogen has very high bond-dissociation energy, high ionization potential, and a negative electron affinity. Proton affinity, though positive, is comparatively small, smaller than, for example, that of methane ($5.3 \pm 0.3$ eV). Dinitrogen is a very weak base, and does not interact even with strong acids. It should be noted, however, that the bond-dissociation energy does not by itself explain the inertness of dinitrogen. The triple-bond-dissociation energy of acetylene (230 kcal mol$^{-1}$)

**Table 1.** Some physical constants and characteristics of dinitrogen.

| | |
|---|---|
| Interatomic distance | 1.095 Å |
| Ionization potential | 15.58 eV |
| N≡N dissociation energy | 225 kcal mol$^{-1}$ |
| Vibration frequency (gas) | 2231 cm$^{-1}$ |
| Electron affinity | $-1.8$ eV |
| Proton affinity | 5.12 eV |
| Singlet-triplet energy | 6.17 eV |
| Solubility in water | $1.7 \times 10^{-3}$ cm$^3$/cm$^3$ |
| Solubility in benzene | $1.11 \times 10^{-1}$ cm$^3$/cm$^3$ |

is approximately the same as for dinitrogen, and for carbon monoxide it is even higher (256 kcal mol$^{-1}$), but both $C_2H_2$ and CO are usually chemically much more reactive than dinitrogen.

The electronic configuration of dinitrogen can be represented as $(1\sigma_s)^2(1\sigma_u)^2(2\sigma_g)^2(2\sigma_u)^2(1\pi_u)^4(3\sigma_g)^2$. The highest occupied orbital $3\sigma_g$ is higher than $1\pi_u$ because of the mixing of s and p orbitals. Both occupied orbitals $3\sigma_g$ and $1\pi_u$ are strongly bonding; the energy level of $3\sigma_g$ is $-15.6$ eV; for $1\pi_u$ it is $-17.1$ eV. The lowest unoccupied orbital $1\pi_g*$ is strongly antibonding: $+7.3$ eV.

Important conclusions about the reasons for the inertness of dinitrogen can be drawn by considering the energies required for consecutive cleavage of the triple bonds in the $N_2$ molecule. Dissociation of the first of the three bonds requires more than 100 kcal mol$^{-1}$, almost half the total triple bond energy. For acetylene, the first bond of the three to be broken is, in contrast, the weakest (53 kcal mol$^{-1}$). This sharp difference is the main reason for their different reactivity. The strength of this first bond in $N_2$ (compared with acetylene) is evidently largely a consequence of high energy level of diazene; this is because of the repulsion of the two unshared electron pairs and the electron pair of the remaining $\pi$-bond in $N_2H_2$ after cleavage of one of the $\pi$-bonds of $N_2$; in acetylene, the electron pairs of two $\sigma$-bonds with H atoms (as distinct from the two unshared pairs on $N_2$) are less strongly repelled by electrons of the $\pi$-bond.

This feature of dinitrogen is an intrinsic characteristic of the molecule, rather than a reflection of the nature of species reacting with it. Thus, the energy of the $N_2$ triplet state (which can be regarded as a $N_2$ molecule with one of $\pi$-bonds cleaved) is much higher (6.17 eV) than the corresponding triplet state of acetylene (ca 3.7 eV).

This difference results in different thermodynamics of one- and two-electron reactions of acetylene and dinitrogen—often these reactions are thermodynamically favorable for acetylene but forbidden for dinitrogen.

For example, the hydrogenation of acetylene (and ethylene) by one molecule of $H_2$ is strongly exothermic ($\Delta H°$ for the reaction $C_2H_2 + H_2 \rightarrow C_2H_4$ is $-42$ kcal mol$^{-1}$) whereas the corresponding reaction for dinitrogen, $N_2 + H_2 \rightarrow N_2H_2$, is strongly endothermic ($\Delta H°$ is $+51$ and $+56$ kcal mol$^{-1}$ for *trans*- and *cis*-diazene respectively). Even addition of a hydrogen atom, which for acetylene is 41 kcal mol$^{-1}$ exothermic, for dinitrogen (the reaction $N_2 + H^• \rightarrow {}^•N_2H$) is ca 9 kcal mol$^{-1}$ endothermic. A radical chain reaction of dihydrogen with dinitrogen initiated by addition of H atom to $N_2$ is, therefore impossible, as is catalytic hydrogenation via diazene as an intermediate (and this is probably the main reason for the absence of activity of typical hydrogenation catalysts towards dinitrogen).

Consecutive electron transfer to the $N_2$ molecule with simultaneous addition of protons reflects these thermodynamics peculiarities of dinitrogen (Figure 1).

One- and two-electron transfer with simultaneous addition of protons

$$N_2 + e + H^+ = {}^•N_2H$$

$$N_2 + 2e + 2H^+ = N_2H_2$$

corresponds to negative redox potentials and requires much stronger reducing

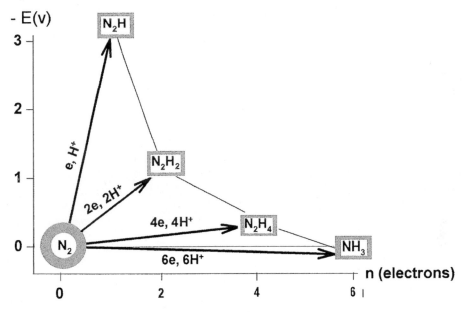

**Figure 1.** Redox potentials of one-, two-, four-, and six-electron reductions of dinitrogen.

agents than dihydrogen. Strong reducing agents such as lithium metal can react with $N_2$ to form nitrides, but such reactions are very unlikely in protic media because the conditions favor a simpler reaction with protons to form $H_2$:

$$2H^+ + 2e \rightarrow H_2$$

At the same time consideration of Figure 1 reveals the possibility of using milder reducing agents which are perhaps sufficiently stable in protic media.

The second and the third bonds cleaved in $N_2$ are, in contrast, very weak—ca 60 kcal mol$^{-1}$ each. Four- and six-electron reductions to hydrazine and ammonia in aqueous solution thus correspond to redox potentials of $-0.36$ and $+0.55$ V, respectively, and require considerably weaker reductants than for one- and two-electron reductions.

If, therefore, we have a metal cluster capable of accepting, e.g., four electrons one by one and then of reacting with $N_2$ to form a hydrazine derivative, further protonation might lead to the formation of hydrazine without intermediate formation of not only the $N_2H$ radical but also the $N_2H_2$ molecule. Such a cluster can then function as a catalyst for $N_2$ reduction to hydrazine or ammonia by comparatively weak one-electron reducing agent sufficiently stable in protic media. The mechanism of four-electron reduction of dinitrogen was proposed ca 30 years ago [1] and helped the discovery of catalysts of $N_2$ reduction in protic media. The first step in such a four-electron or even six-electron mechanism of dinitrogen reduction must always be formation of an intermediate complex with a transition metal compound.

The properties of dinitrogen and systems reducing $N_2$ are considered in more detail in the author's recent book (Ref. [2] and references therein).

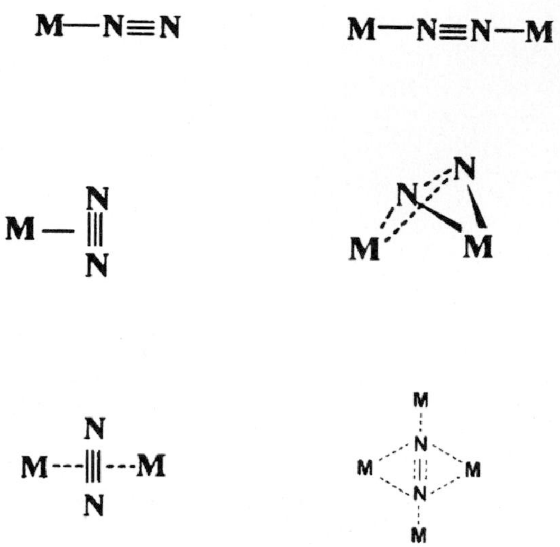

$$\text{M—N}\equiv\text{N} \qquad\qquad \text{M—N}\equiv\text{N—M}$$

$$\text{M —}\ \overset{\text{N}}{\underset{\text{N}}{|||}}$$

$$\text{M---}\overset{\text{N}}{\underset{\text{N}}{|||}}\text{---M}$$

Figure 2. Typical structures of dinitrogen complexes with transition metal compounds (M = metal + other ligands).

## 6.3 Dinitrogen Complexes with Transition Metals: Possible Catalysts for N₂ Reduction

Since 1965, when the first dinitrogen complex with ruthenium(II) was discovered [3a] $N_2$ complexes with several transition metal compounds have been prepared. As with other unsaturated molecules, in these complexes dinitrogen acts both as electron donor and electron acceptor, the latter properties being more pronounced. Different kinds of complex are obtained experimentally and considered theoretically (e.g. see review [3b]).

Examples of the structures of $N_2$ complexes are shown in Figure 2.

It is important for further reduction of the complexes that the dinitrogen ligand is a much stronger base than free dinitrogen and can be protonated. Addition of protons strongly increases the electron-accepting properties of coordinated dinitrogen and facilitates electron transfer from metals M to $N_2$, leading eventually to reduction of dinitrogen to $N_2H_4$ or $NH_3$. (Below 'M' denotes a metal atom with corresponding ligands.)

Linear complexes of the types M–N≡N and M–N≡N–M were discovered first, and are generally more stable and more thoroughly studied than other types of complex. Their relative stability is, apparently, because the highest occupied molecular orbital ($3\sigma_g$) is a $\sigma$-orbital. Also, in mononuclear M–N≡N complexes as distinct from edge-on $\text{M}-\overset{\text{N}}{\underset{\text{N}}{|||}}$ complexes both $\pi$-bonds of dinitrogen are involved in binding with M. Other types of complex might, however, be important as intermediates in a reduction process.

Let us now consider the participation if metal–dinitrogen complexes in $N_2$ reduction. We may safely postulate that the stronger the M–N bonds in the complex the weaker is the NN bond, i.e. the more activated is dinitrogen. If we want to avoid thermodynamically unfavorable diazene, $N_2H_2$, we must have at least four available d-electrons entering both $\pi^*$-orbitals of coordinated dinitrogen. From consideration of experimental and theoretical results obtained for protonation and reduction of dinitrogen in stable dinitrogen complexes with transition metal compounds the following general conclusions can be made.

## 6.4  Mononuclear Complexes M–N≡N

In such complexes dinitrogen usually remains chemically inert, although it can occasionally be reduced to hydrazine or ammonia by protonation by acids:

$$M\text{–}N{\equiv}N + 2H^+ \rightarrow M{=}N\text{–}NH_2 \xrightarrow{H^+} M^{n+4} + N_2H_4$$

$$M\text{–}N{\equiv}N + 3H^+ \rightarrow M{\equiv}N + NH_3 \xrightarrow{H^+} M^{n+6} + 2NH_3$$

It is apparent that four or even six electrons should be available on M and on reduction its oxidation state should be changed by four or six units. Initially, therefore, M must be in a very low oxidation state, and it is difficult to imagine mononuclear dinitrogen complexes as *intermediates* not only in catalytic $N_2$ reduction but even in stoichiometric reduction of dinitrogen if these complexes have to be formed from low-valent M and $N_2$ in the presence of an acid. Actually specially prepared mononuclear complexes containing dinitrogen can occasionally be protonated, as was found by Chatt et al. [3b], and this can be coupled with the reduction of $N_2$, e.g. tungsten(0) dinitrogen complex reacts with sulfuric acid in methanol, forming tungsten(VI):

$$cis\text{-}[W(N_2)(PR_3)_4] + H_2SO_4 \rightarrow N_2 + 2NH_3$$

$$+W(VI)\text{products} + 4[PR_3H]HSO_4$$

With the similar molybdenum complex the yield of ammonia reaches only ca 0.7 mol per metal atom, the remainder being evolved as free dinitrogen, with molybdenum being oxidized to molybdenum(III). Apparently, however, no such complexes have been observed as intermediates in dinitrogen reduction in protic media.

## 6.5  Linear Dinuclear Complexes M–N≡N–M

Protonation of dinitrogen with $N_2$ reduction is, in principle, more favorable thermodynamically in dinuclear complexes than as described in Section 6.4, because the oxidation state of metals increase by two units only in the formation of hydrazine and by three units in the formation of ammonia. Consideration of the molecular orbitals of such complexes leads to the conclusion that in linear dinuclear complexes only $d^2$–$d^4$ electronic configurations correspond to sufficient activation of $N_2$, $d^3$ being the optimum (e.g. see Ref. [2]).

Increasing the number of electrons reduces the activation of $N_2$, because the electrons occupy the orbitals which are bonding with respect to the NN bond, and actually stabilize it. In agreement with this prediction dinitrogen is sufficiently activated to be reduced by protonation by dinuclear complexes of titanium(II), zirconium(II), niobium(III), tantalum(III), molybdenum(IV), and tungsten(IV), whereas it is not reduced by protonation by certain $d^6$–$d^{10}$ complexes, such as those of molybdenum(0), ruthenium(II), or rhodium(I). Apparently dinuclear complexes M–N≡N–M in which M has the $d^3$ electronic configuration can be intermediates in dinitrogen reduction in protic media, particularly if they represent part of polynuclear complexes (vide infra).

In di- and polynuclear complexes with $N_2$ perpendicular orientation of NN in a

complex $M\text{--}\overset{\displaystyle N}{\underset{\displaystyle N}{|||}}\text{--}M$ seems to correspond to more activated $N_2$ molecules than the

linear coordination $M \cdots N{\equiv}N \cdots M$


## 6.6 Polynuclear Dinitrogen Complexes


As the number of metal atoms in a polynuclear dinitrogen complex is increased, weaker reducing agents can be used to achieve the same level of the substrate activation, because with sufficient metal atoms surrounding dinitrogen the oxidation state of metal atoms can increase by one unit only and the complexes can have sufficient electron capacity to afford four- or even six-electron reduction with comparatively small energy requirements. Few polynuclear complexes have yet been isolated and generalizations can be made only with care. Nevertheless, up to six metal atoms can be bound to dinitrogen, each with an M–N bond, for example the complex with gold [3c]. Presumably in non-linear dinitrogen complexes there is no definite limitation to the number of d-electrons present on the metal atoms for the activation of bound dinitrogen; it must, however, also be taken into account that there might be an optimum value of this number for $N_2$ activation.

It should be mentioned that consecutive two-electron reduction of dinitrogen is still usually considered likely in the literature, with diazene, $N_2H_2$, as an intermediate but stabilized by complexation with transition metals. At the same time there is no doubt that both $\pi$-bonds of $N_2$ can be used simultaneously for four d-electrons (entering two degenerate $\pi_g{}^*$-orbitals) from one, two, or several metal atoms forming a complex with dinitrogen, and separate cleavage of each $\pi$-bond seems to be unnecessary, more so because it is thermodynamically unfavorable. Therefore, formation of, for example, a hydrazine derivative and subsequently hydrazine itself:

$$M^n-N{\equiv}N-M^n \xrightarrow{H^+} M{=}N-\overset{+}{N}(H){=}M \xrightarrow{H^+} 2M^{n+2} + N_2H_4$$

can be regarded as a genuine four-electron reduction.

## 6.7 Polynuclear Complex as Optimum Catalyst for $N_2$ Reduction

In principle protic media favor *catalytic* dinitrogen reduction more than aprotic media, because solvent molecules can protonate the intermediate product forming $N_2H_4$ or $NH_3$ and liberating the catalyst for new cycles of dinitrogen activation and reduction. There is, however, an obstacle in protic media which has already been mentioned: strong reducing agents can react with the solvent to form $H_2$ instead of reacting with $N_2$.

We can try to visualize an optimum catalyst for dinitrogen reduction in protic media taking into account the considerations presented earlier. An optimum catalyst might be defined as that functioning with the weakest, yet thermodynamically permitted, reducing agent. The catalyst is expected to be a polynuclear cluster capable to coordinate dinitrogen. At least four electrons of the cluster should be available to enter both antibonding $\pi_g{}^*$ orbitals of dinitrogen. Metal atoms should be bound by flexible bridges to enable adjustment to changes of distances in the process of dinitrogen reduction. The simplest proposal for catalytic mechanism of dinitrogen reduction would be the formation of a polynuclear complex able to reduce $N_2$ to $N_2H_4$ or $NH_3$. The formation of such a complex will, presumably, involve several steps of intermolecular one-electron transfer from the external electron donor to the oxidized state of the complex, after that the complex will reduce $N_2$ independently in multi-electron manner, while transforming itself again into its initial oxidized state.

This mechanism looks simple and reasonable, but it is *not* optimum, at least for such a 'difficult' substrate as dinitrogen. Indeed, to reduce the catalyst to the state which then is able to reduce $N_2$ we need a reducing agent which is a stronger one-electron reductant than the cluster catalyst in each step of its reduction including, for example, the final difficult one-electron step to form the highly reduced state of the catalyst capable of coordinating and reducing $N_2$. The reducing agent might be less strong and the catalytic complex more suitable for dinitrogen activation if intermolecular electron transfer from the external reducing agent and intra-molecular electron transfer from the catalytic complex to the coordinated dinitrogen occur simultaneously with 'reductive coaction' of the external electron donor. It might be expected that if the catalyst and the external reducing agent function in such a coupled manner in the most difficult steps of substrate reduction we shall have more optimum catalytic system. Electrons from the external reducing agent should be transferred to the catalyst coordinating $N_2$ together with protons simultaneously or immediately one after the other to avoid accumulation of negative or positive charges which would destabilize the catalyst.

Protons can go directly to the dinitrogen ligand if it is sufficiently basic to accept $H^+$ from the proton donor; alternatively it might temporarily form a hydride, by addition to metal, or protonate another ligand (as S or O atoms) bound to metal in the vicinity of dinitrogen.

Each electron transferred to the cluster will shift electrons from the cluster to coordinated $N_2$, which is mainly an electron-accepting ligand. Therefore, coordination of dinitrogen will increase the positive charge on the metal atoms and facilitate further electron transfer from the external donor. Protonation of dinitrogen will increase the positive charge further and in this its turn will facilitate electron transfer from the external reducing agent. At some stage the protonated dinitrogen ligand will be transformed to a hydrazine or ammonia derivative. This step might be rate-controlling, because the oxidation state of several, perhaps three, metal atoms will be increased by one unit each. It might require further electron transfer from the external reductant (to make the total number of electrons equal to at least four) and this reductant must, therefore, be situated at sufficiently short distance from the catalyst during the entire process of substrate reduction.

This mechanism with *coupled* one-electron intermolecular electron transfer from the external donor and intramolecular multi-electron transfer from the catalyst to coordinated $N_2$ is, presumably, more efficient than the simpler mechanism considered above with one-electron and multi-electron transfers separated in time. In this mechanism the strongest reductant, which is of necessity the external reducing agent, is used for *direct* reduction of the substrate, whereas for consecutive one-electron and multi-electron processes its reducing power is used only to prepare the reduced form of the catalyst.

## 6.8 Electron Transfer in Biological Nitrogen Fixation

Biological nitrogen fixation involves reduction of dinitrogen catalyzed by a metalloenzyme nitrogenase. As has already been mentioned, there has recently been considerable progress in the study of biological nitrogen fixation, and the mechanism of electron transfer in the process of $N_2$ reduction is becoming ever more clear. Several excellent reviews describing the structure and functions of nitrogenase and the mechanism of biological dinitrogen reduction have recently been published [4–6]. Here only main features of this process will be described. The first observed product of the reduction is ammonia; this is subsequently used in amino acid synthesis. Nitrogenase comprises two separable proteins: Fe protein and MoFe protein. Fe protein is a dimer of identical subunits with a molecular mass of 60 kDa. It contains one $[Fe_4S_4]$ cluster and is a specific one-electron donor for MoFe protein. MoFe protein is a tetramer of two types of subunit, $\alpha_2\beta_2$, with a total molecular mass of 240 kDa. It contains two atoms of Mo, 32 atoms of Fe, and the same number of $S^{2-}$ ions per molecule. These atoms are grouped together in two types of cluster—so called P-clusters and iron–molybdenum cofactors (FeMoco). In the presence of the two proteins (Fe protein and MoFe protein) dinitrogen is reduced to ammonia, the stoichiometry under optimum conditions is:

$$N_2 + 8e + 8\,H^+ + 16\,MgATP \rightarrow 2NH_3 + H_2 + 16MgADP + 16HPO_4{}^{2-}$$

Thus $N_2$ reduction is coupled with ATP hydrolysis; it is apparent that two ATP molecules are hydrolyzed on transfer of each electron. One mole of dihydrogen is produced per mole of fixed dinitrogen even at saturating $N_2$ pressures. These results are obtained with molybdenum nitrogenase [4], containing iron molybdenum cofactor as the active site for dinitrogen and other nitrogenase substrates. As well as molybdenum nitrogenase two other 'alternative' nitrogenases are known [5]; these do not contain molybdenum. One contains vanadium instead of molybdenum, the other contains only iron as a transition metal and is called 'iron-only' nitrogenase. (A new nitrogenase functioning at elevated temperature was recently described in a series of articles [7]. This nitrogenase is very different from the traditional ones and it is too early to make any conclusions concerning the mechanism of its action.)

All three nitrogenases have much in common in their structures and functioning, although there are also some differences in detail. Molybdenum nitrogenase seems to be more stable and more efficient than the vanadium-containing enzyme and, particularly, iron-only nitrogenase. With the Mo-containing nitrogenase only one $H_2$ is produced at saturating dinitrogen pressures, according to the stoichiometry presented above, whereas the two others produce several dihydrogen molecules. Other molecules containing triple bonds can be reduced in the presence of nitrogenase, e.g. acetylene produces ethylene, HCN gives $CH_3NH_2$, nitriles (RCN) afford $RCH_3 + NH_3$ etc. Thus even for simple reactions at least two electrons are necessary for the reduction. Important results were obtained recently by Rees et al. [8] (see also review [6]) who determined the molecular structure of Fe protein and MoFe protein by X-ray analysis. The molecular structure of metal-containing clusters was also elucidated. These results contributed much to the understanding of the electron-transfer mechanism in nitrogenase function, although some details of this mechanism are still not quite clear. There are two locking sites for Fe protein at MoFe protein and two pairs of P-clusters and FeMo cofactors. Each set of clusters functions separately (the distance between two FeMoco is 50 Å).

The first electron acceptor from the external donor is the $Fe_4S_4$ cluster situated between two subunits of Fe protein. From $Fe_4S_4$ an electron travels first to a P-cluster, then to FeMo cofactor, and finally to an activated substrate molecule. According to the X-ray data the shortest distance from the $Fe_4S_4$ cluster of the Fe protein to the P cluster of the MoFe protein is ca 18 Å, that from P cluster to FeMo cofactor is ca 14 Å, and the distance from $Fe_4S_4$ to FeMoco is ca 32 Å. The P-cluster lies between $Fe_4S_4$ and FeMoco. Thus, electron transfer in nitrogenase can be presented as:

$$\text{ext.donor} \xrightarrow{e} \text{Fe}_4\text{S}_4 \xrightarrow{e} \text{P-cluster} \xrightarrow{e} \text{FeMoco} \xrightarrow{e} \text{substrate}$$

The distances $18\,\text{Å}$ and $14\,\text{Å}$ inside the protein complex are typical of electron transfer in biological systems. The reduced Fe protein with two bound MgATP binds to the MoFe protein and a single electron is transferred from the Fe protein to the MoFe protein. This process is coupled to the hydrolysis of two ATP.

For the process in vivo flavodoxin is the external electron donor. For isolated nitrogenase, dithionite $S_2O_4{}^{2-}$ is traditionally used as electron donor. If dithionite is the external donor, the oxidized iron protein with 2MgADP (after the electron is transferred) then dissociates from the MoFe protein. This dissociation, which initially seemed necessary for enzyme function, is, however, apparently a result of the salt effect of dithionite, the concentration of which for effective electron transfer must be sufficiently large.

When photo-electron transfer is used with a dye, such as eosin or dibromofluorescein in combination with NADH (and the dithionite concentration is low, ca $4 \times 10^{-4}$ M), no dissociation of proteins is detected. Kinetic laser spectroscopy helped Syrtsova et al. [9] to follow electron transfer from the Fe protein to the MoFe protein and it was shown that, unlike the situation with dithionite, Fe protein in the complex with MoFe protein could undergo reduction by the photodonor as efficiently as in the free state in solution and electron transfer proceeds in the complex of two proteins without dissociation.

According to X-ray analysis the P-cluster which functions as intermediate electron acceptor from $Fe_4S_4$ cluster of the Fe protein contains eight iron atoms and seven inorganic sulfur atoms (Figure 3). The P-cluster can be considered as two bridged $Fe_4S_4$ and $Fe_4S_3$ clusters and therefore presumably it can be a donor of at least two electrons transferred to FeMo cofactor. When both $Fe_4S_n$ parts of the P-cluster are fully reduced the P-cluster must be a comparatively strong electron donor. Moreover spectroscopic studies indicate that all eight iron atoms can be in ferrous state with a full charge of $-4$ [10]. Presumably the state of the most reduced form of the P-cluster is still unknown, but this result confirms its strong electron donor properties even if no more electrons are accepted by P-cluster in the process of substrate reduction until it gives up one electron to the FeMo cofactor.

Iron molybdenum cofactor, FeMoco was isolated from MoFe protein for the first time in 1977 by Shah and Brill [11]. Since then the cofactor, which naturally attracted much interest as the probable active site of the enzyme, has been thoroughly studied. Although the structure of isolated cofactor remains unknown X-ray studies by Rees et al. have enabled construction of the internal structure of the protein.

It is a polynuclear complex and consists of two cubane fragments $Fe_4S_3$ and $Fe_3MoS_3$ connected by three $S^{2-}$ bridges (Figure 3). The 1:7:9 ratio for Mo:Fe:S in the proposed structure is in agreement with analytical data for FeMoco for the isolated cofactor (Mo:Fe:S = 1:6–8:8–9)

The cluster is connected with the protein matrix through cysteine side groups (Cys 275), a ligand of Fe1 and histidine (His 442) which is a ligand of Mo. These Fe

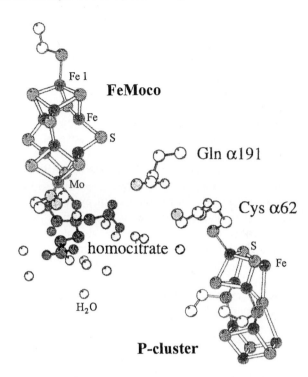

**Figure 3.** FeMo cofactor, P-cluster and their environment of the MoFe protein from *Azotobacter vinelandii* [6]. This is corrected to take into account that the P-cluster is of the 8Fe–7S type, and not of the 8Fe–8S type as previously believed).

and Mo atoms are on the third-order axis a distance of 7.5 Å from each other. In the structure Mo has octahedral coordination and is connected, in addition to three sulfur atoms, to an N atom of histidine and two O atoms of hydroxyl and carboxyl groups of a homocitrate molecule which was also found to be present in FeMo cofactor when it is isolated from the protein. The tetrahedral coordination found for Fe1 and octahedral for Mo are typical of coordination of these atoms both in model compounds and iron–sulfur proteins. The main peculiarity of the model is that six of the seven iron atoms in the cofactor have a triangular-pyramidal coordination, and are thus coordinatively unsaturated.

The Fe–Fe distances in the cluster were found to be 2.4–2.6 Å, which indicates the existence of some bonding between them. This bonding, evidently stabilizing the cluster, would not, however, prevent coordination of a substrate molecule; therefore, although there is no direct evidence, the substrates, including dinitrogen, are coordinated by several, possibly four or even all six iron atoms when they are activated by FeMoco.

The absence of a large number of bridges binding FeMoco and protein explains why the cofactor can be easily extracted from the protein without loss of its catalytic activity when it is reset again on apoprotein (protein matrix deprived of the cofactor).

The role of molybdenum (if there is no direct Mo–N bonding in the complex with the substrate) is not quite clear, but it can be speculated that $Mo^{IV}$ with $d^2$ electronic configuration present in the cofactor reduces the number of electrons (compared with $Fe^{II}$) without much changing its redox potential, and this might help activate dinitrogen (if there is certain optimum number of electrons for $N_2$ activation). Presumably it also changes the polarity of the complex and this might strengthen the activation of bound dinitrogen.

EPR studies of the isolated cofactor in the state reduced by dithionite, FeMoco (s-r) (i.e. semi-reduced) show that it retains the characteristic signal with spin 3/2 observed for the so-called 'M center' (which is in fact FeMoco inside MoFe protein). The broadening of the signal of the isolated FeMoco, compared with the signal of the M center, is attributed to a small change in the ligand sphere of the cofactor upon isolation. In the presence of added thiophenol the EPR signal returns to its form inside the protein apparently because S group of the thiophenol becomes the ligand of the Fe1 atom (instead of cysteine 275 SH group). The cyclic voltammogram of the cofactor on the glass carbon electrode in DMF shows that in addition to the process FeMoco(ox) → FeMoco(s-r) at −0.32 V there is the reduction at −1.1 V with the corresponding oxidation at −0.9 V. These waves are ascribed by the authors [12] to the pair FeMoco(s-r)/FeMoco(red).

When the Marcus theory of electron transfer was applied to electron transfer in nitrogenase, it was found that non-adiabatic theory is not applicable. It was therefore concluded that an event associated with either MgATP binding or hydrolysis acts to gate electron transfer between Fe protein and MoFe protein [13a]. The role of MgATP might be explained by the necessity to transfer electrons against the thermodynamic potential in the process of electron transfer from the $Fe_4S_4$ cluster of the Fe protein to the P-cluster of the MoFe protein to create particularly strong electron donor for FeMoco, although the mechanism of its action is not quite clear.

Recent investigations of electron transfer in the tight complex combining the Fe protein isolated from the bacterium *Clostridium pasteurianum* (Cp2) with the MoFe protein isolated from another bacterium *Azotobacter vinelandii* (Av1) demonstrated that primary electron transfer occurs within this complex even in the absence of MgATP and that the latter accelerates electron transfer more than 10 000-fold [13b]. Analysis of this effect shows that MgATP apparently shifts the redox potential of Fe protein ($E_0$) to a more negative value, therefore changing $\Delta E$ (difference in redox potentials of the electron donor and acceptor); this change alone is not, however, likely to be sufficient for the observed nucleotide-dependent rate acceleration. The important observation of this work is that Cp2Av1 is unable to transfer more than one electron to MoFe protein even when MgATP is present. This explains why Cp2Av1 complex is inactive towards any substrates of nitrogenase (even $H_2$ is not formed) because at least two electrons are required, even for such substrates as acetylene to produce ethylene, and protons to produce dihydrogen.

This supports the view that two electrons are transferred to P-cluster and the second electron requires probably very precise complementarity of the two proteins unattainable in the complex Cp2Av1. Perhaps the transfer of this second electron corresponds to particularly strongly negative value of redox potential.

In conclusion, it can be said that enzymatic reduction of dinitrogen requires substrate coordination on the polynuclear FeMo cofactor. To ensure catalytic reduction of the substrate the reduced cofactor is situated at a distance typical of electron transfer (14 Å) from the P-cluster, which contains at least two electrons capable of being transferred to FeMoco. MgATP hydrolysis helps to increase the rate of electron transfer to the P-cluster and at least one electron is transferred from the $Fe_4S_4$ cluster against the thermodynamic potential presumably increasing the electron-donor properties of the P-cluster for subsequent transfer of electrons to MoFe cofactor. In their reduced states both are sufficiently strong reducing agents with redox potentials presumably more negative than $-1$ V. Using electrochemical terminology the P-cluster can be regarded as a 'nano-electrode' with the catalyst, FeMoco bound to it. It might be thought that the mechanism of catalysis for dinitrogen reduction is close to that for the 'optimum catalyst' considered above. The P-cluster and FeMoco might function in the coupled way in the most difficult stages of dinitrogen reduction in the coordination sphere of FeMoco. Accordingly it was shown that P-clusters are oxidized in a certain step when dinitrogen is reduced by nitrogenase [13a].

## 6.9 N₂ Reduction in Aprotic Media

Enzymatic dinitrogen reduction involves an electron donor, e.g. flavodoxin in vivo or dithionite in vitro (which we now know can be replaced by photoelectron donor), a system of electron transfer, and a site of dinitrogen activation, where the $N_2$ molecule is bound and subsequently reduced to ammonia with participation of protons. As we have seen, this site, evidently FeMoco, is itself a well organized molecular system which, in combination with the P-cluster, reduces $N_2$ by a mechanism of coupled one-electron and multi-electron transfer. If we can use a stronger reducing agent then the system for $N_2$ activation and reduction presumably need not be so well organized. In aprotic media we can use very strong reducing agents, although the reaction naturally cannot produce hydrazine or ammonia because of the lack of protons. At least stoichiometric reaction can be expected to break the NN bond and to form nitrides from intermediate $N_2$ complexes and then to produce ammonia upon subsequent addition of an acid.

Until the 1960s, however, dinitrogen was considered so chemically inert that no attempts had been made by investigators working with transition metal compounds in the presence of strong reducing agents to verify the possibility of $N_2$ reduction— dinitrogen gas was used as an inert atmosphere for conducting various reactions.

In 1964 Vol'pin and Shur published their paper [14] describing the results of dinitrogen reduction in aprotic media. The reaction turned out to be quite general. In the presence of different transition metal compounds, for example $CrCl_3$, $MoCl_5$, $WCl_6$, $FeCl_3$, and $TiCl_4$, when $N_2$ under high pressure (100–150 atm) was reacted with strong reducing agents, for example $LiAlH_4$, $EtMgBr$, or $i\text{-}Bu_3Al$, ammonia was produced on acid decomposition of the products formed. The system $Cp_2TiCl_2 + EtMgBr$ ($Cp = \pi\text{-}C_5H_5$) in ether was found to be even more active towards $N_2$—the yield of $NH_3$ (after hydrolysis) was 67 mol % with respect to Ti compound even at atmospheric pressure and nearly quantitative at elevated $N_2$ pressures. The number of transition metal compounds and possible reductants was later greatly increased by Vol'pin et al. and also by other groups of investigators.

Apart from the compounds already mentioned, vanadium, manganese, and cobalt chlorides, tetra-alkoxy derivatives of titanium, acetylacetonates of V, Cr, Mo, Mn, and Ni, Cp derivatives of Zr and Nb, and triphenyl phosphine complexes of Ti and Fe were found to be active. Later lanthanide complexes were included in the list of dinitrogen-reducing systems, the most effective being compounds of samarium and yttrium.

The transition metals of groups IV, V and VI, particularly Ti, V, Cr, Mo, and W, have the strongest $N_2$-reducing capacity. Ti compounds are particularly active. In the first row of transition metals, the ammonia yields decrease generally from left to right, in line with the decreasing stability of the metal nitrides. Co and Ni compounds are usually of low or no activity. Palladium, copper and platinum complexes have no activity in any system tested.

Yields of $NH_3$ (after hydrolysis) depend on the reducing agents. They include such strong reductants as organometallic compounds ($RMgX$, $RLi$, and $R_3Al$), metal hydrides ($LiAlH_4$ and $LiH$), free metals, and aromatic radical ions. Catalytic hydrogenation was not observed, but occasionally $H_2$ increased the yield of $NH_3$ (after hydrolysis) in combination with other reducing agents.

Catalytic reduction was, however, occasionally observed. $TiCl_4$ can catalyze the reduction of $N_2$ by aluminum metal in the presence of $AlBr_3$ at 130 °C and $N_2$ pressure of 100 atm. Yields of ammonia (after hydrolysis) usually vary from 0.01 to 1 mol mol$^{-1}$ transition metal compound, indicating that two transition metal atoms participate in the reduction of one $N_2$ molecule, although in some systems (particularly with very strong reductants) up to 2 mol $NH_3$ are produced per mol transition metal.

## 6.10 Dinitrogen Reduction in Protic Media

As has already been mentioned protic media are attractive for $N_2$ reduction, in particular because hydrazine or ammonia can be directly produced and the search for catalytic reduction is more promising. There is, however, an important obstacle —the strong reducing agent can be unstable in protic surroundings, readily producing dihydrogen:

$$2H^+ + 2e^- \rightarrow H_2$$

The choice of the systems active to $N_2$ in protic solutions must, therefore, naturally be more limited than in aprotic media.

The first reproducible results demonstrating efficient dinitrogen reduction with participation of protons were reported in 1970 [15]. Table 2 lists systems reducing $N_2$ in the presence of water or methyl alcohol. They are mainly based on $V^{II}$ and

**Table 2.** Systems reducing $N_2$ in protic media ($p_{N2} \approx 100$ atm).

| M | Reductant | $T(^{\circ}C)$ | Products | Yield (mol mol$^{-1}$ M$^{-1}$) |
|---|---|---|---|---|
| Ti$^{II}$ (d$^2$) | Na(Hg) | 20 | $N_2H_4$, $NH_3$ | 0.01 |
| V$^{III}$ (d$^3$) | V(OH)$_2$ + Mg(OH)$_2$ | | | |
| | pH 14.3 | 20 | $N_2H_4$, $NH_3$ | 0.65 |
| | pH 12 | 20 | $NH_3$ | 0.35 |
| | V$^{II}$ + catechol, pH 10.5 | 20 | $NH_3$ | 0.75 |
| Mo$^{III}$ (d$^3$) | Ti(OH)$_3$ | 60 | $N_2H_4$, $NH_3$ | 1 |
| | Ti(OH)$_3$ + Mg(OH)$_2$ | 110 | $N_2H_4$, $NH_3$ | 170 |
| | Cr(OH)$_2$ | 90 | $N_2H_4$, $NH_3$ | 0.80 |
| | without Mo | 90 | | 0.015 |
| | Na(Hg) ($p = 1$ atm) | 20 | $N_2H_4$, $NH_3$ | 1700 |
| | Eu(Hg) ($p = 1$ atm) | 20 | $N_2H_4$, $NH_3$ | 26 |
| Nb$^{III}$ (d$^2$) | Nb(OH)$_3$ | 35 | $N_2H_4$, $NH_3$ | 0.09 |
| Ta$^{III}$ (d$^2$) | Ta(OH)$_3$ | 35 | $N_2H_4$ | 0.02 |

Mo$^{III}$ compounds, although later Nb$^{III}$, Ta$^{III}$ and Ti$^{II}$ compounds were also included. It is clear that all the systems found so far are based on d$^2$ and d$^3$ electronic configurations. They are active in alkaline solution and include mainly hydroxides, i.e. the systems are heterogeneous. There is also a unique family of homogeneous systems based on solutions of vanadium(II) complexes with catechols. One of the simplest and most effective heterogeneous systems is mixed V$^{II}$–Mg$^{II}$ hydroxide. The reduction of $N_2$ occurs in aqueous or alcohol suspensions of freshly prepared hydroxide formed by adding excess alkali to the solution of a mixture of VCl$_2$ and MgCl$_2$. The reaction produces hydrazine and ammonia at high rates at room and lower temperatures and atmospheric $N_2$ pressure.

At high concentrations of alkali (pH 13–14) and high $N_2$ pressures hydrazine is mainly produced according to the stoichiometry:

$$4V(OH)_2 + N_2 + 4H_2O \rightarrow 4V(OH)_3 + N_2H_4$$

At low alkali concentrations (pH 8–12) ammonia is produced directly from dinitrogen without intermediate formation of free hydrazine:

$$6V(OH)_2 + N_2 + 6H_2O \rightarrow 6V(OH)_3 + 2NH_3$$

Freshly prepared mixed hydroxide contains vanadium(II) clusters reactive towards dinitrogen. Some indirect evidence indicates that the number of vanadium ions in the clusters activating dinitrogen approaches four or six. For example, introduction of other ions, such as V$^{3+}$ inhibits $N_2$ reduction and quantitative analysis of the V$^{3+}$ inhibition effect leads to the conclusion that tetramers are the likely species; tetramers are also suggested by analysis of ethane and ethylene formation in the reduction of acetylene.

Kinetic analysis of $N_2$ reduction shows that the reaction can be regarded as pseudo-homogeneous. Analysis reveals Michaelis–Menten dependence of the reaction rate on dinitrogen pressure. This enables estimation of the enthalpy of dinitrogen complex formation and the activation energy—$\Delta H = -4$ kcal mol$^{-1}$ and $E = 8.4$ kcal mol$^{-1}$. These values indicate that the intermediate complex is very unstable and highly reactive. Mixed $V^{II}$–$Mg^{II}$ hydroxide is one system that disproves generally accepted views of dinitrogen as very inert molecule. $N_2$ can be very active in systems that are neither strong bases or acids nor very strong reductants or oxidants. The important condition for high chemical reactivity of $N_2$ is the possibility of accepting simultaneously at least four electrons by intramolecular electron transfer in an intermediate polynuclear complex including at least four atoms that are electron donors.

## 6.11 Soluble Complexes of Vanadium (II)

The exact structure of the intermediate vanadium(II) cluster in the mixed $V^{II}$–$Mg^{II}$ hydroxide and in other hydroxides active towards $N_2$ is, of course, difficult to elucidate, and it was important to find a homogeneous system reducing $N_2$ in protic media. This was realized in 1972 with complexes of $V^{II}$ with catechols or substituted aromatic diols and triols which were found to reduce $N_2$ efficiently in homogeneous water and alcohol solutions [16].

With unsubstituted catechol complexes in water the reaction proceeds in the pH range 8.5 to 13.5 with maximum yields at ca pH 10.5. For substituted catechols the pH range corresponding to the most active complexes reacting with $N_2$ differs noticeably (Table 3). Later the number of catecholate $V^{II}$ complexes active towards dinitrogen was greatly increased but the family of the $V^{II}$ complexes with aromatic diols and their relationship towards $N_2$ remain quite unique—no other complexes have yet been found that reduce dinitrogen in protic media, homogeneously, with reasonably high yields of products.

The reaction of $N_2$ with $V^{II}$ catechol complexes in methanol occurs at room temperature and atmospheric $N_2$ pressure. The yield of $NH_3$, which is the only reaction product, reaches ca 50 % relative to $V^{II} \rightarrow V^{III}$ oxidation. In aqueous so-

**Table 3.** Dinitrogen reduction by vanadium(II) complexes with aromatic ligands (L) (room temperature, $[V^{II}] = 0.05$ M, $[L] = 0.5$ M).

| L | Solvent | $p_{N2}$ (atm) | Time (h) | Yield $NH_3$ (%) | [NaOH]/[L][a] or pH[b] |
|---|---------|----------------|----------|------------------|------------------------|
| Catechol | $CH_3OH$ | 1 | 0.03 | 47 | 1.14[a] |
| Catechol | $CH_3OH$ | 15 | 0.17 | 75 | 1.14[a] |
| Catechol | $H_2O$ | 100 | 1.5 | 60 | 10.5[b] |
| Gallic acid | $H_2O$ | 100 | 4.0 | 48 | 12.9[b] |
| Pyrogallol | $H_2O$ | 100 | 4.0 | 60 | 13.5[b] |

lutions such yields are observed only at elevated dinitrogen pressures. In the absence of $N_2$ a parallel reaction of $V^{II}$ oxidation by the solvent protons proceeds with $H_2$ evolution. As $N_2$ pressure increases the yields of $NH_3$ increase and those of $H_2$ decrease. But even at the elevated pressure the yield of $NH_3$ does not exceed 75 % and does not increase when the $N_2$ pressure is increased further; 25 % of the electrons are used for $H_2$ formation. Thus, analogous to enzymatic reduction with Mo nitrogenase, the stoichiometry of the reaction at saturating $N_2$ pressure corresponds to the equation:

$$8V^{2+} + N_2 + 8H_2O \rightarrow 8V^{3+} + 2NH_3 + H_2 + 8OH^-$$

This equation including $H_2$ evolution *coupled* with the reduction of $N_2$ reflects the polynuclear character of the intermediate complex.

Acetylene is quantitatively reduced by $V^{II}$ catecholate to ethylene with *cis*-dideuteroethylene formed selectively from $C_2D_2$, also similarly to enzymatic reduction of $C_2D_2$ by nitrogenase. Distinct from $N_2$ reduction, the pH range for the reduction of acetylene is much broader than for dinitrogen (from ca pH 5 to concentrated alkali solution). Kinetic studies of the oxidation of vanadium(II) catecholate complex by dinitrogen have led to the reaction equation:

$$-d[V^{II}]/dt = k_1[V^{II}]^2[N_2] + k_2[V^{II}]^{0.5}$$

The first term of the equation corresponds to the reduction of $N_2$ to $NH_3$ and the second to the parallel and *independent* evolution of $H_2$ formed from the solvent protons.

$N_2H_4$ which could be the intermediate in the reaction is easily and quantitatively reduced to $NH_3$ in $V^{II}$–catechol solutions. When $N_2$ reduction is stopped at its initial stages by addition of an acid or an oxidant ($VOSO_4$) a small quantity of $N_2H_4$ is found. A kinetic study of the reduction of specially added hydrazine revealed, however, that the rate constant is at least two orders of magnitude smaller than necessary for free hydrazine to be the intermediate producing ammonia. The results show that $N_2H_4$ is formed by decomposition of an intermediate by acid or oxidant; the intermediate, not being free hydrazine, seems similar to a vanadium hydrazine derivative. Similar results indicating intermediate $N_2^{4-}$ derivative formation were obtained in enzymatic dinitrogen reduction. The second order of vanadium concentration in the rate equation for $N_2$ reduction, and at the same time the one-half order for $H_2$ formation, are indicative of the complex polynuclear structure. It is possible, for example, to assume that the active complex contains four vanadium atoms and two complexes are needed to reduce $N_2$ to $NH_3$, whereas a dinuclear complex produced $H_2$, and this dinuclear complex is formed in equilibrium dissociation of the tetramer into two dimers. In this case $H_2$ formation coupled with $N_2$ reduction could be explained assuming that after ammonia formation by the octamer, for which six vanadium atoms are required, a dimer of $V^{II}$ is formed and this readily produces dihydrogen.

EPR spectra observed for the active complex at a pH of ca 10.5 led to the conclusion that the complex is trinuclear and that ammonia is formed when the trinu-

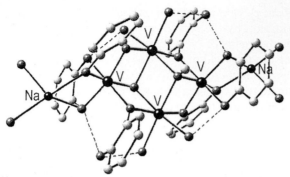

$[V_4(\mu_3\text{-}OCH_3)2L_4(LH)_2\cdot 2CH_3OH]Na_2\cdot 4CH_3OH$

L =

**Figure 4.** The X-ray structure of vanadium di-*tert*-butyl catecholate complex [18].

clear complex coordinating $N_2$ reacts with another trinuclear complex, not containing $N_2$, to produce two molecules of $NH_3$ by use of six vanadium electrons [17]. Although this mechanism did not explain the one-half order for (independent) formation of $H_2$ it was accepted in the paper [17] and in later reviews (e.g. Ref. [2]).

New important information about the structure of the intermediate complex was obtained from the X-ray studies of $V^{II}$ complex with di-*tert*-butylcatecholate [18]. The complex, the structure of which is presented in Figure 4, contains four vanadium ions; two are divalent and two trivalent, but they are indistinguishable, thus their oxidation state is 2.5 and the complex can be regarded as existing in a *semi-reduced* state, similar to the s-r state of FeMoco in the presence of dithionite, which is not sufficiently reduced to activate dinitrogen.

The tetrameric vanadium complex contains also two sodium ions. Each is coordinated by two oxygen atoms of methanol molecules and by two oxygen atoms of different catechol molecules. Because of this coordination, oxygen atoms are $\mu_2$-bridging Na and V atoms. One of the three independent catechol molecules participates in the coordination of the vanadium atoms by one O only whereas the second atom is protonated and does not participate in the metal coordination. It is apparent that the complex

```
      V
   V     V
      V
```

contains the core with two $\mu_3$-bridges formed by methanol OR groups. When all vanadium ions are in the $V^{II}$ state in the form of the complex active to $N_2$ this

$\mu_3$-bridging OR group is probably loosely bound to three $V^{2+}$ ions, and this core containing four $V^{2+}$ ions most probably becomes the site activating $N_2$ in a complex of the type:

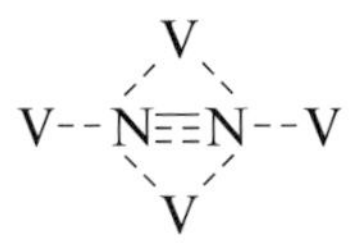

```
            V
          /   \
   V-- N≡≡N --V
          \   /
            V
```

Ultimate intramolecular electron transfer from $4V^{2+}$ to $N_2$ will produce the hydrazido derivative, particularly on protonation of one of the four nitrogen atoms:

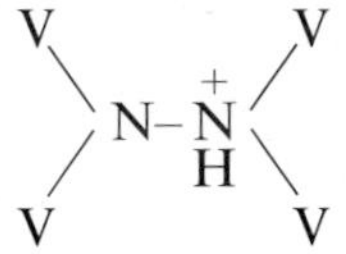

```
   V          V
     \       /
      > N– N <
     /    |   \
   V      H     V
```

These findings prompted us to reconsider the results of EPR spectra and to make more profound analysis of kinetic data for $N_2$ reduction by catecholate vanadium(II) complexes. The conclusion was that a tetra-vanadium structure for the complexes is in a better agreement with the EPR spectra than a trinuclear structure; kinetic results also confirmed the tetranuclear structure of the complex [19]. We therefore regard the tetranuclear structure as confirmed for unsubstituted and substituted catecholate $V^{II}$ complexes, at least for those so far investigated.

Thus the conclusion is that tetranuclear complexes are likely to be the active species in the reduction of $N_2$ to hydrazido derivatives, and then to hydrazine, by reaction with solvent in protic media, whereas octanuclear complexes (or transition states) are needed to transform $N_2^{4-}$ species into two ammonia molecules.

## 6.12 Catalytic Dinitrogen Reduction

A catalytic system for dinitrogen requires the presence of a sufficiently strong reducing agent that would reduce the oxidized form of the catalytic complex (after reaction with $N_2$) to the initial reduced state. For example if we could reduce vanadium(III) complexes formed in $N_2$ reduction by vanadium(II) we would be able to create a catalytic system functioning with the chosen reductant. Finding such a strong reducing agent which is able to reduce $V^{III}$ to $V^{II}$ in *alkaline solutions* is, however, apparently difficult and $H_2$ is formed instead.

More likely candidates as catalysts from the data of Table 2 are molybdenum(III) complexes which activate $N_2$ to reduction only in the presence of a stronger reducing agent, such as $Ti(OH)_3$, $Cr(OH)_2$, or $Ta(OH)_3$. Presumably these strong reductants take part in the process, transferring electrons to molybdenum atoms coordinating dinitrogen. Molybdenum(III) activates $N_2$ in a di- or polynuclear complex forming an intermediate of the type $Mo={=}N{\equiv}N{=}=Mo$ which is unable to produce hydrazine or ammonia readily at the expense of $Mo^{III}$ oxidation to $Mo^V$ or $Mo^{VI}$. The reaction might become thermodynamically more favorable if a stronger reducing agent situated nearby simultaneously transfers electrons, reducing molybdenum to its initial $Mo^{III}$ state. Therefore to reduce dinitrogen catalytically we might try to construct a system containing a bulk reducing agent which adsorbs the catalytic $Mo^{III}$ complex at the surface and is an electroconductive material, in order to use the electrons from remotely situated atoms. For electrochemical reduction electrons from the cathode can be used if the catalytic complex is adsorbed at the surface of the electrode.

The first catalytic system based on this approach for $N_2$ reduction in protic media was realized with $Mo^{III}$ as the catalyst and $Ti^{III}$ hydroxide as the reductant [15]. With pure $Ti(OH)_3$ the yields of products of $N_2$ reduction in the presence of $Mo^{III}$ formed in the process of co-precipitation of both metal hydroxides (or hydroxo complexes) by addition of alkali reaches only equimolar amounts with respect to molybdenum even at elevated temperatures and $N_2$ pressures. In the presence of salts of some other metals, e.g. $Mg^{2+}$, $Ca^{2+}$, or $Sr^{2+}$, in solution before the addition of alkali, the yields increase and the system becomes catalytic. The effect of magnesium salts was particularly pronounced. Yields of hydrazine and ammonia are increased and at sufficiently high temperatures and pressures reach several hundred turnovers per molybdenum. The highest yields are obtained when the ratio of $Mg^{2+}$ to $Ti^{3+}$ is 1:2. At this ratio a compound of formula $MgTi_2O_4$ is produced and forms fine crystals, as was revealed by X-ray analysis and electron microscopy. The catalytic effect of $Mo^{III}$ is observed when the complexes of molybdenum(III) are adsorbed on the surface of $MgTi_2O_4$. Accordingly, whereas for stoichiometric reduction with mixed hydroxides it is essential to have both metals present in homogeneous solution before addition of alkali, for catalytic reduction $MgTi_2O_4$ can be formed initially without molybdenum. After subsequent addition of the molybdenum compound catalytic activity towards $N_2$ is observed.

The magnesium–titanium compound $MgTi_2O_4$ has semiconducting properties, and evidently the molybdenum complex with coordinated dinitrogen forms an electron trap on the surface. Dihydrogen is rapidly evolved in the absence of $N_2$, particularly at high temperatures. Introduction of molybdenum even in very small amounts compared with titanium strongly inhibits $H_2$ evolution in the presence of $N_2$. Evidently Mo complexes are adsorbed at the active sites of $MgTiO_4$, preventing electron flow to the $H_2$ evolution centers. The dependence of the reaction rate on $N_2$ pressure is Michaelis in form, which enables estimation of the equilibrium constant for dinitrogen complex formation and the rate constant for $N_2$ reduction in the complex. The temperature dependence of the constants gives a value of $-7$ kcal mol$^{-1}$ and $-17$ e.u. for the enthalpy and entropy of complex formation; the activation energy, $E$, for $N_2$ reduction in the complex is 20 kcal mol$^{-1}$ [20].

It can be mentioned that a titanium(III) compound was recently found [21] to be an efficient electron donor for the $Fe_4S_4$ cluster in Fe protein of nitrogenase and can therefore be used in $N_2$ enzymatic reduction.

## 6.13  Catalytic Dinitrogen Reduction by Amalgams

The mechanism proposed for the catalytic reduction of $N_2$ by an electroconductive reductant has prompted the use of amalgams as reducing agents. Amalgams such as those formed by sodium and europium are sufficiently strong reductants (redox potentials $-1.84$ and $-1.4$ V, respectively) and at the same time they are reasonably stable in contact with water or alcohol. The amalgams can be prepared electrochemically by use of a mercury cathode and passing electron current through the solution containing metal ions; they can, therefore, be used for the electrochemical reduction of dinitrogen.

Introduction of the sodium amalgam has increased the yield of the reduction products, $N_2H_4$ and $NH_3$, in the catalytic system based on $Mo^{III}$ and $MgTi_2O_4$ described above. Obviously the amalgam transfers electrons through $MgTi_2O_4$ to the $Mo^{III}$ catalytic complex and prevents titanium(III) oxidation. It is clear that if conditions could be found under which the catalytic complex were bound directly to the surface of the amalgam, an electron-conducting material such as $MgTi_2O_4$ would be unnecessary.

We found these conditions by introducing surface-active materials to a solution of a specially prepared catalytic complex containing molybdenum and stabilized by magnesium ions [22]. The structure of the complex which was isolated from the solution in its oxidized form is presented on Figure 5 [23]. When it is reduced to the $Mo^{III}$ state the complex becomes an active catalyst of dinitrogen reduction to hydrazine and ammonia by sodium amalgam. Phospholipid (phosphatidylcholine)

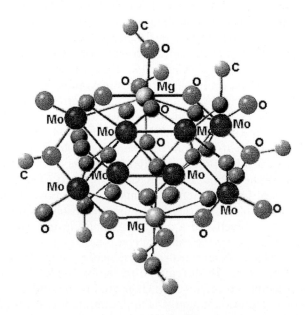

**Figure 5.** The X-ray structure of the anionic part of the octamolybdenum complex [23].

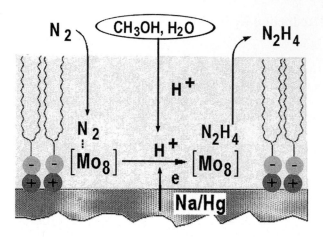

**Figure 6.** Proposed mechanism of dinitrogen reduction catalyzed by octamolybdenum(III) complex incorporated in the phospholipid film on the amalgam surface.

and also polyvinyl alcohol, both surface-active materials, were found to strongly increase the reaction rates and the products yields.

The yields were found also to increase in the presence of phosphines, particularly trimethyl or tributyl phosphine. After all the improvements of the catalyst and reaction conditions the system became by far the most active of known non-biological catalytic systems for the reduction of dinitrogen at ambient temperature and pressure. The specific activity (the rate of $N_2$ reduction per mole of the complex) reached and even exceeded that of nitrogenase. Up to 1000 turnovers relative to the molybdenum complex can be observed at atmospheric pressure and more than 10 000 turnovers at elevated $N_2$ pressures.

The role of the surface-active materials, e.g. phosphatidylcholine, is to form a thin film on the surface of the amalgam, with the catalyst incorporated and thus bound to the surface of the electron donor (Figure 6). With phosphatidylcholine the film is kept at the surface because of attachment of the positive phospholipid heads to the negatively charged surface. Electrons are transferred from the amalgam to the catalytic complex in the process of catalytic reduction. Similarly to the function of the P-cluster in nitrogenase the electron donor participates not only in the initial reduction of molybdenum in the catalytic complex to the $Mo^{III}$ state but also during subsequent steps of substrate reduction. We can therefore speak of 'reductive coaction' of the donor in the process of $N_2$ reduction, according to the mechanism described in previous sections.

If the complex is reduced to the $Mo^{III}$ state and there is no external electron donor in the system (the amalgam is removed) there is no reduction of dinitrogen, even stoichiometric, confirming the mechanism in which the reducing agent participates directly, in a coupled manner, in the reduction of the substrate which is coordinated by the catalytic complex.

More evidence for the participation of the reductant in the catalytic process is the dependence of the rate on the redox potential of the reductant. The reduction rate for the europium amalgam (redox potential $-1.4$ V compared with $-1.84$ V for sodium amalgam) is approximately two orders of magnitude smaller than for

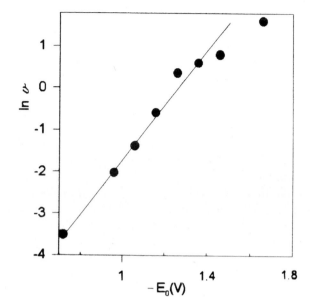

**Figure 7.** Dependence on induced potential of the rate of formation of ethylene by reduction of acetylene [25]. The catalyst: octamolybdenum cluster ($8.5 \times 10^{-6}$ M); the reductant: Zn amalgam (2 %, $w/w$); the cocatalyst: thiophenol ($4.7 \times 10^{-3}$ M); $T = 20\,^{\circ}$C; $p_{(C2H2)} = 100$ mm Hg.

Na(Hg). The dependence of the rate of reaction for acetylene reduction by the Zn amalgam on the induced potential is shown in Figure 7. It is apparent that the dependence follows the Tafel law: $\ln v = A + \alpha E$ (here $v$ is the rate and $E$ is the induced potential).

This dependence was also found for electrochemical reduction of dinitrogen. As has been mentioned, the structure of the catalytic complex presented in Figure 4 corresponds to its oxidized form ($Mo^V$ and $Mo^{VI}$). Unfortunately no crystals suitable for X-ray analysis were obtained for the $Mo^{III}$ state of the complex, but there is evidence that in the reduced state the complex still contains 8Mo and 2Mg. It is possible that on reduction of the complex to the $Mo^{III}$ state the structure becomes less rigid and some coordination sites become open to coordinate $N_2$.

## 6.14 FeMoco as Catalyst for the Reduction of Nitrogenase Substrates at Amalgam Surfaces

As we have already seen, FeMo cofactor of nitrogenase is a polynuclear complex of composition $Fe_7MoS_9$ (homocitrate), and all the available evidence implies it is the active site at which dinitrogen and other nitrogenase substrates are activated and reduced. Yet despite its first isolation in 1977 the catalytic activity of FeMoco was detected only in 1997, i.e. 20 years later.

One possible reason for this is that, according to the mechanism described above, the cofactor must function in a coupled manner with electron transfer from the

P-cluster which is situated close to FeMoco. Presumably attempts to use it as a catalyst outside the MoFe protein had been made earlier with such reductants as $NaBH_4$ which functions as a separate entity only during collisions, i.e. comparatively rarely. As has already been mentioned, in FeMoco iron atoms probably coordinate dinitrogen activating it to further reduction. No iron complexes have yet been synthesized which function as catalysts for $N_2$ reduction in protic media, although several iron complexes reducing dinitrogen in *aprotic* media are known (e.g. Ref. [2]). Presumably such complexes will be prepared in the future, but meanwhile in the light of the catalytic mechanism described above it was natural to investigate the catalytic activity of the FeMo cofactor itself on the surface of multi-electron donors such as amalgams.

It was, indeed, found that FeMo cofactor in solution in DMF and NMF is an active catalyst for reduction of acetylene and other nitrogenase substrates by zinc, europium, and sodium amalgams [24, 25]. The reaction proceeds on the surface of the electron donor, thiophenol functions as a cocatalyst, not only as the proton donor but evidently also the cofactor ligand connecting it, in particular, with the surface of the amalgam. Some of the properties of the isolated cofactor, for example formation of ethylene and ethane from acetylene, inhibiting action of carbon monoxide, dihydrogen evolution, etc. are similar to those of the cofactor inside the protein, in particular, when the cofactor is bound to proteins of $N_2$-fixing organisms different from that from which FeMoco was isolated. Dinitrogen was found to inhibit acetylene reduction by europium amalgam in the presence of FeMoco, the inhibitive effect being *quantitatively identical* with that observed in the reduction catalyzed by nitrogenase (Figure 8) [26]. The conclusion can therefore be drawn that the catalytic action of FeMoco is similar to that of synthetic molybdenum(III) complexes bound to the surface of the amalgam. To be catalytically active FeMoco

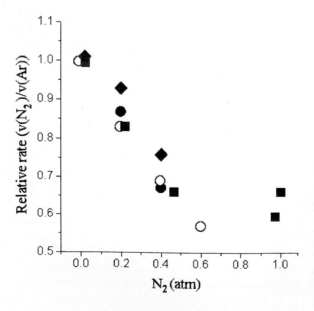

**Figure 8.** Inhibition of acetylene reduction by dinitrogen in protein and non-protein systems with FeMoco as catalyst, $p_{(C2H2)} = 10$ mm Hg. ■, isolated FeMoco with Eu amalgam as a reductant [26]; ●, nitrogenase from *A. vinelandii* (wild type) [27]; ◆, mutant DJ540 [28]; ○, wild-type MoFe protein [28].

must be bound to a multi-electron donor (such as amalgam or cathode in electro-chemical reduction) and to be open to an acid with a certain pK which ensures synchronous protonation and electron transfer. To reduce dinitrogen it seems to be necessary to have a sufficiently strong acid to protonate the substrate but not to destroy the catalyst, which is less stable than synthetic catalysts in the presence of acids.

If both conditions are fulfilled (the presence of a reducing agent in close vicinity to FeMoco and a proton-donating substance of an appropriate p$K$) FeMoco can be catalytically active outside the MoFe protein.

In conclusion, it may be said that to achieve catalytic reduction of dinitrogen in both enzymatic and purely chemical systems in protic media it is essential to unite a one-electron reductant and a polynuclear catalyst activating dinitrogen in a single system such that they function in a coupled manner—one-electron transfer from the external reducing agent must be coupled with poly-electron transfer inside the poly-nuclear catalyst containing the substrate. Hopefully further development of this field along the lines of this mechanism will lead to new active catalytic systems re-ducing dinitrogen.

## References

1. Likhtenshtein, G. I. and Shilov, A. E., *Zh. Fiz. Khim.*, 44, 849, **1970**.
2. Shilov, A. E., *Metal Complexes in Biomimetic Chemical Reactions*, CRC Press, Boca Raton, New York, **1997**.
3. (a) Allen, A. D. and Senoff, C. V., *Chem. Commun.*, 621, **1965**; (b) Hidai, M. and Mizobe, Y., *Chem. Rev.*, 95, 1115, **1995**; (c) Chatt, J., Dilworth, J. R., and Richards, R. L., *Chem. Rev.*, 78, 589, **1978**; (d) Shan, H., Yang, Y., James, A. J. and Sharp P. R., *Science*, 275, 1460, **1997**.
4. Burgess, B. K. and Lowe, D. J., *Chem. Rev.*, 96, 2983, **1996**.
5. Eady, R. R., *Chem. Rev.*, 96, 3013, **1996**.
6. Howard, J. B., and Rees, D. C., *Chem. Rev.*, 96, 2965, **1996**.
7. Gadkari, D., Mörsdorf, G. and Meyer, O., *J. Bacteriol.*, 174, 6840, **1992**.
8. Kim, J. and Rees, D. C., *Nature*, 360, 553, **1992**.
9. Druzhinin, S. Yu., Syrtsova, L. A., Khramov, A. V., Moravsky, A. P. and Shkodina, N. I., *Biochemistry* (Moscow), 61, 480, **1996**.
10. Surerus, K. K., Hendrich, M. P., Christie, P. D., Rottgardt, D., Orme-Johnson, W. H. and Münk, E., *J. Am. Chem. Soc.*, 114, 8579, **1992**.
11. Shah, V. K. and Brill, W. J., *Proc. Natl. Acad. Sci USA*, 74, 3249, **1977**.
12. Schultz, F. A., Greller, S. F., Burgess, B. K., Lough, S. and Newton, W. E., *J. Am. Chem. Soc.*, 107, 5364, **1985**.
13. (a) Lanzilotta, W. N., Parker, V. D. and Seefeldt, *Biochemistry*, 37, 399, **1998**; (b) Chan, J. M., Ryle, M. J. and Seefeldt, *J. Biol. Chem.*, 274, 17593, **1999**.
14. Vol'pin and Shur, V. B., *Dokl. Akad. Nauk SSSR*, 156, 1102, **1964**.
15. (a) Denisov, N. T., Shuvalov, V. F., Shuvalova, N. I., Shilova, A. K. and Shilov, A. E., *Kinet. Katal.*, 11, 813, **1970**; (b) Shilov, A. E., Denisov, N. T., Efimov, O. N., Shuvalov, V. F., Shu-valova, N. I. and Shilova, A. K., *Nature*, 231, 469, **1971**.
16. Nikonova, L. A., Ovcharenko, A. G., Efimov, O. N, Avilov, V. A. and Shilov, A. E., *Kinet. Katal.*, 13, 1602, **1972**.
17. Luneva, N. P., Moravsky, A. P. and Shilov, A. E., *Nouv. J. Chim.*, 5, 21, **1981**.
18. Luneva, N. P., Mironova, S. A., Shilov, A. E., Antipin, M. Yu., Struchkov, Yu. T., *Angew. Chem. Int. Ed. Engl.*, 32, 1178, **1993**.
19. Dzhabiev, T. S., Mironova, S. A. and Shilov, A. E., *Kinet. Katal.*, 40, 844, **1999**.

20. Denisov, N. T. and Shuvalova, N. I., *React. Kinet. Catal. Lett.*, 4, 431, **1976**.
21. Angove, H. C., Yoo, S. J., Burgess, B. K. and Münck, E., *J. Am. Chem. Soc.*, 119, 8730, **1997**.
22. Didenko, L. P., Gavrilov, A. B., Shilova, A. K., Strelets, V. V., Tsarev, V. N., Shilov, A. E., Machaev, V. D., Banerjee, A. K. and Pospišil, L., *Nouv. J. Chim.*, 10, 584, **1986**.
23. Antipin, M. Yu., Struchkov, Yu. T., Shilov, A. E. and Shilova, A. K., *Gazz. Chim. Ital.*, 123, 265, **1993**.
24. Bazhenova, T. A., Bazhenova, M. A., Petrova, G. N. and Shilov, A. E., *Kinet. Katal.*, 38, 319, **1997**.
25. Bazhenova, T. A., Bazhenova, M. A., Mironova, S. A., Petrova, G. N., Shilova, A. K., Shuvalova, N. I. and Shilov, A. E., *Inorg. Chim. Acta*, 270, 221, **1998**.
26. Bazhenova, T. A., Bazhenova, M. A., Petrova, G. N. and Shilov, A. E., *Kinet. Katal.*, #6, **1999** (in print).
27. Rivera-Ortiz, J. M., and Burris, R. M., *J. Bacteriol.*, 123, 537, **1975**.
28. Kim, C.-H., Newton, W. E. and Dean, D. R., *Biochemistry*, 34, 2798, **1995**.

# 7 Transition-metal Complexes as Models of the Active Sites of Hydrogenases*

*Cameron E. Forde and Robert H. Morris*

## 7.1 Hydrogen Chemistry

### 7.1.1 Introduction

Controlling the movement of electrons and protons is vital to life and a variety of biologic systems have evolved with this purpose. Certain enzymes have the ability to separate the protons and electrons of dihydrogen and to harness the energy of this molecule to the synthesis of more complex molecules. Dihydrogen is the simplest molecule and yet the enzymes used to activate it (or to produce it) are amazingly complex. Part of the complexity of hydrogenase enzymes extends from the wealth of oxidation and protonation states of the active site metal cluster. This is due in part to the variety of interactions between dihydrogen and transition metals. While the ability of certain metals to evolve dihydrogen when exposed to acid or even water has been known for a long time, the ability of metals to form hydrides and subsequently the existence of 'non-classical' hydrides or dihydrogen complexes of the metals are more recent discoveries. Long before chemists proposed that dihydrogen forms complexes with transition metals, enzymes had been making use of transition metals to activate dihydrogen.

The chemistry of model systems of hydrogenase enzymes has developed along with the study of the enzymes. Developments in the characterization of the enzyme have spurred the development of model systems that provide a better understanding of the function of the enzyme. Ultimately, a complete understanding of the chemistry of the hydrogenase enzymes will require the synthesis of both areas of study. In this review we consider the work that has been done to characterize the enzyme, focussing on the areas that are amenable for study using model systems. We then

---

*Portions of this work appeared in the Ph. D. thesis of Cameron E. Forde, University of Toronto, 1997.

consider the model systems that shed light on the spectroscopic properties, the function and the structure of the hydrogenase enzymes. The work on theoretical models is then presented with a view toward a fuller understanding of the enzymes and the development of better models. In order to place the chemistry of hydrogenase enzymes and model systems in the context of known metal hydride and dihydrogen chemistry, we present a brief summary of these fields first.

### 7.1.2 Metal Hydride Chemistry

The properties of metal hydrides vary a great deal and depend to a large extent upon the metal. For example, the interactions between hydrides and alkali or alkaline earth elements tend to be ionic in nature while more covalent interactions are observed between hydride ligands of later elements (e.g. tin). For the transition metals the M–H bond strength increases going from 3d to 4d to 5d metals. The M–H bond strength is also influenced by the *trans* ligand. That is, the M–H bond is destabilized by high *trans*-influence ligands such as hydride or alkyl. The reactivity of metal hydrides can be described in terms of their ability to transfer a hydride (hydricity) [1] or their ability to transfer a proton (acidity) [2]. The hydricity is higher for the early transition elements with electron-donating ligands.

Transition metal hydride complexes are often highly fluxional. This feature is often reported for polyhydride complexes. An example of this is the report of the solid state and solution structures of the series of the iron group metals $[MH_3L_4]^+$ where L is either $PMe_3$ or $PEt_3$ [3]. These complexes can adopt a six-coordinate dihydrogen hydride or a seven coordinate trihydride conformation. The dihydrogen hydride conformation can exist in *cis* or *trans* configurations while the trihydride conformation can adopt a tricapped tetrahedral or a classical *cis*-trihydride arrangement. Variable temperature NMR spectra were used to determine the activation parameters for the fluxional processes. The relative stabilities of these different geometries for iron and ruthenium have been determined using ab initio calculations [4, 5].

Protonation of a metal hydride is one of the routes by which metal dihydrogen complexes can be formed. Addition of a proton to a hydride is a more facile process than the addition of a proton to a metal. This is due to the greater electronic rearrangement required to protonate a metal center than that required to protonate a hydride ligand. This feature of metal hydrides is a kinetic effect: the ultimate fate of the nascent dihydrogen complex is determined by thermodynamics. Depending upon the relative stabilities, the dihydrogen ligand can oxidize the metal center forming a dihydride, remain coordinated to the metal, or detach from the metal as $H_2(g)$.

### 7.1.3 Metal Dihydrogen Chemistry

The physical properties of dihydrogen are well understood. Free hydrogen gas has a bond length of 0.74 Å, with a bond strength of 103 kcal $mol^{-1}$. The potential of proton reduction is set at 0 V for the normal hydrogen electrode (NHE) under standard conditions (pH 0, 25 °C). This potential is pH dependent and a potential of $-414$ mV is calculated for pH 7 according to the Nernst equation. This equation also reveals that as the concentration of dihydrogen decreases from saturation the potential will rise. There is thus a range of potentials over which dihydrogen can be formed depending upon the conditions.

The first stable dihydrogen complex reported was $W(H_2)(CO)_3(PPr^i_3)_2$ [6]. Several hundred stable dihydrogen complexes have been reported since this initial report and several reviews have appeared covering the synthesis, properties and reactivity of dihydrogen complexes [7–9]. The majority of the transition metals form stable dihydrogen complexes. Of interest with respect to the chemistry of hydrogenase enzymes are the dihydrogen complexes of nickel and of iron. There are numerous iron dihydrogen complexes but only one nickel complex has been reported. The nickel(0) complex $Ni(H_2)(CO)_3$ is formed by photolysis of $Ni(CO)_4$ in a hydrogen matrix at low temperature [10]. If the hydrogenase enzymes make use of nickel dihydrogen interactions then much of the chemistry of these enzymes has no precedence in the current literature. We attribute the lack of stable nickel dihydrogen complexes to the reduced ability of nickel to participate in stabilizing $\pi$-back donation with the dihydrogen ligand. Iron is better able to form stable dihydrogen complexes as its d-orbitals are higher in energy and able to stabilize the interaction with dihydrogen. More iron dihydrogen chemistry is discussed later in the section on modeling.

There are two components of the bonding interaction between a dihydrogen ligand and a transition metal. The interaction between dihydrogen and a transition metal is described as having $\sigma(H_2)$-to-$\sigma$(metal) and $\pi$(metal)-to-$\sigma^*(H_2)$ components. The extent of the $\sigma$-donation is controlled by a variety of factors including the metal and the oxidation state as well as the ancillary ligands, especially the ligand *trans* to the $H_2$ ligand. The $\pi$-back-donation contribution confers stability, though homolytic cleavage results from too much back bonding. A positive charge on the metal enhances the $\sigma$-interaction with dicationic complexes with dihydrogen *trans* to a $\pi$-acid ligand being particularly strongly bonded [11].

The H–H bond length increases upon coordination to a metal center. The H–H separation in a dihydrogen complex can be determined by a number of techniques. Neutron diffraction yields this value directly, though there is a need to correct for the vibrational motion of the ligand [12]. X-ray crystallography is a less useful technique because of the poor X-ray scattering characteristics of the dihydrogen ligand. An NMR method is commonly used to correlate the spin–lattice relaxation time with the H–H bond length [13, 14]. The motion of the dihydrogen ligand has an effect on the relaxation time and two H–H separations are calculated for each of the two regimes depending upon whether the rotation of the dihydrogen moiety is greater or less than the spectrometer frequency. It is not uncommon for the rotational frequency of the dihydrogen ligand to be greater than 500 MHz! Another NMR technique to probe the H–H separation relies on the determination of the coupling constant between hydrogen and deuterium observed in the analogous HD complex. The H–D coupling constant decreases with a linear correlation from that of the free HD gas (43.2 Hz [15]) as the separation between the nuclei increases. These methods are quite useful for characterizing diamagnetic model complexes but have limited applicability to the characterization of any $H_2$ ligands that might be present in an enzyme containing paramagnetic centers.

The acidity of a dihydrogen molecule increases upon coordination. The $pK_a$ decreases from 35 for free $H_2$ gas (we find that in THF it is greater than 46 with $pK_a$ $HPCy_3^+$ at 9.7) as a function of the metal and its oxidation state as well as the ancillary ligands. A very wide range of $pK_a$ values has been reported for transition-metal dihydrogen complexes. There are now several examples of dihydrogen complexes with $pK_a$ values below zero. The acidity of dihydrogen complexes of the $d^6$ transition metals can be estimated [16]. This method is applicable to the dihydrogen complexes of iron(II). The predictive powers of this method demonstrate that the properties of a dihydrogen ligand are influenced by the ancillary ligands. This is a direct result of the ability of ancillary ligands to influence the electrochemistry of the metal center.

The effect of the metal center on the properties of the dihydrogen ligand has been probed for the series of complexes trans-$[M(H_2)H(R_2PCH_2CH_2PR_2)_2]^+$ (M = Fe, Ru, Os; R = Ph, Et) [17]. These dihydrogen complexes are formed by protonation of the dihydride precursor compounds. The stretching frequency of the terminal hydride ($v_{M-H}$) is found to increase down the period while the H–H separation in the dihydrogen ligand is approximately the same for iron and ruthenium but is greater for osmium. The lability of the dihydrogen ligand as judged by the H/D exchange properties increases from Os to Fe to Ru.

The ligand *trans* to the dihydrogen ligand is found to have a profound effect on the properties of the dihydrogen ligand. An early report compared the effect of chloride versus hydride in the series trans-$[Ru(H_2)X(R_2PCH_2CH_2PR_2)_2]^+$ (R = Ph, Et) [18] The H–H separation in the dihydrogen ligand is longer when the *trans* ligand is chloride than hydride. This is attributed to the fact that chloride has a weak ligand field. Therefore the $t_{2g}$ $\pi$-bonding d-electrons are higher in energy and back-donate more strongly into the $\sigma^*$ orbital of the dihydrogen ligand.

An experimental report of the effect of the ligand *trans* to dihydrogen confirmed these theoretical results. The acidity of a series of ruthenium and osmium complexes, trans-$[M(H_2)XL_2]^{n+}$ (L = $Ph_2P(CH_2)_3PPh_2$; X = CO, Cl, H) has been determined [19]. The complex $[Ru(H_2)(CO)(dppp)_2]^{2+}$ is highly acidic with an estimated $pK_a$ of $-6$.

The effects of the *trans* ligand on the acidity of some dihydrogen complexes of ruthenium and osmium with the general formula trans-$[M(H_2)L(H_2PCH_2CH_2PH_2)_2]^{n+}$ have been quantified using density functional theory [20]. The ligand, L, affects the stability of the dihydrogen ligand as well as the stability of the conjugate base hydride, both of which influence the acidity of the dihydrogen complex. Ligands which destabilize the dihydrogen ligand while stabilizing the conjugate base hydride complex will lead to the most acidic dihydrogen complexes. The effect can be broken down in terms of the role of the ligand as a sigma donor and either a $\pi$-acceptor or $\pi$-donor. Strong sigma donor ligands (like hydride or alkyl) destabilize the conjugate base hydride and result in less acidic dihydrogen complexes. Strong $\pi$-donating ligands stabilize the dihydrogen ligand and destabilize the conjugate base hydride resulting in a less acidic dihydrogen complex. The most acidic dihydrogen complexes result when the *trans* ligand is a strong $\pi$-acceptor ligand. Such a ligand stabilizes the conjugate base hydride and destabilizes the dihydrogen ligand. This result is of importance to the chemistry of hydrogenases as the strong $\pi$-acceptor ligands CN and CO are implicated as discussed later.

The effects of the ancillary ligand, L, on the $M-H_2$ bond dissociation energy in the chromium, molybdenum and tungsten complexes, $ML(CO)_4(H_2)$, have been determined using ab initio calculations [21]. In this report it was found that the stability of the dihydrogen ligand increases when L is a strong $\sigma$-donor ligand while strong $\pi$-accepting ligands weaken the metal–dihydrogen bond. Conversely, the degree of activation of the dihydrogen ligand increases when the number of $\pi$-accepting ligands is reduced.

The dihydrogen ligand is often considered to be a highly labile ligand that interacts only transiently with metals. This perception is due in part to the properties of the Kubas complex that is only stable in an atmosphere of dihydrogen. Indeed, there are many other examples of highly labile dihydrogen ligands. However, this is not to say that all dihydrogen ligands are readily displaced from metal centers. The substitution of dihydrogen from *trans*-$[Fe(H_2)H(dppe)_2]^+$ by a neutral ligand (MeCN, PhCN, or DMSO) is an example of a dihydrogen ligand which is difficult to displace [22]. In this case the kinetic and activation parameters support a mechanism in which one arm of one of a chelating diphosphine ligand opens up a coordination site for attack of the neutral ligand.

## 7.2 Hydrogenase Enzymes

### 7.2.1 Introduction to Hydrogenase Enzymes

Hydrogenases (EC 1.12.2.1) are enzymes that catalyze the two-electron oxidation of dihydrogen as well as the reduction of protons to form $H_2$. A given hydrogenase enzyme is optimized either for $H_2$ uptake or $H_2$ production, though there is also an example of a bidirectional hydrogenase (*Clostridium pasteurianum* hydrogenase I). The function of a given hydrogenase enzyme ($H_2$ uptake as opposed to $H_2$ forming) is determined by the physiological conditions like pH and factors inherent to the enzyme such as the co-factors that are present. These enzymes are found in lower life forms (archaebacteria, bacteria, and algae) and exist in both soluble and membrane-bound (both periplasmic and cytosolic) forms. With the exception of the $H_2$-producing methylenetetrahydromethanopterin dehydrogenase [23], all hydrogenases are metalloenzymes with redox-active iron–sulfur clusters and an active site metal cluster where $H_2$ is either consumed or formed. The metal content is used to subdivide the hydrogenase enzymes into three categories: iron-only [Fe], iron–nickel [FeNi] and iron–nickel–selenium [FeNiSe]-hydrogenases. These three categories of hydrogenase enzymes are distinct, though the [FeNiSe]-hydrogenase enzymes exhibit more resemblance to the [FeNi]-hydrogenases than they do to the [Fe]-hydrogenase enzymes. The majority of hydrogenase enzymes are of the [FeNi] variety and these have also been the most studied. The hydrogenase enzymes have a number of interesting properties that have been modeled in a number of ways. In order to put the model chemistry into context we will present the properties of these enzymes in this section before considering the model systems in the following section.

Several recent reviews cover the model chemistry of [FeNi]-hydrogenases. The role of nickel in the active site of [FeNi]-hydrogenases has been reviewed [24]. This report concludes that a redox active role for the nickel ion in [FeNi]-hydrogenase enzymes is not supported by the current model chemistry. The involvement of cysteinate ligands and/or the iron center is proposed to account for the redox activity of [FeNi]-hydrogenase enzymes. The nickel thiolate model chemistry is the subject of another review that raises the possibility of the involvement of nickel(I) hydrido complexes in the catalytic cycle of [FeNi]-hydrogenase [25]. The chemistry of nickel thiolate model complexes has been reviewed in order to elucidate the preferred oxidation state and coordination chemistry of nickel with relevance to the active site of [FeNi]-hydrogenases [26]. Nickel is known to take on a variety of roles in biology and these roles have also been reviewed [27–29].

The hydrogenase enzymes are very complex and a number of techniques have been used to probe the properties of these enzymes. Part of the complexity of hydrogenase enzymes is due to the different redox and protonation states in which the enzymes can exist. The presence of a variety of protonation and redox states is to be expected from enzymes whose function is to turn protons into dihydrogen (or vice versa). The existence of multiple redox and protonation states greatly complicates studies of the enzyme. The different redox states of the enzyme were first identified in relation to the different EPR signals that are observed. In the next section we consider the EPR evidence about the active sites of [Fe]-, [FeNi]- and [FeNiSe]-hydrogenase enzymes. A system of nomenclature to describe these various redox states will be developed in this section. We then present results of the oxidative and reductive titrations that have been reported to characterize the different redox states. Another useful technique used to characterize hydrogenase enzymes is infrared spectroscopy. A number of signals are observed for these enzymes that are not normally associated with protein samples. This section is rounded out with a description of some of the other techniques that have been used to explore the properties of the active sites of hydrogenases. After having presented the spectroscopic characterization we present the experiments that have been reported on the characterization of the function of these enzymes. We end our discussion of hydrogenase enzymes with the solid state structural characterization of the [FeNi]-, [FeNiSe]- and [Fe]-hydrogenase enzymes.

## 7.2.2 Spectroscopic Characterization of Hydrogenase Enzymes

### Electron paramagnetic resonance

One of the techniques that have been used extensively to characterize hydrogenase enzymes is electron paramagnetic resonance. There are several paramagnetic

**Table 1.** Some examples of the EPR parameters observed for the various states of [FeNi]-hydrogenase enzymes from different sources.

| Organism | State | $g$ Values | Refs |
|---|---|---|---|
| *D. gigas* | A | 2.31, 2.23, 2.02 | 31, 32 |
| | B | 2.33, 2.16, 2.02 | |
| | C | 2.19, 2.16, 2.02 | |
| *C. vinosum* | A | 2.34, 2.16, 2.01 | 33 |
| | B | 2.32, 2.24, 2.01 | |
| *D. vulgaris* (Hildenborough) | A | 2.31, 2.23, 2.02 | 31 |
| | B | 2.33, 2.16, 2.02 | |
| | C | 2.19, 2.16, 2.02 | |
| *D. vulgaris* (Hildenborough) cytoplasmic | A | 2.32, 2.23, 2.0 | 34 |
| | B | 2.33, 2.16, 2.0 | |
| | C | 2.19, 2.14, 2.01 | |
| *T. roseopersicina* | A | 2.32, 2.23, 2.01 | 35 |
| | B | 2.33, 2.17, 2.01 | |
| | C | 2.19, 2.14, 2.01 | |

centers in the various metal-containing hydrogenases. Of interest to this work are the EPR signals associated with the active site. The active sites of the hydrogenase enzymes exist in different redox and protonation states, some of which give rise to EPR signals. Different EPR signals are observed for the [FeNi]-, [FeNiSe]-, and [Fe]-hydrogenases. There are also other states that are not redox active and in the case of the [FeNi]-hydrogenases, we will present these states in the discussion of the infrared properties below.

The active sites of [FeNi]-hydrogenases give rise to a number of interesting EPR spectra. There are three EPR active states labeled **A**, **B** and **C**. Each of these states produces a rhombic EPR spectrum. Data for a variety of different [FeNi]-hydrogenases are collected in Table 1. These signals have been assigned to the Ni ion on the basis of [61]Ni-labeling studies [30].

The [FeNi]-hydrogenase from *Desulfovibrio gigas* was studied by electron spin echo envelope modulation (ESEEM) spectroscopy in the **A** and **C** states [36]. The two states differ in their solvent accessibility as determined by exchange with solvent deuterons. The **A** state is solvent inaccessible while the **C** state is solvent accessible. Coupling to a [14]N nucleus is observed in both states. The solvent accessibility of these two states has been confirmed by ENDOR spectroscopy. This study reveals two types of exchangeable proton species in the **C** state [37]. One of these protons interacts less strongly with the paramagnetic center and is consistent with a bound water or hydroxide ligand. The second proton species interacts more strongly and is not consistent with assignment as a hydride. This ligand is possibly a dihydrogen ligand or an agostic interaction. Oxidation of the **C** state produces the **B** state which retains only one exchangeable proton species. This species has a smaller coupling consistent with a water or hydroxide ligand. A comparison of these three states using Q-band pulsed ENDOR on [57]Fe and natural-abundance Fe [FeNi]-hydrogenases from both *D. gigas* and *D. desulfuricans* reveals that the non-protein bridge present in the **A** state is broken in the other two states [38]. This work also assigns the oxidation states of the active site metal ions in the **A** state as [Fe(II)Ni(III)] where the iron ion is low spin, $S = 0$.

The active sites of [Fe]-hydrogenases exhibit different EPR spectra than the [FeNi]-hydrogenases. An early study of [Fe]-hydrogenase enzymes from *Desulfovibrio vulgaris* (Hildenborough), *Megasphaera elsdenii*, and two from *Clostridium pasteurianum* revealed that these enzymes exhibit signals due to [4Fe-4S]$^+$ clusters in their reduced state and more complex signals when oxidized [39]. These authors conclude that the active site of all four have $S$ greater than 0 and integer, which in the oxidized state is exchange-coupled to an $S = 1/2$ species, most likely a low spin Fe(III) center.

The active sites of the [Fe]-hydrogenases from *Megasphaera elsdenii* and *Desulfovibrio vulgaris* (Hildenborough) have been investigated with one- and two-dimensional pulsed EPR spectroscopy [40]. The presence of a nitrogen-containing ligand was detected in the active sites of both enzymes. Unusual quadrupole values suggest a non-protein source for these nitrogen-containing ligands, consistent with cyanide ligands. The presence of an imidazole ring from a histidine residue is also suggested based on these EPR measurements, and this residue is likely part of a proton shuttle.

**Infra-red absorption spectroscopy**

The infrared spectra of hydrogenase enzymes exhibit absorbances in a region that is not normally associated with protein samples. This region, 1900–2100 cm$^{-1}$ is more commonly associated with molecules with triple bonds. As we will see later this absorption can be attributed to carbonyl and cyanide ligands at the active site. In this section we consider the evidence from infra-red absorption spectroscopy as it pertains to the active sites of hydrogenase enzymes. Once again, we present the work on [FeNi]-hydrogenases and then the [Fe]-hydrogenases.

Studies of the infrared spectra carbon monoxide inhibited forms of [FeNi]-hydrogenase revealed that the native protein exhibits absorbances at energies not normally associated with protein samples. These absorbances in the 1900–2100 cm$^{-1}$ region are also observed in [Fe]-hydrogenase samples. This absorption is attributed to the cyanide and carbonyl ligands at the active site metal centers. Cyanide and carbon monoxide are not commonly found as ligands to metal centers in metalloproteins. The different redox states of the enzymes give rise to different absorption spectra.

Exposure of the [FeNi]-hydrogenase from *Chromatium vinosum* to carbon monoxide produces a band in the infrared spectrum at 2060 cm$^{-1}$ [41]. A shift to 2017 cm$^{-1}$ is observed when $^{13}$CO is employed. This band is in addition to the three bands at 2082, 2069 and 1929 cm$^{-1}$ that are attributed to the native cyanide and carbonyl ligands. The two weak bands in the region 2040–2100 cm$^{-1}$ are separated by 12–16 cm$^{-1}$ and the separation of the average of these weak bands from the strong band shifts according to the state of the enzyme [42]. Labeling studies, using

**Table 2.** Infrared absorption wavenumbers in the range of 1900–2100 cm$^{-1}$ associated with various states of the active site of the [FeNi]-hydrogenase enzymes from *C. vinosum* and *D. gigas*.

| State | $v_{CN}$(sym) | $v_{CN}$(asym) | $v_{CO}$ | Ref. |
|---|---|---|---|---|
| *Chromatium vinosum* | | | | |
| A/B | 2093 | 2081 | 1944 | 41, 42 |
| C | 2088 | 2076 | 1950 | |
| R | 2075 | 2060 | 1936 | |
| *Desulfovibrio gigas* | | | | |
| A | 2093 | 2083 | 1947 | 44, 45 |
| SU | 2099 | 2089 | 1950 | |
| B | 2090 | 2079 | 1946 | |
| SI' | 2069 | 2055 | 1914 | |
| SI" | 2085 | 2075 | 1934 | |
| C | 2086 | 2073 | 1952 | |
| R | 2073 | 2060 | 1940 | |

$^{15}$N or $^{13}$C, and chemical analyzes have implicated one carbon monoxide molecule and two cyanide groups [43].

In the case of the [FeNi]-hydrogenase of *Desulfovibrio gigas* IR absorption in the 1900–2100 cm$^{-1}$ region has been assigned for each of the redox-active states [44]. These data are collected in Table 2. This study revealed the presence of an EPR silent state labeled **SU** that exists as a mixture of two protonation states. The changes in IR properties have been used to determine the electrochemical potentials required for the changes in redox state. These data are presented in the following section.

The unusual IR properties of [FeNi]-hydrogenases are shared by the [Fe]-hydrogenases [46]. Analysis of the spectra of the [Fe]-hydrogenase of *D. vulgaris* obtained under a variety of redox conditions strongly indicate that [Fe]-hydrogenases contain a low-spin Fe ion in the active site with one CN group and one CO molecule as intrinsic, non-protein ligands [47].

**Electrochemical titrations**

Reductive titrations of the [FeNi]-hydrogenase from *Chromatium vinosum* were performed by variation of the H$_2$-partial pressure [48]. Changes in the redox state of the active site were monitored by loss of the EPR signal. The reduction potential of the active site exhibits a pH dependence of $-60$ mV (pH unit)$^{-1}$. This dependence suggests that the reduction of the active site is coupled with protonation.

The electrochemical potentials of different states of the [FeNi]-hydrogenase enzyme from *D. gigas* have been probed using oxidative titrations at pH 8.0 [49, 50]. In addition to the redox events of the iron–sulfur clusters the data are best fit to a model involving four different active site redox states between 0 and $-400$ mV. The potentials of these four redox states are collected in Table 3.

**Table 3.** Mid-point reduction potentials (mV relative to the NHE) reported for [FeNi]-hydrogenase enzymes under the given conditions.

| Organism | A–SU | B–SI | SI–C | C–R | pH/$T$ | Ref. |
|---|---|---|---|---|---|---|
| *C. vinosum* | −115 | −115 | nr | nr | 8.0/30 | 48 |
| *D. gigas* | nr | −150 | −330 | −370 | 8.0/nr | 50 |
| *D. gigas* | −210 | −135 | −365 | −430 | 7.7/40 | 44 |

nr = not reported

The periplasmic [Fe]-hydrogenase from *Desulfovibrio vulgaris* (Hildenborough) exists in two different catalytic forms: as isolated the protein is $O_2$-insensitive; upon reduction the protein becomes active and $O_2$-sensitive [51]. EPR-monitored redox titrations reveal a single reduction of the active site at a potential of −307 mV, just above the onset potential of $H_2$ production.

**Other techniques**

Two techniques that are particularly useful in the characterization of hydrogenase enzymes are Mössbauer and X-ray absorption spectroscopies. At low temperature (4.2 K), the [3Fe–xS] cluster exhibits a paramagnetic Mössbauer spectrum typical for oxidized [3Fe–xS] clusters. At higher temperatures (greater than 20 K), the paramagnetic spectrum collapses into a quadrupole doublet with parameters magnitude of delta EQ magnitude of =0.7 ± 0.06 mm s$^{-1}$ and delta = 0.36 ± 0.06 mm s$^{-1}$, typical of high-spin Fe(III). The observed isomer shift is slightly larger than those observed for the three-iron clusters in *D. gigas* ferredoxin II [52] and in *Azotobacter vinelandii* ferredoxin I [53] and may indicate a different iron coordination environment. When *D. gigas* hydrogenase is poised at potentials lower than −80 mV (relative to the normal hydrogen electrode), the [3Fe–xS] cluster is reduced and becomes EPR-silent. The Mössbauer data indicate that the reduced [3Fe–xS] cluster remains intact, i.e. it does not interconvert into a [4Fe–4S] cluster. Also, the electronic properties of the reduced [3Fe–xS] cluster suggest that it is magnetically isolated from the other paramagnetic centers [54].

X-ray absorption spectroscopy is a useful technique to probe the ligand environment of the nickel ion at the active site. Early EXAFS experiments using the NiK-edge on the [FeNi]-hydrogenase from *Thiocapsa roseopersicina* in the **C** state suggested the involvement of an unusual nickel–iron cluster [55]. When these studies were extended to compare the three EPR active states (**A**, **B** and **C**) and two EPR silent states (**SI** and **R**) no significant change in the edge energy is observed [56]. The data are best fit to a model involving six-coordinate nickel with a mixture of N/O- and S-donor ligands. Similar results are obtained for the [FeNi]-hydrogenase from *Alcaligenes eutrophus* in the as-isolated **A** state [57]. Upon reduction a model with four S-donor ligands with a Ni–S distance of 2.19 Å is appropriate. The incorporation of an iron atom in accordance with the crystallographic results (vide infra) improves the fit of the model to the data. However, the EXAFS technique is limited in that it is difficult to distinguish between S and Fe scattering atoms at distances greater than 2.4 Å.

### 7.2.3 Hydrogenase Function

The function of a hydrogenase enzyme is either to produce $H_2$ or to oxidize $H_2$. The [FeNi]-hydrogenases, in general, function as $H_2$-uptake enzymes, though there are examples of [FeNi]-hydrogenases that evolve $H_2$ (e.g. *Pyrococcus furiosus*). Another generalization that can be made is that [Fe]-hydrogenase enzymes are 10 to 100 times as active as the [FeNi]-hydrogenase enzymes. However this greater activity comes at the price of greater oxygen sensitivity. While [FeNi]-hydrogenases are deactivated to varying extents by exposure to oxygen, [Fe]-hydrogenases are irreversibly deactivated by oxygen. The [FeNiSe]-hydrogenases are much less sensitive to exposure to oxygen.

Different patterns in the pH dependence of hydrogenase activity have been observed with enzymes purified from different species of *Desulfovibrio*. With the cytoplasmic hydrogenase from *Desulfovibrio baculatus* strain 9974, the pH optima in $H_2$ production and uptake were respectively 4.0 and 7.5 with a higher activity in production than in uptake. This contrasts with the periplasmic hydrogenase from *Desulfovibrio vulgaris* (Hildenborough). In this case the highest exchange activity was near pH 5.5. The periplasmic hydrogenase from *Desulfovibrio gigas* has the same pH optimum in the exchange (7.5–8.0) as in the $H_2$ uptake. The ratio of $H_2$ to HD production is greater than one for the first of these enzymes and below one for the latter two enzymes [58]. The absolute activities cannot be compared due to the effect of enzyme concentration on both the $H_2$-uptake and $H_2$-evolution assays [59].

### 7.2.4 Structural Characterization of Hydrogenase Enzymes

The structural characterization of hydrogenase enzymes has clarified much of the mystery that had surrounded the active sites of these enzymes. The report of the crystal structure of the [FeNi]-hydrogenase from *D. gigas* [60] revealed that the active site of this enzyme contained nickel and a second metal later identified as iron (PDB 1frv and 2frv) [45]. The nickel ion is coordinated by four cysteine residues. Two of these residues form a bridge to the iron ion. The second reported structure also contains a third bridging ligand that these authors suggest is most likely hydroxide. No other protein-based ligands coordinate these metals, though the iron ion has three other ligands that were modeled as water in the original structure. The later structure model used two cyanide ligands and one carbon monoxide ligand.

A second [FeNi]-hydrogenase structure, from *Desulfovibrio vulgaris* (Miyazaki F), revealed a similar active site cluster (PDB 1h2a) [61]. Like the structure of the *D. gigas* enzyme, the active site cluster consists of a nickel ion coordinated by four cysteine residues, two of which form a bridge to the iron ion. The iron ion has three nonprotein ligands that are proposed to be the diatomics S=O, CO and CN molecules. In this case the third bridging atom is proposed to be sulfide as hydrogen sulfide (0.37 equiv.) is released on incubation under $H_2$ in the presence of cytochrome c3 or methylviologen [62]. The X-ray crystal structure of the reduced form of *D. vulgaris* (Miyazaki F) [FeNi]-hydrogenase reveals an overall structure that is very similar to that of the oxidized form, with the exception that the third monatomic bridge is absent, leaving this site unoccupied [63]. The coordinates of one other structure, the [FeNi]-hydrogenase enzyme from *D. fructosovorans*, have been deposited (PDB 1frf).

The single crystal X-ray diffraction structure of the [FeNiSe]-hydrogenase from *Desulfomicrobium baculatum* in the reduced, active form has also been reported [64]. The active site metal cluster is similar to those of the [FeNi]-hydrogenases reported with some interesting differences. The most obvious difference is the presence of a selenocysteine residue in place of one of the terminal cysteine residues. The reduced active site of *D. baculatum* has a nickel–iron distance that is 0.4 Å shorter than in the oxidized *D. gigas* enzyme. Like the reduced *D. vulgaris* enzyme, the bridging ligand is absent in the reduced *D. baculatum* hydrogenase.

The structures of two [Fe]-hydrogenase enzymes have also been reported. The structure of the [Fe]-hydrogenases from *Clostridium pasteurianum* reveals that the active site H-cluster of the enzyme is composed of a typical [4Fe–4S] cubane bridged via a cysteine thiolate to a binuclear Fe center (PDB 1feh) [65]. The structure of the [Fe]-hydrogenase from *Desulfovibrio desulfuricans* reveals this same arrangement (PDB 1hfc) [66]. These two crystal structure reports differ in their assignment of the bridging ligands of the two iron atoms of the binuclear cluster. The structure of the binuclear cluster from *C. pasteurianum* is reported to have two sulfides and one carbon monoxide as bridging ligands. In place of the sulfide bridges the binuclear cluster from *D. desulfuricans* is modeled with a bridging 1,3-propanedithiol molecule. The iron atoms are modeled as having octahedral coordination geometries with carbonyl or cyanide molecules. The presence of an iron moiety with CO and CN ligands in all types of hydrogenases suggests that this functional unit is incorporated for its ability to activate dihydrogen [66].

The structure of the CO inhibited form of the [Fe]-hydrogenase from *Clostridium pasteurianum* has also been reported (PDB 1c4c) [67]. The structural characterization indicates the addition of a single molecule of carbon monoxide. Carbon monoxide binds to an iron atom of the binuclear iron center at the site of a terminally bound water molecule in the native state (PDB 1c4a).

# 7.3 Modeling with Transition-metal Complexes

## 7.3.1 Introduction

We have shown in the previous section that there are a number of unusual properties of hydrogenase enzymes. In this section we present the work that has appeared in the literature toward modeling these unusual properties. As we will see there are a number of cases where the development of model systems has led to a greater understanding of the enzyme. That is, after all, the goal of making model complexes. In the following presentation of model systems we will focus on the ways in which the model chemistry provides an improved understanding of the hydrogenase enzymes.

The presentation of the modeling of the hydrogenase enzymes is divided into four sections based on the aspect of the hydrogenase enzyme that is being modeled; spectroscopic, structural, functional and theoretical modeling. Spectroscopic models attempt to reproduce the unusual spectroscopic features that characterize the hydrogenase enzymes. These include EPR, IR and electrochemical properties. The structural aspects of the enzyme active site have been modeled through attempts to synthesize a complex or cluster with similar coordination environments as the enzyme active site. This type of modeling is usually assessed based upon the ability to reproduce metrical parameters such bond lengths or angles. The abilities of the enzymes to convert protons to dihydrogen (or vice versa) and to catalyze the exchange of $H^+/H_2$ comprise the functional aspects of the enzyme system that have been modeled with transition metal complexes. Theoretical techniques have also been employed to better understand the chemistry of hydrogenase enzymes. In the following sections we present the work on each of these types of modeling systems.

## 7.3.2 Spectroscopic Modeling

Many spectroscopic models have focussed on reproducing the unusual EPR properties of the different paramagnetic states of the active sites of the hydrogenase enzymes. Model chemistry should be instrumental in determining the coordination geometry and oxidation state of the paramagnetic center. As we mentioned above, the nickel ion has been implicated as the source of the EPR signals of the active site and model chemistry has focussed on monomeric nickel complexes.

The Ni(III) complexes of *N*-mercaptoacetylglycyl-L-histidine and *N*-mercaptoacetylglycylglycylglycine clearly show the rhombic EPR pattern and *g* values similar to the Ni(III) chromophore of hydrogenases [68]. These results suggested that the Ni(III) center of hydrogenases contains one cysteine sulfur coordination as equatorial ligand in a tetragonal geometry. In addition, an axial nitrogen ligand and a sulfur-rich Ni(III) site, as in an S4 donor set, were thought to be ruled out.

A number of model systems have attempted to reproduce the anomalous IR spectra of [FeNi]-hydrogenases. Initial efforts focussed on cyanide, carbon monoxide and nitric oxide complexes as these ligands are known to absorb in this portion of the infrared spectrum [69].

Some iron and nickel cyanide and carbonyl complexes have been reported as models of the [FeNi]-hydrogenase enzymes. The preparation and structures of the trigonal bipyramidal nickel and iron complexes with the tetradentate ligands tris(2-phenylthiol)phosphine (PS3) and tris(3-phenyl-2-thiophenyl)phosphine (PS3*) have been reported [70, 71]. The nickel carbonyl complex $[Ni(PS3*)(CO)]^{2-}$ exhibits $v_{CO}$ at 2029 cm$^{-1}$ compared with the value of 1940 cm$^{-1}$ for the iron carbonyl complex $[Fe(PS3*)(CO)]^{2-}$. Both of these complexes lose CO upon oxidation. The use of cyanide in place of carbon monoxide allows for the preparation of both $[Fe^{II}(PS3)(CN)]^{2-}$ and $[Fe^{III}(PS3*)(CN)]^-$ complexes. The IR properties of

**Table 4.** Selected IR properties of some carbonyl and cyanide complexes of nickel and iron with the PS3 and PS3* ligands [70, 71].

| Complex | $v_{CN}$ | $v_{CO}$ |
|---|---|---|
| $[Ni^{II}(PS3^*)(CO)]^{2-}$ | na | 2029 |
| $[Fe^{II}(PS3^*)(CO)]^{2-}$ | na | 1940 |
| $[Fe^{II}(PS3)(CN)]^{2-}$ | 2070 | na |
| $[Fe^{III}(PS3^*)(CN)]^{-}$ | 2094 | na |
| $[Fe^{II}(PS3)(CN)(CO)]^{2-}$ | 2079 | 1904 |
| $[Fe^{III}(PS3)(CN)(CO)]^{-}$ | 2108 | 2006 |

these complexes are collected in Table 4. Addition of carbon monoxide to $[Fe^{II}(PS3)(CN)]^{2-}$ produces the six-coordinate complex $[Fe^{II}(PS3)(CN)(CO)]^{2-}$ where the cyanide ligand remains *trans* to the phosphine. This latter complex can be oxidized electrochemically and increases in both $v_{CN}$ and $v_{CO}$ are observed.

The work of Darensbourg et al. has superseded these early model systems. A series of model compounds with the core unit $Fe(CO)_2(CN)$ or $Fe(CO)(CN)_2$ were found to reproduce the unique IR absorption spectra of [FeNi]-hydrogenases very well. The IR spectrum of the [FeNi]-hydrogenase enzyme from *D. gigas* in the **A** state exhibits bands at 1947, 2093, and 2083 cm$^{-1}$. The IR spectrum of the iron(II) model complex K $[CpFe(CO)(CN)_2]$ in acetonitrile exhibits absorption bands at 1949, 2094, and 2088 cm$^{-1}$ which are assigned to the $v_{CO}$, the symmetric $v_{CN}$ and the asymmetric $v_{CN}$, respectively. The energies and peak-widths at half-maximum of this absorption are sensitive to the oxidation state of the iron center, to the medium and to the counter-ion. Polar media produce broad bands with peak width at half-maximum of the $v_{CO}$ band of 17 cm$^{-1}$ in water. The use of non-polar solvents is required to achieve the narrow (4 cm$^{-1}$) peak width at half-maximum observed for the enzyme.

In spite of their structural and amino acid sequence differences, Fe-only and Ni-containing hydrogenases achieve the same catalytic reactions. A chemical modification of histidine residues using a highly specific reagent (pentaammineruthenium II) has been carried out on *Desulfovibrio vulgaris* Hildenborough Fe-hydrogenase and *Desulfovibrio desulfuricans* Norway Ni–Fe–Se-hydrogenase. The preliminary results suggest the existence of a general mechanism involving histidine residues in the two groups of hydrogenases. These residues may be part of the histidine-containing motive shown to be present in both Fe- and Ni–Fe-hydrogenase sequences by Hydrophobic Cluster Analysis. This analysis suggests a functional role for the small subunit of *Desulfovibrio vulgaris* Hildenborough [Fe]-hydrogenase [72].

### 7.3.3 Functional Modeling

As mentioned in the introduction, the addition of acid, or in some cases even water (e.g. the alkali metals), to metals can result in the evolution of hydrogen. The potential role of nickel dihydrogen complexes in metal-containing hydrogenases was proposed early in the history of hydrogenase literature [73]. The report of a nickel(II) complex that produces dihydrogen electrochemically was later reported [74]. This was followed by a report of the production of dihydrogen by protonation of a nickel(I) complex [75]. However, the key to modeling the active sites of hydrogenase enzymes requires the ability to do more than evolve dihydrogen. Ideally what is needed in a successful model is the ability to not only evolve dihydrogen but also to activate dihydrogen. In this section we present the work on the model systems that mimic the function of hydrogenase enzymes.

One interesting model system incorporates nickel(II) into the rubredoxin protein scaffold [76]. Rubredoxins are electron-transport proteins with a redox-active mononuclear iron center coordinated by four cysteine residues. The nickel-substituted rubredoxin evolves $H_2$ and HD when exposed to $D_2$. This system is much less active than the native enzymes with activities less than an order-of-magnitude below [FeNi]- and [FeNiSe]-hydrogenases. The ratio of $H_2$ to $H_2 + HD$ initially produced is used as an indicator of the mechanism. The Ni–rubredoxin model system has ratios in the range 0.45–0.60. Values of this ratio around 0.3 are indicative of heterolytic cleavage of the $D_2$ unit while values closer to 1 support a homolytic activation mechanism. Like the hydrogenase enzymes, the model system is inhibited by carbon monoxide. The concentration of CO required for 50 % inhibition is similar to that required for [Fe]-hydrogenase from *D. vulgaris* but much lower than that required for 50 % inhibition of the [FeNi]-hydrogenase enzymes.

Model systems that couple $H_2$ oxidation with the reduction of a cofactor have been reported. The complex Cp*$L_2$RuH catalyzes the reduction of NAD [77] or methyl viologen [78] under dihydrogen. This system demonstrates that a ruthenium model system can catalyze the activation of hydrogen without the benefit of the protein scaffold.

An iron(II) hydride model system presents the possibility that protonation can occur at the cyanide ligand [79]. The site of protonation in the complexes *trans*-FeH(CN)($R_2PCH_2CH_2PR_2$)$_2$ is controlled by the nature of the substituents R of the chelating diphosphine ligands. When the substituents are ethyl groups the site of protonation is the hydride ligand and the stable dihydrogen complex *trans*-[Fe($H_2$)(CN)($Et_2PCH_2CH_2PEt_2$)$_2$]$^+$ is formed. When the substituents are phenyl groups the site of protonation is at cyanide forming the hydrogen isocyanide ligand in the complex *trans*-[Fe(H)(CNH)($Ph_2PCH_2CH_2PPh_2$)$_2$]$^+$. This variation in reactivity is attributed to the greater electron-donating ability of the alkyl-substituted phosphine ligands. The presence of more electron-donating ethyl groups results in a more basic hydride and thus a less acidic dihydrogen complex. Presumably the less electron-donating phenyl groups result in a dihydrogen complex that is more acidic than the corresponding hydrogen isocyanide ligand.

Two very acidic iron(II) dihydrogen complexes have been reported that have either carbonyl or hydrogen isocyanide ligands [11]. The complexes *trans*-[Fe($H_2$)(L)($Ph_2PCH_2CH_2PPh_2$)$_2$]$^{2+}$ (L = CO, CNH) are prepared by addition of triflic acid to the hydrido complexes *trans*-[FeH(CO)($Ph_2PCH_2CH_2PPh_2$)$_2$]$^+$ or *trans*-FeH(CN)($Ph_2PCH_2CH_2PPh_2$)$_2$. The short H–H separations in these dihydrogen complexes, as determined by NMR methods, suggest that there is very little back donation from the metal to the $H_2$ unit. Some degree of back-donation has been thought to be necessary for the stabilization of the metal–dihydrogen interaction.

The analogous ruthenium and osmium complexes [80, 81] demonstrate similar reactivity. The series of complexes *trans*-[M(H$_2$)(CN)L$_2$]$^+$ and *trans*-[M(H$_2$)(CNH)L$_2$]$^{2+}$, where M = Fe, Ru, Os and L = bidentate phosphine, are prepared by protonation of the hydrido precursor complexes using triflic acid. The ruthenium complexes are the least stable and the dihydrogen ligand dissociates and is replaced by the triflate anion. The exposure of *trans*-[Ru(OTf)(CNH)L$_2$](OTf) to 1 atm H$_2$ results in the elimination of triflic acid and produces a mixture of *trans*-[Ru(H$_2$)(CN)L$_2$]$^+$ and *trans*-[Ru(H)(CNH)L$_2$]$^+$ with the ratio of these two complexes being controlled by the electronic nature of the bidentate phosphine ligands. As mentioned above for the analogous iron complexes, the more electron-donating phosphine donor ligands stabilize the formation of the dihydrogen complex while the hydrido analog is favored by less electron-donating phosphine ligands.

One of the intriguing features of hydrogenase enzymes is their ability both to activate and to produce dihydrogen at neutral pH. This feature suggests that there is a hydrido species that can be protonated at neutral pH, or in other words a dihydrogen species with a p$K_a$ near 7. The ability of metals to lower the p$K_a$ of dihydrogen from 35 in the gas was discussed in the section on dihydrogen chemistry. There are other sites of potential protonation that must be considered. Two examples of such sites are the cysteine thiolates and the cyanide ligands at the active site. Protonation of ligands in the coordination sphere alters the properties of the active site. For example the protonation of thiolato donors results in an increase in the lability of this ligand [82].

### 7.3.4 Structural Modeling

Several reports have been published of complexes that model the binuclear active sites of [FeNi]-, [FeNiSe]- and [Fe]-hydrogenases. Most of these efforts have focussed on reproducing the metrical parameters that have been reported for the corresponding enzyme. Initially efforts centered on modeling a tetrathiolato nickel center thought to be the active site of [FeNi]-hydrogenases. The revelation that the active site is an iron–nickel heterodimer has prompted revision of the models. Much of the nickel thiolate model chemistry through 1995 has been reviewed [24–26] and in the following section we will focus on recent results that model the known binuclear active sites. Efforts to model the heterodimeric nickel–iron center will be presented first followed by the di-iron model complexes.

A nickel–iron complex is formed from the reaction of the nickel(II) complex of the tetradentate ligand *N,N'*-diethyl-3,7-diazanonane-1,9-dithiolate with Fe(CO)$_2$(NO)$_2$ [83]. The carbonyl ligands on the iron are displaced and the thiolato donors assume a bridging mode leaving two terminal nitrosyl ligands about the four-coordinate iron center. X-ray crystallography revealed that the dimer has a short Fe–Ni separation of 2.797(1) Å and Fe–N–O angles of 167.0(5) and 174.6(5)°. The infrared spectrum exhibits two intense bands attributed to the symmetric and asymmetric stretching modes of the nitrosyl ligands at 1663 and 1624 cm$^{-1}$.

A binuclear nickel–iron complex is formed by the reaction of [Fe(CO)-N(CH$_2$CH$_2$S)$_3$]$^-$ with NiCl$_2$(dppe) under an atmosphere of carbon monoxide [84]. One of the chloride ligands is displaced from nickel and two of the thiolato donors adopt bridging modes while the iron becomes six-coordinate by addition of one CO ligand. The dimer is characterized by a long Fe–Ni separation of 3.308 Å, much longer than that found reported for the hydrogenase enzyme of *D. gigas* [60]. The IR spectrum exhibits intense absorptions at 1944 and 2000 cm$^{-1}$ which is assigned to the axial and equatorial CO ligands, respectively.

Two recent reports present the displacement of carbon monoxide ligands by addition of cyanide to iron carbonyl dimers. In the first report [85] the addition of cyanide to the known complexes [Fe(CO)$_3$]$_2$($\mu$-SMe)$_2$ (see Ref. [86].) or [Fe(CO)$_3$]$_2$($\mu^2$-SCH$_2$CH$_2$CH$_2$S) (see Ref. [87].) results in the formation of [Fe(CN)(CO)$_2$]$_2$($\mu$-SMe)$_2$$^{2-}$ or [Fe(CN)(CO)$_2$]$_2$($\mu^2$-SCH$_2$CH$_2$CH$_2$S)$^{2-}$, respectively. The synthesis and structure of the latter complex is also the subject of the second report [88]. Attempts to prepare a monocyano-substituted complex were unsuccessful and the use of excess cyanide lead to the formation of the same dicyano complex. X-ray crystallographic studies of (Ph$_4$P)$_2$ [Fe(CN)(CO)$_2$]$_2$($\mu$-SMe)$_2$ were foiled by the co-crystallization of isomers where the methyl thiolate ligands are disposed to create a mixture of diequatorial and equatorial, axial isomers. These isomers are in equilibrium and the latter isomer is favored four to one. X-ray crystallographic studies of (Et$_4$N)$_2$ [Fe(CN)(CO)$_2$]$_2$($\mu^2$-SCH$_2$CH$_2$CH$_2$S) reveal a short Fe–Fe separation 2.517 Å. This distance is within bonding range and shorter than the 2.6 Å separation observed in the two protein structures.

### 7.3.5 Theoretical Modeling

The crystallographic studies have provided the theoretical studies with the fodder to make models to better understand the chemistry of the active sites of [FeNi]- and [Fe]-hydrogenases. The theoretical models start with the coordinates from the X-ray crystal structures and use density functional theory (DFT) to handle the complexity of these transition-metal systems. The geometries are optimized and the energies calculated using some level of theory. The catalytic cycle is then explored by preparing a number of models using assumptions about what species are involved in the catalytic cycle. In the presentation of the theoretical models that follows we will present the assumptions of each model system and the implications about the chemistry of the active site that are gleaned.

In one model system, that uses DFT and the B3LYP level of theory, the dinuclear cluster is assumed to be neutral in all states [89]. The active site coordinates are taken from the crystal structure of the [FeNi]-hydrogenase from *D. gigas* [60] and two cyanide ligands and one carbonyl ligand are added. The cysteine residues are modeled as either HS$^-$ or H$_2$S. The bridging ligands are modeled as HS$^-$ since H$_2$S results in dissociation of these ligands. This model has reasonable metal thiolate distances (2.8–3.0 Å) and a bent structure with a Fe–S–Ni angle close to 90°. The terminal cysteine residues are modeled in the same manner: as two thiolates, two thiols, or one of each. Taken together with the assumption of cluster neutrality, each of these states requires a different set of oxidation states for the two metals. The model with two terminal HS ligands uses Ni(III) and Fe(III) oxidation states to maintain overall neutrality. This model is ruled out as the reaction with $H_2$ is calculated to be far too exothermic. The other two models more faithfully reproduce the experimental results. The model with two terminal $H_2S$ ligands requires Ni(II)–Fe(II) oxidation states for the cluster while the mixed $H_2S$/HS-model needs Ni(III)–Fe(II) or Ni(II)–Fe(III) oxidation states. Of these two models, the mixed ligand system reproduces the experimental energetics. In this model only the iron is found as a site for $H_2$ binding. The catalytic cycle suggested by these calculations involves transfer of a proton from the iron-bound dihydrogen to a bridging HS-ligand. This newly formed $H_2S$ bridge dissociates from the nickel center and one of the cyanide ligands forms a Fe–CN–Ni bridge. A mechanism where this proton is transferred to a protein residue cannot be ruled out. Throughout this mechanism of $H_2$ oxidation the formal oxidation states of the metals do not change. This apparent contradiction is due to the formal assignment of hydride ligands as H whereas thiol hydrogens are treated as $H^+$. Thus the two electrons from $H_2$ oxidation are allotted to the hydride ligand without changing the oxidation state of the metals. One of the merits of this model is the low energy barrier for the $H_2$ activation step that is in good agreement with the experimental value. The catalytic cycle proceeds with a hydride transfer from iron to nickel.

A second model of [FeNi]-hydrogenase, that also uses DFT with the B3LYP level of theory, uses the infrared spectroscopy results as a check of the quality of the model in a given state [90]. This test makes use of the observation that the C–O stretching frequencies correlate well with the C–O bond length [91]. Since the C–O stretching frequency is sensitive to changes in the electron density of the iron center, it reflects the electronic distribution. In this manner the ability of the model to reproduce the electronic state is assessed. In this model system the cysteine residues are modeled as methyl thiolate ligands and the charge on the metal cluster is not restricted to neutrality. One of the main features of this model is the ability of the thiolate bridges to fold when one of the terminal thiolates is protonated. The mechanism starts with one of these bent structures. The **SI** state exists as a mixture of protonation states which in this model are $[(MeS)_2Ni(\mu\text{-}MeS)_2Fe(CN)_2(CO)]^{2-}$ and $[(MeS)(MeSH)Ni(\mu\text{-}MeS)_2Fe(CN)_2(CO)]^-$. The latter complex reacts with $H_2$ and again the iron is found to be the site of $H_2$ activation. The mechanism proceeds by oxidation of the nickel(II) ion. The dihydrogen unit is split heterolytically to form $(MeSH)_2Ni(\mu\text{-}H)(\mu\text{-}MeS)_2Fe(CN)_2(CO)$. This cluster loses a proton from a terminal methane thiol ligand followed by a rearrangement where the hydride ligand is transferred as a proton to regenerate the methane thiol ligand and form nickel(I). The oxidation of the nickel(I) center and loss of a proton from one of the methane thiol ligands recreates the initial species of the catalytic cycle.

In a third theoretical model the [FeNi]-hydrogenase from *D. gigas* is modeled using a hybrid density functional theory–molecular mechanics (DFT/MM) method [92]. In this model approximately 30 atoms of the active site (including the four cysteine residues) are modeled with DFT while molecular mechanics is used for the rest of the atoms within a 7 Å radius (about 300 atoms). The next shell includes about 10 000 protein and solvent atoms within 27 Å of the active site whose positions are not optimized. A 45 Å radius shell of water molecules rounds out the model. The inclusion of the protein and water matrix provides an improvement over the previous models that use a vacuum medium. This is reflected by the observation that the calculated C–O stretching frequencies are consistently lower when the protein and solvent matrix are excluded. The active states of the enzyme are modeled with a bridging hydride ligand replacing the bridging oxide of the inactive form. Thus the model of the active site in the **SI** state can be represented as $[(Cys)_2Ni(\mu\text{-}H)(\mu\text{-}Cys)_2Fe(CN)_2(CO)]^{3-}$. Unlike the previous two models, nickel is invoked as the site of dihydrogen attack in this model, though an exhaustive search for alternative binding sites was not performed. The attack of dihydrogen on the nickel ion results in heterolytic activation with the formation of a nickel hydride and a protonated cysteine residue.

A DFT model of the [Fe]-hydrogenase from *C. pasteurianum* has also appeared [93]. This model uses $[(MeS)(CO)(CN)Fe(\mu\text{-}S)_2(\mu\text{-}CO)Fe(CO)(CN)]^z$, where $z$ is the overall charge on the cluster. One of the unique features of this model is that the two bridging sulfides are redox active, being represented as either two sulfide, $S^{2-}$, ions or as one persulfide, $S_2^{2-}$, ion, a difference of two electrons. A similar property has not been proposed for the bridging cysteines in the [FeNi]-hydrogenase models. The bridging carbonyl ligand is found to have a low energy barrier to shifting to coordination to the distal iron. This distal iron atom is also a site of dihydrogen coordination, though the binding is weak. There is a low energy path to transfer of a proton to one of the bridging sulfides.

## 7.4 Conclusions

Much information on the nature of the active sites of hydrogenase enzymes has been gleaned from studies on the enzymes as well as from studies of model systems. The information we have presented here demonstrates how each of these areas of study has contributed towards illuminating the chemistry of the hydrogenase enzymes. While the enzyme studies reveal much about the possible mechanisms by which hydrogenase enzymes might operate, it is the modeling studies that will provide the information required for a full understanding of these complex enzymatic systems. Modeling studies reveal much of the hidden workings of the enzymes. The techniques are complementary. Studies of the hydrogenase enzymes provide the scaffold around which modeling studies can build a viable mechanism of dihydrogen activation or production.

A number of generalities can be drawn from the similarities that are observed between the three classes of enzymes, [Fe]-, [FeNi]-, and [FeNiSe]-hydrogenases. The similarities in the structure of the active sites are the most pertinent for the purposes of modeling experiments. The three classes of enzymes share a common iron center with two cyanide ligands and one carbon monoxide ligand. This moiety was first proposed from experiments on the enzyme and confirmed by model studies. This $[Fe(CN)_2(CO)]$ moiety is connected to a second metal ion which is either nickel or iron, depending on the class of hydrogenase enzyme. This second metal is potentially redox active and is bridged to the dicyano carbonyl iron center via two cysteine residues. This second metal center has two terminal ligands that are either cysteine or selenocysteine. These similarities suggest a common mechanism of $H_2$ activation for all three classes of hydrogenase enzyme. Given that no model nickel compound has been reported to activate dihydrogen we are left to conclude that it is the $Fe(CN)_2(CO)$ unit which is responsible for $H_2$-activation. The Ni-61 EPR studies would then suggest that the second metal site could act as a hydride reservoir. For now this is merely speculation though it does account for some of the interesting properties of the enzymes.

In addition to the common features of the three classes of hydrogenase it is instructive to examine the differences that exist between them. One of the most obvious differences is the function of the different classes of hydrogenase. The [Fe]-hydrogenase enzymes are primarily $H_2$-activating while the [FeNi]- and [FeNiSe]-hydrogenase enzymes are $H_2$-producing. This is an over-simplification as there are hydrogenase enzymes that are able both to form $H_2$ as well as to activate $H_2$. It does, however, suggest that the enzymes have evolved with a preference for the metal which better enables it to perform the task of either activating or producing dihydrogen.

# References

1. T. Y. Cheng, B. S. Brunschwig, R. M. Bullock, *J. Am. Chem. Soc. 120* (**1998**) 13121–13137.
2. S. S. Kristjánsdóttir, J. R. Norton, in A. Dedieu (Ed.): *Transition Metal Hydrides: Recent Advances in Theory and Experiment*, VCH, New York **1992**.
3. D. G. Gusev, R. Hübener, P. Burger, O. Orama, H. Berke, *J. Am. Chem. Soc. 119* (**1997**) 3716–3731.
4. F. Maseras, N. Koga, K. Morokuma, *Organometallics 13* (**1994**) 4008–4016.
5. F. Maseras, M. Duran, A. Lledós, J. Bertrán, *J. Am. Chem. Soc. 113* (**1991**) 2879–2884.
6. G. J. Kubas, R. R. Ryan, B. I. Swanson, P. J. Vergamini, H. J. Wasserman, *J. Am. Chem. Soc. 106* (**1984**) 451–452.
7. G. J. Kubas, *Acc. Chem. Res. 21* (**1988**) 120–128.
8. P. G. Jessop, R. H. Morris, *Coord. Chem. Rev. 121* (**1992**) 155–284.
9. D. M. Heinekey, W. J. J. Oldham, *Chem. Rev. 93* (**1993**) 913–926.
10. R. L. Sweany, M. A. Polito, A. Moroz, *Organometallics 8* (**1989**) 2305–2308.
11. C. E. Forde, S. E. Landau, R. H. Morris, *J. Chem. Soc., Dalton Trans.* (**1997**) 1663–1664.
12. P. A. Maltby, M. Schlaf, M. Steinbeck, A. J. Lough, R. H. Morris, W. T. Klooster, T. F. Koetzle, R. C. Srivastava, *J. Am. Chem. Soc. 118* (**1996**) 5396–5407.
13. D. G. Hamilton, R. H. Crabtree, *J. Am. Chem. Soc. 110* (**1988**) 4126–4133.
14. M. T. Bautista, K. A. Earl, P. A. Maltby, R. H. Morris, C. T. Schweitzer, A. Sella, *J. Am. Chem. Soc. 110* (**1988**) 7031–7036.
15. P. E. Bloyce, A. J. Rest, I. Whitwell, W. A. G. Graham, R. Holmes-Smith, *J. Chem. Soc., Chem. Commun.* (**1988**) 846–848.
16. R. H. Morris, *Inorg. Chem. 31* (**1992**) 1471–1478.
17. M. T. Bautista, E. P. Cappellani, S. D. Drouin, R. H. Morris, C. T. Schweitzer, A. Sella, J. Zubkowski, *J. Am. Chem. Soc. 113* (**1991**) 4876–4887.
18. B. Chin, A. J. Lough, R. H. Morris, C. T. Schweitzer, C. D'Agostino, *Inorg. Chem. 33* (**1994**) 6278–6288.
19. E. Rocchini, A. Mezzetti, H. Rüegger, U. Burckhardt, V. Gramlich, A. Del Zotto, P. Martinuzzi, P. Rigo, *Inorg. Chem. 36* (**1997**) 711–720.

20. Z. Xu, I. Bytheway, G. Jia, Z. Lin, *Organometallics 18* (**1999**) 1761–1766.
21. S. Dapprich, G. Frenking, *Organometallics 15* (**1996**) 4547–4551.
22. M. G. Basallote, J. Durán, M. J. Fernandez-Trujillo, G. González, M. A. Máñez, M. Martínez, *Inorg. Chem. 37* (**1998**) 1623–1628.
23. C. Zirngibl, W. Van Dongen, B. Schworer, R. Von Bunau, M. Richter, A. Klein, R. K. Thauer, *Eur. J. Biochem. 208* (**1992**) 511–520.
24. M. J. Maroney, *Comments Inorg. Chem. 17* (**1995**) 347–375.
25. C. M. Goldman, P. K. Mascharak, *Comments Inorg. Chem. 18* (**1995**) 1–25.
26. J. C. Fontecilla-Camps, *J. Biol. Inorg. Chem. 1* (**1996**) 91–98.
27. J. R. J. Lancaster, *The Bioinorganic Chemistry of Nickel*, VCH, New York **1988**.
28. S. W. Ragsdale, *Curr. Opin. Chem. Biol. 2* (**1998**) 208–215.
29. M. J. Maroney, *Curr. Opin. Chem. Biol. 3* (**1999**) 188–199.
30. J. J. Moura, I. Moura, B. H. Huynh, H. J. Kruger, M. Teixeira, R. C. DuVarney, D. V. DerVartanian, A. V. Xavier, H. D. Peck, Jr., J. LeGall, *Biochem. Biophys. Res. Commun. 108* (**1982**) 1388–1393.
31. M. Teixeira, I. Moura, A. V. Xavier, B. H. Huynh, D. V. DerVartanian, H. D. Peck, Jr., J. LeGall, J. J. Moura, *J. Biol. Chem. 260* (**1985**) 8942–8950.
32. R. Cammack, D. S. Patil, V. M. Fernandez, *Biochem. Soc. Trans. 13* (**1985**) 572–578.
33. J. W. van der Zwaan, S. P. Albracht, R. D. Fontijn, E. C. Slater, *FEBS Lett 179* (**1985**) 271–7.
34. C. V. Romao, I. A. Pereira, A. V. Xavier, J. LeGall, M. Teixeira, *Biochem. Biophys. Res. Commun. 240* (**1997**) 75–79.
35. R. Cammack, C. Bagyinka, K. L. Kovacs, *Eur. J. Biochem. 182* (**1989**) 357–362.
36. A. Chapman, R. Cammack, C. E. Hatchikian, J. McCracken, J. Peisach, *FEBS Lett. 242* (**1988**) 134–138.
37. C. Fan, M. Teixera, J. J. G. Moura, I. Moura, B.-H. Huynh, J. Le Gall, H. D. J. Peck, B. M. Hoffman, *J. Am. Chem. Soc. 113* (**1991**) 20–24.
38. J. E. Huyett, M. Carepo, A. Pamplona, R. Franco, I. Moura, J. J. G. Moura, B. M. Hoffman, *J. Am. Chem. Soc. 119* (**1997**) 9291–9292.
39. M. W. Adams, M. K. Johnson, I. C. Zambrano, L. E. Mortenson, *Biochimie 68* (**1986**) 35–42.
40. P. J. van Dam, E. J. Reijerse, W. R. Hagen, *Eur. J. Biochem. 248* (**1997**) 355–361.
41. K. A. Bagley, C. J. Van Garderen, M. Chen, E. C. Duin, S. P. Albracht, W. H. Woodruff, *Biochemistry 33* (**1994**) 9229–9236.
42. K. A. Bagley, E. C. Duin, W. Roseboom, S. P. Albracht, W. H. Woodruff, *Biochemistry 34* (**1995**) 5527–5535.
43. A. J. Pierik, W. Roseboom, R. P. Happe, K. A. Bagley, S. P. Albracht, *J. Biol. Chem. 274* (**1999**) 3331–3337.
44. A. L. de Lacey, E. C. Hatchikian, A. Volbeda, M. Frey, J. C. Fontecilla-Camps, V. M. Fernandez, *J. Am. Chem. Soc. 119* (**1997**) 7181–7189.
45. A. Volbeda, E. Garcia, C. Piras, A. L. Delacey, V. M. Fernandez, E. C. Hatchikian, M. Frey, J. C. Fontecilla-Camps, *J. Am. Chem. Soc. 118* (**1996**) 12989–12996.
46. T. M. van der Spek, A. F. Arendsen, R. P. Happe, S. Yun, K. A. Bagley, D. J. Stufkens, W. R. Hagen, S. P. Albracht, *Eur. J. Biochem. 237* (**1996**) 629–634.
47. A. J. Pierik, M. Hulstein, W. R. Hagen, S. P. Albracht, *Eur. J. Biochem. 258* (**1998**) 572–578.
48. J. M. Coremans, C. J. van Garderen, S. P. Albracht, *Biochim. Biophys. Acta 1119* (**1992**) 148–156.
49. D. P. Barondeau, L. M. Roberts, P. A. Lindahl, *J. Am. Chem. Soc. 116* (**1994**) 3442–3448.
50. L. M. Roberts, P. A. Lindahl, *Biochemistry 33* (**1994**) 14339–14350.
51. A. J. Pierik, W. R. Hagen, J. S. Redeker, R. B. Wolbert, M. Boersma, M. F. Verhagen, H. J. Grande, C. Veeger, P. H. Mutsaers, R. H. Sands, et al., *Eur. J. Biochem. 209* (**1992**) 63–72.
52. B. H. Huynh, J. J. G. Moura, I. Moura, T. A. Kent, J. LeGall, A. V. Xavier, E. Munck, *J. Biol. Chem. 255* (**1980**) 3242–3244.
53. M. H. Emptage, T. A. Kent, B. H. Huynh, J. Rawlings, W. H. Orme-Johnson, E. Munck, *J. Biol. Chem. 255* (**1980**) 1793–1796.
54. B. H. Huynh, D. S. Patil, I. Moura, M. Teixeira, J. J. Moura, D. V. DerVartanian, M. H. Czechowski, B. C. Prickril, H. D. Peck, Jr., J. LeGall, *J. Biol. Chem. 262* (**1987**) 795–800.
55. M. J. Maroney, G. J. Colpas, C. Bagyinka, N. Baidya, P. K. Mascharak, *J. Am. Chem. Soc. 113* (**1991**) 3962–3972.

56. C. Bagyinka, J. P. Whitehead, M. J. Maroney, *J. Am. Chem. Soc. 115* (**1993**) 3576–3585.
57. Z. Gu, J. Dong, C. B. Allan, S. B. Choudhury, R. Franco, J. J. G. Moura, I. Moura, J. LeGall, A. E. Przybyla, W. Roseboom, S. P. J. Albracht, M. J. Axley, R. A. Scott, M. J. Maroney, *J. Am. Chem. Soc. 118* (**1996**) 11155–11165.
58. P. A. Lespinat, Y. Berlier, G. Fauque, M. Czechowski, B. Dimon, J. Le Gall, *Biochimie 68* (**1986**) 55–61.
59. A. Der, C. Bagyinka, T. Pali, K. L. Kovacs, *Anal. Biochem. 150* (**1985**) 481–486.
60. A. Volbeda, M. H. Charon, C. Piras, E. C. Hatchikian, M. Frey, J. C. Fontecilla-Camps, *Nature 373* (**1995**) 580–587.
61. Y. Higuchi, T. Yagi, N. Yasuoka, *Structure 5* (**1997**) 1671–1680.
62. Y. Higuchi, T. Yagi, *Biochem. Biophys. Res. Commun. 255* (**1999**) 295–259.
63. Y. Higuchi, H. Ogata, K. Miki, N. Yasuoka, T. Yagi, *Structure Fold Des. 7* (**1999**) 549–556.
64. E. Garcin, X. Vernede, E. C. Hatchikian, A. Volbeda, M. Frey, J. C. Fontecilla-Camps, *Structure Fold Des. 7* (**1999**) 557–566.
65. J. W. Peters, W. N. Lanzilotta, B. J. Lemon, L. C. Seefeldt, *Science 282* (**1998**) 1853–1858.
66. Y. Nicolet, C. Piras, P. Legrand, C. E. Hatchikian, J. C. Fontecilla-Camps, *Structure Fold Des. 7* (**1999**) 13–23.
67. B. J. Lemon, J. W. Peters, *Biochemistry 38* (**1999**) 12969–12973.
68. Y. Sugiura, J. Kuwahara, T. Suzuki, *Biochem. Biophys. Res. Commun. 115* (**1983**) 878–881.
69. K. Nakamoto, *Infrared and Raman Spectra of Inorganic and Coordination Compounds*, Wiley, New York **1997**.
70. D. H. Nguyen, H. F. Hsu, M. Millar, S. A. Koch, C. Achim, E. L. Bominaar, E. Munck, *J. Am. Chem. Soc. 118* (**1996**) 8963–8964.
71. H. F. Hsu, S. A. Koch, C. V. Popescu, E. Munck, *J. Am. Chem. Soc. 119* (**1997**) 8371–8372.
72. I. Mus-Veteau, F. Guerlesquin, *Biochem. Biophys. Res. Commun. 201* (**1994**) 128–134.
73. R. H. Crabtree, *Inorg. Chim. Acta 125* (**1986**) L7–L8.
74. L. L. Efros, H. H. Thorp, G. W. Brudvig, R. H. Crabtree, *Inorg. Chem. 31* (**1992**) 1722–1724.
75. T. L. James, L. Cai, M. C. Muetterties, R. H. Holm, *Inorg. Chem. 35* (**1996**) 4148–4161.
76. P. Saint-Martin, P. A. Lespinat, G. Fauque, Y. Berlier, J. LeGall, I. Moura, M. Teixera, A. V. Xavier, J. J. G. Moura, *Proc. Natl. Acad. Sci. (USA) 85* (**1988**) 9378–9380.
77. R. T. Hembre, S. McQueen, *J. Am. Chem. Soc. 116* (**1994**) 2141–2142.
78. R. T. Hembre, J. S. McQueen, V. W. Day, *J. Am. Chem. Soc. 118* (**1996**) 798–803.
79. P. I. Amrhein, S. D. Drouin, C. E. Forde, A. J. Lough, R. H. Morris, *J. Chem. Soc., Chem. Commun.* (**1996**) 1665–1666.
80. T. P. Fong, A. J. Lough, R. H. Morris, A. Mezzetti, E. Rocchini, P. Rigo, *J. Chem. Soc., Dalton Trans.* (**1998**) 211–213.
81. T. P. Fong, C. E. Forde, A. J. Lough, R. H. Morris, P. Rigo, E. Rocchini, T. Stephan, *J. Chem. Soc., Dalton Trans.* (**1999**) 4475–4486.
82. T. Y. Bartucz, A. Golombek, A. J. Lough, P. A. Maltby, R. H. Morris, R. Ramachandran, M. Schlaf, *Inorg. Chem. 37* (**1998**) 1555–1562.
83. F. Osterloh, W. Saak, D. Haase, S. Pohl, *Chem. Commun.* (**1997**) 979–980.
84. S. C. Davies, D. J. Evans, D. L. Hughes, S. Longhurst, J. R. Sanders, *Chem. Commun.* (**1999**) 1935–1936.
85. M. Schmidt, S. M. Contakes, T. B. Rauchfuss, *J. Am. Chem. Soc. 121* (**1999**) 9736–9737.
86. R. B. King, *J. Am. Chem. Soc. 84* (**1962**) 2460.
87. A. Winter, L. Zsolnai, G. Huttner, *Z. Naturforsch. 37b* (**1982**) 1430–1436.
88. A. Le Cloirec, S. P. Best, S. Borg, S. C. Davies, D. J. Evans, D. L. Hughes, C. J. Pickett, *Chem. Commun.* (**1999**) 2285–2286.
89. M. Pavlov, P. E. M. Siegbahn, M. R. A. Blomberg, R. H. Crabtree, *J. Am. Chem. Soc. 120* (**1998**) 548–555.
90. S. Niu, L. M. Thomson, M. B. Hall, *J. Am. Chem. Soc. 121* (**1999**) 4000–4007.
91. S. L. Morrison, J. J. Turner, *J. Mol. Struct. 317* (**1994**) 39–47.
92. P. Amara, A. Volbeda, J. C. Fontecilla-Camps, M. J. Field, *J. Am. Chem. Soc. 121* (**1999**) 4468–4477.
93. I. Dance, *Chem. Commun.* (**1999**) 1655–1656.

# 8 Biomimetic Electron-transfer Chemistry of Porphyrins and Metalloporphyrins

*Shunichi Fukuzumi and Hiroshi Imahori*

## 8.1 Introduction

Porphyrins contain an extensively conjugated $\pi$ system and a highly delocalized $\pi$ system such as this is suitable for efficient electron-transfer reactions, because the uptake or release of electrons results in minimal changes of structure and solvation on electron transfer [1]. Numerous metal ions can tightly bind with porphyrin macrocyclic ligands to afford metalloporphyrins which can undergo metal-centered electron-transfer reactions and porphyrin ligand-centered reactions [1]. Therefore, a variety of redox states can be achieved for metalloporphyrins through multi-electron oxidation or reduction processes [1]. Metalloporphyrins can form a metal–carbon $\sigma$-bond to afford organometallic porphyrins [2]. The strength of such binding is significantly altered, depending on the redox state of metalloporphyrins [2]. Consequently, cleavage or formation of the metal–carbon bond can be finely tuned by changing the redox state. Thus, the electron transfer oxidation or reduction of metalloporphyrins and organometallic porphyrins is often accompanied by the subsequent formation or cleavage of the bond, leading to a rich chemistry for metalloporphyrins and organometallic porphyrins [1, 2]. Such formation of reactive intermediates associated with electron-transfer reactions of metalloporphyrins has played an important role in the catalytic function of metalloporphyrins not only in chemical but also in biological processes [3–5]. Axial ligands can also bind or coordinate to the metal to attenuate the electron-transfer reactions of metalloporphyrins including the site of electron transfer, i.e. metal-centered or the porphyrin ligand-centered electron transfer [1]. The use of photoexcited states of porphyrins and metalloporphyrins (i.e., in the singlet or triplet excited state) further widens the scope of electron-transfer reactions, because photoexcitation significantly enhances both the electron-acceptor or -donor properties of photoexcited porphyrins and metalloporphyrins. The rich redox properties of porphyrins and metalloporphyrins both in the ground and excited states have rendered them essential components of biological electron transport systems including photosynthesis and respiration [6–15].

Because the electron-transfer processes of porphyrins and metalloporphyrins have been reviewed elsewhere [1, 16, 17], this chapter focuses on mechanistic aspects of chemical processes associated with the electron-transfer reactions of metalloporphyrins and organometallic porphyrins. When an electron-transfer process is followed by subsequent bond-breaking and bond-forming, the overall reaction often becomes the same as electrophile–nucleophile reactions in which electron shift and bond formation occur in a concerted manner via a polar mechanism. Such electron transfer or alternative polar (concerted) processes have been among the most central propositions in reaction mechanisms [18–26]. Because both processes involve the formation of a significant amount of charge in the transition state, it has been difficult to differentiate between electron transfer and polar mechanisms on the basis of the classical approach of electronic and substitution effects. Advances in time-resolved spectroscopy, including laser flash photolysis and a low-temperature stopped-flow technique, have, however, provided the means of monitoring fast electron-transfer reactions and subsequent chemical reactions, by enabling detection of unstable reactive intermediates produced in electron-transfer reactions.

In the first part of this review, recent advances in the mechanistic aspects of reactions of high-valent metalloporphyrin intermediates produced by the oxidation of metalloporphyrins and organometallic porphyrins are highlighted in relation to the enzymatic reaction mechanisms of biological oxidation of substrates including water. This section is followed by an overview of the reverse process, i.e. the four-electron reduction of $O_2$ to water catalyzed by cytochrome c oxidases and model complexes. Finally biomimetic electron-transfer systems of porphyrins and metalloporphyrins in organized media are presented in relation to practical endeavors such as the design of molecular-scale photonic devices and solar-energy conversion systems.

## 8.2 High-valent Intermediates of Heme Enzymes

### 8.2.1 P-450 Catalytic Mechanisms

Cytochromes P-450, the CO complexes of which have a characteristic absorption maximum at 450 nm, use molecular oxygen to catalyze monooxygenation of versatile organic compounds such as hydrocarbons, sulfides, and amines with reducing agents (Eq. 1) [27–33].

$$RH + O_2 + 2H^+ + 2e^- \xrightarrow[P\text{-}450]{} ROH + H_2O \qquad (1)$$

For decades extensive efforts have been devoted to the elucidation of the molecular mechanisms of oxygen activation and oxidation reactions. The following mechanism is generally assumed for dioxygen activation and monooxygenation (Scheme 1) [27–33].

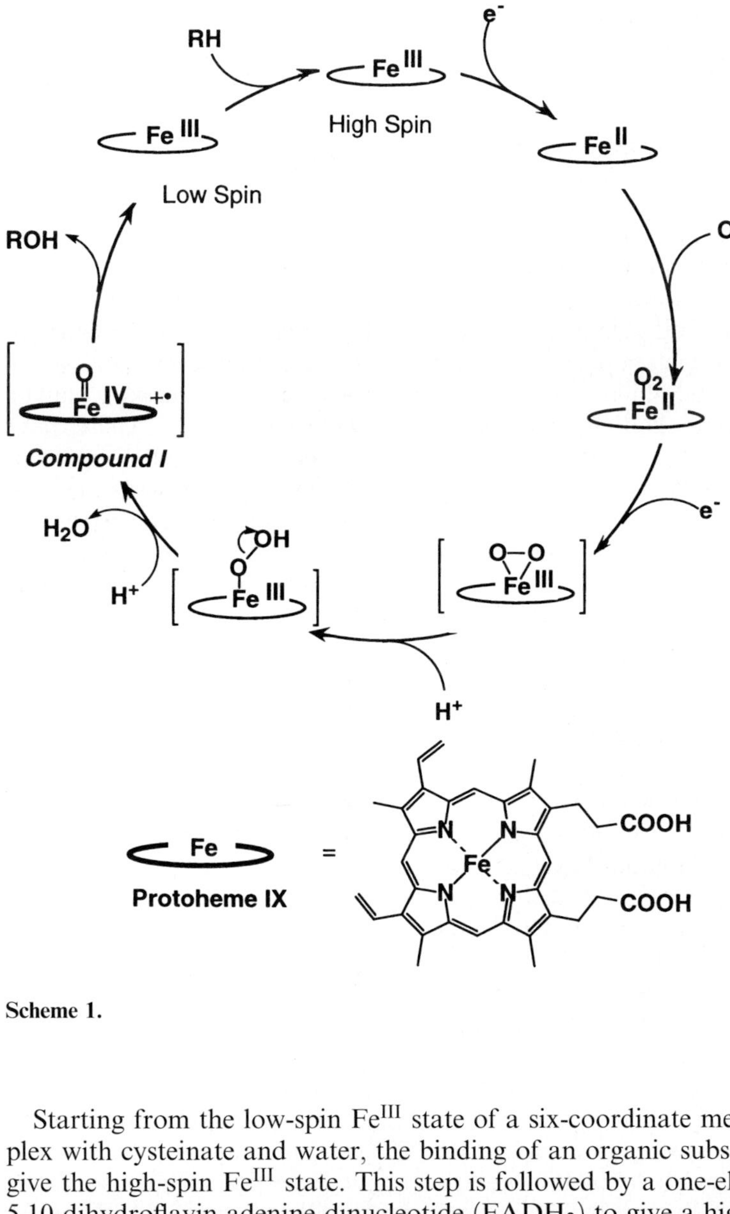

**Scheme 1.**

Starting from the low-spin $Fe^{III}$ state of a six-coordinate metalloporphyrin complex with cysteinate and water, the binding of an organic substrate (RH) occurs to give the high-spin $Fe^{III}$ state. This step is followed by a one-electron reduction via 5,10-dihydroflavin adenine dinucleotide ($FADH_2$) to give a high-spin $Fe^{II}$ complex which can bind oxygen. After a second one-electron reduction to a high-spin $Fe^{II}$ complex, a low-spin peroxo $Fe^{III}$ complex is formed. Two protons are then added to produce a high-valent iron–oxo porphyrin $\pi$ radical cation, $[(P)Fe^{IV}=O]^{\cdot+}$ (or a formally pentavalent iron–oxo complex $[(P)Fe^{V}=O]$) by release of one water molecule, thereby cleaving the O–O bond. Formal oxygen transfer from $[(P)Fe^{IV}=O]^{\cdot+}$ to RH occurs to yield the monooxygenated product, accompanied by regeneration of the starting low-spin $Fe^{III}$ state (Scheme 1). The high-valent iron–oxo porphyrin

$([(P)Fe^{IV}=O]^{\cdot+}$ or $(P)Fe^{V}=O)$ is the reactive intermediate in the monooxygenation step.

The use of a synthetic model system has provided valuable mechanistic insights into the molecular catalytic mechanism of P-450. Groves et al. [34]. were the first to report cytochrome P-450-type activity in a model system comprising iron *meso*-tetraphenylporphyrin chloride [(TPP)FeCl] and iodosylbenzene (PhIO) as an oxidant which can oxidize the $Fe^{III}$ porphyrin directly to $[(TPP)Fe^{IV}=O]^{\cdot+}$ in a 'shunt' pathway. Thus, (TPP)FeCl and other metalloporphyrins can catalyze the monooxygenation of a variety of substrates by PhIO [35–40], hypochlorite salts [41, 42], *p*-cyano-*N,N*-dimethylaniline *N*-oxide [43–46], percarboxylic acids [47–50] and hydroperoxides [51, 52]. Catalytic activity was, however, rapidly reduced because of the destruction of the metalloporphyrin during the catalytic cycle [34–52]. When (TPP)FeCl was immobilized on the surface of silica or silica–alumina, catalytic reactivity and catalytic lifetime both increased significantly [53]. There have been several reports of supported catalysts based on such metalloporphyrins adsorbed or covalently bound to polymers [54–56]. Catalyst lifetime was also significantly improved by use of iron porphyrins such as *meso*-tetramesitylporphyrin chloride [(TMP)FeCl] and iron *meso*-tetrakis(2,3,4,5,6-pentafluorophenyl)porphyrin chloride [(TPFPP)FeCl], which resist oxidative destruction, because of steric and electronic effects and thereby act as efficient catalysts of P-450 type reactions [57–65].

## 8.2.2 Reaction Pathways of $[(P)Fe^{IV}=O]^{\cdot+}$

In the catalytic cycle in Scheme 1, details of the reaction between the high-valent iron–oxo porphyrin $\pi$ radical cation $[(P)Fe^{IV}=O]^{\cdot+}$ and a substrate have yet to be clarified [66]. Three possibilities have survived after extensive studies of the mechanisms of the reactions of high-valent iron–oxo porphyrins, in particular the dealkylation of amines by cytochrome P-450: (**A**) a sequential electron–proton–electron transfer; (**B**) direct hydrogen transfer then electron transfer; and (**C**) electron transfer followed by hydrogen transfer (Scheme 2) [67–74].

It was believed that a sequential electron–proton–electron transfer mechanism would result in small isotope effects and that a hydrogen transfer mechanism would result in large isotope effects. Because small isotope effects were observed for amine dealkylation by cytochrome P-450, a sequential electron–proton–electron transfer mechanism was proposed for the reaction [67, 75–77]. The magnitude of a single isotope effect provides only limited information about a reaction mechanism, however, and isotope effects might be logically consistent with any type of C–H cleavage reaction.

To distinguish between paths **A** and **B** in Scheme 2, Dinnocenzo and coworkers [78–80] determined intramolecular isotope effects for deprotonation from the radical cations (path **A**) and for hydrogen-atom abstraction from a series of substituted *N,N*-dimethylanilines (path **B**); these were compared with isotope effects for *N*-demethylation of the same series of *N,N*-dimethylanilines. The deprotonation isotope-effect profile for the radical cations of *N,N*-dimethylanilines were determined by monitoring electron-transfer reactions from *N,N*-dimethylanilines to $[Fe(phen)_3]^{3+}$

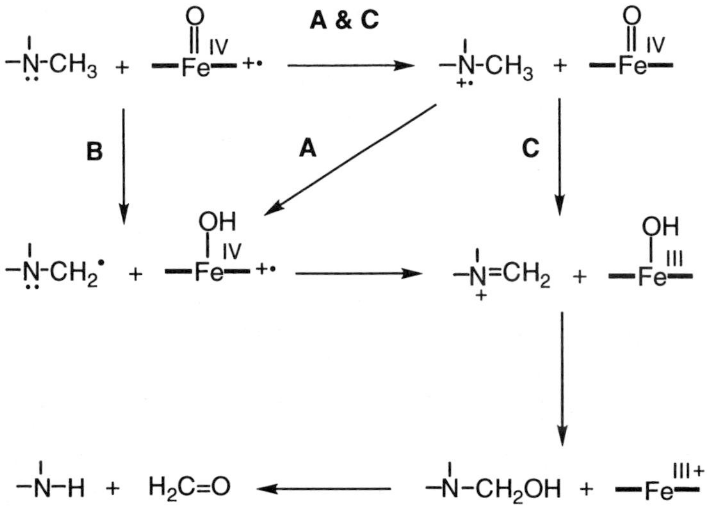

**Scheme 2.**

(phen = 1,10-phenanthroline), a well-known outer-sphere one-electron oxidant [81], in the presence of pyridine as base. The mechanism of such electron-transfer reactions has been well established in electron-transfer reactions of NADH analogs with $[Fe(phen)_3]^{3+}$ in the presence of pyridine [82–84], and can be applied to the reaction of *N,N*-dimethylanilines as shown in Scheme 3.

The rate of initial electron transfer from *N,N*-dimethylaniline to $[Fe(phen)_3]^{3+}$ is diffusion-limited. This is followed by the rate-determining proton transfer from the radical cation to pyridine to give the deprotonated $\alpha$-amino radical which is rapidly oxidized by a second equivalent of $[Fe(phen)_3]^{3+}$ to yield the product iminium ion. Kinetic isotope effects $(k_H/k_D)$ for the proton transfer were determined from the $d_3/d_0$ ratios of the products derived from *p*-substituted *N*-methyl-*N*-trideuteromethylanilines. The $k_H/k_D$ value first increases and then decreases with increasing $pK_a$ of *p*-substituted *N,N*-dimethylaniline. Such a bell-shaped isotope effect profile is typical of proton-transfer reactions [82, 85]. The maximum $k_H/k_D$ value is determined as 8.8 which is much larger than the corresponding value for the demethylation of the same substrate by cytochrome P-450 (2.6) [79].

The isotope effect profile for hydrogen atom abstraction from the same series of *p*-substituted *N,N*-dimethylanilines was determined by using the *tert*-butoxyl radical $(t\text{-BuO}^{\bullet})$ as the hydrogen-atom abstracting agent; this species is known to abstract hydrogen atoms from the $\alpha$-carbon atoms of amines [86]. The *tert*-butoxyl radical was generated by the coumarin-sensitized photoreduction of the *N-tert*-butoxy-4-phenylpyridinium ion [87, 88]. A small, steady decrease in the $k_H/k_D$ value for hydrogen-atom abstraction from *p*-substituted *N,N*-dimethylanilines is observed when the $pK_a$ of the radical cations is increased. This is quite different from the bell-shaped isotope-effect profile for the proton-transfer reactions. Such differences demonstrate that isotope profiles can serve as sensitive probes for distinguishing

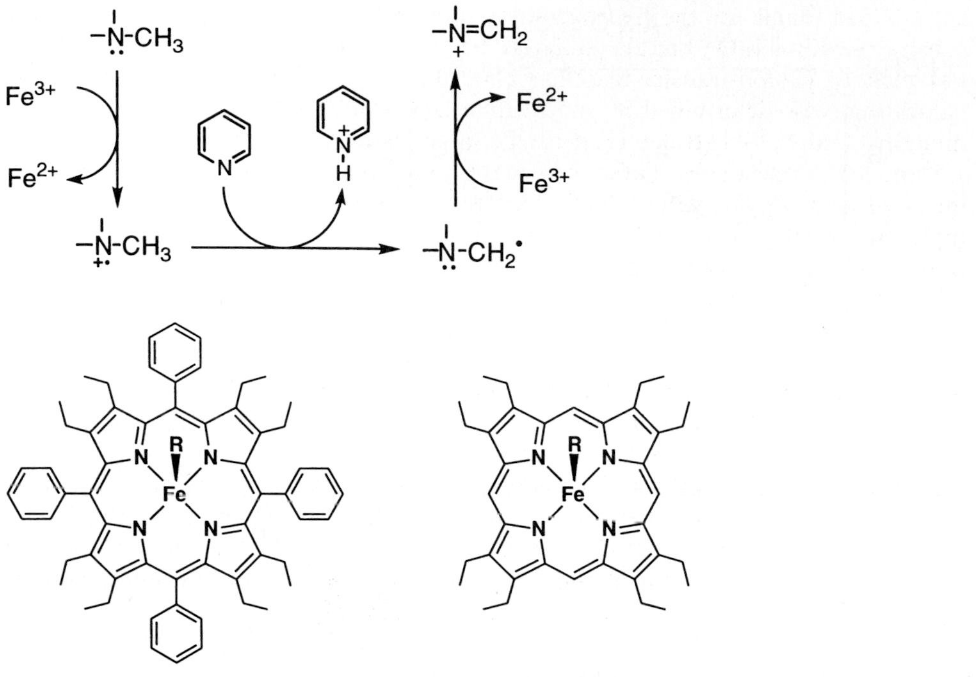

**(OETPP)Fe(R)**

R = $C_6H_5$, 3,5-$C_6F_2H_3$,
2,4,6-$C_6F_3H_2$, or $C_6F_5$

**(OEP)Fe(R)**

R = $C_6H_5$, 2,4,6-$C_6F_3H_2$
or 2,3,5,6-$C_6F_4H$

**Scheme 3**

P-450 amine dealkylation mechanisms. Because $k_H/k_D$ values for $N$-demethylation of $p$-substituted $N,N$-dimethylanilines by cytochrome P-450 are nearly identical with those for the hydrogen abstraction reactions of $t$-BuO$^{\bullet}$, it was proposed that P-450 reacted by a direct hydrogen-atom-abstraction mechanism (**B**) rather than a sequential electron–proton–electron-transfer mechanism (**A**) [78, 80]. Identical relationships between $k_H/k_D$ values for reactions of $t$-BuO$^{\bullet}$ and P-450 has been expanded for other substrates including $p$-xylene, toluene, benzyl alcohol, and tertiary trialkylamine [78]. It has thus been suggested that all these P-450 reactions proceed by a common hydrogen-atom transfer mechanism [78].

For a synthetic P-450 model system, however, an opposite conclusion was drawn from similar analysis of kinetic isotope effects for $N$-demethylation of $p$-substituted $N,N$-dimethylanilines by PhIO, catalyzed by iron *meso*-tetrakis(2,3,4,5,6-penta-fluorophenyl)porphyrin chloride, (TPFPP)FeCl [89]. Intramolecular $k_H/k_D$ values obtained from the $d_3/d_0$ ratios of the products derived from 4-X-$N$-methyl-$N$-tri-deuteromethylanilines [2.0 (X = NO$_2$), 2.0 (X = CN), 2.6 (X = Br), 3.1 (X = H), 3.2 (X = Me), and 3.3 (X = OMe)], decreased regularly on going from electron-donating to electron-withdrawing substituents. Such a trend is the complete opposite to that found for the hydrogen-atom-abstraction reactions of some of these substrates with $t$-BuO$^{\bullet}$, but also different from the bell-shaped isotope-effect profile obtained for proton-transfer reactions (vide supra) [89]. The intermolecular $k_H/k_D$ values were also determined by competitive experiments with 4-X-substituted $N,N$-dimethyl- and $N,N$-bis(trideuteromethyl)anilines, with X = H, Br, and OMe, as 1.6, 1.5 and 1.9, respectively. These values are quite different from the corresponding intramolecular $k_H/k_D$ values. Such a difference between intramolecular and inter-molecular kinetic isotope effects implies a multi-step rather than single-step reaction. An electron transfer pathway followed by hydrogen transfer (**C** in Scheme 2) is consistent with these results.

### 8.2.3 Reorganization Energies for Electron-transfer Reactions of High-valent Iron Porphyrins

The electron transfer pathway has been examined by analyzing relative reactivities of 4-X-substituted $N,N$-dimethylanilines, determined by competitive kinetics; the reactivities span a wide reactivity range of 25 (from X = NO$_2$ to X = MeO) and can be fitted to the Rehm–Welter equation [90] for electron-transfer reactions (Eq. 2):

$$\Delta G^{\neq} = (\Delta G^{\circ}{}_{et}/2) + [(\Delta G^{\circ}{}_{et}/2)^2 + (\Delta G^{\neq}{}_0)^2]^{1/2} \tag{2}$$

where $\Delta G^{\circ}{}_{et}$ is the free energy change for electron transfer and $\Delta G^{\neq}{}_0$ is the intrinsic barrier that represents the activation free energy when the driving force of electron transfer is zero, i.e., $\Delta G^{\neq} = \Delta G^{\neq}{}_0$ at $\Delta G^{\circ}{}_{et} = 0$ [90]. The $\Delta G^{\neq}$ values are related to the rate constant of electron transfer ($k_{et}$) as given by Eq. 3, where $Z$ is the collision frequency (i.e., $1 \times 10^{11}$ M$^{-1}$ s$^{-1}$).

$$\Delta G^{\neq} = 2.3RT \log(Z/k_{et}) \tag{3}$$

The $k_{et}$ values can be calculated from the $\Delta G^{\circ}{}_{et}$ and $\Delta G^{\neq}{}_0$ values by use of Eqs 2 and 3. The $\Delta G^{\circ}{}_{et}$ value is obtained from the one-electron oxidation potential of an electron donor ($E^{\circ}{}_{ox}$) and the one-electron reduction potential of an acceptor ($E^{\circ}{}_{red}$), because $\Delta G^{\circ}{}_{et} = F(E^{\circ}{}_{ox} - E^{\circ}{}_{red})$.

The reorganization energy of electron transfer ($\lambda$) is defined as $4\Delta G^{\neq}{}_0$. The $\lambda$ value was evaluated as 47 kcal mol$^{-1}$ for the postulated electron transfer from $N,N$-dimethylanilines to [(TPFPP)Fe$^{IV}$=O]$^{\cdot+}$ by fitting the data to Eqs 2 and 3. Essentially the same $\lambda$ value can be obtained by fitting the data to the Marcus equation (Eq. 4) [91] except for the region where the rate becomes close to a diffusion-limited value [81, 92].

$$\Delta G^{\neq} = (\lambda/4)[1 + (\Delta G^{\circ}{}_{et}/\lambda)]^2 \tag{4}$$

Although such a large $\lambda$ value is rather unusual in electron-transfer reactions of metalloporphyrins [1], the electron transfer kinetics for generation of iron(IV) por-

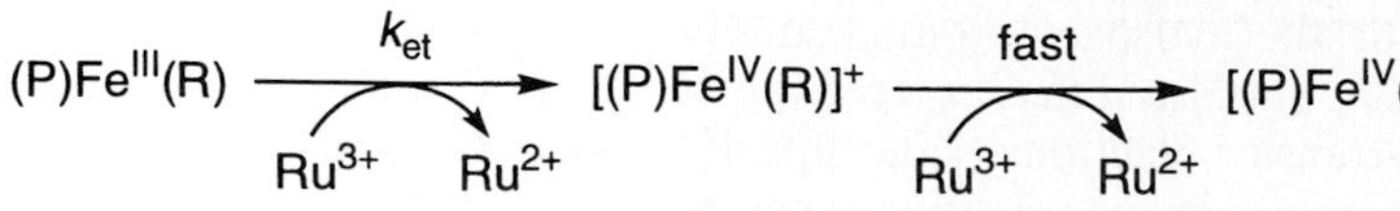

$$(P)Fe^{III}(R) \xrightarrow[\;Ru^{3+}\quad Ru^{2+}\;]{k_{et}} [(P)Fe^{IV}(R)]^+ \xrightarrow[\;Ru^{3+}\quad Ru^{2+}\;]{fast} [(P)Fe^{IV}(R)]^{2+\bullet}$$

**Scheme 4.**

phyrins or iron(IV) porphyrin $\pi$ radical cations indicated a large reorganization energy was required for electron-transfer reactions involving high-valent metalloporphyrins [93]. The investigated compounds are represented as (P)Fe(R) where $P^{2-}$ = the dianion of 2,3,7,8,12,13,17,18-octaethyl-5,10,15,20-tetraphenylporphyrin (OETPP) and R = $C_6H_5$, 3,5-$C_6F_2H_3$, 2,4,6-$C_6F_3H_2$, or $C_6F_5$, or $P^{2-}$ = the dianion of 2,3,7,8,12,13,17,18-octaethylporphyrin (OEP) and R = $C_6H_5$, 2,4,6-$C_6F_3H_2$, or 2,3,5,6-$C_6F_4H$.

The first one-electron transfer from (P)Fe(R) to $[Ru(bpy)_3]^{3+}$ (bpy = 2,2'-bipyridine) leads to an Fe(IV) $\sigma$-bonded complex, $[(P)Fe^{IV}(R)]^+$ [94], and occurs at a rate which is much lower than that of the second one-electron transfer from $[(P)Fe^{IV}(R)]^+$ to $[Ru(bpy)_3]^{3+}$ to give $[(P)Fe^{IV}(R)]^{2+\bullet}$ (Scheme 4).

Metal-centered oxidation is therefore kinetically more difficult than the porphyrin macrocycle oxidation. The reorganization energies ($\lambda$ in kcal mol$^{-1}$) for the metal-centered oxidation of (OETPP)Fe$^{III}$(R) to [(OETPP)Fe$^{IV}$(R)]$^+$ (R = $C_6H_5$, 3,5-$C_6F_2H_3$, and 2,4,6-$C_6F_3H_2$), (OETPP)Fe$^{III}$($C_6F_5$) to [(OETPP)Fe$^{IV}$($C_6F_5$)]$^+$, and (OEP)Fe$^{III}$(R) to [(OEP)Fe$^{IV}$(R)]$^+$ are 54, 49, and 42 kcal mol$^{-1}$, respectively [93]. Each is significantly larger than reorganization energies for the porphyrin-centered oxidations involving the same series of compounds, i.e. the second electron transfer of (P)Fe(R) for the ligand-centered oxidation, ca 24 kcal mol$^{-1}$, which is comparable with the $\lambda$ value for free base porphyrins [95]. Thus, a large reorganization energy is required for the electron-transfer oxidation of (P)Fe$^{III}$(R) to [(P)Fe$^{IV}$(R)]$^{\bullet+}$. It is interesting to note that the $\lambda$ value (49 kcal mol$^{-1}$) for the electron transfer oxidation of (OETPP)Fe$^{III}$($C_6F_5$) to [(OETPP)Fe$^{IV}$($C_6F_5$)]$^+$ is approximately the same as the $\lambda$ value (47 kcal mol$^{-1}$) for the postulated electron-transfer reduction of [(TPFPP)Fe$^{IV}$=O]$^{\bullet+}$ by *N,N*-dimethylanilines (vide infra) [89].

## 8.2.4 Electron-transfer Pathway of [(P)Fe$^{IV}$=O]$^{\bullet+}$

To clarify the mechanism of reaction of P-450, it is crucial to characterize the reactive intermediates in the rate-determining step. Definitive evidence for an electron-transfer mechanism (**C** in Scheme 2) for the *N*-demethylation of *N,N*-dimethylanilines has been obtained by direct observation of the reduction of the high-valent species responsible for P-450 catalysis [96]. For peroxidase, an oxoferryl porphyrin $\pi$-radical cation, compound **I** ([(P)Fe$^{IV}$=O]$^{\bullet+}$), has been well characterized as the species equivalent to the proposed active intermediate of P-450 [97–103]. Compound **I** of horseradish peroxidase (HRP) can be readily generated by chemical oxidation of HRP [100–103]. The involvement of the electron-transfer process of compound **I** in the oxidation of several amines catalyzed by HRP was

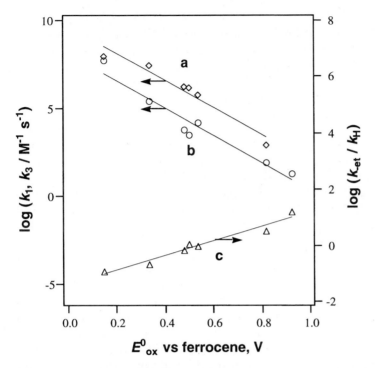

**Figure 1.** Dependence of kinetic values on the oxidation potential of DMA ($E^\circ_{ox}$). (a) ($\Diamond$): bimolecular rate constant of HRP compound **I** with DMA, $k_1$. (b) ($\bigcirc$): bimolecular rate constant of [(TMP)Fe$^{IV}$=O]$^{\cdot+}$ with DMA, $k_3$. (c) ($\triangle$): $k_{-et}/k_H$ determined by use of Eq. 6 [96].

demonstrated by the ESR detection of the corresponding aminium radicals [97–99]. Reactions of compound **I** with 4-X-substituted *N,N*-dimethyl- and *N,N*-bis (trideuteromethyl)anilines (DMA) were monitored spectroscopically by use of a double-mixing stopped-flow technique [96]. When HRP compound **I** and the amines were mixed under stopped-flow conditions in a buffer solution (pH 7.0) at 273 K, rapid formation of compound **II** [(P)Fe$^{IV}$=O] ($k_1$) and the subsequent relatively slow conversion ($k_2$) to the resting ferric state were observed directly. Both $\log k_1$ and $\log k_2$ increase with decreasing $E^\circ_{ox}$ as the electron-transfer process becomes thermodynamically more favorable. A linear correlation between $\log k_1$ and $E^\circ_{ox}$ is shown in Figure 1a. No kinetic isotope effects are observed. Thus, the rate-determining step is clearly electron transfer from DMA to HRP compound **I** and compound **II**.

According the Marcus theory of outer-sphere electron-transfer (Eq. 4), the slope of a linear correlation between $\log k_1$ and $E^\circ_{ox}$ can be derived as given by Eq. 5.

$$\partial(\log k_1)/\partial(E^\circ_{ox}) = -\tfrac{1}{2}(1/2.3RT)(1 + \Delta G^\circ_{et}/\lambda) \tag{5}$$

The slope of the linear correlation between $\log k_1$ and $E^\circ_{ox}$ ($-7.65$) is slightly less negative than the value ($-9.2$) expected from Eq. 5 at $\Delta G^\circ_{et} = 0$

$[\partial(\log k_1)/\partial(E^{\circ}_{ox}) = -(1/2)(1/2.3RT) = -9.2$ at 273 K]. This means that electron transfer from DMA to HRP compound **I** is slightly exergonic, because $\partial(\log k_1)/\partial(E^{\circ}_{ox}) > -(1/2)(1/2.3RT) = -9.2$ at 273 K when $\Delta G^{\circ}_{et} < 0$. This is consistent with the highly positive one-electron reduction potential of HRP compound **I** ($E^{\circ}_{red} = 0.898$ V relative to the NHE, or 0.662 V relative to the SCE) [104]. The unknown value of the reorganization energy ($\lambda$) for electron-transfer reactions from *N,N*-dimethylanilines to HRP compound **I** and the experimental errors involved in estimating the redox potentials in water have precluded complete fit of the rate constants to values calculated from Eqs 3 and 4, although a linear correlation between $\log k_1$ and $E^{\circ}_{ox}$ for a wide range of $E^{\circ}_{ox}$ indicates that a large reorganization energy is required for the electron-transfer reduction of compound **I**, i.e. $|\Delta G^{\circ}_{et}/\lambda| \ll 1$ in Eq. 5 as postulated for electron transfer from *N,N*-dimethylanilines to $[(TPFPP)Fe^{IV}=O]^{\bullet+}$ (vide supra) [89]. A large reorganization energy ($\lambda = 61.9$ kcal mol$^{-1}$) [104] has, in fact, been derived from the electron self-exchange rate between HRP compound **I** and compound **II** ($4.4 \times 10^{-1}$ M$^{-1}$ s$^{-1}$) by applying the Marcus theory of electron transfer (Eq. 4) [91] to the rate of electron transfer from ferrocyanide to HRP compound **I** ($8 \times 10^5$ M$^{-1}$ s$^{-1}$) [105]. An even larger reorganization energy ($\lambda = 78.0$ kcal mol$^{-1}$) [104] was derived from the electron self-exchange rate between HRP compound **II** and ferric HRP ($4.9 \times 10^{-4}$ M$^{-1}$ s$^{-1}$) [104]. The extremely large $\lambda$ value (78.0 kcal mol$^{-1}$) for the metal-centered electron-exchange between HRP compound **II** (Fe$^{IV}$) and ferric HRP (Fe$^{III}$) is consistent with the large $\lambda$ value (83 kcal mol$^{-1}$) derived from the reaction between $[(OEP)Fe^{IV}(R)]^+$ and $(OEP)Fe^{III}(R)$ [93]. The large $\lambda$ value for electron self-exchange between HRP compound **I** and compound **II** ($\lambda = 78.0$ kcal mol$^{-1}$) [104] suggests a significant contribution of the metal-centered reduction of $[(P)Fe^{IV}=O]^{\bullet+}$ to give compound **II**, which should have $(P)Fe^{III}–O^{\bullet}$ character as another resonance form of $(P)Fe^{IV}=O$. This system can be used as an authentic case for the electron-transfer reactivity of compound **I** with *N,N*-dimethylanilines which can be compared with the reactivities of compound **I** of P-450 and of synthetic analogs with the same substrates.

Although characterization of compound **I** of P-450 has remained poor [106], compound **I** of a synthetic model, i.e., $[(TMP)Fe^{IV}=O]^{\bullet+}$ has been well characterized spectrophotometrically [107]. $[(TMP)Fe^{IV}=O]^{\bullet+}$ was prepared by reaction of $(TMP)Fe^{III}(OH)$ with *m*-chloroperoxybenzoic acid (*m*CPBA) in CH$_2$Cl$_2$ at 223 K [193]. The reactions of $[(TMP)Fe^{IV}=O]^{\bullet+}$ with 4-X-substituted *N,N*-dimethylanilines (DMA) in CH$_2$Cl$_2$ at 223 K were monitored by UV–Vis spectral changes of $[(TMP)Fe^{IV}=O]^{\bullet+}$ and found to afford $(TMP)Fe^{III}$ without observation of any intermediates [96]. There is linear correlation between the rate constants ($k_3$) and $E^{\circ}_{ox}$ (Figure 1b), similar to the correlation observed for $\log k_1$ and $E^{\circ}_{ox}$ (Figure 1a). The parallel relationship between these two plots indicates that electron transfer from DMA to $[(TMP)Fe^{IV}=O]^{\bullet+}$ plays an important role in the rate-determining step for the reaction of the P-450 model complex with DMA. Kinetic isotope effects are, however, clearly observed for reactions of $[(TMP)Fe^{IV}=O]^{\bullet+}$ with 4-X-substituted *N,N*-dimethyl- and *N,N*-bis(trideuteromethyl)anilines. The $k_{3H}/k_{3D}$ value increases with increasing $E^{\circ}_{ox}$ from a value close to unity (1.3 for X = OMe) to a large value of 5.9 for X = NO$_2$ [96]. More importantly, there is a large difference between the intermolecular kinetic isotope effects ($k_{3H}/k_{3D}$) and intramolecular isotope effects

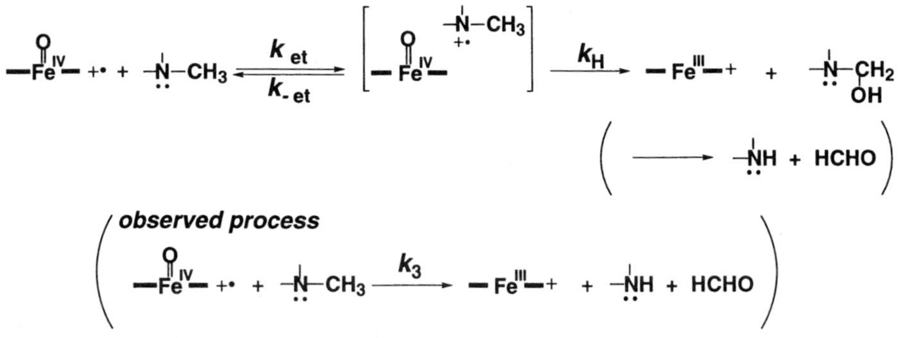

**Scheme 5.**

$(k_H/k_D)$ derived from product analysis (e.g., $k_{3H}/k_{3D} = 1.3$, $k_H/k_D = 3.9$ for $X = OMe$) [96]. The small $k_H/k_D$ values compared with those for proton-transfer reactions of DMA$^{\cdot+}$ [79] indicate that a proton-transfer step is unlikely to be involved. Thus, the rate-determining step might involve a hydrogen atom-transfer step after the initial electron transfer, as shown in Scheme 5 (mechanism **C** in Scheme 2).

If this is so, the product isotope effect reflects the hydrogen-transfer step ($k_H$) from the radical cation of *N,N*-dimethylaniline (DMA$^{\cdot+}$) to (TMP)Fe$^{IV}$=O in competition with the back-electron-transfer step ($k_{-et}$). With decreasing one-electron oxidation potential ($E^\circ_{ox}$) of DMA, equivalent to the one-electron reduction potential ($E^\circ_{red}$) of DMA$^{\cdot+}$, the rate of back-electron-transfer from (TMP)Fe$^{IV}$=O to DMA$^{\cdot+}$ decreases, becoming slower than the hydrogen-transfer step, at which point the electron-transfer step from DMA to [(TMP)Fe$^{IV}$=O]$^{\cdot+}$ becomes the rate-determining step. This might be the reason why the kinetic isotope effect is close to unity for $X = OMe$ ($k_{3H}/k_{3D} = 1.3$) whereas the product isotope effect directly reflects the hydrogen-transfer step ($k_H/k_D = 3.9$ for $X = OMe$). For $X = NO_2$ the hydrogen-transfer step might be the rate-determining step, because of the fast back-electron-transfer whereby the $k_{3H}/k_{3D}$ value (5.9) becomes nearly the same as the $k_H/k_D$ value (6.2). According to Scheme 5, a quantitative relationship can be derived between the kinetic isotope effects ($k_{3H}/k_{3D}$) and the product isotope effects ($k_H/k_D$); this is given by Eq. 6 [96].

$$k_{3H}/k_{3D} = (k_H/k_D)\{(k_D/k_H) + (k_{-et}/k_H)\}/\{1 + (k_{-et}/k_H)\} \qquad (6)$$

When forward electron transfer is the rate-determining step, $k_{-et}/k_H \ll 1$, Eq. 6 is reduced to $k_{3H}/k_{3D} = 1$. On the other hand, when $k_{-et}/k_H \gg 1$, Eq. 6 is reduced to $k_{3H}/k_{3D} = k_H/k_D$. Thus, the difference between $k_H/k_D$ and $k_{3H}/k_{3D}$ should decrease with increasing $k_{-et}/k_H$. The $k_{-et}/k_H$ value can be obtained from the $k_{3H}/k_{3D}$ and $k_H/k_D$ values by use of Eq. 7, which is derived from Eq. 7 [96]. The $k_{-et}$ value for back-electron-transfer from (TMP)Fe$^{IV}$=O to DMA$^{\cdot+}$:

$$k_{-et}/k_H = \{(k_{3H}/k_{3D}) - 1\}/\{(k_H/k_D) - (k_{3H}/k_{3D})\} \qquad (7)$$

is expected to increase with increasing $E^\circ_{ox}$ for DMA, which corresponds to $E^\circ_{red}$ of DMA$^{\cdot+}$, as the free energy change of back-electron-transfer decreases. The hydrogen atom-transfer rate constant ($k_H$) might be rather insensitive to $E^\circ_{ox}$ compared with the back-electron-transfer rate constant ($k_{-et}$). In fact, the $k_{-et}/k_H$ value increases with increasing $E^\circ_{ox}$, as shown in Figure 1c [96]. Thus comparison of the rate constants for reactions of [(TMP)Fe$^{IV}$=O]$^{\cdot+}$ and HRP compound **I** with DMA and analyses of kinetic and product isotope effects clearly indicates that *N*-demethylation of DMA by [(TMP)Fe$^{IV}$=O]$^{\cdot+}$ proceeds via electron transfer from DMA to [(TMP)Fe$^{IV}$=O]$^{\cdot+}$ followed by hydrogen-atom transfer from DMA$^{\cdot+}$ to (TMP)Fe$^{IV}$=O in competition with back-electron-transfer (**C** in Scheme 2). This competition is responsible for the difference between the kinetic and product isotope effects which vary depending on the ratio of the rate constant for back-electron-transfer to that for the hydrogen transfer. The contribution of metal-centered reduction of [(TMP)Fe$^{IV}$=O]$^{\cdot+}$ by DMA would give an iron(III)-oxyl radical character [(TMP)Fe$^{III}$–O$^\cdot$] (vide supra), which would facilitate hydrogen atom-abstraction from DMA$^{\cdot+}$.

### 8.2.5 Electron Transfer Compared with Direct Oxygen Transfer

Comparison of the rate constants of [(TMP)Fe$^{IV}$=O]$^{\cdot+}$ and HRP compound **I** was further extended to reactions with a series of sulfides [108]. HRP is known to convert thioanisole to the corresponding sulfoxide [109], although peroxidases typically catalyze two sequential one-electron oxidations such as one-electron oxidation of phenol derivatives to phenoxy radicals [110]. The yield of the sulfoxide from the stoichiometric reaction of HRP compound **I** with thioanisole is only 25 $\pm$ 5 %. The sulfoxidation involves oxygen-transfer from an oxoferryl species to sulfide, because $^{18}$O in H$_2$$^{18}$O$_2$ has been shown to be incorporated into the product sulfoxide [108, 111]. The initial rapid conversion of compound **I** to compound **II** ($k_1$) by thioanisole is followed by further reduction to the ferric resting state, as is found for reactions with DMA [108, 112, 113]. A linear correlation between $\log k_1$ and $E^\circ_{ox}$ for the reactions of HRP compound **I** with thioanisoles (Figure 2) is readily combined with the relationship for DMA (Figure 1a) into a single, common relationship (Eq. 8) [108].

$$\log k_1 = -10.5 E^\circ_{ox} + 14.6 \tag{8}$$

Such a unified correlation strongly indicates that the reduction of HRP compound **I** by thioanisoles proceeds via electron transfer rather than direct oxygen transfer. Electron transfer from the sulfide to HRP compound **I** in the protein cage might be followed by two competitive processes: (i) oxygen rebound to afford the sulfoxide; and (ii) diffusion of a sulfenium radical from the protein cage to enable the observation of HRP compound **II** as shown in Scheme 6 [108].

In contrast with the reactions of HRP compound **I** in Figure 2, there is no unified correlation for the reactions of a synthetic model [(TMP)Fe$^{IV}$=O]$^{\cdot+}$ with DMA and thioanisoles in plots of $\log k_{obs}$ against $E^\circ_{ox}$ (Figure 3) [108].

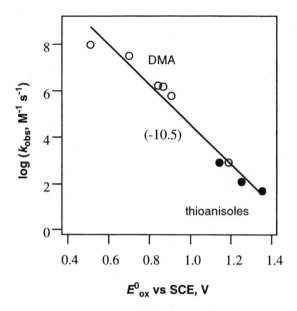

**Figure 2.** Plots of bimolecular rate constants for the reaction of HRP compound **I** with DMAs ($\bigcirc$) and thioanisoles ($\bullet$) against the oxidation potentials of DMAs and thioanisoles ($E^{\circ}_{ox}$). The slope of the line is shown in parentheses [108].

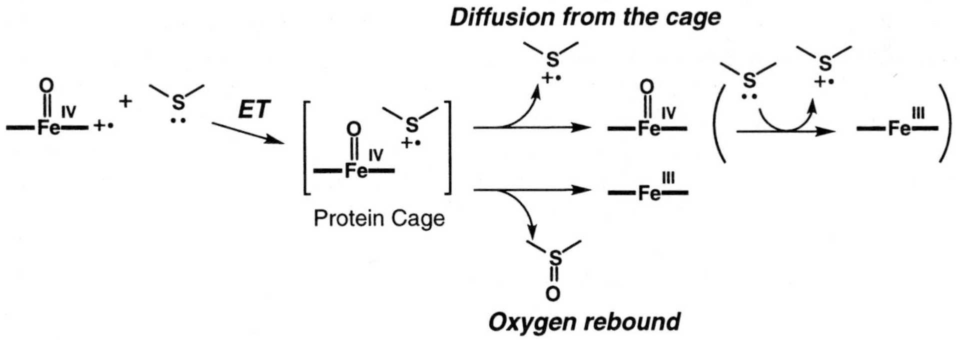

**Diffusion from the cage**

**Protein Cage**

**Oxygen rebound**

**Scheme 6.**

Values of $k_{obs}$ for reactions of $[(TMP)Fe^{IV}=O]^{\bullet+}$ with thioanisoles are at least by two-orders of magnitude larger than those for electron transfer from DMA to $[(TMP)Fe^{IV}=O]^{\bullet+}$. In addition, $k_{obs}$ values of thioanisoles are much less sensitive to $E^{\circ}_{ox}$ values of thioanisoles compared to the large slope observed for the linear correlation between $\log k_1$ and $E^{\circ}_{ox}$ for electron transfer from DMA to $[(TMP)Fe^{IV}=O]^{\bullet+}$. This indicates that the reduction of $[(TMP)Fe^{IV}=O]^{\bullet+}$ by thio-anisoles proceeds via direct oxygen transfer as shown in Scheme 7 rather than the electron-transfer/oxygen-rebound pathway in Scheme 6 [108].

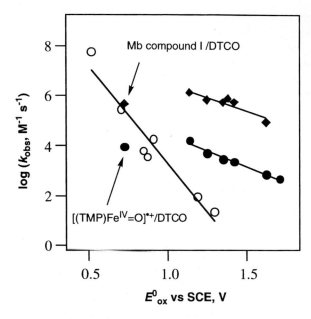

**Figure 3.** Plots of bimolecular rate constants for the reaction of compound **I** against the oxidation potentials, $E°_{ox}$, of DMA and sulfides: $[(TMP)Fe^{IV}=O]^{\bullet+}$ with DMAs (○) and sulfides (●), His64Ser Mb compound 1 with sulfides (◆) [108].

**Direct Oxygen Transfer**

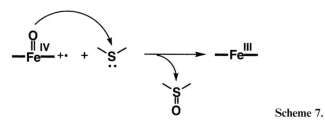

Scheme 7.

The reactions of compound **I** of a sperm whale myoglobin His64Ser mutant (His64Ser Mb) [114] were also examined to compare the reactivity of His64Ser Mb compound **I** toward thioanisoles with that of the synthetic model $[(TMP)Fe^{IV}=O]^{\bullet+}$ [108]. His64Ser Mb compound **I** is readily produced by reaction with *m*CPBA. There is a parallel relationship between the plots for $[(TMP)Fe^{IV}=O]^{\bullet+}$ and His64Ser Mb compound **I**, both of which are far above the electron transfer correlation for the reduction of HRP compound **I** by DMA and thioanisoles [108]. This indicates that the reduction of His64Ser Mb compound **I** also proceeds via direct oxygen transfer in Scheme 7 rather than the electron-transfer/oxygen-rebound pathway (Scheme 6) [108].

Comparison of the linear correlation between $\log k_{obs}$ and $E°_{ox}$ for direct oxygen transfer from $[(TMP)Fe^{IV}=O]^{\bullet+}$ to sulfides with that for the electron transfer from DMA to $[(TMP)Fe^{IV}=O]^{\bullet+}$ in Figure 3 suggests that the relationship for direct oxygen transfer might be included in the relationship for electron transfer when a sulfide with a lower $E^\circ_{ox}$ value is employed for the reaction with $[(TMP)Fe^{IV}=O]^{·+}$. In such a reaction electron transfer from the sulfide to $[(TMP)Fe^{IV}=O]^{·+}$ could become preferable to direct oxygen-transfer and an alternation of the reaction mechanism from direct oxygen transfer to electron transfer would occur, depending on the $E^\circ_{ox}$ value. This supposition was demonstrated by use of 1,5-dithiacyclooctane (DTCO), the $E^\circ_{ox}$ value of which is much lower than those of thioanisoles [108]. The $k_1$ value for the reduction of $[(TMP)Fe^{IV}=O]^{·+}$ by DTCO becomes even smaller than the value expected from the electron-transfer correlation between $\log k_1$ and $E^\circ_{ox}$ in Figure 3 [108]. Such a small $k_{obs}$ value for DTCO compared with the $k_{obs}$ value for DMA at the same $E^\circ_{ox}$ value is consistent with the larger reorganization energy ($\lambda$) expected for electron transfer oxidation producing a $\sigma$ radical cation (DTCO$^{·+}$) than the $\lambda$ values of $\pi$ radical cations (DMA$^{·+}$) [18]. Thus reduction of $[(TMP)Fe^{IV}=O]^{·+}$ by DTCO might proceed via electron transfer from DTCO to $[(TMP)Fe^{IV}=O]^{·+}$ rather than direct oxygen transfer.

Alteration of the reaction mechanism from direct oxygen transfer to electron transfer becomes more evident in the reduction of His64Ser Mb compound **I** by DTCO, because rapid formation of compound **II** was clearly observed; this was followed by the slow conversion of compound **II** to the ferric state [108]. The $k_{obs}$ value for electron transfer from DTCO to His64Ser Mb compound **I** agrees with a linear correlation for reactions of $[(TMP)Fe^{IV}=O]^{·+}$ with DMA via electron transfer (see Figure 3) [108].

In summary, the reduction of HRP compound **I** and $[(TMP)Fe^{IV}=O]^{·+}$ by thioanisoles proceeds via electron transfer whereas the sulfoxidation of thioanisoles with $[(TMP)Fe^{IV}=O]^{·+}$ and His64Ser Mb compound **I** occurs via direct oxygen transfer. When thioanisoles are replaced by a much stronger reductant (DTCO), however, the reactions of both $[(TMP)Fe^{IV}=O]^{·+}$ and His64Ser Mb compound **I** proceed via electron transfer. The heme group of Mb has been shown to be located close to the protein surface [29, 115–117]; it is therefore more accessible for substrates than the heme group of HRP. Thus reactions of His64Ser Mb compound **I** and synthetic model $[(TMP)Fe^{IV}=O]^{·+}$ can proceed via direct oxygen transfer which requires strong interaction between compound **I** and the substrates rather than electron transfer which can occur at a longer distance for the reactions of HRP compound **I**. When a strong electron donor such as DTCO is employed, the electron-transfer pathway can become energetically preferable to direct oxygen transfer; it can proceed at a long separation distance before interacting strongly with the reaction center.

## 8.2.6 'Agostic' Interaction

In the hydroxylation of unactivated C–H bonds in hydrocarbons by cytochromes P-450, a hydrogen atom is thought to be abstracted from the substrate C–H bond by a high-valent iron–oxo intermediate, forming a carbon radical which captures an hydroxyl radical from iron (Scheme 8) [27, 118, 119].

This is called an oxygen-rebound mechanism. Large intramolecular isotope effects ($k_H/k_D = 11$) for substrates with CHD groups [120, 121], C–H bond selectivi-

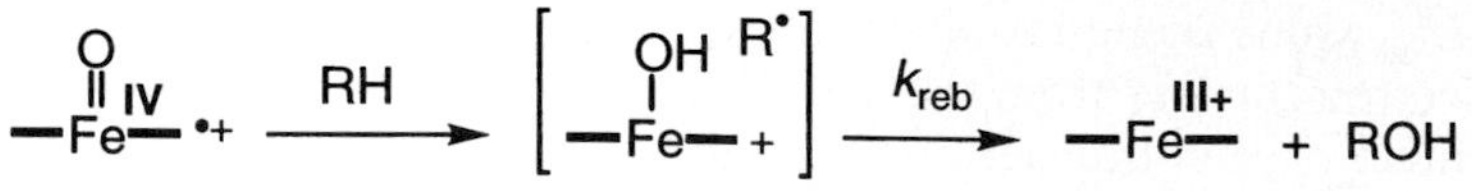

$$\underset{IV}{\overset{O}{\overset{\|}{-Fe-}}}{}^{\bullet+} \quad \xrightarrow{\quad RH \quad} \quad \left[ \overset{OH}{\underset{+}{\overset{|}{-Fe-}}} R^{\bullet} \right] \quad \xrightarrow{\quad k_{reb} \quad} \quad \underset{III+}{-Fe-} \; + \; ROH$$

**Scheme 8.**

ties consistent with those for radical reactions [122] (tertiary > secondary > primary) [123, 124], and loss of stereochemistry in the hydroxylations of norbornane [121, 125] suggest the hydrogen transfer mechanism (Scheme 8). This mechanism was further supported by the detection of rearranged products from carbon radicals in several P-450 oxidations [126]. When, however, the rate constant for the oxygen-rebound step in hydroxylations ($k_{reb}$ in Scheme 8) by a rat liver P-450 isozyme was determined by use of a radical clock substrate [127] widely varying results were obtained ($1.4 \times 10^{10}$–$1.3 \times 10^{13}$ s$^{-1}$) [128–130]. When a hypersensitive radical clock (*trans,trans*-2-*tert*-butoxy-3-phenylcyclopropylmethane) was used to determine the rate constant more accurately it was found that most of the rearranged products were derived from a cationic intermediate produced during the course of the hydroxylation reaction; it was suggested that this process is the origin of the wide variation in the rate constant [131]. The radical species in the hydroxylation reaction was found to have a very short lifetime (70 fs) and it cannot, therefore, be a true intermediate but is part of a reacting ensemble [131]. The small amount of the radical rearrangement product indicates that the rebound step is non-synchronous but concerted with C–H bond cleavage leading to C–O bond formation [131]. The extremely short lifetime of the carbon radical also indicates that the oxygen atom is within bonding distance of carbon at the instant of hydrogen abstraction. This means that a 'side-on' approach of oxygen to the C–H bond is favored over the linear C–H–O arrangement of conventional hydrogen-atom abstraction. In addition, the regioselectivity in the reaction of methylcubane with *t*-BuO$^{\bullet}$ has been shown to be quite different from that with cytochrome P-450 enzymes [132]. This is consistent with a concerted mechanism involving 'side-on' approach to the C–H bond of the substrate.

Such a 'side-on' approach is reminiscent of a three-center two-electron bond, referred to as an agostic interaction [133–136]. An 'agostic' substrate–catalyst complex has recently been proposed as a key intermediate before the oxygen-transfer step in the hydroxylation of cyclohexane by PhIO with (TPFPP)FeCl, as shown in Scheme 9 [137].

Substrates such as H$_2$ which can form strong agostic complexes are expected to inhibit the oxidation of more weakly bound substrates by sequestering the active oxidant (Scheme 9). In fact, the oxidation of cyclohexane was inhibited by the presence of H$_2$ [137]. On removal of the H$_2$ from the reaction system, the rate of cyclohexanol production returns to its original value in the absence of H$_2$, demonstrating that the inhibition is reversible. The pressure at which the rate was half that of the uninhibited reaction ($P_{1/2}$) was 1.4 atm at 0 °C [137]. When H$_2$ was replaced by D$_2$, the $P_{1/2}$ value became smaller (1.0 atm) [137]. Such an inverse isotope effect can be compared with the value (0.78) reported for H$_2$/D$_2$ binding to a tungsten

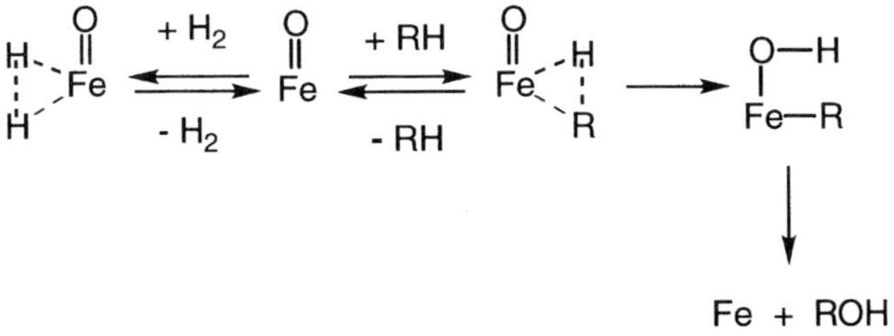

**Scheme 9.**

complex [138]. Methane also inhibits the oxygenation of cyclohexane, possibly because of formation of a complex with $[(TPFPP)Fe^{IV}=O]^{\cdot+}$. It has been shown that even heptane can coordinate to iron(II) with a double A-frame porphyrin and that the metal-to-carbon distances determined by X-ray analysis are 2.5–2.8 Å, typical of moderate to weak agostic interaction [139]. Results from density-functional calculations on iron porphyrin complexes of methane and ethane are indicative of significant charge-transfer character of the complexes (e.g. 0.23 $e$ for ethane) in which the alkane acts as an electron donor [139]. Thus, a charge-transfer-type interaction might play an important role in the 'agostic' substrate-catalyst complex in Scheme 9. Although the importance of charge-transfer complexes as intermediates in a variety of redox reactions has been well documented [83, 140–145], the 'agostic' substrate–catalyst complex proposed in Scheme 9 has yet to be characterized [137].

### 8.2.7 Site of Electron Transfer in Compound I

Electron-transfer reactions between cytochrome c and cytochrome c peroxidase have been studied extensively because of the well-characterized structures and biophysical properties of the reactants [146–150]. It is well known that the resting ferric form of cytochrome c peroxidase is oxidized by hydrogen peroxide to compound **I**, which contains an oxyferryl heme moiety in which the iron atom has a formal oxidation state of 4+ [146–150]. The other is a porphyrin $\pi$ radical cation or organic radical $(R^{\cdot+})$ localized on an amino acid residue of Trp-191 [151–154]; this is formed by transfer of the oxidized equivalent to the amino acid side chain [150]. The site of electron transfer in the reduction of compound **I** has been controversial and two forms of compound **II** have been identified, $(P)Fe^{IV}=O$ containing the oxyferryl heme Fe(IV) [155–158] and $[(P)Fe^{III}]^{\cdot2+}$ containing Fe(III) and the porphyrin $\pi$ radical cation which oxidizes the amino acid side-chain to produce an organic radical $[(P)Fe^{III+}, R^{\cdot+}]$ [159–165] as shown in Scheme 10.

These two pathways have been shown to be affected by ionic strength ($\mu$) [166]. At low ionic strength (below $\mu = 50$ mм) the Fe(IV) site of compound **I** is preferentially reduced, followed by reduction of the radical site in the slow phase of the reaction [166]. At high ionic strength (e.g. $\mu = 200$ mм), the radical site is preferentially reduced, giving a compound **II** that retains the Fe(IV) site [166]. The switch of the site of electron transfer between the Fe(IV) site and the radical site seems to

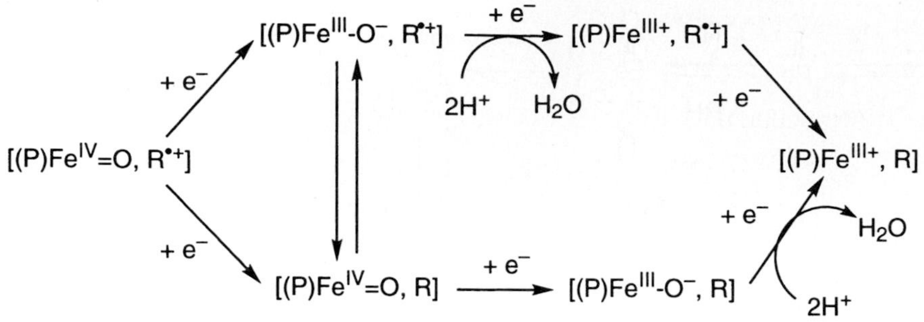

$$[(P)Fe^{III}\text{-}O^-, R^{\bullet+}] \xrightarrow{+ e^-} [(P)Fe^{III+}, R^{\bullet+}]$$

$[(P)Fe^{IV}=O, R^{\bullet+}]$

$[(P)Fe^{III+}, R]$

$2H^+ \quad H_2O$

$$[(P)Fe^{IV}=O, R] \xrightarrow{+ e^-} [(P)Fe^{III}\text{-}O^-, R]$$

$+ e^- \qquad H_2O \qquad 2H^+$

**Scheme 10.**

be controlled by the orientation of the two proteins in the complex, which depends on ionic strength [147]. In yeast cytochrome c-peroxidase complex crystallized at high ionic strength, a short electron transfer pathway from heme to Trp-191 becomes available; there is more intimate contact between cytochrome c and the peroxidase for the complex crystallized at low ionic strength [147]. It was suggested that at low ionic strength the orientation was dominated by electrostatic forces and that at high ionic strength the orientation was dominated by the hydrophobic interaction [166].

### 8.2.8 Photoinduced Electron-transfer Oxidation

The fast oxidation processes of HRP and its functional model peptides [microperoxidase-8 (MP8)] generating the corresponding compound **I** ([(P)Fe$^{IV}$=O]$^{\bullet+}$) and compound **II** [(P)Fe$^{IV}$=O] have been successfully monitored by use of laser-flash photolysis [167, 168]. The photocatalytic system for generation of HRP compound **I** and compound **II** is shown in Scheme 11 [168].

The MLCT (metal to ligand charge transfer) excited state of [Ru(bpy)$_3$]$^{2+}$ ([Ru(bpy)$_3$]$^{2+}$*) is produced by irradiation with visible light; electron transfer from [Ru(bpy)$_3$]$^{2+}$* to the Co$^{III}$ complex gives [Ru(bpy)$_3$]$^{3+}$ [168]. The oxidizing power of [Ru(bpy)$_3$]$^{3+}$ is sufficient to oxidize the ferric resting state of HRP ([(P)Fe$^{III}$]$^+$) at the ligand center to give the porphyrin $\pi$ radical cation ([(P)Fe$^{III}$]$^{\bullet 2+}$). This step is followed by the slower step of conversion of [(P)Fe$^{III}$]$^{\bullet 2+}$ to compound **II**, (P)Fe$^{IV}$=O, accompanied by deprotonation of the coordinated water [168].

The kinetics at 430 nm after a laser flash of the photocatalytic system (pH 8.5) in Scheme 11 are biphasic—a fast reaction on a millisecond time-scale because of formation of [(P)Fe$^{III}$]$^{\bullet 2+}$ was followed by a much slower reaction on a second time-scale because of the conversion of [(P)Fe$^{III}$]$^{\bullet 2+}$ to compound **II**, (P)Fe$^{IV}$=O [168]. In general, ferric porphyrins have ligand-centered one-electron oxidation potentials at which ferric porphyrins are oxidized to ferric porphyrin $\pi$ radical cations; these are higher than oxidation potentials for metal-centered one-electron oxidation to ferryl porphyrins [102, 170]. Despite the smaller (by ca 0.3 eV) driving force for ligand-centered oxidation than for metal-centered oxidation [102, 170], ligand-centered oxidation of HRP occurs before metal-centered oxidation (Scheme 11). This is be-

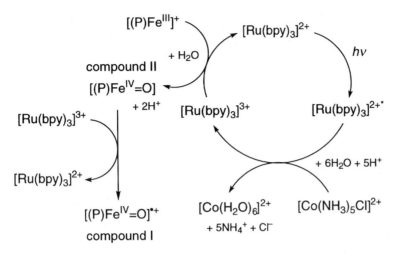

**Scheme 11.**

cause ligand-centered oxidation requires a much smaller reorganization energy for electron transfer and it is, therefore, kinetically favored—as seen in the electron-transfer oxidation of (P)Fe(R) [93]. Only the first process requires $[Ru(bpy)_3]^{3+}$, which was reduced to $[Ru(bpy)_3]^{2+}$. The formation of $[Ru(bpy)_3]^{2+}$ was readily monitored at 450 nm. The rate of formation of $[(P)Fe^{III}]^{\cdot 2+}$ shows first-order dependence on the $[Ru(bpy)_3]^{3+}$ concentration (Eq. 9), whereas conversion to compound **II** is first-order with respect to $[(P)Fe^{III}]^{\cdot 2+}$ concentration (Eq. 10). The rate constant ($k_{et1}$) for electron transfer from $[(P)Fe^{III}]^+$ to $[Ru(bpy)_3]^{3+}$ was determined as $2.5 \times 10^7$ $M^{-1}$ $s^{-1}$ at pH 10.3 [168]. The rate constant ($k_{et2}$) for generation of $(P)Fe^{IV}=O$ at pH 10.8 ($95 \pm 7$ $s^{-1}$) is over 20 times larger than that at lower pH ($4.1 \pm 0.9$ $s^{-1}$) [168].

$$d[[(P)Fe^{III}]^{\cdot 2+}]/dt = k_1[[(P)Fe^{III}]^+][[Ru(bpy)_3]^{3+}] \tag{9}$$

$$d[(P)Fe^{IV}=O]/dt = k_2[[(P)Fe^{III}]^{\cdot 2+}] \tag{10}$$

In acidic solutions ferric MP8 is also rapidly oxidized by $[Ru(bpy)_3]^{3+}$ to a ferric porphyrin $\pi$ radical cation; the rate constant $k_{et1} = 5.6 \times 10^9$ $M^{-1}$ $s^{-1}$ which is much larger than the value for HRP [167]. In alkaline solutions, the oxidation product is ferryl MP8 which is represented as a generalized form, $(P)Fe^{IV}=O$. In this instance, however, transformation of $[(P)Fe^{III}]^{\cdot 2+}$ to $(P)Fe^{IV}=O$ was too fast to be detected [167].

In contrast with HRP, P-450 cannot be oxidized efficiently by external redox agents such as $[Ru(bpy)_3]^{3+}$, because heme is deeply buried in P-450 [169]. By tethering a $[Ru(bpy)_3]^{2+}$ photosensitizer to a protein substrate through a hydrocarbon chain, however, the heme porphyrin $\pi$ radical cation or the $Fe^{IV}$ species which corresponds to compound **II** has been detected successfully upon laser photolysis of the $[Ru(bpy)_3]^{2+}$-tethered P-450 in the presence of $[Co(NH_3)_5Cl]^{2+}$ [169].

HRP compound **II** is further oxidized by electron transfer with $[Ru(bpy)_3]^{3+}$ to afford compound **I** (Scheme 11) [168]. Compound **II** was generated spontaneously after photolysis of the HRP-$[Ru(bpy)_3]^{2+}$-$[Co(NH_3)_5Cl]^{2+}$ system [168]. The rate of formation of compound **I** obeys second-order kinetics, being first order in each reactant, as given by Eq. 11 [168].

$$d[[(P)Fe^{IV}=O]^{\cdot+}]/dt = k_{et}[(P)Fe^{IV}=O][[Ru(bpy)_3]^{3+}] \tag{11}$$

The rate constant for electron transfer from HRP compound **II** to $[Ru(bpy)_3]^{3+}$ is nearly independent of pH. Although the $k_{et}$ value ($1.1 \times 10^8$ $M^{-1}$ $s^{-1}$) is slightly larger than the value for formation of $[(P)Fe^{III}]^{\cdot2+}$ ($2.5 \times 10^7$ $M^{-1}$ $s^{-1}$) at pH 10.3, the $k_{et}$ value ($1.1 \times 10^8$ $M^{-1}$ $s^{-1}$) is significantly smaller than the value ($5.6 \times 10^9$ $M^{-1}$ $s^{-1}$) for the porphyrin ligand-centered oxidation of ferric MP8 by $[Ru(bpy)_3]^{3+}$, despite the higher driving force for electron transfer from HRP compound **II** to $[Ru(bpy)_3]^{3+}$ ($\Delta G^\circ_{et} = -8.1$ kcal mol$^{-1}$ obtained from the $E^\circ_{ox}$ value of HRP compound **I**, which is equal to $E^\circ_{red}$ of HRP compound **II** [104], and the $E^\circ_{red}$ value of $[Ru(bpy)_3]^{3+}$) [171]. When the Marcus theory (Eq. 4) [91] is applied to the $k_{et}$ value ($1.1 \times 10^8$ $M^{-1}$ $s^{-1}$), the $\lambda$ value is evaluated as $\lambda = 30.2$ kcal mol$^{-1}$. If the reorganization energy for electron self-exchange between $[Ru(bpy)_3]^{3+}$ and $[Ru(bpy)_3]^{2+}$ is neglected [93], a large $\lambda$ value (60.4 kcal mol$^{-1}$) is obtained for the electron self-exchange between HRP compound **I** and compound **II**, because the $\lambda$ value for the cross reaction is equal to the average of the $\lambda$ value for electron self-exchange for each reactant. This value is quite consistent with a large reorganization energy ($\lambda = 61.9$ kcal mol$^{-1}$) [104] derived from the electron-transfer rate from ferrocyanide to HRP compound **I** (vide supra). Such a large $\lambda$ value for electron self-exchange between HRP compound **I** and compound **II** and the slower rate than that for the purely ligand-centered oxidation of ferric MP8, despite the larger driving force, is indicative of a significant contribution from the metal-centered oxidation of HRP compound **II** rather than the porphyrin ligand-centered oxidation. This means that HRP compound **II**, formally $(P)Fe^{IV}=O$ should contain a significant amount of $(P)Fe^{III}-O^\cdot$ character, as discussed in the section on P-450 reaction mechanisms (vide supra).

Stepwise conversion of $[(P)Fe^{III}]^{\cdot2+}$ to $(P)Fe^{IV}=O$ has been clearly shown by cofactor reconstitution and use of myoglobin (Mb) bearing a ruthenium complex $[Ru(bpy)_3-Mb]$ (see Scheme 12) [172].

The reconstitution of chemically modified heme with apo-Mb has been well established [173–176]. The one-electron reduction potential of ferryl-Mb ($Fe^{IV}$-heme) was reported as 0.896 V relative to the NHE [177] which is less positive than the one-electron oxidation potential of $[Ru(bpy)_3]^{2+}$ (1.25 V). Thus, generation of ferryl-Mb by electron transfer from the met-Mb (ferric state) moiety to the $Ru^{3+}$ moiety might be thermodynamically favorable. An appropriate amount of a sacrificial electron acceptor such as $[Co(NH_3)_3Cl]^{2+}$ which can quench the $Ru^{2+}$ excited state oxidatively was, however, required to produce the $Ru^{3+}$ state in competition with the reductive quenching of the $Ru^{2+}$ excited state by the met-Mb [178, 179]. No direct electron transfer from the met-Mb (ferric sate) moiety to the $Ru^{2+}$ excited state occurs, because this process is thermodynamically disfavored [178, 179].

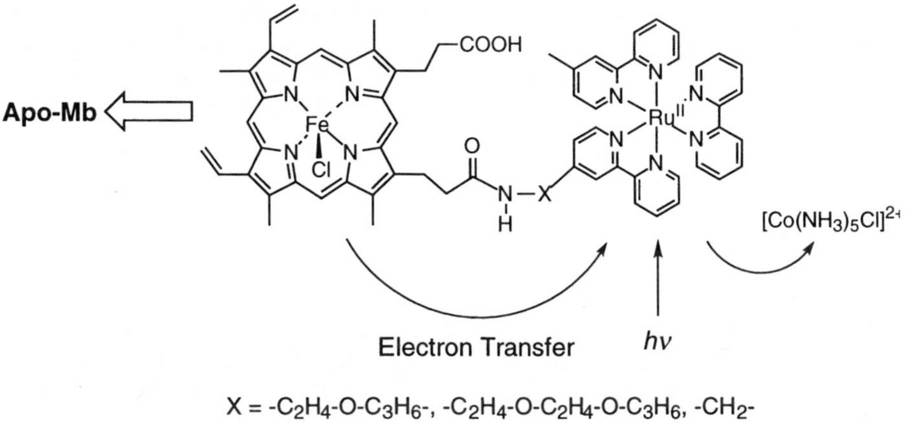

Apo-Mb

[Co(NH$_3$)$_5$Cl]$^{2+}$

Electron Transfer    *hv*

X = -C$_2$H$_4$-O-C$_3$H$_6$-, -C$_2$H$_4$-O-C$_2$H$_4$-O-C$_3$H$_6$, -CH$_2$-

**Scheme 12.**

Steady-state irradiation with visible light of a deaerated aqueous solution containing Ru(bpy)$_3$-met-Mb in the presence of [Co(NH$_3$)$_5$Cl]$^{2+}$ at pH = 7.5 results in the formation of ferric-Mb [172] which has a strong absorption maximum at 408 nm with weak peaks at 505 and 630 nm [180]. No reaction occurred at pH < 6. Laser-flash photolysis study has revealed that the Ru$^{2+}$ excited state is oxidatively quenched by [Co(NH$_3$)$_5$Cl]$^{2+}$ (3 mM) within 0.04 μs in competition with the reductive quenching by met-Mb. Simple bleaching of a Soret band region as a result of formation of the ferric porphyrin π radical cation [181] was detected 1 μs after the laser pulse; this rapid reaction was followed by the appearance of a new peak, from ferryl-Mb, at 426 nm. The rate constants for the first and second stages were determined as $7.1 \times 10^5$ s$^{-1}$, and $4.0 \times 10^4$ s$^{-1}$, respectively [172]. The rate constant for electron transfer from the met-Mb to the Ru$^{3+}$ state (first stage) is independent of pH whereas that for transformation from the ferric porphyrin π radical cation to the ferryl-Mb (second stage) is dependent on pH and the value increased from $4.0 \times 10^4$ s$^{-1}$ at pH 7.5 to $2.0 \times 10^5$ at pH 9.0, as expected from the concomitant deprotonation of coordinated water to generate the ferryl-Mb [172].

### 8.2.9 Oxygen-evolving Complex

Four-electron oxidation of water to oxygen in photosystem II (PSII) is catalyzed by the oxygen-evolving complex (OEC) which contains a tetranuclear manganese cluster as the active site of water oxidation in photosynthetic organisms [182–184]. Extensive efforts have been devoted to developing synthetic models of OEC [185–189], but only rarely have the model complexes actually catalyzed water oxidation [190–195]. The first example of a manganese-based homogeneous catalyst for the oxidation of water is the dimanganese complex of cofacial dimers of triphenylporphyrin linked by an *o*-phenylene bridge, as shown in Scheme 13 [190].

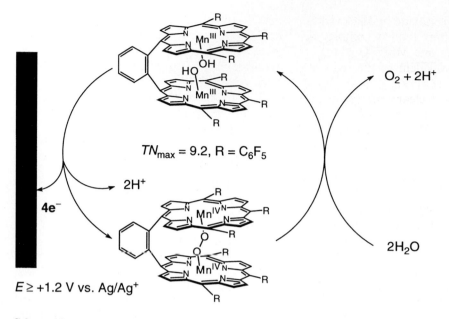

$TN_{max} = 9.2$, R = $C_6F_5$

$E \geq +1.2$ V vs. Ag/Ag$^+$

**Scheme 13.**

Anodic oxidation of an aqueous acetonitrile solution in the absence of the manganese porphyrin dimers did not result in evolution of oxygen at potentials up to +2.3 V (relative to Ag/Ag$^+$), whereas electrochemical oxidation in the presence of the dimer resulted in oxygen evolution at potentials between +1.2 and +2.0 V with a faradaic efficiency of 5–17 % [190]. The maximum turnover number (mol O$_2$/mol catalyst) was 9.2 when a manganese porphyrin dimer with R = C$_6$F$_5$ in Scheme 13 was employed as a catalyst at 2.0 V in the presence of *n*-Bu$_4$NOH [190]. It was confirmed that the rate of oxygen evolution was proportional to catalyst concentration and, on the basis of a labeling experiment using H$_2$$^{18}$O, that the O$_2$ evolved came from H$_2$O [190]. Although the catalytic mechanism remains to be clarified, it is proposed that OH$^-$ might coordinate to each Mn ion of the dimer inside the cavity and that four-electron oxidation at the electrode gives Mn$^{IV}$=O dimer in which the O–O bond is formed efficiently to afford an Mn$^{IV}$–O–O–Mn$^{IV}$ complex because of the proximity of two Mn ions. The peroxy bridge could then be replaced by an OH$^-$ ion, evolving O$_2$ and accompanied by regeneration of the Mn$^{III}$ dimer (Scheme 13) [190].

Similar manganese porphyrin dimers linked by different spacer molecules had catalase activity [196]. The activity was highest when the Mn–Mn separation was ca 4 Å, in agreement with the Mn–Mn separation in the hypothetical Mn$^{IV}$–O–O–Mn$^{IV}$ complex [196]. The high-valent Mn porphyrin dimer has been prepared by the oxidation of the Mn$^{III}$ dimer with *m*CPBA. The Mn$^{IV}$ dimer thus formed was stable for hours at temperatures ranging from −78 to −20 °C and it was detected by ESI MS (electrospray ionization mass spectroscopy) [197]. Cyclooctene was oxidized by *m*CPBA (1.1 equivalent for each Mn ion) with the Mn$^{III}$ dimer to yield the epoxide in 41 % yield on the basis of the amount of *m*CPBA [197]. After the reaction, the Mn complex was recovered in the Mn$^{IV}$ state [197]. This indicates that only one Mn$^{IV}$=O in the dimer was effective in the epoxidation and that another equivalent of the oxidant was consumed in conversion of the Mn$^{III}$ to the Mn$^{IV}$ dimer [197].

The first clear example of an O–O bond-forming reaction involving a di-$\mu$-oxo dimanganese complex has recently been reported for the Mn(III)–Mn(IV) complex, [H$_2$O(terpy)Mn(O)$_2$Mn(terpy)OH$_2$](NO$_3$)$_3$ (terpy is 2,2':6',2"-terpyridine) [194]. The X-ray structure was well-characterized and revealed a di-$\mu$-oxo core with dimensions typical of such complexes (2.7 Å for the Mn–Mn distance) [194]. An ESR spectrum of the di-$\mu$-oxo dimanganese complex contained 16 lines characteristic of Mn(III)–Mn(IV) mixed valence dimers [198]. Water is oxidized to O$_2$ by NaClO in the presence of the di-$\mu$-oxo dimanganese complex. An $^{18}$O-labeling experiment indicated the participation of Mn$^{IV}$=O species in the water oxidation, as shown in Scheme 14. The first step is the formation of a Mn$^V$=O, in which the O atom originates from $^{16}$OCl$^-$. The terminal oxo ligand can rapidly exchange with an $^{18}$O-labeled hydroxide or aqua ligand on the other Mn ion of the di-$\mu$-oxo dimanganese complex. The $^{16}$O-hydroxide/aqua ligand can then exchange rapidly with the solvent, to produce a complex in which where both Mn sites are $^{18}$O labeled, to yield $^{18}$O$_2$ (Scheme 14) [194].

Unfortunately, however, the catalytic reactions end with the complete conversion of the Mn to permanganate after ca 6 h [194]. The detailed mechanism for the catalytic oxidation of water remains to be clarified.

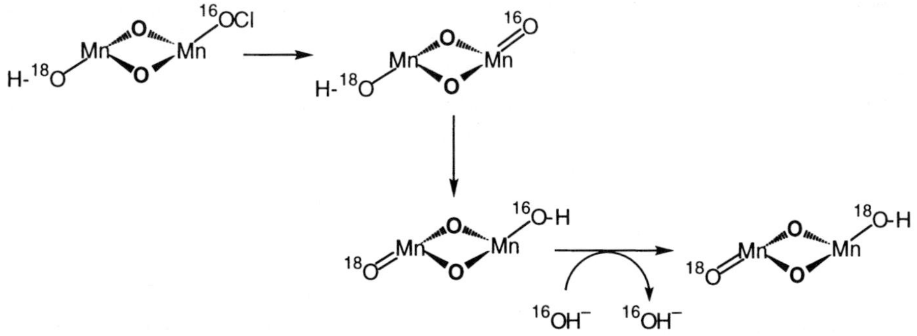

**Scheme 14.**

## 8.3 Catalytic Reduction of Oxygen

The reverse reaction of the oxidation of water to oxygen described above is four-electron reduction of O$_2$ to H$_2$O by cytochrome c oxidases; this is the key step in the respiratory storage of metabolic energy as ATP (adenosine triphosphate) [199–203]. The most important feature of cytochrome c oxidases is that they do not release one-electron or two-electron reduced products ($O_2^-$ and $H_2O_2$, respectively) both of which would be toxic and devastating to the cell [199–203]. To understand the catalytic mechanism of the four-electron reduction of $O_2$ considerable efforts have long been made to prepare synthetic analogs of the active center of cytochrome c oxidases [204–207]. Studies on cytochrome c oxidases and the model complexes have recently been stimulated by the publication of X-ray structures of cytochrome c oxidases [208–212]. Cytochrome c oxidase has a heteronuclear heme $a_3$/$Cu_B$ metal center in which a myoglobin-type iron center and a copper atom ($Cu_B$) coordinate to three imidazoles from histidine residues on the distal side [208–212]. A Cu(II) binding site has recently been constructed successfully by use of cytochrome c peroxidase mutant (Arg48His, Trp51His, His52Ala, Ser81His) from yeast mitochondria [213]. Two other metal sites, a coordinately saturated heme and a dinuclear $Cu_2$ center are involved in storing and transferring the electrons during the four-electron reduction of $O_2$ [208–212]. Although many issues still remain to be re solved, particularly the exact role of Cu in the catalytic cycle, ongoing research on cytochrome c oxidase model systems has provided valuable mechanistic insight into the catalytic mechanism of the four-electron reduction of $O_2$ as described below.

### 8.3.1 Two-electron Reduction

Even simple metalloporphyrins such as [(TPP)M]$^+$ where M = Co, Fe and Mn can enable reduction of $O_2$ to $H_2O$ by ferrocene derivatives (Fc) used as an electron donor in the presence of $HClO_4$ in acetonitrile [214, 215]. However, the rate-determining step is the two-electron reduction of $O_2$ to $H_2O_2$ (Scheme 15) [214, 125]. Nonetheless, the catalytic effect of [(TPP)M]$^+$ for the reduction of $O_2$ by Fc is remarkable, because the oxidation of ferrocene by $O_2$ hardly occurred in the presence of $HClO_4$ without [(TPP)M]$^+$ [216]. The rate-determining step for the [(TPP)M]$^+$-catalyzed two-electron reduction of $O_2$ to $H_2O_2$ has been shown to be the initial electron transfer from Fc to [(TPP)M]$^+$ as shown in Scheme 15 [215]. The reduced metalloporphyrin, (TPP)M, is rapidly oxidized by acid-catalyzed electron-transfer reduction of $O_2$ by Fc, regenerating [(TPP)M]$^+$ via formation of the putative hydroperoxo complex (Scheme 15) [215].

The rate constants ($k_{et}$) of electron transfer from Fc to [(TPP)M]$^+$ agree well with those evaluated in light of the Marcus equations [91] for outer-sphere electron transfer (Eq. 4) [215]. Such agreement clearly demonstrates that electron transfer from Fc to [(TPP)M]$^+$ in Scheme 15 proceeds via an outer-sphere pathway. In contrast to this, the $k_{et}$ value of the acid-catalyzed electron transfer from (TPP)Co to $O_2$ is $10^9$-fold larger than that expected from an outer-sphere electron transfer [215]. Such huge enhancement of the observed rate relative to that calculated for outer-sphere electron transfer indicates the strong inner-sphere nature of acid-catalyzed electron transfer from (TPP)Co to $O_2$; this should result in formation of the hydroperoxo complex, [(TPP)CoO$_2$H]$^+$ (Scheme 15, M = Co). Other metalloporphyrins (M = Fe and Mn) can also act as efficient catalysts of the reduction of

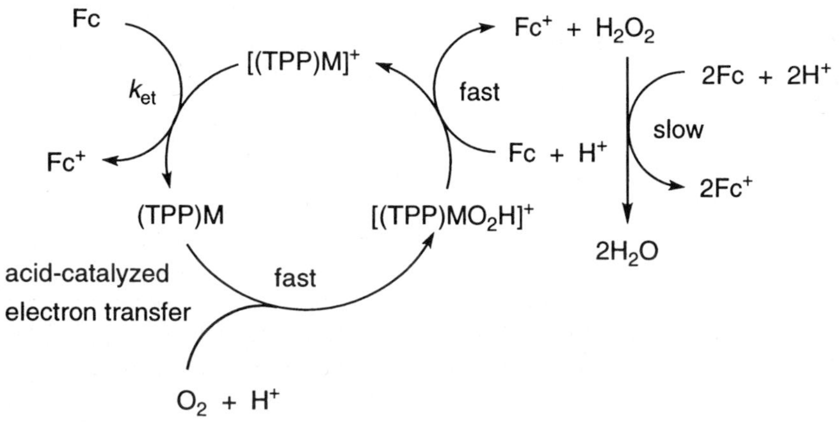

**Scheme 15.**

$O_2$ by Fc in the presence of $HClO_4$ in acetonitrile. The inner-sphere nature of the acid-catalyzed electron transfer from (TPP)M to $O_2$ is essential to the catalytic cycle. Otherwise $[(TPP)M]^+$ would act solely as an electron mediator and no acceleration of the overall electron transfer from Fc to $O_2$ would be achieved.

An NADH analog, 10-methyl-9,10-dihydroacridine (AcrH$_2$) can also reduce $O_2$ to $H_2O$ in the presence of a catalytic amount of $[(TPP)M]^+$ (M = Co, Fe) in the presence of $HClO_4$ in MeCN [217, 218]. In this instance also the two-electron reduction of $O_2$ to $H_2O_2$ is the rate-determining step. The catalytic two-electron reduction of $O_2$ by AcrH$_2$ also proceeds via electron transfer from AcrH$_2$ to $[(TPP)M]^+$ in the presence of $O_2$ and $HClO_4$ to produce a radical pair (AcrH$_2^{\bullet+}$ $[(TPP)CoO_2H]^{\bullet+}$) followed by hydrogen transfer from AcrH$_2^{\bullet+}$ to $[(TPP)MO_2H]^{\bullet+}$ to yield 10-methylacridinium ion (AcrH$^+$) and $H_2O_2$, accompanied by regeneration of $[(TPP)M]^+$ (Scheme 16) [218].

The kinetic isotope effect $(k_H/k_D = 7.1)$ observed when AcrH$_2$ is replaced by AcrD$_2$ is ascribed to this hydrogen-transfer step [218].

To achieve the four-electron reduction of $O_2$ to $H_2O$ without releasing $H_2O_2$ another metal ion such as Cu ion in cytochrome c oxidase is required to stabilize the two-electron reduced peroxy species by forming a bimetallic peroxo complex. This

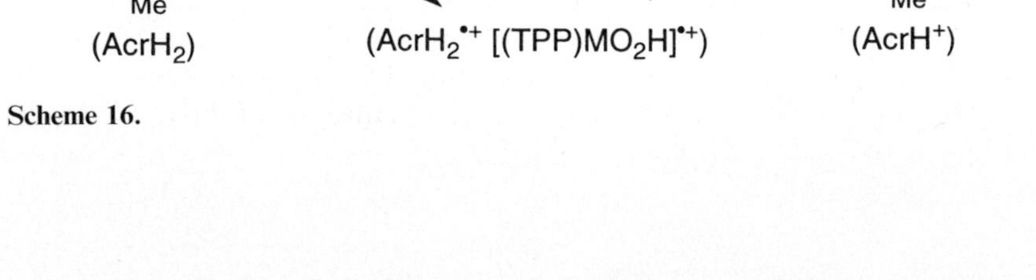

**Scheme 16.**

would make the release of peroxide unfavorable, resulting in the direct four-electron reduction of $O_2$ to $H_2O$. The initial electron transfer to $O_2$ is also accelerated by the interaction of $O_2^-$ with a metal ion which can act as a Lewis acid. Such catalysis of metal ions to accelerate the electron-transfer reduction of a variety of substrates, including oxygen, has been well documented [219, 220]. The existence of a strong interaction between $O_2^-$ and $Sc^{3+}$ has been demonstrated by observation of the ESR spectrum of the $O_2^-$–$Sc^{3+}$ complex [221]. The $O_2^-$–$Sc^{3+}$ complex is generated by photoinduced electron transfer from the dimeric 1-benzyl-1,4-dihydronicotinamide [(BNA)$_2$], which can act as a unique two-electron donor [222, 223], to $O_2$ [221]. The clear eight-line isotropic spectrum at the center is ascribed to the superhyperfine coupling of $O_2^-$ with the 7/2 nuclear spin of the scandium nucleus ($a_{Sc} = 3.82$ G) [221]. The isotropic $g$ value (2.0165) is appreciably smaller than the average value (2.030) [224] of the principal three $g$ components of $O_2^-$ at 77 K, being consistent with spin delocalization to the scandium nucleus, as demonstrated by observation of the superhyperfine coupling. The two inequivalent $a(^{17}O)$ values (21 and 14 G) observed for the $O_2^-$–$Sc^{3+}$ complex containing $^{17}O$ are fully consistent with 'end-on' coordination form of $^{\cdot}O–O^-$–$Sc^{3+}$ in which the electron spin is more localized at the terminal oxygen [221].

## 8.3.2 Four-electron Reduction

Several dimeric metalloporphyrins [225–230] have been used to achieve four-electron reduction of $O_2$ to $H_2O$. The binding of $O_2$ to the metal center inside the cavity of two porphyrins seems to be essential for the four-electron reduction of $O_2$ to $H_2O$ as shown in Scheme 17 [230].

When $O_2$ binds the Co(II) center of the bisporphyrin inside the cavity to produce the *endo* complex (path **a**), the superoxo complex interacts strongly with the Sc(III) ion, leading to the further reduction to $H_2O$ via the corresponding peroxo complex in which the interaction with Sc(III) becomes much stronger. Heterolytic cleavage

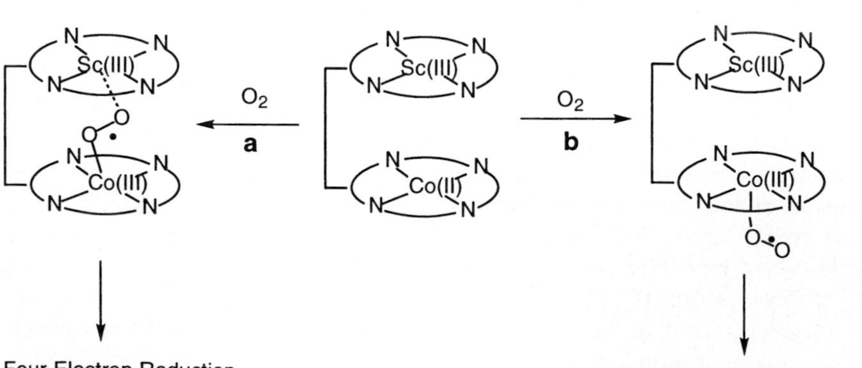

Four-Electron Reduction                                    Two-Electron Reduction

**Scheme 17.**

of the O–O bond can then occur, followed by a further two-reduction to yield $H_2O$. When the Co(II) center binds $O_2$ outside the cavity, however, there is no interaction of the superoxo complex with the Sc(III) ion (path **b**). Then protonation predominates, followed by the further one-electron reduction to yield $H_2O_2$.

Several heterodinuclear Fe–Cu or Co–Cu complexes that closely resemble the native enzyme active sites have recently been prepared to elucidate the catalytic mechanism of cytochrome c oxidases [231–245]. The use of a covalently attached axial ligand seems essential to achieve efficient electroreduction of $O_2$ to $H_2O$ [235, 238, 241–243]. The closest structural analogs of the heme $a_3/Cu_B$ active site of cytochrome c oxidases reported so far are Fe–Cu complexes (**1** and **2**) in which the Cu coordination site is provided by three imidazole ligands [242]. These biomimetic model complexes afford clean electroreduction of $O_2$ to $H_2O$ over a wide range of pH with no leakage of $H_2O_2$ [243].

The acetamide linkages in **1** and **2** provide enough flexibility to the distal imidazole ligands, and they are short enough to keep Cu–Fe distance within the 4.5–5.5 Å range, but not to coordinate to the iron atom [242]. Oxygen binds irreversibly with **1** or **2** to form a putative peroxo-$Fe^{III}(O^{2-})Cu^{II}$ intermediate, which is too unstable to be characterized [243].

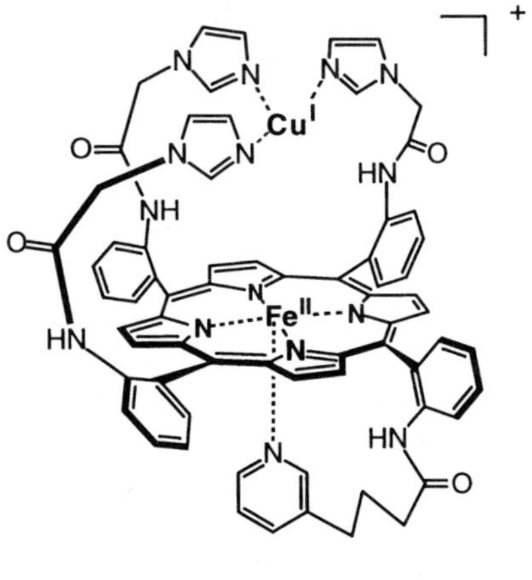

1

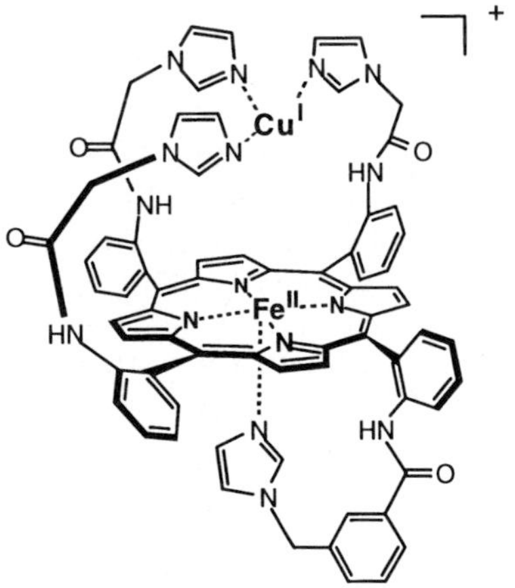

2

Extensive efforts have been made to identify the active oxygen intermediates during the catalytic reduction of $O_2$ to $H_2O$ by cytochrome c oxidases, although the characterization of the active oxygen intermediates remains to be fully understood [246–248]. A stable peroxo $Fe^{III}–Cu^{II}$ complex (**4**) has been reported to be formed by the reaction of a $Fe^{II}–Cu^{I}$ complex (**3**) with $O_2$ (Scheme 18); this complex was characterized spectroscopically [237].

The ESI mass spectrum in MeCN gives a peak at $m/z = 1094$ and the isotope pattern expected from a monocation **4** [237]. The observed Raman band indicates a peroxo O–O stretching vibration at 803 $cm^{-1}$ which is shifted to 759 $cm^{-1}$ on use of $^{18}O_2$ ($\Delta^{16}O_2/^{18}O_2 = 44$ $cm^{-1}$) [237, 247]. The peroxo complex **4** is ESR silent,

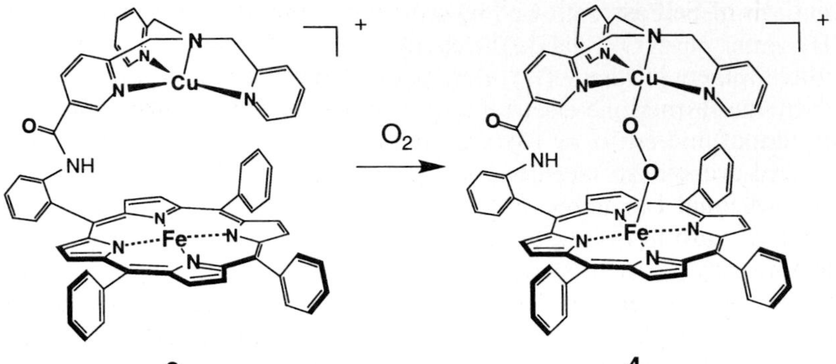

**Scheme 18.**

which is indicative of strong magnetic coupling between the two metals [237]. These results suggest that the peroxo ligand in **4** is bridging in end-on fashion, as shown in Scheme 18, rather than side-on fashion [237]. However, a side-on coordination mode is also possible for other Cu–Fe bimetallic peroxide complexes with different ligands [241, 244]. The magnetic properties and reactivities of Cu–Fe bimetallic peroxide complexes have been shown to be altered drastically by subtle changes of ligand architecture [244]. The electron-transfer properties of this new type of $O_2$ adduct remain to be clarified.

## 8.4 Electron Transfer in Organized Media

When porphyrins and metalloporphyrins are placed with suitable redox couples in organized media, biomimetic electron-transfer systems involving the redox-active porphyrins can be constructed to achieve catalytic redox reaction and vectorial electron transport [249–254]. The biomimetic examples include Langmuir–Blodgett films [255–265], lipid bilayer membranes [266–284], micelles [285–292], vesicles [293–298], self-assembled monolayers [299–339], and others [340, 341]. The co-valent attachment of a porphyrin to a donor and/or an acceptor enhances the electronic coupling between the porphyrin and the redox couples through the bridge, with the result that intramolecular electron transfer predominates over an intermolecular electron transfer. Thus, incorporation of the porphyrin-containing dyads and the more sophisticated systems into the organized media seems to be promising for the realization of molecule-based catalysis and artificial photosynthetic materials in which the occurrence of efficient electron transfer is the heart of the strategy. This section focuses on self-assembled monolayers of porphyrin-containing systems, which have been developed intensively during the recent years, although a variety of important studies from other areas is also reported.

### 8.4.1 Electrocatalysis of Self-assembled Monolayers of Porphyrins

Self-assembled monolayers have recently attracted much attention as a new methodology for molecular assembly [249, 342]. They enable highly organized chemical binding of molecules of interest to the surfaces of, e.g., metals, semiconductors, and insulators. The well-ordered structure of self-assembled monolayers is in sharp contrast with conventional Langmuir–Blodgett films and lipid bilayer membranes in terms of stability, uniformity, and manipulation. Functional molecules can be arranged unidirectionally at the molecular level on substrates when substituents which will self-assemble on the substrates are attached to a terminal of the molecules. The wide variety of examples reported to date include porphyrins and metalloporphyrins in self-assembled monolayers [299–339].

One of the first reported self-assembled monolayers of porphyrins is cobalt(II)-porphyrins **5** and **6** and the analog free-base porphyrins **7** and **8**, in which the number and location of the thiol-containing tails are varied systematically [300, 314].

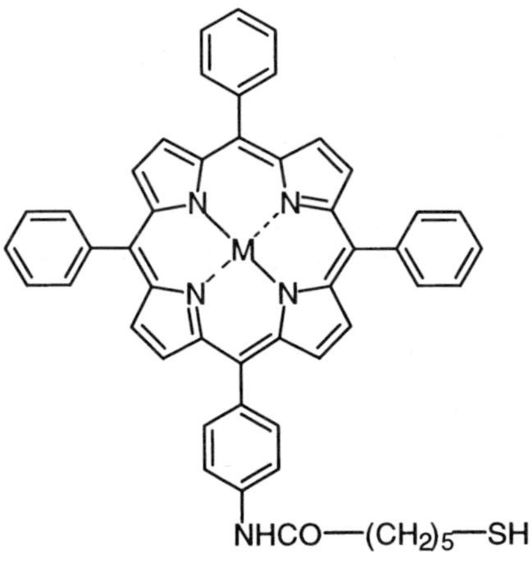

NHCO—(CH$_2$)$_5$—SH

**5: M=Co**
**7: M=H$_2$**

HS                SH

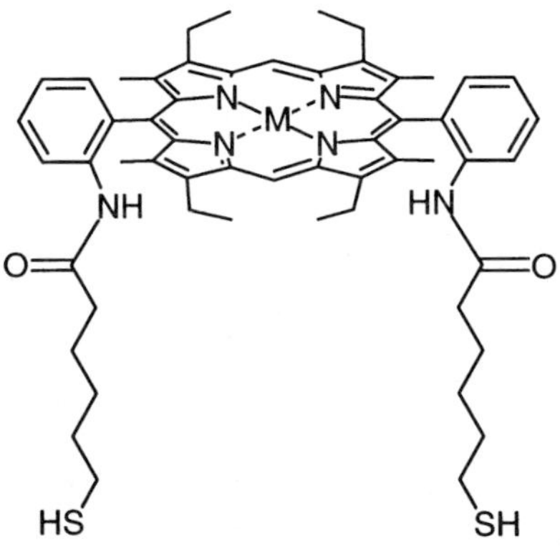

**6: M=Co**
**8: M=H$_2$**

Electrochemical data indicate that self-assembled monolayers of **5** and **6** catalyze the two-electron reduction of O$_2$ to H$_2$O$_2$. The monolayer from **6** is a more effective electrocatalyst for the reduction of O$_2$ than that from **5** [300]. The different reactivity results from different interfacial architecture; this is confirmed by infrared, X-ray photoelectron, and visible spectroscopic measurements [300] which revealed coplanar, inclined π–π stacking of the porphyrin ring in the monolayer of **5** and head-to-tail orientation of the porphyrin ring in the monolayer of **6**. Treatment of the monolayer of **8** with Co(OAc)$_2$ in methanol resulted in electrocatalytic activity in the reduction of O$_2$ [300]. In contrast, a monolayer of **7** treated similarly failed to catalyze dioxygen reduction [300], although treatment of a mixed monolayer of **7** and CH$_3$(CH$_2$)$_3$SH with Co(OAc)$_2$ results in electrocatalytic activity similar to that of **6**.

A tetraphenylporphyrin derivative with four thiol moieties **9** and a similar bis(cobalt) cofacial diporphyrin with three thiol moieties **10**, prepared independently at the same time, electrocatalytically reduce $O_2$ on a variety of gold surfaces [301, 316].

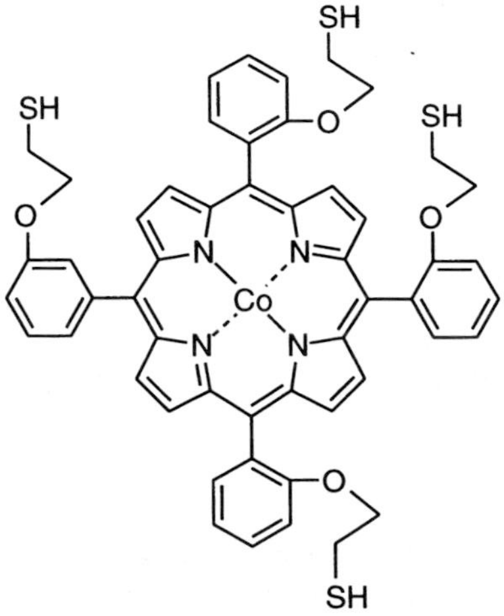

**9** 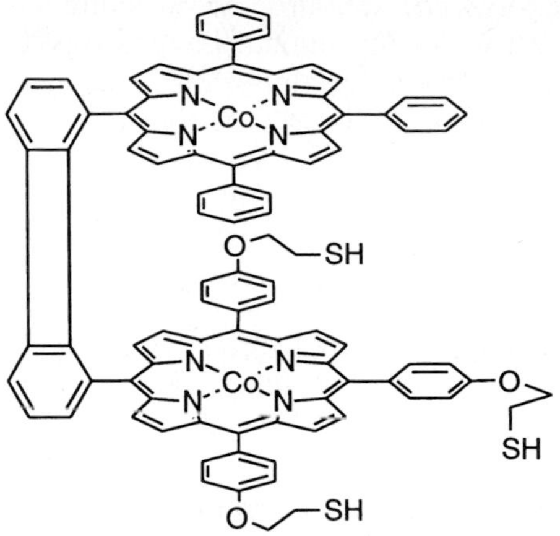          **10**

Cyclic voltammetric and X-ray photoelectron spectroscopic measurements showed the formation of nearly one monolayer where the porphyrin rings with thiol groups are coplanar to the electrode surface through two or three S–Au linkages per porphyrin [301, 316]. The value ($\Gamma$) of surface coverage for **9** was determined to be $7.2$–$8.6 \times 10^{-11}$ mol $cm^{-2}$ ($190$–$230\,\text{Å}^2$ molecule$^{-1}$). This value is quite consistent with a preference for orientation of the porphyrin ring coplanar with the gold surface ($262\,\text{Å}^2$ molecule$^{-1}$). Self-assembled monolayers of **9** and **10** afforded two-electron reduction of $O_2$ to $H_2O_2$. This is in sharp contrast with the electrocatalytic activity of **10** chemisorbed on an edge-plane pyrolytic graphite electrode in which the four-electron reduction of $O_2$ was achieved [225–227, 343–346]. These results indicate that the edge-plane graphite surface plays a crucial role in biomimetic catalysis of the four-electron reduction of $O_2$.

Self-assembled monolayers of metalloporphyrins **11–16**, similar to **5** and **7**, have also been constructed on gold surfaces [307, 335]. The $\Gamma$ value of surface coverage was estimated to be $3.8 \times 10^{-10}$ mol $cm^{-2}$ ($44\,\text{Å}^2$ molecule$^{-1}$), suggesting closely packed structures of the porphyrin with perpendicular orientation to the gold surface [335]. The insertion of the metal ion into a self–assembled monolayer of free base porphyrin **17** on the gold surface has been performed by refluxing the metal ion solution in which the modified gold electrode was immersed [335].

The extent of metal insertion is strongly dependent on the type of metal ion. Cobalt (II), nickel (II), and zinc (II) are readily incorporated into the porphyrin ring, whereas manganese (II), iron (II), and copper (II) are difficult to insert under the same experimental conditions [335]. Self-assembled monolayers of **12** and **13** have electro-catalytic activity in the reduction of $O_2$ to $H_2O$ and $H_2O_2$, respectively.

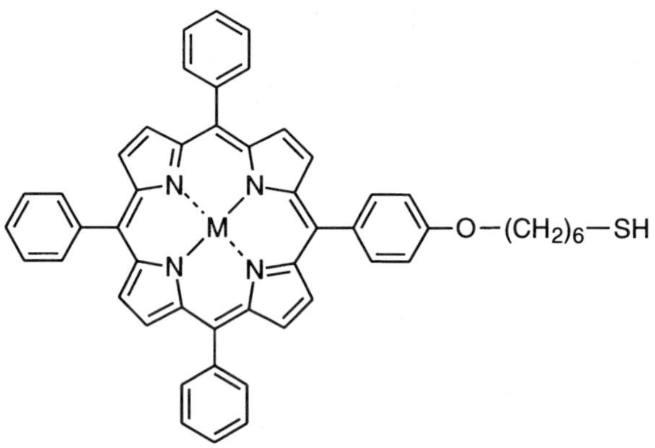

**11:** M=Mn(II); **12:** M=Fe(II); **13:** M=Co(II); **14:** M=Ni(II); **15:** M=Cu(II); **16:** M=Zn(II); **17:** M=$H_2$

Although mechanistic aspects of electron transfer in these systems remains ambiguous, catalysts self-assembled on the gold surface are amenable to the surface analysis as compared with those immobilized on conventional electrodes. More detailed studies including elaborate molecular design are required to elucidate the relationship between the interfacial structure and the electrocatalytic function.

### 8.4.3 Photoinduced Electron Transfer of Metalloporphyrins in Self-assembled Monolayers

Photoinduced electron-transfer reactions on gold electrodes modified with porphyrins and metalloporphyrins have been extensively studied to enable the development of molecular devices for solar-energy conversion and of photonic sensors [302, 305, 309–312, 318–320, 323, 325, 327–329, 331–334, 336, 338, 339]. Systematic studies on the structure and photoelectrochemical properties of the self-assembled monolayer of porphyrins **18** on a gold electrode have been reported to clarify the effects of the spacer length [302, 325, 338]. In the molecular design of **18**, six *t*-butyl groups were introduced into *meso*-phenyl rings of the porphyrin moiety to increase solubility in organic solvents and to suppress the quenching of the excited states of the porphyrins in the monolayers because of the porphyrin aggregation (Scheme 19) [347].

The structure of the self-assembled monolayers was investigated by use of ultraviolet–visible absorption spectroscopy in transmission mode, cyclic voltammetry, ultraviolet–visible ellipsometry, and fluorescence spectroscopy [338]. These experiments showed that as the length of the spacing methylene chain was increased the self-assembled monolayers tended to form highly ordered structures on the gold electrode, with surface coverage reaching $1.5 \times 10^{-10}$ mol cm$^{-2}$ (110 Å$^2$ molecule$^{-1}$) [338]. Adjacent porphyrin rings take J-aggregate-like partially stacked struc-

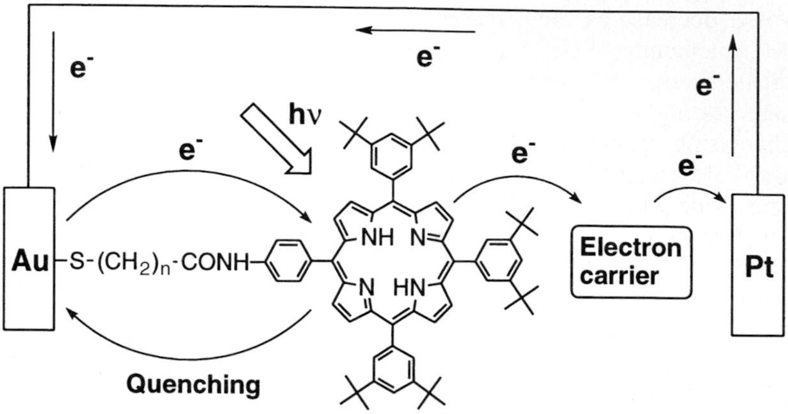

**18:** n=1-7,10,11

**Scheme 19.**

tures in the monolayer [338]. The porphyrin ring plane in the monolayer with an even number ($n = 2, 4, 6, 10$) of methylene spacers ($-(CH_2)_n-$) is tilted significantly to the gold surface, whereas porphyrins with odd numbers ($n = 1, 3, 5, 7, 11$) of methylene spacers adopt nearly perpendicular orientation to the gold surface [338].

Photoelectrochemical studies of gold electrodes modified with **18** were performed in an argon-saturated $Na_2SO_4$ aqueous solution containing methyl viologen as an electron carrier and using a platinum wire counter electrode and a Ag/AgCl reference electrode [338]. An increase in the cathodic photocurrent was observed on increasing the negative bias (from 700 mV to $-200$ mV) relative to the gold electrode [338]. This indicates that the vectorial photocurrent flows from the gold electrode to the counter electrode through the monolayer and the electrolyte. The quantum yield for photocurrent generation increases in a zigzag fashion with increasing spacer length up to $n = 6$ (ca 0.3 %) and then starts decreasing with a further increase in the spacer length. Such dependence of the quantum yield on spacer length can be explained by competition between electron transfer and energy transfer of the singlet excited state of the porphyrin as shown in Scheme 19.

Photoirradiation of the modified electrode results in electron transfer from the singlet excited state of the porphyrin ($E^{\circ}_{ox} = -0.80$ V ($n = 11$) relative to Ag/AgCl) to the methyl viologen ($E^{\circ}_{red} = -0.62$ V relative to Ag/AgCl) or $O_2$ ($E^{\circ}_{red} = -0.48$ V relative to Ag/AgCl). The reduced electron carrier diffuses to release an electron to the platinum electrode, whereas the resultant porphyrin radical cation ($+1.10$ V relative to Ag/AgCl) captures an electron from the gold electrode, generating the cathodic current flow. There is, however, a competitive deactivation pathway in the excited porphyrin, via energy transfer quenching by the gold electrode, judging from the extremely short fluorescence lifetimes of the porphyrins (ca 10–40 ps) on the gold surface compared with those (ca 1–10 ns) on quartz or semiconductor surfaces [347–351]. Electronic coupling between the porphyrin and the electrode would decrease as the number of methylene spacers was increased, leading to the less quenching of the excited porphyrin by the electrode via energy transfer through the spacer [352–356]. The rate of electron transfer from the gold electrode to the resulting porphyrin radical cation should, on the other hand, decrease with increasing spacer length. Thus, the offset effect of suppression of quenching by the gold electrode and reduced electron-transfer between the porphyrin and the electrode with increasing methylene spacer length is responsible for the nonlinear dependence of the quantum yield on the spacer length (vide supra). These results clearly show that optimization of each process is vital to achieving highly efficient sequential electron transfer by organized molecular systems used to mimic vectorial electron flow in photosynthesis.

The surface plasmon resonance technique has been applied to the characterization of the monolayer thickness of a self-assembled monolayer and to measurement of protein interactions with a self-assembled monolayer on the metal surface [357, 358]. Surface plasmon excitation has also been used as an effective excitation source to cause photoinduced electron transfer from the singlet excited state of the porphyrin in a self-assembled monolayer to electron carriers such as methyl viologen and $O_2$ in the electrolyte solution, leading to photocurrent generation [309, 327, 328, 336]. The surface plasmon was generated with p-polarized 632.8 nm light at an incident angle of 73° on a porphyrin (**19**)-modified gold surface of BK-7 right-angle prism by an attenuated total reflection method and by use of a Kretschmann–Raether configuration as shown in Scheme 20 [327, 328, 336].

The photoelectrochemical measurements were conducted in an oxygen-saturated 0.1 M aqueous solution of $Na_2SO_4$ using the **19**-modified gold working electrode, a platinum counter electrode, and an Ag/AgCl reference electrode [328]. The intensity of the cathodic photocurrent under the surface plasmon excitation is enhanced by a factor of ca 6 compared with direct photoirradiation [325, 338]. Such an enhancement of the photocurrent indicates that it is produced via the generation of surface plasmon and excitation of the porphyrin immobilized on the gold surface. These

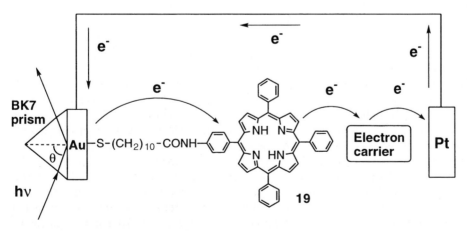

**Scheme 20.**

results demonstrate that the surface plasmon resonance can be used to develop molecule-based photonic materials.

### 8.4.4 Self-assembled Monolayers of Porphyrin-containing Dyads

The importance and complexity of electron-transfer reactions in photosynthesis have inspired many synthetic chemists to design and prepare donor–acceptor linked systems that mimic the highly efficient multistep electron-transfer process [17, 334]. Because porphyrins and metalloporphyrins are essential components in photosynthetic electron transfer, they have been frequently employed in donor–acceptor linked molecules [17, 334]. Some covalently linked porphyrin-containing arrays, e.g. triads, tetrads, and pentads, produce a long-lived, charge-separated state with a high quantum yield, demonstrating that a sequential electron transfer is a pivotal strategy for the construction of artificial photosynthetic systems [17, 334]. Efficient conversion of light to photocurrents or chemical energies via the charge-separated state has, however, often been hampered by the poor unidirectional electron flow in covalently linked molecules incorporated into artificial membranes such as lipid bilayers and Langmuir–Blodgett membranes and/or organized at electrodes [258, 259, 261–265, 268, 271, 294, 295].

The spherical shape of buckminsterfullerene ($C_{60}$) containing a large number of $\pi$-electrons makes this carbon allotrope an ideal component for construction of efficient electron-transfer systems. Recent advances in synthetic methodology have made it possible to link $C_{60}$ with porphyrins [17, 323, 334]. The first porphyrin–fullerene dyad (**20**) in a self-assembled monolayer was reported in 1996 [309, 318, 332]. Because sulfides are relatively stable compared with thiols and disulfides, a methylthio group is attached to the end of a porphyrin ring whereas the opposite end is a $C_{60}$ moiety in **20**. The structures of the self-assembled monolayer formed on a gold electrode were studied by spectroscopic methods including X-ray photoelectron spectroscopy, ultraviolet–visible spectroscopic ellipsometry, ultraviolet–visible absorption spectroscopy, and Fourier transform infrared spectroscopy, as well as electrochemical methods [309, 318, 332]. The experimental data revealed that the porphyrin-$C_{60}$ molecules were tilted and nearly parallel to the gold surface, leading to the formation of loosely packed structures ($\Gamma = 1.6$–$3.7 \times 10^{-11}$ mol cm$^{-2}$). Photoelectrochemical measurements were conducted in an argon-saturated aqueous solution of $Na_2SO_4$, containing methyl viologen as an electron carrier, and using the modified gold electrode as a working electrode and a platinum counter electrode, as shown in Scheme 21 [309, 318, 332].

Under short-circuit conditions the cathodic photocurrent was observed for the photoelectrochemical cell; the quantum yield was ca 0.5 % [309, 318, 332]. The photocurrent intensity in the free-base porphyrin-$C_{60}$ system was enhanced fivefold compared with that in the corresponding free-base porphyrin monomer system. Such enhancement of the photocurrent demonstrates that $C_{60}$ acts an effective mediator in a sequential electron-transfer process. The photocurrent intensity in the free-base porphyrin-$C_{60}$ cell was ca one order of magnitude larger than that in the zinc porphyrin-$C_{60}$ cell [309, 318, 332]. Two different electron-transport mecha-

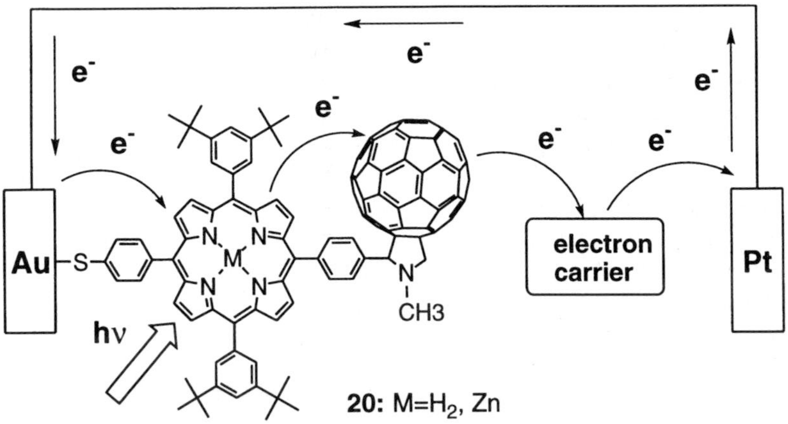

**Scheme 21.**

nisms were proposed for photocurrent generation in self-assembled monolayers of porphyrin-$C_{60}$ systems—ion-pair formation for the zinc porphyrin-$C_{60}$ cell and exciplex formation for the free-base porphyrin-$C_{60}$ cell. The shorter lifetime in the charge-separated state of the former leads to a poor generation of the photocurrent, whereas the longer lifetime in the exciplex of the latter results in a pronounced increase of the photocurrent [309, 318, 332].

The dyad **21** consisting of a porphyrin linked to a thiol-appended benzoquinone as an acceptor was also prepared to construct a self-assembled monolayer system [310, 320]. The amount of **21** adsorbed, ($\Gamma$), was evaluated as $6.2 \times 10^{-11}$ mol $cm^{-2}$ ($270 \, \text{Å}^2$ molecule$^{-1}$) from the amount of charge of the cyclic voltammogram. This occupied area of **21** doesn't agree with that expected for orientation of the porphyrin ring vertical to the electrode ($90 \, \text{Å}^2$ molecule$^{-1}$), indicating the rather poor packing of the molecule at the gold electrode. Photoelectrochemical measurements were performed in a nitrogen-saturated aqueous solution of $Na_2SO_4$ (pH 4.5), containing 50 mM EDTA as electron donor and 5 mM methyl viologen as electron acceptor, and using the modified gold working electrode, a platinum wire counter electrode, and an Ag/AgCl reference electrode as shown in Scheme 22.

Anodic and cathodic photocurrents were observed when the applied potential was more positive and negative than +200 mV which corresponds to the first reduction potential of the quinone moiety in the monolayer, respectively [359]. The potential dependence of the photocurrents as a function of pH in the electrolyte solution correlates well with the shift of the redox potential of the quinone as a function of pH [359]. When the applied potential is more positive than the redox potential of the quinone, photoinduced electron transfer occurs from the singlet excited state of the porphyrin to the gold electrode through the quinone, followed by a hole shift from the resultant porphyrin radical cation to EDTA, leading to the generation of anodic electron flow [359]. On the other hand, when the applied potential is more negative than the redox potential of the quinone moiety, photoinduced electron transfer takes place from the singlet excited state of the porphyrin

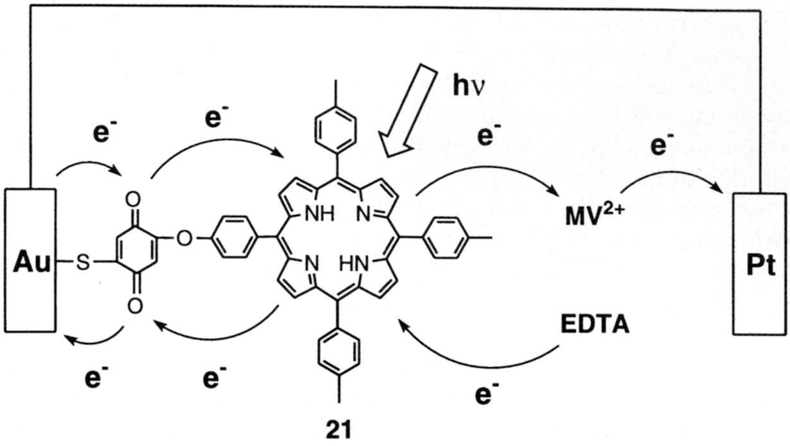

**21**

$MV^{2+}$: methylviologen; EDTA: ethylenediaminetetraacetic acid

**Scheme 22.**

to the methyl viologen, followed by a successive hole shift from the porphyrin radical cation to the gold electrode through the quinone moiety, producing the anodic electron flow [359].

Ferrocene acting as an efficient electron donor has also been linked with a porphyrin ring to construct the highly ordered structure of self-assembled monolayers of ferrocene–porphyrin dyad with an alkanethiol tail **22–24**, as shown in Scheme 23 [319, 329, 331].

The amounts of **22–24** adsorbed were estimated as $2.3 \times 10^{-10}$ mol cm$^{-2}$ (71 Å$^2$ molecule$^{-1}$) for all the monolayers of **22–24** from the amount of charge in anodic peaks of the ferrocene moieties. The occupied area is close to that expected for ori-

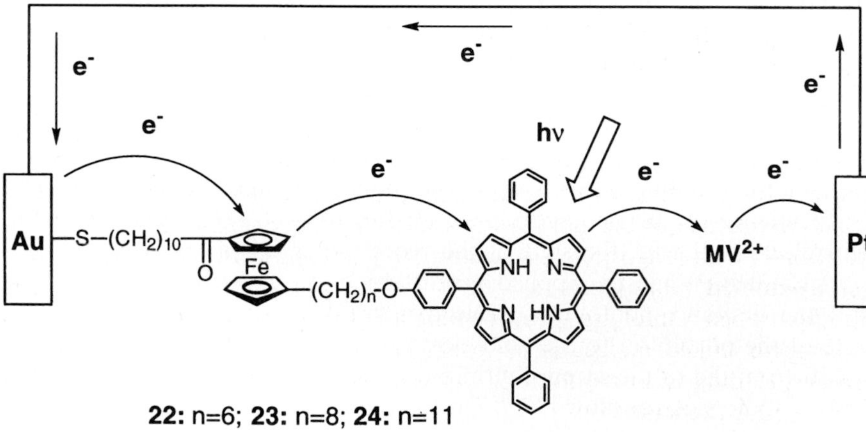

**22**: n=6; **23**: n=8; **24**: n=11

**Scheme 23.**

entation of the porphyrin ring perpendicular to the electrode $(50 \, \text{Å}^2 \, \text{molecule}^{-1})$ rather than that for vertical orientation $(250 \, \text{Å}^2 \, \text{molecule}^{-1})$. Angle-resolved X-ray photoelectron spectroscopic studies on the monolayer of **23** confirmed that alkyl chains both between the porphyrin and the ferrocene and between the ferrocene and the thiol group were aligned with a tilt angle of 30° to the gold surface, with perpendicular orientation of the porphyrin plane [329]. The cathodic photocurrent was observed when the gold electrode modified with monolayers of **22–24** was irradiated with monochromatic light of 430 nm at an applied potential of $-200$ mV in a 0.1 M aqueous $NaClO_4$ solution containing 5 mM methyl viologen as an acceptor [331]. Quantum yields based on the absorbed photons were determined as 4.0, 11, and 12 % for photoelectrochemical cells of monolayers of **22**, **23** and **24**, respectively [331]. The photocurrent is generated via photoinduced electron transfer from the singlet excited state of the porphyrin to methyl viologen, then by successive hole transfer from the porphyrin to the gold electrode through the ferrocene moiety (Scheme 23) [331]. Photocurrent generation is limited by the efficiency of intermolecular electron-transfer quenching of the singlet excited state of the porphyrin by methyl viologen in competition with facile energy transfer quenching by the gold electrode, as described for the **18** monolayer system (vide supra). In addition, strong aggregation of the porphyrins without bulky substituents around the porphyrin moiety (**22–24**) in the monolayer might accelerate the non-radiative decay of the singlet excited state of the porphyrin. The flexible methylene spacer between the porphyrin and the ferrocene moieties does not enable comparison of the photodynamics in solutions with those in the monolayer on the gold surface; this is in sharp contrast with photoelectrochemical cells of the monolayers prepared from porphyrin-$C_{60}$ dyad [309, 318, 332] and triad systems (vide infra) [333, 334, 339].

### 8.4.5 Self-assembled Monolayers of Porphyrin-containing Triads

A highly ordered self-assembled monolayer system of porphyrin-containing dyads has been extended to accommodate porphyrin-containing triads (**25** and **26**) [333, 334, 339]. The alkanethiol-attached triads **25** and **26** involve a linear array of ferrocene (Fc), porphyrin (P), and $C_{60}$ as shown in Scheme 24. The energy gradient of each redox state in the triad is in the order: $Fc-{}^1P^*-C_{60} > Fc-P^{\bullet+}-C_{60}{}^{\bullet-} > Fc^+-P-C_{60}{}^{\bullet-}$. This makes it possible to achieve sequential electron transfer within the triad. The use of $C_{60}$ as an electron acceptor in the triad has enabled the attainment of a long lifetime of the charge separated state and high quantum efficiency, as a consequence of the small reorganization energy of $C_{60}$ [323, 334, 360–373]. Alkanethiols with long methylene chains ($n \geq 10$) are known to form densely packed self-assembled monolayers at a gold surface by use of the S–Au linkage [249, 342]. Thus, when a thiol group with a long alkyl chain is introduced at the end of donor–acceptor linked molecules, it would be arranged unidirectionally at the gold electrodes, leading to formation of a uniform self-assembled monolayer with a thickness of ca 50 Å (Scheme 24).

The fluorescence lifetime measurements of **25** and **26** in solution suggest that the quenching of the singlet excited porphyrin by the attached $C_{60}$ is a major deactiva-

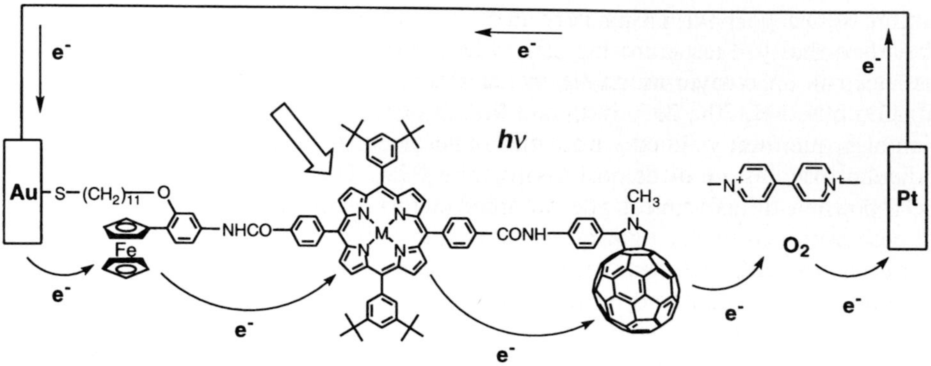

**25:** M=H$_2$; **26:** M=Zn

**Scheme 24.**

tion pathway in both triads [339]. The time-resolved transient spectra of the triads in different solvents revealed that initial photoinduced electron transfer or partial charge-transfer from $^1$P* to C$_{60}$ occurred in the triads, followed by rapid charge shift from the ferrocene to P$^{\bullet+}$ to produce Fc$^+$–P–C$_{60}$$^{\bullet-}$ [369]. The zinc complex affords a lifetime of 40 ns and 15.6 µs in benzene and DMF, respectively [369]. The overall quantum yield for formation of Fc$^+$–P–C$_{60}$$^{\bullet-}$ depends on the solvent used. The quantum yield of the zinc complex is 65 % in benzonitrile, based on the absorption of both the porphyrin and C$_{60}$ [369]. Given the relative absorption ratio of the porphyrin and the C$_{60}$, the quantum yield relative to $^1$ZnP* exceeds 80 %. The amount of **25** adsorbed at the gold electrode was estimated as $1.9 \times 10^{-10}$ mol cm$^{-2}$ ($= 86 \,\text{Å}^2$ molecule$^{-1}$), which is comparable with those for well-ordered self-assembled monolayers of porphyrins [319, 325, 331, 338] and C$_{60}$ [374–376] at gold electrodes. This indicates that the triad molecules are well-packed with nearly perpendicular orientation to the gold surface.

Photoelectrochemical experiments using the triad-modified gold electrode, a platinum wire, and an Ag/AgCl electrode were conducted in the presence of electron carriers such as oxygen and/or methyl viologen [333, 334, 339]. The cathodic photocurrent was detected under irradiation of the modified gold electrode with monochromatic light of 438.5 nm. Increasing cathodic photocurrent was observed with increasing negative bias on the gold electrode. This indicates that the direction of the electron flow is from the gold electrode to the counter electrode through the electrolyte. When the applied potential was more negative than +500 mV, which corresponds to the oxidation potential of the ferrocene, the photocurrent intensity increased drastically. Thus, the photocurrent efficiency is controlled by the rate of electron-transfer between the gold electrode and the ferrocene [319, 331]. The action spectra of the cells agree with the absorption spectra of **25** and **26** on the gold electrode. Such agreement indicates clearly that the porphyrin is a major photoactive species in photocurrent generation. The quantum yields for photoelectrochemical cells of **25** and **26** under the optimum conditions using an oxygen-saturated solution in the presence of methyl viologen were found to be 25 % and 20 %, respectively. This is the highest value (25 %) among donor–acceptor linked molecules on monolayer-modified metal electrodes and in artificial membranes [258, 259, 268, 271, 294, 295, 309, 310, 318, 319, 325, 331, 332, 377].

The high quantum yield may be achieved via photoinduced electron transfer from the singlet excited state of the porphyrin to the $C_{60}$, then hole-shift from the resulting porphyrin radical cation to the ferrocene to produce the charge-separated state, $Fc^+–P–C_{60}^{\bullet-}$ in the monolayer (Scheme 24). The fast electron transfer to $C_{60}$, because of the small reorganization energy can compete well with deactivation as a result of the gold electrode [325, 338]. Electron carriers such as oxygen ($E^\circ_{red} = -0.48$ V for $O_2/O_2^{\bullet-}$) [378] and/or methyl viologen ($E^\circ_{red} = -0.62$ V for $MV^{+2}/MV^{\bullet+}$) are reduced by the $C_{60}^{\bullet-}$ moiety of $Fc^+–P–C_{60}^{\bullet-}$ to produce the radical anions which eventually give an electron to the counter electrode. The electron-transfer rate from the gold electrode to $Fc^+$ in $Fc^+–P–C_{60}^{\bullet-}$ is, on the other hand, affected by the potential applied to the gold electrode. Thus the rate of electron-transfer from the gold electrode to $Fc^+$ increases as the applied potential is reduced, leading to an increase in the photocurrent.

## 8.5 Summary

This review has focused on recent advances in the electron-transfer chemistry of porphyrins and metalloporphyrins, which is relevant to their biological role. Electron-transfer reactions of porphyrins and metalloporphyrins have been studied for long time and they are now relatively well understood on the basis of fundamental redox properties such as reorganization energies and the one-electron redox potentials in the light of the Marcus theory of electron transfer [1]. However, reaction mechanisms involving reactive intermediates such as high-valent metalloporphyrins produced in electron-transfer reactions of metalloporphyrins have remained incompletely understood, as discussed in this review. Axial coordination sites available in metalloporphyrins play an essential role in the construction of useful catalytic systems for biomimetic reactions and energy conversion, because simple combination of outer-sphere electron-transfer reactions would not increase overall electron-transfer rates. Strong inner-sphere interaction between substrates and metalloporphyrins results in the formation of a variety of substrate-bound intermediates; in particular new types of bimetallic active oxygen porphyrin complexes acting as reactive intermediates during the four-electron reduction of oxygen have been explored rapidly. Recent advances in self-assembled monolayer systems of porphyrins and metalloporphyrins have provided a promising means of developing biomimetic electron-transfer systems (in which highly ordered, well-designed architectures act as efficient catalysts) and organic solar cells. In each the search for new and better catalytic systems involving porphyrins and metalloporphyrins continues to be guided by beautifully constructed natural enzymatic systems.

## Acknowledgment

The authors are deeply indebted to the work of all collaborators and coworkers whose names are listed in the references of this chapter (in particular, Prof. K. M. Kadish, Prof. Y. Watanabe and Prof. Y. Sakata). S.F. acknowledges continuous support of his study on electron-transfer chemistry by a Grant-in-Aid from the Ministry of Education, Science, Culture and Sports, Japan.

## References

1. S. Fukuzumi, *The Porphyrin Handbook*, Vol. 8 (Eds.: K. M. Kadish, K. Smith, R. Guilard), Academic Press, San Diego, CA, **2000**, pp. 115–152.
2. R. Guilard, K. M. Kadish, *Chem. Rev.* **1988**, *88*, 1121.
3. D. Mansuy, P. Battioni, in *Metalloporphyrins in Catalytic Oxidations* (Ed.: R. A. Sheldon), Marcel Dekker, New York, **1994**, pp. 99–132.
4. D. Mansuy, *Pure Appl. Chem.* **1987**, *59*, 759.
5. D. Mansuy, *Coord. Chem. Rev.* **1993**, *125*, 129.
6. A. R. Battersby, *Science* **1994**, *264*, 1551.
7. J. Deisenhofer, H. Michel, *Angew. Chem. Int. Ed. Engl.* **1989**, *28*, 829.
8. G. von Jagow, W. D. Engel, *Angew. Chem. Int. Ed. Engl.* **1980**, *19*, 659.
9. W. Kaim, B. Schwederski, *Bioinorganic Chemistry: Inorganic Elements in the Chemistry of Life*, Wiley, New York, **1994**.
10. R. E. Dickerson, R. Timkovich, *Enzymes* (3rd Ed.), **1975**, *11*, 397.
11. M. A. Cusanovich, in *Bioorganic Chemistry*, Vol. 4 (Ed.: E. E. van Tamelen), Academic Press, New York, **1978**.
12. E. Margoliash, in *Electron Transport and Oxygen Utilization* (Ed.: C. Ho), Elsevier, New York, **1982**.
13. H. Sigel, A. Sigel, Eds., *Electron-transfer reactions in Metalloproteins, in Metal Ions in Biological Systems*, Vol. 27, Marcel Dekker, New York, **1991**.
14. S. S. Isied, *Prog. Inorg. Chem.* **1984**, *32*, 443.
15. T. E. Meyer, M. D. Kamen, *Adv. Protein Chem.* **1982**, *35*, 105.
16. K. M. Kadish, E. van Caemelbeck, G. Royal, *The Porphyrin Handbook*, Vol. 8 (Eds: K. M. Kadish, K. Smith, R. Guilard), Academic Press, San Diego, CA, **2000**, pp. 1–114.
17. D. Gust, T. A. Moore, *The Porphyrin Handbook*, Vol. 8 (Eds.: K. M. Kadish, K. Smith, R. Guilard), Academic Press, San Diego, CA, **2000**, pp. 153–190.
18. L. Eberson, *Electron-transfer reactions in Organic Chemistry*, Springer, Tokyo, **1987**.
19. D. Astruc, *Electron-Transfer and Radical Processes in Transition-Metal Chemistry*, VCH, New York, **1995**.
20. J. F. Bunnett, *Acc. Chem. Res.* **1992**, *25*, 2.
21. M. Chanon, M. L. Tobe, *Angew. Chem. Int. Ed. Engl.* **1982**, *21*, 1.
22. S. Fukuzumi, in *Advances in Electron-transfer chemistry*, Vol. 2 (Ed.: P. S. Mariano), JAI Press, Greenwich, CT, **1992**, pp. 65–175.
23. M. Patz, S. Fukuzumi, *J. Phys. Org. Chem.* **1997**, *10*, 129.
24. J. K. Kochi, *Angew. Chem. Int. Ed. Engl.* **1988**, *27*, 1227.
25. E. C. Ashby, *Acc. Chem. Res.* **1988**, *21*, 414.
26. M. Chanon, M. Rajzmann, F. Chanon, *Tetrahedron* **1990**, *46*, 6193.
27. P. R. Ortiz de Montellano, *Cytochrome P450. Structure, Mechanism, and Biochemistry* (2nd Ed.), Plenum Publishing Corporation, New York, **1995**.
28. F. P. Guengerich, T. L. Macdonald, *Acc. Chem. Res.* **1984**, *17*, 9.
29. P. R. Ortiz de Montellano, *Acc. Chem. Res.* **1987**, *20*, 289.
30. F. P. Guengerich, *J. Biol. Chem.* **1991**, *266*, 10019.
31. D. Dolphin, T. G. Traylor, L. Y. Xie, *Acc. Chem. Res.* **1997**, *30*. 251.

32. D. Ostovic, T. C. Bruice, *Acc. Chem. Res.* **1992**, *25*, 314.
33. D. Mansuy, *Comp. Biochem. Phys. C* **1998**, *121*, 5.
34. J. T. Groves, T. E. Nemo, R. S. Myers, *J. Am. Chem. Soc.* **1979**, *101*, 1032.
35. J. T. Groves, T. E. Nemo, *J. Am. Chem. Soc.* **1983**, *105*, 5786.
36. J. T. Groves, R. S. Myers, *J. Am. Chem. Soc.* **1983**, *105*, 5791.
37. J. T. Groves, D. V. Subramanian, *J. Am. Chem. Soc.* **1984**, *106*, 2177.
38. J. P. Collman, T. Kodadek, S. A. Raybuck, J. I. Brauman, L. M. Papazian, *J. Am. Chem. Soc.* **1985**, *107*, 4343.
39. C. M. Dicken, F.-L. Lu, M. W. Nee, T. C. Bruice, *J. Am. Chem. Soc.* **1985**, *107*, 5776.
40. T. Mori, T. Santa, T. Higuchi, T. Mashino, M. Hirobe, *Chem. Pharm. Bull.* **1993**, *41*, 292.
41. J. P. Collman, J. I. Brauman, B. Meunier, T. Hayashi, T. Kodadek, S. A. Raybuck, *J. Am. Chem. Soc.* **1985**, *107*, 2000.
42. B. Meunier, E. Guilmet, M.-E. De Carvalho, R. Poilblanc, *J. Am. Chem. Soc.* **1984**, *106*, 6668.
43. M. W. Nee, T. C. Bruice, *J. Am. Chem. Soc.* **1982**, *104*, 6123.
44. M. F. Powell, E. F. Pai, T. C. Bruice, *J. Am. Chem. Soc.* **1984**, *106*, 3277.
45. D. C. Heimbrook, R. I. Murray, K. D. Egeberg, S. G. Sligar, M. W. Nee, T. C. Bruice, *J. Am. Chem. Soc.* **1984**, *106*, 1514.
46. C. M. Dicken, T. C. Woon, T. C. Bruice, *J. Am. Chem. Soc.* **1986**, *108*, 1636.
47. J. T. Groves, Y. Watanabe, T. J. McMurry, *J. Am. Chem. Soc.* **1983**, *105*, 4489.
48. L.-C. Yuan, T. C. Bruice, *J. Am. Chem. Soc.* **1985**, *107*, 512.
49. W. A. Lee, T. C. Bruice, *J. Am. Chem. Soc.* **1985**, *107*, 513.
50. L.-C. Yuan, T. C. Bruice, *J. Am. Chem. Soc.* **1986**, *108*, 1643.
51. H. J. Ledon, P. Durbut, F. Varescon, *J. Am. Chem. Soc.* **1981**, *103*, 3601.
52. D. Mansuy, P. Battioni, J.-P. Renaud, *J. Chem. Soc. Chem. Commun.* **1984**, 1255.
53. S. Fukuzumi, S. Mochizuki, T. Tanaka, *Isr. J. Chem.* **1987/1988**, *28*, 29.
54. J. R. Lindsay-Smith, in *Metalloporphyrins in Catalytic Oxidations* (Ed.: R. A. Sheldon), Marcel Dekker, New York, **1994**, pp. 325–361.
55. F. Bedioui, *Coord. Chem. Rev.* **1995**, *144*, 39.
56. P. Battioni, E. Cardin, M. Louloudi, B. Schöllhorn, G. A. Spyroulias, D. Mansuy, T. G. Traylor, *Chem. Commun.* **1996**, 2037.
57. B. Meunier, *Chem. Rev.* **1992**, *92*, 1411.
58. P. E. Ellis, Jr., J. E. Lyons, *Coord. Chem. Rev.* **1990**, *105*, 181.
59. G.-X. He, R. D. Arasasingham, G.-H. Zhang, T. C. Bruice, *J. Am. Chem. Soc.* **1991**, *113*, 9828.
60. P. Battioni, J. P. Renaud, J. F. Bartoli, M. Reina- Artiles, M. Fort, D. Mansuy, *J. Am. Chem. Soc.* **1988**, *110*, 8462.
61. A. Thellend, P. Battioni, D. Mansuy, *J. Chem. Soc. Chem. Commun.* **1994**, 1035.
62. O. Brigaud, P. Battioni, D. Mansuy, C. Giessner- Prettre, *New J. Chem.* **1992**, *16*, 1031.
63. C. K. Chang, F. Ebina, *J. Chem. Soc. Chem. Commun.* **1981**, 778.
64. J. F. Bartoli, O. Brigaud, P. Battioni, D. Mansuy, *J. Chem. Soc. Chem. Commun.* **1991**, 440.
65. M. W. Grinstaff, M. G. Hill, J. A. Labinger, H. B. Gray, *Science* **1994**, *264*, 1311.
66. W.-D. Woggon, H. Fretz, in *Adcances in Detailed Reaction Mechanism*, Vol. 2 (Ed.: J. M. Coxon), JAI Press, Greenwich, CT, **1992**, pp. 111–147.
67. G. T. Miwa, J. S. Walsh, G. L. Kedderis, P. F. Hollenberg, *J. Biol. Chem.* **1983**, *258*, 14445.
68. P. G. Wislocki, G. T. Miwa, A. Y. H. Lu, in *Enzymatic Basis of Detoxification*, Vol. 1 (Ed.: W. B. Jakoby), Academic Press, New York, **1980**, p 135.
69. R. E. White, M. J. Coon, *Ann. Rev. Biochem.* **1980**, *49*, 315.
70. F. P. Guengerich, T. L. Macdonald, *Acc. Chem. Res.* **1984**, *17*, 9.
71. J. H. Dawson, M. Sono, *Chem. Rev.* **1987**, *87*, 1255.
72. T. D. Porter, M. J. Coon, *J. Biol. Chem.* **1991**, *266*, 13469.
73. P. F. Hollenberg, *FASEB J.* **1992**, *6*, 686.
74. F. P. Guengerich, T. L. Macdonald, in *Advances in Electron-transfer chemistry*, Vol. 3 (Ed.: P. S. Mariano), JAI press, Greenwich, CT, **1993**, p. 191.
75. F. P. Guengerich, C.-H. Yun, T. L. MacDonald, *J. Biol. Chem.* **1996**, *271*, 27321.
76. J. R. L. Smith, D. N. Mortimer, *J. Chem. Soc. Chem. Comun.* **1985**, 64.

77. J. R. L. Smith, D. N. Mortimer, *J. Chem. Soc. Perkin Trans. 2* **1986**, 1743.
78. J. I. Manchester, J. P. Dinnocenzo, L. Higgins, J. P. Jones, *J. Am. Chem. Soc.* **1997**, *119*, 5069.
79. S. B. Karki, J. P. Dinnocenzo, J. P. Jones, K. R. Korzekwa, *J. Am. Chem. Soc.* **1995**, *117*, 3657.
80. J. P. Dinnocenzo, S. B. Karki, J. P. Jones, *J. Am. Chem. Soc.* **1993**, *115*, 7111.
81. S. Fukuzumi, C. L. Wong, J. K. Kochi, *J. Am. Chem. Soc.* **1980**, *102*, 2928.
82. S. Fukuzumi, Y. Kondo, T. Tanaka, *J. Chem. Soc. Perkin Trans. 2* **1984**, 673.
83. S. Fukuzumi, K. Koumitsu, K. Hironaka, T. Tanaka, *J. Am. Chem. Soc.* **1987**, *109*, 305.
84. S. Fukuzumi, Y. Tokuda, T. Kitano, T. Okamoto, J. Otera, *J. Am. Chem. Soc.* **1993**, *115*, 8960.
85. R. P. Bell, *The Proton in Chemistry* (2nd Ed.), Cornell University Press, Ithaca, NY, **1973**, Chapter 12.
86. D. Griller, J. A. Howard, P. R. Marriott, J. C. Scaiano, *J. Am. Chem. Soc.* **1981**, *103*, 619.
87. I. Wölfle, J. Lodaya, B. Sauerwein, G. B. Schuster, *J. Am. Chem. Soc.* **1992**, *114*, 9304.
88. T. M. Bockman, K. Y. Lee, J. K. Kochi, *J. Chem. Soc. Perkin Trans. 2* **1992**, 1581.
89. E. Baciocchi, O. Lanzalunga, A. Lapi, L. Manduchi, *J. Am. Chem. Soc.* **1998**, *120*, 5783.
90. A. Rehm, A. Weller, *Isr. J. Chem.* **1970**, *8*, 259.
91. R. A. Marcus, *Annu. Rev. Phys. Chem.* **1964**, *15*, 155.
92. F. Scandola, V. Balzani, *J. Am. Chem. Soc.* **1979**, *101*, 6142.
93. S. Fukuzumi, I. Nakanishi, K. Tanaka, T. Suenobu, A. Tabard, R. Guilard, E. Van Caemelbecke, K. M. Kadish, *J. Am. Chem. Soc.* **1999**, *121*, 785.
94. K. M. Kadish, E. Van Caemelbecke, E. Gueletii, S. Fukuzumi, K. Miyamoto, T. Suenobu, A. Tabard, R. Guilard, *Inorg. Chem.* **1998**, *37*, 1759.
95. S. Marguet, P. Hapiot, P. Neta, *J. Phys. Chem.* **1994**, *98*, 7136.
96. Y. Goto, Y. Watanabe, S. Fukuzumi, J. P. Jones, J. P. Dinnocenzo, *J. Am. Chem. Soc.* **1998**, *120*, 10762.
97. B. W. Griffin, P. L. Ting, *Biochemistry* **1978**, *17*, 2206.
98. J. Van der Zee, D. R. Duling, R. P. Mason, T. E. Eling, *J. Biol. Chem.* **1989**, *264*, 19828.
99. H. B. Dunford, J. S. Stillman, *Coord. Chem. Rev.* **1987**, *19*, 187.
100. P. George, *Science* **1953**, *117*, 220.
101. R. R. Fergusson, *J. Am. Chem. Soc.* **1956**, *78*, 741.
102. Y. Hayashi, I. Yamazaki, *J. Biol. Chem.* **1979**, *254*, 9101.
103. Z. S. Farhangrazi, B. R. Copeland, T. Nakayama, T. Amachi, I. Yamazaki, L. S. Powers, *Biochemistry* **1994**, *33*, 5647.
104. Z. S. Farhangrazi, M. E. Fossett, L. S. Powers, W. R. Ellis, Jr. *Biochemistry* **1995**, *34*, 2866.
105. B. B. Hasinoff, H. B. Dunford, *Biochemistry* **1970**, *9*, 4930.
106. T. Egawa, H. Shimada, Y. Ishimura, *Biochem. Biophys. Res. Commun.* **1994**, *201*, 1464.
107. J. T. Groves, Y. Watanabe, *J. Am. Chem. Soc.* **1988**, *110*, 8443.
108. Y. Goto, T. Matsui, S. Ozaki, Y. Watanabe, S. Fukuzumi, *J. Am. Chem. Soc.* **1999**, *121*, 9497.
109. S. Kobayashi, M. Nakano, T. Goto, T. Kimura, A. P. Schaap, *Biochem. Biophys. Res. Commun.* **1986**, *135*, 166.
110. L. P. Candeias, L. K. Folkes, P. Wardman, *Biochemistry* **1997**, *36*, 7081.
111. D. R. Doerge, N. M. Cooray, M. E. Brewster, *Biochemistry* **1991**, *30*, 8960.
112. U. Perez, H. B. Dunford, *Biochim. Biophys. Acta.* **1990**, *1038*, 98.
113. U. Perez, H. B. Dunford, *Biochemistry* **1990**, *29*, 2757.
114. T. Matsui, S. Ozaki, Y. Watanabe, *J. Biol. Chem.* **1997**, *272*, 32735.
115. M. Gajhede, D. J. Schuller, A. Henriksen, A. T. Smith, T. L. Poulos, *Nat. Struc. Biol.* **1997**, *4*, 1932.
116. M. L. Quillin, R. M. Arduini, J. S. Olson, G. N. Phillips, Jr. *J. Mol. Biol.* **1993**, *234*, 140.
117. P. R. Ortiz de Montellano, D. E. Kerr, *Biochemistry* **1985**, *24*, 1147.
118. E.-D. Wogon, H. Fretz, in *Advances in Detailed Reaction Mechanisms*, Vol. 2 (Ed., J. M. Coxon), JAI Press, Greenwich, CT, pp. 111–147.
119. F. P. Guengerich, T. L. Macdonald, *FASEB J.* **1990**, *4*, 2453.
120. L. M. Hjelmeland, L. Aronow, J. R. Trudell, *Biochem. Biophys. Res. Commun.* **1977**, *76*, 541.

121. J. T. Groves, G. A. McClusky, R. E. White, M. J. Coon, *Biochem. Biophys. Res. Commun.* **1978**, *81*, 154.
122. J. Fossey, D. Lefort, M. Massoudi, J.-Y. Nedelec, J. Sorba, *Can. J. Chem.* **1985**, *63*, 678.
123. J. T. Groves, T. E. Nemo, *J. Am. Chem. Soc.* **1983**, *105*, 6243.
124. D. Bouy-Debec, O. Brigaud, P. Leduc, P. Battioni, D. Mansy, *Gazz. Chim. Ital.* **1996**, *126*, 233.
125. M. H. Gelb, D. C. Heimbrook, P. Malkonen, S. G. Sligar, *Biochemisty* **1982**, *21*, 370.
126. P. R. Ortiz de Montellano, R. A. Stearns, *J. Am. Chem. Soc.* **1987**, *109*, 3415.
127. D. Griller, K. U. Ingold, *Acc. Chem. Res.* **1980**, *13*, 317.
128. J. K. Atkinson, K. U. Ingold, *Biochemistry* **1993**, *32*, 9209.
129. J. K. Atkinson, P. F. Hollenberg, K. U. Ingold, C. C. Johnson, M.-H. Le Tadic, M. Newcomb, D. A. Putt, *Biochemistry* **1994**, *33*, 10630.
130. M. Newcomb, M.-H. Le Tadic, D. A. Putt, P. F. Hollenberg, *J. Am. Chem. Soc.* **1995**, *117*, 3312.
131. M. Newcomb, M-H. Le Tadic, D. L. Chestney, E. S. Roberts, P. F. Hollenberg, *J. Am. Chem. Soc.* **1995**, *117*, 12085.
132. S.-Y. Choi, P. E. Eaton, P. F. Hollenberg, K. E. Liu, S. J. Lippard, M. Newcomb, D. A. Putt, S. P. Upadhyaya, Y. Xiong, *J. Am. Chem. Soc.* **1996**, *118*, 6547.
133. M. Brookhart, M. L. H. Green, L. L. Wong, *Prog. Inorg. Chem.* **1988**, *36*, 1.
134. D. H. R. Barton, D. Doller, *Acc. Chem. Res.* **1992**, *25*, 504.
135. R. H. Crabtree, *Angew. Chem. Int. Ed. Engl.* **1993**, *32*, 789.
136. C. Hall, R. N. Perutz, *Chem. Rev.* **1996**, *96*, 3125.
137. J. P. Collman, A. S. Chien, T. A. Eberspacher, J. I. Brauman, *J. Am. Chem. Soc.* **1998**, *120*, 425.
138. B. R. Bender, G. J. Kubas, L. H. Jones, B. I. Swanson, J. Eckert, K. B. Capps, C. D. Hoff, *J. Am. Chem. Soc.* **1997**, *119*, 9179.
139. D. R. Evans, T. Drovetskaya, R. Bau, C. A. Reed, P. D. W. Boyd, *J. Am. Chem. Soc.* **1997**, *119*, 3633.
140. S. Fukuzumi, K. Mochida, J. K. Kochi, *J. Am. Chem. Soc.* **1979**, *101*, 5961.
141. S. Fukuzumi, J. K. Kochi, *J. Am. Chem. Soc.* **1980**, *102*, 2141.
142. S. Fukuzumi, J. K. Kochi, *J. Am. Chem. Soc.* **1980**, *102*, 7290.
143. S. Fukuzumi, J. K. Kochi, *J. Am. Chem. Soc.* **1981**, *103*, 7240.
144. S. Fukuzumi, N. Nishizawa, T. Tanaka, *J. Org. Chem.* **1984**, *49*, 3571.
145. K. M. Zaman, S. Yamamoto, N. Nishimura, J. Maruta, S. Fukuzumi, *J. Am. Chem. Soc.* **1994**, *116*, 12099.
146. H. R. Bosshard, H. Anni, T. Yonetani, in *Peroxidases in Chemistry and Biology*, Vol. II (Eds.: J. Everse, K. E. Everse, M. B. Grisham), CRC Press, Boca Raton, FL, **1990**, pp. 51–84.
147. H. Pelletier, J. Kraut, *Science* **1992**, *258*, 1748.
148. A. F. W. Coulson, J. E. Erman, T. Yonetani, *J. Biol. Chem.* **1971**, *246*, 917.
149. P. S. Ho, B. M. Hoffman, C. H. Kang, E. Margoliash, *J. Biol. Chem.* **1983**, *258*, 4356.
150. J. E. Erman, L. B. Vitello, J. M. Mauro, J. Kraut, *Biochemistry* **1989**, *28*, 7992.
151. M. Sivaraja, D. B. Goodin, M. Smith, B. M. Hoffman, *Science* **1989**, *245*, 738.
152. L. A. Fishel, M. F. Farnum, J. M. Mauro, M. A. Miller, J. Kraut, Y. Liu, X.-L. Tan, C. P. Scholes, *Biochemistry* **1991**, *30*, 1986.
153. J. M. Mauro, L. A. Fishel, J. T. Hazzard, T. E. Meyer, G. Tollin, M. A. Cusanovich, J. Kraut, *Biochemistry* **1988**, *27*, 6243.
154. C. P. Scholes, Y. Liu, L. A. Fishel, M. F. Farnum, J. M. Mauro, J. Kraut, *Isr. J. Chem.* **1989**, *29*, 85.
155. L. Geren, S. Hahm, B. Durham, F. Millett, *Biochemistry* **1991**, *30*, 9450.
156. S. Hahm, B. Durham, F. Millett, *Biochemistry* **1992**, *31*, 3472.
157. K. Wang, H. Mei, L. Geren, M. A. Miller, A. Saunders, X. Wang, J. L. Waldner, G. J. Pielak, B. Durham, F. Millett, *Biochemistry* **1996**, *35*, 15107.
158. S. Hahm, L. Geren, B. Durham, F. Millett, *J. Am. Chem. Soc.* **1993**, *115*, 3372.
159. J. T. Hazzard, T. L. Poulos, G. Tollin, *Biochemistry* **1987**, *26*, 2836.
160. J. T. Hazzard, S. J. Moench, J. E. Erman, J. D. Satterlee, G. Tollin, *Biochemistry* **1988**, *27*, 2002.

161. M. A. Miller, J. T. Hazzard, J. M. Mauro, S. L. Edwards, P. C. Simons, G. Tollin, J. Kraut, *Biochemistry* **1988**, *27*, 9081.
162. G. Tollin, J. T. Hazzard, *Arch. Biochem. Biophys.* **1991**, *287*, 1.
163. J. T. Hazzard, G. Tollin, *J. Am. Chem. Soc.* **1991**, *113*, 8956.
164. F. E. Summers, J. E. Erman, *J. Biol. Chem.* **1988**, *263*, 14267.
165. J. E. Erman, D. S. Kang, K. L. Kim, F. E. Summers, A. L. Matthis, L. B. Vitello, *Mol. Cryst. Liq. Cryst.* **1991**, *194*, 253.
166. M. R. Nuevo, H.-H. Chu, L. B. Vitello, J. E. Erman, *J. Am. Chem. Soc.* **1993**, *115*, 5873.
167. D. W. Low, J. R. Winkler, H. B. Gray, *J. Am. Chem. Soc.* **1996**, *118*, 117.
168. J. Berglund, T. Pascher, J. R. Winkler, H. B. Gray, *J. Am. Chem. Soc.* **1997**, *119*, 2464.
169. J. J. Wilker, I. J. Dmochowski, J. H. Dawson, J. R. Winkler, H. B. Gray, *Angew. Chem. Int. Ed. Engl.* **1999**, *38*, 90.
170. T. C. Bruice, *Acc. Chem. Res.* **1991**, *24*, 243.
171. M. Z. Hoffman, F. Bolleta, L. Moggi, G. L. Hug, *J. Phys. Chem. Ref. Data* **1989**, *18*, 219.
172. I. Hamachi, S. Tsukiji, S. Shinkai, S. Oishi, *J. Am. Chem. Soc.* **1999**, *121*, 5500.
173. Y.-Z. Hu, S. Tsukiji, S. Shinkai, S. Oishi, I. Hamachi, *J. Am. Chem. Soc.* **2000**, *122*, 241.
174. I. Hamachi, S. Tanaka, S. Shinkai, *J. Am. Chem. Soc.* **1993**, *115*, 10458.
175. I. Hamachi, Y. Tajiri, S. Shinkai, *J. Am. Chem. Soc.* **1994**, *116*, 7437.
176. I. Hamachi, Y. Tajiri, T. Nagase, S. Shinkai, *Chem. Eur. J.* **1997**, *3*, 1025.
177. B. He, R. Sinclair, B. R. Copeland, R. Makino, L. S. Powers, I. Yamazaki, *Biochemistry* **1996**, *35*, 2413.
178. I. Hamachi, S. Tanaka, S. Shinkai, *J. Am. Chem. Soc.* **1993**, *115*, 10458.
179. I. Hamachi, S. Tanaka, S. Tsukiji, S. Shinkai, S. Oishi, *Inorg. Chem.* **1998**, *37*, 4380.
180. T. Yonetani, H. Schleyer, *J. Biol. Chem.* **1967**, *242*, 1974.
181. P. Gans, G. Buisson, E. Duée, J.-C. Marchon, B. S. Erler, W. F. Scholz, C. A. Reed, *J. Am. Chem. Soc.* **1986**, *108*, 1223.
182. V. K. Yachandra, K. Sauer, M. P. Klein, *Chem. Rev.* **1996**, *96*, 2927.
183. R. J. Debus, *Biochim. Biophys. Acta* **1992**, *1102*, 269.
184. B. A. Diner, G. T. Babcock, in *Oxygenic Photosynthesis: The Light Reactions* (Eds.: D. R. Ort, C. F. Yocum), Kluwer Academic Publishers, Dordrecht, The Netherlands, **1996**, p 213.
185. K. Wieghardt, *Angew. Chem. Int. Ed. Engl.* **1994**, *33*, 725.
186. G. Christou, *Acc. Chem. Res.* **1989**, *22*, 328.
187. J. Limburg, V. A. Szalai, G. W. Brudvig, *J. Chem. Soc. Dalton Trans.* **1999**, 1353.
188. W. Rüttinger, G. C. Dismukes, *Chem. Rev.* **1997**, *97*, 1.
189. R. Manchanda, G. W. Brudvig, R. H. Crabtree, *Coord. Chem. Rev.* **1995**, *144*, 1.
190. S. W. Gersten, G. J. Samuels, T. J. Meyer, *J. Am. Chem. Soc.* **1982**, *104*, 4029.
191. Y. Naruta, M. Sasayama, T. Sasaki, *Angew. Chem. Int. Ed. Engl.* **1994**, 33, 1839.
192. D. Geselowitz, T. J. Meyer, *Inorg. Chem.* **1990**, *29*, 3894.
193. J. A. Halfen, S. Mahapatra, E. C. Wilkinson, S. Kaderli, V. G. Young, Jr., L. Que, Jr., A. D. Zuberbuhler, W. B. Tolman, *Science* **1996**, *271*, 1397.
194. J. Limburg, J. S. Vrettos, L. M. Liable-Sands, A. L. Rheingold, R. H. Crabtree, G. W. Brudvig, *Science* **1999**, *283*, 1524.
195. J. Limburg, G. W. Brudvig, R. H. Crabtree, *J. Am. Chem. Soc.* **1997**, *119*, 2761.
196. Y. Naruta, M. Sasayama, *J. Chem. Soc. Chem. Commun.* **1994**, 2667.
197. K. Ichihara, Y. Naruta, *Chem. Lett.* **1998**, 185.
198. H. H. Thorp, G. W. Brudvig, *New J. Chem.* **1991**, *15*, 479.
199. L. Stryer, *Biochemistry* (4th Ed.), Freeman, New York, **1995**.
200. G. Palmer, *J. Bioenerg. Biomembr.* **1993**, *25*, 145.
201. B. G. Malmström, *Chem. Rev.* **1990**, *90*, 1247.
202. G. T. Babcock, M. Wikström, *Nature* **1992**, *356*, 301.
203. S. Ferguson-Miller, G. T. Babcock, *Chem. Rev.* **1996**, *96*, 2889.
204. J. P. Collman, F. C. Anson, S. Bencosme, A. Chong, T. Collins, P. Denisevich, E. Exitt, T. Geiger, J. A. Ibers, G. Jameson, C. Konai, K. Meier, R. Oakley, R. Pettman, E. Schmittou, J. Sessler, in *Molecular Engineering: The Design and Synthesis of Catalysis for the Rapid 4-Electron Reduction of Molecular Oxygen to Water* (Eds.: J. P. Collman et al.), Pergamon, U. K., **1981**, pp. 29–45.

205. J. P. Collman, P. S. Wagenknecht, J. E. Hutchison, *Angew. Chem. Int. Ed. Engl.* **1994**, *33*, 1537.
206. S. J. Lippard, *Science* **1993**, *261*, 699.
207. K. D. Karlin, *Science* **1993**, *261*, 701.
208. S. Iwata, C. Ostermeier, B. Ludwig, H. Michel, *Nature* **1995**, *376*, 660.
209. T. Tsukihara, H. Aoyama, E. Yamashita, T. Tomizaki, H. Yamaguchi, K. Shinzawa-Itoh, R. Nakashima, R. Yaono, S. Yoshikawa, *Science* **1995**, *269*, 1069.
210. T. Tsukihara, H. Aoyama, E. Yamashita, T. Tomizaki, H. Yamaguchi, K. Shinzawa-Itoh, R. Nakashima, R. Yaono, S. Yoshikawa, *Science* **1996**, *272*, 1136.
211. S. Yoshikawa, K. Shinzawa-Itoh, R. Nakashima, R. Yaono, E. Yamashita, N. Inoue, M. Yao, M. J. Fei, C. P. Libeu, T. Mizushima, H. Yamaguchi, T. Tomizaki, T. Tsukihara, *Science* **1998**, *280*, 1723.
212. C. Ostermeier, A. Harrenga, U. Ermler, H. Michel, *Proc. Natl. Acad. Sci. U.S.A.* **1997**, *94*, 10547.
213. J. A. Sigman, B. C. Kwok, A. Gengenbach, Y. Lu, *J. Am. Chem. Soc.* **1999**, *121*, 8949.
214. S. Fukuzumi, S. Mochizuki, T. Tanaka, *Chem. Lett.* **1989**, 27.
215. S. Fukuzumi, S. Mochizuki, T. Tanaka, *Inorg. Chem.* **1989**, *28*, 2459.
216. S. Fukuzumi, K. Ishikawa, T. Tanaka, *Chem. Lett.* **1986**, 1.
217. S. Fukuzumi, S. Mochizuki, T. Tanaka, *J. Chem. Soc. Chem. Commun.* **1989**, 391.
218. S. Fukuzumi, S. Mochizuki, T. Tanaka, *Inorg. Chem.* **1990**, *29*, 653.
219. S. Fukuzumi, *Bull. Chem. Soc. Jpn.* **1997**, *70*, 1.
220. S. Fukuzumi and S. Itoh, in *Advances in Photochemistry*, Vol. 25 (Eds. D. C. Neckers, D. H. Volman, G. von Bünau), Wiley, New York, **1999**, pp. 107–172.
221. S. Fukuzumi, M. Patz, T. Suenobu, Y. Kuwahara, S. Itoh, *J. Am. Chem. Soc.* **1999**, *121*, 1605.
222. S. Fukuzumi, T. Suenobu, M. Patz, T. Hirasaka, S. Itoh, M. Fujitsuka, O. Ito, *J. Am. Chem. Soc.* **1998**, *120*, 8060.
223. M. Patz, Y. Kuwahara, T. Suenobu, S. Fukuzumi, *Chem. Lett.* **1997**, 567.
224. R. N. Bagchi, A. M. Bond, F. Scholz, R. Stösser, *J. Am. Chem. Soc.* **1989**, *111*, 8270.
225. R. R. Durand, Jr., C. S. Bencosme, J. P. Collman, F. C. Anson, *J. Am. Chem. Soc.* **1983**, *105*, 2710.
226. J. P. Collman, P. Denisevich, Y. Konai, M. Marrocco, K. Koval, F. C. Anson, *J. Am. Chem. Soc.* **1980**, *102*, 6027.
227. C. K. Chang, H.-Y. Liu, I. Abdalmuhdi, *J. Am. Chem. Soc.* **1984**, *106*, 2725.
228. J. P. Collman, C. S. Bencosme, R. R. Durand, Jr., R. P. Kreh, F. C. Anson, *J. Am. Chem. Soc.* **1983**, *105*, 2699.
229. Y. Le Mest, M. L'Her, J. P. Collman, N. H. Hendricks, L. McElwee-White, *J. Am. Chem. Soc.* **1986**, *108*, 533.
230. R. Guilard, S. Brandes, C. Tardieux, A. Tabard, M. L'Her, C. Miry, P. Gouerec, Y. Knop, J. P. Collman, *J. Am. Chem. Soc.* **1995**, *117*, 11721.
231. A. Nanthakumar, M. S. Nasir, K. D. Karlin, N. Ravi, B. H. Huynh, *J. Am. Chem. Soc.* **1992**, *114*, 6564.
232. S. C. Lee, R. H. Holm, *J. Am. Chem. Soc.* **1993**, *115*, 5833.
233. J. P. Collman, P. C. Herrmann, B. Boitrel, X. Zhang, T. A. Eberspacher, L. Fu, J. Wang, D. L. Rousseau, E. R. Williams, *J. Am. Chem. Soc.* **1994**, *116*, 9783.
234. T. Sasaki, Y. Naruta, *Chem. Lett.* **1995**, 663.
235. J. P. Collman, L. Fu, P. C. Herrmann, X. Zhang, *Science* **1997**, *275*, 949.
236. A. Nnthakumar, S. Fox, N. N. Murthy, K. D. Karlin, *J. Am. Chem. Soc.* **1997**, *119*, 3898.
237. T. Sasaki, N. Nakamura, Y. Naruta, *Chem. Lett.* **1998**, 351.
238. J. P. Collman, L. Fu, P. C. Herrmann, Z. Wang, M. Rapta, M. Bröring, R. Schwenninger, B. Boitrel, *Angew. Chem. Int. Ed. Engl.* **1998**, *37*, 3397.
239. F. Tani, Y. Matsumoto, Y. Tachi, T. Sasaki, Y. Naruta, *Chem. Commun.* **1998**, 1731.
240. B. S. Lim, R. H. Holm, *Inorg. Chem.* **1998**, *37*, 4898.
241. J. P. Collman, *Inorg. Chem.* **1997**, 36, 5145.
242. J. P. Collman, R. Schwenninger, M. Rapta, M. Bröring, L. Fu, *Chem. Commun.* **1999**, 137.

243. J. P. Collman, M. Rapta, M. Bröring, L. Aptova, R. Schwenninger, B. Boitrel, L. Fu, M. L'Her, *J. Am. Chem. Soc.* **1999**, *121*, 1387.
244. M.-A. Kopf, Y.-M. Neuhold, A. D. Zuberbühler, K. D. Karlin, *Inorg. Chem.* **1999**, *38*, 3093.
245. R. A. Ghiladi, T. D. Ju, D.-H. Lee, P. Moënne-Loccoz, S. Kaderli, Y.-M. Neuhold, A. D. Zuberbühler, A. S. Woods, R. J. Cotter, K. D. Karlin, *J. Am. Chem. Soc.* **1999**, *121*, 9885.
246. R. B. Gennis, *Proc. Natl. Acad. Sc. U.S.A.* **1998**, *95*, 12747.
247. T. Kitagawa, T. Ogura, *Prog. Inorg. Chem.* **1997**, *45*, 431.
248. R. B. Gennis, *Biochim. Biophys. Acta* **1998**, *1365*, 241.
249. A. Ulman, *An Introduction to Ultrathin Organic Films*, Academic Press, San Diego, **1991**.
250. S. V. Lymar, V. N. Parmon, K. I. Zamarev, In *Photoinduced Electron Transfer III*, (Ed.: J. Mattay), Springer, Berlin, **1991**, pp. 1–66.
251. M. A. Fox, In *Photoinduced Electron Transfer III*, (Ed.: J. Mattay), Springer, Berlin, **1991**, pp. 67–102.
252. I. Willner, B. Willner, In *Photoinduced Electron Transfer III*, (Ed.: J. Mattay), Springer, Berlin, **1991**, pp. 153–218.
253. J. K. Hurst, In *Kinetics and Catalysis in Microheterogeneous Systems*, (Eds.: M. Grätzel, K. Kalyanasunderam), Marcel Dekker, New York, **1991**, pp. 183–226.
254. J. N. Robin, D. J. Cole-Hamilton, *Chem. Soc. Rev.* **1991**, *20*, 49.
255. K. Naito, A. Miura, M. Azuma, *Thin Solid Films* **1992**, *210–211*, 268.
256. S. Isoda, S. Nishikawa, S. Ueyama, Y. Hanazato, H. Kawakubo, M. Maeda, *Thin Solid Films* **1992**, *210–211*, 290.
257. Y. Nishikawa, S.-i. Fukui, M.-a. Kakimoto, Y. Imai, K. Nishiyama, M. Fujihira, *Thin Solid Films* **1992**, *210–211*, 296.
258. X. D. Wang, B. W. Zhang, J. W. Bai, Y. Cao, X. R. Xiao, J. M. Xu, *J. Phys. Chem.* **1992**, *96*, 2886.
259. Y. Cao, B. W. Zhang, W. Y. Qian, X. D. Wang, J. W. Bai, X. R. Xiao, J. G. Jia, J. W. Xu, *Sol. Energy Mater. Solar Cells* **1995**, *38*, 139.
260. T.-H. Tran-Thi, T. Fournier, A. Y. Sharonov, N. Tkachenko, H. Lemmetyinen, P. Grenier, K.-D. Truong, D. Houde, *Thin Solid Films* **1996**, *273*, 8.
261. H. Yonemura, K. Ohishi, T. Matsuo, *Chem. Lett.* **1996**, 661.
262. H. Hosono, T. Tani, I. Uemura, *Chem. Commun.* **1996**, 1893.
263. K. Liang, K.-Y. Law, D. G. Whitten, *J. Phys. Chem. B* **1997**, *101*, 540.
264. H. Hosono, M. Kaneko, *J. Chem. Soc., Faraday Trans.* **1997**, *97*, 1313.
265. N. V. Tkachenko, A. Y. Tauber, P. H. Hynninen, A. Y. Sharonov, H. Lemmetyinen, *J. Phys. Chem. A* **1999**, *103*, 3657.
266. S. W. Feldberg, G. H. Armen, J. A. Bell, C. K. Chang, C.-B. Wang, *Biophys. J.* **1981**, *34*, 149.
267. E. Bienvenue, P. Seta, A. Hofmanová, C. Gavach, M. Momenteau, *J. Electroanal. Chem.* **1984**, *162*, 275.
268. P. Seta, E. Bienvenue, A. L. Moore, P. Mathis, R. V. Bensasson, P. A. Liddell, P. J. Pessiki, A. Joy, T. A. Moore, D. Gust, *Nature* **1985**, *316*, 653.
269. T. J. Dannhauser, M. Nango, N. Oku, K. Anzai, P. A. Loach, *J. Am. Chem. Soc.* **1986**, *108*, 5865.
270. M. Woodle, J. W. Zhang, D. Mauzerall, *Biophys. J.* **1987**, *52*, 577.
271. Y. Sakata, H. Tatemitsu, E. Bienvenue, P. Seta, *Chem. Lett.* **1988**, 1625.
272. P. Seta, E. Bienvenue, P. Maillard, M. Momenteau, *Photochem. Photobiol.* **1989**, *49*, 537.
273. M. Nango, A. Mizusawa, T. Miyake, J. Yoshinaga, *J. Am. Chem. Soc.* **1990**, *112*, 1640.
274. K. C. Hwang, D. Mauzerall, *J. Am. Chem. Soc.* **1992**, *114*, 9705.
275. M. Nango, H. Kryu, P. A. Loach, *J. Chem. Soc., Chem. Commun.* **1988**, 697.
276. J. T. Groves, S. B. Ungashe, *J. Am. Chem. Soc.* **1990**, *112*, 7796.
277. P. J. Clapp, B. Armitage, P. Roosa, D. F. O'Brien, *J. Am. Chem. Soc.* **1994**, *116*, 9166.
278. J. T. Groves, G. D. Fate, J. Lahiri, *J. Am. Chem. Soc.* **1994**, *116*, 5477.
279. J. Lahiri, G. D. Fate, S. B. Ungashe, J. T. Groves, *J. Am. Chem. Soc.* **1996**, *118*, 2347.
280. K. Iida, M. Nango, M. Matsutaka, M. Yamaguchi, K. Sato, Kazumasa, Tanaka, K. Akimoto, K. Yamashita, K. Tsuda, Y. Kurono, *Langmuir* **1996**, *12*, 450.
281. M. Nango, K. Iida, M. Yamaguchi, K. Yamashita, K. Tsuda, A. Mizusawa, T. Miyake, A. Masuda, J. Yoshinaga, *Langmuir* **1996**, *12*, 1981.

282. T. Komatsu, K. Yamada, E. Tsuchida, U. Siggel, C. Böttcher, J.-H. Fuhrhop, *Langmuir* **1996**, *12*, 6242.
283. M. Nango, T. Hikita, T. Nakano, T. Yamada, M. Nagata, Y. Kurono, T. Ohtsuka, *Langmuir* **1998**, *14*, 407.
284. K. Sun, D. Mauzerall, *J. Phys. Chem. B* **1998**, *102*, 6440.
285. M. P. Pileni, A. M. Braun, M. Grätzel, *J. Photochem. Photobiol.* **1979**, *31*, 423.
286. M. P. Pileni, *Chem. Phys. Lett.* **1980**, *75*, 540.
287. M.-P. Pileni, M. Grätzel, *J. Phys. Chem.* **1980**, *84*, 1822.
288. P.-A. Brugger, P. P. Infelta, A. M. Braun, M. Grätzel, *J. Am. Chem. Soc.* **1981**, *103*, 320.
289. R. H. Schmehl, D. G. Whitten, *J. Phys. Chem.* **1981**, *85*, 3473.
290. P. Brochette, M. P. Pileni, *Nouv. J. Chim.* **1985**, *9*, 551.
291. S. M. B. Costa, J. M. F. M. Lopes, M. J. T. Martins, *J. Chem. Soc., Faraday Trans. 2* **1986**, *82*, 2371.
292. P. Brochette, T. Zemb, P. Mathis, M.-P. Pileni, *J. Phys. Chem.* **1987**, *91*, 1444.
293. J. K. Hurst, D. H. P. Thompson, J. S. Connolly, *J. Am. Chem. Soc.* **1987**, *109*, 507.
294. G. Steinberg-Yfrach, P. A. Liddell, S.-C. Hung, A. L. Moore, D. Gust, T. A. Moore, *Nature* **1997**, *385*, 239.
295. G. Steinberg-Yfrach, J.-L. Rigaud, E. N. Durantini, A. L. Moore, D. Gust, T. A. Moore, *Nature* **1998**, *392*, 479.
296. S. V. Lymar, R. F. Khairutdinov, V. A. Soldatenkova, J. K. Hurst, *J. Phys. Chem. B* **1998**, *102*, 2811.
297. R. F. Khairutdinov, J. K. Hurst, *J. Phys. Chem. B* **1998**, *102*, 6663.
298. R. F. Khairutdinov, J. K. Hurst, *Nature* **1999**, *402*, 509.
299. D. K. Luttrull, J. Graham, J. A. DeRose, D. Gust, T. A. Moore, S. Lindsay, *Langmuir* **1992**, *8*, 765.
300. J. Zak, H. Yuan, M. Ho, L. K. Woo, M. D. Porter, *Langmuir* **1993**, *9*, 2772.
301. J. E. Hutchison, T. A. Postlethwaite, R. W. Murray, *Langmuir* **1993**, *9*, 3277.
302. T. Akiyama, H. Imahori, Y. Sakata, *Chem. Lett.* **1994**, 1447.
303. D. Li, L. W. Moore, B. I. Swanson, *Langmuir* **1994**, *10*, 1177.
304. G. A. Schick, Z. Sun, *Thin Solid Films*, **1994**, *248*, 86.
305. T. R. E. Simpson, D. A. Russell, I. Chambrier, M. J. Cook, A. B. Horn, S. C. Thorpe, *Sens. Actuators, B* **1995**, *29*, 353.
306. I. Chambrier, M. J. Cook, D. A. Russell, *Synthesis* **1995**, 1283.
307. K. Shimazu, M. Takechi, H. Fujii, M. Suzuki, H. Saiki, T. Yoshimura, K. Uosaki, *Thin Solid Films* **1996**, *273*, 250.
308. W. Han, S. Li, S. M. Lindsay, D. Gust, T. A. Moore, A. L. Moore, *Langmuir* **1996**, *12*, 5742.
309. T. Akiyama, H. Imahori, A. Ajavakom, Y. Sakata, *Chem. Lett.* **1996**, 907.
310. T. Kondo, T. Ito, S. Nomura, K. Uosaki, *Thin Solid Films* **1996**, *284–285*, 652.
311. T. R. E. Simpson, M. J. Cook, M. C. Petty, S. C. Thorpe, D. A. Russell, *Analyst* **1996**, *121*, 1501.
312. M. J. Cook, R. Hersans, J. McMurdo, D. A. Russell, *J. Mater. Chem.* **1996**, *6*, 149.
313. L.-H. Guo, G. McLendon, H. Razafitrimo, Y. Gao, *J. Mater. Chem.* **1996**, *6*, 369.
314. H. Yuan, L. K. Woo, *J. Porphyrin Phthalocyanine* **1997**, *1*, 189.
315. T. R. E. Simpson, D. J. Revell, M. J. Cook, D. A. Russell, *Langmuir* **1997**, *13*, 460.
316. J. E. Hutchison, T. A. Postlethwaite, C.-h. Chen, K. W. Hathcock, R. S. Ingram, W. Ou, R. W. Linton, R. W. Murray, *Langmuir* **1997**, *13*, 2143.
317. E. Katz, I. Willner, *Langmuir* **1997**, *13*, 3364.
318. H. Imahori, T. Azuma, K. Ushida, M. Takahashi, T. Akiyama, M. Hasegawa, T. Okada, Y. Sakata, *SPIE* **1997**, *3142*, 104.
319. K. Uosaki, T. Kondo, X.-Q. Zhang, M. Yanagida, *J. Am. Chem. Soc.* **1997**, *119*, 8367.
320. T. Kondo, M. Yanagida, S.-i. Nomura, T. Ito, K. Uosaki, *J. Electroanal. Chem.* **1997**, *438*, 121.
321. D. W. J. McCallien, P. L. Burn, H. L. Anderson, *J. Chem. Soc., Perkin Trans. 1* **1997**, 2581.
322. M. J. Crossley, J. K. Prashar, *Tetrahedron Lett.* **1997**, *38*, 6751.
323. H. Imahori and Y. Sakata, *Adv. Mater.* **1997**, *9*, 537.
324. D. L. Pillound, C. C. Moser, K. S. Reddy, P. L. Dutton, *Langmuir* **1998**, *14*, 4809.

325. H. Imahori, H. Norieda, S. Ozawa, K. Ushida, H. Yamada, T. Azuma, K. Tamaki, Y. Sakata, *Langmuir*, **1998**, *14*, 5335.
326. D. A. Offord, S. B. Sachs, M. S. Ennis, T. A. Eberspacher. J. H. Griffin, C. E. D. Chidsey, J. P. Collman, *J. Am. Chem. Soc.* **1998**, *120*, 4478.
327. A. Ishida, Y. Sakata, T. Majima, *Chem. Commun.* **1998**, 57.
328. A. Ishida, Y. Sakata, T. Majima, *Chem. Lett.* **1998**, 267.
329. M. Yanagida, T. Kanai, X.-Q. Zhang, T. Kondo, K. Uosaki, *Bull. Chem. Soc. Jpn.* **1998**, *71*, 2555.
330. F. D. Cruz, K. Driaf, C. Berthier, J.-M. Lameille, F. Armand, *Thin Solid Films* **1999**, *349*, 155.
331. T. Kondo, T. Kanai, K. Iso-o, K. Uosaki, *Z. Phys. Chem.* **1999**, *212*, 23.
332. H. Imahori, S. Ozawa, K. Ushida, M. Takahashi, T. Azuma, A. Ajavakom, T. Akiyama, M. Hasegawa, S. Taniguchi, T. Okada, Y. Sakata, *Bull. Chem. Soc. Jpn.* **1999**, *72*, 485.
333. H. Imahori, H. Yamada, S. Ozawa, K. Ushida, Y. Sakata, *Chem. Commun.* **1999**, 1165.
334. H. Imahori, Y. Sakata, *Eur. J. Org. Chem.* **1999**, 2445.
335. N. Nishimura, M. Ooi, K. Shimazu, H. Fujii, K. Uosaki, *J. Electroanal. Chem.* **1999**, *473*, 75.
336. A. Ishida, T. Majima, *Chem. Commun.* **1999**, 1299.
337. D. T. Gryko, C. Clausen, J. S. Lindsey, *J. Org. Chem.* **1999**, *64*, 8635.
338. H. Imahori, H. Norieda, Y. Nishimura, I. Yamazaki, K. Higuchi, N. Kato, T. Motohiro, H. Yamada, T. Tamaki, M. Arimura, Y. Sakata, *J. Phys. Chem. B*, **2000**, *104*, 1253.
339. H. Imahori, H. Yamada, Y. Nishimura, I. Yamazaki, Y. Sakata, *J. Phys. Chem. B*, **2000**, *104*, 2099.
340. J. Kiwi, M. Grätzel, *J. Phys. Chem.* **1980**, *84*, 1503.
341. U. Resch, S. M. Hubig, M. A. Fox, *Langmuir* **1991**, *7*, 2923.
342. A. Ulman, *Chem. Rev.* **1996**, *96*, 1533.
343. H. Y. Liu, M. J. Weaver, C.-B. Wang, C. K. Chang, *J. Electroanal. Chem.* **1983**, *145*, 439.
344. H. Y. Liu, I. Abdalmuhdi, C. K. Chang, F. C. Anson, *J. Phys. Chem.* **1985**, *89*, 665.
345. J. P. Collman, K. Kim, *J. Am. Chem. Soc.* **1986**, *108*, 7847.
346. J. P. Collman, L. L. Chang, D. A. Tyvoll, *Inorg. Chem.* **1995**, *34*, 1311.
347. M. Anikin, N. V. Tkachenko, H. Lemmetyinen, *Langmuir* **1997**, *13*, 3002.
348. H. A. Dick, J. R. Bolton, G. Picard, G. Munger, R. M. Leblanc, *Langmuir* **1988**, *4*, 133.
349. D. Gust, T. A. Moore, A. L. Moore, D. K. Luttrull, J. M. DeGraziano, N. J. Boldt, M. V. Auweraer, F. C. De Schryver, *Langmuir* **1991**, *7*, 1483.
350. B. Choudhury, A. C. Weedon, J. R. Bolton, *Langmuir* **1998**, *14*, 6192.
351. B. Choudhury, A. C. Weedon, J. R. Bolton, *Langmuir* **1998**, *14*, 6199.
352. R. R. Chance, A. Prock, R. Silbey, *Adv. Chem. Phys.* **1978**, *37*, 1.
353. D. H. Waldeck, A. P. Alivisatos, C. B. Harris, *Surface Sci.* **1985**, *158*, 103.
354. X.-L. Zhou, X.-Y. Zhu, J. M. White, *Acc. Chem. Res.* **1990**, *23*, 327.
355. G. Cnosse, K. E. Drabe, D. A. Wiersma, *J. Chem. Phys.* **1993**, *98*, 5276.
356. W. L. Barnes, *J. Mod. Opt.* **1998**, *45*, 661.
357. D. G. Hanken, R. M. Corn, *Anal. Chem.* **1995**, *67*, 3767.
358. G. B. Sigal, C. Bamdad, A. Barberis, J. Strominger, Jr., G. M. Whitesides, *Anal. Chem.* **1996**, *68*, 490.
359. Y. Sato, M. Fujita, F. Mizutani, K. Uosaki, *J. Electroanal. Chem.* **1996**, *409*, 145.
360. H. Imahori, K. Hagiwara, T. Akiyama, S. Taniguchi, T. Okada, Y. Sakata, *Chem. Lett.* **1995**, 265.
361. H. Imahori, Y. Sakata, *Chem. Lett.* **1996**, 199.
362. H. Imahori, K. Hagiwara, T. Akiyama, M. Aoki, S. Taniguchi, T. Okada, M. Shirakawa, Y. Sakata, *J. Am. Chem. Soc.* **1996**, *118*, 11771.
363. H. Imahori, K. Hagiwara, T. Akiyama, M. Aoki, S. Taniguchi, T. Okada, M. Shirakawa, Y. Sakata, *Chem. Phys. Lett.* **1996**, *263*, 545.
364. Y. Sakata, H. Imahori, H. Tsue, S. Higashida, T. Akiyama, E. Yoshizawa, M. Aoki, K. Yamada, K. Hagiwara, S. Taniguchi, T. Okada, *Pure Appl. Chem.* **1997**, *69*, 1951.
365. H. Imahori, K. Yamada, M. Hasegawa, S. Taniguchi, T. Okada, Y. Sakata, *Angew. Chem. Int. Ed. Engl.* **1997**, *36*, 2626.
366. S. Higashida, H. Imahori, T. Kaneda, Y. Sakata, *Chem. Lett.* **1998**, 605.

367. K. Tamaki, H. Imahori, Y. Nishimura, I. Yamazaki, A. Shimomura, T. Okada, Y. Sakata, *Chem. Lett.* **1999**, 227.
368. K. Tamaki, H. Imahori, Y. Nishimura, I. Yamazaki, Y. Sakata, *Chem. Commun.* **1999**, 625.
369. M. Fujitsuka, O. Ito, H. Imahori, K. Yamada, H. Yamada, Y. Sakata, *Chem. Lett.* **1999**, 721.
370. K. Yamada, H. Imahori, Y. Nishimura, I. Yamazaki, Y. Sakata, *Chem. Lett.* **1999**, 895.
371. H. Imahori, K. Tamaki, H. Yamada, K. Yamada, Y. Sakata, Y. Nishimura, I. Yamazaki, M. Fujitsuka, O. Ito, *Carbon*, **2000**, *38*, 1599.
372. C. Luo, D. M. Guldi, H. Imahori, K. Tamaki, Y. Sakata, *J. Am. Chem. Soc.* **2000**, *122*, 6535.
373. N. V. Tkachenko, C. Guenther, H. Imahori, K. Tamaki, Y. Sakata, S. Fukuzumi, H. Lemmetyinen, *Chem. Phys. Lett.* **2000**, *326*, 344.
374. C. A. Mirkin, W. B. Caldwell, *Tetrahedron* **1996**, *52*, 5113.
375. H. Imahori, T. Azuma, S. Ozawa, H. Yamada, K. Ushida, A. Ajavakom, H. Norieda, Y. Sakata, *Chem. Commun.* **1999**, 557.
376. H. Imahori, T. Azuma, A. Ajavakom, H. Norieda, H. Yamada, Y. Sakata, *J. Phys. Chem. B* **1999**, *103*, 7233.
377. M. Fujihira, *Mol. Cryst. Liq. Cryst.* **1990**, *183*, 59.
378. *In Standard Potentials in Aqueous Solution* (Eds.: A. J. Bard, R. Rarsons, J. Jordan), Marcel Dekker, New York, **1985**.

# 9 ESR Spectroscopy of Inorganic and Organometallic Radicals

*Wolfgang Kaim*

## 9.1 Introduction: The Information Accessible from ESR

Electron spin resonance (ESR) or, more generally, electron paramagnetic resonance (EPR) [1], is a physical method of analysis which, although restricted to paramagnetic species, is a powerful tool not only for the identification of such systems but also for kinetic investigations and, above all, for the study of their electronic structures. Information can come from the often temperature-dependent line-widths of the ESR signals (which are related to mobility and relaxation behavior), from the $g$ factors (the equivalents of NMR 'chemical shifts'), from electron–nuclear hyperfine coupling (which might require electron–nuclear double resonance, ENDOR, as a more sophisticated method of analysis), and from zero-field splitting parameters, $D$, or exchange coupling values, $J$, for two-spin (triplet) or multi-spin systems.

It is assumed that the interested reader of this chapter has a basic knowledge of the theory and instrumentation associated with this particular variety of magnetic resonance, including perhaps a general introduction to the field of organic radicals. In keeping with the handbook character of this series, the following discussion will focus on special aspects of this method in inorganic and organometallic chemistry with emphasis on practical guidelines and references for the experimentalist, as suggested by own experience and illustrated by selected recent examples; a complementary excellent monograph on radical processes specifically in transition metal chemistry has recently appeared [2].

## 9.2 Inorganic and Organometallic Radicals: Definition, Generation and Peculiarities

The ESR spectroscopy of inorganic and organometallic systems with main group or transition element [2, 3] paramagnetic centers is distinguished from the study of organic radicals [4] by this method:

1) by the large number of isotopes available from the periodic table for electron–nuclear hyperfine interaction [5];
2) by the large spin–orbit coupling effects from the heavier elements (which affect both line-widths and $g$ factors) [6];
3) by the possible use of s, p, d or f orbitals for the primary accommodation of spin; and
4) by possible occurrence of spin-delocalized spatial arrangements ('clusters') like, e.g., near-octahedral $[B_6X_6]^{\bullet-}$ [7], which are quite different from the familiar planar $\pi$ systems in organic radical compounds such as the formally analogous ions $[C_6X_6]^{\bullet-}$ (X = halogen) [8].

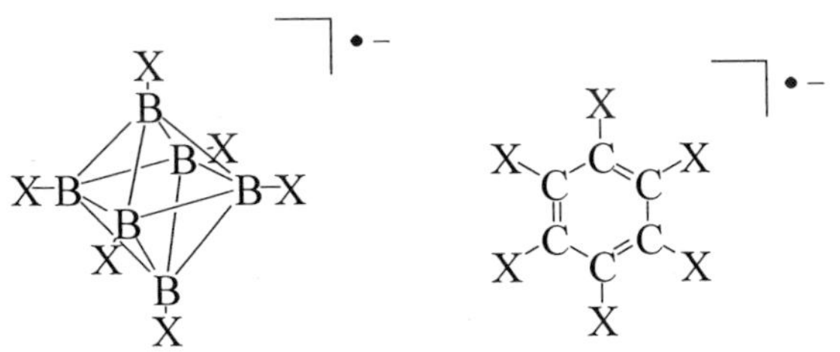

Although inorganic or organometallic radical species may well consist of a single paramagnetic center with at least one unpaired s, p, d or f electron, there are also oligonuclear compounds with the ESR-detectable spin distributed over more than one 'inorganic' atom. Such phenomena are quite familiar in the field of oligomers (e.g. the 'platinum blues' $[Pt^{2.x}L_n]_4$) or paramagnetic clusters of transition metals such as $(CoCp)_3(\mu_3\text{-CPh})_2$ [9]; they also occur, however, in the area of main-group elements as exemplified by ESR-characterized species such as $[R_2Al–AlR]^{\bullet}$ [10], $[R_4Ga_4]^{\bullet-}$ [11], $[B_6X_6]^{\bullet-}$ [7], or $[B_9X_9]^{\bullet-}$ [12] (R = alkyl, X = halogen), all with Group 13 element-centered spins.

In the context of inorganic and organometallic chemistry the term 'radical' needs a qualifying statement. Whereas paramagnetic main-group element compounds and most paramagnetic d block organometallic complexes with significant covalent bonding are usually referred to as radicals, the classical, i.e. non-organometallic transition metal coordination compounds [2, 3] e.g. of copper(II) or manganese(II) are usually not labeled correspondingly. Thus, the compounds $(RX)Mn(CO)_2(C_5R'_5)$ have been regarded as 'radical' species [13] although the spin resides mainly on the metal (low-spin $Mn^{II}$ configuration **A** instead of the radical alternative **B** in Eq. 1), depending to some extent on the electronic effect of R [14]. In contrast, more conventional low-spin manganese(II) complexes such as $[(ON)Mn(CN)_5]^{2-}$ are not generally referred to as radicals [15]. The main reason for this practice lies in the often analogous reactivity (e.g. H-abstraction [16]) of organometallic 'radicals' in comparison with typical organic radicals such as $^{\bullet}CH_3$.

$$(RX^-)Mn^{II}(CO)_2(C_5R'_5) \leftrightarrow (RX^{\bullet})Mn^{I}(CO)_2(C_5R'_5) \qquad (1)$$

$$\mathbf{A} \qquad\qquad\qquad\qquad\qquad \mathbf{B}$$

X = NH, S, Se; R' = H, CH$_3$; R = aryl, alkyl

Paramagnetic coordination compounds where the spin resides *predominantly* on one or more of the organic (or inorganic) ligands are quite common, especially complexes of radical anions [17] such as *o*-semiquinones **1** or other negatively charged chelates such as tetrapyrrole macrocycles **2** [18].

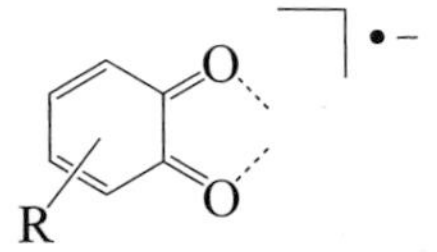

**1** 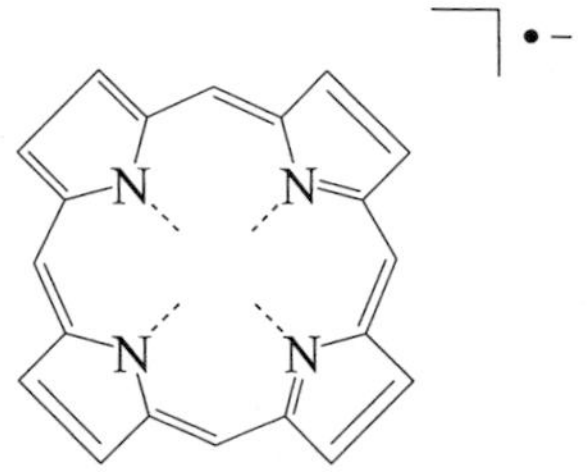                    **2**

(⋯⋯ : potential coordinative bond)

Neutral radicals (e.g. NO˙ [19] or *N*-methylpyrazinium **3** [20]) and even cation radicals such as **4** [21] can also bind metal complex fragments to form ESR-detectable species.

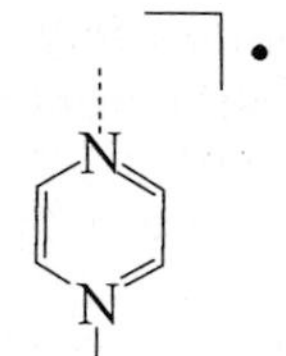

**3** 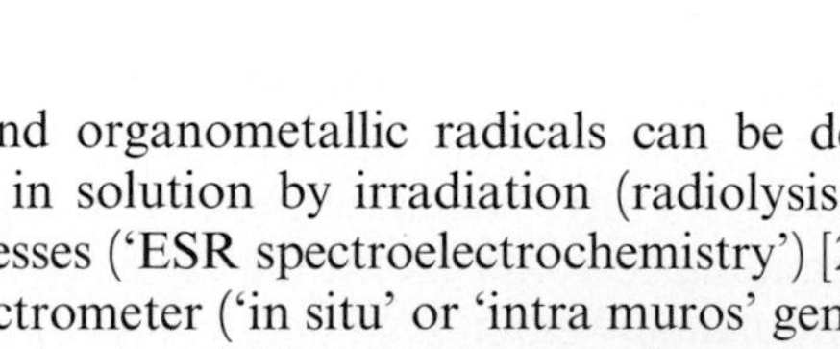                    **4**

Inorganic and organometallic radicals can be deliberately generated for ESR measurements in solution by irradiation (radiolysis, photolysis) [22] or by use of electrode processes ('ESR spectroelectrochemistry') [23] inside the microwave cavity of an ESR spectrometer ('in situ' or 'intra muros' generation), and they can occur as intermediates (transients) [2] or final 'escape' products of thermally or light-induced electron transfer reactions [24, 25]. The use of suitable oxidants or reductants for chemical radical generation in the organometallic field has been excellently reviewed by Connelly and Geiger [26].

The following discussion summarizes important criteria and parameters for the application of ESR in inorganic and organometallic chemistry, illustrated by selected examples.

## 9.3 Selected ESR Parameters

### 9.3.1 Concentration and Lifetime of Radicals

Many of the frequently invoked radicals in inorganic or organometallic chemistry, e.g. the thermally or photochemically generated species $\cdot M(CO)_5$, $M = Mn$ or Re [27], are too short-lived to be observed in the necessary useful concentrations of about $c > 10^{-7}$ M by conventional ESR methods under standard conditions. Although continuous wave (CW) ESR is a well-established detection method for kinetic techniques such as stopped or continuous flow [28], this method is limited to lifetimes of ca $\tau > 10^{-4}$ s for the paramagnetic intermediates. To detect and study species which would be shorter-lived under standard conditions the following techniques have been developed; they are illustrated by examples involving an important elementary process in organometallic chemistry, viz. homolytic $M-C_{sp^3}$ bond dissociation.

**Irradiation of immobilized species in matrices by use of electrons, UV–Vis, X- or $\gamma$-rays and direct observation of thus stabilized radicals under static conditions [22]**

Many short-lived small inorganic radicals and unusual transition metal centers have been generated by photolysis or radiolysis methods in matrices [22]. (The matrix can be a glassy frozen solvent or a solid, e.g. zeolite or other oxide). This method is, however, also suitable for investigation of rather complex systems such as alkyl cobalamins and related '$B_{12}$' species [29, 30]. Irradiation with visible light is even sufficient to cleave the crucial $Co-C_{alkyl}$ bond in these macrocyclic compounds to yield $Co^{II}$ products and the less persistent alkyl radical intermediates; both can be detected by ESR in low-temperature matrices (Eq. 2) [30].

$$[Co^{III}]-CH_2R \quad \xrightarrow[77\ K]{h\nu} \quad [Co^{II}] + \cdot CH_2R \tag{2}$$

$$CD_3OD$$

[Co] = cobalt corrin complex

**Trapping of short-lived radical intermediates as generated by photolytic, thermal or chemical processes with the help of NO-containing 'spin traps' such as alkyl nitrones or nitroso compounds [31]**

This popular approach, illustrated by Eq. 3 [32] for an organometallic example, must be carefully checked with respect to the question whether the resulting stable and thus ESR-observable species (e.g. a nitroxide radical complex) is perhaps just a product from a preformed 'charge-transfer' complex between the substrate and the spin trap.

$$(N^\wedge N)PtMe_4 \xrightarrow[\text{240 K}]{h\nu} Me^\bullet + [(N^\wedge N)PtMe_3]^\bullet$$

$$Et_2O \quad \text{(not directly ESR detectable)}$$

$$\xrightarrow{+\textit{tert.-}BuNO} \quad (N^\wedge N)PtMe_3 \tag{3}$$
$$|$$
$$\textit{tert.-}BuNO^\bullet$$

$N^\wedge N = \alpha\text{-diimine}$

$$a(^{14}N) = 1.58 \text{ mT}$$

$$a(^{195}Pt) = 4.0 \text{ mT}$$

**Pulsed ESR methods**

Pulsed ESR methods (FT-ESR) have been increasingly developed during the last decade for use in direct time-resolved studies [33] of transient species. Although this technique is excellently suited to the investigation of typical inorganic and organometallic reactivity such as the light-induced metal/$sp^3$-carbon bond homolysis (Eq. 4) [34], the corresponding metal-containing radical products such as $[(N^\wedge N)Re(CO)_3]^\bullet$ ($N^\wedge N = \alpha$-diimine ligand) may not be directly detectable owing to unfavorable relaxation properties (cf. Section 9.3.2).

$$[(N^\wedge N)Re(CO)_3(CH_3)] \xrightarrow[\text{toluene}]{h\nu} [(N^\wedge N)Re(CO)_3]^\bullet + {}^\bullet CH_3 \tag{4}$$

$N^\wedge N = 4,4'\text{-dimethyl-2,2'-bipyridine}$

This organometallic radical product could, however, be detected by irradiation in glassy frozen matrices [34]. The alkyl radicals, as very short-lived metal-free products, could be observed and analyzed by FT-ESR with respect to CIDEP (chemically induced dynamic electron polarization) effects [34].

**9.3.2 Line-width and Temperature**

Whereas almost all organic radicals give rather narrow ESR signals in solution, with widths of individual hyperfine lines well below 1 Gauss (0.1 mT) [4], signals in the spectra of inorganic and organometallic radicals can be severely broadened, sometimes beyond ESR detectability even at very low temperatures. In the absence of special (e.g. dynamic) effects, this phenomenon is usually a result of strongly shortened relaxation times after excitation by microwave radiation. The enhanced relaxation of inorganic or organometallic radicals is generally favored by the presence of heavy atoms with their large spin–orbit coupling constants [6] and the numerous available states, often close-lying, involving contributions from s, p, d or f atomic orbitals of one or more of the 'inorganic' atoms. Di- and polynuclear radical compounds (clusters) and high symmetry species ($T_d$, $O_h$) with degenerate spin-bearing orbitals can have particularly pronounced effects for the very same reason—close-lying excited states from which rapid relaxation to the ground state can occur.

To detect such radicals by ESR, low temperatures must be routinely employed, and even then otherwise (e.g. structurally or magnetically) well-characterized species can remain 'ESR-silent' at 77 K ($[B_6I_6]^{\bullet-}$ [8], $[Fe(C_5H_5)_2]^+$ [35a], $[(bpy)_2Ru(\mu\text{-}OC(R)NNC(R)O)Ru(bpy)_2]^{3+}$ [35b]) or 4 K ($Re^{VI}S_4{}^{2-}$ [36a], $[(N^\wedge N)Pt^{III}R_2]^+$ ($N^\wedge N$ = $\alpha$-diimine) [36b], $[In_4R_4]^{\bullet-}$ [11], $[(bpy)_2Os(\mu\text{-}OC(R)NNC(R)O)Os(bpy)_2]^{3+}$ [35c]). Obviously, such effects can occur with main-group element-, transition metal- and even ligand-centered spins [37]. NMR of paramagnetic substances can then sometimes be employed as a complementary technique [38]; lowering the symmetry, e.g. by coordination of metal complex fragments to $ReS_4{}^{2-}$ is another alternative [36a, c].

In contrast, disappearance of ESR signals when the temperature is *lowered* is usually a sign of intermolecular association, e.g. to form a spin–spin coupled dimer with a singlet ground state. Triplet states might, however, then be thermally accessible, or the triplet may even become the ground state of the system (cf. Section 9.3.5).

Decreasing ESR line-width and improved spectral resolution at higher temperatures is usually a sign of enhanced mobility in solution resulting in better averaged anisotropic contributions from hyperfine and $g$ tensors; ESR spectroscopy at X-band frequency is distinguished by a characteristic time-scale of ca $10^{-8}$ s (cf. Section 9.3.6). Inorganic and organometallic radicals are often severely affected by selective 'anisotropic line-broadening' [4] because of large $A$ or $g$ anisotropy and/or slow tumbling (long rotational correlation times) in solution; the higher masses of inorganic elements and the resulting larger moments of inertia often contribute to increased ESR line-widths as exemplified by the series of compounds $[R_2M\text{–}MR]^{\bullet}$, M = Al, Ga [10, 39b, c], or **6** [39a].

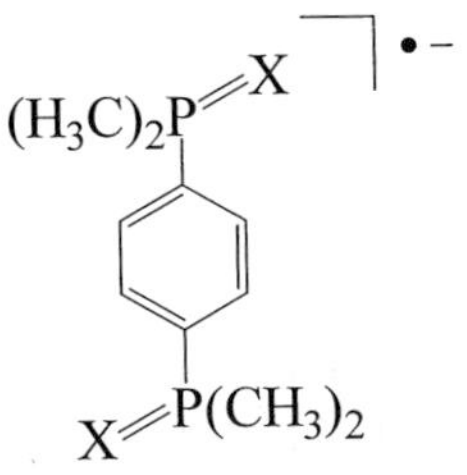

X = O, S, Se

**6**

### 9.3.3 The *g* Factor

According to the ESR resonance condition, $h\nu_r = g\beta H_r$, the $g$ value is the proportionality factor between the resonance frequency $\nu_r$ and the resonance field $H_r$.

The $g$ factor thus has the same role as the absolute chemical shift in NMR spectroscopy. In a purely electron spin-governed ('spin only') situation, the $g$ factor would have the value $g_e = 2.0023$ as calculated and measured for the free electron [1], although even small contributions from excited states with non-zero orbital angular momentum through spin–orbit coupling will affect this value, especially if nuclei with high spin–orbit coupling constants [6] contribute significantly. Thus easily detectable deviations, $\Delta g$, of the isotropic $g$ factor from the free electron value are observed for inorganic and organometallic radicals, even if the spin is still largely (>95 %) ligand-centered; a deviation such as $g_{iso} = 2.0093$ for the $Cu^I$– semiquinone complex **7** [40] translates to $\Delta g = 0.0070$, a relative shift of about 3500 ppm.

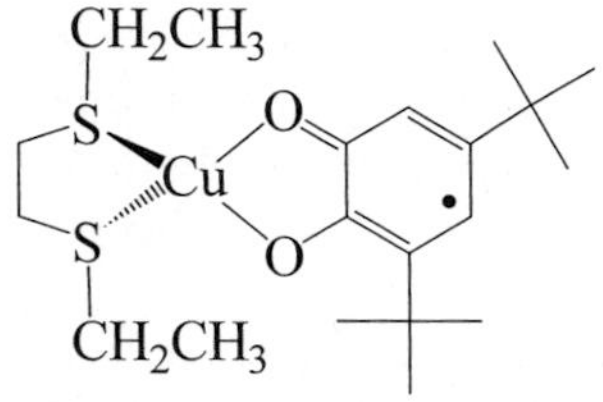

**7**

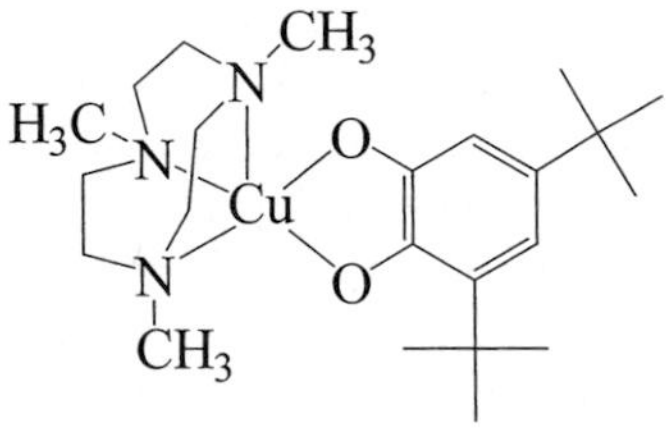

**8**

For metal-centered spin, this deviation can be substantially higher. Thus, for the $Cu^{II}$/catecholate compound **8** $g_{iso} = 2.113$, which corresponds to a shift of ca $5.5 \times 10^4$ ppm [40].

The sign of the deviation is indicative of the frontier orbital situation—according to Stone's approximation (Eq. 5) [41]:

$$g = g_e - \frac{2}{3}\sum_i \sum_n \sum_{kj} \frac{\langle\psi_0|\xi_k\mathbf{L}_{ik}\delta_k|\psi_n\rangle\langle\psi_n|\mathbf{L}_{ij}\delta_j|\psi_0\rangle}{E_n - E_0} = g_e + \Delta g \qquad (5)$$

$\xi$: spin orbit coupling constant

$\mathbf{L}$: angular momentum operator

$E_0$: energy of singly occupied molecular orbital (SOMO)

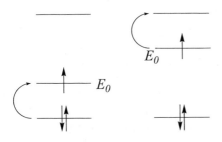

dominant term:
$E_0 > E_n$
$g > g_e$

dominant term:
$E_0 < E_n$
$g < g_e$

**Scheme 1.**

A deviation to higher values than $g_e$ is to be expected if $E_0 > E_n$, i.e. if the highest lying fully occupied MO lies closer to the SOMO than the lowest empty MO. Such is the case for almost all low-spin $d^5$ and $d^9$ systems [2] (cf. **8**) or for the radical complex **7** [40].

The other alternative, $E_0 < E_n$, with the lowest empty MO lying closer to the SOMO than the highest occupied MO leads to lower $g$ factors than $g_e$ as is typical for $d^1$ systems [2] and, e.g., for **9** [42].

$\overset{\bullet\,+}{\left[N\text{---}Ru^{II}\text{---}N\right]}$

$\overset{\frown}{N\quad N}$ : bpy and related ligands

**9**

The effect of increasing spin–orbit coupling constants on the deviation $\Delta g$ of $g$ from $g_e$ is evident, e.g., for Group-6 carbonylmetal(0) complexes $[(OC)_n M]_k (A^{\bullet-})$ of heterocyclic anion radical ligands where the greatest such deviation—positive or negative—is invariably observed for the tungsten homolog [43]. Such correlations are, however, only valid for paramagnetic systems with the inorganic nuclei in chemically comparable environments. An example where the $4d^5$ compound has less $g$ anisotropy $g_1 - g_3$ than a $3d^5$ analog is described below [44a].

In efforts to mimic the function of heterobimetallic hydrogenases and provide potential molecular catalysts for fuel cell technology the Fe,Ru-heterobimetallic hydride complex $[Cp^*RuH(dppf)]$, dppf = 1,1′-bis(diphenylphosphino)ferrocene, has been reported [44b] to catalyze the elementary reaction $H_2 \rightarrow 2H^+ + 2e^-$. It was suggested [44b] that the crucial oxidation involves the ruthenium center and not

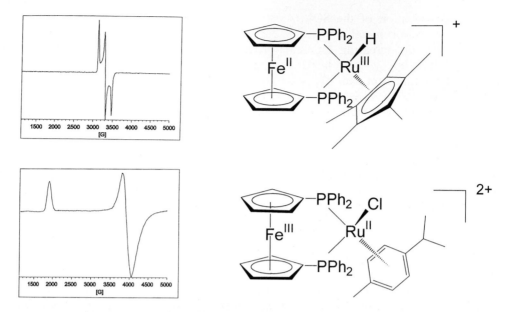

**Figure 1.** Comparison of iron and ruthenium oxidation in 1,1′-bis(diphenylphosphino)ferrocene–ruthenium(II) complexes (ESR spectra [44a] from electrolytically oxidized solutions of precursors [44b, c] in THF at 4 K): Oxidation of [Cp*RuH(dppf)], dppf = 1,1′-bis(diphenylphosphino)ferrocene, and [(Cym)RuCl(dppf)](PF$_6$), Cym = *p*-cymene, occurs at ruthenium for the former compound but yields a ferrocenium species for the latter.

the iron of the ferrocene ligand. Figure 1 illustrates the EPR evidence that there is a viable alternative Fe$^{II}$Ru$^{III}$ rather than Fe$^{III}$Ru$^{II}$ by describing a similar species, the recently structurally characterized [44c] complex [(Cym)RuCl(dppf)](PF$_6$) with a neutral arene (*p*-cymene) instead of a cyclopentadienide and a chloride instead of a hydride co-ligand at ruthenium.

Both complexes are reversibly oxidized but the resulting ESR signals in glassy frozen solution differ significantly (Figure 1). Whereas the organometallic low-symmetry low-spin ruthenium(III) center in [Cp*RuH(dppf)]$^+$ has relatively small $g$ anisotropy with $g_1 = 2.187$, $g_2 = 2.088$ and $g_3 = 1.991$ ($g_1 - g_3 = 0.196$), the ferrocenium systems with their special symmetry and d orbital splitting are typically distinguished by broad ESR signals, often observable only at 4 K [35a], with large $g$ anisotropy. Complex [(Cym)RuCl(dppf)]$^{2+}$ is thus properly formulated as a ferrocenium/Ru$^{II}$ system with $g_1 = 3.612$ and $g_{2,3} = 1.765$ ($g_1 - g_3 = 1.847$) [44a].

In just the same way as the hyperfine coupling, the $g$ factor has tensor structure [1] which can be studied in detail by examining single crystals of radicals [3b, 45]. In many instances, however, the intermediate character of such species allows for investigation in the powder state only or in glassy frozen solutions which then yield up to three separated $g$ factor components, according to the symmetry of the SOMO [3]. The splitting of these $g$ components, including the total $g$ anisotropy, symmetry, and positions of the individual components is valuable ESR information for the identification of the SOMO as has long been known for classical para-magnetic transition metal complexes, e.g. in metal d [3] or f [46] orbital analysis. The influence of increasing spin–orbit coupling constants from heavier elements on the $g$ anisotropy $g_1$–$g_3$ is nicely illustrated by the series of chalcogen-based 19-electron triatomic bent radicals [47a] $O_3^{\cdot-}$ ($g_1 - g_3 = 0.0161$), $SeO_2^{\cdot-}$ ($g_1 - g_3 = 0.0342$), and the new $TeS_2^{\cdot-}$ ($g_1 - g_3 = 0.0719$) [47b].

Large line-widths, sometimes in conjunction with insufficiently resolved hyperfine structure, can preclude the determination of individual $g$ factor components close to $g = 2$ under conventional ESR conditions (X band, 9.5 GHz, 330 mT). As in NMR spectroscopy, an increase in the field and frequency is then appropriate via the Q band (35 GHz) to the W band (95 GHz) or even higher frequencies and magnetic fields [48]. An application of such high-field ESR for the inorganic radical **10** with sizeable metal participation has been reported [49a].

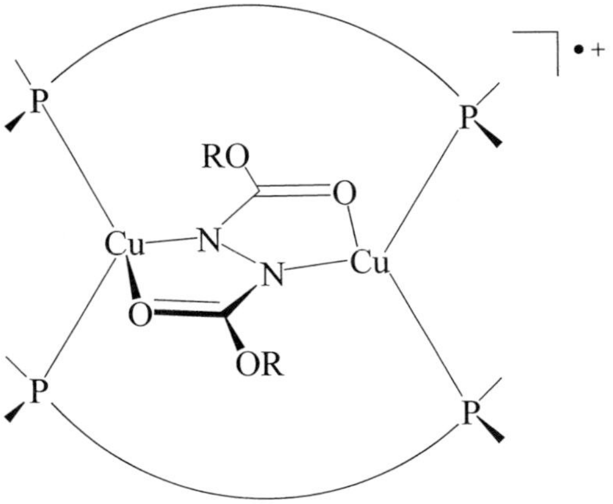

**10**

The rather large $^{63,65}$Cu hyperfine coupling of about 1.7 mT [49b] from two copper nuclei in the stable, crystalline [49c] blue complex **10** has raised the question whether these centers are still correctly described as $Cu^I$ or whether they have some $Cu^{II}$ character. Unfortunately, the extensive hyperfine coupling [49b] from two $^{63,65}$Cu, two $^{14}$N and four $^{31}$P nuclei caused complete broadening of the X band ESR signal in glassy frozen solution. Increasing the frequency by a factor of 25 to 245 GHz, however, revealed three well-separated $g$ components [49a] the values of which ruled out any significant contribution from copper(II) states.

### 9.3.4 Electron–Nuclear Hyperfine Coupling

In principle, ESR hyperfine coupling between the unpaired electron and the various nuclei (isotopes) of the radical species is the most informative data from ESR, both for identification of the species and for elucidation of the electronic structure. Although, ideally, all nuclei with non-zero nuclear spins couple to a certain extent with the unpaired electron in a molecule and thus reveal the nature of the singly occupied molecular orbital (SOMO), there might be severe limitations affecting this kind of information.

1) To begin with, the intrinsic line-width might be too large for the resolution of the hyperfine structure for the reasons given in Section 9.3.2 or because of dynamic phenomena (e.g. electron hopping [42, 50]; Section 9.3.6). Graphical analysis of only partially resolved ESR hyperfine structure may be facilitated by mathematical derivatization, e.g. of the usually recorded first derivative to the second derivative spectrum [45] (whereby mere turning points become maxima or minima), or by computer-assisted simulation (spectrum synthesis) with estimated line-widths and line-shapes. Computer simulation is not only necessary when very many lines (see point 4, below) from several different coupling nuclei overlap, it may also be appropriate when large line-widths and unusual nuclear spin situations combine to yield ESR spectra of unfamiliar appearance. One such example are the various manifestations of 'two-line' patterns as observed [51, 86], e.g., for the dominant coupling of one rhenium atom with the unpaired electron. It arises through the cancellation (by overlap) of the innermost lines of the sextet expected for the hyperfine interaction with $^{185,187}$Re (100 % total natural abundance, $I = 5/2$ and very similar nuclear magnetic moments for both isotopes). Increasing the line-width by lowering the temperature illustrates this effect experimentally (Figure 2).

2) Secondly, the isotopic properties [5] of the nuclei of interest might not be suitable because of absent nuclear spin ($I = 0$: e.g. $^{12}$C, $^{18}$O, $^{32}$S, $^{56}$Fe) or the low natural abundance of isotopes with $I \neq 0$ ($^{13}$C: 1.1 %, $I = 1/2$; $^{17}$O: 0.038 %, $I = 5/2$; $^{33}$S: 0.75 %, $I = 3/2$; $^{57}$Fe: 2.15 %, $I = 1/2$). Manipulation of the natural isotopic composition, e.g. by as selective enrichment, is then appropriate. The enrichment may involve metal isotopes such as $^{95}$Mo (15.9 %, $I = 5/2$) or $^{61}$Ni (1.1 %, $I = 3/2$) in biological probes [52] (where this method might even analytically establish the presence of corresponding trace elements), or, e.g., $^{13}$C nuclei in radical ligands, as illustrated by compound **11** with its very different participation of axial and equatorial carbonyl ligands in the spin distribution [53, 64b].

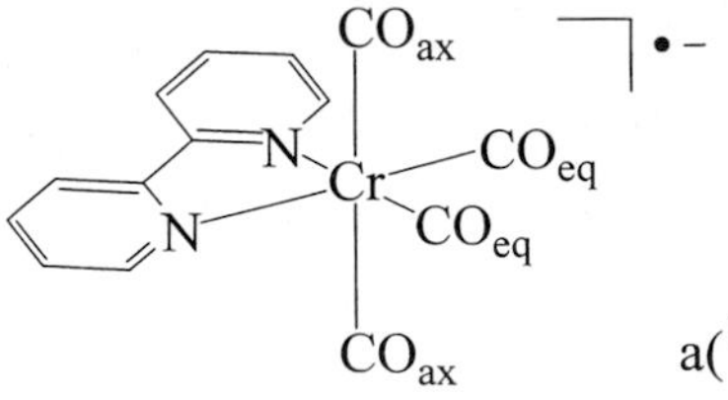

$$a(^{13}C_{eq}) < 0.01 \text{ mT}$$

$$a(^{13}C_{ax}) = 0.601 \text{ mT}$$

**11**

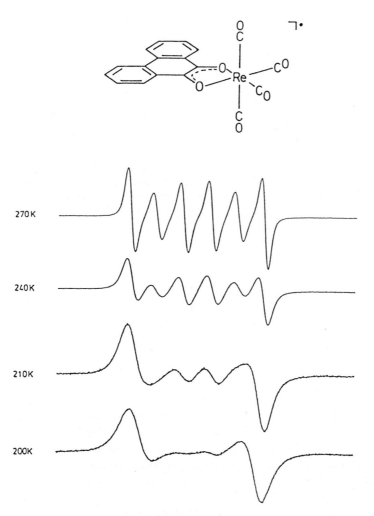

**Figure 2.** Temperature dependence of the ESR spectrum of tetracarbonyl(phenanthrenesemiquinone)rhenium(I) complex [(pq)Re(CO)$_4$]$^•$ in toluene (cf. K. A. M. Creber, T. I. Ho, C. Depew, D. Weir, J. K. S. Wan, *Can. J. Chem.* **1982**, *60*, 1504). The $^{185,187}$Re sextet hyperfine coupling dominates, yet increasing line-width at lower temperatures causes significant distortion of the sextet coupling as a result of overlapping of the innermost lines.

Similarly, de-enrichment or isotope exchange may also improve the detectability of ESR hyperfine structure and permit the assignment of coupling constants. Isotopically pure $^{65}$Cu instead of the naturally occurring $^{63,65}$Cu isotope mixture has been used in bioinorganic studies and H/D exchange ($^2$D: $I = 1$, $\mu_H/\mu_D = 6.5$) is frequently employed in, e.g., transition metal hydride chemistry [54].

3) A further limitation even in otherwise favorable circumstances is the very low

nuclear magnetic moment $\mu$ of certain isotopes such as $^{101}$Rh (100 %, $I = 1/2$) [55a] or $^{191,193}$Ir (100 %, $I = 3/2$) which then results in small and frequently undetectable hyperfine coupling, even with high spin density on the nucleus. As an example, distinctly increased $g$ anisotropy and relaxation rates are observed for electrogenerated $M^I/M^{II}$ mixed-valent systems $[(Cp^*M(\mu\text{-dip})MCp^*]^+$ (M = Rh, Ir; dip = 2,5-diiminopyrazine ligand) in comparison with the related radicals $[Cp^*ClM(\mu\text{-dip})MClCp^*]^{\cdot+}$; hyperfine information (which would settle the question of localized or delocalized valences) could not, however, be obtained because of the unfavorable isotopic situation [55b, c].

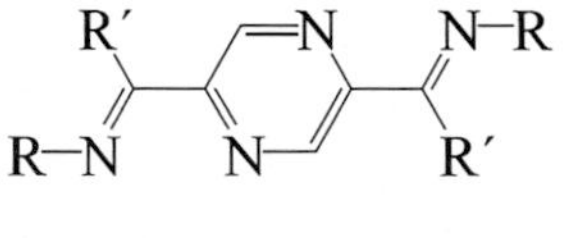

dip

Conversely, a fairly large inorganic isotope coupling constant such as 1.8 mT for $^{69,71}$Ga may suggest an unusual oxidation state (here mononuclear Ga$^{II}$) [56] in a compound **12** where in fact a regular diamagnetic Ga$^{III}$ center binds one radical anion and one dianionic $\alpha$-diimine ligand [57].

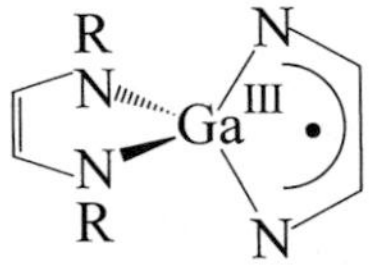

**12**

In a similar manner, the platinum(I) oxidation state has sometimes been invoked [58] in complexes $[(N^\wedge N)PtL_n]^k$ ($N^\wedge N$ = $\alpha$-diimine ligand) because of large $^{195}$Pt isotope coupling where in fact the Pt$^{II}$ center is bound to a one-electron reduced $N^\wedge N$ ligand [36b, 59]. True Pt$^I$ species with 5d$^9$ configuration are quite rare and reactive and have very large coupling constants and $g$ anisotropies [60].

4) The multiplicative connection in Eq. 6 for the total number of hyperfine lines can rapidly lead to so many overlapping individual lines that analysis from the ESR experiment alone becomes impossible and an unresolved, inhomogeneously broadened signal results.

$$N = \prod_i (2n_i I_i + 1) \tag{6}$$

N: number of hyperfine lines

$n_i$: number of equivalent nuclei in set i

$I$: nuclear spin of the isotopes in set i

Fortunately, many nuclei do not contribute to hyperfine splitting in any given radical because of absent nuclear spin, low magnetic moment, or marginal spin density; there are, however, exceptions. For instance, the symmetrical delocalization of spin over all boron and halogen atoms in the stable, near-octahedral cluster radical anion $B_6I_6^{\cdot-}$ [7] results in a total number of 25 606 theoretical lines involving the nuclei $^{10}B$ (19.8 %, $I = 3$), $^{11}B$ (80.2, $I = 3/2$), $^{127}I$ (100 %, $I = 5/2$), according to a statistical treatment for the seven isotopic combinations of the chemically symmetrical species $[^{10}B_n{}^{11}B_{6-n}{}^{127}I_6]^{\cdot-}$, $n = 0$–6. In such cases, or for low-symmetry systems with many *different* coupling constants [21b], use of the electron–nuclear double resonance (ENDOR) technique reduces the number of lines, because each coupling constant is represented by only two lines on the radiofrequency scale [61]. Because of the required saturation of the ESR transition and rather specific requirements for the relaxation rates [61], which are different for each nucleus, the ENDOR method is not always straightforward with inorganic or organometallic radicals in solution; examples are, however, known for main group element radicals such as **13** ($^{10}B$, $^{11}B$, $^{27}Al$ heteronuclear ENDOR) [62] and transition metal containing species [21b, 31b, 63] such as **14** [63b].

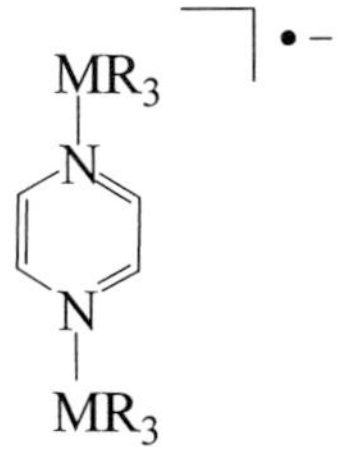

**13**

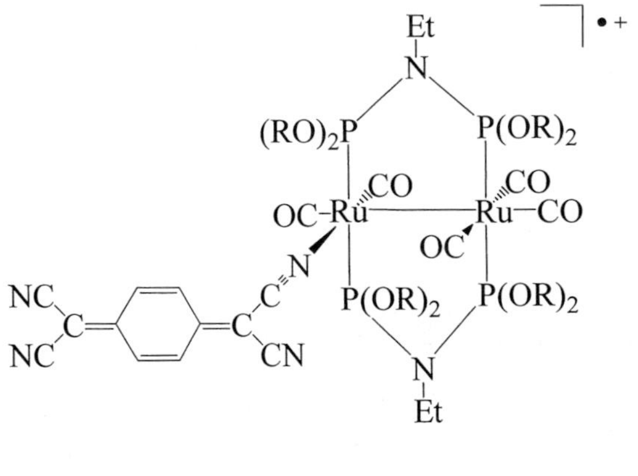

**14**

Pulsed ENDOR methods have been developed to obtain maximum hyperfine information [33a, 55a].

The occurrence of hyperfine coupling for a particular isotope indicates non-zero spin density at that nucleus in accordance with the Fermi contact term (which includes the electron wave function evaluated at the nucleus) [1]. The analysis in terms of spin density then requires at least a comparison with the (calculated [5]) isotropic hyperfine splitting constant $a_o$; ideally, results from increasingly available open-shell quantum-chemical calculation procedures [64] are employed.

The hyperfine structure may also indicate, in a rather elementary way, the symmetry and thus intramolecular electron exchange situation of radicals. For instance, organometallic mixed-valent species $(R_5C_5)(OC)_2Mn^{1.5}(\mu\text{-}L^-)Mn^{1.5}(CO)_2(C_5R_5)$, $L = RO^-$ or imidazolate (**15**), exhibit their delocalized ground state (involving rapid intramolecular electron exchange) through the coupling of one unpaired electron with two *equivalent* nuclei $^{55}Mn$ (100 %, $I = 5/2$) [65]. The resulting eleven lines are spaced with a coupling constant roughly one half of that for corresponding mononuclear species $(R_5C_5)(OC)_2Mn^{II}(L^-)$ [65]. In contrast, dinuclear $Mn^{III}Mn^{IV}$ model complexes for the oxygen-evolving manganese clusters in photosystem II clearly show 16 lines with an intensity distribution that indicates two different coupling constants for the two Mn centers, signifying slow intramolecular electron exchange and a localized mixed-valent situation [66].

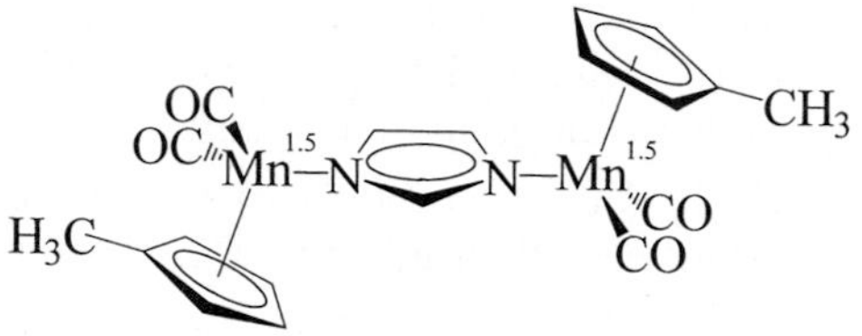

**15**

Similar metal–metal spin delocalization could be established directly by ESR for dimolybdenum systems **16** [67] or $[(Tp^*)(X)ClMo(\mu\text{-}L)MoCl(X)(Tp^*)]^x$,

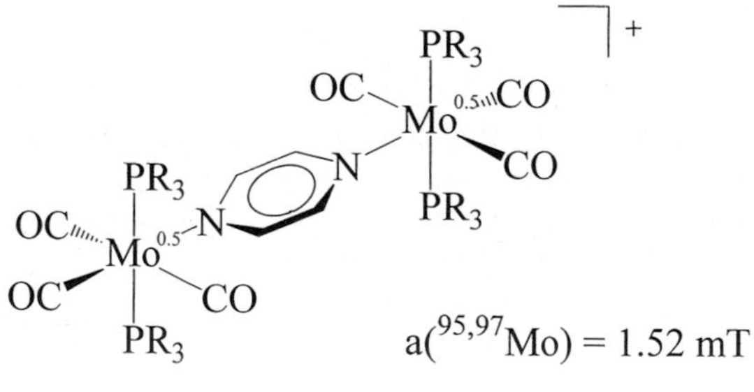

$a(^{95,97}Mo) = 1.52$ mT

**16**

$Tp^* =$ tris(pyrazolyl)borate derivative, $L =$ bispyridine ($X = NO$) or bisphenolate ($X = O$) bridging ligands [68]. Because of the presence of only 25.5 % as hyperfine coupling-active nuclei *Mo ($^{95}Mo$ with 15.9 % and $^{97}Mo$ with 9.6 %, both with $I = 5/2$), the delocalization is evident from the approximately halved hyperfine coupling constants (as compared with mononuclear $Mo^I$ species) and from the relative ESR intensities of lines corresponding to molecules with *Mo–*Mo, *Mo–Mo and Mo–Mo isotope combinations [67, 68].

### 9.3.5 Zero-field Splitting and Exchange Coupling

The presence of two spins in a triplet [70, 71] or biradical [72] species results in additional ESR features, viz. fine structure effects as a result of electron spin–electron spin interactions [69]. The phenomena observed and results derived therefrom are best discussed for simple examples, viz., the interaction of two equivalent centers in structurally well-characterized molecules.

A particularly simple situation exists in titanium(III) or vanadium(IV) dimers with $d^1$ configured centers. Titanium(III) dimers with sandwich structures have been studied in respect of ESR and magnetic effects of $d^1$–$d^1$ interaction [70]. As an example, neutral and ionic titanacarborane dimers with $Ti^{III}$–$Ti^{III}$ distances of ca 3.7 Å afforded well-resolved triplet ESR spectra [70], especially, with the support of a commercial 'dual mode' ESR cavity which, in the parallel polarization ESR mode, $B_z \| B_1$, enhances the observability of otherwise forbidden $\Delta M_S = \pm 2$ transitions over the regular $\Delta M_S = \pm 1$ transitions. Analysis of both features provides zero-field splitting parameters $D$ (axial) and $E$ (rhombic), $g$ factor components, and spin–spin distances which agree with the crystallographically determined features and permit a discussion of through-bond and through-space interaction [70].

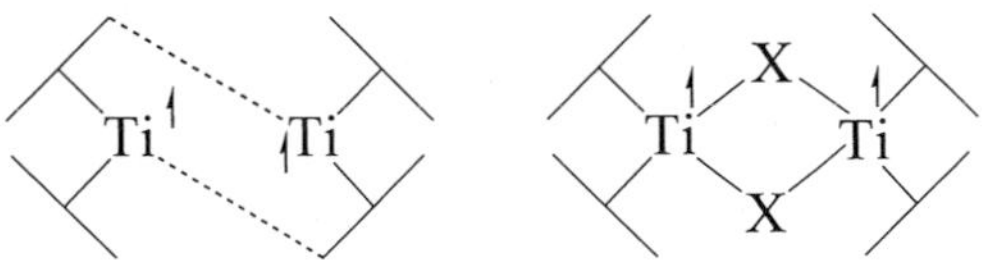

Inspection of the half-field region, especially with the dual mode cavity, may reveal triplet ESR features and thus spin–spin interaction in coordination chemistry even when such effects are not necessarily expected. The tendency of inorganic and organometallic compounds to associate via coordinate or other bonds can be detected through corresponding triplet ESR signals, as illustrated by the crystallographically characterized copper(II) complex **8** [79] which seems to form a dimer with ESR half-field features on cooling in a frozen glassy solution of THF [79].

Weak coupling of organometallic biradicals such as the hydrogen-bridged dimer of $(\eta^7\text{-}C_7H_7)V(\eta^5\text{-}C_5H_4COOH)$ can be detected in solution via simulation of ESR spectra [72]. Rather small exchange coupling constants $|J| < 1$ cm$^{-1}$ may thus be established, providing a complementary technique to NMR and susceptibility measurements.

### 9.3.6 Dynamic Effects

In common with all spectroscopic methods, ESR is characterized by a certain 'characteristic time-scale' as determined by the typical energy involved in the transition. For ESR at X band frequency, this window lies between ca $10^{-7}$ and $10^{-10}$ s, i.e. in the range of intermediate-to-slow molecular rotations and vibrations. The processes resulting in dynamic ESR effects might thus involve, e.g., conformational changes, as frequently encountered in the ESR spectroscopy of organic radicals [4]. Such structural changes also play an important role in inorganic and organo-metallic chemistry [73]; the changes in electron and atom location (i.e. coordination) are, however, more typical and will be illustrated here by examples.

The exchange of one unpaired electron between two or more equivalent or nearly equivalent sites can be monitored by ESR spectroscopy. A classical example is the 'hopping' of an added or charge transfer-generated electron between the three $\alpha$-diimine chelate ligands of $[Ru(bpy)_3]^{n+}$ (**9**) and related complexes.

**Scheme 2**

In contrast with the short-lived MLCT-excited states $*[Ru(bpy)_3]^{2+}$ [74], the electrogenerated $[Ru(bpy)_3]^{\cdot+}$ can be conveniently studied by temperature-dependent ESR [42, 50a, b]. Analysis of the broad unresolved lines of this and related species reveals electron hopping between the three equivalent ligand sites with activation energies of approximately 10 kJ mol$^{-1}$. Similar slow vs. fast exchange phenomena were reported for the 'extended atom' [75] species **17** which contains an encapsulated Na$^+$ ion and mobile spins in the bpy-containing periphery.

Characteristically, more unsymmetrical analogs $[Ru(L)(bpy)_2]^{\cdot+}$ with one better accepting ligand L give much narrower, often resolved, ESR signal lines, because of the localization of spin on L [42].

For the remaining alternative $[Ru(L)_2(bpy)]^{\cdot+}$ in that series, e.g. with L = 2,2'-azobispyridine (abpy), the hopping behavior between the two L sites was found

**17**

[50c] to depend characteristically on the stereochemistry of the complex which can exist in three different configurations [50c].

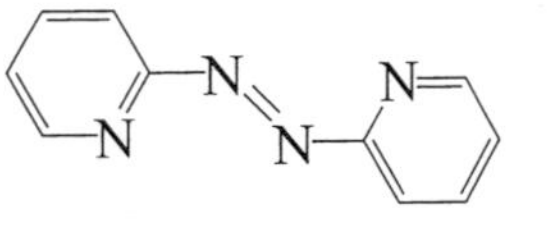

abpy

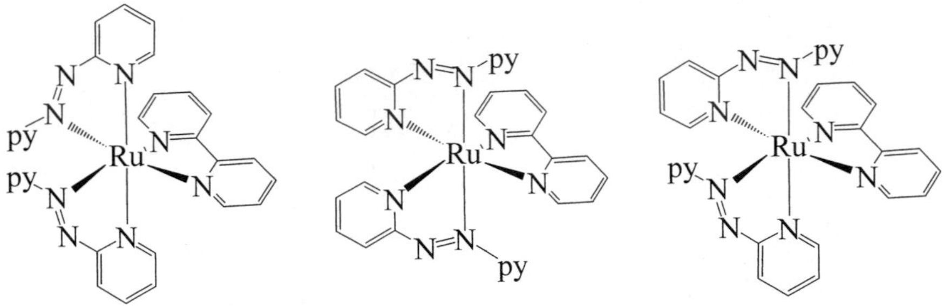

In contrast, the complex $[Cu^I(N^\wedge N)_2]^\cdot$ gave no ESR signs for rapid electron hopping between the two $N^\wedge N = 2,9$-diphenyl-1,10-phenanthroline ligands—probably as a result of the nearly orthogonal arrangement of the ligand $\pi$ systems [77].

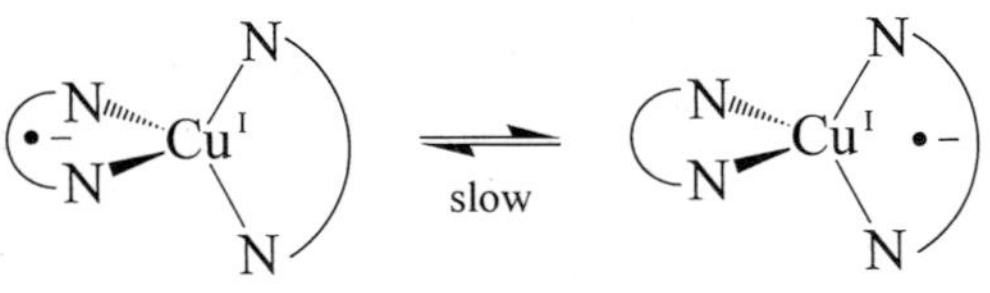

In special circumstances the metal and a ligand can compete for the spin in a paramagnetic complex. As this alternative involves oxidation-state changes, it is referred to as valence tautomerization or redox isomerization [78]. Such behavior is observed for *o*-semiquinone complexes of cobalt and manganese [78]; recently, a copper(I)-semiquinone–copper(II)-catecholate equilibrium system (7) of biochemical interest has been analyzed by temperature-dependent ESR [79].

An equilibrium such as (7) between two paramagnetic species with sufficiently different ESR signals can be analyzed quantitatively [79] by double integration, i.e. from the conventionally recorded first derivative signal to the zeroeth derivative spectrum and then further to the first integrated form. The procedure requires a careful definition of the baseline, especially for broader spectra.

Changes in the coordination mode of paramagnetic complexes can sometimes be easily followed by ESR spectroscopy. For instance, the alternative binding (**18**, **19**) of the Group-6 metal fragments $(OC)_5M$ to either the N or the chalcogen atom of

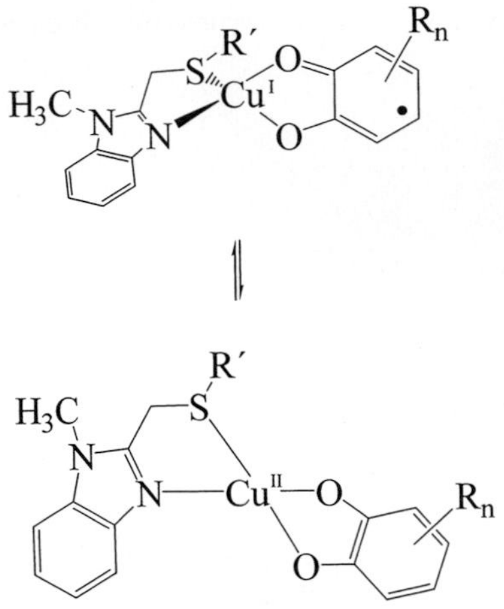

(7)

reduced 2,1,3-benzochalcogenodiazoles can be evident from the symmetry of the ESR hyperfine splitting in the radical anion ligand [80].

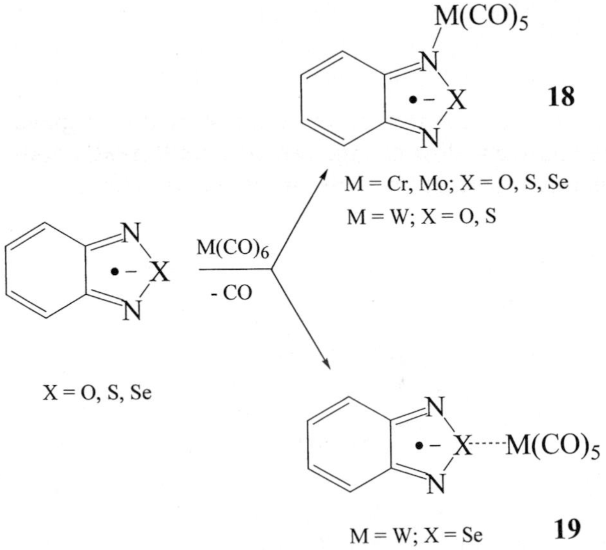

An electron transfer-sensitive $\sigma/\pi$ coordination alternative of substituted olefins to metal centers has been demonstrated [81] in the system $[(NC)_2C=C(CN)_2W(CO)_5]^n$ (**20**) where the $\pi$ complex with $n = 0$ (olefin coordination) changes reversibly to an ESR detectable $\sigma$ complex (nitrile coordination) after one-electron reduction ($n = -1$). Even in the absence of a detailed spin population analysis the low symmetry of the anionic complex is evident from at least three distinctly different $^{14}N$ coupling constants in the ESR spectrum [81].

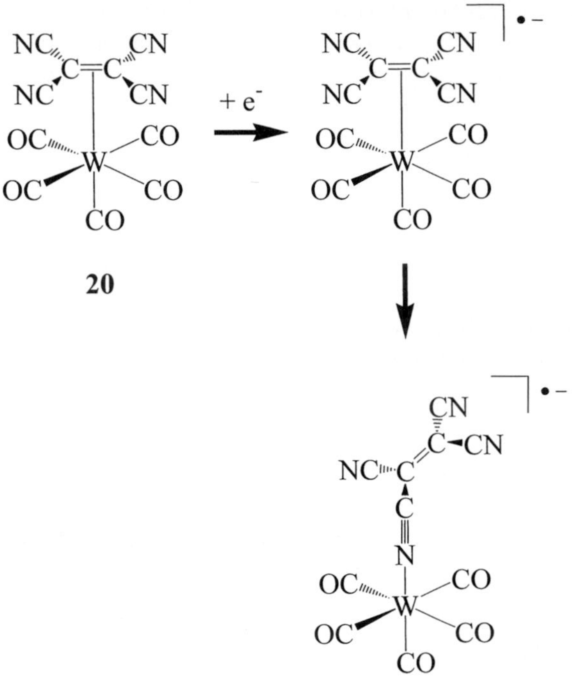

**20**

Examples of ESR-detected coordination changes do not have to involve organic radical ligands. For instance, the relatively slow changeover in a multimodal copper(II) catenate could be conveniently monitored by ESR spectroscopy [82].

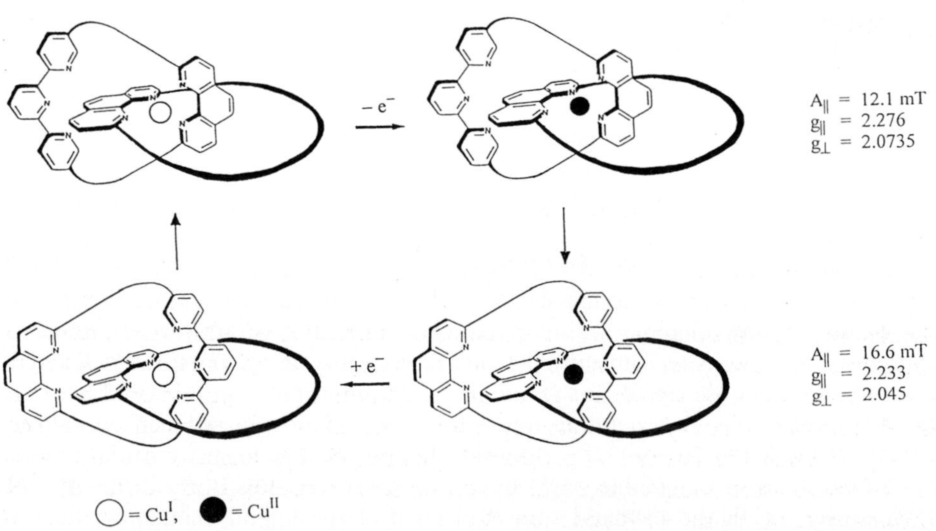

$A_\parallel$ = 12.1 mT
$g_\parallel$ = 2.276
$g_\perp$ = 2.0735

$A_\parallel$ = 16.6 mT
$g_\parallel$ = 2.233
$g_\perp$ = 2.045

$\bigcirc$ = Cu$^I$,  $\bullet$ = Cu$^{II}$

The electrochemically triggered molecular rearrangement involves oxidation from a four-coordinate $Cu^I$ to an intermediate four-coordinate $Cu^{II}$ species which has a distinctly different ESR spectrum (higher $g$, lower $a(^{63,65}Cu)$; measured in frozen solution) than the more stable five-coordinate $Cu^{II}$ alternative. The latter is formed by fairly slow rotation of interlocking macrocyclic rings; it reverts to the four-coordinate arrangement on re-reduction [82].

In favorable circumstances, ESR spectroscopy is also capable of detecting coordination changes which might have otherwise gone unnoticed. As an example, the reduction of the complex **21** produces a well-resolved ESR spectrum which shows only two equivalent $^{31}P$ nuclei (100 %, $I = 1/2$) interacting with the unpaired electron [83]. Apparently, two triorganophosphane ligands are lost after the reduction because of the high electron density created at the $d^6$-configured metal centers [83] (pentacoordination [84]).

### 9.3.7 ESR Information and Chemical Reactivity

ESR Spectroscopy is not only superbly suited to the analysis of metal centered [3], metal/ligand mixed [35b] or ligand centered [85] singly occupied molecular orbitals; in favorable circumstances some ESR parameters can even be directly correlated with chemical reactivity. Although such correlations are obvious if ESR-monitored metal oxidation-state changes are involved, even small but variable amounts, $\delta$, of unpaired electron density transmitted from radical ligands to potentially reactive metal centers can cause significant reactivity variations which are mirrored by ESR data. As an example, the rhenium halide bond labilization in reduced complexes $[(N^\wedge N)Re(CO)_3X]^{\cdot-}$ (**22**; $N^\wedge N = \alpha$-diimine ligand, $X$ = halogen) correlates with the $^{185,187}Re$ (and, if detectable, $^{14}N$) hyperfine splittings [86]. Both kinds of ESR data quantify the ligand-to-metal spin transfer $\delta$ at the ligand–metal interface in such $18+\delta$ valence electron compounds [86–88]; this, in turn, determines the sub-stitutional labilization of the axial and partially [51, 86] charge-accepting ligand.

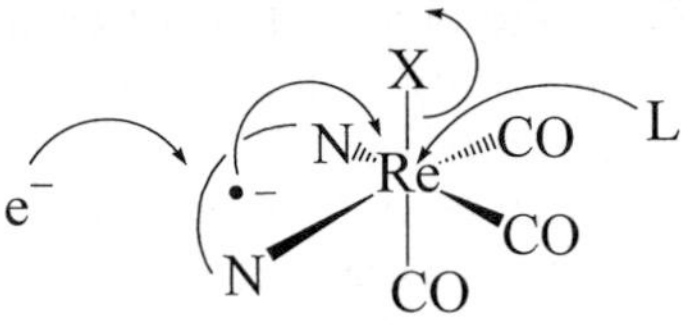

$$X = Cl^-, Br^-$$

$$L = CH_3CN, PPh_3,$$
$$CN^-, CO_2$$

**22**

A related interesting phenomenon is the electron reservoir behavior of coordination compounds [89]. For instance, compounds $[(A)(C_nR_n)M^kX]^+$ with acceptor-bound organometallic reaction centers can accommodate one first added electron in the acceptor ligand A before a second added electron combines with the stored one to effect a chemically productive two-electron process at the metal (Eq. 8) [90].

$$[(A)(C_nR_n)M^kX]^+ \xrightarrow[E]{+e^-} [(A^{\cdot-})(C_nR_n)M^kX]^{\cdot}$$
$$\text{ESR detectable}$$

$$\xrightarrow[EC]{+e^-} [(A)(C_nR_n)M^{k-2}] + X^- \qquad (8)$$

$$M = Rh, \; Ir, \; k = +III, \; n = 5; \; M = Os, \; k = +II, \; n = 6; \; X = \text{halogen};$$

$$A = \text{acceptor, e.g. } \alpha\text{-diimine}$$

In the synthetically attractive field of electron transfer catalysis (ETC) [2, 91], ESR data of radical intermediates can correlate with the variable efficiency of processes such as the ETC-catalyzed substitution (9) by substantiating the amount of ligand-to-metal electron transfer in the ground state [87].

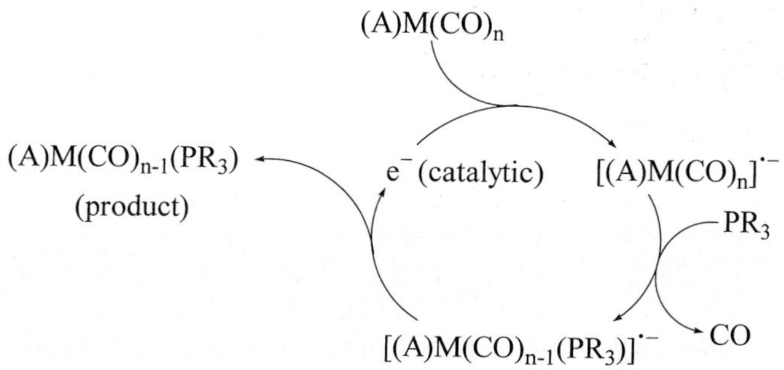

$$(9)$$

requirement for ETC:
$E_{red} [(A)M(CO)_{n-1}(PR_3)]$
$< E_{red} [(A)M(CO)_n];$
$A = $ pyridine derivative, $\alpha$-diimine

### 9.3.8 Concluding Remarks

The previous chapters and examples were meant to familiarize the reader with current conventional ESR methodology and remove any barriers to the application of this kind of spectroscopy in electron-transfer studies. Unavoidably, not all aspects of inorganic ESR could be treated properly; for instance, both the ESR spectroscopy of bioinorganic systems (e.g. heme species or iron–sulfur clusters) or materials chemistry (e.g. defect centers, radicals in zeolites) are vast and entirely independent research areas. Nevertheless, the molecular species presented here might serve as introductory examples to illustrate the versatility of this particular kind of magnetic resonance.

### Acknowledgments

Support for the author's ESR research has been continuously provided by the Deutsche Forschungsgemeinschaft (DFG) and Volkswagenstiftung. In particular, support for the recent graduate college 'Magnetic Resonance' at the University of Stuttgart is gratefully acknowledged. For assistance with the processing of the manuscript my special thanks are due to Mrs Angela Winkelmann and Dr. Brigitte Schwederski.

### References

1. J. A. Weil, J. R. Bolton, J. E. Wertz, *Electron Paramagnetic Resonance*, Wiley, New York, **1994**.
2. D. Astruc, *Electron Transfer and Radical Processes in Transition-Metal Chemistry*, VCH, New York, **1995**.
3. (a) B. A. Goodman, J. B. Raynor, *Adv. Inorg. Chem. Radiochem.* **1970**, *13*, 135; (b) F. E. Mabbs, D. Collison, *Electron Paramagnetic Resonance of d Transition Metal Compounds*, Elsevier, Amsterdam, **1992**; (c) J. R. Pilbrow, *Transition Ion Electron Paramagnetic Resonance*, Clarendon Press, Oxford, **1990**; (d) P. H. Rieger in *Organometallic Radical Processes* (W. C. Trogler, Ed.), Elsevier, Amsterdam, **1990**, p. 270.
4. F. Gerson, *High Resolution E.S.R. Spectroscopy*, VCH, Weinheim, **1970**.
5. J. A. Weil, J. R. Bolton, J. E. Wertz, *Electron Paramagnetic Resonance*, Wiley, New York, **1994**, pp. 534.
6. J. A. Weil, J. R. Bolton, J. E. Wertz, *Electron Paramagnetic Resonance*, Wiley, New York, **1994**, pp. 532.
7. (a) V. Lorenzen, W. Preetz, F. Baumann, W. Kaim, *Inorg. Chem.* **1998**, *37*, 4011; (b) M. Wanner, W. Kaim, V. Lorenzen, W. Preetz, *Z. Naturforsch.* **1999**, *54b*, 1103.
8. M. Ballester, J. Castaner, J. Riera, J. Pujadas, O. Armet, C. Onrubia, J.A. Rio, *J. Org. Chem.* **1984**, *49*, 770.
9. (a) O. Renn, A. Albinati, B. Lippert, *Angew. Chem.* **1990**, *102*, 71; *Angew. Chem. Int. Ed. Engl.* **1990**, *29*, 84; (b) W. E. Geiger, N. G. Connelly, *Adv. Organomet. Chem.* **1985**, *24*, 87; (c) D. Astruc, *Electron Transfer and Radical Processes in Transition-Metal Chemistry*, VCH, New York, **1995**, p. 250.
10. N. Wiberg, T. Blank, W. Kaim, B. Schwederski, G. Linti, *Eur. J. Inorg. Chem.* **2000**, 1475.
11. A. Haaland, K.-G. Martinsen, H. V. Volden, W. Kaim, E. Waldhör, W. Uhl, U. Schütz, *Organometallics* **1996**, *15*, 1146.

12. H. Binder, R. Kellner, K. Vaas, M. Hein, F. Baumann, M. Wanner, R. Winter, W. Kaim, W. Hönle, Y. Grin, U. Wedig, M. Schultheiss, R.K. Kremer, H.G. von Schnering, O. Groeger, G. Engelhardt, *Z. Anorg. Allg. Chem.* **1999**, *625*, 1059.

13. (a) D. Sellmann, J. Müller, P. Hofmann, *Angew. Chem.* **1982**, *94*, 708; *Angew. Chem. Int. Ed. Engl.* **1982**, *21*, 691; (b) D. Sellmann, J. Müller, *J. Organomet. Chem.* **1985**, *281*, 249; (c) A. Winter, G. Huttner, L. Zsolnai, P. Kroneck, M. Gottlieb, *Angew. Chem.* **1984**, *96*, 986; *Angew. Chem. Int. Ed. Engl.* **1984**, *23*, 975; (d) A. Winter, G. Huttner, M. Gottlieb, I. Jibril, *J. Organomet. Chem.* **1985**, *286*, 317.

14. (a) R. Gross, W. Kaim, *Angew. Chem.* **1985**, *97*, 869; *Angew. Chem. Int. Ed. Engl.* **1985**, *24*, 856; (b) R. Gross und W. Kaim, *Inorg. Chem.* **1987**, *26*, 3596.

15. Cf. p. 283 in ref. 3a).

16. D. Astruc, *Electron Transfer and Radical Processes in Transition-Metal Chemistry*, VCH, New York, **1995**, p. 358.

17. (a) W. Kaim, *Coord. Chem. Rev.* **1987**, *76*, 187; (b) W. Kaim, M. Moscherosch, *Coord. Chem. Rev.* **1994**, *129*, 157; (c) D. Astruc, *Electron Transfer and Radical Processes in Transition-Metal Chemistry*, VCH, New York, **1995**, p. 234.

18. (a) D. Astruc, *Electron Transfer and Radical Processes in Transition-Metal Chemistry*, VCH, New York, **1995**, p. 216.

19. F. Baumann, W. Kaim, L. M. Baraldo, L. D. Slep, J. A. Olabe, J. Fiedler, *Inorg. Chim. Acta* **1999**, *285*, 129.

20. E. Waldhör, W. Kaim, J. A. Olabe, L. D. Slep, J. Fiedler, *Inorg. Chem.* **1997**, *36*, 2969.

21. (a) W. Matheis, W. Kaim, *J. Chem. Soc., Faraday Trans.* **1990**, *86*, 3337; (b) W. Matheis, J. Poppe, W. Kaim, S. Zalis, *J. Chem. Soc., Perkin Trans. 2* **1994**, 1923.

22. P. W. Atkins, M. C. R. Symons, *The Structure of Inorganic Radicals*, Elsevier, Amsterdam, **1967**, p. 36.

23. (a) P. T. Kissinger, W. R. Heineman, *Laboratory Techniques in Electroanalytical Chemistry*, 2nd. Ed., Marcel Dekker, New York, **1996**, p. 930; (b) A. J. Bard, L. R. Faulkner, *Electrochemical Methods*, Wiley, New York, **1980**, p. 618.

24. (a) W. Kaim, *Top. Curr. Chem.* **1994**, *169*, 231; (b) S. Hasenzahl, W. Kaim, T. Stahl, *Inorg. Chim. Acta* **1994**, *225*, 23; (c) M. Kaupp, H. Stoll, H. Preuss, W. Kaim, T. Stahl, G. van Koten, E. Wissing, W. J. J. Smeets, A. L. Spek, *J. Am. Chem. Soc.* **1991**, *113*, 5606.

25. (a) W. Kaim, in *Electron and Proton Transfer in Chemistry and Biology*, (Eds.: A. Müller, E. Diemann, W. Junge, H. Ratajczak), Elsevier, Amsterdam, **1992**, p. 45; (b) W. Kaim, B. Olbrich-Deussner, in *Organometallic Radical Processes*, (Ed.: W. C. Trogler), Elsevier, Amsterdam, **1990**, p. 173.

26. N. G. Connelly, W. E. Geiger, *Chem. Rev.* **1996**, *96*, 877.

27. (a) T. Kobayashi, K. Yasufuku, J. Iwai, H. Yesaka, H. Noda, H. Ohtani, *Coord. Chem. Rev.* **1985**, *64*, 1; (b) M. C. R. Symons, R. L. Sweany, *Organometallics* **1982**, *1*, 834.

28. P. W. Atkins, M. C. R. Symons, *The Structure of Inorganic Radicals*, Elsevier, Amsterdam, **1967**, p. 41.

29. C. D. Garr, R.G. Finke, *Inorg. Chem.* **1993**, *32*, 4414.

30. (a) D. N. R. Rao, M. C. R. Symons, *J. Chem. Soc. Chem. Commun.* **1982**, 954; (b) M. C. R. Symons, *NATO ASI Ser., Ser. C* **1989**, 257; (c) K. Warncke, J. C. Schmidt, S.-C. Ke, *J. Am. Chem. Soc.* **1999**, *121*, 10522.

31. (a) D. Rehorek, *Chem. Soc. Rev.* **1991**, *20*, 341; (b) E. Samuel, D. Caurant, C. Gourier, Ch. Elschenbroich, K. Agbaria, *J. Am. Chem. Soc.* **1998**, *120*, 8088.

32. A. Klein, S. Hasenzahl, W. Kaim, *J. Chem. Soc., Perkin Trans. 2* **1997**, 2573.

33. (a) A. Schweiger, *Angew. Chem.* **1991**, *103*, 223; *Angew. Chem. Int. Ed. Engl.* **1991**, *30*, 265; (b) H. van Willigen, P. R. Levstein, M. H. Ebersole, *Chem. Rev.* **1993**, *93*, 173.

34. C. J. Kleverlaan, D. J. Stufkens, I. P. Clark, M. W. George, J. J. Turner, D. M. Martino, H. van Willigen, A. Vlcek, Jr., *J. Am. Chem. Soc.* **1998**, *120*, 10871.

35. (a) Ch. Elschenbroich, E. Bilger, R. D. Ernst, D. R. Wilson, M. S. Kralik, *Organometallics* **1985**, *4*, 2068; (b) V. Kasack, W. Kaim, H. Binder, J. Jordanov, E. Roth, *Inorg. Chem.* **1995**, *34*, 1924; (c) W. Kaim, V. Kasack, *Inorg. Chem.* **1990**, *29*, 4696.

36. (a) R. Schäfer, W. Kaim, M. Moscherosch, M. Krejcik, *J. Chem. Soc., Chem. Commun.* **1992**, 834; (b) W. Kaim, A. Klein, *Organometallics* **1995**, *14*, 1176; (c) W. Kaim, R. Schäfer,

F. Hornung, M. Krejcik, J. Fiedler, S. Zalis in *Transition Metal Sulphides—Chemistry and Catalysis*, NATO ASI Series, (T. Weber, Ed.) Kluwer Academic Publishers, Dordrecht, **1998**, p. 37.

37. J. Poppe, M. Moscherosch, W. Kaim, *Inorg. Chem.* **1993**, *32*, 2640.
38. (a) Y. Ohsawa, M. K. DeArmond, K. W. Hanck, C. G. Moreland, *J. Am. Chem. Soc.* **1985**, *107*, 5383; (b) G. N. La Mar, Horrocks, Jr., W. DeWitt, R. H. Holm (Eds.), *NMR of Paramagnetic Molecules*, Academic Press, New York, **1973**; (c) N. Juranic, *Coord. Chem. Rev.* **1989**, *96*, 253.
39. (a) W. Kaim, *Z. Naturforsch.* **1981**, *36b*, 150; (b) W. Uhl, U. Schütz, W. Kaim, E. Waldhör, *J. Organomet. Chem.* **1995**, *501*, 79; (c) N. Wiberg, K. Amelunxen, H. Nöth, H. Schwenk, W. Kaim, A. Klein, T. Scheiring, *Angew. Chem.* **1997**, *109*, 1258; *Angew. Chem. Int. Ed. Engl.* **1997**, *36*, 1213.
40. J. Rall, W. Kaim, *J. Chem. Soc., Faraday Trans.* **1994**, *90*, 2905.
41. A. J. Stone, *Mol. Phys.* **1964**, *7*, 311; (b) H. Fischer in *Free Radicals*, Vol. II (Ed.: J. K. Kochi), Wiley, New York, **1973**, p. 452.
42. W. Kaim, S. Ernst, V. Kasack, *J. Am. Chem. Soc.* **1990**, *112*, 173.
43. W. Kaim, *Inorg. Chem.* **1984**, *23*, 3365.
44. (a) T. Sixt, J. Fiedler, W. Kaim, *Inorg. Chem. Commun.* **2000**, *3*, 80; (b) R. T. Hembre, J. S. McQueen, V. W. Day, *J. Am. Chem. Soc.* **1996**, *118*, 798; (c) S. B. Jensen, S. J. Rodger, M. D. Spicer, *J. Organomet. Chem.* **1998**, *556*, 151.
45. J. H. MacNeil, A. C. Chiverton, S. Fortier, M. C. Baird, R. C. Hynes, A. J. Williams, K. F. Preston, T. Ziegler, *J. Am. Chem. Soc.* **1991**, *115*, 9834.
46. (a) M. Yamaga, N. Kodama, T. Yosida, B. Henderson, K. Kindo, *J. Phys. Condens. Matter* **1997**, *9*, 9639; (b) E. D. Thoma, H. Shields, Y. Zhang, B. C. McCollum, R. T. Williams, *J. Luminscence* **1997**, *71*, 93.
47. (a) P. W. Atkins, M. C. R. Symons, *The Structure of Inorganic Radicals*, Elsevier, Amsterdam, **1967**, pp. 119 and 142; (b) A. Pfitzner, F. Baumann, W. Kaim, *Angew. Chem.* **1998**, *110*, 2057; *Angew. Chem. Int. Ed. Engl.* **1998**, *37*, 1955.
48. A.-L. Barra, L-C. Brunel, D. Gatteschi, L. Pardi, R. Sessoli, *Acc. Chem. Res.* **1998**, *31*, 460.
49. (a) A.-L. Barra, L.-C. Brunel, F. Baumann, M. Schwach, M. Moscherosch, W. Kaim, *J. Chem. Soc., Dalton Trans.* **1999**, 3855; (b) W. Kaim, M. Moscherosch, *J. Chem. Soc., Faraday Trans.* **1991**, *87*, 3185; (c) M. Moscherosch, J. S. Field, W. Kaim, S. Kohlmann, M. Krejcik, *J. Chem. Soc. Dalton Trans.* **1993**, 211.
50. (a) D. E. Morris, K. W. Hanck, M. K. De Armond, *Inorg Chem.* **1983**, *105*, 3032; (b) D. E. Morris, K. W. Hanck, M. K. De Armond, *J. Electroanal. Chem.* **1983**, *149*, 115; (c) M. Heilmann, F. Baumann, W. Kaim, J. Fiedler, *J. Chem. Soc., Faraday Trans.* **1996**, *92*, 4227.
51. (a) W. Kaim, S. Kohlmann, *Chem. Phys. Lett.* **1987**, *139*, 365; (b) W. Kaim, S. Kohlmann, *Inorg. Chem.* **1990**, *29*, 2909.
52. (a) R. C. Bray, L. S. Meriwether, *Nature (London)* **1966**, *212*, 467; (b) S. P. J. Albracht, *Biochim. Biophys. Acta* **1994**, *1188*, 167.
53. A. Vlcek, F. Baumann, W. Kaim, F.-W. Grevels, F. Hartl, *J. Chem. Soc., Dalton Trans.* **1998**, 215.
54. D. Astruc, *Electron Transfer and Radical Processes in Transition-Metal Chemistry*, VCH, New York, **1995**, p. 217.
55. (a) H. Schönberg, S. Boulmaâz, M. Wörle, L. Liesum, A. Schweiger, H. Grützmacher, *Angew. Chem.* **1998**, *110*, 1492; *Angew. Chem. Int. Ed. Engl.* **1998**, *37*, 1423; (b) S. Berger, A. Klein, M. Wanner, J. Fiedler, W. Kaim, *Inorg. Chem.* **2000**, *39*, 2516; (c) W. Kaim, S. Berger, S. Greulich, R. Reinhardt, J. Fiedler, *J. Organomet. Chem.* **1999**, *582*, 153.
56. (a) F. G. N. Cloke, G. R. Hanson, M. J. Henderson, P. B. Hitchcock, C. L. Raston, *J. Chem. Soc. Chem. Commun.* **1989**, 1002; (b) F. G. N. Cloke, C. I. Dalby, M. J. Henderson, P. B. Hitchcock, C. H. L Kennard, R. M. Lamb, C. L. Raston, *J. Chem. Soc. Chem. Commun.* **1990**, 1394.
57. W. Kaim, W. Matheis, *J. Chem. Soc., Chem. Commun.* **1991**, 597.
58. F. P. Fanizzi, G. Natile, M. Lanfranchi, A. Tiripicchio, F. Laschi, P. Zanello, *Inorg. Chem.* **1996**, *35*, 3173.

59. (a) P.S. Braterman, J.-I. Song, F.M. Wimmer, S. Wimmer, W. Kaim, A. Klein, R.D. Peacock, *Inorg. Chem.* **1992**, *31*, 5084; (b) A. Klein, W. Kaim, E. Waldhör, H.-D. Hausen, *J. Chem. Soc., Perkin Trans. 2* **1995**, 2121; (c) A. Klein, E. J. L. McInnes, T. Scheiring, S. Zalis, *J. Chem. Soc. Faraday Trans.* **1998**, *94*, 2979; (d) E. J. L. McInnes, R. D. Farley, S. A. Macgregor, K. J. Taylor, L. J. Yellowlees, C. C. Rowlands, *J. Chem. Soc. Faraday Trans.* **1998**, *94*, 2985.

60. T. Schmauke, E. Möller, E. Roduner, *J. Chem. Soc. Chem.Commun.* **1998**, 2589 and references cited.

61. H. Kurreck, B. Kirste, W. Lubitz, *Electron Nuclear Double Resonance Spectroscopy of Radicals in Solution*, VCH, New York, **1988**.

62. (a) W. Kaim, W. Lubitz, *Angew. Chem.* **1983**, *95*, 915; *Angew. Chem. Int. Ed. Engl.* **1983**, *22*, 892; *Angew. Chem. Suppl.* **1983**, 1209; (b) H. Kurreck, B. Kirste, W. Lubitz, *Electron Nuclear Double Resonance Spectroscopy of Radicals in Solution*, VCH, New York, **1988**, p. 103.

63. (a) W. Kaim, S. Ernst, *J. Phys. Chem.* **1986**, *90*, 5010; (b) S.E. Bell, J.S. Field, R.I. Haines, M. Moscherosch, W. Matheis, W. Kaim, *Inorg. Chem.* **1992**, *31*, 3269.

64. (a) V. Barone, in *Recent Advances in Density Functional Methods, Part I* (D. O. Chong, Ed.), World Scientific Publ. Co., Singapore, **1996**; (b) S. Zalis, C. Daniel, A. Vlcek, Jr., *J. Chem. Soc. Dalton Trans.* **1999**, 3081.

65. R. Gross, W. Kaim, *Inorg. Chem.* **1986**, *25*, 4865.

66. (a) K. Wieghardt, *Angew. Chem.* **1989**, *101*, 1179; *Angew. Chem. Int. Ed. Engl.* **1989**, *28*, 1153; (b) G. C. Dismukes, in *Mixed Valency Systems: Applications in Chemistry, Physics and Biology*, NATO ASI Ser., Ser. C, Vol. 343 (K. Prassides, Ed.), Kluwer Academic Publishers, Dordrecht, **1991**, p. 137.

67. W. Bruns, W. Kaim, E. Waldhör, M. Krejcik, *J. Chem. Soc., Chem. Commun.* **1993**, 1868.

68. J. A. McCleverty, M. D. Ward, *Acc. Chem. Res.* **1998**, *31*, 842.

69. J. A. Weil, J. R. Bolton, J. E. Wertz, *Electron Paramagnetic Resonance*, Wiley, New York, **1994**, p. 151.

70. N. S. Hosmane, Y. Wang, H. Zhang, K.-J. Lu, J. A. Maguire, T. G. Gray, K. A. Brooks, E. Waldhör, W. Kaim, R. K. Kremer, *Organometallics* **1997**, *16*, 1365 and references cited.

71. (a) Ch. Elschenbroich, J. Heck, *J. Am. Chem. Soc.* **1979**, *101*, 6773; (b) Ch. Elschenbroich, J. Heck, *Angew. Chem.* **1981**, *93*, 278; *Angew. Chem. Int. Ed. Engl.* **1981**, *20*, 267; (c) Ch. Elschenbroich, A. Bretschneider-Hurley, J. Hurley, A. Behrendt, W. Massa, S. Wocadlo, E. Reijerse, *Inorg. Chem.* **1995**, *34*, 743.

72. Ch. Elschenbroich, O. Schiemann, O. Burghaus, K. Harms, *J. Am. Chem. Soc.* **1997**, *119*, 7452.

73. W. E. Geiger, *Prog. Inorg. Chem.* **1985**, *33*, 275.

74. H. Yersin, W. Humbs, J. Strasser, *Coord. Chem. Rev.* **1997**, *159*, 325.

75. (a) L. Echegoyen, A. DeCian, J. Fischer, J.-M. Lehn, *Angew. Chem.* **1991**, *103*, 884; *Angew. Chem. Int. Ed. Engl.* **1991**, *30*, 838; (b) L. Echegoyen, E. Perez-Cordero in *Transition Metals in Supramolecular Chemistry* (L. Fabbrizzi, A. Poggi, Eds.), Kluwer Academic Publishers, Dordrecht, **1994**, p. 115.

76. J. Fees, H.-D. Hausen, W. Kaim, *Z. Naturforsch.* **1995**, *50b*, 15.

77. A.F. Stange, E. Waldhör, M. Moscherosch, W. Kaim, *Z. Naturforsch.* **1995**, 50b, 115.

78. (a) O.-S. Jung, D. H. Jo, Y.-A. Lee, Y. S. Sohn, C. G. Pierpont, *Angew. Chem.* **1996**, *108*, 1796; *Angew. Chem. Int. Ed. Engl.* **1996**, *35*, 1694; (b) D. M. Adams, B. Li, J. D. Simon, D. N. Hendrickson, *Angew. Chem.* **1995**, *107*, 1580; *Angew. Chem. Int. Ed. Engl.* **1995**, *34*, 1481; (c) A. S. Attia, C. G. Pierpont, *Inorg. Chem.* **1995**, *34*, 1172.

79. J. Rall, M. Wanner, M. Albrecht, F. M. Hornung, W. Kaim, *Chem. Eur. J.* **1999**, *5*, 2802.

80. W. Kaim, V. Kasack, *Angew. Chem.* **1982**, *94*, 712; *Angew. Chem. Int. Ed. Engl.* **1982**, *21*, 700.

81. T. Roth, W. Kaim, *Inorg. Chem.* **1992**, *31*, 1930.

82. F. Baumann, A. Livoreil, W. Kaim, J.-P. Sauvage, *J. Chem. Soc., Chem. Commun.* **1997**, 35.

83. W. Bruns, W. Kaim, E. Waldhör, M. Krejcik, *Inorg. Chem.* **1995**, *34*, 663.

84. (a) D. J. Darensbourg, K. K. Klausmeyer, J. H. Reibenspies, *Inorg. Chem.* **1995**, *34*, 4676; (b) F. Hartl, A. Vlcek, Jr., L. A. deLearie, C. G. Pierpont, *Inorg. Chem.* **1990**, *29*, 1073.

85. (a) J. Fees, W. Kaim, M. Moscherosch, W. Matheis, J. Klima, M. Krejcik, S. Zalis, *Inorg. Chem.* **1993**, *32*, 166; (b) S. Ernst, C. Vogler, A. Klein, W. Kaim, S. Zalis, *Inorg. Chem.* **1996**, *35*, 1295; (c) E. Waldhör, M. M. Zulu, S. Zalis, W. Kaim, *J. Chem. Soc., Perkin Trans. 2* **1996**, 1197.

86. A. Klein, C. Vogler, W. Kaim, *Organometallics* **1996**, *15*, 236.
87. (a) B. Olbrich-Deussner, W. Kaim, *J. Organomet. Chem.* **1988**, *340*, 71; (b) B. Olbrich-Deussner, W. Kaim, *J. Organomet. Chem.* **1989**, *361*, 335.
88. (a) D. M. Schut, K. J. Keana, D. R. Tyler, P. H. Rieger, *J. Am. Chem. Soc.* **1995**, *117*, 8939; (b) D. Astruc, *Electron Transfer and Radical Processes in Transition-Metal Chemistry*, VCH, New York, **1995**, p. 207.
89. (a) M.-H. Desbois, D. Astruc, J. Guillin, J.-P. Mariot, F. Varret, *J. Am. Chem. Soc.* **1985**, *107*, 52; (b) D. Astruc, *Electron Transfer and Radical Processes in Transition-Metal Chemistry*, VCH, New York, **1995**, p. 337.
90. W. Kaim, R. Reinhardt, J. Fiedler, *Angew. Chem.* **1997**, *109*, 2600; *Angew. Chem. Int. Ed. Engl.* **1997**, *36*, 2493. See also F. Baumann, A. Stange, W. Kaim, *Inorg. Chem. Commun.* **1998**, *1*, 305.
91. D. Miholova, A. A. Vlcek, *J. Organomet. Chem.* **1985**, *279*, 317.